THERMAL ENGINEERING

(*Including* : **Thermodynamics, Heat Engines and Non-Conventional Power Generation**)

For Engineering Students preparing for B.E./B.Tech. ; AMIE-Section B (India) ; GATE and UPSC (Engg. Services) Examinations.

SI UNITS

By

Er. R.K. RAJPUT

M.E. (Heat Power Engg.) *Hons.–Gold Medallist* ; Grad. (*Mech. Engg. & Elect. Engg.*) ;
M.I.E. (*India*) ; M.S.E.S.I. ; M.I.S.T.E. ; C.E. (*India*)

Recipient of : *"Best Teacher (Academic) Award"*
"Distinguished Author Award"
"Jawahar Lal Nehru Memorial Gold Medal"
for an outstanding research paper
(Institution of Engineers—India)

Principal (*Formerly*):

- ***Thapar Polytechnic College***
- ***Punjab College of Information Technology***

PATIALA

LAXMI PUBLICATIONS (P) LTD

(An ISO 9001:2015 Company)

BENGALURU • CHENNAI • GUWAHATI • HYDERABAD • JALANDHAR
KOCHI • KOLKATA • LUCKNOW • MUMBAI • RANCHI • NEW DELHI
BOSTON (USA) • NAIROBI (KENYA)

THERMAL ENGINEERING

Printed and bound in India

Typeset at Goswami Associates, Delhi

First Edition: 1993, *Second Edition:* 1999, *Third Edition:* 2001, *Fourth Edition:* 2004, *Fifth Edition:* 2005, *Sixth Edition:* 2006
Seventh Edition: 2009, *Eighth Edition:* 2010, *Reprint:* 2011, 2012, *Ninth Edition:* 2013, *Reprint:* 2014, *Tenth Edition:* 2015, Edition : 2018

ISBN 978-81-318-0804-7

Branches

Bengaluru	080-26 75 69 30	
Chennai	044-24 34 47 26,	24 35 95 07
Guwahati	0361-254 36 69,	251 38 81
Hyderabad	040-27 55 53 83,	27 55 53 93
Jalandhar	0181-222 12 72	
Kochi	0484-237 70 04,	405 13 03
Kolkata	033-22 27 43 84	
Lucknow	0522-220 99 16	
Mumbai	022-24 93 12 61	
Ranchi	0651-220 44 64	

PUBLISHED IN INDIA BY

Laxmi Publications (P) Ltd.

(An ISO 9001:2015 Company)
113, GOLDEN HOUSE, DARYAGANJ,
NEW DELHI - 110002, INDIA
Telephone : 91-11-4353 2500, 4353 2501
Fax : 91-11-2325 2572, 4353 2528
www.laxmipublications.com info@laxmipublications.com

C—15630/018/01
Printed at: IPP, Noida

TO ALMIGHTY

Contents

Chapters *Pages*

Chapters *Pages*

Chapters *Pages*

Chapters *Pages*

Chapters *Pages*

Chapters *Pages*

Chapters *Pages*

Preface to The Tenth Edition

I am pleased to present the **Tenth Edition** *(Two Colours)* of this standard treatise on large size paper to enable the reader to handle the book comfortably. The warm reception which the previous editions and reprints of this book have enjoyed all over India and abroad, has been a matter of much satisfaction to me.

In this edition a **new chapter 30** has been added with the following *Contents:*

Chapter 30: 'GATE' and 'UPSC' Examinations' Questions (Latest-selected) with Answers/Solutions

(A) Conventional Questions with Answers/Solutions.

(B) Multiple-Choice Questions with Answers and "Explanations".

Miscellaneous Objective Type Questions with "Answers"

It is hoped that the book in this comprehensive and complete form will prove to be of immense use for the students preparing for **'Universities'** and **'Competitive' Examinations.**

The author's thanks are due to his wife Ramesh Rajput for extending all cooperation during preparation and proof reading of this revised manuscript.

Any suggestions for the improvement of this book will be thankfully acknowledged and incorporated in the next edition.

Er. R.K. Rajput
(Author)

Preface to The First Edition

On the subject of **"Thermal Engineering"** several books are available in the market but most of them either lack in proper subject treatment or the solved examples are small in number and their solutions are too difficult to understand. The purpose of writing this book is to bridge these gaps and to present such a book to the students which should be easy to understand and assimilate comprehensive treatment of the subject matter in simple, lucid and direct language.

The book comprises of 26 Chapters, dealing with '**Thermodynamics**', '**Heat Engines**' and '**Non-Conventional Power Generation**'.

Each chapter contains adequate text supported by simple and self explanatory figures, large number of Worked Examples both in SI and MKS units, Highlights, Theoretical Questions and several Unsolved Examples.

The book will prove to be of great help to the students preparing for engineering degree, A.M.I.E. (India), diploma and other competitive examinations.

The author's grateful thanks are due to Smt. Ramesh Rajput (wife) for extending full co-operation during the preparation of the manuscript.

The author is also thankful to Mr. Rakesh Syal for drawing neat diagrams for this book.

In the end, the author wishes to express his grateful thanks to M/s Laxmi Publications for taking extra pains in printing the book at a rapid pace and very systematically.

Although every care has been taken to make the book free of errors both in text as well as in solved examples, yet the author shall feel obliged if errors present are brought to his notice. Constructive criticism of the book will be warmly received.

Er. R.K. Rajput
(Author)

Foreword

It is a pleasure for me to write these few words on Shri R.K. Rajput's latest book on "Thermal Engineering".

Written by a highly experienced and gifted teacher of our pace-setting Polytechnic, this book deals with Thermodynamics and Heat Engines besides having an additional chapter on Non-Conventional Power Generation. I have no doubt that it will prove extremely useful for students of engineering institutions in our country.

Among the several distinctive features of this book, special mention may be made of the following : Development of the subject in a logical and coherent manner, neat and simple diagrams to explain the text, large number of worked examples as well as unsolved examples for practice, numerous objective-type questions and the consistent use of SI units.

As is well known to his peers, the author of this volume has written a number of popular books earlier on various engineering subjects for the benefit of students in Polytechnics and Engineering Colleges. As is to be expected, a great deal of systematic and meticulous work has gone into the writing of this book.

I wish the author and his book all success.

Professor Dr. T.R. ANANTHARAMAN
Director, Thapar Institute of Engineering
and Technology and Thapar Corporate
Research and Development Centre
PATIALA

Introduction to SI Units and Conversion Factors

A. INTRODUCTION TO SI UNITS

SI, the international system of units are divided into three classes :

1. Base units
2. Derived units
3. Supplementary units.

From the scientific point of view division of SI units into these classes is to a certain extent arbitrary, because it is not essential to the physics of the subject. Nevertheless the General Conference, considering the advantages of a single, practical, world-wide system for international relations, for teaching and for scientific work, decided to base the international system on a choice of six well-defined units given in Table 1 below :

Table 1. SI Base Units

Quantity	*Name*	*Symbol*
length	metre	m
mass	kilogram	kg
time	second	s
electric current	ampere	A
thermodynamic temperature	kelvin	K
luminous intensity	candela	cd
amount of substance	mole	mol

The second class of SI units contains derived units, *i.e.,* units which can be formed by combining base units according to the algebraic relations linking the corresponding quantities. Several of these algebraic expressions in terms of base units can be replaced by special names and symbols can themselves be used to form other derived units.

Derived units may, therefore, be classified under three headings. Some of them are given in Tables 2, 3 and 4.

Table 2. Examples of SI Derived Units Expressed in Terms of Base Units

Quantity	*SI Units*	
	Name	*Symbol*
area	square metre	m^2
volume	cubic metre	m^3
speed, velocity	metre per second	m/s
acceleration	metre per second squared	m/s^2
wave number	1 per metre	m^{-1}
density, mass density	kilogram per cubic metre	kg/m^3
concentration (of amount of substance)	mole per cubic metre	mol/m^3
activity (radioactive)	1 per second	s^{-1}
specific volume	cubic metre per kilogram	m^3/kg
luminance	candela per square metre	cd/m^2

Table 3. SI Derived Units with Special Names

Quantity	*SI Units*			
	Name	*Symbol*	*Expression in terms of other units*	*Expression in terms of SI base units*
frequency	hertz	Hz	—	s^{-1}
force	newton	N	—	$m.kg.s^{-2}$
pressure	pascal	Pa	N/m^2	$m^{-1}.kg.s^{-2}$
energy, work, quantity of heat power	joule	J	N.m	$m^2.kg.s^{-2}$
radiant flux quantity of electricity	watt	W	J/S	$m^2.kg.s^{-3}$
electric charge	coulomb	C	A.s	s.A
electric tension, electric potential	volt	V	W/A	$m^2.kg.s^{-3}.A^{-1}$
capacitance	farad	F	C/V	$m^{-2}.kg^{-1}.s^4$
electric resistance	ohm	Ω	V/A	$m^2.kg.s^{-3}.A^{-2}$
conductance	siemens	S	A/V	$m^{-2}.kg^{-1}.s^3.A^2$
magnetic flux	weber	Wb	V.S.	$m^2.kg.s^{-2}.A^{-1}$
magnetic flux density	tesla	T	Wb/m^2	$kg.s^{-2}.A^{-1}$
inductance	henry	H	Wb/A	$m^2.kg.s^{-2}.A^{-2}$
luminous flux	lumen	lm	—	cd.sr
illuminance	lux	lx	—	m^{-2}.cd.sr

Table 4. Examples of SI Derived Units Expressed by Means of Special Names

Quantity	*SI Units*		
	Name	*Symbol*	*Expression in terms of SI base units*
dynamic viscosity	pascal second	Pa-s	$m^{-1}.kg.s^{-1}$
moment of force	metre newton	N.m	$m^2.kg.s^{-2}$
surface tension	newton per metre	N/m	$kg.s^{-2}$
heat flux density, irradiance	watt per square metre	W/m^2	$kg.s^{-3}$
heat capacity, entropy	joule per kelvin	J/K	$m^2.kg.s^{-2}.K^{-1}$
specific heat capacity, specific entropy	joule per kilogram kelvin	J/(kg.K)	$m^2.s^{-2}.K^{-1}$
specific energy	joule per kilogram	J/kg	$m^2.s^{-2}$
thermal conductivity	watt per metre kelvin	W/(m.K)	$m.kg.s^{-3}.K^{-1}$
energy density	joule per cubic metre	J/m^3	$m^{-1}.kg.s^{-2}$
electric field strength	volt per metre	V/m	$m.kg.s^{-3}.A^{-1}$
electric charge density	coulomb per cubic metre	C/m^3	m^{-3}.s.A
electric flux density	coulomb per square metre	C/m^2	m^{-2}.s.A
permitivity	farad per metre	F/m	$m^{-3}.kg^{-1}.s^4$
current density	ampere per square metre	A/m^2	—
magnetic field strength	ampere per metre	A/m	—
permeability	henry per metre	H/m	$m.kg.s^{-2}.A^{-2}$
molar energy	joule per mole	J/mol	$m^2.kg.s^{-2}mol^{-1}$
molar heat capacity	joule per mole kelvin	J/(mol.K)	$m^2.kg.s^{-2}.K^{-1}.mol^{-1}$

The SI units assigned to third class called "Supplementary units" may be regarded either as base units or as derived units. Refer to Tables 5 and 6.

Table 5. SI Supplementary Units

Quantity	*SI Units*	
	Name	*Symbol*
plane angle	radian	rad
solid angle	steradian	sr

Table 6. Examples of SI Derived Units Formed by Using Supplementary Units

Quantity	*SI Units*	
	Name	*Symbol*
angular velocity	radian per second	rad/s
angular acceleration	radian per second squared	rad/s^2
radiant intensity	watt per steradian	W/sr
radiance	watt per square metre steradian	$W\text{-}m^{-2}.sr^{-1}$

Table 7. SI Prefixes

Factor	*Prefix*	*Symbol*	*Factor*	*Prefix*	*Symbol*
10^{12}	tera	T	10^{-1}	deci	d
10^{9}	giga	G	10^{-2}	centi	c
10^{6}	mega	M	10^{-3}	milli	m
10^{3}	kilo	k	10^{-6}	micro	μ
10^{2}	hecto	h	10^{-9}	nano	n
10^{1}	deca	da	10^{-12}	pico	p
			10^{-15}	fasnto	f
			10^{-18}	atto	a

B. CONVERSION FACTORS

1. Force :

1 newton = $kg\text{-}m/sec^2$ = 0.012 kgf

1 kgf = 9.81 N

2. Pressure :

1 bar = 750.06 mm Hg = 0.9869 atm

= 10^5 N/m^2 = 10^3 $kg/m\text{-}sec^2$

1 N/m^2 = 1 pascal = 10^{-5} bar

= 10^{-2} $kg/m\text{-}sec^2$

1 atm = 760 mm Hg = 1.03 kgf/cm^2

= 1.01325 bar = 1.01325 × 10^5 N/m^2

3. Work, Energy or Heat :

1 joule = 1 newton metre = 1 watt-sec

= 2.7778 × 10^{-7} kWh = 0.239 cal

= 0.239 × 10^{-3} kcal

1 cal = 4.184 joule = 1.1622 × 10^{-6} kWh

1 kcal = 4.184 × 10^3 joule = 427 kgf-m

= 1.1622 × 10^{-3} kWh

1 kWh = 8.6042 × 10^5 cal = 860 kcal

= 3.6 × 10^6 joule

$$1 \text{ kgf-m} = \left(\frac{1}{427}\right) \text{kcal} = 9.81 \text{ joules}$$

4. Power :

1 watt = 1 joule/sec = 0.860 kcal/h

1 h.p. = 75 m kgf/sec = 0.1757 kcal/sec

= 735.3 watt

1 kW = 1000 watts = 860 kcal/h

5. Specific heat :

1 kcal/kg-°K = 0.4184 joules/kg-K

6. Thermal conductivity :

1 watt/m-K = 0.8598 kcal/h-m-°C

1 kcal/h-m-°C = 1.163 watt/m-K

= 1.163 joules/s-m-K.

7. Heat transfer co-efficient :

1 $watt/m^2\text{-}K$ = 0.8598 $kcal/m^2\text{-}h\text{-}°C$

1 $kcal/m^2\text{-}h\text{-}°C$ = 1.163 $watt/m^2\text{-}K$.

C. IMPORTANT ENGINEERING CONSTANTS AND EXPRESSIONS

Engineering constants and expressions	*M.K.S. system*	*SI Units*
1. Value of g_0	9.81 kg-m/kgf-sec^2	1 kg-m/N-sec^2
2. Universal gas constant	848 kgf-m/kg mole-°K	848 × 9.81 = 8314 J/kg-mole-K ($\because$ 1 kgf-m = 9.81 joules)
3. Gas constant (R)	29.27 kgf-m/kg-°K (for air)	$\frac{8314}{29}$ = 287 joules/kg-K for air
4. Specific heats (for air)	c_v = 0.17 kcal/kg-°K c_p = 0.24 kcal/kg-°K	c_v = 0.17 × 4.184 = 0.71128 kJ/kg-K c_p = 0.24 × 4.184 = 1 kJ/kg-K
5. Flow through nozzle-Exit velocity (C_2)	$91.5\sqrt{U}$, where U is in kcal	$44.7\sqrt{U}$, where U is in kJ
6. Refrigeration 1 ton	= 50 kcal/min	= 210 kJ/min
7. **Heat transfer** The Stefan-Boltzmann Law is given by :	$Q = \sigma T^4$ kcal/m^2-h where $\sigma = 4.9 \times 10^{-8}$ kcal/h-m^2-°K^4	$Q = \sigma T^4$ watts/m^2-h where $\sigma = 5.67 \times 10^{-8}$ W/m^{2}K^4

Nomenclature

A	area
b	steady-flow availability function
C	velocity
°C	temperature on the celsius (or centigrade) scale
c	specific heat
c_p	specific heat at constant pressure
c_v	specific heat at constant volume
C_p	molar heat at constant pressure
C_v	molar heat at constant volume
D, d	bore ; diameter
E	emissive power ; total energy
e	base of natural logarithms
g	gravitational acceleration
H	enthalpy
h	specific enthalpy ; heat transfer co-efficient
h_f	specific enthalpy of saturated liquid (fluid)
h_{fg}	latent heat
h_g	specific enthalpy of saturated vapour ; gases
K	temperature on kelvin scale (*i.e.*, celsius absolute, compressibility)
k	thermal conductivity, blade velocity co-efficient
L	stroke
M	molecular weight
m	mass
$\dot{m}$	rate of mass flow
N	rotational speed
n	polytropic index, number of moles ; number of cylinders
P	power
p	absolute pressure
p_m	mean effective pressure
p_i	indicated mean effective pressure
p_b	brake mean effective pressure, back pressure
Q	heat, rate of heat transfer
q	rate of heat transfer per unit area
R	gas constant ; thermal resistance ; radius ; total expansion ratio in compound steam engines
R_0	universal gas constant
r	radius, expansion ratio, compression ratio
S	entropy
s	specific entropy
T	absolute temperature ; torque
t	temperature
U	internal energy ; overall heat transfer co-efficient
u	specific internal energy
V	volume
v	specific volume
W	work ; rate of work transfer ; brake load ; weight
w	specific weight ; velocity of whirl
x	dryness fraction ; length.

Greek Symbols

α	absorptivity
γ	ratio of specific heats, c_p/c_v
$\in$	emissivity ; effectiveness
η	efficiency
θ	temperature difference, angle
ρ	density
σ	Stefan-Boltzmann constant
ϕ	relative humidity, angle.

Kinetic Theory of Gases and Basic Concepts of Thermodynamics

1

1.1. INTRODUCTION TO KINETIC THEORY OF GASES

The kinetic theory of gases deals with the behaviour of molecules constituting the gas. According to this theory, the molecules of all gases are in continuous motion. As a result of this they possess kinetic energy which is transferred from molecule to molecule during their collision. The energy so transferred produces a change in the velocity of individual molecules.

The complete phenomenon of molecular behaviour is quite complex. The *assumptions* are therefore made to simplify the application of theory of an ideal gas.

Assumptions :

1. The molecules of gases are assumed to be rigid, perfectly elastic solid spheres, identical in all respects such as mass, form etc.
2. The mean distance between molecules is very large compared to their own dimensions.
3. The molecules are in state of random motion moving in all directions with all possible velocities and gas is said to be in state of molecular chaos.
4. The collisions between the molecules are perfectly elastic and there are no intermolecular forces of attraction or repulsion. This means that energy of gas is all kinetic.
5. The number of molecules in a small volume is very large.
6. The time spent in collision is negligible, compared to the time during which the molecules are moving independently.
7. Between collisions, the molecules move in a straight line with uniform velocity because of frictionless motion between molecules. The distance between two collisions is called '*free path*' of the molecule, the average distance travelled by a molecule between successive collision is known as '*mean free path*'.
8. The volume of molecule is so small that it is negligible compared to total volume of the gas.

Pressure exerted by an Ideal Gas

Let us consider a quantity of gas to be contained in a cubical vessel of side l with perfectly elastic wall and N represent the very large number of molecules in the vessel. Now let us consider a molecule which may be assumed to have a velocity C_1 in a certain direction. The velocity can be resolved into three components u_1, v_1, w_1 parallel to three co-ordinate axes X, Y and Z which are again assumed parallel to the sides of the cube as shown in Fig. 1.1.

Thus, $C_1^2 = u_1^2 + v_1^2 + w_1^2$.

Let this molecule having mass m strike wall surface $ABCD$ of the cube with velocity u_1. Since the collision is perfectly elastic, the molecule will rebound from this surface with the same velocity u_1. Therefore,

The momentum of the molecule before it strikes the face $ABCD = mu_1$

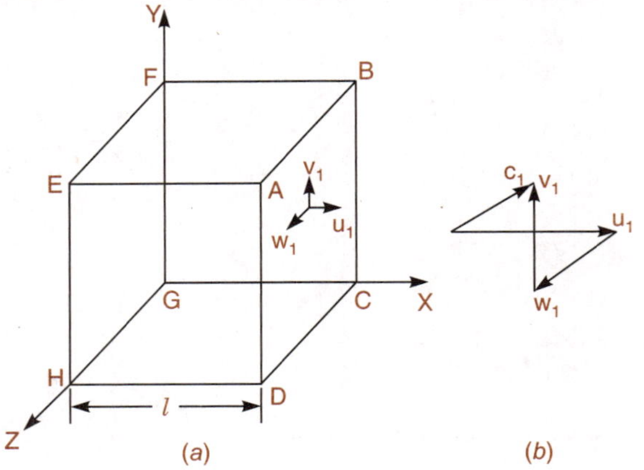

Fig. 1.1

The momentum of the molecule after impact $= -mu_1$.

Hence change of momentum at each impact in direction normal to the surface

$$ABCD = mu_1 - (-mu_1) = 2mu_1$$

After striking the surface *ABCD*, the molecule rebounds and travels back to the face *EFGH*, collides with it and travels back again to the face *ABCD* covering $2l$ distance. This means molecule covers $2l$ distance to hit the same face again. Hence the time taken by the same molecule to strike the same face *ABCD* again is $\frac{2l}{u_1}$.

Therefore, the rate of change of momentum for one molecule of the gas $= \frac{2mu_1}{\frac{2l}{u_1}} = \frac{mu_1^2}{l}$

According to Newton's second law of motion the rate of change of 'momentum is the force'. If F_1 is the force due to one molecule, then $F_1 = \frac{mu_1^2}{l}$

Similarly, then force F_2 due to the impact of another molecule having velocity C_2 whose components are u_2, v_2, w_2 is given by $F_2 = \frac{mu_2^2}{l}$

Hence total force F_x on the face *ABCD* due to impact of *N* molecules is given by $F_x = \frac{m}{l}(u_1^2 + u_2^2 + \ldots\ldots u_N^2)$

Since the pressure (p) is the force per unit area, hence pressure exerted on the wall *ABCD* is given by

$$p_x = \frac{F_x}{l^2} = \frac{m}{l^3}(u_1^2 + u_2^2 + \ldots\ldots u_N^2)$$

Similarly, if p_y and p_z represent the pressures on other faces which are perpendicular to the *Y* and *Z*-axes respectively, we have

$$p_y = \frac{m}{l^3}(v_1^2 + v_2^2 + \ldots\ldots v_N^2)$$

and $$p_z = \frac{m}{l^3}(w_1^2 + w_2^2 + \ldots\ldots w_N^2)$$

Since pressure exerted by the gas is the same in all directions, *i.e.*, $p_x = p_y = p_z$ the average pressure p of the gas is given by

$$p = \frac{p_x + p_y + p_z}{3}$$

or $$p = \frac{m}{l^3}[(u_1^2 + v_1^2 + w_1^2) + (u_2^2 + v_2^2 + w_2^2) + \ldots\ldots(u_N^2 + v_N^2 + w_N^2)$$

But $$C_1^2 = (u_1^2 + v_1^2 + w_1^2)$$
$$C_2^2 = (u_2^2 + v_2^2 + w_2^2) \text{ and so on}$$
$$l^3 = V = \text{volume of gas } (m^3)$$

$$\therefore \quad p = \frac{1}{3}\frac{m}{v}(C_1^2 + C_2^2 + C_3^2 + \ldots\ldots C_N^2)$$

or $$p = \frac{1}{3}\frac{m}{v}N\overline{C}^2 \qquad \ldots(1.1)$$

where $\overline{C}^2 = \left(\frac{C_1^2 + C_2^2 + C_3^2 + \ldots\ldots C_N^2}{N}\right)$ known as mean square velocity

or $$\overline{C} = \sqrt{\frac{C_1^2 + C_2^2 + C_3^2 + \ldots\ldots C_N^2}{N}}$$

where $\overline{C}$ is called the root *mean square velocity* of the molecules and equal to the square root of the mean of square of velocities of individual molecules which is evidently not the same as mean of velocities of different molecules

$$\left[i.e., C_{mean} = \frac{C_1 + C_2 + C_3 + \ldots\ldots C_N}{N}\right]$$

or $$pV = \frac{1}{3}m\,N\overline{C}^2 \qquad \ldots(1.2)$$

This *equation is the fundamental equation of kinetic theory of gases* and is often referred to as ***kinetic equation of gases.***

Equation (1.2) may be written as

$$pV = 2/3 \times 1/2\ m\ N\overline{C}^2$$

where $\frac{1}{2}mN\overline{C}^2$ is the *average transmission or linear kinetic energy of the system of particles.*

Equation (1.1) can be written as

$$p = 1/3\ \rho\ \overline{C}^2 \qquad \ldots(1.3)$$

where ρ is the density. $\left[\because \quad \rho = \frac{mN}{V}, i.e., \frac{\text{Total mass}}{\text{Total volume}}\right]$

This equation expresses the pressure which any volume of gas exerts in terms of its density under the prevailing conditions and its mean square molecular speed.

From equations (1.2) and (1.3),

$$\overline{C} = \sqrt{\frac{3p}{\rho}} = \sqrt{\frac{3pV}{mN}}$$

Kinetic interpretation of temperature :

If V_{mol} is the volume occupied by a gram molecule of a gas and N_0 is the number of moles in one gram molecule of gas,

$$M = \text{molecular weight} = mN_0. \quad ...(i)$$

Since $p\ V_{mol} = R_0T$ ......Molar gas equation ...(*ii*)

From equations (1.2) and (*ii*),

$$1/3\ m\ N_0\ \bar{C} = R_0T$$

R_0 = Universal gas constant

$$\text{or, } 2/3 \times \tfrac{1}{2}m\ N_0\ \bar{C}^2 = R_0T$$

N_0 = Avogadro's number

$$\text{or, } \tfrac{1}{2}\ m\bar{C}^2 = 3/2\ KT \quad ...(1.4)$$

$\dfrac{R_0}{N_0} = K$ (Boltzman's constant)

(*i.e.*, K.E. per molecule = 3/2 *KT*)

$$\text{or, } \quad \bar{C} = \sqrt{\frac{3KT}{m}}$$

$$\text{or, } \quad \bar{C} = \sqrt{\frac{3R_0T}{M}}$$

$$\because \frac{K}{m} = \frac{R_0}{N_0\,m} = \frac{R_0}{M}$$

$$\text{or, } \quad \bar{C} = \sqrt{3RT} \quad ...(1.5)$$

$$\because R = \frac{R_0}{M}$$

where R is characteristic gas constant.

From equation (1.4) it is seen that *temperature is a measure of the average kinetic energy of translation possessed by molecule.* It is known as the ***kinetic interpretation of temperature.*** Hence, the absolute temperature of a gas is proportional to the mean translational kinetic energy of the molecules it consists. If the temperature is fixed, then the average K.E. of the molecules remains constant despite encounters.

1.2. DEFINITION OF THERMODYNAMICS

Thermodynamics may be *defined* as follows :

Thermodynamics is an axiomatic science which deals with the relations among heat, work and properties of system which are in equilibrium. It describes state and changes in state of physical systems.

Or

Thermodynamics is *the science of the regularities governing processes of energy conversion.*

Or

Thermodynamics is *the science that deals with the interaction between energy and material systems.*

Thermodynamics, basically entails *four laws* or axioms known as Zeroth, First, Second and Third law of thermodynamics.

- The ***First law*** throws light on *concept of internal energy*.
- The ***Zeroth law*** deals with *thermal equilibrium* and establishes a *concept of temperature*.
- The ***Second law*** indicates the limit of *converting heat into work* and introduces the *principle of increase of entropy*.
- The ***Third law*** defines the *absolute zero of entropy*.

These laws are based on experimental observations and have no *mathematical proof.* Like all physical laws, these laws are based on *logical reasoning*.

1.3. THERMODYNAMIC SYSTEMS

1.3.1. System, Boundary and Surroundings

System. A system is *a finite quantity of matter or a prescribed region of space* (Refer to Fig. 1.2).

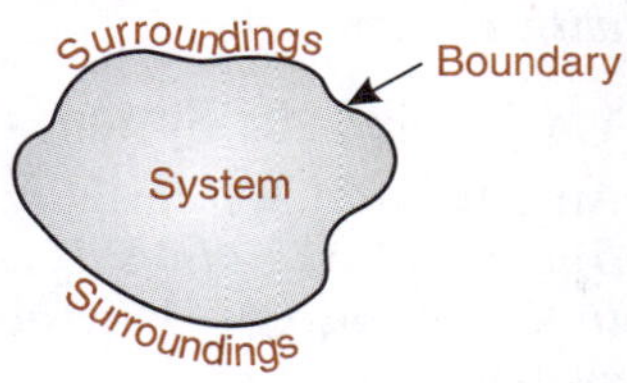

Fig. 1.2. The system.

Boundary. The *actual or hypothetical envelope enclosing the system* is the boundary of the system. The boundary may be fixed or it may move, as and when a system containing a gas is compressed or expanded. The boundary may be *real* or *imaginary.* It is not difficult to envisage a real boundary but an example of imaginary boundary would be one drawn around a system consisting of the fresh mixture about to enter the cylinder of an I.C. engine together with the remanants of the last cylinder charge after the exhaust process (Refer to Fig. 1.3).

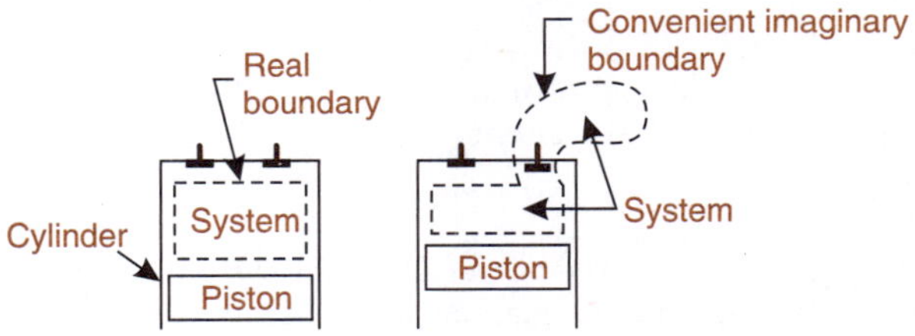

Fig. 1.3. The real and imaginary boundaries.

1.3.2. Closed System

Refer to Fig. 1.4. *If the boundary of the system is impervious to the flow of matter, it is called* a ***closed system.*** An *example* of this system is mass of gas or vapour contained in an engine cylinder, the boundary of which is drawn by the cylinder walls, the cylinder head and piston crown. Here the *boundary is continuous and no matter may enter or leave.*

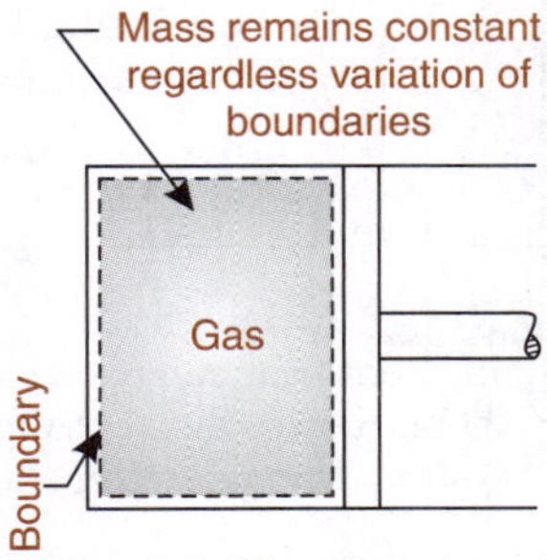

Fig. 1.4. Closed system.

1.3.3. Open System

Refer to Fig. 1.5. An open system is one in which *matter flows into or out of the system*. Most of the engineering systems are open.

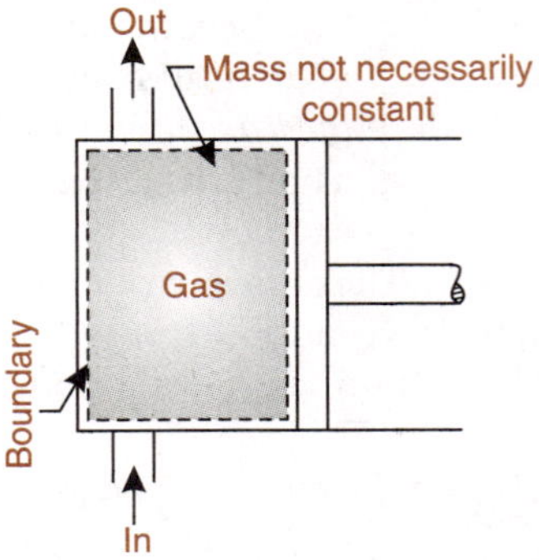

Fig. 1.5. Open system.

1.3.4. Isolated System

An isolated system is that system *which exchanges neither energy nor matter with any other system or with environment*.

1.3.5. Adiabatic System

An adiabatic system is one *which is thermally insulated from its surroundings*. It can, however, *exchange work with its surroundings*. If it does not, it becomes an isolated system.

Phase. A phase is a quantity of matter which is homogeneous throughout in chemical composition and physical structure.

1.3.6. Homogeneous System

A system which consists of a single phase is termed as *homogeneous system*. Examples : Mixture of air and water vapour, water plus nitric acid and octane plus heptane.

1.3.7. Heterogeneous System

A system which consists of two or more phases is called a *heterogeneous system*. Examples : Water plus steam, ice plus water and water plus oil.

1.4. MACROSCOPIC AND MICROSCOPIC POINTS OF VIEW

Thermodynamic studies are undertaken by the following two different approaches :

1. Macroscopic approach—(*Macro* mean *big* or *total*)
2. Microscopic approach—(*Micro* means *small*)

These approaches are discussed (in a comparative way) below :

S. No.	*Macroscopic approach*	*Microscopic approach*
1.	In this approach a certain quantity of matter is considered *without* taking into account the events occurring at molecular level. In other words this approach to thermodynamics is concerned with *gross or overall behaviour*. This is known as *classical thermodynamics*.	The approach considers that the system is made up of a very large number of discrete particles known as *molecules*. These molecules have different velocities and energies. The values of these energies are constantly changing with time. This approach to thermodynamics which is concerned directly with the *structure of the matter* is known as *statistical thermodynamics*.
2.	The analysis of macroscopic system requires simple mathematical formulae.	The behaviour of the system is found by using statistical methods as the number of molecules is very large. So advanced statistical and mathematical methods are needed to explain the changes in the system.
3.	The values of the properties of the system are their average values. For example, consider a sample of a gas in a closed container. The *pressure* of the gas is the average value of the pressure exerted by millions of individual molecules. Similarly the *temperature* of this gas is the average value of translational kinetic energies of millions of individual molecules. These properties like *pressure* and *temperature* can be measured very easily. *The changes in properties can be felt by our senses.*	The properties like *velocity, momentum*, impulse, kinetic energy, force of impact etc. which describe the molecule *cannot be easily measured by instruments. Our senses cannot feel them.*
4.	In order to describe a system only a few properties are needed.	Large number of variables are needed to describe a system. So the approach is complicated.

Note. Although the macroscopic approach seems to be different from microscopic one, there exists a relation between them. Hence when both the methods are applied to a particular system, they give the same result.

1.5. PURE SUBSTANCE

A pure substance is one that has a homogeneous and invariable chemical composition even though there is a change of phase. In other words, it is a system which is (*a*) homogeneous in composition, (*b*) homogeneous in chemical aggregation. Examples : Liquid, water, mixture of liquid water and steam, mixture of ice and water. The mixture of liquid air and gaseous air is not a pure substance.

1.6. THERMODYNAMIC EQUILIBRIUM

A system is in *thermodynamic equilibrium* if the temperature and pressure at all points are same; there should be no velocity gradient; the chemical equilibrium is also necessary. Systems under temperature and pressure equilibrium but not under chemical equilibrium are sometimes said to be in metastable equilibrium conditions. *It is only under thermodynamic equilibrium conditions that the properties of a system can be fixed.*

Thus for attaining a state of *thermodynamic equilibrium* the following *three* types of equilibrium states must be achieved :

1. **Thermal equilibrium.** The temperature of the system does not change with time and has same value at all points of the system.

2. **Mechanical equilibrium.** There are no unbalanced forces within the system or between the surroundings. The pressure in the system is same at all points and does not change with respect to time.

3. **Chemical equilibrium.** No chemical reaction takes place in the system and the chemical composition which is same throughout the system does not vary with time.

1.7. PROPERTIES OF SYSTEMS

A property of a system is a characteristic of the system which depends upon its state, but not upon how the state is reached. There are two sorts of property :

1. **Intensive properties.** These properties *do not depend on the mass of the system. Examples* : Temperature and pressure.

2. **Extensive properties.** These properties *depend on the mass of the system. Example* : Volume. Extensive properties are often divided by mass associated with them to obtain the intensive properties. For example, if the volume of a system of mass m is V, then the specific volume of matter within the system is $\frac{V}{m} = v$, which is an intensive property.

1.8. STATE

***State** is the condition of the system at an instant of time as described or measured by its properties. Or each unique condition of a system is called a **state**.*

It follows from the definition of state that each property has a single value at each state. Stated differently, all properties are *state* or *point functions*. Therefore, all properties are identical for identical states.

On the basis of the above discussion, we can determine if a given variable is *property* or not by applying the following *tests* :

— *A variable is a property, if and only if, it has a single value at each equilibrium state.*

— A variable is a property, if and only if, the change in its value between any two prescribed equilibrium states is single-valued.

Therefore, *any variable whose change is fixed by the end states is a property*.

1.9. PROCESS

A process occurs when the system undergoes a change in a state or an energy transfer at a steady state. A process may be *non-flow* in which a fixed mass within the defined boundary is undergoing a change of state. *Example* : A substance which is being heated in a closed cylinder undergoes a ***non-flow process*** (Fig. 1.4). *Closed systems undergo non-flow processes.* A process may be a ***flow process** in which mass is entering and leaving through the boundary of an open system*. In a steady flow process (Fig. 1.5) mass is crossing the boundary from surroundings at entry, and an equal mass is crossing the boundary at the exit so that the total mass of the system remains constant. In an open system it is necessary to take account of the work delivered from the surroundings to the system at entry to cause the mass to enter, and also of the work delivered from the system at surroundings to cause the mass to leave, as well as any heat or work crossing the boundary of the system.

Quasi-static process. Quasi means *'almost'*. A quasi-static process is also called a *reversible process. This process is a succession of equilibrium states and infinite slowness is its characteristic feature.*

1.10. CYCLE

Any process or series of processes whose end states are identical is termed a **cycle**. The processes through which the system has passed can be shown on a state diagram, but a complete section of the path requires in addition a statement of the heat and work crossing the boundary of the system. Fig. 1.6 shows such a cycle in which a system commencing at condition '1' changes in pressure and volume through a path 123 and returns to its initial condition '1'.

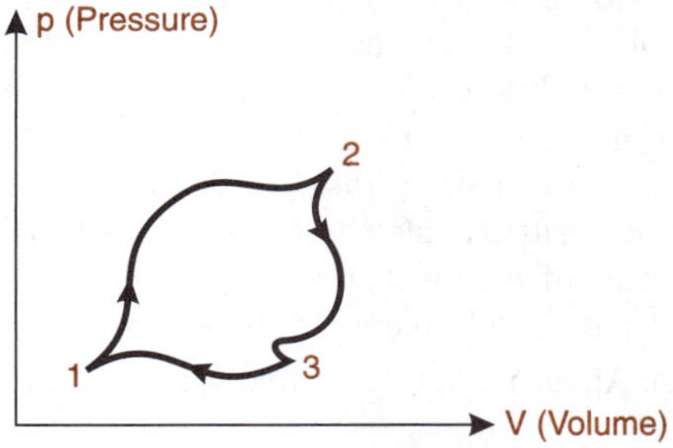

Fig. 1.6. Cycle of operations.

1.11. POINT FUNCTION

When two properties locate a point on the graph (co-ordinate axes) then those properties are called as **point function.**

Examples. Pressure, temperature, volume etc.

$$\int_1^2 dV = V_2 - V_1 \text{ (an } \textit{exact differential}\text{).}$$

1.12. PATH FUNCTION

There are certain quantities which cannot be located on a graph by a *point* but are given by the *area* or so, on that graph. In that case, the area on the graph, pertaining to the particular process, *is a function of the path of the process.* Such quantities are called **path functions.**

Examples. Heat, work etc.

Heat and work are *inexact differentials*. Their change cannot be written as difference between their end states.

Thus $\int_1^2 \delta Q \neq Q_2 - Q_1$ and is shown as ${}_1Q_2$ or Q_{1-2}

Similarly $\int_1^2 \delta W \neq W_2 - W_1$, and is shown as ${}_1W_2$ or W_{1-2}

Note. The operator δ is used to denote inexact differentials and operator *d* is used to denote exact differentials.

1.13. TEMPERATURE

- *The temperature is a thermal state of a body which distinguishes a hot body from a cold body.* The temperature of a body is *proportional to the stored molecular energy i.e.,* the average molecular kinetic energy of the molecules in a system. (A particular molecule does not have a temperature, it has energy. The gas as a system has temperature).
- Instruments for measuring *ordinary temperatures* are known as ***thermometers*** and those for measuring *high temperatures* are known as ***pyrometers.***
- It has been found that a gas will not occupy any volume at a certain temperature. This temperature is known as *absolute zero temperature*. The temperatures measured with absolute zero as basis are called *absolute temperatures*. Absolute temperature is stated in degrees centigrade. The point of absolute temperature is found to occur at 273.15°C below the freezing point of water.

Then, Absolute temperature = Thermometer reading in °C + 273.15.

Absolute temperature can also be represented in degrees kelvin, denoted by K (SI unit).

1.14. ZEROTH LAW OF THERMODYNAMICS

- **'Zeroth law of thermodynamics'** *states that if two systems are each equal in temperature to a third, they are equal in temperature to each other.*

Example. Refer to Fig. 1.7. System '1' may consist of a mass of gas enclosed in a rigid vessel fitted with a pressure gauge. If there is no change of pressure when this system is brought into contact with system '2' a block of iron, then the two systems are equal in temperature (assuming that the systems 1 and 2 do not react each other chemically or electrically). Experiment reveals that if system '1' is brought into contact with a third system '3' again with no change of properties then systems '2' and '3' will show no change in their properties when brought into contact provided they do not react with each other chemically or electrically. Therefore, '2' and '3' must be in equilibrium.

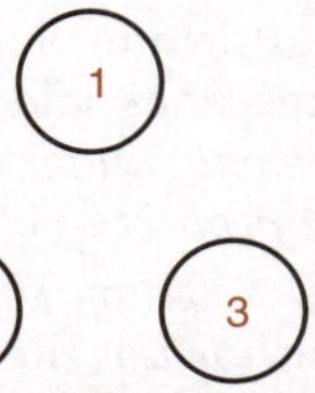

Fig. 1.7. Zeroth law of thermodynamics.

- This law was enunciated by R.H. Fowler in the year 1931. However, since the first and second laws already existed at that time, it was designated as ***zeroth law*** so that it *precedes* the first and second laws *to form a logical sequence.*

1.15. THE THERMOMETER AND THERMOMETRIC PROPERTY

1.15.1. Introduction

- The zeroth law of thermodynamics provides the basis for the measurement of temperature. It enables us to compare temperatures of two bodies '1' and '2' with the help of a third body '3' and say that the temperature of '1' is the same as the temperature of '2' without actually bringing '1' and '2' in thermal contact. In practice, body '3' in the zeroth law is called the **thermometer**. It is brought into thermal equilibrium with a set of standard temperature of a body '2', and is thus calibrated. Later, when any other body '1' is brought in thermal communication with the thermometer, we say that the body '1' has attained equality of temperature with the thermometer, and hence with body '2'. This way, the body '1' has the temperature of body '2' given for example by, say the height of mercury column in the thermometer '3'.

- The height of mercury column in a thermometer, therefore, becomes a **thermometric property**.

There are other methods of temperature measurement which utilize various other properties of materials, that are functions of temperature, as thermometric properties.

Six different kinds of thermometers, and the names of the corresponding thermometric properties employed are given below :

Thermometer	*Thermometric property*
1. Constant volumes gas	Pressure (p)
2. Constant pressure gas	Volume (V)
3. Alcohol or mercury-in-glass	Length (L)
4. Electric resistance	Resistance (R)
5. Thermocouple	Electromotive force (E)
6. Radiation (pyrometer)	Intensity of radiation (I or J)

1.15.2. Measurement of Temperature

Temperature can be depicted as a *thermal state which depends upon the internal or molecular energy of the body.*

1.15.2.1. Temperature Measuring Instruments

These instruments may be classified in two broad categories :

1. **Non-electrical methods :**

(*i*) By using change in volume of a liquid when its temperature is changed.

(*ii*) By using change in pressure of a gas when its temperature is changed.

(*iii*) By using changes in the vapour pressure when the temperature is changed.

2. **Electrical method :**

(*i*) By thermocouples.

(*ii*) By change in resistance of material with change in temperature.

(*iii*) By comparing the colours of filament and the object whose temperature is to be found out.

(*iv*) By ascertaining the energy received by radiation.

The **thermometers** may also be classified as follows :

1. ***Expansion thermometers***

(*i*) Liquid-in-glass thermometers

(*ii*) Bimetallic thermometers.

2. ***Pressure thermometers***

(*i*) Vapour pressure thermometers

(*ii*) Liquid-filled thermometers

(*iii*) Gas-filled thermometers.

3. Thermocouple thermometers

4. Resistance thermometers

5. Radiation pyrometers

6. Optical pyrometers.

1. Expansion Thermometers :

The expansion thermometers make use of the differential expansion of two different substances. Thus in liquid-in-glass thermometers, it is the difference in expansion of liquid and the containing glass. And in bimetallic thermometers, the indication is due to the difference in expansion of the two solids. These thermometers are discussed below :

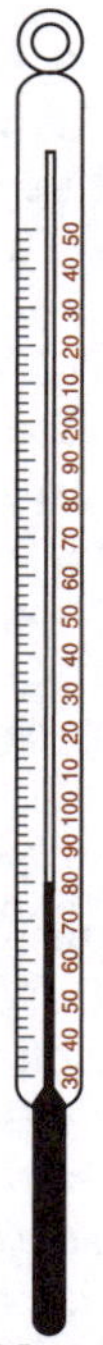

Fig. 1.8. Mercury-in-glass thermometer.

(*i*) **Liquid-in-glass thermometer.** This is a very familiar type of thermometer. The mercury or other liquid fills the glass bulb and extends into the bore of the glass stem. Mercury is the most suitable liquid and is used from – 38.9°C (melting point) to about 600°C. The thermometers employed in the laboratory have the scale engraved directly on the glass stem. A usual type of mercury-in-glass thermometer is shown in Fig. 1.8. An expansion bulb is usually provided at the top of the stem to allow room for expansion of mercury, in case the thermometer is subjected to temperature above its range. The upper limit for mercury-in-glass thermometers is about 600°C. As the upper limit is far above the boiling point of mercury, some inert gas *i.e.,* nitrogen is introduced above the mercury to prevent boiling.

Pentane, ethyl alcohol and toluene are the other liquids which can be used for liquid-in-glass thermometers. Since these liquids are normally colourless a dye is added to facilitate reading. These liquids have a low freezing point as shown below and are suitable for low temperature thermometers.

Liquid	*Boiling point*	*Freezing point*
Pentane	36°C	– 130°C
Ethyl alcohol	78°C	– 100°C
Toluene	110°C	– 92°C

(*ii*) **Bimetallic thermometers.** In a bimetallic thermometer differential expansion of bimetallic strips is used to indicate the temperature. It has the advantage over the liquid-in-glass thermometer, that it is *less fragile* and is *easier to read*. In this type of thermometer two flat strips of different metals are placed side by side and are welded together. Many different metals can be used for this purpose. Generally one is a *low expanding metal* and the other is *high expanding metal*. The bimetal strip is coiled in the form of a spiral or helix. *Due to rise in temperature, the* *curvature of the strip changes. The differential expansion of a strip causes the pointer to move on the dial of the thermometer.*

2. Pressure Thermometers :

In pressure thermometers liquids, gases and vapours can all be used. The principle on which they work is quite simple. The fluid is confined in a closed system. In this case the *pressure is a function of the temperature*, so that when the fluid is heated, the pressure will rise. And the temperature can be indicated by Bourdon type pressure gauge. In general, the thermometer consists of a bulb which contains bulk of the fluid. The bulb is placed in the region whose temperature is required. A capillary tube connects the bulb to a Bourdon tube, which is graduated with a temperature scale.

Pressure thermometers are discussed below :

(*i*) **Vapour pressure thermometer.** A schematic diagram of a vapour pressure thermometer is shown in Fig. 1.9. When the bulb containing the fluid is installed in the region whose temperature is required, some of the fluid vapourizes, and increases the vapour pressure. This change of pressure is indicated on the Bourdon tube. The relation between temperature and vapour pressure of a volatile liquid is of the *exponential form*. Therefore, the scale of a vapour pressure thermometer will not be linear.

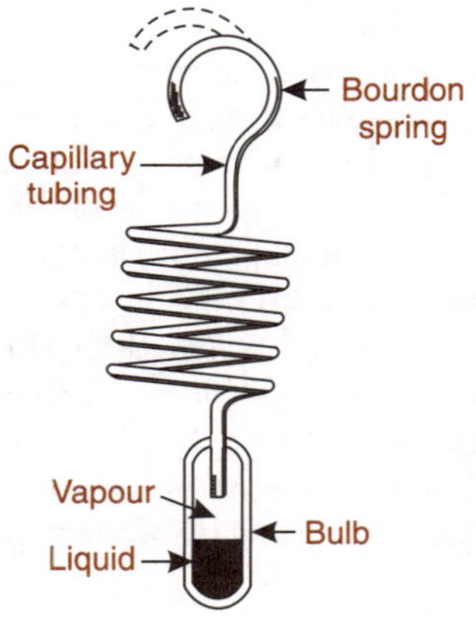

Fig. 1.9. Vapour pressure thermometer.

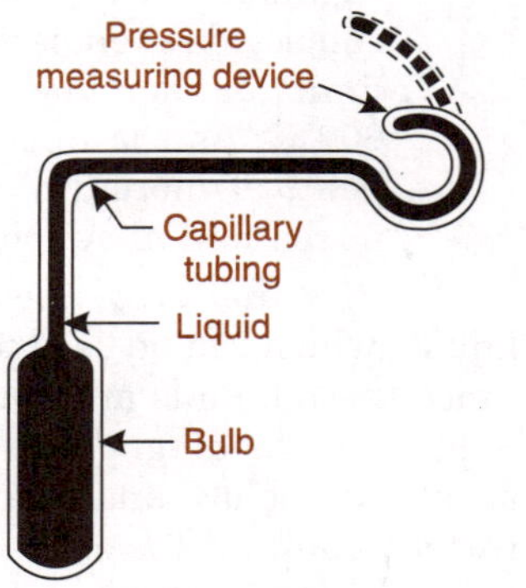

Fig. 1.10. Liquid-filled thermometer.

(*ii*) **Liquid-filled thermometer.** A liquid-filled thermometer is shown in Fig. 1.10. In this case, the *expansion of the liquid causes the pointer to move in the dial*. Therefore liquids having high co-efficient of expansion should be used. In practice many liquids *e.g.,* mercury, alcohol, toluene and glycerine have been successfully used. The operating pressure varies from about 3 to 100 bar. These type of thermometers could be used for a temperature upto 650°C in which mercury could be used as the liquid.

In actual design, the internal diameter of the capillary tube and Bourdon tube is, made much smaller than that of the bulb. This is because the capillary tube is subjected to a temperature which is quite different from that of the bulb. Therefore, to minimise the effect of variation in temperature to which the capillary tube is subjected, the *volume of the bulb is made as large as possible as compared with the volume of the capillary*. However, large volume of bulb tends to increase time lag, therefore, a *compensating device is usually built into the recording or indicating mechanism, which compensates the variations in temperature of the capillary and Bourdon tubes.*

(*iii*) **Gas-filled thermometers.** The temperature range for gas thermometer is practically the same as that of liquid filled thermometer. The gases used in the gas thermometers are *nitrogen* and *helium*. Both these gases are chemically inert, have good values for their co-efficient of expansion and have low specific heats. The construction of this type of thermometer is more or less the same as mercury-thermometer in which Bourdon spring is used. The errors are also compensated likewise. The only difference in this case is that bulb is made much larger than used in liquid-filled thermometers. For good performance the volume of the bulb should be made at least 8 times than that of the rest of the system.

These thermometers are generally used for pressures below 35 bar.

3. Thermocouple Thermometers :

For higher range of temperature *i.e.,* above 650°C, filled thermometers are unsuitable. For higher range of temperature, thermocouples and pyrometers are used.

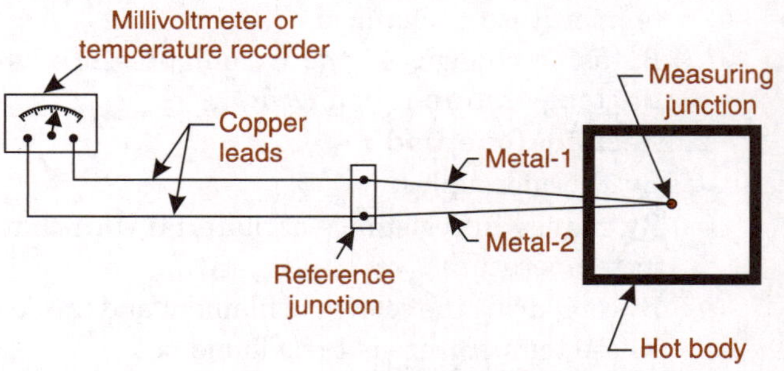

Fig. 1.11. Thermocouple.

In its simplest form a thermocouple consists of *two dissimilar metals or alloys which develop e.m.f. when the reference and measuring junctions are at different temperatures*. The reference junction or cold junction is usually maintained at some constant temperature, such as 0°C. Fig. 1.11, shows a simple circuit of a thermocouple and the temperature measuring device. In many industrial installations the instruments are equipped with automatic compensating devices for temperature changes of the reference junction, thus eliminating the necessity of maintaining this junction at constant temperature.

Table 1.1 gives the composition, useful temperatures range and temperature versus e.m.f. relationship for some commercial thermocouples.

Table 1.1. Composition, Useful Temperature Range and e.m.f. Produced for Some Thermocouples

S.No.	*Thermocouple*	*Composition*	*Temperature (°C)*		*Thermoelectric power*		*Remarks*
			Useful range	*Max.*	*°C*	*Millivolt*	
1.	Platinum vs Platinum-rhodium	Pure platinum vs Pt + 10 or 13% Rh	400 to 1450	1700	0 500 1000 1500	0.0 4.219 9.569 15.498	Used for high temperature measurements
2.	Chromel vs alumel	90% Ni + 10% Cr vs 95% Ni + 5% (Al + Sn) Mn	– 200 to 1200	1450	– 200 0 300 600 900 1200	– 5.75 0.0 12.21 24.90 37.36 48.85	High resistance to oxidation
3.	Iron vs constantan	Pure iron vs 45–60% Cu + 55–40% Ni	– 200 to 750	1000	– 200 0 300 600 900	– 8.27 0.0 16.59 33.27 52.29	—
4.	Copper vs constantan	Pure copper vs Cu–Ni constantan	– 200 to 350	600	– 200 0 200 400	– 5.539 0.0 9.285 20.865	Not suitable in air due to excessive oxidation

4. Resistance Thermometers :

The fact that the electrical resistance of the metals increases with temperature is made use of in resistance thermometers which are purely electrical in nature. *A resistance thermometer is used for precision measurements below 150°C.*

A simple resistance thermometer consists of a *resistance element* or *bulb, electrical loads* and *a resistance measuring or recording instrument*. The resistance element (temperature sensitive element) is usually supplied by the manufacturers with its protecting tube and is ready for electrical connections. The resistance of the metal used as resistance element should be reproducible at any given temperature. *The resistance is reproducible if the composition or physical properties of the metal do not change with temperature. For this purpose platinum is preferred.* A *platinum resistance the rmometer* can measure temperatures to within ± 0.01°C. However, because of high cost of platinum, nickel and copper are used as resistance elements for industrial purposes for low temperatures. The fine resistance wire is wound in a spiral form on a mica frame. The delicate coil is then enclosed in a porcelain or quartz tube. The change of resistance of this unit can be measured by instruments such as Wheatstone bridge, potentiometer or galvanometer.

Advantages :

The resistance thermometers possess the following *advantages* over other devices :

1. A resistance thermometer is very accurate for low ranges below 150°C.
2. It requires no reference junction like thermocouples and as such is more effective at room temperature.
3. The distance between the resistance element and the recording element can be made much larger than is possible with pressure thermometers.
4. It resists corrosion and is physically stable.

Disadvantages :

1. The resistance thermometers cost more.
2. They suffer from time lag.

5. Radiation Pyrometers :

A device which measures the total intensity of radiation emitted from a body is called ***radiation pyrometer.***

The elements of a total radiation pyrometer are illustrated in Fig. 1.12. It collects the radiation from an object (hot body) whose temperature is required. A mirror is used to focus this radiation on a thermocouple. This energy which is concentrated on the thermocouple raises its temperature, and in turn generates an e.m.f. This e.m.f. is then measured either by the galvanometer or potentiometer method. Thus *rise of temperature is a function of the amount of radiation emitted from the object.*

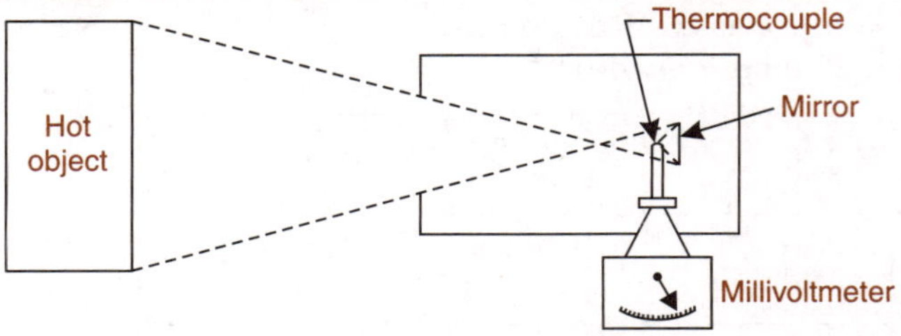

Fig. 1.12. A schematic diagram of radiation pyrometer.

Advantages of the pyrometers

1. The temperatures of moving objects can be measured.
2. A higher temperature measurement is possible than that possible by thermocouples etc.
3. The average temperatures of the extended surface can be measured.
4. The temperature of the objects which are not easily accessible can be measured.

6. Optical Pyrometers :

An optical pyrometer works on the principle *that matters glow above 480°C and the colour of visible radiation is proportional to the temperature of the glowing matter*. The amount of light radiated from the glowing matter (solid or liquid) is measured and employed to determine the temperature.

Fig. 1.13 shows a *disappearing filament pyrometer*.

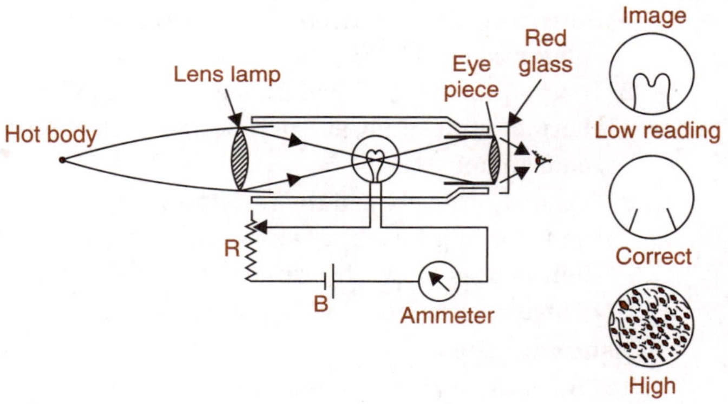

Fig. 1.13. Optical pyrometer.

Operation :

- The optical pyrometer is sighted at the hot body and focused.
- In the beginning filament will appear dark as compared to the background which is bright (being hot).
- By varying the resistance (R) in the filament circuit more and more current is fed into it, till filament becomes equally bright as the background and hence disappears.
- The current flowing in the filament at this stage is measured with the help of an ammeter which is calibrated directly in terms of temperature.
- If the filament current is further increased, the filament appears brighter as compared to the background which then looks dark.
- An optical pyrometer can measure temperatures ranging from 700 to 4000°C.

Table 1.2 gives the summary of temperature range of different instruments.

Table 1.2. Summary of Temperature Range of Different Instruments

Types of instruments	*Liquid in glass thermometer*	*Bimetallic thermometer*	*Pressure thermometers*			*Thermocouples*				*Resistance thermometers*			*Radiation pyrometers*	*Optical pyrometers*
	Mercury		Vapour pressure	Liquid filled	Gas filled	Iron-Constantan	Copper-Constantan	Chromel-Alumel	Platinum-Platinum-Rhodium	Copper	Nickel	Platinum	–	–
Low temperature limit	– 39°C	– 75°C	– 250°C	– 87°C	– 269°C	– 196°C	– 250°C	–184°C	0°C	– 140°C	– 184°C	– 184°C	About room temp.	760°C
High temperature limit	600°C	540°C	316°C	650°C	540°C	760°C	350°C	1260°C	1540°C	120°C	316°C	760°C	As high as desired	As high as desired
Remark	Low price ; Fairly accurate ; Easily broken	Easier to read than Liquid-in-glass thermometers	Normally scale non-linear	To be compensated for variation in ambient temperatures	Requires large bulb ; Linear scale	Widely used in Industry	Constantan is 60% Cu 40% Ni	Chromel is 20% ; Cr 80% ; Ni-Alumel is 2% Al, 98% Ni,	Expensive	Very sensitive. Suitable for narrow range instruments			–	–

1.15.3. The International Practical Temperature Scale

For the calibration of thermometric instruments the Seventh General Conference on Weight and Measures held in 1927 formulated a convenient scale known as the *International Practical Temperature Scale*. It was revised at Thirteenth General Conference in 1968. It consists of *reproducible reference temperatures or primary fixed points* defined by a number of pure substances with assigned values of temperatures determined with precision on ideal or perfect gas temperature scale as given in Table 1.3.

It is stated here that :

— The *triple point* represents an equilibrium state between solid, liquid and vapour phases of a substance.

— *Normal boiling point* is the temperature at which the substance boils at standard atmospheric pressure of 760 mm Hg.

— *Normal freezing point* is the solidification or the melting point temperature of the substance at standard atmospheric pressure.

Table 1.3. Fixed Points of the International Practical Temperature Scale of 1968

Equilibrium state	*Assigned value of temperature*	
	T, K	*t°C*
1. Triple point of hydrogen	13.81	– 259.34
2. Boiling point of hydrogen at 33.306 kPa	17.042	– 266.108
3. Normal boiling point of hydrogen	20.28	– 252.87
4. Normal boiling point of neon	27.102	– 246.048
5. Triple point of oxygen	54.361	– 218.789
6. Normal boiling point of oxygen	90.188	– 182.962
7. Triple point of water	273.16	0.01
8. Normal boiling point of water	373.15	100.00
9. Normal freezing point of antimony (antimony point)	630.74	357.59
10. Normal freezing point of zinc (zinc point)	692.73	419.58
11. Normal freezing point of silver (silver point)	1235.08	961.93
12. Normal freezing point of gold (gold point)	1337.58	1064.43

Based on the available method of measurement, the whole temperature scale may be divided into four ranges. The equations for interpolation for each range are as follows :

1. From – 259.34°C (triple point of hydrogen) to 0°C :

A *platinum resistance thermometer of a standard design* is used and a polynomial of the following form is fitted between the resistance of the wire R_t and temperature t

$$R_t = R_0 (1 + At + Bt^2 + Ct^3) \quad ...(1.6)$$

where R_0 = resistance at the ice point.

2. From 0°C to 630.74°C (Antimony point) :

- It is also based on *platinum resistance thermometer.*
- The diameter of the platinum wire must lie between 0.05 and 0.2 mm.

3. From 630.74°C to 1064.43°C (Gold point) :

- It is based on *standard platinum versus platinum-rhodium thermocouple.*
- Following equation between e.m.f. E and temperature t is employed :

$$E = a + bt + ct^2 \quad ...(1.7)$$

4. Above 1064.43°C :

- It is *based on the intensity of radiation J_T at temperature T emitted by a black body at a wavelength λ in the visible spectrum and by comparing this to the intensity of radiation J_{Au} at the same wavelength emitted by a black body at the gold point.*
- The temperature is calculated from Planck's equation for black body radiation

$$\frac{J_T}{J_{Au}} = \frac{\exp\left(\frac{C_2}{\lambda T_{Au}}\right) - 1}{\exp\left(\frac{C_2}{\lambda T}\right) - 1} \quad ...(1.8)$$

where C_2 = 0.01438 in °C, and λ = wavelength in metres.

Following points are worth noting for gas thermometers :

- *The gas thermometers are never used for the measurement of temperatures.* However, *they are ideal when used for calibration for establishing the ideal gas temperature scale, and for establishing a standard because of precision, reproducible results, and their reading being independent of the thermometric substance used.*
- *The gas thermometers can be used only for temperatures upto which gases do not liquify.*

Method in use before 1954 :

- Celsius and Fahrenheit scales are the two commonly used scales for the measurement of temperature. Symbols C and F are respectively used to denote the readings on these two scales. Until 1954 the temperature scales were based on two fixed points : (*i*) the *steam point* (boiling point of water at standard atmospheric pressure), and (*ii*) the *ice point* (freezing point of water).

The fixed points for these temperature scales are :

Temperature	Celsius scale	Fahrenheit scale
Steam point	100	212
Ice point	0	32
Interval	100	180

- The relation between a particular value C on celsius scale and F on Fahrenheit scale is found to be as mentioned below :

$$\frac{C}{100} = \frac{F-32}{180} \quad \text{or} \quad \frac{C}{5} = \frac{F-32}{9} \qquad ...(1.9)$$

- Further the relation between a temperature difference Δt_F on Fahrenheit scale and Δt_C on celsius scale is

$$\Delta t_F = \frac{180}{100}\Delta t_C = \frac{9}{5}\Delta t_C = 1.8\ \Delta t_C \qquad ...(1.10)$$

- The use of two fixed points was found unsatisfactory and later *abandoned* because of the following *reasons* :

(*i*) It is difficult to achieve equilibrium between pure ice and air-saturated water (since when ice melts, it surrounds itself only with pure water and prevents intimate contact with air-saturated water).

(*ii*) There is extreme sensitiveness of the steam point to the change in pressure.

Method in use after 1954 :

It was suggested by Kelvin that a single fixed point only was necessary to establish a temperature. He pointed out that *triple point of water* (the state at which ice, liquid water and water vapour coexist in equilibrium) could be used as the single point. The tenth CGPM, in 1954, adopted this fixed point, and value was set at 0.01°C or 273.16 K in the Kelvin scale thus established. Correspondingly, the ice point of 0°C on the Celsius scale becomes equal to 273.15 K on the Kelvin scale. Celsius and Kelvin scales are distinguished by using distinct symbols t and T, the relation between these two is then given by :

$$T(K) = t(°C) + 273.15 \qquad ...(1.11)$$

1.15.4. Ideal Gas

From experimental observations it has been established that an ideal gas (to a good approximation) behaves according to the simple equation

$$pV = mRT \qquad ...(1.12)$$

where p, V and T are the pressure, volume and temperature of gas having mass m and R is a constant for the gas known as its ***gas constant.***

Eqn. (1.12) can be written as

$$pv = RT \qquad ...(1.13)$$

(where v = V/m)

In reality there is no gas which can be qualified as an ideal or perfect gas. However *all gases tend to ideal or perfect gas behaviour at all temperatures as their pressure approaches* ***zero pressure.***

For two states of the gas, eqn. (1.12) can be written as:

$$\frac{p_1V_1}{T_1} = \frac{p_2V_2}{T_2} \quad \text{or} \quad \frac{T_2}{T_1} = \frac{p_2}{p_1} \times \frac{V_2}{V_1} \qquad ...(1.14)$$

With the help of this eqn. (1.14), the temperatures can be measured or compared.

1.16. PRESSURE

1.16.1. Definition of Pressure

Pressure is defined as a *force per unit area*. Pressures are exerted by gases, vapours and liquids. The instruments that we generally use, however, record pressure as the difference between two pressures. Thus, it is the *difference between the pressure exerted by a fluid of interest and the ambient atmospheric pressure.* Such devices indicate the pressure either above or below that of the atmosphere. When it is *above the atmospheric pressure*, it is termed *gauge pressure* and is *positive*. When it is *below* atmospheric, it is *negative* and is known as *vacuum*. Vacuum readings are given in millimetres of mercury or millimetres of water below the atmosphere.

It is necessary to establish an absolute pressure scale which is independent of the changes in atmospheric pressure. A pressure of absolute zero can exist only in complete vacuum. *Any pressure measured above the absolute zero of pressure* is termed an '*absolute pressure*'.

A schematic diagram showing the *gauge pressure, vacuum pressure* and the *absolute pressure* is given in Fig. 1.14.

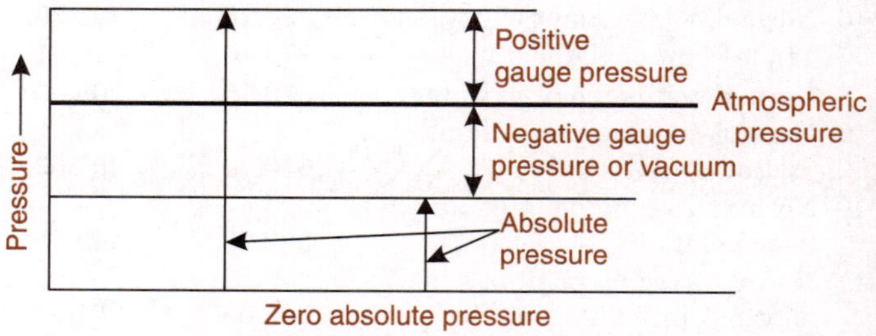

Fig. 1.14. Schematic diagram showing gauge, vacuum and absolute pressures.

Mathematically :

(*i*) Absolute pressure = Atmospheric pressure + Gauge pressure

$$p_{abs.} = p_{atm.} + p_{gauge}.$$

(*ii*) Vacuum pressure = Atmospheric pressure – Absolute pressure.

Vacuum is defined as the *absence of pressure*. A *perfect vacuum* is obtained when *absolute pressure is zero*, at this instant *molecular momentum is zero.*

Atmospheric pressure is measured with the help of barometer.

1.16.2. Unit for Pressure

The fundamental SI unit of pressure is N/m^2 (sometimes called *pascal*, Pa) or bar. 1 bar = 10^5 N/m^2 = 10^5 Pa.

Standard atmospheric pressure = 1.01325 bar = 0.76 m (or 760 mm) Hg.

Low pressures are often expressed in terms of mm of water or mm of mercury. This is an abbreviated way of saying that the pressure is such that which will support a liquid column of stated height.

1.16.3. Types of Pressure Measurement Devices

The pressure may be measured by means of indicating gauges or recorders. These instruments may be mechanical, electro-mechanical, electrical or electronic in operation.

1. **Mechanical instruments.** These instruments may be classified into following two groups :

— The *first group* includes those instruments in which the *pressure* measurement is made by *balancing an unknown force with a known force*.

— The *second group* includes those employing *quantitative deformation of an elastic member for pressure measurement*.

2. **Electro-mechanical instruments.** These instruments usually *employ a mechanical means for detecting the pressure and electrical means for indicating or recording the detected pressure*.

3. **Electronic instruments.** Electronic pressure measuring instruments normally depend on some physical change that can be detected and indicated or recorded electronically.

1.16.4. Mechanical-type Instruments

The mechanical-type instruments are classified as follows :

1. **Manometer gauges**
   (*i*) U-tube manometer
   (*ii*) Cistern manometer
   (*iii*) Micro-manometer etc.
2. **Pressure gauges**
   (*i*) Bourdon tube pressure gauge
   (*ii*) Diaphragm gauge
   (*iii*) Vacuum gauge.

1.16.4.1. Liquid manometers

Low pressures are generally determined by *manometers* which employ liquid columns. It is difficult and costly to construct manometers to measure high pressures, as otherwise the liquid column will become unwieldy and temperature corrections will also be difficult. Their use is, therefore, *restricted to low pressures only*, and for such purposes they are *quite accurate*.

The liquids commonly employed for manometers are mercury and water. *Mercury is used for high and water for low pressures*. For this purpose a liquid is suitable if it has a low viscosity, so that it can adjust itself quickly, and also a low co-efficient of thermal expansion, so that density changes with temperature are minimum.

1. *U-tube manometer :*

A U-tube manometer is in the form of U-tube and is made of glass. When no pressure is applied, the height of the liquid in the two legs is the same. The pressure is then applied to one leg, whilst the other is open to the atmosphere. Under this pressure the liquid will *sink* in this leg and will *rise* in the other. As the other leg is open to the air, therefore, the pressure on this side is known, and is barometric. Now the pressure applied to the first leg can be calculated. This is explained with reference to Fig. 1.15. This consists of a water manometer.

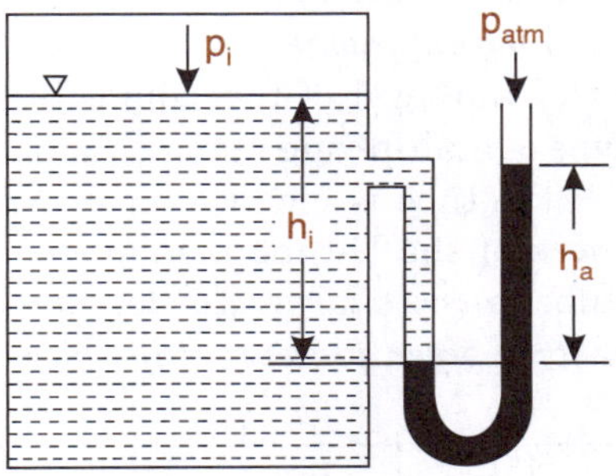

Fig. 1.15. Principle of U-tube manometer.

Considering equilibrium condition, we have

$$p_{\text{atm}} + w_a h_a = p_i + w_i h_i$$

$$\therefore \quad p_i = p_{\text{atm}} + w_a h_a - w_i h_i$$

where $p_{\text{atm.}}$ = Atmospheric pressure,

p_i = Pressure over water surface in the container,

h_a = Height of liquid in U-tube manometer,

h_i = Difference between water surface and lower surface of the liquid in manometer,

w_a = Specific weight of liquid,

w_i = Specific weight of water.

The U-tube manometer shown in Fig. 1.16 is of the simplest form. However, readings have to be taken *at two different places*. Moreover, the *deflection of the two columns may not be the same*. To avoid this difficulty cistern or well type manometer is used.

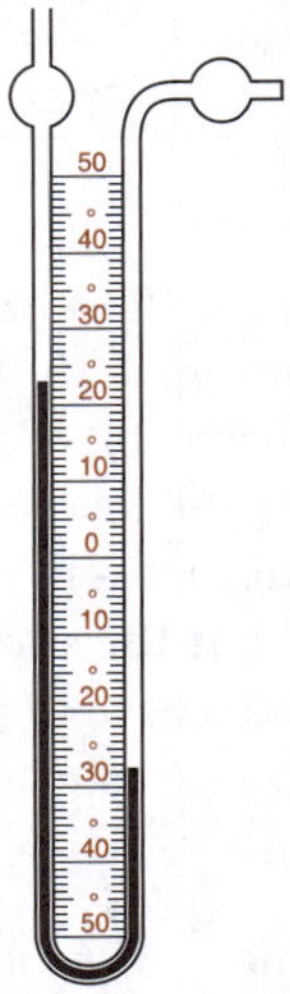

Fig. 1.16. U-tube manometer.

2. *Cistern manometer :*

Fig. 1.17 shows a cistern manometer. The mercury reservoir A is made large enough so that change of level in the reservoir is negligible. This form of manometer is generally used for measuring pressures above atmospheric. In this case, *only one reading of the level in the column is required*. However, a zero setting is necessary.

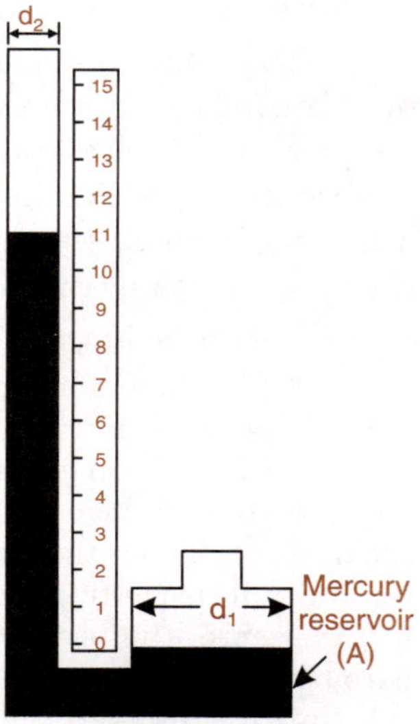

Fig. 1.17. Cistern manometer.

3. *Micro-manometer :*

The U-tube manometer (discussed above) is not very suitable for measuring very low pressures. Therefore, for such purposes, precision types are required. They are called *multiplying* or *micro-manometers*, because they multiply the movement of the level of the liquid. By far the most widely used type of multiplying manometer is the ***inclined manometer.*** If the tube is inclined as shown in Fig. 1.18 the sensitiveness of the U-tube manometer is increased. The inclined tube *causes a larger displacement of the liquid along the tube for a given pressure difference*.

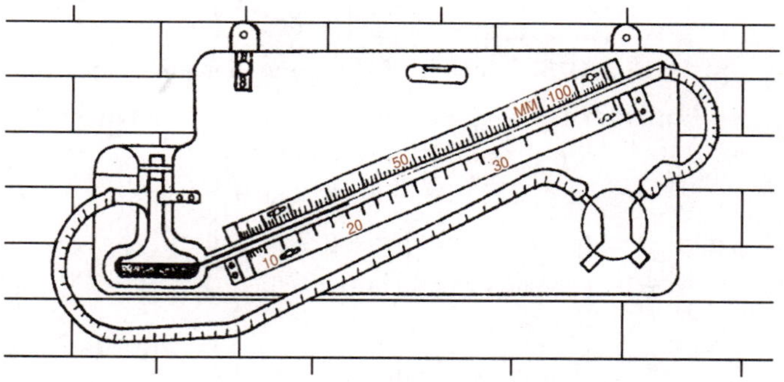

Fig. 1.18. Inclined manometer.

The principle of the inclined manometer is explained in Fig. 1.19. If pressure $p_1 = p_2$, then the level of liquid is shown by LM. However, when p_1 is slightly greater than p_2, the level is the reservoir sinks by h_2, whilst level in the tube rises by a greater distance h_1 as shown in the diagram. If h is the vertical distance between the two surfaces due to difference of pressure, then

$$h = h_1 + h_2$$

also $\quad h_1 = d \sin \theta$

and $\quad h_2 \times A = d \times a \quad \text{or} \quad h_2 = d \times \dfrac{a}{A}$

where A = Area of cross-section of the reservoir, and

a = Area of cross-section of the inclined tube.

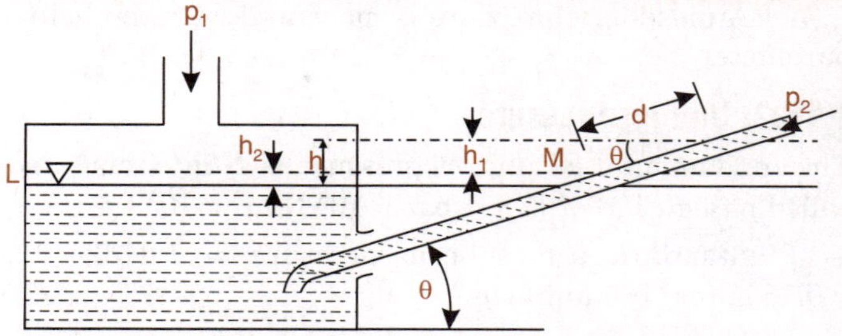

Fig. 1.19. Principle of inclined manometer.

Also the pressure difference Δp *i.e.*, $(p_1 - p_2)$

$$= hw = \left(d \times \sin \theta + d \times \frac{a}{A}\right) w = dw\left(\sin \theta + \frac{a}{A}\right)$$

where w is the specific weight of the liquid.

The sensitiveness of the instrument can be varied by *changing the slope of the inclined tube*. The position of the inclined tube is so arranged that $\left(\sin \theta + \dfrac{a}{A}\right)$ is round figure.

The multiplication factor of the gauge is

$$\frac{d}{h} = \frac{1}{\sin \theta + \dfrac{a}{A}}$$

Thus the multiplication factor depends on θ and $\dfrac{a}{A}$.

The *smaller the values of* θ *and* $\dfrac{a}{A}$, *the greater the multiplication factor*.

Advantages and disadvantages of manometers :

Advantages. The manometer claims the following *advantages :*

1. Relatively inexpensive.
2. Easy to fabricate.
3. Requires very little maintenance.
4. Good accuracy and sensitivity.
5. Their sensitivity can be changed by changing manometric fluids.
6. Particularly suitable to low pressures and low differential pressures.

Disadvantages. The *disadvantages* of manometers are as follows :

1. Unsuitable for recording.
2. Generally large, bulky and fragile.
3. Their calibration is affected by changes in gravitational force and density of fluids and their calibration changes with altitude and temperature.
4. Surface tension of manometric fluid creates a capillary effect and possible hysteresis.
5. A particular manometer can be used only for measurement of a particular fluid/fluids.

6. Meniscus height has to be determined by accurate means to ensure improved accuracy.

Elastic pressure elements

Elastic pressure elements or mechanical type of transducers are used for measurement of very high pressures upto about 700 MN/m^2. There are *three* main types of pressure elements.

(*a*) Bourdon tube
(*b*) Diaphragms
(*c*) Bellows gauges

The action of these mechanical transducers depends upon the *displacement caused by the pressure*. The displacement produced may actuate a pointer whose deflection may be direct measure of the pressure applied or the displacement is measured with the help of a secondary transducer which is electrical in nature. The output of the secondary transducer which is electrical in format is a measure of the pressure applied.

(*a*) **Bourdon tubes/elements :**

Fig. 1.20 shows Bourdon tube configurations.

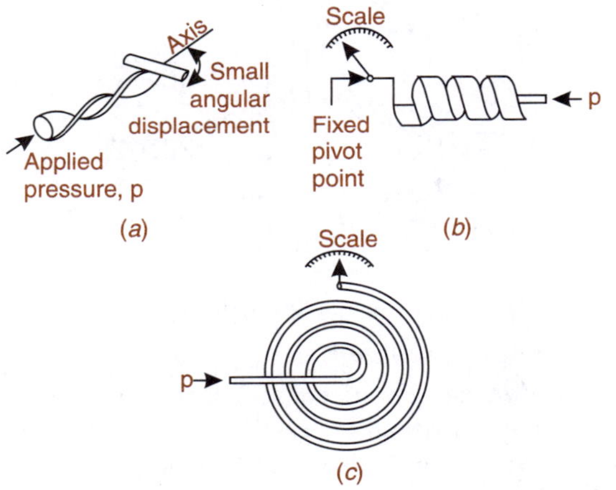

Fig. 1.20. (*a*) Twisted tube, (*b*) Helical, (*c*) Spiral.

Advantages. The Bourdon tube element has the following *advantages* :

1. Simple in construction and cheap.
2. Available in several different ranges.
3. Capability to measure gauge, absolute and differential pressures.
4. The sensitivities of Bourdon tube may be changed by changes in their dimensions.
5. Excellent sensitivity.
6. Simple and straight forward calibration with dead weight tester.
7. Easily adapted to strain, capacitance, magnetic and other electrical transducers.

Disadvantages :

1. Susceptibility to shock and vibration.
2. Inherent hysteresis and slow response to pressure changes.
3. Unsuitable for low pressure applications.

(*b*) **Diaphragm elements :**

Fig. 1.21 shows basic diaphragm types.

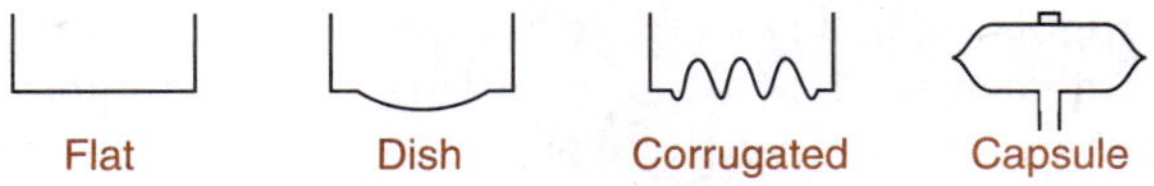

Fig. 1.21. Basic diaphragm types.

Advantages :

1. Small in size and moderately priced.
2. Wide linear range.
3. Can withstand high over pressures and under pressures.
4. Small hysteresis.
5. Can be used for measurement of absolute and differential pressures as also vacuum.

Disadvantages :

1. Need protection against shocks and vibrations.
2. Cannot be used to measure high pressures.
3. Difficult to repair.

(*c*) **Bellow gauges/elements :**

Fig. 1.22 shows some bellow gauges.

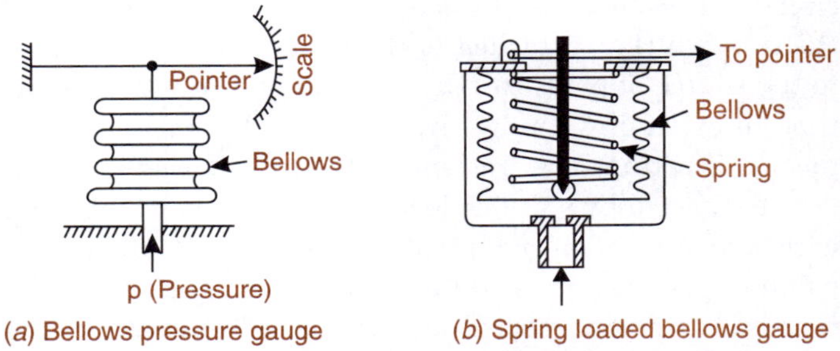

Fig. 1.22. Bellow gauges.

Advantages :

1. Simple and rugged construction.
2. Useful for measurement of low and medium pressures.
3. Moderate cost.
4. Can be used for measurement of absolute, gauge and differential pressures.

Disadvantages :

1. Need spring for accurate characterisation.
2. Greater hysteresis and zero drift problems.
3. Unsuitable for transient measurements due to longer relative motion and mass.
4. Requires compensation for ambient temperature changes.

1.16.4.2. Important types of pressure gauges

The manometers and U-tubes (discussed earlier) are suitable for comparatively *low pressures*. For high pressures they become unnecessarily larger even when they are filled with heavy liquids. Therefore for measuring medium and high pressures, we make use of elastic pressure gauges. They employ different forms of elastic systems such as tubes, diaphragms or bellows etc. to measure the pressure. The elastic deformation of these elements is used to show the effect of pressure. Since these elements are deformed within the elastic limit only, therefore these gauges are sometimes called *elastic gauges.* Sometimes they are also called *secondary instruments,* which implies that they must be calibrated by comparison with primary instruments such as manometer etc.

Some of the important types of these gauges are enumerated and discussed below :

1. Bourdon tube type pressure gauge
2. Diaphragm gauge
3. Vacuum gauge.

1. *Bourdon tube type pressure gauge :*

A Bourdon type tube pressure gauge is used for measuring *high as well as low pressures.* A simple form of this gauge is shown in Fig. 1.23. In this case the pressure element consists of a metal tube of approximately *elliptical cross-section.* This tube is bent in the form of a segment of a circle and responds to pressure changes. When one end of the tube which is attached to the gauge case, is connected to the source of pressure, the internal pressure causes the tube to expand, whereby circumferential stress *i.e.,* hoop tension is set up. The free end of the tube moves and is in turn connected by suitable levers to a rack, which engages with a small pinion mounted on the same spindle as the pointer. Thus the pressure applied to the tube causes the rack and pinion to move. The pressure is indicated by the pointer over a dial which can be graduated in a suitable scale.

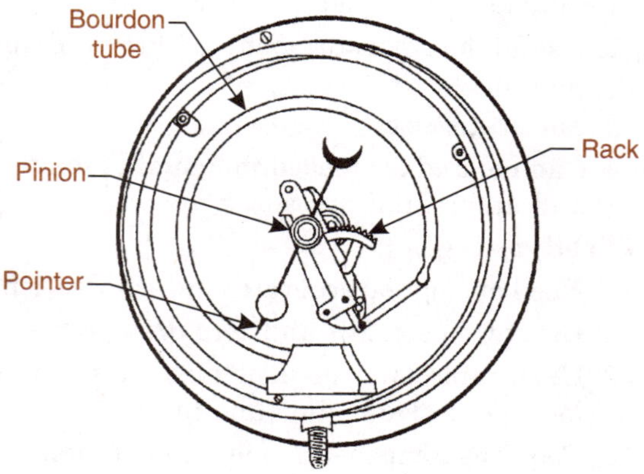

Fig. 1.23. Bourdon tube pressure gauge.

The Bourdon tubes are generally made of *bronze* or *nickel steel*. The former is generally used for low pressures and the latter for high pressures.

Depending upon the purpose for which they are required Bourdon tube gauges are made in different forms, some of them are :

(*i*) ***Compound Bourdon tube*** ...... used for measuring pressures both above and below atmospheric.

(*ii*) ***Double Bourdon tube*** ...... used where vibrations are encountered.

2. *Diaphragm gauge :*

This type of gauge employs a metallic disc or diaphragm instead of a bent tube. This disc or diaphragm is used for *actuating* the indicating device.

Refer to Fig. 1.24. When pressure is applied on the lower side of the diaphragm, it is deflected upward. This movement of the diaphragm is transmitted to a rack and pinion. The latter is attached to the spindle of needle moving on a graduated dial. The dial can again be graduated in a suitable scale.

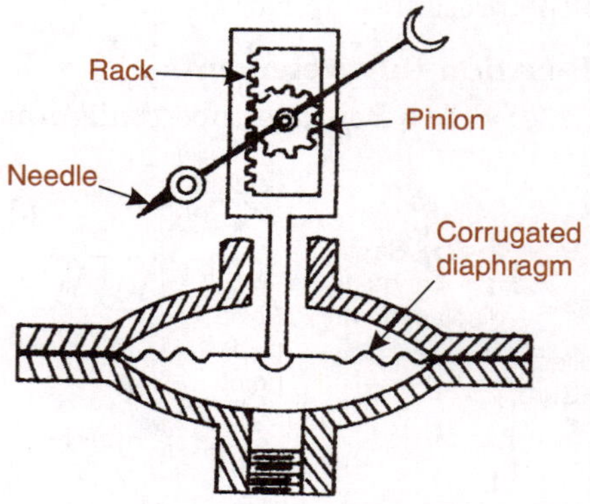

Fig. 1.24. Principle of diaphragm gauge.

3. *Vacuum gauge :*

Bourdon gauges discussed earlier can be used to measure vacuum instead of pressure. Slight changes in the design are required for this purpose. Thus, in this case, the tube be *bent inward instead of outward* as in pressure gauge. *Vacuum gauges* are graduated in millimetres of mercury below atmospheric pressure. In such cases, therefore, absolute pressure in millimetres of mercury is the difference between barometer reading and vacuum gauge reading.

Vacuum gauges are used to measure the vacuum in the *condensers* etc. If there is leakage, the vacuum will drop.

The pressure gauge installation require the following *considerations* :

(*i*) Flexible copper tubing and compression fittings are recommended for most installations.

(*ii*) The installation of a gauge cock and tee in the line close to the gauge is recommended because it permits the gauge to be removed for testing or replacement without having to shut down the system.

(*iii*) Pulsating pressures in the gauge line are not required.

(*iv*) The gauge and its connecting line is filled with an inert liquid and as such liquid seals are provided. Trapped air at any point of gauge lines may cause serious errors in pressure reading.

1.17. SPECIFIC VOLUME

The *specific volume* of a system is the volume occupied by the unit mass of the system. The symbol used is v and units are ; for example, m^3/kg. The symbol V will be used for volume. (Note that specific volume is *reciprocal of density*).

Example 1.1. *Convert the following readings of pressure to kPa assuming that barometer reads 760 mm of Hg.*

(*i*) *80 cm of Hg* (*ii*) *30 cm Hg vacuum*

(*iii*) *1.35 m H_2O gauge* (*iv*) *4.2 bar.*

Solution. Assuming density of Hg,

$$\rho_{Hg} = 13.596 \times 1000 \text{ kg/m}^3$$

Pressure of 760 mm of Hg will be

$$= \rho \times g \times h = 13.596 \times 1000 \times 9.806 \times \frac{760}{1000}$$

$= 101325$ Pa $= 101.325$ kPa.

(*i*) **Pressure of 80 cm of Hg**

$$= \frac{800}{760} \times 101.325 = \mathbf{106.65 \ kPa.} \quad \textbf{(Ans.)}$$

(*ii*) **30 cm Hg vacuum**

$= 76 - 30 = 46$ cm of Hg absolute.

Pressure due to 46 cm of Hg

$$= \frac{460}{760} \times 101.325 = \mathbf{61.328 \ kPa.} \quad \textbf{(Ans.)}$$

(*iii*) **Pressure due to 1.35 m H_2O gauge**

$= 1000 \times 9.806 \times 1.35 = 13238$ Pa

$=$ **13.238 kPa. (Ans.)**

(*iv*) **4.2 bar**

$= 4.2 \times 10^2$ kPa $=$ **420 kPa. (Ans.)**

Note. Pressure of 1 atmosphere = 760 mm of Hg = 101325 N/m^2.

The above values are standard. To get this value we have to use $\rho_{Hg} = 13596$ kg/m³ and $g = 9.806$ m/s². When we use $\rho_{Hg} = 13600$ kg/m³ and $g = 9.81$ m/s², we get $p_{atm.} = 101396$ N/m² which is slightly different from 101325 N/m². It is recommended that for pressure of 1 atm. the value 101325 N/m² should be used.

Example 1.2. *On a piston of 10 cm diameter a force of 1000 N is uniformly applied. Find the pressure on the piston.*

Solution. Diameter of the piston, $d = 10$ cm (= 0.1 m)

Force applied on the piston, $F = 1000$ N

∴ Pressure on the piston,

$$p = \frac{\text{Force}}{\text{Area}} = \frac{F}{A} = \frac{1000}{\pi/4 \times (0.1)^2}$$

$= 127307$ N/m² $=$ **127.307 kN/m². (Ans.)**

Example 1.3. *A tube contains an oil of specific gravity 0.9 to a depth of 120 cm. Find the gauge pressure at this depth (in kN/m²).*

Solution. Specific gravity of oil = 0.9

Depth of oil in the tube, $h = 120$ cm = (1.2 m)

We know that $p = wh$

$= \rho.g.h$, ρ being the mass density

$= (0.9 \ \rho_w) \times g \times h$, ρ_w being mass density of water

$$\left[\text{Specific gravity} = \frac{\rho}{\rho_w}\right]$$

$= 0.9 \times 1000 \times 9.81 \times 1.2$ N/m²

$= 10594.8$ N/m² $=$ **10.595 kN/m². (Ans.)**

Example 1.4. *A vacuum recorded in the condenser of a steam power plant is 740 mm of Hg. Find the absolute pressure in the condenser in Pa. The barometric reading is 760 mm of Hg.*

Solution. Vacuum recorded in the condenser

= 740 mm of Hg

Barometric reading = 760 mm of Hg

We know that,

Absolute pressure in the condenser

= Barometric reading – vacuum in the condenser

= 760 – 740 = 20 mm of Hg

$= 20 \times 133.4$ N/m² $\quad (\because$ 1 mm of Hg $= 133.4$ N/m²$)$

$= 2668$ N/m² $=$ **2668 Pa. (Ans.)**

Example 1.5. *A vessel of cylindrical shape is 50 cm in diameter and 75 cm high. It contains 4 kg of a gas. The pressure measured with manometer indicates 620 mm of Hg above atmosphere when barometer reads 760 mm of Hg. Determine :*

(*i*) *The absolute pressure of the gas in the vessel in bar.*

(*ii*) *Specific volume and density of the gas.*

Solution. Diameter of the vessel, $d = 50$ cm (= 0.5 m)

Height of the vessel, $h = 75$ cm (= 0.75 m)

Mass of gas in the vessel, $m = 4$ kg

Manometer reading

= 620 mm of Hg above atmosphere

Barometer reading = 760 mm of Hg

Now, volume of the vessel

$$= \frac{\pi}{4} d^2 \times h = \frac{\pi}{4} \times (0.5)^2 \times (0.75) = 0.147 \text{ m}^3.$$

(*i*) **Total pressure in the vessel**

= 760 + 620 = 1380 mm of Hg

$= 1380 \times 133.4$ N/m²

$[\because$ 1 mm of Hg $= 133.4$ N/m²$]$

$= 1.841 \times 10^5$ N/m² $=$ **1.841 bar. (Ans.)**

$[\because$ 1 bar $= 10^5$ N/m²$]$

(*ii*) **Specific volume**

$$= \frac{0.147}{4} = \mathbf{0.03675\ m^3/kg.}\ \textbf{(Ans.)}$$

$$\textbf{Density} = \frac{4}{0.147} = \mathbf{27.21\ kg/m^3.}\ \textbf{(Ans.)}$$

Example 1.6. *In a pipe line the pressure of gas is measured with a mercury manometer having one limb open to the atmosphere (Fig. 1.25). If the difference in the height of mercury in the two limbs is 550 mm, calculate the gas pressure.*

Given : Barometric reading = 761 mm of Hg

Acceleration due to gravity = 9.79 m/s²

Density of mercury = 13640 kg/m³.

Solution. At the plane *LM*, we have

$$p = p_0 + \rho gh$$

Now, $p_0 = \rho g h_0$

where h_0 = barometric height ; ρ = density of mercury ; p_0 = atmospheric pressure

Therefore, $p = \rho g h_0 + \rho gh = \rho g (h_0 + h)$

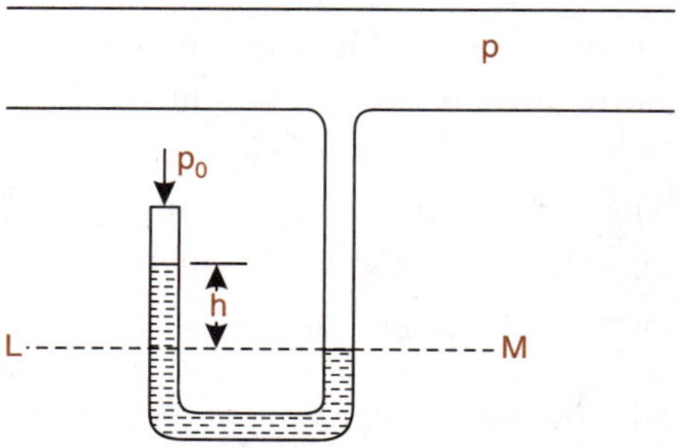

Fig. 1.25

$$= 13640 \times 9.79 \left(\frac{761}{1000} + \frac{550}{1000}\right)$$

$$= 13640 \times 9.79\ (0.761 + 0.55)$$

$$= 175.065 \times 10^3\ \text{N/m}^2 = 175.065\ \text{kPa}$$

$$= \mathbf{1.75\ bar.}\ \textbf{(Ans.)}$$

Example 1.7. *A U-tube mercury manometer with one arm open to atmosphere is used to measure pressure in a steam pipe. The level of mercury in open arm is 97.5 mm greater than that in the arm connected to the pipe. Some of steam in the pipe condenses in the manometer arm connected to the pipe. The height of this column is 34 mm. The atmospheric pressure is 760 mm of Hg. Find the absolute pressure of steam.* **(Poona University, Nov. 1997)**

Solution. Equating the pressure in mm of Hg on both arms above the line *XX* (Fig. 1.26), we get

$$p_{abs.} + p_{water} = p_{Hg} + p_{atm.}$$

Now, $p_{water} = \dfrac{34}{13.6} = 2.5$ mm of Hg

$\therefore$ $p_{abs} + 2.5 = 97.5 + 760$

or $p_{abs} = 97.5 + 760 - 2.5$

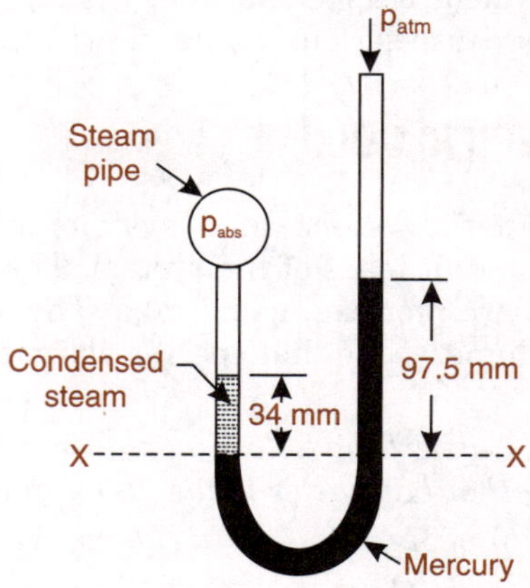

Fig. 1.26

= 855 mm of Hg

$= 855 \times p_{Hg} \times g \times 10^{-5}$ bar

$$= \frac{855}{1000}(m) \times (13.6 \times 1000)\ (\text{kg/m}^3) \times 9.81 \times 10^{-5}$$

$= \mathbf{1.1407\ bar.}$ **(Ans.)**

Example 1.8. *A U-tube manometer is connected to a gas pipe. The level of the liquid in the manometer arm open to the atmosphere is 170 mm lower than the level of the liquid in the arm connected to the gas pipe. The liquid in the manometer has specific gravity of 0.8. Find the absolute pressure of the gas if the manometer reads 760 mm of Hg.* **(Poona University)**

Solution. Equating pressure on both arms above the line *XX* (Fig. 1.27), we get

$$p_{gas} + p_{liquid} = p_{atm.} \quad ...(i)$$

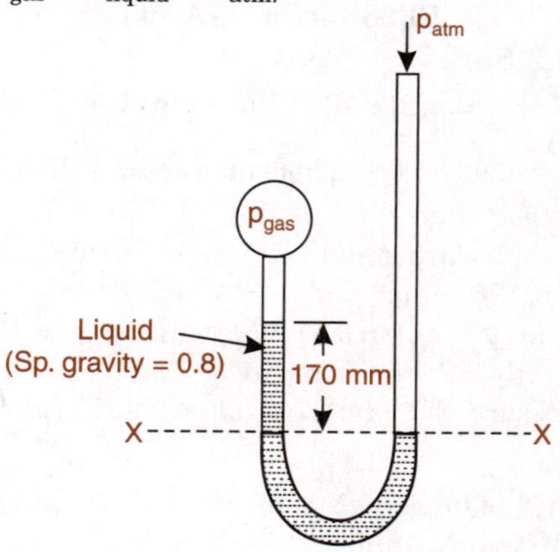

Fig. 1.27

Now, $p_{liquid} = \rho.g.h$ ...(*ii*)

$$= (0.8 \times 1000) \times 9.81 \times \frac{170}{1000}$$

$= 1334.16\ \text{N/m}^2 = 0.0133416$ bar

$p_{atm.} = 760$ mm of Hg $= 1.01325$ bar

Substituting these value is eqn. (*i*) above, we have

$$p_{gas} + 0.0133416 = 1.01325$$

$$\therefore \quad \mathbf{p_{gas} = 0.9999\ bar.} \quad \textbf{(Ans.)}$$

Example 1.9. *Estimate the mass of a piston that can be supported by a gas entrapped under the piston in a 200 mm diameter vertical cylinder when a manometer indicates a difference of 117 mm of Hg column for the gas pressure.*

(Delhi University)

Solution. Refer to Fig. 1.28.

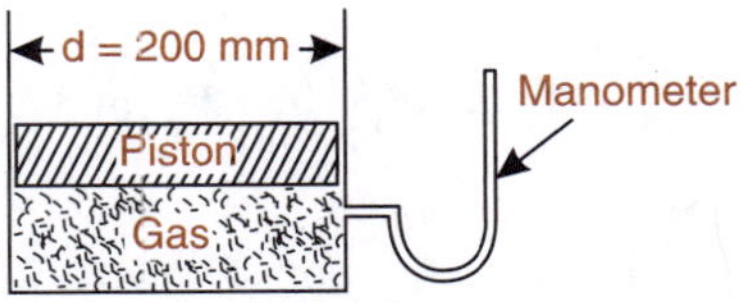

Fig. 1.28

Let m = mass of the piston, kg.

p = pressure of the gas
= 117 mm of Hg column

Dia. of vertical cylinder, d = 200 mm

Now, downward force = $m.g$ ...(*i*)

and upward force = $p \times \pi/4\ d^2$ ...(*ii*)

Equating eqns. (*i*) and (*ii*), we get

$$m.g = p \times \pi/4\ d^2$$

$$m \times 9.81 = \left(13.6 \times 1000 \times 9.81 \times \frac{117}{1000}\right) \times \frac{\pi}{4} \times \left(\frac{200}{1000}\right)^2$$

$$(\because \quad p = \rho gh)$$

$$\therefore \quad \mathbf{m = 49.989\ kg.} \quad \textbf{(Ans.)}$$

1.18. REVERSIBLE AND IRREVERSIBLE PROCESSES

Reversible process. *A reversible process (also sometimes known as quasi-static process) is one which can be stopped at any stage and reversed so that the system and surroundings are exactly restored to their initial states.*

This process has the following *characteristics* :

1. It must pass through the same states on the reversed path as were initially visited on the forward path.

2. This process when undone will leave no history of events in the surroundings.

3. It must pass through a continuous series of equilibrium states.

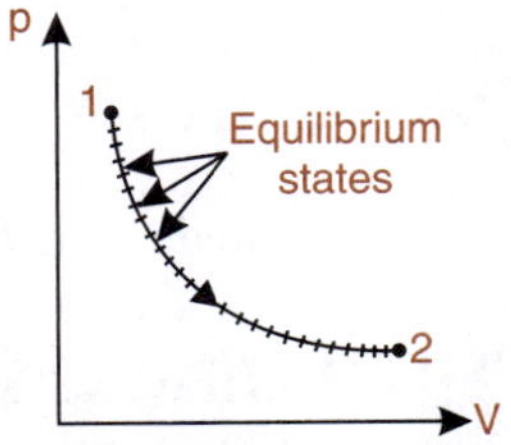

Fig. 1.29. Reversible process.

No real process is truely reversible but some processes may approach reversibility, to close approximation.

Examples. Some examples of nearly reversible processes are :

(*i*) Frictionless relative motion.
(*ii*) Expansion and compression of spring.
(*iii*) Frictionless adiabatic expansion or compression of fluid.
(*iv*) Polytropic expansion or compression of fluid.
(*v*) Isothermal expansion or compression.
(*vi*) Electrolysis.

Irreversible process. An *irreversible process* is one *in which heat is transferred through a finite temperature.*

Examples.

(*i*) Relative motion with friction
(*ii*) Combustion
(*iii*) Diffusion
(*iv*) Free expansion
(*v*) Throttling
(*vi*) Electricity flow through a resistance
(*vii*) Heat transfer
(*viii*) Plastic deformation.

An irreversible process is usually represented by a dotted (or discontinuous) line joining the end states to indicate that the intermediate states are indeterminate (Fig. 1.30).

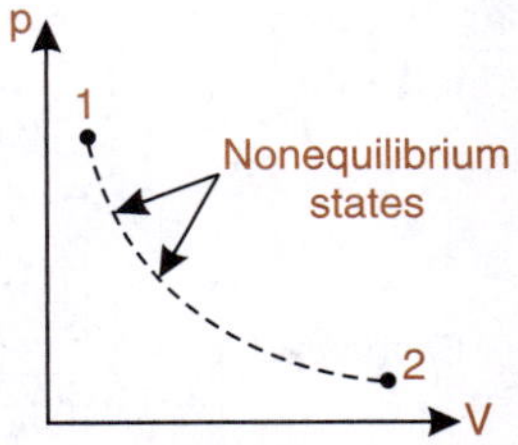

Fig. 1.30. Irreversible process.

Irreversibilities are of *two types* :

1. **External irreversibilities.** These are associated with *dissipating effects outside the working fluid.*

Example. *Mechanical friction occurring during a process due to some external source.*

2. **Internal irreversibilities.** These are associated with *dissipating effects within the working fluid.*

Example. *Unrestricted expansion of gas, viscosity and inertia of the gas.*

1.19. ENERGY, WORK AND HEAT

1.19.1. Energy

Energy is a general term embracing *energy in transition* and *stored energy*. The stored energy of a substance may be in the forms of *mechanical energy* and *internal energy* (other forms of stored energy may be chemical energy and electrical energy). Part of the stored energy may take the form of either potential energy (which is the gravitational energy due to height above a chosen datum line) or kinetic energy due to velocity. The balance part of the energy is known as *internal energy*. In a ***non-flow process*** *usually there is no* *change of potential or kinetic energy and hence change of mechanical energy will not enter the calculations*. In a ***flow process***, however, *there may be changes in both potential and kinetic energy and these must be taken into account while considering the changes of stored energy*. ***Heat and work*** are *the forms of energy in transition*. These are the only forms in which energy can cross the boundaries of a system. *Neither heat nor work can exist as stored energy.*

1.19.2. Work and Heat

Work

Work is said to be done when a *force moves through a distance*. If a part of the boundary of a system undergoes a displacement under the action of a pressure, the work done W is the product of the force (pressure × area), and the distance it moves in the direction of the force. Fig. 1.31 (*a*) illustrates this with the conventional piston and cylinder arrangement, the heavy line defining the boundary of the system. Fig. 1.31 (*b*) illustrates another way in which work might be applied to a system. A force is exerted by the paddle as it changes the momentum of the fluid, and since this force moves during rotation of the paddle work is done.

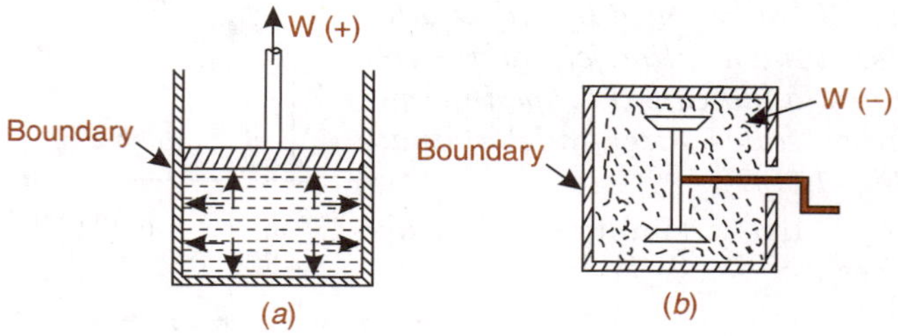

Fig. 1.31

Work is a transient quantity which only appears at the boundary while a change of state is taking place within a system. Work is 'something' which appears at the boundary when a system changes its state due to the movement of a part of the boundary under the action of a force.

Sign convention

- If the work is done *by* the system *on* the surroundings, *e.g.,* when a fluid expands pushing a piston outwards, the work is said to be *positive*.

  i.e., *Work output of the system = + W*

- If the work is done *on* the system *by* the surroundings, *e.g.,* when a force is applied to a rotating handle, or to a piston to compress a fluid, the work is said to be *negative*.

  i.e., *Work input to system = – W*

Heat

Heat (denoted by the symbol Q), may be, defined in an analogous way to work as follows :

"Heat is 'something' which appears at the boundary when a system changes its state due to a difference in temperature between the system and its surroundings".

Heat, like work, is a transient quantity which only appears at the boundary while a change is taking place within the system.

It is apparent that neither δW or δQ are exact differentials and therefore any integration of the elemental quantities of work or heat which appear during a change from state 1 to state 2 must be written as

$$\int_1^2 \delta W = W_{1-2} \text{ or } {}_1W_2 \text{ (or } W\text{), and}$$

$$\int_1^2 \delta Q = Q_{1-2} \text{ or } {}_1Q_2 \text{ (or } Q\text{)}$$

Sign convention :

If the heat flows *into* a system *from* the surroundings, the quantity is said to be *positive* and, conversely, if heat flows *from* the system to the surroundings it is said to be *negative*.

In other words :

Heat received by the system = + Q

Heat rejected or given up by the system = – Q.

Comparison of Work and Heat

Similarities :

(*i*) Both are *path functions and inexact differentials*.

(*ii*) Both are boundary phenomenon *i.e.,* both are recognized at the boundaries of the system as they cross them.

(*iii*) Both are associated with a process, not a state. Unlike properties, work or heat has no meaning at a state.

(*iv*) Systems possess energy, but not work or heat.

Dissimilarities :

(*i*) In heat transfer temperature difference is required.

(*ii*) In a stable system there cannot be work transfer, however, there is no restriction for the transfer of heat.

(*iii*) The sole effect external to the system could be reduced to rise of a weight but in the case of a heat transfer other effects are also observed.

1.20. REVERSIBLE WORK

Let us consider an ideal frictionless fluid contained in a cylinder above a piston as shown in Fig. 1.32 (*a*). *Assume that the pressure and temperature of the fluid are uniform and that there is no friction between the piston and the cylinder walls.*

Let A = Cross-sectional area of the piston,
p = Pressure of the fluid at any instant,
$(p - dp) A$ = Restraining *force exerted* by the surroundings on the piston, and
dl = The distance moved by the piston under the action of the *force exerted*.

Then work done by the fluid on the piston is given by force times the distance moved,

i.e., Work done by the fluid

$= (pA) \times dl = pdV$

(where dV = a small increase in volume)

Or considering unit mass

Work done = pdv

(where v = specific volume)

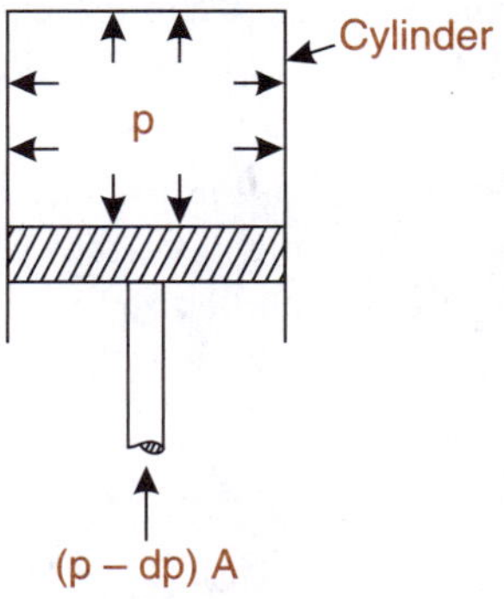

Fig. 1.32 (*a*)

This is only *true* when (*a*) *the process is frictionless and* (*b*) *the difference in pressure between the fluid and its surroundings during the process is infinitely small*. Hence when a reversible process takes place between state 1 and state 2, we have

$$\text{Work done by the unit mass of fluid} = \int_1^2 pdv$$

When a fluid undergoes a reversible process a series of state points can be joined up to form a line on a diagram of properties. The work done by the fluid during any reversible process is therefore given by the area under the line of process plotted on a p-v diagram [Fig. 1.32 (*b*)].

i.e., Work done = Shaded area on Fig. 1.32 (*b*)

$$= \int_1^2 pdv.$$

When p can be expressed in terms of v then the integral, $\int_1^2 pdv$, can be evaluated.

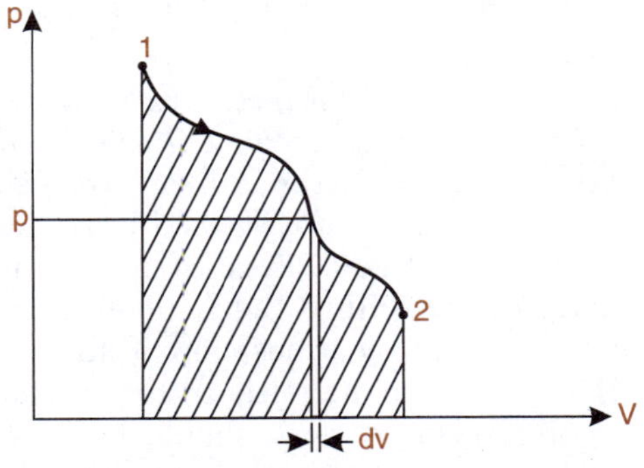

Fig. 1.32 (*b*)

Example 1.10. *An artificial satellite revolves round the earth with a relative velocity of 800 m/s. If acceleration due to gravity is 9 m/s² and gravitational force is 3600 N, calculate its kinetic energy.*

Solution. Relatively velocity of satellite,

$$v = 800 \text{ m/s}$$

Acceleration due to gravity, $g = 9 \text{ m/s}^2$

Gravitational force, $m.g = 3600$ N

$\therefore$ Mass, $m = \frac{3600}{g} = \frac{3600}{9} = 400$ kg.

Kinetic energy $= \frac{1}{2} mv^2 = \frac{1}{2} \times 400 \times (800)^2$ J

$= 128 \times 10^6$ J or **128 MJ. (Ans.)**

Example 1.11. *The specific heat capacity of the system during a certain process is given by*

$$c_n = (0.4 + 0.004\ T)\ kJ/kg°C.$$

If the mass of the gas is 6 kg and its temperature changes from 25°C to 125°C find :

(*i*) *Heat transferred ;*

(*ii*) *Mean specific heat of the gas.*

Solution. Mass of the gas, m = 6 kg

Change in temperature of the gas = 25°C to 125°C

(*i*) **Heat transferred, Q :**

We know that heat transferred is given by,

$$Q = \int m\, c_n\, dT = 6 \int_{25}^{125} (0.4 + 0.004\, T)\, dT$$

$$= 6 \left[0.4\, T + 0.04 \left(\frac{T^2}{2} \right) \right]_{25}^{125}$$

$$= 6[0.4\,(125 - 25) + 0.002\,(125^2 - 25^2)]$$

$$= 6(40 + 30) = \textbf{420 kJ.} \quad \textbf{(Ans.)}$$

(*ii*) **Mean specific heat of the gas, c_n :**

$$Q = m.c_n.dT$$

i.e., $420 = 6 \times c_n \times (125 - 25)$

$\therefore$ $\mathbf{c_n} = \frac{420}{6 \times 100}$ = **0.7 kJ/kg°C. (Ans.)**

Example 1.12. *A temperature scale of certain thermometer is given by the relation*

$$t = a \ln p + b$$

where a and b are constants and p is the thermometric property of the fluid in the thermometer. If at the ice point and steam point the thermometric properties are found to be 1.5 and 7.5 respectively what will be the temperature corresponding to the thermometric property of 3.5 on Celsius scale. **(Roorkee University)**

Solution. $t = a \ln p + b$ ...(Given)

On Celsius scale :

Ice point = 0°C, and

Steam point = 100°C

∴ From given conditions, we have

$$0 = a \ln 1.5 + b \quad ...(i)$$

and $$100 = a \ln 7.5 + b \quad ...(ii)$$

i.e., $$0 = a \times 0.4054 + b \quad ...(iii)$$

and $$100 = a \times 2.015 + b \quad ...(iv)$$

Subtracting (*iii*) from (*iv*), we get

$$100 = 1.61a \quad \text{or} \quad a = 62.112$$

Substituting this value in eqn. (*iii*), we get

$$b = -0.4054 \times 62.112 = -25.18$$

∴ When $p = 3.5$ the value of temperature is given by

$$t = 62.112 \ln (3.5) - 25.18 = \mathbf{52.63°C.} \quad \textbf{(Ans.)}$$

Example 1.13. *A thermocouple with test junction at t°C on gas thermometer scale and reference junction at ice point gives the e.m.f. as:*

$$e = 0.20\, t - 5 \times 10^{-4} t^2 \text{ mV}.$$

The millivoltmeter is calibrated at ice and steam points. What will be the reading on this thermometer where the gas thermometer reads 70°C ?

Solution. $e = 0.20\, t - 5 \times 10^{-4} t^2$ mV ...(Given)

At ice point : When $t = 0°C$, $e = 0$

At steam point : When $t = 100°C$,

$$e = 0.20 \times 100 - 5 \times 10^{-4} \times (100)^2 = 15 \text{ mV}$$

Now, when $t = 70°C$

$$e = 0.20 \times 70 - 5 \times 10^{-4} \times (70)^2 = 11.55 \text{ mV}$$

∴ When the gas thermometer reads 70°C the thermocouple will read

$$t = \frac{100 \times 11.55}{15} = \mathbf{77°C.} \quad \textbf{(Ans.)}$$

☞**Example 1.14.** *Comment whether the following quantities can be called as properties or not* :

(*i*) $\int pdV$, (*ii*) $\int Vdp$, *and*

(*iii*) $\int pdV + \int Vdp.$

Solution. (*i*) $\int pdV$:

p is a function of *V* and integral can only be evaluated if relation between *p* and *V* is known. It is thus an *inexact differential* and hence **not a property.**

(*ii*) $\int Vdp$:

It is **not a property** for the *same reason* as mentioned in (*i*).

(*iii*) $\int pdV + \int Vdp$:

$$\int pdV + \int Vdp = \int pdV + Vdp = \int d(pV) = pV.$$

Thus the integral can be evaluated without knowing the relation between *p* and *V*. It is an *exact differential* and hence **it is a property.**

Example 1.15. *Gas from a cylinder of compressed helium is used to inflate an inelastic flexible balloon, originally folded completely flat, to a volume 0.6 m³. If the barometer reads 760 mm Hg, what is the amount of work done upon the atmosphere by the balloon ? Sketch the system before and after the process.*

Solution. Refer to Fig. 1.33. The firm line B_1 shows the boundary of the system before the process, and dotted line B_2 shows the boundary after the process.

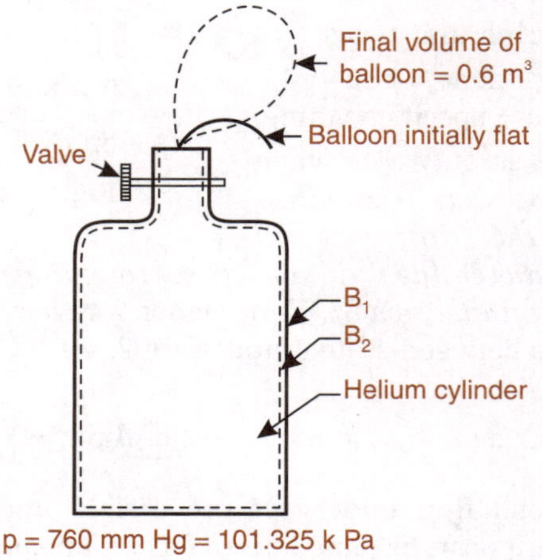

Fig. 1.33

The displacement work,

$$W_d = \int_{\text{cylinder}} pdV + \int_{\text{balloon}} pdV = 0 + \int_{\text{balloon}} pdV$$

$$= 101.325 \times 0.6 \quad [\because \; dV = 0.6 \text{ m}^3]$$

$$= \mathbf{60.795 \text{ kJ.}} \quad \textbf{(Ans.)}$$

This is a positive work, because the *work is done by the system*. Work done by the atmosphere is – 60.795 kJ. Since the wall of the cylinder is rigid there is no *pdV*-work involved in it.

It is assumed that the pressure in the balloon is atmospheric at all times, since the balloon fabric is light, inelastic and unstressed. If the balloon were elastic and stressed during the filling process, the work done by the gas would be *greater* than 60.795 kJ by an amount equal to the work done in stretching the balloon, although the displacement work done by atmosphere is still – 60.795 kJ. However, if the system includes both the gas and the balloon, the displacement work should be 60.795 kJ, as estimated above.

Example 1.16. *Determine the work done by the air which enters into an evacuated vessel from atmosphere when the valve is opened. The atmospheric pressure is 1.013 bar and 1.5 m³ of air at atmospheric condition enters into the vessel.*

Solution. Fig. 1.34 shows the initial and final condition of the system.

No work is done by the boundary in contact with the vessel as the boundary does not move. Work is done by the external boundary at constant pressure.

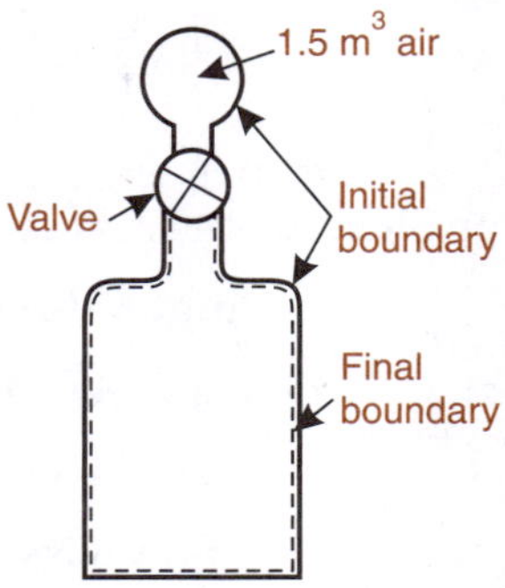

Fig. 1.34

$$\therefore \quad W = \int_{V_1}^{V_2} p\,dV = \int_{1.5}^{0} p\,dV$$

$[\because V_1 = 1.5 \text{ m}^3 \text{ and } V_2 = 0]$

$= p(0 - 1.5) = 1.013 \times 10^5 \times (-1.5)$

$= -1.5195 \times 10^5$ J = – **151.95 kJ.** **(Ans.)**

Since the free air boundary is contracting, the work done by the system is negative, and the surroundings do positive work upon the system.

Example 1.17. *A piston and cylinder machine containing a fluid system has a stirring device as shown in Fig. 1.35. The piston is frictionless, and it is held down against the fluid due to atmospheric pressure of 101.3 kPa. The stirring device is turned 9500 revolutions with an average torque against the fluid of 1.25 Nm. Meanwhile the piston of 0.65 m diameter moves out 0.6 m. Find the net work transfer for the system.*

Solution. Refer to Fig. 1.35.

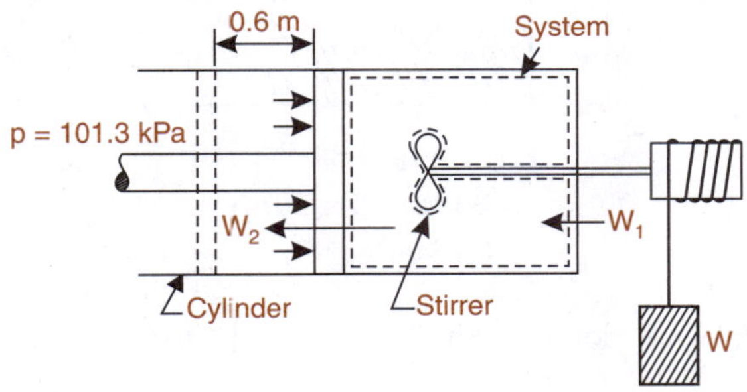

Fig. 1.35

Work done by *the stirring device* upon the system,

$$W_1 = 2\pi NT$$

where, T = Torque = 1.25 Nm

N = Number of revolutions = 9500

$W_1 = 2\pi \times 9500 \times 1.25 = 74622$ Nm = 74.622 kJ

This is *negative work* for the system.

Work done *by the system* upon the surroundings

$$W_2 = (pA).\ L$$

where, p = Pressure = 101.3 kPa

A = Area of the piston

$= \pi/4 \times (0.65)^2 = 0.3318$ m², and

L = Distance moved by the piston = 0.6 m

$W_2 = 101.3 \times 0.3318 \times 0.6 = 20.167$ kJ

This is a *positive work for the system.*

Hence, the network transfer for the system

$W_{net} = W_1 + W_2$

$= -74.622 + 20.167 =$ **– 54.455 kJ.** **(Ans.)**

Example 1.18. *A diesel engine piston which has an area of 45 cm² moves 5 cm during part of suction stroke. 300 cm³ of fresh air is drawn in from the atmosphere. The pressure in the cylinder during suction stroke is 0.9 × 10⁵ N/m² and the atmospheric pressure is 1.013 × 10⁵ N/m². The difference between the suction and atmospheric pressure is accounted for flow resistance in the suction pipe and inlet valve. Find the network done during the process.*

Solution. Area of diesel engine piston

$= 45 \text{ cm}^2 = 45 \times 10^{-4} \text{ m}^2$

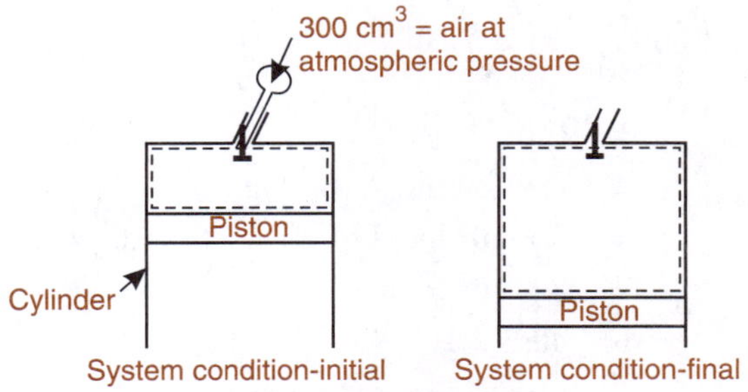

Fig. 1.36

Amount of fresh air drawn in from atmosphere

$= 300 \text{ cm}^3 = 300 \times 10^{-6} \text{ m}^3$

The pressure inside the cylinder during suction stroke

$= 0.9 \times 10^5 \text{ N/m}^2$

Atmospheric pressure $= 1.013 \times 10^5 \text{ N/m}^2$

Initial and final conditions of the system are shown in Fig. 1.36.

Net work done = Work done by free air boundary + work done on the piston

The work done by the free air = –ve because *boundary contracts*

The work done by the cylinder on the piston = +ve because the *boundary expands*

$$\therefore \quad \textbf{Net work done} = \int_{\text{Piston}} pdV + \int_{\substack{\text{Free air}\\ \text{boundary}}} pdV$$

$$= \left[0.9 \times 10^5 \times 45 \times 10^{-4} \times \frac{5}{100} - 1.013 \times 10^5 \times 300 \times 10^{-6}\right]$$

$= [20.25 - 30.39] =$ **– 10.14 Nm or J.** **(Ans.)**

Example 1.19. *The properties of a closed system change following the relation between pressure and volume as pV = 3.0, where p is in bar V is in m³. Calculate the work done when the pressure increases from 1.5 bar to 7.5 bar.*

Solution. Initial pressure, $p_1 = 1.5$ bar

Final pressure, $p_2 = 7.5$ bar

Relation between p and V, $pV = 3.0$

Work done, W :

The work done during the process is given by

$$W = \int_{V_1}^{V_2} p dV$$

$$V_1 = \frac{3.0}{p_1} = \frac{3.0}{1.5} = 2 \text{ m}^3$$

$$V_2 = \frac{3.0}{p_1} = \frac{3.0}{1.5} = 0.4 \text{ m}^3$$

$$\therefore \quad W = 10^5 \int_2^{0.4} \frac{3.0}{V} dV \text{ Nm}$$

$[\because$ 1 bar = 10^5 N/m$^2]$

$$= 10^5 \times 3.0 \left[\log_e V\right]_2^{0.4}$$

$$= 10^5 \times 3.0(\log_e 0.4 - \log_e 2)$$

$$= -3 \times 10^5 \log_e (2/0.4) = -3 \times 10^5 \times \log_e 5$$

$$= -3 \times 10^5 \times 1.61 \text{ Nm}$$

$$= -4.83 \times 10^5 \text{ Nm}$$

$= -4.83 \times 10^5$ J = **– 483 kJ. (Ans.)**

Example 1.20. *To a closed system 150 kJ of work is supplied. If the initial volume is 0.6 m^3 and pressure of the system changes as $p = 8 - 4V$, where p is in bar and V is in m^3, determine the final volume and pressure of the system.*

Solution. Amount of work supplied to a closed system

= 150 kJ

Initial volume = 0.6 m^3

Pressure-volume relationship, $p = 8 - 4V$

The work done during the process is given by

$$W = \int_{V_1}^{V_2} p dV$$

$$= 10^5 \int_{0.6}^{V_2} (8 - 4V) dV$$

$$= 10^5 \left[8V - 4 \times \frac{V^2}{2}\right]_{0.6}^{V}$$

$$= 10^5[8(V_2 - 0.6) - 2(V_2^2 - 0.6^2)]$$

$$= 10^5[8V_2 - 4.8 - 2V_2^2 + 0.72]$$

$$= 10^5[8V_2 - 2V_2^2 - 4.08] \text{ Nm or J}$$

But this work is equal to -150×10^3 J as this work is supplied to the system.

$$\therefore \quad -150 \times 10^3 = 10^5[8V_2 - 2V_2^2 - 4.08]$$

or $$2V_2^2 - 8V_2 + 2.58 = 0$$

$$V_2 = \frac{8 \pm \sqrt{64 - 4 \times 2 \times 2.58}}{4} = \frac{8 \pm 6.585}{4} = 0.354 \text{ m}^3$$

Positive sign is incompatible with the present problem, therefore it is not considered.

$\therefore$ **Final volume, V_2 = 0.354 m^3. (Ans.)**

and, **final pressure,** $p_2 = 8 - 4V$

$$= 8 - 4 \times 0.354 = 6.584 \text{ bar}$$

= **6.584 × 10^5 N/m^2 or Pa. (Ans.)**

REVERSIBLE WORK

Example 1.21. *A fluid at a pressure of 3 bar, and with specific volume of 0.18 m^3/kg, contained in a cylinder behind a piston expands reversibly to a pressure of 0.6 bar according to a law, $p = \frac{C}{v^2}$, where C is a constant. Calculate the work done by the fluid on the piston.*

Solution. Refer to Fig. 1.37.

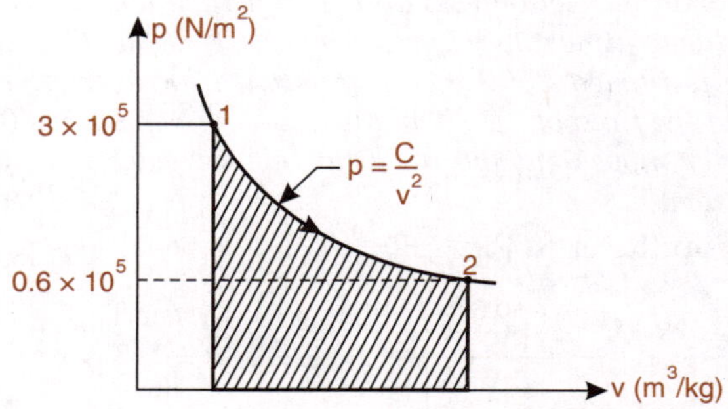

Fig. 1.37

$$p_1 = 3 \text{ bar} = 3 \times 10^5 \text{ N/m}^2$$

$$v_1 = 0.18 \text{ m}^3\text{/kg}$$

Work done = Shaded area = $\int_1^2 p\, dv$

i.e., Work done, $W = \int_1^2 \frac{C}{v^2} , dv = C\int_1^2 \frac{dv}{v^2} = C\left|\frac{v^{-2+1}}{-2+1}\right|_{v_1}^{v_2}$

$$= C\left[-v^{-1}\right]_{v_1}^{v_2} = C\left[-\frac{1}{v}\right]_{v_1}^{v_2} = C\left[\frac{1}{v_1} - \frac{1}{v_2}\right] \quad ...(i)$$

Also $C = pv^2 = p_1 v_1^2$

$$= 3 \times 0.18^2 = 0.0972 \text{ bar (m}^3\text{/kg)}^2$$

and $$v_2 = \sqrt{\frac{C}{p_2}} = \sqrt{\frac{0.0972}{0.6}} = 0.402 \text{ m}^3\text{/kg}$$

Substituting the values of C, v_1 and v_2 in eqn. (*i*), we get

Work done, $W = 0.0972 \times 10^5 \left[\frac{1}{0.18} - \frac{1}{0.402}\right]$ Nm/kg

= **29840 Nm/kg. (Ans.)**

Example 1.22. *A cylinder contains 1 kg of a certain fluid at an initial pressure of 20 bar. The fluid is allowed to expand reversibly behind a piston according to a law pV^2 = constant until the volume is doubled. The fluid is then cooled reversibly at constant pressure until the piston regains its original position ; heat is then supplied reversibly with the piston firmly locked in position until the pressure rises to the original value of 20 bar. Calculate the net work done by the fluid, for an initial volume of 0.05 m^3.*

Solution. Refer to Fig. 1.38.

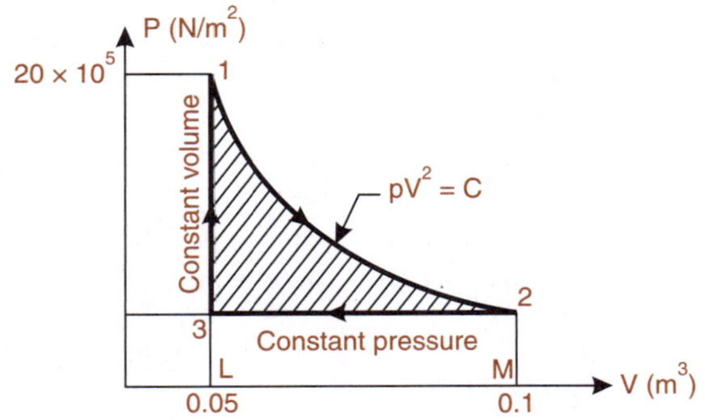

Fig. 1.38

Mass of fluid, m = 1 kg

$$p_1 = 20 \text{ bar} = 20 \times 10^5 \text{ N/m}^2$$

$$V_1 = 0.05 \text{ m}^3$$

Considering the process 1–2

$$p_1V_1^2 = p_2V_2^2$$

$$\therefore \quad p_2 = p_1\left(\frac{V_1}{V_2}\right)^2 = 20\left(\frac{V_1}{2V_1}\right)^2 \quad [\because \; V_2 = 2V_1 \text{ (given)}]$$

$$= \frac{20}{4} = 5 \text{ bar}$$

Work done by the fluid from 1 to 2

$$= \text{Area 12 ML1} = \int_1^2 p\,dV$$

i.e.,

$$W_{1-2} = \int_{v_1}^{v_2} \frac{C}{V^2}\,dV,$$

where $C = p_1V_1^2 = 20 \times 0.05^2$ bar m^6

$$\therefore \quad W_{1-2} = 10^5 \times 20 \times 0.0025\left[-\frac{1}{V}\right]_{0.05}^{0.1}$$

$$= 10^5 \times 20 \times 0.0025\left(\frac{1}{0.05} - \frac{1}{0.1}\right)$$

$$= 50000 \text{ Nm}$$

Work done on fluid from 2 to 3

$$= \text{Area 32ML3} = p_2(V_2 - V_3)$$

$$= 10^5 \times 5 \times (0.1 - 0.05) = 25000 \text{ Nm}$$

Work done during the process 3-1

= 0, because piston is locked in position (*i.e., Volume remains constant*)

$\therefore$ **Net work done by the fluid**

= Enclosed area 1231 = 50000 – 25000

= **25000 Nm. (Ans.)**

HIGHLIGHTS

1. *Thermodynamics* is an axiomatic science which deals with the relations among heat, work and properties of systems which are in equilibrium. It basically entails four laws or axioms known as *Zeroth, First, Second* and *Third* law of thermodynamics.
2. A *system* is a finite quantity of matter or a prescribed region of space.

   A system may be a *closed, open* or *isolated* system.
3. A *phase* is a quantity of matter which is homogeneous throughout in chemical composition and physical structure.
4. A *homogeneous system* is one which consists of a *single phase*.
5. A *heterogeneous system* is one which consists of *two or more phases*.
6. A *pure substance* is one that has a homogeneous and invariable chemical composition even though there is a change of phase.
7. A system is in *thermodynamic equilibrium* if temperature and pressure at all points are same ; there should be no *velocity gradient*.
8. A *property of a system* is a characteristic of the system which depends upon its state, but not upon how the state is reached.

   Intensive properties do not depend on the mass of the system.

   Extensive properties depend on the mass of the system.
9. *State* is the condition of the system at an instant of time as described or measured by its properties. Or each unique condition of a system is called a state.
10. A *process* occurs when the system undergoes a change in state or an energy transfer takes place at a steady state.
11. Any process or series of processes whose end states are identical is termed a *cycle*.
12. The *pressure* of a system is the force exerted by the system on unit area of boundaries. Vacuum is defined as the absence of pressure.
13. A *reversible process* is one which can be stopped at any stage and reversed so that the system and surroundings are exactly restored to their initial states.

    An *irreversible process* is one in which heat is transferred through a finite temperature.

14. Zeroth law of termodynamics states that if two systems are each equal in temperature to a third, they are equal in temperature to each other.

15. Infinite slowness is the characteristic feature of a quasi-static process. A quasi-static process is a succession of equilibrium states. It is also called a reversible process.

OBJECTIVE TYPE QUESTIONS

Choose the Correct Answer :

1. A definite area or space where some thermodynamic process takes place is known as
(*a*) thermodynamic system (*b*) thermodynamic cycle
(*c*) thermodynamic process (*d*) thermodynamic law.

2. An open system is one in which
(*a*) heat and work cross the boundary of the system, but the mass of the working substance does not
(*b*) mass of working substance crosses the boundary of the system but the heat and work do not
(*c*) both the heat and work as well as mass of the working substances cross the boundary of the system
(*d*) neither the heat and work nor the mass of the working substances cross the boundary of the system.

3. An isolated system
(*a*) is a specified region where transfer of energy and/or mass take place
(*b*) is a region of constant mass and only energy is allowed to cross the boundaries
(*c*) cannot transfer either energy or mass to or from the surroundings
(*d*) is one in which mass within the system is not necessarily constant
(*e*) none of the above.

4. In an extensive property of a thermodynamic system
(*a*) extensive heat is transferred
(*b*) extensive work is done
(*c*) extensive energy is utilised
(*d*) all of the above
(*e*) none of the above.

5. Which of the following is an intensive property of a thermodynamic system ?
(*a*) Volume (*b*) Temperature
(*c*) Mass (*d*) Energy.

6. Which of the following is the extensive property of a thermodynamic system ?
(*a*) Pressure (*b*) Volume
(*c*) Temperature (*d*) Density.

7. When two bodies are in thermal equilibrium with a third body they are also in thermal equilibrium with each other. This statement is called
(*a*) Zeroth law of thermodyamics
(*b*) First law of thermodynamics
(*c*) Second law of thermodynamics
(*d*) Kelvin Planck's law.

8. The temperature at which the volume of a gas becomes zero is called
(*a*) absolute scale of temperature
(*b*) absolute zero temperature
(*c*) absolute temperature
(*d*) none of the above.

9. The value of one bar (in SI units) is equal to
(*a*) 100 N/m^2 (*b*) 1000 N/m^2
(*c*) 1×10^4 N/m^2 (*d*) 1×10^5 N/m^2
(*e*) 1×10^6 N/m^2.

10. The absolute zero pressure will be
(*a*) when molecular momentum of the system becomes zero
(*b*) at sea level
(*c*) at the temperature of – 273 K
(*d*) under vacuum conditions
(*e*) at the centre of the earth.

11. Absolute zero temperature is taken as
(*a*) – 273°C (*b*) 273°C
(*c*) 237°C (*d*) – 373°C.

12. Which of the following is *correct* ?
(*a*) Absolute pressure = gauge pressure + atmospheric pressure
(*b*) Gauge pressure = absolute pressure + atmospheric pressure
(*c*) Atmospheric pressure = absolute pressure + gauge pressure
(*d*) Absolute pressure = gauge pressure – atmospheric pressure.

13. The unit of energy in SI units is
(*a*) Joule (J) (*b*) Joule metre (Jm)
(*c*) Watt (W) (*d*) Joule/metre (J/m).

14. One watt is equal to
(*a*) 1 Nm/s (*b*) 1 N/min
(*c*) 10 N/s (*d*) 100 Nm/s
(*e*) 100 Nm/m.

15. One joule (J) is equal to
(*a*) 1 Nm (*b*) kNm
(*d*) 10 Nm/s (*d*) 10 kNm/s.

16. The amount of heat required to raise the temperature of 1 kg of water through 1°C is called
(*a*) specific heat at constant volume
(*b*) specific heat at constant pressure
(*c*) kilo calorie (*d*) none of the above.

17. The heating and expanding of a gas is called
(*a*) thermodynamic system (*b*) thermodynamic cycle
(*c*) thermodynamic process (*d*) thermodynamic law.

18. A series of operations, which take place in a certain order and restore the initial condition is known as
(*a*) reversible cycle (*b*) irreversible cycle
(*c*) thermodynamic cycle (*d*) none of the above.

19. The condition for the reversibility of a cycle is:
(*a*) the pressure and temperature of the working substance must not differ, appreciably, from those of the surroundings at any stage in the process
(*b*) all the processes, taking place in the cycle of operation, must be extremely slow
(*c*) the working parts of the engine must be friction free
(*d*) there should be no loss of energy during the cycle of operation
(*e*) all of the above (*f*) none of the above.

20. In an irreversible process, there is a:
(*a*) loss of heat (*b*) no loss of heat
(*c*) gain of heat (*d*) no gain of heat.

21. The main cause of the irreversibility is:
(*a*) mechanical and fluid friction
(*b*) unrestricted expansion
(*c*) heat transfer with a finite temperature difference
(*d*) all of the above (*e*) none of the above.

22. According to kinetic theory of heat
(*a*) temperature should rise during boiling
(*b*) temperature should fall during freezing
(*c*) at low temperature all bodies are in solid state
(*d*) at absolute zero there is absolutely no vibration of molecules
(*e*) none of the above.

23. A system comprising a single phase is called a
(*a*) closed system
(*b*) open system
(*c*) isolated system
(*d*) homogeneous system
(*e*) heterogeneous system.

ANSWERS

1. (*a*)	**2.** (*c*)	**3.** (*c*)	**4.** (*e*)
5. (*b*)	**6.** (*b*)	**7.** (*a*)	**8.** (*b*)
9. (*d*)	**10.** (*a*)	**11.** (*a*)	**12.** (*a*)
13. (*a*)	**14.** (*a*)	**15.** (*a*)	**16.** (*c*)
17. (*b*)	**18.** (*c*)	**19.** (*e*)	**20.** (*a*)
21. (*d*)	**22.** (*d*)	**23.** (*d*).	

THEORETICAL QUESTIONS

1. Define a thermodynamic system. Differentiate between open system, closed system and an isolated system.
2. How does a homogeneous system differ from a heterogeneous system ?
3. What do you mean by a pure substance ?
4. Explain the following terms :
(*i*) State, (*ii*) Process, and
(*iii*) Cycle.
5. Explain briefly zeroth law of thermodynamics.
6. What is a quasi-static process ?
7. What do you mean by 'reversible work' ?

UNSOLVED EXAMPLES

1. A cylindrical vessel of 60 cm diameter and 80 cm height contains 3.2 kg of a gas. The pressure measured with manometer indicates 60 cm of Hg above atmosphere when barometer reads 760 mm of Hg. Find : (*i*) The absolute pressure of the gas in the vessel in bar, and (*ii*) Specific volume and density of gas.
[**Ans.** (*i*) 1.81 bar ; (*ii*) 0.09 m^3/kg ; 11.11 kg/m^3]
2. A force of 1600 N is applied uniformly on a piston of 8 cm diameter. Determine the pressure on the piston.
[**Ans.** 319 kN/m^2]
3. An oil of specific gravity 0.8 is contained in a tube to a depth of 80 cm. Determine the gauge pressure at this depth in kN/m^2. [**Ans.** 6.275 kN/m^2]
4. A vacuum recorded in the condenser of a steam power plant is 745 mm of Hg. Find the absolute pressure in the condenser in Pa. The barometer reading is 760 mm of Hg. [**Ans.** 2262 Pa]
5. The pressure of gas in a pipe line is measured with a mercury manometer having one limb open to the atmosphere. If the difference in the height of mercury in the two limbs is 562 mm, calculate the gas pressure. The barometer reads 761 mm Hg, the acceleration due to gravity is 9.79 m/s^2, and the density of mercury is 13640 kg/m^3. [**Ans.** 1.77 bar]
6. A turbine is supplied with steam at a gauge pressure of 1.4 MPa. After expansion in the turbine the steam flows into a condenser which is maintained at a vacuum of 710 mm Hg. The barometric pressure is 772 mm Hg. Express the inlet and exhaust steam pressures in pascals (absolute). Take the density of mercury as 13600 kg/m^3.
[**Ans.** 1.503 MPa ; 8.27 kPa]
7. Gas from a bottle of compressed helium is used to inflate an inelastic flexible balloon, originally folded completely flat to a volume of 0.5 m^3. If the barometer reads 760 mm Hg, what is the amount of work done upon the atmosphere by the balloon ? [**Ans.** 50.66 kJ]
8. A rubber balloon (flexible) is to be filled with hydrogen from a storage cylinder which contains hydrogen at 5 bar pressure until it has a volume of 1 m^3. The atmospheric pressure may be taken as 1.013×10^2 N/m^2. Determine the work done by the system comprising the hydrogen initially in the bottle. [**Ans.** 101.3 kJ]
9. When the valve of the evacuated bottle is opened, atmospheric air rushes into it. If the atmospheric pressure is 101.325 kPa, and 1.2 m^3 of air (measured at atmospheric conditions) enters into the bottle, calculate the work done by the air. [**Ans.** – 60.8 kJ]

10. A pump forces 1.2 m^3/min of water horizontally from an open well to a closed tank where the pressure is 0.9 MPa. Compute the work the pump must do upon the water in an hour just to force the water into the tank against the pressure. [**Ans.** 15972 kJ]

11. 14.5 litres of gas at 172 MN/m^2 is expanded at constant pressure until its volume becomes 130.5 litres. Determine the work done by the gas. [**Ans.** 199.5 kJ]

12. 1 kg of a fluid is compressed reversibly according to a law $pv = 0.25$, where p is in bar and v is in m^3/kg. The final volume is 1/4 of the initial volume. Calculate the work done on the fluid and sketch the process on a *p-v* diagram. [**Ans.** 34660 Nm]

13. A gas having a volume of 0.05 m^3 and pressure of 6.9 bar expands reversibly in a cylinder behind a piston according to law pv = constant until the volume is 0.08 m^3. Calculate the work done by the gas. Also sketch the process on a *p-v* diagram. [**Ans.** 15300 Nm]

14. A certain fluid at 10 bar is contained in a cylinder behind a piston, the initial volume being 0.05 m^3. Calculate the work done by the fluid when it expands reversibly, (*a*) At constant pressure to final volume of 0.2 m^3 ; (*b*) According to linear law to final volume of 0.2 m^3 and a final pressure of 2 bar ; (*c*) According to a law pV = constant to a final volume of 0.1 m^3 ; (*d*) According to law pV^3 = constant to a final volume of 0.06 m^3. Sketch all processes on *p-V* diagram. [**Ans.** 150000 ; 90000 ; 34700 ; 7640 Nm]

15. A fluid undergoes the following processes :

(*i*) Heated reversibly at a constant pressure of 1.05 bar until it has a specific volume of 0.1 m^3/kg.

(*ii*) It is then compressed reversibly according to a law pv = constant to a pressure of 4.2 bar.

(*iii*) It is then allowed to expand reversibly according to a law $pv^{1.3}$ = constant.

(*iv*) Finally it is heated at constant volume back to initial conditions.

The work done in the constant pressure process is 515 Nm and the mass of fluid present is 0.2 kg. Calculate the net work done on or by the fluid in the cycle and sketch the cycle on a *p-v* diagram. [**Ans.** – 422 Nm]

Properties of Pure Substances

2

2.1. DEFINITION OF THE PURE SUBSTANCE

A pure substance is a system which is (*i*) *homogeneous in composition*, (*ii*) *homogeneous in chemical aggregation*, and (*iii*) *invariable in chemical aggregation*.

— "*Homogeneous in composition*" means that the composition of each part of the system is the *same* as the composition of *every other part*. "*Composition* means the relative proportions of the chemical elements into which the sample can be analysed. It does not matter how these elements are combined.

For example in Fig. 2.1 system (*a*), comprising steam and water, is homogeneous in composition, since chemical analysis would reveal that hydrogen and oxygen atoms are presents in the ratio 2 : 1 whether the sample be taken from the steam or from the water. The same is true of system (*b*) containing uncombined hydrogen and oxygen gas in the atomic ratio 2 : 1 in the upper part, and water in the lower part. System (*c*) however, is not homogeneous in composition, for the hydrogen and oxygen are present in the ratio 1 : 1 in the upper part, but in the ratio 2 : 1 (as water) in the lower part.

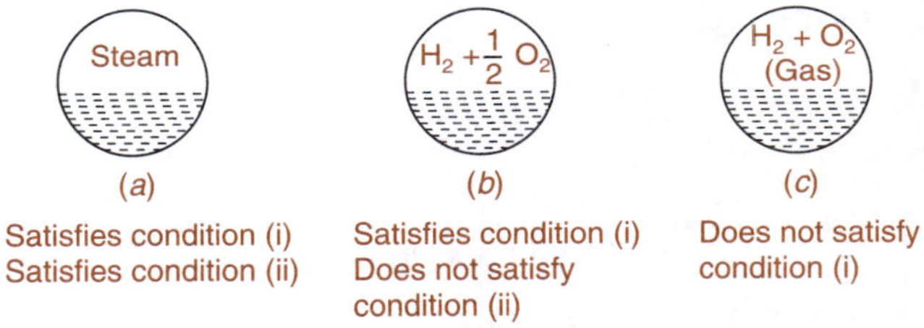

Fig. 2.1. Illustrating the definition of a pure substance.

— "*Homogeneous in chemical aggregation*" means that the chemical elements must be combined chemically in the same way in all parts of the system. Consideration of Fig. 2.1 again shows that the system (*a*) satisfies this condition also ; for steam and water consist of identical molecules. System (*b*) on the other hand is not homogeneous in chemical aggregation since in the upper part of the system the hydrogen and oxygen are not combined chemically (individual atoms of H and O are not uniquely associated), whereas in the lower part of the system the hydrogen and oxygen are combined to form water.

Note however that a uniform mixture of steam, hydrogen gas, and oxygen gas would be regarded as homogeneous in both composition and chemical aggregation whatever the relative proportions of the components.

— "*Invariable in chemical aggregation*" means that the state of chemical combination of the system does not change with *time* (condition (*ii*) referred to variation with *position*). Thus a mixture of hydrogen and oxygen, which changed into steam during the time that the system was under consideration, would not be a pure substance.

2.2. PHASE CHANGE OF A PURE SUBSTANCE

Let us consider 1 kg of liquid water at a temperature of 20°C in a cylinder fitted with a piston, which exerts on the water a constant pressure of one atmosphere (1.0132 bar) as shown in Fig. 2.2 (*i*).

— As the water is heated slowly its temperature rises until the temperature of the liquid water becomes 100°C. During the process of heating, the *volume slightly increases* as indicated by the line 1–2 on the temperature-specific volume diagram (Fig. 2.3). The piston starts moving upwards.

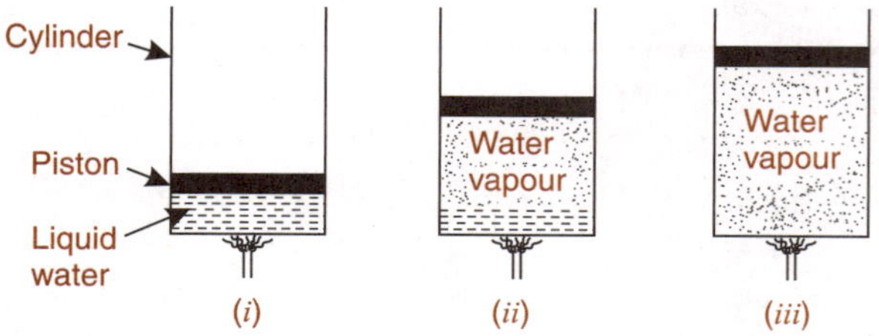

Fig. 2.2. Phase change of water at constant pressure from liquid to vapour phase.

— If the heating of the liquid, after it attains a temperature of 100°C, is continued it *undergoes a change in phase*. A portion of the liquid water changes into vapour as shown in Fig. 2.2 (*ii*). This

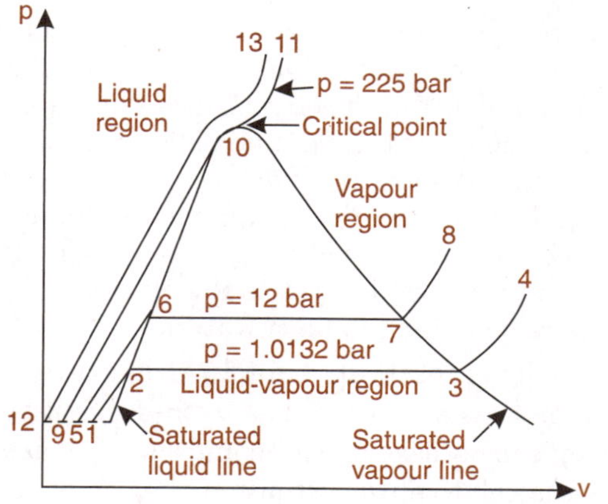

Fig. 2.3

state is described by the line 2–3 in Fig. 2.3. The amount of heat required to convert the liquid water completely into vapour under this condition is called the *heat of vaporisation*. The temperature at which vaporisation takes place at a given pressure is called the *saturation temperature* and the given pressure is called the *saturation pressure*.

During the process represented by the line 2–3 (Fig. 2.3) the volume increases rapidly and piston moves upwards as shown in Fig. 2.2 (*iii*).

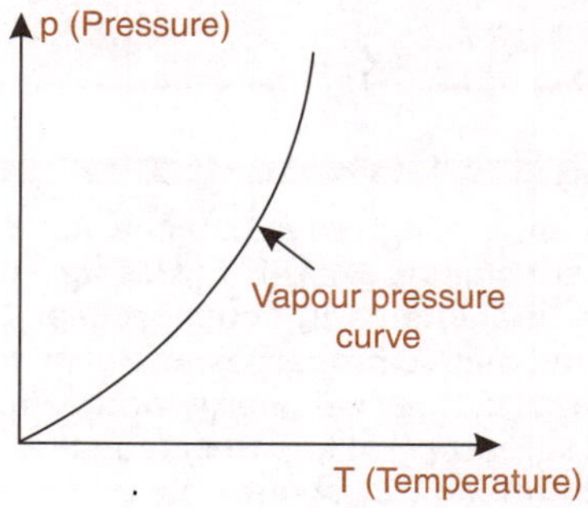

Fig. 2.4. Vapour pressure curve for water.

For a pure substance, definite relationship exists between the saturation pressure and saturation temperature as shown in Fig. 2.4, the curve so obtained is called *vapour pressure curve*.

— It may be noted that if the temperature of the liquid water on cooling becomes lower than the saturation temperature for the given pressure, the liquid water is called a *sub-cooled liquid*. The point '1' (in Fig. 2.3) illustrates this situation, when the liquid water is cooled under atmospheric pressure to a temperature of 20°C, which is below the saturation temperature (100°C).

— Further, at point '1' the temperature of liquid is 20°C and corresponding to this temperature, the saturation pressure is 0.0234 bar, which is lower than the pressure on the liquid water, which is 1 atmosphere. Thus the pressure on the liquid water is greater than the saturation pressure at a given temperature. In this condition, the liquid water is known as the *compressed liquid*.

The term *compressed liquid* or *sub-cooled liquid* is used to distinguish it from *saturated liquid*. *All points in the liquid region indicate the states of the compressed liquid.*

— When all the liquid has been evaporated completely and heat is further added, the *temperature of the vapour increases*. The curve 3-4 in Fig. 2.3 describes the process. When the temperature increases above the saturation temperature (in this case 100°C), the vapour is known as the *superheated vapour* and the temperature at this state is called the *superheated temperature*. There is *rapid increase in volume* and the piston moves upwards [Fig. 2.2 (*iii*)].

The difference between the superheated temperature and the saturation temperature at the given pressure is called the *degree of superheat*.

— If the above mentioned heating process is repeated at different pressures a number of curve similar to 1-2-3-4 are obtained. Thus, if the heating of the liquid water in the piston cylinder arrangement takes place under a constant pressure of 12 bar with an initial temperature of 20°C until the liquid water is converted into superheated steam, then curve 5-6-7-8 will represent the process.

— In the above heating process, it may be noted that, as the pressure increases the *length of constant temperature vaporisation gets reduced*.

From the heating process at a constant pressure of 225 bar represented by the curve 9-10-11 in Fig. 2.3, it can be seen that there is *no constant temperature vaporisation line*. The specific volume of the saturated liquid and of the saturated vapour is the same, *i.e.*, $v_f = v_g$. Such a state of the substance is called the *critical state*. The parameters like temperature, pressure, volume, etc., at such a state are called *critical parameters*.

— The curve 12-13 (Fig. 2.3) *represents a constant pressure heating process, when the pressure is greater than the critical pressure*. At this state, *the liquid water is directly converted into superheated steam*. As there is no definite point at which the liquid water changes into superheated steam, it is generally called *liquid water when the temperature is less than the critical temperature and superheated steam when the temperature is above the critical temperature*.

2.3. p-T (Pressure-Temperature) DIAGRAM FOR A PURE SUBSTANCE

If the vapour pressure of a solid is measured at various temperatures until the *triple point* is reached and then that of the liquid is measured until the critical point is reached, the result when plotted on a *p-T* diagram appears as in Fig. 2.5.

If the substance at the triple point is compressed until there is no vapour left and the pressure on the resulting mixture of liquid and solid is increased, the temperature will have to be changed for equilibrium to exist between the solid and the liquid.

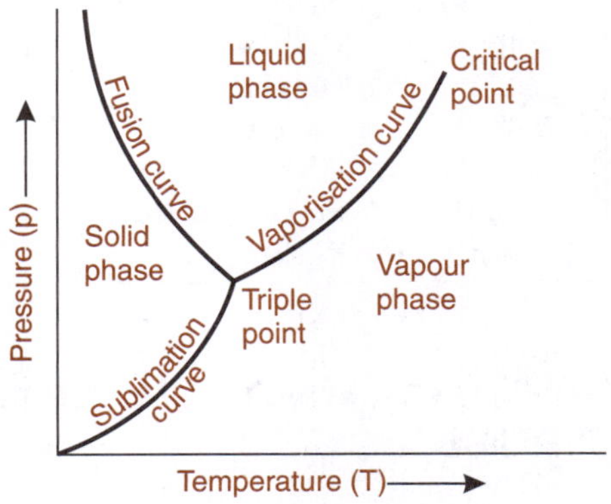

Fig. 2.5. *p-T* diagram for a substance such as water.

Measurements of these pressures and temperatures give rise to a third curve on the *p-T* diagram, starting at the triple point and continuing indefinitely.

The points representing the coexistence of (*i*) *solid* and *vapour* lie on the '*sublimation curve*', (*ii*) *liquid* and *vapour* lie on the '*vaporisation curve*', (*iii*) *liquid* and *solid* lie on the '*fusion curve*'. In the particular case of *water*, the sublimation curve is called the *frost line*, the vaporisation curve is called the *steam line*, and the fusion curve is called the *ice line*.

The slopes of sublimation and the vaporisation curves for all substances are *positive*. The slope of the fusion curve, however may be positive or negative. The fusion curve of *most substances* have a *positive slope*. Water is one of the important exceptions.

Triple point

The triple point is merely the point of intersection of sublimation and vapourisation curves. It must be understood that only on *p-T* diagram is the triple point represented by a *point*. On *p-V* diagram it is a *line*, and on a *U-V* diagram it is a *triangle*.

— The pressure and temperature at which all three phases of a pure substance coexist may be measured with the apparatus that is used to measure vapour pressure.

— Triple-point data for some interesting substances are given in Table 2.1.

Table 2.1. Triple-point Data

S. No.	Substance	Temp., K	Pressure, mm Hg
1.	Hydrogen (normal)	13.96	54.1
2.	Deuterium (normal)	18.63	128
3.	Neon	24.57	324
4.	Nitrogen	63.18	94
5.	Oxygen	54.36	1.14
6.	Ammonia	195.40	45.57
7.	Carbon dioxide	216.55	3.880
8.	Sulphur dioxide	197.68	1.256
9.	Water	273.16	4.58

2.4. p-V-T (Pressure-Volume-Temperature) SURFACE

A detailed study of the heating process reveals that the temperature of the solid rises and then during the change of phase from solid to liquid (or solid to vapour) the temperature remains constant. This phenomenon is common to all phase changes. Since the temperature is constant, pressure and temperature are not independent properties and cannot be used to specify state during a change of phase.

The combined picture of change of pressure, specific volume and temperature may be shown on a three dimensional state model. Fig. 2.6 illustrates the equilibrium states for a pure substance which expands on fusion. Water is an example of a substance that exhibits this phenomenon.

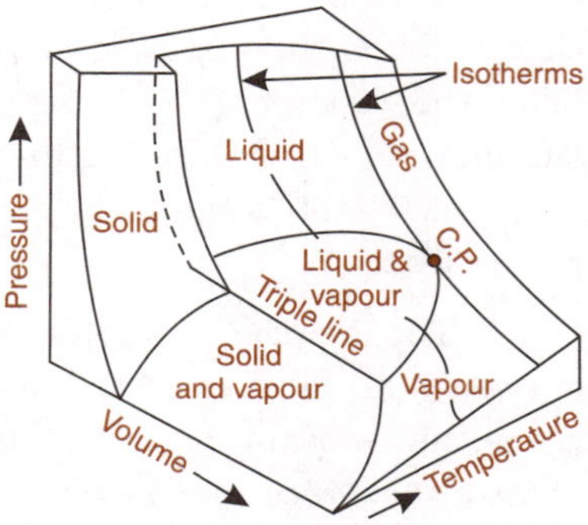

Fig. 2.6. A pressure-volume-temperature (*p-V-T*) surface.

All the equilibrium states lie on the surface of the model. States represented by the space above or below the surface are not possible. It may be seen that the *triple point appears as a line in this representation*. The point *C.P.* is called the *critical point* and no liquid phase exists at temperatures above the isotherms through this point. The term evaporation is meaningless in this situation.

At the critical point the temperature and pressure are called the critical temperature and the critical pressure respectively and when the temperature of a substance is above the critical value, it is called a gas. It is not possible to cause a phase change in a gas unless the temperature is lowered to a value less than the critical temperature. Oxygen and nitrogen are examples of gases that have critical temperatures below normal atmospheric temperature.

2.5. PHASE CHANGE TERMINOLOGY AND DEFINITIONS

Suffices : Solid i
Liquid f
Vapour g

Phase change	Name	Process	Process suffix
1. Solid-liquid	Fusion	Freezing, melting	*if*
2. Solid-vapour	Sublimation	Frosting, defrosting	*ig*
3. Liquid-vapour	Evaporation	Evaporating, Condensing	*fg*

Triple point. The only state at which the solid, liquid and vapour phases *coexist in equilibrium*.

Critical point (*C.P.*). The *limit of distinction between a liquid and vapour*.

Critical pressure. The pressure at the critical point.

Critical temperature. The temperature at the critical point.

Gas. A vapour whose *temperature is greater than the critical temperature*.

Liquid-vapour terms : Refer to Fig. 2.7.

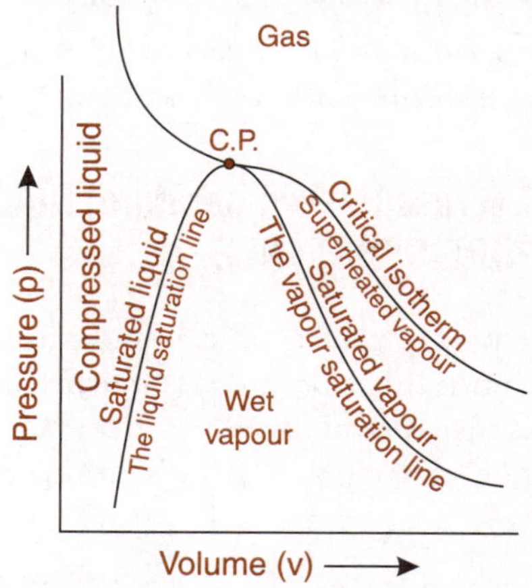

Fig. 2.7. Phase change terminology.

Saturation temperature. The phase change temperature corresponding to the saturation pressure. Sometimes called the *boiling temperature*.

Saturation pressure. The phase change pressure.

Compressed liquid. Liquid whose temperature is *lower than the saturation temperature*. Sometimes called a *sub-cooled liquid*.

Saturated liquid. Liquid at the saturation temperature corresponding to the saturation pressure. That is liquid about to commence evaporating, represented by the point f on a diagram.

Saturated vapour. A term including wet and dry vapour.

Dry (saturated) vapour. Vapour which has just completed evaporation. The pressure and temperature of the vapour are the saturation values. Dry vapour is represented by a point g on a diagram.

Wet vapour. The mixture of saturated liquid and dry vapour during the phase change.

Superheated vapour. Vapour whose temperature is greater than the saturation temperature corresponding to the pressure of the vapour.

Degree of superheat. The term used for the numerical amount by which the temperature of a superheated vapour exceeds the saturation temperature.

2.6. PROPERTY DIAGRAMS IN COMMON USE

Besides *p-V* diagram which is useful because *pressure and volume are easily visualised* and the *T-s* chart which is used in *general thermodynamic work*, there are other charts which are of *practical use for particular applications*. The *specific enthalpy-specific entropy chart is used for steam plant work and the pressure-specific enthalpy chart is used in refrigeration work*. Sketches of these charts are shown in Fig. 2.8. These charts are drawn for H_2O (water and steam) and represent the correct shape of the curves for this substance.

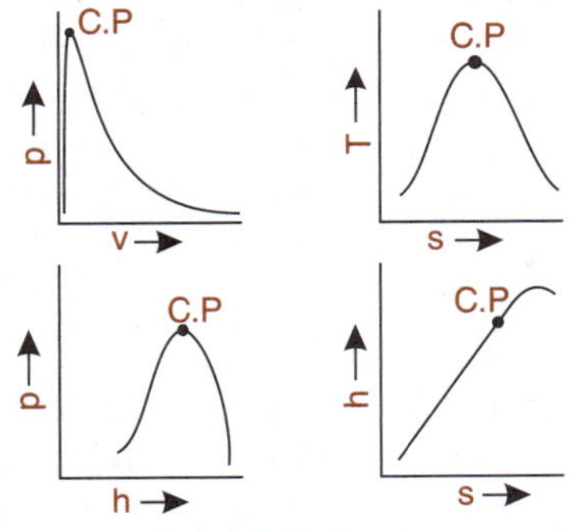

Fig. 2.8

2.7. FORMATION OF STEAM

The process of formation of steam is discussed in detail in the following few pages :

Consider a cylinder fitted with a piston which can move freely upwards and downwards in it. Let, for the sake of simplicity, there be 1 kg of water at 0°C with volume v_f m^3 under the piston [Fig. 2.9 (*i*)]. Further let the piston is loaded with load *W* to ensure heating at constant pressure. Now if the heat is imparted to water, a rise in temperature will be noticed and this rise will continue till boiling point is reached. The temperature at which water starts boiling depends upon the pressure and as such for *each pressure* (under which water is heated) *there is a different boiling point*. This boiling temperature is known as the temperature of formation of steam or *saturation temperature*.

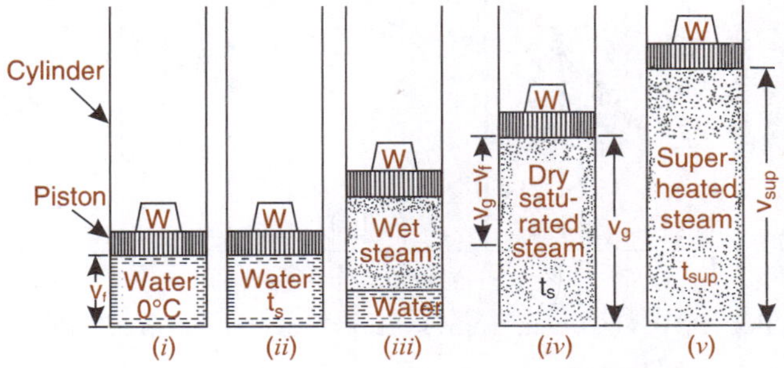

Fig. 2.9. Formation of steam.

It may be noted during heating up to boiling point that there will be slight increase in volume of water due to which piston moves up and hence work is obtained as shown in Fig. 2.9 (*ii*). This work, however, is so *small* that is can be *neglected*.

Now, if supply of heat to water is continued it will be noticed that rise of temperature after the boiling point is reached *nil* but piston starts moving upwards which indicates that there is increase is volume which is only possible if steam formation occurs. The heat being supplied does not show any rise of temperature but changes water into vapour state (steam) and is known as *latent heat* or *hidden heat*. So long as the steam is in contact with water, it is called *wet steam* [Fig. 2.9 (*iii*)] and if heating of steam is further progressed [as shown in Fig. 2.9 (*iv*)] such that all the water particles associated with steam are evaporated, the steam so obtained is called *dry and saturated steam*. If v_g m^3 is the volume of 1 kg of dry and saturated steam then work done on the piston will be

$$p(v_g - v_f) \quad ...(2.1)$$

where p is the constant pressure (due to weight '*W*' on the piston).

Again, if supply of heat to the dry and saturated steam is continued at constant pressure there will be increase in temperature and volume of steam. The steam so obtained is called *superheated steam* and it *behaves like a perfect gas*. This phase of steam formation is illustrated in Fig. 2.9 (*v*).

Fig. 2.10 shows the graphical representation of formation of steam.

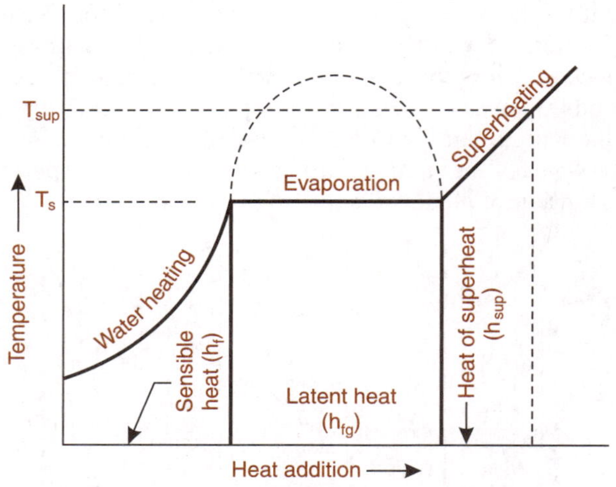

Fig. 2.10. Graphical representation of formation of steam.

2.8. IMPORTANT TERMS RELATING TO STEAM FORMATION

1. Sensible heat of water (h_f). *It is defined as the quantity of heat absorbed by 1 kg of water when it is heated from 0°C (freezing point) to boiling point.* It is also called total heat (or enthalpy) of water or liquid heat invariably. It is reckoned from 0°C where sensible heat is taken as zero. If 1 kg of water is heated from 0°C to 100°C the sensible heat added to it will be 4.18 × 100 = 418 kJ but if water is at say 20°C initially then sensible heat added will be 4.18 × (100 – 20) = 334.4 kJ. This type of heat is denoted by letter h_f and its value can be directly read from the steam tables.

Note. The value of specific heat of water may be taken as 4.18 kJ/kg K at low pressures but at high pressures it is different from this value.

2. Latent heat or hidden heat (h_{fg}). *It is the amount of heat required to convert water at a given temperature and pressure into steam at the same temperature and pressure.* It is expressed by the symbol h_{fg} and its value is available from steam tables. *The value of latent heat is not constant and varies according to pressure variation.*

3. Dryness fraction (x). The term dryness fraction is related with wet steam. *It is defined as the ratio of the mass of actual dry steam to the mass of steam containing it.* It is usually expressed by the symbol '*x*' or '*q*'.

If m_s = Mass of dry steam contained in steam considered, and

m_w = Weight of water particles in suspension in the steam considered,

Then,
$$x = \frac{m_s}{m_s + m_w} \qquad ...(2.2)$$

Thus if in 1 kg of wet steam 0.9 kg is the dry steam and 0.1 kg water particles then $x = 0.9$.

Note. No steam can be completely dry and saturated, so long as it is in contact with the water from which it is being formed.

4. Total heat or enthalpy of wet steam (h). *It is defined as the quantity of heat required to convert 1 kg of water at 0°C into wet steam at constant pressure.* It is the sum of total heat of water and the latent heat and this sum is also called **enthalpy.**

In other words,
$$h = h_f + xh_{fg} \qquad ...(2.3)$$

If steam is dry and saturated, then $x = 1$ and $h_g = h_f + h_{fg}$.

5. Superheated steam. When steam is heated after it has become dry and saturated, it is called superheated steam and the process of heating is called *superheating. Superheating is always carried out at constant pressure*. The additional amount of heat supplied to the steam during superheating is called as *'Heat of superheat'* and can be calculated by using the specific heat of superheated steam at constant pressure (c_{ps}), the value of which varies from 2.0 to 2.1 kJ/kg K depending upon pressure and temperature.

If T_{sup}., T_s are the temperatures of superheated steam in *K* and wet or dry steam, then $(T_{sup} - T_s)$ is called '*degree of superheat*'.

The total heat of superheated steam is given by
$$h_{sup} = h_f + h_{fg} + c_{ps}\,(T_{sup} - T_s) \qquad ...(2.4)$$

Superheated steam behaves like a gas and therefore it follows the gas laws. The value of *n* for this type of steam is 1.3 and the law for the adiabatic expansion is $pv^{1.3}$ = constant.

The ***advantages*** obtained by using '*superheated*' steam are as follows :

(*i*) By superheating steam, its heat content and hence its capacity to do work is increased without having to increase its pressure.

(*ii*) Superheating is done in a superheater which obtains its heat from waste furnace gases which would have otherwise passed uselessly up the chimney.

(*iii*) High temperature of superheated steam results in an increase in thermal efficiency.

(*iv*) Since the superheated steam is at a temperature above that corresponding to its pressure, it can be considerably cooled during expansion in an engine before its temperature falls below that at which it will condense and thereby become wet. Hence, heat losses due to condensation of steam on cylinder walls etc., are avoided to a great extent.

6. Volume of wet and dry steam. If the steam has dryness fraction of x, then 1 kg of this steam will contain x kg of dry steam and $(1 - x)$ kg of water. If v_f is the volume of 1 kg of water and v_g is the volume of 1 kg of perfect dry steam (also known as specific volume), then volume of 1 kg of wet steam = volume of dry steam + volume of water.

$$= xv_g + (1 - x)v_f \quad ...(2.5)$$

Note. The volume of v_f at low pressures is very small and is generally neglected. Thus is general, the volume of 1 kg of wet steam is given by, xv_g and density $\frac{1}{xv_g}$ kg/m³.

$$= xv_g + v_f - xv_f$$
$$= v_f + x(v_g - v_f)$$
$$= v_f + xv_{fg} \quad ...[2.5\ (a)]$$
$$= v_f + xv_{fg} + v_{fg} - v_{fg}$$
$$= (v_f + v_{fg}) - (1 - x)\, v_{fg}$$
$$= v_g - (1 - x)v_{fg} \quad ...[2.5\ (b)]$$

7. Volume of superheated steam. As superheated steam behaves like a perfect gas its volume can be found out in the same way as the gases.

If v_g = Specific volume of dry steam at pressure p,

T_s = Saturation temperature in K,

T_{sup} = Temperature of superheated steam in K, and

v_{sup} = Volume of 1 kg of superheated steam at pressure p,

Then
$$\frac{p \cdot v_g}{T_s} = \frac{p \cdot v_{sup}}{T_{sup}}$$

or
$$v_{sup} = \frac{v_g T_{sup}}{T_s} \quad ...(2.6)$$

2.9. THERMODYNAMIC PROPERTIES OF STEAM AND STEAM TABLES

In engineering problem, for any fluid which is used as working fluid, the six basic thermodynamic properties required are : p (pressure), T (temperature), v (volume), u (internal energy), h (enthalpy) and s (entropy). These properties must be known at different pressure for analysing the thermodynamic cycles used for work producing devices. The values of these properties are determined theoretically or experimentally and are tabulated in the form of tables which are known as '*Steam Tables*'. The properties of wet steam are then computed from such tabulated data. Tabulated values are also available for superheated steam. It may be noted that *steam has only one saturation temperature at each pressure*.

Following are the thermodynamic properties of steam which are tabulated in the form of table :

p = Absolute pressure (bar or kPa) ;
t_s = Saturation temperature (°C) ;
h_f = Enthalpy of saturated liquid (kJ/kg) ;
h_{fg} = Enthalpy or latent heat of vaporisation (kJ/kg) ;
h_g = Enthalpy of saturated vapour (steam) (kJ/kg) ;
s_f = Entropy of saturated liquid (kJ/kg K) ;
s_{fg} = Entropy of vaporisation (kJ/kg K) ;
s_g = Entropy of saturated vapour (steam) (kJ/kg K) ;
v_f = Specific volume of saturated liquid (m³/kg) ;
v_g = Specific volume of saturated vapour (steam) (m³/kg).

Also, $h_{fg} = h_g - h_f$ ...... Change of enthalpy during evaporation

$s_{fg} = s_g - s_f$ ...... Change of entropy during evaporation

$v_{fg} = v_g - v_f$ ...... Change of volume during evaporation.

The above mentioned properties at different pressures are tabulated in the form of tables as under :

The internal energy of steam ($u = h - pv$) is also tabulated in some steam tables.

STEAM TABLES

Absolute pressure bar, p	*Temperature °C* t_s	*Specific enthalpy kJ/kg* h_f	h_{fg}	h_g	*Specific entropy kJ/kg K* s_f	s_{fg}	s_g	*Specific volume m³/kg* v_f	v_g
1.0	99.6	417.5	2257.9	2675.4	1.3027	6.0571	7.3598	0.001043	1.6934
50.0	263.9	1154.9	1639.7	2794.2	2.9206	3.0529	5.9735	0.001286	0.00394
100.0	311.1	1408.0	1319.7	2727.7	3.3605	2.2593	5.6198	0.001452	0.01811

2.10. EXTERNAL WORK DONE DURING EVAPORATION

When water is evaporated to form *saturated* steam, its volume increases from v_f to v_g at a constant pressure, and thus external work is done by steam due to increase in volume. The energy for doing the work is obtained during the absorption of latent heat. This work is called **external work of evaporation** and is given by $p(v_g - v_f)$.

i.e., External work of evaporation $= p(v_g - v_f)$ ...(2.7)

As at low pressure v_f is very small and hence neglected, work of evaporation is

$$p \cdot v_g \quad \text{...(2.8)}$$

In case of *wet* steam with dryness fraction x, work of evaporation will be

$$pxv_g \quad \text{...(2.9)}$$

2.11. INTERNAL LATENT HEAT

The latent heat consists of true latent heat and the work of evaporation. This true latent heat is called the *internal latent heat* and may also be found as follows :

$$\text{Internal latent heat} = h_{fg} - \frac{pv_g}{J} \quad \text{...(2.10)}$$

J = 1 in SI units.

2.12. INTERNAL ENERGY OF STEAM

It is defined as the actual energy stored in the steam. As per previous articles, the total heat of steam is sum of sensible heat, internal latent heat and the external work of evaporation. Work of evaporation is not stored in the steam as it is utilised in doing external work. Hence the internal energy of steam could be found by subtracting work of evaporation from the total heat.

In other words,

$$h = \frac{pv_g}{J} + u,$$

where u is internal energy of 1 kg of steam at pressure p

or
$$u = h - \frac{pv_g}{J}$$

In case of wet steam with dryness fraction 'x'

$$u = h - \frac{pxv_g}{J} \quad \text{...(2.11)}$$

and if steam is superheated to a volume of v_{sup} per kg.

$$h_{sup} = h_f + h_{fg} + c_{ps}(T_{sup} - T_s)$$

and
$$u = h_{sup} - \frac{p \cdot v_{sup}}{J} \quad \text{...(2.12)}$$

2.13. ENTROPY OF WATER

Note. For definition of entropy phase refer Article 4.12).

Consider 1 kg of water being heated from temperature T_1 to T_2 at constant pressure. The change in entropy will be given by,

$$ds = \frac{dQ}{T} = c_{pw} \cdot \frac{dT}{T}$$

Integrating both sides, we get

$$\int_{s_1}^{s_2} ds = \int_{T_1}^{T_2} c_{pw} \frac{dT}{T}$$

$$s_2 - s_1 = c_{pw} \log_e \frac{T_2}{T_1} \quad \text{...(2.13)}$$

If 0°C is taken as datum, then entropy of water per kg at any temperature T above this datum will be

$$s_f = c_{pw} \log_e \frac{T}{273} \quad \text{...(2.14)}$$

2.14. ENTROPY OF EVAPORATION

The change of entropy (ds) is given by,

$$ds = \frac{dQ}{T}$$

or $s_2 - s_1 = \dfrac{Q}{T}$, where Q is the heat absorbed.

When water is evaporated to steam completely the heat absorbed is the latent heat and this heat goes into water without showing any rise of temperature.

Then $Q = h_{fg}$

and
$$s_{evap.} = \frac{h_{fg}}{T_s} \quad \text{...(2.15)}$$

However, in case of wet steam with dryness fraction x the evaporation will be partial and heat absorbed will be xh_{fg} per kg of steam. The change of entropy will be $\dfrac{xh_{fg}}{T_s}$.

2.15. ENTROPY OF WET STEAM

The total entropy of wet steam is the sum of entropy of water (s_f) and entropy of evaporation (s_{fg}).

In other words, $s_{wet} = s_f + \dfrac{xh_{fg}}{T_s}$ ...(2.16)

where s_{wet} = Total entropy of wet steam,

s_f = Entropy of water, and

$\dfrac{xh_{fg}}{T_s}$ = Entropy of evaporation.

If steam is dry and saturated, *i.e.*, $x = 1$, then

$$s_g = s_f + \frac{h_{fg}}{T_s} \quad \text{...(2.17)}$$

2.16. ENTROPY OF SUPERHEATED STEAM

Let 1 kg of dry saturated steam at T_s (saturation temperature of steam) be heated to T_{sup}. If specific heat at constant pressure is c_{ps}, then change of entropy during superheating at constant pressure p

$$= c_{ps} \log_e \left(\frac{T_{sup}}{T_s}\right).$$

Total entropy of superheated steam above the freezing point of water.

s_{sup} = Entropy of dry saturated steam + change of entropy during superheating

$$= s_f + \frac{h_{fg}}{T_s} + c_{ps} \log_e \left(\frac{T_{sup}}{T_s}\right) = s_g + c_{ps} \log_e \left(\frac{T_{sup}}{T_s}\right) \quad ...(2.18)$$

2.17. ENTHALPY-ENTROPY (h-s) CHART OR MOLLIER DIAGRAM

Dr. Mollier, in 1904, conceived the idea of plotting total heat against entropy, and his diagram is more widely used than any other entropy diagram, since the work done on vapour cycles can be scaled from this diagram directly *as a length ; whereas on T-s diagram it is represented by an area.*

A sketch of the *h-s* chart is shown in Fig. 2.11.

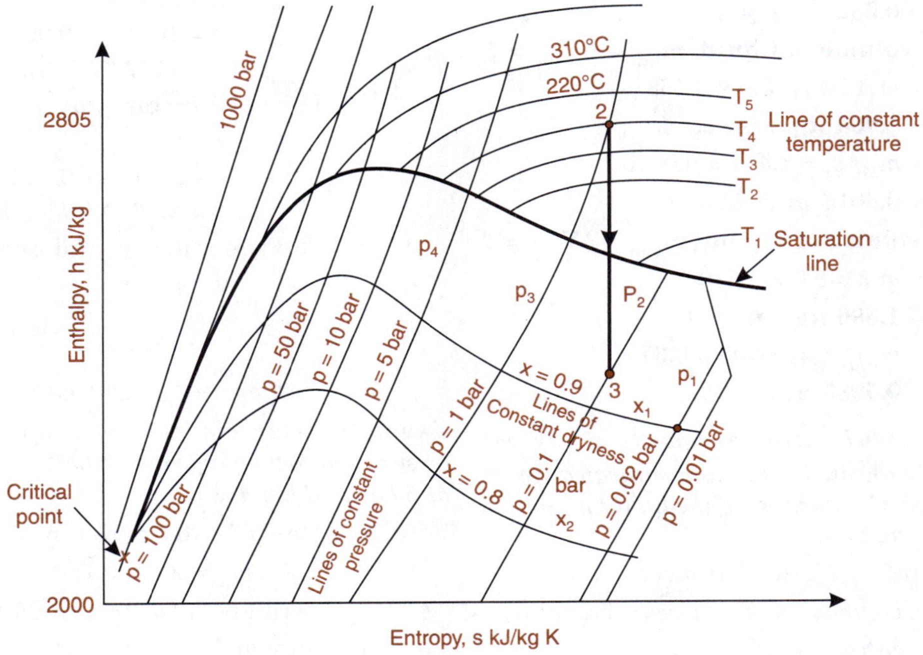

Fig. 2.11. Enthalpy-entropy (*h-s*) chart.

— Lines of constant pressure are indicated by p_1, p_2 etc., lines of constant temperature by T_1, T_2, etc.

— Any two independent properties which appear on the chart are sufficient to define the state (*e.g.,* p_1 and x_1 define state 1 and h can be read off the vertical axis).

— In the superheat region, pressure and temperature can define the state (*e.g.,* p_3 and T_4 define the state 2, and h_2 can be read off).

— A line of constant entropy between two state points 2 and 3 defines the properties at all points during an *isentropic process* between the two states.

Example 2.1. *Calculate the dryness fraction (quality) of steam which has 1.5 kg of water in suspension with 50 kg of steam.*

Solution. Mass of dry steam, m_s = 50 kg

Mass of water in suspension, m_w = 1.5 kg

∴ **Dryness fraction,** x

$$= \frac{\text{Mass of dry steam}}{\text{Mass of dry steam + Mass of water in suspension}}$$

$$= \frac{m_s}{m_s + m_w} = \frac{50}{50 + 1.5} = \mathbf{0.971.} \quad \textbf{(Ans.)}$$

☞ **Example 2.2.** *A vessel having a volume of 0.6 m³ contains 3.0 kg of liquid water and water vapour mixture in equilibrium at a pressure of 0.5 MPa. Calculate :*

(*i*) *Mass and volume of liquid ;*

(*ii*) *Mass and volume of vapour.*

Solution. Volume of the vessel, $V = 0.6 \text{ m}^3$

Mass of liquid water and water vapour, $m = 3.0$ kg

Pressure, $p = 0.5$ MPa = 5 bar

Thus, specific volume, $v = \dfrac{V}{m} = \dfrac{0.6}{3.0} = 0.2 \text{ m}^3\text{/kg}$

At 5 bar : From steam tables,

$$v_{fg} = v_g - v_f = 0.375 - 0.00109 = 0.3739 \text{ m}^3\text{/kg}$$

We know that, $v = v_g - (1 - x)\, v_{fg}$, where x = quality of the vapour.

$$0.2 = 0.375 - (1 - x) \times 0.3739$$

$$\therefore \quad (1 - x) = \frac{(0.375 - 0.2)}{0.3739} = 0.468$$

or $\quad x = 0.532$

(*i*) **Mass and volume of liquid, $m_{liq.}$ = ? $V_{liq.}$ = ?**

$$\mathbf{m_{liq.}} = m(1 - x) = 3.0 \times 0.468 = \mathbf{1.404 \text{ kg.}} \quad \textbf{(Ans.)}$$

$$\mathbf{V_{liq.}} = m_{liq.}\, v_f = 1.404 \times 0.00109 = \mathbf{0.0015 \text{ m}^3.} \quad \textbf{(Ans.)}$$

(*ii*) **Mass and volume of vapour, $m_{vap.}$ = ? $V_{vap.}$ = ?**

$$\mathbf{m_{vap.}} = m.x = 3.0 \times 0.532 = \mathbf{1.596 \text{ kg.}} \quad \textbf{(Ans.)}$$

$$\mathbf{V_{vap.}} = m_{vap.}\, v_g = 1.596 \times 0.375 = \mathbf{0.5985 \text{ m}^3.} \quad \textbf{(Ans.)}$$

☞ **Example 2.3.** *A vessel having a capacity of 0.05 m³ contains a mixture of saturated water and saturated steam at a temperature of 245°C. The mass of the liquid present is 10 kg. Find the following :*

(*i*) *The pressure,* (*ii*) *The mass,*

(*iii*) *The specific volume,* (*iv*) *The specific enthalpy,*

(*v*) *The specific entropy, and*

(*vi*) *The specific internal energy.*

Solution. From steam tables, corresponding to 245°C :

p_{sat} = 36.5 bar, $v_f = 0.001239 \text{ m}^3\text{/kg}$, $v_g = 0.0546 \text{ m}^3\text{/kg}$

h_f = 1061.4 kJ/kg, h_{fg} = 1740.2 kJ/kg

s_f = 2.7474 kJ/kg K, s_{fg} = 3.3585 kJ/kg K.

(*i*) **The pressure = 36.5 bar** (or 3.65 MPa). **(Ans.)**

(*ii*) **The mass, m :**

Volume of liquid, $V_f = m_f v_f$

$$= 10 \times 0.001239 = 0.01239 \text{ m}^3$$

Volume of vapour,

$$V_g = 0.05 - 0.01239 = 0.03761 \text{ m}^3$$

∴ Mass of vapour,

$$m_g = \frac{V_g}{v_g} = \frac{0.03761}{0.0546} = 0.688 \text{ kg}$$

∴ The total mass of mixture,

$$\mathbf{m} = m_f + m_g = 10 + 0.688 = \mathbf{10.688 \text{ kg.}} \quad \textbf{(Ans.)}$$

(*iii*) **The specific volume, v :**

Quality of the mixture,

$$x = \frac{m_g}{m_g + m_f} = \frac{0.688}{0.688 + 10} = 0.064$$

$$\therefore \quad \mathbf{v} = v_f + x v_{fg}$$

$$= 0.001239 + 0.064 \times (0.0546 - 0.001239) \quad (\because \; v_{fg} = v_g - v_f)$$

$$= \mathbf{0.004654 \text{ m}^3\text{/kg.}} \quad \textbf{(Ans.)}$$

(*iv*) **The specific enthalpy, h :**

$$\mathbf{h} = h_f + x h_{fg} = 1061.4 + 0.064 \times 1740.2 = \mathbf{1172.77 \text{ kJ/kg.}} \quad \textbf{(Ans.)}$$

(*v*) **The specific entropy, s :**

$$\mathbf{s} = s_f + x s_{fg} = 2.7474 + 0.064 \times 3.3585 = \mathbf{2.9623 \text{ kJ/kg K.}} \quad \textbf{(Ans.)}$$

(*vi*) **The specific internal energy, u :**

$$\mathbf{u} = h - pv$$

$$= 1172.77 - \frac{36.5 \times 10^5 \times 0.004654}{1000}$$

$$= \mathbf{1155.78 \text{ kJ/kg.}} \quad \textbf{(Ans.)}$$

Example 2.4. *Determine the amount of heat, which should be supplied to 2 kg of water at 25°C to convert it into steam at 5 bar and 0.9 dry.*

Solution. Mass of water to be converted to steam,

$$m_w = 2 \text{ kg}$$

Temperature of water, t_w = 25°C

Pressure and dryness fraction of steam

= 5 bar, 0.9 dry

At 5 bar : From steam tables,

h_f = 640.1 kJ/kg ; h_{fg} = 2107.4 kJ/kg

Enthalpy of 1 kg of steam (above 0°C)

$$h = h_f + x h_{fg} = 640.1 + 0.9 \times 2107.4 = 2536.76 \text{ kJ/kg}$$

Sensible heat associated with 1 kg of water

$$= m_w \times c_{pw} \times (t_w - 0) = 1 \times 4.18 \times (25 - 0) = 104.5 \text{ kJ}$$

Net quantity of heat to be supplied per kg of water

$$= 2536.76 - 104.5 = 2432.26 \text{ kJ}$$

Total amount of heat to be supplied

$$= 2 \times 2432.26 = \mathbf{4864.52 \text{ kJ.}} \quad \textbf{(Ans.)}$$

Example 2.5. *What amount of heat would be required to produce 4.4 kg of steam at a pressure of 6 bar and temperature of 250°C from water at 30°C ? Take specific heat for superheated steam as 2.2 kJ/kg K.*

Solution. Mass of steam to be produced, $m = 4.4$ kg

Pressure of steam, $p = 6$ bar

Temperature of steam, $t_{sup} = 250°C$

Temperature of water $= 30°C$

Specific heat of steam, $c_{ps} = 2.2$ kJ/kg

At 6 bar, 250°C : From steam tables,

$t_s = 158.8°C$, $h_f = 670.4$ kJ/kg, $h_{fg} = 2085$ kJ/kg

Enthalpy of 1 kg superheated steam reckoned from 0°C,

$$h_{sup} = h_f + h_{fg} + c_{ps}(T_{sup} - T_s)$$
$$= 670.4 + 2085 + 2.2(250 - 158.8)$$
$$= 2956 \text{ kJ}$$

Amount of heat already with 1 kg of water

$$= 1 \times 4.18 \times (30 - 0) = 125.4 \text{ kJ}$$

Net amount of heat required to be supplied per kg

$$= 2956 - 125.4 = 2830.6 \text{ kJ}$$

Total amount of heat required

$$= 4.4 \times 2830.6 = \mathbf{12454.6\ kJ.} \quad \textbf{(Ans.)}$$

☞ **Example 2.6.** *Determine the mass of 0.15 m³ of wet steam at a pressure of 4 bar and dryness fraction 0.8. Also calculate the heat of 1 m³ of steam.*

Solution. Volume of wet steam, $v = 0.15$ m³

Pressure of wet steam, $p = 4$ bar

Dryness fraction, $x = 0.8$

At 4 bar. From steam tables,

$v_g = 0.462$ m³/kg, $h_f = 604.7$ kJ/kg, $h_{fg} = 2133$ kJ/kg

$$\therefore \quad \text{Density} = \frac{1}{xv_g} = \frac{1}{0.8 \times 0.462} = 2.7056 \text{ kg/m}^3$$

Mass of 0.15 m³ of steam

$$= 0.15 \times 2.7056 = \mathbf{0.4058\ kg.} \quad \textbf{(Ans.)}$$

Total heat of 1 m³ of steam which has a mass of 2.7056 kg

$$= 2.7056\ h$$

(where h is the total heat of 1 kg of steam)

$$= 2.7056\ (h_f + xh_{fg})$$
$$= 2.7056(604.7 + 0.8 \times 2133)$$
$$= \mathbf{6252.9\ kJ.} \quad \textbf{(Ans.)}$$

Example 2.7. *1000 kg of steam at a pressure of 16 bar and 0.9 dry is generated by a boiler per hour. The steam passes through a superheater via boiler stop valve where its temperature is raised to 380°C. If the temperature of feed water is 30°C, determine :*

(*i*) *The total heat supplied to feed water per hour to produce wet steam.*

(*ii*) *The total heat absorbed per hour in the superheater.*

Take specific heat for superheated steam as 2.2 kJ/kg K.

Solution. Mass of steam generated, $m = 1000$ kg/h

Pressure of steam, $p = 16$ bar

Dryness fraction, $x = 0.9$

Temperature of superheated steam,

$$T_{sup} = 380 + 273 = 653 \text{ K}$$

Temperature of feed water = 30°C

Specific heat of superheated steam,

$$c_{ps} = 2.2 \text{ kJ/kg K.}$$

At 16 bar. From steam tables,

$t_s = 201.4°C$ ($T_s = 201.4 + 273 = 474.4$ K) ;

$h_f = 858.6$ kJ/kg ; $h_{fg} = 1933.2$ kJ/kg

(*i*) *Heat supplied to feed water per hour* to produce wet steam is given by :

$$H = m\ [(h_f + xh_{fg}) - 1 \times 4.18 \times (30 - 0)]$$
$$= 1000\ [(858.6 + 0.9 \times 1933.2) - 4.18 \times 30]$$
$$= 1000(858.6 + 1739.88 - 125.4)$$
$$= \mathbf{2473.08 \times 10^3\ kJ.} \quad \textbf{(Ans.)}$$

(*ii*) *Heat absorbed by superheater per hour*

$$= m[(1 - x)\ h_{fg} + c_{ps}\ (T_{sup} - T_s)]$$
$$= 1000[(1 - 0.9) \times 1933.2 + 2.2\ (653 - 474.4)]$$
$$= 1000(193.32 + 392.92)$$
$$= \mathbf{586.24 \times 10^3\ kJ.} \quad \textbf{(Ans.)}$$

Example 2.8. *Using steam tables, determine the mean specific heat for superheated steam :*

(*i*) *at 0.75 bar, between 100°C and 150°C ;*

(*ii*) *at 0.5 bar, between 300°C and 400°C.*

Solution. (*i*) **At 0.75 bar.** From steam tables ;

At 100°C, $h_{sup} = 2679.4$ kJ/kg

At 150°C, $h_{sup} = 2778.2$ kJ/kg

$$\therefore \quad 2778.2 = 2679.4 + c_{ps}\ (150 - 100)$$

i.e.,
$$c_{ps} = \frac{2778.2 - 2679.4}{50} = \mathbf{1.976.} \quad \textbf{(Ans.)}$$

(*ii*) **At 0.5 bar.** From steam tables ;

At 300°C, $h_{sup} = 3075.5$ kJ/kg

At 400°C, $h_{sup} = 3278.9$ kJ/kg

$$\therefore \quad 3278.9 = 3075.5 + c_{ps}\ (400 - 300)$$

i.e.,
$$c_{ps} = \frac{3278.9 - 3075.5}{100} = \mathbf{2.034.} \quad \textbf{(Ans.)}$$

☞ **Example 2.9.** *A pressure cooker contains 1.5 kg of saturated steam at 5 bar. Find the quantity of heat which must be rejected so as to reduce the quality to 60% dry. Determine the pressure and temperature of the steam at the new state.*

Solution. Mass of steam in the cooker = 1.5 kg

Pressure of steam, p = 5 bar

Initial dryness fraction of steam, $x_1 = 1$

Final dryness fraction of steam, $x_2 = 0.6$

Heat to be rejected :

Pressure and temperature of the steam at the new state :

At 5 bar. From steam tables,

$t_s = 151.8°C$; $h_f = 640.1$ kJ/kg ;

$h_{fg} = 2107.4$ kJ/kg ; $v_g = 0.375$ m³/kg

Thus, the volume of pressure cooker

$= 1.5 \times 0.375 = 0.5625$ m³

Internal energy of steam per kg at initial point 1,

$$u_1 = h_1 - p_1 v_1$$

$$= (h_f + h_{fg}) - p_1 v_{g_1} \qquad (\because v_1 = v_{g1})$$

$$= (640.1 + 2107.4) - 5 \times 10^5 \times 0.375 \times 10^{-3}$$

$$= 2747.5 - 187.5 = 2560 \text{ kJ/kg}$$

Also, $V_1 = V_2$ (V_2 = volume at final condition)

i.e., $0.5625 = 1.5[(1 - x_2)\, v_{f2} + x_2 v_{g2}]$

$= 1.5\, x_2 v_{g2}$ ($\because$ v_{f2} is negligible)

$= 1.5 \times 0.6 \times v_{g2}$

$$\therefore \quad v_{g2} = \frac{0.5625}{1.5 \times 0.6} = 0.625 \text{ m}^3/\text{kg}.$$

From steam tables corresponding to 0.625 m³/kg,

$p_2 \simeq$ **2.9 bar**, t_s = **132.4°C**, $h_f = 556.5$ kJ/kg,
$h_{fg} = 2166.6$ kJ/kg

Internal energy of steam per kg, at final point 2,

$$u_2 = h_2 - p_2 v_2$$

$$= (h_{f_2} + x_2 h_{fg_2}) - p_2 x v_{g2} \qquad (\because v_2 = x v_{g2})$$

$$= (556.5 + 0.6 \times 2166.6) - 2.9 \times 10^5 \times 0.6 \times 0.625 \times 10^{-3}$$

$$= 1856.46 - 108.75 = 1747.71 \text{ kJ/kg}.$$

Heat transferred at constant volume per kg

$= u_2 - u_1 = 1747.71 - 2560 = -812.29$ kJ/kg

Thus, **total heat transferred**

$= -812.29 \times 1.5 =$ **– 1218.43 kJ. (Ans.)**

Negative sign indicates that heat has been **rejected.**

☞ **Example 2.10.** *A spherical vessel of 0.9 m³ capacity contains steam at 8 bar and 0.9 dryness fraction. Steam is blown off until the pressure drops to 4 bar. The valve is then closed and the steam is allowed to cool until the pressure falls to 3 bar. Assuming that the enthalpy of steam in the vessel remains constant during blowing off periods, determine :*

(i) The mass of steam blown off ;

(ii) The dryness fraction of steam in the vessel after cooling ;

(iii) The heat lost by steam per kg during cooling.

Solution. Capacity of the spherical vessel, $V = 0.9$ m³

Pressure of the steam, $p_1 = 8$ bar

Dryness fraction of steam, $x_1 = 0.9$

Pressure of steam after blow off, $p_2 = 4$ bar

Final pressure of steam, $p_3 = 3$ bar.

(*i*) **The mass of steam blown off :**

The mass of steam in the vessel

$$m_1 = \frac{V}{x_1 v_{g_1}} = \frac{0.9}{0.9 \times 0.24} = 4.167 \text{ kg}$$

($\because$ At 8 bar : $v_g = 0.24$ m³/kg)

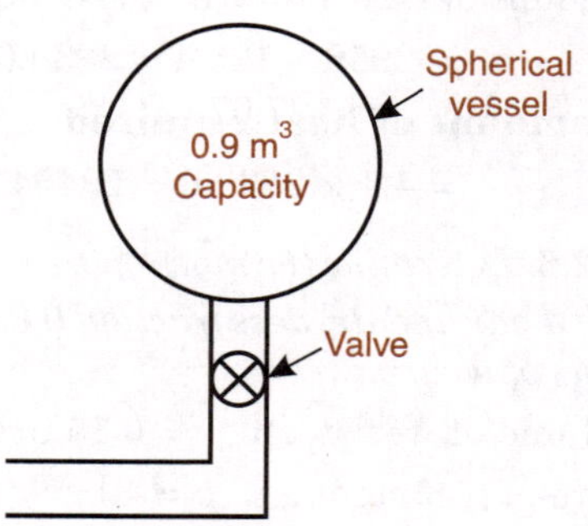

Fig. 2.12

The enthalpy of steam before blowing off (per kg)

$= h_{f_1} + x_1 h_{fg_1} = 720.9 + 0.9 \times 2046.5$ ...... at pressure *8 bar*

$= 2562.75$ kJ/kg

Enthalpy before blowing off = Enthalpy after blowing off

$\therefore$ $2562.75 = (h_{f_2} + x_2 h_{fg_2})$ at pressure 4 bar

$= 604.7 + x_2 \times 2133$ ...... at pressure *4 bar*

$$\therefore \quad x_2 = \frac{2562.75 - 604.7}{2133} = 0.918$$

Now the mass of steam in the vessel after blowing off,

$$m_2 = \frac{0.9}{0.918 \times 0.462} = 2.122 \text{ kg}$$

[$v_{g_2} = 0.462$ m³/kg.......at 4 bar]

Mass of steam blown off,

$m = m_1 - m_2 = 4.167 - 2.122$

$=$ **2.045 kg. (Ans.)**

(*ii*) **Dryness fraction of steam in the vessel after cooling, x_3 :**

As it is constant volume cooling

$\therefore \quad x_2 v_{g_2}$ (at 4 bar) $= x_3 v_{g_3}$ (at 3 bar)

$0.918 \times 0.462 = x_3 \times 0.606$

$$\therefore \quad x_3 = \frac{0.918 \times 0.462}{0.606} = \mathbf{0.699.} \quad \textbf{(Ans.)}$$

(*iii*) **Heat lost during cooling :**

Heat lost during cooling $= m(u_3 - u_2)$, where u_2 and u_3 are the internal energies of steam before starting cooling or after blowing and at the end of the cooling.

$$\therefore \quad u_2 = h_2 - p_2 x_2 v_{g_2} = (h_{f_2} + x_2 h_{fg2}) - p_2 x_2 v_{g2}$$

$$= (604.7 + 0.918 \times 2133) - 4 \times 10^5 \times 0.918 \times 0.462 \times 10^{-3}$$

$$= 2562.79 - 169.65 = 2393.14 \text{ kJ/kg}$$

$$u_3 = h_3 - p_3 x_3 v_{g_3} = (h_{f_3} + x_3 h_{fg3}) - p_3 x_3 v_{g3}$$

$$= (561.4 + 0.669 \times 2163.2) - 3 \times 10^5 \times 0.699 \times 0.606 \times 10^{-3}$$

$$= 2073.47 - 127.07 = 1946.4 \text{ kJ/kg}$$

$\therefore$ Heat transferred during cooling

$$= 2.045\ (1946.4 - 2393.14) = -913.6 \text{ kJ.}$$

i.e., **Heat lost during cooling = 913.6 kJ. (Ans.)**

Example 2.11. *If a certain amount of steam is produced at a pressure of 8 bar and dryness fraction 0.8. Calculate :*

(*i*) *External work done during evaporation.*

(*ii*) *Internal latent heat of steam.*

Solution. Pressure of steam, p = 8 bar

Dryness fraction, x = 0.8

At 8 bar. From steam tables,

$v_g = 0.240 \text{ m}^3/\text{kg}$, $h_{fg} = 2046.5$ kJ/kg

(*i*) **External work done during evaporation**

$$= pxv_g = 8 \times 10^5 \times 0.8 \times 0.24 \text{ Nm}$$

$$= \frac{8 \times 10^5 \times 0.8 \times 0.24}{10^3} = \mathbf{153.6 \ kJ.} \quad \textbf{(Ans.)}$$

(*ii*) **Internal latent heat**

$= xh_{fg}$ – external work done

$= 0.8 \times 2046.5 - 153.6$

$=$ **1483.6 kJ. (Ans.)**

Example 2.12. *A quantity of steam at 10 bar and 0.85 dryness occupies 0.15 m³. Determine the heat supplied to raise the temperature of the steam to 300°C at constant pressure and percentage of this heat which appears as external work.*

Take specific heat of superheated steam as 2.2 kJ/kg K.

Solution. Pressure of steam, $p_1 = p_2$ = 10 bar

Dryness fraction, x_1 = 0.85

Volume of steam, $V_1 = 0.15 \text{ m}^3$

Final temperature of steam, t_{sup_2} = 300°C

Specific heat of superheated steam, c_{ps} = 2.2 kJ/kg K

$$\text{Mass of steam} = \frac{V_1}{x_1 v_{g_1}} = \frac{0.15}{0.85 \times 0.194} = 0.909 \text{ kg}$$

$(\because$ At 10 bar : $v_g = 0.194 \text{ m}^3/\text{kg})$

Heat supplied per kg of steam

$$= (1 - x_1)\, h_{fg_1} + c_{ps}\,(300 - 179.9)$$

$$= (1 - 0.85)2013.6 + 2.2 \times 120.1$$

$$\left[\because \; t_s = 179.9°\text{C... at 10 bar, and } h_{fg} = 2013.6 \text{ kJ/kg}\right]$$

$$= 566.26 \text{ kJ/kg}$$

Total heat supplied

$$= 0.909 \times 566.26 = \mathbf{514.7 \ kJ.} \quad \textbf{(Ans.)}$$

External work done during this process

$$= p\,(v_{sup_2} - x_1 v_{g_1}) \qquad [\because p_1 = p_2 = p]$$

$$= 10 \times 10^5 \left[\left(v_{g_1} \times \frac{T_{sup_2}}{T_{s_1}}\right) - x_1 v_{g_1}\right] \times 10^{-3}$$

$$\left[\because \; \frac{v_{g_1}}{T_{s_1}} = \frac{v_{sup_2}}{T_{sup_2}} \; i.e., \; v_{sup_2} = \frac{v_{g_1} \times T_{sup_2}}{T_{s_1}}\right]$$

$$= 10 \times 10^5 \left[0.194 \times \frac{(300 + 273)}{(179.9 + 273)} - 0.85 \times 0.194\right] \times 10^{-3}$$

$$= \frac{10 \times 10^5}{10^3}\ (0.245 - 0.165) = 80 \text{ kJ/kg}$$

$\therefore$ **Percentage of total heat supplied (per kg) which appears as external work**

$$= \frac{80}{566.26} = 0.141 = \mathbf{14.1\%.} \quad \textbf{(Ans.)}$$

Example 2.13. *Find the specific volume, enthalpy and internal energy of wet steam at 18 bar, dryness fraction 0.85.*

Solution. Pressure of steam, p = 18 bar

Dryness fraction, x = 0.85

From steam tables corresponding to 18 bar pressure :

h_f = 884.6 kJ/kg, h_{fg} = 1910.3 kJ/kg,

$v_g = 0.110 \text{ m}^3/\text{kg}$, u_f = 883 kJ/kg, u_g = 2598 kJ/kg.

(*i*) **Specific volume of wet steam,**

$$v = xv_g = 0.85 \times 0.110 = \mathbf{0.0935 \ m^3/kg.} \quad \textbf{(Ans.)}$$

(*ii*) **Specific enthalpy of wet steam,**

$$h = h_f + xh_{fg} = 884.6 + 0.85 \times 1910.3$$

$$= \mathbf{2508.35 \ kJ/kg.} \quad \textbf{(Ans.)}$$

(*iii*) **Specific internal energy of wet steam,**

$$u = (1 - x)u_f + xu_g$$
$$= (1 - 0.85) \times 883 + 0.85 \times 2598$$
$$= \mathbf{2340.75\ kJ/kg.} \quad \textbf{(Ans.)}$$

Example 2.14. *Find the dryness fraction, specific volume and internal energy of steam at 7 bar and enthalpy 2550 kJ/kg.*

Solution. Pressure of steam, $p = 7$ bar

Enthalpy of steam, $h = 2550$ kJ

From steam tables corresponding to 7 bar pressure :

$h_f = 697.1$ kJ/kg, $h_{fg} = 2064.9$ kJ/kg,

$v_g = 0.273$ m³/kg, $u_f = 696$ kJ/kg, $u_g = 2573$ kJ/kg.

(*i*) **Dryness fraction, x :**

At 7 bar, $h_g = 2762$ kJ/kg, hence since the actual enthalpy is given as 2550 kJ/kg, the steam must be in the wet vapour state.

Now, using the equation,

$$h = h_f + xh_{fg}$$

$$\therefore \quad 2550 = 697.1 + x \times 2064.9$$

i.e.,
$$x = \frac{2550 - 697.1}{2064.9} = 0.897$$

Hence, **dryness fraction = 0.897. (Ans.)**

(*ii*) **Specific volume of wet steam,**

$$v = xv_g = 0.897 \times 0.273$$
$$= \mathbf{0.2449\ m^3/kg.} \quad \textbf{(Ans.)}$$

(*iii*) **Specific internal energy of wet steam,**

$$u = (1 - x)u_f + xu_g$$
$$= (1 - 0.897) \times 696 + 0.897 \times 2573$$
$$= \mathbf{2379.67\ kJ/kg.} \quad \textbf{(Ans.)}$$

Example 2.15. *Steam at 120 bar has a specific volume of 0.01721 m³/kg, find the temperature, enthalpy and the internal energy.*

Solution. Pressure of steam, $p = 120$ bar

Specific volume, $v = 0.01721$ m³/kg

(*i*) **Temperature :**

First it is necessary to decide whether the steam is wet, dry saturated or superheated.

At 120 bar, $v_g = 0.0143$ m³/kg, which is less than the actual specific volume of 0.01721 m³/kg, and hence the steam is **superheated.**

From the superheat tables at 120 bar, the specific volume is 0.01721 m³/kg at a temperature of **350°C. (Ans.)**

(*ii*) **Enthalpy :**

From the steam tables the specific enthalpy at 120 bar, 350°C,

$$h = \mathbf{2847.7\ kJ/kg.} \quad \textbf{(Ans.)}$$

(*iii*) **Internal energy :**

To find internal energy, using the equation,

$$u = h - pv$$
$$= 2847.7 - \frac{120 \times 10^5 \times 0.01721}{10^3}$$
$$= \mathbf{2641.18\ kJ/kg.} \quad \textbf{(Ans.)}$$

Example 2.16. *Steam at 140 bar has an enthalpy of 3001.9 kJ/kg, find the temperature, the specific volume and the internal energy.*

Solution. Pressure of steam, $p = 140$ bar

Enthalpy of steam, $h = 3001.9$ kJ/kg

(*i*) **Temperature :**

At 140 bar, $h_g = 2642.4$ kJ, which is less than the actual enthalpy of 3001.9 kJ/kg, and hence the steam is **superheated.**

From superheat tables at 140 bar, $h = 3001.9$ kJ/kg at a temperature of **400°C. (Ans.)**

(*ii*) The **specific volume,**

$$v = \mathbf{0.01722\ m^3/kg.} \quad \textbf{(Ans.)}$$

(*iii*) The **internal energy** (specific),

$$u = h - pv = 3001.9 - \frac{140 \times 10^5 \times 0.01722}{10^3}$$
$$= \mathbf{2760.82\ kJ/kg.} \quad \textbf{(Ans.)}$$

☞ **Example 2.17.** *Calculate the internal energy per kg of superheated steam at a pressure of 10 bar and a temperature of 300°C. Also find the change of internal energy if this steam is expanded to 1.4 bar and dryness fraction 0.8.*

Solution. At 10 bar, 300°C. From steam tables for superheated steam.

$$h_{sup} = 3051.2 \text{ kJ/kg } (T_{sup} = 300 + 273 = 573 \text{ K})$$

and corresponding to 10 bar (from tables of dry saturated steam)

$$T_s = 179.9 + 273 = 452.9 \text{ K} \,;\, v_g = 0.194 \text{ m}^3\text{/kg}$$

To find $v_{sup.}$, using the relation,

$$\frac{v_g}{T_s} = \frac{v_{sup}}{T_{sup}}$$

$$\therefore \quad v_{sup} = \frac{v_g \times T_{sup}}{T_s} = \frac{0.194 \times 573}{452.9} = 0.245 \text{ m}^3\text{/kg.}$$

Internal energy of superheated steam at 10 bar,

$$u_1 = h_{sup} - pv_{sup}$$
$$= 3051.2 - 10 \times 10^5 \times 0.245 \times 10^{-3}$$
$$= \mathbf{2806.2\ kJ/kg.} \quad \textbf{(Ans.)}$$

At 1.4 bar. From steam tables ;

$h_f = 458.4$ kJ/kg, $h_{fg} = 2231.9$ kJ/kg ;

$v_g = 1.236$ m³/kg

Enthalpy of wet steam (after expansion)

$$h = h_f + xh_{fg}$$
$$= 458.4 + 0.8 \times 2231.9 = 2243.92 \text{ kJ.}$$

Internal energy of this steam,

$$u_2 = h - pxv_g$$
$$= 2243.92 - 1.4 \times 10^5 \times 0.8 \times 1.236 \times 10^{-3}$$
$$= 2105.49 \text{ kJ}$$

Hence change of internal energy per kg

$$u_2 - u_1 = 2105.49 - 2806.2$$
$$= \mathbf{-700.7 \text{ kJ.}} \quad \textbf{(Ans.)}$$

Negative sign indicates **decrease** in internal energy.

Example 2.18. *Find the internal energy of 1 kg of steam at 20 bar when*

(i) it is superheated, its temperature being 400°C ;

(ii) it is wet, its dryness being 0.9.

Assume superheated steam to behave as a perfect gas from the commencement of superheating and thus obeys Charle's law. Specific heat for steam = 2.3 kJ/kg K.

Solution. Mass of steam = 1 kg

Pressure of steam, p = 20 bar

Temperature of superheated steam

= 400°C (T_{sup} = 400 + 273 = 673 K)

Dryness fraction, x = 0.9

Specific heat of superheated steam, c_{ps} = 2.3 kJ/kg K

(*i*) **Internal energy of 1 kg of superheated steam :**

At 20 bar. From steam tables,

t_s = 212.4°C ; h_f = 908.6 kJ/kg ;

h_{fg} = 1888.6 kJ/kg, v_g = 0.0995 m³/kg

Now,
$$h_{sup} = h_f + h_{fg} + c_{ps}(T_{sup} - T_s)$$
$$= 908.6 + 1888.6 + 2.3(400 - 212.4)$$
$$= 3228.68 \text{ kJ/kg}$$

Also,
$$h_{sup} = u + p \,.\, v_{sup}$$

or
$$u = h_{sup} - p \,.\, v_{sup}$$

The value of v_{sup} can be found out by Charle's law

$$\frac{v_g}{T_g} = \frac{v_{sup}}{T_{sup}}$$

$$\therefore \quad v_{sup} = \frac{v_g \times T_{sup}}{T_s} = \frac{0.0995 \times 673}{(212.4 + 273)}$$
$$= 0.1379 \text{ m}^3\text{/kg}$$

Hence internal energy,

$$u = 3228.68 - 20 \times 10^5 \times 0.1379 \times 10^{-3}$$
$$= \mathbf{2952.88 \text{ kJ/kg.}} \quad \textbf{(Ans.)}$$

(*ii*) **Internal energy of 1 kg of wet steam :**

$$h = h_f + xh_{fg} = 908.6 + 0.9 \times 1888.6$$
$$= 2608.34 \text{ kJ/kg}$$

Again $h = u + p \,.\, x \,.\, v_g$

$$\therefore \quad u = h - p \,.\, x \,.\, v_g$$
$$= 2608.34 - 20 \times 10^5 \times 0.9 \times 0.0995 \times 10^{-3}$$
$$= 2429.24 \text{ kJ/kg}$$

Hence internal energy = 2429.24 kJ/kg. (Ans.)

Example 2.19. *Two boilers one with superheater and other without superheater are delivering equal quantities of steam into a common main. The pressure in the boilers and main is 20 bar. The temperature of steam from a boiler with a superheater is 350°C and temperature of the steam in the main is 250°C.*

Determine the quality of steam supplied by the other boiler. Take c_{ps} = 2.25 kJ/kg.

Solution. Boiler B₁. 20 bar, 350°C :

Enthalpy,
$$h_1 = h_{g_1} + c_{ps}(T_{sup} - T_s)$$
$$= 2797.2 + 2.25(350 - 212.4)$$
$$= 3106.8 \text{ kJ/kg} \quad \text{...}(i)$$

Boiler B_2. 20 bar (temperature not known) :

$$h_2 = h_{f_2} + x_2 h_{fg_2}$$
$$= (908.6 + x_2 \times 1888.6) \text{ kJ/kg} \quad \text{...}(ii)$$

Main. 20 bar, 250°C.

Total heat of 2 kg of steam in the steam main

$$= 2[h_g + c_{ps}(T_{sup} - T_s)]$$
$$= 2[2797.2 + 2.25(250 - 212.4)]$$
$$= 5763.6 \text{ kJ} \quad \text{...}(iii)$$

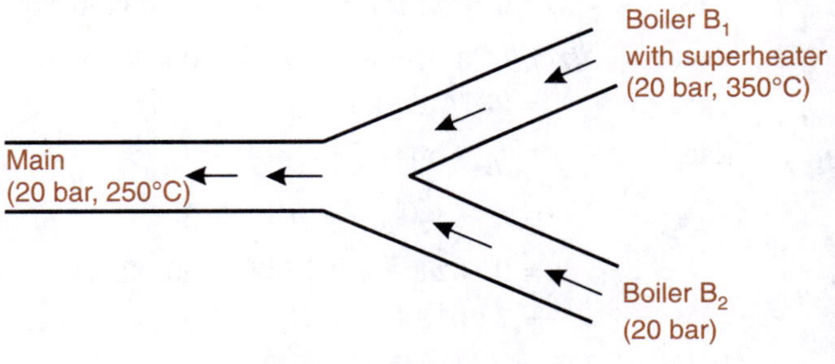

Fig. 2.13

Equating (*i*) and (*ii*) with (*iii*), we get

$$3106.8 + 908.6 + x_2 \times 1888.6 = 5763.6$$
$$4015.4 + 1888.6x_2 = 5763.6$$

$$\therefore \quad x_2 = \frac{5763.6 - 4015.4}{1888.6} = 0.925$$

Hence, **quality of steam supplied by the other boiler = 0.925. (Ans.)**

Example 2.20. *Determine the entropy of 1 kg of wet steam at a pressure of 6 bar and 0.8 dry, reckoned from freezing point (0°C).*

Solution. Mass of wet steam, m = 1 kg

Pressure of steam, p = 6 bar

Dryness fraction, x = 0.8

At 6 bar. From steam tables,

$t_s = 158.8°C,\ h_{fg} = 2085$ kJ/kg

Entropy of wet steam is given by

$$s_{wet} = c_{pw} \log_e \frac{T_s}{273} + \frac{xh_{fg}}{T_s}$$

(where c_{pw} = specific heat of water)

$$= 4.18 \log_e \left(\frac{158.8 + 273}{273}\right) + \frac{0.8 \times 2085}{(158.8 + 273)}$$

$= 1.9165 + 3.8700 = 5.7865$ kJ/kg K

Hence, **entropy of wet steam**

= 5.7865 kJ/kg K. (Ans.)

Example 2.21. *Steam enters an engine at a pressure 10 bar absolute and 400°C. It is exhausted at 0.2 bar. The steam at exhaust is 0.9 dry. Find :*

(i) Drop in enthalpy ; (ii) Change in entropy.

Solution. Initial pressure of steam, $p_1 = 10$ bar

Initial temperature of steam, $t_{sup} = 400°C$

Final pressure of steam, $p_2 = 0.2$ bar

Final condition of steam, $x_2 = 0.9$

At 10 bar, 400°C. From steam tables,

$h_{sup} = 3263.9$ kJ/kg ; $s_{sup} = 7.465$ kJ/kg K

i.e., $h_1 = h_{sup} = 3263.9$ kJ/kg

and $s_1 = s_{sup} = 7.465$ kJ/kg K

At 0.2 bar. From steam tables,

$h_f = 251.5$ kJ/kg ; $h_{fg} = 2358.4$ kJ/kg ;

$s_f = 0.8321$ kJ/kg K ; $s_g = 7.9094$ kJ/kg K

Also, $h_2 = h_{f_2} + x_2 h_{fg_2} = 251.5 + 0.9 \times 2358.4$

$= 2374$ kJ/kg.

Also, $s_2 = s_{f_2} + x_2 s_{fg_2}$

$= s_{f_2} + x_2(s_{g_2} - s_{f_2})$

$= 0.8321 + 0.9(7.9094 - 0.8321)$

$= 7.2016$ kJ/kg K

Hence, (*i*) **Drop in enthalpy,**

$= h_1 - h_2 = 3263.9 - 2374$

= 889.9 kJ/kg. (Ans.)

(*ii*) **Change in entropy**

$= s_1 - s_2 = 7.465 - 7.2016$

= 0.2634 kJ/kg K (decrease). (Ans.)

Example 2.22. *Find the entropy of 1 kg of superheated steam at a pressure of 12 bar and a temperature of 250°C. Take specific heat of superheated steam as 2.1 kJ/kg K.*

Solution. Mass of steam, $m = 1$ kg

Pressure of steam, $p = 12$ bar

Temperature of steam, $T_{sup} = 250 + 273 = 523$ K

Specific heat of superheated steam, $c_{ps} = 2.1$ kJ/kg K

At 12 bar. From steam tables,

$T_s = 188 + 273 = 461$ K, $h_{fg} = 1984.3$ kJ/kg

∴ Entropy of 1 kg of superheated steam,

$$s_{sup} = c_{pw} \log_e \frac{T_s}{273} + \frac{h_{fg}}{T_s} + c_{ps} \log_e \frac{T_{sup}}{T_s}$$

$$= 4.18 \log_e \left(\frac{461}{273}\right) + \frac{1984.3}{461} + 2.1 \times \log_e \left(\frac{523}{461}\right)$$

$= 2.190 + 4.304 + 0.265$

= 6.759 kJ/kg. (Ans.)

Example 2.23. *A piston-cylinder contains 3 kg of wet steam at 1.4 bar. The initial volume is 2.25 m³. The steam is heated until its temperature reaches 400°C. The piston is free to move up or down unless it reaches the stops at the top. When the piston is up against the stops the cylinder volume is 4.65 m³. Determine the amount of work and heat transfer to or from steam.* **(U.P.S.C.)**

Solution. Initial volume per kg of steam

$$= \frac{2.25}{3} = 0.75 \text{ m}^3/\text{kg}$$

Specific volume of steam at 1.4 bar

$= 1.2363$ m³/kg

Dryness fraction of initial steam

$$= \frac{0.75}{12363} = 0.607$$

At 1.4 bar, the enthalpy of 3 kg of steam

$= 3\left[h_f + xh_{fg}\right] = 3\ [458.4 + 0.607 \times 2231.9]$

$= 5439.5$ kJ

At 400°C, volume of steam per kg

$$= \frac{4.65}{3} = 1.55 \text{ m}^3/\text{kg}$$

At 400°C, when $v_{sup} = 1.55$ m³/kg, from steam tables,

Pressure of steam = 2.0 bar

Saturation temperature

$= 120.2°C,\ h = 3276.6$ kJ/kg

Degree of superheat

$= t_{sup} - t_s = 400 - 120.2 = 279.8°C$

Enthalpy of superheated steam at 2.0 bar,

400°C = 3 × 3276.6 = 9829.8 kg

Heat *added during the process*

$= 9829.8 - 5439.5 =$ **4390.3 kJ. (Ans.)**

Internal energy of 0.607 dry steam at 1.4 bar

$= 3 \times 1708 = 5124$ kJ.

Internal energy of superheated steam at 2 bar, 400°C

$= 3(h_{sup} - pv) = 3(3276.6 - 2 \times 10^2 \times 1.55)$

$= 8899.8$ kJ (∵ 1 bar = 10^2 kPa)

Change in internal energy

$= 8899.8 - 5124 = 3775.8$ kJ

Hence, **work done**

= 4390.3 – 3775.8 = **614.5 kJ.** **(Ans.)**

($\because$ $W = Q - \Delta U$)

2.18. DETERMINATION OF DRYNESS FRACTION OF STEAM

The dryness fraction of steam can be measured by using the following *calorimeters* :

1. Tank or bucket calorimeter
2. Throttling calorimeter
3. Separating and throttling calorimeter.

2.18.1. Tank or Bucket Calorimeter

The dryness fraction of steam can be found with the help of tank calorimeter as follows :

A known mass of steam is passed through a known mass of water and steam is completely condensed. The heat lost by steam is equated to heat gained by the water.

Fig. 2.14 shows the arrangement of this calorimeter.

The steam is passed through the sampling tube into the bucket calorimeter containing *a known* mass of water.

The weights of calorimeter with water before mixing with steam and after mixing the steam are obtained by weighing.

The temperature of water before and after mixing the steam are measured by *mercury thermometer*.

The pressure of steam passed through the sampling tube is measured with the help of *pressure gauge*.

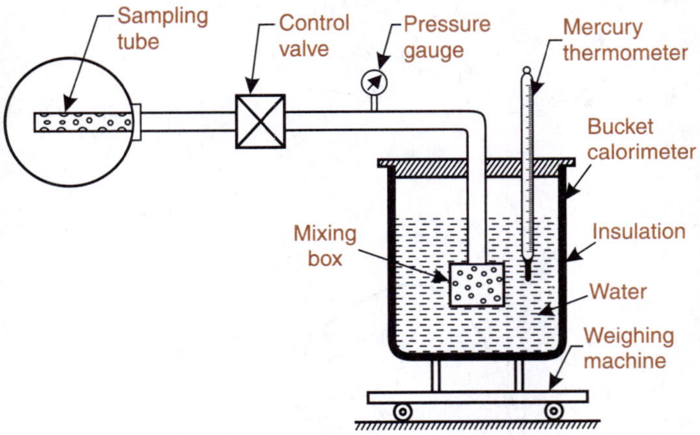

Fig. 2.14. Tank or bucket calorimeter.

Let, p_s = Gauge pressure of steam (bar),

p_a = Atmospheric pressure (bar),

t_s = Saturation temperature of steam known from steam table at pressure $(p_s + p_a)$,

h_{fg} = Latent heat of steam,

x = Dryness fraction of steam,

c_{pw} = Specific heat of water,

c_{pc} = Specific heat of calorimeter,

m_c = Mass of calorimeter, kg,

m_{cw} = Mass of calorimeter and water, kg,

$m_w = (m_{cw} - m_c)$ = Mass of water in calorimeter, kg,

m_{cws} = Mass of calorimeter, water and condensed steam, kg,

$m_s = (m_{cws} - m_{cw})$ = Mass of steam condensed in calorimeter, kg,

t_{cw} = Temperature of water and calorimeter before mixing the steam, °C, and

t_{cws} = Temperature of water and calorimeter after mixing the steam, °C.

Neglecting the losses and *assuming* that the *heat lost by steam is gained by water and calorimeter*, we have

$$(m_{cws} - m_{cw})\,[xh_{fg} + c_{pw}\,(t_s - t_{cws})] = (m_{cw} - m_c)c_{pw}\,(t_{cws} - t_{cw}) + m_c\,c_{pc}\,(t_{cws} - t_{cw})$$

$$\therefore\ m_s[xh_{fg} + c_{pw}\,(t_s - t_{cws})] = (t_{cws} - t_{cw})\,[(m_{cw} - m_c)(c_{pw} + m_c c_{pc})] \quad ...(2.19)$$

or $$m_s[xh_{fg} + c_{pw}\,(t_s - t_{cws})] = (t_{cws} - t_{cw})(m_w c_{pw} + m_c c_{pc})$$

The $m_c c_{pc}$ is known as *water equivalent of calorimeter*.

The value of dryness fraction '*x*' can be found by solving the above equation.

The value of dryness fraction found by this method involves some *inaccuracy* since losses due to convection and radiation are *not* taken into account.

The calculated value of dryness fraction neglecting losses is *always less* than the actual value of the dryness.

Example 2.24. *Steam at a pressure of 5 bar passes into a tank containing water where it gets condensed. The mass and temperature in the tank before the admission of steam are 50 kg and 20°C respectively. Calculate the dryness fraction of steam as it enters the tank if 3 kg of steam gets condensed and resulting temperature of the mixture becomes 40°C. Take water equivalent of tank as 1.5 kg.*

Solution. Pressure of steam, p = 5 bar

Mass of water in the tank = 50 kg

Initial temperature of water = 20°C

Amount of steam condensed, m_s = 3 kg

Final temperature after condensation of steam = 40°C

Water equivalent of tank = 1.5 kg

Dryness fraction of steam, x :

At 5 bar. From steam tables,

h_f = 640.1 kJ/kg ; h_{fg} = 2107.4 kJ/kg

Total mass of water, m_w = mass of water in the tank + water equivalent of tank

$= 50 + 1.5 = 51.5$ kg

Also, heat lost by steam = heat gained by water

$$m_s\,[(h_f + xh_{fg}) - 1 \times 4.18\,(40 - 0)] = m_w[1 \times 4.18\,(40 - 20)]$$

or $\quad 3[(640.1 + x \times 2107.4) - 4.18 \times 40] = 51.5 \times 4.18 \times 20$

or $\quad 3(472.9 + 2107.4x) = 4305.4$

or $\quad 472.9 + 2107.4x = 1435.13$

$$\therefore \quad x = \frac{1435.13 - 472.9}{2107.4} = 0.456.$$

Hence, **dryness fraction of steam = 0.456.** (Ans.)

Example 2.25. *Steam at a pressure of 1.1 bar and 0.95 dry is passed into a tank containing 90 kg of water at 25°C. The mass of tank is 12.5 kg and specific heat of metal is 0.42 kJ/kg K. If the temperature of water rises to 40°C after the passage of the steam, determine the mass of steam condensed. Neglect radiation and other losses.*

Solution. Pressure of steam, $p = 1.1$ bar

Dryness fraction of steam, $x = 0.95$

Mass of water in the tank $= 90$ kg

Initial temperature of water in the tank $= 25°C$

Mass of tank $= 12.5$ kg

Specific heat of metal $= 0.42$ kJ/kg K

Final temperature of water $= 40°C$.

Mass of steam condensed, m_s :

Since the radiation losses are neglected,

$\therefore$ Heat lost by steam = Heat gained by water

or
$$m_s[(h_f + xh_{fg}) - 1 \times 4.18\,(40 - 0)] = m[1 \times 4.18(40 - 25)]$$

But $\quad m = m_1 + m_2$

where, m_1 = Mass of cold water in the vessel before steam supply, and

m_2 = Water equivalent of vessel

$= 0.42 \times 12.5 = 5.25$ kg

At 1.1 bar. From steam tables,

$h_f = 428.8$ kJ/kg ; $h_{fg} = 2250.8$ kJ/kg

$\therefore \quad m_s\,[(428.8 + 0.95 \times 2250.8) - 1 \times 4.18 \times 40] = (90 + 5.25)\,[1 \times 4.18 \times (40 - 25)]$

$m_s\,[2567.06 - 167.20] = 95.25 \times 62.7$

i.e., $\quad 2399.86 m_s = 5972.17$

$\therefore \quad m_s = 2.488$ kg

Hence, **mass of steam condensed = 2.488 kg.** (Ans.)

2.18.2. Throttling Calorimeter

The dryness fraction of wet steam can be determined by using a throttling calorimeter which is illustrated diagrammatically in Fig. 2.15.

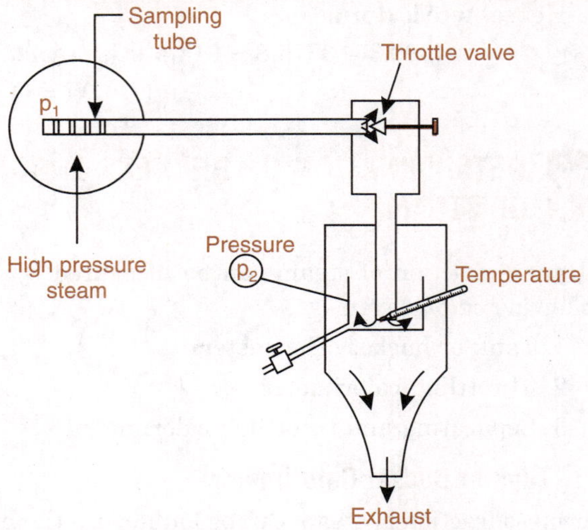

Fig. 2.15. Throttling calorimeter.

The steam to be sampled is taken from the pipe by means of suitable positioned and dimensioned sampling tube. It passes into an insulated container and is throttled through an orifice to atmospheric pressure. Here the temperature is taken and the steam ideally should have about 5.5 K of superheat.

The throttling process is shown on *h-s* diagram in Fig. 2.16 by the line 1–2. If steam initially wet is throttled through a sufficiently large pressure drop, then the steam at state 2 will become superheated. State 2 can then be defined by the *measured pressure and temperature*. The enthalpy, h_2 can then be found and hence

$$h_2 = h_1 = (h_{f_1} + x_1 h_{fg_1}) \text{ at } p_1$$

[where $h_2 = h_{f_2} + h_{fg_2} + c_{ps}\,(T_{sup_2} - T_{s_2})$]

$$\therefore \quad x_1 = \frac{h_2 - h_{f_1}}{h_{fg_1}} \qquad ...(2.20)$$

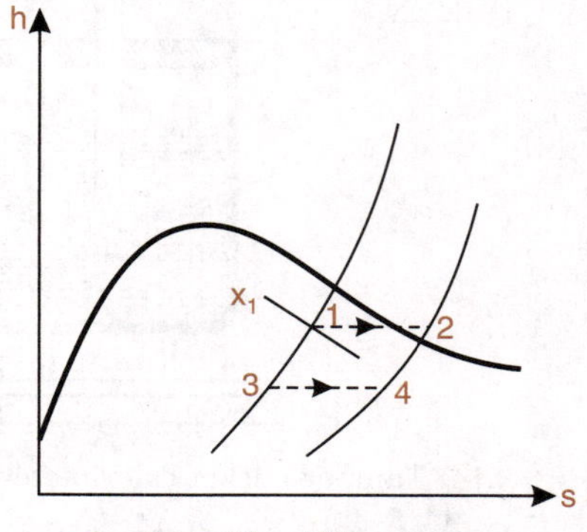

Fig. 2.16. Throttling process.

Hence, the dryness fraction is determined and state 1 is defined.

Example 2.26. *A throttling calorimeter is used to measure the dryness fraction of the steam in the steam main which has steam flowing at a pressure of 8 bar. The steam after* *passing through the calorimeter is at 1 bar pressure and 115°C.*

Calculate the dryness fraction of the steam in the main. Take c_{ps} = 2.1 kJ/kg K.

Solution. Condition of steam *before throttling* :

$$p_1 = 8 \text{ bar}, x_1 = ?$$

Condition of steam *after throttling* :

$$p_2 = 1 \text{ bar}, t_2 = t_{sup_2} = 115°\text{C}$$

As throttling is a *constant enthalpy* process

$$\therefore \quad h_1 = h_2$$

i.e., $$h_{f_1} + x_1 h_{gf_1} = h_{f_2} + h_{fg_2} + c_{p_s}(T_{\text{sup}_2} - T_{s_2})$$

$$[\because T_{sup_2} = 115 + 273 = 388 \text{ K}$$

$$T_{s_2} = 99.6 + 273 = 372.6 \text{ K (at 1 bar)}]$$

$$720.9 + x_1 \times 2046.5 = 417.5 + 2257.9 + 2.1(388 - 372.6)$$

$$720.9 + 2046.5\, x_1 = 2707.7$$

$$\therefore \quad x_1 = \frac{2707.7 - 720.9}{2046.5} = 0.97$$

Hence, **dryness fraction of steam in the main = 0.97. (Ans.)**

2.18.3. Separating and Throttling Calorimeter

If the steam whose dryness fraction is to be determined is *very wet* then throttling to atmospheric pressure *may not be sufficient to ensure superheated steam at exit*. In this case it is *necessary to dry the steam partially, before throttling*. This is done by *passing the steam sample from the main through a separating calorimeter* as shown in Fig. 2.17. The steam is made to change direction suddenly, and the water, being denser than the dry steam is separated out. The quantity of water which is separated out (m_w) is measured at the separator, the steam remaining, which *now has a higher dryness fraction*, is passed through the *throttling calorimeter*. With the combined separating and throttling calorimeter it is *necessary* to condense the steam after throttling and measure the amount of condensate (m_s). If a throttling calorimeter only is sufficient, there is no need to measure condensate, the pressure and temperature measurements at exit being sufficient.

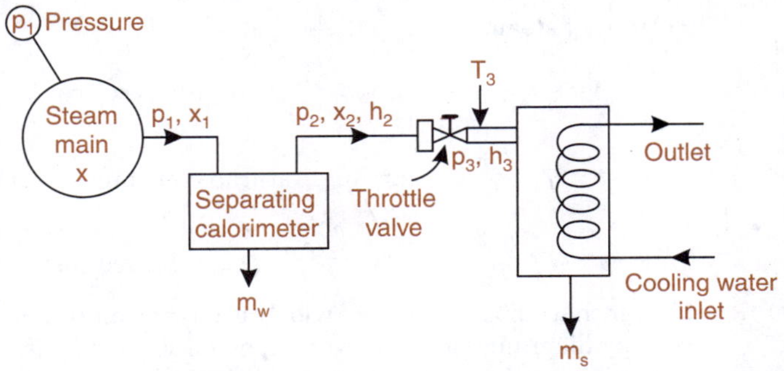

Fig. 2.17. Separating and throttling calorimeter.

Dryness fraction at 2 is x_2, therefore, the mass of dry steam leaving the separating calorimeter is equal to $x_2 m_s$ and this must be the mass of dry vapour in the sample drawn from the main at state 1.

Hence fraction in main,

$$x_1 = \frac{\text{Mass of dry vapour}}{\text{Total mass}} = \frac{x_2\, m_s}{m_w + m_s}.$$

The dryness fraction, x_2, can be determined as follows :

$$^*h_3 = h_2 = h_{f_2} + x_2 h_{fg_2} \text{ at } p_2$$

$$[^* h_3 = h_{f_3} + h_{fg_3} + c_{ps}(T_{sup_3} - T_{s_3}) \text{ at pressure } p_3]$$

or, $$x_2 = \frac{h_3 - h_{f_2}}{h_{fg_2}}$$

The values of h_{f_2} and h_{fg_2} are read from steam tables at pressure p_2. The pressure in the separator is small so that p_1 is approximately equal to p_2.

Example 2.27. *The following observations were taken with a separating and a throttling calorimeter arranged in series :*

Water separated = 2 kg, steam discharged from the throttling calorimeter = 20.5 kg, temperature of steam after throttling = 110°C, initial pressure = 12 bar abs., barometer = 760 mm of Hg, final pressure = 5 mm of Hg.

Estimate the quality of steam supplied.

Solution. Quantity of water separated out, m_w = 2 kg

Steam (condensate) discharged from the throttling calorimeter, m_s = 20.5 kg

Temperature of steam after throttling, t_{sup} = 110°C

Initial pressure of steam, p_1 = 12 bar abs.

Final pressure of steam, p_3 = 760 + 5 = 765 mm

$$= \frac{765}{1000} \times 1.3366 \quad (\because \ 1 \text{ m Hg} = 1.3366 \text{ bar})$$

$$\simeq 1 \text{ bar}$$

From steam tables :

At $p_1 = p_2$ = **12 bar :** h_f = 798.4 kJ/kg,

h_{fg} = 1984.3 kJ/kg

At p_3 = **1 bar :** t_s = 99.6°C, h_f = 417.5 kJ/kg,

h_{fg} = 2257.9 kJ/kg

t_{sup} = 110°C (given)

Also, $$h_3 = h_2$$

$$(h_{f_3} + h_{fg_3}) + c_{ps}(T_{\text{sup}_3} - T_{s_3}) = hf_2 + x_2\, h_{fg_2}$$

Taking c_{ps} = 2 kJ/kg K, we get

$$417.5 + 2257.9 + 2[(110 + 273) - (99.6 + 273)]$$

$$= 798.4 + x_2 \times 1984.3$$

$$2696.2 = 798.4 + 1984.3\, x_2$$

$$\therefore \quad x_2 = \frac{2696.2 - 798.4}{1984.3} = 0.956$$

Now, **quality of steam supplied,**

$$x_1 = \frac{x_2 m_s}{m_w + m_s} = \frac{0.956 \times 20.5}{2 + 20.5} = \mathbf{0.87.} \quad \textbf{(Ans.)}$$

☞ **Example 2.28.** *The following data were obtained in a test on a combined separating and throttling calorimeter :*

Pressure of steam sample = 15 bar, pressure of steam at exit = 1 bar, temperature of steam at the exit = 150°C, discharge from separating calorimeter = 0.5 kg/min, discharge from throttling calorimeter = 10 kg/min.

Determine the dryness fraction of the sample steam.

Solution. Pressure of steam sample, $p_1 = p_2 = 15$ bar

Pressure of steam at the exit, $p_3 = 1$ bar

Temperature of steam at the exit, $t_{sup_3} = 150°C$

Discharge from separating calorimeter,

$$m_w = 0.5 \text{ kg/min}$$

Discharge from throttling calorimeter,

$$m_s = 10 \text{ kg/min}$$

From steam tables :

At $p_1 = p_2 =$ **15 bar :** $h_{f_2} = 844.7$ kJ/kg,

$$h_{fg_2} = 1945.2 \text{ kJ/kg}$$

At $p_3 =$ **1 bar and 150°C :** $h_{sup_3} = 2776.4$ kJ/kg

Also, $h_2 = h_3$

$$h_{f_2} + x_2 h_{fg_2} = h_{sup_3}$$

$$844.7 + x_2 \times 1945.2 = 2776.4$$

$$\therefore \quad x_2 = \frac{2776.4 - 844.7}{1945.2} = 0.993$$

Now, **quality of steam supplied,**

$$x_1 = \frac{x_2 m_s}{m_s + m_w} = \frac{0.993 \times 10}{10 + 0.5} = \mathbf{0.946.} \quad \textbf{(Ans.)}$$

HIGHLIGHTS

1. A pure substance is a system which is (*i*) homogeneous in composition, (*ii*) homogeneous in chemical aggregation, (*iii*) invariable in chemical aggregation.
2. The triple point is merely the point of intersection of sublimation and vapourisation curves. It must be understood that only on *p-T* diagram is the triple point represented by a point. On *p-V* diagram it is a line, and on a *U-V* diagram it is a triangle.
3. Steam as a vapour does not obey laws of perfect gases unless and until it is highly in super dry condition.
4. *Dryness fraction* is the ratio of the mass of actual dry steam to the mass of steam containing it.

   i.e., Dryness fraction $= \dfrac{m_s}{m_s + m_w}$

   where, m_s = Mass of dry steam contained in steam considered

   m_w = Mass of water particles in suspension in the steam considered.
5. Superheated steam behaves like a gas and therefore, it follows gas laws. The law for adiabatic expansion is $pv^{1.3} = C$.
6. External work of evaporation $= p(v_g - v_f)$

   Internal latent heat $= h_{fg} - \dfrac{pv_g}{J}$

   Internal energy of steam, u :

   (*i*) For wet steam : $u = h - \dfrac{p.x.v_g}{J}$

   (*ii*) For superheated steam : $u = h_{sup} - \dfrac{p \cdot v_{sup}}{J}$

   (J = 1, in SI units)
7. Entropy of water when heated from temperature T_1 to T_2 ;

   $$s_2 - s_1 = c_{pw} \log_e \frac{T_2}{T_1}$$

   If 0°C is taken as the datum then entropy of water at any temperature T, above this datum will be

   $$s_f = c_{pw} \log_e \frac{T}{273}$$

   Entropy of evaporation :

   $$s_{evap.} = \frac{h_{fg}}{T_s}$$

   ...(when water is evaporated to steam completely)

   $$= \frac{xh_{fg}}{T_s}$$

   ...(when water is evaporated partially and dryness fraction of steam is x)

   Entropy of steam :

   $$s_{wet} = s_f + \frac{xh_{fg}}{T_s} \quad \text{...[wet steam } (x < 1)]$$

   $$s_g = s_f + \frac{h_{fg}}{T_s} \quad \text{...[dry and saturated steam } (x = 1)]$$

   $$s_{sup} = s_f + \frac{h_{fg}}{T_s} + c_{ps} \log_e \frac{T_{sup}}{T_s} \quad \text{...(superheated steam)}$$
8. Mollier chart/diagram is more widely used than any other entropy diagram, since the work done on vapour cycles can be scaled from this diagram directly as a length, whereas on *T-s* diagram it is represented by an area.

9. *Different processes :*

(*i*) Constant volume heating or cooling

$$x_1 v_{g_1} = x_2 v_{g_2}$$

$$x_1 v_{g_1} = v_{sup_2} = v_{g_2} \cdot \frac{T_{sup_2}}{T_{s_2}}$$

(*ii*) Constant pressure heating or cooling

$$Q = h_2 - h_1$$

(*iii*) Isentropic expansion (non-flow process)

$$W = (u_1 - u_2) \text{ and } s_1 = s_2$$

(*iv*) Throttling

$$h_{f_1} + x_1 h_{fg_1} = h_{f_2} + x_2 h_{fg_2} \quad \text{...(For wet condition)}$$

$$= h_{f_2} + h_{fg_2} + c_{ps} (T_{sup} - T_{s_2})$$

...(For superheated condition)

10. Dryness fraction of steam can determined by the following methods :

(*i*) Bucket calorimeter

(*ii*) Throttling calorimeter

(*iii*) Separating and throttling calorimeter.

OBJECTIVE TYPE QUESTIONS

Choose the Correct Answer :

1. (*a*) Specific volume of water decreases on freezing

(*b*) Boiling point of water decreases with increasing pressure

(*c*) Specific volume of CO_2 increases on freezing

(*d*) Freezing temperature of water decreases with increasing pressure.

2. (*a*) The slope of vaporisation curve is always negative

(*b*) The slope of vaporisation curve is always positive

(*c*) The slope of sublimation curve in negative for all pure substances

(*d*) The slope of fusion curve is positive for all pure substances.

3. (*a*) The process of passing from liquid to vapour is condensation

(*b*) An isothermal line is also a constant pressure line during wet region

(*c*) Pressure and temperature are independent during phase change

(*d*) The term dryness fraction is used to describe the fraction by mass of liquid in the mixture of liquid water and water vapour.

4. The latent heat of vaporisation at critical point is

(*a*) less than zero (*b*) greater than zero

(*c*) equal to zero (*d*) none of the above.

5. (*a*) Critical point involves equilibrium of solid and vapour phases

(*b*) Critical point involves equilibrium of solid and liquid phases

(*c*) Critical point involves equilibrium of solid, liquid and vapour phases

(*d*) Triple point involves equilibrium of solid, liquid and vapour phases.

6. With the increase in pressure

(*a*) boiling point of water increases and enthalpy of evaporation increases

(*b*) boiling point of water increases and enthalpy of evaporation decreases

(*c*) boiling point of water decreases and enthalpy of evaporation increases.

7. With increase in pressure

(*a*) enthalpy of dry saturated steam increases

(*b*) enthalpy of dry saturated steam decreases

(*c*) enthalpy of dry saturated steam remains same

(*d*) enthalpy of dry saturated steam first increases and then decreases.

8. Dryness fraction of steam is defined as

(*a*) mass of water vapour in suspension/(mass of water vapour in suspension + mass of dry steam)

(*b*) mass of dry steam/mass of water vapour in suspension

(*c*) mass of dry steam/(mass of dry steam + mass of water vapour in suspension)

(*d*) mass of water vapour in suspension/mass of dry steam.

9. The specific volume of water when heated at 0°C

(*a*) first increases and then decreases

(*b*) first decreases and then increases

(*c*) increases steadily

(*d*) decreases steadily.

10. Only throttling calorimeter is used for measuring

(*a*) very low dryness fraction upto 0.7

(*b*) very high dryness fraction upto 0.98

(*c*) dryness fraction of only low pressure steam

(*d*) dryness fraction of only high pressure steam.

11. Heat of superheated steam is given by

(*a*) $h_{sup} = h_f + h_{fg} + c_{ps} \log_e \frac{T_{sup}}{T_s}$

(*b*) $h_{sup} = h_f + xh_{fg}$

(*c*) $h_{sup} = h_f + h_{fg}$

(*d*) $h_{sup} = h_f + xh_{fg} + c_{ps} \log_e \frac{T_s}{273}$.

12. Volume of wet steam (per kg) with dryness fraction x is given by

(*a*) $x^3 v_g$ (*b*) xv_f

(*c*) $x^2(v_g - v_f)$ (*d*) $x^2 v_g$

(*e*) none of the above.

13. Internal latent heat is given by

(*a*) $h_{fg} - \frac{pv_g}{J}$ (*b*) $h_g - \frac{pv_g}{J}$

(*c*) $h_{sup} - \frac{pv_f}{J}$ (*d*) $h_{fg} + \frac{pv_g}{J}$

(*e*) none of the above.

14. Entropy of 1 kg of water at T K is given by

(a) $c_{pw} \log_e \frac{T}{273}$ (b) $c_{pw} \log_e \frac{T_2}{T_1}$

(c) $c_{pw} \log_{10} \frac{T}{273}$ (d) $c_{pw} \log_e \frac{T_2}{273}$

(e) none of the above.

15. Entropy of wet steam (1 kg) is given by

(a) $s_f + \frac{xh_{fg}}{T_s}$ (b) $s_g + \frac{xh_{fg}}{T_s}$

(c) $s_f + \frac{h_{fg}}{T_s}$ (d) $s_f + c_{ps} \log_e \frac{T_{sup}}{T_s}$

(e) none of the above.

16. In throttling process

(a) $h_1^2 = h_2$ (b) $h_1 = h_2$

(c) $h_1 = h_2 + \frac{h_{fg}}{T_s}$ (d) $h_2 = h_1 + \frac{h_{fg}}{T_s}$

(e) none of the above.

17. In isentropic process

(a) $W = 2(u_2 - u_1)$ (b) $W = (u_2 - u_1)^2$

(c) $W = u_2 - u_1$ (d) $W = (u_2 - u_1)^{1/2}$

(e) none of the above.

ANSWERS

1. (d)	2. (a)	3. (b)	4. (c)
5. (d)	6. (b)	7. (b)	8. (c)
9. (b)	10. (b)	11. (a)	12. (e)
13. (a)	14. (a)	15. (a)	16. (b)
17. (c).			

THEORETICAL QUESTIONS

1. What is a pure substance ?
2. Draw and explain a p-T (pressure-temperature) diagram for a pure substance.
3. What is a triple point ?
4. Explain with a neat diagram p-V-T surface.
5. Does wet steam obey laws of perfect gases ?
6. Describe the process of formation of steam and give its graphical representation also.
7. Explain the following terms relating to steam formation :
   (i) Sensible heat of water,
   (ii) Latent heat of steam,
   (iii) Dryness fraction of steam,
   (iv) Enthalpy of wet steam, and
   (v) Superheated steam.
8. What advantages are obtained if superheated steam is used in steam prime movers ?
9. What do you mean by the following :
   (i) Internal latent heat
   (ii) Internal energy of steam
   (iii) External work of evaporation
   (iv) Entropy of evaporation
   (v) Entropy of wet steam
   (vi) Entropy of superheated steam.
10. Write a short note on Mollier chart.
11. Draw a neat sketch of throttling calorimeter and explain how dryness fraction of steam is determined ; clearly explain its limitations.
12. Describe with a neat sketch a separating-throttling calorimeter for measuring the dryness fraction of steam.

UNSOLVED EXAMPLES

1. Find the specific volume, enthalpy and internal energy of wet steam at 18 bar, dryness fraction 0.9.
   [**Ans.** 0.0994 m^3/kg ; 2605.8 kJ/kg ; 2426.5 kJ/kg]
2. Find the dryness fraction, specific volume and internal energy of steam at 7 bar and enthalpy 2600 kJ/kg.
   [**Ans.** 0.921 ; 0.2515 m^3/kg, 2420 kJ/kg]
3. Steam at 110 bar has a specific volume of 0.0196 m^3/kg, find the temperature, the enthalpy and the internal energy. [**Ans.** 350°C ; 2889 kJ/kg ; 2673.4 kJ/kg]
4. Steam at 150 bar has an enthalpy of 3309 kJ/kg, find the temperature, the specific volume and the internal energy.
   [**Ans.** 500°C ; 0.02078 m^3/kg ; 2997.3 kJ/kg]
5. Steam at 19 bar is throttled to 1 bar and the temperature after throttling is found to be 150°C. Calculate the initial dryness fraction of the steam. [**Ans.** 0.989]
6. Find the internal energy of one kg of steam at 14 bar under the following conditions :
   (i) When the steam is 0.85 dry ;
   (ii) When steam is dry and saturated ; and
   (iii) When the temperature of steam is 300°C. Take c_{ps} = 2.25 kJ/kg K.
   [**Ans.** (i) 2327.5 kJ/kg ; (ii) 2592.5 kJ/kg ; (iii) 2784 kJ/kg]
7. Calculate the internal energy of 0.3 m^3 of steam at 4 bar and 0.95 dryness. If this steam is superheated at constant pressure through 30°C, determine the heat added and change in internal energy.
   [**Ans.** 2451 kJ/kg ; 119 kJ ; 107.5 kJ/kg]
8. Water is supplied to the boiler at 15 bar and 80°C and steam is generated at the same pressure at 0.9 dryness. Determine the heat supplied to the steam in passing through the boiler and change in entropy.
   [**Ans.** 2260.5 kJ/kg ; 4.92 kJ/kg K]
9. A cylindrical vessel of 5 m^3 capacity contains wet steam at 1 bar. The volume of vapour and liquid in the vessel are 4.95 m^3 and 0.05 m^3 respectively. Heat is transferred to the vessel until the vessel is filled with saturated vapour. Determine the heat transfer during the process.
   [**Ans.** 104.93 MJ]
10. A pressure cooker contains 1.5 kg of steam at 5 bar and 0.9 dryness when the gas was switched off. Determine the quantity of heat rejected by the pressure cooker when the pressure in the cooker falls to 1 bar. [**Ans.** – 2355 kJ]

11. A vessel of spherical shape having a capacity of 0.8 m^3 contains steam at 10 bar and 0.95 dryness. Steam is blown off until the pressure drops to 5 bar. The valve is then closed and the steam is allowed to cool until the pressure falls to 4 bar. Assuming that the enthalpy of steam in the vessel remains constant during blowing off periods, determine :

(*i*) The mass of steam blown-off,

(*ii*) The dryness fraction of steam in the vessel after cooling, and

(*iii*) The heat lost by steam per kg during cooling.

[**Ans.** (*i*) 2.12 kg ; (*ii*) 0.78 ; (*iii*) – 820 kJ]

12. Two boilers one with superheater and other without superheater are delivering equal quantities of steam into a common main. The pressure in the boilers and the main is 15 bar. The temperature of the steam from a boiler with a superheater is 300°C and temperature of the steam in the main is 200°C. Determine the quality of steam supplied by the other boiler. [**Ans.** 0.89]

13. A tank of capacity 0.5 m^3 is connected to a steam pipe through a valve which carries steam at 14 bar and 300°C. The tank initially contains steam at 3.5 bar and saturated condition. The valve in the line connecting the tank is opened and the steam is allowed to pass into the tank until the pressure in the tank becomes 14 bar.

Determine the mass of steam that entered into the tank.

[**Ans.** 1.565 kg]

First Law of Thermodynamics

3

3.1. INTERNAL ENERGY

It is the heat energy stored in a gas. If a certain amount of heat is supplied to a gas the result is that temperature of gas may increase or volume of gas may increase thereby doing some external work or both temperature and volume may increase ; but it will be decided by the conditions under which the gas is supplied heat. *If during heating of the gas the temperature increases its internal energy will also increase.*

Joule's law of internal energy states that internal energy of a perfect gas is a *function of temperature only.* In other words, internal energy of a gas is dependent on the temperature change only and is not affected by the change in pressure and volume.

We do not know how to find the absolute quantity of internal energy in any substance ; however, what is needed in engineering is the change of internal energy (ΔU).

3.2. LAW OF CONSERVATION OF ENERGY

In the early part of nineteenth century the scientists developed the concept of energy and hypothesis that it can be neither created nor destroyed ; this came to be known as the *law of the conservation of energy*. The first law of thermodynamics is merely one statement of this general law/principle with particular reference to heat energy and mechanical energy *i.e.,* work.

3.3. FIRST LAW OF THERMODYNAMICS

It is observed that when a system is made to undergo a complete cycle then network is done *on* or *by* the system. Consider a cycle in which network is done by the system. Since energy cannot be created, this mechanical energy must have been supplied from some source of energy. Now the system has been returned to its initial state : Therefore, its *intrinsic* energy is unchanged, and hence the mechanical energy has not been provided by the system itself. The only other energy involved in the cycle is the heat which was supplied and rejected in various processes. Hence, by the law of conservation of energy, the network done by the system is equal to the net heat supplied to the system. The First Law of Thermodynamics can, therefore, be stated as follows :

"When a system undergoes a thermodynamic cycle then the net heat supplied to the system from the surroundings is equal to network done by the system on its surroundings".

or $$\oint dQ = \oint dW$$

where $\oint$ represents the sum for a complete cycle.

The first law of Thermodynamics *cannot be proved analytically, but experimental evidence has repeatedly confirmed its validity*, and since no phenomenon has been shown to contradict it, the first law is accepted as a *law of nature.* It may be remarked that no restriction was imposed which limited the application of first law to reversible energy transformation. Hence, the first law applies to reversible as well as irreversible transformations : For non-cyclic process, a more general formulation of first law of thermodynamics is required. A new concept which involves a term called *internal energy* fulfills this need.

— The First Law of Thermodynamics may also be stated as follows :

"Heat and work are mutually convertible but since energy can neither be created nor destroyed, the total energy associated with an energy conversion remains constant".

Or

— **"No machine can produce energy without corresponding expenditure of energy, *i.e.*, it is impossible to construct a perpetual motion machine of first kind".**

Fig. 3.1 shows the experiment for checking first law of thermodynamics.

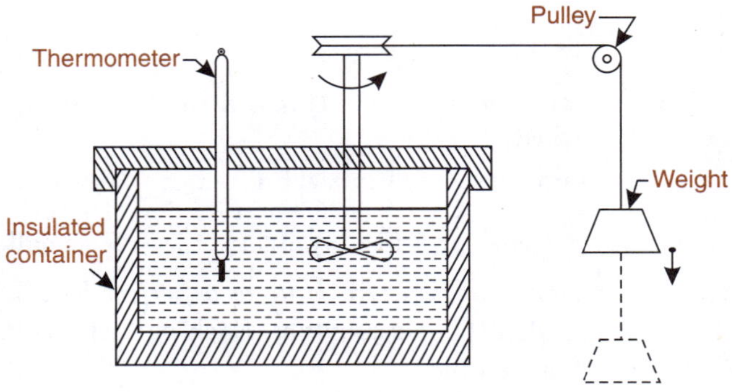

Fig. 3.1. Heat and work.

The work input to the paddle wheel is measured by the fall of weight, while the corresponding temperature rise of liquid in the insulated container is measured by the thermometer. It is already known to us from experiments on heat transfer that temperature rise can also be produced by heat transfer. The experiments show : (*i*) A definite quantity of work is always required to accomplish the same temperature rise obtained with a unit amount of heat. (*ii*) Regardless of whether the temperature of liquid is raised by work transfer or heat transfer, the liquid can be returned by heat transfer in opposite direction to the identical state from which it started. The above results lead to the inference that *work and heat* are different forms of something more general, which is called *energy*.

— It can be stated as an invariable experience that whenever a physical system passes through a complete cycle the algebraic sum of the work transfers during the cycle $\oint dW$ bears a definite ratio to the algebraic sum of the heat transfers during the cycle, $\oint dQ$. This may be expressed by the equation,

$$\oint dW = J \oint dQ \qquad ...(3.1)$$

where J is the proportionality constant and is known as *Mechanical Equivalent of heat*. In S.I. units its value is unity, *i.e.*, 1 Nm/J.

3.4. APPLICATION OF FIRST LAW TO A PROCESS

When a process is executed by a system, the *change in stored energy of the system is numerically equal to the net heat interactions minus the network interaction during the process.*

$$\therefore \quad E_2 - E_1 = Q - W$$

$$\therefore \quad \Delta E = Q - W \qquad [\text{or } Q = \Delta E + W]$$

or

$$\int_1^2 d(Q - W) = \Delta E = E_2 - E_1 \qquad ...(3.2)$$

where E represents the *total internal energy.*

If the electric, magnetic and chemical energies are *absent* and changes in potential and kinetic energy for a *closed system* are neglected, the above equation can be written as

$$\int_1^2 d(Q - W) = \Delta U = U_2 - U_1 \qquad ...(3.3)$$

$$\therefore \quad Q - W = \Delta U = U_2 - U_1 \qquad ...(3.4)$$

Generally, when heat is added to a system its temperature rises and external work is performed due to increase in volume of the system. *The rise in temperature is an indication of increase of internal energy.*

Heat *added* to the system will be considered as *positive* and the heat *removed* or *rejected*, from the system, as *negative.*

3.5. ENERGY—A PROPERTY OF SYSTEM

Consider a system which changes its state from state 1 to state 2 by following the path L, and returns from state 2 to state 1 by following the path M (Fig. 3.2). So the system undergoes a cycle. Writing the first law for path L

$$Q_L = \Delta E_L + W_L \qquad ...(3.5)$$

and for path M

$$Q_M = \Delta E_M + W_M \qquad ...(3.6)$$

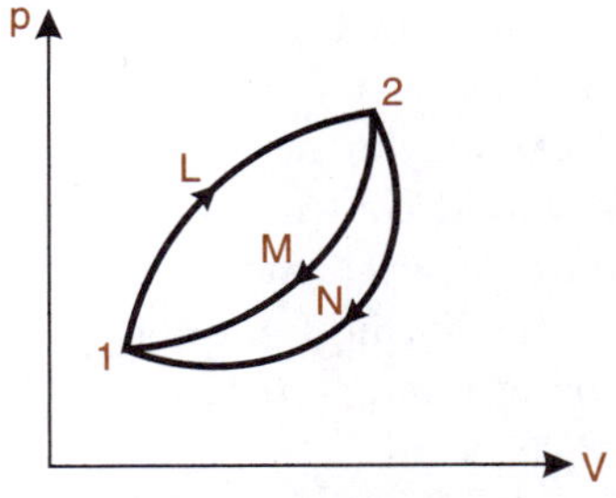

Fig. 3.2. Energy—a property of system.

The processes L and M together constitute a cycle, for which

$$\oint dW = \oint dQ$$

$$W_L + W_M = Q_L + Q_M$$

or $$Q_L - W_L = W_M - Q_M \quad ...(3.7)$$

From equations (3.5), (3.6) and (3.7), it yields

$$\Delta E_L = -\Delta E_M \quad ...(3.8)$$

Similarly, had the system returned from state 2 to state 1 by following the path N instead of path M

$$\Delta E_L = -\Delta E_N \quad ...(3.9)$$

From equations (3.8) and (3.9),

$$\Delta E_M = \Delta E_N \quad ...(3.10)$$

Thus, it is seen that the *change in energy between two states of a system is the same, whatever path the system may follow in undergoing that change of state.* If some arbitrary value of energy is assigned to state 2, the value of energy at state 1 is fixed independent of the path the system follows. Therefore, *energy has a definite value for every state of the system.* Hence, it is *a point function and a property of the system.*

3.6. PERPETUAL MOTION MACHINE OF THE FIRST KIND—PMM 1

The first law of thermodynamics states the general principle of the conservation of energy. Energy is neither created nor destroyed, but only gets transformed from one form to another. *There can be no machine which would continuously supply mechanical work without some form of energy disappearing simultaneously* (Fig. 3.3). Such a *fictitious machine* is called a ***perpetual motion machine of the first kind,*** or in brief, PMM 1. A PMM 1 is thus ***impossible.***

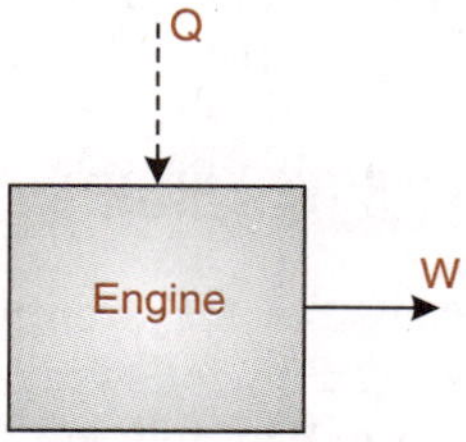

Fig. 3.3. A PPM 1.

— The converse of the above statement is also true, *i.e.*, there can be no machine which would continuously consume work without some other form of energy appearing simultaneously (Fig. 3.4).

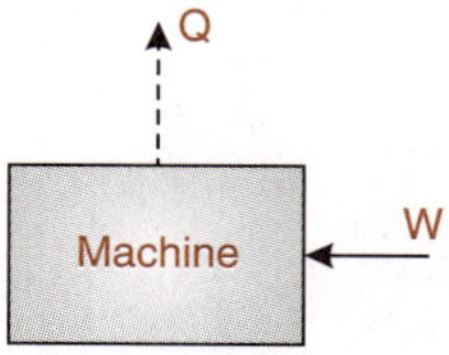

Fig. 3.4. The converse of PMM 1.

3.7. ENERGY OF AN ISOLATED SYSTEM

An isolated system is one in which there is no interaction of the system with the surroundings.

For an isolated system,

$$dQ = 0, dW = 0$$

The first law of thermodynamics gives

$$dE = 0 \quad \text{or} \quad E = \text{constant}$$

The energy of an isolated system is always constant.

3.8. THE PERFECT GAS

3.8.1. The Characteristic Equation of State

— At temperatures that are considerably in excess of critical temperature of a fluid, and also at very low pressure, the vapour of fluid tends to obey the equation

$$\frac{pv}{T} = \text{constant} = R$$

In practice, no gas obeys this law rigidly, but many gases tend towards it.

An imaginary ideal gas which obeys this law is called a *perfect gas*, and the equation $\frac{pv}{T} = R$, is called the *characteristic equation of a state of a perfect gas.* The constant R is called the *gas constant*. Each perfect gas has a different gas constant.

Units of R are Nm/kg K or kJ/kg K.

Usually, the characteristic equation is written as

$$pv = RT \quad ...(3.11)$$

or for m kg, occupying V m^3

$$pV = mRT \quad ...(3.12)$$

— The characteristic equation in *another form*, can be derived by using kilogram-mole as a unit.

The *kilogram-mole* is defined as a quantity of a gas equivalent to M kg of the gas, where M is the molecular weight of the gas (*e.g.*, since the molecular weight of oxygen is 32, then 1 kg mole of oxygen is equivalent to 32 kg of oxygen).

As per definition of the kilogram-mole, for m kg of a gas, we have

$$m = nM \quad ...(3.13)$$

where n = number of moles.

Note. Since the standard of mass is the kg, kilogram-mole will be written simply as mole.

Substituting for m from eqn. (3.13) in eqn. (3.12) gives

$$pV = nMRT$$

or $$MR = \frac{pV}{nT}$$

According to *Avogadro's hypothesis* the volume of 1 mole of any gas is the same as the volume of 1 mole of any other gas, when the gases are at the same temperature and pressure. Therefore, $\frac{V}{n}$ is the same for all gases at the same value of p and T. That is the quantity $\frac{pV}{nT}$ is a *constant* for all gases. This constant is called *universal gas constant*, and is given the symbol, R_0.

i.e., $$MR = R_0 = \frac{pV}{nT}$$

or $$pV = nR_0T \qquad ...(3.14)$$

Since $MR = R_0$, then

$$R = \frac{R_0}{M} \qquad ...(3.15)$$

It has been found experimentally that the volume of 1 mole of any perfect gas at 1 bar and 0°C is approximately 22.71 m^3.

Therefore, from eqn. (3.14),

$$R_0 = \frac{pV}{nT} = \frac{1 \times 10^5 \times 22.71}{1 \times 273.15}$$

$$= 8314.3 \text{ Nm/mole K}$$

Using eqn. (3.15), the gas constant for any gas can be found when the molecular weight is known.

Example. For oxygen which has a molecular weight of 32, the gas constant

$$R = \frac{R_0}{M} = \frac{8314}{32} = 259.8 \text{ Nm/kg K.}$$

3.8.2. Specific Heats

— The specific heat of a solid or liquid is usually defined as the *heat required to raise unit mass through one degree temperature rise.*

— For small quantities, we have

$$dQ = mcdT$$

where m = mass,

c = specific heat, and

dT = temperature rise.

For a gas there are an infinite number of ways in which heat may be added between any two temperatures, and hence a *gas could have an infinite number of specific heats*. However, only two specific heats for gases are defined.

Specific heat at constant volume, c_v

and *Specific heat at constant pressure,* c_p.

We have

$dQ = m\,c_p\,dT$ For a reversible non-flow process at *constant pressure* ...(3.16)

and $dQ = m\,c_v\,dT$ For a reversible non-flow process at *constant volume* ...(3.17)

The values of c_p and c_v, for a perfect gas, are constant for anyone gas at all pressures and temperatures. Hence, integrating eqns. (3.16) and (3.17), we have

Flow of heat in a reversible constant pressure process

$$= mc_p\,(T_2 - T_1) \qquad ...(3.18)$$

Flow of heat in a reversible constant volume process

$$= mc_v\,(T_2 - T_1) \qquad ...(3.19)$$

In case of *real gases, c_p* and *c_v vary with temperature*, but a suitable *average value* may be used for most practical purposes.

3.8.3. Joule's Law

Joule's law states as follows :

"The internal energy of a perfect gas is a function of the absolute temperature only."

i.e., $$u = f(T)$$

To evaluate this function let 1 kg of a perfect gas be heated at constant volume.

According to non-flow energy equation,

$$dQ = du + dW$$

$dW = 0$, since volume remains constant

$$\therefore \quad Q = du$$

At constant volume for a perfect gas, from eqn. (3.17), for 1 kg

$$dQ = c_v dT$$

$$\therefore \quad dQ = du = c_v dT$$

and integrating $u = c_v\,T + K$, K being constant.

According to Joule's law $u = f(T)$, which means that internal energy varies linearly with absolute temperature. Internal energy can be made zero at any arbitrary reference temperature. For a perfect gas it can be assumed that $u = 0$ when $T = 0$, hence constant K is zero.

i.e., Internal energy, $u = c_v\,T$ for a perfect gas ...(3.20)

or For mass m, of a perfect gas

Internal energy, $U = mc_v\,T$ ...(3.21)

For a perfect gas, in any process between states 1 and 2, we have from eqn. (3.21)

Gain in internal energy,

$$U_2 - U_1 = mc_v\,(T_2 - T_1) \qquad ...(3.22)$$

Eqn. (3.22) gives the gains of internal energy for a perfect gas between two states *for any process, reversible* or *irreversible.*

3.8.4. Relationship between Two Specific Heats

Consider a perfect gas being heated at constant pressure from T_1 to T_2.

According to non-flow equation,

$$Q = (U_2 - U_1) + W$$

Also for a perfect gas,

$$U_2 - U_1 = mc_v\,(T_2 - T_1)$$

$$Q = mc_v\,(T_2 - T_1) + W$$

In a constant pressure process, the work done by the fluid,

$$W = p(V_2 - V_1)$$

$$= mR(T_2 - T_1) \qquad \left[\begin{array}{l} \because \; p_1V_1 = mRT_1 \\ p_2V_2 = mRT_2 \\ p_1 = p_2 = p \text{ in this case} \end{array}\right]$$

On substituting

$$Q = mc_v (T_2 - T_1) + mR (T_2 - T_1)$$
$$= m(c_v + R) (T_2 - T_1)$$

But for a constant pressure process,

$$Q = mc_p (T_2 - T_1)$$

By equating the two expressions, we have

$$m(c_v + R)(T_2 - T_1) = mc_p(T_2 - T_1)$$

$$\therefore \quad c_v + R = c_p$$

or $$c_p - c_v = R \qquad ...(3.23)$$

Dividing both sides by c_v, we get

$$\frac{c_p}{c_v} - 1 = \frac{R}{c_v}$$

$$\therefore \quad c_v = \frac{R}{\gamma - 1} \qquad ...[3.23\ (a)]$$

(where $\gamma = c_p/c_v$)

Similarly, dividing both sides by c_p, we get

$$c_p = \frac{\gamma R}{\gamma - 1} \qquad ...[3.23\ (b)]$$

$$\left[\begin{array}{l} \text{In M.K.S. units}: c_p - c_v = \dfrac{R}{J}; \\ c_v = \dfrac{R}{J(\gamma - 1)}, \; c_p = \dfrac{\gamma R}{(\gamma - 1)J} \\ \text{In S.I. units the value of } J \text{ is unity.} \end{array}\right]$$

3.8.5. Enthalpy

— One of the fundamental quantities which occur invariably in thermodynamics is the sum of internal energy (u) and pressure volume product (pv). This sum is called **Enthalpy** (h).

i.e., $$h = u + pv \qquad ...(3.24)$$

— The enthalpy of a fluid is the property of the fluid, since it consists of the sum of a property and the product of the two properties. Since enthalpy is a property like internal energy, pressure, specific volume and temperature, it can be introduced into any problem whether the process is a flow or a non-flow process.

The total enthalpy of mass, m, of a fluid can be

$$H = U + pV, \text{ where } H = mh.$$

For a **perfect gas,**

Referring equation (3.24),

$$h = u + pv$$
$$= c_vT + RT \qquad [\because \; pv = RT]$$
$$= (c_v + R)T$$
$$= c_pT \qquad [\because \; c_p = c_v + R]$$

i.e., $$h = c_pT$$

and $$H = mc_pT.$$

(Note that, since it has been assumed that $u = 0$ at $T = 0$, then $h = 0$ at $T = 0$).

3.8.6. Ratio of Specific Heats

The ratio of specific heat at constant pressure to the specific heat at constant volume is given the symbol γ (gamma).

i.e., $$\gamma = \frac{c_p}{c_v} \qquad ...(3.25)$$

Since $c_p = c_v + R$, it is clear that c_p must be *greater* than c_v for any perfect gas. It follows, therefore, that the ratio, $\frac{c_p}{c_v} = \gamma$ is *always greater than unity.*

In general, the approximate values of γ are as follows :

For *monoatomic* gases such as *argon, helium* = 1.6.

For *diatomic* gases such as *carbon monoxide, hydrogen, nitrogen* and *oxygen* = 1.4.

For triatomic gases such as *carbon dioxide* and *sulphur dioxide* = 1.3.

For some hydrocarbons the value of γ is quite low.

[*e.g.*, for ethane γ = 1.22, and for isobutane γ = 1.11]

3.9. APPLICATION OF FIRST LAW OF THERMODYNAMICS TO NON-FLOW OR CLOSED SYSTEM

1. Reversible Constant Volume Process (v = constant)

In a constant volume process the working substance is contained in a rigid vessel, hence the boundaries of the system are immovable and no work can be done on or by the system, other than paddle-wheel work input. It will be

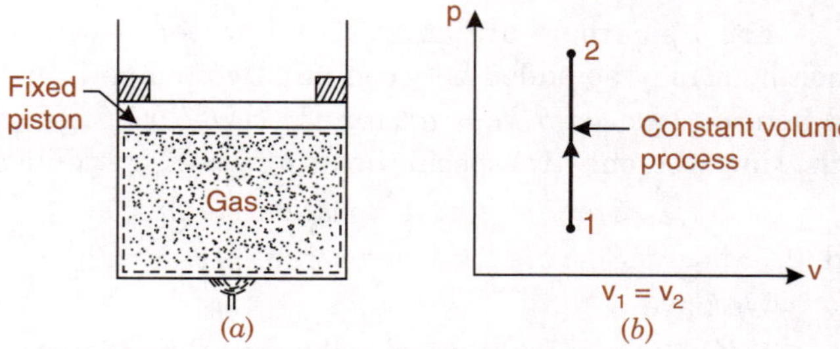

Fig. 3.5. Reversible constant volume process.

assumed that '*constant volume*' implies zero work unless stated otherwise.

Fig. 3.5 shows the system and states before and after the heat addition at constant volume.

Considering mass of the working substance *unity* and applying first law of thermodynamics to the process

$$Q = (u_2 - u_1) + W \quad ...(3.26)$$

The work done $W = \int_1^2 pdv = 0$ as $dv = 0$.

$$\therefore \quad Q = (u_2 - u_1) = c_v(T_2 - T_1) \quad ...[3.27\ (a)]$$

where c_v = Specific heat at constant volume.

For mass, m, of working substance

$$Q = U_2 - U_1 = mc_v(T_2 - T_1) \quad ...[3.27\ (b)]$$

$[\because \ mu = U]$

2. Reversible Constant Pressure Process (p = constant)

It can be seen from Fig. 3.5 (*b*) that when the boundary of the system is *inflexible* as in a constant volume process, then the pressure rises when heat is supplied. Hence, for a constant pressure process, the boundary must move against an external resistance as heat is supplied ; for instance a gas [Fig. 3.6 (*a*)] in a cylinder behind a piston can be made to undergo a constant pressure process. Since the *piston is pushed through a certain distance* by the force exerted by the gas, then the work is done by the gas on its surroundings.

Fig. 3.6 shows the system and states before and after the heat addition at constant pressure.

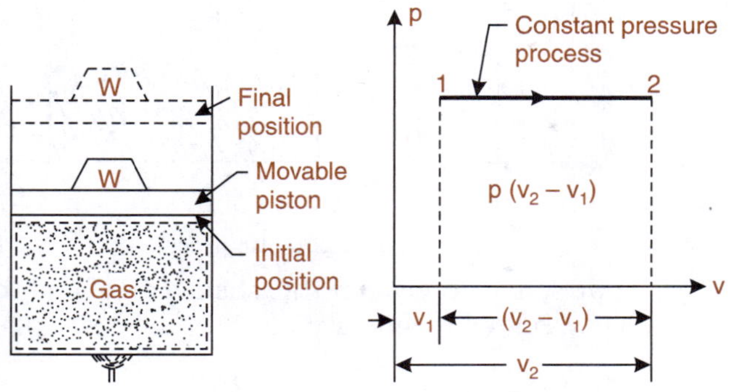

(*a*) (*b*)

Fig. 3.6. Reversible constant pressure process.

Considering unit mass of working substance and applying first law of thermodynamics to the process

$$Q = (u_2 - u_1) + W$$

The work done, $W = \int_1^2 pdv = p(v_2 - v_1)$

$$\therefore \quad Q = (u_2 - u_1) + p(v_2 - v_1) = u_2 - u_1 + pv_2 - pv_1$$
$$= (u_2 + pv_2) - (u_1 + pv_1) = h_2 - h_1$$

$[\because \ h = u + pv]$

or $$Q = h_2 - h_1 = c_p\,(T_2 - T_1) \quad ...(3.28)$$

where h = Enthalpy (specific), and

c_p = Specific heat at constant pressure.

For mass, m, of working substance

$$Q = H_2 - H_1 = mc_p\,(T_2 - T_1) \quad ...[3.28\ (a)]$$

$[\because \ mh = H]$

3. Reversible Temperature or Isothermal Process ($p.v$ = constant, T = constant)

A process at a constant temperature is called an isothermal process. When a working substance in a cylinder behind a piston expands from a high pressure to a low pressure there is a tendency for the temperature to fall. In an isothermal expansion heat must be added continuously in order to keep the temperature at the initial value. Similarly in an isothermal compression heat must be removed from the working substance continuously during the process.

Fig. 3.7 shows the system and states before and after the heat addition at constant temperature.

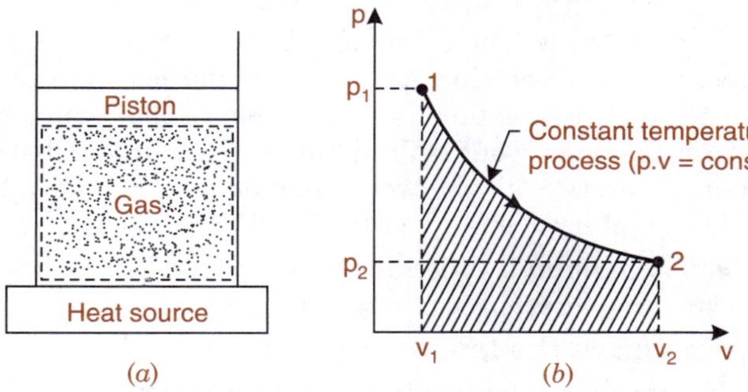

(*a*) (*b*)

Fig. 3.7. Reversible isothermal process.

Considering unit mass of working substance and applying first law to the process

$$Q = (u_2 - u_1) + W$$
$$= c_v\,(T_2 - T_1) + W$$
$$= 0 + W \quad [\because \ T_2 = T_1]$$

The work done, $W = \int_1^2 pdv$

In this case pv = constant or $p = \dfrac{C}{v}$

(where C = constant)

$$\therefore \quad W = \int_{v_1}^{v_2} C\frac{dv}{v} = C[\log_e v]_{v_1}^{v_2} = C\log_e\frac{v_2}{v_1}$$

The constant C can either be written as p_1v_1 or as p_2v_2, since

$$p_1v_1 = p_2v_2 = \text{constant, } C$$

i.e., $W = p_1v_1 \log_e \dfrac{v_2}{v_1}$ per unit mass of working substance

or $W = p_2v_2 \log_e \dfrac{v_2}{v_1}$ per unit mass of working substance

$$\therefore \quad Q = W = p_1 v_1 \log_e \frac{v_2}{v_1} \quad \text{...(3.29)}$$

For mass, m, of the working substance

$$Q = p_1 V_1 \log_e \frac{V_2}{V_1} \quad \text{...[3.29 (a)]}$$

or
$$Q = p_1 V_1 \log_e \frac{p_1}{p_2} \quad \left[\because \frac{V_2}{V_1} = \frac{p_1}{p_2}\right] \quad \text{...[3.29 (b)]}$$

4. Reversible Adiabatic Process (pv^γ = constant)

An **adiabatic process** *is one in which no heat is transferred to or from the fluid during the process.* Such a process can be reversible or irreversible. The reversible adiabatic non-flow process will be considered in this section.

Considering unit mass of working substance and applying first law to the process

$$Q = (u_2 - u_1) + W$$
$$O = (u_2 - u_1) + W$$

or $W = (u_1 - u_2)$ for any adiabatic process ...(3.30)

Eqn. (3.30) is true for an adiabatic process whether the process is reversible or not. In an adiabatic expansion, the work done W by the fluid is at the expense of a reduction in the internal energy of the fluid. Similarly in an adiabatic compression process all the work done on the fluid goes to increase the internal energy of the fluid.

For an adiabatic process to take place, perfect thermal insulation for the system must be available.

To derive the law pv$^\gamma$ = constant :

To obtain a law relating p and v for a reversible adiabatic process let us consider the non-flow energy equation in differential form,

$$dQ = du + dW$$

For a *reversible process*

$$dW = pdv$$

$$\therefore \quad dQ = du + pdv = 0$$

(Since for an adiabatic process $Q = 0$)

Also for a perfect gas

$$pv = RT \text{ or } p = \frac{RT}{v}$$

Hence substituting,

$$du + \frac{RTdv}{v} = 0$$

Also $\quad u = c_v T$ or $du = c_v dT$

$$\therefore \quad c_v dT + \frac{RTdv}{v} = 0$$

Dividing both sides by T, we get

$$c_v \frac{dT}{T} + \frac{Rdv}{v} = 0$$

Integrating

$$c_v \log_e T + R \log_e v = \text{constant}$$

Substituting $T = \dfrac{pv}{R}$

$$c_v \log_e \frac{pv}{R} + R \log_e v = \text{constant}$$

Dividing throughout both sides by c_v

$$\log_e \frac{pv}{R} + \frac{R}{c_v} . \log_e v = \text{constant}$$

Again $\quad c_v = \dfrac{R}{(\gamma - 1)}$ or $\dfrac{R}{c_v} = \gamma - 1$

Hence substituting

$$\log_e \frac{pv}{R} + (\gamma - 1) \log_e v = \text{constant}$$

$$\therefore \quad \log_e \frac{pv}{R} + \log_e v^{\gamma - 1} = \text{constant}$$

$$\log_e \frac{pv \times v^{\gamma - 1}}{R} = \text{constant}$$

i.e.,
$$\log_e \frac{pv^\gamma}{R} = \text{constant}$$

i.e.,
$$\frac{pv^\gamma}{R} = e^{\text{constant}} = \text{constant}$$

or
$$pv^\gamma = \text{constant} \quad \text{...(3.31)}$$

Expression for work W :

A reversible adiabatic process for a perfect gas is shown on a *p-v* diagram in Fig. 3.8 (*b*).

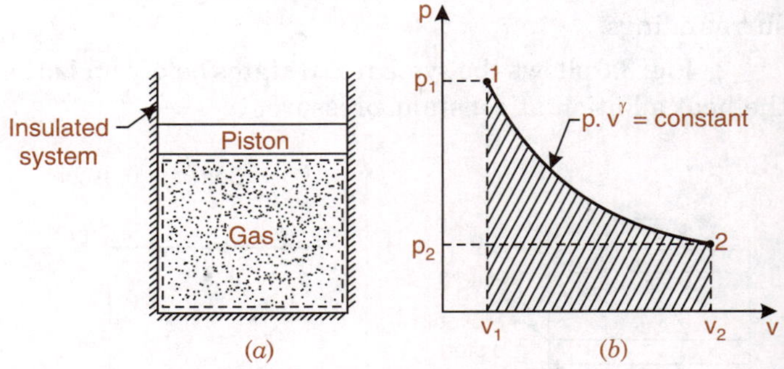

Fig. 3.8. Reversible adiabatic process.

The work done is given by the shaded area, and this area can be evaluated by integration.

i.e.,
$$W = \int_{v_1}^{v_2} p\, dv$$

Therefore, since pv^γ = constant, C, then

$$W = \int_{v_1}^{v_2} C \frac{dv}{v^\gamma} \quad \left[\because p = \frac{C}{v^\gamma}\right]$$

i.e.,
$$W = C \int_{v_1}^{v_2} \frac{dv}{v^\gamma} = C \left| \frac{v^{-\gamma + 1}}{-\gamma + 1} \right|_{v_1}^{v_2}$$

$$= C \left(\frac{v_2^{-\gamma + 1} - v_1^{-\gamma + 1}}{1 - \gamma} \right) = C \left(\frac{v_1^{-\gamma + 1} - v_2^{-\gamma + 1}}{\gamma - 1} \right)$$

The constant in this equation can be written as $p_1 v_1^{\gamma}$ or as $p_2 v_2^{\gamma}$. Hence,

$$W = \frac{p_1 v_1^{\gamma} v_1^{-\gamma+1} - p_2 v_2^{\gamma} v_2^{-\gamma+1}}{\gamma - 1} = \frac{p_1 v_1 - p_2 v_2}{\gamma - 1}$$

i.e., $$W = \frac{p_1 v_1 - p_2 v_2}{\gamma - 1} \quad ...(3.32)$$

or $$W = \frac{R(T_1 - T_2)}{\gamma - 1} \quad ...(3.33)$$

Relationship between T and v, and T and p :

By using equation $pv = RT$, the relationship between T and v, and T and p, may by derived as follows :

i.e., $$pv = RT$$

$$\therefore \quad p = \frac{RT}{v}$$

Putting this value in the equation pv^{γ} = constant

$$\frac{RT}{v} . v^{\gamma} = \text{constant}$$

i.e., $$Tv^{\gamma-1} = \text{constant} \quad ...(3.34)$$

Also $v = \frac{RT}{p}$; hence substituting in equation pv^{γ} = constant

$$p\left(\frac{RT}{p}\right)^{\gamma} = \text{constant}$$

$$\therefore \quad \frac{T^{\gamma}}{p^{\gamma-1}} = \text{constant}$$

or $$\frac{T}{(p)^{\frac{\gamma-1}{\gamma}}} = \text{constant} \quad ...(3.35)$$

Therefore, for a reversible adiabatic process for a perfect gas between states 1 and 2, we can write :

From eqn. (3.31),

$$p_1 v_1^{\gamma} = p_2 v_2^{\gamma} \quad \text{or} \quad \frac{p_2}{p_1} = \left(\frac{v_1}{v_2}\right)^{\gamma} \quad ...(3.36)$$

From eqn. (3.34),

$$T_1 v_1^{\gamma-1} = T_2 v_2^{\gamma-1} \quad \text{or} \quad \frac{T_2}{T_1} = \left(\frac{v_1}{v_2}\right)^{\gamma-1} \quad ...(3.37)$$

From eqn. (3.35),

$$\frac{T_1}{(p_1)^{\frac{\gamma-1}{\gamma}}} = \frac{T_2}{(p_2)^{\frac{\gamma-1}{\gamma}}} \quad \text{or} \quad \frac{T_2}{T_1} = \left(\frac{p_2}{p_1}\right)^{\frac{\gamma-1}{\gamma}} \quad ...(3.38)$$

From eqn. (3.30), the work done in an adiabatic process per kg of gas is given by $W = (u_1 - u_2)$. The gain in internal energy of a perfect gas is given by equation :

$$u_2 - u_1 = c_v (T_2 - T_1) \quad \text{(for 1 kg)}$$

$$\therefore \quad W = c_v (T_1 - T_2)$$

Also, we know that

$$c_v = \frac{R}{\gamma - 1}$$

Hence substituting, we get

$$W = \frac{R(T_1 - T_2)}{\gamma - 1}$$

Using equation, $pv = RT$

$$W = \frac{p_1 v_1 - p_2 v_2}{\gamma - 1}$$

This is the same expression obtained before as eqn. (3.32).

5. Polytropic Reversible Process (pv^n = constant)

It is found that many processes in practice approximate to a reversible law of form pv^n = constant, where n is a constant. Both vapours and perfect gases obey this type of law closely in many non-flow processes. Such processes are *internally reversible*.

We know that for any reversible process,

$$W = \int p\, dv$$

For a process in pv^n = constant, we have

$$p = \frac{C}{v^n}, \text{ where } C \text{ is a constant}$$

$$\therefore \quad W = C\int_{v_1}^{v_2} \frac{dv}{v^n} = C\left|\frac{v^{-n+1}}{-n+1}\right| = C\left(\frac{v_2^{-n+1} - v_1^{-n+1}}{-n+1}\right)$$

i.e., $$W = C\left(\frac{v_1^{-n+1} - v_2^{-n+1}}{n-1}\right) = \frac{p_1 v_1^n v_1^{-n+1} - p_2 v_2^n v_2^{-n+1}}{n-1}$$

(since the constant C, can be written as $p_1 v_1^n$ or as $p_2 v_2^n$)

i.e., Work done, $$W = \frac{p_1 v_1 - p_2 v_2}{n - 1} \quad ...(3.39)$$

or $$W = \frac{R(T_1 - T_2)}{n - 1} \quad ...(3.40)$$

Eqn. (3.39) is true for any working substance undergoing a reversible polytropic process. It follows also that for any polytropic process, we can write

$$\frac{p_2}{p_1} = \left(\frac{v_1}{v_2}\right)^n \quad ...(3.41)$$

The following relations can be derived (following the same procedure as was done under reversible adiabatic process)

$$\frac{T_2}{T_1} = \left(\frac{v_1}{v_2}\right)^{n-1} \quad ...(3.42)$$

$$\frac{T_2}{T_1} = \left(\frac{p_2}{p_1}\right)^{\frac{n-1}{n}} \quad ...(3.43)$$

Heat transfer during polytropic process (for perfect gas pv = RT) :

Using non-flow energy equation, the heat flow/transfer during the process can be found,

i.e., $$Q = (u_2 - u_1) + W$$

$$= c_v(T_2 - T_1) + \frac{R(T_1 - T_2)}{n - 1}$$

i.e., $$Q = \frac{R(T_1 - T_2)}{n - 1} - c_v (T_1 - T_2)$$

Also $$c_v = \frac{R}{(\gamma - 1)}$$

On substituting,

$$Q = \frac{R}{n - 1}(T_1 - T_2) - \frac{R}{(\gamma - 1)}(T_1 - T_2)$$

i.e., $$Q = R(T_1 - T_2)\left(\frac{1}{n-1} - \frac{1}{\gamma - 1}\right)$$

$$= \frac{R(T_1 - T_2)(\gamma - 1 - n + 1)}{(\gamma - 1)(n - 1)} = \frac{R(T_1 - T_2)(\gamma - n)}{(\gamma - 1)(n - 1)}$$

$$\therefore \quad Q = \frac{(\gamma - n)}{(\gamma - 1)} \frac{R(T_1 - T_2)}{(n - 1)}$$

or $$Q = \left(\frac{\gamma - n}{\gamma - 1}\right) W \quad \left[\because W = \frac{R(T_1 - T_2)}{(n - 1)}\right] \quad ...(3.44)$$

In a polytropic process, the *index n depends only on the heat and work quantities* during the process. The various processes considered earlier are special cases of polytropic process for a perfect gas. For example,

When $n = 0$ pv° = constant *i.e.,* p = constant

When $n = \infty$ pv^∞ = constant

or $p^{1/\infty} v$ = constant, *i.e.,* v = constant

When $n = 1$ pv = constant, *i.e.,* T = constant

[since $(pv)/T$ = constant for a perfect gas]

When $n = \gamma$ pv^γ = constant, *i.e.,* reversible adiabatic

This is illustrated on a *p-v* diagram in Fig. 3.9.

(*i*) State 1 to state *A* is *constant pressure cooling* ($n = 0$).

(*ii*) State 1 to state *B* is *isothermal compression* ($n = 1$).

(*iii*) State 1 to state *C* is *reversible adiabatic compression* ($n = \gamma$).

(*iv*) State 1 to state *D* is *constant volume heating* ($n = \infty$).

Similarly,

(*i*) State 1 to state *A′* is *constant pressure heating* ($n = 0$).

(*ii*) State 1 to state *B′* is *isothermal expansion* ($n = 1$).

(*iii*) State 1 to state *C′* is *reversible adiabatic expansion* ($n = \gamma$).

(*iv*) State 1 to state *D′* is *constant volume cooling* ($n = \infty$).

It may be noted that, since γ is always greater than unity, than process 1 to *C* must lie between processes 1 to *B* and 1 to *D* ; similarly, process 1 to *C′* must lie between processes 1 to *B′* and 1 to *D′*.

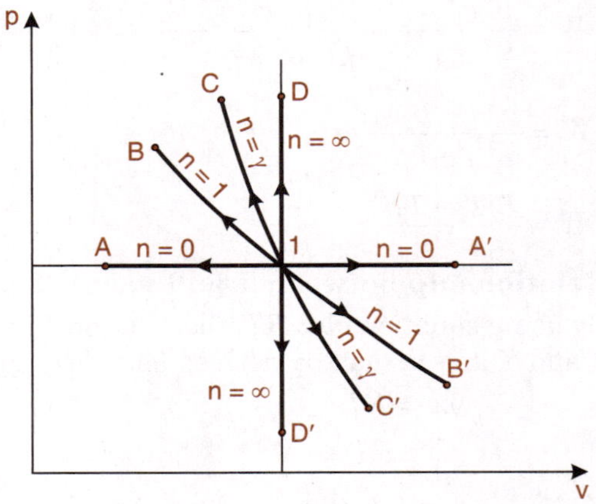

Fig. 3.9

6. Free Expansion

Consider two vessels 1 and 2 interconnected by a short pipe with a valve *A*, and perfectly thermally insulated (Fig. 3.10). Initially let the vessel 1 be filled with a fluid at a certain pressure, and let 2 be completely evacuated. When the valve *A* is opened the fluid in 1 will expand rapidly to fill both vessels 1 and 2. The pressure finally will be lower than the initial pressure in vessel 1. This is known as *free or unresisted expansion*. The process is *highly irreversible* ; since the fluid is eddying continuously during the process. Now applying first law of thermodynamics (or non-flow energy equation) between the initial and final states,

$$Q = (u_2 - u_1) + W$$

In this process, no work is done on or by the fluid, since the boundary of the system does not move. No heat flows to or from the fluid since the system is well lagged. The process is therefore, *adiabatic but irreversible*.

i.e., $$u_2 - u_1 = 0 \quad \text{or} \quad u_2 = u_1$$

In a free expansion, therefore, the internal energy initially equals the initial energy finally.

For a perfect gas,

$$u = c_v T$$

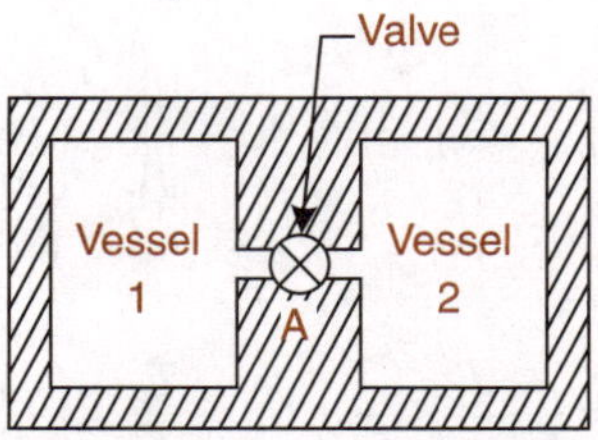

Fig. 3.10. Free expansion.

$\therefore$ For a free expansion of a perfect gas,

$$c_v T_1 = c_v T_2 \quad i.e., \; T_1 = T_2$$

That is, for a perfect gas undergoing a free expansion, the initial temperature is equal to the final temperature.

Table 3.1. Summary of Processes for Perfect Gas (Unit mass)

Process	*Index* n	*Heat added*	$\int_1^2 pdv$	*p, v, T relations*	*Specific heat, c*
Constant pressure	$n = 0$	$c_p(T_2 - T_1)$	$p(v_2 - v_1)$	$\frac{T_2}{T_1} = \frac{v_2}{v_1}$	c_p
Constant volume	$n = \infty$	$c_v(T_2 - T_1)$	0	$\frac{T_1}{T_2} = \frac{p_1}{p_2}$	c_v
Constant temperature	$n = 1$	$p_1v_1 \log_e \frac{v_2}{v_1}$	$p_1v_1 \log_e \frac{v_2}{v_1}$	$p_1v_1 = p_2v_2$	∞
Reversible adiabatic	$n = \gamma$	0	$\frac{p_1v_1 - p_2v_2}{\gamma - 1}$	$p_1v_1^{\gamma} = p_2v_2^{\gamma}$ $\frac{T_2}{T_1} = \left(\frac{v_1}{v_2}\right)^{\gamma - 1} = \left(\frac{p_2}{p_1}\right)^{\frac{\gamma - 1}{\gamma}}$	0
Polytropic	$n = n$	$c_n(T_2 - T_1)$ $= c_v\left(\frac{\gamma - n}{1 - n}\right) \times (T_2 - T_1)$ $= \frac{\gamma - n}{\gamma - 1} \times$ work done (non-flow)	$\frac{p_1v_1 - p_2v_2}{n - 1}$	$p_1v_1^{n} = p_2v_2^{n}$ $\frac{T_2}{T_1} = \left(\frac{v_1}{v_2}\right)^{n - 1} = \left(\frac{p_2}{p_1}\right)^{\frac{n - 1}{n}}$	$c_n = c_v\left(\frac{\gamma - n}{1 - n}\right)$

Note. Equations must be used keeping dimensional consistence.

Example 3.1. *In an internal combustion engine, during the compression stroke the heat rejected to the cooling water is 50 kJ/kg and the work input is 100 kJ/kg.*

Calculate the change in internal energy of the working fluid stating whether it is a gain or loss.

Solution. Heat rejected to the cooling water,

$$Q = -50 \text{ kJ/kg}$$

(–ve sign since heat is rejected)

Work input, $W = -100$ kJ/kg

(–ve sign since work is supplied to the system)

Using the relation, $Q = (u_2 - u_1) + W$

$$-50 = (u_2 - u_1) - 100$$

or $u_2 - u_1 = -50 + 100 = 50$ kJ/kg

Hence, **gain in internal energy = 50 kJ/kg. (Ans.)**

Example 3.2. *In an air motor cylinder the compressed air has an internal energy of 450 kJ/kg at the beginning of the expansion and an internal energy of 220 kJ/kg after expansion. If the work done by the air during the expansion is 120 kJ/kg, calculate the heat flow to and from the cylinder.*

Solution. Internal energy at beginning of the expansion,

$$u_1 = 450 \text{ kJ/kg}$$

Internal energy after expansion,

$$u_2 = 220 \text{ kJ/kg}$$

Work done by the air during expansion,

$$W = 120 \text{ kJ/kg}$$

Heat flow, Q :

Using the relation, $Q = (u_2 - u_1) + W$

$\therefore$ $Q = (220 - 450) + 120$

$= -230 + 120 = -110$ kJ/kg

Hence, **heat rejected by air = 110 kJ/kg. (Ans.)**

Example 3.3. *0.3 kg of nitrogen gas at 100 kPa and 40°C is contained in a cylinder. The piston is moved compressing nitrogen until the pressure becomes 1 MPa and temperature becomes 160°C. The work done during the process is 30 kJ.*

Calculate the heat transferred from the nitrogen to the surroundings.

c_v for nitrogen = 0.75 kJ/kg K.

Solution. Mass of nitrogen, $m = 0.3$ kg

Temperature before compression = 40°C or 313 K

Temperature after compression = 160°C or 433 K

The work done during the compression process,

$$W = -30 \text{ kJ}$$

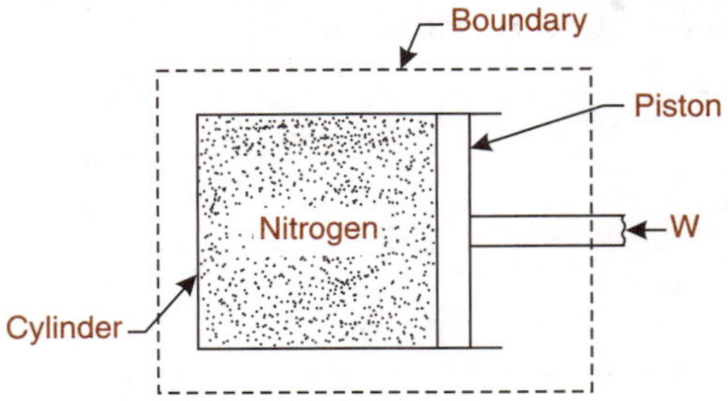

Fig. 3.11

According to first law of thermodynamics,

$$Q = \Delta U + W = (U_2 - U_1) + W$$
$$= mc_v (T_2 - T_1) + W$$
$$= 0.3 \times 0.75(433 - 313) - 30 = -3 \text{ kJ}$$

Hence, **heat 'rejected' during the process**

= 3 kJ. (Ans.)

Note. Work, W has been taken –ve because it has been supplied from outside.

Example 3.4. *When a stationary mass of gas was compressed without friction at constant pressure its initial state of 0.4 m³ and 0.105 MPa was found to change to final state of 0.20 m³ and 0.105 MPa. There was a transfer of 42.5 kJ of heat from the gas during the process.*

How much did the internal energy of the gas change ?

Solution.

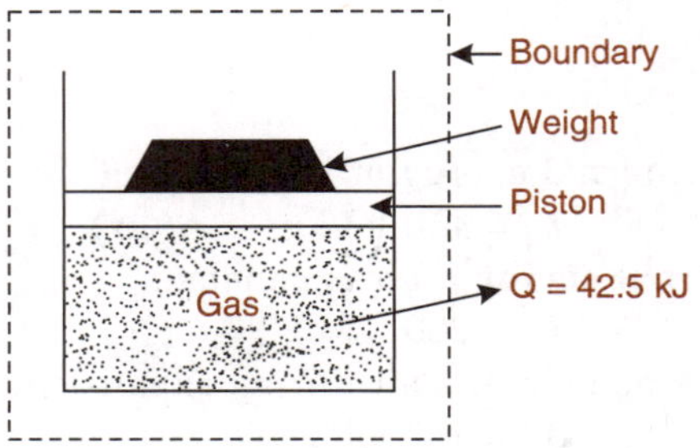

Fig. 3.12

Initial state

Pressure of gas, $p_1 = 0.105$ MPa

Volume of gas, $V_1 = 0.4$ m³

Final state

Pressure of gas, $p_2 = 0.105$ MPa

Volume of gas, $V_2 = 0.20$ m³

Process used : *Constant pressure*

Heat transferred, $Q = -42.5$ kJ

(–ve sign indicates that heat is rejected)

Change in internal energy, $\Delta U = U_2 - U_1$:

First law for a stationary system in a process gives

$$Q = \Delta U + W$$

or $$Q_{1-2} = (U_2 - U_1) + W_{1-2} \quad ...(i)$$

Here $W_{1-2} = \int_{V_1}^{V_2} pdV = p(V_2 - V_1)$

$= 0.105(0.20 - 0.40)$ MJ $= -21$ kJ

$[\because \ 1 \text{ MJ} = 10^3 \text{ kJ}]$

Substituting this value of W_{1-2} in equation (i), we get

$$-42.5 = (U_2 - U_1) - 21$$

$\therefore \quad U_2 - U_1 = -42.5 + 21 = -21.5$ kJ

Hence **'decrease' in internal energy**

= 21.5 kJ. (Ans.)

Example 3.5. *A container is divided into compartments by a partition. The container is completely insulated so that there is no heat transfer. One portion contains gas at temperature T_1 and pressure p_1 while the other portion also has the same gas but at temperature T_2 and pressure p_2.*

How will the First Law of Thermodynamics conclude the result if partition is removed ?

Solution. Refer to Fig. 3.13.

According to First Law of Thermodynamics,

$$\delta Q = \delta U + \delta W$$

When partition removed, $\delta Q = 0$

$\delta W = 0$

$\therefore \quad \delta U = 0.$

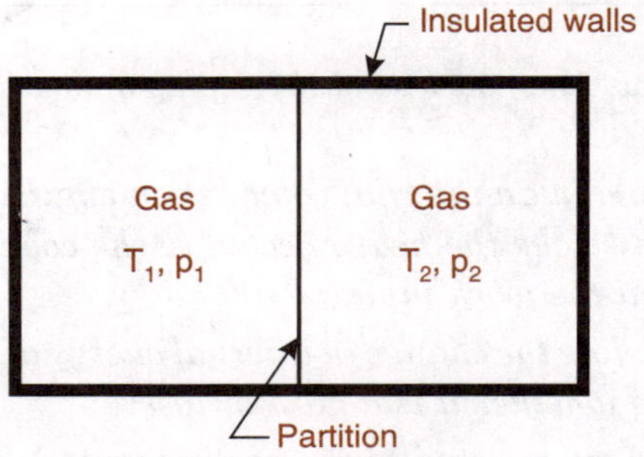

Fig. 3.13

Conclusion. *There is conservation of internal energy.*

Example 3.6. *Air enters a compressor at 10⁵ Pa and 25°C having volume of 1.8 m³/kg and is compressed to 5 × 10⁵ Pa isothermally.*

Determine : (i) Work done ;

(ii) Change in internal energy ; and

(iii) Heat transferred.

Solution. Initial pressure of air, $p_1 = 10^5$ Pa

Initial temperature of air, $T_1 = 25 + 273$

$= 298$ K

Final pressure of air, $p_2 = 5 \times 10^5$ Pa

Final temperature of air, $T_2 = T_1 = 298$ K

(isothermal process)

Since, it is a closed steady state process, we can write down the first law of thermodynamics as,

$$Q = (u_2 - u_1) + W \text{ ......per kg}$$

(*i*) *For isothermal process* :

$$W_{1-2} = \int_1^2 p \,.\, dv = p_1 v_1 \log_e \left(\frac{p_1}{p_2}\right)$$

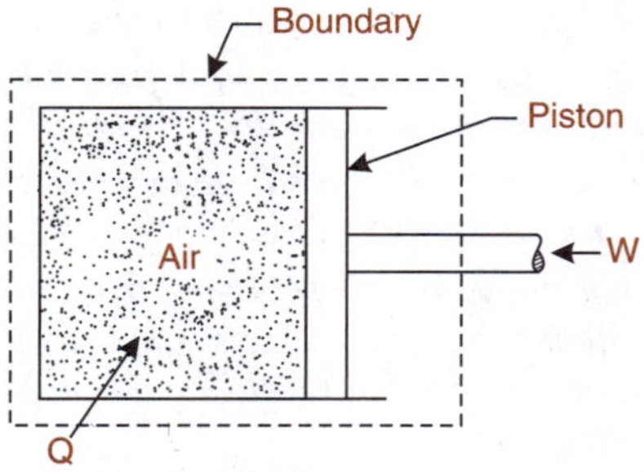

Fig. 3.14

as $\qquad p_1 v_1 = p_2 v_2$ for isothermal process

$$\therefore \qquad W_{1-2} = -10^5 \times 1.8 \log_e \left(\frac{1 \times 10^5}{5 \times 10^5}\right)$$

$$= -2.897 \times 10^5 = -289.7 \text{ kJ/kg.}$$

(– ve sign indicates that the work is supplied to the air)

∴ **Work done on the air = 289.7 kJ/kg. (Ans.)**

(*ii*) Since temperature is constant,

$$\therefore \qquad u_2 - u_1 = 0$$

∴ **Change in internal energy = zero. (Ans.)**

(*iii*) Again, $Q_{1-2} = (u_2 - u_1) + W$

$$= 0 + (-289.7) = -289.7 \text{ kJ}$$

(– ve sign indicates that heat is lost from the system to the surroundings)

∴ **Heat transferred = 289.7 kJ/kg. (Ans.)**

Example 3.7. *A cylinder containing the air comprises the system. Cycle is completed as follows :*

(*i*) *82000 Nm of work is done by the piston on the air during compression stroke and 45 kJ of heat are rejected to the surroundings.*

(*ii*) *During expansion stroke 100000 Nm of work is done by the air on the piston.*

Calculate the quantity of heat added to the system.

Solution. Refer to Fig. 3.15.

Compression stroke. Process 1–2 :

Work done by the piston *on the air,*

$$W_{1-2} = -82000 \text{ Nm } (= -82 \text{ kJ})$$

Heat rejected to the system, $Q_{1-2} = -45$ kJ

Now, $\quad Q_{1-2} = (U_2 - U_1) + W$

$$-45 = (U_2 - U_1) + (-82)$$

$$\therefore \quad (U_2 - U_1) = 37 \text{ kJ} \qquad ...(i)$$

Expansion stroke. Process 2–1 :

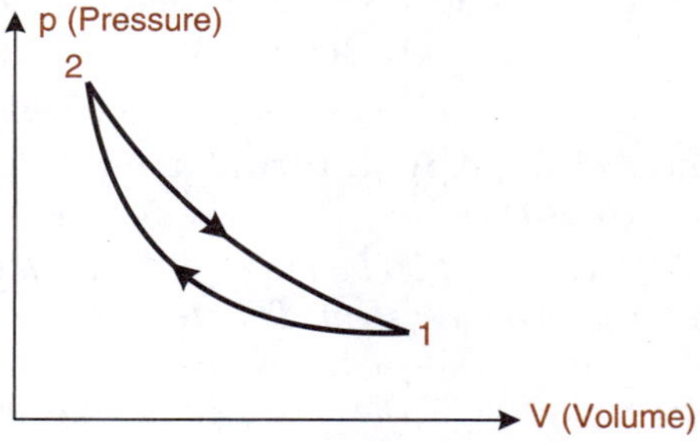

Fig. 3.15

Work done by air on the piston,

$$W_{2-1} = 100000 \text{ Nm } (= 100 \text{ kJ})$$

Now, $\quad Q_{2-1} = (U_1 - U_2) + W$

$$= -37 + 100 \text{ kJ} = 63 \text{ kJ}$$

Hence, **quantity of heat added to the system = 63 kJ. (Ans.)**

☞**Example 3.8.** *A tank containing air is stirred by a paddle wheel. The work input to the paddle wheel is 9000 kJ and the heat transferred to the surroundings from the tank is 3000 kJ.*

Determine : (*i*) *Work done ;*

(*ii*) *Change in internal energy of the system.*

Solution. Refer to Fig. 3.16.

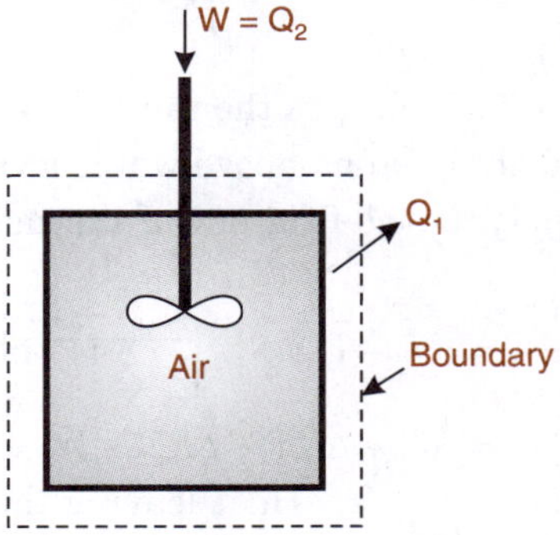

Fig. 3.16

Work input to the paddle wheel = 9000 kJ

Heat transferred to the surroundings from the tank = 3000 kJ

As it is a closed system, the first law of thermodynamics can be written as

$$U_1 - Q + W = U_2 \qquad ...(i)$$

The *work* enters into the tank in the form of energy only so this should be considered as *heat input.*

$$\therefore \qquad Q = Q_1 - Q_2$$

$$= 3000 - 9000 = -6000 \text{ kJ.}$$

(*i*) Since volume does not change (being constant volume process)

$\therefore$ Work done, $W = 0$. **(Ans.)**

Putting the value of $W = 0$ in equation (*i*), we get

(*ii*) $U_1 - (-6000) + 0 = U_2$

$\therefore$ $U_2 - U_1 = 6000$ kJ

Hence, **change in internal energy (increase) = 6000 kJ. (Ans.)**

Example 3.9. *A stone of 20 kg mass and a tank containing 200 kg water comprise a system. The stone is 15 m above the water level initially. The stone and water are at the same temperature initially. If the stone falls into water, then determine ΔU, ΔPE, ΔKE, Q and W, when*

(*i*) *the stone is about to enter the water,*

(*ii*) *the stone has come to rest in the tank, and*

(*iii*) *the heat is transferred to the surroundings in such an amount that the stone and water come to their initial temperature.*

Solution. Refer to Fig. 3.17.

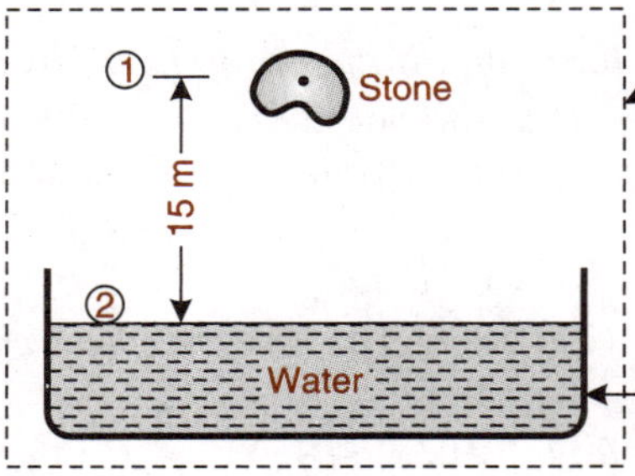

Fig. 3.17

Mass of stone = 20 kg

Mass of water in the tank = 200 kg

Height of stone above water level = 15 m

Applying the first law of thermodynamics,

$$Q = (U_2 - U_1) + m\left[\frac{C_2^2 - C_1^2}{2}\right] + mg\,(Z_2 - Z_1) + W$$

$$= \Delta U + \Delta KE + \Delta PE + W \quad ...(1)$$

Here Q = Heat leaving the boundary.

(*i*) When the stone is about to enter the water,

$Q = 0,\ W = 0,\ \Delta U = 0$

$\because$ $-\Delta KE = \Delta PE = mg\,(Z_2 - Z_1)$

$= 20 \times 9.81\,(0 - 15) = -2943$ J

$\therefore$ $\Delta KE = 2943$ J

and $\Delta PE = $ **– 2943 J. (Ans.)**

(*ii*) When the stone dips into the tank and comes to rest

$Q = 0,\ W = 0,\ \Delta KE = 0$

Substituting these values in eqn. (1), we get

$0 = \Delta U + 0 + \Delta PE + 0$

$\therefore$ $\Delta U = -\Delta PE = -(-2943) =$ **2943 J. (Ans.)**

This shows that the internal energy (temperature) of the system increases.

(*iii*) When the water and stone come to their initial temperature,

$W = 0,\ \Delta KE = 0$

Substituting these values in eqn. (1), we get

$\therefore$ $Q = -\Delta U =$ **– 2943 J. (Ans.)**

The *negative* sign shows that the *heat is lost from the system to the surroundings.*

Example 3.10. *When a system is taken from state l to state m, in Fig. 3.18, along path lqm, 168 kJ of heat flows into the system, and the system does 64 kJ of work :*

(*i*) *How much will be the heat that flows into the system along path lnm if the work done is 21 kJ ?*

(*ii*) *When the system is returned from m to l along the curved path, the work done on the system is 42 kJ. Does the system absorb or liberate heat, and how much of the heat is absorbed or liberated ?*

(*iii*) *If $U_l = 0$ and $U_n = 84$ kJ, find the heat absorbed in the processes ln and nm.*

Solution. Refer to Fig. 3.18.

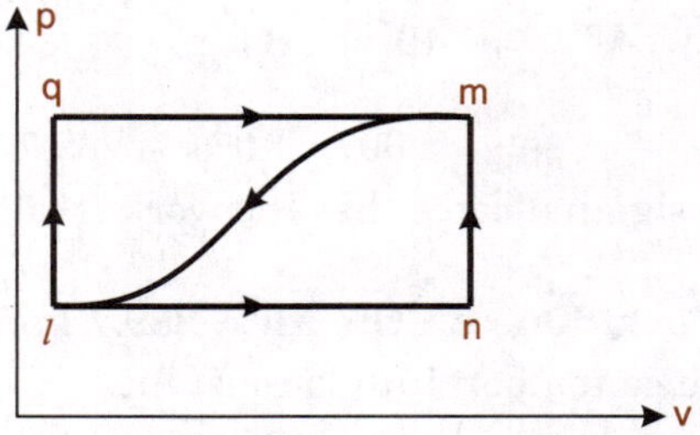

Fig. 3.18

$Q_{l-q-m} = 168$ kJ

$W_{l-q-m} = 64$ kJ

We have, $Q_{l-q-m} = (U_m - U_l) + W_{l-q-m}$

$168 = (U_m - U_l) + 64$

$\therefore$ $U_m - U_l =$ **104 kJ. (Ans.)**

(*i*) $Q_{l-n-m} = (U_m - U_l) + W_{l-n-m}$

$= 104 + 21 =$ **125 kJ. (Ans.)**

(*ii*) $Q_{m-l} = (U_l - U_m) + W_{m-l}$

$= -104 + (-42)$

$=$ **– 146 kJ. (Ans.)**

The system liberates 146 kJ.

(*iii*) $W_{l-n-m} = W_{l-n} + W_{n-m} = W_{l-m} = 21$ kJ

[$\because$ $W_{n-m} = 0$, since volume does not change.]

$\therefore$ $Q_{l-n} = (U_n - U_l) + W_{l-n}$

$= (84 - 0) + 21 =$ **105 kJ. (Ans.)**

Now $Q_{l-m-n} = 125 \text{ kJ} = Q_{l-n} + Q_{n-m}$

$\therefore$ $Q_{n-m} = 125 - Q_{l-n}$

$= 125 - 105 =$ **20 kJ. (Ans.)**

Example 3.11. *In a system, executing a non-flow process, the work and heat per degree change of temperature are given by*

$$\frac{dW}{dT} = 200 \text{ W-s/°C and } \frac{dQ}{dT} = 160 \text{ J/°C}$$

What will be the change of internal energy of the system when its temperature changes from

$$T_1 = 55°C \text{ to } T_2 = 95°C \text{ ?}$$

Solution. Initial temperature, $T_1 = 55°C$;

Final temperature, $T_2 = 95°C$

$$\frac{dW}{dT} = 200 \text{ W-s/°C} ; \frac{dQ}{dT} = 160 \text{ J/°C}.$$

Change of internal energy :

Now, $\frac{dW}{dT} = 200$ W-s/°C

$$\therefore \quad W = \int_{T_1}^{T_2} 200\, dT = \Big| 200\, T \Big|_{T_1}^{T_2} = \Big| 200\, T \Big|_{55}^{95}$$

$$= 200\,(95 - 55) = 8000 \text{ W-s} = 8000 \text{ J}$$

[∵ 1 W-s = 1 J]

Also, $\frac{dQ}{dT} = 160$ J/°C

$$\therefore \quad Q = \int_{T_1}^{T_2} 160\, dT = \Big| 160\, T \Big|_{T_1}^{T_2}$$

$$= \Big| 160\, T \Big|_{55}^{95} = 160\,(95 - 55) = 6400 \text{ J}$$

Applying the first law of thermodynamics to the given non-flow system,

$$Q = \Delta U + W$$

$$\therefore \quad 6400 = \Delta U + 8000$$

or $\Delta U = -1600$ J = **– 1.6 kJ. (Ans.)**

The –ve sign indicates that there is *decrease* in internal energy.

Example 3.12. *A fluid system, contained in a piston and cylinder machine, passes through a complete cycle of four processes. The sum of all heat transferred during a cycle is – 340 kJ. The system completes 200 cycles per min.*

Complete the following table showing the method for each item, and compute the net rate of work output in kW.

Process	*Q (kJ/min)*	*W (kJ/min)*	*ΔE (kJ/min)*
1–2	*0*	*4340*	—
2–3	*42000*	*0*	—
3–4	*– 4200*	—	*– 73200*
4–1	—	—	—

Solution. Sum of all heat transferred during the cycle

= – 340 kJ.

Number of cycles completed by the system

= 200 cycles/min.

Process 1–2 :

$$Q = \Delta E + W$$
$$0 = \Delta E + 4340$$
$$\therefore \quad \Delta E = -4340 \text{ kJ/min.}$$

Process 2–3 :

$$Q = \Delta E + W$$
$$42000 = \Delta E + 0$$
$$\Delta E = 42000 \text{ kJ/min.}$$

Process 3–4 :

$$Q = \Delta E + W$$
$$-4200 = -73200 + W$$
$$\therefore \quad W = 69000 \text{ kJ/min.}$$

Process 4–1 :

$$\sum_{\text{cycle}} Q = -340 \text{ kJ}$$

The system completes 200 cycles/min

$$\because \quad Q_{1-2} = Q_{2-3} + Q_{3-4} + Q_{4-1}$$
$$= -340 \times 200 = -68000 \text{ kJ/min}$$

or $0 + 42000 + (-4200) + Q_{4-1} = -68000$

$$Q_{4-1} = -105800 \text{ kJ/min.}$$

Now, $\oint dE = 0$, since cyclic integral of any property is zero.

$$\Delta E_{1-2} + \Delta E_{2-3} + \Delta E_{3-4} + \Delta E_{4-1} = 0$$
$$-4340 + 42000 + (-73200) + \Delta E_{4-1} = 0$$
$$\therefore \quad \Delta E_{4-1} = 35540 \text{ kJ/min.}$$
$$\therefore \quad W_{4-1} = Q_{4-1} - \Delta E_{4-1}$$
$$= -105800 - 35540 = -141340 \text{ kJ/min}$$

The completed table is given below :

Process	*Q(kJ/min)*	*W(kJ/min)*	*Δ E(kJ/min)*
1–2	0	4340	– 4340
2–3	42000	0	42000
3–4	– 4200	69000	– 73200
4–1	– 105800	– 141340	35540

Since $\sum_{\text{cycle}} Q = \sum_{\text{cycle}} W$

Rate of work output = – 68000 kJ/min

$$= -\frac{68000}{60} \text{ kJ/s or kW}$$

= **1133.33 kW. (Ans.)**

Example 3.13. *The power developed by a turbine in a certain steam plant is 1200 kW. The heat supplied to the steam in the boiler is 3360 kJ/kg, the heat rejected by the system to cooling water in the condenser is 2520 kJ/kg and the feed pump work required to pump the condensate back into the boiler is 6 kW.*

Calculate the steam flow round the cycle in kg/s.

Solution. The power developed by the turbine = 1200 kW

The heat supplied to the steam in the boiler

= 3360 kJ/kg

The heat rejected by the system to cooling water

= 2520 kJ/kg

Feed pump work = 6 kW

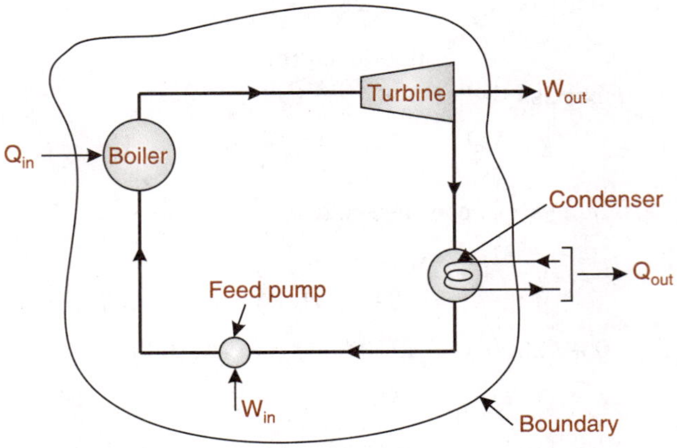

Fig. 3.19

Fig. 3.19 shows the cycle. A boundary is shown which encompasses the entire plant. Strictly, this boundary should be thought of as encompassing the working fluid only.

$$\oint dQ = 3360 - 2520 = 840 \text{ kJ/kg}$$

Let the system flow be in kg/s.

$$\therefore \quad \oint dQ = 840\ \dot{m} \text{ kJ/s}$$

$$\oint dW = 1200 - 6 = 1194 \text{ kJ/s}$$

$$\text{But} \quad \oint dQ = \oint dW$$

i.e., $\quad 840\ \dot{m} = 1194$

$$\therefore \quad \dot{m} = \frac{1194}{840} = 1.421 \text{ kg/s}$$

∴ **Steam flow round the cycle**

= 1.421 kg/s. (Ans.)

Example 3.14. *A closed system of constant volume experiences a temperature rise of 25°C when a certain process occurs. The heat transferred in the process is 30 kJ. The specific heat at constant volume for the pure substance comprising the system is 1.2 kJ/kg°C, and the system contains 2.5 kg of this substance. Determine :*

(*i*) *The change in internal energy ;*

(*ii*) *The work done.*

Solution. (*i*) Temperature rise, $(T_2 - T_1) = 25°C$

The heat transferred in the process, $Q = 30$ kJ

Specific heat at constant volume, $c_v = 1.2$ kJ/kg°C

Mass of the substance, $m = 2.5$ kg

$$\text{Now,} \quad \Delta U = m \int_{T_1}^{T_2} c_v \, dT$$

$$= 2.5 \int_{T_1}^{T_2} 1.2 \, dT = 3.0 \times (T_2 - T_1)$$

$$= 3.0 \times 25 = \mathbf{75\ kJ}$$

Hence, **the change in internal energy is 75 kJ. (Ans.)**

(*ii*) According to the first law of thermodynamics,

$$Q = \Delta U + W$$

$$\therefore \quad 30 = 75 + W$$

$$\therefore \quad W = 30 - 75 = -45 \text{ kJ}$$

Hence, **the work done = – 45 kJ. (Ans.)**

It may be observed that *even though the volume is constant the work is not zero.* Clearly, the process is **irreversible.**

Example 3.15. *A system receives 50 kJ of heat while expanding with volume change of 0.14 m³ against an atmosphere of 1.2 × 10⁵ N/m². A mass of 90 kg in the surroundings is also lifted through a distance of 5.5 metres.*

(*i*) *Find the change in energy of the system.*

(*ii*) *The system is returned to its initial volume by an adiabatic process which requires 110 kJ of work. Find the change in energy of the system.*

(*iii*) *For the combined processes of (i) and (ii) determine the change in energy of the system.*

Solution. Heat received by the system,

$Q = 50$ kJ

Change in volume $\quad \Delta V = 0.14\ m^3$

Pressure $\quad = 1.2 \times 10^5\ N/m^2$

Mass lifted in the surroundings = 90 kg

Distance through which lifted = 5.5 m

Work done during adiabatic process = – 110 kJ.

(*i*) $\quad Q = \Delta E + W \quad$ ...(*i*)

Now, $\quad W = p.\Delta V + W_{net}$

$$= \left(\frac{1.2 \times 10^5 \times 0.14}{1000}\right) + \left(\frac{90 \times 5.5 \times 9.8}{1000}\right) \text{kJ}$$

$$= 16.8 + 4.85 = 21.65 \text{ kJ}$$

But [from (*i*)], $\Delta E = Q - W$

$= 50 - 21.65 =$ **28.35 kJ. (Ans.)**

(*ii*) Since the process is adiabatic,

$$Q = 0$$

and $\Delta E = -W$

$= -(-110) =$ **110 kJ. (Ans.)**

(*iii*) Change in internal energy,

$$\Delta E = Q - W$$

$= 50 - [(-110) + 21.65]$

$=$ **138.35 kJ. (Ans.)**

Example 3.16. *A fluid system undergoes a non-flow frictionless process following the pressure-volume relation as* $p = \frac{5}{V} + 1.5$, *where p is in bar and V is in* m^3. *During the process the volume changes from 0.15* m^3 *to 0.05* m^3 *and the system rejects 45 kJ of heat. Determine :*

(*i*) *Change in internal energy ;*

(*ii*) *Change in enthalpy.*

Solution. Pressure-volume relation : $p = \frac{5}{V} + 1.5$

Initial volume, $V_1 = 0.15 \text{ m}^3$

Final volume, $V_2 = 0.05 \text{ m}^3$

Heat *rejected* by the system, $Q = -45 \text{ kJ}$

Work done is given by,

$$W = \int_1^2 p\,.\,dV = \int_{V_1}^{V_2} \left(\frac{5}{V} + 1.5\right) dV$$

$$= \int_{0.15}^{0.05} \left(\frac{5}{V} + 1.5\right) dV$$

$$= \left[5 \log_e \frac{V_2}{V_1} + 1.5\,(V_2 - V_1)\right]_{0.15}^{0.05} \times 10^5 \text{ Nm}$$

$$= 10^5 \left[5 \log_e \frac{0.05}{0.15} + 1.5\,(0.05 - 0.15)\right]$$

$= 10^5(-5.49 - 0.15)$ Nm

$= -5.64 \times 10^5$ Nm

$= -5.64 \times 10^5$ J [$\because$ 1 Nm = 1 J]

$= -564$ kJ.

(*i*) Applying the first law energy equation,

$$Q = \Delta U + W$$

$$-45 = \Delta U + (-564)$$

$\therefore$ **ΔU = 519 kJ. (Ans.)**

This shows that the internal energy is **increased.**

(*ii*) Change in enthalpy,

$$\Delta H = \Delta U + \Delta\,(pV)$$

$$= 519 \times 10^3 + (p_2V_2 - p_1V_1)$$

$$p_1 = \frac{5}{V_1} + 1.5 = \frac{5}{0.15} + 1.5 = 34.83 \text{ bar}$$

$= 34.83 \times 10^5 \text{ N/m}^2$

$$p_2 = \frac{5}{V_2} + 1.5 = \frac{5}{0.05} + 1.5$$

$= 101.5 \text{ bar} = 101.5 \times 10^5 \text{ N/m}^2$

$\therefore$ $\Delta H = 519 \times 10^3 + (101.5 \times 10^5 \times 0.05 - 34.83 \times 10^5 \times 0.15)$

$= 519 \times 10^3 + 10^3(507.5 - 522.45)$

$= 10^3(519 + 507.5 - 522.45) = 504$ kJ

$\therefore$ **Change in enthalpy = 504 kJ. (Ans.)**

Example 3.17. *The following equation gives the internal energy of a certain substance*

$$u = 3.64\,pv + 90$$

where u is kJ/kg, p is in kPa and v is in m^3/kg.

A system composed of 3.5 kg of this substance expands from an initial pressure of 500 kPa and a volume of 0.25 m^3 *to a final pressure 100 kPa in a process in which pressure and volume are related by* $pv^{1.25}$ = *constant.*

(*i*) *If the expansion is quasi-static, find Q, ΔU and W for the process.*

(*ii*) *In another process, the same system expands according to the same pressure-volume relationship as in part (i), and from the same initial state to the same final state as in part (i), but the heat transfer in this case is 32 kJ. Find the work transfer for this process.*

(*iii*) *Explain the difference in work transfer in parts (i) and (ii).*

Solution. Internal energy equation : $u = 3.64\,pv + 90$

Initial volume, $V_1 = 0.25 \text{ m}^3$

Initial pressure, $p_1 = 500$ kPa

Final pressure, $p_2 = 100$ kPa

Process : $pv^{1.25}$ = constant.

(*i*) Now, $u = 3.64\,pv + 90$

$$\Delta u = u_2 - u_1$$

$= 3.64\,(p_2v_2 - p_1v_1)$ ...per kg

$\therefore$ $\Delta U = 3.64\,(p_2V_2 - p_1V_1)$ ...for 3.5 kg

Now, $p_1V_1^{1.25} = p_2V_2^{1.25}$

$$V_2 = V_1 \left(\frac{p_1}{p_2}\right)^{1/1.25} = 0.25 \left(\frac{500}{100}\right)^{1/1.25}$$

$= 0.906 \text{ m}^3$

$\therefore$ $\Delta U = 3.64\,(100 \times 10^3 \times 0.906 - 500 \times 10^3 \times 0.25)$ J [$\because$ 1 Pa = 1 N/m²]

$= 3.64 \times 10^5\,(0.906 - 5 \times 0.25)$ J

$= -3.64 \times 10^5 \times 0.344 \text{ J} = -125.2$ kJ

i.e., **ΔU = – 125.2 kJ. (Ans.)**

For a **quasi-static process**

$$W = \int p dV = \frac{p_1V_1 - p_2V_2}{n-1}$$

$$= \frac{(500 \times 10^3 \times 0.25 - 100 \times 10^3 \times 0.906)}{(1.25 - 1)}$$

$$= \frac{125 - 90.6}{0.25} \text{ kJ} = 137.6 \text{ kJ}$$

$\therefore \quad Q = \Delta U + W$

$= -125.2 + 137.6 = 12.4$ kJ

i.e., **Q = 12.4 kJ. (Ans.)**

(*ii*) Here $Q = 32$ kJ

Since the end states are the same, ΔU would remain the same as in (*i*)

$\therefore \quad W = Q - \Delta U = 32 - (-125.2)$

= 157.2 kJ. (Ans.)

(*iii*) The work in (*ii*) is **not equal** to $\int p\, dV$ since the process is **not quasi-static.**

☞**Example 3.18.** *The properties of a system, during a reversible constant pressure non-flow process at p = 1.6 bar, changed from $v_1 = 0.3$ m³/kg, $T_1 = 20°C$ to $v_2 = 0.55$ m³/kg, $T_2 = 260°C$. The specific heat of the fluid is given by*

$$c_p = \left(1.5 + \frac{75}{T+45}\right) \text{ kJ/kg°C, where } T \text{ is in °C.}$$

Determine : (i) Heat added/kg ; (ii) Work done/kg ; (iii) Change in internal energy/kg ; (iv) Change in enthalpy/kg.

Solution. Initial volume, $v_1 = 0.3$ m³/kg

Initial temperature, $T_1 = 20°C$

Final volume, $v_2 = 0.55$ m³/kg

Final temperature, $T_2 = 260°C$

Constant pressure, $p = 1.6$ bar

Specific heat at constant pressure,

$$c_p = \left(1.5 + \frac{75}{T+45}\right) \text{ kJ/kg°C}$$

(*i*) The **heat added** per kg of fluid is given by

$$Q = \int_{T_1}^{T_2} c_p\, dT = \int_{20}^{260} \left(1.5 + \frac{75}{T+45}\right) dT$$

$$= \Big| 1.5\,T + 75 \log_e (T+45) \Big|_{20}^{260}$$

$$= 1.5\,(260 - 20) + 75 \times \log_e \left(\frac{260+45}{20+45}\right)$$

$= 475.94$ kJ

$\therefore$ **Heat added = 475.94 kJ/kg. (Ans.)**

(*ii*) The **work done** per kg of fluid is given by

$$W = \int_{v_1}^{v_2} p dv = p(v_2 - v_1)$$

$= 1.6 \times 10^5 (0.55 - 0.3)$ Nm

$= 40 \times 10^3$ J = 40 kJ

$\therefore$ **Work done = 40 kJ/kg. (Ans.)**

(*iii*) **Change in internal energy,**

$\Delta u = Q - W = 475.94 - 40$

= 435.94 kJ/kg. (Ans.)

(*iv*) **Change in enthalpy,** (for **non-flow process**)

$\Delta h = Q =$ **475.94 kJ/kg. (Ans.)**

Example 3.19. *1 kg of gaseous CO_2 contained in a closed system undergoes a reversible process at constant pressure. During this process 42 kJ of internal energy is decreased. Determine the work done during the process.*

Take $c_p = 840$ J/kg°C and $c_v = 600$ J/kg°C.

Solution. Mass CO_2, $m = 1$ kg

Decrease in internal energy, $\Delta u = -42$ kJ $= -42 \times 10^3$ J

Specific heat at constant pressure, $c_p = 840$ J/kg°C

Specific heat at constant volume, $c_v = 600$ J/kg°C

Let, initial temperature of CO_2 $= T_1$

Final temperature of CO_2 $= T_2$

Now change in internal energy,

$$\Delta U = m \times c_v (T_2 - T_1)$$

$$-42 \times 10^3 = 1 \times 600 (T_2 - T_1)$$

$$\therefore \quad T_2 - T_1 = -\frac{42 \times 10^3}{600} = -70°C$$

The heat supplied or rejected,

$Q = mc_p (T_2 - T_1)$

$= 1 \times 840 \times (-70)$

$= -58800$ J or -58.8 kJ

Applying first law to the process,

$Q = \Delta U + W$

$-58.8 = -42 + W$ or $W = -16.8$ kJ

$\therefore$ **Work done during the process**

= – 16.8 kJ. (Ans.)

☞**Example 3.20.** *A fluid is contained in a cylinder by a spring-loaded, frictionless piston so that the pressure in the fluid is a linear function of the volume (p = a + bV). The internal energy of the fluid is given by the following equation*

$$U = 42 + 3.6\,pV$$

where U is in kJ, p in kPa, and V in cubic metre. If the fluid changes from an initial state of 190 kPa, 0.035 m³ to a final state of 420 kPa, 0.07 m³, with no work other than that done on the piston, find the direction and magnitude of the work and heat transfer.

Solution. Relation between pressure and volume,

$$p = a + bV.$$

Equation of internal energy : $U = 42 + 3.6pV$

Initial pressure, $p_1 = 190$ kPa

Initial volume, $V_1 = 0.035\ m^3$

Final pressure, $p_2 = 420$ kPa

Final volume, $V_2 = 0.07\ m^3$

The change in internal energy of the fluid during the process

$$U_2 - U_1 = (42 + 3.6p_2V_2) - (42 + 3.6p_1V_1)$$
$$= 3.6(p_2V_2 - p_1V_1)$$
$$= 3.6(4.2 \times 10^5 \times 0.07 - 1.9 \times 10^5 \times 0.035)\ \text{J}$$
$$= 360(4.2 \times 0.07 - 1.9 \times 0.035)\ \text{kJ}$$
$$= 81.9\ \text{kJ}$$

Now, $p = a + bV$

$$190 = a + b \times 0.035 \quad ...(i)$$
$$420 = a + b \times 0.07 \quad ...(ii)$$

Subtracting (*i*) from (*ii*), we get

$$230 = 0.035\ b \text{ or } b = \frac{230}{0.035} = 6571\ \text{kN/m}^5$$

and $a = -40\ \text{kN/m}^2$

Work transfer involved during the process

$$W_{1-2} = \int_{V_1}^{V_2} p dV = \int_{V_1}^{V_2} (a + bV)\, dV$$
$$= a(V_2 - V_1) + b\left(\frac{V_2^2 - V_1^2}{2}\right)$$
$$= (V_2 - V_1)\left[a + \frac{b}{2}(V_1 + V_2)\right]$$
$$= (0.07 - 0.035)\left[-40\ \text{kN/m}^2 + \frac{6571}{2}\ \text{kN/m}^5\ (0.035 + 0.07)\right]$$
$$= 10.67\ \text{kJ}$$

∴ **Work done by the system = 10.67 kJ. (Ans.)**

Heat transfer involved,

$$Q_{1-2} = (U_2 - U_1) + W_{1-2} = 81.9 + 10.67 = 92.57\ \text{kJ.}$$

92.57 kJ of heat flow into the system during the process. (Ans.)

Example 3.21. *90 kJ of heat are supplied to a system at a constant volume. The system rejects 95 kJ of heat at constant pressure and 18 kJ of work is done on it. The system is brought to original state by adiabatic process. Determine :*

(i) The adiabatic work ;

(ii) The values of internal energy at all end states if initial value is 105 kJ.

Solution. Refer to Fig. 3.20.

Heat supplied at constant volume = 90 kJ

Heat rejected at constant pressure = – 95 kJ

Work done *on* the system = – 18 kJ

Initial value of internal energy, $U_l = 105$ kJ

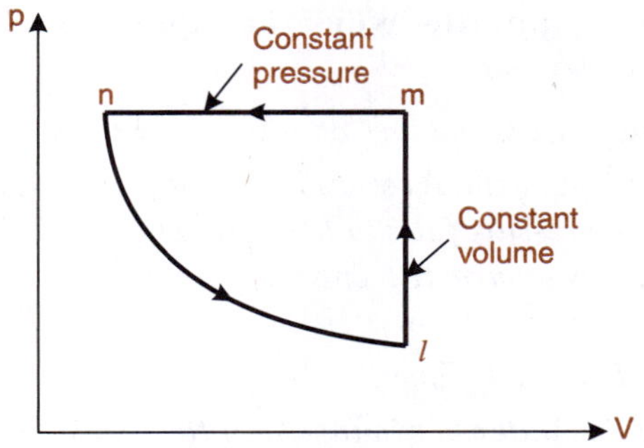

Fig. 3.20

Process *l–m* (*constant volume*) :

$$W_{l-m} = 0$$
$$Q_{l-m} = 90 = U_m - U_l$$
$$\therefore \quad U_m = U_l + 90 = 105 + 90 = 195\ \text{kJ}$$

Process *m–n* (*constant pressure*) :

$$Q_{m-n} = (U_n - U_m) + W_{m-n}$$
$$-95 = (U_n - U_m) - 18$$
$$\therefore \quad U_n - U_m = -77\ \text{kJ}$$
$$\therefore \quad U_n = 195 - 77 = 118\ \text{kJ}$$

$Q_{n-l} = 0$ being adiabatic process

$$\therefore \quad \oint \delta Q = 90 - 95 = -5\ \text{kJ}$$

and
$$\oint \delta W = -18 + W_{n-l} = -5$$
$$\therefore \quad W_{n-l} = -5 + 18 = 13\ \text{kJ}$$

Hence, $\mathbf{W_{n-l} = 13\ kJ\ ;\ U_l = 105\ kJ\ ;}$
$\mathbf{U_m = 195\ kJ\ ;\ U_n = 118\ kJ.}$ **(Ans.)**

Example 3.22. *A movable frictionless piston closes a fully insulated cylinder on one side and offers a constant resistance during its motion. A paddle work is drawn into the cylinder and does work on the system.*

Prove that the paddle work is equal to change in enthalpy.

Solution. Refer to Fig. 3.21.

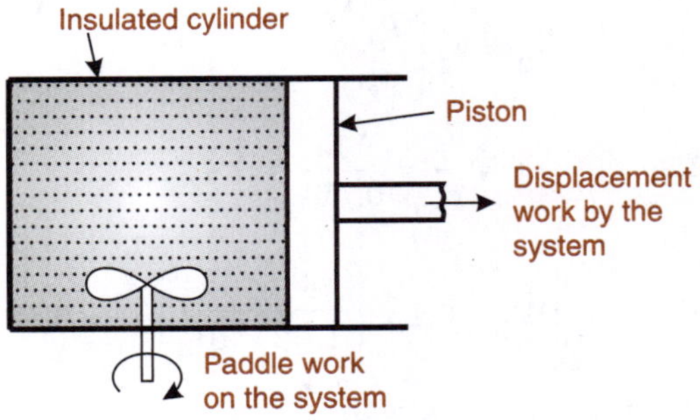

Fig. 3.21

$$Q = W_{paddle} = \Delta U + p\Delta V = \Delta U + \Delta(pV)$$
$$= \Delta(U + pV) = \Delta H$$

Hence, **paddle work is equal to change in enthalpy. (Ans.)**

Example 3.23. *0.2 m^3 of air at 4 bar and 130°C is contained in a system. A reversible adiabatic expansion takes place till the pressure falls to 1.02 bar. The gas is then heated at constant pressure till enthalpy increases by 72.5 kJ. Calculate :*

(i) The work done ;

(ii) The index of expansion, if the above processes are replaced by a single reversible polytropic process giving the same work between the same initial and final states.

Take c_p = 1 kJ/kg K, c_v = 0.714 kJ/kg K.

Solution. Refer to Fig. 3.22.

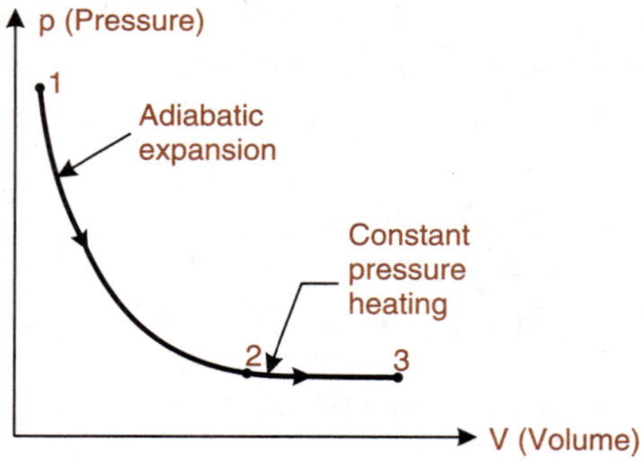

Fig. 3.22

Initial volume, $V_1 = 0.2\ m^3$

Initial pressure, $p_1 = 4\ bar = 4 \times 10^5\ N/m^2$

Initial temperature, $T_1 = 130 + 273 = 403\ K$

Final pressure after adiabatic expansion,

$$p_2 = 1.02\ bar = 1.02 \times 10^5\ N/m^2$$

Increase in enthalpy during constant pressure process = 72.5 kJ.

(*i*) **Work done :**

Process 1–2 : *Reversibe adiabatic process* :

$$p_1 V_1^{\gamma} = p_2 V_2^{\gamma}$$

$$V_2 = V_1 \left(\frac{p_1}{p_2}\right)^{\frac{1}{\gamma}}$$

Also, $\gamma = \dfrac{c_p}{c_v} = \dfrac{1}{0.714} = 1.4$

$$\therefore \quad V_2 = 0.2 \times \left(\frac{4 \times 10^5}{1.02 \times 10^5}\right)^{\frac{1}{1.4}} = 0.53\ m^3$$

Also, $\dfrac{T_2}{T_1} = \left(\dfrac{p_2}{p_1}\right)^{\frac{\gamma - 1}{\gamma}}$

$$\therefore \quad T_2 = T_1 \left(\frac{p_2}{p_1}\right)^{\frac{\gamma - 1}{\gamma}}$$

$$= 403 \left(\frac{1.02 \times 10^5}{4 \times 10^5}\right)^{\frac{1.4 - 1}{1.4}} = 272.7\ K$$

Mass of the gas,

$$m = \frac{p_1 V_1}{R T_1} \qquad [\because \ pV = mRT]$$

where, $R = (c_p - c_v) = (1 - 0.714)$ kJ/kg K

$= 0.286$ kJ/kg K = 286 J/kg K or 286 Nm/kg K

$$\therefore \quad m = \frac{4 \times 10^5 \times 0.2}{286 \times 403} = 0.694\ kg.$$

Process 2–3. *Constant pressure* :

$$Q_{2-3} = mc_p (T_3 - T_2)$$
$$72.5 = 0.694 \times 1 \times (T_3 - 272.7)$$

$$\therefore \quad T_3 = \frac{72.5}{0.694} + 272.7 = 377\ K$$

Also, $\dfrac{V_2}{T_2} = \dfrac{V_3}{T_3}$ or $\dfrac{0.53}{272.7} = \dfrac{V_3}{377}$

$$\therefore \quad V_3 = \frac{0.53 \times 377}{272.7} = 0.732\ m^3$$

Work done by the path 1-2-3 is given by

$$W_{1-2-3} = W_{1-2} + W_{2-3}$$

$$= \frac{p_1V_1 - p_2V_2}{\gamma - 1} + p_2 (V_3 - V_2)$$

$$= \frac{4 \times 10^5 \times 0.2 - 1.02 \times 10^5 \times 0.53}{1.4 - 1} + 1.02 \times 10^5 (0.732 - 0.53)$$

$$= \frac{10^5(4 \times 0.2 - 1.02 \times 0.53)}{0.4} + 1.02 \times 10^5 (0.732 - 0.53)$$

$$= 64850 + 20604 = 85454\ Nm\ or\ J$$

Hence, **total work done = 85454 Nm or J. (Ans.)**

(*ii*) **Index of expansion, n :**

If the work done by the polytropic process is the same,

$$W_{1-2-3} = W_{1-3} = \frac{p_1V_1 - p_3V_3}{n - 1}$$

$$85454 = \frac{4 \times 10^5 \times 0.2 - 1.02 \times 10^5 \times 0.732}{(n - 1)} = \frac{5336}{n - 1}$$

$$\therefore \quad n = \frac{5336}{85454} + 1$$

i.e., $n = 1.062$

Hence, **value of index = 1.062. (Ans.)**

Example 3.24. *The following is the equation which connects u, p and v for several gases*

$$u = a + bpv$$

where a and b are constants. Prove that for a reversible adiabatic process,

$$pv^{\gamma} = constant, \text{ where } \gamma = \frac{b+1}{b}.$$

Solution. Consider a unit mass.

For a reversible adiabatic process, first law gives

$$0 = du + pdv$$

$$\therefore \quad \frac{du}{dv} = -p \quad ...(i)$$

Also, $u = a + bpv$

$$\therefore \quad \frac{du}{dv} = \frac{d(a+bpv)}{dv} = bv\frac{dp}{dv} + bp$$

$$= b\left(p + v \cdot \frac{dp}{dv}\right) \quad ...(ii)$$

Equating (*i*) and (*ii*), we get

$$b\left(p + v \cdot \frac{dp}{dv}\right) = -p$$

$$bp + b \cdot v \cdot \frac{dp}{dv} = -p$$

$$bp + p + bv \cdot \frac{dp}{dv} = 0$$

$$p(b+1) + bv \cdot \frac{dp}{dv} = 0$$

Multiplying both sides by $\frac{dv}{bpv}$, we get

$$\left(\frac{b+1}{b}\right)\frac{dv}{v} + \frac{dp}{p} = 0$$

or

$$\frac{dp}{p} + \left(\frac{b+1}{b}\right)\frac{dv}{v} = 0$$

$$d(\log_e p) + \left(\frac{b+1}{b}\right) d(\log_e v) = 0$$

Also, $\frac{b+1}{b} = \gamma$ ...(Given)

$$\therefore \quad d(\log_e p) + \gamma d(\log_e v) = 0$$

Integrating, we get pv^{γ} = constant.

Example 3.25. *A 15 cm diameter vertical cylinder, closed by a piston contains a combustible mixture at a temperature of 30°C. The piston is free to move and its weight is such that the mixture pressure is 3 bar. Upper surface of the piston is exposed to the atmosphere. The mixture is ignited. As the reaction proceeds, the piston moves slowly upwards and heat transfer to the surroundings takes place. When the reaction is complete and the contents have been reduced to the initial temperature of 30°C, it is found that the piston has moved upwards a distance of 8.5 cm and the magnitude of heat transfer is 4 kJ. Evaluate :*

(*i*) *The work ;*

(*ii*) *Decrease in internal energy of the system.*

Solution. Diameter of vertical cylinder, d = 15 cm (or 0.15 m)

Temperature of combustible mixture

= 30°C (or 303 K)

Pressure of the mixture

= 3 bar = 3×10^5 N/m^2

Upward displacement of the system

= 8.5 cm (or 0.085 m)

Magnitude of heat transfer, $Q = -4$ kJ ...(*i*)

(*i*) Work done by the system, $W = \int pdv$

$$= 3 \times 10^5 \int dv$$

$$[\because \quad p = \text{constant} = 3 \times 10^5 \text{ N/m}^2]$$

$$= 3 \times 10^5 \left[\frac{\pi}{4} \times (0.15)^2 \times 0.085\right] \text{Nm}$$

= 450.62 Nm or J = 0.4506 kJ

$\therefore$ W = 0.4506 kJ.

(*ii*) By first law of thermodynamics,

$$Q = \Delta U + W$$

$$-4 = \Delta U + 0.4506$$

$$\therefore \quad \Delta U = -4.4506 \text{ kJ}$$

$\therefore$ **Decrease in internal energy**

= 4.4506 kJ. (Ans.)

Example 3.26. *A housewife, on a warm summer day, decides to beat the heat by closing the windows and doors in the kitchen and opening the refrigerator door. At first she feels cool and refreshed, but after a while the effect begins to wear off.*

Evaluate the situation as it relates to First Law of Thermodynamics, considering the room including the refrigerator as the system.

Solution. Initially, the temperature of air in the room falls when it communicates with the cool refrigerator with its door open. This makes the housewife feel cool.

Considering the room and its contents as the system, and assuming the walls, windows and doors non-conducting, we find, $Q = 0$.

To operate the refrigerator, *electricity* is supplied from outside and hence *external work W is done on the system*.

Applying the first law to the system,

$$Q = \Delta U + W$$

$$0 = \Delta U + (-W)$$

$$\therefore \quad \Delta U = W$$

The right hand side is a positive figure indicating the increase in energy of the system with time. As the energy is increasing the temperature of air increases and hence the effect of coolness gradually begins to wear off.

It may be pointed out here that in this case the energy rise manifests itself in a rise in temperature.

Example 3.27. *A cylinder contains 0.45 m^3 of a gas at 1×10^5 N/m^2 and 80°C. The gas is compressed to a volume of 0.13 m^3, the final pressure being 5×10^5 N/m^2. Determine :*

(i) The mass of gas ;

(ii) The value of index 'n' for compression ;

(iii) The increase in internal energy of the gas ;

(iv) The heat received or rejected by the gas during compression.

Take γ = 1.4, R = 294.2 J/kg°C.

Solution. Initial volume of gas, $V_1 = 0.45 \text{ m}^3$

Initial pressure of gas, $p_1 = 1 \times 10^5 \text{ N/m}^2$

Initial temperature, $T_1 = 80 + 273 = 353 \text{ K}$

Final volume after compression, $V_2 = 0.13 \text{ m}^3$

The final pressure, $p_2 = 5 \times 10^5 \text{ N/m}^2$.

(*i*) To find mass '*m*' using the relation

$$m = \frac{p_1V_1}{RT_1} = \frac{1 \times 10^5 \times 0.45}{294.2 \times 353}$$

= **0.433 kg. (Ans.)**

(*ii*) To find index '*n*' using the relation

$$p_1V_1^n = p_2V_2^n$$

or

$$\left(\frac{V_1}{V_2}\right)^n = \frac{p_2}{p_1}$$

$$\left(\frac{0.45}{0.13}\right)^n = \left(\frac{5 \times 10^5}{1 \times 10^5}\right) = 5$$

or $(3.46)^n = 5$

Taking log on both sides, we get

$$n \log_e 3.46 = \log_e 5$$

$$n = \log_e 5/\log_e 3.46 = \mathbf{1.296.} \quad \textbf{(Ans.)}$$

(*iii*) In a polytropic process,

$$\frac{T_2}{T_1} = \left(\frac{V_1}{V_2}\right)^{n-1} = \left(\frac{0.45}{0.13}\right)^{1.296-1} = 1.444$$

$\therefore \quad T_2 = 353 \times 1.444 = 509.7 \text{ K}$

Now, increase in internal energy,

$$\Delta U = mc_v (T_2 - T_1)$$

$$= 0.433 \times \frac{R}{(\gamma - 1)} (T_2 - T_1) \quad \left[\because c_v = \frac{R}{(\gamma - 1)}\right]$$

$$= 0.433 \times \frac{294.2}{(1.4 - 1)\, 1000} (509.7 - 353)$$

= **49.9 kJ. (Ans.)**

(*iv*) $Q = \Delta U + W$

Now,

$$W = \frac{p_1V_1 - p_2V_2}{n - 1} = \frac{mR(T_1 - T_2)}{n - 1}$$

$$= \frac{0.433 \times 294.2(353 - 509.7)}{1.296 - 1}$$

= – 67438 Nm or – 67438 J

= – 67.44 kJ

$\therefore \quad Q = 49.9 + (-67.44) = -17.54 \text{ kJ}$

$\therefore$ **Heat rejected = 17.54 kJ. (Ans.)**

Example 3.28. *Air at 1.02 bar, 22°C, initially occupying a cylinder volume of 0.015 m^3, is compressed reversibly and adiabatically by a piston to a pressure of 6.8 bar. Calculate :*

(i) The final temperature ;

(ii) The final volume ;

(iii) The work done.

Solution. Initial pressure, $p_1 = 1.02$ bar

Initial temperature, $T_1 = 22 + 273 = 295$ K

Initial volume, $V_1 = 0.015 \text{ m}^3$

Final pressure, $p_2 = 6.8$ bar

Law of compression : $pv^\gamma = C$

(*i*) **Final temperature :**

Using the relation,

$$\frac{T_2}{T_1} = \left(\frac{p_2}{p_1}\right)^{\frac{\gamma-1}{\gamma}}$$

$$\frac{T_2}{295} = \left(\frac{6.8}{1.02}\right)^{\frac{1.4-1}{1.4}} \quad [\because \gamma \text{ for air} = 1.4]$$

$$\therefore \quad T_2 = 295 \left(\frac{6.8}{1.02}\right)^{\frac{1.4-1}{1.4}} = 507.24 \text{ K}$$

i.e., **Final temperature** = 507.24 – 273

= **234.24°C. (Ans.)**

(*ii*) **Final volume :**

Using the relation,

$$p_1V_1^\gamma = p_2V_2^\gamma$$

$$\frac{p_1}{p_2} = \left(\frac{V_2}{V_1}\right)^\gamma \quad \text{or} \quad \frac{V_2}{V_1} = \left(\frac{p_1}{p_2}\right)^{\frac{1}{\gamma}}$$

$$\therefore \quad V_2 = V_1 \times \left(\frac{p_1}{p_2}\right)^{\frac{1}{\gamma}}$$

$$= 0.015 \times \left(\frac{1.02}{6.8}\right)^{\frac{1}{1.4}} = 0.00387 \text{ m}^3$$

i.e., **Final volume = 0.00387 m³. (Ans.)**

(*iii*) **Work done :**

Now, work done *on the air*,

$$W = \frac{mR(T_1 - T_2)}{(\gamma - 1)} \quad ...(i)$$

where m is the mass of air and is found by the following relation,

$$pV = mRT$$

$$\therefore \quad m = \frac{p_1V_1}{RT_1} = \frac{1.02 \times 10^5 \times 0.015}{0.287 \times 10^3 \times 295}$$

$[\because \; R \text{ for air} = 0.287 \times 10^3]$

$$= 0.01807 \text{ kg}$$

$$\therefore \quad W = \frac{0.01807 \times 0.287 \times 10^3 (295 - 507.24)}{(1.4 - 1)}$$

$$= -2751 \text{ J or } -2.751 \text{ kJ}$$

i.e., **Work done = 2.751 kJ. (Ans.)**

(– ve sign indicates that work is done on the air).

Example 3.29. *0.44 kg of air at 180°C expands adiabatically to three times its original volume and during the process, there is a fall in temperature to 15°C. The work done during the process is 52.5 kJ. Calculate c_p and c_v.*

Solution. Refer to Fig. 3.23.

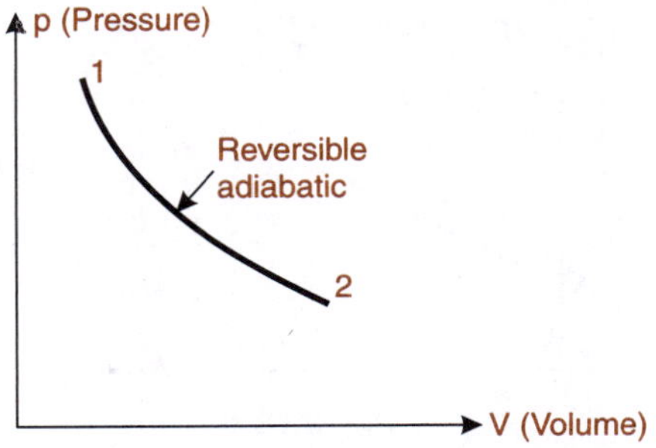

Fig. 3.23

Mass of air, $m = 0.44$ kg

Initial temperature, $T_1 = 180 + 273 = 453$ K

$$\text{Ratio} = \frac{V_2}{V_1} = 3$$

Final temperature, $T_2 = 15 + 273 = 288$ K

Work done during the process, $W_{1-2} = 52.5$ kJ

$$\boldsymbol{c_p = ?, c_v = ?}$$

For adiabatic process, we have

$$\frac{T_2}{T_1} = \left(\frac{V_1}{V_2}\right)^{\gamma - 1}$$

$$\frac{288}{453} = \left(\frac{1}{3}\right)^{\gamma - 1} \quad \text{or} \quad 0.6357 = (0.333)^{\gamma - 1}$$

or Taking log on both sides, we get

$$\log_e (0.6357) = (\gamma - 1) \log_e (0.333)$$

$$-0.453 = (\gamma - 1) \times (-1.0996)$$

$$\therefore \quad \gamma = \frac{0.453}{1.0996} + 1 = 1.41$$

Also, $\dfrac{c_p}{c_v} = \gamma = 1.41$

Work done during adiabatic process,

$$W_{1-2} = \frac{mR(T_1 - T_2)}{\gamma - 1}$$

$$\therefore \quad 52.5 = \frac{0.44 \, R(453 - 288)}{(1.41 - 1)}$$

$$\therefore \quad R = \frac{52.5 \, (1.41 - 1)}{0.44 \, (453 - 288)} = 0.296$$

$$\therefore \quad c_p - c_v = 0.296 \qquad [\because \; R = c_p - c_v]$$

Also $\dfrac{c_p}{c_v} = 1.41$ or $c_p = 1.41 \, c_v$

$$\therefore \; 1.41 \, c_v - c_p = 0.296$$

or $c_v = $ **0.722 kJ/kg K. (Ans.)**

and $c_p = $ **1.018 kJ/kg K. (Ans.)**

☞ **Example 3.30.** *1 kg of ethane (perfect) gas is compressed from 1.1 bar, 27°C according to a law $pV^{1.3}$ = constant, until the pressure is 6.6 bar. Calculate the heat flow to or from the cylinder walls.*

Given : Molecular weight of ethane = 30,

c_p = 1.75 kJ/kg K.

Solution. Mass of ethane gas, $m = 1$ kg

Initial pressure, $p_1 = 1.1$ bar

Initial temperature, $T_1 = 27 + 273 = 300$ K

Final pressure, $p_2 = 6.6$ bar

Law of compression, $pV^{1.3} = C$

Quantity of heat transferred, Q :

Now, characteristic gas constant,

$$R = \frac{\text{Universal gas constant } (R_0)}{\text{Molecular weight } (M)}$$

$$= \frac{8314}{30} = 277.13 \text{ Nm/kg K} = 277.31 \text{ J/kg K}$$

$$= 0.277 \text{ kJ/kg K}$$

Also $c_p - c_v = R$

$$\therefore \quad c_v = c_p - R = 1.75 - 0.277 = 1.473 \text{ kJ/kg K}$$

$$\gamma = \frac{c_p}{c_v} = \frac{1.75}{1.473} = 1.188$$

In case of a polytropic process,

$$\frac{T_2}{T_1} = \left(\frac{p_2}{p_1}\right)^{\frac{n-1}{n}} = \left(\frac{6.6}{1.1}\right)^{\frac{1.3-1}{1.3}} = 1.5119$$

$$\therefore \quad T_2 = 300 \times 1.5119 = 453.6 \text{ K}$$

Now, work done,

$$W = \frac{R \, (T_1 - T_2)}{n - 1} = \frac{0.277(300 - 453.6)}{1.3 - 1}$$

$$= -141.8 \text{ kJ/kg}$$

To find heat flow, using the relation,

$$Q = \left(\frac{\gamma - n}{\gamma - 1}\right) W = \left(\frac{1.188 - 1.3}{1.188 - 1}\right) \times -141.8$$

$$= + 84.5 \text{ kJ/kg}$$

i.e., **Heat supplied = 84.5 kJ/kg. (Ans.)**

Example 3.31. *Fig. 3.24 shows a cylinder of 8 cm inside diameter having a piston loaded with a spring (stiffness = 150 N/cm of compression). The initial pressure, volume and temperature of air in the cylinder are 3 × 10⁵ N/m², 0.000045 m³ and 20°C respectively. Determine the amount of heat added to the system so that piston moves by 3.5 cm.*

Assume c_v = 0.71 kJ/kg K and R = 0.287 kJ/kg K.

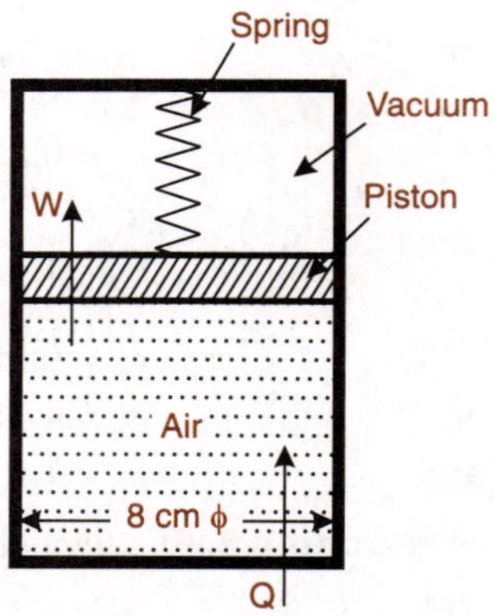

Fig. 3.24

Solution. Insider diameter of the cylinder

= 8 cm

Stiffness of the spring, S = 150 N/cm

Initial pressure of air,

$$p_1 = 3 \times 10^5 \text{ N/m}^2 \text{ or } 30 \text{ N/cm}^2$$

Initial volume of air,

$$V_1 = 0.000045 \text{ m}^3 = 45 \text{ cm}^3$$

Initial temperature of air,

$$T_1 = 20 + 273 = 293 \text{ K}$$

Specific heat at constant volume,

$$c_v = 0.71 \text{ kJ/kg K}$$

Characteristic constant for air,

$$R = 0.287 \text{ kJ/kg K}$$

Refer to Fig. 3.25.

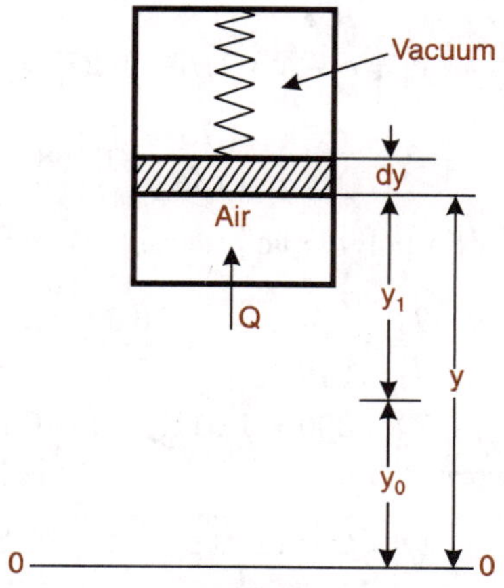

Fig. 3.25

Let, oo = An arbitrary datum from which the position of the lower face of the piston is to be measured,

y = Distance of the lower face of the piston,

$y = y_0$, when spring length is its free length, and

p = Pressure of air within the cylinder when $y = y_0$.

Now, force balance for the piston is given by

$$Ap = S(y - y_0) \quad \text{...}(i)$$

where, A = The area of the piston, and

S = Stiffness of the spring.

With heat transfer to the air, let the pressure inside the cylinder increase by dp forcing the piston to move upward by distance dy. Now the force balance for the piston is

$$A(p + dp) = S(y + dy - y_0) \quad \text{...}(ii)$$

From eqns. (*i*) and (*ii*), we have

$$Adp = Sdy \quad \text{...}(iii)$$

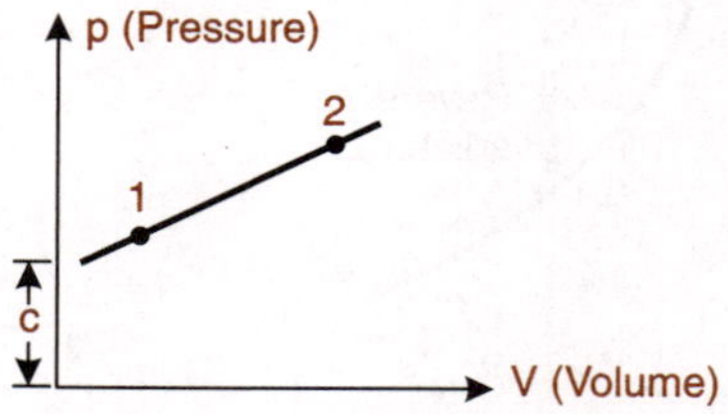

Fig. 3.26

The increase in volume dV of the gas for the piston displacement is given by

$$dV = Ady \quad \text{...}(iv)$$

$$\therefore \quad dp = \frac{S}{A^2} dy \quad \text{...}(v)$$

$$\therefore \quad p = \frac{S}{A^2} V + C \quad \text{...}(vi)$$

The *p-V* relationship for the process is a straight line (Fig. 3.26) having a slope of $\frac{S}{A^2}$ and pressure axis intercept of C. The value of C can be found out from the knowledge of pressure and volume at any state point.

Now, substituting the values of p_1, V_1, A in eqn. (*vi*), we get

$$p = \frac{150}{\left(\frac{\pi}{4} \times 8^2\right)^2} V + C$$

or

$$p = 0.0594\, V + C \quad \text{...}(vii)$$

where p is in N/cm² and V is in cm³.

$$\therefore \quad p_1 = 0.0594\, V_1 + C$$

$$30 = 0.0594 \times 45 + C$$

$$\therefore \quad C = 27.33$$

Hence, p-V relationship for the process is,

$$p = 0.0594\ V + 27.33 \qquad ...(viii)$$

During the process the piston is moved by a distance of 3.5 cm.

This increases the volume of gas by

$$3.5 \times A^2 = 3.5 \times \left(\frac{\pi}{4} \times 8^2\right) = 175.9\ \text{cm}^3$$

Hence, the final volume of air,

$$V_2 = 45 + 175.9 = 220.9\ \text{cm}^3$$

Substituting this value in equation (*viii*), we get

$$p(= p_2) = 0.0594 \times 220.9 + 27.33$$
$$= 40.45\ \text{N/cm}^2$$

The work done W during the process is given by

$$W = \int p dV = \int_{p_1}^{p_2} \frac{A^2}{S} p dp$$
$$= \frac{A^2}{S}\left(\frac{p_2^2 - p_1^2}{2}\right)$$
$$= \frac{A^2}{S}\left(\frac{p_2 + p_1}{2}\right)\left(\frac{p_2 - p_1}{2}\right)$$
$$= \frac{A^2}{S}\left(\frac{p_2 + p_1}{2}\right)\frac{S}{A^2}(V_2 - V_1)$$

or
$$W = \left(\frac{p_2 + p_1}{2}\right)(V_2 - V_1) \qquad ...(ix)$$
$$= \text{(Mean pressure)} \times \text{(Change in volume)}$$
$$W = \left(\frac{40.45 + 30}{2}\right) \times (220.9 - 45)$$
$$= 6196\ \text{N-cm or } 61.96\ \text{Nm}$$

It may be noted that work done does not cross the system boundary when spring and cylinder are considered system.

Now, to find T_2, using the relation,

$$\frac{p_1V_1}{T_1} = \frac{p_2V_2}{T_2}$$

$$\therefore \quad T_2 = \frac{p_2V_2T_1}{p_1V_1} = \frac{40.45 \times 220.9 \times 293}{30 \times 45} = 1939.3\ \text{K}$$

Also, $m = \dfrac{p_1v_1}{R_1T_1} = \dfrac{30 \times 45}{(0.287 \times 10^3) \times 293} = 0.0001605$ kg

Now, change in internal energy,

$$\Delta U = m \times c_v \times (T_2 - T_1)$$
$$= 0.0001605 \times 0.71 \times (1939.3 - 293)$$
$$= 0.1876\ \text{kJ}$$

According to first law,

$$Q_{1-2} = \Delta U + W$$
$$= 0.1876 + 61.96 \times 10^{-3} = 0.2495\ \text{kJ}$$

∴ **Amount of heat added to the system = 0.2495 kJ. (Ans.)**

3.10. APPLICATION OF FIRST LAW TO STEADY FLOW PROCESS

Steady Flow Energy Equation (S.F.E.E.)

In many practical problems, the rate at which the fluid flows through a machine or piece of apparatus is constant. This type of flow is called *steady flow.*

Assumptions :

The following *assumptions* are made in the system analysis :

(*i*) The mass flow through the system remains constant.

(*ii*) Fluid is uniform in composition.

(*iii*) The only interaction between the system and surroundings are work and heat.

(*iv*) The state of fluid at any point remains constant with time.

(*v*) In the analysis only potential, kinetic and flow energies are considered.

Fig. 3.27 shows a schematic flow process for an open system. An open system is one in which both mass and energy may cross the boundaries. A wide interchange of energy may take place within an open system. Let the system be an automatic engine with the inlet manifold at the first state point and exhaust pipe as the second point. There would be an interchange of chemical energy in the fuel, kinetic energy of moving particles, internal energy of gas and heat transferred and shaft work within the system. From Fig. 3.27 it is obvious that if there is no variation of flow of mass or energy with time across the boundaries of the system the steady flow will prevail. The conditions may pass through the cyclic or non-cyclic changes within the

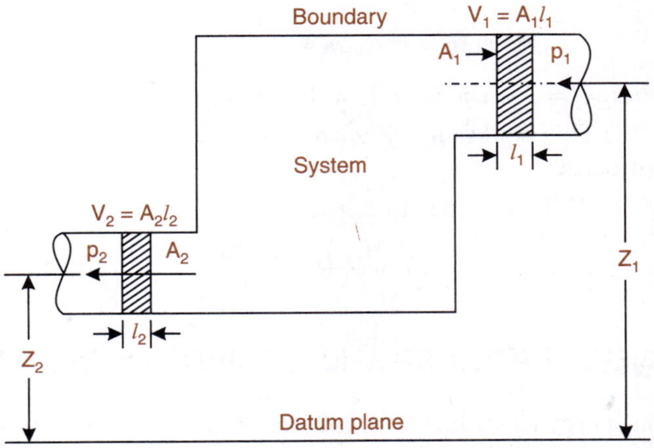

Fig. 3.27

system. As a result the mass entering the system equals the mass leaving, also energy entering the system equals energy leaving.

The steady flow equation can be expressed as follows :

$$u_1 + \frac{C_1^2}{2} + Z_1 g + p_1 v_1 + Q = u_2 + \frac{C_2^2}{2} + Z_2 g + p_2 v_2 + W \quad ...(3.45)$$

$$(u_1 + p_1 v_1) + \frac{C_1^2}{2} + Z_1 g + Q = (u_2 + p_2 v_2) + \frac{C_2^2}{2} + Z_2 g + W$$

$$h_1 + \frac{C_1^2}{2} + Z_1 g + Q = h_2 + \frac{C_2^2}{2} + Z_2 g + W \quad [\because \; h = u + pv]$$

If Z_1 and Z_2 are neglected, we get

$$h_1 + \frac{C_1^2}{2} + Q = h_2 + \frac{C_2^2}{2} + W \quad ...[3.45\,(a)]$$

where, Q = Heat supplied (or entering the boundary) per kg of fluid,

W = Work done by (or work coming out of the boundary) 1 kg of fluid,

C = Velocity of fluid ,

Z = Height above datum,

p = Pressure of the fluid,

u = Internal energy per kg of fluid, and

pv = Energy required for 1 kg of fluid.

This equation is applicable to any medium in any steady flow. It is applicable not only to rotary machines such as centrifugal fans, pumps and compressors but also to reciprocating machines such as steam engines.

In a steady flow the rate of mass flow of fluid at any section is the same as at any other section. Consider any section of cross-sectional area A, where the fluid velocity is C, the rate of volume flow past the section is CA. Also, since mass flow is volume flow divided by specific volume,

$$\therefore \quad \text{Mass flow rate, } \dot{m} = \frac{CA}{v} \quad ...(3.46)$$

(where v = *Specific volume* at the section)

This equation is known as the **continuity of mass equation**.

With reference to Fig. 3.27.

$$\therefore \quad \dot{m} = \frac{C_1 A_1}{v_1} = \frac{C_2 A_2}{v_2} \quad ...[3.46\,(a)]$$

3.11. ENERGY RELATIONS FOR FLOW PROCESS

The energy equation (m kg of fluid) for a steady flow system is given as follows :

$$m\left(u_1 + \frac{C_1^2}{2} + Z_1 g + p_1 v_1\right) + Q = m\left(u_2 + \frac{C_2^2}{2} + Z_2 g + p_2 v_2\right) + W$$

i.e.,
$$Q = m\left[(u_2 - u_1) + (Z_2 g - Z_1 g) + \left(\frac{C_2^2}{2} - \frac{C_1^2}{2}\right) + (p_2 v_2 - p_1 v_1)\right] + W$$

i.e.,
$$Q = m\left[(u_2 - u_1) + g(Z_2 - Z_1) + \left(\frac{C_2^2 - C_1^2}{2}\right) + (p_2 v_2 - p_1 v_1)\right] + W$$

$$= \Delta U + \Delta PE + \Delta KE + \Delta\,(pv) + W$$

where
$$\Delta U = m\,(u_2 - u_1)$$
$$\Delta PE = mg\,(Z_2 - Z_1)$$
$$\Delta KE = m\left(\frac{C_2^2 - C_1^2}{2}\right)$$
$$\Delta pv = m(p_2 v_2 - p_1 v_1)$$

$$\therefore \quad Q - \Delta U = [\Delta PE + \Delta KE + \Delta(pV) + W] \quad ...(3.47)$$

For non-flow process,

$$Q = \Delta U + W = \Delta U + \int_1^2 pdV$$

i.e.,
$$Q - \Delta U = \int_1^2 p.dV \quad ...(3.48)$$

The internal energy is a function of temperature only and it is a point function. Therefore, for the same two temperatures, change in internal energy is the same whatever may be the process, non-flow, or steady flow, reversible or irreversible.

For the same value of Q transferred to non-flow and steady flow process and for the same temperature range, we can equate the values of eqns. (3.47) and (3.48) for $(Q - \Delta U)$.

$$\therefore \quad \int_1^2 p.dV = \Delta\,PE + \Delta\,KE + \Delta\,(pV) + W \quad ...(3.49)$$

where, W = Work transfer in flow process

and $\int_1^2 p.dV$ = Total change in mechanical energy of reversible steady flow process.

Property Relations for Energy Equations

We know that $\quad h = u + pv$

Differentiating above equation

$$dh = du + pdv + vdp$$

But $\quad dQ = du + p.dv \quad$ (as per first law applied to closed system)

or $\quad du = dQ - p.dv$

Substituting this value of du in the above equation, we get

$$dh = dQ - p.dv + pdv + vdp$$
$$= dQ + vdp$$
$$\therefore \quad vdp = dh - dQ$$
$$\therefore \quad -\int_1^2 vdp = Q - \Delta h \quad \text{...(3.50)}$$

where $-\int_1^2 vdp$ represents on a p-v diagram the area behind 1–2 as shown in Fig. 3.28 (b).

The eqn. (3.47) for a unit mass flow can be written as

$$dQ = d(PE) + d(KE) + du + d(pv) + dW$$

Substituting the value of $dQ = du + p.dv$ in the above equation, we get

$$du + pdv = d(PE) + d(KE) + du + pdv + vdp + dW$$
$$\therefore \quad -vdp = d(PE) + d(KE) + dW$$
$$\therefore \quad -\int_1^2 vdp = \Delta\, PE + \Delta\, KE + W \quad \text{...[3.50 (a)]}$$

If $\Delta PE = 0$ (as in most of thermodynamic systems)

$$-\int_1^2 vdp = \Delta\, KE + W \quad \text{...[3.50 (b)]}$$

If $W = 0$, the area behind the curve represents $\Delta\, KE$ and if $\Delta\, KE = 0$, area behind the curve represents W which is shaft work.

$-\int_1^2 vdp$ is a *positive quantity and represents work done by the system.*

If $\Delta\, PE = 0$ and $W = 0$, then

$-\int_1^2 vdp = \Delta\, KE$, this is applicable in case of a *nozzle.*

i.e., $\int_1^2 vdp = \dfrac{C^2}{2}$ in the case of a nozzle.

If $\Delta\, PE = 0$ and $\Delta\, KE = 0$, as in case of a *compressor*,

$$-\int_1^2 vdp = W$$

or $W = \int_1^2 vdp$ in the case of a *compressor*.

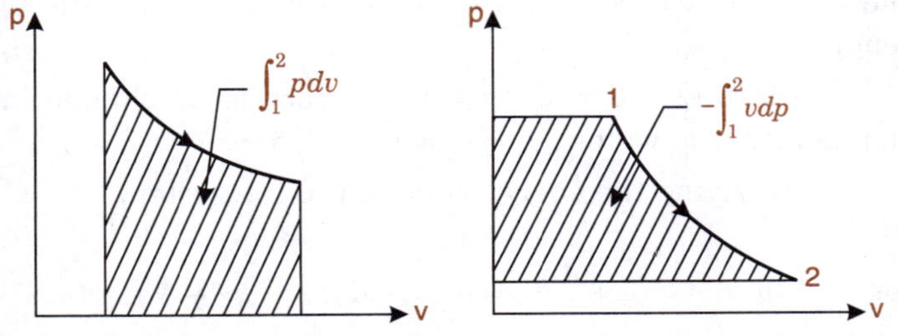

(a) Work done in non-flow process. (b) Work done in flow process.

Fig. 3.28. Representation of work on p-v diagram.

The integral $\int_1^2 pdv$ and $\int_1^2 vdp$ are shown in Fig. 3.28 (a) and (b).

The work done during *non-flow process* is given by

$$\int_1^2 pdv = Q - \Delta u \quad \text{...[3.50 (c)]}$$

For isothermal process, we have

$$\Delta\, u = 0 \text{ and } \Delta h = 0.$$

Substituting these values in equations (3.50) and [3.50 (c)]

$$-\int_1^2 vdp = Q \text{ and } \int_1^2 pdv = Q$$
$$\therefore \quad \int_1^2 pdv = -\int_1^2 vdp$$

The above equation indicates that the *area under both curves is same for an isothermal process.*

Note. In all the above equations 'v' represents volume per unit mass as mass flow is considered unity.

Now let us find out expressions for work done for different flow processes as follows :

(i) **Steady flow constant pressure process :**

$$W = -\int_1^2 v\,.\,dp = 0 \quad [\because \;\; dp = 0] \quad \text{...(3.51)}$$

(ii) **Steady flow constant volume process :**

$$W = -\int_1^2 Vdp = -V(p_2 - p_1) = V(p_1 - p_2)$$

i.e.,
$$W = V(p_1 - p_2) \quad \text{...(3.52)}$$

(iii) **Steady flow constant temperature process :**

The constant temperature process is represented by

$$pV = p_1V_1 = p_2V_2 = C \text{ (constant)}$$

$$\therefore \quad W = -\int_1^2 Vdp$$
$$= -\int_1^2 \frac{C}{p}\, dp \quad \left[\because \; V = \frac{C}{p}\right]$$
$$= -C\int_1^2 \frac{dp}{p} = -C\,\big|\log_e p\big|_1^2$$
$$= -C\,\log_e \frac{p_2}{p_1} = C\,\log_e \frac{p_1}{p_2}$$

i.e.,
$$W = p_1V_1 \log_e\left(\frac{p_1}{p_2}\right) \quad \text{...(3.53)}$$

Now substituting the values of W in the equation (3.49), considering unit mass flow :

(a) The energy equation for *constant pressure flow process*

$$dQ = \Delta\, PE + \Delta\, KE + \Delta\, h$$
$$= \Delta\, h \text{ (if } \Delta PE = 0 \text{ and } \Delta\, KE = 0).$$

(*b*) The energy equation for constant volume flow process

$$dQ = -\int_1^2 vdp + \Delta PE + \Delta KE + \Delta u + pdv + vdp$$
$$= \Delta PE + \Delta KE + \Delta u$$
$$\left[\because\ pdv = 0 \text{ and } v\,.\,dp = \int_1^2 vdp\right]$$

$\therefore \quad dQ = \Delta u$ (if $\Delta PE = 0$ and $\Delta KE = 0$)

3.12. ENGINEERING APPLICATIONS OF STEADY FLOW ENERGY EQUATION (S.F.E.E.)

3.12.1. Water Turbine

Refer to Fig. 3.29. In a water turbine, water is supplied from a height. The potential energy of water is converted into kinetic energy when it enters into the turbine and part of it is converted into useful work which is used to generate electricity.

Considering centre of turbine shaft as *datum*, the energy equation can be written as follows :

$$\left(u_1 + p_1v_1 + Z_1g + \frac{C_1^2}{2}\right) + Q = \left(u_2 + p_2v_2 + Z_2g + \frac{C_2^2}{2}\right) + W$$

In this case,

$$Q = 0$$
$$\Delta u = u_2 - u_1 = 0$$
$$\therefore \quad v_1 = v_2 = v$$
$$Z_2 = 0$$

$$\therefore \left(p_1v + Z_1g + \frac{C_1^2}{2}\right) = \left(p_2v + Z_2g + \frac{C_2^2}{2}\right) + W \quad ...(3.54)$$

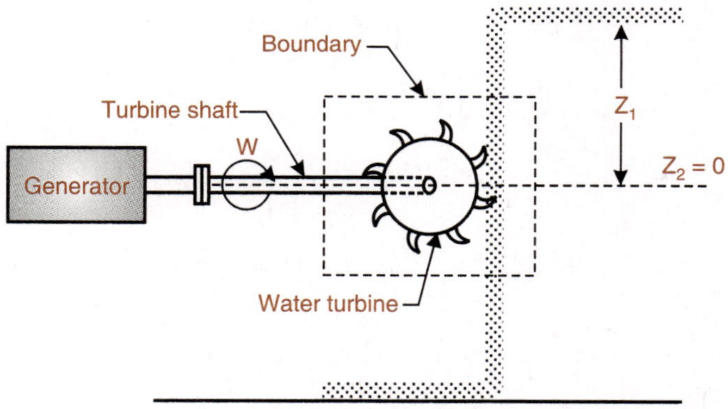

Fig. 3.29. Water turbine.

W is *positive* because work is done by the system (or work comes out of the boundary).

3.12.2. Steam or Gas Turbine

In a steam or gas turbine steam or gas is passed through the turbine and part of its energy is converted into work in the turbine. This output of the turbine runs a generator to produce electricity as shown in Fig. 3.30. The steam or gas leaves the turbine at lower pressure or temperature.

Applying energy equation to the system.

Here, $\quad Z_1 = Z_2$ (*i.e.*, $\Delta Z = 0$)

$$h_1 + \frac{C_1^2}{2} - Q = h_2 + \frac{C_2^2}{2} + W \quad ...(3.55)$$

The sign of *Q* is *negative* because heat is *rejected* (or comes out of the boundary).

The sign of *W* is *positive* because work is done by the system (or work comes out of the boundary).

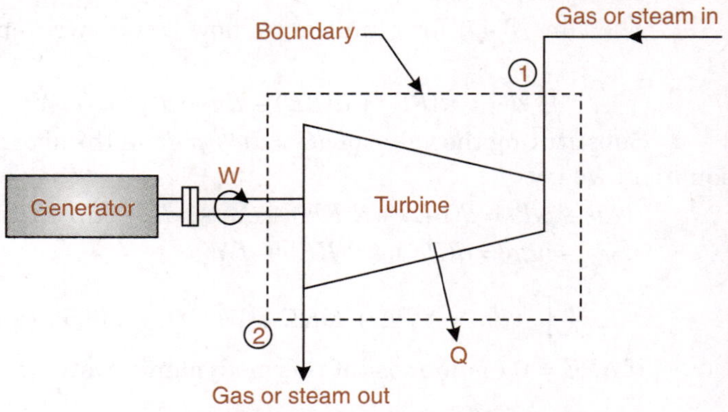

Fig. 3.30. Steam or gas turbine.

3.12.3. Centrifugal Water Pump

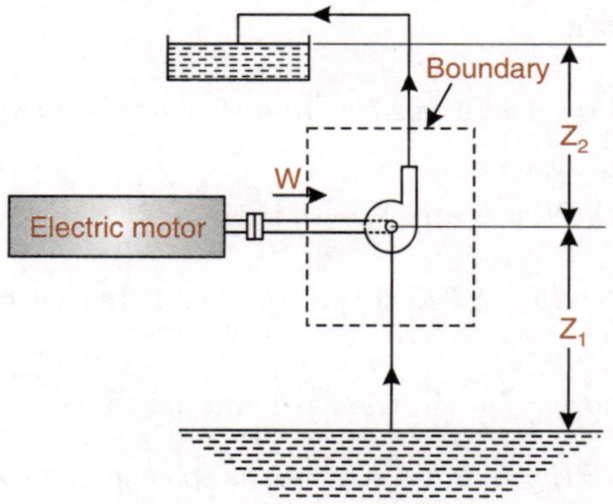

Fig. 3.31. Centrifugal water pump.

A centrifugal water pump draws water from a lower level and pumps to higher level as shown in Fig. 3.31. Work is required to run the pump and this may be supplied from an external source such as an electric motor or a diesel engine.

Here $Q = 0$ and $\Delta u = 0$ as there is no change in temperature of water ; $v_1 = v_2 = v$.

Applying the energy equation to the system

or $$p_1v_1 + Z_1g + \frac{C_1^2}{2} = p_2v_2 + Z_2g + \frac{C_2^2}{2} - W \quad ...(3.56)$$

The sign of *W* is *negative* because *work is done on the system* (or *work enters the boundary*).

3.12.4. Centrifugal Compressor

Refer to Fig. 3.32. A centrifugal compressor compresses air and supplies the same at moderate pressure and in large quantity.

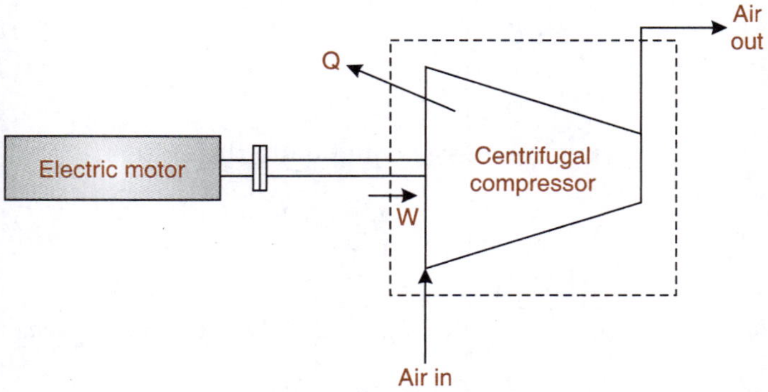

Fig. 3.32. Centrifugal compressor.

Applying energy equation to the system (Fig. 3.32)

$\Delta Z = 0$ (generally taken)

$$\left(h_1 + \frac{C_1^2}{2}\right) - Q = \left(h_2 + \frac{C_1^2}{2}\right) - W$$

The Q is taken as *negative* as heat is *lost* from the system and W is taken as *negative* as work is *supplied* to the system.

or $$\left(h_1 + \frac{C_1^2}{2}\right) - Q = \left(h_2 + \frac{C_2^2}{2}\right) - W \quad ...(3.57)$$

3.12.5. Reciprocating Compressor

Refer to Fig. 3.33. The reciprocating compressor draws in air from atmosphere and supplies the air at a considerable higher pressure in small quantities (compared with centrifugal compressor). The reciprocating compressor can be considered as steady flow system *provided the control volume includes the receiver which reduces the fluctuations of flow considerably.*

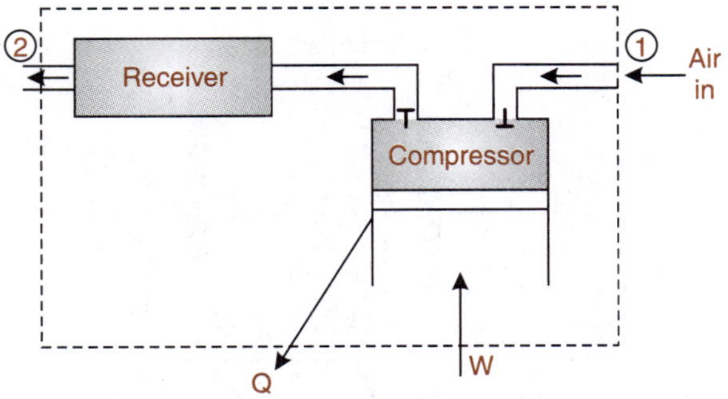

Fig. 3.33. Reciprocating compressor.

Applying energy equation to the system, we have :

$\Delta PE = 0$ and $\Delta KE = 0$ since these changes are negligible compared with other energies.

$$\therefore \quad h_1 - Q = h_2 - W \quad ...(3.58)$$

3.12.6. Boiler

A boiler transfers heat to the incoming water and generates the steam. The system is shown in Fig. 3.34.

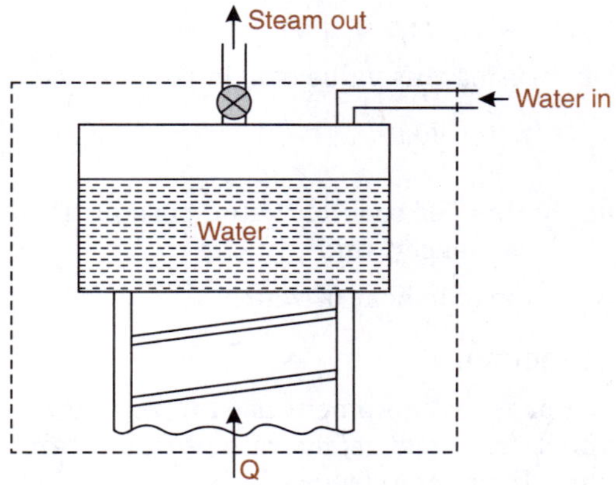

Fig. 3.34. Boiler.

For this system,

$$\Delta Z = 0 \text{ and } \Delta\left(\frac{C_2^2}{2}\right) = 0$$

$W = 0$ since neither any work is developed nor absorbed.

Applying energy equation to the system

$$h_1 + Q = h_2 \quad ...(3.59)$$

3.12.7. Condenser

The condenser is used to condense the steam in case of steam power plant and condense the refrigerant vapour in the refrigeration system using water or air as cooling medium. Fig. 3.35 shows the system.

For this system :

$\Delta PE = 0$, $\Delta KE = 0$ (as their values are very small compared with enthalpies)

$W = 0$ (since neither any work is developed nor absorbed)

Using energy equation to steam flow

$$h_1 - Q = h_2 \quad ...[3.60\ (a)]$$

where Q = Heat lost by 1 kg of steam passing through the condenser.

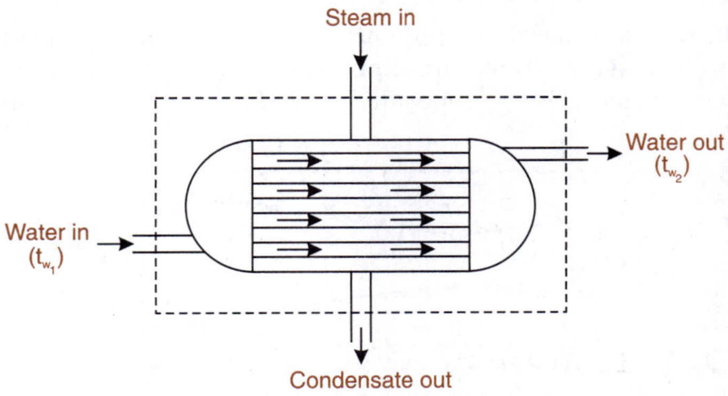

Fig. 3.35. Condenser.

Assuming there are no other heat interactions except the heat transfer between steam and water, then

Q = Heat gained by water passing through the condenser

$$= m_w (h_{w2} - h_{w1}) = m_w c_w (t_{w2} - t_{w1})$$

Substituting this value of Q in eqn. [3.60 (*a*)], we get

$$h_1 - h_2 = m_w (h_{w2} - h_{w1}) = m_w c_w (t_{w2} - t_{w1}) \quad ...[3.60 (b)]$$

where, m_w = Mass of cooling water passing through the condenser, and

c_w = Specific heat of water.

3.12.8. Evaporator

An evaporator is an equipment used in refrigeration plant to carry heat from the refrigerator to maintain the low temperature. Here the refrigerant liquid is passed through the evaporator and it comes out as vapour absorbing its latent heat from the surroundings of the evaporator. Fig. 3.36 shows the system. For this system

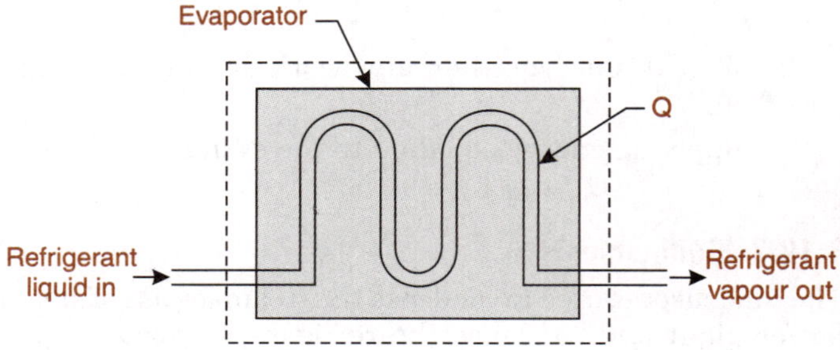

Fig. 3.36. Evaporator.

$$\Delta PE = 0, \Delta KE = 0$$

$W = 0$ [$\because$ No work is absorbed or supplied]

Applying the energy equation to the system

$$h_1 + Q = h_2 \quad ...(3.61)$$

Q is taken as +ve because heat *flows from the surroundings to the system* as the temperature in the system is lower than the surroundings.

3.12.9. Steam Nozzle

In case of a nozzle as the enthalpy of the fluid decreases and pressure drops simultaneously the flow of fluid is *accelerated*. This is generally used to convert the part of the energy of steam into *kinetic energy of steam* supplied to the turbine.

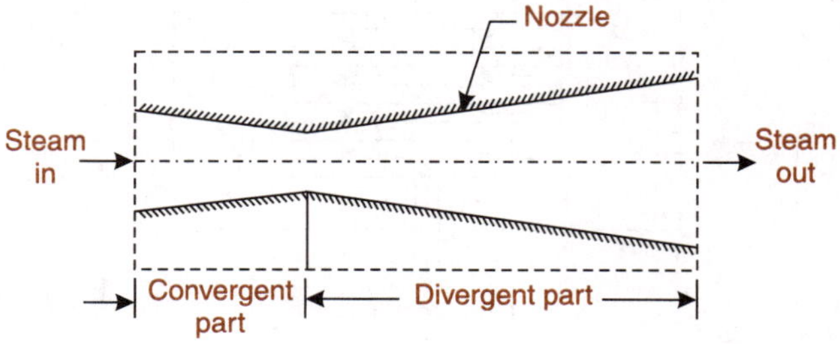

Fig. 3.37. Steam nozzle.

Fig. 3.37 shows a commonly used convergent-divergent nozzle.

For this system,

$$\Delta PE = 0$$
$$W = 0$$
$$Q = 0$$

Applying the energy equation to the system,

$$h_1 + \frac{C_1^2}{2} = h_2 + \frac{C_2^2}{2}$$

or $$\frac{C_2^2}{2} - \frac{C_1^2}{2} = h_1 - h_2 \quad \text{or} \quad C_2^2 - C_1^2 = 2(h_1 - h_2)$$

or $$C_2^2 = C_1^2 + 2(h_1 - h_2)$$

$$\therefore \quad C_2 = \sqrt{C_1^2 + 2(h_1 - h_2)} \quad ...(3.62)$$

where velocity C is in m/s and enthalpy h in joules.

If $C_1 << C_2$, then

$$C_2 = \sqrt{2(h_1 - h_2)} \quad ...[3.63 (a)]$$

$$\therefore \quad C_2 = \sqrt{2\Delta h}\,. \quad ...[3.63 (b)]$$

3.13. THROTTLING PROCESS AND JOULE-THOMSON POROUS PLUG EXPERIMENT

Throttling process involves the *passage of a higher pressure fluid through a narrow constriction. The effect is the reduction in pressure and increase in volume.* This process is *adiabatic* as no heat flows from and to the system, but it is *not reversible*. It is *not an isentropic process*. The *entropy* of the fluid actually *increases*.

Such a process occurs in a flow through a *porous plug*, a *partially closed valve* and a *very narrow orifice*. The porous plug is shown in Fig. 3.38.

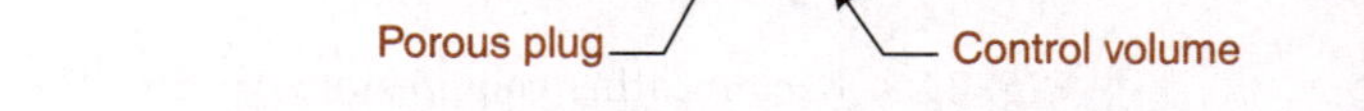

Fig. 3.38. The Joule-Thomson porous plug experiment.

In this system,

$Q = 0$ ($\because$ System is isolated)

$W = 0$ ($\because$ There is no work interaction)

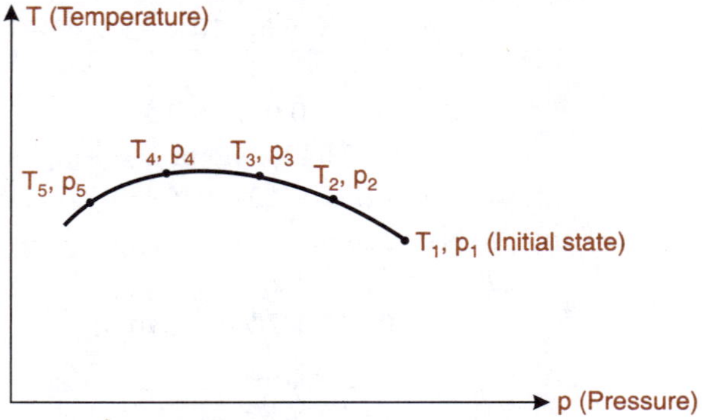

Fig. 3.39. Constant enthalpy curve.

$\Delta PE = 0$ ($\because$ Inlet and outlet are at the same level)

$\Delta KE = 0$ ($\because$ Kinetic energy does not change significantly)

Applying the energy equation to the system

$$h_1 = h_2$$

This shows that *enthalpy remains constant during adiabatic throttling process.*

The throttling process is commonly used for the following purposes :

(*i*) For determining the condition of steam (dryness fraction).

(*ii*) For controlling the speed of the turbine.

(*iii*) Used in refrigeration plant for reducing the pressure of the refrigerant before entry into the evaporator.

Throttling process frequently encountered in practice was investigated by Joule and Thomson (Lord Kelvin) in their famous *porous plug experiment* (Fig. 3.38). A stream of gas at pressure p_1 and temperature T_1 is forced continuously through a *porous plug* in a tube from which it emerges at a lower pressure p_2 and temperature T_2. The whole apparatus is *thermally insulated.*

In this process (as earlier stated)

$$h_1 = h_2$$

Whether the temperature and internal energy change in a throttling process depends on whether the fluid behaves as an ideal gas or not. Since the enthalpy of an ideal gas is a function of temperature alone, it follows that

$$T_1 = T_2 \text{ for (throttling process)}_{ideal\ gas} \quad \text{...(3.64)}$$

and, therefore, $\quad u_1 = u_2$

For an *ideal gas,* therefore, the *throttling* process takes place at

(*i*) constant enthalpy,

(*ii*) constant temperature, and

(*iii*) constant internal energy.

The *enthalpy of a real gas is not a function of temperature alone.* In this case

$$T_1 \neq T_2 \quad \text{...(3.65)}$$

Also since the *pv* product may be different before and after throttling, the change in internal energy is not zero, as it is in *free expansion*, but is given by

$$u_2 - u_1 = p_1v_1 - p_2v_2 \quad \text{...(3.66)}$$

Joule-Thomson and Joule Co-efficients

When a real gas undergoes a throttling process a change in temperature takes place. Let us perform a series of the experiments on the same gas, keeping p_1 and T_1 constant, by varying the pressure downstream of the plug to various values p_2, p_3, p_4 etc. After throttling let T_1, T_2, T_3, T_4 etc., be the corresponding temperatures. Now if a graph is plotted between p and T (Fig. 3.39), a smooth curve drawn through these points will be a curve of *constant enthalpy* because $h_1 = h_2 = h_3 = h_4$ etc.

It may be noted that this curve does *not* represent the process executed by the gas in passing through the plug, since the process is *irreversible* and the gas does not pass through a sequence of equilibrium states.

The *slope* of a constant enthalpy line or a *p-T* diagram at a particular state may be *positive, zero* or *negative* value. The slope is called ***Joule-Thomson co-efficient,*** μ and is given by

$$\mu = \left(\frac{\partial T}{\partial p}\right)_h \quad \text{...(3.67)}$$

$$= 0 \text{ for ideal gas.}$$

If we carry out other series of experiments similar to described above starting from different initial states, we can obtain a family of constant enthalpy curves as shown in Fig. 3.40. The states where μ = 0 are called '***inversion states***' and locus of these states is called the ***inversion curve***.

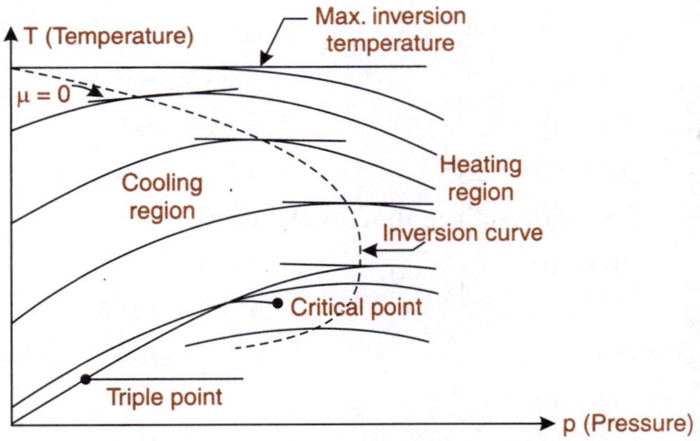

Fig. 3.40. Inversion curve.

The region inside the inversion curve is the cooling region since μ is *positive*, and temperature falls with fall in pressure.

The *region outside the inversion curve is the heating region* since μ is *negative* and temperature rises with fall in pressure.

Cooling can take place only if the initial temperature before throttling is below the *maximum inversion temperature.* This temperature is about $5T_c$.

The *maximum inversion temperatures* of some gases are given below :

(*i*) He = 24 K (*ii*) H_2 = 195 K

(*iii*) Air = 603 K (*iv*) N_2 = 261 K

(*v*) A = 732 K (*vi*) CO_2 = 1500 K.

The free expansion process is also a Joule (not Joule-Thomson). The Joule co-efficient is defined by

$$\text{Joule co-efficient} = \eta = -\left(\frac{\partial T}{\partial v}\right)_u \quad ...(3.68)$$

For free expansion of gases the experimental data obtained is limited. From the data available it appears that η is positive (*i.e.*, cooling accompanies a fall in pressure or increase in specific volume).

Note. The throttling process is used in the liquification of gases, where μ is positive.

STEADY FLOW SYSTEMS

Example 3.32. *10 kg of fluid per minute goes through a reversible steady flow process. The properties of fluid at the inlet are : p_1 = 1.5 bar, ρ_1 = 26 kg/m³, C_1 = 110 m/s and u_1 = 910 kJ/kg and at the exit are p_2 = 5.5 bar, ρ_2 = 5.5 kg/m³, C_2 = 190 m/s and u_2 = 710 kJ/kg. During the passage, the fluid rejects 55 kJ/s and rises through 55 metres. Determine :*

(*i*) *The change in enthalpy* (Δ*h*) ;

(*ii*) *Work done during the process* (*W*).

Solution. Flow of fluid = 10 kg/min

Properties of fluid at the **inlet** :

Pressure, p_1 = 1.5 bar = 1.5 × 10⁵ N/m²

Density, ρ_1 = 26 kg/m³

Velocity, C_1 = 110 m/s

Internal energy, u_1 = 910 kJ/kg

Properties of the fluid at the **exit** :

Pressure, p_2 = 5.5 bar = 5.5 × 10⁵ N/m²

Density, ρ_2 = 5.5 kg/m³

Velocity, C_2 = 190 m/s

Internal energy, u_2 = 710 kJ/kg

Heat *rejected* by the fluid, Q = 55 kJ/s

Rise is elevation of fluid = 55 m.

(*i*) **The change in enthalpy,**

$$\Delta h = \Delta u + \Delta(pv) \quad ...(i)$$

$$\Delta(pv) = \frac{p_2 v_2 - p_1 v_1}{1}$$

$$= \frac{p_2}{\rho_2} - \frac{p_1}{\rho_1} = \frac{5.5 \times 10^5}{5.5} - \frac{1.5 \times 10^5}{26}$$

$$= 1 \times 10^5 - 0.0577 \times 10^5$$

$$= 10^5 \times 0.9423 \text{ Nm or J} = 94.23 \text{ kJ}$$

$$\Delta u = u_2 - u_1 = (710 - 910) = -200 \text{ kJ/kg}$$

Substituting the value in eqn. (*i*), we get

$$\Delta h = -200 + 94.23$$

$$= -\mathbf{105.77\ kJ/kg.} \quad \textbf{(Ans.)}$$

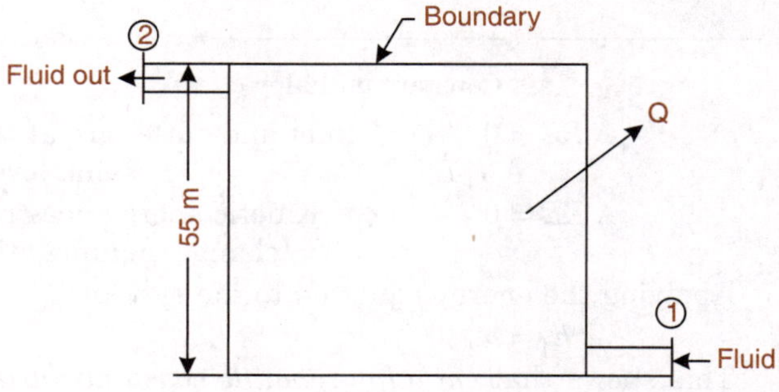

Fig. 3.41

(*ii*) The steady flow equation for unit mass flow can be written as

$$Q = \Delta KE + \Delta PE + \Delta h + W$$

where Q is the heat transfer per kg of fluid

$$Q = 55 \text{ kJ/s} = \frac{55 \text{ kJ/s}}{\frac{10}{60} \text{ kg/s}} = 55 \times 6 = 330 \text{ kJ/kg}$$

Now, $$\Delta KE = \frac{C_2^2 - C_1^2}{2} = \frac{(190)^2 - (110)^2}{2} \text{ Nm or J}$$

$$= 12000 \text{ J or } 12 \text{ kJ/kg}$$

$$\Delta PE = (Z_2 - Z_1)\,g = (55 - 0) \times 9.81 \text{ Nm or J}$$

$$= 539.5 \text{ J or} \approx 0.54 \text{ kJ/kg}$$

Substituting the value in steady flow equation,

$$-330 = 12 + 0.54 - 105.77 + W$$

or $$W = -236.77 \text{ kJ/kg.}$$

Work done per second

$$= -236.77 \times \frac{10}{60} = -39.46 \text{ kJ/s}$$

$$= -\mathbf{39.46\ kW.} \quad \textbf{(Ans.)}$$

Example 3.33. *In a gas turbine unit, the gases flow through the turbine is 15 kg/s and the power developed by the turbine is 12000 kW. The enthalpies of gases at the inlet and outlet are 1260 kJ/kg and 400 kJ/kg respectively, and the velocity of gases at the inlet and outlet are 50 m/s and 110 m/s respectively. Calculate :*

(*i*) *The rate at which heat is rejected to the turbine, and*

(*ii*) *The area of the inlet pipe given that the specific volume of the gases at the inlet is 0.45 m³/kg.*

Solution. Rate of flow of gases, $\dot{m} = 15$ kg/s

Volume of gases at the inlet, $v = 0.45$ m³/kg

Power developed by the turbine, $P = 12000$ kW

$\therefore$ Work done, $W = \dfrac{12000}{15}$

$= 800$ kJ/kg

Enthalpy of gases at the inlet, $h_1 = 1260$ kJ/kg

Enthalpy of gases at the oulet, $h_2 = 400$ kJ/kg

Velocity of gases at the inlet, $C_1 = 50$ m/s

Velocity of gases at the outlet, $C_2 = 110$ m/s.

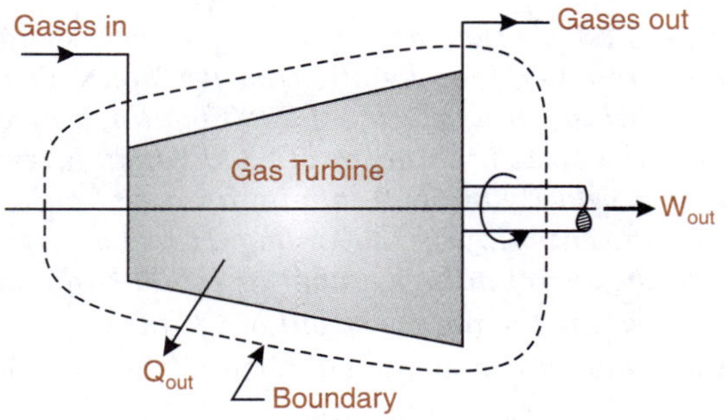

Fig. 3.42

(*i*) **Heat rejected, Q :**

Using the flow equation,

$$h_1 + \frac{C_1^2}{2} + Q = h_2 + \frac{C_2^2}{2} + W \qquad ...(i) \quad [\because \; Z_1 = Z_2]$$

Kinetic energy at inlet

$$= \frac{C_1^2}{2} = \frac{50^2}{2} \text{ m}^2/\text{s}^2$$

$= 1250$ Nm/kg $= 1.25$ kJ/kg

Kinetic energy at outlet

$$= \frac{C_1^2}{2} = \frac{110^2}{2 \times 1000} = 6.05 \text{ kJ/kg}$$

Substituting these values in eqn. (*i*), we get

$$1260 + 1.25 + Q = 400 + 6.05 + 800$$

$$\therefore \quad Q = -55.2 \text{ kJ/kg}$$

i.e., **Heat rejected** $= + 55.2$ kJ/kg

$= 55.2 \times 15$ kJ/s $=$ **828 kW. (Ans.)**

(*ii*) **Inlet area, A :**

Using the relation,

$$\dot{m} = \frac{CA}{v}$$

$$\therefore \quad A = \frac{v\dot{m}}{C} = \frac{0.45 \times 15}{50} = \mathbf{0.135\,m^2}. \quad \textbf{(Ans.)}$$

☞**Example 3.34.** *In an air compressor air flows steadily at the rate of 0.5 kg/s through an air compressor. It enters the compressor at 6 m/s with a pressure of 1 bar and a specific volume of 0.85 m³/kg and leaves at 5 m/s with a pressure of 7 bar and a specific volume of 0.16 m³/kg. The internal energy of the air leaving is 90 kJ/kg greater than that of the air entering. Cooling water in a jacket surrounding the cylinder absorbs heat from the air at the rate of 60 kJ/s. Calculate :*

(*i*) *The power required to drive the compressor ;*

(*ii*) *The inlet and output pipe cross-sectional areas.*

Solution. Air flow rate through the compressor,

$\dot{m} = 0.5$ kg/s

Velocity of air at the inlet to compressor,

$C_1 = 6$ m/s

Velocity of air at the outlet of compressor,

$C_2 = 5$ m/s

Pressure of air at the inlet to the compressor,

$p_1 = 1$ bar

Pressure of air at outlet to the compressor,

$p_2 = 7$ bar

Specific volume of air at inlet to the compressor,

$v_1 = 0.85$ m³/kg

Specific volume of air at outlet to the compressor,

$v_2 = 0.16$ m³/kg

Difference of internal energy at the outlet and inlet of the compressor,

$(u_2 - u_1) = 90$ kJ/kg

Heat rejected by air (to cooling water),

$$Q = -\frac{60}{0.5} = -120 \text{ kJ/kg.}$$

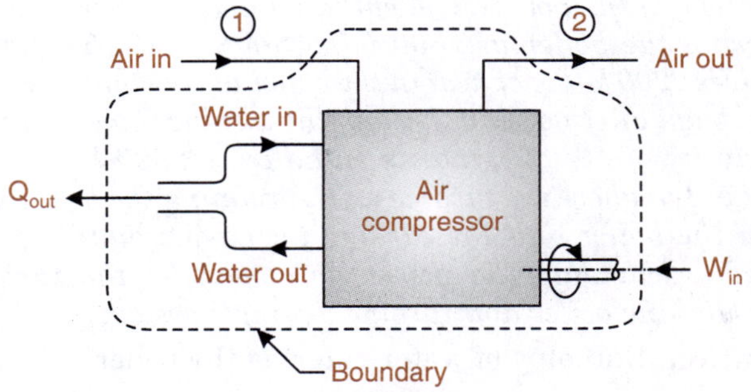

Fig. 3.43

(*i*) **Power required to drive the compressor :**

Using the steady flow energy equation,

$$u_1 + \frac{C_1^2}{2} + p_1v_1 + Q = u_2 + \frac{C_2^2}{2} + p_2v_2 + W$$

$$\therefore \quad W = (u_1 - u_2) + \left(\frac{C_1^2}{2} - \frac{C_2^2}{2}\right) + (p_1v_1 - p_2v_2) + Q$$

$$= -90 + \frac{1}{1000}\left(\frac{6^2}{2} - \frac{5^2}{2}\right)$$

$$+ \frac{10^5}{1000}(1 \times 0.85 - 7 \times 0.16) + (-120)$$

$= -90 + 0.0055 - 27 - 120 = -237$ kJ/kg (app.).

(Note that the change in kinetic energy is negligibly small in comparison with the other terms).

i.e., Work input required

$$= 237 \text{ kJ/kg} = 237 \times 0.5 \text{ kJ/s} = 118.5 \text{ kW}$$

Hence, **power required to drive the compressor = 118.5 kW. (Ans.)**

(*ii*) **Inlet and outlet pipe cross-sectional areas, A_1 and A_2 :**

Using the relation,

$$\dot{m} = \frac{CA}{v}$$

$$\therefore \quad A_1 = \frac{\dot{m}v_1}{C_1} = \frac{0.5 \times 0.85}{6} \text{ m}^2 = 0.0708 \text{ m}^2$$

i.e., Inlet pipe cross-sectional area,

A_1 = **0.0708 m². (Ans.)**

$$\text{Again,} \quad A_2 = \frac{\dot{m}v_2}{C_2} = \frac{0.5 \times 0.16}{5} \text{ m}^2 = 0.016 \text{ m}^2.$$

i.e., Outlet pipe cross-sectional area,

A_2 = **0.016 m². (Ans.)**

Note. In this example, the steady flow energy equation has been used, despite the fact the compression consists of : suction of air ; compression in a closed cylinder ; and discharge of air. The steady flow equation can be used because the cycle of processes takes place many times in a minute, and therefore, average effect is steady flow of air through the machine.

Example 3.35. *In a steam plant, 1 kg of water per second is supplied to the boiler. The enthalpy and velocity of water entering the boiler are 800 kJ/kg and 5 m/s. The water receives 2200 kJ/kg of heat in the boiler at constant pressure. The steam after passing through the turbine comes out with a velocity of 50 m/s, and its enthalpy is 2520 kJ/kg. The inlet is 4 m above the turbine exit. Assuming the heat losses from the boiler and the turbine to the surroundings are 20 kJ/s, calculate the power developed by the turbine. Consider the boiler and turbine as single system.*

Solution. Enthalpy of water entering the boiler,

$$h_1 = 800 \text{ kJ/kg}$$

Velocity of water entering the boiler,

$$C_1 = 5 \text{ m/s}$$

Enthalpy of steam at the outlet of the turbine,

$$h_2 = 2520 \text{ kJ/kg}$$

Velocity of steam at the outlet of the turbine,

$$C_2 = 50 \text{ m/s}$$

Elevation difference, $(Z_1 - Z_2) = 4$ m

Net heat added to the water in the boiler,

$$Q = 2200 - 20 = 2180 \text{ kJ/kg}$$

Power developed by the turbine :

Using the flow equation,

$$h_1 + \frac{C_1^2}{2} + Z_1 g + Q = h_2 + \frac{C_2^2}{2} + Z_2 g + W$$

$$\therefore \quad W = (h_1 - h_2) + \left(\frac{C_1^2}{2} - \frac{C_2^2}{2}\right) + (Z_1 - Z_2)\,g + Q$$

$$= (800 - 2520) + \frac{1}{1000}\left[\frac{5^2}{2} - \frac{50^2}{2}\right] + \frac{4 \times 9.81}{1000} + 2180$$

$$= -1720 + \frac{1}{1000}(12.5 - 1250) + \frac{39.24}{1000} + 2180$$

$$= -1720 - 1.2375 + 0.03924 + 2180$$

$$= 458.8 \text{ kJ/kg} = 458.8 \text{ kJ/s} = 458.8 \text{ kW}$$

Hence, **power developed by the turbine = 458.8 kW. (Ans.)**

Example 3.36. *A turbine, operating under steady-flow conditions, receives 4500 kg of steam per hour. The steam enters the turbine at a velocity of 2800 m/min, an elevation of 5.5 m and a specific enthalpy of 2800 kJ/kg. It leaves the turbine at a velocity of 5600 m/min, an elevation of 1.5 m and a specific enthalpy of 2300 kJ/kg. Heat losses from the turbine to the surroundings amout to 16000 kJ/h.*

Determine the power output of the turbine.

Solution. Quantity of steam supplied to the turbine,

$$m = 4500 \text{ kg/h}$$

Steam velocity at the entrance to the turbine,

$$C_1 = 2800 \text{ m/min}$$

Elevation at the entrance,

$$Z_1 = 5.5 \text{ m}$$

Specific enthalpy at the entrance,

$$h_1 = 2800 \text{ kJ/g}$$

Steam velocity at the exit,

$$C_2 = 5600 \text{ m/min}$$

Elevation at the exit,

$$Z_2 = 1.5 \text{ m}$$

Specific enthalpy at the exit,

$$h_2 = 2300 \text{ kJ/kg}$$

Heat losses from the turbine to the surroundings,

$$Q = -16000 \text{ kJ/h}$$

Applying the steady flow energy equation at entry (1) and exit (2)

$$m\left(h_1 + \frac{C_1^2}{2} + Z_1 g\right) + Q = m\left(h_2 + \frac{C_2^2}{2} + Z_2 g\right) + W$$

$$\therefore \quad Q - W = m\left[(h_2 - h_1) + \left(\frac{C_2^2 - C_1^2}{2}\right) + (Z_2 - Z_1)g\right]$$

$$-\frac{16000}{3600} - W = \frac{4500}{3600}\left[(2300 - 2800) + \left\{\frac{\left(\frac{5600}{60}\right)^2 - \left(\frac{2800}{60}\right)^2}{2 \times 1000}\right\} + \frac{(1.5 - 5.5) \times 9.81}{1000}\right]$$

$$-4.44 - W = 1.25\,(500 + 3.26 - 0.039)$$

or $$W = 633.44 \text{ kJ/s}$$

$\therefore$ **Power output of the turbine**

= 633.44 kW. (Ans.)

Example 3.37. *Steam at a 6.87 bar, 205°C, enters in an insulated nozzle with a velocity of 50 m/s. It leaves at a pressure of 1.37 bar and a velocity of 500 m/s.*

Determine the final enthalpy of steam.

Solution. Pressure of steam at the entrance, $p_1 = 6.87$ bar

The velocity with which steam enters the nozzle,

$$C_1 = 50 \text{ m/s}$$

Pressure of steam at the exit, $p_2 = 1.37$ bar

Velocity of steam at the exit, $C_2 = 500$ m/s.

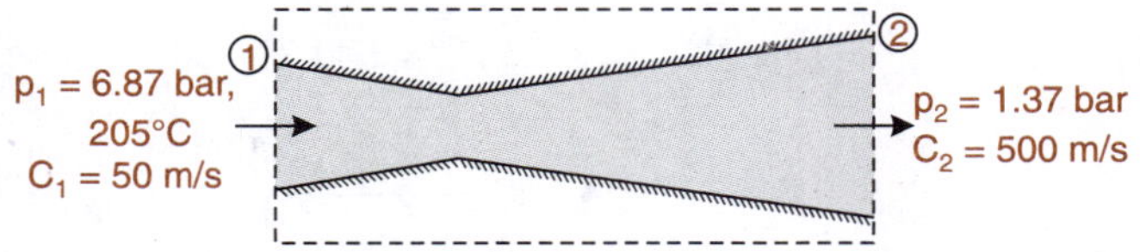

Fig. 3.44

The steady flow energy equation is given by

$$h_1 + \frac{C_1^2}{2} + Z_1 g + Q = h_2 + \frac{C_2^2}{2} + Z_2 g + W \qquad ...(i)$$

Considering the nozzle as an open system, it is evident that :

— there is *no work transfer* across the boundary of the system (*i.e.*, $W = 0$)

— there is *no heat transfer* because the nozzle is insulated (*i.e.*, $Q = 0$).

— the change in potential energy is negligible since there is no significant difference in elevation between the entrance and exit of the nozzle [*i.e.*, $(Z_2 - Z_1)\,g = 0$].

Thus eqn. (*i*) reduces to

$$h_1 + \frac{C_1^2}{2} = h_2 + \frac{C_2^2}{2}$$

$$\therefore \quad (h_2 - h_1) + \frac{C_2^2 - C_1^2}{2} = 0$$

From steam table corresponding to 6.87 bar,

$$h_1 = 2850 \text{ kJ/kg}$$

$$\therefore \quad (h_2 - 2850) + \frac{(500)^2 - (50)^2}{2 \times 1000} = 0$$

or $$h_2 - 2850 + 123.75 = 0 \text{ or } h_2 = 2726.25 \text{ kJ}$$

Hence final enthalpy of steam

= 2726.25 kJ. (Ans.)

Example 3.38. *The working fluid, in a steady flow process flows at a rate of 220 kg/min. The fluid rejects 100 kJ/s passing through the system. The conditions of the fluid at inlet and outlet are given as : $C_1 = 320$ m/s, $p_1 = 6.0$ bar, $u_1 = 2000$ kJ/kg, $v_1 = 0.36$ m³/kg and $C_2 = 140$ m/s, $p_2 = 1.2$ bar, $u_2 = 1400$ kJ/kg, $v_2 = 1.3$ m³/kg. The suffix 1 indicates the condition at inlet and 2 indicates at outlet of the system.*

Determine the power capacity of the system in MW.

The change in potential energy may be neglected.

Solution. Refer to Fig. 3.45.

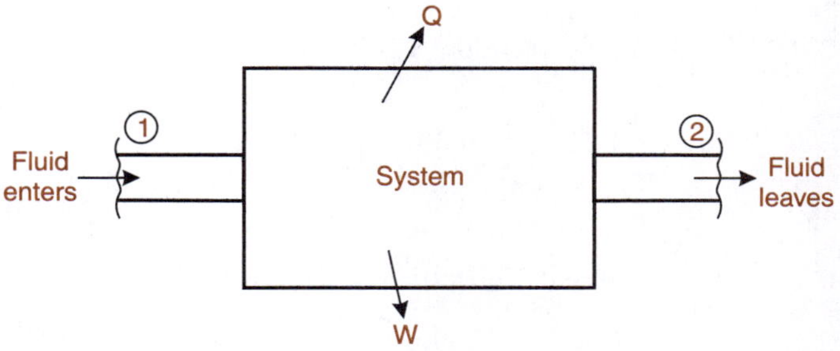

Fig. 3.45

Conditions of the fluid at point 1 :

Velocity, $C_1 = 320$ m/s

Pressure, $p_1 = 6.0$ bar $= 6 \times 10^5$ N/m²

Internal energy, $u_1 = 2000$ kJ/kg

Specific volume, $v_1 = 0.36$ m³/kg.

Conditions of the fluid at point 2 :

Velocity, $C_2 = 140$ m/s

Pressure, $p_2 = 1.2$ bar $= 1.2 \times 10^5$ N/m²

Internal energy, $u_2 = 1400$ kJ/kg

Specific volume, $v_2 = 1.3$ m³/kg

Heat rejected by the fluid, $Q = 100$ kJ/s (–).

Power capacity of the system :

Applying the energy equation at '1' and '2', we get

$$m\left[u_1 + p_1v_1 + \frac{C_1^2}{2} + Z_1 g\right] \pm Q$$

$$= m\left[u_1 + p_2v_2 + \frac{C_2^2}{2} + Z_2 g\right] \pm W$$

Taking –ve sign for Q as the system *rejects* heat and +ve sign for W as the system *develops work*.

$$m\left[u_1 + p_1v_1 + \frac{C_1^2}{2} + Z_1 g\right] - Q$$

$$= m\left[u_2 + p_2v_2 + \frac{C_2^2}{2} + Z_2 g\right] + W$$

$$\therefore \; W = m\left[(u_1 - u_2) + (p_1v_1 - p_2v_2) + \left(\frac{C_1^2 - C_2^2}{2}\right)\right] - Q\,.$$

$$[\because \quad (Z_1 - Z_2)\,g = 0]$$

In the above equation :
— the mass flow is in kg/s
— velocity in m/s
— internal energy in J/kg
— pressure in N/m^2
— specific volume m^3/kg
— the value of Q is in J/s

Then the unit of W will be J/s.

$$\therefore \quad W = \frac{220}{60}\left[(2000 - 1400) \times 10^3 + 10^5 (6 \times 0.36 - 1.2 \times 1.3) + \left(\frac{320^2 - 140^2}{2}\right)\right] - 100 \times 10^3$$

$$= \frac{220}{60}[600 \times 10^3 + 10^5 \times 0.6 + 41.4 \times 10^3] - 100 \times 10^3$$

$$= \frac{220}{60}[600 \times 10^3 + 60 \times 10^3 + 41.4 \times 10^3] - 100 \times 10^3$$

$= 10^3 \times 2471.8$ J/s [$\because$ 1 kJ = 10^3 J]

= 2471.8 kJ/s or kW = 2.4718 MW

Hence, **power capacity of the system**
= 2.4718 MW. (Ans.)

☞**Example 3.39.** *A stream of gases at 7.5 bar, 750°C and 140 m/s is passed through a turbine of a jet engine. The stream comes out of the turbine at 2.0 bar, 550°C and 280 m/s. The process may be assumed adiabatic. The enthalpies of gas at the entry and exit of the turbine are 950 kJ/kg and 650 kJ/kg of gas respectively.*

Determine the capacity of the turbine if the gas flow is 5 kg/s.

Solution. Refer to Fig. 3.46.

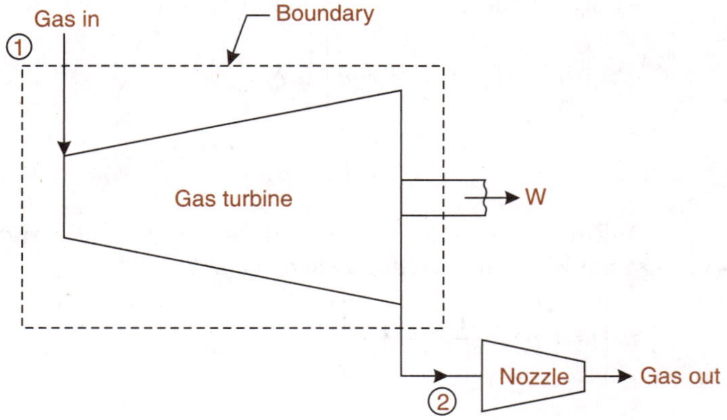

Fig. 3.46

Conditions at '1' :

Pressure, p_1 = 7.5 bar = 7.5 × 10^5 N/m^2, 750°C
Velocity, C_1 = 140 m/s
Enthalpy, h_1 = 950 kJ/kg

Conditions at '2' :

Pressure, p_2 = 2.0 bar = 2 × 10^5 N/m^2, 550°C
Velocity, C_2 = 280 m/s
Enthalpy, h_2 = 650 kJ/kg
Gas flow, m = 5 kg/s.

Capacity of the turbine :

Considering the flow of gas as 1 kg and neglecting the change in potential energy, we can write the steady flow energy equation for the turbine as

$$h_1 + \frac{C_1^2}{2} \pm Q = h_2 + \frac{C_2^2}{2} \pm W$$

$Q = 0$ as the system is adiabatic and W should be taken as +ve since it develops work.

$$\therefore \quad h_1 + \frac{C_1^2}{2} = h_2 + \frac{C_2^2}{2} + W$$

$$\therefore \quad W = (h_1 - h_2) + \frac{C_1^2 - C_2^2}{2}$$

$$= 10^3 (950 - 650) + \frac{140^2 - 280^2}{2}$$

$= 10^3 \times 300 - 29.4 \times 10^3$

$= 270.6 \times 10^3$ J/kg = 270.6 kJ/kg.

Power capacity of the turbine

$= \dot{m}W = 5 \times 270.6 = 1353$ kJ/s

= 1353 kW. (Ans.)

☞**Example 3.40.** *12 kg of air per minute is delivered by a centrifugal air compressor. The inlet and outlet conditions of air are C_1 = 12 m/s, p_1 = 1 bar, v_1 = 0.5 m^3/kg and C_2 = 90 m/s, p_2 = 8 bar, v_2 = 0.14 m^3/kg. The increase in enthalpy of air passing through the compressor is 150 kJ/kg and heat loss to the surroundings is 700 kJ/min.*

Find : (i) Motor power required to drive the compressor ;

(ii) Ratio of inlet to outlet pipe diameters.

Assume that inlet and discharge lines are at the same level.

Solution. Quantity of air delivered by the compressor,

$$m = \frac{12}{60} = 0.2 \text{ kg/s.}$$

Conditions of air at the inlet 1 :

Velocity, C_1 = 12 m/s
Pressure, p_1 = 1 bar = 1 × 10^5 N/m^2
Specific volume, v_1 = 0.5 m^3/kg

Conditions of air at the outlet 2 :

Velocity, C_2 = 90 m/s
Pressure, p_2 = 8 bar = 8 × 10^5 N/m^2
Specific volume, v_2 = 0.14 m^3/kg

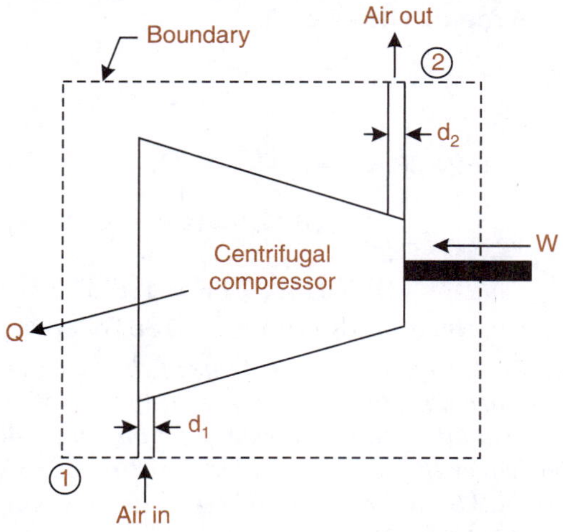

Fig. 3.47

Increase in enthalpy of air passing through the compressor,

$$(h_2 - h_1) = 150 \text{ kJ/kg}$$

Heat lost to the surroundings,

$$Q = -\ 700 \text{ kJ/min} = -\ 11.67 \text{ kJ/s}.$$

(*i*) **Motor power required to drive the compressor :**

Applying energy equation to the system,

$$m\left(h_1 + \frac{C_1^2}{2} + Z_1 g\right) + Q = m\left(h_2 + \frac{C_2^2}{2} + Z_2 g\right) + W$$

Now $\quad Z_1 = Z_2 \quad$ (given)

$$\therefore \quad m\left(h_1 + \frac{C_1^2}{2}\right) + Q = m\left(h_2 + \frac{C_2^2}{2}\right) + W$$

$$W = m\left[(h_1 - h_2) + \frac{C_1^2 - C_2^2}{2}\right] + Q$$

$$= 0.2\left[-\ 150 + \frac{12^2 - 90^2}{2 \times 1000}\right] + (-\ 11.67)$$

$$= -\ 42.46 \text{ kJ/s} = -\ 42.46 \text{ kW}$$

$\therefore$ **Motor power required (or work done on the air) = 42.46 kW. (Ans.)**

(*ii*) **Ratio of inlet to outlet pipe diameter, $\frac{d_1}{d_2}$:**

The mass of air passing through the compressor is given by

$$m = \frac{A_1 C_1}{v_1} = \frac{A_2 C_2}{v_2}$$

$$\therefore \quad \frac{A_1}{A_2} = \frac{C_2}{C_1} \times \frac{v_1}{v_2} = \frac{90}{12} \times \frac{0.5}{0.14} = 26.78$$

$$\therefore \quad \left(\frac{d_1}{d_2}\right)^2 = 26.78 \quad \text{or} \quad \frac{d_1}{d_2} = 5.175$$

Hence, ratio of inlet to outlet pipe diameters = 5.175. (Ans.)

Example 3.41. *In a test of water cooled air compressor, it is found that the shaft work required to drive the compressor is 175 kJ/kg of air delivered and the enthalpy of air leaving is 70 kJ/kg greater than that entering and that the increase in enthalpy of circulating water is 92 kJ/kg.*

Compute the amount of heat transfer to the atmosphere from the compressor per kg of air.

Solution. Refer to Fig. 3.48.

Shaft work required to drive the compressor,

$$W = -\ 175 \text{ kJ/kg}$$

Increase in enthalpy of air passing through the compressor, $(h_2 - h_1) = 70$ kJ/kg

Increase in enthalpy of circulating water,

$$Q_{water} = -\ 92 \text{ kJ/kg}$$

Amount of heat transferred to atmosphere, $Q_{atm.} = ?$

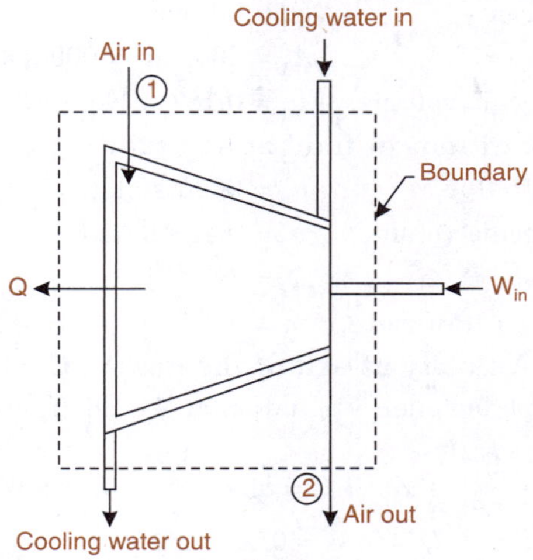

Fig. 3.48

Applying steady-flow energy equation at '1' and '2', we get

$$h_1 + \frac{C_1^2}{2} + Z_1 g + Q = h_2 + \frac{C_2^2}{2} + Z_2 g + W$$

or
$$Q = (h_2 - h_1) + \left(\frac{C_2^2 - C_1^2}{2}\right) + (Z_2 - Z_1)g + W$$

Assuming change in P.E. and K.E. to be negligible.

$$\therefore \quad Q = (h_2 - h_1) + W = 70 + (-\ 175) = -\ 105 \text{ kJ}$$

But $\quad Q = Q_{atm} + Q_{water} \quad$ or $\quad -105 = Q_{atm} + (-\ 92)$

$$\therefore \quad Q_{atm} = -\ 13 \text{ kJ/kg.}$$

Thus heat transferred to atmosphere = 13 kJ/kg. (Ans.)

Example 3.42. *At the inlet to a certain nozzle the enthalpy of fluid passing is 2800 kJ/kg, and the velocity is 50 m/s. At the discharge end the enthalpy is 2600 kJ/kg. The nozzle is horizontal and there is negligible heat loss from it.*

(*i*) *Find the velocity at exit of the nozzle.*

(*ii*) *If the inlet area is 900 cm² and the specific volume at inlet is 0.187 m³/kg, find the mass flow rate.*

(*iii*) *If the specific volume at the nozzle exit is 0.498 m³/kg, find the exit area of nozzle.*

Solution. Refer to Fig. 3.49.

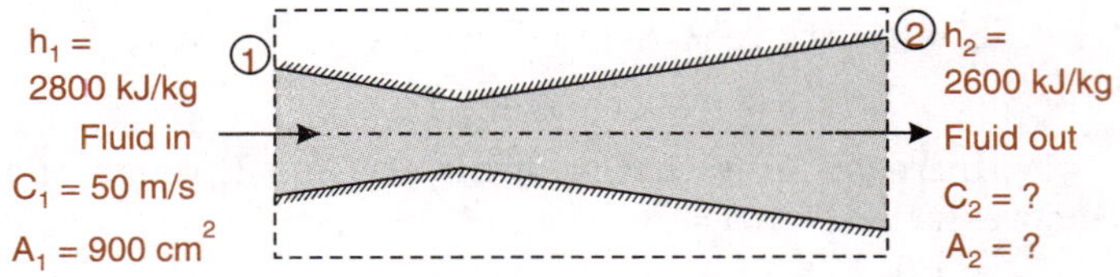

Fig. 3.49

Conditions of fluid at inlet (1) :

Enthalpy, $h_1 = 2800$ kJ/kg

Velocity, $C_1 = 50$ m/s

Area, $A_1 = 900 \text{ cm}^2 = 900 \times 10^{-4} \text{ m}^2$

Specific volume, $v_1 = 0.187 \text{ m}^3/\text{kg}$

Conditions of fluid at exit (2) :

Enthalpy, $h_2 = 2600$ kJ/kg

Specific volume, $v_2 = 0.498 \text{ m}^3/\text{kJ}$

Area, $A_2 = ?$

Mass flow rate, $\dot{m} = ?$

(*i*) **Velocity at exit of the nozzle, C_2 :**

Applying energy equation at '1' and '2', we get

$$h_1 + \frac{C_1^2}{2} + Z_1 g + Q = h_2 + \frac{C_2^2}{2} + Z_2 g + W$$

Here $Q = 0, W = 0, Z_1 = Z_2$

$$\therefore \quad h_1 + \frac{C_1^2}{2} = h_2 + \frac{C_2^2}{2}$$

$$\frac{C_2^2}{2} = (h_1 - h_2) + \frac{C_1^2}{2}$$

$$= (2800 - 2600) \times 1000 + \frac{50^2}{2}$$

$$= 201250 \text{ Nm}$$

$\therefore \quad C_2^2 = 402500$

$\therefore \quad$ **C_2 = 634.4 m/s. (Ans.)**

(*ii*) **Mass flow rate $\dot{m}$:**

By continuity equation,

$$\dot{m} = \frac{AC}{v} = \frac{A_1 C_1}{v_1}$$

$$= \frac{900 \times 10^{-4} \times 50}{0.187} \text{ kg/s} = 24.06 \text{ kg/s}$$

$\therefore$ **Mass flow rate = 24.06 kg/s. (Ans.)**

(*iii*) **Area at the exit, A_2 :**

Now, $$\dot{m} = \frac{A_2 C_2}{v_2}$$

$$24.06 = \frac{A_2 \times 634.4}{0.498}$$

$$\therefore \quad A_2 = \frac{24.06 \times 0.498}{634.4}$$

$$= 0.018887 \text{ m}^2 = 188.87 \text{ cm}^2$$

Hence, **area at the exit = 188.87 cm². (Ans.)**

Example 3.43. *In one of the sections of the heating plant in which there are no pumps enters a steady flow of water at a temperature of 50°C and a pressure of 3 bar (h = 240 kJ/kg). The water leaves the section at a temperature of 35°C and at a pressure of 2.5 bar (h = 192 kJ/kg). The exit pipe is 20 m above the entry pipe.*

Assuming change in kinetic energy to be negligible, evaluate the heat transfer from the water per kg of water flowing.

Solution. Refer to Fig. 3.50.

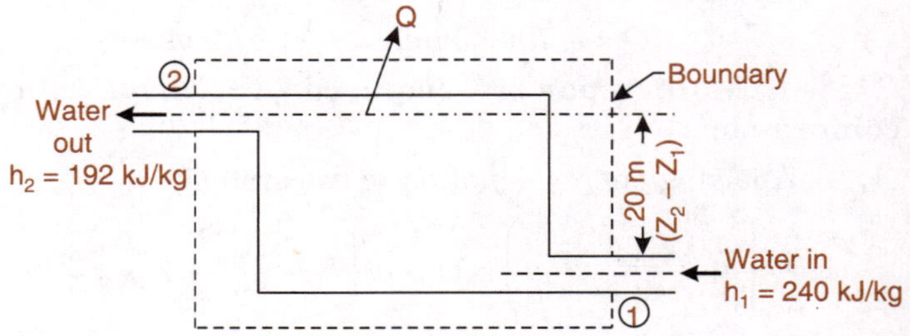

Fig. 3.50

Enthalpy at '1', $h_1 = 240$ kJ/kg

Enthalpy at '2', $h_2 = 192$ kJ/kg

$Z_2 - Z_1 = 20$ m

Applying steady flow energy equation,

$$h_1 + \frac{C_1^2}{2} + Z_1 g + Q = h_2 + \frac{C_2^2}{2} + Z_2 g + W$$

$$Q = (h_2 - h_1) + \left(\frac{C_2^2 - C_1^2}{2}\right) + (Z_2 - Z_1) g + W$$

Here $W = 0$ (no pumps)

$$\frac{C_2^2 - C_1^2}{2} = 0 \quad \text{(given)}$$

$$\therefore \quad Q = (192 - 240) + \frac{20 \times 9.81}{1000} = -47.8 \text{ kJ/kg}$$

$\therefore$ **Heat transfer from water/kg = 47.8 kJ/kg. (Ans.)**

Example 3.44. *The gas leaving the turbine of a gas turbine jet engine flows steadily into the engine jet pipe at a temperature of 900°C, a pressure of 2 bar and a velocity of 300 m/s relative to the pipe. Gas leaves the pipe at a temperature of 820°C and a pressure of 1.1 bar. Heat transfer* *from the gas is negligible. Using the following data evaluate the relative velocity of gas leaving the jet pipe. For the gas at t = 820°C, h = 800 kJ/kg and at 910°C, 915 kJ/kg.*

Solution. Pressure at entry to the engine jet pipe,

$$p_1 = 2 \text{ bar}$$

Velocity relative to the pipe, C_1 = 300 m/s

Heat transfer from gas, $Q = 0$

At temperature, t_1 = 910°C, h_1 = 915 kJ/kg

At temperature, t_2 = 820°C, h_2 = 800 kJ/kg

Relative velocity of gas leaving the jet pipe, C_2 :

Steady flow energy equation is given by :

$$h_1 + \frac{C_1^2}{2} + Z_1g + Q = h_2 + \frac{C_2^2}{2} + Z_2g + W$$

$$Q = 0$$

$$W = 0$$

$$Z_1 = Z_2 \text{ (assumed)}$$

$$h_1 + \frac{C_1^2}{2} = h_2 + \frac{C_2^2}{2}$$

$$\frac{C_2^2}{2} = (h_1 - h_2) + \frac{C_1^2}{2}$$

$$= (915 - 800) \times 1000 + \frac{300^2}{2}$$

$$\therefore \quad C_2^2 = 320000 \quad \text{or} \quad C_2 = 565.7 \text{ m/s.}$$

Hence relative velocity of gas leaving the jet pipe = 565.7 m/s. (Ans.)

Example 3.45. *A centrifugal pump delivers 50 kg of water per second. The inlet and outlet pressures are 1 bar and 4.2 bar respectively. The suction is 2.2 m below the centre of the pump and delivery is 8.5 m above the centre of the pump. The suction and delivery pipe diameters are 20 cm and 10 cm respectively.*

Determine the capacity of the electric motor to run the pump.

Solution. Refer to Fig. 3.51.

Quantity of water delivered by the pump,

$$m_w = 50 \text{ kg/s}$$

Inlet pressure, p_1 = 1 bar = 1 × 10^5 N/m²

Outlet pressure, p_2 = 4.2 bar = 4.2 × 10^5 N/m²

Suction-below the centre of the pump = 2.2 m

Delivery-above the centre of the pump = 8.5 m

Diameter of suction pipe, d_1 = 20 cm = 0.2 m

Diameter of delivery pipe, d_2 = 10 cm = 0.1 m

Capacity of electric motor :

Steady flow energy equation is given by

$$m_w\left(u_1 + p_1v_1 + \frac{C_1^2}{2} + Z_1g\right) + Q$$

$$= m_w\left(u_2 + p_2v_2 + \frac{C_2^2}{2} + Z_2g\right) + W \qquad ...(i)$$

Considering the datum from suction 1, as shown

$$Z_1 = 0, Z_2 = 8.5 + 2.2 = 10.7 \text{ m}$$

$$u_2 - u_1 = 0 \;;\; Q = 0$$

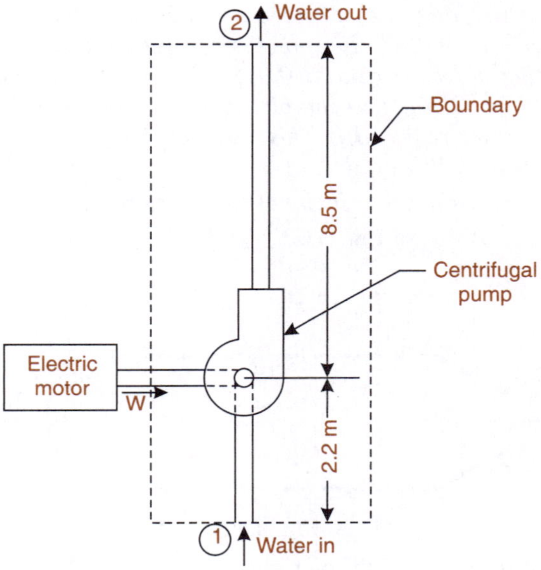

Fig. 3.51

Thus eqn. (*i*) reduces to

$$W = m_w\left[(p_1v_1 - p_2v_2) + (Z_1 - Z_2)g + \left(\frac{C_1^2 - C_2^2}{2}\right)\right] \qquad ...(ii)$$

As water is incompressible fluid.

$$\therefore \quad v_2 = v_1 = v = \frac{1}{\rho} = \frac{1}{1000}$$

The mass flow through inlet and exit pipe is given by

$$m_w = \frac{\pi}{4} \times d_1^2 \times C_1 \times \rho = \frac{\pi}{4} \times d_2^2 \times C_2 \times \rho$$

as $\rho_1 = \rho_2 = \rho$ (for water)

$$\therefore \quad 50 = \frac{\pi}{4} \times (0.2)^2 \times C_1 \times 1000$$

$$\therefore \quad C_1 = \frac{50 \times 4}{\pi \times (0.2)^2 \times 1000} = 1.59 \text{ m/s}$$

and

$$C_2 = \frac{50 \times 4}{\pi \times (0.1)^2 \times 1000} = 6.37 \text{ m/s}$$

Substituting the values in eqn. (*ii*), we get

$$W = 50\left[\left(1\times 10^5 \times \frac{1}{1000} - 4.2\times 10^5 \times \frac{1}{1000}\right) + (0 - 10.7)\times 9.81 + \left(\frac{1.59^2 - 6.37^2}{2}\right)\right]$$

$$= 50[-\ 320 - 104.96 - 19.02]$$

$$= 22199 \text{ J/s or } 22.199 \text{ kJ/s} \simeq 22.2 \text{ kW.}$$

Hence capacity of electric motor = **22.2 kW. (Ans.)**

Example 3.46. *During flight, the air speed of a turbojet engine is 250 m/s. Ambient air temperature is – 14°C. Gas temperature at outlet of nozzle is 610°C. Corresponding enthalpy values for air and gas are respectively 250 and 900 kJ/kg. Fuel air ratio is 0.0180. Chemical energy of fuel is 45 MJ/kg. Owing to incomplete combustion 6% of chemical energy is not released in the reaction. Heat loss from the engine is 21 kJ/kg of air.*

Calculate the velocity of the exhaust jet.

Solution. Refer to Fig. 3.52.

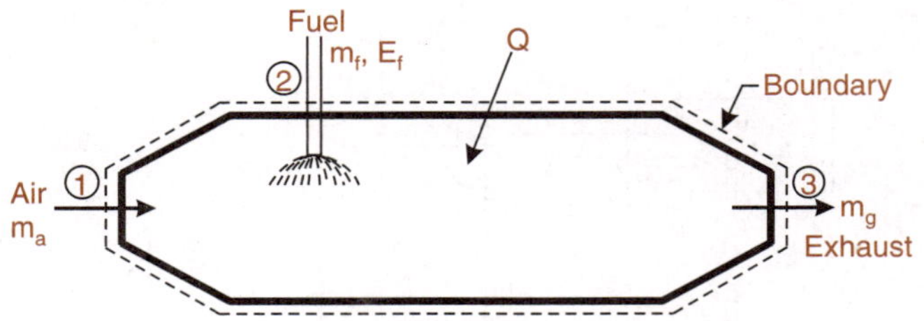

Fig. 3.52

Air speed of turbojet engine, $C_a = 250$ m/s

Ambient air temperature = – 14°C

Gas temperature at outlet of nozzle = 610°C

Enthalpy of air, $h_a = 250$ kJ/kg

Enthalpy of gas, $h_g = 900$ kJ/kg

Fuel air ratio = 0.0180

[If, mass of air, $m_a = 1$ kg, then mass of fuel, $m_f = 0.018$ kg and mass of gas $= 1 + .018 = 1.018$ kg]

Chemical energy of the fuel = 45 MJ/kg.

Heat loss from the engine, $Q = 21$ kJ/kg of air

Velocity of the exhaust gas jet, C_g :

Energy equation for turbojet engine is given by,

$$m_a\left(h_a + \frac{C_a^2}{2}\right) + m_f E_f + Q = m_g\left(h_g + \frac{C_g^2}{2} + E_g\right)$$

$$1\left(250 + \frac{250^2}{2\times 1000}\right) + 0.018 \times 45 \times 10^3 + (-\ 21)$$

$$= 1.018\left[900 + \frac{C_g^2}{2\times 1000} + 0.06 \times \frac{0.018}{1.018}\times 45\times 10^3\right]$$

$$281.25 + 810 - 21 = 1.018\left(900 + \frac{C_g^2}{2000} + 47.74\right)$$

$$1070.25 = 1.018\left(947.74 + \frac{C_g^2}{2000}\right)$$

$$\therefore \quad C_g = 455.16 \text{ m/s}$$

Hence, **velocity of exhaust gas jet = 455.16 m/s. (Ans.)**

☞**Example 3.47.** *Air at a temperature of 20°C passes through a heat exchanger at a velocity of 40 m/s where its temperature is raised to 820°C. It then enters a turbine with same velocity of 40 m/s and expands till the temperature falls to 620°C. On leaving the turbine, the air is taken at a velocity of 55 m/s to a nozzle where it expands until the temperature has fallen to 510°C. If the air flow rate is 2.5 kg/s, calculate :*

(*i*) *Rate of heat transfer to the air in the heat exchanger ;*

(*ii*) *The power output from the turbine assuming no heat loss ;*

(*iii*) *The velocity at exit from the nozzle, assuming no heat loss.*

Take the enthalpy of air as $h = c_p t$, where c_p is the specific heat equal to 1.005 kJ/kg°C and t the temperature.

Solution. Refer to Fig. 3.53.

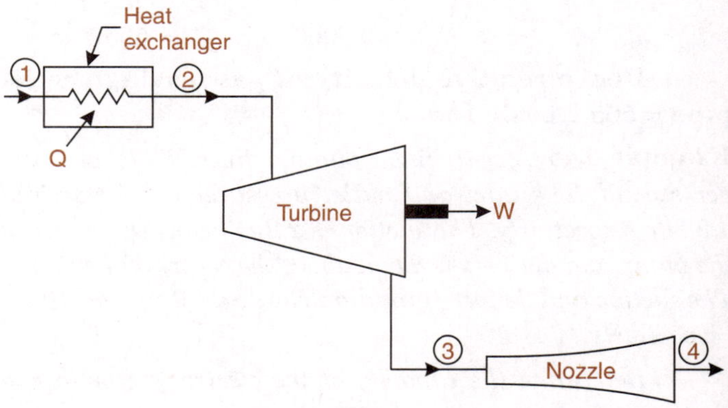

Fig. 3.53

Temperature of air, $t_1 = 20$°C

Velocity of air, $C_1 = 40$ m/s.

Temperature of air after passing the heat exchanger, $t_2 = 820$°C

Velocity of air at entry to the turbine, $C_2 = 40$ m/s

Temperature of air after leaving the turbine, $t_3 = 620$°C

Velocity of air at entry to nozzle, $C_3 = 55$ m/s

Temperature of air after expansion through the nozzle, $t_4 = 510$°C

Air flow rate, $\dot{m} = 2.5$ kg/s..

(*i*) **Heat exchanger :**

Rate of heat transfer :

Energy equation is given as,

$$m\left(h_1 + \frac{C_1^2}{2} + Z_1 g\right) + Q_{1-2} = m\left(h_2 + \frac{C_2^2}{2} + Z_2 g\right) + W_{1-2}$$

Here, $Z_1 = Z_2$, C_1, $C_2 = 0$, $W_{1-2} = 0$

$\therefore \quad mh_1 + Q_{1-2} = mh_2$

or $\quad Q_{1-2} = m(h_2 - h_1)$

$= mc_p (t_2 - t_1) = 2.5 \times 1.005\,(820 - 20)$

$= 2010$ kJ/s.

Hence, **rate of heat transfer = 2010 kJ/s. (Ans.)**

(*ii*) **Turbine :**

Power output of turbine :

Energy equation for turbine gives

$$m\left(h_2 + \frac{C_2^2}{2}\right) = m\left(h_3 + \frac{C_3^2}{2}\right) + W_{2-3}$$

$$[\because \quad Q_{2-3} = 0, Z_1 = Z_2]$$

$$\therefore \quad W_{2-3} = m\left(h_2 + \frac{C_2^2}{2}\right) - m\left(h_3 + \frac{C_3^2}{2}\right)$$

$$= m\left[(h_2 - h_3) + \left(\frac{C_2^2 - C_3^2}{2}\right)\right]$$

$$= m\left[c_p\,(t_2 - t_3) + \frac{C_2^2 - C_3^2}{2}\right]$$

$$= 2.5\left[1.005\,(820 - 620) + \frac{(40)^2 - (55)^2}{2 \times 1000}\right]$$

$= 2.5\,[201 + 0.7125]$

$= 504.3$ kJ/s or 504.3 kW

Hence, **power output of turbine = 504.3 kW. (Ans.)**

(*iii*) **Nozzle :**

Velocity at exit from the nozzle :

Energy equation for nozzle gives,

$$h_3 + \frac{C_3^2}{2} = h_4 + \frac{C_4^2}{2} \quad [\because \quad W_{3-4} = 0, Q_{3-4} = 0, Z_1 = Z_2]$$

$$\frac{C_4^2}{2} = (h_3 - h_4) + \frac{C_3^2}{2} = c_p(t_3 - t_4) + \frac{C_3^2}{2}$$

$$= 1.005(620 - 510) + \frac{55^2}{2 \times 1000}$$

$= 112.062 \times 10^3$ J

$\therefore \quad C_4 = 473.4$ m/s.

Hence, **velocity at exit from the nozzle = 473.4 m/s. (Ans.)**

3.14. HEATING-COOLING AND EXPANSION OF VAPOURS

The basic energy equations for non-flow and flow processes are *also valid for vapours*.

$\therefore$ When $\Delta KE = 0$ and $\Delta PE = 0$

$dQ = du + p.dv$ ......for *non-flow process.*

$dQ = dh - v.dp$ ......for *flow process.*

The various processes using vapour are discussed below :

1. Constant Volume Heating or Cooling. The constant volume heating process is represented on *p-v*, *T-s* and *h-s* diagram as shown in Fig. 3.54 (*a*), (*b*), (*c*) respectively. It is assumed that the steam is in wet condition before heating at pressure p_1, becomes superheated after heating and pressure increases from p_1 to p_2.

Since the mass of steam, *m*, remains constant during the heating process,

$$\therefore \quad m = \frac{V}{x_1 v_{g_1}} = \frac{V}{v_{sup_2}}$$, where *V* is the total constant volume of steam

Also $$\frac{v_{sup_2}}{v_{g_2}} = \frac{T_{sup_2}}{T_{s_2}}$$

$$\therefore \quad \frac{T_{sup_2}}{T_{s_2}} = \frac{x_1 v_{g_1}}{v_{g_2}} \qquad ...(3.69)$$

v_{g_2} and T_{s_2} can be found from the steam tables corresponding to pressure p_2 and then T_{sup_2} can be calculated by using the above equation. When the final condition is known, the change in all other properties can be found easily.

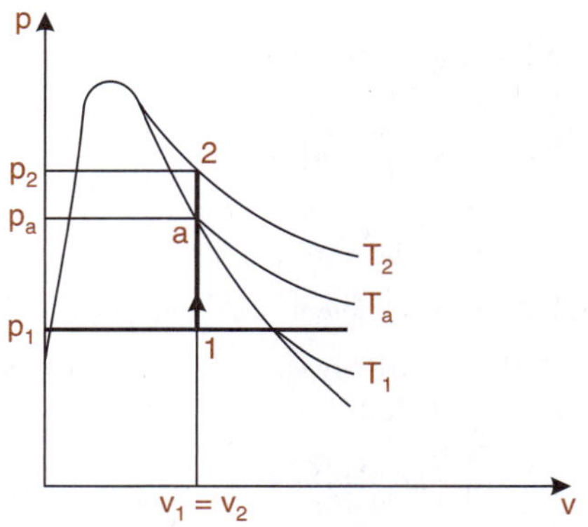

(*a*)

If after cooling the condition of steam remains wet, then the mass fraction be obtained as follows :

$$\frac{V}{x_1 v_{g_1}} = \frac{V}{x_2 v_{g_2}}$$

$$x_2 = \frac{x_1 v_{g_1}}{v_{g_2}} \quad ...(3.70)$$

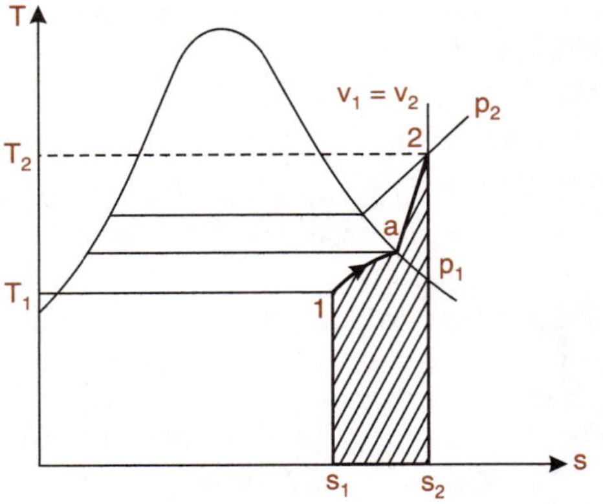

(b)

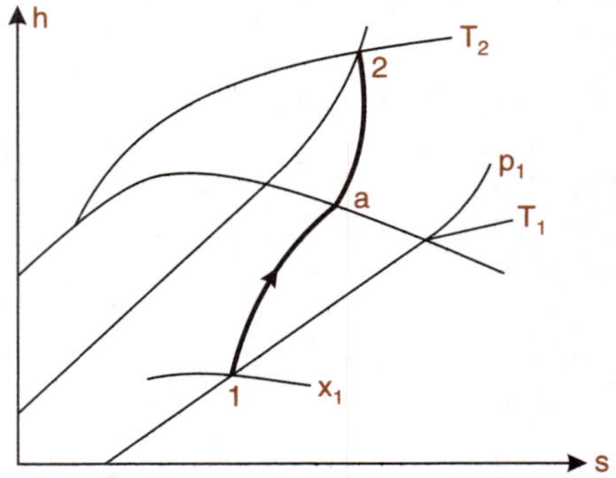

(c)

Fig. 3.54. Constant volume process.

where v_{g_2} can be found from the steam tables corresponding to pressure p_2.

Applying the first law of thermodynamics, we have

$$Q = \Delta u + \int_1^2 p \, . \, dv$$

$$= \Delta u \text{ as } \int_1^2 pdv = 0$$

$$= u_2 - u_1$$

i.e., $$Q_1 = [h_2 - p_2 v_{sup_2}] - [h_1 - p_1(x_1 v_{g_1})] \quad ...(3.71)$$

In case the condition of steam remains wet after heating, then

$$Q = (u_2 - u_1) = [h_2 - p_2(x_2 v_{g_2})] - [h_1 - p_1(x_1 v_{g_1})] \quad ...(3.72)$$

In the *cooling process,* the same equations are used except that the suffixes 1, 2 are interchanged.

☞**Example 3.48.** *A rigid cylinder of volume 0.028 m³ contains steam at 80 bar and 350°C. The cylinder is cooled until the pressure is 50 bar. Calculate :*

(*i*) *The state of steam after cooling ;*

(*ii*) *The amount of heat rejected by the steam.*

Solution. Volume of rigid cylinder = 0.028 m³

Pressure of steam before cooling, p_1 = 80 bar

Temperature of steam before cooling = 350°C

Pressure of steam after cooling, p_2 = 50 bar

Steam at 80 bar and 350°C is in a superheated state, and the specific volume from tables is 0.02995 m³/kg. Hence the mass of steam in the cylinder is given by

$$m = \frac{0.028}{0.02995} = 0.935 \text{ kg}$$

Internal energy at state 1, (*80 bar, 350°C*),

$$u_1 = h_1 - p_1 v_1$$

$$= 2987.3 - \frac{80 \times 10^5 \times 0.02995}{10^3}$$

or $$u_1 = 2747.7 \text{ kJ/kg.}$$

(*i*) **State of steam after cooling :**

At state 2, p_2 = 50 bar and v_2 = 0.02995 m³/kg, therefore, steam is wet, and dryness fraction is given by,

$$x_2 = \frac{v_2}{v_{g_2}} = \frac{0.02995}{0.0394} = 0.76.$$

(*ii*) **Heat rejected by the steam :**

Internal energy at state 2 (50 bar),

$$u_2 = (1 - x_2)\, u_{f_2} + x_2 u_{g_2}$$

$$= (1 - 0.76) \times 1149 + 0.76 \times 2597$$

$$= 2249.48 \text{ kJ/kg}$$

At constant volume,

$$Q = U_2 - U_1 = m(u_2 - u_1)$$

$$= 0.935(2249.48 - 2747.7) = -465.5 \text{ kJ}$$

i.e., **Heat rejected = 465.5 kJ. (Ans.)**

Fig. 3.55 shows the process drawn on *T-s* diagram, the shaded area representing the heat rejected by the system.

2. Constant Pressure Heating or Cooling. Fig. 3.56 (*a*), (*b*) and (*c*) shows the constant pressure heating process on *p-v*, *T-s* and *h-s* diagrams respectively.

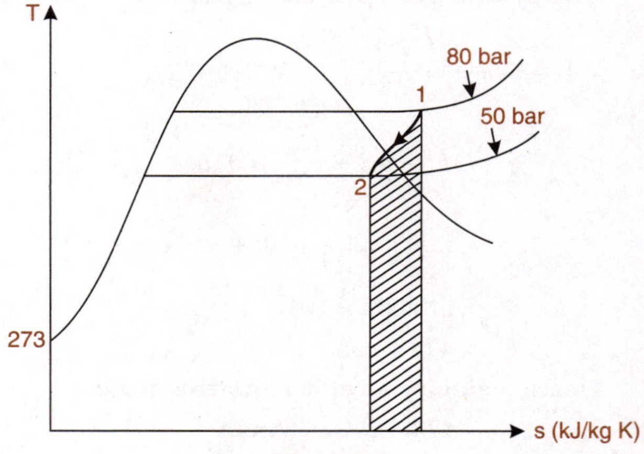

Fig. 3.55

Generation of steam in the boilers is an example of constant pressure heating.

Applying first law of thermodynamics, we have

$$Q = \Delta u + \int_1^2 p \, . \, dv$$

$$= (u_2 - u_1) + p(v_2 - v_1)$$

$$= (u_2 + pv_2) - (u_1 + pv_1) = h_2 - h_1$$

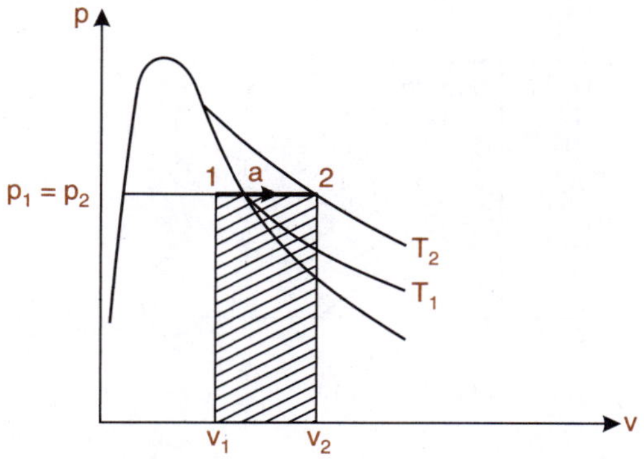

(a)

If the initial condition of steam is wet and final condition is superheated, then

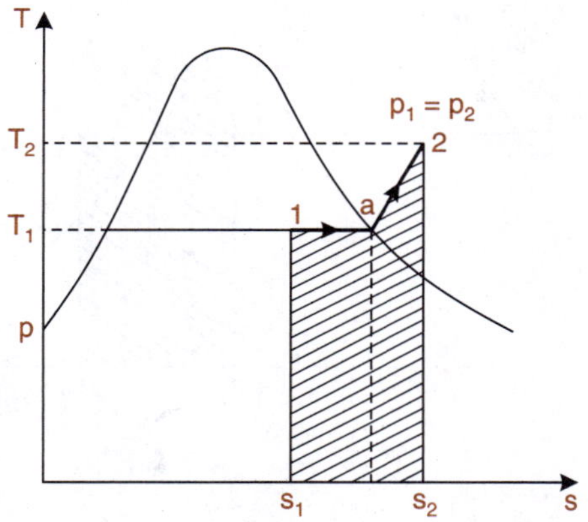

(b)

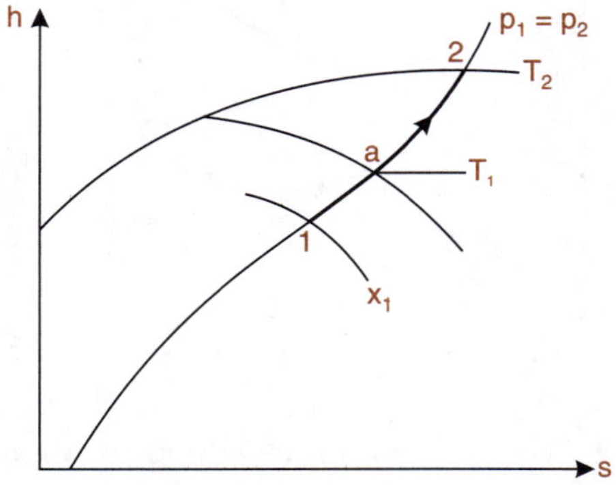

(c)

Fig. 3.56. Constant pressure process.

$$Q = (u_2 + pv_{sup_2}) - (u_1 + p \, . \, x_1 \, v_{g_1})$$

$$= (h_2 - h_1) \quad ...(3.73)$$

here h_1 and h_2 are the actual enthalpies of steam per kg before and after heating.

The heat added during the constant pressure process is equal to the change in enthalpy of steam during the process. When the steam is wet before heating and becomes superheated after heating the work done,

$$W = p \, (v_{sup_2} - x_1 v_{g_1}) \quad ...(3.74)$$

Example 3.49. *0.08 kg of dry steam is heated at a constant pressure of 2 bar until the volume occupied is 0.10528 m³. Calculate :*

(*i*) *Heat supplied ;*

(*ii*) *Work done.*

Solution. Mass of steam, $m = 0.08$ kg

Pressure of steam, $p = 2$ bar

Volume occupied after heating = 0.10528 m³

Initially the steam is dry saturated at 2 bar, hence

$$h_1 = h_g \text{ (at 2 bar)} = 2706.3 \text{ kJ/kg}$$

Finally the steam is at 2 bar and the specific volume is given by

$$v_2 = \frac{0.10528}{0.08} = 1.316 \text{ m}^3\text{/kg}$$

Hence the steam is *superheated finally* (since the value of v_g at 2 bar = 0.885 m³/kg). From superheat tables at 2 bar and 1.316 m³/kg the temperature of steam is 300°C, and the enthalpy, h_2 = 3071.8 kJ/kg.

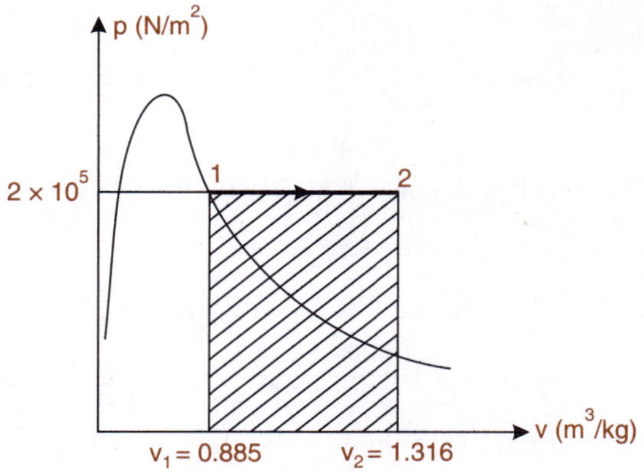

Fig. 3.57

(*i*) **Heat supplied :**

Heat supplied, $Q = H_2 - H_1 = m(h_2 - h_1)$

$= 0.08(3071.8 - 2706.3)$

$=$ **29.24 kJ. (Ans.)**

(*ii*) **Work done :**

The process is shown on a *p-v* diagram in Fig. 3.57. The work done is given by the shaded area

i.e., $W = p(v_2 - v_1)$ Nm/kg

Here $v_1 = v_g$ at 2 bar = 0.885 m³/kg

and $v_2 = 1.316$ m³/kg

$\therefore$ $W = 2 \times 10^5 (1.316 - 0.885)$

$= 2 \times 10^5 \times 0.431$ Nm/kg

Now work done by the total mass of steam (0.08 kg) present

$= 0.08 \times 2 \times 10^5 \times 0.431 \times 10^{-3}$ kJ

= **6.896 kJ. (Ans.)**

Example 3.50. *1 kg of steam at 8 bar, entropy 6.55 kJ/kg K, is heated reversibly at constant pressure until the temperature is 200°C. Calculate the heat supplied, and show on a T-s diagram the area which represents the heat flow.*

Solution. Mass of steam, m = 1 kg

Pressure of steam, p = 8 bar

Entropy of steam (at 8 bar), s = 6.55 kJ/kg K

Temperature after heating = 200°C

At 8 bar, s_g = 6.66 kJ/kg K, hence steam is *wet*, since the actual entropy, s, is less than s_g.

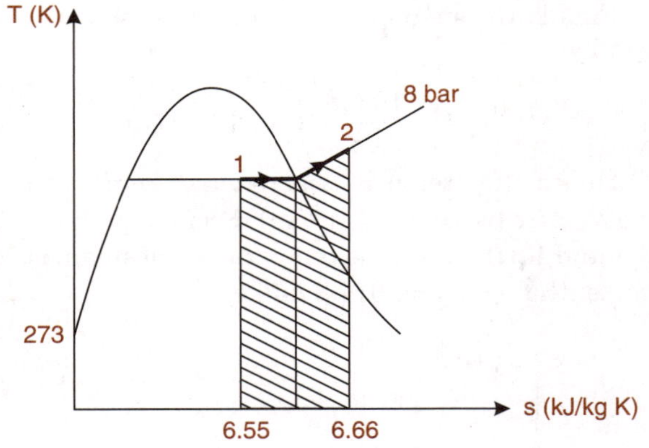

Fig. 3.58

To find the dryness fraction x_1, using the relation,

$$s_1 = s_{f_1} + x_1 s_{fg_1}$$

$$6.55 = 2.0457 + x_1 \times 4.6139$$

$$\therefore \quad x_1 = \frac{6.55 - 2.0457}{4.6139} = 0.976$$

Now, *initial enthalpy* (at 8 bar),

$$h_1 = h_{f_1} + x_1 h_{fg_1}$$

$$= 720.9 + 0.976 \times 2046.5$$

$$= 2718.28 \text{ kJ/kg}$$

Final enthalpy, h_2 : At state 2 the steam is at 200°C at 8 bar and is therefore, superheated.

From superheated tables, h_2 = 2839.3 kJ/kg

Now, $Q = h_2 - h_1 = 2839.3 - 2718.28$

$= 121.02$ kJ/kg

i.e., **Heat supplied = 121.02 kJ/kg. (Ans.)**

The *T-s* diagram showing the process is given in Fig. 3.58, the shaded area representing the heat flow.

3. Constant Temperature or Isothermal Expansion. Fig. 3.59 (*a*), (*b*) and (*c*) shows the constant temperature or isothermal expansion on *p-v*, *T-s* and *h-s* diagrams respectively.

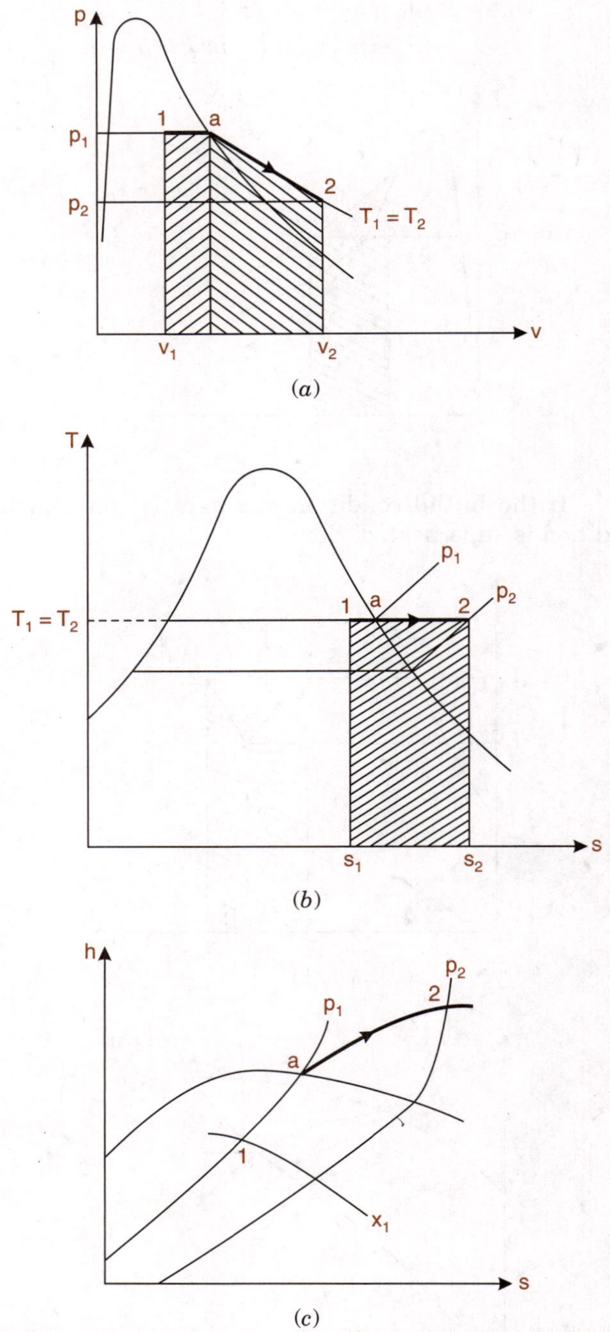

Fig. 3.59. Constant temperature or isothermal expansion.

In the wet region, the constant temperature process is also a constant pressure process during evaporation and as well as condensation. When the steam becomes saturated it behaves like a gas and constant temperature process in superheated region becomes hyperbolic (pv = constant).

When the wet steam is heated at constant temperature till it becomes *dry and saturated,* then the heat transfer (Q) is given by :

$$Q = h_2 - h_1$$

and work done, $W = p_1(v_{g_2} - x_1 v_{g_1})$

$$= pv_{g_1}(1 - x_1)$$

[$\because$ $v_{g_2} = v_{g_1}$ as pressure remains constant during this process]

This process is limited to *wet region only.*

Hyperbolic process (pv = constant) is also an isothermal process in the superheat region as the steam behaves like a gas in this region. The work done during the hyperbolic expansion in a *non-flow system* is given by

$$W = \int_1^2 pdv = \int_1^2 \frac{C}{v} dv = C \log_e \left(\frac{v_2}{v_1}\right)$$

$$= p_1 v_1 \log_e \left(\frac{v_2}{v_1}\right) \quad ...(3.75)$$

where v_1 and v_2 are the specific volumes of steam before and after expansion.

Applying first law of energy equation,

$$Q = \Delta u + \int_1^2 p \cdot dv$$

$$= (u_2 - u_1) + p_1 v_1 \log_e \left(\frac{v_2}{v_1}\right)$$

$$= (h_2 - p_2 v_2) - (h_1 - p_1 v_1) + p_1 v_1 \log_e \left(\frac{v_2}{v_1}\right)$$

Since $p_1 v_1 = p_2 v_2$

$$\therefore \quad Q = (h_2 - h_1) + p_1 v_1 \log_e \frac{v_2}{v_1} \quad ...(3.76)$$

☞**Example 3.51.** *Steam at 7 bar and dryness fraction 0.95 expands in a cylinder behind a piston isothermally and reversibly to a pressure of 1.5 bar. The heat supplied during the process is found to be 420 kJ/kg. Calculate per kg :*

(i) The change of internal energy ;

(ii) The change of enthalpy ;

(iii) The work done.

Solution. Initial pressure of steam,

$$p_1 = 7 \text{ bar} = 7 \times 10^5 \text{ N/m}^2$$

Final pressure of steam,

$$p_2 = 1.5 \text{ bar} = 1.5 \times 10^5 \text{ N/m}^2$$

Heat supplied during the process, Q = 420 kJ/kg.

The process is shown in Fig. 3.60. The saturation temperature corresponding to 7 bar is 165°C. Therefore, the steam is *superheated at the state 2.*

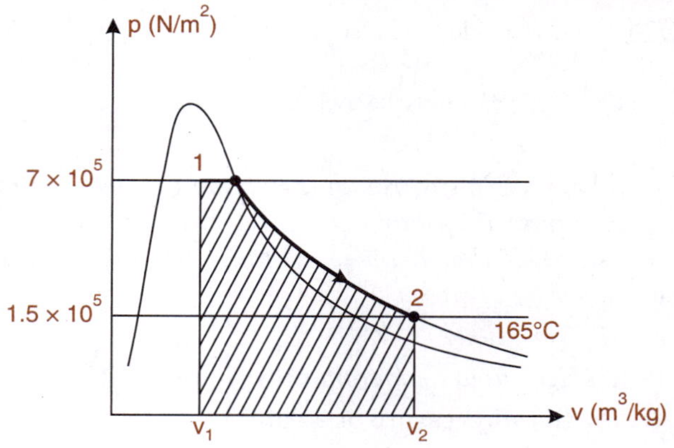

Fig. 3.60

(i) **Change of internal energy :**

The internal energy at state 1 is found by using the relation :

$$u_1 = (1 - x)\, u_f + x u_g$$
$$= (1 - 0.95)\, 696 + (0.95 \times 2573)$$
$$\therefore \quad u_1 = 2479.15 \text{ kJ/kg}$$

Interpolating from superheat tables at 1.5 bar and 165°C, we have

$$u_2 = 2580 + \frac{15}{50}(2856 - 2580)$$
$$= 2602.8 \text{ kJ/kg}$$

$\therefore$ **Gain in internal energy,**

$$u_2 - u_1 = 2602.8 - 2479.15$$
$$= \mathbf{123.65 \text{ kJ/kg.}} \quad \textbf{(Ans.)}$$

(ii) **Change of enthalpy :**

Enthalpy at state 1 (7 bar),

$$h_1 = h_{f_1} + x_1 h_{fg_1}$$

At 7 bar. h_f = 697.1 kJ/kg and h_{fg} = 2064.9 kJ/kg

$$\therefore \quad h_1 = 697.1 + 0.95 \times 2064.9 = 2658.75 \text{ kJ/kg}$$

Interpolating from superheat tables at 1.5 bar and 165°C, we have

$$h_2 = 2772.6 + \frac{15}{50}(2872.9 - 2772.6) = 2802.69 \text{ kJ/kg}$$

$\therefore$ Change of enthalpy

$$= h_2 - h_1 = 2802.69 - 2658.75$$
$$= \mathbf{143.94 \text{ kJ/kg.}} \quad \textbf{(Ans.)}$$

(iii) **Work done :**

From non-flow energy equation,

$$Q = (u_2 - u_1) + W$$
$$\therefore \quad W = Q - (u_2 - u_1) = 420 - 123.65$$
$$= 296.35 \text{ kJ/kg}$$

i.e., **Work done by the steam**

$$= \mathbf{296.35 \text{ kJ/kg.}} \quad \textbf{(Ans.)}$$

Note. The work done is also given by the area in Fig. 3.60 $\left(\int_{v_1}^{v_2} pdv\right)$, this can only be evaluated graphically.

Example 3.52. *In a steam engine cylinder the steam expands from 5.5 bar to 0.75 bar according to a hyperbolic law, pv = constant. If the steam is initially dry and saturated, calculate per kg of steam :*

(i) *Work done ;*

(ii) *Heat flow to or from the cylinder walls.*

Solution. Initial pressure of steam,

$$p_1 = 5.5 \text{ bar} = 5.5 \times 10^5 \text{ N/m}^2$$

Initial condition of steam, $x_1 = 1$

Final pressure of steam,

$$p_2 = 0.75 \text{ bar} = 0.75 \text{ bar} \times 10^5 \text{ N/m}^2$$

At 5.5 bar, $v_1 = v_g = 0.3427 \text{ m}^3\text{/kg}$

Also $p_1v_1 = p_2v_2$

$$\therefore \quad v_2 = \frac{p_1v_1}{p_2} = \frac{5.5 \times 0.3427}{0.75} = 2.513 \text{ m}^3\text{/kg}$$

At 0.75 bar, $v_g = 2.217 \text{ m}^3\text{/kg}$.

Since $v_2 > v_g$ (at 0.75 bar), therefore, the steam is *superheated* at state 2.

Interpolating from superheat tables at 0.75 bar, we have

$$u_2 = 2510 + \left(\frac{2.513 - 2.271}{2.588 - 2.271}\right)(2585 - 2510)$$

$$= 2510 + \frac{0.242}{0.317} \times 75 = 2567.25 \text{ kJ/kg.}$$

For dry saturated steam at 5.5 bar

$$u_1 = u_g = 2565 \text{ kJ/kg}$$

Hence, gain in internal energy

$$= u_2 - u_1 = 2567.25 - 2565 = 2.25 \text{ kJ/kg}$$

The process is shown on a *p-v* diagram in Fig. 3.61, the shaded area representing the work done.

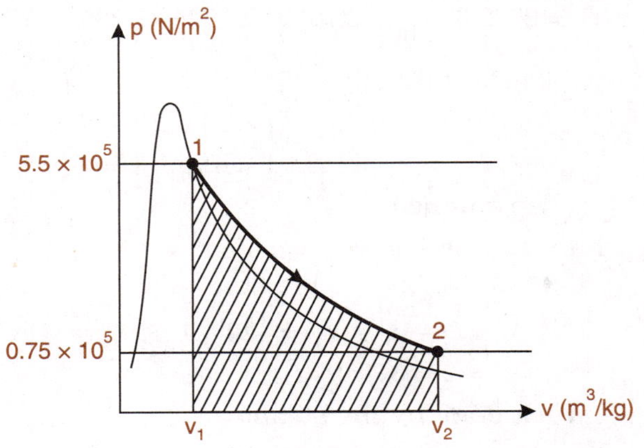

Fig. 3.61

Now, $W = \int_{v_1}^{v_2} pdv$

$$= \int_{v_1}^{v_2} \left(\frac{\text{constant}}{v}\right) dv$$

$$\left[\because \ pv = \text{constant, and } p = \frac{\text{constant}}{v}\right]$$

$$= \text{constant}\left[\log_e v\right]_{v_1}^{v_2}$$

The constant is either p_1v_1 or p_2v_2

i.e., $W = 5.5 \times 10^5 \times 0.3427 \times \log_e \frac{p_1}{p_2}$

$$\left[\because \ p_1v_1 = p_2v_2 \quad \text{or} \quad \frac{v_2}{v_1} = \frac{p_1}{p_2}\right]$$

$$= 5.5 \times 10^5 \times 0.3427 \times \log_e\left(\frac{5.5}{0.75}\right)$$

$$= 375543 \text{ Nm/kg.}$$

Using non-flow energy equation, we get

$$Q = (u_2 - u_1) + W$$

$$= 2.25 + \frac{375543}{10^3} = 378 \text{ kJ/kg}$$

i.e., **Heat supplied = 378 kJ/kg. (Ans.)**

Example 3.53. *Dry saturated steam at 100 bar expands isothermally and reversibly to a pressure of 10 bar. Calculate per kg of steam :*

(i) *The heat supplied ;* (ii) *The work done.*

Solution. Initial pressure of steam, $p_1 = 100$ bar

Final pressure of steam, $p_2 = 10$ bar

The process is shown in Fig. 3.62, the shaded area representing the heat supplied.

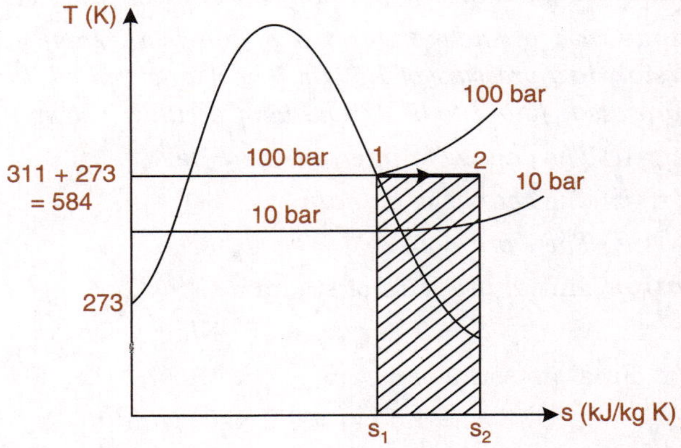

Fig. 3.62

At 100 bar, dry saturated : From steam tables,

$$s_1 = s_g = 5.619 \text{ kJ/kg K and } t_{s_1} = 311°\text{C}$$

At 10 bar and 311°C the steam is superheated, hence interpolating

$$s_2 = 7.124 + \left(\frac{311 - 300}{350 - 300}\right)(7.301 - 7.124)$$

or $$s_2 = 7.163 \text{ kJ/kg K.}$$

(*i*) **Heat supplied :**

Now, heat supplied, Q = shaded area = $T(s_2 - s_1)$

$= 584(7.163 - 5.619)$

$=$ **901.7 kJ/kg. (Ans.)**

(*ii*) **Work done :**

To find work done, applying non-flow energy equation,

$$Q = (u_2 - u_1) + W \quad \text{or} \quad W = Q - (u_2 - u_1)$$

From steam tables at 100 bar, dry saturated,

$$u_1 = u_g = 2545 \text{ kJ/kg}$$

At 10 bar 311°C, interpolating,

$$u_2 = 2794 + \left(\frac{311 - 300}{350 - 300}\right)(2875 - 2794)$$

i.e., $$u_2 = 2811.8 \text{ kJ/kg}$$

Then, $W = Q - (u_2 - u_1)$

$= 901.7 - (2811.8 - 2545) = 634.9$ kJ/kg

Hence, **work done by the steam**

= 634.9 kJ/kg. (Ans.)

4. Reversible Adiabatic or Isentropic Process. Fig. 3.63 (*a*), (*b*) and (*c*) shows the isentropic process on *p-v*, *T-s* and *h-s* diagrams respectively.

Let us consider that the process is non-flow reversible adiabatic. Now applying first law energy equation, we have

$$Q = \Delta u + \int_1^2 p\,.dv = (u_2 - u_1) + W$$

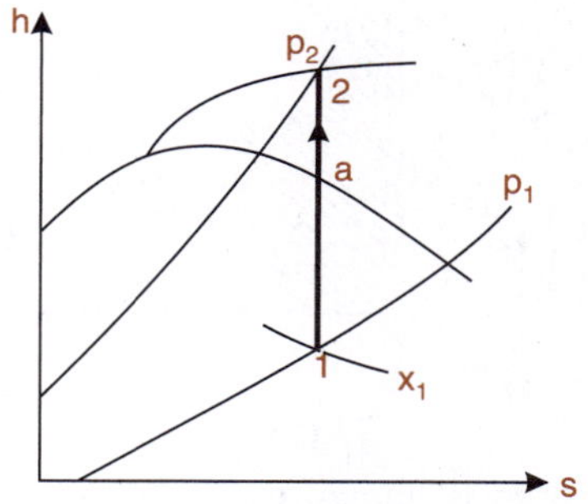

(*c*)

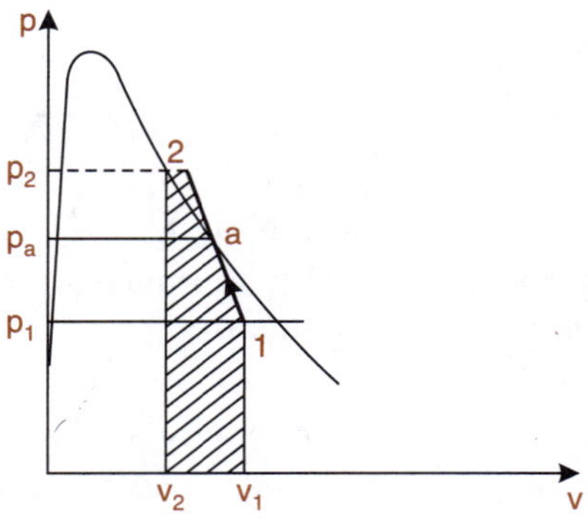

(*a*)

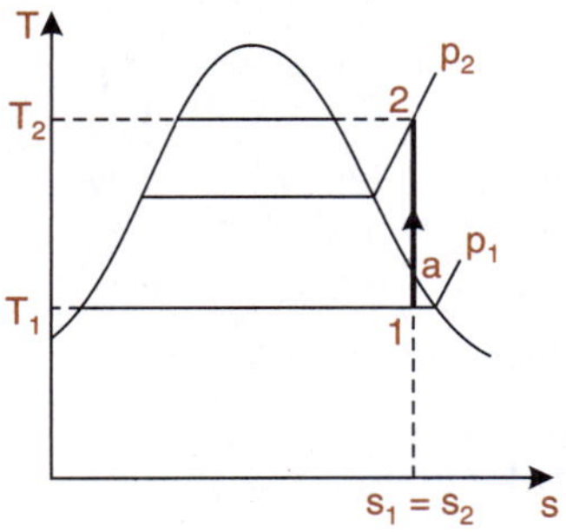

(*b*)

Fig. 3.63. Reversible adiabatic or isentropic process.

As for adiabatic process, $Q = 0$

$\therefore$ $$W = (u_1 - u_2) \quad \text{...(3.77)}$$

In case the process is steady flow reversible adiabatic, then first law energy equation can be written as

$$u_1 + p_1v_1 + Q = u_2 + p_2v_2 + W$$

where v_1 and v_2 are the specific volumes of steam before and after executing the process.

$\therefore$ $$h_1 + 0 = h_2 + W \quad (\because \; Q = 0)$$

$\therefore$ $$W = (h_1 - h_2) \quad \text{...(3.78)}$$

Example 3.54. *1 kg of steam at 120 bar and 400°C expands reversibly in a perfectly thermally insulated cylinder behind a piston until the pressure is 38 bar and the steam is then dry saturated. Calculate the work done by the steam.*

Solution. Mass of steam, $m = 1$ kg

Initial pressure of steam, $p_1 = 120$ bar $= 120 \times 10^5$ N/m²

Initial temperature of steam, $t_1 = 400$°C.

Final pressure of steam, $p_2 = 38$ bar

From superheat tables, at *120 bar and 400°C*

$$h_1 = 3051.3 \text{ kJ/kg and } v_1 = 0.02108 \text{ m}^3\text{/kg}$$

Now, using the equation :

$$u = h - pv$$

$\therefore$ $$u_1 = 3051.3 - \frac{120 \times 10^5 \times 0.02108}{10^3}$$

$= 2798.34$ kJ/kg

Also, $u_1 = u_g$ at 38 bar = 2602 kJ/kg.

Since the cylinder is perfectly thermally insulated then no heat flows to or from the steam during the expansion, the process therefore is adiabatic.

$\therefore$ **Work done by the steam,**

$W = u_1 - u_2$

$= 2798.34 - 2602$

$=$ **196.34 kJ/kg. (Ans.)**

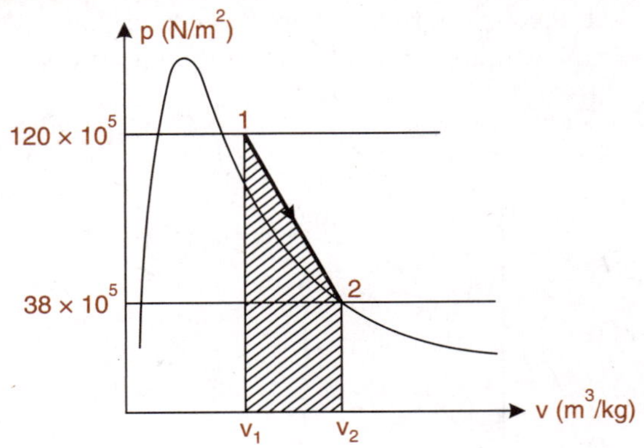

Fig. 3.64

The process is shown on p-v diagram in Fig. 3.64, the shaded area representing the work done.

5. Polytropic process. In this process, the steam follows the law pv^n = constant. This process on p-v, T-s and h-s diagrams is shown in Fig. 3.65 (*a*), (*b*) and (*c*).

The work done during this process is given by

$$W = \left(\frac{p_1v_1 - p_2v_2}{n-1}\right) \text{ Nm/kg}$$

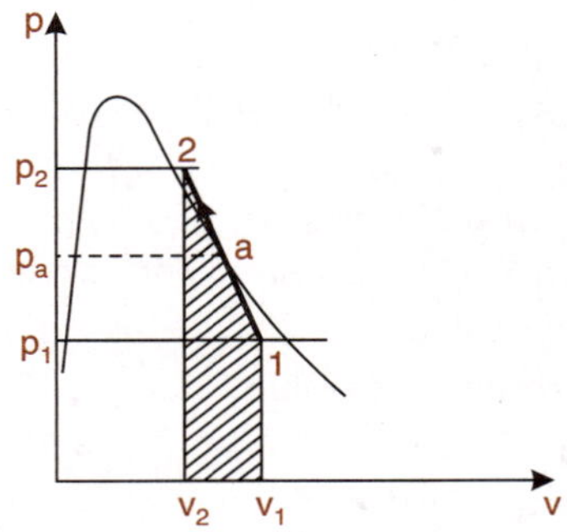

(*a*)

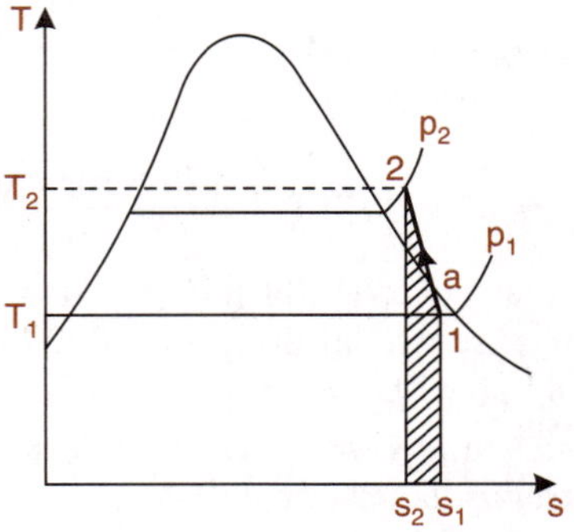

(*b*)

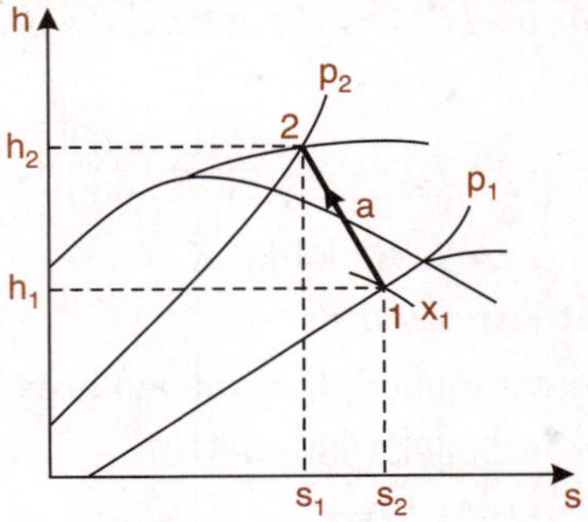

(*c*)

Fig. 3.65. Polytropic process.

Applying the first law energy equation to non-flow process, we have

$$Q = \Delta u + W$$

$$= (u_2 - u_1) + \left(\frac{p_1v_1 - p_2v_2}{n-1}\right)$$

$$= (h_2 - p_2v_2) - (h_1 - p_1v_1) + \left(\frac{p_1v_1 - p_2v_2}{n-1}\right)$$

$$= (h_2 - h_1) + (p_1v_1 - p_2v_2)\left(1 + \frac{1}{n-1}\right)$$

$$= (h_2 - h_1) + \frac{n}{n-1}(p_1v_1 - p_2v_2) \quad ...(3.79)$$

In adiabatic process $Q = 0$ and if $\Delta s \neq 0$ then the process behaves like *adiabatic process* and *not isentropic*. Such a process with steam will be a particular case of the law pv^n = constant. The index n in this case will be that particular index which will satisfy the condition :

$$Q = 0$$

$$\therefore \quad 0 = \Delta u + W$$

$$\therefore \quad W = -\Delta u = -(u_2 - u_1) = (u_1 - u_2)$$

i.e.,
$$W = (u_1 - u_2) \quad ...(3.80)$$

Adiabatic process (not reversible) is also a polytropic process with an index n. The appropriate value of n for adiabatic compression of steam are

$n = 1.13$ for wet steam

$n = 1.3$ for superheated steam

When the initial condition and end condition are *both in wet region* then $p_1v_1^{\,n} = p_2\,v_2^{\,n}$ reduces to :

$$p_1(x_1v_{g_1})^n = p_2(x_2v_{g_2})^n.$$

As p_1, x_1, n and p_2 are specified the value of x_2 can be calculated.

When the end condition is *superheated*, then

$$p_1(x_1v_{g_1})^n = p_2(v_{sup_2})^n.$$

Solving for v_2, then using

$$\frac{v_2}{v_{g_2}} = \frac{T_{sup_2}}{T_{s_2}}$$

T_{sup_2} can be calculated. Knowing T_{s_2} and T_{sup} all properties at the end condition can be calculated.

☞**Example 3.55.** *In a steam engine the steam at the beginning of the expansion process is at 7 bar, dryness fraction 0.98 and expansion follows the law $pv^{1.1}$ = constant, down to a pressure of 0.34 bar. Calculate per kg of steam :*

(*i*) *The work done during expansion ;*

(*ii*) *The heat flow to or from the cylinder walls during the expansion.*

Solution. Initial pressure of steam,

$$p_1 = 7 \text{ bar} = 7 \times 10^5 \text{ N/m}^2$$

Dryness fraction, $x_1 = 0.98$

Law of expansion, $pv^{1.1} = \text{constant}$

Final pressure of steam, $p_2 = 0.34$ bar $= 0.34 \times 10^5$ N/m².

At 7 bar : $v_g = 0.273$ m³/kg

$$\therefore \quad v_1 = x_1 v_g = 0.98 \times 0.273 = 0.267 \text{ m}^3/\text{kg}$$

Also, $p_1 v_1^n = p_2 v_2^n$

i.e.,
$$\frac{v_2}{v_1} = \left(\frac{p_1}{p_2}\right)^{1/n}$$

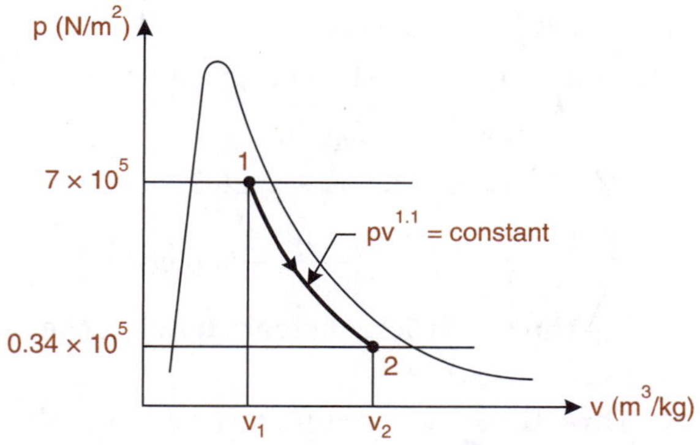

Fig. 3.66

$$\therefore \quad \frac{v_2}{0.267} = \left(\frac{7}{0.34}\right)^{\frac{1}{1.1}}$$

or
$$v_2 = 0.267\left(\frac{7}{0.34}\right)^{\frac{1}{1.1}} = 4.174 \text{ m}^3/\text{kg}.$$

(*i*) **Work done by the steam during the process :**

$$W = \frac{p_1v_1 - p_2v_2}{n-1} = \frac{7 \times 10^5 \times 0.267 - 0.34 \times 10^5 \times 4.174}{(1.1 - 1)}$$

$$= \frac{10^5}{0.1}(1.869 - 1.419) = 10^5 \times 4.5 \text{ Nm/kg}$$

i.e., **Work done** $= \dfrac{10^5 \times 4.5}{10^3}$ **= 450 kJ/kg. (Ans.)**

At 0.34 bar : $v_{g2} = 4.65$ m³/kg, therefore, steam is *wet* at state 2 (since $v_2 < v_{g2}$).

Now, $v_2 = x_2 v_{g2}$, where x_2 = dryness fraction at pressure p_2 (0.34 bar)

$$4.174 = x_2 \times 4.65 \quad \text{or} \quad x_2 = \frac{4.174}{4.65} = 0.897$$

The expansion is shown on a *p-v* diagram shown in Fig. 3.66, the area under 1–2 represents the work done per kg of steam.

(*ii*) **Heat transferred :**

Internal energy of steam at initial state 1 per kg,

$$u_1 = (1 - x_1)u_{f1} + x_1 u_{g_1}$$
$$= (1 - 0.98)\, 696 + 0.98 \times 2573$$
$$= 2535.46 \text{ kJ/kg}$$

Internal energy of steam at final state 2 per kg,

$$u_2 = (1 - x_2)\, u_{f2} + x_2 u_{g2}$$
$$= (1 - 0.897)\, 302 + 0.897 \times 2472$$
$$= 2248.49 \text{ kJ/kg}$$

Using the non-flow energy equation,

$$Q = (u_2 - u_1) + W$$
$$= (2248.49 - 2535.46) + 450$$
$$= 163.03 \text{ kJ/kg}$$

i.e., **Heat supplied = 163.03 kJ/kg. (Ans.)**

Example 3.56. *Steam enters a steam turbine at a pressure of 15 bar and 350°C with a velocity of 60 m/s. The steam leaves the turbine at 1.2 bar and with a velocity of 180 m/s. Assuming the process to be reversible adiabatic, determine the work done per kg of steam flow through the turbine.*

Neglect the change in potential energy.

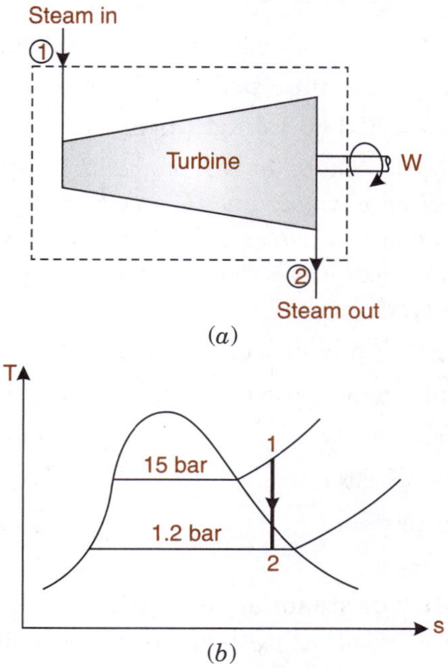

Fig. 3.67

Solution. Initial pressure of steam, $p_1 = 15$ bar

Initial temperature of steam, $t_1 = t_{sup} = 350°C$

Initial velocity of steam, $C_1 = 60$ m/s

Final pressure, $p_2 = 1.2$ bar

Final velocity, $C_2 = 180$ m/s

Process of expansion : *Reversible adiabatic*

As the process is reversible adiabatic, it will be represented by a vertical line on *T-s* diagram by 1–2 as it is also constant entropy process.

The condition at point '2' can be calculated by equating the entropy at point '1' and point '2',

i.e., $s_1 = s_2$ .....per kg of steam

$$7.102 = s_{f_2} + x_2(s_{g_2} - s_{f_2})$$
$$= 1.3609 + x_2(7.2984 - 1.3609)$$
$$\therefore \quad x_2 = \frac{7.102 - 1.3609}{7.2984 - 1.3609} = 0.967$$
$$h_2 = h_{f_2} + x_2 h_{fg_2}$$
$$= 439.4 + 0.967 \times 2244.1$$
$$= 2609.44 \text{ kJ/kg}$$

h_1 (at 15 bar and 350°C) = 3147.5 kJ/kg

Applying the first law energy equation for steady flow process,

$$h_1 + \frac{C_1^2}{2} = h_2 + \frac{C_2^2}{2} + W$$

i.e.,
$$W = (h_1 - h_2) + \left(\frac{C_1^2 - C_2^2}{2}\right)$$
$$= 3147.5 - 2609.44 + \left(\frac{60^2 - 180^2}{2 \times 10^3}\right)$$
$$= 3147.5 - 2609.44 - 14.4 = 523.66 \text{ kJ/kg.}$$

Hence **work done per kg of steam**

= 523.66 kJ/kg. (Ans.)

Example 3.57. *Steam at 10 bar and 200°C enters a convergent divergent nozzle with a velocity of 60 m/s and leaves at 1.5 bar and with a velocity of 650 m/s. Assuming that there is no heat loss, determine the quality of the steam leaving the nozzle.*

Solution. Initial pressure of steam, $p_1 = 10$ bar

Initial temperature of steam, $t_1 = t_{sup} = 200°C$

Initial velocity, $C_1 = 60$ m/s

Final velocity, $C_2 = 650$ m/s

Final pressure, $p_2 = 1.5$ bar

Heat loss = nil

Quality of steam at the outlet :

It is a steady-state non-work developing system. Applying the steady flow energy equation to the process, we get

$$h_1 + \frac{C_1^2}{2} = h_2 + \frac{C_2^2}{2} \qquad (\because \quad Q = 0, W = 0)$$
$$\therefore \quad h_2 = h_1 + \left(\frac{C_1^2 - C_2^2}{2}\right)$$

At 10 bar, 250°C : $h_1 = 2827.9$ kJ/kg (from steam tables)

$$\therefore \quad h_2 = 2827.9 + \left[\frac{60^2 - 650^2}{2 \times 10^3}\right] = 2618.45 \text{ kJ/kg}$$

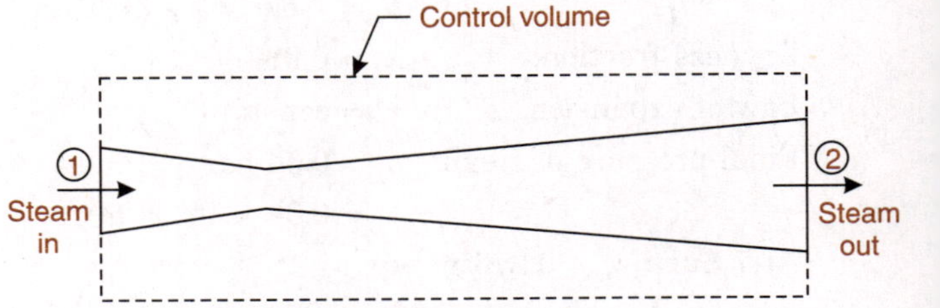

Fig. 3.68

As the enthalpy and pressure of steam at the exit of the nozzle are known, we can find out quality of steam,

$$h_{g_2} \text{ (at 1.5 bar)} = 2693.4 \text{ kJ/kg}$$

As $h_2 < h_{g_2}$, the steam is *wet*.

The enthalpy of wet steam is given by

$$h_2 = h_{f_2} + x_2 h_{fg_2}$$
$$2618.45 = 467.1 + x_2 \times 2226.2$$
$$\therefore \quad x_2 = \frac{2618.45 - 467.1}{2226.2} = 0.966.$$

Hence **the condition of steam leaving the nozzle is 96.6% dry. (Ans.)**

6. Throttling. A flow of fluid is said to be throttled when there is some *restriction to the flow*, when the velocities before and after the restriction are either equal or negligibly small, and when there is a *negligible heat loss to the surroundings.*

The restriction to the flow can be :

(*i*) partly open valve

(*ii*) an orifice or

(*iii*) any other sudden reduction in the cross-section of the flow.

An example of throttling is shown in Fig. 3.69. It is represented on *T-s* and *h-s* diagrams as shown in Figs. 3.70 and 3.71 respectively. The fluid (say steam) flowing steadily along a well-lagged pipe, passes through an orifice at section *X*. Since the pipe is well-lagged it can be assumed that no heat flows to or from the fluid.

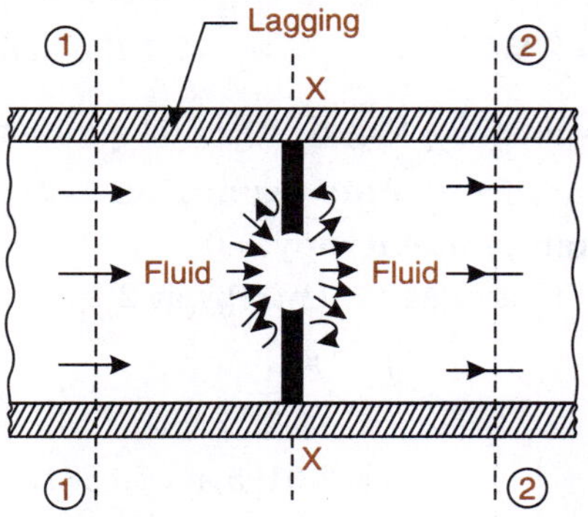

Fig. 3.69. Throttling.

Applying flow equation between any two sections of the flow, we have

$$h_1 + \frac{C_1^2}{2} + Q = h_2 + \frac{C_2^2}{2} + W$$

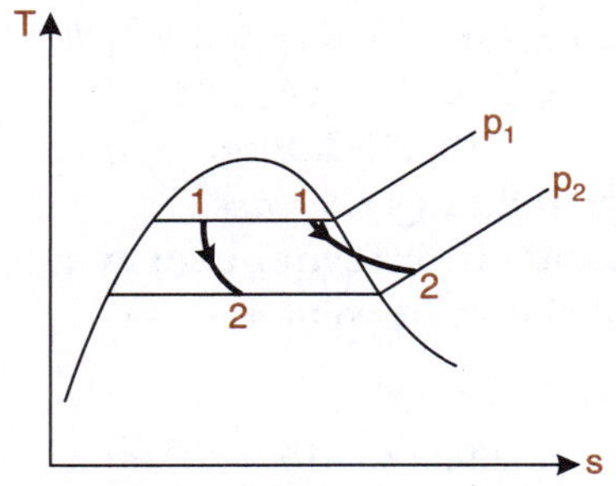

Fig. 3.70. *T-s* diagram.

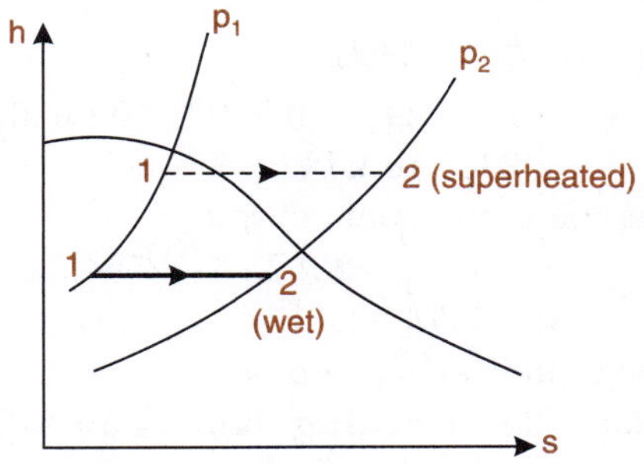

Fig. 3.71. *h-s* diagram.

Now since $Q = 0$, and $W = 0$, then

$$h_1 + \frac{C_1^2}{2} = h_2 + \frac{C_2^2}{2}$$

When the velocities C_1 and C_2 are small, or when C_1 is very nearly equal to C_2, then the K.E. terms may be neglected.

Then $\quad h_1 = h_2 \quad$ ...(3.81)

i.e., For a throttling process :

Initial enthalpy = Final enthalpy.

The process is adiabatic but highly irreversible because of the eddying of the fluid around the orifice at X. Between sections 1 and *X* the enthalpy drops and K.E. increases as the fluid accelerates through the orifice. Between sections *X* and 2 the *enthalpy increases as K.E. is destroyed by fluid eddies.*

During throttling *pressure always falls.*

The throttling process is *used* for the following *purposes* :

1. To *determine the dryness fraction of steam.*
2. To *control the speed of the engine and turbine.*
3. To *reduce* the pressure and temperature of the liquid refrigerant from the condenser condition to evaporator condition in a *refrigeration system.*

☞**Example 3.58.** *Steam at 18 bar is throttled to 1 bar and the temperature after throttling is found to be 150°C. Calculate the initial dryness fraction of the steam.*

Solution. Pressure of steam before throttling, p_1 = 18 bar

Pressure of steam after throttling = 1 bar

Temperature after throttling = 150°C

Initial dryness fraction, x_1 :

From superheat tables at **1 bar and 150°C,** we have

$$h_2 = 2776.4 \text{ kJ/kg}$$

Then for throttling, $h_1 = h_2 = 2776.4$

But $\quad h_1 = h_{f_1} + x_1 h_{fg_1}$

At 18 bar : h_f = 884.6 kJ/kg, h_{fg} = 1910.3 kJ/kg

$\therefore \quad 2776.4 = 884.6 + x_1 \times 1910.3$

or $\quad x_1 = \dfrac{2776.4 - 884.6}{1910.3} = 0.99$

i.e., **Initial, dryness fraction = 0.99. (Ans.)**

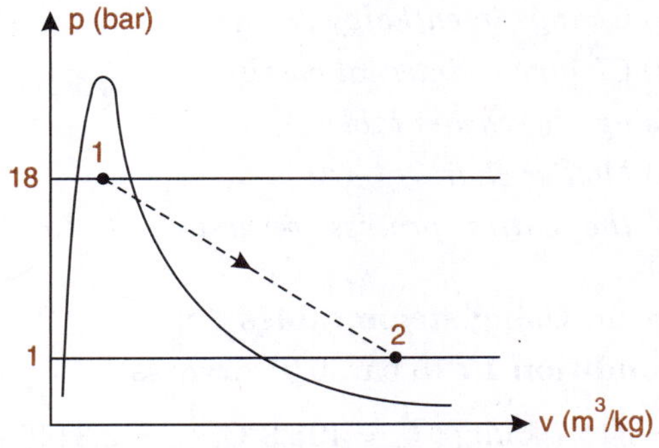

Fig. 3.72

The process is shown on a *p-v* diagram in Fig. 3.72. States 1 and 2 are fixed, but the intermediate states are indeterminate ; the process must be drawn dotted, as shown. No work is done during the process, and the area under the line 1–2 is **not equal to work done.**

Example 3.59. *Steam at 10 bar and 0.9 dryness fraction is throttled to a pressure of 2 bar. Determine the exit condition of steam using* **Mollier chart.**

Solution. Refer to Fig. 3.73.

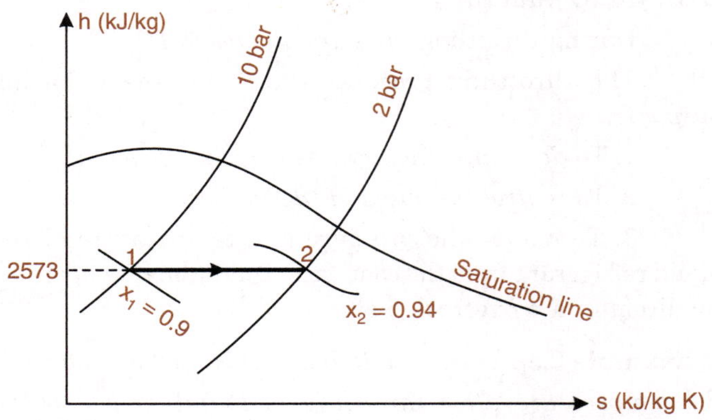

Fig. 3.73

Locate the point '1' at an intersection of the 10 bar pressure line and 0.9 dryness fraction line.

Throttling is a constant enthalpy line so draw a line parallel to *X*-axis till it cuts the 2 bar line and locate the point 2. The dryness fraction of steam at point 2 is 0.94.

(The total enthalpy before and after throttling = 2573 kJ/kg)

Hence **exit condition of steam = 0.94. (Ans.)**

Note. This process occurs during the control of flow of steam supplied to a turbine to take care of the varying load.

Example 3.60. *Steam initially at a pressure of 15 bar and 0.95 dryness expands isentropically to 7.5 bar and is then throttled until it is just dry. Determine per kg of steam :*

(*i*) *Change in entropy ;*

(*ii*) *Change in enthalpy ;*

(*iii*) *Change in internal energy.*

Using : (*a*) *Steam tables*

(*b*) *Mollier chart.*

Is the entire process reversible ? Justify your statement.

Solution. (*a*) **Using steam tables**

Condition 1 : 15 bar, 0.95 dryness

h_{f_1} = 844.7 kJ/kg ; t_{s_1} = 198.3°C, s_{f_1} = 2.3145 kJ/kg K,

s_{g_1} = 6.4406 kJ/kg K, v_{g_1} = 0.132 m³/kg

$$h_1 = h_{f_1} + x_1 h_{fg_1} = 844.7 + 0.95 \times 1945.2 = 2692.64 \text{ kJ/kg}$$

$$s_1 = s_{f_1} + x_1 (s_{g_1} - s_{f_1}) = 2.3145 + 0.95(6.4406 - 2.3145) = 6.2343 \text{ kJ/kg K.}$$

Condition 2 : 7.5 bar

h_{f_2} = 709.3 kJ/kg, t_{s_2} = 167.7°C,

h_{fg_2} = 2055.55 kJ/kg, s_{f_2} = 2.0195 kJ/kg K

s_{g_2} = 6.6816 kJ/kg K, v_{g_2} = 0.255 m³/kg.

Consider **isentropic expansion 1–2 :**

(*i*) **Change in entropy** = 0

i.e., Entropy at 1 = entropy at 2

$\therefore$ $s_1 = s_2$

$$6.2343 = s_{f_2} + x_2 (s_{g_2} - s_{f_2}) = 2.0195 + x_2(6.6816 - 2.0195)$$

$$\therefore \quad x_2 = \frac{6.2343 - 2.0195}{6.6816 - 2.0195} = 0.9$$

Now, enthalpy at point 2,

$$h_2 = h_{f_2} + x_2 h_{fg_2} = 709.3 + 0.9 \times 2055.55 = 2559.29 \text{ kJ/kg.}$$

(*ii*) **Change in enthalpy** $= h_2 - h_1$

= 2559.29 – 2692.64

= **– 133.35 kJ/kg. (Ans.)**

(–ve sign indicates *decrease*).

(*iii*) **Change in internal energy :**

Internal energy at point 1,

$$u_1 = h_1 - p_1 x_1 v_{g_1} = 2692.64 - 15 \times 10^5 \times 0.95 \times 0.132 \times 10^{-3} = 2504.54 \text{ kJ/kg}$$

Internal energy at point 2,

$$u_2 = h_2 - p_2 x_2 v_{g_2} = 2559.29 - 7.5 \times 10^5 \times 0.9 \times 0.255 \times 10^{-3} = 2387.16 \text{ kJ/kg}$$

$\therefore$ Change in internal energy

$$= u_2 - u_1 = 2387.16 - 2504.54 = -117.38 \text{ kJ/kg}$$

(–ve sign indicates *decrease*)

Consider the **throttling expansion 2–3 :**

Entropy at point 2,

$$s_2 = (s_1) = 6.2343 \text{ kJ/kg K}$$

Entropy at point 3,

$$s_3 = s_{f_3} + x_3 (s_{g_3} - s_{f_3})$$

The pressure at point 3 can be read from *h-s* chart (p_3 = 0.06 bar) and the corresponding values of sf_3 and h_{fg_3} from steam tables.

Condition 3 : At 0.06 bar, x_3 = 1. From steam tables,

s_{f_3} = 0.521 kJ/kg K, s_{g_3} = 8.330 kJ/kg K

$\therefore$ s_3 = 0.521 + 1 × (8.330 – 0.521) = 8.330 kJ/kg K

Change in entropy $= s_3 - s_2$

= 8.330 – 6.2343 = 2.0957 kJ/kg K

Change in enthalpy = 0

i.e., $h_2 = h_3$

Change in internal energy = 0

i.e., $u_3 = u_2$

Combining the results obtained from isentropic and throttling expansion, we get during the entire process :

(*i*) **Change in entropy**

= 2.0957 kJ/kg K (increase). (Ans.)

(*ii*) **Change in enthalpy**

= 133.35 kJ/kg K (decrease). (Ans.)

(*iii*) **Change in internal energy**

= 117.38 kJ/kg (decrease). (Ans.)

Only the expansion of steam from point 1 to 2 (*i.e., isentropic expansion*) is *reversible* because of unresisted flow whereas the expansion from point 2 to point 3 (*i.e., throttling expansion*) is *irreversible* because of frictional resistance to flow. *Increase of entropy also shows that expansion from point 2 to point 3 is irreversible.*

(*b*) **Using Mollier chart.**

Refer to Fig. 3.74.

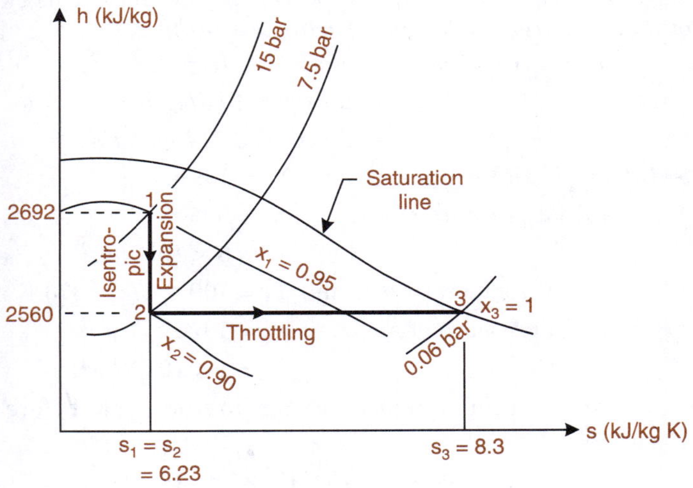

Fig. 3.74

- Locate point 1 at an intersection of 15 bar pressure line and 0.95 dryness fraction line.
- Draw vertical line from point 1 intersecting 7.5 bar pressure line at point 2. Line 1–2 represents *isentropic expansion.*
- From point 2 draw a horizontal line intersecting at the saturation line at point 3. Line 2–3 then represents *throttling expansion.*

From *Mollier chart* :

$h_1 = 2692$ kJ/kg, $h_2 = 2560$ kJ/kg

$s_1 = s_2 = 6.23$ kJ/kg, $s_3 = 8.3$ kJ/kg K

$\therefore$ (*i*) **Change in entropy** $= s_3 - (s_1 \text{ or } s_2)$

$= 8.3 - 6.23$

= 2.07 kJ/kg K (increase). (Ans.)

(*ii*) **Change in enthalpy** $= h_2$ (or h_3) $- h_1$

$= 2560 - 2692 = -132$ kJ/kg

= 132 kJ/kg (decrease). (Ans.)

3.15. UNSTEADY FLOW PROCESSES

In engineering practice, the variable flow process applications are as common as the steady flow process. The *rate of energy and mass transfer into and out of the control volume are not same in the case of unstable* (or variable or transient) *flow process.*

Following two cases only will be discussed :

1. Filling a tank.
2. Emptying a tank or tank discharge.

1. Filling a tank :

Let m_1 = Initial mass of fluid,

p_1 = Initial pressure,

v_1 = Initial specific volume,

T_1 = Initial temperature,

u_1 = Initial specific internal energy,

and m_2 = Final mass of fluid,

p_2 = Final pressure,

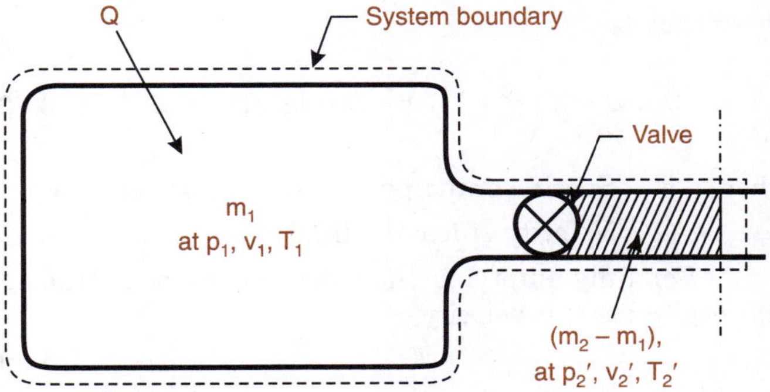

Fig. 3.75

v_2 = Final specific volume,

T_2 = Final temperature,

u_2 = Final specific internal energy,

Also, let p' = Entering fluid pressure,

v' = Entering fluid specific volume,

T' = Entering fluid temperature,

C' = Entering fluid velocity,

u' = Entering specific internal energy of fluid, and

h' = Entering specific enthalpy of fluid.

The quantity of fluid entering

$= m_2 - m_1$

Energy of entering fluid

$$= (m_2 - m_1)\left(u' + p'v' + \frac{C'^2}{2}\right) \quad ...(3.82)$$

$$= (m_2 - m_1)\left(h' + \frac{C'^2}{2}\right) \quad ...(3.83)$$

If Q = Heat transferred into the control volume, we have

$$(m_2 - m_1)\left(h' + \frac{C'^2}{2}\right) + Q = m_2u_2 - m_1u_1 \quad ...(3.84)$$

When the tank is fully insulated and thus no heat transfer takes place,

$$Q = 0$$

and $$(m_2 - m_1)\left(h' + \frac{C'^2}{2}\right) = m_2u_2 - m_1u_1 \quad ...(3.85)$$

Also, if the tank is empty initially and fully insulated for heat transfer,

$$m_1 = 0$$

Thus $$h' + \frac{C'^2}{2} = u_2 \quad ...(3.86)$$

Also, if kinetic energy in the pipe line is neglected

$$h' = u_2 \quad ...(3.87)$$

2. Emptying a tank :

Analogous to the filling of the tank, the equation can be written as

$$(m_1 - m_2)\left(h' + \frac{C'^2}{2}\right) - Q = m_1u_1 - m_2u_2 \quad ...(3.88)$$

where h' = Specific enthalpy of leaving fluid, and

C' = Velocity of leaving fluid.

For fully emptying the tank and no heat transfer and negligible exit velocity,

$$h' = u_1 \quad ...(3.89)$$

Example 3.61. *An air receiver of volume 5.5 m³ contains air at 16 bar and 42°C. A valve is opened and some air is allowed to blow out to atmosphere. The pressure of the air in the receiver drops rapidly to 12 bar when the valve is then closed.*

Calculate the mass of air which has left the receiver.

Solution. Initial volume of air, $V_1 = 5.5 \text{ m}^3$

Initial pressure of air, $p_1 = 16$ bar

Initial temperature of air, $T_1 = 42 + 273 = 315$ K

Final volume of air, $V_2 = V_1 = 5.5 \text{ m}^3$

Final pressure of air, $p_2 = 12$ bar

Mass of air which left the receiver :

Mass of air in the *initial* condition,

$$m_1 = \frac{p_1V_1}{RT_1} = \frac{16 \times 10^5 \times 5.5}{(0.287 \times 10^3) \times 315} = 97.34 \text{ kg.}$$

Assuming that the mass in the receiver undergoes a reversible adiabatic process, then

$$\frac{T_2}{T_1} = \left(\frac{p_2}{p_1}\right)^{\frac{\gamma-1}{\gamma}}$$

$$\frac{T_2}{315} = \left(\frac{12}{16}\right)^{\frac{1.4-1}{1.4}} = \left(\frac{12}{16}\right)^{0.286}$$

or $$T_2 = 315 \times \left(\frac{12}{16}\right)^{0.286} = 290 \text{ K}$$

Now mass of air in the receiver in *final* condition,

$$m_2 = \frac{p_2V_2}{RT_2} = \frac{12 \times 10^5 \times 5.5}{(0.287 \times 10^3) \times 290} = 79.3 \text{ kg.}$$

∴ **Mass of air which left the receiver,**

$m = m_1 - m_2 = 97.34 - 79.3 =$ **18.04 kg. (Ans.)**

Example 3.62. *A 1.6 m³ tank is filled with air at a pressure of 5 bar and a temperature of 100°C. The air is then let off to the atmosphere through a valve. Assuming no heat transfer, determine the work obtainable by utilising the kinetic energy of the discharge air to run a frictionless turbine.*

Take : Atmospheric pressure = 1 bar ;

c_p for air = 1 kJ/kg K ;

c_v for air = 0.711 kJ/kg K.

Solution. Initial volume of air, $V_1 = 1.6 \text{ m}^3$

Initial pressure of air, $p_1 = 5$ bar $= 5 \times 10^5 \text{ N/m}^2$

Initial temperature of air, $T_1 = 100 + 273 = 373$ K

Final pressure of air, $p_2 = 1$ bar $= 1 \times 10^5 \text{ N/m}^2$

Now, initial quantity of air in the tank before discharge,

$$m_1 = \frac{p_1V_1}{RT_1} = \frac{5 \times 10^5 \times 1.6}{(0.287 \times 10^3) \times 373} = 7.47 \text{ kg.}$$

Assuming that system undergoes a reversible adiabatic expansion

$$\frac{T_2}{T_1} = \left(\frac{p_2}{p_1}\right)^{\frac{\gamma-1}{\gamma}}$$

where T_2 is the final temperature of air in the tank.

$$\therefore \quad \frac{T_2}{373} = \left(\frac{1}{5}\right)^{\frac{1.4-1}{1.4}} = 0.631$$

$T_2 = 373 \times 0.631 = 235.4$ K (*i.e.*, finally in the line)

The final quantity of air remaining in the tank is

$$m_2 = \frac{p_2V_2}{RT_2} = \frac{1 \times 10^5 \times 1.6}{(0.287 \times 10^3) \times 235.4} = 2.368 \text{ kg.}$$

With $Q = 0$, kinetic energy is found from,

$$(m_1 - m_2)\left(h' + \frac{C'^2}{2}\right) = m_1u_1 - m_2u_2$$

or $$(m_1 - m_2)\, h' + (m_1 - m_2)\,\frac{C'^2}{2} = m_1u_1 - m_2u_2$$

∴ Kinetic energy,

$$(m_1 - m_2)\,\frac{C'^2}{2} = (m_1u_1 - m_2u_2) - (m_1 - m_2)\, h'$$
$$= m_1c_vT_1 - m_2c_vT_2 - (m_1 - m_2)\, c_pT_2$$
$$= 7.47 \times 0.771 \times 373 - 2.368 \times 0.711 \times 235.4 - (7.47 - 2.368) \times 1 \times 235.4$$
$$= 2148.24 - 396.33 - 1201$$
$$= \mathbf{550.9\ kJ.}\ \textbf{(Ans.)}$$

☞**Example 3.63.** *A frictionless piston is free to move in a closed cylinder. Initially there is 0.035 m³ of oxygen at 4.5 bar, 60°C on one side of the piston and 0.07 m³ of methane at 4.5 bar and – 12°C on the other side. The cylinder walls and piston may be regarded as perfect thermal insulators but the oxygen may be heated electrically. Heating takes place so that the volume of oxygen doubles. Find :*

(i) *Final state condition ;*

(ii) *Work done by the piston ;*

(iii) *Heat transferred to oxygen.*

Treat both gases as perfect and take :

For oxygen $c_p = 0.88$ *kJ/kg K,* $R = 0.24$ *kJ/kg K*

For methane $c_p = 1.92$ *kJ/kg K,* $R = 0.496$ *kJ/kg K.*

Solution. For oxygen :

Initial volume, $V_1 = 0.035\ m^3$

Initial pressure, $p_1 = 4.5$ bar

Initial temperature, $T_1 = 60 + 273 = 333$ K

For methane :

Initial volume, $V_1 = 0.07\ m^3$

Final volume, $V_2 = 0.035\ m^3$

Initial pressure, $p_1 = 4.5$ bar

Initial temperature of methane,

$$T_1 = -12 + 273 = 261\ \text{K}.$$

For Methane :

$$c_p = R \times \frac{\gamma}{\gamma - 1} \quad \text{or} \quad 1.92 = 0.496\left(\frac{\gamma}{\gamma - 1}\right)$$

or, $$\frac{1.92}{0.496} = \frac{\gamma}{\gamma - 1} \quad \text{or} \quad 1.92\,(\gamma - 1) = 0.496\,\gamma$$

$$\therefore \quad \gamma = \frac{1.92}{(1.92 - 0.496)} = 1.348 \text{ say } 1.35$$

For Oxygen :

$$c_v = c_p - R = 0.88 - 0.24 = 0.64\ \text{kJ/kg K}.$$

(i) According to problem ; for *methane*

pV^γ = constant holds good

$$\therefore \quad p_1V_1^\gamma = p_2V_2^\gamma$$

$$p_2 = p_1 \cdot \left(\frac{V_1}{V_2}\right)^\gamma$$
$$= 4.5\,(2)^{1.35} = \mathbf{11.47\ bar.}\ \textbf{(Ans.)}$$

Also, $$\frac{p_1V_1}{T_1} = \frac{p_2V_2}{T_2}$$

or $$T_2 = \frac{p_2V_2T_1}{p_1V_1} = \frac{11.47 \times 0.035 \times 261}{4.5 \times 0.07}$$
$$= \mathbf{332.6\ K.}\ \textbf{(Ans.)}$$

∴ Work done

$$= \frac{p_1V_1 - p_2V_2}{\gamma - 1}$$
$$= \frac{4.5 \times 10^5 \times 0.07 - 11.47 \times 10^5 \times 0.035}{(1.35 - 1)}\ \text{J}$$
$$= \frac{10^5\,(4.5 \times 0.07 - 11.47 \times 0.035)}{0.35 \times 1000}\ \text{kJ}$$
$$= -24.7\ \text{kJ}\ (\textit{done on the methane})$$

(ii) **The piston will be in virtual equilibrium and hence zero work is effected by the piston. (Ans.)**

(iii) Work done by oxygen = work done on methane and expansion of oxygen is effected in the system

$$\therefore \quad W_{oxygen} = +24.7\ \text{kJ}$$

and $$Q = (U_2 - U_1) + W$$

Amount of oxygen present

$$= \frac{p_1V_1}{RT_1} = \frac{4.5 \times 10^5 \times 0.035}{0.24 \times 1000 \times 333} = 0.197\ \text{kg}$$

and $$T_2 = \frac{p_2V_2}{p_1V_1} \times T_1 = \frac{11.47 \times 0.07 \times 333}{4.5 \times 0.035}$$
$$= \mathbf{1697.5\ K.}\ \textbf{(Ans.)}$$

(As the piston is free, the final pressure of oxygen and methane will be **same**).

$$\therefore \quad Q = (U_2 - U_1) + W = mc_v\,(T_2 - T_1) + W$$
$$= 0.197 \times 0.64\,(1697.5 - 333) + 24.7$$
$$= \mathbf{196.7\ kJ.}\ \textbf{(Ans.)}$$

ADDITIONAL/TYPICAL WORKED EXAMPLES

Example 3.64. *Derive the equation of state for a perfect gas.*

Solution. To derive the equation of state for a perfect gas let us consider a *unit mass* of a perfect gas to change its state in the following two successive process (Fig. 3.76) :

(i) Process 1–2′ at constant pressure, and

(ii) Process 2′–2 at constant temperature,

For the *process* 1–2′, applying Charle's law, we get

$$\frac{v_1}{T_1} = \frac{v_2'}{T_2'}$$

and, since $T_2' = T_2$, we may write

$$\frac{v_1}{T_1} = \frac{v_2'}{T_2'} \qquad \text{...}(i)$$

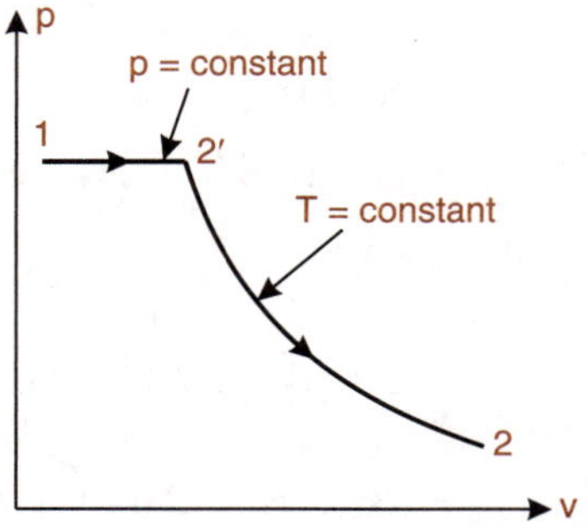

Fig. 3.76. Formulation of equation of state of a perfect gas.

For the *process* 2′–2, using Boyle's law, we have

$$p_2'v_2' = p_2v_2$$

and, since $\quad p_2' = p_1$

$$p_1v_2' = p_2v_2$$

i.e., $\quad v_2' = \dfrac{p_2v_2}{p_1} \quad$ ...(*ii*)

Substituting the value of v_2' from eqn. (*ii*) in eqn. (*i*), we get

$$\frac{v_1}{T_1} = \frac{p_2v_2}{p_1T_2}, \quad \text{or,} \quad \frac{p_1v_1}{T_1} = \frac{p_2v_2}{T_2}$$

i.e., $\quad \dfrac{pv}{T} = \text{constant} \quad$ ...(1)

The magnitude of this constant depends upon the particular gas and it is denoted by R, where R is called the **specific gas constant.** Then

$$\frac{pv}{T} = R$$

The equation of the state for a perfect gas is thus given by the equation

$$pv = RT \quad \text{...(2)}$$

or, for m kg, occupying V m^3,

$$pV = mRT \quad \text{...(3)}$$

If the mass is chosen to be numerically equal to the molecular weight of the gas then 1 mole of the gas has been considered, *i.e.*, 1 kg mole of oxygen is 32 kg oxygen, or 1 kg mole of hydrogen is 2 kg hydrogen.

The equation may be written as

$$pV_0 = MRT \quad \text{...(4)}$$

where, V_0 = Molar volume, and

M = Molecular weight of the gas.

Avogadro discovered that V_0 is the same for all gases at the same pressure and temperature and therefore it may be seen that MR = a constant ; R_0 and thus

$$pV_0 = R_0T \quad \text{...(5)}$$

R_0 is called the **molar or universal gas constant** and its value is **8.3143 kJ/kg mol K.**

If there are n moles present then the ideal gas equation may be written as

$$pV = nR_0T \quad \text{...(6)}$$

where, V is the volume occupied by n moles at pressure p and temperature T.

Example 3.65. *0.1 m^3 of an ideal gas at 300 K and 1 bar is compressed adiabatically to 8 bar. It is then cooled at constant volume and further expanded isothermally so as to reach the condition from where it started. Calculate :*

(*i*) *Pressure at the end of constant volume cooling.*

(*ii*) *Change in internal energy during constant volume process.*

(*iii*) *Network done and heat transferred during the cycle. Assume*

$$c_p = 14.3 \text{ kJ/kg K and } c_v = 10.2 \text{ kJ/kg K.}$$

Solution. *Given :* $V_1 = 0.1$ m^3 ; $T_1 = 300$ K ; $p_1 = 1$ bar ;

$c_p = 14.3$ kJ/kg K ; $c_v = 10.2$ kJ/kg K.

Refer to Fig. 3.77.

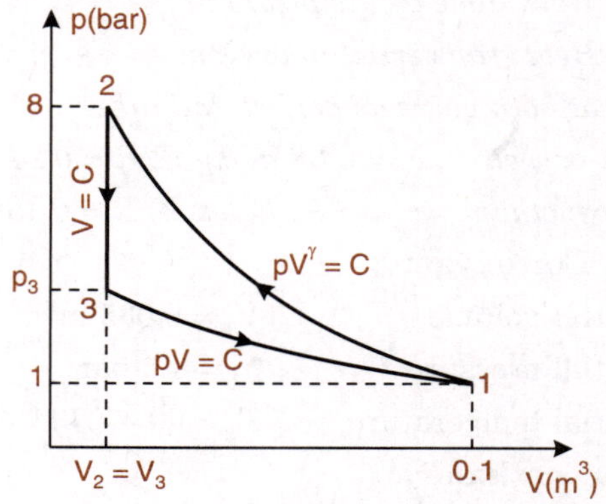

Fig. 3.77

(*i*) **Pressure at the end of constant volume cooling, p_3 :**

$$\gamma = \frac{c_p}{c_v} = \frac{14.3}{10.2} = 1.402$$

Characteristic gas constant,

$$R = c_p - c_v = 14.3 - 10.2 = 4.1 \text{ kJ/kg K}$$

Considering *process 1–2*, we have :

$$p_1V_1^{\gamma} = p_2V_2^{\gamma}$$

$$V_2 = V_1 \times \left(\frac{p_1}{p_2}\right)^{\frac{1}{\gamma}}$$

$$= 0.1 \times \left(\frac{1}{8}\right)^{\frac{1}{1.402}} = 0.0227 \text{ m}^3$$

Also, $\quad \dfrac{T_2}{T_1} = \left(\dfrac{p_2}{p_1}\right)^{\frac{\gamma-1}{\gamma}} = \left(\dfrac{8}{1}\right)^{\frac{1.402-1}{1.402}} = 1.815$

or $\quad T_2 = T_1 \times 1.815 = 300 \times 1.815 = 544.5$ K

Considering *process 3–1*, we have

$$p_3V_3 = p_1V_1$$

$$\therefore \quad p_3 = \frac{p_1V_1}{V_3} = \frac{1 \times 0.1}{0.0227} = \textbf{4.4 bar.} \quad \textbf{(Ans.)}$$

$$(\because \quad V_3 = V_2)$$

(*ii*) **Change in internal energy during constant volume process, $(U_3 - U_2)$:**

$$\text{Mass of gas, } m = \frac{p_1V_1}{RT_1} = \frac{(1 \times 10^5) \times 0.1}{(4.1 \times 1000) \times 300}$$

$$= 0.00813 \text{ kg}$$

$\therefore$ Change in internal energy during *constant volume process 2–3,*

$$U_3 - U_2 = mc_v(T_3 - T_2)$$

$$= 0.00813 \times 10.2\,(300 - 544.5) \qquad (\because \quad T_3 = T_1)$$

$$= \textbf{–20.27 kJ.} \quad \textbf{(Ans.)}$$

(–ve sign means *decrease* in internal energy)

- During constant volume cooling process, temperature and hence internal energy is *reduced*. This *decrease in internal energy equals to heat flow to surroundings since work done is zero.*

(*iii*) **Network done and heat transferred during the cycle :**

$$W_{1-2} = \frac{p_1V_1 - p_2V_2}{\gamma - 1} = \frac{mR(T_1 - T_2)}{\gamma - 1}$$

$$= \frac{0.00813 \times 4.1(300 - 544.5)}{1.402 - 1} = -\,20.27 \text{ kJ}$$

$W_{2-3} = 0$ ... since volume remains constant

$$W_{3-1} = p_3V_3 \log_e\left(\frac{V_1}{V_3}\right) = p_1V_1 \log_e\left(\frac{p_3}{p_1}\right)$$

$$= (1 \times 10^5) \times 0.1 \times \log_e\left(\frac{4.4}{1}\right) \quad (\because \; p_3V_3 = p_1V_1)$$

$$= 14816 \text{ Nm (or J) or } 14.82 \text{ kJ}$$

$\therefore$ **Network done** $= W_{1-2} + W_{2-3} + W_{3-1}$

$= (-\,20.27) + 0 + 14.82$

$=$ **– 5.45 kJ**

–ve sign indicates that work has been done *on the system.* **(Ans.)**

For a cyclic process : $\oint \delta Q = \oint \delta W$

$\therefore$ Heat transferred during the complete cycle

$=$ **– 5.45 kJ**

–ve sign means heat has been *rejected i.e., lost from the system.* **(Ans.)**

Example 3.66. *0.15 m³ of an ideal gas at a pressure of 15 bar and 550 K is expanded isothermally to 4 times the initial volume. It is then cooled to 290 K at constant volume and then compressed back polytropically to its initial state.*

Calculate the network done and heat transferred during the cycle.

Solution. *Given :* $V_1 = 0.15 \text{ m}^3$; $p_1 = 15$ bar ;

$$T_1 = T_2 = 550 \text{ K} \,;\, \frac{V_2}{V_1} = 4 \,;\, T_3 = 290 \text{ K}$$

Refer to Fig. 3.78.

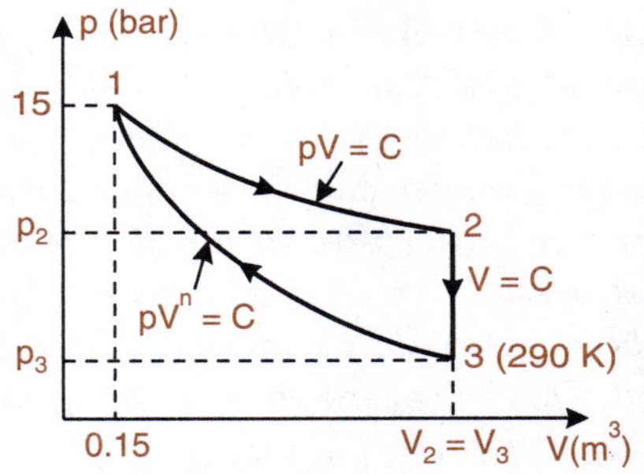

Fig. 3.78

Considering the *isothermal process* 1–2, we have

$$p_1V_1 = p_2V_2 \quad \text{or} \quad p_2 = \frac{p_1V_1}{V_2}$$

or,
$$p_2 = \frac{15 \times 0.15}{(4 \times 0.15)} = 3.75 \text{ bar}$$

$$\text{Work done, } W_{1-2} = p_1V_1 \log_e\left(\frac{V_2}{V_1}\right)$$

$$= (15 \times 10^5) \times 0.15 \times \log_e(4)$$

$$= 311916 \text{ J} = 311.9 \text{ kJ}$$

Considering *constant volume process 2–3*, we get

$$V_2 = V_3 = 4 \times 0.15 = 0.6 \text{ m}^3$$

$$\frac{p_2}{T_2} = \frac{p_3}{T_3}$$

or,
$$p_3 = p_2 \times \frac{T_3}{T_2} = 3.75 \times \frac{290}{550} = 1.98 \text{ bar}$$

$W_{2-3} = 0$ ... since volume remains constant

Consider *polytropic process 3–1 :*

$$p_3V_3^{\,n} = p_1V_1^{\,n} \quad \text{or} \quad \frac{p_1}{p_3} = \left(\frac{V_3}{V_1}\right)^n$$

Taking log on both sides, we get

$$\log_e(p_1/p_3) = n \log_e(V_3/V_1)$$

or,
$$n = \frac{\log_e(p_1/p_3)}{\log_e(V_3/V_1)} = \frac{\log_e(15/1.98)}{\log_e(4)} = 1.46$$

$$W_{3-1} = \frac{p_3V_3 - p_1V_1}{n - 1}$$

$$= \frac{1.98 \times 10^5 \times 0.6 - 15 \times 10^5 \times 0.15}{(1.46 - 1)}$$

$= -230869$ J or -230.87 kJ

$\therefore$ *Network done* $= W_{1-2} + W_{2-3} + W_{3-1}$

$= 311.9 + 0 + (-230.87)$

$=$ **81.03 kJ. (Ans.)**

For a cyclic process, $\oint \delta Q = \oint \delta W$

$\therefore$ Heat *transferred during the cycle*

$=$ **81.03 kJ. (Ans.)**

Example 3.67. *A system consisting of 1 kg of an ideal gas at 5 bar pressure and 0.02 m³ volume executes a cyclic process comprising the following three distinct operations : (i) Reversible expansion to 0.08 m³ volume, 1.5 bar pressure, presuming pressure to be a linear function of volume (p = a + bV), (ii) Reversible cooling at constant pressure and (iii) Reversible hyperbolic compression according to law pV = constant. This brings the gas back to initial conditions.*

(i) Sketch the cycle on p-V diagram.

(ii) Calculate the work done in each process starting whether it is done on or by the system and evaluate the net cyclic work and heat transfer.

Solution. *Given :* $m = 1$ kg ; $p_1 = 5$ bar ; $V_1 = 0.02$ m³ ;

$V_2 = 0.08$ m³ ; $p_2 = 1.5$ bar.

(*i*) **p-V diagram :** *p-V* diagram of the cycle is shown in Fig. 3.79.

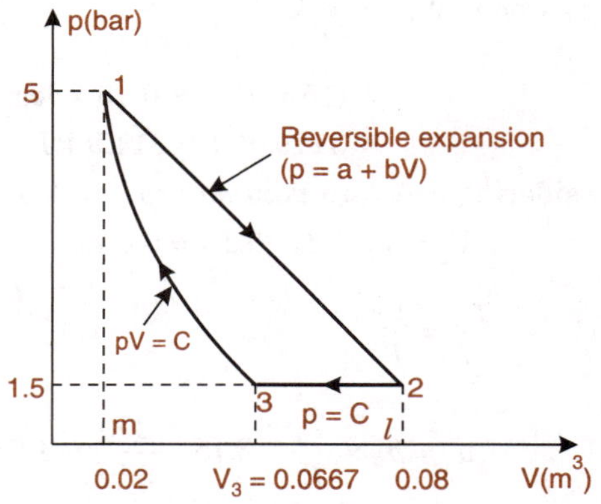

Fig. 3.79. *p-V* diagram.

(*ii*) **Work done and heat transfer :**

- *Process 1–2* (*Linear law*) :

$$p = a + bV \qquad ...(Given)$$

The values of constants a and b can be determined from the values of pressure and volume at the state points 1 and 2.

$$5 = a + 0.02b \qquad ...(i)$$

$$1.5 = a + 0.08b \qquad ...(ii)$$

From (*i*) and (*ii*) we get, $b = -58.33$ and $a = 6.167$

$$W_{1-2} = \int_1^2 pdV = \int_1^2 (a + bV)\, dV$$

$$= \int_1^2 (6.167 - 58.33V)\, dV$$

$$= 10^5 \left| 6.167\,V - 58.33 \times \frac{V^2}{2} \right|_{0.02}^{0.08}$$

$$= 10^5 \left| 6.167\,(0.08 - 0.02) - 58.33 \times \frac{(0.08^2 - 0.02^2)}{2} \right| \times 10^{-3} \text{ kJ} = \mathbf{19.5 \text{ kJ}}$$

This is work done **by the system. (Ans.)**

[*Alternatively :* W_{1-2} = Area under the process line 1–2
= Area of trapezium 1–2–*l*–*m*

$$= \left[\frac{5 + 1.5}{2} \times 10^5\right] \times (0.08 - 0.02)$$

= 19.5 kJ]

- *Process 2–3* (*constant pressure*) :

$$p_3 = p_2 = 1.5 \text{ bar}$$

The volume V_3 can be worked out from the hyperbolic compression 3–1, as follows :

$$p_1V_1 = p_3V_3$$

or $$V_3 = \frac{p_1V_1}{p_3} = \frac{5 \times 0.02}{1.5} = 0.0667 \text{ m}^3$$

$\therefore$ $$W_{2-3} = p_2(V_3 - V_2)$$

$$= 1.5 \times 10^5\,(0.0667 - 0.08) \times 10^{-3} \text{ kJ}$$

$$= -1.995 \text{ kJ}$$

- *Process 3–1* (*hyperbolic process*) :

$$W_{3-1} = p_3V_3 \log_e \left(\frac{V_1}{V_3}\right)$$

$$= (10^5 \times 1.5) \times 0.0667 \log_e \left(\frac{0.02}{0.0667}\right) \times 10^{-3} \text{ kJ}$$

$= \mathbf{-12.05 \text{ kJ.}}$

This is the work **done on the system. (Ans.)**

Network done, $W_{net} = W_{1-2} + W_{2-3} + W_{3-1}$

$= 19.5 + (-1.995) + (-12.05)$

$=$ **5.445 kJ. (Ans.)**

Heat transferred during the complete cycle,

$$\oint \delta Q = \oint \delta W = \mathbf{5.455 \text{ kJ.}} \quad \textbf{(Ans.)}$$

Example 3.68. *5 m³ of air at 2 bar, 27°C is compressed up to 6 bar pressure following pv^1.3 = constant. It is subsequently expanded adiabatically to 2 bar. Considering the two processes to be reversible, determine the network. Also plot the processes on T-s diagram.*

Solution. Refer to Fig. 3.80.

Given : $V_1 = 5 \text{ m}^3$; $p_1 = 2$ bar ; $T_1 = 27 + 273 = 300$ K ;

$p_2 = 6$ bar ; $p_3 = 2$ bar

Network :

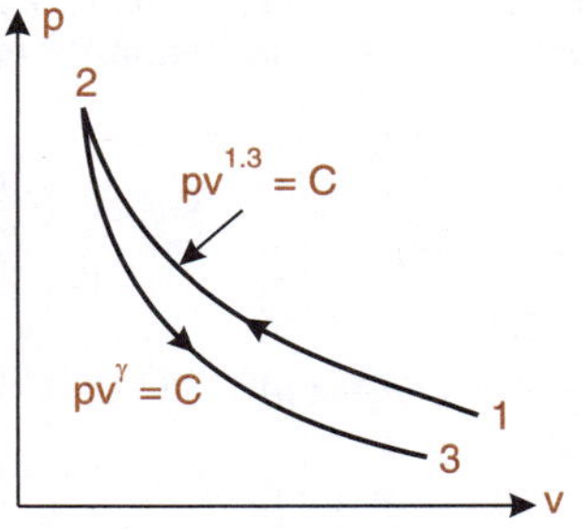

Fig. 3.80. p-V diagram.

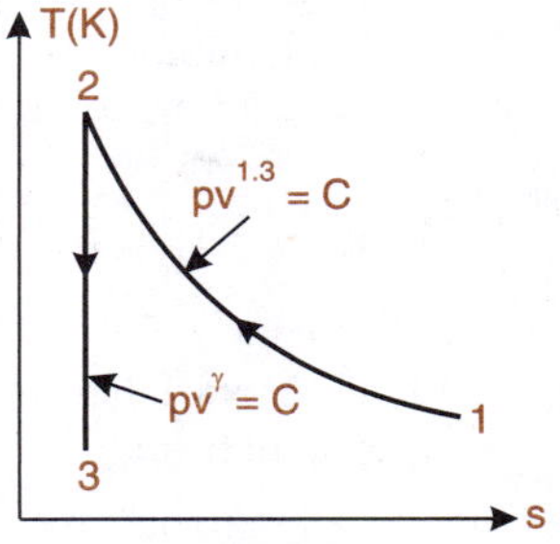

Fig. 3.81. T-s diagram.

Mass of air, $m = \dfrac{p_1V_1}{RT_1} = \dfrac{2 \times 10^5 \times 5}{287 \times 300} = 11.61$ kg.

Considering polytropic compression process 1–2, we have

$$\frac{T_2}{T_1} = \left(\frac{p_2}{p_1}\right)^{\frac{n-1}{n}}, \text{ or, } \frac{T_2}{300} = \left(\frac{6}{2}\right)^{\frac{1.3-1}{1.3}} \text{ or } T_2 = 386.5 \text{ K.}$$

Considering isentropic process 2–3, we get

$$\frac{T_2}{T_3} = \left(\frac{p_2}{p_3}\right)^{\frac{\gamma-1}{\gamma}} = \left(\frac{p_2}{p_1}\right)^{\frac{\gamma-1}{\gamma}} = \left(\frac{6}{2}\right)^{\frac{14-1}{1.4}} = 1.369$$

$(\because \quad p_3 = p_1)$

$$\therefore \quad T_3 = \frac{T_2}{1.369} = \frac{386.5}{1.369} = 282.3 \text{ K}$$

Now, work done during polytropic *compression* 1–2,

$$W_{1-2} = \frac{mR(T_1 - T_2)}{n-1} = \frac{11.61 \times 0.287(300 - 386.5)}{1.3 - 1}$$

$= -\,960.7$ kJ

and, work done during *adiabatic expansion* 2–3

$$W_{2-3} = \frac{mR(T_2 - T_3)}{\gamma - 1}$$

$$= \frac{11.61 \times 0.287(386.5 - 282.3)}{1.4 - 1} = 868 \text{ kJ}$$

$\therefore$ Network done $= W_{1-2} + W_{2-3}$

$= -\,960.7 + 868 = -\,92.7$ kJ

Hence **network done on the air = 92.7 kJ. (Ans.)**

The process plotted on *T-s* diagram is shown in Fig. 3.81.

Example 3.69. *A rigid cylinder containing 0.004 m³ of nitrogen at 1 bar and 300 K is heated reversibly until temperature becomes 400 K. Determine :*

(i) The heat supplied.

(ii) The entropy change.

Assume nitrogen to be perfect gas (molecular mass = 28) and take γ = 1.4.

Solution. *Given :* $V_1 = 0.004 \text{ m}^3$; $p_1 = 1$ bar ; $T_1 = 300$ K ;

$T_2 = 400$ K ; M for $N_2 = 28$; $\gamma = 1.4$.

(*i*) **The heat supplied :**

$$\text{Gas constant } R = \frac{R_0 \text{ (Universal gas constant)}}{M \text{ (Molecular mass)}}$$

$$= \frac{8.314}{28} = 0.297 \text{ kJ/kg K}$$

$$\text{Mass, } m = \frac{p_1V_1}{RT_1} = \frac{(1 \times 10^5) \times 0.004}{(0.297 \times 1000) \times 300}$$

$= 0.00449$ kg

$$c_v = \frac{R}{\gamma - 1} = \frac{0.297}{1.4 - 1} = 0.742 \text{ kJ/kg K}$$

$\therefore$ Heat supplied $= m\, c_v(T_2 - T_1)$

$= 0.00449 \times 0.742(400 - 300)$

$=$ **0.333 kJ. (Ans.)**

(*ii*) **The entropy change :**

*The entropy change, $S_2 - S_1 = m\, c_v \log_e \left(\dfrac{T_2}{T_1}\right)$

$$= 0.00449 \times 0.742 \times \log_e \left(\frac{400}{300}\right)$$

$=$ **9.584 × 10⁻⁴ kJ/kg K. (Ans.)**

HIGHLIGHTS

1. *Internal energy* is the heat energy stored in a gas. The internal energy of a perfect gas is a function of *temperature* only.
2. First law of thermodynamics states :
   — Heat and work are mutually convertible but since energy can neither be created nor destroyed, the total energy associated with an energy conversion remains constant.

   Or

   — No machine can produce energy without corresponding expenditure of energy, *i.e.*, it is impossible to construct a perpetual motion machine of first kind.

   First law can be expressed as follows :

   $Q = \Delta E + W$

*Refer to Art. 4.17.

$Q = \Delta U + W$ ... if electric, magnetic, chemical energies are absent and changes in potential and kinetic energies are neglected.

3. There can be no machine which would continuously supply mechanical work without some form of energy disappearing simultaneously. Such a fictitious machine is called a perpetual motion machine of the first kind, or in brief, PMM1. A PMM1 is thus impossible.
4. The energy of an isolated system is always constant.
5. In case of
   (*i*) **Reversible constant volume process**
   (v = constant)
   $\Delta u = c_v(T_2 - T_1)$; $W = 0$; $Q = c_v (T_2 - T_1)$
   (*ii*) **Reversible constant pressure process**
   (p = constant)
   $\Delta u = c_v(T_2 - T_1)$; $W = p(v_2 - v_1)$; $Q = c_p (T_2 - T_1)$
   (*iii*) **Reversible temperature or isothermal process**
   (pv = constant)
   $$\Delta u = 0,\ W = p_1V_1 \log_e r,\ Q = W$$
   where r = expansion or compression ratio.
   (*iv*) **Reversible adiabatic process** (pv^γ = constant)
   $$\pm \Delta u = \mp W = \frac{R(T_1 - T_2)}{\gamma - 1} ; Q = 0 ;$$
   $$\frac{T_2}{T_1} = \left(\frac{v_1}{v_2}\right)^{\gamma - 1} = \left(\frac{p_2}{p_1}\right)^{\frac{\gamma - 1}{\gamma}}$$
   (*v*) **Polytropic reversible process** (pv^n = constant)
   $$\Delta u = c_v (T_2 - T_1) ; W = \frac{R(T_1 - T_2)}{n - 1} ; Q = \Delta u + W ;$$
   $$\text{and } \frac{T_2}{T_1} = \left(\frac{v_1}{v_2}\right)^{n - 1} = \left(\frac{p_2}{p_1}\right)^{\frac{n - 1}{n}} \text{ and } Q = \left(\frac{\gamma - n}{n - 1}\right) \times W.$$
6. Steady flow equation can be expressed as follows :
   $$u_1 + \frac{C_1^2}{2} + Z_1g + p_1v_1 + Q$$
   $$= u_2 + \frac{C_2^2}{2} + Z_2g + p_2v_2 + W \quad ...(i)$$
   or $h_1 + \dfrac{C_1^2}{2} + Q = h_2 + \dfrac{C_2^2}{2} + W$, neglecting Z_1 and Z_2 ...(*ii*)

   where, Q = Heat supplied per kg of fluid ;
   W = Work done by 1 kg of fluid ;
   C = Velocity of fluid ;
   Z = Height above datum ;
   p = Pressure of the fluid ;
   u = Internal energy per kg of fluid ;
   pv = Energy required per kg of fluid.

   This equation is applicable to any medium in any steady flow.
7. During adiabatic *throttling process* enthalpy remains constant. The slope of a constant enthalpy line on a p-T diagram is called Joule-Thomson co-efficient, μ.
8. In unsteady-flow processes, the rates at which mass and energy enter the control volume may not be the same as the rate of flow of mass and energy moving out of the control volume. The filling of a tank is an example of unsteady flow process.

OBJECTIVE TYPE QUESTIONS

Choose the Correct Answer :

1. If all the variables of a stream are independent of time it is said to be in
   (*a*) steady flow (*b*) unsteady flow
   (*c*) uniform flow (*d*) closed flow
   (*e*) constant flow.
2. A control volume refers to
   (*a*) a fixed region in space (*b*) a specified mass
   (*c*) an isolated system (*d*) a reversible process only
   (*e*) a closed system.
3. Internal energy of a perfect gas depends on
   (*a*) temperature, specific heats and pressure
   (*b*) temperature, specific heats and enthalpy
   (*c*) temperature, specific heats and entropy
   (*d*) temperature only.
4. In reversible polytropic process
   (*a*) true heat transfer occurs
   (*b*) the entropy remains constant
   (*c*) the enthalpy remains constant
   (*d*) the internal energy remains constant
   (*e*) the temperature remains constant.
5. An isentropic process is always
   (*a*) irreversible and adiabatic
   (*b*) reversible and isothermal
   (*c*) frictionless and irreversible
   (*d*) reversible and adiabatic
   (*e*) none of the above.
6. The network done per kg of gas in a polytropic process is equal to
   (*a*) $p_1v_1 \log_e \dfrac{v_2}{v_1}$ (*b*) $p_1 (v_1 - v_2)$
   (*c*) $p_2 \left(v_2 - \dfrac{v_1}{v_2}\right)$ (*d*) $\dfrac{p_1v_1 - p_2v_2}{n - 1}$
   (*e*) $\dfrac{p_2v_1 - p_2v_2}{n - 1}$.
7. Steady flow occurs when
   (*a*) conditions do not change with time at any point
   (*b*) conditions are the same at adjacent points at any instant
   (*c*) conditions change steadily with the time
   (*d*) $\left(\dfrac{\partial v}{\partial t}\right)$ is constant.

8. A reversible process requires that
(*a*) there be no heat transfer
(*b*) Newton's law of viscosity be satisfied
(*c*) temperature of system and surroundings be equal
(*d*) there be no viscous or coulomb friction in the system
(*e*) heat transfer occurs from surroundings to system only.

9. The first law of thermodynamics for steady flow
(*a*) accounts for all energy entering and leaving a control volume
(*b*) is an energy balance for a specified mass of fluid
(*c*) is an expression of the conservation of linear momentum
(*d*) is primarily concerned with heat transfer
(*e*) is restricted in its application to perfect gases.

10. The characteristic equation of gases $pV = mRT$ holds good for
(*a*) monoatomic gases (*b*) diatomic gases
(*c*) real gases (*d*) ideal gases
(*e*) mixture of gases.

11. A gas which obeys kinetic theory perfectly is known as
(*a*) monoatomic gas (*b*) diatomic gas
(*c*) real gas (*d*) pure gas
(*e*) perfect gas.

12. Work done in a free expansion process is
(*a*) zero (*b*) minimum
(*c*) maximum (*d*) positive
(*e*) negative.

13. Which of the following is not a property of the system ?
(*a*) Temperature (*b*) Pressure
(*c*) Specific volume (*d*) Heat
(*e*) None of the above.

14. In the polytropic process equation pv^n = constant, if $n = 0$, the process is termed as
(*a*) constant volume (*b*) constant pressure
(*c*) constant temperature (*d*) adiabatic
(*e*) isothermal.

15. In the polytropic process equation pv^n = constant, if n is infinitely large, the process is termed as
(*a*) constant volume (*b*) constant pressure
(*c*) constant temperature (*d*) adiabatic
(*e*) isothermal.

16. The processes or systems that do not involve heat are called
(*a*) isothermal processes (*b*) equilibrium processes
(*c*) thermal processes (*d*) steady processes
(*e*) adiabatic processes.

17. In a reversible adiabatic process the ratio (T_1/T_2) is equal to
(*a*) $\left(\frac{p_1}{p_2}\right)^{\frac{\gamma-1}{\gamma}}$ (*b*) $\left(\frac{v_1}{v_2}\right)^{\frac{\gamma-1}{\gamma}}$
(*c*) $(v_1 v_2)^{\frac{\gamma-1}{2\gamma}}$ (*d*) $\left(\frac{v_2}{v_1}\right)^{\gamma}$.

18. In isothermal process
(*a*) temperature increases gradually
(*b*) volume remains constant
(*c*) pressure remains constant
(*d*) enthalpy change is maximum
(*e*) change in internal energy is zero.

19. During throttling process
(*a*) internal energy does not change
(*b*) pressure does not change
(*c*) entropy does not change
(*d*) enthalpy does not change
(*e*) volume change is negligible.

20. When a gas is to be stored, the type of compression that would be ideal is
(*a*) isothermal (*b*) adiabatic
(*c*) polytropic (*d*) constant volume
(*e*) none of the above.

21. If a process can be stopped at any stage and reversed so that the system and surroundings are exactly restored to their initial states, it is known as
(*a*) adiabatic process (*b*) isothermal process
(*c*) ideal process (*d*) frictionless process
(*e*) energyless process.

22. The state of a substance whose evaporation from its liquid state is complete, is known as
(*a*) vapour (*b*) perfect gas
(*c*) air (*d*) steam.

23. In SI units, the value of the universal gas constant is
(*a*) 0.8314 J/mole/K (*b*) 8.314 J/mole/K
(*c*) 83.14 J/mole/K (*d*) 831.4 J/mole/K
(*e*) 8314 J/mole/K.

24. When the gas is heated at constant pressure, the heat supplied
(*a*) increases the internal energy of the gas
(*b*) increases the temperature of the gas
(*c*) does some external work during expansion
(*d*) both (*b*) and (*c*)
(*e*) none of the above.

25. The gas constant (R) is equal to the
(*a*) sum of two specific heats
(*b*) difference of two specific heats
(*c*) product of two specific heats
(*d*) ratio of two specific heats.

26. The heat absorbed or rejected during a polytropic process is
(*a*) $\left(\frac{\gamma-n}{\gamma-1}\right)$ × work done (*b*) $\left(\frac{\gamma-n}{\gamma-1}\right)^2$ × work done
(*c*) $\left(\frac{\gamma-n}{\gamma-1}\right)^{1/2}$ × work done (*d*) $\left(\frac{\gamma-n}{\gamma-1}\right)^3$ × work done.

ANSWERS

1. (*a*)	**2.** (*a*)	**3.** (*d*)	**4.** (*a*)
5. (*d*)	**6.** (*d*)	**7.** (*a*)	**8.** (*d*)
9. (*a*)	**10.** (*c*)	**11.** (*e*)	**12.** (*a*)
13. (*d*)	**14.** (*b*)	**15.** (*a*)	**16.** (*e*)

17. (*a*) **18.** (*e*) **19.** (*d*) **20.** (*a*)
21. (*c*) **22.** (*b*) **23.** (*e*) **24.** (*d*)
25. (*b*) **26.** (*a*).

THEORETICAL QUESTIONS

1. Define 'internal energy' and prove that it is a property of a system.
2. Explain the First Law of Thermodynamics as referred to closed systems undergoing a cyclic change.
3. State the First Law of Thermodynamics and prove that for a non-flow process, it leads to the energy equation $Q = \Delta U + W$.
4. What is the mechanical equivalent of heat ? Write down its value when heat is expressed in kJ and work is expressed in Nm.
5. What do you mean by "Perpetual motion machine of first kind-PMM 1" ?
6. Why only in constant pressure non-flow process, the enthalpy change is equal to heat transfer ?
7. Prove that the rate of change of heat interchange per unit change of volume when gas is compressed or expanded is given by $\frac{\gamma - n}{\gamma - 1} \times \frac{pdv}{J}$.
8. Write down the general energy equation for steady flow system and simplify when applied for the following systems :
   (*i*) Centrifugal water pump
   (*ii*) Reciprocating air compressor
   (*iii*) Steam nozzle (*iv*) Steam turbine
   (*v*) Gas turbine.
9. Explain clearly the difference between a non-flow and a steady flow process.
10. For isothermal flow and non-flow steady processes, prove that

$$\int_1^2 pdv = -\int_1^2 v \, . dp$$

Also state the assumptions made.

UNSOLVED EXAMPLES

Closed Systems

1. In a cyclic process, heat transfers are + 14.7 kJ, – 25.2 kJ, – 3.56 kJ and + 31.5 kJ. What is the network for this cyclic process ? **[Ans.** 17.34 kJ]
2. A domestic refrigerator is loaded with food and the door closed. During a certain period the machine consumes 1 kWh of energy and the internal energy of the system drops by 5000 kJ. Find the net heat transferred in the system. **[Ans.** – 8.6 MJ]
3. 1.5 kg of liquid having a constant specific heat of 2.5 kJ/kg°C is stirred in a *well-insulated* chamber causing the temperature to rise by 15°C. Find :
   (*i*) Change in internal energy, and
   (*ii*) Work done for the process.
   [Ans. (*i*) 56.25 kJ, (*ii*) $W = -56.25$ kJ]
4. A system is composed of a stone having a mass of 10 kg and a bucket containing 100 kg of water. Initially the stone and water are at the same temperature, the stone then falls into the water. Determine ΔU, ΔKE, ΔPE, ΔQ and ΔW for the following cases :
   (*i*) At the instant the stone is about to enter the water.
   (*ii*) Just after the stone comes to rest in the bucket.
   [Ans. (*i*) $\Delta Q = \Delta W = \Delta E = 0$, $\Delta KE = 4.184$ kJ, $\Delta PE = -4.184$ kJ ; (*ii*) $\Delta Q = 0$, $\Delta W = 0$, $\Delta KE = 0$, $\Delta U = +4.184$ kJ, $\Delta PE = -4.184$ kJ]
5. A closed system of constant volume experiences a temperature rise of 20°C when a certain process occurs. The heat transferred in the process is 18 kJ. The specific heat at constant volume for the pure substance comprising the system is 1.2 kJ/kg°C, and the system contains 2 kg of this substance. Determine the change in internal energy and the work done. **[Ans.** $\Delta U = 48$ kJ ; $W = -30$ kJ]
6. A stationary mass of gas is compressed without friction from an initial state of 2 m^3 and 2×10^5 N/m^2 to a final state of 1 m^3 and 2×10^5 N/m^2, the pressure remaining the same. There is a transfer of 360 kJ of heat from the gas during the process. How much does the internal energy of the gas change ? **[Ans.** $\Delta U = -160$ kJ]
7. The internal energy of a certain substance is given by the following equation :

$$u = pv + 84$$

where u is given in kJ/kg, p is in kPa and v is in m^3/kg.
A system composed of 3 kg of this substance expands from an initial pressure of 500 kPa and a volume of 0.22 m^3 to a final pressure 100 kPa in a process in which pressure and volume are related by $pv^{1.2}$ = constant.
   (*i*) If the expansion is quasi-static, find Q, ΔU and W for the process.
   (*ii*) In another process the same system expands according to the same pressure-volume relationship as in part (*i*) and from the same initial state to the same final state as in part (*i*) but the heat transfer in this case is 30 kJ. Find the work transfer for this process.
   (*iii*) Explain the difference in work transfer in parts (*i*) and (*ii*).
   [Ans. (*i*) $\Delta U = -91$ kJ, $W = 127.5$ kJ, $Q = 36.5$ kJ (*ii*) $W = 121$ kJ, (*iii*) The work in (*ii*) is not equal to $\int pdV$ since the process is not quasi-static.]
8. A fluid is contained in a cylinder by a spring-loaded, frictionless piston so that the pressure in the fluid is linear function of the volume ($p = a + bV$). The internal energy of the fluid is given by the following equation:

$$U = 34 + 3.15\, pV$$

where U is in kJ, p in kPa and V in cubic metre. If the fluid changes from an initial state of 170 kPa, 0.03 m^3 to a final state of 400 kPa, 0.06 m^3, with no work other than that done on the piston, find the direction and magnitude of the work and heat transfer.

[**Ans.** $W_{1-2} = 10.35$ kJ ; $Q_{1-2} = 69.85$ kJ (heat flows into the system during the process)]

9. A piston cylinder arrangement has a gas in the cylinder space. During a constant pressure expansion to a larger volume the work effect for the gas are 1.6 kJ, the heat added to the gas and cylinder arrangement is 3.2 kJ and the friction between the piston and cylinder wall amounts to 0.24 kJ. Determine the change in internal energy of the entire apparatus (gas, cylinder, piston).

[**Ans.** 1.84 kJ]

10. A system receives 42 kJ of heat while expanding with volume change of 0.123 m^3 against an atmosphere of 12 N/cm^2. A mass of 80 kg in the surroundings is also lifted through a distance of 6 metres.

(*i*) Find the change in energy of the system.

(*ii*) The system is returned to its initial volume by an adiabatic process which requires 100 kJ of work. Find the change in energy of system.

(*iii*) Determine the total change in energy of the system.

[**Ans.** (*i*) 22.54 kJ, (*ii*) 100 kJ, (*iii*) 122.54 kJ]

11. A thermally insulated battery is being discharged at atmospheric pressure and constant volume. During a 1 hour test it is found that a current of 50 A and 2 V flows while the temperature increases from 20°C to 32.5°C. Find the change in internal energy of the cell during the period of operation. [**Ans.** – 36 × 10^4 J]

12. In a certain steam plant the turbine develops 1000 kW. The heat supplied to the steam in the boiler is 2800 kJ/kg, the heat received by the system from cooling water in the condenser is 2100 kJ/kg and the feed pump work required to pump the condensate back into the boiler is 5 kW. Calculate the steam flow round the cycle in kg/s.

[**Ans.** 1.421 kg/s]

13. In the compression stroke of an internal-combustion engine the heat rejected to the cooling water is 45 kJ/kg and the work input is 90 kJ/kg. Calculate the change in internal energy of the working fluid stating whether it is a gain or a loss. [**Ans.** 45 kJ/kg (gain)]

14. 85 kJ of heat are supplied to a system at constant volume. The system rejects 90 kJ of heat at constant pressure and 20 kJ of work is done on it. The system is brought to its original state by adiabatic process. Determine the adiabatic work. Determine also the value of internal energy at all end states if initial value is 100 kJ.

[**Ans.** $W = 15$ kJ ; $U_1 = 100$ kJ, $U_2 = 185$ kJ ; $U_3 = 115$ kJ]

15. A closed system undergoes a reversible process at a constant pressure process of 3.5 bar and its volume changes from 0.15 m^3 to 0.06 m^3. 25 kJ of heat is rejected by the system during the process. Determine the change in internal energy of the system.

[**Ans.** 6.5 kJ (increase)]

16. An air compressor takes in air at 10^5 Pa and 27°C having volume of 1.5 m^3/kg and compresses it to 4.5 × 10^5 Pa. Find the work done, heat transfer and change in internal energy if the compression is isothermal.

[**Ans.** – 225 kJ ; – 225 kJ ; $\Delta U = 0$]

17. A cylinder fitted with piston contains 0.2 kg of N_2 at 100 kPa and 30°C. The piston is moved compressing N_2 until the pressure becomes 1 MPa and temperature becomes 150°C. The work done during the process is 20 kJ. Determine the heat transferred from N_2 to the surroundings. Take $c_v = 0.75$ kJ/kg K for N_2. [**Ans.** – 2 kJ]

18. A closed system consisting of 1 kg of gaseous CO_2 undergoes a reversible process at constant pressure causing a decrease of 30 kJ in internal energy. Determine the work done during the process. Take c_p = 840 J/kg°C and c_v = 600 J/kg°C. [**Ans.** – 12 kJ]

19. The specific heat at constant pressure of one kg fluid undergoing a non-flow constant pressure process is given by

$$c_p = \left[2.5 + \frac{40}{T + 20}\right] \text{ kg/kg°C}$$

where T is in °C.

The pressure during the process is maintained at 2 bar and volume changes from 1 m^3 to 1.8 m^3 and temperature changes from 50°C to 450°C. Determine :

(*i*) Heat added (*ii*) Work done

(*iii*) Change in internal energy

(*iv*) Change in enthalpy.

[**Ans.** (*i*) 1076 kJ ; (*ii*) 160 kJ ; (*iii*) 916 kJ ; (*iv*) 1076 kJ]

20. 1 kg of nitrogen (molecular weight 28) is compressed reversibly and isothermally from 1.01 bar, 20°C to 4.2 bar. Calculate the work done and the heat flow during the process. Assume nitrogen to be a perfect gas.

[**Ans.** $W = 124$ kJ/kg ; $Q = -124$ kJ/kg]

21. Air at 1.02 bar, 22°C, initially occupying a cylinder volume of 0.015 m^3, is compressed reversibly and adiabatically by a piston to a pressure of 6.8 bar. Calculate :

(*i*) The final temperature (*ii*) The final volume

(*iii*) The work done on the mass of air in the cylinder.

[**Ans.** (*i*) 234.5°C; (*ii*) 0.00388 m^3 ; (*iii*) 2.76 kJ]

22. 1 kg of a perfect gas is compressed from 1.1 bar, 27°C according to a law $pv^{1.3}$ = constant, until the pressure is 6.6. bar. Calculate the heat flow to or from the cylinder walls,

(*i*) When the gas is ethane (molecular weight 30), which has

$$c_p = 1.75 \text{ kJ/kg K.}$$

(*ii*) When the gas is argon (molecular weight 40), which has

$$c_p = 0.515 \text{ kJ/kg K.}$$

[**Ans.** (*i*) 84.5 kJ/kg, (*ii*) –59.4 kJ/kg]

23. 1 kg of air at 1 bar, 15°C is compressed reversibly and adiabatically to a pressure of 4 bar. Calculate the final temperature and the work done on the air.

[**Ans.** 155°C ; 100.5 kJ/kg]

24. A certain perfect gas is compressed reversibly from 1 bar, 17°C to a pressure of 5 bar in a perfectly thermally insulated cylinder, the final temperature being 77°C. The work done on the gas during the compression is 45 kJ/kg. Calculate γ, c_v, R and the molecular weight of the gas.
[**Ans.** 1.132 ; 0.75 kJ/kg K ; 0.099 kJ/kg K ; 84]

25. 1 kg of air at 1.02 bar, 20°C is compressed reversibly according to a law $pv^{1.3}$ = constant, to a pressure of 5.5 bar. Calculate the work done on the air and heat flow to or from the cylinder walls during the compression.
[**Ans.** 133.5 kJ/kg ; – 33.38 kJ/kg]

26. 0.05 kg of carbon dioxide (molecular weight 44), occupying a volume of 0.03 m^3 at 1.025 bar, is compressed reversibly until the pressure is 6.15 bar. Calculate final temperature, the work done on the CO_2, the heat flow to or from the cylinder walls,
(*i*) When the process is according to law $pv^{1.4}$ = constant,
(*ii*) When the process is isothermal,
(*iii*) When the process takes place in a perfectly thermally insulated cylinder.
Assume CO_2 to be a perfect gas, and take γ = 1.3.
[**Ans.** 270°C ; 5.138 kJ ; 1.713 kJ ; 52.6°C ; 5.51 kJ ; – 5.51 kJ ; 219°C ; 5.25 kJ ; 0 kJ]

27. Oxygen (molecular weight 32) is compressed reversibly and polytropically in a cylinder from 1.05 bar, 15°C to 4.2 bar in such a way that one-third of the work input is rejected as heat to the cylinder walls. Calculate the final temperature of the oxygen.
Assume oxygen to be a perfect gas and take c_v = 0.649 kJ/kg K. [**Ans.** 113°C]

28. A cylinder contains 0.5 m^3 of a gas at 1 × 10^5 N/m^2 and 90°C. The gas is compressed to a volume of 0.125 m^3, the final pressure being 6 × 10^5 N/m^2. Determine :
(*i*) The mass of gas.
(*ii*) The value of index '*n*' for compression.
(*iii*) The increase in internal energy of gas.
(*iv*) The heat received or rejected by the gas during compression.
(γ = 1.4, R = 294.2 Nm/kg°C).
[**Ans.** 0.468 kg ; 1.292 ; 62.7 kJ ; –22.67 kJ]

Steady Flow Systems

29. 12 kg of a fluid per minute goes through a reversible steady flow process. The properties of fluid at the inlet are p_1 = 1.4 bar, ρ_1 = 25 kg/m^3, C_1 = 120 m/s and u_1 = 920 kJ/kg and at the exit are p_2 = 5.6 bar, ρ_2 = 5 kg/m^3, C_2 = 180 m/s and u_2 = 720 kJ/kg. During the passage, the fluid rejects 60 kJ/s and rises through 60 metres. Determine : (*i*) the change in enthalpy (Δh) and (*ii*) work done during the process (W).
[**Ans.** (*i*) Δh = – 93.6 kJ/kg ; (*ii*) W = – 44.2 kW]

30. In the turbine of a gas turbine unit the gases flow through the turbine is 17 kg/s and the power developed by the turbine is 14000 kW. The enthalpies of the gases at inlet and outlet are 1200 kJ/kg and 360 kJ/kg respectively, and the velocities of the gases at inlet and outlet are 60 m/s and 150 m/s respectively. Calculate the rate at which the heat is rejected from the turbine. Find also the area of the inlet pipe given that the specific volume of the gases at inlet is 0.5 m^3/kg.
[**Ans.** 119.3 kW (heat rejected) ; 0.142 m^3]

31. Air flows steadily at the rate of 0.4 kg/s through an air compressor, entering at 6 m/s with a pressure of 1 bar and a specific volume of 0.85 m^3/kg, and leaving at 4.5 m/s with a pressure of 6.9 bar and a specific volume of 0.16 m^3/kg. The internal energy of air leaving is 88 kJ/kg greater than that of the air entering. Cooling water in a jacket surrounding the cylinder absorbs heat from the air at the rate of 59 kJ/s. Calculate the power required to drive the compressor and the inlet and outlet pipe cross-sectional areas. [**Ans.** 104.4 kW ; 0.057 m^2 ; 0.014 m^2]

32. A turbine operating under steady flow conditions receives steam at the following state : pressure 13.8 bar ; specific volume 0.143 m^3/kg ; internal energy 2590 kJ/kg ; velocity 30 m/s. The state of the steam leaving the turbine is : pressure 0.35 bar ; specific volume 4.37 m^3/kg ; internal energy 2360 kJ/kg ; velocity 90 m/s. Heat is lost to the surroundings at the rate of 0.25 kJ/s. If the rate of steam flow is 0.38 kg/s, what is the power developed by the turbine ?
[**Ans.** 102.8 kW]

33. A nozzle is a device for increasing the velocity of a steadily flowing stream of fluid. At the inlet to a certain nozzle the enthalpy of the fluid is 3025 kJ/kg and the velocity is 60 m/s. At the exit from the nozzle the enthalpy is 2790 kJ/kg. The nozzle is horizontal and there is negligible heat loss from it.
(*i*) Find the velocity at the nozzle exit.
(*ii*) If the inlet area is 0.1 m^2 and specific volume at inlet is 0.19 m^3/kg, find the rate of flow of fluid.
(*iii*) If the specific volume at the nozzle exit is 0.5 m^3/kg, find the exit area of the nozzle.
[**Ans.** (*i*) 688 m/s ; (*ii*) 31.6 kg/s ; (*iii*) 0.0229 m^2]

34. A gas flows steadily through a rotary compressor. The gas enters the compressor at a temperature of 16°C, a pressure of 100 kPa, and an enthalpy of 391.2 kJ/kg. The gas leaves the compressor at a temperature of 245°C, a pressure of 0.6 MPa and an enthalpy of 534.5 kJ/kg. There is no heat transfer to or from the gas as it flows through the compressor.
(*i*) Evaluate the external work done per unit mass of gas assuming the gas velocities at entry and exit to be negligible.
(*ii*) Evaluate the external work done per unit mass of gas when the gas velocity at entry is 80 m/s and that at exit is 160 m/s.
[**Ans.** (*i*) 143.3 kJ/kg; (*ii*) 152.9 kJ/kg]

35. A turbine, operating under steady-flow conditions, receives 5000 kg of steam per hour. The steam enters the turbine at a velocity of 3000 m/min, an elevation of 5 m and a specific enthalpy of 2787 kJ/kg. It leaves the turbine at a velocity of 6000 m/min, an elevation of 1 m and a specific enthalpy of 2259 kJ/kg. Heat losses from the turbine to the surroundings amount to 16736 kJ/h.
Determine the power output of the turbine.
[**Ans.** 723 kW]

36. In a steady flow process, the working fluid flows at a rate of 240 kg/min. The fluid rejects 120 kJ/s passing through

the system. The conditions of fluid at inlet and outlet are given as : $C_1 = 300$ m/s, $p_1 = 6.2$ bar, $u_1 = 2100$ kJ/kg, $v_1 = 0.37$ m^3/kg and $C_2 = 150$ m/s, $p_2 = 1.3$ bar, $u_2 = 1500$ kJ/kg, $v_2 = 1.2$ m^3/kg. The suffix 1 indicates the conditions at inlet and 2 indicates at outlet of the system. Neglecting the change in potential energy, determine the power capacity of the system in MW. [**Ans.** 2.7086 MW]

37. Steam enters a turbine at 20 m/s and specific enthalpy of 3000 kJ/kg and leaves the turbine at 40 m/s and specific enthalpy of 2500 kJ/kg. Heat lost to the surroundings is 25 kJ/kg of steam as the steam passes through the turbine. If the steam flow rate is 360000 kg/h, determine the output from the turbine in MW. [**Ans.** 47.44 MW]

38. A stream of gases at 7.5 bar, 800°C and 150 m/s is passed through a turbine of a jet engine. The stream comes out of the turbine at 2.0 bar, 600°C and 300 m/s. The process may be assumed adiabatic. The enthalpies of gas at the entry and exit of the turbine are 960 kJ/kg and 700 kJ/kg gas respectively.

Determine the capacity of the turbine if the gas flow is 4 kg/s. [**Ans.** 905 kW]

39. In a steam power plant 1.5 kg of water is supplied per second to the boiler. The enthalpy and velocity of water entering into the boiler are 800 kJ/kg and 10 m/s. Heat at the rate of 2200 kJ per kg of water is supplied to the water. The steam after passing through the turbine comes out with a velocity of 50 m/s and enthalpy of 2520 kJ/kg. The boiler inlet is 5 m above the turbine exit. The heat loss from the boiler is 1800 kJ/min and from the turbine 600 kJ/min.

Determine the power capacity of the turbine, considering boiler and turbine as single unit. [**Ans.** 678 kW]

40. 15 kg of air per minute is delivered by a centrifugal compressor. The inlet and outlet conditions of air are : $C_1 = 10$ m/s, $p_1 = 1$ bar, $v_1 = 0.5$ m^3/kg and $C_2 = 80$ m/s, $p_2 = 7$ bar, $v_2 = 0.15$ m^3/kg. The increase in enthalpy of air passing through the compressor is 160 kJ/kg, and heat loss to the surroundings is 720 kJ/min. Assuming that inlet and discharge lines are at the same level, find :

(*i*) Motor power required to drive the compressor.

(*ii*) Ratio of inlet to outlet pipe diameter.

$$\left[\textbf{Ans. } (i)\ 52.78 \text{ kW} ; (ii)\ \frac{d_1}{d_2} = 5.16\right]$$

41. A centrifugal air compressor used in gas turbine receives air at 100 kPa and 300 K and it discharges air at 400 kPa and 500 K. The velocity of air leaving the compressor is 100 m/s. Neglecting the velocity at the entry of the compressor, determine the power required to drive the compressor if the mass flow rate is 15 kg/s.

Take c_p (air) = 1 kJ/kg K and assume that there is no heat transfer from the compressor to the surroundings. [**Ans.** 3075 kW]

42. In a water cooled compressor 0.5 kg of air is compressed per second. A shaft input of 60 kW is required to run the compressor. Heat lost to the cooling water is 30 per cent of input and 10 per cent of the input is lost in bearings and other frictional effects. Air enters the compressor at 1 bar and 20°C. Neglecting the changes in kinetic energy and potential energy, determine the exit air temperature. Take c_p = 1 kJ/kg°C air.

Consider steady flow process. [**Ans.** 92°C]

43. Steam at 7 bar and 200°C enters an insulated convergent divergent nozzle with a velocity of 60 m/s. It leaves the nozzle at a pressure of 1.4 bar and enthalpy of 2600 kJ/kg. Determine the velocity of the steam at exit. [**Ans.** 701 m/s]

44. A petrol engine develops 50 kW brake power. The fuel and air-flow rates are 10 kg and 107 kg/h. The temperature of fuel-air mixture entering the engine is 20°C and temperature of gases leaving the engine is 500°C. The heat transfer rate from the engine to the jacket cooling water is 50 kJ/s and that to the surroundings is 10 kJ/s.

Evaluate the increase in the specific enthalpy of the mixture as it flows through the engine. [**Ans.** – 110 kJ/s]

45. A compressor takes air at 100 kN/m^2 and delivers the same at 550 kN/m^2. The compressor discharges 16 m^3 of free air per minute. The densities of air at inlet and exit are 1.25 kg/m^3 and 5 kg/m^3. The power of the motor driving the compressor is 40 kW. The heat lost to the cooling water circulated around the compressor is 30 kJ/kg of air passing through the compressor.

Neglecting changes in P.E. and K.E. determine the change in specific internal energy. [**Ans.** 60 kJ/kg]

46. A centrifugal pump operating under steady flow conditions delivers 3000 kg of water per minute at 20°C. The suction pressure is 0.8 bar and delivery pressure is 3 bar. The suction pipe diameter is 15 cm and discharge pipe diameter is 10 cm. Find the capacity of the drive motor.

Neglect the change in internal energy and assume that the suction and discharge are at same level. [**Ans.** 11.8 kW]

47. 60 kg of water is delivered by a centrifugal pump per second. The inlet and outlet pressures are 1 bar and 4 bar respectively. The suction is 2 m below the centre of the pump and delivery is 8 m above the centre of the pump. Determine the capacity of the electric motor to run the pump. The suction and delivery pipe diameters are 20 cm and 10 cm and respectively. [**Ans.** 27.15 kW]

48. The air speed of a turbojet engine in flight is 270 m/s. Ambient air temperature is –15°C. Gas temperature outlet of the nozzle is 600°C. Corresponding enthalpy values for air and gas are respectively 260 and 912 kJ/kg. Fuel air ratio is 0.0190. Chemical energy of the fuel is 44.5 MJ/kg. Owing to incomplete combustion 5% of the chemical energy is not released in the reaction. Heat loss from the engine is 21 kJ/kg of air.

Calculate the velocity of exhaust jet. [**Ans.** 560 m/s]

49. Air at a temperature of 15°C passes through a heat exchanger at a velocity of 30 m/s, where its temperature is raised to 800°C. It then enters a turbine with the same velocity of 30 m/s and expands until the temperature falls to 650°C. On leaving the turbine, the air is taken at a velocity of 60 m/s to a nozzle where it expands until the temperature has fallen to 500°C. If the air flow rate is 2 kg/s, calculate (*i*) the rate of heat transfer to the air, (*ii*) the power output from the turbine assuming no heat loss, and (*iii*) the velocity at exit from nozzle, assuming no heat loss.

Take the enthalpy of air as $h = c_p t$, where c_p is the specific heat equal to 1.005 kJ/kg°C and t the temperature.

[**Ans.** (*i*) 1580 kJ/s ; (*ii*) 298.8 kW ; (*iii*) 554 m/s]

Vapour (Steam)

50. 0.05 kg of steam is heated at a constant pressure of 2 bar until the volume occupied is 0.0658 m^3. Calculate the heat supplied and work done. [**Ans.** 18.25 kJ ; 4.304 kJ]

51. Steam at 7 bar and dryness fraction 0.9 expands in a cylinder behind a piston isothermally and reversibly to a pressure of 1.5 bar. Calculate the change of internal energy and the change of enthalpy per kg of steam. The heat supplied during the process is found to be 400 kJ/kg. Calculate the work done per kg of steam.

[**Ans.** 217.5 kJ/kg (gain) ; 245.7 kJ/kg ; 182.5 kJ/kg]

52. 1 kg of steam at 100 bar and 375°C expands reversibly in a perfectly thermally insulated cylinder behind a piston until pressure is 38 bar and the steam is then saturated. Calculate the work done by the steam.

[**Ans.** 169.7 kJ/kg]

53. In a steam engine the steam at the beginning of the expansion process is at 7 bar, dryness fraction 0.95, and the expansion follows the law $pv^{1.1}$ = constant, down to a pressure of 0.34 bar. Calculate the work done per kg of steam during the expansion, and the heat flow per kg of steam to or from the cylinder walls during the expansion.

[**Ans.** 436 kJ/kg ; 155.6 kJ/kg (heat supplied)]

54. Steam at 19 bar is throttled to 1 bar and the temperature after throttling is found to be 150°C. Calculate the initial dryness fraction of the steam. [**Ans.** 0.989]

55. 1 kg of steam at 7 bar, entropy 6.5 kJ/kg K, is heated reversibly at constant pressure until the temperature is 250°C. Calculate the heat supplied, and show on a *T-s* diagram the area which represents the heat flow.

[**Ans.** 283 kJ/kg]

56. 1 kg of steam at 20 bar, dryness fraction 0.9, is heated reversibly at constant pressure to a temperature of 300°C. Calculate the heat supplied and change of entropy and show the process on a *T-s* diagram, indicating the area which represents the heat flow.

[**Ans.** 415 kJ/kg ; 0.8173 kJ/kg K]

57. Steam at 0.05 bar, 100°C is to be condensed completely by a reversible constant pressure process.

Calculate the heat to be removed per kg of steam and the change of entropy. Sketch the process on a *T-s* diagram and shade in the area which represents the heat flow.

[**Ans.** 2550 kJ/kg ; 8.292 kJ/kg K]

58. 0.05 kg of steam at 10 bar, dryness fraction 0.84, is heated reversibly in a rigid vessel until the pressure is 20 bar. Calculate the change of entropy and the heat supplied. Show the area which represents the heat supplied on a *T-s* diagram. [**Ans.** 0.0704 kJ/kg K ; 36.85 kJ]

59. 1 kg of steam undergoes a reversible isothermal process from 20 bar and 250°C to a pressure of 30 bar. Calculate the heat flow, stating whether it is supplied or rejected and sketch the process on a *T-s* diagram.

[**Ans.** –135 kJ/kg]

60. Steam at 5 bar, 250°C, expands isentropically to a pressure of 0.7 bar. Calculate the final condition of steam.

[**Ans.** 0.967]

61. Steam expands reversibly in a cylinder behind a piston from 6 bar dry saturated, to a pressure of 0.65 bar. Assuming that the cylinder is perfectly thermally insulated, calculate the work done during the expansion per kg of steam. Sketch the process on a *T-s* diagram.

[**Ans.** 323.8 kJ/kg]

62. A steam engine receives steam at 4 bar, dryness fraction 0.8, and expands it according to a law $pv^{1.05}$ = constant to a condenser pressure of 1 bar. Calculate the change of entropy per kg of steam during the expansion, and sketch the process on a *T-s* diagram. [**Ans.** 0.381 kJ/kg K]

63. Steam at 15 bar is throttled to 1 bar and a temperature of 150°C. Calculate the initial dryness fraction and the change of entropy. Sketch the process on a *T-s* diagram and state the assumptions made in the throttling process.

[**Ans.** 0.992, 1.202 kJ/kg K]

64. Steam enters a turbine at 70 bar, 500°C and leaves at 2 bar in a dry saturated state. Calculate the isentropic efficiency and effectiveness of the process. Neglect changes of kinetic and potential energy and assume that the process is adiabatic.

The atmospheric temperature is 17°C. [**Ans.** 84.4% ; 88%]

65. Steam at 10 bar and 250°C expands until the pressure becomes 2.75 bar. The dryness fraction of the steam at the end of expansion is 0.95. Determine the change in internal energy. [**Ans.** – 273 kJ/kg]

66. Calculate the quantity of heat required to form 2.5 kg of dry steam at 11 bar from water at 20°C. Also determine the amount of heat removed at constant pressure to cause the steam to become 0.95 dry. Calculate the specific volume at the respective conditions.

[**Ans.** 6740 kJ ; 250 kJ ; 0.1775 m^3/kg ; 0.167 m^3/kg]

67. Steam at 10 bar and 0.95 dryness is available. Determine the final condition of steam in each of the following operations :

(*i*) 160 kJ of heat is removed at constant pressure ;

(*ii*) It is cooled at constant volume till the temperature inside falls to 140°C.

(*iii*) Steam expands isentropically in a steam turbine developing 300 kJ of work per kg of steam when the exit pressure of steam is 0.5 bar.

[**Ans.** (*i*) 0.874 ; (*ii*) 0.367 ; (*iii*) 0.882]

68. Calculate the internal energy of 0.3 m^3 of steam at 4 bar and 0.95 dryness. If this steam is superheated at constant pressure through 30°C, determine the heat added and change in internal energy.

[**Ans.** 2451 kJ/kg ; 119 kJ ; 107.5 kJ/kg]

69. 1 kg of water at 30°C and 1 bar is heated at constant pressure until it becomes saturated vapour. Determine the change in volume, and internal energy during the process. [**Ans.** 1.694 m^3/kg (app.) ; 2380.6 kJ/kg]

70. Water is supplied to the boiler at 15 bar and 80°C and steam is generated at the same pressure at 0.9 dryness. Determine the heat supplied to the steam in passing through the boiler and change in entropy.
[**Ans.** 2260.5 kJ/kg ; 4.92 kJ/kg K]

71. A cylindrical vessel of 5 m^3 capacity contains wet steam at 1 bar. The volume of vapour and liquid in the vessel are 4.95 m^3 and 0.05 m^3 respectively. Heat is transferred to the vessel until the vessel is filled with saturated vapour. Determine the heat transfer during the process.
[**Ans.** 104.93 MJ]

72. A closed vessel of 0.5 m^3 capacity contains dry saturated steam at 3.5 bar. The vessel is cooled until the pressure is reduced to 2 bar. Calculate :
(*i*) The mass of steam in the vessel,
(*ii*) Final dryness fraction of the steam, and
(*iii*) The amount of heat transferred during the process.
[**Ans.** (*i*) 0.955 kg ; (*ii*) 0.582 ; (*iii*) – 828 kJ]

73. A closed vessel of 0.3 m^3 capacity contains steam at 8 bar and 200°C ;
(*i*) Determine the mass of the steam in the vessel.
(*ii*) The vessel is cooled till the steam becomes just dry and saturated. What will be the pressure of the steam in the vessel at this stage ?
(*iii*) The vessel is further cooled till the temperature drops to 158.85°C. Determine the pressure and condition of the steam.
[**Ans.** (*i*) 1.2 kg ; (*ii*) 7.362 bar ; (*iii*) 6 bar, 0.826]

74. 0.5 kg of steam at 4 bar is contained in a cylinder fitted with a piston. The initial volume of steam is 0.1 m^3. Heat is transferred to the steam at constant pressure until the temperature becomes 300°C. Determine the heat transfer and work done during the process. [**Ans.** 771 kJ ; 91 kJ]

75. A quantity of steam at 13 bar and 0.8 dryness occupies 0.1 m^3. Determine the heat supplied to raise the temperature of the steam to 250°C at constant pressure and percentage of this heat which appears as external work. Take specific heat for superheated steam as 2.2 kJ/kg K. [**Ans.** 423 kJ/kg ; 15.3%]

76. A certain quantity of dry and saturated steam at 1.5 bar occupies initially a volume of 2.32 m^3. It is compressed until the volume is halved :
(*i*) Isothermally,
(*ii*) As per the law pv = constant, determine the final condition of steam in each case.

Also determine the heat rejected during the isothermal compression process. [**Ans.** (*i*) 0.5, 2226.5 kJ ; (*ii*) 0.956]

77. Steam enters a turbine at a pressure of 10 bar and 300°C with a velocity of 50 m/s. The steam leaves the turbine at 1.5 bar and with a velocity of 200 m/s. Assuming the process to be reversible adiabatic and neglecting the change in potential energy, determine the work done per kg of steam flow through the turbine.
[**Ans.** 375.55 kJ/kg]

78. Steam at 10 bar and 300°C passing through a convergent divergent nozzle expands reversibly and adiabatically till the pressure falls to 2 bar. If the velocity of steam entering into the nozzle is 50 m/s, determine the exit velocity of the steam. [**Ans.** 832 m/s]

Unsteady Flow Processes

79. An air receiver of volume 6 m^3 contains air at 15 bar and 40.5°C. A valve is opened and some air is allowed to blow out to atmosphere. The pressure of the air in the receiver drops rapidly to 12 bar when the valve is then closed. Calculate the mass of air which has left the receiver.
[**Ans.** 14.7 kg]

80. The internal energy of air is given, at ordinary temperatures, by $u = u_0 + 0.718t$
where u is in kJ/kg, u_0 is any arbitrary value of u at 0°C, kJ/kg and t is temperature in °C.

Also for air, $pv = 0.287 (t + 273)$, where p is in kPa and v is in m^3/kg.

(*i*) An evacuated bottle is fitted with a valve through which air from the atmosphere, at 760 mm Hg and 25°C, is allowed to flow slowly to fill the bottle. If no heat is transferred to or from the air in the bottle, what will its temperature be when the pressure in the bottle reaches 760 mm Hg ?
(*ii*) If the bottle initially contains 0.03 m^3 of air at 400 mm Hg and 25°C, what will the temperature be when the pressure in the bottle reaches 760 mm of Hg ? [**Ans.** (*i*) 144.2°C ; (*ii*) 71.6°C]

Second Law of Thermodynamics and Entropy

4

4.1. LIMITATIONS OF FIRST LAW OF THERMODYNAMICS AND INTRODUCTION TO SECOND LAW

It has been observed that *energy can flow* from a system in the form of *heat* or *work*. The first law of thermodynamics sets no limit to the amount of the total energy of a system which can be caused to flow out as work. A limit is imposed, however, as a result of the principle enunciated in the second law of thermodynamics which states that heat will flow naturally from one energy reservoir to another at a lower temperature, but not in opposite direction without assistance. This is very important because a heat engine operates between two energy reservoirs at different temperatures.

Further the first law of thermodynamics *establishes equivalence between the quantity of heat used and the mechanical work but does not specify the conditions under which conversion of heat into work is possible, neither the direction in which heat transfer can take place*. This gap has been *bridged* by the second law of thermodynamics.

4.2. PERFORMANCE OF HEAT ENGINES AND REVERSED HEAT ENGINES

Refer to Fig. 4.1 (*a*). A *heat engine* is used to produce the maximum work transfer from a given positive heat transfer. The measure of success is called the *thermal efficiency* of the engine and is defined by the ratio :

$$\text{Thermal efficiency, } \eta_{th} = \frac{W}{Q_1} \quad ...(4.1)$$

where, W = Net work transfer from the engine, and

Q_1 = Heat transfer to engine.

For a *reversed heat engine* [Fig. 4.1 (*b*)] acting as a *refrigerator* when the purpose is to achieve the maximum heat transfer from the cold reservoir, the measure of success is called the *co-efficient of performance* (C.O.P.). It is defined by the ratio :

$$\text{Co-efficient of performance, } (\text{C.O.P.})_{ref.} = \frac{Q_2}{W} \quad ...(4.2)$$

where, Q_2 = Heat transfer *from cold reservoir*, and

W = The network transfer to the refrigerator.

For a **reversed heat engine** [Fig. 4.1 (*b*)] acting as a *heat pump*, the measure of success is again called the *co-efficient of performance*. It is defined by the ratio :

$$\text{Co-efficient of performance, } (\text{C.O.P.})_{heat\ pump} = \frac{Q_1}{W} \quad ...(4.3)$$

where, Q_1 = Heat transfer *to hot reservoir*, and

W = Net work transfer to the heat pump.

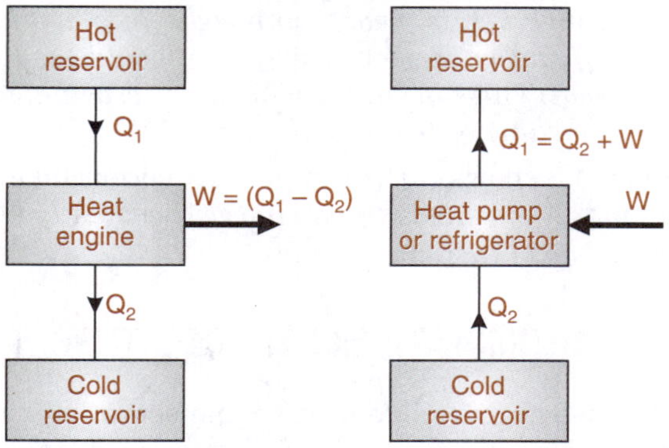

(*a*) Heat engine (*b*) Heat pump or refrigerator

Fig. 4.1

In all the above three cases application of the first law gives the relation $Q_1 - Q_2 = W$, and this can be used to rewrite the expressions for thermal efficiency and co-efficient of performance solely in terms of the heat transfers.

$$\eta_{th} = \frac{Q_1 - Q_2}{Q_1} \quad ...(4.4)$$

$$(\text{C.O.P.})_{ref} = \frac{Q_2}{Q_1 - Q_2} \quad ...(4.5)$$

$$(\text{C.O.P.})_{heat\ pump} = \frac{Q_1}{Q_1 - Q_2} \quad ...(4.6)$$

It may be seen that η_{th} is *always less than unity* and $(C.O.P.)_{heat\ pump}$ is *always greater than unity*.

4.3. REVERSIBLE PROCESSES

A reversible process should fulfil the following *conditions* :

1. The process should not involve friction of any kind.
2. Heat transfer should not take place with finite temperature difference.
3. The energy transfer as heat and work during the forward process should be identically equal to energy transfer as heat and work during the reversal of the process.
4. There should be no free or unrestricted expansion.
5. There should be no mixing of the fluids.
6. The process must proceed in a series of equilibrium states.

Some examples of *ideal reversible processes* are :

(*i*) Frictionless adiabatic expansion or compression ;
(*ii*) Frictionless isothermal expansion or compression ;
(*iii*) Condensation and boiling of liquids.

Some examples of *irreversible processes are* :

(*i*) Combustion process ;
(*ii*) Mixing of two fluids ;
(*iii*) All processes involving friction ;
(*iv*) Flow of electric current through a resistance ;
(*v*) Heat flow from a higher temperature to lower temperature.

Reversible processes are preferred because the devices which produce work such as engines and turbines, reversible process of the working fluid delivers *more work* than the corresponding irreversible processes. Also in case of fans, compressors, refrigerators and pumps *less power input* is required when *reversible processes* are used in place of corresponding irreversible ones.

In thermodynamic analysis concept of reversibility, though hypothetical, is very important because a reversible process is the *most efficient process*. Only reversible processes can be truely represented on property diagrams. Thermodynamic reversibility can only be approached but can *never* be achieved. Thus the main task of the engineer is to design the system which will evolve approximate reversible processes.

4.4. STATEMENTS OF SECOND LAW OF THERMODYNAMICS

The second law of thermodynamics has been enunciated meticulously by Clausius, Kelvin and Planck in slightly different words although both statements are basically identical. Each statement is based on an *irreversible process*. The *first considers transformation of heat between two thermal reservoirs* while the *second considers the transformation of heat into work*.

4.4.1. Clausius Statement

"It is impossible for a self acting machine working in a cyclic process unaided by any external agency, to convey heat from a body at a lower temperature to a body at a higher temperature".

In other words, heat of, itself, cannot flow from a colder to a hotter body.

4.4.2. Kelvin-Planck Statement

"It is impossible to construct an engine, which while operating in a cycle produces no other effect except to extract heat from a single reservoir and do equivalent amount of work".

Although the Clausius and Kelvin-Planck statements appear to be different, they are really equivalent in the sense that a *violation of either statement implies violation of other*.

4.4.3. Equivalence of Clausius Statement to the Kelvin-Planck Statement

Refer to Fig. 4.2. Consider a higher temperature reservoir T_1 and low temperature reservoir T_2. Fig. 4.2 shows a heat pump which requires no work and transfers an amount of Q_2 from a low temperature to a higher temperature reservoir (in violation of the Clausius statement). Let an amount of heat Q_1 (greater than Q_2) be transferred from high temperature reservoir to heat engine which develops a net work, $W = Q_1 - Q_2$ and rejects Q_2 to the low temperature reservoir. Since there is no heat interaction with the low temperature, it can be eliminated. The combined system of the heat engine and heat pump acts then like a heat engine exchanging heat with a single reservoir, which is the violation of the Kelvin-Planck statement.

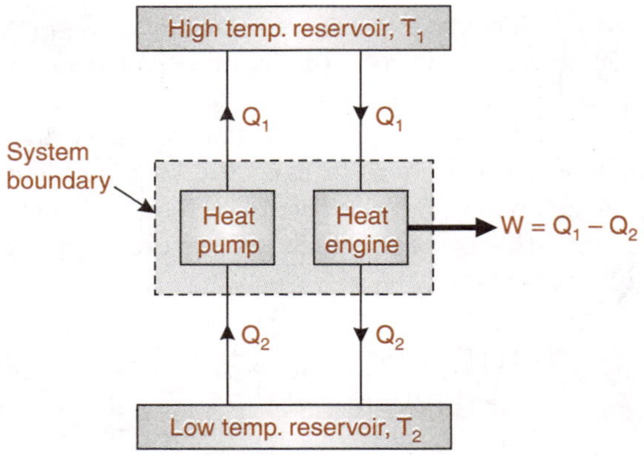

Fig. 4.2. Equivalence of Clausius statement to Kelvin-Planck statement.

4.5. PERPETUAL MOTION MACHINE OF THE SECOND KIND

— A machine which violates the first law of thermodynamics is called the perpetual motion machine of the first kind (PMM1). Such a machine creates its own energy from nothing and *does not exist*.

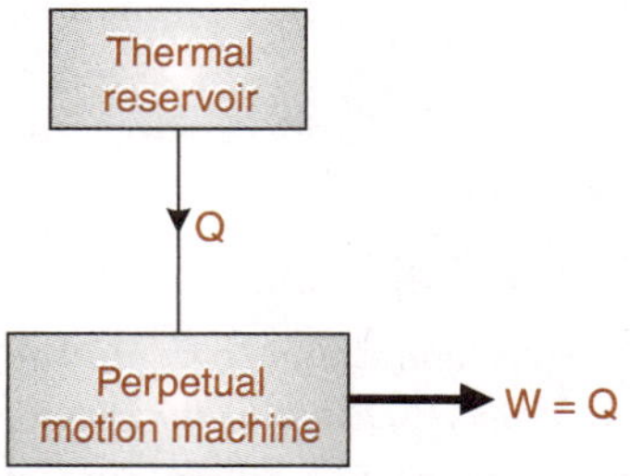

Fig. 4.3. Perpetual motion machine of second kind (PMM2).

— Without violating the first law, a machine can be imagined which would continuously absorb heat from a single thermal *reservoir and would convert this heat completely into work. The efficiency* of such a machine would be *100 per cent*. This machine is called the *perpetual motion machine of the second kind (PMM2)*.

Fig. 4.3 shows the perpetual motion machine of the second kind. *A machine of this kind will evidently violates the second law of thermodynamics.*

4.6. THERMODYNAMIC TEMPERATURE

Take the case of reversible heat engine operating between two reservoirs. Its thermal efficiency is given by the eqn. (4.4),

$$\eta_{th} = \frac{Q_1 - Q_2}{Q_1} = 1 - \frac{Q_2}{Q_1}$$

The temperature of a reservoir remains uniform and fixed irrespective of heat transfer. This means that reservoir has only one property defining its state and the heat transfer from a reservoir is some function of that property, *temperature*. Thus $Q = \phi(K)$, where K is the temperature of reservoir. The choice of the function is universally accepted to be such that the relation,

$$\frac{Q_1}{Q_2} = \frac{\phi(K_1)}{\phi(K_2)} \text{ becomes } \frac{Q_1}{Q_2} = \frac{T_1}{T_2} \quad ...(4.7)$$

where T_1 and T_2 are the thermodynamic temperatures of the reservoirs. Zero thermodynamic temperature (that temperature to which T_2 tends, as the heat transfer Q_2 tends to zero) has never been attained and *one form of third law of thermodynamics is the statement* :

"The temperature of a system cannot be reduced to zero in a finite number of processes".

After establishing the concept of a zero thermodynamic temperature, a reference reservoir is chosen and assigned a numerical value of temperature. Any other thermodynamic temperature may now be defined in terms of reference value and the heat transfers that would occur with reversible engine,

$$T = T_{ref.} \frac{Q}{Q_{ref.}} \quad ...(4.8)$$

The determination of thermodynamic temperature cannot be made in this way as it is not possible to build a reversible engine. Temperatures are determined by the application of thermodynamic relations to other measurements.

The SI unit of thermodynamic temperature is the kelvin (K). The relation between thermodynamic temperature and celsius scale, which is in common use is :

Thermodynamic temperature = Celsius temperature + 273.15°.

The kelvin unit of thermodynamic temperature is the fraction $\frac{1}{273.15}$ of thermodynamic temperature of 'Triple point' of water.

4.7. CLAUSIUS INEQUALITY

When a reversible engine uses more than two reservoirs the third or higher numbered reservoirs will not be equal in temperature to the original two. Consideration of expression for efficiency of the engine indicates that for maximum efficiency, all the heat transfer should take place at maximum or minimum reservoir temperatures. Any intermediate reservoir used will, therefore, lower the efficiency of the heat engine. Practical engine cycles often involve continuous changes of temperature during heat transfer. A relationship among processes in which these sort of changes occur is necessary. The ideal approach to a cycle in which temperature continually changes is to consider the system to be in communication with a large number of reservoirs in procession. Each reservoir is considered to have a temperature differing by a small amount from the previous one. In such a model it is possible to imagine that each reservoir is replaced by a reversible heat engine in communication with standard reservoirs at same temperature T_0. Fig. 4.4 shows one example to this substitution.

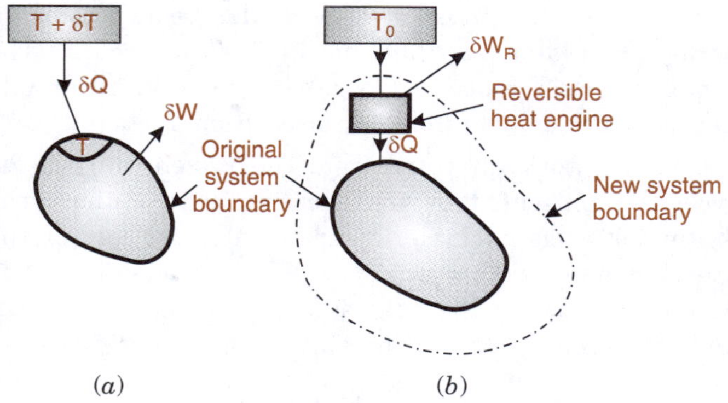

Fig. 4.4. The clausius inequality.

The system to which the heat transfer is effected is neither concerned with the source of energy it receives nor with the method of transfer, save that it must be reversible. Associated with the small heat transfer dQ to the original system is a small work transfer dW and for this system the first law gives

$$\sum_{\text{cycle}} (\delta Q - \delta W) = 0 \qquad \text{...(4.9)}$$

Now consider the engine replacing the reservoirs and apply the second law to the new system in Fig. 4.4 (*b*). If the new system is not a perpetual motion machine of second kind, no positive work transfer is possible with a single reservoir.

Therefore, $$\sum_{\text{cycle}} (\delta W - \delta W_R) \le 0 \qquad \text{...(4.10)}$$

But by the definition of thermodynamic temperature in equation (4.8)

$$\frac{\delta W_R}{\delta Q} = \frac{\delta Q_0 - \delta Q}{\delta Q} = \frac{T_0 - T}{T} \qquad \text{...(4.11)}$$

and by combination of eqns. (4.9), (4.10) and (4.11)

$$T_0 \sum_{\text{cycle}} \left(\frac{\delta Q}{T}\right) \le 0 \text{ but } T_0 \ne 0 \text{ and therefore ;}$$

$$\sum_{\text{cycle}} \left(\frac{\delta Q}{T}\right) \le 0 \qquad \text{...(4.12)}$$

This is known as **Clausius inequality**.

Let us now consider the case of a reversible engine for which

$$\sum_{\text{cycle}} \left(\frac{\delta Q}{T}\right) \le 0,$$

reverse the engine and for the reversible heat pump obtained it is possible to develop the expression,

$$-\sum_{\text{cycle}} \left(\frac{\delta Q}{T}\right) \le 0$$

The *negative sign indicates that the heat transfers have all reversed in direction when the engine was reversed.* This means that for the same machine we have two relations which are only satisfied if in the reversible case,

$$\sum_{\text{cycle}} \left(\frac{\delta Q}{T}\right) \le 0 \qquad \text{...(4.13)}$$

For a reversible case, as the number of reservoirs used tends to infinity, the limiting value of the summation will be

$$\sum_{\text{cycle}} \left(\frac{\delta Q}{T}\right) = 0$$

In words, the Clausius inequality may be expressed as follows :

"When a system performs a reversible cycle, then

$$\sum_{\text{cycle}} \left(\frac{\delta Q}{T}\right) = 0,$$

but when the cycle is not reversible

$$\sum_{\text{cycle}} \left(\frac{\delta Q}{T}\right) < 0\text{''}.$$

4.8. CARNOT CYCLE

The cycle was first suggested by a French engineer Sadi Carnot in 1824 which works on reversible cycle and is known as *Carnot cycle*.

Any fluid may be used to operate the Carnot cycle (Fig. 4.5) which is performed in an engine cylinder the head of which is supposed alternatively to be perfect conductor or a perfect insulator of a heat. Heat is caused to flow into the cylinder by the application of high temperature energy source to the cylinder head during expansion, and to flow from the cylinder by the application of a lower temperature energy source to the head during compression.

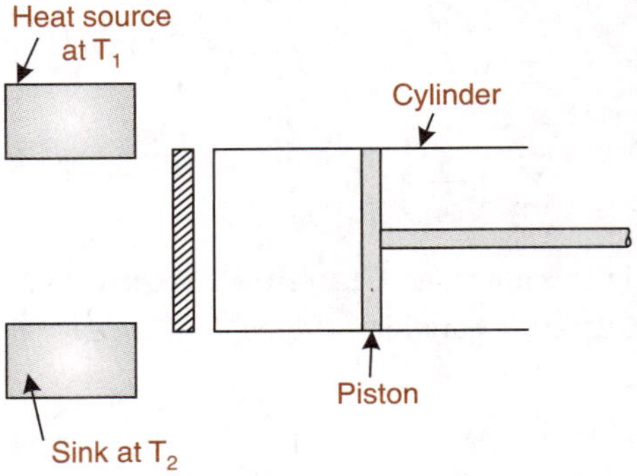

(*a*)

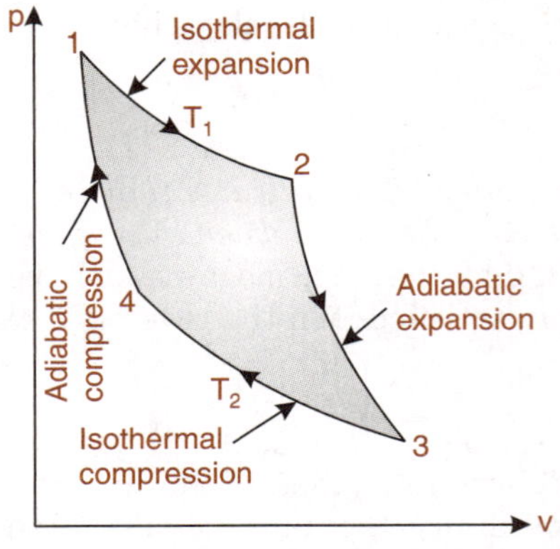

(*b*) Carnot engine cycle

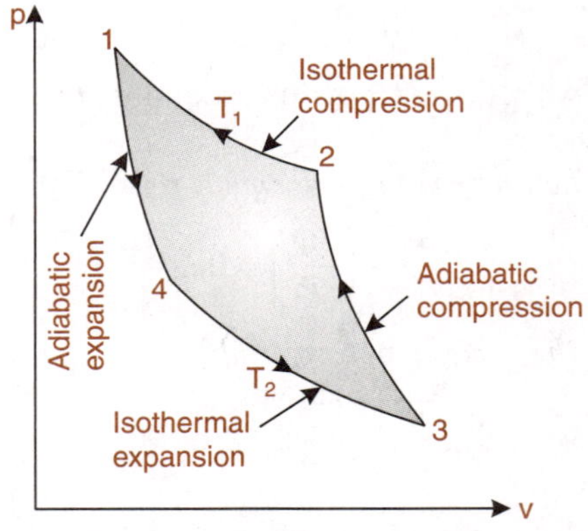

(*c*) Carnot heat pump cycle

Fig. 4.5

The ***assumptions*** made for describing the working of the Carnot engine are as follows :

(*i*) The piston moving in a cylinder does not develop any friction during motion.

(*ii*) The walls of piston and cylinder are considered as perfect insulators of heat.

(*iii*) The cylinder head is so arranged that it can be a perfect heat conductor or perfect heat insulator.

(*iv*) The transfer of heat does not affect the temperature of source or sink.

(*v*) Working medium is a perfect gas and has constant specific heat.

(*vi*) Compression and expansion are reversible.

Following are the *four stages* of Carnot cycle :

Stage 1. (Process 1-2). Hot energy source is applied. Heat Q_1 is taken in whilst the fluid expands isothermally and reversibly at constant high temperature T_1.

Stage 2. (Process 2-3). The cylinder becomes a perfect insulator so that no heat flow takes place. The fluid expands adiabatically and reversibly whilst temperature falls from T_1 to T_2.

Stage 3. (Process 3-4). Cold energy source is applied. Heat Q_2 flows from the fluid whilst it is compressed isothermally and reversibly at constant lower temperature T_2.

Stage 4. (Process 4-1). Cylinder head becomes a perfect insulator so that no heat flow occurs. The compression is continued adiabatically and reversibly during which temperature is raised from T_2 to T_1.

The work delivered from the system during the cycle is represented by the enclosed area of the cycle. Again for a closed cycle, according to first law of the thermodynamics the work obtained is equal to the difference between the heat supplied by the source (Q_1) and the heat rejected to the sink (Q_2).

$$\therefore \qquad W = Q_1 - Q_2$$

Also, thermal efficiency,

$$\eta_{th} = \frac{\text{Work done}}{\text{Heat supplied by the source}} = \frac{Q_1 - Q_2}{Q_1}$$

$$= 1 - \frac{Q_2}{Q_1} \left(= 1 - \frac{T_2}{T_1} \right)$$

$$\left[\begin{array}{r} \because \quad Q_1 = m\, c_p\, T_1 \\ Q_2 = m\, c_p\, T_2 \\ \text{where, } m = \text{mass of fluid.} \end{array} \right]$$

Such an engine since it consists entirely of reversible processes, can operate in the reverse direction so that it follows the cycle shown in Fig. 4.5 (*b*) and operates as a **heat pump.** Q_2 is being taken in at the lower temperature T_2 during the isothermal expansion (process 4-3) and heat Q_1 is being rejected at the upper temperature T_1 (process 2-1). Work W will be needed to drive the pump. Again, the enclosed area represents this work which is exactly equal to that flowing from it when used as engine.

The Carnot cycle *cannot be performed in practice* because of the following **reasons** :

1. It is imposible to perform a frictionless process.

2. It is impossible to transfer the heat without temperature potential.

3. Isothermal process can be achieved only if the piston moves very slowly to allow heat transfer so that the temperature remains constant. Adiabatic process can be achieved only if the piston moves as fast as possible so that the heat transfer is negligible due to very short time available. The isothermal and adiabatic processes take place during the same stroke therefore the piston has to move very slowly for part of the stroke and it has to move very fast during remaining stroke. This variation of motion of the piston during the same stroke is not possible.

4.9. CARNOT'S THEOREM

"It states that of all engines operating between a given constant temperature source and a given constant temperature sink, none has a higher efficiency than a reversible engine".

Refer to Fig. 4.6.

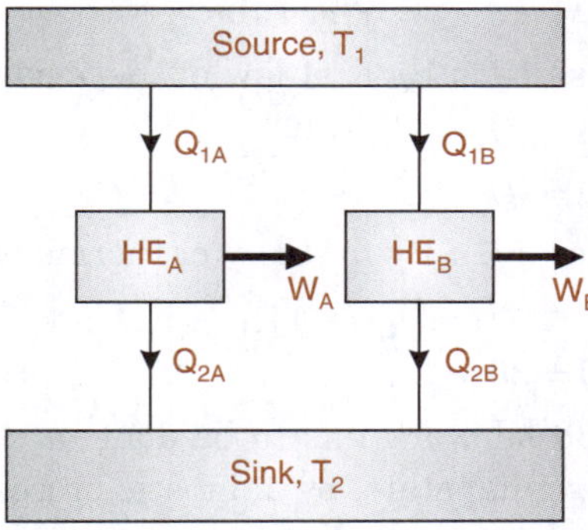

Fig. 4.6. Two cyclic heat engines HE_A and HE_B operating between the same source and sink, of which HE_B is reversible.

HE_A and HE_B are the two engines operating between the given source at temperature T_1 and the given sink at temperature T_2.

Let HE_A be *any* heat engine and HE_B be *any reversible* heat engine. We have to prove that efficiency of HE_B is more than that of HE_A. Let us assume that $\eta_A > \eta_B$. Let the rates of working of the engines be such that

$$Q_{1A} = Q_{1B} = Q_1$$

Since $\quad \eta_A > \eta_B$

$$\frac{W_A}{Q_{1A}} > \frac{W_B}{Q_{1B}}$$

$\therefore \quad W_A > W_B$

Now, let HE_B be reversed. Since HE_B is a reversible heat engine, the magnitudes of heat and work transfer quantities will remain the same, but their directions will be reversed, as shown in Fig. 4.7. Since $W_A > W_B$, some part of W_A (equal to W_B) may be fed to drive the reversed heat engine $\exists H_B$. Since $Q_{1A} = Q_{1B} = Q_1$, the heat discharged by $\exists H_B$ may be supplied to HE_A. The source may, therefore, be eliminated (Fig. 4.8). The net result is that HE_A and $\exists H_B$ together constitute a heat engine which, operating in a cycle produces net work $W_A - W_B$ while exchanging heat with a single reservoir at T_2. This violates the Kelvin-Planck statement of the second law. Hence the assumption that $\eta_A > \eta_B$ is wrong.

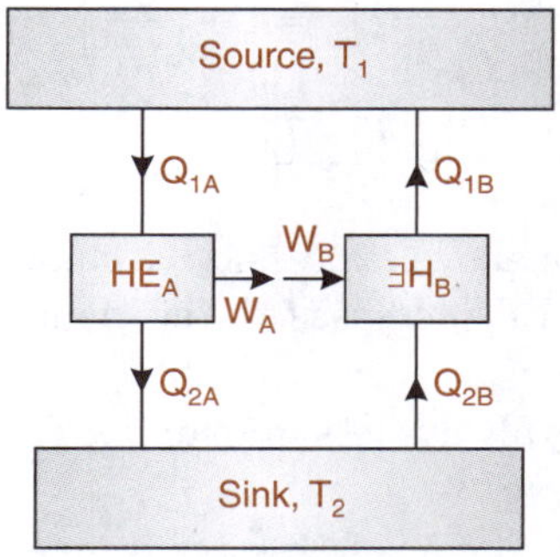

Fig. 4.7. HE_B is reversed.

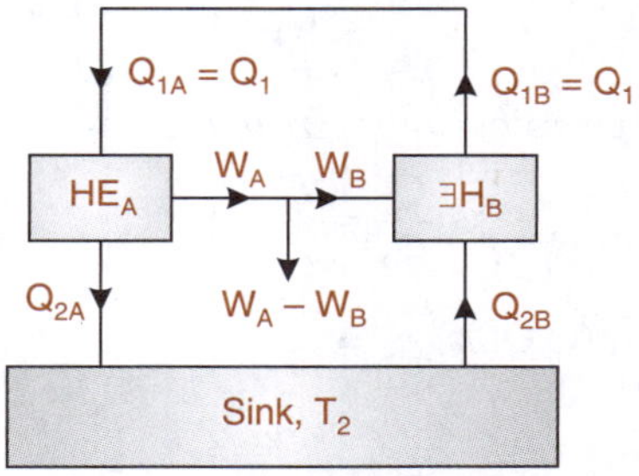

Fig. 4.8. HE_A and $\exists H_B$ together violate the Kelvin-Planck statement.

$\therefore \quad \eta_B \ge \eta_A.$

4.10. COROLLARY OF CARNOT'S THEOREM

"The efficiency of all reversible heat engines operating between the same temperature levels is the same".

Refer to Fig. 4.6. Let both the heat engines HE_A and HE_B be *reversible*. Let us assume $\eta_A > \eta_B$. Similar to the procedure outlined in the Art. 4.9, if HE_B is reversed to run say, as a heat pump using some part of the work output (W_A) of engine HE_A, we see that the combined system of heat pump HE_B and engine HE_A, becomes a *PMM*2. So η_A cannot be greater than η_B. Similary, if we assume $\eta_B > \eta_A$ and reverse the engine HE_A, we observe that η_B cannot be greater than η_A

$\therefore \quad \eta_A = \eta_B.$

Since the efficiencies of all reversible engines operating between the same heat reservoirs are the same, *the efficiency of a reversible engine is independent of the nature or amount of the working substance undergoing the cycle.*

4.11. EFFICIENCY OF THE REVERSIBLE HEAT ENGINE

The efficiency of a reversible heat engine in which heat is received solely at T_1 is found to be

$$\eta_{rev.} = \eta_{max} = 1 - \left(\frac{Q_2}{Q_1}\right)_{rev.} = 1 - \frac{T_2}{T_1}$$

or
$$\eta_{rev.} = \frac{T_1 - T_2}{T_1}$$

From the above expression, it may be noted that as T_2 decreases and T_1 increases, the efficiency of the reversible cycle increases.

Since η is always less than unity, T_2 is always greater than zero and +ve.

The C.O.P. of a refrigerator is given by

$$(\text{C.O.P.})_{ref.} = \frac{Q_2}{Q_1 - Q_2} = \frac{1}{\frac{Q_1}{Q_2} - 1}$$

For a *reversible refrigerator*, using

$$\frac{Q_1}{Q_2} = \frac{T_1}{T_2}$$

$$(\text{C.O.P.})_{rev.} = \frac{1}{\frac{T_1}{T_2} - 1}$$

$$\therefore \quad [(\text{C.O.P.})_{ref.}]_{rev.} = \frac{T_2}{T_1 - T_2} \quad ...(4.14)$$

Similarly, for a *reversible heat pump*

$$[(\text{C.O.P.})_{heat\ pump}]_{rev.} = \frac{T_1}{T_1 - T_2} \quad ...(4.15)$$

Example 4.1. *A heat engine receives heat at the rate of 1500 kJ/min and gives an output of 8.2 kW. Determine :*

(i) The thermal efficiency ;

(ii) The rate of heat rejection.

Solution. Heat received by the heat engine,

$$Q_1 = 1500 \text{ kJ/min} = \frac{1500}{60} = 25 \text{ kJ/s}$$

Work output,

$$W = 8.2 \text{ kW} = 8.2 \text{ kJ/s.}$$

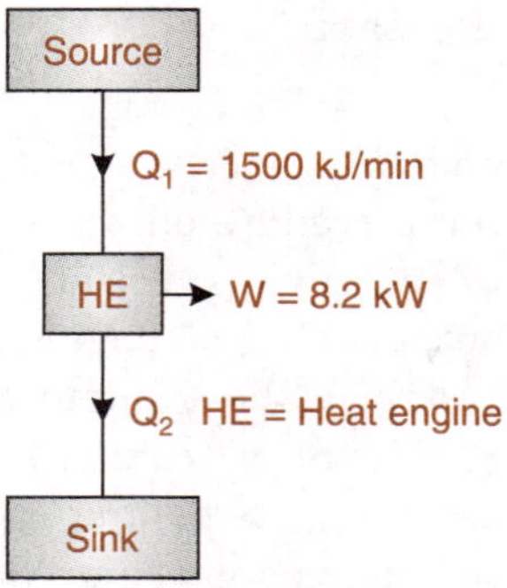

Fig. 4.9

(*i*) **Thermal efficiency,** $\eta_{th} = \frac{W}{Q_1}$

$$= \frac{8.2}{25} = 0.328 = 32.8\%$$

Hence, **thermal efficiency = 32.8%. (Ans.)**

(*ii*) **Rate of heat rejection,**

$$Q_2 = Q_1 - W = 25 - 8.2 = 16.8 \text{ kJ/s}$$

Hence, **the rate of heat rejection = 16.8 kJ/s. (Ans.)**

☞**Example 4.2.** *During a process a system receives 30 kJ of heat from a reservoir and does 60 kJ of work. Is it possible to reach initial state by an adiabatic process ?*

Solution. Heat received by the system = 30 kJ

Work done = 60 kJ

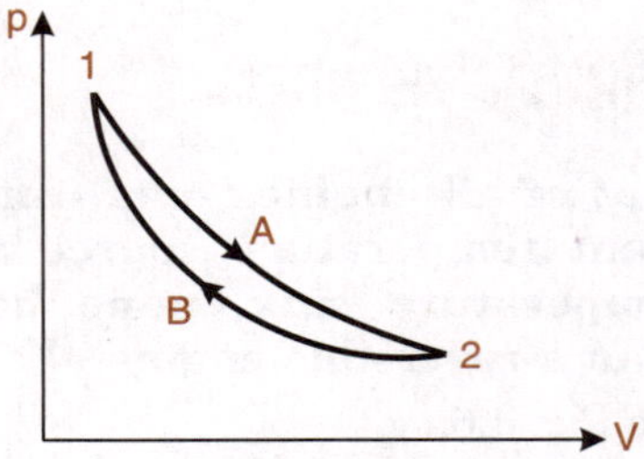

Fig. 4.10

Process 1–2 : By first law of thermodynamics,

$$Q_{1-2} = (U_2 - U_1) + W_{1-2}$$

$$30 = (U_2 - U_1) + 60 \qquad \therefore \quad (U_2 - U_1) = -30 \text{ kJ.}$$

Process 2–1 : By first law of thermodynamics,

$$Q_{2-1} = (U_1 - U_2) + W_{2-1}$$

$$\therefore \quad 0 = 30 + W_{2-1} \qquad \therefore \quad W_{2-1} = -30 \text{ kJ.}$$

Thus 30 kJ work has to be done *on the system* to restore it to original state, by adiabatic process.

Example 4.3. *Find the co-efficient of performance and heat transfer rate in the condenser of a refrigerator in kJ/h which has a refrigeration capacity of 12000 kJ/h when power input is 0.75 kW.*

Solution. Refer to Fig. 4.11.

Refrigeration capacity,

$$Q_2 = 12000 \text{ kJ/h}$$

Power input,

$$W = 0.75 \text{ kW} (= 0.75 \times 60 \times 60 \text{ kJ/h})$$

Co-efficient of performance, C.O.P. :

Heat transfer rate :

$$(\text{C.O.P.})_{refrigerator} = \frac{\text{Heat absorbed at lower temperature}}{\text{Work input}}$$

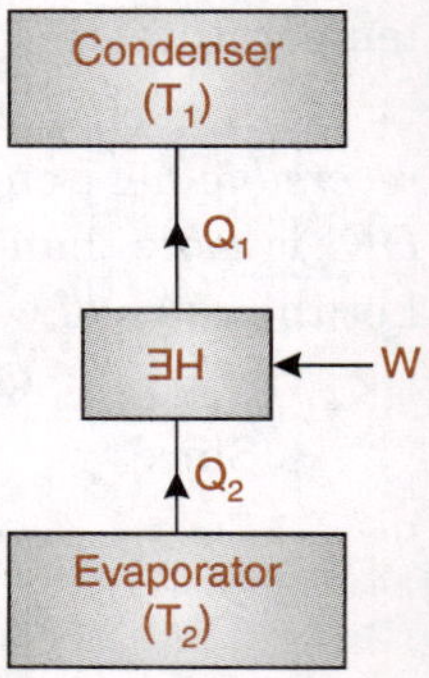

Fig. 4.11

$\therefore \quad \text{C.O.P.} = \dfrac{Q_2}{W} = \dfrac{12000}{0.75 \times 60 \times 60} = 4.44$

Hence **C.O.P. = 4.44. (Ans.)**

Hence transfer rate in condenser = Q_1

According to the first law

$Q_1 = Q_2 + W = 12000 + 0.75 \times 60 \times 60 = 14700$ kJ/h

Hence, **heat transfer rate = 14700 kJ/h. (Ans.)**

Example 4.4. *A domestic food refrigerator maintains a temperature of –12°C. The ambient air temperature is 35°C. If heat leaks into the freezer at the continuous rate of 2 kJ/s determine the least power necessary to pump this heat out continuously.*

Solution. Freezer temperature,

$T_2 = -12 + 273 = 261$ K

Ambient air temperature,

$T_1 = 35 + 273 = 308$ K

Rate of heat leakage into the freezer = 2 kJ/s

Least power required to pump the heat :

The refrigerator cycle removes heat from the freezer at the same rate at which heat leaks into it (Fig. 4.12).

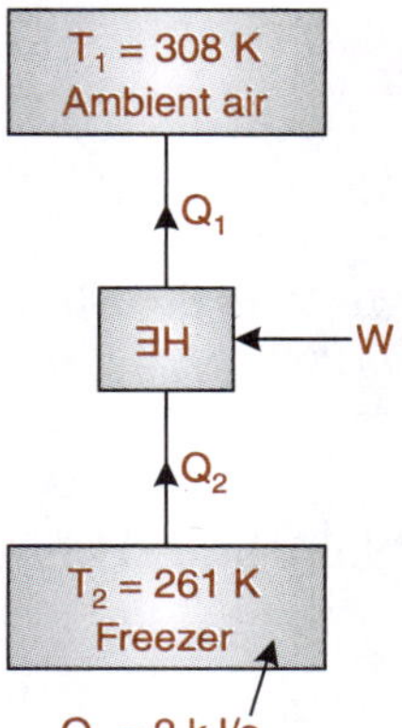

Fig. 4.12

For minimum power requirement

$$\frac{Q_2}{T_2} = \frac{Q_1}{T_1}$$

$\therefore \quad Q_1 = \dfrac{Q_2}{T_2} \times T_1 = \dfrac{2}{261} \times 308 = 2.36$ kJ/s

$\therefore \quad W = Q_1 - Q_2$

$= 2.36 - 2 = 0.36$ kJ/s $= 0.36$ kW

Hence, *least power required to pump the heat continuously*

= 0.36 kW. (Ans.)

Example 4.5. *A house requires 2 × 10⁵ kJ/h for heating in winter. Heat pump is used to absorb heat from cold air outside in winter and send heat to the house. Work required to operate the heat pump is 3 × 10⁴ kJ/h. Determine :*

(*i*) *Heat abstracted from outside ;*

(*ii*) *Co-efficient of performance.*

Solution. (*i*) Heat requirement of the house, Q_1 (or heat rejected)

$= 2 \times 10^5$ kJ/h

Work required to operate the heat pump,

$W = 3 \times 10^4$ kJ/h

Now, $\quad Q_1 = W + Q_2$

where Q_2 is the heat abstracted from outside.

$\therefore \quad 2 \times 10^5 = 3 \times 10^4 + Q_2$

Thus $\quad Q_2 = 2 \times 10^5 - 3 \times 10^4$

$= 200000 - 30000 = 170000$ kJ/h

Hence, **heat abstracted from outside = 170000 kJ/h. (Ans.)**

(*ii*) $\text{(C.O.P.)}_{heat\ pump} = \dfrac{Q_1}{Q_1 - Q_2}$

$= \dfrac{2 \times 10^5}{2 \times 10^5 - 170000} = 6.66$

Hence, **co-efficient of performance = 6.66. (Ans.)**

Note. If the heat requirements of the house were the same but this amount of heat had to be abstracted from the house and rejected out, *i.e., cooling of the house in summer*, we have

$$\text{(C.O.P.)}_{refrigerator} = \frac{Q_2}{Q_1 - Q_2} = \frac{Q_2}{W} = \frac{170000}{3 \times 10^4} = \mathbf{5.66}$$

Thus the same device has two values of C.O.P. depending upon the objective.

Example 4.6. *What is the highest possible theoretical efficiency of a heat engine operating with a hot reservoir of furnace gases at 2100°C when the cooling water available is at 15°C ?*

Solution. Temperature of furnace gases,

$T_1 = 2100 + 273 = 2373$ K

Temperature of cooling water,

$T_2 = 15 + 273 = 288$ K

Now, $\eta_{max} (= \eta_{carnot}) = 1 - \dfrac{T_2}{T_1}$

$= 1 - \dfrac{288}{2373} = \mathbf{0.878 \text{ or } 87.8\%}$. **(Ans.)**

Note. It should be noted that a system in practice operating between similar temperatures (*e.g.*, a steam generating plant) would have a thermal efficiency of about 30%. The discrepency is *due to irreversibility in the actual plant*, and *also because of deviations from the ideal Carnot cycle made for various practical reasons.*

Example 4.7. *A Carnot cycle operates between source and sink temperatures of 250°C and – 15°C. If the system receives 90 kJ from the source, find :*

(*i*) *Efficiency of the system ;*

(*ii*) *The net work transfer ;*

(*iii*) *Heat rejected to sink.*

Solution. Temperature of source, $T_1 = 250 + 273 = 523$ K

Temperature of sink, $T_2 = -15 + 273 = 258$ K

Heat received by the system, $Q_1 = 90$ kJ

(*i*) **Efficiency of the system,**

$$\eta_{\text{carnot}} = 1 - \frac{T_2}{T_1} = 1 - \frac{258}{523}$$

$= 0.506 = \mathbf{50.6\%}$. **(Ans.)**

(*ii*) **The net work transfer,**

$$W = \eta_{carnot} \times Q_1 \qquad \left[\because \ \eta_{carnot} = \frac{W}{Q_1}\right]$$

$= 0.506 \times 90 = \mathbf{45.54\ kJ}$. **(Ans.)**

(*iii*) **Heat rejected to the sink,**

$$Q_2 = Q_1 - W \qquad [\because \ W = Q_1 - Q_2]$$

$= 90 - 45.54 = \mathbf{44.46\ kJ}$. **(Ans.)**

Example 4.8. *An inventor claims that his engine has the following specifications :*

Temperature limits	...... *750°C and 25°C*
Power developed	...... *75 kW*
Fuel burned per hour	...... *3.9 kg*
Heating value of the fuel	...... *74500 kJ/kg*

State whether his claim is valid or not.

Solution. Temperature of source, $T_1 = 750 + 273 = 1023$ K

Temperature of sink, $T_2 = 25 + 273 = 298$ K

We know that the thermal efficiency of Carnot cycle is the maximum between the specified temperature limits.

$$\text{Now,} \quad \eta_{carnot} = 1 - \frac{T_2}{T_1} = 1 - \frac{298}{1023}$$

$= 0.7086$ or 70.86%

The actual thermal efficiency claimed,

$$\eta_{thermal} = \frac{\text{Work done}}{\text{Heat supplied}} = \frac{75 \times 1000 \times 60 \times 60}{3.9 \times 74500 \times 1000}$$

$= 0.9292$ or 92.92%.

Since $\eta_{thermal} > \eta_{carnot}$, therefore claim of the inventor is **not valid** (or possible). **(Ans.)**

Example 4.9. *A cyclic heat engine operates between a source temperature of 1000°C and a sink temperature of 40°C. Find the least rate of heat rejection per kW net output of the engine.*

Solution. Temperature of source,

$$T_1 = 1000 + 273 = 1273 \text{ K}$$

Temperature of sink,

$$T_2 = 40 + 273 = 313 \text{ K}$$

Least rate of heat rejection per kW net output :

For a reversible heat engine, the rate of heat rejection will be minimum (Fig. 4.13)

$$\eta_{max} = \eta_{rev.} = 1 - \frac{T_2}{T_1}$$

$$= 1 - \frac{313}{1273} = 0.754$$

$$\text{Now} \quad \frac{W_{net}}{Q_1} = \eta_{max} = 0.754$$

$$\therefore \quad Q_1 = \frac{W_{net}}{0.754} = \frac{1}{0.754} = 1.326 \text{ kW}$$

$$\text{Now} \quad Q_2 = Q_1 - W_{net} = 1.326 - 1 = 0.326 \text{ kW}$$

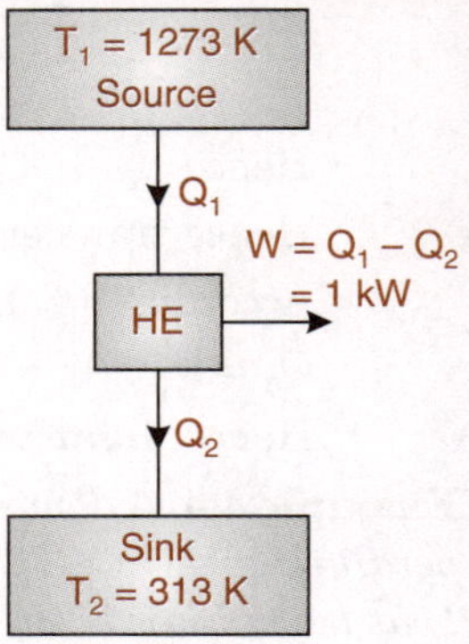

Fig. 4.13

Hence, **the least rate of heat rejection = 0.326 kW. (Ans.)**

Example 4.10. *A fish freezing plant requires 40 tons of refrigeration. The freezing temperature is – 35°C while the ambient temperature is 30°C. If the performance of the plant is 20% of the theoretical reversed Carnot cycle working within the same temperature limits, calculate the power required.*

Given : 1 ton of refrigeration = 210 kJ/min.

Solution. Cooling required = 40 tons = 40 × 210

= 8400 kJ/min

Ambient temperature, $T_1 = 30 + 273 = 303$ K

Freezing temperature, $T_2 = -35 + 273 = 238$ K

Performance of plant

= 20% of the theoretical reversed Carnot cycle

$$(\text{C.O.P.})_{refrigerator} = \frac{T_2}{T_1 - T_2} = \frac{238}{303 - 238} = 3.66$$

$\therefore$ Actual C.O.P = 0.20 × 3.66 = 0.732

Now work needed to produce cooling of 40 tons is calculated as follows :

$$(\text{C.O.P.})_{actual} = \frac{\text{Cooling reqd.}}{\text{Work needed}}$$

$$\Rightarrow \quad 0.732 = \frac{8400}{W}$$

or

$$W = \frac{8400}{0.732} \text{ kJ/min} = 191.25 \text{ kJ/s} = 191.25 \text{ kW}$$

Hence, **power required = 191.25 kW. (Ans.)**

Example 4.11. *Source 1 can supply energy at the rate of 12000 kJ/min at 320°C. A second source 2 can supply energy at the rate of 120000 kJ/min at 70°C. Which source (1 or 2) would you choose to supply energy to an ideal reversible heat engine that is to produce large amount of power if the temperature of the surroundings is 35°C ?*

Solution. Source 1 :

Rate of supply of energy = 12000 kJ/min

Temperature, $T_1 = 320 + 273 = 593$ K.

Source 2 :

Rate of supply of energy = 120000 kJ/min

Temperature, $T_1 = 70 + 273 = 343$ K

Temperature of the surroundings,

$$T_2 = 35 + 273 = 308 \text{ K}$$

Let the Carnot engine be working in the two cases with the two source temperatures and the single sink temperature. The efficiency of the cycle will be given by :

$$\eta_1 = 1 - \frac{T_2}{T_1} = 1 - \frac{308}{593} = 0.4806 \quad \text{or} \quad 48.06\%$$

$$\eta_2 = 1 - \frac{T_2}{T_1} = 1 - \frac{308}{343} = 0.102 \quad \text{or} \quad 10.2\%$$

∴ The work delivered in the two cases is given by

$$W_1 = 12000 \times 0.4806 = 5767.2 \text{ kJ/min}$$

and $W_2 = 120000 \times 0.102 = 12240$ kJ/min.

Thus, choose **source 2. (Ans.)**

Note. The source 2 is selected even though *efficiency* in this case is *lower*, because the *criterion for selection is the larger output.*

Example 4.12. *A reversible heat engine operates between two reservoirs at temperatures 700°C and 50°C. The engine drives a reversible refrigerator which operates between reservoirs at temperatures of 50°C and – 25°C. The heat transfer to the engine is 2500 kJ and the net work output of the combined engine refrigerator plant is 400 kJ.*

(i) Determine the heat transfer to the refrigerant and the net heat transfer to the reservoir at 50°C ;

(ii) Reconsider (i) given that the efficiency of the heat engine and the C.O.P. of the refrigerator are each 45 per cent of their maximum possible values.

Solution. Refer to Fig. 4.14.

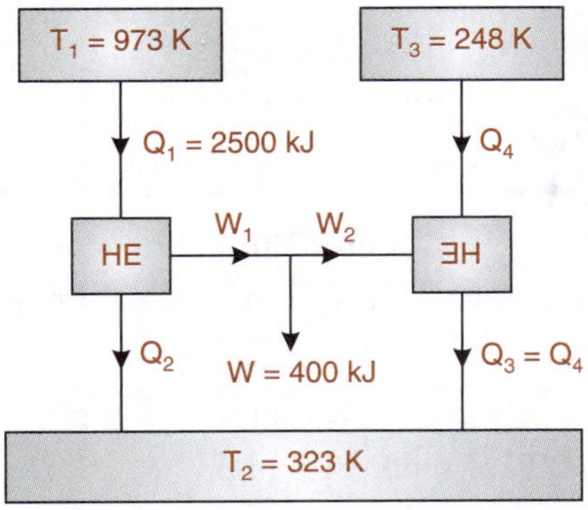

Fig. 4.14

Temperature, $T_1 = 700 + 273 = 973$ K

Temperature, $T_2 = 50 + 273 = 323$ K

Temperature, $T_3 = -25 + 273 = 248$ K

The heat transfer to the heat engine, $Q_1 = 2500$ kJ

The network output of the combined engine refrigerator plant,

$$W = W_1 - W_2 = 400 \text{ kJ.}$$

(*i*) Maximum efficiency of the heat engine cycle is given by

$$\eta_{max} = 1 - \frac{T_2}{T_1} = 1 - \frac{323}{973} = 0.668$$

Again, $\frac{W_1}{Q_1} = 0.668$

∴ $W_1 = 0.668 \times 2500 = 1670$ kJ

$$(\text{C.O.P.})_{max} = \frac{T_3}{T_2 - T_3} = \frac{248}{323 - 248} = 3.306$$

Also, C.O.P. $= \frac{Q_4}{W_2} = 3.306$

Since, $W_1 - W_2 = W = 400$ kJ

$$W_2 = W_1 - W = 1670 - 400 = 1270 \text{ kJ}$$

∴ $Q_4 = 3.306 \times 1270 = 4198.6$ kJ

$$Q_3 = Q_4 + W_2 = 4198.6 + 1270 = 5468.6 \text{ kJ}$$

$$Q_2 = Q_1 - W_1 = 2500 - 1670 = 830 \text{ kJ.}$$

Heat rejection to the 50°C reservoir

$= Q_2 + Q_3 = 830 + 5468.6$

$=$ **6298.6 kJ. (Ans.)**

(*ii*) Efficiency of actual heat engine cycle,

$$\eta = 0.45\ \eta_{max} = 0.45 \times 0.668 = 0.3$$

∴ $W_1 = \eta \times Q_1 = 0.3 \times 2500 = 750$ kJ

∴ $W_2 = 750 - 400 = 350$ kJ

C.O.P. of the actual refrigerator cycle,

$$\text{C.O.P.} = \frac{Q_4}{W_2} = 0.45 \times 3.306 = 1.48$$

∴ $Q_4 = 350 \times 1.48 =$ **518 kJ. (Ans.)**

$Q_3 = 518 + 350 = 868$ kJ

$Q_2 = 2500 - 750 = 1750$ kJ

Heat rejected to 50°C reservoir

$= Q_2 + Q_3 = 1750 + 868 =$ **2618 kJ. (Ans.)**

Example 4.13. (*i*) *A reversible heat pump is used to maintain a temperature of 0°C in a refrigerator when it rejects the heat to the surroundings at 25°C. If the heat removal rate from the refrigerator is 1440 kJ/min, determine the C.O.P. of the machine and work input required.*

(*ii*) *If the required input to run the pump is developed by a reversible engine which receives heat at 380°C and rejects heat to atmosphere, then determine the overall C.O.P. of the system.*

Solution. Refer to Fig. 4.15 (*a*).

(*i*) Temperature, $T_1 = 25 + 273 = 298$ K

Temperature, $T_2 = 0 + 273 = 273$ K

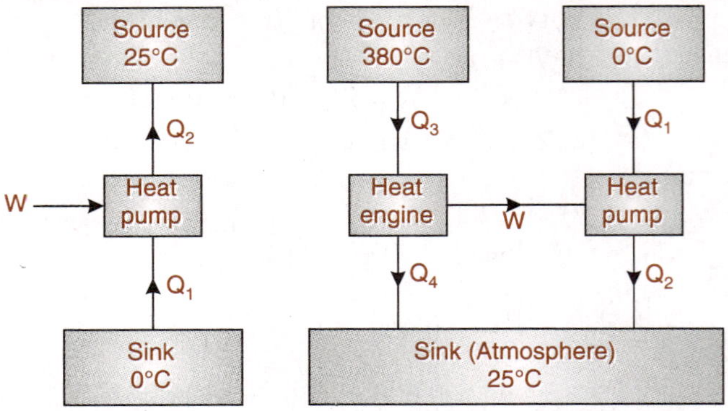

(*a*) Single system (*b*) Combined system

Fig. 4.15

Heat removal rate from the refrigerator,

$$Q_1 = 1440 \text{ kJ/min} = 24 \text{ kJ/s}$$

Now, co-efficient of performance, for reversible heat pump,

$$\textbf{C.O.P.} = \frac{T_1}{T_1 - T_2} = \frac{298}{(298 - 273)}$$

$$= \mathbf{11.92.} \ \textbf{(Ans.)}$$

$$\therefore \ (\text{C.O.P.})_{ref.} = \frac{T_2}{T_1 - T_2} = \frac{273}{298 - 273} = 10.92$$

$$\text{Now,} \quad 10.92 = \frac{Q_1}{W} = \frac{24}{W}$$

$$\therefore \quad W = 2.2 \text{ kW}$$

i.e., **Work input required = 2.2 kW. (Ans.)**

$$Q_2 = Q_1 + W = 24 + 2.2 = 26.2 \text{ kJ/s}$$

(*ii*) Refer to Fig. 4.15 (*b*).

The overall C.O.P. is given by,

$$\text{C.O.P.} = \frac{\text{Heat removed from the refrigerator}}{\text{Heat supplied from the source}}$$

$$= \frac{Q_1}{Q_3} \quad ...(i)$$

For the reversible engine, we can write

$$\frac{Q_3}{T_3} = \frac{Q_4}{T_4}$$

or $$\frac{Q_4 + W}{T_3} = \frac{Q_4}{T_4}$$

or $$\frac{Q_4 + 2.2}{(380 + 273)} = \frac{Q_4}{(25 + 273)}$$

or $$\frac{Q_4 + 2.2}{653} = \frac{Q_4}{298}$$

or $$298(Q_4 + 2.2) = 653\, Q_4$$

or $$Q_4(653 - 298) = 298 \times 2.2$$

or $$Q_4 = \frac{298 \times 2.2}{(653 - 298)} = 1.847 \text{ kJ/s}$$

$$\therefore \quad Q_3 = Q_4 + W = 1.847 + 2.2$$

$$= 4.047 \text{ kJ/s}$$

Substituting this value in eqn. (*i*), we get

$$\textbf{C.O.P.} = \frac{24}{4.047} = \mathbf{5.93.} \ \textbf{(Ans.)}$$

If the purpose of the system is to supply the heat to the sink at 25°C, then

$$\textbf{Overall C.O.P.} = \frac{Q_2 + Q_4}{Q_3} = \frac{26.2 + 1.847}{4.047}$$

$$= \mathbf{6.93.} \ \textbf{(Ans.)}$$

Example 4.14. *An ice plant working on a reversed Carnot cycle heat pump produces 15 tonnes of ice per day. The ice is formed from water at 0°C and the formed ice is maintained at 0°C. The heat is rejected to the atmosphere at 25°C. The heat pump used to run the ice plant is coupled to a Carnot engine which absorbs heat from a source which is maintained at 220°C by burning liquid fuel of 44500 kJ/kg calorific value and rejects the heat to the atmosphere. Determine :*

(*i*) *Power developed by the engine ;*

(*ii*) *Fuel consumed per hour.*

Take enthalpy of fusion of ice = 334.5 kJ/kg.

Solution. (*i*) Fig. 4.16 shows the arrangement of the system.

Amount of ice produced per day = 15 tonnes.

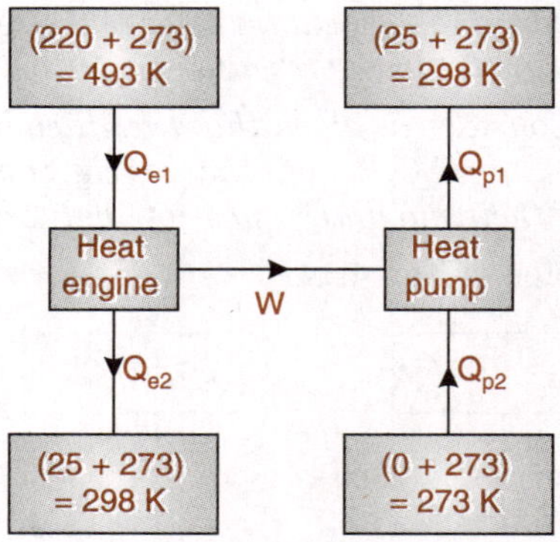

Fig. 4.16

∴ The amount of heat removed by the heat pump,

$$Q_{p_2} = \frac{15 \times 1000 \times 334.5}{24 \times 60} = 3484.4 \text{ kJ/min}$$

C.O.P. of the heat pump

$$= \frac{Q_{p_2}}{W} = \frac{273}{298 - 273}$$

$$\therefore \quad W = Q_{p_2} \times \frac{298 - 273}{273} = 3484.4 \times \frac{25}{273}$$

$$= 319.08 \text{ kJ/min}$$

This work must be developed by the Carnot engine,

$$W = \frac{319.08}{60} = 5.3 \text{ kJ/s} = 5.3 \text{ kW}$$

Thus **power developed by the engine**

$$= \mathbf{5.3\ kW.} \quad \textbf{(Ans.)}$$

(*ii*) The efficiency of Carnot engine is given by

$$\eta_{carnot} = \frac{W}{Q_{e_1}} = 1 - \frac{298}{493} = 0.396$$

$$\therefore \quad Q_{e_1} = \frac{W}{0.396} = \frac{5.3}{0.396} = 13.38 \text{ kJ/s}$$

$$\therefore \quad Q_{e_1 (per\ hour)} = 13.38 \times 60 \times 60 = 48168 \text{ kJ}$$

$\therefore$ **Quantity of fuel consumed/hour**

$$= \frac{48168}{44500} = \mathbf{1.082\ kg/h.} \quad \textbf{(Ans.)}$$

Example 4.15. *Two Carnot engines work in series between the source and sink temperatures of 550 K and 350 K. If both engines develop equal power determine the intermediate temperature.*

Solution. Fig. 4.17 shows the arrangement of the system.

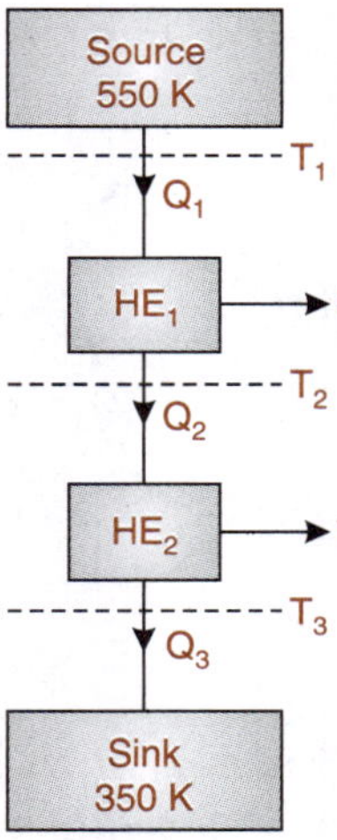

Fig. 4.17

Temperature of the source,

$$T_1 = 550 \text{ K}$$

Temperature of the sink,

$$T_3 = 350 \text{ K}$$

Intermediate temperature, T_2 :

The efficiencies of the engines HE_1 and HE_2 are given by

$$\eta_1 = \frac{W}{Q_1} = \frac{T_1 - T_2}{T_1} = \frac{W}{Q_2 + W} \quad ...(i)$$

$$\eta_2 = \frac{W}{Q_2} = \frac{T_2 - T_3}{T_2} = \frac{W}{Q_3 + W} \quad ...(ii)$$

From eqn. (*i*), we get

$$W = (Q_2 + W)\left(\frac{T_1 - T_2}{T_1}\right)$$

$$\therefore \quad W\left[1 - \left(\frac{T_1 - T_2}{T_1}\right)\right] = Q_2\left(\frac{T_1 - T_2}{T_1}\right)$$

$$\therefore \quad W\left(\frac{T_2}{T_1}\right) = Q_2\left(\frac{T_1 - T_2}{T_1}\right)$$

$$\therefore \quad W = Q_2\left(\frac{T_1 - T_2}{T_2}\right) \quad ...(iii)$$

From eqn. (*ii*), we get

$$W = Q_2\left(\frac{T_2 - T_3}{T_2}\right) \quad ...(iv)$$

Now from eqns. (*iii*) and (*iv*), we get

$$T_1 - T_2 = T_2 - T_3$$

$$2T_2 = T_1 + T_3 = 550 + 350$$

$$\therefore \quad T_2 = 450 \text{ K}$$

Hence, **intermediate temperature**

$$= \mathbf{450\ K.} \quad \textbf{(Ans.)}$$

Example 4.16. *A Carnot heat engine draws heat from a reservoir at temperature T_1 and rejects heat to another reservoir at temperature T_3. The Carnot forward cycle engine drives a Carnot reversed cycle engine or Carnot refrigerator which absorbs heat from reservoir at temperature T_2 and rejects heat to a reservoir at temperature T_3. If the high temperature T_1 = 600 K and low temperature T_2 = 300 K, determine :*

(*i*) *The temperature T_3 such that heat supplied to engine Q_1 is equal to the heat absorbed by refrigerator Q_2.*

(*ii*) *The efficiency of Carnot engine and C.O.P. of Carnot refrigerator.*

Solution. Refer to Fig. 4.18.

Temperature, T_1 = 600 K

Temperature, T_2 = 300 K

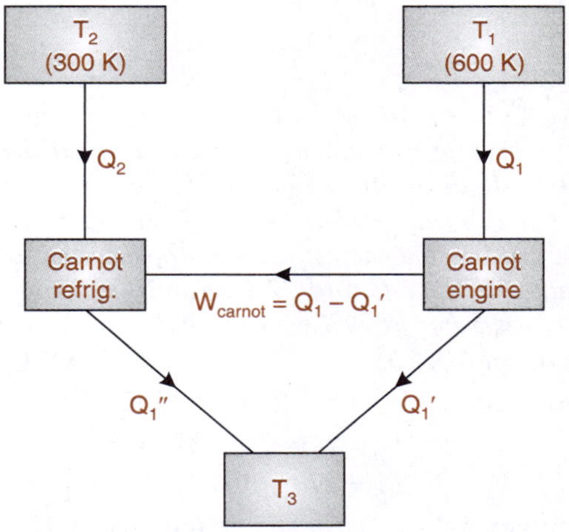

Fig. 4.18

Efficiency of Carnot engine,

$$\eta_{carnot\ engine} = \frac{Q_1 - Q_1'}{Q_1} = \frac{T_1 - T_3}{T_1}$$

$$= \frac{\text{Work of Carnot engine}}{\text{Heat supplied to the Carnot engine}}$$

$$= \frac{W_{carnot}}{Q_1}$$

or

$$W_{carnot} = Q_1\left(\frac{T_1 - T_3}{T_1}\right) \quad ...(i)$$

Also,

$$\text{C.O.P.}_{(carnot\ refrigerator)} = \frac{Q_2}{Q_1'' - Q_2} = \frac{T_2}{T_3 - T_2}$$

$$= \frac{\text{Heat absorbed}}{W_{carnot}} = \frac{Q_2}{W_{carnot}}$$

or $$W_{carnot} = Q_2\left(\frac{T_3 - T_2}{T_2}\right) \quad ...(ii)$$

(*i*) **Temperature, T_3 :**

From eqns. (*i*) and (*ii*), we get

$$Q_1\left(\frac{T_1 - T_3}{T_1}\right) = Q_2\left(\frac{T_3 - T_2}{T_2}\right)$$

$$\therefore \quad \frac{Q_2}{Q_1} = \frac{T_2}{T_1}\left(\frac{T_1 - T_3}{T_3 - T_2}\right)$$

or $$\frac{Q_2}{Q_1} = 1 = \frac{300}{600}\left(\frac{600 - T_3}{T_3 - 300}\right)$$

or $$600 - T_3 = 2(T_3 - 300)$$

$$600 - T_3 = 2T_3 - 600 \quad \text{or} \quad T_3 = 400 \text{ K}$$

Hence, **temperature, T_3 = 400 K. (Ans.)**

(*ii*) **Efficiency of Carnot engine and C.O.P. of Carnot refrigerator :**

$$\eta_{carnot\ engine} = \frac{T_1 - T_3}{T_1} = \frac{600 - 400}{600}$$

$$= 0.3333 = \mathbf{33.33\%.} \quad \textbf{(Ans.)}$$

$$\text{C.O.P.}_{refrigerator} = \frac{T_2}{T_3 - T_2} = \frac{300}{400 - 300} = \mathbf{3.} \quad \textbf{(Ans.)}$$

Example 4.17. *A heat pump working on a reversed carnot cycle takes in energy from a reservoir maintained at 5°C and delivers it to another reservoir where temperature is 77°C. The heat pump derives power for its operation from a reversible engine operating within the higher and lower temperatures of 1077°C and 77°C. For 100 kJ/kg of energy supplied to reservoir at 77°C, estimate the energy taken from the reservoir at 1077°C.* **(U.P.S.C., 1994)**

Solution. *Given :* $T_3 = 5 + 273 = 278$ K ;

$T_2 = T_4 = 77 + 273 = 350$ K ;

$T_1 = 273 + 1077 = 1350$ K ;

Energy taken from the reservoir at 1077°C, Q_1 :

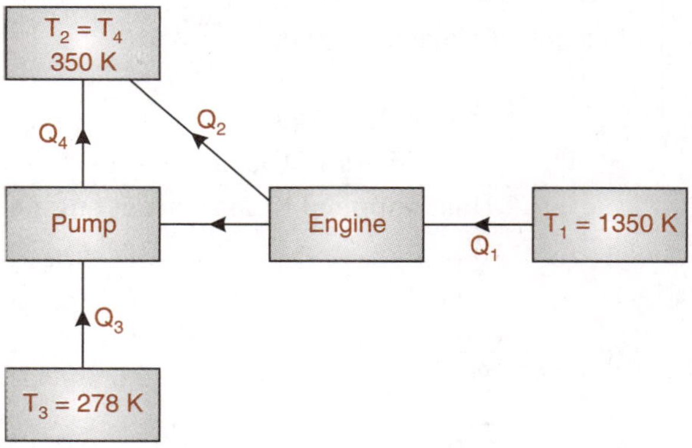

Fig. 4.19

For reversible engine, $$\eta = \frac{Q_1 - Q_2}{Q_1} = \frac{T_1 - T_2}{T_1} \quad ...(i)$$

or $$1 - \frac{Q_2}{Q_1} = 1 - \frac{T_2}{T_1}$$

$$\therefore \quad \frac{Q_1}{Q_2} = \frac{T_1}{T_2}$$

For reversible heat pump,

$$\text{C.O.P.} = \frac{Q_4}{Q_4 - Q_3} = \frac{T_4}{T_4 - T_3} \quad ...(ii)$$

Since work for running the pump is being supplied by the engine

$$\therefore \quad Q_1 - Q_2 = Q_4 - Q_3$$

or $$\frac{Q_1}{T_1}(T_1 - T_2) = \frac{Q_4}{T_4}(T_4 - T_3) \quad \text{[From (i) and (ii)]}$$

$$\therefore \quad \frac{Q_1}{Q_4} = \frac{T_1}{T_4}\left(\frac{T_4 - T_3}{T_1 - T_2}\right)$$

$$= \frac{1350}{350}\left(\frac{350 - 278}{1350 - 350}\right) = 0.278$$

or $$Q_4 = \frac{Q_1}{0.278} = 3.6\ Q_1$$

and $$Q_2 = \frac{T_2}{T_1} \times Q_1 = \frac{350}{1350} Q_1 = 0.259\ Q_1$$

$$\therefore \quad Q_4 + Q_2 = (3.6 + 0.259)\ Q_1 = 100$$

$$\therefore \quad \mathbf{Q_1} = \frac{100}{3.6 + 0.259} = \mathbf{25.9\ kJ.} \quad \textbf{(Ans.)}$$

CLAUSIUS INEQUALITY

☞**Example 4.18.** *300 kJ/s of heat is supplied at a constant fixed temperature of 290°C to a heat engine. The heat rejection takes place at 8.5°C. The following results were obtained :*

(*i*) *215 kJ/s are rejected.*

(*ii*) *150 kJ/s are rejected.*

(*iii*) *75 kJ/s are rejected.*

Classify which of the result report a reversible cycle or irreversible cycle or impossible results.

Solution. Heat supplied at 290°C = 300 kJ/s

Heat rejected at 8.5°C : (*i*) 215 kJ/s, (*ii*) 150 kJ/s, (*iii*) 75 kJ/s.

Applying *Clausius inequality* to the cycle or process, we have :

(*i*) $$\sum_{cycle} \frac{\delta Q}{T} = \frac{300}{290 + 273} - \frac{215}{8.5 + 273}$$

$$= 0.5328 - 0.7637 = -0.2309 < 0.$$

$\therefore$ **Cycle is irreversible. (Ans.)**

(ii) $\sum_{\text{cycle}} \frac{\delta Q}{T} = \frac{300}{290 + 273} - \frac{150}{8.5 + 273}$

$= 0.5328 - 0.5328 = 0$

∴ **Cycle is reversible. (Ans.)**

(iii) $\sum_{\text{cycle}} \frac{\delta Q}{T} = \frac{300}{290 + 273} - \frac{75}{8.5 + 273}$

$= 0.5328 - 0.2664 = 0.2664 > 0.$

This **cycle is impossible** by second law of thermodynamics, *i.e., Clausius inequality*. **(Ans.)**

Example 4.19. *A steam power plant operates between boiler temperature of 160°C and condenser temperature of 50°C. Water enters the boiler as saturated liquid and steam leaves the boiler as saturated vapour. Verify the Clausius inequality for the cycle.*

Given : Enthalpy of water entering boiler = 687 kJ/kg.

Enthalpy of steam leaving boiler = 2760 kJ/kg

Condenser pressure = 0.124 × 10⁵ N/m².

Solution. Boiler temperature, $T_1 = 160 + 273 = 433$ K

Condenser temperature, $T_2 = 50 + 273 = 323$ K

From steam tables :

Enthalpy of water entering boiler, $h_{f_1} = 687$ kJ/kg

Enthalpy of steam leaving boiler, $h_2 = 2760$ kJ/kg

Condenser pressure $= 0.124 \times 10^5$ N/m²

Boiler pressure $= 6.18 \times 10^5$ N/m²

......(corresponding to 160°C)

Enthalpy of vapour leaving the turbine,

$h_3 = 2160$ kJ/kg

(assuming isentropic expansion)

Enthalpy of water leaving the condenser,

$h_{f4} = 209$ kJ/kg

Now Q_{boiler}, $Q_1 = h_2 - h_{f1} = 2760 - 687 = 2073$ kJ/kg

and $Q_{condenser}$, $Q_2 = h_{f4} - h_3 = 209 - 2160$

$= -1951$ kJ/kg

∴ $\sum_{\text{cycle}} \frac{\delta Q}{T} = \frac{Q_1}{T_1} + \frac{Q_2}{T_2} = \frac{2073}{433} + \left(\frac{-1951}{323}\right)$

$= -1.25$ kJ/kg K

< 0. ...... **Proved.**

☞**Example 4.20.** *In a power plant cycle, the temperature range is 164°C to 51°C, the upper temperature being maintained in the boiler where heat is received and the lower temperature being maintained in the condenser where heat is rejected. All other processes in the steady flow cycle are adiabatic. The specific enthalpies at various points are given in Fig. 4.20.*

Verify the Clausius Inequality.

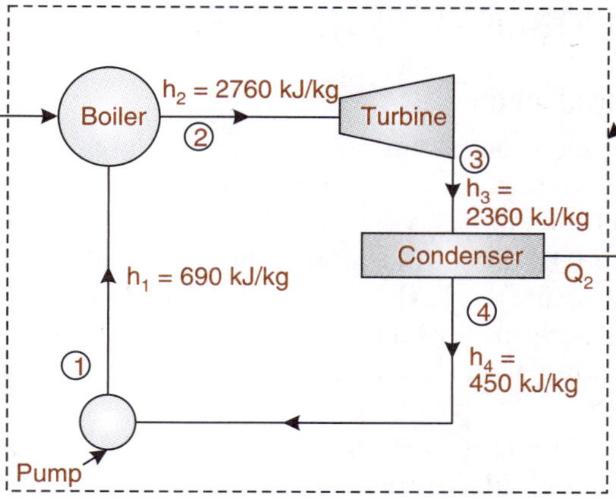

Fig. 4.20

Solution. Temperature maintained in boiler,

$T_1 = 164 + 273 = 437$ K

Temperature maintained in condenser,

$T_2 = 51 + 273 = 324$ K

Heat transferred in the boiler per kg of fluid,

$Q_1 = h_2 - h_1 = 2760 - 690 = 2070$ kJ/kg

Heat transferred out at the condenser per kg of fluid,

$Q_2 = h_4 - h_3 = 450 - 2360 = -1910$ kJ/kg

Since there is no transfer of heat at any other point, we have per kg

$$\sum_{\text{cycle}} \frac{\delta Q}{T} = \frac{Q_1}{T_1} + \frac{Q_2}{T_2} = \frac{2070}{437} + \left(\frac{-1910}{324}\right)$$

$= 4.737 - 5.895$

$= -1.158$ kJ/kg K < 0.

The Clausius Inequality is proved. The steady flow cycle is obviously irreversible.

If the cycle is reversible between the same temperature limits and the heat supplied at higher temperature is same, the heat rejected can be calculated as follows :

$$\eta_{reversible} = 1 - \frac{T_2}{T_1} = 1 - \frac{324}{437} = 0.2586 \text{ or } 25.86\%$$

∴ Heat rejected per kg is given by

$Q_2 = (1 - 0.2586) \times Q_1$

$= (1 - 0.2586) \times 2070 = 1534.7$ kJ/kg

$$\sum_{\text{cycle}} \frac{\delta Q}{T} = \frac{2070}{437} - \frac{1534.7}{324} = 4.73 - 4.73 = 0$$

i.e., $$\sum_{\text{cycle}} \frac{\delta Q}{T} = \frac{Q_{added}}{T_{source}} = \frac{Q_{rejected}}{T_{sink}} = 0$$

Thus **Clausius Equality sign for a reversible engine is verified.**

4.12. ENTROPY

4.12.1. Introduction

In heat engine theory, the term entropy plays a vital role and leads to important results which by other methods can be obtained much more laboriously.

It may be noted that all heat is not equally valuable for converting into work. Heat that is supplied to a substance at high temperature has a greater possibility of conversion into work than heat supplied to a substance at a lower temperature.

"***Entropy** is a function of a quantity of heat which shows the possibility of conversion of that heat into work. The increase in entropy is small when heat is added at a high temperature and is greater when heat addition is made at a lower temperature. Thus for maximum entropy, there is minimum availability for conversion into work and for minimum entropy there is maximum availability for conversion into work*".

4.12.2. Entropy—a Property of a System

Refer to Fig. 4.21. Let us consider a system undergoing a reversible process from state 1 to state 2 along path L and then from state 2 to the original state 1 along path M. Applying the Clausius theorem to this reversible cyclic process, we have

$$\oint_R \frac{\delta Q}{T} = 0$$

(where the subscript designates a reversible cycle)

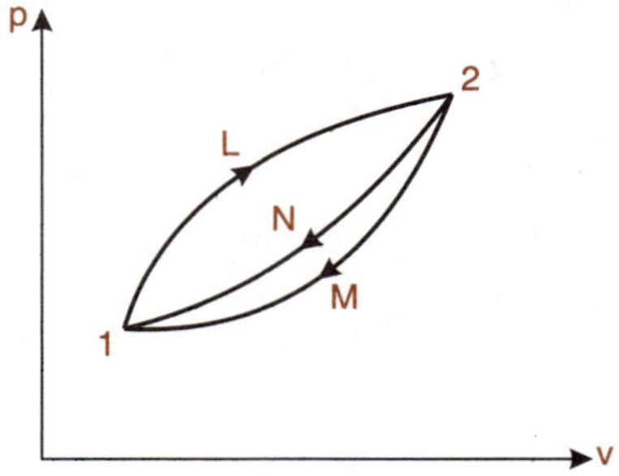

Fig. 4.21. Reversible cyclic process between two fixed end states.

Hence when the system passes through the cycle 1-L-2-M-1, we have

$$\int_{1(L)}^{2} \frac{\delta Q}{T} + \int_{2(M)}^{1} \frac{\delta Q}{T} = 0 \quad ...(4.16)$$

Now consider another reversible cycle in which the system changes from state 1 to state 2 along path L, but returns from state 2 to the original state 1 along a different path N. For this reversible cyclic process, we have

$$\int_{1(L)}^{2} \frac{\delta Q}{T} + \int_{2(N)}^{1} \frac{\delta Q}{T} = 0 \quad ...(4.17)$$

Subtracting equation (4.17) from equation (4.16), we have

$$\int_{2(M)}^{1} \frac{\delta Q}{T} - \int_{2(N)}^{1} \frac{\delta Q}{T} = 0$$

or

$$\int_{1}^{2(M)} \frac{\delta Q}{T} = \int_{1}^{2(N)} \frac{\delta Q}{T}$$

As no restriction is imposed on paths L and M, except that they must be reversible, the *quantity* $\frac{\delta Q}{T}$ *is a function of the initial and final states of the system and is independent of the path of the process. Hence it represents a property of the system.* This property is known as the **"entropy".**

4.12.3. Change of Entropy in a Reversible Process

Refer to Fig. 4.21.

Let S_1 = Entropy at the initial state 1, and

S_2 = Entropy at the final state 2.

Then, the change in entropy of a system, as it undergoes a change from state 1 to 2, becomes

$$S_2 - S_1 = \int_1^2 \left(\frac{\delta Q}{T}\right)_R \quad ...(4.18)$$

Lastly, if the two equilibrium states 1 and 2 are infinitesimal near to each other, the integral sign may be omitted and $S_2 - S_1$ becomes equal to dS.

Hence equation (4.18) may be written as

$$dS = \left(\frac{\delta Q}{T}\right)_R \quad ...(4.19)$$

where dS is an exact differential.

Thus, from equation (4.19), we find that the change of entropy in a reversible process is equal to $\frac{\delta Q}{T}$. *This is the mathematical formulation of the second law of thermodynamics.*

Equation (4.19) indicates that when an *inexact* differential δQ is divided by an integrating factor T during a reversible process, it becomes an *exact differential*.

The ***third law of thermodynamics*** states "*When a system is at zero absolute temperature, the entropy of system is zero*".

It is clear from the above law that the absolute value of entropy corresponding to a given state of the system could be determined by integrating $\left(\frac{\delta Q}{T}\right)_R$ between the state at absolute zero and the given state. *Zero entropy, however, means the absence of all molecular, atomic, electronic and nuclear disorders.*

As it is not practicable to get data at zero absolute temperature, the change in entropy is calculated either between two known states or by selecting some convenient point at which the entropy is given an arbitrary value of zero. For steam, the reference point at which the entropy is given an arbitrary value of zero is 0°C and for refrigerants like ammonia, Freon-12, carbon dioxide etc. the reference point is – 40°C, at which the entropy it taken as zero.

Thus, in practice we can *determine the change in entropy and not the absolute value of entropy.*

4.13. ENTROPY AND IRREVERSIBILITY

We know that change in entropy in a *reversible process* is equal to $\left(\frac{\delta Q}{T}\right)_R$ [eqn. (4.19)]. Let us now find the change in entropy in an *irreversible process.*

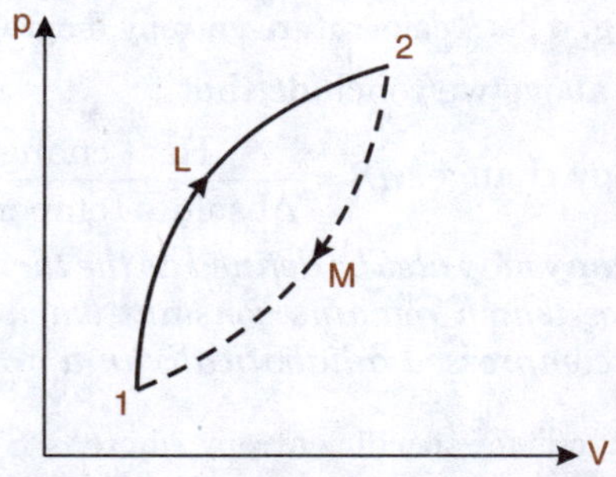

Fig. 4.22. Entropy change for an irreversible process.

Consider a closed system undergoing a change from state 1 to state 2 by a reversible process 1-*L*-2 and returns from state 2 to the initial state 1 by an irreversible process 2-*M*-1 as shown in Fig. 4.22 on the thermodynamic coordinates, pressure and volume.

Since entropy is a thermodynamic property, we can write

$$\oint dS = \int_{1(L)}^{2} (dS)_R + \int_{2(M)}^{1} (dS)_I = 0 \quad ...(4.20)$$

(Subscript *I* represents the irreversible process).

Now for a reversible process, from eqn. (4.19), we have

$$\int_{1(L)}^{2} (dS)_R = \int_{1(L)}^{2} \left(\frac{\delta Q}{T}\right)_R \quad ...(4.21)$$

Substituting the value of $\int_{1(L)}^{2} (dS)_R$ in eqn. (4.20), we get

$$\int_{1(L)}^{2} \left(\frac{\delta Q}{T}\right)_R + \int_{2(M)}^{1} (dS)_I = 0 \quad ...(4.22)$$

Again, since in eqn. (4.20) the processes 1-*L*-2 and 2-*M*-1 together form an irreversible cycle, applying Clausius equality to this expression, we get

$$\oint \frac{\delta Q}{T} = \int_{1(L)}^{2} \left(\frac{\delta Q}{T}\right)_R + \int_{2(M)}^{1} \left(\frac{\delta Q}{T}\right)_I < 0 \quad ...(4.23)$$

Now subtracting eqn. (4.23) from eqn. (4.22), we get

$$\int_{2(M)}^{1} (dS)_I > \int_{2(M)}^{1} \left(\frac{\delta Q}{T}\right)_I$$

which for infinitesimal changes in states can be written as

$$(dS)_I > \int \left(\frac{\delta Q}{T}\right)_I \quad ...(4.24)$$

Eqn. (4.24) states that the *change in entropy in an irreversible process is greater than* $\frac{\delta Q}{T}$.

Combining eqns. (4.23) and (4.24), we can write the equation in the general form as :

$$dS \geq \frac{\delta Q}{T} \quad ...(4.25)$$

where *equality sign stands for the reversible process and inequality sign stands for the irreversible process.*

It may be noted here that the *effect of irreversibility is always to increase the entropy of the system.*

Let us now consider an *isolated system.* We know that in an isolated system, matter, work or heat cannot cross the boundary of the system. Hence according to first law of thermodynamics, the internal energy of the system will remain constant.

Since for an isolated system, $\delta Q = 0$, from eqn. (4.25), we get

$$(dS)_{isolated} \geq 0 \quad ...(4.26)$$

Eqn. (4.26) states that the *entropy of an isolated system either increases or remains constant.* This is a corollary of the second law. It explains the *principle of increase in entropy.*

4.14. CHANGE IN ENTROPY OF THE UNIVERSE

We know that the entropy of an isolated system either increase or remains constant, *i.e.,*

$$(dS)_{isolated} \geq 0$$

By including any system and its surrounding within a single boundary, as shown in Fig. 4.23, an isolated system can be formed. The combination of the

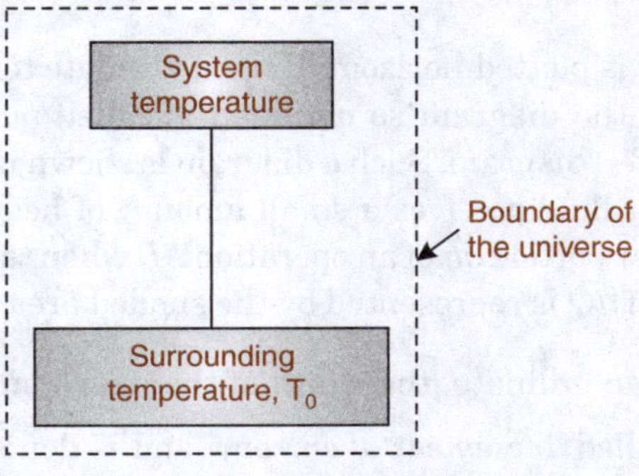

Fig. 4.23. Entropy change of universe.

system and the surroundings within a single boundary is sometimes called the **Universe.** Hence, applying the principle of increase in entropy, we get

$$(dS)_{universe} \geq 0$$

where $(dS)_{universe} = (dS)_{system} + (dS)_{surroundings}$.

In the combined closed system consider that a quantity of heat δQ is transferred from the system at temperature T to the surroundings at temperature T_0. Applying eqn. (4.24) to this process, we can write

$$(dS)_{system} > -\frac{\delta Q}{T}$$

(–ve sign indicates that heat is transferred from the system).

Similarly, since an amount of heat δQ is absorbed by the surroundings, for a reversible process, we can write

$$(dS)_{surroundings} = \frac{\delta Q}{T_0}$$

Hence, the total change in entropy for the combined system

$$(dS)_{system} + (dS)_{surroundings} \geq -\frac{\delta Q}{T} + \frac{\delta Q}{T_0}$$

or $$(dS)_{universe} \geq dQ\left(-\frac{1}{T} + \frac{1}{T_0}\right)$$

The same result can be obtained in the case of an open system.

For both closed and open systems, we can write

$$(dS)_{universe} \geq 0 \quad ...(4.27)$$

Eqn. (4.27) states that the process involving the interaction of a system and the surroundings takes place only if the net entropy of the combined system increases or in the limit remains constant. *Since all natural processes are irreversible, the entropy is increasing continually.*

The entropy attains its maximum value when the system reaches a stable equilibrium state from a non-equilibrium state. This is the state of maximum disorder and is one of maximum thermodynamic probability.

4.15. TEMPERATURE-ENTROPY DIAGRAM

If entropy is plotted-horizontally and absolute temperature vertically the diagram so obtained is called *temperature-entropy* (*T-s*) diagram. Such a diagram is shown in Fig. 4.24. If working fluid receives a small amount of heat dQ in an elementary portion ab of an operation AB when temperature is T, and if dQ is represented by the shaded area of which T is the mean ordinate, the width of the figure must be $\frac{dQ}{T}$. This is called '*increment of entropy*' and is denoted by dS. The total heat received by the operation will be given by the area under the curve AB and $(S_B - S_A)$ will be corresponding increase of entropy.

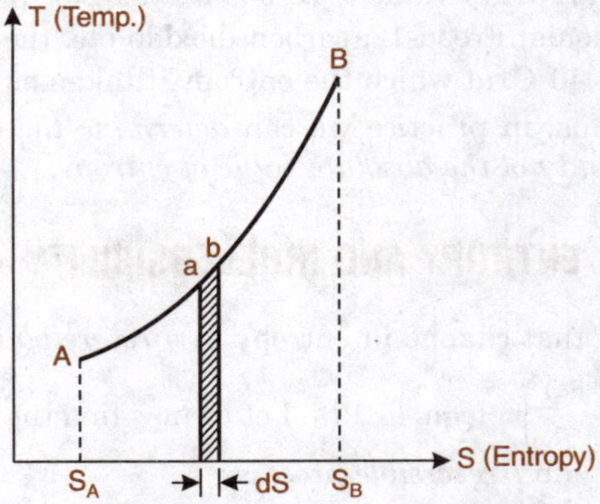

Fig. 4.24. Temperature-entropy diagram.

From above we conclude that :

$$\text{Entropy change, } dS = \frac{\text{Heat change } (Q)}{\text{Absolute temperature } (T)}.$$

"***Entropy*** *may also be defined as the thermal property of a substance which remains constant when substance is expanded or compressed adiabatically in a cylinder*".

Note. '*s*' stands for specific entropy whereas '*S*' means total entropy (*i.e.*, $S = ms$).

4.16. CHARACTERISTICS OF ENTROPY

The characteristics of entropy in a summarised form are given below :

1. It increases when heat is supplied irrespective of the fact whether temperature changes or not.
2. It decrease when heat is removed whether temperature changes or not.
3. It remains unchanged in all adiabatic frictionless processes.
4. It increases if temperature of heat is lowered without work being done as in a throttling process.

4.17. ENTROPY CHANGES FOR A CLOSED SYSTEM

4.17.1. General Case for Change of Entropy of a Gas

Let 1 kg of gas at a pressure p_1, volume v_1, absolute temperature T_1 and entropy s_1, be heated such that its final pressure, volume, absolute temperature and entropy are p_2, v_2, T_2 and s_2 respectively. Then by law of conservation of energy,

$$dQ = du + dW$$

where, dQ = Small change of heat,

du = Small internal energy, and

dW = Small change of work done (pdv).

Now, $dQ = c_v dT + pdv$

Dividing both sides by T, we get

$$\frac{dQ}{T} = \frac{c_v dT}{T} + \frac{pdv}{T}$$

But $\frac{dQ}{T} = ds$

and as $pv = RT$

$$\therefore \quad \frac{p}{T} = \frac{R}{v}$$

Hence $ds = \frac{c_v dT}{T} + R\frac{dv}{v}$

Integrating both sides, we get

$$\int_{s_1}^{s_2} ds = c_v \int_{T_1}^{T_2} \frac{dT}{T} + R\int_{v_1}^{v_2} \frac{dv}{v}$$

or $$(s_2 - s_1) = c_v \log_e \frac{T_2}{T_1} + R \log_e \frac{v_2}{v_1} \quad ...(4.28)$$

This expression can be reproduced in the following way :

According to the gas equation, we have

$$\frac{p_1 v_1}{T_1} = \frac{p_2 v_2}{T_2} \quad \text{or} \quad \frac{T_2}{T_1} = \frac{p_2}{p_1} \times \frac{v_2}{v_1}$$

Substituting the value of $\frac{T_2}{T_1}$ in eqn. (4.28), we get

$$s_2 - s_1 = c_v \log_e \frac{p_2}{p_1} \times \frac{v_2}{v_1} + R \log_e \frac{v_2}{v_1}$$

$$= c_v \log_e \frac{p_2}{p_1} + c_v \log_e \frac{v_2}{v_1} + R \log_e \frac{v_2}{v_1}$$

$$= c_v \log_e \frac{p_2}{p_1} + (c_v + R) \log_e \frac{v_2}{v_1}$$

$$= c_v \log_e \frac{p_2}{p_1} + c_p \log_e \frac{v_2}{v_1}$$

$$\therefore \quad s_2 - s_1 = c_v \log_e \frac{p_2}{p_1} + c_p \log_e \frac{v_2}{v_1} \quad ...(4.29)$$

Again, from gas equation,

$$\frac{p_1 v_1}{T_1} = \frac{p_2 v_2}{T_2} \quad \text{or} \quad \frac{v_2}{v_1} = \frac{p_1}{p_2} \times \frac{T_2}{T_1}$$

Putting the value of $\frac{v_2}{v_1}$ in eqn. (4.28), we get

$$(s_2 - s_1) = c_v \log_e \frac{T_2}{T_1} + R \log_e \frac{p_1}{p_2} \times \frac{T_2}{T_1}$$

$$= c_v \log_e \frac{T_2}{T_1} + R \log_e \frac{p_1}{p_2} + R \log_e \frac{T_2}{T_1}$$

$$= (c_v + R) \log_e \frac{T_2}{T_1} - R \log_e \frac{p_2}{p_1}$$

$$= c_p \log_e \frac{T_2}{T_1} - R \log_e \frac{p_2}{p_1}$$

$$\therefore \quad s_2 - s_1 = c_p \log_e \frac{T_2}{T_1} - R \log_e \frac{p_2}{p_1}. \quad ...(4.30)$$

4.17.2. Heating a Gas at Constant Volume

Refer to Fig. 4.25. Let 1 kg of gas be heated at constant volume and let the change in entropy and absolute temperature be from s_1 to s_2 and T_1 to T_2 respectively.

Then $Q = c_v(T_2 - T_1)$

Differentiating to find small increment of heat dQ corresponding to small rise in temperature dT.

$$dQ = c_v dT$$

Dividing both sides by T, we get

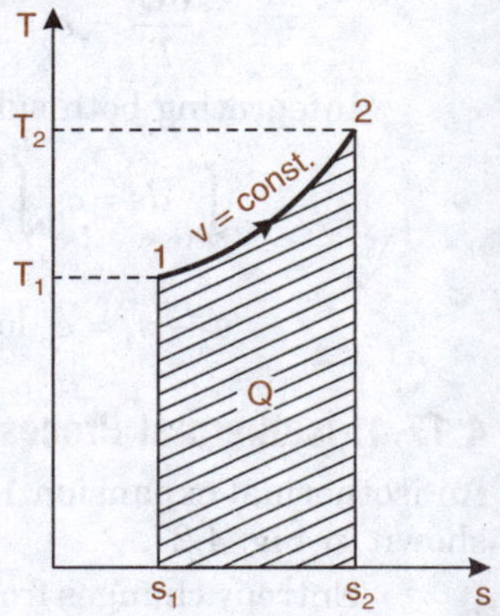

Fig. 4.25. *T-s* diagram : Constant volume process.

$$\frac{dQ}{T} = c_v \cdot \frac{dT}{T}$$

or $$ds = c_v \cdot \frac{dT}{T}$$

Integrating both sides, we get

$$\int_{s_1}^{s_2} ds = c_v \int_{T_1}^{T_2} \frac{dT}{T}$$

or $$s_2 - s_1 = c_v \log_e \frac{T_2}{T_1} \quad ...(4.31)$$

4.17.3. Heating a Gas at Constant Pressure

Refer to Fig. 4.26. Let 1 kg of gas be heated at constant pressure, so that its absolute temperature changes from T_1 to T_2 and entropy s_1 to s_2.

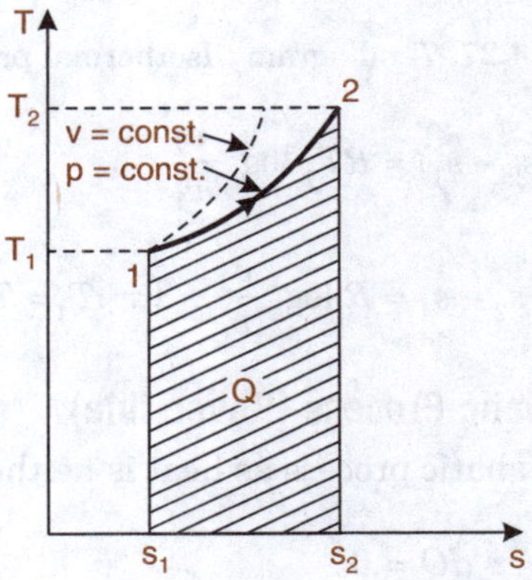

Fig. 4.26. *T-s* diagram : Constant pressure process.

Then, $Q = c_p(T_2 - T_1)$.

Differentiating to find small increase in heat, dQ of this gas when the temperature rise is dT.

$$dQ = c_p \, . \, dT$$

Dividing both sides by T, we get

$$\frac{dQ}{T} = c_p . \frac{dT}{T} \quad \text{or} \quad ds = c_p . \frac{dT}{T}$$

Integrating both sides, we get

$$\int_{s_1}^{s_2} ds = c_p \int_{T_1}^{T_2} \frac{dT}{T}$$

$$s_2 - s_1 = c_p \log_e \frac{T_2}{T_1} \quad \text{...(4.32)}$$

4.17.4. Isothermal Process

An isothermal expansion 1–2 at constant temperature T is shown in Fig. 4.27.

Entropy changes from s_1 to s_2 when gas absorbs heat during expansion. The heat taken by the gas is given by the area under the line 1–2 which also represents the work done during expansion. In other words, $Q = W$.

But $$Q = \int_{s_1}^{s_2} T ds = T(s_2 - s_1)$$

and $$W = p_1 v_1 \log_e \frac{v_2}{v_1}$$

$$= RT_1 \log_e \frac{v_2}{v_1} \text{ per kg of gas } [\because \; p_1 v_1 = RT_1]$$

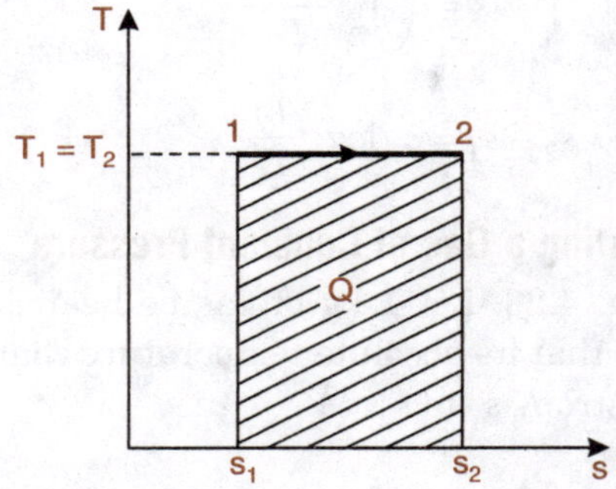

Fig. 4.27. *T-s* diagram : Isothermal process.

$$\therefore \quad T(s_2 - s_1) = RT_1 \log_e \frac{v_2}{v_1}$$

or $$s_2 - s_1 = R \log_e \frac{v_2}{v_1} . \quad [\because \; T_1 = T_2 = T] \quad \text{...(4.33)}$$

4.17.5. Adiabatic Process (Reversible)

During an adiabatic process as heat is neither supplied nor rejected,

$$dQ = 0$$

or $$\frac{dQ}{dT} = 0 \quad \text{or} \quad ds = 0 \quad \text{...(4.34)}$$

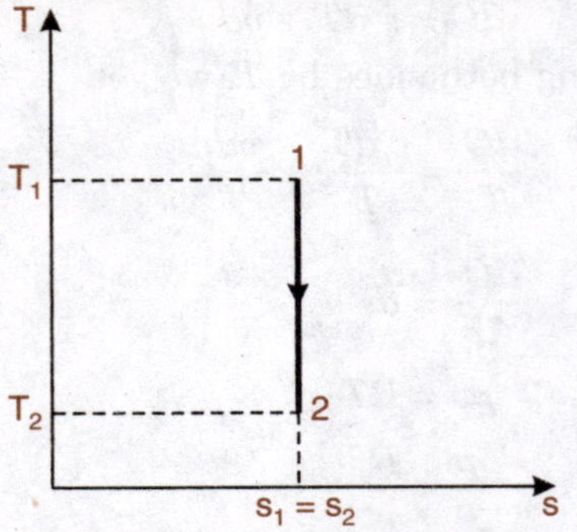

Fig. 4.28. *T-s* diagram : Adiabatic process.

This shows that there is no change in entropy and hence it is known as *isentropic process.*

Fig. 4.28 represents an adiabatic process. It is a vertical line (1–2) and therefore area under this line is nil ; hence heat supplied or rejected and entropy change is zero.

4.17.6. Polytropic Process

Refer to Fig. 4.29.

The expression for 'entropy change' in polytropic process (pv^n = constant) can be obtained from eqn. (4.28)

i.e., $$s_2 - s_1 = c_v \log_e \frac{T_2}{T_1} + R \log_e \frac{v_2}{v_1}$$

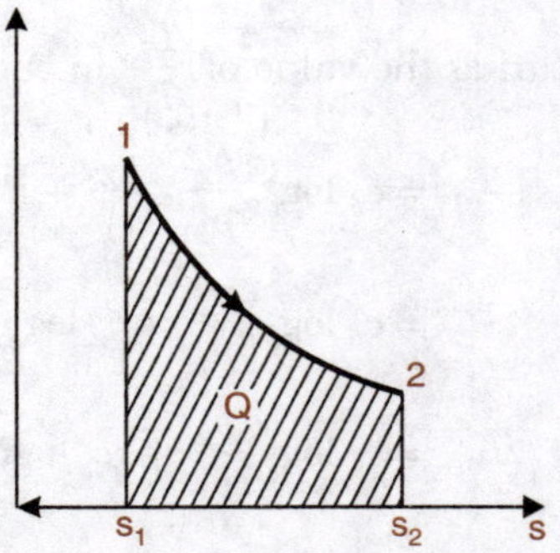

Fig. 4.29. *T-s* diagram : Polytropic process.

Also $$p_1 v_1^n = p_2 v_2^n$$

or $$\frac{p_1}{p_2} = \left(\frac{v_2}{v_1}\right)^n \quad \text{...(i)}$$

Again, as $$\frac{p_1 v_1}{T_1} = \frac{p_2 v_2}{T_2}$$

or $$\frac{p_1}{p_2} = \frac{v_2}{v_1} \times \frac{T_1}{T_2} \quad \text{...(ii)}$$

From (*i*) and (*ii*), we get

$$\left(\frac{v_2}{v_1}\right)^n = \frac{v_2}{v_1} \times \frac{T_1}{T_2}$$

or $$\left(\frac{v_2}{v_1}\right)^{n-1} = \frac{T_1}{T_2} \quad \text{or} \quad \frac{v_2}{v_1} = \left(\frac{T_1}{T_2}\right)^{\frac{1}{n-1}}$$

Substituting the value of $\frac{v_2}{v_1}$ in eqn. (4.28), we get

$$s_2 - s_1 = c_v \log_e \frac{T_2}{T_1} + R \log_e \left(\frac{T_1}{T_2}\right)^{\frac{1}{n-1}}$$

$$= c_v \log_e \frac{T_2}{T_1} + R \left(\frac{1}{n-1}\right) \log_e \frac{T_1}{T_2}$$

$$= c_v \log_e \frac{T_2}{T_1} - R \left(\frac{1}{n-1}\right) \log_e \frac{T_2}{T_1}$$

$$= c_v \log_e \frac{T_2}{T_1} - (c_p - c_v) \times \left(\frac{1}{n-1}\right) \log_e \frac{T_2}{T_1} \qquad [\because R = c_p - c_v]$$

$$= c_v \log_e \frac{T_2}{T_1} - (\gamma \,.\, c_v - c_v) \times \left(\frac{1}{n-1}\right) \log_e \frac{T_2}{T_1} \qquad [\because c_p = \gamma \,.\, c_v]$$

$$= c_v \left[1 - \left(\frac{\gamma - 1}{n-1}\right)\right] \log_e \frac{T_2}{T_1}$$

$$= c_v \left[\frac{(n-1) - (\gamma - 1)}{(n-1)}\right] \log_e \frac{T_2}{T_1}$$

$$= c_v \left(\frac{n - 1 - \gamma + 1}{n-1}\right) \log_e \frac{T_2}{T_1}$$

$$= c_v \,.\, \left(\frac{n-\gamma}{n-1}\right) \log_e \frac{T_2}{T_1} \text{ per kg of gas}$$

$$\therefore \quad s_2 - s_1 = c_v \left(\frac{n-\gamma}{n-1}\right) \log_e \frac{T_2}{T_1} \text{ per kg of gas} \qquad ...(4.35)$$

4.17.7. Approximation for Heat Absorbed

The curve LM shown in Fig. 4.30 is obtained by heating 1 kg of gas from initial state L to final state M. Let temperature during heating increases from T_1 to T_2. Then heat absorbed by the gas will be given by the area (shown shaded) under curve LM.

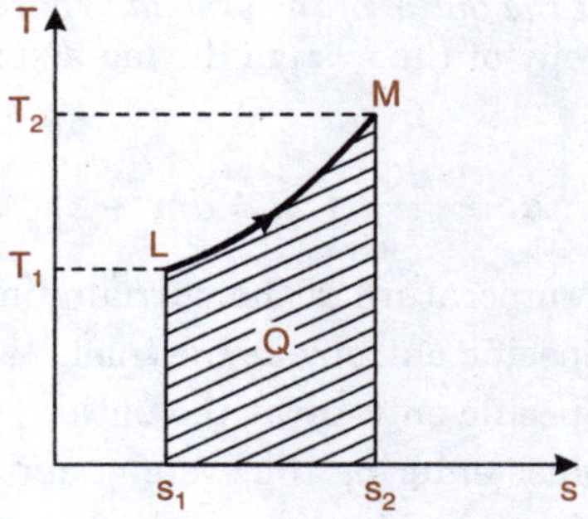

Fig. 4.30

As the curve on T-s diagram which represents the heating of the gas, usually has very slight curvature, it can be assumed a straight line for a small temperature range. Then,

$$\text{Heat absorbed} = \text{Area under the curve } LM$$

$$= (s_2 - s_1) \left(\frac{T_1 + T_2}{2}\right) \qquad ...(4.36)$$

In other words, heat absorbed *approximately* equals the product of change of entropy and mean absolute temperature.

Table 4.1. Summary of Formulae

S. No.	*Process*	*Change of entropy (per kg)*
1.	General case	(*i*) $c_v \log_e \frac{T_2}{T_1} + R \log_e \frac{v_2}{v_1}$ (in terms of T and v) (*ii*) $c_v \log_e \frac{p_2}{p_1} + c_v \log_e \frac{v_2}{v_1}$ (in terms of p and v) (*iii*) $c_p \log_e \frac{T_2}{T_1} - R \log_e \frac{p_2}{p_1}$ (in terms of T and p)
2.	Constant volume	$c_v \log_e \frac{T_2}{T_1}$
3.	Constant pressure	$c_p \log_e \frac{T_2}{T_1}$
4.	Isothermal	$R \log_e \frac{v_2}{v_1}$
5.	Adiabatic	Zero
6.	Polytropic	$c_v \left(\frac{n-\gamma}{n-1}\right) \log_e \frac{T_2}{T_1}$

4.18. ENTROPY CHANGES FOR AN OPEN SYSTEM

In an open system, as compared with closed system, there is additional change of entropy due to the mass crossing the boundaries of the system. *The net change of entropy of a system due to mass transport is equal to the difference between the product of the mass and its specific entropy at the inlet and at the outlet of the system.* Therefore, the total change of entropy of the system during a small interval is given by

$$dS \geq \frac{dQ}{T_0} + \Sigma s_i \,.\, dm_i - \Sigma s_0 \,.\, dm_0$$

where, T_0 = Temperature of the surroundings,
s_i = Specific entropy at the inlet,
s_0 = Specific entropy at the outlet,
dm_i = Mass entering the system, and
dm_0 = Mass leaving the system.
(Subscripts i and 0 refer to inlet and outlet conditions)

The above equation in general form can be written as

$$dS \geq \frac{dQ}{T_0} + \Sigma\, s \,.\, dm \qquad ...(4.37)$$

In equation (4.37) entropy flow *into* the system is considered *positive* and entropy *out-flow* is considered *negative*. The *equality sign is applicable to reversible process* in which the heat interactions and mass transport to and from the system is accomplished reversibly. The *inequality sign is applicable to irreversible processes.*

If equation (4.37) is divided by dt, then it becomes a rate equation and is written as

$$\frac{dS}{dt} \geq \frac{1}{T_0} \cdot \frac{dQ}{dt} + \sum s \,.\, \frac{dm}{dt} \qquad ...(4.38)$$

In a steady-state, steady flow process, the rate of change of entropy of the system $\left(\frac{dS}{dt}\right)$ becomes zero.

$$\therefore \qquad 0 \geq \frac{1}{T_0} \frac{dQ}{dt} + \sum s \,.\, \frac{dm}{dt}$$

or

$$\frac{1}{T_0} \dot{Q} + \Sigma s \,.\, \dot{m} \leq 0 \qquad ...(4.39)$$

where $\qquad \dot{Q} = \frac{dQ}{dt}$ and $\dot{m} = \frac{dm}{dt}$.

For *adiabatic steady flow process*, $\dot{Q} = 0$

$$\Sigma s \,.\, \dot{m} \leq 0 \qquad ...(4.40)$$

If the process is *reversible adiabatic*, then

$$\Sigma s \,.\, \dot{m} = 0 \qquad ...(4.41)$$

4.19. THE THIRD LAW OF THERMODYNAMICS

- The third law of thermodynamics is stated as follow :

  "The entropy of all perfect crystalline solids is zero at absolute zero temperature".

- The third law of thermodynamics, often referred to as *Nernst Law*, provides the basis for the calculation of absolute entropies of substances.

According to this law, if the entropy is zero at $T = 0$, the absolute entropy $s_{ab.}$ of a substance at any temperature T and pressure p is expressed by the expression.

$$s_{ab} = \int_0^{T_s = T_{f1}} c_{ps} \frac{dT}{T} + \frac{h_{sf}}{T_s} + \int_{T_s}^{T_{f2} = T_g} c_{pf} \frac{dT}{T} + \frac{h_{fg}}{T_g} \int_{T_g}^{T} c_{pg} \frac{dT}{T} \qquad ...(4.42)$$

where $T_s = T_{f1} = T_{sf} = T_{sat}$ ...... for fusion,
$T_{f2} = T_g = T_{fg} = T_{sat}$ ...... for vaporisation
c_{ps}, c_{pf}, c_{pg} = Constant pressure specific heats for solids, liquids and gas,
h_{sf}, h_{fg} = Latent heats of fusion and vaporisation.

Thus by putting $s = 0$ at $T = 0$, one may integrate zero kelvin and standard state of 278.15 K and 1 atm., and find the entropy difference.

- Further, it can be shown that the entropy of a crystalline substance at $T = 0$ is not a function of pressure, *viz.*,

$$\left(\frac{\partial s}{\partial p}\right)_{T=0} = 0$$

However, *at temperatures above absolute zero, the entropy is a function of pressure also.* The absolute entropy of a substance at 1 atm pressure can be calculated using eqn. (4.42) ; for pressures different from 1 atm, necessary corrections have to be applied.

ENTROPY

Example 4.21. *An iron cube at a temperature of 400°C is dropped into an insulated bath containing 10 kg water at 25°C. The water finally reaches a temperature of 50°C at steady state. Given that the specific heat of water is equal to 4186 J/kg K. Find the entropy changes for the iron cube and the water. Is the process reversible ? If so why ?* **(GATE)**

Solution. *Given* : Temperature of iron cube
= 400°C = 673 K
Temperature of water = 25°C = 298 K
Mass of water = 10 kg
Temperature of water and cube after equilibrium
= 50°C = 323 K

Specific heat of water, c_{pw} = 4186 J/kg K

Entropy changes for the iron cube and the water :

Is the process reversible ?

Now, Heat lost by iron cube = Heat gained by water

$$m_i\, c_{pi}\,(673 - 323) = m_w\, c_{pw}\,(323 - 298) = 10 \times 4186\,(323 - 298)$$

$$\therefore \quad m_i\, c_{pi} = \frac{10 \times 4186\,(323 - 298)}{(623 - 323)} = 2990$$

where, m_i = Mass of iron, kg, and

c_{pi} = Specific heat of iron, J/kg K

Entropy of iron at 673 K

$$= m_i c_{pi} \ln\left(\frac{673}{273}\right) = 2990 \ln\left(\frac{673}{273}\right)$$

$$= 2697.8 \text{ J/K} \qquad \text{[Taking 0°C as datum]}$$

Entropy of water at 298 K

$$= m_w\, c_{pw} \ln\left(\frac{298}{273}\right) = 10 \times 4186 \ln\left(\frac{298}{273}\right) = 3667.8 \text{ J/K}$$

Entropy of iron at 323 K

$$= 2990 \times \ln\left(\frac{323}{273}\right) = 502.8 \text{ J/K}$$

Entropy of water at 323 K

$$= 10 \times 4186 \ln\left(\frac{323}{273}\right) = 7040.04 \text{ J/K}$$

Changes in entropy of iron

$$= 502.8 - 2697.8 = -2195 \text{ J/K}$$

Change in entropy of water

$$= 7040.04 - 3667.8 = 3372.24 \text{ J/K}$$

Net change in entropy

$$= 3372.24 - 2195 = 1177.24 \text{ J/K}$$

Since $\Delta S > 0$ hence the process is **irrevesible. (Ans.)**

Example 4.22. *An ideal gas is heated from temperature T_1 to T_2 by keeping its volume constant. The gas is expanded back to its initial temperature according to the law pv^n = constant. If the entropy change in the two processes are equal, find the value of n in terms of the adiabatic index γ.* **(U.P.S.C.)**

Solution. Change in entropy during constant volume process

$$= m\, c_v \ln\left(\frac{T_2}{T_1}\right) \qquad ...(i)$$

Change in entropy during polytropic process (pv^n = constant)

$$= m\, c_v \left(\frac{\gamma - n}{n - 1}\right) \ln\left(\frac{T_2}{T_1}\right) \qquad ...(ii)$$

For the same entropy, equating (*i*) and (*ii*), we have

$$\frac{\gamma - n}{n - 1} = 1, \quad \text{or} \quad (\gamma - n) = (n - 1) \quad \text{or} \quad 2n = \gamma + 1$$

$$\therefore \quad \mathbf{n} = \frac{\gamma + 1}{2}. \quad \textbf{(Ans.)}$$

Example 4.23. *Air at 20°C and 1.05 bar occupies 0.025 m³. The air is heated at constant volume until the pressure is 4.5 bar, and then cooled at constant pressure back to original temperature. Calculate :*

(*i*) *The net heat flow from the air.*

(*ii*) *The net entropy change.*

Sketch the process on T-s diagram.

Solution. The processes are shown on a *T-s* diagram in Fig. 4.31.

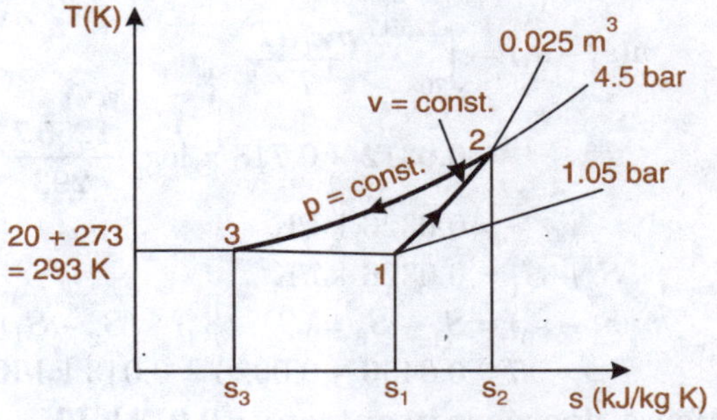

Fig. 4.31

For air :

Temperature, T_1 = 20 + 273 = 293 K

Volume, $V_1 = V_3$ = 0.025 m³

Pressure, p_1 = 1.05 bar = 1.05 × 10⁵ N/m²

Pressure, p_2 = 4.5 bar = 4.5 × 10⁵ N/m².

(*i*) **Net heat flow :**

For a perfect gas (corresponding to point 1 of air),

$$m = \frac{p_1 V_1}{RT_1} = \frac{1.05 \times 10^5 \times 0.025}{0.287 \times 10^3 \times 293} = 0.0312 \text{ kg}$$

For a perfect gas at *constant volume*,

$$\frac{p_1}{T_1} = \frac{p_2}{T_2}$$

$$\frac{1.05}{293} = \frac{4.5}{T_2} \quad \text{or} \quad T_2 = \frac{4.5 \times 293}{1.05} = 1255.7 \text{ K.}$$

At **constant volume,**

$$Q = mc_v\,(T_2 - T_1) = 0.0312 \times 0.718\,(1255.7 - 293)$$

i.e., $\quad Q_{1-2}$ = 21.56 kJ.

Also, at **constant pressure,**

$$Q = m \times c_p \times (T_3 - T_2) = 0.0312 \times 1.005\,(293 - 1255.7)$$

i.e., $\quad Q_{2-3}$ = – 30.18 kJ

$\therefore$ **Net heat flow** $= Q_{1-2} + Q_{2-3}$

$= 21.56 + (-30.18) = -8.62$ kJ

i.e., **Heat rejected = 8.62 kJ. (Ans.)**

(*ii*) **Net entropy change :**

Referring to Fig. 4.31.

Net decrease in entropy,

$$S_1 - S_2 = (S_2 - S_3) - (S_2 - S_1)$$

At **constant pressure,** $dQ = mc_p\, dT$, hence

$$m(s_2 - s_3) = \int_{293}^{1255.7} \frac{mc_p dT}{T}$$

$$= 0.0312 \times 1.005 \times \log_e \frac{1255.7}{293}$$

i.e., $S_2 - S_3 = 0.0456$ kJ/K

At **constant volume,** $dQ = mc_v\, dT$, hence

$$m(s_2 - s_1) = \int_{293}^{1255.7} \frac{mc_v dT}{T}$$

$$= 0.0312 \times 0.718 \times \log_e \frac{1255.7}{293}$$

$= 0.0326$ kJ/K

i.e., $S_2 - S_1 = 0.0326$ kJ/K

$\therefore$ $m(s_1 - s_3) = S_1 - S_3 = (S_2 - S_3) - (S_2 - S_1)$

$= 0.0456 - 0.0326 = 0.013$ kJ/K

Hence, **decrease in entropy = 0.013 kJ/K. (Ans.)**

Note that since *entropy is a property*, the decrease in entropy is given by $S_1 - S_3$, is *independent of the process undergone between states 1 and 3.*

Example 4.24. *0.04 m³ of nitrogen contained in a cylinder behind a piston is initially at 1.05 bar and 15°C. The gas is compressed isothermally and reversibly until the pressure is 4.8 bar. Calculate :*

(*i*) *The change of entropy,*

(*ii*) *The heat flow, and*

(*iii*) *The work done.*

Sketch the process on a p-v and T-s diagram.

Assume nitrogen to act as a perfect gas. Molecular weight of nitrogen = 28.

Solution. Refer to Fig. 4.32.

Initial pressure, $p_1 = 1.05$ bar $= 1.05 \times 10^5$ N/m²

Initial volume, $V_1 = 0.04$ m³

Temperature, $T_1 = 15 + 273 = 288$ K

Final pressure, $p_2 = 4.8$ bar $= 4.8 \times 10^5$ N/m²

Final temperature, $T_2 = T_1 = 288$ K.

The process is shown on a *p-v* and a *T-s* diagram in Figs. 4.32 (*a*) and 4.32 (*b*) respectively. The shaded area in Fig. 4.32 (*a*) represents **work input**, and the shaded area on Fig. 4.32 (*b*) represents **heat rejected.**

Characteristic gas constant,

$$R = \frac{\text{Universal gas constant, } R_0}{\text{Molecular weight, } M}$$

$$= \frac{8314}{28} = 297 \text{ Nm/kg K}$$

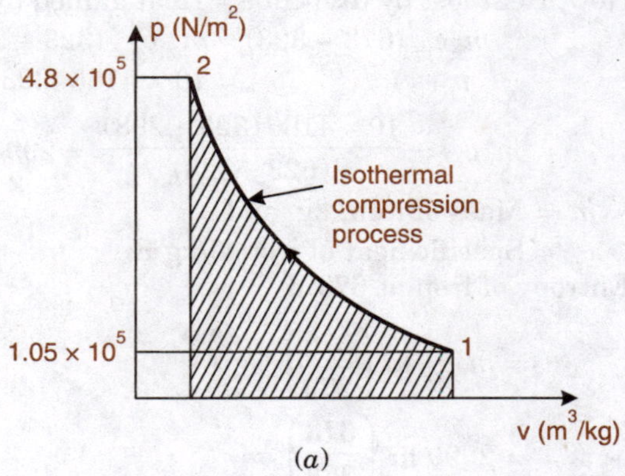

(*a*)

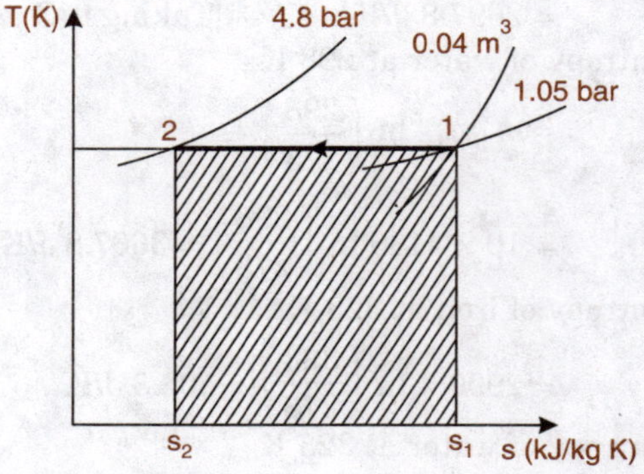

(*b*)

Fig. 4.32

Now, using characteristic gas equation (to find mass '*m*' of nitrogen), we have

$$p_1 V_1 = mRT_1$$

$$m = \frac{p_1 V_1}{RT_1} = \frac{1.05 \times 10^5 \times 0.04}{297 \times 288} = 0.0491 \text{ kg}$$

(*i*) **The change of entropy,**

$$S_2 - S_1 = mR \log_e \frac{p_1}{p_2}$$

$$= 0.0491 \times \frac{297}{10^3} \log_e \left(\frac{1.05}{4.8}\right)$$

i.e., $S_2 - S_1 = -0.02216$ kJ/K.

$\therefore$ **Decrease in entropy, $S_1 - S_2$**

= 0.02216 kJ/K. (Ans.)

(*ii*) **Heat rejected** = Shaded area on Fig. 4.32 (*b*)

$= T(S_1 - S_2) = 288 \times 0.02216$

= 6.382 kJ. (Ans.)

(*iii*) For an isothermal process for a perfect gas,

$$W = Q = 6.382 \text{ kJ}$$

Hence, **the work done on air = 6.382 kJ. (Ans.)**

Example 4.25. *1 kg of gas enclosed in an isolated box of volume v_1, temperature T_1 and pressure p_1 is allowed to expand freely till volume increases to $v_2 = 2v_1$.*

Determine the change in entropy.

Take R for gas as 287 kJ/kg K.

Solution. During the process of free expansion in an isolated box,

$$\Delta U = 0, W = 0 \text{ and } Q = \Delta U + W = 0$$

The process is represented by dotted line on p-v diagram as shown in Fig. 4.33 (a), where $v_2 = 2v_1$.

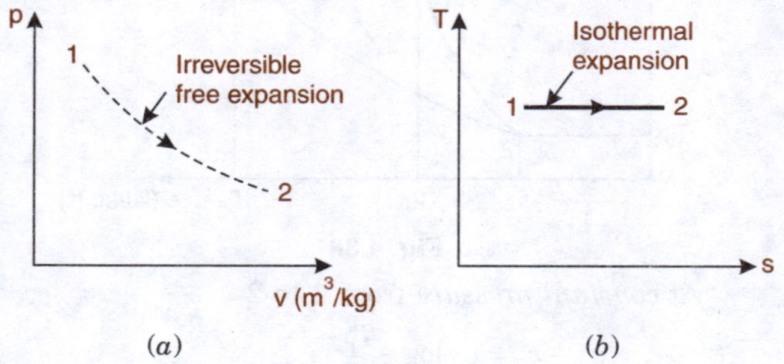

Fig. 4.33

To calculate the entropy change, assume that the irreversible free expansion process is replaced by a reversible isothermal process as temperature in free expansion remains constant, in such a way that the volume increases to double of its original as shown in Fig. 4.33 (b). As the work is developed by the system and heat is given to the system at constant temperature, during isothermal reversible system then as per first law of thermodynamics :

$$\Delta U = 0, Q = W$$

i.e.,

$$Q = \int_{v_1}^{v_2} p \, . dv$$

$$= \int_{v_1}^{v_2} \frac{RT}{v} . dv \qquad \left[\because pv = RT \text{ and } p = \frac{RT}{v} \right]$$

$$= RT \log_e \frac{v_2}{v_1}$$

$$\therefore \quad \frac{Q}{T} = R \log_e \frac{v_2}{v_1}$$

But this is the expression for change in entropy of the system. Entropy being the property of the system, its change is same whether it is reversible or irreversible process.

$\therefore$ For the given process,

$$\Delta s = R \log_e \left(\frac{v_2}{v_1} \right)$$

$$= 287 \log_e (2) \qquad [\because v_2 = 2v_1 \text{ (given)}]$$

$$= 198.9 \text{ kJ/kg K}$$

Hence **change in entropy**

= 198.9 kJ/kg K. (Ans.)

Example 4.26. *0.04 kg of carbon dioxide (molecular weight = 44) is compressed from 1 bar, 20°C, until the pressure is 9 bar, and the volume is then 0.003 m³. Calculate the change of entropy.*

Take c_p for carbon dioxide as 0.88 kJ/kg K, and assume carbon dioxide to be a perfect gas.

Solution. Mass of carbon dioxide, $m = 0.04$ kg

Molecular weight, $M = 44$

Initial pressure, $p_1 = 1$ bar $= 1 \times 10^5$ N/m²

Initial temperature, $T_1 = 20 + 273 = 293$ K

Final pressure, $p_2 = 9$ bar

Final volume, $V_2 = 0.003$ m³

c_p for carbon dioxide $= 0.88$ kJ/kg K

Change of entropy :

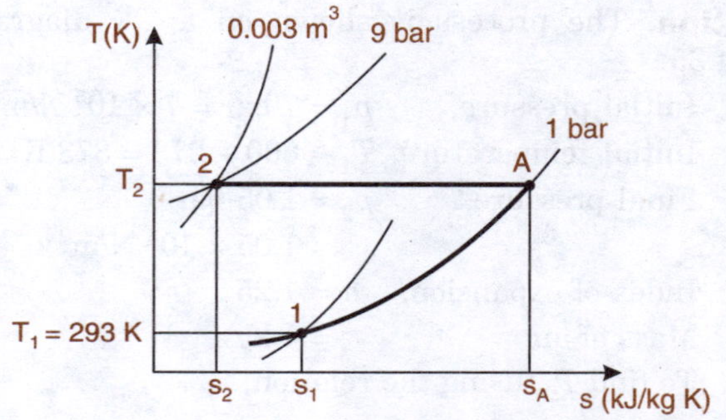

Fig. 4.34

Characteristics gas constant,

$$R = \frac{R_0}{M} = \frac{8314}{44} = 189 \text{ Nm/kg K}$$

To find T_2, using the relation,

$$p_2V_2 = mRT_2$$

$$\therefore \quad T_2 = \frac{p_2V_2}{mR} = \frac{9 \times 10^5 \times 0.003}{0.04 \times 189} = 357 \text{ K}$$

Now, $s_A - s_2 = R \log_e \dfrac{p_2}{p_1} = \dfrac{189}{10^3} \log_e \left(\dfrac{9}{1} \right)$

$= 0.4153$ kJ/kg K

Also at constant pressure from 1 to A

$$s_A - s_1 = c_p \log_e \frac{T_2}{T_1} = 0.88 \log_e \left(\frac{357}{293} \right)$$

$= 0.1738$ kJ/kg K

Then $(s_1 - s_2) = (s_A - s_2) - (s_A - s_1)$

$= 0.4153 - 0.1738 = 0.2415$ kJ/kg K

Hence for 0.04 kg of carbon dioxide **decrease in entropy,**

$$S_1 - S_2 = m(s_1 - s_2) = 0.04 \times 0.2415$$

$=$ **0.00966 kJ/K. (Ans.)**

Note. In short, the change of entropy can be found by using the following relation :

$$(s_2 - s_1) = c_p \log_e \frac{T_2}{T_1} - R \log_e \frac{p_2}{p_1}$$

$$= 0.88 \log_e \left(\frac{357}{293}\right) - \frac{189}{10^3} \log_e \left(\frac{9}{1}\right)$$

$$= 0.1738 - 0.4153 = -0.2415 \text{ kJ/kg K}$$

$$\therefore \quad S_2 - S_1 = m(s_2 - s_1) = 0.04 \times (-0.2415)$$

$$= -0.00966 \text{ kJ/K}$$

(–ve sign means *decrease* in entropy)

or $\quad S_1 - S_2 = 0.00966$ kJ/K.

Example 4.27. *Calculate the change of entropy of 1 kg of air expanding polytropically in a cylinder behind a piston from 7 bar and 600°C to 1.05 bar. The index of expansion is 1.25.*

Solution. The process is shown on a *T-s* diagram in Fig. 4.35.

Initial pressure, $p_1 = 7 \text{ bar} = 7 \times 10^5 \text{ N/m}^2$

Initial temperature, $T_1 = 600 + 273 = 873$ K

Final pressure, $p_2 = 1.05$ bar $= 1.05 \times 10^5 \text{ N/m}^2$

Index of expansion, $n = 1.25$

Mass of air $= 1$ kg

To find T_2, using the relation,

$$\frac{T_2}{T_1} = \left(\frac{p_2}{p_1}\right)^{\frac{n-1}{n}}$$

$$\therefore \quad \frac{T_2}{873} = \left(\frac{1.05}{7}\right)^{\frac{1.25-1}{1.25}}$$

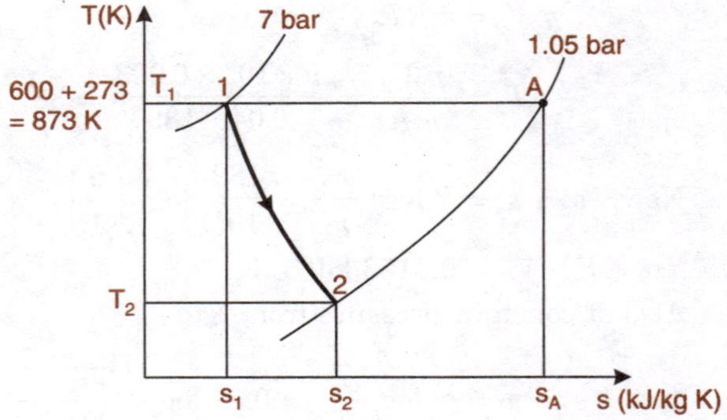

Fig. 4.35

or $$T_2 = 873 \times \left(\frac{1.05}{7}\right)^{\frac{0.25}{1.25}}$$

$$= 873 \times (0.15)^{0.2} = 597.3 \text{ K}.$$

Now replace the process 1 to 2 by processes, 1 to A and A to 2.

Then at *constant temperature* from 1 to A,

$$s_A - s_1 = R \log_e \frac{v_2}{v_1} = R \log_e \frac{p_1}{p_2}$$

$$= 0.287 \log_e \left(\frac{7}{1.05}\right) = 0.544 \text{ kJ/kg K}.$$

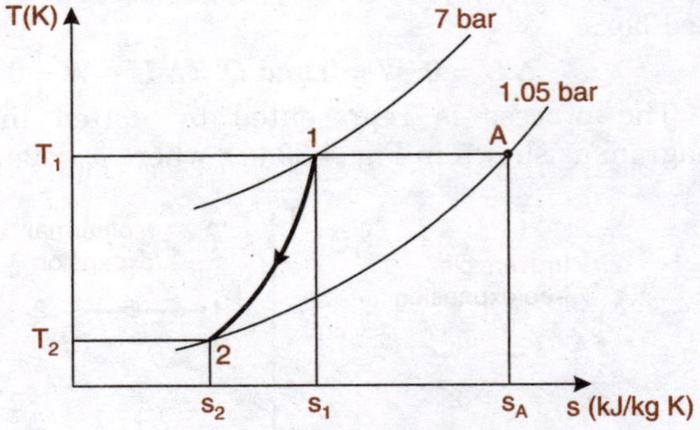

Fig. 4.36

At *constant pressure* from A to 2

$$s_A - s_2 = c_p \log_e \frac{T_1}{T_2}$$

$$= 1.005 \log_e \frac{873}{597.3} = 0.3814 \text{ kJ/kg K}$$

Then $\quad s_2 - s_1 = 0.544 - 0.3814 = 0.1626$ kJ/kg K

i.e., **Increase in entropy = 0.1626 kJ/kg K. (Ans.)**

Note that if in this problem $s_A - s_2$ happened to be greater than $s_A - s_1$, this would mean that s_1 was greater than s_2, and the process should appear as shown in Fig. 4.36.

Note. The change of entropy can also be found by using the following relation :

$$s_2 - s_1 = c_v \left(\frac{n - \gamma}{n - 1}\right) \log_e \frac{T_2}{T_1}$$

$$= 0.718 \left(\frac{1.25 - 1.399}{1.25 - 1}\right) \log_e \left(\frac{597.3}{873}\right)$$

$$\left[\because \ \gamma = \frac{c_p}{c_v} = \frac{1.005}{0.718} = 1.399\right]$$

$$= 0.718 \times (-0.596) \times (-0.3795)$$

$$= 0.1626 \text{ kJ/kg K (increase).}$$

Example 4.28. *In an air turbine the air expands from 7 bar and 460°C to 1.012 bar and 160°C. The heat loss from the turbine can be assumed to be negligible.*

(*i*) *Show that the process is irreversible ;*

(*ii*) *Calculate the change of entropy per kg of air.*

Solution. Refer to Fig. 4.37.

Initial pressure, $p_1 = 7 \text{ bar} = 7 \times 10^5 \text{ N/m}^2$

Initial temperature, $T_1 = 460 + 273 = 733$ K

Final pressure, $p_2 = 1.012$ bar $= 1.012 \times 10^5 \text{ N/m}^2$

Final temperature, $T_2 = 160 + 273 = 433$ K

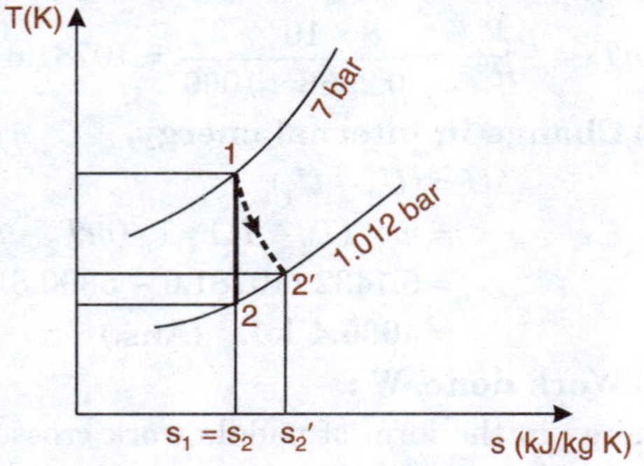

Fig. 4.37

(*i*) **To prove that the process is irreversible :**

Since the heat loss is negligible, the process is ***adiabatic.***

For a reversible adiabatic process for a perfect gas, using the following equation, we have :

$$\frac{T_2}{T_1} = \left(\frac{p_2}{p_1}\right)^{\frac{\gamma-1}{\gamma}}$$

$$\frac{T_2}{733} = \left(\frac{1.012}{7}\right)^{\left(\frac{1.4-1}{1.4}\right)}$$

$$\therefore \quad T_2 = 733 \times \left(\frac{1.012}{7}\right)^{\frac{0.4}{1.4}}$$

$$= 733 \times (0.1446)^{0.286} = 421.6 \text{ K}$$

$$= 421.6 - 273 = 148.6°\text{C}.$$

But the actual temperature is 160°C at the pressure of 1.012 bar, *hence the process is irreversible.* **Proved.**

(*ii*) **Change of entropy per kg of air :**

The change of entropy $s_2' - s_1$, can be found by considering a reversible constant pressure process between 2 and 2′.

$$\therefore \quad s_2' - s_2 = c_p \log_e \frac{T_2'}{T_2} = 1.005 \log_e \frac{433}{421.6}$$

$$= 0.02681 \text{ kJ/kg K}$$

i.e., **Increase of entropy, $s_2' - s_1$**

= 0.02681 kJ/kg K. (Ans.)

Example 4.29. *A fluid undergoes a reversible adiabatic compression from 4 bar, 0.3 m³ to 0.08 m³ according to the law, $pv^{1.25}$ = constant.*

Determine : (i) Change in enthalpy ;

(ii) Change in internal energy ;

(iii) Change in entropy ; (iv) Heat transfer ;

(v) Work transfer.

Solution. Refer to Fig. 4.38.

Initial volume, $V_1 = 0.3$ m³

Initial pressure, $p_1 = 4$ bar $= 4 \times 10^5$ N/m²

Final volume, $V_2 = 0.08$ m³

Law of compression : $pv^{1.25}$ = constant.

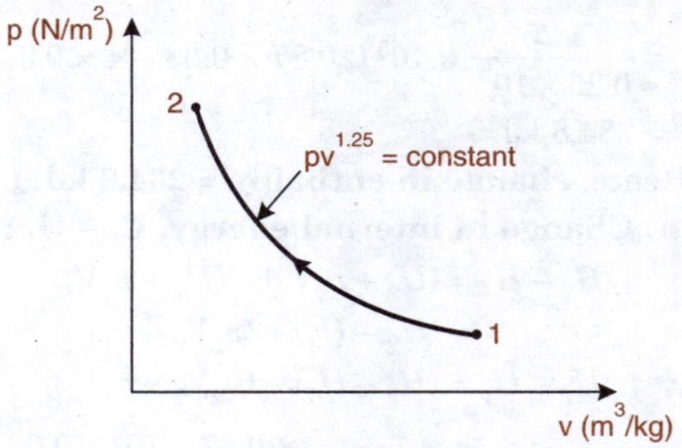

Fig. 4.38

For reversible adiabatic process,

$$p_1V_1^n = p_2V_2^n$$

or

$$\frac{p_2}{p_1} = \left(\frac{V_1}{V_2}\right)^n$$

$$\therefore \quad p_2 = p_1 \times \left(\frac{V_1}{V_2}\right)^n$$

$$= 4 \times \left(\frac{0.3}{0.08}\right)^{1.25} = 20.87 \text{ bar.}$$

(*i*) **Change in enthalpy, $H_2 - H_1$:**

We know that, $$\int_{H_1}^{H_2} dH = \int_{p_1}^{p_2} V dp \quad ...(i)$$

Also $$p_1V_1^n = pV^n$$

$$\therefore \quad V = \left(\frac{p_1V_1^n}{p}\right)^{1/n}$$

Substituting this value of V in eqn. (*i*), we get

$$\int_{H_1}^{H_2} dH = \int_{p_1}^{p_2} \left[\left(\frac{p_1V_1^n}{p}\right)\right]^{1/n} dp = (p_1V_1^n)^{1/n} \int_{p_1}^{p_2} p^{-1/n}\, dp$$

$$= (p_1V_1^n)^{1/n} \left[\frac{p^{-\frac{1}{n}+1}}{-\frac{1}{n}+1}\right]_{p_1}^{p_2}$$

$$= [(p_1V_1^n)]^{1/n} \left[\frac{p_2^{\left(1-\frac{1}{n}\right)} - p_1^{\left(1-\frac{1}{n}\right)}}{1-\frac{1}{n}}\right]$$

$$= (p_1V_1^n)^{1/n} \times \frac{n}{(n-1)}\left[(p_2)^{\left(1-\frac{1}{n}\right)} - (p_1)^{\left(1-\frac{1}{n}\right)}\right]$$

$$= \frac{n(p_2V_2 - p_1V_1)}{(n-1)} \qquad [\because \; p_1V_1^n = p_2V_2^n]$$

$$= \frac{1.25}{(1.25-1)\times 10^3}[20.87 \times 10^5 \times 0.08 - 4 \times 10^5 \times 0.3] \text{ kJ}$$

$$= \frac{1.25}{0.25 \times 10^3} \times 10^5 (20.87 \times 0.08 - 4 \times 0.3) \text{ kJ}$$

$= 234.8$ kJ.

Hence, **change in enthalpy = 234.8 kJ. (Ans.)**

(*ii*) **Change in internal energy, $U_2 - U_1$:**

$$H_2 - H_1 = (U_2 + p_2V_2) - (U_1 + p_1V_1)$$
$$= (U_2 - U_1) + (p_2V_2 - p_1V_1)$$
$$\therefore \quad U_2 - U_1 = (H_2 - H_1) - (p_2V_2 - p_1V_1)$$
$$= 234.8 - \left(\frac{20.87 \times 10^5 \times 0.08 - 4 \times 10^5 \times 0.3}{10^3}\right) \text{kJ}$$
$$= 234.8 - 46.96 = 187.84 \text{ kJ.}$$

Hence, **change in internal energy = 187.84 kJ. (Ans.)**

(*iii*) **Change in entropy,** $S_2 - S_1 = 0$. **(Ans.)**

(*iv*) **Heat transfer,** $Q_{1-2} = 0$. **(Ans.)**

(*v*) **Work transfer, W_{1-2} :**

$$Q_{1-2} = (U_2 - U_1) + W_{1-2}$$
$$\therefore \quad Q_{1-2} = Q_{1-2} - (U_2 - U_1)$$
$$= 0 - 187.84 = -187.84 \text{ kJ}$$

Hence, **work done on the fluid = 187.84 kJ. (Ans.)**

☞**Example 4.30.** *An insulated cylinder of volume capacity 4 m³ contains 20 kg of nitrogen. Paddle work is done on the gas by stirring it till the pressure in the vessel gets increased from 4 bar to 8 bar. Determine :*

(*i*) *Change in internal energy,*
(*ii*) *Work done,*
(*iii*) *Heat transferred, and*
(*iv*) *Change in entropy.*

Take for nitrogen : c_p = 1.04 kJ/kg K, and c_v = 0.7432 kJ/kg K.

Solution. Pressure, $p_1 = 4 \text{ bar} = 4 \times 10^5 \text{ N/m}^2$

Pressure, $p_2 = 8 \text{ bar} = 8 \times 10^5 \text{ N/m}^2$

Volume, $V_1 = V_2 = 4 \text{ m}^3$

and it is constant for both end states.

Now, $$\frac{p_1}{T_1} = \frac{p_2}{T_2}$$

or $$\frac{T_2}{T_1} = \frac{p_2}{p_1} = \frac{8 \times 10^5}{4 \times 10^5} = 2$$

Also, $$R = c_p - c_v = 1.04 - 0.7432 = 0.2968 \text{ kJ/kg K.}$$

The mass of the gas in the cylinder is given by

$$m = \frac{pV}{RT} \quad \text{or} \quad mT = \frac{pV}{R}$$

$$\therefore \quad mT_1 = \frac{p_1V_1}{R} = \frac{4 \times 10^5 \times 4}{0.2968 \times 1000} = 5390.8 \text{ kg K}$$

and $$mT_2 = \frac{p_2V_2}{R} = \frac{8 \times 10^5 \times 4}{0.2968 \times 1000} = 10781.6 \text{ kg K.}$$

(*i*) **Change in internal energy,**

$$\Delta U = (U_2 - U_1)$$
$$= mc_v (T_2 - T_1) = c_v (mT_2 - mT_1)$$
$$= 0.7432 (10781.6 - 5390.8)$$
$$= \mathbf{4006.4 \text{ kJ.}} \quad \textbf{(Ans.)}$$

(*ii*) **Work done, W :**

Energy in the form of paddle work crosses into the system, but there is no change in system boundary or *pdv* work is absent. No heat is transferred to the system. We have

$$Q_{1-2} = (U_2 - U_1) + W_{1-2}$$

But $$Q_{1-2} = 0$$

$$\therefore \quad W_{1-2} = -(U_2 - U_1)$$
$$= \mathbf{-4006.4 \text{ kJ or kNm.}} \quad \textbf{(Ans.)}$$

(*iii*) **Heat transferred,** $Q_{1-2} = 0$. **(Ans.)**

(*iv*) **Change in entropy,**

$$S_2 - S_1 = mc_v \log_e \frac{T_2}{T_1}$$

For constant volume process

$$= 20 \times 0.7432 \log_e 2$$
$$= \mathbf{10.3 \text{ kJ/K.}} \quad \textbf{(Ans.)}$$

Example 4.31. *A certain gas has a specific heat at constant volume of 1.25 kJ/kg K. When it is expanded reversibly and adiabatically from a specific volume of 0.0624 m³/kg and a temperature of 530 K to a specific volume of 0.186 m³/kg its temperature falls by 165 K. When it is expanded into an evacuated space from the same initial condition to the same final specific volume its temperature falls only by 25 K.*

Find the change in entropy in each of the adiabatic processes.

Solution. Refer to Fig. 4.39.

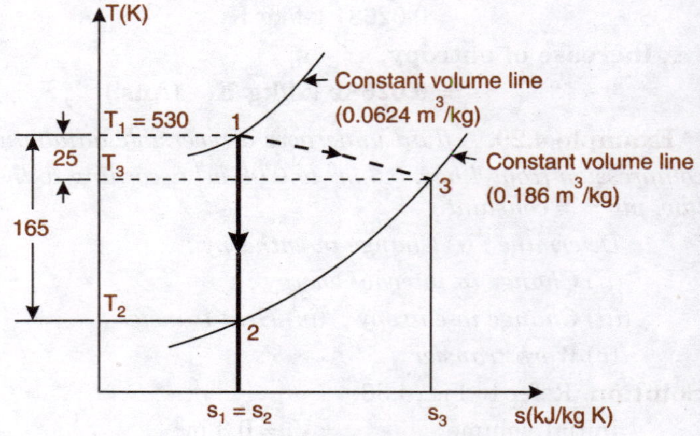

Fig. 4.39

Specific heat of gas at constant volume,

$$c_v = 1.25 \text{ kJ/kg K}$$

Initial specific volume, $v_1 = 0.0624$ m³/kg
Initial temperature, $T_1 = 530$ K
Final specific volume, $v_2 = 0.186$ m³/kg
Temperature fall when expanded reversibly and adiabatically = 165 K
Temperature fall when expanded into an evacuated space = 25 K

Change in entropy :

Path 1-2 : *Reversible adiabatic process.*

Change in entropy, $(s_2 - s_1) = 0$.

Path 1-3 : *Adiabatic process such that*

$$v_3 = 0.186 \text{ m}^3/\text{kg} \ (= v_2)$$

(States 2 and 3 lie on the same constant volume line on *T-s* diagram)

and $\quad T_1 - T_3 = 25$ K.

Change in entropy during this adiabatic process $= s_3 – s_1$.

To calculate $(s_3 – s_1)$ a reversible path has to be selected joining 3 and 1.

This is achieved by selecting the reversible adiabatic path 1-2 and the reversible constant volume process 2-3.

$$s_3 - s_1 = (s_3 - s_2) + (s_2 - s_1)$$
$$= (s_3 - s_2) + 0 = (s_3 - s_2)$$
$$= c_v \log_e \frac{T_3}{T_2} = 1.25 \log_e \left(\frac{530-25}{530-165}\right)$$
$$= 1.25 \log_e \left(\frac{505}{365}\right)$$

$= \mathbf{0.4058\ kJ/kg\ K.}$ **(Ans.)**

Example 4.32. *A heat pump operates between two identical bodies which are at temperature T_1 and cools one of the bodies to a temperature T_2 $(T_2 < T_1)$. Prove that for this operation the minimum work required by the heat pump is given by*

$$W = c_p\left(\frac{T_1^2}{T_2^{\ 2}} + T_2 - 2T_1\right)$$

where c_p is the specific heat which is same for both the bodies.

Solution. The arrangement is shown in Fig. 4.40.

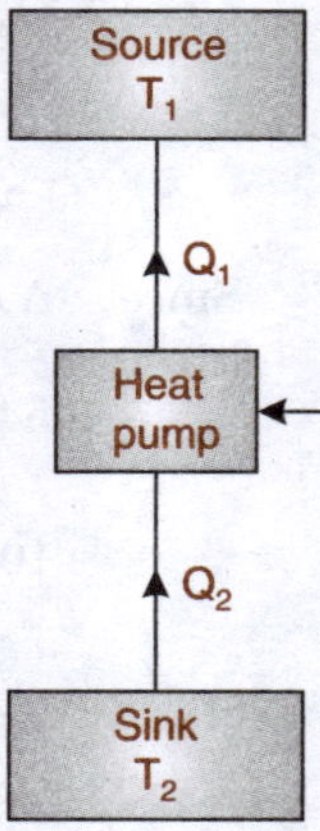

Fig. 4.40

For the minimum work absorbed by the heat pump, the heat pump must be reversed Carnot cycle engine and the required condition is

$$\oint\left(\frac{dQ}{T}\right) = 0$$

For infinitely small changes, we can write

$$c_p \frac{dT_1}{T_1} + c_p \frac{dT_2}{T_2} = 0$$

If T_1' is the final temperature of the high level reservoir, then the above equation can be written as

$$c_p \int_{T_1}^{T_1'}\left(\frac{dT_1}{T_1}\right) + c_p \int_{T_1}^{T_2}\left(\frac{dT_2}{T_2}\right) = 0$$

$$\therefore \quad \log_e\left(\frac{T_1'}{T_1}\right) + \log_e\left(\frac{T_2}{T_1}\right) = 0 = \log_e(1)$$

$$\therefore \quad \log_e\left(\frac{T_1' T_2}{T_1^{\ 2}}\right) = \log_e(1)$$

$$\therefore \quad T_1' = \frac{T_1^2}{T_2}$$

Now the work given to the heat pump
= Heat rejected at higher level temperature
– heat picked up at lower level temperature

$$\therefore \quad W = c_p \int_{T_1}^{T_1'} (dT) - c_p \int_{T_2}^{T_1} (dT)$$
$$= c_p\,[(T_1' - T_1) - (T_1 - T_2)]$$
$$= c_p\,(T_1' + T_2 - 2T_1)$$

Now substituting the value of T_1' in the above equation in terms of T_1 and T_2

$$W = c_p\left[\frac{T_1^2}{T_2} + T_2 - 2T_1\right].$$ **Proved.**

☞**Example 4.33.** *The connections of a reversible engine to three sources at 500 K, 400 K and 300 K are shown in Fig. 4.41. It draws 1500 kJ/min of energy from the source at 800 K and develops 200 kJ/min of work.*

(i) Determine the heat interactions with the other two sources of heat.

(ii) Evaluate the entropy change due to each heat interaction with the engine.

(iii) Total entropy change during the cycle.

Solution. Refer to Fig. 4.41.

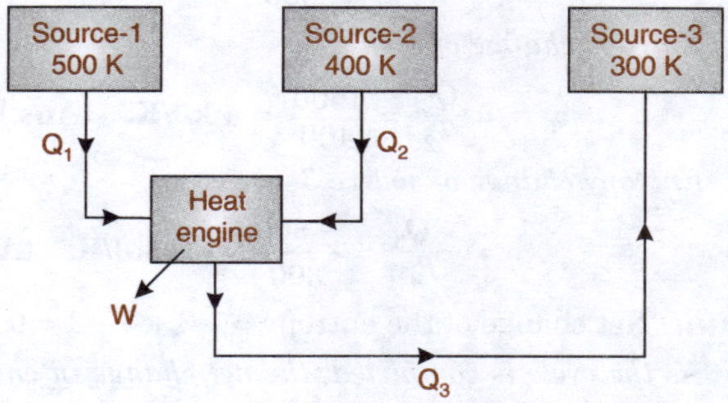

Fig. 4.41

Temperature of source 1 = 500 K
Temperature of source 2 = 400 K
Temperature of source 3 = 300 K

Heat energy drawn from source 1,

$$Q_1 = 1500 \text{ kJ/min}$$

Work developed, $W = 200$ kJ/min.

(*i*) The direction of heat flow from source 1 is known as given in the problem. Assume that the quantities of heats Q_2 and Q_3 are taken from heat sources and their directions are arbitrarily chosen.

For the cyclic operation of the engine

$$\oint \left(\frac{dQ}{T}\right) = 0$$

$$\therefore \quad \frac{Q_1}{T_1} + \frac{Q_2}{T_2} - \frac{Q_3}{T_3} = 0$$

and $$Q_1 + Q_2 - Q_3 = W$$

$$\therefore \quad \frac{1500}{500} + \frac{Q_2}{400} - \frac{Q_3}{300} = 0 \quad ...(i)$$

and $$1500 + Q_2 - Q_3 = 200 \quad ...(ii)$$

Solving eqns. (*i*) and (*ii*), we get

$$Q_2 = -1600 \text{ kJ/min and } Q_3 = -300 \text{ kJ/min.}$$

The above values indicate that the direction of Q_2 and Q_3 are reversed. Since Q_2 should be +v and Q_3 also must be +ve but both are –ve therefore, their assumed directions should be reversed. The arrangement is shown in Fig. 4.42.

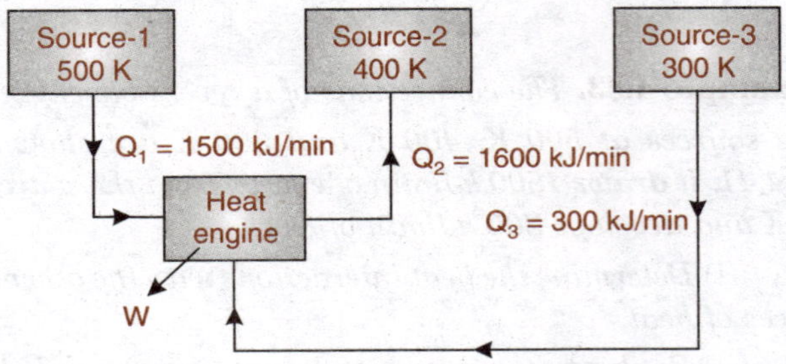

Fig. 4.42

(*ii*) *Entropy change of source 1*

$$= \frac{-Q_1}{T_1} = \frac{-1500}{500} = \mathbf{-3\ kJ/K.} \quad \textbf{(Ans.)}$$

Entropy change of sink 2

$$= \frac{Q_2}{T_2} = \frac{1600}{400} = \mathbf{4\ kJ/K.} \quad \textbf{(Ans.)}$$

Entropy change of source 3

$$= \frac{-Q_3}{T_3} = \frac{-300}{300} = \mathbf{-1\ kJ/K.} \quad \textbf{(Ans.)}$$

(*iii*) Net change of the entropy = – 3 + 4 – 1 = 0

As *the cycle is completed, the net change in entropy must be zero because entropy is a property.*

It may be observed from the new arrangement that the engine takes heat from source 1 and source 3 and rejects to source 2, only *i.e.*, the equipment does both *a heat engine and a heat pump function.*

Example 4.34. *The heat capacity of a system at constant volume is given by $C_v = ZT^2$, where $Z = 0.045$ J/K³*

A system is originally at 250 K, and a thermal reservoir at 125 K is available. Determine the maximum amount of work that can be recovered as the system is cooled down to the temperature of the reservoir.

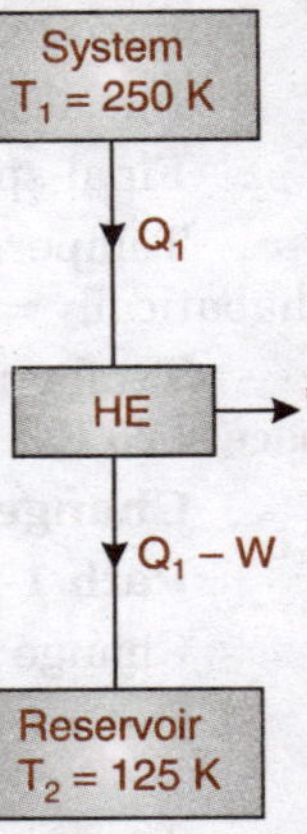

Fig. 4.43

Solution. Refer to Fig. 4.43.

Initial temperature of the system,

$$T_1 = 250 \text{ K}$$

Temperature of the reservoir,

$$T_2 = 125 \text{ K}$$

Heat capacity of the system at constant volume is

$$C_v = ZT^2 = 0.045\, T^2$$

> The product of mass and specific heat (*mc*) is called the *heat capacity* of the substance. The capital letter C, C_p or C_v is used for heat capacity

Heat removed from the system

$$Q_1 = \int_{T_1}^{T_2} C_v\, dT = \int_{250}^{125} 0.045\, T^2\, dT$$

$$= 0.045 \left[\frac{T^3}{3}\right]_{250}^{125} = \frac{0.045}{3}(125^3 - 250^3)$$

$$= -205.08 \times 10^3 \text{ J}$$

$$(\Delta S)_{system} = \int_{250}^{125} C_v \frac{dT}{T} = \int_{250}^{125} 0.045\, T^2 \frac{dT}{T}$$

$$= 0.045 \int_{250}^{125} T dT = 0.045 \left[\frac{T^2}{2}\right]_{250}^{125}$$

$$= \frac{0.045}{2}(125^2 - 250^2) = -1054.7 \text{ J/K}$$

$$(\Delta S)_{reservoir} = \frac{Q_1 - W}{T_{reservoir}} = \frac{205.08 \times 10^3 - W}{125} \text{ J/K}$$

$$(\Delta S)_{\text{working fluid in } HE} = 0$$

$$\therefore \quad (\Delta S)_{universe} = (\Delta S)_{system} + (\Delta S)_{reservoir}$$

$$= -1054.7 + \frac{205.08 \times 10^3 - W}{125}$$

Since $(\Delta S)_{universe} \geq 0$

$$\therefore \quad -1054.7 + \frac{205.08 \times 10^3 - W}{125} \geq 0$$

or $$-1054.7 + 1640.64 - \frac{W}{125} \geq 0$$

or $$585.94 - \frac{W}{125} \geq 0$$

or $$585.94 \geq \frac{W}{125} \quad \text{or} \quad \frac{W}{125} \leq 585.94$$

i.e., $$W_{(max.)} = 585.94 \times 125 = \mathbf{73.24\ kJ.} \quad \textbf{(Ans.)}$$

Example 4.35. *In an insulated duct air is flowing steadily. The pressure and temperature measurements of the air at two stations A and B are given below :*

Station	*Pressure*	*Temperature*
A	*140 kPa*	*60°C*
B	*110 kPa*	*15°C*

Establish the direction of the flow of the air in the duct.

For air assume that :

$$c_p = 1.005 \text{ kJ/kg K}$$

$$h = c_p T \text{ and } \frac{v}{T} = \frac{0.287}{p}$$

where p, v and T are pressure (in kPa), volume (in m³/kg) and temperature (in K) respectively.

Solution. From property relation,

$$Tds = dh - vdp$$

$$ds = \frac{dh}{T} - \frac{vdp}{T}$$

For two states at A and B the entropy changes of the system is given by

$$\int_{s_A}^{s_B} ds = \int_{T_A}^{T_B} \frac{c_p dT}{T} - \int_{p_A}^{p_B} 0.287 \frac{dp}{p}$$

$$\therefore \quad s_B - s_A = 1.005 \log_e \frac{T_B}{T_A} - 0.287 \log_e \frac{110}{140}$$

$$= 1.005 \log_e \left(\frac{15 + 273}{60 + 273}\right) - 0.287 \log_e \frac{110}{140}$$

$$= - 0.1459 + 0.0692 = - 0.0767 \text{ kJ/kg K}$$

$$(\Delta s)_{\text{system}} = - 0.0767 \text{ kJ/kg K}$$

Since the duct is insulated, $(\Delta s)_{\text{surroundings}} = 0$

$$\therefore \quad (\Delta s)_{\text{universe}} = - 0.0767 \text{ kJ/kg K.}$$

This is impossible.

So, **the flow is from B to A. (Ans.)**

Example 4.36. *3 kg of water at 80°C is mixed with 4 kg of water at 15°C in an isolated system. Calculate the change of entropy due to mixing process.*

Solution. Fig. 4.44 shows the isolated system before mixing. When barrier is removed, the water from two compartments mix each other. Let t_m is the final equilibrium temperature after mixing.

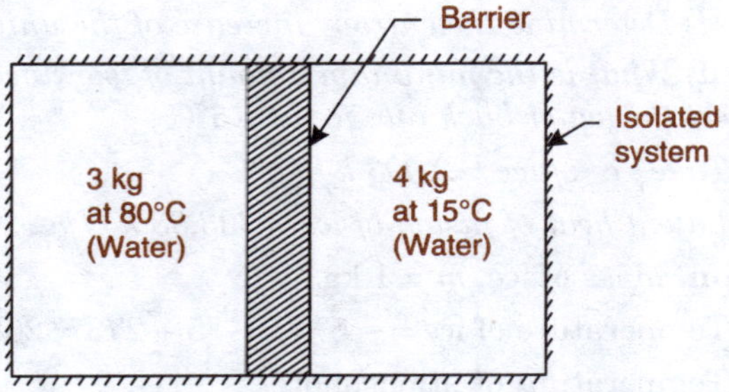

Fig. 4.44

Applying first law of thermodynamics to the isolated system :

Total energy before mixing

= Total energy after mixing

$$\therefore \quad 3c_{pw}(80 - 0) + 4c_{pw}(15 - 0) = 7\, c_{pw}\,(t_m - 0)$$

[c_{pw} = Specific heat of water at constant pressure]

or $\quad 240c_{pw} + 60c_{pw} = 7c_{pw}\, t_m$

or $\quad 240 + 60 = 7\, t_m$

$$\therefore \quad t_m = \frac{300}{7} = 42.85°\text{C}$$

Initial entropy of the system,

$$= 3c_{pw} \log_e \left(\frac{80 + 273}{273}\right) + 4c_{pw} \log_e \left(\frac{15 + 273}{273}\right)$$

$$= 0.7709c_{pw} + 0.2139\, c_{pw} = 0.9848\, c_{pw}$$

Final entropy of the system

$$= (3 + 4)\, c_{pw} \log_e \left(\frac{42.85 + 273}{273}\right) = 1.0205\, c_{pw}$$

Net change in entropy,

$$\Delta S = \text{Final entropy} - \text{Initial entropy}$$

$$= 1.0205c_{pw} - 0.9848\, c_{pw} = 0.0357\, c_{pw}$$

$$= 0.0357 \times 4.187 \text{ kJ/K} \quad [\because \; c_{pw} = 4.187 \text{ kJ/kg K}]$$

$$= 0.1495 \text{ kJ/K}$$

Hence, **net change in entropy**

= 0.1495 kJ/K. (Ans.)

Example 4.37. *A mass 'm' of fluid at temperature T_1 is mixed with an equal mass of the same fluid at T_2. Prove that the resultant change of entropy of the universe is $2\,mc\,\frac{(T_1 + T_2)/2}{\sqrt{T_1 T_2}}$ and also prove that it is always positive.*

Solution. Mean temperature of the mixture $= \left(\frac{T_1 + T_2}{2}\right)$.

Thus change in entropy is given by

$$\Delta S = mc \int_{T_1}^{(T_1 + T_2)/2} \frac{dT}{T} - mc \int_{(T_1 + T_2)/2}^{T_2} \frac{dT}{T}$$

$$= mc \log_e \left(\frac{T_1 + T_2}{2T_1}\right) - mc \log_e \left(\frac{2T_2}{T_1 + T_2}\right)$$

$$= mc \log_e \left(\frac{T_1 + T_2}{2T_1}\right) + mc \log_e \left(\frac{T_1 + T_2}{2T_2}\right)$$

$$= mc \log_e \frac{(T_1 + T_2)^2}{4T_1 T_2}$$

$$= mc \log_e \left[\frac{T_1 + T_2}{2\sqrt{T_1 T_2}}\right]^2 = 2\, mc \log_e \left[\frac{T_1 + T_2}{2\sqrt{T_1 T_2}}\right]$$

$$= 2\, mc \log_e \left[\frac{(T_1 + T_2)/2}{\sqrt{T_1 T_2}}\right]$$

i.e., **Resultant change of entropy of universe**

$$= 2\,mc\,\log_e \left[\frac{(T_1 + T_2)/2}{\sqrt{T_1 T_2}}\right] \quad \text{...Proved}$$

The arithmetic mean $\frac{T_1 + T_2}{2}$ is greater than geometric mean $\sqrt{T_1 T_2}$.

Therefore, $\log_e \frac{(T_1 + T_2)/2}{\sqrt{T_1 T_2}}$ is +ve.

Thus, **the entropy of the universe increases.**

☞**Example 4.38.** (*a*) *One kg of water at 0°C is brought into contact with a heat reservoir at 90°C. When the water has reached 90°C, find :*

(*i*) *Entropy change of water ;*
(*ii*) *Entropy change of the heat reservoir ;*
(*iii*) *Entropy change of the universe.*

(*b*) *If water is heated from 0°C to 90°C by first bringing it in contact with a reservoir at 40°C and then with a reservoir at 90°C, what will the entropy change of the universe be ?*

(*c*) *Explain how water might be heated from 0°C to 90°C with almost no change in the entropy of the universe.*

Solution. Mass of water, $m = 1$ kg

Temperature of water,
$$T_1 = 0 + 273 = 273 \text{ K}$$
Temperature of the heat reservoir,
$$T_2 = 90 + 273 = 363 \text{ K.}$$

(*a*) Refer to Fig. 4.45. Water is being heated through a finite temperature difference. The entropy of water would increase and that of the reservoir would decrease so that net entropy change of the water (system) and the reservoir together would be positive definite. *Water is being heated, irreversibly,* and to find the entropy change of water, we have to assume a reversible path between the end states, which are at equilibrium.

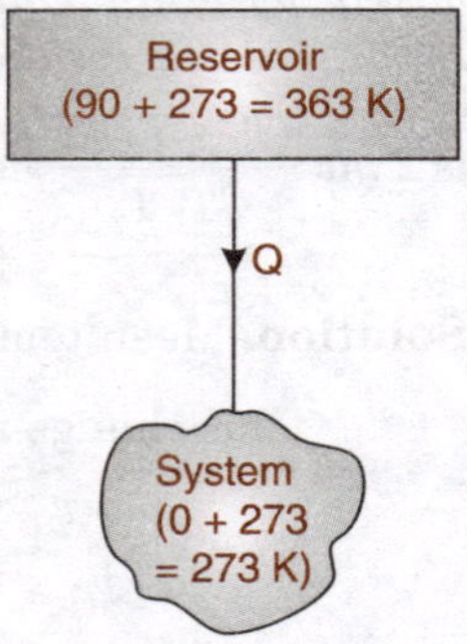

Fig. 4.45

(*i*) *Entropy change of water,*

$$(\Delta s)_{water} = \int_{T_1}^{T_2} \frac{mc\,dT}{T} = mc \log_e \frac{T_2}{T_1}$$

(c = Specific heat of water)

$$= 1 \times 4.187 \times \log_e \frac{363}{273}$$

= 1.193 kJ/kg K. (Ans.)

(*ii*) The temperature of the reservoir remains *constant* irrespective of the amount of the heat withdrawn.

Amount of heat absorbed by the system from the reservoir,

$$Q = 1 \times 4.187 \times (363 - 273) = 376.8 \text{ kJ}$$

∴ *Entropy change of the reservoir,*

$$(\Delta s)_{reservoir} = -\frac{Q}{T} = -\frac{376.8}{363} = \mathbf{-1.038\ kJ/K.} \quad \textbf{(Ans.)}$$

(–ve sign indicates decrease in entropy).

(*iii*) *Entropy change of the universe,*

$$(\Delta s)_{universe} = (\Delta s)_{water\ (or\ system)} + (\Delta s)_{reservoir}$$
$$= 1.193 + (-1.038) = \mathbf{0.155\ kJ/K.} \quad \textbf{(Ans.)}$$

(*b*) The heating of water is being carried out in two stages, first from 0°C (273 K) to 40°C (*i.e.*, 313 K) by bringing in contact with a reservoir at 40°C (313 K), and then from 40°C (313 K) to 90°C (363 K) by bringing in contact with a second reservoir at 90°C (363 K).

$$(\Delta s)_{water} = \int_{273}^{313} mc\frac{dT}{T} + \int_{313}^{363} mc\frac{dT}{T}$$
$$= 1 \times 4.187 \left(\log_e \frac{313}{273} + \log_e \frac{363}{313}\right)$$
$$= 4.187\,(0.1367 + 0.1482) = 1.1928 \text{ kJ/K.}$$

$$(\Delta s)_{res.\ I} = -\frac{1 \times 4.187 \times (313 - 273)}{313} = -0.535 \text{ kJ/K}$$

$$(\Delta s)_{res.\ II} = -\frac{1 \times 4.187 \times (363 - 313)}{363} = -0.576 \text{ kJ/K}$$

$$\therefore \quad (\Delta s)_{univ.} = (\Delta s)_{water} + (\Delta s)_{res.\ I} + (\Delta s)_{res.\ II}$$
$$= 1.1928 + (-0.535) + (-0.576)$$
$$= 0.0818 \text{ kJ/K.}$$

i.e., **Entropy change of universe**
= 0.0818 kJ/K. (Ans.)

(*c*) The entropy change of universe would be *less and less*, if the water is heated in *more and more stages*, by bringing the water in contact successively with *more and more heat reservoirs,* each succeeding reservoir being at a *higher temperature* than the preceding one. When water is heated in infinite steps, by bringing in contact with an infinite number of reservoirs in succession, so that at any instant the temperature difference between the water and the reservoir in contact is infinitesimally small, then the entropy change of the universe would be *zero* and the water would be *reversibly* heated.

Example 4.39. *1 kg of ice at – 5°C is exposed to the atmosphere which is at 25°C. The ice melts and comes into thermal equilibrium.*

(*i*) *Determine the entropy increase of the universe.*

(*ii*) *What is the minimum amount of work necessary to convert the water back into ice at – 5°C ?*

Take : c_p *of ice = 2.093 kJ/kg°C*

Latent heat of fusion of ice = 333.33 kJ/kg.

Solution. Mass of ice, $m = 1$ kg

Temperature of ice = – 5°C (= – 5 + 273 = 268 K)

Temperature of atmosphere

= 25°C (= 25 + 273 = 298 K)

Heat absorbed by ice from the atmosphere (Fig. 4.46)

= Heat absorbed in solid phase + latent heat + heat absorbed in liquid phase

$= 1 \times 2.093 \times [0 - (-5)] + 1 \times 333.33 + 1 \times 4.187 \times (25 - 0)$

$= 10.46 + 333.33 + 104.67$

$= 448.46$ kJ.

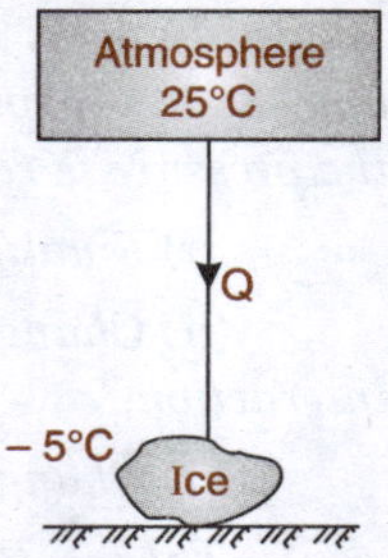

Fig. 4.46

(*i*) **Entropy increase of the universe,** $(\Delta s)_{universe}$ **:**

Entropy change of the atmosphere,

$$(\Delta s)_{atm.} = -\frac{Q}{T} = -\frac{448.46}{298} = -1.5049 \text{ kJ/K}$$

Entropy change of system (ice) as it gets heated from – 5°C to 0°C,

$$(\Delta s_I)_{system} = \int_{268}^{273} mc_p \frac{dT}{T}$$

$$= 1 \times 2.093 \log_e \frac{273}{268} = 0.0386 \text{ kJ/K}$$

Entropy change of the system as ice melts at 0°C to become water at 0°C.

$$(\Delta s_{II})_{system} = \frac{333.33}{273} = 1.2209 \text{ kJ/K}$$

Entropy change of water as it gets heated from 0°C to 25°C

$$(\Delta s_{III})_{system} = \int_{273}^{298} mc_p \frac{dT}{T}$$

$$= 1 \times 4.187 \log_e \left(\frac{298}{273}\right) = 0.3668 \text{ kJ/K}$$

Total entropy change of ice as it melts into water

$$(\Delta s)_{total} = \Delta s_I + \Delta s_{II} + \Delta s_{III}$$

$$= 0.0386 + 1.2209 + 0.3668$$

$$= 1.6263 \text{ kJ/K}$$

Then temperature-entropy diagram for the system as ice at – 5°C converts to water at 25°C is shown in Fig. 4.47.

∴ **Entropy increase of the universe,**

$$(\Delta s)_{univ.} = (\Delta s)_{system} + (\Delta s)_{atm.}$$

$$= 1.6263 + (-1.5049)$$

$= \mathbf{0.1214}$ **kJ/K. (Ans.)**

(*ii*) **Minimum amount of work necessary to convert the water back into ice at – 5°C,** $\mathbf{W_{min.}}$ **:**

To convert 1 kg of water at 25°C to ice at – 5°C, 448.46 kJ of heat have to be removed from it, and the system has to be brought from state 4 to state 1 (Fig. 4.47). A refrigerator cycle, as shown in Fig. 4.48, is assumed to accomplish this. The entropy change of the system would be the same, *i.e.*, $s_4 - s_1$, with the only difference that its sign will be negative, because heat is removed from the system (Fig. 4.47).

$$(\Delta s)_{system} = s_1 - s_4 \ (negative)$$

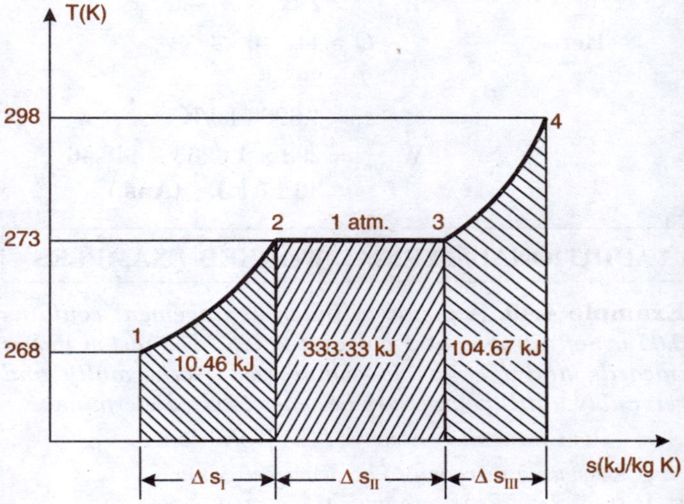

Fig. 4.47

The entropy change of the working fluid in the refrigerator would be zero, since it is operating in a cycle, *i.e.*,

$$(\Delta s)_{refrigerator} = 0.$$

The entropy change of the atmosphere (*positive*),

$$(\Delta s)_{atm.} = \frac{Q + W}{T}$$

∴ Entropy change of the universe

$$(\Delta s)_{univ.} = (\Delta s)_{system} + (\Delta s)_{refrigerator} + (\Delta s)_{atm.}$$

$$= (s_1 - s_4) + 0 + \frac{Q + W}{T} = (s_1 - s_4) + \frac{Q + W}{T}$$

By the principle of increase of entropy

$$(\Delta s)_{univ. \text{ or isolated system}} \geq 0$$

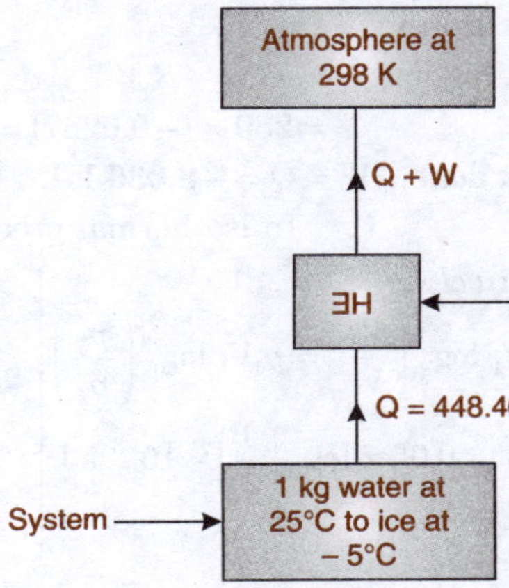

Fig. 4.48

$$\therefore \quad \left[(s_1 - s_4) + \frac{Q+W}{T} \right] \geq 0$$

$$\therefore \quad \frac{Q+W}{T} \geq (s_4 - s_1)$$

$$W \geq T(s_4 - s_1) - Q$$

$$W_{(min)} = T(s_4 - s_1) - Q$$

Here $Q = 448.46$ kJ

$T = 298$ K

$s_4 - s_1 = 1.6263$ kJ/K

$$\therefore \quad \mathbf{W}_{(min)} = 298 \times 1.6263 - 448.46$$

$= \mathbf{36.17}$ **kJ. (Ans.)**

ADDITIONAL/TYPICAL WORKED EXAMPLES

Example 4.40. *A piston-cylinder arrangement contains 0.05 m³ of nitrogen at 1 bar and 280 K. The piston moves inwards and the gas is compressed isothermally and reversibly until the pressure becomes 5 bar. Determine :*

(i) Change in entropy. (ii) Work done.

Assume nitrogen to be a perfect gas.

Solution. *Given :* $V_1 = 0.05$ m³ ; $p_1 = 1$ bar ; $T_1 = 280$ K ; $p_2 = 5$ bar.

(*i*) **Change in entropy, *($S_2 - S_1$) :**

Gas constant, $R = \dfrac{R_0}{M} = \dfrac{8.314}{28} = 0.297$ kJ/kg K

Mass of the gas, $m = \dfrac{p_1V_1}{RT_1} = \dfrac{(1 \times 10^5) \times 0.05}{(0.297 \times 1000) \times 280}$

$= 0.06$ kg

$\therefore$ Change in entropy

$$S_2 - S_1 = mR \log_e \left(\frac{p_1}{p_2} \right)$$

$$= 0.06 \times 0.297 \log_e \left(\frac{1}{5} \right)$$

$= \mathbf{-0.0287}$ **kJ/K. (Ans.)**

(*ii*) **Work done :**

Heat interaction, $Q = T(S_2 - S_1)$

$= 280 \times (-0.0287) = -8.036$ kJ

$\therefore$ Work done, $W = Q = \mathbf{-8.036}$ **kJ. (Ans.)**

($\because$ In isothermal process, $W = Q$)

Alternatively :

$$W = p_1V_1 \log_e \left(\frac{V_2}{V_1} \right) = p_1V_1 \log_e \left(\frac{p_1}{p_2} \right)$$

$$= 1 \times 10^5 \times 0.05 \times \log_e \left(\frac{1}{5} \right) \times 10^{-3} \text{ kJ}$$

$= 8.04$ kJ

*Refer to Art. 4.17.

Example 4.41. *1 kg of air initially at 8 bar pressure and 380 K expands polytropically ($pv^{1.2}$ = constant) until the pressure is reduced to one-fifth value. Calculate :*

(i) Final specific volume and temperature.

(ii) Change of internal energy, work done and heat interaction.

(iii) Change in entropy.

Take : $R = 0.287$ kJ/kg K and $\gamma = 1.4$.

Solution. *Given :* $m = 1$ kg ; $p_1 = 8$ bar ; $T_1 = 380$ K ; Law of expansion : $pv^{1.2}$ = constant ;

$$p_2 = \frac{p_1}{5} = \frac{8}{5} = 1.6 \text{ bar} ; R = 0.287 \text{ kJ/kg K} ; \gamma = 0.4.$$

(*i*) **Final specific volume and temperature, $\mathbf{v_2, T_2}$:**

$$p_1v_1 = RT_1$$

or, $v_1 = \dfrac{RT_1}{p_1} = \dfrac{(0.287 \times 10^3) \times 380}{8 \times 10^5}$

$= \mathbf{0.1363}$ **m³/kg. (Ans.)**

Also, $p_1v_1^n = p_2v_2^n$, or, $v_2 = v_1 \times \left(\dfrac{p_1}{p_2} \right)^{\frac{1}{n}}$

$$\therefore \quad v_2 = 0.1363 \times \left(\frac{8}{1.6} \right)^{\frac{1}{1.2}}$$

$= \mathbf{0.5211}$ **m³/kg. (Ans.)**

Again,

$$\frac{T_2}{T_1} = \left(\frac{p_2}{p_1} \right)^{\frac{n-1}{n}}, \quad \text{or,} \quad \frac{T_2}{380} = \left(\frac{1}{5} \right)^{\frac{1.2-1}{1.2}}$$

$$\therefore \quad T_2 = 290.6 \text{ K}$$

[*Alternatively* : $T_2 = \dfrac{p_2v_2}{R} = \dfrac{1.6 \times 10^5 \times 0.5211}{287}$ $= 290.6$ K]

(*ii*) **Change of internal energy, work done and heat interaction :**

Change of internal energy

$$\mathbf{u_2 - u_1} = c_v(T_2 - T_1) = \frac{R}{\gamma - 1}(T_2 - T_1)$$

$$= \frac{0.287}{1.4 - 1}(290.6 - 380)$$

$= \mathbf{-64.14}$ **kJ/kg** *(Decrease)*. **(Ans.)**

Work done, $\mathbf{W_{1\text{-}2}} = \dfrac{p_1v_1 - p_2v_2}{n-1} = \dfrac{R(T_1 - T_2)}{n-1}$

$$= \frac{0.287(380 - 290.6)}{1.2 - 1}$$

$= \mathbf{128.29}$ **kJ/kg** *(Work done by air)*. **(Ans.)**

Heat interaction, $Q_{1-2} = (u_2 - u_1) + W$
$= -64.14 + 128.29$
$=$ **64.15 kJ/kg** *(Heat received)*. **(Ans.)**

$$\left[\begin{aligned} &\textit{Alternatively}: Q = \frac{\gamma - n}{\gamma - 1} \times W \\ &= \frac{1.4 - 1.2}{1.4 - 1} \times 128.29 = 64.15 \text{ kJ/kg} \end{aligned}\right]$$

(*iii*) **Change in entropy, $(s_2 - s_1)$:**

$$s_2 - s_1 = c_v \log_e \frac{T_2}{T_1} + R \log_e \frac{v_2}{v_1}$$

$$= \frac{0.287}{1.4 - 1} \times \log_e \left(\frac{290.6}{380}\right) + 0.287 \log_e \left(\frac{0.5211}{0.1363}\right)$$

$$= -0.192 + 0.385$$

$=$ **0.193 kJ/kg K** *(increase)*. **(Ans.)**

Example 4.42. (*a*) *Show that approximate change of entropy during a polytropic process equals the quantity of heat transferred divided by the mean absolute temperature.*

(*b*) *One kg of air at 290 K is compressed in a cylinder according to the polytropic law $pv^{1.3}$ = constant. If the compression ratio is 16, calculate the entropy change of air during the compression process stating whether it is an increase or decrease.*

What would be the percentage error if the entropy change is calculated by dividing the quantity of heat exchanged by the mean absolute temperature during the process ?

Take $\gamma = 1.4$ and $c_v = 0.718$ kJ/kg K.

Solution. (*a*) In Fig. 4.49 curve 1-2 represents the polytropic process ($pv^n = c$) from state point 1 to state point 2. The area under the process curve 1-2 on *T-S* diagram represents the heat transferred during the process. The slope of the curve 1-2 is usually *small* and can be considered to be a straight line (shotted dotted).

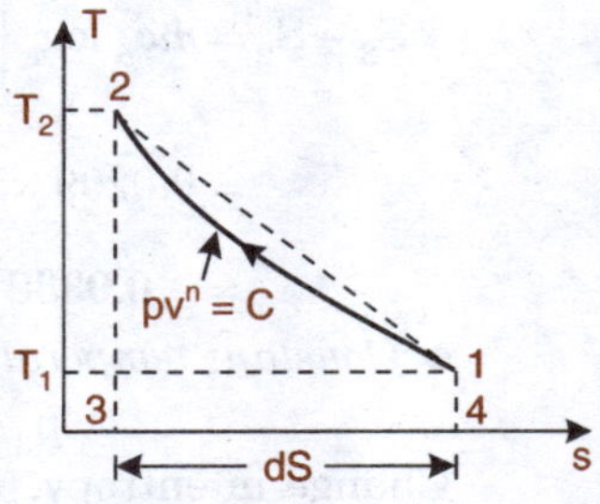

Fig. 4.49

Heat transferred = Area of trapezium 1-2-3-4
= Base × mean ordinate

$$= dS \times \left(\frac{T_1 + T_2}{2}\right)$$

= Entropy change × mean absolute temperature during the process

or, $$\text{Entropy change} = \frac{\text{Heat transferred}}{\text{Mean absolute temperature}}$$

(*b*) *Given* : $m = 1$ kg ; $T_1 = 290$ K ; $pv^{1.3}$ = constant ; $r = 16$; $\gamma = 1.4$; $c_v = 0.718$ kJ/kg K

For a polytropic process : $\dfrac{T_2}{T_1} = \left(\dfrac{v_1}{v_2}\right)^{n-1}$

or, $$T_2 = 290 \times (16)^{1.3-1} = 666.2 \text{ K}$$

Now, $$s_2 - s_1 = c_v \left(\frac{n - \gamma}{n - 1}\right) \log_e \left(\frac{T_2}{T_1}\right) \text{ ... per kg}$$

$$= 0.718 \left(\frac{1.3 - 1.4}{1.3 - 1}\right) \log_e \left(\frac{666.2}{290}\right)$$

$=$ **– 0.199 kJ/kg K. (Ans.)**

The –ve sign indicates *decrease* in entropy.

Heat transferred during the process is given by,

$$Q = \frac{\gamma - n}{\gamma - 1} \times W = \frac{\gamma - n}{\gamma - 1} \times \frac{R(T_1 - T_2)}{n - 1}$$

$$= c_v \left(\frac{\gamma - n}{\gamma - 1}\right)(T_1 - T_2) \text{ ...per kg}$$

$$= 0.718 \left(\frac{1.4 - 1.3}{1.3 - 1}\right)(290 - 666.2)$$

$$= -90.04 \text{ kJ/kg} \qquad \left(\because c_v = \frac{R}{\gamma - 1}\right)$$

Mean absolute temperature,

$$T_{\text{mean}} = \frac{T_1 + T_2}{2} = \frac{290 + 666.2}{2} = 478.1 \text{ K}$$

Approximate change of entropy

$$= \frac{Q}{T_{\text{mean}}} = \frac{-90.04}{478.1} = -0.188 \text{ kJ/kg K}$$

$$\therefore \quad \%\text{age error} = \frac{0.199 - 0.188}{0.199} \times 100$$

$=$ **5.53%. (Ans.)**

- The approximate value of entropy change is lower, because in the relation $Q = T_{av} \times dS$ actual value of heat transferred is substituted instead of approximate value (*i.e.*, Area under the straight line) which is higher.

Example 4.43. *1.2 m³ of air is heated reversibly at constant pressure from 300 K to 600 K, and is then cooled reversibly at constant volume back to initial temperature. If the initial pressure is 1 bar, calculate :*

(*i*) *The net heat flow.*

(*ii*) *The overall change in entropy.*

Represent the processes on T-S plot.

Take $c_p = 1.005$ kJ/kg K and $R = 0.287$ kJ/kg K

Solution. *Given* : $V_1 = 1.2$ m³ ; $p_1 = p_2$; $T_1 = 300$ K ; $T_2 = 600$ K ; $p_1 = 1$ bar ; $c_p = 1.005$ kJ/kg K ; $R = 0.287$ kJ/kg K

Fig. 4.50 shows the *T-S* plot of the processes.

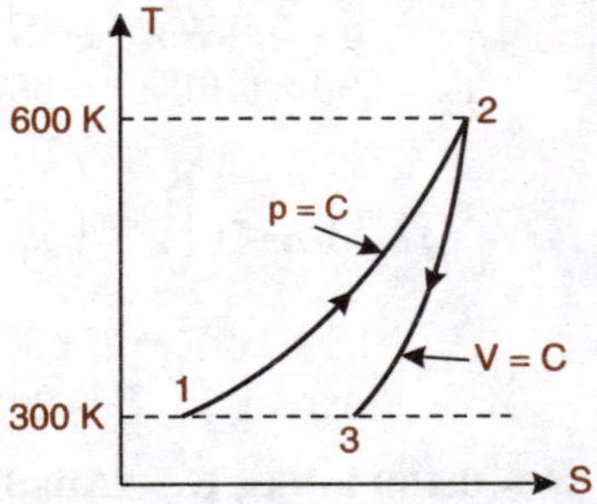

Fig. 4.50

(*i*) **The net heat flow, Q :**

$$\text{Mass of air, } m = \frac{p_1V_1}{RT_1} = \frac{1\times10^5 \times 1.2}{(0.287\times1000)\times300} = 1.394 \text{ kg}$$

$$Q = Q_{1\text{-}2} + Q_{2\text{-}3}$$
$$= mc_p(T_2 - T_1) + mc_v(T_3 - T_2)$$
$$= mc_p(T_2 - T_1) + mc_v(T_1 - T_2) \quad ...(\because T_1 = T_3)$$
$$= m(T_2 - T_1)(c_p - c_v) = m(T_2 - T_1) \times R$$
$$= 1.394(600 - 300) \times 0.287 = \mathbf{120\ kJ.} \quad \textbf{(Ans.)}$$

(*ii*) **The overall change in entropy :**

Entropy change during *constant pressure process 1-2*,

$$S_2 - S_1 = mc_p \log_e\left(\frac{T_2}{T_1}\right)$$
$$= 1.394 \times 1.005 \log_e\left(\frac{600}{300}\right) = 0.9711 \text{ kJ/K}$$

Entropy change during *constant volume process 2-3* ;

$$S_3 - S_2 = mc_v \log_e\left(\frac{T_3}{T_2}\right) = m(c_p - R)\log_e\left(\frac{T_1}{T_2}\right)$$
$$= 1.394 \times (1.005 - 0.287)\log_e\left(\frac{300}{600}\right)$$
$$= -0.6938 \text{ kJ/K}$$

$\therefore$ Overall change of entropy

$$= (S_2 - S_1) + (S_3 - S_2)$$
$$= 0.9771 + (-0.6938) = \mathbf{0.2833\ kJ/K.} \quad \textbf{(Ans.)}$$

Example 4.44. *A closed system contains air at a pressure 1 bar, temperature 300 K and volume 0.018 m³. This system undergoes a thermodynamic cycle consisting of the following three processes in series : (i) Constant volume heat addition till pressure becomes 5 bar, (ii) Constant pressure cooling, and (iii) Isothermal heating to initial state.*

Represent the cycle on T-s and p-V plots and evaluate the change in entropy for each process.

Take c_p = 0.718 kJ/kg K and R = 0.287 kJ/kg K.

Solution. *Given :* p_1 = 1 bar ; T_1 = 300 K ; V_1 = 0.018 m³ ; p_2 = 5 bar ; c_v = 0.718 kJ/kg K ; R = 0.287 kJ/kg K.

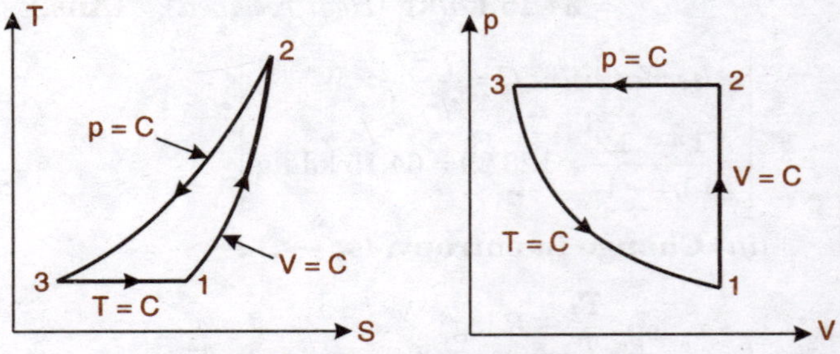

Fig. 4.51. *T-s* and *p-V* diagrams.

$$\text{Mass of air, } m = \frac{p_1V_1}{RT_1} = \frac{(1\times10^5)\times0.018}{(0.287\times1000)\times300} = 0.0209 \text{ kg}$$

Refer to Fig. 4.51.

- *Constant volume process 1-2 :*

$$\frac{p_1}{T_1} = \frac{p_2}{T_2} \quad \text{or} \quad T_2 = T_1 \times \frac{p_2}{p_1} = 300 \times \frac{5}{1} = 1500 \text{ K}$$

$\therefore$ Change in entropy, $\mathbf{S_2 - S_1} = mc_v \log_e\left(\frac{T_2}{T_1}\right)$

$$= 0.0209 \times 0.718 \times \log_e\left(\frac{1500}{300}\right)$$
$$= \mathbf{0.0241\ kJ/K.} \quad \textbf{(Ans.)}$$

- *Constant pressure process 2-3 :*

$$T_3 = T_1 = 300 \text{ K}$$

Now, change in entropy,

$$\mathbf{S_3 - S_2} = mc_p \log_e\left(\frac{T_3}{T_2}\right) = m(c_v + R)\log_e\left(\frac{T_3}{T_2}\right)$$
$$= 0.0209 \times (0.718 + 0.287) \times \log_e\left(\frac{300}{1500}\right)$$
$$= \mathbf{-\,0.0338\ kJ/K.} \quad \textbf{(Ans.)}$$

- *Constant temperature (isothermal) process 3-1 :*

$$p_3 = p_2 = 5 \text{ bar}$$

Change in entropy,

$$\mathbf{S_1 - S_3} = mR \log_e\left(\frac{p_3}{p_1}\right)$$
$$= 0.0209 \times 0.287 \times \log_e\left(\frac{5}{1}\right)$$
$$= \mathbf{0.00965\ kJ/K.} \quad \textbf{(Ans.)}$$

Example 4.45. *Derive expressions for entropy change with variable specific heat.*

Solution. Let us assume that the specific heats of a gas vary with temperature according to the linear relations :

$$c_p = a + kT, \quad \text{and} \quad c_v = b + kT$$

where, *a*, *b* and *k* = Constants, and *T* = Temperature, *K*.

For *unit mass* of gas,

$$Q = T\,ds = du + pdv$$
$$= c_v dT + pdv$$

or,
$$ds = c_v \frac{dT}{T} + \frac{pdv}{T} = c_v \frac{dT}{T} + R\frac{dv}{v}$$
$$(\because \quad pv = RT)$$

Now,
$$R = c_p - c_v = (a + kT) - (b + kT) = a - b$$

$$\therefore \quad ds = (b + kT)\frac{dT}{T} + (a - b)\frac{dv}{v}$$

This is the differential form of entropy change.

Integrating both sides between limits 1 and 2, we get

$$s_2 - s_1 = b \log_e\left(\frac{T_2}{T_1}\right) + k(T_2 - T_1) + (a - b)\log_e\left(\frac{v_2}{v_1}\right) \quad ...(1)$$

For the entropy change the following expressions can be obtained by suitable manipulations to *eqn. (1)* :

1. *Expression for entropy change in terms of* **temperature** *only* :

$$\frac{T_2}{T_1} = \left(\frac{v_1}{v_2}\right)^{n-1}$$

or,
$$\log_e \frac{T_2}{T_1} = (n-1)\log_e\left(\frac{v_1}{v_2}\right) = -(n-1)\log_e\left(\frac{v_2}{v_1}\right)$$

$$\therefore \quad s_2 - s_1 = b\log_e\left(\frac{T_2}{T_1}\right) + k(T_2 - T_1)$$
$$-\left(\frac{a-b}{n-1}\right)\log_e\left(\frac{T_2}{T_1}\right) \quad ...[\text{From eqn. (1)}]$$

or,
$$\mathbf{s_2 - s_1} = \left(b - \frac{a-b}{n-1}\right)\log_e\left(\frac{T_2}{T_1}\right) + k(T_2 - T_1) \quad ...(i)$$

2. *Expression for entropy change in terms of* **pressure, volume and temperature.**

From eqn. (1), we have

$$s_2 - s_1 = b\log_e\left(\frac{T_2}{T_1}\right) + k(T_2 - T_1) + a\log_e\left(\frac{v_2}{v_1}\right) - b\log_e\left(\frac{v_2}{v_1}\right)$$
$$= a\log_e\left(\frac{v_2}{v_1}\right) + b\log_e\left(\frac{T_2}{T_1}\times\frac{v_1}{v_2}\right) + k(T_2 - T_1)$$

or,
$$\mathbf{s_2 - s_1} = a\log_e\left(\frac{v_2}{v_1}\right) + b\log_e\left(\frac{p_2}{p_1}\right) + k(T_2 - T_1) \quad ...(ii)$$

3. *Expression for entropy change in terms of* **pressure and temperature only.**

Again, from eqn. (1), we have

$$s_2 - s_1 = b\log_e\left(\frac{T_2}{T_1}\right) + k(T_2 - T_1) + a\log_e\left(\frac{v_2}{v_1}\right) - b\log_e\left(\frac{v_2}{v_1}\right)$$
$$= a\log_e\left(\frac{T_2}{T_1}\times\frac{p_1}{p_2}\right) + b\log_e\left(\frac{T_2}{T_1}\times\frac{v_1}{v_2}\right) + k(T_2 - T_1)$$
$$= a\log_e\left(\frac{T_2}{T_1}\right) - a\log_e\left(\frac{p_2}{p_1}\right) + b\log_e\left(\frac{p_2}{p_1}\right) + k(T_2 - T_1)$$

or,
$$\mathbf{s_2 - s_1} = a\log_e\left(\frac{T_2}{T_1}\right) + (b - a)\log_e\left(\frac{p_2}{p_1}\right) + k(T_2 - T_1) \quad ...(iii)$$

• Derivation of the formula $T^b v^{a-b} e^{kT}$ = constant for the *adiabatic expansion of gas* :

We know that, $ds = (a - b)\dfrac{dv}{v} + (b + kT)\dfrac{dT}{T}$

$$\left.\begin{aligned} s_2 - s_1 &= a\log_e v - b\log_e v + b\log_e T + kT \\ &= a\log_e v + b\log_e\left(\frac{T}{v}\right) + kT \\ &= a\log_e v + b\log_e\left(\frac{p}{R}\right) + kT \\ &= a\log_e v + \log_e\left(\frac{p}{a-b}\right)^b + kT \end{aligned}\right\} \begin{aligned} &= 0 \text{ for} \\ &\text{adiabatic} \\ &\text{expansion} \end{aligned}$$

This gives : $v^a p^b e^{kT}$ = constant

$p v^{a-b} e^{kT}$ = constant

$T^b v^{a-b} e^{kT}$ = constant

These expressions can be obtained by taking kT on right-side and taking the antilog of the resulting expressions.

Example 4.46. *Determine the entropy change of 4 kg of a perfect gas whose temperature varies from 127°C to 227°C during a constant volume process. The specific heat varies linearly with absolute temperature and is represented by the relation :*

$$c_v = (0.48 + 0.0096\,T)\ \text{kJ/kg K}.$$

Solution. *Given :* $m = 4$ kg ; $T_1 = 127 + 273 = 400$ K ;

$T_2 = 227 + 273 = 500$ K ;

$c_v = (0.48 + 0.0096\,T)$ kJ/kg K.

Entropy variation for a constant volume process is given by :

$$dS = mc_v\frac{dT}{T}, \quad \text{or,} \quad dS = 4 \times (0.48 + 0.0096T)\frac{dT}{T}$$

Integrating both sides, we get

$$S_2 - S_1 = 4 \times 0.48 \int_{T_1}^{T_2} \frac{dT}{T} + 4 \times 0.0096 \int_{T_1}^{T_2} dT$$
$$= 1.92\log_e\left(\frac{T_2}{T_1}\right) + 0.0384\,(T_2 - T_1)$$

$$= 1.92 \log_e \left(\frac{500}{400}\right) + 0.0384(500 - 400)$$

$$= 4.268 \text{ kJ/K}$$

i.e., $\mathbf{S_2 - S_1 = 4.268 \text{ kJ.}}$ **(Ans.)**

Example 4.47. *The specific heats of a gas vary linearly with absolute temperature according to the following relations :*

$$c_p = (0.85 + 0.00025\, T) \text{ kJ/kg K, and}$$

$$c_v = (0.56 + 0.00025\, T) \text{ kJ/kg K}$$

If the entropy of the gas at 1 bar pressure and 273 K is zero, find the entropy of the gas at 25 bar and 750 K temperature.

Solution. *Given :* $c_p = (0.85 + 0.00025\, T)$ kJ/kg K ;

$c_v = (0.56 + 0.00025\, T)$ kJ/kg K ;

$p_1 = 1$ bar ; $T_1 = 273$ K ; $p_2 = 25$ bar ; $T_2 = 750$ K.

We know that,

$$ds = c_v \frac{dT}{T} + \frac{p}{T} dv = c_v \frac{dT}{T} + R \frac{dv}{v}$$

Integrating both sides, we get

$$s_2 - s_1 = \int c_v \frac{dT}{T} + R \log_e \left(\frac{v_2}{v_1}\right)$$

$$= \int c_v \frac{dT}{T} + (c_p - c_v) \log_e \left(\frac{p_1}{p_2} \times \frac{T_2}{T_1}\right)$$

$$= \int \left(\frac{0.56}{T} + 0.00025\right) dT + 0.29 \log_e \left(\frac{1}{25} \times \frac{750}{273}\right)$$

$$= \left[0.56 \log_e \left(\frac{T_2}{T_1}\right) + 0.00025\,(T_2 - T_1)\right] - 0.6404$$

$$= 0.56 \log_e \left(\frac{750}{273}\right) + 0.00025(750 - 273) - 0.6404$$

$$= 0.0448 \text{ kJ/kg K}$$

i.e., $\mathbf{s_2 - s_1 = 0.0448 \text{ kJ/kg K.}}$ **(Ans.)**

Example 4.48. *An insulated vessel of 0.5 m³ capacity is divided by a rigid conducting diaphragm into two chambers A and B, each having a capacity of 0.25 m³. Chamber A contains air at 1.4 bar pressure and 290 K temperature and the corresponding parameters for air in chamber B are 4.2 bar and 440 K. Calculate :*

(i) *Final equilibrium temperature,*

(ii) *Final pressure on each side of the diaphragm, and*

(iii) *Entropy change of system.*

For air take $c_v = 0.715$ *kJ/kg K and* $R = 0.287$ *kJ/kg K.*

Solution. *Given :* $V_A = 0.25 \text{ m}^3$; $p_{Ai} = 1.4$ bar ; $T_{Ai} = 290$ K ; $V_B = 0.25 \text{ m}^3$; $p_{Bi} = 4.2$ bar ; $T_{Bi} = 440$ K ; $c_v = 0.715$ kJ/kg K.

(i) **Final equilibrium temperature,** $\mathbf{T_f}$ **:**

Mass of air,

$$m_A = \frac{p_{Ai} V_A}{R T_{Ai}} = \frac{(1.4 \times 10^5) \times 0.25}{(0.287 \times 1000) \times 290} = 0.4205 \text{ kg}$$

$$m_B = \frac{p_{Bi} V_B}{R T_{Bi}} = \frac{(4.2 \times 10^5) \times 0.25}{(0.287 \times 1000) \times 440} = 0.8315 \text{ kg}$$

Let T_f be the final equilibrium temperature (K). Since the diaphragm is conducting.

Heat gained by air in chamber A = Heat lost by air in chamber B

$$m_A\, c_v\, (T_f - 290) = m_B\, c_v (440 - T_f)$$

or, $$0.4205 \times (T_f - 290) = 0.8315(440 - T_f)$$

or, $$0.4205\, T_f - 121.94 = 365.86 - 0.8315\, T_f$$

$\therefore$ $\mathbf{T_f = 389.6 \text{ K.}}$ **(Ans.)**

(ii) **Final pressure on each side of the diaphragm :** $\mathbf{p_{Af}}$ **;** $\mathbf{p_{Bf}}$**.**

$$p_{Af} = \frac{1.4 \times 389.6}{290} = \mathbf{1.88 \text{ bar.}} \quad \textbf{(Ans.)}$$

$$p_{Bf} = \frac{4.2 \times 389.6}{440} = \mathbf{3.72 \text{ bar.}} \quad \textbf{(Ans.)}$$

(iii) **Entropy change of the system :**

Entropy change for chamber A

$$= m_A\, c_v \log_e \left(\frac{T_f}{T_{Ai}}\right)$$

$$= 0.4205 \times 0.715 \times \log_e \left(\frac{389.6}{290}\right)$$

$$= 0.0888 \text{ kJ/K}$$

Entropy change for chamber B

$$= m_B\, c_v \log_e \left(\frac{T_f}{T_{Bi}}\right)$$

$$= 0.8315 \times 0.715 \times \log_e \left(\frac{389.6}{440}\right)$$

$$= -0.0723 \text{ kJ/K}$$

$\therefore$ Net change of entropy

$$= 0.0888 + (-0.0723)$$

$$= \mathbf{0.0165 \text{ kJ/K.}} \quad \textbf{(Ans.)}$$

HIGHLIGHTS

1. **Clausius statement :**

   "It is impossible for a self-acting machine working in a cyclic process, unaided by any external agency, to convey heat from a body at a lower temperature to a body at a higher temperature."

   Kelvin-Planck statement :

   "It is impossible to construct an engine, which while operating in a cycle produces no other effect except to extract heat from a single reservoir and do equivalent amount of work".

   Although above statements of second law of thermodynamic appear to be different, they are really equivalent in the sense that violation of either statement implies violation of other.

2. **Perpetual motion machine of second kind** (*PMM*2) is that imaginary machine which would continuously absorb heat from a single thermal reservoir and convert this heat completely into work. The efficiency of such a machine would be 100%.

3. Clausius inequality is given by:

$$\sum_{\text{Cycle}} \left(\frac{\delta Q}{T}\right) \le 0$$

   "When a system performs a reversible cycle, then

$$\sum_{\text{Cycle}} \left(\frac{\delta Q}{T}\right) = 0,$$

   but when the cycle is not reversible

$$\sum_{\text{Cycle}} \left(\frac{\delta Q}{T}\right) < 0\text{''}.$$

4. **'Entropy'** is a function of a quantity of heat which shows the possibility of conversion of that heat into work. The increase in entropy is small when heat is added at a high temperature and is greater when heat addition is made at lower temperature. Thus for maximum entropy, there is a minimum availability for conversion into work and for minimum entropy there is maximum availability for conversion into work.

5. **Entropy changes for a closed system (per kg) :**

   (*i*) *General case* :

   (*a*) $c_v \log_e \frac{T_2}{T_1} + R \log_e \frac{v_2}{v_1}$ (in terms of T and v)

   (*b*) $c_v \log_e \frac{p_2}{p_1} + c_p \log_e \frac{v_2}{v_1}$ (in terms of p and v)

   (*c*) $c_p \log_e \frac{T_2}{T_1} - R \log_e \frac{p_2}{p_1}$ (in terms of T and p)

   (*ii*) *Constant volume* : $c_v \log_e \frac{T_2}{T_1}$

   (*iii*) *Constant pressure* : $c_p \log_e \frac{T_2}{T_1}$

   (*iv*) *Isothermal* : $R \log_e \frac{v_2}{v_1}$ (*v*) *Adiabatic* : Zero

   (*vi*) *Polytropic* : $c_v \left(\frac{n-\gamma}{n-1}\right) \log_e \frac{T_2}{T_1}$.

6. **Entropy change for an open system**

$$dS \ge \frac{dQ}{T_0} + \Sigma s_i \,.\, dm_i - \Sigma s_0 \,.\, dm_0$$

   where, T_0 = Temperature of the surroundings.

   Subscripts i and 0 refer to inlet and outlet conditions.

OBJECTIVE TYPE QUESTIONS

Choose the Correct Answer :

1. Second law of thermodynamics defines
   (*a*) heat (*b*) work
   (*c*) enthalpy (*d*) entropy
   (*e*) internal energy.
2. For a reversible adiabatic process, the change in entropy is
   (*a*) zero (*b*) minimum
   (*c*) maximum (*d*) infinite
   (*e*) unity.
3. For any reversible process, the change in entropy of the system and surroundings is
   (*a*) zero (*b*) unity
   (*c*) negative (*d*) positive
   (*e*) infinite.
4. For any irreversible process the net entropy change is
   (*a*) zero (*b*) positive
   (*c*) negative (*d*) infinite
   (*e*) unity.
5. The processes of a Carnot cycle are
   (*a*) two adiabatic and two constant volume
   (*b*) one constant volume and one constant pressure and two isentropics
   (*c*) two adiabatics and two isothermals
   (*d*) two constant volumes and two isothermals
   (*e*) two isothermals and two isentropics.
6. Isentropic flow is
   (*a*) irreversible adiabatic flow
   (*b*) ideal fluid flow (*c*) perfect gas flow
   (*d*) frictionless reversible flow
   (*e*) reversible adiabatic flow.
7. In a Carnot engine, when the working substance gives heat to the sink
   (*a*) the temperature of the sink increases
   (*b*) the temperature of the sink remains the same
   (*c*) the temperature of the source decreases
   (*d*) the temperatures of both the sink and the source decrease
   (*e*) changes depend on the operating conditions.

8. If the temperature of the source is increased, the efficiency of the Carnot engine

(*a*) decreases (*b*) increases

(*c*) does not change

(*d*) will be equal to the efficiency of a practical engine

(*e*) depends on other factors.

9. The efficiency of an ideal Carnot engine depends on

(*a*) working substance

(*b*) on the temperature of the source only

(*c*) on the temperature of the sink only

(*d*) on the temperatures of both the source and the sink

(*e*) on the construction of engine.

10. The efficiency of a Carnot engine using an ideal gas as the working substance is

(*a*) $\dfrac{T_1 - T_2}{T_1}$ (*b*) $\dfrac{T_1}{T_1 - T_2}$

(*c*) $\dfrac{T_1 T_2}{T_1 - T_2}$ (*d*) $\dfrac{T_1 - T_2}{T_1 T_2}$

(*e*) $\dfrac{T_2(T_1 - T_2)}{T_1(T_1 + T_2)}$.

11. In a reversible cycle, the entropy of the system

(*a*) increases (*b*) decreases

(*c*) does not change

(*d*) first increases and then decreases

(*e*) depends on the properties of working substance.

12. A frictionless heat engine can be 100% efficient only if its exhaust temperature is

(*a*) equal to its input temperature

(*b*) less than its input temperature

(*c*) 0°C (*d*) 0°K

(*e*) – 100°C.

13. Kelvin-Planck's law deals with

(*a*) conservation of energy (*b*) conservation of heat

(*c*) conservation of mass

(*d*) conversion of heat into work

(*e*) conversion of work into heat.

14. Which of the following statements is *correct* according to Clausius statement of second law of thermodynamics ?

(*a*) It is impossible to transfer heat from a body at a lower temperature to a body at a higher temperature

(*b*) It is impossible to transfer heat from a body at a lower temperature to a body at a higher temperature, without the aid of an external source.

(*c*) It is possible to transfer heat from a body at a lower temperature to a body at a higher temperature by using refrigeration cycle

(*d*) None of the above.

15. According to Kelvin-Planck's statement of second law of thermodynamics

(*a*) It is impossible to construct an engine working on a cyclic process, whose sole purpose is to convert heat energy into work

(*b*) It is possible to construct an engine working on a cyclic process, whose sole purpose is to convert the heat energy into work

(*c*) It is impossible to construct a device which while working in a cyclic process produces no effect other than the transfer of heat from a colder body to a hotter body

(*d*) When two dissimilar metals are heated at one end and cooled at the other, the e.m.f. developed is proportional to the difference of their temperatures at the two end.

(*e*) None of the above.

16. The property of a working substance which increases or decreases as the heat is supplied or removed in a reversible manner is known as

(*a*) enthalpy (*b*) internal energy

(*c*) entropy (*d*) external energy.

17. The entropy may be expressed as a function of

(*a*) pressure and temperature

(*b*) temperature and volume

(*c*) heat and work (*d*) all of the above

(*e*) none of the above.

18. The change of entropy, when heat is absorbed by the gas is

(*a*) positive (*b*) negative

(*c*) positive or negative.

19. Which of the following statements is *correct* ?

(*a*) The increase in entropy is obtained from a given quantity of heat at a low temperature

(*b*) The change in entropy may be regarded as a measure of the rate of the availability of heat for transformation into work

(*c*) The entropy represents the maximum amount of work obtainable per degree drop in temperature

(*d*) All of the above.

20. The condition for the reversibility of a cycle is

(*a*) the pressure and temperature of working substance must not differ, appreciably from those of the surroundings at any stage in the process

(*b*) all the processes taking place in the cycle of operation, must be extremely slow

(*c*) the working parts of the engine must be friction free

(*d*) there should be no loss of energy during the cycle of operation

(*e*) all of the above.

21. In an irreversible process there is a

(*a*) loss of heat (*b*) no loss of work

(*c*) gain of heat (*d*) no gain of heat.

22. The main cause for the irreversibility is

(*a*) mechanical and fluid friction

(*b*) unrestricted expansion

(*c*) heat transfer with a finite temperature difference

(*d*) all of the above.

23. The efficiency of the Carnot cycle may be increased by
(*a*) increasing the highest temperature
(*b*) decreasing the highest temperature
(*c*) increasing the lowest temperature
(*d*) decreasing the lowest temperature
(*e*) keeping the lowest temperature constant.

24. Which of the following is the *correct* statement ?
(*a*) All the reversible engines have the same efficiency
(*b*) All the reversible and irreversible engines have the same efficiency
(*c*) Irreversible engines have maximum efficiency
(*d*) All engines are designed as reversible in order to obtain maximum efficiency.

ANSWERS

1. (*d*)	**2.** (*a*)	**3.** (*a*)	**4.** (*b*)
5. (*e*)	**6.** (*e*)	**7.** (*b*)	**8.** (*b*)
9. (*d*)	**10.** (*a*)	**11.** (*c*)	**12.** (*d*)
13. (*d*)	**14.** (*b*)	**15.** (*e*)	**16.** (*c*)
17. (*a*)	**18.** (*a*)	**19.** (*d*)	**20.** (*e*)
21. (*a*)	**22.** (*d*)	**23.** (*d*)	**24.** (*a*).

THEORETICAL QUESTIONS

1. State the limitations of first law of thermodynamics.
2. What is the difference between a heat engine and a reversed heat engine ?
3. Enumerate the conditions which must be fulfilled by a reversible process. Give some examples of ideal reversible processes.
4. What is an irreversible process ? Give some examples of irreversible processes.
5. Give the following statements of second law of thermodynamics.
(*i*) Clausius statement
(*ii*) Kelvin-Planck statement.
6. Define heat engine, refrigerator and heat pump.
7. What is the perpetual motion machine of the second kind ?
8. What do you mean by 'Thermodynamic temperature' ?
9. What do you mean by 'Clausius inequality' ?
10. Describe the working of a Carnot cycle.
11. Derive an expression for the efficiency of the reversible heat engine.
12. What do you mean by the term 'Entropy' ?
13. What are the characteristics of entropy ?
14. Prove that entropy is a property of a system.
15. Derive an expression for the change in entropy of the universe.
16. What is a temperature-entropy diagram ?
17. Derive expressions for entropy changes for a closed system in the following cases :
(*i*) General case for change of entropy of a gas
(*ii*) Heating a gas at constant volume
(*iii*) Heating a gas at constant pressure
(*iv*) Polytropic process.
18. Give an expression for entropy changes for an open system.

UNSOLVED EXAMPLES

Heat Engines and Reversible Heat Engines

1. A cyclic heat engine operates between a source temperature of 800°C and a sink temperature of 30°C. What is the least rate of heat rejection per kW net output of the engine ? [**Ans.** 0.392 kW]

2. A domestic food freezer maintains a temperature of – 15°C. The ambient air temperature is 30°C. If heat leaks into the freezer at the continuous rate of 1.75 kJ/s what is the least power necessary to pump this heat out continuously? [**Ans.** 0.31 kW]

3. A reversible heat engine operates between two reservoirs at temperatures of 600°C and 40°C. The engine drives a reversible refrigerator which operates between reservoirs at temperatures of 40°C and – 20°C. The heat transfer to the heat engine is 2000 kJ and the net work output for the combined engine refrigerator is 360 kJ. (*i*) Calculate the heat transfer to the refrigerant and the net heat transfer to the reservoir at 40°C. (*ii*) Reconsider (*i*) given that the efficiency of the heat engine and the C.O.P. of the refrigerator are each 40 per cent of their maximum possible values. [**Ans.** Heat rejection to 40°C reservoir (*i*) 5539 kJ ; (*ii*) 1899.6 kJ]

4. A heat engine is supplied heat at the rate of 1700 kJ/min and gives an output of 9 kW. Determine the thermal efficiency and the rate of heat rejection. [**Ans.** 31.76% ; 9.333 kJ/s]

5. What is the highest possible theoretical efficiency of a heat engine operating with a hot reservoir of furnace gases at 2000°C when the cooling water available is at 10°C ? [**Ans.** 87.54%]

6. A Carnot cycle operates between source and sink temperatures of 260°C and – 17.8°C. If the system receives 100 kJ from the source, find (*i*) efficiency of the system, (*ii*) the net work transfer, (*iii*) heat rejected to the sink. [**Ans.** 52.2% ; 52.2 kJ ; 47.8 kJ]

7. Source *A* can supply energy at a rate of 11000 kJ/min at 320°C. A second source *B* can supply energy at a rate of 110000 kJ/min at 68°C. Which source *A* or *B*, would you choose to supply energy to an ideal reversible engine that is to produce large amount of power if the temperature of the surroundings is 40°C ? [**Ans.** Source *B*]

8. A fish freezing plant requires 50 tons of refrigeration. The freezing temperature is – 40°C while the ambient temperature is 35°C. If the performance of the plant is 15% of the theoretical reversed Carnot cycle working within the same temperature limits, calculate the power required. [**Ans.** 375 kW]
Take 1 ton = 210 kJ/ min.

Clausius Inequality

9. A heat engine is supplied with 278 kJ/s of heat at a constant fixed temperature of 283°C and the heat rejection takes place at 5°C. The following results were reported : (*i*) 208 kJ/s are rejected, (*ii*) 139 kJ/s are rejected, (*iii*) 70 kJ/s are rejected.
Classify which of the results report a reversible cycle or irreversible cycle or impossible results.
[**Ans.** (*i*) Irreversible; (*ii*) Reversible; (*iii*) Impossible]

Entropy

10. Air at 15°C and 1.05 bar occupies 0.02 m^3. The air is heated at constant volume until the pressure is 4.2 bar, and then cooled at constant pressure back to the original temperature. Calculate the net heat flow to or from the air and the net entropy change.

Sketch the process on a *T-s* diagram.

[**Ans.** Heat rejected = 6.3 kJ, decrease in entropy of air = 0.0101 kJ/K]

11. 0.03 m^3 of nitrogen contained in a cylinder behind a piston is initially at 1.05 bar and 15°C. The gas is compressed isothermally and reversibly until the pressure is 4.2 bar. Calculate the change of entropy, the heat flow, and the work done, and sketch the process on a *p-v* and *T-s* diagrams. Assume nitrogen to act as a perfect gas. Molecular weight of nitrogen = 28.

[**Ans.** 0.01516 kJ/K (decrease) ; 4.37 kJ (heat rejected) ; 4.37 kJ]

12. Calculate the change of entropy of 1 kg of air expanding polytropically in a cylinder behind a piston from 6.3 bar and 550°C to 1.05 bar. The index of expansion is 1.3.

[**Ans.** 0.1 kJ/kg K (increase)]

13. 0.05 kg of carbon dioxide (molecular weight = 44) is compressed from 1 bar, 15°C, until the pressure is 8.3 bar, and the volume is then 0.004 m^3. Calculate the change of entropy. Take c_p for carbon dioxide as 0.88 kJ/kg K, and assume carbon dioxide to be a perfect gas.

[**Ans.** 0.0113 kJ/K (decrease)]

14. In an air turbine the air expands from 6.8 bar and 430°C to 1.013 bar and 150°C. The heat loss from the turbine can be assumed to be negligible. Show that the process is irreversible, and calculate the change of entropy per kg of air. [**Ans.** 0.0355 kJ/kg K (increase)]

15. (*i*) One kg of water at 273 K is brought into contact with a heat reservoir at 373 K. When the water has reached 373 K, find the entropy change of the water of the heat reservoir, and of the universe.

(*ii*) If water is heated from 273 K to 373 K by first bringing it in contact with a reservoir at 323 K and then with reservoir at 373 K, what will the entropy change of the universe be ?

[**Ans.** (*i*) 0.183 kJ/K ; (*ii*) 0.098 kJ/K]

16. One kg of ice at – 5°C is exposed to the atmosphere which is at 20°C. The ice melts and comes into thermal equilibrium with the atmosphere.

(*i*) Determine the entropy increase of the universe.

(*ii*) What is the minimum amount of work necessary to convert the water back into ice at – 5°C ? c_p of ice is 2.093 kJ/kg °C and the latent heat of fusion of ice is 333.3 kJ/kg.

[**Ans.** (*i*) 0.0949 kJ/K (increase); (*ii*) 28.5 kJ]

17. A system has a heat capacity at constant volume $c_v = AT^2$, where A = 0.042 J/K^3. The system is originally at 200 K and a thermal reservoir at 100 K is available. What is the maximum amount of work that can be recovered as the system is cooled down to the temperature of the reservoir ? [**Ans.** 35 kJ]

18. A fluid undergoes a reversible adiabatic compression from 0.5 MPa, 0.2 m^3 to 0.05 m^3 according to the law, $pv^{1.3}$ = constant. Determine the change in enthalpy, internal energy and entropy, and the heat transfer and work transfer during the process.

[**Ans.** 223.3 kJ ; 171.77 kJ, ; zero ; zero ; – 171.77 kJ]

19. A rigid cylinder containing 0.006 m^3 of nitrogen (molecular weight 28) at 1.04 bar, 15°C, is heated reversibly until the temperature is 90°C. Calculate the change of entropy and the heat supplied. Sketch the process on *T-s* diagram. Take the isentropic index, γ, for nitrogen as 1.4, and assume that nitrogen is a perfect gas. [**Ans.** 0.00125 kJ/K ; 0.407 kJ]

20. 1 m^3 of air is heated reversibly at constant pressure from 15°C to 300°C, and is then cooled reversibly at constant volume back to the initial temperature. The initial pressure is 1.03 bar. Calculate the net heat flow and overall change of entropy, and sketch the process on a *T-s* diagram. [**Ans.** 101.5 kJ ; 0.246 kJ/K]

21. 1 kg of air is allowed to expand reversibly in a cylinder behind a piston in such a way that the temperature remains constant at 260°C while the volume is doubled. The piston is then moved in, and heat is rejected by the air reversibly at constant pressure until the volume is the same as it was initially. Calculate the net heat flow and the overall change of entropy. Sketch the processes on a *T-s* diagram. [**Ans.** – 161.9 kJ/kg ; – 0.497 kJ/kg K]

22. 1 kg of air at 1.013 bar, 17°C, is compressed according to a law $pv^{1.3}$ = constant, until the pressure is 5 bar. Calculate the change of entropy and sketch the process on a *T-s* diagram, indicating the area, which represents the heat flow. [**Ans.** – 0.0885 kJ/kg K]

23. 0.06 m^3 of ethane (molecular weight 30), at 6.9 bar and 60°C, is allowed to expand isentropically in a cylinder behind a piston to a pressure of 1.05 bar and a temperature of 107°C. Calculate γ, R, c_p, c_v for ethane, and calculate the work done during the expansion. Assume ethane to be a perfect gas.

The same mass of ethane at 1.05 bar, 107°C, is compressed to 6.9 bar according to a law $pv^{1.4}$ = constant. Calculate the final temperature of the ethane and the heat flow to or from the cylinder walls during the compression. Calculate also the change of entropy during the compression, and sketch both processes on *p-v* and *T-s* diagrams. [**Ans.** 1.219 ; 0.277 kJ/kg K ; 1.542 kJ/kg K ; 1.265 kJ/kg K ; 54.2 kJ ; 378°C ; 43.4 kJ ; 0.0867 kJ/K]

24. In a reversible process the rate of heat transfer to the system per unit temperature rise is given by $\frac{dQ}{dT}$ = 0.5 kJ/°C. Find the change in entropy of the system if its temperature rises from 500 K to 800 K.

[**Ans.** 0.235 kJ/kg°C]

25. In a reversible cycle 100 kJ of heat is received at 500 K ; then an adiabatic expansion occurs to 400 K, at which temperature 50 kJ of heat is received, then a further adiabatic expansion to 300 K at which temperature 100 kJ of heat is rejected :

(*i*) Find the change in entropy which occurs as the system is restored to its initial state in the remainder of the cycle.

(*ii*) If during the remainder of the cycle heat is transferred only at 400 K, how much heat is transferred and in what direction ? [**Ans.** 0.008 kJ/K ; + 3.2 kJ]

26. 1 kg of air is compressed according to the law $pv^{1.25}$ = constant from 1.03 bar and 15°C to 16.45 bar. Calculate the change in entropy. [**Ans.** 0.255 kJ/kg K]

27. A quantity of gas (mean molecular weight 36.2) is compressed according to the law pv^n = constant, the initial pressure and volume being 1.03 bar and 0.98 m^3 respectively. The temperature at the start of compression is 17°C and at the end it is 115°C. The amount of heat rejected during compression is 3.78 kJ, c_p = 0.92. Calculate :

(*i*) Value of *n*, (*ii*) Final pressure,

(*iii*) Change in entropy.

[**Ans.** (*i*) 1.33 ; (*ii*) 1.107 bar ; (*iii*) 0.228 kJ/kg K]

Availability and Irreversibility

5

5.1. AVAILABLE AND UNAVAILABLE ENERGY

There are many forms in which an energy can exist. But even under ideal conditions all these forms cannot be converted completely into work. This indicates that energy has two parts :

— *Available part.*

— *Unavailable part.*

'***Available energy***' is the *maximum portion of energy which could be converted into useful work by ideal processes which reduce the system to a dead state* (a state in equilibrium with the earth and its atmosphere). Because there can be only one value for maximum work which the system alone could do while descending to its dead state, it follows immediately that *'Available energy' is a property*.

A system which has a pressure difference from that of surroundings, work can be obtained from an expansion process, and if the system has a different temperature, heat can be transferred to a cycle and work can be obtained. But when the temperature and pressure becomes equal to that of the earth, transfer of energy ceases, and although the system contains internal energy, this energy is ***unavailable.***

Summarily available energy denote, the latent capability of energy to do work, and in this sense it can be applied to energy in the system or in the surroundings.

The theoretical maximum amount of work which can be obtained from a system at any state p_1 and T_1 when operating with a reservoir at the constant pressure and temperature p_0 and T_0 is called '**availability**'.

5.2. AVAILABLE ENERGY REFERRED TO A CYCLE

The ***available energy (A.E.)*** or the available part of the energy supplied is the maximum work output obtainable from a certain heat input in a cyclic heat engine (Fig. 5.1). The minimum energy that has to be rejected to the sink by the second law is called the ***unavailable energy (U.E.),*** or the unavailable part of the energy supplied.

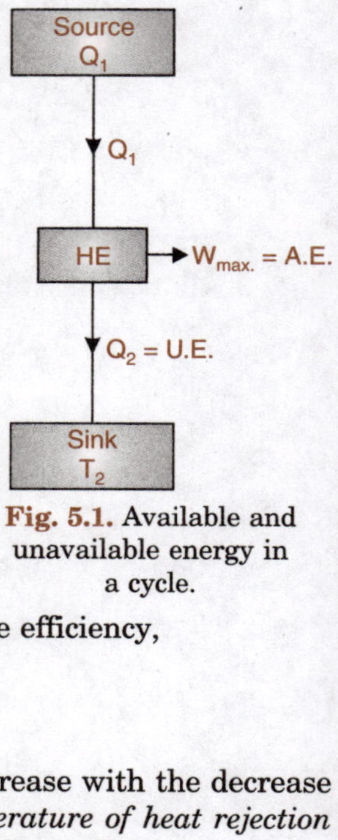

Fig. 5.1. Available and unavailable energy in a cycle.

$$\therefore \quad Q_1 = \text{A.E.} + \text{U.E.}$$

or $$W_{max} = \text{A.E.} = Q_1 - \text{U.E.}$$

For the given values of the source temperature T_1 and sink temperature T_2, the reversible efficiency,

$$\eta_{rev.} = 1 - \frac{T_2}{T_1}$$

For a given T_1, $\eta_{rev.}$ will increase with the decrease of T_2. *The lowest practicable temperature of heat rejection is the temperature of the surroundings, T_0.*

$$\therefore \quad \eta_{max} = 1 - \frac{T_0}{T_1}$$

and $$W_{max} = \left(1 - \frac{T_0}{T_1}\right) Q_1$$

Consider a finite process *l-m*, in which heat is supplied reversibly to a heat engine (Fig. 5.2). Taking an elementary cycle, if dQ_1 is the heat received by the engine reversibly at T_1,

$$\text{Then } dW_{max} = \frac{T_1 - T_0}{T_1} dQ_1$$

$$= dQ_1 - \frac{T_0}{T_1} dQ_1 = \text{A.E.}$$

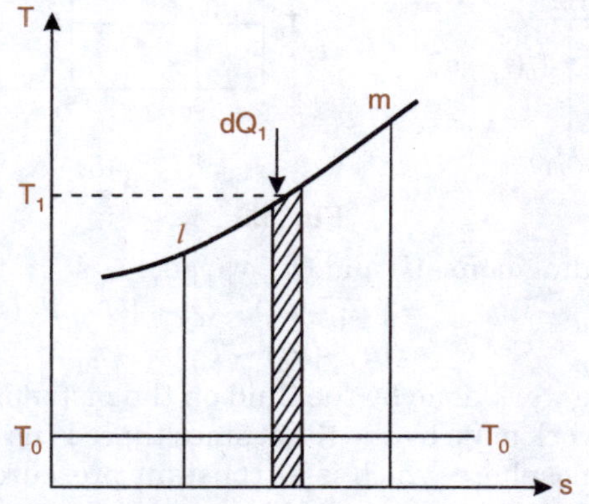

Fig. 5.2. Availability of energy.

For the heat engine receiving heat for the whole process *l-m*, and rejecting heat at T_0

$$\int_l^m dW_{max} = \int_l^m dQ_1 - \int_l^m \frac{T_0}{T_1} dQ_1$$

$$\therefore \quad W_{max} = \text{A.E.} = Q_{l-m} - T_0 (s_l - s_m) \quad \text{...(5.1)}$$

or unavailable energy, $\text{U.E.} = Q_{l-m} - W_{max}$

or $\text{U.E.} = T_0 (s_l - s_m)$

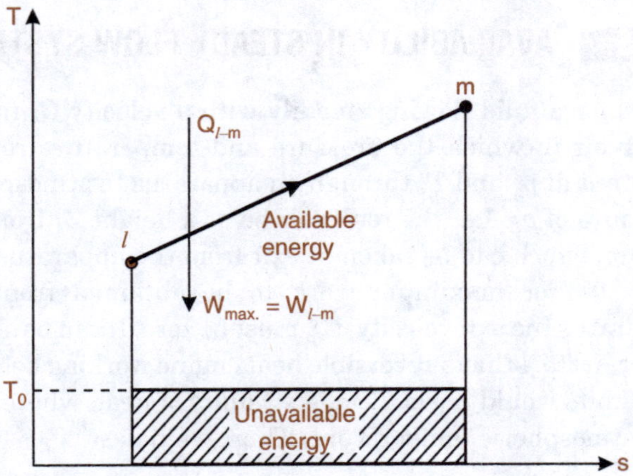

Fig. 5.3. Unavailable energy by the second law of thermodynamics.

Thus unavailable energy is the product of the lowest temperature of heat rejection, and the change of entropy of the system during the process of supplying heat (Fig. 5.3).

5.3. DECREASE IN AVAILABLE ENERGY WHEN HEAT IS TRANSFERRED THROUGH A FINITE TEMPERATURE DIFFERENCE

When transfer of heat takes place through a finite temperature difference, there is a decrease in the availability of energy so transferred. Consider a reversible heat engine operating between temperatures T_1 and T_0 (Fig. 5.4). Then

$$Q_1 = T \,.\, \Delta s \,;$$

$$Q_2 = T_0 \, \Delta s \,;$$

and $\quad W = \text{A.E.} = [T_1 - T_0] \, \Delta s.$

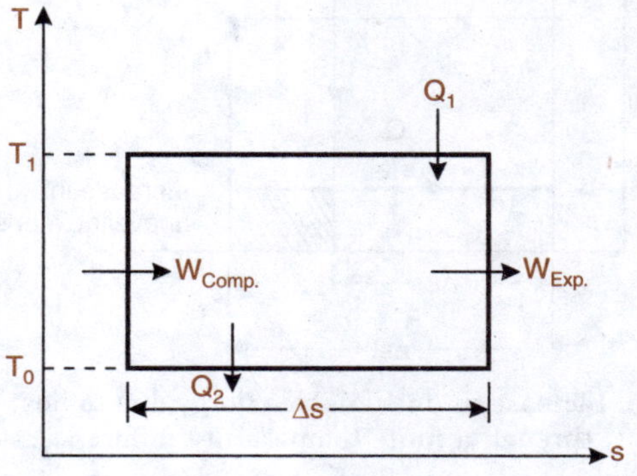

Fig. 5.4. Carnot cycle.

Assume that heat Q_1 is transferred through a finite temperature difference from the reservoir or source at T_1 to the engine absorbing heat at T_1', lower than T_1 (Fig. 5.4). The availability of Q_1 as received by the engine at T_1' can be found by allowing the engine to operate reversibly in a cycle between T_1' and T_0 receiving Q_1 and rejecting Q_2'.

Now, $\quad Q_1 = T_1 \Delta s = T_1' \Delta s'$

$\because \quad T_1 > T_1'$

$\therefore \quad \Delta s' > \Delta s$

$Q_2 = T_0 \, \Delta s$

$Q_2' = T_0 \, \Delta s'$

$\because \quad \Delta s' > \Delta s$

$\therefore \quad Q_2' > Q_2$

$\therefore \quad W' = Q_1 - Q_2' = T_1' \, \Delta s' - T_0 \, \Delta s'$

and $\quad W = Q_1 - Q_2 = T_1 \, \Delta s - T_0 \, \Delta s$

$\therefore \quad W' < W$, because $Q_2' > Q_2$

The *loss of available energy* due to irreversible heat transfer through finite temperature difference between the source and the working fluid during the heat addition process is given as :

$$W - W' = Q_2' - Q_2$$

$$= T_0 \, (\Delta s' - \Delta s)$$

i.e., Decrease in available energy, A.E.

$$= T_0 (\Delta s' - \Delta s) \quad ...(5.2)$$

Thus the decrease in A.E. is the product of the lowest feasible temperature of heat rejection and the additional entropy change in the system while receiving heat irreversibly, compared to the case of reversible heat transfer from the same source. *The greater is the temperature difference* $(T_1 - T_1')$, *the greater is the heat rejection* Q_2' *and the greater will be the unavailable part of the energy supplied* (Fig. 5.5).

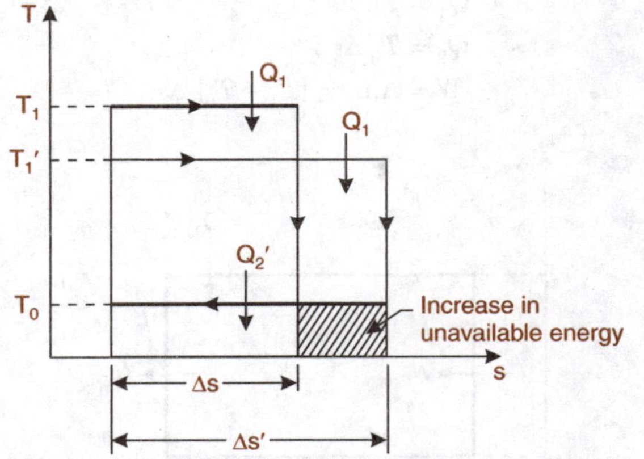

Fig. 5.5. Increase in unavailable energy due to heat transfer through a finite temperature difference.

Energy is said to be *degraded* each time it flows through a finite temperature difference. That is, why the second law of thermodynamics is sometimes called the *law of the degradation of energy*, and energy is said to *'run down hill'*.

5.4. AVAILABILITY IN NON-FLOW SYSTEMS

Let us consider a system consisting of a fluid in a cylinder behind a piston, the fluid expanding reversibly from initial condition of p_1 and T_1 to final atmospheric conditions of p_0 and T_0. Imagine also that the system works in conjunction with a reversible heat engine which receives heat reversibly from the fluid in the cylinder such that the working substance of the heat engine follows the cycle *01L0* as shown in Fig. 5.6, where $s_1 = s_L$ and $T_0 = T_L$ (the only possible way in which this could occur would be if an infinite number of reversible heat engines were arranged in parallel, each operating on a Carnot cycle, each one receiving heat at a different constant temperature and each one rejecting heat at T_0). The work done by the engine is given by :

$$W_{engine} = \text{Heat supplied} - \text{Heat rejected}$$
$$= Q - T_0 (s_1 - s_0) \quad ...(i)$$

The heat supplied to the engine is equal to the heat rejected by the fluid in the cylinder. Therefore, for the fluid in the cylinder undergoing the process 1 to 0, we have

$$-Q = (u_0 - u_1) + W_{fluid}$$

i.e.,
$$W_{fluid} = (u_1 - u_0) - Q \quad ...(ii)$$

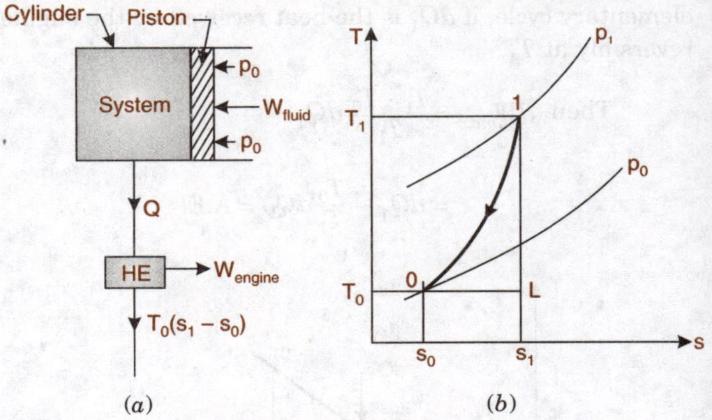

Fig. 5.6

Adding eqns. (*i*) and (*ii*), we get

$$W_{fluid} + W_{engine} = [(u_1 - u_0) - Q] + [Q - T_0 (s_1 - s_0)]$$
$$= (u_1 - u_0) - T_0 (s_1 - s_0)$$

The work done by the fluid on the piston is less than the total work done by the fluid, since there is no work done on the atmosphere which is at constant pressure p_0

i.e., Work done on atmosphere $= p_0 (v_0 - v_1)$

Hence, *maximum work available*

$$= (u_1 - u_0) - T_0 (s_1 - s_0) - p_0 (v_0 - v_1)$$

Note. When a fluid undergoes a complete cycle then the net work done on the atmosphere is zero.

$$W_{max} = (u_1 + p_0 v_1 - T_0 s_1) - (u_0 + p_0 v_0 - T_0 s_0) \quad ...(5.3)$$

$$\therefore \quad W_{max} = a_1 - a_0 \quad ...[5.3\ (a)]$$

The property, $a = u + p_0 v - T_0 s$ (per unit mass) is called the **non-flow availability function.**

5.5. AVAILABILITY IN STEADY FLOW SYSTEMS

Consider a fluid flowing steadily with a velocity C_1 from a reservoir in which the pressure and temperature remain constant at p_1 and T_1 through an apparatus to atmospheric pressure of p_0. Let the reservoir be at a height Z_1 from the datum, which can be taken at exit from the apparatus, *i.e.*, $Z_0 = 0$. For maximum work to be obtained from the apparatus the exit velocity, C_0, must be zero. It can be shown as for Art. 5.4 that a reversible heat engine working between the limits would reject $T_0 (s_1 - s_0)$ units of heat, where T_0 is the atmospheric temperature. Thus, we have

$$W_{max} = \left(h_1 + \frac{C_1^2}{2} + Z_1 g\right) - h_0 - T_0 (s_1 - s_0)$$

In several thermodynamic systems the kinetic and potential energy terms are negligible

i.e.,
$$W_{max} = (h_1 - T_0 s_1) - (h_0 - T_0 s_0)$$
$$= b - b_0$$

The property, $b = h - T_0s$ (per unit mass) is called the **steady-flow availability function.**

[In the equation $b = h - T_0s$; the function 'b' (like the function 'a') is a composite property of a system and its environment ; this is also known as ***Keenan function***].

Note 1. The alternative names for availability and unavailable quantity $T_0\Delta s$ are ***energy*** and ***anergy*** respectively.

2. The only difference between $a = u + p_0v - T_0s$ function and $b = (h - T_0s) = (u + pv - T_0s)$ function is in pressure only.

5.6. HELMHOLTZ AND GIBBS FUNCTIONS

The work done in a non-flow reversible system (per unit mass) is given by :

$$W = Q - (u_0 - u_1)$$
$$= T.ds - (u_0 - u_1)$$
$$= T(s_0 - s_1) - (u_0 - u_1)$$

i.e., $$W = (u_1 - Ts_1) - (u_0 - Ts_0) \quad ...(5.4)$$

The term $(u - Ts)$ is known as *Helmholtz function*. This gives maximum possible output when the heat Q is transferred at constant temperature and is the case with a very large source.

If work against atmosphere is equal to $p_0(v_0 - v_1)$, then the maximum work available,

$$W_{max} = W - \text{work against atmosphere}$$
$$= W - p_0(v_0 - v_1)$$
$$= (u_1 - Ts_1) - (u_0 - Ts_0) - p_0(v_0 - v_1)$$
$$= (u_1 + p_0v_1 - Ts_1) - (u_0 + p_0v_0 - Ts_0)$$
$$= (h_1 - Ts_1) - (h_0 - Ts_0)$$

i.e., $$W_{max} = g_1 - g_0 \quad ...(5.5)$$

where $g = h - T.s$ is known as ***Gibb's function or free energy function.***

The maximum possible available work when system changes from 1 to 2 is given by

$$W_{max} = (g_1 - g_0) - (g_2 - g_0) = g_1 - g_2 \quad ...(5.6)$$

Similarly, for *steady flow system* the maximum work available is given by

$$W_{max} = (g_1 - g_2) + (\text{KE}_1 - \text{KE}_2) + (\text{PE}_1 - \text{PE}_2) \quad ...(5.7)$$

where K.E. and P.E. represent the kinetic and potential energies.

It may be noted that Gibb's function $g = (h - Ts)$ is a property of the system where availability function $a = (u + p_0v - T_0s)$ is a composite property of the system and surroundings.

Again,
$$a = u + p_0v - T_0s$$
$$b = u + pv - T_0s$$
$$g = u + pv - Ts$$

When state 1 proceeds to *dead state* (zero state)

$$a = b = g.$$

5.7. IRREVERSIBILITY

The actual work which a system does is always *less* than the idealized reversible work, and the difference between the two is called the *irreversibility of the process.*

Thus, Irreversibility, $I = W_{max} - W \quad ...(5.8)$

This is also sometimes referred to as '***degradation***' or '***dissipation***'.

For a *non-flow process* between the equilibrium states, when the system exchanges heat only with environment, irreversibility (*per unit mass*),

$$i = [(u_1 - u_2) - T_0(s_1 - s_2)] - [(u_1 - u_2) + Q]$$
$$= T_0(s_2 - s_1) - Q$$
$$= T_0(\Delta s)_{system} + T_0(\Delta s)_{surr.}$$

i.e., $$i = T_0[(\Delta s)_{system} + (\Delta s)_{surr.}] \quad ...(5.9)$$

$\therefore$ $$i \geq 0$$

Similarly, for *steady flow-process*

$$i = W_{max} - W \text{ (per unit mass)}$$
$$= \left[\left(b_1 + \frac{C_1^2}{2} + gZ_1\right) - \left(b_2 + \frac{C_2^2}{2} + gZ_2\right)\right]$$
$$- \left[\left(h_1 + \frac{C_1^2}{2} + gZ_1\right) - \left(h_2 + \frac{C_2^2}{2} + gZ_2\right) + Q\right]$$
$$= T_0(s_2 - s_1) - Q$$
$$= T_0(\Delta s)_{system} + T_0(\Delta s)_{surr.}$$

i.e., $$i = T_0(\Delta s_{system} + \Delta s_{surr.})$$

The same expression for irreversibility applies to both flow and non-flow processes.

The quantity $T_0(\Delta s_{system} + \Delta s_{surr.})$ represents (per unit mass) an increase in unavailable energy (or energy).

5.8. EFFECTIVENESS

***Effectiveness** is defined as the ratio of actual useful work to the maximum useful work.* The useful output of a system is given by the increase of availability of the surroundings.

Effectiveness, $\in$

$$= \frac{\text{Increase of availability of surroundings}}{\text{Loss of availability of the system}} \quad ...(5.10)$$

For a compression or heating process the effectiveness is given by

$$\in = \frac{\text{Increase of availability of the system}}{\text{Loss of availability of the surroundings}}$$

or $$\in = \frac{W_{useful}}{W_{max.\,useful}} \quad ...[5.10\,(a)]$$

The *effectiveness of an actual process is always less than unity*. Thus effectiveness of a process is the measure of the extent to which advantage has been taken of an opportunity to obtain useful work.

Example 5.1. *One kg of air is compressed polytropically from 1 bar pressure and temperature of 300 K to a pressure of 6.8 bar and temperature of 370 K. Determine the irreversibility if the sink temperature is 293 K. Assume $R = 0.287\ kJ/kg\ K$, $c_p = 1.004\ kJ/kg\ K$ and $c_v = 0.716\ kJ/kg/K$.* **(U.P.S.C.)**

Solution. Irreversibility $I = W_{max} - W_{act}$

$-W_{max}$ = Change in internal energy − T_0 × Change in entropy

or $-W_{max} = (u_2 - u_1) - T_0(s_2 - s_1) = W_{rev}$

or $-W_{max} = c_v(T_2 - T_1) - T_0[c_p \ln (T_2/T_1) - R \ln (p_2/p_1)]$

$= 0.716(370 - 300) - 293 \times [1.005 \ln (370/300) - 0.287 \ln (6.8/1)]$

or $W_{max} = -149.53$ kJ/kg $= W_{rev}$

(negative sign indicates that work is done on air)

The index of compression '*n*' is given by

$$\frac{T_2}{T_1} = \left(\frac{p_2}{p_1}\right)^{[(n-1)/n]}$$

or $$\frac{n-1}{n} = \frac{\ln (T_2/T_1)}{\ln (p_2/p_1)} = \frac{\ln (370/300)}{\ln (6.8/1)}$$

or $n = 1.123$

$$W_{actual} = \frac{mR(T_1 - T_2)}{n-1} = \frac{1 \times 0.287(300 - 370)}{1.123 - 1}$$

$= -163.33$ kJ/kg

$I = W_{rev} - W_{act} = -149.53 - (-163.33)$

$=$ **13.8 kJ/kg.** **(Ans.)**

Example 5.2. *A system at 500 K receives 7200 kJ/min from a source at 1000 K. The temperature of atmosphere is 300 K. Assuming that the temperatures of system and source remain constant during heat transfer find out :*

(*i*) *The entropy produced during heat transfer ;*

(*ii*) *The decrease in available energy after heat transfer.*

Solution. Refer to Fig. 5.7.

Temperature of source, $T_1 = 1000$ K

Temperature of system, $T_2 = 500$ K

Temperature of atmosphere, $T_0 = 300$ K

Heat received by the system, $Q = 7200$ kJ/min.

(*i*) **Net change of entropy :**

Change in entropy of the *source* during heat transfer

$$= \frac{-Q}{T_1} = \frac{-7200}{1000} = -7.2 \text{ kJ/min-K}$$

Change in entropy of the *system* during heat transfer

$$= \frac{Q}{T_2} = \frac{7200}{500} = 14.4 \text{ kJ/min-K}$$

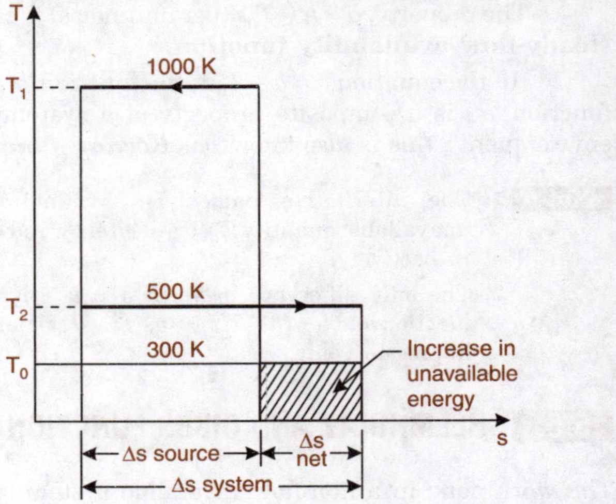

Fig. 5.7

The net change of entropy, ΔS

$= -7.2 + 14.4 =$ **7.2 kJ/min-K.** **(Ans.)**

(*ii*) **Decrease in available energy :**

Available energy with source

$= (1000 - 300) \times 7.2 = 5040$ kJ

Available energy with the system

$= (500 - 300) \times 14.4 = 2880$ kJ

∴ **Decrease in available energy**

$= 5040 - 2880 =$ **2160 kJ.** **(Ans.)**

[Also, increase in available energy

$= T_0(S_2 - S_1) = T_0\ \Delta S$

$= 300 \times 7.2 = 2160$ kJ]

Example 5.3. *8 kg of air at 650 K and 5.5 bar pressure is enclosed in a closed system. If the atmosphere temperature and pressure are 300 K and 1 bar respectively, determine :*

(*i*) *The availability if the system goes through the ideal work producing process.*

(*ii*) *The availability and effectiveness if the air is cooled at constant pressure to atmospheric temperature without bringing it to complete dead state. Take c_v = 0.718 kJ/kg K ; c_p = 1.005 kJ/kg K.*

Solution. Mass of air, $m = 8$ kg

Temperature, $T_1 = 650$ K

Pressure, $p_1 = 5.5$ bar

Atmospheric pressure, $p_0 = 1$ bar

Atmospheric temperature, $T_0 = 300$ K

For air : $c_v = 0.718$ kJ/kg K ; $c_p = 1.005$ kJ/kg K.

(*i*) Change in available energy (for bringing the system to dead state),

$= m[(u_1 - u_0) - T_0 \Delta s]$

Also $$\Delta s = c_v \log_e \left(\frac{T_1}{T_0}\right) + R \log_e \frac{v_1}{v_0}$$

Using the ideal gas equation,

$$\frac{p_1 v_1}{T_1} = \frac{p_0 v_0}{T_0}$$

$$\therefore \quad \frac{v_0}{v_1} = \frac{p_1}{p_0} \cdot \frac{T_0}{T_1} = \frac{5.5}{1} \times \frac{300}{650} = 2.54$$

$$\therefore \quad \Delta s = 0.718 \log_e \left(\frac{650}{300}\right) + 0.287 \log_e \left(\frac{1}{2.54}\right)$$

$$= 0.555 + (-0.267) = 0.288 \text{ kJ/kg K}$$

$\therefore$ Change in available energy

$$= m[(u_1 - u_0) - T_0 \Delta s] = m[c_v(T_1 - T_0) - T_0 \Delta s]$$

$$= 8[0.718(650 - 300) - 300 \times 0.288]$$

$$= 1319.2 \text{ kJ}$$

Loss of availability per unit mass during the process

$$= p_0 (v_0 - v_1) \text{ per unit mass}$$

Total loss of availability $= p_0(V_0 - V_1)$

But $V_1 = \dfrac{mRT_1}{p_1} = \dfrac{8 \times 287 \times 650}{5.5 \times 10^5}$

$$= 2.713 \text{ m}^3 \qquad \left[\because pV = mRT \text{ or } V = \frac{mRT}{p}\right]$$

and $\quad V_0 = 2.54 \times 2.713 = 6.891 \text{ m}^3$

$\therefore$ **Loss of availability**

$$= \frac{1 \times 10^5}{10^3}(6.891 - 2.713) = \mathbf{417.8 \text{ kJ.}} \quad \textbf{(Ans.)}$$

(*ii*) Heat transferred during cooling (constant pressure) process

$$= m.\, c_p (T_1 - T_0)$$

$$= 8 \times 1.005 (650 - 300) = 2814 \text{ kJ}$$

Change in entropy during cooling

$$\Delta s = mc_p \log_e \left(\frac{T_1}{T_0}\right)$$

$$= 8 \times 1.005 \times \log_e \left(\frac{650}{300}\right) = 6.216 \text{ kJ/K}$$

Unavailable energy $= T_0 \Delta S$

$$= 300 \times 6.216 = 1864.8 \text{ kJ}$$

Available energy $= 2814 - 1864.8$

$= \mathbf{949.2 \text{ kJ.}}$ **(Ans.)**

Effectiveness, $\quad \in = \dfrac{\text{Available energy}}{\text{Change in available energy}}$

$$= \frac{949.2}{1319.2} = \mathbf{0.719.} \quad \textbf{(Ans.)}$$

☞**Example 5.4.** *In a power station, the saturated steam is generated at 200°C by transferring the heat from hot gases in a steam boiler. Find the increase in total entropy of the combined system of gas and water and increase in unavailable energy due to irreversible heat transfer. The gases are cooled from 1000°C to 500°C and all the heat from gases goes to water. Assume water enters the boiler at saturated condition and leaves as saturated steam.*

Take : c_{pg} (for gas) = 1.0 kJ/kg K, h_{fg} (latent heat of steam at 200°C) = 1940.7 kJ/kg.

Atmospheric temperature = 20°C.

Obtain the results on the basis of 1 kg of water.

Solution. Refer to Fig. 5.8.

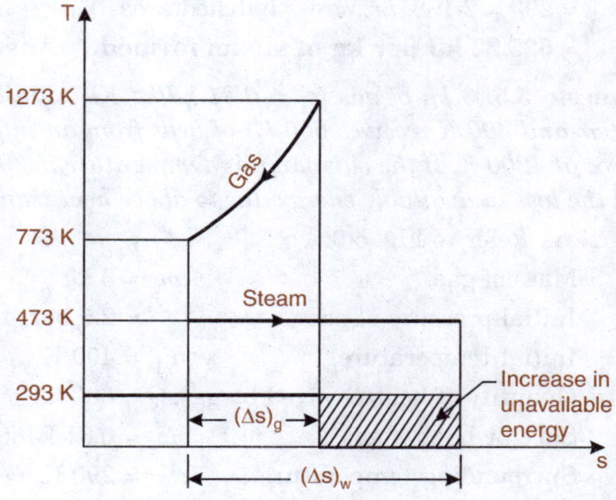

Fig. 5.8

Temperature of saturation steam $= 200 + 273$
$= 473$ K

Initial temperature of gases $= 1000 + 273$
$= 1273$ K

Final temperature of gases $= 500 + 273$
$= 773$ K

For gases : $\quad c_{pg} = 1$ kJ/kg K

Latent heat of steam of 200°C saturation temperature, $\quad h_{fg} = 1940.7$ kJ/kg

Atmospheric temperature $= 20 + 273$
$= 293$ K

Heat lost by gases = Heat gained by 1 kg saturated water when it is converted to steam at 200°C.

$$\therefore \quad m_g c_{pg} (1273 - 773) = 1940.7$$

[where m_g = mass of gases, c_{pg} = specific heat of gas at constant pressure]

i.e., $\quad m_g = \dfrac{1940.7}{1.0 \times (1273 - 773)} = 3.88$ kg

Change of entropy of m_g kg of gas,

$$(\Delta S)_g = m_g \, c_{pg} \log_e \left(\frac{773}{1273}\right)$$

$$= 3.88 \times 1.0 \times \log_e \left(\frac{773}{1273}\right) = -1.935 \text{ kJ/K}$$

Change of entropy of water (per kg) when it is converted into steam,

$$(\Delta s)_w = \frac{h_{fg}}{T_s} = \frac{1940.7}{(200 + 273)} = 4.103 \text{ kJ/kg K.}$$

Net change in entropy due to heat transfer

$= -1.935 + 4.103 =$ **2.168 kJ/K. (Ans.)**

Increase in unavailable energy due to heat transfer

$= 293 \times 2.168$, *i.e.,* cross hatched area

= **635.22 kJ per kg of steam formed. (Ans.)**

Example 5.5. *3 kg of gas (c_v = 0.81 kJ/kg K) initially at 2.5 bar and 400 K receives 600 kJ of heat from an infinite source at 1200 K. If the surrounding temperature is 290 K, find the loss in available energy due to above heat transfer.*

Solution. Refer to Fig. 5.9.

Mass of gas, $m_g = 3$ kg

Initial pressure of gas = 2.5 bar

Initial temperature, $T_1' = 400$ K

Quantity of heat received by gas, $Q = 600$ kJ

Specific heat of gas, $c_v = 0.81$ kJ/kg K

Surrounding temperature = 290 K

Temperature of infinite source, $T_1 = 1200$ K

Heat received by the gas is given by,

$$Q = m_g c_v (T_2' - T_1')$$

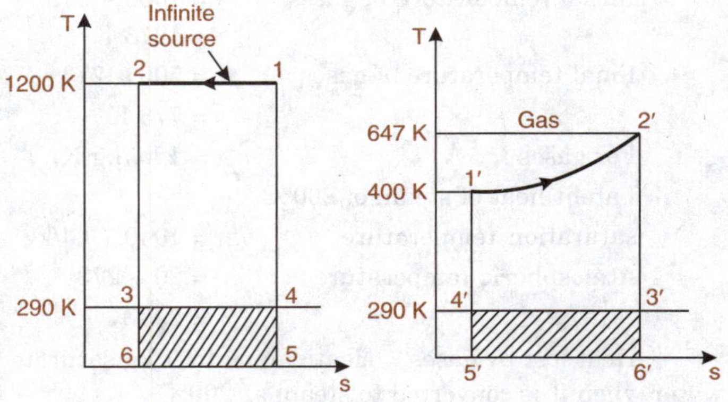

Fig. 5.9

$$600 = 3 \times 0.81 (T_2' - 400)$$

$$\therefore \quad T_2' = \frac{600}{3 \times 0.81} + 400 = 646.9 \text{ K say } 647 \text{ K}$$

Available energy with the source

= area 1-2-3-4-1

$$= (1200 - 290) \times \frac{600}{1200} = 455 \text{ kJ}$$

Change in entropy of the gas

$$= m_g c_v \log_e \left(\frac{T_2'}{T_1'}\right)$$

$$= 3 \times 0.81 \times \log_e \left(\frac{647}{400}\right) = 1.168 \text{ kJ/K}$$

Unavailability of the gas

= area 3′- 4′- 5′- 6′- 3′

$= 290 \times 1.168 = 338.72$ kJ

Available energy with the gas

$= 600 - 338.72 = 261.28$ kJ

∴ **Loss in available energy due to heat transfer**

$= 455 - 261.28 =$ **193.72 kJ. (Ans.)**

☞**Example 5.6.** *Calculate the unavailable energy in 60 kg of water at 60°C with respect to the surroundings at 6°C, the pressure of water being 1 atmosphere.*

Solution. Refer to Fig. 5.10.

Mass of water, $m = 60$ kg

Temperature of water, $T_1 = 60 + 273 = 333$ K

Temperature of surroundings, $T_0 = 6 + 273 = 279$ K

Pressure of water, $p = 1$ atm.

If the water is cooled at a constant pressure of 1 atm. from 60°C to 6°C the heat given up may be used as a *source for a series of Carnot engines each using the surroundings as a sink.* It is assumed that the amount of energy received by any engine is small relative to that in the source and temperature of the source does not change while heat is being exchanged with the engine.

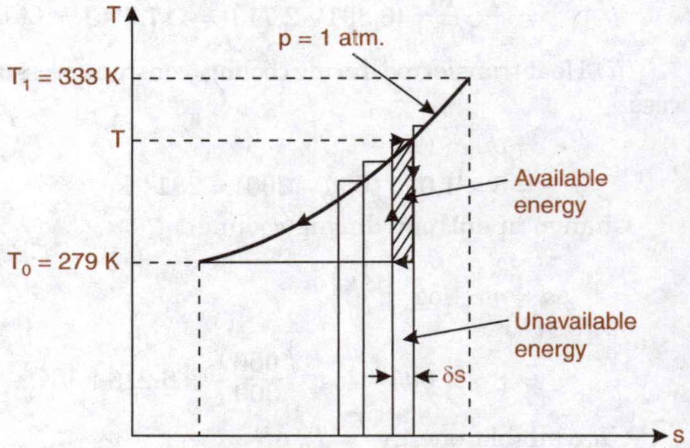

Fig. 5.10

Consider that the source has fallen to temperature T, at which level there operates a Carnot engine which takes in heat at this temperature and rejects heat at $T_0 = 279$ K. If δs is the entropy change of water, the work obtained is

$$\delta W = -m(T - T_0)\, \delta s$$

where δs is negative.

$$\therefore \quad \delta W = -60\,(T - T_0)\, \frac{c_p \delta T}{T}$$

$$= -60\, c_p \left(1 - \frac{T_0}{T}\right) \delta T$$

With a very great number of engines in the series, the total work (maximum) obtainable when the water is cooled from 333 K to 279 K would be

$$W_{max} = \text{Available energy}$$

$$= -\lim. \sum_{333}^{279} 60\, c_p \left(1 - \frac{T_0}{T}\right) \delta T$$

$$= \int_{279}^{333} 60\, c_p \left(1 - \frac{T_0}{T}\right) dT$$

$$= 60\, c_p \left[(333 - 279) - 279 \log_e \left(\frac{333}{279}\right)\right]$$

$$= 60 \times 4.187\,(54 - 49.36) = 1165.7 \text{ kJ}$$

Also, $Q_1 = 60 \times 4.187 \times (333 - 279)$

$= 13565.9$ kJ

$\therefore$ **Unavailable energy** $= Q_1 - W_{max}$

$= 13565.9 - 1165.7 =$ **12400.2 kJ. (Ans.)**

Example 5.7. *15 kg of water is heated in an insulated tank by a churning process from 300 K to 340 K. If the surrounding temperature is 300 K, find the loss in availability for the process.*

Solution. Mass of water, $m = 15$ kg

Temperature, $T_1 = 340$ K

Surrounding temperature, $T_0 = 300$ K

Specific heat of water, $c_p = 4.187$ kJ/kg K

Loss in availability :

Work added during churning

= Increase in enthalpy of the water

$= 15 \times 4.187 \times (340 - 300) = 2512.2$ kJ

Now the energy in the water = 2512.2 kJ

The availability out of this energy is given by

$$m[(u_1 - u_0) - T_0\, \Delta s]$$

where $\Delta s = c_p \log_e \left(\frac{T_1}{T_0}\right)$

$\therefore$ $\Delta s = 4.187 \log_e \left(\frac{340}{300}\right) = 0.524$ kJ/kg K

$\therefore$ Available energy

$= m\,[c_v\,(T_1 - T_0) - T_0\, \Delta s]$

$= 15\,[4.187\,(340 - 300) - 300 \times 0.524]$

$= 158.7$ kJ

$\therefore$ **Loss in availability**

$= 2508 - 158.7 =$ **2349.3 kJ. (Ans.)**

This shows that *conversion of work into heat is highly irreversible process* (since out of 2512.2 kJ of work energy supplied to increase the temperature, only 158.7 kJ will be available again for conversion into work).

Example 5.8. *5 kg of air at 550 K and 4 bar is enclosed in a closed system.*

(i) Determine the availability of the system if the surrounding pressure and temperature are 1 bar and 290 K respectively.

(ii) If the air is cooled at constant pressure to the atmospheric temperature, determine the availability and effectiveness.

Solution. Mass of air, $m = 5$ kg

Temperature, $T_1 = 550$ K

Pressure, $p_1 = 4 \text{ bar} = 4 \times 10^5 \text{ N/m}^2$

Temperature, $T_2 = T_0 = 290$ K

Pressure, $p_2 = p_0 = 1 \text{ bar} = 1 \times 10^5 \text{ N/m}^2$.

(*i*) **Availability of the system :**

Availability of the system is

$= m[(u_1 - u_0) - T_0(s_1 - s_0)]$

$= m[c_v(T_1 - T_0) - T_0\, \Delta s]$

$$\Delta s = c_p \log_e \frac{T_1}{T_0} - R \log_e \left(\frac{p_1}{p_0}\right)$$

$$= 1.005 \log_e \left(\frac{550}{290}\right) - 0.287 \log_e \left(\frac{4}{1}\right)$$

$= 0.643 - 0.397 = 0.246$ kJ/kg K

$\therefore$ **Availability of the system**

$= m\,[c_v\,(T_1 - T_0) - T_0 \Delta s]$

$= 5[0.718\,(550 - 290) - 290 \times 0.246]$

$=$ **576.7 kJ. (Ans.)**

(*ii*) Heat transferred during cooling

$Q = m \times c_p \times (T_1 - T_0)$

$= 5 \times 1.005 \times (550 - 290)$

$= 1306.5$ kJ ...... heat lost by the system

Change of entropy during cooling

$$\Delta S = m \times c_p \times \log_e \left(\frac{T_1}{T_0}\right)$$

$$= 5 \times 1.005 \times \log_e \left(\frac{550}{290}\right) = 3.216 \text{ kJ/K}$$

Unavailable portion of this energy

$= T_0 \,.\, (\Delta S) = 290 \times 3.216 = 932.64$ kJ

$\therefore$ **Available energy**

$= 1306.5 - 932.64 =$ **373.86 kJ. (Ans.)**

Effectiveness,

$$\in = \frac{\text{Available energy}}{\text{Availability of the system}} = \frac{373.86}{576.7}$$

$=$ **0.648 or 64.8%. (Ans.)**

Example 5.9. *Air at the rate of 25 kg/min is compressed in a centrifugal air compressor from 1 bar to 2 bar. The temperature increases from 15°C to 100°C during* *compression. Determine actual and minimum power required to run the compressor. The surrounding air temperature is 15°C.*

Neglect the heat interaction between the compressor and surroundings and changes in potential and kinetic energy.

Take for air, c_p = 1.005 kJ/kg K, R = 0.287 kJ/kg K.

Solution. Rate of flow of air, m = 25 kg/min.

Initial pressure, $p_1 = 1$ bar

Final pressure, $p_2 = 2.0$ bar

Initial temperature, $T_1 = T_0 = 15 + 273 = 288$ K

Final temperature, $T_2 = 100 + 273 = 373$ K.

Applying energy equation to compressor,

$$W_{actual} = h_2 - h_1 \quad [\text{as } Q = 0, \Delta PE = 0, \Delta KE = 0]$$

$$= c_p (T_2 - T_1) = 1.005 (373 - 288)$$

$$= 85.4 \text{ kJ/kg}$$

Total actual work done/min

$$= 25 \times 85.4 = 2135 \text{ kJ/min}$$

$$= \frac{2135}{60} = 35.58 \text{ kJ/s} = 35.58 \text{ kW}$$

The *minimum work required is given by the increase in availability of the air stream.*

$$W_{min} = b_2 - b_1 = (h_2 - h_1) - T_0 (s_2 - s_1)$$

$$s_2 - s_1 = c_p \log_e \left(\frac{T_2}{T_1}\right) - R \log_e \left(\frac{p_2}{p_1}\right) \quad \text{...per unit mass}$$

$$= 1.005 \log_e \left(\frac{373}{288}\right) - 0.287 \log_e \left(\frac{2.0}{1}\right)$$

$$= 0.2599 - 0.1989 = 0.061 \text{ kJ/kg K}$$

$$\therefore W_{min} = (h_2 - h_1) - T_0(s_2 - s_1)$$

$$= 85.4 - 288 \times 0.061 = 67.8 \text{ kJ/kg}$$

$\therefore$ **Minimum work required**

$$\frac{25 \times 67.8}{60} = 28.25 \text{ kJ/s} = \mathbf{28.25 \text{ kW}}. \quad \textbf{(Ans.)}$$

Example 5.10. *1 kg of oxygen at 1 bar and 450 K is mixed with 1 kg of hydrogen at the same temperature and pressure by removing the diaphragm as shown in Fig. 5.11. Determine the loss in availability if the surrounding temperature is 290 K.*

Assume that the system is fully isolated.

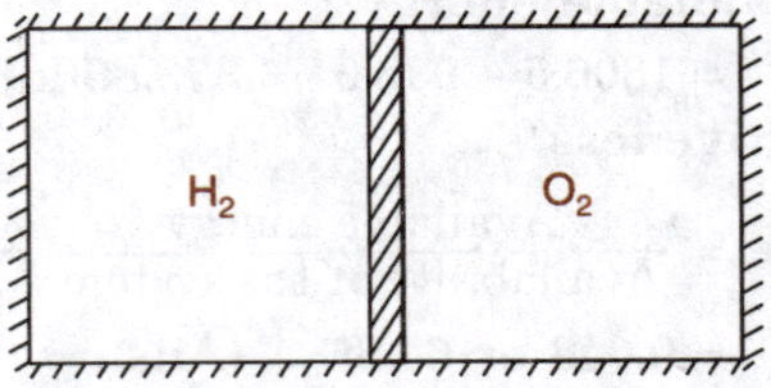

Fig. 5.11

Solution. Mass of oxygen, $m_{O_2} = 1$ kg

Mass of hydrogen, $m_{H_2} = 1$ kg

Pressure, $p = 1 \text{ bar} = 1 \times 10^5 \text{ N/m}^2$

Temperature, $T_{O_2} = T_{H_2} = 450$ K

Surrounding temperature = 290 K

Characteristic gas constant of O_2,

$$R_{O_2} = \frac{R_0}{M_{O_2}} = \frac{8314}{32} = 259.6 \text{ J/kg K}$$

Now to find volume of O_2 using the relation,

$$pv = mRT$$

$$v = \frac{mRT}{p} = \frac{1 \times 259.6 \times 450}{1 \times 10^5} = 1.168 \text{ m}^3$$

i.e., $v_{O_2} = 2.336 \text{ m}^3$

Characteristic gas constant of H_2,

$$R_{H_2} = \frac{R_0}{M_{H_2}} = \frac{8314}{2} = 4157 \text{ J/kg K}$$

Volume of H_2, $v_{H_2} = \dfrac{1 \times 4157 \times 450}{1 \times 10^5} = 18.706 \text{ m}^3$

Total volume after mixing

$$= v_{O_2} + v_{H_2} = 1.168 + 18.706$$

$$= 19.874 \text{ m}^3$$

The partial pressure of each gas changes after the mixing even though the temperature is the same due to increase in volume.

Change in entropy of oxygen

$$= R_{O_2} \log_e \frac{v_2}{v_1}$$

$$= 259.6 \log_e \left(\frac{19.874}{1.168}\right) = 735.7 \text{ J/K}$$

Change in entropy of hydrogen

$$= R_{H_2} \log_e \frac{v_2}{v_1} = 4157 \log_e \left(\frac{19.874}{18.706}\right)$$

$$= 251.78 \text{ J/K}$$

Net change in entropy,

$$\Delta S = 735.7 + 251.78 = 987.48 \text{ J/K}$$

Loss in availability

$$= T_0 \Delta S = 290 \times 987.48 \text{ J}$$

$$= \frac{290 \times 987.48}{10^3} \text{ kJ} = 286.36 \text{ kJ}$$

i.e., **Loss in availability = 286.36 kJ. (Ans.)**

Example 5.11. *Calculate the decrease in available energy when 20 kg of water at 90°C mixes with 30 kg of water at 30°C, the pressure being taken as constant and the temperature of the surroundings being 10°C.*

Take c_p of water as 4.18 kJ/kg K.

Solution. Temperature of surrounding,

$$T_0 = 10 + 273 = 283 \text{ K}$$

Specific heat of water, c_p = 4.18 kJ/kg K

The available energy of a system of mass m, specific heat c_p, and at temperature T, is given by,

$$\text{Available energy, A.E.} = mc_p \int_{T_0}^{T} \left(1 - \frac{T_0}{T}\right) dT$$

Now, available energy of 20 kg of water at 90°C,

$$(\text{A.E.})_{20\text{ kg}} = 20 \times 4.18 \int_{(10+273)}^{(90+273)} \left(1 - \frac{283}{T}\right) dT$$

$$= 83.6 \left[(363 - 283) - 283 \log_e\left(\frac{363}{283}\right)\right]$$

$$= 83.6\,(80 - 70.45) = 798.38 \text{ kJ}$$

Available energy of 30 kg of water at 30°C,

$$(\text{A.E.})_{30\text{ kg}} = 30 \times 4.18 \int_{(10+273)}^{(30+273)} \left(1 - \frac{283}{T}\right) dT$$

$$= 30 \times 4.18 \left[(303 - 283) - 283 \log_e\left(\frac{303}{283}\right)\right]$$

$$= 125.4\,(20 - 19.32) = 85.27 \text{ kJ}$$

Total available energy,

$$(\text{A.E.})_{\text{total}} = (\text{A.E.})_{20\text{ kg}} + (\text{A.E.})_{30\text{ kg}}$$

$$= 798.38 + 85.27 = 883.65 \text{ kJ}$$

If t°C is the final temperature after mixing, then

$$20 \times 4.18 \times (90 - t) = 30 \times 4.18\,(t - 30)$$

or

$$20(90 - t) = 30\,(t - 30)$$

$$\therefore \quad t = \frac{20 \times 90 + 30 \times 30}{20 + 30} = 54°\text{C}$$

Total mass after mixing = 20 + 30 = 50 kg

Available energy of 50 kg of water at 54°C

$$(\text{A.E.})_{50\text{ kg}} = 50 \times 4.18 \left[(327 - 283) - 283 \log_e\left(\frac{327}{283}\right)\right]$$

$$= 209\,(44 - 40.89) = 649.99 \text{ kJ}$$

$\therefore$ **Decrease in available energy due to mixing**

= Total energy before mixing

– Total energy after mixing

= 883.65 – 649.99 = **233.66 kJ. (Ans.)**

☞ **Example 5.12.** *In an heat exchanger (parallel flow type) waters enter at 50°C and leaves at 70°C while oil (specific gravity = 0.82, specific heat = 2.6 kJ/kg K) enters at 240°C and leaves at 90°C. If the surrounding temperature is 27°C determine the loss in availability on the basis of one kg of oil per second.*

Solution. Refer to Fig. 5.12.

Inlet temperature of water, T_{w_1} = 50°C = 323 K

Outlet temperature of water, T_{w_2} = 70°C = 343 K

Inlet temperature of oil, T_{0_1} = 240°C = 513 K

Outlet temperature of oil, T_{0_2} = 90°C = 363 K

Specific gravity of oil = 0.82

Specific heat of oil = 2.6 kJ/kg K

Surrounding temperature, T_0 = 27 + 273 = 300 K.

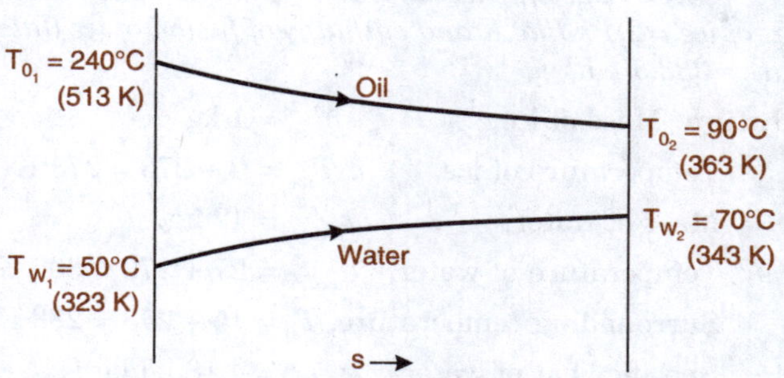

Fig. 5.12

Loss in availability :

Consider one kg of oil.

Heat lost by oil = Heat gained by water

$$m_0 \times c_{p0} \times (T_{0_1} - T_{0_2}) = m_w \times c_{pw} \times (T_{w_2} - T_{w_1})$$

where c_{p0} = Specific heat of oil (2.6 kJ/kg K),

c_{pw} = Specific heat of water (4.18 kJ/kg K), and

m_0 = Mass of oil (= 1 kg),

m_w = Mass of water (= ?)

$$\therefore \quad 1 \times 2.6 \times (513 - 363) = m_w \times 4.18 \times (343 - 323)$$

or

$$390 = 83.6\, m_w \text{ or } m_w = 4.66 \text{ kg}$$

Entropy change of water

$$= m_w\, c_{pw} \log_e \frac{T_{w_2}}{T_{w_1}}$$

$$= 4.66 \times 4.18 \times \log_e\left(\frac{343}{323}\right) = 1.17 \text{ kJ/K}$$

Entropy change of oil

$$= m_0 c_{p0} \log_e \left(\frac{T_{0_2}}{T_{0_1}}\right) = 1 \times 2.6 \log_e\left(\frac{363}{513}\right)$$

$$= -\,0.899 \text{ kJ/K}$$

Change in availability of water

$$= m_w[c_{pw}(T_{w_2} - T_{w_1})] - T_0\,(\Delta S)_w$$

$$= 4.66[(4.18\,(343 - 323)] - 300 \times 1.17$$

$$= 38.57 \text{ kJ}$$

+ve sign indicates an *increase in availability*

Change in availability of oil

$$= m_0[c_{p0}(T_{0_2} - T_{0_1})] - T_0\,(\Delta S)_0$$

$$= 1[2.6(363 - 513)] - 300 \times (-\,0.899)$$

$$= -\,120.3 \text{ kJ/K}$$

$\therefore$ **Loss in availability**

= – 120.3 + 38.57 = – **81.73 kJ. (Ans.)**

(–ve sign indicates the **loss**).

Example 5.13. *1 kg of ice at 0°C is mixed with 12 kg of water at 27°C. Assuming the surrounding temperature as 15°C, calculate the net increase in entropy and unavailable energy when the system reaches common temperature :*

Given : Specific heat of water = 4.18 kJ/kg K ; specific heat of ice = 2.1 kJ/kg K and enthalpy of fusion of ice (latent heat) = 333.5 kJ/kg.

Solution. Mass of ice, $m_{ice} = 1$ kg

Temperature of ice, $T_{ice} = 0 + 273 = 273$ K

Mass of water, $m_{water} = 12$ kg

Temperature of water, $T_{water} = 27 + 273 = 300$ K

Surrounding temperature, $T_0 = 15 + 273 = 288$ K

Specific heat of water = 4.18 kJ/kg K

Specific heat of ice = 2.1 kJ/kg K

Latent heat of ice = 333.5 kJ/kg

Let T_c = *common temperature* when heat flows between ice and water stops.

Heat lost by water = Heat gained by ice

i.e., $12 \times 4.18(300 - T_c) = 4.18(T_c - 273) + 333.5$

or $15048 - 50.16T_c = 4.18T_c - 1141.14 + 333.5$

or $54.34\, T_c = 15855.64$

$\therefore \quad T_c = 291.8$ K or 18.8°C.

Change of entropy of water

$$= 12 \times 4.18 \log_e \left(\frac{291.8}{300}\right) = -1.39 \text{ kJ/K}$$

Change of entropy of ice

$$= 1 \times 4.18 \log_e \left(\frac{291.8}{273}\right) + \frac{333.5}{273} = 1.499 \text{ kJ/K}$$

Net change of entropy,

$$\Delta S = -1.39 + 1.499 = 0.109 \text{ kJ/K}$$

Hence, **net increase in entropy**

= 0.109 kJ/K. (Ans.)

Increase in unavailable energy

$= T_0 \Delta S = 288 \times 0.109 =$ **31.39 kJ. (Ans.)**

☞**Example 5.14.** *A vapour, in a certain process, while condensing at 400°C, transfers heat to water at 200°C. The resulting steam is used in a power cycle which rejects heat at 30°C.*

What is the fraction of the available energy in the heat transferred from the process vapour at 400°C that is lost due to the irreversible heat transfer at 200°C ?

Solution. Refer to Fig. 5.13.

Temperature of vapour, $T_1 = 400 + 273 = 673$ K

Temperature of water, $T_2 = 200 + 273 = 473$ K

Temperature at which heat is rejected,

$T_0 = 30 + 273 = 303$ K.

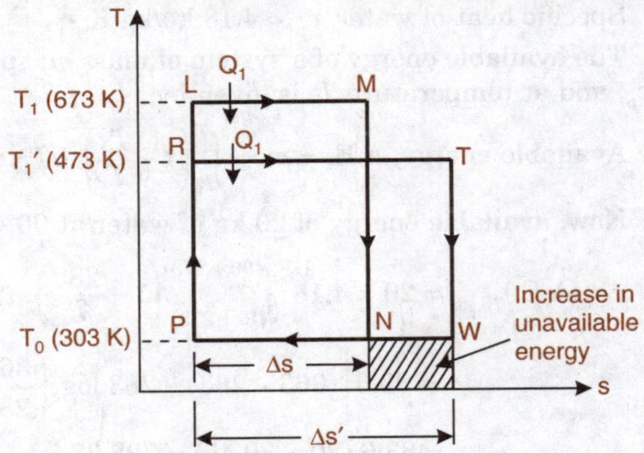

Fig. 5.13

LMNP (Fig. 5.13) would have been the power cycle, if there was no temperature difference between the vapour condensing and the vapour evaporating, and the area under *NP* would have been the unavailable energy. *RTWP* is the power cycle when the vapour condenses at 400°C and the water evaporates at 200°C. The unavailable energy becomes the area under *PW*. Therefore, the increase in unavailable energy due to irreversible heat transfer is represented by the area under *NW*.

Now, $Q_1 = T_1 \Delta s = T_1' \Delta s'$

$$\frac{\Delta s'}{\Delta s} = \frac{T_1}{T_1'}$$

W = Work done in cycle *LMNP*

$= (T_1 \;\; T_0)\, \Delta s$ ...per unit mass

W' = Work done in cycle *RTWP*

$= (T_1 - T_0)\, \Delta s'$ ...per unit mass

The fraction of energy that becomes unavailable due to irreversible heat transfer,

$$\frac{W - W'}{W} = \frac{(T_1 - T_0)\Delta s - (T_1 - T_0)\Delta s'}{(T_1 - T_0)\Delta s} = \frac{T_0\,(\Delta s' - \Delta s)}{(T_1 - T_0)\Delta s}$$

$$= \frac{T_0\left(\dfrac{\Delta s'}{\Delta s} - 1\right)}{(T_1 - T_0)} = \frac{T_0\left(\dfrac{T_1}{T_1'} - 1\right)}{(T_1 - T_0)}$$

$$= \frac{T_0\,(T_1 - T_1')}{T_1'\,(T_1 - T_0)} = \frac{303(673 - 473)}{473(673 - 303)} = 0.346$$

Hence **the fraction of energy that becomes unavailable = 0.346** or **34.6%. (Ans.)**

Example 5.15. *A liquid is heated at approximately constant pressure from 20°C to 80°C by passing it through tubes which are immersed in a furnace. The furnace temperature is constant at 1500°C. Calculate the effectiveness of the heating process when the atmospheric temperature is 15°C.*

Take specific heat of liquid as 6.3 kJ/kg K.

Solution. Initial temperature of fluid,

$T_1 = 20 + 273 = 293$ K

Final temperature of fluid,

$T_2 = 80 + 273 = 353$ K

Temperature of the furnace,

$$T_f = 1500 + 273 = 1773 \text{ K}$$

Atmospheric temperature,

$$T_0 = 15 + 273 = 288 \text{ K}$$

Specific heat of liquid,

$$c_{pl} = 6.35 \text{ kJ/kg K}$$

Increase of availability of the liquid

$$= b_2 - b_1 = (h_2 - h_1) - T_0(s_2 - s_1)$$

i.e., $$b_2 - b_1 = c_{pl}\,(T_2 - T_1) - T_0 \times c_{pl} \log_e \frac{T_2}{T_1}$$

$$= 6.3\,(353 - 293) - 288 \times 6.3 \times \log_e \left(\frac{353}{293}\right)$$

$$= 39.98 \text{ kJ/kg}$$

Now, the heat rejected by the furnace = Heat supplied to the liquid, $(h_2 - h_1)$.

If this quantity of heat were supplied to a heat engine operating on the Carnot cycle its thermal efficiency would be,

$$\eta_{th} = \left(1 - \frac{T_0}{T_f}\right) = \left(1 - \frac{288}{1773}\right) = 0.837 \text{ (or 83.7\%)}$$

$\therefore$ Work which could be obtained from a heat engine

= Heat supplied × Thermal efficiency

i.e., Possible work of heat engine = $(h_2 - h_1) \times 0.837$

The possible work from a heat engine is a measure of the loss of availability of the furnace.

$\therefore$ Loss of availability of surroundings

$$= (h_2 - h_1) \times 0.837 = c_{pl}\,(T_2 - T_1) \times 0.837$$

$$= 6.3\,(353 - 293) \times 0.837 = 316.38 \text{ kJ/kg}$$

Then, *effectiveness of the heating process,*

$$\in = \frac{\text{Increase of availability of the liquid}}{\text{Loss of availability of surroundings}}$$

$$= \frac{39.98}{316.38} = \mathbf{0.1263 \text{ or } 12.63\%}. \quad \textbf{(Ans.)}$$

Note. *The very low value of effectiveness reflects the irreversibility of the transfer of heat through a large temperature difference.* If the furnace temperature were much lower then process would be much more effective, although the heat transferred to the liquid would remain the same.

Example 5.16. *Air at 20°C is to be heated to 50°C by mixing it in steady flow with a quantity of air at 100°C. Assuming that the mixing process is adiabatic and neglecting changes in kinetic and potential energy, calculate :*

(i) The ratio of mass flow of air initially at 100°C to that initially at 20°C.

(ii) The effectiveness of heating process, if the atmospheric temperature is 20°C.

Solution. (*i*) Let, x = ratio of mass flows.

Stream 1 = air at 20°C ($T_1 = 20 + 273 = 293$ K)

Stream 2 = air at 100°C ($T_2 = 100 + 273 = 373$ K)

Stream 3 = air at 50°C ($T_3 = 50 + 273 = 323$ K)

If, c_p = Specific heat of air constant pressure

Then $$c_pT_1 + xc_pT_2 = (1 + x)c_pT_3$$

or $$c_pT_1 + xc_pT_2 = c_pT_3 + xc_pT_3$$

or $$xc_p(T_2 - T_3) = c_p(T_3 - T_1)$$

i.e., $$xc_p(373 - 323) = c_p(323 - 293)$$

$\therefore$ $$x = \frac{30}{50} = \mathbf{0.6.} \quad \textbf{(Ans.)}$$

(*ii*) Let the system considered be a stream of air of unit mass, heated from 20°C to 50°C.

Increase of availability of system

$$= b_3 - b_1 = (h_3 - h_1) - T_0(s_3 - s_1)$$

$$= c_p(T_3 - T_1) - T_0(s_3 - s_1)$$

$$= 1.005(323 - 293) - 293(s_3 - s_1)$$

$$[\because \; T_0 = 20 + 273 = 293 \text{ K}]$$

Also, $$s_3 - s_1 = c_p \log_e \frac{T_3}{T_1}$$

$$= 1.005 \log_e \frac{323}{293} = 0.0979 \text{ kJ/kg K}$$

$\therefore$ *Increase of availability of system*

$$= 1.005 \times 30 - 293 \times 0.0979 = 1.465 \text{ kJ/kg.}$$

The system, which is the air being heated, is 'surrounded' by the air stream being cooled. Therefore, the loss of availability of the surroundings is given by, $x(b_2 - b_3)$.

i.e., *Loss of availability of surroundings*

$$= x[(h_2 - h_3) - T_0(s_2 - s_3)]$$

$$= 0.6[c_p(T_2 - T_3) - T_0(s_2 - s_3)]$$

$$= 0.6\left[1.005\,(373 - 323) - 293 \times 1.005 \log_e \left(\frac{373}{323}\right)\right]$$

$$= 0.6[50.25 - 42.38] = 4.722 \text{ kJ/kg}$$

$\therefore$ *Effectiveness*

$$= \frac{\text{Increase of availability of system}}{\text{Loss of availability of surroundings}}$$

$$= \frac{1.465}{4.722} = \mathbf{0.31} \text{ or } \mathbf{31\%}. \quad \textbf{(Ans.)}$$

The *low* figure for the effectiveness is an indication of the *highly irreversible nature of the mixing process.*

☞**Example 5.17.** *2.5 kg of air at 6 bar, 90°C expands adiabatically in a closed system until its volume is doubled and its temperature becomes equal to that of the surroundings which is at 1 bar, 5°C. For this process determine :*

(*i*) *The maximum work ;*

(*ii*) *The change in availability ;*

(*iii*) *The irreversibility.*

For air take : $c_v = 0.718$ *kJ/kg K,* $R = 0.287$ *kJ/kg K.*

Solution. Mass of air, $m = 2.5$ kg

Initial pressure of air,

$$p_1 = 6 \text{ bar} = 6 \times 10^5 \text{ N/m}^2$$

Ratio of final to initial volume, $\dfrac{V_2}{V_1} = 2$

Initial temperature of air,

$$T_1 = 90 + 273 = 363 \text{ K}$$

Final pressure of air,

$$p_2 = 1 \text{ bar} = 1 \times 10^5 \text{ N/m}^2$$

Final temperature of air,

$$T_2 = T_0 = 5 + 273 = 278 \text{ K}$$

From the property relation

$$TdS = dU + pdV$$

or

$$dS = \frac{dU}{T} + \frac{pdV}{T}$$

$$dS = \frac{mc_v dT}{T} + \frac{mRdV}{V}$$

$$\left[\because u = c_v\, dT \text{ and } pV = mRT \text{ or } \frac{p}{T} = \frac{mR}{V}\right]$$

$\therefore$ The entropy change of air between the initial and final states is

$$\int_1^2 dS = \int_1^2 \frac{mc_v dT}{T} + \int_1^2 \frac{mRdV}{V}$$

or

$$S_2 - S_1 = mc_v \log_e \frac{T_2}{T_1} + mR \log_e \frac{V_2}{V_1}$$

(*i*) **The maximum work, W_{max} :**

Also, $W_{max} = (U_1 - U_2) - T_0(S_1 - S_2)$

$$= m\left[c_v(T_1 - T_2) + T_0\left(c_v \log_e \frac{T_2}{T_1} + R \log_e \frac{v_2}{v_1}\right)\right]$$

$$= 2.5\left[0.718(363 - 278) + 278\left(0.718 \log_e\left(\frac{278}{363}\right) + 0.287 \log_e 2\right)\right]$$

$$= 2.5[61.03 + 278(-0.1915 + 0.1989)] = 157.7 \text{ kJ}$$

Hence, **maximum work = 157.7 kJ. (Ans.)**

(*ii*) **The change in availability :**

The change in availability is given by,

$$A_1 - A_2 = (U_1 - U_2) - T_0(S_1 - S_2) + p_0(V_1 - V_2)$$

$$= W_{max} + p_0(V_1 - V_2)$$

$$= 157.7 + p_0(V_1 - 2V_1)$$

$$= 157.7 - P_0V_1 \qquad \left[\begin{matrix} p_1V_1 = mRT_1 \\ \because V_1 = \dfrac{mRT_1}{p_1} \end{matrix}\right]$$

$$= 157.7 - \frac{1 \times 10^5}{10^3}\left[\frac{2.5 \times (0.287 \times 1000) \times 363}{6 \times 10^5}\right]$$

$$= 114.29 \text{ kJ}$$

Hence **change in availability = 114.29 kJ. (Ans.)**

(*iii*) **The irreversibility, I :**

The irreversibility is given by

$$I = W_{max.\ useful} - W_{actual}$$

From the first law of thermodynamics,

$$W_{actual} = Q - \Delta U = -\Delta U = U_1 - U_2$$

$$[\because Q = 0 \text{ ... adiabatic process}]$$

$$I = (U_1 - U_2) - T_0(S_1 - S_2) - (U_1 - U_2)$$

$$= T_0(S_2 - S_1)$$

$$= T_0(\Delta S)_{system}$$

For adiabatic process, $(\Delta S)_{surr.} = 0$

$$I = T_0\left[mc_v \log_e \frac{T_2}{T_1} + mR \log_e \frac{v_2}{v_1}\right]$$

$$= 278 \times 2.5\left[0.718 \log_e\left(\frac{278}{363}\right) + 0.287 \log_e 2\right]$$

$$= 695(-0.1915 + 0.1989) = 5.143 \text{ kJ}$$

Hence, **the irreversibility = 5.143 kJ. (Ans.)**

☞**Example 5.18.** *In a turbine the air expands from 7 bar, 600°C to 1 bar, 250°C. During expansion 9 kJ/kg of heat is lost to the surroundings which is at 1 bar, 15°C. Neglecting kinetic energy and potential energy changes, determine per kg of air :*

(*i*) *The decrease in availability ;*

(*ii*) *The maximum work ;*

(*iii*) *The irreversibility.*

For air, take : $c_p = 1.005$ *kJ/kg K,* $h = c_pT$*, where* c_p *is constant.*

Solution. Mass of air considered = 1 kg

Pressure, $p_1 = 7 \text{ bar} = 7 \times 10^5 \text{ N/m}^2$

Temperature, $T_1 = 600 + 273 = 873$ K

Pressure, $p_2 = 1 \text{ bar} = 1 \times 10^5 \text{ N/m}^2$

Temperature, $T_2 = 250 + 273 = 523$ K

Surrounding temperature, $T_0 = 15 + 273 = 288$ K

Heat lost to the surroundings during expansion,

$$Q = 9 \text{ kJ/kg.}$$

(*i*) From the property relation,

$$TdS = dH - Vdp$$

$$dS = \frac{dH}{T} - \frac{Vdp}{T} = \frac{m.c_p dT}{T} - \frac{mRdp}{p}$$

or $\int_1^2 dS = \int_1^2 \frac{m.c_p dT}{T} - \int_1^2 \frac{mRdp}{p}$

or $S_2 - S_1 = mc_p \log_e \frac{T_2}{T_1} - mR \log_e \frac{p_2}{p_1}$ $\left\{ \because h = c_pT, dh = c_p dT;\ dH = mc_p dT \text{ and } pV = mRT \text{ or } V = \frac{mRT}{p} \right\}$

For 1 kg of air

$$s_2 - s_1 = c_p \log_e \frac{T_2}{T_1} - R \log_e \frac{p_2}{p_1}$$

Now, the change in availability is given by

$$b_1 - b_2 = (h_1 - T_0 s_1) - (h_2 - T_0 s_2)$$
$$= (h_1 - h_2) - T_0(s_1 - s_2)$$
$$= c_p (T_1 - T_2) - T_0 \left(R \log_e \frac{p_2}{p_1} - c_p \log_e \frac{T_2}{T_1} \right)$$
$$= 1.005(873 - 523) - 288 \left[0.287 \log_e \left(\frac{1}{7}\right) - 1.005 \log_e \left(\frac{523}{873}\right) \right]$$
$$= 351.75 - 288(- 0.5585 + 0.5149)$$
$$= 364.3 \text{ kJ/kg}$$

i.e., **Decrease in availability = 364.3 kJ/kg. (Ans.)**

(*ii*) **The maximum work,**

W_{max} = Change in availability

= **364.3 kJ/kg. (Ans.)**

(*iii*) From steady flow energy equation,

$$Q + h_1 = W + h_2$$
$$W = (h_1 - h_2) + Q = c_p (T_1 - T_2) + Q$$
$$= 1.005(873 - 523) + (- 9) = 342.75 \text{ kJ/kg}$$

The irreversibility,

$$I = W_{max} - W$$
$$= 364.3 - 342.75 = \mathbf{21.55 \text{ kJ/kg.}} \text{ (Ans.)}$$

[Alternatively, $I = T_0 (\Delta S_{system} + \Delta S_{surr.})$

$$= 288 \left[1.005 + \log_e \left(\frac{523}{873}\right) - 0.287 \log_e \left(\frac{1}{7}\right) + \frac{9}{288} \right]$$
$$= 288 [- 0.5149 + 0.5585 + 0.03125] = 21.55 \text{ kJ/kg.}]$$

Example 5.19. *1 kg of air undergoes a polytropic compression from 1 bar and 290 K to 6 bar and 400 K. If the temperature and pressure of the surroundings are 290 K and 1 bar respectively, determine :*

(*i*) *The irreversibility ;*

(*ii*) *The effectiveness.*

Take for air : c_p = 1.005 kJ/kg K, c_v = 0.718 kJ/kg K, R = 0.287 kJ/kg K.

Solution. Mass of air, $m = 1$ kg

Initial temperature, $T_1 = T_0 = 290$ K

Final temperature, $T_2 = 400$ K

Initial pressure, $p_1 = p_0 = 1$ bar

Final pressure, $p_2 = 6$ bar.

(*i*) **The irreversibility, I :**

We know that reversible work,

$W_{rev.}$ = Change in internal energy − T_0 × Change in entropy

$$= (u_2 - u_1) - T_0 \left[c_p \log_e \left(\frac{T_2}{T_1}\right) - R \log_e \left(\frac{p_2}{p_1}\right) \right]$$
$$= c_v (T_2 - T_1) - T_0 \left[c_p \log_e \left(\frac{T_2}{T_1}\right) - R \log_e \left(\frac{p_2}{p_1}\right) \right]$$
$$= 0.718 (400 - 290) - 290 \left[1.005 \log_e \left(\frac{400}{290}\right) - 0.287 \log_e \left(\frac{6}{1}\right) \right]$$
$$= 78.98 - 290 (0.3232 - 0.5142)$$
$$= 134.37 \text{ kJ/kg } (-)$$

[–ve sign indicates that work is done on the air]

Actual work is given by

$$W_{actual} = \frac{mR(T_1 - T_2)}{n - 1} = \frac{R(T_1 - T_2)}{n - 1} \text{ as } m = 1 \text{ kg.}$$

The index n for the compression is given by :

$$\frac{T_2}{T_1} = \left(\frac{p_2}{p_1}\right)^{\frac{n-1}{n}}$$

$$\therefore \quad \frac{400}{290} = \left(\frac{6}{1}\right)^{\frac{n-1}{n}} \quad \text{or} \quad 1.379 = (6)^{\frac{n-1}{n}}$$

$$\log_e 1.379 = \left(\frac{n-1}{n}\right) \log_e 6$$

or
$$0.3213 = 1.7917 \left(\frac{n-1}{n}\right)$$

$$\therefore \quad \frac{n-1}{n} = \frac{0.3213}{1.7917} = 0.1793 \quad \text{or} \quad n = 1.218$$

$$\therefore \quad W_{actual} = \frac{0.287 (290 - 400)}{(1.218 - 1)} = - 144.8 \text{ kJ}$$

(–ve sign means that the work is done on the air)

Now, **irreversibility** = $W_{rev} - W_{actual}$

= − 134.37 − (− 144.8) = **10.43 kJ. (Ans.)**

(*ii*) **The effectiveness, ∈ :**

$$\in = \frac{W_{rev.}}{W_{actual}} = \frac{-134.37}{-144.8}$$

= **0.928** or **92.8%. (Ans.)**

Example 5.20. *A flywheel whose moment of inertia is 0.62 kg m² rotates at a speed 2500 r.p.m. in a large heat insulated system, the temperature of which is 20°C.*

(*i*) *If the K.E. of the flywheel is dissipated as frictional heat at the shaft bearings which have a water equivalent of 1.9 kg, find the rise in the temperature of the bearings when the flywheel has come to rest.*

(ii) Calculate the greatest possible amount of the above heat which may be returned to the flywheel as high-grade energy, showing how much of the original K.E. is now unavailable. What would be the final r.p.m. of the flywheel, if it is set in motion with this available energy ?

Solution. Moment of inertia of the flywheel,

$$I = 0.62 \text{ kg m}^2$$

Initial angular velocity of the flywheel,

$$\omega_1 = \frac{2\pi N_1}{60} = \frac{2\pi \times 2500}{60} = 261.8 \text{ rad/s.}$$

Temperature of insulated system,

$$T_0 = 20 + 273 = 293 \text{ K}$$

Water equivalent of shaft bearings = 1.9 kg

(*i*) Initial available energy of the flywheel,

$$(\text{K.E.})_{initial} = \frac{1}{2} I\omega_1^2 = \frac{1}{2} \times 0.62 \times (261.8)^2$$

$$= 2.12 \times 10^4 \text{ Nm} = 21.2 \text{ kJ.}$$

When this K.E. is dissipated as frictional heat, if Δt is the temperature rise of the bearings, we have

Water equivalent of bearings × rise in temperature

= 21.2

i.e., $(1.9 \times 4.18)\,\Delta t = 21.2$ or $\Delta t = \frac{21.2}{1.9 \times 4.18} = 2.67°\text{C}$

Hence, **rise in temperature of bearings = 2.67°C. (Ans.)**

∴ **Final temperature of the bearings = 20 + 2.67 = 22.67°C.**

(ii) The maximum amount of energy which may be returned to the flywheel as high-grade energy is,

$$\text{A.E.} = 1.9 \times 4.18 \int_{293}^{295.67} \left(1 - \frac{293}{T}\right) dT$$

$$= 1.9 \times 4.18 \left[(295.67 - 293) - 293 \log_e \left(\frac{295.67}{293}\right)\right]$$

= **0.096 kJ. (Ans.)**

The amount of energy rendered unavailable is

$$\text{U.E.} = (\text{A.E.})_{\text{initial}} - (\text{A.E.})_{\text{returnable as high grade energy}}$$

$$= 21.2 - 0.096 = 21.1 \text{ kJ.}$$

Since the amount of energy returnable to the flywheel is 0.096 kJ, if ω_2 is the final angular velocity, and the flywheel is set in motion with this energy, then

$$0.096 \times 10^3 = \frac{1}{2} \times 0.62 \times \omega_2^2$$

$$\therefore \quad \omega_2^2 = \frac{0.096 \times 10^3 \times 2}{0.62} = 309.67$$

or $\omega_2 = 17.59$ rad/s.

If N_2 is the final r.p.m. of the flywheel, then

$$\omega_2 = \frac{2\pi N_2}{60} \quad \text{or} \quad N_2 = \frac{17.59 \times 60}{2\pi} = 168 \text{ r.p.m.}$$

i.e., **Final r.p.m. of the flywheel = 168 r.p.m. (Ans.)**

☞**Example 5.21.** *The air, in a steady flow, enters the system at a pressure of 8 bar and 180°C with a velocity of 80 m/s and leaves at 1.4 bar and 20°C with a velocity of 40 m/s. The temperature of the surroundings is 20°C and pressure is 1 bar. Determine :*

(i) Reversible work and actual work assuming the process to be adiabatic ;

(ii) Irreversibility and effectiveness of the system on the basis of 1 kg of air flow.

Take for air : c_p = 1.005 kJ/kg K ; R = 0.287 kJ/kg K.

Solution. Initial pressure of air, $p_1 = 8$ bar

Initial temperature of air, $T_1 = 180 + 273 = 453$ K

Final pressure of air, $p_2 = 1.4$ bar

Final temperature, $T_2 = T_0 = 20 + 273 = 293$ K

Surroundings' pressure, $p_0 = 1$ bar

Mass of air = 1 kg

Initial velocity of air, $C_1 = 80$ m/s

Final velocity of air, $C_2 = 40$ m/s.

(*i*) **Reversible work and actual work :**

Availability of air at the inlet

$$= (h_1 - h_0) - T_0 (s_1 - s_0) + \frac{C_1^2}{2}$$

$$= c_p (T_1 - T_0) - T_0 (s_1 - s_0) + \frac{C_1^2}{2}$$

$$(s_1 - s_0) = c_p \log_e \left(\frac{T_1}{T_0}\right) - R \log_e \left(\frac{p_1}{p_0}\right)$$

$$= 1.005 \log_e \left(\frac{453}{293}\right) - 0.287 \log_e \left(\frac{8}{1}\right)$$

$$= 0.437 - 0.596 = -0.159 \text{ kJ/kg K}$$

∴ Availability of air at the inlet

$$= 1.005\,(453 - 293) - 293\,(-0.159) + \frac{80^2}{2 \times 10^3}$$

$$= 160.8 + 46.58 + 3.2 = 210.58 \text{ kJ}$$

Availability at the exit

$$= (h_2 - h_0) - T_0 (s_2 - s_0) + \frac{C_2^2}{2}$$

$$= -T_0 (s_2 - s_0) + \frac{C_2^2}{2}$$

as $h_2 = h_0$ because $T_2 = T_0 = 293$ K

Now $s_2 - s_0 = -R \log_e \left(\frac{p_2}{p_0}\right)$

$$= -0.287 \log_e \left(\frac{1.4}{1}\right) = -0.09656 \text{ kJ/kg K}$$

∴ Availability at the exit

$$= -293\,(-0.09656) + \frac{40^2}{2 \times 10^3} = 29.09 \text{ kJ/kg}$$

Reversible/theoretical work which must be available,

$W_{rev} = 210.58 - 29.09 =$ **181.49 kJ. (Ans.)**

Actual work developed can be calculated by using the energy equation for adiabatic steady flow process as follows :

$$W_{actual} = (h_1 - h_2) + \left(\frac{C_1^2 - C_2^2}{2}\right)$$

$$= c_p\,(T_1 - T_2) + \left(\frac{C_1^2 - C_2^2}{2}\right)$$

$$= 1.005(453 - 293) + \left(\frac{80^2 - 40^2}{2 \times 10^3}\right)$$

$= 160.8 + 2.4 =$ **163.2 kJ/kg. (Ans.)**

(*ii*) **Irreversibility and effectiveness :**

Irreversibility, $I = W_{rev} - W_{actual}$

$= 181.49 - 163.2$

$=$ **18.29 kJ/kg. (Ans.)**

Effectiveness, $\in = \frac{W_{actual}}{W_{rev}} = \frac{163.2}{181.49}$

$=$ **0.899** or **89.9%. (Ans.)**

Example 5.22. *Steam expands adiabatically in a turbine from 20 bar, 400°C to 4 bar, 250°C. Calculate :*

(*i*) *The isentropic efficiency of the process ;*

(*ii*) *The loss of availability of the system assuming an atmospheric temperature of 20°C ;*

(*iii*) *The effectiveness of the process ;*

The changes in K.E. and P.E. may be neglected.

Solution. Initial pressure of steam, $p_1 = 20$ bar

Initial temperature of steam, $t_1 = 400°C$

Final pressure of steam, $p_2 = 4$ bar

Final temperature of steam, $t_2 = 250°C$

Atmospheric temperature, $= 20°C$ (= 293 K).

Initial state 1 : 20 bar, 400°C ; From steam tables,

$h_1 = 3247.6$ kJ/kg ; $s_1 = 7.127$ kJ/kg K

Final state 2 : 4 bar 250°C ; From steam tables,

$h_2' = 2964.2$ kJ/kg, $s_2' = 7.379$ kJ/kg K

The process is shown as 1 to 2′ in Fig. 5.14

$s_1 = s_2 = 7.127$ kJ/kg K

Hence, interpolating,

$$h_2 = 2752.8 + \left(\frac{7.127 - 6.930}{7.171 - 6.930}\right)(2860.5 - 2752.8)$$

$$= 2752.8 + \frac{0.197}{0.241} \times 107.7 = 2840.8 \text{ kJ/kg.}$$

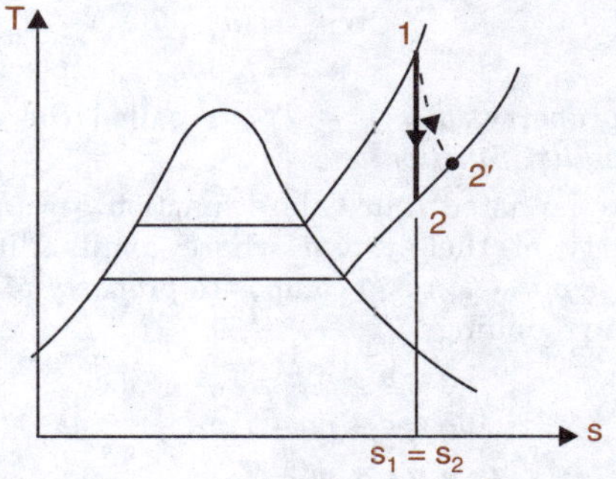

Fig. 5.14

(*i*) **Isentropic efficiency :**

$$\eta_{isen.} = \frac{\text{Actual work output}}{\text{Isentropic work}}$$

$$= \frac{h_1 - h_2'}{h_1 - h_2} = \frac{3247.6 - 2964.2}{3247.6 - 2840.8}$$

$= \frac{283.4}{406.8} =$ **0.6966** or **69.66%. (Ans.)**

(*ii*) **Loss of availability :**

Loss of availability $= b_1 - b_2$

$= h_1 - h_2' + T_0\,(s_2' - s_1')$

$= 3247.6 - 2964.2 + 293\,(7.379 - 7.127)$

$= 283.4 + 73.83 =$ **357.23 kJ/kg. (Ans.)**

(*iii*) **Effectiveness :**

$$\text{Effectiveness, } \in = \frac{W}{b_1 - b_2'} = \frac{h_1 - h_2'}{b_1 - b_2'}$$

$$= \frac{3247.6 - 2964.2}{357.9}$$

$=$ **0.7918** or **79.18%. (Ans.)**

HIGHLIGHTS

1. '*Available energy*' is the maximum portion of the energy which could be converted into useful work by ideal processes which reduce the system to a dead state.
2. The theoretical maximum amount of work which can be obtained from a system at any state p_1 and T_1 when operating with a reservoir at the constant pressure and temperature p_0 and T_0 is called '*availability*'.
3. Energy is said to be *degraded* each time it flows through a finite temperature difference. That is, why the second law of thermodynamics is sometimes called the *law of*

the degradation of energy, and energy is said to '*run down hill*'.

4. In non-flow systems :

Maximum work available,

$$W_{max} = (u_1 - u_0) - T_0(s_1 - s_0) - p_0(v_0 - v_1)$$
$$= (u_1 + p_0v_1 - T_0s_1) - (u_0 + p_0v_0 - T_0s_0)$$
$$= a_1 - a_0 \quad \text{...per unit mass}$$

The property $a = u + p_0v - T_0s$ is called the *non-flow availability function*.

5. In steady-flow systems :

Maximum work available,

$$W_{max} = (h_1 - T_0s_1) - (h_0 - T_0s_0)$$
$$= b - b_0 \quad \text{...per unit mass}$$

The property, $b = h - T_0s$ is called the *steady-flow availability function*.

6. It may be noted that Gibb's function $g = (h - Ts)$ is a property of the system where availability function $a = u + p_0v - T_0s$ is a composite property of the system and surroundings.

Again, $a = u + p_0v - T_0s$

$b = u + pv - T_0s$

$g = u + pv - Ts$

When state 1 proceeds to dead state (zero state)

$a = b = g.$

7. The actual work which a system does is always less than the idealized reversible work, and the difference between the two is called the *irreversibility of the process*. This is also sometimes referred to as *degradation* or *dissipation*.

Effectiveness is defined as the ratio of actual useful work to the maximum useful work.

OBJECTIVE TYPE QUESTIONS

Choose the Correct Answer :

1. Gibb's function is expressed as
   (*a*) $(u + pv - Ts)$ (*b*) $(u + pv - Tds)$
   (*c*) $(u + pdv - Tds)$ (*d*) $(u + pv - sdT)$.
2. Availability function is expressed as
   (*a*) $a = (u + p_0v - T_0s)$ (*b*) $a = (u + p_0dv + T_0ds)$
   (*c*) $a = (du + p_0dv - T_0ds)$ (*d*) $a = (u + p_0v + T_0s)$.
3. To increase work capacity of energy transferred by heat transfer from high temperature to low temperature
   (*a*) lower temperature should be lowered keeping temperature difference same
   (*b*) higher temperature should be increased keeping temperature difference same
   (*c*) temperature difference should be increased
   (*d*) temperature difference should be decreased.
4. Helmholtz function is expressed as
   (*a*) $(u - Ts)$ (*b*) $(h - Ts)$
   (*c*) $(- sdT + vdp)$ (*d*) $(u + pv)$.
5. If a heat source at temperature T_1 transfers heat to a system at temperature $T_2 (T_1 > T_2)$, state which of the following statements is *not true* ?
   (*a*) Δ_{system} decreases (*b*) Δ_{source} decreases
   (*c*) $(\Delta_{system} + \Delta_{source})$ decreases
   (*d*) $(\Delta_{system} + \Delta_{source})$ increases.

ANSWERS

1. (*a*) 2. (*a*) 3. (*d*) 4. (*a*)
5. (*d*).

THEORETICAL QUESTIONS

1. Explain the concept of available and unavailable energy. When does the system become dead ?
2. Define the term 'availability'.
3. Is the availability function same for a non-flow and a flow process ?
4. Define availability function and find the relationship between availability function and change in availability.
5. How are the concepts of entropy and unavailable energy related to each other ?
6. Derive an expression for availability in non-flow systems.
7. Derive an expression for availability in steady flow systems.
8. Differentiate between availability function and Gibb's energy function.
9. Derive an expression for decrease in available energy when heat is transferred through a finite temperature difference.
10. Derive a general expression for irreversibility in (*i*) non-flow process, (*ii*) steady flow process.
11. What is the effectiveness of a system and how does it differ from efficiency ?

UNSOLVED EXAMPLES

1. A system receives 10000 kJ of heat at 500 K from a source at 1000 K. The temperature of the surroundings is 300 K. Assuming that the temperature of the system and source remains constant during heat transfer, find :
   (*i*) The entropy production due to above mentioned heat transfer ;
   (*ii*) Decrease in available energy.
   [**Ans.** (*i*) 10 kJ/K ; (*ii*) 3000 kJ]
2. In a power station, saturated steam is generated at 252°C by transferring heat from the hot gases generated in the combustion chamber. The gases are cooled from 1100°C to 550°C during transferring the heat for steam generation. Determine the increase in total entropy of the combined system of gas and steam and increase in unavailable energy on the basis of one kg of steam generated. Assume water enters the boiler at saturated condition and leaves as saturated steam.
   [**Ans.** 1.99 kJ/K ; 597 kJ/kg of steam formed]
3. Air at 15°C is to be heated to 40°C by mixing it in steady flow with a quantity of air at 90°C. Assuming that the mixing process is adiabatic and neglecting changes in kinetic and potential energy, calculate the ratio of the mass flow of air initially at 90°C to that initially at 15°C.

Calculate also the effectiveness of the heating process, if the atmospheric temperature is 15°C.

[**Ans.** 0.5, 0.327 or 32.7%]

4. A liquid of specific heat 6.3 kJ/kg K is heated at approximately constant pressure from 15°C to 70°C by passing it through tubes which are immersed in a furnace. The furnace temperature is constant at 1400°C.

Calculate the effectiveness of the heating process when the atmospheric temperature is 10°C.

[**Ans.** 0.121 or 12.1%]

5. 500 kJ of heat from an infinite source at 1000 K is supplied to 2 kg of gas initially at 2 bar and 350 K in a closed tank. Find the loss in available energy due to above heat transfer. Take : c_v (gas) = 0.8 kJ/kg K and surrounding temperature = 300 K. [**Ans.** 157.5 kJ]

6. In an heat exchanger of parallel flow type, water enters at 60°C and leaves at 80°C while oil of specific gravity 0.8 enters at 250°C and leaves at 100°C. The specific heat of oil is 2.5 kJ/kg K and surrounding temperature is 300 K. Determine the loss in availability on the basis of one kg of oil flow per second. [**Ans.** – 59.9 kJ]

7. 1 kg of ice at 0°C is mixed with 10 kg of water at 30°C. Determine the net increase in the entropy and unavailable energy when the system reaches common temperature. Assume that surrounding temperature is 10°C. Take, specific heat of water = 4.18 kJ/kg K ; specific heat of ice = 2.1 kJ/kg K ; latent heat of ice = 333.5 kJ/kg. [**Ans.** 0.114 kJ/K, 32.30 kJ]

8. 1000 kJ of heat leaves hot gases at 1400°C from a fire box of a boiler and goes to steam at 250°C. Atmospheric temperature is 20°C. Divide the energy into available and unavailable portions :

(*i*) As it leaves the hot gases.

(*ii*) As it enters the steam.

[**Ans.** (*i*) 825 kJ, 175 kJ ; (*ii*) 440 kJ, 580 kJ]

9. In a certain process, a vapour, while condensing at 420°C, transfers heat to water evaporating at 250°C. The resulting steam is used in a power cycle which rejects heat at 35°C. What is the fraction of the available energy in the heat transferred from the process vapour at 420°C that is lost due to the irreversible heat transfer at 250°C.

[**Ans.** 0.26]

10. In a steam boiler, hot gases from a fire transfer heat to water which vapourizes at constant temperature. In certain case, the gases are cooled from 1100°C to 550°C while the water evaporates at 220°C. The specific heat of gases is 1.005 kJ/kg K, and the latent heat of water at 220°C is 1858.5 kJ/kg. All the heat transferred from the gases goes to the water. How much does the total entropy of the combined system of gas and water increase as a result of irreversible heat transfer ? Obtain the result on the basis of 1 kg of water evaporated. If the temperature of the surroundings is 30°C find the increase in unavailable energy due to irreversible heat transfer.

[**Ans.** 2.045 kJ/K, 620 kJ]

11. Calculate the unavailable energy in 40 kg of water at 75°C with respect to the surroundings at 5°C, the pressure being 1 atmosphere. [**Ans.** 10420 kJ]

12. Calculate the decrease in available energy when 25 kg of water at 95°C mixes with 35 kg of water at 35°C, the pressure being taken as constant and the temperature of the surroundings being 15°C.

Take c_p of water = 4.18 kJ/kg K. [**Ans.** 281.81 kJ]

13. 2 kg of air at 5 bar, 80°C expands adiabatically in a closed system until its volume is doubled and its temperature becomes equal to that of the surroundings which is at 1 bar, 5°C. For this process, determine :

(*i*) The maximum work ;

(*ii*) The change in availability ;

(*iii*) The irreversibility.

For air take : c_v = 0.718 kJ/kg K, $u = c_vT$, where c_v is constant and $pV = mRT$, where p is in bar, V volume in m^3, m mass in kg, R is constant equal to 0.287 kJ/kg K, and T temperature in K.

[**Ans.** (*i*) 122.72 kJ ; (*ii*) 82.2 kJ ; (*iii*) 15.2 kJ]

14. One kg of air at a pressure p, and temperature 900 K is mixed with 1 kg of air at the same pressure and 500 K. Determine the loss in availability if the surrounding temperature is 300 K. [**Ans.** 54 kJ]

15. 10 kg of water is heated in an insulated tank by a churning process from 300 K to 350 K. Determine the loss in availability for the process if the surrounding temperature is 300 K. [**Ans.** 1968 kJ]

16. A closed system contains 10 kg of air at 600 K and 5 bar. Determine the availability of the system if the surrounding pressure and temperature are 1 bar and 300 K respectively.

If the air is cooled at constant pressure to the atmospheric temperature, determine the availability and effectiveness.

[**Ans.** 1464 kJ ; 921 kJ ; 0.63]

17. In a turbine air expands from 5 bar, 520°C to 1 bar, 300°C. During expansion 10 kJ/kg of heat is lost to the surroundings which is at 0.98 bar, 20°C. Neglecting kinetic and potential energy changes, determine per kg of air :

(*i*) The decrease in availability ;

(*ii*) The maximum work ;

(*iii*) The irreversibility.

For air take : c_p = 1.005 kJ/kg K ; $h = c_pT$, where c_p is constant.

[**Ans.** (*i*) 260.7 kJ/kg ; (*ii*) 260.7 kJ/kg ; (*iii*) 49.6 kJ/kg]

18. A centrifugal air compressor compresses air at the rate of 20 kg/min from 1 bar to 2 bar. The temperature increases from 20°C to 120°C during the compression. Determine actual and minimum power required to run the compressor. The surrounding air temperature is 20°C. Neglect the heat interaction between the compressor and surroundings and changes in potential and kinetic energy. [**Ans.** 24.2 kW]

19. 1 kg of air is compressed polytropically from 1 bar and 300 K to 7 bar and 380 K. Determine the irreversibility and effectiveness, assuming temperature and pressure as 300 K and 1 bar.

Take for air : c_p = 1.005 kJ/kg K, c_v = 0.718 kJ/kg K, R = 0.287 kJ/kg K. [**Ans.** 1.25 kJ ; 0.924]

20. The moment of inertia of a flywheel is 0.54 kg-m^2 and it rotates at a speed of 3000 r.p.m. in a large heat insulated system, the temperature of which is 15°C. If the kinetic energy of the flywheel is dissipated as frictional heat at the shaft bearings which have a water equivalent of 2 kg, find the rise in the temperature of the bearings when the flywheel has come to rest.

Calculate the greatest possible amount of the above heat which may be returned to the flywheel as high-grade energy, showing how much of the original kinetic energy is now unavailable. What would be the final r.p.m. of the flywheel, if it is set in motion with this available energy ?

[**Ans.** 3.19°C ; 0.1459 kJ ; 222 r.p.m.]

21. In a steady flow air enters the system at a pressure of 10 bar and 200°C with a velocity of 100 m/s and leaves at 1.5 bar and 25°C with a velocity of 50 m/s. The temperature of the surroundings is 25°C and pressure is 1 bar. Determine reversible work and actual work assuming the process to be adiabatic.

Determine also the irreversibility and effectiveness of the system on the basis of one kg of air flow.

Take for air : c_p = 1 kJ/kg K ; R = 287 J/kg K.

[**Ans.** 200.65 kJ ; 178.75 kJ/kg, 21.90 kJ/kg ; 0.894]

Thermodynamic Relations

6

6.1. General aspects. 6.2. Fundamentals of partial differentiation. 6.3. Some general thermodynamic relations. 6.4. Entropy equations (Tds equations). 6.5. Equations for internal energy and enthalpy. 6.6. Measurable quantities—Equation of state—Co-efficient of expansion and compressibility— Specific heats—Joule-Thomson co-efficient. 6.7. Clausius-Claperyon equation. *Highlights—Objective Type Questions—Exercises.*

6.1. GENERAL ASPECTS

In this chapter, some important thermodynamic relations are deduced ; principally those which are useful when tables of properties are to be compiled from limited experimental data, those which may be used when calculating the work and heat transfers associated with processes undergone by a liquid or solid. It should be noted that the relations only apply to a substance in the solid phase when the stress, *i.e.*, the pressure, is uniform in all directions ; if it is not, a single value for the pressure cannot be alloted to the system as a whole.

Eight properties of a system, namely *pressure* (*p*), *volume* (*v*), *temperature* (*T*), *internal energy* (*u*), *enthalpy* (*h*), *entropy* (*s*), *Helmholtz function* (*f*) and *Gibbs function* (*g*) have been introduced in the previous chapters. *h*, *f* and *g* are sometimes referred to as **thermodynamic potentials.** Both *f* and *g* are useful when considering chemical reactions, and the former is of fundamental importance in statistical thermodynamics. The Gibbs function is also useful when considering processes involving a change of phase.

Of the above eight properties only the first three, *i.e.*, *p*, *v* and *T* are directly measurable. We shall find it convenient to introduce other combination of properties which are relatively easily measurable and which, together with measurements of *p*, *v* and *T*, enable the values of the remaining properties to be determined. These combinations of properties might be called **'thermodynamic gradients'** ; they are all defined as the rate of change of one property with another while a third is kept constant.

6.2. FUNDAMENTALS OF PARTIAL DIFFERENTIATION

Let three variables are represented by *x*, *y* and *z*. Their functional relationship may be expressed in the following forms :

$$f(x, y, z) = 0 \quad ...(i)$$

$$x = x(y, z) \quad ...(ii)$$

$$y = y(x, z) \quad ...(iii)$$

$$z = z(x, y) \quad ...(iv)$$

Let *x* is a function of two independent variables *y* and *z*

$$x = x(y, z) \quad ...(6.1)$$

Then the differential of the dependent variable *x* is given by

$$dx = \left(\frac{\partial x}{\partial y}\right)_z dy + \left(\frac{\partial x}{\partial z}\right)_y dz \quad ...(6.2)$$

where *dx* is called an **exact differential.**

If $\left(\frac{\partial x}{\partial y}\right)_z = M$ and $\left(\frac{\partial x}{\partial z}\right)_y = N$

Then $dx = Mdy + Ndz$ ...(6.3)

Partial differentiation of *M* and *N* with respect to *z* and *y*, respectively, gives

$$\frac{\partial M}{\partial z} = \frac{\partial^2 x}{\partial y \partial z} \quad \text{and} \quad \frac{\partial N}{\partial y} = \frac{\partial^2 x}{\partial z \partial y}$$

or

$$\frac{\partial M}{\partial z} = \frac{\partial N}{\partial y} \quad ...(6.4)$$

dx is a perfect differential when eqn. (6.4) is satisfied for any function *x*.

Similarly if $y = y(x, z)$ and $z = z(x, y)$ ...(6.5)

then from these two relations, we have

$$dy = \left(\frac{\partial y}{\partial x}\right)_z dx + \left(\frac{\partial y}{\partial z}\right)_x dz \quad ...(6.6)$$

$$dz = \left(\frac{\partial z}{\partial x}\right)_y dx + \left(\frac{\partial z}{\partial y}\right)_x dy \quad ...(6.7)$$

$$dy = \left(\frac{\partial y}{\partial x}\right)_z dx + \left(\frac{\partial y}{\partial z}\right)_x \left[\left(\frac{\partial z}{\partial x}\right)_y dx + \left(\frac{\partial z}{\partial y}\right)_x dy\right]$$

$$= \left[\left(\frac{\partial y}{\partial x}\right)_z + \left(\frac{\partial y}{\partial z}\right)_x \left(\frac{\partial z}{\partial x}\right)_y\right] dx + \left(\frac{\partial y}{\partial z}\right)_x \left(\frac{\partial z}{\partial y}\right)_x dy$$

$$= \left[\left(\frac{\partial y}{\partial x}\right)_z + \left(\frac{\partial y}{\partial z}\right)_x \left(\frac{\partial z}{\partial x}\right)_y\right] dx + dy$$

or $$\left(\frac{\partial y}{\partial x}\right)_z + \left(\frac{\partial y}{\partial z}\right)_x \left(\frac{\partial z}{\partial x}\right)_y = 0$$

or $$\left(\frac{\partial y}{\partial z}\right)_x \left(\frac{\partial z}{\partial x}\right)_y = -\left(\frac{\partial y}{\partial x}\right)_z$$

or $$\left(\frac{\partial x}{\partial y}\right)_z \left(\frac{\partial z}{\partial x}\right)_y \left(\frac{\partial y}{\partial z}\right)_x = -1 \quad ...(6.8)$$

In terms of p, v and T, the following relation holds good

$$\left(\frac{\partial p}{\partial v}\right)_T \left(\frac{\partial T}{\partial p}\right)_v \left(\frac{\partial v}{\partial T}\right)_p = -1 \quad ...(6.9)$$

6.3. SOME GENERAL THERMODYNAMIC RELATIONS

The first law applied to a closed system undergoing a reversible process states that

$$dQ = du + pdv$$

According to second law,

$$ds = \left(\frac{dQ}{T}\right)_{rev.}$$

Combining these equations, we get

$$Tds = du + pdv$$

or $$du = Tds - pdv \quad ...(6.10)$$

The properties h, f and g may also be put in terms of T, s, p and v as follows :

$$dh = du + pdv + vdp = Tds + vdp$$

Helmholtz free energy function,

$$df = du - Tds - sdT \quad ...(6.11)$$

$$= -pdv - sdT \quad ...(6.12)$$

Gibb's free energy function,

$$dg = dh - Tds - sdT = vdp - sdT \quad ...(6.13)$$

Each of these equations is a *result of the two laws of thermodynamics.*

Since du, dh, df and dg are the *exact differentials,* we can express them as

$$du = \left(\frac{\partial u}{\partial s}\right)_v ds + \left(\frac{\partial u}{\partial v}\right)_s dv,$$

$$dh = \left(\frac{\partial h}{\partial s}\right)_p ds + \left(\frac{\partial h}{\partial p}\right)_s dp,$$

$$df = \left(\frac{\partial f}{\partial v}\right)_T dv + \left(\frac{\partial f}{\partial T}\right)_v dT,$$

$$dg = \left(\frac{\partial g}{\partial p}\right)_T dp + \left(\frac{\partial g}{\partial T}\right)_p dT.$$

Comparing these equations with (6.10) to (6.13) we may equate the corresponding co-efficients. For example, from the two equations for du, we have

$$\left(\frac{\partial u}{\partial s}\right)_v = T \quad \text{and} \quad \left(\frac{\partial u}{\partial v}\right)_s = -p$$

The complete group of such relations may be summarised as follows :

$$\left(\frac{\partial u}{\partial s}\right)_v = T = \left(\frac{\partial h}{\partial s}\right)_p \quad ...(6.14)$$

$$\left(\frac{\partial u}{\partial v}\right)_s = -p = \left(\frac{\partial f}{\partial v}\right)_T \quad ...(6.15)$$

$$\left(\frac{\partial h}{\partial p}\right)_s = v = \left(\frac{\partial g}{\partial p}\right)_T \quad ...(6.16)$$

$$\left(\frac{\partial f}{\partial T}\right)_v = -s = \left(\frac{\partial g}{\partial T}\right)_p \quad ...(6.17)$$

Also, $$\left(\frac{\partial T}{\partial v}\right)_s = -\left(\frac{\partial p}{\partial s}\right)_v \quad ...(6.18)$$

$$\left(\frac{\partial T}{\partial p}\right)_s = \left(\frac{\partial v}{\partial s}\right)_p \quad ...(6.19)$$

$$\left(\frac{\partial p}{\partial T}\right)_v = \left(\frac{\partial s}{\partial v}\right)_T \quad ...(6.20)$$

$$\left(\frac{\partial v}{\partial T}\right)_p = -\left(\frac{\partial s}{\partial p}\right)_T \quad ...(6.21)$$

The equations (6.18) to (6.21) are known as **Maxwell relations.**

It must be emphasised that eqns. (6.14) to (6.21) do not refer to a process, but simply express relations between properties which must be satisfied when any system is in a state of equilibrium. Each partial differential co-efficient can itself be regarded as a property of state. The state may be defined by a point on a three dimensional surface, the surface representing all possible states of stable equilibrium.

6.4. ENTROPY EQUATIONS (Tds Equations)

Since entropy may be expressed as a function of any other two properties, *e.g.* temperature T and specific volume v,

$$s = f(T, v)$$

i.e., $$ds = \left(\frac{\partial s}{\partial T}\right)_v dT + \left(\frac{\partial s}{\partial v}\right)_T dv$$

or $$Tds = T\left(\frac{\partial s}{\partial T}\right)_v dT + T\left(\frac{\partial s}{\partial v}\right)_T dv \quad ...(6.22)$$

But for a reversible constant volume change

$$dq = c_v (dT)_v = T(ds)_v$$

or $$c_v = T\left(\frac{\partial s}{\partial T}\right)_v \quad ...(6.23)$$

But, $$\left(\frac{\partial s}{\partial v}\right)_T = \left(\frac{\partial p}{\partial T}\right)_v \quad \text{[Maxwell's eqn. (6.20)]}$$

Hence, substituting in eqn. (6.22), we get

$$Tds = c_v dT + T\left(\frac{\partial p}{\partial T}\right)_v dv \quad ...(6.24)$$

This is known as the *first form of entropy equation* or the **first Tds equation.**

Similarly, writing

$$s = f(T, p)$$

$$Tds = T\left(\frac{\partial s}{\partial T}\right)_p dT + T\left(\frac{\partial s}{\partial p}\right)_T dp \quad ...(6.25)$$

where $$c_p = T\left(\frac{\partial s}{\partial T}\right)_p \quad ...(6.26)$$

Also $$\left(\frac{\partial s}{\partial p}\right)_T = -\left(\frac{\partial v}{\partial T}\right)_p \quad \text{[Maxwell's eqn. (6.21)]}$$

whence, substituting in eqn. (6.25)

$$Tds = c_p dT - T\left(\frac{\partial v}{\partial T}\right)_p dp \quad ...(6.27)$$

This is known as the *second form of entropy equation* or the **second Tds equation.**

6.5. EQUATIONS FOR INTERNAL ENERGY AND ENTHALPY

(*i*) Let $u = f(T, v)$

$$du = \left(\frac{\partial u}{\partial T}\right)_v dT + \left(\frac{\partial u}{\partial v}\right)_T dv = c_v\, dT + \left(\frac{\partial u}{\partial v}\right)_T dv \quad ...(6.28)$$

To evaluate $\left(\frac{\partial u}{\partial v}\right)_T$ let $u = f(s, v)$

Then $$du = \left(\frac{\partial u}{\partial s}\right)_v ds + \left(\frac{\partial u}{\partial v}\right)_s dv$$

or $$\left(\frac{\partial u}{\partial v}\right)_T = \left(\frac{\partial u}{\partial s}\right)_v \left(\frac{\partial s}{\partial v}\right)_T + \left(\frac{\partial u}{\partial v}\right)_s$$

But $$\left(\frac{\partial u}{\partial s}\right)_v = T, \left(\frac{\partial s}{\partial v}\right)_T = \left(\frac{\partial s}{\partial T}\right)_v, \left(\frac{\partial u}{\partial v}\right)_s = -p$$

Hence $$\left(\frac{\partial u}{\partial v}\right)_T = T\left(\frac{\partial p}{\partial T}\right)_v - p \quad ...(6.29)$$

This is sometimes called the **energy equation.**

From equation (6.28), we get

$$du = c_v dT + \left\{T\left(\frac{\partial p}{\partial T}\right)_v - p\right\} dv \quad ...(6.30)$$

(*ii*) To evaluate dh we can follow similar steps as under

$$h = f(T, p)$$

$$dh = \left(\frac{\partial h}{\partial T}\right)_p dT + \left(\frac{\partial h}{\partial p}\right)_T dp$$

$$= c_p dT + \left(\frac{\partial h}{\partial p}\right)_T dp \quad ...(6.31)$$

To find $\left(\frac{\partial h}{\partial p}\right)_T$; let $h = f(s, p)$

Then, $$dh = \left(\frac{\partial h}{\partial s}\right)_p ds + \left(\frac{\partial h}{\partial p}\right)_s dp$$

$$\left(\frac{\partial h}{\partial p}\right)_T = \left(\frac{\partial h}{\partial s}\right)_p \left(\frac{\partial s}{\partial p}\right)_T + \left(\frac{\partial h}{\partial p}\right)_s$$

But, $$\left(\frac{\partial h}{\partial s}\right)_p = T, \left(\frac{\partial s}{\partial p}\right)_T = -\left(\frac{\partial v}{\partial p}\right)_p, \left(\frac{\partial h}{\partial p}\right)_s = v$$

Hence
$$\left(\frac{\partial h}{\partial p}\right)_T = v - T\left(\frac{\partial v}{\partial T}\right)_p \quad ...(6.32)$$

From eqn. (6.31), we get

$$dh = c_p\, dT + \left\{v - T\left(\frac{\partial v}{\partial T}\right)_p\right\} dp \quad ...(6.33)$$

6.6. MEASURABLE QUANTITIES

Out of eight thermodynamic properties, as earlier stated, only *p, v* and *T* are directly measurable. Let us now examine the information that can be obtained from measurements of these primary properties, and then see what other easily measurable quantities can be introduced.

The following will be discussed :

(*i*) Equation of state

(*ii*) Co-efficient of expansion and compressibility

(*iii*) Specific heats

(*iv*) Joule-Thomson co-efficient.

6.6.1. Equation of State

Let us imagine a series of experiments in which the volume of a substance is measured over a range of temperatures while the pressure is maintained constant, this being repeated for various pressures. The results might be represented graphically by a three-dimensional surface, or by a family of constant.pressure lines on a *v-T* diagram. It is useful if an equation can be found to express the relation between *p, v* and *T*, and this can always be done over a limited range of states. No single equation will hold for all phases of a substance, and usually more than one equation is required even in one phase if the accuracy of the equation is to match that of the experimental results. Equations relating *p, v* and *T* are called *equations of state* or *characteristic equations.* Accurate equations of state are usually complicated, a typical form being

$$pv = A + \frac{B}{v} + \frac{C}{v^2} + ...$$

where *A, B, C*, ... are functions of temperature which differ for different substances.

An equation of state of a particular substance is an empirical result, and *it cannot be deduced from the laws of thermodynamics.* Nevertheless the general form of the equation may be predicted from hypotheses about the microscopic structure of matter. This type of prediction has been developed to a high degree of precision for gases, and to a lesser extent for liquids and solids. The simplest postulates about the molecular structure of gases lead to the concept of the perfect gas which has the equation of state $pv = RT$. Experiments have shown that the behaviour of real gases at low pressure with high temperature agrees well with this equation.

6.6.2. Co-efficient of Expansion and Compressibility

From *p-v-T* measurements, we find that an equation of state is not the only useful information which can be obtained. When the experimental results are plotted as a series of constant pressure lines on a *v-T* diagrams, as in Fig. 6.1 (*a*), the slope of a constant pressure line at any given state is $\left(\frac{\partial v}{\partial T}\right)_p$. If the gradient is divided by the volume at that state, we have a value of a property of the substance called its co-efficient of cubical expansion β. That is,

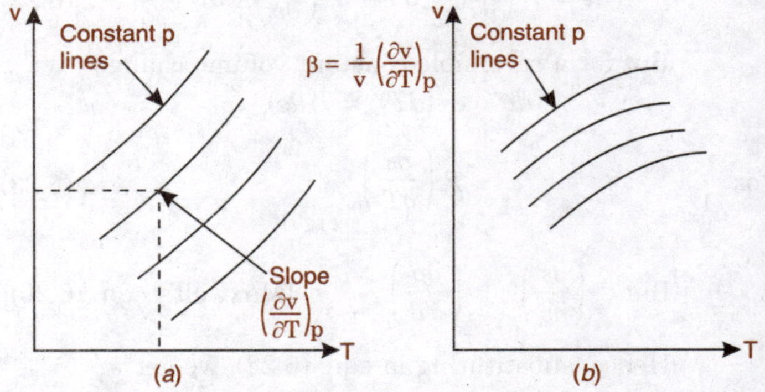

Fig. 6.1. Determination of co-efficient of expansion from *p-v-T* data.

$$\beta = \frac{1}{v}\left(\frac{\partial v}{\partial T}\right)_p \quad ...(6.34)$$

Value of β can be tabulated for a range of pressures and temperatures, or plotted graphically as in Fig. 6.1 (*b*). For solids and liquids over the normal working range of pressure and temperature, the variation of β is small and can often be neglected. In tables of physical properties β is usually quoted as an average value over a small range of temperature, the pressure being atmospheric. This average co-efficient may be symbolised by β and it is defined by

$$\beta = \frac{v_2 - v_1}{v_1 (T_2 - T_1)} \quad ...(6.35)$$

Fig. 6.2 (*a*) can be replotted to show the variation of volume with pressure for various constant values of temperature. In this case, the gradient of a curve at any state is $\left(\frac{\partial v}{\partial p}\right)_T$. When this gradient is divided by the volume at that state, we have a property known as the *compressibility K of the substance.* Since this gradient is always *negative, i.e., the volume of a substance always decreases* with *increase of pressure* when the *temperature is constant*, the compressibility is usually made a positive quantity by defining it as

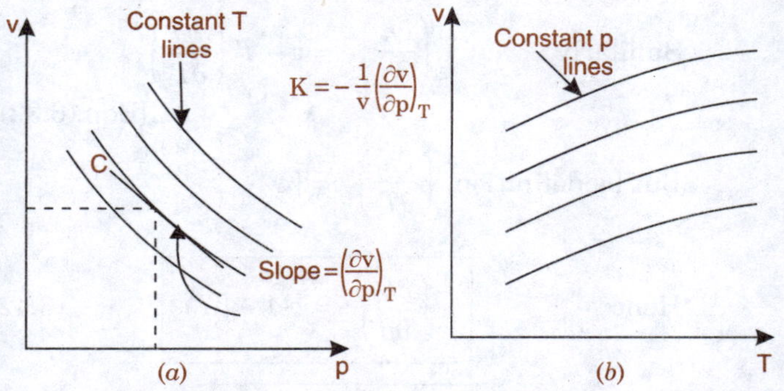

Fig. 6.2. Determination of compressibility from *p-T* data.

$$K = -\frac{1}{v}\left(\frac{\partial v}{\partial p}\right)_T \qquad ...(6.36)$$

K can be regarded as a constant for many purposes for solids and liquids. In tables of properties it is often quoted as an average a value over a small range of pressure at atmospheric temperature, *i.e.,*

$$\overline{K} = -\frac{v_2 - v_1}{v_1(p_2 - p_1)}$$

When β and K are known, we have

$$\left(\frac{\partial p}{\partial T}\right)_v\left(\frac{\partial T}{\partial v}\right)_p\left(\frac{\partial v}{\partial p}\right)_T = -1$$

Since $\left(\frac{\partial v}{\partial T}\right)_p = \beta v$ and $\left(\frac{\partial v}{\partial p}\right)_T = -Kv$,

$$\left(\frac{\partial p}{\partial T}\right)_v = \frac{\beta}{K}, \qquad ...(6.37)$$

When the equation of state is known, the co-efficient of cubical expansion and compressibility can be found by differentiation. For a perfect gas, for example, we have

$$\left(\frac{\partial v}{\partial T}\right)_p = \frac{R}{p} \quad \text{and} \quad \left(\frac{\partial v}{\partial p}\right)_T = \frac{RT}{p^2}$$

Hence $\beta = \frac{1}{v}\left(\frac{\partial v}{\partial T}\right)_p = \frac{R}{pv} = \frac{1}{T}$,

and $K = -\frac{1}{v}\left(\frac{\partial v}{\partial p}\right)_T = \frac{RT}{p^2 v} = \frac{1}{p}$.

6.6.3. Specific Heats

Following are the three differential co-efficients which can be relatively easily determined experimentally :

Consider the first quantity $\left(\frac{\partial u}{\partial T}\right)_v$. During a process at constant volume, the first law informs us that an increase of internal energy is equal to heat supplied. If a calorimetric experiment is conducted with a known mass of substance at constant volume, the quantity of heat Q required to raise the temperature of unit mass by ΔT may be measured. We can then write : $\left(\frac{\Delta u}{\Delta T}\right)_v = \left(\frac{Q}{\Delta T}\right)_v$. The quantity obtained this way is known as the mean *specific heat at constant volume* over the temperature range ΔT. It is found to vary with the conditions of the experiment, *i.e.,* with the temperature range and the specific volume of the substance. As the temperature range is reduced the value approaches that of $\left(\frac{\partial u}{\partial T}\right)_v$, and the *true specific heat at constant volume* is defined by $c_v = \left(\frac{\partial u}{\partial T}\right)_v$. This is a property of the substance and in general its value varies with the state of the substance, *e.g.,* with temperature and pressure.

According to first law of thermodynamics the heat supplied is equal to the increase of enthalpy during a reversible constant pressure process. Therefore, a calorimetric experiment carried out with a substance at constant pressure gives us, $\left(\frac{\Delta h}{\Delta T}\right)_p = \left(\frac{Q}{\Delta T}\right)_p$ which is the mean specific heat at constant pressure. As the range of temperature is made infinitesimally small, this becomes the rate of change of enthalpy with temperature at a particular state defined by T and p, and this is true *specific heat at constant pressure* defined by $c_p = \left(\frac{\partial h}{\partial T}\right)_p$. c_p also varies with the state, *e.g.,* with pressure and temperature.

The description of experimental methods of determining c_p and c_v can be found in texts on physics. When solids and liquids are considered, it is not easy to measure c_v owing to the stresses set up when such a substance is prevented from expanding. However, a relation between c_p, c_v, β and K can be found as follows, from which c_v may be obtained if the remaining three properties have been measured.

The First Law of Thermodynamics, for a reversible process states that

$$dQ = du + p\,dv$$

Since we may write $u = \phi(T, v)$, we have

$$du = \left(\frac{du}{\partial T}\right)_v dT + \left(\frac{\partial u}{\partial v}\right)_T dv$$

$$\therefore \quad dQ = \left(\frac{\partial u}{\partial T}\right)_v dT + \left\{p + \left(\frac{\partial u}{\partial v}\right)_T\right\} dv$$

$$= c_v\,dT + \left\{p + \left(\frac{\partial u}{\partial v}\right)_T\right\} dv$$

This is true for any reversible process, and so, for a reversible constant pressure process,

$$dQ = c_p(dT)_p = c_v(dT)_p + \left\{p + \left(\frac{\partial u}{\partial v}\right)_T\right\}(dv)_p$$

Hence $c_p - c_v = \left\{p + \left(\frac{\partial u}{\partial v}\right)_T\right\}\left(\frac{\partial v}{\partial T}\right)_p$

Also $\left(\frac{\partial p}{\partial T}\right)_v = \left(\frac{\partial s}{\partial v}\right)_T = \frac{1}{T}\left\{p + \left(\frac{\partial u}{\partial v}\right)_T\right\}$,

and therefore $c_p - c_v = T\left(\frac{\partial p}{\partial T}\right)_v\left(\frac{\partial v}{\partial T}\right)_p$

Now, from eqns. (6.34) and (6.37), we have

$$c_p - c_v = \frac{\beta^2 T v}{K} \quad \text{...(6.38)}$$

Thus at any state defined by T and v, c_v can be found if c_p, β and K are known for the substance at that state. The values of T, v and K are always positive and, although β may sometimes be negative (*e.g.*, between 0° and 4°C water contracts on heating at constant pressure), β^2 is always positive. It follows that c_p *is always greater than* c_v.

The other expressions for c_p and c_v can be obtained by using the equation (6.14) as follows :

Since $c_v = \left(\frac{\partial u}{\partial T}\right)_v = \left(\frac{\partial u}{\partial s}\right)_v\left(\frac{\partial s}{\partial T}\right)_v$

We have $c_v = T\left(\frac{\partial s}{\partial T}\right)_v$ ...(6.39)

Similarly, $c_p = \left(\frac{\partial h}{\partial T}\right)_p = \left(\frac{\partial h}{\partial s}\right)_p\left(\frac{\partial s}{\partial T}\right)_p$

Hence, $c_p = T\left(\frac{\partial s}{\partial T}\right)_p$ ...(6.40)

Alternative Expressions for Internal Energy and Enthalpy

(*i*) Alternative expressions for equations (6.29) and (6.32) can be obtained as follows :

$$\left(\frac{\partial u}{\partial v}\right)_T = T\left(\frac{\partial p}{\partial T}\right)_v - p \quad \text{...[Eqn.(6.29)]}$$

But, $\left(\frac{\partial p}{\partial T}\right)_v\left(\frac{\partial T}{\partial v}\right)_p\left(\frac{\partial v}{\partial p}\right)_T = -1$

or $\left(\frac{\partial p}{\partial T}\right)_v = -\left(\frac{\partial v}{\partial T}\right)_p\left(\frac{\partial p}{\partial v}\right)_T = +\frac{\beta v}{Kv} = \frac{\beta}{K}$

Substituting in eqn. (6.29), we get

$$\left(\frac{\partial u}{\partial v}\right)_T = T\frac{\beta}{K} - p \quad \text{...(6.41)}$$

Thus, $$du = c_v dT + \left(\frac{T\beta}{K} - p\right) dv \quad \text{...[6.28 (a)]}$$

Similarly, $\left(\frac{\partial h}{\partial p}\right)_T = v - T\left(\frac{\partial v}{\partial T}\right)_p$ ...[Eqn.(6.32)]

But by definition, $\left(\frac{\partial u}{\partial T}\right)_p = \beta v$

Hence $$\left(\frac{\partial h}{\partial p}\right)_T = v(1 - \beta T) \quad \text{...(6.42)}$$

Thus $$dh = c_p\, dT + v(1 - \beta T)\, dp \quad \text{...[6.31 (a)]}$$

(*ii*) Since $u = h - pv$

or $\left(\frac{\partial u}{\partial p}\right)_T = \left(\frac{\partial h}{\partial p}\right)_T - p\left(\frac{\partial v}{\partial p}\right)_T - v$

$= v - v\beta T + pKv - v$

Hence, $\left(\frac{\partial u}{\partial p}\right)_T = pKv - v\beta T$ ...(6.43)

6.6.4. Joule-Thomson Co-efficient

Let us consider the partial differential co-efficient $\left(\frac{\partial T}{\partial p}\right)_h$.

We know that if a fluid is flowing through a pipe, and the pressure is reduced by a throttling process, the enthalpies on either side of the restriction may be equal.

The throttling process is illustrated in Fig. 6.3 (*a*). The velocity increases at the restriction, with a consequent decrease of enthalpy, but this increase of kinetic energy is dissipated by friction, as the eddies die down after restriction. The steady-flow energy equation implies that the enthalpy of the fluid is restored to its initial value if the flow is adiabatic and if the velocity before restriction is equal to that downstream of it. These conditions are very nearly satisfied in the following experiment which is usually referred to as the Joule-Thomson experiment.

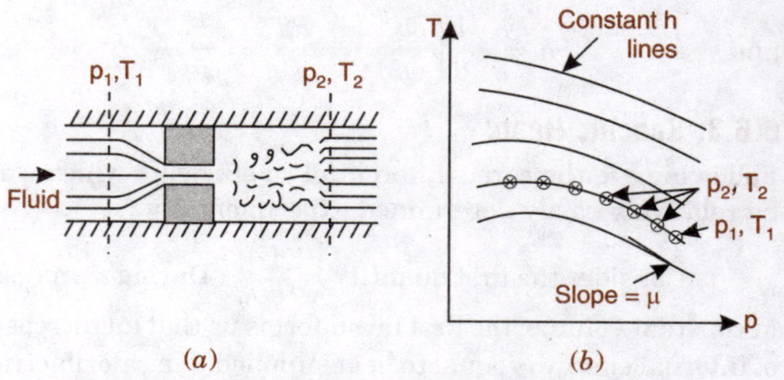

Fig. 6.3. Determination of Joule-Thomson co-efficient.

Through a *porous plug* (inserted in a pipe) a fluid is allowed to flow steadily from a high pressure to a low pressure. The pipe is well lagged so that any heat flow to or from the fluid is negligible when steady conditions have been reached. Furthermore, the velocity of the flow is kept low, and any difference between the kinetic energy upstream and downstream of the plug is negligible. *A porous plug is used because the local increase of directional kinetic energy, caused by the restriction, is rapidly converted to random molecular energy by viscous friction in fine passages of the plug. Irregularities in the flow die out in a very short distance* downstream of the plug, and temperature and pressure measurements taken there will be values for the fluid in *a state of thermodynamic equilibrium*.

By keeping the upstream pressure and temperature constant at p_1 and T_1, the downstream pressure p_2 is reduced in steps and the corresponding temperature T_2 is measured. The fluid in the successive states defined by the values of p_2 and T_2 must always have the same value of the enthalpy, namely the value of the enthalpy corresponding to the state defined by p_1 and T_1. From these results, points representing equilibrium states of the same enthalpy can be plotted on a *T-s* diagram, and joined up to form a curve of constant enthalpy. The curve does *not* represent the throttling process itself, which is irreversible. During the actual process, the fluid undergoes first a decrease and then an increase of enthalpy, and no single value of the specific enthalpy can be ascribed to all elements of the fluid. If the experiment is repeated with different values of p_1 and T_1, a family of curves may be obtained (covering a range of values of enthalpy) as shown in Fig. 6.3 (*b*).

The slope of a curve [Fig. 6.3 (*b*)] at any point in the field is a function only of the state of the fluid, it is the Joule-Thomson co-efficient μ, defined by $\mu = \left(\frac{\partial T}{\partial p}\right)_h$. The change of temperature due to a throttling process is small and, if the fluid is a gas, it may be an increase or decrease. At any particular pressure there is a temperature, the *temperature of inversion*, above which a gas can never be cooled by a throttling process.

Both c_p and μ, as it may be seen, are defined in terms of *p*, *T* and *h*. The third partial differential co-efficient based on these three properties is given as follows :

$$\left(\frac{\partial h}{\partial p}\right)_T \left(\frac{\partial p}{\partial T}\right)_h \left(\frac{\partial T}{\partial h}\right)_p = -1$$

Hence
$$\left(\frac{\partial h}{\partial p}\right)_T = -\mu c_p \quad \text{...(6.44)}$$

μ may be expressed in terms of c_p, *p*, *v* and *T* as follows :

The property relation for *dh* is $dh = T\,ds + v\,dp$

From second *Tds* equation, we have

$$Tds = c_p\,dT - T\left(\frac{\partial v}{\partial T}\right)_p dp$$

$$\therefore \quad dh = c_p\,dT - \left[T\left(\frac{\partial v}{\partial T}\right)_p - v\right] dp \quad \text{...(6.45)}$$

For a constant enthalpy process $dh = 0$. Therefore,

$$0 = (c_p\,dT)_h + \left[\left\{v - T\left(\frac{\partial v}{\partial T}\right)_p\right\} dp\right]_h$$

or
$$(c_p\,dT)_h = \left[\left\{T\left(\frac{\partial v}{\partial T}\right)_p - v\right\} dp\right]_h$$

$$\therefore \quad \mu = \left(\frac{\partial T}{\partial p}\right)_h = \frac{1}{c_p}\left[T\left(\frac{\partial v}{\partial T}\right)_p - v\right] \quad \text{...(6.46)}$$

For an ideal gas, $pv = RT$; $v = \frac{RT}{p}$

or
$$\left(\frac{\partial v}{\partial T}\right)_p = \frac{R}{p} = \frac{v}{T}$$

$$\therefore \quad \mu = \frac{1}{c_p}\left(T \times \frac{v}{T} - v\right) = 0.$$

Therefore, if an ideal gas is throttled, there will not be any change in temperature.

Let $h = f(p, T)$

Then
$$dh = \left(\frac{\partial h}{\partial p}\right)_T dp + \left(\frac{\partial h}{\partial T}\right)_p dT \quad \text{...(6.47)}$$

But $\left(\frac{\partial h}{\partial T}\right)_p = c_p$

$$\therefore \quad dh = \left(\frac{\partial h}{\partial p}\right)_T dp + c_p\,dT$$

For throttling process, $dh = 0$

$$\therefore \quad 0 = \left(\frac{\partial h}{\partial p}\right)_T \left(\frac{\partial p}{\partial T}\right)_h + c_p \quad \text{...(6.48)}$$

or
$$c_p = -\frac{1}{\mu}\left(\frac{\partial h}{\partial p}\right)_T \quad \text{...(6.49)}$$

$\left(\frac{\partial h}{\partial p}\right)_T$ is known as the *constant temperature co-efficient.*

6.7. CLAUSIUS-CLAPERYON EQUATION

Clausius-Claperyon equation is a relationship between the *saturation pressure*, *temperature*, the *enthalpy of evaporation*, and the *specific volume* of the *two phases* involved. This equation provides a basis for calculations of properties in a two-phase region. It gives the *slope* of a curve separating the two phases in the *p-T* diagram.

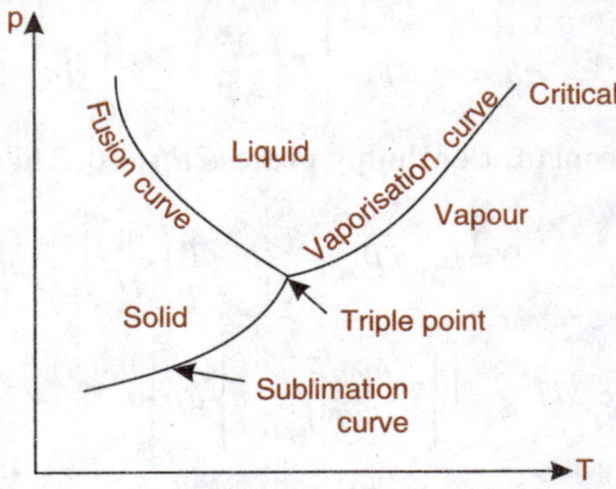

Fig. 6.4. *p-T* diagram.

The Clausius-Claperyon equation can be derived in different ways. The method given below involves the use of the Maxwell relation [eqn. (6.20)]

$$\left(\frac{\partial p}{\partial T}\right)_v = \left(\frac{\partial s}{\partial v}\right)_T$$

Let us consider the change of state from saturated liquid to saturated vapour of a pure substance which takes place at constant temperature. During the evaporation, the pressure and temperature are independent of volume.

$$\therefore \quad \left(\frac{dp}{dT}\right) = \frac{s_g - s_f}{v_g - v_f}$$

where s_g = Specific entropy of saturated vapour,

s_f = Specific entropy of saturated liquid,

v_g = Specific volume of saturated vapour, and

v_f = Specific volume of saturated liquid.

Also, $s_g - s_f = s_{fg} = \dfrac{h_{fg}}{T}$

and $v_g - v_f = v_{fg}$

where s_{fg} = Increase in specific entropy,

v_{fg} = Increase in specific volume, and

h_{fg} = Latent heat added during evaporation at saturation temperature T.

$$\therefore \quad \frac{dp}{dT} = \frac{s_g - s_f}{v_g - v_f} = \frac{s_{fg}}{v_{fg}} = \frac{h_{fg}}{T\,.v_{fg}} \quad ...(6.50)$$

This is known as **Clausius-Claperyon or Claperyon equation** for *evaporation of liquids*. The derivative $\dfrac{dp}{dT}$ is the slope of vapour pressure versus temperature curve. Knowing this slope and the specific volume v_g and v_f from experimental data, we can determine the enthalpy of evaporation, $(h_g - h_f)$ which is relatively difficult to measure accurately.

Eqn. (6.50) is also valid for the change from a solid to liquid, and from solid to a vapour.

At very low pressures, if we assume $v_g \simeq v_{fg}$ and the equation of the vapour is taken as $pv = RT$, then eqn. (6.50) becomes

$$\frac{dp}{dT} = \frac{h_{fg}}{Tv_g} = \frac{h_{fg}\,p}{RT^2} \quad ...(6.51)$$

or

$$h_{fg} = \frac{RT^2}{p}\frac{dp}{dT} \quad ...(6.52)$$

Eqn. (6.52) may be used to obtain the enthalpy of vapourisation. This equation can be rearranged as follows :

$$\frac{dp}{p} = \frac{h_{fg}}{R} \cdot \frac{dT}{T^2}$$

Integrating the above equation, we get

$$\int \frac{dp}{p} = \frac{h_{fg}}{R} \int \frac{dT}{T^2}$$

$$\ln \frac{p_2}{p_1} = \frac{h_{fg}}{R}\left[\frac{1}{T_1} - \frac{1}{T_2}\right] \quad ...(6.53)$$

Knowing the vapour pressure p_1 at temperature T_1 we can find the vapour pressure p_2 corresponding to temperature T_2 from eqn. (6.53).

From eqn. (6.50), we see that the *slope of the vapour pressure curve is always +ve*, since $v_g > v_f$ and h_{fg} is always +ve. Consequently, the vapour pressure of any simple compressible substance increases with temperature.

— It can be shown that the slope of the *sublimation curve is also +ve for any pure substance.*

— However, the slope of the *melting curve could be +ve or –ve.*

— *For a substance that contracts on freezing, such as water, the slope of the melting curve will be negative.*

☞**Example 6.1.** *For a perfect gas, show that*

$$c_p - c_v = \left[p + \left(\frac{\partial u}{\partial v}\right)_T\right]\left(\frac{\partial v}{\partial T}\right)_p = pv\beta + v\beta\left(\frac{\partial u}{\partial v}\right)_T$$

where β *is the co-efficient of cubical/volume expansion.*

Solution. The first law of thermodynamics applied to a closed system undergoing a reversible process states as follows :

$$dQ = du + pdv \quad ...(i)$$

As per second law of thermodynamics,

$$ds = \left(\frac{dQ}{T}\right)_{rev.} \quad ...(ii)$$

Combining these equations (*i*) and (*ii*), we have

$$Tds = du + pdv$$

Also, since $h = u + pv$

$$\therefore \quad dh = du + pdv + vdp = Tds + vdp$$

Thus, $Tds = du + pdv = dh - vdp$

Now, writing relation for u taking T and v as independent, we have

$$du = \left(\frac{\partial u}{\partial T}\right)_v dT + \left(\frac{\partial u}{\partial v}\right)_T dv$$

$$= c_v\, dT + \left(\frac{\partial u}{\partial v}\right)_T dv$$

Similarly, writing relation for h taking T and p as independent, we have

$$dh = \left(\frac{\partial h}{\partial T}\right)_p dT + \left(\frac{\partial h}{\partial p}\right)_T dp$$

$$= c_p\, dT + \left(\frac{\partial h}{\partial p}\right)_T dp$$

In the equation for Tds, substituting the value of du and dh, we have

$$c_v\, dT + \left(\frac{\partial u}{\partial v}\right)_T dv + pdv = c_p\, dT + \left(\frac{\partial h}{\partial p}\right)_T dp - vdp$$

or,
$$c_v\, dT + \left[p + \left(\frac{\partial u}{\partial v}\right)_T\right] dv = c_p\, dT - \left[v - \left(\frac{\partial h}{\partial p}\right)_T\right] dp$$

Since the above equation is true for any process, therefore, it will also be true for the case when $dp = 0$ and hence

$$(c_p - c_v)\,(dT)_p = \left[p + \left(\frac{\partial u}{\partial v}\right)_T\right](dv)_p$$

or,
$$(c_p - c_v) = \left[p + \left(\frac{\partial u}{\partial v}\right)_T\right]\left(\frac{\partial v}{\partial T}\right)_p$$

By definition,
$$\beta = \frac{1}{v}\left(\frac{\partial v}{\partial T}\right)_p$$

$\therefore$ The above equation becomes,

$$c_p - c_v = \left[p + \left(\frac{\partial u}{\partial v}\right)_T\right] v\beta$$

$$= pv\beta + v\beta\left(\frac{\partial u}{\partial v}\right)_T \qquad \textbf{Proved.}$$

☞**Example 6.2.** *Find the value of co-efficient of volume expansion β and isothermal compressibility K for a van der Waals' gas obeying*

$$\left(p + \frac{a}{v^2}\right)(v - b) = RT.$$

Solution. van der Waals equation is

$$\left(p + \frac{a}{v^2}\right)(v - b) = RT$$

Rearranging this equation, we can write

$$p = \frac{RT}{v - b} - \frac{a}{v^2}$$

Now for β we require $\left(\frac{\partial u}{\partial T}\right)_p$. This can be found by writing the cyclic relation,

$$\left(\frac{\partial v}{\partial T}\right)_p \left(\frac{\partial T}{\partial p}\right)_v \left(\frac{\partial p}{\partial v}\right)_T = -1$$

Hence,
$$\left(\frac{\partial v}{\partial T}\right)_p = -\frac{\left(\frac{\partial p}{\partial T}\right)_v}{\left(\frac{\partial p}{\partial v}\right)_T}$$

From the van der Waals equation,

$$\left(\frac{\partial p}{\partial T}\right)_v = \frac{R}{v - b}$$

Also,
$$\left(\frac{\partial p}{\partial v}\right)_T = -\frac{RT}{(v - b)^2} + \frac{2a}{v^3}$$

Hence,
$$\beta = \frac{1}{v}\left(\frac{\partial v}{\partial T}\right)_p = \frac{1}{v}\left[-\frac{\left(\frac{\partial p}{\partial T}\right)_v}{\left(\frac{\partial p}{\partial v}\right)_T}\right]$$

or,
$$\beta = \frac{1}{v}\left[-\frac{\frac{R}{v - b}}{\frac{-RT}{(v - b)^2} + \frac{2a}{v^3}}\right] = \frac{\mathbf{Rv^2\,(v - b)}}{\mathbf{RTv^3 - 2a(v - b)^2}}. \quad \textbf{(Ans.)}$$

Also,
$$K = -\frac{1}{v}\left(\frac{\partial v}{\partial p}\right)_T = -\frac{1}{v}\left[\frac{1}{\frac{2a}{v^3} - \frac{RT}{(v - b)^2}}\right]$$

$$= \frac{\mathbf{v^2\,(v - b)^2}}{\mathbf{RTv^3 - 2a(v - b)^2}}. \quad \textbf{(Ans.)}$$

Example 6.3. *Prove that the internal energy of an ideal gas is a function of temperature alone.*

Solution. The equation of state for an ideal gas is given by

$$p = \frac{RT}{v}$$

But
$$\left(\frac{\partial u}{\partial v}\right)_T = T\left(\frac{\partial p}{\partial T}\right)_v - p \qquad \text{[Eqn. (6.29)]}$$

$$= T\,\frac{R}{v} - p = p - p = 0.$$

Thus, if the temperature remains constant, there is no change in internal energy with volume (and therefore also with pressure). Hence internal energy (u) is a function of temperature (T) alone. **...Proved.**

Example 6.4. *Prove that specific heat at constant volume (c_v) of a van der Waals' gas is a function of temperature alone.*

Solution. The van der Waals equation of state is given by,

$$p = \frac{RT}{v - b} - \frac{a}{v^2}$$

or $$\left(\frac{\partial p}{\partial T}\right)_v = \frac{R}{v-b}$$

or $$\left(\frac{\partial^2 p}{\partial T^2}\right)_v = 0$$

Now $$\left(\frac{dc_v}{dv}\right)_T = T\left(\frac{\partial^2 p}{\partial T^2}\right)_v$$

Hence $$\left(\frac{\partial c_v}{\partial v}\right)_T = 0$$

Thus c_v of a van der Waals gas is independent of volume (and therefore of pressure also). Hence it is a function of temperature alone.

☞**Example 6.5.** *Determine the following when a gas obeys van der Waals' equation,*

$$\left(p + \frac{a}{v^2}\right)(v-b) = RT$$

(*i*) *Change in internal energy ;*
(*ii*) *Change in enthalpy ;*
(*iii*) *Change in entropy.*

Solution. (*i*) **Change in internal energy :**

The change in internal energy is given by

$$du = c_v dT + \left[T\left(\frac{\partial p}{\partial T}\right)_v - p\right] dv$$

But, $$\left(\frac{\partial p}{\partial T}\right)_v = \left[\frac{\partial}{\partial T}\left\{\frac{RT}{v-b} - \frac{a}{v^2}\right\}\right]_v = \frac{R}{v-b}$$

$$\therefore \quad \int_1^2 du = c_v \int_1^2 dT + \int_1^2 \left[T\left(\frac{R}{v-b}\right) - p\right] dv$$

$$= c_v \int_1^2 dT + \int_1^2 \left[T\left(\frac{R}{v-b}\right) - \left\{\frac{RT}{v-b} - \frac{a}{v^2}\right\}\right] dv$$

$$= c_v \int_1^2 dT + \int_1^2 \left[\frac{RT}{v-b} - \frac{RT}{v-b} + \frac{a}{v^2}\right] dv$$

$$= c_v \int_1^2 dT + \int_1^2 \frac{a}{v^2} . dv$$

$$\therefore \quad u_2 - u_1 = \mathbf{c_v(T_2 - T_1) + a\left(\frac{1}{v_1} - \frac{1}{v_2}\right)}. \quad \textbf{(Ans.)}$$

(*ii*) **Change in enthalpy :**

The change in enthalpy is given by

$$dh = c_p dT + \left[v - T\left(\frac{\partial v}{\partial T}\right)_p\right] dp$$

$$\left(\frac{\partial h}{\partial p}\right)_T = 0 + v - T\left(\frac{\partial v}{\partial T}\right)_p \quad ...(1)$$

Let us consider $p = f(v, T)$

$$\therefore \quad dp = \left(\frac{\partial p}{\partial v}\right)_T dv + \left(\frac{\partial p}{\partial T}\right)_v dT$$

$$\therefore \quad (dp)_T = \left(\frac{\partial p}{\partial v}\right)_T dv + 0 \quad \text{as } dT = 0 \quad ...(2)$$

From equation (1),

$$(dh)_T = \left[v - T\left(\frac{\partial v}{\partial T}\right)_p\right](dp)_T.$$

Substituting the value of $(dp)_T$ from eqn. (2), we get

$$(dh)_T = \left[v - T\left(\frac{\partial v}{\partial T}\right)_p\right]\left(\frac{\partial p}{\partial v}\right)_T dv$$

$$= \left[v\left(\frac{\partial p}{\partial v}\right)_T - T\left(\frac{\partial v}{\partial T}\right)_p\left(\frac{\partial p}{\partial v}\right)_T\right] dv \quad ...(3)$$

Using the cyclic relation for p, v, T which is

$$\left(\frac{\partial v}{\partial T}\right)_p\left(\frac{\partial T}{\partial p}\right)_v\left(\frac{\partial p}{\partial v}\right)_T = -1$$

$$\therefore \quad \left(\frac{\partial v}{\partial T}\right)_p\left(\frac{\partial p}{\partial v}\right)_T = -\left(\frac{\partial p}{\partial T}\right)_v$$

Substituting this value in eqn. (3), we get

$$(dh)_T = \left[v\left(\frac{\partial p}{\partial v}\right)_T + T\left(\frac{\partial p}{\partial T}\right)_v\right] dv \quad ...(4)$$

For Van der Waals equation

$$\left(\frac{\partial p}{\partial v}\right)_T = \frac{\partial}{\partial v}\left[\left(\frac{RT}{v-b}\right) - \frac{a}{v^2}\right]_T$$

$$= -\frac{RT}{(v-b)^2} + \frac{2a}{v^3} \quad ...(5)$$

$$\left(\frac{\partial p}{\partial T}\right)_v = \frac{\partial}{\partial T}\left[\left(\frac{RT}{v-b} - \frac{a}{v^2}\right)\right]_v = \frac{R}{v-b} \quad ...(6)$$

Substituting the values of eqns. (5) and (6) in equation (1), we get

$$(dh)_T = \left[v\left\{-\frac{RT}{(v-b)^2} + \frac{2a}{v^3}\right\} + T\left(\frac{R}{v-b}\right)\right] dv$$

$$\therefore \quad \int_1^2 (dh)_T = -RT\int_1^2 \frac{v}{(v-b)^2} dv + 2a\int_1^2 \frac{dv}{v^2} + RT\int_1^2 \frac{dv}{(v-b)}$$

$$\therefore \quad (h_2 - h_1)_T = -RT\left[\log_e\left(\frac{v_2 - b}{v_1 - b}\right) - b\left\{\frac{1}{v_2 - b} - \frac{1}{v_1 - b}\right\}\right] - 2a\left(\frac{1}{v_2} - \frac{1}{v_1}\right) + RT\log_e\left(\frac{v_2 - b}{v_1 - b}\right)$$

$$= \mathbf{bRT}\left[\frac{1}{(v_2 - b)} - \frac{1}{(v_1 - b)}\right] - 2a\left[\frac{1}{v_2} - \frac{1}{v_1}\right].$$ **(Ans.)**

(*iii*) **Change in entropy :**

The change in entropy is given by:

$$ds = c_p \frac{dT}{T} + \left(\frac{\partial p}{\partial T}\right)_v . dv$$

For van der Waals equation,

$$\left(\frac{\partial p}{\partial T}\right)_v = \frac{R}{v - b} \quad \text{...as per eqn. (6)}$$

$$\therefore \quad ds = c_v \frac{dT}{T} + \frac{R}{v - b} dv$$

$$\therefore \quad \int_1^2 ds = c_v \int_1^2 \left[\frac{dT}{T}\right] + R\int_1^2 \frac{dv}{(v - b)}$$

$$\therefore \quad \mathbf{s_2 - s_1 = c_v \log_e \left[\frac{T_2}{T_1}\right] + R \log_e \left[\frac{v_2 - b}{v_1 - b}\right].}$$ **(Ans.)**

Example 6.6. *The equation of state in the given range of pressure and temperature is given by*

$$v = \frac{RT}{p} - \frac{C}{T^3}$$

where C is constant.

Derive an expression for change of enthalpy and entropy for this substance during an isothermal process.

Solution. The general equation for finding dh is given by

$$dh = c_p\, dT + \left[v - T\left(\frac{\partial v}{\partial T}\right)_p\right] dp$$

$$\int_1^2 dh = 0 + \int_1^2 \left[\left\{v - T\left(\frac{\partial v}{\partial T}\right)_p\right\}\right]_T$$

as $dT = 0$ for isothermal change.

From the given equation of state, we have

$$\left(\frac{\partial v}{\partial T}\right)_p = \frac{R}{p} + \frac{3C}{T^4} \quad \text{...}(i)$$

$$\therefore \quad h_2 - h_1 = \int_1^2 \left[\left\{\left(\frac{RT}{p} - \frac{C}{T^3}\right) - \frac{RT}{p} - \frac{3C}{T^3}\right\} dp\right]_T$$

$$= \left[\int_1^2 \left(-\frac{4C}{T^3}\right) dp\right]_T = -\frac{4C}{T^3}\left[(p_2 - p_1)\right]_T$$

The general equation for finding ds is given by:

$$ds = c_p \frac{dT}{T} - \left(\frac{\partial v}{\partial T}\right)_p dp$$

$$\int_1^2 ds = \left[-\int_1^2 \left(\frac{\partial v}{\partial T}\right)_p dp\right]_T$$

as $dT = 0$ for isothermal change.

Substituting the value from eqn. (*i*), we get

$$(s_2 - s_1) = \left[\int_1^2 -\left(\frac{R}{p} + \frac{3C}{T^4}\right) dp\right]_T$$

$$= \mathbf{-R \log_e \left(\frac{p_2}{p_1}\right) - \left(\frac{3C}{T^4}\right)(p_2 - p_1).}$$ **(Ans.)**

Example 6.7. *For a perfect gas obeying $pv = RT$, show that c_v and c_p are independent of pressure.*

Solution. Let $s = f(T, v)$

$$\text{Then} \quad ds = \left(\frac{\partial s}{\partial T}\right)_v dT + \left(\frac{\partial s}{\partial v}\right)_T dv$$

$$\text{Also} \quad u = f(T, v)$$

$$\text{Then} \quad du = \left(\frac{\partial u}{\partial T}\right)_v dT + \left(\frac{\partial u}{\partial v}\right)_T dv$$

$$= c_v\, dT + \left(\frac{\partial u}{\partial v}\right)_T dv$$

$$\text{Also,} \quad du = Tds - pdv$$

$$Tds - pdv = c_v\, dT + \left(\frac{\partial u}{\partial v}\right)_T dv$$

$$ds = c_v \frac{dT}{T} + \frac{1}{T}\left[\left(\frac{\partial u}{\partial v}\right)_T + p\right] dv$$

Equating the co-efficients of dT in the two equations of ds, we have

$$\frac{c_v}{T} = \left(\frac{\partial s}{\partial T}\right)_v$$

$$c_v = T\left(\frac{\partial s}{\partial T}\right)_v$$

$$\left(\frac{\partial c_v}{\partial v}\right)_T = T\frac{\partial^2 s}{\partial T \partial v}$$

From eqn. (6.20),

$$\left(\frac{\partial s}{\partial v}\right)_T = \left(\frac{\partial p}{\partial T}\right)_v$$

$$\frac{\partial^2 s}{\partial v \partial T} = \left(\frac{\partial^2 p}{\partial T^2}\right)_v$$

$$\left(\frac{\partial c_v}{\partial v}\right)_T = T\left(\frac{\partial^2 p}{\partial T^2}\right)_v$$

$$\text{Also} \quad p = \frac{RT}{v} \quad \text{...(Given)}$$

$$\left(\frac{\partial p}{\partial T}\right)_v = \frac{R}{v}$$

$$\left(\frac{\partial^2 p}{\partial T^2}\right)_v = 0 \quad \text{or} \quad \left(\frac{\partial c_v}{\partial v}\right)_T = 0$$

This shows that c_v *is a function of T alone*, or c_v *is independent of pressure.*

Also, $c_p = T\left(\frac{\partial s}{\partial T}\right)_p$

$$\left(\frac{\partial c_p}{\partial p}\right)_T = T\frac{\partial^2 s}{\partial T \partial p}$$

From eqn. (6.21),

$$\left(\frac{\partial s}{\partial p}\right)_T = -\left(\frac{\partial v}{\partial T}\right)_p$$

$$\frac{\partial^2 s}{\partial p \partial T} = -\left(\frac{\partial^2 v}{\partial T^2}\right)_p$$

$$\left(\frac{\partial c_p}{\partial p}\right)_T = -T\left(\frac{\partial^2 v}{\partial T^2}\right)_p$$

Again, $v = \frac{R}{p}$ ...(Given)

$$\left(\frac{\partial v}{\partial T}\right)_p = \frac{R}{p}$$

and $\left(\frac{\partial^2 v}{\partial T^2}\right)_p = 0$; $\left(\frac{\partial c_p}{\partial p}\right)_T = 0$

This shows that c_p *is a function of T alone*, or c_p *is independent of pressure*.

Example 6.8. *Using the first Maxwell equation, derive the remaining three.*

Solution. The *first Maxwell relation* is as follows :

$$\left(\frac{\partial T}{\partial v}\right)_s = -\left(\frac{\partial p}{\partial s}\right)_v \quad ...(i) \text{ [Eqn. (6.18)]}$$

(1) Using the cyclic relation

$$\left(\frac{\partial T}{\partial v}\right)_s \cdot \left(\frac{\partial v}{\partial s}\right)_T \cdot \left(\frac{\partial s}{\partial T}\right)_v = -1$$

$$\therefore \quad \left(\frac{\partial s}{\partial v}\right)_T = -\left(\frac{\partial T}{\partial v}\right)_s \cdot \left(\frac{\partial s}{\partial T}\right)_v \quad ...(ii)$$

Substituting the value from eqn. (*i*) in eqn. (*ii*), we get

$$\left(\frac{\partial s}{\partial v}\right)_T = \left(\frac{\partial p}{\partial s}\right)_v \cdot \left(\frac{\partial s}{\partial T}\right)_v \quad ...(iii)$$

Using the chain rule,

$$\left(\frac{\partial p}{\partial s}\right)_v \cdot \left(\frac{\partial s}{\partial T}\right)_v \cdot \left(\frac{\partial T}{\partial p}\right)_v = 1 \quad ...(iv)$$

Substituting the value of eqn. (*iv*) in eqn. (*iii*), we get

$$\left(\frac{\partial s}{\partial v}\right)_T = \left(\frac{\partial p}{\partial T}\right)_v$$

This is *Maxwell Third relation.*

(2) Again using the cyclic relation

$$\left(\frac{\partial s}{\partial p}\right)_v \cdot \left(\frac{\partial p}{\partial v}\right)_s \cdot \left(\frac{\partial v}{\partial s}\right)_p = -1$$

$$\therefore \quad \left(\frac{\partial v}{\partial s}\right)_p = -\left(\frac{\partial p}{\partial s}\right)_v \cdot \left(\frac{\partial v}{\partial p}\right)_s \quad ...(v)$$

Substituting the value from eqn. (*i*) into eqn. (*v*)

$$\left(\frac{\partial v}{\partial s}\right)_p = \left(\frac{\partial T}{\partial v}\right)_s \cdot \left(\frac{\partial v}{\partial p}\right)_s \quad ...(vi)$$

Again using the chain rule,

$$\left(\frac{\partial T}{\partial v}\right)_s \cdot \left(\frac{\partial v}{\partial p}\right)_s \cdot \left(\frac{\partial p}{\partial T}\right)_s = 1$$

Substituting the value of (*vi*) into (*v*), we get

$$\left(\frac{\partial v}{\partial s}\right)_p = \left(\frac{\partial T}{\partial p}\right)_s$$

This is *Maxwell second relation.*

(3) $\left(\frac{\partial v}{\partial T}\right)_p \cdot \left(\frac{\partial T}{\partial p}\right)_v \cdot \left(\frac{\partial p}{\partial v}\right)_T = -1$

$$\left(\frac{\partial v}{\partial T}\right)_p = -\left(\frac{\partial p}{\partial T}\right)_v \cdot \left(\frac{\partial v}{\partial p}\right)_T$$

$$= -\left(\frac{\partial p}{\partial s}\right)_v \left(\frac{\partial s}{\partial T}\right)_v \left(\frac{\partial s}{\partial p}\right)_T \left(\frac{\partial v}{\partial s}\right)_T$$

Substituting the value from eqn. (*i*), we get

$$\left(\frac{\partial v}{\partial T}\right)_p = \left(\frac{\partial T}{\partial v}\right)_s \left(\frac{\partial s}{\partial T}\right)_v \left(\frac{\partial s}{\partial p}\right)_T \left(\frac{\partial v}{\partial s}\right)_T$$

$$= \left\{\left(\frac{\partial T}{\partial v}\right)_s \cdot \left(\frac{\partial v}{\partial s}\right)_T \cdot \left(\frac{\partial s}{\partial T}\right)_v\right\} \left(\frac{\partial s}{\partial p}\right)_T$$

$$= -\left(\frac{\partial s}{\partial p}\right)_T$$

$$\therefore \quad \left(\frac{\partial v}{\partial T}\right)_p = -\left(\frac{\partial s}{\partial p}\right)_T$$

This is *Maxwell fourth relation.*

Example 6.9. *Derive the following relations :*

(*i*) $u = a - T\left(\frac{\partial a}{\partial T}\right)_v$ (*ii*) $h = g - T\left(\frac{\partial g}{\partial T}\right)_p$

(*iii*) $c_v = -T\left(\frac{\partial^2 a}{\partial T^2}\right)_v$ (*iv*) $c_p = -T\left(\frac{\partial^2 g}{\partial T^2}\right)_p$

where a = *Helmholtz function (per unit mass), and*

g = *Gibbs function (per unit mass).*

Solution. (*i*) Let $a = f(v, T)$

Then $da = \left(\frac{\partial a}{\partial v}\right)_T dv + \left(\frac{\partial a}{\partial T}\right)_v dT$

Also $\quad da = -pdv - sdT$

Comparing the co-efficients of dT, we get

$$\left(\frac{\partial a}{\partial T}\right)_v = -s$$

Also $\quad a = u - Ts$

or $\quad u = a + Ts = a - T\left(\frac{\partial a}{\partial T}\right)_v$

Hence $\quad \mathbf{u = a - T\left(\frac{\partial a}{\partial T}\right)_v}$. **(Ans.)**

(*ii*) Let $\quad g = f(p, T)$

Then $\quad dg = \left(\frac{\partial g}{\partial p}\right)_T dp + \left(\frac{\partial g}{\partial T}\right)_p dT$

Also $\quad dg = vdp - sdT$

Comparing the co-efficients of dT, we get

$$\left(\frac{\partial g}{\partial T}\right)_p = -s$$

Also $\quad h = g + Ts = g - T\left(\frac{\partial g}{\partial T}\right)_p$

Hence $\quad \mathbf{h = g - T\left(\frac{\partial g}{\partial T}\right)_p}$. **(Ans.)**

(*iii*) From eqn. (6.23), we have

$$c_v = T\left(\frac{\partial s}{\partial T}\right)_v \quad ...(i)$$

Also $\quad \left(\frac{\partial a}{\partial T}\right)_v = -s$

or $\quad \left(\frac{\partial s}{\partial T}\right)_v = -\left(\frac{\partial^2 a}{\partial T^2}\right)_v \quad ...(ii)$

From eqns. (*i*) and (*ii*), we get

$$\mathbf{c_v = -T\left(\frac{\partial^2 a}{\partial T^2}\right)_v}. \quad \textbf{(Ans.)}$$

(*iv*) From eqn. (6.26), we have

$$c_p = T\left(\frac{\partial s}{\partial T}\right)_p \quad ...(i)$$

Also $\quad \left(\frac{\partial g}{\partial T}\right)_p = -s$

or $\quad \left(\frac{\partial s}{\partial T}\right)_p = -\left(\frac{\partial^2 g}{\partial T^2}\right)_p \quad ...(ii)$

From eqns. (*i*) and (*ii*), we get

$$\mathbf{c_p = -T\left(\frac{\partial^2 g}{\partial T^2}\right)_p}. \quad \textbf{(Ans.)}$$

Example 6.10. *Find the expression for ds in terms of dT and dp.*

Solution. Let $\quad s = f(T, p)$

Then $\quad ds = \left(\frac{\partial s}{\partial T}\right)_p . dT + \left(\frac{\partial s}{\partial p}\right)_T dp$

As per Maxwell relation (6.21)

$$\left(\frac{\partial s}{\partial p}\right)_T = -\left(\frac{\partial v}{\partial T}\right)_p$$

Substituting this in the above equation, we get

$$ds = \left(\frac{\partial s}{\partial T}\right)_p dT - \left(\frac{\partial v}{\partial T}\right)_p . dp \quad ...(i)$$

The enthalpy is given by

$$dh = c_p dT = Tds + vdp$$

Dividing by dT at constant pressure

$$\left(\frac{\partial h}{\partial T}\right)_p = c_p = T\left(\frac{\partial s}{\partial T}\right)_p + 0$$

(as $dp = 0$ when pressure is constant)

Now substituting this in eqn. (*i*), we get

$$ds = c_p \frac{dT}{T} - \left(\frac{\partial s}{\partial T}\right)_p . dp \quad ...(ii)$$

But $\quad \beta = \frac{1}{v}\left(\frac{\partial v}{\partial T}\right)_p$

Substituting this in eqn. (*ii*), we get

$$ds = c_p \frac{dT}{T} - \beta v dp \quad \textbf{(Ans.)}$$

Example 6.11. *Derive the following relations :*

(*i*) $\left(\frac{\partial T}{\partial p}\right)_s = \frac{Tv\beta}{c_p}$ $\quad$ (*ii*) $\left(\frac{\partial T}{\partial v}\right)_s = -\frac{T\beta}{c_v K}$.

where β = *Co-efficient of cubical expansion, and*
K = *Isothermal compressibility.*

Solution. (*i*) Using the Maxwell relation (6.19), we have

$$\left(\frac{\partial T}{\partial p}\right)_s = \left(\frac{\partial v}{\partial s}\right)_p = \left(\frac{\partial v}{\partial T}\right)_p \left(\frac{\partial T}{\partial s}\right)_p$$

Also $\quad c_p = T\left(\frac{\partial s}{\partial T}\right)_p$

From eqn. (6.34),

$$\beta = \frac{1}{v}\left(\frac{\partial v}{\partial T}\right)_p$$

$$\left(\frac{\partial T}{\partial p}\right)_s = \frac{\beta v T}{c_p}$$

i.e., $\quad \mathbf{\left(\frac{\partial T}{\partial p}\right)_s = \frac{Tvb}{c_p}}$. **(Ans.)**

(*ii*) Using the Maxwell relation (6.18)

$$\left(\frac{\partial T}{\partial v}\right)_s = -\left(\frac{\partial p}{\partial s}\right)_v = -\left(\frac{\partial p}{\partial T}\right)_v \left(\frac{\partial T}{\partial s}\right)_v$$

Also $\quad c_v = T\left(\frac{\partial s}{\partial T}\right)_v \quad$ [Eqn. (6.23)]

$$K = -\frac{1}{v}\left(\frac{\partial v}{\partial p}\right)_T \quad \text{[Eqn. (6.36)]}$$

Then, $\left(\frac{\partial T}{\partial v}\right)_s = -\frac{T}{c_v}\left(\frac{\partial p}{\partial T}\right)_v$

Also, $\left(\frac{\partial p}{\partial v}\right)_T\left(\frac{\partial v}{\partial T}\right)_p\left(\frac{\partial T}{\partial p}\right)_v = -1$

i.e., $\left(\frac{\partial p}{\partial T}\right)_v = -\left(\frac{\partial p}{\partial v}\right)_T\left(\frac{\partial v}{\partial T}\right)_p = -\left(\frac{1}{-vK}\right)\beta v = \frac{\beta}{K}$

$\therefore$ $\left(\frac{\partial T}{\partial v}\right)_s = \frac{-\mathbf{Tb}}{\mathbf{c_v K}}$. **(Ans.)**

☞**Example 6.12.** *Derive the third Tds equation*

$$Tds = c_v\left(\frac{\partial T}{\partial p}\right)_v dp + c_p\left(\frac{\partial T}{\partial v}\right)_p dv$$

and also show that this may be written as :

$$Tds = \frac{c_v}{\beta} Kdp + \frac{c_p}{\beta v} dv.$$

Solution. Let $s = f(p, v)$

Then $ds = \left(\frac{\partial s}{\partial p}\right)_v dp + \left(\frac{\partial s}{\partial v}\right)_p dv$

or $Tds = T\left(\frac{\partial s}{\partial p}\right)_v dp + T\left(\frac{\partial s}{\partial v}\right)_p dv$

$$= T\left(\frac{\partial s}{\partial T}\right)_v\left(\frac{\partial T}{\partial p}\right)_v + T\left(\frac{\partial s}{\partial T}\right)_p\left(\frac{\partial T}{\partial v}\right)_p dv$$

But $\left(\frac{\partial s}{\partial T}\right)_v = \frac{c_v}{T}$ and $\left(\frac{\partial s}{\partial T}\right)_p = \frac{c_p}{T}$

Hence, $\mathbf{Tds = c_v\left(\frac{\partial T}{\partial p}\right)_v dp + c_p\left(\frac{\partial T}{\partial v}\right)_p dv}$ **...Proved.**

Also, $\left(\frac{\partial T}{\partial p}\right)_v = \frac{-1}{\left(\frac{\partial p}{\partial v}\right)_T\left(\frac{\partial v}{\partial T}\right)_p}$

$$= -\left(\frac{\partial T}{\partial v}\right)_p\left(\frac{\partial v}{\partial p}\right)_T = \frac{K}{\beta}$$

and $\left(\frac{\partial T}{\partial v}\right)_p = \frac{1}{\beta v}$

Substituting these values in the above *Tds* equation, we get

$$\mathbf{Tds = \frac{c_v K}{\beta} dp + \frac{c_p}{\beta v} dv} \quad \textbf{...Proved.}$$

Example 6.13. *Using Maxwell relation derive the following Tds equation*

$$Tds = c_p\, dT - T\left(\frac{\partial v}{\partial T}\right)_p dp. \quad \textbf{(U.P.S.C.)}$$

Solution. $s = f(T, p)$

$$Tds = T\left(\frac{\partial s}{\partial T}\right)_p dT + T\left(\frac{\partial s}{\partial p}\right)_T dp \quad ...(i)$$

where $c_p = T\left(\frac{\partial s}{\partial T}\right)_p$

Also, $\left(\frac{\partial s}{\partial p}\right)_T = -\left(\frac{\partial v}{\partial T}\right)_p$ ...... Maxwell relation

Substituting these in eqn. (*i*), we get

$$T_{ds} = \mathbf{c_p\, dT - T\left(\frac{\partial v}{\partial T}\right)_p dp.} \quad \textbf{(Ans.)}$$

Example 6.14. *Derive the following relations:*

$$\left(\frac{\partial T}{\partial v}\right)_u = \frac{T\left(\frac{\partial p}{\partial T}\right)_v - p}{c_v}.$$

Solution. $\left(\frac{\partial T}{\partial v}\right)_u$ can be expressed as follows :

$$\left(\frac{\partial T}{\partial v}\right)_u = \frac{-\left(\frac{\partial T}{\partial u}\right)_v}{\left(\frac{\partial v}{\partial u}\right)_T} = -\frac{\left(\frac{\partial u}{\partial v}\right)_T}{\left(\frac{\partial u}{\partial T}\right)_v}$$

Also $Tds = du + pdv$

or $du = Tds - pdv$

or $\left(\frac{\partial u}{\partial v}\right)_T = T\left(\frac{\partial s}{\partial v}\right)_T - p\left(\frac{\partial u}{\partial v}\right)_T$

or $\left(\frac{\partial u}{\partial v}\right)_T = T\left(\frac{\partial s}{\partial v}\right)_T - p$ ...(*i*)

or $\left(\frac{\partial u}{\partial T}\right)_v = T\left(\frac{\partial s}{\partial T}\right)_v$ ...(*ii*)

Dividing eqn. (*i*) by eqn. (*ii*), we get

$$\left(\frac{\partial T}{\partial v}\right)_u = \frac{T\left(\frac{\partial s}{\partial v}\right)_T - p}{T\left(\frac{\partial s}{\partial T}\right)_v} \quad ...(iii)$$

Also $c_v = T\left(\frac{\partial s}{\partial T}\right)_v$

and $\left(\frac{\partial s}{\partial v}\right)_T = \left(\frac{\partial p}{\partial T}\right)_v$ ... Maxwell relation

Substituting these value in eqn. (*iii*), we get

$$\left(\frac{\partial T}{\partial v}\right)_u = \frac{T\left(\frac{\partial p}{\partial T}\right)_v - p}{c_v} \quad \textbf{...Proved.}$$

Example 6.15. *Prove that for any fluid*

$$(i)\ \left(\frac{\partial h}{\partial v}\right)_T = v\left(\frac{\partial p}{\partial v}\right)_T + T\left(\frac{\partial p}{\partial T}\right)_v$$

$$(ii)\ \left(\frac{\partial h}{\partial p}\right)_T = v - T\left(\frac{\partial v}{\partial T}\right)_p$$

Show that for a fluid obeying van der Waal's equation,

$$p = \frac{RT}{v-b} - \frac{a}{v^2}$$

where R, a and b are constants

$$h \text{ (enthalpy)} = \frac{RTb}{v-b} - \frac{2a}{v} + f(T)$$

where f(T) is arbitrary.

Solution. (*i*) We know that

$$ds = \frac{c_v}{T}dT + \left(\frac{\partial p}{\partial T}\right)_v dv \quad \text{...[Eqn. (6.24)]}$$

Also $\quad dh = Tds + vdp$

$$= T\left[\frac{c_v}{T}dT + \left(\frac{\partial p}{\partial T}\right)_v dv\right] + vdp$$

i.e., $\quad dh = c_v dT + T\left(\frac{\partial p}{\partial T}\right)_v + dv + vdp$

Putting $dT = 0$, we get

$$\left(\frac{\partial \mathbf{h}}{\partial \mathbf{v}}\right)_\mathbf{T} = \mathbf{T}\left(\frac{\partial \mathbf{p}}{\partial \mathbf{T}}\right)_\mathbf{v} + \mathbf{v}\left(\frac{\partial \mathbf{p}}{\partial \mathbf{v}}\right)_\mathbf{T} \quad \text{...Proved.}$$

(*ii*)
$$\left(\frac{\partial h}{\partial p}\right)_T = \left(\frac{\partial h}{\partial v}\right)_T\left(\frac{\partial v}{\partial p}\right)_T$$

$$= \left[T\left(\frac{\partial p}{\partial T}\right)_v + v\left(\frac{\partial p}{\partial v}\right)_T\right]\left(\frac{\partial v}{\partial p}\right)_T$$

i.e.,
$$\left(\frac{\partial h}{\partial p}\right)_T = T\left(\frac{\partial p}{\partial T}\right)_v\left(\frac{\partial v}{\partial p}\right)_T + v \quad \text{...}(i)$$

Also
$$\left(\frac{\partial p}{\partial T}\right)_v\left(\frac{\partial v}{\partial p}\right)_T = -\left(\frac{\partial v}{\partial T}\right)_p$$

∴ Eqn. (*i*) becomes

$$\left(\frac{\partial \mathbf{h}}{\partial \mathbf{p}}\right)_\mathbf{T} = \mathbf{v} - \mathbf{T}\left(\frac{\partial \mathbf{v}}{\partial \mathbf{T}}\right)_\mathbf{p} \quad \text{...Proved.}$$

Now
$$p = \frac{RT}{v-b} - \frac{a}{v^2}$$

$$\left(\frac{\partial p}{\partial v}\right)_T = -\frac{RT}{(v-b)^2} + \frac{2a}{v^3}$$

and
$$\left(\frac{\partial p}{\partial T}\right)_v = \frac{R}{v-b}$$

∴
$$\left(\frac{\partial h}{\partial v}\right)_T = v\left[\frac{-RT}{(v-b)^2} + \frac{2a}{v^3}\right] + T\left(\frac{R}{v-b}\right)$$

$$= -\frac{RTv}{(v-b)^2} + \frac{2a}{v^2} + \frac{RT}{v-b}$$

$$= \frac{-RTv}{(v-b)^2} + \frac{RT}{v-b} + \frac{2a}{v^2}$$

$$= \frac{-RTv + RT(v-b)}{(v-b)^2} + \frac{2a}{v^2}$$

$$= \frac{-RTv + RTv - RTb}{(v-b)^2} + \frac{2a}{v^2}$$

i.e.,
$$\left(\frac{\partial h}{\partial v}\right)_T = \frac{-RTb}{(v-b)^2} + \frac{2a}{v^2}$$

or
$$\mathbf{h} = \frac{\mathbf{RTb}}{\mathbf{v-b}} - \frac{\mathbf{2a}}{\mathbf{v}} + \mathbf{f(T)} \quad \text{...Proved.}$$

This shows h depends on T and v.

Example 6.16. *Derive the following relations :*

$$(i)\ \left(\frac{\partial h}{\partial p}\right)_T = v - T\left(\frac{\partial v}{\partial T}\right)_p = -c_p\left(\frac{\partial T}{\partial p}\right)_h$$

$$(ii)\ \left(\frac{\partial u}{\partial v}\right)_T = T\left(\frac{\partial p}{\partial T}\right)_v - p$$

with the aid of eqn. (ii) show that

$$\left(\frac{\partial u}{\partial p}\right)_T = -T\left(\frac{\partial v}{\partial T}\right)_p - p\left(\frac{\partial v}{\partial p}\right)_T$$

The quantity $c_p\left(\frac{\partial T}{\partial p}\right)_h$ *is known as Joule-Thomson cooling effect. Show that this cooling effect for a gas obeying the equation of state* $(v-b) = \frac{RT}{p} - \frac{C}{T^2}$ *is equal to* $\left(\frac{3C}{T^2}\right) - b$.

Solution. (*i*) We know that

$$\left(\frac{\partial h}{\partial p}\right)_T = -\mu c_p \quad \text{...[Eqn. (6.44)]}$$

Also
$$\mu = \frac{1}{c_p}\left[T\left(\frac{\partial v}{\partial T}\right)_p - v\right] \quad \text{...[Eqn. (6.46)]}$$

∴
$$\left(\frac{\partial h}{\partial p}\right)_T = -\left[T\left(\frac{\partial v}{\partial T}\right)_p - v\right] = v - T\left(\frac{\partial v}{\partial T}\right)_p$$

... Proved.

Also
$$\mu = \left(\frac{\partial T}{\partial p}\right)_h$$

∴
$$\left(\frac{\partial h}{\partial p}\right)_T = -c_p\left(\frac{\partial T}{\partial p}\right)_h.$$

(*ii*) Let $\quad u = f(T, v)$

$$du = \left(\frac{\partial u}{\partial T}\right)_v dT + \left(\frac{\partial u}{\partial v}\right)_T dv$$

$$= c_v\, dT + \left(\frac{\partial u}{\partial v}\right)_T dv \quad \text{...}(i)$$

Also $\quad du = Tds - pdv$

Substituting the value of Tds [from eqn. (6.24)], we get

$$du = c_v\, dT + T\left(\frac{\partial p}{\partial T}\right)_v dv - pdv$$

$$= c_v\, dT + \left[T\left(\frac{\partial p}{\partial T}\right)_v - p\right] dv \quad ...(ii)$$

From eqns. (i) and (ii), we get

$$\left(\frac{\partial u}{\partial v}\right)_T = T\left(\frac{\partial p}{\partial T}\right)_v - p \quad \text{...Proved.}$$

Also $$\left(\frac{\partial u}{\partial p}\right)_T = \left(\frac{\partial u}{\partial v}\right)_T \left(\frac{\partial v}{\partial p}\right)_T$$

or, $$\left(\frac{\partial u}{\partial p}\right)_T = \left(\frac{\partial v}{\partial p}\right)_T \left[T\left(\frac{\partial p}{\partial T}\right)_v - p\right]$$

or, $$\left(\frac{\partial u}{\partial p}\right)_T = T\left(\frac{\partial p}{\partial T}\right)_v \left(\frac{\partial v}{\partial p}\right)_T - p\left(\frac{\partial v}{\partial p}\right)_T$$

...Proved.

We know that $$\left(\frac{\partial p}{\partial v}\right)_T \left(\frac{\partial v}{\partial T}\right)_p \left(\frac{\partial T}{\partial p}\right)_v = -1$$

or, $$\left(\frac{\partial p}{\partial T}\right)_v \left(\frac{\partial v}{\partial p}\right)_T = -\left(\frac{\partial v}{\partial T}\right)_p$$

Also, $$\left(\frac{\partial u}{\partial p}\right)_T = -T\left(\frac{\partial v}{\partial T}\right)_p - p\left(\frac{\partial v}{\partial p}\right)_T$$

... already proved.

and $$\mu = \frac{1}{c_p}\left[T\left(\frac{\partial u}{\partial T}\right)_p - v\right] \quad \text{... [Eqn. (6.46)]}$$

Now, $$v - b = \frac{RT}{p} - \frac{C}{T^2} \quad \text{... [Given]}$$

$$\left(\frac{\partial v}{\partial T}\right)_p = \frac{R}{p} + \frac{2C}{T^3}$$

Substituting this value in the expression of μ above, we get:

$$\mu = \frac{1}{c_p}\left[T\left(\frac{R}{p} + \frac{2C}{T^3}\right) - v\right]$$

or, $$\mu c_p = T\left(\frac{R}{p} + \frac{2C}{T^3}\right) - \frac{RT}{p} + \frac{C}{T^2} - b = \frac{3C}{T^2} - b$$

or, $$c_p\left(\frac{\partial T}{\partial p}\right)_h = \frac{3C}{T^2} - b \quad \text{... Proved.}$$

Example 6.17. *The pressure on the block of copper of 1 kg is increased from 20 bar to 800 bar in a reversible process maintaining the temperature constant at 15°C. Determine the following :*

(i) Work done on the copper during the process,
(ii) Change in entropy,
(iii) The heat transfer,
(iv) Change in internal energy, and
(v) $(c_p - c_v)$ for this change of state.

Given : β (volume expansitivity) = 5 × 10^{-5}/K, K (thermal compressibility) = 8.6 × 10^{-12} m^2/N and v (specific volume) = 0.114 × 10^{-3} m^3/kg.

Solution. (i) **Work done on the copper, W :**

Work done during isothermal compression is given by

$$W = \int_1^2 pdv$$

The isothermal compressibility is given by

$$K = -\frac{1}{v}\left(\frac{\partial v}{\partial p}\right)_T$$

$$\therefore \quad dv = -K(v.dp)_T$$

$$\therefore \quad W = -\int_1^2 pKv.dp = -vK\int_1^2 pdp$$

Since v and K remain essentially constant

$$\therefore \quad W = -\frac{vK}{2}(p_2^2 - p_1^2)$$

$$= -\frac{0.114 \times 10^{-3} \times 8.6 \times 10^{-12}}{2}[(800 \times 10^5)^2 - (20 \times 10^5)^2]$$

$$= -\frac{0.114 \times 8.6 \times 10^{-15}}{2} \times 10^{10}[(800)^2 - (20)^2]$$

$$= -\frac{0.114 \times 8.6 \times 10^{-5}}{2}(640000 - 400)$$

= **– 3.135 J/kg. (Ans.)**

The *negative sign indicates that the work is done on the copper block.*

(ii) **Change in entropy :**

The change in entropy can be found by using the following Maxwell relation :

$$\left(\frac{\partial s}{\partial p}\right)_T = -\left(\frac{\partial v}{\partial T}\right)_p = -\frac{v}{v}\left(\frac{\partial v}{\partial T}\right)_p = -v\beta$$

as $$\frac{1}{v}\left(\frac{\partial v}{\partial T}\right)_p = \beta$$

$$\therefore \quad (ds)_T = -v\beta\,(dp)_T$$

Integrating the above equation, assuming v and β remaining constant, we get

$$\mathbf{s_2 - s_1} = -v\beta\,(p_2 - p_1)_T$$
$$= -0.114 \times 10^{-3} \times 5 \times 10^{-5}[800 \times 10^5 - 20 \times 10^5]$$
$$= -0.114 \times 10^{-3} \times 5\,(800 - 20)$$

= **– 0.446 J/kg K. (Ans.)**

(iii) **The heat transfer, Q :**

For a reversible isothermal process, the heat transfer is given by :

$$\mathbf{Q} = T(s_2 - s_1) = (15 + 273)(-0.4446)$$

= **– 128 J/kg. (Ans.)**

(*iv*) **Change in internal energy, du :**

The change in internal energy is given by :

$$\mathbf{du} = Q - W$$
$$= -128 - (-3.135)$$
$$= \mathbf{-124.8\ J/kg.}\quad \textbf{(Ans.)}$$

(*v*) $\mathbf{c_p - c_v}$ **:**

The difference between the specific heat is given by :

$$\mathbf{c_p - c_v} = \frac{\beta^2 T v}{K} \quad \text{... [Eqn. (6.38)]}$$

$$= \frac{(5 \times 10^{-5})^2 \times (15 + 273) \times 0.114 \times 10^{-3}}{8.6 \times 10^{-12}}$$

$$= \mathbf{9.54\ J/kg\ K.}\quad \textbf{(Ans.)}$$

Example 6.18. *Using Clausius-Claperyon's equation, estimate the enthalpy of vapourisation. The following data is given :*

At 200°C : $v_g = 0.1274\ m^3/kg$; $v_f = 0.001157\ m^3/kg$;

$$\left(\frac{dp}{dT}\right) = 32\ kPa/K.$$

Solution. Using the equation

$$\left(\frac{dp}{dT}\right) = \frac{h_{fg}}{T_s(v_g - v_f)}$$

where, h_{fg} = Enthalpy of vapourisation.

Substituting the various values, we get

$$32 \times 10^3 = \frac{h_{fg}}{(200 + 273)(0.1274 - 0.001157)}$$

$$\therefore \quad \mathbf{h_{fg}} = 32 \times 10^3\ (200 + 273)(0.1274 - 0.001157)\ \text{J}$$

$$= 1910.8 \times 10^3\ \text{J/kg}$$

$$= \mathbf{1910.8\ kJ/kg.}\quad \textbf{(Ans.)}$$

Example 6.19. *An ice skate is able to glide over the ice because the skate blade exerts sufficient pressure on the ice that a thin layer of ice is melted. The skate blade then glides over this thin melted water layer. Determine the pressure an ice skate blade must exert to allow smooth ice skate at – 10°C.*

The following data is given for the range of temperatures and pressures involved :

$h_{fg(ice)} = 334\ kJ/kg$; $v_{liq.} = 1 \times 10\ m^3/kg$;

$v_{ice} = 1.01 \times 10^3\ m^3/kg$.

Solution. Since it is a problem of phase change from solid to liquid, therefore, we can use Clausius-Claperyon equation given below :

$$\frac{dp}{dT} = \frac{h_{fg}}{v_{fg}} \cdot \frac{1}{T}$$

Multiplying both the sides by dT and integrating, we get

$$\int_{p_1}^{p_2} dp = \frac{h_{fg}}{v_{fg}} \int_{T_1}^{T_2} \frac{dT}{T}$$

or $$(p_2 - p_1) = \frac{h_{fg}}{v_{fg}} \log_e \left(\frac{T_2}{T_1}\right) \quad \text{...(i)}$$

But at $p_1 = 1$ atm., $t_1 = 0°C$

Thus, $p_1 = 1.013$ bar, $T_1 = 0 + 273 = 273$ K

$p_2 = ?$, $T_2 = -10 + 273 = 263$ K

Substituting these values in eqn. (*i*), we get

$$(p_2 - 1.013 \times 10^5) = \frac{334 \times 10^3}{(1 - 1.01)} \times \log_e \left(\frac{263}{273}\right)$$

$$= \frac{334 \times 10^3}{0.01} \times \log_e \left(\frac{273}{263}\right)$$

$$= 12.46 \times 10^5\ \text{N/m}^2$$

or $$p_2 = 12.46 \times 10^5 + 1.013 \times 10^5$$

$$= 13.47 \times 10^5\ \text{N/m}^2 \text{ or } \mathbf{13.47\ bar.}\quad \textbf{(Ans.)}$$

This pressure is considerably high. It can be achieved with ice skate blade by having only a small portion of the blade surface in contact with the ice at any given time. If *the temperature drops lower than – 10°C, say – 15°C, then it is not possible to generate sufficient pressure to melt the ice and conventional ice skating will not be possible.*

Example 6.20. *For mercury, the following relation exists between saturation pressure (bar) and saturation temperature (K) :*

$$\log_{10} p = 7.0323 - 3276.6/T - 0.652 \log_{10} T$$

Calculate the specific volume v_g *of saturation mercury vapour at 0.1 bar.*

Given that the latent heat of vapourisation at 0.1 bar is 294.54 kJ/kg.

Neglect the specific volume of saturated mercury liquid.

Solution. Latent heat of vaporisation,

$$h_{fg} = 294.54\ \text{kJ/kg (at 0.1 bar)} \quad \text{...(given)}$$

Using Clausius-Claperyon equation

$$\frac{dp}{dT} = \frac{h_{fg}}{v_{fg}T} = \frac{h_{fg}}{(v_g - v_f)T} \quad \text{...(i)}$$

Since v_f is neglected, therefore eqn. (*i*) becomes

$$\frac{dp}{dT} = \frac{h_{fg}}{v_g T}$$

Now, $$\log_{10} p = 7.0323 - \frac{3276.6}{T} - 0.652 \log_{10} T$$

Differentiating both sides, we get

$$\frac{1}{2.302p} \cdot \frac{dp}{dT} = \frac{3276.6}{T^2} - \frac{0.652}{2.302\,T}$$

$$\frac{dp}{dT} = 2.302 \times 3276.6 \times \frac{p}{T^2} - 0.652\ \frac{p}{T} \quad \text{...(ii)}$$

From eqns. (*i*) and (*ii*), we have

or $$\frac{h_{fg}}{v_g T} = 2.302 \times 3276.6 \times \frac{p}{T^2} - 0.652\ \frac{p}{T} \quad \text{...(iii)}$$

We know that

$$\log_{10} p = 7.0323 - \frac{3276.6}{T} - 0.652 \log_{10} T \quad \text{...(Given)}$$

At $p = 0.1$ bar,

$$\log_{10} (0.1) = 7.0332 - \frac{3276.6}{T} - 0.652 \log_{10} T$$

$$-1 = 7.0323 - \frac{3276.6}{T} - 0.652 \log_{10} T$$

or $$0.652 \log_{10} T = 8.0323 - \frac{3276.6}{T}$$

or $$\log_{10} T = 12.319 - \frac{5025.4}{T}$$

Solving by hit and trial method, we get

$$T = 523 \text{ K}$$

Substituting this value in eqn. (*iii*), we get

$$\frac{294.54 \times 10^3}{v_g \times 523} = 2.302 \times 3276.6 \times \frac{0.1 \times 10^5}{(523)^2} - 0.652 \times \frac{0.1 \times 10^5}{523}$$

$$\frac{563.17}{v_g} = 275.75 - 12.46$$

i.e., $\mathbf{v_g = 2.139}$ **m³/kg. (Ans.)**

HIGHLIGHTS

1. Maxwell relations are given by

$$\left(\frac{\partial T}{\partial v}\right)_s = -\left(\frac{\partial p}{\partial s}\right)_v \; ; \; \left(\frac{\partial T}{\partial p}\right)_s = \left(\frac{\partial v}{\partial s}\right)_p$$

$$\left(\frac{\partial p}{\partial T}\right)_v = \left(\frac{\partial s}{\partial v}\right)_T \; ; \; \left(\frac{\partial v}{\partial T}\right)_p = -\left(\frac{\partial s}{\partial p}\right)_T .$$

2. The specific heat relations are

$$c_p - c_v = \frac{vT\beta^2}{K} \; ; c_v = T\left(\frac{\partial s}{\partial T}\right)_v \; ; c_p = T\left(\frac{\partial s}{\partial T}\right)_p .$$

3. Joule-Thomson co-efficient is expressed as

$$\mu = \left(\frac{\partial T}{\partial p}\right)_h .$$

4. Entropy equations (*Tds* equations) :

$$Tds = c_v dT + T\left(\frac{\partial p}{\partial T}\right)_v dv \quad \text{...(1)}$$

$$Tds = c_p dT - T\left(\frac{\partial v}{\partial T}\right)_p dp \quad \text{...(2)}$$

5. Equations for *internal energy* and *enthalpy* :

$$\left(\frac{\partial u}{\partial v}\right)_T = T\left(\frac{\partial p}{\partial T}\right)_v - p \quad \text{...(1)}$$

$$du = c_v dT + \left\{T\left(\frac{\partial p}{\partial T}\right)_v - p\right\} dv \quad \text{...[1 (a)]}$$

$$\left(\frac{\partial h}{\partial p}\right)_T = v - T\left(\frac{\partial v}{\partial T}\right)_p \quad \text{...(2)}$$

$$dh = c_p dT + \left\{v - T\left(\frac{\partial v}{\partial T}\right)_p\right\} dp \quad \text{...[2 (a)]}$$

OBJECTIVE TYPE QUESTIONS

Choose the Correct Answer :

1. The specific heat at constant pressure (c_p) is given by

(*a*) $c_p = T\left(\frac{\partial s}{\partial T}\right)_p$ (*b*) $c_p = T\left(\frac{\partial T}{\partial s}\right)_p$

(*c*) $c_p = T\left(\frac{\partial v}{\partial T}\right)_p$ (*d*) $c_p = T\left(\frac{\partial v}{\partial T}\right)_p$.

2. The specific heat relation is

(*a*) $(c_p - c_v) = \frac{vT\beta^2}{K}$ (*b*) $(c_p - c_v) = \frac{vTK}{\beta^2}$

(*c*) $(c_p - c_v) = \frac{pTK}{\beta^2}$ (*d*) $(c_p - c_v) = \frac{v^2T\beta}{K}$.

3. The relation of internal energy is :

(*a*) $du = \left(\frac{K}{\beta}c_v\right) dp + \left(\frac{c_p}{v\beta} - p\right) dv$

(*b*) $du = \left(\frac{K}{\beta}c_v\right) dp + \left(\frac{c_p}{v\beta} + p\right) dv$

(*c*) $du = \left(\frac{K}{\beta}c_p\right) dp + \left(\frac{c_p}{v\beta} - v\right) dv$

(*d*) $du = \left(\frac{K}{\beta}c_p\right) dp + \left(\frac{c_v}{v\beta} - p\right) dv$.

4. *Tds* equation is :

(*a*) $Tds = c_v dT + \frac{T\beta}{K} dv$ (*b*) $Tds = c_p dT - \frac{T\beta}{K} dv$

(*c*) $Tds = c_v dT + \frac{TK}{\beta} dv$ (*d*) $Tds = c_v dT + \frac{T\beta}{K} dp$.

ANSWERS

1. (*a*) 2. (*a*) 3. (*a*) 4. (*a*).

EXERCISES

1. Define the co-efficient of :
   (*i*) Volume expansion
   (*ii*) Isothermal compressibility
   (*iii*) Adiabatic compressibility.
2. Derive the Maxwell relations and explain their importance in thermodynamics.
3. Show that the equation of state of a substance may be written in the form

$$\frac{dv}{v} = -Kdp + \beta dT.$$

4. A substance has the volume expansivity and isothermal compressibility :

$$\beta = \frac{1}{T} \; ; K = \frac{1}{p}$$

Find the equation of state. $\left[\textbf{Ans. } \frac{pv}{T} = \text{constant} \right]$

5. For a perfect gas, show that the difference in specific heats is $c_p - c_v = \frac{R}{T}$.

6. For the following given differential equations,

$$du = Tds - pdv$$

and $$dh = Tds + vdp$$

prove that for perfect gas equation,

$$\left(\frac{\partial u}{\partial p}\right)_T = 0 \quad \text{and} \quad \left(\frac{\partial h}{\partial p}\right)_T = 0.$$

7. Using the cyclic equation, prove that

$$\left(\frac{\partial p}{\partial T}\right)_v = \frac{\beta}{KT} .$$

8. Prove that the change in entropy is given by

$$ds = \frac{c_v}{T}\left[\frac{KT}{\beta} . dp + \frac{c_p}{\beta v}\right] dv.$$

9. Deduce the following thermodynamic relations :

(*i*) $\left(\frac{\partial h}{\partial p}\right)_T = v - T\left(\frac{\partial v}{\partial T}\right)_p = -c_p\left(\frac{\partial T}{\partial p}\right)_h$

(*ii*) $\left(\frac{\partial u}{\partial v}\right)_T = T\left(\frac{\partial p}{\partial T}\right)_v - p.$

10. Show that for a van der Waals gas

$$c_p - c_v = \frac{R}{1 - 2a\,(v - b)^2 / RTv^3} .$$

11. A gas obeys $p(v - b) = RT$, where b is positive constant. Find the expression for the Joule-Thomson co-efficient of this gas. Could this gas be cooled effectively by throttling ?

12. The pressure on the block of copper of 1 kg is increased from 10 bar to 1000 bar in a reversible process maintaining the temperature constant at 15°C. Determine :

(*i*) Work done on the copper during the process

(*ii*) Change in entropy

(*iii*) The heat transfer

(*iv*) Change in internal energy

(*v*) $(c_p - c_v)$ for this change of state.

The following data may be assumed :

Volume expansivity (β) = 5×10^{-5}/K

Isothermal compressibility (K) = 8.6×10^{-12} m²/N

Specific volume (v) = 0.114×10^{-3} m³/kg

[**Ans.** (*i*) – 4.9 J/kg ; (*ii*) – 0.57 J/kg K ; (*iii*) – 164 J/kg ; (*iv*) – 159.1 J/kg ; (*v*) 9.5 J/kg K]

Ideal and Real Gases

7

7.1. INTRODUCTION

An *'ideal gas'* is defined as a *gas having no forces of intermolecular attraction*. The gases which *follow the gas laws at all ranges of pressures and temperatures* are considered as *"ideal gases"*. However, *'real gases' follow these laws at low pressures or high temperatures or both*. This is because the forces of attraction between molecules *tend to be very small* at reduced pressures and elevated temperatures.

An *ideal gas* obeys the law $pv = RT$. The specific heat capacities *are not constant but are functions of temperature*. A *perfect gas* obeys the law $pv = RT$ and has *constant specific heat capacities*.

A perfect gas is well suited to mathematical manipulation and is therefore a most useful model to use for analysis of practical machinery which uses real gases as a working substance.

In *reality there is no ideal or perfect gas*. At a very low pressure and at a very high temperature, real gases like hydrogen, oxygen, nitrogen, helium etc. behave nearly the same way as perfect gases. These gases are called *semi-perfect* or *permanent gases*. The term semi-perfect has the implication that the behaviour of the gases are nearly the same as that of a perfect gas. The term *'permanent'* was used for these gases by earlier chemists who thought that these gases did not change their phase (*i.e.*, did not condense to a liquid state). Hence they are called permanent gases. There is no gas which does not change phase, and there is no permanent gas in the real sense. However, these gases can be changed into a liquid phase only if they are subjected to a great decrease in temperature and increase in pressure.

All gases behave in nearly in a similar way, especially at pressures considerably lower than the critical pressure, and at temperatures above the critical temperature. *The relation between the independent properties, such as pressure, specific volume and temperature for a pure substance* is known as the *'equation of state'*. For engineering calculations, the equation of state for perfect gases can be used for real gases so long as the pressures are well below their critical pressure and the temperatures are above the critical temperature.

7.2. THE EQUATION OF STATE FOR A PERFECT GAS

Boyle's law. It states that *volume of a given mass of a perfect gas varies inversely as the absolute pressure when temperature is constant.*

If p is the absolute pressure of the gas and V is the volume occupied by the gas, then

$$V \propto \frac{1}{p}$$

or pV = Constant, so long as the temperature is constant ...(7.1)

Fig. 7.1 shows the graphical representation of Boyle's law. The curves are rectangular hyperbolas asymptotic to the p-v axis. Each curve corresponds to a different temperature. For any two points on the curve,

$$\frac{p_1}{p_2} = \frac{V_2}{V_1} \qquad ...(7.2)$$

Charle's law. It states that *if any gas is heated at constant pressure, its volume changes directly as its absolute temperature.*

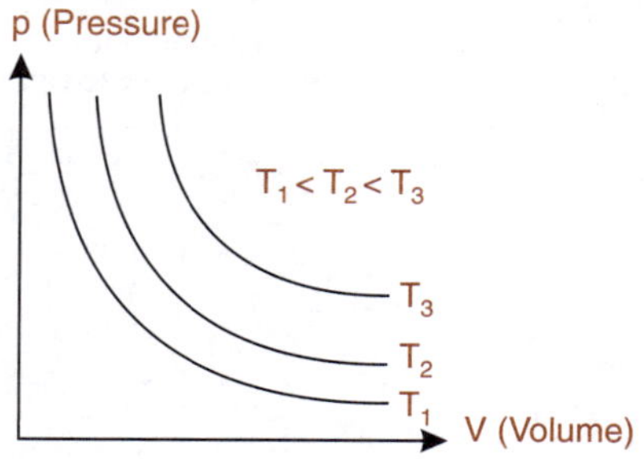

Fig. 7.1. *p-V* relation of a perfect gas at constant temperature.

In other words, $V \propto T$

or $\frac{V}{T}$ = Constant, so long as pressure is constant ...(7.3)

If a gas changes its volume from V_1 to V_2 and absolute temperature from T_1 to T_2 without any change of pressure, then

$$\frac{V_1}{T_1} = \frac{V_2}{T_2} \quad ...(7.4)$$

Fig. 7.2 gives the graphical representation of Charle's law.

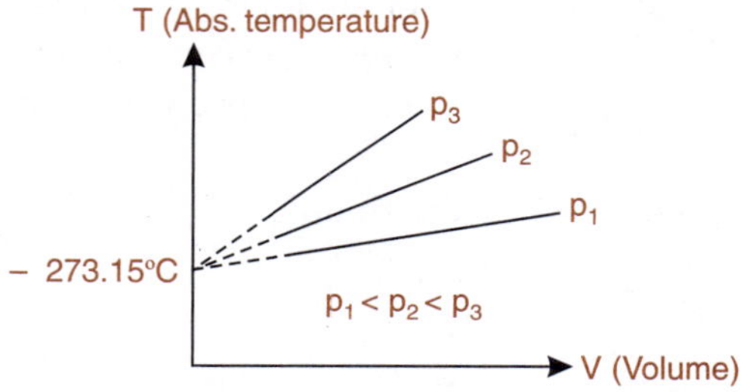

Fig. 7.2. *T-V* relation of a perfect gas constant pressure.

To derive the *equation of state* for a perfect gas let us consider a *unit mass* of a perfect gas to change its state in the following two successive processes (Fig. 7.3)

(*i*) Process 1-2′ at constant pressure, and

(*ii*) Process 2′-2 at constant temperature.

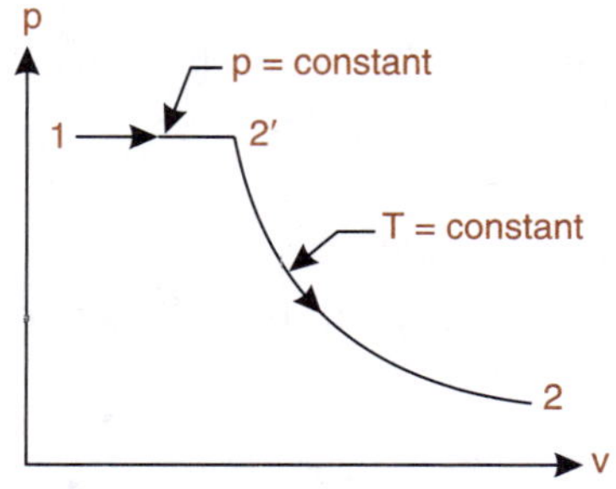

Fig. 7.3. Formulation of equation of state of a perfect gas.

For the *process 1-2′*, applying Charle's law

$$\frac{v_1}{T_1} = \frac{v_2'}{T_2'}$$

and since $T_2' = T_2$, we may write

$$\frac{v_1}{T_1} = \frac{v_2'}{T_2} \quad ...(i)$$

For the *process 2′-2*, using Boyle's law

$$p_2'v_2' = p_2v_2$$

and since $p_2' = p_1$

$$p_1v_2' = p_2v_2$$

i.e., $$v_2' = \frac{p_2v_2}{p_1} \quad ...(ii)$$

Substituting the value of v_2' from eqn. (*ii*) in eqn. (*i*), we get

$$\frac{v_1}{T_1} = \frac{p_2v_2}{p_1T_2}$$

or $$\frac{p_1v_1}{T_1} = \frac{p_2v_2}{T_2}$$

i.e., $$\frac{pv}{T} = \text{constant} \quad ...(7.5)$$

The magnitude of this constant depends upon the particular gas and it is denoted by R, where R is called the **specific gas constant**. Then

$$\frac{pv}{T} = R$$

The equation of the state for a perfect gas is thus given by the equation

$$pv = RT \quad ...(7.6)$$

or for m kg, occupying V m^3,

$$pV = mRT \quad ...(7.7)$$

If the mass is chosen to be numerically equal to the molecular weight of the gas then 1 mole of the gas has been considered, *i.e.*, 1 kg mole of oxygen is 32 kg oxygen, or 1 kg mole of hydrogen is 2 kg hydrogen.

The equation may be written as

$$pV_0 = MRT \quad ...(7.8)$$

where V_0 = Molar volume, and

M = Molecular weight of the gas.

Avogadro discovered that V_0 *is the same for all gases at the same pressure and temperature* and therefore it may be seen that MR = a constant ; R_0 and thus

$$pV_0 = R_0T \quad ...(7.9)$$

R_0 is called the **molar or universal gas constant** and its value is **8.3143 kJ/kg mol K.**

If there are n moles present then the ideal gas equation may be written as

$$pV = nR_0T \quad ...(7.10)$$

where V is the volume occupied by n moles at pressure p and temperature T.

7.3. p-v-T SURFACE OF AN IDEAL GAS

The equation of state of an ideal gas is a relationship between the variables pressure (p), volume (v) and temperature (T). On plotting these variables along three mutually perpendicular axes, we get a surface which represents the equation of state ($pv = RT$). Such a surface is called *p-v-T surface*. These surfaces represent the fundamental properties of a substance and *provide a tool to study the thermodynamic properties and processes of that substance.* Fig. 7.4 shows a portion of a p-v-T surface for an *ideal gas. Each point on this surface represents an equilibrium state and a line on the surface represents a process*. The Fig. 7.4 also shows the constant pressure, constant volume and constant temperature lines.

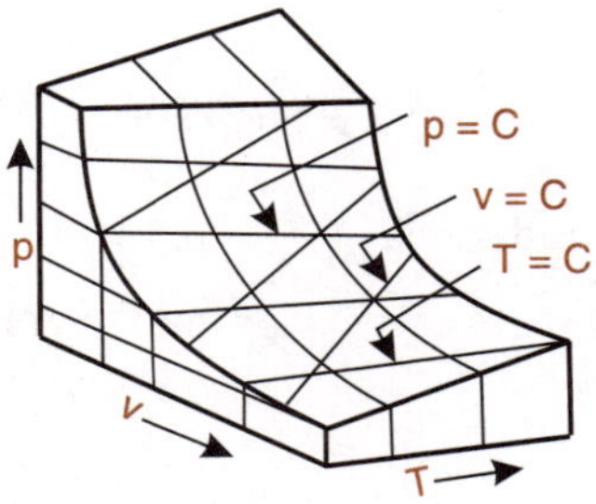

Fig. 7.4

7.4. INTERNAL ENERGY AND ENTHALPY OF A PERFECT GAS

Joule's Law. *Joule's law states that the specific internal energy of a gas depends only on the temperature of the gas and is independent of both pressure and volume.*

i.e., $u = f(T)$

Joule concluded this result from a series of experiments conducted with an apparatus similar to the one shown in Fig. 7.5.

— Two tanks connected by a valve were submerged in a bath of water.

— Initially one tank was evacuated and the other was filled with air under high pressure.

— A thermometer was placed in the water bath.

— After the tank and water had attained the same temperature, the valve between the two tanks was opened to pass air slowly from high pressure tank to the evacuated tank. Time was allowed for equilibrium to be attained.

Joule observed that there was *no change in temperature of water during or after the process*. Since there was no change in the temperature of water, he concluded that there *was no heat transfer between air and water i.e.,* $\delta Q = 0$. And since there was no work during the process, *i.e.,* $\delta W = 0$, from the first law of thermodynamics, $\delta Q = dE + \delta W$, Joule *concluded that change in internal energy of the air is zero, i.e., $dE = 0$.*

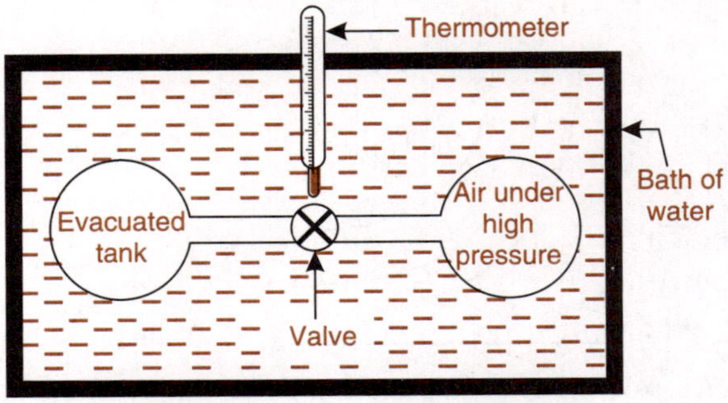

Fig. 7.5. Apparatus for demonstration of Joule's law.

Again, since both pressure and volume changed during the process, he remarked that *internal energy was a function only of temperature ; since during the process temperature did not change, the internal energy remained constant*.

Later on when experiments were conducted with more refined instruments, it was found that there was a *very small change in temperature of water,* indicating that *for real gases internal energy was not a function of temperature alone.* However, at low pressure and high temperature where real gases behave like semi-perfect gases and where the equation of state for a semi-perfect gas, $pv = RT$, is sufficiently accurate, Joule's law holds equally good in that range.

From definition of enthalpy,

$$h = u + pv$$

Also $pv = RT$

$$\therefore \quad h = u + RT \quad \text{...(7.11)}$$

Since u is a function of temperature only, h is a function of temperature,

$$i.e., \quad h = f(T) \quad \text{...(7.12)}$$

7.5. SPECIFIC HEAT CAPACITIES OF AN IDEAL GAS

The specific heat capacity at constant volume of any substance is defined by

$$c_v = \left(\frac{\partial u}{\partial T}\right)_v$$

It may be seen that as Joule's law for an ideal gas states $u = f(T)$, then

$$c_v = \frac{du}{dT} \quad \text{...(7.13)}$$

Since $h = u + pv$, Boyle's law, $pV = f(T)$ and Joule's law $u = f(T)$ together show, $h = f(T)$ and by similar argument to the above it may be seen that :

$$c_p = \frac{dh}{dT} \quad \text{...(7.14)}$$

Further as $h = u + pv$, then $h = u + RT$ and by differentiation

$$\frac{dh}{dT} = \frac{du}{dT} + R$$

Substitution from eqns. (7.13) and (7.14) gives,

$$c_p = c_v + R \quad i.e., \quad c_p - c_v = R \qquad ...(7.15)$$

If expressed in terms of molar quantities then eqn. (7.15) becomes

$$C_p - C_v = R_0 \qquad ...(7.16)$$

where C_p and C_v are *molar specific heat capacities.*

Equations for specific heat capacities of ideal gases

Since both u and h are functions of temperature, the equations to c_p and c_v must also be functions of temperature. They are usually expressed in a form :

$$c_p = a + KT + K_1T^2 + K_2T^3 \qquad ...(7.17)$$

$$c_v = b + KT + K_1T^2 + K_2T^3 \qquad ...(7.18)$$

where a, b, K, K_1 and K_2 are constants. Values of specific enthalpy etc. are then obtained by integration.

7.6. REAL GASES

It has been observed that when experiments are performed at relatively low pressures and temperatures most of the real gases obey Boyle's and Charle's laws quite closely. But the actual behaviour of real gases at elevated pressures and at low temperatures deviates considerably.

The ideal gas equation $pv = RT$ can be derived analytically using the kinetic theory of gases by making the following **assumptions** :

(*i*) A finite volume of gas contains large number of molecules.

(*ii*) The collision of molecules with one another and with the walls of the container are perfectly elastic.

(*iii*) The molecules are separated by large distances compared to their own dimensions.

(*iv*) The molecules do not exert forces on one another except when they collide.

As long as the above assumptions are valid the behaviour of a real gas approaches closely that of an ideal gas.

7.7. VAN DER WAALS' EQUATION

Van der Waals' equation (for a real gas) may be written as :

$$\left(p + \frac{a}{v^2}\right)(v - b) = RT \qquad ...[7.19\ (a)]$$

The constants a and b are specific constants and depend upon the type of the fluid considered, 'v' represents the volume per unit mass and R is the gas constant.

If the volume of one mole is considered then the above equation can be written as:

$$\left(p + \frac{a}{\bar{v}^{2}}\right)(\bar{v} - b) = R_0T \qquad ...[7.19\ (b)]$$

The units of p, $\bar{v}$, T, R, a and b are as follows :

p (N/m^2), $\bar{v}$ (m^3/kg-mol), T (K) and R = 8314 Nm/kg-mol K, a [Nm^4/(kg-mol)^2], b (m^3/kg mol).

Table 7.1. Constants of van der Waals' Equation

S. No.	*Substance*	*a* *Nm⁴/(kg-mol)²*	*b* *m³/kg-mol*
1.	Hydrogen (H_2)	25105	0.0262
2.	Oxygen (O_2)	139250	0.0314
3.	Carbon dioxide (CO_2)	362850	0.0423
4.	Helium (He)	3417620	0.0228
5.	Air	135522	0.0362
6.	Water (H_2O) vapour	551130	0.0300
7.	Mercury (Hg) vapour	2031940	0.0657

Van der Waals equation was proposed in 1873 for the gaseous and liquid states of a fluid, and accounts qualitatively for many important properties, but quantitatively it fails in many particulars.

The characteristic equation for a perfect gas is obtained by neglecting the finite size of the molecules. If this be taken into account it is obvious that the equation must be modified, for the distance travelled by a molecule between two successive encounters will be less than if the molecules were point spheres. Let the average distance traversed by a molecule between two successive encounters be denoted by λ, the mean free path. In Fig. 7.6 suppose L and M to be the two molecules of diameter 'd' at a distance λ apart. If these molecules were to impinge along the line of centres the path moved over would be less by an amount 'd' than if the molecules were point spheres. Now all the encounters between molecules are not direct, so their mean free paths will be lessened by an amount kd, where k is a fraction. That is, the mean free path is diminished in the ratio $(\lambda - kd) : \lambda$ or $\left(1 - \frac{kd}{\lambda}\right) : 1$.

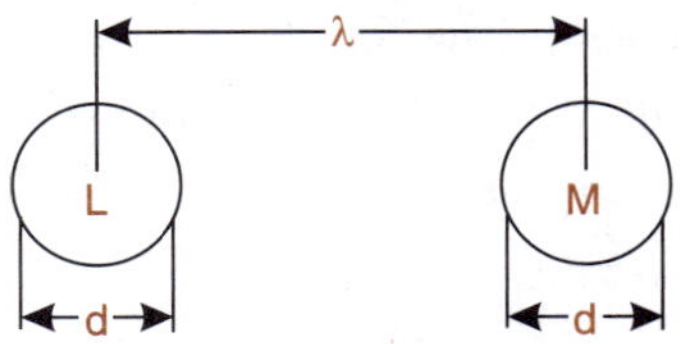

Fig. 7.6

If the mean free path is lessened in this ratio, the encounters per second will be increased in the ratio $1 : 1 - \frac{kd}{\lambda}$. But the pressure of the gas depends upon the encounters per second with the wall of the containing vessel. Hence the new pressure is given by

$$p = \frac{1}{3}\rho\overline{C}^2 \cdot \frac{1}{1 - \dfrac{kd}{\lambda}} \qquad ...(7.20)$$

where ρ is the density and $\overline{C}$ is the average velocity.

The mean free path is inversely proportional to the density of the gas, for if the volume were halved, *i.e.*, the density doubled, there would be twice as many molecules in the same space, and therefore any molecule would only have to travel approximately half as far before encountering another molecule. Hence writing v for $\frac{1}{\rho}$ and $\frac{b}{v}$ for $\frac{kd}{\lambda}$ in eqn. (7.20), we get

$$pv\left(1 - \frac{b}{v}\right) = \frac{\overline{C}^2}{3} = RT$$

or $$p(v - b) = RT \qquad ...(7.21)$$

Next consider the *forces of cohesion* which act between a molecule and those surrounding it. When the molecule is sufficiently far removed from the surface of the gas in all directions the resultant of these cohesives forces are equally probable, as the individual forces are varying continuously as the surrounding molecules change their positions. Hence if the resultant is averaged over a sufficient length of time the aggregate force will be nil. *This is not true, however, when the molecule is near the surface*. Let the force from each molecule be resolved into normal and tangential components. All directions for the resultant in the tangential plane are equally likely, but the resultant normal component is most often directed inwards. Averaged over a sufficient length of time the total resultant force will therefore be a normal force always directed inwards. Thus the average effect of the cohesive forces is the same as if there was a permanent field of force acting at and near the surface. This field of force can be regarded as exerting a pressure p_1 over the boundary of the gas. The pressure is proportional to the number of molecules per unit area near the boundary surface and to the normal component of the force. Both of these factors are proportional to the density, so p_1 will be proportional to the square of the density.

i.e., $$p_1 = a\rho^2 \qquad ...(7.22)$$

where a is a constant.

Hence the molecules are not deflected by impact alone on reaching the boundary, but as the total result of their impact and of the action of the supposed field of force. That is, their change of momentum may be supposed to be produced by a total pressure $p + p_1$ instead of by the simple pressure p.

Hence eqn. (7.21) now becomes :

$$(p + p_1)(v - b) = RT,$$

or $$\left(p + \frac{a}{v^2}\right)(v - b) = RT$$

by substitution from p_1 from (7.22) and replacing ρ^2 by $\frac{1}{v^2}$.

Evaluation of constants *a* and *b*

The general form of the isothermals for carbon dioxide given by van der Waals' equation is shown in Fig. 7.7. These curves are obtained from the equation,

$$\left(p + \frac{0.00874}{v^2}\right)(v - 0.0023) = \frac{1.00646}{273}\,T \qquad ...(7.23)$$

where the unit of pressure is the atmosphere, and the unit of volume that of the gas at 0°C under one atmosphere pressure.

van der Waals' equation being a cubic in v has three roots which may be either all real, or two imaginary and one real, as imaginary roots always occur in pairs. In Fig. 7.7, the 40°C isothermal corresponds to the first condition, and the other isothermals to the latter. There is one isothermal

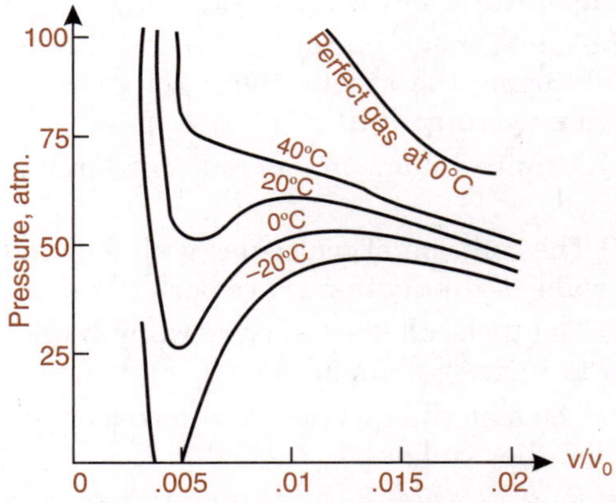

Fig. 7.7. Van der Waals' Isothermal for CO_2.

where there are three real coincident roots at a point of inflexion. All the isothermals for temperatures higher than that corresponding to the isothermal with three real coincident roots have no horizontal tangent, and all those lower have a maximum and minimum. Consequently this curve is identified with the critical isothermal. The temperature of the critical isothermal is obtained in the following manner. Equation (7.19) may be written

$$v^3 - \left(b + \frac{RT}{p}\right)v^2 + \frac{av}{p} - \frac{ab}{p} = 0 \qquad ...(7.24)$$

$$\left(p + \frac{a}{v^2}\right)(v - b) = RT$$

$$= pv - pb + \frac{a}{v^2} \times v - \frac{a}{v^2} \times b - RT = 0$$

$$= pv - pb + \frac{a}{v} - \frac{ab}{v^2} - RT = 0$$

Multiplying both sides by $\frac{v^2}{p}$, we get

$$pv \times \frac{v^2}{p} - pb \times \frac{v^2}{p} + \frac{a}{v} \times \frac{v^2}{p} - \frac{ab}{v^2} \times \frac{v^2}{p} - \frac{RT}{p}v^2 = 0$$

$$v^3 - \left(b + \frac{RT}{p}\right)v^2 + \frac{av}{p} - \frac{ab}{p} = 0$$

Now at the critical point, as the three roots are equal, the equation must be of the form :

$$(v - v_c)^3 = 0 \qquad ...(7.25)$$

where the suffix c denotes conditions at the critical point. For the critical point equation (7.24) becomes

$$v^3 - \left(b + \frac{RT_c}{p_c}\right)v^2 + \frac{av}{p_c} - \frac{ab}{p_c} = 0 \qquad ...(7.26)$$

Equations (7.25) and (7.26) are identical, hence equating coefficients

$$3v_c = b + \frac{RT_c}{p_c},$$

$$3v_c^2 = \frac{a}{p_c},$$

$$v_c^3 = \frac{ab}{p_c},$$

and from these by a simple reduction, we have

$$\left.\begin{aligned} v_c &= 3b \\ p_c &= \frac{a}{27b^2} \\ T_c &= \frac{8}{27} \cdot \frac{a}{bR} \end{aligned}\right\} \qquad ...(7.27)$$

From these equations it follows that *the critical volume, pressure and temperature are all completely determined by the constants of equation (7.19).*

The equation (7.27) indicates the critical constants for a particular gas and leads to the following results :

The values of a and b are also given by

$$a = 3p_c v_c^2 = \frac{9}{8} RT_c v_c = \frac{27}{64} \cdot \frac{R^2T_c^2}{p_c} \qquad ...(i)$$

$$b = \frac{v_c}{3} = \frac{RT_c}{8p_c} \qquad ...(ii)$$

and $$R = \frac{8}{3} \frac{p_c v_c}{T_c} \qquad ...(iii)$$

Using the values of a, b and R in equation (7.23), and substituting in (7.26), we have for carbon dioxide

p_c = 61.2 atmospheres,

T_c = 305.3 K or 32.2°C.

It is frequently assumed that the approximate agreement between the calculated and experimental values of the critical temperature for carbon dioxide is a sufficient verification of Van der Waals' theory, but the constant b cannot be calculated with the required degree of accuracy from Regnault's experiments to make this an adequate test of the theory.

Also from equation (7.27), we have

$$\frac{p_c v_c}{RT_c} = \frac{3}{8} = 0.375$$

whereas experiment shows that about 0.27 as the average value of this ratio, varying considerably, however, from gas to gas.

The Reduced Equation

When the pressure, volume and temperature of the fluid are expressed as fractions of the critical pressure, volume and temperature the reduced form of van der Waals' equation is obtained. Thus, writing

$$p = ep_c = \frac{ea}{27b^2},$$

$$v = nv_c = 3nb,$$

$$T = mT_c = \frac{8}{27} \cdot \frac{ma}{bR}$$

and substituting these values in eqn. (7.19), this reduces to

$$\left(e + \frac{3}{n^2}\right)(3n - 1) = 8m$$

In this "*reduced*" equation the three constants which characterised a particular fluid have disappeared. The equation is accordingly true of any substance which satisfies an equation of the van der Waals type, and the form of the curves connecting e, n and m is the same for all these substances. Thus we see that two substances, the behaviour of each of which is represented by van der Waals' equation, will be in corresponding states when the pressure, volume and temperature are the same multiples of their critical values.

This *theorem of corresponding states*, enunciated by van der Waals, was tested by Amagat and found to be approximately true for a large number of fluids. The theorem of corresponding states is not unique to the equation of van der Waals. Any equation of state giving a critical point and having not more than three constants will serve equally well to give a reduced equation, in which the constants peculiar to anyone fluid disappear, and therefore become the basis of the theorem of corresponding states.

It must be remembered in applying the theorem that the accuracy of results deduced by its aid cannot be greater than the accuracy with which the original equation represents the behaviour of the fluids under consideration.

Amagat's Experiments

As per Amagat's experiments Van der Waals' equation accounts for the variation of the product pv with increasing pressure as follows.

Writing equation (7.19) in the form

$$pv = \frac{RTv}{v-b} - \frac{a}{v},$$

and differentiating with respect to p, keeping T constant, we have

$$\left(\frac{d(pv)}{dp}\right)_T = \left\{\frac{a}{v^2} - \frac{RTb}{(v-b)^2}\right\}\left(\frac{dv}{dp}\right)_T \quad ...(7.28)$$

Since the condition for a minimum on any isothermal is

$$\left(\frac{d(pv)}{dp}\right)_T = 0,$$

the right-hand side of equation (7.28) must vanish at this point. Now $\left(\frac{dv}{dp}\right)_T$ is *never zero,* so we have as the condition for a *minimum* :

$$\frac{RTb}{(v-b)^2} = \frac{a}{v^2} \quad \text{or} \quad RT \cdot \frac{b}{a} = \left(1 - \frac{b}{v}\right)^2 \quad ...(7.29)$$

This equation shows that the *volume at which the minimum value of pv occurs on any isothermal gradually increases as the temperature is raised.*

To find the locus of minima the temperature T must be eliminated from equation (7.29) by substitution from the original equation. Thus from equation (7.19)

$$RT = \left(p + \frac{a}{v^2}\right)(v - b),$$

and substituting this in equation (7.28), we have

$$\frac{b}{a}\left(p + \frac{a}{v^2}\right)(v-b) = \left(1 - \frac{b}{v}\right)^2,$$

which reduces to $v = b\left(\frac{pv^2}{a} + 2\right)$

Multiply each side of this equation by p, and put $pv = y$ and $p = x$, and we obtain

$$y = b\left(\frac{y^2}{a} + 2x\right) \quad \text{or} \quad y(a - by) = 2abx$$

The above expression gives the locus of minima and is a parabola with axis parallel to the x-axis as shown in Fig. 7.8.

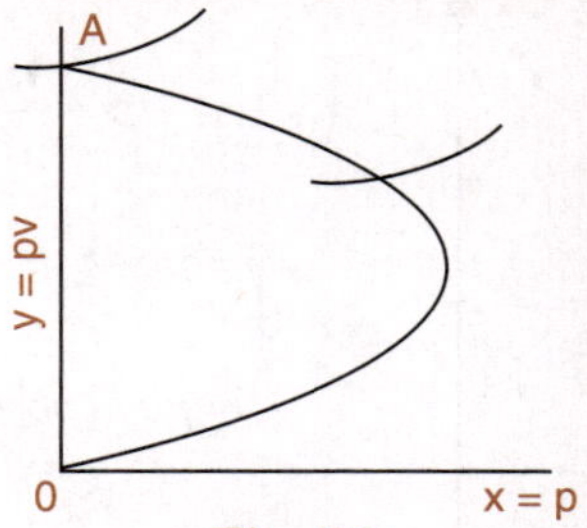

Fig. 7.8

Consider the isothermal which goes through the point A.

Here $x = 0$ and $y = \frac{a}{b}$.

Writing Van der Waals' equation in terms of x and y, we have

$$\left(1 + \frac{ax}{y^2}\right)(y - bx) = RT,$$

and substituting the coordinates of the point A

$$RT = \frac{a}{b}$$

or

$$T = \frac{a}{bR} \quad ...(7.30)$$

For temperatures above that given by equation (7.30) the minima lie in the region of negative pressure, so an Amagat isothermal for a temperature equal to or greater than $\frac{a}{bR}$ will slope upwards along its whole length for increasing values of p, but for a temperature less than $\frac{a}{bR}$ the isothermals first dip to a minimum and then rise.

Using the result from equation (7.27)

$$T_c = \frac{8}{27}\frac{a}{bR},$$

we see that the limiting temperature for an isothermal to show a *minimum* is

$$T = \frac{27}{8} \cdot T_c$$

The reason for Amagat finding no dip in the isothermals for hydrogen is now apparent. The critical temperature is 35 K, and therefore the limiting temperature above which minima do not occur is $\frac{27}{8} \times 35 = 118.1$ K or – 155°C, and all Amagat's experiments were conducted between 0°C and 100°C.

The Cooling Effect

The most gases show an inversion of the cooling effect at a certain temperature. The equation of van der Waals indicates at what temperature this occurs.

We have

$$\left(p + \frac{a}{v^2}\right)(v - b) = RT \quad \text{[From eqn. (7.15)]}$$

Keeping p constant and differentiating with respect to T, we get

$$\left\{p - \frac{a}{v^2} + \frac{2ab}{v^3}\right\}\left(\frac{dv}{dT}\right)_p = R$$

or

$$\left(\frac{dv}{dT}\right)_p = \frac{R}{\left\{p - \frac{a}{v^2} + \frac{2ab}{v^3}\right\}}$$

Substituting this value of $\left(\frac{dv}{dT}\right)_p$ in the equation

$c_p\mu = T\left(\frac{dv}{dT}\right)_p - v$ (where μ is a measure of cooling effect), we get

$$c_p\mu = \frac{RT}{\left\{p - \frac{a}{v^2} + \frac{2ab}{v^3}\right\}} - v$$

and substituting for RT from equation (7.19) this reduces to

$$c_p\mu = \frac{-bp + \frac{2a}{v} - \frac{3ab}{v^2}}{p - \frac{a}{v^2} + \frac{2ab}{v^3}}$$

The denominator of this expression is *always positive*, since it is $R\left(\frac{dT}{dv}\right)_p$. Hence the cooling effect, μ, is *positive* if

$$bp < \frac{2a}{v} - \frac{3ab}{v^2} \qquad \text{...(7.31)}$$

and *negative* if

$$bp > \frac{2a}{v} - \frac{3ab}{v^2} \qquad \text{...(7.32)}$$

and *inversion occurs* when

$$bp = \frac{2a}{v} - \frac{3ab}{v^2}$$

or

$$p = \frac{a}{b}\left(\frac{2}{v} - \frac{3b}{v^2}\right) \qquad \text{...(7.33)}$$

In order to get the temperature of inversion this equation must be combined with the original equation. Thus

$$\frac{2a}{b}\left(1 - \frac{b}{v}\right)^2 = RT \qquad \text{...(7.34)}$$

Since v is necessarily always greater than b, it will be seen that as *v increases so also does the temperature of inversion*.

The form of curve given by equation (7.33) is shown in Fig. 7.9. The pressure is zero when $v = \frac{3b}{2}$, or infinity. These values of v determine the limiting values of the temperature of inversion, as it is only between these limits that p is positive. Substituting these limits of v in equation (7.34) the limiting values of the temperature of inversion are $\frac{2}{9}\frac{a}{bR}$ and $\frac{2a}{bR}$, or from (7.27), $\frac{3}{4}T_c$ and $\frac{27}{4}T_c$.

The equation (7.33) being quadratic there are two values of v for a constant value of p at which inversion occurs, as may also be seen by reference to Fig. 7.9. Consequently by equation (7.34) there are two temperatures for a constant value of p at which inversion occurs. As the temperature increases through the lower of these values the change is from a heating to a cooling effect, and as it increases through the higher of these values the change is from a cooling to a heating effect.

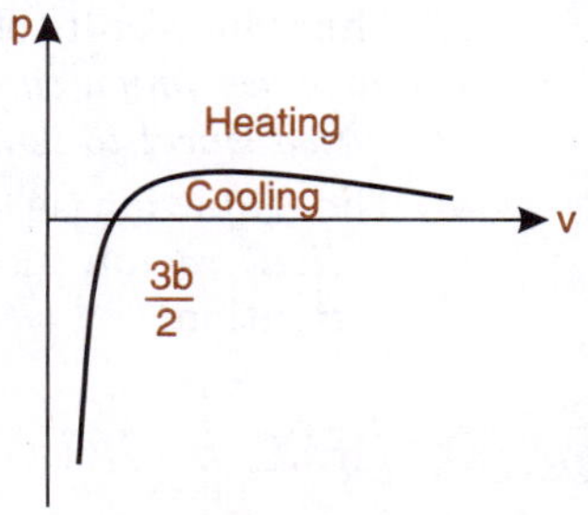

Fig. 7.9

The inversion will occur when the maximum value of p is $\frac{a}{3b^2}$, when $v = 3b$. For any value of p less than this there is a cooling effect provided the condition of the substance is represented by a point inside the area enclosed by the curve and the axis of volume, Fig. 7.9, and for any greater value of p there is a heating effect as indicated by equations (7.31) and (7.32) respectively.

Let us take the case of *hydrogen*. In the experiments of Joule and Thomson the pressure used was 4.7 atmospheres. The critical temperature and pressure are 35 K and 15 atmospheres. From equation (7.33) we can find the values of $\frac{b}{v}$ corresponding to the pressure used by Joule and Thomson, and by substitution in equation (7.34) find the two temperatures at which inversion occurs at this pressure. Equation (7.33) can be written as :

$$p = 27p_c\left[2\frac{b}{v} - 3\left(\frac{b}{v}\right)^2\right]$$

Hence

$$\frac{b}{v} = \frac{2 \pm \sqrt{4 - \frac{12p}{27\,p_c}}}{6} = 0.6608 \text{ or } 0.0058$$

by substitution of the above values for p and p_c.

Writing equation (7.34) in the form

$$T = \frac{27}{4}T_c\left(1 - \frac{b}{v}\right)^2,$$

we have by substitution for $\frac{b}{v}$: T = 233.5 K or 27.2 K
that is, below – 245.9°C there would be a *heating effect*, between – 245.9°C and – 39.6°C *a cooling effect*, and above – 39.6°C a heating effect. Thus **Van der Waals' equation qualitatively accounts for the heating effect observed at ordinary temperatures.**

Limitations of Van der Waals' Equation

Van der Waals' equation under actual condition becomes *invalid* as discussed below :

— The values of a and b (which are assumed to be constant) are found to vary with temperature. Thus the results obtained from the equation are *incorrect when the variation of a and b is large with respect to temperature.*

— The equation is *not accurate enough in the critical region* and it is also obvious from its derivation.

7.8. VIRIAL EQUATION OF STATE

The virial (a Latin word used for force which refers to interaction forces between molecules) equation of state may be expressed as follows :

$$\frac{pv}{RT} = A_0 + A_1 p + A_2 p^2 + A_3 p^3 + \ldots\ldots \quad \ldots(7.35)$$

or

$$\frac{pv}{RT} = B_0 + \frac{B_1}{v} + \frac{B_2}{v^2} + \frac{B_3}{v^3} + \ldots\ldots \quad \ldots(7.36)$$

where A_0, A_1, ... and B_0, B_1, ... are called the *virial co-efficients which are functions of temperature only.*

— The virial equation can be used only for gases at low and medium densities.

— The *advantage* of virial equation is that the *virial co-efficients can be determined from experimental p-v-T data.*

7.9. BEATTIE-BRIDGEMAN EQUATION

Beattie-Bridgeman equation is expressed as follows :

$$p = \frac{R_0 T(1-e)}{(\bar{v})^2}(\bar{v} + B) - \frac{A}{(\bar{v})^2} \quad \ldots(7.37)$$

where p = pressure

$$A = A_0\left(1 - \frac{a}{\bar{v}}\right)$$

$$B = B_0\left(1 - \frac{b}{\bar{v}}\right)$$

and $$e = \frac{c}{\bar{v}T^3}$$

The factors A_0, a, B_0, b and c are constants whose values for different gases are given in Table 7.2.

— This equation is normally used for substances at pressures less than critical pressure.

— The equation is accurate enough when the volumes involved are greater than twice the critical volume.

— The equation fits the data of fourteen gases down to the critical point and over a wide range of pressure within ± 0.5% error. However, it is *inaccurate near critical point.*

Table 7.2. Constants of the Beattie-Bridgeman Equation of State

Gas	A_0	a	B_0	b	$c \times 10^{-4}$
Hydrogen (H_2)	20.0117	– 0.00506	0.02096	– 0.04359	0.0504
Oxygen (O_2)	151.0857	0.02562	0.04624	0.004208	4.80
Carbon dioxide (CO_2)	507.2836	0.07132	0.10476	0.07235	66.0
Helium (He)	2.1886	0.05984	0.01400	0.0	0.0040
Air	131.8441	0.01931	0.04611	– 0.001101	4.34
Nitrogen	136.2315	0.02617	0.05046	– 0.00691	4.20
Argon	130.7802	0.02328	0.03931	0.0	5.99

7.10. REDUCED PROPERTIES

The ratios of pressure, temperature and specific volume of a real gas to the corresponding critical values are called the **reduced properties**.

$$p_r = \frac{p}{p_c},\ T_r = \frac{T}{T_c},\ v_r = \frac{v}{v_c} \quad \ldots(7.38)$$

Table 7.3. Critical Constants

Substance	*Pressure (p_c) bar*	*Temperature (T_c) K*
Air	37.69	132.5
Argon	48.64	151.0
Carbon dioxide	73.86	304.2
Carbon monoxide	34.96	133.0
Helium	2.29	5.3
Hydrogen	12.97	33.3
Nitrogen	33.94	126.2
Oxygen	50.76	154.8
Water	228.59	647.15
Ethane	48.84	305.5
Ethylene	51.17	282.4
Methane	46.41	191.1
Propane	42.55	370.0

Table 7.4. Properties of Gases

Gas	*Molecular weight (M)*	c_p *(kJ/kg K)*	c_v *(kJ/kg K)*	$R = c_p - c_v$ *(kJ/kg K)*	$\gamma = \frac{c_p}{c_v}$	$Z_c = \frac{p_c \bar{v}}{R_0 T_c}$
Air	28.97	1.005	0.718	0.287	1.4	0.284
Oxygen	32	0.920	0.660	0.260	1.4	0.307
Nitrogen	28	1.046	0.754	0.292	1.39	0.291
Hydrogen	2	14.40	10.40	4.0	1.38	0.304
Carbon monoxide	28	1.046	0.754	0.292	1.39	—
Carbon dioxide	44	0.840	0.650	0.190	1.3	0.274
Water	18	—	—	0.462	—	0.230
Methane	16	2.22	1.70	0.520	1.3	—
Sulphur dioxide	64	0.796	0.67	0.126	1.19	0.268
Ammonia	17	—	—	0.488	—	—

7.11. LAW OF CORRESPONDING STATES

If any two gases have equal values of reduced pressure and reduced temperature, then they have same values of reduced volume ; i.e., $v_R = f(T_r, p_r)$ for all gases and the function is the same.

This law is most accurate in the *vicinity of the critical point*.

7.12. COMPRESSIBILITY CHART

The *compressibility factor* (Z) of any gas is a function of only two properties, usually temperature and pressure, so that $Z = f(T_r, p_r)$ except near the critical point. The value of Z for any real gas may be less or more than unity, depending on pressure and temperature conditions of the gas.

The general compressibility chart is plotted with Z versus p_r for various values of T_r. This is constructed by plotting the known data of one or more gases and can be used for any gas. Such a chart is shown in Fig. 7.10. This chart gives best results for the regions well removed from the critical state for all gases.

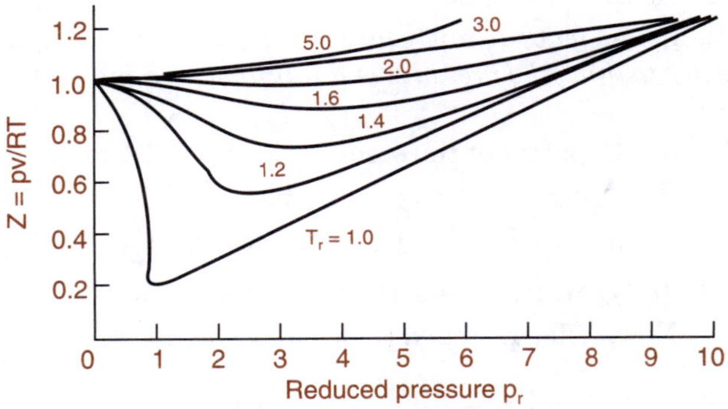

Fig. 7.10. Generalised compressibility chart.

IDEAL GASES

Example 7.1. *The volume of a high altitude chamber is 40 m³. It is put into operation by reducing pressure from 1 bar to 0.4 bar and temperature from 25°C to 5°C.*

How many kg of air must be removed from the chamber during the process ? Express this mass as a volume measured at 1 bar and 25°C.

Take R = 287 J/kg K for air.

Solution. $V_1 = 40\ m^3$ $\quad V_2 = 40\ m^3$

$p_1 = 1$ bar $\quad p_2 = 0.4$ bar

$T_1 = 25 + 273 = 298$ K $\quad T_2 = 5 + 273 = 278$ K

kg of air to be removed :

Assuming nitrogen to be a perfect gas,

$$p_1V_1 = m_1RT_1 \qquad \left(\text{or } m_1 = \frac{p_1V_1}{RT_1}\right)$$

$$p_2V_2 = m_2RT_2 \qquad \left(\text{or } m_2 = \frac{p_2V_2}{RT_2}\right)$$

Mass of air removed during the process

$$= (m_1 - m_2) \text{ kg}$$

$$(m_1 - m_2) = \frac{p_1V_1}{RT_1} - \frac{p_2V_2}{RT_2}$$

$$= \frac{1}{R}\left(\frac{p_1V_1}{T_1} - \frac{p_2V_2}{T_2}\right)$$

$$= \frac{1}{287}\left[\frac{(1 \times 10^5) \times 40}{298} - \frac{(0.4 \times 10^5) \times 40}{278}\right]$$

$= \mathbf{26.71\ kg.}$ **(Ans.)**

Volume of this mass of gas at 1 bar and 25°C is given by

$$V = \frac{mRT}{p} = \frac{26.71 \times 287 \times 298}{1 \times 10^5}$$

$= \mathbf{22.84\ m^3.}$ **(Ans.)**

Example 7.2. *A steel flask of 0.04 m³ capacity is to be used to store nitrogen at 120 bar, 20°C. The flask is to be protected against excessive pressure by a fusible plug which will melt and allow the gas to escape if the temperature rises too high.*

(i) How many kg of nitrogen will the flask hold at the designed conditions ?

(ii) At what temperature must the fusible plug melt in order to limit the pressure of a full flask to a maximum of 150 bar ?

Solution. Capacity of the steel flask, $V = 0.04\ m^3$

Pressure, $p = 120$ bar

Temperature, $T = 20 + 273 = 293$ K

(*i*) **kg of nitrogen the flask can hold :**

Now, R for nitrogen (molecular weight, $M = 28$)

$$= \frac{R_0}{M} = \frac{8314}{28} = 296.9 \text{ J/kg K}$$

Assuming nitrogen to be a perfect gas, we get

Mass of nitrogen in the flask at designed condition

$$= m = \frac{pV}{RT} = \frac{120 \times 10^5 \times 0.04}{296.9 \times 293}$$

$= \mathbf{5.51\ kg.}$ **(Ans.)**

(*ii*) **Temperature at which fusible plug should melt, t :**

When the fusible plug is about to melt

$$p = 150 \text{ bar} ;\ V = 0.04\ m^3 ;\ m = 5.51 \text{ kg}$$

Therefore, temperature t at which fusible plug must melt is given by:

or $$T = \frac{pV}{mR} = \frac{150 \times 10^5 \times 0.04}{5.51 \times 296.9} = 366.7 \text{ K}$$

or $\mathbf{t} = 366.7 - 273 = \mathbf{93.7°C.}$ **(Ans.)**

Example 7.3. *A balloon of spherical shape 6 m in diameter is filled with hydrogen gas at a pressure of 1 bar abs. and 20°C. At a later time, the pressure of gas is 94 per cent of its original pressure at the same temperature :*

(i) What mass of original gas must have escaped if the dimensions of the balloon is not changed ?

(ii) Find the amount of heat to be removed to cause the same drop in pressure at constant volume.

Solution. Diameter of the spherical balloon = 6 m

Pressure of hydrogen gas,

$$p_1 = 1 \text{ bar abs.}$$

Temperature of hydrogen gas,

$$T_1 = 20°C \text{ or } 293 \text{ K}$$

At a later time pressure of the gas,

$$p_2 = 0.94p_1 \text{ at } 293 \text{ K.}$$

(*i*) **Mass of original gas escaped :**

$$\Delta m = m_1 - m_2$$

[where m_1 and m_2 are the initial and final masses of the gas]

$$= \frac{p_1V_1}{RT_1} = \frac{p_2V_2}{RT_2} = \frac{V_1}{RT_1}(p_1 - p_2)$$

$[\because\ V_1 = V_2,\ T_1 = T_2 \text{ and } p_2 = 0.94p_1]$

$$= \frac{V_1}{RT_1}(p_1 - 0.94p_1) = \frac{p_1V_1}{RT_1}(1 - 0.94)$$

$\therefore$ **%age mass escaped** $= \dfrac{\Delta m}{m_1} \times 100$

$$= \frac{\frac{p_1V_1}{RT_1}(1 - 0.94)}{\frac{p_1V_1}{RT_1}} = \mathbf{6\%.} \text{ } \textbf{(Ans.)}$$

(*ii*) **Amount of heat to be removed :**

Using the gas equation,

$$\frac{p_1V_1}{T_1} = \frac{p_2V_2}{T_2}$$

or $$\frac{p_1}{T_1} = \frac{0.94p_1}{T_2} \quad (\because \; V_1 = V_2 \text{ and } p_2 = 0.94p_1)$$

$\therefore$ $$T_2 = 0.94T_1 = 0.94 \times 293$$
$$= 275.4 \text{ K or } 2.42°\text{C}$$

The heat to be removed is given by

$$Q = mc_v(T_1 - T_2)$$

where $$m = \frac{p_1V_1}{RT_1} = \frac{1 \times 10^5 \times \frac{4}{3}\pi \times 3^3}{\frac{8314}{2} \times 293} = 9.28 \text{ kg}$$

$$\left[\begin{array}{l} \because \quad MR = 8314 \\ \therefore \quad R = \dfrac{8314}{2} \\ \text{as } M \text{ for } H_2 = 2 \end{array}\right]$$

$c_v = 10400$ J/kg K for H_2

$\therefore$ Q (heat to be removed)

$= 9.28 \times 10400\ (293 - 275.4) =$ **1.69 MJ. (Ans.)**

Example 7.4. *A vessel of capacity 3 m*^{3} *contains 1 kg mole of* N_2 *at 90°C.*

(*i*) *Calculate pressure and the specific volume of the gas.*

(*ii*) *If the ratio of specific heats is 1.4, evaluate the values of* c_p *and* c_v.

(*iii*) *Subsequently, the gas cools to the atmospheric temperature of 20°C ; evaluate the final pressure of gas.*

(*iv*) *Evaluate the increase in specific internal energy, the increase in specific enthalpy, increase in specific entropy and magnitude and sign of heat transfer.*

Solution. Mass of N_2, $m = 1$ kg mole *i.e.,* 28 kg

Capacity of the vessel, $V_1 = 3$ m^3

Temperature, $T_1 = 90 + 273 = 363$ K

(*i*) **Pressure (p_1) and specific volume (v_1) of the gas :**

Using the relation

$$p_1V_1 = mRT_1$$

$$p_1 \times 3 = 28 \times \left(\frac{8314}{28}\right) \times 363$$

$$\left[\because \; R = \frac{R_0}{M} = \frac{8314}{28}\right]$$

$\therefore$ $\mathbf{p_1 = 1005994}$ **J/m² or 10.06 bar. (Ans.)**

Specific volume, $$\mathbf{v_1} = \frac{V_1}{m} = \frac{3}{28}$$

$= \mathbf{0.107}$ **m³/kg. (Ans.)**

(*ii*) $\mathbf{c_p = ?,\ c_v = ?}$

$$\frac{c_p}{c_v} = 1.4 \text{ (given)} \quad ...(i)$$

But $$c_p - c_v = R = \frac{8314}{28} \quad ...(ii)$$

Solving for c_p and c_v between eqns. (*i*) and (*ii*)

$\mathbf{c_p = 1.039}$ **kJ/kg K ;**

$\mathbf{c_v = 0.742}$ **kJ/kg K. (Ans.)**

(*iii*) **Final pressure of the gas after cooling to 20°C :**

Initially	*After cooling*
$p_1 = 10.06$ bar	$p_2 = ?$
$V_1 = 3$ m^3	$V_2 = 3$ m^3
$T_1 = 363$ K	$T_2 = 20 + 273 = 293$ K

Now, $$\frac{p_1V_1}{T_1} = \frac{p_2V_2}{T_2}$$

or $$\frac{p_1}{T_1} = \frac{p_2}{T_2} \quad (\text{as } V_1 = V_2)$$

$\therefore$ $$\mathbf{p_2} = \frac{p_1T_2}{T_1} = \frac{10.06 \times 293}{363}$$

$=$ **8.12 bar. (Ans.)**

(*iv*) **Δu, Δh, Δs, Q :**

For a perfect gas,

Increase in specific internal energy

$$\mathbf{\Delta u} = c_v(T_2 - T_1) = 0.742(293 - 363)$$
$= \mathbf{-\ 51.94}$ **kJ/kg. (Ans.)**

Increase in specific enthalpy,

$$\mathbf{\Delta h} = c_p(T_2 - T_1) = 1.039(293 - 363)$$
$= \mathbf{-\ 72.73}$ **kJ/kg. (Ans.)**

Increase in specific entropy,

$$\mathbf{\Delta s} = c_v \log_e\left(\frac{T_2}{T_1}\right) + R \log_e\left(\frac{v_2}{v_1}\right)$$

But, $v_1 = v_2$

$\therefore$ $$\Delta s = c_v \log_e\left(\frac{T_2}{T_1}\right) = 0.742 \log_e\left(\frac{293}{363}\right)$$

$= \mathbf{-\ 0.1589}$ **kJ/kg K. (Ans.)**

Now, $Q = \Delta u + W$

Here $W = 0$ as change in volume is zero

$\therefore$ $Q = \Delta u$

$\therefore$ **Heat transfer, Q** $= -\ 51.94$ kJ/kg

$= -\ 51.94 \times 28$

$= \mathbf{-\ 1454.32}$ **kJ. (Ans.)**

Example 7.5. (*a*) *1 kg of air at a pressure of 8 bar and a temperature of 100°C undergoes a reversible polytropic process following the law* $pv^{1.2}$ *= constant. If the final pressure is 1.8 bar, determine :*

(*i*) *The final specific volume, temperature and increase in entropy ;*

(*ii*) *The work done and the heat transfer.*

Assume $R = 0.287$ *kJ/kg K and* $\gamma = 1.4$.

(*b*) *Repeat (a) assuming the process to be irreversible and adiabatic between end states.*

Solution. (*a*) Mass of air, $m = 1$ kg

Pressure, $p_1 = 8$ bar

Temperature, $T_1 = 100 + 273 = 373$ K

The law followed : $pv^{1.2}$ = constant

Final pressure, $p_2 = 1.8$ bar

Characteristic gas constant, $R = 0.287$ kJ/kg K

Ratio of specific heats, $\gamma = 1.4$

(*i*) $\mathbf{v_2}$, $\mathbf{T_2}$ **and** $\mathbf{\Delta s}$ **:**

Assuming air to be a perfect gas,

$$p_1 v_1 = RT_1$$

$$\therefore \quad v_1 = \frac{RT_1}{p_1} = \frac{(0.287 \times 1000) \times 373}{8 \times 10^5}$$

$$= 0.1338 \text{ m}^3/\text{kg}$$

Also, $p_1 v_1^{1.2} = p_2 v_2^{1.2}$

or
$$\frac{v_2}{v_1} = \left(\frac{p_1}{p_2}\right)^{1/1.2}$$

or
$$v_2 = v_1 \left(\frac{p_1}{p_2}\right)^{1/1.2} = 0.1338 \left(\frac{8}{1.8}\right)^{1/1.2}$$

$$= 0.4637 \text{ m}^3/\text{kg}$$

i.e., Final specific volume,

$\mathbf{v_2 = 0.4637}$ **m³/kg. (Ans.)**

Again, $p_2 v_2 = RT_2$

$$T_2 = \frac{p_2 v_2}{R} = \frac{1.8 \times 10^5 \times 0.4637}{(0.287 \times 1000)} = 290.8 \text{ K}$$

i.e., Final temperature,

$\mathbf{t_2}$ = 290.8 – 273 = **17.8°C. (Ans.)**

Increase in entropy Δs is given by,

$$\Delta s = c_v \log_e \left(\frac{T_2}{T_1}\right) + R \log_e \left(\frac{v_2}{v_1}\right)$$

But
$$\gamma = \frac{c_p}{c_v} = 1.4 \text{ (given)} \quad ...(i)$$

and
$$c_p - c_v = R \text{ (= 0.287 kJ/kg K for air)} \quad ...(ii)$$

Solving for c_v between (*i*) and (*ii*),

$$c_v = 0.717 \text{ kJ/kg K}$$

$$\therefore \quad \Delta s = 0.717 \log_e \left(\frac{290.8}{373}\right) + 0.287 \log_e \left(\frac{0.4637}{0.1338}\right)$$

$$= -0.1785 + 0.3567 = 0.1782 \text{ kJ/kg K}$$

i.e., Increase in entropy, $\mathbf{\Delta s}$ = **0.1782 kJ/kg K. (Ans.)**

(*ii*) **Work done and heat transfer :**

The work done in a polytropic process is given by,

$$W = \frac{p_1 v_1 - p_2 v_2}{n - 1} = \frac{R(T_1 - T_2)}{n - 1}$$

$$= \frac{0.287\,(373 - 290.8)}{(1.2 - 1)} = 117.96 \text{ kJ/kg}$$

i.e., **Work done = 117.96 kJ/kg. (Ans.)**

Heat transfer, $Q = \Delta u + W$

where $\Delta u = c_v(T_2 - T_1)$

$$= 0.717\,(290.8 - 373) = -58.94 \text{ kJ/kg}$$

$$\therefore \quad Q = -58.94 + 117.96 = 59.02 \text{ kJ/kg}$$

Hence **heat transfer = 59.02 kJ/kg. (Ans.)**

(*b*) (*i*) Though the process is assumed now to be irreversible and adiabatic, the end states are given to be the same as in (*a*). Therefore, **all the properties at the end of the process are the same as in (a). (Ans.)**

(*ii*) As the process is adiabatic, Q (heat transfer) **= 0. (Ans.)**

$\Delta u = \Delta u$ in (*a*)

Applying first law for this process

$$Q = \Delta u + W$$

$$0 = \Delta u + W$$

or
$$W = -\Delta u$$

$$= -(-58.94) = 58.94$$

$\therefore$ **Work done = 58.94 kJ/kg. (Ans.)**

Example 7.6. *Two spheres each 2.5 m in diameter are connected to each other by a pipe with a valve as shown in Fig. 7.11. One sphere contains 16 kg of air and other 8 kg of air when the valve is closed. The temperature of air in both sphere is 25°C. The valve is opened and the whole system is allowed to come to equilibrium conditions. Assuming there is no loss or gain of energy, determine the pressure in the spheres when the system attains equilibrium.*

Neglect the volume of the pipe.

Solution. Volume of each sphere

$$= \frac{4}{3}\pi R^3 = \frac{4}{3}\pi \times \left(\frac{2.5}{2}\right)^3 = 8.18 \text{ m}^3$$

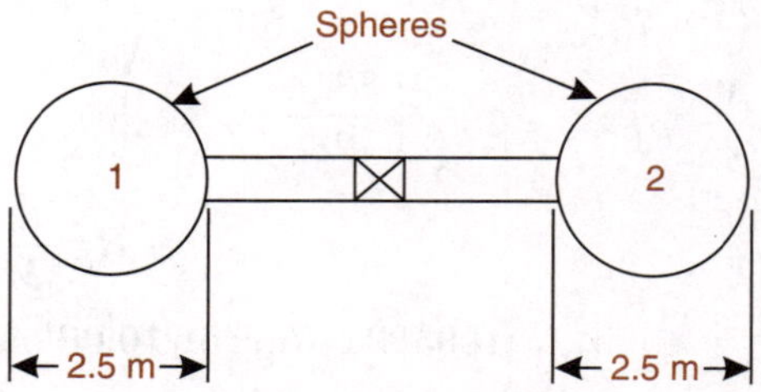

Fig. 7.11

The temperature in both spheres is same (25°C)

i.e., $T_1 = T_2 = 25 + 273 = 298$ K

As no energy exchange occurs, the temperature reached after equilibrium is 298 K.

Mass of air in sphere 1, $m_1 = 16$ kg

Mass of air in sphere 2, $m_2 = 8$ kg

After opening the valve

Total volume, $V = 8.18 + 8.18 = 16.36 \text{ m}^3$

Total mass, $m = m_1 + m_2$

$= 16 + 8 = 24$ kg

Now using characteristic gas equation

$$pV = mRT$$

$$\therefore \quad p = \frac{mRT}{V} = \frac{24 \times 287 \times 298}{16.36}$$

$= 1.255 \times 10^5 \text{ N/m}^2$ or 1.255 bar

Hence **pressure in the spheres when the system attains equilibrium**

= 1.255 bar. (Ans.)

Example 7.7. *CO_2 flows at a pressure of 10 bar and 180°C into a turbine, located in a chemical plant, and there it expands reversibly and adiabatically to a final pressure of 1.05 bar. Calculate the final specific volume, temperature and increase in entropy. Neglect changes in velocity and elevation.*

If the mass flow rate is 6.5 kg/min. evaluate the heat transfer rate from the gas and the power delivered by the turbine.

Assume CO_2 to be a perfect gas and $c_v = 0.837$ kJ/kg K.

Solution. *At entry to turbine* — *At exit of turbine*

Pressure, $p_1 = 10$ bar — Pressure, $p_2 = 1.05$ bar

Temperature, $T_1 = 180 + 273 = 453$ K

Since the expansion is reversible and adiabatic, therefore, the equation pv^γ = constant is applicable.

$$\therefore \quad p_1 v_1^{\gamma} = p_2 v_2^{\gamma} \quad ...(i)$$

Eliminating v_1 and v_2 using the perfect gas equation

$$v = \frac{RT}{p}$$

We can write equation (*i*) as

$$\frac{T_1}{T_2} = \left(\frac{p_1}{p_2}\right)^{(\gamma - 1)/\gamma}$$

$$\therefore \quad \frac{453}{T_2} = \left(\frac{10}{1.05}\right)^{(\gamma - 1)/\gamma} \quad ...(ii)$$

$c_v = 0.837$ kJ/kg K (given)

$$R = \frac{R_0}{M} = \frac{8.314}{44}$$

(Molecular weight of $CO_2 = 44$)

$= 0.1889$ kJ/kg K

Also $c_p - c_v = R$

$\therefore \quad c_p - 0.837 = 0.1889$

$c_p = 1.0259$ kJ/kg K

$$\therefore \quad \gamma = \frac{c_p}{c_v} = \frac{1.0259}{0.837} = 1.23$$

Substituting for γ in equation (*ii*)

$$\frac{453}{T_2} = \left(\frac{10}{1.05}\right)^{(1.23 - 1)/1.23}$$

$\therefore \quad T_2 = 297$ K

Final temperature

$= 297 - 273 =$ **24°C. (Ans.)**

$$p_2 v_2 = RT_2$$

$\therefore \quad 1.05 \times 10^5 \times v_2 = (0.1889 \times 1000) \times 297$

$$\therefore \quad v_2 = \frac{(0.1889 \times 1000) \times 297}{1.05 \times 10^5}$$

$= 0.5343 \text{ m}^3\text{/kg}$

i.e., **Final specific volume**

= 0.5343 m³/kg. (Ans.)

As the process is reversible and adiabatic

$$\Delta s = 0$$

i.e., **Increase in entropy = 0. (Ans.)**

Since the process is adiabatic, therefore, **heat transfer rate from turbine = 0. (Ans.)**

Applying steady flow energy equation (S.F.E.E.) on unit time basis,

$$\dot{m}\left[h_1 + \frac{C_1^2}{2} + Z_1\right] + \dot{Q} = \dot{m}\left[h_2 + \frac{C_2^2}{2} + Z_2\right] + W$$

By data changes in velocity and elevation are negligible, and $Q = 0$.

$\therefore$ S.F.E.E. reduces to

i.e., $W = \dot{m}(h_1 - h_2)$

$= \dot{m}\, c_p(T_1 - T_2)$

$$\left[as \frac{dh}{dT} = c_p, h_1 - h_2 = c_p (T_1 - T_2)\right]$$

$$= \frac{6.5}{60} \times 1.0259 (453 - 297) = 17.34 \text{ kW}$$

Hence power delivered by the turbine

= 17.34 kW. (Ans.)

Example 7.8. *A certain quantity of air initially at a pressure of 8 bar and 280°C has a volume of 0.035 m³. It undergoes the following processes in the following sequence in a cycle :*

(a) Expands at constant pressure to 0.1 m³,

(b) Follows polytropic process with n = 1.4, and

(c) A constant temperature process (which completes the cycle).

Evaluate the following :

(i) The heat received in the cycle ;

(ii) The heat rejected in the cycle ;

(iii) Efficiency of the cycle.

Solution. Fig. 7.12 shows the cycle on p-V and T-s planes.

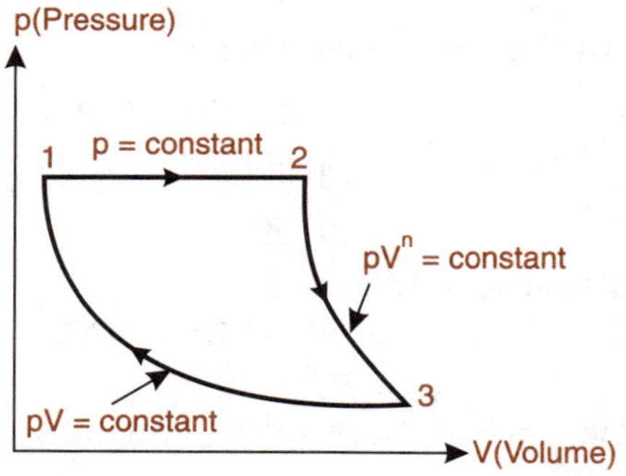

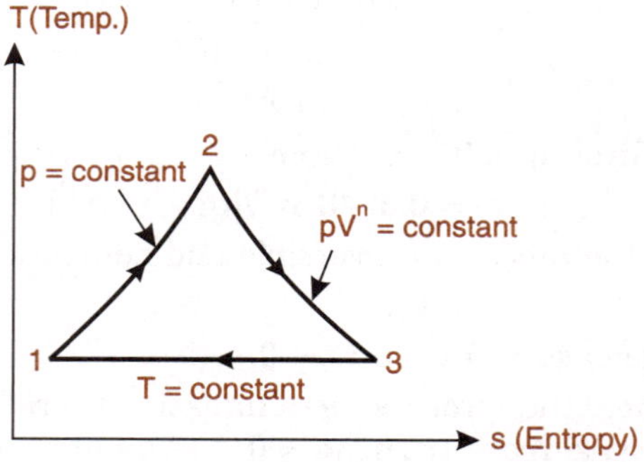

Fig. 7.12

Pressure, $p_1 = 8$ bar

Volume, $V_1 = 0.035$ m^3

Temperature, $T_1 = 280 + 273 = 553$ K

Pressure, $p_2 = 8$ bar $(= p_1)$

Volume, $V_2 = 0.1$ m^3

Index, $n = 1.4$

To find mass of air, use the relation

$$p_1V_1 = mRT_1$$

$$\therefore \quad m = \frac{p_1V_1}{RT_1} = \frac{8 \times 10^5 \times 0.035}{287 \times 553} = 0.1764 \text{ kg}$$

From $p_2V_2 = mRT_2$

$$T_2 = \frac{p_2V_2}{mR} = \frac{8 \times 10^5 \times 0.1}{0.1764 \times 287} = 1580 \text{ K}$$

Also, $p_2V_2^{1.4} = p_3V_3^{1.4}$

and $$\frac{T_2}{T_3} = \left(\frac{p_2}{p_3}\right)^{(1.4-1)/1.4}$$

But $T_3 = T_1$ as 1 and 3 are on an isothermal line.

$$\therefore \quad \frac{1580}{553} = \left(\frac{8}{p_3}\right)^{0.4/1.4}$$

$$2.857 = \left(\frac{8}{p_3}\right)^{0.2857}$$

$$p_3 = \frac{8}{(2.857)^{1/0.2857}} = \frac{8}{(2.857)^{3.5}} = 0.2 \text{ bar}$$

Now, $p_3V_3 = mRT_3$

$$\therefore \quad 0.2 \times 10^5 \times V_3 = 0.1764 \times 287 \times 553$$

$$\therefore \quad V_3 = \frac{0.1764 \times 287 \times 553}{0.2 \times 10^5} = 1.399 \text{ m}^3$$

(*i*) **The heat received in the cycle :**

Applying first law to the constant pressure *process 1-2*,

$$Q = \Delta U + W$$

$$W = \int_1^2 pdV \quad \text{(as the process is reversible)}$$

$$= p(V_2 - V_1)$$

$$= 8 \times 10^5 (0.1 - 0.035)$$

$= 52000$ J or 52 kJ (work done by air)

$$\therefore \quad Q = m \times c_v(T_2 - T_1) + 52$$

$$= 0.1764 \times 0.71(1580 - 553) + 52 = 180.6 \text{ kJ}$$

i.e., Heat received = 180.6 kJ

Applying first law to reversible polytropic *process 2-3*

$$Q = \Delta U + W$$

But $$W = \frac{p_2V_2 - p_3V_3}{n-1} = \frac{mR(T_2 - T_3)}{n-1}$$

$$= \frac{0.1764 \times 0.287\,(1580 - 553)}{1.4 - 1}$$

$= 129.98$ kJ (work done **by air**)

$$\therefore \quad Q = mc_v(T_3 - T_2) + 129.98$$

$$= 0.1764 \times 0.71\,(553 - 1580) + 129.98$$

$= -128.6 + 129.98 = 1.354$ kJ (heat received)

$\therefore$ **Total heat received in the cycle**

$= 180.6 + 1.354 =$ **181.954 kJ. (Ans.)**

(*ii*) **The heat rejected in the cycle :**

Applying first law to reversible isothermal *process 3-1*,

$$Q = \Delta U + W$$

$$W = p_3V_3 \log_e\left(\frac{V_1}{V_3}\right)$$

$$= 0.2 \times 10^5 \times 1.399 \times \log_e\left(\frac{0.035}{1.399}\right) \times 10^{-3}$$

$= -103.19$ kJ (work done **on** the air)

$$\therefore \quad Q = mc_v(T_1 - T_3) + W$$

$= 0 - 103.19 = -103.19$ kJ $\quad (\because T_1 = T_3)$

Hence heat rejected in the cycle

= 103.19 kJ. (Ans.)

(*iii*) **Efficiency of the cycle, η_{cycle} :**

$$\eta_{cycle} = \frac{\text{Heat received} - \text{Heat rejected}}{\text{Heat received}}$$

$$= \frac{181.954 - 103.19}{181.954} = 0.433 \text{ or } 43.3\%$$

i.e., **Efficiency of the cycle = 43.3%. (Ans.)**

REAL GASES

Example 7.9. *One kg of CO_2 has a volume of 1 m³ at 100°C. Compute the pressure by*

(*i*) *Van der Waals' equation*

(*ii*) *Perfect gas equation.*

Solution. (*i*) **Using Van der Waals' equation :**

Molar specific volume, $\bar{v} = 1 \times 44 = 44$ m³/kg-mol ($\because$ M for $CO_2 = 44$)

Temperature, $T = 100 + 273 = 373$ K

The values of a and b for CO_2 (from Table 7.1)

$a = 362850$ Nm⁴/(kg-mol)²

and $b = 0.0423$ m³/kg-mol

$R_0 = 8314$ Nm/kg-mol K

van der Waals' equation is written as

$$\left(p + \frac{a}{\bar{v}^2}\right)(\bar{v} - b) = R_0 T$$

or

$$p = \left(\frac{R_0 T}{\bar{v} - b} - \frac{a}{\bar{v}^2}\right)$$

Substituting the values in the above equation, we get

$$\therefore \quad \mathbf{p} = \frac{8314 \times 373}{44 - 0.0423} - \frac{362850}{44^2}$$

$= 70548 - 187$

= 70361 N/m² or 0.7036 bar. (Ans.)

(*ii*) **Using perfect gas equation :**

$$p\bar{v} = R_0 T$$

$$\therefore \quad \mathbf{p} = \frac{R_0 T}{\bar{v}} = \frac{8314 \times 373}{44}$$

= 70480 N/m² or 0.7048 bar. (Ans.)

Example 7.10. *A container of 3 m³ capacity contains 10 kg of CO_2 at 27°C. Estimate the pressure exerted by CO_2 by using :*

(*i*) *Perfect gas equation*

(*ii*) *Van der Waals' equation*

(*iii*) *Beattie Bridgeman equation.*

Solution. Capacity of the container, $V = 3$ m³

Mass of CO_2, $m = 10$ kg

Temperature of CO_2, $T = 27 + 273 = 300$ K

Pressure exerted by CO_2, p :

(*i*) **Using perfect gas equation :**

Characteristic gas constant,

$$R = \frac{R_0}{M} = \frac{8314}{44} = 188.95 \text{ Nm/kg K (for } CO_2)$$

Using perfect gas equation

$$pV = mRT$$

$$\therefore \quad \mathbf{p} = \frac{mRT}{V} = \frac{10 \times 188.95 \times 300}{3}$$

= 188950 N/m² or 1.889 bar. (Ans.)

(*ii*) **Using Van der Waals' equation :**

$$\left(p + \frac{a}{\bar{v}^2}\right)(\bar{v} - b) = R_0 T$$

$$p = \frac{R_0 T}{\bar{v} - b} - \frac{a}{\bar{v}^2}$$

From Table 7.1

For CO_2 : $a = 362850$ Nm⁴/(kg-mol)²

$b = 0.0423$ m³/(kg-mol)

$\bar{v}$ = Molar specific volume

$$= \frac{3 \times 44}{10} = 13.2 \text{ m}^3\text{/kg-mol}$$

Now substituting the values in the above equation, we get

$$p = \frac{8314 \times 300}{13.2 - 0.0423} - \frac{362850}{(13.2)^2}$$

$= 189562 - 2082.5$

= 187479.5 N/m² or 1.875 bar. (Ans.)

(*iii*) **Using Beattie Bridgeman equation :**

$$p = \frac{R_0 T(1 - e)}{(\bar{v})^2}(\bar{v} + B) - \frac{A}{\bar{v}^2}$$

where p = pressure, $A = A_0\left(1 - \frac{a}{\bar{v}}\right)$, $B = B_0\left(1 - \frac{b}{\bar{v}}\right)$

and $e = \frac{c}{\bar{v}T^3}$

From Table 7.2

$A_0 = 507.2836$, $a = 0.07132$

$B_0 = 0.10476$, $b = 0.07235$

$C = 66 \times 10^4$

$$\therefore \quad A = 507.2836\left(1 - \frac{0.07132}{13.2}\right) = 504.5$$

$$B = 0.10476\left(1 - \frac{0.07235}{13.2}\right) = 0.1042$$

$$C = \frac{66 \times 10^4}{13.2 \times (300)^3} = 0.001852$$

Now substituting the various values in the above equation, we get

$$\mathbf{p} = \frac{8314 \times 300(1 - 0.001852)}{(13.2 + 0.1042) - (13.2)^2} - \frac{504.5}{(13.2)^2}$$

$= 190093 - 2.89 \simeq 1.9 \times 10^5$ N/m²

= 1.9 bar. (Ans.)

Example 7.11. *One kg-mol of oxygen undergoes a reversible non-flow isothermal compression and the volume decreases from 0.2 m³/kg to 0.08 m³/kg and the initial temperature is 60°C. If the gas obeys Van der Waals' equation find :*

(i) The work done during the process

(ii) The final pressure.

Solution. The Van der Waals' equation is written as

$$\left(p + \frac{a}{\bar{v}^2}\right)(\bar{v} - b) = R_0 T$$

where p = pressure of the gas ; a, b = constants ;

$\bar{v}$ = molar volume ; R_0 = universal gas constant

From Table 7.1

For O_2 : $a = 139250$ Nm⁴/(kg-mol)²

$b = 0.0314$ m³/kg-mol

and $R_0 = 8314$ Nm/kg-mol K

$\bar{v}_1 = 0.2 \times 32 = 6.4$ m³/kg-mol

$\bar{v}_2 = 0.08 \times 32 = 2.56$ m³/kg-mol.

(*i*) **Work done during the process :**

The work done per kg mole of O_2 is given by

$$W = \int_1^2 p \,.\, d\bar{v} = \int_1^2 \left[\left(\frac{R_0 T}{\bar{v} - b}\right) - \frac{a}{\bar{v}^2}\right] d\bar{v}$$

$$= R_0 T \Big[\log_e (\bar{v} - b)\Big]_{\bar{v}_1}^{\bar{v}_2} + \left[\frac{a}{\bar{v}}\right]_{\bar{v}_1}^{\bar{v}_2}$$

$$= R_0 T \left[\log_e \left(\frac{\bar{v}_2 - b}{\bar{v}_1 - b}\right)\right] + \left[a\left(\frac{1}{\bar{v}_2} - \frac{1}{\bar{v}_1}\right)\right]$$

$$= 8314 \times (60 + 273) \left[\log_e \left(\frac{2.56 - 0.0314}{6.4 - 0.0314}\right)\right] + \left[139250\left(\frac{1}{2.56} - \frac{1}{6.4}\right)\right]$$

$= -2557359 + 32636$

= – 2524723 Nm/kg-mol. (Ans.)

(*ii*) **The final pressure, $\mathbf{p_2}$:**

$$p_2 = \frac{R_0 T}{\bar{v}_2 - b} - \frac{a}{\bar{v}_2^2}$$

$$= \frac{8314 \times 333}{2.56 - 0.0314} - \frac{139250}{(2.56)^2}$$

= 1073651 N/m² or 10.73 bar. (Ans.)

Example 7.12. *If the values for reduced pressure and compressibility factor for ethylene are 20 and 1.25 respectively, compute the temperature.*

Solution. Reduced pressure, $p_r = 20$

Compressibility factor, $Z = 1.25$

Temperature, T = ?

From the generalised compressibility chart on $Z - p_r$ co-ordinates corresponding to $p_r = 20$ and $Z = 1.25$, $T_r = 8.0$.

Now, since $T = T_c \, T_r$

∴ $\mathbf{T} = 282.4 \times 8.0$

[From Table 7.3, $T_c = 282.4$ K]

= 2259.2 K. (Ans.)

Example 7.13. *Calculate the density of N_2 at 260 bar and 15°C by using the compressibility chart.*

Solution. Pressure, $p = 260$ bar

Temperature, $T = 15 + 273 = 288$ K

Density, ρ = ?

For $\mathbf{N_2}$ (from Table 7.3) :

$p_c = 33.94$ bar

$T_c = 126.2$ K

Now, $p_r = \dfrac{p}{p_c} = \dfrac{260}{33.94} = 7.6$

and $T_r = \dfrac{T}{T_c} = \dfrac{288}{126.2} = 2.28$

From the compressibility chart for $p_r = 7.6$ and $T_r = 2.28$, $Z \simeq 1.08$

Also $Z = \dfrac{pv}{RT} = \dfrac{p}{\rho RT}$,

where ρ stands for density

or $$\rho = \frac{p}{ZRT} = \frac{260 \times 10^5}{1.08 \times \frac{8314}{28} \times 288}$$

= 281.5 kg/m³. (Ans.)

Example 7.14. *What should be the temperature of 1.3 kg of CO_2 gas in a container at a pressure of 200 bar to behave as an ideal ?*

Solution. Pressure, $p = 200$ bar

Temperature, T = ?

For $\mathbf{CO_2}$ (from Table 7.3)

$p_c = 73.86$ bar

$T_c = 304.2$ K

As the gas behaves like an ideal gas, $Z = 1$

$$p_r = \frac{p}{p_c} = \frac{200}{73.86} = 2.7$$

From compressibility chart for $Z = 1$, $p_r = 2.7$

$$T_r = 2.48$$

$$\therefore \quad \mathbf{T} = T_r \times T_c$$

$$= 2.48 \times 304.2 = \mathbf{754.4\ K.} \quad \textbf{(Ans.)}$$

Example 7.15. *A spherical shaped balloon of 12 m diameter contains H_2 at 30°C and 1.21 bar. Find the mass of H_2 in the balloon using real gas equation.*

Solution. Diameter of spherical balloon = 12 m

∴ Volume, $V = 4/3\ \pi \times (6)^3 = 904.78\ m^3$

Temperature, $T = 30 + 273 = 303$ K

Pressure, $p = 1.21$ bar

Mass of H_2 in the balloon, m :

For H_2 (from Table 7.3)

$$p_c = 12.97 \text{ bar}$$

$$T_c = 33.3 \text{ K}$$

Now, $$p_r = \frac{p}{p_c} = \frac{1.21}{12.97} = 0.093$$

$$T_r = \frac{T}{T_c} = \frac{303}{33.3} = 9.1$$

From compressibility charge, corresponding to $p_r = 0.093$ and $T_r = 9.1$

$$Z \simeq 1$$

(This indicates that the gas having higher critical pressure and lower critical temperature behaves like an ideal gas at normal pressure and temperature conditions.)

Also, $pV = ZmRT$

or $$\mathbf{m} = \frac{pV}{ZRT} = \frac{1.21 \times 10^5 \times 904.78}{1 \times \left(\frac{8314}{2}\right) \times 303}$$

$$= \mathbf{86.9\ kg.} \quad \textbf{(Ans.)}$$

Example 7.16. *Determine the value of compressibility factor at critical point (Z_{cp}) for the Van der Waals' gas.*

Solution. Refer to Fig. 7.13.

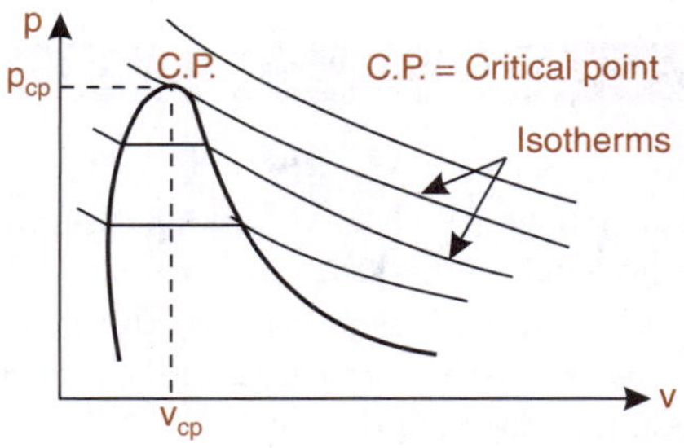

Fig. 7.13

From the isotherms plotted on p-v diagram in Fig. 7.13 it can be seen that the critical isotherms has an inflection point, whose *tangent is horizontal* at the *critical point*.

$$\left(\frac{\partial p_c}{\partial v}\right)_{cp} = 0 \text{ and } \left|\frac{\partial^2 p}{\partial v^2}\right|_{cp} = 0$$

The Van der Waal's equation at the critical point is

$$p_{cp} = \frac{R_0 T_{cp}}{\bar{v}_{cp} - b} - \frac{a}{\bar{v}_{cp}^{\,2}} \quad ...(i)$$

As T_{cp} is constant

$$\left(\frac{\partial p_{cp}}{\partial v_{cp}}\right) = -\frac{R_0 T_{cp}}{(\bar{v}_{cp} - b)^2} + \frac{2a}{\bar{v}_{cp}^{\,3}} = 0 \quad ...(ii)$$

$$\left(\frac{\partial^2 p_{cp}}{\partial v_{cp}^{\,2}}\right) = \frac{2R_0 T_{cp}}{(\bar{v}_{cp} - b)^3} - \frac{6a}{(\bar{v}_{cp})^4} = 0 \quad ...(iii)$$

$\left(\frac{3}{\bar{v}_{cp}}\right) \times (ii) + (iii)$ gives

$$-\frac{3R_0 T}{\bar{v}_{cp}(\bar{v}_{cp} - b)^2} + \frac{2R_0 T}{(\bar{v}_{cp} - b)^3} = 0$$

or $$\frac{3}{\bar{v}_{cp}} = \frac{2}{(\bar{v}_{cp} - b)}$$

or $$\bar{v}_{cp} = 3b$$

Substituting for b in (ii), we get

$$-\frac{R_0 T_{cp}}{[\bar{v}_{cp} - (1/3)\bar{v}_{cp}]^2} + \frac{2a}{(\bar{v}_{cp})^3} = 0$$

$$\therefore \quad a = \frac{9}{8} R_0 T_{cp} \bar{v}_{cp}$$

Substituting for a and b in (i), we get

$$p_{cp} = \frac{R_0 T_{cp}}{\bar{v}_{cp} - (1/3)\bar{v}_{cp}} - \frac{(9/8) R_0 T_{cp} \bar{v}_{cp}}{\bar{v}_{cp}^{\,2}}$$

$$\therefore \quad \frac{p_{cp} \bar{v}_{cp}}{R_0 T_{cp}} = \frac{1}{(2/3)} - \frac{(9/8)}{1}$$

But $$\frac{p_{cp} \bar{v}_{cp}}{R_0 T_{cp}} = Z_{cp}$$

$$\therefore \quad \mathbf{Z_{cp}} = \frac{3}{2} - \frac{9}{8} = \mathbf{\frac{3}{8}}. \quad \textbf{(Ans.)}$$

HIGHLIGHTS

1. An '*ideal gas*' is defined as a gas having no forces of intermolecular attraction. It obeys the law $pv = RT$. The specific heat capacities are *not constant* but are functions of temperature.
   A '*perfect gas*' obeys the law $pv = RT$ and has *constant* specific heat capacities.
2. The relation between the independent properties, such as pressure, specific volume and temperature for a pure substance is known as '*equation of state*'.
3. Each point on a *p-v-T* surface represents an equilibrium state and a line on the surface represents a process.
4. *Joule's law* states that the specific internal energy of a gas depends only on the temperature of the gas and is independent of both pressure and volume.
5. Van der Waals' equation may be written as:

$$\left(p + \frac{a}{v^2}\right)(v - b) = RT$$

   where a and b are constants for the particular fluid and R is the gas constant.

OBJECTIVE TYPE QUESTIONS

Choose the Correct Answer :

1. (*a*) A perfect gas does not obey the law $pv = RT$
   (*b*) A perfect gas obeys the law $pv = RT$ and has constant specific heat
   (*c*) A perfect gas obeys the law $pv = RT$ but have variable specific heat capacities.
2. Boyle's law states that, when temperature is constant, the volume of a given mass of a perfect gas
   (*a*) varies directly as the absolute pressure
   (*b*) varies inversely as the absolute pressure
   (*c*) varies as square of the absolute pressure
   (*d*) does not vary with the absolute pressure.
3. Charle's law states that if any gas is heated at constant pressure, its volume
   (*a*) changes directly as its absolute temperature
   (*b*) changes inversely as its absolute temperature
   (*c*) changes as square of the absolute temperature
   (*d*) does not change with absolute temperature.
4. The equation of the state per kg of a perfect gas is given by
   (*a*) $p^2v = RT$ (*b*) $pv = RT$
   (*c*) $pv^2 = RT$ (*d*) $p^2v^2 = RT$.
   where p, v, R and T are the pressure, volume, characteristic gas constant and temperature of the gas respectively.
5. The equation of state of an ideal gas is a relationship between the variables :
   (*a*) pressure and volume
   (*b*) pressure and temperature
   (*c*) pressure, volume and temperature
   (*d*) none of the above.
6. Joule's law states that the specific internal energy of a gas depends only on
   (*a*) the pressure of the gas
   (*b*) the volume of the gas
   (*c*) the temperature of the gas
   (*d*) none of the above.
7. Equation for specific heat at constant pressure of an ideal gas is given by
   (*a*) $c_p = a + KT + K_1T^2 + K_2T^3$
   (*b*) $c_p = a + KT^2 + K_1T^3 + K_2T^4$
   (*c*) $c_p = a + KT^2 + K_1T^4 + K_2T$
   (*d*) $c_p = a + KT^2 + K_1T^3 + K_2T^2$.
   where a, K, K_1 and K_2 are constants.
8. Van der Waals' equation may be written as:
   (*a*) $\left(p + \frac{a}{v}\right)(v - b) = RT$ (*b*) $\left(p + \frac{a}{v^2}\right)(v - b) = RT$
   (*c*) $\left(p + \frac{a}{v^2}\right)(v^2 - b) = RT$ (*d*) $\left(p + \frac{a}{v^2}\right)(v^2 - b) = RT^2$.

ANSWERS

1. (*b*)	**2.** (*b*)	**3.** (*a*)	**4.** (*b*)
5. (*c*)	**6.** (*c*)	**7.** (*a*)	**8.** (*b*).

THEORETICAL QUESTIONS

1. What is an ideal gas ?
2. What is the difference between an ideal and a perfect gas ?
3. What are semi-perfect or permanent gases ?
4. Define 'Equation of state'.
5. State Boyle's and Charle's laws and derive an equation of the state for a perfect gas.
6. What is a *p-v-T* surface ? Draw a portion of such a surface.
7. Derive the relationship between the two principal specific heats and characteristic gas constant for a perfect gas.
8. Write a short note on Van der Waals' equation.

UNSOLVED EXAMPLES

IDEAL GASES

1. A vessel of 0.03 m^3 capacity contains gas at 3.5 bar pressure and 35°C temperature. Determine the mass of the gas in the vessel. If the pressure of this gas is increased to 10.5 bar while the volume remains constant, what will be the temperature of the gas ?
   For the gas take R = 290 J/kg K. [**Ans.** 0.118 kg, 650°C]
2. The tyre of an automobile contains a certain volume of air at a gauge pressure of 2 bar and 20°C. The barometer reads 75 cm of Hg. The temperature of air in the tyre rises to 80°C due to running of automobile for two hours. Find the new pressure in the tyre.

Assume that the air is an ideal gas and tyre does not stretch due to heating. **[Ans.** 2.62 bar]

3. A tank made of metal is designed to bear an internal gauge pressure of 7 bar. The tank is filled with a gas at a pressure of 5.5 bar abs., and 15°C. The temperature in the tank may reach to 50°C when the tank stands in the sun.
(*i*) If the tank does not expand with temperature, will the design pressure be exceeded on a day when atmospheric pressure is 1 bar and air in the tank reaches 50°C when exposed to hot sun.
(*ii*) What temperature would have to be reached to raise the air pressure to the design limit ?
[Ans. (*i*) 6.16 bar, (*ii*) 147°C]

4. A vessel of spherical shape is 1.5 m in diameter and contains air at 40°C. It is evacuated till the vacuum inside the vessel is 735 mm of mercury. Determine :
(*i*) The mass of air pumped out ;
(*ii*) If the tank is then cooled to 10°C what is the pressure in the tank ?
The barometer reads 760 mm of mercury. Assume that during evacuation, there is no change in temperature of air. **[Ans.** (*i*) 1.91 kg, (*ii*) 3 kPa]

5. A balloon of spherical shape is 8 m in diameter and is filled with hydrogen at a pressure of 1 bar abs. and 15°C. At a later time, the pressure of gas is 95 per cent of its original pressure at the same temperature.
(*i*) What mass of original gas must have escaped if the dimensions of the balloon are not changed ?
(*ii*) Find the amount of heat to be removed to cause the same drop in pressure at constant volume.
[Ans. (*i*) 5 per cent, (*ii*) 3.26 MJ]

6. Find the molecular weight and gas constant for the gas whose specific heats are as follows :
c_p = 1.967 kJ/kg K, c_v = 1.507 kJ/kg K.
[Ans. 180.461 kJ/kg K]

7. A constant volume chamber of 0.3 m^3 capacity contains 1 kg of air at 20°C. Heat is transferred to the air until its temperature is 200°C. Find :
(*i*) Heat transferred ;
(*ii*) Change in entropy and enthalpy.
[Ans. (*i*) 128.09 kJ, (*ii*) 0.339 kJ/kg K, 180.8 kJ]

8. 1 kg of air at 20°C occupying a volume of 0.3 m^3 undergoes a reversible constant pressure process. Heat is transferred to the air until its temperature is 200°C. Determine :
(*i*) The work and heat transferred.
(*ii*) The change in internal energy, enthalpy and entropy.
[Ans. (*i*) 51.5 kJ, 180.8 kJ ; (*ii*) 128.09 kJ, 180.8 kJ, 0.479 kJ/kg K]

9. A balloon of spherical shape, 10 m in diameter is filled with hydrogen at 20°C and atmospheric pressure. The surrounding air is at 15°C and barometer reads 75 mm of Hg. Determine the load lifting capacity of the balloon.
[Ans. 587.2 kg]

10. Air expands in a cylinder in a reversible adiabatic process from 13.73 bar to 1.96 bar. If the final temperature is to be 27°C, what would be the initial temperature ?
Also calculate the change in specific enthalpy, heat and work transfers per kg of air.
[Ans. 524 K, 224.79 kJ/kg, zero, 160.88 kJ/kg]

11. 1 kg mole of N_2 is contained in a vessel of volume 2.5 m^3 at 100°C.
(*i*) Find the mass, the pressure and the specific volume of the gas.
(*ii*) If the ratio of the specific heats is 1.4, evaluate the values of c_p and c_v.
(*iii*) Subsequently, the gas cools to the atmospheric temperature of 30°C, evaluate the final pressure of the gas.
(*iv*) Evaluate the increase in specific internal energy, the increase in specific enthalpy, increase in specific entropy and magnitude and sign of heat transfer.
[Ans. (*i*) 28 kg, 12.37 bar, 0.089 m^3/kg ;
(*ii*) c_p = 1.038 kJ/kg K, c_v = 0.745 kJ/kg K ;
(*iii*) 10.22 bar ; (*iv*) – 52.16 kJ/kg,
– 72.67 kJ/kg, – 0.1536 kJ/kg K, 1465.1 kJ]

12. The pressure and volume of a gas, during a process, change from 1 bar absolute and 2 m^3 respectively to 6 bar absolute and 0.4 m^3 respectively. During the process the increase in the enthalpy of the gas is 200 kJ. Taking c_v = 10.4 kJ/kg K, determine c_p, R and ΔU.
[Ans. 13 kJ/kg K, 2.6 kJ/kg K, 160 kJ]

13. 1 kg of air at 27°C is heated reversibly at constant pressure until the volume is doubled and then heated reversibly at constant volume until the pressure is doubled. For the total path find :
(*i*) The work ; (*ii*) Heat transfer ;
(*iii*) Change of entropy.
[Ans. (*i*) 86.14 kJ, (*ii*) 728.36 kJ, (*iii*) 1.186 kJ/kg K]

14. A mass of air initially at 260°C and a pressure of 6.86 bar has a volume of 0.03 m^3. The air is expanded at constant pressure to 0.09 m^3, a polytropic process with n = 1.5 is then carried out, followed by a constant temperature process which completes the cycle. All processes are reversible. Find (*i*) The heat received and rejected in the cycle, (*ii*) The efficiency of the cycle.
Show the cycle on *p-v* and *T-s* planes.
[Ans. (*i*) 143.58 kJ, – 20.3 kJ ; (*ii*) 38.4%]

REAL GASES

15. One kg-mol of oxygen undergoes a reversible non-flow isothermal compression and the volume decreases from 0.15 m^3/kg to 0.06 m^3/kg and the initial temperature is 50°C. The gas obeys Van der Waals' equation during the compression. Find :
(*i*) The work done during the process ;
(*ii*) The final pressure.
[Ans. (*i*) – 6706500 Nm/kg-mol, (*ii*) 13.825 bar]

16. Determine the compressibility factor for O_2 at (*i*) 100 bar and 70°C and (*ii*) at 5 bar and 30°C.
[Ans. (*i*) 0.71, (*ii*) 0.98]

17. Determine the pressure of air at 205°C having a specific volume of 0.00315 m^3/kg by means of :
(*i*) Ideal gas equation ;
(*ii*) Van der Waals' equation ;
(*iii*) Beattie-Bridgeman equation.
[Ans. (*i*) 435.7 bar, (*ii*) 557.3 bar, (*iii*) 525.8 bar]

Gases and Vapour Mixtures

8

8.1. Introduction. 8.2. Dalton's law and Gibbs-Dalton law. 8.3. Volumetric analysis of a gas mixture. 8.4. The apparent molecular weight and gas constant. 8.5. Specific heats of a gas mixture. 8.6. Adiabatic mixing of perfect gases. 8.7. Gas and vapour mixtures. *Highlights—Objective Type Questions—Theoretical Questions—Unsolved Examples.*

8.1. INTRODUCTION

- A *pure substance* is defined as *a substance having a constant and uniform chemical composition*. A homogeneous mixture of gases which do not react with one another may, therefore, be considered a pure substance. For example, air is a homogeneous mixture of nitrogen, oxygen and traces of other substances like argon, helium, carbon dioxide, etc., and as they do not react with one another, air is regarded a pure substance. The properties of such a mixture can be determined and tabulated just like those of any other pure substance. The properties of air and some combustion products have been determined and tabulated in gas tables. But it is not possible to determine the properties of the unlimited number of mixtures possible, the properties of the mixtures are determined from the properties of the constituent gases.
- In this chapter the mixtures to be considered are those composed of perfect gases, and perfect gases and vapours. The properties of such mixtures are important in combustion calculations. Air and water vapour mixtures are considered later in the chapter with reference to surface condensers, but for moist atmospheric air there is a special nomenclature and this is considered in a separate chapter on *Psychrometrics*.

8.2. DALTON'S LAW AND GIBBS-DALTON LAW

Dalton's law

Let us consider a closed vessel of volume V at temperature T, which contains a mixture of perfect gases at a known pressure. If some of the mixture were removed, then the pressure would be *less* than the initial value. If the gas removed were the full amount of one of the constituents then the reduction in pressure would be equal to the contribution of that constituent to the initial total pressure. Each constituent contributes to the total pressure by an amount which is known as the *partial pressure* of the constituent.

The *relationship between the partial pressures of the constituents* is expressed by Dalton's law, as follows :

— *The pressure of a mixture of gases is equal to the sum of the partial pressures of the constituents.*

— *The partial pressure of each constituent is that pressure which the gas would exert if it occupied alone that volume occupied by the mixtures at the same temperature.*

This is expressed diagrammatically in Fig. 8.1. The gases A and B, originally occupying volume V at temperature T are mixed in the third vessel which is of the *same volume* and is *at the same temperature*.

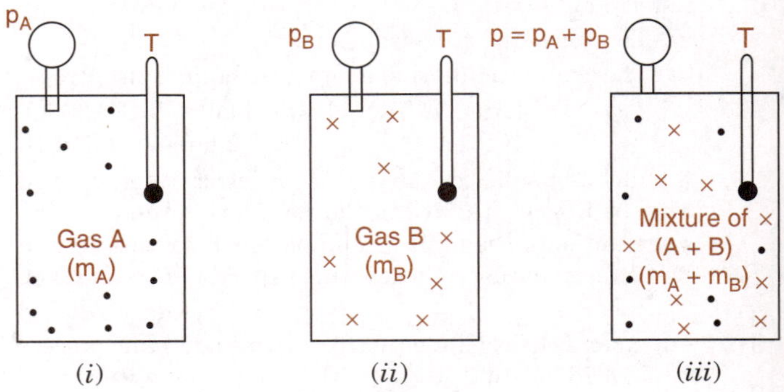

Fig. 8.1

By the *consideration of mass*,

$$m = m_A + m_B \qquad ...(8.1)$$

By *Dalton's law*, $p = p_A + p_B$ ...(8.2)

Dalton's law is based on experiment and is found to be obeyed more accurately by gas mixtures at low pressures. As shown in Fig. 8.1 each occupant occupies the whole vessel. The example given in Fig. 8.1 and relationship in eqns. (8.1) and (8.2) refer to a mixture of two gases, but the law can be extended to any number of gases,

i.e., $$m = m_A + m_B + m_C + ... \text{ or } m = \Sigma\, m_i \qquad ...(8.3)$$

where m_i = Mass of a constituent.

Similarly $p = p_A + p_B + p_C + ...$ or $p = \Sigma p_i$ ...(8.4)

where p_i = The partial pressure of a constituent.

Gibbs-Dalton law

Dalton's law was re-formulated by Gibbs to include a second statement on the properties of mixtures. The combined statement is known as the Gibbs-Dalton law, and is as follows :

— *The internal energy, enthalpy, and entropy of a gaseous mixture are respectively equal to the sums of the internal energies, enthalpies, and entropies, of the constituents.*

— *Each constituent has that internal energy, enthalpy and entropy, which it could have if it occupied alone that volume occupied by the mixture at the temperature of the mixture.*

This statement leads to the following equations :

$$mu = m_A u_A + m_B u_B + ... \text{ or } mu = \Sigma\, m_i u_i \qquad ...(8.5)$$

and $$mh = m_A h_A + m_B h_B + ... \text{ or } mh = \Sigma\, m_i h_i \qquad ...(8.6)$$

and $$ms = m_A s_A + m_B s_B + ... \text{ or } ms = \Sigma\, m_i s_i \qquad ...(8.7)$$

— ***Properties of air.*** The properties of air which is the most common mixture are given below :

Constituent	*Molecular weight*	*Chemical symbol*	*Volumetric analysis %*	*Gravimetric analysis %*
Oxygen	31.999	O_2	20.95	23.14
Nitrogen	28.013	N_2	78.09	75.53
Argon	39.948	Ar	0.93	1.28
Carbon dioxide	44.01	CO_2	0.03	0.05

Mean molecular weight of air = 28.96

Gas constant $R = 0.287$ kJ/kg K

For *approximate* calculations the air is said to be composed of oxygen and 'atmospheric nitrogen'.

Constituent	*Molecular weight*	*Volumetric analysis %*	*Gravimetric analysis %*
Oxygen	32	21	23.3
Atmospheric nitrogen	28	79	76.7
Nitrogen/ Oxygen	—	3.76 : 1	3.29 : 1

Note. Volumetric analysis is the analysis by **volume** ; gravimetric analysis is the analysis by **weight or mass**.

8.3. VOLUMETRIC ANALYSIS OF A GAS MIXTURE

It is usual practice to quote the analysis of a mixture by volume as this is the most convenient for practical determinations. In Art. 10.15, the analysis of exhaust or flue gases by means of the Orsat apparatus is described. The volume of the gas sample is measured at atmospheric pressure, and the temperature is held constant by means of a water jacket round the gas sample. The constituents are absorbed chemically one by one, and the remainder of the sample is measured after each absorption ; the difference in volume gives the partial volume occupied by the constituent in the mixture.

Let us consider a volume V of a gaseous mixture at a temperature T, consisting of three constituents A, B and C [Fig. 8.2 (*a*)]. Let us further assume that each of the constituents is compressed to a pressure p equal to the total pressure of the mixture, and let the *temperature remain constant*. The partial volumes then occupied by the constituents will be V_A, V_B and V_C.

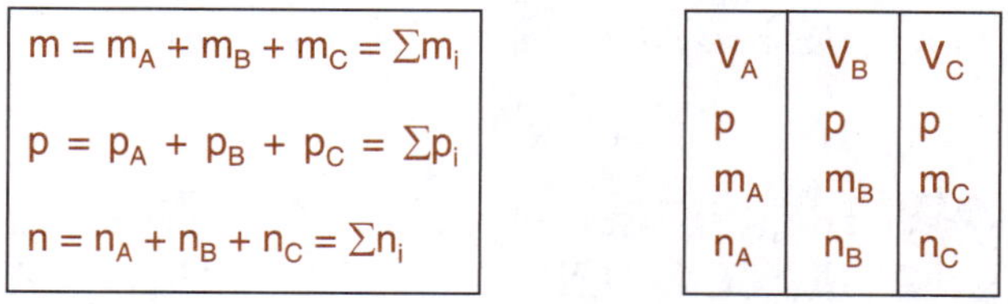

(*a*) (*b*)

Fig. 8.2

Now using the eqn. $pV = mRT$, we get:

$$m_A = \frac{p_A V}{R_A T} \qquad \text{... referring to Fig. 8.2 } (a)$$

and, $$m_A = \frac{p V_A}{R_A T} \qquad \text{... referring to Fig. 8.2 } (b)$$

Now equating the two values for m_A, we have

$$\frac{p_A V}{R_A T} = \frac{p V_A}{R_A T} \quad i.e., \quad p_A V = p V_A$$

or $$V_A = \frac{p_A}{p} V$$

In general, therefore,

$$V_i = \frac{p_i}{p} V \quad ...(8.8)$$

i.e.,
$$\Sigma V_i = \sum \frac{p_i V}{p} = \frac{V}{p} \Sigma p_i$$

Now from eqn. (8.4), $p = \Sigma\, p_i$, therefore,

$$\Sigma\, V_i = V \quad ...(8.9)$$

Thus, the *volume of a mixture of gases is equal to the sum of the volumes of the individual constituents when each exists alone at the pressure and temperature of the mixture.*

This is the statement of another empirical law, *the law of partial volumes*, sometimes called *Amagat's law* or *Leduc's law*.

— The analysis of mixtures, oftenly, is simplified if it is carried out in moles. The mole is given by the equation

$$n = \frac{m}{M}$$

where, n = Number of moles,
m = Mass of gas, and
M = Molecular weight.

According to *Avogadro's law*, the *number of moles of any gas is proportional to the volume of the gas at a given pressure and temperature.* Referring to Fig. 8.2 (*a*), the volume V contains n moles of the mixture at p and T. In Fig. 8.2 (*b*), the gas A occupies a volume V_A at p and T, and this volume contains n_A moles. Similarly there are n_B moles of gas B in volume V_B and n_C moles of gas C in volume V_C.

From eqn. (8.9), $\Sigma\, V_i = V$

or $V_A + V_B + V_C = V$

$\therefore$ The *total number of moles in the vessel must equal the sum of the moles of the individual constituents,*

$$n = n_A + n_B + n_C = \Sigma\, n_i \quad ...(8.10)$$

8.4. THE APPARENT MOLECULAR WEIGHT AND GAS CONSTANT

The Apparent Molecular Weight

In a gas mixture if a gas occupies a total volume of V at a temperature T, then from the definition of partial pressure and equation $pV = nR_0T$, we have

$$p_i V = n_i R_0 T \quad ...(8.11)$$

(where R_0 is the *universal gas constant*)

$\therefore$ $\Sigma\, p_i V = \Sigma\, n_i R_0 T$

i.e., $\Sigma\, p_i = R_0 T\, \Sigma\, n_i$

Also $p = \Sigma\, p_i$ [From eqn. (8.4)]

$\therefore$ $pV = R_0 T\, \Sigma\, n_i$

Also $n = \Sigma\, n_i$ [From eqn. (8.10)]

$\therefore$ $pV = nR_0T$

The *mixture therefore acts as a perfect gas,* and *this is the characteristic equation for mixture.*

An apparent molecular weight is defined by the equation

$$M = \frac{m}{n}$$

where, m = Mass of the mixture, and
n = Number of moles of mixture.

The Gas Constant

The *apparent gas constant* (similarly as above) is defined by the equation

$$R = \frac{R_0}{M}$$

It can be assumed that a mixture of perfect gases obeys all the perfect gas laws.

In order to determine the gas constant for the mixture in terms of the gas constants of the constituents let us consider the equation $pV = mRT$ both for the mixture and for a constituent as follows

$pV = mRT$

and $p_i V = m_i R_i T$

Then $\Sigma\, p_i V = \Sigma\, m_i R_i T$

$\therefore$ $V\, \Sigma\, p_i = T\, \Sigma\, m_i R_i$

Also $p = \Sigma\, p_i$

$\therefore$ $pV = T\, \Sigma\, m_i R_i$

or $pV = mRT = T\, \Sigma\, m_i R_i$

i.e., $mR = \Sigma\, m_i R_i$

or
$$R = \Sigma\, \frac{m_i}{m} R_i \quad ...(8.12)$$

where, $\frac{m_i}{m}$ = mass fraction of a constituent.

— From equation (8.11), $p_i V = n_i R_0 T$, and combining this with eqn. (8.8) $\left(V_i = \frac{p_i}{p} V\right)$ applied to the mixture (*i.e.,* $pV = nR_0T$), we have

$$\frac{p_i V}{pV} = \frac{n_i R_0 T}{n R_0 T}$$

or
$$\frac{p_i}{p} = \frac{n_i}{n} \quad ...(8.13)$$

On c ombining this with eqn. (8.8), we get

$$\frac{p_i}{p} = \frac{n_i}{n} = \frac{V_i}{V} \quad ...(8.14)$$

This means that the *molar analysis is identical with the volumetric analysis,* and *both are equal to the ratio of the partial pressure to the total pressure.*

— The *apparent molecular weight* can be also be determined by the following method.

Let us apply characteristic equation to each constituent and to mixture, we have

$$m_i = \frac{p_i V}{R_i T}$$

$$m = \frac{pV}{RT}$$

Also $\quad m = \Sigma\, m_i \quad$ [From eqn. (8.3)]

$$\therefore \quad \frac{pV}{RT} = \sum \frac{p_i V}{R_i T}$$

$$\therefore \quad \frac{p}{R} = \sum \frac{p_i}{R_i}$$

Using the relation $R = \frac{R_0}{M}$, and substituting, we have

$$\frac{pM}{R_0} = \sum \frac{p_i M_i}{R_0}$$

or $$pM = \sum p_i M_i$$

i.e., $$M = \sum \frac{p_i}{p} M_i \quad ...(8.15)$$

Now using eqn. (8.14), we have

$$M = \sum \frac{V_i}{V} M_i \quad ...(8.16)$$

and $$M = \sum \frac{n_i}{n} M_i \quad ...(8.17)$$

Alternately

$$p = \Sigma\, p_i = p_A + p_B + ...\, p_i$$

Also $\quad pV = mRT$

Similarly

$$p_A V = m_A R_A T$$

$$p_B V = m_B R_B T$$

$$\therefore \quad pV = p_A V + p_B V + ...$$

or $$mRT = m_A R_A T + m_B R_B T + ...$$

$$\therefore \quad mR = m_A R_A + m_B R_B + ...$$

But $\quad R = \frac{R_0}{M},\ R_A = \frac{R_0}{M_A},\ R_2 = \frac{R_0}{M_B}$

Substituting this in the above equation, we get

$$m\frac{R_0}{M} = m_A \frac{R_0}{M_A} + m_B \frac{R_0}{M_B} + ...$$

$$\therefore \quad \frac{1}{M} = \frac{m_A}{m} \cdot \frac{1}{M_A} + \frac{m_B}{m} \cdot \frac{1}{M_B} + ...$$

$$= \frac{m_{fA}}{M_A} + \frac{m_{fB}}{M_B} + ...$$

where m_{fA}, m_{fB} etc. are the *mass fractions* of the constituents.

$$\therefore \quad \frac{1}{M} = \sum \frac{m_{fi}}{M_i}$$

$$\therefore \quad M = \frac{1}{\sum \frac{m_{fi}}{M_i}} \quad ...(8.18)$$

8.5. SPECIFIC HEATS OF A GAS MIXTURE

— As per Gibbs-Dalton law, the internal energy of a mixture of gases is given by

$$mu = \Sigma\, m_i u_i \quad ...(i) \quad \text{[From eqn. (8.5)]}$$

Also $\quad u = c_v T \quad ...(ii) \quad$ [From perfect gas equation]

Now from (*i*) and (*ii*), we have

$$mc_v T = \Sigma\, m_i c_{vi} T$$

$$\therefore \quad mc_v = \Sigma\, m_i c_{vi}$$

or $$c_v = \sum \frac{m_i}{m} c_{vi} \quad ...(8.19)$$

Similarly from equations,

$$mh = \Sigma\, m_i h_i$$

and $\quad h = c_p T$, we get

$$mc_p T = \Sigma\, m_i c_{pi} T$$

$$\therefore \quad m_{cp} = \Sigma\, m_i c_{pi}$$

or $$c_p = \sum \frac{m_i}{m} c_{pi} \quad ...(8.20)$$

From eqns. (8.18) and (8.19),

$$c_p - c_v = \sum \frac{m_i}{m} c_{pi} - \sum \frac{m_i}{m} c_{vi}$$

$$= \sum \frac{m_i}{m} (c_{pi} - c_{vi})$$

Also $\quad c_{pi} - c_{vi} = R_i$, therefore,

$$c_p - c_v = \sum \frac{m_i}{m} R_i$$

Also from eqn. (8.12), $R = \sum \frac{m_i}{m} R_i$, therefore for the mixture

$$c_p - c_v = R$$

The following equations can be applied to a mixture of gases

$$\gamma = \frac{c_p}{c_v}; \quad c_v = \frac{R}{\gamma - 1}; \quad c_p = \frac{\gamma R}{\gamma - 1}$$

It should be borne in mind that γ *must be determined from the eqn.* $\gamma = \frac{c_p}{c_v}$*; there is no weighted mean expression as there is for R,* c_v *and* c_p.

— In problems on mixtures it is often convenient to work in moles and the specific heats can be expressed in terms of the mole. These are known as *molar heats*, and are denoted by C_p and C_v.

Molar heats are defined as follows :

$$C_p = Mc_p \quad \text{and} \quad C_v = Mc_v \qquad ...(8.21)$$

But $\quad c_p - c_v = R$

$\therefore \quad C_p - C_v = Mc_p - Mc_v = M(c_p - c_v) = MR$

But $\quad MR = R_0$

$$\therefore \quad C_p - C_v = R_0 \qquad ...(8.22)$$

Also $\quad U = mc_vT = \dfrac{mMc_vT}{M}$

Again, $\quad \dfrac{m}{M} = n \quad \text{and} \quad Mc_v = C_v$

$$\therefore \quad U = nC_vT \qquad ...(8.23)$$

$$\text{Similarly,} \quad H = nC_pT \qquad ...(8.24)$$

By the Gibbs-Dalton law,

$$U = \Sigma U_i \quad \text{and} \quad H = \Sigma H_i$$

$$\therefore \quad nC_vT = \Sigma n_iC_{vi}T \quad \text{and} \quad nC_pT = \Sigma n_iC_{pi}T$$

i.e.,
$$C_v = \sum \frac{n_i}{n} C_{vi} \qquad ...(8.25)$$

and
$$C_p = \sum \frac{n_i}{n} C_{pi}. \qquad ...(8.26)$$

8.6. ADIABATIC MIXING OF PERFECT GASES

— Fig. 8.3 shows two gases A and B separated from each other in a closed vessel by a thin diaphragm. If the diaphragm is removed or punctured then the gases mix and each then occupies the total volume, behaving as if the other gas were not present. This process is equivalent to a *free expansion* of each gas, and is *irreversible*. The process can be simplified by the assumption that it is adiabatic ; this means that the vessel is perfectly thermally insulated and there will therefore be an *increase in entropy of the system.*

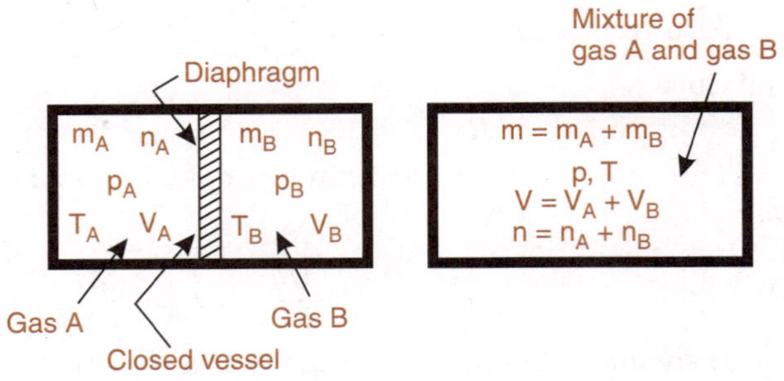

Fig. 8.3

In a free *expansion process, the internal energy initially is equal to the internal energy finally.* In this case, from eqn. (8.23),

$$U_1 = n_AC_{vA}\,T_A + n_BC_{vB}T_B$$

and
$$U_2 = (n_AC_{vA} + n_BC_{vB})T$$

If this result is extended to any number of gases, we have

$$U_1 = \Sigma n_iC_{vi}T_i \quad \text{and} \quad U_2 = T\,\Sigma n_iC_{vi}$$

Then $\quad U_1 = U_2$

i.e., $\quad \Sigma n_iC_{vi}T_i = T\,\Sigma n_iC_{vi}$

i.e.,
$$T = \frac{\Sigma n_iC_{vi}T_i}{\Sigma n_iC_{vi}} \qquad ...(8.27)$$

— When two streams of fluid meet to form a *common stream in steady flow*, they give *another form of mixing* (Fig. 8.4).

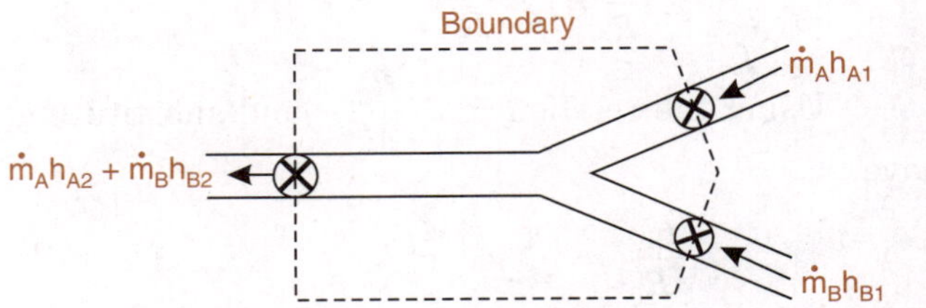

Fig. 8.4

Applying steady-flow energy equation to the *mixing section* (neglecting changes in kinetic and potential energy), we get

$$\dot{m}_Ah_{A_1} + \dot{m}_Bh_{B_1} + Q = \dot{m}_Ah_{A_2} + \dot{m}_Bh_{B_2} + W$$

In case of *adiabatic flow* : $Q = 0$, and also $W = 0$ in this case

$$\therefore \quad \dot{m}_Ah_{A_1} + \dot{m}_Bh_{B_1} = \dot{m}_Ah_{A_2} + \dot{m}_Bh_{B_2}$$

Also $h = c_pT$, hence,

$$\dot{m}_Ac_{pA}T_A + \dot{m}_Bc_{pB}T_B = \dot{m}_Ac_{pA}T + \dot{m}_Bc_{pB}T$$

For any number of gases this becomes

$$\Sigma\, \dot{m}_ic_{pi}T_i = T\,\Sigma\, \dot{m}_ic_{pi}$$

i.e.,
$$T = \frac{\Sigma\, m_ic_{pi}T_i}{\Sigma\, m_ic_{pi}} \qquad ...(8.28)$$

Also, $\quad C_p = Mc_p \quad \text{and} \quad M = m/n$

$\therefore \quad nC_p = mc_p$

Hence,
$$T = \frac{\Sigma n_iC_{pi}T_i}{\Sigma n_iC_{pi}} \qquad ...(8.29)$$

Eqns. (8.28) and (8.29) represent *one condition which must be satisfied in an adiabatic mixing process of perfect gas in steady flow.* In a particular problem some other information must be known (*e.g.,* specific volume or the final pressure) before a complete solution is possible.

8.7. GAS AND VAPOUR MIXTURES

Fig. 8.5 (*i*) shows a vessel of fixed volume which is maintained at a constant temperature. The vessel is *evacuated* and the absolute pressure is therefore *zero.*

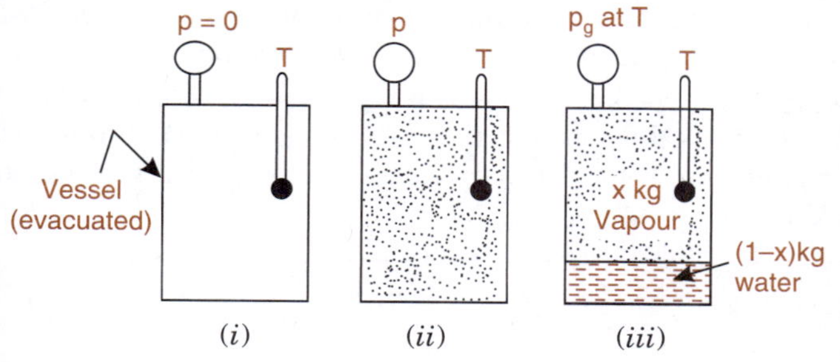

Fig. 8.5

In Fig. 8.5 (*ii*) a small quantity of water is introduced into the vessel and it evaporated to occupy the whole volume. For a small quantity of water introduced, the pressure in the vessel will be *less* than the saturation pressure corresponding to the temperature of the vessel. At this condition of pressure and temperature the vessel will be occupied by *superheated vapour*. As more water is introduced the pressure increases and the water continues to evaporate until such a condition is reached that the *volume can hold no water*. Any additional water introduced into the vessel after this will not evaporate but will exist as water, the condition being as in Fig. 8.5 (*iii*), which shows the vapour in contact with its liquid per kg of water introduced, the vessel can be thought of as containing either $(1 - x)$ kg of water plug x kg of dry saturated vapour, or as containing 1 kg of wet steam of dryness fraction x.

The *temperature remains constant during the whole process of evaporation*. If the temperature is now raised by the addition of heat, then more vapour will evaporate and the pressure in the vessel will increase. Eventually the *vessel will contain a superheated vapour as before, but at a higher pressure and temperature.*

In Fig. 8.5 the vessel is considered to be initially evacuated, but the water would evaporate in exactly the same way, if the vessel contains a gas or a mixture of gases. As stated in the Gibbs-Dalton law, each constituent behaves as if it occupies the whole vessel at the temperature of the vessel. When a little water is sprayed into a vessel containing a gas mixture, then the vapour forced will exert the saturation pressure corresponding to the temperature of the vessel, and this is the partial pressure of the vapour in the mixture.

In case there is a *saturated vapour in a mixture*, then the partial pressure of the vapour can be found from tables at the temperature of the mixture. This means that a saturated vapour obeys the Gibbs-Dalton law ; *this is only a good approximation at low values of the total pressure.*

Mixtures of air and water vapour have been considered in this chapter ; in chapter 9 moist atmospheric air (*i.e.,* a mixture of dry air and water vapour) has been considered as a separate topic-*Psychrometrics.*

Example 8.1. *A vessel of 0.35 m³ capacity contains 0.4 kg of carbon monoxide (molecular weight = 28) and 1 kg of air at 20°C. Calculate :*

(*i*) *The partial pressure of each constituent,*

(*ii*) *The total pressure in the vessel, and*

The gravimetric analysis of air may be taken as 23.3% oxygen (molecular weight = 32) and 76.7% nitrogen (molecular weight = 28).

Solution. Capacity of the vessel, $V = 0.35$ m³

Mass of carbon monoxide $= 0.4$ kg

Mass of air $= 1$ kg

Temperature, $T = 20°C$ or 293 K

Mass of oxygen present in 1 kg of air

$$= \frac{23.3}{100} \times 1 = 0.233 \text{ kg}$$

Mass of nitrogen present in 1 kg of air

$$= \frac{76.7}{100} \times 1 = 0.767 \text{ kg}$$

But, characteristic gas constant,

$$R = \frac{R_0}{M} \quad ...(i)$$

where, R_0 = Universal gas constant (= 8.314 kJ/kg K), and
M = Molecular weight.

Also, the characteristic gas equation is given by

$$pV = mRT \quad ...(ii)$$

$$\therefore \quad pV = \frac{mR_0T}{M} \quad ...(iii)$$

i.e.,

$$p = \frac{mR_0T}{MV} \quad ...(iv)$$

Hence, for a constituent,

$$p_i = \frac{m_iR_0T}{M_iV} \quad ...(v)$$

Substituting the values, we get the partial pressures as follows :

(*i*) **Partial pressures :**

For O_2, $$p_{O_2} = \frac{0.233 \times (8.314 \times 10^3) \times 293}{32 \times 0.35 \times 10^5}$$

$= \mathbf{0.5068}$ **bar. (Ans.)**

For N_2, $$p_{N_2} = \frac{0.767 \times (8.314 \times 10^3) \times 293}{28 \times 0.35 \times 10^5}$$

$= \mathbf{1.9065}$ **bar. (Ans.)**

For CO, $$p_{CO} = \frac{0.40 \times (8.314 \times 10^3) \times 293}{28 \times 0.35 \times 10^5}$$

$= \mathbf{0.9943}$ **bar. (Ans.)**

(*ii*) **Total pressure in the vessel, p :**

$$p = \Sigma p_i = p_{O_2} + p_{N_2} + p_{CO}$$

$$= 0.5068 + 1.9065 + 0.9943$$

$$= \mathbf{3.4076\ bar.}$$ **(Ans.)**

Example 8.2. *The gravimetric analysis of air and other data are as follows :*

Constituent	*Percentage*	*Molecular weight*
Oxygen	*23.14*	*32*
Nitrogen	*75.53*	*28*
Argon	*1.28*	*40*
Carbon dioxide	*0.05*	*44*

Calculate : (i) Gas constant for air ; (ii) Apparent molecular weight.

Solution. Using the relation,

$$R = \frac{R_0}{M} \quad ...(i)$$

$$R_{O_2} = \frac{8.314}{32} = 0.2598 \text{ kJ/kg K}$$

$$R_{N_2} = \frac{8.314}{28} = 0.2969 \text{ kJ/kg K}$$

$$R_{Ar} = \frac{8.314}{40} = 0.2078 \text{ kJ/kg K}$$

$$R_{CO_2} = \frac{8.314}{44} = 0.1889 \text{ kJ/kg K}$$

(*i*) **Gas constant for air :**

Now using the equation,

$$R = \sum \frac{m_i}{m} R_i, \text{ we have} \quad ...(ii)$$

$$R = 0.2314 \times 0.2598 + 0.7553 \times 0.2969 + 0.0128 \times 0.2078 + 0.0005 \times 0.1889$$

$$= 0.0601 + 0.2242 + 0.00266 + 0.00009$$

$$= 0.2871 \text{ kJ/kg K}$$

Hence, gas constant for air

$= \mathbf{0.2871\ kJ/kg\ K.}$ **(Ans.)**

(*ii*) **Apparent molecular weight :**

Now from eqn. (*i*), we have

$$M = \frac{8.314}{0.2871} = 28.96$$

Hence apparent molecular weight = **28.96. (Ans.)**

Example 8.3. *Following is the gravimetric analysis of air :*

Constituent	*Percentage*
Oxygen	*23.14*
Nitrogen	*75.53*
Argon	*1.28*
Carbon dioxide	*0.05*

Calculate the analysis by volume and the partial pressure of each constituent when the total pressure is 1 bar.

Solution. We know that the analysis by volume V_i/V, is the same as the mole fraction n_i/n. Also $n_i = m_i/M_i$; therefore considering 1 kg of mixture and using a tabular method, we have

Constituent	m_i	M_i	$n_i = \frac{m_i}{M_i}$	$\frac{n_i}{n} \times 100\% = \frac{V_i}{V} \times 100\%$
Oxygen	0.2314	32	0.00723	$\frac{0.00723 \times 100}{0.03453}$ = **20.94%. (Ans.)**
Nitrogen	0.7553	28	0.02697	$\frac{0.02697 \times 100}{0.03453}$ = **78.10%. (Ans.)**
Argon	0.0128	40	0.00032	$\frac{0.00032 \times 100}{0.03453}$ = **0.93%. (Ans.)**
Carbon dioxide	0.0005	44	0.00001	$\frac{0.00001 \times 100}{0.03453}$ = **0.03%. (Ans.)**
			$n = \Sigma n_i$ = 0.03453	

Also, $$\frac{p_i}{p} = \frac{V_i}{V} = \frac{n_i}{n}$$

$\therefore$ $$p_i = \frac{n_i}{n} p, \text{ hence}$$

For O_2, $p_{O_2} = 0.2094 \times 1 = \mathbf{0.2094\ bar.}$ **(Ans.)**

For N_2, $p_{N_2} = 0.7810 \times 1 = \mathbf{0.7810\ bar.}$ **(Ans.)**

For Ar, $p_{Ar} = 0.0093 \times 1 = \mathbf{0.0093\ bar.}$ **(Ans.)**

For CO_2, $p_{CO_2} = 0.0003 \times 1 = \mathbf{0.0003\ bar.}$ **(Ans.)**

☞**Example 8.4.** *A vessel contains at 1 bar and 20°C a mixture of 1 mole of CO_2 and 4 moles of air. Calculate for the mixture :*

(*i*) *The masses of CO_2, O_2 and N_2, and the total mass ;*

(*ii*) *The percentage carbon content by mass ;*

(*iii*) *The apparent molecular weight and the gas constant for the mixture ;*

(*iv*) *The specific volume of the mixture.*

The volumetric analysis of air can be taken as 21% oxygen and 79% nitrogen.

Solution. The pressure in the vessel, p = 1 bar

Temperature in the vessel, $T = 20 + 273 = 293$ K

No. of moles of CO_2 = 1 mole

No. of moles of air = 4 mole

From equation, $n_i = \left(\frac{V_i}{V}\right)n$, we have

$$n_{O_2} = 0.21 \times 4 = 0.84$$

$$n_{N_2} = 0.79 \times 4 = 3.16$$

(*i*) From equation, $m_i = n_i M_i$, we have

$$m_{CO_2} = 1 \times 44 = \mathbf{44\ kg.} \quad \textbf{(Ans.)}$$

$$m_{O_2} = 0.84 \times 32 = \mathbf{26.88\ kg.} \quad \textbf{(Ans.)}$$

and $m_{N_2} = 3.16 \times 28 = \mathbf{88.48\ kg.}$ **(Ans.)**

The *total mass*, $m = m_{CO_2} + m_{O_2} + m_{N_2}$

$= 44 + 26.88 + 88.48$

$= \mathbf{159.36\ kg.}$ **(Ans.)**

(*ii*) Since the molecular weight of carbon is 12, therefore, there are 12 kg of carbon present for every mole of CO_2,

i.e., **Percentage carbon in mixture**

$$= \frac{12 \times 100}{159.36} = \mathbf{7.53\%\ by\ mass.} \quad \textbf{(Ans.)}$$

(*iii*) From equation $n = \Sigma\, n_i$, we have

$$n = n_{CO_2} + n_{O_2} + n_{N_2}$$

$$= 1 + 0.84 + 3.16 = 5.0$$

Now using the equation $M = \sum \frac{n_i}{n} M_i$, we have

$$M = \frac{1}{5} \times 44 + \frac{0.84}{5} \times 32 + \frac{3.16}{5} \times 28$$

$$= 8.8 + 5.376 + 17.696 = 31.872$$

i.e., **Apparent molecular weight = 31.872. (Ans.)**

From equation, $R = \frac{R_0}{M}$, we have

$$R = \frac{8.314}{31.872} = 0.2608 \text{ kJ/kg K}$$

i.e., **Gas constant for the mixture**

= 0.2608 kJ/kg K. (Ans.)

(*iv*) To find specific volume of the mixture, v using the relation :

$$pv = RT$$

$$v = \frac{RT}{p} = \frac{0.2608 \times 10^3 \times 293}{1 \times 10^5}$$

$$= 0.7641 \text{ m}^3\text{/kg}$$

i.e., **Specific volume of the mixture**

= 0.7641 m³/kg. (Ans.)

Example 8.5. *A mixture of hydrogen (H_2) and oxygen (O_2) is to be made so that the ratio of H_2 to O_2 is 2 : 1 by volume. If the pressure and temperature are 1 bar and 25°C respectively, calculate :*

(*i*) *The mass of O_2 required ;*

(*ii*) *The volume of the container.*

Solution. Pressure, $p = 1$ bar

Temperature, $T = 25 + 273 = 298$ K

Ratio of H_2 to O_2 = 2 : 1 by volume.

(*i*) **The mass of O_2 required :**

Let the mass of O_2 per kg of H_2 = x kg

Now, $n_i = \frac{m_i}{M_i}$

$$\therefore \quad n_{H_2} = \frac{1}{2} = 0.5$$

and $n_{O_2} = \frac{x}{32}$

From equation, $\frac{V_i}{V} = \frac{n_i}{n}$, we have

$$\frac{V_{H_2}}{V_{O_2}} = \frac{n_{H_2}}{n_{O_2}} \quad \text{and} \quad \frac{V_{H_2}}{V_{O_2}} = 2 \text{ (given)}$$

$$\therefore \quad \frac{0.5}{x/32} = 2 \qquad \therefore \quad x = \frac{32 \times 0.5}{2} = 8$$

i.e., Mass of O_2 per kg of H_2 = **8 kg. (Ans.)**

(*ii*) **The volume of the container, V :**

The total number of moles in the vessel per kg of H_2 is

$$n = n_{H_2} + n_{O_2}$$

$$= 0.5 + \frac{x}{32} = 0.5 + \frac{8}{32} = 0.75$$

Now using the relation,

$$pV = nR_0T$$

$$\therefore \quad V = \frac{nR_0T}{p} = \frac{0.75 \times (8.314 \times 10^3)}{1 \times 10^5} \times 298$$

$$= 18.58 \text{ m}^3$$

i.e., **The volume of the container**

= 18.58 m³. (Ans.)

Example 8.6. *A gaseous mixture of composition by volume, 78% H_2 and 22% CO is contained in a vessel. It is desired that the mixture should be made in proportion 52% H_2 and 48% CO by removing some of the mixture and adding some CO. Calculate per mole of mixture the mass of mixture to be removed, and mass of CO to be added.*

Assume that the pressure and temperature in the vessel remain constant during the procedure.

Solution. Composition of mixture by volume :

$$H_2 = 78\%, \ CO = 22\%$$

Final composition desired : H_2 = 52%, CO = 48%

Since the pressure and temperature remain constant, then the number of moles in the vessel remain the same throughout.

$\therefore$ Moles of mixture removed = Moles of CO added.

Let x kg of mixture be removed and y kg of CO be added.

For the mixture, $M = \sum \frac{V_i}{V} M_i$

$\therefore$ $M = 0.78 \times 2 + 0.22 \times 28 = 7.72$

Also from equation, $n = \frac{m}{M}$, we have

Moles of mixture removed

$$= \frac{x}{7.72} = \text{moles of CO added} = \frac{y}{28}$$

From equation, $\frac{V_i}{V} = \frac{n_i}{n}$, we have

Moles of H_2 in the mixture removed

$$= 0.78 \times \frac{x}{7.72} = 0.101\,x$$

and Moles of H_2 initially = 0.78 × 1 = 0.78

Hence, Moles of H_2 remaining in vessel

$$= 0.78 - 0.101\,x$$

But 1 mole of the new mixture is 52% H_2 and 48% CO, therefore

$$0.78 - 0.101\,x = 0.52$$

$\therefore$ $0.101\,x = 0.26$ or $x = 2.57$

i.e., **Mass of mixture removed = 2.57 kg. (Ans.)**

Also since $\frac{x}{7.72} = \frac{y}{28}$

$\therefore$ $y = \frac{28}{7.72} \times x = \frac{28}{7.72} \times 2.57 = 9.32$ kg

i.e., **Mass of CO added = 9.32 kg. (Ans.)**

☞**Example 8.7.** *In an engine cylinder a gas has a volumetric analysis of 13% CO_2, 12.5% O_2, and 74.5% N_2. The temperature at the beginning of expansion is 950°C and the gas mixture expands reversibly through a volume ratio of 8 : 1, according to the law $pv^{1.2}$ = constant. Calculate per kg of gas :*

(i) The work done ;

(ii) The heat flow ;

(iii) Change of entropy per kg of mixture.

The values of c_p for the constituents CO_2, O_2 and N_2 are 1.235 kJ/kg K, 1.088 kJ/kg K and 1.172 kJ/kg K respectively.

Solution. From equation $m_i = n_i M_i$, the conversion from volume fraction to mass fraction is as follows :

Consider 1 mole of the mixture.

Constituent	n_i	M_i	$m_i = n_i M_i$	$\frac{m_i}{m}$ = *Fraction by mass*
CO_2	0.13	44	5.72	$\frac{5.72}{30.58} = 0.187$
O_2	0.125	32	4.00	$\frac{4.0}{30.58} = 0.131$
N_2	0.745	28	20.86	$\frac{20.86}{30.58} = 0.682$
			$\Sigma m_i = 30.58$	

Now using the equation

$$c_p = \sum \frac{m_i}{m} c_{pi}$$

$\therefore$ $c_p = 0.187 \times 1.235 + 0.131 \times 1.088 + 0.682 \times 1.172$

$= 0.231 + 0.1425 + 0.799$

$= 1.1725$ kJ/kg K

From equations, $R = \sum \frac{m_i}{m} R_i$ and $R_i = \frac{R_0}{M_i}$, we have

$$R = 0.187 \times \frac{8.314}{44} + 0.131 \times \frac{8.314}{32} + 0.682 \times \frac{8.314}{28}$$

$= 0.0353 + 0.0340 + 0.2025$

$= 0.2718$ kJ/kg K

From equation, $c_p - c_v = R$, we have

$c_v = 1.1725 - 0.2718 = 0.9$ kJ/kg K

(*i*) **The work done, W :**

$$W = \frac{R(T_1 - T_2)}{n - 1}.$$

T_2 can be found by using the equation

$$\frac{T_2}{T_1} = \left(\frac{v_1}{v_2}\right)^{n-1} = \left(\frac{1}{8}\right)^{1.2-1} = 0.659$$

$\therefore$ $T_2 = (950 + 273) \times 0.659 = 805.9$ K

$\therefore$ $W = \frac{0.2718(1223 - 805.9)}{1.2 - 1}$

$=$ **566.8 kJ/kg. (Ans.)**

(*ii*) **The heat flow, Q :**

Also from equation,

$u_2 - u_1 = c_v(T_2 - T_1)$ ... for 1 kg

$= 0.9(805.9 - 1223) = -375.39$ kJ/kg

Now from the non-flow energy equation,

$Q = (u_2 - u_1) + W$

$= -375.39 + 566.8 = 191.41$ kJ/kg

i.e., **Heat supplied = 191.41 kJ/kg. (Ans.)**

(*iii*) **Change of entropy per kg of mixture :**

Refer to Fig. 8.6. The change of entropy between state 1 and state 2 can be found by imagining the process replaced by two other processes 1 to A and A to 2.

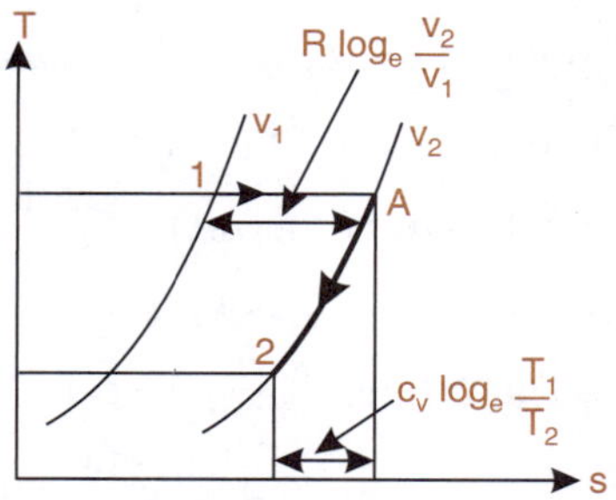

Fig. 8.6

For *isothermal process* 1-*A*

$$s_A - s_1 = R \log_e \frac{v_2}{v_1}$$

$$= 0.2718 \log_e 8 = 0.565 \text{ kJ/kg K}$$

For *constant volume process A*-2 :

$$s_A - s_2 = c_v \int_2^A \frac{dT}{T} = c_v \log_e \frac{T_1}{T_2}$$

$$= 0.9 \log_e \frac{1223}{805.9}$$

i.e., $s_A - s_2 = 0.375$ kJ/kg K

Then by subtraction,

$$s_2 - s_1 = 0.565 - 0.375 = 0.19 \text{ kJ/kg K}$$

i.e., **Change of entropy per kg of mixture**

= 0.19 kJ/kg K. (Ans.)

Example 8.8. *The following is the volumetric analysis of a producer gas : CO = 28%, H_2 = 13%, CH_4 = 4%, CO_2 = 4%, N_2 = 51%. The values of C_p for the constituents CO, H_2, CH_4, CO_2 and N_2 are 29.27 kJ/mole K, 28.89 kJ/mole K, 35.8 kg/mole K, 37.22 kJ/mole K, 29.14 kJ/mole K respectively. Calculate the values of C_p, C_v, c_p and c_v for the mixture.*

Solution. Using the relation

$$C_p = \sum \frac{n_i}{n} C_{pi}$$

$\therefore$ $C_p = 0.28 \times 29.27 + 0.13 \times 28.89 + 0.04 \times 35.8 + 0.04 \times 37.22 + 0.51 \times 29.14$

i.e., $\mathbf{C_p}$ **= 29.733 kJ/mole K. (Ans.)**

Also $C_p - C_v = R_0$

$\therefore$ $\mathbf{C_v} = C_p - R_0 = 29.733 - 8.314$

= 21.419 kJ/mole K. (Ans.)

To find the molecular weight, using the equation :

$$M = \sum \frac{n_i}{n} M_i$$

$$= 0.28 \times 28 + 0.13 \times 2 + 0.04 \times 16 + 0.04 \times 44 + 0.51 \times 28$$

$$= 24.78$$

Now, $\mathbf{c_p} = \frac{C_p}{M} = \frac{29.733}{24.78}$

= 1.199 kJ/kg K. (Ans.)

and $\mathbf{c_v} = \frac{C_v}{M} = \frac{21.419}{24.78}$

= 0.864 kJ/kg K. (Ans.)

Example 8.9. *The analysis by weight of a perfect gas mixture at 20°C and 1.3 bar is 10% O_2, 70% N_2, 15% CO_2 and 5% CO. For a reference state of 0°C and 1 bar determine :*

(*i*) *Partial pressures of the constituents ;*

(*ii*) *Gas constant of mixture.*

Solution. Using the relation, $M = \dfrac{1}{\sum \dfrac{m_{f_i}}{M_i}}$

where, M = Molecular weight of the mixture ;

m_{f_i} = Mass fraction of a constituent, ;

M_i = Molecular weight of a constituent.

$$M = \frac{1}{\frac{(m_f)_{O_2}}{M_{O_2}} + \frac{(m_f)_{N_2}}{M_{N_2}} + \frac{(m_f)_{CO_2}}{M_{CO_2}} + \frac{(m_f)_{CO}}{M_{CO}}}$$

$$= \frac{1}{\frac{0.1}{32} + \frac{0.7}{28} + \frac{0.15}{44} + \frac{0.05}{28}}$$

$$= \frac{1}{0.03332} = 30 \text{ kg/kg mole}$$

Considering 1 kg of the mixture

$$n = \frac{m}{M} = \frac{1}{30} = 0.033$$

Now mole fraction $x_i = \frac{n_i}{\Sigma n_i} = \frac{n_i}{n} = \frac{m_i / M_i}{n}$

$\therefore$ $x_{O_2} = \frac{(0.1/32)}{0.033} = 0.094$;

$x_{N_2} = \frac{(0.7/28)}{0.033} = 0.757$

$x_{CO_2} = \frac{(0.15/44)}{0.033} = 0.103$;

$x_{CO} = \frac{(0.05/28)}{0.033} = 0.054$

(*i*) **Partial pressures of the constituents :**

$\mathbf{p_{O_2}} = x_{O_2} \times p = 0.094 \times 1.3$

= 0.132 bar. (Ans.)

$\mathbf{p_{N_2}} = x_{N_2} \times p = 0.757 \times 1.3$

= 0.984 bar. (Ans.)

$$p_{CO_2} = x_{CO_2} \times p = 0.103 \times 1.3$$
$$= \mathbf{0.1334\ bar.} \quad \textbf{(Ans.)}$$
$$p_{CO} = x_{CO} \times p = 0.054 \times 1.3$$
$$= \mathbf{0.0702\ bar.} \quad \textbf{(Ans.)}$$

(*ii*) **Gas constant of mixture, R_{mix} :**

$$R_{mix} = \frac{R_0}{M} = \frac{8.314}{30}$$
$$= \mathbf{0.277\ kJ/kg\ K.} \quad \textbf{(Ans.)}$$

Example 8.10. *A mixture of ideal gases consists of 4 kg of nitrogen and 6 kg of carbon dioxide at a pressure of 4 bar and a temperature of 20°C. Find :*

(i) The mole fraction of each constituent,
(ii) The equivalent molecular weight of the mixture,
(iii) The equivalent gas constant of the mixture,
(iv) The partial pressures and partial volumes,
(v) The volume and density of the mixture, and
(vi) The c_p and c_v of the mixture.

If the mixture is heated at constant volume to 50°C, find the changes in internal energy, enthalpy and entropy of the mixture. Find the changes in internal energy, enthalpy and entropy of the mixture if the heating is done at constant pressure.

Take γ : for CO_2 = 1.286 and for N_2 = 1.4.

Solution. (*i*) **The mole fraction of each constituent :**

Since mole fraction $\quad x_i = \dfrac{n_i}{\Sigma n_i}$

$$\therefore \quad x_{N_2} = \frac{\frac{4}{28}}{\frac{4}{28} + \frac{6}{44}} = \frac{0.1428}{0.1428 + 0.1364}$$
$$= \frac{0.1428}{0.2792} = \mathbf{0.511.} \quad \textbf{(Ans.)}$$
$$x_{CO_2} = \frac{\frac{6}{44}}{\frac{4}{28} + \frac{6}{44}} = \frac{0.1364}{0.1428 + 0.1364}$$
$$= \mathbf{0.488.} \quad \textbf{(Ans.)}$$

(*ii*) **The equivalent molecular weight of the mixture, M :**

$$M = 0.511 \times 28 + 0.488 \times 44$$
$$= \mathbf{35.78\ kg/kg\ mole.} \quad \textbf{(Ans.)}$$

(*iii*) **The equivalent gas constant of the mixture, R_{mix} :**

Total mass, $\quad m = m_{N_2} + m_{CO_2} = 4 + 6 = 10$ kg

$$\therefore \quad R_{mix} = \frac{m_{N_2} R_{N_2} + m_{CO_2} R_{CO_2}}{m}$$
$$= \frac{4 \times \left(\frac{8.314}{28}\right) + 6 \times \left(\frac{8.314}{44}\right)}{10} \quad \left[\because \ R = \frac{R_0}{M}\right]$$
$$= \mathbf{0.232\ kJ/kg\ K.} \quad \textbf{(Ans.)}$$

(*iv*) **The partial pressures and partial volumes :**

$$p_{N_2} = x_{N_2} \times p = 0.511 \times 4$$
$$= \mathbf{2.044\ bar.} \quad \textbf{(Ans.)}$$
$$p_{CO_2} = x_{CO_2} \times p = 0.488 \times 4$$
$$= \mathbf{1.952\ bar.} \quad \textbf{(Ans.)}$$
$$V_{N_2} = \frac{m_{N_2} R_{N_2} T}{p} = \frac{4 \times \frac{8.314}{28} \times 293 \times 10^3}{4 \times 10^5}$$
$$= \mathbf{0.87\ m^3.} \quad \textbf{(Ans.)}$$
$$V_{CO_2} = \frac{m_{CO_2} R_{CO_2} T}{p} = \frac{6 \times \frac{8.314}{44} \times 293 \times 10^3}{4 \times 10^5}$$
$$= \mathbf{0.83\ m^3.} \quad \textbf{(Ans.)}$$

(*v*) **The volume and density of the mixture :**

Total volume of the mixture,

$$V = \frac{mRT}{p} = \frac{m_{N_2} R_{N_2} T}{p_{N_2}} = \frac{m_{CO_2} R_{CO_2} T}{p_{CO_2}}$$
$$\therefore \quad V = \frac{10 \times 0.232 \times 293 \times 10^3}{4 \times 10^5}$$
$$= \mathbf{1.699\ m^3.} \quad \textbf{(Ans.)}$$

Density of the mixture,

$$\rho_{mix} = \rho_{N_2} + \rho_{CO_2} = \frac{m}{V}$$
$$= \frac{10}{1.699} = \mathbf{5.88\ kg/m^3.} \quad \textbf{(Ans.)}$$

(*vi*) **c_p and c_v of the mixture :**

$$c_{p_{N_2}} - c_{v_{N_2}} = R_{N_2}$$
$$\therefore \quad c_{v_{N_2}} = \frac{R_{N_2}}{\gamma - 1} = \frac{8.314}{28(1.4 - 1)} = 0.742 \text{ kJ/kg K.}$$

and $\quad c_{p_{N_2}} = 1.4 \times 0.742 = 1.039$ kJ/kg K

Now, $\quad c_{v_{CO_2}} = \dfrac{R_{CO_2}}{\gamma - 1} = \dfrac{8.314}{44(1.286 - 1)} \quad [\because \gamma \text{ for } CO_2 = 1.286]$

$$= 0.661 \text{ kJ/kg K}$$

and $\quad c_{p_{CO_2}} = 1.286 \times 0.661 = 0.85$ kJ/kg K

For the mixture :

$$c_p = \frac{m_{N_2} c_{p_{N_2}} + m_{CO_2} c_{p_{CO_2}}}{m_{N_2} + m_{CO_2}}$$

$$= \frac{4 \times 1.039 + 6 \times 0.85}{(4 + 6)}$$

$= \mathbf{0.925\ kJ/kg\ K.}$ **(Ans.)**

$$c_v = \frac{m_{N_2} c_{v_{N_2}} + m_{CO_2} c_{v_{CO_2}}}{m_{N_2} + m_{CO_2}}$$

$$= \frac{4 \times 0.742 + 6 \times 0.661}{(4 + 6)}$$

$= \mathbf{0.693\ kJ/kg\ K.}$ **(Ans.)**

When the mixture is heated at constant volume :

Change in internal energy,

$U_2 - U_1 = mc_v(T_2 - T_1)$

$= 10 \times 0.693(50 - 20) = \mathbf{207.9\ kJ.}$ **(Ans.)**

Change in entropy,

$H_2 - H_1 = mc_p(T_2 - T_1)$

$= 10 \times 0.925(50 - 20) = \mathbf{277.5\ kJ.}$ **(Ans.)**

Change in entropy,

$$S_2 - S_1 = mc_v \log_e \frac{T_2}{T_1} + mR \log_e \frac{V_2}{V_1}$$

$$= mc_v \log_e \frac{T_2}{T_1} \qquad (\because V_1 = V_2)$$

$$= 10 \times 0.693 \times \log_e \left(\frac{273 + 50}{273 + 20}\right)$$

$= \mathbf{0.675\ kJ/kg\ K.}$ **(Ans.)**

When the mixture is heated at constant pressure :

If the mixture is heated at constant pressure *ΔU and ΔH will remain the same.*

The **change in entropy** will be

$$S_2 - S_1 = mc_p \log_e \frac{T_2}{T_1} - mR \log_e \frac{p_2}{p_1}$$

$$= mc_p \log_e \frac{T_2}{T_1} \qquad (\because p_1 = p_2)$$

$$= 10 \times 0.925 \log_e \left(\frac{323}{293}\right)$$

$= \mathbf{0.902\ kJ/kg\ K.}$ **(Ans.)**

Example 8.11. *A vessel of 1.8 m³ capacity contains oxygen at 8 bar and 50°C. The vessel is connected to another vessel of 3.6 m³ capacity containing carbon monoxide at 1 bar and 20°C. A connecting valve is opened and the gases mix adiabatically. Calculate :*

(i) The final temperature and pressure of the mixture ;

(ii) The change of entropy of the system.

Take : For oxygen C_v = 21.07 kJ/mole K.

For carbon monoxide C_v = 20.86 kJ/mole K.

Solution. Using the relation, $n = \dfrac{pV}{R_0T}$

$$\therefore \quad n_{O_2} = \frac{8 \times 10^5 \times 1.8}{(8.314 \times 10^3) \times 323} = 0.536$$

(where T_{O_2} = 50 + 273 = 323 K)

and
$$n_{CO} = \frac{1 \times 10^5 \times 3.6}{(8.314 \times 10^3) \times 293} = 0.1478$$

(where T_{CO} = 20 + 273 = 293 K).

(*i*) **Final temperature (T) and pressure (p) of the mixture :**

Before mixing :

$U_1 = \Sigma\, n_i C_{vi} T_i$

$= 0.536 \times 21.07 \times 323 + 0.1478 \times 20.86 \times 293$

i.e., $U_1 = 4551.15$ kJ

After mixing :

$U_2 = T\, \Sigma\, n_i C_{vi}$

$= T\,(0.536 \times 21.07 + 0.1478 \times 20.86)$

i.e., $U_2 = 14.37\, T$

For **adiabatic mixing,**

$U_1 = U_2$

$\therefore \quad 4551.15 = 14.37\, T$

$$\therefore \quad T = \frac{4551.15}{14.37} = 316.7 \text{ K}$$

$\therefore$ Temperature of the mixture

$= 316.7 - 273 = \mathbf{43.7°C.}$ **(Ans.)**

Now $p = \dfrac{nR_0T}{V}$

$$\therefore \quad p = \frac{(0.536 + 0.1478) \times 8.314 \times 10^3 \times 316.7}{(1.8 + 3.6) \times 10^5}$$

$= 3.33$ bar

i.e., Pressure after mixing = **3.33 bar. (Ans.)**

(*ii*) **Change of entropy of the system :**

Change of entropy of the system = change of entropy of the O_2 + change of entropy of CO

...... Gibbs-Dalton law

Referring to Fig. 8.7, the change of entropy of O_2 can be calculated by replacing the process undergone by the oxygen by the two processes 1 to *A* and *A* to 2.

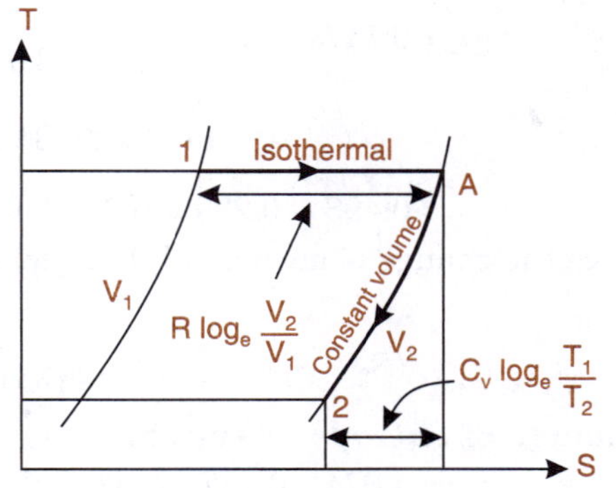

Fig. 8.7

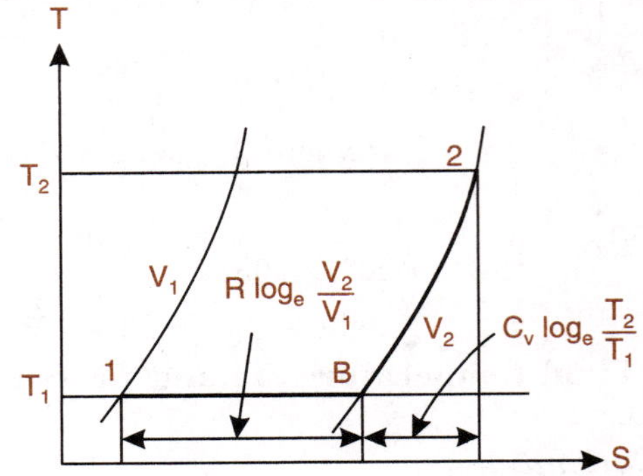

Fig. 8.8

For an *isothermal process* 1-*A* :

$$S_A - S_1 = mR \log_e \frac{V_A}{V_1}$$

or,
$$S_A - S_1 = nR_0 \log_e \frac{V_A}{V_1}$$

$$= 0.536 \times 8.314 \times \log_e \frac{5.4}{1.8}$$

$$= 4.896 \text{ kJ/K}$$

For *constant volume process A*-2 :

$$S_A - S_2 = mc_v \int_2^A \frac{dT}{T} = mc_v \log_e \frac{T_1}{T_2}$$

i.e.,
$$S_A - S_2 = nC_v \log_e \frac{T_1}{T_2}$$

$$= 0.536 \times 21.07 \times \log_e \frac{323}{316.7}$$

$$= 0.222 \text{ kJ/K}$$

$$\therefore \quad S_2 - S_1 = (S_A - S_1) - (S_A - S_2)$$
$$= 4.896 - 0.222 = 4.674 \text{ kJ/K}$$

Referring to Fig. 8.8, the change of entropy of CO can be found in a similar way to the above,

i.e.,
$$S_2 - S_1 = (S_B - S_1) + (S_2 - S_B)$$

$$\therefore \quad S_2 - S_1 = nR_0 \log_e \frac{V_B}{V_1} + nC_v \log_e \frac{T_2}{T_1}$$

$$= 0.1478 \times 8.314 \times \log_e \frac{5.4}{3.6}$$

$$+ 0.1478 \times 20.86 \log_e \frac{316.7}{293}$$

$$= 0.498 + 0.239 = 0.737 \text{ kJ/K}$$

Hence the change of entropy of the whole system is given by

$$(S_2 - S_1)_{system} = (S_2 - S_1)_{O_2} + (S_2 - S_1)_{CO}$$

i.e., **Change of entropy of system**

$$= 4.674 + 0.737 = \mathbf{5.411 \text{ kJ/K.}} \quad \textbf{(Ans.)}$$

☞Example 8.12. *Two vessels, A and B, both containing nitrogen, are connected by a valve which is opened to allow the contents to mix and achieve an equilibrium temperature of 30°C. Before mixing the following information is known about the gases in the two vessels.*

Vessel	*Pressure (p)*	*Temp. (t)*	*Contents*
A	*16 bar*	*55°C*	*0.6 kg mole*
B	*6.4 bar*	*25°C*	*3.6 kg*

(a) Calculate : (i) The final equilibrium pressure ;

(ii) The amount of heat transferred to the surroundings ;

(b) If the vessel had been perfectly insulated, calculate :

(i) The final temperature ;

(ii) The final pressure.

Assume γ = 1.4.

Solution. Refer to Fig. 8.9.

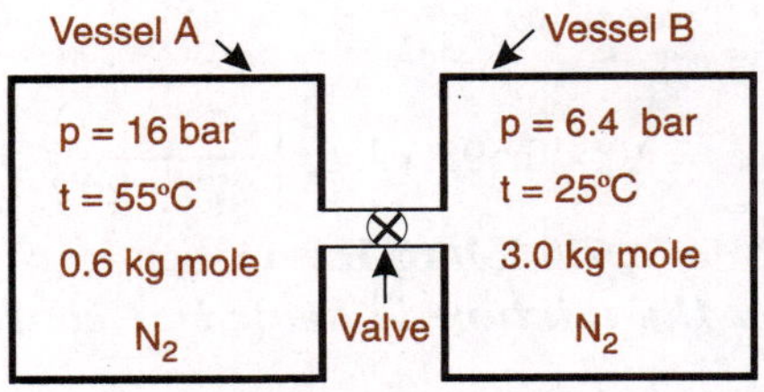

Fig. 8.9

Vessel A :

$$p_A V_A = n_A R_0 T_A$$

(where V_A = volume of vessel *A*)

$$16 \times 10^5 \times V_A = 0.6 \times 8.314 \times 10^3 \times (55 + 273)$$

$$\therefore \quad V_A = 1.022 \text{ m}^3$$

The mass of gas in vessel *A*

$$m_A = n_A M_A = 0.6 \times 28 = 16.8 \text{ kg}$$

Characteristic gas constant *R* of nitrogen

$$R = \frac{8.314}{28} = 0.297 \text{ kJ/kg K}$$

Vessel B :

$$p_B V_B = m_B R T_B$$

$$6.4 \times 10^5 \times V_B = 3.0 \times 0.297 \times 10^3 \times (25 + 273)$$

$$\therefore \quad V_B = 0.415 \text{ m}^3$$

Total volume of *A* and *B*

$$V = V_A + V_B = 1.022 + 0.415 = 1.437 \text{ m}^3$$

Total mass of gas

$$m = m_A + m_B = 16.8 + 3 = 19.8 \text{ kg}$$

Final temperature after mixing

$$T = 30 + 273 = 303 \text{ K}$$

(*a*) (*i*) **Final equilibrium pressure, p :**

$$pV = mRT$$

or $p \times 1.437 = 19.8 \times 0.297 \times 303 \times 10^3$

$= 12.4 \times 10^5 \text{ N/m}^2$

i.e., **p = 12.4 bar. (Ans.)**

Also $c_v = \dfrac{R}{\gamma - 1} = \dfrac{0.297}{1.4 - 1} = 0.743$ kJ/kg K.

(*ii*) **Amount of heat transferred, Q :**

As there is no work transfer, the amount of heat transfer,

Q = change of internal energy

$= U_2 - U_1$

Measuring the internal energy above the datum of absolute zero (at $T = 0$ K, $u = 0$ kJ/kg).

Internal energy U_1 (before mixing)

$= m_A c_v T_A + m_B c_v T_B$

$= 16.8 \times 0.743 \times 328 + 3.0 \times 0.743 \times 298$

$= 4758.5$ kJ

Final internal energy U_2 (after mixing)

$= mc_v T = 19.8 \times 0.743 \times 303 = 4457.5$ kJ

$\therefore$ $\mathbf{Q} = U_2 - U_1 = 4457.5 - 4758.5$

$=$ **–301 kJ. (Ans.)**

(*b*) **If the vessel were insulated :**

(*i*) **Final temperature, $\mathbf{t_2}$:**

If the vessel were insulated

$$Q = U_2 - U_1 = 0$$

i.e., $U_1 = U_2$

$$m_A c_v T_A + m_B c_v T_B = mc_v T$$

$$\therefore \quad T = \frac{m_A c_v T_A + m_B c_v T_B}{mc_v} = \frac{m_A T_A + m_B T_B}{m}$$

$$= \frac{16.8 \times 328 + 3.0 \times 298}{19.8} = 323.5 \text{ K}$$

i.e., **t** = 323.5 – 273 = **50.5°C. (Ans.)**

(*ii*) **Final pressure, p :**

$$pV = mRT$$

or $p = \dfrac{mRT}{V} = \dfrac{19.8 \times 0.297 \times 10^3 \times 323.5}{1.437}$

$= 13.24 \times 10^5 \text{ N/m}^2$

i.e., **p = 13.24 bar. (Ans.)**

Example 8.13. *Calculate the increase in entropy when 3 kg of O_2 at 50°C are mixed with 9 kg of N_2 at the same temperature. The initial pressure of each constituent is 11 bar and is the same as that of the mixture.*

Solution. $x_{O_2} = \dfrac{p_{O_2}}{p} = \dfrac{3/32}{3/32 + 9/28} = 0.225$

$$x_{N_2} = \frac{p_{N_2}}{p} = \frac{9/28}{3/32 + 9/28} = 0.774$$

Increase of entropy due to diffusion

$$\Delta S = -\, m_{O_2} R_{O_2} \log_e \frac{p_{O_2}}{p} - m_{N_2} R_{N_2} \log_e \frac{p_{N_2}}{p}$$

$$= -3 \times \left(\frac{8.314}{32}\right) \log_e 0.225 - 9 \times \left(\frac{8.314}{28}\right) \times \log_e 0.774$$

$= 1.1626 + 0.6846$

$=$ **1.8472 kJ/kg K. (Ans.)**

Example 8.14. *2.5 kg of N_2 at 15 bar and 40°C is contained a rigid vessel. Adequate quantity of O_2 is added to increase the pressure to 20 bar while the temperature remains constant at 40°C.*

Calculate the mass of O_2 added.

Solution. $m_{N_2} = 2.5$ kg, $M_{N_2} = 28$

Initial pressure, $p_{N_2} = 15$ bar

Final pressure, $p_{\text{total}} \; (= p_{N_2} + p_{O_2}) = 20$ bar

Temperature = 40°C

In this process, the number of nitrogen moles remains *constant* at

$$n_{N_2} = \frac{m_{N_2}}{M_{N_2}} = \frac{2.5}{28} = 0.0893$$

After the oxygen is added, $p_{N_2} = 15$ bar, since the volume remains unchanged

$$p_{O_2} = p_{\text{total}} - p_{N_2} = 20 - 15 = 5 \text{ bar}$$

Now $$p_{N_2} = \frac{n_{N_2} R_0 T}{V} \quad ...(i)$$

and $$p_{O_2} = \frac{n_{O_2} R_0 T}{V} \quad ...(ii)$$

Dividing (*ii*) by (*i*), we get

$$\frac{p_{O_2}}{p_{N_2}} = \frac{n_{O_2}}{n_{N_2}}$$

$$\therefore \quad n_{O_2} = \frac{p_{O_2} \times n_{N_2}}{p_{N_2}}$$

$$= \frac{5 \times 10^5 \times 0.0893}{15 \times 10^5} = 0.0297$$

$\therefore$ **Mass of $\mathbf{O_2}$ added,**

$$m_{O_2} = n_{O_2} \times M_{O_2}$$

$= 0.0297 \times 32 =$ **0.9504 kg. (Ans.)**

Example 8.15. *Given that air consists of 21% oxygen and 79% nitrogen by volume. Determine :*

(i) The moles of nitrogen per mole of oxygen ;

(ii) The partial pressure of oxygen and nitrogen if the total pressure is atmosphere ;

(iii) The kg of nitrogen per kg of mixture.

Solution. Let $n_{O_2} = 1.0$ and V = volume of air

so that $V_{O_2} = 0.21V$; $V_{N_2} = 0.79V$

Let V contain $n = n_{O_2} + n_{N_2}$ moles of air at p and T.

(*i*) **Moles of N_2 per mole of O_2 :**

Now $pV_{O_2} = n_{O_2} R_0 T$ ...(*i*)

and $pV_{N_2} = n_{N_2} R_0 T$ ...(*ii*)

Dividing (*i*) by (*ii*), we get

$$\frac{V_{O_2}}{V_{N_2}} = \frac{n_{O_2}}{n_{N_2}}$$

$$\therefore \quad \mathbf{n_{N_2}} = \frac{n_{O_2} \times V_{N_2}}{V_{O_2}}$$

$$= \frac{1 \times 0.79V}{0.21V} = \mathbf{3.76\ moles.} \quad \textbf{(Ans.)}$$

(*ii*) $\mathbf{p_{O_2}}$ **and** $\mathbf{p_{N_2}}$ **:**

Now $x_{O_2} = \dfrac{n_{O_2}}{4} = \dfrac{p_{O_2}}{p}$ $(n = n_{O_2} + n_{N_2})$

(where p = total pressure = 1 atm.)

$$\therefore \quad \mathbf{p_{O_2}} = \frac{n_{O_2}}{n} \times p = \frac{1}{(1 + 3.76)} \times 1$$

$$= \mathbf{0.21\ atm.} \quad \textbf{(Ans.)}$$

Again, $x_{N_2} = \dfrac{n_{N_2}}{n} = \dfrac{p_{N_2}}{p}$

$$\therefore \quad \mathbf{p_{N_2}} = \frac{n_{N_2}}{n} \times p$$

$$= \frac{3.76}{1 + 3.76} \times 1 = \mathbf{0.79\ atm.} \quad \textbf{(Ans.)}$$

(*iii*) **The kg of nitrogen per kg of mixture :**

$$\frac{m_{N_2}}{m_{N_2} + m_{O_2}} = \frac{n_{N_2} M_{N_2}}{n_{N_2} M_{N_2} + n_{O_2} M_{O_2}}$$

$$= \frac{3.76 \times 28}{3.76 \times 28 + 1 \times 32}$$

$$= \mathbf{0.77\ kg\ N_2/kg\ mix.} \quad \textbf{(Ans.)}$$

Example 8.16. *Air (N_2 = 77%, O_2 = 23% by weight) at 25°C and 12 bar is contained in a vessel of capacity 0.6 m³. Some quantity of CO_2 is forced into the vessel so that the temperature remains at 25°C but the pressure rises to 18 bar.*

Find the masses of O_2, N_2 and CO_2 in the cylinder.

Solution. Volume of the vessel, $V = 0.6\ m^3$

Temperature (constant), $T = 25 + 273 = 298$ K

Initial pressure = 12 bar

Final pressure = 18 bar

Now $pV = nR_0T$

or
$$n = \frac{pV}{R_0T} = \frac{12 \times 10^5 \times 0.6}{8.314 \times 10^3 \times 298} = 0.29$$

Also
$$R_{air} = \left(\frac{m_{O_2}}{M_{O_2}} + \frac{m_{N_2}}{M_{N_2}}\right)\frac{R_0}{m} \qquad (m = n \times M)$$

Considering 100 kg of air

$$m_{O_2} = 23 \text{ kg}, \ m_{N_2} = 77 \text{ kg}$$

$$\therefore \quad R_{air} = \left(\frac{23}{32} + \frac{77}{28}\right) \times \frac{8.314}{100} = 0.288 \text{ kJ/kg K}$$

$$M_{air} = \frac{R_0}{R_{air}} = \frac{8.314}{0.288} = 28.87$$

Now $pV = mRT$

$$\therefore \quad m = \frac{pV}{RT}$$

$$= \frac{12 \times 10^5 \times 0.6}{0.288 \times 298 \times 10^3} = 8.39 \text{ kg of air}$$

Mass of O_2, $m_{O_2} = 0.23 \times 8.39 = \mathbf{1.93\ kg.}$ **(Ans.)**

Mass of N_2, $m_{N_2} = 0.77 \times 8.39 = \mathbf{6.46\ kg.}$ **(Ans.)**

After adding CO_2 in the vessel :

$p_{N_2} + p_{O_2} = 12$ bar ... before adding CO_2

$p_{CO_2} + (p_{N_2} + p_{O_2}) = 18$ bar ... after adding CO_2

or $p_{CO_2} + 12 = 18$

$p_{CO_2} = 6$ bar

Now
$$p_{CO_2} = \frac{m_{CO_2} R_{CO_2} T_{CO_2}}{V}$$

$$\therefore \quad m_{CO_2} = \frac{p_{CO_2}}{R_{CO_2} T_{CO_2}} = \frac{6 \times 10^5 \times 0.6}{\left(\frac{8.314}{44}\right) \times 298 \times 10^3}$$

$$= 6.39 \text{ kg}$$

i.e., **Mass of CO_2 = 6.39 kg. (Ans.)**

Example 8.17. *A vessel of 6 m³ capacity contains two gases A and B in proportion of 45 per cent and 55 per cent respectively at 30°C. If the value of R for the gases is 0.288 kJ/kg K and 0.295 kJ/kg K and if the total weight of the mixture is 2 kg, calculate :*

(i) The partial pressure ; (ii) The total pressure ;

(iii) The mean value of R for the mixture.

Solution. Capacity of the vessel, $V = 6 \text{ m}^3$

%age content of gas $A = 45\%$

%age content of gas $B = 55\%$

R for gas A, $R_A = 0.288$ kJ/kg K

R for gas B, $R_B = 0.295$ kJ/kg K

Total weight of the mixture $= 2$ kg

Temperature, $T = 30 + 273 = 303$ K.

(*i*) **The partial pressures, p_A, p_B :**

$$m_A = 2 \times 0.45 = 0.9 \text{ kg}$$

$$m_B = 2 \times 0.55 = 1.1 \text{ kg}$$

Now, $p_A V = m_A R_A T_A$

$$\therefore \quad \mathbf{p_A} = \frac{m_A R_A T_A}{V}$$

$$= \frac{0.9 \times 0.288 \times 303 \times 10^3}{6 \times 10^5}$$

$= \mathbf{0.13\ bar.}$ **(Ans.)**

and $$\mathbf{p_B} = \frac{m_B R_B T_B}{V} = \frac{1.1 \times 0.295 \times 303 \times 10^3}{6 \times 10^5}$$

$= \mathbf{0.164\ bar.}$ **(Ans.)**

(*ii*) **The total pressure, p :**

$$p = p_A + p_B$$

$= 0.13 + 0.164 = \mathbf{0.294\ bar.}$ **(Ans.)**

(*iii*) **The mean value of R for the mixture, R_m :**

$$R_m = \frac{m_A R_A + m_B R_B}{m_A + m_B}$$

$$= \frac{0.9 \times 0.288 + 1.1 \times 0.295}{(0.9 + 1.1)}$$

$= 0.292$ kJ/kg K

i.e., **Mean value of R for the mixture**

= 0.292 kJ/kg K. (Ans.)

Example 8.18. *The pressure and temperature of mixture of 4 kg of O_2 and 6 kg of N_2 are 4 bar and 27°C respectively. For the mixture determine the following :*

(*i*) *The mole fraction of each component ;*

(*ii*) *The average molecular weight ;*

(*iii*) *The specific gas constant ;*

(*iv*) *The volume and density ;*

(*v*) *The partial pressures and partial volumes.*

Solution. Mass of oxygen, $m_{O_2} = 4$ kg

Mass of nitrogen, $m_{N_2} = 6$ kg

Pressure, $p = 4$ bar

Temperature, $T = 27 + 273 = 300$ K.

(*i*) **The mole fraction of each component :**

$$n = \frac{m}{M}$$

$$\therefore \quad n_{O_2} = \frac{4}{32} = 0.125 \text{ and } n_{N_2} = \frac{6}{28} = 0.214$$

Now $$\mathbf{x_{O_2}} = \frac{0.125}{0.125 + 0.214} = \mathbf{0.3687.} \textbf{ (Ans.)}$$

and $$\mathbf{x_{N_2}} = \frac{0.214}{0.125 + 0.214} = \mathbf{0.6313.} \textbf{ (Ans.)}$$

(*ii*) **The average molecular weight, M :**

$$M = \frac{n_{O_2} M_{O_2} + n_{N_2} \times M_{N_2}}{n_{O_2} + n_{N_2}}$$

$$= \frac{0.125 \times 32 + 0.214 \times 28}{0.125 + 0.214} = 29.475$$

i.e., **Average molecular weight = 29.475. (Ans.)**

(*iii*) **The specific gas constant, R :**

$$\mathbf{R} = \frac{R_0}{M}$$

$$= \frac{8.314}{29.475} = \mathbf{0.282\ kJ/kg\ K.} \textbf{ (Ans.)}$$

(*iv*) **The volume and density :**

$pV = mRT$ for mixture

$$\mathbf{V} = \frac{mRT}{p} = \frac{(4 + 6) \times 0.282 \times 10^3 \times 300}{4 \times 10^5}$$

$= \mathbf{2.115\ m^3.}$ **(Ans.)**

Density, $\rho = \rho_{O_2} + \rho_{N_2}$

$$\rho_{O_2} = \frac{m_{O_2}}{V} = \frac{4}{2.115} = 1.891 \text{ kg/m}^3$$

$$\rho_{N_2} = \frac{m_{N_2}}{V} = \frac{6}{2.115} = 2.837 \text{ kg/m}^3$$

$\therefore \quad \rho = 1.891 + 2.837$

$= \mathbf{4.728\ kg/m^3.}$ **(Ans.)**

(*v*) **The partial pressures and partial volumes :**

$$p_{O_2} V = n_{O_2} R_0 T$$

$$\therefore \quad \mathbf{p_{O_2}} = \frac{n_{O_2} R_0 T}{V}$$

$$= \frac{0.125 \times 8.314 \times 10^3 \times 300}{2.115 \times 10^5}$$

$= \mathbf{1.474\ bar.}$ **(Ans.)**

and $\mathbf{p_{N_2}} = 4 - 1.474 = \mathbf{2.526\ bar.}$ **(Ans.)**

$\mathbf{V_{O_2}} = x_{O_2} V = 0.3687 \times 2.115$

$= \mathbf{0.779\ m^3.}$ **(Ans.)**

$\mathbf{V_{N_2}} = x_{N_2} V = 0.6313 \times 2.115$

$= \mathbf{1.335\ m^3.}$ **(Ans.)**

Example 8.19. *4 kg of carbon dioxide at 40°C and 1.4 bar are mixed with 8 kg of nitrogen at 160°C and 1.0 bar to form a mixture at a final pressure of 0.7 bar. The process occurs adiabatically in a steady flow apparatus. Calculate :*

(i) The final temperature of the mixture ;

(ii) The change in entropy.

Take value of c_p : for CO_2 = 0.85 kJ/kg K and N_2 = 1.04 kJ/kg K.

Solution. Refer to Fig. 8.10.

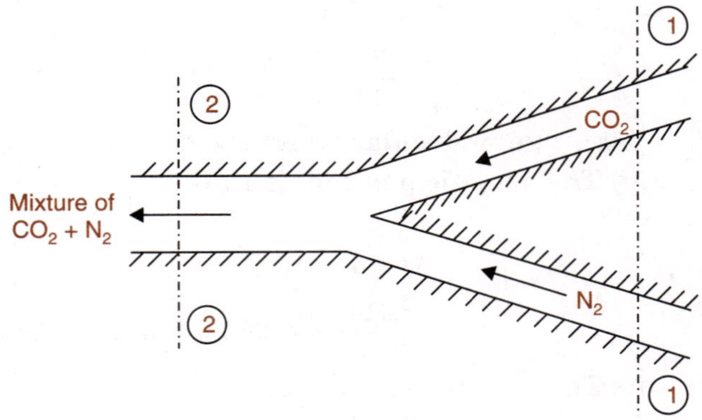

Fig. 8.10

(*i*) **Final temperature, T_2 :**

In this process, $W = 0$, $Q = 0$

∴ The steady flow equation may be written as

$$H_1 = H_2$$

i.e., $$(mh_1)_{CO_2} + (mh_1)_{N_2} = (mh_2)_{\text{mixture}}$$

$$= (mh_2)_{CO_2} + (mh_2)_{N_2}$$

or $$[m(h_1 - h_2)]_{CO_2} + [m(h_1 - h_2)]_{N_2} = 0$$

or $$[mc_p(T_1 - T_2)]_{CO_2} + [mc_p(T_1 - T_2)]_{N_2} = 0$$

or $$4 \times 0.85(40 - T_2) + 8 \times 1.04(160 - T_2) = 0$$

or $$136 - 3.4T_2 + 1331.2 - 8.32T_2 = 0$$

or $$1467.2 - 11.72T_2 = 0$$

∴ $$T_2 = 125.2°C = \mathbf{398.2\ K.}$$ **(Ans.)**

(*ii*) **Change in entropy :**

Now $$n_{CO_2} = \frac{4}{44} = 0.0909$$

$$n_{N_2} = \frac{8}{28} = 0.2857$$

∴ $$n = n_{CO_2} + n_{N_2}$$

$$= 0.0909 + 0.2857 = 0.3766$$

Again, $$\frac{p_{(CO_2)_2}}{p_2} = x_{CO_2}$$

[p_2 = pressure of the mixture]

∴ $$p_{(CO_2)_2} = 0.7 \times \frac{0.0909}{0.3766} = 0.17 \text{ bar}$$

Similarly, $$p_{(N_2)_2} = 0.7 \times \frac{0.2857}{0.3766} = 0.53 \text{ bar}$$

∴ Change in entropy, ΔS

$$= \left[mc_p \log_e \frac{T_2}{T_1} - mR \log_e \frac{p_2}{p_1} \right]_{CO_2} + \left[mc_p \log_e \frac{T_2}{T_1} - mR \log_e \frac{p_2}{p_1} \right]_{N_2}$$

$$= 4\left[0.85 \log_e \frac{398.2}{313} - \frac{8.314}{44} \log_e \frac{0.17}{1.4}\right] + 8\left[1.04 \log_e \frac{398.2}{433} - \frac{8.314}{28} \log_e \frac{0.53}{1.0}\right]$$

$$= 4(0.2046 + 0.3984) + 8(-0.0871 + 0.1885)$$

$$= 3.2232 \text{ kJ/K}$$

i.e., **Change in entropy = 3.2232 kJ/K. (Ans.)**

Example 8.20. *An insulated vessel containing 1 mole of oxygen at a pressure of 2.5 bar and a temperature of 293 K is connected through a valve to a second insulated rigid vessel containing 2 mole nitrogen at a pressure of 1.5 bar and a temperature of 301 K. The valve is opened and adiabatic mixing takes place. Assuming that oxygen and nitrogen are perfect gases calculate the entropy change in the mixing process.*

Assume the following specific heats at constant volume :

$$c_{v(O_2)} = 0.39 \text{ kJ/kg K}$$

$$c_{v(N_2)} = 0.446 \text{ kJ/kg K.}$$

Solution. Consider the system within the boundary of Fig. 8.11.

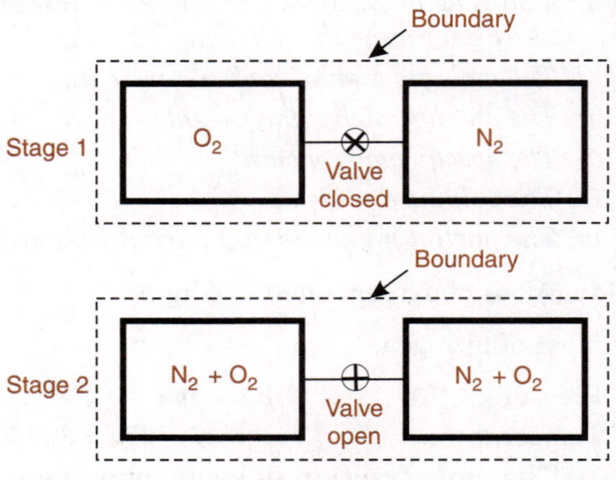

Fig. 8.11

In the process : $Q = 0$, $W = 0$, hence by the first law for a non-flow process $\Delta U = 0$. Let T_2 be the final temperature of the mixture.

$$\therefore \quad \left[m_{O_2} c_{v(O_2)} T_{O_2} + m_{N_2} c_{v(N_2)} T_{N_2}\right]$$

$$= \left[m_{O_2} c_{v(O_2)} + m_{N_2} c_{v(N_2)}\right] T_2$$

or $\quad [(1 \times 32) \times 0.39 \times 293 + (2 \times 28) \times 0.446 \times 301]$

$$= (1 \times 32 \times 0.39 + (2 \times 28) \times 0.446] T_2$$

or $$T_2 = \frac{32 \times 0.39 \times 293 + 56 \times 0.446 \times 301}{32 \times 0.39 + 56 \times 0.446}$$

$$= \frac{11174.4}{37.45} = 298.4 \text{ K}$$

Now entropy change in the mixing process is given by

$$\Delta S = m_{O_2} \Delta_{s(O_2)} + m_{N_2} \Delta_{s(N_2)}$$

where $$\Delta_s = c_v \log_e \frac{T_2}{T_1} + R \log_e \frac{V_2}{V_1}$$

Now to find initial and final volumes of O_2 and N_2 using the relation,

$$V = \frac{nR_0T}{p}$$

$\therefore$ The initial volume of O_2

$$= \frac{1 \times 8.314 \times 293 \times 10^3}{2.5 \times 10^5} = 9.74 \text{ m}^3$$

and Initial volume of N_2

$$= \frac{2 \times 8.314 \times 301 \times 10^3}{1.5 \times 10^5} = 33.4 \text{ m}^3$$

Final volume of the mixture

$$= 9.74 + 33.4 = 43.14 \text{ m}^3$$

$$\text{Thus } \Delta S = \left[(1 \times 32)\left\{0.39 \times \log_e \frac{298.4}{293} + \frac{8.314}{32} \log_e \frac{43.14}{9.74}\right\}\right]$$

$$+ \left[(2 \times 28)\left\{0.446 \times \log_e \frac{298.4}{301} + \frac{8.314}{28} \log_e \frac{43.14}{33.4}\right\}\right]$$

$$= 12.60 + 4.04 = 16.64 \text{ kJ}$$

i.e., **Entropy change in the mixing process = 16.64 kJ. (Ans.)**

Example 8.21. *A tank of capacity 0.45 m³ is insulated and is divided into two sections through a partition. One section initially contains H_2 at 3 bar and 130°C and has a volume of 0.3 m³ and the other section initially holds N_2 at 6 bar and 30°C. The gases are then allowed to mix after removing the adiabatic partition. Determine :*

(i) The temperature of the equilibrium mixture ;

(ii) The pressure of the mixture ;

(iii) The change in entropy for each component and total value.

Assume : $c_{v(N_2)} = 0.744$ *kJ/kg K,*

$c_{v(H_2)} = 10.352$ *kJ/kg K*

$c_{p(N_2)} = 1.041$ *kJ/kg K,*

$c_{p(H_2)} = 14.476$ *kJ/kg K.*

Solution. Total capacity of the tank, $V = 0.45$ m³

$V_{H_2} = 0.3$ m³ ; $\quad T_{H_2} = 130 + 273 = 403$ K

$p_{H_2} = 3$ bar ; $\quad V_{N_2} = 0.15$ m³

(*i.e.,* 0.45 – 0.3 = 0.15 m³)

$p_{N_2} = 6$ bar ; $\quad T_{N_2} = 30 + 273 = 303$ K.

(i) Temperature of equilibrium mixture, T_2 :

Now $p_{H_2} V_{H_2} = m_{H_2} R_{H_2} T_{H_2}$

$$\therefore \quad m_{H_2} = \frac{3 \times 10^5 \times 0.3}{\left(\frac{8.314}{2}\right) \times 403 \times 10^3} = 0.0537 \text{ kg}$$

$$\left[\because R = \frac{R_0}{M}\right]$$

and $$p_{N_2} V_{N_2} = m_{N_2} R_{N_2} T_{N_2}$$

$$\therefore \quad m_{N_2} = \frac{6 \times 10^5 \times 0.15}{\left(\frac{8.314}{28}\right) \times 303 \times 10^3} = 1.0 \text{ kg}$$

According to the first law for a closed system

$$\Delta U = 0$$

i.e., $$U_{\text{initial}} = U_{\text{final}}$$

or $$m_{H_2} c_{v(H_2)} T_{(H_2)_1} + m_{N_2} c_{v(N_2)} T_{(N_2)_1} = m_{H_2} c_{v(H_2)} T_{(H_2)_2} + m_{N_2} c_{v(N_2)} T_{(N_2)_2}$$

or $$m_{H_2} c_{v(H_2)} \left[T_{(H_2)_2} - T_{(H_2)_1}\right] + m_{N_2} c_{v(N_2)} \left[T_{(N_2)_2} - T_{(N_2)_1}\right] = 0$$

or $$0.0537 \times 10.352 \left[T_{(H_2)_2} - 403\right] + 1.0 \times 0.744 \left[T_{(N_2)_2} - 303\right] = 0$$

$$T_{(H_2)_2} = T_{(N_2)_2} = T_2$$

or $$0.556(T_2 - 403) + 0.744(T_2 - 303) = 0$$

or $$0.556T_2 - 224 + 0.744T_2 - 225.4 = 0$$

$$\therefore \quad T_2 = 345.7 \text{ K}$$

i.e., Temperature of the mixture = **345.7 K. (Ans.)**

(ii) Pressure of the mixture, p_2 :

Now $$p_{(H_2)_2} V = m_{H_2} R_{H_2} T_2$$

$$\therefore \quad p_{(H_2)_2} = \frac{0.0537 \times \left(\frac{8.314}{2}\right) \times 345.7 \times 10^3}{0.45 \times 10^5} = 1.71 \text{ bar}$$

Similarly $p_{(N_2)_2} = \dfrac{1.0 \times \left(\dfrac{8.314}{28}\right) \times 345.7 \times 10^3}{0.45 \times 10^5}$

$= 2.28$ bar

$\therefore \quad \mathbf{p_2} = p_{(H_2)_2} + p_{(N_2)_2}$

$= 1.71 + 2.28 = \mathbf{3.99\ bar.}$ **(Ans.)**

(*iii*) **Change in entropy :**

Now $(\Delta S)_{H_2} = m\left(c_p \log_e \dfrac{T_2}{T_1} - R \log_e \dfrac{p_2}{p_1}\right)_{H_2}$

$= 0.0537\left[14.476 \log_e \dfrac{345.7}{403} - \dfrac{8.314}{2} \log_e \dfrac{1.71}{3}\right]$

= 0.00626 kJ/K. (Ans.)

$(\Delta S)_{H_2} = m\left(c_p \log_e \dfrac{T_2}{T_1} - R \log_e \dfrac{p_2}{p_1}\right)$

$= 1.0\left[1.041 \log_e \dfrac{345.7}{303} - \dfrac{8.314}{28} \log_e \dfrac{2.28}{6}\right]$

= 0.424 kJ/K. (Ans.)

$\therefore \quad \Delta S = (\Delta S)_{H_2} + (\Delta S)_{N_2} = 0.00626 + 0.424$

$= 0.43026$ kJ/K

i.e., **Total change in entropy**

= 0.43026 kJ/K. (Ans.)

Example 8.22. *A perfect gas mixture consists of 4 kg of N_2 and 6 kg of CO_2 at a pressure of 4 bar and a temperature of 25°C. Calculate c_v and c_p of the mixture.*

If the mixture is heated at constant volume to 50°C, find the change in internal energy, enthalpy and entropy of the mixture.

Take : $c_{v(N_2)} = 0.745$ *kJ/kg K,*

$c_{v(CO_2)} = 0.653$ *kJ/kg K*

$c_{p(N_2)} = 1.041$ *kJ/kg K,*

$c_{p(CO_2)} = 0.842$ *kJ/kg K.*

Solution. $m_{N_2} = 4$ kg, $m_{CO_2} = 6$ kg, $p_{mix} = 4$ bar

$T_1 = 25 + 273 = 298$ K, $T_2 = 50 + 273 = 323$ K

$\mathbf{c_{v(mix)} = ?,\ c_{p(mix)} = ?}$

Using the relation,

$(m_{N_2} + m_{CO_2})c_{v(mix)} = m_{N_2} c_{v(N_2)} + m_{CO_2} c_{v(CO_2)}$

$(4 + 6)\ c_{v(mix)} = 4 \times 0.745 + 6 \times 0.653$

$\therefore \quad \mathbf{c_{v(mix)}} = \dfrac{4 \times 0.745 + 6 \times 0.653}{4 + 6}$

= 0.6898 kJ/kg K. (Ans.)

Similarly, $\mathbf{c_{p(mix)}} = \dfrac{4 \times 1.041 + 6 \times 0.842}{4 + 6}$

= 0.9216 kJ/kg K. (Ans.)

Change in internal energy, ΔU :

$\mathbf{\Delta U} = [mc_v(T_2 - T_1)]_{mix}$

$= (4 + 6) \times 0.6898(323 - 298)$

= 172.45 kJ. (Ans.)

Change in enthalpy, ΔH :

$\mathbf{\Delta H} = [mc_p(T_2 - T_1)]_{mix}$

$= (4 + 6) \times 0.9216(323 - 298)$

= 230.4 kJ. (Ans.)

Change in entropy, ΔS :

$(\Delta s) = \left(c_v \log_e \dfrac{T_2}{T_1} + R \log_e \dfrac{v_2}{v_1}\right)$

$= \left(c_v \log_e \dfrac{T_2}{T_1}\right) \quad (\because v_2 = v_1)$

$\therefore \quad (\Delta s)_{N_2} = \left(c_v \log_e \dfrac{T_2}{T_1}\right)_{N_2}$

and $(\Delta s)_{CO_2} = \left(c_v \log_e \dfrac{T_2}{T_1}\right)_{CO_2}$

Hence,

$\mathbf{\Delta S} = \left(mc_v \log_e \dfrac{T_2}{T_1}\right)_{N_2} + \left(mc_v \log_e \dfrac{T_2}{T_1}\right)_{CO_2}$

$= 4 \times 0.745 \log_e \dfrac{323}{298} + 6 \times 0.653 \log_e \dfrac{323}{298}$

= 0.5557 kJ/K. (Ans.)

Note. ΔS may also be found out as follows :

$\Delta S = (m_{N_2} + m_{CO}) \ c_{v(mix)} \log_e \dfrac{T_2}{T_1}$

$= (4 + 6) \times 0.6898 \log_e \dfrac{323}{298} = 0.5557$ kJ/K.

HIGHLIGHTS

1. According to Dalton's law :
   (*i*) The pressure of a mixture of gases is equal to the sum of the partial pressures of the constituents.
   (*ii*) The partial pressure of each constituent is that pressure which the gas would exert if it occupied alone that volume occupied by the mixture at the same temperature.
2. According to Gibbs-Dalton law :
   (*i*) The internal energy, enthalpy and entropy of a gaseous mixture are respectively equal to the sum of the internal energies, enthalpies and entropies of the constituents.
   (*ii*) Each constituent has that internal energy, enthalpy and entropy, which it would have if it occupied alone that volume occupied by the mixture at the temperature of the mixture.
3. The characteristic equation for mixture is given as :
$$pV = nR_0T$$
where n = Number of moles of mixture, and
R_0 = Universal gas constant.
4. Molecular weight (M) may be found out by using the following relations :
$$M = \sum \frac{n_i}{n} M_i \quad \text{and} \quad M = \frac{1}{\sum \frac{m_{fi}}{M_i}}$$
where $m_f = \frac{m_i}{m}$ = mass fraction of a constituent.
5. The following condition must be satisfied in an adiabatic mixing process of perfect gas in steady flow :
$$T = \frac{\Sigma m_i c_{pc} T_i}{\Sigma m_i c_{pi}} = \frac{\Sigma n_i C_{pi} T_i}{\Sigma n_i C_{pi}} .$$

OBJECTIVE TYPE QUESTIONS

Choose the Correct Answer :

1. In an ideal gas the partial pressure of a component is
   (*a*) inversely proportional to the square of the mole fraction
   (*b*) directly proportional to the mole fraction
   (*c*) inversely proportional to the mole fraction
   (*d*) equal to the mole fraction.
2. The value of the universal gas constant is
   (*a*) 8.314 J/kg K (*b*) 83.14 kJ/kg K
   (*c*) 848 kJ/kg K (*d*) 8.314 kJ/kg K.
3. In a mixture of gases, the partial pressure p_i of any constituent gas can be found by
   (*a*) $p_i = n_i RT/V$ (*b*) $p_i = m_i RT/V_i$
   (*c*) $p_i = n_i R_0 T/V_i$ (*d*) $p_i = n_i R_0 T/V$
   where, R = Characteristic gas constant,
   R_0 = Universal gas constant,
   V = Volume of the mixture,
   V_i = Volume of the ith constituent, and
   T = Temperature of the mixture.
4. Mole fraction of a component of gas mixture is equal to
   (*a*) $1/f$ (*b*) f^2
   (*c*) f (*d*) f/p
   where, f = Volume fraction, and
   p = Pressure of the mixture.
5. In a gaseous mixture the specific volume of each component is given by
   (*a*) V/m (*b*) V_i/m_i
   (*c*) V/m_i (*d*) none of the above.
   where, V = Volume of the mixture,
   V_i = Volume of the ith component,
   m = Mass of mixture, and
   m_i = Mass of the ith component.

ANSWERS

1.	(*b*)	2.	(*d*)	3.	(*d*)
4.	(*c*)	5.	(*c*).		

THEORETICAL QUESTIONS

1. Define the following terms :
   (*i*) Partial pressure (*ii*) Mole fraction
   (*iii*) Volume fraction of a gas constituent in a mixture.
2. Explain briefly Dalton's law and Gibbs-Dalton law.
3. State and explain Amagat's law or Leduc's law.
4. Prove that the molar analysis is identical with the volumetric analysis, and both are equal to the ratio of the partial pressure to the total pressure.
5. Prove the following relation
$$M = \frac{\Sigma n_i M_i}{\Sigma n_i} \quad \frac{1}{\sum \frac{m_{fi}}{M_i}}$$
where, M = Molecular weight of the mixture,
n_i = Number of moles of an any constituent,
m_{fi} = Mass fraction of the constituent, and
M_i = Molecular weight of the constituent.

UNSOLVED EXAMPLES

1. 0.45 kg of carbon monoxide (28) and 1 kg of air at 15°C are contained in a vessel of volume 0.4 m^3. Calculate the partial pressure of each constituent and the total pressure in the vessel. The gravimetric analysis of air is to be taken as 23.3% oxygen (32) and 76.7% nitrogen (28).
   [**Ans.** p_{O_2} = 0.4359 bar ; p_{N_2} = 1.64 bar, p_{CO} = 0.962 bar]
2. A mixture of CO and O_2 is to be prepared in the proportion of 7 kg to 4 kg in a vessel of 0.3 m^3 capacity. If the temperature of the mixture is 15°C, determine the pressure to which the vessel is subjected. If the temperature is raised to 40°C, what will then be the pressure in the vessel ? [**Ans.** 29.9 bar, 32.5 bar]

3. Assuming that air may be treated as a mixture of ideal gases which has a mass composition 23.2% oxygen and 76.8% nitrogen, find the gas constant and apparent molecular weight of air. [**Ans.** 0.288 kJ/kg K, 28.86]

4. A mixture of 1 mole CO_2 and 3.5 moles of air is contained in a vessel at 1 bar and 15°C. The volumetric analysis of air can be taken as 21% oxygen and 79% nitrogen. Calculate for the mixture : (*i*) The masses of CO_2, O_2 and N_2, and the total mass. (*ii*) The percentage carbon content by mass. (*iii*) The apparent molecular weight and the gas constant for the mixture. (*iv*) The specific volume of the mixture. [**Ans.** (*i*) 44 kg, 23.55 kg, 77.5 kg ; (*ii*) 8.27% ; (*iii*) 32.2, 0.2581 kJ/kg K ; (*iv*) 0.7435 m^3/kg]

5. A mixture of H_2 and O_2 is to be made so that the ratio of H_2 to O_2 is 2 to 1 by volume. Calculate the mass of O_2 required and the volume of the container, per kg of H_2, if the pressure and temperature are 1 bar and 15°C respectively. [**Ans.** 8 kg, 17.96 m^3]

6. A vessel contains a gaseous mixture of composition by volume, 80% H_2 and 20% CO. It is desired that the mixture should be made in proportion 50% H_2 and 50% CO by removing some of the mixture and adding some CO. Calculate per mole of mixture the mass of mixture to be removed, and the mass of CO to be added. The pressure and temperature in the vessel remain constant during the procedure. [**Ans.** 2.7 kg, 10.5 kg]

7. The gas in an engine cylinder has a volumetric analysis of 12% of CO_2, 11.5% O_2 and 76.5% N_2. The temperature at the beginning of expansion is 1000°C and the gas mixture expands reversibly through a volume ratio of 7 : 1, according to a law $pv^{1.25}$ = constant. Calculate the work done and the heat flow per kg of gas. The values of c_p for the constituents are as follows : c_p for CO_2 = 1.235 kJ/kg K ; c_p for O_2 = 1.088 kJ/kg K ; c_p for N_2 = 1.172 kJ/kg K.
[**Ans.** 536.3 kJ/kg ; 96 kJ/kg (heat supplied)]

8. A producer gas has the following volumetric analysis : 29% CO, 12% H_2, 3% CH_4, 4% CO_2, 52% N_2. Calculate the values of C_p, C_v, c_p and c_v for the mixture. The values of C_p for the constituents are as follows : for CO C_p = 29.27 kJ/mole K ; for H_2 C_p = 28.89 kJ/mole K ; for CH_4 C_p = 35.8 kJ/mole K ; for CO_2 C_p = 37.22 kJ/mole K ; for N_2 C_p = 29.14 kJ/mole K.
[**Ans.** 29.676 kJ/mole K, 21.362 kJ/mole K ; 1.178 kJ/kg K ; 0.8476 kJ/kg K]

9. An exhaust gas is analysed and is found to contain, by volume, 78% N_2, 12% CO_2 and 10% O_2. What is the corresponding gravimetric analysis ? Calculate the mass of mixture per mole, and the density if the temperature is 550°C and the total pressure is 1 bar.
[**Ans.** 72% N_2, 17.3% CO_2, 10.6% O_2 ; 30.28 kg/mole ; 0.442 kg/m^3]

10. A vessel of 3 m^3 capacity contains a mixture of nitrogen and carbon dioxide, the analysis by volume showing equal quantities of each. The temperature is 15°C and the total pressure is 3.5 bar. Determine the mass of each constituent. [**Ans.** 6.14 kg N_2 ; 9.65 kg CO_2]

11. A mixture is made up of 25% N_2 35% O_2, 20% CO_2 and 20% CO by volume. Calculate : (*i*) The molecular weight of the mixture. (*ii*) C_p and C_v for the mixture. (*iii*) γ for the mixture. (*iv*) The partial pressure of each constituent when the total pressure is 1.5 bar. (*v*) The density of the mixture at 1.5 bar and 15°C.
[**Ans.** (*i*) 32.6 ; (*ii*) 30.9, 22.53 kJ/mole K ; (*iii*) 1.37 ; (*iv*) 0.375, 0.525, 0.3, 0.3 bar ; (*v*) 2.04 kg/m^3]

12. A mixture of ideal gases consists of 3 kg of nitrogen and 5 kg of carbon dioxide at a pressure of 3 bar and a temperature of 20°C. Find (*i*) the mole fraction of each constituent, (*ii*) the equivalent molecular weight of the mixture, (*iii*) the equivalent gas constant of the mixture, (*iv*) the partial pressures and partial volumes, (*v*) the volume and density of the mixture, and (*vi*) the c_p and c_v of the mixture.
If the mixture is heated at constant volume to 40°C, find the changes in internal energy, enthalpy and entropy of the mixture. Find the changes in internal energy, enthalpy and entropy of the mixture if the heating is done at constant pressure.
Take γ : for CO_2 = 1.286 and N_2 = 1.4.
[**Ans.** (*i*) 0.485, 0.515 ; (*ii*) 36.25 kg/kg mole ; (*iii*) 0.229 kJ/kg K ; (*iv*) 1.455 bar, 1.545 bar ; 0.87 m^3, 0.923 m^3 ; (*v*) 1.79 m^3, 4.46 kg/m^3 ; (*vi*) 0.92 kJ/kg K, 0.69 kJ/kg K ; 110.4 kJ, 147.2 kJ, 0.368 kJ/kg K ; 0.49 kJ/kg K]

13. A vessel of 1.5 m^3 capacity contains oxygen at 7 bar and 40°C. The vessel is connected to another vessel of 3 m^3 capacity containing carbon monoxide at 1 bar and 15°C. A connecting valve is opened and the gases mix adiabatically. Calculate : (*i*) The final temperature and pressure of the mixture. (*ii*) The change in entropy of the system.
Assume : For oxygen C_v = 21.07 kJ/mole K
For carbon monoxide C_v = 20.86 kJ/mole K.
[**Ans.** (*i*) 34°C, 3 bar ; (*ii*) 4.108 kJ/K]

14. Two vessels both containing nitrogen, are connected by a valve which is opened to allow the contents to mix and achieve an equilibrium temperature of 27°C. Before mixing the following information is known about the gases in the two vessels.

Vessel A	*Vessel B*
p = 15 bar	p = 6 bar
t = 50°C	t = 20°C
Contents = 0.5 kg mole	Contents = 2.5 kg

Calculate the final equilibrium pressure and amount of heat transferred to the surroundings. If the vessel had been perfectly insulated, calculate the final temperature and pressure which would have been reached.
Take γ = 1.4.
[**Ans.** 11.68 bar, –226.2 kJ, 45.5°C, 12.4 bar]

15. Find the increase in entropy when 2 kg of oxygen at 60°C are mixed with 6 kg of nitrogen at the same temperature. The initial pressure of each constituent is 10.3 bar and is the same as that of the mixture.
[**Ans.** 1.2314 kJ/kg K]

16. A gas mixture contains hydrogen, nitrogen and carbon monoxide in equal molar proportions. Treating the components of the mixture as perfect gases, determine the gas constant and the ratio of specific heats for the mixture. Calculate the work required to compress 1 kg of the mixture in reversible adiabatic flow through a

pressure ratio of 2 from an initial temperature of 15°C when kinetic and potential energy changes are negligible.
[**Ans.** 0.43 kJ/kg K, 1.4, – 82.76 kJ]

17. 2 kg of nitrogen at 20 bar and 38°C is contained in a rigid vessel. A sufficient quantity of oxygen is added to increase the pressure to 26.66 bar while the temperature remains constant at 38°C.

Calculate the mass of oxygen added. [**Ans.** 0.761 kg]

18. Air (N_2 = 77%, O_2 = 23% by weight) at 20°C and 10 bar is contained in a vessel of capacity of 0.5 m^3. Some quantity of carbon dioxide is forced into the vessel so that the temperature remains at 20°C but the pressure rises to 15 bar. Find the masses of oxygen, nitrogen and carbon dioxide in the cylinder. The universal gas constant is 8.3143 kJ/kg K. [**Ans.** 1.35 kg, 4.54 kg, 4.51 kg]

19. A vessel of 5 m^3 capacity contains a mixture of two gases in proportion of 40 per cent and 60 per cent at 20°C. If the value of R for the gases is 0.287 kJ/kg K and 0.294 kJ/kg K and if the total weight of the mixture is 1.5 kg, calculate :

(*i*) The partial pressure ; (*ii*) Total pressure ;
(*iii*) Mean value of R for the mixture.

[**Ans.** (*i*) 0.1 bar ; (*ii*) 0.155 bar ; (*iii*) 0.291 kJ/kg K]

20. A mixture consisting of 6 kg of O_2 and 9 kg of N_2 has a pressure of 3 bar and temperature of 20°C. For the mixture determine the following :

(*i*) The mole fraction of each component ;
(*ii*) The average molecular weight ;
(*iii*) The specific gas constant ;
(*iv*) The volume and density ;
(*v*) The partial pressures and partial volumes.

[**Ans.** (*i*) 0.3684, 0.6315 ; (*ii*) 29.475 ; (*iii*) 0.282 kJ/kg K ; (*iv*) 4.13 m^3, 3.629 kg/m^3 ; (*v*) 1.1 bar, 1.894 bar ; 1.52 m^3, 2.61 m^3]

21. A closed vessel of 0.1 m^3 capacity contained air at 1 bar pressure and 27°C. Hydrogen was added and the total pressure in the vessel was raised to 1.2 bar at the same temperature. Find the weight of oxygen, nitrogen and hydrogen finally in the vessel and their respective partial pressures.

Air contains 77% N_2 by weight. Take the value of universal gas constant as 8.3143 kJ/kg K and molecular weight of air as 29.

[**Ans.** m_{O_2} = 0.0267 kg, m_{N_2} = 0.0893 kg, m_{H_2} = 0.001603 kg]

22. A vessel of capacity 0.3 m^3 is insulated and divided into two sections by a partition. One section is 0.2 m^3 in volume and initially contains H_2 at 2 bar and 127°C. The remaining section initially holds N_2 at 4 bar and 27°C. The adiabatic partition is then removed, and the gases are allowed to mix. Determine :

(*i*) The temperature of the equilibrium mixture,
(*ii*) The pressure of the mixture, and
(*iii*) The change in entropy for each component and total value.

$c_{v(N_2)}$ = 0.744 kJ/kg K, $c_{v(H_2)}$ = 10.352 kJ/kg K

$c_{p(N_2)}$ = 1.041 kJ/kg K, $c_{p(N_2)}$ = 14.476 kJ/kg K.

[**Ans.** (*i*) 342.6 K ; (*ii*) 2.661 bar ; (*iii*) $(\Delta S)_{H_2}$ = 0.00235 kJ/K ; $(\Delta S)_{N_2}$ = 0.1908 kJ/K ; ΔS = 0.19315 kJ/K]

23. 3 kg of N_2 and 5 kg of CO_2 at a pressure of 3 bar and a temperature of 20°C comprise a perfect gas mixture. Calculate c_v and c_p of the mixture.

If the mixture is heated at constant volume to 40°C, find the change in internal energy, enthalpy and entropy of the mixture.

Take : $c_{v(N_2)}$ = 0.7448 and $c_{p(N_2)}$ = 1.0416 kJ/kg K

$c_{v(CO_2)}$ = 0.6529 and $c_{p(CO_2)}$ = 0.8418 kJ/kg K.

[**Ans.** 0.6873 kJ/kg K, 0.9167 kJ/kg K ; 109.96 kJ, 146.67 kJ, 0.363 kJ/K]

Psychrometrics

9

9.1. CONCEPT OF PSYCHROMETRY AND PSYCHROMETRICS

Air comprises of fixed gases principally, nitrogen and oxygen with an admixture of water vapour in varying amounts. In atmospheric air water is always present and its relative weight averages less than 1% of the weight of atmospheric air in temperate climates and less than 3% by weight under the most extreme natural climatic conditions, it is nevertheless one of most important factors in human comfort and has significant effects on many materials. Its effect on human activities is in fact altogether disproportionate to its relative weights. *The art of measuring* the *moisture content of air is termed* **"psychrometry"**. *The science which investigates the thermal properties of moist air, considers the measurement and control of the moisture content of air, and studies the effect of atmospheric moisture on material and human comfort may properly be termed* **"psychrometrics"**.

9.2. DEFINITIONS

Some of the more important definitions are given below :

1. **Dry air.** The international joint committee on Psychrometric Data has adopted the following exact composition of air expressed in mole fractions (Volumetric) Oxygen 0.2095, Nitrogen 0.7809, Argon 0.0093, Carbon dioxide 0.0003. Traces of rare gases are neglected. Molecular weight of air for all air conditioning calculations will be taken as 28.97. Hence the gas constant,

$$R_{air} = \frac{8.3143}{28.97} = 0.287 \text{ kJ/kg K}$$

Dry air is never found in practice. Air always contains some moisture. Hence the common designation "air" *usually means moist* air. The term *'dry air'* is used to indicate the *water free contents* of air having any degree of moisture.

2. **Saturated air.** Moist air is said to be saturated when its condition is such that it can co-exist in natural equilibrium with an associated condensed moisture phase presenting a flat surface to it. For *a given temperature, a given quantity of air can be saturated with a fixed quantity of moisture.* At higher temperatures, it requires a larger quantity of moisture to saturate it. At saturation, vapour pressure of moisture in air corresponds to the saturation pressure given in steam tables corresponding to the given temperature of air.

3. **Dry-bulb temperature (DBT).** It *is the temperature of air as registered by an ordinary thermometer* (t_{db}).

4. **Wet-bulb temperature (WBT).** *It is the temperature registered by a thermometer when the bulb is covered by a wetted wick and is exposed to a current of rapidly moving air* (t_{wb}).

5. **Adiabatic saturation temperature.** *It is the temperature at which the water or ice can saturate air by evaporating adiabatically into it.* It is numerically equivalent to the measured wet bulb temperature (as corrected, if necessary for radiation and conduction) (t_{wb}).

6. **Wet-bulb depression.** *It is the difference between dry-bulb and wet bulb temperatures* ($t_{db} - t_{wb}$).

7. **Dew point temperature (DPT).** *It is the temperature to which air must be cooled at constant pressure in order to cause condensation of any of its water vapour*. It is equal to steam table saturation temperature corresponding to the actual partial pressure of water vapour in the air (t_{dp}).

8. **Dew point depression.** *It is the difference between the dry bulb and dew point temperatures* $(t_{db} - t_{dp})$.

9. **Specific humidity (Humidity ratio).** *It is the ratio of the mass of water vapour per unit mass of dry air in the mixture of vapour and air*, it is generally expressed as grams of water per kg of dry air. For a given barometric pressure it is a function of dew point temperature alone.

10. **Relative humidity (RH), (ϕ).** *It is the ratio of the partial pressure of water vapour in the mixture to the saturated partial pressure at the dry bulb temperature, expressed as percentage.*

11. **Sensible heat.** *It is the heat that changes the temperature of a substance when added to or abstracted from it.*

12. **Latent heat.** *It is the heat that does not affect the temperature but changes the state of substance when added to or abstracted from it.*

13. **Enthalpy.** *It is the combination energy which represents the sum of internal and flow energy in a steady flow process.* It is determined from an arbitrary datum point for the air mixture and is expressed as kJ per kg of dry air (h).

Note. When air is *saturated* DBT, WBT, DPT are *equal*.

9.3. PSYCHROMETRIC RELATIONS

Pressure

Dalton's law of partial pressure is employed to determine the pressure of a mixture of gases. This law *states that the total pressure of a mixture of gases is equal to the sum of partial pressures which the component gases would exert if each existed alone in the mixture volume at the mixture temperature.*

Precise measurements made during the last few years indicate that this law as well as Boyle's and Charle's laws are only approximately correct. Modern tables of atmospheric air properties are based on the correct versions.

For calculating partial pressure of water vapour in the air many equations have been proposed, probably Dr. Carrier's equation is most widely used.

$$p_v = (p_{vs})_{wb} - \frac{[p_t - (p_{vs})_{wb}](t_{db} - t_{wb})}{1527.4 - 1.3\, t_{wb}} \quad ...(9.1)$$

where p_v = Partial pressure of water vapour,

p_{vs} = Partial pressure of water vapour when air is fully saturated,

p_t = Total pressure of moist air,

t_{db} = Dry bulb temperature (°C), and

t_{wb} = Wet bulb temperature (°C).

Specific humidity W :

$$\text{Specific humidity} = \frac{\text{Mass of water vapour}}{\text{Mass of dry air}}$$

or

$$W = \frac{m_v}{m_a}$$

Also,

$$m_a = \frac{p_a V}{R_a T} \quad ...(9.2)$$

and

$$m_v = \frac{p_v \times V}{R_v \times T} \quad ...(9.3)$$

where p_a = Partial pressure of dry air,

p_v = Partial pressure of water vapour,

V = Volume of mixture,

R_a = Characteristic gas constant for dry air, and

R_v = Characteristic gas constant for water vapour.

From equations (9.2) and (9.3)

$$W = \frac{p_v \times V}{R_v \times T} \times \frac{R_a T}{p_a V} = \frac{R_a}{R_v} \times \frac{p_v}{p_a}$$

But

$$R_a = \frac{R_0}{M_a}$$

$$\left(= \frac{8.3143}{28.97} = 0.287 \text{ kJ/kg K in SI units} \right)$$

$$R_v = \frac{R_0}{M_v}$$

$$\left(= \frac{8.3143}{18} = 0.462 \text{ kJ/kg K in SI units} \right)$$

where R_0 = Universal gas constant,

M_a = Molecular weight of air, and

M_v = Molecular weight of water vapour.

$$\therefore \quad W = \frac{0.287}{0.462} \cdot \frac{p_v}{p_a} = 0.622 \frac{p_v}{p_t - p_v}$$

i.e.,

$$W = 0.622 \frac{p_v}{p_t - p_v} \quad ...(9.4)$$

The masses of air and water vapour in terms of specific volumes are given by expression as

$$m_a = \frac{V}{v_a} \quad \text{and} \quad m_v = \frac{V}{v_v}$$

where v_a = Specific volume of dry air, and

v_v = Specific volume of water vapour.

$$W = \frac{v_a}{v_v} \quad ...(9.5)$$

Degree of saturation (μ) :

Degree of saturation

$$= \frac{\text{Mass of water vapour associated with unit mass of dry air}}{\text{Mass of water vapour associated with saturated unit mass of dry saturated air}}$$

i.e., $$\mu = \frac{W}{W_s} \quad ...(9.6)$$

where, W_s = Specific humidity of air when air is fully saturated

$$\therefore \quad \mu = \frac{0.622\left(\frac{p_v}{p_t - p_v}\right)}{0.622\left(\frac{p_{vs}}{p_t - p_{vs}}\right)} = \frac{p_v(p_t - p_{vs})}{p_{vs}(p_t - p_v)}$$

$$= \frac{p_v}{p_{vs}}\left[\frac{\left(1 - \frac{p_{vs}}{p_t}\right)}{\left(1 - \frac{p_v}{p_t}\right)}\right] \quad ...(9.7)$$

where p_{vs} = Partial pressure of water vapour when air is fully saturated (p_{vs} can be calculated from steam tables corresponding to the dry bulb temperature of the air).

Relative humidity (RH), ϕ :

Relative humidity,

$$\phi = \frac{\text{Mass of water vapour in a given volume}}{\text{Mass of water vapour in the same volume if saturated at the same temp.}}$$

$$= \frac{m}{m_{vs}} = \frac{\frac{p_v T}{R_v T}}{\frac{p_{vs} T}{R_v T}} = \frac{p_v}{p_{vs}} \quad ...(9.8)$$

Inserting the value of equation (9.8) into equation (9.7), we get

$$\mu = \phi\left[\frac{1 - \frac{p_{vs}}{p_v}}{1 - \frac{p_v}{p_t}}\right] = \phi\left[\frac{1 - \frac{p_{vs}}{p_t}}{1 - \frac{\phi p_{vs}}{p_t}}\right] = \phi\left(\frac{p_t - p_{vs}}{p_t - \phi\, p_{vs}}\right)$$

$$\therefore \quad \phi(p_t - p_{vs}) = \mu(p_t - \phi\, p_{vs})$$

or $$\phi(p_t - p_{vs} + \mu p_{vs}) = \mu p_t$$

$$\therefore \quad \phi = \frac{\mu p_t}{p_t - p_{vs} + \mu p_{vs}} = \frac{\mu}{1 - (1 - \mu)\frac{p_{vs}}{p_t}} \quad ...(9.9)$$

Since $p_{vs} << p_t$

$\therefore$ $\phi \simeq \mu$

Inserting the value of the p_v from equation (9.4) into equation (9.8), we get

$$\phi = \frac{p_a W}{0.622} \times \frac{1}{p_{vs}} = 1.6\, W\, \frac{p_a}{p_{vs}} \quad ...(9.10)$$

Note **1.** Relative humidity as compared to specific humidity plays a vital role in comfort air-conditioning and industrial air-conditioning. *Relative humidity signifies the absorption capacity of air.* If initial relative humidity of air is less it will absorb more moisture.

2. W, μ and ϕ cannot be conveniently measured as they require measurement of p_v and p_{vs}. The value of p_v can be obtained from the measurement of the wet bulb temperature and the value of p_{vs} can be calculated from steam tables corresponding to given air temperature.

Enthalpy of moist air

It is the sum of enthalpy of dry air and enthalpy of water vapour associated with dry air. It is expressed in kJ/kg of dry air

$$h = h_{\text{air}} + W \,.\, h_{\text{vapour}}$$
$$= c_p t_{db} + W \,.\, h_{\text{vapour}}$$

where h = Enthalpy of mixture/kg of dry air,

h_{air} = Enthalpy of 1 kg of dry air,

h_{vapour} = Enthalpy of 1 kg of vapour obtained from steam tables,

W = Specific humidity in kg/kg of dry air, and

c_p = Specific heat of dry air normally assumed as 1.005 kJ/kg K.

Also $h_{\text{vapour}} = h_g + c_{ps}\,(t_{db} - t_{dp})$

where h_g = Enthalpy of saturated steam at dew point temperature,

and c_{ps} = 1.88 kJ/kg K.

$$\therefore \quad h = c_p t_{db} + W[h_g + c_{ps}(t_{db} - t_{dp})] \quad ...(9.11)$$
$$= (c_p + c_{ps} W)\, t_{db} + W(h_g - c_{ps} t_{dp})$$
$$= c_{pm} t_{db} + W(h_g - c_{ps} t_{dp}) \quad ...[9.11(a)]$$

where $c_{pm} = (c_p + c_{ps} W)$ is the *specific heat of humid air or humid specific heat.*

The value of c_{pm} is taken as 1.021 kJ/kg dry air per K. It is the heat capacity of (1 + W) kg of moisture per kg of dry air.

$h_{\text{vapour}} \simeq h_g$ at dry bulb temperature. So,

$$h = c_p t_{db} + W\, h_g. \quad ...(9.12)$$

However, a better approximation is given by the following relationship :

$$h_{\text{vapour}} = 2500 + 1.88 t_{db} \text{ kJ/kg of water vapour} \quad ...[9.12\,(a)]$$

where t_{db} is dry bulb temperature in °C, and the datum state is liquid water at 0°C.

$$\therefore \quad h = 1.005\, t_{db} + W(2500 + 1.88\, t_{db}) \text{ kJ/kg dry air.} \quad ...[9.12\,(b)]$$

Adiabatic saturation process

In an insulated chamber when unsaturated air flows over a long sheet of water (Fig. 9.1), the water evaporates, and the specific humidity of the air increases. As the

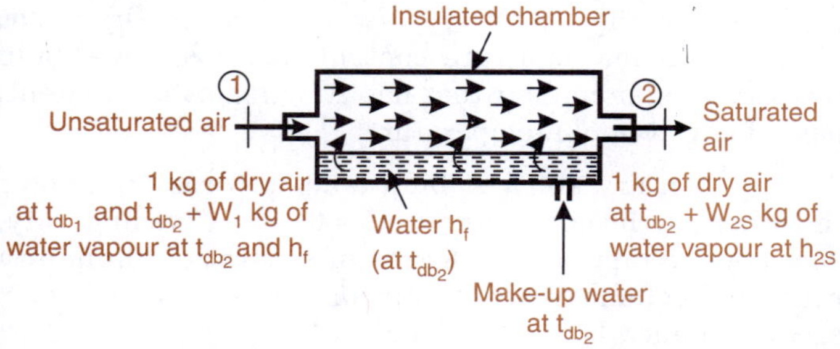

Fig. 9.1. Adiabatic saturation process.

evaporation takes place both the air and water are cooled. *The process continues until the energy transferred from the air to the water is equal to the energy required to vapourise the water.* When this point is reached, thermal equilibrium exists with respect to water, air and water vapour, and consequently the air is saturated. The *equilibrium temperature is called the* ***adiabatic saturation temperature*** *or the* ***thermodynamic wet bulb temperature***. The make-up water is introduced at *this temperature* to make the water level constant. The '*adiabatic*' cooling process is shown in Fig. 9.2 for the vapour in the air-vapour mixture. Although the total pressure of the mixture is constant, the partial pressure of the vapour increases, and in the saturated state corresponds to the adiabatic saturation temperature. The vapour is initially at *DBT* t_{db_1} and is cooled adiabatically to *DBT* t_{db_2} which is equal to the adiabatic saturation t_{wb_2}. The adiabatic saturation temperature and wet bulb temperatures are taken to be equal for all practical purposes. The wet bulb temperature lies between the dry bulb temperature and dew point temperature.

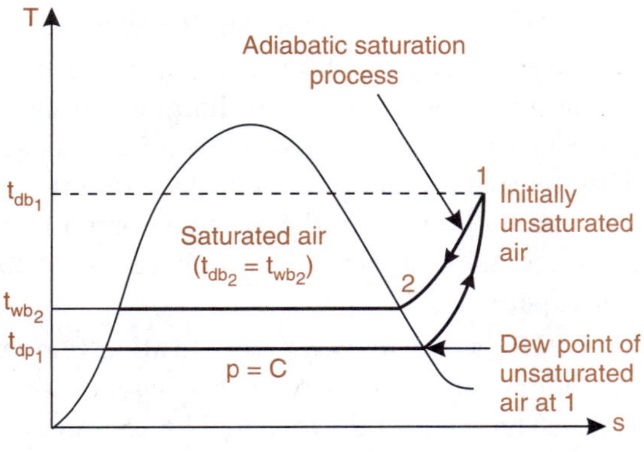

Fig. 9.2. Adiabatic cooling process.

Let us now apply the first law to the entire process. Considering the process to be steady state steady flow, neglecting changes in kinetic and potential energies, we have

$$h_1 + (W_{2s} - W_1)h_{f_2} = h_{2s} \quad \text{...(9.13)}$$

The quantities W_{2s}, h_{2s} and h_{f_2} are the functions of temperature t_{db_2}. The term $(W_{2s} - W_1)h_{f_2}$ is quite small and if this term is neglected, it can be seen that the enthalpy remains constant in adiabatic saturation.

Equation (9.13) may be rewritten as

$$h_1 - W_1 h_{f_2} = h_{2s} - W_{2s} h_{f_2}$$

The inlet term can be generalized and the expression can be written as follows :

$$\Sigma = h_{2s} - W_{2s}h_{f_2} = h_1 - W_1 h_{f_2} = h_x - W_x h_{f_2} \quad \text{...(9.14)}$$

This means that sigma function (Σ) as defined by the equation, is constant for any wet bulb temperature.

Also $\quad h_1 = h_{2s} - (W_{2s} - W_1)\, h_{f_2} \quad$ ...(9.15)

Equation (9.15) indicates that the enthalpy of an air-water vapour mixture is equal to the enthalpy of saturated air at the same wet bulb temperature, less small correction term $(W_{2s} - W_1)\, h_{f_2}$. This correction term is called ***enthalpy deviation.***

$$h_1 = h_{\text{air (1)}} + W_1\, h_{\text{vapour (1)}} \quad \text{...(9.16)}$$

$$h_{2s} = h_{\text{air (2)}} + W_{2s}\, h_{\text{vapour (2)}} \quad \text{...(9.17)}$$

or $\quad h_{\text{air (1)}} + W_1 h_{g1} + (W_{2s} - W_1)h_{f_2} = h_{\text{air (2)}} + W_{2s} h_{g2}$

$$c_p t_{db_1} + W_1 h_{g1} + (W_{2s} - W_1)h_{f_2} = c_p t_{db_2} + W_{2s} h_{g2} \quad \text{...(9.18)}$$

After arranging, we get

$$W_1 = \frac{c_p(t_{db_2} - t_{db_1}) + W_{2s}(h_{g_2} - h_{f_2})}{h_{g_1} - h_{f_2}}$$

or

$$W_1 = \frac{c_p(t_{db_2} - t_{db_1}) + W_{2s} h_{g_2}}{h_{g_1} - h_{f_2}} \quad \text{...[9.18(a)]}$$

Note. The wet bulb temperature is not a property of moist air as it is influenced by heat and mass transfer rates. Thus in psychrometric equations and psychrometric charts where the wet bulb temperature appears, *it is always the thermodynamic wet bulb temperature that is considered.*

9.4. PSYCHROMETERS

A psychrometer is a device which is used for measuring dry bulb and wet bulb temperatures simultaneously.

The psychrometers may be classified as follows :

1. Laboratory psychrometer
2. Sling psychrometer
3. Aspirating psychrometer
4. Continuous recording psychrometer.

The description of a sling psychrometer is given below :

Refer to Fig. 9.3. The sling psychrometer consists of two thermometers mounted on a base plate. The one with the 'sock' is wet-bulb thermometer ; the other is dry-bulb. The wet bulb exists below the dry-bulb. This is done purposely so that 'sock' can be dipped in water without wetting the dry-bulb. The handle of the frame helps for rotating the psychrometer to produce necessary air motion. As the psychrometer is rotated it provides necessary air velocity over the thermometer. Fast movement of air past the 'sock' is necessary to bring the air at temperature t_{db} always in immediate contact with the wet 'sock'. The temperature spread between dry bulb and wet bulb readings depends upon the *amount of moisture in the air*. *Dry air*, or air that has low moisture content has a low wet bulb temperature ; humid air that has a high moisture content, has a high wet-bulb temperature.

When dry and wet bulb temperatures are known the other psychrometric properties like relative humidity, dew point temperature, degree of saturation, humidity ratio, and volume of air-vapour mixture per kg of dry air are determined by calculations.

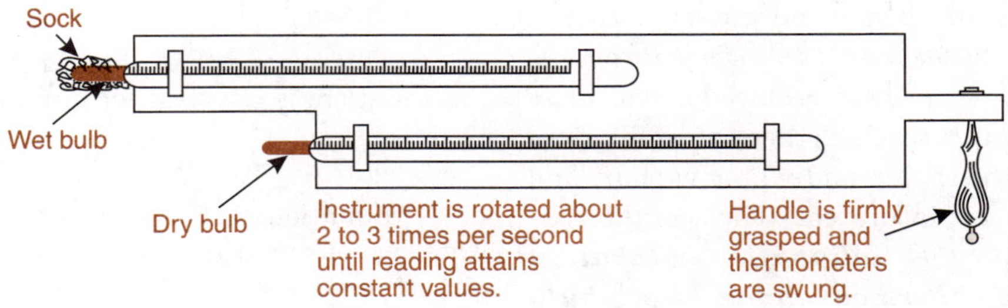

Fig. 9.3. Sling psychrometer.

9.5. PSYCHROMETRIC CHARTS

The psychrometric charts are prepared to represent graphically all the necessary moist air properties used for air conditioning calculations. The values are based on actual measurements verified for thermodynamic consistency.

For psychrometric charts the most convenient co-ordinates are dry bulb temperature of air vapour mixture as the abcissa and moisture content (kg/kg of dry air) or water vapour pressure as the ordinate. Depending upon whether the humidity contents is abcissa or ordinate with temperature co-ordinate, the charts are generally classified as Mollier chart and Carrier chart. Carrier chart having t_{db} as the abcissa and W as the ordinate finds a wide application. The chart is constructed as under :

1. *The dry bulb temperature* (ºC) of unit mass of dry air for different humidity contents or humidity ratios are indicated by *vertical lines drawn parallel to the ordinate.*

2. *The mass of water vapour* in kg (or grams) per kg of dry air is drawn parallel to the abcissa for different values of dry bulb temperature. It is the *major vertical scale of the chart.*

3. *Pressure of water vapour in mm of mercury* is shown in the scale at left and is the absolute pressure of steam.

4. *Dew point temperatures* are temperatures corresponding to the boiling points of water at low pressures of water vapour and are shown in the scale on the *upper curved line.* The dew points for different low pressures are read on *diagonal co-ordinates.*

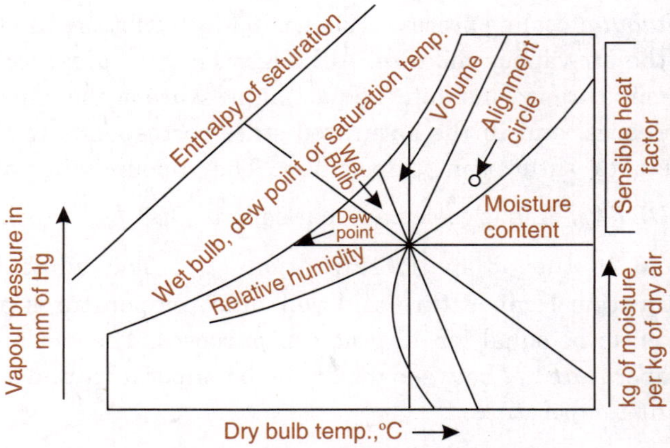

Fig. 9.4. Skeleton psychrometric chart.

5. Constant relative humidity lines in percent are indicated by marking off vertical distances between the saturation line or the upper curved line and the base of the chart. The relative humidity curve depicts quantity (kg) of moisture actually present in the air as a percentage of the total amount possible at various dry bulb temperatures and masses of vapour.

6. *Enthalpy or total heat* at saturation temperature in kJ/kg of dry air is shown by a diagonal system of co-ordinates. The scale on the diagonal line is separate from the body of the chart and is indicated above the saturation line.

7. *Wet bulb temperatures* are shown on the diagonal co-ordinates coinciding with heat co-ordinates. The scale of wet bulb temperatures is shown on the saturation curve. The diagonals run downwards to the right at an angle of 30º to the horizontal.

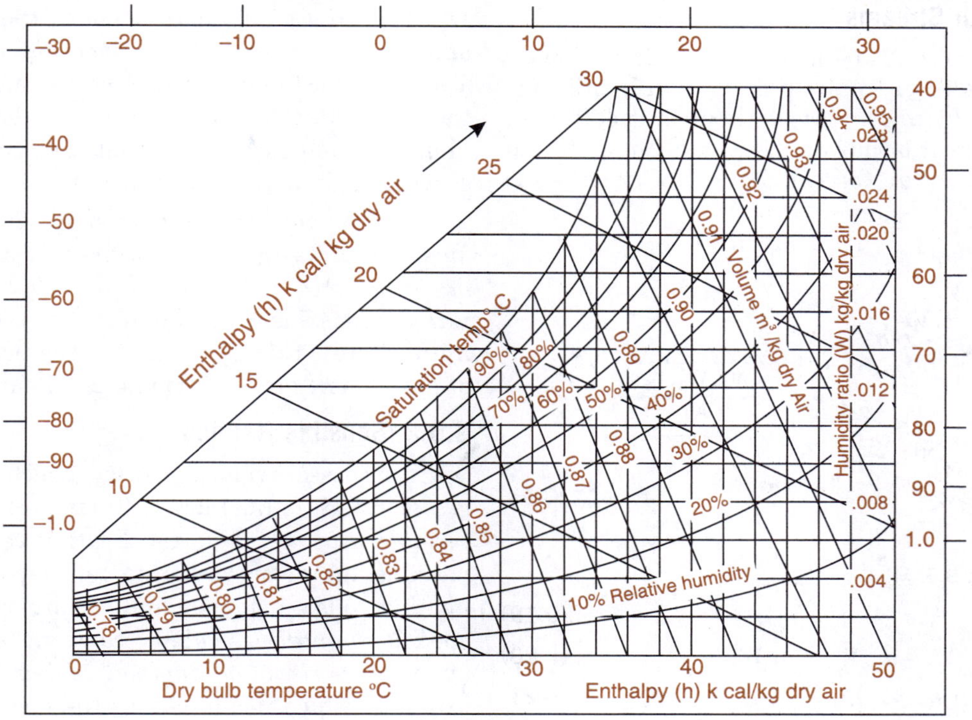

Fig. 9.5. Carrier chart.

8. The volume of air vapour mixture per kg of dry air (specific volume) is also indicated by a set of diagonal co-ordinates but at an angle of 60° with the horizontal.

The other properties of air vapour mixtures can be determined by using formulae (already discussed).

In relation to the psychrometric chart, these terms can quickly indicate many things about the condition of air, for example :

1. If dry bulb and wet bulb temperatures are known, the relative humidity can be read from the chart.

2. If the dry bulb and relative humidity are known, the wet bulb temperature can be determined.

3. If wet bulb temperature and relative humidity are known, the dry bulb temperature can be found.

4. If wet bulb and dry bulb temperatures are known, the dew point can be found.

5. If wet bulb and relative humidity are known, dew point can be read from the chart.

6. If dry-bulb and relative humidity are known, dew point can be found.

7. The quantity (kg) of moisture in air can be determined from any of the following combinations :

(*i*) Dry bulb temperature and relative humidity ;

(*ii*) Dry bulb temperature and dew point ;

(*iii*) Wet bulb temperature and relative humidity ;

(*iv*) Wet bulb temperature and dew point temperature ;

(*v*) Dry bulb temperature and wet bulb temperature ; and

(*vi*) Dew point temperature alone.

Figs. 9.4 and 9.5 show the skeleton psychrometric chart and lines on carrier chart respectively.

9.6. PSYCHROMETRIC PROCESSES

In order to condition air to the conditions of human comfort or of the optimum control of an industrial process required, certain processes are to be carried out on the outside air available. The processes affecting the *psychrometric properties of air are called* ***psychrometric processes.*** These processes involve mixing of air streams, heating, cooling, humidifying, dehumidifying, adiabatic saturation and mostly the combinations of these.

The important psychrometric processes are enumerated and explained in the following text :

1. Mixing of air streams
2. Sensible heating
3. Sensible cooling
4. Cooling and dehumidification
5. Cooling and humidification
6. Heating and dehumidification
7. Heating and humidification.

9.6.1. Mixing of Air Streams

Refer to Figs. 9.6 and 9.7. Mixing of several air streams is the process which is very frequently used in air conditioning. This mixing normally takes place without the addition or rejection of either heat or moisture, *i.e., adiabatically and at constant total moisture content. Thus we can write the following equations* :

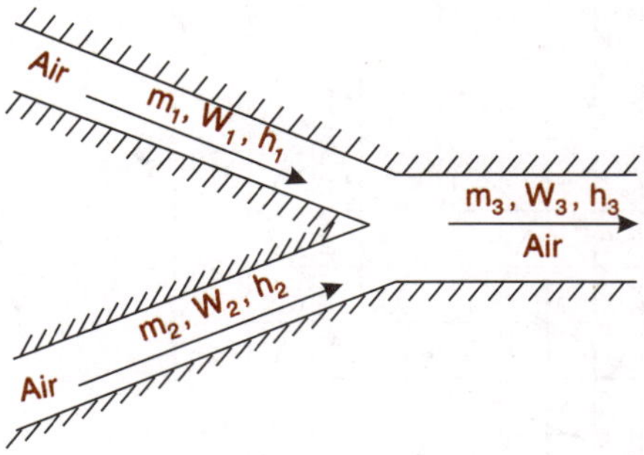

Fig. 9.6. Mixing of air streams.

$$m_1 + m_2 = m_3 \quad ...(9.19)$$

$$m_1W_1 + m_2W_2 = m_3W_3 \quad ...(9.20)$$

$$m_1h_1 + m_2h_2 = m_3h_3 \quad ...(9.21)$$

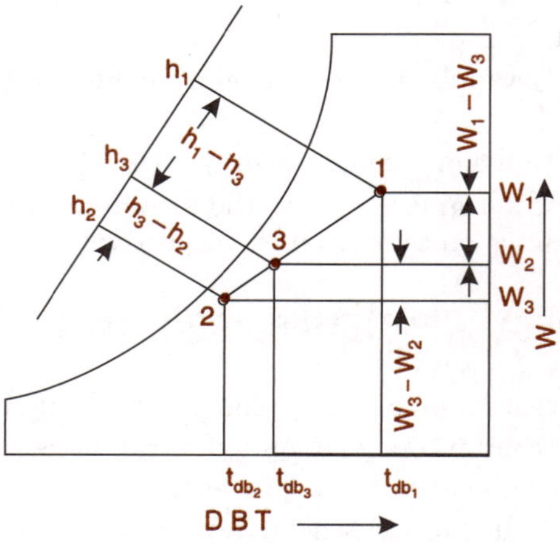

$$\frac{m_1}{m_2} = \frac{W_3 - W_2}{W_1 - W_3} = \frac{h_3 - h_2}{h_1 - h_3}$$

Fig. 9.7

Rearranging of last two equations gives the following :

$$m_1(W_1 - W_3) = m_2(W_3 - W_2)$$

$$m_1(h_1 - h_3) = m_2(h_3 - h_2)$$

or

$$\frac{m_1}{m_2} = \frac{W_3 - W_2}{W_1 - W_3} = \frac{h_3 - h_2}{h_1 - h_3}$$

where m = Mass of dry air, W = Specific humidity, h = Enthalpy — at particular state points.

On the psychrometric chart, the specific humidity and enthalpy scales are linear, ignoring enthalpy deviations. Therefore, the final state 3 lies on a straight line connecting the initial states of the two streams before mixing, and the final state 3 divides this line into two parts that are in the same ratio as were the two masses of air before mixing.

If the air quantities are known in volume instead of mass units, it is generally sufficiently accurate to units of m^3 or m^3/min. in the mixing equations. The inaccuracy introduced is due to the difference in specific volume at two initial states. This difference in densities is small for most of the comfort air conditioning problems.

9.6.2. Sensible Heating

When air passes over a dry surface which is at a temperature *greater* than its (air) dry bulb temperature, it undergoes *sensible heating*. Thus the heating can be achieved by passing the air over heating coil like electric resistance heating coils or steam coils. During such a process, the specific *humidity remains constant but the dry bulb temperature rises and approaches that of the surface.* The extent to which it approaches the mean effective surface temperature of the coil is conveniently expressed in terms of the equivalent **by-pass factor.**

The by-pass factor (BF) for the process is defined as the ratio of the difference between the mean surface temperature of the coil and leaving air temperature to the difference between the mean surface temperature and the entering air temperature. Thus on Fig. 9.8, air at temperature t_{db_1}, passes over a heating coil with an average surface temperature t_{db_3} and leaves at temperature t_{db_2}.

The by-pass factor is expressed as follows :

$$BF = \frac{t_{db_3} - t_{db_2}}{t_{db_3} - t_{db_1}} \quad ...(9.22)$$

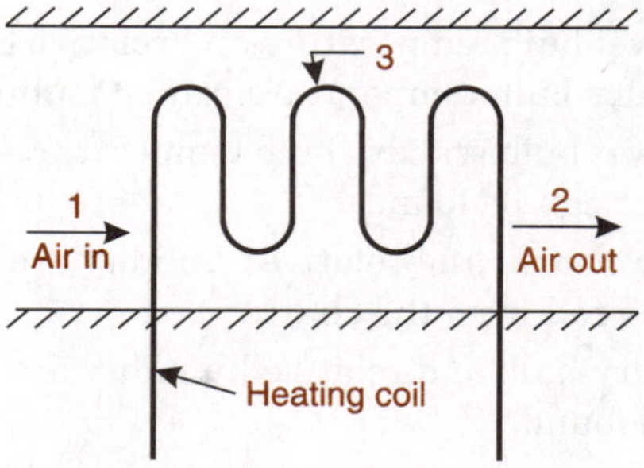

Fig. 9.8. Sensible heating.

Or in terms of lengths on the chart (Fig. 9.9) it is $\frac{\text{length } 2\text{-}3}{\text{length } 1\text{-}3}$. *The value of the by-pass factor is a function of coil design and velocity.* The heat added to the air can be obtained directly from the entering and leaving enthalpies

$(h_2 - h_1)$ or it can be obtained from the humid specific heat multiplied by the temperature difference $(t_{db_2} - t_{db_1})$.

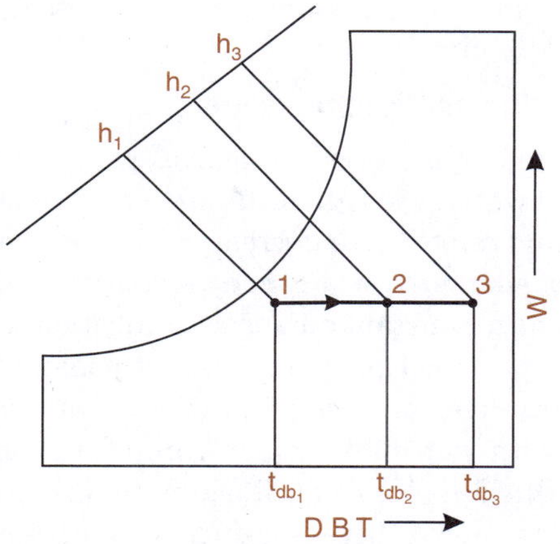

Fig. 9.9

In a complete air conditioning system the *preheating* and *reheating* of air are among the familar examples of sensible heating.

Note. 'By-pass factor' can be considered to represent the fraction of air which *does not* come into contact with coil surface.

9.6.3. Sensible Cooling

Refer to Fig. 9.10. Air undergoes *sensible cooling* whenever it passes over a surface that is at a temperature less than the *dry bulb* temperature of the *air but greater than the dew point temperature*. Thus sensible cooling can be achieved by passing the air over cooling coil like *evaporating coil of the refrigeration cycle or secondary brine coil*. During the process, *the specific humidity remains constant* and *dry bulb temperature decreases*, approaching the mean effective surface temperature. On a psychrometric chart the process will appear as a horizonal line 1–2 (Fig. 9.11), where point 3 represents the effective surface temperature. For this process :

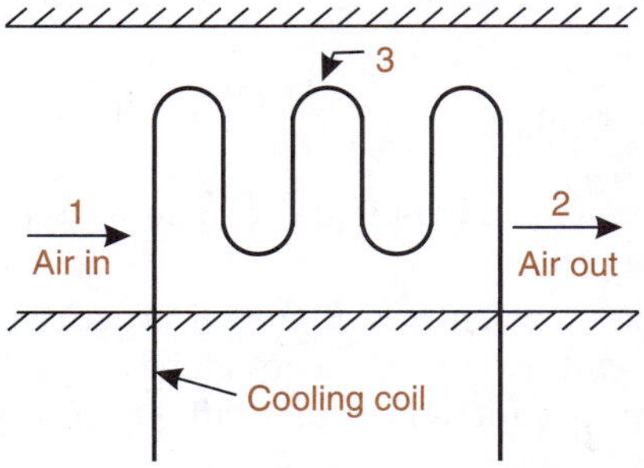

Fig. 9.10. Sensible cooling.

$$\text{By-pass factor BF} = \frac{t_{db_2} - t_{db_3}}{t_{db_1} - t_{db_3}} \quad ...(9.23)$$

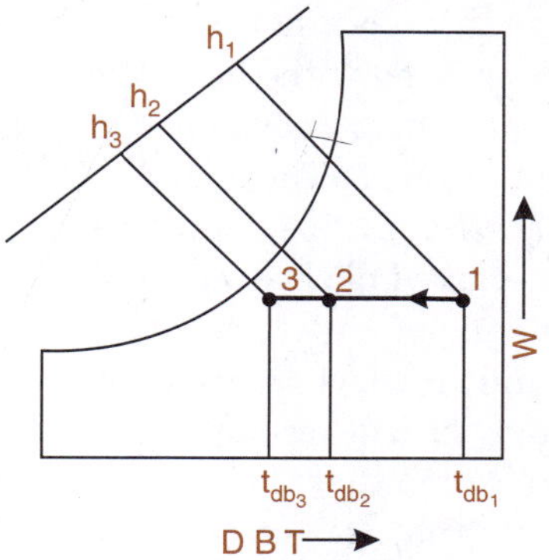

Fig. 9.11

The heat removed from air can be obtained from the enthalpy difference $(h_1 - h_2)$ or from humid specific heat multiplied by the temperature difference $(t_{db_1} - t_{db_2})$.

9.6.4. Cooling and Dehumidification

Refer to Fig. 9.12. Whenever air is made to pass over a surface or through a spray of water that is at a *temperature less than the dew point temperature of the air, condensation of some of the water vapour in air will occur simultaneously with the sensible cooling process.* Any air that comes into

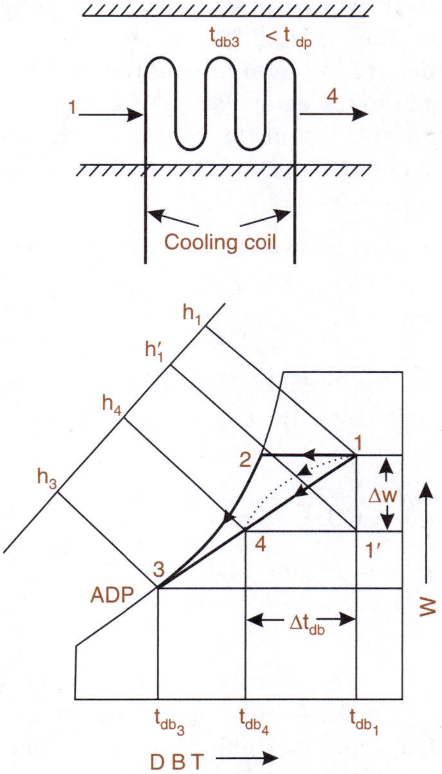

Fig. 9.12. Cooling and dehumidification.

sufficient contact with the cooling surface will be reduced in temperature to the mean surface temperature along a path such as 1-2-3 in Fig. 9.12, with condensation and therefore dehumidification occurring between points

2 and 3. The air that does not contact the surface will be finally cooled by mixing with the portion that did, and the final state point will somewhere on the straight line connecting points 1 and 3. The actual path of air during the path will not be straight line shown but will be something similarly to the curved dashed line 1–4. It will result from a continuous mixing of air which is connecting a particular part of the coil and air which is by passing it. It is convenient, however to analyse the problem with the straight line shown, and to assume that the final air state results from the mixing of air that has completely by passed the coil with air that has been cooled to the mean effective surface temperature. If there is enough contact between air and surface for all the air to come to the mean surface temperature, the process is one of zero by pass. In any practical system, complete saturation is not obtained and final state will be a point such as 4 in Fig. 9.12 with an equivalent by pass factor equal to $\frac{\text{length 3-4}}{\text{length 3-1}}$. For processes involving condensation, the effective surface temperature, *e.g.* t_{db_3} in Fig. 9.12 is called '**apparatus dew point**' (ADP). The final state point of air passing through a cooling and dehumidifying apparatus is in effect a mixture condition that results from mixing the fraction of the air, which is equal to the equivalent by-pass factor (BF) and is at initial state point and the remaining fraction which is equal to one minus by pass factor (1–BF) and is saturated at the apparatus dew point (ADP).

Total heat removed from the air is given by

$$Q_t = h_1 - h_4 = (h_1 - h_1') + (h_1' - h_4)$$
$$= Q_L + Q_S$$

where, Q_L = Latent heat removed $(h_1 - h_1')$, and

Q_S = Sensible heat removed $(h_1' - h_4)$

The ratio $\frac{Q_S}{Q_L}$ *is called sensible heat factor* (SHF) Or *sensible heat ratio* (SHR)

$$\therefore \quad \text{SHF} = \frac{Q_S}{Q_L + Q_S} \quad \text{...(9.24)}$$

The ratio fixes the slope of the line 1–4 on the psychrometric chart. Sensible heat factor slope lines are given on the psychrometric chart. If the initial condition and SHF are known for the given process, then the process line can be drawn through the given initial condition at a slope given by SHF on the psychrometric chart.

The capacity of the cooling coil in *tonnes* of refrigeration is given by,

$$\text{Capacity in TR} = \frac{m_a(h_1 - h_4) \times 60}{14000}, \quad \text{...(9.25)}$$

where, m_a = mass of air, kg/min and h = enthalpy in kJ/kg of air.

9.6.5. Cooling and Humidification

If unsaturated air is passed through a spray of continuously recirculated water, the specific humidity will increase while the dry bulb temperature decreases. This is the process of ***adiabatic saturation* or *evaporative cooling.*** This process is one of constant adiabatic-saturation temperature and for all practical purposes, one of constant wet bulb temperature. The process is illustrated as path 1-2 on Fig. 9.13, with wet bulb temperature of air being that of point 3, which is also equilibrium temperature of the recirculated water if there is sufficient contact between air and spray, the air will leave at a condition very close to that of point 3. The concept of equivalent by pass can be applied to this process but another term is more used to describe the performance of a humidifying apparatus. It is the **'*saturating*'** or **'*humidifying efficiency*'** which is *defined as the ratio of dry-bulb temperature decrease to the entering wet bulb depression usually expressed as percentage.* Thus, from Fig. 9.13, the saturating efficiency is:

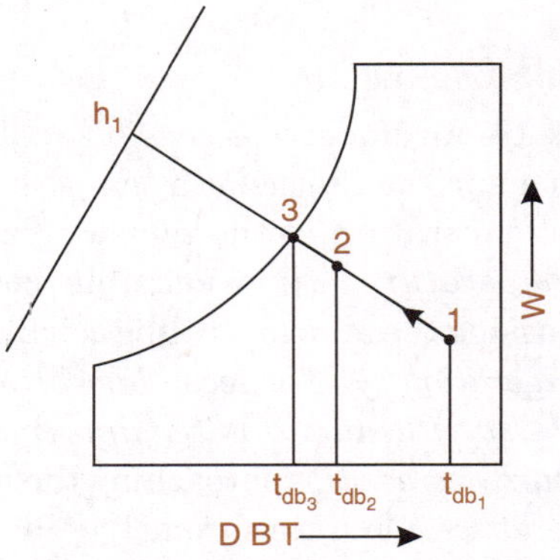

Fig. 9.13. Cooling and humidification.

$$\%\ \eta_{sat} = \left(\frac{t_{db_1} - t_{db_2}}{t_{db_1} - t_{db_3}}\right) \times 100 \quad \text{...(9.26)}$$

As a fraction, it is equal to one minus the by pass factor for the process.

This adiabatic process, for all practical purposes, is line of constant *enthalpy.* The moisture added can be obtained from the increase in specific humidity.

9.6.6. Heating and Dehumidification

If air is passed over a solid absorbent surface or through a liquid absorbent spray simultaneous heating and dehumidification is accompanied. In either case the dehumidification results from adsorbent or absorbent having a lower water vapour pressure than air. Moisture is condensed out of the air, and consequently the latent heat of condensation is liberated, causing sensible heating of air. If these were the only energies involved, the process would be the inverse of the adiabatic saturation process. There is, however, an additional energy absorbed or liberated by the active material, termed the *heat of adsorption* or *absorption*. For the solid adsorbents used commercially, such as silica gel or activated alumina, and for the more common liquid absorbents, such as solutions of organic salts or inorganic compounds like ethylene, glycol, heat is involved and results in additional sensible heating. Thus the path lies above a constant wet bulb line on the psychrometric chart such as path 1-2 in Fig. 9.14.

9.6.7. Heating and Humidification

If air is passed through a humidifier which has heated water sprays instead of simply recirculated spray, the air is humidified and may be heated, cooled or unchanged in temperature. In such a process the *air increases in specific humidity and the* enthalpy, and the *dry bulb temperature will increase or decrease according to the initial temperature of the air and that of the spray*. If sufficient water is supplied relative to the mass flow of air, the air will approach saturation at water temperature. Examples of such processes are shown on Fig. 9.15.

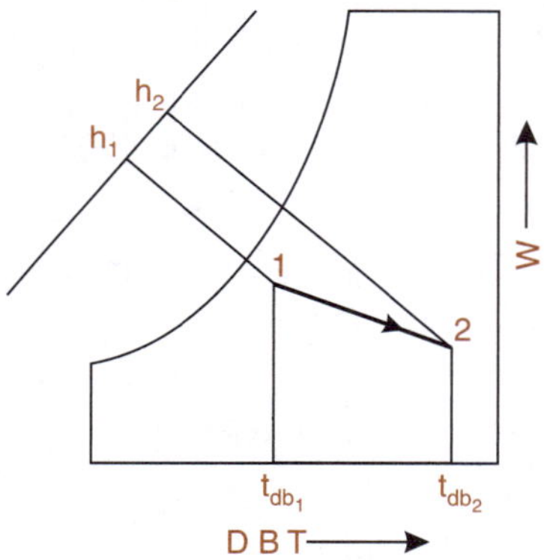

Fig. 9.14. Heating and dehumidification.

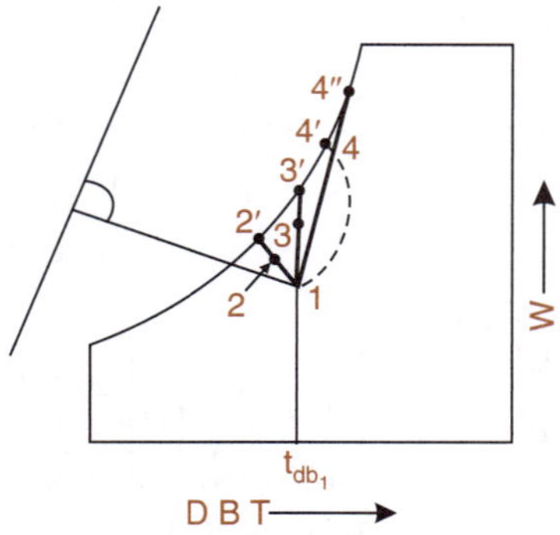

Fig. 9.15. Heating and humidification.

Process 1-2 : It denotes the cases in which the temperature of the heated spray water is less than the air DBT.

Process 1-3 : It denotes the cases in which the temperature is *equal to the air DBT*.

Process 1-4 : It denotes the cases in which a spray temperature is *greater than air DBT*.

As in the case of adiabatic saturation, the degree to which the process approaches saturation can be expressed in terms of the by-pass factor or a saturating efficiency.

If the water rate relative to the air quantity is smaller, the water temperature will drop significantly during the process. The resultant process will be a curved line such as the dashed 1-4 where 4 represents the leaving water temperature.

Note. It is possible to accomplish heating and humidification by evaporation from an open pan of heated water, or by direct injection of heated water or steam. The latter is more common. The process line for it is of little value because the process is essentially an instantaneous mixing of steam and the air. The final state point of the air can be found, however by making a humidity and enthalpy balance for the process. The solution of such a problem usually involves cut-and-try procedure.

Example 9.1. *The atmospheric conditions are ; 20°C and specific humidity of 0.0095 kg/kg of dry air. Calculate the following :*

(*i*) *Partial pressure of vapour*

(*ii*) *Relative humidity*

(*iii*) *Dew point temperature.*

Solution. Dry bulb temperature, t_{db} = 20°C

Specific humidity, W = 0.0095 kg/kg of dry air

(*i*) **Partial pressure of vapour, p_v :**

The specific humidity is given by

$$W = \frac{0.622\, p_v}{p_t - p_v}$$

$$0.0095 = \frac{0.622\, p_v}{1.0132 - p_v}$$

$$0.0095(1.0132 - p_v) = 0.622\, p_v$$

$$0.009625 - 0.0095\, p_v = 0.622\, p_v$$

$$p_v = \mathbf{0.01524\ bar.} \quad \textbf{(Ans.)}$$

(*ii*) **Relative humidity φ :**

Corresponding to 20°C, from steam tables,

$$p_{vs} = 0.0234 \text{ bar}$$

∴ **Relative humidity,**

$$\phi = \frac{p_v}{p_{vs}} = \frac{0.01524}{0.0234} = \mathbf{0.65\ or\ 65\%.} \quad \textbf{(Ans.)}$$

(*iii*) **Dew point temperature, t_{dp} :**

The dew point temperature is the saturation temperature of water vapour at a pressure of 0.01524 bar,

t_{dp} [from steam tables by interpolation]

$$= 13 + \frac{(14-13)}{(0.01598-0.0150)} \times [0.01524 - 0.0150]$$

$$= 13 + \frac{0.00024}{0.00098} = \mathbf{13.24°C.} \quad \textbf{(Ans.)}$$

Example 9.2. *The air supplied to a room of a building in winter is to be at 17°C and have a relative humidity of 60%. If the barometric pressure is 1.01325 bar, find : (i) The specific humidity ; (ii) The dew point under these conditions.*

Solution. Dry bulb temperature, $t_{db} = 17°C$

Relative humidity, $\phi = 60\%$

Barometric or total pressure, $p_t = 1.01325$ bar

(*i*) **Specific humidity, W :**

Corresponding to 17°C, from steam tables,

$$p_{vs} = 0.0194 \text{ bar}$$

Also, $\phi = \dfrac{p_v}{p_{vs}}$

i.e., $0.6 = \dfrac{p_v}{0.0194}$

$\therefore$ $p_v = 0.6 \times 0.0194 = 0.01164$ bar.

Specific humidity, W

$$= \frac{0.622\, p_v}{p_t - p_v} = \frac{0.622 \times 0.01164}{1.01325 - 0.01164}$$

$$= \mathbf{0.00723\ kg/kg\ of\ dry\ air.} \quad \textbf{(Ans.)}$$

(*ii*) **Dew point temperature, t_{dp} :**

If the air is cooled at constant pressure the vapour will begin to condense at the saturation temperature corresponding to 0.01164 bar. By interpolation from steam tables, the dew point temperature, t_{dp} is then

$$\mathbf{t_{dp}} = 9 + (10-9) \times \frac{0.01164 - 0.01150}{0.01230 - 0.01150}$$

$$= \mathbf{9.18°C.} \quad \textbf{(Ans.)}$$

Example 9.3. *0.004 kg of water vapour per kg of atmospheric air is removed and temperature of air after removing the water vapour becomes 20°C. Determine :*

(i) Relative humidity (ii) Dew point temperature.

Assume that condition of atmospheric air is 30°C and 55% R.H. and pressure is 1.0132 bar.

Solution. Corresponding to 30°C, from steam tables,

$$p_{vs} = 0.0425 \text{ bar}$$

$\therefore$ Relative humidity (R.H.),

$$\phi = \frac{p_v}{p_{vs}}$$

i.e., $0.55 = \dfrac{p_v}{0.0425}$

$\therefore$ $p_v = 0.02337$ bar.

Also the specific humidity,

$$W = \frac{0.622\, p_v}{p_t - p_v} = \frac{0.622 \times 0.02337}{1.0132 - 0.02337}$$

$$= 0.01468 \text{ kg/kg of dry air.}$$

The specific humidity after removing 0.004 kg of water vapour becomes,

$$0.01468 - 0.004 = 0.01068 \text{ kg/kg of dry air}$$

and the temperature t_{db} is given as 20°C.

The partial pressure of water vapour, p_v, at this condition can be calculated as follows :

$$W = \frac{0.622\, p_v}{p_t - p_v}$$

$$0.01068 = \frac{0.622\, p_v}{1.0132 - p_v}$$

or $0.01068\,(1.0132 - p_v) = 0.622\, p_v$

or $0.01082 - 0.01068\, p_v = 0.622\, p_v$

$0.6327\, p_v = 0.01082$

$\therefore$ $p_v = 0.0171$ bar

Corresponding to 20°C, from steam tables,

$$p_{vs} = 0.0234 \text{ bar.}$$

(*i*) **Relative humidity, ϕ**

$$= \frac{p_v}{p_{vs}} = \frac{0.0171}{0.0234}$$

$$= \mathbf{0.73\ or\ 73\%.} \quad \textbf{(Ans.)}$$

(*ii*) **Dew point temperature, t_{dp} :**

Corresponding to 0.0171 bar, from steam tables,

$$\mathbf{t_{dp} = 15°C.} \quad \textbf{(Ans.)}$$

Example 9.4. *The sling psychrometer in a laboratory test recorded the following readings :*

Dry bulb temperature = 35°C

Wet bulb temperature = 25°C.

Calculate the following :

(i) Specific humidity

(ii) Relative humidity

(iii) Vapour density in air

(iv) Dew point temperature

(v) Enthalpy of mixture per kg of dry air

Take atmospheric pressure = 1.0132 bar.

Solution. For finding the partial pressure of vapour, using the equation :

$$p_v = (p_{vs})_{wb} - \frac{[p_t - (p_{vs})_{wb}](t_{db} - t_{wb})}{1527.4 - 1.3\, t_{wb}}$$

Corresponding to 25°C (from steam tables),

$$(p_{vs})_{wb} = 0.0317 \text{ bar}$$

Substituting the values in the above equation, we get

$$p_v = 0.0317 - \frac{[1.0132 - 0.0317](35 - 25)}{1527.4 - 1.3 \times 25}$$

$= 0.0317 - 0.0065 = 0.0252$ bar.

(*i*) **Specific humidity,**

$$W = \frac{0.622\, p_v}{p_t - p_v} = \frac{0.622 \times 0.0252}{(1.0312 - 0.0252)}$$

= **0.01586 kg/kg of dry air. (Ans.)**

(*ii*) **Relative humidity,** $\phi = \frac{p_v}{p_{vs}} = \frac{0.0252}{0.0563}$

[p_{vs} = 0.0563 bar corresponding to 35°C, from steam tables]

= **0.447** or **44.7%. (Ans.)**

(*iii*) **Vapour density :**

From characteristic gas equation

$$p_v V_v = m_v R_v T_v$$

$$p_v = \frac{m_v}{V_v} R_v T_v = \rho_v R_v T_v$$

[where ρ_v = vapour density,

$$R_v = \frac{\text{Universal gas constant}}{\text{Molecular weight of } H_2O} = \frac{8314.3}{18}\Bigg]$$

$$\therefore \quad 0.0252 \times 10^5 = \rho_v \times \frac{8314.3}{18} \times (273 + 35)$$

$$\therefore \quad \rho_v = \frac{0.0252 \times 10^5 \times 18}{8314.3 \times 308}$$

= **0.0177 kg/m³. (Ans.)**

(*iv*) **Dew point temperature, t_{dp} :**

Corresponding to 0.0252 bar, from steam tables (by interpolation),

$$t_{dp} = 21 + (22 - 21) \times \frac{(0.0252 - 0.0249)}{(0.0264 - 0.0249)}$$

= **21.2°C. (Ans.)**

(*v*) **Enthalpy of mixture per kg of dry air, h :**

$$h = c_p t_{db} + W h_{vapour}$$

$$= 1.005 \times 35 + 0.01586\, [h_g + 1.88\, (t_{db} - t_{dp})]$$

$$= 35.175 + 0.01586\, [2565.3 + 1.88\, (35 - 21.2)]$$

(where h_g = 2565.3 kJ/kg corresponding to 35°C t_{db})

= **76.27 kJ/kg of dry air. (Ans.)**

Example 9.5. Adiabatic mixing : *One kg of air at 35°C DBT and 60% R.H. is mixed with 2 kg of air at 20°C DBT and 13°C dew-point temperature. Calculate the specific humidity of the mixture.*

Solution. For the air at 35°C DBT and 60% R.H. :

Corresponding to 35°C, from steam tables,

$$p_{vs} = 0.0563 \text{ bar}$$

Relative humidity,

$$\phi = \frac{p_v}{p_{vs}}$$

$$\therefore \quad p_v = \phi\, p_{vs} = 0.6 \times 0.0563 = 0.0338 \text{ bar}$$

$$W = \frac{0.622\, p_v}{p_t - p_v} = \frac{0.622 \times 0.0338}{1.0132 - 0.0338}$$

= 0.0214 kg/kg of dry air

Corresponding to 0.0338 bar, from steam tables,

$$t_{dp} = 26 + (27 - 26)\frac{(0.0338 - 0.0336)}{(0.0356 - 0.0336)}$$

= 26.1°C

Enthalpy, $h = c_p t_{db} + W h_{vapour}$

$$= 1.005\, t_{db} + W\, [h_g + 1.88\, (t_{db} - t_{dp})]$$

$$= 1.005 \times 35 + 0.0214\, [2565.3 + 1.88\, (35 - 26.1)]$$

= 90.43 kJ/kg of dry air.

For the air at 20°C DBT and 13°C dew-point temperature :

p_v is the vapour pressure corresponding to the saturation pressure of steam at 13°C.

$$\therefore \quad p_v = 0.0150 \text{ bar}$$

$$W = \frac{0.622\, p_v}{p_t - p_v} = \frac{0.622 \times 0.015}{1.0132 - 0.015}$$

= 0.00935 kg/kg of dry air

Enthalpy, $h = c_p t_{db} + W h_{vapour}$

$$= 1.005 \times 20 + 0.00935\, [h_g + 1.88\, (t_{db} - t_{dp})]$$

$$= 20.1 + 0.00935\, [2538.1 + 1.88\, (20 - 13)]$$

= 43.95 kJ/kg of dry air

Now enthalpy per kg of moist air

$$= \frac{1}{3}\left[\frac{90.43}{1.0214} + \frac{43.95 \times 2}{1.00935}\right] = 58.54 \text{ kJ/kg of moist air}$$

Mass of vapour/kg of moist air

$$= \frac{1}{3}\left[\frac{0.0214}{1.0214} + \frac{0.00935 \times 2}{1.00935}\right] = 0.01316 \text{ kg/kg of moist air}$$

Specific humidity of mixture

$$= \frac{0.01316}{1 - 0.01316} = \mathbf{0.01333 \text{ kg/kg of dry air.}} \quad \textbf{(Ans.)}$$

Example 9.6. Sensible heating : *90 m³ of air per minute at 20°C and 75% R.H. is heated until its temperature becomes 30°C. Calculate :*

(*i*) *R.H. of the heated air.*

(*ii*) *Heat added to air per minute.*

Solution. ***For air at 20°C and 75% R.H. :***

p_{vs} = 0.0234 bar (from steam tables, at 20°C)

$$p_v = \phi \times p_{vs} = 0.75 \times 0.0234 = 0.01755 \text{ bar}$$

$$t_{dp} = 15 + (16 - 15)\frac{(0.01755 - 0.017)}{(0.0182 - 0.017)} \simeq 15.5°\text{C}$$

$$W_1 = \frac{0.622\, p_v}{p_t - p_v} = \frac{0.622 \times 0.01755}{1.0132 - 0.01755}$$
$= 0.0109$ kg/kg of dry air

Enthalpy, $h_1 = c_p t_{db} + W h_{vapour}$
$= 1.005 \times 20 + 0.0109\, [h_g + 1.88\,(t_{db} - t_{dp})]$
$= 1.005 \times 20 + 0.0109\, [2538.1 + 1.88(20 - 15.5)]$
$= 47.85$ kJ/kg of dry air

(*i*) **Relative humidity of heated air :**

For air at 30°C DBT :

Since the saturation pressure of water vapour at 30°C is higher than the saturation pressure of water vapour at 20°C so it is *sensible heating,* where p_v is same after heating.

$\therefore$ Relative humidity, ϕ

$$= \frac{p_v}{p_{vs}} = \frac{0.01755}{0.0425} = 0.412 \text{ or } 41.2\%$$

($p_{vs} = 0.0425$ bar, corresponding to 30°C)

i.e., Relative humidity of heated air = **41.2%. (Ans.)**

(*ii*) **Heat added to air per minute :**

Enthalpy, $h_2 = c_p t_{db} + W h_{vapour}$
$= 1.005 \times 30 + 0.0109 \times [h_g + 1.88\,(t_{db} - t_{dp})]$
$= 1.005 \times 30 + 0.0109\, [2556.3 + 1.88\,(30 - 15.5)]$
$= 58.31$ kJ/kg of dry air

Mass of dry air in 90 m^3 of air supplied

$$m_a = \frac{pV}{RT} = \frac{(p_t - p_v)V}{RT}$$
$$= \frac{(1.0132 - 0.01755) \times 10^5 \times 90}{287 \times (273 + 20)}$$
$= 106.5$ kg/min.

Amount of heat added per minute
$= 106.5\,(h_2 - h_1)$
$= 106.5\,(58.31 - 47.85) \simeq$ **1114 kJ. (Ans.)**

Example 9.7. Sensible cooling : *40 m^3 of air at 35°C DBT and 50% R.H. is cooled to 25°C DBT maintaining its specific humidity constant. Determine :*

(i) Relative humidity (R.H.) of cooled air ;

(ii) Heat removed from air.

Solution. ***For air at 35°C DBT and 50% R.H. :***

$p_{vs} = 0.0563$ bar (At 35°C, from steam tables)

$$\phi = \frac{p_v}{p_{vs}}$$

$\therefore \quad p_v = \phi \times p_{vs} = 0.5 \times 0.0563 = 0.02815$ bar

$$W_1 = \frac{0.622\, p_v}{p_t - p_v} = \frac{0.622 \times 0.02815}{1.0132 - 0.02815}$$
$= 0.0177$ kg/kg of dry air

$$h_1 = c_p\, t_{db_1} + W_1\, [h_{g_1} + 1.88\,(t_{db_1} - t_{dp_1}]$$

$t_{dp_1} \simeq 23$°C (corresponding to 0.02815 bar)

$\therefore \quad h_1 = 1.005 \times 35 + 0.0177\, [2565.3 + 1.88\,(35 - 23)]$
$= 80.98$ kJ/kg of dry air

For air at 25°C DBT :

(*i*) **R.H. of cooled air :**

Since the specific humidity remains constant the vapour pressure in the air remains constant.

$$\phi = \frac{p_v}{p_{vs}} = \frac{0.02815}{0.0317} = 0.888 \text{ or } 88.8\%$$

i.e., Relative humidity of the cooled air = **88.8%. (Ans.)**

(*ii*) **Heat removed from air :**

$$h_2 = c_p t_{db_2} + W_2\, [h_{g_2} + 1.88\,(t_{db_2} - t_{dp_2})]$$
$= 1.005 \times 25 + 0.0177\, [2547.2 + 1.88\,(25 - 23)]$
$= 70.27$ kJ/kg of dry air.

To find mass of dry air (m_a), using the relation :

[$\because W_1 = W_2 = 0.0177$ kg/kg of dry air
$t_{dp_2} = t_{dp_1} = 23$°C since p_v does not change]

$$p_a v_a = m_a R_a T_a$$

$$\therefore \quad m_a = \frac{p_a v_a}{R_a T_a}$$
$$= \frac{(1.0132 - 0.02815) \times 10^5 \times 40}{287 \times (273 + 35)} = 44.57 \text{ kg}$$

$\therefore$ Heat removed from 40 m^3 of air
$= m_a\,(h_1 - h_2) = 44.57\,(80.98 - 70.27)$
= **477.3 kJ. (Ans.)**

Example 9.8. Cooling and dehumidification : *120 m^3 of air per minute at 35°C DBT and 50% relative humidity is cooled to 20°C DBT by passing through a cooling coil.*

Determine the following :

(i) Relative humidity of out coming air and its wet bulb temperature.

(ii) Capacity of cooling coil in tonnes of refrigeration.

(iii) Amount of water vapour removed per hour.

Solution. ***For the air at 35°C DBT and 50% R.H. :***

$p_{vs} = 0.0563$ bar (At 35°C, from steam tables)
$p_v = \phi \times p_{vs} = 0.5 \times 0.0563 = 0.02815$ bar.

$$W_1 = \frac{0.622\, p_v}{p_t - p_v} = \frac{0.622 \times 0.02815}{1.0132 - 0.02815}$$
$= 0.0177$ kg/kg of dry air.

$$h_1 = c_p\, t_{db_1} + W_1\, [h_{g_1} + 1.88\,(t_{db_1} - t_{dp_1})]$$

$t_{dp_1} \approx 23$°C (Corresponding to 0.02815 bar).

$\therefore \quad h_1 = 1.005 \times 35 + 0.0177\, [2565.3 + 1.88\,(35 - 23)]$
$= 80.98$ kJ/kg of dry dir.

For the air at 20°C. As the saturation vapour pressure at 20°C is 0.0234 bar, less than the vapour pressure 0.02815 bar at 35°C, so that *condensation takes place* and air *will be saturated at 20°C*.

(*i*) ∴ **Relative humidity of exit air is 100 per cent. (Ans.)**

Since the air is saturated, wet bulb temperature is equal to dry bulb temperature = 20°C. (Ans.)

$$\therefore \quad p_v = p_{vs} = 0.0234 \text{ bar.}$$

$$W_2 = \frac{0.622\, p_v}{p_t - p_v} = \frac{0.622 \times 0.0234}{(1.0132 - 0.0234)}$$

$$= 0.0147 \text{ kg/kg of dry air}$$

$$h_2 = c_p\, t_{db_2} + W_2\, [h_{g_2} + 1.88\,(t_{db_2} - t_{dp_2})]$$

$$= 1.005 \times 20 + 0.0147\, [2538.1 + 1.88\,(20 - 20)]$$

$$[\because \text{ When air is saturated } t_{db} = t_{dp}]$$

$$= 57.41 \text{ kJ/kg of dry air}$$

The weight of water vapour removed per kg of dry air

$$= 0.0177 - 0.0147 = 0.003 \text{ kg/kg of dry air}$$

Heat removed per kg of dry air

$$= h_1 - h_2 = 80.98 - 57.41$$

$$= 23.57 \text{ kJ/kg of dry air}$$

Mass of dry air passing per minute

$$m_a = \frac{p_a V_a}{R_a T_a} = \frac{(1.0132 - 0.02815) \times 10^5 \times 120}{287 \times (35 + 273)}$$

$$= 133.7 \text{ kg/min}$$

(*ii*) **Capacity of the cooling coil in tonnes of refrigeration**

$$= \frac{m_a\,(h_1 - h_2)}{14000}$$

$$= \frac{133.7 \times 23.57 \times 60}{14000} = \mathbf{13.5\ TR.}$$ **(Ans.)**

(*iii*) **Amount of water vapour removed per hour**

$$= m_a\,(W_1 - W_2) \times 60$$

$$= 133.7\,(0.0177 - 0.0147) \times 60$$

$$= \mathbf{24.066\ kg/h.}$$ **(Ans.)**

Example 9.9. Adiabatic humidification : *150 m³ of air per minute is passed through the adiabatic humidifier. The condition of air at inlet is 35°C DBT and 20 per cent relative humidity and the outlet condition is 20°C DBT and 15°C WBT.*

Determine the following :

(*i*) *Dew point temperature*

(*ii*) *Relative humidity of the exit air*

(*iii*) *Amount of water vapour added to the air per minute.*

Solution. ***For air at 35°C DBT and 20% relative humidity.***

$$p_{vs} = 0.0563 \text{ bar (At 35°C from steam tables)}$$

$$p_v = \phi \times p_{vs} = 0.2 \times 0.0563 = 0.01126 \text{ bar}$$

$$W_1 = \frac{0.622\, p_v}{p_t - p_v} = \frac{0.622 \times 0.01126}{1.0132 - 0.01126}$$

$$= 0.00699 \text{ kg/kg of dry air}$$

(*i*) The dew point temperature of air which is the saturation temperature of steam corresponding to the pressure 0.01126 bar is

$$8 + (9 - 8) \times \frac{(0.01126 - 0.01072)}{(0.01150 - 0.01072)} = 8.7°\text{C}$$

i.e., **Dew point temperature = 8.7°C. (Ans.)**

(*ii*) **Relative humidity of the exit air :**

For air at 20°C *DBT* and 15°C *WBT*.

$$p_v = (p_{vs})_{wb} - \frac{[p_t - (p_{vs})_{wb}]\,(t_{db} - t_{wb})}{1527.4 - 1.3\, t_{wb}}$$

$$= 0.0170 - \frac{[1.0132 - 0.0170]\,(20 - 15)}{1527.4 - 1.3 \times 15}$$

$$= 0.0137 \text{ bar}$$

$$W_2 = \frac{0.622\, p_v}{p_t - p_v} = \frac{0.622 \times 0.0137}{(1.0132 - 0.0137)}$$

$$= 0.00852 \text{ kg/kg of dry air}$$

Relative humidity

$$= \frac{p_v}{p_{vs}} = \frac{0.0137}{0.0234} = \mathbf{0.585 \text{ or } 58.5\%.}$$ **(Ans.)**

($\because$ p_{vs} = 0.0234 bar, corresponding to 20°C, from steam tables)

The *dew point temperature of air* which is the saturation temperature of steam corresponding to 0.0137 bar is **11°C** (from steam tables). **(Ans.)**

The amount of water vapour per kg of dry air

$$= W_2 - W_1 = 0.00852 - 0.00699$$

$$= 0.00153 \text{ kg}$$

The mass of dry air in 150 m³ of mixture

$$m_a = \frac{p_a V_a}{R_a T_a} = \frac{(1.0132 - 0.01126) \times 10^5 \times 150}{287 \times (35 + 273)}$$

$$= 170 \text{ kg}$$

(*iii*) **The amount of water vapour added to air per minute**

$$= m_a\,(W_2 - W_1) = 170 \times 0.00153$$

$$= \mathbf{0.26\ kg/min.}$$ **(Ans.)**

Example 9.10. Adiabatic saturation process : *An air-water vapour mixture enters an adiabatic saturation chamber at 28°C and leaves at 18°C, which is the adiabatic saturation temperature. The pressure remains constant at 1.0 bar.*

Determine the relative humidity and humidity ratio of the inlet mixture.

Solution. The specific humidity at the *exit*

$$W_{ss} = \frac{0.622\, p_s}{p_t - p_s} = \frac{0.622 \times 0.0206}{(1.00 - 0.0206)}$$

$$= 0.01308 \text{ kg/kg of dry air}$$

The specific humidity at the inlet [equation (9.18)]

$$W_1 = \frac{c_p\,(t_{db_2} - t_{db_1}) + W_{2s}\,(h_{g_2} - h_{f_2})}{h_{g_1} - h_{f_2}}$$

$$= \frac{1.005\,(18 - 28) + 0.01308\,(2534.4 - 75.6)}{(2552.6 - 75.6)}$$

$$= \frac{42.211}{2477} = 0.01704 \text{ kg/kg of dry air}$$

$$W_1 = \frac{0.622\,p_{v_1}}{p_t - p_{v_1}}$$

$$\therefore \quad 0.01704 = \frac{0.622\,p_{v_1}}{1.00 - p_{v_1}}$$

or $\quad 0.01704\,(1.00 - p_{v_1}) = 0.622\,p_{v_1}$

or $\quad 0.01704 - 0.01704\,p_{v_1} = 0.622\,p_{v_1}$

or $\quad 0.0639\,p_{v_1} = 0.01704$

$\therefore \quad p_{v_1} = 0.02666$ bar

$\therefore$ **Relative humidity**

$$= \frac{p_{v_1}}{p_{s_1}} = \frac{0.02666}{0.03782} = \mathbf{0.7 \text{ or } 70\%.} \quad \textbf{(Ans.)}$$

Example 9.11. *An air-water vapour mixture enters an air-conditioning unit at a pressure of 1.0 bar. 38°C DBT, and a relative humidity of 75%. The mass of dry air entering is 1 kg/s. The air-vapour mixture leaves the air-conditioning unit at 1.0 bar, 18°C, 85% relative humidity. The moisture condensed leaves at 18°C. Determine the heat transfer rate for the process.*

Solution. $t_{db_1} = 38°C$, R.H., $\phi_1 = 75\%$

$t_{tb_2} = 18°C$, R.H., $\phi_2 = 85\%$

The flow diagram and the process are shown in Fig. 9.16 (*a*) and (*b*) respectively.

At 38°C:

From steam tables :

$p_{vs} = 0.0663$ bar, $h_{g_1} = 2570.7$ kJ/kg

$\therefore \quad p_v = \phi \times p_{vs} = 0.75 \times 0.0663 = 0.0497$ bar

$$W_1 = \frac{0.622 \times 0.0497}{1.0 - 0.0497} = 0.0325 \text{ kg/kg of dry air}$$

At 18°C:

From steam tables :

$p_{vs} = 0.0206$ bar, $h_{g_2} = 2534.4$ kJ/kg

$h_{f_2} = 75.6$ kJ/kg

$p_v = 0.85 \times 0.0206 = 0.01751$ bar

$$W_2 = \frac{0.622 \times 0.01751}{1 - 0.01751} = 0.01108 \text{ kg/kg of dry air}$$

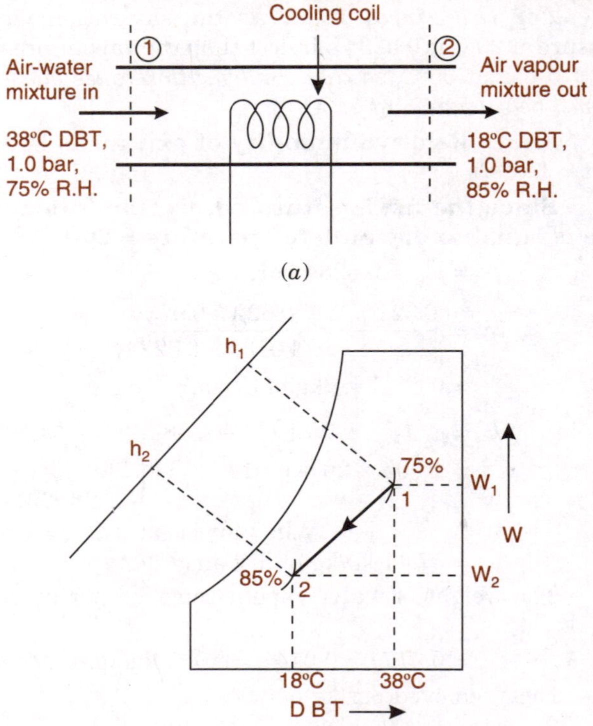

(*a*)

(*b*)

Fig. 9.16

Heat transfer rate,

$$q = (W_2 h_{g_2} - W_1 h_{g_1}) + c_p\,(t_{db_2} - t_{db_1}) + (W_1 - W_2)\,h_{f_2}$$

$$= (0.01108 \times 2534.4 - 0.0325 \times 2570.7) + 1.005\,(18 - 38) + (0.0325 - 0.01108) \times 75.6$$

$$= -\,55.46 - 20.1 + 1.62$$

$$= \mathbf{-\,73.94 \text{ kJ/kg of dry air.}} \quad \textbf{(Ans.)}$$

Example 9.12. Evaporative Cooler : *Atmospheric air at 38°C and 25% relative humidity passes through an evaporator cooler. If the final temperature of air is 18°C, how much water is added per kg of dry air and what is the final relative humidity ?*

Solution. At 38°C :

$p_{vs} = 0.0663$ bar, $h_{g_1} = 2570.7$ kJ/kg

and $\quad p_v = \phi \times p_{vs} = 0.25 \times 0.0663 = 0.01657$ bar

At 18°C :

$h_{g_2} = 2534.4$ kJ/kg, $p_{vs} = 0.0206$ bar

$$\therefore \quad W_1 = \frac{0.622 \times 0.01657}{1.0132 - 0.01657}$$

$= 0.01037$ kg/kg of dry air

Since enthalpy remains constant during the process

$$\therefore \quad c_p t_{db_1} + W_1 h_{g_1} = c_p t_{db_2} + W_2 h_{g_2}$$

$$1.005 \times 38 + 0.01034 \times 2570.7 = 1.005 \times 18 + W_2 \times 2534.4$$

$(\because$ At 18°C, $h_{g_2} = 2534.4$ kJ/kg)

i.e., $$W_2 = \frac{1.005 \times 38 + 0.01034 \times 2570.7 - 1.005 \times 18}{2534.4}$$

$= 0.01842$ kg/kg of dry air

$\therefore$ **Amount of water added**

$= W_2 - W_1 = 0.01842 - 0.01034$

$=$ **0.00808 kg/kg of dry air. (Ans.)**

$$\text{Also, } 0.00808 = \frac{0.622\, p_{v_2}}{1.0132 - p_{v_2}}$$

or $\quad 0.00808\,(1.0132 - p_{v_2}) = 0.622\, p_{v_2}$

$0.00818 - 0.00808\, p_{v_2} = 0.622\, p_{v_2}$

$\therefore \quad p_{v_2} = 0.01298$ bar

$\therefore$ **Final relative humidity**

$$= \frac{0.01298}{0.0206} = \mathbf{0.63 \text{ or } 63\%.} \quad \textbf{(Ans.)}$$

Example 9.13. *Saturated air at 3°C is required to be supplied to a room where the temperature must be held at 22°C with a relative humidity of 55%. The air is heated and then water at 10°C is sprayed to give the required humidity. Determine :*

(i) The mass of spray water required per m³ of air at room conditions.

(ii) The temperature to which the air must be heated.

Neglect the fan power. Assume that the total pressure is constant at 1.0132 bar.

Solution. The flow diagram is shown in Fig. 9.17 (*a*) and the processes are shown in Fig. 9.17 (*b*).

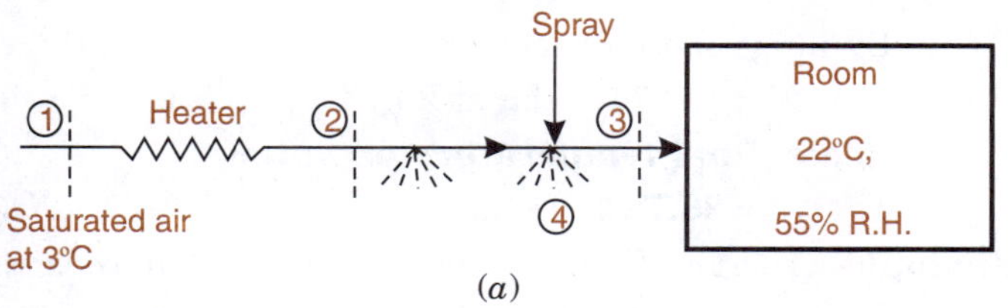

(*a*)

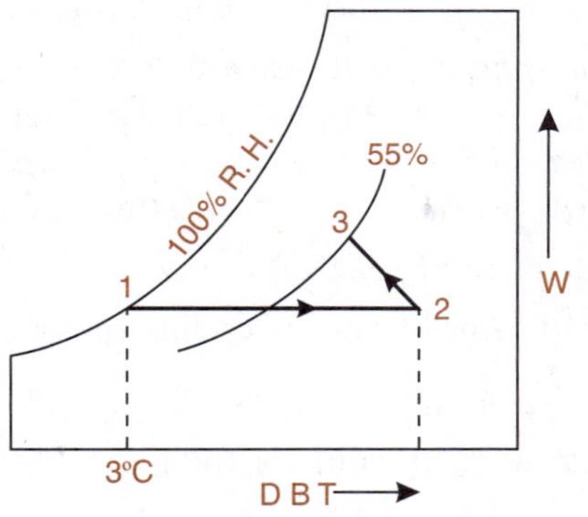

(*b*)

Fig. 9.17

(*i*) **Mass of spray water required:**

At 22°C:

From steam tables :

$p_{vs} = 0.0264$ bar

$$\phi_3 = \frac{p_{v_3}}{p_{vs_3}} = \frac{p_{vs_3}}{0.0264} = 0.55$$

$\therefore \quad p_{v_3} = 0.55 \times 0.0264 = 0.01452$ bar

$$\therefore \quad W_3 = \frac{0.622 \times 0.01452}{(1.0132 - 0.01452)}$$

$= 0.00904$ kg/kg of dry air

At 3°C:

From steam tables :

$p_{vs} = 0.0076$ bar

$$\phi_1 = \frac{p_{v_1}}{p_{vs_1}} = 1.00$$

$\therefore \quad p_{v_1} = p_{vs_1} = 0.0076$ bar

$$W_1 = \frac{0.622 \times 0.0076}{1.0132 - 0.0076}$$

$= 0.0047$ kg/kg of dry air

$W_3 - W_1 = 0.00904 - 0.0047$

$= 0.00434$ kg/kg of dry air

$$v_{a_3} = \frac{R_a T_3}{p_{a_3}} = \frac{287 \times (273 + 22)}{(1.0132 - p_{v_3}) \times 10^5} = \frac{287 \times 295}{0.9987 \times 10^5}$$

$= 0.847$ m³/kg of dry air

$$\text{Spray water } = \frac{0.00434}{0.847}$$

$=$ **0.005124 kg moisture/m³. (Ans.)**

(*ii*) **Temperature to which the air must be heated t_{db_2} :**

Now, $\quad h_2 + (W_3 - W_2)\, h_4 = h_3$

$[c_p t_{db_2} + W_2 h_{\text{vapour (2)}}] + (W_3 - W_2) h_2$

$= c_p t_{db_3} + W_3 h_{\text{vapour (3)}}$

$\therefore \quad c_p (t_{db_3} - t_{db_2}) + W_3 h_{\text{vapour (3)}}$

$- W_2 h_{\text{vapour (2)}} - (W_3 - W_2) h_4 = 0$

From the steam tables at $p_v = 0.01452$ bar :

$h_g = 2524$ kJ/kg

and $\quad t_{dp} = t_{sat} = 12.5$°C

$\therefore \quad 1.005(22 - t_{db_2}) + 0.00904[2524 + 1.88(22 - 12.5)]$

$- 0.0047\,[2524 + 1.88(t_{db_2} - 12.5)]$

$- (0.00434 \times 4.187 \times 10) = 0$

$22.11 - 1.005\, t_{db_2} + 22.97 - 11.86$

$- 0.0088\, t_{db_2} + 0.11 - 0.1817 = 0$

$1.014\, t_{db_2} = 33.148$

$\therefore \quad \mathbf{t_{db_2} = 32.7°C.} \quad$ **(Ans.)**

Example 9.14. Cooling tower : *A small-size cooling tower is designed to cool 5.5 litres of water per second, the inlet temperature of which is 44°C. The motor-driven fan induces 9 m³/s of air through the tower and the power absorbed is 4.75 kW. The air entering the tower is at 18°C, and has a relative humidity of 60%. The air leaving the tower can be assumed to be saturated and its temperature is 26°C. Calculate :*

(i) The amount of cooling water (make-up) required per second.

(ii) The final temperature of the water.

Assume that the pressure remains constant throughout the tower at 1.013 bar.

Solution. The cooling tower is shown diagrammatically in Fig. 9.18.

(*i*) **Make-up water required :**

At 18°C $p_{vs} = 0.0206$ bar,

$\therefore \quad p_v = \phi \times p_{vs} = 0.6 \times 0.0206 = 0.01236$ bar

$\therefore \quad p_{a_1} = 1.013 - 0.01236 = 1.00064$ bar

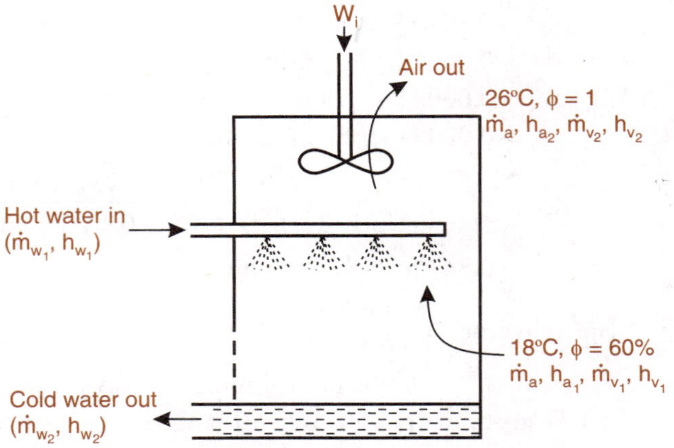

Fig. 9.18

Then,

$$\dot{m}_a = \frac{10^5 \times 1.00064 \times 9}{(0.287 \times 10^3) \times (18 + 273)} = 10.78 \text{ kg/s}$$

and
$$\dot{m}_{v_1} = \frac{10^5 \times 0.01236 \times 9}{(0.4618 \times 10^3) \times (18 + 273)} = 0.0828 \text{ kg/s}$$

(Script '*v*' denotes vapour and the script '*a*' denotes the air).

At exit at 26°C,

$p_{vs} = 0.0336$ bar and $\phi = 1$

$\therefore \quad p_v = p_{vs} = 0.0336$ bar

$$W_2 = \frac{0.622\, p_v}{p_t - p_v} = \frac{0.622 \times 0.0336}{1.013 - 0.0336} = 0.02133 \text{ kg.}$$

But
$$W = \frac{m_v}{m_a}$$

$\therefore \quad \dot{m}_{v_2} = 0.02133 \times 10.78 = 0.23$ kg/s

Hence, **make-up water required**

$= 0.23 - 0.0828 =$ **0.1472 kg/s. (Ans.)**

(*ii*) **Final temperature of the water :**

Also, $\quad \dot{m}_{w_1} = 5.5 \times 1 = 5.5$ kg/s

and $\quad \dot{m}_{w_2} = \dot{m}_{w_1} -$ (make-up water)

$= 5.5 - 0.1472 = 5.353$ kg/s

Applying the steady flow energy equation and neglecting changes in kinetic energy and potential energy, we have

$$W_i + \dot{m}_{w_1} h_{w_1} + \dot{m}_{a_1} h_{a_1} + \dot{m}_{v_1} h_{v_1} = \dot{m}_{a_2} h_{a_2} + \dot{m}_{v_2} h_{v_2} + \dot{m}_{w_2} h_{w_2}$$

Now, W_i (*i.e.*, work input) = 4.75 kW = 4.75 kJ/s.

Evaluating the enthalpies from a datum of 0°C, we have :

$h_{w_1} = h_f$ at 44°C = 184.3 kJ/kg,

$h_{a_1} = 1.005\,(18 - 0) = 18.09$ kJ/kg,

$h_{v_1} = 2519.7 + 1.88\,(18 - 10) = 2534.74$ kJ/kg.

[Corresponding to $p_v = 0.01236$ bar, $t_s = t_{dp} \simeq 10°C$ *i.e.*, the vapour is superheated]

$h_{v_1} = h_g$ at 26°C = 2549 kJ/kg

$h_{a_2} = 1.005\,(26 - 0) = 26.13$ kJ/kg.

Then, substituting, we get

$4.75 + 5.5 \times 184.3 + 10.78 \times 18.09 + 0.0828 \times 2534.74$

$= 10.78 \times 26.13 + 0.23 \times 2549 + 5.353 \times h_{w_2}$

or $\quad 5.353\, h_{w_2} = 1423.28 - 867.95 = 555.33$

or $\quad h_{w_2} = 103.74$ kJ/kg.

By interpolation,

$h_f = 103.74$ kJ/kg at 26.7°C.

Hence, **final temperature of water**

= 26.7°C. (Ans.)

Example 9.15. *A cooling tower used in power plant consists of 10 big fans, $\dot{m}_{water}$ = 1000 kg/min. It is cooled from 35°C to 30°C. Atmospheric conditions are 35°C DBT, 25°C WBT. Air leaves the tower at 30°C, 90% RH. Find out the quantity of air handled per fan per hour and the quantity of make-up water required per hour.* **(Punjab University)**

Solution. Refer to Fig. 9.19.

Heat absorbed from the cooling tower

$= \dot{m}_{water} \times c \times \Delta T$

$= (1000 \times 60) \times 4.186 \times (35 - 30)$

$= 1.256 \times 10^6$ kJ/h

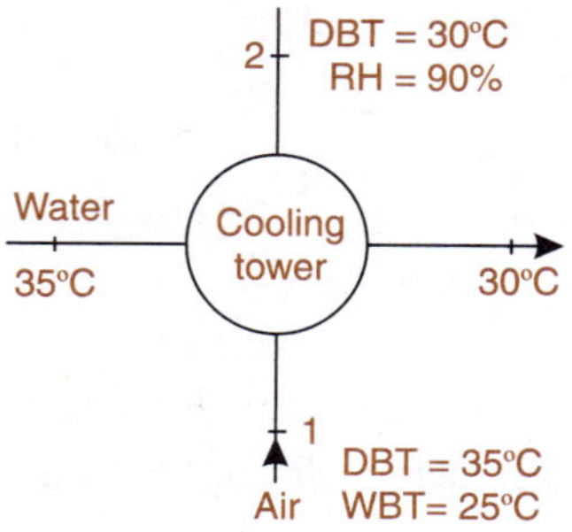

Fig. 9.19

From psychrometric chart, we have:

At 35°C DBT and 25°C WBT :

h_1 = 76.5 kJ/kg ; W_1 = 0.016 kg/kg of air

At 30°C and 90% RH :

h_2 = 92.5 kJ/kg ; W_2 = 0.0246 kg/kg of air

Heat gained by air = Heat lost by water

$$\dot{m}_{air}\ (h_2 - h_1) = 1.256 \times 10^6$$

$$\therefore \quad \text{Mass of air, } \dot{m}_{air} = \frac{1.256 \times 10^6}{(h_2 - h_1)}$$

$$= \frac{1.256 \times 10^6}{(92.5 - 76.5)} = 78.5 \times 10^3 \text{ kg/h}$$

∴ **Quantity of air handled per fan**

$$= \frac{78.5 \times 10^3}{10} = \textbf{7850 kg/h.} \quad \textbf{(Ans.)}$$

Quantity of make-up water

$= \dot{m}_{air}\ (W_2 - W_1)$

$= 78.5 \times 10^3\ (0.0246 - 0.016)$

= **675.1 kg/h.** **(Ans.)**

SOLUTIONS USING PSYCHROMETRIC CHARTS

Example 9.16. *The following data pertain to an air-conditioning system :*

Unconditioned space DBT = 30°C

Unconditioned space WBT = 22°C

Cold air duct supply surface temperature = 14°C.

Determine : (i) Dew point temperature.

(ii) Whether or not condensation will form on the duct.

Solution. Refer to Fig. 9.20.

(*i*) To determine the dew point temperature for the given conditions, find the intersection of 30°C *DBT* and 22°C *WBT* and move horizontally (as shown by the arrow) to the dew point temperature scale. The **dew point (t_{dp}) is 18.6°C.** **(Ans.)**

(*ii*) Since the duct temperature (14°C) is less than t_{dp} (18.6°C) therefore *moisture will condense on the duct surface.* **(Ans.)**

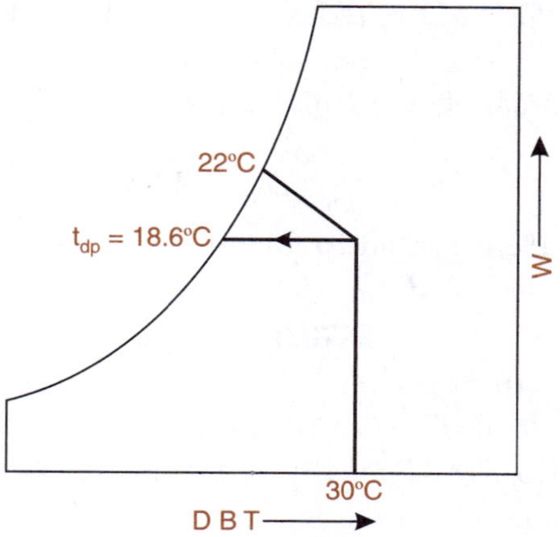

Fig. 9.20

Example 9.17. *200 m³ of air per minute at 15°C DBT and 75% R.H. is heated until its temperature is 25°C.*

Find : (i) R.H. of heated air.

(ii) Wet bulb temperature of heated air.

(iii) Heat added to air per minute.

Solution. Refer to Fig. 9.21.

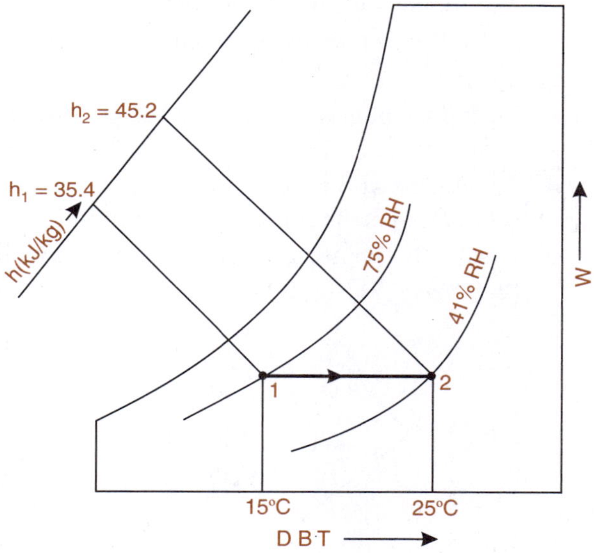

Fig. 9.21

- Locate point 1 on the psychrometric chart on intersection of 15°C *DBT* and 75% R.H. lines.
- Through point 1 draw a horizontal line to cut 25°C *DBT* line and get point 2.
- Read the following values from the psychrometric chart :

h_1 = 35.4 kJ/kg

h_2 = 45.2 kJ/kg

v_{s_1} = 0.8267 m³/kg.

(*i*) **R.H. of heated air** (read from chart)

= **41%.** **(Ans.)**

(*ii*) **WBT of heated air** (read from chart)

$= \mathbf{16.1°C}$. **(Ans.)**

(*iii*) Mass of air circulated per min.,

$$m_a = \frac{200}{0.8267} = 241.9 \text{ kg.}$$

$\therefore$ **Heat added to air/min.**

$$= m_a (h_2 - h_1) = 241.9 (45.2 - 35.4)$$

$$= \mathbf{2370.6 \ kJ.} \quad \textbf{(Ans.)}$$

Example 9.18. *It is required to design an air-conditioning plant for a small office room for following winter conditions :*

Outdoor conditions ...... 14°C DBT and 10°C WBT

Required conditions ...... 20°C DBT and 60% R.H.

Amount of air circulation ...... 0.30 m³/min./person.

Seating capacity of office ...... 60.

The required condition is achieved first by heating and then by adiabatic humidifying.

Determine the following :

(i) Heating capacity of the coil in kW and the surface temperature required if the by-pass factor of coil is 0.4.

(ii) The capacity of the humidifier.

Solve the problem by using psychrometric chart.

Solution. Refer to Fig. 9.22.

- Locate the points '1' and '3' on the psychrometric chart.
- Draw a constant enthalpy line through '3' and constant specific humidity line through '1'.
- Locate the point '2' where the above two lines intersect.

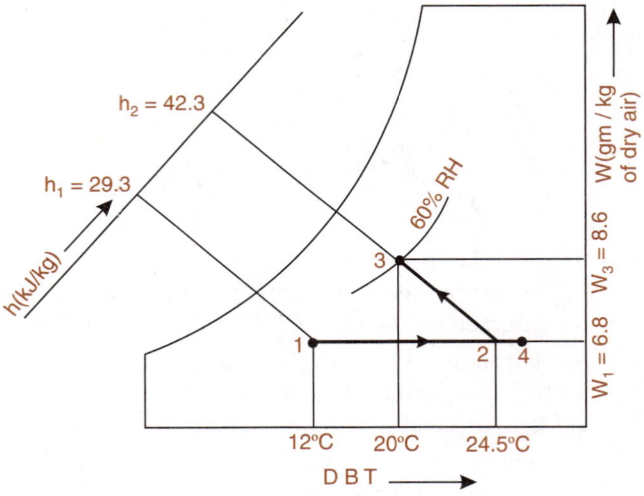

Fig. 9.22

From the psychrometric chart :

$$h_1 = 29.3 \text{ kJ/kg}, \ h_2 = h_3 = 42.3 \text{ kJ/kg}$$

$$t_{db_2} = 24.5\text{°C}, \ v_{s_1} = 0.817 \text{ m}^3\text{/kg}$$

The mass of air circulated per minute,

$$m_a = \frac{0.30 \times 60}{0.817} = 22.03 \text{ kg/min.}$$

(*i*) **Heating capacity of the heating coil**

$$= m_a(h_2 - h_1) = 22.03 \,(42.3 - 29.3)$$

$$= 286.4 \text{ kJ/min.}$$

$$= \mathbf{4.77 \ kJ/s \ or \ 4.77 \ kW.} \quad \textbf{(Ans.)}$$

The by-pass factor (*BF*) of heating coil is given by :

$$BF = \frac{t_{db_4} - t_{db_2}}{t_{db_4} - t_{db_1}}$$

$$0.4 = \frac{t_{db_4} - 24.5}{t_{db_4} - 12}$$

$$\therefore \quad 0.4\,(t_{db_4} - 12) = t_{db_4} - 24.5$$

i.e., t_{db_4} *(coil surface temperature)* = **32.8°C. (Ans.)**

(*ii*) **The capacity of the humidifier**

$$= \frac{m_a\,(W_3 - W_1)}{1000} \times 60 \text{ kg/h}$$

$$= \frac{22.03\,(8.6 - 6.8)}{1000} \times 60 = \mathbf{2.379 \ kg/h.} \quad \textbf{(Ans.)}$$

Example 9.19. *It is required to design an air-conditioning system for an industrial process for the following hot and wet summer conditions :*

Outdoor conditions ...... 32°C DBT and 65% R.H.

Required air inlet conditions ...... 25°C DBT and 60% R.H.

Amount of free air circulated ...... 250 m³/min.

Coil dew temperature ...... 13°C.

The required condition is achieved by first cooling and dehumidifying and then by heating.

Calculate the following :

(i) The cooling capacity of the cooling coil and its by-pass factor.

(ii) Heating capacity of the heating coil in kW and surface temperature of the heating coil if the by-pass factor is 0.3.

(iii) The mass of water vapour removed per hour.

Solve this problem with the use of psychrometric chart.

Solution. Refer to Fig. 9.23.

- Locate the points '1', '5' and '3' as shown on psychrometric chart.
- Join the line 1–5.
- Draw constant specific humidity line through '3' which cuts the line 1–5 at point '2'. The point '2' is located in this way.

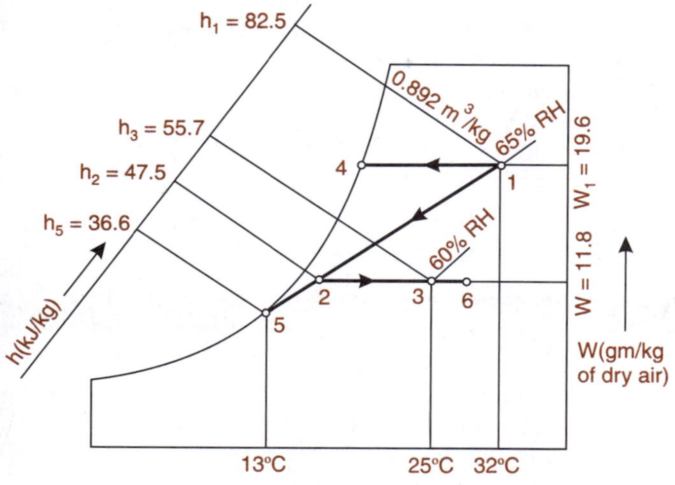

Fig. 9.23

From psychrometric chart :

$h_1 = 82.5$ kJ/kg, $h_2 = 47.5$ kJ/kg

$h_3 = 55.7$ kJ/kg, $h_5 = 36.6$ kJ/kg

$W_1 = 19.6$ gm/kg, $W_3 = 11.8$ gm/kg

$t_{db_2} = 17.6$°C, $v_{s_1} = 0.892$ m³/kg.

The mass of air supplied per minute,

$$m_a = \frac{250}{0.892} = 280.26 \text{ kg/min.}$$

(*i*) **The capacity of the cooling coil**

$$= \frac{m_a (h_1 - h_2) \times 60}{14000}$$

$$= \frac{280.26\,(82.5 - 47.5) \times 60}{14000}$$

= **42.04 TR. (Ans.)**

The *by-pass factor of the cooling coil* is given by :

$$BF = \frac{h_2 - h_5}{h_1 - h_5} = \frac{47.5 - 36.6}{82.5 - 36.6}$$

= **0.237. (Ans.)**

(*ii*) **The heating capacity of the heating coil**

$$= m_a (h_3 - h_2) = 280.26\,(55.7 - 47.5)$$

$$= 2298.13 \text{ kJ/min} = \frac{2298.13}{60} \text{ kJ/s}$$

= **38.3 kW. (Ans.)**

The by-pass factor of the *heating coil* is given by

$$BF = \frac{t_{db_6} - t_{db_3}}{t_{db_6} - t_{db_2}}$$

$$0.3 = \frac{t_{db_6} - 25}{t_{db_6} - 17.6}$$

$\therefore$ $t_{db_6} = 28.2$°C.

Hence, surface temperature of heating coil = 28.2°C. (Ans.)

(*iii*) **The mass of water vapour removed per hour**

$$= \frac{280.26\,(W_1 - W_3) \times 60}{1000}$$

$$= \frac{280.26\,(19.6 - 11.8)}{1000} \times 60$$

= **131.16 kg/h. (Ans.)**

HIGHLIGHTS

1. 'Air-conditioning' is the simultaneous control of temperature, humidity, motion and purity of the atmosphere in a confined space.
2. 'Psychrometry' is an art of measuring moisture content of air.

   The science which investigates the thermal properties of moist air, considers the measurement and control of the moisture content of air, and studies the effects of atmospheric moisture on material and human comfort may properly be termed 'Psychrometrics'.
3. When air is saturated *DBT, WBT, DPT* are equal.
4. *Dalton's law of partial pressure* states, that the total pressure of a mixture of gases is equal to the sum of partial pressure which the component gases would exert if each existed alone in the mixture volume at the mixture temperature.
5. Specific humidity, $W = \dfrac{0.622\, p_v}{p_t - p_v}$.
6. Degree of saturation (μ)

   $$= \frac{\text{Mass of water vapour associated with unit mass of dry air } (W)}{\text{Mass of water vapour associated with saturated unit mass of dry air } (W_s)} \cdot$$
7. Relative humidity, $\phi = \dfrac{p_v}{p_{vs}}$.
8. A 'Psychrometer' is a device which is used for measuring dry bulb and wet-bulb temperatures simultaneously.
9. The processes which affect the psychrometric properties of air are called '*psychrometric processes*'.

   Important psychrometric processes are :

   (*i*) Mixing of air streams
   (*ii*) Sensible heating
   (*iii*) Sensible cooling
   (*iv*) Cooling and dehumidification
   (*v*) Cooling and humidification
   (*vi*) Heating and dehumidification
   (*vii*) Heating and humidification.

OBJECTIVE TYPE QUESTIONS

Choose the Correct Answer :

1. In an unsaturated air the state of a vapour is
   (*a*) wet (*b*) superheated
   (*c*) saturated (*d*) unsaturated.
2. For saturated air
   (*a*) Wet bulb depression is zero
   (*b*) Wet bulb depression is positive
   (*c*) Wet bulb depression is negative
   (*d*) Wet bulb depression can be either positive or negative.
3. Which one of the following statements is *correct* ?
   (*a*) Dew point temperature can be measured with the help of thermometer
   (*b*) Dew point temperature is the saturation temperature corresponding to the partial pressure of the water vapour in moist air.
   (*c*) Dew point temperature is the same as the thermodynamic wet bulb temperature.
   (*d*) For saturated air, dew point temperature is less than the wet bulb temperature.
4. During sensible heating of moist air, enthalpy
   (*a*) increases (*b*) decreases
   (*c*) remains constant (*d*) none of the above.
5. During sensible cooling, wet bulb temperature
   (*a*) decreases (*b*) increases
   (*c*) remains constant (*d*) can decrease or increase.
6. Which one of the following statements is *correct* ?
   (*a*) Evaporative cooling and sensible cooling is the same
   (*b*) Evaporative cooling is a cooling and humidification process
   (*c*) Evaporative cooling is a cooling and dehumidification process
   (*d*) Evaporative cooling is not effective for hot and dry climates.
7. An air washer can work as a
   (*a*) filter only (*b*) humidifier only
   (*c*) dehumidifier only (*d*) all of the above.
8. The relative humidity, during sensible heating,
   (*a*) can increase or decrease (*b*) increases
   (*c*) decreases (*d*) remains constant.
9. The vapour pressure, during sensible heating of moist air,
   (*a*) increases (*b*) decreases
   (*c*) can increase or decrease
   (*d*) remains constant.
10. The relative humidity, during heating and humidification,
    (*a*) increases (*b*) decreases
    (*c*) may increase or decrease
    (*d*) remains constant.
11. The relative humidity, during cooling and dehumidification of moist air
    (*a*) increases (*b*) decreases
    (*c*) can increase or decrease
    (*d*) remains constant.

ANSWERS

1. (*b*)	**2.** (*a*)	**3.** (*b*)	**4.** (*a*)
5. (*a*)	**6.** (*b*)	**7.** (*d*)	**8.** (*b*)
9. (*d*)	**10.** (*a*)	**11.** (*c*).	

THEORETICAL QUESTIONS

1. Define the following terms :
   (*i*) Saturated air (*ii*) Dry bulb temperature
   (*iii*) Dew point temperature
   (*iv*) Relative humidity
   (*v*) Specific humidity.
2. State 'Dalton's law of partial pressure'.
3. Derive the following relations :
   (*i*) Specific humidity, $W = 0.622 \dfrac{p_v}{p_t - p_v}$.
   (*ii*) Degree of saturation, $\mu = \dfrac{p_v}{p_s}\left[\dfrac{\left(1 - \dfrac{p_s}{p_t}\right)}{\left(1 - \dfrac{p_v}{p_t}\right)}\right]$.
4. Explain briefly with a neat sketch a 'sling psychrometer'.
5. Describe briefly any two of the following processes :
   (*i*) Sensible heating
   (*ii*) Cooling and dehumidification
   (*iii*) Heating and humidification
   (*iv*) Heating and dehumidification.
6. Write a short note on 'by-pass factor'.

UNSOLVED EXAMPLES

1. The atmospheric conditions are 30°C and specific humidity of 0.0215 kg/kg of air. Determine :
   (*i*) Partial pressure of air (*ii*) Relative humidity
   (*iii*) Dew point temperature.
   Atmospheric pressure = 756 mm Hg.
   [**Ans.** (*i*) 14.89 mm of Hg, (*ii*) 46.8%, (*iii*) 17°C]
2. A mixture of air and water vapour at 1 bar and 25°C has a dew point temperature of 15°C. Determine the relative humidity and specific humidity.
   [**Ans.** 53.8%, 0.01078 kg/kg of dry air]
3. An air-water vapour mixture at 1.24 bar has a temperature of 38°C and relative humidity of 60%. Calculate the kg-mass of water vapour per kg of air and per kg of mixture. Also find the dew point.
   [**Ans.** 0.0203 kg/kg of dry air, 0.0199 kg/kg of mixture, 28.5°C]
4. In a house, the temperature of the windows on a day in winter is 5°C. When the temperature in the room is 23°C, and the barometric pressure is 748.8 mm Hg, what would be maximum relative humidity that could be maintained in the room without condensation on the window panes ?

Under these conditions, find the partial pressure of the water vapour and air, the specific humidity, and the density of the mixture.

[**Ans.** 30.7%, 0.00872 bar, 0.989 bar, 0.00548 kg/kg of dry air, 1.164 kg/m^3]

5. Atmospheric air enters a heater at 4.5°C and 60% relative humidity and leaves at a temperature of 21°C. Calculate :

(*i*) The heat supplied to the air ;

(*ii*) The final relative humidity.

[**Ans.** 16.8 kJ/kg, 29.2%]

6. The air supplied to a room of building in winter is to be at 17°C and have a relative humidity of 60%. If the barometric pressure is 1.01325 bar, calculate the specific humidity. What would be the dew point under these conditions ? [**Ans.** 0.00723 kg/kg of dry air, 9.18°C]

7. If air at the condition of example 6, is passed at the rate of 0.5 m^3/s over a cooling coil which is at a temperature of 6°C, calculate the amount of vapour which will be condensed. Assume that the barometric pressure is the same as in example 6, and that the air leaving the coil is saturated. [**Ans.** 3.082 kg/h]

8. An air and water vapour mixture at 1 bar and 26.7°C has a specific humidity of 0.0085. Determine the percentage saturation. [**Ans.** 37.7%]

9. A mixture of air and water vapour at 1.013 bar and 16°C has a dew point of 5°C. Determine the relative and specific humidities. [**Ans.** 48%, 0.0054 kg/kg of dry air]

10. Atmospheric air at a pressure of 760 mm Hg has a temperature of 32°C and a percentage saturation as determined from a psychrometric chart of 52%. Calculate

(*i*) The partial pressure of the vapour and the dry air

(*ii*) The specific humidity

(*iii*) The dew point

(*iv*) The density of the mixture.

[**Ans.** (*i*) 0.0247 bar, 0.988 bar, (*ii*) 0.01556, (*iii*) 20.9°C, (*iv*) 1.147 kg/m^3]

11. In a laboratory test, a psychrometer recorded 36°C DBT and 30°C WBT. Calculate :

(*i*) Vapour pressure (*ii*) Relative humidity

(*iii*) Specific humidity (*iv*) Degree of saturation

(*v*) Dew point temperature

(*vi*) Enthalpy of the mixture.

[**Ans.** (*i*) 0.0385 bar, (*ii*) 64.5%, (*iii*) 0.025 kg/kg dry air, (*iv*) 0.63, (*v*) 28°C, (*vi*) 99.2 kJ]

12. The pressure and temperature of the air in a room is 1 bar and 28°C. If the relative humidity is found to be 30 per cent, determine :

(*i*) The partial pressure of the water vapour and dew point,

(*ii*) The specific volume of each constituent, and

(*iii*) The specific humidity.

[**Ans.** (*i*) 0.0378 bar, 8.8°C ; (*ii*) v_{air} = 0.874 m^3/kg, v_{vap} = 122.7 m^3/kg ; (*iii*) 0.00712 kg/kg dry air]

13. 100 m^3 of air per minute at 35°C DBT and 60% relative humidity is cooled to 20°C DBT by passing through a cooling coil. Find the following :

(*i*) Capacity of cooling coil in kJ/h

(*ii*) Amount of water vapour removed per hour, and

(*iii*) Relative humidity of air coming out and its wet-bulb temperature.

[**Ans.** (*i*) 1037088 kJ/h, (*ii*) 465.36 kg/h, (*iii*) 100%, 20°C]

14. Atmospheric air at 38°C and 40 per cent relative humidity is to be cooled and dehumidified to a state of saturated air at 10°C. The mass rate of flow of atmospheric air entering the dehumidifier is 45.4 kg/h. Neglecting any pressure drop, determine :

(*i*) The mass of water removed ;

(*ii*) The quantity of heat removed.

[**Ans.** (*i*) 0.397 kg/h, (*ii*) 2287 kJ/h]

15. 1 kg of air at 24°C and a relative humidity of 70% is to be mixed adiabatically in a steady state, steady flow device with 1 kg of air at 16°C and a relative humidity of 10%. Assuming that the mixing is to be carried out at a constant pressure of 1.0 atm, determine the temperature and relative humidity of the stream leaving the device.

[**Ans.** 19.5°C, 50%]

16. An air-water vapour mixture enters an adiabatic saturator at 30°C and leaves at 20°C, which is the adiabatic saturation temperature. The pressure remains constant at 1 bar. Determine the relative humidity and the humidity ratio of the inlet mixture.

[**Ans.** 39.8%, 0.0107 kg/kg dry air]

Fuels and Combustion (Including Chemical Thermodynamics)

10

10.1. INTRODUCTION

- Fuel may be *chemical* or *nuclear*. Here we shall consider briefly *chemical fuels* only.

A chemical fuel is a substance which releases heat energy on combustion. The principal combustible elements of each fuel are *carbon* and *hydrogen*. Though *sulphur* is a combustible element too but its presence in the fuel is considered to be *undesirable*.

- In **chemical thermodynamics** the study of systems involving chemical reactions is an important topic. *A chemical reaction may be defined as the rearrangement of atoms due to redistribution of electrons*. In a chemical reaction the terms, **reactants** and the **products** are frequently used. '*Reactants*' comprise of initial constituents which start the reaction while '*products*' comprise of *final constituents* which are formed by the chemical reaction. Although the basic principles which will be discussed in this chapter apply to any chemical reaction, here main attention will be focused on an important type of chemical reaction—**combustion**.

10.2. CLASSIFICATION OF FUELS

Fuels can be classified according to whether :

1. They occur in nature called **primary fuels** or are prepared called **secondary fuels** ;

2. They are in solid, liquid or gaseous state. The detailed classification of fuels can be given in a summary form as follows :

Type of fuel	*Natural (Primary)*	*Prepared (Secondary)*
Solid	Wood	Coke
	Peat	Charcoal
	Lignite coal	Briquettes
Liquid	Petroleum	Gasoline
		Kerosene
		Fuel oil
		Alcohol
		Benzol
		Shale oil
Gaseous	Natural gas	Petroleum gas
		Producer gas
		Coal gas
		Coke-oven gas
		Blast furnace gas
		Carburetted gas
		Sewer gas

10.3. SOLID FUELS

Coal. Its main constituents are carbon, hydrogen, oxygen, nitrogen, sulphur, moisture and ash. Coal passes through different stages during its formation from vegetation. These stages are enumerated and discussed below :

Plant debris—Peat—Lignite—Brown coal—Sub-bituminous coal—Bituminous coal—Semi-bituminous coal—Semi-anthracite coal—Anthracite coal—Graphite.

Peat. It is the first stage in the formation of coal from wood. It contains huge amount of moisture and therefore it is dried for about 1 to 2 months before it is put to use. It is used as a domestic fuel in Europe and for power generation in Russia. In India it does not come in the categories of good fuels.

Lignites and brown coals. These are intermediate stages between peat and coal. They have a woody or often a clay like appearance associated with high moisture, high ash and low heat contents. Lignites are usually amorphous in character and impose transport difficulties as they break easily. They burn with a smoky flame. Some of this type are suitable for local use only.

Bituminous coal. It burns with long yellow and smoky flames and has high percentages of volatile matter. The average calorific value of bituminous coal is about 31350 kJ/kg. It may be of two types, namely *caking* or *noncaking*.

Semi-bituminous coal. It is softer than the anthracite. It burns with a very small amount of smoke. It contains 15 to 20 per cent volatile matter and has a tendency to break into small sizes during storage or transportation.

Semi-anthracite. It has less fixed carbon and less lustre as compared to true anthracite and gives out longer and more luminous flames when burnt.

Anthracite. It is very hard coal and has a shining black lustre. It ignites slowly unless the furnace temperature is high. It is non-caking and has high percentage of fixed carbon. It burns either with very short blue flames or without flames. The calorific value of this fuel is high to the tune of 35500 kJ/kg and as such is *very suitable for steam generation*.

Wood charcoal. It is obtained by destructive distillation of wood. During the process the volatile matter and water are expelled. The physical properties of the residue (charcoal), however depends upon the rate of heating and temperature.

Coke. It consists of carbon, mineral matter with about 2% sulphur and small quantities of hydrogen, nitrogen and phosphorus. It is solid residue left after the destructive distillation of certain kinds of coals. It is smokeless and clear fuel and can be produced by several processes. It is *mainly used in blast furnace* to produce heat and at the same time to reduce the iron ore.

Briquettes. These are prepared from fine coal or coke by compressing the material under high pressure.

10.4. LIQUID FUELS

The chief source of liquid fuels is *petroleum* which is obtained from wells under the earth's crust. These fuels have proved *more advantageous in comparison to solid fuels* in the following respects.

Advantages :

1. Require less space for storage.
2. Higher calorific value.
3. Easy control of consumption.
4. Staff economy.
5. Absence of danger from spontaneous combustion.
6. Easy handling and transportation.
7. Cleanliness.
8. No ash problem.
9. Non-deterioration of the oil in storage.

Petroleum. There are different opinions regarding the origin of petroleum. However, now it is accepted that petroleum has originated probably from organic matter like fish and plant life etc., by bacterial action or by their distillation under pressure and heat. It consists of a mixture of gases, liquids and solid hydrocarbons with small amounts of nitrogen and sulphur compounds. In India, the main sources of Petroleum are Assam and Gujarat.

Heavy fuel oil or crude oil is imported and then refined at different refineries. The refining of crude oil supplies the most important product called *petrol*. Petrol can also be made by polymerization of refinery gases.

Other liquid fuels are kerosene, fuels oils, colloidal fuels and alcohol.

10.5. GASEOUS FUELS

Natural gas. The main constituents of natural gas are *methane* (CH_4) and *ethane* (C_2H_6). It has calorific value nearly 21000 kJ/m^3. Natural gas is used alternately or simultaneously with oil for internal combustion engines.

Coal gas. Mainly consists of *hydrogen, carbon monoxide* and *hydrocarbons*. It is prepared by carbonisation of coal. It finds its use in boilers and sometimes used for commercial purposes.

Coke-oven gas. It is obtained during the production of coke by heating the bituminous coal. The volatile content of coal is driven off by heating and major portion of this gas is utilised in heating the ovens. This gas *must be thoroughly filtered before using in gas engines.*

Blast furnace gas. It is obtained from smelting operation in which air is forced through layers of coke and iron ore, the example being that of pig iron manufacture where this gas is produced as by product and contains about 20% carbon monoxide (CO). After filtering it may be blended with richer gas or used in gas engines directly. The heating value of this gas is very low.

Producer gas. It results from the partial oxidation of coal, coke or peat when they are burnt with an insufficient quantity of air. It is produced in specially designed retorts.

It has low heating value and in general is suitable for large installations. It is also used in steel industry for firing open hearth furnaces.

Water or illuminating gas. It is produced by blowing steam into white hot coke or coal. The decomposition of steam takes place liberating free hydrogen, and oxygen in the steam combines with carbon to form carbon monoxide according to the reaction :

$$C + H_2O \longrightarrow CO + H_2$$

The gas composition varies as the hydrogen content if the coal is used.

Sewer gas. It is obtained from sewage disposal vats in which fermentation and decay occur. It consists of mainly marsh gas (CH_4) and is collected at large disposal plants. It works as a fuel for gas engines which in turn drive the plant pumps and agitators.

Gaseous fuels are becoming popular because of following *advantages* they possess.

Advantages :

1. Better control of combustion.
2. Much less excess air is needed for complete combustion.
3. Economy in fuel and more efficiency of furnace operation.
4. Easy maintenance of oxidizing or reducing atmosphere.
5. Cleanliness.
6. No problem of storage if the supply is available from public supply line.
7. The distribution of gaseous fuels even over a wide area is easy through the pipe lines and as such handling of the fuel is altogether eliminated.
8. Gaseous fuels give economy of heat and produce higher temperatures (as they can be preheated in regenerative furnances and thus heat from hot flue gases can be recovered).

10.6. BASIC CHEMISTRY

Before considering combustion problems it is necessary to understand the construction and use of chemical formulae. This involves elementary concepts which are discussed below briefly.

Atoms. It is not possible to divide the chemical elements *indefinitely*, and the *smallest particle which can take part in a chemical change* is called an **'atom'**. If an atom is split as in nuclear reaction, the *divided atom does not retain the original chemical properties*.

Molecules. It is rare to find elements to exist naturally as single atom. Some elements have atoms which exist in pairs, each pair forming a molecule (*e.g.* oxygen), and the atoms of each molecule are held together by stronger *inter-atomic forces*. The isolation of a molecule of oxygen would be tedious, but possible; the isolation of an atom of oxygen would be a different prospect. The molecules of some substances are formed by the mating up of atoms of different elements. For example, water has a molecule which consists of two atoms of hydrogen and one atom of oxygen. The atoms of different elements have different masses and these values are important when a quantitative analysis is required. The actual masses are infinitesimally small, and the ratios of the masses of atoms are used. These ratios are indicated by **atomic weight** quoted on a scale which defines the atomic weight of oxygen as 16.

The symbols and molecular weights of some important elements, compounds and gases are given in Table 10.1.

Table 10.1. Symbols and Molecular weights

Elements / Compounds / Gases	*Molecule*		*Atom*	
	Symbol	*Molecular weight*	*Symbol*	*Molecular weight*
Hydrogen	H_2	2	H	1
Oxygen	O_2	32	O	16
Nitrogen	N_2	28	N	14
Carbon	C	12	C	12
Sulphur	S	32	S	32
Water	H_2O	18	—	—
Carbon monoxide	CO	28	—	—
Carbon dioxide	CO_2	44	—	—
Sulphur dioxide	SO_2	64	—	—
Marsh gas (Methane)	CH_4	16	—	—
Ethylene	C_2H_4	28	—	—
Ethane	C_2H_6	30	—	—

10.7. COMBUSTION EQUATIONS

- In a combustion chamber proportionate masses of air and fuel enter where the chemical reaction takes place, and then the combustion products pass to the exhaust. By the conservation of mass the mass flow remains constant (*i.e.*, total mass of *products* = total mass of *reactants*), but the reactants are chemically different from the products, and the products leave at a higher temperature. The *total number of atoms of each element concerned in the combustion remains constant, but the atoms are rearranged into groups having different chemical properties.* This information is expressed in the chemical equation which shows (*i*) the reactants and the products of combustion, (*ii*) the relative quantities of the reactants and products. The two sides of the equation must be **consistent**, each having the same number of atoms of each element involved.

- The oxygen supplied for combustion is *usually* provided by *atmospheric air*, and it is necessary to use accurate and consistent analysis of air by *mass* and by *volume*. It is usual in combustion calculations to take air as 23.3% O_2, 76.7% N_2 by mass, and 21% O_2, 79% N_2 by volume. The small traces of other gases in dry air are included in nitrogen, which is sometimes called *'atmospheric nitrogen'*.

Some important *combustion equations* are given below :

1. Combustion of hydrogen

$$2H_2 + O_2 = 2H_2O \quad ...(10.1)$$

The above equation of combustion of hydrogen tell us that :

(*i*) Hydrogen reacts with water to form steam or water.

(*ii*) Two molecules of hydrogen react with one molecule of oxygen to give two molecules of steam or water,

i.e., 2 volumes H_2 + 1 volume $O_2 \longrightarrow$ 2 volumes H_2O

The H_2O may be liquid or a vapour depending on whether the product has been cooled sufficiently to cause condensation.

The *proportions by mass* are obtained by using *atomic weights* as follows :

$$2H_2 + O_2 \longrightarrow 2H_2O$$

$\therefore$ $2(2 \times 1) + 2 \times 16 \longrightarrow 2(2 \times 1 + 16)$

i.e., $4 \text{ kg } H_2 + 32 \text{ kg } O_2 \longrightarrow 36 \text{ kg } H_2O$

or $\mathbf{1 \text{ kg } H_2 + 8 \text{ kg } O_2 \longrightarrow 9 \text{ kg } H_2O}$

The same proportions are obtained by writing the equation (10.1) as: $H_2 + \frac{1}{2} O_2 \longrightarrow H_2O$, and this is sometimes done.

It will be noted from equation (10.1) that the total volume of the *reactants* is

2 volumes H_2 + 1 volume O_2 = 2 volumes H_2O.

The total volume of the *product* is only 2 volumes. There is therefore a *volumetric contraction on combustion*.

Since the oxygen is accompanied by nitrogen if air is supplied for the combustion, then this nitrogen should be included in the equation. As nitrogen is *inert* as far as chemical reaction is concerned, it will appear on both sides of the equation.

With one mole of oxygen there are 79/21 moles of nitrogen, hence equation (10.1) becomes,

$$2H_2 + O_2 + \tfrac{79}{21} N_2 \longrightarrow 2H_2O + \tfrac{79}{21} N_2 \quad ...(10.2)$$

2. Combustion of carbon

(*i*) ***Complete combustion of carbon to carbon dioxide***

$$C + O_2 \longrightarrow CO_2 \quad ...(10.3)$$

and including the nitrogen,

$$C + O_2 + \tfrac{79}{21} N_2 \longrightarrow CO_2 + \tfrac{79}{21} N_2 \quad ...(10.4)$$

By volume :

0 volume C + 1 volume O_2 + $\frac{79}{21}$ volumes N_2

$\longrightarrow$ 1 volume CO_2 + $\frac{79}{21}$ volumes N_2 ...[10.4 (*a*)]

The volume of carbon is written as zero since the volume of solid is negligible in comparison with that of a gas.

By mass :

$$12 \text{ kg C} + (2 \times 16) \text{ kg } O_2 + \tfrac{79}{21} (2 \times 14) \text{ kg } N_2$$

$$\longrightarrow (12 + 2 \times 16) \text{ kg } CO_2 + \tfrac{79}{21} (2 \times 14) N_2$$

i.e., $12 \text{ kg C} + 32 \text{ kg } O_2 + 105.3 \text{ kg } N_2 \longrightarrow 44 \text{ kg } CO_2 + 105.3 \text{ kg } N_2$

or $1 \text{ kg C} + \frac{8}{3} \text{ kg } O_2 + \frac{105.3}{12} \text{ kg } N_2$

$\longrightarrow \frac{11}{3} \text{ kg } CO_2 + \frac{105.3}{12} \text{ kg } N_2.$ ...[10.4 (*b*)]

(*ii*) ***The incomplete combustion of carbon.*** The incomplete combustion of carbon occurs when there is an insufficient supply of oxygen to burn the carbon completely to carbon dioxide.

$$2C + O_2 \longrightarrow 2CO \quad ...(10.5)$$

and including the nitrogen,

$$2C + O_2 + \frac{79}{21} N_2 \longrightarrow 2CO + \frac{79}{21} N_2 \quad ...(10.6)$$

By mass :

$$(2 \times 12) \text{ kg C} + (2 \times 16) \text{ kg } O_2 + \frac{79}{21} (2 \times 14) \text{ kg } N_2$$

$$\longrightarrow 2(12 + 16) \text{ kg CO} + \frac{79}{21} (2 \times 14) \text{ kg } N_2$$

i.e., $24 \text{ kg C} + 32 \text{ kg } O_2 + 105.3 \text{ kg } N_2 \longrightarrow 56 \text{ kg CO} + 105.3 \text{ kg } N_2$

or $1 \text{ kg C} + \frac{4}{3} \text{ kg } O_2 + \frac{105.3}{24} \text{ kg } N_2$

$\longrightarrow \frac{7}{3} \text{ kg CO} + \frac{105.3}{24} \text{ kg } N_2$ ...[10.6 (*a*)]

If a further supply of oxygen is available then the combustion can continue to completion,

$$2CO + O_2 + \tfrac{79}{21} N_2 \longrightarrow 2CO_2 + \tfrac{79}{21} N_2 \quad ...(10.7)$$

By mass :

$$56 \text{ kg } CO + 32 \text{ kg } O_2 + \frac{79 \times 28}{21} \text{ kg } N_2$$

$$\longrightarrow 88 \text{ kg } CO_2 + \frac{79 \times 28}{21} \text{ kg } N_2$$

or

$$1 \text{ kg } CO + \frac{4}{7} \text{ kg } O_2 + \frac{105.3}{56} \text{ kg } N_2$$

$$\longrightarrow \frac{11}{7} \text{ kg } CO_2 + \frac{105.3}{56} \text{ kg } N_2. \quad ...[10.7\ (a)]$$

10.8. THEORETICAL AIR AND EXCESS AIR

The *minimum amount of air* that supplies sufficient oxygen for the complete combustion of all the carbon, hydrogen, and any other elements in the fuel that may oxidise is called the **"theoretical air".** When complete combustion is achieved with theoretical air, the products contain no oxygen.

In practice, it is found that complete combustion is not likely to be achieved unless the amount of air supplied is somewhat greater than the theoretical amount. Thus 150 per cent theoretical air means that air actually supplied is 1.5 times the theoretical air.

The complete combustion of methane with minimum amount of theoretical air and 150 per cent theoretical air respectively is written as :

$$CH_4 + 2O_2 + 2\left(\frac{79}{21}\right) N_2 \longrightarrow CO_2 + 2H_2O + 2\left(\frac{79}{21}\right) N_2 \quad ...(10.8)$$

$$CH_4 + 2(1.5)\, O_2 + 2\left(\frac{79}{21}\right) (1.5)\, N_2$$

$$\longrightarrow CO_2 + 2H_2O + O_2 + 3\left(\frac{29}{21}\right) N_2 \quad ...(10.9)$$

(with 150 per cent theoretical air)

The amount of air actually supplied may also be expressed in terms of per cent excess air. The excess air is the amount of air supplied over and above the theoretical air. Thus 150 per cent theoretical air is equivalent to 50 per cent excess air.

Note. For complete combustion of fuel we need air. As per theoretical basis there is a minimum amount of air which is required by the fuel to burn completely, *but always, air in excess is used because whole of air supplied for combustion purposes does not come in contact with the fuel completely and as such portion of fuel may be left unburnt*. But if a large quantity of excess air is used it exercises a *cooling effect on combustion process which however can be avoided by preheating the air. The weight of excess air supplied can be determined from the weight of oxygen which is left unused.* The amount of excess air supplied varies with the type of fuel and the firing conditions. It may approach a value of 100% but modern practice is to use 25% to 50% excess air.

10.9. STOICHIOMETRIC AIR FUEL (A/F) RATIO

Stoichiometric (or chemically correct) *mixture* of air and fuel is one that contains *just sufficient oxygen for complete combustion of the fuel.*

A *weak mixture* is one which has an *excess of air*.

A *rich mixture* is one which has a *deficiency* of air.

The percentage of excess air is given as under :

%age excess air

$$= \frac{\text{Actual A/F ratio} - \text{Stoichiometric A/F ratio}}{\text{Stoichiometric A/F ratio}} \quad ...(10.10)$$

(where A and F denote *air* and *fuel* respectively)

The ratios are expressed as follows :

For *gaseous fuels*	*By volume*
For solid and liquid fuels	*By mass*

For *boiler* plant the mixture is usually greater than 20% weak ; for *gas turbines* it can be as much as 300% weak. *Petrol engines* have to meet various conditions of load and speed, and operate over a wide range of mixture strength. The following definition is used :

$$\text{Mixture strength} = \frac{\text{Stoichiometric A/F ratio}}{\text{Actual A/F ratio}} \quad ...(10.11)$$

The working value range between 80% (weak) and 120% (rich).

Note. The reciprocal of the air fuel ratio is called the *fuel-air (F/A) ratio.*

10.10. AIR-FUEL RATIO FROM ANALYSIS OF PRODUCTS

When analysis of combustion products is known air-fuel ratio can be calculated by the following methods :

1. **Fuel composition known**
   (*i*) Carbon balance method
   (*ii*) Hydrogen balance method
   (*iii*) Carbon-hydrogen balance method.
2. **Fuel composition unknown**
   (*i*) Carbon-hydrogen balance method.

1. **Fuel composition known**

(*i*) *Carbon balance method.* When the fuel composition is known, the carbon balance method is quite accurate if *combustion takes place with excess air and when free (solid) carbon is not present in the products*. It may be noted that the Orsat analysis will not determine the quantity of solid carbon in the products.

(*ii*) *Hydrogen balance method*. This method is used when solid carbon is suspected to be present.

(*iii*) *Carbon-hydrogen balance method*. This method may be employed *when there is some uncertainty about the nitrogen percentage reported by the Orsat analysis.*

2. **Fuel composition unknown**

When the fuel composition is not known the carbon-hydrogen balance method has to be employed.

10.11. HOW TO CONVERT VOLUMETRIC ANALYSIS TO WEIGHT ANALYSIS ?

The conversion of volumetric analysis to weight analysis involves the following steps :

1. Multiply the volume of each constituent by its molecular weight.
2. Add all these weights and then divide each weight by the total of all and express it as percentage.

10.12. HOW TO CONVERT WEIGHT ANALYSIS TO VOLUMETRIC ANALYSIS ?

1. Divide the weight of each constituent by its molecular weight.
2. Add up these volumes and divide each volume by the total of all and express it as a percentage.

10.13. WEIGHT OF CARBON IN FLUE GASES

The weight of carbon contained in one kg of flue or exhaust gas can be calculated from the amounts of CO_2 and CO contained in it.

In eqn. [10.4 (*b*)], it was shown that 1 kg of carbon produces 11/3 kg of CO_2 when completely burnt. Hence 1 kg of CO_2 will contain 3/11 kg of carbon.

In eqn. [10.6 (*a*)], it can be seen that 1 kg of carbon produces 7/3 kg of CO, hence 1 kg CO contains 3/7 kg of carbon.

Therefore, weight of carbon per kg of fuel

$$= \left(\frac{3}{11}CO_2 + \frac{3}{7}CO\right)$$

where CO_2 and CO are the quantities of carbon dioxide and carbon monoxide present in 1 kg of flue or exhaust gas.

10.14. WEIGHT OF FLUE GASES PER KG OF FUEL BURNT

Due to supply of air, the weight of flue gas or exhaust gas is always more than that of fuel burnt. The actual weight of dry flue gases can be obtained by comparing the weight of carbon present in the flue gases with the weight of carbon in the fuel, since there is no loss of carbon during the combustion process. As the analysis of the exhaust gases is volumetric, so this *must first be reduced to weight analysis.*

Also, *total weight of carbon in one kg of flue gas is*

$$= \left(\frac{3}{11}CO_2 + \frac{3}{7}CO\right)$$

$\therefore$ The *weight of flue gas/kg of fuel burnt*

$$= \frac{\text{Weight of carbon in one kg of fuel}}{\text{Weight of carbon in one kg of flue gas}}.$$

10.15. ANALYSIS OF EXHAUST AND FLUE GAS

The combustion products are mainly gaseous. When a sample is taken for analysis it is usually cooled down to a temperature which is below the saturation temperature of the steam present. The *steam content* is therefore *not included* in the analysis, which is then quoted as the *analysis of the dry products*. Since the products are gaseous, it is usual to quote the analysis by *volume*. *An analysis which includes the steam in the exhaust is called a wet analysis.*

Practical analysis of combustion products :

The most common means of analysis of the combustion products is the ***Orsat apparatus*** which is described below :

Construction. An Orsat's apparatus consists of the following :

(*i*) A burette

(*ii*) A gas cleaner

(*iii*) Four absorption pipettes 1, 2, 3, 4.

The pipettes are interconnected by means of a manifold fitted with cocks S_1, S_2, S_3 and S_4 and contain different chemicals to absorb carbon dioxide (CO_2), carbon monoxide (CO) and oxygen (O_2). Each pipette is also fitted with a number of small glass tubes which provide a greater amount of surface. These tubes are wetted by the absorbing agents and are exposed to the gas under analysis. The measuring burrette is surrounded by a *water jacket* to prevent, changes in temperature and density of the gas. The pipettes 1, 2, 3, 4 contain the following chemicals :

***Pipette 1* :** Contains '*KOH*' (caustic soda) *to absorb* CO_2 (carbon dioxide)

***Pipette 2* :** Contains *an alkaline solution of 'pyrogallic acid' to absorb* O_2 (oxygen)

***Pipette 3, 4* :** Contain an *acid solution of 'cuprous chloride' to absorb CO* (carbon monoxide)

Furthermore the apparatus has a *levelling bottle* and a three way cock to connect the apparatus either to gases or to atmosphere.

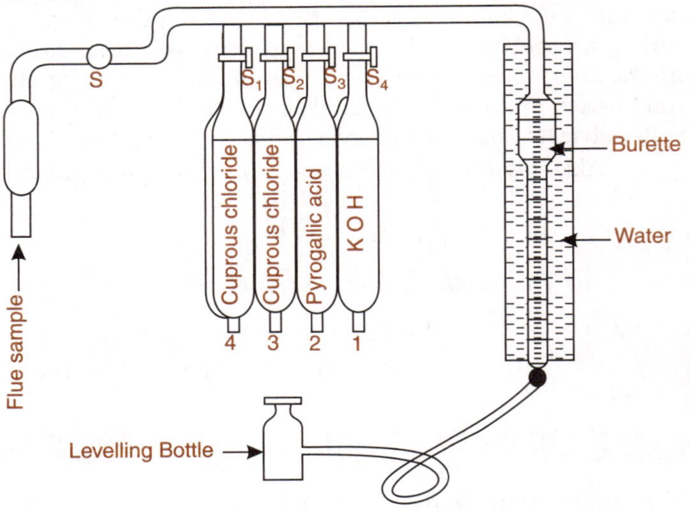

Fig. 10.1. Orsat's apparatus.

Procedure. 100 cm^3 of gas whose analysis is to be made is drawn into the bottle by lowering the levelling bottle. The stop cock S_4 is then opened and the whole flue gas is forced to pipette 1. The gas remains in this pipette for sometime and most of the carbon dioxide is absorbed. The levelling bottle is then lowered to allow the chemical to come to its original level. The volume of gas thus absorbed is read on the scale of the measuring bottle. The flue gas is then forced through the pipette 1 for a number of times to ensure that the whole of the CO_2 is absorbed. Further, the remaining flue gas is then forced to the pipette 2 which contains pyrogallic acid to absorb whole of O_2. The reading on the measuring burette will be the sum of volume of CO_2 and O_2. The oxygen content can then be found out by subtraction. Finally, as before, the sample of gas is forced through the pipettes 3 and 4 to absorb carbonmonoxide completely.

The amount of nitrogen in the sample can be determined by subtracting from total volume of gas the sum of CO_2, CO and O_2 contents.

Orsat apparatus *gives an analysis of the dry products of combustion*. Steps may have been taken to remove the steam from the sample by condensing, but as the sample is collected over water it becomes saturated with water. The resulting analysis is nevertheless a true analysis of the dry products. This is because the volume readings are taken at a constant temperature and pressure, and the partial pressure of the vapour is constant. This means that the sum of the partial pressures of the remaining constituents is constant. The vapour then occupies the same proportion of the total volume at each measurement. Hence the vapour does not affect the result of the analysis.

Note. Quantitatively the dry product analysis can be used to calculate A/F ratio. This method of obtaining the A/F ratio is not so reliable as direct measurement of air consumption and fuel consumption of the engine. More caution is required when analysing the products of consumption of a *solid fuel* since some of the products do not appear in the flue gases (*e.g.* ash and unburnt carbon). The residual solid must be analysed as well in order to determine the carbon content, if any. With an engine using *petrol or diesel fuel* the exhaust may include unburnt particles of carbon and this quantity will not appear in the analysis. The exhaust from internal combustion engines may contain also some CH_4 and H_2 due to incomplete combustion. Another piece of equipment called the **Heldane apparatus** measures the CH_4 content as well as CO_2, O_2 and CO.

10.16. INTERNAL ENERGY AND ENTHALPY OF FORMATION

The first law of thermodynamics can be applied to any system. Non-flow and steady-flow energy equations deducted from this law must be applicable to systems undergoing combustion processes.

It has been proved experimentally that the energy released, when a unit mass of a fuel undergoes complete combustion, depends on the *temperature at which the process is carried out*. Thus such quantities quoted are related to temperature. Now it will be shown that if the energy released by a fuel at one temperature is known then it can be calculated at other temperatures.

The process of combustion is defined as taking place from reactants at a state identified by the reference temperature T_0 and another property, either pressure or volume, to products at the same state.

Let U_{R_0} = Internal energy of the *reactants* (which is a mixture of fuel and air) at T_0,

U_{P_0} = Internal energy of *products* of combustion at T_0,

U_{R_1} = Internal energy of *reactants* at temperature T_1,

U_{P_1} = Internal energy of *products* at temperature T_1,

U_{R_2} = Internal energy of *reactants* at temperature T_2,

U_{P_2} = Internal energy of *products* at temperature T_2,

ΔU_0 = Constant volume heat of combustion,

Q = Heat transferred to the surroundings during the process, and

W = Work obtained during combustion process.

Analysis for a non-flow process involving combustion at 'constant volume' :

When the combustion process is carried out at *constant volume* then the non-flow energy equation, $Q = (U_2 - U_1) + W$, can be applied to give

$$Q = (U_{P_0} - U_{R_0}) \qquad ...(10.12)$$

where, $W = 0$ for constant volume combustion,

$$U_1 = U_{R_0}, \text{ and}$$

$$U_2 = U_{P_0}.$$

The internal energy change is independent of the path between the two states and depends only on the initial and final values and is given by the quantity Q. This is illustrated in Fig. 10.2 and *property diagram* of Fig. 10.3. The heat so transferred is called the *internal energy of combustion* at T_0 (or *constant volume heat of combustion*), and is denoted by ΔU_0. Thus,

$$\Delta U_0 = U_{P_0} - U_{R_0} \qquad ...(10.13)$$

ΔU_0 is a *negative quantity* since the internal energy of the reactants includes the potential chemical energy and heat is transferred *from* the system.

It may be noted that in case of real constant volume combustion processes the initial and final temperatures will not be same as T_0 (reference temperature). The change in internal energy, for analytical purposes, between reactants at state 1 to products at state 2 can be considered in the following *three steps* (stages) :

(*i*) The change for the reactants from state 1 to T_0.

(*ii*) The constant volume combustion process from reactants to products at T_0.

(*iii*) The change for the products from T_0 to state 2.

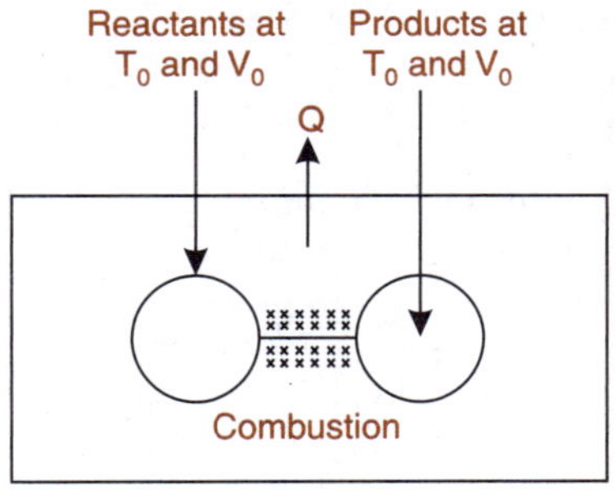

Fig. 10.2

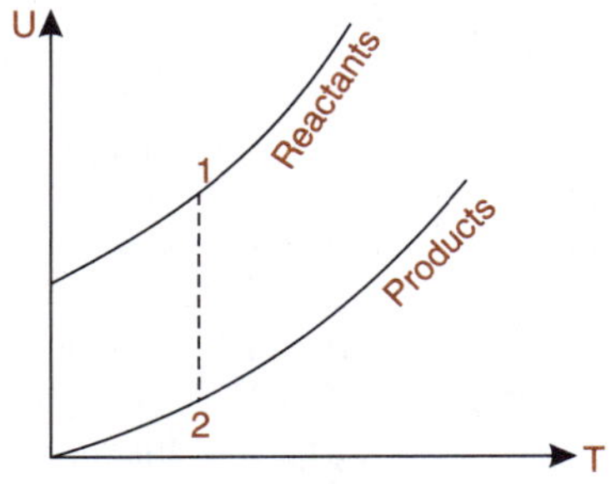

Fig. 10.3

The entire process can be thought of as taking place in piston-cylinder device as shown in Fig. 10.4.

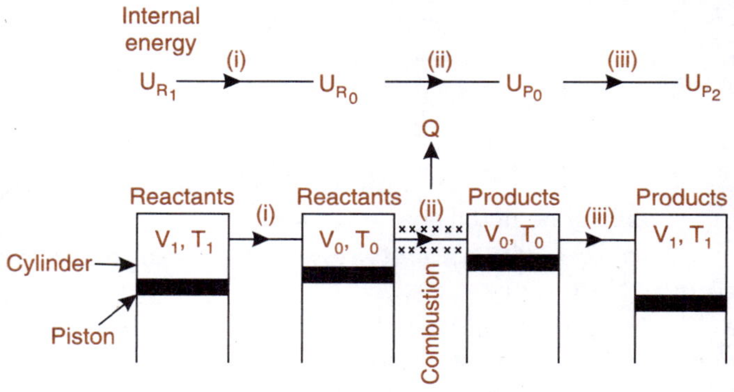

Fig. 10.4

Thus $U_2 - U_1$, the change in internal energy between states 1 and 2, can be written as $(U_{P_2} - U_{R_1})$ to show the chemical changes involved and this can be further expanded for analytical purposes as follows :

$$U_{P_2} - U_{R_1} = (U_{P_2} - U_{P_0}) + (U_{P_0} - U_{R_0}) + (U_{R_0} - U_{R_1})$$

$$i.e., \; U_{P_2} - U_{R_1} = \underbrace{(U_{P_2} - U_{P_0})}_{\substack{\text{Products} \\ (iii)}} + \underset{(ii)}{\Delta U_0} + \underbrace{(U_{R_0} - U_{R_1})}_{\substack{\text{Reactants} \\ (i)}} \qquad ...(10.14)$$

The values of $(U_{R_0} - U_{R_1})$ and $(U_{P_2} - U_{P_0})$ can be calculated from the following relations :

$$U_{R_0} - U_{R_1} = \sum_R n_i (u_{i_0} - u_{i_1}) \qquad ...(10.15)$$

where, u_i = Tabulated value of the internal energy for any constituent at the required temperature T_0 or T_1 in heat unit per mole

n_i = Number of moles of the constituent, and

$\sum_R$ = Summation for all the constituents of the reactants denoted by i.

If *mass* base is used for tabulated values or calculation, then

$$U_{R_0} - U_{R_1} = \sum_R m_i (u_{i_0} - u_{i_1}) \qquad ...(10.16)$$

where, u_i = Internal energy per unit mass.

The above expression in terms of the *specific heats* (average values for the required temperature range) may be written as

$$U_{R_0} - U_{R_1} = \sum_R m_i c_{vi} (T_0 - T_1) = (T_0 - T_1) \sum_R m_i c_{vi} \qquad ...(10.17)$$

For *products*, similar expressions may be written as :

$$U_{P_2} - U_{P_0} = \sum_P n_i (u_{i_2} - u_{i_0}) \; ... \textit{ on mole basis}$$

$$U_{P_2} - U_{P_0} = \sum_P m_i (u_{i_2} - u_{i_0}) \; ... \textit{ on mass basis}$$

$$U_{P_2} - U_{P_0} = \sum_P m_i c_{vi}(T_2 - T_0)$$

$$= (T_2 - T_0) \sum_P m_i c_{vi}$$

... in terms of *mean specific heats*

It may be noted that $n_i C_{vi} = m_i c_{vi}$

Analysis for a steady flow or 'constant pressure' combustion process :

In such an analysis the *changes in enthalpy (H)* are important. An analysis carried out as above will give the following expressions :

$$H_{P_2} - H_{R_1} = \underbrace{(H_{P_2} - H_{P_0})}_{\text{Products}} + \Delta H_0 + \underbrace{(H_{R_0} - H_{R_1})}_{\text{Reactants}} \quad ...(10.18)$$

where, $\Delta H_0 = H_{P_0} - H_{R_0}$, and is always *negative* ...(10.19)

[ΔH_0 = *Enthalpy of combustion* at T_0 or the *constant pressure heat of combustion* at T_0]

Expressions for change of enthalpy of reactants and products :

Reactants :

$$H_{R_0} - H_{R_1} = \sum_R n_i (h_{i_0} - h_{i_1}) \quad \text{... on mole basis} \quad ...(10.20)$$

$$H_{R_0} - H_{R_1} = \sum_R m_i (h_{i_0} - h_{i_1}) \quad \text{... on mass basis} \quad ...(10.21)$$

$$H_{R_0} - H_{R_1} = \sum_R m_i c_{pi}(T_0 - T_1)$$

$$= (T_0 - T_1) \sum_R m_i c_{pi}$$

... in terms of *mean specific heats*

Products : $$H_{P_2} - H_{P_0} = \sum_P n_i (h_{i_2} - h_{i_0})$$

... *on mole basis* ...(10.22)

$$H_{P_2} - H_{P_0} = \sum_P m_i (h_{i_2} - h_{i_0})$$

... *on mass basis* ...(10.23)

$$H_{P_2} - H_{P_0} = \sum_P m_i c_{pi}(T_2 - T_0)$$

$$= (T_2 - T_0) \sum_P m_i c_{pi}$$

... in terms of *mean specific heats* ...[10.23 (*a*)]

It may be noted that $n_i C_{pi} = m_i c_{pi}$

From the definition of the enthalpy of a perfect gas

$$H = U + pV = U + nR_0T$$

So if we are concerned only with gaseous mixtures in the reaction then for products and reactants

$$H_{P_0} = U_{P_0} + n_P R_0 T_0$$

and $$H_{R_0} = U_{R_0} + n_R R_0 T_0$$

where n_P and n_R are the moles of products and reactants respectively and the temperature is the reference temperature T_0.

Thus, using eqns. (10.13) and (10.19), we have

$$\Delta H_0 = \Delta U_0 + (n_P - n_R) R_0 T_0 \quad ...(10.24)$$

If there is no change in the number of moles during the reaction or if the reference temperature is absolute zero, then ΔH_0 and ΔU_0 will be equal.

10.17. ENTHALPY OF FORMATION (ΔH_f)

A combustion reaction is a particular kind of chemical reaction in which products are formed from reactants with the release or absorption of energy as heat is transferred to as from the surroundings. In some substances like hydrocarbon fuels which are many in number and complex in structure the *heat of reaction* or *combustion* may be calculated on the basis of known values of the enthalpy of formation, ΔH_f of the constituent of the reactants and products at the temperature T_0 (reference temperature). *The enthalpy of formation (ΔH_f) is the increase in enthalpy when a compound is formed from its constituent elements in their natural form and in a standard state.* The standard state is 25°C, and 1 atm. pressure, but it must be borne in mind that not all substances can exist in natural form, *e.g.* H_2O cannot be a vapour at 1 atm. and 25°C.

The expression of a particular reaction, for calculation purposes, may be given as :

$$\Delta H_0 = \sum_P n_i \Delta H_{f_i} - \sum_R n_i \Delta H_{f_i} \quad ...(10.25)$$

Typical values of ΔH_f for different substances at 25°C (298 K) in **kJ/mole** are given below :

S. No.	*Substance*	*Formula*	*State*	ΔH_f
1.	Oxygen	O	gas	249143
		O_2	gas	zero
2.	Water	H_2O	liquid	– 285765
			vapour	– 241783
3.	Carbon	C	gas	714852
			diamond	1900
			graphite	zero
4.	Carbon monoxide	CO	gas	– 111508
5.	Carbon dioxide	CO_2	gas	– 393443
6.	Methane	CH_4	gas	– 74855
7.	Methyl alcohol	CH_3OH	vapour	– 240532
8.	Ethyl alcohol	C_2H_5OH	vapour	– 281102
9.	Ethane	C_2H_6	gas	– 83870
10.	Ethene	C_2H_4	gas	51780
11.	Propane	C_3H_8	gas	– 102900
12.	Butane	C_4H_{10}	gas	– 125000
13.	Octane	C_8H_{18}	liquid	– 247600

10.18. CALORIFIC OR HEATING VALUES OF FUELS

The "calorific value or heating value" of the fuel is defined as the energy liberated by the complete oxidation of a unit mass or volume of a fuel. It is expressed in kJ/kg for solid and liquid fuels and kJ/m^3 for gases.

If a fuel contains hydrogen water will be formed as one of the products of combustion. If this *water is condensed, a large amount of heat will be released* than if the water exists in the vapour phase. For this reason *two heating values* are defined ; the *higher or gross heating value* and the *lower* or *net heating value*.

The *higher heating value*, HHV, is obtained when the *water formed by combustion is completely condensed.*

The *lower heating value*, LHV, is obtained when the water formed by combustion exists *completely in the vapour phase*.

Thus : $(\text{HHV})_p = (\text{LHV})_p + m\, h_{fg}$ ...(10.26)

$(\text{HHV})_v = (\text{LHV})_v + m(u_g - u_f)$ ...(10.27)

where m = Mass of water formed by combustion,

h_{fg} = Enthalpy of vaporisation of water, kJ/kg,

u_g = Specific internal energy of vapour, kJ/kg, and

u_f = Specific internal energy of liquid, kJ/kg.

In almost all practical cases, the water vapour in the products is vapour, the lower value is the one *which usually applies*.

10.19. DETERMINATION OF CALORIFIC OR HEATING VALUES

The calorific value of fuels can be determined either from chemical analysis or in the laboratory.

10.19.1. Solid and Liquid Fuels

Dulong's formula. Dulong suggested a formula for the calculation of the calorific value of the solid or liquid fuels from their chemical composition which is as given below.

Gross calorific value

$$\text{or H.H.V.} = \frac{1}{100}\left[33800\,\text{C} + 144000\left(\text{H} - \frac{\text{O}}{8}\right) + 9270\,\text{S}\right] \text{kJ/kg} \quad ...(10.28)$$

where C, H, O and S are carbon, hydrogen, oxygen and sulphur in percentages respectively in 100 kg of fuel. In the above formula the oxygen is assumed to be in combination with hydrogen and only extra surplus hydrogen supplies the necessary heat.

Laboratory method (Bomb calorimeter)

The calorific value of solid and liquid fuels is determined in the laboratory by '*Bomb calorimeter*'. It is so named because its shape resembles that of a bomb. Fig. 10.5 shows the schematic sketch of a bomb calorimeter.

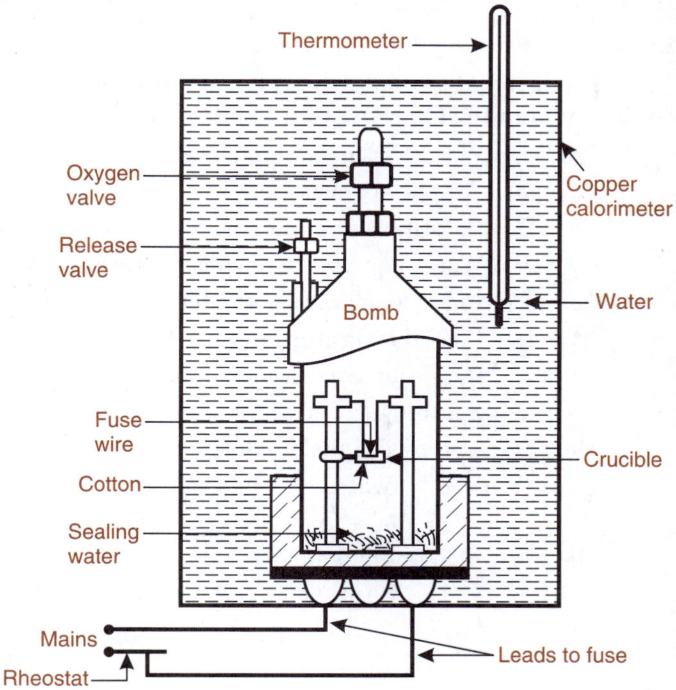

Fig. 10.5. Bomb calorimeter.

The *calorimeter* is made of austenitic steel which provides considerable resistance to corrosion and enables it to withstand high pressure. In the calorimeter is a *strong cylindrical bomb* in which combustion occurs. The bomb has *two valves* at the top. One supplies oxygen to the bomb and other releases the exhaust gases. A *crucible* in which a weighted quantity of fuel sample is burnt is arranged between the *two electrodes* as shown in Fig. 10.5. The calorimeter is fitted with *water jacket* which surrounds the bomb. To reduce the losses due to radiation, calorimeter is further provided with a jacket of water and air. A stirrer for keeping the temperature of water uniform and a thermometer to measure the temperature up to an accuracy of 0.001°C are fitted through the lid of the calorimeter.

Procedure. To start with, about 1 gm of fuel sample is accurately weighed into the crucible and a fuse wire (whose weight is known) is stretched between the electrodes. It should be ensured that *wire is in close contact with the fuel.* To absorb the combustion products of sulphur and nitrogen 2 ml of water is poured in the bomb. Bomb is then supplied with pure oxygen through the valve to an amount of 25 atmosphere. The bomb is then placed in the weighed quantity of water, in the calorimeter. The stirring is started after making necessary electrical connections, and when the thermometer indicates a steady temperature fuel is fired and temperature readings are recorded after 1/2 minute intervals until maximum temperature is attained. The bomb is then removed ; the pressure slowly released through the exhaust valve and the contents of the bomb are carefully weighed for further analysis.

The heat released by the fuel on combustion is absorbed by the surrounding water and the calorimeter.

From the above data the calorific value of the fuel can be found in the following way :

Let w_f = Weight of fuel sample (kg),
w = Weight of water (kg),
C = Calorific value (higher) of the fuel (kJ/kg),
w_e = Water equivalent of calorimeter (kg),
t_1 = Initial temperature of water and calorimeter,
t_2 = Final temperature of water and calorimeter,
t_c = Radiation corrections, and
c = Specific heat of water.

Heat released by the fuel sample = $w_f \times C$

Heat received by water and calorimeter

$$= (w_w + w_e) \times c \times [(t_2 - t_1) + t_c].$$

Heat lost = Heat gained

$$\therefore \quad w_f \times C = (w + w_e) \times c \times [(t_2 - t_1) + t_c]$$

i.e.,
$$C = \frac{(w + w_e) \times c \times [(t_2 - t_1) + t_c]}{w_f} \quad ...(10.29)$$

[Value of c is 4.18 in SI units and unity in MKS units.]

Note 1. Corrections pertain to the heat of oxidation of fuse wire, heat liberated as a result of formation of sulphuric and nitric acids in the bomb itself.

2. It should be noted that *bomb calorimeter measures the higher or gross calorific value because the fuel sample is burnt at a constant volume in the bomb.* Further the bomb calorimeter will measure the H.C.V. directly if the bomb contains adequate amount of water before firing to saturate the oxygen. Any water formed from combustion of hydrogen will, therefore, be condensed.

The procedure of determining calorific values of *liquid fuels* is similar to that described above. However, if the liquid fuel sample is volatile, it is weighed in a glass bulb and broken in a tray just before the bomb is closed. In this way the loss of volatile constituents of fuels during weighing operation is prevented.

Example 10.1. *The following particulars refer to an experimental determination of the calorific value of a sample of coal containing 88% C and 4.2% H_2. Weight of coal = 0.848 gm, weight of fuse wire 0.027 gm, of calorific value 6700 J/gm, weight of water in the calorimeter = 1950 gm, water equivalent of calorimeter = 380 gm, observed temperature rise = 3.06°C, cooling correction = + 0.017°C.*

Find the higher and lower calorific values of the coal.

Solution. Percentage of *carbon* in coal = 88%

Percentage of *hydrogen* in coal = 4.2%

Weight of coal, w_f = 0.848 gm

Weight of fuse wire, w_{fw} = 0.027 gm

Weight of water in the calorimeter, w = 1950 gm

Water equivalent of calorimeter, w_e = 380 gm

Observed temperature rise $(t_2 - t_1)$ = 3.06°C

Cooling correction,

t_c = + 0.017°C

∴ Corrected temperature rise

$= (t_2 - t_1) + t_c$

$= 3.06 + 0.017 = 3.077$°C

Calorific value of fuse wire

= 6700 J/gm

Heat received by water

$= (w + w_e) \times 4.18 \times [(t_2 - t_1) + t_c]$

$= (1950 + 380) \times 4.18 \times 3.077$

= 29968 J

Heat given out by fuse wire

$= w_{fw} \times$ calorific value

$= 0.027 \times 6700 = 180.9$ J

Heat produced due to combustion of fuel

$= 29968 - 180.9 = 29787.1$ J

∴ *Higher calorific value of fuel, H.C.V.*

$$= \frac{29787}{0.848} = 35126 \text{ J/gm}$$

= **35126 kJ/kg. (Ans.)**

Steam produced per kg of coal

$= 9 \times 0.042 = 0.378$ kg

Lower calorific value of coal, L.C.V.

$=$ H.C.V. $- 2465 \times 0.378$

$= 35126 - 931.7$

= **34194.3 kJ/kg. (Ans.)**

10.19.2. Gaseous Fuels

The calorific value of gaseous fuels can be determined by ***Junker's gas calorimeter.***

Fig. 10.6 illustrates *Junker's gas calorimeter*. Its principle is some what similar to Bomb calorimeter; in respect that heat evolved by burning the gas is taken away by the water. In its simplest construction it consists of a combustion chamber in which the gas is burnt (in a gas burner). A water jacket through which a set of tubes called flues pass surrounds this chamber. Thermometers are incorporated at different places (as shown in Fig. 10.6) to measure the temperatures.

Procedure. A metered quantity of gas whose calorific value is to be determined is supplied to the gas burner via a gas meter which records its volume and a gas pressure regulator which measures the pressure of the gas by means of a manometer. When the gas burns the hot products of combustion travel upwards in the chamber and then downwards through the flues and finally escape to the atmosphere through the outlet. The temperature of the escaping gas is recorded by the thermometer fitted at the exit and this temperature should be as close to room temperature as possible so that entire heat of combustion is absorbed by water. The cold water enters the calorimeter near the bottom and leaves near the top. Water which is formed by condensation of steam is collected in a pot.

The quantity of gas used during the experiment is accurately measured by the meter and temperature of ingoing and outgoing water are indicated by the thermometers. From the above data the calorific value of the gas can be calculated.

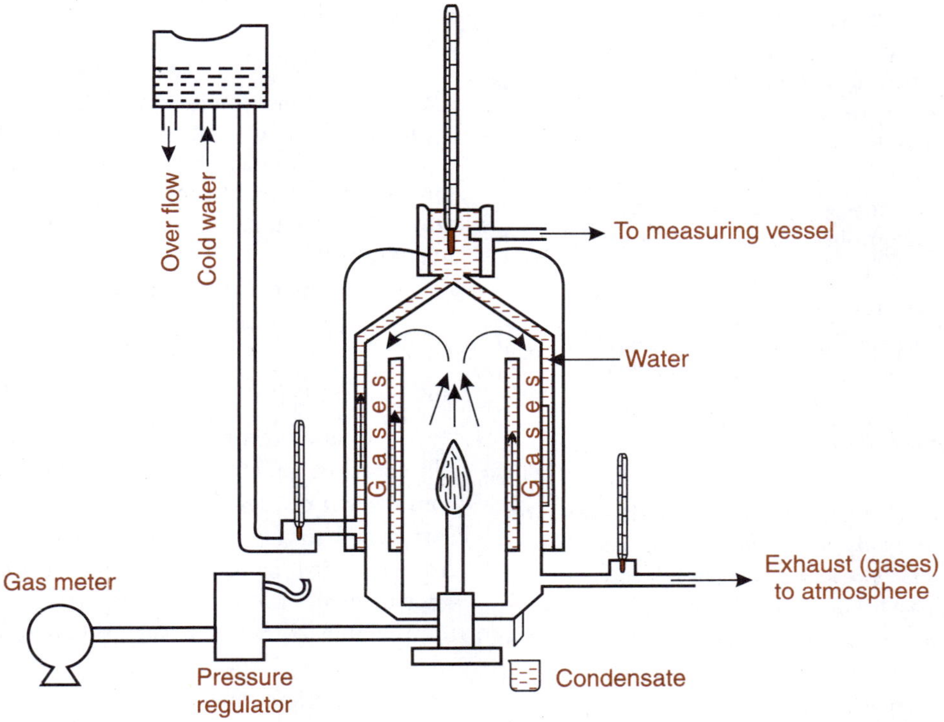

Fig. 10.6. Junker's gas calorimeter.

Note **1.** The volume of gas used during the experiment should be converted to volume at S.T.P. (standard temperature and pressure *i.e.*, 15°C, 760 mm respectively).

2. The gross calorific value is obtained by dividing the heat given out by corrected volume of gas. The net or lower calorific value of the gas is obtained by subtracting from total heat the heat associated with condensed water (which is obtained by multiplying the weighed condensate by latent heat of vaporisation of water.).

Example 10.2. *Following results were obtained when a sample of gas was tested by Junker's gas calorimeter :*

Gas burnt in the calorimeter = 0.08 m³, Pressure of gas supply = 5.2 cm of water, Barometer = 75.5 cm of Hg. Temperature of gas = 13°C, Weight of water heated by gas = 28 kg, Temperature of water at inlet = 10°C, Temperature of water at outlet = 23.5°C, Steam condensed = 0.06 kg.

Determine the higher and lower calorific values per m³ of the gas at a temperature of 15°C and barometric pressure of 76 cm of Hg.

Solution. The volume of the gas is measured at a temperature of 13°C and pressure of 5.2 cm of water. Let us reduce this volume to S.T.P. by using the general gas equation,

$$\frac{p_1V_1}{T_1} = \frac{p_2V_2}{T_2}$$

$$p_1 = 75.5 + \left(\frac{5.2}{13.6}\right) = 75.882 \text{ cm of Hg} ;$$

$$T_1 = 273 + 13 = 286 \text{ K}$$

$$V_1 = 0.08 \text{ m}^3 ; p_2 = 76 \text{ cm of Hg}$$

$$V_2 = ? ; T_2 = 273 + 15 = 288 \text{ K}$$

$$\therefore \quad \frac{75.882 \times 0.08}{286} = \frac{76 \times V_2}{288}$$

i.e.,

$$V_2 = \frac{75.882 \times 0.08 \times 288}{286 \times 76} = 0.0804 \text{ m}^3$$

Heat received by water

$$= 28 \times 4.18 \times (23.5 - 10) = 1580 \text{ kJ}$$

Higher calorific value of fuel

$$= \frac{1580}{0.08} = \mathbf{19750 \text{ kJ/m}^3}. \quad \textbf{(Ans.)}$$

Amount of water vapour formed (*i.e.*, steam condensed)

per m³ of gas burnt $= \dfrac{0.06}{0.08} = 0.75$ kg

Lower calorific value, L.C.V.

$= \text{H.C.V.} - 2465 \times 0.75$

$= 19750 - 1848.7 =$ **17901.3 kJ/kg. (Ans.)**

10.20. ADIABATIC FLAME TEMPERATURE

In a given combustion process, that takes place *adiabatically* and with no work or changes in kinetic or potential energy involved, *the temperature of the products is referred to as the 'adiabatic flame temperature'*. With the assumptions of no work and no changes in kinetic or potential energy, this is the *maximum temperature* that can be achieved for the given reactants because any heat transfer from the reacting substances and any incomplete combustion would tend to lower the temperature of the products.

The following points are worthnoting :

(*i*) *The maximum temperature achieved through adiabatic complete combustion varies with the type of reaction and per cent of theoretical air supplied.*

An *increase in the air-fuel* ratio will effect a *decrease in the maximum temperature*.

(*ii*) For a given fuel and given pressure and temperature of the reactants, the *maximum adiabatic flame temperature that can be achieved is with a 'stoichiometric' mixture.*

(*iii*) The adiabatic flame temperature can be *controlled* by the *amount of excess air* that is used. This is important, for example, in *gas turbines*, where the maximum permissible temperature is determined by *metallurgical considerations* in the turbine, and *close control of the temperature of the products is essential*.

10.21. CHEMICAL EQUILIBRIUM

The calculation of the *adiabatic flame temperature* is based, in part, on the assumption that the reaction goes to *completion*. Owing to *dissociation*, complete conversion of the reactants to the products is not accomplished. As a consequence of the failure to achieve complete conversion of the reactants, the maximum reaction temperature cannot attain the level of the theoretical adiabatic flame temperature.

The combination of CO and O_2 produces CO_2 together with a release of energy. In an adiabatic system no heat is transferred to the surroundings, hence the temperature of the mixture of the products and reacting substances rises rapidly. As the mixture temperature increases to higher levels the rate of dissociation of the CO_2 becomes increasingly more pronounced. Since the dissociation of CO_2 requires absorption of energy, *a condition is reached where the rate of evolution and the rate of absorption of energy are in balance. At this point no further increase in temperature can be observed* and the reaction is in *chemical equilibrium*. For this condition

$$C + O_2 \rightleftharpoons CO_2$$

At each temperature of the equilibrium mixture the substances participating in the reaction exist in unique proportions. For the combustion of CO the right-hand side of the equation

$$CO + \tfrac{1}{2} O_2 = (1 - x)\, CO_2 + x\, CO + \tfrac{x}{2} O_2 \quad ...(10.30)$$

represents the distribution of the equilibrium products resulting from a reaction between CO and O_2. In this equation x denotes the fraction of dissociated CO_2. At low temperatures the fraction $(1 - x)$ approaches unity while at high temperatures $(1-x)$ shows a substantial reduction in magnitude.

For the combustion of H_2 with O_2

$$H_2 + \tfrac{1}{2} O_2 \rightleftharpoons H_2O$$

and

$$H_2 + \tfrac{1}{2} O_2 = (1 - x)\, H_2O + x\, H_2 + \tfrac{x}{2} O_2 \quad ...(10.31)$$

It is essential to distinguish between the effects of dissociation and the losses resulting from incomplete combustion of fuel. Incomplete combustion, which may be attributed to a number of factors, results in a discharge from the system of combustible substances. *Dissociation*, on the other hand, is of *transient nature*. Usually any appreciable degree of dissociation extends over a very short time interval at the highest level of temperature attained in the reaction. The gaseous products are likely to be discharged from the system at a temperature that is indicative of a low degree of dissociation. *For example, dissociation does not influence the heating value determined in a fuel calorimeter*. Although the maximum temperature attained in the calorimeter is limited by chemical equilibrium, the combustion process moves to completion with the decrease in the temperature of the products. The reduction in temperature is a result of heat transfer to the jacket water. Dissociation of the products is negligible at room temperature, which is essentially the calorimeter reaction temperature.

The temperature of the products discharged from the combustion chamber of the gas turbine power plant is limited to approximately 870°C by introduction of a *large quantity of excess air*. Absorption of energy in the water walls of a boiler furnace limits the outlet gas temperature to approximately 1100°C. The quantity of dissociated products at temperatures ranging upward to 1100°C is not appreciable. In the cylinder of I.C. engine, considerably higher maximum temperatures—that is, in excess of 1100°C are attained, hence in the analysis of this thermal power system consideration must be given to the effects of dissociation. Of particular significance is the effect of reduced maximum temperature on the system availability. As a result of heat transfer and work performed by the gaseous medium the products are discharged from the system at a temperature below the level at which an appreciable degree of dissociation is observed.

The proportions of the dissociated products in chemical equilibrium at temperature T are established from the *equilibrium constant*. The evaluation of the equilibrium constant is achieved in accordance with the analysis presented by Van't Hoff.

10.22. ACTUAL COMBUSTION ANALYSIS

In evaluating the performance of an actual combustion process a number of different parameters can be defined depending on the nature of the process and the system considered. The combustion efficiency in a gas turbine for instance can be defined as:

$$\eta_{combustion} = \frac{(F/A)_{ideal}}{(F/A)_{actual}} \quad ...(10.32)$$

where $(F/A)_{ideal}$ = Fuel-air ratio required for adiabatic and complete combustion and in which the products would attain the adiabatic flame temperature.

In case of a *steam generator (boiler)*

$$\eta_{steam\ generator} = \frac{\text{Heat transferred to steam / kg fuel}}{\text{Higher heating value of the fuel}} \quad ...(10.33)$$

In case of an *internal combustion engine,*

$$\eta_{thermal} = \frac{W_{actual}}{\text{Heating value}} \quad ...(10.34)$$

Example 10.3. *A coal sample gave the following analysis by weight, Carbon 85 per cent, Hydrogen 6 per cent, Oxygen 6 per cent, the remainder being incombustible. Determine minimum weight of air required per kg of coal for chemically correct composition.*

Solution.

Element, wt. (kg)	*O_2 required (kg)*
C = 0.85	$0.85 \times \frac{8}{3} = 2.27$
H_2 = 0.06	$0.06 \times 8 = 0.48$
O_2 = 0.06	----------
	Total O_2 = 2.75

Weight of O_2 to be supplied

= Wt. of O_2 needed

– Wt. of O_2 already present in fuel

= 2.75 – 0.06 = 2.69 kg

Weight of air needed

$$= 2.69 \times \frac{100}{23} = \mathbf{11.70\ kg.} \quad \textbf{(Ans.)}$$

Example 10.4. *The percentage composition of sample of liquid fuel by weight is, C = 84.8 per cent, and H_2 = 15.2 per cent. Calculate (i) the weight of air needed for the combustion of 1 kg of fuel ; (ii) the volumetric composition of the products of combustion if 15 per cent excess air is supplied.*

Solution.

Element, wt. (kg)	*O_2 used (kg)*	*Dry products (kg)*
C = 0.848	$0.848 \times \frac{8}{3} = 2.261$	$\frac{0.848 \times 11}{3} = 3.109\ (CO_2)$
H_2 = 0.152	$0.152 \times 8 = 1.216$	
	Total O_2 = 3.477	

(*i*) **Minimum weight of air needed for combustion**

$$= \frac{3.477 \times 100}{23} = \mathbf{15.11\ kg.} \quad \textbf{(Ans.)}$$

Excess air supplied

$$= \frac{15.11 \times 15}{100} = 2.266 \text{ kg}$$

Wt. of oxygen in excess air

$$= \frac{2.266 \times 23}{100} = 0.521 \text{ kg}$$

Total air supplied for combustion

= Minimum air + Excess air

= 15.11 + 2.266 = 17.376 kg

∴ Wt. of nitrogen (N_2) in flue gases

$$= \frac{17.376 \times 77}{100} = 13.38 \text{ kg.}$$

(*ii*) To get **volumetric composition of the product of combustion** let us use tabular method.

Name of gas	*Weight (x)*	*Molecular weight (y)*	*Proportional volume (z)* $= \frac{(x)}{(y)}$	*Percentage volume* $= \frac{(z)}{\Sigma(z)} \times 100$
CO_2	3.109	44	0.0707	**12.51 per cent. (Ans.)**
O_2	0.521	32	0.0163	**2.89 per cent. (Ans.)**
N_2	13.38	28	0.4780	**84.60 per cent. (Ans.)**
			$\Sigma z = 0.5650$	

Example 10.5. *Percentage volumetric analysis of a sample of flue gases of a coal fired boiler gave CO_2 = 10.4 ; CO = 0.2 ; O_2 = 7.8 and N_2 = 81.6 (by difference). Gravemetric percentage analysis of coal was C = 78, H_2 = 6, O_2 = 3 and incombustible = 13. Estimate :*

(i) Weight of dry flue gases per kg of fuel.

(ii) Weight of excess air per kg of fuel.

Solution.

Element, wt. (kg)	O_2 *reqd. (kg)*	*Dry products (kg)*
C = 0.78	$0.78 \times \frac{8}{3} = 2.08$	$0.78 \times \frac{11}{3} = 2.86\ (CO_2)$
H_2 = 0.06	$0.06 \times 8 = 0.48$	
O_2 = 0.03		
	Total O_2 = 2.56	

Minimum wt. of air needed for combustion

$$= (2.56 - 0.03) \times \frac{100}{23} = 11 \text{ kg.}$$

(*i*) **Weight of dry flue gases per kg of fuel :**

To determine the wt. of flue gases per kg of fuel let us use tabular method to convert volumetric analysis to analysis by weight.

Name of gas	*Volume per* m^3 *of flue gas (x)*	*Molecular weight (y)*	*Relative volume* $z = x \times y$	*Weight per kg of flue gas* $= \frac{z}{\Sigma z}$
CO_2	0.104	44	4.576	0.1525
CO	0.002	28	0.056	0.0019
N_2	0.816	28	22.848	0.7616
O_2	0.078	32	2.496	0.0832
			Σz = 29.976 (say 30)	

Amount of carbon present per kg of gases

= Amount of carbon in 0.1525 kg of CO_2

+ Amount of carbon present in 0.0019 kg of CO

$$= \frac{3}{11} \times 0.1525 + \frac{3}{7} \times 0.0019$$

= 0.0416 + 0.0008 = 0.0424 kg.

Also carbon in the fuel = 0.78 kg.

∴ *Weight of dry flue gas per kg of fuel*

$$= \frac{\text{Weight of carbon in 1 kg of fuel}}{\text{Weight of carbon in 1 kg of flue gas}} = \frac{0.78}{0.0424}$$

= **18.4 kg. (Ans.)**

(*ii*) **Weight of excess air per kg of fuel :**

Weight of excess oxygen per kg of flue gas

$$= 0.0832 - \frac{4}{7} \times .0019 \quad \left[\begin{array}{c} 2CO + O_2 = 2CO_2 \\ 56 \quad 32 \quad 88 \\ 1\text{ kg} + \frac{4}{7}\text{ kg} = \frac{11}{7}\text{ kg} \end{array}\right]$$

= 0.0832 – 0.0011 (allowing for unburnt carbon monoxide)

= 0.0821 kg.

Weight of excess O_2 per kg of fuel

= 18.4 × 0.0821 = 1.51 kg

∴ *Weight of excess air per kg of fuel*

$$= \frac{1.51 \times 100}{23} = \mathbf{6.56\ kg.}\ \textbf{(Ans.)}$$

Example 10.6. *A single cylinder was supplied with a gas having the following percentage volumetric analysis ; CO = 5,* CO_2 *= 10,* H_2 *= 50,* CH_4 *= 25,* N_2 *= 10. The percentage volumetric analysis of dry gases was* CO_2 *= 8,* O_2 *= 6 and* N_2 *= 86. Determine the air-fuel ratio by volume.*

Solution. Combustion equations are :

$$2H_2 + O_2 = 2H_2O$$

1 vol + 1/2 vol = 1 vol

$$2CO + O_2 = 2CO_2$$

1 vol + 1/2 vol = 1 vol

$$CH_4 + 2O_2 = CO_2 + 2H_2O$$

1 vol + 2 vol = 1 vol + 2 vol.

Gas	*Vol* (m^3)	O_2 *needed* (m^3)	*Products* (m^3)	
			CO_2	N_2
CO	0.05	0.025	0.05	—
CO_2	0.10	—	0.10	—
H_2	0.50	0.25	—	—
CH_4	0.25	0.50	0.25	—
N_2	0.10	—	—	0.10
Total	1.0	0.775	0.4	0.10

Volume of air required

$= 0.775 \times 100/21 = 3.69\ m^3$

Volume of nitrogen in the air

$= 3.69 \times 79/100 = 2.92\ m^3$

Dry combustion products of 1 m^3 of gases (V) contain 0.4 m^3 of CO_2 + 0.1 m^3 of N_2 (as given in the table) + 2.92 m^3 of N_2 (from air supplied for complete combustion) = 3.42 m^3.

Excess air supplied

$$= \frac{O_2 \times V}{(21 - O_2)} = \frac{6.0 \times 3.42}{(21 - 6)} = \frac{20.52}{15} = 1.37\ m^3$$

Total quantity of air supplied

$= 3.69 + 1.37 = 5.06\ m^3$.

Air fuel ratio

$$= \frac{\text{Volume of air}}{\text{Volume of fuel}} = \frac{5.06}{1} = \mathbf{5\ (say).}\ \textbf{(Ans.)}$$

☞**Example 10.7.** *The following is the ultimate analysis of a sample of petrol by weight :*

Carbon = 85 per cent ; Hydrogen = 15 per cent.

Calculate the ratio of air to petrol consumption by weight if the volumetric analysis of the dry exhaust gas is :

CO_2 = 11.5 per cent ; CO = 1.2 per cent ; O_2 = 0.9 per cent ; N_2 = 86 per cent.

Also find percentage excess air.

Solution.

Name of gas	*Volume per m^3 of flue gas (x)*	*Molecular weight (y)*	*Relative weight $z = x \times y$*	*Weight per kg of flue gas $= \frac{z}{\Sigma z}$*
CO_2	0.115	44	5.06	0.1700
CO	0.012	28	0.336	0.0113
O_2	0.009	32	0.288	0.0096
N_2	0.86	28	24.08	0.8091
			Σz = 29.76	

∴ Weight of carbon per kg of flue gas

= Weight of carbon in 0.17 kg of CO_2 + Weight of carbon in 0.0113 kg of CO

$$= \frac{3}{11} \times 0.17 + \frac{3}{7} \times 0.0113 = 0.0512 \text{ kg}$$

∴ Weight of dry flue gas per kg of fuel

$$= \frac{0.85}{0.0512} = 16.6 \text{ kg}$$

Vapour of combustion

= 9 × 0.15 = 1.35 kg

Total weight of gas

= 16.6 + 1.35 = 17.95 kg per kg of fuel

∴ Air supplied

= (17.95 – 1) = 16.95 kg/kg of fuel

∴ *Ratio of air to petrol*

= **16.95 : 1. (Ans.)**

Stoichiometric air

$$= \left[\left(0.85 \times \frac{8}{3}\right) + (0.15 \times 8)\right] \times \frac{100}{23}$$

= 15.07 kg per kg of fuel

∴ Excess air

= 16.95 – 15.07 = 1.88 kg

∴ *Percentage excess air*

$$= \frac{1.88}{15.07} \times 100 = \mathbf{12.47\%}. \quad \textbf{(Ans.)}$$

☞**Example 10.8.** *A sample of fuel has the following percentage composition : Carbon = 86 per cent ; Hydrogen = 8 per cent ; Sulphur = 3 per cent ; Oxygen = 2 per cent ; Ash = 1 per cent.*

For an air-fuel ratio of 12 : 1, calculate :

(i) Mixture strength as a percentage rich or weak.

(ii) Volumetric analysis of the dry products of combustion.

Solution.

Element, wt. (kg)	*O_2 reqd. (kg)*
C = 0.86	$0.86 \times \frac{8}{3} = 2.29$
H_2 = 0.08	0.08 × 8 = 0.64
S = 0.03	$0.03 \times \frac{1}{1} = 0.03$
O_2 = 0.02	
	Total O_2 = 2.96

Weight of oxygen to be supplied per kg of fuel

= 2.96 – 0.02 = 294 kg.

Weight of minimum air required for complete combustion $= \frac{2.94 \times 100}{23} = 12.78$ kg

Hence "*correct*" fuel-air ratio $= \frac{1}{12.78} : 1$

But actual ratio is $\frac{1}{12} : 1$.

(*i*) **Mixture strength**

$$= \frac{12.78}{12} \times 100 = 106.5\%$$

This show that mixture is **6.5% rich. (Ans.)**

Deficient amount of air = 12.78 – 12 = 0.78 kg

Amount of air saved by burning 1 kg of C to CO instead of CO_2

$$= \text{Oxygen saved} \times \frac{100}{23}$$

$$= \left[\frac{8}{3}(CO_2) - \frac{4}{3}(CO)\right] \times \frac{100}{23} = 5.8 \text{ kg}$$

Hence $\frac{0.78}{5.8}$ = 0.134 kg of carbon burns to CO and as such 0.86 – 0.134 = 0.726 kg of carbon burns to CO_2.

$$\therefore \text{ CO formed} = 0.134 \times \frac{7}{3} = 0.313 \text{ kg}$$

$$CO_2 \text{ formed} = 0.726 \times \frac{11}{3} = 2.662 \text{ kg}$$

N_2 supplied = 12 × 0.77 = 9.24 kg

SO_2 formed = 0.03 × 2 = 0.06 kg.

(*ii*) **The percentage composition of dry flue gases** is given as below :

Dry products	*Weight (kg)* x	*Molecular weight* y	*Proportional volume* $z = \frac{x}{y}$	*Percentage volume* $\frac{z}{\Sigma z} \times 100$
CO	0.313	28	0.0112	**2.78 per cent (Ans.)**
CO_2	2.662	44	0.0605	**15.03 per cent (Ans.)**
N_2	9.24	28	0.3300	**81.97 per cent (Ans.)**
SO_2	0.06	64	0.0009	**0.22 per cent (Ans.)**
			$\Sigma z = 0.4026$	

Example 10.9. *A fuel ($C_{10}H_{22}$) is burnt using an air fuel ratio of 13 : 1 by weight. Determine the complete volumetric analysis of the products of combustion, assuming that the whole amount of hydrogen burns to form water vapour and there is neither any free oxygen nor any free carbon. The carbon burns to CO_2 and CO.*

Air contains 77% of nitrogen and 23% of oxygen by weight.

Solution. Combustion equation is :

$$2C_{10}H_{22} + 31O_2 = 20CO_2 + 22H_2O$$

$$2 \times 142 + 31 \times 32 = 20 \times 44 + 22 \times 18$$

or $$284 + 992 = 880 + 396$$

$\therefore$ Air required for complete combustion

$$= \frac{992 \times 100}{284 \times 23} = 15.2 \text{ kg/kg of fuel}$$

Air actually supplied

$= 13$ kg/kg of fuel

$\therefore$ Deficiency of air

$= 15.2 - 13 = 2.2$ kg/kg of fuel

Also 1 kg of C requires $\frac{4}{3} \times \frac{100}{23} = 5.8$ kg of less air to burn to CO instead of CO_2.

Hence $\frac{2.2}{5.8} = 0.379$ kg C is burnt to CO ;

and $\left(\frac{12 \times 10}{142}\right) - 0.379 = 0.466$ kg of C is burnt to CO_2.

Weight of CO_2 formed

$$= 0.466 \times \frac{11}{3} = 1.708 \text{ kg}$$

Weight of CO formed

$$= 0.379 \times \frac{7}{3} = 0.884 \text{ kg}$$

Weight of H_2O formed

$$= \left(\frac{22}{142}\right) \times 9 = 1.394 \text{ kg}$$

Weight of N_2 from air

$= 13 \times 0.77 = 10.01$ kg.

The percentage composition of products of combustion

Products of combustion	*Weight* x	*Molecular weight* y	*Proportional volume* $z = \frac{x}{y}$	*Percentage* $= \frac{z}{\Sigma z} \times 100$
CO_2	1.708	44	0.0388	**7.678 per cent (Ans.)**
CO	0.884	28	0.0316	**6.254 per cent (Ans.)**
H_2O	1.394	18	0.0774	**15.317 per cent (Ans.)**
N_2	10.01	28	0.3575	**70.750 per cent (Ans.)**
			$\Sigma z = 0.5053$	

Example 10.10. *Find an expression for the weight of air supplied per kg of fuel when carbon content of the fuel and the volumetric analysis of the fuel gas is known.*

Show that carbon burnt to CO per kg of fuel is $\frac{(CO)(C)}{CO + CO_2}$, where CO and CO_2 represent percentage monoxide and carbon dioxide in dry flue gas by volume and C represents percentage by weight of carbon that is actually burnt.

Solution. Let C = Percentage of carbon, by weight, in the fuel burnt,

CO_2 = Percentage by volume of carbon dioxide in the dry flue gas,

CO = Percentage by volume of carbon monoxide in the dry flue gas, and

N_2 = Percentage by volume of nitrogen in the dry flue gas.

Consider 100 m³ of the gas.

The actual weight of CO_2

$= CO_2 \times a \times b \times 44$ in 100 m³ of gas

The actual weight of CO

$= CO \times a \times b \times 28$ in 100 m³ of gas

The actual weight of N_2

$= N_2 \times a \times b \times 28$ in 100 m³ of gas

The molecular weight of CO = 28

The molecular weight of N_2 = 28

Total mass of carbon per 100 m^3 of flue gas

$$= [a \times b \times 44 \times CO_2] \times \frac{12}{44} + [a \times b \times 28 \times CO] \times \frac{12}{28}$$

The reaction producing CO_2 and CO are :

$$C + O_2 = CO_2$$
$$12 \quad 32 \quad 44$$

$$\therefore \quad C \text{ per kg of } CO_2 = \frac{12}{44}$$

Also

$$2C + O_2 = 2CO$$
$$24 \quad 32 \quad 56$$

$$\therefore \quad C \text{ per kg of CO} = \frac{12}{28}$$

All the carbon available in carbon dioxide and carbon monoxide comes from carbon present in the fuel. Also it can be safely assumed that all nitrogen comes from air (*this analysis being not applicable to the fuels containing nitrogen*).

Total carbon in the flue gas

$= a \times b \times 12\ (CO_2 + CO)$ per kg of carbon in the flue gas

$$= \frac{a \times b \times 28 \times N_2}{a \times b \times 12(CO + CO_2)} \qquad ...(i)$$

For every kg of fuel supplied for combustion $\frac{C}{100}$ kg of carbon pass out of the flue, whence the weight of N_2 per kg of fuel is :

$$\frac{28 N_2}{12(CO + CO_2)} \times \frac{C}{100} \text{ kg} \qquad ...(ii)$$

If all this nitrogen has come from the atmosphere, then air supplied per kg of fuel

$$= \frac{28 \times N_2 \times C \times 100}{12(CO + CO_2) \times 100 \times 77} \qquad ...(iii)$$

(since air contains 77% nitrogen by weight)

$\therefore$ *Air per kg of fuel*

$$= \frac{N_2 \times C}{\frac{100 \times 77 \times 12}{28 \times 100}(CO + CO_2)}$$

$$= \mathbf{\frac{N_2 \times C}{33(CO + CO_2)}}. \quad \textbf{(Ans.)}$$

It has been shown above that the total carbon of the flue gas is given by $a \times b \times 12\ (CO_2 + CO)$ and that in CO by $a \times b \times 12$ CO, whence the proportion of carbon burnt to CO is given by

$$\frac{a \times b \times 12 \times CO}{a \times b \times 12(CO + CO_2)} = \frac{CO}{CO + CO_2}$$

and if C is the proportion of carbon actually burnt per kg of fuel, then

Carbon burnt to CO per kg of fuel

$$= \frac{(CO)(C)}{CO + CO_2}. \quad \textbf{Proved.}$$

☞ **Example 10.11.** *The following results were obtained in a trial on a boiler fitted with economiser :*

	CO_2	CO	O_2	N_2
Analysis of gas entering the economiser	*8.3*	*0*	*11.4*	*80.3*
Analysis of gas leaving the economiser	*7.9*	*0*	*11.5*	*80.6*

(*i*) *Determine the air leakage into the economiser if the carbon content of the fuel is 80 per cent.*

(*ii*) *Determine the reduction in temperature of the gas due to air leakage if atmospheric temperature is 20°C and flue gas temperature is 410°C. Ash collected from ash pan is 15 per cent by weight of the fuel fired.*

Take : c_p for air = 1.005 kJ/kg K and c_p for flue gas = 1.05 kJ/kg K.

Solution. (*i*) Air supplied $= \frac{N_2 \times C}{33(CO_2 + CO)}$

Air supplied on the basis of conditions at entry to the economiser

$$= \frac{80.3 \times 80}{33(8.3 + 0)} = 23.45 \text{ kg}$$

Air applied on the basis of conditions at exit

$$= \frac{80.6 \times 80}{33(7.9 + 0)} = 24.73 \text{ kg}$$

$\therefore$ *Air leakage* = 24.73 – 23.45

= **1.28 kg of air per kg of fuel. (Ans.)**

(*ii*) For each kg of fuel burnt, the ash collected is 15% *i.e.*, 0.15 kg.

$\therefore$ Weight of fuel passing up the chimney

= 1 – 0.15 = 0.85 kg

$\therefore$ Total weight of products

= Weight of air supplied per kg of fuel + Weight of fuel passing through chimney per kg of fuel

= 23.45 + 0.85 = 24.3 kg

Heat in flue gases per kg of coal

= Weight of flue gases × Specific heat × Temperature rise above 0°C

= 24.3 × 1.05 × (410 – 0) = 10461 kJ

Heat in leakage air

= Weight of leakage air × Specific heat × Temperature rise of air above 0°C

= 1.28 × 1.005 × (20 – 0) = 25.73 kJ.

We can still consider, in the mixture, the gas and the air as separate and having their own specific heats, but sharing a common temperature t.

For heat balance :

$$(1.005 \times 1.28 + 24.3 \times 1.05)\, t = 10461 + 25.73$$

$$26.8\, t = 10486.73$$

$$\therefore \quad t = 391.3°C$$

$\therefore$ *Fall in temperature as a result of the air leakage into the economiser*

$= 410 - 391.3 =$ **18.7°C. (Ans.)**

AIR-FUEL RATIO AND ANALYSIS OF PRODUCTS OF COMBUSTION

Example 10.12. *The chemical formula for alcohol is C_2H_6O. Calculate the stoichiometric air/fuel ratio by mass and the percentage composition of the products of combustion per kg of C_2H_6O.* **(GATE)**

Solution. Chemical equation for complete combustion of given fuel can be written as follows :

$$C_2H_6O + 3O_2 \longrightarrow 2CO_2 + 3H_2O$$
$$(1 \times 46) + (3 \times 32) = (2 \times 44) + (3 \times 18)$$

For complete combustion of 1 kg of C_2H_6O, oxygen required

$$= \frac{3 \times 32}{46} = 2.087 \text{ kg of oxygen}$$

$$= 2.087 \times \frac{100}{23} = 9.074 \text{ kg of air}$$

A : F ratio for complete combustion

= **9.074 : 1. Ans.**

Also 46 kg of fuel produces products of combustion = 88 + 54 = 142 kg

[$\because$ 1 kg of fuel produces $\frac{142}{46}$ = 3.087 kg of products of combustion (*i.e.*, CO_2 and H_2O)]

Hence $\mathbf{CO_2}$ produced by fuel

$$= \frac{88}{46} \times 100 = \mathbf{1.913} \text{ or } \mathbf{191.3\%}. \text{ (Ans.)}$$

$\mathbf{H_2O}$ produced by fuel

$$= \frac{54}{46} \times 100 = \mathbf{1.174} \text{ or } \mathbf{117.4\%}. \text{ (Ans.)}$$

Example 10.13. *Calculate the amount of theoretical air required for the combustion of 1 kg of acetylene (C_2H_2) to CO_2 and H_2O.*

Solution. For combustion of acetylene (C_2H_2) the stoichiometric equation is written as

$$C_2H_2 + x\, O_2 \longrightarrow a\, CO_2 + b\, H_2O \quad ...(i)$$

Balancing the *carbon atoms* on both sides of the combustion eqn. (*i*), we get

$$2C = a\,C \quad i.e., \quad a = 2$$

Now balancing *hydrogen atoms* on both sides, we get

$$2H = 2bH$$
$$b = 1$$

Thus, eqn. (*i*) becomes

$$C_2H_2 + x\, O_2 \longrightarrow 2CO_2 + H_2O$$

Now, balancing *oxygen atoms* in the above equation

$$2x = 2 \times 2 + 1 = 5 \quad i.e., \quad x = 2.5$$

Hence, the final combustion eqn. (*i*) is

$$C_2H_2 + 2.5O_2 \longrightarrow 2CO_2 + H_2O \quad ...(ii)$$

Thus, for combustion of C_2H_2 in air, we get

$$C_2H_2 + 2.5O_2 + 2.5\left(\frac{79}{21}\right) N_2 \longrightarrow 2CO_2 + H_2O + 2.5\left(\frac{79}{21}\right) N_2$$

On a mass basis, this becomes

$$(2 \times 12 + 2 \times 1)\, C_2H_2 + 2.5\,(2 \times 16)\, O_2 + 2.5\left(\frac{79}{21}\right)(2 \times 14)\, N_2$$
$$\longrightarrow 2\,(12 + 2 \times 16)\, CO_2 + (1 \times 2 + 1 \times 16)\, H_2O + 2.5\left(\frac{79}{21}\right)(2 \times 14)\, N_2$$

$$26 \text{ kg } C_2H_2 + 80 \text{ kg } O_2 + 263.3\, N_2 \longrightarrow 88 \text{ kg } CO_2 + 18 \text{ kg } H_2O + 263.3 \text{ kg } N_2$$

or

$$1 \text{ kg } C_2H_2 + 3.076 \text{ kg } O_2 + 10.12 \text{ kg } N_2 \longrightarrow 3.38 \text{ kg } CO_2 + 0.69 \text{ kg } H_2O + 10.12 \text{ kg } N_2$$

i.e., Amount of air = 3.076 + 10.12

= 13.196 kg of air per kg of C_2H_2

Hence amount of theoretical air required for combustion of 1 kg acetylene

= 13.196 kg. (Ans.)

Example 10.14. *Determine the gravimetric analysis of the products of complete combustion of acetylene with 200 per cent stoichiometric air.*

Solution. The stoichiometric air equation (Example 10.3) is written as :

$$C_2H_2 + 2.5\, O_2 + 2.5\left(\frac{79}{21}\right) N_2 \longrightarrow 2CO_2 + H_2O + 2.5\left(\frac{79}{21}\right) N_2$$

If *200 per cent* stoichiometric air is used, the combustion equation for C_2H_2 becomes

$$C_2H_2 + (2)\,(2.5)\,O_2 + (2)\,(2.5)\left(\frac{79}{21}\right) N_2$$

$$\longrightarrow 2CO_2 + H_2O + (2)\,(2.5)\left(\frac{79}{21}\right) N_2 + (1)\,(2.5)\,O_2$$

$$(2 \times 12 + 2 \times 1) \text{ kg } C_2H_2 + (2)\,(2.5)\,(2 \times 16) \text{ kg } O_2 + (2)\,(2.5)\left(\frac{79}{21}\right)(2 \times 14) \text{ kg } N_2$$

$$\longrightarrow 2\,(12 + 2 \times 16) \text{ kg } CO_2 + (2 \times 1 + 16) \text{ kg } H_2O + (2)\,(2.5)\left(\frac{79}{21}\right)(2 \times 14) \text{ kg } N_2 + (1)\,(2.5)\,(2 \times 16) \text{ kg } O_2$$

or $\quad 26 \text{ kg } C_2H_2 + 160 \text{ kg } O_2 + 526.6 \text{ kg } N_2 \longrightarrow 88 \text{ kg } CO_2 + 18 \text{ kg } H_2O + 526.6 \text{ kg } N_2 + 80 \text{ kg } O_2$

or $\quad 1 \text{ kg } C_2H_2 + 6.15 \text{ kg } O_2 + 20.25 \text{ kg } N_2 \longrightarrow 3.38 \text{ kg } CO_2 + 0.69 \text{ kg } H_2O + 20.25 \text{ kg } N_2 + 3.07 \text{ kg } O_2$

Thus for 1 kg of fuel, the products will consist of

$CO_2 = 3.38$ kg

$H_2O = 0.69$ kg

$O_2 = 3.07$ kg

$N_2 = 20.25$ kg

Total mass of products $= 27.39$ kg

$\therefore$ Mass fractions are :

$$CO_2 = \frac{3.38}{27.39} = 0.123$$

$$H_2O = \frac{0.69}{27.39} = 0.025$$

$$O_2 = \frac{3.07}{27.39} = 0.112$$

$$N_2 = \frac{20.25}{27.39} = 0.739.$$

Hence the gravimetric analysis of the complete combustion is :

CO_2 = 12.3%, H_2O = 2.5%, O_2 = 11.2%, N_2 = 73.9%. (Ans.)

Example 10.15. *Calculate the theoretical air-fuel ratio for the combustion of octane, C_8H_{18}. The combustion equation is :*

$$C_8H_{18} + 12.5\,O_2 + 12.5\left(\frac{79}{21}\right) N_2 \longrightarrow 8CO_2 + 9H_2O + 12.5\left(\frac{79}{21}\right) N_2.$$

Solution. The air-fuel ratio on a mole basis is :

$$A/F = \frac{12.5 + 12.5\left(\frac{79}{21}\right)}{1} = 59.5 \text{ mol air/mol fuel}$$

The theoretical air-fuel ratio on a mass basis is found by introducing the molecular weight of the air and fuel

$$A/F = \frac{59.5(28.97)}{(8 \times 12 + 1 \times 18)}$$

$= \mathbf{15.08}$ **kg air/kg fuel. (Ans.)**

Example 10.16. *One kg of octane (C_8H_{18}) is burned with 200% theoretical air. Assuming complete combustion determine :*

(i) Air-fuel ratio

(ii) Dew point of the products at a total pressure 100 kPa.

Solution. The equation for the combustion of C_8H_{18} with theoretical air is :

$$C_8H_{18} + 12.5\,O_2 + 12.5\left(\frac{79}{21}\right) N_2 \longrightarrow 8CO_2 + 9H_2O + 12.5\left(\frac{79}{21}\right) N_2$$

For *200% theoretical air* the combustion equation would be

$$C_8H_{18} + (2)\,(12.5)\,O_2 + (2)\,(12.5)\left(\frac{79}{21}\right) N_2 \longrightarrow 8CO_2 + 9H_2O + (1)\,(12.5)\,O_2 + (2)\,(12.5)\left(\frac{79}{21}\right) N_2$$

Mass of fuel $= (1)\,(8 \times 12 + 1 \times 18) = 114$ kg/mole

Mass of air $= (2)\,(12.5)\left(1 + \frac{79}{21}\right) 28.97$

$= 3448.8$ kg/mole of fuel

***(i)* Air-fuel ratio :**

$$\text{Air-fuel ratio, } A/F = \frac{\text{Mass of air}}{\text{Mass of fuel}} = \frac{3448.8}{114} = 30.25$$

i.e., **A/F = 30.25. (Ans.)**

***(ii)* Dew point of the products, t_{dp} :**

Total number of moles of products

$$= 8 + 9 + 12.5 + (2)\,(12.5)\left(\frac{79}{21}\right)$$

$= 123.5$ moles/mole fuel

Mole fraction of $H_2O = \frac{9}{123.5} = 0.0728$

Partial pressure of $H_2O = 100 \times 0.0728 = 7.28$ kPa

The saturation temperature corresponding to this pressure is 39.7°C which is also the dew-point temperature.

Hence t_{dp} = 39.7°C. **(Ans.)**

Note. The water condensed from the products of combustion usually contains some dissolved gases and therefore may be quite *corrosive*. For this reason the products of combustion are often *kept above the dew point until discharged to the atmosphere*.

Example 10.17. *One kg of ethane (C_2H_6) is burned with 90% of theoretical air. Assuming complete combustion of hydrogen in the fuel determine the volumetric analysis of the dry products of combustion.*

Solution. The complete combustion equation for C_2H_6 is written as :

$$C_2H_6 + 3.5O_2 \longrightarrow 2CO_2 + 3H_2O$$

The combustion equation for C_2H_6 for 90% theoretical air is written as :

$$C_2H_6 + (0.9)(3.5)\,O_2 + (0.9)(3.5)\left(\frac{79}{21}\right) N_2$$

$$\longrightarrow a\,CO_2 + b\,CO + 3H_2O + (0.9)(3.5)\left(\frac{79}{21}\right) N_2$$

By balancing *carbon atoms* on both the sides, we get

$$2 = a + b \qquad ...(i)$$

By balancing *oxygen atoms* on both the sides, we get

$$(0.9)(3.5)(2) = 2a + b + 3 \qquad ...(ii)$$

Substituting the value of b (= 2 – a) from eqn. (*i*) in eqn. (*ii*), we get

$$(0.9)(3.5)(2) = 2a + 2 - a + 3$$

$$6.3 = a + 5$$

$$\therefore \quad a = 1.3$$

and $$b = 2 - a = 2 - 1.3 = 0.7$$

Thus the combustion equation becomes :

$$C_2H_6 + (0.9)(3.5)\,O_2 + (0.9)(3.5)\left(\frac{79}{21}\right) N_2$$

$$\longrightarrow 1.3CO_2 + 0.7CO + 3H_2O + (0.9)(3.5)\left(\frac{79}{21}\right) N_2$$

Total number of moles of *dry products* of combustion

$$= 1.3 + 0.7 + (0.9)(3.5)\left(\frac{79}{21}\right)$$

$$= 1.3 + 0.7 + 11.85 = 13.85 \text{ moles/mole of fuel}$$

Volumetric analysis of dry products of combustion is as follows :

$$CO_2 = \frac{1.3}{13.85} \times 100 = \mathbf{9.38\%}. \quad \textbf{(Ans.)}$$

$$CO = \frac{0.7}{13.85} \times 100 = \mathbf{5.05\%}. \quad \textbf{(Ans.)}$$

$$N_2 = \frac{11.85}{13.85} \times 100 = \mathbf{85.56\%}. \quad \textbf{(Ans.)}$$

☞ **Example 10.18.** *Methane (CH_4) is burned with atmospheric air. The analysis of the products on a 'dry' basis is as follows :*

CO_2 = 10.00%, O_2 = 2.37%, CO = 0.53%, N_2 = 87.10%.

(*i*) *Determine the combustion equation ;*

(*ii*) *Calculate the air-fuel ratio ;*

(*iii*) *Per cent theoretical air.*

Solution. (*i*) **Combustion equation :**

From the analysis of the products, the following equation can be written, keeping in mind that this analysis is on a *dry basis*.

$$x\,CH_4 + y\,O_2 + z\,N_2 \longrightarrow 10.0\,CO_2 + 0.53\,CO + 2.37\,O_2 + a\,H_2O + 87.1\,N_2$$

To determine all the unknown co-efficients let us find balance for each of the elements.

Nitrogen balance : $z = 87.1$

Since all the nitrogen comes from the air,

$$\frac{z}{y} = \frac{79}{21}\ ; \ y = \frac{87.1}{(79/21)} = 23.16$$

Carbon balance :

$$x = 10.00 + 0.53 = 10.53$$

Hydrogen balance :

$$a = 2x = 2 \times 10.53 = 21.06$$

Oxygen balance. All the unknown co-efficients have been solved for, and in this case the oxygen balance provides a check on the accuracy. Thus, y can also be determined by an oxygen balance

$$y = 10.00 + \frac{0.53}{2} + 2.37 + \frac{21.06}{2} = 23.16$$

Substituting these values for x, y, z and a, we have,

$$10.53\,CH_4 + 23.16\,O_2 + 87.1\,N_2 \longrightarrow 10.0\,CO_2 + 0.53\,CO + 2.37\,O_2 + 21.06\,H_2O + 87.1\,N_2$$

Dividing both sides by 10.53, we get the **combustion equation** per mole of fuel as :

$$\mathbf{CH_4 + 2.2\,O_2 + 8.27\,N_2 \longrightarrow 0.95\,CO_2 + 0.05\,CO + 2H_2O + 0.225\,O_2 + 8.27\,N_2.} \quad \textbf{(Ans.)}$$

(*ii*) **Air-fuel ratio A/F :**

The air-fuel ratio on a *mole basis* is :

2.2 + 8.27 = **10.47 moles air/mole fuel.** **(Ans.)**

The air-fuel ratio on a *mass basis* is found by introducing the molecular weights

$$\textbf{A/F} = \frac{10.47 \times 28.97}{(12 + 1 \times 4)} = \textbf{18.96 kg air/kg fuel.} \quad \textbf{(Ans.)}$$

The theoretical air-fuel ratio is found by writing the combustion equation for theoretical air,

$$CH_4 + 2O_2 + 2\left(\frac{79}{21}\right) N_2 \longrightarrow CO_2 + 2H_2O + (2)\left(\frac{79}{21}\right) N_2$$

$$\textbf{A/F}_{\textbf{theor}} = \frac{\left[2 + (2)\left(\frac{79}{21}\right)\right] 28.97}{(12 + 1 \times 4)}$$

$$= \textbf{17.24 kg air/kg fuel.} \quad \textbf{(Ans.)}$$

(*iii*) **Per cent theoretical air :**

Per cent theoretical air

$$= \frac{18.96}{17.24} \times 100 = \textbf{110\%.} \quad \textbf{(Ans.)}$$

Example 10.19. *The gravimetric analysis of a sample of coal is given as 82% C, 10% H_2 and 8% ash. Calculate :*

(*i*) *The stoichiometric A/F ratio ;*

(*ii*) *The analysis of the products by volume.*

Solution. (*i*) **The stoichiometric A/F ratio :**

1 kg of coal contains 0.82 kg C and 0.10 kg H_2.

$\therefore$ 1 kg of coal contains $\frac{0.82}{12}$ moles C and $\frac{0.10}{2}$ moles H_2

Let the oxygen required for complete combustion

= x moles

Then the nitrogen supplied with the oxygen

$$= x \times \frac{79}{21} = 3.76x \text{ moles}$$

For 1 kg of coal the combustion equation is therefore as follows :

$$\frac{0.82}{12} C + \frac{0.10}{2} H_2 + x\ CO_2 + 3.76x\ N_2 \longrightarrow a\ CO_2 + b\ H_2O + 3.76\ x\ N_2$$

Then, **Carbon balance :**

$$\frac{0.82}{12} = a \qquad \therefore \quad a = 0.068 \text{ moles}$$

Hydrogen balance :

$$2 \times \frac{0.10}{2} = 2b \qquad \therefore \quad b = 0.05 \text{ moles}$$

Oxygen balance :

$$2x = 2a + b \quad \therefore \quad x = \left(\frac{2 \times 0.068 + 0.05}{2}\right) = 0.093 \text{ moles}$$

The mass of 1 mole of oxygen is 32 kg, therefore, the mass of O_2 supplied per kg of coal = 32 × 0.093 = 2.976 kg

i.e., *Stoichiometric A/F ratio* $= \frac{2.976}{0.233}$ = **12.77.** **(Ans.)**

(where air is assumed to contain 23.3% O_2 and 76.7% N_2 by mass)

(*ii*) **The analysis of products by volume :**

Total moles of products

$= a + b + 3.76x$

$= 0.068 + 0.05 + 3.76 \times 0.093 = 0.467$ moles

Hence *the analysis of the products* is

$$\textbf{CO}_2 = \frac{0.068}{0.467} \times 100 = \textbf{14.56\%.} \quad \textbf{(Ans.)}$$

$$\textbf{H}_2 = \frac{0.05}{0.467} \times 100 = \textbf{10.7\%.} \quad \textbf{(Ans.)}$$

$$\textbf{N}_2 = \frac{(3.76 \times 0.093)}{0.467} \times 100 = \textbf{74.88\%.} \quad \textbf{(Ans.)}$$

☞**Example 10.20.** *Calculate the stoichiometric air-fuel ratio for the combustion of a sample of dry anthracite of the following composition by mass :*

Carbon (C)	*= 88 per cent*
Hydrogen (H_2)	*= 4 per cent*
Oxygen (O_2)	*= 3.5 per cent*
Nitrogen (N_2)	*= 1 per cent*
Sulphur (S)	*= 0.5 per cent*
Ash	*= 3 per cent*

If 30 per cent excess air is supplied determine :

(*i*) *Air-fuel ratio ;*

(*ii*) *Wet dry analysis of the products of combustion by volume.*

Solution. Stoichiometric air-fuel (A/F) ratio :

In case of a fuel with several constituents a *tabular method* is advisable, as shown below. Each constituent is taken separately and the amount of oxygen required for complete combustion is found from the chemical equation. The oxygen in the fuel is included in the column headed '*oxygen required*' as a *negative quantity*.

Composition	Mass per kg of coal	Combustion equation	Oxygen required per kg of coal	Products per kg of coal
C	0.88	$C + O_2 \longrightarrow CO_2$	$0.88 \times \frac{32}{12} = 2.346$ kg	$0.88 \times \frac{44}{12} = 3.23$ kg CO_2
		12 kg + 32 kg $\longrightarrow$ 44 kg		
H_2	0.04	$2H_2 + O_2 \longrightarrow 2H_2O$	$0.04 \times 8 = 0.32$ kg	$0.04 \times 9 = 0.36$ kg H_2O
		1 kg + 8 kg $\longrightarrow$ 9 kg		
O_2	0.035	—	— 0.035 kg	—
N_2	0.01	—	—	0.01 kg N_2
S	0.005	$S + O_2 \longrightarrow SO_2$	$0.005 \times \frac{32}{32} = 0.005$ kg	$0.005 \times \frac{64}{32} = 0.01$ kg SO_2
		32 kg + 32 kg $\longrightarrow$ 64 kg		
Ash	0.03	—	—	—
			Total $O_2 = 2.636$ kg	

From table :

O_2 required per kg of coal $= 2.636$ kg

$\therefore$ Air required per kg of coal $= \frac{2.636}{0.233} = 11.31$ kg

(where air is assumed to contain 23.3% O_2 *by mass*)

N_2 associated with this air

$= 0.767 \times 11.31 = 8.67$ kg

$\therefore$ Total N_2 in products

$= 8.67 + 0.01 = 8.68$ kg

The stoichiometric A/F ratio = 11.31/1. (Ans.)

When 30 per cent excess air is used :

(*i*) **Actual A/F ratio :**

Actual A/F ratio $= 11.31 + 11.31 \times \frac{30}{100}$

$=$ **14.7/1. (Ans.)**

(*ii*) **Wet and dry analyses of products of combustion by volume :**

As per actual A/F ratio, N_2 supplied

$= 0.767 \times 14.7 = 11.27$ kg

Also O_2 supplied

$= 0.233 \times 14.7 = 3.42$ kg

(where air is assumed to contain $N_2 = 76.7\%$ and $O_2 = 23.3$)

In the products then, we have

$N_2 = 11.27 + 0.01 = 11.28$ kg

and excess $O_2 = 3.42 - 2.636 = 0.78$ kg

The products are entered in the following table and the analysis by volume is obtained :

— In column 3 the percentage by mass is given by the mass of each product divided by the total mass of 15.66 kg.

— In column 5 the moles per kg of coal are given by equation $n = \frac{m}{M}$. The total of column 5 gives the total moles of wet products per kg of coal, and by subtracting the moles of H_2O from this total, the total moles of dry products is obtained as 0.5008.

Product	*Mass/kg coal*	*% by mass*	*M*	*Moles/kg coal*	*% by vol. wet*	*% by vol. dry*
1	*2*	*3*	*4*	*5*	*6*	*7*
CO_2	3.23	20.62	44	0.0734	14.10	14.66
H_2O	0.36	2.29	18	0.0200	3.84	—
SO_2	0.01	0.06	64	0.0002 (say)	0.04	0.04
O_2	0.78	4.98	32	0.0244	4.68	4.87
N_2	11.28	72.03	28	0.4028	77.34	80.43
	15.66 kg			Total wet = 0.5208	100.00	100.00
				$-H_2O$ = 0.0200		**(Ans.)**
				Total dry = 0.5008		

— Column 6 gives the proportion of each constituent of column 5 expressed as a percentage of the total moles of the wet products.

— Similarly column 7 gives the percentage by volume of the dry products.

Example 10.21. *The following analysis relate to coal gas :*

$H_2 = 50.4$ *per cent* $CO = 17$ *per cent*

$CH_4 = 20$ *per cent* $C_4H_8 = 2$ *per cent*

$O_2 = 0.4$ *per cent* $N_2 = 6.2$ *per cent*

$CO_2 = 4$ *per cent.*

(*i*) *Calculate the stoichiometric A/F ratio.*

(*ii*) *Find also the wet and dry analyses of the products of combustion if the actual mixture is 30 per cent weak.*

Solution. The example is solved by a tabular method ; a specimen calculation is given below :

For CH_4 :

$$CH_4 + 2O_2 \longrightarrow CO_2 + 2H_2O$$

i.e., 1 mole CH_4 + 2 moles $O_2 \longrightarrow$ 1 mole CO_2 + 2 moles H_2O

There are 0.2 moles of CH_4 per mole of the coal gas, hence

$$0.2 \text{ moles } CH_4 + 0.2 \times 2 \text{ moles } O_2 \longrightarrow 0.2 \text{ moles } CO_2 + 0.2 \times 2 \text{ moles } H_2O$$

$\therefore$ O_2 required for the CH_4 in the coal gas = 0.4 moles per mole of coal gas.

The oxygen in the fuel (0.004 moles) is included in column 4 as a *negative quantity*.

Product	*Moles/mole fuel*	*Combustion equation*	*O_2 moles/ mole fuel*	*Products CO_2*	*H_2O*
1	2	3	4	5	6
H_2O	0.504	$2H_2 + O_2 \longrightarrow 2H_2O$	0.252	—	0.504
CO	0.17	$2CO + O_2 \longrightarrow 2CO_2$	0.085	0.17	—
CH_4	0.20	$CH_4 + 2O_2 \longrightarrow CO_2 + 2H_2O$	0.400	0.20	0.40
C_4H_8	0.02	$C_4H_8 + 6O_2 \longrightarrow 4CO_2 + 4H_2O$	0.120	0.08	0.08
O_2	0.004	—	– 0.004	—	—
N_2	0.062	—	—	—	—
CO_2	0.04	—	—	0.04	—
			Total = 0.853	0.49	0.984

(*i*) **Stoichiometric A/F ratio :**

$$\text{Air required} = \frac{0.853}{0.21} = 4.06 \text{ moles/mole of fuel}$$

(where air is assumed to contain 21% O_2 by volume)

$\therefore$ **Stoichiometric A/F ratio**

= 4.06/1 by volume. (Ans.)

(*ii*) **Wet and dry analyses of the products of combustion if the actual mixture is 30% weak :**

Actual A/F ratio with 30% weak mixture

$$= 4.06 + \frac{30}{100} \times 4.06 = 1.3 \times 4.06 = 5.278/1$$

Associated $N_2 = 0.79 \times 5.278 = 4.17$ moles/mole fuel

Excess oxygen $= 0.21 \times 5.278 - 0.853 = 0.255$ moles

Total moles of N_2 in products $= 4.17 + 0.062 = 4.232$ moles/mole fuel.

Analysis by volume of wet and dry products :

Product	*Moles/mole fuel*	*% by vol. (dry)*	*% by vol. (wet)*
CO_2	0.490	9.97	8.31
H_2O	0.984	—	16.68
O_2	0.255	5.19	4.32
N_2	4.170	84.84	70.69
	Total wet = 5.899 – H_2O = 0.984 Total dry = 4.915	100.00	100.00 **(Ans.)**

Example 10.22. *Find the stoichiometric air-fuel ratio for the combustion of ethyl alcohol (C_2H_6O), in a petrol engine. Calculate the air-fuel ratios for the extreme mixture strengths of 80 per cent and 130 per cent. Determine also the wet and dry analyses by volume of the exhaust gas for each mixture strength.*

Solution. The equation for combustion of ethyl alcohol is as follows :

$$C_2H_6O + 3O_2 \longrightarrow 2CO_2 + 3H_2O$$

— Since there are two atoms of carbon in each mole of C_2H_6O then there must be two moles of CO_2 in the products, giving two atoms of carbon on each side of the equation.

— Similarly, since there are six atoms of hydrogen in each mole of ethyl alcohol then there must be three moles of H_2O in the products, giving six atoms of hydrogen on each side of the equation.

— Then balancing the atoms of oxygen, it is seen that there are (2 × 2 + 3) = 7 atoms on the right hand side of the equation, hence seven atoms must appear on the left hand side of the equation. There is one atom of oxygen in ethyl alcohol, therefore a further six atoms of oxygen must be supplied, and hence three moles of oxygen are required as shown.

Since the O_2 is supplied as air, the associated N_2 must appear in the equation,

i.e., $$C_2H_6O + 3O_2 + 3 \times \frac{79}{21} N_2 \longrightarrow 2CO_2 + 3H_2O + 3 \times \frac{79}{21} N_2$$

One mole of fuel has a mass of (2 × 12 + 1 × 6 + 16) = 46 kg. Three moles of oxygen have a mass of (3 × 32) = 96 kg.

$$\therefore \quad O_2 \text{ required per kg of fuel} = \frac{96}{46} = 2.09 \text{ kg}$$

∴ **Stoichiometric A/F ratio**

$$= \frac{2.09}{0.233} = \mathbf{8.96/1.} \quad \textbf{(Ans.)}$$

Considering a mixture strength of 80% :

$$\text{Now, mixture strength} = \frac{\text{Stoichiometric A/F ratio}}{\text{Actual A/F ratio}}$$

i.e., $$0.8 = \frac{8.96/1}{\text{Actual A/F ratio}}$$

$$\therefore \quad \textbf{Actual A/F ratio} = \frac{8.96}{0.8} = \mathbf{11.2/1.} \quad \textbf{(Ans.)}$$

This means that 1/0.8 or 1.25 times as much air is supplied as is necessary for complete combustion. The exhaust will therefore contain 0.25 stoichiometric oxygen.

i.e., $$C_2H_6O + 1.25\left(3O_2 + 3 \times \frac{79}{21} N_2\right)$$

$$\longrightarrow 2CO_2 + 3H_2O + 0.25 \times 3O_2 + 1.25 \times 3 \times \frac{79}{21} N_2$$

i.e., The products are :

2 moles CO_2 + 3 moles H_2O + 0.75 moles O_2 + 14.1 moles N_2

The total moles = 2 + 3 + 0.75 + 14.1 = 19.85

Hence wet analysis is :

$$CO_2 = \frac{2}{19.85} \times 100 = \mathbf{10.08\%.} \quad \textbf{(Ans.)}$$

$$H_2O = \frac{3}{19.85} \times 100 = \mathbf{15.11\%.} \quad \textbf{(Ans.)}$$

$$O_2 = \frac{0.75}{19.85} \times 100 = \mathbf{3.78\%.} \quad \textbf{(Ans.)}$$

$$N_2 = \frac{14.1}{19.85} \times 100 = \mathbf{71.03\%.} \quad \textbf{(Ans.)}$$

The total dry moles = 2 + 0.75 + 14.1 = 16.85

Hence dry analysis is :

$$CO_2 = \frac{2}{16.85} \times 100 = \mathbf{11.87\%.} \quad \textbf{(Ans.)}$$

$$O_2 = \frac{0.75}{16.85} \times 100 = \mathbf{4.45\%.} \quad \textbf{(Ans.)}$$

$$N_2 = \frac{14.1}{16.85} \times 100 = \mathbf{83.68\%.} \quad \textbf{(Ans.)}$$

Considering a mixture strength of 130% :

$$\text{Now,} \quad 1.3 = \frac{\text{Stoichiometric ratio}}{\text{Actual A/F ratio}}$$

$$\therefore \quad \textbf{Actual A/F ratio} = \frac{8.96}{1.3} = \mathbf{6.89/1.} \quad \textbf{(Ans.)}$$

This means that $\frac{1}{1.3}$ or 0.769 of the stoichiometric air is supplied. *The combustion cannot be complete, as the necessary oxygen is not available. It is usual to assume that all hydrogen is burned to H_2O, since hydrogen atoms have a greater affinity for oxygen than carbon atoms. The carbon in the fuel will burn to CO and CO_2, but the relative proportions have to be determined.*

Let, a = Number of moles of CO_2 in the products,

and b = Number of moles of CO in the products.

Then the combustion equation is as follows :

$$C_2H_6O + 0.769\left(3O_2 + 3 \times \frac{79}{21} N_2\right)$$

$$\longrightarrow a\, CO_2 + b\, CO + 3H_2O + 0.769 \times 3 \times \frac{79}{21} N_2$$

To find a and b a balance of carbon and oxygen atoms can be made,

i.e., ***Carbon balance :***

$$2 = a + b \quad ...(i)$$

and ***Oxygen balance :***

$$1 + 2 \times 0.769 \times 3 = 2a + b + 3$$

or $$2.614 = 2a + b \quad ...(ii)$$

From eqns. (*i*) and (*ii*), we get

$$a = 0.614,\ b = 1.386$$

i.e., The products are : 0.614 moles CO_2 + 1.386 moles CO + 3 moles H_2O + 8.678 moles N_2

The total moles = 0.614 + 1.386 + 3 + 8.678 = 13.678.

Hence wet analysis is :

$$CO_2 = \frac{0.614}{13.678} \times 100 = \mathbf{4.49\%}. \quad \text{(Ans.)}$$

$$CO = \frac{1.386}{13.678} \times 100 = \mathbf{10.13\%}. \quad \text{(Ans.)}$$

$$H_2O = \frac{3}{13.678} \times 100 = \mathbf{21.93\%}. \quad \text{(Ans.)}$$

$$N_2 = \frac{8.678}{13.678} \times 100 = \mathbf{63.45\%}. \quad \text{(Ans.)}$$

The total dry moles = 0.614 + 1.386 + 8.678 = 10.678

Hence dry analysis is :

$$CO_2 = \frac{0.614}{10.678} \times 100 = \mathbf{5.75\%}. \quad \text{(Ans.)}$$

$$CO = \frac{1.386}{10.678} \times 100 = \mathbf{12.98\%}. \quad \text{(Ans.)}$$

$$N_2 = \frac{8.678}{10.678} \times 100 = \mathbf{81.27\%}. \quad \text{(Ans.)}$$

Example 10.23. *For the stoichiometric mixture of example 10.22 calculate :*

(*i*) *The volume of the mixture per kg of fuel at a temperature of 50°C and a pressure of 1.013 bar.*

(*ii*) *The volume of the products of combustion per kg of fuel after cooling to a temperature of 130°C at a pressure of 1 bar.*

Solution. As before,

$$C_2H_6O + 3O_2 + 3 \times \frac{79}{21} N_2 \longrightarrow 2CO_2 + 3H_2O + 3 \times \frac{79}{21} N_2$$

$$\therefore \quad \text{Total moles reactants} = 1 + 3 + 3 \times \frac{79}{21} = 15.3$$

From equation,

$$pV = nR_0T$$

$$V = \frac{nR_0T}{p} = \frac{15.3 \times 8.314 \times 10^3 \times (50 + 273)}{1.013 \times 10^5}$$

$$= 405.6 \text{ m}^3\text{/mole of fuel}$$

In 1 mole of fuel there are (2 × 12 + 6 + 16) = 46 kg

(*i*) ∴ **Volume of reactants per kg of fuel**

$$= \frac{405.6}{46} = \mathbf{8.817\ m^3}. \quad \text{(Ans.)}$$

When the products are cooled to 130°C the H_2O exists as steam, since the temperature is well above the saturation temperature corresponding to the partial pressure of the H_2O. This must be so since the saturation temperature corresponding to the *total* pressure is 99.6°C, and the *saturation temperature decreases with pressure*. The total moles of the products is

$$= \left(2 + 3 + 3 \times \frac{79}{21}\right) = 16.3$$

From equation,

$$pV = nR_0T$$

$$V = \frac{nR_0T}{p} = \frac{16.3 \times 8.314 \times 10^3 \times (130 + 273)}{1 \times 10^5}$$

$$= 546.14 \text{ m}^3\text{/mole of fuel.}$$

(*ii*) ∴ **Volume of products per kg of fuel**

$$= \frac{546.14}{46} = \mathbf{11.87\ m^3}. \quad \text{(Ans.)}$$

☞**Example 10.24.** *The following is the composition of coal gas supplied to a gas engine :*

H_2 = 50.6 per cent ; CO = 10 per cent ; CH_4 = 26 per cent ; C_4H_8 = 4 per cent ; O_2 = 0.4 per cent ; CO_2 = 3 per cent ; N_2 = 6 per cent.

If the air-fuel ratio is 7/1 by volume, calculate the analysis of the dry products of combustion. It can be assumed that the stoichiometric A/F ratio is less than 7/1.

Solution. Since it is given that the actual *A/F ratio is greater than the stoichiometric,* therefore it follows that *excess air has been supplied*. The products will therefore consist of CO_2, H_2O, O_2 and N_2.

The combustion equation can be written as follows :

$$0.506H_2 + 0.1CO + 0.26CH_4 + 0.04C_4H_8 + 0.004O_2 + 0.03CO_2 + 0.06N_2 + 0.21 \times 7O_2 + 0.79 \times 7N_2 \longrightarrow a\,CO_2 + b\,H_2O + c\,O_2 + d\,N_2$$

Then,

Carbon balance **:**

$0.1 + 0.26 + 4 \times 0.04 + 0.03 = a \quad \therefore \quad a = 0.55$

Hydrogen balance **:**

$2 \times 0.506 + 4 \times 0.26 + 8 \times 0.04 = 2b \quad \therefore \quad b = 1.186$

Oxygen balance **:**

$0.1 + 2 \times 0.004 + 2 \times 0.03 + 0.21 \times 7 \times 2 = 2a + b + 2c \quad \therefore \quad c = 0.411$

Nitrogen balance **:**

$2 \times 0.06 + 2 \times 0.79 \times 7 = 2d \quad \therefore \quad d = 5.59$

∴ Total moles of dry products = 0.55 + 0.411 + 5.59 = 6.65

Then analysis by volume is :

$$CO_2 = \frac{0.55}{6.55} \times 100 = \mathbf{8.39\%}. \quad \text{(Ans.)}$$

$$O_2 = \frac{0.411}{6.55} \times 100 = \mathbf{6.27\%}. \quad \text{(Ans.)}$$

$$N_2 = \frac{5.59}{6.55} \times 100 = \mathbf{85.34\%}. \quad \text{(Ans.)}$$

Example 10.25. *The following is the analysis (by weight) of a chemical fuel :*

Carbon = 60 per cent ; Hydrogen = 20 per cent ; Oxygen = 5 per cent ; Sulphur = 5 per cent and Nitrogen = 10 per cent.

Find the stoichiometric amount of air required for complete combustion of this fuel.

Solution. On the basis of 100 kg fuel let us assume an equivalent formula of the form :

$$C_aH_bO_cN_dS_e$$

From the given analysis by weight, we can write

$12a = 60$ or $a = 5$

$1b = 20$ or $b = 20$

$16c = 5$ or $c = 0.3125$

$14d = 10$ or $d = 0.7143$

$32e = 5$ or $e = 0.1562$

Then the formula of the fuel can be written as

$$C_5H_{20}O_{0.3125}N_{0.7143}S_{0.1562}$$

The combustion equation is :

$$C_5H_{20}O_{0.3125}N_{0.7143}S_{0.1562} + x\,O_2 + x\left(\frac{79}{21}\right)N_2 \longrightarrow p\,CO_2 + q\,H_2O + r\,SO_2 + s\,N_2$$

Then,

Carbon balance :

$5 = p$ $\therefore$ $p = 5$

Hydrogen balance :

$20 = 2q$ $\therefore$ $q = 10$

Sulphur balance :

$0.1562 = r$ $\therefore$ $r = 0.1562$

Oxygen balance :

$$0.3125 + 2x = (2p + q + 2r)$$

i.e.,

$$x = p + \frac{q}{2} + r - \frac{0.3125}{2}$$

$$= 5 + \frac{10}{2} + 0.1562 - \frac{0.3125}{2} = 9.99$$

Nitrogen balance :

$$0.7143 + 2x \times \frac{79}{21} = 2s$$

$$\therefore \quad s = \frac{0.7143}{2} + x \times \frac{79}{21}$$

$$= \frac{0.7143}{2} + 9.99 \times \frac{79}{21} = 37.94$$

Hence, the combustion equation is written as follows :

$$C_5H_{20}O_{0.3125}N_{0.7143}S_{0.1562} + 9.99O_2 + 9.99\left(\frac{79}{21}\right)N_2 \longrightarrow 5CO_2 + 10H_2O + 0.1562SO_2 + 37.94N_2$$

$\therefore$ **Stoichiometric air required**

$$= \frac{9.92 \times 32 + 9.99 \times \left(\frac{79}{21}\right) \times 28}{100}$$

= 13.7 kg/kg of fuel. (Ans.)

Note. This example can also be solved by tabular method as explained in example 10.20.

Example 10.26. *A sample of fuel has the following percentage composition by weight :*

Carbon = 84 per cent *Hydrogen = 10 per cent*

Oxygen = 3.5 per cent *Nitrogen = 1.5 per cent*

Ash = 1 per cent

(i) Determine the stoichiometric air-fuel ratio by mass.

(ii) If 20 per cent excess air is supplied, find the percentage composition of dry flue gases by volume.

Solution. (*i*) **Stoichiometric air-fuel ratio :**

On the basis of 100 kg of fuel let us assume an equivalent formula of the form :

$$C_aH_bO_cN_d$$

From the given analysis by weight, we can write

$12a = 84$ *i.e.,* $a = 7$

$1b = 10$ *i.e.,* $b = 10$

$16c = 3.5$ *i.e.,* $c = 0.218$

$14d = 1.5$ *i.e.,* $d = 0.107$

The formula of fuel is

$$C_7H_{10}O_{0.218}N_{0.107}$$

The combustion equation is written as

$$C_7H_{10}O_{0.218}N_{0.107} + x\,O_2 + x\left(\frac{79}{21}\right)N_2 \longrightarrow p\,CO_2 + q\,H_2O + r\,N_2$$

Then,

Carbon balance :

$7 = p$ *i.e.,* $p = 7$

Hydrogen balance :

$10 = 2q$ *i.e.,* $q = 5$

Oxygen balance :

$$0.218 + 2x = (2p + q)$$

or $0.218 + 2x = 2 \times 7 + 5$ *i.e.,* $x = 9.39$

Nitrogen balance :

$$0.107 + 2x\left(\frac{79}{21}\right) = 2r$$

or $0.107 + 2 \times 9.39 \times \frac{79}{21} = 2r$ *i.e.,* $r = 35.4$

Hence the combustion equation becomes

$$C_7H_{10}O_{0.218}N_{0.107} + 9.39O_2 + 9.39\left(\frac{79}{21}\right)N_2 \longrightarrow 7CO_2 + 5H_2O + 35.4N_2$$

$\therefore$ **Stoichiometric A/F ratio**

$$= \frac{9.39 \times 32 + 9.39 \times \frac{79}{21} \times 28}{100}$$

$= \mathbf{12.89.}$ **(Ans.)**

(*ii*) **Percentage composition of dry flue gases by volume with 20 per cent excess air :**

If 20 per cent excess air is used, the combustion equation becomes

$$C_7H_{10}O_{0.218}N_{0.107} + (1.2)(9.39)\,O_2 + (1.2)(9.39)\left(\frac{79}{21}\right)N_2$$
$$\longrightarrow 7CO_2 + 5H_2O + (0.2)(9.39)\,O_2 + (1.2)(35.4)\,N_2$$

Total number of moles of dry products of combustion

$$n = 7 + (0.2)(9.39) + (1.2)(35.4)$$
$$= 7 + 1.878 + 42.48 = 51.358$$

$\therefore$ **Percentage composition of dry flue gases by volume is as follows :**

$$CO_2 = \frac{7}{51.358} \times 100 = \mathbf{13.63\%}. \quad \textbf{(Ans.)}$$

$$O_2 = \frac{1.878}{51.358} \times 100 = \mathbf{3.66\%}. \quad \textbf{(Ans.)}$$

$$N_2 = \frac{42.48}{51.358} \times 100 = \mathbf{82.71\%}. \quad \textbf{(Ans.)}$$

Example 10.27. *Orsat analysis of the products of combustion of a hydrocarbon fuel of unknown composition is as follows :*

Carbon dioxide (CO_2)	*= 8%*
Carbon monoxide (CO)	*= 0.5%*
Oxygen (O_2)	*= 6.3%*
Nitrogen (N_2)	*= 85.2%*

Determine the following :

(*i*) *Air-fuel ratio ;*

(*ii*) *Per cent theoretical air required for combustion.*

Solution. From the given Orsat analysis the combustion equation is written as follows :

$$a\,C + b\,H + c\,O_2 + \left(\frac{79}{21}\right)c\,N_2 = 8CO_2 + 0.5CO + 6.3O_2 + x\,H_2O + 85.2N_2$$

Then,

Carbon balance :

$$a = 8 + 0.5 = 8.5 \qquad i.e., \qquad a = 8.5$$

Nitrogen balance :

$$\frac{79}{21}c = 85.2 \qquad i.e., \qquad c = 22.65$$

Oxygen balance :

$$c = 8 + \frac{0.5}{2} + 6.3 + \frac{x}{2}$$

or $\quad 22.65 = 8 + 0.25 + 6.3 + \frac{x}{2} \qquad i.e., \quad x = 16.2$

Hydrogen balance :

$$b = 2x = 2 \times 16.2 = 32.4 \qquad i.e., \quad b = 32.4$$

(*i*) **Air-fuel ratio :**

The air supplied per 100 moles of dry products is

$$= 22.65 \times 32 + \left(\frac{79}{21}\right) \times 22.65 \times 28$$
$$= 3110.6 \text{ kg}$$

$\therefore$ **Air-fuel ratio** $= \frac{3110.6}{8.5 \times 12 + 32.4 \times 1}$

= **23.1 kg of air/kg of fuel. (Ans.)**

(*ii*) **Per cent theoretical air required for combustion :**

Mass fraction of carbon

$$= \frac{12 \times 8.5}{12 \times 8.5 + 32.4 \times 1} = 0.759$$

Mass fraction of hydrogen

$$= \frac{32.4 \times 1}{12 \times 8.5 + 32.4} = 0.241$$

Considering 1 kg of fuel, the *air required for complete combustion* is

$$= \left[0.759 \times \left(\frac{8}{3}\right) \times \frac{100}{23.3}\right] + \left[0.241 \times 8 \times \frac{100}{23.3}\right]$$
$$= 16.96 \text{ kg}$$

$\therefore$ **Per cent theoretical air required for combustion** $= \frac{23.1}{16.96} \times 100 = \mathbf{136.2\%}.$ **(Ans.)**

Example 10.28. *The following is the volumetric analysis of the dry exhaust from an internal combustion engine :*

CO_2 = 8.9% ; CO = 8.2% ; H_2 = 4.3% ; CH_4 = 0.5% and N_2 = 78.1%.

If the fuel used is octane (C_8H_{18}) determine air-fuel ratio on mass basis :

(*i*) *By a carbon balance.* (*ii*) *By a hydrogen-oxygen balance.*

Solution. (*i*) As per analysis of dry products, the combustion equation is written as

$$a\,C_8H_{18} + 78.1N_2 + 78.1\left(\frac{21}{79}\right)O_2 \longrightarrow 8.9CO_2 + 8.2CO + 4.3H_2 + 0.5CH_4 + 78.1N_2 + x\,H_2O$$

Carbon balance :

$$8a = 8.9 + 8.2 + 0.5 = 17.6 \; i.e., \; a = 2.2$$

$\therefore$ **Air-fuel (A/F) ratio**

$$= \frac{78.1 \times 28 + 78.1 \times \frac{21}{79} \times 32}{2.2(8 \times 12 + 1 \times 18)}$$

$$= \frac{2186.8 + 664.3}{250.8} = \frac{2851.1}{250.8} = \mathbf{11.37.} \quad \textbf{(Ans.)}$$

(*ii*) In this case the combustion equation is written as

$$a\ C_8H_{18} + b\ O_2 + b\left(\frac{79}{21}\right)N_2 \longrightarrow 8.9CO_2 + 8.2CO$$

$$+ 4.3H_2 + 0.5CH_4 + b\left(\frac{79}{21}\right)N_2 + x\ H_2O$$

Carbon balance :

$$8a = 8.9 + 8.2 + 0.5 = 17.6$$

i.e., $a = 2.2$

Hydrogen balance :

$$18a = 4.3 \times 2 + 0.5 \times 4 + 2x$$

or $18 \times 2.2 = 8.6 + 2 + 2x$

i.e., $x = 14.5$

Oxygen balance :

$$2b = 8.9 \times 2 + 8.2 + x$$

or $2b = 17.8 + 8.2 + 14.5$ *i.e.,* $b = 20.25$

$\therefore$ **Air-fuel (A/F) ratio**

$$= \frac{(20.25 \times 32) + (20.25)\left(\frac{79}{21}\right) \times 28}{2.2(8 \times 12 + 1 \times 18)}$$

$$= \frac{2781}{250.8} = \mathbf{11.09.} \quad \textbf{(Ans.)}$$

☞ **Example 10.29.** *The exhaust from an engine running on benzole was measured with the help of Orsat apparatus. Orsat analysis showed a CO_2 content of 12%, but no CO. Assuming that the remainder of the exhaust contains only oxygen and nitrogen, calculate the air-fuel ratio of the engine.*

The ultimate analysis of benzole is C = 88% and H_2 = 12%.

Solution. 1 kg of fuel, consisting of 0.88 kg C and 0.12 kg H_2, can be written as 0.88/2 moles C and 0.12/2 moles H_2. Therefore, considering 1 mole of *dry exhaust gas* (D.E.G.) we can write the combustion equation as follows :

$$X\left(\frac{0.88}{12}C + \frac{0.12}{2}H_2\right) + Y\,O_2 + \frac{79}{21}\,Y\,N_2$$

$$\longrightarrow 0.12CO_2 + a\ O_2 + (0.88 - a)\ N_2 + b\ H_2O$$

[Let the D.E.G. contain a moles of O_2.
The moles of CO_2 in 1 mole of D.E.G. are 0.12.
Therefore the D.E.G. contains $(1 - a - 0.12)$
$= (0.88 - a)$ moles of N_2.]

where, X = Mass of fuel per mole D.E.G.,
Y = Moles of O_2 per mole D.E.G.,
a = Moles of *excess* O_2 per mole D.E.G., and
b = Moles of H_2O per mole D.E.G.

Now,

Carbon balance :

$$\frac{0.88}{12}X = 0.12 \quad \therefore \quad X = 1.636$$

Hydrogen balance :

$$0.06X = b \quad \therefore \quad b = 0.06 \times 1.636 = 0.098$$

Oxygen balance :

$$2Y = 2 \times 0.12 + 2a + b$$

or $2Y = 0.24 + 2a + 0.098 \quad \therefore \quad Y = 0.169 + a$

Nitrogen balance :

$$\frac{79}{21}Y = (0.88 - a) \quad \therefore \quad Y = 0.234 - 0.266a$$

Equating the expressions for Y gives

$$0.234 - 0.266a = 0.169 + a \quad \therefore \quad a = 0.0513$$

i.e., $Y = 0.169 + 0.0513 = 0.2203$

$\therefore$ O_2 supplied $= 0.2203 \times 32$ kg/mole D.E.G.

i.e., Air supplied $= \frac{0.2203 \times 32}{0.233}$

$= 30.26$ kg/mole D.E.G.

Since $X = 1.636$, then, the fuel supplied per mole D.E.G. is 1.636 kg

$\therefore$ **A/F ratio** $= \frac{30.26}{1.636} =$ **18.5/1. (Ans.)**

Example 10.30. *The analysis of the dry exhaust from an internal combustion engine is as follows :*

Carbon dioxide (CO_2) = 15 per cent
Carbon monoxide (CO) = 3 per cent
Methane (CH_4) = 3 per cent
Hydrogen (H_2) = 1 per cent
Oxygen (O_2) = 2 per cent
Nitrogen (N_2) = 76 per cent

Calculate the proportions by mass of carbon to hydrogen in the fuel, assuming it to be a pure hydrocarbon.

Solution. Let 1 kg of fuel contain x kg of carbon (C) and y kg hydrogen (H_2). Then considering 1 mole of D.E.G. and introducing X and Y, we can write

$$X\left(\frac{x\,C}{12} + \frac{y}{2}H_2\right) + Y\,O_2 + \frac{79}{21}\,Y\,N_2$$

$$\longrightarrow 0.15CO_2 + 0.03CO + 0.03CH_4 + 0.01H_2 + 0.02O_2 + a\ H_2O + 0.76N_2$$

Then,

Nitrogen balance :

$$\frac{79}{21}Y = 0.76 \quad \therefore \quad Y = 0.202$$

Oxygen balance :

$$Y = 0.15 + \frac{0.03}{2} + 0.02 + \frac{a}{2}$$

or $\quad 0.202 = 0.15 + 0.015 + 0.02 + \frac{a}{2} \quad \therefore \quad a = 0.034$

Carbon balance :

$$\frac{Xx}{12} = 0.15 + 0.03 + 0.03 \quad \therefore \quad Xx = 2.52 \quad ...(i)$$

Hydrogen balance :

$$\frac{Xy}{2} = 2 \times 0.03 + 0.01 + a = 0.06 + 0.01 + 0.034$$

$\therefore \quad Xy = 0.208 \quad ...(ii)$

Dividing equations (*i*) and (*ii*), we get

$$\frac{Xx}{Xy} = \frac{2.52}{0.208} \quad \text{or} \quad \frac{x}{y} = 12.1$$

i.e., **Ratio of C to H_2 in fuel** $= \frac{x}{y} = \mathbf{\frac{12.1}{1}}$. **(Ans.)**

INTERNAL ENERGY AND ENTHALPY OF COMBUSTION

Example 10.31. *ΔH_0 (enthalpy of combustion at reference temperature T_0) for benzene vapour (C_6H_6) at 25°C is – 3301000 kJ/mole with the H_2O in the liquid phase. Calculate ΔH_0 for the H_2O in the vapour phase.*

Solution. If H_2O remains as a vapour the heat transferred to the surroundings will be less than when the vapour condenses by the amount due to the change in enthalpy of the vapour during condensation at the reference temperature.

$$\Delta H_0 \text{ (vapour)} = \Delta H_0 \text{ (liquid)} + m_s\, h_{fgo}$$

where, m_s = Mass of H_2O formed, and

h_{fgo} = Change in enthalpy of steam between saturated liquid and saturated vapour at the reference temperature T_0

= 2441.8 kJ at 25°C

For the reaction :

$$C_6H_6 + 7.5O_2 \longrightarrow 6CO_2 + 3H_2O$$

3 moles of H_2O are formed on combustion of 1 mole of C_6H_6 ; 3 moles of H_2O

$$= 3 \times 18 = 54 \text{ kg } H_2O$$

$\therefore$ **ΔH_0 (vapour)** $= -\ 3301000 + 54 \times 2441.8$

$= \mathbf{-\ 3169143\ kJ/mole.}$ **(Ans.)**

Example 10.32. *Calculate ΔU_0 in kJ/kg for the combustion of benzene (C_6H_6) vapour at 25°C given that ΔH_0 = – 3169100 kJ/mole and the H_2O is in the vapour phase.*

Solution. Given : $\Delta H_0 = -\ 3169100$ kJ

The combustion equation is written as

$$C_6H_6 + 7.5O_2 \longrightarrow 6CO_2 + 3H_2O \text{ (vapour)}$$

$$n_R = 1 + 7.5 = 8.5,\ n_P = 6 + 3 = 9$$

Using the relation,

$$\Delta U_0 = \Delta H_0 - (n_P - n_R)R_0T_0$$

$= -\ 3169100 - (9 - 8.5) \times 8.314 \times (25 + 273)$

$= -\ 3169100 - 1239 = -\ 3170339$ kJ/mole

(It may be noted that ΔU_0 is negligibly different from ΔH_0)

1 mole of $C_6H_6 = 6 \times 12 + 1 \times 6 = 78$ kg

$$\therefore \quad \mathbf{\Delta U_0} = \frac{-\ 3170339}{78}$$

$= \mathbf{-\ 40645\ kJ/kg.}$ **(Ans.)**

Example 10.33. *ΔH_0 for CO at 60°C is given as – 285200 kJ/mole. Calculate ΔH_0 at 2500°C given that the enthalpies of gases concerned in kJ/mole are as follows :*

Gas	*60°C*	*2500°C*
CO	*9705*	*94080*
O	*9696*	*99790*
CO_2	*10760*	*149100*

Solution. The reaction equation is given by

$$CO + \tfrac{1}{2} O_2 \longrightarrow CO_2$$

Refer to Fig. 10.7.

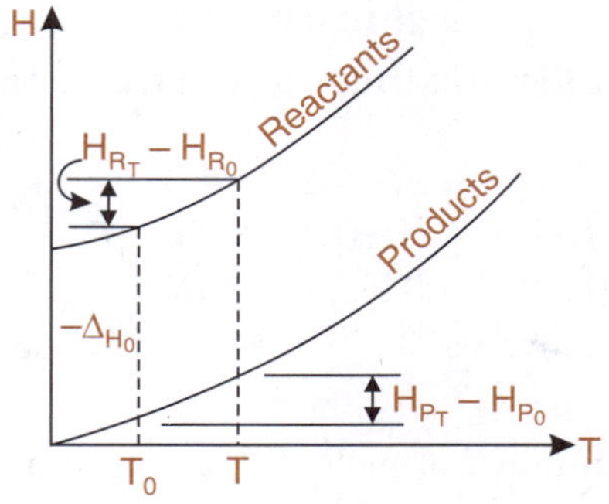

Fig. 10.7

It can be seen from the property diagram of Fig. 10.7 that the enthalpy of combustion at temperature T, ΔH_T can be obtained from ΔH_0 and T_0 by the relationship

$$-\ \Delta H_T = -\ \Delta H_0 + (H_{R_T} - H_{R_0}) - (H_{P_T} - H_{P_0}) \quad ...(i)$$

where $\quad H_{R_T} - H_{R_0}$ = increase in enthalpy of the *reactants* from T_0 to T

and $\quad H_{P_T} - H_{P_0}$ = increase in enthalpy of the *products* from T_0 to T.

Now, from the given data, we have

$$H_{R_0} = 1 \times 9705 + \tfrac{1}{2} \times 9696 = 14553 \text{ kJ}$$

$$H_{R_T} = 1 \times 94080 + \tfrac{1}{2} \times 99790 = 143975 \text{ kJ}$$

$$H_{P_0} = 1 \times 10760 = 10760 \text{ kJ}$$

$$H_{P_T} = 1 \times 149100 \text{ kJ} = 149100 \text{ kJ}$$

Using equation (*i*), we get

$-\ \Delta H_T = +\ 285200 + (143975 - 14553)$

$- (149100 - 10760)$

$= 285200 + 129422 - 138340 = 276282$

$\therefore \quad \mathbf{\Delta H_T = -\ 276282\ kJ/mole.}$ **(Ans.)**

HEATING VALUES OF FUELS

Example 10.34. *The lower heating value of propane at constant pressure and 25°C is 2044009 kJ per kg mole. Find the higher heating value at constant pressure and at constant volume.*

Solution. (*i*) **Higher heating value at constant pressure, $(HHV)_P$:**

The combustion reaction for propane is written as

$$C_3H_8 + 5O_2 = 3CO_2 + 4H_2O$$

Now $(HHV)_p = (LHV)_p + mh_{fg}$

where, $(HHV)_p$ = Higher heating value at constant pressure,

LHV = Lower heating value,

m = Mass of water formed by combustion

= 4 × 18 = 72 kg per kg mole, and

h_{fg} = Latent heat of vaporisation at given temperature per unit mass of water

= 2443 kJ/kg at 25°C.

∴ $\mathbf{(HHV)_p} = 2044009 + 72(2442)$

= **2219833 kJ/kg. (Ans.)**

(*ii*) **Higher heating value at constant volume, $(HHV)_v$:**

Now $(\Delta U) = \Delta H - \Delta n R_0 T$

or $-(HHV)_v = -(HHV)_p - \Delta n R_0 T$

or $(HHV)_v = (HHV)_p + \Delta n R_0 T$

where R_0 = universal gas constant = 8.3143 kJ/kg mol K

$\Delta n = n_P - n_R$

$\left[\begin{array}{l} n_P = \text{number of moles of gaseous products} \\ n_R = \text{number of moles of gaseous reactants} \end{array}\right]$

Now, the reaction for higher heating value is

$$C_3H_8 + 5O_2 = 3CO_2 + 4H_2O \text{ (liquid)}$$

$$\Delta n = 3 - (1 + 5) = -3$$

∴ $\mathbf{(HHV)_v} = 2219905 - 3(8.3143)(25 + 273)$

= **2212472 kJ/kg. (Ans.)**

Example 10.35. *Calculate the lower heating value of gaseous octane at constant volume if $(\Delta U)_{25°C} = -5494977$ kJ for the reaction :*

$$[C_8H_{18}\text{ (gas)} + 12.5O_2\text{ (gas)}] = [8CO_2\text{ (gas)} + 9H_2O\text{ (liquid)}].$$

Solution. The given value of ΔU corresponds to the *higher heating value* at constant volume because the water in the products is in *liquid phase*.

$(HHV)_v$ = 5494977 kJ/kg

$(LHV)_v = (HHV)_v - m(u_g - u_f)$

m = 9 × 18 = 162 kg/kg mole C_8H_{18}

$(u_g - u_f)$ = 2305 kJ/kg at 25°C

∴ $\mathbf{(LHV)_v}$ = 5494977 – 162(2305)

= **5121567 kJ/kg. (Ans.)**

Example 10.36. *Calculate the lower and higher heating values at constant pressure per kg of mixture at 25°C, for the stoichiometric mixtures of :*

(*i*) *Air and benzene vapour (C_6H_6), and*

(*ii*) *Air and octane vapour (C_8H_{18}).*

Given that the enthalpies of combustion at 25°C are :

C_6H_6 = – 3169500 kJ/mole

C_8H_{18} = – 5116200 kJ/mole

Both the above figures are for the case where the water in the products is in the vapour phase.

Solution. (*i*) **Air and benzene vapour :**

For benzene, the combustion equation is as follows :

$$C_6H_6(g) + 7.5O_2(g) + 7.5\left(\frac{79}{21}\right)N_2(g)$$

$$= 6CO_2(g) + 3H_2O(g) + 7.5\left(\frac{79}{21}\right)N_2(g)$$

Since the water in the products is in *vapour phase*, therefore, the given value of enthalpy of combustion corresponds to the *lower heating* value at constant pressure.

i.e., $(LHV)_p$ = 3169500 kJ/mole

$(LHV)_v$ per kg of mixture

$$= \frac{3169500}{(12 \times 6 + 6 \times 1) + (7.5 \times 32) + 7.5\left(\frac{79}{21}\right)(28)}$$

$$= \frac{3169500}{78 + 240 + 790} = \mathbf{2861\ kJ/kg.}\ \textbf{(Ans.)}$$

Now, $(HHV)_p = (LHV)_p + mh_{fg}$

where $(HHV)_p$ = Higher heating value at constant pressure,

$(LHV)_p$ = Lower heating value at constant pressure,

m = Mass of water formed by combustion.

= 3 × 18 = 54 kg/kg mole of fuel, and

h_{fg} = Latent heat of vapourisation at given temperature per unit mass of water

= 2442 kJ/kg at 25°C.

∴ $(HHV)_p$ = 3169500 + 54 × 2442

= 3301368 kJ/mole

Thus, **$(HHV)_p$ per kg of mixture**

$$= \frac{3301368}{78 + 240 + 790} = \mathbf{2980\ kJ/kg.}\ \textbf{(Ans.)}$$

(*ii*) **Air and octane vapour :**

$(LHV)_p$ = 5116200 kJ/mole of C_8H_{18}

For octane, the combustion equation is written as follows :

$$C_8H_{18}(g) + 12.5O_2(g) \longrightarrow 8CO_2(g) + 9H_2O(g) + 12.5\left(\frac{79}{21}\right)N_2(g)$$

$(LHV)_p$ per kg of mixture

$$= \frac{5116200}{(12 \times 8 + 18 \times 1) + 12.5 \times 32 + 12.5 \times \frac{79}{21} \times 28}$$

$$= \frac{5116200}{114 + 400 + 1317} = \textbf{2794 kJ/kg.} \quad \textbf{(Ans.)}$$

$$(HHV)_p = (LHV)_p + mh_{fg}$$

$$m = 9 \times 18 = 162 \text{ kJ/kg mole of fuel}$$

$$\therefore \quad (HHV)_p = 5116200 + 162 \times 2442 = 5511804$$

Hence, **$(HHV)_p$ per kg of mixture**

$$= \frac{5511804}{114 + 400 + 1317} = \textbf{3010 kJ/kg.} \quad \textbf{(Ans.)}$$

☞**Example 10.37.** *The higher heating value of kerosene at constant volume whose ultimate analysis is 88% carbon and 12% hydrogen, was found to be 45670 kJ/kg. Calculate the other three heating values.*

Solution. Combustion of 1 kg of fuel produces the following products :

$$CO_2 = \frac{44}{12} \times 0.88 = 3.23 \text{ kg}$$

$$H_2O = \frac{18}{2} \times 0.12 = 1.08 \text{ kg}$$

At 25°C : $(u_g - u_f)$ *i.e.,* $u_{fg} = 2304$ kJ/kg

$h_{fg} = 2442$ kJ/kg

(*i*) **$(LHV)_v$:**

$$(LHV)_v = (HHV)_v - m(u_g - u_f)$$

$$= 45670 - 1.08 \times 2304 = 43182 \text{ kJ/kg}$$

Hence **$(LHV)_v$ = 43182 kJ/kg. (Ans.)**

(*ii*) **$(HHV)_p$, $(LHV)_p$:**

The combustion equation is written as follows :

$$1 \text{ mole fuel} + \frac{x}{32} O_2 \longrightarrow \frac{3.23}{44} CO_2 + \frac{1.08}{18} H_2O$$

i.e.,
$$\frac{x}{32} = \frac{3.23}{44} + \frac{1.08}{18 \times 2}$$

or
$$x = 3.31 \text{ kg}$$

i.e., 1 kg fuel + 3.31 kg O_2 = $3.23CO_2 + 1.08H_2O$

Also, $\Delta H = \Delta U + \Delta n R_0 T$

i.e., $-(HHV)_p = -(HHV)_v + \Delta n R_0 T$

or $(HHV)_p = (HHV)_v - \Delta n R_0 T$

where $\Delta n = n_p - n_R$

$$= \left(\frac{3.23}{44} - \frac{3.31}{32}\right)$$

$$\begin{bmatrix} n_p = \text{number of moles of gaseous products} \\ n_R = \text{number of moles of gaseous reactants} \end{bmatrix}$$

Since in case of higher heating value, H_2O will appear in liquid phase

$$\textbf{(HHV)}_\textbf{p} = 45670 - \left(\frac{3.23}{44} - \frac{3.31}{32}\right) \times 8.3143 \times (25 + 273)$$

$$= \textbf{45744 kJ/kg.} \quad \textbf{(Ans.)}$$

$$\textbf{(LHV)}_\textbf{p} = (HHV)_p - 1.08 \times 2442$$

$$= 45774 - 1.08 \times 2442$$

$$= \textbf{43107 kJ/kg.} \quad \textbf{(Ans.)}$$

HIGHLIGHTS

1. A *chemical reaction* may be defined as the rearrangement of atoms due to redistribution of electrons. *'Reactants'* comprise of initial constituents which start the reaction while *'products'* comprise of final constituents which are formed by the chemical reaction.
2. A chemical fuel is a substance which releases heat energy on combustion.
3. The total number of atoms of each element concerned in the combustion remains constant, but the atoms are rearranged into groups having different chemical properties.
4. The amount of excess air supplied varies with the type of the fuel and the firing conditions. It may approach a value of 100 per cent but modern practice is to use 25% to 50% excess air.
5. *Stoichiometric* (or chemically correct) *mixture* of air and fuel is one that contains just sufficient oxygen for complete combustion of the fuel.
6. Mixture strength $= \dfrac{\text{Stoichiometric A/F ratio}}{\text{Actual A/F ratio}}$.
7. When analysis of combustion products is known air-fuel ratio can be calculated by the following methods :
   (*a*) Fuel composition known
   (*i*) Carbon balance method
   (*ii*) Hydrogen balance method
   (*iii*) Carbon hydrogen balance method
   (*b*) Fuel composition unknown
   (*i*) Carbon hydrogen balance method.
8. The most common means of analysis of the combustion products is Orsat apparatus.
9. The *enthalpy of formation* (ΔH_f) is the increase in enthalpy when a compound is formed from its constituent elements in their natural form and in a standard state. The standard state is 25°C and 1 atm. pressure (but it must be borne in mind that not all substances can exist in natural form, *e.g.* H_2O cannot be a vapour at 1 atm. and 25°C).
10. $(HHV)_p = (LHV)_p + m\,h_{fg}$
    $(HHV)_v = (LHV)_v + m(u_g - u_f)$

where HHV = Higher heating value,

LHV = Lower heating value,

m = Mass of water formed by combustion,

h_{fg} = Enthalpy of vapourisation of water, kJ/kg,

u_g = Specific internal energy of vapour, kJ/kg, and

u_f = Specific internal energy of liquid, kJ/kg.

11. In a given combustion process, that takes place adiabatically and with no work or changes in kinetic or potential energy involved, the temperature of the products is referred to as the '*adiabatic flame temperature*'.
12. For a given fuel and given pressure and temperature of the reactants, the maximum adiabatic flame temperature that can be achieved is with a 'stoichiometric' mixture.

OBJECTIVE TYPE QUESTIONS

Choose the Correct Answer :

1. The smallest particle which can take part in a chemical change is called
(*a*) atom (*b*) molecule
(*c*) electron (*d*) compound.
2. A chemical fuel is a substance which releases ...... on combustion.
(*a*) chemical energy (*b*) heat energy
(*c*) sound energy (*d*) magnetic energy.
3. The most important solid fuel is
(*a*) wood (*b*) charcoal
(*c*) coal (*d*) all of the above.
4. For each mole of oxygen, number of moles of nitrogen required for complete combustion of carbon are
(*a*) 20/21 (*b*) 2/21
(*c*) 77/21 (*d*) 79/21.
5. Modern practice is to use ...... excess air.
(*a*) 5 to 10 per cent (*b*) 15 to 20 per cent
(*c*) 20 to 25 per cent (*d*) 25 to 50 per cent.
6. Stoichiometric air-fuel ratio by mass for combustion of petrol is.
(*a*) 5 (*b*) 10
(*c*) 12 (*d*) 15.05.
7. An analysis which includes the steam in the exhaust is called
(*a*) dry analysis (*b*) wet analysis
(*c*) dry and wet analysis (*d*) none of the above.
8. The Orsat apparatus gives
(*a*) volumetric analysis of the dry products of combustion
(*b*) gravimetric analysis of the dry products of combustion
(*c*) gravimetric analysis of products of combustion including H_2O
(*d*) volumetric analysis of products of combustion including H_2O.
9. In the Orsat apparatus KOH solution is used to absorb
(*a*) carbon monoxide (*b*) carbon dioxide
(*c*) oxygen (*d*) none of the above.
10. Enthalpy of formation is defined as enthalpy of compounds at
(*a*) 25°C and 10 atmospheres (*b*) 25°C and 1 atmosphere
(*c*) 0°C and 1 atmosphere
(*d*) 100°C and 1 atmosphere.
11. Bomb calorimeter is used to find the calorific value of ...... fuels.
(*a*) solid (*b*) gaseous
(*c*) solid and gaseous (*d*) none of the above.
12. When the fuel is burned and the water appears in the vapour phase, the heating value of fuel is called
(*a*) enthalpy of formation (*b*) lower heating value
(*c*) higher heating value (*d*) none of the above.
13. Heat released in a reaction at constant pressure is called
(*a*) entropy change (*b*) enthalpy of reaction
(*c*) internal energy of reaction
(*d*) none of the above
(*e*) all of the above.
14. When the fuel is burned and water is released in the liquid phase, the heating value of fuel is called
(*a*) higher heating value (*b*) lower heating value
(*c*) enthalpy of formation (*d*) none of the above.
15. Choose the *correct* statement :
(*a*) Number of atoms of each constituent are not conserved in a chemical reaction.
(*b*) The mass of all the substances on one side of the equation may not be equal to the mass of all the substances on the other side.
(*c*) The number of atoms of each constituent are conserved in a chemical reaction.
(*d*) The number of moles of the reactants in a chemical equation are equal to the number of moles of the products.

ANSWERS

1. (*a*)	2. (*b*)	3. (*c*)	4. (*d*)
5. (*d*)	6. (*d*)	7. (*b*)	8. (*a*)
9. (*b*)	10. (*b*)	11. (*a*)	12. (*b*)
13. (*b*)	14. (*a*)	15. (*c*).	

THEORETICAL QUESTIONS

1. What is chemical thermodynamics ?
2. What is a chemical fuel ?
3. What are primary fuels ? List some important primary fuels.
4. What are secondary fuels ? List some important secondary fuels.
5. Write a short note on 'excess air'.
6. What do you mean by stoichiometric air-fuel (A/F) ratio ?
7. Enumerate the methods by which air-fuel ratio can be calculated when analysis of combustion products is known.
8. How is analysis of exhaust and flue gas carried out ?
9. Derive relations for internal energy and enthalpy of reaction.

10. What is enthalpy of formation (ΔH_f) = ?
11. Define heating value of fuel.
12. What is the difference between higher heating value (HHV) and lower heating value (LHV) of the fuel ?
13. Describe with the help of neat sketches the following calorimeters used for the determination of heating values :
    (*i*) Bomb calorimeter (*ii*) Junkers gas calorimeter.
14. What is 'adiabatic flame temperature' ?
15. Write a short note on chemical equilibrium.

UNSOLVED EXAMPLES

1. A fuel has the following composition by weight : Carbon = 86% ; hydrogen = 11.75% and oxygen = 2.25%. Calculate the theoretical air supply per kg of fuel, and the weight of products of combustion per kg of fuel. [**Ans.** 13.98 kg ; 4.21 kg]
2. The volumetric analysis of a fuel gas is : CO_2 = 14% ; CO = 1% ; O_2 = 5% and N_2 = 80%. Calculate the fuel gas composition by weight. [**Ans.** 20.24% ; 0.93% ; 5.25% ; 73.58%]
3. The ultimate analysis of a dry coal burnt in a boiler gauge C = 84% ; H_2 = 9% and incombustibles 7% by weight. Determine the weight of dry flue gases per kg of coal burnt, if volumetric combustion of the flue gas is :
   CO_2 = 8.75%, CO = 2.25% ; O_2 = 80% and N_2 = 81%. [**Ans.** 18.92 kg]
4. During a trial in a boiler, the dry flue gas analysis by volume was obtained as CO_2 = 13%, CO = 0.3%, O_2 = 6%, N_2 = 80.7%. The coal analysis by weight was reported as C = 62.4%, H_2 = 4.2%, O_2 = 4.5%, moisture = 15% and ash 13.9%. Estimate :
   (*a*) Theoretical air required to burn 1 kg of coal.
   (*b*) Weight of air actually supplied per kg of coal.
   (*c*) The amount of excess air supplied per kg of coal burnt. [**Ans.** (*a*) 8.5 kg ; (*b*) 11.5 kg ; (*c*) 3 kg]
5. A steam boiler uses pulverised coal in the furnace. The ultimate analysis of coal (by weight) as received is : C = 78% ; H_2 = 3% ; O_2 = 3% ; ash 10% and moisture 5%. Excess air supplied is 30%. Calculate the weight of air to be supplied and weight of gaseous product formed per kg of coal burnt.
   [**Ans.** 13 kg ; CO_2 = 2.86 kg ; H_2 = 0.27 kg ; excess O_2 = 0.69 kg and N_2 = 9.81 kg per kg of coal]
6. The percentage composition by mass of a crude oil is given as follows : C = 90% ; H_2 = 3.3% ; O_2 = 3%, N_2 = 0.8% ; S = 0.9% and remaining incombustible. If 50% excess air is supplied find the percentage of dry exhaust gases found by volume.
   [**Ans.** CO_2 = 12.7% ; SO_2 = 0.05% ; O_2 = 7%, N_2 = 80.25%]
7. In a boiler trial, the analysis of the coal used is as follows : C = 20%, H_2 = 4.5%, O_2 = 7.5%, remainder—incombustible matter.
   The dry flue gas has the following composition by volume : CO_2 = 8.5%, CO = 1.2%, N_2 = 80.3%, O_2 = 10%. Determine :
   (*i*) Minimum weight of air required per kg of coal.
   (*ii*) Percentage excess air. [**Ans.** (*i*) 3.56 kg, (*ii*) 63.2%]
8. The ultimate analysis of a sample of petrol by weight is : Carbon 0.835 ; hydrogen 0.165. Calculate the ratio of air to petrol consumption by weight, if the volumetric analysis of the dry exhaust gas is : Carbon dioxide 12.1 ; Carbon monoxide 1.1 ; Oxygen 0.8 ; Nitrogen 85.4 per cent. Also find the percentage excess air. [**Ans.** 16.265 : 1 ; 5.6%]
9. The percentage composition of a sample of fuel was found to be C = 85%, H_2 = 9%, S = 3%, O_2 = 1.5%, Ash = 1.5%. For an air-fuel ratio of 12 : 1. Calculate : (*i*) The mixture strength as a percentage rich or weak. (*ii*) The volumetric analysis of the dry products of combustion.
   [**Ans.** 8.83% rich ; CO_2 = 12.16%, CO = 2.54%, N_2 = 84.83%, SO_2 = 0.47%]
10. Determine the gravimetric analysis of the products of complete combustion of acetylene (C_2H_2) with 125 per cent stoichiometric air.
    [**Ans.** CO_2 = 19.5%, H_2O = 3.9%, O_2 = 4.4%, N_2 = 72.2%]
11. One kg of ethane (C_2H_6) is burned with 80% of theoretical air. Assuming complete combustion of the hydrogen in the fuel determine the volumetric analysis of the dry products of combustion. [**Ans.** CO_2 = 4.8%, CO = 11.2%, N_2 = 84%]
12. The gravimetric analysis of a sample of coal is given as 80% C, 12% H_2 and 8% ash. Calculate the stoichiometric A/F ratio and the analysis of the products by volume.
    [**Ans.** CO_2 = 13.6%, H_2 = 12.2%, N_2 = 74.2%]
13. Calculate the stoichiometric air-fuel ratio for the combustion of a sample of dry anthracite of the following composition by mass :
    C = 90 per cent ; H_2 = 3 per cent ; N_2 = 1 per cent ; Sulphur = 0.5 per cent ; ash = 3 per cent.
    If 20 per cent excess air is supplied determine :
    (*i*) Air-fuel ratio
    (*ii*) Wet analysis of the products of combustion by volume.
    [**Ans.** 11.25/1 (*i*) 13.5/1 ; (*ii*) CO_2 = 16.3%, H_2O = 0.03%, SO_2 = 3.51%, N_2 = 80.3%]
14. The following is the analysis of a supply of coal gas :
    H_2 = 49.4 per cent ; CO = 18 per cent ; CH_4 = 20 per cent ; C_4H_8 = 2 per cent ; O_2 = 0.4 per cent ; N_2 = 6.2 per cent ; CO_2 = 4 per cent.
    (*i*) Calculate the stoichiometric A/F ratio.
    (*ii*) Find also the wet and dry analyses of the products of combustion if the actual mixture is 20 per cent weak.
    [**Ans.** (*i*) 4.06/1 by volume ; (*ii*) Wet analysis :
    CO_2 = 9.0%, H_2O = 17.5%, O_2 = 3.08%, N_2 = 70.4%.
    Dry analysis : CO_2 = 10.9%, O_2 = 3.72%, N_2 = 85.4%]
15. Find the stoichiometric air-fuel ratio for the combustion of ethyl alcohol (C_2H_6O), in a petrol engine. Calculate the air-fuel ratios for the extreme mixture strengths of 90% and 120%. Determine also the wet

and dry analyses by volume of the exhaust gas for each mixture strength.

[**Ans.** 8.96/1 ; 9.95/1 ; 7.47/1, Wet analysis : CO_2 = 11.2%, H_2O = 16.8%, O_2 = 1.85%, N_2 = 70.2%
Dry analysis : CO_2 = 13.45%, O_2 = 2.22%, N_2 = 84.4%
Wet analysis : CO_2 = 6.94%, CO = 6.94%, H_2 = 20.8%, N_2 = 65.3%
Dry analysis : CO_2 = 8.7%, CO = 8.7%, N_2 = 82.5%]

16. The chemical analysis of a fuel by weight is as follows :
Carbon = 50 per cent ; Hydrogen = 25 per cent ; Oxygen = 10 per cent ; Sulphur = 5 per cent and Nitrogen = 10 per cent.
Find the stoichiometric amount of air required for complete combustion of this fuel. [**Ans.** 14.26 kg]

17. The percentage composition of a fuel by weight is as follows :
Carbon = 89.3 per cent ; Hydrogen = 5 per cent ; Oxygen = 4.2 per cent ; Nitrogen = 1.5 per cent and the remainder ash. Determine the stoichiometric air-fuel ratio by mass.
If 30 per cent excess air is supplied, find the percentage composition of dry flue gases by volume.
[**Ans.** 11.74 ; CO_2, = 14.3%, O_2 = 4.9%, N_2 = 80.8%]

18. Orsat analysis of the products of combustion of hydrocarbon fuel of unknown composition is as follows :
Carbon dioxide (CO_2) = 9%
Carbon monoxide (CO) = 0.6%
Oxygen (O_2) = 7.3%
Nitrogen (N_2) = 83.1%
Determine the following :
(*i*) Air-fuel ratio
(*ii*) Per cent theoretical air required for combustion.
[**Ans.** (*i*) 22.1, (*ii*) 146.2%]

19. An Orsat analysis of the exhaust from an engine running on benzole showed a CO_2 content of 15 per cent, but no CO. Assuming that the remainder of the exhaust contains only oxygen and nitrogen, calculate the air-fuel ratio of the engine.
The ultimate analysis of benzole is C = 90 per cent and H_2 = 10%. [**Ans.** 15.2/1]

20. The analysis of the dry exhaust from and internal-combustion engine gave :
CO_2 = 12 per cent ; CO = 2 per cent ; CH_4 = 4 per cent ; H_2 = 1 per cent ; O_2 = 4.5 per cent and the remainder nitrogen.
Calculate the proportions by mass of carbon to hydrogen in the fuel, assuming it to be a pure hydrocarbon.
[**Ans.** 7.35/1]

21. The following is the percentage analysis by mass of a fuel :
Hydrogen (H_2) = 10 per cent
Oxygen (O_2) = 2 per cent
Sulphur (S) = 1 per cent
Nitrogen (N_2) = 3 per cent
Determine the following :
(*i*) The amount of air required to completely burn 1 kg of this fuel.
(*ii*) The products of combustion as a percentage by mass.
[**Ans.** 13.17 kg ; CO_2 = 21.7% ; H_2O = 6.35% ; SO_2 = 0.141% ; N_2 = 71.75%]

22. An Orsat analysis of the products of combustion resulting from the burning in air of a hydrocarbon fuel yielded the following :
Carbon dioxide (CO_2) = 12.2% Oxygen (O_2) = 1.1%
Carbon monoxide (CO) = 0.5% Nitrogen (N_2) = 86.2%
Determine : (*i*) The mass fraction of carbon in the fuel.
(*ii*) Air-fuel ratio.
(*iii*) Per cent of air theoretically needed for complete combustion.
[**Ans.** (*i*) 80.4%, (*ii*) 16.57/1, (*iii*) 103.9%]

23. If the higher heating value at constant pressure (C_6H_6) at 25°C is 3298354 kJ/kg mole, determine its lower calorific value at constant pressure.
[**Ans.** 316647 kJ/mole]

24. The lower heating value of propane at constant pressure and 25°C is 2042055 kJ/kg mole. Find the higher heating value at constant pressure and at constant volume.
[**Ans.** 2217816 kJ/kg mole, 2210333 kJ/kg mole]

25. The higher heating value of kerosene at constant volume whose ultimate analysis is 86% carbon and 14% hydrogen, was found to be 46890 kJ/kg. Calculate the other three heating values.
[**Ans.** $(LHV)_v$ = 43987 kJ/kg ; $(HHV)_p$ = 46977 kJ/kg ; $(LHV)_p$ = 43900 kJ/kg]

11 Steam Generators

11.1. INTRODUCTION

In simple words a boiler may be defined as *a closed vessel in which steam is produced from water by combustion of fuel*.

According to American Society of Mechanical Engineers (A.S.M.E.) a *'steam generating unit'* is defined as :

"A combination of apparatus for producing, furnishing or recovering heat together with the apparatus for transferring the heat so made available to the fluid being heated and vapourised".

The steam generated is employed for the following *purposes* :

(*i*) For generating power in steam engines or steam turbines.

(*ii*) In the textile industries for sizing and bleaching etc., and many other industries like sugar mills ; chemical industries.

(*iii*) For heating the buildings in cold weather and for producing hot water for hot water supply.

The *primary requirements* of steam generators or boilers are :

(*i*) The water must be contained safely.

(*ii*) The steam must be safely delivered in desired condition (as regards its pressure, temperature, quality and required rate).

11.2. CLASSIFICATION OF BOILERS

The boilers may be classified as follows :

1. Horizontal, Vertical or Inclined

If the axis of the boiler is horizontal, the boiler is called as *horizontal*, if the axis is vertical, it is called *vertical* boiler and if the axis is inclined it is known as *inclined boiler*. *The parts of a horizontal boiler can be inspected and repaired easily but it occupies more space. The vertical boiler occupies less floor area.*

2. Fire Tube and Water Tube

In the fire tube boilers, the hot gases are inside the tubes and the water surrounds the tubes. Examples : *Cochran, Lancashire* and *Locomotive boilers*.

In the water tube boilers, the water is inside the tubes and hot gases surround them. Examples : *Babcock* and *Wilcox, Stirling, Yarrow boiler etc.*

3. Externally Fired and Internally Fired

The boiler is known as externally fired if the fire is outside the shell. Examples : *Babcock and Wilcox boiler, Stirling boiler etc.*

In case of internally fired boilers, the furnace is located inside the boiler shell. Examples : *Cochran, Lancashire boiler etc.*

4. Forced Circulation and Natural Circulation

In *forced circulation* type of boilers, the circulation of water is done *by a forced pump*. Examples : *Velox, Lamont, Benson boiler etc.*

In *natural circulation* type of boilers, circulation of water in the boiler takes place due to *natural convention* currents produced by the application of heat. Examples : *Lancashire, Babcock* and *Wilcox boiler* etc.

5. High Pressure and Low Pressure Boilers

The boilers which produce steam at *pressures of 80 bar and above* are called *high pressure boilers*. Examples : *Babcock* and *Wilcox, Velox, Lamont, Benson boilers.*

The boilers which produce steam at pressure *below 80 bar* are called low *pressure boilers.* Examples : *Cochran, Cornish, Lancashire and Locomotive boilers.*

6. Stationary and Portable

Primarily, the boilers are classified as either stationary (*land*) or mobile (*marine* and *locomotive*).

— Stationary boilers are used for power plant-steam, for central station utility power plants, for plant process steam etc.

— Mobile boilers or portable boilers include locomotive type, and other small units for temporary use at sites (just as in small coal-field pits).

7. Single Tube and Multi-tube Boilers

The fire tube boilers are classified as single tube and multi-tube boilers, depending upon whether the fire tube is one or more than one. The examples of the former type are cornish, simple vertical boiler and rest of the boilers are multi-tube boilers.

11.3. COMPARISON BETWEEN 'FIRE-TUBE' AND 'WATER-TUBE' BOILERS

S. No.	Particulars	Fire-tube boilers	Water-tube boilers
1.	*Position of water and hot gases*	Hot gases inside the tubes and water outside the tubes.	Water inside the tubes and hot gases outside the tubes.
2.	*Mode of firing*	Generally internally fired.	Externally fired.
3.	*Operating pressure*	Operating pressure limited to 16 bar.	Can work under as high pressure as 100 bar.
4.	*Rate of steam production*	Lower	Higher.
5.	*Suitability*	Not suitable for large power plants.	Suitable for large power plants.
6.	*Risk on bursting*	Involves lesser risk on explosion due to lower pressure.	Involves more risk on bursting due to high pressure.
7.	*Floor area*	For a given power it occupies more floor area.	For a given power it occupies less floor-area.
8.	*Construction*	Difficult	Simple
9.	*Transportation*	Difficult	Simple
10.	*Shell diameter*	Large for same power	Small for same power
11.	*Chances of explosion*	Less	More
12.	Treatment of water	Not so necessary	More necessary
13.	*Accessibility of various parts*	Various parts not so easily accessible for cleaning, repair and inspection.	Various parts are more accessible.
14.	*Requirement of skill*	Require less skill for efficient and economic working.	Require more skill and careful attention.

11.4. SELECTION OF A BOILER

While selecting a boiler the following *factors* should be considered :

1. The working pressure and quality of steam required (*i.e.*, whether wet or dry or superheated).
2. Steam generation rate.
3. Floor area available.
4. Accessibility for repair and inspection.
5. Comparative initial cost.
6. Erection facilities.
7. The portable load factor.
8. The fuel and water available.
9. Operating and maintenance costs.

11.5. ESSENTIALS OF A GOOD STEAM BOILER

A good boiler should possess the following *features* :

1. The boiler should produce the maximum weight of steam of the required quality at minimum expenses.
2. Steam production rate should be as per requirements.
3. It should be absolutely reliable.
4. It should occupy minimum space.
5. It should be light in weight.
6. It should be capable of quick starting.
7. There should be an easy access to the various parts of the boiler for repairs and inspection.
8. The boiler components should be transportable without difficulty.
9. The installation of the boiler should be simple.
10. The tubes of the boiler should not accumulate soot or water deposits and should be sufficiently strong to allow for wear and corrosion.
11. The water and gas circuits should be such as to allow minimum fluid velocity (for low frictional losses).

11.6. BOILER TERMS

Shell. The shell of a boiler consists of one or more steel plates bent into a cylindrical form and riveted or welded together. The shell ends are closed with the end plates.

Setting. The primary function of setting is to confine heat to the boiler and form a passage for gases. It is made of brickwork and may form the wall of the furnace and the combustion chamber. It also provides support in some types of boilers (*e.g.*, Lancashire boilers).

Grate. It is the platform in the furnace upon which fuel is burnt and it is made of cast iron bars. The bars are so arranged that air may pass on to the fuel for combustion. The area of the grate on which the fire rests in a coal or wood fired boiler is called *grate surface*.

Furnace. It is a chamber formed by the space above the grate and below the boiler shell, in which combustion takes place. It is also called a *fire-box*.

Water space and steam space. The volume of the shell that is occupied by the water is termed *water space* while the entire shell volume less the water and tubes (if any) space is called *steam space*.

Mountings. The items such as stop valve, safety valves, water level gauges, fusible plug, blow-off cock, pressure gauges, water level indicator etc., are termed as *mountings* and a *boiler cannot work safely without them*.

Accessories. The items such as superheaters, economisers, feed pumps etc., are termed as *accessories* and they form integral part of the boiler. They *increase the efficiency of the boiler*.

Water level. The level at which water stands in the boiler is called *water level*. The space above the water level is called *steam space*.

Foaming. Formation of steam bubbles on the surface of boiler water due to high surface tension of the water.

Scale. A deposit of medium to extreme hardness occurring on water heating surfaces of a boiler because of an undesirable condition in the boiler water.

Blowing off. The removal of the mud and other impurities of water from the lowest part of the boiler (where they usually settle) is termed as *'blowing off'*. This is accomplished with the help of a blow off cock or valve.

Lagging. Blocks of asbestos or magnesia insulation wrapped on the outside of a boiler shell or steam piping.

Refractory. A heat insulation material, such as fire brick or plastic fire clay, used for such purposes as lining combustion chambers.

11.7. FIRE TUBE BOILERS

The various fire tube boilers are described as follows :

11.7.1. Simple Vertical Boiler

Refer to Fig. 11.1. It consists of a cylindrical shell, the greater portion of which is full of water (which surrounds the fire box also) and remaining is the steam space. At the bottom of the fire box is grate on which fuel is burnt and the ash from it falls in the ash pit.

The fire box is provided with two cross tubes. This increases the heating surface and the circulation of water. The cross tubes are fitted inclined. This ensures efficient circulation of water. At the ends of each cross tube are provided hand holes to give access for cleaning these tubes. The combustion gases after heating the water and thus converting it into steam escape to the atmosphere through the chimney. Man hole, is provided to clean the interior of the boiler and exterior of the combustion chamber and chimney. The various mountings shown in Fig. 11.1 are (*i*) Pressure gauge, (*ii*) Water level gauge or indicator, (*iii*) Safety valve, (*iv*) Steam stop valve, (*v*) Feed check valve, and (*vi*) Man hole.

Flow of combustion gases and circulation of water in water jackets are indicated by arrows in Fig. 11.1.

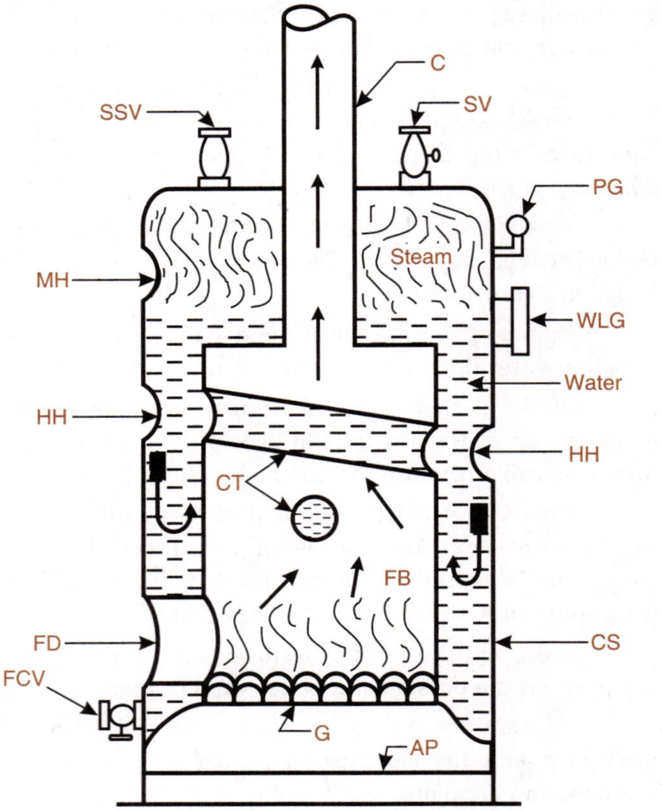

CS = Cylindrical shell
MH = Man hole
CT = Cross tubes
G = Grate
PG = Pressure gauge
SV = Safety valve
WLG = Water level gauge
C = Chimney
HH = Hand hole
FD = Fire door
FB = Fire box
AP = Ash pit
SSV = Steam stop valve
FCV = Feed check valve

Fig. 11.1. Simple vertical boiler.

The rate of production in such a boiler normally does not exceed 2500 kg/hr and pressure is normally limited to 7.5 to 10 bar.

A simple vertical boiler is self-contained and can be transported easily.

11.7.2. Cochran Boiler

It is *one of the best* types of vertical multi-tubular boiler, and has a number of horizontal fire tubes.

Dimensions, working pressure, capacity, heating surface and efficiency are given below :

Shell diameter	......	2.75 m
Height	......	5.79 m
Working pressure	......	6.5 bar (max. pressure = 15 bar)
Steam capacity	......	3500 kg/hr (max. capacity = 4000 kg/hr)
Heating surface	......	120 m^2
Efficiency	......	70 to 75% (depending on the fuel used)

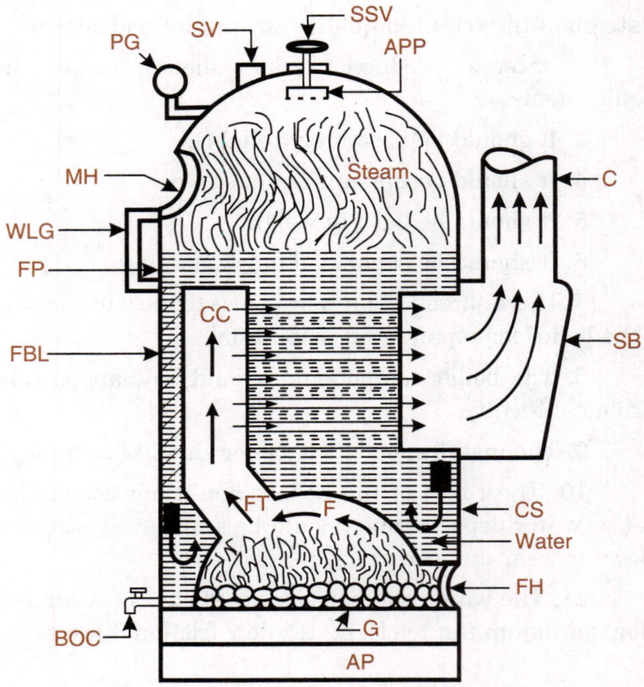

CS = Cylindrical shell
CC = Combustion chamber
FBL = Fir brick lining
F = Furnace (dome shaped)
BOC = Blow off cock
SSV = Steam stop valve
APP = Antipriming pipe
PG = Pressure gauge
FT = Flue tube
SB = Smoke box
C = Chimney
FH = Fire hole
G = Grate
AP = Ash pit
SV = Safety valve
MH = Man hole
WLG = Water level gauge

Fig. 11.2. Cochran boiler.

Cochran boiler consists of a cylindrical shell with a dome shaped top where the space is provided for steam. The furnace is one piece construction and is seamless. Its crown has a hemispherical shape and thus provides maximum volume of space. The fuel is burnt on the grate and ash is collected and disposed of from ash pit. The gases of combustion produced by burning of fuel enter the combustion chamber through the flue tube and strike against fire brick lining which directs them to pass through number of horizontal tubes, being surrounded by water. After which the gases escape to the atmosphere through smoke box and chimney. A number of hand-holes are provided around the outer shell for cleaning purposes.

The various boiler mountings shown in Fig. 11.2 are : (*i*) Water level gauge, (*ii*) Safety valve, (*iii*) Steam stop valve, (*iv*) Blow off cock, (*v*) Man hole and, (*vi*) Pressure gauge.

The path of combustion of gases and circulation of water are shown by arrows in Fig. 11.2.

11.7.3. Cornish Boiler

This form of boiler was first adopted by Trevithick, the Cornish engineer, at the time of introduction of high-pressure steam to the early Cornish engine, and is still used.

The *specifications* of Cornish boiler are given below :

No. of flue tubes	......	One
Diameter of the shell	......	1.25 to 1.75 m
Length of the shell	......	4 to 7 m
Pressure of the steam	......	10.5 bar
Steam capacity	......	6500 kg/h.

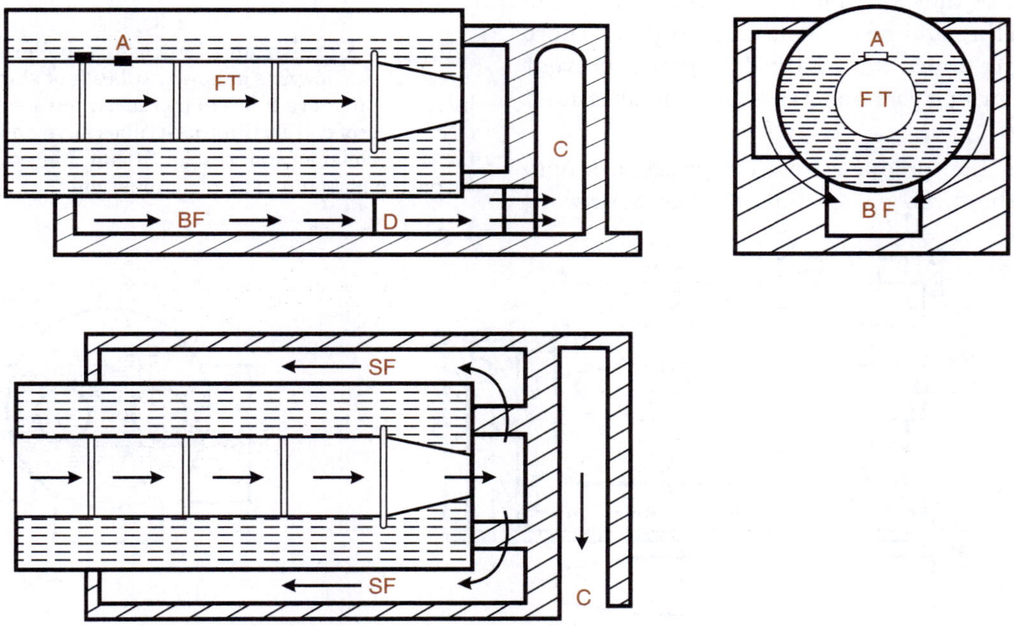

A = Fusible plug
BF = Bottom flue
D = Damper
FT = Flue tube
SF = Side flue
C = Passage to chimney

Fig. 11.3. Cornish boiler.

Refer to Fig. 11.3. It consists of a cylindrical shell with flat ends through which passes a smaller flue tube containing the furnace. The products of combustion pass from the fire grate forward over the brickwork bridge to the end of the furnace tube ; they then return by the two side flues to the front end of the boiler, and again pass to the back end of a flue along the bottom of the boiler to the chimney.

The various boiler mountings which are used on this boiler are : (*i*) Steam stop valve, (*ii*) Pressure gauge, (*iii*) Water gauge, (*iv*) Fusible plug, (*v*) Blow off cock, (*vi*) High steam low water safety valve, (*vii*) Feed check valve and (*viii*) Man hole.

The advantage possessed by this type of boiler is that the sediment contained in the water falls to the bottom, where the plates are not brought into contact with the hottest portion of the furnace gases. The reason for carrying the product of combustion first through the side flues, and lastly through the bottom flue, is because the gases, having parted with much of their heat by the time they reach the bottom flue, are less liable to unduly heat the plates in the bottom of the boiler, where the sediment may have collected.

11.7.4. Lancashire Boiler

This boiler is *reliable,* has *simplicity of design, ease of operation* and *less operating and maintenance costs.* It is commonly used in *sugar-mills* and *textile industries* where alongwith the power steam and steam for the process work is also needed. In addition this boiler is used where larger reserve of water and steam are needed.

The *specifications* of Lancashire boiler are given below :

Diameter of the shell	......	2 to 3 m
Length of the shell	......	7 to 9 m
Maximum working pressure	......	16 bar
Steam capacity	......	9000 kg/h
Efficiency	......	50 to 70%

Refer to Fig. 11.4. The Lancashire boiler consists of a cylindrical shell inside which two large tubes are placed. The shell is constructed with several *rings* of cylindrical from and it is placed horizontally over a brickwork which forms several channels for the flow of hot gases. These two tubes are also constructed with several rings of cylindrical form. They pass from one and of the shell to the other and are covered with water. The furnace is placed at the front end of each tube and they are known as furnace tubes. The coal is introduced through the fire hole into the grate. There is low brickwork fire bridge at the back of the gate to prevent the entry of the burning coal and ashes into the interior of the furnace tubes.

The combustion products from the grate pass upto the back end of the furnace tubes and then in downward direction. Thereafter they move through the bottom channel or bottom flue upto the front end of the boiler where they are divided and pass upto the side flues. Now they move along the two side flues and come to the chimney flue from where they lead to the chimney. To control the flow of hot gases to the chimney, dampers (in the form of sliding doors) are provided. As a result the flow of air to the grate can be controlled. The various mountings used on the boiler are shown in Fig. 11.4.

Note. In Cornish and Lancashire boilers, conical shaped cross tubes known as galloway tubes (not shown) may be fitted inside the furnace tubes to increase their heating surfaces and circulation of water. But these tubes have now become absolete for their considerable cost of fitting. Moreover, they cool the furnace gases and retard combustion.

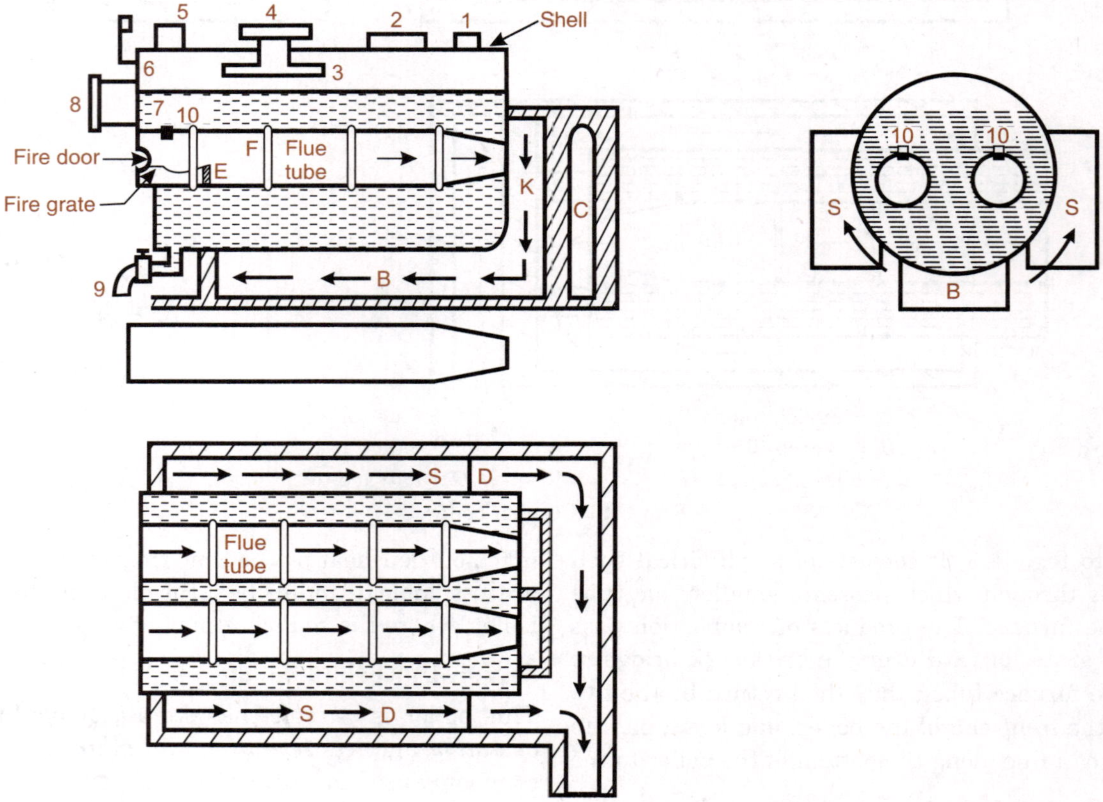

B = Bottom flue
C = Chimney
D = Dampers
E = Fire bridge
F = Flue tube
K = Main flue
S = Side flue

1. High steam low water safety valve
2. Man hole
3. Antipriming pipe
4. Steam stope valve
5. Safety valve
6. Pressure gauge
7. Feed check valve
8. Water gauge
9. Blow down cock
10. Fusible plug

Fig. 11.4. Lancashire boiler.

11.7.5. Locomotive Boiler

It is mainly employed in locomotives though it may also be used as a stationary boiler. It is compact and its capacity for steam production is quite high for its size as it can raise large quantity of steam rapidly.

Dimensions and the specifications of the locomotive boilers (made at Chitranjan works in India) are given below :

Barrel diameter	......	2.095 m
Length of the barrel	......	5.206 m
Size of the tubes (superheater)	......	14 cm
No. of superheater tubes	......	38
Size of ordinary tubes	......	5.72 cm
No. of ordinary tubes	......	116
Steam capacity	......	9000 kg/h
Working pressure	......	14 bar
Grate Area	......	4.27 m^2
Coal burnt/hr	......	1600 kg
Heating surface	......	271 m^2
Efficiency	......	70%

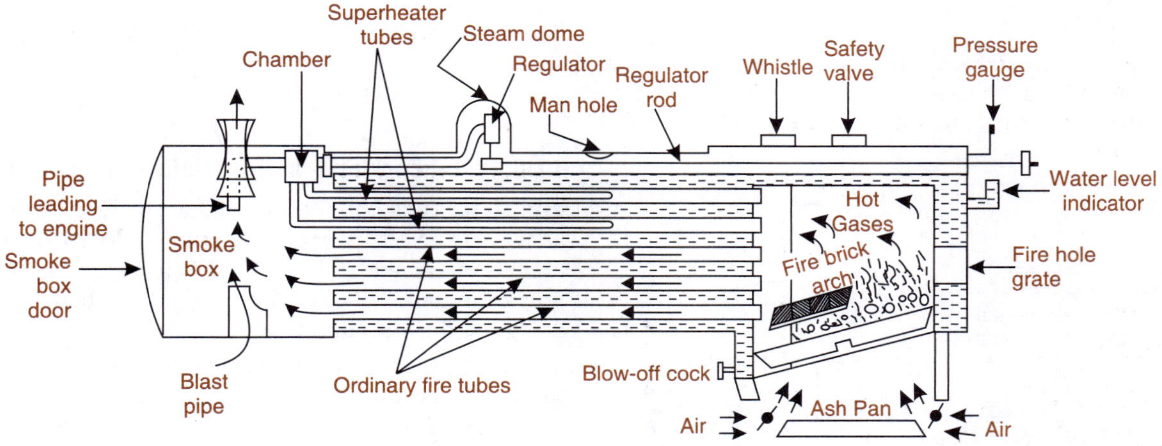

Fig. 11.5. Locomotive boiler.

Refer to Fig. 11.5. The locomotive boiler consists of a cylindrical barrel with a rectangular fire box at one end and a smoke box at the other end. The coal is introduced through the fire hole into the grate which is placed at the bottom of the fire box. The hot gases which are generated due to burning of the coal are deflected by an arch of fire bricks, so that walls of the fire box may be heated properly. The fire box is entirely surrounded by water except for the fire hole and the ash pit which is situated below the fire box which is fitted with dampers at its front and back ends. The dampers control the flow of air to the grate. The hot gases pass from the fire box to the smoke box through a series of fire tubes and then they are discharged into the atmosphere through the chimney. The fire tubes are placed inside the barrel. Some of these tube are of larger diameter and the others of smaller diameter. *The superheater tubes are placed inside the fire tubes of larger diameter*. The heat of the hot gases is transmitted into the water through the heating surface of the fire tubes. The steam generated is collected over the water surface.

A dome shaped chamber known as *steam dome* is fitted on the upper part of the barrel, from where the steam flows through a steam pipe into the chamber. The flow of steam is regulated by means of a regulator. From the chamber it passes through the superheater tubes and returns to the superheated steam chamber (not shown) from which it is led to the cylinders through the pipes, one to each cylinder.

In this boiler natural draught cannot be obtained because it requires a very high chimney which cannot be provided on a locomotive boiler since it has to run on rails. Thus some artificial arrangement has to be used to produce a correct draught. As such the draught here is *produced by exhaust steam* from the cylinder which is discharged through the blast pipe to the chimney. When the *locomotive is standing and no exhaust steam is available from the engine fresh steam from the boiler is used for the purpose.*

The various boiler mountings include :

Safety valves, pressure gauge, water level indicator, fusible plug, man hole, blow-off cock and feed check valve.

A locomotive boiler entails the following merits and demerits :

Merits :

1. High steam capacity.
2. Low cost of construction.
3. Portability.
4. Low installation cost.
5. Compact.

Demerits :

1. There are chances to corrosion and scale formation in the water legs due to the accumulation of sediments and the mud particles.
2. It is difficult to clean some water spaces.
3. Large flat surfaces need bracing.
4. It cannot carry high overloads without being damaged by overheating.
5. There are practical constructional limits for pressure and capacity which do not meet requirements.

11.7.6. Scotch Boiler

The scotch type marine boiler is probably the *most popular* boiler for steaming capacities upto about 1000 kg/hr and pressure of about 17 bar. It is of compact size and occupies small floor space.

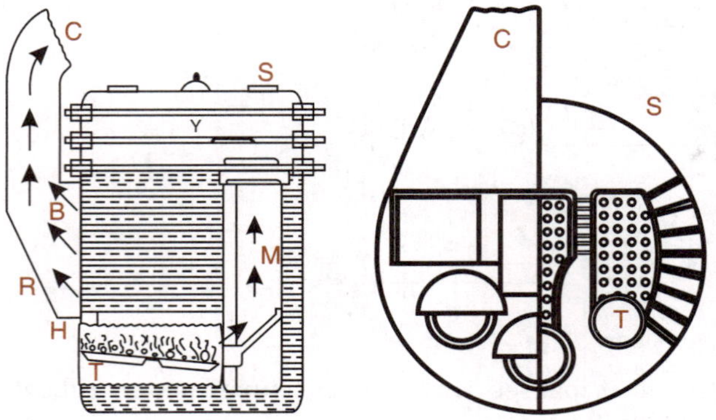

T = Furnace tube
H = Fire hole
M = Combustion chamber
B = Smoke box
R = Smoke box door
C = Chimney
S = Shell
Y = Bar stays

Fig. 11.6. Scotch boiler.

Fig. 11.6 shows a single ended scotch type marine boiler. It consists of a cylindrical shell in which are incorporated one to four cylindrical, corrugated steel furnaces. The furnaces are internally fired and surrounded by water. A combustion chamber is located at the back end of the furnace and is also surrounded by water. Usually each furnace has its own combustion chamber. A nest of fire tubes run from the front tube plate to the back tube plate.

The hot gases produced due to burning of fuel move to the combustion chambers (by means of the draught). Then they travel to the smoke box through the fire tubes and finally leave the boiler *via* uptake and the chimney.

In a double ended scotch boiler furnaces are provided at each end. They look like single ended boilers placed back to back. A double ended boiler for same evaporation capacity, is cheaper and occupies less space as compared to single ended boiler.

11.8. WATER TUBE BOILERS

The types of water tube boilers are given below :

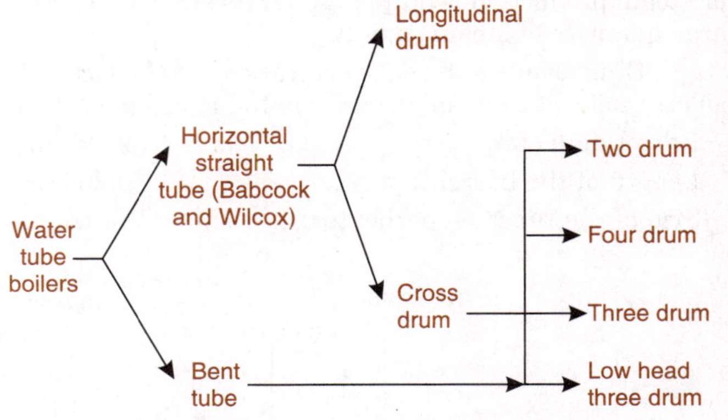

11.8.1. Babcock and Wilcox Water-tube Boiler

The water tube boilers are used exclusively, when pressure above 10 bar and capacity in excess of 7000 kg of steam per hour is required. Babcock and Wilcox water-tube boiler is an example of horizontal straight tube boiler and may be designed for stationary or marine purposes.

The *particulars* (dimensions, capacity etc.) relating to this boiler are given below :

Diameter of the drum	......	1.22 to 1.83 m
Length	......	6.096 to 9.144 m
Size of the water tubes	......	7.62 to 10.16 cm
Size of superheater tubes	......	3.84 to 5.71 cm
Working pressure	......	40 bar (max.)
Steaming capacity	......	40000 kg/h (max.)
Efficiency	......	60 to 80%

Fig. 11.7 shows a Babcock and Wilcox boiler with longitudinal drum. It consists of a drum connected to a series of front end and rear end header by short riser tubes. To these headers are connected a series of inclined water tubes of solid drawn mild steel.

The angle of inclination of the water tubes to the horizontal is about 15° or more. A hand hole is provided in the header in front of each tube for cleaning and inspection of tubes. A feed valve is provided to fill the drum and inclined tubes with water the level of which is indicated by the water level indicator. Through the fire door the fuel is supplied to grate where it is burnt. The hot gases are forced to move upwards between the tubes by baffle plates provided. The water from the drum flows through the inclined tubes *via* downtake header and goes back into the shell in the form of water and steam *via* uptake header. The steam gets collected in the steam space of the drum. The steam then enters through the antipriming pipe and flows in the superheater tubes where it is further heated and is finally taken out through the main stop valve and supplied to the engine when needed.

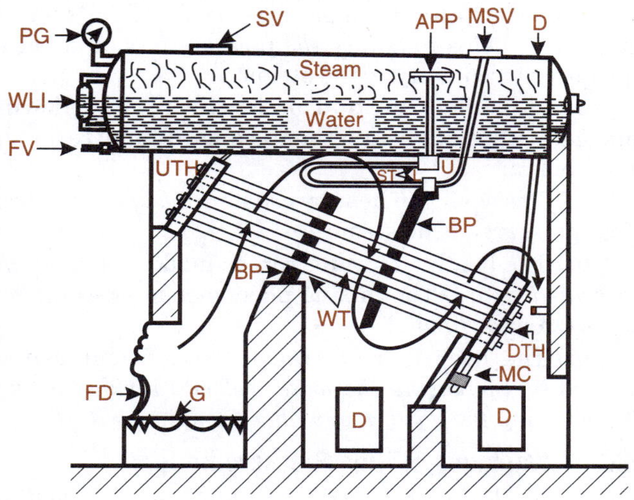

D = Drum
DTH = Down take header
WT = Water tubes
BP = Baffle plates
D = Doors
G = Grate
FD = Fire door
MC = Mud collector
WLI = Water level indicator
PG = Pressure gauge
ST = Superheater tubes
SV = Safety valve
MSV = Main stop valve
APP = Antipriming pipe
L = Lower junction box
U = Upper junction box
FV = Feed valve

Fig. 11.7. Babcock and Wilcox boiler.

At the lowest point of the boiler is provided a mud collector to remove the mud particles through a blow-down-cock.

The entire boiler except the furnace are hung by means of metallic slings or straps or wrought iron girders supported on pillars. This arrangement enables the drum and the tubes to expand or contract freely. The brickwork around the boiler encloses the furnace and the hot gases.

The various mountings used on the boiler are shown in Fig. 11.7.

A Babcock Wilcox water tube boiler with cross draw differs from longitudinal drum boiler in a way that how drum is placed with reference to the axis of the water tubes of the boiler. The longitudinal drum restricts the number of tubes that can be connected to one drum circumferentially and limits the capacity of the boiler. In the cross drum there is no limitation of the number of connecting tubes.

The pressure of steam in case of cross drum boiler may be as high as 100 bar and steaming capacity upto 27000 kg/h.

11.8.2. Stirling Boiler

Stirling water tube boiler is an example of *bent tube* boiler. The main elements of a bent type water tube boiler are essentially drum or drums and headers connected by bent tubes. For large central power stations these boilers are very popular. They have steaming capacities as high as 50000 kg/h and pressure as high as 60 bar.

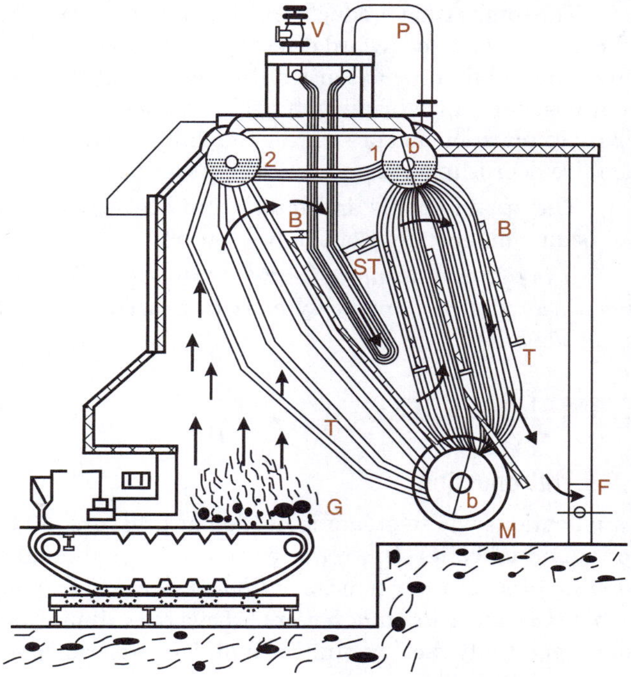

V = Stop valve
P = Steam pipe
b = Water baffle
B = Baffle wall
ST = Superheater tubes
T = Water tubes
G = Grate
M = Mud drum

Fig. 11.8. Stirling boiler.

Fig. 11.8 shows a small-sized stirling water tube boiler. It consists of two upper drums known as *steam drums* and a lower drum known as *mud or water drum*. The steam drums are connected to mud drum by banks of bent tubes. The steam and water space of the steam drums are interconnected with each other, so that balance of water and steam may be obtained. For carrying out cleaning operation a man hole at one end of each drum is provided. The feed water from the economiser (not shown) is delivered to the steam drum-1 which is fitted with a baffle. The baffle deflects the water to move downwards into the drum. The water flows from the drum 1 to the mud drum through the rearmost water tubes at the backside. So the mud particles and other impurities will move to the mud drum, where these particles may be deposited. As this drum is not subjected to high temperature, so the impurities may not cause harm to the drum. The blow-off cock blows off the impurities. The baffle provided at the mud drum deflects the pure water to move upwards to the drum 1 through the remaining half of the water tubes at the back. The water also flows from it to the drum 2 through the water tubes which are just over the furnace. So they attain a higher temperature than the remaining portion of the boiler and a major portion of evaporation takes place in these tubes. The steam is taken from the drum 1 through a steam pipe and then it passes through the superheater tubes where the steam is superheated. Finally the steam moves to the stop valve from where it can be supplied for further use.

The combustion products ensuing from the grate move in the upward and downward directions due to the brickwall baffles and are finally discharged through the chimney into the atmosphere. Fire brick arch gets incandescent hot and helps in combustion and preventing the chilling of the furnace when fire door is opened and cold air rushes in.

The steam drums and mud drum are supported on steel beams independent of the brick-work.

It is lighter and more flexible than the straight tube boilers. But it is comparatively more difficult to clean and inspect the bent tubes.

11.9. HIGH PRESSURE BOILERS

11.9.1. Introduction

In applications where steam is needed at pressure, 30 bar, and individual boilers are required to raise less than about 30000 kg of steam per hour, *shell boilers are considerably cheaper than the water tube boilers.* Above these limits, shell boilers (generally factory built) are difficult to transport if not impossible. There are no such limits to water tube boilers. These can be site erected from easily transportable parts, and moreover the pressure parts are of smaller diameter and therefore can be thinner. The geometry can be varied to suit a wide range of situations and furnace is not limited to cylindrical form. Therefore, *water tube boilers are generally preferred for high pressure and high output whereas shell boilers for low pressure and low output.*

The modern high pressure boilers employed for power generation are for steam capacities 30 to 650 tonnes/h and above with a pressure upto 160 bar and maximum steam temperature of about 540°C.

11.9.2. Unique Features of the High Pressure Boilers

Following are the unique features of high pressure boilers :

1. Method of water circulation
2. Type of tubing
3. Improved method of heating.

1. Method of water circulation. The circulation of water through the boiler may be *natural circulation* due to density difference or *forced circulation.* In all modern high pressure boiler plants, the water circulation is maintained with the help of *pump* which forces the water through the boiler plant. The *use of natural circulation is limited to sub-critical boilers due to its limitations.*

2. Type of tubing. In most of the high pressure boilers, the water circulated through the tubes and their external surfaces are exposed to the flue gases. In water tube boilers, if the flow takes place through one continuous tube, the large pressure drop takes place due to friction. This is *considerably reduced by arranging the flow to pass through parallel system of tubing.* In most of the cases, several sets of the tubings are used. This type of arrangement helps to *reduce the pressure loss,* and better control over the quality of the steam.

3. Improved method of heating. The following improved methods of heating may be used to increase the heat transfer :

(*i*) The saving of heat by *evaporation of water* above critical pressure of the steam.

(*ii*) The heating of water can be made by *mixing the superheated steam.* The mixing phenomenon *gives highest heat transfer co-efficient.*

(*iii*) The overall heat transfer co-efficient can be increased by *increasing the water velocity inside the tube and increasing the gas velocity above sonic velocity.*

11.9.3. Advantages of High Pressure Boilers

The following are the advantages of high pressure boilers.

1. In high pressure boilers pumps are used to maintain forced circulation of water through the tubes of the boiler. This *ensures positive circulation of water and increases evaporative capacity of the boiler and less number of steam drums will be required.*
2. The *heat of combustion is utilised more efficiently* by the use of small diameter tube in large number and in multiple circuits.
3. *Pressurised combustion* is used which *increases rate of firing of fuel* thus increasing the rate of heat release.
4. Due to compactness less floor space is required.
5. The tendency of scale formation is eliminated due to high velocity of water through the tubes.
6. All the parts are uniformly heated, therefore the danger of overheating is reduced and thermal stress problem is simplified.
7. The differential expansion is reduced due to uniform temperature and this reduces the possibility of gas and air leakages.
8. The components can be arranged horizontally as high head required for natural circulation is eliminated using forced circulation. There is a greater flexibility in the components arrangement.
9. The steam can be raised quickly to meet the variable load requirements without the use of complicated control devices.
10. The *efficiency of plant is increased upto 40 to 42 per cent* by using high pressure and high temperature steam.
11. A very rapid start from cold is possible if an external supply of power is available. Hence the boiler can be used for carrying peak loads or stand by purposes with hydraulic station.
12. Use of high pressure and high temperature steam is economical.

11.9.4. LaMont Boiler

This boiler works on a forced circulation and the circulation is maintained by a centrifugal pump, driven by a steam turbine using steam from the boiler. For emergency an electrically-driven pump is also fitted.

Fig. 11.9 shows a LaMont steam boiler. The feed water passes through the economiser to the drum from which it is drawn to the circulation pump. The pump delivers the feed water to the tube evaporating section which in turn sends a mixture of steam and water to the drum. The steam in the drum is then drawn through the superheater. The superheated steam so obtained is then supplied to the prime mover.

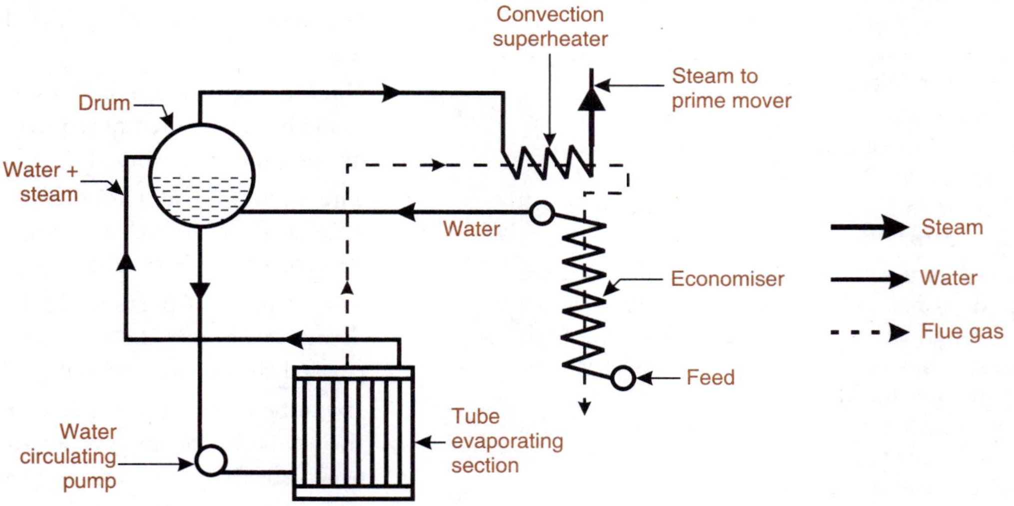

Fig. 11.9. LaMont boiler.

These boilers have been built to generate of 45 to 50 tonnes of superheated steam at a pressure of 130 bar and at a temperature of 500°C.

11.9.5. Loeffler Boiler

In a LaMont boiler the major difficulty experienced is the deposition of salt and sediment on the inner surfaces of the water tubes. The deposition reduces the heat transfer and ultimately the generating capacity. This further increases the danger of overheating the tubes due to salt deposition as it has high thermal resistance. This difficulty was solved in Loeffler boiler by *preventing the flow* of water into the boiler tubes.

This boiler also makes use of forced circulation. Its novel principle is *the evaporating of the feed water by means of superheated steam from the superheater, the hot gases from the furnace being primarily used for superheating purposes.*

Fig. 11.10 shows a diagrammatic view of a Loeffler boiler. The high pressure feed pump draws water through the economiser (or feed water heater) and deliver it into the evaporating drum. The steam circulating pump draws saturated steam from the evaporating drum and passes it through radiant and convective superheaters where steam is heated to required temperature. From the superheater about one-third of the superheated steam passes to the prime mover (turbine) the remaining two-thirds passing through the water in the evaporating drum in order to evaporate feed water.

This boiler can carry higher salt concentrations than any other type and is more compact than indirectly heated boilers having natural circulation. These qualities fit it for land or sea transport power generation.

Loeffler boilers with generating capacity of 100 tonnes/h and operating at 140 bar are already commissioned.

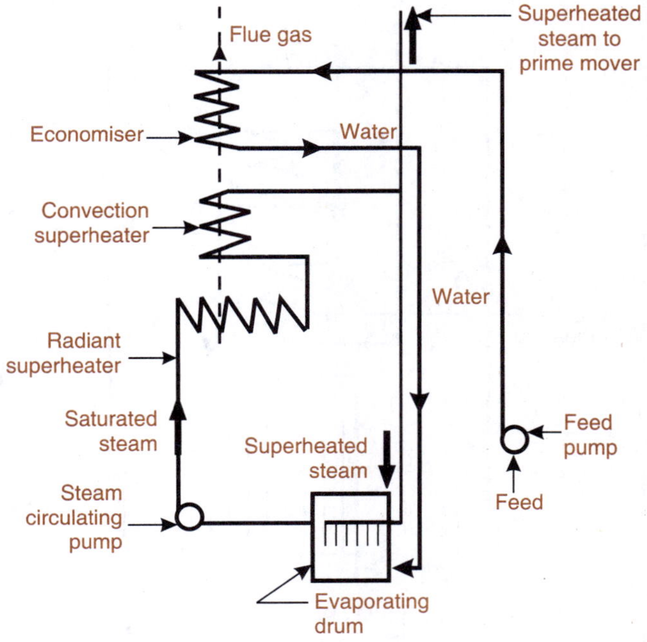

Fig. 11.10. Loeffler boiler.

11.9.6. Benson Boiler

In the LaMont boiler, the main difficult experienced *is the formation and attachment of bubbles on the inner surfaces of the heating tubes. The attached bubbles to the tube surfaces reduce the heat flow and steam generation as it offers high thermal resistance than water film. Benson in 1922 argued that if the boiler pressure was raised to critical pressure (225 atm.), the steam and water have the same density and therefore, the danger of bubble formation can be easily eliminated.* The first high pressure Benson boiler was put into operation in 1927 in West Germany.

This boiler too makes use of forced circulation and uses oil as fuel. It chief novel *principle is that it eliminates the latent heat of water by first compressing the feed to a pressure of 235 bar, it is then above the critical pressure and its latent heat is zero.*

Fig. 11.11 shows a schematic diagram of a Benson boiler. This boiler does not use any drum. The feed water after circulation through the economic tubes flows through the radiant parallel tube section to evaporate partly. The steam water mixture produced then moves to the transit section where this mixture is converted into steam. The steam is now passed through the convection superheater and finally supplied to the prime mover.

Boilers having as high as 650°C temperature of steam had been put into service. The maximum working pressure obtained so far from commercial Benson boiler is 500 atm. The Benson boilers of 150 tonnes/h generating capacity are in use.

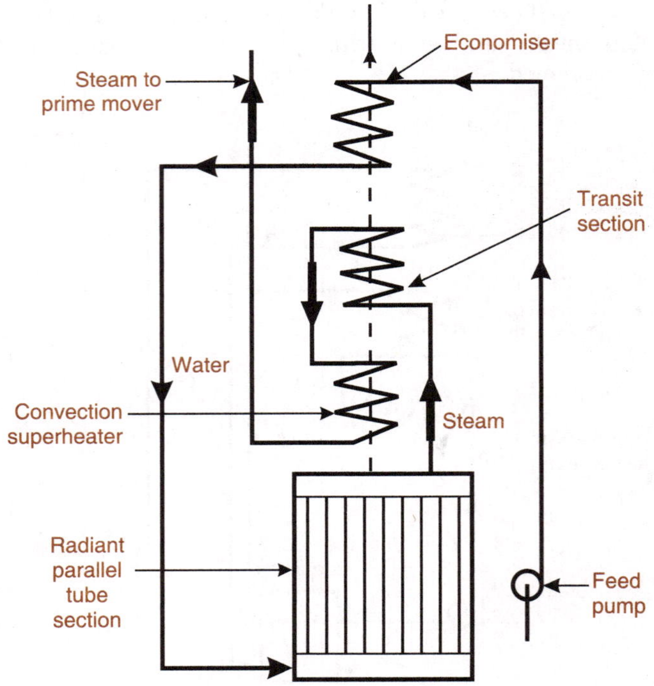

Fig. 11.11. Benson boiler.

Advantages of a Benson Boiler

The Benson boiler possesses the following *advantages* :

1. It can be erected in a comparatively smaller floor area.
2. The total weight of a Benson boiler is 20% less than other boilers, since there are no drums. This also reduces the cost of the boiler.
3. It can be started very quickly because of welded joints.
4. Natural convection boilers require expansion joints but these are not required for Benson boiler as the pipes are welded.
5. The furnace walls of the boiler can be more efficiently protected by using smaller diameter and closed pitched tubes.
6. The transfer of parts of the boiler is easy as no drums are required and majority of the parts are carried to the site without pre-assembly.
7. It can be operated most economically by varying the temperature and pressure at partial loads and overloads. The desired temperature can also be maintained constant at any pressure.
8. The blow-down losses of the boiler are hardly 4% of natural circulation boiler of the same capacity.
9. Explosion hazards are not severe as it consists of only tubes of small diameter and has very little storage capacity.
10. The superheater in a Benson boiler is an integral part of forced circulation system, therefore no special starting arrangement for superheater is required.

11.9.7. Velox Boiler

It is a well known fact that when the gas velocity exceeds the sound-velocity, the heat is transferred from the gas at a much higher rate than rates achieved with sub-sonic flow. The advantage of this theory is taken to effect the large heat transfer from a smaller surface area in this boiler.

This boiler makes use of **pressurised combustion.**

The gas turbine drives the axial flow compressor which raises the incoming air from atmosphere pressure to furnace pressure. The combustion gases after heating the water and steam flow through the gas turbine to the atmosphere. The feed water after passing through the economiser is pumped by a water circulating pump to the tube evaporating section. Steam separated in steam separating section flows to the superheater, from there it moves to the prime mover.

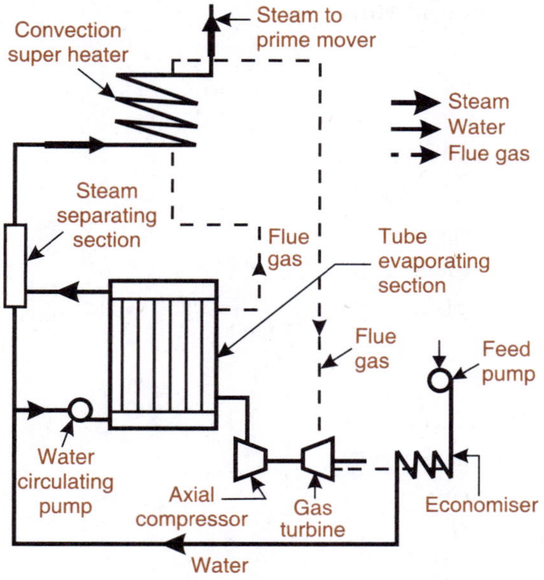

Fig. 11.12. Velox boiler.

The size of the Velox boiler is limited to 100 tonnes/h because 600 B.H.P. is required to run the air compressor at this output. The power developed by the gas turbine is not sufficient to run the compressor and therefore some power from external source must be supplied.

Advantages

1. The boiler is very compact and has greater flexibility.
2. Very high combustion rates are possible.
3. It can be quickly started.
4. Low excess air is required as the pressurised air is used and the problem of draught is simplified.

11.9.8. Super-critical Boilers

A large number of steam generating plants are designed between working ranges of 125 atm. and 510°C to 300 atm. and 660°C ; these are basically characterised as *sub-critical and super-critical.*

Usually *a sub-critical boiler* consists of three distinct section as *preheater* (economiser), *evaporator* and *superheater*.

A *super-critical boiler* requires only *preheater* and *superheater.*

The constructional layout of both the above types of boilers is, however, practically identical.

These days it has become a rule to use *super-critical boilers above 300 MW capacity units.*

The super-critical boilers claim the following *advantages* over critical type :

1. Large heat transfer rates.
2. Owing to less heat capacity of the generator the pressure level is more stable and therefore gives better response.
3. Because of absence of two phase mixture the problems of erosion and corrosion are minimised.
4. More adaptable to load fluctuations (because of great ease of operation, simplicity and flexibility).
5. The turbo-generators connected to super-critical boilers can generate peak loads by changing the pressure of operation.
6. Higher thermal efficiency.

Presently, 246 atm. and 538°C are used for unit size above 500 MW capacity plants.

11.9.9. Supercharged Boiler

In a supercharged boiler, the combustion is carried out under pressure in the combustion chamber by supplying the compressed air. The exhaust gases from the combustion chamber are used to run the gas turbine as they are exhausted to high pressure. The gas turbine runs the air compressor to supply the compressed air to the combustion chamber.

Advantages :

1. Owing to very high overall heat transfer co-efficient the heat transfer surface required is hardly 20 to 25% of the heat transfer surface of a conventional boiler.
2. The part of the gas turbine output can be used to drive other auxiliaries.
3. Small heat storage capacity of the boiler plant gives better response to control.
4. Rapid start of the boiler is possible.
5. Comparatively less number of operators are required.

11.10. COMBUSTION EQUIPMENT FOR STEAM BOILERS

11.10.1. General Aspects

The combustion equipment is a component of the steam generator. Since the source of heat is the combustion of a fuel, a working unit must have whatever equipment is necessary to receive the fuel and air, proportioned to each other and to the boiler steam demand, mix, ignite and perform any other special combustion duties, such as distillation of volatile from coal prior to ignition.

Fluid fuels are handled by *burners* ; solid lump fuels by *stokers.*

In boiler plants hand firing on grates is practically unheard of now-a-days in new plants, although there are many small industrial plants still in service with hand firing.

Since so many different principles are used in combustion equipment, Table 11.1 gives the more important manufactured types of stokers and burners.

Table 11.1. Important Types of Stokers and Burners

- Fluid fuels (Burners)
  - Gas
    - Multiple jet
    - Fan mix
    - Pre-mix
  - Oil
    - Pressure atomizing
      - Multiple burner
      - Replaceable tip
      - Wide range tip
    - Steam atomizing
      - Outside mix
      - Inside mix
    - Rotary cup
    - Vaporiser
      - Wick
      - Hot plate
  - Crushed and finely sized coal
    - Pulverized coal
      - Short flame
      - Long flame
      - Tangential
    - High turbulence furnace (cyclone)
- Lump coal stokers
  - Overfeed
    - Spreader stoker
      - Mechanical throw
        - Overthrow
        - Underthrow
          - Dump grate
          - Travelling grate
      - Jet throw
        - Air jet
        - Steam jet
    - Conveyor stoker
      - Travelling grate
      - Chain grate
        - Natural draft
        - Forced draft
  - Underfeed
    - Horizontal retort
      - Single
      - Twin
        - Ram feed
        - Screw feed
          - Grates stationary
          - Grates agitated
    - Sloping retort—Multiple report, large capacity.

The fuels are mainly bituminous coal, fuel oil, and natural gas mentioned in order of importance. All are composed of hydrocarbons, and coal has, as well much fixed carbon and little sulphur. To burn these fuels to the desired and products, CO_2 and H_2O, requires (*i*) *air in sufficient proportions,* (*ii*) *a good mixing of the fuel and air,* (*iii*) *a turbulence or relative motion between fuel and air.* The combustion equipment *must fulfil these requirements and, in addition, be capable of close regulation of rate of firing the fuel, for boilers ordinarily operate on variable load.* Coal-firing equipment *must also have a means for holding and discharging the ash residue.*

The basic requirements of combustion equipment are :

1. Thorough mixing of fuel and air.
2. Optimum fuel-air ratios leading to most complete combustion possible maintained over full load range.
3. Ready and accurate response of rate fuel feed to load demand (usually as reflected in boiler steam pressure).
4. Continuous and reliable ignition of fuel.
5. Practical distillation of volatile components of coal followed by adequate action in items 1 and 4 above.
6. Adequate control over point of formation and accumulation of ash, when coal is the fuel.

Natural gas is used as a boiler fuel in gas well regions where fuel is relatively cheap and coal sources comparatively distant. The transportation of natural gas overland to supply cities with domestic and industrial heat has made the gas in the well more valuable and the gas fired steam generator more difficult to justify in comparison with coal, on fuel cost alone. Cleanliness and convenience in use are other criteria of selection, but more decisive in small plants in central power stations.

Transportation costs add less to the delivery price of oil than gas ; also fuel oil may be stored in tanks at a reasonable cost, whereas gas cannot. Hence although fuel oil is usually more costly than coal per kg of steam generated, many operators select fuel oil burners rather than stokers because of the simplicity and cleanliness of storing and transporting the fuel from storage to burner.

11.10.2. Burning of Coal

The two most commonly used methods for the burning of coal are :

1. ***Stoker firing***
2. ***Pulverised fuel firing.***

The selection of one of the above methods depends upon the following *factors* :

(*i*) Characteristics of the coal available.
(*ii*) Capacity of the boiler unit.
(*iii*) Load fluctuations.
(*iv*) Station load factor.
(*v*) Reliability and efficiency of the various types of combustion equipment available.

11.10.2.1. Stoker Firing

A ***stoker*** *is a power operated fuel feeding mechanism and grate.*

Stoker firing (as compared to hand firing) entails the following *advantages* and *disadvantages*.

Advantages :

1. A cheaper grade of fuel can be used.
2. A higher efficiency attained.
3. A greater flexibility of operations assured.
4. Less smoke produced.
5. Generally less building space is necessary.
6. Can be used for small or large boiler units.
7. Very reliable, and maintenance charges are reasonably low.
8. Practically immune from explosions.
9. Reduction in auxiliary plant.

Disadvantages :

1. Construction is complicated.
2. In case of very large units the initial cost may be rather higher than with pulverised fuel.
3. There is always a certain amount of loss of coal in the form of riddlings through the grates.
4. Sudden variations in the steam demand cannot be met to the same degree.
5. Troubles due to slagging and clinkering of combustion chamber walls are experienced.
6. Banking and stand by losses are always present.

Classification of stoker firing

Automatic stokers are classified as follows :

1. ***Overfeed stokers***
2. ***Underfeed stokers.***

In the case of *overfeed stokers,* the coal is fed into the grate above the point of air admission and in case of *underfeed stokers*, the coal is admitted into the furnace below the point of air admission. The difference is made clear in Fig. 11.13.

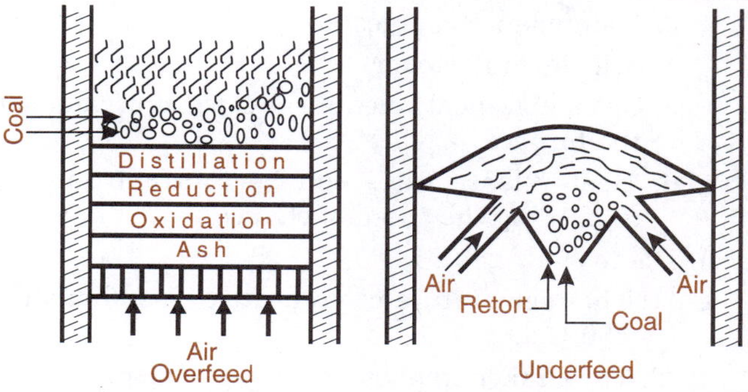

Fig. 11.13. Difference between underfeed and overfeed stoker firing.

1. Overfeed stokers

These types of stokers are used for *large capacity boiler installation where the coal is burned without pulverisation.* The overfeed stokers are of mainly two types : (*i*) Travelling grate stoker, and (*ii*) Spreader stoker.

Travelling grate stoker. The travelling stoker may be chain grate type or bar grate type. These two differ only in the details of grate construction.

Fig. 11.14 shows a *chain grate stoker.* The speed of the stoker is 12.5 cm to 50 cm per minute. An index plate with pointer shows the coal bed thickness at all times. This can be regulated either by adjusting the opening of the fuel gate or by the speed control of the stoker driving motor. The air is admitted from the underside of the grate which is divided into several compartments each connected to an air duct. The grate should be saved from being overheated. For this, the coal should have sufficient ash content which will form a layer on the grate. Since there is practically no agitation of the fuel bed, *non-coking coals are best suited for chain grate stokers.*

The rate of burning with this stoker is 200 to 300 kg per m^2 per hour when forced draught is used.

The advantages and disadvantages of chain grate stoker are listed below :

Advantages :

1. Simple in construction.
2. Low initial cost.

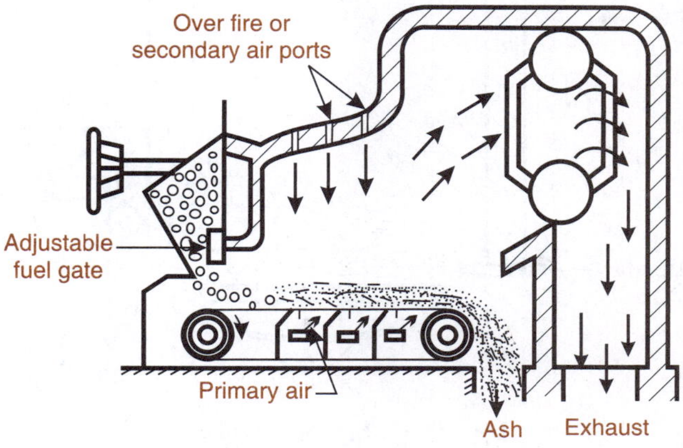

Fig. 11.14. Chain grate stoker.

3. Maintenance charges low.
4. Self-cleaning stoker.
5. Gives high heat release rates per unit volume of the furnace.
6. Heat release rates can be controlled just by controlling the speed of chain.

Disadvantages :

1. The temperature of preheated air is limited as 180°C.
2. The clinker troubbs are very common.
3. Ignition arches are required.
4. There is always some loss of coal in the form of fine particles carried with the ashes.
5. This cannot be used for high capacity boilers (200 tonnes/h) or more.

Spreader stoker

Refer to Fig. 11.15. In a spreader stoker the coal is not fed into the furnace by means of grate. The function of the grate is only to support a bed of ash and move it out of the furnace. From the coal hopper, coal is fed into the path of a rotor by means of a conveyor, and is thrown into the furnace by the rotor and is burnt in suspension. The air for combustion is supplied through the holes in the grate. Overfire air or secondary air to create turbulence and supply oxygen for the thorough combustion of coal is supplied through nozzles located directly above the ignition arch. Unburnt coal and ash are deposited on the grate which can be moved periodically to remove ash out of furnace.

Spreader stokers can be used for boiler capacities from 70000 kg per hr of steam to about 140000 kg per hr.

Advantages :

1. A wide variety of coal can be burnt.
2. The clinkering difficulties are reduced even with coals which have high clinkering tendencies.
3. The use of high temperature preheated air is possible.
4. Operation cost is considerably low.

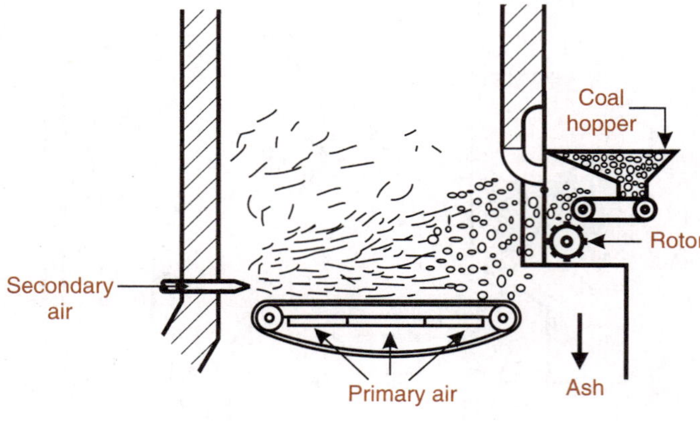

Fig. 11.15. Spreader stoker.

Disadvantages :

1. It is always difficult to operate spreader with varying sizes of coal with varying moisture content.
2. Fly-ash is much more.

2. Underfeed stokers

Refer to Fig. 11.16. In these stokers, the coal is fed into the furnace below the point of air admission. Coal from the hopper is pushed into the retort by means of reciprocating plunger. When coal gets heated, all the volatiles in it are distilled and when coal reaches the zone of active combustion, it is in the form of coke and ash. The ash discharge plates are at the back of the furnace and by the time coal is pushed down on to those, all the combustion has been completed. Air is admitted into the furnace through holes in the sides of the retort. The coal is continuously agitated by the plunger and also by three pusher plates along the bottom of the retort. Due to this, the fuel bed remains porous and free from clinkers.

Underfeed stokers are suitable for *non-clinkering, high volatile coals having coking properties and low ash content*.

Advantages :

1. Give higher thermal efficiency compared with chain grate stokers.
2. Combustion rate is considerably higher.
3. The grate is self cleaning.
4. Different variety of coals can be used.
5. Grate bars, tuyeres and retorts are not subjected to high temperature as they remain always in contact with fresh coal.
6. Smokeless operation is possible even at very light load.

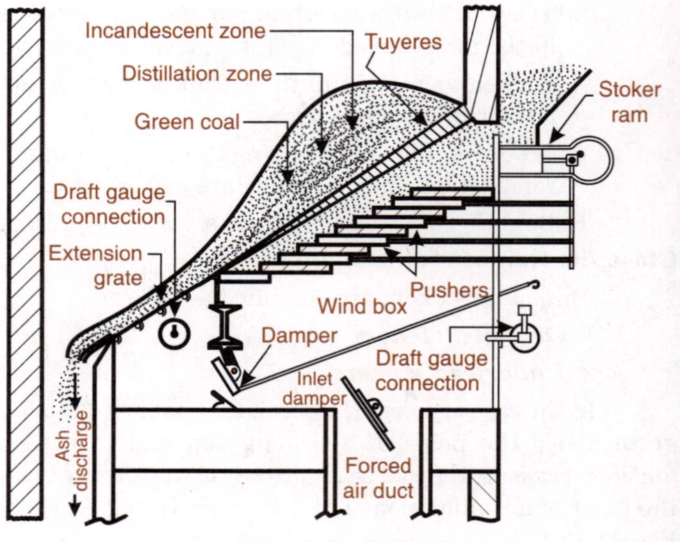

Fig. 11.16. Underfeed stokers.

Disadvantages :

1. High initial cost.
2. Requires large building space.
3. The clinker troubles are usually present.
4. Low grade fuels with high ash content cannot burn economically.

11.10.2.2. Pulverised Fuel Firing

In a pulverised fuel firing system *the coal is reduced to a fine powder with the help of grinding mill and then projected into the combustion chamber with the help of hot air current.* The amount of air required (known as secondary air) to complete the combustion is supplied separately to the combustion chamber. The resulting turbulence in the combustion chamber helps for uniform mixing of fuel and air and thorough combustion. *The amount of air which is used to carry the coal and to dry it before entering into the combustion chamber is known as 'primary air'* and the *amount of air which is supplied separately for completing the combustion is known as 'secondary air'*. The efficiency of pulverised fuel firing system mostly depends upon the size of the powder. The fineness of the coal should be such as 70% of it would pass through a 200 mesh sieve and 90% through 50 mesh sieve.

Several modern thermal power plants use pulverised fuel system when the available coal is cheap and is not suitable for stoker firing.

Advantages :

1. Any grade of coal can be used since coal is powdered before use.
2. The rate of feed of the fuel can be regulated properly resulting in fuel economy.
3. Since there is almost complete combustion of the fuel there is increased rate of evaporation and higher boiler efficiency.
4. Greater capacity to meet peak loads.
5. The system is practically free from sagging and clinkering troubles.
6. No standby losses due to banked fires.
7. Practically no ash handling troubles.
8. No moving parts in the furnace subjected to high temperatures.
9. The external heating surfaces are free from corrosion.
10. This system works successfully with or in combination with gas and oil.

Disadvantages :

1. High capital cost.
2. Lot of fly ash in the exhaust which makes the removing of fine dust uneconomical.
3. The possibilities of explosion are more as coal burns like a gas.
4. The maintenance of furnace brick work is costly.
5. Special equipment is needed to start this system.
6. The skilled operators are required.

Coal burners. A coal burner fires the pulverised coal along with primary air into the furnace. The secondary air is admitted separately below the burner, around the burner or elsewhere in the furnace. Ignition takes place by means of radiation and flame propagation from the fuel already burning in the furnace.

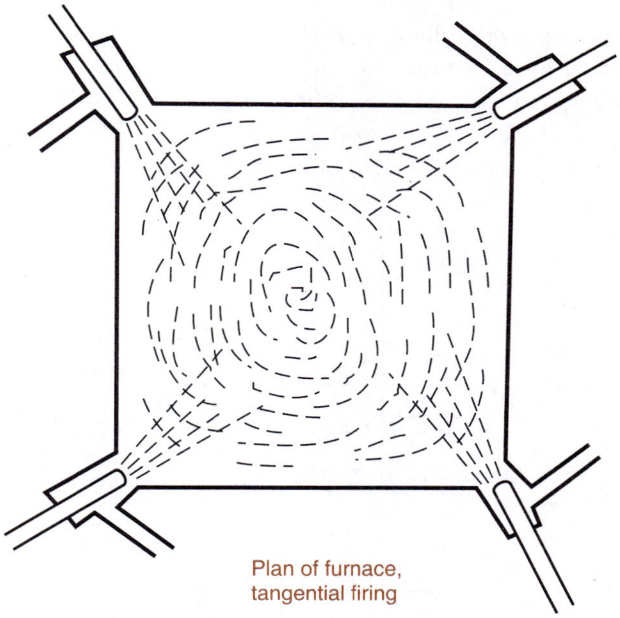

Fig. 11.17. Tangential firing.

Coal burner may be classified according to their design and by their arrangement in the furnace as discussed below :

In *"opposite firing"* burners are placed on the opposite walls of the furnace and they fire directly against each other. This results in intimate mixing of the fuel and the air.

In *"cross firing"*, the burners fire in the vertical directions and in horizontal direction and the fuel streams intersect with each other.

In *"tangential firing"*, the burners are placed in the corners of the furnace and they send horizontal streams of air and fuel *tangent to an imaginary circle in the centre of the furnace.* This results in *intense turbulence* and thorough mixing of the fuel and air. All the fuel and air nozzles can be tilted 24° above and below the horizontal.

Fig. 11.17 shows the plan view of the tangential firing.

Cyclone burners. In such burners *crushed coal is used and not the pulverised fuel*. This is done to *eliminate pulverisers and to reduce the fly ash difficulties.* From the feeder the crushed coal (max. 6.5 mm) and the primary air enter with a vortex motion at the centre of the cyclone. The secondary air admitted separately aids in the vortex motion. The fuel is quickly burned and ash in the form of molten slag drains down the inner wall of the cyclone. Hot flue gases with 10 to 20% of the ash in the coal in the form of fly ash enter the furnace. Due to centrifugal action, most of this fly ash is thrown against the walls of the furnace and is drained away along with the molten film of slag. Thus the fuel gases leaving the furnace are quite clean to flow through the rest of heat exchanger passages. This results in *better heat transfer, good combustion*. Also, *less furnace cleaning is required and fly ash trouble is greatly reduced.*

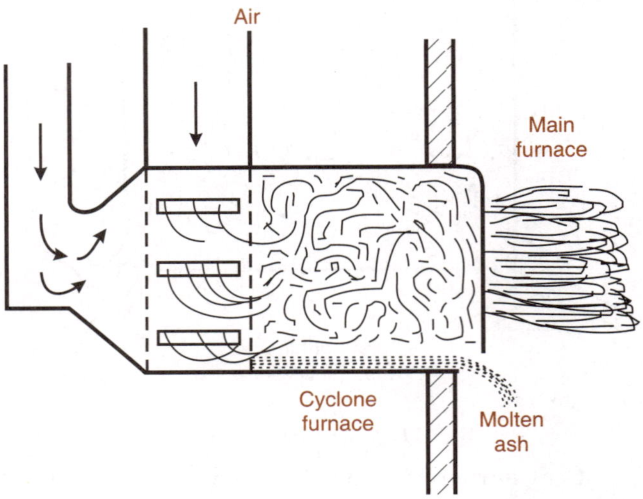

Fig. 11.18. Cyclone burner.

The *merits* of the *cyclone burner* over the other types are listed below :

1. Simplified coal crushing equipments can be used instead of costly pulverise mills.
2. Excess air required can be reduced to 15% minimum using forced draught fan.
3. The cyclone furnace can use low grade fuels, reduces the size of the steam generator and limits the fly ash emission so that excessive furnace cleaning and precipitations are not required.

Oil Burning

Fig. 11.19 shows the simple method of oil burning.

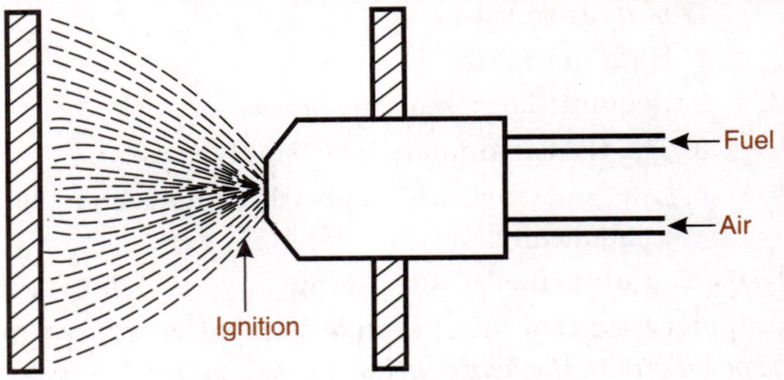

Fig. 11.19. Principle of oil firing.

The functions of the oil burner are to mix the fuel and air in the proper proportion and to prepare the fuel for combustion. Following are the two ways to achieve it :

1. The oil may be vaporised or gasified by heating within the burner.
2. The oil may be atomised by the burner and its vaporisation occurs in the combustion space.

Vaporising burners find little use as there are limited range of fuels they can handle. In the second arrangement, the atomisation of the oil is done in three basic ways :

(*i*) The oil is broken into small droplets by using air or steam under pressure.

(*ii*) Forcing oil under pressure through a suitable nozzle.

(*iii*) Tearing an oil film into drops by centrifugal force.

Gas burning. Gas burning is much simpler as the fuel is ready for combustion and requires no preparation. The remaining parts of the job *i.e.,* proportioning, mixing and burning can be achieved in many ways. The most simple and familiar gas burner is the atmospheric burner. In this the momentum of the incoming low pressure gas stream is used to draw in or aspirate the air needed for combustion. Gas and air together pass through a tube leading to the burner ports, mixing in the process. The mixture burns at the ports or the openings in the burner head. Secondary air is drawn into the flame from the surrounding atmosphere. Larger counterparts of this type having ring or sectional burning heads with many ports are used to fire small boilers.

Example 11.1. *Sketch and describe the working of a once-through boiler. Discuss its special features.* **(U.P.S.C.)**

Solution.

- The **once-through boiler** or steam generator is also called the *forced circulation, Benson or universal pressure boiler* because it is *applicable to all temperature and pressures, although economically it is suited to large sizes and pressures in the high subcritical and supercritical range.*

- In a once through boiler, in contrast to the drum type (Fig. 11.20), the feedwater goes through the economiser, furnace walls, and superheat sections, changing sequentially to saturated water, saturated steam, and superheated steam in *one continuous pass* (Fig. 11.21). *No steam drum is required to separate saturated steam from boiling water and no water circulation takes place.* Reheat of steam after it is expanded in the high-pressure turbine is accomplished by a *reheater* in the usual manner.

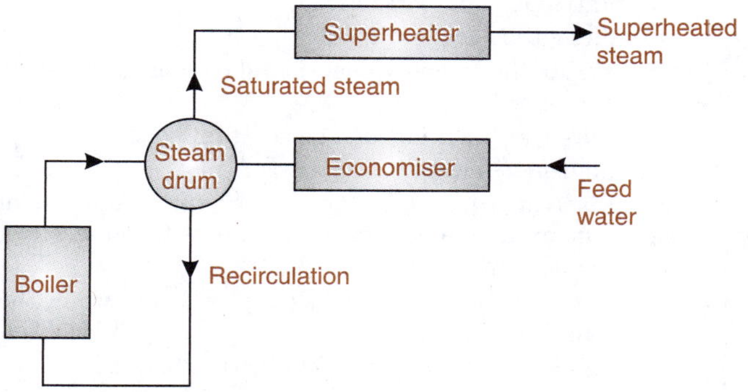

Drum type steam generator

Fig. 11.20

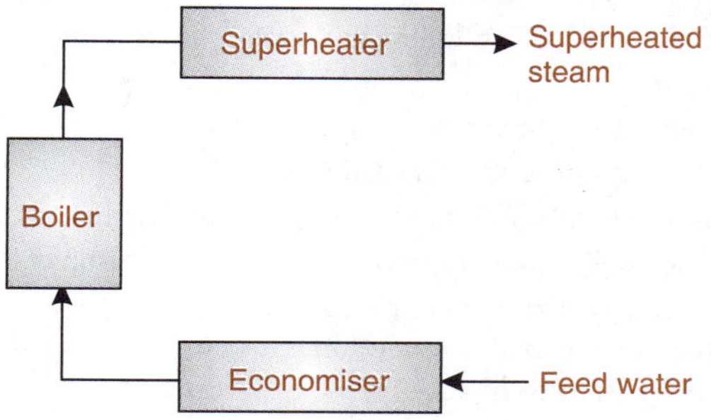

Once through boiler or steam generator

Fig. 11.21

- Because of the once-through mode of operation, *very high purity feed water is required.*
- Fig. 11.21 shows a typical once-through steam generator.
- The *once-through boiler in the only type suited to supercritical pressure operation* (above 221 bar, for steam) because the latent heat of vaporisation at and beyond critical pressure is zero, and liquid and vapour are one and the same, so no separation is drum is possible or necessary.
- *While particularly applicable to supercritical pressure, once-through steam generators are used economically for high pressure and sub-critical steam.*

HIGHLIGHTS

1. A **'boiler'** is defined as a closed vessel in which steam is produced from water by combustion of fuel.
   A **'steam generating unit'** is a combination of apparatus for producing, furnishing or recovering heat together with the apparatus for transferring the heat so made available to the fluid being heated and vaporised.
2. **Fire tube boilers.** In these boilers the hot gases are inside the tubes and water surrounds the tubes.
   Examples : Cochran, Lancashire and Locomotive boilers.
3. **Water tube boilers.** In these boilers the water is inside the tubes and hot gases surround them.
   Examples : Babcock and Wilcox, Stirling, Yarrow boiler etc.
4. **High pressure boilers.** The modern high pressure boilers employed for power generation have steam capacities 30 to 650 tonnes/h and above with a pressure upto 160 bar and minimum steam temperature of about 540°C.
   Examples : LaMont, Loeffler, Benson, Velox boilers.
5. A large number of steam generating plants are designed between working ranges of 125 atm. and 510°C to 300 atm. and 660°C, these are basically characterised as *sub-critical* and *super-critical*.
6. In a super-charged boiler, the combustion is carried out under pressure in the combustion chamber by supplying the compressed air.
7. 'Burning of coal' can be carried out by :
   (*i*) Stoker firing
   (*ii*) Pulverised fuel firing.
   Automatic stokers are classified as :
   (*i*) Overfeed stokers
   (*ii*) Underfeed stokers.
   In a **'pulverised fuel firing system'** the coal is reduced to a fine powder with the help of grinding mill and then projected into the combustion chamber with the help of hot air current.

OBJECTIVE TYPE QUESTIONS

Say 'Yes' or 'No'

1. A closed vessel in which steam is produced from water by combustion of fuel is called a 'boiler'.
2. There is no difference between 'steam generating unit' and 'boiler'.
3. It is very difficult to inspect and repair the parts of a horizontal boiler.

4. If the axis of the boiler is vertical, it is called a vertical boiler.
5. Locomotive boiler is a water tube boiler.
6. Stirling boiler is a fire tube boiler.
7. Babcock and Wilcox is a water tube boiler.
8. Velox boiler uses a forced pump for circulation of the fluid.
9. In Lancashire boiler the circulation of water takes place due to natural convection currents produced by the application of heat.
10. Benson boiler is a high pressure boiler.
11. Velox boiler is a low pressure boiler.
12. Cornish boiler is a single tube boiler.
13. Fire-tube boilers are not suitable for large power plants.
14. For a given power, water tube boilers, occupy more floor area.
15. The construction of a fire tube boiler is difficult.
16. There are more chances of explosion in water tube boilers.
17. A fire tube boiler has a higher rate of steam production.
18. A simple vertical boiler is self contained and can be transported easily.
19. A locomotive boiler is mainly employed in locomotives though it may also be used as a stationary boiler.
20. Stirling water tube boiler is an example of *bent tube* boiler.
21. Water tube boilers are generally preferred for low pressure and low output.
22. A velox boiler makes use of pressurised combustion.
23. A stoker is a power operated fuel feeding mechanism and grate.
24. Overfeed stokers are used for large capacity boiler installation where the coal is burned without pulverisation.
25. In underfeed stokers the coal is fed into furnace above the point of air admission.
26. In pulverised fuel firing the ash handling troubles are very acute.
27. Cyclone burners use pulverised fuel.

ANSWERS

1. Yes	2. No	3. No	4. Yes
5. No	6. No	7. Yes	8. Yes
9. Yes	10. Yes	11. No	12. Yes
13. Yes	14. No	15. Yes	16. Yes
17. No	18. Yes	19. Yes	20. Yes
21. No	22. Yes	23. Yes	24. Yes
25. No	26. No	27. No.	

THEORETICAL QUESTIONS

1. Define 'a boiler' and 'a steam generating unit'.
2. State the uses of steam produced by the boilers.
3. List the primary requirements of steam generators.
4. How are boilers classified ?
5. Give the comparison between 'Fire-tube and Water-tube' boilers.
6. State the differences between the following boilers :
   (*i*) Externally fired and internally fired.
   (*ii*) Forced circulation and natural circulation.
   (*iii*) High pressure and low pressure.
   (*iv*) Stationary and portable.
   (*v*) Single tube and multi-tube.
7. Enumerate the factors which should be considered while selecting a boiler.
8. What are the essentials of a good steam boiler ?
9. Explain the following boiler terms :
   Shell, setting, grate, furnace, water space and steam space, mountings, accessories, water level, foaming, scale, blowing off, lagging, refractory.
10. Explain with the help of neat diagrams any two of the following fire-tube boilers :
    (*i*) Simple vertical boiler. (*ii*) Cochran boiler.
    (*iii*) Cornish boiler. (*iv*) Lancashire boiler.
    (*v*) Locomotive boiler. (*vi*) Scotch boiler.
11. Give the construction and working of the following water tube boilers :
    (*i*) Babcock and Wilcox boiler
    (*ii*) Stirling boiler.
12. Explain the unique features of the high pressure boilers.
13. List the advantages of high pressure boilers.
14. Explain with neat sketches the construction and working of any two of the following high pressure boilers :
    (*i*) LaMont boiler (*ii*) Loeffler boiler
    (*iii*) Benson boiler (*iv*) Velox boiler.
15. Write short notes on the following :
    (*i*) Supercharged boilers (*ii*) Supercritical boilers.
16. What are the basic requirements of combustion equipment ?
17. Explain briefly the following methods of burning of coal :
    (*i*) Stoker firing (*ii*) Pulverised fuel firing.
18. Explain briefly the following :
    (*i*) Travelling grate stoker (*ii*) Spreader stokers.
19. What is the difference between 'overfeed stokers and underfeed stokers' ?
20. Write a short note on 'pulverised fuel firing'.
21. What are the advantages of 'Pulverised fuel firing' over 'stoker firing' ?
22. Write a short note on coal burners.

Boiler Mountings and Accessories

12

12.1. Introduction. 12.2. Boiler mountings—Water level indicator—Pressure gauge—Safety valves—High steam and low water safety valve—Fusible plug—Blow-off cock—Feed check valve—Junction or stop valve. **12.3. Accessories**—Feed pumps—Injector—Economiser—Air preheater—Superheater—Steam separator—Steam trap. *Highlights—Objective Type Questions—Theoretical Questions.*

12.1. INTRODUCTION

Boiler Mountings. These are different fittings and devices which are *necessary for the operation and safety of a boiler*. Usually these devices are *mounted over boiler shell*.

In accordance with the Indian boiler regulation the following *mountings* should be fitted to the boilers.

- Two safety valves
- Two water level indicators
- A pressure gauge
- A steam stop valve
- A feed check valve
- A blow-off cock
- An attachment for inspector's test gauge
- A man hole
- Mud holes or sight holes.

Boilers of Lancashire and Cornish type should be fitted with a *high pressure and low water safety valve.*

All land boilers should have a *fusible plug* in each furnace.

Boiler Accessories. These are *auxiliary plants required for steam boilers for their proper operation and for the increase of their efficiency.* Commonly used boiler *accessories* are :

- Feed pumps
- Injector
- Economiser
- Air preheater
- Superheater
- Steam separator
- Steam trap.

12.2. BOILER MOUNTINGS

The various boiler mountings are discussed as follows :

12.2.1. Water Level Indicator

The function of a water level indicator is to indicate the level of water in the boiler constantly. It is also called *water gauge.* Normally two water level indicators are fitted at the front end of every boiler. Where the boiler drum is situated at considerable height from the floor, the water gauge is often inclined to make the water level visible from any position. When the water being heated in the boiler transforms into steam the level of water in the boiler shell goes on decreasing. For the proper working of the boiler, the water must be kept at safe-level. If the water level falls below the safe level and the boiler goes on producing steam without the addition of feed water, great *damage like crack and leak can occur to the parts of the boiler which get uncovered from water.* This can result in the *stoppage of steam generation and boiler operation.*

Fig. 12.1 shows a Hopkinson's water gauge. It is a common form of glass tube water-level gauge. *A* is the front end plate of the boiler. *F* is a very hard glass tube indicating water level and is connected to the boiler plate through stuffing boxes in hollow gun metal castings (*B*, *C*) having flanges *X*, *Y* for bolting the plate.

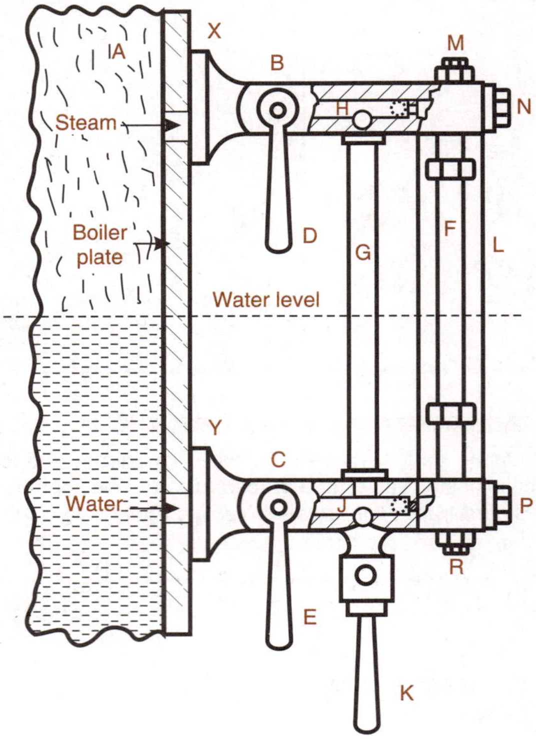

A = End plate of boiler
B and C = Hollow gun metal castings
D and E = Cocks
F = Gauge glass
G = Hollow metal column
H and J = Two balls
K = Drain cock
L = Guard glass
M, N, P, R = Screwed caps
X, Y = Flanges

Fig. 12.1. Water level indicator.

For controlling the passage of steam and water cocks D and E are provided. When these cocks are opened the water stands in the glass tube at the same level as in the boiler. K is the drain cock to blow out water at intervals so as not to allow any sediments to accumulate. Upper and lower stuffing boxes are connected by a hollow metal column G. Balls J and H rest in the position shown in the normal working of the gauge. When the glass tube breaks due to rush of water in the bottom passage the balls move to dotted positions and shut off the water and steam. Then the cocks D and E can be safely closed and broken glass tube replaced. M, N, P and R are screwed caps for internal cleaning of the passage after dismantling. L is the guard glass ; it is tough and does not give splinters on breaking. Thus when the gauge glass breaks, and this guard glass which normally will hold flying pieces, also gives way, the pieces will not fly one and *hurt* the attendant.

12.2.2. Pressure Gauge

The function of a pressure gauge is to measure the pressure exerted inside the vessel. The gauge is usually mounted on the front top of the shell or the drum. It is usually constructed to indicate upto double the maximum working pressure. Its dial is graduated to read pressures in kgf/cm^2 (or bar) gauge (*i.e.*, above atmospheric). There are two types of pressure gauges : (*i*) Bourdon tube pressure gauge and (*ii*) Diaphragm type pressure gauge. A pointer, which rotates over a circular graduated scale, indicates the pressure.

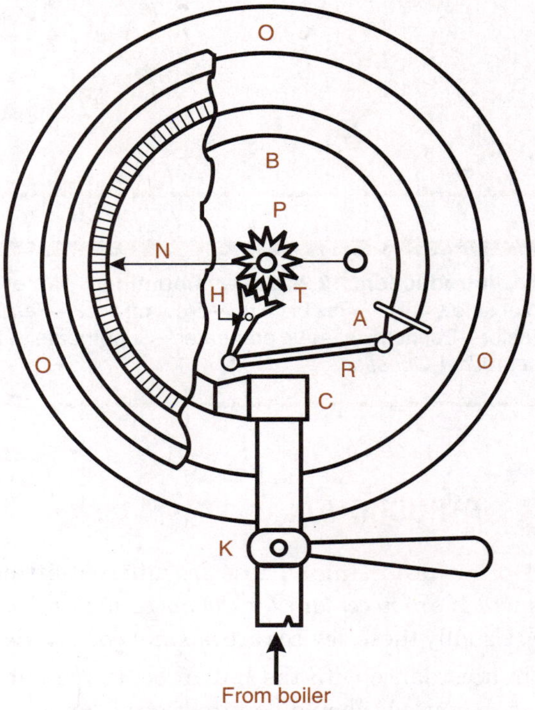

Fig. 12.2. Bourdon pressure gauge.

A pressure gauge is known as *compound pressure gauge* if it is designed in such a fashion so as to measure pressures above and below the atmosphere on the same dial.

Fig. 12.2 shows a Bourdon pressure gauge (single tube) a common type of pressure gauge used. The essential feature of this gauge is the elliptical spring tube which is made of a special quality of bronze and is solid drawn. One end A is closed by a plug and the other is connected with a block C, the block is connected with a syphon tube (which is full of condensed water). The steam pressure forces the water from the syphon tube into elliptical tube and this causes the tube to become circular is cross-section. As the tube is fixed at C, the other end A moves outwards. This outward movement is magnified by the rod R and transmitted to toothed sector T. This toothed sector is hinged at the point H and meshes with the pinion P fixed to the spindle of the pointer N. Thus the pointer moves and registers the pressure on a graduated dial.

The movement of the free end of the elliptical tube *is proportional to the difference between external and internal pressure on the tube.* Since the outside pressure on the tube is atmospheric, the movement of the free end is a measure of the boiler pressure above atmospheric *i.e.*, *gauge pressure.*

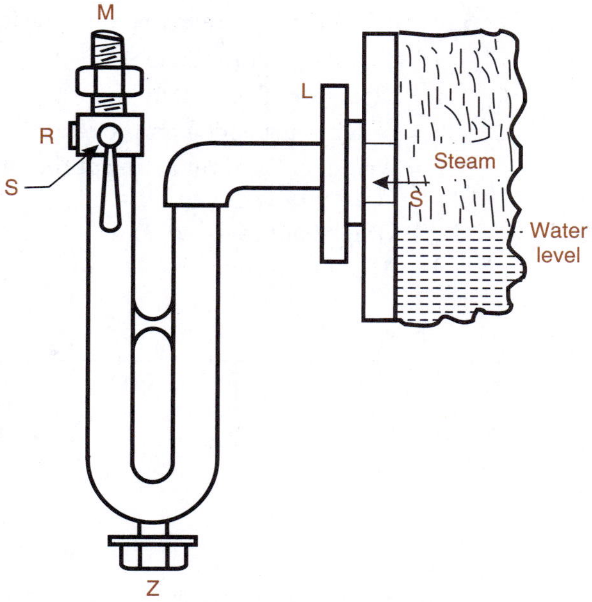

L = Flange
M = Connection pressure gauge
Z = Plug for cleaning syphon
R = Plug for testing pressure gauge
S = Three way cock

Fig. 12.3. U-tube syphon.

Fig. 12.3 shows a U-tube syphon which connects the gauge to the boiler. The U-tube syphon is connected to the steam space of the boiler and contains condensed steam which enters the gauge tube. The condensed water transmits pressure to the gauge, and at the same time prevents steam from entering the pressure gauge. In case steam passes into the gauge tube it will expand the tube and reading obtained will be false. Furthermore metal may be affected. Plug R is used for connecting the inspector's standard gauge and testing accuracy of boiler pressure gauge while in service. Plug Z is employed for cleaning the syphon. Three way cock S is used for either connecting the boiler pressure gauge to steam space or inspector's pressure gauge to the steam space.

Note. *The double-tube Bourdon gauge is more rigid than the single tube and more suitable for locomotive and portable boilers.*

12.2.3. Safety Valves

The function of a safety valve is to release the excess steam when the pressure of steam inside the boiler exceeds the rated pressure. As soon as the pressure of steam inside the boiler exceeds the rated pressure the safety valve automatically opens and excess steam rushes out into the atmosphere till the pressure drops down to the normal value. A safety valve is generally mounted on the top of the shell.

As per boiler regulations every boiler must be fitted at least with two safety valves.

The various types of safety valves are enumerated and discussed as follows :

1. Dead weight safety valve.
2. Lever safety valve.
3. Spring loaded safety valve.
4. High steam and low water safety valve.

12.2.3.1. Dead Weight Safety Valve

Fig. 12.4 shows a dead weight safety valve. A is the vertical cast iron pipe through which steam pressure acts. B is the bottom flange directly connected to seating block on the boiler shell communicating to the steam space. V is the gun metal valve and VS is the gun metal valve seat. D is another cast iron pipe for discharge of excess steam from the boiler. W are the weights in the form of cylindrical disc of cast iron. WC is the weight carrier carrying the weights W. The cover plate C covers these weights. The steam pressure acts in the upward direction and is balanced by the force of the dead weights W. The total dead-weights consist of the sum of the weights W, weight of the valve V, weight of the weight carrier and weight of the cover plate C.

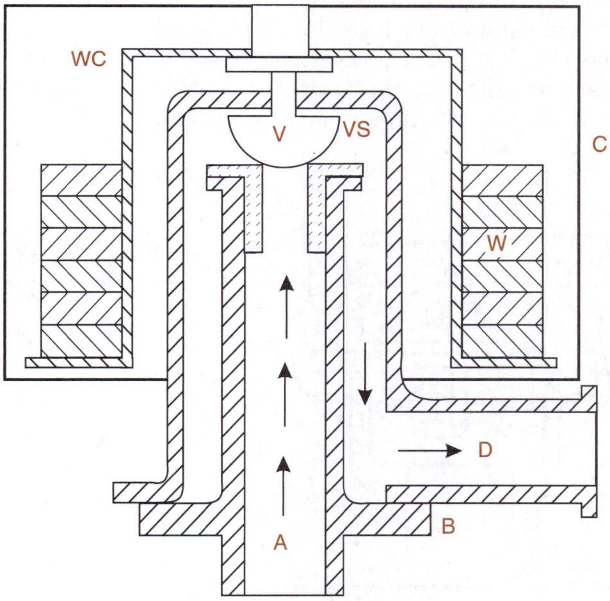

A = Cast iron pipe
B = Bottom flange
V = Gun metal valve
W = Cast iron weights
WC = Weight carrier
D = Discharge pipe
VS = Gun metal valve seat
C = Cover plate

Fig. 12.4. Dead weight safety valve.

When the steam pressure is greater than the working pressure it lifts the valve with its weights. So the steam escapes from the boiler and the steam pressure thereby decreases.

Merits of dead weight safety valve :

1. Simplicity of design.
2. Gives quite a satisfactory performance during operation.
3. It cannot be easily tempered from the pressure adjustment view-point.

Demerits :

1. Unsuitable for use on any boiler *where extensive vibration and movement are experienced* (*e.g.*, locomotive and marine work).
2. It is not suitable for high pressure boilers because a large amount of weight is required to balance the steam pressure.

Uses. It is mainly used for *low pressures, low capacity, stationary boilers* of the Cornish and Lancashire types.

12.2.3.2. Lever Safety Valve

Refer to Fig. 12.5. It consists of a lever and weight *W*. The valve (made of gun metal) rests on the valve seat (gun metal) which is screwed into the valve body ; the valve seat can be replaced if required. The valve body is fitted on the boiler shell. One end of the lever is hinged while at the other is suspended a weight *W*. The strut presses against the valve on seat against the steam pressure below the valve. The slotted lever guide allows vertical movement to the lever.

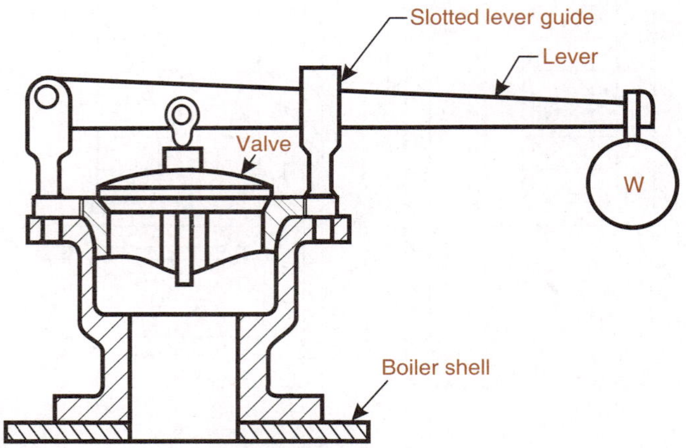

Fig. 12.5. Lever safety valve.

When the steam pressure becomes greater than the normal working pressure, the valve is lifted with the lever and the weight. Consequently, the steam escapes through the passages between the valve and seat and the steam pressure decreases.

The disadvantages of this valve is that it admits of being tempered with, and the effect of a small addition to the weight is magnified considerably in its action on the valve.

Fig. 12.6 shows the loading arrangement on the lever.

Let p = Steam pressure (gauge),
d = Diameter of the valve,
W = Weight suspended on the lever,
W_l = Weight of the lever acting at the centre of gravity G,
W_v = Weight of the valve, and
A = Area of the valve.

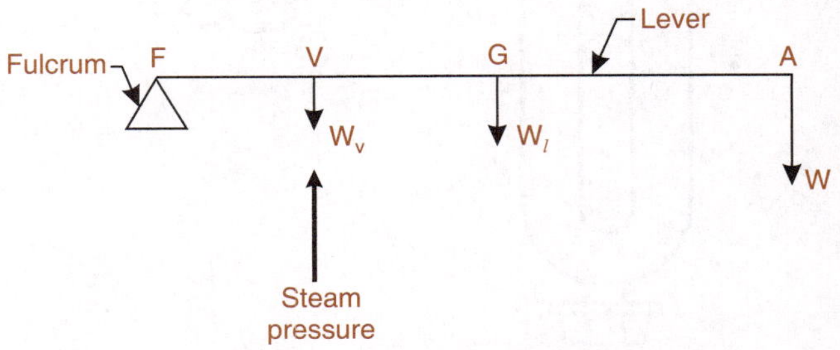

Fig. 12.6. Loading arrangement on the lever.

Taking moments about the fulcrum F, we get

$$W \times AF + W_l \times GF + W_v \times VF = p \times a \times VF,$$

where $a = \frac{\pi}{4} d^2$.

From the above equation we can find the weight W or length of lever for a given pressure of steam (p).

12.2.3.3. Spring Loaded Safety Valve

For locomotives and marine engines both the lever and dead-weight types are unsuitable for abvious reasons, and the valve must be spring loaded, as such valve is *unaffected by vibration or deviations from the vertical.*

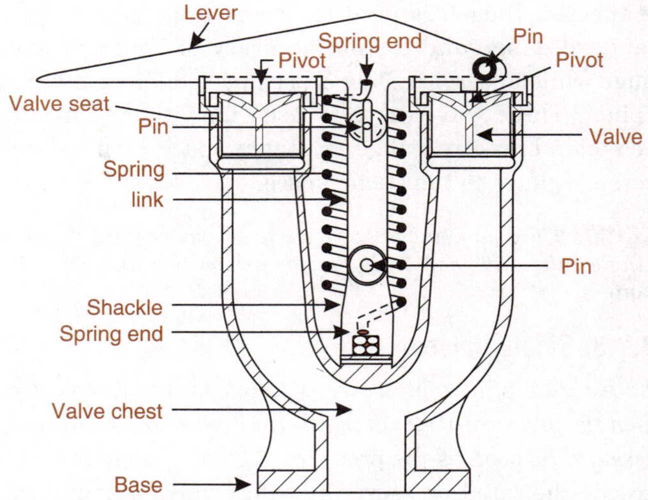

Fig. 12.7. Ramsbottom spring loaded safety valve.

Fig. 12.7 illustrates what is known as Ramsbottom spring loaded safety valve. It consists of two separate valves and seatings having one lever, bearing on the two valves, and loaded by a spring, the spring being placed between the valves. The tension on the spring can be adjusted by the nuts. By pulling or raising the lever the operator/driver can relieve the pressure from either valve separately, and ascertain it is not sticking on the seating.

One *disadvantage* of the spring-loaded safety valve is that *the load on the valve increases as the valve lifts, so that pressure required just to lift the valve is less than that required to open it fully.* From this reason is some cases *it is arranged that the area acted on by the steam is greater when the valve is open than the valve is closed.*

12.2.4. High Steam and Low Water Safety Valve

The high steam and low water safety valve serves the following *purposes : (i) The steam automatically escapes out when the level of water falls below a certain level. (ii) It automatically discharges the excess steam when the pressure of the steam rises above a certain pressure.* This is a single device in which two valves are combined in one to serve the above mentioned purposes.

It is generally used on Lancashire or Cornish boiler. It *cannot* be used on *mobile boilers.*

Fig. 12.8 shows the details of Hopkinson's high steam and low water safety valve. It consists of valve V resting on the valve seat VS and the valve U loaded with the weights W rests on the valve V. Now, when the steam pressure rises above the rated pressure of the boiler, the valve V is uplifted alongwith the valve U and the steam escapes out. Therefore, in case when the steam pressure exceeds the rated pressure, top valve acts as a lever safety valve as shown in Fig. 12.8. F_1 is the fulcrum, W_1 is the weight suspended on one end of the lever L_1.

The most important arrangement is lever L_2 with the fulcrum F_2. On one end is attached to a float E, usually made of tile and on the other end is fixed a balance weight W. When the float E is submersed in water, the lever L_2 is balanced about fulcrum F_2. When the water level falls below a certain level, the weight of the float E increases and it produces a swing towards the right so that the knife edge K lifts the spindle C which opens the valve U and thus the steam rushes out. This escape of steam *acts as a warning for the boiler attendant as it produces sufficient noise.*

12.2.5. Fusible Plug

The function of a fusible plug is to protect the boiler against damage due to overheating for low water level. It is fitted on the fire box crown plate or over the combustion chamber at its appropriate place.

A common from of fusible plug is illustrated in Fig. 12.9. It consists of a hollow gun metal body screwed into the fire box crown. The body has a hexagonal flange to tighten it into the shell. A gun metal plug having a hexagonal flange is screwed into the gun metal body. There is another hollow gun metal plug separated from the metal plug by an annulus of fusible metal. The fusible metal is protected from fire by flange on the hollow gun metal plug.

Under normal condition when the water-level in the boiler shell is normal, the fusible plug is fully submerged under water. In this case, the heat from the fusible plug is being conducted to water which keeps the fusible metal at an almost constant temperature and *below its melting point.* But when the water level falls below the fusible plug, it gets uncovered from water and is exposed to steam. The heat conduction from the fusible plug to steam is very little compared with that to water. Hence fusible plug becomes overheated and it melts with the result that the hollow gun metal *plug falls down making a hole. The steam and water being under pressure immediately rush to fire box and extinguish the fire.*

The fusible plugs should generally be renewed after a period of about two years as they are liable to become defective over a long period of use (because they are subjected to heat on one side and scale deposits on the other).

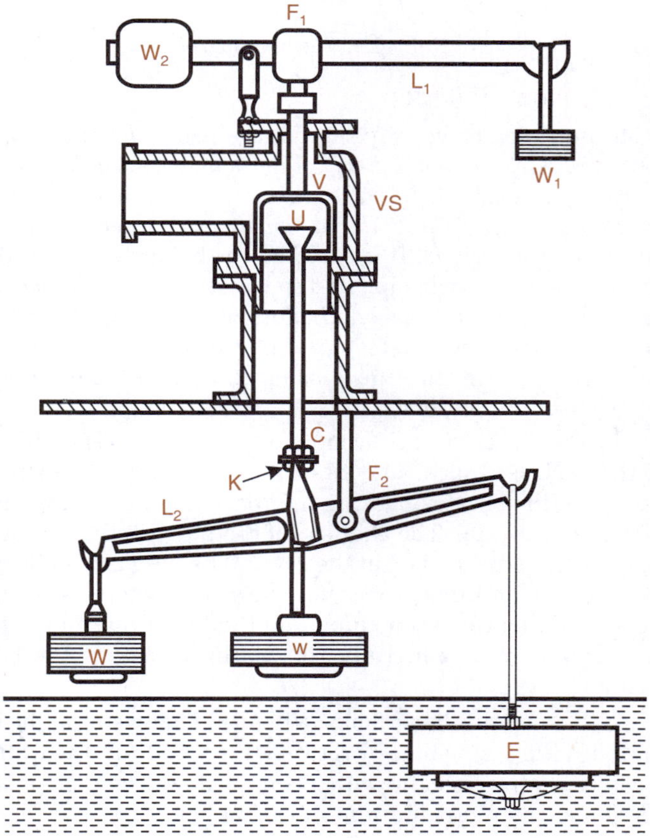

W_1 = Main weight
W_2 = Counter weight
L_1, L_2 = Levers
V = Valve
U = Hemispherical valve
VS = Valve seat
F_1 = Fulcrum (upper lever L_1)
F_2 = Fulcrum (lower level L_2)
E = Float
K = Knife edge
C = Spindle
w = Dead weight
W = Balance weight

Fig. 12.8. High steam and low water safety valve.

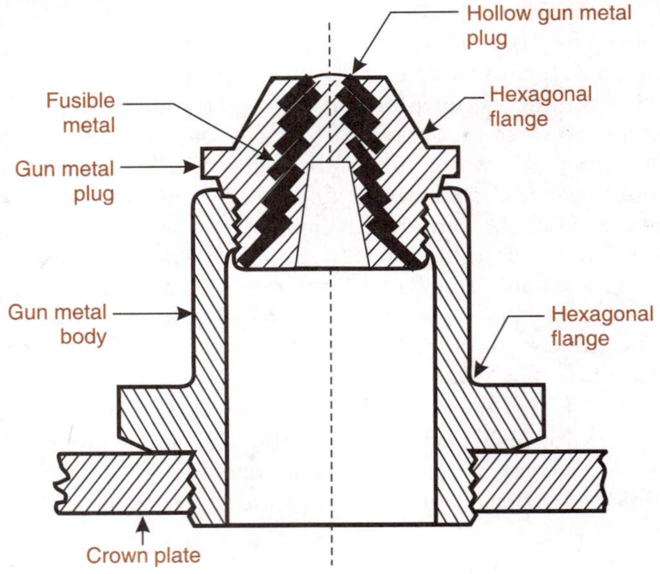

Fig. 12.9. Fusible plug.

12.2.6. Blow-off Cock

A blow-off cock or valve performs the *two functions* : (*i*) *It may discharge a portion of water when the boiler is in operation to blow out mud, scale or sediments perodically.* (*ii*) *It may empty the boiler when necessary for cleaning, inspection and repair.* It is fitted on the boiler shell directly or to a short branch pipe at the *lowest part of the water space*. When more than one boilers are working and they drain in the same waste pipe line, an isolating valve is necessary to prevent the discharge of one boiler, from entering into the other.

Fig. 12.10 shows a common type of plug. The plug *P* of the cock is conical and fits into the casing *C* which is packed with asbestos packing in grooves round the top and bottom of the plug. The shank *S* of the plug passes through a gland and stuffing box in the cover. The plug is held down by a yoke *Y* and two studs (not shown). *A* are the vertical slots for fixing the box spanner, on the top of the yoke. The plug spindle *S* is generally rotated by means of the box spanner.

The plug *P* has a hole. When this hole is brought in line with the casing hole by rotating the spindle *S*, the water flows out of it. And the water cannot flow when the solid portion of the plug is infront of casing hole.

12.2.7. Feed Check Valve

The function of a feed check valve is to control the supply of water to the boiler and to prevent the escaping of water from the boiler when the pump pressure is less or pump is stopped. The feed check valve is fitted in the water space of the boiler *slightly below* the normal level of the water.

Fig. 12.10 shows a common design of a feed check valve. It consists of a check valve *CV* which moves automatically up and down under the pressure of water on its gun metal seat. *FV* is the feed check valve which can be raised or lowered on its gun metal seat, thereby opening the delivery passage and its opening controls simultaneously lift of the check valve *CV*. *F* is the flange which is bolted to the front end of the boiler shell at a point from which an internally perforated pipe leads the feed water and distributes it near the working level of the water in the boiler. The feed check valve *CV* is operated by the handwheel *H* to control the supply of water to the boiler to maintain the water level constant.

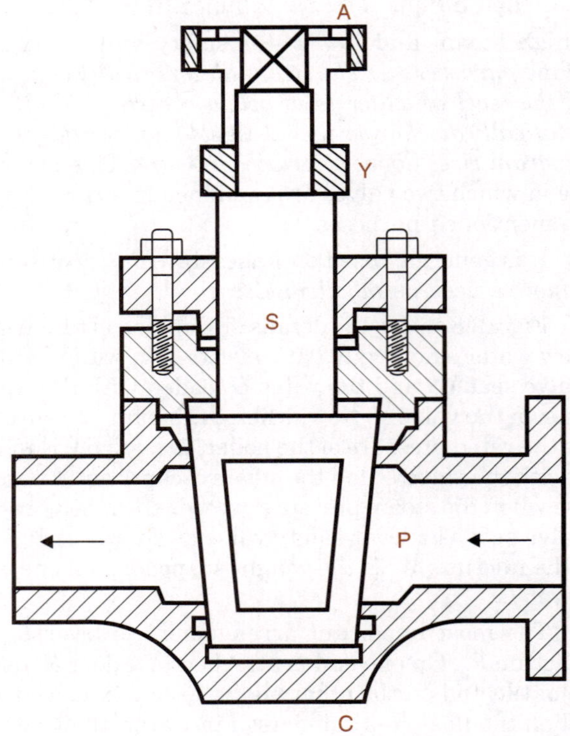

C = Casing *A* = Vertical slots
S = Shank *Y* = Yoke
P = Plug

Fig. 12.10. Blow-off cock.

During normal operation the feed valve may be lifted due to the pressure of water from the feed pump and water may be fed into the boiler. When the pump stops working, the pump pressure is less than the boiler pressure, the valve may be closed to its seat automatically due to the pressure of water from the boiler.

It is very important to check that the non-return valve (*CV*) is in good condition as it is subjected to automatic upward and downward movement due to the fluctuations in the feed water pressures. The valve *CV* can be inspected for repair by fully closing the valve *FV*. If the boiler is in operation it should be seen that there is sufficient water in the boiler before such work is undertaken.

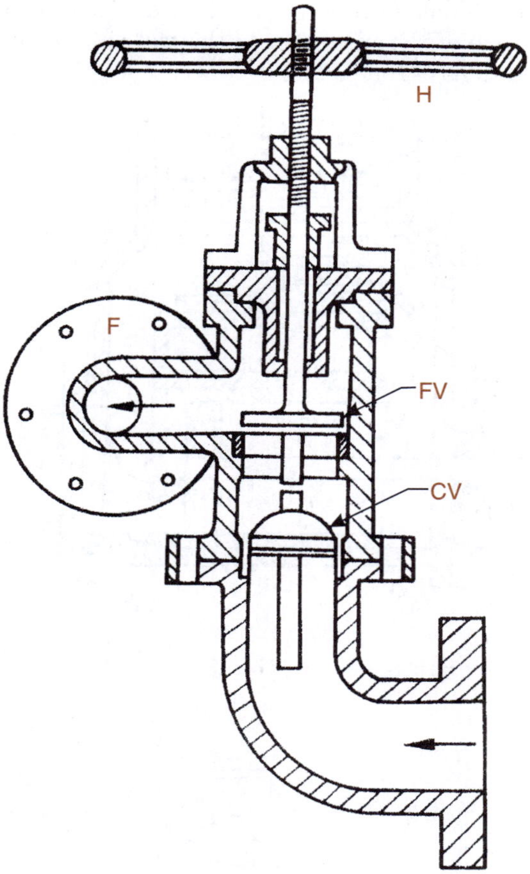

FV = Feed check valve *F* = Flange
CV = Check valve *H* = Hand wheel

Fig. 12.11. Feed check valve.

12.2.8. Junction or Stop Valve

A *junction valve* is a valve which is *placed directly over a boiler* and connected to a steam pipe which carries steam to the engine. If a valve is placed in the steam pipe leading steam to the engine and placed near the engine, it is usually termed as **stop valve.** Junction valve and stop valve are essentially the same, the *larger sizes are called junction valves* and the *smaller sizes stop valves.* The *function of the stop valve* or *junction valve is to regulate the flow of steam from one steam pipe to the other or from the boiler to the steam pipe.*

The common type of stop valve is shown in Fig. 12.12. It consists of a valve *V* which is attached to the valve spindle *VS*. The spindle is connected to hand wheel *H* and passes through a screwed portion like the nut and through a gland or stuffing box *G* to prevent leakage of steam.

On turning the hand wheel *H* the spindle is raised or lowered depending upon the sense of rotation of wheel (clockwise for lowering and anticlockwise for raising). The passage of steam flow on opening is shown by arrows.

The size of the valve is designed by the pipe diameter it connects (e.g., 5 cm, 7 cm and 10 cm stop valve).

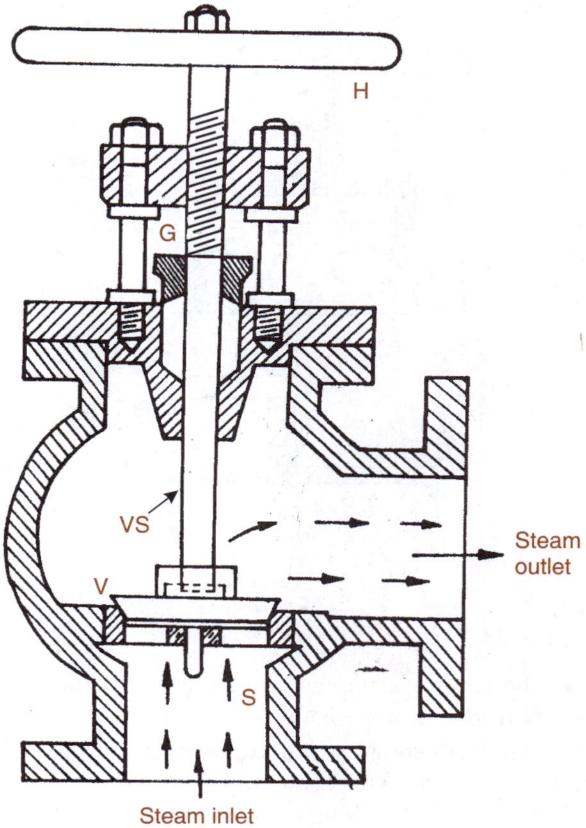

H = Hand wheel *V* = Valve
VS = Valve spindle *S* = Valve seat
G = Gland

Fig. 12.12. Stop valve.

Note. In a locomotive boiler the supply of steam is regulated by means of a regulator which is placed inside the boiler shell.

12.3. ACCESSORIES

Commonly used accessories are discussed as follows :

12.3.1. Feed Pumps

The feed pump is a pump which is used to *deliver feed water to the boiler.* It is desirable that the quantity of water supplied should be *at least equal to that evaporated and supplied to the engine.* Two types of pumps which are commonly used as feed pumps are (*i*) reciprocating pump and (*ii*) rotary pump.

The reciprocating pump consists of a pump cylinder and a piston. Inside the cylinder reciprocates a piston which displaces water. The reciprocating pump may be of two types :

1. Single acting pump
2. Double acting pump

In a single acting pump the water is displaced by one side of the piston only and so the water is discharged in alternate strokes.

In a double acting pump, the water is discharged in each stroke of the piston since the water is displaced by both the sides of the piston.

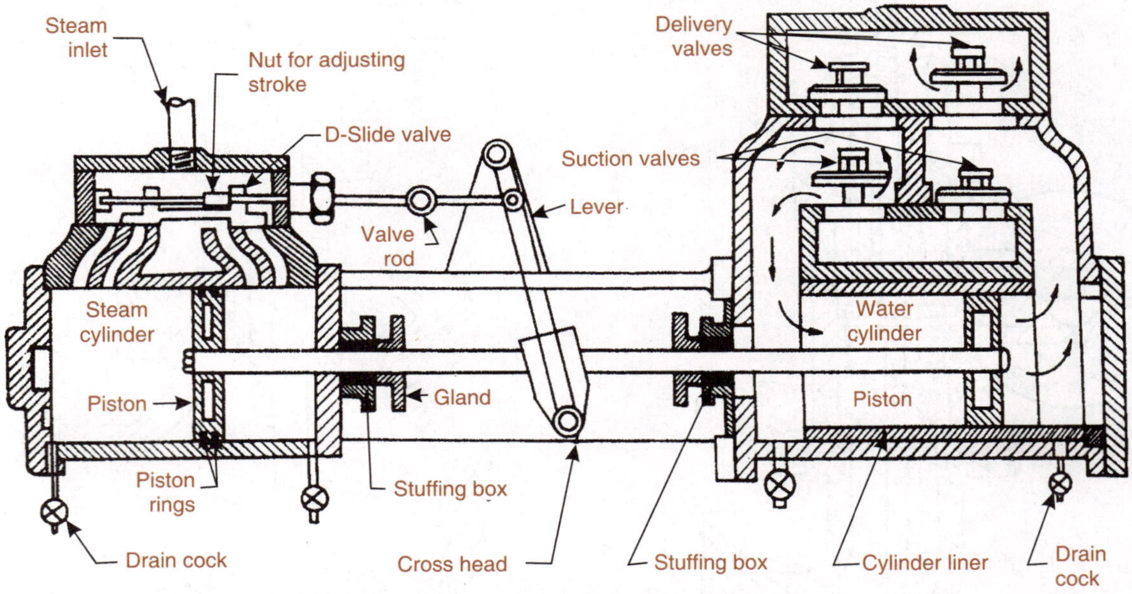

Fig. 12.13. Feed pump.

The reciprocating feed pumps are continuously run by steam from the same boiler to which water is to be fed.

Fig. 12.13 shows a *duplex direct acting steam pump.* Here there are two single steam cylinders placed side by side. Slide valves distribute the steam in each cylinder. The slide valve in each cylinder steam chest is operated by the crosshead on the piston rod of the opposite cylinder, through an arrangement of rods and rocker arms. The feed pump is generally double acting. On each side of the pump plunger there are suction and discharge valves. The pumps work alternately and consequently continuous flow of water is maintained. *Double feed pump is commonly employed for medium size boilers.*

Rotary feed pumps are of centrifugal type and are commonly run either by a small steam turbine or by an electric motor. A rotary pump consists of a casing and a rotating element known as impeller which is fitted over a shaft. It utilises the centrifugal force of the rotating impeller for pumping the liquid from one place to the other.

12.3.2. Injector

The function of an injector is to feed water into the boiler. It is commonly employed for vertical and locomotive boilers and does not find its application in large capacity high pressure boilers. It is also used where the *space is not available for the installation of a feed pump.*

In an injector the water is delivered to the boiler by steam pressure ; the *kinetic energy of steam is used to increase the pressure and velocity of the feed water.*

Fig. 12.14 shows an injector. It consists of a spindle P, a steam cone S, a combining cone K, a delivery cone D,

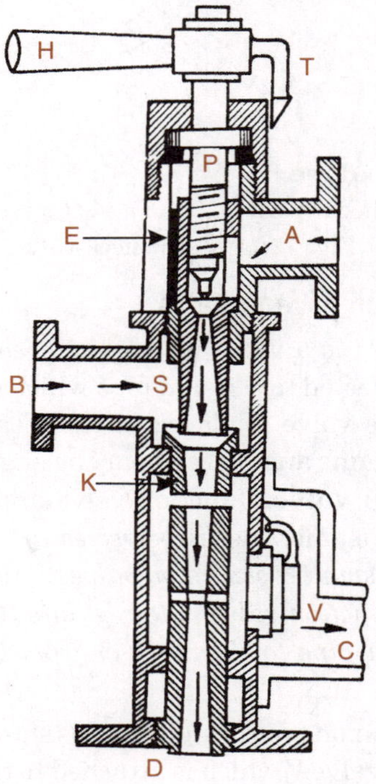

S = Steam cone
K = Combining cone
D = Delivery cone
P = Spindle
A = Steam pipe
B = Water pipe
E = Key
H = Handle
T = Pointer
V = Valve
C = Overflow pipe

Fig. 12.14. Injector.

and a handle *H*, with a pointer *T*. The spindle's upper end is provided with a handle while the lower end serves the purpose of a valve. The pointer on the handle indicates the 'shut' and 'open' position of the valve. The lower part of the spindle has a screw which works in a nut which is integral part of the steam cone. The key *E* checks the rotation of steam cone. With the rotation of the handle steam cone moves up or down and consequently the valve controls the steam flow through the steam cone. The steam enters through the steam pipe *A*, while the feed water enters through the water pipe *B*. The flow of water is also regulated due to sliding motion of the steam cone by its lower end. The water mixes with the steam at the combining cone where it is condensed. The mixture then passes through the delivery cone and there its kinetic energy is converted into pressure energy. The final pressure must be greater than the steam pressure of boiler otherwise water will not enter into the boiler. The excess water finds its way through the overflow pipe.

Advantages :

1. Low initial cost.
2. Simplicity.
3. Compactness.
4. Absence of dynamic parts.
5. Thermal efficiency very high (about 99%).
6. Ease of operation.

Disadvantages :

1. Pumping efficiency is low.
2. It cannot force very hot water.
3. Irregularity of operation under extreme variation in steam pressure.

Note. An injector is more efficient than a feed pump because all the heat in the operating steam is returned to boiler and in addition to performing the work of a pump, the injector acts as a feed water heater. But when a large quantity of feed water is involved (*e.g.*, marine and large installations) feed pumps are employed because they have greater reliability and require lesser amount of attention.

12.3.3. Economiser

An economiser is a device in which the waste heat of the flue gases is utilised for heating the feed water.

Economiser are of the two types : (*i*) *Independent type*, and (*ii*) *Integral type*. *Former* is installed in chamber apart from the boiler setting. The chamber is situated at the passage of the flow of the flue gases from the boiler or boiler to the chimney. *Latter* is a part of the boiler heating surface and is installed within the boiler setting.

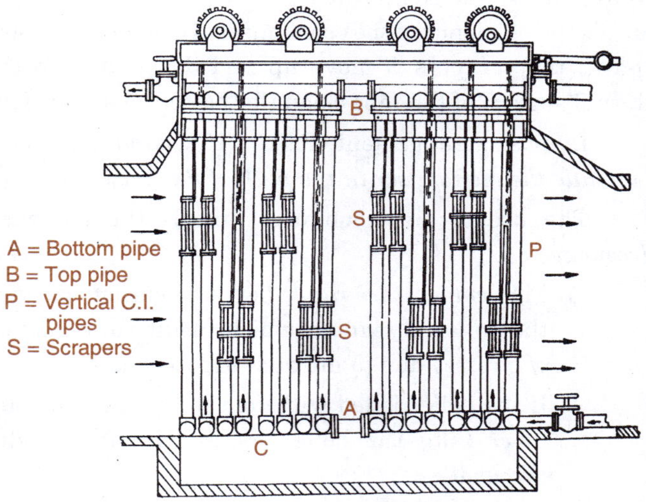

A = Bottom pipe *P* = Vertical C.I. pipes
B = Top pipe *S* = Scrapers

Fig. 12.15. Economiser.

Fig. 12.15 shows an independent type vertical tube economiser (called Green's economiser). It is employed for boilers of medium pressure range upto about 25 bar. It consists of a large number of vertical cast iron pipes *P* which are connected with two horizontal pipes, one at the top and the other at the bottom. *A* is the bottom pipe through which the feed water is pumped into the economiser. The water comes into the top pipe *B* from the bottom pipe (*via* vertical pipes) and finally flows to the boiler. The flue gases move around the pipes in the direction opposite to the flow of water. Consequently, heat transfer through the surfaces of the pipes takes place and water is thereby heated.

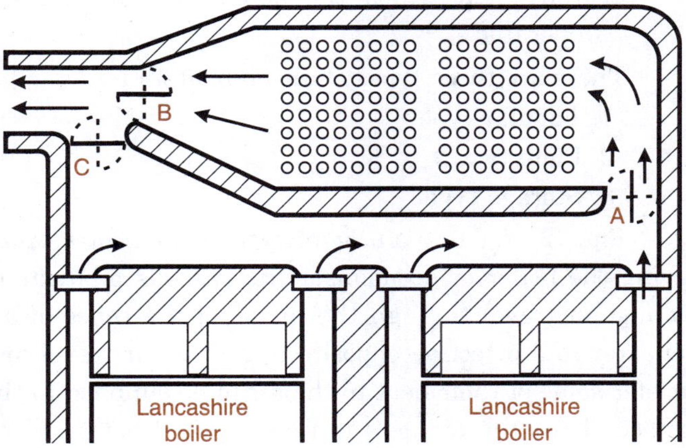

Fig. 12.16. By-pass arrangement of flues.

A blow-off cock is provided at the back end of vertical pipes to remove sediments deposited in the bottom boxes. The soot of the flue gases which gets deposited on the pipes reduces the efficiency of the economiser. To prevent the soot deposit, the scrapers *S* move up and down to keep the external surface of the pipe clean (for better heat transfer).

By-pass arrangement (Fig. 12.16) *enables to isolate or include the economiser in the path of flue gases.*

The use of an economiser entails the following *advantages* :

1. The *temperature range* between various parts of the boiler is *reduced* which results in *reduction of stresses* due to unequal expansion.
2. If the boiler is fed with cold water it may result in *chilling* the boiler metal. Hot feed water *checks* it.
3. *Evaporative capacity* of the boiler is *increased.*
4. *Overall efficiency* of the plant is *increased.*

12.3.4. Air Preheater

The function of the air pre-heater is to increase the temperature of air before it enters the furnace. It is generally *placed after the economiser* ; so the flue gases pass through the economiser and then to the air preheater.

An air-preheater consists of plates or tubes with hot gases on one side and air on the other. It preheats the air to be supplied to the furnace. *Preheated air accelerates* the *combustion and facilitates the burning of coal.*

Degree of preheating depends on :

(*i*) Type of fuel,

(*ii*) Type of fuel burning equipment, and

(*iii*) Rating at which the boiler and furnace are operated.

There are three types of air preheaters :

1. Tubular type
2. Plate type
3. Storage type.

Fig. 12.17 shows a *tubular type air preheater*. After leaving the boiler or economiser the gaseous products of combustion travel through the inside of the tubes of air preheater in a direction opposite to that of air travel and transfer some of their heat to the air to be supplied to the furnace. Thus the air gets initially heated before being supplied to the furnace. The gases reverse their direction near the bottom of the air heater, and a soot hopper is fitted to the bottom of air heater casing to collect soot.

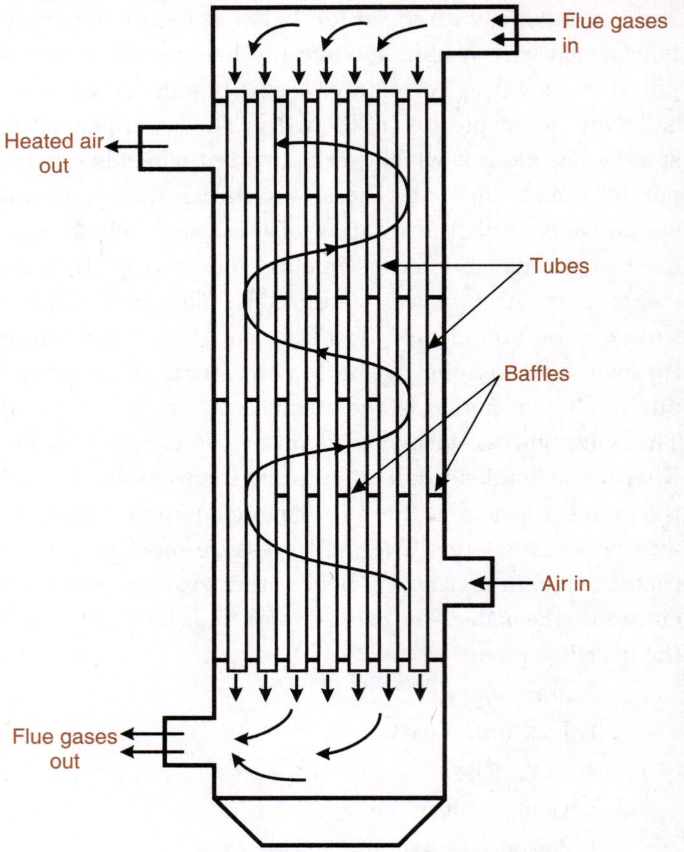

Fig. 12.17. Tubular type air preheater.

In the *plate type air preheater* the air absorbs heat from the hot gases being swept through the heater at high velocity on the opposite side of a plate.

Fig. 12.18 shows a self explanatory sketch of a storage type air preheater (heat exchanger).

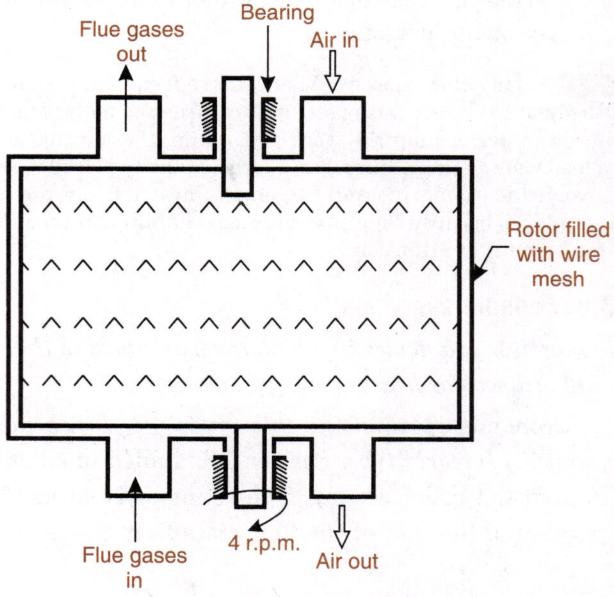

Fig. 12.18. Storage type air preheater.

Finally the gases escape to the atmosphere through the stack (chimney). The temperature of the gases leaving the stack should be kept as low as possible so that there is minimum loss of heat to the stack.

Note. Storage type air preheaters are employed widely in *larger plants*.

12.3.5. Superheater

The function of a superheater is to increase the temperature of the steam above its saturation point. The superheater is very important accessory of a boiler and can be used both on fire-tube and water-tube boilers. The small boilers are not commonly provided with a superheater.

Superheated steam has the following ***advantages*** :

(*i*) *Steam consumption* of the engine or turbine is *reduced*.

(*ii*) *Losses due to condensation* in the cylinders and the steam pipes are *reduced.*

(*iii*) *Erosion of turbine blade is eliminated.*

(*iv*) *Efficiency* of the steam plant is *increased.*

Superheaters are located in the path of the furnace gases so that heat is recovered by the superheater from the hot gases.

There are two types of superheaters :

1. Convective superheater
2. Radiant superheater.

Convective superheater *makes use of heat in flue gases whereas a* ***radiant superheater*** *is placed in the furnace and wall tubes receives heat from the burning fuel through radiant process. The radiant type of superheater is generally used where a high amount of superheat temperature is required.*

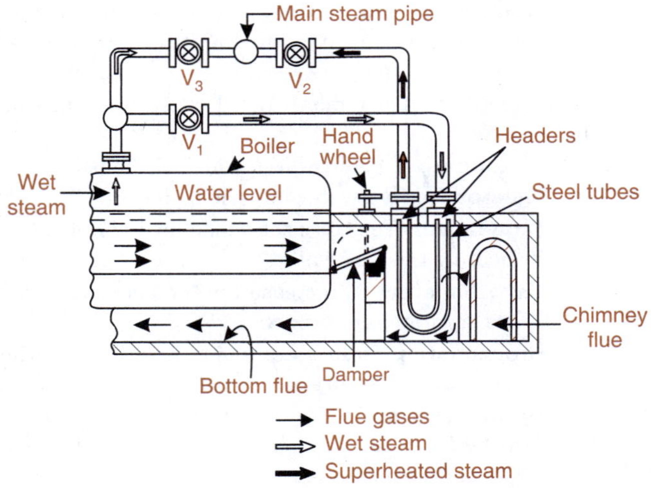

Fig. 12.19. Sugden's superheater.

Fig. 12.19 shows Sugden's superheater installed in a Lancashire boiler. It consists of two steel headers to which are attached solid drawn 'U' tubes of steel. These tubes are arranged in groups of four and one pair of the headers generally carries ten of these groups or total of forty tubes. The steam from the boiler enters and leaves the headers as shown by the arrows. Fig. 12.19 also shows how the steam pipes may be arranged so as to pass the steam through the superheater or direct to the main steam pipe. When the steam is taken from the boiler direct to the main steam pipe, the valves V_1 and V_2 are closed and V_3 is opened ; when the steam is passed through the superheater *i.e.,* when the superheater is in action the valve V_3 is closed the valve V_1 and V_2 are opened.

The path of the gases is controlled by the *damper* which is operated by the hand wheel.

12.3.6. Steam Separator

The steam available from a boiler may be either wet, dry ; or superheated ; but in many cases there will be loss of heat from it during its passage through the steam pipe from the boiler to the engine tending to produce wetness. The use of wet steam in an engine or turbine is uneconomical besides involving some risk ; hence it is usual to endeavour to separate any water that may be present from the steam before the latter enters the engine. This is accomplished by the use of a *steam separator*. Thus *the function of a steam separator is to remove the entrained water particles from the steam conveyed to the steam engine or turbine.* It is installed as close to the steam engine as possible on the main steam pipe from the boiler.

According to the principle of operation the steam separators are classified as follows :

1. Impact or baffle type
2. Reverse current type
3. Centrifugal type.

Fig. 12.20 shows *baffle plate steam separator.* The steam enters the flange *A* and flows down. In its passage it

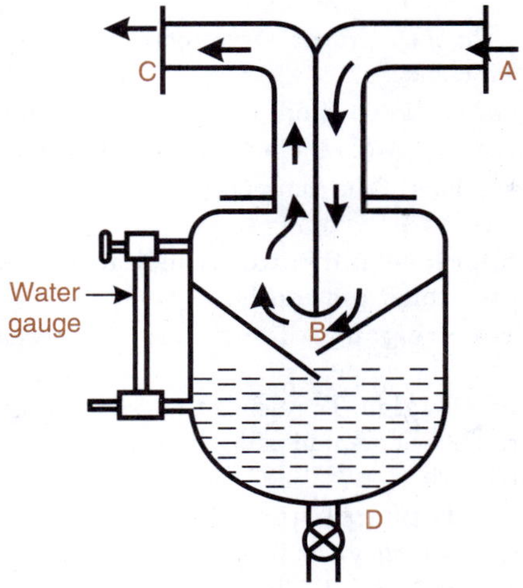

A, C = Flanges *D* = Drain pipe
B = Baffles

Fig. 12.20. Baffle plate steam separator.

strikes the baffles B ; as a result it gets deflected, but water particles having greater density and greater inertia fall to the bottom of the separator. The drier steam discharges through the flange C. To see the level of water collected a water gauge is provided. The water collected in the vessel is removed at intervals through the drain pipe D.

12.3.7. Steam Trap

The function of a steam trap is to drain away automatically the condensed steam from the steam pipes, steam jackets and steam separators without permitting any steam to escape.

The steam traps are classified as follows :

1. Expansion or thermostatic type
2. Bucket or float type.

Fig. 12.21 shows an expansion steam trap (sirius type). A hollow spring tube B of nickel steel contains a liquid which becomes a gas at temperatures above the lowest that the steam is likely to have. The water enters at A, and, being lower in temperature than the steam, it causes the tube B to contracted and the valve is opened. When all the water is discharged and steam enters the trap, the increased temperature converts the liquid in the tube B into the gas, the curved tube tends to straighten itself and consequently the valve V is closed.

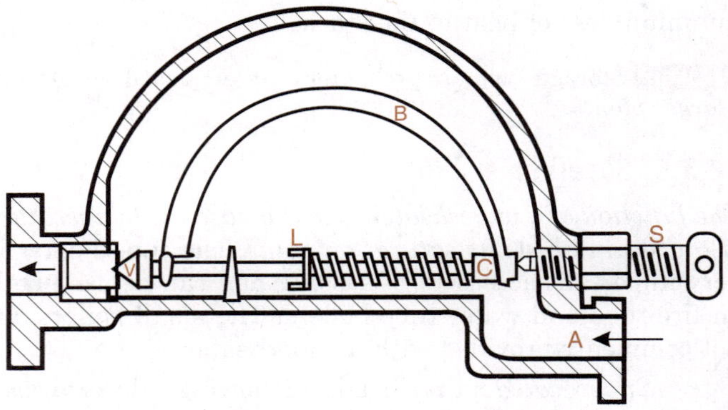

A = Water entry S = Screw
B = Hollow spring tube V = Valve
C = End of the tube L = Lugs

Fig. 12.21. Expansion steam trap.

By adjusting the screw S the trap can be arranged to discharge either continuously or intermittently. The end C of the tube is held against the adjusting screw by the spring, one end of which presses against the lugs L fixed to the casing.

HIGHLIGHTS

1. *Boiler mountings* are different fittings and devices which are necessary for operation and safety of a boiler.
   Examples. Pressure gauge, safety valves, water level indicator, steam stop valve, feed check valve, blow-off cock etc.
2. *Boiler accessories* are auxiliary plants required for steam boilers for their proper operation and for the increase of their efficiency.
   Examples. Feed pumps, injector, economiser, air preheater, superheater, steam separator and steam trap.
3. The functions of the commonly used *boiler mountings* are as follows :
   (*i*) **Water level indicator.** To indicate the level of water in the boiler constantly.
   (*ii*) **Pressure gauge.** To measure the pressure exerted inside the vessel.
   (*iii*) **Safety valve.** To release the excess steam when the pressure of steam inside the boiler exceeds the rated pressure.
   (*iv*) **Fusible plug.** To protect the boiler against damage due to overheating for low water level.
   (*v*) **Blow-off cock.** To discharge a portion of water when the boiler is in operation to blow out mud, scale or sediments periodically or to empty the boiler when necessary for cleaning, inspection and repair.
   (*vi*) **Feed check valve.** To control the supply of water to the boiler and to prevent the escaping of water from the boiler when the pump pressure is less or pump is stopped.
   (*vii*) **Junction or stop valve.** To regulate the flow of steam from one steam pipe to the other or from boiler to the steam pipe.
4. The functions of commonly used *accessories* are as follows :
   (*i*) **Feed pump.** To deliver feed water to the boiler.
   (*ii*) **Injector.** To feed water into the boiler.
   (*iii*) **Economiser.** To utilise the waste heat of flue gases for heating the feed water.
   (*iv*) **Air preheater.** To increase the temperature of air before it enters the furnace.
   (*v*) **Superheater.** To increase the temperature of the steam above its saturation point.
   (*vi*) **Steam separator.** To remove the entained water particles from the steam conveyed to the steam engine or steam turbine.
   (*vii*) **Steam trap.** To drain away automatically the condensed steam from the steam pipes, steam jackets and steam separators without permitting any steam to escape.

OBJECTIVE TYPE QUESTIONS

Say 'Yes' or 'No' :

1. A boiler can function without boiler mountings.
2. Boiler mountings means boiler accessories.
3. All land boilers should have a fusible plug in each furnace.
4. The efficiency of a boiler decreases with the use of accessories.
5. Water level indicator is used to indicate the level of water in the boiler constantly.
6. Pressure gauge measures absolute pressure of steam in the boiler.
7. A pressure gauge is known as 'compound pressure gauge' if it is designed in such a fashion so as to measure pressures above and below the atmosphere on the same dial.
8. The double tube Bourdon gauge is less rigid than the single tube and more suitable for locomotive and portable boilers.
9. Only one safety valve should be used on each boiler.
10. Dead weight safety valve is quite simple in design.
11. Dead weight safety valve is mainly used for low pressures, low capacity, stationary boilers of Cornish and Lancashire types.
12. A spring loaded safety valve is most suitable for locomotives and marine engines.
13. High steam and low water safety valve is most suitable for mobile boilers.
14. A fusible plug is used to protect the boiler against damage due to overheating for low water level.
15. A blow-off cock is used to blow-off steam from the boiler if the pressure of the steam becomes excessive.
16. Feed check valve is used to regulate the flow of steam from one steam pipe to the other.
17. A stop valve is used to control the supply of water to the boiler.
18. A feed pump is a pump which is used to deliver feed water to the boiler.
19. Double feed pump is commonly employed for medium size boilers.
20. An injector injects feed water into the boiler.
21. An injector is less efficient than a feed pump.
22. An economiser preheats the air before it is supplied to the boiler.
23. An air preheater is generally placed after the economiser.
24. The function of a superheater is to increase the temperature of the steam above its saturation point.
25. The function of a steam separator is the same as that of a steam trap.

ANSWERS

1. No	2. No	3. Yes	4. No
5. Yes	6. No	7. Yes	8. No
9. No	10. Yes	11. Yes	12. Yes
13. No	14. Yes	15. No	16. No
17. No	18. Yes	19. Yes	20. Yes
21. No	22. No	23. Yes	24. Yes
25. No.			

THEORETICAL QUESTIONS

1. What is the function of boiler mountings ? Can a boiler work without mountings ?
2. How do accessories differ from mountings ?
3. Enumerate the various accessories normally used in a steam generating plant.
4. What is the function of a safety valve ? State the minimum number of safety valves to be used on a boiler.
5. What is a fusible plug and state where it is located in a boiler ?
6. Explain with neat sketches any three of the following mountings :
   (*i*) Water level indicator (*ii*) Pressure gauge
   (*iii*) Feed check valve (*iv*) Blow-off cock
   (*v*) High steam and low water safety valve
   (*vi*) Junction or stop valve.
7. Explain with neat sketches any two of the following boiler accessories :
   (*i*) Injector (*ii*) Superheater
   (*iii*) Air preheater (*iv*) Economiser.
8. What is a steam trap ? Explain with a neat sketch expansion type of steam trap.
9. What is the function of a steam separator ? Discuss with a neat sketch anyone type of steam separators.

13 Draught

13.1. DEFINITION AND CLASSIFICATION OF DRAUGHT

The small pressure difference which causes a flow of gas to take place is termed as a **draught.** The function of the draught, in case of a boiler, is to force air to the fire and to carry away the gaseous products of combustion. In a boiler furnace proper combustion takes place only when sufficient quantity of air is supplied to the burning fuel.

The draught may be classified as :

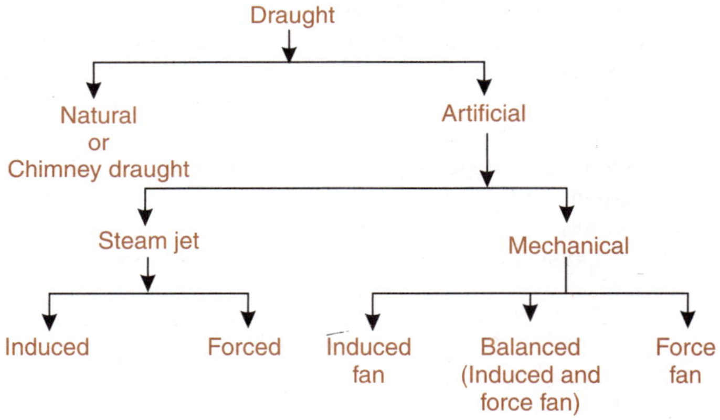

13.2. NATURAL DRAUGHT—CHIMNEY

Natural draught is obtained by the use of a chimney. The chimney in a boiler installation performs one or more of the following functions : (*i*) *It produces the draught whereby the air and gas are forced through the fuel bed, furnace, boiler passes and settings* ; (*ii*) *It carries the* products of combustion to such a height before discharging them that they will not be objectionable or injurious to surroundings. A *chimney* is vertical tubular structure built either of masonry, concrete or steel. *The draught produced by the chimney is due to the density difference between the column of hot gases inside the chimney and the cold air outside.*

Fig. 13.1 shows a diagrammatic arrangement of a chimney of height 'H' metres above the grate.

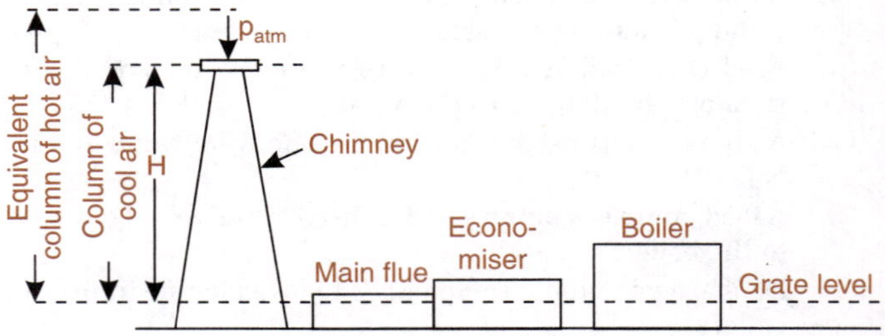

Fig. 13.1

We have $p_1 = p_a + \rho_g \cdot gH$

where, p_1 = Pressure at the grate level (Chimney side),

p_a = Atmospheric pressure at chimney top,

$\rho_g \cdot gH$ = Pressure due to the column of hot gas of height H metres, and

ρ_g = Average mass density of hot gas.

Similarly, $p_2 = p_a + \rho_a \cdot gH$

where, p_2 = Pressure acting on the grate on the *open side,*

$\rho_a \cdot gH$ = Pressure exerted by the column of cold air outside the chimney of height H metres, and

ρ_a = Mass density of air outside the chimney.

∴ Net pressure difference causing the flow through the combustion chamber,

$$\Delta p = p_2 - p_1 = (\rho_a - \rho_g)\, gH \qquad ...(13.1)$$

This *difference of pressure causing the flow of gases is known as '**static draught**'*. Its value is small and is generally *measured by a water manometer*.

It may be noted that this pressure difference in chimney is generally *less* than 12 mm of water.

13.3. CHIMNEY HEIGHT AND DIAMETER

Let us assume that the volume of products of combustion is equal to the volume of air supplied both reduced to the same temperature and pressure conditions.

Let, m_a = Mass of air supplied per kg of fuel,

$m_a + 1$ = Mass of chimney gases,

T_a = Absolute temperature of atmosphere, and

T_g = Average absolute temperature of chimney gases.

Also, $$\frac{\text{Mass of hot gases}}{\text{Mass of air}} = \frac{m_a+1}{m_a},$$

temperature and pressure being same.

Now, $$\rho_a = \frac{p}{RT_a} = \frac{1.01325 \times 10^5}{287} \cdot \frac{1}{T_a}$$

$$= 353 \cdot \frac{1}{T_a} \quad ...(13.2)$$

and, $$\rho_g = \frac{p}{RT_g} \cdot \left(\frac{m_a+1}{m_a}\right)$$

$$= \frac{1.033 \times 10^5}{287} \cdot \frac{1}{T_g} \cdot \left(\frac{m_a+1}{m_a}\right)$$

$$= 353 \cdot \frac{1}{T_g} \cdot \left(\frac{m_a+1}{m_a}\right) \quad ...(13.3)$$

Inserting the values of ρ_a and ρ_g into eqn. (13.1), we get

$$\Delta p = 353\, gH \left[\frac{1}{T_g} - \frac{1}{T_g} \cdot \left(\frac{m_a+1}{m_a}\right)\right] \text{(N/m}^2\text{)} \quad ...(13.4)$$

Assuming that the draught pressure Δp produced is equivalent to H_1 metres height of burnt gases, we have :

$$\Delta p = \rho_g \cdot gH_1 = 353 \left(\frac{m_a+1}{m_a}\right) \cdot \frac{1}{T_g}\, g\, H_1 \quad ...(13.5)$$

Equating eqns. (13.4) and (13.5), we get

$$353 \left(\frac{m_a+1}{m_a}\right) \frac{1}{T_a}\, g.\, H_1 = 353\, g\, H \left[\frac{1}{T_a} - \frac{1}{T_g}\left(\frac{m_a+1}{m_a}\right)\right]$$

$$\frac{353}{T_a}\left(\frac{m_a+1}{m_a}\right) g\, H_1 = \frac{353}{T_g}\, g\, H - \frac{353}{T_g}\, g\, H \left(\frac{m_a+1}{m_a}\right)$$

$$\frac{353}{T_a}\left(\frac{m_a+1}{m_a}\right) g\, H_1 = \frac{353}{T_g}\left(\frac{m_a+1}{m_a}\right) g\, H \left[\left(\frac{m_a}{m_a+1}\right)\frac{T_g}{T_a} - 1\right]$$

$$\therefore \quad H_1 = H \left[\left(\frac{m_a}{m_a+1}\right)\frac{T_g}{T_a} - 1\right] \quad ...(13.6)$$

Due to losses at various sections along the path of the flue gas, the actual draught available is *always less* than that given by the eqn. (13.4).

If h_w is the height, in mm of a column of water which will produce the pressure Δp, then

$$h_w = 353\, H \left[\frac{1}{T_a} - \frac{1}{T_g}\left(\frac{m_a+1}{m_a}\right)\right] \quad ...(13.7)$$

(1 mm of water = 9.81 Pa)

The height h_w would be shown by the use of a U-tube manometer.

Note. The formula as expressed by eqn. (13.7) is used for numerical calculation work only.

Chimney diameter :

Assuming no loss, the velocity of the gases passing through the chimney is given by

$$C = \sqrt{2gH_1}$$

If the pressure loss in the chimney is equivalent to a hot-gas column of h' metres, then

$$C = \sqrt{2g\,(H_1 - h')} = 4.43\sqrt{H_1 - h'}$$

$$= 4.43\sqrt{H_1}\sqrt{1 - \frac{h'}{H_1}} = K\sqrt{H_1} \quad ...(13.8)$$

where, $$K = 4.43\sqrt{1 - \frac{h'}{H_1}}$$

The value of K : 0.825 .... For *brick chimneys*, and
1.1 .... For *steel chimneys*

The mass of the gases flowing through any cross-section of the chimney is given by

$$\dot{m}_g = \rho_g \,.\, A \,.\, C \text{ kg/s}$$

or $$\dot{m}_g = \rho_g \cdot \frac{\pi}{4} D^2 .\, C$$

$$\therefore \quad D^2 = \frac{\dot{m}_g}{\rho_g . C} \times \frac{4}{\pi}$$

or $$D = 1.128\sqrt{\frac{\dot{m}_g}{\rho_g . C}} \quad ...(13.9)$$

13.4. CONDITION FOR MAXIMUM DISCHARGE THROUGH A CHIMNEY

The chimney draught is most effective when the maximum weight of hot gases is discharged in a given time, and it will be shown that this occurs when the absolute temperature of the chimney gases bears a certain relation to the absolute temperature of the outside air.

We know that the velocity of gas through the chimney, assuming the losses to be negligible, is given by

$$C = \sqrt{2gH_1}, \quad \text{where } h' = 0$$

Inserting the value of H_1 from eqn. (13.6)

$$C = \sqrt{2gH\left[\left(\frac{m_a}{m_a+1}\right)\frac{T_g}{T_a} - 1\right]} \quad ...(13.10)$$

The density of the hot gas is given by

$$\rho_g = \frac{p}{RT_g} \quad ...(13.11)$$

The mass of gas discharged per second,

$$m_g = A \times C \,.\, \rho_g$$

Inserting the values of C and ρ_g from eqns. (13.10) and (13.11), we get

$$m_g = A\sqrt{2gH\left[\left(\frac{m_a}{m_a+1}\right)\frac{T_g}{T_a} - 1\right]}\ \frac{p}{RT_g}$$

$$m_g = \frac{K}{T_g}\sqrt{\left(\frac{m_a}{m_a+1}\right)\frac{T_g}{T_a} - 1} \quad ...(13.12)$$

where constant $K = \dfrac{A \times p \times \sqrt{2gH}}{R}$

The value of m_g will be *maximum*, if

$\dfrac{dm_g}{dT_g} = 0$ as T_a and m_a are fixed quantities

$$\therefore \quad \frac{d}{dT_g}\left[\frac{K}{T_g}\sqrt{\left(\frac{m_a}{m_a+1}\right)\frac{T_g}{T_a} - 1}\right] = 0$$

or,
$$\frac{d}{dT_g}\left[\frac{1}{T_g}\sqrt{\left(\frac{m_a}{m_a+1}\right)\frac{T_g}{T_a} - 1}\right] = 0$$

or,
$$\frac{d}{dT_g}\left[\frac{(ZT_g - 1)^{1/2}}{T_g}\right] = 0,$$

where $Z = \dfrac{m_a}{m_a+1} \cdot \dfrac{1}{T_a}$

or,
$$\frac{d}{dT_g}\left[(ZT_g - 1)^{1/2} \times T_g^{-1}\right] = 0$$

or, $(Z \,.\, T_g - 1)^{1/2} \,.\, (-1)(T_g)^{-2} + T_g^{-1} \times \frac{1}{2}(ZT_g - 1)^{-1/2} \times Z = 0$

or,
$$\frac{-(ZT_g - 1)^{1/2}}{T_g^{\,2}} + \frac{Z}{2T_g(ZT_g - 1)^{1/2}} = 0$$

or,
$$\frac{-2(ZT_g - 1) + ZT_g}{2(T_g)^3(ZT_g - 1)^{1/2}} = 0$$

or,
$$-2(ZT_g - 1) + ZT_g = 0$$

or,
$$-2ZT_g + 2 + ZT_g = 0$$

or,
$$ZT_g = 2$$

or,
$$\frac{m_a}{m_a+1} \cdot \frac{T_g}{T_a} = 2$$

i.e.,
$$\frac{T_g}{T_a} = 2\left(\frac{m_a+1}{m_a}\right) \quad ...(13.13)$$

Thus we see that the *absolute temperature of the chimney gases bears a certain ratio to the absolute temperature of the outside air.*

Putting the value of $\dfrac{T_g}{T_a}$ in eqn. (13.6), we get

$$(H_1)_{max} = H\left[\left(\frac{m_a}{m_a+1}\right) \times 2\left(\frac{m_a+1}{m_a}\right) - 1\right]$$

$$= H(2 - 1) = H$$

i.e.,
$$(H_1)_{max} = H \quad ...(13.14)$$

The draught in mm of water column for maximum discharge can be evaluated by inserting the value of T_g/T_a in eqn. (13.7)

$$\therefore \quad (h_w)_{max} = 353\,H\left(\frac{1}{T_a} - \frac{1}{2T_a}\right)$$

$$= \frac{176.5H}{T_a} \text{ mm of water} \quad ...(13.15)$$

13.5. EFFICIENCY OF A CHIMNEY

The temperature of the flue gases leaving a chimney, in case of *natural draught,* is higher than that of flue gases leaving it in case of *artificial draught* system because a certain minimum temperature is needed to produce a given draught with the given height of a chimney. As far as steam generation is concerned, in case of a natural draught system the heat carried away by flue gases is more due to *higher* flue gas temperature. This indicates that the draught is created at the cost of thermal efficiency of the boiler plant installation since a portion of the heat carried away by flue gases to produce the required draught could have been used either in heating the air going to furnace or in heating the feed water going to boiler ; thereby improving the thermal efficiency of the installation. Let,

T' = Absolute temperature of flue gases leaving the chimney to create the draught of h_w mm of water,

T'' = Absolute temperature of flue gases leaving the chimney in case of artificial draught of h_w mm of water, and

c_p = Mean specific heat of flue gases.

The extra heat carried away by 1 kg of flue gas due to higher temperature required to produce the natural draught

$= c_p (T' - T'')$, since T' is *greater* than T''.

The draught pressure produced by the natural draught system in height of hot gases column,

$$H_1 = H\left[\left(\frac{m_a}{m_a+1}\right) \times \frac{T_g}{T_a} - 1\right] \text{ metre head.}$$

The maximum energy this head would give to 1 kg of flue gas which is at the expense of extra heat carried away from the boiler plant

$$= H\left[\left(\frac{m_a}{m_a+1}\right) \times \frac{T_g}{T_a} - 1\right] \times g \text{ Nm/kg (or J/kg)}$$

$$= \frac{H}{J}\left[\left(\frac{m_a}{m_a+1}\right) \times \frac{T_g}{T_a} - 1\right] \times g \times 10^{-3} \text{ kJ/kg}$$

∴ Efficiency of chimney,

$$\eta_{ch} = \frac{\frac{H}{J}\left[\left(\frac{m_a}{m_a+1}\right) \times \frac{T_g}{T_a} - 1\right] \times g \times 10^{-3} \text{ (kJ/kg)}}{c_p (T' - T'') \text{ (kJ/kg)}} \quad ...(13.16)$$

In the equation, $T' = T_g$.

(Is S.I. units J = 1)

The efficiency of a chimney is proportional to the height but even for a very tall chimney the efficiency will be *less than 1%* and thus we see that chimney is *very inefficient as an instrument for creating draught.*

13.6. DRAUGHT LOSSES

The loss in a draught may be due to the *reasons* mentioned below :

(*a*) The frictional resistance offered by the flues and gas passages to the flow of flue gases.

(*b*) Loss near the bends in the gas flow circuit.

(*c*) Loss due to friction head in equipments like grate, economiser, superheater etc.

(*d*) Loss due to imparting velocity to the flue gases.

The loss in draught in a chimney is 20 per cent of the total draught produced by it.

13.7. ARTIFICIAL DRAUGHT

In the boiler installations of today the total static draught required may vary from 30 to 350 mm of water column.

It may not be possible to build a chimney high enough to produce draught of such a large magnitude. To meet this requirement artificial draught system should be used. It may be a *mechanical draught* or a *steam jet draught.* The former is used for central power stations and many other boiler installations while the latter is employed for small installations and in locomotives.

13.7.1. Forced Draught

In a mechanical draught system, the draught is produced by a fan. In a forced draught system, a blower or a fan is installed *near or at the base of the boiler* to force the air through the cool bed and other passages through the furnace, flues, air preheater, economiser etc. It is a *positive pressure draught. The enclosure for the furnace etc., has to be very highly sealed so that gases from the furnace do not leak out in the boiler house.*

13.7.2. Induced Draught

In this system a fan or blower is located *at or near the base of the chimney.* The pressure over the fuel bed is reduced below that of the atmosphere. By creating a partial vacuum in the furnace and flues, the products of combustion are drawn from the main flue and they pass up the chimney. *This draught is used usually when economisers and air preheaters are incorporated in the system.* The draught is similar in action to the natural draught.

13.7.3. Balanced Draught

It is a *combination of the forced and induced draught systems.* In this system the *forced draught fan overcomes the resistance in the air preheater and chain grate stoker while the induced draught fan overcomes draught losses through boiler, economiser, air preheater and connecting flues.*

The *forced draught* entails following *advantages* over *induced draught* :

1. Forced draught fan does not require water cooled bearings.
2. Tendency to air leak into the boiler furnace is reduced.
3. No loss due to in rush of cold air through the furnace doors when they are opened for fire and cleaning fires.
4. Fan size and power required for the same draught are 1/5 to 1/2 of that required from an induced draught fan installation because forced draught fan handles cold air.

13.7.4. Advantages of Mechanical Draught

The mechanical draught possesses the following *advantages* :

1. Easy control of combustion and evaporation.
2. Increase in evaporative power of a boiler.
3. Improvement in the efficiency of the plant.
4. Reduced chimney height.
5. Prevention of smoke.

6. Capability of consuming low grade fuel.
7. Low grade fuel can be used as the intensity of artificial draught is high.
8. The fuel consumption per H.P. due to artificial draught is 15% less than that for natural draught.
9. The fuel burning capacity of grate is 200 to 300 kg/m^2-h with mechanical draught whereas it is hardly 50 to 100 kg/m^2-h with natural draught.

13.7.5. Power Required to Drive Fan

Let p = Draught, P_a,
h_f = Draught produced by the fan, mm,
V = Volumetric flow rate of combustion air at fan conditions, m^3/h, and
η_f = Fan efficiency.

Power required

$$= \frac{pV}{\eta_f} = \frac{\rho g h_f V}{\eta_f}$$

$$= \frac{1000 \times g \times h_f V}{1000 \times \eta_f \times 3600} \text{ watts}$$

$$= \frac{g h_f V}{\eta_f \times 1000 \times 3600} \text{ kW} \quad ...(13.17)$$

$$= 2.725 \times 10^{-6} \frac{hV}{\eta_f} \text{ kW} \quad ...(13.18)$$

(*i*) **Forced draught (F.D.) fan power, P_{FD} :**

Let M = Quantity of fuel burnt per hour,
m_a = Mass of air supplied, kg/kg of fuel,
T_a = Temperature of atmospheric air, and
T_o = Temperature at N.T.P.

Volume of air at T_o,

$$V_o = \frac{m_a M}{\rho}$$

$$\therefore \quad V = \frac{T_a V_o m_a M}{273} \text{ m}^3\text{/h}$$

Substituting in eqn. (13.17), we get

Power of F.D. fan,

$$P_{FD} = \frac{2.725 \times 10^{-6}}{273} \left(\frac{h V_o m_a M T_a}{\eta_f} \right) \text{ kW}$$

$$= 0.998 \times 10^{-8} \left(\frac{h V_o m_a M T_a}{\eta_f} \right) \text{ kW} \quad ...(13.19)$$

(*ii*) **Induced draught (I.D.) fan power, P_{ID} :**

Let m = Mass of air supplied per kg of fuel,
Then $m_a + 1$ = Mass of the products of combustion.
At the same temperature

$$\frac{\rho_g}{\rho_a} = \frac{m_a + 1}{m_a}$$

But $$\rho_a = \frac{273}{T_g} \times \frac{1}{V_o}$$

$$\therefore \quad \rho_g = \frac{273}{T_g} \times \frac{1}{V_o}$$

Volume handled by I.D. fan

$$= \frac{\text{Total mass handled by I.D. fan}}{\text{Density of gases}}$$

$$= \frac{M(m_a + 1)}{\rho_g} = \frac{M(m_a + 1) T_g V_o m_a}{273 (m_a + 1)}$$

$$= \frac{M m_a T_g V_o}{273}$$

Substituting in eqn. (13.18), we get

Power of I.D. fan,

$$P_{ID} = \frac{2.725 \times 10^{-6} \times h M m_a T_g V_o}{273 \times \eta_f}$$

$$= 0.998 \times 10^{-8} \frac{h V m_a M T_g}{\eta_f} \quad ...(13.20)$$

Assuming *same efficiency*, for the *same draught* (neglecting leakage), we have

$$\frac{\text{Power of I.D. fan}}{\text{Power of F.D. fan}} = \frac{T_g}{T_a} \quad ...(13.21)$$

13.7.6. Steam Jet Draught

Steam jet draught is a simple and easy method of producing artificial draught. It may be forced or induced type depending upon where the steam jet to produce draught is located. If the steam jet is *directed into the smoke box* near the stack, the air is *induced* through the flues, the grate and ash pit to the smoke box. If the jet is *located before the grate,* air is *forced* through the fuel bed, furnace and flues to the chimney.

The steam jet draught entails the following *advantages* :

(*i*) Very simple and economical.
(*ii*) Occupies minimum space.
(*iii*) Requires very little attention.
(*iv*) In forced type steam jet draught, the steam keeps the fire bars cool and prevents the adhering of clinker to the fire bars.
(*v*) Several classes of low grade fuels can be used with this system.

WORKED EXAMPLES

Example 13.1. *Calculate the height of chimney required to produce a draught equivalent to 1.7 cm of water if the flue gas temperature is 270°C and ambient temperature is 22°C and minimum amount of air per kg of fuel is 17 kg.*

Solution. Draught in mm of H_2O,

$$h_w = 1.7 \text{ cm} = 17 \text{ mm}$$

Flue gas temperature,

$$T_g = 270 + 273 = 543 \text{ K}$$

Ambient temperature,

$$T_a = 22 + 273 = 295 \text{ K}$$

Minimum amount of air per kg of fuel, $m_a = 17$ kg

Height of chimney, H :

Using the relation,

$$h_w = 353\,H\left[\frac{1}{T_a} - \frac{1}{T_g}\left(\frac{m_a+1}{m_a}\right)\right] \text{ mm of } H_2O$$

$$17 = 353\,H\left(\frac{1}{295} - \frac{1}{543} \times \frac{18}{17}\right) \text{ mm of } H_2O$$

$$= 353\,H\,(0.00339 - 0.00195)$$

$$= 0.508\,H \text{ or } H = 33.46 \text{ m}$$

Hence, *height of chimney* = **33.46 m. (Ans.)**

Example 13.2. *Calculate the mass of flue gases flowing through the chimney when the draught produced is equal to 1.9 cm of water. Temperature of flue gases is 290°C and ambient temperature is 20°C. The flue gases formed per kg of fuel burnt are 23 kg. Neglect the losses and take the diameter of the chimney as 1.8 m.*

Solution. Draught in mm of water,

$$h_w = 1.9 \text{ cm} = 19 \text{ mm}$$

Temperature of flue gases,

$$T_g = 290 + 273 = 563 \text{ K}$$

Ambient temperature,

$$T_a = 20 + 273 = 293 \text{ K}$$

Flue gases formed per kg of fuel burnt,

$$(m_a + 1) = 23 \text{ kg}$$

Diameter of the chimney, $D = 1.8$ m

Mass of flue gases, m_g :

Using the relation,

$$H_1 = H\left[\left(\frac{m_a}{m_a+1} \times \frac{T_g}{T_a}\right) - 1\right]$$

where H_1 is the head in terms of gas column.

Also $$h_w = 353\,H\left[\frac{1}{T_a} - \frac{1}{T_g} \times \left(\frac{m_a+1}{m_a}\right)\right]$$

$$19 = 353\,H\left(\frac{1}{293} - \frac{1}{563} \times \frac{23}{22}\right)$$

$$\left(\begin{array}{ll}\because & m_a + 1 = 23 \text{ kg (given)} \\ i.e., & m_a = 22 \text{ kg}\end{array}\right)$$

$$19 = 353\,H\,(0.00341 - 0.00185)$$

$$= 0.548\,H$$

or $$H = 34.67 \text{ m}$$

$$\therefore \quad H_1 = 34.67\left(\frac{22}{23} \times \frac{563}{293} - 1\right)$$

$$= 29.05 \text{ m of air.}$$

Velocity of gases,

$$C = \sqrt{2gH_1} = \sqrt{2 \times 9.81 \times 29.05}$$

$$= 23.87 \text{ m/s.}$$

Now, mass of flue gases,

$$m_g = A \times C \times \rho_g$$

where, $$\rho_g = 353\left(\frac{m_a+1}{m_a}\right) \cdot \frac{1}{T_g}$$

$$= 353 \times \frac{23}{22} \times \frac{1}{563} = 0.655 \text{ kg/m}^3$$

$$\therefore \quad m_g = \pi/4 \times 1.8^2 \times 23.87 \times 0.655$$

$$= \mathbf{39.8 \text{ kg/s.}} \text{ (Ans.)}$$

Hence *mass of flue gases passing through the chimney* = **39.8 kg/s. (Ans.)**

☞**Example 13.3.** *A chimney of height 32 m is used for producing a draught of 16 mm of water. The temperatures of ambient air and flue gases are 27°C and 300°C respectively. The coal burned in the combustion chamber contains 81% carbon, 5% moisture and remaining ash. Neglecting losses and assuming the value of burnt products equivalent to the volume of air supplied and complete combustion of fuel find the percentage of excess air supplied.*

Solution. Height of the chimney, $H = 32$ m

Draught in mm of water = 16 mm

Ambient air temperature, $T_a = 27 + 273 = 300$ K

Temperature of flue gases, $T_g = 300 + 273 = 573$ K

Percentage of carbon in the fuel = 81%

Percentage of excess air supplied :

Using the relation,

$$h_w = 353\,H\left[\frac{1}{T_a} - \frac{1}{T_g}\left(\frac{m_a+1}{m_a}\right)\right] \text{ mm of water}$$

$$16 = 353 \times 32\left[\frac{1}{300} - \frac{1}{573}\left(\frac{m_a+1}{m_a}\right)\right]$$

$$\frac{16}{353 \times 32} = \frac{1}{300} - \frac{1}{573}\left(\frac{m_a+1}{m_a}\right)$$

$$\frac{m_a+1}{m_a} = 573\left(\frac{1}{300} - \frac{16}{353 \times 32}\right)$$

$$= 573\,(0.003333 - 0.001416) = 1.0984$$

$$\therefore \quad m_a = 10.16 \text{ kg/kg of fuel}$$

Again, $$C + O_2 = CO_2$$

$$12 + 32 = 44$$

Thus 1 kg of carbon requires $\frac{32}{12} = \frac{8}{3}$ kg of oxygen

and $\frac{8}{3} \times \frac{100}{23}$ = 11.6 kg of air. One kg of fuel contains only 0.81 carbon.

∴ Air required for complete combustion

= 11.6 × 0.81 = 9.396 kg/kg of fuel.

Hence, *Percentage excess air supplied*

$$= \frac{10.16 - 9.396}{9.396} \times 100 = \mathbf{8.13\%.} \quad \textbf{(Ans.)}$$

Example 13.4. *To provide a natural draught a chimney of height 16 m is used. Calculate (i) The draught in mm of water when the temperature of chimney gases is such that the mass of the gases discharged is maximum, (ii) If the temperature of flue gases does not exceed 350°C find air supplied per kg of fuel, when discharge is maximum. Take ambient temperature as 20°C.*

Solution. Height of chimney, H = 16 m

Temperature of flue gases, T_g = 350 + 273 = 623 K

Temperature of ambient air, T_a = 20 + 273 = 293 K

(*i*) **Draught, $h_{w(max)}$:**

Using the relation,

$$h_{w(max)} = \frac{176.5\,H}{T_a} \text{ mm of water}$$

$$= \frac{176.5 \times 16}{(273 + 20)} = \mathbf{9.638\ mm.} \quad \textbf{(Ans.)}$$

(*ii*) **Air supplied, m_a :**

The condition for maximum discharge is given by

$$\frac{T_g}{T_a} = 2\left(\frac{m_a + 1}{m_a}\right)$$

$$\frac{(350 + 273)}{(20 + 273)} = 2\left(\frac{m_a + 1}{m_a}\right)$$

$$\therefore \quad \frac{m_a + 1}{m_a} = 1.063$$

or m_a = **15.87 kg/kg of fuel. (Ans.)**

☞ **Example 13.5.** *With a chimney of height 45 metres, the temperature of flue gases with natural draught was 370°C. The same draught was developed by induced draught fan and the temperature of the flue gases was 150°C. Mass of the flue gases formed is 25 kg per kg of coal fired. The boiler house temperature is 35°C. Assuming c_p = 1.004 kJ/kg K for the flue gases determine the efficiency of the chimney.*

Solution. Height of chimney, H = 45 m

Temperature of flue gases with natural draught,

$$T_{g_1} = 370 + 273 = 643 \text{ K}$$

Temperature of flue gases with induced draught fan,

$$T_{g_2} = 150 + 273 = 423 \text{ K}$$

Mass of flue gases formed per kg of coal fired

$$(m_a + 1) = 25 \text{ kg}$$

Boiler house temperature,

$$T_a = 35 + 273 = 308 \text{ K}$$

Specific heat of flue gases,

$$c_p = 1.004 \text{ kJ/kg K}$$

Efficiency of chimney :

Let H_1 be the column of flue gases equivalent of draught produced by artificial draught fan in metre head.

$$\text{Then } H_1 = H\left[\left(\frac{m_a}{m_a + 1}\right)\frac{T_g}{T_a} - 1\right]$$

$$= 45\left(\frac{24}{25} \times \frac{643}{308} - 1\right) \quad \left(\because \begin{matrix} m_a + 1 = 25 \text{ kg} \\ \text{and} \quad m_a = 24 \text{ kg}\end{matrix}\right)$$

= 45.18 metre head

= 45.18 × 9.81 Nm/kg

= 45.18 × 9.81 × 10^{-3} kNm/kg

= 0.443 kJ/kg

Additional heat lost to natural draught per kg mass of the gases

$$= c_p\,(T_{g_1} - T_{g_2}) = 1.004\,(643 - 423)$$

= 220.88 kJ/kg

$$\therefore \quad \eta_{chimney} = \frac{0.443}{220.88} \times 100 = \mathbf{0.2\%.} \quad \textbf{(Ans.)}$$

☞**Example 13.6.** *Determine the height and diameter of the chimney used to produce a draught for a boiler which has an average coal consumption of 1800 kg/h and flue gases formed per kg of coal fired are 14 kg. The pressure losses through the system are given below :*

Pressure loss in fuel bed = 7 mm of water, pressure loss in boiler flues = 7 mm of water, pressure loss in bends = 3 mm of water, pressure loss in chimney = 3 mm of water.

Pressure head equivalent to velocity of flue gases passing through the chimney = 1.3 mm of water.

The temperatures of ambient air and flue gases are 35°C and 310°C respectively.

Assume actual draught is 80% of theoretical.

Solution. Average coal consumption

= 1800 kg/h

Flue gases formed per kg of coal fired

= 14 kg

Temperature of ambient air,

$$T_a = 35 + 273 = 308 \text{ K}$$

Temperature of flue gases,

$$T_g = 310 + 273 = 583 \text{ K.}$$

Height of chimney, H :

Draught required is equivalent to overcome the losses and velocity head

$$= 7 + 7 + 3 + 3 + 1.3 = 21.3 \text{ mm of water.}$$

Actual draught to be produced,

$$h_w = \frac{21.3}{0.8} = 26.62 \text{ mm of water}$$

$$h_w = 353\, H \left(\frac{1}{T_a} - \frac{1}{T_g} \cdot \frac{m_a + 1}{m_a}\right)$$

$$26.62 = 353\, H \left(\frac{1}{308} - \frac{1}{583} \times \frac{14}{13}\right)$$

$$= H\,(1.146 - 0.652)$$

$$= 0.494\, H \text{ or } H = 53.88 \text{ m}$$

Diameter of the chimney, D :

$$\rho_g = \frac{353}{T_g}\left(\frac{m_a + 1}{m_a}\right)$$

$$= \frac{353}{583} \times \frac{14}{13} = 0.652 \text{ kg/m}^3$$

Flue gases formed per second

$$= \frac{1800 \times 14}{3600} = 7 \text{ kg}$$

$$m_g = A \times C \times \rho_g \quad ...(i)$$

But $C = \sqrt{2gH_1}$

where H_1 is the equivalent velocity expressed in m of gas

$$H_1 \rho_g = h_w \rho_w$$

where h_w is the water head equivalent to velocity head responsible for giving velocity to the gas,

$$\therefore \quad H_1 = \frac{h_w \rho_w}{\rho_g} = \frac{1.3 \times 1000}{1000 \times 0.652} = 1.993 \text{ m}$$

$$\therefore \quad C = \sqrt{2 \times 9.81 \times 1.993} = 6.25 \text{ m/s}$$

Substituting this value in eqn. (*i*), we get

$$7 = \frac{\pi}{4} D^2 \times 6.25 \times 0.625$$

or

$$D^2 = \frac{7 \times 4}{\pi \times 6.25 \times 0.652}$$

$$\therefore \quad D = \mathbf{1.478 \text{ m.}} \quad \textbf{(Ans.)}$$

Example 13.7. *How much air is used per kg of coal burnt in a boiler having chimney of 32.3 m height to create a draught of 19 mm of water column when the temperature of flue gases in the chimney is 370°C and the temperature of the boiler house is 29.5°C ?* **(AMIE)**

Solution. Height of the chimney, $H = 32.3$ m

Draught of water, $h_w = 19$ mm

Temperature of flue gases in the chimney,

$$T_g = 370 + 273 = 643 \text{ K}$$

Ambient temperature,

$$T_a = 29.5 + 273 = 302.5 \text{ K}$$

Air used per kg of coal burnt, m_a :

We know that

$$h_w = 353\, H \left[\frac{1}{T_a} - \frac{1}{T_g}\left(\frac{m_a + 1}{m_a}\right)\right]$$

$$19 = 353 \times 32.3 \left[\frac{1}{302.5} - \frac{1}{643}\left(\frac{m_a + 1}{m_a}\right)\right]$$

or

$$\frac{19}{353 \times 32.3} = \frac{1}{302.5} - \frac{1}{643}\left(\frac{m_a + 1}{m_a}\right)$$

$$0.001666 = 0.003305 - 0.001555\left(\frac{m_a + 1}{m_a}\right)$$

$$0.001555\left(\frac{m_a + 1}{m_a}\right) = 0.003305 - 0.001666 = 0.001639$$

$$\frac{m_a + 1}{m_a} = \frac{0.001639}{0.001555} = 1.054$$

or

$$1 + \frac{1}{m_a} = 1.054$$

or

$$m_a = \mathbf{18.52 \text{ kg/kg of coal.}} \quad \textbf{(Ans.)}$$

Example 13.8. *A chimney has a height of 24 m. The ambient temperature is 25°C. Temperature of flue gases passing through the chimney is 300°C. If the air-flow through the combustion space is 20 kg/kg of fuel burned, find the following :*

(i) The theoretical draught in mm of water ;

(ii) Velocity of the flue gases passing through the chimney if 50% of the theoretical draught is lost in friction at the grate and passage. **(M.U.)**

Solution. (*i*) **Theoretical draught** :

Using the relation,

$$h_w = 353\, H \left[\frac{1}{T_a} - \frac{1}{T_g}\,\frac{m_a + 1}{m_a}\right]$$

where, h_w = Head in mm of water

H = Height of the chimney = 24 m

T_a = Absolute temperature of atmosphere (= 25 + 273 = 298 K)

T_g = Absolute temperature of chimney gases (= 273 + 300 = 573 K)

m_a = Mass of air supplied per kg of fuel.

Substituting the values, we get

$$h_w = 353 \times 24 \left[\frac{1}{298} - \frac{1}{573} \times \left(\frac{20 + 1}{20}\right)\right]$$

$$= \mathbf{12.9 \text{ mm of } H_2O.} \quad \textbf{(Ans.)}$$

(*ii*) **Velocity of flue gases :**

The equivalent gas head,

$$H_1 = H\left[\frac{m_a}{m_a+1}\times\frac{T_g}{T_a}-1\right]$$

$$= 24\left[\frac{20}{20+1}\times\frac{573}{298}-1\right] = 19.95 \text{ m}$$

Available head

$$= 0.5 \times 19.95 = 9.975 \text{ m}$$

∴ *Velocity of the flue gases*

$$= \sqrt{2gH_1} = \sqrt{2\times 9.81\times 9.975}$$

$$= \mathbf{13.99 \ m/s.} \quad \textbf{(Ans.)}$$

Example 13.9. *Calculate power of a motor required to drive a fan which maintains a draught of 54 mm of water under the following conditions for (i) induced draught fan, (ii) forced draught fan :*

Temperature of flue gases leaving the boiler in each case	*= 240°C*
Temperature of air in the boiler house	*= 20°C*
Air supplied per kg of fuel in each case	*= 18.5 kg*
Mass of coal burnt per hour	*= 1820 kg*
Efficiency of the fan	*= 82 per cent.*

Solution. Temperature of gases, $T_g = 240 + 273 = 513$ K

Temperature of air, $T_a = 20 + 273 = 293$ K

Mass of air used, $m_a = 18.5$ kg

Mass of coal used, $M = 1820$ kg

Draught produce by the fan, $h_w = 54$ mm of water

Efficiency of each fan, $\eta_f = 82$ per cent.

(*i*) **Power of a motor required to drive induced draught fan, P_{ID} :**

Using the relation :

$$P_{ID} = \frac{0.998\times 10^{-8} hV_o m_a M T_g}{\eta_f} \text{ kW}$$

$$= \frac{0.998\times 10^{-8}\times 54\times 0.7734\times 18.5\times 1820\times 513}{0.82}$$

(where $V_o = 0.7734 \text{ m}^3$ at 0°C and 760 mm of Hg)

$$= \mathbf{8.78 \ kW.} \quad \textbf{(Ans.)}$$

(*ii*) **Power of a motor required to drive forced draught fan, P_{FD} :**

$$P_{FD} = \frac{0.998\times 10^{-8} hV_o m_a M T_a}{\eta_f} \text{ kW}$$

$$= \frac{0.998\times 10^{-8}\times 54\times 0.7734\times 18.5\times 1820\times 293}{0.82}$$

$$= \mathbf{5.014 \ kW.} \quad \textbf{(Ans.)}$$

HIGHLIGHTS

1. The small pressure difference which causes a flow of gas to take place is termed as a *draught*.
2. *Natural* draught is obtained by the use of a *chimney*. The draught produced by the chimney is due to density difference between the column of hot gases inside the chimney and the cold air outside.
3. $h_w = 353\, H\left[\frac{1}{T_a}-\frac{1}{T_g}\left(\frac{m_a+1}{m_a}\right)\right]$.
4. The draught pressure produced by the natural draught system in height of hot gases column,

$$H_1 = H\left[\left(\frac{m_a}{m_a+1}\right)\times\frac{T_g}{T_a}-1\right].$$

5. Diameter (D) of the chimney is given by,

$$D = 1.128\sqrt{\frac{\dot{m}_g}{\rho_g . C}}$$

where, $\dot{m}_g$ = Mass of gases flowing through any cross-section of the chimney,

ρ_g = Density of gases, and

C = Velocity of gases passing through the chimney.

6. Condition for maximum discharge through a chimney is :

$$\frac{T_g}{T_a} = 2\left(\frac{m_a+1}{m_a}\right).$$

7. Efficiency of chimney,

$$\eta_{\text{chimney}} = \frac{\frac{H}{J}\left[\left(\frac{m_a}{m_a+1}\right)\times\frac{T_g}{T_a}-1\right]}{c_p\,(T'-T'')}$$

where, T' = Absolute temperature of flue gases leaving the chimney to create the draught of h_w mm of water, and

T'' = Absolute temperature of flue gases leaving the chimney in case of artificial draught of h_w mm of water.

8. **Power required to drive fan :**

Power of F.D. fan $= 0.998 \times 10^{-8}\left(\frac{hV_o\, m_a M T_a}{\eta_f}\right)$ kW

Power of I.D. fan $= 0.998 \times 10^{-8}\left(\frac{hV_o\, m_a M T_g}{\eta_f}\right)$ kW.

9. Balanced draught is a combination of the forced and induced draught system.

OBJECTIVE TYPE QUESTIONS

Choose the Correct Answer :

1. The draught which a chimney produces is called
   (*a*) induced draught (*b*) natural draught
   (*c*) forced draught (*d*) balanced draught.
2. The draught produced by steel chimney as compared to that produced by brick chimney for the same height is
   (*a*) less (*b*) more
   (*c*) same (*d*) may be more or less.
3. In a boiler installation the natural draught is produced
   (*a*) due to the fact that furnace gases being light go through the chimney giving place to cold air from outside to rush in.
   (*b*) due to the fact that pressure at the grate due to cold column is higher than the pressure at chimney base due to hot column.
   (*c*) due to the fact that at the chimney top the pressure is more than its environmental pressure.
   (*d*) all of the above.
4. The draught produced, for a given height of the chimney and given mean temperature of chimney gases
   (*a*) decreases with increase in outside air temperature
   (*b*) increases with increase in outside air temperature
   (*c*) remains the same irrespective of outside air temperature
   (*d*) may increase or decrease with increase in outside air temperature.
5. The draught produced by chimney of given height at given outside temperature
   (*a*) decreases if the chimney gas temperature increases
   (*b*) increases if the chimney gas temperature increases
   (*c*) remains same irrespective of chimney gas temperature
   (*d*) may increase or decrease.
6. For forced draught system, the function of chimney is mainly
   (*a*) to produce draught to accelerate the combustion of fuel
   (*b*) to discharge gases high up in the atmosphere to avoid hazard
   (*c*) to reduce the temperature of the hot gases discharged
   (*d*) none of the above.
7. Artificial draught is produced by
   (*a*) induced fan (*b*) forced fan
   (*c*) induced and forced fan (*d*) all of the above.
8. The draught in locomotive boilers is produced by
   (*a*) forced fan (*b*) chimney
   (*c*) steam jet (*d*) only motion of locomotive.
9. For the same draught produced the power of induced draught fan as compared to forced draught fan is
   (*a*) less (*b*) more
   (*c*) same (*d*) not predictable.
10. Artificial draught is produced by
    (*a*) air fans (*b*) steam jet
    (*c*) fan or steam jet (*d*) all of the above.
11. The artificial draught normally is designed to produce
    (*a*) less smoke (*b*) more draught
    (*c*) less chimney gas temperature
    (*d*) all of the above.
12. For the induced draught the fan is located
    (*a*) near bottom of chimney (*b*) near bottom of furnace
    (*c*) at the top of the chimney (*d*) anywhere permissible.
13. The pressure at the furnace is minimum in case of
    (*a*) forced draught system (*b*) induced draught system
    (*c*) balanced draught system (*d*) natural draught system.
14. For maximum discharge of hot gases through the chimney the height of hot gas column producing draught is
    (*a*) twice the height of chimney
    (*b*) equal to the height of chimney
    (*c*) half the height of chimney
    (*d*) none of the above.
15. The efficiency of chimney is approximately
    (*a*) 80% (*b*) 40%
    (*c*) 20% (*d*) 0.25%.
16. In balanced draught system the pressure at force fan inlet
    (*a*) is greater than pressure at chimney outlet
    (*b*) is less than pressure at chimney outlet
    (*c*) approximately same as that at chimney outlet.

ANSWERS

1.	(*b*)	**2.**	(*b*)	**3.**	(*b*)	**4.**	(*a*)
5.	(*b*)	**6.**	(*b*)	**7.**	(*d*)	**8.**	(*c*)
9.	(*b*)	**10.**	(*d*)	**11.**	(*d*)	**12.**	(*a*)
13.	(*c*)	**14.**	(*b*)	**15.**	(*d*)	**16.**	(*c*).

THEORETICAL QUESTIONS

1. What is the function of a boiler chimney ?
2. Why is there no chimney in the case of a locomotive boiler ?
3. What do you understand by the term "boiler draught" ?
4. What are the limitations of chimney draught ?
5. Define the chimney efficiency and find out expression for the same.
6. What are the various types of draughts used in usual practice ?
7. What are the advantages of artificial draught over natural draught ?
8. What do you understand by steam jet draught ? Where is it generally employed ?
9. Derive an expression for maximum discharge rate of gases through the chimney for a given height of the chimney.

UNSOLVED EXAMPLES

1. Calculate the quantity of air supplied per kg of fuel burnt in the combustion chamber of a boiler when the required draught of 1.85 cm of water is produced by a chimney of 32 m height. The temperatures of the flue gases and ambient air recorded are 370°C and 30°C respectively. [**Ans.** 15.3 kg]

2. Determine the draught produced in cm of water by a chimney of 50 metres height when the temperature of the flue gases passing through the chimney is such that the mass of flue gases discharge is maximum in a given time. The ambient air temperature is 20°C. [**Ans.** 30.2 mm]

3. A boiler is provided with a chimney of 24 m height. The ambient temperature is 25°C. The temperature of flue gases passing through the chimney is 300°C. If the air flow through the combustion chamber is 20 kg/kg of fuel burnt, find (*i*) the theoretical draught in cm of water, (*ii*) velocity of the flue gases passing through the chimney if 50% of the theoretical draught is lost in friction at grate and passage. [**Ans.** 12.9 mm of water, 14 m/s]

4. Estimate the mass of flue gases flowing through the chimney as per data given below :

Draught produced	= 20 mm of water column
Temperature of the flue gases	= 573 K
Ambient temperature	= 303 K
The mass of air used	= 19 kg per kg of fuel burnt
Diameter of the chimney	= 2 m.

Neglect the losses. [**Ans.** 50.11 kg/s]

5. A 30 metres high chimney discharges flue gases at 357°C, when the outside temperature is 27°C. Air-fuel ratio of 16 is required to burn the coal on the grate. Determine :
   (*i*) The draught in mm of water column.
   (*ii*) The draught produced in terms of height of column of flue gases in metres.
   (*iii*) Volume of flue gases passing through the chimney per second if 1360 kg of coal is burnt per hour over the grate.
   (*iv*) The base diameter of the chimney if the velocity of the flue gases at the base of the chimney is given by

$$H_1 = K\frac{C^2}{2}, \text{ where the value of } K = 1.627.$$

[**Ans.** (*i*) 17.44 mm of water column, (*ii*) 29.29 m of flue gases, (*iii*) 38835.12 kg/h, (*iv*) 1.513 m]

6. A 40 metres high chimney is discharging flue gases at 623 K, when the ambient temperature is 303 K. The quantity of air supplied is 13 kg of air per kg of fuel burnt. Determine :
   (*i*) Draught produced in mm of water column.
   (*ii*) Equivalent draught in metres of hot gas column.
   (*iii*) Efficiency of the chimney, if the minimum temperature of artificial draught is 423 K. The mean specific heat of flue gases is 1.005 kJ/kg K.
   (*iv*) Percentage of heat spent in natural draught system, if the net calorific value of fuel supplied be 30600 kJ/kg.
   (*v*) Temperature of chimney gases for maximum discharge in a given time, and what would be the draught produced correspondingly ?

[**Ans.** (*i*) 22.19 mm of water, (*ii*) 36.369 m, (*iii*) 0.177%, (*iv*) 9.19%, (*v*) 652.6 K, 23.3 mm of water, 40 m]

7. Calculate the power of a motor required to drive a fan which maintains a draught of 54 mm of water under the following conditions for (*i*) induced draught fan, (*ii*) forced draught fan :
   Temperature of flue gases leaving the boiler in each case = 200°C, temperature of air in the boiler house = 20°C, air supplied per kg of fuel = 18.5 kg, mass of coal burnt per hour = 1820 kg.
   Efficiency of fan may be assumed as 80 per cent.

[**Ans.** (*i*) 6.91 kW, (*ii*) 4.28 kW]

14 Performance of Steam Generators

14.1. EVAPORATIVE CAPACITY

The evaporative capacity of a boiler may be expressed in terms of :

(*i*) kg of steam/h

(*ii*) kg of steam/h/m^2 of heating surface

(*iii*) kg of steam/kg of fuel fired.

14.2. EQUIVALENT EVAPORATION

Generally the output or evaporative capacity of the boiler is given as kg of water evaporated per hour but as different boilers generate steam at different pressures and temperatures (from feed water at different temperatures) and as such have different amounts of heat ; the number of kg of water evaporated per hour in no way provides the exact means for comparison of the performance of the boilers. Hence to compare the evaporative capacity or performance of different boilers working under different conditions it becomes imperative to provide a common base so that water be supposed to be evaporated under standard conditions. The *standard conditions* adopted are : *Temperature of feed water 100°C and converted into dry and saturated steam at 100°C.* As per these standard conditions 1 kg of water at 100°C necessitates 2257 kJ (539 kcal in MKS units) to get converted to steam at 100°C.

Thus the **equivalent evaporation** *may be defined as the amount of water evaporated from water at 100°C to dry and saturated steam at 100°C.*

Consider a boiler generating m_a kg of steam per hour at a pressure p and temperature T.

Let h = Enthalpy of steam per kg under the generating conditions.

$$\begin{bmatrix} h = h_f + h_{fg} \text{ ...... Dry saturated steam at pressure } p \\ h = h_f + xh_{fg} \text{ ...... Wet steam with dryness fraction } x \text{ at pressure } p \\ h = h_f + h_{fg} + c_p\,(T_{sup} - T_s) \text{ ...... Superheated steam at pressure } p \text{ and temperature } T_{sup} \end{bmatrix}$$

h_{f_1} = Specific enthalpy of water at a given feed temperature.

Then heat gained by the steam from the boiler per unit time

$$= m_a\,(h - h_{f_1})$$

The equivalent evaporation (m_e) from the definition is obtained as

$$m_e = \frac{m_a\,(h - h_{f_1})}{h_{fg}} = \frac{m_a\,(h - h_{f_1})}{2257} \qquad \text{...(14.1)}$$

The evaporation rate of the boiler is also sometimes given in terms of *kg of steam / kg of fuel*.

The presently accepted standard of expressing the capacity of a boiler is in terms of the total heat added per hour.

14.3. FACTOR OF EVAPORATION

It is defined as *the ratio of heat received by 1 kg of water under working conditions to that received by 1 kg of water evaporated from and at 100°C*. It is denoted by F_e.

$$\therefore \quad \text{Factor of evaporation, } F_e = \frac{h - h_{f_1}}{2257} \qquad \text{...(14.2)}$$

14.4. BOILER EFFICIENCY

'Boiler efficiency' is the *ratio of heat actually utilised in generation of steam to the heat supplied by the fuel in the same period.*

i.e., Boiler efficiency $= \dfrac{m_a (h - h_{f_1})}{C}$ ...(14.3)

where, m_a = Mass of water actually evaporated into steam per kg of fuel at the working pressure, and

C = Calorific value of the fuel in kJ/kg.

If the boiler, economiser, and superheater are considered as a single unit, then the *boiler efficiency is termed as overall efficiency of the boiler plant.*

The following are the *factors* on which the boiler efficiency depends :

1. Fixed factors.
2. Variable factors.

1. Fixed Factors. These are :

(*i*) *Boiler design.* It includes the arrangement and effectiveness of the heating surfaces, the shape and volume of the furnace, the arrangement of flues, the arrangement of steam and water circulation.

(*ii*) *Heat recovery equipment.* It includes the economiser, superheater, air preheater and feed water heater.

(*iii*) *Built in losses.* It includes the heat transfer properties of the settings and construction materials, flue gas and ash heat losses.

(*iv*) Rated rate of firing, the furnace volume and heating surface.

(*v*) Properties and characteristics of fuel burnt.

2. Variable Factors. These are :

(*i*) Actual firing rate.

(*ii*) Fuel condition as it is fired.

(*iii*) The conditions of heat absorbing surfaces.

(*iv*) Excess air fluctuations.

(*v*) Incomplete combustion and combustibles in the refuse.

(*vi*) Change in draught from the rated, due to atmospheric conditions.

(*vii*) Humidity and temperature of the combustion air.

14.5. HEAT LOSSES IN A BOILER PLANT

The following *heat losses* occur in a boiler plant :

1. Heat lost to flue gases.
2. Heat lost due to incomplete combustion.
3. Heat lost due to unburnt fuel.
4. Convection and radiation losses.

1. Heat lost to flue gases. The flue gases contain dry products of combustion as well as the steam generated due to combustion of hydrogen in the fuel.

Heat lost through *dry flue gases*,

$$Q_g = m_g\, c_{pg}\, (T_g - T_a) \quad ...(14.4)$$

where, m_g = Mass of gases formed per kg of coal,

c_{pg} = Specific heat of gases,

T_g = Temperature of the gases, and

T_a = Temperature of air entering the combustion chamber of boiler.

Heat carried away by the steam in flue gases,

$$Q_s = m_{s_1}\, (h_{s_1} - h_{f_1}) \quad ...(14.5)$$

where, m_{s_1} = Mass of steam formed per kg of fuel due to the combustion of H_2 in the fuel,

h_{f_1} = Enthalpy of water at boiler house temperature, and

h_{s_1} = Enthalpy of steam at the gas temperature and partial pressure of steam vapour.

The heat lost to flue gases can be reduced by passing the flue gases through the economiser and air preheater.

2. Heat lost due to incomplete combustion. The combustion is said to be incomplete if the carbon burns to CO instead of CO_2. One kg of carbon releases 10120 kJ of heat if it burns to CO whereas it can release 33800 kJ/kg if it burns to CO_2.

∴ Heat loss due to incomplete combustion of 1 kg of carbon

= 33800 – 10120 = 23680 kJ

If CO is present in the flue gases it indicates that combustion of fuel is incomplete. If the percentages of CO and CO_2 in flue gases by volume are known, then carbon burnt to CO instead of CO_2 per kg of fuel is given by

$$\text{Mass of carbon burnt to CO} = \frac{\text{CO} \times \text{C}}{\text{CO}_2 + \text{CO}} \quad ...(14.6)$$

where CO and CO_2 are expressed as % by volume in flue gases and C as the fraction of carbon in one kg of fuel.

∴ Heat lost due to incomplete combustion of carbon per kg of fuel

$$= \frac{\text{CO} \times \text{C}}{\text{CO}_2 + \text{CO}} \times 23680 \text{ kJ/kg of fuel.}$$

This loss (due to incomplete combustion) can be reduced by supplying excess quantity of air and giving a turbulent motion to the air before it enters the furnace in order to help the mixing process.

3. Heat lost due to unburnt fuel. If m_{f_1} is the mass of unburnt fuel per kg of fuel used and C is the calorific value of the fuel, then heat lost due to unburnt fuel,

$$Q = m_{f_1} \times \text{C} \quad ...(14.7)$$

In case of solid fuels this loss cannot be completely avoided.

4. **Convection and radiation losses.** As the hot surfaces of the boiler are exposed to the atmosphere, therefore heat is lost to atmosphere by convection and radiation.

The loss of heat due to convection and radiation losses

= Heat released per kg of fuel – total of the heat losses given by eqns. (14.4), (14.5), (14.6) and (14.7).

These losses can be reduced by providing heat insulation on the boiler surface.

WORKED EXAMPLES

Example 14.1. *An oil fuel with a lower calorific value of 44700 kJ is burnt in a boiler with air-fuel ratio as 20 : 1. Neglecting ash, calculate the maximum temperature attained in the furnace of the boiler. Assume that whole of the heat of combustion is given to the products of combustion and their average specific heat is 1.08. Take boiler room temperature as 38°C.*

Solution. L.C.V. of fuel = 44700 kJ

Air-fuel ratio = 20 : 1

Average specific heat, c_{pg} = 1.08 kJ/kg K

Boiler room temperature, $T_1 = 38 + 273 = 311$ K

Maximum furnace temperature attained, T_2 :

Since the whole of heat is taken by the gases, therefore,

Heat of combustion = Heat of gases

i.e., $1 \times 44700 = m_g \times c_{pg} \times (T_2 - T_1)$ (m_g = mass of gases)

$= (20 + 1) \times 1.08\,(T_2 - 311)$

$$\therefore \quad T_2 = \frac{1 \times 44700}{(20+1) \times 1.08} + 311 = 2282 \text{ K}$$

i.e., T_2 **= 2282 K or 2009°C. (Ans.)**

Example 14.2. *The steam used by the turbine is 5.4 kg/kWh at a pressure of 50 bar and a temperature of 350°C. The efficiency of boiler is 82 per cent with feed water at 150°C.*

(i) How many kg of 28100 kJ coal are required/kWh ?

(ii) If the cost of coal/tonne is ₹500, what is fuel cost/kWh ?

Solution. Mass of steam used, m_s = 5.4 kg/kWh

Pressure of steam, p = 50 bar

Temperature of steam, t_{sup} = 350°C

Boiler efficiency = 82%

Feed water temperature = 150°C

Calorific value of coal, C = 28100 kJ

Cost of coal/tonne = ₹ 500

(*i*) **Coal required, kg/kWh :**

Boiler efficiency is given by, $\eta_{boiler} = \dfrac{m_s\,(h - h_{f_1})}{m_f \times C}$ ...(*i*)

where, m_f is the mass of fuel used, kg/kWh

At 45 bar and 350°C. From steam tables,

h_{sup} = 3068.4 kJ/kg

h_{f_1} (at 150°C) = 1 × 4.18 × (150 – 0) = 627 kJ/kg

Putting these values in eqn. (*i*), we get

$$0.82 = \frac{5.4\,(3068.4 - 627)}{m_f \times 28100}$$

$$\therefore \quad m_f = \frac{5.4\,(3068.4 - 627)}{0.82 \times 28100}$$

= 0.572 kg/kWh. (Ans.)

(*ii*) **Fuel cost/kWh :**

The cost of fuel (coal)/kWh = m_f in tonnes/kWh × cost/tonne

$$= \frac{0.572}{1000} \times 500 \times 100$$

= 28.6 paise/kWh. (Ans.)

Example 14.3. *In a boiler test 1250 kg of coal are consumed in 24 hours. The mass of water evaporated is 13000 kg and the mean effective pressure is 7 bar. The feed water temperature was 40°C, heating value of coal is 30000 kJ/kg. The enthalpy of 1 kg of steam at 7 bar is 2570.7 kJ. Determine :*

(i) Equivalent evaporation per kg of coal ;

(ii) Efficiency of the boiler.

Solution.

Quantity of coal consumed in 24 hours = 1250 kg

Mass of water evaporated = 13000 kg

Mean effective pressure of steam = 7 bar

Feed water temperature = 40°C

Enthalpy of steam at 7 bar = 2570.7 kJ/kg

Heating value of coal, C = 30000 kJ/kg

(*i*) **Equivalent evaporation per kg of coal, m_e :**

Mass of water actually evaporated per kg of fuel,

$$m_a = \frac{13000}{1250} = 10.4 \text{ kg}$$

Heat required to produce 1 kg of steam

$= h - h_{f_1} = 2570.7 - 1 \times 4.18 \times (40 - 0)$

= 2403.5 kJ

Now, $$m_e = \frac{m_a\,(h - h_{f_1})}{h_{fg}}$$ [Refer eqn. (14.1)]

$$= \frac{10.4 \times 2403.5}{2257} = \textbf{11.075 kg.}$$ **(Ans.)**

(*ii*) **Efficiency of boiler, η_{boiler} :**

$$\eta_{boiler} = \frac{m_a (h - h_{f_1})}{C} = \frac{10.4 \times 2403.5}{30000}$$

$= \mathbf{0.833}$ **or 83.3%. (Ans.)**

Example 14.4. *The following readings were obtained during a boiler trial of 6 hours duration.*

Mean steam pressure = 12 bar ; mass of steam generated = 40000 kg ; mean dryness fraction = 0.85 ; mean feed water temperature = 30°C, coal used = 4000 kg. Calorific value of coal = 33400 kJ/kg. Calculate :

(*i*) *Factor of equivalent evaporation ;*

(*ii*) *Equivalent evaporation from and at 100°C ;*

(*iii*) *Efficiency of the boiler.*

Solution. Mean steam pressure, $p = 12$ bar

Mass of steam generated = 4000 kg

Mean dryness fraction, $x = 0.85$

Mean feed water temperature = 30°C

Coal used = 4000 kg

Calorific value of coal, $C = 33400$ kJ/kg

From steam tables, *corresponding to 12 bar :*

$h_f = 798.4$ kJ/kg, $h_{fg} = 1984.3$ kJ/kg

Now, $h = h_f + x\,h_{fg} = 798.4 + 0.85 \times 1984.3$

$= 2485.05$ kJ/kg

Heat of feed water, $h_{f_1} = 1 \times 4.18 \times (30 - 0)$

$= 125.4$ kJ/kg

Total net heat given to produce 1 kg of steam

$= h - h_{f_1} = 2485.05 - 125.4$

$= 2359.65$ kJ/kg.

(*i*) **Factor of equivalent evaporation, F_e :**

$$F_e = \frac{h - h_{f_1}}{2257} = \frac{2359.65}{2257}$$

$= \mathbf{1.045}$. **(Ans.)**

(*ii*) **Equivalent evaporation from and at 100°C, m_e :**

$$m_e = \frac{m_a (h - h_{f_1})}{2257}$$

But, $m_a = \dfrac{40000}{4000}$

$= 10$ kg/kg of fuel

$\therefore \quad m_e = \dfrac{10 \times 2359.65}{2257}$

$= \mathbf{10.45}$ **kg of steam/kg of coal. (Ans.)**

(*iii*) **Efficiency of boiler, η_{boiler} :**

$$\eta_{boiler} = \frac{m_a (h - h_{f_1})}{C}$$

$$= \frac{10 \times 2359.65}{33400}$$

$= \mathbf{0.7065}$ **or 70.65%. (Ans.)**

Example 14.5. *A steam generator evaporates 18000 kg/h of steam at 12.5 bar and a quality of 0.97 from feed water at 105°C, when coal is fired at the rate of 2040 kg/h. If the higher calorific value of the coal is 27400 kJ/kg, find :*

(*i*) *The heat rate of boiler in kJ/h ;*

(*ii*) *The equivalent evaporation ;*

(*iii*) *The thermal efficiency.*

Solution.

Steam generated, $m = 18000$ kg/h

Steam pressure, $p = 12.5$ bar

Quality of steam, $x = 0.97$

Feed water temperature = 105°C

Rate of coal firing, $m_f = 2040$ kg/h

Higher calorific value (H.C.V.) of coal, $C = 27400$ kJ/kg

(*i*) **Heat rate of boiler :**

At 12.5 bar : From steam tables,

$h_f = 806.7$ kJ/kg, $h_{fg} = 1977.4$ kJ/kg

$\therefore \quad h = h_f + x h_{fg} = 806.7 + 0.97 \times 1977.4$

$= 2724.78$ kJ/kg

h_{f_1} (heat of feed water) $= 1 \times 4.18 \times (105 - 0)$

$= 438.9$ kJ/kg

Heat rate of the boiler = Heat supplied per hour

$= m(h - h_{f_1}) = 18000\,(2724.78 - 438.9)$

$= \mathbf{4.1146 \times 10^7}$ **kJ/h. (Ans.)**

(*ii*) **Equivalent evaporation, m_e :**

$$m_e = \frac{m_a (h - h_{f_1})}{2257}$$

where, $m_a = \dfrac{18000}{2040} = 8.823$ kg/kg of fuel

$\therefore \quad m_e = \dfrac{8.823\,(2724.78 - 438.9)}{2257}$

$= \mathbf{8.936}$ **kg of steam/kg of fuel. (Ans.)**

(*iii*) **Thermal efficiency $\eta_{thermal}$:**

$$\eta_{thermal} = \frac{m (h - h_{f_1})}{m_f \times C}$$

$$= \frac{18000\,(2724.78 - 438.9)}{2040 \times 27400}$$

$= \mathbf{0.7361}$ **or 73.61%. (Ans.)**

Example 14.6. *The following data refer to a boiler plant consisting of an economiser, a boiler and a superheater.*

Mass of water evaporated per hour = 5940 kg, mass of coal burnt per hour = 675 kg, L.C.V. of coal = 31600 kJ / kg, pressure of steam at boiler stop valve = 14 bar, temperature of feed water entering the economiser = 32°C, temperature of feed water leaving the economiser = 115°C, dryness fraction of steam leaving the boiler and entering superheater = 0.96, temperature of steam leaving the superheater = 260°C, specific heat of superheated steam = 2.33. Determine :

(*i*) *Percentage of heat in coal utilised in economiser, boiler and superheater ;*

(*ii*) *Overall efficiency of boiler plant.*

Solution.

Mass of water evaporated = 5940 kg/h

Mass of coal burnt = 675 kg/h

Lower calorific value of coal = 31600 kJ/kg

Pressure of steam at boiler stop valve, p_1 = 14 bar

Temperature of feed water entering the economiser, t_{e_1} = 32°C

Temperature of feed water leaving the economiser, t_{e_2} = 115°C

Dryness fraction of steam entering superheater = 0.96

Temperature of steam leaving the superheater, t_{sup} = 260°C

Specific heat of superheated steam, c_p = 2.3 kJ/kg K

Heat utilised by 1 kg of feed water in *economiser*

$$h_{f_1} = 1 \times 4.18 \times (t_{e_2} - t_{e_1})$$
$$= 1 \times 4.18 \times (115 - 32)$$
$$= 346.9 \text{ kJ/kg}$$

Heat utilised in *boiler* per kg of feed water

$$h_{\text{boiler}} = (h_f + xh_{fg}) - h_{f_1}$$

At 14 bar pressure. From steam tables,

$$t_s = 195°\text{C},\ h_f = 830.1 \text{ kJ/kg},\ h_{fg} = 1957.7 \text{ kJ/kg}$$

$$\therefore \quad h_{\text{boiler}} = (830.1 + 0.96 \times 1957.7) - 346.9$$
$$= 2362.6 \text{ kJ/kg}$$

Heat utilised in *superheater* by 1 kg of feed water,

$$h_{\text{superheater}} = (1 - x)\, h_{fg} + c_p\,(T_{sup} - T_s)$$
$$= (1 - 0.96) \times 1957.7 + 2.3\,(260 - 195)$$
$$= 78.3 + 149.5 = 227.8 \text{ kJ/kg}$$

Also, mass of water evaporated/hour/kg of coal burnt

$$= \frac{5940}{675} = 8.8 \text{ kg}$$

(*i*) **Percentage of heat utilised in economiser**

$$= \frac{346.9}{31600} \times 8.8 \times 100 = \mathbf{9.66\%}. \quad \textbf{(Ans.)}$$

Percentage of heat utilised in boiler

$$= \frac{2362.6}{31600} \times 8.8 \times 100 = \mathbf{65.7\%}. \quad \textbf{(Ans.)}$$

Percentage of heat utilised in superheater

$$= \frac{227.8}{31600} \times 8.8 \times 100 = \mathbf{6.34\%}. \quad \textbf{(Ans.)}$$

(*ii*) **Overall efficiency of boiler plant,** η_{overall} :

Total heat absorbed in kg of water

$$= h_{f_1} + h_{\text{boiler}} + h_{\text{superheater}}$$
$$= 346.9 + 2362.6 + 227.8 = 2937.3 \text{ kJ/kg}$$

$$\eta_{\text{overall}} = \frac{8.8 \times 2937.3}{31600}$$
$$= \mathbf{0.8179 \text{ or } 81.79\%}. \quad \textbf{(Ans.)}$$

Example 14.7. *The following observations were made during the trial of a boiler plant consisting of a battery of 6 Lancashire boilers and an economiser :*

Calorific value of coal/kg = 29915 kJ

Mass of feed water per kg of dry coal = 9.1 kg

Equivalent evaporation from and at 100°C per kg of dry coal = 9.6 kg

Temperature of feed water to economiser = 12°C

Temperature of feed water to boiler = 105°C

Air temperature = 13°C

Temperature of the flue gases entering economiser = 370°C

Mass of flue gases entering the economiser = 18.2 kg/kg of coal

Mean specific heat of flue gases = 1.046 kJ/kg°C

Find :

(*i*) *The efficiency of the boiler alone.*

(*ii*) *The efficiency of the economiser alone.*

(*iii*) *The efficiency of the whole boiler plant.*

Solution. Boiler efficiency

Heat supplied for steam generation

$$= 9.6 \times 2257 = 21667 \text{ kJ}$$

(*i*) **Boiler efficiency**
$$= \frac{21667}{29915}$$
$$= \mathbf{0.724 \text{ or } 72.4\%}. \quad \textbf{(Ans.)}$$

(*ii*) **Efficiency of economiser**:

Heat in the flue gases, per kg of dry coal entering economiser

$= 18.2 \times 1.046 \times (370 - 13)$
$= 6796.3$ kJ

Heat utilised in economiser

$= 9.1 \times 4.184 \times (105 - 12)$
$= 3540.9$ kJ

$\therefore$ **Efficiency of economiser** $= \dfrac{3540.9}{6796.3}$

$= \mathbf{0.521 \text{ or } 52.1\%}$. **(Ans.)**

(*iii*) **Boiler plant efficiency:**

Total heat utilisation

$= 21667 + 3540.9$
$= 25207.9$ kJ

$\therefore$ **Boiler plant efficiency** $= \dfrac{25207.9}{299.15}$

$= \mathbf{0.842 \text{ or } 84.2\%}$. **(Ans.)**

Example 14.8. *A boiler produces 200 kg of dry and saturated steam per hour at 10 bar and feed water is heated by an economiser to a temperature of 110°C. 225 kg of coal of a calorific value of 30100 kJ/kg are fired per hour. If 10% of coal remains unburnt, find the thermal efficiency of the boiler and boiler and grate combined.*

Solution.

Rate of production of steam $= 2000$ kg/h
Quality of steam, $x = 1$
Steam pressure, $p = 10$ bar
Feed water temperature rise $= 110°C$
Rate of coal firing $= 225$ kg/h
Calorific value of coal $= 30100$ kJ/kg
Percentage of coal unburnt $= 10\%$
From steam tables, corresponding to 10 bar : $h = h_g$
$= 2776.2$ kJ/kg

Heat contained in 1 kg of feed water before entering the boiler,

$$h_{f_1} = 1 \times 4.18 \times (110 - 0) = 459.8 \text{ kJ}$$

Total heat given to produce 1 kg of steam in boiler

$$= h - h_{f_1} = 2776.2 - 459.8 = 2316.4 \text{ kJ/kg}$$

Mass of coal actually burnt $= 225 \times \dfrac{90}{100} = 202.5$ kg

Mass of steam produced per kg of coal (actually burnt), $m_a = \dfrac{2000}{202.5} = 9.87$ kg

Thermal efficiency of the boiler

$$= \frac{m_a (h - h_{f_1})}{C} = \frac{9.87\,(2776.2 - 459.8)}{30100}$$

$= \mathbf{0.759 \text{ or } 75.9\%}$. **(Ans.)**

Thermal efficiency of boiler and grate combined

$$= \frac{\frac{2000}{225}(2776.2 - 459.8)}{30100}$$

$= \mathbf{0.684 \text{ or } 68.4\%}$. **(Ans.)**

☞**Example 14.9.** *A boiler generates 7.5 kg of steam per kg of coal burnt at a pressure of 11 bar, from feed water having a temperature of 70°C. The efficiency of boiler is 75% and factor of evaporation 1.15, specific heat of steam at constant pressure is 2.3.*

Calculate :

(*i*) *Degree of superheat and temperature of steam generated ;*

(*ii*) *Calorific value of coal in kJ/kg ;*

(*iii*) *Equivalent evaporation in kg of steam per kg of coal.*

Solution.

Steam generated per kg of coal, $m_a = 7.5$ kg
Steam pressure, $p = 11$ bar
Temperature of feed water $= 70°C$
Efficiency of boiler $= 75\%$
Factor of evaporation, $F_e = 1.15$
Specific heat of steam, $c_{ps} = 2.3$ kJ/kg K

(*i*) **Degree of superheat and temperature of steam generated :**

At 11 bar. From steam tables :

$h_f = 781.1$ kJ/kg, $h_{fg} = 1998.5$ kJ/kg,
$t_s = 184.1°C$ ($T_s = 273 + 184.1 = 457.1$ K)

Factor of evaporation,

$$F_e = \frac{[\{h_f + h_{fg} + c_{ps}(T_{sup} - T_s)\} - h_{f_1}]}{2257}$$

$$1.15 = \frac{[781.1 + 1998.5 + 2.3\,(T_{sup} - 457.1) - 1 \times 4.18 \times (70 - 0)]}{2257}$$

$$= \frac{2779.6 + 2.3\,(T_{sup} - 457.1) - 292.6}{2257}$$

$$= \frac{2487 + 2.3\,(T_{sup} - 457.1)}{2257}$$

$\therefore$ $(T_{sup} - 457.1) = (1.15 \times 2257 - 2487)/2.3$
$= 47.2$

i.e., $T_{sup} = \mathbf{504.3 \text{ K.}}$ **(Ans.)**

Degree of superheat

$= (T_{sup} - T_s) = 504.3 - 457.1$
$= \mathbf{47.2°C.}$ **(Ans.)**

(*ii*) **Calorific value of coal, C :**

Boiler efficiency

$$= \frac{m_a (h - h_{f_1})}{C}$$

$$0.75 = \frac{m_a [\{h_f + h_{fg} + c_{ps} (T_{sup} - T_s)\} - h_{f_1}]}{C}$$

$$0.75 = \frac{7.5 [\{781.1 + 1998.5 + 2.3 (504.3 - 457.1) - 1 \times 4.18 \times (70 - 0)]}{C}$$

$$= \frac{7.5 (2888.16 - 292.6)}{C} = \frac{19466.7}{C}$$

$$\therefore \quad C = \frac{19466.7}{0.75} = 25955 \text{ kJ/kg}$$

i.e., *Calorific value of coal* = **25955 kJ/kg. (Ans.)**

(*iii*) **Equivalent evaporation m_e :**

$$m_e = \frac{m_a (h - h_{f_1})}{2257} = \frac{7.5 (2888.16 - 292.6)}{2257}$$

= **8.625 kg. (Ans.)**

Example 14.10. *During the trial of water tube boiler the following data were obtained :*

Steam pressure,	*p = 13 bar*
Degree of superheat	*= 77°C*
Temperature of feed water	*= 85°C*
Water evaporated	*= 3000 kg per hour*
Coal fired	*= 410 kg per hour*
Ash	*= 40 kg per hour*
Percentage of combustible in ash	*= 9.6%*
Moisture in coal	*= 4.5%*
Calorific value of dry coal/kg	*= 30500 kJ/kg.*

Determine :

(i) The efficiency of the boiler plant including superheater.

(ii) The efficiency of the boiler and furnace combined.

Take specific heat of superheated steam = 2.1 kJ/kg K.

Solution. *At 13 bar.* From steam tables :

t_s = 191.6°C, h_f = 814.7 kJ/kg, h_{fg} = 1970.7 kJ/kg

$$h = h_{sup} = h_f + h_{fg} + c_{ps} (T_{sup} - T_s)$$
$$= 814.7 + 1970.7 + 2.1 \times 77 = 2947.1 \text{ kJ/kg}$$

Also $h_{f_1} = 1 \times 4.18 \times (85 - 0) = 355.3$ kJ/kg

$\therefore$ Total heat supplied to produce 1 kg of steam

$$= h_{sup} - h_{f_1} = 2947.1 - 355.3 = 2591.8 \text{ kJ/kg.}$$

(*i*) **Efficiency of boiler plant including superheater**

$$= \frac{m_a (h_{sup} - h_{f_1})}{C}$$

$$\left[\text{Dry coal} = 410 - 410 \times \frac{4.5}{100} = 391.55 \text{ kg}\right]$$

$$= \frac{\frac{3000}{391.55} (2947.1 - 355.3)}{30500}$$

= **0.651 or 65.1%. (Ans.)**

(*ii*) **Efficiency of the boiler and furnace combined**

Combustible in ash per hour $= 40 \times \frac{9.6}{100} = 3.84$ kg

The combustible present in ash is practically carbon and its value may be taken as 33860 kJ/kg.

Heat actually supplied per hour

= Heat of dry coal – heat of combustible in the ash

= 391.55 × 30500 – 3.84 × 33860 = 11812253 kJ

Heat usefully utilised in the boiler per hour

$= 3000(h_{sup} - h_{f_1}) = 3000 (2947.1 - 355.3)$

= 7775400 kJ.

Efficiency of the boiler and furnace combined

$$= \frac{\text{Heat usefully utilised in boiler per hour}}{\text{Heat actually supplied per hour}}$$

$$= \frac{7775400}{11812253} = \textbf{0.658 or 65.8\%.} \quad \textbf{(Ans.)}$$

Example 14.11. *The following data were taken during the test on a boiler for a period of one hour :*

Steam generated = 5000 kg ; coal burnt = 700 kg, calorific value of coal = 31402 kJ/kg, quality of steam = 0.92. If the boiler pressure is 1.2 MPa and the feed water temperature is 45°C, find the boiler equivalent evaporation and the efficiency. **(N.U.)**

Solution.

Rate of steam production	= 5000 kg/h
Rate of coal firing	= 700 kg/h
Calorific value of coal,	C = 31402 kJ/kg
Quality of steam,	x = 0.92
Boiler pressure,	p = 1.2 MPa or 12 bar
Feed water temperature	= 45°C

Equivalent evaporation, m_e :

From steam tables, corresponding to 12 bar :

h_f = 798.4 kJ/kg, h_{fg} = 1984.3 kJ/kg

Now, $h = h_f + xh_{fg} = 798.4 + 0.92 \times 1984.3$

= 2623.96 kJ/kg

Heat of feed water,

$$h_{f_1} = 1 \times 4.18 \times 45 = 188.1 \text{ kJ/kg}$$

∴ Equivalent evaporation,

$$m_e = \frac{m_a (h - h_{f_1})}{2257}$$

where, $$m_a = \frac{5000}{700} = 7.143 \text{ kg/kg of coal}$$

∴ $$m_e = \frac{7.143 (2623.96 - 188.1)}{2257}$$

= **7.709 kg of steam/kg of coal. (Ans.)**

Boiler efficiency, η_{boiler} :

$$\eta_{boiler} = \frac{m_a (h - h_{f_1})}{C} = \frac{7.143 (2623.96 - 188.1)}{31402}$$

= **0.554 or 55.4%. (Ans).**

Example 14.12. *A steam generator delivers steam at 100 bar, 500°C (enthalpy, h = 3373.7 kJ/kg). The feed water inlet temperature is 160°C (h = 677 kJ/kg). The enthalpies of saturated liquid and saturated vapour at 100 bar are 1407.65 and 2724.7 kJ/kg respectively. The steam generation rate is 100000 kg/h and the steam generator efficiency is 88%. Estimate :*

(*i*) *The fuel burning rate in kg/h, if calorific value of fuel is 21 MJ/kg ;*

(*ii*) *The percentage of total heat absorbed in the economiser, evaporator and superheater.*

Assume that only latent heat is absorbed in the evaporator and neglect any pressure drop.

(Roorkee University)

Solution. Enthalpy of steam (at 100 bar, 500°C),

$$h_{sup} = 3373.7 \text{ kJ/kg}$$

Enthalpy of feed water (at inlet temperature 160°C),

$$h_{f_1} = 677 \text{ kJ/kg}$$

Enthalpy of saturated liquid at 100 bar,

$$h_f = 1407.65 \text{ kJ/kg}$$

Enthalpy of saturated vapour at 100 bar,

$$h_g = 2724.7 \text{ kJ/kg}$$

Rate of steam generation, m_s = 100000 kg/h

Efficiency of steam generator, $\eta_{gen.}$ = 88%

Calorific value of fuel,

$$C = 21 \text{ MJ/kg (or } 21 \times 10^3 \text{ kJ/kg)}$$

(*i*) **Fuel burning rate, m :**

$$\eta_{gen.} = \frac{\text{Heat absorbed by steam per hour}}{\text{Heat added by fuel per hour}}$$

$$= \frac{m_s (h_{sup} - h_{f_1})}{m \times C}$$

or $$0.88 = \frac{100000 (3373.7 - 677)}{m \times 21 \times 10^3} = \frac{12841.4}{m}$$

or $$m = \frac{12841.4}{0.88} = \mathbf{14592.5 \text{ kg/h.}} \quad \textbf{(Ans.)}$$

(*ii*) **The percentage of total heat absorbed in economiser, evaporator and superheater :**

Total heat supplied to steam formation = 3373.7 – 677 = 2696.7 kJ/kg

Heat absorbed in economiser

$$= \frac{1407.65 - 677}{2696.7}$$

= **0.2709 or 27.09%. (Ans.)**

Heat absorbed in evaporator

$$= \frac{2724.7 - 1407.65}{2696.7}$$

= **0.4884 or 48.84%. (Ans.)**

Heat absorbed in superheater

$$= \frac{3373.7 - 2724.7}{2696.7}$$

= **0.2407 or 24.07%. (Ans.)**

Example 14.13. *Calculate the efficiency of boiler and economiser having following data :*

(*i*) **Boiler**

Mass of feed water	*= 2060 kg/hour*
Mass of coal supplied	*= 227 kg/hour*
Calorific value of coal	*= 30000 kJ/kg*
Enthalpy of steam produced	*= 2750 kJ/kg*
Enthalpy of feed water	*= 398 kJ/kg*

(*ii*) **Economiser**

Temperature of feed water at inlet	*= 15°C*
Temperature of feed water at exit	*= 95°C*
Atmospheric temperature	*= 18°C*
Temperature of entering flue gases	*= 370°C*
Mass of flue gases	*= 4075 kg / hour.*

(AMIE)

Solution. (*i*) **Efficiency of boiler, η_{boiler} :**

$$\eta_{boiler} = \frac{\text{Mass rate of feed water} \times \begin{Bmatrix} \text{enthalpy} & - & \text{enthalpy of} \\ \text{of steam} & & \text{feed water} \end{Bmatrix}}{\text{Mass rate of coal} \times C.V.}$$

$$= \frac{2060 (2750 - 398)}{227 \times 30000}$$

= **0.7115 or 71.15%. (Ans.)**

(*ii*) **Efficiency of economiser, $\eta_{eco.}$:**

$$\eta_{eco.} = \frac{\text{Mass rate of feed water} \times c_{pw} \times (t_{w,out} - t_{w,in})}{\text{Mass rate of flue gases} \times c_{pg} \times (t_{g,in} - t_{g,out})}$$

$$= \frac{2060 \times 4.187\,(95 - 15)}{4075 \times 1.01 \times (370 - 18)}$$

(Assume c_{pg} = 1.01 kJ/kg°C)

= **0.476 or 47.6%. (Ans.)**

Example 14.14. *In a boiler test the following data were recorded :*

Mean temperature of feed water	*= 50°C*
Mean boiler pressure	*= 5 bar*
Dryness fraction of steam	*= 0.95*
Coal consumption	*= 600 kg/hour*
Calorific value of coal	*= 30400 kJ/kg*
Feed water supplied to boiler	*= 4800 kg/hour*

Determine equivalent evaporation of boiler from and at 100°C. **(P.U.)**

Solution. Given : p = 5 bar, x = 0.95, C = 30400 kJ/kg

From steam tables, corresponding to 5 bar :

h_f = 640.1 kJ/kg, h_{fg} = 2107.4 kJ/kg

Now, $h = h_f + xh_{fg}$

= 640.1 + 0.95 × 2107.4 = 2642.13 kJ/kg

Heat of feed water,

h_{f_1} = 1 × 4.18 × (50 – 0)

= 209 kJ/kg

Equivalent evaporation from and at 100°C, m_e :

$$m_e = \frac{m_a\,(h - h_{f_1})}{2257}$$

But $m_a = \dfrac{4800}{600}$

= 8 kg/kg of fuel

$\therefore$ $m_e = \dfrac{8\,(2642.13 - 209)}{2257}$

= **8.624 kg of steam/kg of coal. (Ans.)**

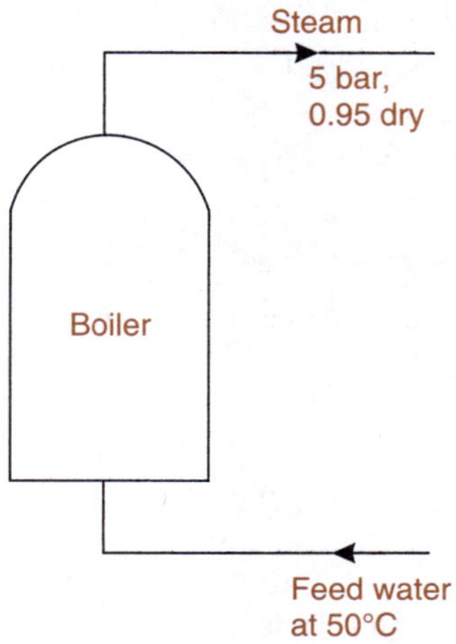

Fig. 14.1

HIGHLIGHTS

1. The *evaporative capacity* of a boiler may be expressed in terms of :
(*i*) kg of steam/h ; (*ii*) kg of steam/h-m² of heating surface ; and (*iii*) kg of steam/kg of fuel fired.
2. *Equivalent evaporation* may be defined as the amount of water evaporated from water at 100°C to dry and saturated steam at 100°C.
i.e., $m_e = \dfrac{m_a\,(h - h_{f_1})}{2257}$.
3. *Factor of evaporation* is defined as the ratio of heat received by 1 kg of water under working conditions to that received by 1 kg of water evaporated from and at 100°C.
i.e., $F_e = \dfrac{h - h_{f_1}}{2257}$.
4. *Boiler efficiency* is the ratio of heat actually utilised in generation of steam to the heat supplied by the fuel in the same period.
i.e., $\eta_{boiler} = \dfrac{m_a(h - h_{f_1})}{C}$.
5. *Heat losses* in a boiler plant are : (*i*) Heat lost to gases ; (*ii*) Heat lost due to incomplete combustion ; (*iii*) Heat lost due to unburnt fuel ; and (*iv*) Convection and radiant loss.

OBJECTIVE TYPE QUESTIONS

Say 'Yes' or 'No'

1. Evaporative capacity of a boiler may be expressed in terms of kg of steam per kg of fuel fired.
2. Equivalent evaporation may be defined as the amount of water evaporated from water at 90°C to dry and saturated steam at 100°C.
3. Factor of evaporation = $\dfrac{h - h_{f_1}}{2557}$.

4. Boiler efficiency

$$= \frac{\text{Heat actually utilised in generation of steam}}{\text{Heat supplied by the fuel}}.$$

5. Boiler efficiency depends only on fixed factors.

ANSWERS

1. Yes 2. No 3. No 4. Yes
5. Yes.

THEORETICAL QUESTIONS

1. What do you mean by 'Evaporative capacity' of a boiler ? How is it expressed ?
2. Define and explain the 'Equivalent evaporation'.
3. Define the term 'Factor of evaporation'.
4. How is 'Boiler efficiency' defined ?
5. Enumerate the heat losses which occur in a boiler plant.

UNSOLVED EXAMPLES

1. 8 kg of steam is produced at 14 bar and 0.95 dryness in a boiler fed with water at 39°C, per kg of coal consumed. Determine the equivalent evaporation from and at 100°C.

[**Ans.** 8.96 kg/kg of coal]

2. A boiler with superheater generates 6000 kg/hour of steam at 15 bar and 0.8 dryness. The boiler exit temperature is 300°C. The feed water temperature is 80°C. The overall efficiency of the plant is 85%. Determine the consumption rate, assuming a calorific value of 30000 kJ/kg. Also find the equivalent evaporation from and at 100°C. What will be the area of superheater surface if the overall heat transfer co-efficient is 450000 kJ/m^2-h ?

[**Ans.** 646 kg/h, 7200 kg/h, 3.86 m^2]

3. A boiler generates 45000 kg/hour of steam at 18 bar the steam temperature being 325°C. The feed water temperature is 49.2°C. The efficiency of the boiler is 80%. When using oil of calorific value 44500 kJ/kg, the steam generated is supplied to a turbine developing 500 kW, and exhausting at 1.8 bar, the dryness of exhaust steam 0.98. Calculate the oil burnt per hour and turbine efficiency. Also find the energy available in the exhaust steam above 49.4°C. [**Ans.** 2448.8 kJ/kg]

4. 5400 kg of steam is produced per hour at a pressure of 7.5 bar in a boiler with feed water at 41.5°C. The dryness fraction of steam at exit is 0.98. The amount of coal burnt per hour is 670 kg of calorific value 31000 kJ/kg. Determine :

(*i*) The boiler efficiency.

(*ii*) Equivalent evaporation.

[**Ans.** 66.2%, 9.12 kg/kg of coal]

5. A boiler with superheater generates 600 kg/hour of steam at 15 bar and 0.98 dryness. The boiler exit temperature is 300°C. The feed water temperature is 80°C. The overall efficiency of the plant is 85%.

(*i*) Determine the coal consumption rate if calorific value of coal is 30000 kJ/kg.

(*ii*) Find the equivalent evaporation from and at 100°C.

(*iii*) What will be the area of superheater surface if the overall heat transfer co-efficient is 450000 kJ/m^2 h ?

[**Ans.** 646 kg/h, 7200 kg/h, 3.86 m^2]

6. During a boiler trial the following observations were made :

Duration of trial = 1 hour ; steam generated = 35500 kg ; steam pressure = 12 bar ; steam temperature = 250°C ; temperature of water entering economiser = 17°C, temperature of water leaving economiser = 77°C ; oil burnt = 3460 kg ; calorific value of oil = 39500 kJ/kg. Calculate :

(*i*) Equivalent evaporation per kg of fuel.

(*ii*) Thermal efficiency of plant.

(*iii*) Percentage heat energy of the fuel energy utilised by the economiser.

[**Ans.** (*i*) 13.02 kg, (*ii*) 74.4%, (*iii*) 6.52%]

7. In a boiler trial of one hour duration the following observations were made :

Steam generated = 5250 kg ; coal burnt = 695 kg ; calorific value of coal = 30200 kJ/kg ; dryness fraction of steam = 0.94 ; rated pressure of the boiler = 12 bar ; temperature of steam leaving the superheater = 240°C ; temperature of hot well = 47°C.

Calculate : (*i*) Equivalent evaporation per kg of fuel without superheater.

(*ii*) Equivalent evaporation per kg of coal with superheater.

(*iii*) Thermal efficiency of the boiler without superheater.

(*iv*) Thermal efficiency of boiler with superheater.

(*v*) Heat supplied by the superheater per hour. Take c_p of steam as 2.184.

[**Ans.** (*i*) 8.26 kg/kg of fuel, (*ii*) 9.04 kg/kg of fuel, (*iii*) 61.7%, (*iv*) 67.53%, (*v*) 1.22 × 10^5 kJ/h]

8. A boiler is provided with an economiser and a superheater. The steam is generated at 21 bar and 0.97 dry. This steam is passed on to a superheater and it leaves the superheater at 250°C. Feed water is supplied to the economiser at 38°C and leaves the economiser at 77°C. The quantity of feed water supplied is 10 kg/kg of fuel which has a calorific value of 32600 kJ/kg. Determine :

(*i*) Efficiency of the entire boiler plant.

(*ii*) Percentage of heat received in the economiser

(*iii*) Percentage of heat received in the superheater.

[**Ans.** (*i*) 88.9%, (*ii*) 5.6%, (*iii*) 3.4%]

9. The following observations were made during a boiler trial :

Mass of feed water = 1520 kg/h ; temperature of feed water = 30°C ; steam pressure = 8.5 bar ; steam temperature = 172.9°C ; dryness fraction of steam = 0.95 ; coal burnt = 200 kg/hour ; calorific value of coal = 27200 kJ/kg ; ash and unburnt coal collected = 16 kg/hour, calorific value of ash and unburnt coal = 2720 kJ/kg ; mass of flue gases = 17.3 kg/kg of coal, temperature of flue gases = 330°C ; boiler room temperature = 18°C ; c_p for gases = 1.006 kJ/kg °C.

Calculate the boiler efficiency. [**Ans.** 70%]

10. Compare the thermal efficiency of two boilers for which the data is given below :

	Boiler 1	*Boiler 2*
Steam pressure	14 bar	14 bar
Steam produced per kg of coal fired	10 kg	14 kg
Quality of steam	0.9 dry	superheated to 240°C
Feed water temperature	27°C	27°C
Calorific value of fuel	34000 kJ/kg	46000 kJ/kg

Specific heat of feed water is 4.18 kJ/kg K and specific heat of steam is 2.1 kJ/kg K.

Which boiler is more efficient ?

[**Ans.** $\eta_{boiler\ 1}$ = 73%, $\eta_{boiler\ 2}$ = 79.5%]

Basic Steam Power Cycles

15

15.1. CARNOT CYCLE

Fig. 15.1 shows a Carnot cycle on *T-s* and *p-V* diagrams. It consists of (*i*) two constant pressure operations (4-1) and (2-3) and (*ii*) two frictionless adiabatics (1-2) and (3-4). These operations are discussed below :

1. **Operation (4-1).** 1 kg of boiling water at temperature T_1 is heated to form wet steam of dryness fraction x_1. Thus heat is absorbed at constant temperature T_1 and pressure p_1 during this operation.

2. **Operation (1-2).** During this operation steam is expanded isentropically to temperature T_2 and pressure p_2. The point '2' represents the condition of steam after expansion.

3. **Operation (2-3).** During this operation heat is rejected at constant pressure p_2 and temperature T_2. As the steam is exhausted it becomes wetter and cooled from 2 to 3.

4. **Operation (3-4).** In this operation the wet steam at '3' is compressed isentropically till the steam regains its original state of temperature T_1 and pressure p_1. Thus cycle is completed.

Refer *T-s* diagram :

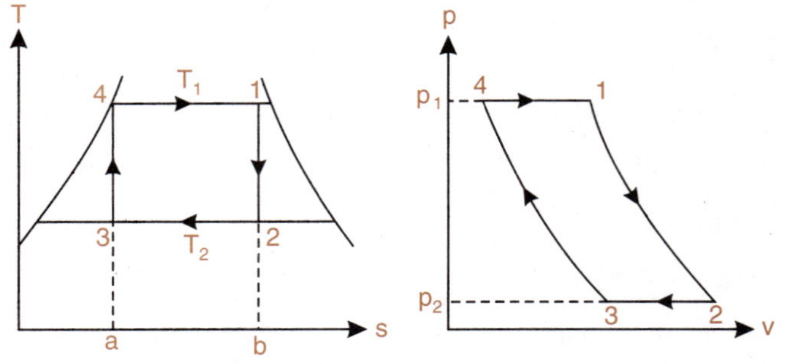

Fig. 15.1. Carnot cycle on *T-s* and *p-v* diagrams.

Heat supplied at constant temperature T_1 [operation (4-1)] = area 4-1-*b*-*a* = $T_1 (s_1 - s_4)$ or $T_1 (s_2 - s_3)$.

Heat rejected at constant temperature T_2 (operation 2-3) = area 2-3-*a*-*b* = $T_2 (s_2 - s_3)$.

Since there is no exchange of heat during isentropic operations (1-2) and (3-4)

$$\text{Net work done} = \text{Heat supplied} - \text{heat rejected}$$
$$= T_1 (s_2 - s_3) - T_2 (s_2 - s_3)$$
$$= (T_1 - T_2)(s_2 - s_3).$$

$$\text{Carnot cycle } \eta = \frac{\text{Work done}}{\text{Heat supplied}}$$

$$= \frac{(T_1 - T_2)(s_2 - s_3)}{T_1 (s_2 - s_3)} = \frac{T_1 - T_2}{T_1} \quad ...(15.1)$$

Limitations of Carnot cycle

Though Carnot cycle is simple (thermodynamically) and has the *highest thermal efficiency* for given values of T_1 and T_2, yet it is *extremely difficult to operate in practice* because of the following *reasons* :

1. It is difficult to compress a wet vapour isentropically to the saturated state as required by the process 3-4.

2. It is difficult to control the quality of the condensate coming out of the condenser so that the state '3' is exactly obtained.

3. The efficiency of the Carnot cycle is greatly affected by the temperature T_1 at which heat is transferred to the working fluid. Since the critical temperature for steam is only 374°C, therefore, if the cycle is to be operated in the *wet region*, the maximum possible temperature is severely limited.

4. The cycle is still more difficult to operate in practice with superheated steam due to the necessity of supplying the superheat at constant temperature instead of constant pressure (as it is customary).

- *In a practical cycle, limits of pressure and volume are far more easily realised than limits of temperature so that at present no practical engine operates on the Carnot cycle, although all modern cycles aspire to achieve it.*

15.2. RANKINE CYCLE

Ranking cycle *is the theoretical cycle on which the steam turbine (or engine) works.*

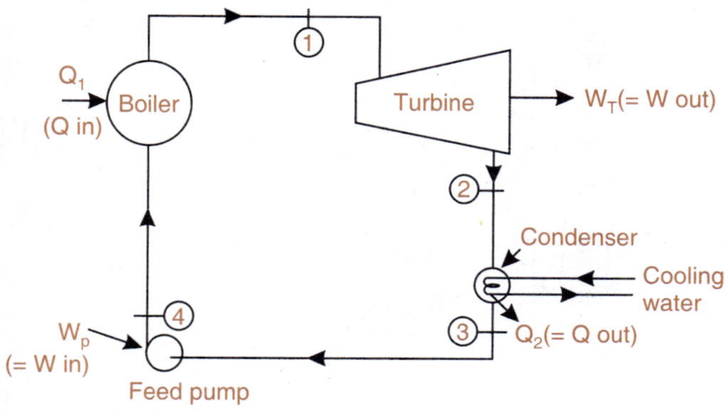

Fig. 15.2. Rankine cycle.

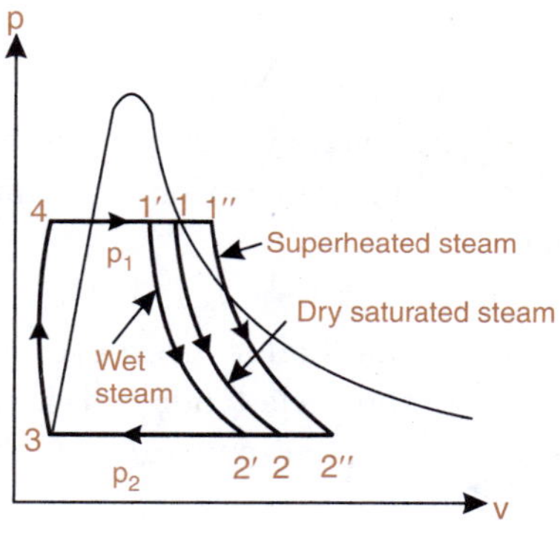

(*a*)

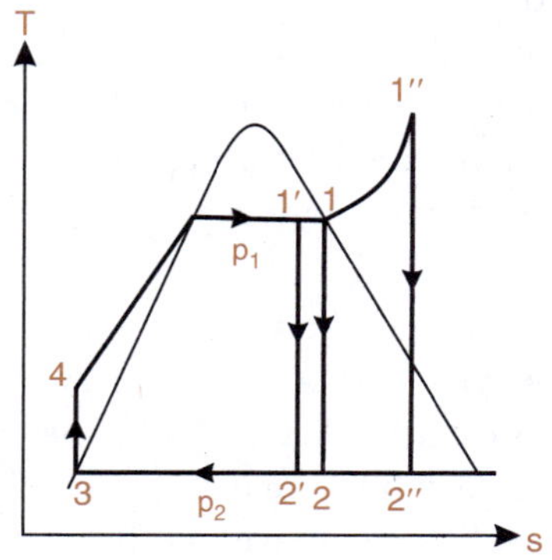

(*b*)

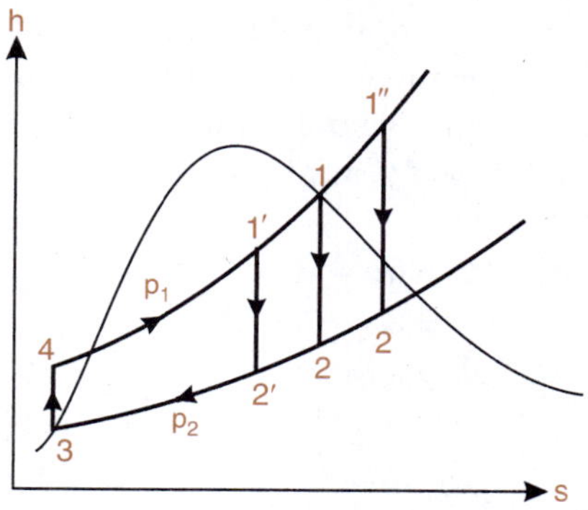

(*c*)

Fig. 15.3. (*a*) *p-v* diagram ; (*b*) *T-s* diagram ; (*c*) *h-s* diagram for Rankine cycle.

The Rankine cycle is shown in Fig. 15.2. It comprises of the following *processes* :

Process 1–2 : Reversible adiabatic expansion in the turbine (or steam engine).

Process 2–3 : Constant-pressure transfer of heat in the condenser.

Process 3–4 : Reversible adiabatic pumping process in the feed pump.

Process 4–1 : Constant-pressure transfer of heat in the boiler.

Fig. 15.3 shows the Rankine cycle on *p-v*, *T-s* and *h-s* diagrams (when the saturated steam enters the turbine, the steam can be wet or superheated also).

Considering 1 kg of fluid :

Applying *steady flow energy equation* (S.F.E.E.) to boiler, turbine, condenser and pump :

(*i*) **For boiler** (as control volume), we get

$$h_{f_4} + Q_1 = h_1$$

$$\therefore \quad Q_1 = h_1 - h_{f_4} \quad \text{...(15.2)}$$

(*ii*) **For turbine** (as control volume), we get

$$h_1 = W_T + h_2, \text{ where } W_T = \text{turbine work}$$

$$\therefore \quad W_T = h_1 - h_2 \quad \text{...(15.3)}$$

(*iii*) **For condenser**, we get

$$h_2 = Q_2 + h_{f_3}$$

$$\therefore \quad Q_2 = h_2 - h_{f_3} \quad \text{...(15.4)}$$

(*iv*) **For the feed pump**, we get

$$h_{f_3} + W_P = h_{f_4}, \quad \text{where } W_P = \text{Pump work}$$

$$\therefore \quad W_P = h_{f_4} - h_{f_3}$$

Now, efficiency of Rankine cycle is given by

$$\eta_{\text{Rankine}} = \frac{W_{\text{net}}}{Q_1} = \frac{W_T - W_P}{Q_1}$$

$$= \frac{(h_1 - h_2) - (h_{f_4} - h_{f_3})}{(h_1 - h_{f_4})} \quad \text{...(15.5)}$$

The feed pump handles liquid water which is incompressible which means with the increase in pressure its density or specific volume undergoes a little change. Using general property relation for reversible adiabatic compression, we get

$$Tds = dh - vdp$$

$\because$ $ds = 0$

$\therefore$ $dh = vdp$

or, $\Delta h = v\,\Delta p$ ...... (Since change in specific volume is negligible)

or, $h_{f_4} - h_{f_3} = v_3\,(p_1 - p_2)$

When p is in bar and v is in m^3/kg, we have

$$h_{f_4} - h_{f_3} = v_3\,(p_1 - p_2) \times 10^5 \text{ J/kg}$$

The feed pump term $(h_{f_4} - h_{f_3})$ being a small quantity in comparison with turbine work, W_T, is usually neglected, *especially when the boiler pressures are low.*

Then, $$\eta_{\text{Rankine}} = \frac{h_1 - h_2}{h_1 - h_{f_4}} \qquad \text{...[15.5 (a)]}$$

Comparison between Rankine Cycle and Carnot Cycle

The following points are worth noting :

(*i*) Between the same temperature limits Rankine cycle provides a higher specific work output than a Carnot cycle, consequently Rankine cycle *requires a smaller steam flow rate resulting in smaller size plant for a given power output.* However, Rankine cycle calls for higher *rates of heat transfer in boiler and condenser.*

(*ii*) Since in Rankine cycle only part of the heat is supplied isothermally at constant higher temperature T_1, therefore, its *efficiency is lower* than that of Carnot cycle. The efficiency of the Rankine cycle will approach that of the Carnot cycle more nearly if the *superheat temperature rise is reduced.*

(*iii*) The advantage of using pump to feed liquid to the boiler instead to compressing a wet vapour is obvious that the *work for compression is very large compared to the pump.*

Fig. 15.4 shows the plots between efficiency and specific steam consumption against boiler pressure for Carnot and ideal Rankine cycles.

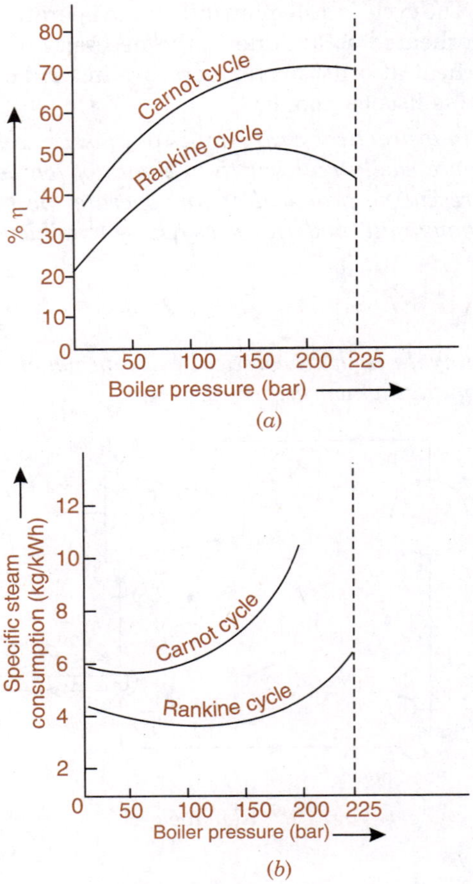

Fig. 15.4

Effect of Operating Conditions on Rankine Cycle Efficiency

The Rankine cycle efficiency can be *improved* by :

(*i*) *Increasing the average temperature at which heat is supplied.*

(*ii*) *Decreasing/reducing the temperature at which heat is rejected.*

This can be achieved by making suitable changes in the conditions of steam generation or condensation, as discussed below :

1. ***Increasing boiler pressure.*** It has been observed that by increasing the boiler pressure (other factors remaining the same) the cycle tends to rise and reaches a maximum value at a boiler pressure of about 166 bar [Fig. 15.5 (*a*)].

2. ***Superheating.*** All other factors remaining the same, if the steam is superheated before allowing it to expand the Rankine cycle efficiency may be increased [Fig. 15.5 (*b*)]. The use of superheated steam also ensures longer turbine blade life because of the absence of erosion from high velocity water particles that are suspended in wet vapour.

3. ***Reducing condenser pressure.*** The thermal efficiency of the cycle can be amply improved by reducing the condenser pressure [Fig. 15.5 (*c*)] (hence by reducing the temperature at which heat is rejected), especially in high vacuums. But the increase in efficiency is obtained at the *increased cost of condensation apparatus.*

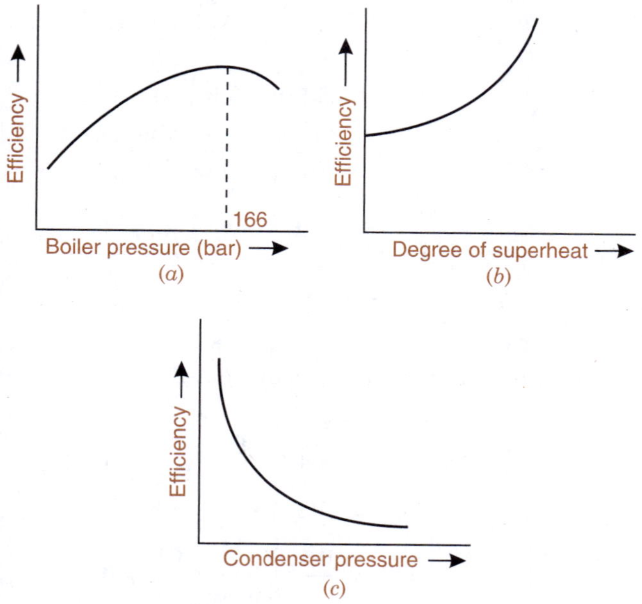

Fig. 15.5. Effect of operating conditions on the thermal efficiency of the Rankine cycle.

The thermal efficiency of the Rankine cycle is also *improved* by the following methods :

(*i*) *By regenerative feed heating.*

(*ii*) *By reheating of steam.*

(*iii*) *By water extraction.*

(*iv*) *By using binary-vapour.*

☞**Example 15.1.** *The following data refer to a simple steam power plant :*

S. No.	*Location*	*Pressure*	*Quality/ Temp.*	*Velocity*
1.	*Inlet to turbine*	*6 MPa (= 60 bar)*	*380°C*	—
2.	*Exit from turbine inlet to condenser*	*10 kPa (= 0.1 bar)*	*0.9*	*200 m/s*
3.	*Exit from condenser and inlet to pump*	*9 kPa (= 0.09 bar)*	*Saturated liquid*	—
4.	*Exit from pump and inlet to boiler*	*7 MPa (= 70 bar)*	—	—
5.	*Exit from boiler*	*6.5 MPa (= 65 bar)*	*400°C*	—
	Rate of steam flow = 10000 kg/h.			

Calculate :

(*i*) *Power output of the turbine.*

(*ii*) *Heat transfer per hour in the boiler and condenser separately.*

(*iii*) *Mass of cooling water circulated per hour in the condenser. Choose the inlet temperature of cooling water 20°C and 30°C at exit from the condenser.*

(*iv*) *Diameter of the pipe connecting turbine with condenser.*

Solution. Refer to Fig. 15.6.

(*i*) **Power output of the turbine, P :**

At 60 bar, 380°C : From steam tables,

$$h_1 = 3043.0 \text{ (at 350°C)} + \frac{3177.2 - 3043.0}{(400 - 350)} \times 30 \text{ ... By interpolation}$$

$$= 3123.5 \text{ kJ/kg}$$

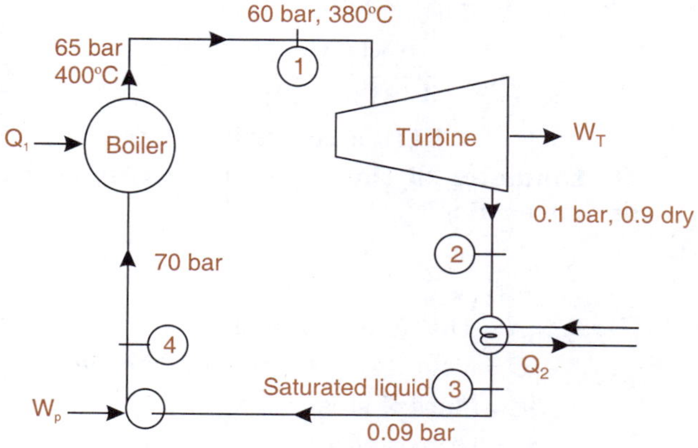

Fig. 15.6

At 0.1 bar :

h_{f_2} = 191.8 kJ/kg,

h_{fg_2} = 2392.8 kJ/kg (from steam tables)

and x_2 = 0.9 (given)

$\therefore\ h_2 = h_{f_2} + x_2 h_{fg_2}$

$= 191.8 + 0.9 \times 2392.8$

$= 2345.3$ kJ/kg

Power output of the turbine = $m_s (h_1 - h_2)$ kW,

where m_s = Rate of steam flow in kg/s and h_1, h_2 = Enthalpy of steam in kJ/kg

$$= \frac{10000}{3600}(3123.5 - 2345.3) = 2162 \text{ kW}$$

Hence power output of the turbine

= 2162 kW. (Ans.)

(*ii*) **Heat transfer per hour in the boiler and condenser :**

At 70 bar : h_{f_4} = 1267.4 kJ/kg

At 65 bar, 400°C :

$$h_a = \frac{3177.2\,(60 \text{ bar}) + 3158.1\,(70 \text{ bar})}{2}$$

$= 3167.6$ kJ/kg ......(By interpolation)

$\therefore$ *Heat transfer per hour in the boiler,*

$Q_1 = 10000\ (h_a - h_{f_4})$ kJ/h

$= 10000\ (3167.6 - 1267.4)$

$= \mathbf{1.9 \times 10^7}$ **kJ/h. (Ans.)**

At 0.09 bar : h_{f_3} = 183.3 kJ/kg

Heat transfer per hour in the condenser,

$$Q_1 = 10000\,(h_2 - h_{f_3})$$
$$= 10000\,(2345.3 - 183.3)$$
$$= \mathbf{2.16 \times 10^7\ kJ/h.}\quad \textbf{(Ans.)}$$

(*iii*) **Mass of cooling water circulated per hour in the condenser, m_w :**

Heat lost by steam = Heat gained by the cooling water

$$Q_2 = m_w \times c_{pw}\,(t_2 - t_1)$$
$$2.16 \times 10^7 = m_w \times 4.18\,(30 - 20)$$
$$\therefore\quad \mathbf{m_w} = \frac{2.16 \times 10^7}{4.18\,(30-20)}$$
$$= \mathbf{1.116 \times 10^7\ kg/h.}\quad \textbf{(Ans.)}$$

(*iv*) **Diameter of the pipe connecting turbine with condenser, d :**

$$\frac{\pi}{4}\,d^2 \times C = m_s\,x_2\,v_{g_2} \quad ...(i)$$

Here, d = Diameter of the pipe (m),
C = Velocity of steam = 200 m/s (given),
m_s = Mass of steam in kg/s,
x_2 = Dryness fraction at '2', and
v_{g_2} = Specific volume at pressure 0.1 bar (= 14.67 m³/kg).

Substituting the various values in eqn. (*i*), we get

$$\frac{\pi}{4}\,d^2 \times 200 = \frac{10000}{3600} \times 0.9 \times 14.67$$
$$d = \left(\frac{10000 \times 0.9 \times 14.67 \times 4}{3600 \times \pi \times 200}\right)^{1/2}$$
$$= \mathbf{0.483\ m\ or\ 483\ mm.}\quad \textbf{(Ans.)}$$

Example 15.2. *In a steam power cycle, the steam supply is at 15 bar and dry and saturated. The condenser pressure is 0.4 bar. Calculate the Carnot and Rankine efficiencies of the cycle. Neglect pump work.*

Solution. Steam supply pressure, p_1 = 15 bar, x_1 = 1

Condenser pressure, p_2 = 0.4 bar

Carnot and Rankine efficiencies :

From steam tables :

At 15 bar : t_s = 198.3°C, h_g = 2789.9 kJ/kg, s_g = 6.4406 kJ/kg K

At 0.4 bar : t_s = 75.9°C, h_f = 317.7 kJ/kg,
h_{fg} = 2319.2 kJ/kg,
s_f = 1.0261 kJ/kg K,
s_{fg} = 6.6448 kJ/kg K

$$T_1 = 198.3 + 273 = 471.3\ K$$
$$T_2 = 75.9 + 273 = 348.9\ K$$

$$\therefore\quad \eta_{carnot} = \frac{T_1 - T_2}{T_1} = \frac{471.3 - 348.9}{471.3}$$
$$= \mathbf{0.259\ or\ 25.9\%.}\quad \textbf{(Ans.)}$$

$$\eta_{Rankine} = \frac{\text{Adiabatic or isentropic heat drop}}{\text{Heat supplied}}$$
$$= \frac{h_1 - h_2}{h_1 - h_{f_2}}$$

where $\quad h_2 = h_{f_2} + x_2\,h_{fg_2} = 317.7 + x_2 \times 2319.2 \quad ...(i)$

Value of x_2 :

As the steam expands isentropically,

$$\therefore\quad s_1 = s_2$$
$$6.4406 = s_{f_2} + x_2$$
$$s_{fg_2} = 1.0261 + x_2 \times 6.6448$$
$$\therefore\quad x_2 = \frac{6.4406 - 1.0261}{6.6448} = 0.815$$
$$\therefore\quad h_2 = 317.7 + 0.815 \times 2319.2$$
$$= 2207.8\ \text{kJ/kg} \quad \text{[From eqn. }(i)]$$

$$\text{Hence,}\quad \eta_{Rankine} = \frac{2789.9 - 2207.8}{2789.9 - 317.7}$$
$$= \mathbf{0.2354\ or\ 23.54\%.}\quad \textbf{(Ans.)}$$

Example 15.3. *In a steam turbine steam at 20 bar, 360°C is expanded to 0.08 bar. It then enters a condenser, where it is condensed to saturated liquid water. The pump feeds back the water into the boiler. Assume ideal processes, find per kg of steam the net work and the cycle efficiency.*

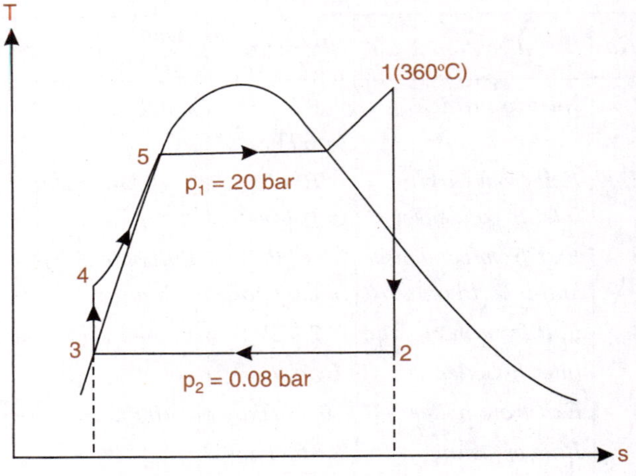

Fig. 15.7

Solution. Boiler pressure, p_1 = 20 bar (360°C)

Condenser pressure, p_2 = 0.08 bar

From steam tables :

At 20 bar (p_1), 360°C : h_1 = 3159.3 kJ/kg
s_1 = 6.9917 kJ/kg K

At 0.08 bar (p_2) :

$$h_3 = h_{f(p_2)} = 173.88 \text{ kJ/kg},$$

$$s_3 = s_{f(p_2)} = 0.5926 \text{ kJ/kg K}$$

$$h_{fg(p_2)} = 2403.1 \text{ kJ/kg},$$

$$s_{g(p_2)} = 8.2287 \text{ kJ/kg K}$$

$$v_{f(p_2)} = 0.001008 \text{ m}^3\text{/kg}$$

$$\therefore \quad s_{fg(p_2)} = 7.6361 \text{ kJ/kg K}$$

Now $s_1 = s_2$

$$6.9917 = s_{f(p_2)} + x_2\, s_{fg(p_2)}$$

$$= 0.5926 + x_2 \times 7.6361$$

$$\therefore \quad x_2 = \frac{0.69917 - 0.5926}{7.6361} = 0.838$$

$$\therefore \quad h_2 = h_{f(p_2)} + x_2\, h_{fg(p_2)}$$

$$= 173.88 + 0.838 \times 2403.1$$

$$= 2187.68 \text{ kJ/kg.}$$

Net work, W_{net} :

$$W_{net} = W_{turbine} - W_{pump}$$

$$W_{pump} = h_{f_4} - h_{f(p_2)} \;(= h_{f_3})$$

$$= v_{f(p_2)}\,(p_1 - p_2)$$

$$= 0.00108 \text{ (m}^3\text{/kg)} \times (20 - 0.08) \times 100 \text{ kN/m}^2$$

$$= 2.008 \text{ kJ/kg}$$

[and $h_{f_4} = 2.008 + h_{f(p_2)} = 2.008 + 173.88$

$= 175.89$ kJ/kg]

$$W_{turbine} = h_1 - h_2 = 3159.3 - 2187.68$$

$$= 971.62 \text{ kJ/kg}$$

$$\therefore \quad W_{net} = 971.62 - 2.008$$

$= \mathbf{969.61}$ **kJ/kg. (Ans.)**

Cycle efficiency, η_{cycle} :

$$Q_1 = h_1 - h_{f_4} = 3159.3 - 175.89$$

$$= 2983.41 \text{ kJ/kg}$$

$$\therefore \quad \eta_{cycle} = \frac{W_{net}}{Q_1} = \frac{969.61}{2983.41}$$

$= \mathbf{0.325}$ **or 32.5%. (Ans.)**

Example 15.4. *A Rankine cycle operates between pressures of 80 bar and 0.1 bar. The maximum cycle temperature is 600°C. If the steam turbine and condensate pump efficiencies are 0.9 and 0.8 respectively, calculate the specific work and thermal efficiency. Relevant steam table extract is given below.*

p(bar)	*t(°C)*	*Specific volume (m^3/kg)*		*Specific enthalpy (kJ/kg)*			*Specific entropy (kJ/kg K)*		
		v_f	v_g	h_f	h_{fg}	h_g	s_f	s_{fg}	s_g
0.1	*45.84*	*0.0010103*	*14.68*	*191.9*	*2392.3*	*2584.2*	*0.6488*	*7.5006*	*8.1494*
80	*295.1*	*0.001385*	*0.0235*	*1317*	*1440.5*	*2757.5*	*3.2073*	*2.5351*	*5.7424*

80 bar, 600°C	*v*	*0.486 m^3/kg*
Superheat	*h*	*3642 kJ/kg*
table	*s*	*7.0206 kJ/kg K*

(GATE)

Solution. Refer to Fig. 15.8

At 80 bar, 600°C :

$$h_1 = 3642 \text{ kJ/kg ;}$$

$$s_1 = 7.0206 \text{ kJ/kg K.}$$

Since $s_1 = s_2$,

$$\therefore \quad 7.0206 = s_{f_2} + x_2\, s_{fg_2}$$

$$= 0.6488 + x_2 \times 7.5006$$

or

$$x_2 = \frac{7.0206 - 0.6488}{7.5006} = 0.85$$

Now, $h_2 = h_{f_2} + x_2\, h_{fg_2}$

$$= 191.9 + 0.85 \times 2392.3$$

$$= 2225.36 \text{ kJ/kg}$$

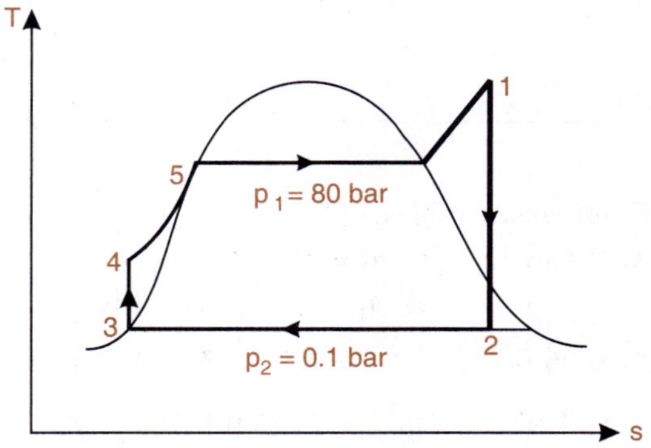

Fig. 15.8

Actual turbine work

$$= \eta_{turbine} \times (h_1 - h_2)$$

$$= 0.9\,(3642 - 2225.36) = 1275 \text{ kJ/kg}$$

Pump work $= v_{f(p_2)}\,(p_1 - p_2)$

$$= 0.0010103\,(80 - 0.1) \times \frac{10^5}{10^3}\ \text{kN/m}^2$$

$$= 8.072\ \text{kJ/kg}$$

Actual pump work $= \dfrac{8.072}{\eta_{pump}} = \dfrac{8.072}{0.8}$

$$= 10.09\ \text{kJ/kg}$$

Specific work

$$(W_{net}) = 1275 - 10.09$$

$$= \mathbf{1264.91\ kJ/kg.}\quad \textbf{(Ans.)}$$

Thermal efficiency $= \dfrac{W_{net}}{Q_1}$

where, $Q_1 = h_1 - h_{f_4}$

But $h_{f_4} = h_{f_3}$ + pump work

$$= 191.9 + 10.09 = 202\ \text{kJ/kg}$$

$\therefore$ Thermal efficiency, $\eta_{th} = \dfrac{1264.91}{3642 - 202}$

$$= \mathbf{0.368\ or\ 36.8\%.}\quad \textbf{(Ans.)}$$

Example 15.5. *A simple Rankine cycle works between pressures 28 bar and 0.06 bar, the initial condition of steam being dry saturated. Calculate the cycle efficiency, work ratio and specific steam consumption.*

Solution.

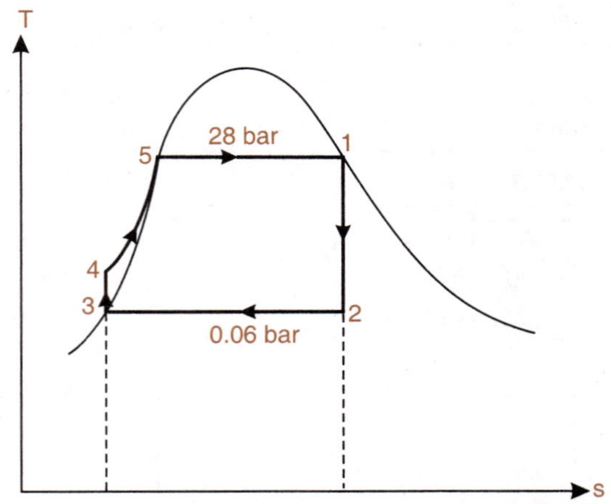

Fig. 15.9

From steam tables,

At 28 bar : $h_1 = 2802$ kJ/kg,

$s_1 = 6.2104$ kJ/kg K

At 0.06 bar : $h_{f_2} = h_{f_3} = 151.5$ kJ/kg,

$h_{fg_2} = 2415.9$ kJ/kg,

$s_{f_2} = 0.521$ kJ/kg K,

$s_{fg_2} = 7.809$ kJ/kg K

$v_f = 0.001$ m³/kg

Considering turbine *process 1–2*, we have :

$$s_1 = s_2$$

$$6.2104 = s_{f_2} + x_2\, s_{fg_2}$$

$$= 0.521 + x_2 \times 7.809$$

$$\therefore\quad x_2 = \frac{6.2104 - 0.521}{7.809} = 0.728$$

$$\therefore\quad h_2 = h_{f_2} + x_2\, h_{fg_2}$$

$$= 151.5 + 0.728 \times 2415.9$$

$$= 1910.27\ \text{kJ/kg}$$

$\therefore$ Turbine work,

$$W_{turbine} = h_1 - h_2$$

$$= 2802 - 1910.27$$

$$= 891.73\ \text{kJ/kg}$$

Pump work,

$$W_{pump} = h_{f_4} - h_{f_3} = v_f\,(p_1 - p_2)$$

$$= \frac{0.001\,(28 - 0.06) \times 10^5}{1000}$$

$$= 2.79\ \text{kJ/kg}$$

$$[\because\ h_{f_4} = h_{f_3} + 2.79 = 151.5 + 2.79 = 154.29\ \text{kJ/kg}]$$

$\therefore$ Net work,

$$W_{net} = W_{turbine} - W_{pump}$$

$$= 891.73 - 2.79 = 888.94\ \text{kJ/kg}$$

Cycle efficiency $= \dfrac{W_{net}}{Q_1} = \dfrac{888.94}{h_1 - h_{f_4}}$

$$= \frac{888.94}{2802 - 154.29}$$

$$= \mathbf{0.3357\ or\ 33.57\%.}\quad \textbf{(Ans.)}$$

Work ratio $= \dfrac{W_{net}}{W_{turbine}} = \dfrac{888.94}{891.73}$

$$= \mathbf{0.997.}\quad \textbf{(Ans.)}$$

Specific steam consumption

$$= \frac{3600}{W_{net}} = \frac{3600}{888.94}$$

$$= \mathbf{4.049\ kg/kWh.}\quad \textbf{(Ans.)}$$

☞**Example 15.6.** *In a Rankine cycle, the steam at inlet to turbine is saturated at a pressure of 35 bar and the exhaust pressure is 0.2 bar. Determine :*

(i) The pump work,

(ii) The turbine work,

(iii) The Rankine efficiency,

(iv) The condenser heat flow,

(v) The dryness at the end of expansion.

Assume flow rate of 9.5 kg/s.

Solution. Pressure and condition of steam, at inlet to the turbine,

$p_1 = 35$ bar, $x = 1$

Exhaust pressure, $p_2 = 0.2$ bar

Flow rate, $\dot{m} = 9.5$ kg/s

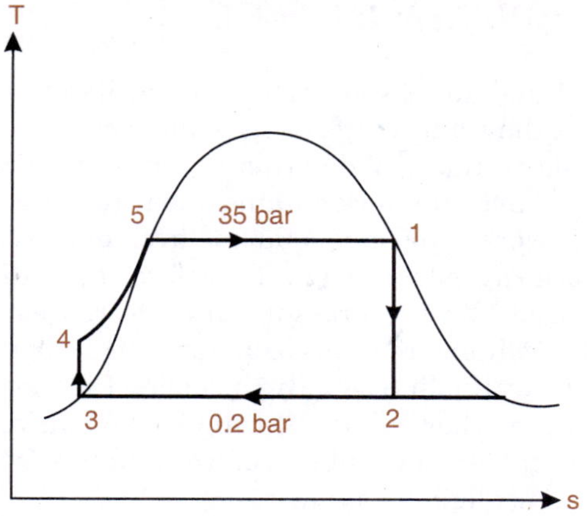

Fig. 15.10

From steam tables :

At 35 bar :

$$h_1 = h_{g_1} = 2802 \text{ kJ/kg},$$

$$s_{g_1} = 6.1228 \text{ kJ/kg K}$$

At 0.26 bar :

$$h_f = 251.5 \text{ kJ/kg},$$

$$h_{fg} = 2358.4 \text{ kJ/kg},$$

$$v_f = 0.001017 \text{ m}^3\text{/kg},$$

$$s_f = 0.8321 \text{ kJ/kg K},$$

$$s_{fg} = 7.0773 \text{ kJ/kg K}.$$

(*i*) **The pump work :**

Pump work $= (p_4 - p_3)$

$v_f = (35 - 0.2) \times 10^5 \times 0.001017$ J

or 3.54 kJ/kg

$$\begin{bmatrix} \text{Also } h_{f_4} - h_{f_3} = \text{Pump work} = 3.54 \\ \therefore \quad h_{f_4} = 251.5 + 3.54 = 255.04 \text{ kJ/kg} \end{bmatrix}$$

Now power required to drive the pump

$= 9.5 \times 3.54$ kJ/s

or **33.63 kW.** **(Ans.)**

(*ii*) **The turbine work :**

$$s_1 = s_2 = s_{f_2} + x_2 \times s_{fg_2}$$

$$6.1228 = 0.8321 + x_2 \times 7.0773$$

$$\therefore \quad x_2 = \frac{6.1228 - 0.8321}{7.0773} = 0.747$$

$$\therefore \quad h_2 = h_{f_2} + x_2\, h_{fg_2}$$

$$= 251.5 + 0.747 \times 2358.4$$

$$= 2013 \text{ kJ/kg}$$

$\therefore$ *Turbine work* $= \dot{m}\,(h_1 - h_2)$

$= 9.5\,(2802 - 2013)$

$=$ **7495.5 kW.** **(Ans.)**

It may be noted that pump work (33.63 kW) is very small as compared to the turbine work (7495.5 kW).

(*iii*) **The Rankine efficiency :**

$$\eta_{\text{rankine}} = \frac{h_1 - h_2}{h_1 - h_{f_2}} = \frac{2802 - 2013}{2802 - 251.5} = \frac{789}{2550.5}$$

$=$ **0.3093 or 30.93%.** **(Ans.)**

(*iv*) **The condenser heat flow :**

The condenser heat flow $= \dot{m}\,(h_2 - h_{f_3})$

$= 9.5\,(2013 - 251.5)$

$=$ **16734.25 kW.** **(Ans.)**

(*v*) **The dryness at the end of expansion, x_2 :**

The dryness at the end of expansion,

$x_2 =$ **0.747 or 74.7%.** **(Ans.)**

Example 15.7. *The adiabatic enthalpy drop across the prime mover of the Rankine cycle is 840 kJ/kg. The enthalpy of steam supplied is 2940 kJ/kg. If the back pressure is 0.1 bar, find the specific steam consumption and thermal efficiency.*

Solution. Adiabatic enthalpy drop, $h_1 - h_2 = 840$ kJ/kg

Enthalpy of steam supplied, $h_1 = 2940$ kJ/kg

Back pressure, $p_2 = 0.1$ bar

From steam tables, corresponding to 0.1 bar :

$$h_f = 191.8 \text{ kJ/kg}$$

Now, $$\eta_{\text{rankine}} = \frac{h_1 - h_2}{h_1 - h_{f_2}} = \frac{840}{2940 - 191.8}$$

$= 0.3056 =$ **30.56%.** **(Ans.)**

Useful work done per kg of steam = 840 kJ/kg

$\therefore$ **Specific steam consumption** $= \dfrac{1}{840}$ kg/s

$= \dfrac{1}{840} \times 3600 =$ **4.286 kg/kWh.** **(Ans.)**

☞**Example 15.8.** *A 35 kW (I.P.) system engines consumes 284 kg/h at 15 bar and 250°C. If condenser pressure is 0.14 bar, determine :*

(*i*) *Final condition of steam ;*

(*ii*) *Rankine efficiency ;*

(*iii*) *Relative efficiency.*

Solution. Power developed by the engine = 35 kW (I.P.)

Steam consumption = 284 kg/h

Condenser pressure = 0.14 bar

Steam inlet pressure = 15 bar, 250°C.

From steam tables :

At 15 bar, 250°C : $h = 2923.3$ kJ/kg, $s = 6.709$ kJ/kg K

At 0.14 bar : $h_f = 220$ kJ/kg, $h_{fg} = 2376.6$ kJ/kg, $s_f = 0.737$ kJ/kg K, $s_{fg} = 7.296$ kJ/kg K (*i*) **Final condition of steam :**

Since steam expands isentropically.

$$\therefore \quad s_1 = s_2 = s_{f_2} + x_2\, s_{fg_2}$$

$$6.709 = 0.737 + x_2 \times 7.296$$

$$\therefore \quad x_2 = \frac{6.709 - 0.737}{7.296}$$

$$= \mathbf{0.818 \simeq 0.82.} \quad \textbf{(Ans.)}$$

$$\therefore \quad h_2 = h_{f_2} + x_2\, h_{fg_2}$$

$$= 220 + 0.82 \times 2376.6$$

$$= 2168.8 \text{ kJ/kg.}$$

(*ii*) **Rankine efficiency :**

$$\eta_{rankine} = \frac{h_1 - h_2}{h_1 - h_{f_2}} = \frac{2923.3 - 2168.8}{2923.3 - 220}$$

$$= \mathbf{0.279 \text{ or } 27.9\%.} \quad \textbf{(Ans.)}$$

(*iii*) **Relative efficiency :**

$$\eta_{thermal} = \frac{\text{I.P.}}{\dot{m}\left(h_1 - h_{f_2}\right)}$$

$$= \frac{35}{\frac{284}{3600}(2923.3 - 220)}$$

$$= 0.1641 \text{ or } 16.41\%$$

$$\eta_{relative} = \frac{\eta_{thermal}}{\eta_{rankine}} = \frac{0.1641}{0.279}$$

$$= \mathbf{0.588 \text{ or } 58.8\%.} \quad \textbf{(Ans.)}$$

Example 15.9. *Calculate the fuel oil consumption required in a industrial steam plant to generate 5000 kW at the turbine shaft. The calorific value of the fuel is 40000 kJ/kg and the Rankine cycle efficiency is 50%. Assume appropriate values for isentropic turbine efficiency, boiler heat transfer efficiency and combustion efficiency.* **(N.U.)**

Solution. Power to be generated at the turbine shaft,

$$P = 5000 \text{ kW}$$

The calorific value of the fuel, C = 40000 kJ/kg

Rankine cycle efficiency, $\eta_{rankine}$ = 50%

Fuel oil combustion, m_f :

Assume : $\eta_{turbine}$ = 90% ; $\eta_{heat\ transfer}$ = 85% ;

$\eta_{combustion}$ = 98%

$$\eta_{rankine} = \frac{\text{Shaft power} / \eta_{turbine}}{m_f \times C \times \eta_{heat\ transfer} \times \eta_{combustion}}$$

or $$0.5 = \frac{(5000/0.9)}{m_f \times 40000 \times 0.85 \times 0.98}$$

$$\therefore \quad m_f = \frac{(5000/0.9)}{0.5 \times 40000 \times 0.85 \times 0.98}$$

$$= \mathbf{0.3335 \text{ kg/s or } 1200.6 \text{ kg/h.}} \quad \textbf{(Ans.)}$$

15.3. MODIFIED RANKINE CYCLE

Figs. 15.11 and 15.12 show the modified Rankine cycle on p-v and T-s diagrams (neglecting pump work) respectively. It will be noted that p-V diagram is very narrow at the toe *i.e.*, point '2'' and the work obtained near to e is very small. In fact this work is too inadequate to overcome friction (due to reciprocating parts) even. Therefore, the adiabatic is terminated at '2' ; the pressure drop decreases suddenly whilst the volume remains constant. This operation is represented by the line 2–3. By this doing the stroke length is reduced ; in other words the cylinder dimensions reduce but at the expense of small loss of work (area 2-3-2') which, however, is negligibly small.

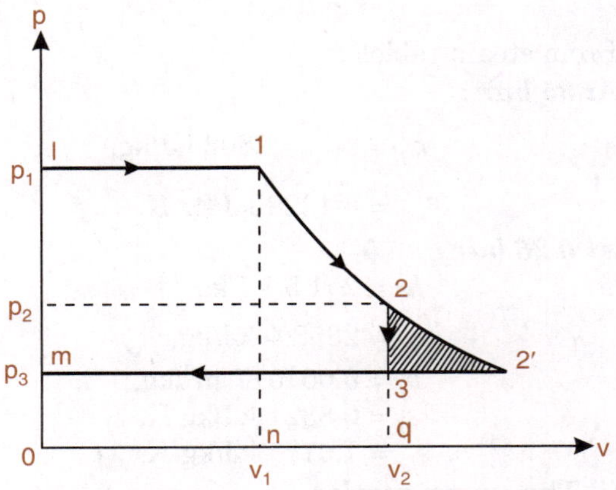

Fig. 15.11. p-v diagram of Modified Rankine Cycle.

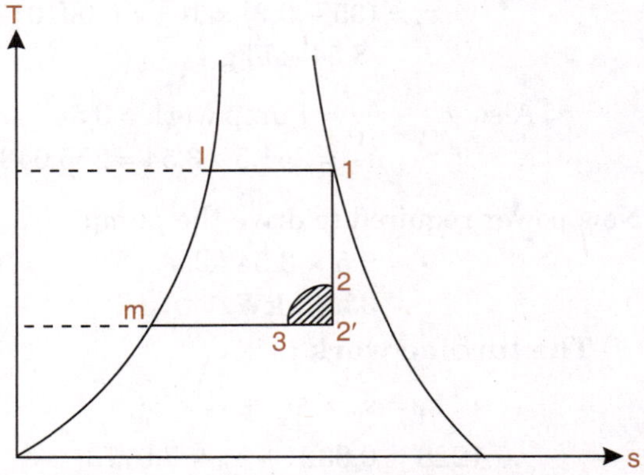

Fig. 15.12. T-s diagram of Modified Rankine cycle.

The work done during the modified Rankine cycle can be calculated in the following way :

Let p_1, v_1, u_1 and h_1 correspond to initial condition of steam at '1'.

p_2, v_2, u_2 and h_2 correspond to condition of steam at '2'.

p_3, h_3 correspond to condition of steam at '3'.

Work done during the cycle/kg of steam

= area l-1-2-3-m

= area 'o-l-1-n' + area '1-2-q-n' – area 'o-m-3-q'

$= p_1v_1 + (u_1 - u_2) - p_3v_2$

Heat supplied $= h_1 - h_{f_3}$

$\therefore$ The modified Rankine efficiency

$$= \frac{\text{Work done}}{\text{Heat supplied}}$$

$$= \frac{p_1v_1 + (u_1 - u_2) - p_3v_2}{h_1 - h_{f_3}} \quad ...(15.6)$$

Alternative method for finding modified Rankine efficiency :

Work done during the cycle/kg of steam

$= \text{area } `l\text{-}1\text{-}2\text{-}3\text{-}m'$

$= \text{area } `l\text{-}1\text{-}2\text{-}s' + \text{area } `s\text{-}2\text{-}3\text{-}m'$

$= (h_1 - h_2) + (p_2 - p_3)\, v_2$

Heat supplied $= h_1 - h_{f_3}$

$$\text{Modified Rankine efficiency} = \frac{\text{Work done}}{\text{Heat supplied}}$$

$$= \frac{(h_1 - h_2) + (p_2 - p_3)\, v_2}{h_1 - h_{f_3}} \quad ...(15.7)$$

Note. Modified Rankine cycle is used for 'reciprocating steam engines' because stroke length and hence cylinder size is reduced with the sacrifice of practically a quite negligible amount of work done.

☞**Example 15.10. (Modified Rankine cycle).** *Steam at a pressure of 15 bar and 300°C is delivered to the throttle of an engine. The steam expands to 2 bar when release occurs. The steam exhaust takes place at 1.1 bar. A performance test gave the result of the specific steam consumption of 12.8 kg/kWh and a mechanical efficiency of 80 per cent. Determine :*

(*i*) *Ideal work or the modified Rankine engine work per kg.*

(*ii*) *Efficiency of the modified Rankine engine or ideal thermal efficiency.*

(*iii*) *The indicated and brake work per kg.*

(*iv*) *The brake thermal efficiency.*

(*v*) *The relative efficiency on the basis of indicated work and brake work.*

Solution. Fig. 15.13 shows the *p-v* and *T-s* diagrams for modified Rankine cycle.

From steam tables :

1. ***At 15 bar, 300°C :***

$h_1 = 3037.6$ kJ/kg, $v_1 = 0.169$ m³/kg,

$s_1 = 6.918$ kJ/kg K.

2. ***At 2 bar :***

$t_{s_2} = 120.2$°C, $h_{f_2} = 504.7$ kJ/kg,

$h_{fg_2} = 2201.6$ kJ/kg,

$s_{f_2} = 1.5301$ kJ/kg K,

$s_{fg_2} = 5.5967$ kJ/kg K,

$v_{f_2} = 0.00106$ m³/kg,

$v_{g_2} = 0.885$ m³/kg.

3. ***At 1.1 bar :***

$t_{s_3} = 102.3$°C, $h_{f_3} = 428.8$ kJ/kg,

$h_{fg_3} = 2250.8$ kJ/kg,

$s_{f_3} = 1.333$ kJ/kg K,

$s_{fg_3} = 5.9947$ kJ/kg K,

$v_{f_3} = 0.001$ m³/kg,

$v_{g_3} = 1.549$ m³/kg.

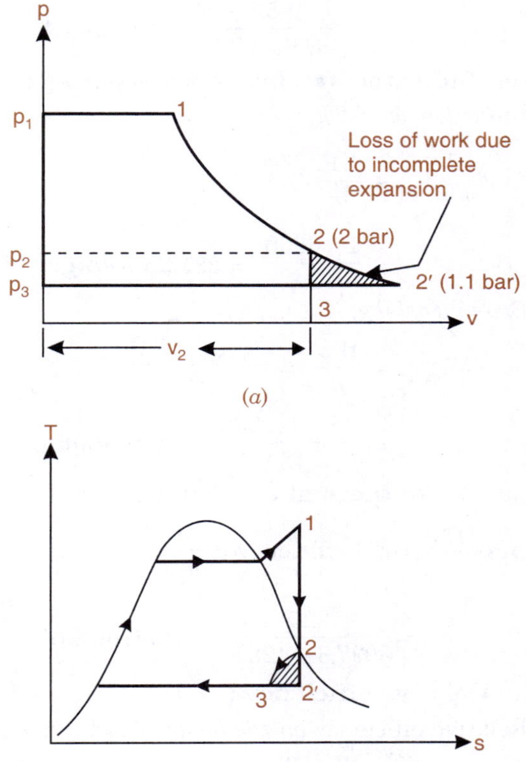

(*a*) (*b*)

Fig. 15.13. *p-v* and *T-s* diagrams.

During isentropic expansion 1–2, we have

$$s_1 = s_2$$

$$6.918 = s_{f_2} + x_2\, s_{fg_2}$$

$$= 1.5301 + x_2 \times 5.5967$$

$$\therefore \quad x_2 = \frac{6.918 - 1.5301}{5.5967} = 0.96.$$

Then $h_2 = h_{f_2} + x_2\, h_{fg_2} = 504.7 + 0.96 \times 2201.6 = 2618.2$ kJ/kg

$$v_2 = x_2\, v_{g_2} + (1 - x_2)\, v_{f_2}$$
$$= 0.96 \times 0.885 + (1 - 0.96) \times 0.00106$$
$$= 0.849 \text{ m}^3/\text{kg}.$$

(*i*) **Ideal work :**

Ideal work or modified Rankine engine work/kg,

$$W = (h_1 - h_2) + (p_2 - p_3)\, v_2$$
$$= (3037.6 - 2618.2) + (2 - 1.1) \times 10^5 \times 0.849/1000$$
$$= 419.4 + 76.41 = \mathbf{495.8 \text{ kJ/kg.}} \quad \textbf{(Ans.)}$$

(*ii*) **Rankine engine efficiency :**

$$\eta_{\text{rankine}} = \frac{\text{Work done}}{\text{Heat supplied}} = \frac{495.8}{(h_1 - h_{f_3})}$$
$$= \frac{495.8}{3037.6 - 428.8} = \mathbf{0.19 \text{ or } 19\%.} \quad \textbf{(Ans.)}$$

(*iii*) **Indicated and brake work per kg :**

Indicated work/kg,

$$W_{\text{indicated}} = \frac{\text{I.P.}}{\dot{m}}$$
$$= \frac{1 \times 3600}{12.8} = \mathbf{281.25 \text{ kJ/kg.}} \quad \textbf{(Ans.)}$$

Brake work/kg,

$$W_{\text{brake}} = \frac{\text{B.P.}}{\dot{m}} = \frac{\eta_{\text{mech.}} \times \text{I.P.}}{\dot{m}}$$
$$= \frac{0.8 \times 1 \times 3600}{12.8} = \mathbf{225 \text{ kJ/kg.}} \quad \textbf{(Ans.)}$$

(*iv*) **Brake thermal efficiency :**

$$\text{Brake thermal efficiency} = \frac{W_{\text{brake}}}{h_1 - h_{f_3}}$$
$$= \frac{225}{3037.6 - 428.8} = \mathbf{0.086 \text{ or } 8.6\%.} \quad \textbf{(Ans.)}$$

(*v*) **Relative efficiency :**

Relative efficiency *on the basis of indicated work*

$$= \frac{\dfrac{W_{\text{indicated}}}{h_1 - h_{f_3}}}{\dfrac{W}{h_1 - h_{f_3}}} = \frac{W_{\text{indicated}}}{W}$$
$$= \frac{281.25}{495.8} = \mathbf{0.567 \text{ or } 56.7\%.} \quad \textbf{(Ans.)}$$

Relative efficiency *on the basis of brake work*

$$= \frac{\dfrac{W_{\text{indicated}}}{(h_1 - h_{f_3})}}{\dfrac{W}{(h_1 - h_{f_3})}} = \frac{W_{\text{brake}}}{W}$$
$$= \frac{225}{495.8} = \mathbf{0.4538 \text{ or } 45.38\%.} \quad \textbf{(Ans.)}$$

Example 15.11. *Superheated steam at a pressure of 10 bar and 400°C is supplied to a steam engine. Adiabatic expansion takes place to release point at 0.9 bar and it exhausts into a condenser at 0.3 bar. Neglecting clearance determine for a steam flow rate of 1.5 kg/s :*

(*i*) *Quality of steam at the end of expansion and the end of constant volume operation.*

(*ii*) *Power developed.*

(*iii*) *Specific steam consumption.*

(*iv*) *Modified Rankine cycle efficiency.*

Solution. Fig. 15.14 shows the *p-v* and *T-s* diagrams for modified Rankine cycle.

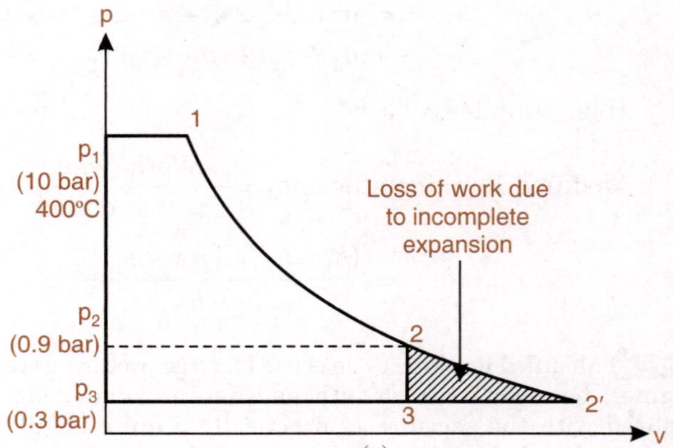

(*a*)

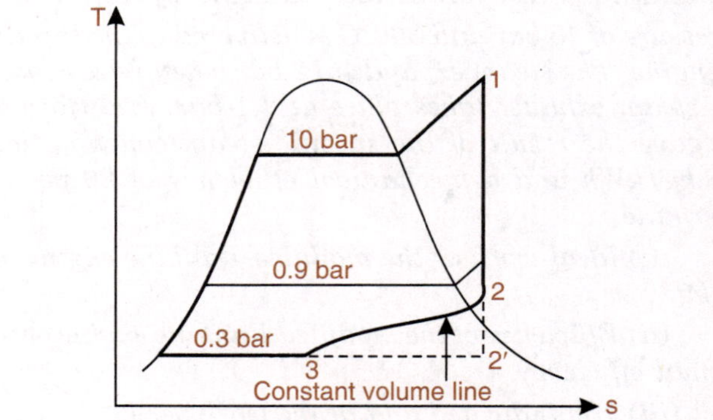

(*b*)

Fig. 15.14. *p-v* and *T-s* diagrams.

From steam tables :

1. At 10 bar, 400°C :

$$h_1 = 3263.9 \text{ kJ/kg}, \; v_1 = 0.307 \text{ m}^3/\text{kg}, \; s_1 = 7.465 \text{ kJ/kg K}$$

2. At 0.9 bar :

$$t_{s_2} = 96.7°\text{C}, \; h_{g_2} = 2670.9 \text{ kJ/kg},$$
$$s_{g_2} = 7.3954 \text{ kJ/kg K},$$
$$v_{g_2} = 1.869 \text{ m}^3/\text{kg}$$

3. At 0.3 bar :

$$h_{f_3} = 289.3 \text{ kJ/kg}, \; v_{g_3} = 5.229 \text{ m}^3/\text{kg}$$

(*i*) **Quality of steam at the end of expansion, T_{sup2} :**

For isentropic expansion 1-2, we have

$$s_1 = s_2$$

$$= s_{g_2} + c_p \log_e \frac{T_{sup\,2}}{T_{s_2}}$$

$$7.465 = 7.3954 + 2.1 \log_e \frac{T_{sup2}}{(96.7 + 273)}$$

$$\left(\frac{7.465 - 7.3954}{2.1}\right) = \log_e \frac{T_{sup2}}{369.7} \text{ or } \log_e \frac{T_{sup2}}{369.7} = 0.0033$$

$$\frac{T_{sup2}}{369.7} = 1.0337 \text{ or } T_{sup2} = 382 \text{ K}$$

or $$t_{sup2} = 382 - 273 = \mathbf{109°C.} \quad \textbf{(Ans.)}$$

$$\therefore \quad h_2 = h_{g_2} + c_{ps}(T_{sup2} - T_{s_2})$$

$$= 2670.9 + 2.1\,(382 - 366.5)$$

$$= 2703.4 \text{ kJ/kg.}$$

Quality of steam at the end of constant volume operation, x_3 :

For calculating v_2 using the relation

$$\frac{v_{g_2}}{T_{s_2}} = \frac{v_2}{T_{sup2}} \quad \text{(Approximately)}$$

$$\frac{1.869}{369.7} = \frac{v_2}{382}$$

or $$v_2 = \frac{1.869 \times 382}{369.7} = 1.931 \text{ m}^3\text{/kg}$$

Also $$v_2 = v_3 = x_3\, v_{g_3}$$

$$1.931 = x_3 \times 5.229$$

or $$\mathbf{x_3} = \frac{1.931}{5.229} = \mathbf{0.37.} \quad \textbf{(Ans.)}$$

(*ii*) **Power developed, P :**

Work done $= (h_1 - h_2) + (p_2 - p_3)\, v_2$

$$= (3263.9 - 2703.4) + \frac{(0.75 - 0.3) \times 10^5 \times 1.931}{1000}$$

$$= 560.5 + 86.9 = 647.4 \text{ kJ/kg}$$

∴ **Power developed** = Steam flow rate × work done (per kg)

$$= 1 \times 647.4 = \mathbf{647.4 \text{ kW.}} \quad \textbf{(Ans.)}$$

(*iii*) **Specific steam consumption, ssc :**

$$\textbf{ssc} = \frac{3600}{\text{Power}} = \frac{1 \times 3600}{647.4}$$

$$= \mathbf{5.56 \text{ kg/kWh.}} \quad \textbf{(Ans.)}$$

(*iv*) **Modified Rankine cycle efficiency, η_{mR} :**

$$\eta_{mR} = \frac{(h_1 - h_2) + (p_2 - p_3)\, v_2}{h_1 - h_{f_3}}$$

$$= \frac{647.4}{3263.9 - 289.3}$$

$$= \mathbf{0.217 \text{ or } 21.7\%.} \quad \textbf{(Ans.)}$$

15.4. REGENERATIVE CYCLE

In the Rankine cycle it is observed that the condensate which is fairly at low temperature has an irreversible mixing with hot boiler water and this results in decrease of cycle efficiency. Methods are, therefore, adopted to heat the feed water from the hot well of condenser irreversibly by interchange of heat within the system and thus improving the cycle efficiency. This heating method is called regenerative feed heat and the cycle is called *regenerative cycle*.

The principle of regeneration can be practically utilised by extracting steam from the turbine at several locations and supplying it to the regenerative heaters. The resulting cycle is known as *regenerative or bleeding cycle*. The heating arrangement comprises of : (*i*) For medium capacity turbines—not more than 3 heaters ; (*ii*) For high pressure high capacity turbines—not more than 5 to 7 heaters ; and (*iii*) For turbines of super critical parameters 8 to 9 heaters. The most advantageous condensate heating temperature is selected depending on the turbine throttle conditions and this determines the number of heaters to be used. The final condensate heating temperature is kept 50 to 60°C below the boiler saturated steam temperature so as to prevent evaporation of water in the feed mains following a drop in the boiler drum pressure. The conditions of steam bled for each heater are so selected that the temperature of saturated steam will be 4 to 10°C higher than the final condensate temperature.

Fig. 15.15 (*a*) shows a diagrammatic layout of a condensing steam power plant in which a surface condenser is used to condense all the steam that is not extracted for feed water heating. The turbine is double extracting and the boiler is equipped with a superheater. The cycle diagram (*T-s*) would appear as shown in Fig. 15.15 (*b*). This arrangement constitutes a *regenerative cycle*.

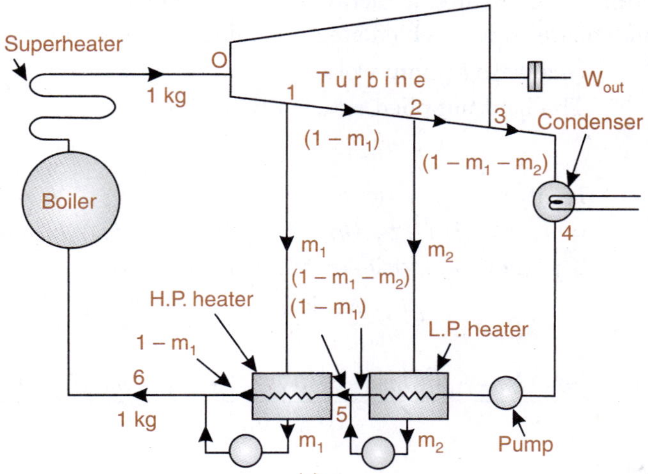

(*a*)

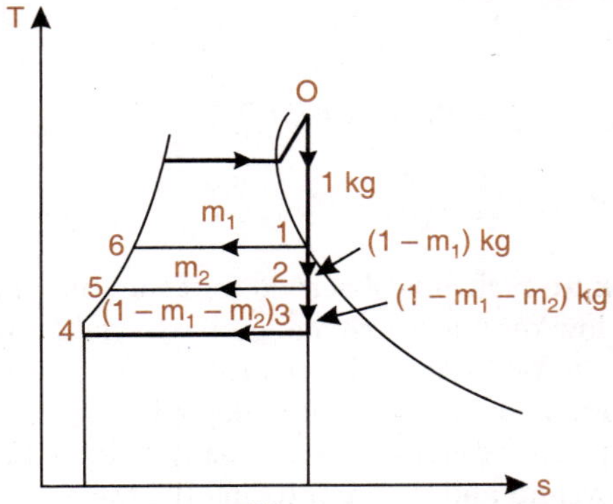

(b)

Fig. 15.15. Regenerative cycle.

Let, m_1 = kg of high pressure (H.P.) steam per kg of steam flow,

m_2 = kg of low pressure (L.P.) steam extracted per kg of steam flow, and

$(1 - m_1 - m_2)$ = kg of steam entering condenser per kg of steam flow.

Energy/Heat balance equation for H.P. heater :

$$m_1 (h_1 - h_{f_6}) = (1 - m_1)(h_{f_6} - h_{f_5})$$

or $$m_1[(h_1 - h_{f_6}) + (h_{f_6} - h_{f_5})] = (h_{f_6} - h_{f_5})$$

or $$m_1 = \frac{h_{f_6} - h_{f_5}}{h_1 - h_{f_5}} \quad ...(15.8)$$

Energy/Heat balance equation for L.P. heater :

$$m_2 (h_2 - h_{f_5}) = (1 - m_1 - m_2)(h_{f_5} - h_{f_3})$$

or $$m_2[(h_2 - h_{f_5}) + (h_{f_5} - h_{f_3})] = (1 - m_1)(h_{f_5} - h_{f_3})$$

or $$m_2 = \frac{(1 - m_1)(h_{f_5} - h_{f_3})}{(h_2 - h_{f_3})} \quad ...(15.9)$$

All enthalpies may be determined ; therefore m_1 and m_2 may be found. The maximum temperature to which the water can be heated is dictated by that of bled steam. The condensate from the bled steam is added to feed water.

Neglecting pump work :

The heat supplied externally in the cycle

$$= (h_0 - h_{f_6})$$

Isentropic work done

$$= m_1 (h_0 - h_1) + m_2 (h_0 - h_2) + (1 - m_1 - m_2)(h_0 - h_3)$$

The thermal efficiency of regenerative cycle is

$$\eta_{thermal} = \frac{\text{Work done}}{\text{Heat supplied}}$$

$$= \frac{m_1 (h_0 - h_1) + m_2 (h_0 - h_2) + (1 - m_1 - m_2)(h_0 - h_3)}{(h_0 - h_{f_6})} \quad ...(15.10)$$

[The work done by the turbine may also be calculated by summing up the products of the steam flow and the corresponding heat drop in the turbine stages.

i.e., Work done $= (h_0 - h_1) + (1 - m_1)(h_1 - h_2) + (1 - m_1 - m_2)(h_2 - h_3)$]

Advantages of Regenerative cycle over Simple Rankine cycle :

1. The heating process in the boiler tends to become reversible.
2. The thermal stresses set up in the boiler are minimised. This is due to the fact that temperature ranges in the boiler are reduced.
3. The thermal efficiency is improved because the average temperature of heat addition to the cycle is increased.
4. Heat rate is reduced.
5. The blade height is less due to the reduced amount of steam passed through the low pressure stages.
6. Due to many extractions there is an improvement in the turbine drainage and it reduces erosion due to moisture.
7. A small size condenser is required.

Disadvantages :

1. The plant becomes more complicated.
2. Because of addition of heaters greater maintenance is required.
3. For given power a large capacity boiler is required.
4. The heaters are costly and the gain in thermal efficiency is not much in comparison to the heavier costs.

Note. In the absence of precise information (regarding actual temperature of the feed water entering and leaving the heaters and of the condensate temperatures) the following assumption should always be made while doing calculations :

1. Each heater is ideal and bled steam just condenses.
2. The feed water is heated to saturation temperature at the pressure of bled steam.
3. Unless otherwise stated the work done by the pumps in the system is considered negligible.
4. There is equal temperature rise in all the heaters (usually 10°C to 15°C).

Example 15.12. *A steam turbine is fed with steam having an enthalpy of 3100 kJ/kg. It moves out of the turbine with an enthalpy of 2100 kJ/kg. Feed heating is done at a pressure of 3.2 bar with steam enthalpy of 2500 kJ/kg. The condensate from a condenser with an enthalpy of 125 kJ/kg enters into the feed heater. The quantity of bled steam is 11200 kg/h. Find the power developed by the turbine. Assume that the water leaving the feed heater is saturated liquid at 3.2 bar and the heater is direct mixing type. Neglect pump work.*

Solution. Arrangement of the components is shown in Fig. 15.16.

At 3.2 bar, $h_{f_2} = 570.9$ kJ/kg.

Consider m kg out of 1 kg is taken to the feed heater (Fig. 15.16).

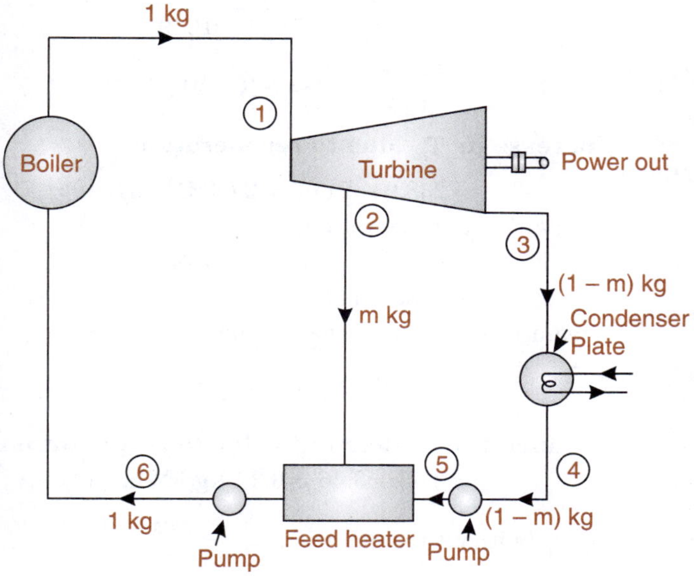

Fig. 15.16

Energy balance for the feed heater is written as :

$$m h_2 + (1 - m)\, h_{f_5} = 1 \times h_{f_2}$$

$$m \times 2100 + (1 - m) \times 125 = 1 \times 570.9$$

$$2100\, m + 125 - 125\, m = 570.9$$

$$1975\, m = 570.9 - 125$$

$$\therefore \quad m = 0.226 \text{ kg}$$

per kg of steam supplied to the turbine

∴ Steam supplied to the turbine per hour

$$= \frac{11200}{0.226} = 49557.5 \text{ kg/h}$$

Net work developed per kg of steam

$$= (h_1 - h_2) + (1 - m)\,(h_2 - h_3)$$

$$= (3100 - 2500) + (1 - 0.226)\,(2500 - 2100)$$

$$= 600 + 309.6 = 909.6 \text{ kJ/kg}$$

∴ Power developed by the turbine

$$= 909.6 \times \frac{49557.5}{3600} \text{ kJ/s}$$

$$= \mathbf{12521.5\ kW.} \quad \textbf{(Ans.)} \qquad (\because\ 1 \text{ kJ/s} = 1 \text{ kW})$$

☞**Example 15.13.** *In a single-heater regenerative cycle the steam enters the turbine at 30 bar, 400°C and the exhaust pressure is 0.10 bar. The feed water heater is a direct contact type which operates at 5 bar. Find :*

(i) The efficiency and the steam rate of the cycle.

(ii) The increase in mean temperature of heat addition, efficiency and steam rate as compared to the Rankine cycle (without regeneration).

Pump work may be neglected.

Solution. Fig. 15.17 shows the flow, *T-s* and *h-s* diagrams.

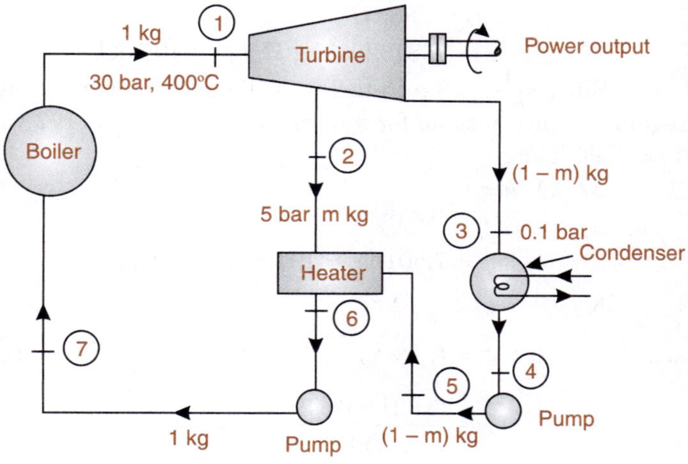

(*a*)

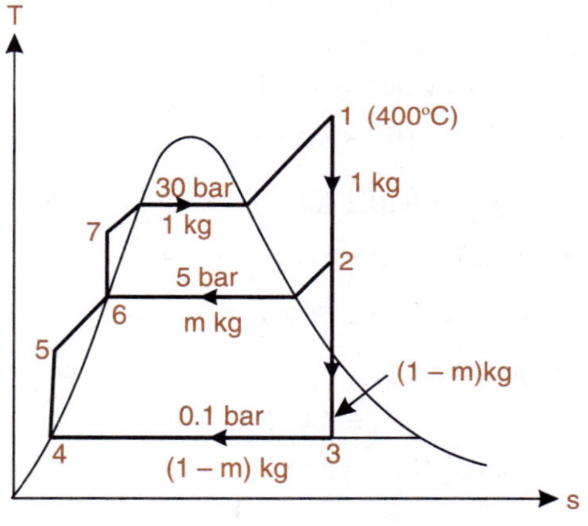

(*b*)

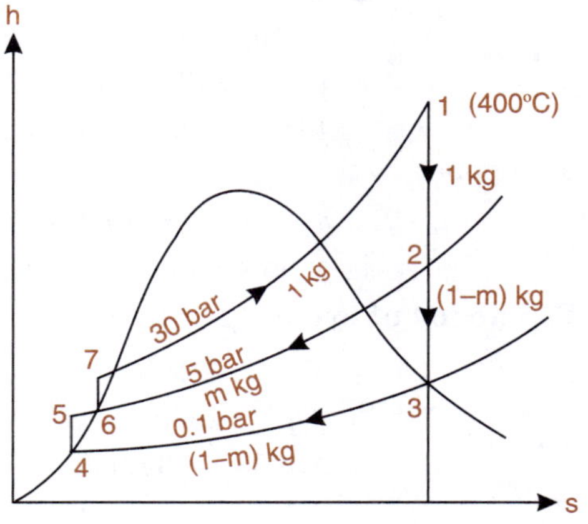

(*c*)

Fig. 15.17

From steam tables :

At 30 bar, 400°C :

h_1 = 3230.9 kJ/kg,

s_1 = 6.921 kJ/kg K = s_2 = s_3,

At 5 bar :

s_f = 1.8604,

s_g = 6.8192 kJ/kg K, h_f = 640.1 kJ/kg

Since $s_2 > s_g$, the state 2 must lie in the superheated region. From the table for superheated steam t_2 = 172°C, h_2 = 2796 kJ/kg.

At 0.1 bar :

s_f = 0.649,

s_{f_g} = 7.501, h_f = 191.8, h_{f_g} = 2392.8

Now, $s_2 = s_3$

i.e., $6.921 = s_{f_3} + x_3\, s_{fg_3} = 0.649 + x_3 \times 7.501$

$$\therefore \quad x_3 = \frac{6.921 - 0.649}{7.501} = 0.836$$

$$\therefore \quad h_3 = h_{f_3} + x_3\, h_{fg_3}$$
$$= 191.8 + 0.836 \times 2392.8$$
$$= 2192.2 \text{ kJ/kg}$$

Since pump work is neglected

$$h_{f_4} = 191.8 \text{ kJ/kg} = h_{f_5}$$

$$h_{f_6} = 640.1 \text{ kJ/kg (at 5 bar)} = h_{f_7}$$

Energy balance for heater gives

$$m\,(h_2 - h_{f_6}) = (1 - m)\,(h_{f_6} - h_{f_5})$$

$$m\,(2796 - 640.1) = (1 - m)\,(640.1 - 191.8)$$
$$= 448.3\,(1 - m)$$
$$2155.9\,m = 448.3 - 448.3\,m$$
$$\therefore \quad m = 0.172 \text{ kg}$$

∴ Turbine work,

$$W_T = (h_1 - h_2) + (1 - m)\,(h_2 - h_3)$$
$$= (3230.9 - 2796)$$
$$+ (1 - 0.172)\,(2796 - 2192.2)$$
$$= 434.9 + 499.9 = 934.8 \text{ kJ/kg}$$

Heat supplied,

$$Q_1 = h_1 - h_{f_6} = 3230.9 - 640.1$$
$$= 2590.8 \text{ kJ/kg.}$$

(*i*) **Efficiency of cycle, η_{cycle} :**

$$\eta_{cycle} = \frac{W_T}{Q_1} = \frac{934.8}{2590.8}$$

= **0.3608 or 36.08%. (Ans.)**

Steam rate $= \dfrac{3600}{934.8}$ = **3.85 kg/kWh. (Ans.)**

(*ii*)
$$T_{m_1} = \frac{h_1 - h_{f_7}}{s_1 - s_7} = \frac{2590.8}{6.921 - 1.8604}$$
$$= 511.9 \text{ K} = 238.9°\text{C}.$$

T_{m_1} (without regeneration)

$$= \frac{h_1 - h_{f_4}}{s_1 - s_4} = \frac{3230.9 - 191.8}{6.921 - 0.649}$$

$$= \frac{3039.1}{6.272} = 484.5 \text{ K} = 211.5°\text{C}.$$

Increase in T_{m_1} due to regeneration

= 238.9 – 211.5 = **27.4°C. (Ans.)**

W_T (without regeneration)

$= h_1 - h_3 = 3230.9 - 2192.2$

= 1038.7 kJ/kg

Steam rate without regeneration

$$= \frac{3600}{1038.7} = 3.46 \text{ kg/kWh}$$

∴ **Increase in steam rate due to regeneration**

= 3.85 – 3.46 = **0.39 kg/kWh. (Ans.)**

$$\eta_{cycle} \text{ (without regeneration)} = \frac{h_1 - h_3}{h_1 - h_{f_4}}$$

$$= \frac{1038.7}{3230.9 - 191.8}$$

= **0.3418 or 34.18%. (Ans.)**

Increase in cycle efficiency due to regeneration

= 36.08 – 34.18 = **1.9%. (Ans.)**

Example 15.14. *Steam is supplied to a turbine at a pressure of 30 bar and a temperature of 400°C and is expanded adiabatically to a pressure of 0.04 bar. At a stage of turbine where the pressure is 3 bar a connection is made to a surface heater in which the feed water is heated by bled steam to a temperature of 130°C. The condensed steam from the feed heater is cooled in a drain cooler to 27°C. The feed water passes through the drain cooler before entering the feed heater. The cooled drain water combines with the condensate in the well of the condenser.*

Assuming no heat losses in the steam, calculate the following :

(i) Mass of steam used for feed heating per kg of steam entering the turbine ;

(ii) Thermal efficiency of the cycle.

Solution. Refer to Fig. 15.18.

From steam tables :

At 3 bar : t_s = 133.5°C,

h_f = 561.4 kJ/kg.

At 0.04 bar : t_s = 29°C,

h_f = 121.5 kJ/kg.

From Mollier chart :

h_0 = 3231 kJ/kg (at 30 bar, 400°C)

h_1 = 2700 kJ/kg (at 3 bar)

h_2 = 2085 kJ/kg (at 0.04 bar).

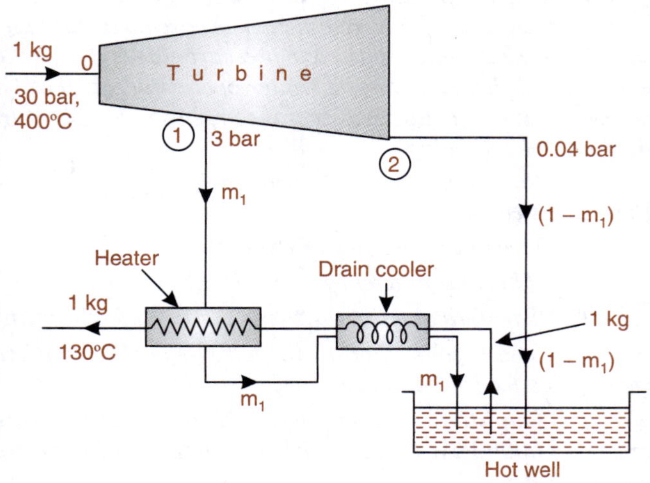

(a)

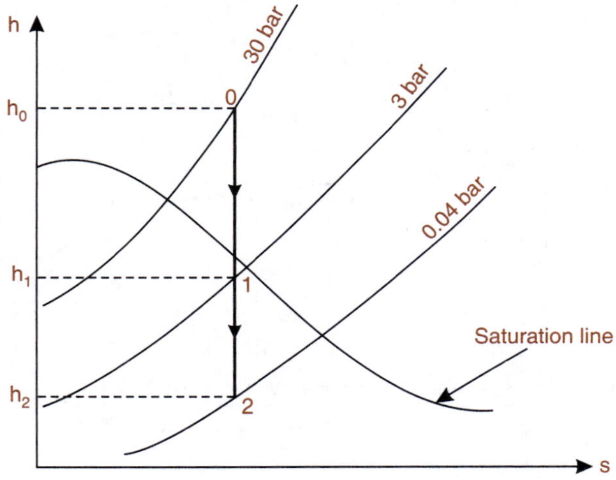

Fig. 15.18

(i) **Mass of steam used, m_1 :**

Heat lost by the steam = Heat gained by water.

Taking the feed-heater and drain-cooler combined, we have :

$$m_1 (h_1 - h_{f_2}) = 1 \times 4.186 (130 - 27)$$

or $$m_1 (2700 - 121.5) = 4.186 (130 - 27)$$

$$\therefore \quad \mathbf{m_1} = \frac{4.186(130 - 27)}{(2700 - 121.5)}$$

= **0.1672 kg. (Ans.)**

(ii) **Thermal efficiency of the cycle :**

Work done per kg of steam

$= 1(h_0 - h_1) + (1 - m_1)(h_1 - h_2)$

$= 1(3231 - 2700) + (1 - 0.1672)(2700 - 2085)$

= 1043.17 kJ/kg

Heat supplied per kg of steam $= h_0 - 1 \times 4.186 \times 130$

= 3231 – 544.18 = 2686.82 kJ/kg.

$$\eta_{\text{Thermal}} = \frac{\text{Work done}}{\text{Heat supplied}} = \frac{1043.17}{2686.82}$$

= **0.3882 or 38.82%. (Ans.)**

Example 15.15. *Steam is supplied to a turbine at 30 bar and 350°C. The turbine exhaust pressure is 0.08 bar. The main condensate is heated regeneratively in two stages by steam bled from the turbine at 5 bar and 1.0 bar respectively. Calculate masses of steam bled off at each pressure per kg of steam entering the turbine and the theoretical thermal efficiency of the cycle.*

Solution. Refer to Fig. 15.19.

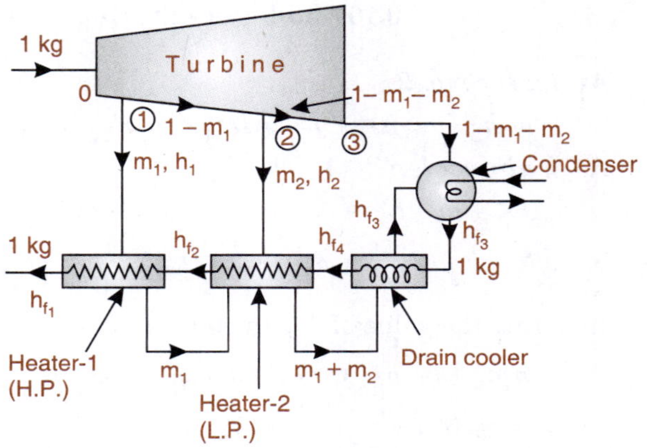

(a)

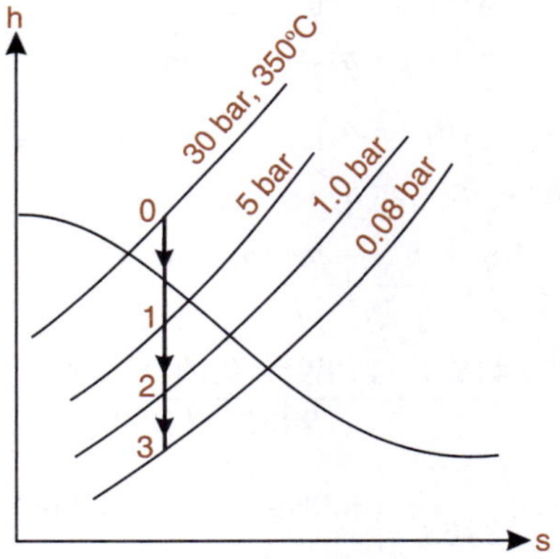

(b)

Fig. 15.19

The following *assumptions* are made :

1. The condensate is heated to the saturation temperature in each heater.

2. The drain water from H.P. (high pressure) heater passes into the steam space of the L.P. (low pressure) heater without loss of heat.

3. The combined drains from the L.P. heater are cooled in a drain cooler to the condenser temperature.

4. The expansion of the steam in the turbine is adiabatic and frictionless.

Enthalpy at 30 bar, 350°C, h_0 = 3115.3 kJ/kg.

After adiabatic expansion (from Mollier chart)

Enthalpy at 5 bar, h_1 = 2720 kJ/kg

Enthalpy at 1.0 bar, h_2 = 2450 kJ/kg

Enthalpy at 0.08 bar, h_3 = 2120 kJ/kg

From steam tables : h_{f_1} = 640.1 kJ/kg (at 5.0 bar)

h_{f_2} = 417.5 kJ/kg (at 1.0 bar)

h_{f_3} = 173.9 kJ/kg (at 0.08 bar)

At heater No. 1 :

$$m_1h_1 + h_{f_2} = m_1h_{f_1} + h_{f_1}$$

$$\mathbf{m_1} = \frac{h_{f_1} - h_{f_2}}{h_1 - h_{f_1}} = \frac{640.1 - 417.5}{2720 - 640.1}$$

= **0.107 kJ/kg of entering steam. (Ans.)**

At heater No. 2 :

$$m_2h_2 + m_1h_{f_1} + h_{f_4} = (m_1 + m_2)h_{f_2} + h_{f_2} \quad ...(i)$$

At drain cooler :

$$(m_1 + m_2)\, h_{f_2} + h_{f_3} = h_{f_4} + (m_1 + m_2)h_{f_3}$$

$$\therefore \quad h_{f_4} = (m_1 + m_2)\,(h_{f_2} - h_{f_3}) + h_{f_3} \quad ...(ii)$$

Inserting the value of h_{f_4} in eqn. (*i*), we get

$$m_2h_2 + m_1h_{f_1} + (m_1 + m_2)\,(h_{f_2} - h_{f_3}) + h_{f_3} = (m_1 + m_2)\,h_{f_2} + h_{f_2}$$

$$m_2h_2 + m_1h_{f_1} + (m_1 + m_2)h_{f_2} - (m_1 + m_2)h_{f_3} + h_{f_3} = (m_1 + m_2)h_{f_2} + h_{f_2}$$

$$m_2h_2 + m_1h_{f_1} - m_1h_{f_3} - m_2h_{f_3} + h_{f_3} = h_{f_2}$$

$$m_2\,(h_2 - h_{f_3}) = (h_{f_2} - h_{f_3}) - m_1(h_{f_1} - h_{f_3})$$

$$\mathbf{m_2} = \frac{(h_{f_2} - h_{f_3}) - m_1\,(h_{f_1} - h_{f_3})}{(h_2 - h_{f_3})} = \frac{(417.5 - 173.9) - 0.107\,(640.1 - 173.9)}{(2450 - 173.9)}$$

$$= \frac{193.7}{2276.1} = \mathbf{0.085\ kJ/kg.} \quad \textbf{(Ans.)}$$

Work done

$$= 1\,(h_0 - h_1) + (1 - m_1)\,(h_1 - h_2) + (1 - m_1 - m_2)\,(h_2 - h_3)$$

$$= 1\,(3115.3 - 2720) + (1 - 0.107)\,(2720 - 2450) + (1 - 0.107 - 0.085)\,(2450 - 2120)$$

$$= 395.3 + 241.11 + 266.64 = 903.05 \text{ kJ/kg}$$

Heat supplied/kg $= h_0 - h_{f_1}$ = 3115.3 − 640.1 = 2475.2 kJ/kg

∴ **Thermal efficiency of the cycle**

$$= \frac{\text{Work done}}{\text{Heat supplied}} = \frac{903.05}{2475.2}$$

= **0.3648 or 36.48%. (Ans.)**

☞Example 15.16. *Steam at a pressure of 20 bar and 250°C enters a turbine and leaves it finally at a pressure of 0.05 bar. Steam is bled off at pressures of 5.0, 1.5 and 0.3 bar. Assuming (i) that the condensate is heated in each heater upto the saturation temperature of the steam in that heater, (ii) that the drain water from each heater is cascaded through a trap into the next heater on the low pressure side of it, (iii) that the combined drains from the heater operating at 0.3 bar are cooled in a drain cooler to condenser temperature, calculate the following :*

(i) Mass of bled steam for each heater per kg of steam entering the turbine

(ii) Thermal efficiency of the cycle,

(iii) Thermal efficiency of the Rankine cycle

(iv) Theoretical gain due to regenerative feed heating,

(v) Steam consumption in kg/kWh with or without regenerative feed heating, and

(vi) Quantity of steam passing through the last stage nozzle of a 50000 kW turbine with and without regenerative feed-heating.

Solution. Refer to Fig. 15.20 (*a*), (*b*).

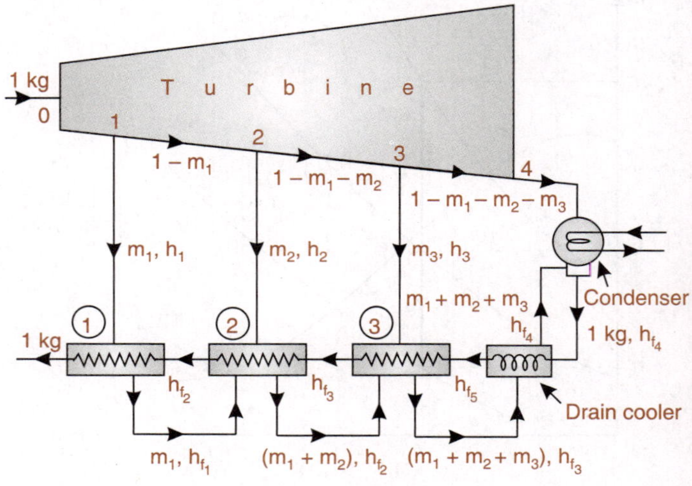

(*a*)

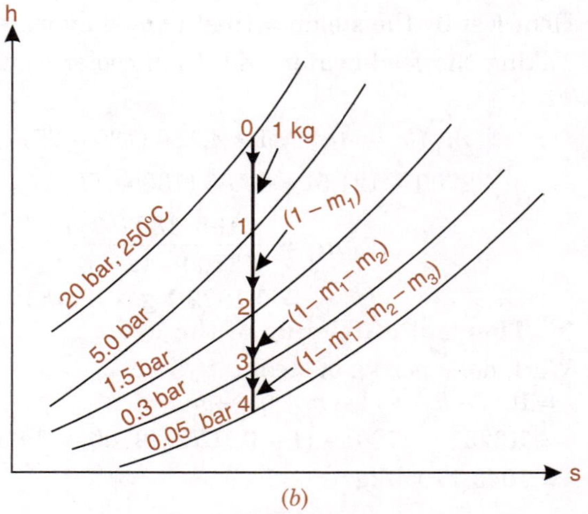

(*b*)

Fig. 15.20

From Mollier Chart :

$h_0 = 2905$ kJ/kg, $h_1 = 2600$ kJ/kg, $h_2 = 2430$ kJ/kg

$h_3 = 2210$ kJ/kg, $h_4 = 2000$ kJ/kg

From steam tables :

At 5 bar : $h_{f_1} = 640.1$ kJ/kg

At 1.5 bar : $h_{f_2} = 467.1$ kJ/kg

At 0.3 bar : $h_{f_3} = 289.3$ kJ/kg

At 0.05 bar : $h_{f_4} = 137.8$ kJ/kg.

(*i*) **Mass of bled steam for each heater per kg of steam :**

Using heat balance equation :

At heater No. 1 :

$$m_1 h_1 + h_{f_2} = m_1 h_{f_1} + h_{f_1}$$

$$\therefore \quad \mathbf{m_1} = \frac{h_{f_1} - h_{f_2}}{h_1 - h_{f_1}} = \frac{640.1 - 467.1}{2600 - 640.1}$$

= 0.088 kJ/kg of entering steam. (Ans.)

At heater No. 2 :

$$m_2 h_2 + h_{f_3} + m_1 h_{f_1} = h_{f_2} + (m_1 + m_2) h_{f_2}$$

$$\mathbf{m_2} = \frac{(h_{f_2} + h_{f_3}) - m_1 (h_{f_1} - h_{f_2})}{(h_2 - h_{f_2})}$$

$$= \frac{(467.1 - 289.3) - 0.088\,(640.1 - 467.1)}{(2430 - 467.1)} = \frac{162.57}{1962.9}$$

= 0.0828 kJ/kg of entering steam. (Ans.)

At heater No. 3 :

$$m_3 h_3 + h_{f_5} + (m_1 + m_2)\, h_{f_2}$$

$$= h_{f_3} + (m_1 + m_2 + m_3)\, h_{f_3} \quad ...(i)$$

At drain cooler :

$$(m_1 + m_2 + m_3)\, h_{f_3} + h_{f_4}$$

$$= h_{f_5} + (m_1 + m_2 + m_3)\, h_{f_4}$$

$$\therefore \quad h_{f_5} = (m_1 + m_2 + m_3)\,(h_{f_3} - h_{f_4}) + h_{f_4} \quad ...(ii)$$

Inserting the value of h_{f_5} in eqn. (*i*), we get

$$m_3 h_3 + (m_1 + m_2 + m_3)\,(h_{f_3} - h_{f_4}) + h_{f_4}$$

$$+ (m_1 + m_2)\, h_{f_2} = h_{f_3} + (m_1 + m_2 + m_3)\, h_{f_3}$$

$$\therefore \quad \mathbf{m_3} = \frac{(h_{f_3} - h_{f_4}) - (m_1 + m_2)\,(h_{f_2} - h_{f_4})}{h_3 - h_{f_4}}$$

$$= \frac{(289.3 - 137.8) - (0.088 + 0.0828)\,(467.1 - 137.8)}{(2210 - 137.8)}$$

$$= \frac{151.5 - 56.24}{2072.2}$$

= 0.046 kJ/kg of entering steam. (Ans.)

Work done/kg (*neglecting pump work*)

$$= (h_0 - h_1) + (1 - m_1)\,(h_1 - h_2) + (1 - m_1 - m_2)\,(h_2 - h_3) + (1 - m_1 - m_2 - m_3)\,(h_3 - h_4)$$

$$= (2905 - 2600) + (1 - 0.088)\,(2600 - 2430) + (1 - 0.088 - 0.0828)\,(2430 - 2210) + (1 - 0.088 - 0.0828 - 0.046)\,(2210 - 2000)$$

$= 305 + 155.04 + 182.42 + 164.47 = 806.93$ kJ/kg

Heat supplied/kg $= h_0 - h_{f_1}$

$= 2905 - 640.1 = 2264.9$ kJ/kg.

(*ii*) **Thermal efficiency of the cycle, $\eta_{Thermal}$:**

$$\eta_{Thermal} = \frac{\text{Work done}}{\text{Heat supplied}} = \frac{806.93}{2264.9}$$

= 0.3563 or 35.63%. (Ans.)

(*iii*) **Thermal efficiency of Rankine cycle, $\eta_{Rankine}$:**

$$h_{Rankine} = \frac{h_0 - h_4}{h_0 - h_{f_4}} = \frac{2905 - 2000}{2905 - 137.8}$$

= 0.327 or 32.7%. (Ans.)

(*iv*) **Theoretical gain due to regenerative feed heating**

$$= \frac{35.63 - 32.7}{35.63} = \mathbf{0.0822 \text{ or } 8.22\%.} \quad \text{(Ans.)}$$

(*v*) **Steam consumption with regenerative feed heating**

$$= \frac{1 \times 3600}{\text{Work done / kg}} = \frac{1 \times 3600}{806.93}$$

= 4.46 kg/kWh. (Ans.)

Steam consumption without regenerative feed heating

$$= \frac{1 \times 3600}{\text{Work done / kg without regeneration}}$$

$$= \frac{1 \times 3600}{h_0 - h_4}$$

$$= \frac{1 \times 3600}{2905 - 2000} = \mathbf{3.97 \text{ kg/kWh.}} \quad \text{(Ans.)}$$

(*vi*) **Quantity of steam passing through the last stage of a 50000 kW turbine with regenerative feed-heating**

$= 4.46\,(1 - m_1 - m_2 - m_3) \times 50000$

$= 4.46\,(1 - 0.088 - 0.0828 - 0.046) \times 50000$

= 174653.6 kg/h. (Ans.)

Same without regenerative arrangement

$= 3.97 \times 50000 =$ **198500 kg/h. (Ans.)**

Example 15.17. *A steam turbine plant developing 120 MW of electrical output is equipped with reheating and regenerative feed heating arrangement consisting of two feed heaters—*

one surface type on H.P. side and other direct contact type on L.P. side. The steam conditions before the steam stop valve are 100 bar and 530°C. A pressure drop of 5 bar takes place due to throttling in valves.

Steam exhausts from the H.P. turbine at 25 bar. A small quantity of steam is bled off at 25 bar for H.P. surface heater for feed heating and the remaining is reheated in a reheater to 550°C and the steam enters at 22 bar in L.P. turbine for further expansion. Another small quantity of steam is bled off at pressure 6 bar for the L.P. heater and the rest of steam expands up to the back pressure of 0.05 bar. The drain from the H.P. heater is led to the L.P. heater and the combined feed from the L.P. heater is pumped to the high-pressure feed heater and finally to the boiler with the help of boiler feed pump.

The component efficiencies are : Turbine efficiency 85%, pump efficiency 90%, generator efficiency 96%, boiler efficiency 90% and mechanical efficiency 95%. It may be assumed that the feed-water is heated up to the saturation temperature at the prevailing pressure in feed heater.
Work out the following :

(i) Sketch the feed heating system and show the process on T-s and h-s diagrams.

(ii) Amounts of steam bled off.

(iii) Overall thermal efficiency of turbo-alternator considering pump work.

(iv) Specific steam consumption in kg/kWh.

(AMIE)

Solution. (*i*) The *schematic arrangement* including feed heating system, and *T-s* and *h-s* diagrams of the process are shown in Figs. 15.21 and 15.22 respectively.

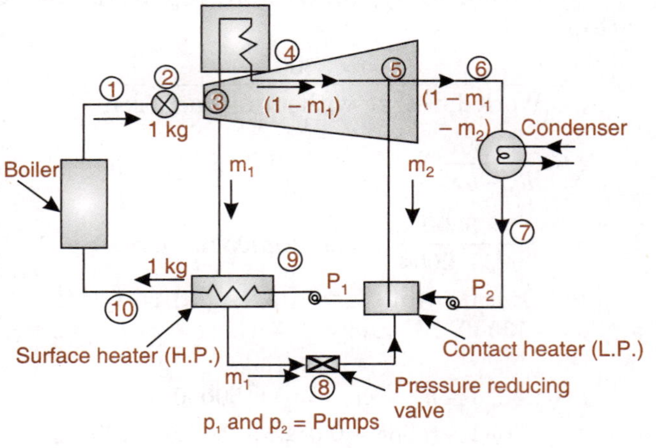

Fig. 15.21

(*ii*) **Amounts of bled off.** The enthalpies at various state points as read from *h-s* diagram/steam tables, in kJ/kg, are :

$h_1 = h_2 = 3460$

$h_3' = 3050$, and $\therefore\ h_3 = 3460 - 0.85(3460 - 3050) = 3111.5$

$h_4 = 3585$

$h_5' = 3140$, and $\therefore\ h_5 = 3585 - 0.85(3585 - 3140) = 3207$

$h_6' = 2335$, and $\therefore\ h_6 = 3207 - 0.85(3207 - 2335) = 2466$

$h_7 = 137.8$ kJ/kg (h_f at 0.05 bar)

$h_8 = h_{10} = 962$ kJ/kg (h_f at 25 bar)

and $h_9 = 670.4$ (h_f at 6 bar).

Enthalpy balance for ***surface heater :***

$m_1h_3 + h_9 = m_1h_8 + h_{10}$, neglecting pump work

or $$m_1 = \frac{h_{10} - h_9}{h_3 - h_8} = \frac{962 - 670.4}{3111.5 - 962} = 0.13566 \text{ kg}$$

Enthalpy balance for ***contact heater :***

$m_2h_5 + (1 - m_1 - m_2)h_7 + m_1h_8 = h_9$, neglecting pump work

or $m_2 \times 3207 + (1 - 0.13566 - m_2) \times 137.8 + 0.13566 \times 962 = 670.4$

or $m_2 = 0.1371$ kg.

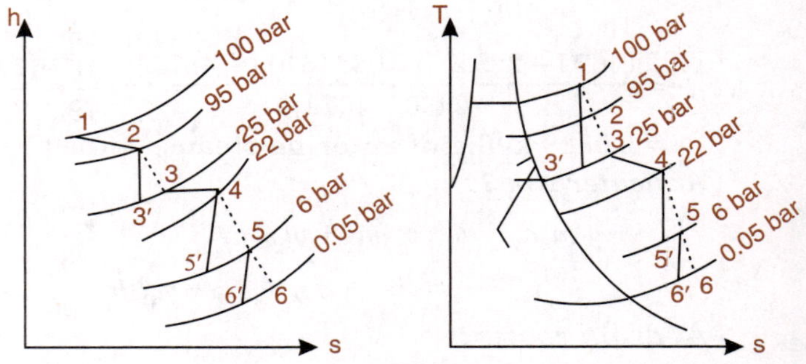

Fig. 15.22

Pump Work. Take specific volume of water as 0.001 m³/kg.

$$(W_{pump})_{L.P.} = (1 - m_1 - m_2)(6 - 0.05) \times 0.001 \times 10^2$$
$$= (1 - 0.13566 - 0.1371) \times 5.95 \times 0.1$$
$$= 0.4327 \text{ kJ/kg.}$$

$$(W_{pump})_{H.P.} = 1 \times (100 - 6) \times 0.001 \times 10^2 = 9.4 \text{ kJ/kg}$$

$$\text{Total pump work (actual)} = \frac{0.4327 + 9.4}{0.9} = 10.925 \text{ kJ/kg}$$

Turbine output (indicated)

$= (h_2 - h_3) + (1 - m_1)(h_4 - h_5) + (1 - m_1 - m_2)(h_5 - h_6)$

$= (3460 - 3111.5) + (1 - 0.13566)(3585 - 3207) + (1 - 0.13566 - 0.1371)(3207 - 2466)$

$= 1214.105$ kJ/kg

Net electrical output

= (Indicated work – Pump work) × $\eta_{mech.}$ × $\eta_{gen.}$

= (1214.105 – 10.925) × 0.9 × 0.96 = 1039.55 kJ/kg

[**Note.** All the above calculations are for 1 kg of main (boiler) flow.]

$\therefore$ Main steam flow rate $= \dfrac{120 \times 10^3 \times 3600}{1039.55}$

$= 4.155 \times 10^5$ kJ/h.

Amounts of bled off are :

(*a*) Surface (high pressure) heater,

= 0.13566 kg/kg of boiler flow

or = $0.13566 \times 4.155 \times 10^5$

i.e., = **5.6367×10^4 kg/h.** **(Ans.)**

(*b*) Direct contact (low pressure) heater

= 0.1371 kg/kg of boiler flow

or = $0.1371 \times 4.155 \times 10^5$

i.e., = **5.697×10^4 kg/h.** **(Ans.)**

(*iii*) **Overall thermal efficiency, $\eta_{overall}$:**

Heat input in boiler

$= \dfrac{h_1 - h_{10}}{\eta_{boiler}}$

$= \dfrac{3460 - 962}{0.9}$

= 2775.6 kJ/kg of boiler flow.

Heat input in reheater

$= \dfrac{h_4 - h_3}{\eta_{boiler}} = \dfrac{3585 - 3111.5}{0.9}$

= 526.1 kJ/kg of boiler flow

$\therefore$ $\eta_{overall} = \dfrac{1039.55}{2775.6 + 526.1} \times 100$

= **31.48%.** **(Ans.)**

(*iv*) **Specific steam consumption :**

Specific steam consumption

$= \dfrac{4.155 \times 10^5}{120 \times 10^3}$

= **3.4625 kg/kWh.** **(Ans.)**

15.5. REHEAT CYCLE

For attaining greater thermal efficiencies when the initial pressure of steam was raised beyond 42 bar it was found that resulting condition of steam after, expansion was increasingly wetter and exceeded in the safe limit of 12 per cent condensation. It, therefore, became necessary to *reheat* the steam after part of expansion was over so that the resulting condition after complete expansion fell within the region of permissible wetness.

The reheating or resuperheating of steam is now universally used when high pressure and temperature steam conditions such as 100 to 250 bar and 500°C to 600°C are employed for throttle. For plants of *still higher pressures and temperatures, a double reheating may be used.*

In actual practice reheat *improves* the cycle efficiency by about 5% for a 85/15 bar cycle. A *second reheat* will give a *much less gain* while the initial cost involved would be so high as to prohibit use of two stage reheat except in case of very high initial throttle conditions. The cost of reheat equipment consisting of boiler, piping and controls may be 5% to 10% more than that of the conventional boilers and this additional expenditure is justified only if gain in thermal efficiency is sufficient to promise a return of this investment. *Usually a plant with a base load capacity of 50000 kW and initial steam pressure of 42 bar would economically justify the extra cost of reheating.*

The improvement in thermal efficiency due to reheat is greatly dependent upon the *reheat pressure* with respect to the original pressure of steam.

Fig. 15.23 shows the reheat pressure selection on cycle efficiency.

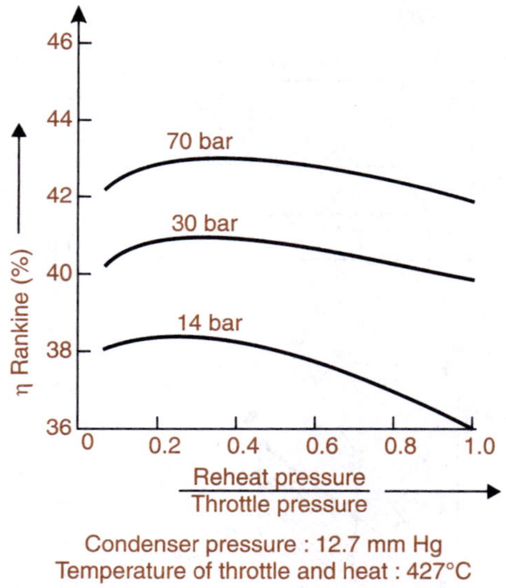

Fig. 15.23. Effect of reheat pressure selection on cycle efficiency.

Fig. 15.24 shows a schematic diagram of a theoretical single-stage reheat cycle. The corresponding representation of ideal reheating process on *T-s* and *h-s* chart is shown in Fig. 15.25 (*a*) and (*b*).

Refer to Fig. 15.24. 5-1 shows the formation of steam in the boiler. The steam as at state point 1 (*i.e.,* pressure p_1 and temperature T_1) enters the turbine and expands isentropically to a certain pressure p_2 and temperature T_2. From this state point 2 the whole of steam is drawn out of the turbine and is reheated in a reheater to a temperature T_3. (Although there is an *optimum pressure* at which the steam should be removed for reheating, if the highest return is to be obtained, yet, for simplicity, the whole steam is removed from the high pressure exhaust, where the pressure is about *one-fifth* of boiler pressure, and after undergoing a 10% pressure drop, in circulating through the

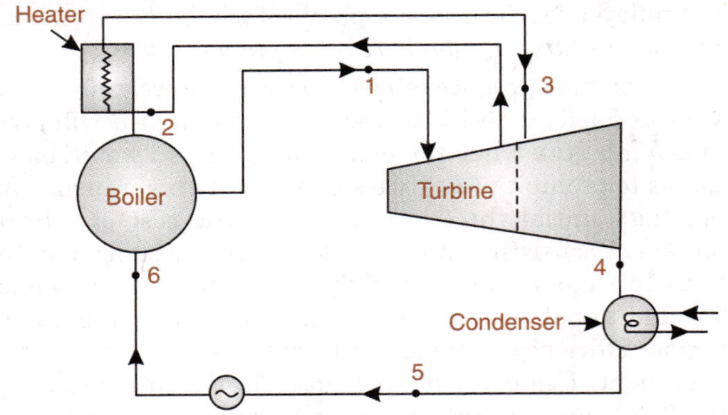

Fig. 15.24. Reheat cycle.

heater, it is returned to intermediate pressure or low pressure turbine). This reheated steam is then readmitted to the turbine where it is expanded to condenser pressure isentropically.

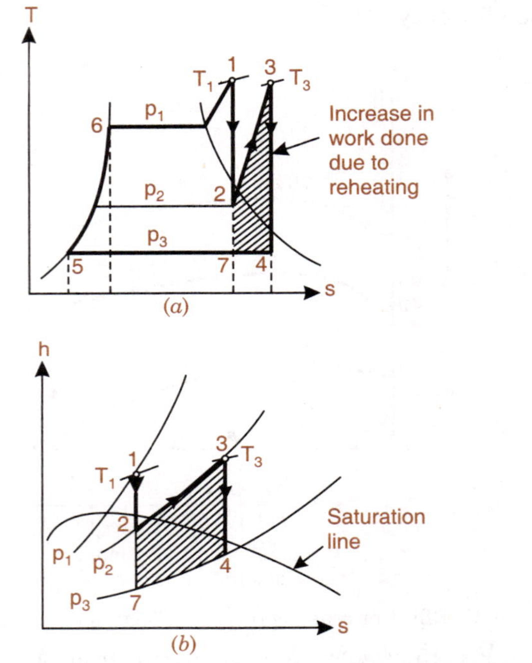

Fig. 15.25. Ideal reheating process on *T-s* and *h-s* chart.

Note. Superheating of steam. The primary object of superheating steam and supplying it to the prime movers is *to avoid too much wetness at the end of expansion*. Use of inadequate degree of superheat in *steam engines would cause greater condensation* in the engine cylinder ; while in case of *turbines* the moisture content of steam would result in *undue blade erosion*. The *maximum wetness in the final condition of steam that may be tolerated without any appreciable harm to the turbine blades is about 12 per cent*. Broadly each *1 per cent* of moisture in steam reduces the efficiency of that part of the turbine in which wet steam passes by 1 per cent to 1.5 per cent and in engines about 2 per cent.

Advantages of superheated steam :

(*i*) Superheating reduces the initial condensation losses in steam engines.

(*ii*) Use of superheated steam results in *improving the plant efficiency* by effecting a *saving in cost of fuel*. This saving may be of the order of 6% to 7% due to first 38°C of superheat and 4% to 5% for next 38°C and so on. This saving results due to the fact that the heat content and consequently the capacity to do work in superheated steam is increased and the quantity of steam required for a given output of power is reduced. Although additional heat has to be added in the boiler there is reduction in the work to be done by the feed pump, the condenser pump and other accessories due to reduction in quantity of steam used. It is estimated that the *quantity of steam may be reduced by 10% to 15% for first 38°C of superheat and somewhat less for the next 38°C of superheat in the case of condensing turbines.*

(*iii*) When a superheater is used in a boiler it helps in *reducing the stack temperatures* by extracting heat from the flue gases before these are passed out of chimney.

Thermal efficiency with *'Reheating'* (neglecting pump work) :

Heat supplied $= (h_1 - h_{f_4}) + (h_3 - h_2)$

Heat rejected $= h_4 - h_{f_4}$

Work done by the turbine

$$= \text{Heat supplied} - \text{heat rejected}$$
$$= (h_1 - h_{f_4}) + (h_3 - h_2) - (h_4 - h_{f_4})$$
$$= (h_1 - h_2) + (h_3 - h_4)$$

Thus, theoretical thermal efficiency of reheat cycle is

$$\eta_{thermal} = \frac{(h_1 - h_2) + (h_3 - h_4)}{(h_1 - h_{f_4}) + (h_3 - h_2)} \quad ...(15.11)$$

If pump work, $W_p = \dfrac{v_f\,(p_1 - p_b)}{1000}$ kJ/kg is considered,

the thermal efficiency is given by :

$$\eta_{thermal} = \frac{[(h_1 - h_4) + (h_3 - h_4)] - W_p}{[(h_1 - h_{f_4}) + (h_3 - h_2)] - W_p} \quad ...(15.12)$$

W_p is usually small and *neglected*.

Thermal efficiency *without reheating* is

$$\eta_{thermal} = \frac{h_1 - h_7}{h_1 - h_{f_4}} \quad (\because \; h_{f_4} = h_{f_7}) \quad ...(15.13)$$

Note 1. The reheater may be incorporated in the walls of the main boiler ; it may be a separately fired superheater or it may be heated by a coil carrying high-pressure superheated steam, this system being analogous to a steam jacket.

2. Reheating should be done at *'optimum pressure'* because if the steam is reheated early in its expansion then the additional *quantity of heat supplied will be small* and thus thermal efficiency gain will be small ; and if the reheating is done at a fairly low pressure, then, although a large amount of additional heat is supplied, the steam will have a high degree of superheat (as is clear from Mollier diagram), thus a large proportion of the heat supplied in the reheating process *will be thrown to waste in the condenser*.

Advantages of 'Reheating' :

1. There is an increased output of the turbine.
2. Erosion and corrosion problems in the steam turbine are eliminated/avoided.
3. There is an improvement in the thermal efficiency of the turbines.
4. Final dryness fraction of steam is improved.
5. There is an increase in the nozzle and blade efficiencies.

Disadvantages :

1. Reheating requires more maintenance.
2. The increase in thermal efficiency is not appreciable in comparison to the expenditure incurred in reheating.

☞**Example 15.18.** *Steam at a pressure of 15 bar and 250°C is expanded through a turbine at first to a pressure of 4 bar. It is then reheated at constant pressure to the initial temperature of 250°C and is finally expanded to 0.1 bar. Using Mollier chart, estimate the work done per kg of steam flowing through the turbine and amount of heat supplied during the process of reheat. Compare the work output when the expansion is direct from 15 bar to 0.1 bar without any reheat. Assume all expansion processes to be isentropic.*

Solution. Refer to Fig. 15.26.

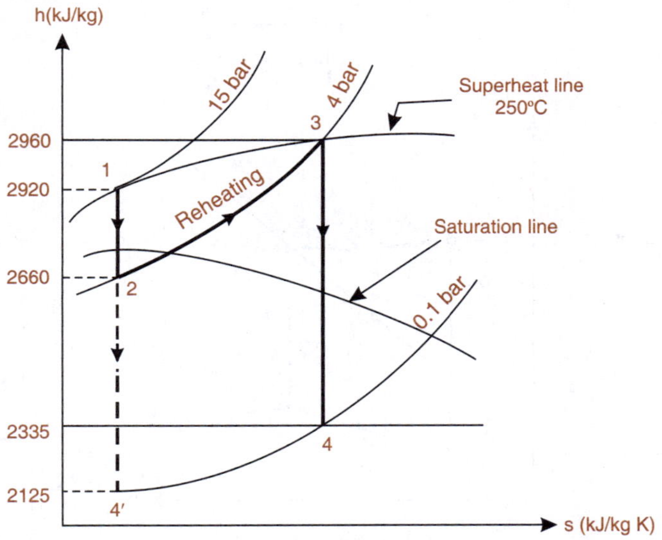

Fig. 15.26

Pressure, $p_1 = 15$ bar ;

$p_2 = 4$ bar ;

$p_4 = 0.1$ bar.

Work done per kg of steam,

$$W = \text{Total heat drop}$$
$$= [(h_1 - h_2) + (h_3 - h_4)] \text{ kJ/kg} \quad ...(i)$$

Amount of heat supplied during process of reheat,

$$h_{reheat} = (h_3 - h_2) \text{ kJ/kg} \quad ...(ii)$$

From Mollier diagram or *h-s* chart,

$h_1 = 2920$ kJ/kg, $h_4 = 2660$ kJ/kg

$h_3 = 2960$ kJ/kg, $h_4 = 2335$ kJ/kg

Now, by putting the values in eqns. (*i*) and (*ii*), we get

$$W = (2920 - 2660) + (2960 - 2335)$$
$$= \mathbf{885 \text{ kJ/kg.}} \quad \textbf{(Ans.)}$$

Hence *work done per kg of steam* = **885 kJ/kg. (Ans.)**

Amount of heat supplied during reheat,

$$h_{reheat} = (2960 - 2660) = \mathbf{300 \text{ kJ/kg.}} \quad \textbf{(Ans.)}$$

If the expansion would have been continuous without reheating *i.e.*, 1 to 4′, the work output is given by

$$W_1 = h_1 - h_4'$$

From Mollier diagram,

$$h_{4'} = 2125 \text{ kJ/kg}$$

$$\therefore \quad W_1 = 2920 - 2125 = \mathbf{795 \text{ kJ/kg.}} \quad \textbf{(Ans.)}$$

Example 15.19. *A steam power plant operates on a theoretical reheat cycle. Steam at boiler at 150 bar, 550°C expands through the high pressure turbine. It is reheated at a constant pressure of 40 bar to 550°C and expands through the low pressure turbine to a condenser at 0.1 bar. Draw T-s and h-s diagrams. Find :*

(*i*) *Quality of steam at turbine exhaust ;*

(*ii*) *Cycle efficiency ;*

(*iii*) *Steam rate in kg/kWh.* **(D.U.)**

Solution. Refer to Figs. 15.27 and 15.28.

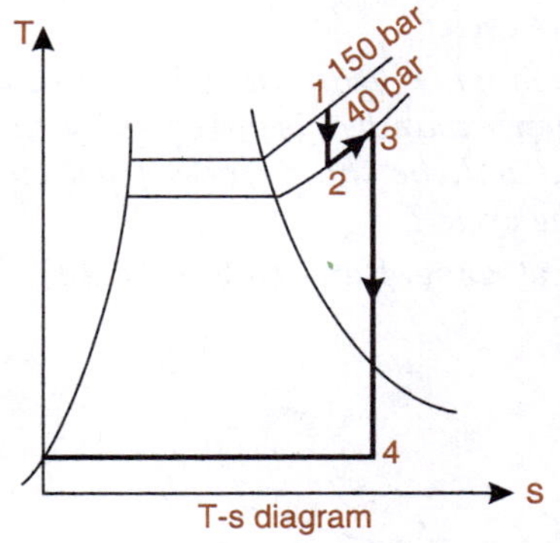

Fig. 15.27

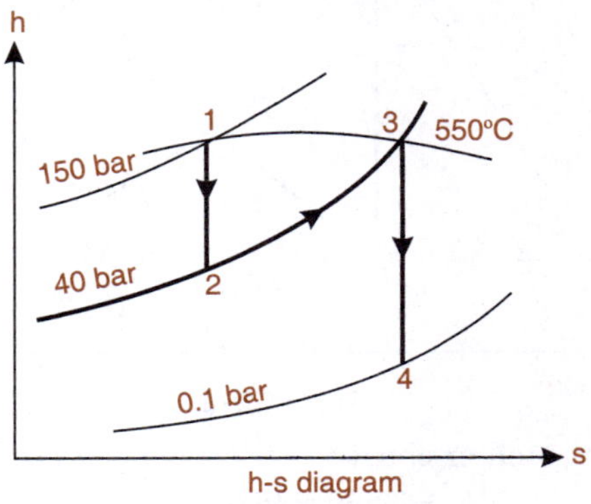

Fig. 15.28

From Mollier diagram (*h-s* diagram) :

$h_1 = 3450$ kJ/kg ; $h_2 = 3050$ kJ/kg ;

$h_3 = 3560$ kJ/kg ; $h_4 = 2300$ kJ/kg

h_{f_4} (from steam tables, at 0.1 bar) = 191.8 kJ/kg

(*i*) **Quality of steam at turbine exhaust, x_4 :**

$x_4 = 0.88$ (From Mollier diagram)

(*ii*) **Cycle efficiency, η_{cycle} :**

$$\eta_{cycle} = \frac{(h_1 - h_2) + (h_3 - h_4)}{(h_1 - h_{f_4}) + (h_3 - h_2)}$$

$$= \frac{(3450 - 3050) + (3560 - 2300)}{(3450 - 191.8) + (3560 - 3050)}$$

$$= \frac{1660}{3768.2} = \mathbf{0.4405 \text{ or } 44.05\%.} \quad \textbf{(Ans.)}$$

(*iii*) **Steam rate in kg/kWh :**

$$\text{Steam rate} = \frac{3600}{(h_1 - h_2) + (h_3 - h_4)}$$

$$= \frac{3600}{(3450 - 3050) + (3560 - 2300)}$$

$$= \frac{3600}{1660} = \mathbf{2.17 \text{ kg/kWh.}} \quad \textbf{(Ans.)}$$

Example 15.20. *A turbine is supplied with steam at a pressure of 32 bar and a temperature of 410°C. The steam then expands isentropically to a pressure of 0.08 bar. Find the dryness fraction at the end of expansion and thermal efficiency of the cycle.*

If the steam is reheated at 5.5 bar to a temperature of 395°C and then expanded isentropically to a pressure of 0.08 bar, what will be the dryness fraction and thermal efficiency of the cycle ?

Solution. First case. Refer to Fig. 15.29.

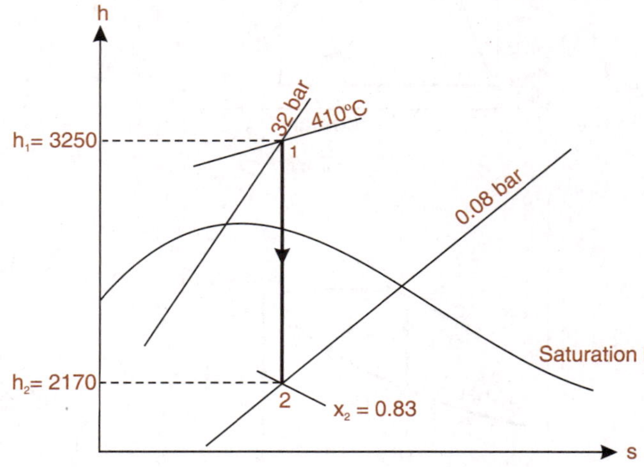

Fig. 15.29

From Mollier chart :

$h_1 = 3250$ kJ/kg

$h_2 = 2170$ kJ/kg

Heat drop (or work done)

$= h_1 - h_2$

$= 3250 - 2170 = 1080$ kJ/kg

Heat supplied $= h_1 - h_{f_2}$

$= 3250 - 173.9$ [$h_{f_2} = 173.9$ kJ/kg at 0.08 bar]

$= 3076.1$ kJ/kg

Thermal efficiency

$$= \frac{\text{Work done}}{\text{Heat supplied}} = \frac{1080}{3076.1}$$

$= \mathbf{0.351 \text{ or } 35.1\%.}$ **(Ans.)**

Exhaust steam condition, x_2

= 0.83 (From Mollier chart). (Ans.)

Second case. Refer to Fig. 15.30 (*b*).

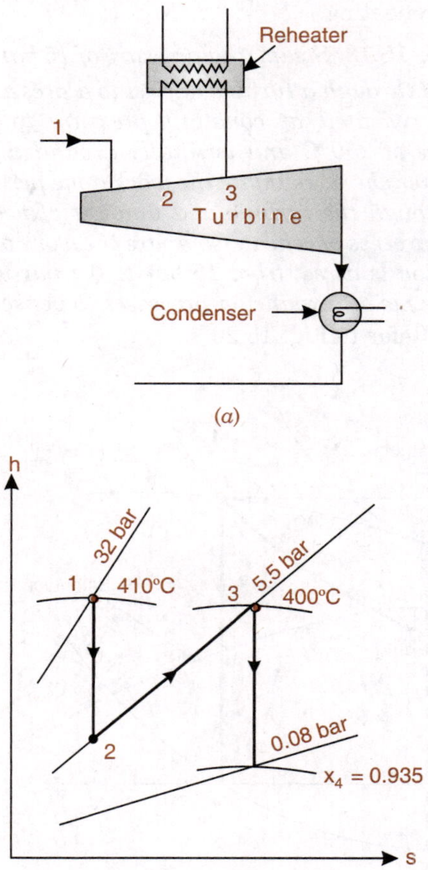

(*a*)

(*b*)

Fig. 15.30

From Mollier chart :

$h_1 = 3250$ kJ/kg ;

$h_2 = 2807$ kJ/kg ;

$h_3 = 3263$ kJ/kg ;

$h_4 = 2426$ kJ/kg.

Work done $= (h_1 - h_2) + (h_3 - h_4)$
$= (3250 - 2807) + (3263 - 2426)$
$= 1280$ kJ/kg

Heat supplied $= (h_1 - h_{f_4}) + (h_3 - h_2)$
$= (3250 - 173.9) + (3263 - 2807)$
$= 3532$ kJ/kg

Thermal efficiency

$$= \frac{\text{Work done}}{\text{Heat supplied}} = \frac{1280}{3532}$$

$= \mathbf{0.362}$ **or 36.2%. (Ans.)**

Condition of steam at the exhaust,

$\mathbf{x_4 = 0.935}$ [*From Mollier chart*]. **(Ans.)**

Example 15.21. (*a*) *How does erosion of turbine blades occur ? State the methods of preventing erosion of turbine blades.*

(*b*) *What do you mean by TTD of a feed water heater ? Draw temperature-path-line diagram of a closed feed water heater used in regenerative feed heating cycle.*

(*c*) *In a 15 MW steam power plant operating on ideal reheat cycle, steam enters the H.P. turbine at 150 bar and 600°C. The condenser is maintained at a pressure of 0.1 bar. If the moisture content at the exit of the L.P. turbine is 10.4%, determine :*

(*i*) *Reheat pressure ;* (*ii*) *Thermal efficiency ;* (*iii*) *Specific steam consumption ; and* (*iv*) *Rate of pump work in kW. Assume steam to be reheated to the initial temperature.* **(AMIE)**

Solution. (*a*) The erosion of the moving blades is *caused by the presence of water particles in (wet) steam in the L.P. stages*. The water particles strike the leading surface of the blades. Such impact, if sufficiently heavy, produces *severe local stresses in the blade material causing the surface metal to fail and flake off*.

The erosion, if any, is more likely to occur in the region where the steam is wettest, *i.e.*, in the last one or two stages of the turbine. Moreover, the water droplets are concentrated in the outer parts of the flow annuals where the velocity of impact is highest.

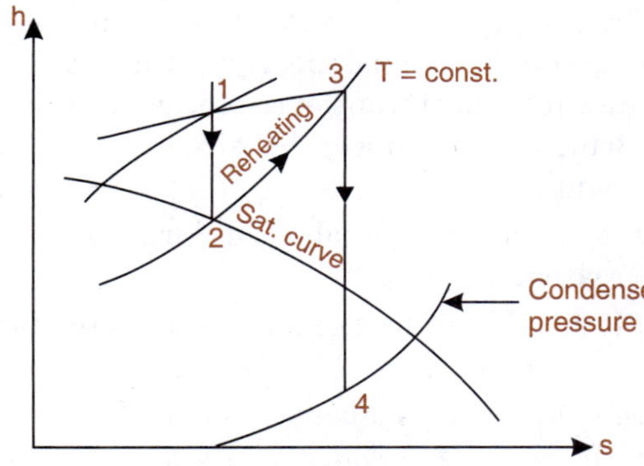

Fig. 15.31

Erosion difficulties due to moisture in the steam may be avoided by reheating (see Fig. 15.31). The whole of steam is taken from the turbine at a suitable point 2, and a further supply of heat is given to it along 2-3 after which the steam is readmitted to the turbine and expanded along 3-4 to condenser pressure.

Erosion *may also be reduced by using steam traps in between the stages to separate moisture from the steam.*

(*b*) TTD means *"Terminal temperature difference"*. It is the *difference between temperatures of bled steam/condensate and the feed water at the two ends of the feed water heater.*

The required temperature-path-line diagram of a closed feed water heater is shown in Fig. 15.32.

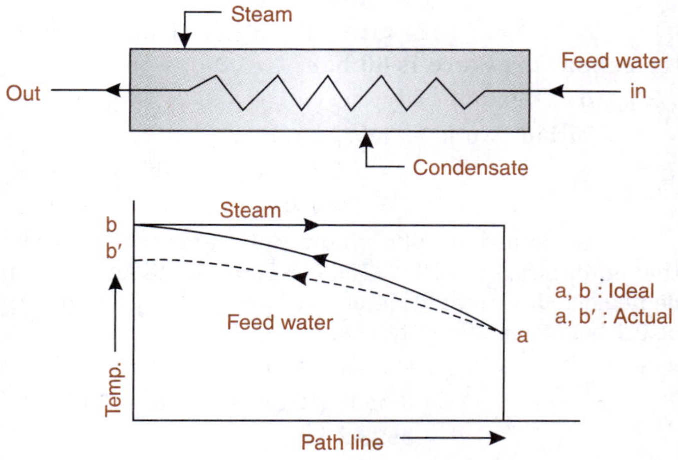

Fig. 15.32

(*c*) The cycle is shown on *T-s* and *h-s* diagrams in Figs. 15.33 and 15.34 respectively. The following values are read from the Mollier diagram :

$h_1 = 3580$ kJ/kg, $h_2 = 3140$ kJ/kg,

$h_3 = 3675$ kJ/kg, and $h_4 = 2335$ kJ/kg

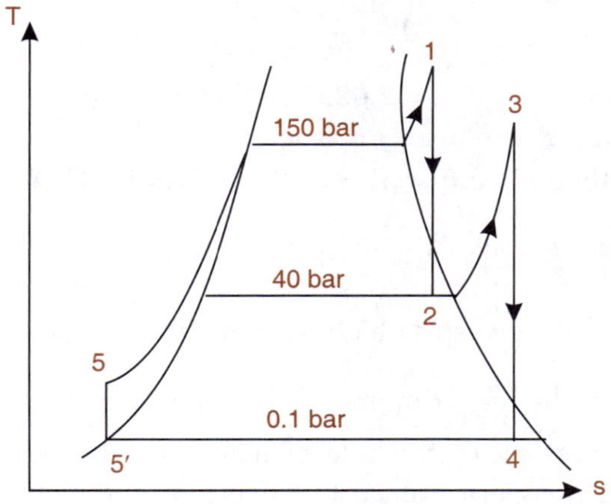

Fig. 15.33

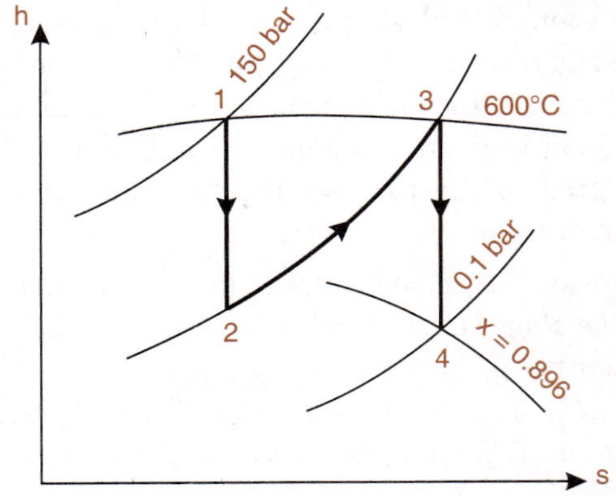

Fig. 15.34

Moisture contents in exit from L.P. turbine = 10.4%

$$x_4 = 1 - 0.104 = 0.896$$

(*i*) **Reheat pressure :** From the Mollier diagram, the **reheat pressure is 40 bar.** **(Ans.)**

(*ii*) **Thermal efficiency, η_{th} :**

Turbine work $= (h_1 - h_2) + (h_3 - h_4)$
$= (3580 - 3140) + (3675 - 2335)$
$= 1780$ kJ/kg.

Assuming specific volume of water = 10^{-3} m^3/kg, the pump work = 10^{-3} (150 – 0.1) = 0.15 kJ/kg, *i.e.*, may be neglected in computing of η_{th}, $h_5 = h_4 = 191.8$ kJ/kg, (h_f at 0.1 bar) from steam tables,

$$Q_{input} = (h_1 - h_5) + (h_3 - h_2)$$
$$= (3580 - 191.8) + (3675 - 3140)$$
$$= 3923.2 \text{ kJ/kg}$$

$$\%\eta_{th} = \frac{1780}{3923.2} \times 100 = \mathbf{45.37\%}.$$ **(Ans.)**

(*iii*) **Specific steam consumption :**

Steam consumption

$$= \frac{15 \times 10^3}{1780} = 8.427 \text{ kg/s}$$

Specific steam consumption

$$= \frac{8.427 \times 3600}{15 \times 10^3}$$

= 2.0225 kg/kWh. **(Ans.)**

(*iv*) **Rate of pump work :**

Rate of pump work = 8.427 × 0.15 = **1.26 kW.** **(Ans.)**

15.6. BINARY VAPOUR CYCLE

Carnot cycle gives the highest thermal efficiency which is given by $\frac{T_1 - T_2}{T_1}$. To approach this cycle in an actual engine it is necessary that whole of heat must be supplied at constant temperature T_1 and rejected at T_2. This can be achieved only by using a vapour in the wet field but not in the superheated. The efficiency depends on temperature T_1 since T_2 is fixed by the natural sink to which heat is rejected. This means that T_1 should be *as large as possible, consistent with the vapour being saturated.*

If we use steam as the working medium the temperature rise is accompanied by rise in pressure and at critical temperature of 374.15°C the pressure is as high as 225 bar which will create many difficulties in design, operation and control. It would be desirable to use some fluid other than steam which has more desirable thermodynamic properties than water. An ideal fluid for this purpose should have a *very high critical temperature combined with low pressure*. Mercury, diphenyl oxide and similar compounds, aluminium bromide and zinc ammonium chloride are fluids which possess the required properties in varying degrees. Mercury is the only working fluid which has been successfully used in practice. It has high critical temperature (588.4°C) and correspondingly low critical pressure (21 bar abs.). *The mercury alone cannot be used as its saturation temperature at atmospheric pressure is high (357°C). Hence* ***binary vapour cycle*** *is generally used to increase the overall efficiency of the plant.* Two fluids (mercury and water) are used in cascade in the binary cycle for production of power.

The *few more properties* required for an ideal binary fluid used in high temperature limit are listed below :

1. It should have high critical temperature at reasonably low pressure.
2. It should have high heat of vaporisation to keep the weight of fluid in the cycle to minimum.
3. Freezing temperature should be below room temperature.
4. It should have chemical stability through the working cycle.
5. It must be non-corrosive to the metals normally used in power plants.
6. It must have an ability to wet the metal surfaces to promote the heat transfer.
7. The vapour pressure at a desirable condensation temperature should be nearly atmospheric which will eliminate requirement of power for maintenance of vacuum in the condenser.
8. After expansion through the primemover the vapour should be nearly saturated so that a desirable heat transfer co-efficient can be obtained which will reduce the size of the condenser required.
9. It must be available in large quantities at reasonable cost.
10. It should not be toxic and, therefore, dangerous to human life.

Although mercury does not have all the required properties, it is more favourable than any other fluid investigated. It is most stable under all operating conditions.

Although, mercury does not cause any corrosion to metals, but it is extremely dangerous to human life, therefore, elaborate precautions must be taken to prevent the escape of vapour. The *major disadvantage* associated with mercury is that it *does not wet surface of the metal and forms a serious resistance to heat flow*. This difficulty can be considerably reduced by adding magnesium and titanium (2 parts in 100000 parts) in mercury.

Thermal properties of mercury :

Mercury fufills practically all the *desirable thermodynamic properties* stated above.

1. Its freezing point is – 3.3°C and boiling point is – 354.4°C at atmospheric pressure.
2. The pressure required when the temperature of vapour is 540°C is only 12.5 bar (app.) and, therefore, heavy construction is not required to get high initial temperature.
3. Its liquid saturation curve is very steep, approaching the isentropic of the Carnot cycle.
4. It has no corrosive or erosive effects upon metals commonly used in practice.
5. Its critical temperature is so far removed from any possible upper temperature limit with existing metals as to cause no trouble.

Some *undesirable properties* of mercury are listed below :

1. Since the latent heat of mercury is quite low over a wide range of desirable condensation temperatures, therefore, several kg of mercury must be circulated per kg of water evaporated in binary cycle.
2. The cost is a considerable item as the quantity required is 8 to 10 times the quantity of water circulated in binary system.
3. Mercury vapour in larger quantities is poisonous, therefore, the system must be perfect and tight.

Fig. 15.35 shows the schematic line diagram of binary vapour cycle using mercury and water as working fluids. The processes are represented on *T-s* diagram as shown in Fig. 15.36.

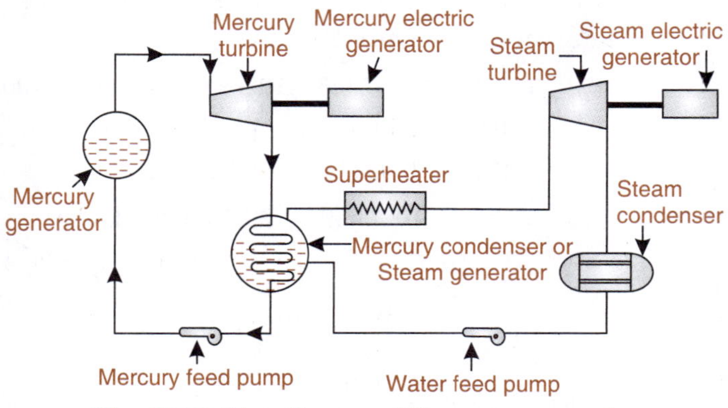

Fig. 15.35. Line diagram of binary vapour cycle.

Analysis of Binary vapour cycle :

h_{hg_1} = Heat supplied per kg of Hg (mercury) vapour formed in the mercury boiler.

h_{hg_2} = Heat lost by one kg of Hg vapour in the mercury condenser.

h_s = Heat given per kg of steam generated in the mercury condenser or steam boiler.

W_{hg} = Work done per kg of Hg in the cycle.

W_s = Work done per kg of steam in the steam cycle.

η_s = Thermal efficiency of the steam cycle.

η_{hg} = Thermal efficiency of the Hg cycle.

m = Mass of Hg in the Hg cycle per kg of steam circulated in the steam cycle.

The heat losses to the surroundings, in the following analysis, are neglected and steam generated is considered one kg and Hg in the circuit is m kg per kg of water in the steam cycle.

Heat supplied in the Hg boiler

$$h_t = m \times h_{hg_1} \quad ...(15.14)$$

Work done in the mercury cycle

$$= m \,.\, W_{hg} \quad ...(15.15)$$

Work done in the steam cycle

$$= 1 \times W_s \quad ...(15.16)$$

Total work done in the binary cycle is given by

$$W_t = m\, W_{hg} + W_s \quad ...(15.17)$$

∴ Overall efficiency of the binary cycle is given by

$$\eta = \frac{\text{Work done}}{\text{Heat supplied}}$$

$$= \frac{W_t}{h_t} = \frac{mW_{hg} + W_s}{mh_{hg_1}} \quad ...(15.18)$$

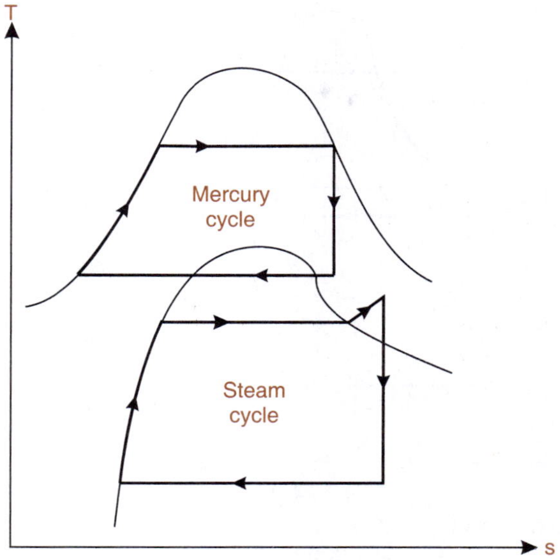

Fig. 15.36. Binary vapour cycle on *T-s* diagram.

Thermal efficiency of the mercury cycle is given by

$$\eta_{hg} = \frac{mW_{hg}}{mh_{hg_1}}$$

$$= \frac{W_{hg}}{h_{hg_1}} = \frac{h_{hg_1} - h_{hg_2}}{h_{hg_1}} = 1 - \frac{h_{hg_2}}{h_{hg_1}} \quad ...(15.19)$$

$$= \frac{mh_{hg_1} - h_s}{mh_{hg_1}} = 1 - \frac{1}{m} \cdot \frac{h_s}{h_{hg_1}} \quad ...(15.20)$$

Heat lost by mercury vapour = Heat gained by steam

$$\therefore \quad m\,h_{hg_2} = 1 \times h_s \quad ...(15.21)$$

Substituting the value of m from eqn. (15.21) into eqn. (15.20), we get

$$\eta_{hg} = 1 - \frac{h_{hg_2}}{h_{hg_1}} \quad ...(15.22)$$

The thermal efficiency of the steam cycle is given by

$$\eta_s = \frac{W_s}{h_s} = \frac{h_{s_1} - h_{s_2}}{h_{s_1}} = \frac{h_{s_1} - h_{s_2}}{mh_{hg_2}} \quad ...(15.23)$$

From the eqns. (15.18), (15.20), (15.21), (15.22) and (15.23), we get

$$\eta = \eta_{hg}\,(1 - \eta_s) + \eta_s \quad ...(15.24)$$

To solve the problems eqns. (15.19), (15.23), (15.24) are used.

In the design of binary cycle, another important problem is the limit of exhaust pressure of the mercury (location of optimum exhaust pressure) which will provide maximum work per kg of Hg circulated in the system and high thermal efficiency of the cycle. It is not easy to decide as number of controlling factors are many.

☞**Example 15.22.** *A binary vapour cycle operates on mercury and steam. Standard mercury vapour at 4.5 bar is supplied to the mercury turbine, from which it exhausts at 0.04 bar. The mercury condenser generates saturated steam at 15 bar which is expanded in a steam turbine to 0.04 bar.*

(i) Determine the overall efficiency of the cycle.

(ii) If 48000 kg/h of steam flows through the steam turbine, what is the flow through the mercury turbine ?

(iii) Assuming that all processes are reversible, what is the useful work done in the binary vapour cycle for the specified steam flow ?

(iv) If the steam leaving the mercury condenser is superheated to a temperature of 300°C in a superheater located in the mercury boiler and if the internal efficiencies of the mercury and steam turbines are 0.84 and 0.88 respectively, calculate the overall efficiency of the cycle. The properties of standard mercury are given below :

p (bar)	*t (°C)*	h_f *(kJ/kg)*	h_g *(kJ/kg)*	s_f *(kJ/kg K)*	s_g *(kJ/kg K)*	v_f *(m³/kg)*	v_g *(m³/kg)*
4.5	*450*	*62.93*	*355.98*	*0.1352*	*0.5397*	79.9×10^{-6}	*0.068*
0.04	*216.9*	*29.98*	*329.85*	*0.0808*	*0.6925*	76.5×10^{-6}	*5.178.*

Solution. The binary vapour cycle is shown in Fig. 15.37.

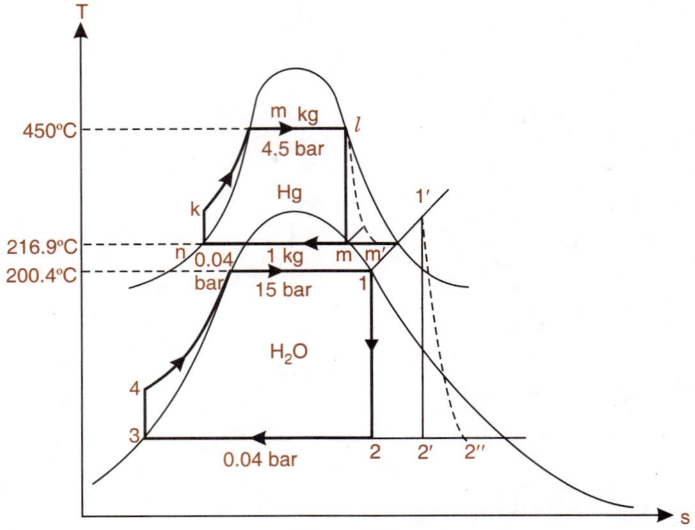

Fig. 15.37

Mercury cycle :

$$h_l = 355.98 \text{ kJ/kg}$$

$$s_l = 0.5397 = s_m = s_f + x_m\, s_{fg}$$

or $$0.5397 = 0.0808 + x_m = (0.6925 - 0.0808)$$

$$\therefore \quad x_m = \frac{(0.5397 - 0.0808)}{(0.6925 - 0.0808)} = 0.75$$

$$h_m = h_f + x_m\, h_{fg}$$
$$= 29.98 + 0.75 \times (329.85 - 29.98)$$
$$= 254.88 \text{ kJ/kg}$$

Work obtained from mercury turbine

$$(W_T)_{Hg} = h_l - h_m = 355.98 - 254.88 = 101.1 \text{ kJ/kg}$$

Pump work in mercury cycle,

$$(W_P)_{Hg} = h_{f_k} - h_{f_n} = 76.5 \times 10^{-6} \times (4.5 - 0.04) \times 100$$
$$= 0.0341 \text{ kJ/kg}$$

$$\therefore \quad W_{net} = 101.1 - 0.0341 \simeq 101.1 \text{ kJ/kg}$$

$$Q_1 = h_l - h_{f_k} = 355.98 - 29.98 = 326 \text{ kJ/kg}$$

$$(\because \; h_{f_n} \simeq h_{f_k})$$

$$\therefore \quad \eta_{\text{Hg cycle}} = \frac{W_{net}}{Q_1} = \frac{101.1}{326}$$
$$= 0.31 \quad \text{or} \quad 31\%.$$

Steam cycle :

At 15 bar : $h_1 = 2789.9$ kJ/kg, $s_1 = 6.4406$ kJ/kg

At 0.04 bar : $h_f = 121.5$ kJ/kg, $h_{f_g} = 2432.9$ kJ/kg,

$s_f = 0.432$ kJ/kg K,

$s_{fg_2} = 8.052$ kJ/kg K,

$v_f = 0.0001$ kJ/kg K

Now, $s_1 = s_2$

$$6.4406 = s_f + x_2\, s_{fg} = 0.423 + x_2 \times 8.052$$

$$\therefore \quad x_2 = \frac{6.4406 - 0.423}{8.052} = 0.747$$

$$h_2 = h_{f_2} + x_2\, h_{f_g}$$

$$= 121.5 + 0.747 \times 2432.9$$

$$= 1938.8 \text{ kJ/kg}$$

Work obtained from steam turbine,

$$(W_T)_{steam} = h_1 - h_2 = 2789.9 - 1938.8 = 851.1 \text{ kJ/kg}$$

Pump work in steam cycle,

$$(W_P)_{steam} = h_{f_4} - h_{f_3} = 0.001\,(15 - 0.04) \times 100$$

$$= 1.496 \text{ kJ/kg} \simeq 1.5 \text{ kJ/kg}$$

or $$h_{f_4} = h_{f_3} + 1.5 = 121.5 + 1.5 = 123 \text{ kJ/kg}$$

$$Q_1 = h_1 - h_{f_4} = 2789.9 - 123 = 2666.9 \text{ kJ/kg}$$

$$(W_{net})_{steam} = 851.1 - 1.5 = 849.6 \text{ kJ/kg}$$

$$\therefore \eta_{steam\ cycle} = \frac{W_{net}}{Q_1} = \frac{849.6}{2666.6} = 0.318 \text{ or } 31.8\%.$$

(*i*) **Overall efficiency of the binary cycle :**

Overall efficiency of the binary cycle

$$= \eta_{Hg_{cycle}} + \eta_{steam\ cycle} - \eta_{Hg_{cycle}} \times \eta_{steam\ cycle}$$

$$= 0.31 + 0.318 - 0.31 \times 0.318 = 0.5294 \text{ or } 52.94\%$$

Hence *overall efficiency of the binary cycle*

= **52.94%. (Ans.)**

$\eta_{overall}$ can also be found out as follows :

Energy balance for a mercury condenser-steam boiler gives :

$$m\,(h_m - h_{f_n}) = 1(h_1 - h_{f_4})$$

where m is the amount of mercury circulating for 1 kg of steam in the bottom cycle

$$\therefore \quad m = \frac{h_1 - h_{f_4}}{h_m - h_{f_n}} = \frac{2666.9}{254.88 - 29.98} = 11.86 \text{ kg}$$

$$(Q_1)_{total} = m\,(h_l - h_{f_k}) = 11.86 \times 326 = 3866.36 \text{ kJ/kg}$$

$$(W_T)_{total} = m\,(h_l - h_m) + (h_1 - h_2)$$

$$= 11.86 \times 101.1 + 851.1 = 2050.1 \text{ kJ/kg}$$

$(W_P)_{total}$ may be *neglected*

$$\eta_{overall} = \frac{W_T}{Q_1} = \frac{2050.1}{3866.36} = 0.53 \text{ or } 53\%.$$

(*ii*) **Flow through mercury turbine :**

If 48000 kg/h of steam flows through the steam turbine, the flow rate of mercury,

$$m_{Hg} = 48000 \times 11.86 = \mathbf{569280\ kg/h.} \quad \textbf{(Ans.)}$$

(*iii*) **Useful work in binary vapour cycle :**

Useful work,

$$(W_T)_{total} = 2050.1 \times 48000 = 9840.5 \times 10^4 \text{ kJ/h}$$

$$= \frac{9840.5 \times 10^4}{3600} = 27334.7 \text{ kW}$$

= **27.33 MW. (Ans.)**

(*iv*) **Overall efficiency under new conditions :**

Considering the efficiencies of turbines, we have :

$$(W_T)_{Hg} = h_l - h_m' = 0.84 \times 101.1$$

$$= 84.92 \text{ kJ/kg}$$

$$\therefore \quad h_m' = h_l - 84.92 = 355.98 - 84.92$$

$$= 271.06 \text{ kJ/kg}$$

$$\therefore \quad m'\,(h_m' - h_n') = (h_1 - h_{f_4})$$

or $$m' = \frac{h_1 - h_{f_4}}{h_m' - h_n'}$$

$$= \frac{2666.9}{271.06 - 29.98} = 11.06 \text{ kg}$$

$$(Q_1)_{total} = m'\,(h_l - h_{f_k}) + 1\,(h_1' - h_1)$$

At 15 bar, 300°C :

$h_g = 3037.6$ kJ/kg,

$s_g = 6.918$ kJ/kg K

$= 11.06 \times 326 + (3037.6 - 2789.9)$

$= 3853.26$ kJ/kg

$$s_1' = 6.918 = s_2' = 0.423 + x_2' \times 8.052$$

$$\therefore \quad x_2' = \frac{6.918 - 0.423}{8.052} = 0.80.$$

$$h_2' = 121.5 + 0.807 \times 2432.9$$

$$= 2084.8 \text{ kJ/kg}$$

$$(W_T)_{steam} = h_1' - h_2'$$

$$= 0.88\,(3037.6 - 2084.8)$$

$$= 838.46 \text{ kJ/kg}$$

$$(W_T)_{total} = 11.06 \times 84.92 + 838.46$$

$$= 1777.67 \text{ kJ/kg}$$

Neglecting pump work,

$$\eta_{overall} = \frac{1777.67}{3853.26} = \mathbf{0.4613 \text{ or } 46.13\%.} \quad \textbf{(Ans.)}$$

ADDITIONAL/TYPICAL WORKED EXAMPLES

Example 15.23. *The following data relate to a regenerative steam power plant generating 22500 kW energy, the alternator directly coupled to steam turbine :*

Condition of steam supplied to the steam turbine	*... 60 bar, 450°C*
Condenser vacuum	*... 707.5 mm*
Pressure at which steam is bled from the steam turbine	*... 3 bar*
Turbine efficiency of each portion of expansion	*... 87 per cent*
Boiler efficiency	*... 86 per cent*
Alternator efficiency	*... 94 per cent*
Mechanical efficiency from turbine to generator	*... 97 per cent*

Neglecting the pump work in calculating the input to the boiler, determine :

(i) The steam bled per kg of steam supplied to the turbine.

(ii) The steam generated per hour if the 9 per cent of the generator output is used to run the pumps.

(iii) The overall efficiency of the plant.

Solution. The schematic arrangement of the steam power plant is shown in Fig. 15.38 (*a*), while the conditions of the fluid passing through the components are represented on *T-s* and *h-s* diagrams as shown in Fig. 15.38 (*b*) and (*c*). The conditions of the fluid entering and leaving the pump are shown by the same point as the rise in temperature due to pump work is neglected.

Given : Power generated = 22500 kW ;

$$p_1 = 60 \text{ bar} \; ; \; t_1 = 450°\text{C} \; ;$$

$$p_2 \, (= p_2{}') = 3 \text{ bar} \; ;$$

$$p_3 \, (= p_3{}') = \frac{760 - 707.5}{760} \times 1.013$$

$$= 0.07 \text{ bar} \; ; \; \eta_{\text{turbine (each portion)}} = 87\% \; ;$$

$$\eta_{\text{boiler}} = 86\% \; ; \; \eta_{\text{alt.}} = 94\%, \; \eta_{\text{mech.}} = 97\%$$

- Locate point 1 corresponding to the conditions : p_1 = 60 bar ; t_1 = 450°C on the *h-s* chart *(Mollier chart)*.

  From *h-s* chart ; we find : h_1 = 3300 kJ/kg.

- Draw vertical line through point 1 till it cuts the 3 bar pressure line, then locate point 2.

$\therefore$ $$h_2 = 2607 \text{ kJ/kg}$$

Now, $$\eta_{\text{turbine}} = 0.87 = \frac{h_1 - h_2{}'}{h_1 - h_2}$$

or, $$0.87 = \frac{3300 - h_2{}'}{3300 - 2607}$$

$\therefore$ $$h_2{}' = 2697 \text{ kJ/kg}$$

- Locate the point 2 on the *h-s* chart as enthalpy and pressure are known and then draw a vertical line through the point 2 till it cuts the 0.07 bar pressure line and then locate the point 3.

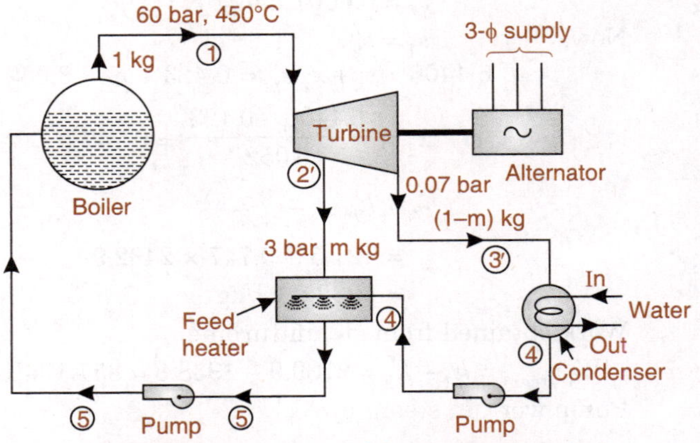

(*a*) Schematic arrangement of the steam power plant

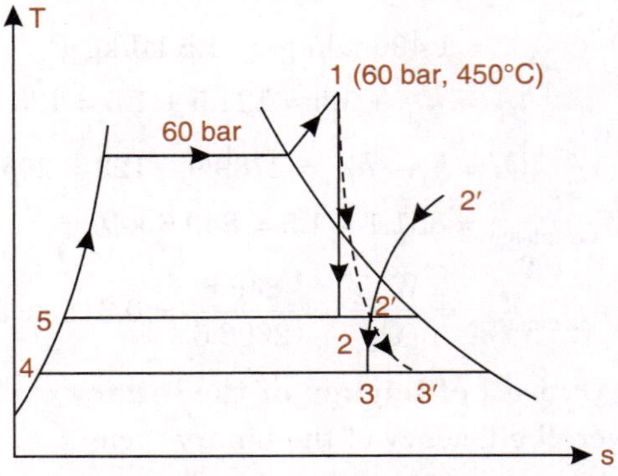

(*b*) T-s diagram (Neglecting pump work)

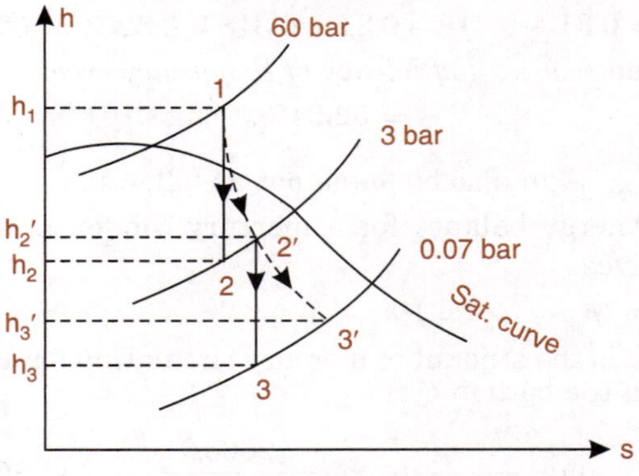

(*c*) h-s diagram

Fig. 15.38

$\therefore$ $$h_3 = 2165 \text{ kJ/kg}$$

Again, $$\eta_{\text{turbine}} = 0.87 = \frac{h_2{}' - h_3{}'}{h_2{}' - h_3}$$

or, $$0.87 = \frac{2697 - h_3{}'}{2697 - 2165}$$

$\therefore$ $$h_3{}' = 2234 \text{ kJ/kg}$$

From steam tables, corresponding to pressures 3 bar and 0.02 bar, the saturated liquid heats at points 4 and 5 are :

$$h_{f4} = 163.4 \text{ kJ/kg} ; h_{f5} = 561.4 \text{ kJ/kg}$$

(*i*) **The steam bled per kg of steam supplied to the turbine, m :**

Considering the *energy balance for feed heater* we have ;

$$m(h_2' - h_{f5}) = (1 - m)(h_{f5} - h_{f4})$$

or, $m(2697 - 561.4) = (1 - m)(561.4 - 163.4)$

or, $2135.6\, m = 398\,(1 - m)$

$\therefore$ m = **0.157 kJ/kg of steam generated. (Ans.)**

(*ii*) **Steam generated per hour :**

Work developed per kg of steam in the turbine

$$= 1(h_1 - h_2') + (1 - m)(h_2' - h_3')$$

$$= (3300 - 2697) + (1 - 0.157)(2697 - 2234)$$

$$= 993.3 \text{ kJ/kg}$$

Actual work developed by the turbine

$$= \frac{22500}{\eta_{alt.} \times \eta_{mech.}} = \frac{22500}{0.94 \times 0.97} = 24676.5 \text{ kW}$$

$\therefore$ Steam generated per hour

$$= \frac{24676.5}{993.3} \times \frac{3600}{1000} \text{ tonnes/h}$$

= **89.43 tonnes/h. (Ans.)**

(*iii*) **The overall efficiency of the plant, $\eta_{overall}$:**

Net power available deducting pump power

$$= 22500\,(1 - 0.09) = 20475 \text{ kW}$$

Heat supplied in the boiler

$$= \frac{89.43 \times 1000\,(h_1 - h_{f5})}{0.86} \text{ kJ/h}$$

$$= \frac{89.43 \times 1000\,(3300 - 561.4)}{0.86 \times 3600} \text{ kW}$$

$$= 79106.3 \text{ kW}$$

$$\therefore \quad \eta_{overall} = \frac{\text{Net power available}}{\text{Heat supplied by the boiler}}$$

$$= \frac{20475}{79106.3} = \mathbf{0.2588 \text{ or } 25.88\%.} \textbf{ (Ans.)}$$

Example 15.24. *A steam power plant of 110 MW capacity is equipped with regenerative as well as reheat arrangement. The steam is supplied at 80 bar and 55°C of superheat. The steam is extracted at 7 bar for feed heating and remaining steam is reheated to 350°C, and then expanded to 0.4 bar in the L.P. stage. Assume indirect type of feed heaters. Determine :*

(*i*) *The ratio of steam bled to steam generated,*

(*ii*) *The boiler generating capacity in tonnes of steam/hour, and*

(*iii*) *Thermal efficiency of the cycle.*

Assume no losses and ideal processes of expansion.

Solution. The schematic arrangement of the plant is shown in Fig. 15.39 (*a*) and the processes are represented on *h-s* chart in Fig. 15.39 (*b*).

Given : Capacity of plant = 110 MW ;

t_1 = 350°C *i.e.*, t_s at 80 bar ≃ 295°C + 55°C = 350°C

$p_2 = p_3$ = 7 bar ; t_3 = 350°C ; p_4 = 0.4 bar

- Locate point 1 corresponding to the condition p_1 = 80 bar and t_1 = 350°C, on the *h-s* chart.
- Locate point 2 by drawing vertical line through point 1 till it cuts the 7 bar pressure line.
- Locate point 3 as the cross point of 7 bar and 350°C temperature line.
- Locate point 4 by drawing vertical line through the point 3 till it cuts the 0.4 bar pressure line.

From *h-s* chart, we find :

h_1 = 2985 kJ/kg ; h_2 = 2520 kJ/kg ;

h_3 = 3170 kJ/kg ; h_4 = 2555 kJ/kg.

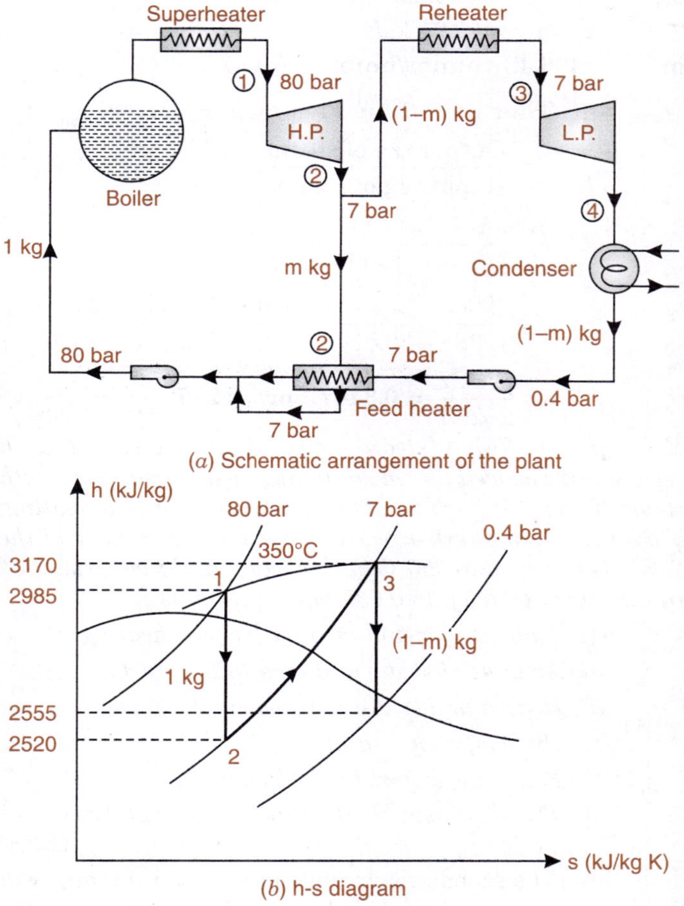

(*a*) Schematic arrangement of the plant

(*b*) h-s diagram

Fig. 15.39

Also, from steam tables, we have :

h_{f2} (at 7 bar) = 697.1 kJ/kg ;

h_{f4} (at 0.4 bar) = 317.7 kJ/kg

(*i*) **The ratio of steam bled to steam generated :**

Consider energy/heat balance of feed heater :

Heat lost by m kg of steam = Heat gained by $(1 - m)$ kg of condensed steam

$$m(h_2 - h_{f2}) = (1 - m)(h_{f2} - h_{f4})$$

$$m(2520 - 697.1) = (1 - m)(697.1 - 317.7)$$

$$1822.9\, m = (1 - m) \times 379.4$$

$$\therefore \quad m = 0.172 \text{ kg}$$

i.e., Amount of steam bled per kg of steam supplied to the turbine = 0.172 kg

$$\therefore \quad \frac{\text{Steam generated}}{\text{Steam bled}} = \frac{1}{0.172} = \mathbf{5.814.} \quad \textbf{(Ans.)}$$

(*ii*) **The boiler generating capacity :**

If m_s is the mass of steam supplied to the power plant per second, then the work developed is given by :

$$m_s(h_1 - h_2) + m_s(1 - m)(h_3 - h_4) = 110 \times 10^3$$

or, $$m_s(2985 - 2520) + m_s(1 - 0.172)(3170 - 2555) = 110 \times 10^3$$

or, $$m_s(465 + 509.22) = 110 \times 10^3$$

$$\therefore \quad m_s = 112.91 \text{ kg/s}$$

or **406.48 tonnes/hour. (Ans.)**

(*iii*) **Thermal efficiency of the cycle, $\eta_{thermal}$:**

$$\eta_{thermal} = \frac{\text{Output / kg of steam}}{\text{Input / kg of steam}}$$

$$= \frac{(h_1 - h_2) + (1 - m)(h_3 - h_4)}{(h_1 - h_{f_2}) + (1 - m)(h_3 - h_2)}$$

$$= \frac{(2985 - 2520) + (1 - 0.172)(3170 - 2555)}{(2985 - 697.1) + (1 - 0.172)(3170 - 2520)}$$

$$= \frac{974.22}{2826.1} = \mathbf{0.3447} \text{ or } \mathbf{34.47\%}. \quad \textbf{(Ans.)}$$

Example 15.25. *A steam power plant equipped with regenerative as well as reheat arrangement is supplied with steam to the H.P. turbine at 80 bar 470°C. For feed heating, a part of steam is extracted at 7 bar and remainder of the steam is reheated to 350°C in a reheater and then expanded in L.P. turbine down to 0.035 bar. Determine :*

(*i*) *Amount of steam bled-off for feed heating,*

(*ii*) *Amount of steam supplied to L.P. turbine,*

(*iii*) *Heat supplied in the boiler and reheater*

(*iv*) *Cycle efficiency, and*

(*v*) *Power developed by the system.*

The steam supplied by the boiler is 50 kg/s.

(B.U.)

Solution. The schematic arrangement of the steam power plant is shown in Fig. 15.40 (*a*) and the processes are represented on *h-s* diagram as shown in Fig. 15.40(*b*).

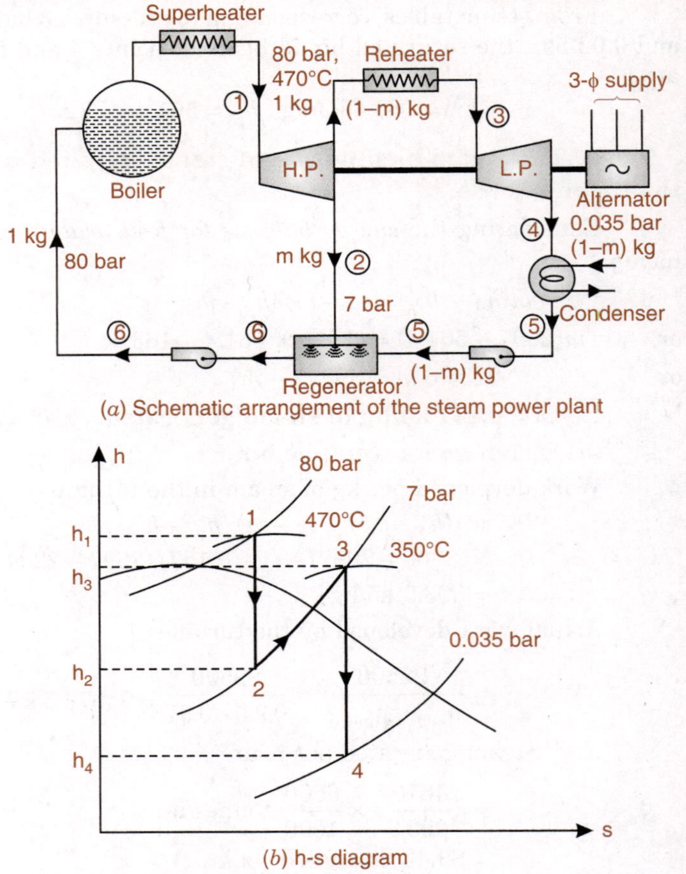

Fig. 15.40

From *h-s* chart and steam tables, we have enthalpies at different points as follows :

h_1 = 3315 kJ/kg ; h_2 = 2716 kJ/kg
h_3 = 3165 kJ/kg ; h_4 = 2236 kJ/kg. } From *h-s* chart

$h_{f6} = h_{f2}$ = 697.1 kJ/kg ;
$h_{f5} = h_{f4}$ = 101.9 kJ/kg } From steam tables.

(*i*) **Amount of steam bled-off for feed heating :**

Considering *energy balance at regenerator, we have :*

Heat lost by steam = Heat gained by water

$$m(h_2 - h_{f6}) = (1 - m)(h_{f6} - h_{f5})$$

or, $$m(h_2 - h_{f2}) = (1 - m)(h_{f2} - h_{f4}) \quad [\because h_{f6} = h_{f2} ; h_{f5} = h_{f4}]$$

or, $$m(2716 - 697.1) = (1 - m)(697.1 - 111.9)$$

or, $$2018.9\, m = 585.2\,(1 - m)$$

$$\therefore \quad m = 0.225 \text{ kg of steam supplied}$$

Hence amount of **steam bled off is 22.5% of steam generated by the boiler. (Ans.)**

(*ii*) **Amount of steam supplied to L.P. turbine :**

Amount of steam supplied to L.P. turbine

= 100 – 22.5

= **77.5% of the steam generated by the boiler. (Ans.)**

(*iii*) **Heat supplied in the boiler and reheater**

Heat supplied in the *boiler* per kg of steam generated

$= h_1 - h_{f6} = 3315 - 697.1 =$ **2617.9 kJ/kg. (Ans.)**

$(\because \; h_{f6} = h_{f2})$

Heat supplied in the *reheater* per kg of steam generated

$= (1 - m)(h_3 - h_2)$

$= (1 - 0.225)(3165 - 2716) =$ **347.97 kJ/kg. (Ans.)**

Total amount of heat supplied by the boiler and reheater per kg of steam generated,

$$Q_s = 2617.9 + 347.97 = 2965.87 \text{ kJ/kg}$$

(*iv*) **Cycle efficiency, η_{cycle} :**

Amount of work done by per kg of steam generated by the boiler,

$W = 1(h_1 - h_2) + (1 - m)(h_3 - h_4)$, Neglecting pump work

$= (3315 - 2716) + (1 - 0.225)(3165 - 2236)$

$\simeq 1319$ kJ/kg

$$\therefore \quad \eta_{cycle} = \frac{W}{Q_s} = \frac{1319}{2965.87}$$

$= \mathbf{0.4447}$ **or 44.47%. (Ans.)**

(*v*) **Power developed by the system :**

Power developed by the system

$= m_s \times W = 50 \times 1319$ kJ/s

$= \dfrac{50 \times 1319}{1000} =$ **65.95 MW. (Ans.)**

Example 15.26. *A steam power plant operates on ideal Rankine cycle using reheater and regenerative feed water heaters. It has one open feed heater. Steam is supplied at 150 bar and 600°C. The condenser pressure is 0.1 bar. Some steam is extracted from the turbine at 40 bar for closed feed water heater and remaining steam is reduced at 40 bar to 600°C. Extracted steam is completely condensed in this closed feed water heater and is pumped to 150 bar before mixing with the feed water heater. Steam for the open feed water heater is bled from L.P. turbine at 5 bar. Determine :*

(*i*) *Fraction of steam extracted from the turbines at each bled heater, and*

(*ii*) *Thermal efficiency of the system.*

Draw the line diagram of the components and represent the cycle on T-s diagram. **(P.U.)**

Solution. The arrangement of the components is shown in Fig. 15.41 (*a*) and the processes are represented on *T-s* diagram as shown in Fig. 15.41 (*b*).

From *h-s* chart and steam tables we have enthalpies at different points as follows :

$h_1 = 3578$ kJ/kg ; $h_2 = 3140$ kJ/kg ;
$h_3 = 3678$ kJ/kg ; $h_4 = 3000$ kJ/kg ;
$h_5 = 2330$ kJ/kg ;
} From *h-s* chart

h_{f1} (at 150 bar) = 1611 kJ/kg
h_{f2} (at 40 bar) = 1087.4 kJ/kg ;
h_{f4} (at 5 bar) = 640.1 kJ/kg ;
$h_{f5} = h_{f6}$ (at 0.1 bar) = 191.8 kJ/kg
} From steam tables

(*i*) **Fraction of steam extracted from the turbines at each bled heater m_1, m_2 :**

Considering *energy balance for closed feed heater,* we have :

$$m_1(h_2 - h_{f2}) = (1 - m_1)(h_{f2} - h_{f4})$$

$$m_1(3140 - 1087.4) = (1 - m_1)(1087.4 - 640.1)$$

or, $\quad 2052.6\, m_1 = (1 - m_1) \times 447.3$

$\therefore$ **m_1 = 0.179 kg/kg of steam supplied by the boiler. (Ans.)**

Considering *energy balance for open feed heater,* we have :

$$m_2(h_4 - h_{f4}) = (1 - m_1 - m_2)(h_{f4} - h_{f6})$$

or, $\quad m_2(h_4 - h_{f4}) = (1 - m_1 - m_2)(h_{f4} - h_{f5})$

$(\because \; h_{f6} = h_{f5})$

or, $m_2(3000 - 640.1) = (1 - 0.179 - m_2)(640.1 - 191.8)$

or, $\quad 2359.9\, m_2 = (0.821 - m_2) \times 448.3$

$= 368.05 - 448.3\, m_2$

$\therefore$ **m_2 = 0.131 kg/kg of steam supplied by boiler. (Ans.)**

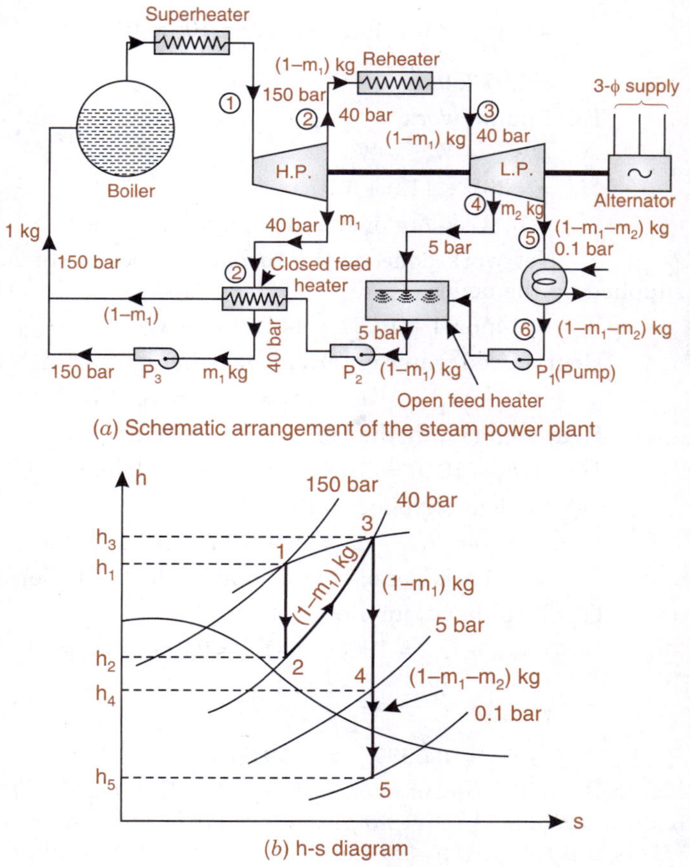

(*a*) Schematic arrangement of the steam power plant

(*b*) h-s diagram

Fig. 15.41

(*ii*) **Thermal efficiency of the system, $\eta_{thermal}$:**

Total work done per kg of steam supplied by the boiler

$$= 1 \times (h_1 - h_2) + (1 - m_1)(h_3 - h_4) + (1 - m_1 - m_2)(h_4 - h_5)$$

$$= (3578 - 3140) + (1 - 0.179)(3678 - 3000) + (1 - 0.179 - 0.131)(3000 - 2330)$$

$$= 438 + 556.64 + 462.3 = 1456.94 \text{ kJ/kg}$$

Work done by the pump P_1

$$W_{P1} = v_{w1} (1 - m_1 - m_2)(5 - 0.1) \times 10^5 \times 10^{-3} \text{ kJ/kg}$$

$$= \frac{1}{1000} (1 - 0.179 - 0.131)(5 - 0.1) \times 10^5 \times 10^{-3}$$

$$= 0.338 \text{ kJ/kg}$$

$$\left[\text{Taking } v_{w1} = v_{w2} = v_{w3} = \frac{1}{1000} \text{ m}^3\text{/kg}\right]$$

Work done by the pump P_2,

$$W_{P2} = v_{w2} (1 - m_1)(150 - 5) \times 10^5 \times 10^{-3} \text{ kJ/kg}$$

$$= \frac{1}{1000} (1 - 0.179)(150 - 5) \times 10^5 \times 10^{-3}$$

$$= 11.9 \text{ kJ/kg}$$

Work done by pump P_3,

$$W_{P3} = v_{w3} \times m_1 \times (150 - 40) \times 10^5 \times 10^{-3}$$

$$= \frac{1}{1000} \times 0.179 (150 - 40) \times 10^5 \times 10^{-3}$$

$$= 1.97 \text{ kJ/kg}$$

Total pump work

$$= W_{P1} + W_{P2} + W_{P3}$$

$$= 0.338 + 11.9 + 1.97$$

$$= 14.21 \text{ kJ/kg of steam supplied by boiler}$$

$\therefore$ Net work done by the turbine per kg of steam supplied by the boiler,

$$W_{net} = 1456.94 - 14.21 = 1442.73 \text{ kJ/kg}$$

Heat of feed water entering the boiler

$$= (1 - m_1) \times 1611 + m_1 \times 1611 = 1611 \text{ kJ/kg}$$

Heat supplied by the boiler per kg of steam,

$$Q_{s1} = h_1 - 1610 = 3578 - 1610 = 1968 \text{ kJ/kg}$$

Q_{s2} = Heat supplied in the reheater

$$= (1 - m_1)(h_3 - h_2) = (1 - 0.179)(3678 - 3140)$$

$$= 441.7 \text{ kJ/kg of steam supplied by the boiler}$$

Q_{st} (Total heat supplied)

$$= Q_{s1} + Q_{s2} = 1968 + 441.7 = 2409.7 \text{ kJ/kg}$$

$$\therefore \eta_{thermal} = \frac{W_{net}}{Q_{st}} = \frac{1442.73}{2409.7}$$

$$= \mathbf{0.5987 \text{ or } 59.87\%.} \quad \textbf{(Ans.)}$$

Example 15.27. *Steam at 70 bar and 450°C is supplied to a steam turbine. After expanding to 25 bar in high pressure stages, it is reheated to 420°C at the constant pressure. Next ; it is expanded in intermediate pressure stages to an appropriate minimum pressure such that part of the steam bled at this pressure heats the feed water to a temperature of 180°C. The remaining steam expands from this pressure to a condenser pressure of 0.07 bar in the low pressure stage. The isentropic efficiency of H.P. stage is 78.5%, while that of the intermediate and L.P. stages is 83% each. From the above data, determine :*

(*i*) *The minimum pressure at which bleeding is necessary.*

(*ii*) *The quantity of steam bled per kg of flow at the turbine inlet.*

(*iii*) *The cycle efficiency.*

Neglect pump work. **(Roorkee University)**

Solution. The schematic arrangement of the plant is shown in Fig. 15.42 (*a*) and the processes are represented on *T-s* and *h-s* diagrams as shown in Fig. 15.42 (*b*) and (*c*) respectively.

(*i*) **The minimum pressure at which bleeding is necessary :**

It would be assumed that the feed water heater is an open heater. Feed water is heated to 180°C. So p_{sat} at 180°C ≃ 10 bar is the pressure at which the heater operates.

Thus, *the pressure at which bleeding is necessary is* **10 bar. (Ans.)**

From the *h-s* chart (Mollier chart), we have :

$$h_1 = 3285 \text{ kJ/kg} \; ; \; h_2 = 2980 \text{ kJ/kg} \; ;$$

$$h_3 = 3280 \text{ kJ/kg} \; ; \; h_4 = 3030 \text{ kJ/kg}$$

$$h_3 - h_4' = 0.83(h_3 - h_4) = 0.83(3280 - 3030)$$

$$= 207.5 \text{ kJ/kg}$$

$$h_4' = h_3 - 207.5 = 3280 - 207.5 = 3072.5 \text{ kJ/kg}$$

$$h_5 = 2210 \text{ kJ/kg}$$

$$h_4' - h_5' = 0.83(h_4' - h_5)$$

$$= 0.83(3072.5 - 2210) \simeq 715.9 \text{ kJ/kg}$$

$$\therefore \quad h_5' = h_4' - 715.9 = 3072.5 - 715.9 = 2356.6 \text{ kJ/kg}$$

From steam tables, we have :

$$h_{f6} = 163.4 \text{ kJ/kg} \; ; \quad h_{f8} = 762.6 \text{ kJ/kg}$$

$$h_1 - h_2' = 0.785(h_1 - h_2) = 0.785(3285 - 2980)$$

$$= 239.4 \text{ kJ/kg}$$

$$\therefore \quad h_2' = h_1 - 239.4 = 3285 - 239.4 = 3045.6 \text{ kJ/kg}$$

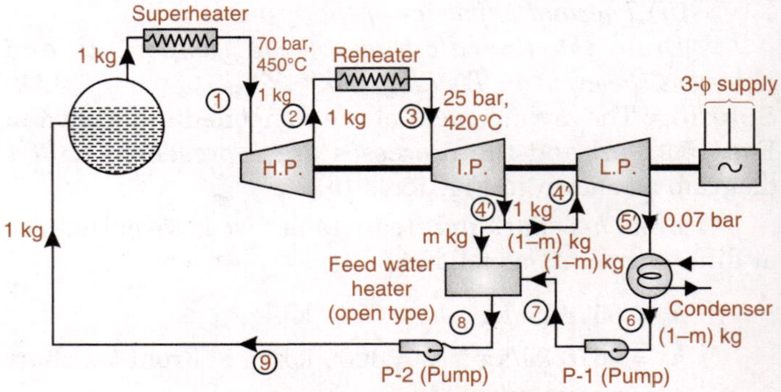

(*a*) Schematic arrangement of the plant

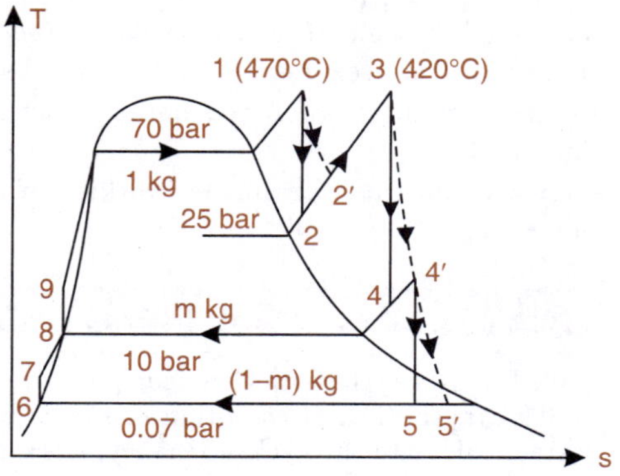

(b) T-s diagram

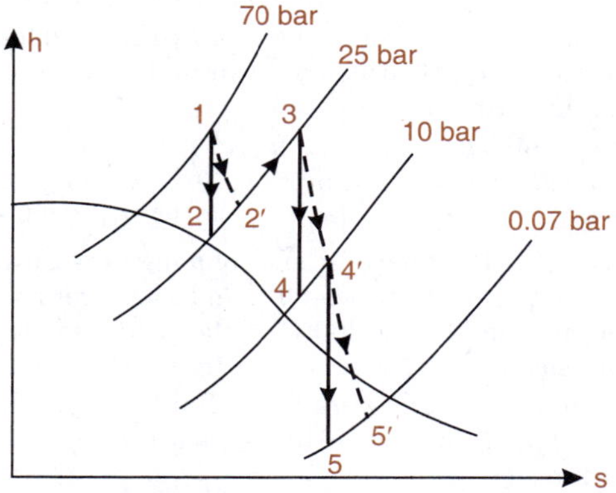

(c) h-s diagram (Pump work not shown)

Fig. 15.42

(*ii*) **The quantity of steam bled per kg of flow at the turbine inlet, m :**

Considering *energy balance for the feed water heater, we have :*

$$m \times h_4' + (1 - m)\, h_{f7} = 1 \times h_{f8}$$

$$m \times 3072.5 + (1 - m) \times 163.4 = 1 \times 762.6$$

$$(\because \quad h_{f7} = h_{f6})$$

$$3072.5\, m + 163.4 - 163.4\, m = 762.6$$

$$\therefore \quad m = \frac{(762.6 - 163.4)}{(3072.5 - 163.4)}$$

= **0.206 kg of steam flow at turbine inlet. (Ans.)**

(*iii*) **Cycle efficiency, η_{cycle} :**

$$\eta_{cycle} = \frac{\text{Work done}}{\text{Heat supplied}}$$

$$= \frac{1(h_1 - h_2) + 1(h_3 - h_4) + (1 - m)(h_4' - h_5')}{(h_1 - h_{f8}) + (h_3 - h_2')}$$

$$= \frac{(3285 - 3045.6) + 207.5 + (1 - 0.206)(715.9)}{(3285 - 762.6) + (3280 - 3045.6)}$$

$$= \frac{1015.3}{2756.8} = \mathbf{0.3683 \text{ or } 36.83\%}. \quad \textbf{(Ans.)}$$

HIGHLIGHTS

1. Carnot cycle efficiency = $\dfrac{T_1 - T_2}{T_1}$.
2. Rankine cycle is the theoretical cycle on which steam primemovers work.

   Rankine efficiency = $\dfrac{h_1 - h_2}{h_1 - h_{f_2}}$.
3. The thermal efficiency of Rankine cycle is increased by
   (*i*) *Increasing the average temperature at which heat is added to the cycle.*
   (*ii*) *Decreasing the average temperature at which heat is rejected to the cycle.*
4. Thermal efficiency of regenerative cycle

$$= \frac{(h_0 - h_1) + (1 - m_1)(h_1 - h_2) + (1 - m_1 - m_2)(h_2 - h_3)}{(h_0 - h_{f_6})}.$$

OBJECTIVE TYPE QUESTIONS

Choose the Correct Answer :

1. Rankine cycle efficiency of a good steam power plant may be in the range of
   (*a*) 15 to 20% (*b*) 35 to 45%
   (*c*) 70 to 80% (*d*) 90 to 95% .
2. Rankine cycle operating on low pressure limit of p_1 and high pressure limit of p_2
   (*a*) has higher thermal efficiency than the Carnot cycle operating between same pressure limits
   (*b*) has lower thermal efficiency than Carnot cycle operating between same pressure limits
   (*c*) has same thermal efficiency as Carnot cycle operating between same pressure limits
   (*d*) may be more or less depending upon the magnitudes of p_1 and p_2.
3. Rankine efficiency of a steam power plant
   (*a*) improves in summer as compared to that in winter
   (*b*) improves in winter as compared to that in summer
   (*c*) is unaffected by climatic conditions
   (*d*) none of the above.
4. Rankine cycle comprises of
   (*a*) two isentropic processes and two constant volume processes
   (*b*) two isentropic processes and two constant pressure processes
   (*c*) two isothermal processes and two constant pressure processes
   (*d*) none of the above.
5. In Rankine cycle the work output from the turbine is given by
   (*a*) change of internal energy between inlet and outlet
   (*b*) change of enthalpy between inlet and outlet

(*c*) change of entropy between inlet and outlet
(*d*) change of temperature between inlet and outlet.

6. Regenerative heating *i.e.,* bleeding steam to reheat feed water to boiler
(*a*) decreases thermal efficiency of the cycle
(*b*) increases thermal efficiency of the cycle
(*c*) does not affect thermal efficiency of the cycle
(*d*) may increase or decrease thermal efficiency of the cycle depending upon the point of extraction of steam.

7. Regenerative cycle thermal efficiency
(*a*) is always greater than simple Rankine thermal efficiency
(*b*) is greater than simple Rankine cycle thermal efficiency only when steam is bled at particular pressure
(*c*) is same as simple Rankine cycle thermal efficiency
(*d*) is always less than simple Rankine cycle thermal efficiency.

8. In a regenerative feed heating cycle, the optimum value of the fraction of steam extracted for feed heating
(*a*) decreases with increase in Rankine cycle efficiency
(*b*) increases with increase in Rankine cycle efficiency
(*c*) is unaffected by increase in Rankine cycle efficiency
(*d*) none of the above.

9. In a regenerative feed heating cycle, the greatest economy is affected
(*a*) when steam is extracted from only one suitable point of steam turbine
(*b*) when steam is extracted from several places in different stages of steam turbine
(*c*) when steam is extracted only from the last stage of steam turbine
(*d*) when steam is extracted only from the first stage of steam turbine.

10. The maximum percentage gain in Regenerative feed heating cycle thermal efficiency
(*a*) increases with number of feed heaters increasing
(*b*) decreases with number of feed heaters increasing
(*c*) remains same unaffected by number of feed heaters
(*d*) none of the above.

ANSWERS

1. (*b*)	**2.** (*a*)	**3.** (*b*)	**4.** (*b*)
5. (*b*)	**6.** (*b*)	**7.** (*a*)	**8.** (*b*)
9. (*b*)	**10.** (*a*).		

THEORETICAL QUESTIONS

1. Explain the various operation of a Carnot cycle. Also represent it on a *T-s* and *p-v* diagrams.
2. Describe the different operations of Rankine cycle. Derive also the expression for its efficiency.
3. State the methods of increasing the thermal efficiency of a Rankine cycle.
4. Explain with the help of neat diagram a 'Regenerative Cycle'. Derive also an expression for its thermal efficiency.
5. State the advantages of regenerative cycle/simple Rankine cycle.
6. Explain with a neat diagram the working of a Binary vapour cycle.

UNSOLVED EXAMPLES

1. A simple Rankine cycle works between pressure of 30 bar and 0.04 bar, the initial condition of steam being dry saturated, calculate the cycle efficiency, work ratio and specific steam consumption.
[**Ans.** 35%, 0.997, 3.84 kg/kWh]

2. A steam power plant works between 40 bar and 0.05 bar. If the steam supplied is dry saturated and the cycle of operation is Rankine, find :
(*i*) Cycle efficiency
(*ii*) Specific steam consumption.
[**Ans.** (*i*) 35.5%, (*ii*) 3.8 kg/kWh]

3. Compare the Rankine efficiency of a high pressure plant operating from 80 bar and 400°C and a low pressure plant operating from 40 bar 400°C, if the condenser pressure in both cases is 0.07 bar. [**Ans.** 0.391 and 0.357]

4. A steam power plant working on Rankine cycle has the range of operation from 40 bar dry saturated to 0.05 bar. Determine :
(*i*) The cycle efficiency (*ii*) Work ratio
(*iii*) Specific fuel consumption.
[**Ans.** (*i*) 34.64%, (*ii*) 0.9957, (*iii*) 3.8 kg/kWh]

5. In a Rankine cycle, the steam at inlet to turbine is saturated at a pressure of 30 bar and the exhaust pressure is 0.25 bar. Determine :
(*i*) The pump work (*ii*) Turbine work
(*iii*) Rankine efficiency (*iv*) Condenser heat flow
(*v*) Dryness at the end of expansion.
Assume flow rate of 10 kg/s.
[**Ans.** (*i*) 30 kW, (*ii*) 7410 kW, (*iii*) 29.2%, (*iv*) 17900 kW, (*v*) 0.763]

6. In a regenerative cycle the inlet conditions are 40 bar and 400°C. Steam is bled at 10 bar in regenerative heating. The exit pressure is 0.8 bar. Neglecting pump work determine the efficiency of the cycle. [**Ans.** 0.296]

7. A turbine with one bleeding for regenerative heating of feed water is admitted with steam having enthalpy of 3200 kJ/kg and the exhausted steam has an enthalpy of 2200 kJ/kg. The ideal regenerative feed water heater is fed with 11350 kg/h of bled steam at 3.5 bar (whose enthalpy is 2600 kJ/h). The feed water (condensate from the condenser) with an enthalpy of 134 kJ/kg is pumped to the heater. It leaves the heater dry saturated at 3.5 bar. Determine the power developed by the turbine.
[**Ans.** 16015 kW]

8. A binary-vapour cycle operates on mercury and steam. Saturated mercury vapour at 4.5 bar is supplied to the mercury turbine, from which it exhaust at 0.04 bar. The mercury condenser generates saturated steam at 15 bar which is expanded in a steam turbine to 0.04 bar.

(*i*) Find the overall efficiency of the cycle.

(*ii*) If 50000 kg/h of steam flows through the steam turbine, what is the flow through the mercury turbine ?

(*iii*) Assuming that all processes are reversible, what is the useful work done in the binary vapour cycle for the specified steam flow ?

(*iv*) If the steam leaving the mercury condenser is superheated to a temperature of 300°C in a superheater located in the mercury boiler, and if the internal efficiencies of the mercury and steam turbines are 0.85 and 0.87 respectively, calculate the overall efficiency of the cycle. The properties of saturated mercury are given below :

p (bar)	*t* (°C)	h_f	h_g	s_f	s_g	v_f	v_g
		(kJ/kg)		(kJ/kg K)		(m^3/kg)	
4.5	450	63.93	355.98	0.1352	0.5397	79.9×10^{-6}	0.068
0.04	216.9	29.98	329.85	0.0808	0.6925	76.5×10^{-3}	5.178

[**Ans.** (*i*) 52.94%, (*ii*) 59.35×10^4 kg/h, (*iii*) 28.49 MW, (*iv*) 46.2%]

Reciprocating Steam Engine

16

16.1. General aspects of heat engines. 16.2. Definition and classification of a reciprocating steam engine. 16.3. Steam engine parts and their description. 16.4. Working of a steam engine. 16.5. Steam engine terminology. 16.6. Hypothetical or theoretical indicator diagram. 16.7. Actual indicator diagram and diagram factor. 16.8. Methods of reducing condensation. 16.9. Mean effective pressure (m.e.p. or p_m). 16.10. Engine indicators—Definition and uses—Type of indicators—Crosby pencil indicator. **16.11. Indicated power (I.P.). 16.12. Brake power (B.P.). 16.13. Efficiencies of steam engine. 16.14. Mass of steam in cylinder. 16.15. Saturation curve and missing quantity. 16.16. Governing of steam engines. 16.17. Valves. 16.18. Heat balance sheet. 16.19. Performance curves**—Worked Examples—Additional/Typical Examples. *Highlights—Objective Type Questions—Theoretical Questions—Unsolved Examples.*

16.1. GENERAL ASPECTS OF HEAT ENGINES

During the last two hundred years men have been building machines capable of converting the chemical energy in fuels into mechanical energy, machines capable of performing mechanical work which may be used to drive other machines.

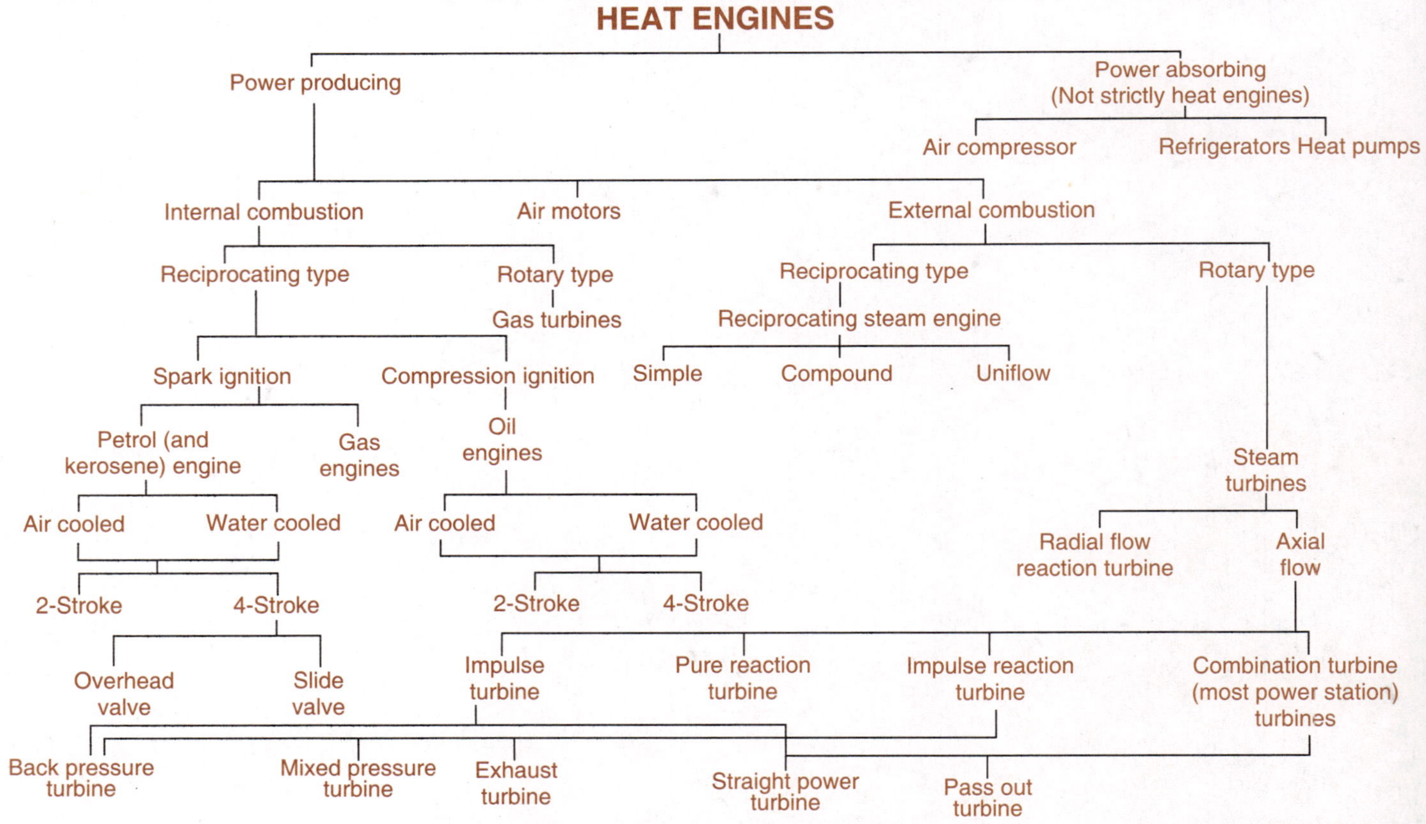

Fig. 16.1. Classification of heat engines.

Such machines are called **prime movers,** and the heat engine is at the present time, the basic prime-mover. It is an *energy converter, and transforms the chemical energy in the fuel supplied to it into mechanical work*. The heat engine is not the only prime-mover, there are many other types. An electric motor, for example, converts the electrical energy brought to it into mechanical work, and/is therefore, a prime mover. The electricity supplied to the motor, however, was most likely produced from a generator driven by a heat engine. The *heat engine* is therefore what might be termed **"a prime mover of the first order",** since it converts energy direct from a natural source into mechanical energy. On the other hand, the electric motor might then be called **"a prime mover of the second order"** since the energy which it converts into work is derived from a natural source through the agency of the other machines.

16.2. DEFINITION AND CLASSIFICATION OF A RECIPROCATING STEAM ENGINE

A reciprocating steam engine is a form of heat engine where *heat energy is converted into mechanical work due to the action of steam upon a reciprocating piston.* A steam boiler supplies steam to the engine and the steam is exhausted to either atmosphere or into the condenser. Where generation of large power is involved steam engines are now being replaced by steam turbines.

The steam engines find their applications in locomotives, ships and in the other fields where *small power* and *low speeds* are required. Its *main advantage* is that engine speed can be *reversed*. However, it has a *low thermal efficiency*.

The reciprocating steam engines are classified as follows :

1. **According to class of service :**

(*i*) Stationary

(*ii*) Marine

(*iii*) Locomotive

(*iv*) Pumping or hoisting (as per the use engine is to be put).

2. **According to speed (rotative) :**

(*i*) High speed (above 300 r.p.m.)

(*ii*) Medium speed (between 125 r.p.m. and 300 r.p.m.)

(*iii*) Low speed (below 125 r.p.m.).

3. **According to arrangement of cylinders :**

(*i*) Vertical

(*ii*) Inclined

(*iii*) Horizontal.

(The common form of vertical engine with the cylinder above the crankshaft is known as a *vertical inverted engine*).

4. **According to type of valve design :**

(*i*) Simple D valve

(*ii*) Balance plate valve

(*iii*) Piston valve

(*iv*) Riding cut-off valve

(*v*) Poppet valve

(*vi*) Corliss valve

(*vii*) Poppet type.

5. **According to method of governing :**

(*i*) Manual or automatic

(*ii*) Throttle or cut-off governing

(*iii*) Centrifugal or inertia type of governor.

6. **According to type of exhaust :**

(*i*) Condensing (meaning an absolute exhaust pressure below atmospheric)

(*ii*) Non-condensing (meaning exhaust pressure at or above atmospheric).

7. **According to approximate type of steam condition :**

(*i*) Saturated or superheated

(*ii*) High or low pressure (although 'high' pressures as known in turbine work is generally well out of the range of pressure in steam engines cylinders).

8. **According to the action of steam upon the piston :**

(*i*) Single acting (steam acts on one side of the piston and the other side is open to atmosphere).

(*ii*) Double acting (steam acts alternately on both sides of the piston).

9. **According to the nature of expansion :**

(*i*) Expansive engines (used for all industrial purposes for the power development etc.).

(*ii*) Non-expansive engines (used in limited cases *e.g.*, direct acting pumps, rolling mills and winches.)

10. **According to the range of expansion of steam :**

(*i*) Simple steam engine (expansion of steam takes place in a single cylinder)

(*ii*) Compound steam engine.

Note. While the reciprocating steam engine and its counter part, the direct-acting steam pump, are rapidly becoming obsolete, a thermodynamic study of the factors affecting their performance is desirable from the view point of fundamental engineering knowledge and from that of learning the limitations to which this type of prime mover can be developed. *Briefly, the steam engine may be said to be characterised by its ruggedness, by its high torque at slow speed, and by its practical inability to expand steam to a pressure lower than 0.14 bar in the cylinder because of the huge volume that would be required.*

16.3. STEAM ENGINE PARTS AND THEIR DESCRIPTION

Fig. 16.2 shows a schematic diagram of a single cylinder, double-acting, non-condensing and horizontal reciprocating steam engine. The various parts are discussed below :

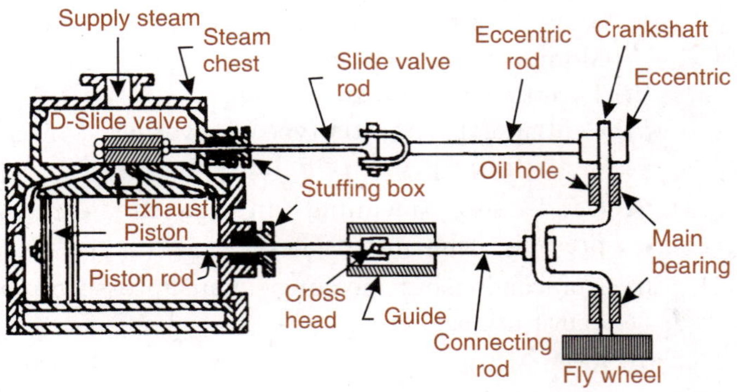

Fig. 16.2. Reciprocating steam engine.

Engine cylinder. The engine cylinder is made of cast-iron and is bored out perfectly true. It is integral with the frame in small engines. The casting of cylinder consists of two chambers : (*i*) cylinder chamber and (*ii*) valve chamber or valve chest. The cylinder is connected at each of its end with the steam chest by means of steam ports. The steam reaches the cylinder after passing through the steam ports. The exhaust port is located midway, the steam-ports which lead the used steam to the atmosphere or the condenser depending upon, whether the steam engine is *non-condensing* or *condensing*. The steam ports should be fairly large in cross-section to admit enough steam into the cylinder during the time the steam port is open otherwise steam will be throttled or wiredrawn.

One of the cylinder ends is fully covered by means of a cover known as **cover end** or **top end** or **front end.** The other end through which the piston rod passes is known as **bottom end** or **back end** or **crank end.** At this back-end cover, the piston rod passes through the *stuffing box* which makes the hole steam-tight and *prevents the leakage of steam*.

The steam engine cylinder is generally *steam jacketed in order to reduce the condensation of steam* as the steam cools during expansion. The condensation of steam is further reduced by covering the cylinder with non-conducting material known as **lagging.** The lagging is then covered with a plain thin sheet to give the cylinder a finished appearance.

In certain cases liners are fitted in the cylinder. The liner is an internal barrel made of steel or cast-iron. When the wear of the cylinder becomes too much, the whole of the cylinder block may have to be replaced which is decidedly a costly affair. In case the liner wears out too much, the replacement of liner alone is much cheaper than the whole cylinder block.

Steam chest. It is reservoir of steam from which the steam is admitted into the cylinder during the admission stroke of the piston. It is cast integral with the cylinder and is closed with a cover known as the steam chest cover through which the slide valve rod passes.

Piston and piston rings. *The function of a piston is to convert the pressure energy of steam into its reciprocating movement*. It moves from end to end of the cylinder. It is made of cast iron, cast steel or forged steel. The piston must form a steam-tight division between the two ends of a cylinder. This is accomplished by putting piston rings (generally made of rectangular section) on the piston. At least two rings should be fitted. The rings are generally of cast-iron. In addition to preventing leakage of steam, piston rings reduce considerably the friction between the piston and walls.

Solid type and hollow or box type pistons are shown in Figs. 16.3 and 16.4 respectively.

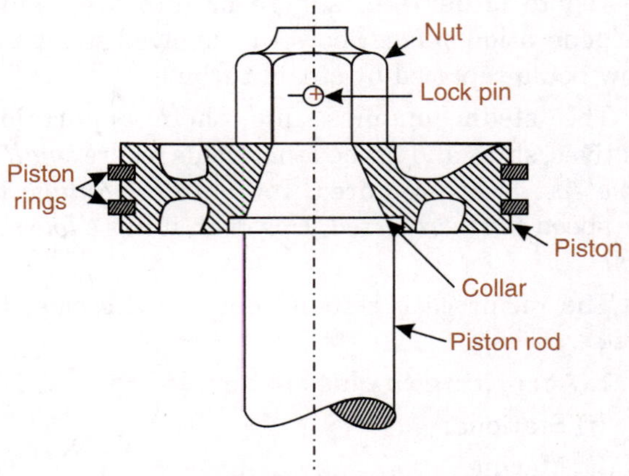

Fig. 16.3. Solid type piston.

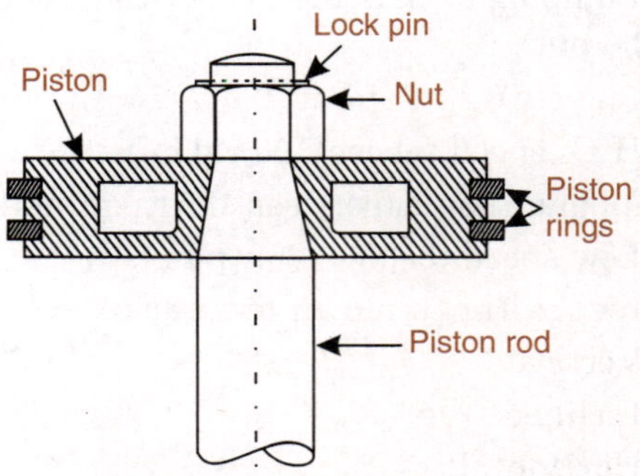

Fig. 16.4. Hollow or box type piston.

Stuffing box. *The function of a stuffing box* (Fig. 16.5) *is to prevent the leakage of fluid where a sliding or rotating* *piece passes through a vessel containing fluid at a certain pressure.* In a steam engine it is placed at the cylinder end and the steam chest end and thus allows the reciprocating movement of the piston rod and the slide valve rod without and leakage of steam. The stuffing box if cast integral with the cylinder is made of cast-iron.

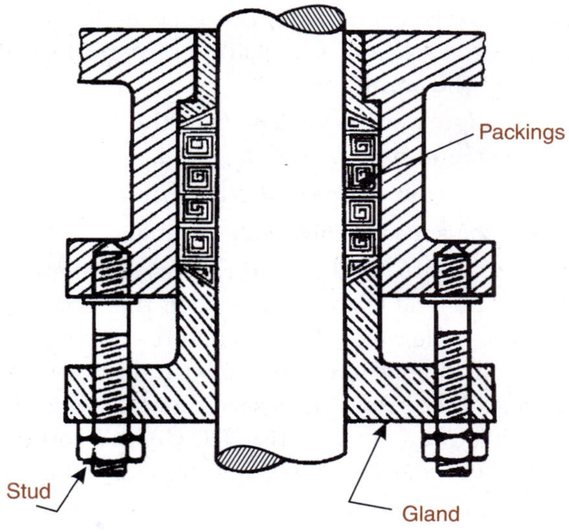

Fig. 16.5. Stuffing box.

The leakage is prevented due to packing between the stuffing box and the gland. The gland is tightened by means of studs and nuts in order to keep the packing in the compressed position. The gland may be made of gun metal or lined with gun-metal bush if the gland is made of cast iron.

There are many types of packings. *Asbestos* is more suitable for superheated steam than greased hemp rope or leather. In case of high pressure engines using superheated steam, soft packings are spoiled under these conditions. So under such circumstances, *metallic packings* consisting of sets of block of white metal are used.

Cross-head and guide. It connects the outer end of the piston rod to the small end of the connecting rod. It is made of wrought iron, forged steel or cast steel. The cross-head slides in between the parallel guides and receives its reciprocating motion from the reciprocating motion of the piston rod. The connecting rod oscillates about its small end in the cross-head and is secured by means of a gudgeon pin. The piston rod is joined to the cross-head either by screwing or by cotter joint or by forging the piston rod to the cross-head end. The guide may be cast with the frame of the cylinder. Small size cross-heads are generally forged as one with the piston rod.

Connecting rod. *The function of a connecting rod is to convert the reciprocating motion of the piston rod and cross-head into circular motion of the crankshaft.* It is made of steel forging of rectangular, *I* or *H* section or circular section. Its motion is oscillating. Its one end is connected to the cross-head by means of a pin known as **gudgeon pin** or **wrist pin** and the other end is connected to the crank by means of a pin known as **crank pin.** The end of the connecting rod which is connected to the crank is known as big end and the end connected to the cross-head is known as small end. The lubrication of the big end bearing may be done by means of "Forced lubrication". The oil under pressure reaches the crank pin by the suitable arrangement. The lubrication of the big end and the small end must be highly satisfactory for the proper running of the engine.

The length of the connecting rod varies from about 2 to 6 times the length of the piston stroke and is measured from the centre of the cross-head pin to the centre of the crank pin.

Crank and the crankshaft. *The function of a crank is to convert the reciprocating motion of the piston into rotary motion of the crankshaft.* The distance between the axis of the crankshaft and crank pin is called the radius of the crank. *Twice the radius of the crank is equal to the stroke of the piston.* The crankshaft after receiving the rotary motion is used to run a variety of systems *e.g.,* generators, pulleys, wheels and pumps etc. It is made of mild steel.

Main bearings. *A bearing is a device which supports the revolving shaft or part of a machine and confines its motion.* The crankshaft is supported by the main bearings. *The portion of the crankshaft rotating in the main bearings is called* a **journal.** The bearings used in the main bearing is of split-type *i.e.,* in two parts, so that where they are required to be replaced, they do not present much difficulty. The main bearings may be lubricated by ring oiling, needle lubrication, wick lubrication, drop lubricator or forced lubrication.

Eccentric. Refer to Fig. 16.6. It converts the rotary motion of the crankshaft into the reciprocating motion of the slide valve. It is fixed on the crankshaft. An eccentric consists of the 'sheave' and the 'strap'. The sheave is a cylindrical grooved pulley fixed on the crankshaft by means of a key. The sheave is hollowed at two places in order to reduce its weight and save the material. The distance between the centre of the sheave and the centre of the crankshaft is called **'eccentricity'.** *Twice the eccentricity is the travel or stroke of the valve or "Throw of eccentric".* The strap is generally in two parts held together by means of bolts. It may be made of cast iron, brass, bronze or steel.

When the sheave revolves the straps move to and fro, conveying this motion to the eccentric rod which is connected to the valve rod by means of a pin. Thus, the necessary reciprocating motion for a slide valve is obtained from the rotary motion of the crankshaft of the engine with the help of the eccentric.

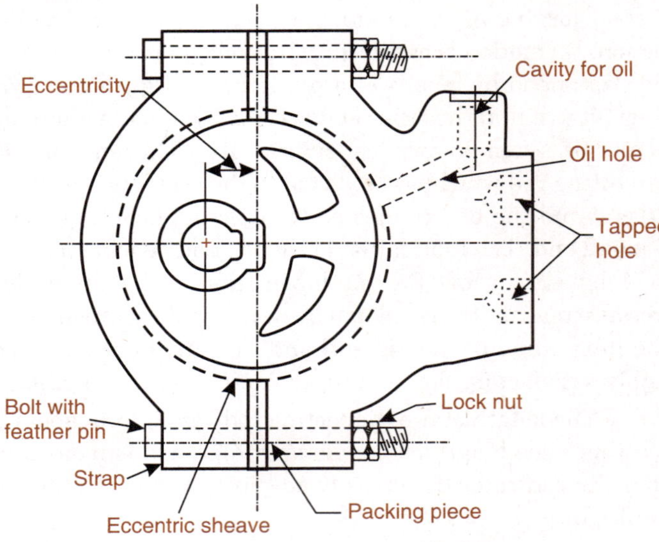

Fig. 16.6. Eccentric.

Flywheel. The flywheel is used to maintain the constant angular motion of the crank during the stroke of the engine. It is made of cast iron. Its shape is like a disc and its weight depends upon the pressure variation on the piston during the cycle. When the engine develops power, the flywheel stores a part of its energy and returns this energy to the crankshaft during the idle period of the cycle. Thus, *a flywheel produces a steady rotation of the crankshaft and avoids the jerks over a cycle.*

Governor. The governor is used to keep the speed of the crankshaft constant over a period when *there is variation of load on the engine. The governor increases or decreases the supply of steam in the engine cylinder as the load increase or decreases in order to maintain the speed of the crankshaft constant*. The steam engine may have throttle governing arrangement also (in which pressure of the steam supplied to the engine is controlled).

16.4. WORKING OF A STEAM ENGINE

Refer to Fig. 16.2. Due to the pressure of the steam over the piston, it reciprocates inside the cylinder. The reciprocating motion of the piston is converted into the circular motion of the crankshaft through the piston rod and the connecting rod. The D-slide valve again gets the reciprocating motion from the crankshaft by means of an eccentric. The circular motion of the crankshaft is converted into the reciprocating motion of the slide valve through the eccentric, eccentric rod and the D-slide valve rod.

16.5. STEAM ENGINE TERMINOLOGY

Cylinder bore (D). It is the inside diameter of the cylinder or the liner.

Piston stroke (L). It is the distance travelled by the piston from the cover end of the cylinder to crank end of the cylinder. The piston makes a forward or out stroke when it moves from the cover end to the crank end and a return or back stroke when it moves from crank end to cover end.

Crank throw or crank radius. It is the distance between the *centre of the crankshaft and the centre of the crank-pin*. This is equal to *half the travel of the piston* (*i.e.*, $r = L/2$). Two strokes of the piston are performed in one revolution of the crankshaft.

Average piston speed. It is the *linear distance travelled by the piston per minute*, and is given by

$$\text{Piston speed} = 2LN \text{ m/min.}$$

where N is the speed of the crank in r.p.m.

Dead centres. The dead centres are the positions of piston at the end of the stroke when the centre lines of the piston rod, the connecting rod and the crank are in the same straight line. There are two such dead centres, one for each end of the stroke. In case of vertical steam engines, they are called top dead centre (T.D.C.) and bottom dead centre (B.D.C.) and inner dead centre (I.D.C.) and outer dead centre (O.D.C.) in case of horizontal engines.

Swept volume (V_s). The volume swept by the piston when it moves from one dead centre position to another dead centre position is called *swept volume.* The crank completes half the revolutions during this period. It is given by

$$V_s = A \times L$$

where A is the area of the piston on which steam acts

$$(= \pi/4\, D^2).$$

It is also called ***piston displacement.***

Clearance. The short distance between the piston and the cylinder cover is called linear piston clearance (L_c). *The volume of space between the piston and the cylinder cover, when the piston is at the end of the stroke, plus the volume of the port leading to this space is called volumetric clearance* (V_c). It is usually expressed as percentage of swept volume.

$$V_c = V_p + \frac{\pi}{4} D^2 . L_c \quad ...(16.1)$$

where, V_p = port volume.

If V_p is neglected, then

$$V_c = \frac{\pi}{4} D^2 L_c \quad ...(16.2)$$

Piston clearance serves the following purposes :

(*i*) *It checks the piston striking cylinder head.*

(*ii*) *It allows small quantity of water to be collected due to steam condensation without excessive rise of pressure.*

(*iii*) *It provides cushioning effect to the fast moving piston when it changes its direction of motion.*

Eccentric throw or eccentricity. It is the *distance between the centre of the eccentric and the centre of the crankshaft.*

Valve travel. The total distance that the valve travels in one direction is called *valve travel. It is equal to twice the throw of the eccentric.*

Back pressure (p_b). The steam pressure which acts on the exhaust side of the piston is known as *back pressure. It reduces the network done on the piston during a stroke.*

16.6. HYPOTHETICAL OR THEORETICAL INDICATOR DIAGRAM

An *indicator diagram* is a plot of the variations of the steam pressure and volume of the steam in the cylinder during the cycle of operations. The hypothetical or theoretical indicator diagram is based upon the following **assumptions :**

(*i*) The steam is admitted to the engine cylinder at boiler pressure and exhausted at condenser pressure.

(*ii*) There is instantaneous opening and closing of the ports.

(*iii*) The expansion is hyperbolic (*i.e.*, pV = constant).

(*iv*) No wire drawing due to restricted valve opening.

(*v*) No compression after exhaust.

(*vi*) No pressure drop due to condensation of steam (due to the difference in temperature of the steam and the cylinder walls).

Fig. 16.7 shows a hypothetical diagram with clearance neglected. The sequence of operations is as follows :

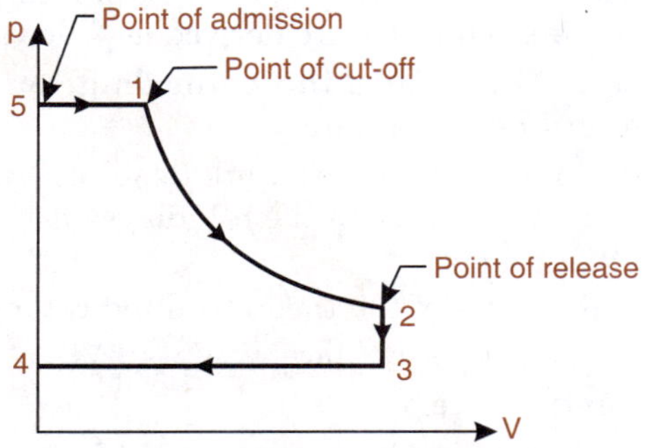

Fig. 16.7. Hypothetical diagram with clearance neglected.

Operation 5–1. The steam is admitted at boiler pressure in the engine cylinder. At point 1, steam supply is cut-off and is known as **point of cut-off.**

Operation 1–2. During the operation steam expands in the engine cylinder to the end of the stroke.

Operation 2–3. At point 2, the exhaust port is opened and the steam is released and the pressure now drops suddenly from 2 to 3 (condenser pressure which is the back pressure on the piston). The point 2 is known as **point of release.**

Operation 3–4. During this operation wet steam is exhausted into the condenser.

At point 4, the piston reaches its extreme position enclosing the clearance volume and fresh steam is admitted into the engine cylinder (point 5) and next cycle repeats.

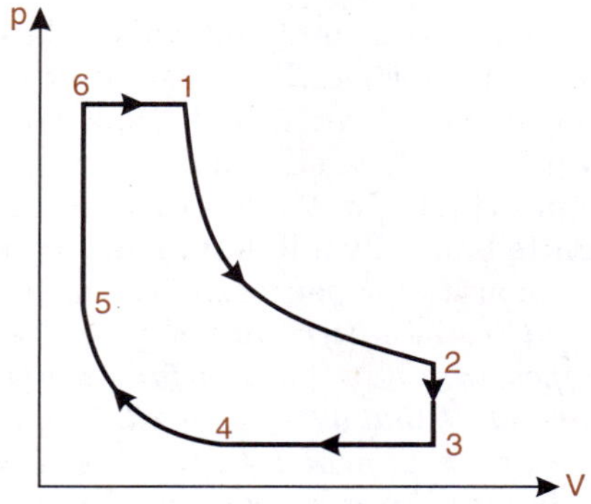

Fig. 16.8. Hypothetical diagram when clearance and compression are considered.

Fig. 16.8 shows hypothetical indicator diagram *when clearance and compression are considered*. The compression starts at 4 and finishes at 5 (which is also point of admission). The admission of steam into the engine cylinder continues upto 1. The curve 1–2 is expansion curve. Release takes place at the point 2 and 3–4 represents the exhaust line.

The expansion curve (1–2) and compression curves (4–5) are assumed to follow the law

$$pV = C \ (i.e., \text{ hyperbolic}).$$

16.7. ACTUAL INDICATOR DIAGRAM AND DIAGRAM FACTOR

The actual indicator diagram is *different* from the theoretical *due to the timings of the valve action and the pressure drop due to friction in the admission port* (Fig. 16.9). *The valve action gives rounding of the diagram since processes cannot be started or terminated instantaneously.* The exhaust process must begin before the end of the stroke to allow the time required for blow-down (*i.e.*, release of pressure). The cushion steam has the practical function of cushioning the reciprocating parts, hence reducing the inertia load on the parts and bearings when the piston changes direction at the end of the stroke.

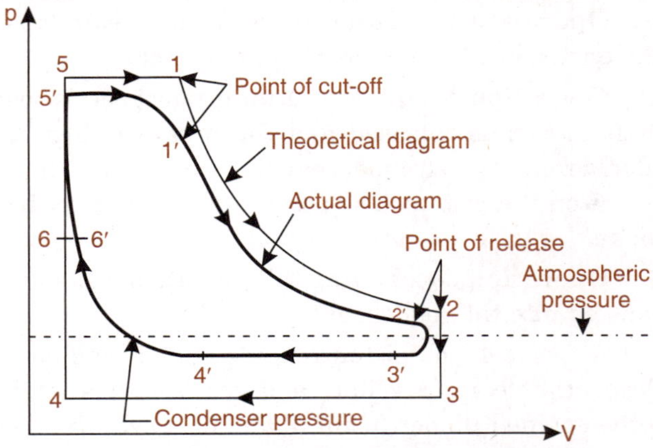

Fig. 16.9. Actual indicator diagram.

Owing to heat transfer which takes place, the actual expansion is not isentropic. The amount of heat transfer depends on the temperature range through which the steam passes between inlet and exhaust, and the mean temperature of cylinder walls. Some condensation occurs in steam admitted initially if it is supplied dry saturated or with a small amount of superheat. *Some of the condensate is re-evaporated in the latter part of the expansion stroke when the temperature of the steam falls below that of the cylinder. The result is that work done is less than that which would be given by the ideal isentropic expansion.* When condensation occurs there is reduction in specific volume and a drop in pressure. The re-evaporation process is the reverse of this, but the recovery is late in the stroke and little use can be made of it. When condensation begins, the rate of heat transfer to the cylinder walls is increased, and this promotes further condensation. The expansion can be made more efficient by limiting the temperature and pressure range through which the steam falls.

'Diagram factor' *is defined as the ratio of the area of actual indicator diagram and the area of the hypothetical indicator diagram.*

Diagram factor

$$= \frac{\text{Area of actual indicator diagram}}{\text{Area of hypothetical indicator diagram}}$$

$$= \frac{p_m \text{ (actual)}}{p_m \text{ (theoretical)}}$$

The diagram factor varies from 0.63 to 0.86, *the higher values being obtained with steam jacketing of the cylinder or by using superheated steam.* The clearance volume in steam engines varies between 1.5% to 15% of swept volume. Within this range the value of diagram factor depends on the particular design, especially of the slide valve and the quality of the manufacture of the engine.

16.8. METHODS OF REDUCING CONDENSATION

To reduce cylinder condensation in a steam engine following methods are employed :

1. **By superheating.** The change in heat content, in the case of a superheated steam, is accompanied by a change in temperature. It has also been found that a reduction in temperature of steam tends to suppress the heat loss. Such things are not applicable to saturated steam. Moreover, wet steam has got better heat transmission capacities and due to its greater density the leakage is more. Considering these facts the method of superheating the steam, so that the moisture can be removed, is used for reducing the cylinder condensation.

2. **By using jacketed cylinder.** The cylinder condensation can also be reduced by using steam jacketed cylinder. But due to its limited surface, temperature differences and the short time available for the flow of heat, it is effective only in the case of slow speed engines.

3. **By compounding the cylinders.** The condensation of steam in the engine cylinder can be reduced by compounding cylinders; because by doing so that range of variation of temperature in each cylinder is reduced.

4. **By increasing the speed of the engine.** The engines which run at higher rotational speeds have small dimensions for developing the same power. Thus, with increased speed the cylinder surface which takes part in heat transfer and time during which heat transfer takes place, both are reduced; consequently condensation is reduced.

5. **By large compression of steam during last part of exhaust.** Finally the cylinder condensation can also be reduced by large compression of steam during last part of exhaust. Due to this process there is an increase in temperature of steam as well as the cylinder wall temperature.

16.9. MEAN EFFECTIVE PRESSURE (m.e.p. or p_m)

It is the constant uniform pressure which if acting on the face of the piston throughout the stroke would have produced the same work area, as obtained under actual working conditions. In simple words, it is the *mean height of the indicator diagram*. The expressions of mean effective pressure for different cases are derived as follows :

Case 1. Theoretical indicator diagram without clearance and compression.

Refer to Fig. 16.10. The work done per cycle by an engine is equal to the area of the *p-V* diagram.

∴ Work done,

W = Area of the theoretical indicator diagram

$$= p_1V_1 + p_1V_1 \log_e V_2/V_1 - p_bV_2$$

Theoretical m.e.p.

$$= \frac{\text{Theoretical work done}}{\text{Stroke volume}}$$

$$= \frac{p_1V_1 + p_1V_1 \log_e V_2/V_1 - p_bV_2}{V_2}$$

$$= \frac{p_1V_1}{V_2}(1 + \log_e V_2/V_1) - p_b$$

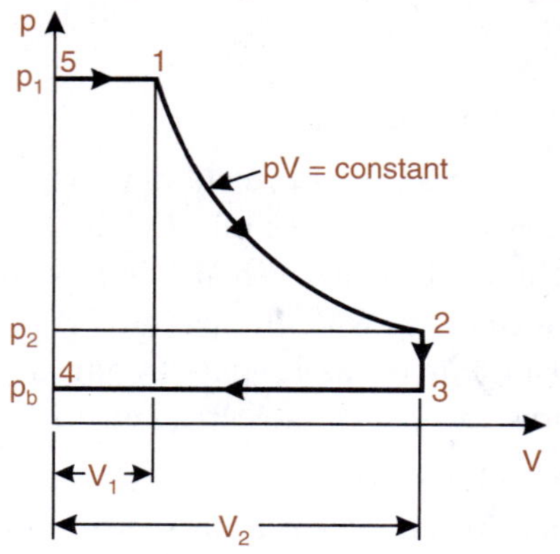

Fig. 16.10

The expansion ratio, r, is defined as $\frac{V_2}{V_1}$, therefore,

Theoretical/Hypothetical m.e.p.,

$$p_m = \frac{p_1}{r}(1 + \log_e r) - p_b \quad ...(16.3)$$

Note. (*i*) The ratio $\frac{V_1}{V_2} = \frac{1}{r}$ = cut-off ratio.

"Cut-off ratio *is the ratio of volume between the points of admission and cut-off to the swept volume*".

(*ii*) A hypothetical steam consumption can be estimated using the specific volume, v, of the inlet steam, and the volume induced per cycle, V_1,

i.e., Steam consumption/cycle = $\frac{V_1}{v}$ kg.

Case 2. Theoretical indicator diagram with clearance.

Refer to Fig. 16.11.

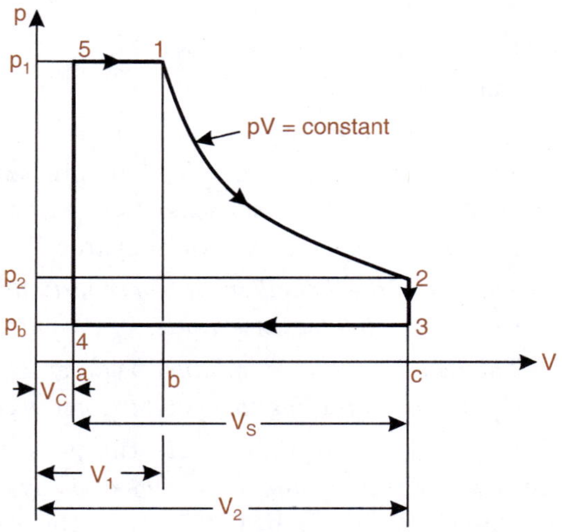

Fig. 16.11

Work done, W = Area of indicator diagram

= Area '512345'

= Area '51ba5' + Area '12cb1' – Area 43ca4

$= p_1(V_1 - V_c) + p_1 V_1 \log_e V_2/V_1 - p_b(V_2 - V_c)$

$[V_c = V_5 = V_4 = \text{Clearance volume}]$

$= p_1(V_1 - V_c) + p_1 V_1 \log_e V_2/V_1 - p_b . V_s \quad ...(16.4)$

$(\because \quad V_2 - V_c = V_s = \text{Swept volume})$

Let c = Ratio of clearance volume to swept volume

$$= V_c / V_s \quad ...(16.5)$$

Also, $V_s = V_2 - V_c \quad \therefore \quad V_2 = V_s + V_c$

Cut-off ratio, $\frac{1}{r} = \frac{V_1 - V_c}{V_s} \quad \therefore \quad V_1 = \frac{V_s}{r} + V_c$

Inserting the value of V_c from eqn. (16.5), we get

$$V_1 = \frac{V_s}{r} + c \,.\, V_s$$

or

$$V_1 = V_s\left(c + \frac{1}{r}\right) \quad ...(16.6)$$

Inserting the values of V_c, V_2 and V_1 in eqn. (16.4), we get

$$W = p_1\left[V_s\left(c + \frac{1}{r}\right) - c\,V_s\right] + p_1\left(c + \frac{1}{r}\right) V_s \log_e\left[\frac{V_s(1 + c)}{V_s\left(c + \frac{1}{r}\right)}\right] - p_b V_s$$

$$= p_1 V_s\left[c + \frac{1}{r} - c\right] + p_1 V_s\left(c + \frac{1}{r}\right)\log_e\left[\frac{1 + c}{c + \frac{1}{r}}\right] - p_b V_s$$

$$= \frac{p_1 V_s}{r} + p_1 V_s\left(c + \frac{1}{r}\right)\log_e\left[\frac{1 + c}{c + \frac{1}{r}}\right] - p_b V_s \quad ...(16.7)$$

$$p_m = \frac{\text{Work done}}{\text{Stroke volume}}$$

$$= \frac{\frac{p_1 V_s}{r} + p_1 V_s\left(c + \frac{1}{r}\right)\log_e\left[\frac{1 + c}{c + \frac{1}{r}}\right] - p_b V_s}{V_s}$$

$$= \frac{p_1}{r} + p_1\left(c + \frac{1}{r}\right)\log_e\left[\frac{1 + c}{c + \frac{1}{r}}\right] - p_b$$

$$\therefore \quad p_m = p_1 \left[\frac{1}{r} + \left(c + \frac{1}{r} \right) \log_e \left(\frac{1+c}{c + \frac{1}{r}} \right) \right] - p_b \quad ...(16.8)$$

Case 3. Theoretical indicator diagram with clearance and compression.

Refer to Fig. 16.12.

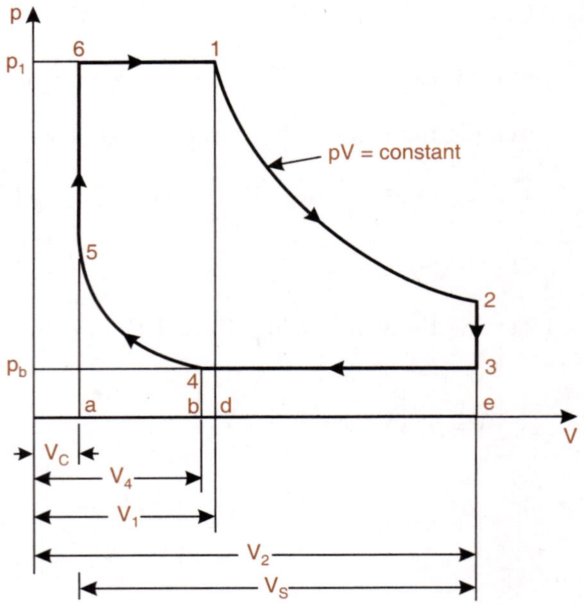

Fig. 16.12

Work done, W = Area of the indicator diagram

= Area '6123456'

= Area '61*da*6' + Area '12*ed*1'

– Area 34*be*3 – Area '45*ab*4'.

Following the same way as discussed in the case 2, we get the following expression for mean effective pressure.

$$p_m = p_1 \left[\frac{1}{r} + \left(c + \frac{1}{r} \right) \log_e \left(\frac{c+1}{c + \frac{1}{r}} \right) \right]$$

$$- p_b \left[(1 - \alpha) + (c + \alpha) \log_e \left(\frac{\alpha + c}{c} \right) \right] \quad ...(16.9)$$

where, $c = \frac{V_c}{V_s}, \frac{1}{r} = \frac{V_1 - V_c}{V_s}$

and, α = ratio of the volume between points of compression and admission to the swept volume V_s

$$= \frac{V_4 - V_c}{V_s} = \frac{V_4}{V_s} - c.$$

16.10. ENGINE INDICATORS

16.10.1. Definition and Uses

An indicator is an instrument in which *p-V* diagram of an engine is produced automatically. Originally, the indicator was invented by James Watt. Although improvement in points of detail have come up with the passage of time, the main features of the instrument as designed by him are substantially retained at present time by indicator makers.

An indicator is *used* :

1. To obtain a diagram from which conclusions may be drawn in respect of the following :
   (*i*) Behaviour of the steam in the cylinder ;
   (*ii*) The promptness of the steam admission ;
   (*iii*) The loss by fall of pressure between the boiler and the cylinder ;
   (*iv*) The loss by wire drawing ;
   (*v*) The extent and character of the expansion ;
   (*vi*) The efficiency of the arrangement for exhaust, including the extent of the back pressure, and
   (*vii*) The amount of compression.
2. To find the mean effective pressure exerted by the steam upon the piston and from there to calculate indicated power of the engine.
3. To determine whether the valves are set correctly by taking diagrams from each end of the cylinder and observing and comparing the respective positions of the points of admission, cut-off, release and compression.

16.10.2. Types of Indicators

Two simple types of indicators are :

1. Pencil indicator
2. Optical indicator.

The *pencil indicator* registers the pressure and volume by means of a *mechanism consisting of links and a spring*. The *essentials* of such an indicator are :

(*i*) *The rise of the 'Pencil' must be proportional to the rise of pressure in the cylinder of the engine.*

(*ii*) *The motion of the drum must be an exact copy of the motion of the engine piston to a reduced scale.*

The *Optical Indicator* registers the pressure and volume on *a photographic plate by means of a ray of light.* The main merit of this indicator is that the *diagram obtained is not affected by the inertia of the moving parts of any mechanism.*

16.10.3. Crosby Pencil Indicator

Fig. 16.13 shows a view of *crosby pencil indicator*. It consists of a small piston and cylinder assembly fitted on the top of the engine cylinder such that the cylinder pressure operates the indicator piston. The indicator piston is loaded with a precisely calibrated spring. The amount of movement of the piston is directly proportional to the engine cylinder pressure and this motion is amplified through a linkage mechanism, so that it is duplicated to a large scale by a pencil/stylus which moves over a drum. The drum executes a rotary motion proportional to the movement of the engine piston, obtained by means of a reducing mechanism attached to the piston rod or cross-head. As a result, the drum rotates back and forth in synchronism with the engine piston, while the stylus moves vertical distances that are proportional to the pressures in the engine cylinder.

The indicator may be placed in communication with either end of the cylinder or with the atmosphere through a three way cock.

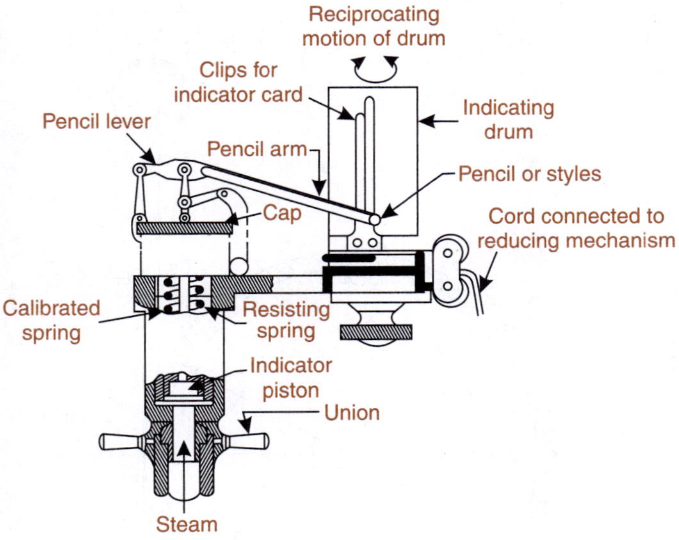

Fig. 16.13. Crosby pencil indicator.

16.11. INDICATED POWER (I.P.)

The ***indicated power*** *is the power developed in the engine cylinder*. It is so named because an indicator is used to obtain *p-V* diagram from which work done by the steam on the piston per stroke can be calculated. The I.P. can be calculated if the following data are known : (*i*) *mean effective pressure acting on the engine piston during the stroke,* (*ii*) *area of the piston,* (*iii*) *length of the stroke* and (*iv*) *number of working strokes/min.*

Let p_{m_1} = Actual mean effective pressure (bar) acting on the piston for cover end side of the engine,

p_{m_2} = Actual mean effective pressure (bar) acting on the piston for crank end side of the engine,

A_1 = Piston area on the cover and side (m^2),

A_2 = Piston area on the crank end side (m^2),

L = Length of the stroke (m), and

N = Speed of the engine (r.p.m.).

Work obtained per stroke for *cover end side*

= Force × distance moved

= (pressure × area) × distance

$= p_{m_1}\ (10^5\ \text{N/m}^2) \times A_1\ (\text{m}^2)\ L(\text{m})\ \text{Nm}$

Therefore, indicated power (*cover end side*)

$= p_{m_1}\ (10^5\ \text{N/m}^2) \times A_1\ (\text{m}^2)\ L(\text{m})\ N\ (\text{r.p.m})\ \text{Nm/min}$

$= p_{m_1} LA_1N \times 10^5\ \text{Nm/min}$

$$= \frac{p_{m_1} LA_1N \times 10^5}{60}\ \text{Nm/s}$$

$$= p_{m_1} LA_1N \times \frac{10}{6}\ \text{kJ/s} = \frac{10\, p_{m_1} LA_1N}{6}\ \text{kW} \quad ...(16.10)$$

[For a *single-acting engine* indicated power,

$$\text{I.P.} = \frac{10\, p_m LAN}{6}\ \text{kW}\Big]$$

In case of a double-acting steam engine, the total indicated power developed is the *sum* of *indicated power on cover end side and indicated power on crank end side.*

Indicated power for *crank end*

$$= \frac{10\, p_{m_2} LA_2N}{6}\ \text{kW} \quad ...(16.11)$$

where, $A_2 = A_1$ – area of piston rod

If D = diameter of piston, and

d = diameter of piston rod

Then, $$A_2 = \frac{\pi}{4} D^2 - \frac{\pi}{4} d^2 = \frac{\pi}{4} (D^2 - d^2)$$

Total indicated power developed by a *double acting steam engine*,

$$\text{I.P.} = \frac{10\, p_{m_1} LA_1N}{6} + \frac{10\, p_{m_2} LA_2N}{6} \quad ...(16.12)$$

If the area of piston rod is *neglected*, then

$$A_1 = A_2 = \frac{\pi}{4} D^2 = A$$

and $$p_{m_1} = p_{m_2} = p_m$$

Then $\text{I.P.} = 2 \times \dfrac{10 p_m LAN}{6}$

$$= \frac{10 p_m LAN}{3} \text{ kW} \quad ...(16.13)$$

Calculation of p_m :

If, A_i = Area of indicator diagram (mm^2),

S_i = Spring scale or spring number (bar/mm),

L_i = Length of indicator diagram (mm),

Then, $$p_m = \frac{\text{Area of indicator diagram} \times \text{spring number}}{\text{Length of the indicator diagram}}$$

i.e., $$p_m = \frac{A_i \times S_i}{L_i} \text{ bar} \quad ...(16.14)$$

16.12. BRAKE POWER (B.P.)

Brake power is *the actual power available from the engine for doing the useful work*. Brake power (B.P.) is always *less* than indicated (I.P.) since a part of power developed in the engine cylinder is used to overcome the *frictional losses* at different moving parts of the engine. The difference between I.P. and B.P. is called F.P. (frictional power).

$$\therefore \quad \text{I.P.} = \text{B.P.} + \text{F.P.} \quad ...(16.15)$$

Brake power of an engine can be determined by a brake of some kind applied to the brake pulley of the engine. This arrangement for determination of B.P. of the engine is known as *dynamometer*. Usually, rope brake dynamometer is used for the purpose.

Rope brake dynamometer. A schematic diagram of rope brake dynamometer is shown in Fig. 16.14. This type of dynamometer is very suitable for measuring B.P. of an engine of moderate size, it is easy to fabricate and inexpensive. It consists of rope wrapped round the brake drum or flywheel keyed to crankshaft of an engine whose B.P. is to be determined. One end of the rope is connected to the spring balance while at the other end is hung a weight *W*. Wooden blocks are incorporated (as shown) to check the rope slipping off the brake drum/flywheel. Since a lot of heat is produced (due to friction) in this arrangement, the rim of the wheel is water cooled.

The arrow shows the direction of rotation of wheel. It is evident from the figure that *W* opposes the rotation of wheel whereas spring balance *S* acts otherwise, so the *net brake load* (which opposes the rotation) is $(W - S)$.

Let W = Dead weight on the rope (newton),

S = Spring balance reading (newton),

D = Diameter of the brake drum/wheel (m),

d = Diameter of the rope (m), and

N = Speed of the engine (r.p.m.)

Net load or frictional force acting on the drum/wheel

$= (W - S)$ newton

The effective radius at which net load acts

$$= \frac{D + d}{2}$$

$\therefore$ Braking torque,

T = Frictional force × radius

$$= (W - S)\left(\frac{D + d}{2}\right) \text{ Nm.}$$

$\therefore$ Power absorbed

= Frictional torque × angle turned in one minute

$$= T \times 2\pi N$$

$$= (W - S)\left(\frac{D + d}{2}\right) \times 2\pi N \text{ Nm/min}$$

$$= \frac{(W - S)\pi(D + d)N}{60} \text{ Nm/s or J.}$$

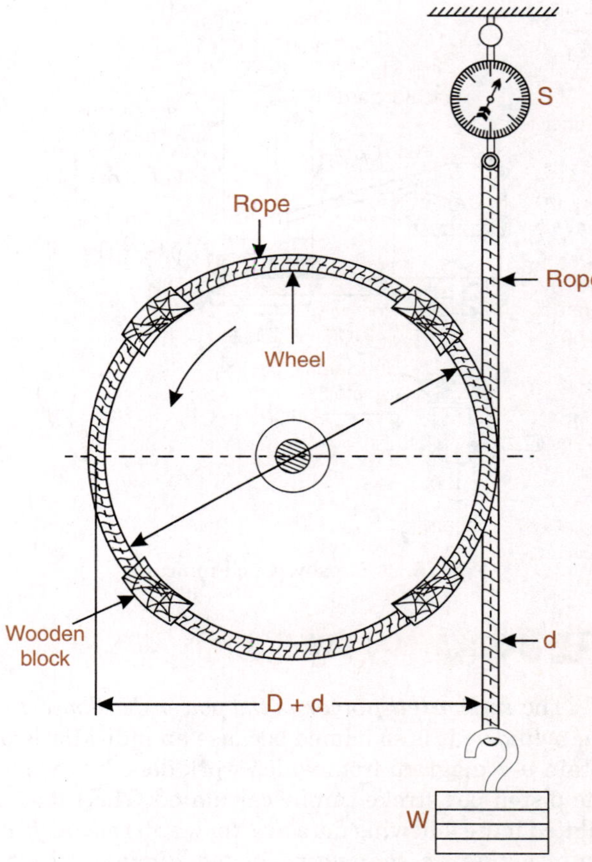

Fig. 16.14. Rope brake dynamometer.

$\therefore$ Brake power, B.P.

$$= \frac{(W - S)\pi(D + d)N}{60 \times 10^3} \text{ kW} \quad ...(16.16)$$

If diameter of the rope '*d*' is neglected, then

$$\text{B.P.} = \frac{(W - S)\,\pi\, DN}{60 \times 10^3}\ \text{kW} \qquad (\because\ d = 0) \quad ...(16.17)$$

Also if T, braking torque is given,

Then $$\text{B.P.} = \frac{2\pi NT}{60 \times 10^3}\ \text{kW.}$$

16.13. EFFICIENCIES OF STEAM ENGINE

1. **Mechanical efficiency** ($\eta_{mech.}$). It is the *ratio of brake power to indicated power.*

i.e., $$\eta_{mech.} = \frac{\text{Brake power}}{\text{Indicated power}} \quad ...(16.18)$$

2. **Thermal efficiency** (η_{th}). It is the *ratio of useful work (heat units) to the heat supplied.*

i.e., $$\eta_{th.} = \frac{\text{Useful work}}{\text{Heat supplied}} \quad ...(16.19)$$

Thermal efficiency on *indicated power (I.P.) basis* is given by :

$$\eta_{th.(i)} = \frac{\text{I.P.}}{\dot{m}_s\,(h_1 - h_{f_2})} \quad ...(16.20)$$

where, $\dot{m}_s$ = Mass of steam supplied (kg/s),

h_1 = Heat entering into the engine in kJ (per kg of steam), and

h_{f_2} = Heat coming out of condenser in kJ (per kg of water).

The eqn. (16.20) can also be expressed as follows :

$$\eta_{th.(i)} = \frac{\text{I.P.} \times 3600}{\dot{m}_s \times 3600\,(h_1 - h_{f_2})} = \frac{3600}{\dfrac{\dot{m}_s \times 3600}{\text{I.P.}} \times (h_1 - h_{f_2})}$$

The term $\dfrac{\dot{m}_s \times 3600}{\text{I.P.}}$ is known as '*specific steam consumption*' (s.s.c.).

In other words, the '*specific steam consumption*' *is the steam flow in kg/h required to develop 1 kW.*

Thermal efficiency on *brake power (B.P.) basis* is given by :

$$\eta_{th.(b)} = \frac{\text{B.P.}}{\dot{m}_s\,(h_1 - h_{f_2})} \quad ...(16.21)$$

3. **Relative efficiency.** *It is the ratio between indicated thermal efficiency and the Rankine efficiency.*

i.e., $$\eta_{\text{Relative}} = \frac{\text{Thermal efficiency (indicated)}}{\text{Rankine efficiency}} \quad ...(16.22)$$

4. **Overall efficiency** ($\eta_{overall}$). It is the *ratio of net output* (*i.e.,* brake work) *of the engine to the input of the boiler.*

i.e., $$\eta_{\text{overall}} = \frac{\text{Net output or brake work}}{\text{Input of the boiler}} = \frac{\text{B.P.}}{\dot{m}_f \times \text{C.V.}} \quad ...(16.23)$$

where $\dot{m}_f$ = Mass of fuel used/sec., and

C.V. = Calorific value of the fuel.

16.14. MASS OF STEAM IN CYLINDER

The total mass of steam in the steam engine cylinder is the *sum of mass of steam admitted per stroke and mass of steam left in the clearance volume from the previous stroke.* The former can be calculated from the mass of steam condensed by the condenser, the latter may be obtained from the *p-V* diagram of the engine.

Then *mass of steam admitted per stroke, m* is given by

$$m = \frac{\text{Mass of steam used per hour}}{\text{Number of strokes per hour}}$$

If m_c = Mass of cushion steam, then total mass of steam in the cylinder after cut-off

= $m_c + m$ (cushion steam acts as a buffer in slowing down the piston at the end of the stroke).

The mass of cushion steam can be calculated from a point on the compression curve of the indicator diagram (Fig. 16.15). In order to do this it is necessary to calibrate the indicator diagram so that the absolute pressure and volume of any point may be read off.

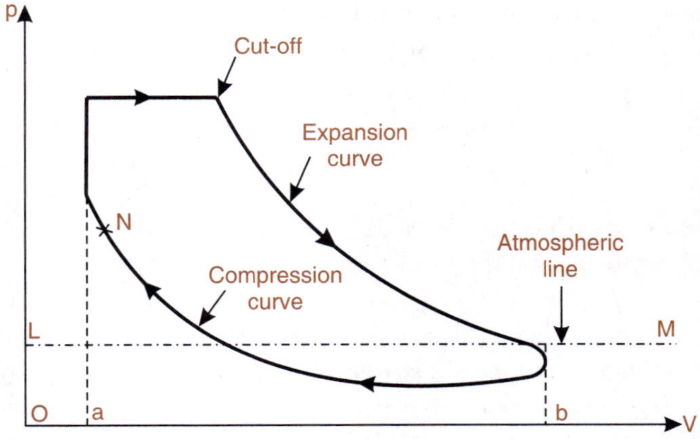

Fig. 16.15

Let *LM* represent the atmospheric line, and let *S* be the pressure (in N/m²) per cm of vertical ordinate. Draw the line *Ob* to represent absolute zero pressure such that : *OL* = atmospheric pressure/S. The pressure may now be marked on the vertical ordinate to the scale of 1 cm = S N/m² starting with zero at '0'.

The length of the diagram represents the stroke volume.

$$\therefore \quad ab = \frac{\pi}{4} D^2 \times L \qquad \begin{bmatrix} D = \text{Diameter of piston (m)} \\ L = \text{Length of stroke (m)} \end{bmatrix}$$

$$\text{Then, volume scale} = \frac{\text{Stroke volume}}{ab}$$

The distance '*oa*' represents the clearance volume to this volume scale :

$$\text{Hence,} \quad oa = \frac{\text{Clearance volume (m}^3)}{\text{Volume scale}}$$

The horizontal line *ob* may now be graduated to the volume scale, starting with its zero at '*o*'.

The absolute pressure and volume at any point may now be read off from this *p-V* diagram.

The mass of cushion steam trapped in the cylinder during the compression period of the back stroke can now be calculated from the compression curve of *p-V* diagram as follows :

Select a point '*N*' on the compression curve, towards the end of the stroke. At this point, the steam may be assumed dry saturated because the compression tends to dry the steam which may be assumed dry saturated at the end of compression stroke. Read the pressure and volume of the steam at *N* from the diagram.

Let p_N = Pressure of steam at point *N* (N/m²), and

V_N = Volume of steam at point *N* (m³).

From steam tables, obtain the specific volume of steam at pressure p_N ; let this volume be v.

Then $\quad V_N = m_C v$

$$\text{or} \qquad m_C = \frac{V_N}{v} \qquad \text{...(16.24)}$$

16.15. SATURATION CURVE AND MISSING QUANTITY

Saturation Curve. The saturation curve is *the curve showing the volume the steam in the cylinder would occupy, during the expansion stroke if the steam is perfectly dry and saturated at all the points.* It is plotted on *p-V* diagram and the wetness of steam can be seen on it at a glance.

Fig. 16.16 shows a calibrated indicator diagram. The total mass of steam in cylinder during expansion stroke $= m_C + m$.

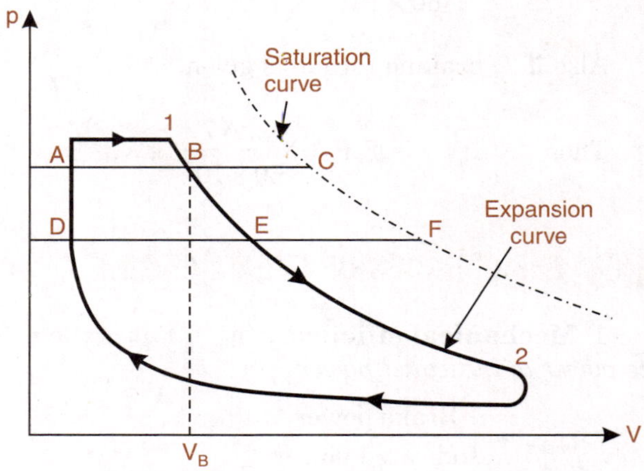

Fig. 16.16. Calibrated indicater diagram.

On the expansion curve, consider any point *B* and read off from the diagram the pressure (p_B) and volume (V_B) at this point. From steam tables, obtain the specific volume v of dry steam at pressure p_B. The volume the steam at *B* would occupy if dry saturated = mv. Let this volume be represented by *AC* to the volume scale of the *p-V* diagram. Then, the point *C* represents the volume the steam at *B* would occupy if *dry saturated*.

∴ Dryness fraction at *B*

$$= \frac{\text{Volume of dry steam in mixture}}{\text{Total volume of steam}} = \frac{AB}{AC}.$$

Following this way, a number of points may be obtained and plotted. The curve passing through these points is known as ***saturation curve*** (because all the points on this line represent the condition of steam dry and saturated).

From this saturation curve, the dryness fraction for all points on expansion curve can be obtained. For instance, the dryness fraction at *E* will be given by :

$$\text{Dryness fraction at } E = \frac{DE}{DF}.$$

From Fig. 16.16, it can be seen that the steam is wet at the beginning of the expansion stroke and becomes drier towards the end. This is owing to the fact that high pressure steam in the initial stage of expansion is hotter than the cylinder walls ; this causes the steam to condense. During the expansion stroke the steam pressure falls and towards the end of the stroke the walls will be hotter than the steam ; consequently the condensed steam re-evaporates and as a result the dryness fraction is improved. The improvement in dryness fraction will not be there if cylinder walls are not jacketed.

Missing quantity. *The missing quantity is the horizontal distance between the actual expansion curve and the saturation curve* (Fig. 16.16). At a pressure p_D, the missing quantity is represented by EF (m^3).

The missing quantity is mainly due to condensation of the steam, but a small amount will be due to leakage past the piston. Due to this missing quantity there is a loss of work represented by the area between the expansion curve and saturation curve.

The missing quantity can be **reduced** in the following ways :

(*i*) By steam jacketing the cylinder walls efficiently.

(*ii*) By reducing the temperature range of the steam during the stroke ; this can be accomplished by *compounding the expansion of steam* in two cylinders instead of allowing the whole pressure to drop to occur in one cylinder.

16.16. GOVERNING OF STEAM ENGINES

A governor is used to keep the speed of engine constant by either controlling the quantity of steam or the pressure of the steam supplied to the engine as per load requirements of the engine.

The following two methods are commonly used for governing the steam engine :

1. Throttle governing, and
2. Cut-off governing.

Throttle governing. In throttle governing, the quantity of steam entering the cylinder is *varied by opening or closing of the throttle valve* under the action of a governor. The indicator diagram showing the effects of throttle governing at various loads is shown in Fig. 16.17.

The method of control by throttling *causes an inherent loss of available energy* from the irreversibility of throttling, and the relatively late cut-off possessed by a purely throttle-controlled engine also causes a loss due to incomplete expansion. On the other hand, since the steam is throttled before it enters the cylinder, it operates between narrow temperature limits within the cylinder and thus *initial condensation is minimised*. Furthermore, when engine is operated on wet steam, the throttling generally tends to dry the steam or even to superheat it slightly, thus combating initial condensation. However, the losses inherent in throttling usually out-weigh its good features. For this reason many automatic engines use cut-off governing.

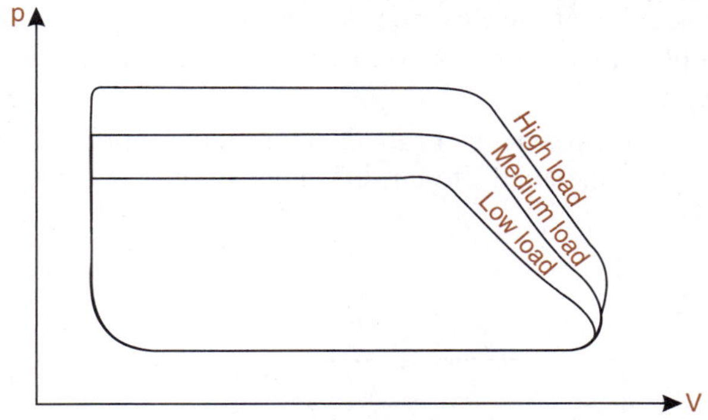

Fig. 16.17. Throttle governing.

Refer to Fig. 16.18. During a test on a throttle governed engine, if the indicated power is varied by altering the load, and the rate of steam consumption is measured for all values of power, it will be found that the steam consumption is a straight line function of the indicated power. This is known as ***Willan's law*** ; *it holds good only for a throttle governed engine running at constant speed.*

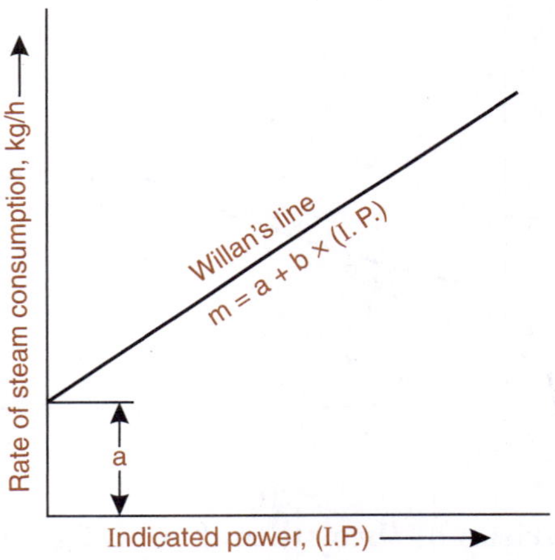

Fig. 16.18

Willan's line is a straight line and may therefore be expressed in the form as follows :

$$m = a + bx. \text{ (I.P.)} \qquad ...(16.25)$$

where, m = Steam consumption in kg/h,

a = Steam consumption at no load in kg/h (constant), and

b = Slope of Willan's line (constant).

Cut-off governing. In cut-off governing, the volume of steam supplied to the engine is altered by *changing the point of cut-off* by a special slide valve working under the control of the centrifugal governor.

In this type of governing for light load (Fig. 16.19), an excellently long expansion can be achieved from full high pressure conditions. The *worst loss here is caused by initial condensation.* Since the cut-off occurs early in the stroke, a large per cent of the entering steam is subjected to the cooling effect of the clearance spaces, the surface-to-volume ratio of which is greater than that of the engine cylinder itself.

The variations of rate of steam consumption against the I.P. developed in cut-off governing are shown in Fig. 16.20.

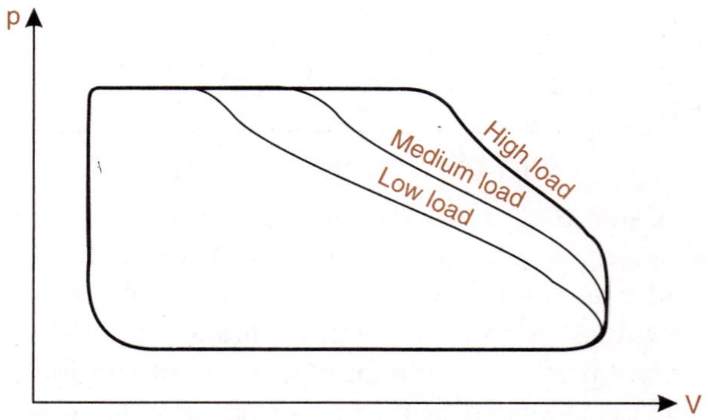

Fig. 16.19. Cut-off governing.

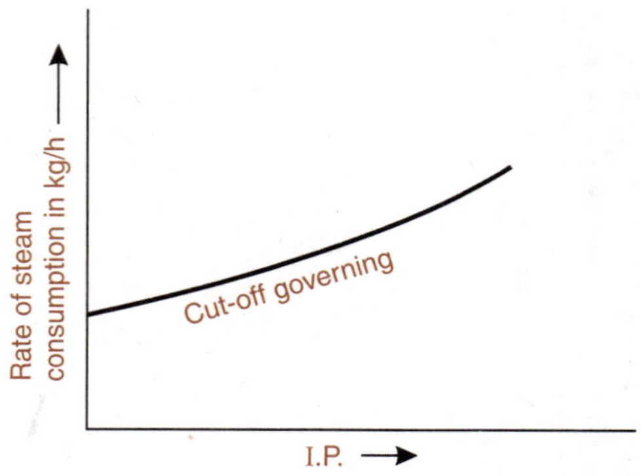

Fig. 16.20

Comparison of throttle and cut-off governing

S. No.	*Particular*	*Throttle governing*	*Cut-off governing*
1.	*Steam consumption per hour at full load.*	Same	Same
2.	*Steam consumption per hour at part load* (on the same engine).	Higher	Lower
3.	*Design and cost.*	Simple and less costly	Complicated and costly
4.	*Thermal efficiency at part load.*	Lower	Higher

16.17. VALVES

For proper operation of a steam engine one of the most essential parts is a valve. Its function is to admit steam to the cylinder at proper time in the stroke, and on the return stroke to open the cylinder to the exhaust and let the steam escape either to the atmosphere or to the condenser. The steam engine works properly if there is proper distribution of steam to the engine cylinder.

The valves may be classified as follows :

1. Sliding valves

(*i*) Simple slide valves,

(*ii*) Balanced slide valves,

(*iii*) Multiported slide valves,

(*iv*) Piston valves, and

(*v*) Meyer expansion valve etc.

2. Non-sliding valves

(*i*) Drop valves, and

(*ii*) Corliss valves.

Some of the important valves are described below :

D-slide valve. The D-slide valve is the most simple and extensively used for distribution of steam in the steam engines. It is so named because its shape is like that letter '*D*'. It is placed on a flat seat over the cylinder with three rectangular ports. The opening at the middle of the cylinder is called the *exhaust port* and the other two openings are called *steam ports.* The exhaust port is larger than the steam-port. The steam is admitted by the two sides of the valve and it is discharged through the inside cavity of the valve.

Refer to Figs. 16.21 and 16.22.

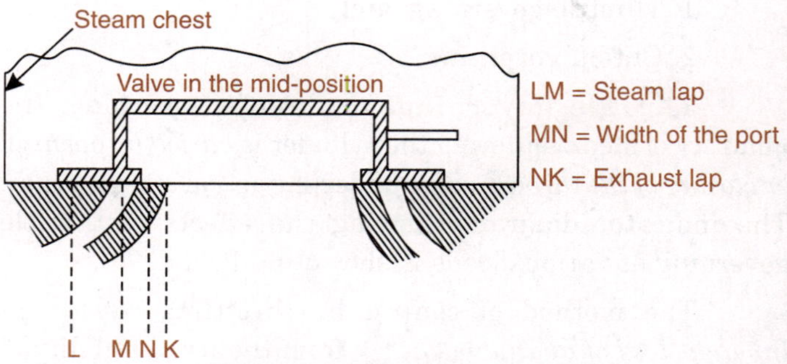

Fig. 16.21. D-slide valve.

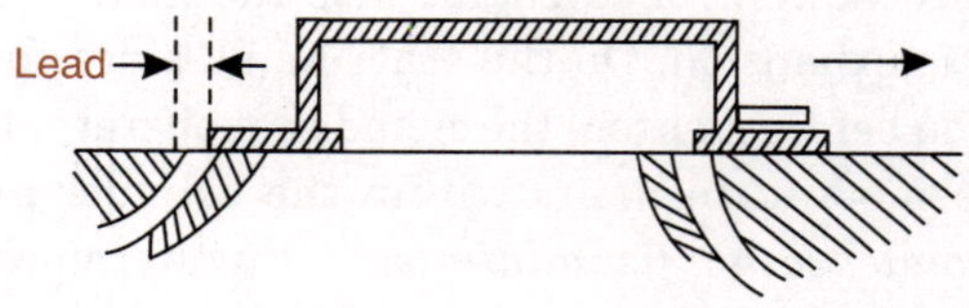

Fig. 16.22. Lead of the slide valve.

The distance travelled by the D-slide valve from one extreme position to another extreme position is called **valve travel.** It is *equal to twice the throw of the eccentric if the valve is connected directly to the eccentric.*

If the valve is half way between the two extreme positions, the position of the valve is known as **mid-position.**

When the valve is at mid-position, it overlaps the steam ports on both sides. This overlapping is called **lap of the valve.** The amount which it overlaps the outside edge of the steam port is called the **outside lap** or **steam lap** (LM) and amount which it overlaps the inside edge is called the **inside lap or exhaust lap** (NK). When the piston is at the beginning of the stroke, a small part of steam port is opened by the slide valve for admission of the steam. This amount of port opening is called **lead of the slide valve** (Fig. 16.22).

Advantages of D-slide valve :

(*i*) Simplicity in construction.

(*ii*) Operating mechanism very simple.

(*iii*) Low maintenance cost.

Disadvantages :

(*i*) There is a wear at the seat due to excessive friction.

(*ii*) The slow opening and closing of steam ports cause wire drawing of the steam, specially at the time of cut-off.

(*iii*) No independent adjustment of valve operations.

(*iv*) D-slide valve requires large clearance volume in comparison with other types of valves.

(*v*) The valve may get distorted due to unequal temperatures on different parts.

Balanced slide valves. Refer to Fig. 16.23. A balanced slide valve is a modification of a D-slide valve where steam pressure on the back of the valve is minimized. Four deep grooves are made over the valve to make the grooves to form a rectangle. Springs are placed inside the grooves over which the metal strips are placed. Due to the action of springs, the strips are pressed outwards against a smooth plate called balance plate. Thus a rectangular area is enclosed at the back of the valve by the metal strips and the balance plate. To provide a free passage between enclosed area and the exhaust steam an opening is made at the middle of the valve.

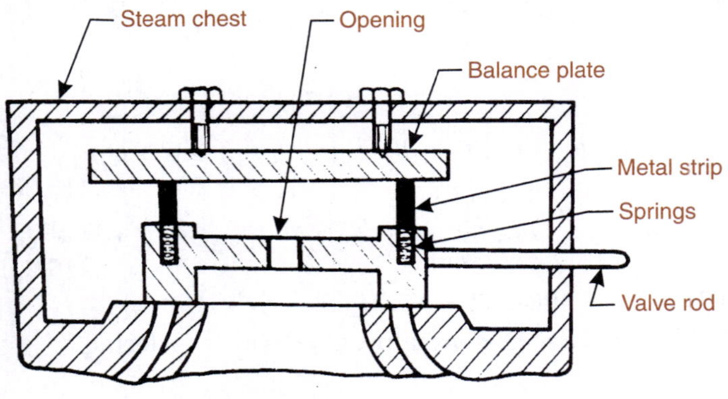

Fig. 16.23. Balanced slide valve.

The pressure exerted on the back of the valve is thus reduced and remains same as the exhaust pressure.

Piston valves. Refer to Fig. 16.24. The piston valve is a particular type of slide valve where friction due to steam pressure is completely minimized. It consists of two pistons connected with a single rod. These valves are made steam tight like an ordinary piston. As the amount of motion is

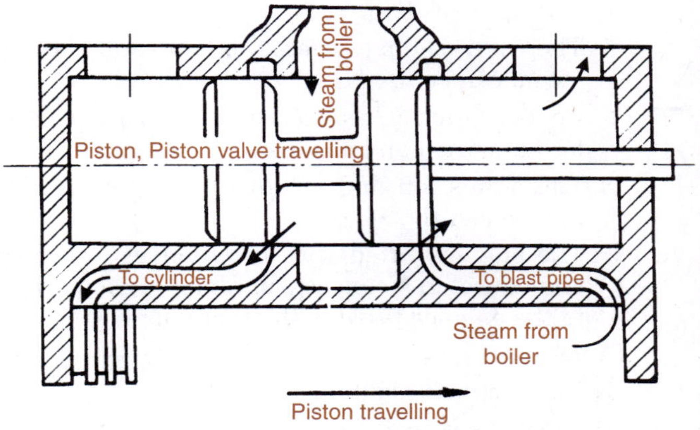

Fig. 16.24. Piston valve.

small as compared with that of the engine piston, the wear is therefore small. When the steam is admitted by the two sides of the valve and is discharged through a central opening, it is called **outside admission valve.** But when the steam is admitted through the central opening and is discharged by the two sides of the valve, it is called **inside admission valve.** The piston valves are generally of inside admission type (Refer to Fig. 16.24) and other slide valves are of outside admission type.

This type of valve is also suitable for high pressure superheated steam. It is commonly used in compound and locomotive engines.

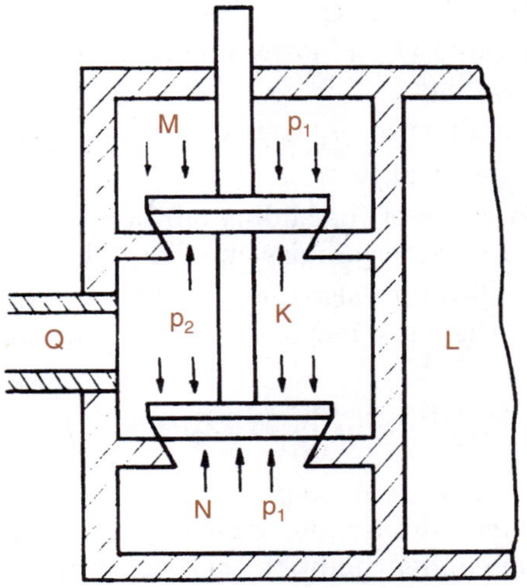

Fig. 16.25. Drop valve.

Drop valve. Fig. 16.25 shows a drop valve. No force except friction and inertia is required to operate it. The pressures on the valve are arranged in such a way so that they oppose each other and the result is a balanced system.

The steam enters the valve at L and passes on to chamber K through M and N and enters the engine cylinder at Q.

These valves are mostly used on *stationary engines.*

Advantages of drop valve :

1. There is no sliding surfaces and thus superheated steam can be used.
2. Condensation is reduced due to separate ports for admission and exhaust.
3. The admission and exhaust can be controlled independently.
4. The engines fitted with drop valves are more economical in steam consumption in comparison to the engines fitted with other valves.

Disadvantages :

1. Initial cost is high.
2. Difficulty is experienced in making both valves to seat simultaneously.

Corliss valve. It is only a modification of the basic slide valve. If a slide valve is coiled about an axis perpendicular to the axis of the piston movement, a Corliss valve would be obtained. A Corliss valve is operated by a wrist plate driven by a single eccentric. The only difficulty with this valve is that two admission valves and two exhaust valves per cylinder are to be provided.

Advantages :

1. Owing to the use of independent admission and exhaust ports, condensation and clearance volume are reduced.
2. It enables independent adjustment of points of admission, cut-off and release.
3. In a horizontal engine, exhaust valve can be located at the lowest position to ensure perfect drainage.
4. Small power is required for operating the valves.

Disadvantages :

1. Because of the sliding surfaces, the valve is not suitable for admission of superheated steam.
2. Unless the valve is very right in bending, the edge of the valve may catch on the edge of the port.

16.18. HEAT BALANCE SHEET

Procedure. For preparing a heat balance sheet for a steam engine cylinder, the engine should be tested over a period of time under conditions of constant load and steam supply. An indicator diagram should be taken and the steam pressure noted, at regular intervals of time ; an account should also be kept of the steam supply to the jackets. The mass of the steam supplied to the cylinder supplied can be obtained from the steam condensed by the condenser.

Analysis. The heat balance for the steam engine is more easily drawn up than that of the internal combustion engine. The working fluid in the steam engine does not undergo any chemical change, and consequently changes in properties may be ascertained with reference to an arbitrary datum.

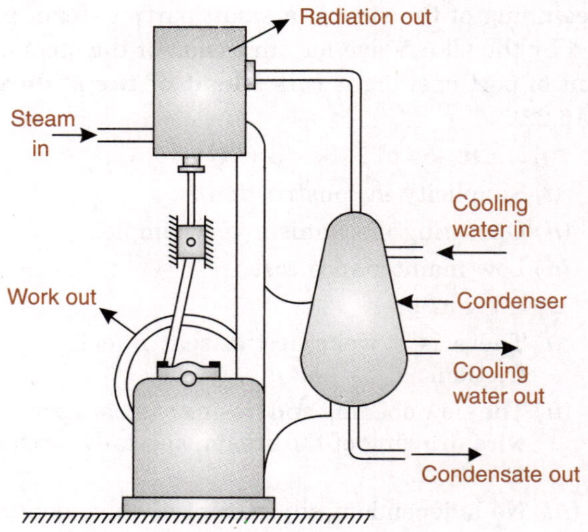

Fig. 16.26

Fig. 16.26 shows, diagrammatically, a steam engine, and the various fluids and energy quantities entering and leaving have been indicated there on. Treating the engine as a flow system as shown in Fig. 16.27, an energy equation may be written as follows :

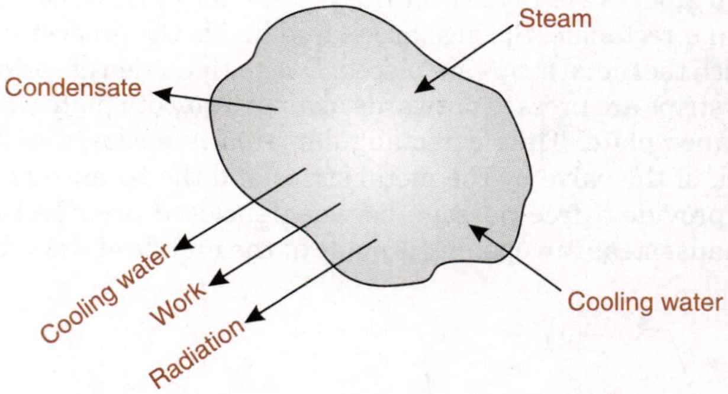

Fig. 16.27

$$m_{s_1} h_{s_1} + m_c h_{c_1} = m_{s_2} h_{s_2} + m_c h_{c_2} + W + Q_R \quad \text{...(16.26)}$$

$$m_{s_1} h_{s_1} = m_{s_2} h_{s_2} + m_c (h_{c_2} - h_{c_1}) + W + Q_R \quad \text{...(16.27)}$$

where, m = Mass, W = Work,

h = Specific enthalpy, Q = Heat transferred

and suffices have the following meanings :

s_1 = Steam supplied,

s_2 = Condensate discharged,

c = Cooling water, R = Radiation,

The equation (16.27) in tabular form is expressed as follows :

Heat in		*Heat out*	
In steam	$m_{s_1} h_{s_1}$	Heat as shaft work	W
		Heat to condensate	$m_{s_2} h_{s_2}$
		Heat to coolant	$m_c (h_{c_2} - h_{c_1})$
		Heat to radiation	Q_R

The value of enthalpy may be obtained from the steam tables. The heat to cooling water is given by $m_c \times c_{pw} \times (t_{out} - t_{in})$ and the heat to radiation obtained by difference. In the absence of leakage, $m_{s_1} = m_{s_2}$.

The heat balance may therefore be written as follows :

Heat in	*Heat out*
Heat in steam	Heat as B.P.
	Heat to condensate.
	Heat to cooling water.
	Heat to radiation, by difference.

It is to be noted that friction terms are absent, since any work done against friction is manifest in heat to the condensate, cooling water and radiation loss.

16.19. PERFORMANCE CURVES

In Fig. 16.28 is shown a set of typical performance curves of a reciprocating steam engine under test conditions.

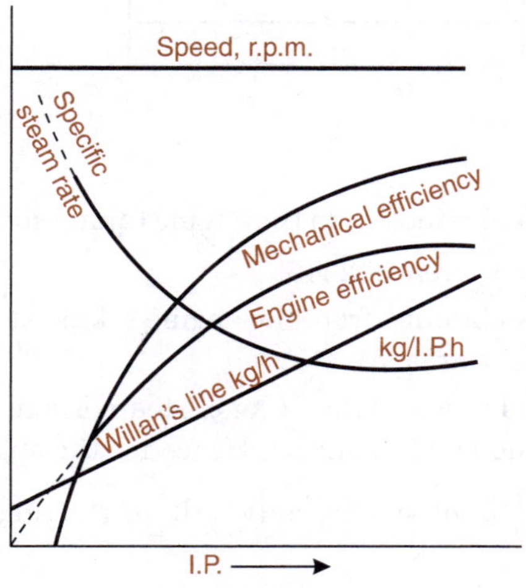

Fig. 16.28

Most of the curves are self-explanatory and should be carefully scrutinised for analysis of the variables involved.

One salient fact is that the speed curve as shown is practically horizontal, with a slight drop as the load is increased ; this shows good governing. The Willan's line as shown is straight ; this line is straight when the governing is by throttling with a fixed cut-off, and indicates a linear relationship between the total steam consumption in kg/h and the power if steam pressure is varied to suit the load. This fact is also of great advantage in predicting the part load steam consumption of steam turbines, most of which use throttle governing. If the steam consumption is plotted for an automatic engine with cut-off governing, the line will not be strictly straight.

WORKED EXAMPLES

Example 16.1. *Steam at 8.8 bar and 0.9 dry is supplied to an engine which expands it adiabatically to the release pressure of 1.2 bar when the pressure falls at constant volume to the exhaust pressure of 0.2 bar. Calculate :*

(i) Steam consumption in kg/kWh ;

(ii) Mean effective pressure ;

(iii) Heat to be removed by the condenser per kg of exhaust steam.

Solution. Steam pressure, $p_1 = 8.8$ bar

Dryness fraction, $x_1 = 0.9$

Pressure, $p_2 = 1.2$ bar

Exhaust pressure, $p_3 = 0.2$ bar

From steam tables :

At p_1 = 8.8 bar. $s_{f_1} = 2.0848$ kJ/kg K,

$s_{g_1} = 6.6269$ kJ/kg K

$h_{f_1} = 738.5$ kJ/kg,

$h_{fg_1} = 2032.8$ kJ/kg

At p_2 = 1.2 bar. $s_{f_2} = 1.3609$ kJ/kg K,

$s_{fg_2} = 7.2984$ kJ/kg K

$h_{f_2} = 439.4$ kJ/kg,

$h_{fg_2} = 2244.1$ kJ/kg,

$v_{g_2} = 1.428$ m³/kg

At p_3 = 0.2 bar. $h_{f_3} = 251.5$ kJ/kg.

Equating entropy at points 1 and 2,

$$s_1 = s_2$$

$$\therefore \quad s_{f_1} + x_1(s_{g_1} - s_{f_1}) = s_{f_2} + x_2 (s_{g_2} - s_{f_2})$$

$$2.0848 + 0.9 (6.6269 - 2.0848) = 1.3069 + x_2 (7.2984 - 1.3609)$$

$$6.1727 = 1.3069 + 5.9375x_2$$

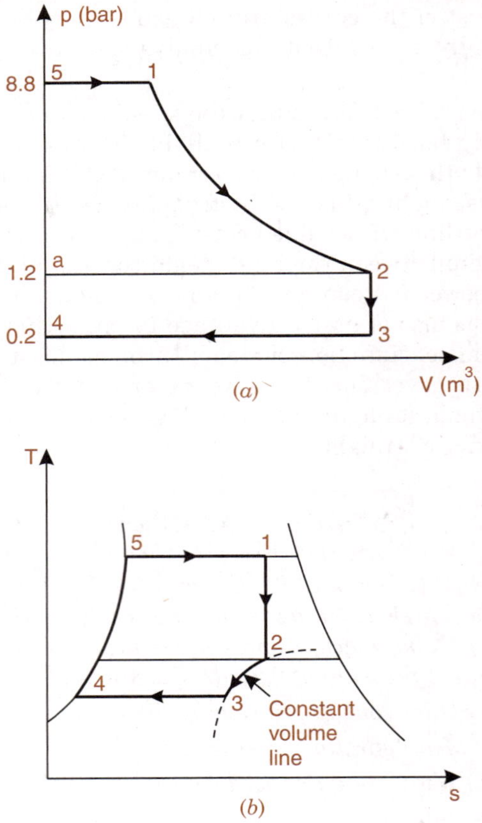

Fig. 16.29

$$\therefore \quad x_2 = \frac{6.1727 - 1.3069}{5.9375} = 0.819$$

$$h_1 = h_{f_1} + x_1 h_{fg_1}$$
$$= 738.5 + 0.9 \times 2032.8 = 2568.02 \text{ kJ/kg}$$
$$h_2 = h_{f_2} + x_2 h_{fg_2}$$
$$= 439.4 + 0.819 \times 2244.1 = 2277.3 \text{ kJ/kg}$$
$$v_2 = x_2 v_{g_2} = 0.819 \times 1.428 = 1.169 \text{ m}^3$$

Total work done

= Area '451234' = Area '512a5' + area 'a234a'

$$= (h_1 - h_2) + (p_2 - p_3) \times v_2$$
$$= (2568.02 - 2277.3) + (1.2 - 0.2) \times 10^5 \times 1.169 \times 10^{-3}$$
$$= 290.72 + 116.9 = 407.62 \text{ kJ/kg.}$$

(*i*) **Steam consumption in kg/kWh :**

$$= \frac{1 \times 3600}{407.62} = \textbf{8.83 kg/kWh. (Ans.)}$$

(*ii*) **Mean effective pressure, p_m :**

$$p_m \times v_2 = 407.62$$

$$\frac{p_m \times 10^5 \times v_2}{10^3} = 407.62$$

$$\therefore \quad p_m = \frac{407.62 \times 10^3}{10^5 \times v_2} = \frac{407.62 \times 10^3}{10^5 \times 1.169}$$

(where p_m is in bar)

= **3.48 bar. (Ans.)**

(*iii*) **Heat to be removed by the condenser :**

Heat to be removed by the condenser

= Heat supplied – Work done

$$= (h_1 - h_{f_3}) - 407.62$$
$$= (2568.02 - 251.5) - 407.62$$

= **1908.9 kJ/kg. (Ans.)**

Example 16.2. *Find the gain in thermal efficiency due to the expansive action of steam considering the work done per kg of steam in two engines receiving dry steam at 17.5 bar and exhausting at 0.7 bar, one of them taking steam for whole stroke and other taking steam for $\frac{1}{5}$th of the stroke.*

Assume negligible clearance and hyperbolic expansion.

Solution. Initial pressure of steam,

$$p_1 = 17.5 \text{ bar}$$

Dryness fraction, $x_1 = 1$

Pressure of steam after expansion,

$$p_3 (= p_b) = 0.7 \text{ bar}$$

Refer to Fig. 16.30.

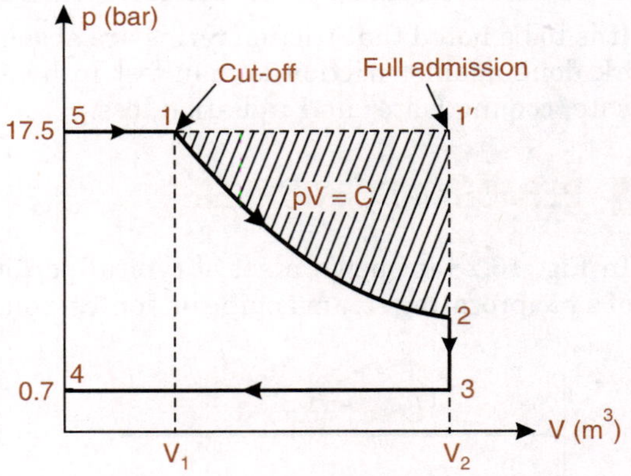

Fig. 16.30

Work obtained from the engine taking steam for $\frac{1}{5}$th of the stroke = area '512345'.

Work obtained from the engine taking steam for the whole stroke = area '51′345'.

Let us consider that 1 kg of steam is supplied to the engine during one full stroke. Hence for the engine taking steam for $\frac{1}{5}$th of stroke, only $\frac{1}{5}$th of the steam will be admitted.

Case 1. Engine taking steam for $\frac{1}{5}$th of stroke :

Considering the *p-V* diagram 51234, we have :

Mean effective pressure,

$$p_m = \frac{p_1}{r}(1 + \log_e r) - p_b$$

Here expansion ratio,

$$r = \frac{V_2}{V_1} = \frac{V_2}{\frac{V_2}{5}} = 5$$

$$\therefore \quad p_m = \frac{17.5}{5}(1 + \log_e 5) - 0.7 = 8.43 \text{ bar}$$

At 17.5 bar. Specific volume, $v = 0.113 \text{ m}^3/\text{kg}$.

Work done by $\frac{1}{5}$th of steam

$$= p_m \times v$$

$$= \frac{8.43 \times 10^5 \times 0.113}{10^3} = 95.26 \text{ kJ/kg.}$$

Work done per kg of steam

$$= 95.26 \times 5 = 476.3 \text{ kJ}$$

Heat supplied per kg

$$= h_1 - h_{f_3}$$

$$\left[\begin{array}{l} \textit{At 17.5 bar. } h_1 = h_{g_1} = 2794.1 \text{ kJ/kg} \\ \textit{At 0.7 bar. } h_{f_3} = 376.8 \text{ kJ/kg.} \end{array}\right]$$

$$= 2794.1 - 376.8$$

$$= 2417.3 \text{ kJ/kg}$$

Thermal efficiency

$$= \frac{\text{Work done}}{\text{Heat supplied}} = \frac{476.3}{2417.3} = 0.197 \text{ or } 19.7\%.$$

Case 2. Engine taking steam for the whole stroke :

Considering the *p-V* diagram 51′ 34.

Work done/kg of steam = area 51′ 345

$$= (p_1 - p_b)\, v_2$$

$$= \frac{(17.5 - 0.7) \times 10^5 \times 0.113}{10^3} = 189.84 \text{ kJ/kg}$$

Heat supplied per kg

$$= h_1 - h_{f_3} = 2794.1 - 376.8$$

$$= 2417.3 \text{ kJ/kg}$$

Thermal efficiency

$$= \frac{189.84}{2417.3} = 0.0785 = 7.85\%$$

Gain in thermal efficiency

$$= 19.7 - 7.85 = \mathbf{11.85\%.} \textbf{ (Ans.)}$$

This gain is an increase in thermal efficiency by more than 150 per cent *due to the expansive working of steam.*

Example 16.3. *The cylinder of a non-condensing steam engine is supplied with steam at 11.5 bar. The clearance volume is $\frac{1}{10}$th of the stroke volume and the cut-off takes place at $\frac{1}{4}$th of the stroke. If the pressure at the end of compression is 5.4 bar, compute the value of mean effective pressure of the steam on the piston. Assume that the expansion and compression are hyperbolic. The back pressure is 1.1 bar.*

Solution. Admission pressure, $p_1 = 11.5$ bar

Clearance volume, $V_c = \frac{1}{10}$th of stroke volume V_s

Cut-off $= \frac{1}{4}$th of the stroke

Pressure at the end of compression, $p_5 = 5.4$ bar

Back pressure, $p_b = p_3 = 1.1$ bar.

Mean effective pressure, p_m :

The *theoretical* mean effective pressure, when clearance and compression are considered, is given by eqn. (16.9) :

$$p_{m(th.)} = p_1\left[\frac{1}{r} + \left(c + \frac{1}{r}\right)\log_e\left(\frac{c+1}{c + \frac{1}{r}}\right)\right]$$

$$- p_b\left[(1 - \alpha) + (c + \alpha)\log_e\left(\frac{\alpha + c}{c}\right)\right] \quad ...(i)$$

$$r = \frac{1}{1/4} = 4 \text{ or } \frac{1}{r} = \frac{1}{4} \text{ and } c = \frac{V_s/10}{V_s} = 0.1.$$

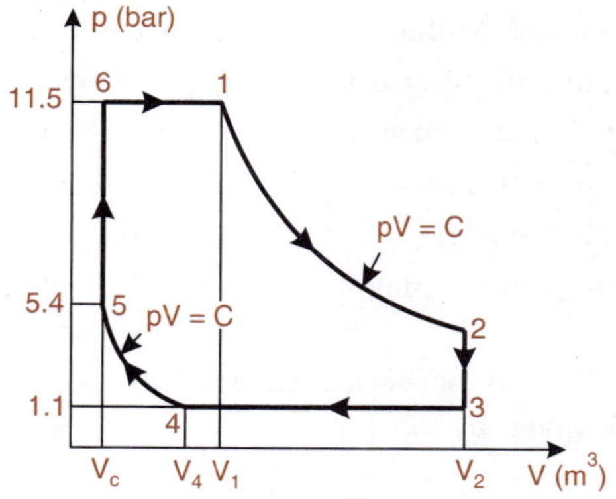

Fig. 16.31

To find α (ratio of volume between points of compression and admission to the swept volume V_s) applying hyperbolic law between the point of beginning and end of compression :

$$p_4V_4 = p_5V_5$$

i.e., $1.1 \times (V_c + \alpha V_s) = 5.4 \times V_c$

$$\therefore \quad \alpha = \frac{5.4V_c - 1.1V_c}{1.1V_s}$$

$$= \frac{5.4V_c - 1.1V_c}{11V_c} = 0.39$$

$$\left[\begin{array}{l} \because\ V_s = V_c \text{ and } p_4 = p_b = 1.1 \text{ bar and} \\ \because\ \alpha = \dfrac{V_4 - V_c}{V_s} \\ \therefore\ V_4 = V_c + \alpha V_s \\ \text{Also } \dfrac{V_c}{V_s} = \dfrac{1}{10} \\ \therefore\ V_s = 10V_c \end{array}\right]$$

Now inserting various values in eqn. (*i*), we get

$$p_{m(th.)} = 11.5\left[\frac{1}{4} + \left(0.1 + \frac{1}{4}\right)\log_e\left(\frac{0.1+1}{0.1+1/4}\right)\right]$$

$$- 1.1\left[(1 - 0.39) + (0.1 + 0.39)\log_e\left(\frac{0.39+0.1}{0.1}\right)\right]$$

$$= 11.5 \times 0.65 - 1.1 \times 1.388 = 5.95 \text{ bar}$$

i.e., *Mean effective pressure of steam* = **5.95 bar. (Ans.)**

Example 16.4. *A double-acting steam engine has a cylinder 200 mm bore by 300 mm stroke and cut-off occurs at 0.4 stroke. The admission and exhaust pressures are 7 and 0.38 bar. If the diagram factor is 0.80, calculate the indicated power of the engine at 200 r.p.m., neglecting the effect of clearance and assuming hyperbolic expansion.*

If, however, clearance volume is 10% of the swept volume, calculate the mean effective pressure, the cut-off remaining at the same point of stroke as before.

Solution. Bore of engine cylinder, D = 200 mm = 0.2 m

Length of stroke, L = 300 mm = 0.3 m

Admission pressure, p_1 = 7 bar

Exhaust pressure, p_b = 0.38 bar

Diagram factor = 0.8

Engine speed, N = 200 r.p.m.

Clearance volume, V_c = 10% of stroke volume V_s

(*i*) The hypothetical/theoretical mean effective pressure is given by

$$p_{m(th.)} = \frac{p_1}{r}(1 + \log_e r) - p_3$$

($p_3 = p_b$ = back pressure)

But $r = \dfrac{V_2}{V_1} = \dfrac{V_2}{0.4V_s} = \dfrac{V_2}{0.4V_2} = 2.5$

($V_s = V_2$ as clearance volume is neglected)

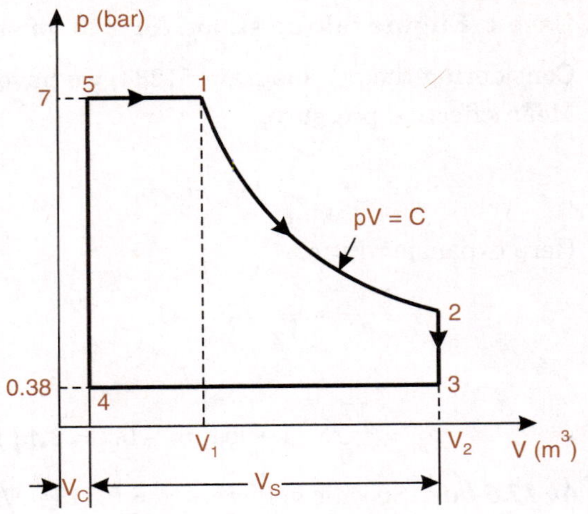

Fig. 16.32

$$\therefore \quad p_{m(th.)} = \frac{7}{2.5}(1 + \log_e 2.5) - 0.38$$

$$= 4.98 \text{ bar}$$

Actual m.e.p.,

$p_{m(act.)}$ = Diagram factor × theoretical m.e.p.

= 0.8 × 4.98 = 3.98 bar

Now, indicated power,

$$\textbf{I.P.} = \frac{10p_mLAN}{3}$$

...since the engine is double-acting

$$= \frac{10 \times 3.98 \times 0.3 \times \pi/4 \times 0.2^2 \times 200}{3}$$

= **25 kW. (Ans.)**

(*ii*) Refer to Fig. 16.32.

The theoretical mean effective pressure, when clearance considered, is given by

$$p_{m(th.)} = p_1\left[\frac{1}{r} + \left(c + \frac{1}{r}\right)\log_e\left(\frac{c+1}{c+\frac{1}{r}}\right)\right] - p_b$$

Here, $r = 2.5$ or $\dfrac{1}{r} = 0.4$ and $c = \dfrac{10}{100} = 0.1$

$$\therefore \quad p_{m(th.)} = 7\left[0.4 + (0.1 + 0.4)\log_e\left(\frac{0.1+1}{0.1+0.4}\right)\right] - 0.38$$

$$= 7(0.4 + 0.394) - 0.38 = 5.18 \text{ bar}$$

Assuming the same diagram factor,

$$p_{m(act.)} = 0.8 \times 5.18 = 4.14 \text{ bar}$$

Hence, *mean effective pressure (actual)*

= **4.14 bar. (Ans.)**

Example 16.5. *A steam engine having swept volume 0.034 m³ and clearance volume 20% of the piston displacement, consumes 2725 kg of steam per hour at 250 r.p.m. The engine is double acting and the pressure of steam near cut-off is 8.8 bar, when piston has traversed 30% of working stroke. If the compression commences at 65% of the return stroke when the steam remaining in the cylinder is dry at 0.14 bar, estimate the mass of cushion steam at this point. Also find the dryness fraction of the steam at cut-off.*

Solution. Swept volume, $V_s = 0.034\ m^3$

Clearance volume, $V_c = 0.2\ V_s$

Steam consumption $= 2725$ kg/h

Engine speed, $N = 250$ r.p.m.

Pressure of steam near cut-off, $p_1 = 8.8$ bar

Exhaust or back pressure, $p_b = 0.14$ bar.

(*i*) **Mass of cushion steam :**

$$V_c = 0.2V_s = 0.2 \times 0.034 = 0.0068\ m^3$$

Total volume $V_2 = V_c + V_s = 0.0068 + 0.034 = 0.0408\ m^3$

$$V_4 = V_c + 0.35V_s = 0.0068 + 0.35 \times 0.034 = 0.0187\ m^3$$

$$V_1 = V_c + 0.3V_s = 0.0068 + 0.3 \times 0.034 = 0.017\ m^3$$

Cylinder feed/stroke

$$= \frac{2725}{60 \times 250 \times 2} = 0.09083 \text{ kg}$$

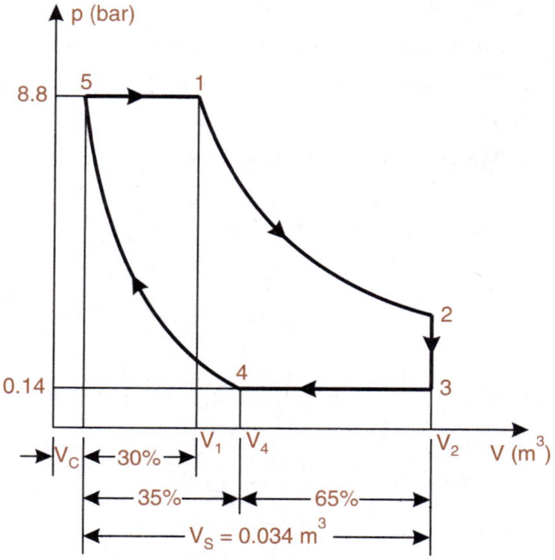

Fig. 16.33

Now, *mass of cushion steam* at point 4 is the mass of 0.0187 m³ dry steam at 0.14 bar.

$$= \frac{0.0187}{10.69} = \mathbf{0.001749\ kg.}\ \textbf{(Ans.)}$$

[From steam tables : Specific volume, v_g of steam at 0.14 bar = 10.69 m³/kg]

(*ii*) **Dryness fraction of steam at cut-off :**

The mass of steam present in the cylinder during expansion

$$= 0.09083 + 0.001749 = 0.09258 \text{ kg}$$

Now, the indicated dry mass at a point of cut-off (point 1)

$$= \frac{0.017}{0.219} = 0.07762 \text{ kg}$$

[Specific volume, v_g of steam at 8.8 bar = 0.219 m³/kg]

∴ *Dryness fraction of steam at cut-off*

$$= \frac{0.07762}{0.09258} = \mathbf{0.838.}\ \textbf{(Ans.)}$$

☞**Example 16.6.** *A single cylinder, double-acting, non-condensing steam engine 200 mm in diameter and 400 mm in stroke develops 30 kW at 100 r.p.m. The clearance is 10% and cut-off is 40% of stroke. The pressure at the point of cut-off is 5 bar. The compression starts at 80% of the stroke during return stroke. The pressure of the steam on compression curve at 90% of the return stroke is 1.5 bar and steam is dry and saturated.*

Calculate the actual and minimum theoretical possible specific steam consumption on I.P. basis. Take the missing quantity of cut-off as 0.0072 kg/stroke.

Solution. Diameter of engine cylinder, $D = 200$ mm $= 0.2$ m

Length of stroke, $L = 400$ mm $= 0.4$ m

Indicated power developed, I.P. $= 30$ kW

Engine speed, $N = 100$ r.p.m.

Clearance volume, $V_c = 0.1\ V_s$

Cut-off $= 0.4\ V_s$

Pressure at the point of cut-off, $p_1 = 5$ bar

Steam pressure on compression curve at 90% of the return stroke,

$$p_s = 1.5 \text{ bar}$$

Missing quantity at cut-off

$$= 0.0072 \text{ kg/stroke.}$$

Refer to Fig. 16.34.

Stroke volume, $V_s = \frac{\pi}{4} D^2 \times L$

$$= \frac{\pi}{4} \times 0.2^2 \times 0.4 = 0.01256\ m^3$$

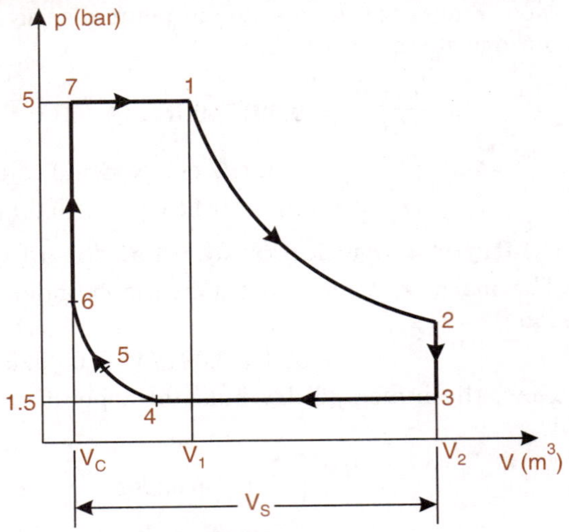

Fig. 16.34

$$V_s = V_c + (1 - 0.9)V_s$$
$$= 0.1\ V_s + 0.1V_s = 0.2V_s$$
$$= 0.2 \times 0.01256 = 0.002512\ m^3$$

Mass of cushion steam/stroke

$$m_c = \frac{V_5}{v_5} = \frac{V_6}{v_{g_5}}$$

$\left[\begin{array}{l}\text{where } v_{g_5} = \text{specific volume} \\ \text{of steam at 1.5 bar ;} \\ \text{dryness fraction } = 1\end{array}\right]$

$$= \frac{0.002512}{1.159} = 0.002167 \text{ kg/stroke}$$

$$V_1 = V_c + 0.4V_s = 0.1V_s + 0.4V_s = 0.5V_s$$
$$= 0.5 \times 0.01256 = 0.00628\ m^3$$

Mass of steam at point of cut-off if it is dry and saturated,

$$m_1 = \frac{V_1}{v_{g_1}}$$

$\left[\begin{array}{l}v_{g_1} = \text{specific volume of steam at 5 bar} \\ \quad = 0.375\ m^3/kg\end{array}\right]$

$$= \frac{0.00628}{0.375} = 0.01674 \text{ kg/stroke}$$

If m_s is the mass of steam supplied/stroke, the missing quantity is given by

$(m_c + m_s) - m_1$ = Missing quantity

$\therefore\quad 0.002167 + m_s - 0.01674 = 0.0072$

or $\quad m_s = 0.02177$ kg/stroke

Mass of steam supplied per hour

$= 0.02177 \times 60 \times 100 \times 2 = 261.24$ kg/h

Specific steam consumption on I.P. basis

$$= \frac{261.24}{30} = \mathbf{8.708\ kg/kWh.}\ \text{(Ans.)}$$

Theoretical minimum possible steam/stroke will be if the steam at cut-off is dry and saturated.

Thus, minimum possible steam supplied/stroke

$= m_1 - m_c = 0.01674 - 0.002167$

$= 0.01457$ kg/stroke

$\therefore$ Minimum steam that may be supplied/hour

$= 0.01457 \times 60 \times 100 \times 2 = 174.84$ kg/h

Hence, *minimum specific steam consumption on I.P. basis*

$$= \frac{174.84}{30} = \mathbf{5.828\ kg/kWh.}\ \text{(Ans.)}$$

Example 16.7. *The following data refer to a single-stage double-acting steam engine :*

High pressure = 7 bar ; low pressure = 1.2 bar ; cut-off and clearance are respectively 40% and 10% of stroke volume ; compression of steam starts at 20% of stroke volume ; specific volume of steam at 1.2 bar is 1.455 m³/kg ; steam consumption = 2700 kg/h ; engine speed = 240 r.p.m. ; quality of steam supplied = 0.9 and amount of jacket steam condensed = 18 kg/h.

Calculate : (i) The m.e.p. of the cycle ;

(ii) The diagram factor ;

(iii) The power out if mechanical efficiency is 0.85 ;

(iv) The missing quantity per cycle at the state during expansion, where the pressure is 5 bar.

Solution. High pressure, $p_1 = 7$ bar

Low pressure, $p_3 = p_b = 1.2$ bar

Clearance volume, $V_c = 0.1\ V_s$

Cut-off $= 0.4\ V_s$

Specific volume of steam at 1.2 bar = 1.455 m³/kg

Steam consumption = 2700 kg/h

Engine speed, $N = 240$ r.p.m.

Quality of steam supplied = 0.9

Amount of jacket steam condensed = 18 kg/h

Mechanical efficiency, $\eta_{mech.} = 0.85$.

Refer to Fig. 16.35.

(*i*) **Mean effective pressure, p_m :**

$$p_m = p_1\left[\frac{1}{r} + \left(c + \frac{1}{r}\right)\log_e\left(\frac{c+1}{c+1/r}\right)\right] - p_b\left[(1-\alpha) + (c+\alpha)\log_e\left(\frac{\alpha + c}{c}\right)\right]$$

Here, $r = \dfrac{1}{0.4}$ or $\dfrac{1}{r} = 0.4, \quad c = \dfrac{V_c}{V_s} = 0.1$

$$\alpha = \frac{V_4 - V_s}{V_s} = \frac{(0.2\ V_s + V_c) - V_c}{V_s}$$

$$= \frac{(0.2\ V_s + 0.1\ V_s) - 0.1V_s}{V_s} = 0.2$$

$$\therefore \quad p_m = 7\left[0.4 + (0.1+0.4)\log_e\left(\frac{0.1+1}{0.1+0.4}\right)\right] - 1.2\left[(1-0.2) + (0.1+0.2)\log_e\left(\frac{0.2+0.1}{0.1}\right)\right]$$

$$= 5.56 - 1.35 = \mathbf{4.21\ bar.}\ \textbf{(Ans.)}$$

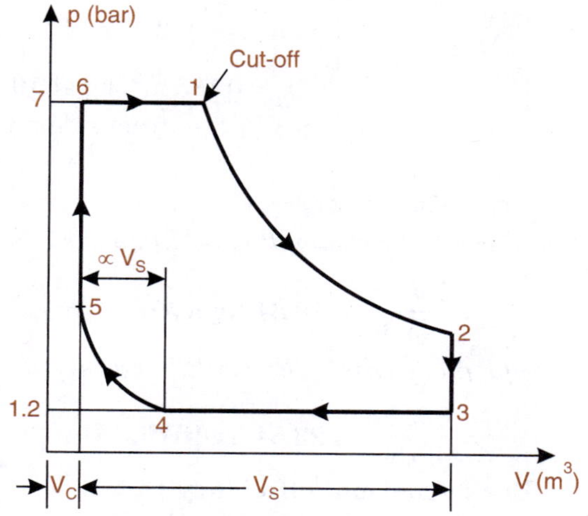

Fig. 16.35

(*ii*) **The diagram factor :**

If the clearance volume is included in the hypothetical or theoretical diagram the mean effective pressure p_m' is given by

$$p_m' = \frac{p_1}{r'}(1 + \log_e r') - p_b$$

$$\frac{1}{r'} = \frac{V_1}{V_s} = \frac{0.4\,V_s + V_c}{V_s} = \frac{0.4\,V_s + 0.1\,V_s}{V_s}$$

$= 0.5$ and $r' = 2$

$\therefore \quad p_m' = 7 \times 0.5\,(1 + \log_e 2) - 1.2 = 4.72$ bar

Hence diagram factor, D.F.

$$= \frac{p_m}{p_m'} = \frac{4.21}{4.72} = \mathbf{0.89.}\ \textbf{(Ans.)}$$

(*iii*) **Power output :**

Since the volume occupied by the liquid is quite small compared with that of the vapour, it may be neglected and specific volume of vapour at cut-off may be written as $x_1 v_g$, x_1 being the quality of vapour, assumed to be 0.9.

Now, mass of steam at cut-off

= Mass of cushion steam + mass admitted/stroke

$= (\alpha V_s + V_c)/1.455 + 2700/(60 \times 2 \times 240)$

$= (0.2\,V_s + 0.1\,V_s)/1.455 + 0.0937$

$= 0.206\,V_s + 0.0937$

Volume of steam at cut-off

$= 0.4\,V_s + 0.1\,V_s$

$= 0.5\,V_s = 0.273 \times 0.9\,(0.206\,V_s + 0.0937)$

$$\therefore \quad V_s = \frac{0.273 \times 0.9 \times 0.0937}{0.5 - 0.273 \times 0.9 \times 0.206}$$

[$\because$ Specific volume at 7 bar, $v_{g_1} = 0.273$ m³/kg]

$$= \frac{0.02302}{0.4493} = 0.0512\ \text{m}^3/\text{stroke}$$

Power output $= p_m\,V_s \times \eta_{mech.}\,(2\,N/60)$

$$= (4.21 \times 10^5 \times 0.0512 \times 10^{-3}) \times 0.85 \times \frac{2 \times 240}{60}$$

$= \mathbf{146.57\ kW.}$ **(Ans.)**

(*iv*) **Missing quantity :**

Volume of steam at 5 bar, V' :

$7 \times 0.5\,V_s = 5 \times V'$ ($\because$ pV = constant)

$$\therefore \quad V' = \frac{7 \times 0.5 \times V_s}{5} = \frac{7 \times 0.5 \times 0.0512}{5}$$

$= 0.03584$ m³

Actual mass of steam

$= 0.206 \times 0.0512 + 0.0937 = 0.1042$ kg

Since the specific volume of dry steam at 5 bar is 0.375 m³/kg, the mass of dry steam occupying a volume of 0.03584 m³ is

$$= \frac{0.03584}{0.375} = 0.09557\ \text{kg}$$

$\therefore$ *Missing quantity*

$= 0.1042 - 0.09557$

$= \mathbf{0.00863\ kg.}$ **(Ans.)**

☞**Example 16.8.** *The following readings were taken during a trial of a single-cylinder double-acting non-condensing steam engine running at 240 r.p.m.*

Cylinder diameter	*250 mm*
Piston rod diameter	*50 mm*
Stroke of the engine	*350 mm*
Cut-off	*30% of the stroke*
Length of the indicator diagram	*53 mm*
Area of the indicator diagram for cover end	*1570 mm²*
Area of the indicator diagram for crank end	*1440 mm²*
Spring number	*0.12 bar/mm*
Circumference of the brake wheel	*5.25 m*
Circumference of the brake rope	*62.5 mm*
Dead load on the brake	*1900 N*
Reading of the spring balance	*190 N*
Pressure of the steam supplied	*10.5 bar*
Dryness fraction of steam supplied	*0.92*

Find : (i) Indicated power (I.P.) ;

(ii) Brake power (B.P.) ;

(*iii*) *Mechanical efficiency ;*

(*iv*) *Specific steam consumption on I.P. and B.P. basis ;*

(*v*) *Brake thermal efficiency.*

Solution. (*i*) **Indicated power (I.P.) :**

$$\text{I.P.} = \frac{10\, p_{m_1} LA_1N}{6} + \frac{10\, p_{m_2} LA_2N}{6} \quad ...(i)$$

$$p_m = \frac{\text{Area of the indicator diagram }(A) \times \text{Spring number }(S)}{\text{Length of the indicator diagram }(L)}$$

$$\therefore \quad p_{m_1} = \frac{A_1S}{L} = \frac{1570 \times 0.12}{53} = 3.55 \text{ bar}$$

and $$p_{m_2} = \frac{A_2S}{L} = \frac{1440 \times 0.12}{53} = 3.26 \text{ bar}$$

$$A_1 = \pi/4\, D^2 = \pi/4 \times 0.25^2 = 0.04908 \text{ m}^2$$

$$A_2 = \pi/4\, (D^2 - d^2)$$

$$= \pi/4\, (0.25^2 - 0.05^2) = 0.04712 \text{ m}^2$$

Inserting these values in eqn. (*i*), we get

$$\text{I.P.} = \frac{10 \times 3.55 \times 0.35 \times 0.04908 \times 240}{6} + \frac{10 \times 3.26 \times 0.35 \times 0.04712 \times 240}{6}$$

$$= 24.4 + 21.5 = \mathbf{45.9\ kW.}$$ **(Ans.)**

(*ii*) **Brake power (B.P.) :**

$$\text{B.P.} = \frac{(W - S)\,\pi(D + d)N}{60 \times 10^3}$$

$$= \frac{(1900 - 190) \times \left(5.25 + \dfrac{62.5}{1000}\right) \times 240}{60 \times 1000}$$

$$= \mathbf{36.34\ kW.}$$ **(Ans.)**

(*iii*) **Mechanical efficiency, η_{mech} :**

$$\left[\because \quad pD = 5.25 \text{ m} \; ; \; \pi d = \frac{62.5}{1000} \text{ m} \quad \text{... Given}\right]$$

$$\eta_{mech} = \frac{\text{B.P.}}{\text{I.P.}} = \frac{36.34}{45.9} = 0.7917 = \mathbf{79.17\%.}$$ **(Ans.)**

(*iv*) **Specific steam consumption :**

Steam consumption per hour, m_s = Steam consumed/revolution × r.p.m. × 60

= [(steam consumption of cover end/stroke + steam consumption of crank end/stroke) × r.p.m. × 60]

$$= \left[\frac{\pi/4\, D^2 \times L}{r} + \frac{\pi/4\,(D^2 - d^2)\, L}{r}\right] \times \frac{\text{r.p.m} \times 60}{\text{specific volume of steam}}$$

where, r = cut-off ratio.

$$\therefore \quad m_s = \frac{\pi}{4} \cdot \frac{L}{r}\, [D^2 + (D^2 - d^2)] \times \frac{240 \times 60}{xv_g}$$

$$= \frac{\pi}{4} \times \frac{0.35}{3.33}\, [0.25^2 + (0.25^2 - 0.05^2)] \times \frac{240 \times 60}{0.92 \times 0.185}$$

$$\left[\because \quad r = \frac{1}{0.3} = 3.33 \text{ and } v_g = 0.185 \text{ m}^3/\text{kg at 10.5 bar from steam tables}\right]$$

$$= 855.56 \text{ kg/h}$$

Specific steam consumption on I.P. basis

$$= \frac{855.56}{45.9} = \mathbf{18.64\ kg/kWh.}$$ **(Ans.)**

Specific fuel consumption on B.P. basis

$$= \frac{855.56}{36.34} = \mathbf{23.54\ kg/kWh.}$$ **(Ans.)**

(*v*) **Brake thermal efficiency :**

Brake thermal efficiency,

$$\eta_{th(B)} = \frac{\text{B.P.}}{\dot{m}_s\,(h_1 - h_{f_2})}$$

Here $$h_1 = h_{f_1} + x_1 h_{fg_1}$$

$$\left[\text{At 10.5 bar, from steam tables : } h_{f_1} = 772 \text{ kJ/kg}, \; h_{fg} = 2005.9 \text{ kJ/kg}\right]$$

$$= 772 + 0.92 \times 2005.9 = 2617.4 \text{ kJ/kg}$$

h_{f_2} = 0, since engine is non-condensing.

$$\therefore \quad \eta_{th(B)} = \frac{36.34}{\dfrac{855.56}{3600} \times 2617.4}$$

$$= \mathbf{0.0584 \text{ or } 5.84\%.}$$ **(Ans.)**

Example 16.9. *Steam is admitted to an engine for 35% of the stroke with a pressure of 7.2 bar, the law of expansion followed is $pV^{1.1}$ = constant. Compression starts at 55% of return stroke and follows the law $pV^{1.2}$ = constant, the clearance volume is 20% of the displacement volume and the back pressure is 660 mm of mercury vacuum when barometer reads 760 mm of mercury. Estimate the mean effective pressure and indicated power of a double-acting engine with cylinder diameter 320 mm, stroke 480 mm and speed 200 r.p.m.*

Solution. Cylinder diameter, D = 320 mm = 0.32 m

Stroke length, L = 480 mm = 0.48 m

Admission steam pressure, p_1 = 7.2 bar

Cut-off $= 0.35\, V_s$

Law of expansion, $pV^{1.1}$ = constant
Law of compression, $pV^{1.2}$ = constant
Clearance volume, $V_c = 0.2\ V_s$
Back pressure, p_b = 660 mm of Hg
Barometer reading = 760 mm of Hg
Engine speed, N = 200 r.p.m.
Refer to Fig. 16.36.

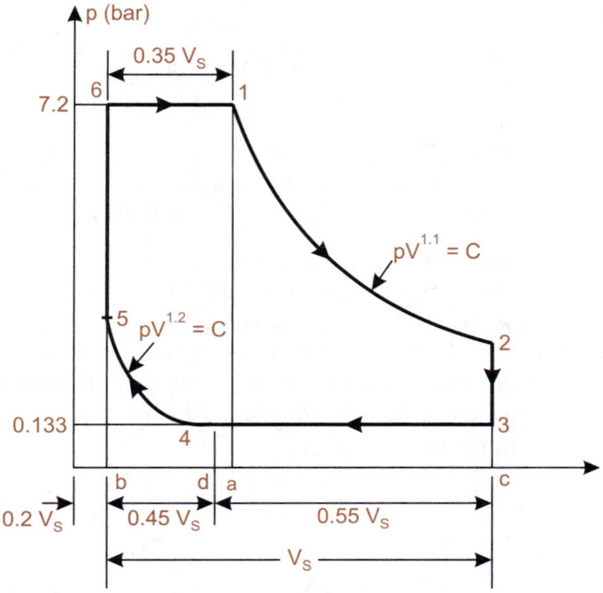

Fig. 16.36

Back pressure, $p_b = \dfrac{760 - 660}{760} \times 1.013 = 0.133$ bar

Considering *expansion curve*,

$$pV^{1.1} = \text{constant}$$

$$p_1V_1^{1.1} = p_2V_2^{1.1}$$

$$p_2 = p_1 \left(\frac{V_1}{V_2}\right)^{1.1} = 7.2\left[\frac{(0.35\ V_s + 0.2\ V_s)}{(V_s + 0.2\ V_s)}\right]^{1.1} = 7.2\left(\frac{0.55}{1.2}\right)^{1.1} = 3.05 \text{ bar}$$

Considering the *compression curve*,

$$pV^{1.2} = \text{constant}$$

$$p_4V_4^{1.2} = p_5V_5^{1.2}$$

$$p_5 = p_4 \times \left(\frac{V_4}{V_2}\right)^{1.2} = 0.133 \times \left[\frac{0.45\ V_s + 0.2\ V_s}{0.2\ V_s}\right]^{1.2} = 0.133 \times \left(\frac{0.65}{0.2}\right)^{1.2} = 0.55 \text{ bar.}$$

Also, **mean effective pressure,**

$$p_m = \frac{\text{Area of the indicator diagram}}{\text{Stroke volume}}$$

$$= \frac{\text{area '61ab6' + area '12ca1' – area '43cd4' – area '54db5'}}{V_s}$$

$$= \frac{\left[(7.2 \times 0.35\ V_s)\left\{\dfrac{7.2 \times 0.55\ V_s - 3.05 \times 1.2\ V_s}{1.1 - 1}\right\}\right]}{V_s} - \frac{(0.133 \times 0.55\ V_s) + \left(\dfrac{0.55 \times 0.2\ V_s - 0.133 \times 0.65\ V_s}{1.2 - 1}\right)}{V_s}$$

$$= \frac{\left[2.52 + \dfrac{3.96 - 3.66}{0.1}\right] - \left[0.073 + \dfrac{0.11 - 0.086}{0.2}\right]}{1}$$

$$= \frac{(2.52 + 3.0) - (0.073 + 0.12)}{1} = 5.33 \text{ bar.}$$

$\therefore$ **Indicated power, I.P.** $= \dfrac{10\ p_m LAN}{3}$ [$\because$ Engine is double-acting]

$$= \frac{10 \times 5.33 \times 0.48 \times \pi/4 \times 0.32^2 \times 200}{3}$$

= 137.17 kW. (Ans.)

Example 16.10. *A steam locomotive is coupled with two single-cylinder, double-acting engine of 350 mm in diameter and 550 mm stroke. The driving wheels are 2.2 m in diameter. The pressure of steam supplied to the engine is 11 bar and dry saturated. The exhaust pressure of steam is 1.1 bar. The maximum cut-off is 80% of the stroke. Find out the tractive effort at 15 km/h speed with maximum cut-off. Assume the diagram factor 0.8 and mechanical efficiency 70%.*

If the resistance to train is 12 N/1000 N at 90 km/h speed, determine the total train load that can be hauled at this speed if the cut-off is 20% of the stroke. Assume diagram factor at this speed as 0.70 and mechanical efficiency 80%.

Assume that the engine is directly coupled to the driving wheel.

Solution. Number of cylinder, $n = 2$

Diameter of a cylinder, D = 350 mm = 0.35 m
Stroke length, L = 550 mm = 0.55 m
Diameter of a driving wheel, D_w = 2.2 m
Pressure of steam supplied, p_1 = 11 bar
Quality of steam, $x_1 = 1$
Exhaust pressure of steam, p_b = 1.1 bar
Maximum cut-off $= 0.8\ V_s$
Diagram factor, $(D.F.)_1 = 0.8$
Mechanical efficiency = 70%

Tractive effort, P at 15 km/h :

Resistance to train = 12 N/1000 N
Speed = 90 km/h
Cut-off = 0.2 V_s
Diagram factor, $(D.F.)_2 = 0.7$
Mechanical efficiency = 80%

(*i*) **Tractive effort, P :**

$$p_{m(th.)} = \frac{p_1}{r}(1 + \log_e r) - p_b$$

$$= \frac{11}{1/0.8}\left(1 + \log_e \frac{1}{0.8}\right) - 1.1 = 9.66 \text{ bar.}$$

$$p_{m(act.)} = \text{Diagram factor } (D.F.)_1 \times p_{m(th.)}$$

$$= 0.8 \times 9.66 = 7.73 \text{ bar}$$

$$\text{Speed of the train} = \frac{\pi D_w N \times 60}{1000}$$

$$= 15 \quad \text{(given)}$$

$$\therefore \quad N = \frac{15 \times 1000}{\pi \times 2.2 \times 60} = 36.2 \text{ r.p.m.}$$

As the engine is directly coupled to the driving wheel, the rotational speed of the engine is equal to the rotational speed of the driving wheel.

$$\therefore \quad \text{I.P.} = \frac{n \times 10 p_m\, LAN}{3}$$

$$= \frac{2 \times 10 \times 7.73 \times 0.55 \times \pi/4 \times 0.35^2 \times 36.2}{3}$$

[∵ $n = 2$ and the engine is double-acting]

$$= 98.7 \text{ kW}$$

$$\therefore \quad \text{B.P.} = \eta_{mech.} \times \text{I.P.} = 0.7 \times 98.7 = 69.1 \text{ kW}$$

B.P. = Tractive effort × speed in metres/sec.

$$69.1 = P \times \left(\frac{15 \times 1000}{60 \times 60}\right), \text{ where } P \text{ is in kN}$$

$$\therefore \quad P = \frac{69.1 \times 3600}{15 \times 1000} = 16.584 \text{ kN.}$$

i.e., *Tractive effort* = **16.584 kN.** **(Ans.)**

(*ii*) **Total train load W_t :**

$$p_{m(th.)} = \frac{p_1}{r}(1 + \log_e r) - p_b$$

$$= \frac{11}{5}(1 + \log_e 5) - 1.1 \qquad \left[\because \ r = \frac{1}{0.2} = 5\right]$$

$$= 4.64 \text{ bar}$$

$$p_{m(act.)} = \text{Diagram factor } (D.F.)_2 \times p_{m(th.)}$$

$$= 0.7 \times 4.64 = 3.25 \text{ bar}$$

$$\text{Speed of the engine} = 36.2 \times \frac{90}{15} = 217.2 \text{ r.p.m.}$$

$$\text{I.P.} \propto p_{m(act.)} N$$

∴ I.P. at net cut-off and new speed is given by

$$\text{I.P.} = 98.7 \times \frac{3.25}{7.73} \times \frac{217.2}{36.2} = 248.98 \text{ kW}$$

$$\text{B.P.} = \eta_{mech.} \times 248.98 = 0.8 \times 248.98 = 199.2 \text{ kW}$$

$$\text{Tractive effort,} \quad P = \frac{\text{B.P.}}{\text{Speed in metre / s}},$$

where P is kN

$$= \frac{199.2}{\left(\frac{90 \times 1000}{60 \times 60}\right)} = 7.968 \text{ kN}$$

∴ *Total train load,* W_t

$$= \frac{7.968\,(\text{kN}) \times 1000\,(\text{N})}{12\,(\text{N})} = \textbf{664 kN.} \quad \textbf{(Ans.)}$$

Example 16.11. *A steam engine is governed by varying the initial steam pressure, used 6 kg/kWh when developing 60 kW and 4.65 kg/kWh when developing 150 kW. If steam supplied is 96% dry at 15 bar and a back pressure of 0.36 bar, determine the steam consumption per hour and indicated thermal efficiency when developing 120 kW.*

Solution. Refer to Fig. 16.37.

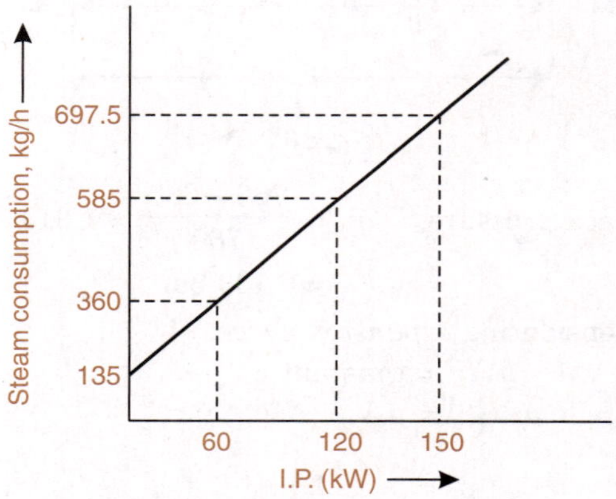

Fig. 16.37

Steam consumption/hour developing 60 kW
$= 60 \times 6 = 360$ kg/h

Steam consumption/hour developing 150 kW
$= 150 \times 4.65 = 697.5$ kg/h

In throttle governing, Willan's law holds good,

i.e., $m = a + b \times \text{I.P.}$

$$360 = a + b \times 60 \quad ...(i)$$

$$697.5 = a + b \times 150 \quad ...(ii)$$

[m = Steam consumption (kg/h)
a = Steam consumption at no-load (kg/h)
b = Slope of Willan's line.]

Subtracting (*i*) from (*ii*), we get

$$337.5 = 90\, b$$

$\therefore \quad b = \dfrac{337.5}{90} = 3.75$

Inserting the value of b in (*i*), we get

$360 = a + 3.75 \times 60$

$\therefore \quad a = 135.$

Steam consumption for developing 120 kW

$= 135 + 3.75 \times 120 =$ **585 kg/h.** **(Ans.)**

Heat supplied/kg $= h_1 - h_{f_2}$

From steam tables :

At 15 bar : $h_{f_1} = 844.7$ kJ/kg,

$h_{fg_1} = 1945.2$ kJ/kg.

At 0.36 bar : $h_{f_2} = 307.1$ kJ/kg.

$\therefore$ Heat supplied/kg

$= (844.7 + 0.96 \times 1945.2) - 307.1$

$= 2405$ kJ/kg.

Indicated thermal efficiency (developing 120 kW)

$$\eta_{th(I)} = \frac{\text{I.P.}}{\dot{m}_s\,(h_1 - h_{f_2})} = \frac{120}{\frac{585}{3600} \times 2405}$$

$= \mathbf{0.3070 \text{ or } 30.7\%}.$ **(Ans.)**

Example 16.12. *The following readings were taken during the test at full load on a single-cylinder, double-acting condensing type throttle governed steam engine :*

Diameter of the cylinder	*400 mm*
Stroke of the engine	*600 mm*
Cut-off	*50% of stroke*
Pressure of steam supplied	*11 bar*
Back pressure	*0.8 bar*
Brake wheel diameter	*4.5 m*
Net load on the brake	*4900 N*
Speed of the engine	*150 r.p.m.*
Diagram factor	*0.82*

(*i*) *Find the indicated power, brake power and mechanical efficiency of the engine at full load.*

(*ii*) *If the load on the brake is reduced to 50% of the full load, find the pressure of steam entering the cylinder. Assume the diagram factor at reduced load as 0.82 and mechanical efficiency 70% of the mechanical efficiency at full load.*

Solution. Theoretical mean effective pressure,

$$p_{m(th.)} = \frac{p_1}{r}(1 + \log_e r) - p_b$$

$$= \frac{11}{2}(1 + \log_e 2) - 0.8 \qquad \left(r = \frac{1}{0.5} = 2\right)$$

$= 8.51$ bar

Actual mean effective pressure,

$$p_{m(act.)} = (\text{D.F.}) \times p_{m(th.)} = 0.82 \times 8.51 = 6.98 \text{ bar}$$

(*i*) $\therefore \quad \text{I.P.} = \dfrac{10\, p_m\, LAN}{3}$

$$= \frac{10 \times 6.98 \times 0.6 \times \pi/4 \times 0.4^2 \times 150}{3}$$

$= \mathbf{263.14 \text{ kW}}.$ **(Ans.)**

$$\text{B.P.} = \frac{(W - S)\pi DN}{60 \times 1000} = \frac{4900 \times \pi \times 4.5 \times 150}{60 \times 1000}$$

$= \mathbf{173.18 \text{ kW}}.$ **(Ans.)**

$$\eta_{mech.} = \frac{\text{B.P.}}{\text{I.P.}} = \frac{173.18}{263.14}$$

$= \mathbf{0.658 \text{ or } 65.8\%}.$ **(Ans.)**

(*ii*) When the load on the brake is reduced to 50% of full-load, B.P. developed at reduced load is given by

$(\text{B.P.})_2 = 173.18 \times 0.5 = 86.6$ kW

$\eta_{mech.}$ at reduced load

$= 0.7 \times 0.658 = 0.46$ or 46%

$\therefore$ $(\text{I.P.})_2$ at reduced load

$$= \frac{(\text{B.P.})_2}{\eta_{mech.}} = \frac{86.6}{0.46} = 188.26 \text{ kW}$$

For governed engine :

$$\frac{(\text{I.P.})_1}{(\text{I.P.})_2} = \frac{p_{m(th.)_1}}{p_{m(th.)_2}}$$

$$\frac{263.14}{188.26} = \frac{8.51}{p_{m(th.)_2}}$$

$$\therefore \quad p_{m(th.)_2} = \frac{8.51 \times 188.26}{263.14} = 6.09 \text{ bar}$$

$$p_{m(th.)_2} = \frac{p_2}{r}(1 + \log_e r) - p_b$$

$$\therefore \quad p_2 = \frac{(p_{m(th.)_2} + p_b) r}{(1 + \log_e r)} = \frac{(6.09 + 0.8) \times 2}{(1 + \log_e 2)}$$

$= 8.14$ bar

i.e., *Pressure of the steam at new load on the brake wheel*

$= \mathbf{8.14 \text{ bar}}.$ **(Ans.)**

☞ **Example 16.13.** *The following observations were made during a trial on a single-cylinder double-acting condensing type steam engine (running at part load).*

Pressure of steam supply	*= 8 bar*
Quantity of condensate available from condenser	*= 95 kg/h*
Quality of steam supplied to the engine	*= dry and saturated*

Vacuum in condenser	= *300 mm of mercury*
Engine speed	= *150 r.p.m.*
m.e.p. on cover end side	= *1.52 bar*
m.e.p. on crank end side	= *1.22 bar*
Diameter of the cylinder	= *200 mm*
Length of the stroke	= *300 mm*
Piston rod diameter	= *38 mm*
Net load on the brake	= *950 N*
Effective diameter of brake wheel	= *0.75 m*
Quantity of cooling water used	= *45 kg/min*
Temperature rise of cooling water	= *18°C*
Condensate temperature	= *55°C*

Determine :

(i) Mechanical efficiency ;

(ii) Brake thermal efficiency ;

(iii) Specific steam consumption on I.P. basis ;

(iv) Specific steam consumption on B.P. basis.

Draw the heat balance sheet.

Solution. $\text{I.P.} = \dfrac{10\, p_{m_1} LA_1 N}{6} + \dfrac{10\, p_{m_2} LA_2 N}{6}$

$$= \frac{10\, LN}{6}(p_{m_1} A_1 + p_{m_2} A_2)$$

$$= \frac{10 \times 0.3 \times 150}{6}\left[1.52 \times \frac{\pi}{4} \times 0.2^2 + 1.22 \times \frac{\pi}{4}(0.2^2 - 0.038^2)\right]$$

$$= 75\ (0.0477 + 0.0369) = 6.345 \text{ kW}$$

$$\text{B.P.} = \frac{(W - S)\,\pi DN}{60 \times 1000} = \frac{950 \times \pi \times 0.75 \times 150}{60 \times 1000}$$

$$= 5.59 \text{ kW}$$

(i) **Mechanical efficiency, $\eta_{mech.}$:**

$$\eta_{mech} = \frac{\text{B.P.}}{\text{I.P.}} = \frac{5.59}{6.345} = \mathbf{0.881 \text{ or } 88.1\%.} \quad \textbf{(Ans.)}$$

(ii) **Brake thermal efficiency, $\eta_{th(B)}$:**

$$\eta_{th(B)} = \frac{\text{B.P.}}{\dot{m}_s (h_1 - h_{f_2})}$$

$$= \frac{5.59}{\frac{95}{3600}(2767.4 - 229.9)}$$

$$\left[\begin{array}{l} \textit{At 8 bar } (p_1) : h_{f_1} = 720.9 \text{ kJ/kg}, \\ \qquad\qquad h_{fg_1} = 2046.5 \text{ kJ/kg} \\ \therefore \quad h_1 = h_{f_1} + x_1 h_{fg_1} \\ \qquad = 720.9 + 1 \times 2046.5 = 2767.4 \text{ kJ/kg} \\ \text{Condensate temperature} = 55°\text{C} \\ \therefore \quad h_{f_2} = 1 \times 4.18 \times (55 - 0) = 229.9 \text{ kJ/kg} \end{array}\right]$$

= **0.0835 or 8.35%. (Ans.)**

(iii) **Specific steam consumption on I.P. basis**

$$= \frac{95}{6.345} = \mathbf{14.97 \text{ kg/kWh.}} \quad \textbf{(Ans.)}$$

(iv) **Specific steam consumption on B.P. basis**

$$= \frac{95}{5.59} = \mathbf{16.99 \text{ kg/kWh.}} \quad \textbf{(Ans.)}$$

$$\text{Heat supplied/min} = \frac{95}{60} \times h_1$$

$$= \frac{95}{60} \times 2767.4 = 4381.7 \text{ kJ/min}$$

Heat equivalent of B.P. = 5.59 × 60 = 335.4 kJ/min

Heat taken away be cooling water

= 45 × 4.18 × 18 = 3385.8 kJ/min

Heat in condensate per minute

$$= \frac{95}{60} \times 4.18 \times (55 - 0) = 364 \text{ kJ/min}$$

Heat unaccounted for (by difference)

= 4381.7 – (335.4 + 3385.8 + 364)

= 296.5 kJ/min

Heat balance sheet on minute basis :

Particulars	*Heat (kJ/min)*	*Percentage (%)*
Heat supplied	*4381.7*	*100*
(i) Heat equivalent of B.P.	335.4	7.65
(ii) Heat carried away by cooling water.	3385.8	77.27
(iii) Heat in condensate.	364.0	8.31
(iv) Heat unaccounted for	296.5	6.76

ADDITIONAL/TYPICAL EXAMPLES

Example 16.14. *A single-cylinder double-acting steam engine with 15 cm bore and 20 cm stroke is to develop 20 kW at 300 r.p.m. with cut-off occurring at 20% of stroke. Back pressure is 0.28 bar. Determine admission pressure if diagram factor is 0.72. Also calculate indicated thermal efficiency if engine receives 222 kg of dry steam per hour. Neglect area of piston rod.* **(AMIE)**

Solution. Engine bore, D = 15 cm = 0.15 m

Stroke length, L = 20 cm = 0.2 m

Speed of the engine, N = 300 r.p.m.

Power developed, I.P. = 20 kW

Diagram factor, D.F. = 0.72

Steam (dry) supplied per hour = 222 kg/h

Cut-off ratio, $\dfrac{V_1}{V_2} = 0.2$

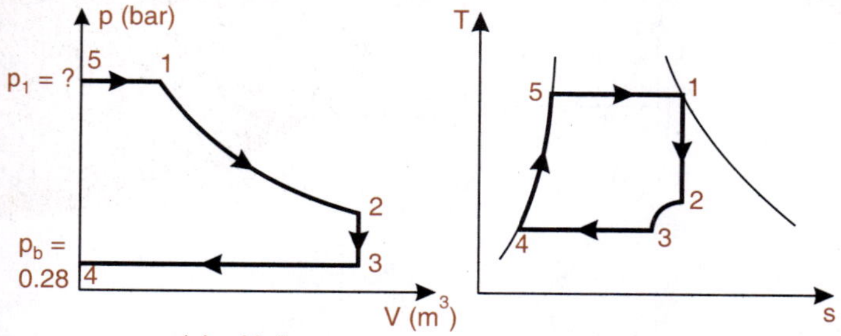

(a) p-V diagram (b) T-s diagram

Fig. 16.38

Admission pressure, p_1 :

Indicated power,

$$\text{I.P.} = \frac{p_m LAN \times 10}{3} \text{ kW} \qquad \text{[Eqn. (16.13)]}$$

$$20 = \frac{p_m \times 0.20 \times \frac{\pi}{4} \times 0.15^2 \times 300 \times 10}{3}$$

$$= 3.534\, p_m$$

$$\therefore \quad p_m = \frac{20}{3.534} = 5.659 \text{ bar}$$

$$(p_m)_{th.} = \frac{p_m}{\text{D.F.}} = \frac{5.659}{0.72} = 7.86 \text{ bar}$$

Also, $(p_m)_{th.} = \frac{p_1}{r} [1 + \log_e r] - p_b$,

where, $r = \left(\text{Expansion ratio} = \frac{V_2}{V_1}\right) = \frac{1}{0.2} = 5$

$$\therefore \quad 7.86 = \frac{p_1}{5} [1 + \log_e 5] - 0.28$$

or $= 0.522\, p_1 - 0.28$

or $p_1 = \frac{(7.86 + 0.28)}{0.522} = $ **15.59 bar. (Ans.)**

Indicated thermal efficiency, $\eta_{th.\,(I)}$:

From steam tables :

$h_1 = 2793.4$ kJ/kg at 15.59 bar ($x = 1$) ;

$h_{f_2} = 282.7$ kJ/kg at 2.8 bar

Heat supplied to steam

$$= \frac{222}{3600} (h_1 - h_{f_2})$$

$$= \frac{222}{3600} (2793.4 - 282.7) = 154.83 \text{ kJ/s}$$

$$\therefore \quad \eta_{th.(I)} = \frac{20}{154.83}$$

= **0.1292 or 12.92%. (Ans.)**

Example 16.15. *A double-acting single-cylinder steam engine runs at 250 r.p.m. and develops 30 kW. The pressure limits of operation are 10 bar and 1 bar. Cut-off is 40% of the stroke. The L/D ratio is 1.25 and the diagram factor is 0.75. Assume dry saturated steam at inlet, hyperbolic expansion and negligible effect of piston rod. Find :*

(i) Mean effective pressure,

(ii) Cylinder dimensions, and

(iii) Indicated thermal efficiency.

(D.U.)

Solution. Speed of the steam engine,

$N = 250$ r.p.m.

Power developed, $P = 30$ kW

Pressure limits of operation : 10 bar (p_1), 1 bar (p_b)

Cut-off ratio, $r = \frac{1}{0.4} = 2.5$

L/D ratio $= 1.25$

Diagram factor, D.F. = 0.75

Condition of steam at inlet to the engine = Dry saturated.

(*i*) **Mean effective pressure, p_m :**

$$p_m = \text{D.F.} \left[\frac{p_1}{r} (1 + \log_e r) - p_b\right]$$

$$= 0.75 \left[\frac{10}{2.5} (1 + \log_e 2.5) - 1\right]$$

$\simeq$ **5.0 bar. (Ans.)**

(*ii*) **Cylinder dimensions, L and D :**

Indicated power, I.P. $= \frac{10\, p_m LAN}{3}$ kW

or $$30 = \frac{10 \times 5.0 \times 1.25 D \times \frac{\pi}{4} D^2 \times 250}{3} = 4090.6\, D^3$$

or $D^3 = \frac{30}{4090.6}$ or Cylinder dia., $D = \left(\frac{30}{4090.6}\right)^{1/3}$

= **0.194 m or 194 mm. (Ans.)**

Length of stroke, $L = 1.25\, D$

$= 1.25 \times 194$

= **242.5 mm. (Ans.)**

(*iii*) **Indicated thermal efficiency, $\eta_{th(I)}$:**

Mean flow rate of steam,

$$\dot{m}_s = \frac{\frac{\pi}{4} D^2 \times L \times \frac{1}{r} \times 2 \times N}{v_g}$$

$$= \frac{\frac{\pi}{4} \times (0.194)^2 \times 0.2425 \times \frac{1}{2.5} \times 2 \times \frac{250}{60}}{0.194}$$

$= 0.1231$ kg/s

[where v_g = specific volume of dry saturated steam at 10 bar = 0.194 m³/kg (from steam tables)]

∴ Indicated thermal efficiency,

$$\eta_{th(I)} = \frac{\text{I.P.}}{\dot{m}_s (h_1 - h_f)}$$

$$= \frac{30}{0.1231(2776.2 - 417.5)}$$

$= \mathbf{0.1033 \text{ or } 10.33\%}$. **(Ans.)**

[where, h_1 = Enthalpy of dry saturated steam at 10 bar = 2776.2 kJ/kg, and

h_f = Enthalpy of water at 1 bar = 417.5 kJ/kg (from steam tables)]

Example 16.16. *Estimate the dryness fraction of steam in a cylinder at 0.7 of the stroke from the following data :*

r.p.m. = 100 ; cut-off = 0.5 stroke ; steam condensed/min = 44.95 kg/min ; clearance = 8% ; swept volume = 0.1062 m³ ; pressure of steam at 0.7 stroke = 4.13 bar ; pressure of steam at 0.8 of return stroke on compression curve = 1.31 bar ; volume of 1 kg of steam at 4.13 bar = 0.438 m³ ; volume of 1 kg of steam at 1.31 bar = 1.296 m³. **(N.U.)**

Solution. Mass of steam used per stroke or cylinder feed

$$= \frac{44.95}{100} = 0.4495 \text{ kg.}$$

Clearance volume

$= 0.08\, V_s = 0.08 \times 0.1062 = 0.008496 \text{ m}^3$

Volume at M on compression curve,

$V_M = 0.008496 + (1 - 0.8) \times 0.1062$

$= 0.02974 \text{ m}^3$

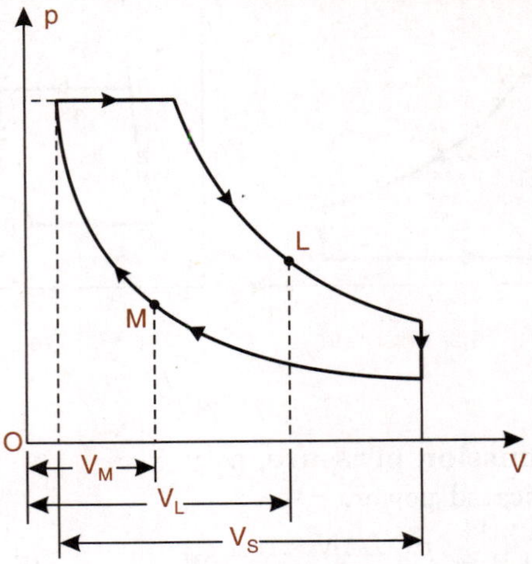

Fig. 16.39

∴ Mass of cushion steam

$$= \frac{V_M}{v} = \frac{0.02974}{1.296} = 0.02295 \text{ kg}$$

∴ Total mass of steam during expansion stroke

$= 0.4495 + 0.02295 = 0.4725$ kg

Volume of steam at L

$= 0.008494 + 0.7 \times 0.1062 = 0.08283 \text{ m}^3$

∴ **Dryness fraction at L**

$$= \frac{\text{Volume of steam in the cylinder at } L}{\text{Volume of steam at } L, \text{ if dry}}$$

$$= \frac{0.08283}{0.4725 \times 0.438} = \mathbf{0.4.} \quad \textbf{(Ans.)}$$

HIGHLIGHTS

1. The **heat engine** is an energy converter, and transforms the chemical energy in the fuel supplied to it into mechanical work.
2. The steam engine may be said to be *characterised* by its ruggedness, by its high torque at slow speed and by its practical inability to expand steam to a pressure lower than 0.14 bar in the cylinder because of the huge volume that would to required.
3. An **indicator diagram** is a plot of the variations of the steam pressure and volume of the steam in the cylinder during the cycle of operations.
4. **'Diagram factor'** is defined as the ratio of the area of actual indicator diagram and the area of the hypothetical diagram.
5. Methods of **reducing condensation** in a steam engine are :
   (*i*) By superheating.
   (*ii*) By using jacketed cylinders.
   (*iii*) By compounding the cylinders.
   (*iv*) By increasing the speed of the engine.
   (*v*) By large compression of steam during last part of exhaust.
6. **Mean effective pressure** (m.e.p. or p_m) is the constant uniform pressure which if acting on the face of the piston throughout the stroke would have produced the same work area, as obtained under actual working conditions.
7. Theoretical m.e.p. (p_m) :
   (*i*) **Without clearance and compression :**

$$p_m = \frac{p_1}{r}(1 + \log_e r) - p_b$$

   (*ii*) **With clearance :**

$$p_m = p_1 \left[\frac{1}{r} + \left(c + \frac{1}{r} \right) \log_e \frac{(c+1)}{\left(c + \frac{1}{r} \right)} \right] - p_b$$

(*iii*) **With clearance and compression :**

$$p_m = p_1\left[\frac{1}{r}+\left(c+\frac{1}{r}\right)\log_e \frac{(c+1)}{\left(c+\frac{1}{r}\right)}\right] - p_b\left[(1-\alpha)+(c+\alpha)\log_e\left(\frac{\alpha+c}{c}\right)\right]$$

where, $c = \frac{V_c}{V_s}$

$\frac{1}{r} = \frac{V_1 - V_c}{V_s}$

and α = Ratio of the volume between points of compression and admission to the swept volume V_s.

8. An **indicator** is an instrument in which *p-V* diagram of an engine is produced automatically. Two simple types of indicators are :
(*i*) Pencil indicator (*ii*) Optical indicator.

9. **Indicated power** is the *power developed* in the engine cylinder. Indicated power is given by

$$\text{I.P.} = \frac{10 p_m LAN}{6} \text{ kW}$$

... for single-acting steam engine

$$= \frac{10 p_m LAN}{3} \text{ kW}$$

... for double-acting steam engine (*neglecting area of piston rod*)

$$= \frac{10 p_{m_1} LA_1 N}{6} + \frac{10 p_{m_2} LA_2 N}{6}$$

... for double-acting steam engine *if area of piston rod is considered.*

10. From an indicator diagram m.e.p. (p_m) can be calculated as follows :

$$p_m = \frac{A_i \times S_i}{L_i} \text{ bar}$$

where, A_i = Area of indicator diagram (mm²)
S_i = Spring scale or spring number (bar/mm)
L_i = Length of indicator diagram (mm).

11. **Brake power** (B.P.) is the actual power available from the engine for doing the useful work.

$$\text{B.P.} = \frac{(W-S)\pi(D+d)N}{60 \times 10^3} \text{ kW}$$

where, $(W - S)$ = Net load or frictional force acting on the drum/wheel (N),
$(D + d)$ = Effective diameter of the drum/where (m), and
N = Engine speed (r.p.m.).

12. **Saturation curve** is the curve showing the volume the steam in the cylinder would occupy, during the expansion stroke if the steam is perfectly dry and saturated at all the points.
Missing quantity is the horizontal distance between the actual expansion curve and the saturation curve.

13. A governor is used to keep the speed of the engine constant by either controlling the quantity of steam or the pressure of the steam supplied to the engine as per load requirements of the engine. The following two methods are commonly used for governing the steam engine :
(*i*) Throttle governing (*ii*) Cut-off governing.

OBJECTIVE TYPE QUESTIONS

Choose the Correct Answer :

1. In a steam engine D-slide valve controls the flow of steam into and out of the steam engine cylinder
(*a*) through one admission port and two exhaust ports
(*b*) through one admission port and one exhaust port
(*c*) through one exhaust port and two admission ports
(*d*) through two exhaust ports and two admission ports.

2. In a steam engine
(*a*) piston rod is connected to cross-head by gudgeon pin
(*b*) connecting rod is connected to crankshaft by gudgeon pin
(*c*) connecting rod is connected to piston by gudgeon pin
(*d*) connecting rod is connected to cross-head by gudgeon pin.

3. For a steam engine the 'diagram factor' is defined as
(*a*) area of hypothetical indicator diagram × area of actual indicator diagram
(*b*) $\frac{\text{area of actual indicator diagram}}{\text{area of hypothetical indicator diagram}}$
(*c*) $\frac{\text{area of hypothetical indicator diagram}}{\text{area of actual indicator diagram}}$
(*d*) none of the above.

4. The function of D-slide valve in steam engine is
(*a*) only to exhaust steam from the cylinder
(*b*) only to admit steam in the cylinder
(*c*) to admit steam and also exhaust steam from the cylinder
(*d*) none of the above.

5. The component that imparts motion to the D-slide valve of steam engine is
(*a*) flywheel (*b*) eccentric
(*c*) crank (*d*) cam.

6. For throttle governing of steam engine
(*a*) rate of steam consumption varies inversely as the speed of the engine
(*b*) rate of steam consumption varies directly as the speed of the engine
(*c*) rate of steam consumption is directly proportional to I.P. of the engine
(*d*) rate of steam consumption is inversely proportional to I.P. of the engine.

7. Willan's line is a curve plotted of steam consumption rate in steam engines versus
(*a*) speed (*b*) I.P.
(*c*) F.P. (*d*) B.P.

ANSWERS

1. (*c*) 2. (*d*) 3. (*b*) 4. (*c*)
5. (*b*) 6. (*b*) 7. (*b*).

THEORETICAL QUESTIONS

1. Define an heat engine. How are heat engines classified ?
2. Define a reciprocating steam engine and state how the reciprocating steam engines are classified.
3. Using neat sketches enumerate and explain the various parts of a steam engine (reciprocating).
4. Explain the following terms as applied to steam engine : Cylinder bore, Piston stroke, Crank throw, Average piston speed, Dead centres, Swept volume and clearance.
5. Define the term 'indicator diagram' and explain how it is obtained.
6. Explain with a neat sketch the construction and working of a Crosby pencil indicator.
7. Define the term 'Diagram factor'.
8. Enumerate and explain briefly the various methods of reducing cylinder condensation in a steam engine.
9. Define the term 'mean effective pressure' and derive its expression in the following cases :
   (*i*) Theoretical indicator diagram without clearance and compression.
   (*ii*) Theoretical indicator diagram with clearance.
   (*iii*) Theoretical indicator diagram with clearance and compression.
10. Explain with a neat sketch a dynamometer commonly used for measuring brake power.
11. Explain the following efficiencies as applied to steam engine :
    (*i*) Mechanical efficiency. (*ii*) Thermal efficiency.
    (*iii*) Relative efficiency. (*iv*) Overall efficiency.
12. Explain clearly what do you mean by 'saturation curve' and 'missing quantity'. Suggest the ways by which missing quantity can be reduced.
13. Explain briefly the following methods of governing :
    (*i*) Throttle governing. (*ii*) Cut-off governing.
14. Explain briefly the following valves :
    (*i*) D-slide valve, (*ii*) Balanced slide valve,
    (*iii*) Piston valve, (*iv*) Drop valve, and
    (*v*) Corliss valve.

UNSOLVED EXAMPLES

1. A double-acting single-cylinder steam engine is required to develop 70 kW at 250 r.p.m. using steam supplied at 11 bar dry saturated, and exhausting into a condenser at 0.85 bar. The cut-off ratio is 0.25 and the stroke/bore ratio is 1.3 : 1. The mechanical efficiency of the engine can be taken as 85%. Calculate on the basis of a hypothetical diagram, the stroke and bore of the engine, assuming the diagram factor to be 0.8. If the brake thermal efficiency of the cycle is 15%, calculate the specific steam consumption of the engine. Neglect the cross-section area of the piston rod. [**Ans.** D = 277 mm, L = 360 mm, 10.07 kg/kWh]
2. A single-cylinder, double-acting steam engine is supplied with steam at 11 bar, dry saturated. The bore is 300 mm and the stroke is 375 mm. The engine runs at 250 r.p.m. with a cut-off ratio of 1/3, and against a back pressure of 0.34 bar. If the diagram factor is 0.8, calculate the I.P. If the mechanical efficiency is 85%, calculate the B.P., and the specific steam consumption in kg/kWh ; the brake thermal efficiency of the cycle is 15%. The enthalpy of the feed water to the boiler is that of water at the saturation temperature corresponding to the condenser pressure. [**Ans.** 130.5 kW ; 110 kW ; 9.68 kWh]
3. A two-cylinder locomotive steam engine is supplied with steam at 12 bar. Cut-off occurs at 40% of the stroke and the back pressure is 12 bar. Diagram factor may be taken as 0.83 and mechanical efficiency 90%. If the length of stroke for both the cylinders is 650 mm, diameter of driving wheels 1.92 m and the engine produces a tractive effort of 22.5 kN at the driving wheels, determine the cylinder bore and the indicated power when engine runs at 72 km per hour. [**Ans.** 330 mm ; 500 kW]
4. A single cylinder double acting steam engine has a bore of 239 mm and a stroke of 299 mm. Steam is admitted at a pressure of 12.5 bar. Cut-off occurs at 30% of stroke and the exhaust pressure is 1.3 bar. The diagram factor is 0.82 and the mechanical efficiency is 78%. The engine runs at 150 r.p.m. Determine the power delivered. [**Ans.** 30 kW]
5. Steam is supplied to a double acting engine running at 120 r.p.m. and developing 37.5 kW at a pressure of 5 bar and is exhausted at a pressure of 0.34 bar. Cut-off occurs at $\frac{1}{3}$rd of stroke. Assuming a diagram factor of 0.7 and a mechanical efficiency of 0.85, determine the swept volume of the cylinder. [**Ans.** 0.05 m^3]
6. A single-cylinder double-acting steam engine develops 75 kW when steam is admitted at 10.35 bar. The cut-off is at $\frac{1}{3}$rd of the stroke and the back pressure is 0.275 bar. The diagram factor is 0.7 and the mechanical efficiency 84%. If the mean piston speed is 240 m/min and the stroke/bore ratio is 1.2, calculate the bore and stroke of the engine. [**Ans.** D = 242 mm, L = 291 mm]
7. Steam at 15.4 bar is supplied to a double acting steam engine of 150 mm diameter and 200 mm stroke running at 300 r.p.m. The cut-off occurs at 20% of the stroke. The exhaust pressure is 0.28 bar. Determine the power developed by the engine if the diagram factor is 0.72. Also calculate the indicated thermal efficiency of the engine if the steam supplied is 222 kg per hour. Neglect the clearance and assume the steam is dry saturated when supplied to the engine. [**Ans.** 20 kW ; 13%]

8. Calculate a suitable admission pressure for a double-acting single-cylinder steam engine of 230 mm bore and 300 mm stroke to develop 27 kW at a speed of 210 r.p.m. The exhaust pressure is 1.3 bar and expansion ratio is 2.5. Assume a diagram factor of 0.8. [**Ans.** 6.25 bar]

9. Dry saturated steam at 9 bar is supplied to a simple-cylinder double-acting engine developing 22.5 kW running at 240 r.p.m. The exhaust pressure is 1.4 bar. Cut-off takes place at 0.4 of stroke. Assuming a diagram factor of 0.8, and stroke to bore ratio of 1.25 and hyperbolic expansion, determine the bore and stroke of the engine.

Calculate also the steam consumption/hour.

[**Ans.** 34 kg]

10. A single-cylinder vertical steam engine is double-acting and runs at 120 r.p.m. It has a bore of 215 mm and a stroke of 305 mm. The piston rod diameter is 37 mm. The indicated area on the head end was 1050 mm^2 and on the piston end was 1020 mm^2. The length of the indicated card area was 76 mm. The spring scale was 0.8 bar/cm. Determine power developed. Also find the power delivered if the mechanical efficiency was 80%. Calculate also the brake torque.

[**Ans.** 4.76 kW, 303 Nm]

11. A single-cylinder double-acting steam engine receives steam at 10 bar and exhausts at 0.55 bar. The dryness of inlet steam is 0.96. The power developed by the engine when running at 210 r.p.m. is 45 kW, with steam consumption of 460 kg/h. The diameter of the engine is 250 mm and the stroke is 375 mm. The expansion ratio is 6, calculate the diagram factor and the indicated thermal efficiency of the engine. [**Ans.** 0.85 ; 14.9%]

12. Following is the data for a single-cylinder, double-acting steam engine :

Piston diameter = 350 mm ; stroke = 500 mm ; clearance volume = 10% of swept volume ; r.p.m. = 200 ; steam supply pressure = 10 bar ; cut-off at 35% of stroke ; law of expansion for steam $pV^{1.22}$ = constant ; exhaust pressure = 0.15 bar ; compression commences at 75% of return stroke ; law of compression for steam, $pV^{1.31}$ = constant ; diagram factor = 0.88.

Determine the indicated power of the engine.

[**Ans.** 196.6 kW]

13. A single-cylinder double-acting steam engine is supplied steam at 12 bar and the steam supply is cut-off at 40% of the stroke. The steam is exhausted from the cylinder at 1.2 bar. The compression commences at 0.85 of the return stroke and the clearance volume is 10% of the displacement volume. Determine : (*i*) the mean effective pressure (*ii*) cylinder dimensions when engine is designed to develop 30 kW at brakes at 120 r.p.m. Take stroke/bore ratio as 1.1. Assume hyperbolic expansion and take mechanical efficiency as 78%. The engine has a mean piston speed of 75 m/min.

[**Ans.** 8.235 bar, D = 284 mm, L = 312.5 mm]

14. The following is the data for a single-cylinder, double-acting, condensing type throttled governed engine, run at full load : Pressure of steam supplied = 10 bar ; back pressure = 0.6 bar ; cylinder diameter = 400 mm ; stroke = 600 mm ; cut-off = 40% of stroke ; brake wheel diameter = 4.2 m ; net load on brake = 5000 N ; r.p.m. = 200 ; diagram factor = 0.82.

Determine :

(*i*) Indicated power,

(*ii*) Brake power,

(*iii*) Mechanical efficiency at full load, and

(*iv*) If the load on the brakes is reduced to 40% of the full load, what will be pressure of the steam entering the cylinder ? Assume diagram factor at full-load as 0.82 and mechanical efficiency as 70% of its value at full load. [**Ans.** (*i*) 283.9 kW, (*ii*) 219.9 kW, (*iii*) 77.46%, (*iv*) 6.05 bar]

15. The following data refer to a test conducted on a single-cylinder, double-acting condensing type steam engine : Pressure of steam supply = 8.5 bar ; temperature of steam as supplied to the engine = 172.29°C ; vacuum in the condenser = 200 mm mercury ; engine speed = 160 r.p.m. ; diameter of cylinder = 230 mm ; length of stroke = 320 mm ; piston rod diameter = 40 mm ; net load on brake = 980 N ; effective brake wheel radius = 450 mm ; rate of cooling water through condenser = 55 kg/min, rise in temperature of cooling water = 17.5°C ; temperature of condensate = 54°C ; condensate coming out of condenser = 97 kg/h.

If the mean effective pressure on cover side is 1.6 bar and that on crank side is 1.25 bar, calculate :

(*i*) Indicated power

(*ii*) Brake power

(*iii*) Mechanical efficiency

(*iv*) Brake thermal efficiency

(*v*) Specific steam consumption based on indicated power. [**Ans.** (*i*) 9.97 kW, (*ii*) 7.39 kW, (*iii*) 74%, (*iv*) 78%, (*v*) 9.729 kg/kWh]

Compound Steam Engines

17

17.1. Introduction. 17.2. Advantages of compound steam engines. 17.3. Classification of compound steam engines. 17.4. Multi-cylinder engines. 17.5. Estimation of cylinder dimensions (compound steam engines). 17.6. Causes of loss of thermal efficiency in compound steam engines. 17.7. The governing of compound steam engines. 17.8. Uniflow steam engine—Worked Examples—Additional/Typical Examples. *Highlights—Objective Type Questions—Theoretical Questions—Unsolved Examples.*

17.1. INTRODUCTION

A simple steam engine is one in which each of the engine cylinder receives steam direct from the boiler, and exhausts into the atmosphere or into a condenser. Compound steam engines are those which have two or more cylinders of successively increasing diameters so arranged that exhaust steam from the first and smallest cylinder is passed forward to do work in a second, and sometimes a third or fourth cylinder, before escaping to the condenser.

'Compounding' is primarily done to secure a *large total ratio of expansion* without the disadvantages attached to a very early cut-off in a single-cylinder. These ***disadvantages*** are :

(*i*) A large range of temperature, accompanied by excessive condensation and re-evaporation effects unless high superheats are used. The range of temperature is not merely the difference between the initial temperature and the exhaust temperature, but includes the fact that the cylinder is heating up during admission and cooling during expansion and exhaust so that with an early cut-off the ratio of the cooling period to the heating period is large.

(*ii*) A cut-off earlier than 30% of the stroke, with an ordinary slide valve, is accompanied by an early release and excessive compression.

(*iii*) When the cut-off is very early the evil effect of clearance is very much accentuated.

(*iv*) With an early cut-off the turning moment on the shaft is more uneven ; as a result a greater flywheel effect is required. This effect, however, is reduced if a number of cylinders and cranks are used.

17.2. ADVANTAGES OF COMPOUND STEAM ENGINES

The compound steam engines claim the following **advantages** over single stage steam engines :

1. Owing to reduction in temperature range per cylinder, the cylinder condensation is minimised.
2. As there is re-evaporation in the early stages of expansion the steam can be expanded still further in the later stages. Consequently, the loss due to condensation in compound steam engines is not cumulative and is generally restricted to low pressure cylinder.
3. The leakage past the valves and piston is minimised since the pressure difference across these components is less.
4. Although the total expansion for the whole engine is large yet the expansion ratio per cylinder is generally not more than four. This permits the use of a simple type of valve gear.
5. The cylinder condensation can be reduced by reheating the steam after expansion in each cylinder.
6. In compound steam engines, the turning moment is more uniform (since the pressure difference is less).
7. In a single stage steam engine if the expansion ratio is large the cylinder has to be made strong enough to withstand the force of high pressure steam. Furthermore it should have large volume to contain low pressure steam.
8. High speeds are possible since mechanical balancing may be made more or less perfect.

9. In compound steam engines since the forces are distributed over more components, the forces in the working parts are reduced.
10. The engine can be started in any position and this quality is of great advantage in marine, locomotive and mining work.
11. If the engine undergoes breakdown it can be modified to run on reduced load ; this is of much help in marine propulsion.
12. For the same power and economy, the cost of the compound engines is less than that of a simple engine.
13. Due to lighter reciprocating parts of an engine, the engine vibrations are reduced.

17.3. CLASSIFICATION OF COMPOUND STEAM ENGINES

The compound steam engines are usually classified as follows :

1. Tandem compound steam engines
2. Cross compound steam engines
   (*i*) Woolfe type compound steam engines
   (*ii*) Receiver type compound steam engines.

1. Tandem compound steam engines (Refer to Fig. 17.1).

In this type of engine, the two cylinders have a common piston rod and are fixed in tandem, working on the same crank. These cylinders may be regarded as having cranks at zero degree to each other.

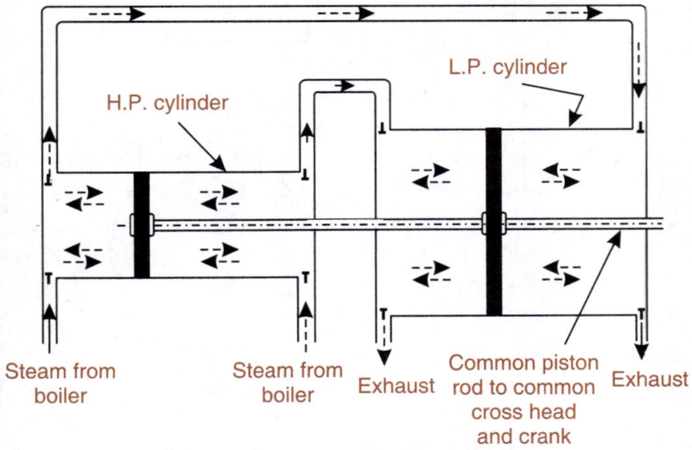

Fig. 17.1. Tandem compound steam engine.

Steam generated in the boiler is supplied to one side of the piston of the high pressure (H.P.) cylinder (shown by solid arrow heads). On the other side of the high pressure piston exhaust takes place simultaneously and this exhaust steam now acts on the piston of the low pressure (L.P.) cylinder (shown by solid arrow heads). The valves should be operated in such a way that there should be continuous admission of steam in the high pressure as well as low pressure cylinder simultaneously. After the steam supply is cut-off in the high pressure cylinder, the steam expands and after the high pressure exhaust steam is admitted to low pressure cylinder upto cut-off point, it further expands to the condenser pressure if the engine is condensing or to the atmosphere for non-condensing engine.

Since usually the steam engines are double-acting, the steam flow takes place during the return stroke as shown by thin arrow heads.

As the cycles of the H.P. and L.P. cylinders are in phase (Fig. 17.2) the maximum turning moment on the crank-shaft, due to each cylinder, will act at the same instant. This is the *disadvantage* of this type of compound engine, as a *large flywheel* is consequently *required. Problems arising out of vibrations and balancing exist in this type of engine.*

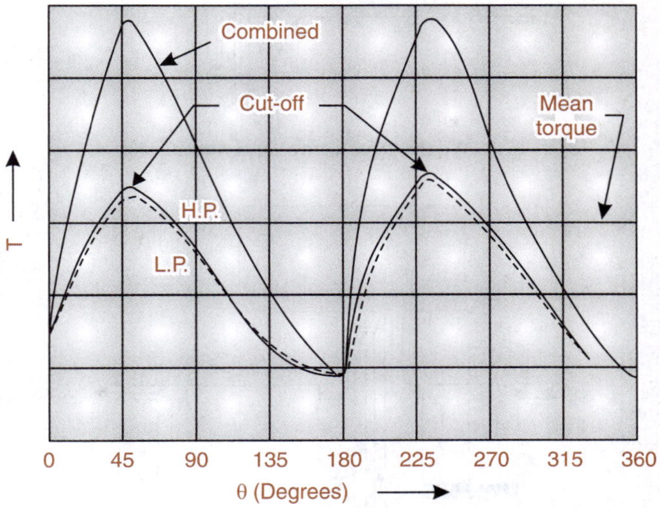

Fig. 17.2

2. Cross compound steam engines

(*i*) **Woolfe type compound steam engines.** Fig. 17.3 shows a schematic diagram of a Woolfe type compound steam engine. In this type of two cylinder compound engine, the cranks of the cylinders are at 180° to each other. The cylinders are arranged side by side, and exhaust steam from H.P. cylinder passes directly into the L.P. cylinder, the expansion is, therefore, continuous during the stroke. As the cranks are at 180° the two cycles are in phase and cause *a large variation in the turning moment* on the crankshaft ; this is the same disadvantage as in the tandem type of compound steam engines.

(*ii*) **Receiver type compound steam engines.** A receiver type compound steam engine is shown in Fig. 17.4. In this type, the steam from one cylinder is not directly discharged into the next cylinder but it is discharged into a chamber known as *'receiver'*. So receiver is nothing but a reservoir of steam from where the steam is admitted into the L.P. cylinder during its admission stroke. In this arrangement, the crank angles are *less than 180°*. In a two cylinder compound engine the angle is 90°.

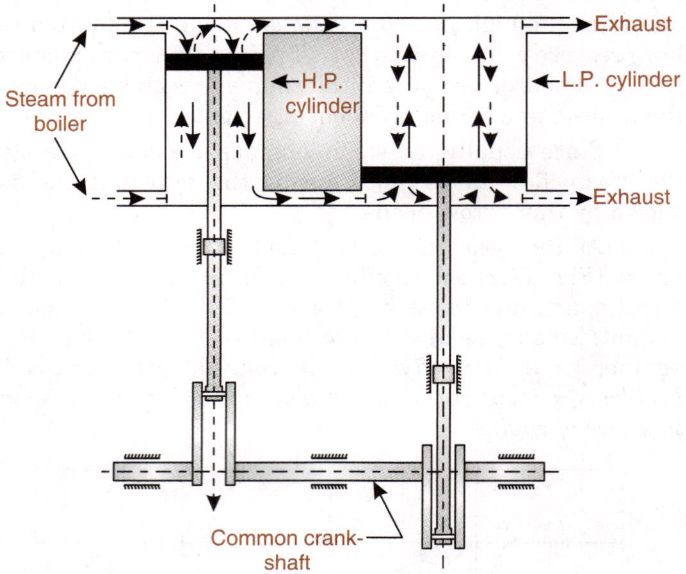

Fig. 17.3. Woolfe type compound steam engine.

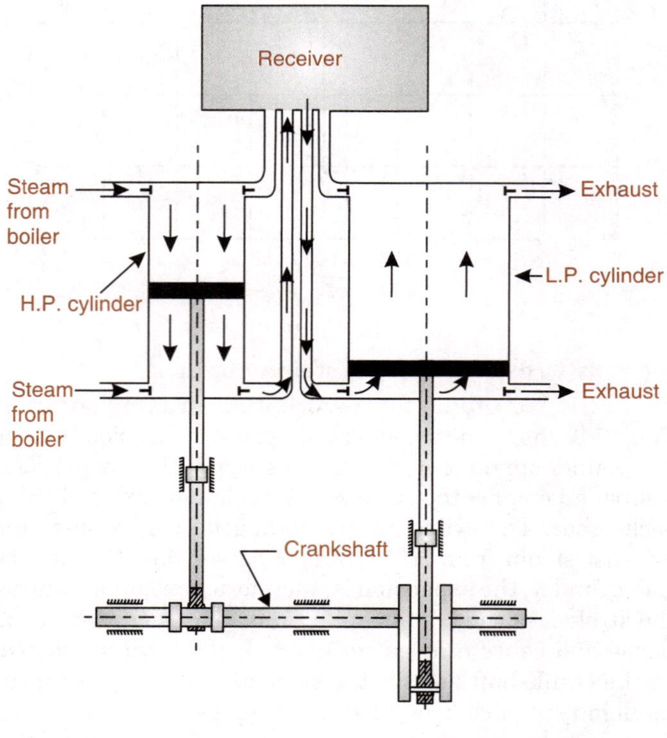

Fig. 17.4. Receiver type compound steam engine.

Fig. 17.5 shows T-θ curve for receiver type compound engine. *The turning moment is more uniform and therefore, a lighter flywheel will be required.*

This type of engine can start in any position. It can also be run at reduced loads, with one cylinder in operation.

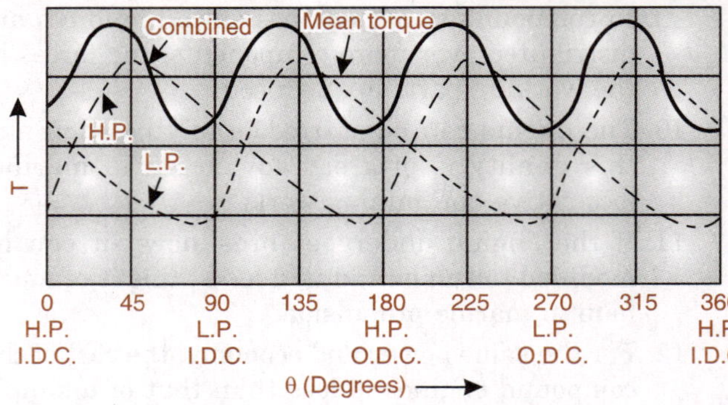

Fig. 17.5. T-θ curve.

There is always an unavoidable pressure drop in the receiver due to condensation of steam but it can be reduced by steam jacketing the receiver. The receiver should be large enough to keep the pressure in it fairly constant ; its volume should be about 1.5 times the H.P. cylinder volume.

According to number of expansion stages, the compound steam engines may also be classified as follows :

1. Double expansion
2. Triple expansion.

1. Double expansion. Refer to Fig. 17.6.

In a double expansion engine, the expansion of steam takes place in two cylinders. The steam is first expanded in high pressure cylinder and then it is discharged into the low pressure cylinder. Finally, it is exhausted into the condenser.

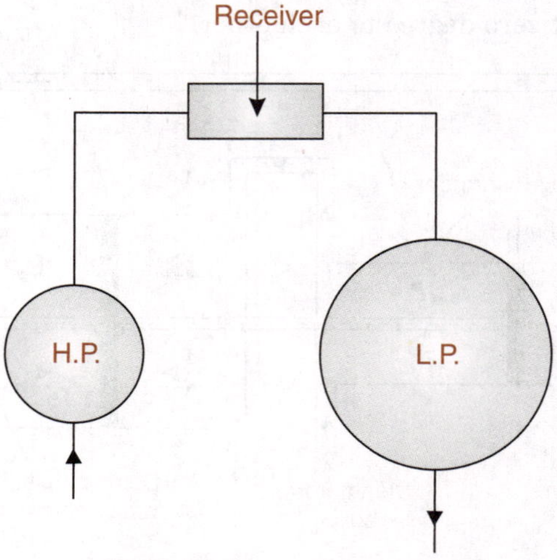

Fig. 17.6. Double expansion engine.

2. Triple expansion. Refer to Fig. 17.7.

In this type of engine, the expansion of steam is completed in three cylinders. The steam from the high pressure cylinder is exhausted into intermediate pressure (I.P.) cylinder and then the steam is discharged into the low pressure cylinder. The steam from the low pressure cylinder is discharged into the condenser.

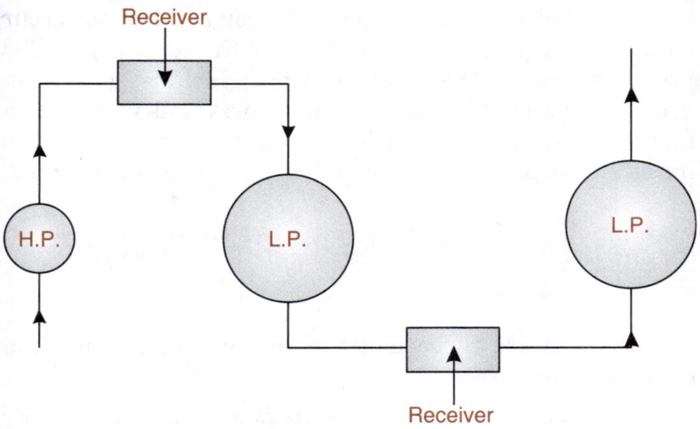

Fig. 17.7. Triple expansion engine.

17.4. MULTI-CYLINDER ENGINES

In a multi-cylinder compound steam engine, the number of cylinders to be provided depends upon the total range of pressure through which the steam is to be expanded.

- Simple engines are usually non-condensing with initial pressure ranging from 6 to 7 bar.
- Compound condensing engines with two cylinders have initial pressures ranging from 7 to 10.5 bar.
- Compound condensing engines with three cylinders (*i.e.,* triple expansion engines) have initial pressures ranging from 10.5 to 14 bar.
- The initial pressure in a quadruple expansion engine (*i.e.,* four cylinder engine) may be from 14 to 21 bar.
- In case of condensing engines the L.P. (release) pressure may vary from 0.7 to 0.85 bar and for non-condensing engines from 1.4 to 1.75 bar.
- In very large engines, for example, when the diameter of L.P. cylinder exceeds 2.5 m, it is usual practice to fit two L.P. cylinders thereby producing a four cylinder triple expansion engine. One H.P. and two L.P. cylinders are used on high speed engines.

17.5. ESTIMATION OF CYLINDER DIMENSIONS (COMPOUND STEAM ENGINES)

The cylinder dimensions, in case of two cylinder compound engine can be determined from the power and the hypothetical indicator diagram. The number of assumptions made will depend on the amount of known data ; there may be :

(*i*) *Equal initial loads/forces on the pistons* ; and

(*ii*) *Equal work in the cylinders.*

Let Fig. 17.8 represent hypothetical indicator diagram for a two-cylinder compound engine, and let there be pressure drop between H.P. cylinder release and the receiver as shown. This pressure drop causes a loss of work represented by the dark area, *in many compound engines this area is a considerable amount and cannot be neglected. Although the pressure drop after release is wasteful, yet it is partly counter-balanced by the drying effect on the steam which it produces.*

Let p_1 = Initial steam pressure (bar),

p_2 = Release pressure in H.P. cylinder (bar),

p_3 = Receiver steam pressure (bar),

p_4 = Steam pressure at release of L.P. cylinder (bar),

p_b = Condenser pressure (bar),

V_1 = Volume at cut-off in H.P. cylinder (m^3),

V_2 = Volume of H.P. cylinder (m^3),

V_3 = Volume at cut-off of L.P. cylinder (m^3),

V_4 = Volume of L.P. cylinder (m^3),

L = Length of stroke (m),

$D_{H.P.}$ = Diameter of H.P. cylinder (m),

$D_{L.P.}$ = Diameter of L.P. cylinder (m),

N = Speed of the engine (r.p.m.), and

$(D.F.)_0$ = Overall diagram factor.

Then, neglecting clearance volume,

H.P. cylinder volume, $V_2 = \frac{\pi}{4} D_{H.P.}^2 L$

L.P. cylinder volume, $V_4 = \frac{\pi}{4} D_{L.P.}^2 L$

Ratio of expansion in H.P. cylinders, $r_{H.P.} = \frac{V_2}{V_1}$

Ratio of expansion in L.P. cylinder, $r_{L.P.} = \frac{V_4}{V_3}$

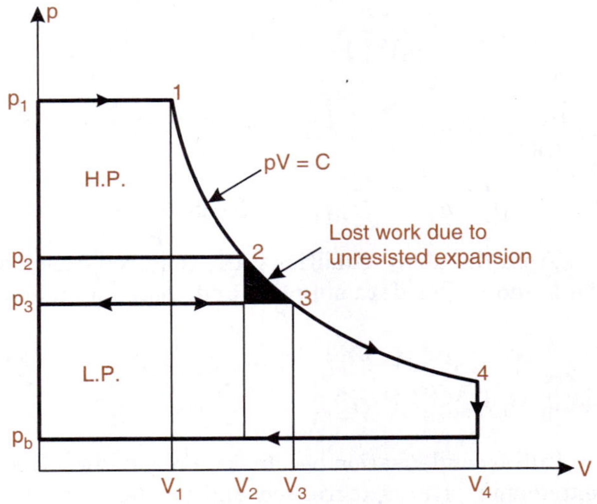

Fig. 17.8. Hypothetical indicator diagram for a two cylinder compound engine.

Ratio of expansion for whole engine, $R = \dfrac{V_4}{V_1}$

As expansion is assumed hyperbolic,

$$p_1V_1 = p_2V_2 = p_3V_3 = p_4V_4$$

Total work done by engine/stroke

$$= (\text{D.F.})_0 \left[p_1V_1 \left(1 + \log_e \frac{V_4}{V_1} \right) - p_bV_4 \right]$$

Total power of engine, neglecting dark area

$$= (\text{D.F.})_0 \left[p_1V_1 \left(1 + \log_e \frac{V_4}{V_1} \right) - p_bV_4 \right] \times \frac{2N \times 10^5}{60 \times 1000} \text{ kW} \quad ...(17.1)$$

($\because$ 1 bar = 10^5 N/m^2)

Area of hypothetical diagram for H.P. cylinder

$$= p_1V_1 \left(1 + \log_e \frac{V_2}{V_1} \right) - p_3V_2 \quad ...(17.2)$$

$$= p_1V_1 \log_e \frac{V_2}{V_1}$$

$$\left[\begin{array}{c} \text{If there is no pressure drop at release} \\ p_1V_1 = p_3V_2 \end{array} \right] \quad ...(17.3)$$

Area of the hypothetical indicator diagram for L.P. cylinder

$$= p_3V_3 \left(1 + \log_e \frac{V_4}{V_3} \right) - p_bV_4.$$

(*i*) ***If work done in each cylinder is same,*** then

$$p_1V_1 \left(1 + \log_e \frac{V_2}{V_1} \right) - p_3V_2 = p_3V_3 \left(1 + \log_e \frac{V_4}{V_3} \right) - p_bV_4$$

$$= \frac{1}{2} \times \left[p_1V_1 \left(1 + \log_e \frac{V_4}{V_1} \right) - p_bV_4 \right] \quad ...(17.4)$$

(*ii*) ***If the initial load/force on the pistons is same,*** then

$$(p_1 - p_3) \frac{\pi}{4} D^2_{\text{H.P.}} = (p_3 - p_b) \frac{\pi}{4} D^2_{\text{L.P.}} \quad ...(17.5)$$

From the above equations, the required dimensions can be found as per data supplied and assumptions made.

17.6. CAUSES OF LOSS OF THERMAL EFFICIENCY IN COMPOUND STEAM ENGINES

Initial condensation becomes increasingly serious as the expansion ratio is increased and as the temperature range across the cylinder becomes greater. The most effective solution adopted in practice is that of compounding, but nevertheless the initial condensation and leakage occur on a reduced scale in each stage of the expansion. The pressure-drop in the receiver, which includes the wire-drawing at the exhaust and inlet ports and valves, is a further loss, resulting in an increase of entropy due to unresisted expansion, and overall available heat is reduced.

17.7. THE GOVERNING OF COMPOUND STEAM ENGINES

There are three methods of governing compound engines. These are :

1. **Throttle governing.** *Reducing the steam supply pressure in the H.P. cylinder* ;
2. **Cut-off governing on H.P. cylinder.** *Varying the point of cut-off in the H.P. cylinder*; and
3. **Cut-off governing on L.P. cylinder.** *Varying the point of cut-off in the L.P. cylinder.*

For simplicity, consider an engine with a cylinder volume ratio of 5 : 2, and with cut-off in the H.P. cylinder at full load of $\frac{1}{2}$. Let the back pressure in the L.P. cylinder be zero.

Fig. 17.9 illustrates the above three methods, the full load diagrams being 1-2-3-4-5-6-1, 1′-2′-3′-4′-5′-6′-1′, 1″-2″-3″- 4″-5″-6″-1″ respectively.

Throttle governing. Refer to Fig. 17.9 (*a*). Suppose the full load pressure 6-1 is reduced to half by throttling, that is to 6-7, the cut-off in each cylinder remaining the same. The light load diagram will become 7-8-10-5-6-7. The high pressure work will be reduced from 1-2-3-7-1 to 7-8-9-11-7 and the low pressure work will be reduced from 7-3-4-5-6-7 to 11-9-10-5-6-11.

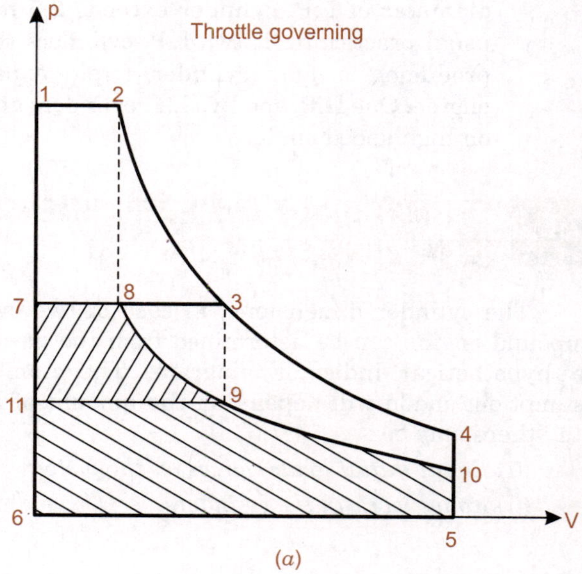

(*a*)

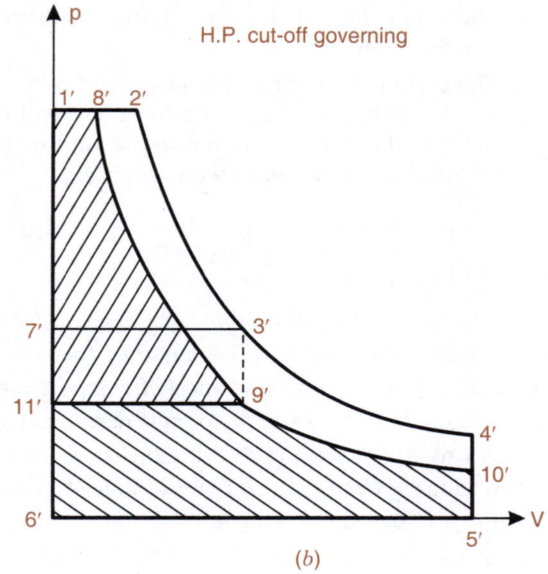

(b)

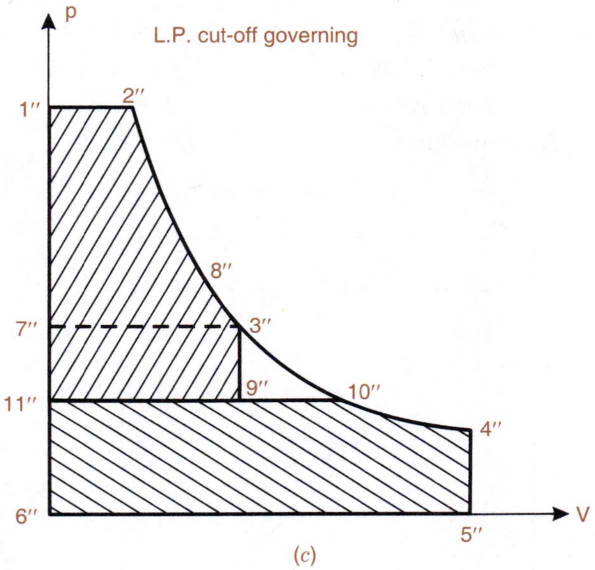

(c)

Fig. 17.9. Governing of compound steam engines.

High pressure cut-off governing. Refer to Fig. 17.9 (*b*). Suppose for the light load the cut-off in high pressure is reduced from $\frac{1}{2}$ to $\frac{1}{4}$, the initial pressure and the cut-off in low pressure remaining constant. The light load diagram will become 1′-8′-10′-5′-6′-1′, the high pressure work becoming 1′-8′-9′-11′-1′, which is not very different from 1′-2′-3′-7′-1′, while the low pressure work is reduced from 7′-3′-4′-5′-6′-7′ to 11′-9′-10′-5′-6′-11′. *Governing by varying the cut-off in the H.P. cylinder will therefore reduce the proportion of work done in the L.P. cylinder at light loads. With condensing engines at very light loads this may cause the average pressure to overcome the back pressure and frictional resistance, thus reducing the efficiency of the engine.*

A comparison of the diagrams Figs. 17.9 (*a*) and 17.9 (*b*) shows that the *high pressure cut-off governing is more economical than throttle governing*. The light load diagrams for the L.P. cylinder, namely 11-9-10-5-6-11 and 11′-9′-10′-5′-6′-11′, are the same in each case, the release pressure 10-5 or 10′-5′ also being the same. Hence the same volume of steam at the same release pressure is exhausted from the engine in each case, whereas the work done with throttle governing is represented by 7-8-10-5-6-7 while the work done with high pressure cut-off governing is represented by the larger area 1′-8′-10′-5′-6′-1′.

Low pressure cut-off governing. Refer to Fig. 17.9 (*c*). Suppose the high pressure cut-off is kept constant at $\frac{1}{2}$, while low pressure cut-off is changed from 2/5 to 3/5. By this change of cut-off the high pressure work is increased from 1″-2″-3″-7″-1″ to 1″-2″-3″-9″-11″-1″ while the low pressure work is decreased from 7″-3″-4″-5″-6″-7″ to 11″-10″-4″-5″-6″-11″. Thus, by *making the cut-off in the L.P. cylinder earlier, the total work done by the engine is only slightly affected, but the proportion of the total work done in the L.P. cylinder is reduced, while the work done in the H.P. cylinder is increased.*

17.8. UNIFLOW STEAM ENGINE

The condensation of steam during admission and part of expansion in the reciprocating steam engine is due to the fact that the walls of the cylinder are relatively cool in comparison with the steam. This low temperature is due to the effect of the cylinder having been filled with low pressure, low temperature steam prior to admission. The 'Uniflow' engine was designed with a view to reducing the effects of initial condensation of the steam, and this *reduction is achieved by virtue of the uni-directional flow of steam in the engine cylinder, so that any particular part of the cylinder wall is maintained at an approximate even temperature*.

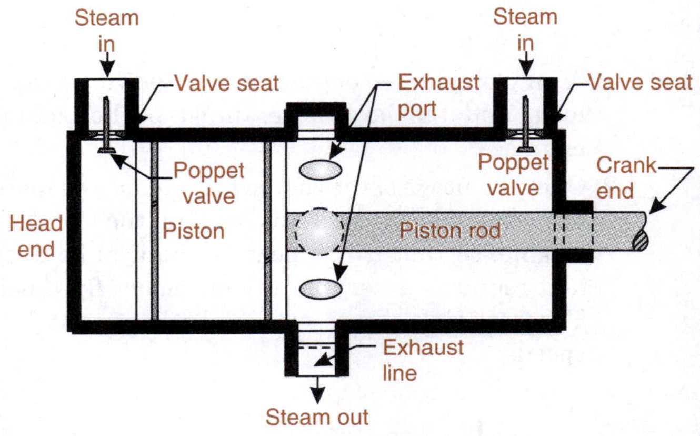

Fig. 17.10. Uniflow steam engine.

A uniflow steam engine is illustrated diagrammatically in Fig. 17.10. It has a cylinder that is long in comparison with its stroke, and has a piston about 45 per cent as long as the cylinder. Steam enters at each end and flows oneway (literally uniflow) towards the middle, where the ring of exhaust ports is uncovered by the piston as it nears the end of stroke. The long piston thus acts as an exhaust valve. On the return stroke, the piston very soon covers the exhaust ports, so that compression (point *L* in Fig. 17.11) starts early and can be carried almost to line pressure. By cleverly using adjustable clearance plugs, an engine of this type can be adjusted to compress the cushion steam to the desired point. Other variations on uniflow type use auxiliary exhaust valves to prohibit excessive compression.

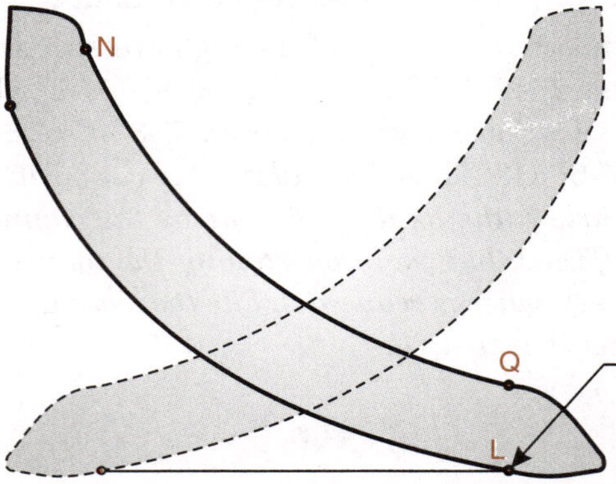

Fig. 17.11. Indicator diagram for uniflow engine.

Advantages :

1. Uniflow engines are fairly simple in design and of robust construction.
2. Since condensation losses are very small the uniflow engine dispenses with the necessity for compounding.
3. Uniflow engines have high efficiency and specific steam consumption is practically constant over its working range of load.
4. Having fewer rubbing parts than the compound engines, it has a higher mechanical efficiency.

Disadvantages :

1. A single cylinder engine of high power is very long, and the large sizes must be horizontal engines.
2. One advantage of the compound engine, the more even turning moment, is lost in the Uniflow Engine, so that it has poor mechanical balance and requires a very large and heavy flywheel. These engines must consequently run at low speeds.
3. There is a loss of net output work due to early compression.
4. The cylinder must be very large and robust as it has to withstand high pressure. The volume of the cylinder is nearly equal to the volume of L.P. cylinder in compound steam engine.

WORKED EXAMPLES

Example 17.1. *A compound engine is to develop 90 kW at 100 r.p.m. Steam is supplied at 7.5 bar and the condenser pressure is 0.21 bar. Assuming hyperbolic expansion and expansion ratio of 15, a diagram factor of 0.72 and neglecting clearance and receiver losses, determine the diameters of the cylinders so that they may develop equal powers. Stroke of each piston = L.P. cylinder diameter.*

Solution. Power to be developed, I.P. = 90 kW

Engine speed, N = 100 r.p.m.

Admission steam pressure, p_1 = 7.5 bar

Condenser pressure, p_b = 0.21 bar

Expansion ratio, R = 15

Diagram factor, D.F. = 0.72.

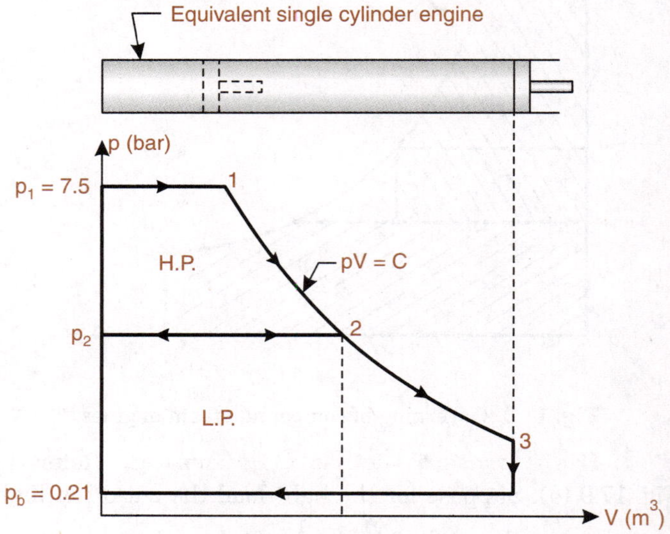

Fig. 17.12

Cylinder diameters, $D_{L.P.}$, $D_{H.P.}$:

Stroke of each piston = L.P. cylinder diameter

$p_{m(\text{actual})}$ referred to L.P. cylinder

$$= \text{D.F.}\left[\frac{p_1}{R}(1+\log_e R) - p_b\right]$$

$$= 0.72\left[\frac{7.5}{15}(1+\log_e 15) - 0.21\right] = 1.18 \text{ bar}$$

Indicated power,

$$\text{I.P.} = \frac{10\, p_{m(actual)}\, LAN}{3}$$

$$= \frac{10 \times 1.18 \times D_{\text{L.P.}} \times \frac{\pi}{4} D^2_{\text{L.P.}} \times 100}{3}$$

$$= 90 = 308.92\; D^3_{\text{L.P.}}$$

$$\therefore \quad D_{\text{L.P.}} = \left(\frac{90}{308.92}\right)^{1/3} = 0.6629 \text{ m} \simeq 663 \text{ mm}$$

i.e., Diameter of low pressure cylinder = **663 mm. (Ans.)**

Work done in H.P. cylinder

$$= p_1V_1 + p_1V_1 \log_e \frac{V_2}{V_1} - p_2V_2$$

But $p_1V_1 = p_2V_2$

$\therefore$ Work done in *H.P. cylinder* $= p_1V_1 \log_e r_{\text{H.P.}}$

$$\left(\because\; r_{\text{H.P.}} = \frac{V_2}{V_1}\right)$$

Work done in L.P. cylinder

$$= p_2V_2 + p_2V_2 \log_e \frac{V_3}{V_2} - p_bV_3$$

$$= p_2V_2 + p_2V_2 \log_e r_{\text{L.P.}} - p_bV_3$$

$$\left(\because\; r_{\text{L.P.}} = \frac{V_3}{V_2}\right)$$

Equating work done in H.P. cylinder to that done in L.P. cylinder, we have :

$$p_1V_1 \log_e r_{\text{H.P.}} = p_2V_2 + p_2V_2 \log_e r_{\text{L.P.}} - p_bV_3$$

$$\therefore \quad p_bV_3 = p_1V_1\,(1 + \log_e r_{\text{L.P.}} - \log_e r_{\text{H.P.}})$$

$$(\because\; p_1V_1 = p_2V_2)$$

or

$$\frac{p_bV_3}{p_1V_1} - 1 = \log_e r_{\text{L.P.}} - \log_e r_{\text{H.P.}}$$

or

$$\log_e \left(\frac{r_{\text{L.P.}}}{r_{\text{H.P.}}}\right) = \frac{p_bV_3}{p_1V_1} - 1$$

But

$$\frac{V_3}{V_1} = R = 15$$

$$\therefore \quad \log_e \left(\frac{r_{\text{L.P.}}}{r_{\text{H.P.}}}\right) = \left(\frac{0.21}{7.5} \times 15 - 1\right)$$

Also

$$r_{\text{L.P.}} = \frac{V_3}{V_2} \text{ and } r_{\text{H.P.}} = \frac{V_2}{V_1}$$

$$\log_e \left(\frac{r_{\text{L.P.}}}{r_{\text{H.P.}}}\right) = \log_e \left(\frac{V_3}{V_2}\right) \times \left(\frac{V_1}{V_2}\right)$$

$$= \log_e \left(\frac{15\,V_1^2}{V_2^2}\right) \quad (\because\; V_3 = 15V_1)$$

$$\therefore \quad \log_e \left(\frac{15\,V_1^2}{V_2^2}\right) = \frac{0.21}{7.5} \times 15 - 1$$

or

$$\log_e \left(\frac{V_2^2}{15\,V_1^2}\right) = 1 - \frac{0.21}{7.5} \times 15 = 0.58$$

or

$$\frac{V_2^2}{15\,V_1^2} = 1.786, \quad \therefore \quad \frac{V_2^2}{V_1^2} = 26.79$$

$\therefore$ Ratio of expansion for H.P. cylinder,

$$r_{\text{H.P.}} = \frac{V_2}{V_1} = (26.79)^{1/2} = 5.176$$

Also

$$\frac{V_3}{V_2} = \frac{V_3}{V_1} \times \frac{V_1}{V_2} = \frac{15}{(V_2/V_1)}$$

$$= \frac{15}{5.176} \simeq 2.9$$

Volume of H.P. cylinder

$$= \frac{\pi}{4} \times \left(\frac{0.663}{\sqrt{2.9}}\right)^2 \times 0.663$$

$$= \frac{\pi}{4}\, D^2_{\text{H.P.}} \times 0.663$$

$$\therefore \quad D_{\text{H.P.}} = \left[\frac{(0.663)^3}{2.9 \times 0.663}\right]^{1/2} = 0.3893 \text{ m}$$

$$= 389 \text{ mm}$$

i.e., Diameter of H.P. cylinder = **389 mm. (Ans.)**

☞**Example 17.2.** *On the basis of the following particulars of a double acting compound engine, calculate :*

(i) The piston diameters,

(ii) Common stroke, and

(iii) The L.P. cut-off if the initial loads on the piston rods are to be equal :

Indicated power developed	=	*309 kW*
Speed	=	*200 r.p.m.*
Steam supply pressure	=	*14 bar*
Exhaust pressure	=	*0.35 bar*
Number of expansions	=	*9*
Ratio of cylinder volumes	=	*4.5 : 1*
Stroke length	=	*0.75 × L.P. piston diameter*
Overall diagram factor	=	*0.62*
Pressure loss in the receiver between the two cylinders	=	*0.2 bar*

Neglect clearance.

Solution. Refer to Fig. 17.13.

(*i*) **Piston diameters :**

$$\frac{V_4}{V_1} = R = 9, \quad \frac{V_4}{V_2} = 4.5$$

Theoretical mean effective pressure of the whole engine,

$$p_{m(th.)} = \frac{p_1}{R}(1 + \log_e R) - p_b$$

$$= \frac{14}{9}(1 + \log_e 9) - 0.35 = 4.623 \text{ bar}$$

Actual mean effective pressure,

$$p_{m(act.)} = \text{Diagram factor} \times p_{m(th.)}$$

$$= 0.62 \times 4.623 = 2.866 \text{ bar}$$

Indicated power,

$$\text{I.P.} = \frac{10\, p_{m(actual)}\, LAN}{3}$$

$$309 = \frac{10 \times 2.866 \times (0.75 \times D_{L.P.}) \times \frac{\pi}{4} \times D^2_{L.P.} \times 200}{3}$$

$(\because \; L = 0.75\, D_{L.P.} \text{.... given})$

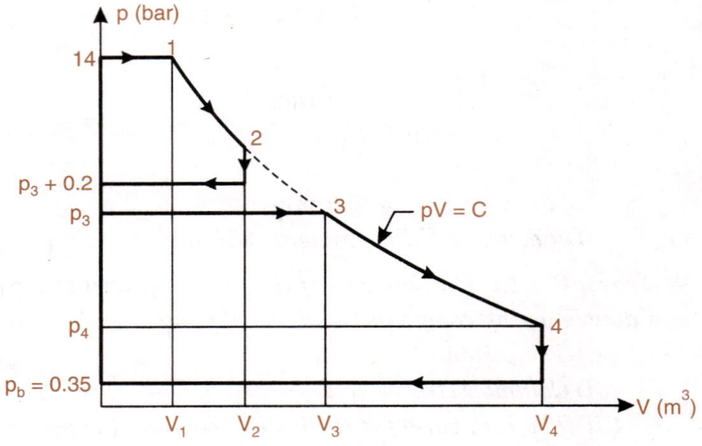

Fig. 17.13

$$\therefore \quad D_{L.P.} = \left(\frac{309 \times 3 \times 4}{10 \times 2.866 \times 0.75 \times \pi \times 200}\right)^{1/3}$$

$$= 0.65 \text{ m or } 650 \text{ mm}$$

i.e., *Diameter of L.P. cylinder* = **650 mm.** **(Ans.)**

Diameter of H.P. cylinder

$$= \frac{650}{\sqrt{4.5}} = \mathbf{306 \text{ mm.}} \quad \textbf{(Ans.)}$$

(*ii*) **Common stroke, L :**

$$L = 0.75 \times D_{L.P.}$$

$$= 0.75 \times 650 = \mathbf{487.5 \text{ mm.}} \quad \textbf{(Ans.)}$$

(*iii*) **Cut-off in L.P. cylinder :**

Let, p_3 = Admission pressure to the L.P. cylinder,

and

$p_3 + 0.2$ = Exhaust pressure of the H.P. cylinder.

Since the initial loads on the two piston rods are to be equal,

$$\therefore \quad [14 - (p_3 + 0.2)]\, V_2 = (p_3 - 0.35)\, V_4$$

or

$$14 - (p_3 + 0.2) = (p_3 - 0.35)\frac{V_4}{V_2} = (p_3 - 0.35) \times 4.5 \qquad \left[\because \frac{V_4}{V_2} = 4.5\right]$$

or

$$14 - p_3 - 0.2 = 4.5p_3 - 1.575 \text{ or } 5.5p_3 = 15.375$$

$$\therefore \quad p_3 = \frac{15.375}{5.5} = 2.795 \text{ bar}$$

As $p_1V_1 = p_4V_4$

$$\therefore \quad p_4 = \frac{p_1V_1}{V_4} = 14 \times \frac{1}{9} = 1.555 \text{ bar}$$

$\therefore$ *Cut-off in the L.P. cylinder*

$$= \frac{V_3}{V_4} = \frac{p_4}{p_3} = \frac{1.555}{2.795} \qquad (\because \; p_3V_3 = p_4V_4)$$

$$= \mathbf{0.556.} \quad \textbf{(Ans.)}$$

☞**Example 17.3.** *A two-cylinder compound, double-acting steam engine is required to develop 62.5 kW (brake) at 350 r.p.m. when supplied with steam at 20 bar and exhausting to a condenser at 0.2 bar. Cut-off ratio in both cylinders to be 0.4. The stroke length for both cylinders is 250 mm. Estimate the cylinder diameters to develop equal work.*

Neglect clearance and assume hyperbolic expansion.

Take, mechanical efficiency = 85% and diagram factor for each cylinder = 0.8.

Solution. Power to be developed, B.P. = 62.5 kW

Engine speed, N = 350 r.p.m.

Steam supply pressure, p_1 = 20 bar

Condenser/back pressure, p_2 = 0.2 bar

Cut-off ratio in both cylinders = 0.4

Stroke length (for both cylinders), L = 250 mm

Mechanical efficiency = 85%
Diagram factor = 0.8 } for each cylinder

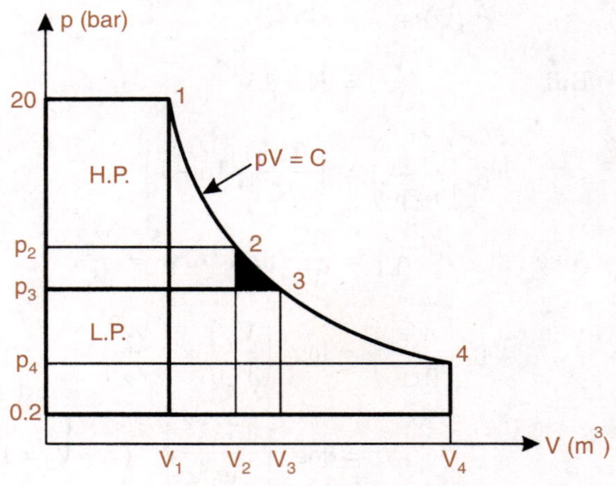

Fig. 17.14

Cylinder diameters :

Refer to Fig. 17.14.

For equal work done in both cylinders,

$$\text{D.F.}\left[p_1V_1\left\{1+\log_e \frac{V_2}{V_1}\right\}-p_3V_2\right]$$

$$= \text{D.F.}\left[p_3V_3\left\{1+\log_e\left(\frac{V_4}{V_3}\right)\right\}-p_bV_4\right]$$

Dividing throughout by V_2, we get

$$p_1\frac{V_1}{V_2}\left[1+\log_e\left(\frac{V_2}{V_1}\right)\right]-p_3$$

$$=p_3\frac{V_3}{V_2}\left[1+\log_e\left(\frac{V_4}{V_3}\right)\right]-p_b\frac{V_4}{V_2}$$

Inserting $\quad p_3V_3 = p_1V_1$

$$p_1\cdot\frac{V_1}{V_2}\left[1+\log_e\left(\frac{V_2}{V_1}\right)\right]-p_3$$

$$=p_1\cdot\frac{V_1}{V_2}\left[1+\log_e\left(\frac{V_4}{V_3}\right)\right]-p_b\cdot\frac{V_4}{V_2}$$

$$\therefore \quad p_1\cdot\frac{V_1}{V_2}\left[\log_e\left(\frac{V_2}{V_1}\right)-\log_e\left(\frac{V_4}{V_3}\right)\right]$$

$$=p_3-p_b\cdot\frac{V_4}{V_2}$$

$$\therefore \quad 20\times 0.4\,[\log_e 2.5-\log_e 2.5]=p_3-p_b\cdot\frac{V_4}{V_2}$$

or
$$p_3-p_b\cdot\frac{V_4}{V_2}=0 \qquad \ldots(i)$$

Considering points of cut-off as on hyperbolic curve

$$p_1V_1 = p_3V_3$$

$$\therefore \quad p_1\times 0.4\,V_2 = p_3\times 0.4\,V_4$$

$$\left[\frac{V_1}{V_2}=\frac{V_3}{V_4}=0.4\ldots\ldots \text{given}\right]$$

$$\therefore \quad \frac{V_4}{V_2}=\frac{p_1}{p_3}=\frac{20}{p_3}$$

Inserting the value of V_4/V_2 in eqn. (*i*), we get

$$p_3-p_b\times\frac{20}{p_3}=0$$

$$\therefore \quad p_3^2 = 20\times p_b = 20\times 0.2 = 4$$

$$\therefore \quad p_3 = 2 \text{ bar}$$

Work developed/stroke in the H.P. cylinder is

$$\frac{62.5\times 60\times 1000}{0.85\times 350\times 2}=\text{D.F.}\left[p_1V_1\left\{1+\log_e\left(\frac{V_2}{V_1}\right)\right\}-p_3V_2\right]$$

$$=\text{D.F.}\times V_2\left[p_1\cdot\frac{V_1}{V_2}\left\{1+\log_e\left(\frac{V_2}{V_1}\right)\right\}-p_3\right]$$

$$= 0.8\times V_2\,[20\times 10^5\times 0.4\,\{1+\log_e 2.5\}-2\times 10^5]$$

$$= 1066426.1\,V_2$$

$$\therefore \quad V_2=\frac{62.5\times 60\times 1000}{0.85\times 350\times 2\times 1066426.1}$$

$$= 0.00591 \text{ m}^3$$

But $\quad V_2 = \pi/4\, D^2_{\text{H.P.}}\times L$

$$0.00591 = \pi/4\times D^2_{\text{H.P.}}\times 0.25$$

$$\therefore \quad D_{\text{H.P.}}=\left(\frac{0.00591\times 4}{\pi\times 0.25}\right)^{1/2}$$

$= 0.173$ m = **173 mm. (Ans.)**

As shown earlier, $\quad \dfrac{V_4}{V_2}=\dfrac{20}{p_3}$

But $\quad \dfrac{V_4}{V_2}=\dfrac{\text{Volume of L.P. cylinder}}{\text{Volume of H.P. cylinder}}$

$$=\frac{\pi/4\,D^2_{\text{L.P.}}\times L}{\pi/4\,D^2_{\text{H.P.}}\times L}=\frac{D^2_{\text{L.P.}}}{D^2_{\text{H.P.}}}=\left(\frac{D_{\text{L.P.}}}{D_{\text{H.P.}}}\right)^2$$

$$\therefore \quad \left(\frac{D_{\text{L.P.}}}{D_{\text{H.P.}}}\right)^2=\frac{20}{p_3}=\frac{20}{2}=10$$

$$\therefore \quad \frac{D_{\text{L.P.}}}{D_{\text{H.P.}}}=\sqrt{10}=3.162$$

i.e., $\quad D_{\text{L.P.}} = 3.162\times D_{\text{H.P.}} = 3.162\times 173$

= **547 mm. (Ans.)**

Example 17.4. *Find the ratio of cylinder diameters for a double-acting condensing steam engine. The steam supplied is at 11.5 bar and the exhaust is at 0.14 bar. Cut-off in each cylinder to be at half stroke. Clearance volume 10% in each case. Total expansion ratio 10, assuming a diagram factor of 75%. Also calculate the indicated power if the steam used per hour was 1090 kg.*

Solution. Steam supply pressure, $\quad p_1$ = 11.5 bar

Exhaust pressure, $\quad p_b$ = 0.14 bar

Cut-off in each cylinder = half stroke

Clearance volume in each case = 10%

Total expansion ratio = 10

Diagram factor, D.F. = 0.75

Steam used = 1090 kg/h

Refer to Fig. 17.15

(*i*) **Ratio of cylinder diameters :**

If the total ratio of expansion is reckoned on the swept volume, then

$$\frac{V_4'}{V_1'} = 10 \quad ...(i)$$

$$\text{Ratio of cylinder volumes} = \frac{V_4'}{V_2'} = \frac{V_4'}{2V_1'} \quad ...(ii)$$

From (*i*) and (*ii*),

$$\text{Ratio of cylinder volumes} = \frac{1}{2} \times 10 = 5$$

$\therefore$ *Ratio of cylinder diameters* = $\sqrt{5}$ = **2.236. (Ans.)**

(*ii*) **Indicated power :**

$$p_m \text{ (with clearance)} = p_1\left(\frac{1}{r} + \left(c + \frac{1}{r}\right)\log_e\left(\frac{c+1}{c+\frac{1}{r}}\right)\right) - p_b$$

To determine p_3 the expansion curve is assumed continuous, since one diagram factor is given, so that $p_1V_1 = p_3V_3$

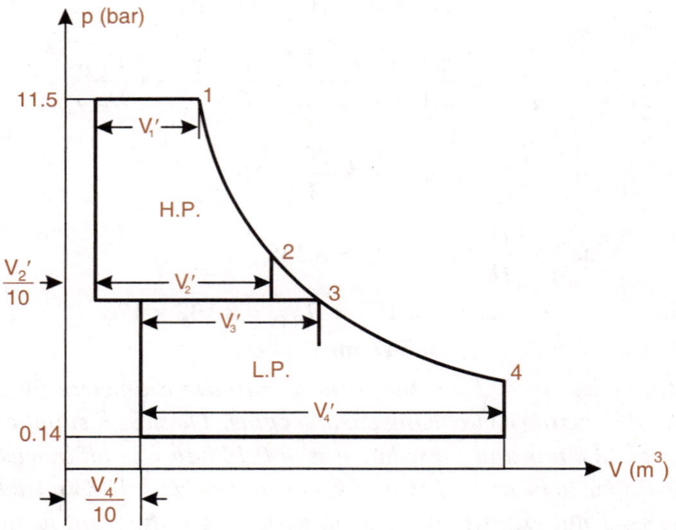

Fig. 17.15

$$\therefore \quad p_3 = p_1 . \frac{V_1}{V_3} = 11.5\left[\frac{\left(\frac{1}{10}+\frac{1}{2}\right)V_2'}{\left(\frac{1}{10}+\frac{1}{2}\right)V_4'}\right]$$

$$= \frac{11.5}{5} = 2.3 \text{ bar}$$

i.e., $p_3 = 2.3$ bar

$$p_m \text{ of H.P.} = 11.5\left[\frac{1}{2} + \left(\frac{1}{10}+\frac{1}{2}\right)\log_e\left\{\frac{\frac{1}{10}+1}{\frac{1}{10}+\frac{1}{2}}\right\}\right] - 2.3$$

$$= 11.5\ [0.5 + 0.6 \log_e 1.833] - 2.3$$

$$= 7.63 \text{ bar}$$

$$p_m \text{ of L.P.} = 2.3\left[\frac{1}{2} + \left(\frac{1}{10}+\frac{1}{2}\right)\log_e\left\{\frac{\left(\frac{1}{10}+1\right)}{\left(\frac{1}{10}+\frac{1}{2}\right)}\right\}\right] - 0.14$$

$$= 1.846 \text{ bar.}$$

Since nothing is said about the dryness after cut-off we must assume the steam dry and saturated, and if we consider 1 kg of steam is used per stroke the specific volume at 11.5 bar is 0.17 m³/kg. The mass of steam remaining in the H.P. cylinder clearance at 2.3 bar is

$$\frac{V_2'}{10 \times 0.777} = \frac{V_2'}{7.77}$$

where 0.777 m³/kg is the specific volume at 2.3 bar.

Total volume at cut-off

$$= \frac{V_2'}{10} + 0.5\,V_2' = 0.6\,V_2'$$

$$\text{But} \quad 0.6\,V_2' = 0.17\left(1 + \frac{V_2'}{7.77}\right)$$

$$= 0.17 + \frac{0.17\,V_2'}{7.77}$$

$$\text{or} \quad 0.6\,V_2' = 0.17 + 0.0218\,V_2'$$

$$\therefore \quad V_2' = \frac{0.17}{(0.6 - 0.0218)} = 0.294 \text{ m}^3$$

$$\text{and} \quad V_4' = 5V_2' = 5 \times 0.294 = 1.47 \text{ m}^3$$

Work done/kg of steam

$$= p_{m(\text{H.P.})}\,V_2' + p_{m(\text{L.P.})} \times V_4'$$

$$= 10^5\,(7.63 \times 0.294 + 1.846 \times 1.47) \text{ Nm}$$

$$= 4.957 \times 10^5 \text{ Nm}$$

$\therefore$ Ideal indicated power,

$$\text{I.P.} = \frac{4.957 \times 10^5 \times 1090}{3600 \times 1000} = 150 \text{ kW}$$

Actual I.P. allowing for a diagram factor of 0.75

= 0.75 × 150 = **112.5 kW. (Ans.)**

Example 17.5. *A triple-expansion engine is required to develop 2940 kW under the following conditions :*

Pressure in the steam chest	*= 15 bar*
Piston speed	*= 210 m/min*
Exhaust pressure	*= 0.15 bar*
Cylinder volume ratios	*= 1 : 2.4 : 7.2*

Total ratio of expansion = 18

Overall diagram factor = 0.62.

Assuming equal initial loading on each piston, determine :

(*i*) *Cylinder diameters*

(*ii*) *Mean receiver pressures*

(*iii*) *Cut-off points in each cylinder.*

Assume hyperbolic expansion and neglect clearance.

Solution. Refer to Fig. 17.16.

Let $V_2 = 1$ unit.

Since $V_2 : V_4 : V_6 = 1 : 2.4 : 7.2$ (given)

$\therefore$ $V_4 = 2.4$ units and $V_6 = 7.2$ units.

Also, total ratio of expansion, $R = V_6/V_1 = 18$

$$\therefore \quad V_1 = \frac{V_6}{18} = \frac{7.2}{18} = 0.4 \text{ units.}$$

(*i*) **Cylinder diameters :**

Mean effective pressure referred to L.P. cylinder,

$$p_m = \text{D.F.} \left[\frac{p_1}{R}(1 + \log_e R) - p_b\right]$$

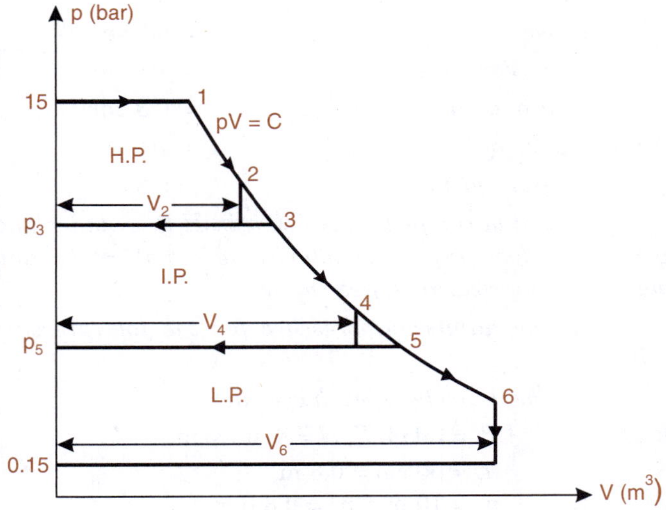

Fig. 17.16

$$= 0.62 \left[\frac{15}{18}(1 + \log_e 18) - 0.15\right] = 1.917 \text{ bar}$$

Indicated power, $\text{I.P.} = 2940 = \dfrac{10\, p_m LAN}{3}$

i.e.,
$$2940 = \frac{10 \times 1.917 \times A \times 105}{3}$$

$$\left[\because \; 2LN = 210 \quad \therefore \; LN = 105 \text{ m/min}\right]$$

$$\therefore \quad A_{\text{L.P.}} = \frac{2940 \times 3}{10 \times 1.917 \times 105} = 4.3818 \text{ m}^2$$

$$D_{\text{L.P.}} = \left(\frac{4.3818 \times 4}{\pi}\right)^{1/2}$$

= **2.362 m or 2362 mm. (Ans.)**

$$D_{\text{H.P.}} = \frac{D_{\text{L.P.}}}{\sqrt{V_6/V_2}} = \frac{2362}{\sqrt{7.2/1.0}}$$

= **880 mm. (Ans.)**

$$D_{\text{I.P.}} = \frac{D_{\text{L.P.}}}{\sqrt{V_6/V_4}} = \frac{2362}{\sqrt{7.2/2.4}}$$

= **1363.7 mm. (Ans.)**

(*ii*) **Mean receiver pressures :**

For equal initial loads on each piston,

$$(p_1 - p_3)\, V_2 = (p_3 - p_5)\, V_4 = (p_5 - p_6) V_6$$

$$(15 - p_3) \times 1 = (p_3 - p_5) \times 2.4 = (p_5 - 0.15) \times 7.2$$

$\therefore$ $15 - p_3 = 7.2\, p_5 - 1.08$

$\therefore$ $p_3 = 16.08 - 7.2\, p_5$ ...(*i*)

Also $2.4\,(p_3 - p_5) = 7.2\, p_5 - 0.15 \times 7.2$

Inserting the value of p_3 from (*i*), we get

$2.4\,[16.08 - 7.2\, p_5 - p_5] = 7.2\, p_5 - 1.08$

$38.592 - 19.68\, p_5 = 7.2\, p_5 - 1.08$

$26.88\, p_5 = 39.672$

$$\therefore \quad p_5 = \frac{39.672}{26.88} = \mathbf{1.476 \text{ bar.}} \quad \textbf{(Ans.)}$$

Inserting the value of p_5 in (*i*), we get

$p_3 = 16.08 - 7.2 \times 1.476 =$ **5.45 bar. (Ans.)**

(*iii*) **Cut-off points in each cylinder :**

$$\textit{H.P. cut-off} = \frac{V_1}{V_2} = \frac{0.4}{1} = \mathbf{0.4.} \quad \textbf{(Ans.)}$$

Now $p_1 V_1 = p_3 V_3$

$15 \times 0.4 = 5.45 \times V_3$

$$\therefore \quad V_3 = \frac{15 \times 0.4}{5.45} = 1.1 \text{ units.}$$

$$\textit{I.P. cut-off} = \frac{V_3}{V_4} = \frac{1.1}{2.4} = \mathbf{0.458.} \quad \textbf{(Ans.)}$$

Again, $p_1 V_1 = p_5 V_5$

$15 \times 0.4 = 1.476 \times V_5$

$$\therefore \quad V_5 = \frac{15 \times 0.4}{1.476} = 4.065 \text{ units}$$

$$\textit{L.P. cut-off} = \frac{V_5}{V_6} = \frac{4.065}{7.2} = \mathbf{0.564.} \quad \textbf{(Ans.)}$$

Example 17.6. *The following observations are recorded during a trial on a jacketed double-acting compound steam engine supplied with dry and saturated steam :*

Steam admission pressure = 6.2 bar

Diameter of H.P. cylinder = 250 mm

Diameter of L.P. cylinder = 375 mm

Stroke	*= 600 mm*
m.e.p. of H.P. cylinder	*= 2.3 bar*
m.e.p. of L.P. cylinder	*= 2.2 bar*
Speed	*= 100 r.p.m.*
Brake torque	*= 4500 Nm*
Receiver pressure	*= 2.6 bar*
Discharge of wet air pump	*= 18 kg/min*
Discharge from jacket drain	*= 2.1 kg/min*
Discharge from receiver drain	*= 1.25 kg/min*
Cooling water from condenser	*= 350 kg/min*
Temperature rise of cooling water	*= 30°C*
Temperature of condensate	*= 50°C*

(i) Draw up a heat balance sheet.

(ii) Find the mechanical and brake thermal efficiencies of engine.

Neglect mass of air carried by steam in condenser.

Solution. Total indicated power of the engine,

$$\text{I.P.} = \frac{10LN}{3}\,[p_{m(\text{H.P.})}A_{\text{H.P.}} + p_{m(\text{L.P.})}A_{\text{L.P.}}]$$

$$= \frac{10 \times 0.6 \times 100}{3}\left[2.3 \times \frac{\pi}{4} \times 0.25^2 + 2.2 \times \frac{\pi}{4} \times 0.375^2\right]$$

= 200(0.1129 + 0.2429) = 71.16 kW

Brake power of the engine,

$$\text{B.P.} = \frac{2\pi NT}{60 \times 1000} = \frac{2\pi \times 100 \times 4500}{60 \times 1000} = 47.12 \text{ kW}$$

Heat equivalent of B.P.

= 47.12 × 60 = 2827.2 kJ/min

Heat in jacket drain

= 2.1 × 676 = 1419.6 kJ/min

(676 kJ/kg is the enthalpy of water at 6.2 bar)

Heat in receiver drain

= 1.25 × 540.9 = 676.1 kJ/min

(540.9 kJ/kg is the enthalpy of water at 2.6 bar)

Heat in condensate

= 18 × 4.18 × 50 = 3762 kJ/min

Heat in cooling water

= 350 × 4.18 × 30 = 43890 kJ/min

Heat supplied to the engine per min.

= (Jacket drain + receiver drain + air pump discharge) × h_1

= (2.1 + 1.25 + 18) × 2756.9 = 58859.8 kJ/min

$$\left[\begin{aligned} &h_1 = 2756.9 \text{ kJ/kg} (= h_g) \text{ at 6.2 bar from steam tables,} \\ &h_{f_2} = 1 \times 4.18 \times 50 = 209 \text{ kJ/kg} \end{aligned}\right]$$

Heat unaccounted for

= 58859.8 – (2827.2 + 1419.6 + 676.1 + 3762 + 43890)

= 6284.9 kJ/min.

Heat balance sheet on minute basis :

	Particulars	*Heat (kJ/min)*	*Percentage (%)*
	Heat supplied	*58859.8*	*100*
(*i*)	Heat equivalent of B.P.	2827.2	4.80
(*ii*)	Heat in cooling water	43890	74.56
(*iii*)	Heat in jacket drain	1419.6	2.41
(*iv*)	Heat in receiver drain	676.1	1.15
(*v*)	Heat in condensate	3762	6.39
(*vi*)	Heat unaccounted for	6284.9	10.67

ADDITIONAL/TYPICAL EXAMPLES

Example 17.7. *The following data refer to a double-acting compound steam engine :*

I.P.	*= 365 kW*
R.P.M.	*= 420*
Stroke	*= 60 cm*
Admission pressure	*= 10 bar*
Back pressure	*= 0.3 bar*
Expansion ratio	*= 10*
Diagram factor	*= 0.8*

Assuming complete expansion in H.P. cylinder and equal initial load, expansion follows the law pV = constant, and neglecting clearance, determine :

(i) The admission pressure for the low pressure cylinder.

(ii) The diameter of each cylinder.

Solution. I.P. = 365 kW, N = 420 r.p.m.,

L = 60 cm = 0.6 m,

p_1 = 10 bar, p_b = 0.3 bar,

$$R = \frac{V_3}{V_1} = 10, \text{ D.F.} = 0.8$$

(*i*), (*ii*) : p_2 = ?, $D_{\text{L.P.}}$ = ?, $D_{\text{H.P.}}$ = ?

Pressure, p_2 ; $D_{\text{L.P.}}$:

$$P_{m(\text{actual})} = \text{D.F.} \times \left[\frac{p_1}{r}(1 + \log_e R) - p_b\right]$$

$$= 0.8 \times \left[\frac{10}{10}(1 + \log_e 10) - 0.3\right]$$

= 2.402 bar

$$\text{I.P.} = \frac{10\, p_{m(\text{actual})} \times LAN}{3}$$

$$365 = \frac{10 \times 2.402 \times 0.6 \times \frac{\pi}{4} \times D^2_{L.P.} \times 420}{3}$$

$$\therefore \quad D^2_{L.P.} = \frac{365 \times 3 \times 4}{10 \times 2.402 \times 0.6 \times \pi \times 420} = 0.2303 \text{ m}^2$$

$$D_{L.P.} = 0.4798 \text{ m} \simeq 0.48 \text{ m or } 480 \text{ mm}$$

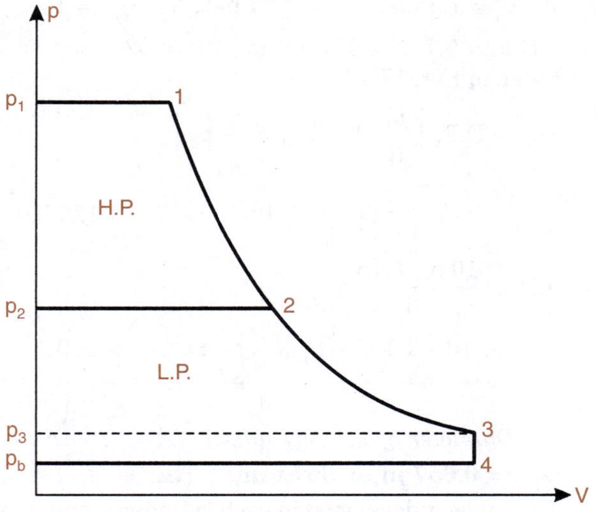

Fig. 17.17

∴ *Diameter of L.P. cylinder,*

$D_{L.P.}$ = **480 mm.** **(Ans.)**

$$p_1V_1 = p_3V_3$$

$$\therefore \quad p_3 = \frac{p_1V_1}{V_3} = 10 \times \frac{1}{10} = 1 \text{ bar} \quad \left[\because \frac{V_1}{V_3} = \frac{1}{10}\right]$$

Also $p_2V_2 = p_3V_3$

$$p_2 \times \frac{\pi}{4} (D_{H.P.})^2 \times L = 1 \times \frac{\pi}{4} (D_{L.P.})^2 \times L$$

or
$$(D_{H.P.})^2 = \frac{0.48^2}{p_2}$$

Since *initial load is same*, therefore,

$$(p_1 - p_2) \times \frac{\pi}{4} \times (D_{H.P.})^2 = (p_2 - p_b) \times \frac{\pi}{4} \times (D_{L.P.})^2$$

$$(10 - p_2) \times \frac{(0.48^2)}{p_2} = (p_2 - 0.3) \times 0.48^2$$

$$(10 - p_2) = p_2 (p_2 - 0.3) = p_2^2 - 0.3\, p_2$$

or
$$p_2^2 + 0.7\, p_2 - 10 = 0$$

or
$$p_2 = \frac{-0.7 \pm \sqrt{0.7^2 + 4 \times 10}}{2} = 2.83 \text{ bar}$$

i.e., p_2 = **2.83 bar.** **(Ans.)**

Diameter of H.P. cylinder, $D_{H.P.}$:

$$D_{H.P.} = \frac{0.48}{\sqrt{p_2}} = \frac{0.48}{\sqrt{2.83}} = 0.285 \text{ or } 285 \text{ mm}$$

i.e., *Diameter of H.P. cylinder,* $D_{H.P.}$ = **285 mm.** **(Ans.)**

Example 17.8. *A compound steam engine is to develop 260 kW, when taking steam at 8.75 bar and exhausting at 0.15 bar abs. The engine speed is 140 r.p.m. and the piston speed is 150 metres/min, the cylinder cut-off ratio is 0.4 and the cylinder volume ratio is 3.7. Allow a diagram factor of 0.83 for the combined cards and determine suitable dimensions of the cylinders if the diagram factor for the H.P. cylinders alone is 0.85 ; determine the separate powers developed in the two cylinders when the L.P. cut-off is arranged to give equal initial loads on the pistons. Assume hyperbolic expansion and neglected clearance effects.*

Solution. p_1 = 8.75 bar, p_b = 0.15 bar,

N = 140 r.p.m.,

Piston speed

$$= 150 \text{ m/min}, \quad \frac{V_1}{V_2} = 0.4, \quad \frac{V_4}{V_2} = 3.7,$$

D.F. = 0.83 ; I.P. = 260 kW

Overall expansion ratio R,

$$\frac{V_4}{V_1} = \frac{V_4}{V_2} \times \frac{V_2}{V_1} = 3.7 \times 2.5 = 9.25$$

If the whole expansion takes place in L.P. cylinder, then,

$$\text{I.P.} = \frac{\text{D.F.} \times 10 \left[\frac{p_1}{R}(1 + \log_e R) - p_b\right] \times V_4 \times N}{6}$$

or
$$260 = \frac{0.83 \times 10 \left[\frac{8.75}{9.25}(1 + \log_e 9.25) - 0.15\right] \times V_4 \times 140}{6}$$

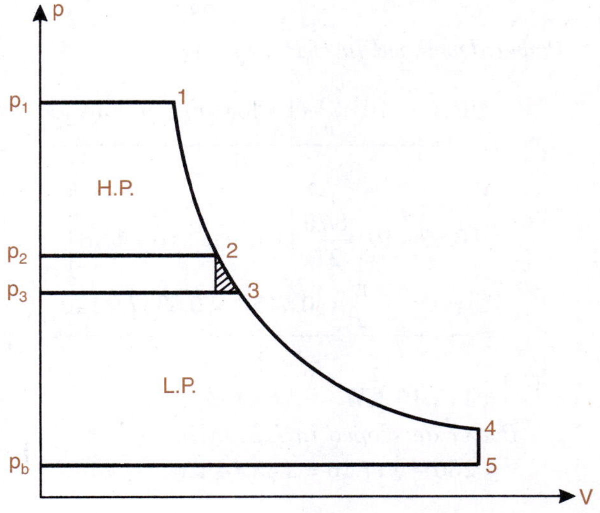

Fig. 17.18

or $$260 = \frac{0.83 \times 10 \times 2.9\, V_4 \times 140}{6}$$

∴ L.P. cylinder volume,

$$V_4 = 0.4629 \text{ m}^3$$

Length of stroke,

$$L = \frac{\text{Piston speed}}{2N} = \frac{150}{2 \times 140}$$

$$= \mathbf{0.5357 \text{ m.}} \quad \textbf{(Ans.)}$$

Also, $\frac{\pi}{4} D^2_{\text{L.P.}} \times 0.5357 = 0.4629$

∴ *Diameter of L.P. cylinder,*

$$D_{\text{L.P.}} \left(\frac{0.4629 \times 4}{\pi \times 0.5357}\right)^{1/2} = \mathbf{1.049 \text{ m.}} \quad \textbf{(Ans.)}$$

Volume of H.P. cylinder

$$= \frac{0.4629}{3.7} = \frac{\pi}{4} D^2_{\text{H.P.}} \times 0.5357$$

∴ *Diameter of H.P. cylinder,*

$$D_{\text{H.P.}} = \mathbf{0.545 \text{ m.}} \quad \textbf{(Ans.)}$$

For *equal initial load,*

$$(p_1 - p_2) \frac{\pi}{4} D^2_{\text{H.P.}} = (p_2 - p_b) \frac{\pi}{4} D^2_{\text{L.P.}}$$

$$(8.75 - p_2) \frac{\pi}{4} \times 0.545^2 = (p_2 - 0.15) \times \frac{\pi}{4} \times (1.049)^2$$

or $$(8.75 - p_2) = (p_2 - 0.15) \times \left(\frac{1.049}{0.545}\right)^2$$

$$= (p_2 - 0.15) \times 3.705$$

or $$8.75 - p_2 = 3.705\, p_2 - 0.556.$$

∴ $$p_2 \simeq 1.98 \text{ bar}$$

Power developed in H.P. cylinder

$$= \frac{\text{D.F.} \times 10 \left[\frac{p_1}{r}(1 + \log_e r) - p_2\right] \times V_2 \times N}{6}$$

$$= \frac{0.85 \times 10 \left[\frac{8.75}{2.5}(1 + \log_e 2.5) - 1.98\right] \times \frac{\pi}{4} \times (0.545)^2 \times 0.5357 \times 140}{6}$$

$$= \mathbf{117.16 \text{ kW.}} \quad \textbf{(Ans.)}$$

∴ *Power developed in L.P. cylinder*

$$= 260 - 117.16 = \mathbf{142.84 \text{ kW.}} \quad \textbf{(Ans.)}$$

Example 17.9. *A two cylinder compound steam engine is to develop 92 kW at 110 r.p.m. Steam is supplied at 7.35 bar abs, and the cylinder pressure is 0.21 bar abs. Stroke of each piston is equal to L.P. cylinder diameter. The total expansion ratio is 15. Allow a diagram factor of 0.7. Assume hyperbolic expansion and neglect clearance and receiver loss. Determine the diameter of the cylinder so that they may develop equal power.*

Solution. $p_1 = 7.35$ bar, $p_b = 0.21$ bar, $R = \frac{V_1}{V_3} = 15$,

D.F. = 0.7, N = 110 r.p.m., I.P. = 92

Refer to Fig. 17.17

$$p_m = \text{D.F.} \left[\frac{p_1}{R}(1 + \log_e R) - p_b\right]$$

$$= 0.7 \left[\frac{7.35}{15}(1 + \log_e 15) - 0.21\right] = 1.125 \text{ bar}$$

$$\text{I.P.} = \frac{10\, p_m\, LAN}{3}$$

$$92 = \frac{10 \times 1.125 \times D_{\text{L.P.}} \times \left(\frac{\pi}{4} D^2_{\text{L.P.}}\right) \times 110}{3}$$

∴ *Diameter of L.P. cylinder,*

$$D_{\text{L.P.}} = \mathbf{0.657 \text{ m or } 657 \text{ mm.}} \quad \textbf{(Ans.)}$$

Since power developed in each cylinder is equal, work done in H.P. cylinder will be equal to the work done in L.P. cylinder.

∴ $$p_1 V_1 \left(1 + \log_e \frac{V_2}{V_1}\right) - p_2 V_2 = \frac{1}{2}\left[p_1 V_1 \left(1 + \log_e \frac{V_3}{V_1}\right) - p_b V_3\right]$$

Dividing both sides by $p_1 V_1$ and considering $p_1 V_1 = p_2 V_2$, we have

or $$\log_e \frac{V_2}{V_1} = \frac{1}{2}\left[1 + \log_e 15 - \frac{0.21}{7.35} \times 5\right]$$

∴ $$\frac{V_2}{V_1} = 5.154$$

∴ Cylinder volume,

$$V_2 = 5.154 \times \frac{\left(\frac{\pi}{4} \times 0.657^2 \times 0.657\right)}{15}$$

$$= \frac{\pi}{4} D^2_{\text{H.P.}} \times 0.657$$

or $$D_{\text{H.P.}} = 0.385 \text{ m or } 385 \text{ mm}$$

i.e., *Diameter of H.P. cylinder* = **385 mm.** **(Ans.)**

HIGHLIGHTS

1. '**Compounding**' is primarily done to secure a large total ratio of expansion.
2. The compound steam engines are usually classified as follows :
   (*i*) Tandem compound steam engines
   (*ii*) Cross compound steam engines
      (*a*) Woolfe type
      (*b*) Receiver type.
3. In a 'double expansion' engine the steam expands in two cylinders.
   In a 'triple expansion' engine the expansion of steam is completed in three cylinders.
4. The following are the methods of governing compound steam engines :
   (*i*) **Throttle governing.** Reducing the steam supply pressure in the H.P. cylinder.
   (*ii*) **Cut-off governing on H.P. cylinder.** Varying the point of cut-off in the H.P. cylinder.
   (*iii*) **Cut-off governing on L.P. cylinder.** Varying the point of cut-off in the L.P. cylinder.
5. Uniflow engines are fairly simple in design and of robust construction.

OBJECTIVE TYPE QUESTIONS

Choose the Correct Answer :

1. The simple steam engine as compared to compound steam engine for the same output has
   (*a*) smaller flywheel
   (*b*) large flywheel
   (*c*) same size flywheel as output is same
   (*d*) same size flywheel as speed is same.
2. The leakage past the piston and initial condensation by compounding the steam engine is
   (*a*) increased
   (*b*) decreased
   (*c*) unaffected
   (*d*) depends on methods of compounding.
3. Woolfe type compound steam engines have
   (*a*) two cranks at 180° phase difference
   (*b*) two cranks at 90° phase difference
   (*c*) only one crank with no phase difference
   (*d*) two cranks with no phase difference.
4. Tandem compound steam engines have
   (*a*) two cranks at 180° phase difference
   (*b*) two cranks at 90° phase difference
   (*c*) only one crank with no phase difference
   (*d*) two cranks with no phase difference.
5. Receiver type compound steam engines have
   (*a*) two cranks at 180° phase difference
   (*b*) two cranks at 90° phase difference
   (*c*) only one crank with no phase difference
   (*d*) two cranks with no phase difference.
6. The compound steam engine that can start in any position is
   (*a*) Tandem type (*b*) Woolfe type
   (*c*) Receiver type (*d*) None of the above.
7. The compound steam engine that requires the smallest flywheel for a given output and given speed is
   (*a*) Tandem type
   (*b*) Woolfe type
   (*c*) Receiver type
   (*d*) Anyone if the speed is same.
8. Governing of compound steam engines is done by
   (*a*) throttling steam to high pressure cylinder
   (*b*) cut-off variation in high pressure cylinder
   (*c*) cut-off variations simultaneously in high pressure and low pressure cylinders.
   (*d*) all of the above methods.
9. The equation used for expansion in compound steam engines namely $pV = C$ signifies
   (*a*) isothermal expansion (*b*) adiabatic expansion
   (*c*) parabolic expansion (*d*) hyperbolic expansion.
10. In a two-cylinder compound steam engine with continuous expansion from initial pressure to back pressure is geometric mean of the initial and back pressure if
    (*a*) work is shared equally by high pressure and low pressure cylinders
    (*b*) the initial loads on the pistons of high pressure cylinder and low pressure cylinder are same
    (*c*) both the cases (*a*) and (*b*)
    (*d*) none of the above.

ANSWERS

1. (*b*)	**2.** (*b*)	**3.** (*a*)	**4.** (*c*)
5. (*b*)	**6.** (*c*)	**7.** (*c*)	**8.** (*d*)
9. (*d*)	**10.** (*c*).		

THEORETICAL QUESTIONS

1. State the advantages of compound steam engines.
2. How are compound steam engines classified ?
3. Explain briefly with neat diagrams the following compound steam engines :
   (*i*) Woolfe type
   (*ii*) Receiver type.
4. Write short note on multi-cylinder engines.
5. Discuss the causes of loss of thermal efficiency in compound steam engines.

6. What different methods are used for governing compound steam engines ? Discuss their relative advantages and disadvantages.
7. Discuss with a neat sketch the construction and working of a uniflow engine. Also mention its merits and demerits.

UNSOLVED EXAMPLES

1. A double acting compound steam engine is supplied with steam at 14 bar and 0.9 dryness. The condenser working pressure is 0.35 bar. The diagram factor referred to the L.P. cylinder is 0.8. The common stroke is 35 cm. The H.P. cylinder bore is 200 mm, while the L.P. cylinder bore is 300 mm. Expansion in the H.P. cylinder is complete. The engine runs at 300 r.p.m. Neglecting clearance and assuming equal initial piston loads, calculate :
   (*i*) The intermediate pressure
   (*ii*) The indicated power output
   (*iii*) The hourly steam consumption at the operating conditions.
   [**Ans.** (*i*) 4.55 bar; (*ii*) 110.5 kW; (*iii*) 1018 kg/h]
2. A two-cylinder double-acting compound steam engine develops 100 kW when running at 240 r.p.m. Dry and saturated steam is supplied to the engine at 16 bar and exhausts from L.P. cylinder occurs at 0.15 bar. Stroke for both the cylinders is 250 mm. Cut-off occurs at 40% of stroke in L.P. as well as H.P. cylinder. The diagram factor for both the cylinders is 0.7.
   Determine the cylinder diameters for the two cylinders if equal work is done by both the cylinders.
   [**Ans.** $D_{L.P.}$ = 327 mm ; $D_{H.P.}$ = 102 mm]
3. A compound steam engine develops 220 kW at a speed of 270 r.p.m., the stroke being 500 mm for both the H.P. and L.P. cylinders. The expansion is hyperbolic. The supply pressure is 12 bar and exhaust pressure is 0.28 bar. The expansion is complete in the H.P. cylinder and the total expansion ratio is 10. Neglecting clearance and assuming equal power development, determine the cylinder diameters. The diagram factor referred to the L.P. cylinder is 0.75.
   [**Ans.** $D_{L.P.}$ = 455 mm ; $D_{H.P.}$ = 312 mm]
4. The diameter and stroke of a L.P. cylinder of a double-acting compound steam engine are each 600 mm. Steam is supplied at a pressure of 11 bar and is exhausted to 0.28 bar. The diagram factor referred to the L.P. cylinder is 0.82. The expansion is complete in the H.P. cylinder. The total expansion ratio in the engine is 8. Assuming equal work developed in each cylinder and engine runs at 240 r.p.m., calculate :
   (*i*) The indicated power
   (*ii*) The intermediate pressure
   (*iii*) The diameter of the H.P. cylinder assuming equal stroke length with that of L.P. cylinder.
   [**Ans.** (*i*) 440 kW; (*ii*) 3.38 bar; (*iii*) $D_{H.P.}$ = 383 mm]
5. A two-cylinder compound steam engine receives steam at a pressure of 8 bar and discharges at 0.2 bar. The cylinder volume ratio is 4. Cut-off in the H.P. and L.P. cylinders occurs at 0.4 and 0.5 of the stroke respectively.
   (*i*) Sketch the hypothetical indicator diagram and insert the pressure at cut-off and release in each cylinder, neglecting clearance.
   (*ii*) Calculate the mean effective pressure in each cylinder.
   (*iii*) Find the ratio of work done in two cylinders.
   [**Ans.** (*ii*) $p_{mH.P.}$ = 4.53 bar, $p_{mL.P.}$ = 1.15 bar; (*iii*) 0.985]
6. Steam is supplied at a pressure of 10.35 bar to an engine having equal strokes 400 mm for the H.P. and L.P. cylinders and having bore diameters of 330 mm and 630 mm for the H.P. and L.P. cylinders. Cut-off for the H.P. cylinder occurs at one-fourth of stroke. The back pressure is 0.21 bar. The power developed by the engine is 120 kW when the mechanical efficiency was 0.82. If the engine runs at 180 r.p.m., determine the diagram factor. Assume that the engine is double-acting.
   [**Ans.** 0.81]
7. For a two-cylinder double-acting compound engine, running at 120 r.p.m., developing 120 kW, the data for H.P. and L.P. cylinders is given below :
   H.P. cylinder : Admission pressure = 13 bar (abs) ; cut-off occurs at 45% of stroke ; clearance = 10% ; diagram factor = 0.8 ; law of expansion = hyperbolic.
   L.P. cylinder : Condenser vacuum = 590 mm Hg ; barometer = 760 mm Hg ; clearance = 12% ; cut-off occurs at 55% of stroke ; diagram factor = 0.8 ; law of expansion = hyperbolic.
   If the length of stroke for both cylinders is equal to L.P. cylinder diameter and the total number of expansion is 10, determine :
   (*i*) Diameter of H.P. cylinder.
   (*ii*) Diameter of L.P. cylinder.
   (*iii*) Length of stroke.
   (*iv*) Ratio of work done in two cylinders.
   [**Ans.** (*i*) 233 mm; (*ii*) 516 mm; (*iii*) 516 mm; (*iv*) 1.029]
8. A two-cylinder compound double-acting steam engine develops 92 kW of power at 110 r.p.m. The steam supply is at 7 bar and exhaust is at 0.21 bar. The overall expansion ratio is 15, and the cylinders develop equal power. The expansion may be assumed to be hyperbolic and expansion in H.P. cylinder is complete. The diagram factor is 0.7 and the common stroke is equal to the diameter of L.P. cylinder. Neglecting clearance, calculate :
   (*i*) Diameter of L.P. cylinder
   (*ii*) Diameter of H.P. cylinder
   (*iii*) Length of stroke.
   [**Ans.** (*i*) 667 mm; (*ii*) 390 mm; (*iii*) 667 mm]
9. The bores of a compound steam engine cylinders are 300 mm and 600 mm while the common stroke is 400 mm. The supply steam is at a pressure of 11 bar and cut-off is at $\frac{1}{3}$rd of stroke in the H.P. cylinder. The back pressure in the L.P. cylinder is 0.32 bar. The details of the indicator card are : for H.P. cylinder ; diagram area 12.5 cm^2 ; spring scale 2.7 bar/cm ; for L.P. cylinder 11.4 cm^2, spring scale 0.8 bar/cm. The length of both diagrams being 7.5 cm. Determine :
   (*i*) The theoretical and actual mean effective pressures referred to the L.P. cylinder.

(*ii*) The overall diagram factor and indicated power if the engine runs at 162 r.p.m.

[**Ans.** (*i*) 2.34 bar. 2.87 bar; (*ii*) D.F. = 0.815 ; 142.5 kW]

10. A compound steam engine is supplied with steam at 8 bar and exhaust occurs at 0.28 bar. The engine runs at 120 r.p.m. and has mean speed of 150 metres/min. Cut-off takes place at 40% of stroke in the H.P. cylinder and at 55% of stroke in L.P. cylinder. If the engine cylinder volume ratio is 3.5 and the power developed by the engine is 360 kW, determine :

(*i*) Cylinder dimensions

(*ii*) L.P. cylinder receiver pressure

(*iii*) Ratio of maximum loads.

Assume diagram factor as 0.8.

[**Ans.** (*i*) $D_{L.P.}$ = 930 mm, $D_{H.P.}$ = 595 mm, Length of stroke = 625 mm; (*ii*) 1.67 bar; (*iii*) 1.385]

11. A double-acting compound steam engine has a common stroke of 400 mm. The supply pressure is 17.25 bar while the exhaust pressure 0.415 bar. The diameter of L.P. cylinder is 450 mm. The ratio of cylinder volumes is 2.5 : 1. The diagram factor referred to the L.P. cylinder is 0.78. The initial piston loads in the cylinders are equal. The engine runs at 270 r.p.m. The total expansion ratio is 9. Neglecting clearance and assuming hyperbolic expansion, calculate :

(*i*) The intermediate pressure

(*ii*) The indicated power

(*iii*) Diameter of H.P. cylinder.

[**Ans.** (*i*) 5.23 bar; (*ii*) 270 kW; (*iii*) 284 mm]

12. The following data relate to a two-cylinder double-acting compound steam engine running at 100 r.p.m.

H.P. cylinder : Diameter = 250 mm, cut-off = 30% of stroke, D.F. = 0.8, clearance = 10% of stroke.

L.P. cylinder : Diameter = 500 mm ; cut-off = 40% of stroke, clearance = 8% of stroke, D.F. = 0.7

The admission and exhaust pressures are 6 bar and 0.30 bar respectively. The common stroke of the engine is 400 mm.

Determine the power output of the engine.

[**Ans.** 27.5 kW]

13. In a triple expansion compound steam engine the steam is supplied at 12.5 bar. All the cylinders have restricted expansion and the cylinder volumes are in the ratio 1 : 2.5 : 6. If overall expansion ratio is 12.5, cut-off in I.P. cylinder occurs at 50% and cut-off in L.P. cylinder occurs at 65% of the stroke, determine :

(*i*) The mean effective pressures of three cylinders.

(*ii*) Loss in power due to restricted expansion in H.P. and I.P. cylinders.

Assume hyperbolic expansion of steam and neglect clearance.

[**Ans.** (*i*) $(p_m)_{H.P.}$ = 5.58 bar, $(p_m)_{I.P.}$ = 2.1 bar, $(p_m)_{L.P.}$ = 0.99 bar; (*ii*) loss in power = 6.7%]

Steam Nozzles

18

18.1. Introduction. 18.2. Steam flow through nozzles—Velocity of steam—Discharge through the nozzle and conditions for its maximum value. **18.3. Nozzle efficiency. 18.4. Supersaturated or metastable expansion of steam in a nozzle. 18.5. General relationship between area, velocity and pressure in nozzle flow. 18.6. Steam injector**—Worked examples. *Highlights—Objective Type Questions—Theoretical Questions—Unsolved Examples.*

18.1. INTRODUCTION

*A **steam nozzle** may be defined as a passage of varying cross-section, through which heat energy of steam is converted to kinetic energy. Its major function is to produce steam jet with high velocity to drive steam turbines.* A turbine nozzle performs *two functions* :

(*i*) It transforms a portion of energy of steam (obtained from steam generating unit) into kinetic energy.

(*ii*) In the impulse turbine it directs the steam jet of high velocity against blades, which are free to move in order to convert kinetic energy into shaft work. In reaction turbines the nozzles which are free to move, discharge high velocity steam. The reactive force of the steam against the nozzle produces motion and work is obtained.

The cross-section of a nozzle at first tapers to a smaller section (*to allow for changes which occur due to changes in velocity, specific volume and dryness fraction as the steam expands*) ; the smallest section being known as ***throat,*** and then it diverges to a large diameter. The nozzle which converges to throat and diverges afterwards is known as ***convergent-divergent*** nozzle (Fig. 18.1). In convergent nozzle there is no divergence after the throat as shown in Fig. 18.2.

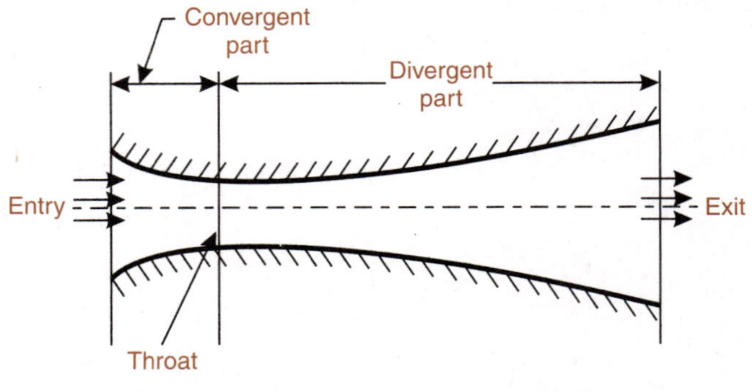

Fig. 18.1. Convergent-divergent nozzle.

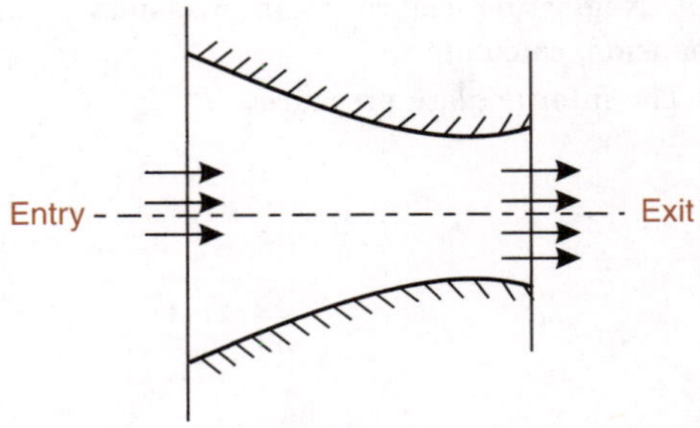

Fig. 18.2. Convergent nozzle.

In a "*convergent-divergent nozzle*", *because of the higher expansion ratio, addition of divergent portion produces steam at higher velocities as compared to a convergent nozzle.*

18.2. STEAM FLOW THROUGH NOZZLES

The steam flow through the nozzle may be assumed as *adiabatic flow* since during the expansion of steam in nozzle neither any heat is supplied nor rejected, work, however, is performed by increasing the kinetic energy of the steam. As the steam passes through the nozzle it loses its pressure as well as the heat. *The work done is equal to the adiabatic heat drop which in turn is equal to Rankine area.*

18.2.1. Velocity of Steam

Steam enters the nozzle with high pressure and low initial velocity (it is so small as compared to the final velocity that it is generally *neglected*) and leaves it with high velocity and low pressure. This is due to the reason that heat energy of steam is *converted* into kinetic energy as it (steam) passes through the nozzle. The final or outlet velocity of steam can be found as follows :

Let C = Velocity of steam at the section considered (m/sec),

h_1 = Enthalpy of steam entering the nozzle,

h_2 = Enthalpy of steam at section considered, and

h_d = Heat drop during expansion of steam is the nozzle = $(h_1 - h_2)$.

Considering 1 kg of steam and flow to be frictionless adiabatic, we have :

Gain in kinetic energy = Adiabatic heat drop

$$\frac{C^2}{2} = h_d$$

$$\therefore \quad C = \sqrt{2 \times 1000\, h_d}, \text{ where } h_d \text{ is in kJ}$$

$$= 44.72\sqrt{h_d} \quad \text{...(18.1)}$$

In practice, there is loss due to friction in the nozzle and its value varies from 10 to 15 per cent of total heat drop. Due to this, total heat drop is minimized. Let heat drop after deducting friction loss be kh_d.

The velocity, $C = 44.72\sqrt{kh_d}$ ...(18.2)

18.2.2. Discharge through the Nozzle and Conditions for its Maximum Value

Let p_1 = Initial pressure of steam,

v_1 = Initial volume of 1 kg of steam at pressure p_1 (m³),

p_2 = Steam pressure at the throat,

v_2 = Volume of 1 kg of steam at pressure p_2 (m³),

A = Cross-sectional area of nozzle at throat (m²), and

C = Velocity of steam (m/s).

The steam flowing through the nozzle follows approximately the equation given below :

$$pv^n = \text{Constant}$$

where, n = 1.135 for *saturated steam,* and

= 1.3 for *superheated steam.*

[For **wet steam,** the value of n can be calculated by Dr. *Zenner's equation,*

$n = 1.035 + 0.1x$, where x is the initial dryness fraction of steam]

Work done per kg of steam during the cycle (Rankine area)

$$= \frac{n}{n-1}(p_1v_1 - p_2v_2)$$

and, Gain in kinetic energy = Adiabatic heat drop

= Work done during Rakine cycle

or

$$\frac{C^2}{2} = \frac{n}{n-1}(p_1v_1 - p_2v_2)$$

$$= \frac{n}{n-1}p_1v_1\left(1 - \frac{p_2v_2}{p_1v_1}\right) \quad \text{...(18.3)}$$

Also $\quad p_1v_1^n = p_2v_2^n$

or

$$\frac{v_2}{v_1} = \left(\frac{p_1}{p_2}\right)^{1/n} \quad \text{...(18.4)}$$

or

$$v_2 = v_1\left(\frac{p_1}{p_2}\right)^{1/n} \quad \text{...(18.5)}$$

Putting the value of v_2/v_1 from eqn. (18.4) in eqn. (18.3), we get

$$\frac{C^2}{2} = \frac{n}{n-1}p_1v_1\left[1 - \frac{p_2}{p_1}\left(\frac{p_1}{p_2}\right)^{1/n}\right]$$

$$= \frac{n}{n-1}p_1v_1\left[1 - \left(\frac{p_2}{p_1}\right)^{1-\frac{1}{n}}\right]$$

$$= \frac{n}{n-1}p_1v_1\left[1 - \left(\frac{p_2}{p_1}\right)^{\frac{n-1}{n}}\right]$$

$$C^2 = 2\left(\frac{n}{n-1}\right)p_1v_1\left[1 - \left(\frac{p_2}{p_1}\right)^{\frac{n-1}{n}}\right]$$

$$C = \sqrt{2\left(\frac{n}{n-1}\right)p_1v_1\left\{1 - \left(\frac{p_2}{p_1}\right)^{\frac{n-1}{n}}\right\}} \quad \text{...(18.6)}$$

If m is the mass of steam discharged in kg/sec.,

Then $$m = \frac{AC}{v_2} \quad \text{...(18.7)}$$

Substituting the value of v_2 from eqn. (18.5) in eqn. (18.7),

$$m = \frac{AC}{v_1\left(\frac{p_1}{p_2}\right)^{1/n}}$$

or $$m = \frac{A}{v_1\left(\frac{p_1}{p_2}\right)^{1/n}}\sqrt{2\left(\frac{n}{n-1}\right)p_1v_1\left[1 - \left(\frac{p_2}{p_1}\right)^{\frac{n-1}{n}}\right]}$$

$$= \frac{A}{v_1}\sqrt{\left[2\left(\frac{n}{n-1}\right)p_1v_1\left\{\left(\frac{p_2}{p_1}\right)^{2/n} - \left(\frac{p_2}{p_1}\right)^{\frac{n-1}{n}}\left(\frac{p_2}{p_1}\right)^{2/n}\right\}\right]}$$

$$= \frac{A}{v_1}\sqrt{\left[2\left(\frac{n}{n-1}\right)p_1v_1\left\{\left(\frac{p_2}{p_1}\right)^{2/n} - \left(\frac{p_2}{p_1}\right)^{\frac{n+1}{n}}\right\}\right]} \quad ...(18.8)$$

It is obvious from above equation that there is only one value of the ratio (called *critical pressure ratio*) p_2/p_1 which will produce the *maximum discharge*. This can be obtained by differentiating 'm' with respect to (p_2/p_1) and equating it to zero.

As other quantities except the ratio p_2/p_1 are constant,

$$\therefore \quad \frac{d}{d\left(\frac{p_2}{p_1}\right)}\left[\left(\frac{p_2}{p_1}\right)^{2/n} - \left(\frac{p_2}{p_1}\right)^{\frac{n+1}{n}}\right] = 0$$

or $$\frac{2}{n}\left(\frac{p_2}{p_1}\right)^{\frac{2}{n}-1} - \left(\frac{n+1}{n}\right)\left(\frac{p_2}{p_1}\right)^{\frac{n+1}{n}-1} = 0$$

or $$\left(\frac{p_2}{p_1}\right)^{\frac{2}{n}-1} = \frac{n+1}{2}\left(\frac{p_2}{p_1}\right)^{1/n}$$

or $$\left(\frac{p_2}{p_1}\right)^{2-n} = \left(\frac{n+1}{2}\right)^n\left(\frac{p_2}{p_1}\right)$$

or $$\left(\frac{p_2}{p_1}\right)^{2-n-1} = \left(\frac{n+1}{2}\right)^n$$

or $$\frac{p_2}{p_1} = \left(\frac{2}{n+1}\right)^{\frac{n}{n-1}} \quad ...(18.9)$$

Hence the discharge through the nozzle will be the *maximum* when critical pressure ratio, *i.e.*,

$$\frac{\text{Throat pressure}}{\text{Inlet pressure}} = \frac{p_2}{p_1} = \left(\frac{2}{n+1}\right)^{\frac{n}{n-1}}$$

For saturated steam : $n = 1.135$

$$\frac{p_2}{p_1} = \left(\frac{2}{1.135+1}\right)^{\frac{1.135}{1.135-1}} = \left(\frac{2}{2.135}\right)^{\frac{1.135}{0.135}} = 0.58$$

For superheated steam : $n = 1.3$

$$\frac{p_2}{p_1} = \left(\frac{2}{1.3+1}\right)^{\frac{1.3}{1.3-1}} = \left(\frac{2}{2.3}\right)^{\frac{1.3}{0.3}} = 0.546$$

Substituting the value of $\frac{p_2}{p_1}$ from eqn. (18.9) into eqn. (18.8), we get the maximum discharge,

$$m_{max} = \frac{A}{v_1}\sqrt{2\left(\frac{n}{n-1}\right)p_1v_1 \times \left[\left\{\left(\frac{2}{n+1}\right)^{\frac{n}{n-1}}\right\}^{\frac{2}{n}} - \left\{\left(\frac{2}{n+1}\right)^{\frac{n}{n-1}}\right\}^{\frac{n+1}{n}}\right]}$$

$$= \frac{A}{v_1}\sqrt{2\left(\frac{n}{n-1}\right)p_1v_1\left[\left(\frac{2}{n+1}\right)^{\frac{2}{n-1}} - \left(\frac{2}{n+1}\right)^{\frac{n+1}{n-1}}\right]}$$

$$= A\sqrt{2\left(\frac{n}{n-1}\right)\frac{p_1}{v_1}\left[\left(\frac{2}{n+1}\right)^{\frac{2}{n-1}} - \left(\frac{2}{n+1}\right)^{\frac{n+1}{n-1}}\right]}$$

$$= A\sqrt{2\left(\frac{n}{n-1}\right)\frac{p_1}{v_1}\left(\frac{2}{n+1}\right)^{\frac{n+1}{n-1}}\left[\left(\frac{2}{n+1}\right)^{\frac{2}{n-1}-\frac{n+1}{n-1}} - 1\right]}$$

$$= A\sqrt{2\left(\frac{n}{n-1}\right)\left(\frac{p_1}{v_1}\right)\left(\frac{2}{n+1}\right)^{\frac{n+1}{n-1}}\left[\left(\frac{2}{n+1}\right)^{\frac{1-n}{n-1}} - 1\right]}$$

$$= A\sqrt{2\left(\frac{n}{n-1}\right)\left(\frac{p_1}{v_1}\right)\left(\frac{2}{n+1}\right)^{\frac{n+1}{n-1}}\left[\left(\frac{2}{n+1}\right)^{-1} - 1\right]}$$

$$= A\sqrt{2\left(\frac{n}{n-1}\right)\left(\frac{p_1}{v_1}\right)\left(\frac{2}{n+1}\right)^{\frac{n+1}{n-1}}\left(\frac{n-1}{2}\right)}$$

i.e., $$m_{max} = A\sqrt{n\left(\frac{p_1}{v_1}\right)\left(\frac{2}{n+1}\right)^{\frac{n+1}{n-1}}} \quad ...(18.10)$$

From the above equation it is evident that *the maximum mass flow depends only on the initial condition of the steam (p_1, v_1) and the throat area* and is *independent of the final pressure of steam i.e.,* at the exit of the nozzle. *The addition of the divergent part of the nozzle after the throat does not affect the discharge of steam passing through the nozzle but it only accelerates the steam leaving the nozzle.*

It may be noted that the discharge through nozzle increases as the pressure at the throat of the nozzle (p_2) decreases, when the supply pressure p_1 is constant. But once the nozzle pressure p_2 reaches the critical value [given by equation (18.9)], the discharge reaches a maximum and after that the throat pressure and mass flow remains constant irrespective of the pressure at the exit.

The velocity of steam at the throat of the nozzle when the discharge is maximum is obtained by substituting the value of $\frac{p_2}{p_1}$ from eqn. (18.9) into eqn. (18.6).

$$C_{max} = \sqrt{2\left(\frac{n}{n-1}\right)p_1v_1\left[1-\left\{\left(\frac{2}{n+1}\right)^{\frac{n}{n-1}}\right\}^{\frac{n-1}{n}}\right]}$$

$$= \sqrt{2\left(\frac{n}{n-1}\right)p_1v_1\left(1-\frac{2}{n+1}\right)}$$

$$= \sqrt{2\left(\frac{n}{n-1}\right)p_1v_1\left(\frac{n-1}{n+1}\right)}$$

i.e., $$C_{max} = \sqrt{2\left(\frac{n}{n+1}\right)p_1v_1} \quad ...(18.11)$$

The above equation indicates that the *velocity is also dependent on the initial conditions of the steam.*

18.3. NOZZLE EFFICIENCY

When the steam flows through a nozzle the final velocity of steam for a given pressure drop is *reduced* due to the following reasons :

(*i*) *The friction between the nozzle surface and steam :*

(*ii*) *The internal friction of steam itself ; and*

(*iii*) *The shock losses.*

Most of these frictional losses occur between the throat and exit in convergent-divergent nozzle. These frictional losses entail the following effects :

(*i*) *The expansion is no more isentropic and enthalpy drop is reduced ;*

(*ii*) *The final dryness friction of steam is increased as the kinetic energy gets converted into heat due to friction and is absorbed by steam ;*

(*iii*) *The specific volume of steam is increased as the steam becomes more dry due to this frictional reheating.*

Fig. 18.3 *represents on Mollier diagram the* ***effect of friction*** *on steam flow through a nozzle.*

The point 1 represents the initial condition of steam which enters the nozzle in a dry saturated state. If the friction is neglected, the expansion of steam from entry to throat is represented by the vertical line 1-2 and that from the throat to the exit by 2-3. Now if the friction were taken into account the heat drop would have been somewhat less

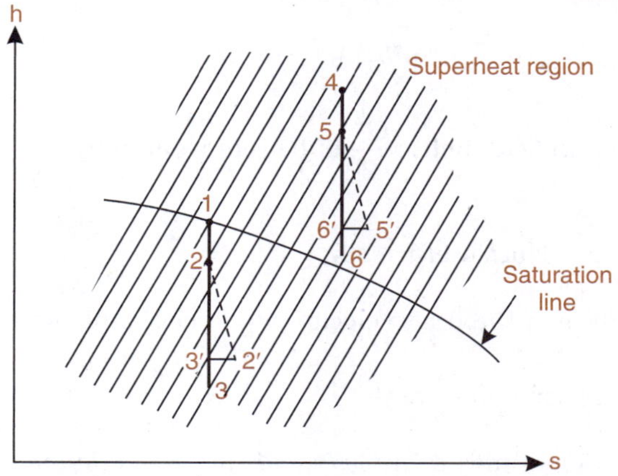

Fig. 18.3. Effect of friction on steam flow through a nozzle.

than 1-3. Let this heat drop be 1-3′. From 3′ draw a horizontal line which cuts the same pressure line on which point 3 lies, at the point 2′ which represents the final condition steam. It may be noted that dryness fraction of steam is **more** at point 2′ than at point 3. Hence *the effect of friction is to improve the quality of steam.* The value of co-efficient '*k*' in the equation for the velocity of expanding steam is given by :

$$k = \frac{\text{Actual heat drop}}{\text{Isentropic heat drop}} = \frac{1-3'}{1-3} = \frac{h_1 - h_3'}{h_1 - h_3}$$

The actual expansion is represented by the curve 1-2-2′ since the *friction occurs mainly between the throat and exit.*

On the other hand, if the steam at entry to nozzle were superheated corresponding to the point 4, the expansion can be represented by the vertical line 4-6 if friction were neglected and by 4-5-5′ if the friction were taken into account. In this case, k = 4-6′/4-6 = $(h_4 - h_6')/(h_4 - h_6)$. The point 5′ represents the final condition of steam. It may be noted that *the friction tends to superheat steam.* Therefore it can be concluded that *friction tends to decrease the velocity of steam and increase the final dryness fraction or superheat the steam.*

The ***nozzle efficiency*** is therefore defined *as the ratio of the actual enthalpy drop to the isentropy enthalpy drop between the same pressures,*

i.e., $$\text{Nozzle efficiency} = \frac{h_1 - h_3'}{h_1 - h_3} \text{ or } \frac{h_4 - h_6'}{h_4 - h_6} \text{ as the case may be.} \quad ...(18.12)$$

If the actual velocity at exit from the nozzle is C_2' and the velocity at exit when the flow is isentropic is C_3, then using the steady flow energy equation, in each case we have

$$h_1 + \frac{C_1^2}{2} = h_3 + \frac{C_3^2}{2} \text{ or } h_1 - h_3 = \frac{C_3^2 - C_1^2}{2}$$

and $h_1 + \frac{C_1^2}{2} = h_2' + \frac{C_2'^2}{2}$ or $h_1 - h_2' = \frac{C_2'^2 - C_1^2}{2}$

$\left[\textbf{In MKS units}: \frac{C^2}{2} \text{ to be represented by } \frac{C^2}{2gJ}\right]$

$\therefore$ Nozzle efficiency $= \frac{C_2'^2 - C_1^2}{C_3^2 - C_1^2}$ ...(18.13)

When the inlet velocity, C_1, is negligibly small then

Nozzle efficiency $= \frac{C_2'^2}{C_3^2}$ ...(18.14)

Sometimes a *"velocity co-efficient" is defined as the ratio of the actual exit velocity to the exit velocity when the flow is isentropic between the same pressures,*

i.e., *Velocity co-efficient* $= \frac{C_2'}{C_3}$ ...(18.15)

It can be seen from eqns. (18.14) and (18.15) that the *"velocity co-efficient" is the square root of the nozzle efficiency, when the inlet velocity is assumed to be negligible.*

18.4. SUPERSATURATED OR METASTABLE EXPANSION OF STEAM IN A NOZZLE

When steam flows through a nozzle, it would normally be expected that the discharge of steam through the nozzle would be slightly less than the theoretical value. But it has been observed during experiments on flow of wet steam that the *discharge is slightly greater than that calculated by the formula.* This phenomenon is explained as follows : The converging part of the nozzle is so short and the steam velocity so high that the molecules of steam have insufficient time to collect and form droplets so that normal condensation does not take place. Such rapid expansion is said to be *metastable* and produces a *supersaturated state.* In this state of supersaturation the steam is undercooled to a temperature less than that corresponding to its pressure ; consequently the density of steam increases and hence the weight of discharge. Prof. Wilson through experiments showed that dry saturated steam, when suddenly expanded *in the absence of dust, does not condense until it density is about 8 times that of the saturated vapour of the same pressure*. This effect is discussed below :

Refer to Fig. 18.4. The point 1 represents initial state of the steam. The steam expands isentropically without any condensation to point 2, 2 being on the superheat constant pressure curve *AB* produced. At point 2 the limit of supersaturation is reached and steam reverts to its normal condition at 3 at the same enthalpy value as 2, and at the same pressure. The steam continues expanding isentropically to a lower pressure to point 4 instead of 4′ which would have been reached if thermal equilibrium had been maintained. Consequently, enthalpy drop is reduced and the condition of the final steam is improved. *The limiting condition of under-cooling at which condensation commences and is assumed to restore conditions of normal thermal equilibrium is called the* ***"Wilson Line"***.

It may be noted that when metastable conditions prevail the h-s chart/diagram should not be used and the expansion must be considered to follow the law $pv^{1.3}$ = C, i.e., with the index of expansion for superheated steam. Thus,

$$\text{Enthalpy drop} = \frac{n}{(n-1)} p_1 v_1 \left[1 - \left(\frac{p_2}{p_2}\right)^{\frac{n-1}{n}}\right]$$

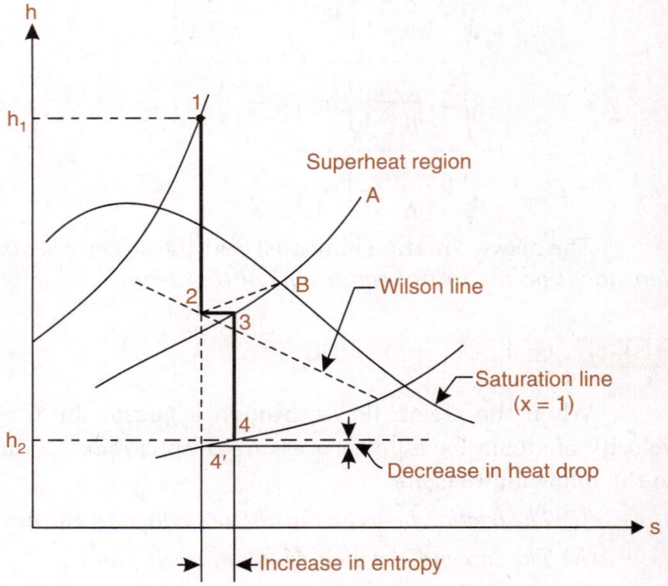

Fig. 18.4. Supersaturated flow of steam.

The relationship, $T_2/T_1 = (p_2/p_1)^{\frac{n-1}{n}}$ may be used to calculate *supercooled temperature.* The *'degree of undercooling' is then the difference between the saturation temperature and the supercooled temperature.*

Effects of supersaturation. In a nozzle in which supersaturation occurs the **effects** may be summarised as follows :

(*i*) There is an increase in the entropy and specific volume of steam.

(*ii*) The heat drop is reduced below that for thermal equilibrium as a consequence the exit velocity of steam is reduced.

(*iii*) Since the condensation does not take place during supersaturated expansion, so the temperature at which the supersaturation occurs will be *less* than the saturation temperature corresponding to the pressure. Therefore, the *density of supersaturated steam will be more than that for the equilibrium conditions which gives the increase in the mass of steam discharged.*

(*iv*) The dryness fraction of steam is improved.

The problems on supersaturated flow cannot be solved by Mollier chart unless Wilson line is drawn on it.

The velocity of steam at the end of expansion is found by using the relation,

$$C_2 = \sqrt{2 \times \left(\frac{n}{n-1}\right) p_1 v_1 \left[1 - \left(\frac{p_2}{p_1}\right)^{\frac{n-1}{n}}\right]}$$

Specific volume, $$v_2 = v_1 \left(\frac{p_2}{p_1}\right)^{1/n}$$

Apparent temperature, $$T_2 = T_1 \left(\frac{p_2}{p_1}\right)^{\frac{n-1}{n}}$$

and $$A_2 = \frac{m \times v_2}{C}.$$

18.5. GENERAL RELATIONSHIP BETWEEN AREA, VELOCITY AND PRESSURE IN NOZZLE FLOW

In Fig. 18.5 is shown a nozzle in which a steady and an isentropic flow is taking place. Let us consider two transverse plane sections at a distance δx apart, assuming that the nozzle is running full and the velocity is uniform across any section.

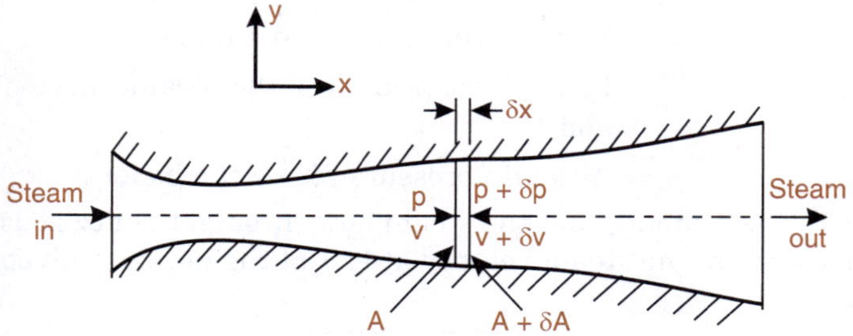

Fig. 18.5

Using continuity equation, we have

$$m = \frac{AC}{v} = \frac{(A + \delta A)(C + \delta C)}{(v + \delta v)}$$

or, $$\frac{v + \delta v}{v} = \frac{(A + \delta A)(C + \delta C)}{AC}$$

or, $$1 + \frac{\delta v}{v} = \left(\frac{A + \delta A}{A}\right)\left(\frac{C + \delta C}{C}\right)$$

$$= \left(1 + \frac{\delta A}{A}\right)\left(1 + \frac{\delta C}{C}\right)$$

or, $$\frac{\delta A}{A} + \frac{\delta C}{C} - \frac{\delta v}{v} = 0$$

and in limits $$\frac{dA}{A} + \frac{dC}{C} - \frac{dv}{v} = 0 \quad ...(18.16)$$

Since the flow is isentropic,

$\therefore$ $$pv^{\gamma} = \text{constant}$$

$\therefore$ $\log_e p + \gamma \log_e v = \log_e K$, where K is constant.

Differentiating this and dividing by pv, we get

$$\frac{dp}{p} + \gamma \cdot \frac{dv}{v} = 0 \quad \text{or} \quad \frac{dv}{v} = -\frac{1}{\gamma} \cdot \frac{dp}{p}$$

Also, from isentropic flow, we have from the momentum equation,

$$CdC = -vdp$$

or, $$\frac{dC}{C} = -\frac{vdp}{C^2} \quad \text{(Dividing both sides by } C^2\text{)}$$

By substituting these values in equation (18.16), we get

$$\frac{dA}{A} + \left(-\frac{vdp}{C^2}\right) - \left(-\frac{1}{\gamma}\frac{dp}{p}\right) = 0$$

or, $$\frac{dA}{A} - \frac{vdp}{C^2} + \frac{1}{\gamma}\frac{dp}{p} = 0$$

$$\frac{dA}{A} = \frac{vdp}{C^2} - \frac{1}{\gamma} \cdot \frac{dp}{p}$$

$$= \frac{1}{\gamma}\frac{dp}{p}\left(\frac{vdp}{C^2} \times \frac{\gamma p}{dp} - 1\right)$$

$$= \frac{1}{\gamma}\frac{dp}{p}\left(\frac{\gamma pv}{C^2} - 1\right)$$

and writing C_s for the sonic velocity at pressure p and specific volume v, we get (as sonic velocity is given by $C_s^2 = \gamma RT = \gamma pv$ in this case)

$$\frac{dA}{A} = \frac{1}{\gamma}\frac{dp}{p}\left(\frac{C_s^2}{C^2} - 1\right) \quad ...(18.17)$$

The ratio of the velocity 'C' to the local sonic velocity 'C_s' is known as the Mach number and is denoted by the letter '*M*'.

$\therefore$ $$\frac{dA}{A} = \frac{1}{\gamma}\frac{dp}{p}\left(\frac{1 - M^2}{M^2}\right) \quad ...(18.18)$$

Equations (18.17) and (18.18) give a useful insight into the changes of nozzle area under certain conditions. These may be explained as follows :

Case I. Accelerated flow, $\frac{dp}{p}$ negative (nozzle), pressure decreases along flow direction.

(*i*) $C < C_s$, $M < 1$. Then $\frac{dA}{A}$ must be *negative*. This corresponds to the convergent part of the nozzle.

As soon as C reaches the value C_s (*i.e.*, $M = 1$), then $\frac{dA}{A} = 0$ and the throat of the nozzle is reached.

(*ii*) $C > C_s$, $M > 1$. Then $\frac{dA}{A}$ must be *positive*. This corresponds to the divergent part of the nozzle.

Type of flow	$\frac{dp}{p}$ *Negative (Nozzle)*	$\frac{dp}{p}$ *Positive (Different)*
1. Subsonic ($M < 1$)	Convergent nozzle	Divergent diffuser
2. Supersonic ($M > 1$)	Divergent nozzle	Convergent diffuser

Fig. 18.6

Case II. Decelerated flow, $\frac{dp}{p}$ positive. (Diffuser)

This applies to diffuser in which the kinetic energy of flow is converted into pressure energy (little application in steam turbine).

(*i*) $C < C_s$, $M < 1$. Here $\frac{dA}{A}$ must be positive, *i.e.*, the diffuser must be of divergent type.

(*ii*) $C > C_s$, $M > 1$. Here $\frac{dA}{A}$ must be negative, *i.e.*, the diffuser must be of the convergent type.

These forms are summarised in Fig. 18.6.

18.6. STEAM INJECTOR

A steam injector is employed to force water into the boiler under pressure. It makes use of the principle of a steam nozzle by which it utilises the kinetic energy of a steam jet for increasing the pressure and velocity of a corresponding quantity of water.

The arrangement of injector for feeding the water from a tank to a boiler is shown in Fig. 18.7. The steam from the boiler is expanded to a high velocity be passing 'it through a convergent nozzle'. The steam jet enters the mixing cone and imparts its momentum to the incoming water supply from the feed water tank (the tank may be above or below the level of the steam injector). The cold water causes the steam to condense. The resulting jet at 2, formed by the steam and water is at atmospheric pressure and has a large velocity. The mixture then enters the delivery pipe at 3 through a diverging cone or diffuser, in which the kinetic energy is reduced and converted into pressure energy. This pressure energy is sufficient to overcome the boiler pressure and lifts the water through H_b. The pressure of water on leaving the delivery pipe must be about *20% higher than the boiler pressure in order to overcome all resistances.* The gap between the mixing cone and diverging cone is provided with an outlet through which any excess water may overflow during starting of steam injector.

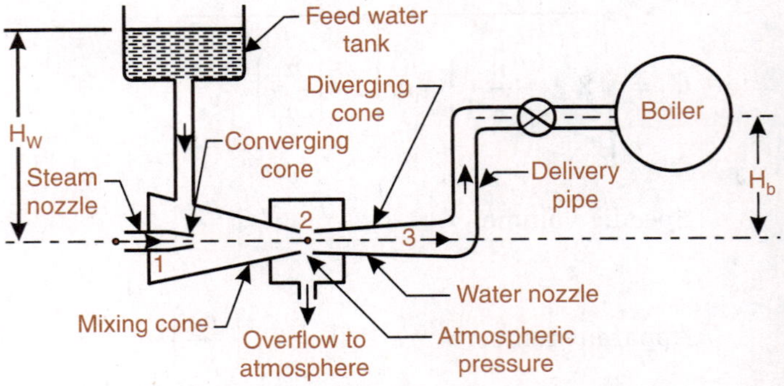

Fig. 18.7. Steam injector.

Let, C_s = Velocity of steam leaving the nozzle,

C_w = Velocity of water at entry to the steam nozzle,

C_m = Velocity of mixture leaving the combined nozzle or entering the water nozzle,

C_{wd} = Velocity of water in the delivery pipe,

m_w = Mass of water drawn from feed water tank per kg of steam supplied to the steam nozzle,

m_s = Mass of steam supplied per sec,

H_w = Water level over the steam injector,

H_b = Boiler inlet level over the steam injector, and

p_b = Absolute pressure inside the boiler.

Assuming that the steam flow through the nozzle is maximum, the steam velocity at exit of the nozzle is given by

$$C_s = \sqrt{2g\left(\frac{n}{n+1}\right)p_1 v_1}$$

$$= 91.5\sqrt{\eta_{\text{nozzle}} \times h_d} \text{ in M.K.S. units}$$

or

$$44.22\sqrt{\eta_{\text{nozzle}} \times h_d} \text{ in S.I. units.}$$

Applying Bernoulli's energy equation at points '3' and '4'

$$0 + \frac{C_m^2}{2g} + \frac{1.033 \times 10^4}{10^3} = H_b + \frac{C^2{}_{wd}}{2g} + \frac{p_b \times 10^4}{10^3}$$

in M.K.S. units.

For all practical purposes, generally, H_b, $\frac{C^2{}_{wd}}{2g}$ and $\frac{1.033 \times 10^4}{10^3}$ are small compared to C_m^2 and may be neglected. The water pressure at entry to the boiler is considered nearly 20% greater than the absolute boiler pressure for overcoming all resistances.

$$\therefore \quad \frac{C_m^2}{2g} = \frac{kp_b \times 10^4}{10^3} \quad ...(18.19)$$

where $k > 1$ and it lies between 1.2 and 1.3 as per assumed or required conditions.

Equating the momentum of water and steam before mixing and after mixing, the following equation may be obtained.

$$\frac{C_s}{g} + \frac{m_w}{g}\sqrt{2gH_w} = \frac{m_w + 1}{g} \times C_m$$

In the above equation, it has been assumed that the mass of water flowing per kg of steam supplied is m_w.

$$C_m = \frac{C_s}{m_w + 1} + \frac{m_w}{m_w + 1} \cdot \sqrt{2gH_w} \quad ...(18.20)$$

If the water tank is below the injector, then

$$C_m = \frac{C_s}{m_w + 1} - \frac{m_w}{m_w + 1}\sqrt{2gH_w} \quad ...(18.21)$$

If H_w is neglected, then eqns. (18.20) or (18.21) become

$$C_m = \frac{C_s}{m_w + 1} \quad ...(18.22)$$

$$\therefore \quad \frac{C_m^2}{2g} = \frac{1}{2g}\left(\frac{C_s}{m_w + 1}\right)^2 \quad ...(18.23)$$

Now equating eqns. (18.19) and (18.23),

$$\frac{1}{2g}\left(\frac{C_s}{m_w + 1}\right)^2 = \frac{kp_b \times 10^4}{10^3}$$

$$\therefore \quad \frac{C_s}{m_w + 1} = \sqrt{20\ gkp_b} \text{ in M.K.S. units.} \quad ...(18.24)$$

From the above equation, the mass of water supplied to the boiler per kg of steam supplied to the steam nozzle can be found.

Area of steam nozzle at the exit is given by

$$A_n = \frac{m_s v}{C_s} \quad ...(18.25)$$

where v is the specific volume of steam at exit.

Area of the discharge end of the combined nozzle or entrance of water nozzle is given by

$$A_{wn} = \frac{m_s(1 + m_w)}{10^3 \times C_m} \quad ...(18.26)$$

It is assumed that all the steam is condensed before coming out of the combined nozzle (mixing cone).

WORKED EXAMPLES

Example 18.1. *(a) Mention the types of nozzles you know. Where are these used ?*

(b) From first principles, prove that maximum discharge per unit area in a steam nozzle at the throat is given by the expression

$$\frac{m_{max}}{A} = \left[2\left(\frac{p_1}{v_1}\right)\left(\frac{2}{n+1}\right)^{\frac{n+1}{n-1}}\right]^{1/2}. \quad \textbf{(U.P.S.C)}$$

Solution. (a) **Types of nozzles :**

1. ***Convergent nozzle.*** It is a smoothly varying cross-sectional area duct which is used for accelerating a steadily flowing fluid. The purpose of this nozzle is to *convert the internal energy of the fluid into the kinetic form.*

2. ***Convergent-divergent nozzle.*** This type of nozzle is a modification of the convergent type where there is a *divergent section which acts as an accelerator for supersonic flow.*

3. ***Steam nozzles.*** This is a special purpose convergent nozzle used in steam turbine for *accelerating the steam at the expense of its pressure.*

4. ***Flow nozzle.*** It is a device used for the measurement of discharge.

(b) Refer to Art. 18.2.2.

Example 18.2. *Dry saturated steam enters a frictionless adiabatic nozzle with negligible velocity at a temperature of 300°C. It is expanded to a pressure of 5000 kPa. The mass flow rate is 1 kg/s. Calculate the exit velocity of steam.*

Properties of Steam

Sat. temp. (°C)	*Sat. pressure (kPa)*	*Enthalpy (kJ/kg)*		*Entropy (kJ/kg/°C)*		*Specific volume (m^3/kg)*	
		Sat. liq.	*Sat. vap*	*Sat. liq.*	*Sat. vap*	*Sat. liq.*	*Sat. vap*
300	8593	1345	2751	3.2552	5.7081	0.0014	0.0216
263.91	5000	1154.5	2794.2	2.9206	5.9735	0.0012	0.0394

(GATE)

Solution. *Given* $h_1 = 2751$ kJ/kg ; $s_1 = 5.7081$ kJ/kg°C

$s_{f_2} = 2.9206$ kJ/kg K ;

$s_{fg_2}\ (s_{g_2} - s_{f_2}) = (5.9735 - 2.9206)$

$h_{f_2} = 1154.5$ kJ/kg ; $h_{g_2} = 2794.2$ kJ/kg.

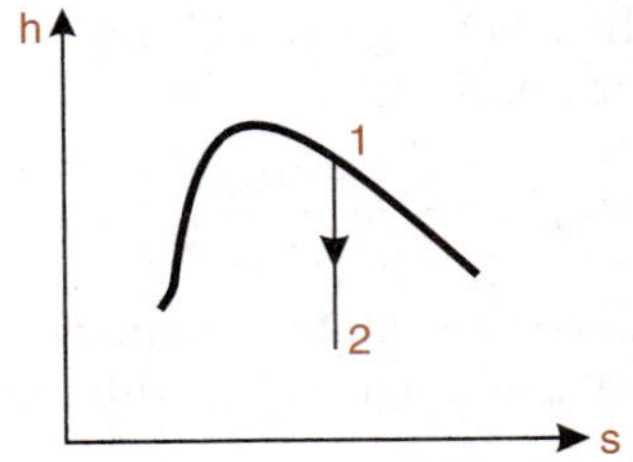

Fig. 18.8

Since the steam expands isentropically, therefore,

$$s_1 = s_2$$

$$5.7081 = 2.9206 + x_2\ (5.9735 - 2.9206)$$

$$\therefore \quad x_2 = \frac{5.7081 - 2.9206}{5.9735 - 2.9206} = 0.913$$

$$h_2 = h_{f_2} + x_2\ h_{fg_2}$$

$$= 1154.5 + 0.913\ (h_{g_2} - h_{fg_2})$$

$$= 1154.5 + 0.913\ (2794.2 - 1154.5)$$

$$= 2651.5 \text{ kJ/kg.}$$

Now, $$\frac{C_1^2}{2} = h_1 - h_2$$

or $$C_1 = \sqrt{2\ (h_1 - h_2)}$$

$$= \sqrt{2\ (2751 - 2651.5) \times 10^3}$$

$$= \mathbf{446.1\ m/s.} \quad \textbf{(Ans.)}$$

Example 18.3. *Steam is expanded in a set of nozzles from 10 bar and 200°C to 5 bar. What type of nozzle is it ? Neglecting the initial velocity find minimum area of the nozzle required to allow a flow of 3 kg/s under the given conditions. Assume that expansion of steam to be isentropic.* **(AMIE)**

Solution. Steam pressure at the entry to the steam nozzles,

$$p_1 = 10 \text{ bar, } 200°\text{C}$$

Steam exit pressure, $p_1 = 5$ bar

We know that, $$\frac{p_2}{p_1} = \left(\frac{2}{n+1}\right)^{\frac{n}{n-1}} = \left[\frac{2}{(1.3+1)}\right]^{\frac{1.3}{0.3}}$$

$$= \left(\frac{2}{2.3}\right)^{4.333} = 0.5457$$

$$\therefore \quad p_2 = p_1 \times 0.5457$$

$$= 10 \times 0.5457 \simeq 5.5 \text{ bar}$$

Since throat pressure (p_2) is greater than the exit pressure, the nozzle used is convergent-divergent nozzle. The minimum area will be at throat, where the pressure is 5.5 bar.

From Mollier chart, $h_1 - h_2 \simeq 120$ kJ/kg

Specific volume, $v \simeq 0.345$ m³/kg

Velocity at the throat, $C_2 = 44.72\sqrt{120}$

$= 489.88$ m/s

Throat area, $$A_2 = \frac{\dot{m}v}{C_2} = \frac{3 \times 0.345}{489.88}$$

$$= \mathbf{0.0021\ m^2.} \quad \textbf{(Ans.)}$$

Example 18.4. *Steam having pressure of 10.5 bar and 0.95 dryness is expanded through a convergent-divergent nozzle and the pressure of steam leaving the nozzle is 0.85 bar. Find the velocity at the throat for maximum discharge conditions. Index of expansion may be assumed as 1.135. Calculate mass rate of flow of steam through the nozzle.*

Solution. The pressure at throat for maximum discharge,

$$p_2 = p_1\left(\frac{2}{n+1}\right)^{\frac{n}{n-1}} = 10.5\left(\frac{2}{1.135+1}\right)^{\frac{1.135}{1.135-1}}$$

$$= 10.5\left(\frac{2}{2.135}\right)^{8.41} = 6.06 \text{ bar}$$

The velocity C_2 at throat for maximum discharge is given by eqn. (18.11)

$$C_2 = \sqrt{2\frac{n}{n+1}p_1v_1}$$

$$= \sqrt{2 \times \frac{1.135}{1.135+1} \times 10.5 \times 10^5\ (0.95 \times 0.185)}$$

$$\simeq 443 \text{ m/s}$$

[C_2 can also be obtained with the help of steam tables or Mollier chart also]

$$p_1v_1^{\,n} = p_2v_2^{\,n}$$

$$10.5(0.95 \times 0.185)^{1.135} = 6.06 \times v_2^{1.135}$$

$$\therefore \quad v_2 = 0.285 \text{ m}^3\text{/kg}$$

Mass flow rate, $$\dot{m} = \frac{A_2C_2}{v_2} = \frac{1 \times 443}{0.285}$$

$$= \mathbf{1554.4\ kg/m^2\ of\ throat\ area.} \quad \textbf{(Ans.)}$$

Example 18.5. *In a steam nozzle, the steam expands from 4 bar to 1 bar. The initial velocity is 60 m/s and the initial temperature is 200°C. Determine the exit velocity if the nozzle efficiency is 92%.*

Solution. Steam pressure at entry to the nozzle,

$$p_1 = 4 \text{ bar, } 200°\text{C}$$

Steam pressure at exit from the nozzle,

$$p_2 = 1 \text{ bar}$$

Initial velocity of steam, $C_1 = 60$ m/s

Nozzle efficiency, $\eta_{nozzle} = 92\%$

Exit velocity, C_2 :

Using steam tables only :

At p_1 = 4 bar, 200°C :

$h_1 = 2860.5$ kJ/kg, $s_1 = 7.171$ kJ/kg

At p_2 = 1 bar :

$h_{f_2} = 417.5$ kJ/kg, $h_{fg_2} = 2257.9$ kJ/kg,

$s_{f_2} = 1.3027$ kJ/kg K $s_{fg_2} = 6.0571$ kJ/kg K

(Refer to Fig. 18.9)

Now, $s_1 = s_2$

$= 7.171 = s_{f_2} + x_2\, s_{fg_2}$

$= 1.3027 + x_2 \times 6.0571$

$$\therefore \quad x_2 = \frac{7.171 - 1.3027}{6.0571} = 0.969$$

$\therefore \quad h_2 = h_{f_2} + x_2\, h_{fg_2}$

$= 417.5 + 0.969 \times 2257.9$

$= 2605.4$ kJ/kg

Enthalpy drop (isentropic)

$= h_1 - h_2$

$= 2860.5 - 2605.4 = 255.1$ kJ/kg

Using Mollier chart :

Refer to Fig. 18.9.

$h_1 = 2860$ kJ/kg

$h_2 = 2605$ kJ/kg

Enthalpy drop (isentropic)

$= h_1 - h_2 = 2860 - 2605 = 255$ kJ/kg

Actual enthalpy drop

$= \eta_{nozzle} \times (h_1 - h_2)$

$= 0.92 \times 255.1 = 234.69$ kJ/kg

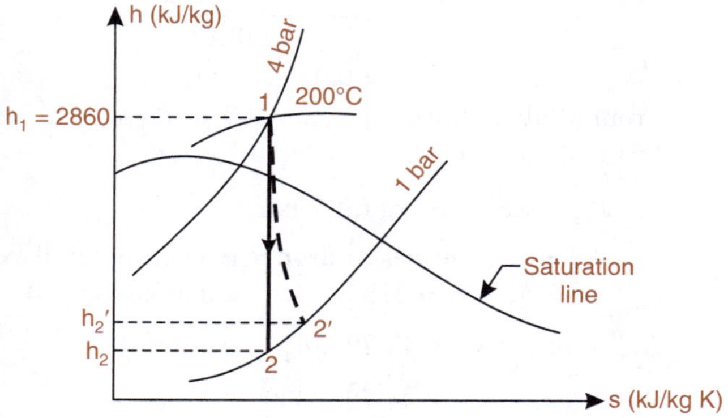

Fig. 18.9

$$\text{Also,} \quad \frac{C_2'^2 - C_1^2}{2} = \text{Enthalpy drop (actual)}$$

i.e., $C_2'^2 - 60^2 = 2 \times 234.69 \times 1000$

$$\therefore \quad C_2' = \sqrt{60^2 + 2 \times 234.69 \times 1000} = 687.7 \text{ m/s}$$

Hence *exit velocity of steam* = **687.7 m/s. (Ans.)**

Example 18.6. *Dry saturated steam enters a steam nozzle at a pressure of 15 bar and is discharged at a pressure of 2.0 bar. If the dryness fraction of discharge steam is 0.96, what will be the final velocity of steam ? Neglect initial velocity of steam.*

If 10% of heat drop is lost in friction, find the percentage reduction in the final velocity.

Solution. Initial pressure of steam, $p_1 = 15$ bar, $x_1 = 1$

Final pressure of steam, $p_2 = 2.0$ bar, $x_2 = 0.96$

From steam tables :

At p_1 = 15 bar, x_1 = 1 :

$h_1 = h_g = 2789.9$ kJ/kg

At p_2 = 2 bar :

$h_{f_2} = 504.7$ kJ/kg,

$h_{fg_2} = 2201.6$ kJ/kg

$h_2 = h_{f_2} + x_2\, h_{fg_2}$

$= 504.7 + 0.96 \times 2201.6 = 2618.2$ kJ/kg

The velocity of steam at discharge from nozzle in S.I. units is given by :

$$C_2 = 44.72\sqrt{h_d} = 44.72\sqrt{(h_1 - h_2)}$$

$$= 44.72\sqrt{(2789.9 - 2618.2)} = 585.9 \text{ m/s}$$

i.e., Final velocity of steam = **585.9 m/s. (Ans.)**

In case 10% of heat drop is lost in friction, nozzle co-efficient

$= 1.0 - 0.1 = 0.9$

Hence the velocity of steam

$= 44.72\sqrt{kh_d}$

$= 44.72\sqrt{0.9(2789.9 - 2618.2)} = 555.9$ m/s

Percentage reduction in velocity

$$= \frac{585.9 - 555.9}{585.9} \times 100 = \textbf{5.12\%}. \quad \textbf{(Ans.)}$$

Example 18.7. *Steam initially dry and saturated is expanded in a nozzle from 15 bar at 300°C to 1.0 bar. If the frictional loss in the nozzle is 12% of the total heat drop calculate the mass of steam discharged when exit diameter of the nozzle is 15 mm.*

Solution. Refer to Fig. 18.10.

Pressure, $p_1 = 15$ bar, 300°C

Pressure, $p_2 = 1.0$ bar

Frictional loss in nozzle = 12%

$\therefore$ Nozzle co-efficient, $k = 1 - 0.12 = 0.88$

Exit diameter of nozzle, $d_2 = 15$ mm

Neglecting the velocity of steam at inlet to the nozzle, the velocity of steam at exit from the nozzle is given by

$$C_2' = 44.72\sqrt{kh_d} = 44.72\sqrt{0.88 \times (h_1 - h_2)}$$

$$= 44.72\sqrt{0.88 \times (3037 - 2515)} = 958.5 \text{ m/s}$$

Dryness fraction of steam at discharge pressure,

$$x_2' = 0.93$$

Specific volume of dry saturated steam at 1.0 bar,

$$v_{g_2} = 1.694 \text{ m}^3/\text{kg}$$

Hence *mass of steam discharged through nozzle per hour*

$$= \frac{A_2 C_2'}{x_2 v_{g_2}} \times 3600$$

$$= \frac{\pi/4 \times (15/1000)^2 \times 958.5}{0.93 \times 1.694} \times 3600$$

= **387 kg/h. (Ans.)**

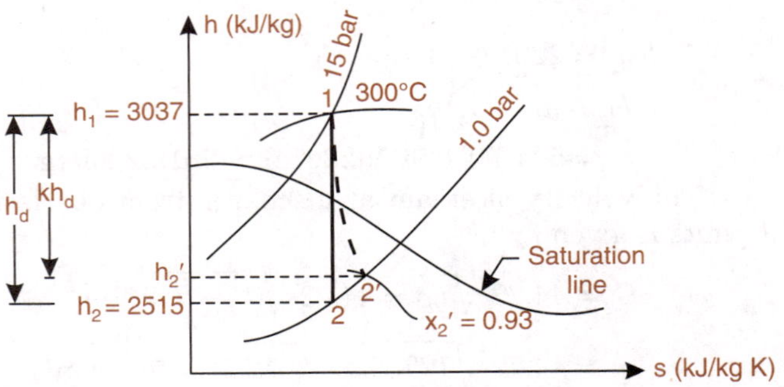

Fig. 18.10

Example 18.8. *Dry saturated steam at a pressure of 11 bar enters a convergent-divergent nozzle and leaves at a pressure of 2 bar. If the flow is adiabatic and frictionless, determine :*

(i) The exit velocity of steam.

(ii) Ratio of cross-section at exit and that at throat.

Assume the index of adiabatic expansion to be 1.135.

Solution. Refer to Fig. 18.11.

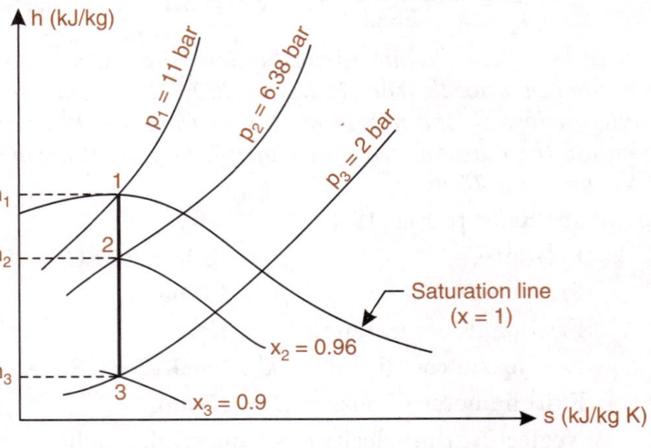

Fig. 18.11

p_1 = 11 bar ; p_3 = 2 bar ;
p_2 = Throat pressure ; n = 1.135

Now,
$$\frac{p_2}{p_1} = \left(\frac{2}{n+1}\right)^{\frac{n}{n-1}}$$

$$= \left(\frac{2}{1.135+1}\right)^{\frac{1.135}{1.135-1}}$$

$$= \left(\frac{2}{2.135}\right)^{\frac{1.135}{0.135}} = 0.58$$

or $p_2 = 0.58 \times p_1 = 0.58 \times 11 = 6.38$ bar.

- From Mollier chart (Fig. 18.11) point 1 is located on the dry saturation line corresponding to 11 bar pressure.
- From '1' vertical line 1–3 is drawn cutting the pressure line 2 bar.
- Point '2' corresponding to throat pressure 6.38 bar is located on the vertical line.

Adiabatic heat drop between inlet and throat,

$$h_d = h_1 - h_2$$
$$= 2780 - 2679 = 101 \text{ kJ/kg}$$
$$x_2 = 0.96$$
$$v_{g_2} = 0.297 \text{ m}^3/\text{kg}$$

Throat velocity, $C_2 = 44.72 \sqrt{h_d} = 44.72 \sqrt{101} = 449.4$ m/s

Also,
$$\dot{m} = \frac{A_2 C_2}{v_2} = \frac{A_2 C_2}{x_2 v_{g_2}}$$

(where $\dot{m}$ = Mass flow rate in kg/s)

or Throat area,
$$A_2 = \frac{\dot{m}\, x_2\, v_{g_2}}{C_2}$$
$$= \frac{\dot{m} \times 0.96 \times 0.297}{449.4}$$
$$= 0.000634\ \dot{m}$$

From Mollier chart, $x_3 = 0.9$

From steam tables,

$v_{g_3} = 0.885$ m³/kg (at 2 bar)

h_d' = Adiabatic heat drop between inlet and exit
$= h_1 - h_3 = 2780 - 2480 = 300$ kJ/kg

Exit velocity, $C_3 = 44.72 \sqrt{h_d'}$
$= 44.72 \sqrt{300}$
= **774.6 m/s. (Ans.)**

Exit area,
$$A_3 = \frac{\dot{m}\, x_3\, v_{g_3}}{C_3} = \frac{\dot{m} \times 0.9 \times 0.885}{774.6}$$
$$= 0.001028\ \dot{m}$$

$$\therefore \text{ Ratio of } \frac{\text{exit area}}{\text{throat area}} = \frac{0.001028\, \dot{m}}{0.000638\, \dot{m}}$$

= **1.62. (Ans.)**

Example 18.9. *The nozzles of a Delaval steam turbine are supplied with dry saturated steam at a pressure of 9 bar. The pressure at the outlet is 1 bar. The turbine has two nozzles with a throat diameter of 2.5 mm. Assuming nozzle efficiency as 90% and that of turbine rotor 35%, find the quality of steam used per hour and the power developed.*

Solution. p_1 = 9 bar, p_3 = 1 bar, p_2 = throat pressure,

Number of nozzles = 2

We know that, $\frac{p_2}{p_1} = \left(\frac{2}{n+1}\right)^{\frac{n}{n-1}}$

Since the steam is dry and saturated, n = 1.135

$$\therefore \quad \frac{p_2}{p_1} = \left(\frac{2}{1.135+1}\right)^{\frac{1.135}{1.135-1}} = 0.58$$

or $\quad p_2 = 0.58\, p_1 = 0.58 \times 9 = 5.22$ bar.

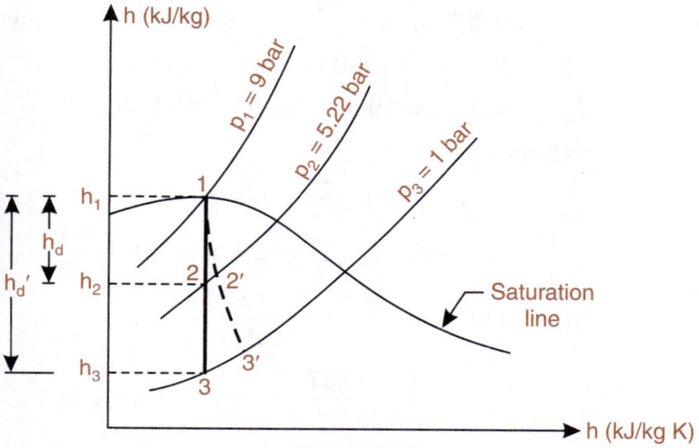

Fig. 18.12

From Mollier chart :

h_1 = 2770 kJ/kg, h_2 = 2670 kJ/kg

h_3 = 2400 kJ/kg

x_2' = 0.96, x_3' = 0.88

Now, $\quad h_d = h_1 - h_2 = 2770 - 2670 = 100$ kJ/kg

$h_d' = h_1 - h_3 = 2770 - 2400 = 370$ kJ/kg

From steam tables :

v_{g_2} = 0.361 m³/kg (at 5.22 bar)

v_{g_3} = 1.694 m³/kg (at 1.0 bar)

Velocity of steam at throat,

$$C_2' = 44.72\sqrt{kh_d}$$

$$= 44.72\sqrt{0.9 \times 100} = 424.2 \text{ m/s}$$

Exit velocity, $\quad C_3' = 44.72\sqrt{kh_{d'}}$

$$= 44.72\sqrt{0.9 \times 370}$$
$$= 816 \text{ m/s}$$

D_2 = 2.5 mm,

$$A_2 = \frac{\pi}{4} D_2^2 \times 2 = \frac{\pi}{4} \times \left(\frac{2.5}{1000}\right)^2 \times 2$$
$$= 9.82 \times 10^{-6} \text{ m}^2$$

Mass of steam used per sec.,

$$\dot{m} = \frac{A_2\, C_2'}{x_2'\, v_{g_2}} = \frac{9.82 \times 10^{-6} \times 424.2}{0.96 \times 0.361}$$
$$= 0.012 \text{ kg/s}$$

Energy supplied by the steam to the wheel per sec

$$= \frac{\dot{m}\, C_3'^2}{2} = \frac{0.012 \times 816^2}{2} = 3995 \text{ W} \simeq 4 \text{ kW}.$$

$\therefore$ Useful work

$$= \eta_{turbine} \times 4 = 0.35 \times 4 = 1.44 \text{ kW}.$$

i.e., *Power developed* = **1.44 kW. (Ans.)**

Example 18.10. *An impulse turbine having a set of 16 nozzles receives steam at 20 bar, 400°C. The pressure of steam at exit is 12 bar. If the total discharge is 260 kg/min and nozzle efficiency is 90%, find the cross-sectional area of the exit of each nozzle. If the steam has a velocity of 80 m/s at entry to the nozzles, find the percentage increase in discharge.*

Solution. Set of nozzles = 16, p_1 = 20 bar, 400°C

Total discharge = 260 kg/min, η_{nozzle} = 90%.

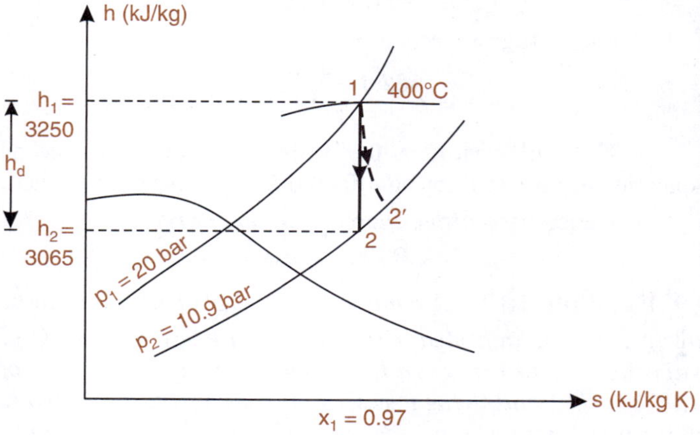

Fig. 18.13

Since the steam supplied to the nozzle is superheated, the throat pressure is given by :

$$\frac{p_2}{p_1} = 0.546$$

$\therefore \quad p_2 = 20 \times p_1 = 20 \times 0.546 = 10.9$ bar

$$\left[\text{For superheated steam} : n = 13 \qquad \frac{p_2}{p_1} = \left(\frac{2}{n+1}\right)^{\frac{n}{n-1}} = \left(\frac{2}{1.3+1}\right)^{\frac{1.3}{1.3-1}} = 0.546\right]$$

Since the pressure is *less* than the exit pressure of steam from nozzle, as such the nozzle is *convergent type*

From Mollier chart :

$h_1 = 3250$ kJ/kg ;

$h_2 = 3065$ kJ/kg ;

$h_d = h_1 - h_2 = 3250 - 3065 = 185$ kJ/kg.

Velocity at exit neglecting initial velocity of steam,

$$C_2' = 44.72\sqrt{kh_d} = 44.72\sqrt{0.9 \times 185} = 577 \text{ m/s}$$

Specific volume at exit,

$v_2' = 0.235$ m³/kg (From Mollier chart)

Area at exit for one nozzle,

$$A_2 = \frac{m \times v_2'}{C_2' \times \text{no. of nozzles}}$$

$$= \frac{260 \times 0.235}{60 \times 577 \times 16} = \mathbf{1.1 \times 10^{-4}\ m^2}. \quad \textbf{(Ans.)}$$

Taking into account the initial velocity of steam as 80 m/s, the velocity of steam at exit, C_2' is calculated as follows :

$$\frac{C_2'^2 - C_1^2}{2} = kh_d$$

i.e., $$C_2'^2 = 2\,kh_d + C_1^2$$

$$= 2 \times 0.9 \times 185 \times 1000 + 80^2 = 339400$$

$\therefore$ $$C_2' = 582.6 \text{ m/s}$$

Percentage increase in velocity

$$= \frac{582.6 - 577}{577} \times 100 = \mathbf{0.97\%.}$$

This will result in 0.97% increase in discharge as specific volume will not be affected by velocity of approach.

Hence *percentage increase in discharge*

$= \mathbf{0.97\%.}$ **(Ans.)**

☞ **Example 18.11.** *A convergent-divergent nozzle is to be designed in which steam initially at 14 bar and 80°C of superheat is to be expanded down to a back pressure of 1.05 bar. Determine the necessary throat and exit diameters of the nozzle for a steam discharge of 500 kg/hour, assuming that the expansion is in thermal equilibrium throughout and friction reheat amounting to 12% of the total isentropic enthalpy drop to be effective in the divergent part of the nozzle.*

Solution. $p_1 = 14$ bar, $t_{sup} - t_s = 80$°C

or $t_{sup} = t_s + 80 = 195 + 80 = 275$°C ;

$p_3 = 1.05$ bar

We know that,

$$\frac{p_2}{p_1} = \left(\frac{2}{n+1}\right)^{\frac{n}{n-1}} = \left(\frac{2}{1.3+1}\right)^{\frac{1.3}{1.3-1}} = 0.546$$

i.e., $$p_2 = p_1 \times 0.546 = 14 \times 0.546 = 7.64 \text{ bar}$$

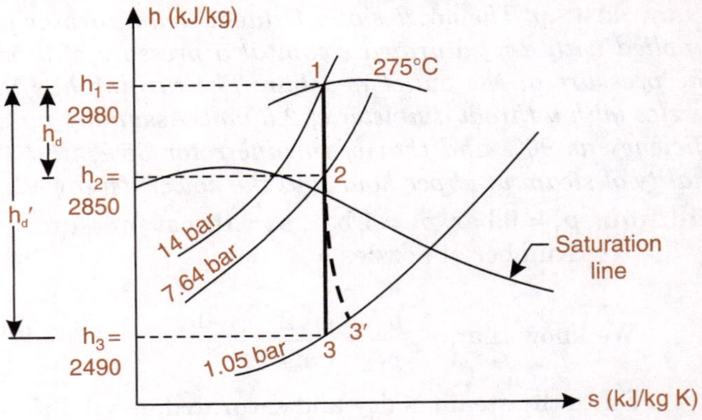

Fig. 18.14

From Mollier chart,

$h_1 = 2980$ kJ/kg, $h_2 = 2850$ kJ/kg

$h_3 = 2490$ kJ/kg, $x_3' = 0.921$

$v_2 = 0.287$ m³/kg (From Mollier chart)

$h_d = h_1 - h_2 = 2980 - 2850 = 130$ kJ/kg

$h_d' = h_1 - h_3 = 2980 - 2490 = 490$ kJ/kg

For throat :

$$C_2 = 44.72\sqrt{h_d}$$

$$= 44.72\sqrt{130} = 509.8 \text{ m/s}$$

Now, $$\dot{m} = \frac{A_2C_2}{v_2} = \frac{A_2 \times 509.8}{0.287}$$

$\therefore$ $$A_2 = \frac{\dot{m} \times 0.287}{509.8}$$

$$= \frac{500 \times 0.287}{3600 \times 509.8} = 7.82 \times 10^{-5} \text{ m}^2$$

i.e., $$\frac{\pi}{4} D_2^2 = 7.82 \times 10^{-5}$$

or $$D_2 = \left(\frac{7.82 \times 10^{-5} \times 4}{\pi}\right)^{1/2}$$

$$= 0.009978 \text{ m or } 9.9 \text{ mm}$$

i.e., *Throat diameter* = **9.9 mm. (Ans.)**

At exit :

$$C_3' = 44.72\sqrt{kh_d'}$$

$$= 44.72\sqrt{(1 - 0.12) \times 490} = 928.6 \text{ m/s}$$

$$v_3' = x_3'\,v_{g_3} = 0.921 \times 1.69 = 1.556 \text{ m}^3/\text{kg}$$

$\therefore$ $$A_3 = \frac{\dot{m}\,v_3'}{C_3'}$$

$$= \frac{500 \times 1.556}{3600 \times 928.6} = 0.0002327 \text{ m}^2$$

i.e., $$\frac{\pi}{4} D_3^2 = 0.0002327$$

or $$D_3 = \left(\frac{0.0002327 \times 4}{\pi}\right)^{1/2}$$
$$= 0.0172 \text{ m or } 17.2 \text{ mm}$$

i.e., *Exit diameter* = **17.2 mm. (Ans.)**

Example 18.12. *Steam at a pressure of 15 bar and dryness fraction 0.97 is discharged through a convergent-divergent nozzle to a back pressure of 0.2 bar. The mass flow rate is 9 kg/kWh. If the power developed is 220 kW, determine :*

(*i*) *Throat pressure.*

(*ii*) *Number of nozzles required if each nozzle has a throat of rectangular cross-section of 4 mm × 8 mm.*

(*iii*) *If 12% of the overall isentropic enthalpy drop reheats by friction the steam in divergent portion find the cross-section of the exit rectangle.*

Solution. p_1 = 15 bar ; x_1 = 0.97 ; p_3 = 0.2 bar

Mass flow rate = 9 kg/kWh ;

Power developed = 220 kW

Throat dimensions = 4 mm × 8 mm

The isentropic index expansion for wet steam

$$n = 1.035 + 0.1\,x_1$$
$$= 1.035 + 0.1 \times 0.97 = 1.132$$

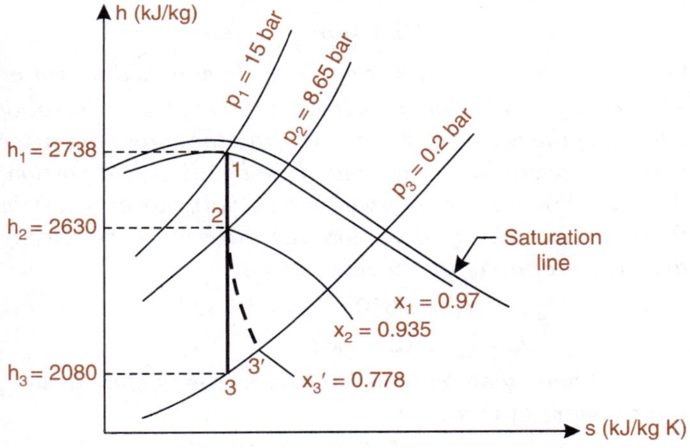

Fig. 18.15

(*i*) **Throat pressure, p_2 :**

$$\frac{p_2}{p_1} = \left(\frac{2}{n+1}\right)^{\frac{n}{n-1}} = \left(\frac{2}{1.132+1}\right)^{\frac{1.132}{1.132-1}}$$
$$= (0.938)^{8.575} = 0.577$$

$\therefore$ $$p_2 = 0.577 \times p_1$$
$$= 0.577 \times 15$$
$$= \mathbf{8.65 \text{ bar.}}$$ **(Ans.)**

(*ii*) **Number of nozzles :**

h_1 = 2738 kJ/kg ; h_2 = 2630 kJ/kg,
h_3 = 2080 kJ/kg

Frictional heating

$$= 0.12\,(h_1 - h_3)$$
$$= 0.12\,(2738 - 2080) = 78.96 \text{ kJ/kg}$$

Thus by frictional reheating the point 3 gets shifted to 3′ on the constant pressure line such that

$$h_3' = h_3 + \text{frictional reheating}$$
$$= 2080 + 78.96 = 2158.96 \text{ kJ/kg}$$

Thus the steam conditions in the nozzle from inlet to exit are shown by 1-2-3′.

Steam consumption,

$$\dot{m} = \frac{9 \times 220}{60 \times 60} = 0.55 \text{ kg/s}$$

Velocity at throat,

$$C_2 = 44.72\sqrt{h_d} = 44.72\sqrt{(h_1 - h_2)}$$
$$= 44.72\sqrt{(2738 - 2630)} = 464.7 \text{ m/s}$$

Total area at throat,

$$A_2 = \frac{\dot{m} \times v_2}{C_2} = \frac{\dot{m} x_2 v_{g_2}}{C_2}$$
$$= \frac{0.55 \times 0.935 \times 0.223}{464.7}$$
$$= 2.467 \times 10^{-4} \text{ m}^2$$

Throat area per nozzle

$$= 4 \times 8 \times 10^{-6} = 0.32 \times 10^{-4} \text{ m}^2$$

$\therefore$ *Number of nozzles*

$$= \frac{2.467 \times 10^{-4}}{0.32 \times 10^{-4}} = \mathbf{7.7 \text{ say } 8.}$$ **(Ans.)**

(*iii*) **Cross-section at the exit :**

Velocity at exit from the nozzle,

$$C_3' = 44.72\sqrt{(h_1 - h_3')}$$
$$= 44.72\sqrt{(2738 - 2158.96)} = 1076 \text{ m/s}$$

$\therefore$ Exit area, $$A_3 = \frac{\dot{m} v_3'}{C_3'} = \frac{\dot{m} x_3' v_{g_3}}{C_3'}$$
$$= \frac{0.55 \times 0.778 \times 7.65}{1076}$$
$$= 0.003042 \text{ m}^2$$

$\therefore$ Area/nozzle $$= \frac{0.003042}{8} = 0.0003802 \text{ m}^2$$

Keeping the same aspect ratio for the rectangle, and let x be the smaller side

$$2x^2 = 0.0003802$$

$$x = \left(\frac{0.0003802}{2}\right)^{1/2} = 0.01378 \text{ m or } 13.78 \text{ mm}$$

Thus exit rectangle is

13.78 mm × 27.56 mm. (Ans.)

Example 18.13. *A Delaval type impulse turbine is to develop 150 kW with a probable consumption of 7.5 kg of steam per kWh with initial pressure being 12 bar and the exhaust 0.15 bar. Taking the diameter at the throat of each nozzle as 6 mm, find the number of nozzles required. Assuming that 10 per cent of the total drop is lost in diverging part of the nozzle, find the diameter at the exit of the nozzle and the quality of steam which is to be fully expanded as it leaves the nozzle.*

Solution. $p_1 = 12$ bar, $p_3 = 0.15$ bar

Assuming the steam to be initially dry and saturated,

$$\frac{p_2}{p_1} = \left(\frac{2}{n+1}\right)^{\frac{n}{n-1}} = \left(\frac{2}{1.135+1}\right)^{\frac{1.135}{0.135}} = 0.58$$

$\therefore \quad p_2 = 12 \times 0.58 = 6.96$ bar

Refer to Fig. 18.16.

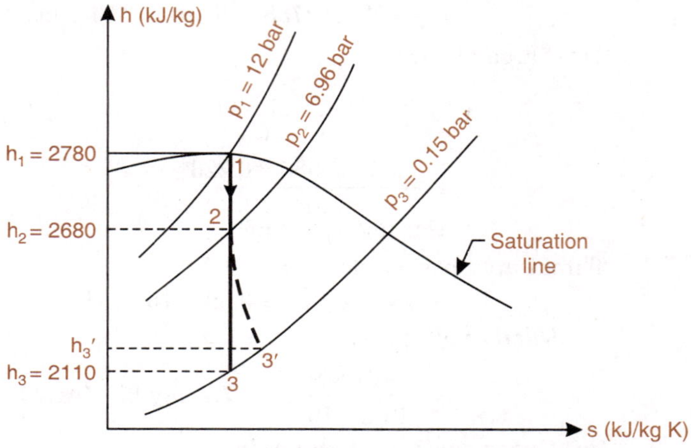

Fig. 18.16

From Mollier chart :

$h_1 = 2780$ kJ/kg ; $h_2 = 2680$ kJ/kg

$x_2 = 0.96$, $h_3 = 2110$ kJ/kg, $x_3' = 0.8$

From steam tables :

$v_{g_2} = 0.274$ m³/kg, $v_{g_3} = 10.022$ m³/kg

The steam consumption of all nozzles per second

$$= \frac{7.5 \times 150}{3600} = 0.3125 \text{ kg}$$

$$C_2 = 44.72\sqrt{h_d} = 44.72\sqrt{(h_1 - h_2)}$$

$$= 44.72\sqrt{(2780 - 2680)} = 447.2 \text{ m/s}$$

$$A_2 = \pi/4\, D_2^2 = \pi/4 \times (6/1000)^2$$
$$= 0.2827 \times 10^{-4} \text{ m}^2$$

$$\dot{m} = \frac{A_2 C_2}{x_2 v_{g_2}}$$

$$= \frac{0.2827 \times 10^{-4} \times 447.2}{0.96 \times 0.274} = 0.04806 \text{ kg/s}$$

$\therefore$ *Number of nozzles required*

$$= \frac{0.3125}{0.04806} = \textbf{6.5 say 7.} \quad \textbf{(Ans.)}$$

At exit :

Quality of steam leaving the nozzle, $x_3' = 0.8$

$$h_d' = h_1 - h_3 = 2780 - 2110 = 670 \text{ kJ/kg}$$

Velocity at the exit,

$$C_3' = 44.72\sqrt{kh_d'}$$

$$= 44.72\sqrt{0.9 \times 670} = 1098 \text{ m/s}$$

Area at the exit,

$$A_3 = \frac{\pi}{4} D_3^2$$

$$\dot{m} = \frac{A_3 C_3'}{v_3'} = \frac{\pi/4\, D_3^2 \times C_3'}{x_3 v_{g_3}}$$

$$\therefore \quad 0.04806 = \frac{\frac{\pi}{4} D_3^2 \times 1098}{0.8 \times 10.022}$$

i.e.,
$$D_3^2 = \frac{0.04806 \times 4 \times 0.8 \times 10.022}{\pi \times 1098}$$

$\therefore \quad D_3 = 0.0211$ m or 21.1 mm

i.e., *Diameter at the exit of the nozzle*

= **21.1 mm. (Ans.)**

☞ **Example 18.14.** *A steam nozzle is supplied steam at 15 bar 350°C and discharges steam at 1 bar. If the diverging portion of the nozzle is 80 mm long and the throat diameter is 6 mm, determine the cone angle of the divergent portion. Assume 12% of the total available enthalpy drop is lost in friction in the divergent portion. Also determine the velocity and temperature of the steam at throat.*

Solution. $p_1 = 15$ bar, 350°C, $p_3 = 1$ bar,

$k = 1 - 0.12 = 0.88$

When steam supplied to the nozzle is *superheated*, the pressure at throat,

$$p_2 = 0.546\, p_1 = 0.546 \times 15 = 8.19 \text{ bar}$$

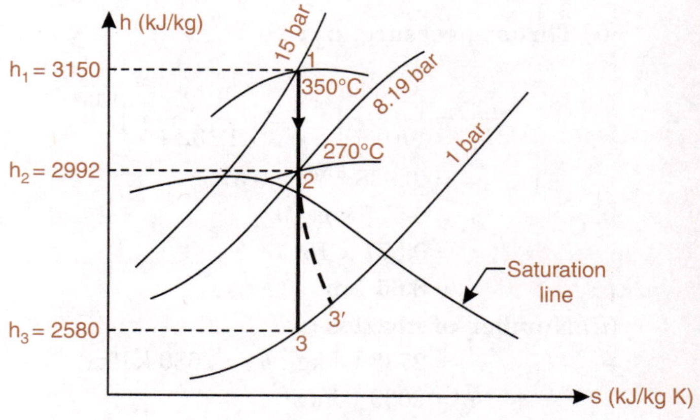

Fig. 18.17

From Mollier chart :

$h_1 = 3150$ kJ/kg, $h_2 = 2992$ kJ/kg

$v_2 = 0.24$ m³/kg, $h_3 = 2580$ kJ/kg

$v_3' = 1.75$ m³/kg, $t_2 = 270°C$

i.e., **Temperature of the steam at throat**

= **270°C. (Ans.)**

Velocity of steam at throat,

$$C_2 = 44.72\sqrt{h_d} = 44.72\sqrt{(h_1 - h_2)}$$

$$= 44.72\sqrt{(3150 - 2992)}$$

$$= \mathbf{562.12\ m/s.}\ \textbf{(Ans.)}$$

From the conditions at nozzle throat, mass flow rate,

$$\dot{m} = \frac{A_2C_2}{v_2}$$

$$= \frac{\pi/4\,(6/1000)^2 \times 562.12}{0.24} = 0.0662 \text{ kg/s}$$

At exit :

$$C_3' = 44.72\sqrt{kh_d'} = 44.72\sqrt{0.88 \times (h_1 - h_3)}$$

$$= 44.72\sqrt{0.88 \times (3150 - 2580)} = 1001.5 \text{ m/s}$$

Exit area of the nozzle,

$$A_3 = \frac{\dot{m} \times v_3'}{C_3'} = \frac{0.0662 \times 1.75}{1001.5} = 0.0001156 \text{ m}^2$$

i.e., $$\frac{\pi}{4}D_3^2 = 0.0001156$$

$$\therefore\ D_3 = \left(\frac{4 \times 0.0001156}{\pi}\right)^{1/2} = 0.012 \text{ m or } 12.1 \text{ mm}$$

If θ be the cone angle of nozzle,

$$\tan\theta = \frac{(12.1 - 6)}{2 \times 80} = 0.03812 \text{ or } \theta = 2°\ 11'$$

Thus, **cone angle = 2 × 2° 11′ = 4° 22′. (Ans.)**

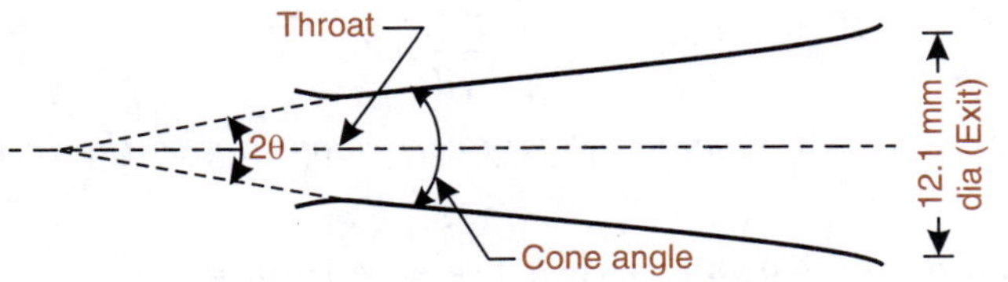

Fig. 18.17 (a)

Example 18.15. *A steam turbine develops 184 kW, with a consumption of 16.45 kg/kWh. The pressure and temperature of the steam entering the nozzle are 11.8 bar and 220°C. The steam leaves the nozzle at 1.18 bar. The diameter of the nozzle at the throat is 7 mm. Find the number of nozzles.*

If 8% of the total enthalpy drop is lost in friction in diverging part of the nozzle, determine the diameter at the exit of the nozzle and exit velocity of the leaving steam.

(D.U.)

Solution. Power developed by the steam turbine = 184 kW

Steam consumption = 16.45 kg/kWh

Steam entry pressure, $p_1 = 11.8$ bar, 220°C

Steam exit pressure, $p_2 = 1.18$ bar

Diameter of the nozzle at the throat = 7 mm = 0.007 m

Total enthalpy drop = 8%

Steam pressure at throat, $p_2 = 0.545 \times p_1$
$= 0.545 \times 11.8$
$= 6.43$ bar

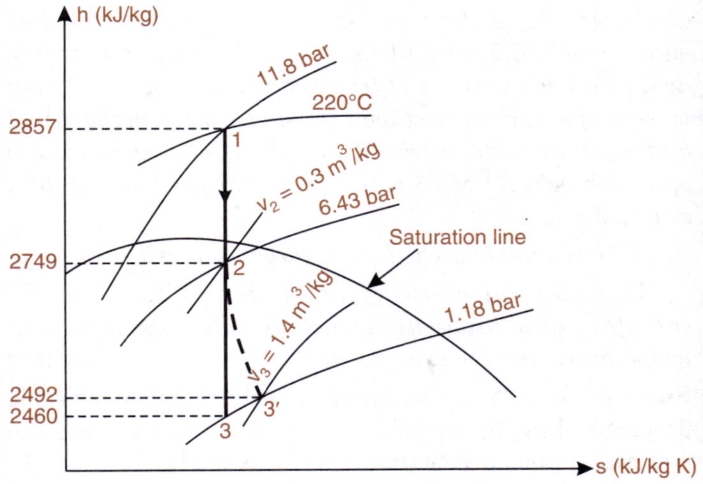

Fig. 18.18

Fig. 18.18 shows the expansion and values of enthalpy and specific volume at various points on Mollier chart.

$$h_3' = 2857 - \left(1 - \frac{8}{100}\right)(2857 - 2460)$$

$$= 2491.7 \text{ kJ/kg}$$

Steam flow, $$\dot{m} = \frac{184 \times 16.45}{3600} = 0.8407 \text{ kg/s}$$

Velocity of steam at throat,

$$C_2 = 44.72\sqrt{(2857 - 2749)} = 464.7 \text{ m/s}$$

$$\dot{m} = \frac{A_2C_2}{v_2} \times N$$

∴ *Number of nozzles,*

$$N = \frac{\dot{m}v_2}{A_2C_2} = \frac{0.8407 \times 0.3}{\pi/4 \times 0.007^2 \times 466.6}$$

= **14.04 or say 14. (Ans.)**

Velocity at exit,

$$C_3' = 44.72\sqrt{2857 - 2492}$$

= **854.4 m/s. (Ans.)**

or $$\dot{m} = \frac{A_3 C_3 \times N}{v_3}$$

$$0.8407 = \frac{\pi/4\, D_3^2 \times 855.4 \times 14}{1.4}$$

$$\therefore \quad D_3^2 = \frac{0.8407 \times 1.4 \times 4}{\pi \times 854.4 \times 14}$$

or $$D_3 = 0.0112 \text{ m} \quad \text{or} \quad 11.2 \text{ mm}$$

Hence *diameter at the exit of the nozzle*

= **11.2 mm. (Ans.)**

Example 18.16. (*a*) *What are the important considerations for selection of the blade material for a steam turbine ?*

(*b*) *A convergent-divergent nozzle is required to discharge 2 kg of steam per second. The nozzle is supplied with steam at 6.9 bar and 180°C and discharge takes place against a back pressure of 0.98 bar. Expansion upto throat is isentropic and the frictional resistance between the throat and exit is equivalent to 62.76 kJ/kg of steam. Taking approach velocity of 75 m/s and throat pressure 3.9 bar, estimate :*

(*i*) *Suitable areas for the throat and exit.*

(*ii*) *Overall efficiency of the nozzle based on the enthalpy drop between the actual inlet pressure, and temperature and the exit pressure.* **(AMIE)**

Solution. (*a*) The selection of material besides desirable properties largely depends on cost. The following are the materials which are generally used for blades :

(*i*) Brass (*ii*) Nickel brass
(*iii*) Phosphor bronze (*iv*) Manganese copper
(*v*) Monel metal (*vi*) Nickel steel
(*vii*) Stainless steel.

(*b*) The expansion and values obtained from Mollier chart are shown in Fig. 18.19.

(*i*) **Areas for throat and exit :**

Between inlet and throat :

$$h_1 + \frac{C_1^2}{2} = h_2 + \frac{C_2^2}{2}$$

$$2790 + \frac{75 \times 75}{2 \times 1000} = 2663 + \frac{C_2^2}{2 \times 1000}$$

$$2792.8 = 2663 + \frac{C_2^2}{2000}$$

$$\therefore \quad C_2 = 509.5 \text{ m/s.}$$

Area at throat, $$\mathbf{A_2} = \frac{\dot{m} v_2}{C_2} = \frac{2 \times 0.45}{509.5}$$

$$= 0.001766 \text{ m}^2$$

= **17.66 cm². (Ans.)**

Between inlet and exit :

$$h_1 + \frac{C_1^2}{2} = h_3' + \frac{C_3'^2}{2}$$

$$2790 + \frac{75 \times 75}{2 \times 1000} = 2498 + \frac{C_3'^2}{2 \times 1000}$$

$$2792.8 = 2498 + \frac{C_3'^2}{2000}$$

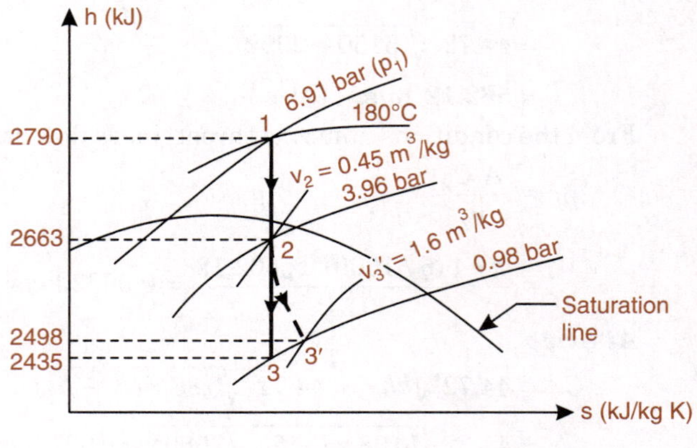

Fig. 18.19

$$\therefore \quad C_3' = 767.8 \text{ m/s}$$

Area at exit, $$A_3 = \frac{\dot{m} v_3'}{C_3'} = \frac{2 \times 1.6}{767.8}$$

= **0.004167 or 41.67 cm². (Ans.)**

(*ii*) **Overall efficiency :**

$$\eta_{overall} = \frac{h_1 - h_3'}{h_1 - h_3} = \frac{2790 - 2498}{2790 - 2435}$$

= **0.8225 or 82.25%. (Ans.)**

Example 18.17. (*a*) *Show that, when critical pressure ratio occurs, the velocity of a compressible fluid at the exit of a convergent nozzle is given by*

$$C_2 = \sqrt{\frac{2}{\gamma + 1}} \cdot a_1$$

where a_1 is the sonic velocity corresponding to the intial conditions. Assume critical pressure ratio $= \left(\frac{2}{\gamma+1}\right)^{\frac{\gamma}{\gamma+1}}$,

where γ is the adiabatic index.

(*b*) *State the factors on which nozzle efficiency depends.*

(*c*) *Determine the throat and exit height of a Delaval nozzle to discharge 27 kg of a perfect gas per minute. The inlet and exit pressures are 480 kPa and 138 kPa respectively. Initial temperature of the gas is 535°C. Nozzle efficiency is 90% and frictional losses occur only after the throat. The molecular weight of the gas is 29 and its adiabatic index = 1.4. Assume square cross-section of the nozzle.*

(Roorkee University)

Solution. (*a*) $\frac{p_2}{p_1} = \left(\frac{2}{\gamma+1}\right)^{\frac{\gamma}{\gamma-1}}$ ...Given

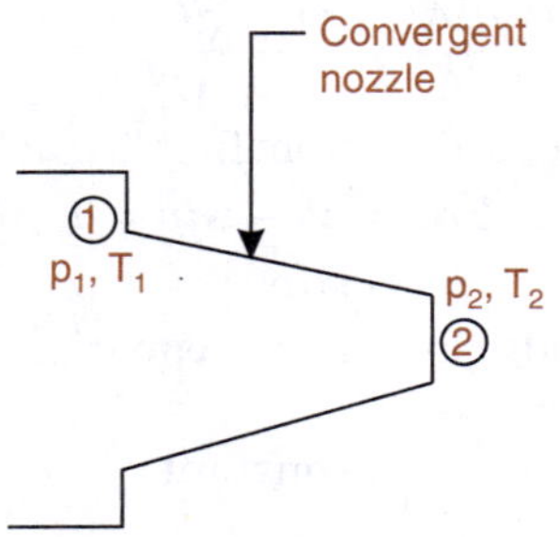

Fig. 18.20

The exit Mach. number is 1, ∴ $T_2 = T^*$

$$\frac{T^*}{T_1} = \left(\frac{p^*}{p_1}\right)^{\frac{\gamma-1}{\gamma}}$$

$$= \left[\left(\frac{2}{\gamma+1}\right)^{\frac{\gamma}{\gamma-1}}\right]^{\frac{\gamma-1}{\gamma}} = \frac{2}{\gamma+1}$$

The exit velocity is sonic velocity,

$$= \sqrt{\gamma R T_2} = \sqrt{\gamma R T^*}$$

$$= \sqrt{\gamma R \frac{2T_1}{\gamma+1}} = \sqrt{\frac{2}{\gamma+1}} \times \sqrt{\gamma R T_1}$$

$$= \sqrt{\frac{2}{\gamma+1}}\, a_1 .$$ **Proved.**

(*b*) **The factors on which nozzle efficiency depends are :**

1. Material of the nozzle.
2. Workmanship of the manufacture of nozzle.
3. Size and shape of the nozzle.
4. Reynolds number of flow.
5. Angle of divergence of divergent portion.
6. Nature of fluid flowing and its state.
7. Turbulence in fluid and its state.

(*c*) *Given :* m_g = 27 kg/min ; p_1 = 480 kPa ;

p_3 = 138 kPa ; T_1 = 535 + 273 = 808 K ;

η_{nozzle} = 90%,

Molecular weight of the gas = 29 ; γ = 1.4.

$$\frac{p_2}{p_1} = \left(\frac{2}{\gamma+1}\right)^{\frac{\gamma}{\gamma-1}}$$

$$= \left(\frac{2}{1.4+1}\right)^{\frac{1.4}{1.4-1}} = 0.528$$

∴ $p_2 = 480 \times 0.528 = 253.44$ kPa

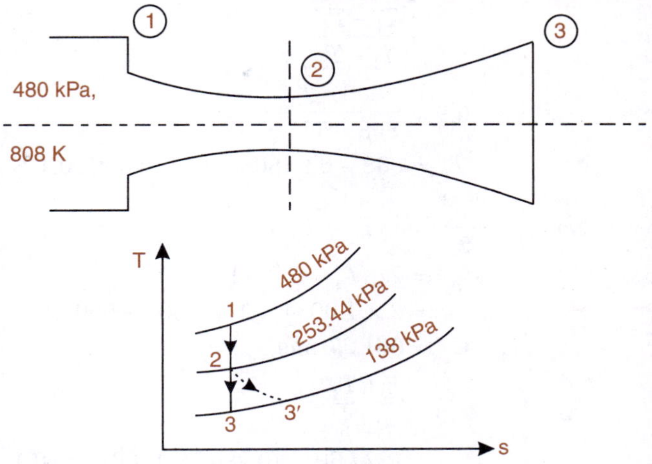

Fig. 18.21

$$\frac{T_2}{T_1} = \frac{2}{\gamma+1} = \frac{2}{1.4+1} = 0.8333$$

∴ $T_2 = 808 \times 0.8333 = 673.3$ K

$$R = \frac{8.314}{29} = 0.2867 \text{ kJ/kg°C}$$

$$c_p = \frac{R\gamma}{\gamma-1} = \frac{0.2867 \times 1.4}{(1.4-1)}$$

$= 1.0034$ kJ/kg°C

Now, $\frac{C_2^2}{2} = c_p(T_1 - T_2)$

or $C_2^2 = 2c_p(T_1 - T_2)$

$= 2 \times 1.0034 \times 1000(808 - 673.3)$

∴ $C_2 = 519.9$ m/s

$$\dot{m}_g = \rho_2 A_2 C_2 = \frac{p_2}{RT_2} A_2 C_2$$

∴ $$A_2 = \frac{\dot{m}_g R T_2}{p_2 C_2}$$

$$= \frac{(27/60)(0.2867 \times 1000) \times 673.3}{(253.44 \times 1000) \times 519.9}$$

$= 6.59 \times 10^{-4}$ m^2

∴ **Height of nozzle** (square section) at **throat**

$= \sqrt{6.59 \times 10^{-4}}$

= 0.0257 m or 25.7 mm. (Ans.)

At the exit : $$\frac{T_3}{T_1} = \left(\frac{p_3}{p_1}\right)^{\frac{\gamma-1}{\gamma}}$$

∴ $$T_3 = T_1 \times \left(\frac{p_3}{p_1}\right)^{\frac{\gamma-1}{\gamma}}$$

$$= 808 \times \left(\frac{138}{480}\right)^{\frac{1.4-1}{1.4}} = 565.9 \text{ K}$$

$$\eta_{nozzle} = \frac{T_1 - T_3'}{T_1 - T_3}$$

or $$0.9 = \frac{808 - T_3'}{808 - 565.9}$$

$\therefore$ $$T_3' = 808 - 0.9\,(808 - 565.9) = 590.1 \text{ K}$$

Now, $$\frac{C_3'^2}{2} = c_p(T_1 - T_3')$$

or $$C_3'^2 = 2c_p(T_1 - T_3')$$
$$= 2 \times 1.0034 \times 1000(808 - 590.1)$$

or $$C_3' = 661.27 \text{ m/s}$$

$\therefore$ $$A_3 = \frac{\dot{m}RT_3'}{p_3C_3'}$$
$$= \frac{(27/60) \times (0.2867 \times 1000) \times 590.1}{(138 \times 1000) \times 661.27}$$
$$= 8.343 \times 10^{-4} \text{ m}^2$$

$\therefore$ **Height of nozzle** (square section) **at the exit**

$$= \sqrt{8.343 \times 10^{-4}}$$
$$= \mathbf{0.02888 \text{ m or } 28.88 \text{ mm.}}$$ **(Ans.)**

Example 18.18. (*a*) *Justify the statement "Nozzles are more efficient than diffusers".*

(*b*) *Derive the following expression for nozzle flow :*

$$\frac{dA}{A} = \frac{1}{\gamma}\frac{dp}{p}\left(\frac{1 - M^2}{M^2}\right),$$ *the symbols having usual meanings.*

(*c*) *It is proposed to design steam nozzles for the following data :*

Initial pressure	*= 30 bar*
Initial temperature	*= 450°C*
Back pressure	*= 6 bar*
Nozzle efficiency	*= 90%*
Initial steam velocity	*= 60 m/s*
Mass flow rate	*= 2 kg/s*

Assuming circular cross-section, calculate the inlet, throat and exit diameter of nozzle. **(AMIE)**

Solution. (*a*) **Nozzles** *are more efficient than diffusers* because in nozzles the *flow is in the direction of decreasing pressure* (favourable pressure gradient). Hence the boundary layer is thin and the frictional losses are less. In **diffusers** the boundary layers are thick due to adverse pressure gradients. This has *greater frictional effects.* Further, the supersonic diffusers have invariably a *normal shock at entry,* even when operating under design conditions. This *further reduces the efficiency.*

(*b*) From continuity equation,

$$\dot{m} = \rho AC = \text{constant} = \frac{AC}{v}$$

Taking logarithms, and the differentials, we have

$$\frac{dA}{A} + \frac{dC}{C} - \frac{dv}{v} = 0$$

or $$\frac{dA}{A} = \frac{dv}{v} - \frac{dC}{C} \quad ...(i)$$

For isentropic flow, from first law of thermodynamics,

$$Tds = dh - vdp = 0$$

or $$dh = vdp \quad ...(ii)$$

From steady flow energy equation

$$h + \frac{C^2}{2} = \text{constant}$$

or $$dh = -C\,dC \quad ...(iii)$$

From (*ii*) and (*iii*), we have

$$dC = -\frac{vdp}{C} \quad ...(iv)$$

For an ideal gas, isentropic relation,

$$pv^\gamma = \text{constant}$$

or $$\frac{dp}{p} + \gamma\frac{dv}{v} = 0$$

or $$\frac{dv}{v} = -\frac{1}{\gamma}\frac{dp}{p} \quad ...(v)$$

Substituting from (*iv*) and (*v*) in (*i*), we get

$$\frac{dA}{A} = -\frac{1}{\gamma}\frac{dp}{p} + \frac{vdp}{C^2}$$
$$= \frac{1}{\gamma}\frac{dp}{p}\left(\frac{pv^\gamma}{C^2} - 1\right)$$
$$= \frac{1}{\gamma}\frac{dp}{p}\left(\frac{\gamma RT}{C^2} - 1\right), \text{ as } pv = RT$$

Further, noting that sonic velocity $C_{sonic} = \sqrt{\gamma RT}$, and $M = Mach\ Number = \frac{C}{C_{sonic}}$, the above relation reduces to

$$\frac{dA}{A} = \frac{1}{\gamma}\frac{dp}{p}\left[\frac{C^2_{sonic}}{C^2} - 1\right] = \frac{1}{\gamma}\frac{dp}{p}\left[\frac{1}{M^2} - 1\right]$$

or $$\frac{dA}{A} = \frac{1}{\gamma}\frac{dp}{p}\left[\frac{1 - M^2}{M^2}\right],$$ the **required derivation.**

(*c*) Refer to Fig. 18.22.

$p_1 = 30$ bar, $p_3 = 6$ bar,

p_2 = throat pressure, $C_1 = 60$ m/s, $\dot{m} = 2$ kg/s.

Inlet : *From Mollier chart :*

$h_1 = 3343$ kJ/kg, $v_1 = 0.1078$ m³/kg

$$h_{01} = h_1 + \frac{C_1^2}{2} = 3343 + \frac{60^2}{2 \times 1000}$$
$$= 3344.8 \text{ kJ/kg} \approx h_1 \qquad [p_{01} \approx p_1 = 30 \text{ bar}]$$

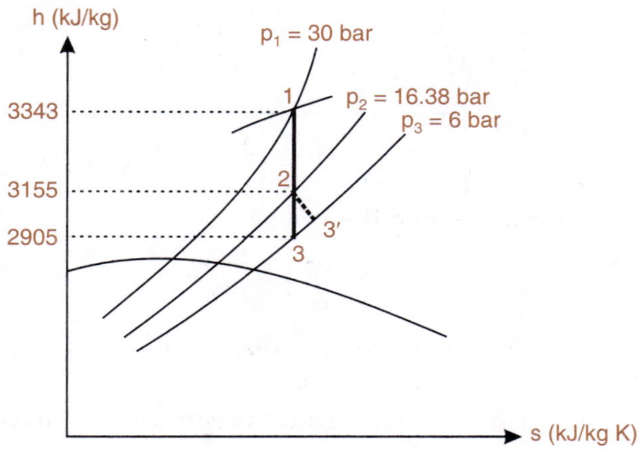

Fig. 18.22

Also, $\dot{m} = \dfrac{A_1C_1}{v_1}$

or $A_1 = \dfrac{\dot{m}v_1}{C_1} = \dfrac{2 \times 0.1078}{60} = 0.003593 \text{ m}^2$

or $\dfrac{\pi}{4}D_1^2 = 0.003593$

∴ $\mathbf{D_1} = \left(\dfrac{0.003593 \times 4}{\pi}\right)^{1/2} = 0.0676 \text{ m}$

= **67.6 mm.** **(Ans.)**

Throat : Assume that frictional losses occur in diverging part only ; *i.e.*, flow upto throat is frictionless.

$$\frac{p_2}{p_1} = 0.546$$

∴ $p_2 = 30 \times 0.546 = 16.38$ bar

$$\left[\begin{array}{l}\text{For superheat steam } n = 1.3 \\ \dfrac{p_2}{p_1} = \left(\dfrac{2}{n+1}\right)^{\frac{n}{n-1}} = \left(\dfrac{2}{1.3+1}\right)^{\frac{1.3}{1.3-1}} = 0.546\end{array}\right]$$

From Mollier chart :

$h_2 = 3155$ kJ/kg, $v_2 = 0.19$ m³/kg

$C_2 = 44.72\sqrt{(3344.8 - 3155)} = 616$ m/s

∴ $A_2 = \dfrac{\dot{m}v_2}{C_2} = \dfrac{2 \times 0.19}{616} = 0.000617 \text{ m}^2$

or $\dfrac{\pi}{4}D_2^2 = 0.000617$

∴ $\mathbf{D_2} = \left(\dfrac{0.000617 \times 4}{\pi}\right)^{1/2}$

= **0.028 m or 28 mm.** **(Ans.)**

Exit : *From Mollier chart* :

$h_3 = 2905$ kJ/kg, $v_3' = 0.4$ m³/kg

$h_3' = h_{01} - \eta_{nozzle}(h_{01} - h_3)$

$= 3344.8 - 0.9(3344.8 - 2905)$

$= 2949$ kJ/kg

∴ $C_3' = 44.72\sqrt{(3344.8 - 2949)} \simeq 890$ m/s

$A_3 = \dfrac{\dot{m}v_3'}{C_3'} = \dfrac{2 \times 0.4}{890} = 0.000899 \text{ m}^2$

or $\dfrac{\pi}{4}D_3^2 = 0.000899.$

or $\mathbf{D_3} = \left(\dfrac{0.000899 \times 4}{\pi}\right)^{1/2}$

= **0.0338 m or 33.8 mm.** **(Ans.)**

Example 18.19. *Determine the throat area, exit area and exit velocity for a steam nozzle to pass a mass flow of 0.2 kg/s when inlet conditions are 10 bar and 250°C and the final pressure is 2 bar. Assume expansion is isentropic and that the inlet velocity is negligible. Use $pv^{1.3}$ constant. Do not calculate from h-s chart.* **(N.U.)**

Solution. Mass of steam flowing through the nozzle,

$\dot{m}_s = 0.2$ kg/s

Inlet pressure, $p_1 = 10$ bar

Inlet temperature, $T_1 = 250 + 273 = 523$ K

Specific volume at 10 bar, 250°C

$= 0.233$ m³/kg

(From steam tables)

Final pressure, $p_3 = 2$ bar

Throat area (A_2), exit area (A_3) and exit velocity (C_3) :

$$\frac{p_2}{p_1} = \left(\frac{2}{n+1}\right)^{\frac{n}{n-1}} = \left(\frac{2}{1.3+1}\right)^{\frac{1.3}{1.3-1}} = 0.5457$$

∴ $p_2 = 0.5457 \times p_1 = 0.5457 \times 10 = 5.457$ bar

At throat :

Velocity $C_2 = \sqrt{\dfrac{2n}{n-1}p_1v_1\left[1 - \left(\dfrac{p_2}{p_1}\right)^{\frac{n-1}{n}}\right]}$

$$= \sqrt{\frac{2 \times 1.3}{1.3-1} \times (10 \times 10^5) \times 0.233\left[1 - \left(\frac{5.457}{10}\right)^{\frac{1.3-1}{1.3}}\right]}$$

$= \sqrt{2019333.3\,(1 - 0.8696)} = 513.15$ m/s

Also $v_2 = v_1\left(\dfrac{p_1}{p_2}\right)^{\frac{1}{n}} = 0.233\left(\dfrac{10}{5.457}\right)^{\frac{1}{1.3}}$

$= 0.3712$ m³/kg

Mass flow rate,

$\dot{m}_s = \dfrac{A_2C_2}{v_2}$ or $0.2 = \dfrac{A_2 \times 513.15}{0.3712}$

or Throat area, $A_2 = \dfrac{0.2 \times 0.3712}{513.15}$
$= \mathbf{1.446 \times 10^{-4}\ m^2}$. **(Ans.)**

At exit :

Velocity, $$C_3 = \sqrt{\frac{2n}{n-1}\, p_1 v_1 \left\{1 - \left(\frac{p_3}{p_1}\right)^{\frac{n-1}{n}}\right\}}$$

$$= \sqrt{\frac{2 \times 1.3}{1.3 - 1} \times (10 \times 10^5) \times 0.233 \left\{1 - \left(\frac{2}{10}\right)^{\frac{1.3-1}{1.3}}\right\}}$$

$$= \sqrt{2019333.33 \times 0.3102} = 791.45 \text{ m/s}$$

$$v_3 = v_1 \left(\frac{p_1}{p_3}\right)^{1/n} = 0.233 \left(\frac{10}{2}\right)^{\frac{1}{1.3}} = 0.8035 \text{ m}^3/\text{kg}$$

Mass flow rate, $\dot{m}_s = \dfrac{A_3 C_3}{v_3}$ or $0.2 = \dfrac{A_3 \times 791.45}{0.8035}$

or Exit area, $A_3 = \dfrac{0.2 \times 0.8035}{791.45}$
$= \mathbf{2.03 \times 10^{-4}\ m^2}$. **(Ans.)**

Example 18.20. *Define critical pressure ratio of a nozzle and discuss why attainment of sonic velocity determines the maximum mass rate of flow through steam nozzle.*

Solution. The critical pressure ratio is given by

$$\frac{p_2}{p_1} = \left(\frac{2}{n+1}\right)^{\frac{n}{n-1}}$$

The velocity at throat

$$C_2 = \sqrt{2\frac{n}{n-1}(p_1 v_1 - p_2 v_2)}$$

[where p_2 and C_2 are the pressure and velocity at the throat]

$$= \sqrt{2\left(\frac{n}{n-1}\right) p_2 v_2 \left[\frac{p_1 v_1}{p_2 v_2} - 1\right]}$$

$$= \sqrt{2\left(\frac{n}{n-1}\right) p_2 v_2 \left[\left(\frac{p_2}{p_1}\right)^{\frac{1-n}{n}} - 1\right]}$$

Substituting the value of $\dfrac{p_2}{p_1}$, we get,

$$C_2 = \sqrt{2\left(\frac{n}{n-1}\right) p_2 v_2 \left[\left\{\left(\frac{2}{n+1}\right)^{\frac{n}{n-1}}\right\}^{\frac{1-n}{n}} - 1\right]}$$

$$= \sqrt{2\left(\frac{n}{n-1}\right) p_2 v_2 \left[\frac{n+1}{2} - 1\right]}$$

$$= \sqrt{2 p_2 v_2 \left(\frac{n}{n-1}\right) \times \left(\frac{n-1}{2}\right)} = \sqrt{n\, p_2 v_2}$$

Velocity of sound is given by

$$a^2 = -v^2 \left(\frac{\partial p}{\partial v}\right)_s$$

For constant entropy process, it is assumed that

$$pv^n = \text{constant}$$

Differentiating this and on substitution, we have,

$$a^2 = npv$$

$$\therefore \quad a = \sqrt{np_2 v_2} = C_2$$

This proves that under the *conditions of maximum discharge the velocity of fluid at throat is equal to the sonic velocity at the throat conditions.*

☞ **Example 18.21.** *In an installation 5.2 kg/s of steam at 30 bar and 350°C is supplied to group of six nozzles in a wheel diameter maintained at 4 bar. Determine :*

(i) The dimensions of the nozzles of rectangular cross-sectional flow area with aspect ratio 3 : 1. The expansion may be considered metastable and friction is neglected ;

(ii) Degree of undercooling and supersaturation ;

(iii) Loss in available heat drop due to irreversibility ;

(iv) Increase in entropy ;

(v) Ratio of mass flow rate with metastable expansion to that if expansion is in thermal equilibrium.

Solution. Refer to Fig. 18.23.

From Mollier chart :

$h_1 = 3115$ kJ/kg ; $\quad h_3 = 2675$ kJ/kg
$v_1 = 0.09$ m³/kg ; $\quad v_3 = 0.46$ m³/kg.

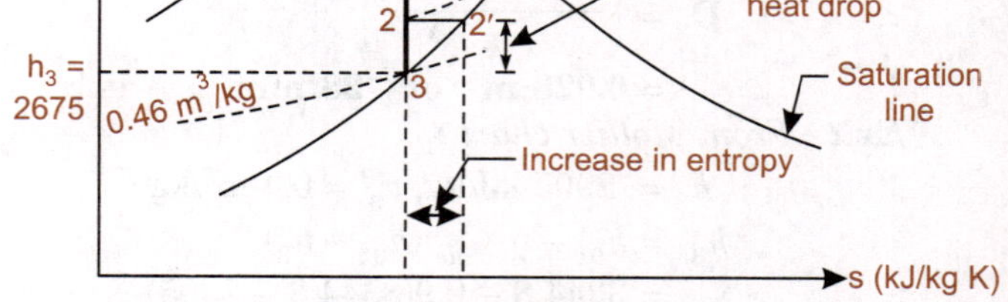

Fig. 18.23

(*i*) **Dimensions of the nozzle :**

For supersaturated steam, the index of expansion is assumed same as for superheated steam, i.e., $n = 1.3$.

Thus isentropic enthalpy drop,

$$(h_1 - h_2) = \frac{n}{n-1} \, \frac{p_1 v_1 \times 10^5}{10^3} \left[1 - \left(\frac{p_2}{p_1}\right)^{\frac{n-1}{n}}\right] \text{ kJ/kg}$$

$$= \frac{1.3}{1.3-1} \times \frac{30 \times 0.09 \times 10^5}{10^3} \left[1 - \left(\frac{4}{30}\right)^{\frac{1.3-1}{1.3}}\right]$$

$$= 1170\,(1 - 0.6282) = 435 \text{ kJ/kg}$$

∴ Velocity at point 2,

$$C_2 = 44.72\sqrt{(h_1 - h_2)}$$

$$= 44.72\sqrt{435} = 932.7 \text{ m/s}$$

$$\text{Also } v_2 = v_1 \left(\frac{p_1}{p_2}\right)^{\frac{1}{n}} = 0.09 \times \left(\frac{30}{4}\right)^{\frac{1}{0.3}} = 0.4239 \text{ m}^3\text{/kg}$$

∴ Mass flow rate,

$$\dot{m} = \frac{A_2 \times C_2}{v_2}$$

$$5.2 = \frac{A_2 \times 932.7}{0.4239}$$

$$\therefore \quad A_2 = \frac{5.2 \times 0.4239}{932.7} = 0.002363 \text{ m}^2$$

Since aspect ratio is 3 : 1, if we assume breadth as x the length will be $3x$ and area of six nozzles will be

$$A_2 = 6 \times 3x \times x$$

$$0.002363 = 18x^2$$

$$\therefore \quad x = \left(\frac{0.002363}{18}\right)^{1/2}$$

$$= \mathbf{0.0114 \text{ m or } 11.4 \text{ mm.}} \quad \textbf{(Ans.)}$$

and *length* $= 3 \times 11.4 =$ **34.2 mm.** **(Ans.)**

(*ii*) **Degree of undercooling and supersaturation :**

Temperature at point 2 is found as follows :

$$\frac{T_2}{T_1} = \left(\frac{p_2}{p_1}\right)^{\frac{n-1}{n}}$$

$$\therefore \quad T_2 = T_1 \times \left(\frac{p_2}{p_1}\right)^{\frac{n-1}{n}}$$

$$= (273 + 350) \times \left(\frac{4}{30}\right)^{\frac{1.3-1}{1.3}}$$

$$= 623 \times 0.628 = 391.2 \text{ K or } 118.2°\text{C}$$

From steam tables saturation temperature at 4 bar $= 143.6°\text{C}$

∴ *Degree of* ***undercooling***

$= 143.6 - 118.2 =$ **25.4°C.** **(Ans.)**

Saturation pressure corresponding to 118.2°C $\simeq 1.9$ bar

∴ *Degree of* ***super saturation***

$$= \frac{4}{1.9} = \mathbf{2.1.} \quad \textbf{(Ans.)}$$

(*iii*) **Loss in available heat drop :**

Isentropic enthalpy drop for expansion under thermal equilibrium conditions as read out from Mollier chart

$$h_1 - h_3 = 3115 - 2675 = 440 \text{ kJ/kg}$$

∴ Loss of available heat drop

$$= 440 - 435 = 5 \text{ kJ/kg.}$$

(*iv*) **Increase in entropy**

$$= \frac{5}{(143.6 + 273)}$$

$$= \mathbf{0.012 \text{ kJ/kg K.}} \quad \textbf{(Ans.)}$$

(*v*) **Ratio of mass flow rate :**

Exit velocity from the nozzle with expansion in thermal equilibrium is given by

$$C_3 = 44.72\sqrt{(h_1 - h_3)}$$

$$= 44.72\sqrt{440} = 938 \text{ m/s}$$

Also specific volume at 4 bar at state point 3 from Mollier chart,

$$v_3 = 0.46 \text{ m}^3\text{/kg}$$

∴ *Mass flow rate for metastable flow*

$$\therefore \quad \frac{\text{Mass flow rate for metastable flow}}{\text{Mass flow rate for thermal equilibrium flow}}$$

$$= \frac{\text{Area of flow} \times C_2}{v_2} \times \frac{v_3}{\text{Area of flow} \times C_3}$$

$$= \frac{v_3 C_2}{v_2 C_3} = \frac{0.46 \times 932.7}{0.4239 \times 938} = \mathbf{1.07.} \quad \textbf{(Ans.)}$$

Example 18.22. *Air enters a convergent nozzle from a reservoir at 2200 kPa and 100°C. If the exit area is 3.25 cm², what is the maximum mass flow rate that this nozzle can handle ? Assume the process to be isentropic and that the air behaves as an ideal gas.* **(D.U.)**

Solution. For maximum mass flow rate, and $\gamma = 1.4$, the pressure at the exit will be critical,

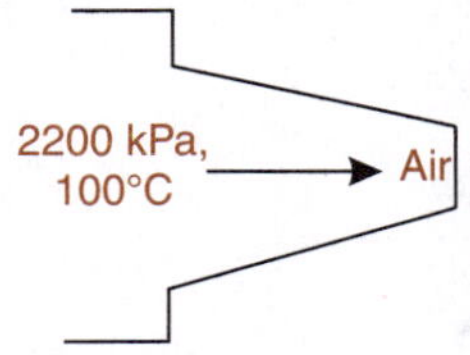

Fig. 18.24

$$p^* = \left(\frac{2}{\gamma + 1}\right)^{[\gamma/(\gamma-1)]}$$

$$p_0 = \left(\frac{2}{1.4 + 1}\right)^{(1.4/1.4.1)} \times p_0$$

$$= 0.528 \times 2200$$

$$\text{and} \quad T^* = T_0 \left(\frac{p^*}{p_0}\right)^{[(\gamma-1)/\gamma]}$$

$$= (100 + 273)\left(\frac{1161.6}{2200}\right)^{(0.4/1.4)} = 310.78 \text{ K}$$

Sonic velocity,

$$C^* = \sqrt{\gamma R T^*}$$

$$= \sqrt{1.4 \times 287 \times 310.78} = 353.37 \text{ m/s}$$

$$\rho^* = \frac{p^*}{RT^*} = \frac{1161.6}{0.287 \times 310.78} = 13.02 \text{ kg/m}^3$$

$$\dot{m} = \rho^* A_e C^* = 13.02 \times \frac{3.25}{10^4} \times 353.37$$

$$= \mathbf{1.495\ kg/s.} \quad \textbf{(Ans.)}$$

Example 18.23. *Air is expanded reversibly and adiabatically in a nozzle from 13 bar and 150°C to a pressure of 6 bar. The inlet velocity of the nozzle is very small and the process occurs under steady state flow conditions. Calculate the exit velocity of the nozzle.* **(U.P.S.C.)**

Solution. *Given :* $p_1 = 13$ bar ; $T_1 = 150 + 273 = 423$ K ; $p_2 = 6$ bar ; $C_1 = 0$

Exit velocity, C_2 :

Applying the energy equation at '1' and '2', we get

$$m\left[h_1 + \frac{C_1^2}{2} + Z_1 g\right] + Q = \left[h_2 + \frac{C_2^2}{2} + Z_2 g\right] + W \quad \text{...}(i)$$

Since the air is expanded reversibly and adiabatically in a nozzle from condition '1' to '2', therefore,

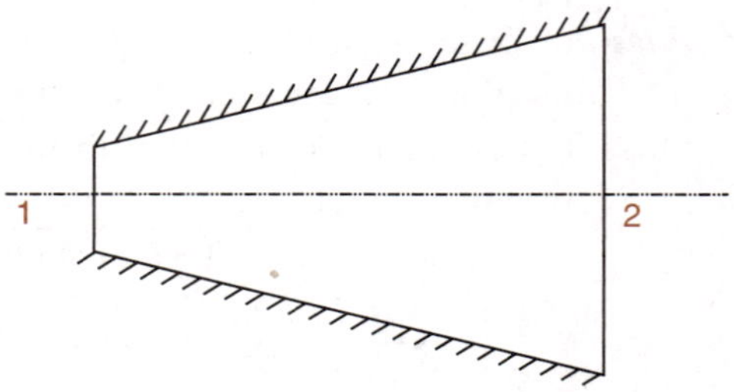

Fig. 18.25

$Q = 0$; Also $W = 0$ and $Z_1 = Z_2$

∴ Eqn. (*i*) reduces to :

$$h_1 + \frac{C_1^2}{2} = h_2 + \frac{C_2^2}{2}$$

But $C_1 = 0$

∴ $C_2^2 = 2(h_1 - h_2)$

or $$C_2 = \sqrt{2(h_1 - h_2)} = \sqrt{2 \times c_p (T_1 - T_2)} \quad \text{...}(ii)$$

Now, $$\frac{T_2}{T_1} = \left(\frac{p_2}{p_1}\right)^{\frac{\gamma-1}{\gamma}} = \left(\frac{6}{13}\right)^{\frac{1.4-1}{1.4}} = 0.8018$$

∴ $T_2 = 423 \times 0.8018 = 339.16$ K

Substituting the values in eqn. (*ii*), we get

$$\mathbf{C_2} = \sqrt{2 \times 1.005 (423 - 339.16)}$$

$$= \mathbf{12.98\ m/s.} \quad \textbf{(Ans.)}$$

Example 18.24. *Air enters a frictionless adiabatic converging nozzle at 10 bar 500 K with negligible velocity. The nozzle discharges to a region at 2 bar. If the exit area of the nozzle is 2.5 cm², find the flow rate of air through the nozzle. Assume for air c_p = 1005 J/kg K and c_v = 718 J/kg K.* **(GATE)**

Solution. Refer to Fig. 18.26.

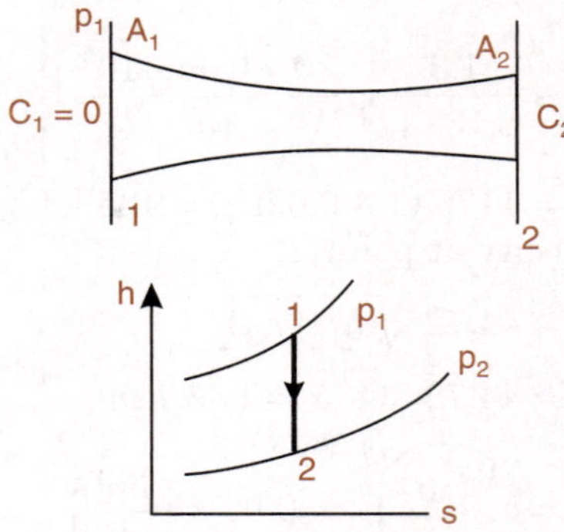

Fig. 18.26

Given : $p_1 = 10$ bar ; $T_1 = 500$ K ; $C_1 = 0$;
$p_2 = 2$ bar ; $A_2 = 2.5$ cm² ;
$c_p = 1005$ J/kg K ; $c_v = 718$ J/kg K.

Flow rate of air, Q :

$$\gamma = \frac{c_p}{c_v} = \frac{1005}{718} = 1.4$$

For the *isentropic process 1-2* in the nozzle,

$$\frac{T_1}{T_2} = \left(\frac{p_1}{p_2}\right)^{\frac{\gamma-1}{\gamma}}$$

or $$\frac{500}{T_2} = \left(\frac{10}{2}\right)^{\frac{1.4-1}{1.4}} = 1.584$$

∴ $$T_2 = \frac{500}{1.584} = 315.6 \text{ K}$$

Now, $$\frac{C_2^2 - C_1^2}{2} = h_1 - h_2$$

or $$C_2^2 - C_1^2 = 2(h_1 - h_2)$$

or $$C_2 = \sqrt{2(h_1 - h_2)} \quad (\because C_1 = 0)$$

$$= \sqrt{2 c_p (T_1 - T_2)}$$

$$= \sqrt{2 \times 1005 (500 - 315.6)}$$

$$= 608.8 \text{ m/s}$$

∴ Flow rate of air, $Q = A_2 C_2$

$$= 2.5 \times 10^{-4} \times 608.8$$

$$= \mathbf{0.1522\ m^3/s.} \quad \textbf{(Ans.)}$$

HIGHLIGHTS

1. A *steam nozzle* may be defined as a passage of varying cross-section, through which heat energy of steam is converted to kinetic energy.
2. The velocity, $C = 44.72\sqrt{kh_d}$ ...... in SI units

   $C = 91.5\sqrt{kh_d}$ ...... in MKS units
3. Critical pressure ratio,

$$\frac{p_2}{p_1} = \left(\frac{2}{n+1}\right)^{\frac{n}{n-1}}$$

...... condition for maximum discharge.

4. The *nozzle efficiency* is defined as the ratio of the actual enthalpy drop to the isentropic enthalpy drop between the same pressures.
5. A steam injector is a device employed to force water into the boiler under pressure.

OBJECTIVE TYPE QUESTIONS

Choose the Correct Answer :

1. For a steam nozzle, if p_1= inlet pressure, p_2 = exit pressure and n is the index of isentropic expansion, the mass flow rate per unit area is maximum if

   (*a*) $\frac{p_2}{p_1} \le \left(\frac{2}{n+1}\right)^{\frac{n-1}{n}}$ (*b*) $\frac{p_2}{p_1} \le \left(\frac{1}{n+1}\right)^{\frac{n}{n+1}}$

   (*c*) $\frac{p_2}{p_1} \le \left(\frac{2}{n+1}\right)^{\frac{n}{n+1}}$ (*d*) $\frac{p_2}{p_1} \le \left(\frac{2}{n+1}\right)^{\frac{n}{n-1}}$.
2. The isentropic expansion of steam through nozzle for the steam initially superheated at inlet is approximated by equation

   (*a*) $pv^{1.3} = C$ (*b*) $pv^{1.125} = C$

   (*c*) $pv^{1.4} = C$ (*d*) $pv = C$.
3. The ratio of exit pressure to inlet pressure for maximum mass flow rate per unit area of steam through a nozzle when steam is initially dry saturated is

   (*a*) 0.6 (*b*) 0.578

   (*c*) 0.555 (*d*) 0.5457.
4. The ratio of exit pressure to inlet pressure for maximum mass flow rate per unit area of steam through nozzle when steam is initially superheated is

   (*a*) 0.555 (*b*) 0.578

   (*c*) 0.5457 (*d*) 0.6.
5. The critical pressure ratio of a convergent nozzle is defined as

   (*a*) the ratio of outlet pressure to inlet pressure of nozzle

   (*b*) the ratio of inlet pressure to outlet pressure of nozzle

   (*c*) the ratio of outlet pressure to inlet pressure only when mass flow rate per unit area is minimum

   (*d*) the ratio of outlet pressure to inlet pressure only when mass flow rate per unit area is maximum.
6. The isentropic expansion of steam through nozzle for the steam initially dry saturated at inlet is approximated by equation

   (*a*) $pv = C$ (*b*) $pv^{1.4} = C$

   (*c*) $pv^{1.3} = C$ (*d*) $pv^{1.135} = C$.
7. The effect of considering friction losses in steam nozzle for the same pressure ratio leads to

   (*a*) increase in exit velocity from the nozzle

   (*b*) decrease in exit velocity from nozzle

   (*c*) no change in exit velocity from nozzle

   (*d*) increase or decrease depending upon the exit quality of steam.
8. The effect of considering friction in steam nozzles for the same pressure ratio leads to

   (*a*) increase in dryness fraction of exit steam

   (*b*) decrease in dryness fraction of exit steam

   (*c*) no change in the quality of exit steam

   (*d*) decrease or increase of dryness fraction of exit steam depending upon inlet quality.

ANSWERS

1. (*d*)	**2.** (*a*)	**3.** (*b*)	**4.** (*c*)
5. (*d*)	**6.** (*d*)	**7.** (*b*)	**8.** (*a*).

THEORETICAL QUESTIONS

1. Define the term 'steam nozzle'. Explain various types of nozzles.
2. State the relation between the velocity of steam and heat during any part of a steam nozzle.
3. Derive an expression for the weight of steam discharged through a nozzle.
4. Define critical pressure ratio for the nozzle of the steam turbine. Obtain analytically its value in terms of the index of expansion.
5. What is the effect of friction on the flow through a steam nozzle ? Explain with the help of *h-s* diagram.
6. What do you mean by a supersaturated flow ? Explain with the help of *h-s* diagram.

UNSOLVED EXAMPLES

1. Dry saturated steam enters a steam nozzle at pressure of 12 bar and is discharged to a pressure of 1.5 bar. If the dryness fraction of a discharged steam is 0.95 what will be the final velocity of steam ? Neglect initial velocity of steam. If 12% of the heat drop is lost in friction, find the percentage reduction in the final velocity. [**Ans.** 633.3 m/s, 6.2%]
2. In a steam nozzle, the steam expands from 3 bar to 1.0 bar. The initial velocity is 90 m/s and initial temperature is 150°C. The nozzle efficiency is 0.95. Determine the exit velocity. [**Ans.** 594 m/s]

3. Steam initially dry and saturated is expanded in a nozzle from 12 bar to 0.95 bar. If the frictional loss in the nozzle is 10% of the total heat drop, calculate the mass of steam discharged when exit diameter of the nozzle is 12 mm. [**Ans.** 224.3 kg/h]

4. The inlet conditions of steam to a convergent-divergent nozzle is 22 bar and 260°C. The exit pressure is 4 bar. Assuming frictionless flow upto the throat and a nozzle efficiency of 85%, determine :
   (*i*) The flow rate for a throat area of 32.2 cm^2.
   (*ii*) The exit area. [**Ans.** (*i*) 11 kg/s ; (*ii*) 60.5 cm^2]

5. In a steam nozzle, dry and saturated steam is expanded from 10 bar to 0.1 bar. Using steam tables, calculate :
   (*i*) Dryness fraction of steam at exit ;
   (*ii*) Heat drop ;
   (*iii*) The velocity of steam at exit from the nozzle when initial velocity is 135 m/s.
   [**Ans.** (*i*) 0.79 ; (*ii*) 694 kJ/kg ; (*iii*) 1185.8 m/s]

6. Determine throat area, exit area and exit velocity for a steam nozzle to pass 0.2 kg/s when the inlet conditions are 12 bar and 250°C and the final pressure is 2 bar. Assume that the expansion is isentropic and that the inlet velocity is negligible. Take $n = 1.3$ for superheated steam. [**Ans.** 1.674 cm^2, 2.015 cm^2, 831.6 m/s]

7. The inlet conditions to a steam nozzle are 10 bar and 250°C. The exit pressure is 2 bar. Assuming isentropic expansion and negligible velocity, determine :
   (*i*) The throat area ;
   (*ii*) The exit velocity ;
   (*iii*) The exit area of the nozzle.
   [**Ans.** (*i*) 1.44 cm^2 ; (*ii*) 795 m/s ; (*iii*) 2.15 cm^2]

8. Dry saturated steam is passed at 7 bar through a convergent-divergent nozzle. The throat cross-sectional area is 4.5 cm^2. Find the mass of steam passing through the nozzle per minute. [**Ans.** 27 kg/min.]

9. Steam enters a nozzle passing a mass flow of 14 kg/s at a pressure of 30 bar and a temperature of 300°C. After expansion to a exit pressure of 5 bar, the exit velocity is 800 m/s.
   (*i*) Determine the nozzle efficiency and the exit area.
   (*ii*) If the losses occur only in the divergent portion, determine the velocity of steam at the throat.
   [**Ans.** (*i*) 80%, 61.2 cm^2 ; (*ii*) 530 m/s]

10. Steam at a pressure of 10 bar and 0.9 dry discharges through a nozzle having throat area of 350 mm^3. If the back pressure is 1.4 bar, find :
   (*i*) Final velocity of the steam.
   (*ii*) Cross-sectional area of the nozzle at exit for maximum discharge.
   [**Ans.** (*i*) 785.2 m/s ; (*ii*) 679 mm^2]

11. A set of 16 nozzles for an impulse turbine receives steam at 16 bar, 300°C. The pressure of steam at exit is 10 bar. If the total discharge is 245 kg/min and nozzle efficiency is 90%, find the cross-sectional area of the exit of each nozzle. If the steam has a velocity of 100 m/s at entry to the nozzles, find the percentage increase in discharge. [**Ans.** 1.256 cm^2 ; 1.94%]

12. A steam nozzle is supplied steam at 7 bar 275°C and it discharges steam at 1 bar. If the diverging portion of the nozzle is 60 mm long and the throat diameter is 6 mm, determine the cone angle of the divergent portion. Assume 10% of the total available enthalpy drop is lost in friction in the divergent portion. Also determine the velocity and temperature of steam at throat. [**Ans.** 2° 8′, 531 m/s, 202.8°C]

13. Steam at a pressure of 10 bar with dryness fraction 0.98 is discharged through a convergent-divergent nozzle at 0.1 bar. The mass flow rate of steam is 10 kg/kWh. If the power developed is 200 kW, determine :
   (*i*) Pressure at the throat.
   (*ii*) Number of nozzles required if each nozzle has a throat of rectangular cross-section of 5 mm × 10 mm, if 10% of the overall isentropic enthalpy drop reheats by friction the steam in the divergent portion.
   [**Ans.** (*i*) 5.8 bar ; (*ii*) 8]

14. Saturated steam at a pressure of 10 bar expands to a pressure of 4 bar isentropically. If the supersaturated flow occurs throughout, determine :
   (*i*) The degree of undercooling.
   (*ii*) The reduction in enthalpy due to supersaturation.
   [**Ans.** (*i*) 50°C; (*ii*) 10 kJ/kg]

15. The dry saturated steam is expanded in a nozzle from pressure of 10 bar to a pressure of 4 bar. If the expansion is supersaturated, find :
   (*i*) The degree of undercooling.
   (*ii*) The degree of supersaturation. [**Ans.** 50°C, 5]

16. Steam at 35 bar and 300°C is supplied to a group of six nozzles. The exit pressure of steam is 8 bar. The rate of flow of steam being 5.2 kg/s. Determine :
   (*i*) The dimensions of the nozzle of rectangular cross-section with aspect ratio of 3 : 1. The expansion may be considered as metastable and friction neglected.
   (*ii*) The degree of undercooling and supersaturation.
   (*iii*) Loss in available heat drop due to irreversibility.
   (*iv*) Increase in entropy.
   (*v*) Ratio of mass flow rate with metastable expansion to thermal expansion.
   [**Ans.** (*i*) 26.7 mm × 8.9 mm ; (*ii*) 35.8°C, 2.58 ; (*iii*) 6.2 kJ/kg ; (*iv*) 0.01398 kJ/kg K ; (*v*) 1.0657]

Steam Turbines

19

19.1. INTRODUCTION

*The **steam turbine** is a prime-mover in which the potential energy of the steam is transformed into kinetic energy, and latter in its turn is transformed into the mechanical energy of rotation of the turbine shaft*. The turbine shaft, directly or with the help of a reduction gearing, is connected with the driven mechanism. Depending on the type of the driven mechanism a steam turbine may be *utilised in most diverse fields of industry, for power generation* and *for transport*. Transformation of the potential energy of steam into the mechanical energy of rotation of the shaft is brought about by different means.

19.2. CLASSIFICATION OF STEAM TURBINES

There are several ways in which the steam turbines may be classified. The most important and common division being with respect to the *action of the steam*, as :

(*a*) Impulse.

(*b*) Reaction.

(*c*) Combination of impulse and reaction.

Other classification are :

1. According to the number of pressure stages:

(*i*) *Single stage turbines with one or more velocity stages usually of small power capacities* ; these turbines are mostly used for driving centrifugal compressors, blowers and other similar machinery.

(*ii*) *Multistage impulse and reaction turbines* ; they are made in a wide range of power capacities varying from small to large.

2. According to the direction of steam flow :

(*i*) *Axial turbines* in which steam flows in a direction parallel to the axis of the turbine.

(*ii*) *Radial turbines* in which steam flows in a direction perpendicular to the axis of the turbine; one or more low-pressure stages in such turbines are made axial.

3. According to the number of cylinders :

(*i*) Single cylinder turbines.

(*ii*) Double cylinder turbines.

(*iii*) Three cylinder turbines.

(*iv*) Four cylinder turbines.

Multi-cylinder turbines which have their rotors mounted on one and the same shaft and coupled to a single generator are known as *single shaft turbines* ; turbines with separate rotor shafts for each cylinder placed parallel to each other are known as **multiaxial turbines.**

4. According to the method of governing :

(*i*) *Turbines with throttle governing* in which fresh steam enters through one or more (depending on the power developed) simultaneously operated throttle valves.

(*ii*) *Turbines with nozzle governing* in which fresh steam enters through two or more consecutively opening regulators.

(*iii*) *Turbines with by pass governing* in which steam turbines besides being fed to the first stage is also directly fed to one, two or even three intermediate stages of the turbine.

5. According to heat drop process :

(*i*) *Condensing turbines with generators* ; in these turbines steam at a pressure less than atmospheric is directed to a condenser ; besides, steam is also extracted from intermediate stages for feed water heating, the number of such extractions usually being from 2–3 to as much 8–9. The latent heat of exhaust steam during the process of condensation is completely lost in these turbines.

(*ii*) *Condensing turbines with one or two intermediate stage extractions* at specific pressures for industrial and heating purposes.

(*iii*) *Back pressure turbines*, the exhaust steam from which is utilised for industrial or heating purposes ; to this type of turbines can also be added (in a relative sense) turbines with deteriorated vacuum, the exhaust steam of which may be used for heating and process purposes.

(*iv*) *Topping turbines* ; these turbines are also of the back pressure type with the difference that the exhaust steam from these turbines is further utilised in medium and low pressure condensing turbines. These turbines, in general, operate at high initial conditions of steam pressure and temperature, and are mostly used during extension of power station capacities, with a view to obtain better efficiencies.

(*v*) *Back pressure turbines with steam extraction from intermediate stages at specific pressure* ; turbines of this type are meant for supplying the consumer with steam of various pressures and temperature conditions.

(*vi*) *Low pressure turbines* in which the exhaust steam from reciprocating steam engines, power hammers, presses, etc., is utilised for power generation purposes.

(*vii*) *Mixed pressure turbines* with two or three pressure stages, with supply of exhaust steam to its intermediate stages.

6. According to steam conditions at inlet to turbine :

(*i*) *Low pressure turbines*, using steam at a pressure of 1.2 to 2 ata.

(*ii*) *Medium pressure turbines*, using steam at pressures of upto 40 ata.

(*iii*) *High pressure turbines*, utilising pressures above 40 ata.

(*iv*) *Turbines of very high pressures*, utilising steam at pressures of 170 ata and higher and temperatures of 550°C and higher.

(*v*) *Turbines of supercritical pressures*, using steam at pressures of 225 ata and above.

7. According to their usage in industry :

(*i*) *Stationary turbines with constant speed of rotation* primarily used for driving alternators.

(*ii*) *Stationary steam turbines* with variable speed meant for driving turbo-blowers, air circulators, pumps, etc.

(*iii*) *Non-stationary turbines with variable speed* ; turbines of this type are usually employed in steamers, ships and railway locomotives.

19.3. ADVANTAGES OF STEAM TURBINE OVER STEAM ENGINES

The following are the *principal advantages of steam turbine over steam engines* :

1. The thermal efficiency of a steam turbine is much higher than that of a steam engine.
2. The power generation in a steam turbine is at a uniform rate, therefore necessity to use a flywheel (as in the case of steam engine) is not felt.
3. Much higher speeds and greater range of speed is possible than in case of a steam engine.
4. In large thermal stations where we need higher outputs, the steam turbines prove very suitable as these can be made in big sizes.
5. With the absence of reciprocating parts (as in steam engine) the balancing problem is minimised.
6. No internal lubrication is required as there are no rubbing parts in the steam turbine.
7. In a steam turbine there is no loss due to initial condensation of steam.
8. It can utilise high vacuum very advantageously.
9. Considerable overloads can be carried at the expense of slight reduction in overall efficiency.

19.4. DESCRIPTION OF COMMON TYPES OF TURBINES

The common types of steam turbines are :

1. Simple impulse turbine.
2. Reaction turbine.

The main difference between these turbines lies *in the way in which the steam is expanded while it moves through them. In the former type steam expands in the nozzles and its pressure does not alter as it moves over the blades while in the latter type the steam expands continuously as it passes over the blades and thus there is gradual fall in the pressure during expansion.*

1. Simple impulse turbines

Fig. 19.1 shows a simple impulse turbine diagrammatically. The top portion of the figure exhibits a longitudinal section through the upper half of the turbine, the middle portion shows one set of nozzles which is followed by a ring of moving blades, while lower part of the diagram indicates approximately changes in pressure and velocity during the flow of steam through the turbine. This turbine is called '*simple*' impulse turbine since the expansion of the steam takes place in *one set of the nozzles*.

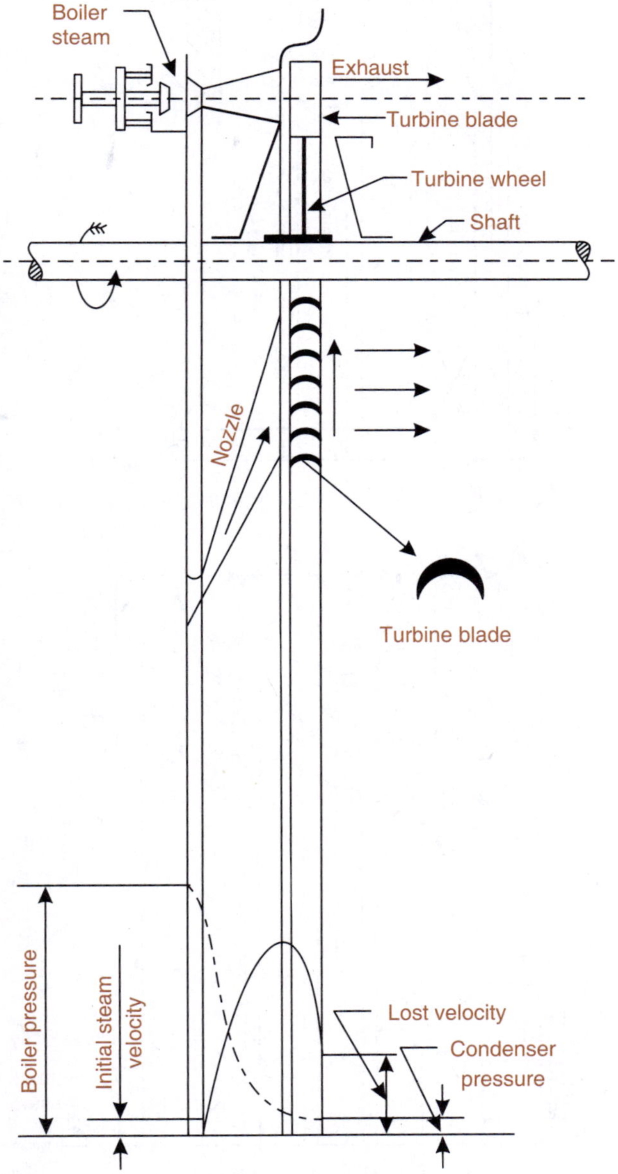

Fig. 19.1. Simple impulse turbine.

As the steam flows through the nozzle its pressure falls from steam chest pressure to condenser pressure (or atmospheric pressure if the turbine is non-condensing). Due to this relatively higher ratio of expansion of steam in the nozzles the steam leaves the nozzle with a very high velocity. Refer to Fig. 19.1, it is evident that the velocity of the steam leaving the moving blades is a large portion of the maximum velocity of the steam when leaving the nozzle. The loss of energy due to this higher exit velocity is commonly called the "**carry over loss**" or "**leaving loss**".

The principal example of this turbine is the well known "**De laval turbine**" and in this turbine the 'exit velocity' or 'leaving velocity' or 'lost velocity' may amount to 3.3 per cent of the nozzle outlet velocity. Also since all the kinetic energy is to be absorbed by one ring of the moving blades only, the velocity of wheel is too high (varying from 25000 to 30000 r.p.m.). This wheel or rotor speed however, can be reduced by different methods (discussed in the following article).

2. Reaction turbine

In this type of turbine, *there is a gradual pressure drop and takes place continuously over the fixed and moving blades*. The function of the fixed blades is (the same as the nozzle) that they alter the direction of the steam as well as allow it expand to a larger velocity. As the steam passes over the moving blades its kinetic energy (obtained due to fall in pressure) is absorbed by them. Fig. 19.2 shows a *three stage* reaction turbine. The changes in pressure and velocity are also shown there in.

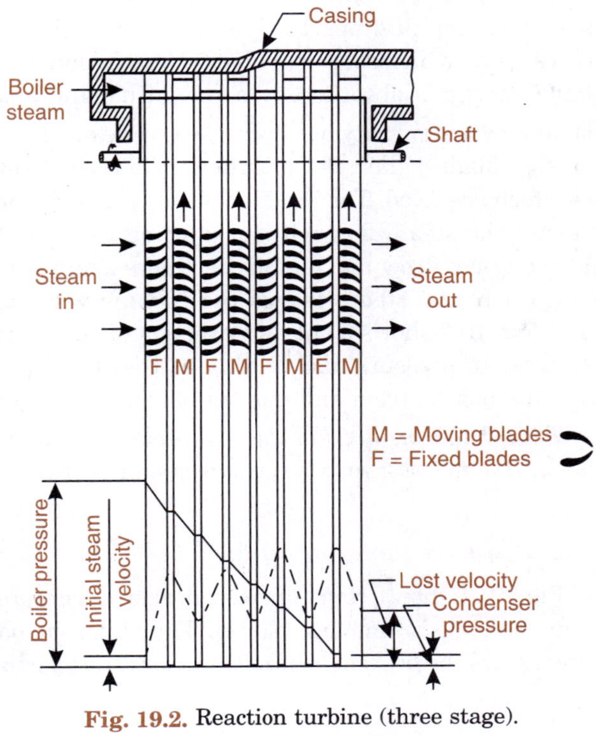

Fig. 19.2. Reaction turbine (three stage).

As the volume of steam increases at lower pressures therefore, the diameter of the turbine must increase after each group of blade rings. It may be noted that in this turbine since the pressure drop per stage is small, therefore the number of stages required is much higher than an impulse turbine of the same capacity.

19.5. METHODS OF REDUCING WHEEL OR ROTOR SPEED

As already discussed under the heading 'simple impulse turbine' that if the steam is expanded from the boiler pressure to condenser pressure in one stage the speed of the rotor becomes *tremendously high which crops up practical complicacies*. There are several methods of reducing this speed to lower value ; all these methods utilise a multiple system of rotor in series, keyed on a common shaft and the steam pressure or jet velocity is absorbed in stages as the steam flows over the blades. This is known as '**compounding**'. The different methods of compounding are :

1. Velocity compounding.
2. Pressure compounding.
3. Pressure velocity compounding.
4. Reaction turbine.

1. Velocity compounding

Steam is expanded through a stationary nozzle from the boiler or inlet pressure to condenser pressure. So the pressure in the nozzle drops, the kinetic energy of the steam increases due to increase in velocity. A portion of this available energy is absorbed by a row of moving blades. The steam (whose velocity has decreased while moving over the moving blades) then flows through the second row of blades which are fixed. The function of these fixed blades is to re-direct the steam flow without altering its velocity to the following next row moving blades where again work is done on them and steam leaves the turbine with a low velocity. Fig. 19.3 shows a cut away section of such a stage and changes in pressure and velocity as the steam passes through the nozzle, fixed and moving blades.

Though this method has the advantage that the initial cost is low due to lesser number of stages yet its efficiency is low.

2. Pressure compounding

Fig. 19.4 shows rings of fixed nozzles incorporated between the rings of moving blades. The steam at boiler pressure enters the first set of nozzles and expands partially. The kinetic energy of the steam thus obtained is absorbed by the moving blades (stage 1). The steam then expands partially in the second set of nozzles where its pressure again falls and the velocity increases ; the kinetic energy so obtained is absorbed by the second ring of moving blades (stage 2). This is repeated in stage 3 and steam finally leaves the turbine at low velocity and pressure. The number of stages (or pressure reductions) depends on the number of rows of nozzles through which the steam must pass.

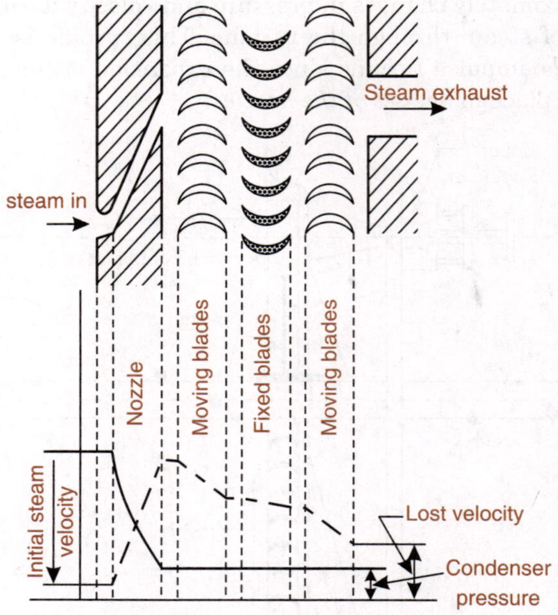

Fig. 19.3. Velocity compounding.

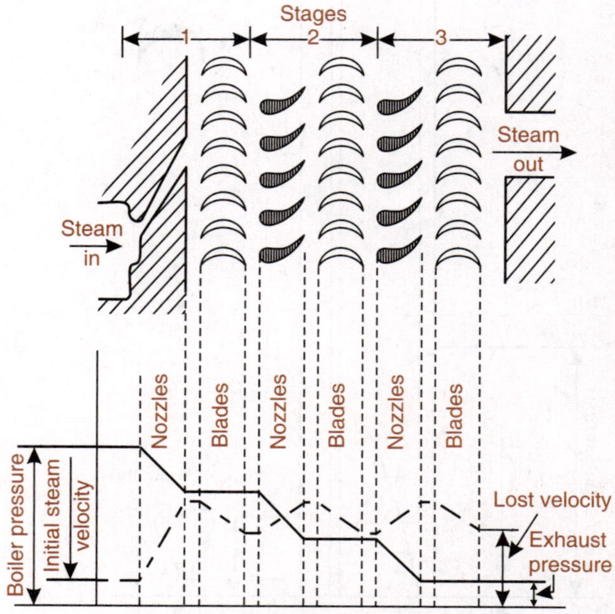

Fig. 19.4. Pressure compounding.

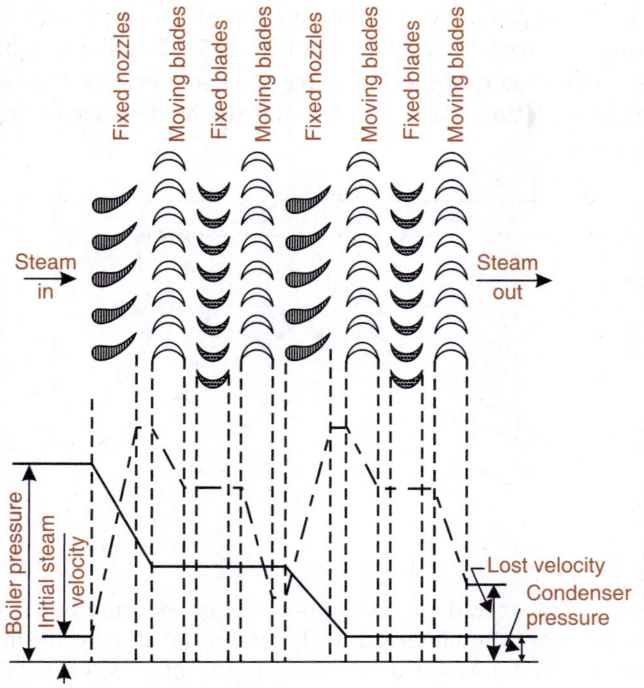

Fig. 19.5. Pressure velocity compounding.

This method of compounding is used in *Rateau and Zoelly turbine. This is most efficient turbine since the speed ratio remains constant but it is expensive owing to a large number of stages.*

3. Pressure velocity compounding

This method of compounding is the combination of two previously discussed method. The total drop in steam pressure is divided into stages and the velocity obtained in each stage is also compounded. The rings of nozzles are fixed at the beginning of each stage and pressure remains constant during each stage. The changes in pressure and velocity are shown in Fig. 19.5.

This method of compounding is used in *Curits and Moore turbine.*

4. Reaction turbine

It has been discussed in Article 19.4.

19.6. DIFFERENCE BETWEEN IMPULSE AND REACTION TURBINES

S. No.	Particulars	Impulse turbine	Reaction turbine
1.	*Pressure drop*	Only in nozzles and not in moving blades.	In fixed blades (nozzles) as well as in moving blades.
2.	*Area of blade channels*	Constant.	Varying (converging type).
3.	*Blades*	Profile type.	Aerofoil type.
4.	*Admission of steam*	Not all round or complete.	All round or complete.
5.	*Nozzles/fixed blades*	Diaphragm contains the nozzle.	Fixed blades similar to moving blades attached to the casing serve as nozzles and guide the steam.
6.	*Power*	Not much power can be developed.	Much power can be developed.
7.	*Space*	Requires less space for same power.	Requires more space for same power.
8.	*Efficiency*	Low.	High.
9.	*Suitability*	Suitable for small power requirements.	Suitable for medium and higher power requirements.
10.	*Blade manufacture*	Not difficult.	Difficult.

19.7. IMPULSE TURBINES

19.7.1. Velocity Diagram for Moving Blade

Fig. 19.6 shows the velocity diagram of a *single stage impulse turbine*.

C_{bl} = Linear velocity of moving blade (m/s).

C_1 = Absolute velocity of steam entering moving blade (m/s).

C_0 = Absolute velocity of steam leaving moving blade (m/s).

C_{w_1} = Velocity of whirl at the entrance of moving blade.
= tangential component of C_1.

C_{w_0} = Velocity of whirl at exit of the moving blade.
= tangential component of C_0.

C_{f_1} = Velocity of flow at entrance of moving blade.
= axial component of C_1.

C_{f_0} = Velocity of flow at exit of the moving blade.
= axial component of C_0.

C_{r_1} = Relative velocity of steam to moving blade at entrance.

C_{r_0} = Relative velocity of steam to moving blade at exit.

α = Angle with the tangent of the wheel at which the steam with velocity C_1 enters. This is also called *nozzle engle.*

β = Angle which the discharging steam makes with the tangent of the wheel at the exit of moving blade.

θ = Entrance angle of moving blade.

ϕ = Exit angle of moving blade.

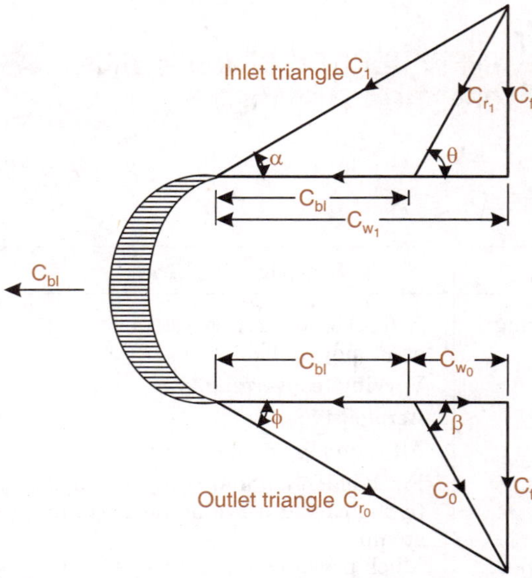

Fig. 19.6. Velocity diagram for moving blade.

The steam jet issuing from the nozzle at a velocity of C_1 impinges on the blade at an angle α. The tangential component of this jet (C_{w_1}) *performs work on the blade*, the axial component (C_{f_1}) however *does no work but causes the steam to flow through the turbine*. As the blades move with a tangential velocity of C_{bl}, the entering steam jet has a relative velocity C_{r_1} (with respect to blade) which makes an angle θ with the wheel tangent. The steam then glides over the blade without any shock and discharges at a relative velocity of C_0 at an angle ϕ with the tangent of the blades. The relative velocity at the inlet (C_{r_1}) is the same as the relative velocity at the outlet (C_{r_0}) if *there is no frictional loss at the blade*. The absolute velocity (C_0) of leaving steam make an angle β to the tangent at the wheel.

To have convenience in solving the problems on turbines it is a common practice to combine the two vector velocity diagrams on a common base which represents the blade velocity (C_{bl}) as shown in Fig. 19.7. This diagram has been obtained by superimposing the inlet velocity diagram on the outlet diagram in order that the blade velocity lines C_{bl} coincide.

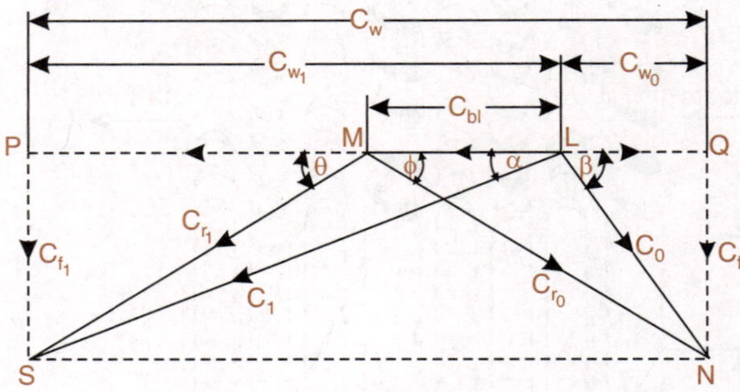

Fig. 19.7

19.7.2. Work done on the Blade

The work done on the blade may be found out from the change of momentum of the steam jet during its flow over the blade. As earlier discussed, it is only the velocity of whirl which performs work on the blade since it acts in its (blade) direction of motion.

From Newton's second law of motion,

Force (tangential) on the wheel

= Mass of steam × acceleration

= Mass of steam/sec. × change of velocity

$$= \dot{m}_s(C_{w_1} - C_{w_0}) \qquad ...(19.1)$$

The value of C_{w_0} is actually negative as the steam is discharged in the *opposite direction* to the blade motion, therefore due consideration should be given to the fact that *values of C_{w_1} and C_{w_0} are to be added while doing the solution of the problem.* (*i.e.*, when $\beta < 90°$)

Work done on blades/sec.

= Force × distance travelled/sec.

$$= \dot{m}_s(C_{w_1} + C_{w_0}) \times C_{bl}$$

$$\text{Power per wheel} = \dot{m}_s(C_{w_1} + C_{w_0}) \times C_{bl}$$

$$= \frac{\dot{m}_s C_w C_{bl}}{1000} \text{ kW} \qquad ...(19.2)$$

$$(\because \ C_w = C_{w_1} + C_{w_0})$$

Blade or diagram efficiency

$$= \frac{\text{Work done the blade}}{\text{Energy supplied to the blade}}$$

$$= \frac{\dot{m}_s(C_{w_1} + C_{w_0}) . C_{bl}}{\frac{\dot{m}\, C_1^2}{2}}$$

$$= \frac{2\, C_{bl}(C_{w_1} + C_{w_0})}{C_1^2} \qquad ...(19.3)$$

If h_1 and h_2 be the total heats before and after expansion through the nozzles, then $(h_1 - h_2)$ is the heat drop through a stage of fixed blades ring and moving blades ring.

$\therefore$ **Stage efficiency,**

$$\eta_{stage} = \frac{\text{Work done on blade per kg of steam}}{\text{Total energy supplied per kg of steam}}$$

$$= \frac{C_{bl}(C_{w_1} + C_{w_0})}{(h_1 - h_2)} \qquad ...(19.4)$$

Now, nozzle efficiency $= \dfrac{C_1^{\,2}}{2(h_1 - h_2)}$

Also η_{stage} = Blade efficiency × nozzle efficiency

$$= \frac{2C_{bl}(C_{w_1} + C_{w_0})}{C_1^{\,2}} \times \frac{C_1^{\,2}}{2(h_1 - h_2)}$$

$$= \frac{C_{bl}(C_{w_1} + C_{w_0})}{(h_1 - h_2)}$$

The **axial thrust** on the wheel is due to *difference* between the velocities of flow at entrance and outlet.

Axial force on the wheel

= Mass of steam × axial acceleration

$$= \dot{m}_s(C_{f_1} - C_{f_0}) \qquad ...(19.5)$$

The *axial force on the wheel must be balanced or must be taken by a thrust bearing.*

Energy converted to heat by blade friction

= loss of kinetic energy during flow over blades

$$= \dot{m}_s(C_{r_1}^{\,2} - C_{r_0}^{\,2}) \qquad ...(19.6)$$

19.7.3. Blade Velocity Co-efficient

In an impulse turbine, if friction is neglected the relative velocity will remain unaltered as it passes over blades. In practice the flow of steam over the blades is resisted by friction. The effect of the friction is to reduce the relative velocity of steam as it passes over the blades. In general there is a loss of 10 to 15 per cent in the relative velocity. Owing to friction in the blades, C_{r_0} is less than C_{r_1} and we may write

$$C_{r_0} = K \,.\, C_{r_1} \qquad ...(19.7)$$

where K is termed a **blade velocity co-efficient.**

19.7.4. Expression for Optimum Value of the Ratio of Blade Speed to Steam Speed (for maximum efficiency) for a Single Stage Impulse Turbine

Refer to Fig. 19.7.

$$C_w = PQ = MP + MQ = C_{r_1} \cos\theta + C_{r_0} \cos\phi$$

$$= C_{r_1} \cos\theta \left[1 + \frac{C_{r_0} \cos\phi}{C_{r_1} \cos\theta}\right]$$

$$= C_{r_1} \cos\theta\,(1 + K \,.\, Z), \text{ where } Z = \frac{\cos\phi}{\cos\theta} \qquad ...(i)$$

Generally, *the angles θ and ϕ are nearly equal for impulse turbine* and hence it can safely be assumed that Z is a constant.

But, $C_{r_1} \cos\theta = MP = LP - LM = C_1 \cos\alpha - C_{bl}$

From eqn. (*i*), $\quad C_w = (C_1 \cos\alpha - C_{bl})(1 + K \,.\, Z)$

We know that,

$$\text{Blade efficiency, } \eta_{bl} = \frac{2C_{bl} \,.\, C_w}{C_1^{\,2}} \qquad ...(ii)$$

$$\eta_{bl} = \frac{2C_{bl}(C_1 \cos\alpha - C_{bl})(1 + KZ)}{C_1^{\,2}}$$

$$= 2(\rho \cos\alpha - \rho^2)(1 + KZ)$$

$$= 2\rho(\cos\alpha - \rho)(1 + KZ) \qquad ...(iii)$$

where $\rho = \dfrac{C_{bl}}{C_1}$ is the ratio of *blade speed to steam speed* and is commonly called as **"Blade speed ratio"**.

For particular impulse turbine α, K and Z may assumed to be constant and from equation (*iii*) it can be seen clearly that η_{bl} depends on the value of ρ only. Hence differentiating (*iii*),

$$\frac{d\eta_{bl}}{d\rho} = 2(\cos\alpha - 2\rho)(1 + KZ)$$

For a maximum or minimum value of η_{bl} this should be zero

$$\cos\alpha - 2\rho = 0, \quad \therefore \quad \rho = \frac{\cos\alpha}{2}$$

Now, $\quad \dfrac{d^2\eta_{bl}}{d\rho^2} = 2(-2)(1 + KZ) = -4(1 + KZ)$

which is a *negative quantity* and thus the value so obtained is the *maximum*.

Optimum value of ratio of blade speed to steam speed is

$$\rho_{opt} = \frac{\cos\alpha}{2} \qquad ...(19.8)$$

Substituting this value of ρ in eqn. (*iii*), we get

$$(\eta_{bl})_{max} = 2 \times \frac{\cos\alpha}{2}\left(\cos\alpha - \frac{\cos\alpha}{2}\right)(1 + K \,.\, Z)$$

$$= \frac{\cos^2\alpha}{2}(1 + KZ) \qquad ...(19.9)$$

It is sufficiently accurate to assume symmetrical blades ($\theta = \phi$) and no friction in fluid passage for the purpose of analysis.

$\therefore \quad Z = 1$ and $K = 1$

$$\therefore \quad (\eta_{bl})_{max} = \cos^2\alpha \qquad ...(19.10)$$

The work done per kg of steam is given by

$$W = (C_{w_1} + C_{w_0})\, C_{bl}$$

Substituting the value of $C_{w_1} + C_{w_0}$ $(= C_w)$

$$W = (C_1 \cos \alpha - C_{bl})(1 + KZ)C_{bl}$$
$$= 2C_{bl}(C_1 \cos \alpha - C_{bl}) \text{ when } K = 1 \text{ and } Z = 1$$

The maximum value of W can be obtained by substituting the value of $\cos \alpha$ from equation (19.8),

$$\cos \alpha = 2\rho = 2 \frac{C_{bl}}{C_1}$$

$$\therefore \quad W_{max} = 2C_{bl}(2C_{bl} - C_{bl}) = 2C_{bl}^2 \quad ...(19.11)$$

It is obvious from the equation (19.8) that the blade velocity should be *approximately* half of absolute velocity of steam jet coming out from the nozzle (fixed blade) for *the maximum work developed per kg of steam or for maximum efficiency. For the other values of blade speed the absolute velocity at outlet from the blade will increase, consequently, more energy will be carried away by the steam and efficiency will decrease.*

For equiangular blades with no friction losses, optimum value of $\frac{C_{bl}}{C_1}$ corresponds to the case, when the outlet absolute velocity is *axial* as shown in Fig. 19.8.

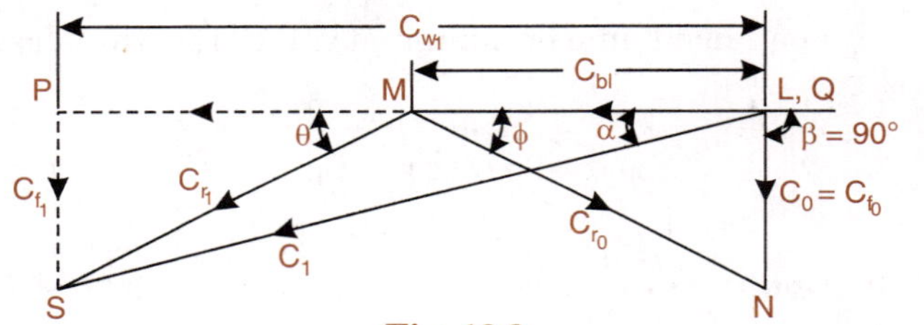

Fig. 19.8

Since the discharge is axial

$$\beta = 90°, \qquad \therefore C_0 = C_{f_0} \text{ and } C_{w_0} = 0.$$

The variations of η_{bl} or work developed per kg of steam with $\frac{C_{bl}}{C_1}$ is shown in Fig. 19.9. This figure shows that :

(*i*) When $\frac{C_{bl}}{C_1} = 0$, the work done becomes zero as the distance travelled by the blade (C_{bl}) is zero, even though the torque on the blade is maximum.

(*ii*) The maximum efficiency is $\cos^2 \alpha$ and maximum work done per kg of steam is $2C^2_{bl}$ when $\frac{C_{bl}}{C_1} = \cos \alpha/2$.

(*iii*) When $\frac{C_{bl}}{C_1} = 1$, the work done is zero as the torque acting on the blade becomes zero even though the distance travelled by the blade is maximum.

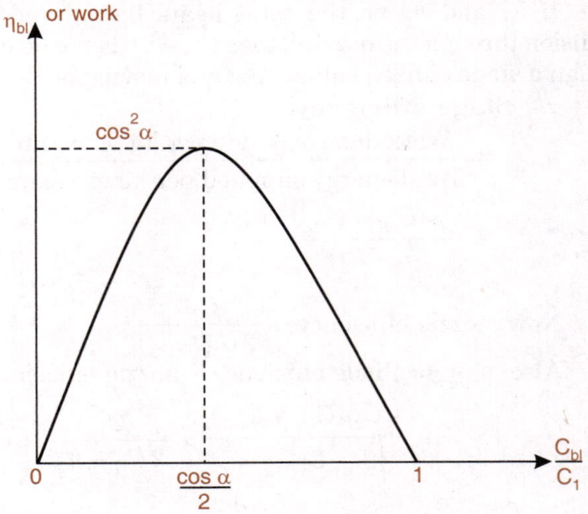

Fig. 19.9

When the high pressure steam is expanded from the boiler pressure to condenser pressure in a single stage of nozzle, the absolute velocity of steam becomes maximum and blade velocity also becomes tremendously high. In such a case, *a velocity compounded stage is used to give lower blade speed ratio* and *better utilization of the kinetic energy of the steam*. The arrangement of velocity compounding has already been dealt with.

Fig. 19.10 shows the velocity diagrams for the first and second row of moving blades of *velocity compounded unit*. The speed and angles are such that the final absolute velocity of the steam leaving the second row is *axial.* With this arrangement, *the K.E. carried by the steam is minimum*, therefore, *the efficiency becomes maximum*.

The velocity of blades (C_{bl}) is same for both the rows since they are mounted on the same shaft.

Consider **first row of moving blades :**

Work done per kg of steam,

$$W_1 = C_{bl}(C_{w_1} + C_{w_2})$$
$$= C_{bl}[C_{r_1} \cos \theta + C_{r_0} \cos \phi]$$

If there is no friction loss and symmetrical blading is used, then

$$C_{r_1} = C_{r_0} \text{ and } \theta = \phi$$

$$\therefore \quad W_1 = C_{bl} \times 2C_{r_1} \cos \theta = 2C_{bl}(C_1 \cos \alpha - C_{bl}) \quad ...(19.12)$$

The magnitude of absolute velocity of steam leaving the first row and entering into the second row of moving blades is same and its direction only is changed.

$$\therefore \quad C_1' = C_0$$

Consider **second row of moving blades :**

Work done per kg, $W_2 = C_{bl} \cdot C'_{w_1}$ as $C'_{w_2} = 0$ because discharge is *axial* and $\beta' = 90°$

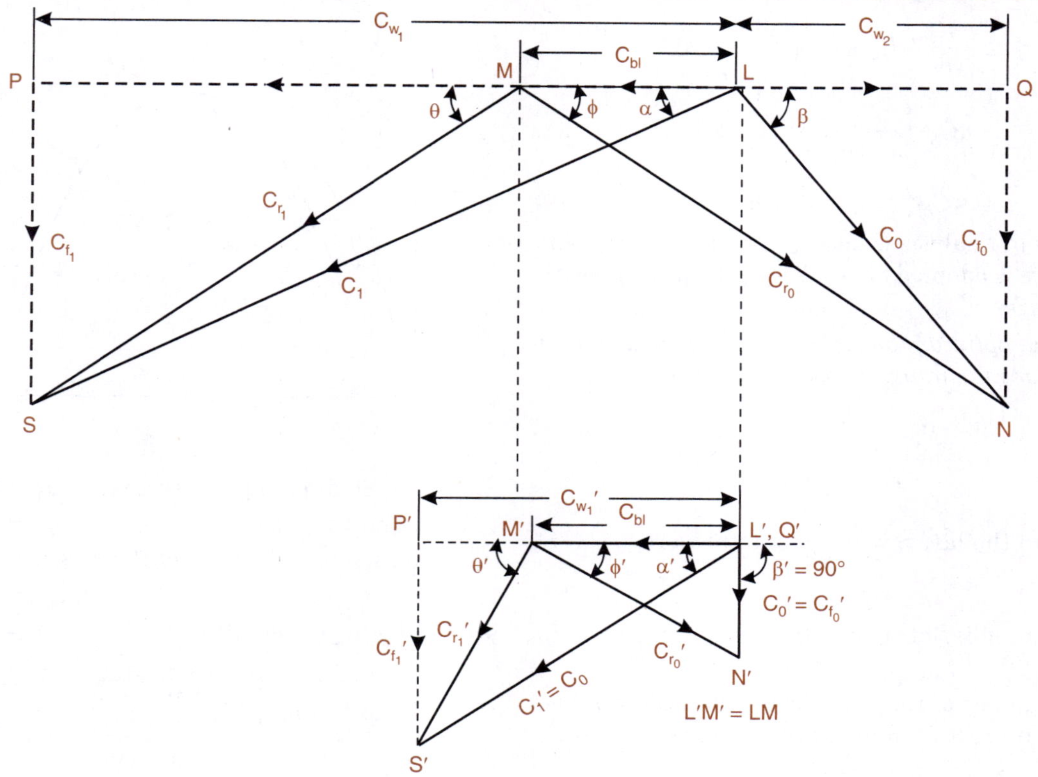

Fig. 19.10

Alternately, $W_2' = C_{bl}\, [C'_{r_1} \cos\theta' + C'_{r_0} \cos\phi']$

For symmetrical blades $\theta' = \phi'$

and, if there is no friction loss, then $C'_{r_1} = C'_{r_0}$

$$\therefore \quad W_2 = 2C_{bl}\, C'_{r_1} \cos\theta'$$

$$= 2C_{bl}\, (C_1' \cos\alpha' - C_{bl}) \qquad ...(19.13)$$

Now α' may be equal to β.

$$\therefore \quad C_1' \cos\alpha' = C_0 \cos\beta = C_{r_0} \cos\phi - C_{bl}$$

$$= C_{r_1} \cos\theta - C_{bl} = (C_1 \cos\alpha - C_{bl}) - C_{bl}$$

$$= C_1 \cos\alpha - 2C_{bl}$$

Substituting the value of $C_1' \cos\alpha'$ in eqn. (19.13), we get

$$W_2 = 2C_{bl}\, [(C_1 \cos\alpha - 2C_{bl}) - C_{bl}]$$

$$= 2C_{bl}\, (C_1 \cos\alpha - 3C_{bl}) \qquad ...(19.14)$$

Total work done per kg of steam passing through both stages is given by

$$W_t = W_1 + W_2$$

$$= 2C_{bl}\, [C_1 \cos\alpha - C_{bl}] + 2C_{bl}\, [C_1 \cos\alpha - 3C_{bl}]$$

$$= 2C_{bl}\, (2C_1 \cos\alpha - 4C_{bl})$$

$$= 4C_{bl}\, (C_1 \cos\alpha - 2C_{bl}) \qquad ...(19.15)$$

The blade efficiency for two stage impulse turbine is given by

$$\eta_{bl} = \frac{W_t}{\frac{C_1^2}{2}} = 4C_{bl}\, [C_1 \cos\alpha_1 - 2C_{bl}] \times \frac{2}{C_1^2}$$

$$= \frac{8C_{bl}}{C_1^2}\, (C_1 \cos\alpha - 2C_{bl}) = 8\, \frac{C_{bl}}{C_1}\left(\cos\alpha - 2\,.\frac{C_{bl}}{C_1}\right)$$

$$= 8\rho\, (\cos\alpha - 2\rho) \qquad ...(19.16)$$

where ρ (velocity ratio) $= \dfrac{C_{bl}}{C_1}$.

The blade efficiency for two stage turbine will be maximum when $\dfrac{d\eta_{bl}}{d\rho} = 0$

$$\therefore \quad \frac{d}{d\rho}\, [8\rho \cos\alpha - 16\rho^2] = 0$$

$$\therefore \quad 8 \cos\alpha - 32\rho = 0$$

From which, $\rho = \dfrac{\cos\alpha}{4}$ ...(19.17)

Substituting this value in eqn. (19.16), we get

$$(\eta_{bl})_{\max} = 8\,.\,\frac{\cos\alpha}{4}\left[\cos\alpha - 2\,.\,\frac{\cos\alpha}{4}\right] = \cos^2\alpha \qquad ...(19.18)$$

The maximum work done per kg of steam is obtained by substituting the value of

$$\rho = \frac{C_{bl}}{C_1} = \frac{\cos\alpha}{4}$$

or $$C_1 = \frac{4C_{bl}}{\cos\alpha} \text{ in the eqn. (19.15).}$$

$$\therefore \quad (W_b)_{max} = 4C_{bl}\left(\frac{4C_{bl}}{\cos\alpha}.\cos\alpha - 2C_{bl}\right)$$

$$= 8C_{bl}^{\,2} \qquad ...(19.19)$$

The present analysis is done for *two stages* only. The similar procedure is adopted for analysing the problem with three or four stages.

In general, optimum blade speed ratio for maximum blade efficiency or maximum work done is given by

$$\rho = \frac{\cos\alpha}{2.n} \qquad ...(19.20)$$

and work done in the last row = $\frac{1}{2^n}$ of total work ...(19.21)

where n is the number of moving/rotating blade rows in series.

As the number of rows increases, the utility of last row decreases. In practice, *more than* two rows are hardly preferred.

19.7.5. Advantages of Velocity Compounded Impulse Turbine

1. Owing to relatively large heat drop, a velocity-compounded impulse turbine requires a comparatively small number of stages.
2. Due to number of stages being small, its cost is less.
3. As the number of moving blades' rows in a wheel increases, the maximum stage efficiency and optimum value of ρ decreases.
4. Since the steam temperature is sufficiently low in a two or three row wheel, therefore, cast iron cylinder may be used. This will cause saving in material cost.

19.7.6. Disadvantages of Velocity Compounded Impulse Turbine

1. It has high steam consumption and low efficiency.
2. In a single row wheel, the steam temperature is high so cast iron cylinder cannot be used due to phenomenon of growth ; cast steel cylinder is used which is costlier than cost iron.

Example 19.1. *A stage of a steam turbine is supplied with steam at a pressure of 50 bar and 350°C, and exhausts at a pressure of 5 bar. The isentropic efficiency of the stage is 0.82 and the steam consumption is 2270 kg/min. Determine the power output of the stage.*

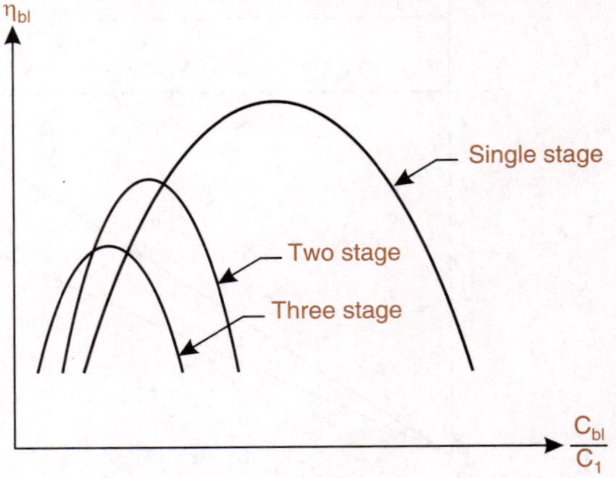

Fig. 19.11

Solution. Steam supply pressure, p_1 = 50 bar, 350°C

Exhaust pressure p_2 = 5 bar

Isentropic efficiency of the stage,

η_{stage} = 0.82

Steam consumption, m_s = 2270 kg/min

Power output of the stage, P :

Refer to Fig. 19.12.

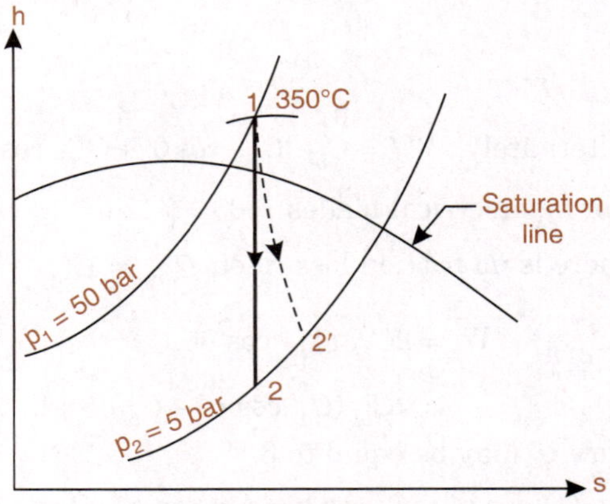

Fig. 19.12

From ***Mollier chart :***

h_1 = 3130.7 kJ/kg of steam

h_2 = 2640 kJ/kg of steam

Isentropic heat drop

$= h_1 - h_2 = 3130.7 - 2640$

$= 490.7$ kJ/kg

Actual heat drop

$= h_1 - h_2'$

But, $$\eta_{isen.\,(stage)} = \frac{h_1 - h_2'}{h_1 - h_2}$$

or $$0.82 = \frac{h_1 - h_2'}{490.7}$$

or $$h_1 - h_2' = 0.82 \times 490.7 = 402.4 \text{ kJ/kg}$$

∴ *Power developed*

$= m_s(h_1 - h_2')$

$= \frac{2270}{60} \times 402.4 \text{ kW} = \mathbf{15224 \text{ kW}}.$ **(Ans.)**

Example 19.2. *In a De-Laval turbine steam issues from the nozzle with a velocity of 1200 m/s. The nozzle angle is 20°, the mean blade velocity is 400 m/s, and the inlet and outlet angles of blades are equal. The mass of steam flowing through the turbine per hour is 1000 kg. Calculate :*

(*i*) *Blade angles.*

(*ii*) *Relative velocity of steam entering the blades.*

(*iii*) *Tangential force on the blades.*

(*iv*) *Power developed.*

(*v*) *Blade efficiency.*

Take blade velocity co-efficient as 0.8.

Solution. Absolute velocity of steam entering the blade, $C_1 = 1200$ m/s

Nozzle blade, $\alpha = 20°$

Mean blade velocity, $C_{bl} = 400$ m/s

Inlet blade angle, θ = Outlet blade angle, ϕ

Blade velocity co-efficient, $K = 0.8$

Mass of steam flowing through the turbine,

$m_s = 1000$ kg/h.

Refer to Fig. 19.13. Procedure of drawing the inlet and outlet triangles (LMS and LMN) respectively is as follows :

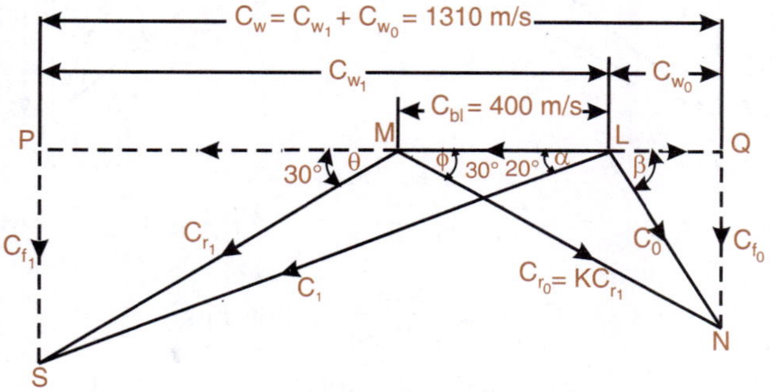

Fig. 19.13

- Select a suitable scale and draw line LM to represent C_{bl}(= 400 m/s).
- At point L make angle of 20° (α) and cut length LS to represent velocity C_1(= 1200 m/s). Join MS. Produce M to meet the perpendicular drawn from S at P. Thus *inlet triangle* is completed.

By measurement : $\theta = 30°$, $C_{r_1} = MS = 830$ m/s

$\theta = \phi = 30°$ (given)

Now, $C_{r_2} = KC_{r_1} = 0.8 \times 830 = 664$ m/s

- At point M make an angle of 30° (ϕ) and cut the length MN to represent C_{r_0} (= 664 m/s). Join LN. Produce L to meet the perpendicular drawn from N at Q. Thus *outlet triangle* is completed.

(*i*) **Blade angles θ, ϕ :**

As the blades are symmetrical (given)

∴ $\theta = \phi = \mathbf{30°}.$ **(Ans.)**

(*ii*) **Relative velocity of steam entering the blades, C_{r_1} :**

$C_{r_1} = MS = \mathbf{830 \text{ m/s}}.$ **(Ans.)**

(*iii*) **Tangential force on the blades :**

Tangential force

$= \dot{m}_s(C_{w_1} + C_{w_0}) = \frac{1000}{60 \times 60}(1310)$

$= \mathbf{363.8 \text{ N}}.$ **(Ans.)**

(*iv*) **Power developed, P :**

$$P = \dot{m}_s(C_{w_1} + C_{w_0})C_{bl}$$

$$= \frac{1000}{60 \times 60} \times \frac{1310 \times 400}{1000} \text{ kW}$$

$= \mathbf{145.5 \text{ kW}}.$ **(Ans.)**

(*v*) **Blade efficiency, η_{bl} :**

$$\eta_{bl} = \frac{2C_{bl}(C_{w_1} + C_{w_0})}{C_1^2} = \frac{2 \times 400 \times 1310}{1200^2}$$

$= \mathbf{72.8\%}.$ **(Ans.)**

Example 19.3. *The velocity of steam exiting the nozzle of the impulse stage of a turbine is 400 m/s. The blades operate close to the maximum blading efficiency. The nozzle angle is 20°. Considering equiangular blades and neglecting blade friction, calculate for a steam flow of 0.6. kg/s, the diagram power and the diagram efficiency.* **(GATE)**

Solution. *Given* :

$C_1 = 400$ m/s, $\alpha = 20°$, $\theta = \phi$; $\dot{m}_s = 0.6$ kg/s.

For maximum blade efficiency,

$$\rho = \frac{C_{bl}}{C_1} = \frac{\cos\alpha}{2}$$

$$\therefore \quad \frac{C_{bl}}{400} = \frac{\cos 20°}{2} \quad \text{or} \quad C_{bl} = 187.9 \text{ m/s}$$

$$C_{w_1} = C_1 \cos\alpha = 400 \cos 20° = 375.9 \text{ m/s}$$

$$C_{f_1} = C_1 \sin\alpha = 400 \sin 20° = 136.8 \text{ m/s}$$

$$\tan\theta = \frac{C_{f_1}}{C_{w_1} - C_{bl}} = \frac{136.8}{375.9 - 187.9} = 0.727$$

$$\therefore \quad \theta = \tan^{-1}(0.727) = 36°$$

Now, $C_{r_1} \sin\theta = C_{f_1}$ or $C_{r_1} = \frac{C_{f_1}}{\sin\theta}$,

$$\therefore \quad C_{r_1} = \frac{136.8}{\sin 36°} = 232.7 \text{ m/s}$$

Now neglecting friction,

$C_{r_0} = C_{r_1} = 232.7$ m/s

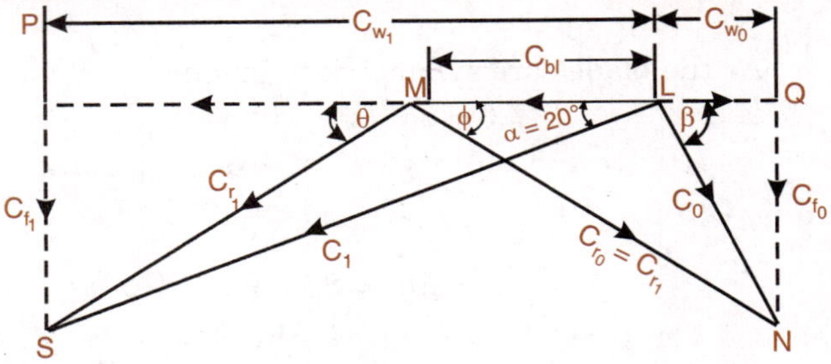

Fig. 19.14

Since the blades are *equiangular*, therefore,

$$\theta = \phi = 36°$$

$$C_{w_0} = C_{r_0} \cos 36° - C_{bl}$$

$$= 237.7 \cos 36° - 187.9 = 0.36 \text{ m/s}$$

$$\therefore \quad C_w = C_{w_1} + C_{w_0} = 375.9 + 0.36$$

$$= 376.26 \text{ m/s}$$

Diagram power,

$$P = \dot{m}_s (C_{w_1} + C_{w_0}) \times C_{bl} \times 10^{-3} \text{ kW}$$

$$= 0.6 \times 376.26 \times 187.9 \times 10^{-3}$$

$$= \mathbf{42.4 \text{ kW.}} \quad \textbf{(Ans.)}$$

Blade or diagram efficiency,

$$\eta_{bl} = \frac{2C_{bl}(C_{w_1} + C_{w_0})}{C_1^2}$$

$$= \frac{2 \times 187.9 \times 376.26}{(400)^2}$$

$$= \mathbf{0.884 \text{ or } 88.4\%} \quad \textbf{(Ans.)}$$

Example 19.4. *A single stage steam turbine is supplied with steam at 5 bar, 200°C at the rate of 50 kg/min. It expands into a condenser at a pressure of 0.2 bar. The blade speed is 400 m/s. The nozzles are inclined at an angle of 20° to the plane of the wheel and the outlet blade angle is 30°. Neglecting friction losses, determine the power developed, blade efficiency, and stage efficiency.* **(U.P.S.C.)**

Solution. *Given* : $p_1 = 5$ bar, 200°C ; $p_2 = 0.2$ bar, $m_s = 50$ kg/min, $C_{bl} = 400$ m/s ; $\alpha = 20°$, $\phi = 30°$; $C_{r_1} = C_{r_0}$ (because friction losses are neglected)

Refer to Fig. 19.15

From steam tables :

At 5 bar, 200°C :

$h_1 = 2855.4$ kJ/kg ; $s_1 = 7.0592$ kJ/kg K

At 0.2 bar :

$h_{f_2} = 251.5$ kJ/kg, $h_{fg_2} = 2358.4$ kJ/kg

$s_{f_2} = 0.832$ kJ/kg K ; $s_{fg} = 7.0773$ kJ/kg K

Since the steam expansion takes place *isentropically*,

$$\therefore \quad s_1 = s_2$$

or $$7.0592 = 0.8321 + x_2 \times 7.0773$$

or $$x_2 = \frac{7.0592 - 0.8321}{7.0773} = 0.88$$

∴ Enthalpy of steam at 0.2 bar,

$$h_2 = h_{f_2} + x_2 h_{fg_2}$$

$$= 251.5 + 0.88 \times 2358.4$$

$$= 2326.9 \text{ kJ/kg}$$

Enthalpy drop

$$= h_1 - h_2 = 2855.4 - 2326.9 = 528.5 \text{ kJ/kg}$$

Velocity of steam entering the blades,

$$C_1 = 44.7\sqrt{h_1 - h_2} = 44.7\sqrt{528.5} \simeq 1028 \text{ m/s}$$

The velocity diagram is shown in Fig. 19.15

Now, $$C_{w_1} = 1028 \cos 20° = 966 \text{ m/s}$$

$$C_{f_1} = 1028 \sin 20° = 351.6 \text{ m/s}$$

$$\tan\theta = \frac{C_{f_1}}{C_1 \cos 20° - C_{bl}}$$

$$= \frac{351.6}{1028 \cos 20° - 400} = 0.6212$$

$$\therefore \quad \theta = \tan^{-1}(0.6212) = 31.85°$$

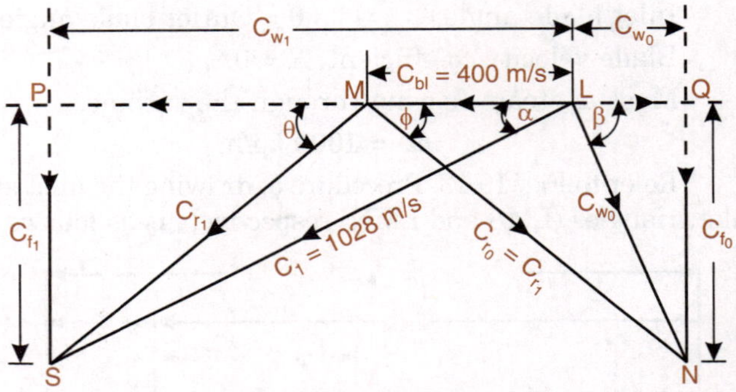

Fig. 19.15

Now, $$C_{r_1} \sin 31.85° = C_{f_1} = 351.6$$

$$\therefore \quad C_{r_1} = \frac{351.6}{\sin 31.85°} = 666 \text{ m/s}$$

$$\therefore \quad C_{w_0} = C_{r_0} \cos 30° - C_{bl} \qquad (\because C_{r_0} = C_{r_1})$$

$$= 666 \cos 30° - 400 \simeq 177 \text{ m/s}$$

Power developed,

$$P = \dot{m}_s (C_{w_1} + C_{w_0})\, C_{bl}$$

$$= \frac{50}{60}(966 + 177) \times 400 \times 10^{-3} \text{ kW}$$

$$= \mathbf{381 \text{ kW.}} \quad \textbf{(Ans.)}$$

Blade efficiency,

$$\eta_{bl} = \frac{2C_{bl}(C_{w_1} + C_{w_0})}{C_1^2}$$

$$= \frac{2 \times 400 \times (966 + 177)}{(1028)^2}$$

$$= \mathbf{0.865 \text{ or } 86.5\%.} \quad \textbf{(Ans.)}$$

Since there are no losses, therefore,

Stage efficiency

= blade efficiency = **86.5%.** **(Ans.)**

Example 19.5. *The following data relate to a single stage impulse turbine :*

Steam velocity	*= 600 m/s ;*
Blade speed	*= 250 m/s*
Nozzle angle	*= 20° ;*
Blade outlet angle	*= 25°.*

Neglecting the effect of friction, calculate the work developed by the turbine for the steam flow rate of 20 kg/s. Also calculate the axial thrust on the bearings.

Solution. Absolute velocity of steam entering the blades, C_1 = 600 m/s

Blade speed, C_{bl} = 250 m/s ; Nozzle angle, α = 20°

Blade outlet angle, ϕ = 25° ; Steam flow rate, $\dot{m}_s$ = 20 kg/s

Refer to Fig. 19.16.

- Triangle *LMS* is drawn with the above data.
- Then angle *LMN i.e.,* ϕ = 25° is drawn such that *NM* = *MS* (because effect of friction is to be neglected *i.e.,* *K* = 1).
- Join *LN* by vector C_0 which represents the velocity of steam at outlet from the wheel,

This completes both inlet and outlet triangles.

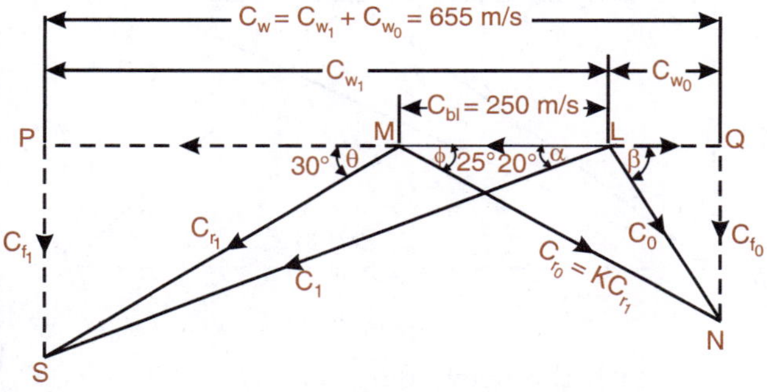

Fig. 19.16

By measurement :

$$C_w = C_{w_1} + C_{w_0} = 655 \text{ m/s} ;$$

$$C_{f_1} = 200 \text{ m/s} ; C_{f_0} = 160 \text{ m/s}.$$

Work developed, W :

$$W = \dot{m}_s(C_{w_1} + C_{w_0})\ C_{bl} = 20 \times 655 \times 250$$

$$= \mathbf{3275000\ Nm/s.}\ \ \textbf{(Ans.)}$$

Axial thrust :

Axial thrust

$$= \dot{m}_s(C_{f_1} - C_{f_0}) = 20(200 - 160)$$

$$= \mathbf{800\ N.}\ \ \textbf{(Ans.)}$$

Example 19.6. *A single row impulse turbine develops 132.4 kW at a blade speed of 175 m/s, using 2 kg of steam per sec. Steam leaves the nozzle at 400 m/s. Velocity coefficient of the blades is 0.9. Steam leaves the turbine blades axially.*

Determine nozzle angle, blade angles at entry and exit, assuming no shock.

Solution. Power developed, P = 132.4 kW

Blade speed, C_{bl} = 175 m/s

Steam used, $\dot{m}_s$ = 2 kg/s

Velocity of steam leaving the nozzle, C_1 = 400 m/s

Blade velocity co-efficient, K = 0.9

Power developed,

$$P = \dot{m}_s \frac{(C_{w_1} + C_{w_0}) \times C_{bl}}{1000} \text{ kW}$$

$$132.4 = \frac{2(C_{w_1} + C_{w_0}) \times 175}{1000}$$

or

$$(C_{w_1} + C_{w_0}) = \frac{132.4 \times 1000}{2 \times 175} = 378 \text{ m/s}$$

C_{w_0} = 0, since the *discharge is axial*.

Construct the velocity diagram as shown in Fig. 19.17.

In this diagram $\dfrac{C_{r_0}}{C_{r_1}}$ = 0.9, β = 90°, since the *discharge is axial* ; and

$$C_{w_1} + C_{w_0} = PL = PQ = 378 \text{ m/s}.$$

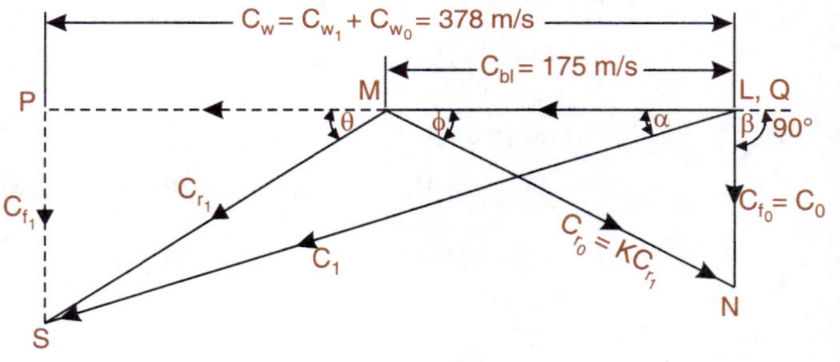

Fig. 19.17

From the diagram (by measurement) :

Nozzle angle,	α = **21°. (Ans.)**
Blade inlet angle,	θ = **36°. (Ans.)**
Blade outlet angle,	ϕ = **32°. (Ans.)**

Example 19.7. *A simple impulse turbine has a mean blade speed of 200 m/s. The nozzles are inclined at 20° to the plane of rotation of the blades. The steam velocity from nozzles is 600 m/s. The turbine uses 3500 kg/h of steam. The absolute velocity at exit is along the axits of the turbine. Determine :*

(i) The inlet and exit angles of the blades.

(ii) The power output of the turbine.

(iii) The diagram efficiency.

(iv) The axial thrust (per kg steam per second).

Assume inlet and outlet angles to be equal.

(U.P.S.C.)

Solution. *Given* : C_{bl} = 200 m/s ; α = 20° ; C_1 = 600 m/s ; m_s = 3500 kg/h ; β = 90° ; $\theta = \phi$.

(i) Inlet and exit angles of the blades, θ, ϕ :

Refer to Fig. 19.18

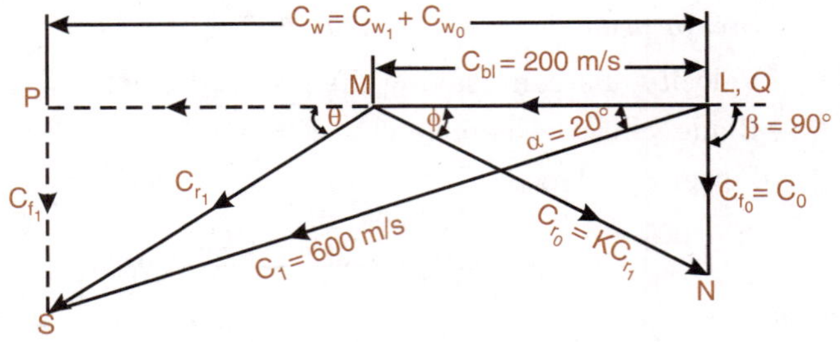

Fig. 19.18

$$C_{f_1} = C_1 \sin 20° = 600 \sin 20° = 205.2 \text{ m/s}$$

$$\tan\theta = \frac{SP}{PM} = \frac{C_{f_1}}{C_1 \cos 20° - C_{bl}}$$

$$= \frac{205.2}{600\cos 20° - 200} = 0.564$$

$\therefore$ $\theta = \tan^{-1}(0.564)$ = **29.4°. (Ans.)**

Also $\theta = \phi$ ...(Given)

$\therefore$ ϕ = **29.4°. (Ans.)**

(ii) The power output of the turbine, P :

$$\mathbf{P} = \dot{m}_s(C_{w_1} + C_{w_0})\, C_{bl}$$

$$= \frac{3500}{3600}(600 \cos 20° + 0) \times 200 \times 10^{-3} \text{ kW}$$

= **109.6 kW. (Ans.)**

(C_{w_0} = 0, since the discharge is axial)

(iii) The blade or diagram efficiency, η_{bl} :

$$\eta_{bl} = \frac{2C_{bl}(C_{w_1} + C_{w_0})}{C_1^2}$$

$$= \frac{2 \times 200(600\cos 20° + 0)}{600^2}$$

= **0.626 or 62.6%. (Ans.)**

(iv) The axial thrust (per kg steam per second) :

The axial thrust per kg per second

$$= 1 \times (C_{f_1} - C_{f_0})\, N$$

where $C_{f_1} = C_1 \sin 20° = 600 \sin 20° = 205.2$ m/s.

Now, $\dfrac{C_{f_0}}{C_{bl}} = \tan 29.4°$

$\therefore$ $C_{f_0} = 200 \times \tan 29.4° = 112.69$ m/s

Substituting the values, we get

Axial thrust (per kg steam per second)

= 1 × (205.2 – 112.69) = **92.51 N. (Ans.)**

☞**Example 19.8.** *Steam with absolute velocity of 300 m/s is supplied through a nozzle to a single stage impulse turbine. The nozzle angle is 25°. The mean diameter of the blade rotor is 1 metre and it has a speed of 2000 r.p.m. Find suitable blade angles for zero axial thrust. If the blade velocity co-efficient is 0.9 and the steam flow rate is 10 kg/s, calculate the power developed.*

Solution. Absolute velocity of steam entering the blade, C_1 = 300 m/s

Nozzle angle,	α = 25°
Mean diameter of the rotor blade,	D = 1 m
Speed of the rotor,	N = 2000 r.p.m.
Blade velocity co-efficient,	K = 0.9
Steam flow rate,	$\dot{m}_s$ = 10 kg/s

Blade angles :

$$\text{Blade speed, } C_{bl} = \frac{\pi DN}{60} = \frac{\pi \times 1 \times 2000}{60} = 105 \text{ m/s.}$$

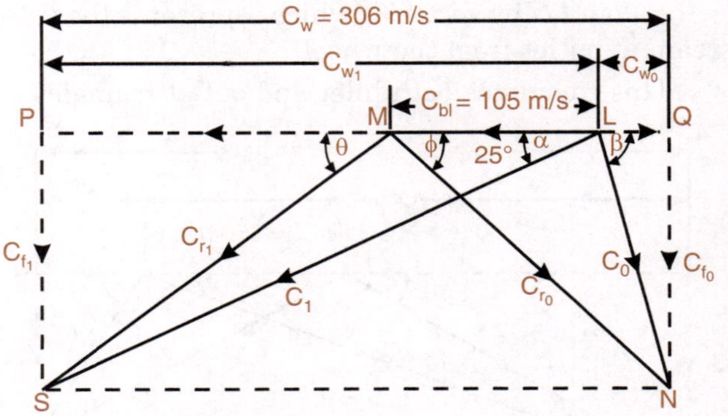

Fig. 19.19

— With the above data (*i.e.*, C_1 = 300 m/s, C_{bl} = 105 m/s and α = 25°) draw triangle LMS (Fig. 19.19). From *S* draw perpendicular *SP* on *LM* produced. Measure C_{r_1}.

— From *S* draw a line *parallel to LP* ($\because C_{f_1} = C_{f_0}$) and from point *M* draw an arc equal to $C_{r_0} (= 0.9\, C_{r_1})$ to get the point of intersection *N*. Complete the triangle *LMN*. From *N* draw perpendicular *NQ* on *PL* produced to get C_{f_0}.

Measure θ and ϕ (the blade angles) from the velocity diagram.

θ = **37°** and ϕ = **42°. (Ans.)**

Power developed, P :

$$P = \frac{\dot{m}_s(C_{w_1} + C_{w_0}) \times C_{bl}}{1000}$$

$$= \frac{10 \times 306 \times 105}{1000} = \mathbf{321.3\ kW.} \quad \textbf{(Ans.)}$$

☞**Example 19.9.** *In an impulse turbine (with a single row wheel) the mean diameter of the blades is 1.05 m and the speed is 3000 r.p.m. The nozzle angle is 18°, the ratio of blade speed to steam speed is 0.42 and the ratio of the relative velocity at outlet from the blades to that at inlet is 0.84. The outlet angle of the blade is to be made 3° less than the inlet angle. The steam flow is 10 kg/s. Draw the velocity diagram for the blades and derive the following :*

(*i*) *Tangential thrust on the blades*
(*ii*) *Axial thrust on the blades*
(*iii*) *Resultant thrust on the blades*
(*iv*) *Power developed in the blades*
(*v*) *Blading efficiency.* **(P.U.)**

Solution. Mean diameter of the blades, $D = 1.05$ m

Speed of the turbine, $N = 3000$ r.p.m.

Nozzle angle, $\alpha = 18°$

Ratio of blade speed to steam speed, $\rho = 0.42$

Ratio, $\frac{C_{r_0}}{C_{r_1}} = 0.84$

Outlet blade angle, $\phi = \theta - 3°$

Steam flow rate $\dot{m}_s = 10$ kg/s

Blade speed,

$$C_{bl} = \frac{\pi DN}{60} = \frac{\pi \times 1.05 \times 3000}{60} = 164.5 \text{ m/s}$$

But $\rho = \frac{C_{bl}}{C_1} = 0.42$ (given)

$$\therefore \quad C_1 = \frac{C_{bl}}{0.42} = \frac{164.5}{0.42} = 392 \text{ m/s}$$

With the data, $C_1 = 392$ m/s ;
$\alpha = 18°$, complete ΔLMS
$\theta = 30°$ (on measurement)
$\therefore \quad \phi = 30° - 3 = 27°.$

Now complete the ΔLMN by taking

$$\phi = 27° \text{ and } C_{r_0} = 0.84\, C_{r_1}.$$

Finally complete the whole diagram as shown in Fig. 19.20.

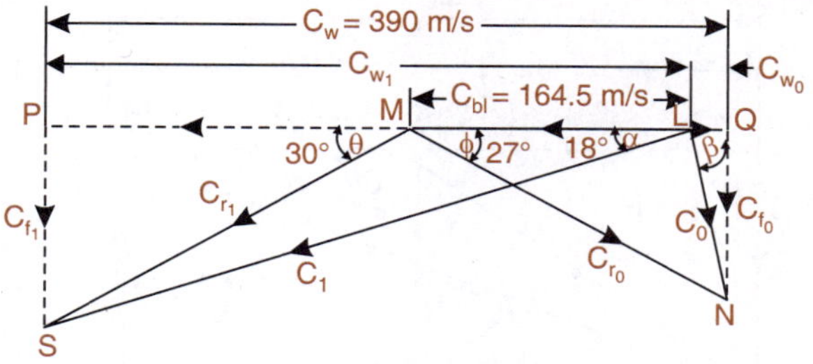

Fig. 19.20

(*i*) **Tangential thrust on the blades :**

Tangential thrust

$$= \dot{m}_s(C_{w_1} + C_{w_0}) = 10 \times 390 = \mathbf{3900\ N.} \quad \textbf{(Ans.)}$$

(*ii*) **Axial thrust :**

$$\text{Axial thrust} = \dot{m}_s(C_{f_1} - C_{f_0}) = 10\,(120 - 95)$$

$$= \mathbf{250\ N.} \quad \textbf{(Ans.)}$$

(*iii*) **Resultant thrust :**

$$\text{Resultant thrust} = \sqrt{(3900)^2 + (250)^2}$$

$$= \mathbf{3908\ N.} \quad \textbf{(Ans.)}$$

(*iv*) **Power developed, P :**

$$P = \frac{\dot{m}_s(C_{w_1} + C_{w_0}) \times C_{bl}}{1000} = \frac{10 \times 390 \times 164.5}{1000}$$

$$= \mathbf{641.55\ kW.} \quad \textbf{(Ans.)}$$

(*v*) **Blading efficiency, η_{bl} :**

$$\eta_{bl} = \frac{2\,C_{bl}\,(C_{w_1} + C_{w_0})}{{C_1}^2} = \frac{2 \times 164.5 \times 390}{392^2}$$

$$= \mathbf{83.5\%.} \quad \textbf{(Ans.)}$$

Example 19.10. *In a stage of impulse reaction turbine provided with single row wheel, the mean diameter of the blades in 1 m. It runs at 3000 r.p.m. The steam issues from the nozzle at a velocity of 350 m/s and the nozzle angle is 20°. The rotor blades are equiangular. The blade friction factor is 0.86. Determine the power developed if the axial thrust on the end bearing of a rotor is 118 N.*

Solution. Mean diameter of the blades, $D = 1$ m

Speed of the turbine, $N = 3000$ r.p.m.

Velocity of steam issuing from the nozzle, $C_1 = 350$ m/s

Nozzle angle, $\alpha = 20°$

Blade angles, $\theta = \phi$

Blade friction factor, $K = 0.86$

Axial thrust $= 118$ N

Power developed, P :

$$\text{Blade, velocity,} \quad C_{bl} = \frac{\pi DN}{60}$$

$$= \frac{\pi \times 1 \times 3000}{60} = 157 \text{ m/s}$$

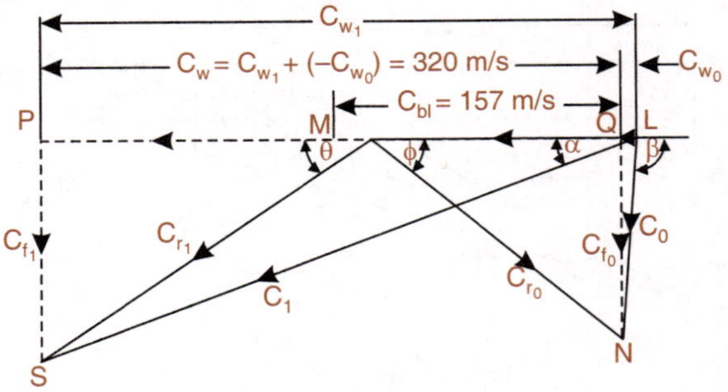

Fig. 19.21

— With the data, C_{bl} = 157 m/s, C_1 = 350 m/s, α = 20°, draw the Δ*LMS* (Fig. 19.21).

By measurement,

$$\theta = 35°$$

Since the blades are equiangular,

$$\theta = \phi = 35°$$

— Now with

$\phi = 35°$ and $C_{r_0} = 0.86\ C_{r_1}$, complete the Δ*LMN*.

On measurement ;

$$C_{f_1} = 120 \text{ m/s}, \ C_{f_0} = 102.5 \text{ m/s}$$

Also, axial thrust $\dot{m}_s(C_{f_1} - C_{f_0}) = 118$

$$\therefore \quad \dot{m}_s = \frac{118}{C_{f_1} - C_{f_0}} = \frac{118}{(120 - 102.5)} = 6.74 \text{ kg/s}$$

Further in this case,

$$C_w = C_{w_1} + C_{w_0} = C_{w_1} + {}^*(-C_{w_0}) = 320 \text{ m/s}$$

(*β > 90°)

Now, *power developed,*

$$P = \frac{\dot{m}_s(C_{w_1} + C_{w_0}) \times C_{bl}}{1000} \text{ kW}$$

$$= \frac{6.74 \times 320 \times 157}{1000} = \mathbf{338.6\ kW.} \quad \textbf{(Ans.)}$$

Example 19.11. *A simple impulse turbine has one ring of moving blades running at 150 m/s. The absolute velocity of steam at exit from the stage is 85 m/s at an angle of 80° from the tangential direction. Blade velocity co-efficient is 0.82 and the rate of steam flowing through the stage is 2.5 kg/s. If the blades are equiangular, determine :*

(i) Blade angles ; (ii) Nozzle angle ;

(iii) Absolute velocity of steam issuing from the nozzle ;

(iv) Axial thrust.

Solution. Blade velocity, C_{bl} = 150 m/s

Absolute velocity of steam at exit from the stage, C_0 = 85 m/s

Angle, β = 80°

Blade velocity co-efficient, $K = \dfrac{C_{r_0}}{C_{r_1}} = 0.82$

Rate of steam flowing through the stage,

$$\dot{m}_s = 2.5 \text{ kg/s}$$

Blades are equiangular, *i.e.*, θ = φ.

— With the above given data velocity triangle for *exit* can be drawn to a suitable scale. From that, value φ = θ can be obtained. Also the value of C_{r_0} can be obtained which helps to get the value of C_{r_1} with the help of given value of '*K*'. With these values having being known the inlet velocity triangle of the velocity diagram can be completed to get the value of C_1, the absolute velocity of steam issuing from the nozzle and value of axial thrust can also be calculated. The Fig. 19.22 gives the velocity diagram of the turbine stage to a suitable scale.

— From the outlet velocity Δ*LMN*

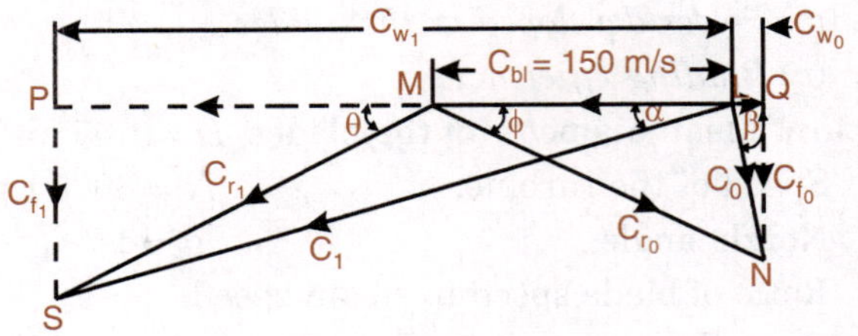

Fig. 19.22

By measurement, C_{r_0} = 186 m/s

$$\therefore \quad C_{r_1} = \frac{C_{r_0}}{K} = \frac{186}{0.82} = 226.8 \text{ m/s}$$

(i) **Blades angles** θ, φ :

By measurement θ = φ = blade angles = **27°. (Ans.)**

(ii) **Nozzle angle,** α :

By measurement ; nozzle angle, α = **16°. (Ans.)**

(iii) **Absolute velocity, C₁ :**

Absolute velocity of steam issuing from the nozzle,

C_1 = **366 m/s (by measurement). (Ans.)**

(iv) **Axial thrust :**

Also, C_{f_0} = 84 m/s, C_{f_1} = 102 m/s — By measurement.

$$\therefore \quad \text{Axial thrust} = \dot{m}_s(C_{f_1} - C_{f_0})$$

$$= 2.5\ (102 - 84) = \mathbf{45\ N.} \quad \textbf{(Ans.)}$$

Example 19.12. *One stage of an impulse turbine consists of a converging nozzle ring and one ring of moving blades. The nozzles are inclined at 22° to the blades whose tip angles are both 35°. If the velocity of steam at exit from nozzle is 660 m/s, find the blade speed so that the steam passes on without shock. Find the diagram efficiency neglecting losses if the blades are run at this speed.* **(U.P.S.C.)**

Solution. *Given :* $\alpha = 22°$; $\theta = \phi = 35°$; $C_1 = 660$ m/s.

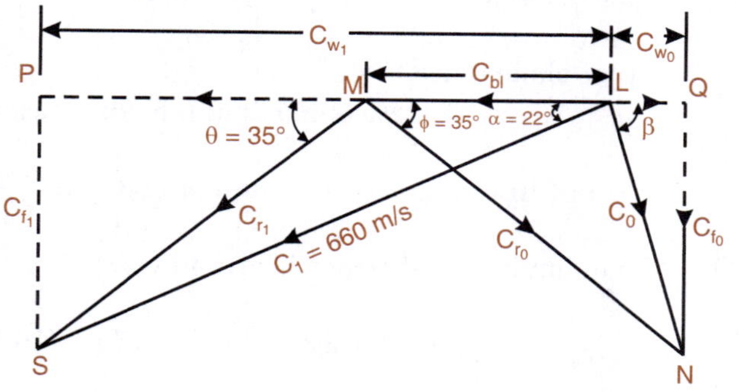

Fig. 19.23

In case of impulse turbine, *maximum blade efficiency,*

$$(\eta_{bl})_{max} = \frac{\cos^2 \alpha}{2}(1 + KZ) \quad ...[\text{Eqn. (19.19)}]$$

where K (= blade velocity co-efficient) = 1,

(∵ Losses are neglected)

$$Z = \frac{\cos \phi}{\cos \theta} = 1$$

(∵ Blades are *equiangular*)

$$\therefore \quad \eta_{bl} = \frac{\cos^2 \alpha}{2}(1 + 1) = \cos^2 \alpha$$

$$= (\cos 22°)^2 = \mathbf{0.86} \text{ or } \mathbf{86\%}. \text{ (Ans.)}$$

Also, $$\rho_{opt.} = \frac{\cos \alpha}{2} \quad ...[\text{Eqn. (19.8)}]$$

or $$\frac{C_{bl}}{C_1} = \frac{\cos 22°}{2} = 0.4636$$

$$\therefore \quad C_{bl} = C_1 \times 0.4636$$

$$= 660 \times 0.4636 \simeq \mathbf{306 \text{ m/s.}} \text{ (Ans.)}$$

☞**Example 19.13.** *In a single stage impulse turbine the mean diameter of the blade ring is 1 metre and the rotational speed is 3000 r.p.m. The steam is issued from the nozzle at 300 m/s and nozzle angle is 20°. The blades are equiangular. If the friction loss in the blade channel is 19% of the kinetic energy corresponding to the relative velocity at the inlet to the blades, what is the power developed in the blading when the axial thrust on the blades is 98 N ?*

Solution.

Mean diameter of the blade ring,

$D = 1$ m

Speed of the turbine,

$N = 3000$ r.p.m.

Absolute velocity of steam issuing from the nozzle,

$C_1 = 300$ m/s

Nozzle angle,

$\alpha = 20°$

Blade angles are equiangular,

$\theta = \phi$

Friction loss in the blade channel

= 19%

i.e., $$C_{r_0} = (1 - 0.19)\, C_{r_1} = 0.81\, C_{r_1}$$

Axial thrust on the blades

= 98 N

Power developed, P :

Blade speed, $$C_{bl} = \frac{\pi DN}{60} = \frac{\pi \times 1 \times 3000}{60} = 157.1 \text{ m/s}$$

Also $\theta = \phi$ (given)

Now, velocity diagram is drawn to a suitable scale as shown in Fig. 19.24.

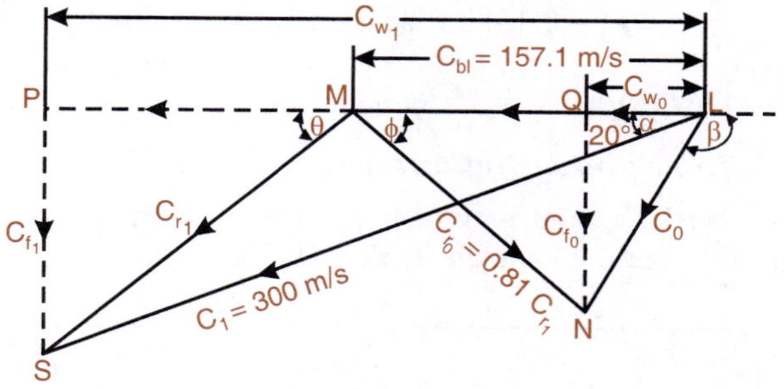

Fig. 19.24

By measurement (from diagram) ;

$$C_{w_1} = 283.5 \text{ m/s} ; \; C_{w_0} = 54 \text{ m/s}$$

$$C_{f_1} = 100.5 \text{ m/s}$$

$$C_{f_0} = 81 \text{ m/s.}$$

Axial thrust $= \dot{m}_s(C_{f_1} - C_{f_0})$

$$98 = \dot{m}_s\,(100.5 - 81)$$

or $$\dot{m}_s = \frac{98}{100.5 - 81} = 5.025 \text{ kg}$$

Power developed,

$$P = \frac{\dot{m}_s[C_{w_1} + {}^*(-C_{w_0})]}{1000}\, C_{bl} \quad (^*\beta > 90°)$$

$$= \frac{5.025\,(283.5 - 54) \times 157.1}{1000} = \mathbf{181.2 \text{ kW.}} \text{ (Ans.)}$$

Example 19.14. *Show that the maximum possible efficiency of a De-Laval steam turbine is 88.3% when nozzle angle is 20°. Deduce the formula used.*

Solution. Maximum possible efficiency, $\eta_{max} = 88.3\%$

Nozzle angle, $\alpha = 20°$

Maximum possible efficiency of a De-Laval turbine (impulse turbine) = $\cos^2 \alpha$, where α is the nozzle angle.

$$\therefore \quad \eta_{max} = \cos^2 20° = (0.9396)^2$$

$$= 0.883 = \mathbf{88.3\%.} \text{ (Proved).}$$

For derivation of the formula used refer to Article 19.7.

Example 19.15. *In a simple impulse turbine the nozzles are inclined at 20° to the direction of motion of the moving blades. The steam leaves the nozzle at 375 m/s. The blade velocity is 165 m/s. Calculate suitable inlet and outlet angles* *for the blades in order that the axial thrust is zero. The relative velocity of steam as it flows over the blades is reduced by 15% by friction. Also, determine the power developed for a flow rate of 10 kg/s.*

Solution. Nozzles, $\alpha = 20°$

Velocity of steam issuing from the nozzles,

$C_1 = 375$ m/s

Blade speed, $C_{bl} = 165$ m/s

Axial thrust = zero *i.e.*, $C_{f_1} = C_{f_0}$

$$\frac{C_{r_0}}{C_{r_1}} = (1 - 0.15) = 0.85, \text{ i.e., 15\% loss due to friction,}$$

steam flow rate, $\dot{m}_s = 10$ kg/s.

Inlet and outlet angles :

With the above given data, draw velocity diagram to a suitable scale as shown in Fig. 19.25.

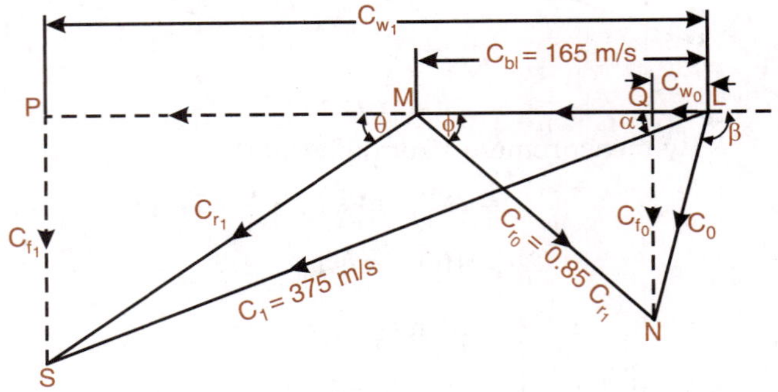

Fig. 19.25

By measurement (from velocity diagram),

$$\left.\begin{aligned}\theta &= \mathbf{35°}\\ \phi &= \mathbf{42°}\\ \beta &= \mathbf{100°}\end{aligned}\right\}. \quad \textbf{(Ans.)}$$

Power developed, P :

Also, $C_{w_1} = 354$ m/s ; $C_{w_0} = 24$ m/s (By measurement)

$\therefore$ *Power developed,*

$$P = \frac{\dot{m}_s\,[C_{w_1} + C_{w_0}] \times C_{bl}}{1000} = \frac{\dot{m}_s\,[C_{w_1} + (-\,C_{w_0})] \times C_{bl}}{1000}$$

$$= \frac{10\,[354 + (-\,24)] \times 165}{1000} = \mathbf{544.5\ kW.} \quad \textbf{(Ans.)}$$

Example 19.16. *In a single stage impulse turbine nozzle angle is 20° and blade angles are equal. The velocity co-efficient for blade is 0.85. Find maximum blade efficiency possible. If the actual blade efficiency is 92% of the maximum blade efficiency, find the possible ratio of blade speed to steam speed.*

Solution. Nozzle angle, $\alpha = 20°$

Blade angles are equal *i.e.*, $\theta = \phi$

Blade velocity co-efficient, $K = 0.85 \left(= \dfrac{C_{r_0}}{C_{r_1}}\right)$

Actual blade efficiency

= 92% of maximum blade efficiency

Ratio of blade speed to steam speed, $\rho = \dfrac{C_{bl}}{C_1}$.

Maximum blade efficiency is given by :

$$(\eta_{bl})_{max} = \frac{\cos^2 \alpha}{2}(1 + KZ) \quad \text{...[Eqn. (19.19)]}$$

$$= \frac{\cos^2 \alpha}{2}(1 + K) \text{ as } Z = \frac{\cos \phi}{\cos \theta} = 1$$

$$(\eta_{bl})_{max} = \frac{\cos^2 20°}{2}(1 + 0.85) = 0.816 \text{ or } 81.6\%$$

The actual efficiency of the turbine

$= 0.92 \times 0.816 = 0.75$

The blade efficiency of a single stage impulse turbine is given by be relation,

$$\eta_{bl} = 2\,(1 + K)(\rho \times \cos \alpha - \rho^2)$$

$$0.75 = 2(1 + 0.85)(\rho \times \cos 20° - \rho^2)$$

$$0.75 = 2 \times 1.85(0.94\,\rho - \rho^2)$$

$$0.203 = 0.94\,\rho - \rho^2$$

$$\rho^2 - 0.94\,\rho + 0.203 = 0$$

$$\therefore \quad \rho = \frac{0.94 \pm \sqrt{(0.94)^2 - 4 \times 0.203}}{2} = \frac{0.94 \pm 0.267}{2}$$

or $\rho = 0.603$ or 0.336

Hence possible ratio, $\rho =$ **0.603 or 0.336. (Ans.)**

☞ **Example 19.17.** *The following data refer to a single stage impulse turbine :*

Isentropic nozzle heat drop = 251 kJ/kg ; nozzle efficiency = 90% ; nozzle angle = 20° ; ratio of blade speed to whirl component of steam speed = 0.5 ; blade velocity co-efficient = 0.9 ; the velocity of steam entering the nozzle = 20 m/s.

Determine : (i) The blade angles at inlet and outlet if the steam enters into the blades without shock and leaves the blades in an axial direction.

(ii) Blade efficiency.

(iii) Power developed and axial thrust if the steam flow is 8 kg/s.

Solution. Isentropic heat drop = 251 kJ/kg

Nozzle efficiency, $\eta_{nozzle} = 90\%$

Nozzle angle, $\alpha = 20°$

Ratio of blade speed to whirl component of steam speed = 0.5

Blade velocity co-efficient, $K = 0.9$

Velocity of steam entering the nozzle = 20 m/s

(*i*) **Blade angles :**

Nozzle efficiency is given by :

$$\eta_{nozzle} = \frac{\text{Useful heat drop}}{\text{Isentropic heat drop}}$$

or $$0.9 = \frac{\text{Useful heat drop}}{251}$$

$\therefore$ Useful heat drop

$$= 0.9 \times 251 = 225.9 \text{ kJ/kg}$$

Applying the energy equation to the nozzle, we get

$$\frac{C_1^2 - 20^2}{2} = 225.9 \times 1000$$

$$\therefore \quad C_1^2 = 2 \times 225.9 \times 1000 + 400 = 452200$$

i.e., $$C_1 = 672.4 \text{ m/s}$$

Other data given : $\alpha = 20°, \frac{C_{bl}}{C_w} = 0.5, K = \frac{C_{r_0}}{C_{r_1}} = 0.9$

and $C_{w_0} = 0$, *as the steam leaves the blades axially.*

$\therefore C_0 = C_{f_0}$

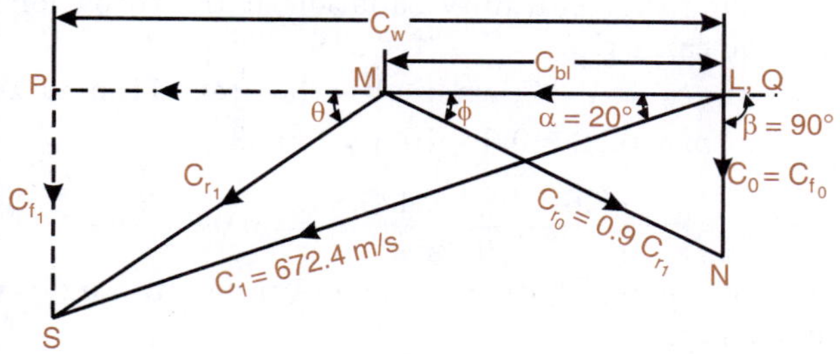

Fig. 19.26

With the above data construct the velocity triangles as follows :

- Select a suitable scale, say 1 cm = 50 m/s.
- Draw a horizontal line through a point L and angle $\alpha = 20°$. Mark the point along LS as

$$LS = C_1 = 672.4 \text{ m} = \frac{672.4}{50} = 13.5 \text{ cm}.$$

- Draw a line through S which is perpendicular to the horizontal line through L and it cuts at the point P. Measure the distance LP = 12.7 cm.

$$\therefore \quad C_{w_1} = C_w = 12.7 \text{ cm}.$$

and $$C_{bl} = 0.5\, C_w = 0.5 \times 12.7 = 6.35 \text{ cm}.$$

- Mark the point M as $LM = C_{bl}$ = 6.35 cm
- Join point MS and complete the inlet velocity triangle LMS.
- Measure MS (C_{r_1}) = 7.7 cm.

$$C_{r_0} = 0.9 \times 7.7 = 6.93 \text{ cm}.$$

- Draw a perpendicular line through point L to the line LM. From M cut an arc of radius 6.93 cm to cut the vertical line through L and mark the point N and join MN which completes the outlet triangle LMN.

Now, find out velocities converting lengths into velocities :

$$C_1 = 672.4 \text{ m/s}$$
$$C_w = 12.7 \times 50 = 635 \text{ m/s}$$
$$C_{bl} = 0.5\, C_w = 0.5 \times 635 = 317.5 \text{ m/s}$$
$$C_{r_1} = 7.7 \text{ cm} = 7.7 \times 50 = 385 \text{ m/s}$$
$$C_{r_0} = 0.9\, C_{r_1} = 0.9 \times 385 = 346.5 \text{ m/s}$$
$$C_{f_1} = 4.45 \text{ cm} = 4.45 \times 50 = 222.5 \text{ m/s}$$
$$C_{f_0} = 2.6 \text{ cm} = 2.6 \times 50 = 130 \text{ m/s}.$$

Blade angles measured from the diagram :

θ = **35°**, ϕ = **22°**. **(Ans.)**

(*ii*) **Blade efficiency, η_{bl} :**

$$\eta_{bl} = \frac{2C_{bl}C_w}{C_1^2} = \frac{2 \times 317.5 \times 635}{(672.4)^2}$$

= **0.89 or 89%. (Ans.)**

(*iii*) **Power developed, P and axial thrust :**

$$P = \frac{\dot{m}_s(C_{w_1} + C_{w_0}) \times C_{bl}}{1000}$$

$$= \frac{8 \times (635 + 0) \times 317.5}{1000} = \mathbf{1612.9\ kW.} \quad \textbf{(Ans.)}$$

Axial thrust $= \dot{m}_s(C_{f_1} - C_{f_0})$

$= 8(222.5 - 130) =$ **740 N. (Ans.)**

☞ **Example 19.18.** *In a single stage steam turbine saturated steam at 10 bar abs. is supplied through a convergent-divergent steam nozzle. The nozzle angle is 20° and the mean blade speed is 400 m/s. The steam pressure leaving the nozzle is 1 bar abs. Find :*

(*i*) *The best blade angles if blades are equiangular.*

(*ii*) *The maximum power developed by the turbine if a number of nozzles used are 5 and area at the throat of each nozzle is 0.6 cm².*

Assume nozzle efficiency 88% and blade friction co-efficient of 0.87.

Solution. Supply steam pressure (to nozzles) = 10 bar abs.

Nozzle angle, $\alpha = 20°$

Mean blade speed, C_{bl} = 400 m/s

Steam pressure leaving the nozzle = 1 bar abs.

Number of nozzles used = 5

Area of throat at each nozzle $= 0.6 \text{ cm}^2$

Nozzle efficiency, $\eta_{nozzle} = 88\%$

Blade friction co-efficient, $K = \dfrac{C_{r_0}}{C_{r_1}} = 0.87.$

The velocity of steam at the outlet of nozzle is found representing the expansion through nozzle on *h-s* chart as shown in Fig. 19.27.

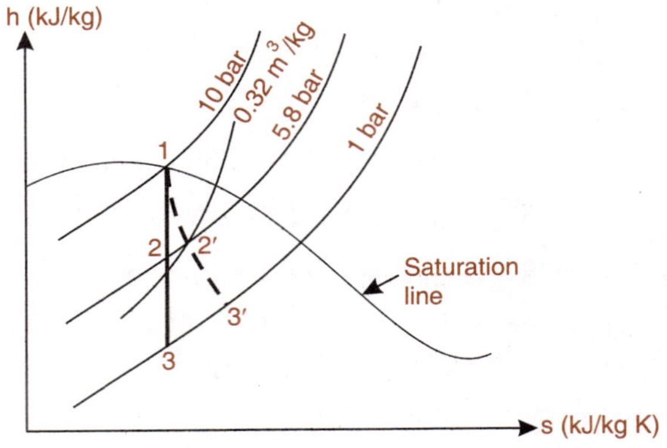

Fig. 19.27

From *h-s* chart,

$$h_1 - h_3 \simeq 402 \text{ kJ/kg}$$

$$\eta_{nozzle} = \frac{h_1 - h_3'}{h_1 - h_3} = 0.88$$

$$\therefore \quad h_1 - h_3' = 0.88 \times 402 = 353.76 \text{ kJ/kg}$$

Also $$\frac{C_3'^2}{2} = h_1 - h_3'$$

or $$C_3' = \sqrt{2(h_1 - h_3')}$$

$$= \sqrt{2 \times 353.76 \times 1000} = 841.14 \text{ m/s}$$

(*i*) Blade angles :

Construct the velocity triangles as per data given as shown in Fig. 19.27.

By measurement, θ (= φ) = **35.5°. (Ans.)**

(*ii*) Maximum power developed, P :

For finding out the maximum power developed by the turbine let us first find out the maximum mass of steam passing through the nozzle.

The required condition for the maximum mass flow through the nozzle is given by

$$\frac{p_2}{p_1} = \left(\frac{2}{n+1}\right)^{\frac{n}{n-1}}$$

where, p_1 = Pressure of steam at inlet of the nozzle,

p_2 = Pressure of steam at the throat of the nozzle,

and n (index of expansion) = 1.135 as steam is saturated.

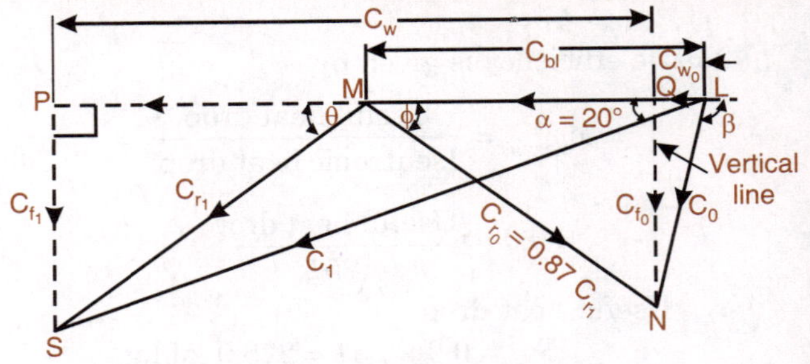

Fig. 19.28

$$\therefore \quad \frac{p_2}{p_1} = \left(\frac{2}{1.135+1}\right)^{\frac{1.135}{0.135}} = 0.58$$

$$\therefore \quad p_2 = 10 \times 0.58 = 5.8 \text{ bar}$$

From *h-s* chart (Fig. 19.27)

$$h_1 - h_2 \simeq 105 \text{ kJ/kg}$$

and $$h_1 - h_2' = 0.88 \times 105 = 92.5 \text{ kJ/kg}$$

v_2' (specific volume at point 2′) $\simeq 0.32 \text{ m}^3/\text{kg}$

The maximum velocity of steam at the throat of the nozzle is given by

$$C = \sqrt{2(h_1 - h_2')} = \sqrt{2 \times 92.4 \times 1000} = 429.88 \text{ m/s}$$

Using the continuity equation at the throat of the nozzle, we can write

$m \,.\, v_2' = A \times C$, where A is the area of the nozzle.

$$\therefore \quad m \times 0.32 = 0.6 \times 10^{-4} \times 429.88$$

$$\therefore \quad m = \frac{0.6 \times 10^{-4} \times 429.88}{0.32} = 0.0806 \text{ kg/s.}$$

Total mass of steam passing through 5 nozzles per second is given by

$$m_t = 0.0806 \times 5 = 0.403 \text{ kg/s}$$

∴ Power developed by the turbine

$$= \frac{m_t \times C_w C_{bl}}{1000} \text{ kW}$$

From velocity diagram,

$$C_w = 750 \text{ m/s (by measurement)}$$

$$\therefore \quad \textit{Power developed} = \frac{0.403 \times 750 \times 400}{1000}$$

$$= \mathbf{120.9 \text{ kW.}} \quad \textbf{(Ans.)}$$

☞ **Example 19.19.** *The first stage of an impulse turbine is compounded for velocity and has two rows of moving blades and one ring of fixed blades. The nozzle angle is 15° and the leaving angles of blades are respectively, first-moving 30°, fixed 20° ; second-moving 30°. The velocity of steam leaving the nozzle is 540 m/s. The friction loss in each blade row is 10% of the relative velocity. Steam leaves the second row of moving blades axially.*

Find : (*i*) *Blade velocity ;*

(*ii*) *Blade efficiency ;*

(*iii*) *Specific steam consumption.*

Solution. Refer to Fig. 19.29.

Nozzle angle, $\alpha = 15°$; $\alpha' = 20°$

$\beta' = 90°$

[since the steam leaves the blades *axially*]

$\phi = \phi' = 30°$

Velocity of steam leaving the nozzle, $C_1 = 540$ m/s

and $\dfrac{C_{r_0}}{C_{r_1}} = 0.9$

$$\frac{C_1'}{C_0} = 0.9 \text{ and } \frac{C_{r_0}'}{C_{r_1}'} = 0.9.$$

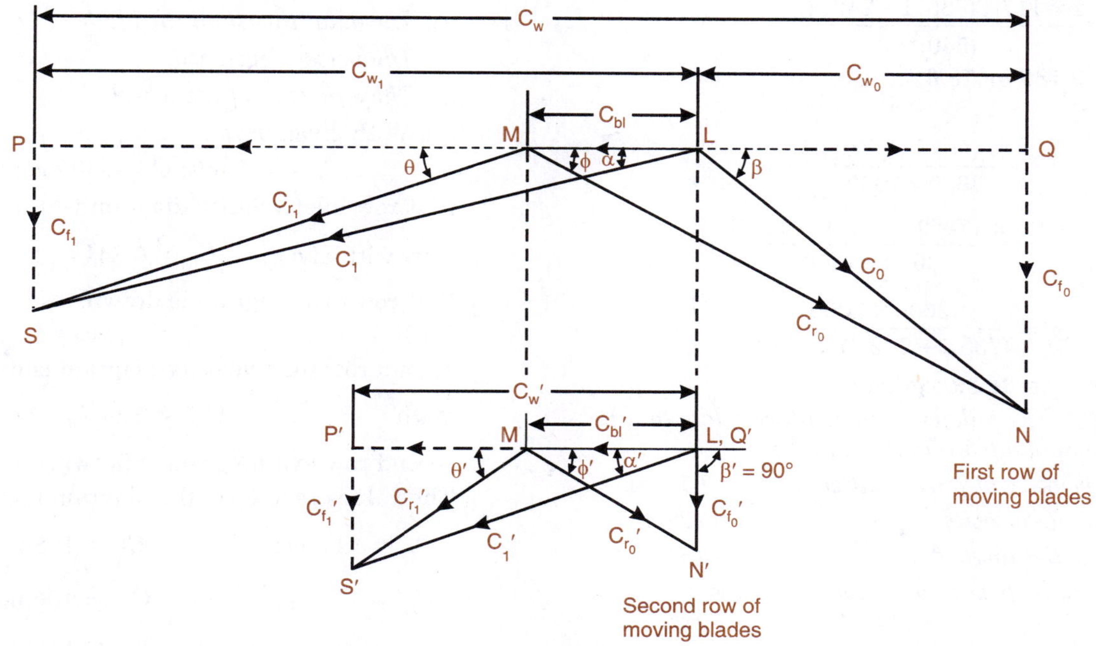

Fig. 19.29

Second row of moving blades :

The velocity triangles should be drawn *starting from the second row of moving blades.*

The procedure is as follows : Refer to Fig. 19.29.

- Draw LM to any convenient scale (say 3 cm) as C_{bl} is *not* known.
- Draw $\phi' = 30°$ and draw perpendicular through the point L (or Q') to LM as $\beta' = 90°$. This meets the line MN' at N'. This completes the outlet triangle LMN'.
- Measure $C_{r_0}' = MN' = 3.5$ cm

$$C_{r_1}' = \frac{C_{r_0}'}{0.9} = \frac{3.5}{0.9} = 3.9 \text{ cm.}$$

- Draw an $\angle\alpha' = 20°$ and draw an arc of radius 3.9 cm with centre at M to cut the line LS' at S'. Join MS'. This completes the inlet velocity $\Delta LMS'$. Measure $LS' = C_1' = 6.5$ cm.

First row of moving blades :

The following steps are involved in drawing velocity triangle for *first row of moving blades.*

- Draw $LM = C_{bl}$ (= 3 cm)

$$C_0 = \frac{C_1'}{0.9} = \frac{6.5}{0.9} = 7.22 \text{ cm.}$$

- Draw $\angle\phi = 30°$, through the point M to the line LM.
- Draw an arc of radius 7.22 cm with centre L. This arc cuts the line LN at point N. Join MN. This completes the outlet ΔLMN.
- Measure, $C_{r_0} = MN = 9.7$ cm

$$MS = C_{r_1} = \frac{C_{r_0}}{0.9} = \frac{9.7}{0.9} = 10.8 \text{ cm}$$

— Draw an $\angle\alpha = 15°$, through a point L. Draw an arc of radius of 10.8 cm with centre at M. This arc cuts the line LS at S. Join MS. This completes the inlet velocity triangle.

Measure LS from the velocity triangle

$$LS = 13.8 \text{ cm} = C_1 = 540 \text{ m/s.}$$

The scale is now calculated from the above.

$$\therefore \quad \text{Scale 1 cm} = \frac{540}{13.8} = 39.1 \text{ m/s}$$

(*i*) **Blade velocity, C_{bl} :**

Measure the following distances from the velocity diagram and convert into velocities :

$C_{bl} = LM = 3$ cm $= 3 \times 39.1 =$ **117.3 m/s. (Ans.)**

(*ii*) **Blade efficiency,** η_{bl}

$$C_w = PQ = 18.8 \text{ cm} = 18.8 \times 39.1 = 735.1 \text{ m/s}$$

$$C_w' = P'Q' = 6.2 \text{ cm} = 6.2 \times 39.1 = 242.4 \text{ m/s}$$

$$= \frac{2C_{bl}(C_w + C_w')}{C_1^2}$$

$$= \frac{2 \times 117.3(735.1 + 242.4)}{(540)^2}$$

$$= \mathbf{0.786 \text{ or } 78.6\%.} \quad \text{(Ans.)}$$

(*iii*) **Specific steam consumption,** m_s **:**

$$1 = \frac{m_s(C_w + C_w')C_{bl}}{3600 \times 1000}$$

$$= \frac{m_s(735.1 + 242.4) \times 117.3}{3600 \times 1000}$$

$$\therefore \quad m_s = \frac{3600 \times 1000}{(735.1 + 242.4) \times 117.3}$$

$$= \mathbf{31.39 \text{ kg/kWh.}} \quad \text{(Ans.)}$$

Example 19.20. *The following particulars relate to a two-row velocity compounded impulse wheel :*

Steam velocity at nozzle outlet = *650 m/s*
Mean blade velocity = *125 m/s*
Nozzle outlet angle = *16°*
Outlet angle, first row of moving blades = *18°*
Outlet angle, fixed guide blades = *22°*
Outlet angle, second row of moving blades = *36°*
Steam flow = *2.5 kg/s*

The ratio of the relative velocity at outlet to that at inlet is 0.84 for all blades.

Determine the following :

(*i*) *The axial thrust on the blades ;*
(*ii*) *The power developed,*
(*iii*) *The efficiency of the wheel.* **(AMIE)**

Solution. With given data,

C_{bl} = 125 m/s, C_1 = 650 m/s, α = 16°,

first row inlet velocity diagram is drawn.

Now with given, $C_{r_0} = 0.84\, C_{r_1}$, ϕ = 18°,

first row exit diagram is drawn.

With $C_1' = C_0$; α' = 22°,

second row inlet velocity diagram is drawn

With $C_{r_0}' = 0.84\, C_{r_1}'$; ϕ' = 36°,

second row exit diagram is drawn.

The values read form the diagram are as follows :

C_{f_1} = 180 m/s, C_{f_0} = 138 m/s,

C_{f_1}' = 122 m/s, C_{f_0}' = 107 m/s,

$C_{w_1} + C_{w_0}$ = 924 m/s, $C_{w_1}' + C_{w_0}'$ = 324 m/s.

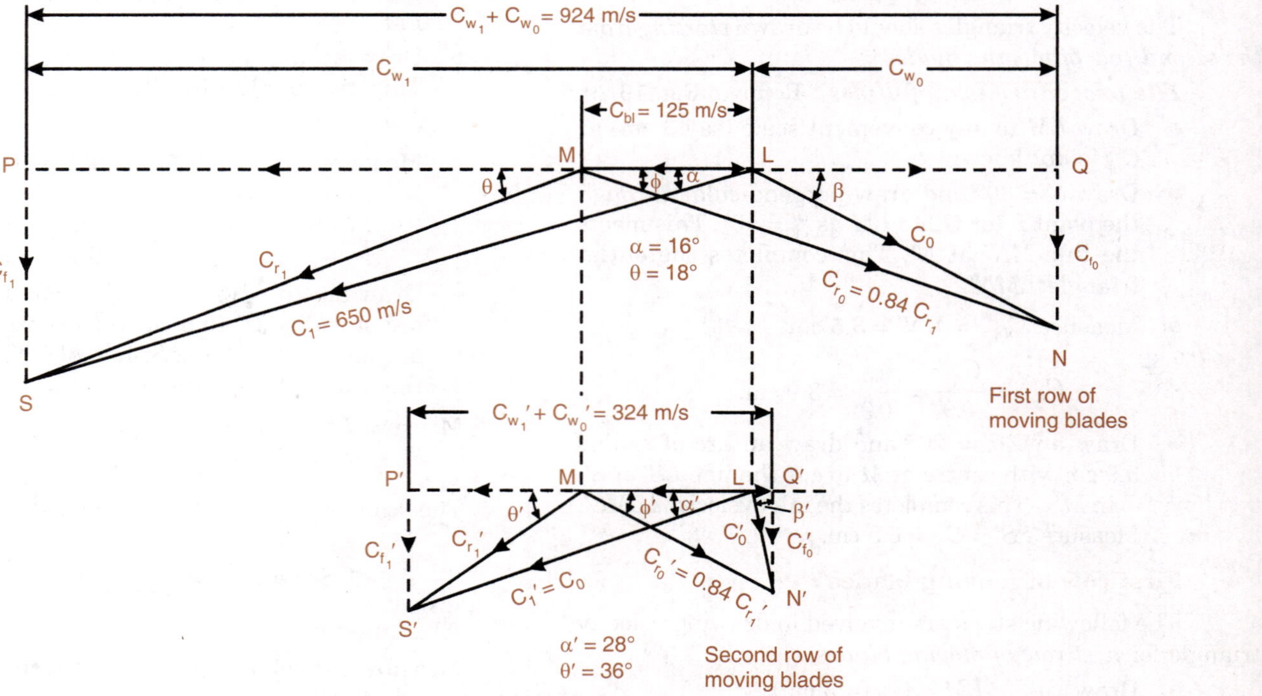

Fig. 19.30

(*i*) **Axial thrust on the blades**

$= \dot{m}_s[(C_{f_1} - C_{f_0}) + (C_{f_1}' - C_{f_0}')]$

$= 2.5\ [(180 - 138) + (122 - 107)]$

$=$ **142.5 N.** **(Ans.)**

(*ii*) **Power developed**

$$= \frac{\dot{m}_s[(C_{w_1} + C_{w_0}) + (C_{w_1}' + C_{w_0}')]C_{bl}}{1000}$$

$$= \frac{2.5(924 + 324) \times 125}{1000} = \textbf{390 kW.} \quad \textbf{(Ans.)}$$

(*iii*) **Efficiency**

$$= \frac{(924 + 324) \times 125}{C_1^2/2} = \frac{(924 + 324) \times 125}{650^2/2}$$

$=$ **0.738 or 73.8%.** **(Ans.)**

Example 19.21. *The first stage of an impulse turbine is compounded for velocity and has two rings of moving blades and one ring of fixed blades. The nozzle angle is 20° and the leaving angles of the blades are respectively as follows :*

First moving 20°, fixed 25° and second moving 30°. Velocity of steam leaving the nozzles is 600 m/sec and the steam velocity relative to the blade is reduced by 10% during the passage through each ring. Find the diagram efficiency and power developed for a steam flow of 4 kg per second. Blade speed may be taken as 125 m/sec. **(M.U.)**

Solution. $C_{bl} = 125$ m/s, $C_1 = 600$ m/s

$\alpha = 20°, \phi = 20°$

$\alpha' = 25°, \phi' = 30°$

$$K = \left(1 - \frac{10}{100}\right) = 0.9$$

$\dot{m}_s = 4$ kg/s

With these values velocity triangles can be drawn (Fig. 19.31).

From diagram (By measurement) :

$C_{w_1} = 565$ m/s, $\quad C_{w_0} = 285$ m/s

$C_{w_1}' = 260$ m/s, $\quad C_{w_0}' = 20$ m/s

Now, $\quad C_w = C_{w_1} + C_{w_0} = 565 + 285 = 850$ m/s

$C_w' = C_{w_1}' + C_{w_0}' = 260 + 20 = 280$ m/s

Power developed

$$= \frac{\dot{m}_s(C_w + C_w')C_{bl}}{1000} = \frac{4 \times (850 + 280) \times 125}{1000}$$

$=$ **565 kW.** **(Ans.)**

Diagram efficiency

$$= \frac{C_{bl}(C_w + C'_w)}{C_1^2/2} = \frac{2 \times 125(850 + 280)}{600^2}$$

$=$ **0.7847 or 78.47%.** **(Ans.)**

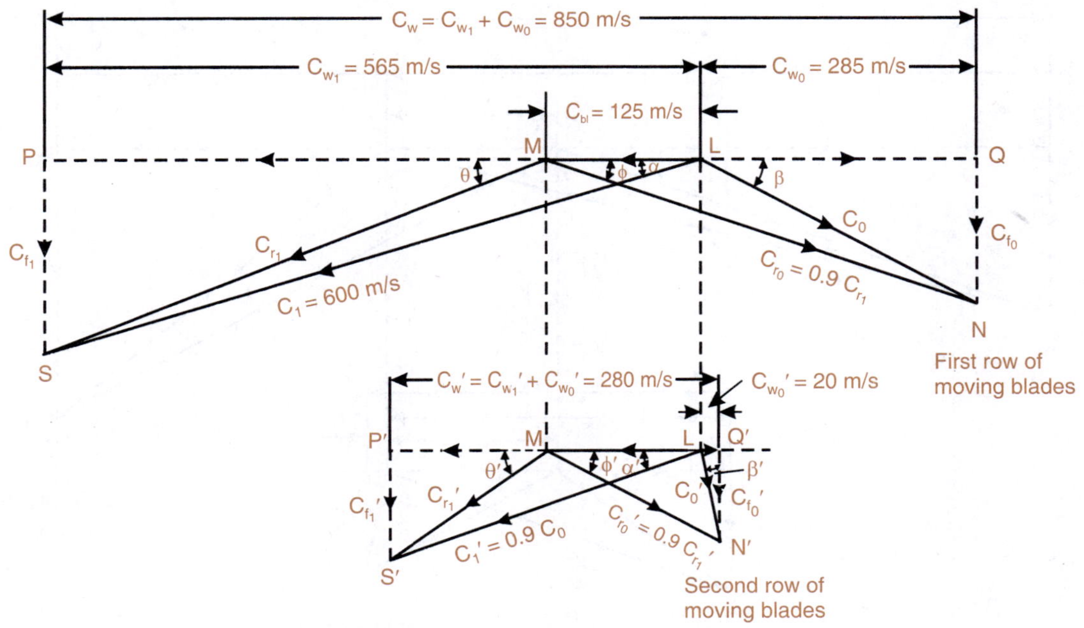

Fig. 19.31

Example 19.22. *The following data relate to a compound impulse turbine having two rows of moving blades and one row of fixed blades in between them.*

The velocity of steam leaving the nozzle	*= 600 m/s*
Blade speed	*= 125 m/s*
Nozzle angle	*= 20°*
First moving blade discharge angle	*= 20°*
First fixed blade discharge angle	*= 25°*
Second moving blade discharge angle	*= 30°*
Friction loss in each ring	*= 10% of relative velocity.*

Find : (i) Diagram efficiency ;

(ii) Power developed for a steam flow of 6 kg/s.

Solution. Refer to Fig. 19.32.

First row of moving blades :

To draw velocity triangles for first row of moving blades the following procedure may be followed :

Select a suitable scale.

- Draw LM = blade velocity (C_{bl}) = 125 m/s.
- Make $\angle MLS$ = nozzle angle, α = 20°.
- Draw LS = velocity of steam leaving the nozzle = 600 m/s.
- Join MS to complete the inlet triangle LMS.
- Make $\angle LMN$ = outlet angle of first moving blades = 20°.

and, cut $MN = 0.9\ MS$, since $K = 0.9$.

— Join LN to complete the outlet ΔLMN.

Second row of moving blades :

The velocity triangles for second row of moving blades may be drawn as follows :

- Draw LM = blade velocity (C_{bl}) = 125 m/s.
- Make $\angle MLS'$ = outlet angle of fixed blade = 25°

and cut $LS' = 0.9\ LN$. $(\because\ K = 0.9)$

- Join MS'. The inlet velocity triangle LMS' is completed.
- Make $\angle LMN'$ = outer angle of second moving blades = 30°

and cut $MN' = 0.9\ MS'$ $(\because\ K = 0.9)$

- Join LN'. The outlet velocity triangle is completed.

The following required data may now be scaled off from the diagram :

$$C_w = C_{w_1} + C_{w_0} = PQ = 845 \text{ m/s}$$

$$C_w' = P'Q' = 280 \text{ m/s}.$$

(*i*) Diagram efficiency,

$$\eta_{bl} = \frac{2C_{bl}(C_w + C_w')}{C_1^2}$$

$$= \frac{2 \times 125(845 + 280)}{(600)^2}$$

$$= \mathbf{0.781 \text{ or } 78.1\%.} \quad \textbf{(Ans.)}$$

(*ii*) Power developed,

$$P = \frac{\dot{m}_s(C_w + C_w')}{1000} C_{bl}$$

$$= \frac{6(845 + 280) \times 125}{1000}$$

$$= \mathbf{843.75 \text{ kW.}} \quad \textbf{(Ans.)}$$

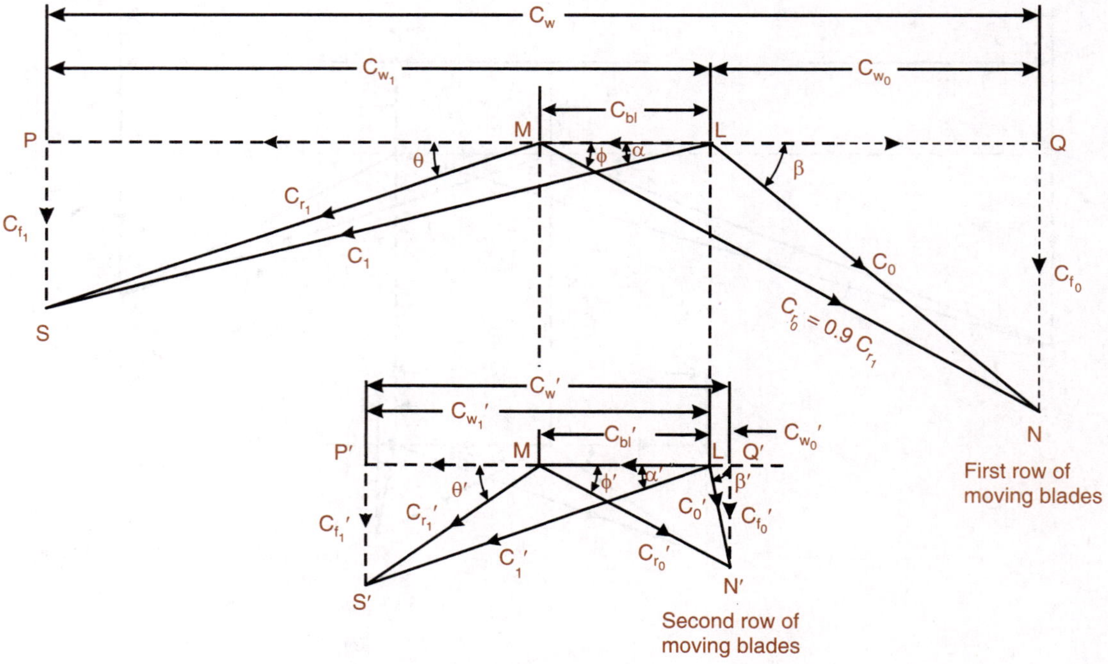

Fig. 19.32

Example 19.23. *The first stage of a turbine is a two-row velocity compounded impulse wheel. The steam velocity at inlet is 600 m/s and the mean blade velocity is 120 m/s. The nozzle angle is 16° and the exit angles for the first-row of moving blades, the fixed blades, and the second row of moving blades are 18°, 21° and 35° respectively.*

(i) Calculate the blade inlet angles for each row.

(ii) Calculate also for each row of moving blades, the driving force and the axial thrust on the wheel for a mass flow of 1 kg/s.

(iii) Calculate the diagram efficiency for the wheel and the diagram power per kg/s steam flow.

(iv) What would be the maximum possible diagram efficiency for the given steam inlet velocity and nozzle angle ?

Take the blade velocity co-efficient as 0.9 for all blades.

Solution. Refer to Fig. 19.33.

$\alpha = 16°, \phi = 18°, C_1 = 600$ m/s, $C_{bl} = 120$ m/s

$\alpha' = 21°, \phi' = 35°, \dot{m}_s = 1$ kg/s,

Blade velocity co-efficient, $K = 0.9$

With the above data velocity triangles can be drawn.

From the diagram (by measurement)

$C_w = C_{w_1} + C_{w_0} = 875$ m/s ;

$C_w' = C_{w_1}' + C_{w_0}' = 294$ m/s

$C_{f_1} = 168$ m/s, $C_{f_0} = 135$ m/s ;

$C_{f_1}' = 106$ m/s, $C_{f_0}' = 97$ m/s.

(i) Blade inlet angles :

First row : $\theta = 20°$ (moving blade)

$\beta = 24.5°$ (fixed blade)

Second row : $\theta' = 34.5°$ (moving blade).

(ii) Driving force :

First row of moving blades

$= \dot{m}_s(C_{w_1} + C_{w_0}) = \dot{m}_s C_w = 1 \times 875 = \mathbf{875\ N}$. **(Ans.)**

Second row of moving blades

$= \dot{m}_s(C_{w_1}' + C_{w_0}') = \dot{m}_s C_w' = 1 \times 294 = \mathbf{294\ N}$. **(Ans.)**

Axial thrust :

First row of moving blades

$= \dot{m}_s(C_{f_1}' + C_{f_0}') = 1 \times (168 - 135) = 33$ N

Second row of moving blades

$= \dot{m}_s(C_{f_1}' - C_{f_0}') = 1 \times (106 - 97) = 9$ N

Total axial thrust

$= 33 + 9 = \mathbf{42\ N\ per\ kg/s}$. **(Ans.)**

(iii) Power developed

$$= \frac{\dot{m}_s(C_w + C_w')\,C_{bl}}{1000}$$

$$= \frac{1 \times (875 + 294) \times 120}{1000}$$

$= \mathbf{140.28\ kW\ per\ kg/s}$. **(Ans.)**

Diagram efficiency

$$= \frac{2C_{bl}(C_w + C_w')}{C_1^2} = \frac{2 \times 120\,(875 + 294)}{(600)^2}$$

$= \mathbf{0.7793\ or\ 77.93\%}$. **(Ans.)**

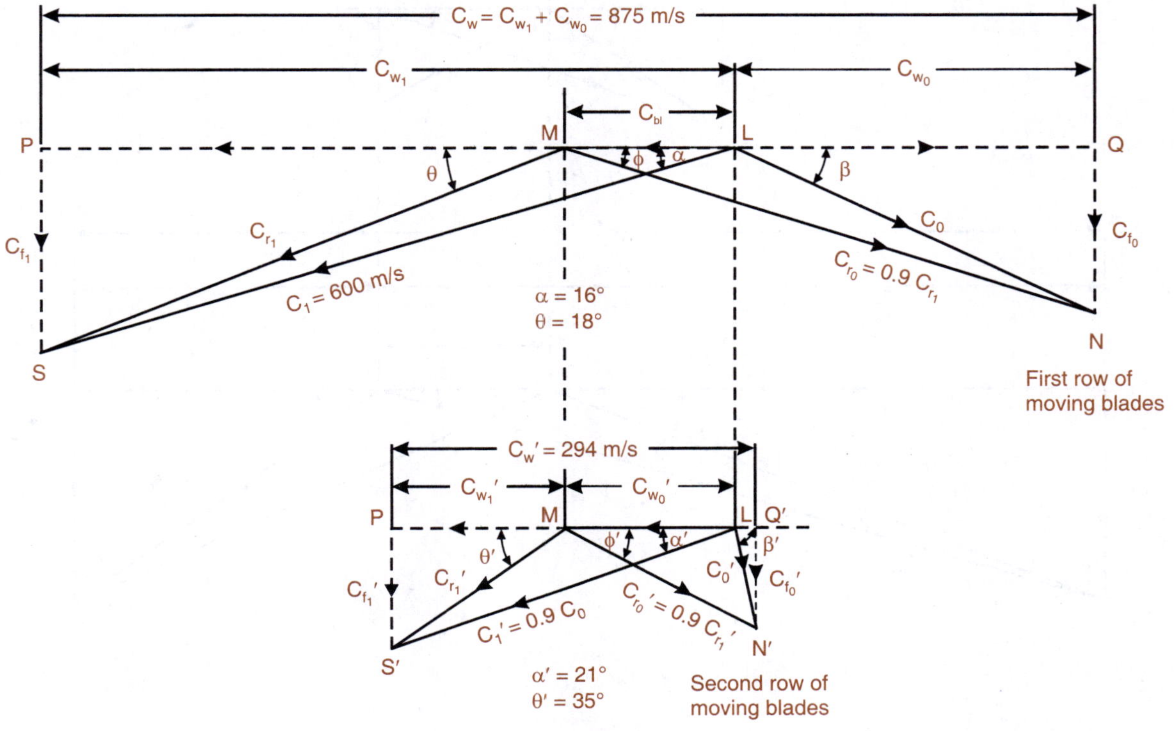

Fig. 19.33

(*iv*) **Maximum diagram efficiency**

$= \cos^2 \alpha = \cos^2 16° =$ **0.924 or 92.4%.** **(Ans.)**

Example 19.24. *An impulse stage of a turbine has two rows of moving blades separated by fixed blades. The steam leaves the nozzles at an angle of 20° with the direction of motion of the blades. The blade exit angles are : 1st moving 30° ; fixed 22° ; 2nd moving 30°.*

If the adiabatic heat drop for the nozzle is 186.2 kJ/kg and the nozzle efficiency 90%, find the blade speed necessary if the final velocity of the steam is to be axial. Assume a loss of 15% in relative velocity for all blade passages. Final also the blade efficiency and the stage efficiency. **(P.U.)**

Solution. Steam velocity, $C_1 = 44.72\sqrt{\eta_n h_d}$

$$= 44.72\sqrt{0.9 \times 186.2} = 579 \text{ m/s}.$$

The velocity diagram *for axial discharge turbine is drawn in reverse direction* (see Fig. 19.34).

- The blade velocity (C_{bl}) *LM* is drawn to any convenient scale.
- As discharge is axial *LN′* is drawn perpendicular to *LM*.
- Knowing the outlet angle of the second moving ring ($\phi' = 30°$), *N′* is located.

MN′ represents relative velocity at outlet (C_{r_0}').

- Relative velocity at inlet to the second moving blade is

$$C_{r_1}' = MS' = \frac{MN'}{0.85}\left(= \frac{C_{r_0}'}{K}\right).$$

- The triangle at inlet to the second moving blades ring *LMS′* is obtained by drawing the discharge angle ∠*MLS′* ($\alpha' = 22°$), where *LS′* (C_1') is the exit velocity of the second blade ring.
- Now, $C_0 = LN = \dfrac{C_1'}{0.85} = \dfrac{LS'}{0.85}$ assuming a loss of 15% (given).
- With compass at centre *L* and radius *LN* arc is drawn and the velocity triangle at the exit of the first moving blade ring *LMN* is completed, knowing the exit angle of the first moving blade ring ($\phi = 30°$).
- With $\alpha = 20°$ and $C_{r_1} = MS = \dfrac{C_{r_0}}{0.85} = \dfrac{MN}{0.85}$ the inlet velocity triangle is completed.

Now *LS* is the absolute velocity from the nozzle. Since this velocity is known, scale can be calculated. It is,

$$= \frac{\text{Absolute velocity at exit from the nozzle}}{\text{Length } LS}$$

and then the blade velocity, etc., can be calculated.

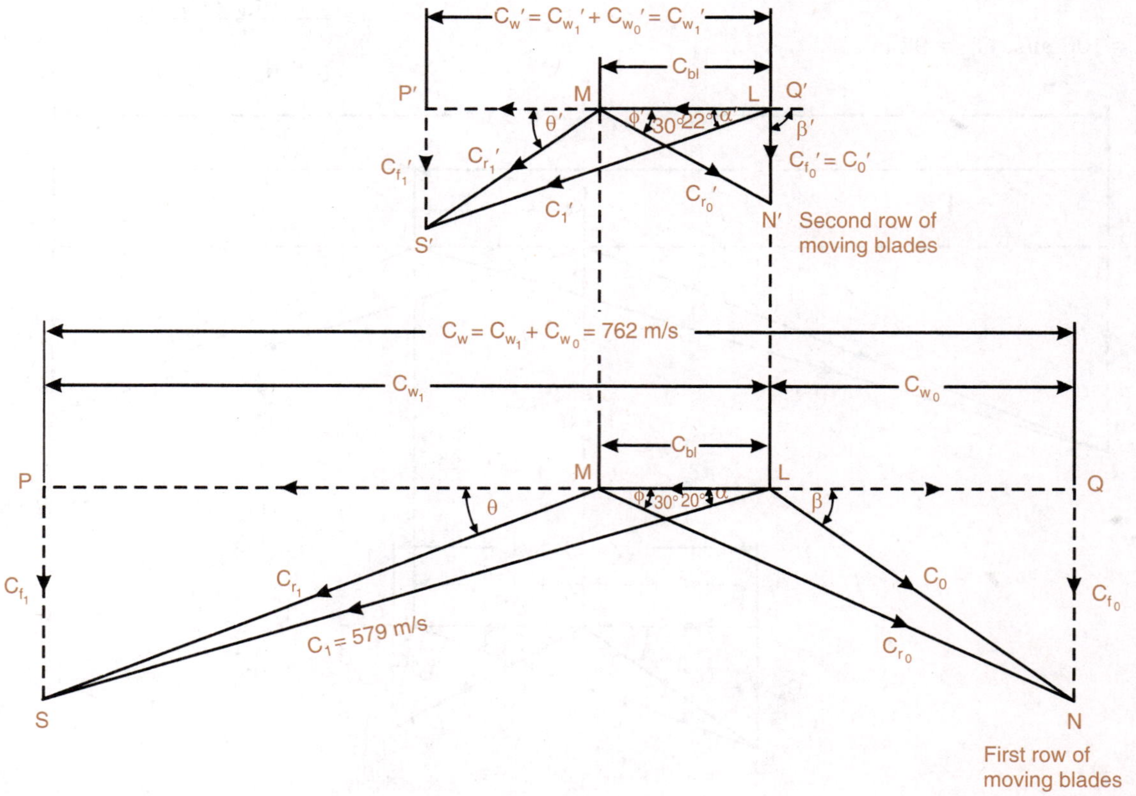

Fig. 19.34

From the diagram :

C_{bl} (= LM) = 117 m/s

$C_{w_1} + C_{w_0}$ = 762 m/s

$C_{w_1}' + C_{w_0}' = C_{w_1}'$ = 234 m/s $\quad (\because \; C_{w_0}' = 0)$

Blade efficiency,

$$\eta_{bl} = \frac{(C_w + C_w') \times C_{bl}}{C_1^2/2} = \frac{(762 + 234) \times 117}{579^2/2}$$

$$= \mathbf{0.6952 \text{ or } 69.52\%.} \quad \textbf{(Ans.)}$$

Stage efficiency,

$$\eta_{stage} = \frac{(C_w + C_w') \times C_{bl}}{h_d} = \frac{(762 + 234) \times 117}{186.2 \times 1000}$$

$$= \mathbf{0.626 \text{ or } 62.6\%.} \quad \textbf{(Ans.)}$$

19.8. REACTION TURBINES

The reaction turbines which are used these days are really **impulse-reaction** turbine. Pure reaction turbines are *not* in general use. The expansion of steam and heat drop occur both in fixed and moving blades.

19.8.1. Velocity Diagram for Reaction Turbine Blade

Fig. 19.35 shows the velocity diagram for reaction turbine blade. In case of an impulse turbine blade the relative velocity of steam either remains constant as the steam glides over the blades or is reduced slightly due to friction. In reaction turbine blades, the steam continuously expands at it flows over the blades. *The effect of the continuous expansion of steam during the flow over the blade is to increase the relative velocity of steam.*

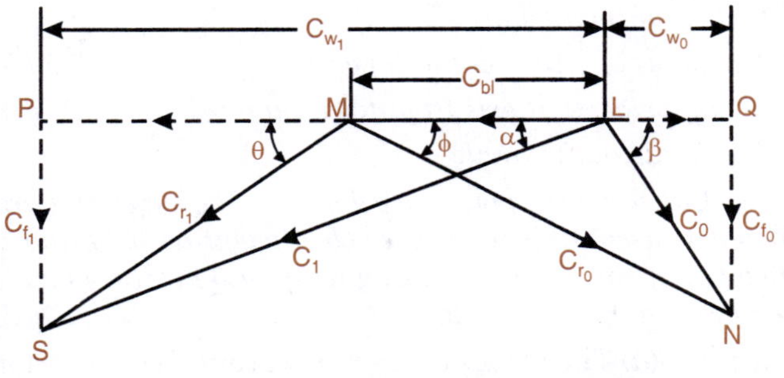

Fig. 19.35. Velocity diagram for reaction turbine blade.

$\therefore \quad C_{r_0} > C_{r_1}$ for *reaction turbine blade.*

$(C_{r_0} \le C_{r_1}$ for *impulse turbine blade*).

19.8.2. Degree of Reaction (R_d)

The ***degree of reaction*** of reaction turbine stage is defined as the *ratio of heat drop over moving blades to the total heat drop in the stage.*

Thus the degree of reaction of reaction turbine is given by,

$$R_d = \frac{\text{Heat drop in moving blades}}{\text{Heat drop in the stage}}$$

$$= \frac{\Delta h_m}{\Delta h_f + \Delta h_m} \text{ as shown in Fig. 19.35.}$$

The heat drop in moving blades is equal to increase in relative velocity of steam passing through the blade.

$$\therefore \quad \Delta h_m = \frac{C_{r_0}^2 - C_{r_1}^2}{2}$$

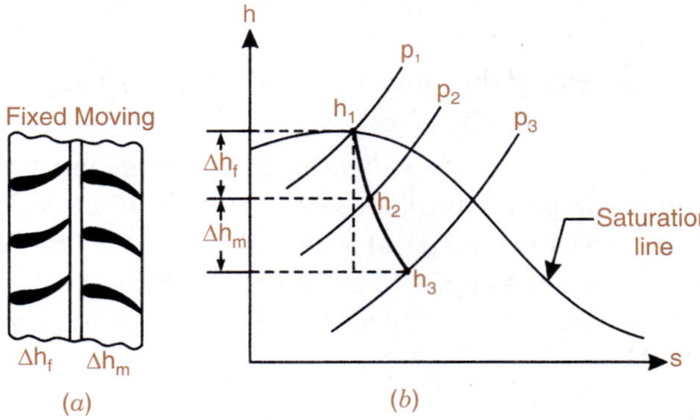

Fig. 19.36

The total heat drop in the stage ($\Delta h_f + \Delta h_m$) is equal to the work done by the steam in the stage and it is given by

$$\Delta h_f + \Delta h_m = C_{bl}\,(C_{w_1} + C_{w_0})$$

$$\therefore \quad R_d = \frac{C_{r_0}^2 - C_{r_1}^2}{2C_{bl}(C_{w_1} + C_{w_0})} \quad \text{...(19.22)}$$

Referring to Fig. 19.36,

$$C_{r_0} = C_{f_0} \operatorname{cosec} \phi \text{ and } C_{r_1} = C_{f_1} \operatorname{cosec} \theta$$

and $\quad (C_{w_1} + C_{w_0}) = C_{f_1} \cot \theta + C_{f_0} \cot \phi$

The velocity of flow generally remains constant through the blades.

$$\therefore \quad C_{f_1} = C_{f_0} = C_f .$$

Substituting the values of C_{r_1}, C_{r_0} and $(C_{w_1} + C_{w_0})$ in eqn. (19.22), we get

$$R_d = \frac{C_f^2 (\operatorname{cosec}^2 \phi - \operatorname{cosec}^2 \theta)}{2\, C_{bl}\, C_f\, (\cot \theta + \cot \phi)}$$

$$= \frac{C_f}{2C_{bl}} \left[\frac{(\cot^2 \phi + 1) - (\cot^2 \theta + 1)}{\cot \theta + \cot \phi} \right]$$

$$= \frac{C_f}{2C_{bl}} \left[\frac{\cot^2 \phi - \cot^2 \theta}{\cot \phi + \cot \theta} \right]$$

$$= \frac{C_f}{2C_{bl}} (\cot \phi - \cot \theta) \quad \text{...(19.23)}$$

If the turbine is designed for 50% reaction ($\Delta h_f = \Delta h_m$), then the eqn. (19.23) can be written as:

$$\frac{1}{2} = \frac{C_f}{2C_{bl}}(\cot\phi - \cot\theta)$$

$$\therefore \quad C_{bl} = C_f(\cot\phi - \cot\theta) \quad ...(19.24)$$

Also C_{bl} can be written as:

$$C_{bl} = C_f(\cot\phi - \cot\beta) \quad ...(19.25)$$

and $$C_{bl} = C_f(\cot\alpha - \cot\theta) \quad ...(19.26)$$

$C_{f_1} = C_{f_0} = C_f$ is assumed in writing the above equations.

Comparing the eqns. (19.24), (19.25), (19.26)

$$\theta = \beta \text{ and } \phi = \alpha$$

which *means that moving blade and fixed blade* must have the same shape if the degree of reaction is 50%. This condition gives symmetrical velocity diagrams. This type of turbine is known as "**Parson's reaction turbine**". Velocity diagram for the blades of this turbine is given in Fig. 19.37.

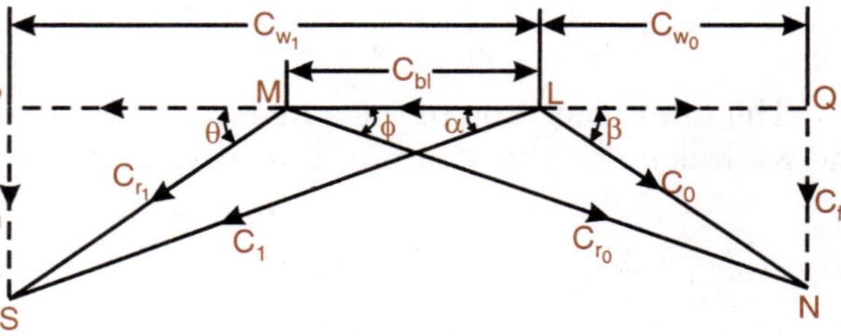

Fig. 19.37

Example 19.25. *Define the term 'degree of reaction' as applied to a steam turbine. Show that for Parson's reaction turbine the degree of reaction is 50%.* **(AMIE)**

Solution. Refer to Fig. 19.38.

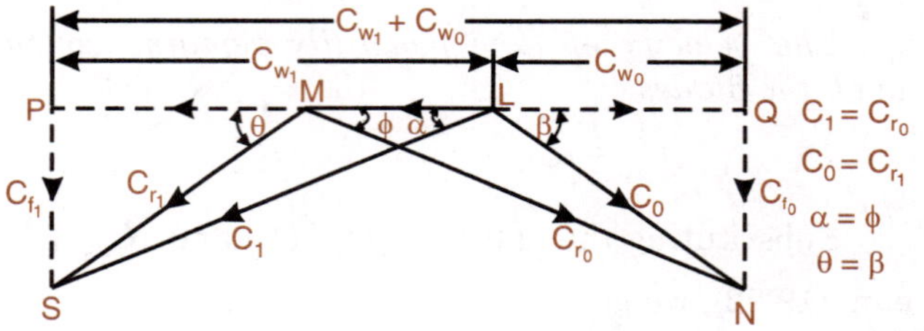

Fig. 19.38

The pressure drop in reaction turbines takes place in both fixed and moving blades. The division generally is given in terms of enthalpy drops. The criterion used is the *degree of reaction*. It is defined as :

$$\frac{\text{Enthalpy drop in rotor blades}}{\text{Total enthalpy drop in stage}} = \frac{\Delta h_m}{\Delta h_f + \Delta h_m}$$

(Refer to Fig. 19.36)

A special case is when the degree of reaction is zero; it means no heat drop in the moving blades. This becomes a case of impulse stage. Other common case is of Parson's turbine which has the same reason for both the fixed and moving blades. The blades are *symmetrical, i.e.,* the exit angle of moving blade is equal to the exit angle of the fixed blade and the inlet angle of the moving blade is equal to the inlet angle of the fixed blade. Since the blades are symmetrical the velocity diagram is also symmetrical. In such a case the degree of reaction is 50%.

Applying the steady flow energy equation to the fixed blades and assuming that the velocity of steam entering the fixed blade is equal to the absolute velocity of steam leaving the previous moving row, we have

$$\Delta h_f = \frac{C_1^2 - C_0^2}{2}$$

Similarly, for the moving blades

$$\Delta h_m = \frac{C_{r_0}^2 - C_{r_1}^2}{2}$$

But $C_1 = C_{r_0}$ and $C_0 = C_{r_1}$

$$\therefore \quad \Delta h_f = \Delta h_m$$

Hence degree of reaction $= \dfrac{\Delta h_m}{\Delta h_m + \Delta h_m} = \dfrac{1}{2}$

This is a proof that **Parson's reaction turbine is a 50% reaction turbine.**

Example 19.26. (*a*) *Explain the functions of the blading of a reaction turbine.*

(*b*) *A certain stage of a Parson's turbine consists of one row of fixed blades and one row of moving blades. The details of the turbine are as below :*

The mean diameter of the blades	*= 68 cm*
R.P.M. of the turbine	*= 3,000*
The mass of steam passing per sec	*= 13.5 kg*
Steam velocity at exit from fixed blades	*= 143.7 m/s*
The blade outlet angle	*= 20°*

Calculate the power developed in the stage and gross efficiency, assuming carry over co-efficient as 0.74 and the efficiency of conversion of heat energy into kinetic energy in the blade channel as 0.92. **(M.U.)**

Solution. (*a*) The blades of reaction turbine has to perform two functions :

1. They change the direction of motion of steam causing change of momentum, responsible for motive force.

2. The blades also act as nozzles causing pressure drop as steam moves in the blade passage.

(*b*) $D = 0.68$ m, $N = 3000$ r.p.m., $\dot{m} = 13.5$ kg/s

$C_{r_0} = 143.7$ m/s, $\phi = 20°$, $\psi = 0.74$, $\eta = 0.92$

Blade velocity, $$C_{bl} = \frac{\pi DN}{60} = \frac{\pi \times 0.68 \times 3000}{60} = 106.8 \text{ m/s}$$

$$C_{w_1} + C_{w_0} = 165 \text{ m/s} \quad \text{(Fig. 19.39)}$$

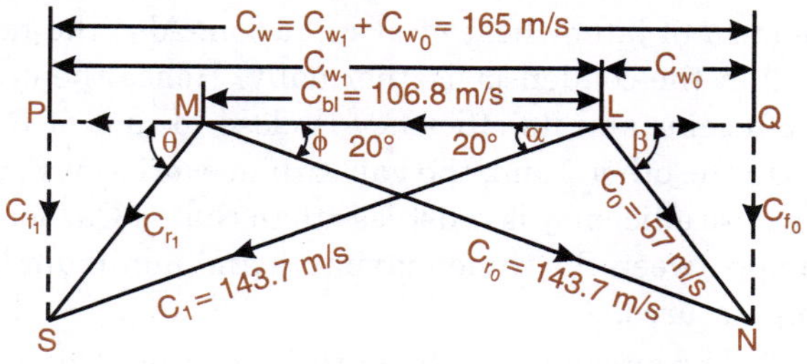

Fig. 19.39

Power developed

$$\dot{m}_s \times \frac{(C_{w_1} + C_{w_0})\, C_{bl}}{1000} = 13.5 \times \frac{165 \times 106.8}{1000}$$

$$= \mathbf{237.89\ kW.} \quad \textbf{(Ans.)}$$

In Parson turbine blades are symmetrical, *i.e.*,

$$\alpha = \phi,\ \theta = \beta$$

$$C_1 = C_{r_0},\ C_{r_1} = C_0$$

Enthalpy drop

$$= 2 \times \frac{C_1^{\,2} - \psi C_0^{\,2}}{2\eta} = 2 \times \left(\frac{143.7^2 - 0.74 \times 57^2}{2 \times 0.92 \times 1000}\right)$$

$= 19.83$ kJ/kg

∴ **Gross efficiency**

$$= \frac{\text{Work done /kg}}{\text{Enthalpy drop /kg}} = \frac{(C_{w_1} + C_{w_0})\, C_{bl}}{19.83 \times 1000}$$

$$= \frac{165 \times 106.8}{19.83 \times 1000} = \mathbf{0.888\ or\ 88.8\%.} \quad \textbf{(Ans.)}$$

Example 19.27. (*a*) *Discuss the factors that influence the erosion of turbine blades. On a sketch mark the portions of the blades more likely to be eroded. Sketch the methods used to prevent erosion of steam turbines blades.*

(*b*) *A reaction turbine running at 360 r.p.m. consumes 5 kg of steam per second. Tip leakage is 10%. Discharge blade tip angle for both moving and fixed blades is 20°. Axial velocity of flow is 0.75 times blade velocity. The power developed by a certain pair is 4.8 kW where the pressure is 2 bar and dryness fraction is 0.95. Find the drum diameter and blades height.* **(U.P.S.C.)**

Solution. (*a*) In the *high pressure* and *intermediate pressure stages* of turbine the pressures and temperatures are high and the blade material should be such that it stands high pressures and temperatures. In the intermediate pressure stages steam is wet therefore, the material should be able to *withstand both corrosion and erosion* due to the presence of water particles. In addition to corrosion and erosion the blades are *also subjected to high centrifugal stresses* as the low pressure stages are longer, therefore, the blade material and its *design* should be such that *it stands corrosion, erosion* and *high centrifugal stresses*.

When the speed is high and moisture exceeds 10 per cent the effect of moisture is most prominent. The most effected portion is the back of the inlet edge of the blade, where either grooves are formed or even some portion breaks away. Due to centrifugal force the water particles tend to concentrate in the outer annulus and their tip speed is greater than the root speed, hence erosion effect is most on tips (Fig. 19.40).

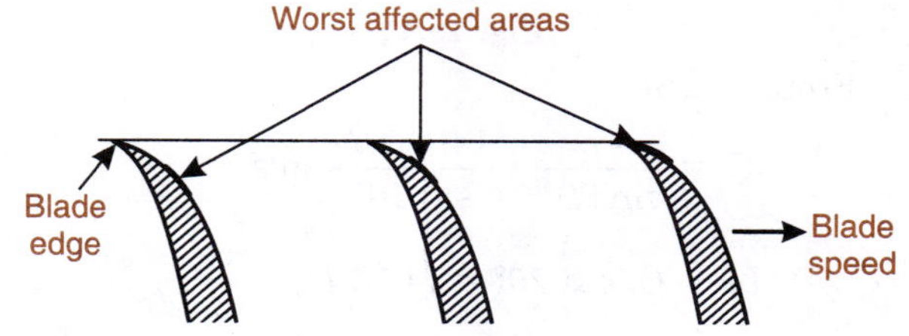

Fig. 19.40

Methods adopted to prevent erosion :

(*i*) By raising the temperature of steam at inlet, so that at exit of turbine the wetness does not exceed 10 per cent.

(*ii*) By adopting reheat cycle ; so that the wetness at exit remains in limit.

(*iii*) Drainage belts are provided on the turbine, so that the water droplets which are on outer periphery, due to centrifugal force are drained. The drained amount is about 25 per cent of total water particles present.

(*iv*) The leading edge of the turbine is provided with a shield of hard material.

— In the method (*i*) difficulties are the limits of temperature a material can withstand.

— Reheat cycle [method (*ii*)] has its own advantages and disadvantages.

— Drainage belts [method (*iii*)] cause structural changes in the turbine casing design, however, bleeding may help.

The most satisfactory solution is *providing tungsten shield. This prolongs the blade life, however, it does not remove the resistance which the water droplets impose on the rotation of the rotor.*

(*b*) Speed,

$$N = 360 \text{ r.p.m.}$$

$$\dot{m}_s = 5\left[1 - \frac{10}{100}(\text{tip leakage})\right] = 4.5 \text{ kg/s}$$

$$\alpha = \phi = 20°,\ C_f = 0.75\ C_{bl}$$

Power developed in a certain pair

$= 4.8$ kW at 2 bar ($x = 0.95$)

Blade velocity,

$$C_{bl}\ (LM) = \frac{\pi DN}{60} = \frac{\pi \times D \times 360}{60} = 18.85\ D \text{ m/s.}$$

$$C_{f_1} = 0.75 \times 18.85\ D = 14.138\ D \text{ m/s}$$

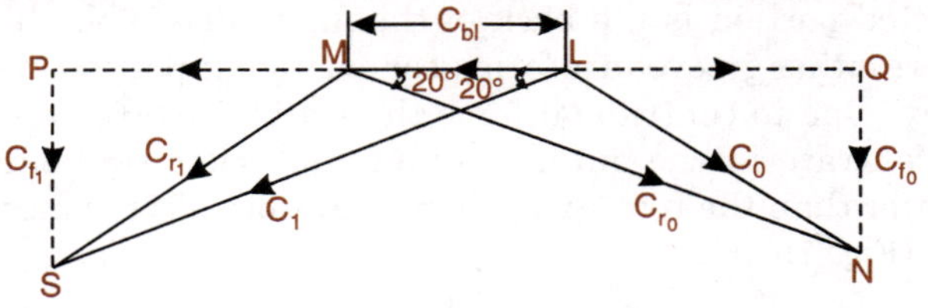

Fig. 19.41

From, Δ *LSP*,

$$C_1 = \frac{C_{f_1}}{\sin 20°} = \frac{14.138\,D}{\sin 20°}\ \text{m/s}$$

and
$$LP = C_1 \cos 20° = 14.13\,D \times \frac{\cos 20°}{\sin 20°}$$
$$= 14.138\,D \cot 20°$$
$$PQ = 2LP - LM$$
$$= (2 \times 14.138\,D \cot 20° - 18.85\,D)\ \text{m/s}$$

$$\text{Power developed} = \frac{\dot{m}_s \times PQ \times LM}{1000}$$

$$4.8 = \frac{4.5 \times (2 \times 14.138\,D \cot 20° - 18.85\,D) \times 18.85\,D}{1000}$$

Solving this equation, we get

Drum diameter, $D =$ **0.98 m. (Ans.)**

From steam tables at 2 bar, $v_g = 0.885\ \text{m}^3/\text{kg}$

$$\text{Mass flow rate} = \frac{\text{Flow area} \times \text{flow velocity}}{\text{Specific volume}}$$

$$4.5 = \frac{\pi D h \times C_{f_1}}{x v_g}$$

$$4.5 = \frac{\pi \times 0.98 \times h \times (14.138 \times 0.98)}{0.95 \times 0.885}$$

$$h = \frac{4.5 \times 0.95 \times 0.885}{\pi \times 0.98 \times 14.138 \times 0.98} = 0.0887\ \text{m}$$

$\therefore$ Blade height = **0.0887 m. (Ans.)**

Example 19.28. *(a) List the advantages of steam turbines over gas turbines.*

(b) Determine the isentropic enthalpy drop in the stage of Parson's reaction turbine which has the following particulars :

Speed = 1500 rpm ; mean diameter of rotor = 1 m ;
stage efficiency = 80% ; speed ratio = 0.7 ;
blade outlet angle = 20°. **(B.U.)**

Solution. (*a*) **Advantages of steam turbines over gas turbines :**

1. The load control in steam turbines is easy simply by throttle governing or cut-off governing. In gas turbines the air-fuel ratio becomes too high, 100 to 150 at part loads. This causes problems to sustain the flame.

2. The steam turbine works on Rankine cycle. In this cycle most of the heat is supplied at constant temperature in the form of latent heat of evaporation. Also the heat is rejected in the condenser isothermally. Hence the cycle is more efficient, and its efficiency is close to that of Carnot cycle. On the other hand, the gas turbine works on Brayton cycle whose efficiency is must less than that of Carnot cycle working between the same maximum and minimum limits of temperatures.

3. The efficiency of steam turbine at part load is not very much reduced. In gas turbines the maximum cycle temperature decreases considerably at part load ; therefore its part load efficiency is considerably low.

4. The blade material for steam turbines is cheap. For gas-turbines the blade material is costly as it is required to sustain considerably high temperatures.

(*b*) For Parson's reaction turbine, the velocity triangles are symmetrical, as shown in Fig. 19.37.

Given : $N = 1500$ r.p.m., $D = 1$ m, $\eta_{stage} = 80\%$;
Speed ratio,

$$\frac{C_{bl}}{C_1} = 0.7,\ \phi = \alpha = 20°$$

$$C_1 = C_{r_0} \text{ and } C_{r_1} = C_0$$

$$C_{bl} = \frac{\pi DN}{60} = \frac{\pi \times 1 \times 1500}{60} = 78.54\ \text{m/s}$$

$$\text{Speed ratio} = 0.7 = \frac{C_{bl}}{C_1}$$

$$\therefore \quad C_1 = \frac{78.54}{0.7} = 112.2\ \text{m/s}$$

$$C_{r_1}^{\,2} = C_1^{\,2} + C_{bl}^{\,2} - 2C_1 C_{bl} \cos \alpha$$
$$= (112.2)^2 + (78.54)^2 - 2 \times 112.2 \times 78.54 \cos 20°$$
$$= 2195.84$$

or $\quad C_{r_1} = 46.86$ m/s

Δh = Actual enthalpy drop for the stage

$$= \tfrac{1}{2}[(C_1^{\,2} - C_0^{\,2}) + (C_{r_0}^{\,2} - C_{r_1}^{\,2})]$$
$$= \tfrac{1}{2}[(C_1^{\,2} - C_{r_1}^{\,2}) + (C_1^{\,2} - C_{r_1}^{\,2})] = C_1^{\,2} - C_{r_1}^{\,2}$$
$$(\because\ C_0 = C_{r_1}\ ;\ C_{r_0} = C_1)$$

or $\quad \Delta h = [(112.2)^2 - (46.84)^2] \times 1/1000$ kJ/kg
$= 10.39$ kJ/kg

$$\text{Isentropic enthalpy drop, } (\Delta h') = \frac{(\Delta h)}{\eta_{stage}} = \frac{10.39}{0.8}$$

= **12.99 kJ/kg. (Ans.)**

19.8.3. Condition for Maximum Efficiency

The condition for *maximum efficiency* is derived by making the following *assumptions* :

(*i*) The degree of reaction is 50%.

(*ii*) The moving and fixed blades are symmetrical.

(iii) The velocity of steam at exit from the preceding stage is same as velocity of steam at the entrance to the succeeding stage.

Refer to Fig. 19.37 (velocity diagram for reaction blade).

Work done per kg of steam,

$$W = C_{bl}\,(C_{w_1} + C_{w_0}) = C_{bl}[C_1 \cos\alpha + (C_{r_0} \cos\phi - C_{bl})]$$

as $\phi = \alpha$ and $C_{r_0} = C_{r_1}$ as per the assumptions

$$\therefore \quad W = C_{bl}\,[2C_1 \cos\alpha - C_{bl}]$$

or

$$W = C_1^2\left[\frac{2C_{bl}\,C_1 \cos\alpha}{C_1^2} - \frac{C_{bl}^2}{C_1^2}\right]$$

$$= C_1^2\,[2\rho\,.\cos\alpha - \rho^2] \qquad ...(19.27)$$

where $\rho = \dfrac{C_{bl}}{C_1}$.

The K.E. supplied to the fixed blade $= \dfrac{C_1^2}{2g}$.

The K.E. supplied to the moving blade $= \dfrac{C_{r_0}^2 - C_{r_1}^2}{2}$.

$\therefore$ Total energy supplied to the stage,

$$\Delta h = \frac{C_1^2}{2} + \frac{C_{r_0}^2 - C_{r_1}^2}{2}$$

as $C_{r_0} = C_1$ for symmetrical triangles.

$$\therefore \quad \Delta h = \frac{C_1^2}{2} + \frac{C_1^2 - C_{r_1}^2}{2}$$

$$= C_1^2 - \frac{C_{r_1}^2}{2} \qquad ...(19.28)$$

Considering the ΔLMS (Fig. 19.35)

$$C_{r_1}^2 = C_1^2 + C_{bl}^2 - 2C_1\,.\,C_{bl}\,.\cos\alpha$$

Substituting this value of $C_{r_1}^2$ is eqn. (19.35), we have

Total energy supplied to the stage

$$\Delta h = C_1^2 - (C_1^2 + C_{bl}^2 - 2C_1\,.\,C_{bl}\,.\cos\alpha)/2$$

$$= (C_1^2 + 2C_1\,C_{bl} \cos\alpha - C_{bl}^2)/2$$

$$= \frac{C_1^2}{2}\left[1 + \frac{2C_{bl}}{C_1}.\cos\alpha - \left(\frac{C_{bl}}{C_1}\right)^2\right]$$

$$= \frac{C_1^2}{2}\left[1 + 2\rho \cos\alpha - \rho^2\right] \qquad ...(19.29)$$

The blade efficiency of the reaction turbine is given by,

$$\eta_{bl} = \frac{W}{\Delta h}$$

Substituting the value of W and Δh from eqns. (19.27) and (19.29), we get

$$\eta_{bl} = \frac{C_1^2\,[2\rho \cos\alpha - \rho^2]}{\dfrac{C_1^2}{2}(1 + 2\rho \cos\alpha - \rho^2)}$$

$$= \frac{2(2\rho \cos\alpha - \rho^2)}{(1 + 2\rho \cos\alpha - \rho^2)} = \frac{2\rho(2 \cos\alpha - \rho)}{(1 + 2\rho \cos\alpha - \rho^2)}$$

$$= \frac{2(1 + 2\rho \cos\alpha - \rho^2) - 2}{(1 + 2\rho \cos\alpha - \rho^2)} = 2 - \frac{2}{1 + 2\rho \cos\alpha - \rho^2} \qquad ...(19.30)$$

The η_{bl} becomes maximum when the value of $(1 + 2\rho \cos\alpha - \rho^2)$ becomes maximum.

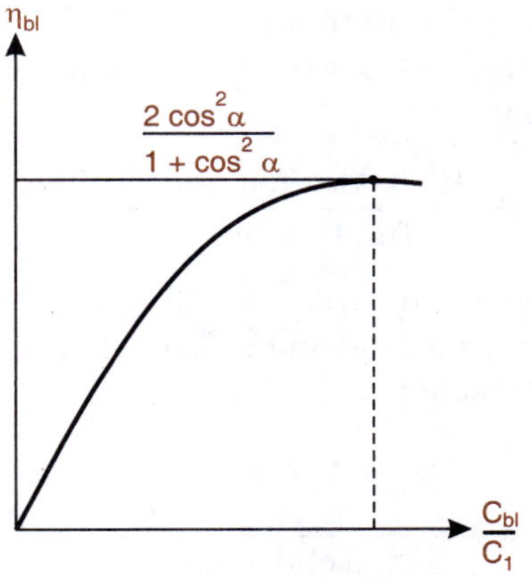

Fig. 19.42

$\therefore$ The required equation is:

$$\frac{d}{d\rho}(1 + 2\rho \cos\alpha - \rho^2) = 0$$

$$2 \cos\alpha - 2\rho = 0$$

$$\therefore \qquad \rho = \cos\alpha \qquad ...(19.31)$$

Substituting the value of ρ from eqn. (19.31) into the eqn. (19.30), the value of *maximum efficiency* is given by,

$$(\eta_{bl})_{max} = 2 - \frac{2}{1 + 2\cos^2\alpha - \cos^2\alpha}$$

$$= 2\left(1 - \frac{1}{1 + \cos^2\alpha}\right) = \frac{2\cos^2\alpha}{1 + \cos^2\alpha}$$

Hence $$(\eta_b)_{max} = \frac{2\cos^2\alpha}{1 + \cos^2\alpha} \qquad ...(19.32)$$

The variation of η_{bl} with blade speed ratio $\left(\dfrac{C_{bl}}{C_1}\right)$ for the reaction stage is shown in Fig. 19.42.

19.9. TURBINES EFFICIENCIES

1. **Blood or diagram efficiency (η_{bl}).** *It is the ratio of work done on the blade per second to the energy entering the blade per second.*

2. **Stage efficiency (η_{stage}).** The stage efficiency covers all the losses in the nozzles, blades, diaphragms and discs that are associated with that stage.

$$\eta_{stage} = \frac{\text{Network done on shaft per stage per kg of steam flowing}}{\text{Adiabatic heat drop per stage}}$$

$$= \frac{\text{Network done on blades} - \text{Disc friction and windage}}{\text{Adiabatic heat drop per stage}}.$$

3. **Internal efficiency ($\eta_{internal}$).** This is equivalent to the stage efficiency when applied to the whole turbine, and is given by :

$$\eta_{internal} = \frac{\text{Heat coverted into useful work}}{\text{Total adiabatic heat drop}}.$$

4. **Overall or turbine efficiency ($\eta_{overall}$).** This efficiency covers internal and external losses ; for example, bearings and steam friction, leakage, radiation etc.

$$\eta_{overall} = \frac{\text{Work delivered at the turbine coupling in heat units per kg of steam}}{\text{Total adiabatic heat drop}}.$$

5. **Net efficiency or efficiency ratio (η_{net}).** It is the ratio

$$\frac{\text{Brake thermal efficiency}}{\text{Thermal efficiency on the Rankine cycle}}$$

Also the actual thermal efficiency

$$= \frac{\text{Heat converted into useful work per kg of steam}}{\text{Total heat in steam at stop valve} - \text{Water heat in exhaust}}.$$

Again, Rankine efficiency

$$= \frac{\text{Adiabatic heat drop}}{\text{Total heat in steam at stop valve} - \text{Water heat in exhaust}}.$$

$$\eta_{net} = \frac{\text{Heat converted into useful work}}{\text{Total adiabatic heat drop}}.$$

Hence $\eta_{net} = \eta_{overall.}$

It is the overall or net efficiency that is meant when the efficiency of a turbine is spoken of without qualification.

19.10. TYPES OF POWER IN STEAM TURBINE PRACTICE

In steam turbine performance the following types of power are generally used :

1. **Adiabatic power (A.P.).** It is the power based on the total internal steam flow and adiabatic heat drop.

2. **Shaft power (S.P.).** It is the actual power transmitted by the turbine.

3. **Rim power (R.P.).** It is the power developed at the rim. It is also called ***blade power.***

Power losses are usually expressed as follows :

(*i*) $(P)_D$ = Power lost in overcoming disc friction.

(*ii*) $(P)_{bw}$ = Power lost in blade windage losses.

Let us consider the case of an impulse turbine. Let $\dot{m}_s$ be the total internal steam flow in kg/s.

Refer to Fig. 19.43. The line (1–2) represents the adiabatic or isentropic expansion of steam in the nozzle from pressure p_1 to p_2. But the actual path of the stage point during expansion in nozzles is shown by (1–3) which takes into account the effect of 'nozzles losses'

Then, $\quad$ A.P. $= \dot{m}_s\ (h_1 - h_2)$ kW $\quad$ ...(19.33)

After expansion in the nozzle the steam enters the blades where the R.P. is developed. Due to blade friction the steam is somewhat reheated and this reheating is shown by (3–4) along the constant pressure p_2 line just for convenience. But in actual practice though the pressure at outlet of the blade is equal to that at the inlet, the pressure in the blade channels is not constant. However, with this simplification ;

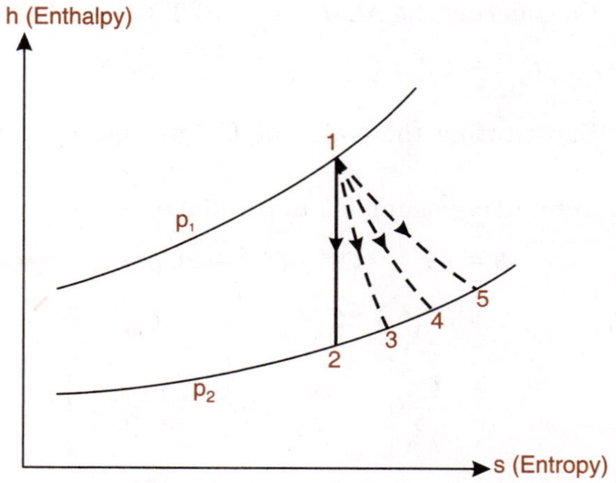

Fig. 19.43

R.P. $= \dot{m}_s\ (h_1 - h_4)$ kW $\quad$ ...(19.34)

4–5 shows the further reheating due to friction and blade windage and these losses are given as:

$(P)_{w.f.} = \dot{m}_s\ (h_5 - h_4)$ kW $\quad$ ...(19.35)

Now points 1 and 5 are the initial and final stage points respectively for a single stage impulse turbine. It, therefore, follows that

$$\text{S.P.} = m_s(h_1 - h_5) \text{ kW.} \quad ...(19.36)$$

REACTION TURBINES

Example 19.29. *The following data refer to a particular stage of a Parson's reaction turbine :*

Speed of the turbine	*= 1500 r.p.m.*
Mean diameter of the rotor	*= 1 metre*
Stage efficiency	*= 80 per cent*
Blade outlet angle	*= 20°*
Speed ratio	*= 0.7*

Determine the available isentropic enthalpy drop in the stage.

Solution. Mean diameter of the rotor, $D = 1$ m

Turbine speed, $N = 1500$ r.p.m.

Blade outlet angle, $\phi = 20°$

Speed ratio, $\rho = \dfrac{C_b}{C_1} = 0.7$

Stage efficiency, $\eta_{stage} = 80\%$

Isentropic enthalpy drop :

Blade speed,

$$C_{bl} = \frac{\pi DN}{60} = \frac{\pi \times 1 \times 1500}{60} = 78.54 \text{ m/s}$$

But $\rho = \dfrac{C_{bl}}{C_1} = 0.7$ (given)

$$\therefore \quad C_1 = \frac{C_{bl}}{0.7} = \frac{75.54}{0.7} = 112.2 \text{ m/s}$$

In Parson's turbine $\alpha = \phi$.

With the above data known, the velocity diagram for the turbine can be drawn to a suitable scale as shown in Fig. 19.44.

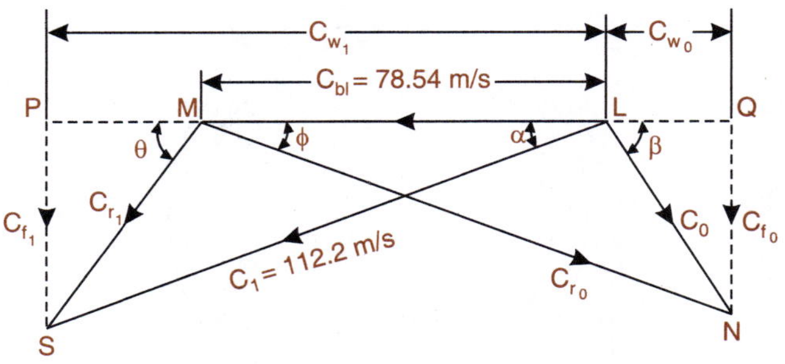

Fig. 19.44

By measurement (from the diagram)

$$C_{w_1} = 106.5 \text{ m/s} \; ; \; C_{w_0} = 27 \text{ m/s}$$

$$\eta_{stage} = \frac{C_{bl}(C_{w_1} + C_{w_0})}{h_d},$$

where h_d = isentropic enthalpy drop.

i.e.,
$$0.8 = \frac{78.54(106.25 + 27)}{h_d \times 1000}$$

$$\therefore \quad h_d = \frac{78.54(106.25 + 27)}{0.8 \times 1000} = 13.08 \text{ kJ}$$

Hence, *isentropic enthalpy drop*

= 13.08 kJ/kg. (Ans.)

Example 19.30. *In a reaction turbine, the blade tips are inclined at 35° and 20° in the direction of motion. The guide blades are of the same shape as the moving blades, but reversed in direction. At a certain place in the turbine, the drum diameter is 1 metre and the blades are 10 cm high. At this place, the steam has a pressure of 1.75 bar and dryness 0.935. If the speed of this turbine is 250 r.p.m. and the steam passes through the blades without shock, find the mass of steam flow and power developed in the ring of moving blades.*

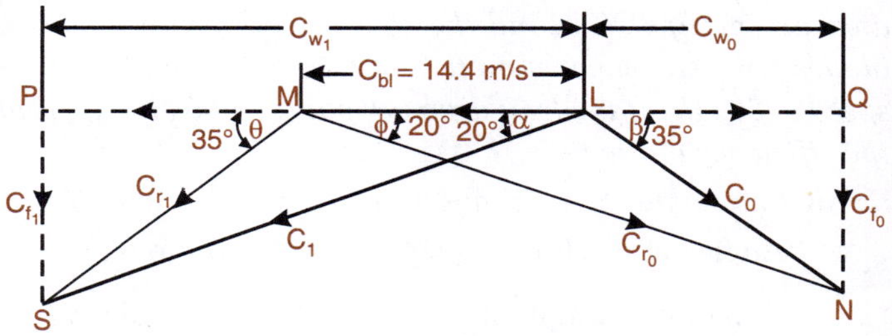

Fig. 19.45

Solution. Refer to Fig. 19.45.

Angles, $\alpha = \phi = 20°$, and $\theta = \beta = 35°$

Mean drum diameter, $D_m = 1 + 0.1 = 1.1$ m

Area of flow $= \pi D_w h$, where h is the height of blade
$= \pi \times 1.1 \times 0.1 = 0.3456 \text{ m}^2$

Steam pressure = 1.75 bar

Dryness fraction of steam, $x = 0.935$

Speed of the turbine, $N = 250$ r.p.m.

Rate of steam flow, $\dot{m}_s$:

Blade speed, $C_{bl} = \dfrac{\pi DN}{60} = \dfrac{\pi \times 1.1 \times 250}{60} = 14.4$ m/s

With the above given data the velocity diagram can be drawn to a suitable scale as shown in Fig. 19.45.

By measurement (from diagram) :

$$C_{w_1} = 30 \text{ m/s} \; ; \; C_{w_0} = 15.45 \text{ m/s} \; ;$$

$$C_{f_1} = C_{f_0} = 10.8 \text{ m/s}$$

From steam tables corresponding to 1.75 bar pressure.

v_g = Specific volume of dry saturated steam
$= 1.004 \text{ m}^3/\text{kg}$

$x = 0.935$ (given)

∴ Specific volume of wet steam

$$xv_g = 0.935 \times 1.004 = 0.938 \text{ m}^3/\text{kg}$$

Mean flow rate is given by :

$$\dot{m}_s = \frac{\text{Area of flow} \times \text{Velocity of flow}}{\text{Specific volume of steam}}$$

$$= \frac{0.3456 \times 10.8}{0.938} = 3.98 \text{ kg/s.}$$

Power developed, P :

$$P = \frac{\dot{m}_s(C_{w_1} + C_{w_0})C_{bl}}{1000} \text{ kW}$$

$$= \frac{3.98(30 + 15.45) \times 14.4}{1000}$$

$$= \mathbf{2.6\ kW.} \quad \textbf{(Ans.)}$$

Example 19.31. *In a reaction turbine, the fixed blades and moving blades are of the same shape but reversed in direction. The angles of the receiving tips are 35° and of the discharging tips 20°. Find the power developed per pair of blades for a steam consumption of 2.5 kg/s, when the blade speed is 50 m/s. If the heat drop per pair is 10.04 kJ/kg, find the efficiency of the pair.*

Solution. Angles of receiving tips, $\theta = \beta = 35°$

Angles of discharging tips, $\alpha = \phi = 20°$

Steam consumption, $\dot{m}_s = 2.5$ kg/s

Blade speed, $C_{bl} = 50$ m/s

Heat drop per pair, $h_d = 10.04$ kJ/kg

Power developed per pair of blades :

Refer to Fig. 19.46.

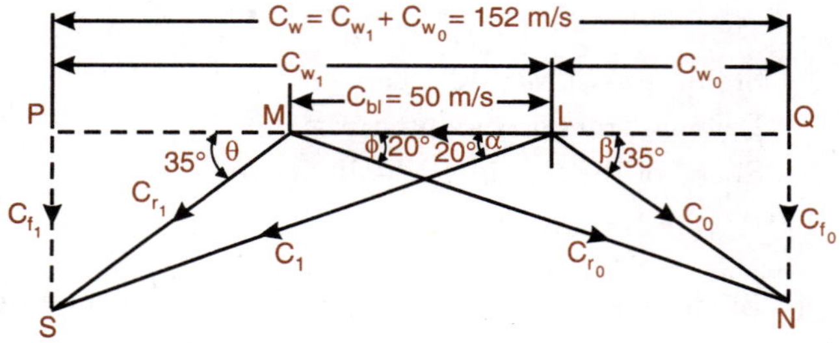

Fig. 19.46

$$NS = PQ = 152 \text{ m/s}$$

Work done per pair per kg of steam

$$= (C_{w_1} + C_{w_0})\ C_{bl} = 152 \times 50 = 7600 \text{ Nm/kg of steam.}$$

$$Power/pair = \frac{\dot{m}_s(C_{w_1} + C_{w_0})\ C_{bl}}{1000}$$

$$= \frac{2.5 \times 7600}{1000} = \mathbf{19\ kW.} \quad \textbf{(Ans.)}$$

Efficiency of the pair :

$$Efficiency = \frac{\text{Work done per pair per kg of steam}}{h_d}$$

$$= \frac{7600}{10.04 \times 1000} = \mathbf{0.757} \text{ or } \mathbf{75.7\%.} \quad \textbf{(Ans.)}$$

☞ **Example 19.32.** *A stage of a turbine with Parson's blading delivers dry saturated steam at 2.7 bar from the fixed blades at 90 m/s. The mean blade height is 40 mm, and the moving blade exit angle is 20°. The axial velocity of steam is 3/4 of the blade velocity at the mean radius. Steam is supplied to the stage at the rate of 9000 kg/h. The effect of the blade tip thickness on the annulus area can be neglected. Calculate :*

(*i*) *The wheel speed in r.p.m. ;*

(*ii*) *The diagram power ;*

(*iii*) *The diagram efficiency ;*

(*iv*) *The enthalpy drop of the steam in this stage.*

Solution. The velocity diagram is shown in Fig. 19.47 (*a*) and the blade wheel annulus is represented in Fig. 19.47 (*b*).

Pressure = 2.7 bar, $x = 1$, $C_1 = 90$ m/s,

$h = 40$ mm $= 0.04$ m

$\alpha = \phi = 20°$, $C_{f_1} = C_{f_0} = 3/4\ C_{bl}$

Rate of steam supply = 9000 kg/h.

(*i*) Wheel speed, N :

$$C_f = 3/4\ C_{bl} = C_1 \sin 20° = 90 \sin 20° = 30.78 \text{ m/s}$$

$$\therefore\ C_{bl} = 30.78 \times 4/3 = 41.04 \text{ m/s}$$

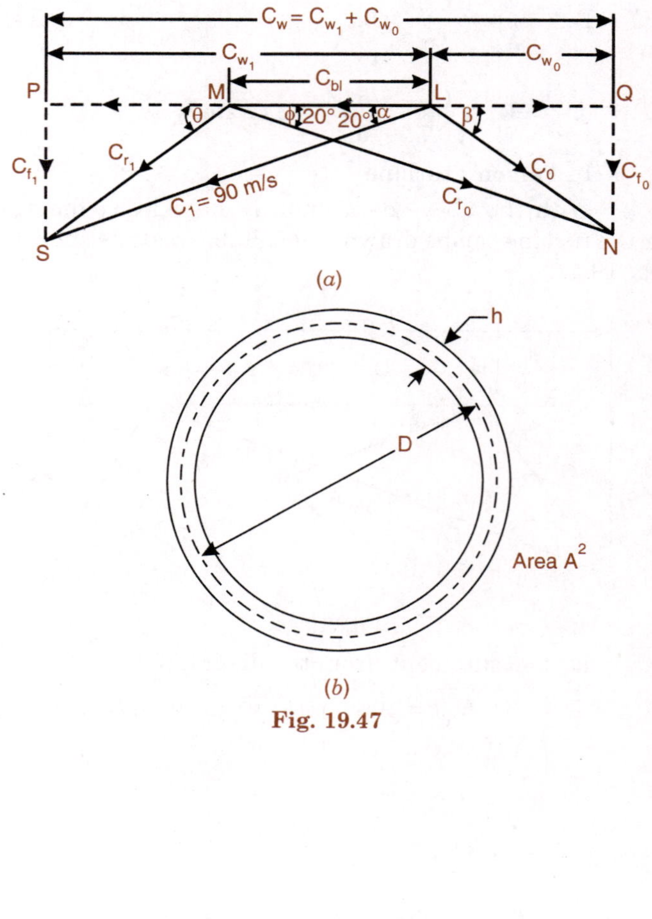

Fig. 19.47

The mass flow of steam is given by : $\dot{m}_s = \dfrac{C_f A}{v}$

(where A is the annulus area, and v is the specific volume of the steam).

In this case, $v = v_g$ at 2.7 bar = 0.6686 m³/kg

$$\therefore \quad \dot{m}_s = \frac{9000}{3600} = \frac{30.78}{0.6686}$$

or $$A = \frac{9000 \times 0.6686}{3600 \times 30.78} = 0.054 \text{ m}^2$$

Now, annulus area, $A = \pi Dh$

(where D is the mean diameter, and h is the mean blade height)

$$\therefore \quad 0.054 = \pi D \times 0.04 \quad \text{or} \quad D = \frac{0.054}{\pi \times 0.04} = 0.43 \text{ m}$$

Also, $$C_{bl} = \frac{\pi DN}{60} \quad \text{or} \quad 41.04 = \frac{\pi \times 0.43 \times N}{60}$$

or $$N = \frac{41.04 \times 60}{\pi \times 0.43} = \textbf{1823 r.p.m.} \quad \textbf{(Ans.)}$$

***(ii)* The diagram power :**

Diagram power = $\dot{m}_s\, C_w\, C_{bl}$

Now, $C_w = 2C_1 \cos \alpha - C_{bl}$

$= 2 \times 90 \times \cos 20° - 41.04 = 128.1$ m/s

$$\therefore \quad \textit{Diagram power} = \frac{9000 \times 128.1 \times 41.04}{3600 \times 1000}$$

$= \textbf{13.14 kW.}$ **(Ans.)**

***(iii)* The diagram efficiency :**

Rate of doing work per kg/s

$= C_w C_{bl} = 128.1 \times 41.04$ Nm/s

Also, energy input to the moving blades per stage

$$= \frac{C_1^2}{2} + \frac{C_{r_0}^2 - C_{r_1}^2}{2} = \frac{C_1^2}{2} + \frac{C_1^2 - C_{r_1}^2}{2}$$

$$= C_1^2 - \frac{C_{r_1}^2}{2} \qquad (\because \; C_{r_0} = C_1)$$

Referring to Fig. 19.47 (*a*), we have

$$C_{r_1}^2 = C_1^2 + C_{bl}^2 - 2C_1 C_{bl} \cos \alpha$$
$$= 90^2 + 41.04^2 - 2 \times 90 \times 41.04 \times \cos 20°$$
$$= 8100 + 1684.28 - 6941.69$$

$$\therefore \quad C_{r_1} = 53.3 \text{ m/s}$$

$$\text{Energy input} = 90^2 - \frac{53.3^2}{2} = 6679.5 \text{ Nm per kg/s}$$

$\therefore$ *Diagram efficiency*

$$= \frac{128.1 \times 41.04}{6679.5} = \textbf{0.787 or 78.7\%.} \quad \textbf{(Ans.)}$$

***(iv)* Enthalpy drop in the stage :**

Enthalpy drop in the moving blades

$$= \frac{C_{r_0}^2 - C_{r_1}^2}{2} = \frac{90^2 - 53.3^2}{2 \times 1000} = 2.63 \text{ kJ/kg}$$

$$(\because \; C_{r_0} = C_{r_1})$$

$\therefore$ *Total enthalpy drop per stage* = 2 × 2.63

= **5.26 kJ/kg.** **(Ans.)**

Example 19.33. *The outlet angle of the blade of a Parson's turbine is 20° and the axial velocity of flow of steam is 0.5 times the mean blade velocity. If the diameter of the ring is 1.25 m and the rotational speed is 3000 r.p.m. determine :*

(i) Inlet angles of blades.

(ii) Power developed if dry saturated steam at 5 bar passes through the blade whose height may be assumed as 6 cm. Neglect the effect of blade thickness.

Solution. Refer to Fig. 19.48.

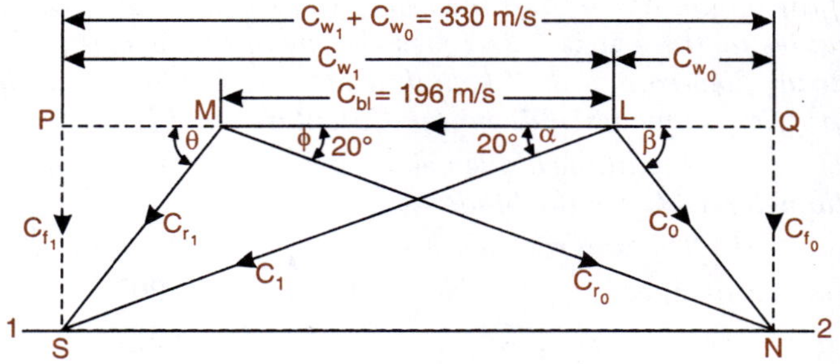

Fig. 19.48

Angles, $\alpha = \phi = 20°$

Axial velocity of flow of steam,

$$C_{f_1} = C_{f_0} = 0.5\, C_{bl} \text{ (blade speed)}$$

Diameter of the ring, $D = 1.25$ m

Rotational speed, $N = 3000$ r.p.m.

$$\text{Blade speed, } C_{bl} = \frac{\pi DN}{60} = \frac{\pi \times 1.25 \times 3000}{60}$$
$$= 196 \text{ m/s}$$

$$\therefore \quad C_{f_1} = C_{f_0} = 0.5 \times 196 = 98 \text{ m/s}$$

Velocity diagram is drawn as follows :

- Takes LM (C_{bl}) = 196 m/s, and $\alpha = \phi = 20°$.
- Draw line 1–2 parallel to LM at a value of 98 m/s (according to scale). The points S and N are thus located on the line 1–2.
- Complete the rest of the diagram as shown in Fig. 19.48.

***(i)* Inlet angles of blades :**

The *inlet angles* (by measurement) are :

$\beta = \theta = $ **55°.** **(Ans.)**

(ii) Power developed, P :

Area of flow is given by,

$A = \pi \times D$ (mean diameter) $\times h$ (height of blade)

Mean flow rate is given by,

$$\dot{m}_s = \frac{\text{Area of flow} \times \text{Velocity of flow}}{\text{Specific volume of steam}} = \frac{\pi D h \times C_f}{v}$$

From steam tables, $v_g = 0.375\ \text{m}^3/\text{kg}$ at 5 bar

$$\therefore\ \dot{m}_s = \frac{\pi \times 1.25 \times \left(\frac{6}{100}\right) \times 98}{0.375} = 61.57\ \text{kg/s}$$

Power developed,

$$P = \frac{\dot{m}_s \times C_w \times C_{bl}}{1000}$$

$$= \frac{61.57 \times 330 \times 196}{1000} = \mathbf{3982.3\ kW.}\ \textbf{(Ans.)}$$

Example 19.34. *A 50% reaction turbine (with symmetrical velocity triangles) running at 400 r.p.m. has the exit angle of the blades as 20° and the velocity of steam relative to the blades at the exit is 1.35 times the mean blade speed. The steam flow rate is 8.33 kg/s and at a particular stage the specific volume is 1.381 m³/kg. Calculate for this stage :*

(i) A suitable blade height, assuming the rotor mean diameter 12 times the blade height, and

(ii) The diagram work. **(N.U.)**

Solution. Speed, $N = 400$ r.p.m. ; $\alpha = 20°$

$$C_{r_0} = C_1 = 1.35\ C_{bl}\ ;\ \dot{m}_s = 8.33\ \text{kg/s}$$

$$v = 1.381\ \text{m}^3/\text{kg}\ ;\ D = 12\ h$$

(i) Blade height, h :

Refer to Fig. 19.49.

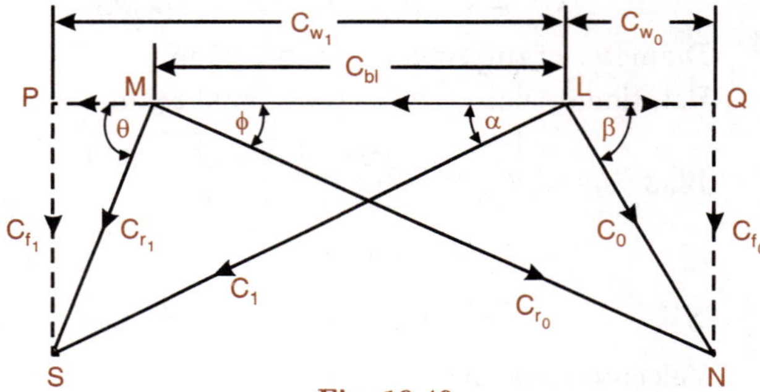

Fig. 19.49

Axial flow velocity,

$$C_{f_1} = C_{f_0} = C_f = C_1 \sin \alpha$$
$$= 1.35\ C_{bl} \sin 20°$$
$$= 0.4617\ C_{bl}$$

Area of flow, $A = \pi D h = \pi D \times \dfrac{D}{12} = \dfrac{\pi D^2}{12}$

Mass flow rate,

$$\dot{m} = \frac{A C_f}{v} \quad \text{or} \quad 8.33 = \frac{A \times 0.4617\ C_{bl}}{1.381}$$

or $$\frac{8.33 \times 1.381}{0.4917} = A \times C_{bl} = \frac{\pi D^2}{12} \times \frac{\pi D N}{60}$$

or $$24.916 = \frac{\pi^2 D^3 \times 400}{720}$$

or $$D^3 = \frac{24.919 \times 720}{\pi^2 \times 400}$$

or $$D = 1.656\ \text{m}$$

$\therefore$ Blade height,

$$h = \frac{D}{12} = \frac{1.656}{12} = \mathbf{0.138\ m\ \ or\ \ 138\ mm.}\ \textbf{(Ans.)}$$

(ii) The diagram work :

Diagram work

$$= \dot{m} \times C_{bl}\ (C_{w_1} + C_{w_0})$$
$$= \dot{m} \times C_{bl}\ (2\ C_1 \cos \alpha - C_{bl})$$
$$= 8.33 \times C_{bl}\ (2 \times 1.35\ C_{bl} \cos 20° - C_{bl})$$
$$= 8.33 \times C_{bl}^2\ (2 \times 1.35 \cos 20° - 1)$$
$$= 8.33 \times \left(\frac{\pi D N}{60}\right)^2 \times 1.537$$
$$= 8.33 \times \left(\frac{\pi \times 1.656 \times 400}{60}\right)^2 \times 1.537$$
$$= \mathbf{15401.2\ W\ \ or\ \ 15.4\ kW.}\ \textbf{(Ans.)}$$

Example 19.35. *300 kg/min of steam (2 bar, 0.98 dry) flows through a given stage of a reaction turbine. The exit angle of fixed blades as well as moving blades is 20° and 3.68 kW of power is developed. If the rotor speed is 360 r.p.m. and tip leakage is 5 per cent, calculate the mean drum diameter and the blade height. The axial flow velocity is 0.8 times the blade velocity.* **(Roorkee University)**

Solution. Rate of flow of steam through the turbine,

$$\dot{m}_s = \frac{300}{60} = 5\ \text{kg/s}$$

Pressure and condition of steam,

$$p = 2\ \text{bar},\ x = 0.98.$$

The exit angles of fixed blades as well as moving blades, $\alpha = \phi = 20°$

Power developed,	$P = 3.68$ kW
Speed of the rotor,	$N = 360$ r.p.m.
Tip leakage	= 5 per cent
Axial flow velocity,	$C_f = 0.8\ C_{bl}$ (blade velocity)

Refer to Fig. 19.49.

Mean drum diameter, D :

Mean blade velocity,

$$C_{bl} = \frac{\pi D N}{60} = \frac{\pi D \times 360}{60} = 18.85\ D\ \text{m/s}$$

Power developed,

$$P = \frac{\dot{m}_s C_{bl} C_w}{1000}$$

or $$3.67 = \frac{(5 \times 0.95) \times 18.85\, D \times C_w}{1000}$$

$$\therefore \quad C_w = \frac{3.67 \times 1000}{(5 \times 0.95) \times 18.85 D} = \frac{40.988}{D}$$

Assuming Parson's reaction turbine, we have

$$C_{f_1} = C_1 \sin \alpha$$

or $$C_1 = \frac{C_{f_1}}{\sin \alpha} = \frac{0.8\, C_{bl}}{\sin \alpha}$$

$$= \frac{0.8 \times 18.85\, D}{\sin 20°} = 44.091\, D$$

$$(C_{f_1} = C_{f_0} = C_f)$$

Also, $C_w = 2\, C_1 \cos \alpha - C_{bl}$

or $$\frac{40.988}{D} = 2 \times 44.091\, D \cos 20° - 18.85\, D$$

$$= 64.01\, D$$

or $$40.988 = 64.01\, D^2$$

or D = **0.8 m or 800 mm. (Ans.)**

Blade height, h :

Mean steam flow rate, $\dot{m}_s = \dfrac{\pi D h C_f}{x v_g}$

or $$(5 \times 0.95) = \frac{\pi \times 0.8 \times h \times C_1 \sin \alpha}{0.98 \times 0.885}$$

$$= \frac{\pi \times 0.8 \times h \times (44.091 D \times \sin 20°)}{0.98 \times 0.885}$$

(At 2 bar : $v_g = 0.885$ kg/m³)

or $$h = \frac{(5 \times 0.95) \times 0.98 \times 0.885}{\pi \times 0.8 \times (44.091 \times 0.80 \times 0.3420)}$$

= **0.1359 m or 135.9 mm. (Ans.)**

Example 19.36. *(a) Why is drum type construction preferred to disc type construction in reaction turbine ?*

(b) Why is partial admission of steam adopted for H.P. impulse stages while full admission is essential for any stage of a reaction turbine ?

(c) In a 50% reaction turbine, the speed of rotation of a blade group is 3000 r.p.m. with mean blade velocity of 120 m/s. The velocity ratio is 0.8 and the exit angle of the blades is 20°. If the mean blade height is 30 mm, calculate the total steam flow rate through the turbine. Neglect the effect of blade edge thickness of the annular area but consider 10% of the total steam flow rate as the tip leakage loss. The mean condition of steam in that blade group is found to be 2.7 bar and 0.95 dry.

(d) What do you mean by once through boilder ?

(AMIE)

Solution. (*a*) The rotor of the turbine can be of drum type or disc type. *Disc type construction is difficult (complicated) to make, but lighter in weight.* Hence the centrifugal stresses are lower at a particular speed. On the other hand drum type construction is simple in construction, and it is easy to attach aerofoil shape blades. Further it is easier to design for tip leakage reduction which is a major problem in reaction turbines. Moreover due to small pressure drop per stage (larger number of stages) in reaction turbines, their rotational speeds are lower and so the *centrifugal stresses are not very high* (even the reaction blades are lighter). Therefore *drum type construction is preferred to disc type in reaction turbines.*

To accommodate increase in specific volume at lower pressures the drum diameter is stepped up which allows greater area without unduly increasing blade height. The *increased drum diameter also increases the torque due to steam pressure.*

(*b*) In **impulse turbines** there is no expansion of steam in moving blades, and the pressure of steam remains constant while flowing over the moving blades. The expansion takes place only in the nozzles at the inlet to the turbine in H.P. stages, or through the fixed blades in the subsequent stages. The nozzles need not occupy the complete circumference. Therefore partial admission of steam is feasible and adopted for H.P. impulse stages.

In **reaction turbines,** *pressure drop is required in the moving blades also. This is not possible with partial admission. Hence full admission is essential for all stages of a reaction turbine.*

(*c*) Refer to Fig. 19.37.

Given : $N = 3000$ r.p.m. ; $\phi = \alpha = 20°$;

$$C_{bl} = 120 \text{ m/s} ; \quad \frac{C_{bl}}{C_1} = 0.8 ;$$

$$\therefore \quad C_1 = \frac{C_{bl}}{0.8} = \frac{120}{0.8} = 150 \text{ m/s}$$

Also $$C_{bl} = \frac{\pi D N}{60}$$

or $$120 = \frac{\pi D N}{60}$$

$$\therefore \quad D = \frac{120 \times 60}{\pi \times 3000} = 0.764 \text{ m.}$$

From steam tables, v_g (at 2.7 bar) = 0.668 m³/kg

$v = 0.95 \times 0.668 = 0.6346$ m³/kg

Flow area $$A = \pi D h = \pi \times 0.764 \times \frac{30}{1000}$$

$$= 0.072 \text{ m}^2$$

Flow velocity $C_f = C_1 \sin \alpha = 150 \sin 20°$

$= 51.3$ m/s $\qquad (C_{f_1} = C_{f_0} = C_f)$

Mass flow rate $$\dot{m} = \frac{A C_f}{v} = \frac{0.072 \times 51.3}{0.6346}$$

$$= 5.82 \text{ kg/s}$$

Accounting for 10 per cent leakage (of total steam flow), the total steam flow rate is

$$\frac{5.82}{0.9} = \mathbf{6.467\ kg/s.}\ \text{(Ans.)}$$

(*d*) *Once through* boiler is a boiler which *does not require any water or steam drum.* It is a *monotube boiler* using about 1.5 kg long tube arranged in the combustion chamber and the furnace. The *economizer, boiler and superheater are in series with no fixed surfaces as separators between the steam and water.*

Benson boiler is an example of once through boiler, operating at supercritical pressure. The tube length to diameter ratio of such a boiler is about 2500. Due to large frictional resistance the feed pressure should be about 1.4 times the boiler pressure.

☞**Example 19.37.** *A twenty-stage Parson turbine receives steam at 15 bar at 300°C. The steam leaves the turbine at 0.1 bar pressure. The turbine has a stage efficiency of 80% and the reheat factor 1.06. The total power developed by the turbine is 10665 kW. Find the steam flow rate through the turbine assuming all stages develop equal power.*

The pressure of steam, at certain stage of the turbine is 1 bar abs., and is dry and saturated. The blade exit angle is 25° and the blade speed ratio is 0.75. Find the mean diameter of the rotor of this stage and also the rotor speed. Take blade height as 1/12th of the mean diameter. The thickness of the blades may be neglected.

Solution. Number of stage = 20

Steam supply pressure = 15 bar, 300°C

Exhaust pressure = 0.1 bar

Stage efficiency of turbine,

η_{stage} = 80%

Reheat factor = 1.06

Total power developed = 10665 kW

Steam pressure at a certain stage

= 1 bar abs., $x = 1$

Blade exit angle = 25°

Blade speed ratio, $\rho = \dfrac{C_{bl}}{C_1} = 0.75$

Height of the blade, $h = \dfrac{1}{12} D$ (mean dia. of rotor)

(*i*) **Steam flow rate, $\dot{m}_s$:**

Refer to Fig. 19.50.

Isentropic drop,

$$(\Delta h)_{isentropic} = h_1 - h_2 = 3040 - 2195 = 845 \text{ kJ/kg}$$

$$\eta_{overall} = \eta_{stage} \times \text{Reheat factor}$$
$$= 0.8 \times 1.06 = 0.848$$

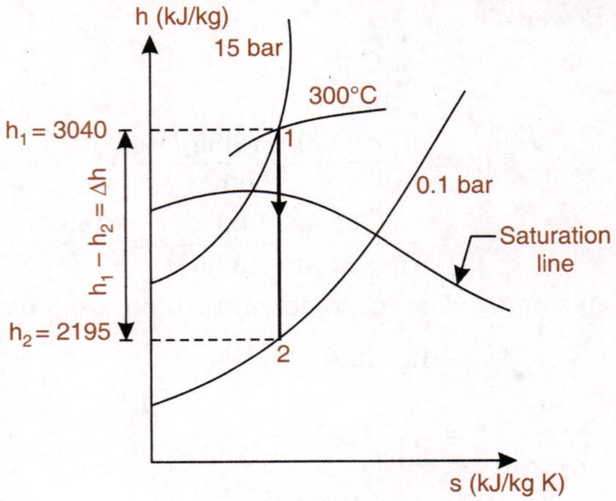

Fig. 19.50

Work done = Actual enthalpy drop

$= (\Delta h)_{isentropic} \times \eta_{overall}$

$= 845 \times 0.848 = 716.56$ kJ/kg

Work done per stage per kg

$$= \frac{716.56}{20} = 35.83 \text{ kJ} \quad ...(i)$$

Also, total power

= No. of stages × $\dot{m}_s$ × work done/kg stage

$\therefore\ 10665 = 20 \times \dot{m}_s \times 35.83$

$$\therefore \quad \dot{m}_s = \frac{10665}{20 \times 35.83} = \mathbf{14.88\ kg/s.}\ \text{(Ans.)}$$

(*ii*) **Mean diameter of rotor, D :**

Rotor speed, N :

Refer to Fig. 19.51.

Work done per kg per stage

$= C_{bl} \times C_w = C_{bl} (2C_1 \cos 25° - C_{bl})$

Also, $\dfrac{C_{bl}}{C_1} = 0.75$ ...(Given)

$$\therefore \quad C_1 = \frac{C_{bl}}{0.75} = 1.33 C_{bl}$$

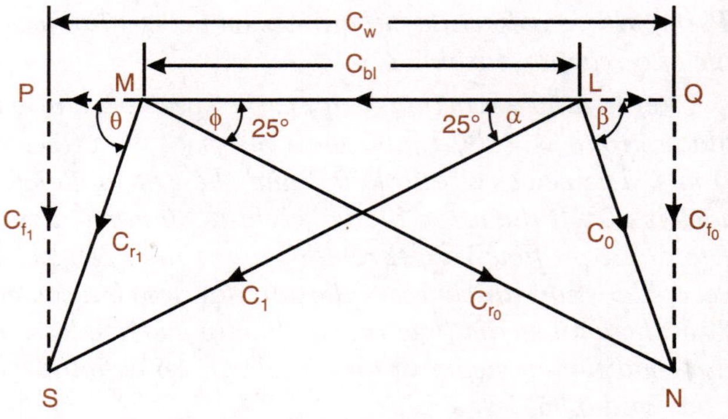

Fig. 19.51

i.e., Work done per kg per stage

$= C_{bl}\,(2 \times 1.33C_{bl} \times 0.906 - C_{bl})$

$= 1.41\; C_{bl}^2$ Nm ...(*ii*)

Equating (*i*) and (*ii*), we get

$1.41\; C_{bl}^2 = 35.83 \times 1000$

$\therefore \quad C_{bl}^2 = \dfrac{35.83 \times 1000}{1.41}$

or $C_{bl} = 159.41$ m/s

$\therefore \quad C_1 = 1.33 \times 159.41 = 212$ m/s

From Fig. 19.51

$C_{f_1} = C_1 \sin \alpha = 212 \sin 25° = 89.59$ m/s

v_g = Specific volume at 1 bar when steam is dry and saturated

= 1.694 m³/kg (from steam tables)

Mass flow rate, $\dot{m}_s = \dfrac{\pi D h C_{f_1}}{v}$

$\therefore \quad 14.88 = \dfrac{\pi \times D \times \left(\dfrac{D}{12}\right) 89.59}{1.694}$

or $D^2 = \dfrac{14.88 \times 1.694 \times 12}{\pi \times 89.59}$

$\therefore$ D = **1.036 m. (Ans.)**

Now, $h = \dfrac{D}{12} = \dfrac{1.036}{12}$

= 0.086 m = **8.6 cm. (Ans.)**

Also, $C_{bl} = \dfrac{\pi D N}{60}$

$\therefore \quad N = \dfrac{C_{bl} \times 60}{\pi D} = \dfrac{159.41 \times 60}{\pi \times 1.036}$

= **2938.7 r.p.m. (Ans.)**

Example 19.38. *The following data relate to a stage of reaction turbine :*

Mean rotor diameter = 1.5 m ; speed ratio = 0.72 ; blade outlet angle = 20° ; rotor speed = 3000 r.p.m.

(i) Determine the diagram efficiency.

(ii) Determine the percentage increase in diagram efficiency and rotor speed if the rotor is designed to run at the best theoretical speed, the exit angle being 20°.

Solution. Mean rotor diameter, $D = 1.5$ m

Speed ratio, $\rho = \dfrac{C_{bl}}{C_1} = 0.72$

Blade outlet angle = 20°

Rotor speed, $N = 3000$ r.p.m.

(This example is solved purely by calculations; Fig. 19.52 is not drawn to scale.)

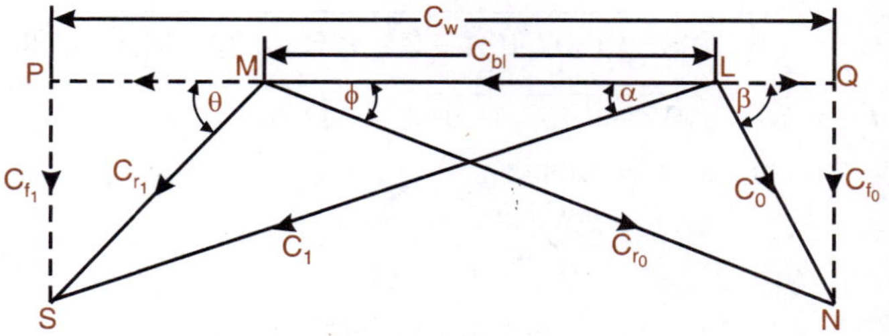

Fig. 19.52

(*i*) Diagram efficiency :

Blade velocity,

$C_{bl} = \dfrac{\pi D N}{60} = \dfrac{\pi \times 1.5 \times 3000}{60} = 235.6$ m/s

Speed ratio, $\rho = \dfrac{C_{bl}}{C_1} = 0.72$

$\therefore \quad C_1 = \dfrac{C_{bl}}{0.72} = \dfrac{235.6}{0.72} = 327.2$ m/s

Assuming that velocity triangles are symmetrical

$\alpha = \phi = 20°$

From the velocity ΔLMS

$C_{r_1}^2 = C_1^2 + C_{bl}^2 - 2C_1 C_{bl} \cos \alpha$

$C_{r_1} = \sqrt{(327.2)^2 + (235.6)^2 - 2 \times 327.2 \times 235.6 \cos 20°}$

$= 100\sqrt{10.7 + 5.55 - 14.48} = 133$ m/s

i.e., $C_{r_1} = 133$ m/s

Work done per kg of steam

$= C_{bl}C_w = C_{bl}(2C_1 \cos \alpha - C_{bl})$

$= 235.6\,(2 \times 327.2 \cos 20° - 235.6) = 89371.3$ Nm.

Energy supplied per kg of steam

$= \dfrac{C_1^2 + C_{r_0}^2 - C_{r_1}^2}{2}$

$= \dfrac{2C_1^2 - C_{r_1}^2}{2} \qquad (\because \; C_1 = C_{r_0})$

$= \dfrac{2 \times (327.2)^2 - (133)^2}{2} = 98215.3$ Nm

$\therefore$ *Diagram efficiency*

$= \dfrac{89371.3}{98215.3}$ = **0.91** or **91%. (Ans.)**

(*ii*) Percentage increase in diagram efficiency:

For the best diagram efficiency (*maximum*), the required condition is

$\rho = \dfrac{C_{bl}}{C_1} = \cos \alpha$

$\therefore \quad C_{bl} = C_1 \cos \alpha = 372.2 \cos 20° = 307.46$ m/s

For this blade speed, the value of C_{r_1} is again calculated by using eqn. (*i*),

$$C_{r_1} = \sqrt{(327.2)^2 + (307.46)^2 - 2 \times 327.2 \times 307.46 \times \cos 20°}$$
$$= 100\sqrt{10.7 + 9.45 - 18.906} = 111.5 \text{ m/s}$$

Diagram efficiency

$$= \frac{2C_{bl}(2C_1 \cos \alpha - C_{bl})}{(C_1^2 + C_{r_0}^2 - C_{r_1}^2)}$$
$$= \frac{2 \times 307.46(2 \times 327.2 \cos 20° - 307.46)}{(327.2)^2 + (327.2)^2 - (111.5)^2}$$
$$= 0.937 \text{ or } 93.7\%$$

Percentage increase in diagram efficiency

$$= \frac{0.937 - 0.91}{0.91}$$
$$= \mathbf{0.0296 \text{ or } 2.96\%.} \text{ (Ans.)}$$

The diagram efficiency for the best speed can also be calculated by using relation:

$$\eta_{blade} = \frac{2\cos^2 \alpha}{1 + \cos^2 \alpha} = \frac{2 \times \cos^2 20°}{1 + \cos^2 20°} = \frac{1.766}{1 + 0.883}$$
$$= \mathbf{0.937 \text{ or } 93.7\%.} \text{ (Ans.)}$$

The **best theoretical speed of the rotor** is given by,

$$C_{bl} = \frac{\pi DN}{60} \text{ or } N = \frac{60\, C_{bl}}{\pi D} = \frac{60 \times 307.46}{\pi \times 1.5}$$
$$= \mathbf{3914.7 \text{ r.p.m.}} \text{ (Ans.)}$$

☞**Example 19.39. (Impulse reaction turbine).** *The following data relate to a stage of an impulse reaction turbine:*

Steam velocity coming out of nozzle = 245 m/s ; nozzle angle = 20° ; blade mean speed = 145 m/s ; speed of the rotor = 300 r.p.m. ; blade height = 10 cm ; specific volume of steam at nozzle outlet and blade outlet respectively = 3.45 m³/kg and 3.95 m³/kg ; Power developed by the turbine = 287 kW ; efficiency of nozzle and blades combinedly = 90% ; carry over co-efficient = 0.82.

Find : (i) The heat drop in each stage ;

(ii) Degree of reaction ;

(iii) Stage efficiency.

Solution. Steam velocity coming out of nozzle,

$C_1 = 245$ m/s

Nozzle angle, $\alpha = 20°$

Blade mean speed, $C_{bl} = 145$ m/s

Speed of the rotor, $N = 3000$ r.p.m.

Blade height, $h = 10 \text{ cm} = 0.1$ m

Specific volume of steam at nozzle outlet or blade inlet,

$$v_1 = 3.45 \text{ m}^3/\text{kg}$$

Specific volume of steam at blade outlet,

$$v_0 = 3.95 \text{ m}^3/\text{kg}$$

Power developed by the turbine = 287 kW

Efficiency of nozzle and blades combinedly = 90%

Carry over co-efficient, $\psi = 0.82$

Blade speed, $C_{bl} = \frac{\pi DN}{60}$

$$\therefore \quad D = \frac{60\, C_{bl}}{\pi N} = \frac{60 \times 145}{\pi \times 3000} = 0.923 \text{ m}$$

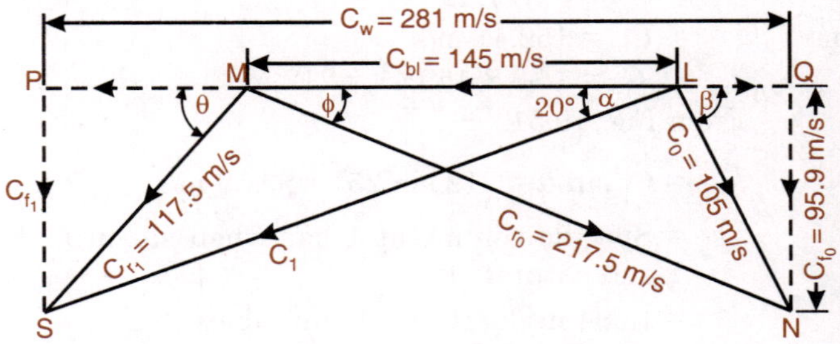

Fig. 19.53

Mass flow rate,

$$\dot{m}_s = \frac{C_{f_1} \times \pi Dh}{v_1} = \frac{C_1 \sin \alpha \times \pi Dh}{v_1}$$
$$= \frac{245 \times \sin 20° \times \pi \times 0.923 \times 0.1}{3.45} = 7.04 \text{ kg/s}$$

Also $\dot{m}_s = \frac{C_{f_0}\, \pi Dh}{v_0}$

$$\therefore \quad C_{f_0} = \frac{\dot{m}_s v_0}{\pi Dh} = \frac{7.04 \times 3.95}{\pi \times 0.923 \times 0.1} = 95.9 \text{ m/s}$$

The power is given by,

$$P = \frac{\dot{m}_s \times C_{bl} \times C_w}{1000}$$
$$287 = \frac{7.04 \times 145 \times C_w}{1000}$$
$$\therefore \quad C_w = \frac{287 \times 1000}{7.04 \times 145} = 281 \text{ m/s}$$

Now draw velocity triangles as follows :

Select a suitable scale (say 1 cm = 25 m/s)

- Draw LM = blade velocity = 145 m/s ; $\angle MLS$ = nozzle angle = 20°
  Join MS to complete the inlet ΔLMS.
- Draw a perpendicular from S which cuts the line through LM at point P.
  Mark the point Q such that $PQ = C_w = 281$ m/s.
- Draw a perpendicular through point Q and the point N as $QN = 95.9$ m/s.
  Join LN and MN to complete the *outlet* velocity triangle.

From the velocity triangles ;

$$C_{r_1} = 117.5 \text{ m/s} ; C_{r_0} = 217.5 \text{ m/s} ; C_0 = 105 \text{ m/s}.$$

(*i*) **Heat drop in each stage, $(\Delta h)_{stage}$:**

Heat drop in *fixed blades* (Δh_f)

$$= \frac{C_1^2 - \psi C_0^2}{2 \times \eta_{nozzle}}$$, where ψ is a carry over co-efficient

$$= \frac{(245)^2 - 0.82 \times (105)^2}{2 \times 0.9 \times 1000} = 28.32 \text{ kJ/kg}$$

Heat drop in *moving blades* (Δh_m)

$$= \frac{C_{r_0}^2 - C_{r_1}^2}{2 \times \eta_{nozzle}} = \frac{(217.5)^2 - (117.5)^2}{2 \times 0.9 \times 1000} = 18.61 \text{ kJ/kg}$$

Total heat drop in a stage,

$$(\Delta h)_{stage} = \Delta h_f + \Delta h_m$$
$$= 28.32 + 18.61 = \mathbf{46.93 \text{ kJ/kg.}} \quad \textbf{(Ans.)}$$

(*ii*) **Degree of reaction, R_d :**

$$R_d = \frac{\Delta h_m}{\Delta h_m + \Delta h_f} = \frac{18.61}{18.61 + 28.32} = \mathbf{0.396.} \quad \textbf{(Ans.)}$$

(*iii*) **Stage efficiency, η_{stage} :**

Work done per kg of steam

$$= \frac{C_{bl} \times C_w}{1000} = \frac{145 \times 281}{1000}$$
$$= 40.74 \text{ kJ/kg of steam}$$

$$\therefore \quad \eta_{stage} = \frac{\text{Work done per kg of steam}}{\text{Total heat drop in a stage}} = \frac{40.74}{46.93}$$
$$= \mathbf{0.868 \text{ or } 86.8\%.} \quad \textbf{(Ans.)}$$

19.11. "STATE POINT LOCUS" AND "REHEAT FACTOR"

The terms *state point locus* and *reheat factor* are discussed below :

State Point Locus

*The **state point** may be defined as that point on h-s diagram which represents the condition of steam at that instant.* Thus knowing the initial condition of steam entering the nozzle of a turbine the initial state point of Fig. 19.43 may be located on *h-s* diagram. If stage efficiency

$$\left[\eta_{stage} = \frac{\text{S.P.}}{\text{A.P.}} = \frac{h_1 - h_5}{h_1 - h_2}\right]$$ be known or assumed the position

of the end state point 5 for a stage be readily obtained. The point 5 now becomes the initial state point for the succeeding stage of the turbine.

Let us now consider a *multistage turbine having four stages*. Refer to Fig. 19.54.

The initial point 1 is located according to the given initial condition. (1-2) is adiabatic expansion in the first stage. h_2' may be calculated from the following relation, $h_1 - h_2' = \eta_{stage}(h_1 - h_2)$. Point 2′ is then located with the value h_2' on p_2 line. Then (2′-3) is drawn showing adiabatic expansion. Point 3′ may be located by finding h_3' from $h_2' - h_3' = \eta_{stage}(h_2' - h_3)$. Proceeding in this way all stage points 3′, 4′ and 5′ may be fixed. The locus passing through these points is called "*State point locus*". The sum of the adiabatic heat drops (1-2) + (2′-3) + (3′-4) + (4′-5) is generally called "*cumulative heat drop*" and is represented as h_{cum}. For the purpose of design the various quantities obtained from the Mollier diagram are set out in some curves form, and these curves are termed "**condition curves**".

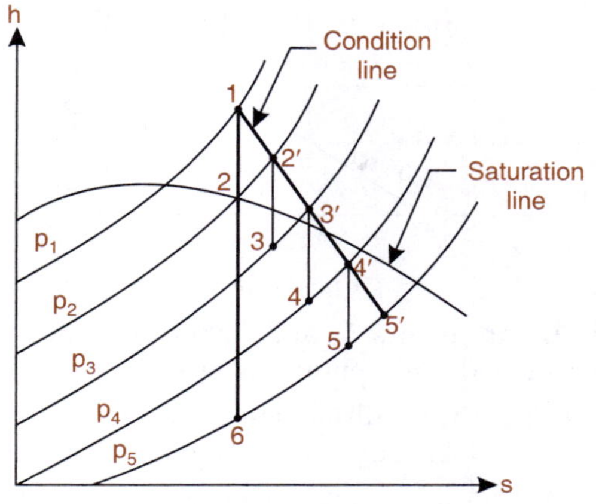

Fig. 19.54

Reheat Factor

Referring to Fig. 19.54, the *adiabatic heat drop* (h_{adi}) from pressure p_1 to final pressure p_5, considering all the stages as one unit, is $(h_1 - h_6)$. It will be found that h_{adi} is less that $h_{cum.}$ The ratio $\frac{h_{cum}}{h_{adi}}$ is termed as ***Reheat factor.***

The value of reheat factor depends on the following factors :

(*i*) Stage efficiency ;

(*ii*) Initial pressure and condition of steam ;

(*iii*) Final pressure.

Example 19.40. *In a three-stage steam turbine steam enters at 35 bar and 400°C and exhausts at 0.05 bar, 0.9 dry. If the work developed per stage is equal, determine :*

(*i*) *Condition of steam at entry to each stage.*

(*ii*) *The stage efficiencies.*

(*iii*) *The reheat factor.*

(*iv*) *Internal turbine efficiency.*

Assume condition line to be straight.

Solution. Initial condition of steam = 35 bar, 400°C

Exhaust condition of steam = 0.05 bar, 0.9 dry

(*i*) **Condition of steam at entry to each stage :**

Refer to Fig. 19.55.

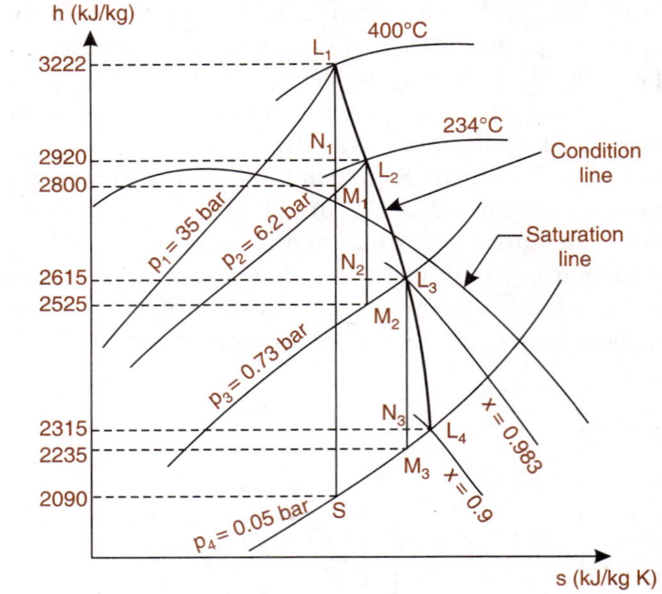

Fig. 19.55

- Locate points L_1 and L_4 corresponding to entry and exhaust conditions of steam.
- Since the condition line is straight (given), points L_1 and L_4 are joined by a straight line.

Heat drop due to expansion from L_1 to L_4 = 3222 – 2315 = 907 kJ/kg

- Since the work developed per stage is equal,

$\therefore$ Useful work/stage

$$= \frac{907}{3} = 302.3 \text{ kJ/kg}$$

i.e., $(h_{L_1} - h_{N_1}) = 302.3$ kJ/kg

Produce horizontal to cut condition line at L_2. L_1N_1 produced cuts the pressure line through L_2 at M_1. Thus L_2, N_1 and M_1 are located.

Proceeding for other stages likewise, we get the following results :

Points	*Pressure*	*Temp. or quality of steam*	
L_2	6.2 bar	234°C	**(Ans.)**
L_3	0.73 bar	0.98 dry	**(Ans.)**

$$h_{L_1M_1} = h_{L_1} - h_{M_1} = 3222 - 2800 = 422 \text{ kJ/kg}$$

$$h_{L_2M_2} = h_{L_2} - h_{M_2} = 2920 - 2525 = 395 \text{ kJ/kg}$$

$$h_{L_3M_3} = h_{L_3} - h_{M_3} = 2615 - 2235 = 380 \text{ kJ/kg.}$$

(*ii*) Stage efficiencies :

Efficiency of stage 1,

$$\eta_1 = \frac{h_{L_1} - h_{N_1}}{h_{L_1} - h_{M_1}} = \frac{302.3}{422}$$

$$= \mathbf{0.7163} \quad \text{or} \quad \mathbf{71.63\%.} \quad \textbf{(Ans.)}$$

Efficiency of stage 2,

$$\eta_2 = \frac{h_{L_2} - h_{N_2}}{h_{L_2} - h_{M_2}} = \frac{302.3}{395}$$

$$= \mathbf{0.7653} \quad \text{or} \quad \mathbf{76.53\%.} \quad \textbf{(Ans.)}$$

Efficiency of stage 3,

$$\eta_3 = \frac{h_{L_3} - h_{N_3}}{h_{L_3} - h_{M_3}} = \frac{302.3}{380}$$

$$= 0.7955 = \mathbf{79.55\%.} \quad \textbf{(Ans.)}$$

(*iii*) Reheat factor :

Reheat factor

$$= \frac{\text{Cumulative drop}}{\text{Isentropic enthalpy drop}} = \frac{h_{L_1M_1} + h_{L_2M_2} + h_{L_3M_3}}{h_{L_1} - h_S}$$

$$= \frac{422 + 395 + 380}{3222 - 2090} = \mathbf{1.057.} \quad \textbf{(Ans.)}$$

(*iv*) Internal turbine efficiency :

Internal turbine efficiency

$$= \frac{h_{L_1} - h_{L_4}}{h_{L_1} - h_S} = \frac{3222 - 2315}{3222 - 2090}$$

$$= \mathbf{0.801} \quad \text{or} \quad \mathbf{80.1\%.} \quad \textbf{(Ans.)}$$

19.12. REHEATING STEAM

Refer to Art. 15.5 (Reheat cycle)

19.13. BLEEDING

***Bleeding** is the process of draining steam from the turbine, at certain points during its expansion, and using this steam for heating the feed water supplied to the boiler.* In this process a small quantity of steam, at certain sections of the turbine, is drained from the turbine and is then circulated around the feed water pipe leading from hot well to the boiler. The steam is thus condensed due to relatively cold water, the heat so lost by steam is transferred to the feed water. The condensed steam then finds its way to hot well.

There is a usual practice in bleeding installations to allow the bled steam to mix with the feed water. The mixture of steam and water then proceeds to the boiler.

By bleeding process hotter water is supplied to the boiler of course at the cost of loss of small amount of turbine work. *Due to this process efficiency is slightly increased but at the same time power developed is also decreased.*

The bleeding process in steam turbines approximates to cascade heating and tends to modify the Rankine cycle to a reversible cycle, thus increasing the efficiency ; but any increase in efficiency due to an approach to the condition of thermodynamic reversibility is accompanied by a decrease in power. Hence it follows that the thermodynamic benefits derived from the process of bleeding are of a *limited character. The ideal Rankine cycle, modified to take into account the effect of bleeding is known as the* ***regenerative cycle.***

Note. Please refer Art. 15.4 (Regenerative cycle) also.

19.14. ENERGY LOSSES IN STEAM TURBINES

The increase in heat energy required for doing mechanical work in actual practice as compared to the theoretical value, in which the process of expansion takes place strictly according to the adiabatic process, is termed as energy loss in a steam turbine.

The losses which appear in an actual turbine may be divided into two following groups :

1. Internal losses. Losses directly connected with the steam conditions while in its flow through the turbine. They may be further classified as :

(*i*) Losses in regulating valves.
(*ii*) Losses in nozzles (guide blades).
(*iii*) Losses in moving blades :
  (*a*) losses due to trailing edge wake ;
  (*b*) impingement losses ;
  (*c*) losses due to leakage of steam through the angular space ;
  (*d*) frictional losses ;
  (*e*) losses due to turning of the steam jet in the blades ;
  (*f*) losses due to shrouding.
(*iv*) Leaving velocity losses (exit velocity).
(*v*) Losses due to friction of disc carrying the blades and windage losses.
(*vi*) Losses due to clearance between the rotor and guide blade discs.
(*vii*) Losses due to wetness of steam.
(*viii*) Losses in exhaust piping etc.

2. External losses. Losses which do not influence the steam conditions. They may be further classified as :

(*i*) Mechanical losses.
(*ii*) Losses due to leakage of steam from the labyrinth gland seals.

19.15. STEAM TURBINE GOVERNING AND CONTROL

The objective of governing is to keep the turbine speed fairly constant irrespective of load.

The principal methods of steam turbine governing are as follows :

1. Throttle governing
2. Nozzle governing
3. By-pass governing
4. Combination of 1 and 2 and 1 and 3.

1. Throttle governing

Throttle governing is the most widely used particularly on *small turbines,* because *its initial cost is less* and *the mechanism is simple.* The object of throttle governing is to *throttle the steam* whenever *there is a reduction of load* compared to economic or design load for maintaining speed and *vice-versa.*

Fig. 19.56 (*a*) shows a simple throttle arrangement. To start the turbine for full load running valve A is opened. The operation of double beat valve B is carried out by an oil servomotor which is controlled by a centrifugal governor. As the steam turbine gains speed the valve B closes to throttle the steam and reduces the supply to the nozzle.

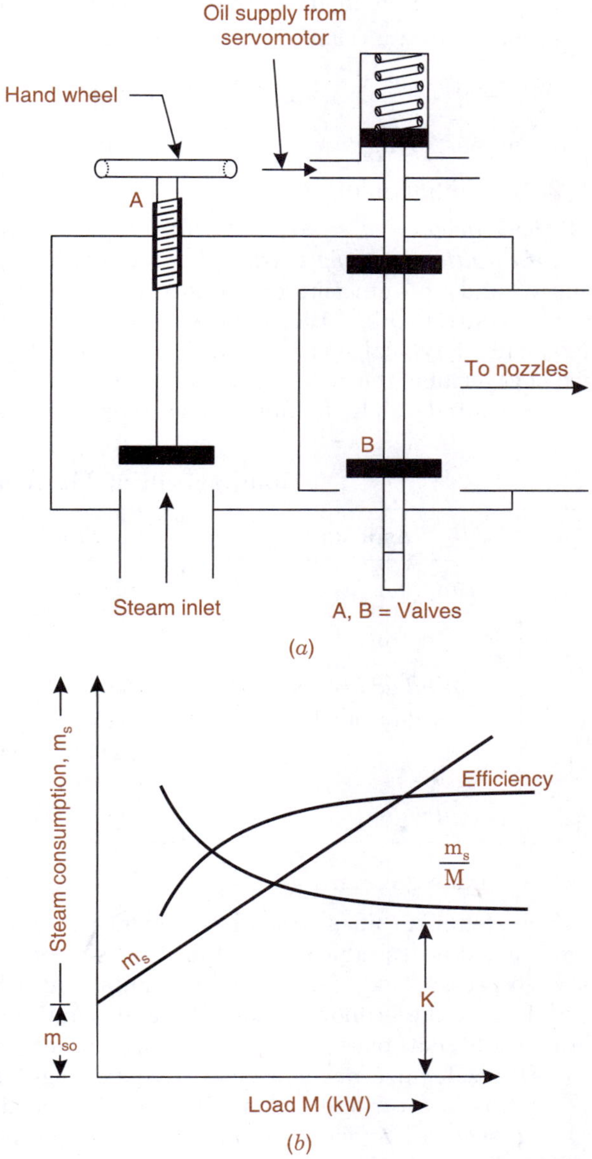

Fig. 19.56. Throttle governing.

For a turbine governed by throttling the relationship between steam consumption and load is given by the well known *Willan's line* as shown in Fig. 19.56 (*b*). Several tests have shown that when a turbine is governed by throttling, the Willan's line is straight. It is expressed as :

$$m_s = KM + m_{s_0} \qquad ...(19.37)$$

where, $\dot{m}_s$ = Steam consumption in kg/h at any load *M*,

m_{s_0} = Steam consumption in kg/h at no load,

m_{s_1} = Steam consumption in kg/h at full load,

M = Any other load in kW,

M_1 = Full load in kW, and

K = Constant.

m_{s_0} varies from about 0.1 to 0.14 times the full load consumption. The eqn. (19.37) can also be written as :

$\frac{m_s}{M} = K + \frac{m_{s_0}}{M}$, where $\frac{m_s}{M}$ is called the steam consumption per kWh.

2. Nozzle governing

The efficiency of a steam turbine is considerably reduced if throttle governing is carried out at low loads. An alternative, and more efficient form of governing is by means of nozzle control. Fig. 19.57 shows a diagrammatic arrangement of typical nozzle control governing. In this method of governing, the nozzles are grouped together 3 to 5 or more groups and supply of steam to each group is controlled by regulating valves. Under full load conditions the valves remain fully open.

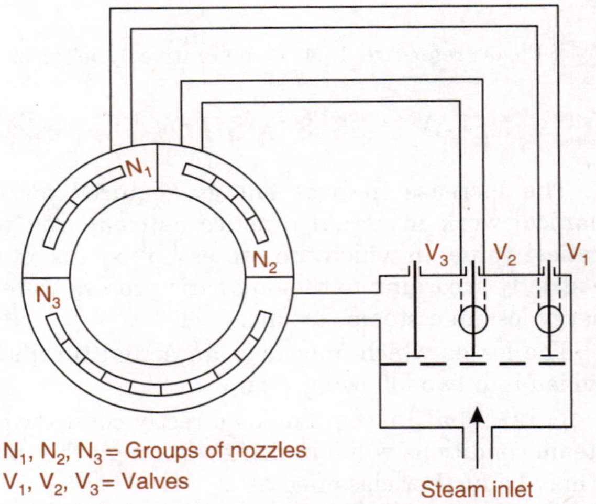

Fig. 19.57. Nozzle governing.

When the load on the turbine becomes more or less than the design value, the supply of steam to a group of nozzles may be varied accordingly so as to restore the original speed.

Nozzle control can only be applied to the first stage of a turbine. It is suitable for *simple impulse turbine* and *larger units which have an impulse stage followed* by an *impulse-reaction turbine.* In pressure compounded impulse turbines, there will be some drop in pressure at entry to second stage when some of the first stage nozzles are cut out.

Comparison of Throttle and Nozzle Control Governing

S. No.	Aspects	Throttle Control	Nozzle Control
1.	*Throttling losses*	Severe	No throttling losses (Actually there are a little throttling losses in nozzles valves which are partially open).
2.	*Partial admission losses*	Low	High.
3.	*Heat drop available*	Lesser	Larger
4.	*Use*	Used in impulse and reaction turbines both.	Used in impulse and also in reaction (if initial stage impulse) turbines.
5.	*Suitability*	Small turbines	Medium and larger turbines.

3. By-pass governing

The steam turbines which are designed to work at economic load it is desirable to have full admission of steam in the high pressure stages. At the maximum load, which is greater than the economic load, the additional steam required could not pass through the first stage since additional nozzles are not available. By-pass regulation allows for this in a turbine which is throttle governed, by means of a second by-pass valve in the first stage nozzle (Fig. 19.58). This valve opens when throttle valve has opened a definite amount. Steam is by-passed through the second valve to a lower stage in the turbine. When by-pass valve operates it is under the control of the turbine governor. The secondary and tertiary supplies of steam in the lower stages increase the work output in these stages, but there is a *loss in efficiency* and a curving of the Willian's line.

In reaction turbines, because of the pressure drop required in the moving blades, nozzles control governing is not possible, and *throttle governing plus by-pass governing, is used.*

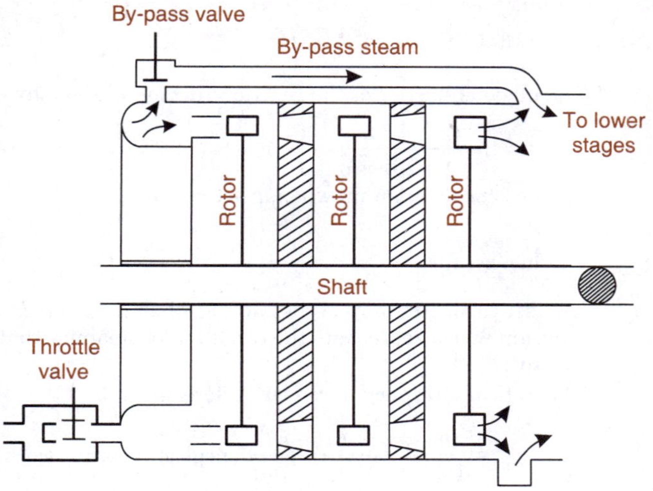

Fig. 19.58. By-pass governing.

19.16. SPECIAL FORMS OF STEAM TURBINES

In many industries such as chemical, sugar refining, paper making, textile etc., where combined use of power and heating and process work is required it is wasteful to generate steam for power and process purposes separately, because about 70 per cent of heat supplied for power purposes will normally be carried away by the cooling water. On the other hand, if the engine or turbine is operated with a normal exhaust pressure then the temperature of the exhaust steam is too low to be of any use for heating process. It would be possible to generate the required power and still have available for process work a large quantity of heat in the exhaust steam, if suitable modification of the initial steam pressure and exhaust pressure is made. Thus in combined power and process plants following type of steam turbines are used : (1) *Back pressure turbines*, and (2) *Steam extraction or pass-out turbines.*

1. Back pressure turbine

In this type of turbine steam at boiler pressure enters the turbine and is exhausted into a pipe. This pipe leads to process plant or other turbine. The back pressure turbine may be used in cases where the power generated (by expanding steam) from economical initial pressure down to the heating pressure is equal to, or greater than, the power requirements. The steam exhausted from the turbine is *usually superheated* and in most cases it is not suitable for process work due to the reasons :

(*i*) It is impossible to control its temperature, and

(*ii*) Rate of the heat transfer from superheated steam to the heating surface is lower than that of saturated steam. Consequently a desuperheater is invariably used. To enhance the power capacity of the existing installation, a high pressure boiler and a back-pressure turbine are added to it. This added high pressure boiler supplies steam to the back pressure turbine which exhausts into the old low pressure turbine.

2. Extraction or pass out turbine

It is found that in several cases the power available from a back pressure turbine (through which the whole of the steam flows) is appreciably less than that required in the factory and this may be due to the following reasons :

(*i*) Small heating or process requirements ;

(*ii*) A relatively high exhaust pressure ; and

(*iii*) A combination of the both.

In such a case it would be possible to install a back-pressure turbine to provide the heating steam and a condensing turbine to generate extra power, but it is possible, and useful, to combine functions of both machines in a single turbine. Such a machine is called **extraction or pass out turbine** and here at some point intermediate between inlet and exhaust some steam is extracted or passed out for process or heating purposes. In this type of turbine a sensitive governor is used which controls the admission of steam to the high pressure section so that regardless of power or process requirements, constant speed is maintained.

Exhaust or low pressure turbine

If an uninterrupted supply of low pressure steam is available (such as from reciprocating steam engines exhaust) it is possible to improve the efficiency of the whole plant by fitting an exhaust or low pressure turbine. The use of exhaust turbine is chiefly made where there are several reciprocating steam engines which work intermittently ; and are non-condensing (*e.g.,* rolling mill and colliery engines). The exhaust steam from these engines is expanded in an exhaust turbine and then condensed. In this turbine some form of heat accumulator is needed to collect the more or less irregular supply of low pressure steam from the non-condensing steam engines and deliver it to the turbine at the rate required. In some cases when the supply of low pressure steam falls below the demand, live steam from the boiler, with its pressure and temperature reduced ; is used to make up the deficiency.

The necessary drop in pressure may be obtained by the use of a reducing valve, or for large flows, more economically by expansion through another turbine. The high pressure and low pressure turbines are sometimes combined on a common spindle and because of two supply pressures this combined unit is known as '**mixed pressure turbine**'.

HIGHLIGHTS

1. The **steam turbine** is a prime mover in which the potential energy of the steam is transformed into kinetic energy, and latter in its turn is transformed into the mechanical energy of rotation of the turbine shaft.
2. The most important **classification** of steam turbines is as follows :
   (*i*) Impulse turbines (*ii*) Reaction turbines
   (*iii*) Combination of impulse and reaction turbines.
3. The main difference between **Impulse** and **Reaction turbines** lies in the way in which steam in expanded while its moves through them. In the former type, steam expands in the nozzle and its pressure does not change as it moves over the blades while in the latter type the steam expands continuously as it passes over the blades and thus there is a gradual fall in pressure during expansion.
4. The different methods of compounding are :
   (*i*) Velocity compounding
   (*ii*) Pressure compounding
   (*iii*) Pressure velocity compounding
   (*iv*) Reaction turbine.
5. Force (tangential) on the wheel $= \dot{m}_s(C_{w_1} + C_{w_0})$ Nm

   Power per wheel $= \dfrac{\dot{m}_s(C_{w_1} + C_{w_0}) \times C_{bl}}{1000}$ kW

   Blade or diagram efficiency, $\eta_{bl} = \dfrac{2C_{bl}(C_{w_1} + C_{w_0})}{C_1^{\,2}}$

   Stage efficiency, $\eta_{\text{stage}} = \dfrac{C_{bl}(C_{w_1} + C_{w_0})}{(h_1 - h_2)}$.
6. The axial thrust on the wheel due to *difference* between the velocities of flow at entrance and outlet.
   Axial force on the wheel $= \dot{m}_s(C_{f_1} - C_{f_0})$.
7. Energy converted to heat by blade friction = Loss of kinetic energy during flow over blades $= \dot{m}_s(C_{r_1}^{\,2} - C_{r_0}^{\,2})$.
8. Optimum value of ratio of blade speed to steam speed is,
   $\rho_{opt.} = \dfrac{\cos\alpha}{2}$.
9. The blade efficiency for *two-stage turbine* will be maximum when, $\rho_{opt.} = \dfrac{\cos\alpha}{4}$.
   In general optimum blade speed ratio for maximum blade efficiency or maximum work done is given by

   $$\rho_{opt.} = \frac{\cos\alpha}{2n}$$

   and the work done in the last row $= \dfrac{1}{2^n}$ of total work,
   where n is the number of moving/rotating blade rows in series.
   In practice more than *two rows are hardly preferred.*
10. The **degree of reaction** of reaction turbine stage is defined as the ratio of heat drop over moving blades to the total heat drop in the stage.
11. The blade efficiency of the reaction turbine is given by

    $$\eta_{bl} = 2 - \frac{2}{1 + 2\rho\cos\alpha - \rho^2};$$

    η_{bl} becomes maximum when, $\rho = \cos\alpha$

    and hence $(\eta_{bl})_{\max} = \dfrac{2\cos^2\alpha}{1 + \cos^2\alpha}$.
12. The **state point** may be defined as that point on *h-s* diagram which represents the condition of steam at that instant.
13. Theoretical efficiency of reheat cycle is given by

    $$\eta_{\text{thermal}} = \frac{(h_1 - h_2) + (h_3 - h_4)}{(h_1 - h_{f_4}) + (h_3 - h_2)}, \text{ neglecting pump work.}$$
14. The principal methods of **steam governing** are as follows:
    (*i*) Throttle governing
    (*ii*) Nozzle governing
    (*iii*) By-pass governing
    (*iv*) Combination of (*i*), (*ii*) and (*iii*).

OBJECTIVE TYPE QUESTIONS

Choose the Correct Answer :

1. In case of impulse steam turbine
   (*a*) there is enthalpy drop in fixed and moving blades
   (*b*) there is enthalpy drop only in moving blades
   (*c*) there is enthalpy drop in nozzles
   (*d*) none of the above.
2. De-Laval turbine is
   (*a*) pressure compounded impulse turbine
   (*b*) velocity compounded impulse turbine
   (*c*) simple single wheel impulse turbine
   (*d*) simple single wheel reaction turbine.
3. The pressure on the two sides of the impulse wheel of a steam turbine
   (*a*) is same
   (*b*) is different
   (*c*) increases from one side to the other side
   (*d*) decreases from one side to the other side.
4. In De-Laval steam turbine
   (*a*) the pressure in the turbine rotor is approximately same as in condenser
   (*b*) the pressure in the turbine rotor is higher than pressure in the condenser
   (*c*) the pressure in the turbine rotor gradually decreases from inlet to exit to condenser
   (*d*) none of the above.
5. In case of reaction steam turbine
   (*a*) there is enthalpy drop both in fixed and moving blades
   (*b*) there is enthalpy drop only in fixed blades
   (*c*) there is enthalpy drop only in moving blades
   (*d*) none of the above.

6. Curtis turbine is
(*a*) reaction steam turbine
(*b*) pressure velocity compounded steam turbine
(*c*) pressure compounded impulse steam turbine
(*d*) velocity compounded impulse steam turbine.

7. Rateau steam turbine is
(*a*) reaction steam turbine
(*b*) velocity compounded impulse steam turbine
(*c*) pressure compounded impulse steam turbine
(*d*) pressure velocity compounded steam turbine.

8. Parson's turbine is
(*a*) pressure compounded steam turbine
(*b*) simple single wheel, impulse steam turbine
(*c*) simple single wheel reaction steam turbine
(*d*) multiwheel reaction steam turbine.

9. Blade or diagram efficiency is given by
(*a*) $\frac{(C_{w_1} \pm C_{w_0})C_{bl}}{C_1}$ (*b*) $\frac{2C_{bl}(C_{w_1} \pm C_{w_0})}{C_1^2}$
(*c*) $\frac{C_{bl}^2}{C_1^2}$ (*d*) $\frac{C_1^2 - C_0^2}{C_1^2}$.

10. Axial thrust on rotor of steam turbine is
(*a*) $\dot{m}_s(C_{f_1} - C_{f_0})$ (*b*) $\dot{m}_s(C_{f_1} - 2C_{f_0})$
(*c*) $\dot{m}_s(C_{f_1} + C_{f_0})$ (*d*) $\dot{m}_s(2C_{f_1} - C_{f_0})$.

11. Stage efficiency of steam turbine is
(*a*) $\eta_{blade}/\eta_{nozzle}$ (*b*) $\eta_{nozzle}/\eta_{blade}$
(*c*) $\eta_{nozzle} \times \eta_{blade}$ (*d*) none of the above.

12. For maximum blade efficiency for single stage impulse turbine
(*a*) $\rho \left(= \frac{C_{bl}}{C_1}\right) = \cos^2 \alpha$ (*b*) $\rho = \cos \alpha$
(*c*) $\rho = \frac{\cos \alpha}{2}$ (*d*) $\rho = \frac{\cos^2 \alpha}{2}$.

13. Degree of reaction as referred to steam turbine is defined as
(*a*) $\frac{\Delta h_f}{\Delta h_m}$ (*b*) $\frac{\Delta h_m}{\Delta h_f}$
(*c*) $\frac{\Delta h_m}{\Delta h_m + \Delta h_f}$ (*d*) $\frac{\Delta h_f}{\Delta h_f + \Delta h_m}$.

14. For Parson's reaction steam turbine, degree of reaction is
(*a*) 75% (*b*) 100%
(*c*) 50% (*d*) 60%.

15. The maximum efficiency for Parson's reaction turbine is given by
(*a*) $\eta_{max} = \frac{\cos \alpha}{1 + \cos \alpha}$ (*b*) $\eta_{max} = \frac{2\cos \alpha}{1 + \cos \alpha}$
(*c*) $\eta_{max} = \frac{2\cos^2 \alpha}{1 + \cos^2 \alpha}$ (*d*) $\eta_{max} = \frac{1 + \cos^2 \alpha}{2\cos^2 \alpha}$.

16. Reheat factor in steam turbines depends on
(*a*) exit pressure only
(*b*) stage efficiency only
(*c*) initial pressure and temperature only
(*d*) all of the above.

17. For multistage steam turbine reheat factor is defined as
(*a*) stage efficiency × nozzle efficiency
(*b*) commulative enthalpy drop × η_{nozzle}
(*c*) $\frac{\text{commulative enthalpy drop}}{\text{isentropic enthalpy drop}}$
(*d*) $\frac{\text{isentropic enthalpy drop}}{\text{cumulative actual enthalpy drop}}$.

18. The value of reheat factor normally varies from
(*a*) 0.5 to 0.6 (*b*) 0.9 to 0.95
(*c*) 1.02 to 1.06 (*d*) 1.2 to 1.6.

19. Steam turbines are governed by the following methods
(*a*) Throttle governing
(*b*) Nozzle control governing
(*c*) By-pass governing
(*d*) All of the above.

20. In steam turbines the reheat factor
(*a*) increases with the increase in number of stages
(*b*) decreases with the increase in number of stages
(*c*) remains same irrespective of number of stages
(*d*) none of the above.

ANSWERS

1.	(*c*)	**2.**	(*c*)	**3.**	(*a*)	**4.**	(*a*)
5.	(*a*)	**6.**	(*d*)	**7.**	(*c*)	**8.**	(*d*)
9.	(*b*)	**10.**	(*a*)	**11.**	(*c*)	**12.**	(*c*)
13.	(*c*)	**14.**	(*c*)	**15.**	(*c*)	**16.**	(*d*)
17.	(*c*)	**18.**	(*c*)	**19.**	(*d*)	**20.**	(*a*).

THEORETICAL QUESTIONS

1. Define a steam turbine and state its fields of application.
2. How are the steam turbines classified ?
3. Discuss the advantages of a steam turbine over the steam engines.
4. Explain the difference between an impulse turbine and a reaction turbine.
5. What do you mean by compounding of steam turbines ? Discuss various methods of compounding steam turbines.
6. What methods are used in reducing the speed of the turbine rotor ?
7. Explain with the help of neat sketch a single-stage impulse turbine. Also explain the pressure and velocity variations along the axial direction.
8. Define the following as related to steam turbines
(*i*) Speed ratio (*ii*) Blade velocity co-efficient
(*iii*) Diagram efficiency (*iv*) Stage efficiency.

9. In case of steam turbines derive expressions for the following :
   (*i*) Force (*ii*) Work done
   (*iii*) Diagram efficiency (*iv*) Stage efficiency
   (*v*) Axial thrust.
10. Derive the expression for maximum blade efficiency in a single-stage impulse turbine.
11. Explain the pressure compounded impulse steam turbine showing pressure and velocity variations along the axis of the turbine.
12. Explain velocity compounded impulse steam turbine showing pressure and velocity variations along the axis of the turbine.
13. Define the term "degree of reaction" used in reaction turbines and prove that it is given by

$$R_d = \frac{C_f}{2C_{bl}} (\cot \phi - \cot \theta) \text{ when } C_{f_1} = C_{f_0} = C_f.$$

Further prove that the moving and fixed blades should have the same shape for a 50% reaction.

14. Prove that the diagram or blade efficiency of a single stage reaction turbine is given by

$$\eta_{bl} = 2 - \frac{2}{1 + 2\rho \cos \alpha - \rho^2}, \text{ where } R_d = 50\% \text{ and } C_{f_1} = C_{f_0}$$

Further prove that maximum blade efficiency is given by $(\eta_{bl})_{max} = \dfrac{2 \cos^2 \alpha}{1 + \cos^2 \alpha}$.

15. Explain 'reheat factor'. Why is its magnitude always greater than unity ?
16. Describe the process and purpose of reheating as applicable to steam flowing through a turbine.
17. State the advantages and disadvantages of reheating steam.
18. Write a short note on 'bleeding of steam turbines'.
19. Enumerate the energy losses in steam turbines.
20. Describe briefly the various methods of 'steam turbine governing'.

UNSOLVED EXAMPLES

Impulse Turbines

1. A steam jet enters the row of blades with a velocity of 380 m/s at an angle of 22° with the direction of motion of the moving blades. If the blade speed is 180 m/s and there is no thrust on the blades, determine the inlet and outlet blade angles. Velocity of steam while passing over the blade is reduced by 10%. Also determine the power developed by turbine when the rate of flow of steam is 1000 kg per minute. [**Ans.** 879 kW]
2. In a simple impulse turbine, the nozzles are inclined at 20° to the direction of motion of moving blades. The steam leaves the nozzles at 375 m/s. The blade speed is 165 m/s. Find suitable inlet and outlet angles for the blades in order that the axial thrust is zero. The relative velocity of steam as it flows over the blades is reduced by 15% by friction. Determine also the power developed for a flow rate of 10 kg/s. [**Ans.** 34°, 41°, 532 kW]
3. In an impulse turbine the nozzles are inclined at 24° to the plane of rotation of the blades. The steam speed is 1000 m/s and blade speed is 400 m/s. Assuming equiangular blades, determine :
   (*i*) Blade angles
   (*ii*) Force on the blades in the direction of motion
   (*iii*) Power developed for a flow rate of 1000 kg/h.
   [**Ans.** (*i*) 39°, (*ii*) 1.135 kN, (*iii*) 113.5 kW]
4. In a De-Laval turbine, the steam issues from the nozzles with a velocity of 850 m/s. The nozzle angle is 20°. Mean blade velocity is 350 m/s. The blades are equiangular. The mass flow rate is 1000 kg/min. Friction factor is 0.8. Determine : (*i*) blade angles, (*ii*) axial thrust on the end bearing, (*iii*) power developed in kW, (*iv*) blade efficiency, (*v*) stage efficiency, if nozzle efficiency is 93%.
   [**Ans.** (*i*) 33°, 33°, (*ii*) 500 N, (*iii*) 4666.7 kW, (*iv*) 77.5%, (*v*) 72.1%]
5. In a single stage impulse turbine the nozzles discharge the steam on to the blades at an angle of 25° to the plane of rotation and the fluid leaves the blades with an absolute velocity of 300 m/s at an angle of 120° to the direction of motion of the blades. If the blades have equal inlet and outlet angles and there is no axial thrust, estimate :
   (*i*) Blade angle
   (*ii*) Power produced per kg/s flow of steam
   (*iii*) Diagram efficiency.
   [**Ans.** (*i*) 36.3°, (*ii*) 144 kW, (*iii*) 0.762]
6. Steam enters the blade row of an impulse turbine with a velocity of 600 m/s at an angle of 25° to the plane of rotation of the blades. The mean blade speed is 255 m/s. The blade angle on the exit side is 30°. The blade friction co-efficient is 10%. Determine :
   (*i*) Work done per kg of steam
   (*ii*) Diagram efficiency
   (*iii*) Axial thrust per kg of steam/s.
   [**Ans.** (*i*) 150.45 kW, (*ii*) 83.6%, (*iii*) 90 N]
7. The nozzles of an impulse turbine are inclined at 22° to the plane of rotation. The blade angles both at inlet and outlet are 36°. The mean diameter of the blade ring is 1.25 m and the steam velocity is 680 m/s. Assuming shockless entry determine : (*i*) The speed of the turbine rotor in r.p.m., (*ii*) the absolute velocity of steam leaving the blades, and (*iii*) the torque on the rotor for a flow rate of 2500 kg/h.
   [**Ans.** (*i*) 4580 r.p.m., (*ii*) 225 m/s, (*iii*) 290.5 Nm]
8. A single-stage steam turbine is provided with nozzles from which steam is released at a velocity of 1000 m/s at an angle of 24° to the direction of motion of blades. The speed of the blades is 400 m/s. The blade angles at inlet and outlet are equal. Find : (*i*) Inlet blade angle, (*ii*) force exerted on the blades in the direction of their motion, (*iii*) power developed in kW for steam flow rate of 40000 kg/h.
   Assume that the steam enters and leaves the blades without shock. [**Ans.** (*i*) 39°, (*ii*) 1135 N, (*iii*) 4540 kW]

9. In a single row impulse turbine the nozzle angle is 30° and the blade speed is 215 m/s. The steam speed is 550 m/s. The blade friction co-efficient is 0.85. Assuming axial exit and a flow rate of 700 kg/h, determine :
   (*i*) Blade angles.
   (*ii*) Absolute velocity of steam at exit.
   (*iii*) The power output of the turbine.
   [**Ans.** (*i*) 46°, 49°, (*ii*) 243 m/s, (*iii*) 19.8 kW]

10. In a steam turbine, steam expands from an inlet condition of 7 bar and 300°C with an isentropic efficiency of 0.9. The nozzle angle is 20°. The stage operates at optimum blade speed ratio. The blade inlet angle is equal to the outlet angle. Determine :
   (*i*) Blade angles.
   (*ii*) Power developed if the steam flow rate is 0.472 kg/s.
   [**Ans.** (*i*) 36°, (*ii*) 75 kW]

11. Steam at 7 bar and 300°C expands to 3 bar in an impulse stage. The nozzle angle is 20°, the rotor blades have equal inlet and outlet angles and the stage operates with the optimum blade speed ratio. Assuming that isentropic efficiency of nozzles is 90% and velocity at entry to the stage is negligible, deduce the blade angles used and the mass flow required for this stage to produce 50 kW.
   [**Ans.** 36°, 0.317 kg/s]

12. In a two-stage velocity compounded steam turbine, the mean blade speed is 150 m/s while the steam velocity as it is issued from the nozzle is 675 m/s. The nozzle angle is 20°. The exit angle of first row moving blade, fixed blade and the second row moving blades are 25°, 25° and 30° respectively. The blade friction co-efficient is 0.9. If the steam flow rate is 4.5 kg/s, determine :
   (*i*) Power output. (*ii*) Diagram efficiency.
   [**Ans.** (*i*) 807 kW, (*ii*) 78.5%]

Reaction Turbines

13. At a particular stage of reaction turbine, the mean blade speed is 60 m/s and the steam pressure is 3.5 bar with a temperature of 175°C. The identical fixed and moving blades have inlet angles of 30° and outlet angles of 20°. Determine :
   (*i*) The blade height if it is $\frac{1}{10}$th of the blade ring diameter, for flow rate of 13.5 kg/s.
   (*ii*) The power developed by a pair.
   (*iii*) Specific enthalpy drop if the stage efficiency is 85%.
   [**Ans.** (*i*) 64 mm, (*ii*) 218 kW, (*iii*) 19.1 kJ/kg]

14. In a stage of impulse reaction turbine operating with 50% degree of reaction, the blades are identical in shape. The outlet angle of the moving blades is 19° and the absolute discharge velocity of steam is 100 m/s in the direction at 100° to the motion of the blades. If the rate of flow of steam through the turbine is 15000 kg/h, calculate the power developed by the turbine in kW. [**Ans.** 327.5 kW]

15. At a stage in a reaction turbine the pressure of steam is 0.34 bar and the dryness 0.95. For a flow rate of 36000 kg/h, the stage develops 950 kW. The turbine runs at 3600 r.p.m. and the velocity of flow is 0.72 times the blade velocity. The outlet angle of both stator and rotor blades is 20°. Determine at this stage :
   (*i*) Mean rotor diameter. (*ii*) Height of blades.
   [**Ans.** (*i*) 0.951 m, (*ii*) 115 mm]

16. In a multi-stage reaction turbine at one of the stages the rotor diameter is 1250 mm and speed ratio 0.72. The speed of the rotor is 3000 r.p.m. Determine : (*i*) The blade inlet angle if the outlet blade angle is 22°, (*ii*) Diagram efficiency, (*iii*) The percentage increase in diagram efficiency and rotor speed if turbine is designed to run at the best theoretical speed.
   [**Ans.** (*i*) 61.5°, (*ii*) 82.2%, (*iii*) 30.47%]

17. In a 50 per cent reaction turbine stage running at 3000 r.p.m., the exit angles are 30° and the inlet angles are 50°. The mean diameter is 1 m. The steam flow rate is 10000 kg/min and the stage efficiency is 85%. Determine :
   (*i*) Power output of the stage.
   (*ii*) The specific enthalpy drop in the stage.
   (*iii*) The percentage increase in the relative velocity of steam when it flows over the moving blades.
   [**Ans.** (*i*) 11.6 MW, (*ii*) 82 kJ/kg, (*iii*) 52.2%]

18. Twelve successive stages of a reaction turbine have blades with effective inlet and outlet angles of 80° and 20° respectively. The mean diameter of the blade row is 1.2 m and the speed or rotation is 3000 r.p.m. Assuming constant velocity of flow throughout, estimate the enthalpy drop per stage.
   For a steam inlet condition of 10 bar and 250°C and an outlet condition of 0.2 bar, estimate the stage efficiency. Assume a reheat factor of 1.04, determine the blade height at a stage where the specific volume is 1.02 m^3/kg. [**Ans.** 40.4 kJ/kg, 70.3%, 57 mm]

20 Steam Condensers

20.1. Introduction. 20.2. Vacuum. 20.3. Organs of a steam condensing plant. 20.4. Classification of condensers—Jet condensers—Surface condensers—Reasons for inefficiency in surface condensers—Comparison between jet and surface condensers. **20.5. Sources of air in condensers. 20.6. Effects of air leakage in a condenser. 20.7. Methods for obtaining maximum vacuum in condensers. 20.8. Vacuum measurement. 20.9. Vacuum efficiency. 20.10. Condenser efficiency. 20.11. Dalton's law of partial pressures. 20.12. Determination of mass of cooling water. 20.13. Heat transmission through walls of tubes of a surface condenser. 20.14. Air pumps. 20.15. Cooling towers**—Worked Examples. *Highlights—Objective Type Questions—Theoretical Questions—Unsolved Examples.*

20.1. INTRODUCTION

A **steam condenser** *is a device or an appliance in which steam condenses and heat released by steam is absorbed by water*. It serves the following purposes :

1. It *maintains a very low back pressure* on the exhaust side of the piston of the steam engine or turbine. Consequently, the steam expands to a greater extent which results in an *increase in available heat energy* for converting into mechanical work. The shaded area in Fig. 20.1. (*i.e.*, area 44′5′5) shows the increase in work obtained by fitting a condenser to a non-condensing engine. *The thermal efficiency of a condensing unit therefore is higher than that of non-condensing unit for the same available steam.*

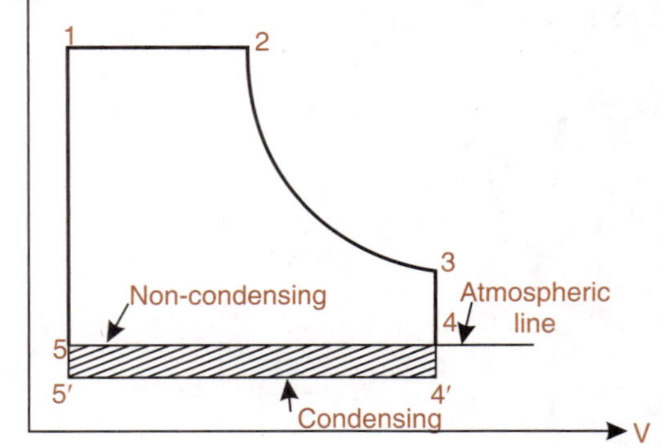

Fig. 20.1

2. It supplies to the boiler pure and hot feed water as the condensed steam which is discharged from the condenser and collected in a hot well, can be used as feed water for the boiler.

20.2. VACUUM

Vacuum is *sub-atmospheric pressure*. It is measured as the *pressure depression below atmospheric*. The condensation of steam in a closed vessel produces a partial vacuum by reason of the great reduction in the volume of the low pressure steam or vapour. The back pressure in steam engine or steam turbine can lowered from 1.013 to 0.2 bar abs. or even less. Since the steam engines are intermittent flow machines and as such cannot take the advantage of a very low vacuum, therefore, for most steam engines the exhaust pressure is about 0.2 to 0.28 bar abs. On the other hand, in steam turbines, which are continuous flow machines, the back pressure may be about 0.025 bar abs.

20.3. ORGANS OF A STEAM CONDENSING PLANT

A steam condensing plant mainly consists of the following *organs/elements* :

1. Condenser (To condense the steam).
2. Supply of cooling (or injection) water.
3. Wet air pump (To remove the condensed steam, the air and uncondensed water vapour and gases

from the condenser ; separate pumps may be used to deal with air and condensate).

4. Hot well (where the condensate can be discharged and from which the boiler feed water is taken).
5. Arrangement for recooling the cooling water in case surface condenser is employed.

20.4. CLASSIFICATION OF CONDENSERS

Mainly, condensers are of two types : (1) Jet condensers, (2) Surface condenser.

In ***jet condensers***, the *exhaust steam and water come in direct contact with each other and temperature of the condensate is the same as that of cooling water leaving the condenser*. The cooling water is usually sprayed into the exhaust steam to cause, rapid condensation.

In ***surface condensers***, *the exhaust steam and water do not come into direct contact*. The steam passes over the outer surface of tubes through which a supply of cooling water is maintained. There may be single-pass or double-pass. In single-pass condensers, the water flows in one direction only through all the tubes, while in two-pass condenser the water flows in one direction through the tubes and returns through the remainder.

A jet condenser is simpler and cheaper than a surface condenser. It should be installed when the cooling water is cheaply and easily made suitable for boiler feed or when a cheap source of boiler and feed water is available. A surface condenser is most commonly used because the condensate obtained is not thrown as a waste but returned to the boiler.

20.4.1. Jet Condensers

These condensers may be classified as :

(*a*) Parallel flow type

(*b*) Counter flow type

(*c*) Ejector type.

Parallel flow and counter flow condensers are further sub-divided into two types : (*i*) Low level type (*ii*) High level type.

In *parallel-flow type* of condenser, both the exhaust steam and cooling water find their entry at the top of the condenser and then flow downwards and condensate and water are finally collected at the bottom.

In *counter-flow type,* the steam and cooling water enter the condenser from opposite directions. Generally, the exhaust steam travels in upward direction and meet the cooling water which flows downwards.

Low level jet condenser (Parallel-flow)

In the Fig. 20.2 is shown a line sketch of a low level parallel flow condenser. The exhaust steam is entering the condenser from the top and cold water is being sprayed on its way. The baffle plate provided in it ensures the proper mixing of the steam and cooling water. An extraction pump at the bottom discharges the condensate to the hot well from where it may be fed to the boiler if the cooling water being used is free from impurities. A separate dry pump may be incorporated to maintain proper vacuum.

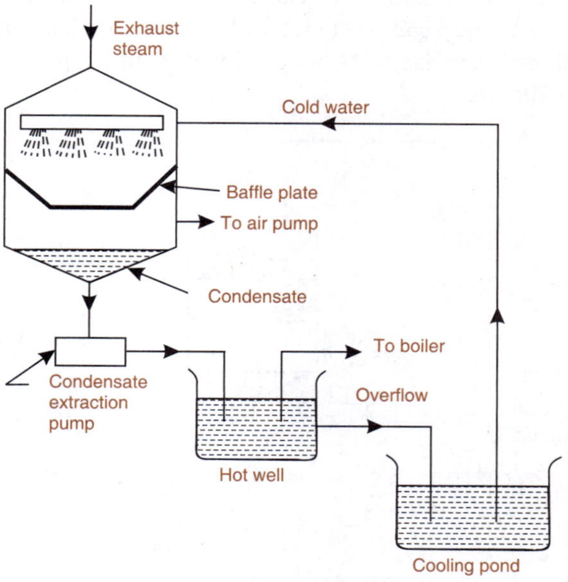

Fig. 20.2. Low level jet condenser (Parallel flow).

Low level jet condenser (Counter-flow)

Refer to Fig. 20.3 *L*, *M* and *N* are the perforated trays which break up water into jets. The steam moving upwards comes in contact with water and gets condensed. The condensate and water mixture is sent to the hot well by means of an extraction pump and the air is removed by an air suction pump provided at the top of the condenser.

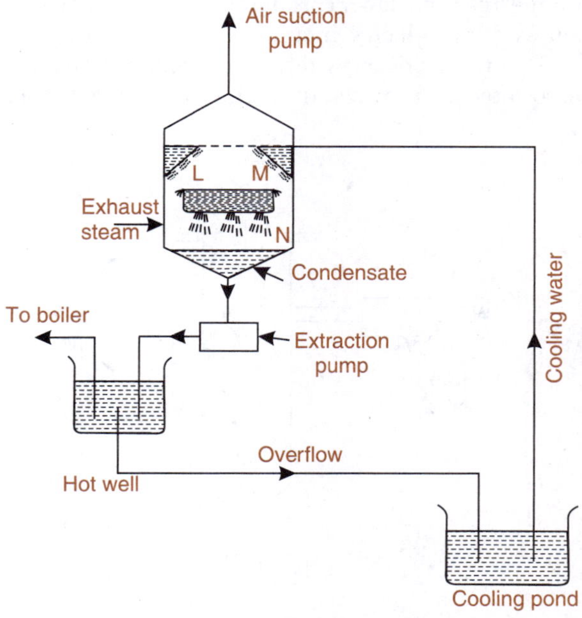

Fig. 20.3. Low level jet condenser (Counter-flow).

High level jet condenser (Counter-flow type).

In Fig. 20.4 is shown a high level counter-flow jet condenser. It is also called ***barometric condenser***. In this case the shell is placed at a height about 10.363 metres above hot well and thus the necessity of providing an extraction pump can be obviated. However provision of own injection pump has to be made if water under pressure is not available.

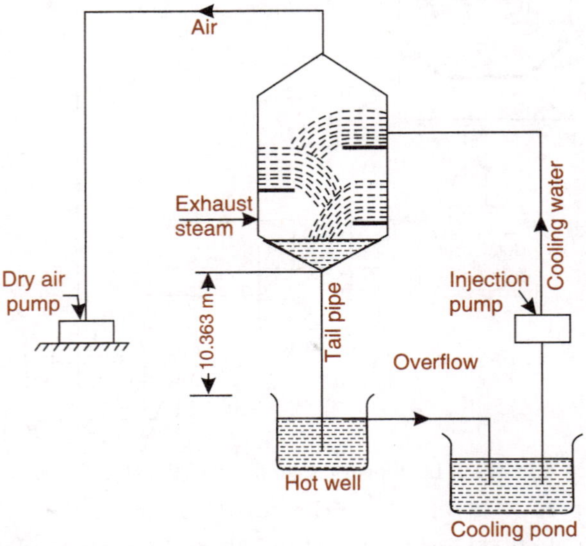

Fig. 20.4. High level jet condenser (Counter-flow type).

Ejector condenser

Fig. 20.5 shows the schematic sketch of an ejector condenser. Here the exhaust steam and cooling water mix in hollow truncated cones. The cold water having a head of about 6 metres flow down through the number of cones and as it moves its velocity increases and drop in pressure results. Due to this decreased pressure exhaust steam along with associated air is drawn through the truncated cones and

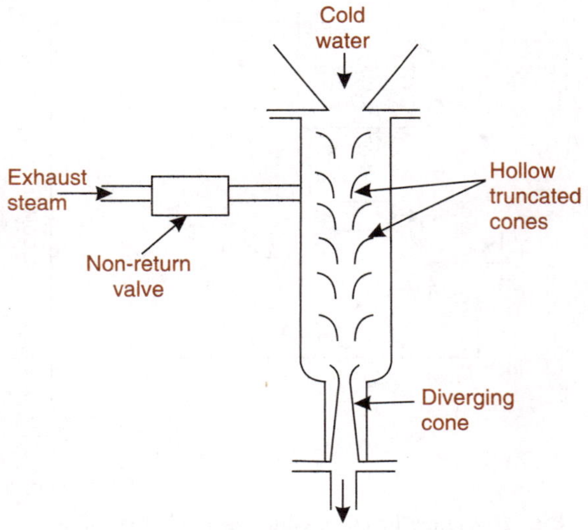

Fig. 20.5. Ejector condenser.

finally lead to diverging cone. In the diverging cone, a portion of kinetic energy gets converted into pressure energy which is more than the atmospheric so that condensate consisting of condensed steam, cooling water and air is discharged into the hot well. The exhaust steam inlet is provided with a non-return valve which does not allow the water from hot well to rush back to the engine in case a failure of cooling water supply to condenser.

20.4.2. Surface Condensers

Most condensers are generally classified on the direction of flow of condensate, the arrangement of the tubing and the position of the condensate extraction pump. The following is the main classification of surface condensers :

(*i*) Down-flow type
(*ii*) Central-flow type
(*iii*) Inverted-flow type
(*iv*) Regenerative type
(*v*) Evaporative type.

(*i*) **Down-flow type**

In Fig. 20.6 is shown a down flow type of surface condenser. It consists of a shell which is generally of cylindrical shape ; though other types are also used. It has cover plates at the ends and furnished with number of parallel brass tubes. A baffle plate partitions the water box into two sections. The cooling water enters the shell at

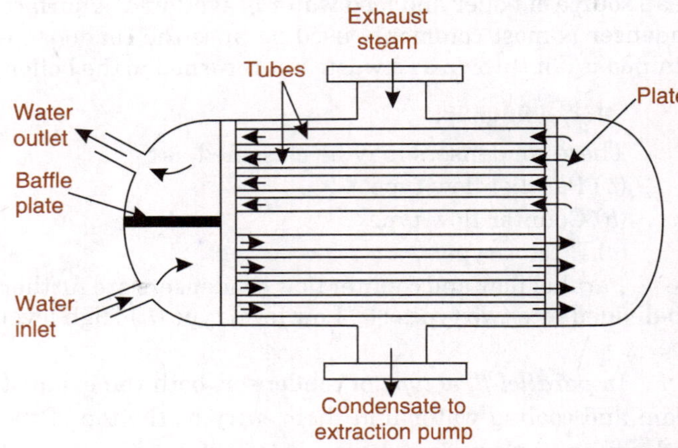

Fig. 20.6. Down flow type.

the lower half section and after travelling through the upper half section comes out through the outlet. The exhaust steam entering shell from the top flows down over the tubes and gets condensed and is finally removed by an extraction pump. Due to the fact that steam flows in a direction right angle to the direction of flow of water, it is also called *cross-surface condenser*.

(*ii*) **Central-flow type**

Refer to Fig. 20.7 (on next page). In this type of condenser, the suction pipe of the air extraction pump is located in the centre of the tubes which results in radial flow of the steam. The better contact between the outer surface of the tubes and steam is ensured, due to large passages the pressure drop of steam is reduced.

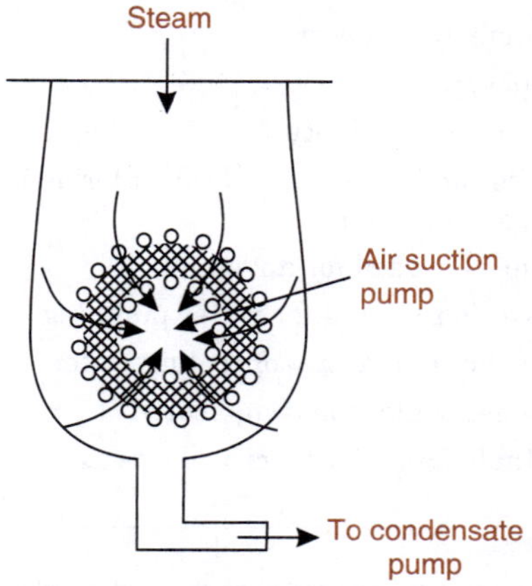

Fig. 20.7. Central flow type.

(iii) **Inverted-flow type**

This type of condenser has the air suction at the top, the steam after entering at the bottom rises up and then again flows down to the bottom of the condenser, by following a path near the outer surface of the condenser. The condensate extraction pump is at the bottom.

(iv) **Regenerative type**

This type is applied to condensers adopting a regenerative method of heating of the condensate. After leaving the tube nest, the condensate is passed through the entering exhaust steam from the steam engine or turbine thus raising the temperature of the condensate, for use as feed water for the boiler.

(v) **Evaporative type**

Fig. 20.8 shows the schematic sketch of an *evaporative condenser*. The underlying principle of this condenser is that when a limited quantity of water is available, its quantity needed to condense the steam can be reduced by causing the circulating water to evaporate under a small partial pressure.

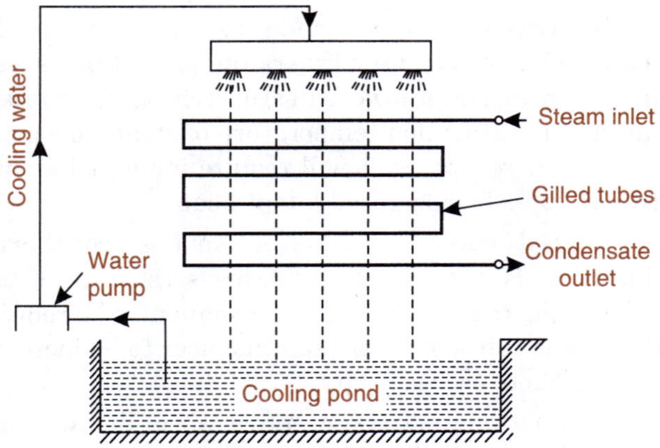

Fig. 20.8. Evaporative condenser.

The exhaust steam enters at the top through gilled pipes. The water pump sprays water on the pipes and descending water condenses the steam. The water which is not evaporated falls into the open tank (cooling pond) under the condenser from which it can be drawn by circulating water pump and used over again. The evaporative condenser is placed in open air and finds its application in small size plants.

20.4.3. Reasons for Inefficiency in Surface Condensers

The various reasons for inefficiency in surface condensers are discussed below :

1. The pressure inside the condenser is less than atmospheric, and in order to obtain the maximum work from unit mass of steam, the pressure should be as low as possible. The pressure in the condenser also depends upon the amount of air. Owing to high vacuum pressure in the condenser it is impossible to prevent air from leaking in through the joints thereby increasing the pressure in the condenser and thus limiting the amount of work done by unit mass of steam in the engine or turbine. Air leakage also results in lowering the partial pressure of steam and temperature. This means that latent heat increases and therefore more cooling water is required and the undercooling of the condensate is likely to be more severe with a resulting lower overall efficiency.
2. One of the main causes of poor performance in surface condensers is the pressure drop which occurs as the steam flows over the tubes ; this pressure drop, by increasing the volume of the steam, tends to destroy the vacuum. The decrease in vacuum results in less amount of work done by unit mass of steam.
3. The heat conduction is through the brass tube walls. This conduction of heat is not perfect and results in less efficiency.
4. On examining the heat balance sheets of steam engine plants, it will be found that more than one-half of the heat supplied by fuel is rejected to the condenser cooling water ; this is chief loss of steam plant and is the cause of its low overall efficiency.
5. Steam entering the condenser with high resistance.
6. Circulating water passing through the condenser with high friction and at a velocity not consistent with high efficiency.
7. Undercooling of condensate.
8. Air extraction from hottest section and with comparatively large amount of water vapour.

20.4.4. Comparison between Jet and Surface Condensers

S.No.	Jet Condenser	Surface Condenser
1.	Cooling water and steam are mixed up.	Cooling water and steam are not mixed up.
2.	Low manufacturing cost.	High manufacturing cost.
3.	Lower upkeep.	Higher upkeep.
4.	Requires small floor space.	Requires large floor space.
5.	The condensate cannot be used as feed water in the boilers unless the cooling water is free from impurities.	Condensate can be reused as feed water as it does not mix with the cooling water.
6.	More power is required for air pump.	Less power is needed for air pump.
7.	Less power is required for water pumping.	More power is required for water pumping.
8.	It requires less quantity of cooling water.	It requires large quantity of cooling water.
9.	The condensing plant is simple.	The condensing plant is complicated.
10.	Less suitable for high capacity plants due to low vacuum efficiency.	More suitable for high capacity plants as vacuum efficiency is high.

20.5. SOURCES OF AIR IN CONDENSERS

The *main sources of air* found in condensers are given below :

1. There is a leakage of air from atmosphere at the joints of the parts which are internally under a pressure less than that of atmosphere. The quantity of air that leaks in can be reduced to a great extent if design and making of the vacuum joints are undertaken carefully.
2. Air is also accompanied with steam from the boiler into which it enters dissolved in feed water. The quantity of air depends upon the treatment which the feed water receives before it enters the boiler. However, the amount of air which enters through this source is relatively small.
3. In jet condensers, a little quantity of air accompanies the injection water (in which it is dissolved).

Note. **1.** In jet condensers, the quantity of air dissolved in injection water is about 0.5 kg/10000 kg of water.

2. In surface condensers of reciprocating steam engines, the air leakage is about 15 kg/10000 kg of steam whereas in surface condensers of well designed and properly maintained steam turbine plants the air leakage is about 5 kg/10000 kg of steam.

In order to check whether there is air leakage in the condenser, the following procedure is adopted :

1. Keep the plant running until the temperature and pressure conditions are steady in the condenser.
2. The steam condenser be isolated by shutting off steam supply and simultaneously closing the condensate and air extraction pumps.

In case there is a leakage the readings of vacuum gauge and thermometer will record a fall.

The following methods are used to check the source of air leakage.

1. Put the steam condenser under air pressure and note its effect on soap water at the points where infiltration is likely to occur.
2. Put the peppráment oil on the suspected joint (when the condenser is operating) and make a check on the peppráment odour in the discharge of air ejector.
3. Large leakages in steam condenser under vacuum can be detected by moving/passing candle flame over possible openings.

20.6. EFFECTS OF AIR LEAKAGE IN A CONDENSER

The following are the *effects of air leakage in a condenser* :

1. **Lowered thermal efficiency.** The leaked air in the condenser results in increased back-pressure on the prime mover which means there is loss of heat drop and consequently thermal efficiency of steam power plant is lowered.

2. **Increased requirement of cooling water.** The leaked air in the condenser lowers the partial pressure of steam which means a lowered saturation temperature of steam. As the saturation temperature of steam lowers, its latent heat increases. So it will require increased amount of cooling water for increased latent heat.

3. **Reduced heat transfer.** Air has poor thermal conductivity. Hence leaked air reduces the rate of heat transfer from the vapour, and consequently it requires surface of the tubes of a surface condenser to be increased for a given condenser capacity.

4. **Corrosion.** The presence of air in the condenser increases the corrosive action.

20.7. METHODS FOR OBTAINING MAXIMUM VACUUM IN CONDENSERS

Following are some of the methods used to obtain *maximum possible vacuum in condensers used in modern steam power plants.*

1. **Air pump.** Air pumps are provided to maintain a desired vacuum in the condenser by extracting the air and other non-condensable gases. They are usually classified as : (*a*) ***Wet air pumps*** which remove a *mixture of condensate and non-condensable gases.* (*b*) ***Dry air pump*** which removes the *air only.*

2. **Steam air ejector.** When a wet air pump (also called extraction pump) is employed then use is made of steam air ejectors to remove air from the mixture. The operation of the ejector consists in utilising the viscous drag of a high velocity steam jet for the ejection of air and other non-condensable gases from a chamber ; it is chiefly used for exhausting the air from steam condensers. In the case of ejectors used for steam plants where a high vacuum pressure is maintained in the condenser, it is necessary to use two, or perhaps three ejectors in series to obtain maximum vacuum.

3. **De-aerated feed water.** The de-aeration of feed water helps both in maintaining better vacuum in the condenser and controlling corrosion of the steel shell and piping of the steam power plant.

4. **Air-tight joints.** The various joints of the steam power plant are rendered air-tight by suitable packing materials etc., at the joints of piping etc., and these are maintained as such by proper inspection from time to time.

20.8. VACUUM MEASUREMENT

The term *vacuum* in case of a condenser means *pressure below atmospheric pressure.* It is generally expressed in mm of mercury. The vacuum is measured by means of a vacuum gauge (Fig. 20.9). Usually for calculation purpose the vacuum gauge reading is corrected to standard barometric reading 760 mm as follows.

Corrected vacuum in mm of Hg

= (760 – absolute pressure in mm of Hg)

= 760 – (actual barometric height – actual vacuum)

Also, 760 mm of Hg = 1.01325 bar

$$\therefore \quad 1 \text{ mm of Hg} = \frac{1.01325}{760} = 0.001333 \text{ bar.}$$

20.9. VACUUM EFFICIENCY

It is defined as the ratio of the actual vacuum to the maximum obtainable vacuum. The latter vacuum is obtained when there is only steam and no air is present in the condenser.

$$\text{Vacuum efficiency} = \frac{\text{Actual vacuum}}{\text{Maximum obtainable vacuum}}$$

$$= \frac{\text{Actual vacuum}}{\text{Barometer pressure} - \text{Absolute pressure of steam}} \quad ...(20.1)$$

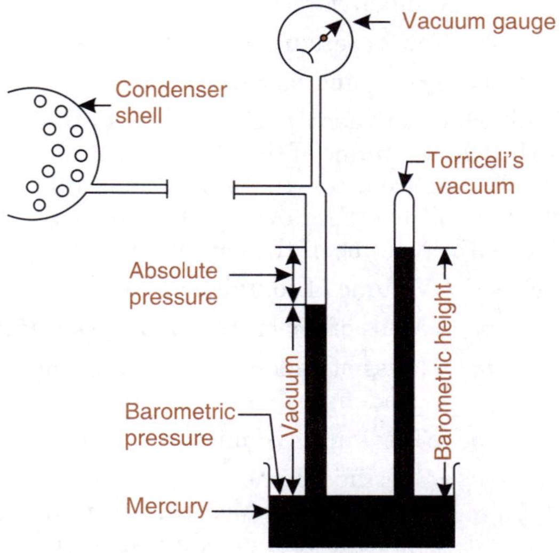

Fig. 20.9. Measurement of vacuum.

Note. In case of the absolute pressure of steam corresponding to the temperature of condensate being equal to the absolute pressure in the condenser, the efficiency would be 100%. Actually some quantity of air is also present in the condenser which may leak in and be accompanied by the entering steam. The vacuum efficiency, therefore, depends on the amount of air removed by the air pump from the condenser.

20.10. CONDENSER EFFICIENCY

It is defined *as the ratio of the difference between the outlet and inlet temperatures of cooling water to the difference between the temperature corresponding to the vacuum in the condenser and inlet temperature of cooling water, i.e.,*

Condenser efficiency

$$= \frac{\text{Rise in temperature of cooling water}}{\begin{bmatrix}\text{Temp. corresponding to} \\ \text{vacuum in the condenser}\end{bmatrix} - \begin{bmatrix}\text{Inlet temp. of} \\ \text{cooling water}\end{bmatrix}}$$

or

$$= \frac{\text{Rise in temperature of cooling water}}{\begin{bmatrix}\text{Temp. corresponding to the} \\ \text{absolute pressure in the condenser}\end{bmatrix} - \begin{bmatrix}\text{Inlet temp. of} \\ \text{cooling water}\end{bmatrix}} \quad ...(20.2)$$

20.11. DALTON'S LAW OF PARTIAL PRESSURES

It states that :

(*i*) In a container in which gas and a vapour are enclosed, the total pressure exerted is the sum of partial pressure of the gas and partial pressure of the vapour at the common temperature.

Let t = Temperature of mixture of air and water vapour in the container in °C

p_a = Partial pressure of air at temperature t,

p_s = Saturation pressure of water vapour at temperature t, and

p = Total pressure in the container

$\therefore \quad p = p_a + p_s$ and $p_a = (p - p_s)$

(*ii*) Each constituent of the mixture in the container occupies the whole volume of the container at its individual pressure. Thus if from the container air is removed, the water vapour will exert its own partial pressure but will occupy the entire volume of the container.

Let V = Volume of container in m^3,

m_a = Mass of air present in the container,

m_s = Mass of vapour (steam) present in the condenser,

m = Total mass of mixture in the condenser,

v_g = Specific volume of saturated water vapour (steam) at temperature $t°$ in m^3, and

v_a = Specific volume of air at temperature $t°$ in m^3.

Now as per Dalton's law, $V = m_a v_a = m_s v_g$

$$\therefore \quad \frac{m_a}{m_s} = \frac{v_g}{v_a} \quad \text{...(20.3)}$$

$\therefore$ Mass of air/m^3 of container

$$\frac{m_a}{V} = \frac{1}{v_a} \quad \text{...(20.4)}$$

and the mass of water vapour (steam) per m^3 of container

$$\frac{m_s}{V} = \frac{1}{v_g} \quad \text{...(20.5)}$$

$\therefore$ Total mass of mixture in the container

$$m = m_s + m_a = m_s\left(1 + \frac{m_a}{m_s}\right)$$

$$= m_s\left(1 + \frac{v_g}{v_a}\right) \quad \text{...(20.6)}$$

Also

$$m = m_a\left(1 + \frac{m_s}{m_a}\right)$$

$$= m_a\left(1 + \frac{v_a}{v_g}\right) \quad \text{...(20.7)}$$

20.12. DETERMINATION OF MASS OF COOLING WATER

Let m_w = Mass of cooling water required in kg/h,

m_s = Mass of steam condensed in kg/h,

t_s = Saturation temperature of steam corresponding to the condenser vacuum in °C,

t_c = Temperature of the condensate leaving the condenser,

t_{w_1} = Temperature of cooling water at inlet in °C,

t_{w_2} = Temperature of cooling water at outlet in °C,

c_{pw} = Specific heat of water at constant pressure,

x = Dryness fraction of steam entering the condenser, and

h_{fg} = Latent heat of 1 kg of steam entering the condenser.

Now, heat lost by steam

$$= m_s[xh_{fg_2} + c_{pw}(t_s - t_c)] \text{ kJ/kg}$$

and heat gained by water $= m_w \times c_{pw}(t_{w_2} - t_{w_1})$

If all heat lost by steam is gained by cooling water, then

$$m_s[x h_{f_g} + c_{pw}(t_s - t_c)] = m_w \times c_{pw}(t_{w_2} - t_{w_1})$$

$$\therefore \quad m_w = \frac{m_s[xh_{fg_2} + c_{pw}(t_s - t_c)]}{c_{pw}(t_{w_2} - t_{w_1})} \text{ kg/h} \quad \text{...(20.8)}$$

Eqn. (20.8) *applies to surface condenser only*.

In a jet condenser, since cooling water and steam mix together, therefore the condensate temperature will be same as that of outlet temperature of cooling water (*i.e.*, $t_c = t_{w_2}$). Thus quantity of cooling water, m_w in case of jet condenser is found by the following expression :

$$m_w = \frac{m_s[xh_{fg_2} + c_{pw}(t_s - t_{w_2})]}{c_{pw}(t_{w_2} - t_{w_1})} \text{ kg/h} \quad \text{...(20.9)}$$

20.13. HEAT TRANSMISSION THROUGH WALLS OF TUBES OF A SURFACE CONDENSER

In case of surface condenser, the *rate of heat transmission varies approximately with square root of the water velocity in tubes*. It thus follows that an increase of heat flow could be obtained by increasing the velocity of flow of water ; but this, in turn, would require a larger amount of energy to circulate the water on account of the corresponding increase in resistance. The following formula is sometimes used for calculating the rate of heat transmission through the walls of the tubes.

Let m_s = Mass of steam used in kg/h,

h = Total heat of 1 kg of steam entering the condenser,

t_m = Mean temperature difference causing heat flow across the tube surface in °C,

t_{w_1} = Temperature of entering cooling water in °C,

t_{w_2} = Temperature of leaving cooling water in °C,
t_p = Temperature of entering steam in °C,
t_c = Temperature of condensate when leaving in °C,
h_{fc} = Total heat of condensate when leaving,
A = Total surface area of condenser tubes in m^2, and
K = Heat transmission co-efficient.

(The value of K must be obtained experimentally for the tubes used and for the cooling water velocity in the tubes; it is a function of both of these factors).

Then, $$K = \frac{m_s (h - h_{f_c})}{t_w A} \quad ...(20.10)$$

where $$t_m = \frac{(t_c - t_{w_1}) - (t_s - t_{w_2})}{\log_e \dfrac{t_c - t_{w_1}}{t_s - t_{w_2}}} \quad ...(20.11)$$

This equation is due to Grashof and gives approximate result only. It does not hold for all types of surface condensers and modifications of the equation have been made to suite particular types. Eqn. (20.11) may be applied to the *contra-flow* conditions.

If the pressure drop in the condenser is nil, $t_s = t_c$ then eqn. (20.11) may be written as

$$t_m = -\frac{t_{w_2} - t_{w_1}}{\log_e \dfrac{(t_s - t_{w_1})}{(t_s - t_{w_2})}} \quad ...(20.12)$$

For a *cross-flow* condition,

$$t_m = \frac{(t_{w_2} - t_{w_1})}{\log_e \left[\dfrac{d}{d - \left(\dfrac{t_{w_2} - t_{w_1}}{t_s - t_c}\right) \log_e \left(\dfrac{t_s - t_{w_1}}{t_s - t_{w_2}}\right)} \right]} \quad ...(20.13)$$

where 'd' is diameter of condenser tubes.

If the value of the heat transmission co-efficient K is known and the value of t_m obtained from eqns. (20.11) (20.12) to (20.13) the necessary area of the heating surface of the tubes can now be obtained from eqn. (20.10).

20.14. AIR PUMPS

The main function which an air pump performs is that it maintains vacuum in the condenser as nearly as possible equal to that corresponding to the exhaust steam temperature, by removing air from the condenser. It may also remove condensate together with air from the condenser.

An air pump which removes the moist air alone is called a ***dry air pump*** *whereas that which removes both air and condensate is called a* ***wet air pump***.

Types of Air Pumps

Air pumps may be classified as follows :

1. Reciprocating piston or bucket pumps
2. Rotary pumps
3. Steam jet air pumps (ejectors)
4. Water jet pumps.

1. Edward's pump

Fig. 20.10 shows Edward's air pump which is a design of reciprocating piston or bucket wet air pump. The feature of this pump is the absence of inaccessible foot and bucket valves. This is effected by *conical end* to the piston and piercing the base of the *liner* with *ports* which communicate with air pump suction pipe down which the condensate gravitates.

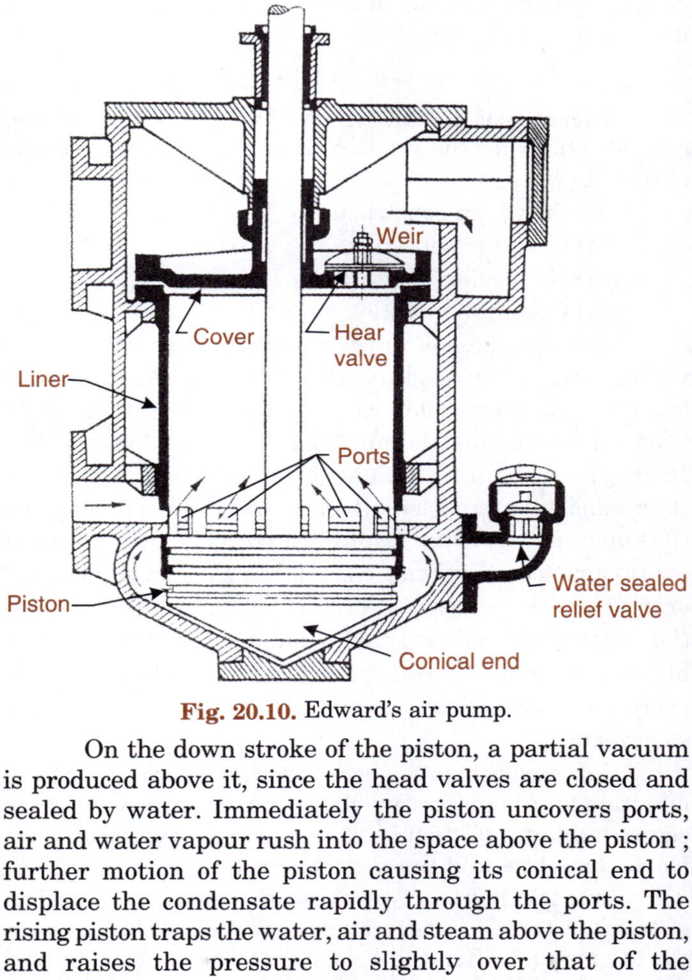

Fig. 20.10. Edward's air pump.

On the down stroke of the piston, a partial vacuum is produced above it, since the head valves are closed and sealed by water. Immediately the piston uncovers ports, air and water vapour rush into the space above the piston ; further motion of the piston causing its conical end to displace the condensate rapidly through the ports. The rising piston traps the water, air and steam above the piston, and raises the pressure to slightly over that of the atmosphere until *headvalves* open and allow the water vapour and air to pass to the waste, and the condensate to gravitate to the hot well over the *weir* which retains sufficient water above *cover* to seal the valves against air leakage. A *water-sealed relief valve* is placed in the base of the cylinder to release the pressure should it, for any reason, exceed atmospheric pressure.

2. **Rotary dry air pump**

The reciprocating air pumps, because of limited speed of operation, become very bulky for higher vacuum or large powers and it is due to this reason that rotary dry air pumps and steam jet ejectors have been developed.

Prof. Maurice Leblance invented a rotary dry air pump which is fairly widely used and resembles one stage of radial flow steam turbine. The revolving vanes project thin film of water, at a velocity of about 39 m/sec, down a collecting cone in which these films act as pistons, the air being entrained between successive sheets of water. Although the pump is charged with the water, it is intended to handle only air, the water and air being discharged through a diverging cone which raises its pressure to slightly greater than atmospheric. The water and air pass on to a slightly elevated tank in which the water is cooled prior to its return to the pump.

3. **Steam-operated air ejector**

Steam-operated air ejectors find a very wide of field of applications for the production of high vacuum because of the following reasons :

(*i*) Simple in construction,

(*ii*) Cheap to construct,

(*iii*) No moving parts, and

(*iv*) Occupy very little space.

The operation of the ejector consists in utilizing the viscous drag of high velocity steam jet for the ejection of air and other incondensable gases from a chamber ; it is chiefly used for exhausting the air from steam condensers. The steam jet flows through an air chamber where it entrains the air and any other gases which are adjacent to its surface ; the kinetic energy of the resulting mixture is then converted to pressure energy by being passed through a diverging cone, or diffuser. The increase of pressure thus obtained enables the mixture to be discharged against a pressure which is higher than that of the entraining chamber. The entraining operation is due to the viscous drag between the air and steam jet.

For steam plants where a high vacuum pressure is maintained it is imperative to use two or three ejectors in series to obtain sufficient increase of pressure in the mixture for its discharge into the atmosphere.

Fig. 20.11 shows **two-stage air injector.** It consists of two injectors in series having a surface cooler between the stages and after the last stage. The function of the surface coolers is to condense the steam used in the ejector. The latent heat thus absorbed by the cooling water is recovered by being transferred to the condensate. The surface coolers also cool the steam and reduce its volume before it is passed on to the next stage. The operating steam for the two stages is controlled by stop valves 'A_1' and 'A_2'. The steam for first stage enters the steam box B_1 and that for the second stage the steam box B_2. The first stage operating steam is expanded through the nozzle '*C*' into the mixing chamber '*D*', where the jet of steam entrains the air and vapour coming from the main condenser through the branch '*E*'. The mixture of steam and air passes on through the diffuser '*F*', in which it receives its initial compression and is then discharged into the first stage cooler G_1, where the steam is condensed. The air passes on through the pipe '*H*' to the second stage ejector in which it is compressed to a little above atmospheric pressure, the operating steam being condensed in the second stage cooler G_2. The cooler consists of a nest of annular tubes through which some of the condensate is passed.

(Refer to Fig. 20.11).

A_1, A_2 = Stop valves ; B_1, B_2 = Steam boxes ; C = Nozzle ; D = Mixing chamber ; E = Branch ; F = Diffuser ; G_1, G_2 = State coolers ; H = Pipe ; U = Pipes ; M_1 = Water box ; L_1, L_2 = Pipes, N = Field tubes ; O = Outer tubes ; P_1, P_2 = Water boxes.

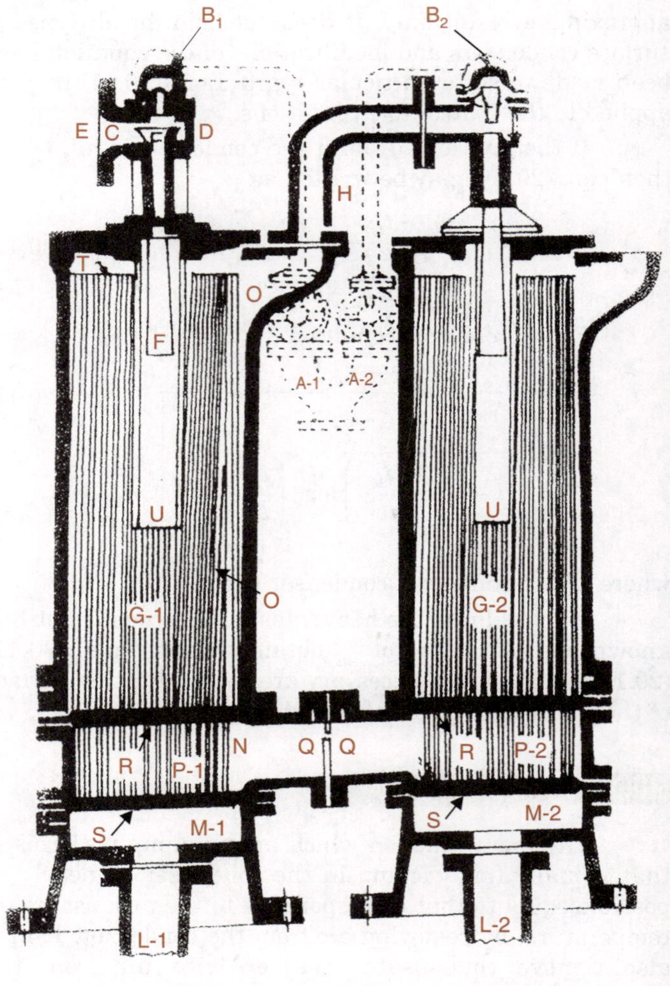

Fig. 20.11. Two-stage air ejector.

The pipes '*U*' direct the mixture of steam and air leaving the diffusers to the bottom of the coolers 'G_1' and

'G_2'. The baffles are provided to cause the steam flow across the nest of tubes. The condensed steam in the first stage cooler, which is under vacuum, is returned to the main condenser through a small '*V*' balance pipe, while the condensed steam in the second stage cooler is drained to the hot well or any other convenient place. The condensate from the main condenser enters the lower box 'M_1' through the pipe 'L_1' and passes upwards through the inner member of the field tubes N, returning downwards through the annuli formed between the inner tubes N and the outer tubes O into the water box P_1, it then passes through the branches Q to the upper water box of second stage cooler, circulates through the tubes in reverse order and leaves the pipe L_2. In the entry water box, a suitable strainer is incorporated.

4. Water jet pumps

A circulating water pump supplies cooling water through a head in the case of jet condensers and circulates the cooling water under pressure in case of surface condensers. The following types of water circulating pumps are employed :

(*i*) Reciprocating type ; (*ii*) Plunger type ;
(*iii*) Centrifugal type ; (*iv*) Propeller type.

20.15. COOLING TOWERS

In power plants, the hot water from condenser is cooled in cooling tower, so that it can be reused in condenser for condensation of steam. *In a cooling tower water is made to trickle down drop by drop so that it comes in contact with the air moving in the opposite direction. As a result of this some water is evaporated and is taken away with air. In evaporation, the heat is taken away from the bulk of water, which is thus cooled.* Factors affecting cooling of water in a cooling tower are :

(*i*) Temperature of air ;
(*ii*) Humidity of air ;
(*iii*) Temperature of hot air ;
(*iv*) Size and height of tower ;
(*v*) Velocity of air entering tower ;
(*vi*) Accessibility of air to all parts of tower ;
(*vii*) Degree of uniformity in descending water ;
(*viii*) Arrangement of plates in tower.

Cooling towers may be classified according to the material of which these are made, *i.e.*, (*i*) *timber*, (*ii*) *concrete* (ferro-concrete, multideck concrete hyperbolic) and (*iii*) *steel duct type*.

Timber towers are rarely used due to following *disadvantages* :

(*i*) Due to exposure to sun, wind, water, etc. ; timber rots easily
(*ii*) Short life ;
(*iii*) High maintenance charges ;
(*iv*) The design generally does not facilitate proper circulation of air ;
(*v*) Limited cooling capacity.

Concrete towers possess the following *advantages* :

(*i*) Large capacity sometimes of the order of 5×10^3 m^3/h ;
(*ii*) Improved draught and air circulation ;
(*iii*) Increased stability under air pressure ;
(*iv*) Low maintenance.

Duct type cooling towers are rarely used in case of modern power plants owing to their small capacity.

The cooling towers require a draught of air for evaporation of water sprayed. The draught may be created by a chimney or the available natural air velocity (*natural draught*) or by fans (*mechanical draught*). The mechanical draught may be *forced* or *induced* depending on the placement of fans.

Induced and *natural draught cooling* towers are shown in Figs. 20.12 and 20.13 respectively.

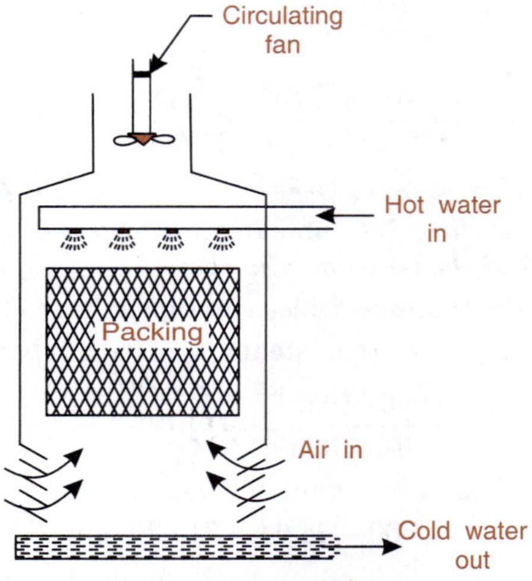

Fig. 20.12. Induced draught cooling tower.

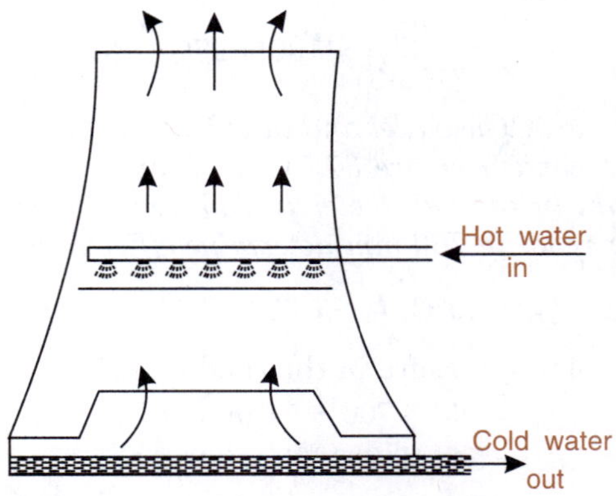

Fig. 20.13. Natural draught cooling tower.

WORKED EXAMPLES

Example 20.1. *If a barometer stands at 760 mm and condenser vacuum is at 710 mm and temperature is 30°C, calculate the mass of air per kg of uncondensed steam.*

Solution. Barometric reading = 760 mm of Hg

Condenser vacuum = 710 mm of Hg

Total absolute pressure in the condenser

= 760 – 710 = 50 mm of Hg

= 50 × 0.001333 = 0.06665 bar

(∵ 1 mm Hg = 0.001333 bar)

Temperature in the condenser = 30°C

Corresponding to this temperature pressure of steam from steam tables

= 0.0425 bar

∴ Pressure exerted by air

= 0.06665 – 0.0425 = 0.024 bar

And specific volume of steam

= 32.89 m³/kg (From steam tables)

Now, $pV = m_a RT$

$$m_a = \frac{pV}{RT}$$

$$= \frac{0.024 \times 10^5 \times 32.89}{287 \times (30 + 273)} = \mathbf{0.908\ kg.} \quad \textbf{(Ans.)}$$

Example 20.2. *Steam enters a condenser at 36°C and with barometer reading 760 mm. If the vacuum of 695 mm is produced find the vacuum efficiency.*

Solution. From steam tables corresponding to 36°C :

Partial pressure of steam, p_s = 0.0595 bar

$$= \frac{0.0595}{0.001333} \text{ mm of Hg} = 44.64 \text{ mm of Hg}$$

∴ Maximum obtainable vacuum

= 760 – 44.64 = 715.36 mm of Hg

$$\therefore \textit{Vacuum efficiency} = \frac{\text{Actual vacuum}}{\text{Max. obtainable vacuum}}$$

$$= \frac{695}{715.36} \times 100 = \mathbf{97\%.} \quad \textbf{(Ans.)}$$

Example 20.3. *The outlet and inlet temperatures of cooling water to a condenser are 37.5°C and 30°C respectively. If the vacuum in the barometer is 706 mm of mercury with barometer reading 760 mm determine efficiency.*

Solution. t_{w_1} = 30°C, t_{w_2} = 37.5°C

Absolute pressure in the condenser

= 760 – 706 = 54 mm of Hg

= 54 × 0.001333 = 0.072 bar

From steam tables corresponding to 0.072 bar,

$t_s \simeq 40°C$

∴ *Condenser efficiency*

$$= \frac{\text{Rise in temp. of cooling water}}{\text{(Temperature corresponding to vacuum in condenser – Inlet temperature of cooling water)}}$$

$$= \frac{(t_{w_2} - t_{w_1})}{(t_s - t_{w_1})} = \frac{37.5 - 30}{40 - 30} = \frac{7.5}{10} = \mathbf{0.75 \text{ or } 75\%.} \quad \textbf{(Ans.)}$$

Example 20.4. *The surface condenser is designed to handle 16000 kg of steam per hour. The steam enters the condenser at 0.09 bar abs. and 0.88 dryness fraction and the condensate leaves the condenser at the corresponding saturation temperature. Determine the rise in cooling water temperature if the cooling water flow rate is 8.96 × 10⁵ kg/hour. Assume that the pressure is constant throughout the condenser.*

Solution. Mass of steam handled per hour = 16000 kg

Cooling water flow per hour,

$m_w = 8.96 \times 10^5$ kg/hour

From steam tables corresponding to dry saturated steam, for given pressure of 0.09 bar of steam, for 1 kg steam;

h_f = 183.3 kJ/kg ; t_s = 43.8°C,

h_{fg} = 2397.7 kJ/kg

Total heat lost by the wet steam of dryness fraction 0.88 in one hour

$= 16000\,(h_f + xh_{fg} - c_{pw} \times t_s)$

= 16000 (183.3 + 0.88 × 2397.7 – 4.184 × 43.8)

= 16000 (183.3 + 2109.9 – 183.26) = 337.59 × 10⁵ kJ

Heat gained by cooling water in one hour

$= 8.96 \times 10^5 \times 4.184 \times (t_{w_2} - t_{w_1})$, $t_{w_2} - t_{w_1}$ being rise of temperature

∴ Heat lost by steam/hour

= Heat gained by water/hour

$$337.59 \times 10^5 = 8.96 \times 10^5\,(t_{w_2} - t_{w_1})$$

$$\therefore \quad (t_{w_2} - t_{w_1}) = \frac{337.59 \times 10^5}{8.96 \times 10^5 \times 4.184} = 8.99°C$$

Hence *rise in cooling water temperature*

= **9°C. (Ans.)**

Example 20.5. *A surface condenser is designed to handle 10000 kg of steam per hour. The steam enters at 0.08 bar abs. and 0.9 dryness and the condensate leaves at the corresponding saturation temperature. The pressure is constant throughout the condenser. Estimate the cooling water flow rate per hour, if the cooling water temperature rise is limited to 10°C.*

Solution. From steam tables.

At 0.08 bar ; t_s = 41.5°C, h_{fg} = 2403.1 kJ/kg

Also m_s = 10000 kg/h ; $(t_{w_2} - t_{w_1})$ = 10°C.

Saturation temperature of condensate, t_c = 41.5°C

Assuming that there is perfect heat transfer in the condenser,

Then, heat lost by the exhaust steam
= heat gained by cooling water.

$$10000\,[xh_{fg} + c_{pw}(t_s - t_c)] = m_w \times c_{pw} \times (t_{w_2} - t_{w_1})$$

or $$10000\,[0.9 \times 2403.1 + 4.184 \times (41.5 - 41.5)] = m_w \times 4.184 \times 10$$

$$\therefore \quad m_w = \frac{10000 \times 0.9 \times 2403.1}{10 \times 4.184}$$

= **516919 kg/h. (Ans.)**

Example 20.6. *A steam jet turbo-generator develops 100 kW using 13.6 kg of steam per kWh. The exhaust steam pressure is 0.14 bar and 680.4 kg of cooling water are passed through the condenser per minute. The inlet and outlet temperatures are respectively 15.6°C and 32.2°C. Estimate the dryness fraction of exhaust steam. Temperature of hot well is 35°C.*

Solution. From steam tables :

At 0.14 bar : $t_s = 52.6°C$; $h_{fg} = 2376.6$ kJ/kg

Also, $m_s = 13.6$ kg/kWh, $t_{w_1} = 15.6°C$,

$t_{w_2} = 32.2°C$

$m_w = 680.4$ kg/min or (680.4×60) kg/h

Hot well temperature, $t_h = 35°C$.

Assuming that there is perfect heat transfer in the condenser,

Then, heat lost by the steam = heat gained by cooling water

$$100 \times m_s\,[xh_{fg} + c_{pw}(t_s - t_h)] = m_w \times c_{pw} \times (t_{w_2} - t_{w_1})$$

or $$100 \times 13.6\,[x \times 2376.6 + 4.184 \times (52.6 - 35.0)] = 680.4 \times 60 \times 4.184 \times (32.2 - 15.6)$$

$$1360\,(2376.6\,x + 73.64) = 2835406$$

$$\therefore \quad x = \frac{2835406 - 1360 \times 73.64}{1360 \times 2376.6} = \frac{2835406 - 100150}{3232176} = 0.846$$

Hence, *dryness fraction of steam* = **0.846. (Ans.)**

Example 20.7. *The volume of condenser which contains 0.144 kg of air with steam is 4.05 m³. Temperature in the condenser is 40°C and there is some water in the condenser. Determine the pressure in the condenser. R for air = 287 joules/kg K.*

Solution. $m_a = 0.144$ kg, $V = 4.05$ m³, $T = 40 + 273 = 313$ K

We know that,

$$p_a V_a = m_a RT$$

$$p_a \times 4.05 = 0.144 \times 287 \times 313$$

$$\therefore \quad p_a = \frac{0.144 \times 287 \times 313}{4.05} \times 10^{-5} \text{ bar} = 0.0319 \text{ bar}$$

Partial pressure of water vapour corresponding to 40°C,

$$p_s = 0.0738 \text{ bar}$$

∴ Pressure in the condenser

$= p_a + p_s = 0.0319 + 0.0738 =$ **0.01057 bar. (Ans.)**

Example 20.8. *The air leakage into the condenser operating in conjunction with a steam turbine is estimated at 0.681 kg per minute. The vacuum near the outlet to the air pump is 710 mm when the barometer reads 760 mm and temperature at this point is 18°C.*

Find : (i) The minimum capacity of air-pump in m³/min ;

(ii) The mass of vapour extracted with the air/min.

Solution. Pressure in the condenser

= (760 – 710) = 50 mm of Hg

= 50 × 0.001333 = 0.0666 bar

Partial pressure of steam at 18°C = 0.0206 bar

Hence partial pressure of air,

$$p_a = 0.0666 - 0.0206 = 0.046 \text{ bar}$$

Mass of air leakage = 0.681 kg/min

$$V = \frac{mRT}{p_a} = \frac{0.681 \times 287 \times (273 + 18)}{0.046 \times 10^5}$$

= **12.36 m³/mm. (Ans.)**

From the Dalton's law of partial pressure, the volume of steam will be same as that of air *i.e.*, 12.36 m³/min.

Mass of steam condensed/min,

$$m_s = \frac{\text{Volume of steam}}{\text{Specific volume of steam at 0.0206 bar}} = \frac{12.36}{65.04} = \textbf{0.19 kg/min.} \quad \textbf{(Ans.)}$$

Example 20.9. *A closed vessel of 0.7 m³ capacity contains saturated water vapour and air at a temperature of 42.7°C and a pressure of 0.13 bar abs. Due to further air leakage into the vessel, the pressure rises to 0.28 bar abs. and temperature falls to 37.6°C. Calculate the mass of air which has leaked in. Take R = 287 J/kg K for air.*

Solution. ***Initial condition :***

From steam tables, partial pressure of steam at 42.7°C,

$$p_s = 0.085 \text{ bar abs.}$$

∴ Partial pressure of air,

$$p_a = 0.13 - 0.085 = 0.045 \text{ bar abs.}$$

Mass of air present *initially* in the vessel of 0.7 m³ capacity,

$$(m_a)_i = \frac{p_a V}{RT_a} = \frac{0.045 \times 10^5 \times 0.7}{287\,(42.7 + 273)} = 0.0347 \text{ kg}$$

Final condition :

From steam tables, partial pressure of steam at 37.6°C,

$$p_s = 0.065 \text{ bar}$$

∴ Partial pressure of air, $p_a = 0.28 - 0.065$
$= 0.215$ bar abs.

Mass of air present *finally* in the vessel of 0.7 m^3 capacity,

$$(m_a)_f = \frac{p_a V}{RT_a} = \frac{0.215 \times 10^5 \times 0.7}{2.87 \times (37.6 + 273)} = 0.1688 \text{ kg}$$

Hence, *air leakage into the vessel* $= (m_a)_f - (m_a)_i$
$= 0.1688 - 0.0347 =$ **0.1341 kg. (Ans.)**

Example 20.10. *In a surface condenser a section of the tubes near to the air pump suction is screened off so that the air is cooled to a temperature below that of the condensate, separate extraction pumps being provided to deal with air and condensate respectively. 5448 kg of steam are condensed per hour and the air leakage is 4.54 kg/h. The temperature of the exhaust steam is 31°C, the temperature of the condensate is 27°C, and the temperature at the air pump suction is 21.1°C. Assuming a constant vacuum throughout the condenser, find :*

(i) The mass of steam condensed per hour in the air cooler ;

(ii) The volume of air in m^3/h to be dealt with by the air pump ;

(iii) The percentage reduction in necessary air pump capacity following the cooling of the air.

Solution. From steam tables, pressure of steam at 31°C,

$$p_s = 0.045 \text{ bar abs.}$$

and partial pressure of steam at air pump suction at 21.1°C,

$$p_s = 0.025 \text{ bar abs.}$$

Partial pressure of air,

$$p_a = 0.045 - 0.025 = 0.02 \text{ bar abs.}$$

Also, at air pump suction ; $T = 21.1 + 273 = 294.1$ K

(*i*) Using characteristic equation of gas for 4.54 kg of air,

$$p_a V = mRT$$

$$V = \frac{mRT}{10^4 \, p_a} = \frac{4.54 \times 287 \times 294.1}{0.02 \times 10^5} = 191.6 \text{ m}^3/\text{h}.$$

As per Dalton's law this is also the volume of the steam mixed with air at the air pump suction.

Specific volume of steam at partial pressure of 0.025 bar = 54.25 m^3/kg

∴ *Mass of steam at air pump suction/hour*

$$= \frac{191.6}{54.25} = \textbf{3.53 kg.} \quad \textbf{(Ans.)}$$

(*ii*) **Volume of air/hour to be dealt**
= 191.6 m^3/h. (Ans.)

(*iii*) Temperature at the pump *without air cooling*,

$$T = 27 + 273 = 300 \text{ K.}$$

Partial pressure of steam at 27°C = 0.0357 bar (From steam tables).

Then, partial pressure of air
$= 0.045 - 0.0357$ bar $= 0.0093$ bar

Again, using the characteristic gas equation to 4.54 kg of air at partial pressure of 0.0093 bar,

$$V = \frac{mRT}{p_a} = \frac{4.54 \times 287 \times 300}{0.0093 \times 10^5} = 420.3 \text{ m}^3/\text{h}.$$

Percentage reduction in air pump capacity due to air cooling

$$= \frac{420.3 - 191.6}{420.3} = \textbf{0.544 or 5.44\%.} \quad \textbf{(Ans.)}$$

☞ **Example 20.11.** *A surface condenser deals with 13625 kg of steam per hour at a pressure of 0.09 bar. The steam enters 0.85 dry and the temperature at the condensate and air extraction pipes is 36°C. The air leakage amounts to 7.26 kg/hour. Determine (i) the surface required if the average heat transmission rate is 3.97 kJ/cm^2 per second ; (ii) the cylinder diameter for the dry air pump, if it is to be single acting at 60 r.p.m. with a stroke to bore ratio of 1.25 and volumetric efficiency of 0.85.* **(M.U.)**

Solution. Corresponding to 0.09 bar (from steam tables)

$$h_f = 183.3 \text{ kJ/kg}, \; h_{fg} = 2397.7 \text{ kJ/kg}$$

Heat extracted/sec. from steam

$$= \frac{13625}{3600} \left[(h_f + x h_{fg}) - h_{f_1} \right]$$

$$= \frac{13625}{3600} [(183.3 + 0.85 \times 2397.7) - 4.184 \times 36]$$

$$= 3.785 [(183.3 + 2038) - 150.62]$$

$$= 7837.5 \text{ kJ/s.}$$

(*i*) **Surface required**

$$= \frac{7837.5}{3.97} = \textbf{1974.18 cm}^2. \quad \textbf{(Ans.)}$$

(*ii*) Quantity of air leakage/min. into the condenser

$$= \frac{7.26}{60} = 0.121 \text{ kg}$$

Partial pressure of steam at 36°C (from steam tables)

$$p_s = 0.0595 \text{ bar}$$

But, $p = p_a + p_s = 0.09$ bar

∴ $p_a = 0.09 - 0.0595 = 0.0305$ bar

Using characteristic gas equation,

$$p_a V = mRT$$

$$0.0305 \times 10^5 \times V = 0.121 \times 287 \times (273 + 36)$$

or $$V = \frac{0.121 \times 287 \times 309}{0.0305 \times 10^5} = 3.52 \text{ m}^3$$

Capacity of air pump/min

$$= \frac{3.52}{0.85} = 4.14 \text{ m}^3 \text{ or } 4.14 \times 10^6 \text{ cm}^3$$

Capacity of the pump/stroke

$$= \frac{4.14 \times 10^6}{60} = 69000 \text{ cm}^3$$

Also $\pi/4 \times d^2 \times 1.25 \times d = 69000$

(where d is the diameter of cylinder for dry air pump)

or $$d^3 = \frac{69000 \times 4}{\pi \times 1.25} = 70282.8$$

or $d \simeq$ **41.3 cm.** **(Ans.)**

☞ **Example 20.12.** *A surface condenser deals with 13000 kg of steam per hour. The leakage air in the system amounts to 1 kg per 2700 kg of steam. The vacuum in the air pump suction is 705 mm of mercury (barometer 760 mm of Hg) and temperature is 34.6°C.*

Determine the discharging capacity of the wet air pump which removes both air and condensate in m³ per minute, taking the volumetric efficiency of the pump as 90%.

If the air pump is single-acting and runs at 60 r.p.m. and piston stroke is 1.25 times the diameter of the pump, find the dimensions of the wet air pump.

Solution. From steam tables, corresponding to 34.6°C,

$$p_s = 0.055 \text{ bar}$$

The combined pressure of steam and air in the condenser,

$$p = 760 - 705 = 55 \text{ mm of Hg}$$
$$= 55 \times 0.001333 = 0.0733 \text{ bar}$$

Partial pressure of air,

$$p_a = p - p_s$$
$$= 0.0733 - 0.055 = 0.0183 \text{ bar}$$

Mass of air leakage in the condenser/min.,

$$m_a = \frac{13000}{2700 \times 60} = 0.0802 \text{ kg.}$$

Volume of air leakage in the condenser/min.,

$$V_a = \frac{m_a R T_a}{p_a} = \frac{0.0802 \times 287 \times (34.6 + 273)}{0.0183 \times 10^5}$$
$$= 3.869 \text{ m}^3$$

Mass of steam condensed/min., $m_s = \dfrac{13000}{60}$ kg

Volume of condensate/min. $= \dfrac{13000}{60 \times 1000} = 0.2167 \text{ m}^3$

[∵ Density of water = 1000 kg/m³]

∴ Volume of mixture (air + condensate) actually discharged/min.

$$= 3.869 + 0.2167 = 4.0857 \text{ m}^3$$

∴ Discharging capacity of the air pump/min.

$$= 4.0857 \times \frac{100}{90} = 4.539 \text{ m}^3$$

∴ Discharging capacity of the air pump/stroke

$$= \frac{4.539}{60} \times 10^6 = 75650 \text{ cm}^3$$

$$\pi/4\, d^2 \times 1.25\, d = 75650$$

[d = dia. of the cylinder for the pump]

$$\boldsymbol{d} = \left(\frac{75650 \times 4}{\pi \times 1.25}\right)^{1/3} = \textbf{42.55 cm.} \quad \textbf{(Ans.)}$$

Piston stroke,

$$\boldsymbol{l} = 1.25\, d = 1.25 \times 42.55 = \textbf{53.2 cm.} \quad \textbf{(Ans.)}$$

Example 20.13. *To check the leakage of air in a condenser, the following procedure is adopted. After ruming the plant to reach the steady conditions the steam supply to the condenser and the air and condensate pump are shut down, thus completely isolating the condenser. The temperature and vacuum readings are noted at shut down and also after a period of 10 minutes. They are 39°C and 685 mm Hg and 28°C and 480 mm Hg respectively. The barometer reads 750 mm Hg. The effective volume of the condenser is 1.5 m³. Determine (i) quantity of air leakage into the condenser during the period of observation ; (ii) the quantity of water vapour condensed during the period.*

Solution. ***At shut down :***

From steam tables, corresponding to t_s = 39°C :

$$p_s = 0.07 \text{ bar} = \frac{0.07}{0.001333} = 52.5 \text{ mm Hg}$$

and $v_g = 20.53 \text{ m}^3/\text{kg}$

The combined pressure of steam and air in the condenser,

$$p = p_a + p_s = 750 - 685 = 65 \text{ mm Hg}$$

$$\therefore \quad p_a = p - p_s = 65 - 52.5 = 12.5 \text{ mm Hg}$$
$$= 12.5 \times 0.001333 = 0.0167 \text{ bar}$$

Now, mass of air in 1.5 m³

$$m_a = \frac{p_a V_a}{R T_a} = \frac{0.0167 \times 10^5 \times 1.5}{287 \times (273 + 39)} = 0.028 \text{ kg}$$

and mass of steam in 1.5 m³

$$m_s = \frac{1.5}{20.53} = 0.073 \text{ kg}$$

After 10 minutes, observed duration,

From steam tables, corresponding to t_s = 28°C :

$$p_s = 0.0378 \text{ bar} = \frac{0.0378}{0.001333}$$
$$= 28.36 \text{ mm Hg}, \; v_g = 36.69 \text{ m}^3/\text{kg}$$

Total pressure in the condenser,

$$p = 750 - 480 = 270 \text{ mm Hg}$$

∴ Air pressure,

$$p_a = p - p_s = 270 - 28.36 = 241.64 \text{ mm Hg}$$
$$= 241.64 \times 0.001333 = 0.322 \text{ bar}$$

Mass of air,

$$m_a = \frac{p_a V_a}{RT_a} = \frac{0.322 \times 10^5 \times 1.5}{287 \times (273 + 28)} = 0.559 \text{ kg.}$$

Mass of steam, $m_s = \frac{1.5}{36.69} = 0.0408$ kg

$\therefore$ **Air leakage** *in 10 minutes period*

$= (0.559 - 0.028) =$ **0.531 kg. (Ans.)**

and **steam condensed** *in 10 minutes period*

$= (0.073 - 0.0408) =$ **0.0322 kg. (Ans.)**

Example 20.14. *A jet condenser is required to condenser 5000 kg of steam per hour. 350 m³ of injection water are used per hour. Initial temperature of the cooling water is 27°C. The volume of air at atmospheric pressure dissolved in injection water is 5% of the volume of water. The amount of air entering the condenser with steam is 1 kg for every 3500 kg of steam. The vacuum in the air pump suction is 686 mm. When the barometer records 760 mm and the temperature of the condensate is 34.6°C, determine the suction capacity of the air-pump in m³/min to remove air and water from the condenser. Assume volumetric efficiency of pump as 85%.*

Solution. Total pressure in the condenser,

$$p = p_a + p_s = (760 - 686) = 74 \text{ mm Hg}$$

$$74 \times 0.001333 = 0.0986 \text{ bar}$$

Partial pressure of steam corresponding to 34.6°C = 0.055 bar (From steam tables)

$\therefore$ Partial pressure of air, $p_a = p - p_s$

$= 0.0986 - 0.055 = 0.0436$ bar

Mass of air entering per minute with steam

$$= \frac{5000}{3500 \times 60} = 0.0238 \text{ kg}$$

Volume of air entering per minute with injection water $= \frac{350 \times 5}{100 \times 60} = 0.292 \text{ m}^3$.

$$\text{Mass of this volume of air} = \frac{p_a V_a}{RT_a}$$

$$= \frac{1.01325 \times 10^5 \times 0.292}{287 \times (273 + 27)} = 0.343 \text{ kg.}$$

Total weight of air entering the condenser per minute

$= 0.0238 + 0.343 = 0.3668$ kg

Now, volume of this air at 0.045 bar and 34.6°C,

$$V = \frac{m_a R T_a}{p_a} = \frac{0.3668 \times 287 \times (273 + 34.6)}{0.0436 \times 10^5}$$

$$= 7.427 \text{ m}^3$$

$$\text{Volume of condensate/min} = \frac{5000}{60 \times 1000} = 0.083 \text{ m}^3$$

$$\text{Volume of injected water/min} = \frac{350}{60} = 5.83 \text{ m}^3$$

Hence, total volume to be handled

$= 7.427 + 0.083 + 5.83 = 13.34 \text{ m}^3$

$\therefore$ Suction capacity of the air pump

$$= \frac{13.34}{0.85} = \mathbf{15.69 \text{ m}^3}. \textbf{ (Ans.)}$$

Example 20.15. *The pressure under the air baffle of a surface condenser is 52 mm of Hg. Temperature of the mixture leaving the cooler suction is 25°C. Assuming available water at 15.5°C, and external water might lower the temperature further to 20°C. Explain the effect of this on the quantity of vapour accompanying the air to the air pump suction.*

Solution. The vapour (steam) under the conditions prevailing in the condenser, is treated as saturated. From steam tables, corresponding to 25°C,

$$p_s = 0.0317 \text{ bar} = \frac{0.0317}{0.001333} = 23.78 \text{ mm Hg};$$

$$v_g = (v_s) = 43.36 \text{ m}^3/\text{kg}$$

Total pressure (air + steam)

$$p = p_a + p_s = 52 \text{ mm of Hg}$$

$$\therefore p_a = p - p_s = 52 - 23.78 = 28.22 \text{ mm Hg}$$

$$= 28.22 \times 0.001333 = 0.0376 \text{ bar}$$

Since $v_a = v_g$ (According to Dalton's law)

$$\therefore v_a = v_g = 43.36 \text{ m}^3/\text{kg}$$

$$\therefore \text{Mass of air, } m_a = \frac{p_a V_a}{RT} = \frac{0.0376 \times 10^5 \times 43.36}{287 \times (273 + 25)}$$

$$= 1.906 \text{ m}^3/\text{kg of vapour}$$

$$m_s = \frac{1}{1.906} = 0.525 \text{ kg of vapour/kg of air}$$

When the temperature is lowered to 20°C with external cooling, we have from the steam tables (corresponding to 20°C),

$$p_s = 0.0234 \text{ bar} = \frac{0.0234}{0.001333} = 17.55 \text{ mm Hg,}$$

$$v_g = 57.48 \text{ m}^3$$

$$\therefore p_a = p - p_s = 52 - 17.55 = 34.55 \text{ mm of Hg}$$

$$= 34.45 \times 0.001333 = 0.046 \text{ bar}$$

and $$m_a = \frac{p_a v_a}{RT} \qquad [\because \; v_a = v_g = 57.79 \text{ m}^3]$$

$$= \frac{0.046 \times 10^5 \times 57.79}{287 \times 293} = 3.16 \text{ kg/kg of vapour}$$

and $$m_s = \frac{1}{3.16} = 0.316 \text{ kg of vapour/kg of air.}$$

From the above calculations, it is evident that the vapour accompanying each kg of air withdrawn is *reduced from 0.51 kg to 0.316 kg. This will influence considerably the ejector capacity and the work done for air pump suction.*

☞ **Example 20.16.** *During a trial on a steam condenser, the following observations were recorded :*

Condenser vacuum	*680 mm Hg*
Barometer reading	*764 mm Hg*
Mean condenser temperature	*36.2°C*
Hot well temperature	*30°C*
Condensate formed per hour	*1780 kg*
Circulating cooling water inlet temperature 20°C	
Circulating cooling water outlet temperature 32°C	
Quantity of cooling water	*1250 kg/min.*

Determine :

(*i*) *Condenser vacuum corrected to standard barometer.*

(*ii*) *Vacuum efficiency.*

(*iii*) *Undercooling of condensate.*

(*iv*) *Condenser efficiency.*

(*v*) *Condition of steam as it enters the condenser.*

(*vi*) *Mass of air present per kg of condensed steam.*

Assume : R for air = *0.287 kJ/kg K*

Specific heat of water = *4.186 kJ/kg/ K.*

(M.U.)

Solution.

(*i*) **Condenser vacuum corrected to standard barometer**

= Standard barometric pressure − (barometric pressure − gauge pressure)

= 760 − (764 − 680) = 676 mm of Hg

But 1 mm of Hg = 1.333×10^{-3} bar

∴ 676 mm of Hg = $676 \times 1.333 \times 10^{-3}$

= **0.9011 bar. (Ans.)**

(*ii*) **Vacuum efficiency :**

From steam tables saturation pressure corresponding to 36.2°C = 0.06 bar.

∴ *Vacuum efficiency*

$$= \frac{\text{Condenser vacuum}}{\text{Barometer reading} - \text{Pressure of steam}}$$

$$= \frac{680 \times 1.333 \times 10^{-3}}{764 \times 1.333 \times 10^{-3} - 0.06}$$

= **0.9458 or 94.58%. (Ans.)**

(*iii*) **Undercooling of condensate :**

= Condensate temperature − Hot well temperature

= 36.2 − 30 = **6.2°C. (Ans.)**

(*iv*) **Condenser efficiency :**

Absolute condenser pressure

= Barometric pressure − Vacuum reading

= 764 − 680 = 84 mm of Hg

= $84 \times 1.333 \times 10^{-3}$ = 0.1119 bar

Saturation temperature corresponding to 0.1119 bar (from steam tables) ≃ 48°C.

∴ *Condenser efficiency*

$$= \frac{\text{Actual cooling water temperature rise}}{\text{Maximum possible temperature rise}}$$

$$= \frac{32 - 20}{48 - 20} = 0.4286 = \mathbf{42.86\%}. \quad \textbf{(Ans.)}$$

(*v*) **Condition of steam entering the condenser:**

Absolute condenser pressure = 0.1119 bar.

For this pressure,

h_f = 200.1 kJ/kg, h_{fg} = 2387.2 kJ/kg

Enthalpy of condensate corresponding to hot well temperature of 30°C

= 125.75 kJ/kg.

Also, heat lost by steam = heat gained by water

$$m_s\,[(h_f + x\,h_{fg}) - h_{\text{hot-well}}] = m_w \times c_{pw} \times (t_{w_2} - t_{w_1})$$

$$= \frac{1780}{60}\,[(200.1 + x \times 2387.2) - 125.75]$$

$$= 1250 \times 4.186 \times (32 - 20)$$

$$(200.1 + x \times 2387.2) - 125.75 = 2116.5$$

Solving, x = 0.855

Hence, *condition of steam entering the condenser*

= **0.855. (Ans.)**

(*vi*) **Mass of air present :**

Specific volume of steam at mean condensate temperature of 36.2°C

= 23.74 m³/kg

Partial pressure of steam at 36.2°C = 0.06 bar

∴ Partial pressure of air

= 0.1119 − 0.06 = 0.0519 bar

∴ *Mass of air present per kg of uncondensed vapour,*

$$m_a = \frac{p_a V}{RT} = \frac{0.0519 \times 10^5 \times 23.74}{0.287 \times 1000 \times (273 + 36.2)}$$

= **1.3884 kg. (Ans.)**

Example 20.17. *A prime mover uses 15000 kg of steam per hour and develops 2450 kW. The steam is supplied at 30 bar and 350°C. The exhaust from the prime mover is condensed at 725 mm Hg when barometer records 755 mm Hg. The condensate temperature from the condenser is 31°C and the rise of temperature of circulating water is from 8°C to 18°C. Determine : (i) The quality of steam entering the condenser, (ii) The quantity of circulating cooling water and the ratio of cooling.*

Assume that no air is present in the condenser and all mechanical drive losses are negligible.

Solution. Quantity of steam used, m_s = 15000 kg/h

Pressure of steam = 30 bar, 350°C

Vacuum reading	= 725 mm Hg
Barometer reading	= 755 mm Hg
Condensate temperature	= 31°C

Rise of temperature of cooling water $t_{w_2} - t_{w_1}$

$= 18 - 8 = 10°C.$

(*i*) Quality of steam entering the condenser, x:

Condenser pressure = 755 – 725 = 30 mm Hg

$= 30 \times 1.333 \times 10^{-3} = 0.0399$ bar

From steam table :

***At 0.0399 bar* :** $h_f = 121.5$ kJ/kg, $h_{fg} = 2432.9$ kJ/kg

***At 30 bar, 350°C* :** $h_g = 3115.3$ kJ/kg

The work done in the turbine = 2450 kW or kJ/s ...(*i*)

Enthalpy drop in the turbine/sec.

$$= \frac{15000}{3600} [3115.3 - (121.5 + x \times 2432.9)] \text{ kJ/s} \quad ...(ii)$$

Since mechanical drive losses are negligible, the expressions (*i*) and (*ii*) are equal.

$$\therefore \quad 2450 = \frac{15000}{3600} [3115.3 - (121.5 + x \times 2432.9)]$$

$$\text{or} \quad \frac{2450 \times 3600}{15000} = 3115.3 - 121.5 - x \times 2432.9$$

$$588 = 3115.3 - 121.5 - 2432.9x$$

$$\therefore \quad x = \frac{3115.3 - 121.5 - 588}{2432.9} = \mathbf{0.988.} \quad \textbf{(Ans.)}$$

(*ii*) Quantity of circulating water, m_w :

Ratio of cooling :

Heat lost by condensing steam = Heat gained by cooling water

$$m_s [h_f + x\, h_{fg} - h_c] = m_w \times c_{pw} \times (t_{w_2} - t_{w_1})$$

$$\text{or} \quad m_w = \frac{m_s [h_f + x\, h_{fg} - h_c]}{c_{pw} \times (t_{w_2} - t_{w_1})}$$

$$= \frac{15000\,[121.5 + 0.988 \times 2432.9 - 4.186 \times 31]}{4.186 \times 10}$$

$= \mathbf{858375}$ **kg/h. (Ans.)**

[where c_{pw} = specific heat of water = 4.186 kJ/kg K]

$\therefore$ *Ratio of cooling*

$$= \frac{m_w}{m_s} = \frac{858375}{15000} = \mathbf{57.225 \text{ kg/kg.}} \quad \textbf{(Ans.)}$$

☞ **Example 20.18.** *A surface condenser is required to deal with 20000 kg of steam per hour, and the air leakage is estimated at 0.3 kg per 1000 kg of steam. The steam enters the condenser dry saturated at 38°C. The condensate is extracted at the lowest point of the condenser at a temperature of 36°C. The condensate loss is made up with water at 7°C. It is required to find the saving in condensate and the saving in heat supplied in the boiler, by fitting a separate air extraction pump which draws the air over an air cooler. Assume that the air leaves the cooler at 27°C. The pressure in the condenser can be assumed to remain constant.*

Solution. The mass of air per kg of steam at entry = 0.3/1000 = 0.0003 kg

***At 38°C* :** Saturation pressure = 0.06624 bar and $v_g = 21.63$ m³/kg (From steam tables)

For 1 kg of steam, the volume is 21.63 m³, and this must be the volume occupied by 0.0003 kg of air when exerting its partial pressure,

i.e., Partial pressure of air

$$= \frac{m_a R_a T}{V} = \frac{0.0003 \times 0.287 \times (273 + 38) \times 10^3}{21.63 \times 10^5}$$

$= 1.2 \times 10^{-5}$ bar

This is negligibly small and may be neglected.

Condensate extraction :

***At 36°C* :** Saturation pressure = 0.0594 bar, v_g = 23.97 m³/kg.

The total pressure in the condenser is 0.06624 bar, hence

$$0.06624 = 0.0594 + p_a$$

$$\therefore \quad p_a = 0.00684 \text{ bar}$$

The mass of air removed per hour is

$$\frac{20000 \times 0.3}{1000} = 6 \text{ kg/h}$$

Hence the volume of air removed per hour is

$$\frac{mRT}{p} = \frac{6 \times 0.287 \times (273 + 36) \times 10^3}{0.00684 \times 10^5} = 778 \text{ m}^3\text{/h}$$

The mass of steam associated with the air removed is therefore given by

$$\frac{778}{23.97} = 32.45 \text{ kg/h.}$$

Separate extraction :

***At 27°C* :** Saturation pressure = 0.03564 bar, v_g = 38.81 m³/kg

The air partial pressure = 0.06624 – 0.03564 = 0.0306 bar

$\therefore$ The volume of air removed

$$= \frac{mRT}{p} = \frac{6 \times 0.287 \times (273 + 27) \times 10^3}{0.0306 \times 10^5} = 168.9 \text{ m}^3\text{/h}$$

$$\therefore \quad \text{Steam removed} = \frac{168.9}{38.81} = 4.35 \text{ kg/h.}$$

Hence, *the saving in condensate by using the separate extraction method*

= 32.45 – 4.35 = **28.1 kg/h. (Ans.)**

The *saving in heat to be supplied in the boiler*

= 28.1 × 4.186 (36 – 7) = **3411 kJ/h. (Ans.)**

Example 20.19. *For the data of example 20.18, calculate :*

(i) The percentage reduction in air pump capacity by using the separate extraction method.

(ii) If the temperature rise of cooling water is 5.5 K, calculate the mass flow of cooling water required.

Solution. Capacity of air pump without air cooler

$= 778 \text{ m}^3/\text{h}$

Capacity of air pump with air cooler $= 168.9 \text{ m}^3/\text{h}$.

$\therefore$ *Percentage reduction in capacity*

$$= \left(\frac{778 - 168.9}{778}\right) \times 100 = \mathbf{78.3\%}. \quad \textbf{(Ans.)}$$

Fig. 20.14 shows the system to be analysed. Let suffixes *s*, *a* and *c* denote steam, air and condensate respectively.

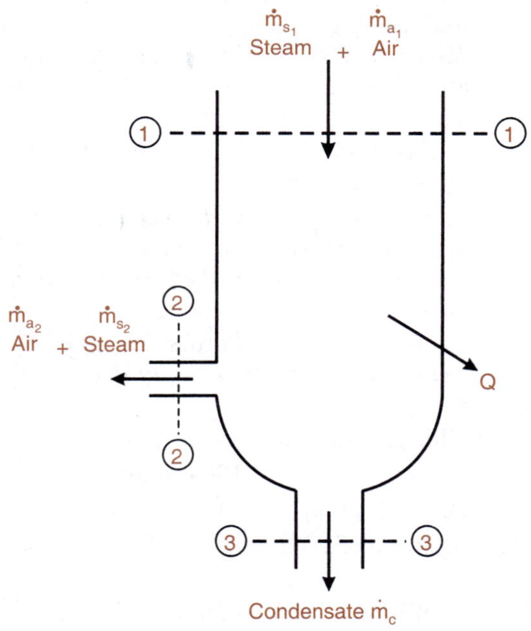

Fig. 20.14

Applying the steady flow energy equation and neglecting changes in kinetic energy, we have

$$Q = (\dot{m}_{s_1} h_{s_1} + \dot{m}_{a_1} h_{a_1}) - (\dot{m}_{s_2} h_{s_2} + \dot{m}_{a_2} h_{a_2}) - \dot{m}_c h_c$$

$$\dot{m}_{a_1} = \dot{m}_{a_2} = 6 \text{ kg/h} \; ; \; \dot{m}_{s_2} = 4.35 \text{ kg/h}$$

$$\dot{m}_c = 20000 - 4.35 = 20000 \text{ kg/h approximately.}$$

Also $\quad h_{a_1} - h_{a_2} = c_p (T_1 - T_2)$

$\therefore \quad Q = 20000 \times 2570.1 + 6 \times 1.005 (38 - 27) - 4.35 \times 2550.3 - 20000 \times 150.9$

[where $\quad h_c = h_f$ at 36°C = 150.9 kJ/kg]

i.e., Heat rejected $= 48.37 \times 10^6$ kJ/h

The *mass of cooling water required* for a 5.5 K rise in temperature

$$= \frac{48.37 \times 10^6}{4.186 \times 5.5} = \mathbf{2.1 \times 10^6 \text{ kg/h.}} \quad \textbf{(Ans.)}$$

Example 20.20. *In a condenser air pump and water pump are separately installed. Steam enters the condenser at 41.5°C and the condensate is removed at 37.6°C. The quantity of air infiltrating into the condenser through various zones is 6 kg/h. Determine :*

(i) The volume of air handled by the air pump.

(ii) The quantity handled by a combined air and condensate pump at 39°C.

Make suitable assumptions and list all such assumptions.

Solution. Pressure of steam corresponding to 41.5°C

= 0.08 bar.

Air pump suction point

Temperature = 37.6°C

Pressure (p_s) corresponding to 37.6°C = 0.065 bar

$\therefore$ Partial pressure of air = 0.08 – 0.065 = 0.015 bar

$\therefore$ *Volume of air,*

$$V_a = \frac{mRT}{p_a} = \frac{6 \times 0.287 \times (273 + 37.6) \times 1000}{0.015 \times 10^5}$$

$$= \mathbf{356.57 \text{ m}^3/\text{h}.} \quad \textbf{(Ans.)}$$

Assumptions :

1. Pressure inside the condenser is uniform.

2. Air is removed at the same temperature as that of condensate.

3. The pressure due to air at the entry of steam is neglected.

If condensate and air to be removed by the same pump, the air is to be removed at 39°C.

Partial pressure of steam at 39°C = 0.07 bar

$\therefore$ Partial pressure of air = 0.08 – 0.07 = 0.01 bar

$$\therefore \quad \textit{Volume of air} = \frac{6 \times 0.287 \times (273 + 39) \times 1000}{0.01 \times 10^5}$$

$$= \mathbf{537.26 \text{ m}^3/\text{h}.} \quad \textbf{(Ans.)}$$

☞ **Example 20.21.** *The following data relate to a two-pass surface condenser :*

Steam condensed	*15400 kg/h*
Temperature of cooling water when it enters the condenser	*15°C*
Temperature of cooling water when it leaves the condenser	*30°C*
The vacuum in the condenser	*675 mm of Hg*
Barometer reading	*750 mm of Hg*
Temperature of the condensate	*32°C*
Quality of exhaust steam	*0.92*
Water velocity in the tubes	*2.6 m/s*
Outside diameter of the tubes	*2.8 cm*
Thickness of the tubes	*0.03 cm*
Heat transfer co-efficient (U)	*3.35 kJ/h/cm²/°C.*

Determine : (i) Area of the tube surface required ;
(ii) Number of tubes ;
(iii) Length of tubes.

Solution. Absolute pressure in the condenser

$= 750 - 675 = 75$ mm Hg $= 75 \times 0.001333 \simeq 0.1$ bar

From steam tables saturation temperature corresponding to 0.1 bar pressure is 45.8°C. (At 0.1 bar ; $h_f = 191.8$ kJ/kg ; $h_{fg} = 2392.8$ kJ/kg).

Mean temperature difference,

$$t_m = \frac{t_{w_2} - t_{w_1}}{\log_e \frac{(t_s - t_{w_1})}{(t_s - t_{w_2})}} = \frac{(30 - 15)}{\log_e \frac{(45.8 - 15)}{(45.8 - 30)}}$$

$$= \frac{15}{\log_e \frac{30.8}{15.8}} = 22.5°\text{C}$$

Heat extracted per kg of steam

$= (h_f + xh_{fg}) - 1 \times 4.184 \times 32$
$= (191.8 + 0.92 \times 2392.8) - 133.9$
$= 2259.3$ kJ/kg.

Now, heat lost by steam = heat gained by water

$15400 \times 2393.2 = m_w \times c_{pw} \times (t_{w_2} - t_{w_1})$
$= m_w \times 4.184 \times (30 - 15)$

∴ Mass of circulating water,

$$m_w = \frac{15400 \times 2259.3}{4.184 \times 15} = 554385 \text{ kg/h.}$$

(i) **Area of tube surface required, A :**

$$A = \frac{Q}{Ut_m}$$

where Q = Heat lost by steam or heat gained by water

$= m_w \times c_{pw} \times (t_{w_2} - t_{w_1})$
$= 554385 \times 4.184 \times (30 - 15)$
$= 34793202$ kJ/h

U = Heat transfer co-efficient
$= 3.35$ kJ/h/cm²/°C

∴ $$A = \frac{34793202}{3.35 \times 22.5} = 461601 \text{ cm}^2.$$

(ii) **Number of tubes :**

$$\text{Total water required/sec.} = \frac{m_w \times 1000}{3600}$$

$$= \frac{554385 \times 1000}{3600}$$

[∵ 1 kg occupies 1000 cm³ volume]

$= 153996$ cm³

Water required/sec./tube

$$= \frac{\pi}{4} \times (2.8 - 2 \times 0.03)^2 \times (2.6 \times 100) = 1533 \text{ cm}^3$$

∴ Number of tubes

$$= \frac{2 \times m_w \times 1000}{3600 \times 1533} = \frac{2 \times 554385 \times 1000}{3600 \times 1533} = 200.$$

(iii) **Length of tubes, l :**

Surface area per tube

$$= \frac{A}{200} = \frac{461601}{200} = 2308$$

i.e., $\pi \times d \times l = 2308$

$\pi \times 2.8 \times l = 2308$

l = **262.4 cm. (Ans.)**

Example 20.22. *At a thermal power station where the boiler pressure is 80 bar and the required flow rate is 130 tonnes/hour, two multistage centrifugal pumps are used in series to pump water from the condenser to the boiler. Each pump is required to produce a head of approximately the total head and run at 1450 r.p.m. If the impellers in stages are identical and the specific speed of each impeller is not less than 15, find :*

(i) The head developed per stage and required number of stages in each pump.

(ii) The required impeller diameters, assuming the speed ratio based on the outer tip diameter to be 0.98 and the shaft power input, if the overall efficiency of each pump is 0.76.

What would you expect if the discharge valve is closed and pump is switched on ? What safety device would you recommend ? **(AMIE)**

Solution. Boiler pressure, $p = 85$ bar ; flow rate $= 130$ tonnes/hour

Speed of each pump, $N = 1450$ r.p.m. ;

Specific speed of each impeller, $N_s = 15$; speed ratio $= 0.98$

Overall efficiency of each pump, $\eta_{op} = 0.76$

(i) **Head developed per stage and no. of stages required in each pump :**

$$\text{Total head} = \frac{p}{w} = \frac{80 \times 10^5}{9810} = 815.5 \text{ m}$$

(∵ For water, $w = 9810$ N/m³)

$$\text{Head per pump} = \frac{815.5}{2} = 407.75 \text{ m}$$

Let n_s = Number of stages per pump,

$$\text{Then, head per stage, } H = \frac{407.75}{n_s}$$

$$\text{Now, specific speed, } N_s = 15 = \frac{N\sqrt{Q}}{H^{3/4}}$$

$$\text{where } Q = \frac{(130 \times 1000) \times 9.81}{9810 \times 3600} \text{ m}^3\text{/s} = 0.0361 \text{ m}^3\text{/s}$$

$$\therefore \quad 15 = \frac{1450 \times \sqrt{0.0361}}{\left(\dfrac{407.75}{n_s}\right)^{3/4}} \quad \text{or} \quad \left(\frac{407.75}{n_s}\right)^{3/4} = \frac{1450 \times \sqrt{0.0361}}{15} = 18.37$$

$$\text{or} \quad \frac{407.75}{n_s} = (18.37)^{4/3} = 48.47 \quad \text{or} \quad n_s = \frac{407.75}{48.47} = 8.41$$

Actual number of stages in each pump = **8. (Ans.)**

Head developed per stage

$$= \frac{407.75}{8} = \mathbf{50.97\ m.} \quad \textbf{(Ans.)}$$

(*ii*) Diameter of each impeller (D) and shaft power input, P :

Speed ratio

$$= 0.98 = \frac{u_2}{\sqrt{2gH}} \quad \text{or} \quad 0.98 = \frac{u_2}{\sqrt{2 \times 9.81 \times 50.97}}$$

$$\text{or} \quad u_2 = 0.98 \times \sqrt{2 \times 9.81 \times 50.97} = 30.99 \text{ m/s}$$

$$\text{Also,} \quad u_2 = \frac{\pi DN}{60} \quad \text{or} \quad 30.99 = \frac{\pi D \times 1450}{60}$$

$$\text{or} \quad D = \frac{30.99 \times 60}{\pi \times 1450} = \mathbf{0.408\ m\ or\ 408\ mm.} \quad \textbf{(Ans.)}$$

If the discharge valve is closed and pump is switched on, *a pressure wave will be reflected from the valve. A relief valve (safety device) may be used on the delivery side of the pump.*

Example 20.23. (*a*) *Discuss in brief the requirements of a good surface condenser.*

(*b*) *A steam condenser is equipped in a steam power plant which handles 15000 kg/h of steam and develops 2.5 MW power. The initial condition of steam entering to turbine is 27 bar and 300°C. The exhaust from the turbine is condensed in the condenser and the vacuum maintained is 72 cm of Hg while the barometer reading is 76 cm of Hg. The temperature of the circulating water is increased from 20°C to 28°C while the condensate is removed at a temperature of 27°C. Work out the following :*

(*i*) *Show the condensing process on T-s and h-s diagrams ;* (*ii*) *Dryness fraction of steam entering the condenser ;* (*iii*) *Mass rate of circulating water and cooling ratio ;* (*iv*) *Degree of undercooling.* **(AMIE)**

Solution. (*a*) There are *two main requirements* of a good surface condenser :

1. Air should be removed from the surface condenser at a cooler section, and it should be *as far as possible dry,* without much water vapour. Humidity ratio or the water vapour quantity contained by unit mass of saturated air decreases with decrease in temperature of air.

2. The *pressure drop of steam in the condenser should be minimum*. Thus, smooth flow of steam should be provided without any unnecessary resistance.

(*b*) (*i*) The **condensing process** (2–3) is shown in Fig. 20.15 on *T-s* and *h-s* diagrams.

The schematic arrangement is shown in Fig. 20.16.

$$p_{cond.} = 76 - 72 = 4 \text{ cm Hg}$$

$$= (13.59 \times 10^3) \times 9.81 \times \frac{4}{100} \times 10^{-5} \text{ bar}$$

$$= 0.0533 \text{ bar}$$

Corresponding to $t_{sat} = 34°C$:

$h_f = 142.5$ kJ/kg and $h_{fg} = 2421$ kJ/kg (from steam tables)

Also $h_1 = 2997$ kJ/kg (at 27 bar and 300°C).

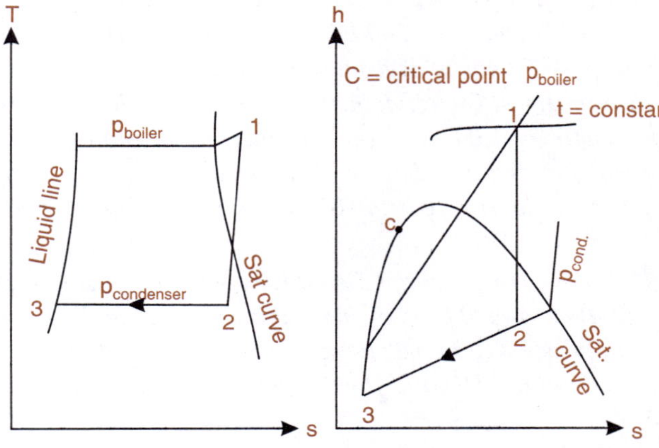

Fig. 20.15

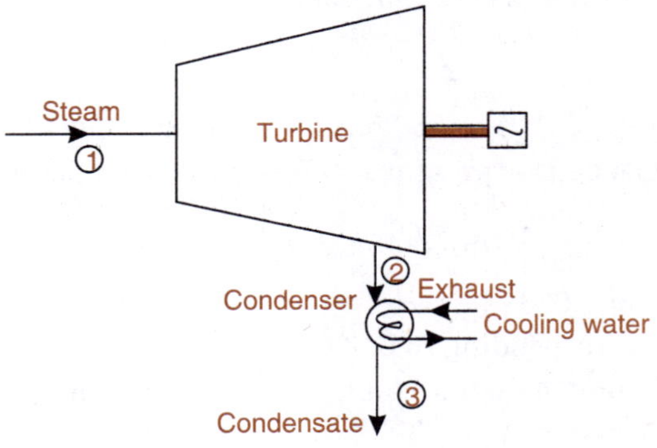

Fig. 20.16

(*ii*) Work per kg

$$= h_1 - h_2 = \frac{(2.5 \times 10^3) \times 3600}{15000} = 600 \text{ kJ/kg}$$

$$h_2 = h_1 - 600 = 2997 - 600 = 2397 \text{ kJ/kg}$$

$$\text{or} \quad 2397 = 142.5 + x_2 \times 2421$$

$\therefore$ **Dryness fraction of steam entering condenser,** x_2 = **0.931. (Ans.)**

(*iii*) **Mass of circulating water, m_w and cooling ratio :**

$h_3 = 113.2$ (h_f at 27°C) kJ/kg

Making *energy balance for condenser,*

Mass rate of cooling water (m_w, kg/hour) × 4.187 × (28 – 20)

$= 15000\,(2397 - 113.2)$

$\therefore \quad m_w = 1.02272 \times 10^6$ kg/h

Cooling ratio

$$= \frac{\text{Cooling water rate}}{\text{Steam rate}} = \frac{1.02272 \times 10^6}{15000}$$

= **68.18.** (It is quite high !). **(Ans.)**

(*iv*) **Degree of undercooling :**

Degree of undercooling = 34 – 27 = **7°C. (Ans.)**

Example 20.24. (*a*) *A 210 MW multistage steam turbine is supplied with steam at 100 bar and 500°C. Vacuum in the condenser is 71 cm of Hg when barometer reads 76 cm of Hg. Assuming stage efficiency of 75% for each stage and reheat factor of 1.04, find the volume rate of cooling water required for the condenser. The rise in cooling water temperature is limited to 10°C and the condensate is undercooled by 2°C.*

(*b*) *A steam turbine discharges 4000 kg of steam per hour at 40°C and 0.85 dry. The estimated air leakage into the condenser is 16 kg per hour. The temperature at the air pump suction is 32°C and the temperature of the condensate is 35°C.*

Find the capacity of the dry air pump if its volumetric efficiency is 80%.

(*c*) *What are the functions of the air cooling section in a surface condenser ?* **(AMIE)**

Solution. (*a*) $\eta_{int} = \eta_{stage} \times$ reheat factor

$= 0.75 \times 1.04 = 0.78$

Condenser pressure = (76 – 71) × 10 = 50 mm Hg

$$= \frac{50}{1000}\,(\text{m}) \times (13.6 \times 1000)\,(\text{kg/m}^3) \times 9.8 \times 10^{-5}$$

= 0.0667 bar

Corresponding to 0.0662 bar $t_s = 38.1$°C.

From Mollier's chart, overall isentropic enthalpy drop, (Δh) is,

$(\Delta h) = h_1 - h_2$

$= 3370 - 2040 = 1330$ kJ/kg

Actual enthalpy drop

$= h_1 - h_2' = (\Delta h) \times \eta_{int}$

$= 1330 \times 0.78 = 1037.4$ kJ/kg

$h_2' = 3370 - 1037.4 = 2332.6$ kJ/kg

h_f at (38.1 – 2°C), *i.e.*, 36.1°C = 151 kJ/kg

$$\text{Steam rate} = \frac{210 \times 10^3}{1037.4} = 202.429 \text{ kg/s}$$

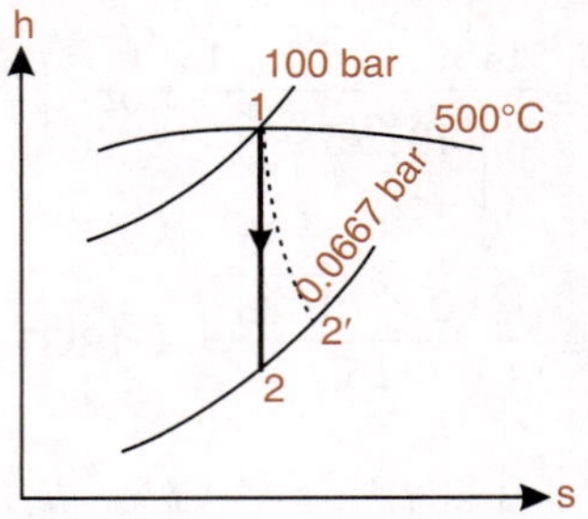

Fig. 20.17

Heat rejected in condenser

= Steam rate × ($h_2' - h_f$)

Heat gained by cooling

$= m_w \times c_{pw} \times (t_{w_2} - t_{w_1})$

$= m_w \times 4.187 \times 10$

$\therefore$ Cooling water rate,

$$\dot{m}_w = \frac{202.429\,(2332.6 - 151)}{4.187 \times 10}$$

= 10547.4 kg/s = **10547.4 l/s. (Ans.)**

(*b*) Specific volume of steam at inlet = xv_g

($\because$ $v_g = 19.52$ m³/kg at 40°C)

$= 0.85 \times 19.52 = 16.59$ m³/kg

Volume rate of steam at inlet,

$$V_1 = \frac{4000 \times 16.59}{60}$$

= 1106 m³/min.

This is also the volume rate of air at inlet.

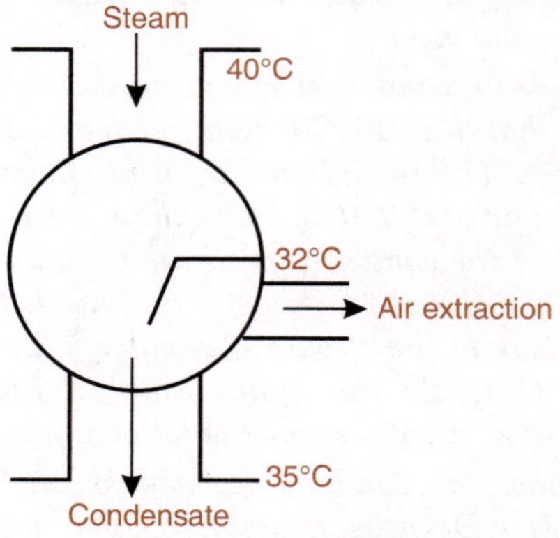

Fig. 20.18

$\therefore$ Partial pressure of air at inlet,

$$p_{a_1} = \frac{m_a R T_1}{V_1}$$

$$= \frac{16}{60} \times \frac{287 \times (273 + 40)}{1106} \times 10^{-3} \text{ kPa}$$

= 0.0216 kPa

Partial pressure of steam at inlet, at 40°C

= 7.38 kPa.

∴ Total condenser pressure

= 7.38 + 0.0216

= 7.406 kPa

Partial pressure of steam at pump inlet (at 32°C)

= 4.76 kPa

∴ Partial pressure of air at pump suction,

$$p_{a_2} = 7.406 - 4.76 = 2.646 \text{ kPa.}$$

∴ Dry air pump capacity,

$$V_2 = \frac{m_a R T_2}{p_{a_2} \times \eta_{vol}}$$

$$= \frac{16}{60} \times \frac{287 \times (273 + 32)}{(2.646 \times 10^3) \times 0.8}$$

= **11.03 m³. (Ans.)**

(*c*) The total pressure in the condenser is the sum of partial pressures of water vapour and air. It remains constant at every section of the condenser, whereas the *partial pressure of steam depends upon the temperature at that section*.

In the condenser, air cooling section is provided near the air extraction pump suction. By circulating relatively cold water at this section the temperature is kept low, thereby *lowering the partial pressure of steam*. This serves ***two functions :***

(*i*) The amount of steam lost to atmosphere (which is otherwise used as pure condensate for boiler feed) is considerably reduced.

(*ii*) The partial pressure of air in this section becomes considerably high, and so its density. For a given rate of air leakage into the condenser, the volume of air handled by the air-extraction pump is *reduced*. This enables the use of a **small** *capacity and power pump*.

HIGHLIGHTS

1. A *steam condenser* is a device or an appliance in which steam condenses and heat released by steam is absorbed by water.
2. *Vacuum* is sub-atmospheric pressure. It is measured as the pressure depression below atmospheric.
3. Condensers are of two types :
   (*i*) *Jet condensers*—exhaust steam and water come in direct contact with each other and the temperature of the condensate is the same as that of cooling water leaving the condenser.
   (*ii*) *Surface condensers*—exhaust steam and water do not come into direct contact. The steam passes over the tubes through which a supply of cooling water is maintained.
4. Jet condensers may be classified as : (*i*) Parallel-flow type, (*ii*) Counter-flow type, and (*iii*) Ejector type.
5. Surface condensers may be classified as : (*i*) Down-flow type, (*ii*) Central-flow type, (*iii*) Inverted-flow type, (*iv*) Regenerative type, and (*v*) Evaporative type.
6. *Vacuum efficiency* is defined as the ratio of the actual vacuum to the maximum obtainable vacuum.
7. *Condenser efficiency* is defined as the ratio of the difference between the outlet and inlet temperatures of cooling water to the difference between the temperature corresponding to the vacuum in the condenser and inlet temperature of cooling water.
8. The quantity of cooling water, *m* is found by the following expression :

$$m_w = \frac{m_s[xh_{fg} + c_{pw}(t_s - t_{w_2})]}{c_{pw}(t_{w_2} - t_{w_1})} \text{ kg/h}$$

...... **Jet condenser.**

$$m_w = \frac{m_s[xh_{fg} + c_{pw}(t_s - t_c)]}{c_{pw}(t_{w_2} - t_{w_1})} \text{ kg/h}$$

......**Surface condenser.**

9. Air pumps may be classified as :
   (*i*) Reciprocating piston or bucket pumps
   (*ii*) Rotary pumps
   (*iii*) Steam jet air pumps (ejectors)
   (*iv*) Wet jet pumps.

OBJECTIVE TYPE QUESTIONS

Choose the Correct Answer :

1. The thermal efficiency of the engine with condenser as compared to without condenser, for a given pressure and temperature of steam, is
   (*a*) higher
   (*b*) lower
   (*c*) same as long as initial pressure and temperature are unchanged
   (*d*) none of the above.
2. In jet type condensers
   (*a*) cooling water passes through tubes and steam surrounds them
   (*b*) steam passes through tubes and cooling water surrounds them
   (*c*) steam and cooling water mix
   (*d*) steam and cooling water do not mix.
3. In a shell and tube surface condenser
   (*a*) steam and cooling water mix to give the condensate

(*b*) cooling water passes through the tubes and steam surrounds them
(*c*) steam passes through the cooling tubes and cooling water surrounds them
(*d*) all of the above varying with situation.

4. In a surface condenser if air is removed, there is
(*a*) fall in absolute pressure maintained in condenser
(*b*) rise in absolute pressure maintained in condenser
(*c*) no change in absolute pressure in the condenser
(*d*) rise in temperature of condensed steam.

5. The cooling section in the surface condenser
(*a*) increases the quantity of vapour extracted alongwith air
(*b*) reduces the quantity of vapour extracted alongwith air
(*c*) does not affect vapour quantity extracted but reduces pump capacity of air extraction pump
(*d*) none of the above.

6. Edward's air pump
(*a*) removes air and also vapour from condenser
(*b*) removes only air from condenser
(*c*) removes only un-condensed vapour from condenser
(*d*) removes air alongwith vapour and also the condensed water from condenser.

7. Vacuum efficiency of a condenser is ratio of
(*a*) $\dfrac{\text{Actual vacuum in condenser with air present}}{\text{Theoretical vacuum in condenser with no air present}}$
(*b*) $\dfrac{\text{Theoretical vacuum in condenser with no air present}}{\text{Actual vacuum in condenser with air present}}$
(*c*) $\dfrac{\text{Partial pressure of vapour + partial pressure of air present}}{\text{Partial pressure of vapour only}}$
(*d*) $\dfrac{\text{Partial pressure of vapour only}}{\text{Partial pressure of vapour + partial pressure of air present}}$

8. In a steam power plant, the function of a condenser is
(*a*) to maintain pressure below atmospheric to increase work output from the prime mover
(*b*) to receive large volumes of steam exhausted from steam prime mover.
(*c*) to condense large volumes of steam to water which may be used again in boiler
(*d*) all of the above.

9. In a regenerative surface condenser
(*a*) there is one pump to remove air and condensate
(*b*) there are two pumps to remove air and condensate
(*c*) there are three pumps to remove air, vapour and condensate
(*d*) there is no pump, the condensate gets removed by gravity.

10. Evaporative type of condenser has
(*a*) steam in pipes surrounded by water
(*b*) water in pipes surrounded by steam
(*c*) either (*i*) and (*ii*)
(*d*) none of the above.

11. Condenser efficiency is defined as
(*a*) $\dfrac{\text{Saturation temperature at condenser pressure}}{\text{Rise in cooling water temperature}}$
(*b*) $\dfrac{\text{Temperature rise of cooling water}}{\text{Saturation temperature corresponding to condenser pressure}}$
(*c*) $\dfrac{\text{Temperature rise of cooling water}}{\text{Saturation temperature corresponding to condenser pressure} - \text{cooling water inlet temperature}}$
(*d*) $\dfrac{\text{Saturation temperature corresponding to condenser pressure}}{\text{Saturation temperature of vapour at its partial pressure in condenser}}$.

ANSWERS

1. (*a*)	2. (*c*)	3. (*b*)	4. (*a*)
5. (*b*)	6. (*d*)	7. (*a*)	8. (*d*)
9. (*b*)	10. (*a*)	11. (*c*).	

THEORETICAL QUESTIONS

1. Define a steam condenser and state its objects.
2. Explain clearly the term 'vacuum'. How is it measured ?
3. State the organs of a steam condensing plant.
4. How will you classify condensers ? In what respects a jet condenser differs from a surface condenser ?
5. Explain briefly the following types of jet condensers :
(*i*) Parallel-flow type (*ii*) Counter-flow type
(*iii*) Ejector type.
6. Classify the surface condensers and explain with neat sketches any two of the following :
(*i*) Down-flow type, (*ii*) Regenerative type,
(*iii*) Evaporative type.
7. Explain the reasons for inefficiency in surface condensers.
8. State the comparison between jet and surface condensers.
9. What are the sources of air in the condensers ?
10. Explain the effects of air leakage in a condenser.
11. Explain the following :
(*i*) Vacuum measurement (*ii*) Vacuum efficiency
(*iii*) Condenser efficiency.
12. State and explain Dalton's law of partial pressures.
13. Derive an expression for determining weight of cooling required in case of a surface condenser.
14. What is the chief function performed by an air pump ? Explain briefly with a neat sketch any one of the following:
(*i*) Bucket pump ; (*ii*) Rotary pump ; (*iii*) Wet jet pump.

UNSOLVED EXAMPLES

1. The reading of a vacuum gauge in a condenser is 714 mm of Hg when barometer is 759 mm. In another case the vacuum is 709 mm of Hg while the barometer is 756 mm. Correct these vacuum gauge readings to the standard barometer of 760 mm.
[**Ans.** 715 mm of Hg, 713 mm of Hg]
2. A vacuum of 635 mm was obtained when barometer reads 750 mm of Hg. Find the corresponding vacuum to a standard barometer of 760 mm. [**Ans.** 647 mm]
3. A vacuum of 68 cm of Hg was obtained with the barometer reading 75 cm of Hg. The temperature of the condensate is 21°C. Correct the vacuum to a standard barometer of 76 cm, and hence determine the partial pressure of air and steam present, and also the weight of air and present per kg of steam. [**Ans.** 69 cm of Hg, 5.136 cm of Hg, 1.864 cm of Hg ; 4.415 kg/kg of steam]
4. Calculate the vacuum efficiency of a condenser from the following data : Vacuum at steam inlet to condenser = 725 mm ; Barometer = 760 mm ; Hot well temperature = 26.4°C. [**Ans.** 98.25%]
5. The observations recorded during the trial on a steam condenser are given below :
Condenser vacuum = 685 mm Hg ; Barometer reading = 765 mm Hg ; Mean condensate temperature = 34°C ; Hot well temperature = 28°C ; Condensate formed per hour = 1750 kg ; Circulating cooling water inlet temperature = 18°C, Circulating cooling water outlet temperature = 30°C ; Quantity of cooling water = 1300 kg/min.
Determine : (*i*) Vacuum efficiency, (*ii*) Undercooling of condensate, (*iii*) Condenser efficiency, (*iv*) Condition of steam as it enters the condenser, (*v*) Mass of air present per kg of uncondensed steam.
Take : *R* for air = 0.287 kJ/kg K.
Specific heat of water = 4.186 kJ/kg K.
[**Ans.** (*i*) 94.47%, (*ii*) 6°C, (*iii*) 37.5%, (*iv*) 0.898, (*v*) 1.613 kg]
6. A surface condenser is fitted with separate air and condensate outlets. A portion of the cooling surface is screened from the incoming steam and the air passes over these screened tubes to the air extraction section and becomes cooled below the condensate temperature. The condenser receives 20000 kg/h of steam dry saturated at 36.2°C. At condensate outlet, the temperature is 34.6°C, and at the air extraction section the temperature is 29°C. The volume of air plus vapour leaving the condenser is 3.8 m^3/min. Assuming constant pressure throughout the condenser, calculate :
(*i*) The mass of air removed per 10000 kg of steam.
(*ii*) The mass of steam condensed in the air cooler per minute.
(*iii*) The heat to be removed per minute by the cooling water.
Neglect the partial pressure of the air inlet to the condenser.
[**Ans.** (*i*) 2.63 kg, (*ii*) 0.492 kg, (*iii*) 807050 kJ]
7. Separate air pump and water pump are installed in a condenser. Steam enters the condenser at 40°C and condensate is removed at 37°C. The quantity of air infiltrating into the condenser through various zones is 5 kg/hour.
(*i*) What will be the volume of air handled by the air pump ?
(*ii*) What will be the quantity handled by a combined air and condensate pump at 38°C.
Make suitable assumptions and list all such assumptions. [**Ans.** 400.8 m^3/h ; 776 m^3/h]
8. In a surface condenser test the following observations were made : Vacuum 70 cm of Hg : barometer 76.5 cm of Hg : mean temperature of condensate 35.82°C ; hot well temperature 30°C ; weight of cooling water 47500 kg/h ; inlet temperature of cooling water 17°C ; outlet temperature of cooling water 32°C ; weight of condensate 1500 kg/h. Calculate : (*i*) The mass of air present per m^3 of condenser volume ; (*ii*) The state of steam entering the condenser, and (*iii*) The vacuum efficiency.
[**Ans.** 0.03145 kg ; 0.812 ; 97.05%]
9. A surface condenser deals with 2724 kg of steam per hour and the air leakage amounts to 2.27 kg per hour. The temperature of the air pump suction is 30°C and vacuum 67 cm of Hg when the barometer reads 74.7 cm of Hg. The volumetric efficiency of the air pump is 70%. Determine the discharging capacity of the air pump in m^3/min., to remove the air and condensed steam.
[**Ans.** 0.805 m^3/min.]
10. A surface condenser deals with 12500 kg of steam per hour. The leakage air in the system amounts to 1 kg per 2500 kg of steam. The vacuum in the air pump suction is 705 mm of mercury (barometer 760 mm of Hg) and the temperature is 34.25°C. Compute the discharging capacity of the wet air pump which removes both air and condensate in m^3/min, taking the volumetric efficiency of the pump as 80%. If the air pump is single-acting and runs at 55 r.p.m. and piston stroke is 1.25 times the diameter of the pump, find the dimensions of the wet air pump. [**Ans.** 4.99 m^3/min ; 45.25 cm, 56.563 cm]
11. A prime mover uses 16000 kg of steam per hour and develops 2600 kW. The steam is supplied at 30 bar and 300°C. The exhaust from the prime mover is condensed at 725 mm Hg (0.984895 bar) when barometer records 755 mm Hg (1.006584 bar). The condensate temperature from the condenser is 30°C and the rise of temperature of circulating water is from 7°C to 17°C. Determine : (*i*) The quality of steam entering the condenser. (*ii*) The quantity of circulating cooling water and the ratio of cooling water.
Assume all mechanical drive losses are negligible and no air is present in the condenser.
[**Ans.** (*i*) 0.9496, (*ii*) 873028 kg/h, 54.56 kg/kg]

Gas Power Cycles

21

21.1. DEFINITION OF A CYCLE

A **cycle** is defined as a *repeated series of operations occurring in a certain order*. It may be repeated by repeating the processes in the same order. The cycle may be of *imaginary perfect engine or actual engine*. The former is called ***ideal cycle*** and the latter ***actual cycle.*** *In ideal cycle all accidental heat losses are prevented and the working substance is assumed to behave like a perfect working substance.*

21.2. AIR STANDARD EFFICIENCY

To compare the effects of different cycles, it is of paramount importance that the effect of the calorific value of the fuel is altogether eliminated and this can be achieved by considering air (which is assumed to behave as a perfect gas) as the working substance in the engine cylinder. *The efficiency of engine using air as the working medium is known as an "**Air standard efficiency**"*. This efficiency is oftenly called ***ideal efficiency.***

The actual efficiency of a cycle is always *less* than the air-standard efficiency of that cycle under ideal conditions. This is taken into account by introducing a new term "***Relative efficiency***" which is defined as :

$$\eta_{\text{relative}} = \frac{\text{Actual thermal efficiency}}{\text{Air standard efficiency}} \quad ...(21.1)$$

The analysis of all air standard cycles is based upon the following *assumptions* :

Assumptions :

1. The gas in the engine cylinder is a *perfect gas i.e.*, it obeys the gas laws and has constant specific heats.
2. The physical constants of the gas in the cylinder are the same as those of air at moderate temperatures *i.e.*, the molecular weight of cylinder gas is 29.

   $c_p = 1.005$ kJ/kg K, $c_v = 0.718$ kJ/kg K.
3. The compression and expansion processes are adiabatic and they take place without internal friction, *i.e.*, these processes are *isentropic*.
4. No chemical reaction takes place in the cylinder. Heat is supplied or rejected by bringing a hot body or a cold body in contact with cylinder at appropriate points during the process.
5. The cycle is considered closed with the same 'air' always remaining in the cylinder to repeat the cycle.

21.3. THE CARNOT CYCLE

This cycle has the *highest possible efficiency* and consists of four simple operations namely,

(*a*) Isothermal expansion
(*b*) Adiabatic expansion
(*c*) Isothermal compression
(*d*) Adiabatic compression.

The condition of the Carnot cycle may be imagined to occur in the following way :

One kg of a air is enclosed in the cylinder which (except at the end) is made of perfect non-conducting material. A source of heat '*H*' is supposed to provide unlimited quantity of heat, non-conducting cover '*C*' and a sump '*S*' which is of infinite capacity so that its temperature remains unchanged irrespective of the fact how much heat is supplied to it. The temperature of source *H* is T_1 and the same is of the working substance. The working substance while rejecting heat to sump '*S*' has the temperature. T_2 *i.e.*, the same as that of sump *S*.

Following are the *four stages* of the Carnot cycle. Refer to Fig. 21.1 (*a*).

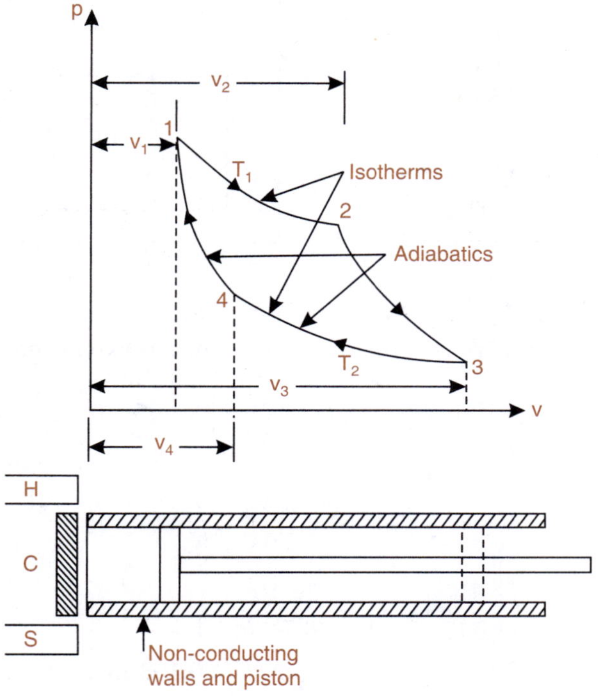

(*a*) Four stages of the carnot cycle

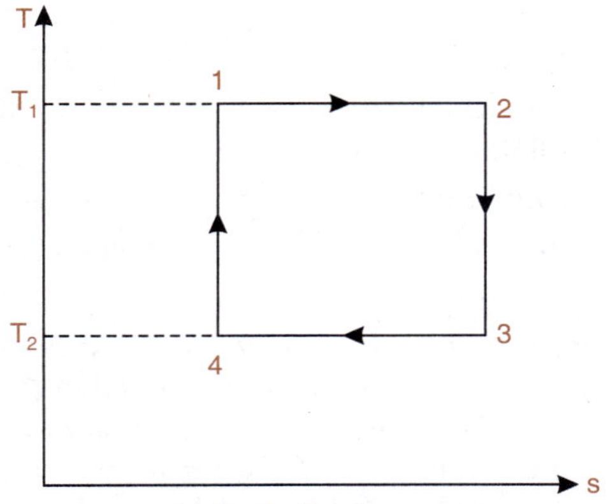

(*b*) *T-s* diagram

Fig. 21.1. Carnot cycle.

Stage (1). Line 1–2 [Fig. 21.1 (*a*)] represents the isothermal expansion which takes place at temperature T_1 when source of heat *H* is applied to the end of cylinder. Heat supplied in this case is given by $RT_1 \log_e r$ and where *r* is the ratio of expansion.

Stage (2). Line 2–3 represents the application of non-conducting cover to the end of the cylinder. This is followed by the adiabatic expansion and the temperature falls from T_1 to T_2.

Stage (3). Line 3–4 represents the isothermal compression which takes place when sump '*S*' is applied to the end of cylinder. Heat is rejected during this operation whose value is given by $RT_2 \log_e r$ where *r*, is the ratio of compression.

Stage (4). Line 4–1 represents repeated application of non-conducting cover and adiabatic compression due to which temperature increases from T_2 to T_1.

It may be noted that ratio of expansion during isotherm 1–2 and ratio of compression during isotherm 3–4 must be equal to get a closed cycle.

Fig. 21.1 (*b*) represents the Carnot cycle on *T-s* coordinates.

Now according to law of conservation of energy,

Heat supplied = Work done + Heat rejected

Work done = Heat supplied – Heat rejected

$$= RT_1 . \log_e r - RT_2 \log_e r$$

Efficiency of cycle

$$= \frac{\text{Work done}}{\text{Heat supplied}} = \frac{R \log_e r (T_1 - T_2)}{RT_1 . \log_e r}$$

$$= \frac{T_1 - T_2}{T_1} \quad \text{...(21.2)}$$

From this equation, it is quite obvious that if temperature T_2 decreases efficiency increases and it becomes 100% if T_2 becomes absolute zero which, of course is impossible to attain. Further more *it is not possible to produce an engine that should work on Carnot's cycle as it would necessitate the piston to travel very slowly during first portion of the forward stroke (isothermal expansion) and to travel more quickly during the remainder of the stroke (adiabatic expansion) which however is not practicable.*

Example 21.1. *A Carnot engine working between 400°C and 40°C produces 130 kJ of work. Determine :*

(*i*) *The engine thermal efficiency.*

(*ii*) *The heat added.*

(*iii*) *The entropy changes during heat rejection process.*

Solution. Temperature, $T_1 = T_2$

$$= 400 + 273 = 673 \text{ K}$$

Temperature,

$$T_3 = T_4 = 40 + 273 = 313 \text{ K}$$

Work produced, $W = 130$ kJ.

(*i*) **Engine thermal efficiency, η_{th} :**

$$\eta_{th.} = \frac{673 - 313}{673} = \mathbf{0.535} \text{ or } \mathbf{53.5\%}. \quad \textbf{(Ans.)}$$

(*ii*) **Heat added :**

$$\eta_{th.} = \frac{\text{Work done}}{\text{Heat added}}$$

i.e.,
$$0.535 = \frac{130}{\text{Heat added}}$$

$$\therefore \quad \text{Heat added} = \frac{130}{0.535} = \mathbf{243\ kJ.} \quad \textbf{(Ans.)}$$

(*iii*) **Entropy change during the heat rejection process, $(S_3 - S_4)$:**

$$\text{Heat rejected} = \text{Heat added} - \text{Work done}$$
$$= 243 - 130 = 113 \text{ kJ}$$

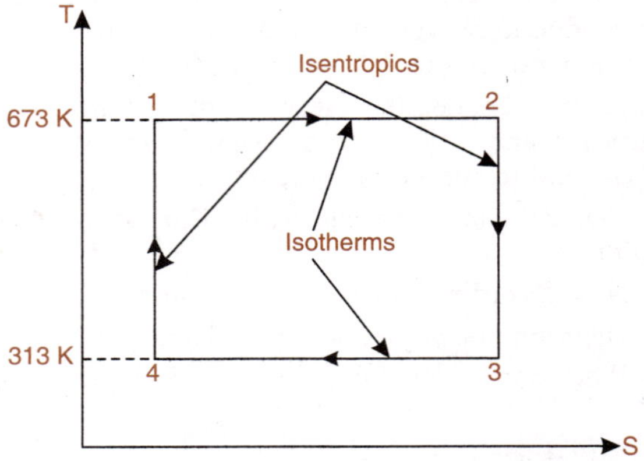

Fig. 21.2

$$\text{Heat rejected} = T_3 (S_3 - S_4) = 113$$

$$\therefore \quad (S_3 - S_4) = \frac{113}{T_3} = \frac{113}{313} = \mathbf{0.361\ kJ/K.} \quad \textbf{(Ans.)}$$

Example 21.2. *0.5 kg of air (ideal gas) executes a Carnot power cycle having a thermal efficiency of 50 per cent. The heat transfer to the air during the isothermal expansion is 40 kJ. At the beginning of the isothermal expansion the pressure is 7 bar and the volume is 0.12 m³. Determine :*

(i) The maximum and minimum temperatures for the cycle in K ;

(ii) The volume at the end of isothermal expansion in m³ ;

(iii) The heat transfer for each of the four processes in kJ.

For air c_v = 0.721 kJ/kg K, and c_p = 1.008 kJ/kg K.

(U.P.S.C.)

Solution. Refer to Fig. 21.3. *Given* : m = 0.5 kg ; η_{th} = 50% ; Heat transferred during isothermal expansion = 40 kJ ;

$$p_1 = 7 \text{ bar}, V_1 = 0.12 \text{ m}^3 ;$$
$$c_v = 0.721 \text{ kJ/kg K} ; c_p = 1.008 \text{ kJ/kg K}.$$

(*i*) **The maximum and minimum temperatures, T_1, T_2 :**

$$p_1V_1 = mRT_1$$
$$7 \times 10^5 \times 0.12 = 0.5 \times 287 \times T_1$$

$\therefore$ Maximum temperature,

$$\mathbf{T_1} = \frac{7 \times 10^5 \times 0.12}{0.5 \times 287} = \mathbf{585.4\ K.} \quad \textbf{(Ans.)}$$

$$\eta_{cycle} = \frac{T_1 - T_2}{T_1} \Rightarrow 0.5 = \frac{585.4 - T_2}{585.4}$$

$\therefore$ Minimum temperature,

$$\mathbf{T_2} = 585.4 - 0.5 \times 585.4 = \mathbf{292.7\ K.} \quad \textbf{(Ans.)}$$

(*ii*) **The volume at the end of isothermal expansion, V_2 :**

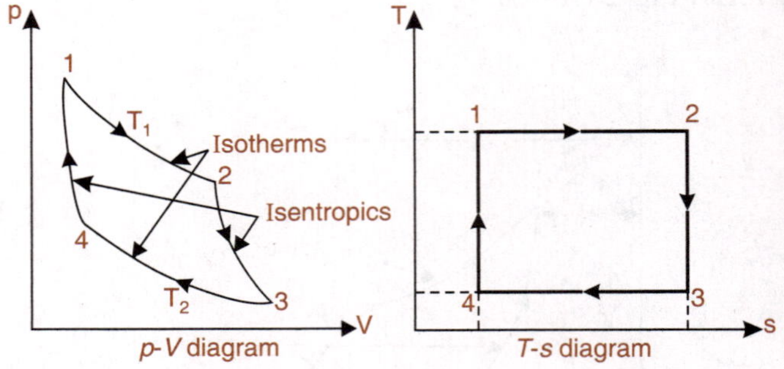

Fig. 21.3. Carnot cycle.

Heat transferred during isothermal expansion

$$= p_1V_1 \ln(r) = mRT_1 \ln\left(\frac{V_2}{V_1}\right) = 40 \times 10^3$$

......(Given)

or
$$0.5 \times 287 \times 585.4 \ln\left(\frac{V_2}{0.12}\right) = 40 \times 10^3$$

or
$$\ln\left(\frac{V_2}{0.12}\right) = \frac{40 \times 10^3}{0.5 \times 287 \times 585.4} = 0.476$$

or
$$\mathbf{V_2} = 0.12 \times (e)^{0.476} = \mathbf{0.193\ m^3.} \quad \textbf{(Ans.)}$$

(*iii*) **The heat transfer for each of the four processes :**

Process	*Classification*	*Heat transfer*
1–2	Isothermal expansion	40 kJ
2–3	Adiabatic reversible expansion	zero
3–4	Isothermal compression	– 40 kJ
4–1	Adiabatic reversible compression	zero.

(Ans.)

☞**Example 21.3.** *In a Carnot cycle, the maximum pressure and temperature are limited to 18 bar and 410°C. The ratio of isentropic compression is 6 and isothermal expansion is 1.5. Assuming the volume of the air at the beginning of isothermal expansion as 0.18 m³, determine :*

(i) The temperature and pressures at main points in the cycle.

(ii) Change in entropy during isothermal expansion.

(iii) Mean thermal efficiency of the cycle.

(*iv*) *Mean effective pressure of the cycle.*

(*v*) *The theoretical power if there are 210 working cycles per minute.*

Solution. Refer to Fig. 21.4.

Maximum pressure, $p_1 = 18$ bar

Maximum temperature, $T_1 = (T_2) = 410 + 273 = 683$ K

Ratio of isentropic (or adiabatic) compression, $\frac{V_4}{V_1} = 6$

Ratio of isothermal expansion, $\frac{V_2}{V_1} = 1.5.$

Volume of the air at the beginning of isothermal expansion, $V_1 = 0.18$ m^3.

(*i*) **Temperatures and pressures at the main points in the cycle :**

For the *isentropic process* 4–1 :

$$\frac{T_1}{T_4} = \left(\frac{V_4}{V_1}\right)^{\gamma - 1} = (6)^{1.4 - 1} = (6)^{0.4} = 2.05$$

$$\therefore \quad T_4 = \frac{T_1}{2.05} = \frac{683}{2.05} = 333.2 \text{ K} = T_3$$

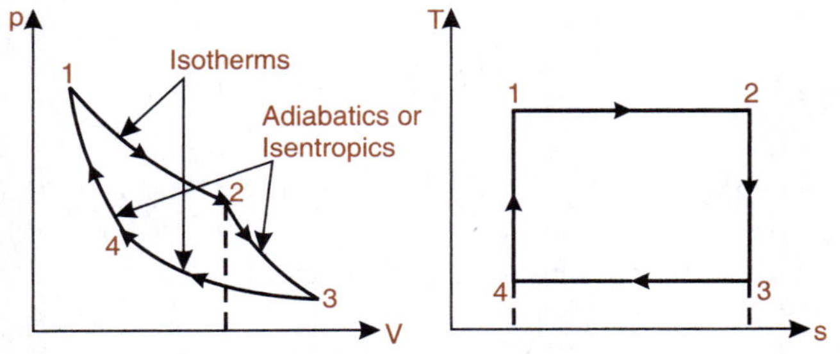

Fig. 21.4

Also,
$$\frac{p_1}{p_4} = \left(\frac{V_4}{V_1}\right)^{\gamma} = (6)^{1.4} = 12.29$$

$$\therefore \quad p_4 = \frac{p_1}{12.29} = \frac{18}{12.29} = 1.46 \text{ bar}$$

For the *isothermal process* 1–2 :

$$p_1V_1 = p_2V_2$$

$$p_2 = \frac{p_1V_1}{V_2} = \frac{18}{1.5} = 12 \text{ bar}$$

For *isentropic process* 2–3, we have :

$$p_2V_2^{\gamma} = p_3V_3^{\gamma}$$

$$p_3 = p_2 \times \left(\frac{V_2}{V_3}\right)^{\gamma} = 12 \times \left(\frac{V_1}{V_4}\right)^{\gamma} \qquad \left[\because \frac{V_4}{V_1} = \frac{V_3}{V_2}\right]$$

$$= 12 \times \left(\frac{1}{6}\right)^{1.4} = \mathbf{0.97 \ bar.} \quad \textbf{(Ans.)}$$

Hence
$p_1 = \mathbf{18\ bar}$, $T_1 = T_2 = 683$ K
$p_2 = \mathbf{12\ bar}$
$p_3 = \mathbf{0.97\ bar}$, $T_3 = T_4 = 333.2$ K
$p_4 = \mathbf{1.46\ bar}$
(Ans.)

(*ii*) **Change in entropy :**

Change in entropy during isothermal expansion,

$$S_2 - S_1 = mR \log_e \left(\frac{V_2}{V_1}\right) = \frac{p_1V_1}{T_1} \log_e \left(\frac{V_2}{V_1}\right)$$

$$\left[\because \quad pV = mRT \quad \text{or} \quad mR = \frac{pV}{T}\right]$$

$$= \frac{18 \times 10^5 \times 0.18}{10^3 \times 683} \log_e (1.5)$$

$$= \mathbf{0.192\ kJ/K.} \quad \textbf{(Ans.)}$$

(*iii*) **Mean thermal efficiency of the cycle :**

Heat supplied, $Q_s = p_1V_1 \log_e \left(\frac{V_2}{V_1}\right)$

$= T_1 (S_2 - S_1)$

$= 683 \times 0.192 = 131.1$ kJ

Heat rejected, $Q_r = p_4V_4 \log_e \left(\frac{V_3}{V_4}\right)$

$= T_4 (S_3 - S_4)$ because increase in entropy during heat addition is equal to decrease in entropy during heat rejection.

$$\therefore \quad Q_r = 333.2 \times 0.192 = 63.97 \text{ kJ}$$

$$\therefore \quad \textit{Efficiency}, \quad \eta = \frac{Q_s - Q_r}{Q_s} = 1 - \frac{Q_r}{Q_s}$$

$$= 1 - \frac{63.97}{131.1}$$

$$= \mathbf{0.512} \quad \text{or} \quad \mathbf{51.2\%.} \quad \textbf{(Ans.)}$$

(*iv*) **Mean effective pressure of the cycle, p_m :**

The mean effective pressure of the cycle is given by

$$p_m = \frac{\text{Work done per cycle}}{\text{Stroke volume}}$$

$$\frac{V_3}{V_1} = 6 \times 1.5 = 9$$

Stroke volume, $V_s = V_3 - V_1 = 9V_1 - V_1$

$= 8V_1 = 8 \times 0.18 = 1.44$ m^3

$$\therefore \quad p_m = \frac{(Q_s - Q_r) \times J}{V_s} = \frac{(Q_s - Q_r) \times 1}{V_s}$$

$$(\because \quad J = 1)$$

$$= \frac{(131.1 - 63.97) \times 10^3}{1.44 \times 10^5} = \mathbf{0.466\ bar.} \quad \textbf{(Ans.)}$$

(*v*) **Power of the engine, P :**

Power of the engine working on this cycle is given by

$P = (131.1 - 63.97) \times (210/60) = \mathbf{234.9\ kW.}$ **(Ans.)**

Example 21.4. *A reversible engine converts one-sixth of the heat input into work. When the temperature of the sink is reduced by 70°C, its efficiency is doubled. Find the temperature of the source and the sink.*

Solution. Let, T_1 = temperature of the source (*K*), and

T_2 = temperature of the sink (*K*)

First case :

$$\frac{T_1 - T_2}{T_1} = \frac{1}{6}$$

i.e., $6T_1 - 6T_2 = T_1$

or $5T_1 = 6T_2$ or $T_1 = 1.2T_2$ ...(*i*)

Second case :

$$\frac{T_1 - [T_2 - (70 + 273)]}{T_1} = \frac{1}{3}$$

$$\frac{T_1 - T_2 + 343}{T_1} = \frac{1}{3}$$

$$3T_1 - 3T_2 + 1029 = T_1$$

$$2T_1 = 3T_2 - 1029$$

$$2 \times (1.2T_2) = 3T_2 - 1029 \quad (\because T_1 = 1.2T_2)$$

$$2.4T_2 = 3T_2 - 1029$$

or $0.6T_2 = 1029$

$\therefore$ $T_2 = \frac{1029}{0.6}$ = **1715 K** or **1442°C. (Ans.)**

and $T_1 = 1.2 \times 1715$ = **2058 K** or **1785°C. (Ans.)**

Example 21.5. *An inventor claims that a new heat cycle will develop 0.4 kW for a heat addition of 32.5 kJ/min. The temperature of heat source is 1990 K and that of sink is 850 K. Is his claim possible ?*

Solution. Temperature of heat source, T_1 = 1990 K

Temperature of sink, T_2 = 850 K

Heat supplied = 32.5 kJ/min

Power developed by the engine, P = 0.4 kW

The *most efficient engine is one that works on Carnot cycle*

$$\eta_{carnot} = \frac{T_1 - T_2}{T_1} = \frac{1990 - 850}{1990} = 0.573 \text{ or } 57.3\%$$

Also, thermal efficiency of the engine,

$$\eta_{th} = \frac{\text{Work done}}{\text{Heat supplied}} = \frac{0.4}{(32.5/60)} = \frac{0.4 \times 60}{32.5}$$

$$= 0.738 \text{ or } 73.8\%$$

which is not feasible as no engine can be more efficient than that working on Carnot cycle.

Hence claims of the inventor is **not true. (Ans.)**

Example 21.6. *An ideal engine operates on the Carnot cycle using a perfects gas as the working fluid. The ratio of the greatest to the least volume is fixed and is x : 1, the lower temperature of the cycle is also fixed, but the volume compression ratio 'r' of the reversible adiabatic compression is variable. The ratio of the specific heats is γ.*

Show that if the work done in the cycle is a maximum then,

$$(\gamma - 1) \log_e \frac{x}{r} + \frac{1}{r^{\gamma - 1}} - 1 = 0.$$

Solution. Refer to Fig. 21.1.

$$\frac{V_3}{V_1} = x \; ; \; \frac{V_4}{V_1} = r$$

During isotherms, since compression ratio = expansion ratio

$$\therefore \quad \frac{V_3}{V_4} = \frac{V_2}{V_1}$$

Also $$\frac{V_3}{V_4} = \frac{V_3}{V_1} \times \frac{V_1}{V_4} = x \times \frac{1}{r} = \frac{x}{r}$$

Work done per kg of the gas

$$= \text{Heat supplied} - \text{Heat rejected}$$

$$= RT_1 \log_e \frac{x}{r} - RT_2 \log_e \frac{x}{r}$$

$$= R(T_1 - T_2) \log_e \frac{x}{r}$$

$$= RT_2 \left(\frac{T_1}{T_2} - 1\right) \log_e \frac{x}{r}$$

But $$\frac{T_1}{T_2} = \left(\frac{V_4}{V_1}\right)^{\gamma - 1} = (r)^{\gamma - 1}$$

$\therefore$ Work done per kg of the gas,

$$W = RT_2 \, (r^{\gamma - 1} - 1) \log_e \frac{x}{r}$$

Differentiating *W* w.r.t. '*r*' and equating to zero

$$\frac{dW}{dr} = RT_2$$

$$\left[(r^{\gamma - 1} - 1)\left\{\frac{r}{x} \times (-xr^{-2})\right\} + \log_e \frac{x}{r}\{(\gamma - 1)r^{\gamma - 2}\}\right] = 0$$

or $$(r^{\gamma - 1} - 1)\left(-\frac{1}{r}\right) + (\gamma - 1) \times r^{\gamma - 2} \log_e \frac{x}{r} = 0$$

or $$-r^{\gamma - 2} + \frac{1}{r} + r^{\gamma - 2} (\gamma - 1) \log_e \frac{x}{r} = 0$$

or $$r^{\gamma - 2}\left\{-1 + \frac{1}{r \cdot r^{\gamma - 2}} + (\gamma - 1) \log_e \frac{x}{r}\right\} = 0$$

or $$-1 + \frac{1}{r \cdot r^{\gamma - 2}} + (\gamma - 1) \log_e \frac{x}{r} = 0$$

$$(\gamma - 1) \log_e \frac{x}{r} + \frac{1}{r^{\gamma - 1}} - 1 = 0.$$ ...... **Proved.**

21.4. CONSTANT VOLUME OR OTTO CYCLE

This cycle is so named as it was conceived by 'Otto'. On this cycle, petrol, gas and many types of oil engines work. It is the standard of comparison for internal combustion engines.

Fig. 21.5 (*a*) and (*b*) shows the theoretical *p-v* diagram and *T-s* diagrams of this cycle respectively.

- The point 1 represents that cylinder is full of air with volume V_1, pressure p_1 and absolute temperature T_1.
- Line 1–2 represents the *adiabatic compression* of air due to which p_1, V_1 and T_1 change to p_2, V_2 and T_2 respectively.
- Line 2–3 shows the *supply of heat* to the air *at constant volume* so that p_2 and T_2 change to p_3 and T_3 (V_3 being the same as V_2).
- Line 3–4 represents the *adiabatic expansion* of the air. During expansion p_3, V_3 and T_3 change to a final value of p_4, V_4 or V_1 and T_4 respectively.
- Line 4–1 shows the *rejection of heat* by air at *constant volume* till original state (point 1) reaches.

Consider **1 kg of air** (working substance) :

Heat supplied at constant volume $= c_v(T_3 - T_2)$.

Heat rejected at constant volume $= c_v(T_4 - T_1)$.

But, work done = Heat supplied – Heat rejected

$$= c_v(T_3 - T_2) - c_v(T_4 - T_1)$$

$$\therefore \quad \text{Efficiency} = \frac{\text{Work done}}{\text{Heat supplied}}$$

$$= \frac{c_v(T_3 - T_2) - c_v(T_4 - T_1)}{c_v(T_3 - T_2)}$$

$$= 1 - \frac{T_4 - T_1}{T_3 - T_2} \qquad \text{...}(i)$$

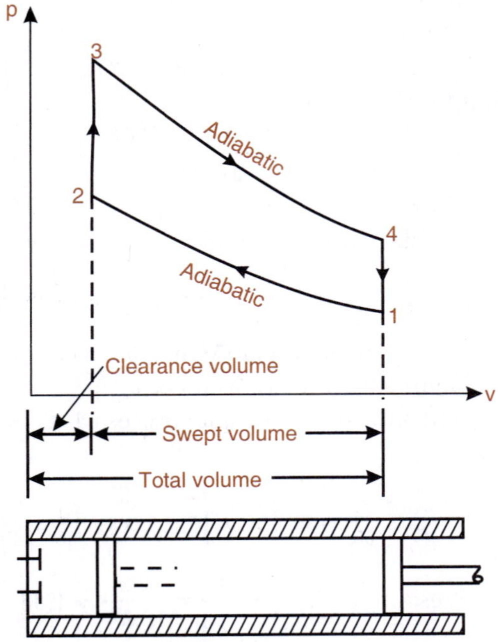

(*a*)

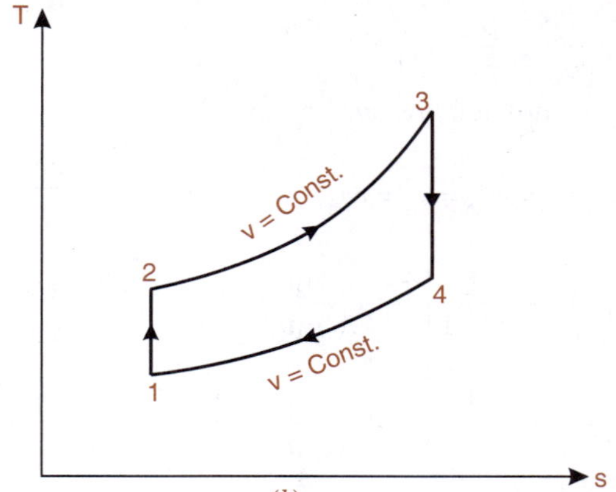

(*b*)

Fig. 21.5

Let compression ratio, $r_c\ (= r) = \dfrac{v_1}{v_2}$

and expansion ratio, $r_e\ (= r) = \dfrac{v_4}{v_3}$

(These *two ratios are same* in this cycle)

As $$\frac{T_2}{T_1} = \left(\frac{v_1}{v_2}\right)^{\gamma - 1}$$

Then, $$T_2 = T_1 \cdot (r)^{\gamma - 1}$$

Similarly, $$\frac{T_3}{T_4} = \left(\frac{v_4}{v_3}\right)^{\gamma - 1}$$

or $$T_3 = T_4 \cdot (r)^{\gamma - 1}$$

Inserting the values of T_2 and T_3 in equation (*i*), we get

$$\eta_{otto} = 1 - \frac{T_4 - T_1}{T_4 \cdot (r)^{\gamma - 1} - T_1 \cdot (r)^{\gamma - 1}}$$

$$= 1 - \frac{T_4 - T_1}{r^{\gamma - 1}(T_4 - T_1)} = 1 - \frac{1}{(r)^{\gamma - 1}} \qquad \text{...(21.3)}$$

This expression is known as the ***air standard efficiency of the Otto cycle.***

It is clear from the above expression that efficiency increases with the increase in the value of r, which means we can have maximum efficiency by increasing r to a considerable extent, but *due to practical difficulties its value is limited to about 8.*

The *net work done per kg* in the Otto cycle can also be expressed in terms of p, v. If p is expressed in bar *i.e.*, 10^5 N/m^2, then work done

$$W = \left(\frac{p_3 v_3 - p_4 v_4}{\gamma - 1} - \frac{p_2 v_2 - p_1 v_1}{\gamma - 1}\right) \times 10^2 \text{ kJ} \qquad \text{...(21.4)}$$

Also $$\frac{p_3}{p_4} = r^{\gamma} = \frac{p_2}{p_1}$$

$$\therefore \quad \frac{p_3}{p_2} = \frac{p_4}{p_1} = r_p$$

where r_p stands for *pressure ratio.*

and $\quad v_1 = rv_2 = v_4 = rv_3 \qquad \left[\because \frac{v_1}{v_2} = \frac{v_4}{v_3} = r\right]$

$$\therefore \quad W = \frac{1}{\gamma - 1}\left[p_4 v_4\left(\frac{p_3 v_3}{p_4 v_4} - 1\right) - p_1 v_1\left(\frac{p_2 v_2}{p_1 v_1} - 1\right)\right]$$

$$= \frac{1}{\gamma - 1}\left[p_4 v_4\left(\frac{p_3}{p_4 r} - 1\right) - p_1 v_1\left(\frac{p_2}{p_1 r} - 1\right)\right]$$

$$= \frac{v_1}{\gamma - 1}\left[p_4 (r^{\gamma - 1} - 1) - p_1 (r^{\gamma - 1} - 1)\right]$$

$$= \frac{v_1}{\gamma - 1}\left[(r^{\gamma - 1} - 1)(p_4 - p_1)\right]$$

$$= \frac{p_1 v_1}{\gamma - 1}\left[(r^{\gamma - 1} - 1)(r_p - 1)\right] \qquad \text{...[21.4 (a)]}$$

Mean effective pressure (p_m) is given by :

$$p_m = \left[\left(\frac{p_3 v_3 - p_4 v_4}{\gamma - 1} - \frac{p_2 v_2 - p_1 v_1}{\gamma - 1}\right) \div (v_1 - v_2)\right] \text{bar} \qquad \text{...(21.5)}$$

$$\text{Also} \quad p_m = \frac{\left[\frac{p_1 v_1}{\gamma - 1}(r^{\gamma - 1} - 1)(r_p - 1)\right]}{(v_1 - v_2)}$$

$$= \frac{\frac{p_1 v_1}{\gamma - 1}[(r^{\gamma - 1} - 1)(r_p - 1)]}{v_1 - \frac{v_1}{r}}$$

$$= \frac{\frac{p_1 v_1}{\gamma - 1}[(r^{\gamma - 1} - 1)(r_p - 1)]}{v_1\left(\frac{r - 1}{r}\right)}$$

i.e.,
$$\mathbf{p_m} = \frac{p_1 r[(r^{\gamma - 1} - 1)(r_p - 1)]}{(\gamma - 1)(r - 1)} \qquad \text{...(21.6)}$$

Example 21.7. *The efficiency of an Otto cycle is 60% and $\gamma = 1.5$. What is the compression ratio ?*

Solution. Efficiency of Otto cycle, $\eta = 60\%$

Ratio of specific heats, $\gamma = 1.5$

Compression ratio, $r = ?$

Efficiency of Otto cycle is given by,

$$\eta_{Otto} = 1 - \frac{1}{(r)^{\gamma - 1}}$$

$$0.6 = 1 - \frac{1}{(r)^{1.5 - 1}}$$

or $\quad \frac{1}{(r)^{0.5}} = 0.4 \quad \text{or} \quad (r)^{0.5} = \frac{1}{0.4} = 2.5 \quad \text{or} \quad r = 6.25$

Hence, *compression ratio* = **6.25. (Ans.)**

Example 21.8. *An engine of 250 mm bore and 375 mm stroke works on Otto cycle. The clearance volume is 0.00263 m³. The initial pressure and temperature are 1 bar and 50°C. If the maximum pressure is limited to 25 bar, find the following :*

(i) The air standard efficiency of the cycle.

(ii) The mean effective pressure for the cycle.

Assume the ideal conditions.

Solution. Bore of the engine, $D = 250 \text{ mm} = 0.25 \text{ m}$

Stroke of the engine, $L = 375 \text{ mm} = 0.375 \text{ m}$

Clearance volume, $V_c = 0.00263 \text{ m}^3$

Initial pressure, $p_1 = 1 \text{ bar}$

Initial temperature, $T_1 = 50 + 273 = 323 \text{ K}$

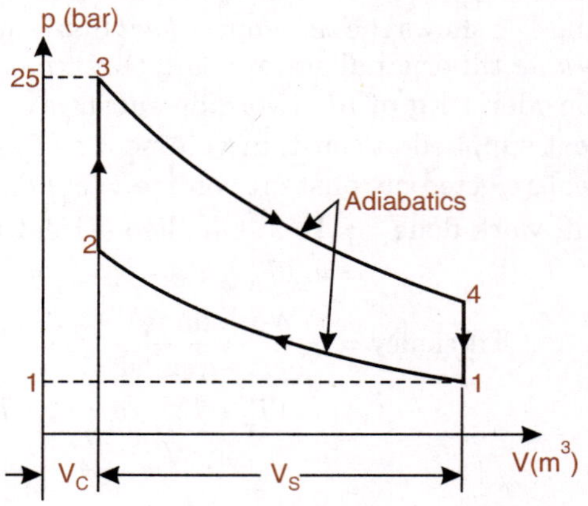

Fig. 21.6

Maximum pressure, $p_3 = 25$ bar

Swept volume,

$$V_s = \pi/4\, D^2 L = \pi/4 \times 0.25^2 \times 0.375 = 0.0184 \text{ m}^3$$

Compression ratio,

$$r = \frac{V_s + V_c}{V_c} = \frac{0.0184 + 0.00263}{0.00263} = 8.$$

***(i)* Air standard efficiency :**

The air standard efficiency of Otto cycle is given by

$$\eta_{Otto} = 1 - \frac{1}{(r)^{\gamma - 1}} = 1 - \frac{1}{(8)^{1.4 - 1}} = 1 - \frac{1}{(8)^{0.4}}$$

$$= 1 - 0.435 = \mathbf{0.565} \quad \text{or} \quad \mathbf{56.5\%}. \quad \textbf{(Ans.)}$$

***(ii)* Mean effective pressure, p_m :**

For adiabatic (or isentropic) process 1–2

$$p_1 V_1^{\gamma} = p_2 V_2^{\gamma}$$

or $\quad p_2 = p_1\left(\frac{V_1}{V_2}\right)^{\gamma} = 1 \times (r)^{1.4} = 1 \times (8)^{1.4} = 18.38 \text{ bar}$

$$\therefore \quad \text{Pressure ratio, } r_p = \frac{p_3}{p_2} = \frac{25}{18.38} = 1.36$$

The mean effective pressure is given by

$$p_m = \frac{p_1 r[(r^{\gamma-1}-1)(r_p-1)]}{(\gamma-1)(r-1)}$$

$$= \frac{1 \times 8\,[\{(8)^{1.4-1}-1\}\,(1.36-1)]}{(1.4-1)\,(8-1)} \quad \text{...[Eqn. (21.6)]}$$

$$= \frac{8\,(2.297-1)(0.36)}{0.4 \times 7} = 1.334 \text{ bar}$$

Hence *mean effective pressure* = **1.334 bar.** **(Ans.)**

Example 21.9. *The minimum pressure and temperature in an Otto cycle are 100 kPa and 27°C. The amount of heat added to the air per cycle is 1500 kJ/kg.*

(i) Determine the pressures and temperatures at all points of the air standard Otto cycle.

(ii) Also calculate the specific work and thermal efficiency of the cycle for a compression ratio of 8 : 1.

Take for air : c_v = 0.72 kJ/kg K, and γ = 1.4.

(GATE)

Solution. Refer to Fig. 21.7. *Given* : p_1 = 100 kPa = 10^5 N/m² or 1 bar ;

T_1 = 27 + 273 = 300 K ; Heat added = 1500 kJ/kg ;

r = 8 : 1 ; c_v = 0.72 kJ/kg ; γ = 1.4.

Consider 1 kg of air.

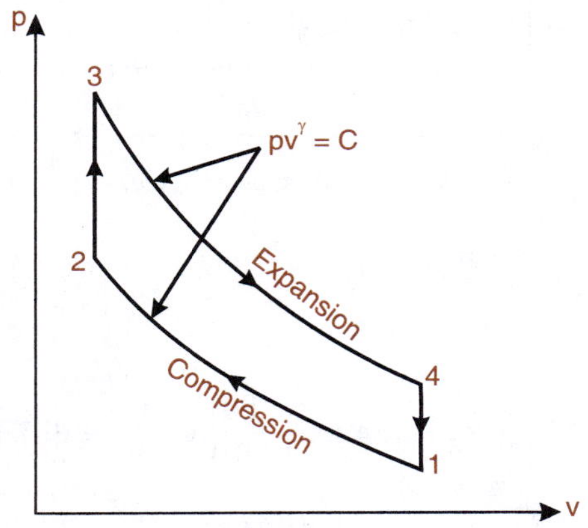

Fig. 21.7

(i) **Pressures and temperatures at all points :**

Adiabatic compression process 1–2 :

$$\frac{T_2}{T_1} = \left(\frac{v_1}{v_2}\right)^{\gamma-1} = (r)^{\gamma-1} = (8)^{1.4-1} = 2.297$$

$\therefore$ $\mathbf{T_2}$ = 300 × 2.297 = **689.1 K.** **(Ans.)**

Also $p_1 v_1^\gamma = p_2 v_2^\gamma$

or $$\frac{p_2}{p_1} = \left(\frac{v_1}{v_2}\right)^{\gamma} = (8)^{1.4} = 18.379$$

$\therefore$ $\mathbf{p_2}$ = 1 × 18.379 = **18.379 bar.** **(Ans.)**

Constant volume process 2–3 :

Heat added during the process,

$$c_v\,(T_3 - T_2) = 1500$$

or $$0.72\,(T_3 - 689.1) = 1500$$

or $$\mathbf{T_3} = \frac{1500}{0.72} + 689.1 = \mathbf{2772.4\ K.} \quad \textbf{(Ans.)}$$

Also, $\dfrac{p_2}{T_2} = \dfrac{p_3}{T_3} \Rightarrow \mathbf{p_3} = \dfrac{p_2 T_3}{T_2}$

$$= \frac{18.379 \times 2772.4}{689.1} = \mathbf{73.94\ bar.} \quad \textbf{(Ans.)}$$

Adiabatic Expansion process 3–4 :

$$\frac{T_3}{T_4} = \left(\frac{V_4}{V_3}\right)^{\gamma-1} = (r)^{\gamma-1} = (8)^{1.4-1} = 2.297$$

$\therefore$ $$\mathbf{T_4} = \frac{T_3}{2.297} = \frac{2772.4}{2.297} = \mathbf{1206.9\ K.} \quad \textbf{(Ans.)}$$

Also, $p_3 v_3^\gamma = p_4 v_4^\gamma \Rightarrow \mathbf{p_4} = p_3 \times \left(\dfrac{v_3}{v_4}\right)^{\gamma}$

$$= 73.94 \times \left(\frac{1}{8}\right)^{1.4} = \mathbf{4.023\ bar.} \quad \textbf{(Ans.)}$$

(ii) **Specific work and thermal efficiency :**

Specific work = Heat added – heat rejected

$= c_v\,(T_3 - T_2) - c_v(T_4 - T_1) = c_v\,[(T_3 - T_2) - (T_4 - T_1)]$

= 0.72 [(2772.4 – 689.1) – (1206.9 – 300)]

= **847 kJ/kg.** **(Ans.)**

Thermal efficiency, $\eta_{th} = 1 - \dfrac{1}{(r)^{\gamma-1}}$

$$= 1 - \frac{1}{(8)^{1.4-1}} = \mathbf{0.5647} \text{ or } \mathbf{56.47\%.} \quad \textbf{(Ans.)}$$

Example 21.10. *An air standard Otto cycle has a volumetric compression ratio of 6, the lowest cycle pressure of 0.1 MPa and operates between temperature limits of 27°C and 1569°C.*

(i) Calculate the temperature and pressure after the isentropic expansion (ratio of specific heats = 1.4).

(ii) Since it is observed that values in (i) are well above the lowest cycle operating conditions, the expansion process was allowed to continue down to a pressure of 0.1 MPa. Which process is required to complete the cycle ? Name the cycle so obtained.

(iii) Determine by what percentage the cycle efficiency has been improved. **(GATE)**

Solution. Refer to Fig. 21.8. *Given* : $\dfrac{v_1}{v_2} = \dfrac{v_4}{v_3} = r = 6$; p_1 = 0.1 MPa = 1 bar ; T_1 = 27 + 273 = 300 K ; T_3 = 1569 + 273 = 1842 K ; γ = 1.4.

(*i*) Temperature and pressure after the isentropic expansion, T_4, p_4 :

Consider 1 kg of air :

For *the compression process 1–2 :*

$$p_1v_1^{\gamma} = p_2v_2^{\gamma} \Rightarrow p_2 = p_1 \times \left(\frac{v_1}{v_2}\right)^{\gamma}$$

$$= 1 \times (6)^{1.4} = 12.3 \text{ bar}$$

Also $$\frac{T_2}{T_1} = \left(\frac{v_1}{v_2}\right)^{\gamma-1} = (6)^{1.4-1} = 2.048$$

$\therefore$ $$T_2 = 300 \times 2.048 = 614.4 \text{ K}$$

For *the constant volume process 2–3 :*

$$\frac{p_2}{T_2} = \frac{p_3}{T_3} \Rightarrow p_3 = \frac{p_2T_3}{T_2} = 12.3 \times \frac{1842}{614.4}$$

$$= 36.9 \text{ bar}$$

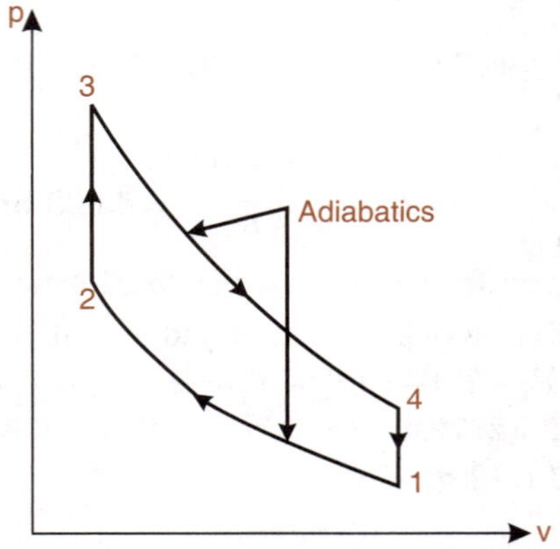

Fig. 21.8

For *the expansion process 3–4 :*

$$\frac{T_3}{T_4} = \left(\frac{v_4}{v_3}\right)^{\gamma-1} = (6)^{1.4-1} = 2.048$$

$\therefore$ $$\mathbf{T_4} = \frac{T_3}{2.048} = \frac{1842}{2.048} = \mathbf{900\ K.} \quad \textbf{(Ans.)}$$

Also $$p_3v_3^{\gamma} = p_4v_4^{\gamma} \Rightarrow p_4 = p_3 \times \left(\frac{v_3}{v_4}\right)^{\gamma}$$

or $$\mathbf{p_4} = 36.9 \times \left(\frac{1}{6}\right)^{1.4} = \mathbf{3\ bar.} \quad \textbf{(Ans.)}$$

(*ii*) Process required to complete the cycle :

Process required to complete the cycle is the *constant pressure scavenging.*

The cycle is called **Atkinson cycle** (Refer to Fig. 21.9).

(*iii*) Percentage improvement/increase in efficiency :

$$\eta_{Otto} = 1 - \frac{1}{(r)^{\gamma-1}} = 1 - \frac{1}{(6)^{1.4-1}}$$

$$= \mathbf{0.5116} \text{ or } \mathbf{51.16\%.} \quad \textbf{(Ans.)}$$

$$\eta_{Atkinson} = \frac{\text{Work done}}{\text{Heat supplied}}$$

$$= \frac{\text{Heat supplied} - \text{Heat rejected}}{\text{Heat supplied}}$$

$$= \frac{c_v(T_3 - T_2) - c_p(T_5 - T_1)}{c_v(T_3 - T_2)}$$

$$= 1 - \frac{c_p(T_5 - T_1)}{c_v(T_3 - T_2)} = 1 - \frac{\gamma(T_5 - T_1)}{(T_3 - T_2)}$$

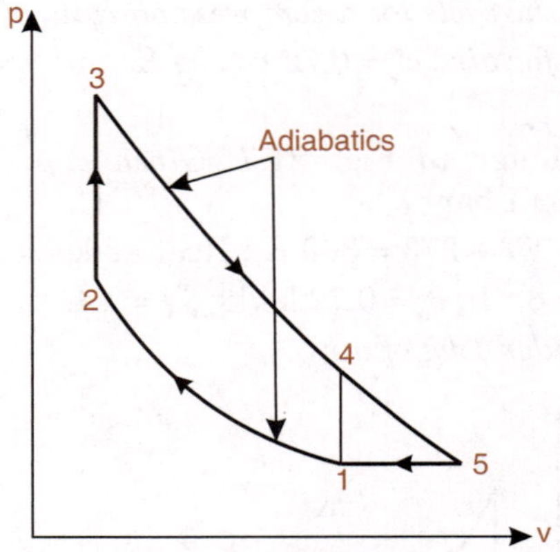

Fig. 21.9. Atkinson cycle.

Now, $$\frac{T_5}{T_3} = \left(\frac{p_5}{p_3}\right)^{\frac{\gamma-1}{\gamma}}$$

or $$T_5 = 1842 \times \left(\frac{1.0}{36.9}\right)^{\frac{1.4-1}{1.4}} = 657 \text{ K}$$

$\therefore$ $$\eta_{Atkinson} = 1 - \frac{1.4(657 - 300)}{(1842 - 614.4)}$$

$$= \mathbf{0.5929} \text{ or } \mathbf{59.29\%.}$$

$\therefore$ **Improvement in efficiency**

$$= 59.29 - 51.16 = \mathbf{8.13\%.} \quad \textbf{(Ans.)}$$

Example 21.11. *A certain quantity of air at a pressure of 1 bar and temperature of 70°C is compressed adiabatically until the pressure is 7 bar in Otto cycle engine. 465 kJ of heat per kg of air is now added at constant volume. Determine:*

(*i*) *Compression ratio of the engine.*

(*ii*) *Temperature at the end of compression.*

(*iii*) *Temperature at the end of heat addition.*

Take for air c_p = 1.0 kJ/kg K, c_v = 0.706 kJ/kg K.

Show each operation on p-V and T-s diagrams.

Solution. Refer to Fig. 21.10.

Initial pressure, p_1 = 1 bar

Initial temperature, T_1 = 70 + 273 = 343 K

Pressure after adiabatic compression,

p_2 = 7 bar

Heat addition at constant volume,

Q_s = 465 kJ/kg of air

Specific heat at constant pressure,

c_p = 1.0 kJ/kg K

Specific heat at constant volume,

c_v = 0.706 kJ/kg K

$$\therefore \quad \gamma = \frac{c_p}{c_v} = \frac{1.0}{0.706} = 1.41$$

(*i*) Compression ratio of engine, r :

According to *adiabatic compression 1–2*

$$p_1V_1^\gamma = p_2V_2^\gamma$$

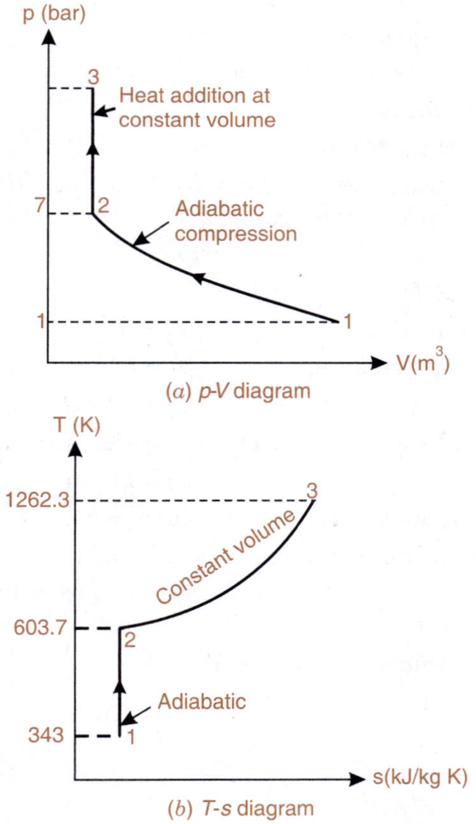

(*a*) *p-V* diagram

(*b*) *T-s* diagram

Fig. 21.10

$$\text{or} \quad \left(\frac{V_1}{V_2}\right)^\gamma = \frac{p_2}{p_1}$$

$$\text{or} \quad (r)^\gamma = \frac{p_2}{p_1} \qquad \left(\because \frac{V_1}{V_2} = r\right)$$

$$\text{or} \quad r = \left(\frac{p_2}{p_1}\right)^{\frac{1}{\gamma}} = \left(\frac{7}{1}\right)^{\frac{1}{1.41}} = (7)^{0.709} = 3.97$$

Hence *compression ratio of the engine* = **3.97. (Ans.)**

(*ii*) Temperature at the end of compression, T_2:

In case of *adiabatic compression 1–2,*

$$\frac{T_2}{T_1} = \left(\frac{V_1}{V_2}\right)^{\gamma - 1} = (3.97)^{1.41 - 1} = 1.76$$

$$\therefore \quad T_2 = 1.76\ T_1 = 1.76 \times 343 = 603.7 \text{ K or } 330.7°\text{C}$$

Hence *temperature at the end of compression* = **330.7°C. (Ans.)**

(*iii*) Temperature at the end of heat addition, T_3:

According to *constant volume heating operation 2–3*

$$Q_s = c_v\,(T_3 - T_2) = 465$$

$$0.706\,(T_3 - 603.7) = 465$$

$$\text{or} \quad T_3 - 603.7 = \frac{465}{0.706}$$

$$\text{or} \quad T_3 = \frac{465}{0.706} + 603.7 = 1262.3 \text{ K or } 989.3°\text{C}$$

Hence *temperature at the end of heat addition* = **989.3°C. (Ans.)**

Example 21.12. *In a constant volume 'Otto cycle', the pressure at the end of compression is 15 times that at the start, the temperature of air at the beginning of compression is 38°C and maximum temperature attained in the cycle is 1950°C. Determine :*

(*i*) *Compression ratio.*

(*ii*) *Thermal efficiency of the cycle.*

(*iii*) *Work done.*

Take γ for air = 1.4.

Solution. Refer to Fig. 21.11.

Initial temperature, T_1 = 38 + 273 = 311 K

Maximum temperature, T_3 = 1950 + 273 = 2223 K.

(*i*) Compression ratio, r :

For *adiabatic compression 1–2,*

$$p_1V_1^\gamma = p_2V_2^\gamma$$

$$\text{or} \quad \left(\frac{V_1}{V_2}\right)^\gamma = \frac{p_2}{p_1}$$

$$\text{But} \quad \frac{p_2}{p_1} = 15 \qquad \text{...(given)}$$

$$\therefore \quad (r)^\gamma = 15 \qquad \left[\because r = \frac{V_1}{V_2}\right]$$

$$\text{or} \quad (r)^{1.4} = 15$$

$$\text{or} \quad r = (15)^{\frac{1}{1.4}} = (15)^{0.714} = 6.9$$

Hence *compression ratio* = **6.9. (Ans.)**

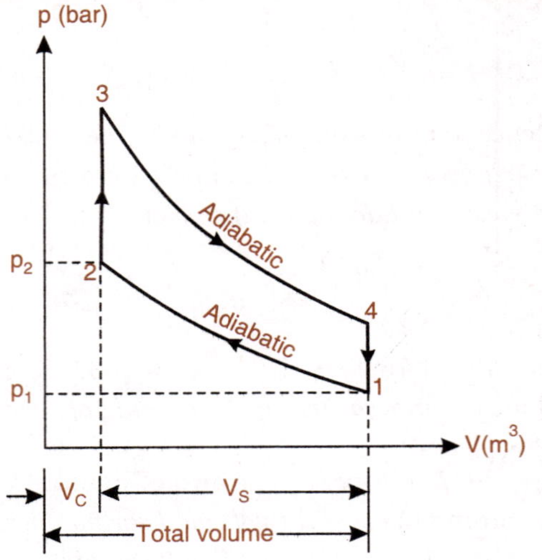

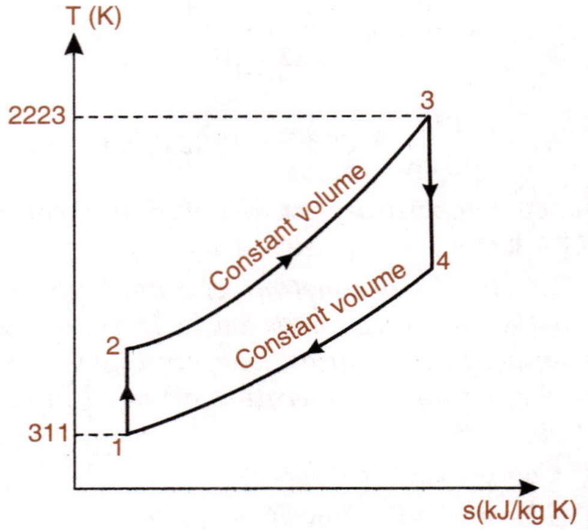

Fig. 21.11

(*ii*) **Thermal efficiency :**

$$\text{Thermal efficiency, } \eta_{th} = 1 - \frac{1}{(r)^{\gamma-1}} = 1 - \frac{1}{(6.9)^{1.4-1}}$$

$$= \mathbf{0.538} \text{ or } \mathbf{53.8\%}. \quad \textbf{(Ans.)}$$

(*iii*) **Work done :**

Again, for *adiabatic compression 1–2,*

$$\frac{T_2}{T_1} = \left(\frac{V_1}{V_2}\right)^{\gamma-1} = (r)^{\gamma-1}$$

$$= (6.9)^{1.4-1} = (6.9)^{0.4} = 2.16$$

or $T_2 = T_1 \times 2.16 = 311 \times 2.16 = 671.7 \text{ K or } 398.7°\text{C}$

For *adiabatic expansion process 3–4*

$$\frac{T_3}{T_4} = \left(\frac{V_4}{V_3}\right)^{\gamma-1} = (r)^{\gamma-1} = (6.9)^{0.4} = 2.16$$

or $$T_4 = \frac{T_3}{2.16} = \frac{2223}{2.16} = 1029 \text{ K or } 756°\text{C}$$

Heat supplied *per kg of air*

$$= c_v(T_3 - T_2) = 0.717(2223 - 671.7)$$

$$= 1112.3 \text{ kJ/kg or air}$$

$$\left[c_v = \frac{R}{\gamma - 1} = \frac{0.287}{1.4 - 1} = 0.717 \text{ kJ/ kg K}\right]$$

Heat rejected per kg of air

$$= c_v(T_4 - T_1) = 0.717(1029 - 311)$$

$$= 514.8 \text{ kJ/kg of air}$$

∴ *Work done per kg of air*

= Heat supplied – heat rejected

= 1112.3 – 514.8

= **597.5 kJ or 597500 Nm. (Ans.)**

☞**Example 21.13.** *An engine working on Otto cycle has a volume of 0.45 m³, pressure 1 bar and temperature 30°C at the beginning of compression stroke. At the end of compression stroke, the pressure is 11 bar. 210 kJ of heat is added at constant volume. Determine :*

(*i*) *Pressures, temperatures and volumes at salient points in the cycle.*

(*ii*) *Percentage clearance.*

(*iii*) *Efficiency.*

(*iv*) *Mean effective pressure.*

(*v*) *Ideal power developed by the engine if the number of working cycles per minute is 210.*

Assume the cycle is reversible.

Solution. Refer to Fig. 21.12

Volume, $V_1 = 0.45 \text{ m}^3$

Initial pressure, $p_1 = 1 \text{ bar}$

Initial temperature, $T_1 = 30 + 273 = 303 \text{ K}$

Pressure at the end of compression stroke,

$p_2 = 11 \text{ bar}$

Heat added at constant volume = 210 kJ

Number of working cycles/min. = 210.

(*i*) **Pressures, temperatures and volumes at salient points :**

For *adiabatic compression 1–2*

$$p_1V_1^{\gamma} = p_2V_2^{\gamma}$$

or $$\frac{p_2}{p_1} = \left(\frac{V_1}{V_2}\right)^{\gamma} = (r)^{\gamma}$$

or $$r = \left(\frac{p_2}{p_1}\right)^{\frac{1}{\gamma}} = \left(\frac{11}{1}\right)^{\frac{1}{1.4}} = (11)^{0.714} = 5.5$$

Also $$\frac{T_2}{T_1} = \left(\frac{V_1}{V_2}\right)^{\gamma-1} = (r)^{\gamma-1} = (5.5)^{1.4-1}$$

$$= 1.977 \simeq 1.98$$

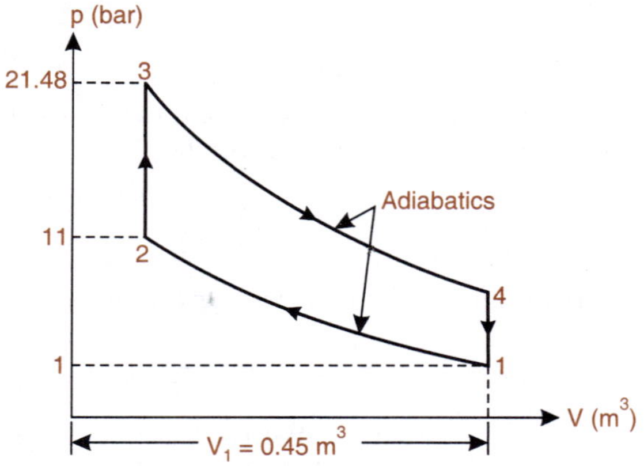

Fig. 21.12

$\therefore \quad T_2 = T_1 \times 1.98 = 303 \times 1.98 = $ **600 K.** **(Ans.)**

Applying gas laws to points 1 and 2

$$\frac{p_1V_1}{T_1} = \frac{p_2V_2}{T_2}$$

$$\therefore \quad V_2 = \frac{T_2}{T_1} \times \frac{p_1}{p_2} \times V_1 = \frac{600 \times 1 \times 0.45}{303 \times 11}$$

$= \mathbf{0.081\ m^3}.$ **(Ans.)**

The heat supplied during the *process 2–3* is given by :

$$Q_s = m\, c_v\, (T_3 - T_2)$$

where $m = \dfrac{p_1V_1}{RT_1} = \dfrac{1 \times 10^5 \times 0.45}{287 \times 303} = 0.517$ kg

$\therefore \quad 210 = 0.517 \times 0.71\, (T_3 - 600)$

or $T_3 = \dfrac{210}{0.517 \times 0.71} + 600 = $ **1172 K.** **(Ans.)**

For the *constant volume process 2–3*

$$\frac{p_3}{T_3} = \frac{p_2}{T_2}$$

$$\therefore \quad p_3 = \frac{T_3}{T_2} \times p_2 = \frac{1172}{600} \times 11$$

$= $ **21.48 bar.** **(Ans.)**

$V_3 = V_2 = \mathbf{0.081\ m^3}.$ **(Ans.)**

For the *adiabatic (or isentropic) process 3–4*

$$p_3V_3^{\gamma} = p_4V_4^{\gamma}$$

$$p_4 = p_3 \times \left(\frac{V_3}{V_4}\right)^{\gamma} = p_3 \times \left(\frac{1}{r}\right)^{\gamma}$$

$$= 21.48 \times \left(\frac{1}{5.5}\right)^{1.4} = \mathbf{1.97\ bar.}\ \textbf{(Ans.)}$$

Also $\dfrac{T_4}{T_3} = \left(\dfrac{V_3}{V_4}\right)^{\gamma - 1} = \left(\dfrac{1}{r}\right)^{\gamma - 1} = \left(\dfrac{1}{5.5}\right)^{1.4 - 1} = 0.505$

$\therefore \quad T_4 = 0.505\, T_3 = 0.505 \times 1172 = $ **591.8 K.** **(Ans.)**

$V_4 = V_1 = \mathbf{0.45\ m^3}.$ **(Ans.)**

(*ii*) **Percentage clearance :**

Percentage clearance

$$= \frac{V_c}{V_s} = \frac{V_2}{V_1 - V_2} \times 100 = \frac{0.081}{0.45 - 0.081} \times 100$$

$= $ **21.95%.** **(Ans.)**

(*iii*) **Efficiency :**

The heat rejected per cycle is given by

$$Q_r = mc_v(T_4 - T_1)$$

$= 0.517 \times 0.71\, (591.8 - 303) = 106$ kJ

The air-standard efficiency of the cycle is given by

$$\eta_{otto} = \frac{Q_s - Q_r}{Q_s} = \frac{210 - 106}{210}$$

$= $ **0.495 or 49.5%.** **(Ans.)**

Alternatively :

$$\eta_{otto} = 1 - \frac{1}{(r)^{\gamma - 1}} = 1 - \frac{1}{(5.5)^{1.4 - 1}}$$

$= $ **0.495 or 49.5%.** **(Ans.)**

(*iv*) **Mean effective pressure, p_m :**

The mean effective pressure is given by

$$p_m = \frac{W \text{ (work done)}}{V_s \text{(swept volume)}} = \frac{Q_s - Q_r}{(V_1 - V_2)}$$

$$= \frac{(210 - 106) \times 10^3}{(0.45 - 0.081) \times 10^5}$$

$= $ **2.818 bar.** **(Ans.)**

(*v*) **Power developed, P :**

Power developed,

$P = $ Work done per second

$= $ Work done per cycle × number of cycles per second

$= (210 - 106) \times (210/60)$

$= $ **364 kW.** **(Ans.)**

Example 21.14. (*a*) *Show that the compression ratio for the maximum work to be done per kg of air in an Otto cycle between upper and lower limits of absolute temperatures T_3 and T_1 is given by*

$$r = \left(\frac{T_3}{T_1}\right)^{1/2(\gamma - 1)}$$

(*b*) *Determine the air-standard efficiency of the cycle when the cycle develops maximum work with the temperature limits of 310 K and 1220 K and working fluid is air. What will be the percentage change in efficiency if helium is used as working fluid instead of air ? The cycle operates between the same temperature limits for maximum work development.*

Consider that all conditions are ideal.

Solution. Refer to Fig. 21.13.

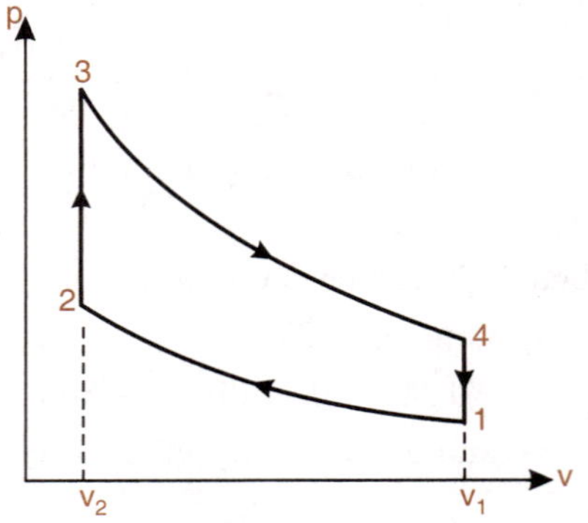

Fig. 21.13

(*a*) The work done per kg of fluid in the cycle is given by

$$W = Q_s - Q_r = c_v (T_3 - T_2) - c_v (T_4 - T_1)$$

But $$\frac{T_2}{T_1} = \left(\frac{v_1}{v_2}\right)^{\gamma - 1} = (r)^{\gamma - 1}$$

$\therefore$ $$T_2 = T_1 \,.\, (r)^{\gamma - 1} \quad ...(i)$$

Similarly, $$T_3 = T_4 \,.\, (r)^{\gamma - 1} \quad ...(ii)$$

$\therefore$ $$W = c_v \left[T_3 - T_1 \,.\, (r)^{\gamma - 1} - \frac{T_3}{(r)^{\gamma - 1}} + T_1\right] \quad ...(iii)$$

This expression is a function of r when T_3 and T_1 are fixed. The value of W will be maximum when,

$$\frac{dW}{dr} = 0.$$

$\therefore$ $$\frac{dW}{dr} = - T_1 \,.\, (\gamma - 1)\, (r)^{\gamma - 2} - T_3 (1 - \gamma)\, (r)^{-\gamma} = 0$$

or $$T_3\, (r)^{-\gamma} = T_1\, (r)^{\gamma - 2}$$

or $$\frac{T_3}{T_1} = (r)^{2(\gamma - 1)}$$

$\therefore$ $$r = \left(\frac{T_3}{T_1}\right)^{1/2(\gamma - 1)}.$$ **...... Proved.**

(*b*) **Change in efficiency :**

For *air* $\gamma = 1.4$

$\therefore$ $$r = \left(\frac{T_3}{T_1}\right)^{1/2(1.4 - 1)} = \left(\frac{1220}{310}\right)^{1/0.8} = 5.54$$

The air-standard efficiency is given by

$$\eta_{otto} = 1 - \frac{1}{(r)^{\gamma - 1}} = 1 - \frac{1}{(5.54)^{1.4 - 1}}$$

$= \mathbf{0.495}$ or **49.5%.** **(Ans.)**

If *helium* is used, then the values of

$c_p = 5.22$ kJ/kg K and $c_v = 3.13$ kJ/kg K

$\therefore$ $$\gamma = \frac{c_p}{c_v} = \frac{5.22}{3.13} = 1.67$$

The compression ratio for maximum work for the temperature limits T_1 and T_3 is given by

$$r = \left(\frac{T_3}{T_1}\right)^{1/2(\gamma - 1)} = \left(\frac{1220}{310}\right)^{1/2(1.67 - 1)} = 2.77$$

The air-standard efficiency is given by

$$\eta_{otto} = 1 - \frac{1}{(r)^{\gamma - 1}} = 1 - \frac{1}{(2.77)^{1.67 - 1}}$$

$= \mathbf{0.495}$ **or 49.5%.**

Hence change in efficiency is nil. **(Ans.)**

Example 21.15. (*a*) *An engine working on Otto cycle, in which the salient points are 1, 2, 3 and 4, has upper and lower temperature limits T_3 and T_1. If the maximum work per kg of air is to be done, show that the intermediate temperature is given by*

$$T_2 = T_4 = \sqrt{T_1 T_3}.$$

(*b*) *If an engine works on Otto cycle between temperature limits 1450 K and 310 K, find the maximum power developed by the engine assuming the circulation of air per minute as 0.38 kg.*

Solution. (*a*) Refer to Fig. 21.13 (Example 21.14).

Using the equation (*iii*) of example 21.14.

$$W = c_v \left[T_3 - T_1 \,.\, (r)^{\gamma - 1} - \frac{T_3}{(r)^{\gamma - 1}} + T_1\right]$$

and differentiating W w.r.t. r and equating to zero

$$r = \left(\frac{T_3}{T_1}\right)^{1/2(\gamma - 1)}$$

$$T_2 = T_1 (r)^{\gamma - 1} \text{ and } T_4 = T_3/(r)^{\gamma - 1}$$

Substituting the value of r in the above equation, we have

$$T_2 = T_1 \left[\left(\frac{T_3}{T_1}\right)^{1/2(\gamma - 1)}\right]^{\gamma - 1} = T_1 \left(\frac{T_3}{T_1}\right)^{1/2} = \sqrt{T_1 T_3}$$

Similarly,

$$T_4 = \frac{T_3}{\left[\left(\frac{T_3}{T_1}\right)^{1/2(\gamma - 1)}\right]^{\gamma - 1}} = \frac{T_3}{\left(\frac{T_3}{T_1}\right)^{1/2}} = \sqrt{T_3 T_1}$$

$\therefore$ $T_2 = T_4 = \sqrt{T_1 T_3}$. **Proved.**

(*b*) **Power developed, P :**

$$\left.\begin{aligned} T_1 &= 310 \text{ K} \\ T_3 &= 1450 \text{ K} \\ m &= 0.38 \text{ kg} \end{aligned}\right\} \quad ....\text{(given)}$$

Work done $$W = c_v \, [(T_3 - T_2) - (T_4 - T_1)]$$

$$T_2 = T_4 = \sqrt{T_1 T_3} = \sqrt{310 \times 1450} = 670.4 \text{ K}$$

$$\therefore \quad W = 0.71\,[(1450 - 670.4) - (670.4 - 310)]$$
$$= 0.71\,(779.6 - 360.4) = 297.6 \text{ kJ/kg}$$

Work done per second

$$= 297.6 \times (0.38/60) = 1.88 \text{ kJ/s}$$

Hence **power developed, P = 1.88 kW.** **(Ans.)**

Example 21.16. *For the same compression ratio, show that the efficiency of Otto cycle is greater than that of Diesel cycle.*

Solution. Refer to Fig. 21.14.

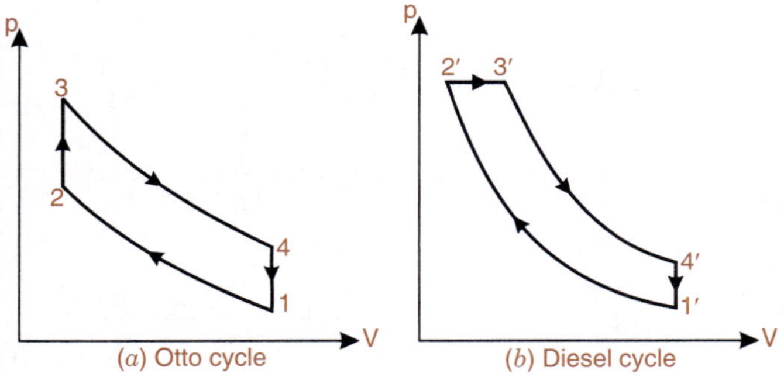

Fig. 21.14

We know that

$$\eta_{otto} = 1 - \frac{1}{(r)^{\gamma - 1}}$$

and

$$\eta_{diesel} = 1 - \frac{1}{(r)^{\gamma - 1}} \times \frac{1}{\gamma}\left\{\frac{\rho^{\gamma} - 1}{\rho - 1}\right\}$$

As the compression ratio is same,

$$\frac{V_1}{V_2} = \frac{V_1'}{V_2'} = r$$

If $\dfrac{V_4'}{V_3'} = r_1$, then cut-off ratio, $\rho = \dfrac{V_3'}{V_2'} = \dfrac{r}{r_1}$

Putting the value of ρ in η_{diesel}, we get

$$\eta_{diesel} = 1 - \frac{1}{(r)^{\gamma - 1}} \times \frac{1}{\gamma}\left[\frac{\left(\frac{r}{r_1}\right)^{\gamma} - 1}{\frac{r}{r_1} - 1}\right]$$

From above equation, we observe

$$\frac{r}{r_1} > 1$$

Let $r_1 = r - \delta$, where δ is a small quantity.

Then

$$\frac{r}{r_1} = \frac{r}{r - \delta} = \frac{r}{r\left(1 - \frac{\delta}{r}\right)}$$

$$= \left(1 - \frac{\delta}{r}\right)^{-1} = 1 + \frac{\delta}{r} + \frac{\delta^2}{r^2} + \frac{\delta^3}{r^3} + \ldots\ldots$$

and

$$\left(\frac{r}{r_1}\right)^{\gamma} = \frac{r^{\gamma}}{r^{\gamma}\left(1 - \frac{\delta}{r}\right)^{\gamma}} = \left(1 - \frac{\delta}{r}\right)^{-\gamma}$$

$$= 1 + \frac{\gamma\delta}{r} + \frac{\gamma(\gamma + 1)}{2!} \cdot \frac{\delta^2}{r^2} + \ldots\ldots$$

$$\therefore \quad \eta_{diesel} = 1 - \frac{1}{(r)^{\gamma - 1}} \times \frac{1}{\gamma}\left[\frac{\frac{\gamma . \delta}{r} + \frac{\gamma(\gamma + 1)}{2!} \cdot \frac{\delta^2}{r^2} + \ldots\ldots}{\frac{\delta}{r} + \frac{\delta^2}{r^2} + \ldots\ldots}\right]$$

$$= 1 - \frac{1}{(r)^{\gamma - 1}}\left[\frac{\frac{\delta}{r} + \frac{\gamma + 1}{2} \cdot \frac{\delta^2}{r^2} + \ldots\ldots}{\frac{\delta}{r} + \frac{\delta^2}{r^2} + \ldots\ldots}\right]$$

The ratio inside the bracket is greater than 1 since the co-efficients of terms δ^2/r^2 is greater than 1 in the numerator. Its means that something more is subtracted in case of diesel cycle than in Otto cycle.

Hence, *for same compression ratio* $\eta_{otto} > \eta_{diesel}$.

21.5. CONSTANT PRESSURE OR DIESEL CYCLE

This cycle was introduced by Dr. R. Diesel in 1897. It differs from Otto cycle in that *heat is supplied at constant pressure instead of at constant volume*. Fig. 21.15 (*a* and *b*) shows the *p-v* and *T-s* diagrams of this cycle respectively.

This cycle comprises of the following ***operations*** :

(*i*) 1–2......*Adiabatic compression*.

(*ii*) 2–3......*Addition of heat at constant pressure*.

(*iii*) 3–4......*Adiabatic expansion*.

(*iv*) 4–1......*Rejection of heat at constant volume*.

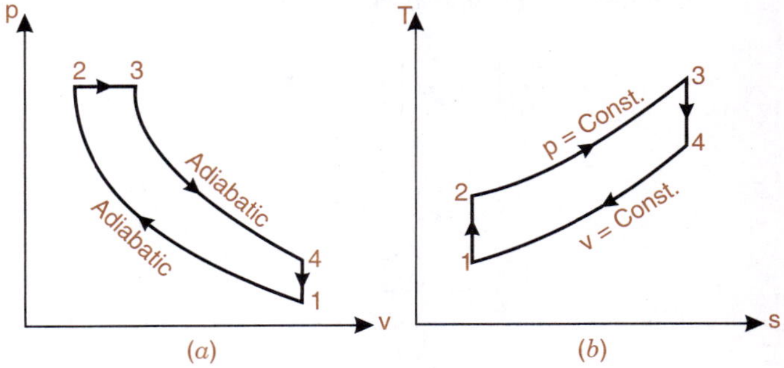

Fig. 21.15

Point 1 represents that the cylinder is full of air. Let p_1, V_1 and T_1 be the corresponding pressure, volume and absolute temperature. The piston then compresses the air adiabatically (*i.e.*, pV^{γ} = constant) till the values become p_2, V_2 and T_2 respectively (at the end of the stroke) at point 2. Heat is then added from a hot body at a constant pressure. During this addition of heat let volume increases from V_2 to V_3 and temperature T_2 to T_3, corresponding to point 3.

This point (3) is called the ***point of cut-off***. The air then expands adiabatically to the conditions p_4, V_4 and T_4 respectively corresponding to point 4. Finally, the air rejects the heat to the cold body at constant volume till the point 1 where it returns to its original state.

Consider 1 kg of air.

Heat supplied at constant pressure $= c_p(T_3 - T_2)$

Heat rejected at constant volume $= c_v(T_4 - T_1)$

Work done $=$ Heat supplied $-$ heat rejected

$$= c_p(T_3 - T_2) - c_v(T_4 - T_1)$$

$$\therefore \quad \eta_{diesel} = \frac{\text{Work done}}{\text{Heat supplied}}$$

$$= \frac{c_p(T_3 - T_2) - c_v(T_4 - T_1)}{c_p(T_3 - T_2)}$$

$$= 1 - \frac{(T_4 - T_1)}{\gamma(T_3 - T_2)} \quad ...(i)$$

$$\left[\because \frac{c_p}{c_v} = \gamma\right]$$

Let compression ratio, $r = \frac{v_1}{v_2}$, and cut-off ratio,

$$\rho = \frac{v_3}{v_2} \; i.e., \; \frac{\text{Volume at cut-off}}{\text{Clearance volume}}$$

Now, during *adiabatic compression 1–2*,

$$\frac{T_2}{T_1} = \left(\frac{v_1}{v_2}\right)^{\gamma - 1} = (r)^{\gamma - 1} \text{ or } T_2 = T_1 \,.\, (r)^{\gamma - 1}$$

During *constant pressure process 2–3*,

$$\frac{T_3}{T_2} = \frac{v_3}{v_2} = \rho \text{ or } T_3 = \rho \,.\, T_2 = \rho \,.\, T_1 \,.\, (r)^{\gamma - 1}$$

During *adiabatic expansion 3–4*

$$\frac{T_3}{T_4} = \left(\frac{v_4}{v_3}\right)^{\gamma - 1} = \left(\frac{r}{\rho}\right)^{\gamma - 1}$$

$$\left(\because \frac{v_4}{v_3} = \frac{v_1}{v_3} = \frac{v_1}{v_2} \times \frac{v_2}{v_3} = \frac{r}{\rho}\right)$$

$$\therefore \quad T_4 = \frac{T_3}{\left(\frac{r}{\rho}\right)^{\gamma - 1}} = \frac{\rho \,.\, T_1 (r)^{\gamma - 1}}{\left(\frac{r}{\rho}\right)^{\gamma - 1}} = T_1 \,.\, \rho^{\gamma}$$

By inserting values of T_2, T_3 and T_4 in eqn. (*i*), we get

$$\eta_{diesel} = 1 - \frac{(T_1 \,.\, \rho^{\gamma} - T_1)}{\gamma\,[\rho \,.\, T_1 \,.\, (r)^{\gamma - 1} - T_1 \,.\, (r)^{\gamma - 1}]}$$

$$= 1 - \frac{(\rho^{\gamma} - 1)}{\gamma (r)^{\gamma - 1} (\rho - 1)}$$

or

$$\eta_{diesel} = 1 - \frac{1}{\gamma (r)^{\gamma - 1}} \left[\frac{\rho^{\gamma} - 1}{\rho - 1}\right] \quad ...(21.7)$$

It may be observed that eqn. (21.7) for efficiency of diesel cycle is different from that of the Otto cycle only in bracketed factor. This factor is always greater than unity, because $\rho > 1$. Hence *for a given compression ratio, the Otto cycle is more efficient*.

The *net work* for diesel cycle can be expressed in terms of *pv* as follows :

$$W = p_2(v_3 - v_2) + \frac{p_3 v_3 - p_4 v_4}{\gamma - 1} - \frac{p_2 v_2 - p_1 v_1}{\gamma - 1}$$

$$= p_2\,(\rho v_2 - v_2) + \frac{p_3 \rho v_2 - p_4 r v_2}{\gamma - 1} - \frac{p_2 v_2 - p_1 r v_2}{\gamma - 1}$$

$$\left[\begin{array}{l} \because \frac{v_3}{v_2} = \rho \quad \therefore \quad v_3 = \rho v_2 \text{ and } \frac{v_1}{v_2} = r \quad \therefore \quad v_1 = r v_2 \\ \text{But } v_4 = v_1 \quad \therefore \quad v_4 = r v_2 \end{array}\right]$$

$$= p_2 v_2\,(\rho - 1) + \frac{p_3\,\rho v_2 - p_4 r v_2}{\gamma - 1} - \frac{p_2 v_2 - p_1 r v_2}{\gamma - 1}$$

$$= \frac{v_2\,[p_2\,(\rho - 1)(\gamma - 1) + p_3 \rho - p_4 r - (p_2 - p_1 r)]}{\gamma - 1}$$

$$= \frac{v_2 \left[p_2\,(\rho - 1)(\gamma - 1) + p_3 \left(\rho - \frac{p_4 r}{p_3}\right) - p_2 \left(1 - \frac{p_1 r}{p_2}\right)\right]}{\gamma - 1}$$

$$= \frac{p_2 v_2\,[(\rho - 1)(\gamma - 1) + \rho - \rho^{\gamma} \,.\, r^{1-\gamma} - (1 - r^{1-\gamma})]}{\gamma - 1}$$

$$\left[\because \frac{p_4}{p_3} = \left(\frac{v_3}{v_4}\right)^{\gamma} = \left(\frac{\rho}{r}\right)^{\gamma} = \rho^{\gamma} r^{-\gamma}\right]$$

$$= \frac{p_1 v_1 r^{\gamma - 1}\,[(\rho - 1)(\gamma - 1) + \rho - \rho^{\gamma} r^{1-\gamma} - (1 - r^{1-\gamma})]}{\gamma - 1}$$

$$\left[\begin{array}{l} \because \frac{p_2}{p_1} = \left(\frac{v_1}{v_2}\right)^{\gamma} \text{ or } p_2 = p_1 \,.\, r^{\gamma} \\ \text{and } \frac{v_1}{v_2} = r \text{ or } v_2 = v_1\, r^{-1} \end{array}\right]$$

$$= \frac{p_1 v_1 r^{\gamma - 1}\,[\gamma(\rho - 1) - r^{1-\gamma}\,(\rho^{\gamma} - 1)]}{(\gamma - 1)} \quad ...(21.8)$$

Mean effective pressure $\mathbf{p_m}$ is given by :

$$p_m = \frac{p_1 v_1 r^{\gamma - 1}\,[\gamma(\rho - 1) - r^{1-\gamma}\,(\rho^{\gamma} - 1)]}{(\gamma - 1) v_1 \left(\frac{r - 1}{r}\right)}$$

or

$$\mathbf{p_m} = \frac{p_1 r^{\gamma}\,[\gamma(\rho - 1) - r^{1-\gamma}\,(\rho^{\gamma} - 1)]}{(\gamma - 1)(r - 1)} \quad ...(21.9)$$

Example 21.17. *A diesel engine has a compression ratio of 15 and heat addition at constant pressure takes place at 6% of stroke. Find the air standard efficiency of the engine.*

Take γ for air as 1.4.

Solution. Refer to Fig. 21.16.

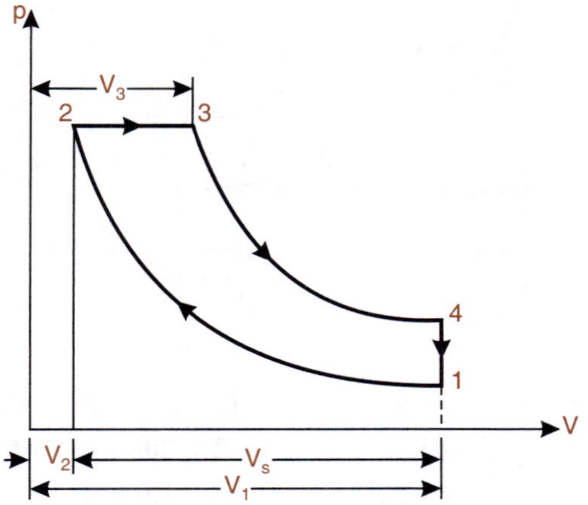

Fig. 21.16

Compression ratio, $r\left(=\dfrac{V_1}{V_2}\right) = 15$

γ for air = 1.4

Air standard efficiency of diesel cycle is given by

$$\eta_{\text{diesel}} = 1 - \frac{1}{\gamma(r)^{\gamma-1}}\left[\frac{\rho^\gamma - 1}{\rho - 1}\right] \quad ...(i)$$

where ρ = cut-off ratio $= \dfrac{V_3}{V_2}$

But $V_3 - V_2 = \dfrac{6}{100} V_s$ (V_s = stroke volume)

$= 0.06 (V_1 - V_2) = 0.06 (15 V_2 - V_2)$

$= 0.84 V_2$ or $V_3 = 1.84 V_2$

$\therefore \quad \rho = \dfrac{V_3}{V_2} = \dfrac{1.84 V_2}{V_2} = 1.84$

Putting the value in eqn. (*i*), we get

$$\eta_{diesel} = 1 - \frac{1}{1.4\,(15)^{1.4-1}}\left[\frac{(1.84)^{1.4} - 1}{1.84 - 1}\right]$$

$= 1 - 0.2417 \times 1.605$

$=$ **0.612** or **61.2%.** **(Ans.)**

Example 21.18. *The stroke and cylinder diameter of a compression ignition engine are 250 mm and 150 mm respectively. If the clearance volume is 0.0004 m³ and fuel injection takes place at constant pressure for 5 per cent of the stroke determine the efficiency of the engine. Assume the engine working on the diesel cycle.*

Solution. Refer to Fig. 21.16.

Length of stroke, $L = 250$ mm $= 0.25$ m

Diameter of cylinder, $D = 150$ mm $= 0.15$ m

Clearance volume, $V_2 = 0.0004$ m³

Swept volume, $V_s = \pi/4\, D^2 L = \pi/4 \times 0.15^2 \times 0.25 = 0.004418$ m³

Total cylinder volume

= Swept volume + clearance volume

= 0.004418 + 0.0004 = 0.004818 m³

Volume at point of cut-off, $V_3 = V_2 + \dfrac{5}{100} V_s$

$= 0.0004 + \dfrac{5}{100} \times 0.004418 = 0.000621$ m³

$\therefore$ Cut-off ratio, $\rho = \dfrac{V_3}{V_2} = \dfrac{0.000621}{0.0004} = 1.55$

Compression ratio,

$$r = \frac{V_1}{V_2} = \frac{V_s + V_2}{V_2} = \frac{0.004418 + 0.0004}{0.0004} = 12.04$$

Hence,

$$\eta_{diesel} = 1 - \frac{1}{\gamma(r)^{\gamma-1}}\left[\frac{\rho^\gamma - 1}{\rho - 1}\right]$$

$$= 1 - \frac{1}{1.4 \times (12.04)^{1.4-1}}\left[\frac{(1.55)^{1.4} - 1}{1.55 - 1}\right]$$

$= 1 - 0.264 \times 1.54 =$ **0.593** or **59.3%.** **(Ans.)**

Example 21.19. *Calculate the percentage loss in the ideal efficiency of a diesel engine with compression ratio 14 if the fuel cut-off is delayed from 5% to 8%.*

Solution. Let the clearance volume (V_2) be unity.

Then, compression ratio, $r = 14$

Now, when the fuel is cut-off at 5%, we have

$$\frac{\rho - 1}{r - 1} = \frac{5}{100} \quad \text{or} \quad \frac{\rho - 1}{14 - 1} = 0.05$$

or $\rho - 1 = 13 \times 0.05 = 0.65$

$\therefore \quad \rho = 1.65$

$$\eta_{\text{diesel}} = 1 - \frac{1}{\gamma(r)^{\gamma-1}}\left[\frac{\rho^\gamma - 1}{\rho - 1}\right]$$

$$= 1 - \frac{1}{1.4 \times (14)^{1.4-1}}\left[\frac{(1.65)^{1.4} - 1}{1.65 - 1}\right]$$

$= 1 - 0.248 \times 1.563 = 0.612$ or 61.2%

When the fuel is cut-off at 8%, we have

$$\frac{\rho - 1}{r - 1} = \frac{8}{100} \quad \text{or} \quad \frac{\rho - 1}{14 - 1} = \frac{8}{100} = 0.08$$

$\therefore \quad \rho = 1 + 1.04 = 2.04$

$$\eta_{diesel} = 1 - \frac{1}{\gamma (r)^{\gamma-1}} \left[\frac{\rho^{\gamma} - 1}{\rho - 1}\right]$$

$$= 1 - \frac{1}{1.4 \times (14)^{1.4-1}} \left[\frac{(2.04)^{1.4} - 1}{2.04 - 1}\right]$$

$$= 1 - 0.248 \times 1.647 = 0.591 \quad \text{or} \quad 59.1\%.$$

Hence percentage loss in efficiency due to delay in fuel cut-off

$$= 61.2 - 59.1 = \mathbf{2.1\%}. \quad \textbf{(Ans.)}$$

Example 21.20. *The mean effective pressure of a Diesel cycle is 7.5 bar and compression ratio is 12.5. Find the percentage cut-off of the cycle if its initial pressure is 1 bar.*

Solution. Mean effective pressure, $p_m = 7.5$ bar

Compression ratio, $r = 12.5$

Initial pressure, $p_1 = 1$ bar

Refer to Fig. 21.15.

The mean effective pressure is given by

$$p_m = \frac{p_1 r^{\gamma} [\gamma(\rho - 1) - r^{1-\gamma}(\rho^{\gamma} - 1)]}{(\gamma - 1)(r - 1)}$$

$$7.5 = \frac{1 \times (12.5)^{1.4} [1.4(\rho - 1) - (12.5)^{1-1.4}(\rho^{1.4} - 1)]}{(1.4 - 1)(12.5 - 1)}$$

$$7.5 = \frac{34.33[1.4\rho - 1.4 - 0.364\rho^{1.4} + 0.364]}{4.6}$$

$$7.5 = 7.46\,(1.4\rho - 1.036 - 0.364\rho^{1.4})$$

$$1.005 = 1.4\rho - 1.036 - 0.364\rho^{1.4}$$

or $\quad 2.04 = 1.4\rho - 0.364\rho^{1.4}$

or $\quad 0.346\rho^{1.4} - 1.4\rho + 2.04 = 0$

Solving by trial and error method, we get

$$\rho = 2.24$$

$$\therefore \quad \%\ cut\text{-}off = \frac{\rho - 1}{r - 1} \times 100$$

$$= \frac{2.24 - 1}{12.5 - 1} \times 100 = \mathbf{10.78\%}. \quad \textbf{(Ans.)}$$

☞**Example 21.21.** *An engine with 200 mm cylinder diameter and 300 mm stroke works on theoretical Diesel cycle. The initial pressure and temperature of air used are 1 bar and 27°C. The cut-off is 8% of the stroke. Determine :*

(i) Pressures and temperatures at all salient points.

(ii) Theoretical air standard efficiency.

(iii) Mean effective pressure.

(iv) Power of the engine if the working cycles per minute are 380.

Assume that compression ratio is 15 and working fluid is air.

Consider all conditions to be ideal.

Solution. Refer to Fig. 21.17.

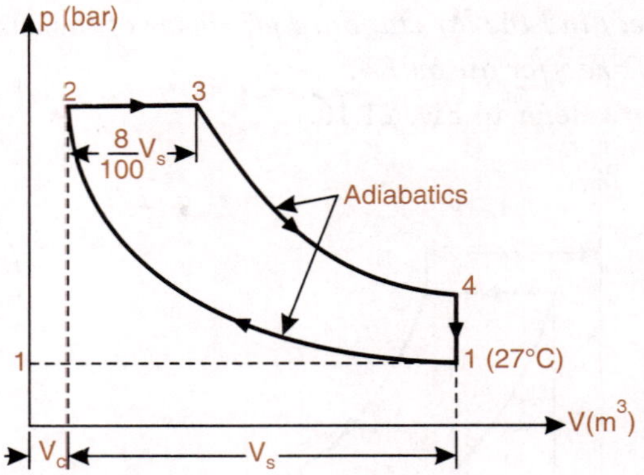

Fig. 21.17

Cylinder diameter, $D = 200$ mm or 0.2 m

Stroke length, $L = 300$ mm or 0.3 m

Initial pressure, $p_1 = 1.0$ bar

Initial temperature, $T_1 = 27 + 273 = 300$ K

Cut-off $= \frac{8}{100} V_s = 0.08\, V_s$

(i) **Pressures and temperatures at salient points :**

Now, stroke volume,

$$V_s = \pi/4\, D^2 L = \pi/4 \times 0.2^2 \times 0.3 = 0.00942 \text{ m}^3$$

$$V_1 = V_s + V_c = V_s + \frac{V_s}{r - 1} \qquad \left[\because\ V_c = \frac{V_s}{r - 1}\right]$$

$$= V_s\left(1 + \frac{1}{r - 1}\right) = \frac{r}{r - 1} \times V_s$$

i.e., $\quad \mathbf{V_1} = \frac{15}{15 - 1} \times V_s = \frac{15}{14} \times 0.00942$

$$= \mathbf{0.0101\ m^3}. \quad \textbf{(Ans.)}$$

Mass of the air in the cylinder can be calculated by using the gas equation,

$$p_1 V_1 = mRT_1$$

$$m = \frac{p_1 V_1}{RT_1} = \frac{1 \times 10^5 \times 0.0101}{287 \times 300} = 0.0117 \text{ kg/cycle}$$

For the *adiabatic (or isentropic) process 1–2*

$$p_1 V_1^{\gamma} = p_2 V_2^{\gamma} \quad \text{or} \quad \frac{p_2}{p_1} = \left(\frac{V_1}{V_2}\right)^{\gamma} = (r)^{\gamma}$$

$$\therefore \quad \mathbf{p_2} = p_1 \cdot (r)^{\gamma} = 1 \times (15)^{1.4}$$

$$= \mathbf{44.31\ bar}. \quad \textbf{(Ans.)}$$

Also, $\quad \frac{T_2}{T_1} = \left(\frac{V_1}{V_2}\right)^{\gamma - 1} = (r)^{\gamma - 1} = (15)^{1.4 - 1} = 2.954$

$\therefore \quad \boldsymbol{T_2} = T_1 \times 2.954 = 300 \times 2.954$
$= \mathbf{886.2\ K.}$ **(Ans.)**

$$\boldsymbol{V_2} = V_c = \frac{V_s}{r-1} = \frac{0.00942}{15-1}$$
$= \mathbf{0.0006728\ m^3.}$ **(Ans.)**

$\boldsymbol{p_2} = p_3 = \mathbf{44.31\ bar.}$ **(Ans.)**

$$\%\ \text{cut-off ratio} = \frac{\rho - 1}{r - 1}$$

$$\frac{8}{100} = \frac{\rho - 1}{15 - 1}$$

i.e., $\rho = 0.08 \times 14 + 1 = 2.12$

$\therefore \quad V_3 = \rho\, V_2 = 2.12 \times 0.0006728$
$= \mathbf{0.001426\ m^3.}$ **(Ans.)**

[V_3 can also be calculated as follows:
$V_3 = 0.08V_s + V_c = 0.08 \times 0.00942 + 0.0006728$
$= 0.001426\ m^3$]

For the *constant pressure process 2–3,*

$$\frac{V_3}{T_3} = \frac{V_2}{T_2}$$

$$\therefore \quad \boldsymbol{T_3} = T_2 \times \frac{V_3}{V_2} = 886.2 \times \frac{0.001426}{0.0006728}$$
$= \mathbf{1878.3\ K.}$ **(Ans.)**

For the *isentropic process 3–4,*

$$p_3V_3^{\gamma} = p_4V_4^{\gamma}$$

$$p_4 = p_3 \times \left(\frac{V_3}{V_4}\right)^{\gamma} = p_3 \times \frac{1}{(7.07)^{1.4}}$$

$$= \frac{44.31}{(7.07)^{1.4}} = \mathbf{2.866\ bar.} \quad \textbf{(Ans.)}$$

$$\left[\because \frac{V_4}{V_3} = \frac{V_4}{V_2} \times \frac{V_2}{V_3} = \frac{V_1}{V_2} \times \frac{V_2}{V_3} = \frac{r}{\rho}, \quad \because V_4 = V_1 = \frac{15}{2.12} = 7.07\right]$$

Also, $$\frac{T_4}{T_3} = \left(\frac{V_3}{V_4}\right)^{\gamma - 1} = \left(\frac{1}{7.07}\right)^{1.4 - 1} = 0.457$$

$\therefore \quad \boldsymbol{T_4} = T_3 \times 0.457 = 1878.3 \times 0.457$
$= \mathbf{858.38\ K.}$ **(Ans.)**

$\boldsymbol{V_4} = V_1 = \mathbf{0.0101\ m^3.}$ **(Ans.)**

(*ii*) Theoretical air standard efficiency :

$$\eta_{diesel} = 1 - \frac{1}{\gamma(r)^{\gamma - 1}}\left[\frac{\rho^{\gamma} - 1}{\rho - 1}\right]$$

$$= 1 - \frac{1}{1.4\,(15)^{1.4 - 1}}\left[\frac{(2.12)^{1.4} - 1}{2.12 - 1}\right]$$

$= 1 - 0.2418 \times 1.663$
$= \mathbf{0.598\ or\ 59.8\%.}$ **(Ans.)**

(*iii*) Mean effective pressure, p_m :

Mean effective pressure of Diesel cycle is given by

$$\boldsymbol{p_m} = \frac{p_1(r)^{\gamma}[\gamma(\rho - 1) - r^{1-\gamma}(\rho^{\gamma} - 1)]}{(\gamma - 1)(r - 1)}$$

$$= \frac{1 \times (15)^{1.4}[1.4(2.12 - 1) - (15)^{1 - 1.4}(2.12^{1.4} - 1)]}{(1.4 - 1)(15 - 1)}$$

$$= \frac{44.31[1.568 - 0.338 \times 1.863]}{0.4 \times 14}$$

$= \mathbf{7.424\ bar.}$ **(Ans.)**

(*iv*) Power of the engine, P :

Work done per cycle

$$= p_mV_s = \frac{7.424 \times 10^5 \times 0.00942}{10^3} = 6.99\ \text{kJ/cycle}$$

Work done per second = Work done per cycle × no. of cycles per second

= 6.99 × 380/60 = 44.27 kJ/s = 44.27 kW

Hence *power of the engine* = **44.27 kW. (Ans.)**

Example 21.22. *The volume ratios of compression and expansion for a diesel engine as measured from an indicator diagram are 15.3 and 7.5 respectively. The pressure and temperature at the beginning of the compression are 1 bar and 27°C.*

Assuming an ideal engine, determine the mean effective pressure, the ratio of maximum pressure to mean effective pressure and cycle efficiency.

Also find the fuel consumption per kWh if the indicated thermal efficiency is 0.5 of ideal efficiency, mechanical efficiency is 0.8 and the calorific value of oil 42000 kJ/kg.

Assume for air : c_p = 1.005 kJ/kg K ; c_v = 0.718 kJ/kg K, γ = 1.4. **(U.P.S.C.)**

Solution. Refer to Fig. 21.18. *Given :* $\frac{V_1}{V_2} = 15.3\ ;\ \frac{V_4}{V_3} = 7.5$

$p_1 = 1$ bar ; $T_1 = 27 + 273 = 300$ K ;
$\eta_{th(I)} = 0.5 \times \eta_{air\text{-}standard}\ ;\ \eta_{mech.} = 0.8$;
$C = 42000$ kJ/kg.

The cycle is shown in Fig. 21.18, the subscripts denote the respective points in the cycle.

Mean effective pressure, p_m :

$$p_m = \frac{\text{Work done by the cycle}}{\text{Swept volume}}$$

Work done = Heat added – heat rejected
Heat added $= mc_p\,(T_3 - T_2)$, and
Heat rejected $= mc_v\,(T_4 - T_1)$

Now assume air as a perfect gas and mass of oil in the air-fuel mixture is negligible and is not taken into account.

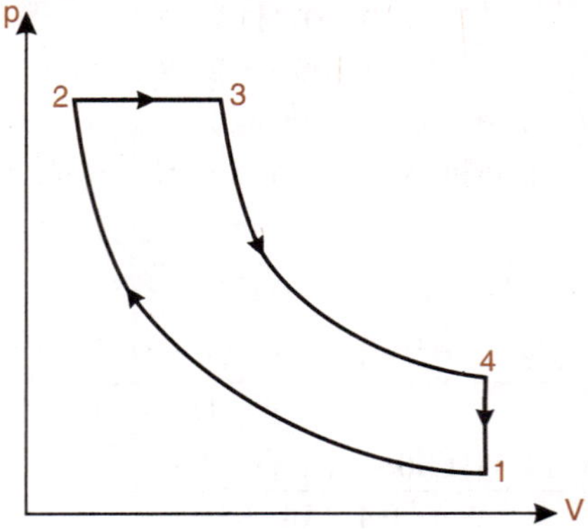

Fig. 21.18. Diesel cycle.

Process 1–2 is an *adiabatic compression process*, thus

$$\frac{T_2}{T_1} = \left(\frac{V_1}{V_2}\right)^{\gamma - 1} \quad \text{or} \quad T_2 = T_1 \times \left(\frac{V_1}{V_2}\right)^{1.4 - 1}$$

(since $\gamma = 1.4$)

or $$T_2 = 300 \times (15.3)^{0.4} = 893.3 \text{ K}$$

Also, $$p_1V_1^\gamma = p_2V_2^\gamma \Rightarrow p_2 = p_1 \times \left(\frac{V_1}{V_2}\right)^\gamma$$

$$= 1 \times (15.3)^{1.4} = 45.56 \text{ bar}$$

Process 2–3 is a *constant pressure process*, hence

$$\frac{V_2}{T_2} = \frac{V_3}{T_3} \Rightarrow T_3 = \frac{V_3T_2}{V_2}$$

$$= 2.04 \times 893.3 = 1822.3 \text{ K}$$

Assume that the volume at point 2 (V_2) is 1 m^3. Thus the mass of air involved in the process,

$$m = \frac{p_2V_2}{RT_2} = \frac{45.56 \times 10^5 \times 1}{287 \times 893.3} = 17.77 \text{ kg}$$

$$\left[\because \frac{V_4}{V_3} = \frac{V_1}{V_3} = \frac{V_1}{V_2} \times \frac{V_2}{V_3} \quad \text{or} \quad \frac{V_3}{V_2} = \frac{V_1}{V_2} \times \frac{V_3}{V_4} = \frac{15.3}{7.5} = 2.04\right]$$

Process 3–4 is an *adiabatic expansion process*, thus

$$\frac{T_4}{T_3} = \left(\frac{V_3}{V_4}\right)^{\gamma - 1} = \left(\frac{1}{7.5}\right)^{1.4 - 1} = 0.4466$$

or $$T_4 = 1822.3 \times 0.4466 = 813.8 \text{ K}$$

$\therefore$ Work done $= mc_p (T_3 - T_2) - mc_v (T_4 - T_1)$

$= 17.77 [1.005 (1822.3 - 893.3)$

$- 0.718 (813.8 - 300)] = 10035$ kJ

$\therefore$ $$\mathbf{p_m} = \frac{\text{Work done}}{\text{Swept volume}}$$

$$= \frac{10035}{(V_1 - V_2)} = \frac{10035}{(15.3V_2 - V_2)} = \frac{10035}{14.3}$$

$= 701.7$ kN/m^2 = **7.017 bar. (Ans.)**

($\because V_2 = 1$ m^3 assumed)

Ratio of maximum pressure to mean effective pressure

$$= \frac{p_2}{p_m} = \frac{45.56}{7.017} = \mathbf{6.49.} \quad \textbf{(Ans.)}$$

Cycle efficiency, η_{cycle} :

$$\eta_{\text{cycle}} = \frac{\text{Work done}}{\text{Heat supplied}}$$

$$= \frac{10035}{mc_p (T_3 - T_2)} = \frac{10035}{17.77 \times 1.005 (1822.3 - 897.3)}$$

= 0.6048 or 60.48%. (Ans.)

Fuel consumption per kWh ; m_f :

$\eta_{th(I)} = 0.5\, \eta_{cycle} = 0.5 \times 0.6048$

$= 0.3024$ or 30.24%

$\eta_{th(B)} = 0.3024 \times 0.8 = 0.242$

Also, $$\eta_{th(B)} = \frac{\text{B.P.}}{m_f \times C} = \frac{1}{\frac{m_f}{3600} \times 42000}$$

$$= \frac{3600}{m_f \times 42000}$$

or $$0.242 = \frac{3600}{m_f \times 42000}$$

or $$\mathbf{m_f} = \frac{3600}{0.242 \times 42000} = \mathbf{0.354 \text{ kg/kWh.}} \quad \textbf{(Ans.)}$$

21.6. DUAL COMBUSTION CYCLE

This cycle (also called the *limited pressure cycle or mixed cycle*) is a combination of Otto and Diesel cycles, in a way, that heat is added partly at constant volume and partly at constant pressure ; *the advantage of which is that more time is available to fuel (which is injected into the engine cylinder before the end of compression stroke) for combustion. Because of lagging characteristics of fuel this cycle is invariably used for diesel and hot spot ignition engines.*

The dual combustion cycle (Fig. 21.19) consists of the following **operations** :

(*i*) 1–2—*Adiabatic compression*

(*ii*) 2–3—*Addition of heat at constant volume*

(*iii*) 3–4—*Addition of heat at constant pressure*

(*iv*) 4–5—*Adiabatic expansion*

(*v*) 5–1—*Rejection of heat at constant volume.*

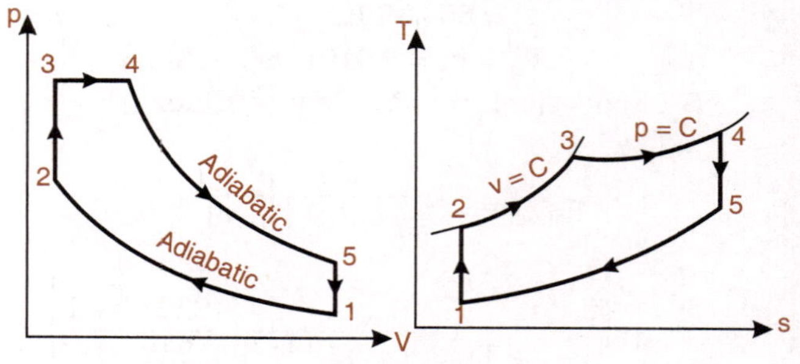

Fig. 21.19

Consider 1 kg of air.

Total heat supplied = Heat supplied during the operation 2–3 + heat supplied during the operation 3–4

$$= c_v(T_3 - T_2) + c_p(T_4 - T_3)$$

Heat rejected during operation 5–1 = $c_v(T_5 - T_1)$

Work done = Heat supplied – heat rejected

$$= c_v(T_3 - T_2) + c_p(T_4 - T_3) - c_v(T_5 - T_1)$$

$$\eta_{dual} = \frac{\text{Work done}}{\text{Heat supplied}}$$

$$= \frac{c_v(T_3 - T_2) + c_p(T_4 - T_3) - c_v(T_5 - T_1)}{c_v(T_3 - T_2) + c_p(T_4 - T_3)}$$

$$= 1 - \frac{c_v(T_5 - T_1)}{c_v(T_3 - T_2) + c_p(T_4 - T_3)}$$

$$= 1 - \frac{c_v(T_5 - T_1)}{(T_3 - T_2) + \gamma(T_4 - T_3)} \quad ...(i)$$

$$\left(\because \gamma = \frac{c_p}{c_v}\right)$$

Compression ratio, $r = \frac{v_1}{v_2}$

During *adiabatic compression process 1–2,*

$$\frac{T_2}{T_1} = \left(\frac{v_1}{v_2}\right)^{\gamma - 1} = (r)^{\gamma - 1} \quad ...(ii)$$

During *constant volume heating process,*

$$\frac{p_3}{T_3} = \frac{p_2}{T_2}$$

or $\frac{T_3}{T_2} = \frac{p_3}{p_2} = \beta$, where β is known as **pressure or explosion ratio**.

or
$$T_2 = \frac{T_3}{\beta} \quad ...(iii)$$

During *adiabatic expansion process,*

$$\frac{T_4}{T_5} = \left(\frac{v_5}{v_4}\right)^{\gamma - 1} = \left(\frac{r}{\rho}\right)^{\gamma - 1} \quad ...(iv)$$

$$\left(\because \frac{v_5}{v_4} = \frac{v_1}{v_4} = \frac{v_1}{v_2} \times \frac{v_2}{v_4} = \frac{v_1}{v_2} \times \frac{v_3}{v_4} = \frac{r}{\rho}, \; \rho \text{ being the cut-off ratio}\right)$$

During *constant pressure heating process,*

$$\frac{v_3}{T_3} = \frac{v_4}{T_4}$$

$$T_4 = T_3 \frac{v_4}{v_3} = \rho \, T_3 \quad ...(v)$$

Putting the value of T_4 in the eqn. (*iv*), we get

$$\frac{\rho T_3}{T_5} = \left(\frac{r}{\rho}\right)^{\gamma - 1} \quad \text{or} \quad T_5 = \rho \, . \, T_3 \, . \left(\frac{\rho}{r}\right)^{\gamma - 1}$$

Putting the value of T_2 in eqn. (*ii*), we get

$$\frac{\frac{T_3}{\beta}}{T_1} = (r)^{\gamma - 1}$$

$$T_1 = \frac{T_3}{\beta} \cdot \frac{1}{(r)^{\gamma - 1}}$$

Now inserting the values of T_1, T_2, T_4 and T_5 in eqn. (*i*), we get

$$\eta_{dual} = 1 - \frac{\left[\rho \, . \, T_3\left(\frac{\rho}{r}\right)^{\gamma - 1} - \frac{T_3}{\beta} \cdot \frac{1}{(r)^{\gamma - 1}}\right]}{\left[\left(T_3 - \frac{T_3}{\beta}\right) + \gamma(\rho T_3 - T_3)\right]}$$

$$= 1 - \frac{\frac{1}{(r)^{\gamma - 1}}\left(\rho^\gamma - \frac{1}{\beta}\right)}{\left(1 - \frac{1}{\beta}\right) + \gamma(\rho - 1)}$$

i.e.,
$$\eta_{dual} = 1 - \frac{1}{(r)^{\gamma - 1}} \cdot \frac{(\beta \, . \, \rho^\gamma - 1)}{[(\beta - 1) + \beta\gamma(\rho - 1)]} \quad ...(21.10)$$

Work done is given by,

$$W = p_3(v_4 - v_3) + \frac{p_4 v_4 - p_5 v_5}{\gamma - 1} - \frac{p_2 v_2 - p_1 v_1}{\gamma - 1}$$

$$= p_3 v_3(\rho - 1) + \frac{(p_4 \rho v_3 - p_5 r v_3) - (p_2 v_3 - p_1 r v_3)}{\gamma - 1}$$

$$= \frac{p_3 v_3(\rho - 1)(\gamma - 1) + p_4 v_3\left(\rho - \frac{p_5}{p_4} r\right) - p_2 v_3\left(1 - \frac{p_1}{p_2} r\right)}{\gamma - 1}$$

Also $\frac{p_5}{p_4} = \left(\frac{v_4}{v_5}\right)^\gamma = \left(\frac{\rho}{r}\right)^\gamma$ and $\frac{p_2}{p_1} = \left(\frac{v_1}{v_2}\right)^\gamma = r^\gamma$

also, $p_3 = p_4$, $v_2 = v_3$, $v_5 = v_1$

$$\therefore \; W = \frac{v_3[p_3(\rho - 1)(\gamma - 1) + p_3(\rho - \rho^\gamma r^{1-\gamma}) - p_2(1 - r^{1-\gamma})]}{(\gamma - 1)}$$

$$= \frac{p_2 v_2[\beta(\rho - 1)(\gamma - 1) + \beta(\rho - \rho^\gamma r^{1-\gamma}) - (1 - r^{1-\gamma})]}{(\gamma - 1)}$$

$$= \frac{p_1(r)^\gamma v_1/r[\beta\gamma(\rho - 1) + (\beta - 1) - r^{1-\gamma}(\beta\rho^\gamma - 1)]}{\gamma - 1}$$

$$= \frac{p_1 v_1 r^{\gamma - 1}[\beta\gamma(\rho - 1) + (\beta - 1) - r^{\gamma - 1}(\beta\rho^\gamma - 1)]}{\gamma - 1} \quad ...(21.11)$$

Mean effective pressure (p_m) is given by,

$$p_m = \frac{W}{v_1 - v_2} = \frac{W}{v_1\left(\frac{r-1}{r}\right)}$$

$$= \frac{p_1 v_1 [r^{1-\gamma}\beta\gamma(\rho-1) + (\beta-1) - r^{1-\gamma}(\beta\rho^\gamma - 1)]}{(\gamma-1)v_1\left(\frac{r-1}{r}\right)}$$

$$\mathbf{p_m} = \frac{p_1(r)^\gamma[\beta(\rho-1) + (\beta-1) - r^{1-\gamma}(\beta\rho^\gamma-1)]}{(\gamma-1)(r-1)}$$

...(21.12)

Example 21.23. *The swept volume of a diesel engine working on dual cycle is 0.0053 m³ and clearance volume is 0.00035 m³. The maximum pressure is 65 bar. Fuel injection ends at 5 per cent of the stroke. The temperature and pressure at the start of the compression are 80°C and 0.9 bar. Determine the air standard efficiency of the cycle. Take γ for air = 1.4.*

Solution. Refer to Fig. 21.20.

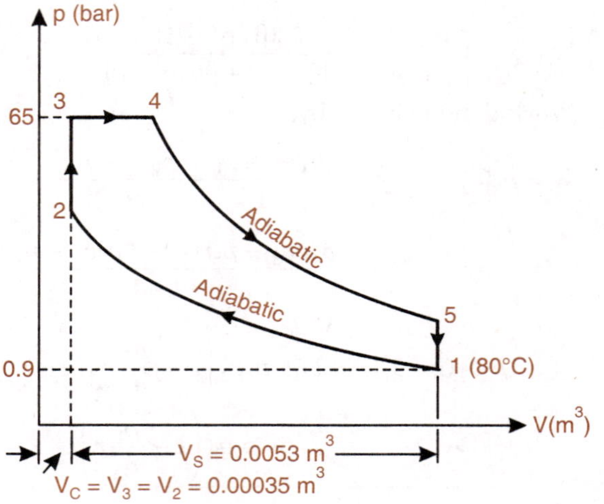

Fig. 21.20

Swept volume, $V_s = 0.0053 \text{ m}^3$

Clearance volume, $V_c = V_3 = V_2 = 0.00035 \text{ m}^3$

Maximum pressure, $p_3 = p_4 = 65$ bar

Initial temperature, $T_1 = 80 + 273 = 353$ K

Initial pressure, $p_1 = 0.9$ bar

$\eta_{dual} = ?$

The efficiency of a dual combustion cycle is given by

$$\eta_{dual} = 1 - \frac{1}{(r)^{\gamma-1}}\left[\frac{\beta.\rho^\gamma - 1}{(\beta-1)+\beta\gamma(\rho-1)}\right] \quad ...(i)$$

Compression ratio,

$$r = \frac{V_1}{V_2} = \frac{V_s + V_c}{V_c} = \frac{0.0053 + 0.00035}{0.00035} = 16.14$$

$[\because V_2 = V_c = \text{Clearance volume}]$

Cut-off ratio,

$$\rho = \frac{V_4}{V_3} = \frac{\frac{5}{100}V_s + V_3}{V_3} = \frac{0.05V_s + V_c}{V_c}$$

$(\because V_2 = V_3 = V_c)$

$$= \frac{0.05 \times 0.0053 + 0.00035}{0.00035} = 1.757 \text{ say } 1.76$$

Also during the *compression operation 1–2,*

$$p_1V_1^\gamma = p_2V_2^\gamma$$

or $$\frac{p_2}{p_1} = \left(\frac{V_1}{V_2}\right)^\gamma = (16.14)^{1.4} = 49.14$$

or $$p_2 = p_1 \times 49.14 = 0.9 \times 49.14 = 44.22 \text{ bar}$$

Pressure or explosion ratio, $\beta = \frac{p_3}{p_2} = \frac{65}{44.22} = 1.47$

Putting the value of r, ρ and β in eqn. (*i*), we get

$$\eta_{dual} = 1 - \frac{1}{(16.14)^{1.4-1}}\left[\frac{1.47 \times (1.76)^{1.4} - 1}{(1.47-1) + 1.47 \times 1.4\,(1.76-1)}\right]$$

$$= 7 - 0.328\left[\frac{3.243 - 1}{0.47 + 1.564}\right]$$

$= \mathbf{0.6383}$ or **63.83%.** **(Ans.)**

Example 21.24. *An oil engine working on the dual combustion cycle has a compression ratio 14 and the explosion ratio obtained from an indicator card is 1.4. If the cut-off occurs at 6 per cent of stroke, find the ideal efficiency. Take γ for air = 1.4.*

Solution. Refer to Fig. 21.19.

Compression ratio, $r = 14$

Explosion ratio, $\beta = 1.4$

If ρ is the cut-off ratio, then

$$\frac{\rho - 1}{r - 1} = \frac{6}{100} \quad \text{or} \quad \frac{\rho-1}{14-1} = 0.06$$

$\therefore \quad \rho = 1.78$

Ideal efficiency is given by

$$\eta_{\text{ideal or dual}} = 1 - \frac{1}{(r)^{\gamma-1}}\left[\frac{(\beta\rho^\gamma - 1)}{(\beta-1)+\beta\gamma(\rho-1)}\right]$$

$$= 1 - \frac{1}{(14)^{1.4-1}}\left[\frac{1.4 \times (1.78)^{1.4} - 1}{(1.4-1) + 1.4 \times 1.4(1.78-1)}\right]$$

$$= 1 - 0.348\left[\frac{3.138 - 1}{0.4 + 1.528}\right] = \mathbf{0.614} \text{ or } \mathbf{61.4\%}. \quad \textbf{(Ans.)}$$

Example 21.25. *The compression ratio for a single-cylinder engine operating on dual cycle is 9. The maximum pressure in the cylinder is limited to 60 bar. The pressure and temperature of the air at the beginning of the cycle are 1 bar and 30°C. Heat is added during constant pressure process upto 4 per cent of the stroke. Assuming the cylinder* *diameter and stroke length as 250 mm and 300 mm respectively, determine :*

(i) The air standard efficiency of the cycle.

(ii) The power developed if the number of working cycles are 3 per second.

Take for air $c_v = 0.71$ kJ/kg K and $c_p = 1.0$ kJ/kg K

Solution. Refer to Fig. 21.21.

Cylinder diameter, $D = 250$ mm $= 0.25$ m

Compression ratio, $r = 9$

Stroke length, $L = 300$ mm $= 0.3$ m

Initial pressure, $p_1 = 1$ bar

Initial temperature, $T_1 = 30 + 273 = 303$ K

Maximum pressure, $p_3 = p_4 = 60$ bar

Cut-off $= 4\%$ of stroke volume

Number of working cycles/sec. = 3.

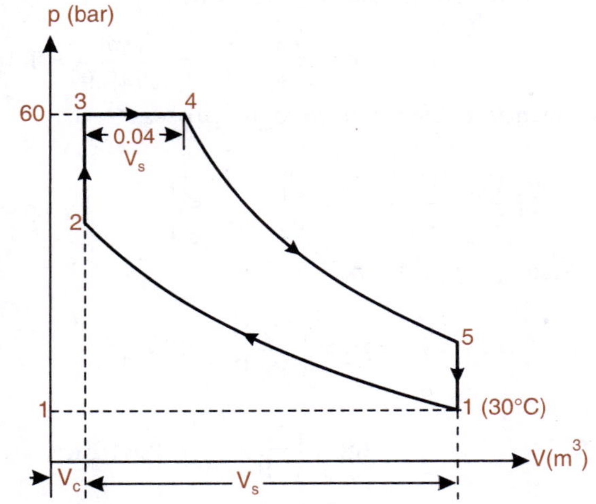

Fig. 21.21

(i) **Air standard efficiency :**

Now, swept volume, $V_s = \pi/4\, D^2 L = \pi/4 \times 0.25^2 \times 0.3$
$= 0.0147$ m^3

Also, compression ratio,

$$r = \frac{V_s + V_c}{V_c}$$

i.e.,
$$9 = \frac{0.0147 + V_c}{V_c}$$

$$\therefore \quad V_c = \frac{0.0147}{8} = 0.0018 \text{ m}^3$$

$$\therefore \quad V_1 = V_s + V_c = 0.0147 + 0.0018 = 0.0165 \text{ m}^3$$

For the *adiabatic (or isentropic) process 1–2,*

$$p_1 V_1^{\gamma} = p_2 V_2^{\gamma}$$

$$p_2 = p_1 \times \left(\frac{V_1}{V_2}\right)^{\gamma} = 1 \times (r)^{\gamma}$$
$$= 1 \times (9)^{1.4} = 21.67 \text{ bar}$$

Also,
$$\frac{T_2}{T_1} = \left(\frac{V_1}{V_2}\right)^{\gamma - 1} = (r)^{\gamma - 1} = (9)^{1.4 - 1}$$
$$= (9)^{0.4} = 2.408$$

$$\therefore \quad T_2 = T_1 \times 2.408 = 303 \times 2.408 = 729.6 \text{ K}$$

For the *constant volume process 2–3,*

$$\frac{T_3}{p_3} = \frac{T_2}{p_2}$$

$$\therefore \quad T_3 = T_2 \cdot \frac{p_3}{p_2} = 729.6 \times \frac{60}{21.67} = 2020 \text{ K}$$

Also,
$$\frac{\rho - 1}{r - 1} = \frac{4}{100} \quad \text{or} \quad 0.04$$

$$\therefore \quad \frac{\rho - 1}{9 - 1} = 0.04 \quad \text{or} \quad \rho = 1.32$$

For the *constant pressure process 3–4,*

$$\frac{V_4}{T_4} = \frac{V_3}{T_3} \quad \text{or} \quad \frac{T_4}{T_3} = \frac{V_4}{V_3} = \rho$$

$$\therefore \quad T_4 = T_3 \times \rho = 2020 \times 1.32 = 2666.4 \text{ K}$$

Also expansion ratio,

$$\frac{V_5}{V_4} = \frac{V_5}{V_2} \times \frac{V_2}{V_4} = \frac{V_1}{V_2} \times \frac{V_3}{V_4} = \frac{r}{\rho}$$

$[\because V_5 = V_1 \text{ and } V_2 = V_3]$

For *adiabatic process 4–5,*

$$\frac{T_5}{T_4} = \left(\frac{V_4}{V_5}\right)^{\gamma - 1} = \left(\frac{\rho}{r}\right)^{\gamma - 1}$$

$$\therefore \quad T_5 = T_4 \times \left(\frac{\rho}{r}\right)^{\gamma - 1}$$

$$= 2666.4 \times \left(\frac{1.32}{9}\right)^{1.4 - 1} = 1237 \text{ K}$$

Also $p_4 V_4^{\gamma} = p_5 V_5^{\gamma}$

$$p_5 = p_4 \cdot \left(\frac{V_4}{V_5}\right)^{\gamma} = 60 \times \left(\frac{r}{\rho}\right)^{\gamma}$$

$$= 60 \times \left(\frac{1.32}{9}\right)^{1.4} = 4.08 \text{ bar}$$

Heat supplied,

$$Q_s = c_v(T_3 - T_2) + c_p(T_4 - T_3)$$
$$= 0.71\,(2020 - 729.6) + 1.0\,(2666.4 - 2020)$$
$$= 1562.58 \text{ kJ/kg}$$

Heat rejected,

$$Q_r = c_v\,(T_5 - T_1)$$
$$= 0.71\,(1237 - 303) = 663.14 \text{ kJ/kg}$$

$$\eta_{\text{air-standard}} = \frac{Q_s - Q_r}{Q_s} = \frac{1562.85 - 663.14}{1562.58}$$

$= \mathbf{0.5756}$ or **57.56%.** **(Ans.)**

(ii) **Power developed by the engine, P :**

Mass of air in the cycle is given by

$$m = \frac{p_1V_1}{RT_1} = \frac{1 \times 10^5 \times 0.0165}{287 \times 303} = 0.0189 \text{ kg}$$

$\therefore$ Work done per cycle $= m(Q_s - Q_r)$

$= 0.0189\ (1562.58 - 663.14) = 16.999$ kJ

Power developed = Work done per cycle × no. of cycles per second

$= 16.999 \times 3 =$ **50.99 say 51 kW.** (Ans.)

☞**Example 21.26.** *In an engine working on Dual cycle, the temperature and pressure at the beginning of the cycle are 90°C and 1 bar respectively. The compression ratio is 9. The maximum pressure is limited to 68 bar and total heat supplied per kg of air is 1750 kJ. Determine :*

(i) Pressure and temperatures at all salient points

(ii) Air standard efficiency

(iii) Mean effective pressure.

Solution. Refer to Fig. 21.22.

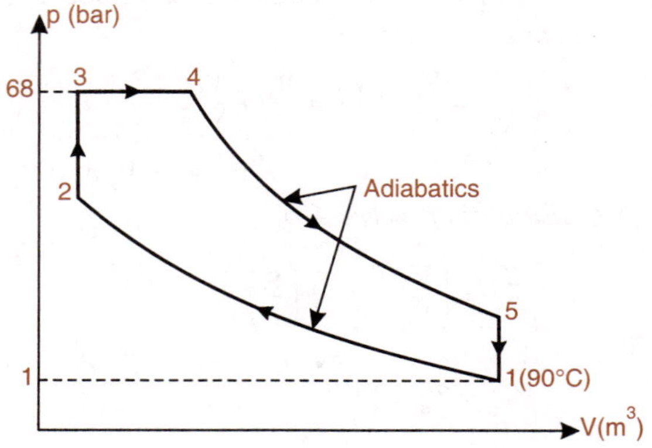

Fig. 21.22

Initial pressure, $p_1 = 1$ bar

Initial temperature, $T_1 = 90 + 273 = 363$ K

Compression ratio, $r = 9$

Maximum pressure, $p_3 = p_4 = 68$ bar

Total heat supplied $= 1750$ kJ/kg

(i) **Pressures and temperatures at salient points :**

For the *isentropic process 1–2,*

$$p_1V_1^{\gamma} = p_2V_2^{\gamma}$$

$$\boldsymbol{p_2} = p_1 \times \left(\frac{V_1}{V_2}\right)^{\gamma} = 1 \times (r)^{\gamma}$$

$= 1 \times (9)^{1.4} =$ **21.67 bar.** (Ans.)

Also, $\dfrac{T_2}{T_1} = \left(\dfrac{V_1}{V_2}\right)^{\gamma - 1} = (r)^{\gamma - 1}$

$= (9)1.4 - 1 = 2.408$

$\therefore$ $T_2 = T_1 \times 2.408$

$= 363 \times 2.408 =$ **874.1 K.** (Ans.)

$\boldsymbol{p_3 = p_4 =}$ **68 bar.** (Ans.)

For the *constant volume process* 2–3,

$$\frac{p_2}{T_2} = \frac{p_3}{T_3}$$

$\therefore$ $\boldsymbol{T_3} = T_2 \times \dfrac{p_3}{p_2} = 874.1 \times \dfrac{68}{21.67}$

= **2742.9 K.** (Ans.)

Heat added at constant volume

$= c_v\ (T_3 - T_2) = 0.71\ (2742.9 - 874.1)$

$= 1326.8$ kJ/kg

$\therefore$ Heat added at constant pressure

= Total heat added – heat added at constant volume

$= 1750 - 1326.8 = 423.2$ kJ/kg

$\therefore$ $c_p(T_4 - T_3) = 423.2$

or $1.0(T_4 - 2742.9) = 423.2$

$\therefore$ $\boldsymbol{T_4 =}$ **3166 K.** (Ans.)

For *constant pressure process* 3–4,

$$\rho = \frac{V_4}{V_3} = \frac{T_4}{T_3} = \frac{3166}{2742.9} = 1.15$$

For *adiabatic (or isentropic) process* 4–5,

$$\frac{V_5}{V_4} = \frac{V_5}{V_2} \times \frac{V_2}{V_4} = \frac{V_1}{V_2} \times \frac{V_3}{V_4} = \frac{r}{\rho} \qquad \left(\because\ \rho = \frac{V_4}{V_3}\right)$$

Also $p_4V_4^{\gamma} = p_5V_5^{\gamma}$

$\therefore$ $p_5 = p_4 \times \left(\dfrac{V_4}{V_5}\right)^{\gamma} = 68 \times \left(\dfrac{\rho}{r}\right)^{\gamma}$

$= 68 \times \left(\dfrac{1.15}{9}\right)^{1.4} =$ **3.81 bar.** (Ans.)

Again, $\dfrac{T_5}{T_4} = \left(\dfrac{V_4}{V_5}\right)^{\gamma - 1} = \left(\dfrac{\rho}{r}\right)^{\gamma - 1}$

$= \left(\dfrac{1.15}{9}\right)^{1.4 - 1} = 0.439$

$\therefore$ $T_5 = T_4 \times 0.439 = 3166 \times 0.439$

= **1389.8 K.** (Ans.)

(ii) **Air standard efficiency :**

Heat rejected during constant volume process 5–1,

$Q_r = c_v(T_5 - T_1) = 0.71(1389.8 - 363) = 729$ kJ/kg

$\therefore$ $\eta_{\text{air-standard}} = \dfrac{\text{Work done}}{\text{Heat supplied}} = \dfrac{Q_s - Q_r}{Q_s}$

$= \dfrac{1750 - 729}{1750}$

= **0.5834** or **58.34%.** (Ans.)

(iii) **Mean effective pressure, $\boldsymbol{p_m}$:**

Mean effective pressure is given by

$$p_m = \frac{\text{Work done per cycle}}{\text{Stroke volume}}$$

or $$p_m = \frac{1}{V_s}\left[p_3(V_4 - V_3) + \frac{p_4V_4 - p_5V_5}{\gamma - 1} - \frac{p_2V_2 - p_1V_1}{\gamma - 1}\right]$$

$V_1 = V_5 = r\,V_c$, $V_2 = V_3 = V_c$, $V_4 = \rho V_c$,
$V_s = (r - 1)\,V_c$

$$\left[\because\; r = \frac{V_s + V_c}{V_c} = 1 + \frac{V_s}{V_c} \quad \therefore\; V_s = (r-1)V_c\right]$$

$$\therefore\; p_m = \frac{1}{(r-1)V_c}\left[p_3(\rho V_c - V_c) + \frac{p_4\rho V_c - p_5 \times rV_c}{\gamma - 1} - \frac{p_2V_c - p_1 rV_c}{\gamma - 1}\right]$$

$r = 9$, $\rho = 1.15$, $\gamma = 1.4$
$p_1 = 1$ bar, $p_2 = 21.67$ bar,
$p_3 = p_4 = 68$ bar, $p_5 = 3.81$ bar

Substituting the above values in the above equation, we get

$$p_m = \frac{1}{(9-1)}\left[68(1.15 - 1) + \frac{68 \times 1.15 - 3.81 \times 9}{1.4 - 1} - \frac{21.67 - 9}{1.4 - 1}\right]$$

$$= \frac{1}{8}(10.2 + 109.77 - 31.67) = 11.04 \text{ bar}$$

Hence, *mean effective pressure* = **11.04 bar. (Ans.)**

Example 21.27. *An I.C. engine operating on the dual cycle (limited pressure cycle) the temperature of the working fluid (air) at the beginning of compression is 27°C. The ratio of the maximum and minimum pressures of the cycle is 70 and compression ratio is 15. The amounts of heat added at constant volume and at constant pressure are equal. Compute the air standard thermal efficiency of the cycle. State three main reasons why the actual thermal efficiency is different from the theoretical value.*

Take γ for air = 1.4. **(U.P.S.C.)**

Solution. Refer to Fig. 21.23. *Given :* $T_1 = 27 + 273 = 300$ K ;

$$\frac{p_3}{p_1} = 70, \quad \frac{v_1}{v_2} = \frac{v_1}{v_3} = 15$$

Air standard efficiency, $\eta_{\text{air-standard}}$:

Consider 1 kg of air.

Adiabatic compression process 1–2 :

$$\frac{T_2}{T_1} = \left(\frac{v_1}{v_2}\right)^{\gamma - 1} = (15)^{1.4 - 1} = 2.954$$

$$\therefore\quad T_2 = 300 \times 2.954 = 886.2 \text{ K}$$

$$\frac{p_2}{p_1} = \left(\frac{v_1}{v_2}\right)^{\gamma} = (15)^{1.4} \Rightarrow p_2 = 44.3\,p_1$$

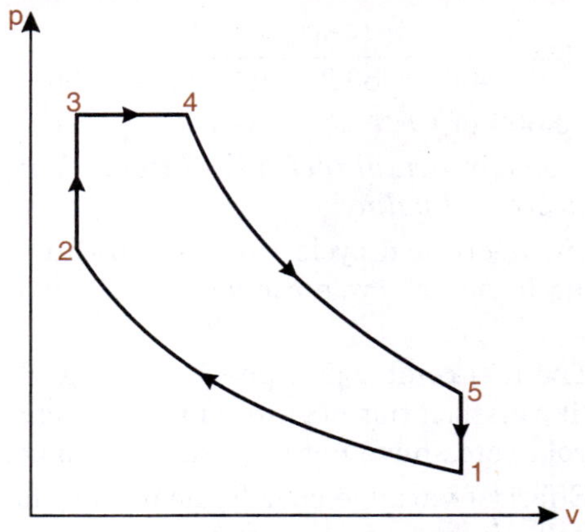

Fig. 21.23. Dual cycle.

Constant pressure process 2–3 :

$$\frac{p_2}{T_2} = \frac{p_3}{T_3}$$

or $$T_3 = T_2 \times \frac{p_3}{p_2} = 886.2 \times \frac{70p_1}{44.3p_1} = 1400 \text{ K}$$

Also, Heat added at constant volume = Heat added at constant pressure ...(Given)

or $$c_v(T_3 - T_2) = c_p(T_4 - T_3)$$

or $$T_3 - T_2 = \gamma(T_4 - T_3)$$

or $$T_4 = T_3 + \frac{T_3 - T_2}{\gamma} = 1400 + \frac{1400 - 886.2}{1.4} = 1767 \text{ K.}$$

Constant volume process 3–4 :

$$\frac{v_3}{T_3} = \frac{v_4}{T_4} \Rightarrow \frac{v_4}{v_3} = \frac{T_4}{T_3} = \frac{1767}{1400} = 1.26$$

Also, $$\frac{v_4}{v_3} = \frac{v_4}{(v_1/15)} = 1.26 \text{ or } v_4 = 0.084\,v_1$$

Also, $v_5 = v_1$

Adiabatic expansion process 4–5 :

$$\frac{T_4}{T_5} = \left(\frac{v_5}{v_4}\right)^{\gamma - 1} = \left(\frac{v_1}{0.084v_1}\right)^{1.4 - 1} = 2.69$$

$$\therefore\quad T_5 = \frac{T_4}{2.69} = \frac{1767}{2.69} = 656.9 \text{ K}$$

$$\therefore\quad \eta_{\text{air-standard}} = \frac{\text{Work done}}{\text{Heat supplied}}$$

$$= \frac{\text{Heat supplied} - \text{Heat rejected}}{\text{Heat supplied}}$$

$$= 1 - \frac{\text{Heat rejected}}{\text{Heat supplied}}$$

$$= 1 - \frac{c_v(T_5 - T_1)}{c_v(T_3 - T_2) + c_p(T_4 - T_3)}$$

$$= 1 - \frac{(T_5 - T_1)}{(T_3 - T_2) + \gamma(T_4 - T_3)}$$

$$= 1 - \frac{(656.9 - 300)}{(1400 - 886.2) + 1.4(1767 - 1400)}$$

= **0.653** or **65.3%.** **(Ans.)**

Reasons for actual thermal efficiency being different from the theoretical value :

1. In theoretical cycle working substance is taken *air* whereas in actual cycle *air with fuel acts as working substance.*

2. The fuel combustion phenomenon and associated problems like dissociation of gases, dilution of charge during suction stroke, etc., have *not* been taken into account.

3. Effect of variable specific heat, heat loss through cylinder walls, inlet and exhaust velocities of air/gas etc., have *not* been taken into account.

Example 21.28. *A Diesel engine working on a dual combustion cycle has a stroke volume of 0.0085 m³ and a compression ratio 15 : 1. The fuel has a calorific value of 43890 kJ/kg. At the end of suction, the air is at 1 bar and 100°C. The maximum pressure in the cycle is 65 bar and air fuel ratio is 21 : 1. Find for ideal cycle the thermal efficiency. Assume $c_p = 1.0$ and $c_v = 0.71$.*

Solution. Refer to Fig. 21.24.

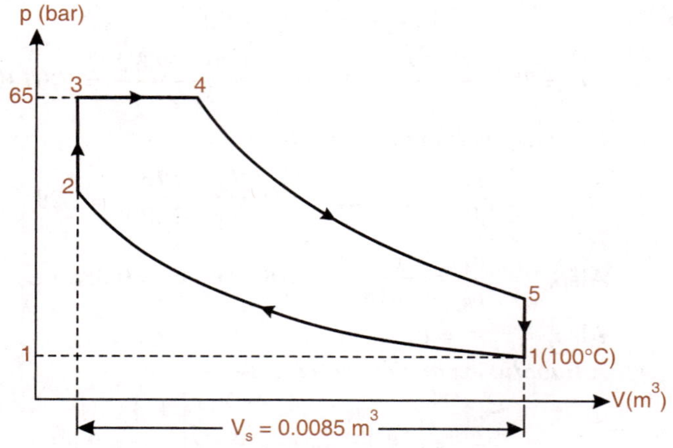

Fig. 21.24

Initial temperature, $T_1 = 100 + 273 = 373$ K

Initial pressure, $p_1 = 1$ bar

Maximum pressure in the cycle, $p_3 = p_4 = 65$ bar

Stroke volume, $V_s = 0.0085$ m³

Air-fuel ratio $= 21 : 1$

Compression ratio, $r = 15 : 1$

Calorific value of fuel, $C = 43890$ kJ/kg

$c_p = 1.0,\ c_v = 0.71$

Thermal efficiency :

$$V_s = V_1 - V_2 = 0.0085$$

and as $r = \frac{V_1}{V_2} = 15$, then $V_1 = 15V_2$

$\therefore \quad 15V_2 - V_2 = 0.0085$

or $\quad 14V_2 = 0.0085$

or $\quad V_2 = V_3 = V_c = \frac{0.0085}{14} = 0.0006$ m³

or $\quad V_1 = 15V_2 = 15 \times 0.0006 = 0.009$ m³

For *adiabatic compression process 1–2,*

$$p_1V_1^{\gamma} = p_2V_2^{\gamma}$$

or $\quad p_2 = p_1 \cdot \left(\frac{V_1}{V_2}\right)^{\gamma} = 1 \times (15)^{1.41}$

$$\left[\gamma = \frac{c_p}{c_v} = \frac{1.0}{0.71} = 1.41\right]$$

$= 45.5$ bar

Also, $\quad \frac{T_2}{T_1} = \left(\frac{V_1}{V_2}\right)^{\gamma - 1} = (r)^{\gamma - 1} = (15)^{1.41 - 1} = 3.04$

$\therefore \quad T_2 = T_1 \times 3.04 = 373 \times 3.04$

$= 1134$ K or 861°C

For *constant volume process 2–3,*

$$\frac{p_2}{T_2} = \frac{p_3}{T_3}$$

or $\quad T_3 = T_2 \times \frac{p_3}{p_2} = 1134 \times \frac{65}{45.5}$

$= 1620$ K or 1347°C

According to characteristic equation of gas,

$$p_1V_1 = mRT_1$$

$\therefore \quad m = \frac{p_1V_1}{RT_1} = \frac{1 \times 10^5 \times 0.009}{287 \times 373}$

$= 0.0084$ kg (air)

Heat added during constant volume process 2–3,

$= m \times c_v (T_3 - T_2)$

$= 0.0084 \times 0.71 (1620 - 1134)$

$= 2.898$ kJ

Amount of fuel added during the constant volume process 2–3,

$$= \frac{2.898}{43890} = 0.000066 \text{ kg}$$

Also as air-fuel ratio is 21 : 1.

$\therefore$ Total amount of fuel added

$$= \frac{0.0084}{21} = 0.0004 \text{ kg}$$

Quantity of fuel added during the process 3–4,

$= 0.0004 - 0.000066 = 0.000334$ kg

$\therefore$ *Heat added during the constant pressure operation 3–4*

$= 0.000334 \times 43890 = 14.66$ kJ

But $\quad (0.0084 + 0.0004)\ c_p (T_4 - T_3) = 14.66$

or $\quad 0.0088 \times 1.0 (T_4 - 1620) = 14.66$

$\therefore \quad T_4 = \frac{14.66}{0.0088} + 1620 = 3286$ K or 3013°C

Again for *process 3–4*,

$$\frac{V_3}{T_3} = \frac{V_4}{T_4} \quad \text{or} \quad V_4 = \frac{V_3 T_4}{T_3} = \frac{0.0006 \times 3286}{1620}$$

$= 0.001217 \text{ m}^3$

For *adiabatic expansion operation 4–5*,

$$\frac{T_4}{T_5} = \left(\frac{V_5}{V_4}\right)^{\gamma - 1} = \left(\frac{0.009}{0.001217}\right)^{1.41 - 1} = 2.27$$

or $$T_5 = \frac{T_4}{2.27} = \frac{3286}{2.27} = 1447.5 \text{ K or } 1174.5°\text{C}$$

Heat rejected during constant volume process 5–1,

$= m\, c_v\, (T_5 - T_1)$

$= (0.00854 + 0.0004) \times 0.71\, (1447.5 - 373)$

$= 6.713$ kJ

Work done = Heat supplied – Heat rejected

$= (2.898 + 14.66) - 6.713 = 10.845$ kJ

∴ *Thermal efficiency,*

$$\eta_{th.} = \frac{\text{Work done}}{\text{Heat supplied}} = \frac{10.845}{(2.898 + 14.66)}$$

$=$ **0.6176** or **61.76%.** **(Ans.)**

☞**Example 21.29.** *The compression ratio and expansion ratio of an oil engine working on the dual cycle are 9 and 5 respectively. The initial pressure and temperature of the air are 1 bar and 30°C. The heat liberated at constant pressure is twice the heat liberated at constant volume. The expansion and compression follow the law $pV^{1.25}$ = constant. Determine :*

(*i*) *Pressures and temperatures at all salient points.*

(*ii*) *Mean effective pressure of the cycle.*

(*iii*) *Efficiency of the cycle.*

(*iv*) *Power of the engine if working cycles per second are 8.*

Assume : Cylinder bore = 250 mm and stroke length = 400 mm.

Solution. Refer to Fig. 21.25.

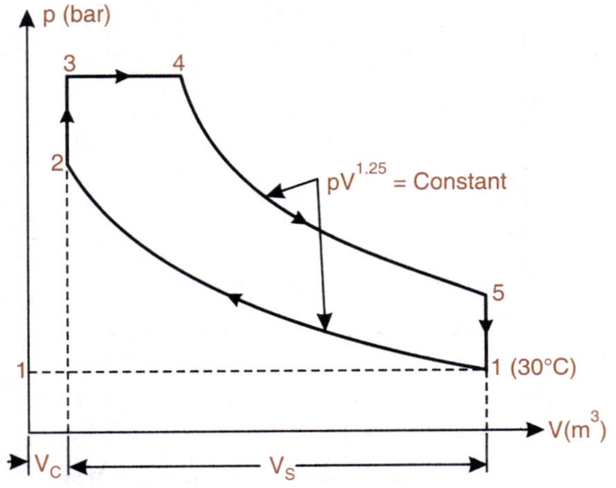

Fig. 21.25

Initial temperature, $T_1 = 30 + 273 = 303$ K

Initial pressure, $p_1 = 1$ bar

Compression and expansion law,

$pV^{1.25}$ = Constant

Compression ratio, $r_c = 9$

Expansion ratio, $r_e = 5$

Number of cycles/sec. $= 8$

Cylinder diameter, $D = 250$ mm $= 0.25$ m

Stroke length, $L = 400$ mm $= 0.4$ m

Heat liberated at constant pressure

= 2 × heat liberated at constant volume

(*i*) **Pressures and temperatures at all salient points :**

For *compression process 1–2,*

$$p_1 V_1^{\,n} = p_2 V_2^{\,n}$$

∴ $$\boldsymbol{p_2} = p_1 \times \left(\frac{V_1}{V_2}\right)^n$$

$= 1 \times (9)^{1.25} =$ **15.59 bar.** **(Ans.)**

Also, $$\frac{T_2}{T_1} = \left(\frac{V_1}{V_2}\right)^{n-1} = (9)^{1.25-1} = 1.732$$

∴ $\boldsymbol{T_2} = T_1 \times 1.732 = 303 \times 1.732$

$=$ **524.8 K** or **251.8°C.** **(Ans.)**

Also, $c_p(T_4 - T_3) = 2 \times c_v(T_3 - T_2)$ ...... (given) ...(*i*)

For *constant pressure process 3–4,*

$$\frac{T_4}{T_3} = \frac{V_4}{V_3} = \rho = \frac{\text{Compression ratio } (r_c)}{\text{Expansion ratio } (r_e)} = \frac{9}{5} = 1.8$$

$$T_4 = 1.8 T_3$$

$$\left[\begin{aligned} \frac{V_5}{V_4} (i.e., r_e) &= \frac{V_5}{V_3} \times \frac{V_3}{V_4} \\ &= \frac{V_1}{V_3} \times \frac{1}{\rho} \\ &= \frac{V_1}{V_2} \times \frac{1}{\rho} = \frac{r_c}{\rho} \\ \therefore \rho &= \frac{r_c}{\frac{V_5}{V_4}} = \frac{r_c}{r_e} \end{aligned}\right]$$

Substituting the values of T_2 and T_4 in the eqn. (*i*), we get

$1.0(1.8T_3 - T_3) = 2 \times 0.71(T_3 - 524.8)$

$0.8T_3 = 1.42(T_3 - 524.8)$

$0.8T_3 = 1.42T_3 - 745.2$

∴ $0.62T_3 = 745.2$

$T_3 =$ **1201.9 K** or **928.9°C.** **(Ans.)**

Also, $$\frac{p_3}{T_3} = \frac{p_2}{T_2} \qquad \text{...... for } \textit{process 2–3}$$

∴ $$\boldsymbol{p_3} = p_2 \times \frac{T_3}{T_2}$$

$= 15.59 \times \dfrac{1201.9}{524.8} =$ **35.7 bar.** **(Ans.)**

$\boldsymbol{p_4} = \boldsymbol{p_3} =$ **35.7 bar.** **(Ans.)**

$$T_4 = 1.8T_3 = 1.8 \times 1201.9$$
$$= \mathbf{2163.4\ K} \text{ or } \mathbf{1890.4°C.} \quad \textbf{(Ans.)}$$

For expansion process 4–5,

$$p_4V_4^n = p_5V_5^n$$

$$p_5 = p_4 \times \left(\frac{V_4}{V_5}\right)^n = p_4 \times \frac{1}{(r_e)^n} = \frac{35.7}{(5)^{1.25}}$$
$$= \mathbf{4.77\ bar.} \quad \textbf{(Ans.)}$$

Also
$$\frac{T_5}{T_4} = \left(\frac{V_4}{V_5}\right)^{n-1} = \frac{1}{(r_e)^{n-1}} = \frac{1}{(5)^{1.25-1}}$$
$$= 0.668$$

$$\therefore \quad T_5 = T_4 \times 0.668 = 2163.4 \times 0.668$$
$$= \mathbf{1445\ K} \text{ or } \mathbf{1172°C.} \quad \textbf{(Ans.)}$$

(*ii*) Mean effective pressure, p_m :

Mean effective pressure is given by

$$p_m = \frac{1}{V_s}\left[p_3(V_4 - V_3) + \frac{p_4V_4 - p_5V_5}{n-1} - \frac{p_2V_2 - p_1V_1}{n-1}\right]$$

$$= \frac{1}{(r_c - 1)}\left[p_3(\rho - 1) + \frac{p_4\rho - p_5r_c}{n-1} - \frac{p_2 - p_1r_c}{n-1}\right]$$

Now, $r_c = \rho$, $\rho = 1.8$, $n = 1.25$,

$p_1 = 1$ bar, $p_2 = 15.59$ bar, $p_3 = 35.7$ bar,

$p_4 = 35.7$ bar, $p_5 = 4.77$ bar

$$\therefore \quad p_m = \frac{1}{(9-1)}\left[35.7(1.8-1) + \frac{35.7 \times 1.8 - 4.77 \times 9}{1.25 - 1} - \frac{15.59 - 1 \times 9}{1.25 - 1}\right]$$

$$= \frac{1}{8}\,[28.56 + 85.32 - 26.36] = 10.94 \text{ bar}$$

Hence *mean effective pressure* = **10.94 bar.** **(Ans.)**

(*iii*) Efficiency of the cycle :

Work done per cycle is given by $W = p_mV_s$

Here, $V_s = \pi/4D^2L = \pi/4 \times 0.25^2 \times 0.4$
$= 0.0196\ m^3$

$$\therefore W = \frac{10.94 \times 10^5 \times 0.0196}{1000} \text{ kJ/cycle} = 21.44 \text{ kJ/cycle}$$

Heat supplied per cycle = mQ_s,

where m is the mass of air per cycle which is given by

$$m = \frac{p_1V_1}{RT_1} \text{ where } V_1 = V_s + V_c = \frac{r_c}{r_c - 1}V_s$$

$$\left[\begin{array}{l} r_c = \dfrac{V_s + V_c}{V_c} = 1 + \dfrac{V_s}{V_c} \text{ or } V_c = \dfrac{V_s}{r_c - 1} \\ \therefore\ V_1 = V_s + \dfrac{V_s}{r_c - 1} = V_s\left(1 + \dfrac{1}{r_c - 1}\right) = \dfrac{r_c}{r_c - 1}V_s \end{array}\right]$$

$$= \frac{9}{9-1} \times 0.0196 = 0.02205\ m^3$$

$$\therefore \quad m = \frac{1 \times 10^5 \times 0.02205}{287 \times 303} = 0.02535 \text{ kg/cycle}$$

$\therefore$ Heat supplied per cycle

$= mQ_s = 0.02535[c_v(T_3 - T_2) + c_p(T_4 - T_3)]$

$= 0.02535[0.71(1201.9 - 524.8) + 1.0(2163.4 - 1201.9)]$

$= 36.56$ kJ/cycle

$$\textit{Efficiency} = \frac{\text{Work done per cycle}}{\text{Heat supplied per cycle}} = \frac{21.44}{36.56}$$
$$= \mathbf{0.5864} \text{ or } \mathbf{58.64\%.} \quad \textbf{(Ans.)}$$

(*iv*) Power of the engine, P :

Power of the engine,

P = Work done per second

= Work done per cycle × no. of cycles/sec.

= 21.44 × 8 = **171.52 kW.** **(Ans.)**

21.7. COMPARISON OF OTTO, DIESEL AND DUAL COMBUSTION CYCLES

Following are the *important variable factors which are used as a basis for comparison of the cycles :*

- Compression ratio.
- Maximum pressure
- Heat supplied
- Heat rejected
- Net work

Some of the above mentioned variables are fixed when the performance of Otto, Diesel and dual combustion cycles is to be compared.

21.7.1. Efficiency Versus Compression Ratio

Fig. 21.26 shows the comparison for the air standard efficiencies of the Otto, Diesel and Dual combustion cycles at various compression ratios and with given cut-off ratio for the Diesel and Dual combustion cycles. It is evident from the Fig. 21.26 that the air standard efficiencies *increase with the increase in the compression ratio. For a given compression ratio Otto cycle is the most efficient while the Diesel cycle is the least efficient.* ($\eta_{otto} > \eta_{dual} > \eta_{diesel}$).

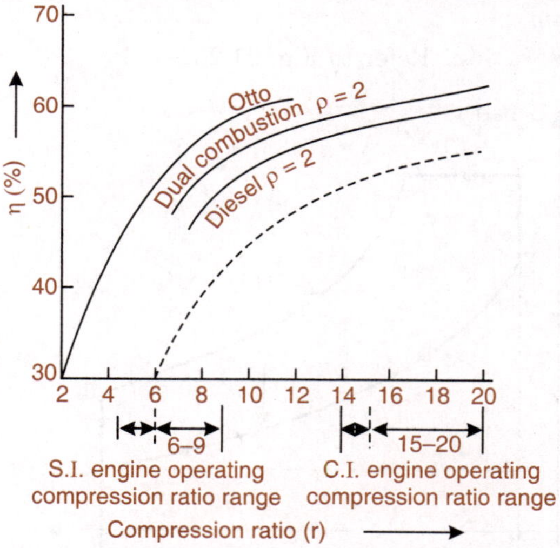

Fig. 21.26. Comparison of efficiency at various compression ratios.

Note. The maximum compression ratio for the petrol engine is limited by detonation. In their respective ratio ranges, the Diesel cycle is more efficient than the Otto cycle.

21.7.2. For the Same Compression Ratio and the Same Heat Input

A comparison of the cycles (Otto, Diesel and Dual) on the *p-v* and *T-s* diagrams for the *same compression ratio and heat supplied* is shown in Fig. 21.27.

We know that, $$\eta = 1 - \frac{\text{Heat rejected}}{\text{Heat supplied}} \quad ...(21.13)$$

Since all the cycles reject their heat at the same specific volume, process line from state 4 to 1, the quantity of *heat rejected from each cycle is represented by the appropriate area under the line 4 to 1 on the T-s diagram.* As is evident from the eqn. (21.13) the cycle which has the least heat rejected will have the highest efficiency. Thus, Otto cycle is the most efficient and Diesel cycle is the least efficient of the three cycles.

i.e., $\eta_{otto} > \eta_{dual} > \eta_{diesel}$.

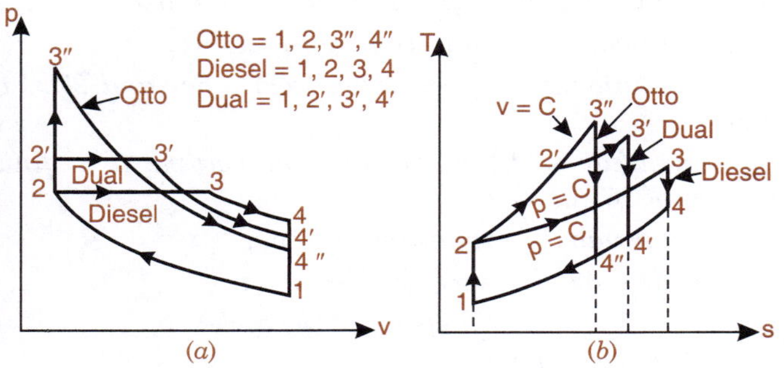

(*a*) (*b*)

Fig. 21.27. (*a*) *p-v* diagram, (*b*) *T-s* diagram.

21.7.3. For Constant Maximum Pressure and Heat Supplied

Fig. 21.28 shows the Otto and Diesel cycles on *p-v* and *T-s* diagrams for constant maximum pressure and heat input respectively.

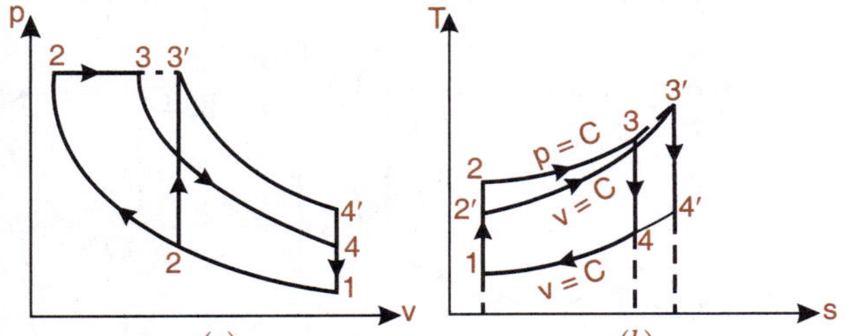

(*a*) (*b*)

Fig. 21.28. (*a*) *p-v* diagram, (*b*) *T-s* diagram.

— For the maximum pressure the points 3 and 3′ must lie on a constant pressure line.

— On *T-s* diagram the heat rejected from the Diesel cycle is represented by the area under the line 4 to 1 and this area is less than the Otto cycle area under the curve 4′ to 1 ; hence the *Diesel cycle is more efficient than the Otto cycle for the condition of maximum pressure and heat supplied.*

Example 21.30. *With the help of p-v and T-s diagram compare the cold air standard otto, diesel and dual combustion cycles for same maximum pressure and maximum temperature.* **(AMIE Summer, 1998)**

Solution. Refer to Fig. 21.29 (*a*), (*b*).

The air-standard Otto, Dual and Diesel cycles are drawn on common *p-v* and *T-s* diagrams for the same maximum pressure and maximum temperature, for the purpose of comparison.

Otto 1–2–3–4–1, Dual 1–2′–3′–3–4–1, Diesel 1–2″–3–4–1 [Fig 21.29 (*a*)].

Slope of constant volume lines on *T-s* diagram is higher than that of constant pressure lines. [Fig. 21.29 (*b*)].

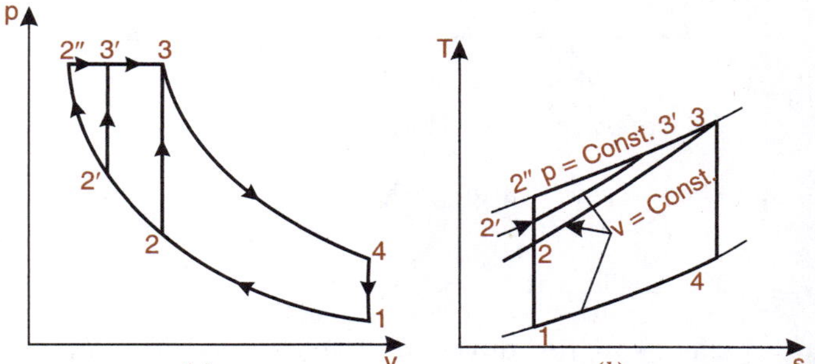

(*a*) (*b*)

Fig. 21.29

Here the otto cycle must be limited to a low compression ratio (r) to fulfil the condition that point 3 (same maximum pressure and temperature) is to be a common state for all the three cycles.

The construction of cycles on *T-s* diagram proves that for the given conditions the heat rejected is same for all the three cycles (area under process line 4–1). Since, by definition,

$$\eta = 1 - \frac{\text{Heat rejected, } Q_r}{\text{Heat supplied, } Q_s} = 1 - \frac{\text{Const.}}{Q_s}$$

the cycle, with greater heat addition will be more efficient. From the *T-s* diagram,

$$Q_{s(diesel)} = \text{Area under } 2''\text{–}3$$
$$Q_{s(dual)} = \text{Area under } 2'\text{–}3'\text{–}3$$
$$Q_{s(otto)} = \text{Area under } 2\text{–}3.$$

It can be seen that, $Q_{s(diesel)} > Q_{s(dual)} > Q_{s(otto)}$

and thus, $\eta_{diesel} > \eta_{dual} > \eta_{otto}$.

21.8. ATKINSON CYCLE

This cycle consists of two adiabatics, a constant volume and a constant pressure process.

p-V diagram of this cycle is shown in Fig. 21.30. It consists of the following *four operations* :

(*i*) 1–2—*Heat rejection at constant pressure*

(*ii*) 2–3—*Adiabatic compression*

(*iii*) 3–4—*Addition of heat at constant volume*

(*iv*) 4–1—*Adiabatic expansion.*

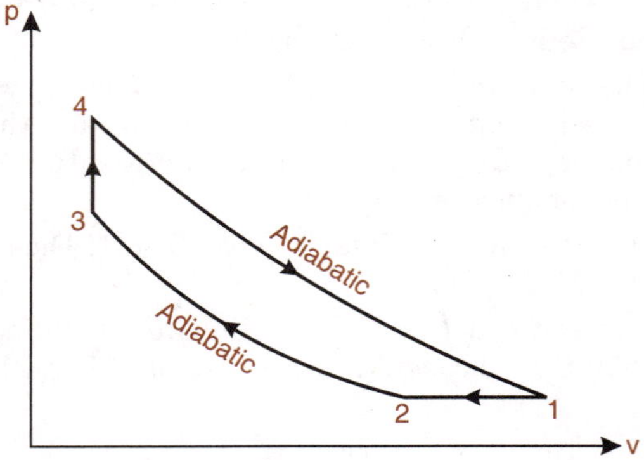

Fig. 21.30

Considering 1 kg of air.

Compression ratio $= \dfrac{v_2}{v_3} = \alpha$

Expansion ratio $= \dfrac{v_1}{v_4} = r$

Heat supplied at constant volume $= c_v(T_4 - T_3)$

Heat rejected $= c_v(T_1 - T_2)$

Work done = Heat supplied – heat rejected

$= c_v(T_4 - T_3) - c_v(T_1 - T_2)$

$$\eta = \frac{\text{Work done}}{\text{Heat supplied}} = \frac{c_v\,(T_4 - T_3) - c_p(T_1 - T_2)}{c_v\,(T_4 - T_3)}$$

$$= 1 - \gamma \,.\, \frac{(T_1 - T_2)}{(T_4 - T_3)} \qquad \text{...}(i)$$

During *adiabatic compression 2–3,*

$$\frac{T_3}{T_2} = \left(\frac{v_2}{v_3}\right)^{\gamma - 1} = (\alpha)^{\gamma - 1}$$

or $$T_3 = T_2\,(\alpha)^{\gamma - 1} \qquad \text{...}(ii)$$

During *constant pressure operation 1–2,*

$$\frac{v_1}{T_1} = \frac{v_2}{T_2}$$

or $$\frac{T_2}{T_1} = \frac{v_2}{v_1} = \frac{\alpha}{r} \qquad \text{...}(iii) \quad \left(\frac{v_2}{v_1} = \frac{v_2}{v_3} \times \frac{v_3}{v_1} = \frac{v_2}{v_3} \times \frac{v_4}{v_1} = \frac{\alpha}{r}\right)$$

During *adiabatic expansion 4–1,*

$$\frac{T_4}{T_1} = \left(\frac{v_1}{v_4}\right)^{\gamma - 1} = (r)^{\gamma - 1}$$

$$T_1 = \frac{T_4}{(r)^{\gamma - 1}} \qquad \text{...}(iv)$$

Putting the value of T_1 in eqn. (*iii*), we get

$$T_2 = \frac{T_4}{(r)^{\gamma - 1}} \cdot \frac{\alpha}{r}$$

$$= \frac{\alpha T_4}{r^{\gamma}} \qquad \text{...}(v)$$

Substituting the value of T_2 in eqn. (*ii*), we get

$$T_3 = \frac{\alpha T_4}{r^{\gamma}}\,(\alpha)^{\gamma - 1} = \left(\frac{\alpha}{r}\right)^{\gamma} . \, T_4$$

Finally putting the values of T_1, T_2 and T_3 in eqn. (*i*), we get

$$\eta = 1 - \gamma \left(\frac{\dfrac{T_4}{r^{\gamma - 1}} - \dfrac{\alpha T_4}{(r)^{\gamma}}}{T_4 - \left(\dfrac{\alpha}{r}\right)^{\gamma} . \, T_4}\right) = 1 - \gamma \left(\frac{r - \alpha}{r^{\gamma} - \alpha^{\gamma}}\right)$$

Hence, air standard efficiency $= 1 - \gamma \,.\, \left(\dfrac{r - \alpha}{r^{\gamma} - \alpha^{\gamma}}\right)$ ...(21.14)

Example 21.31. *A perfect gas undergoes a cycle which consists of the following processes taken in order :*

(*a*) *Heat rejection at constant pressure.*

(*b*) *Adiabatic compression from 1 bar and 27°C to 4 bar.*

(*c*) *Heat addition at constant volume to a final pressure of 16 bar.*

(*d*) *Adiabatic expansion to 1 bar.*

Calculate : (*i*) *Work done/kg of gas.*

(*ii*) *Efficiency of the cycle.*

Take : $c_p = 0.92$, $c_v = 0.75$.

Solution. Refer to Fig. 21.31.

Pressure, $p_2 = p_1 = 1$ bar

Temperature, $T_2 = 27 + 273 = 300$ K

Pressure after adiabatic compression, $p_3 = 4$ bar

Final pressure after heat addition, $p_4 = 16$ bar

For *adiabatic compression 2–3,*

$$\frac{T_3}{T_2} = \left(\frac{p_3}{p_2}\right)^{\frac{\gamma - 1}{\gamma}} = \left(\frac{4}{1}\right)^{\frac{1.22 - 1}{1.22}} = 1.284$$

$$\left[\gamma = \frac{c_p}{c_v} = \frac{0.92}{0.75} = 1.22\right]$$

$\therefore \quad T_3 = T_2 \times 1.284$

$= 300 \times 1.284 = 385.2$ K or 112.2°C

For *constant volume process 3–4,*

$$\frac{p_4}{T_4} = \frac{p_3}{T_3}$$

$$T_4 = \frac{p_4 T_3}{p_3} = \frac{16 \times 385.2}{4} = 1540.8 \text{ K or } 1267.8°\text{C}$$

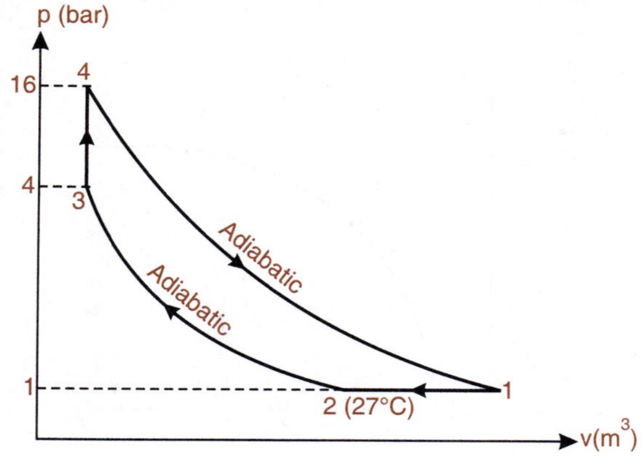

Fig. 21.31

For *adiabatic expansion process 4–1,*

$$\frac{T_4}{T_1} = \left(\frac{p_4}{p_1}\right)^{\frac{\gamma-1}{\gamma}} = \left(\frac{16}{1}\right)^{\frac{1.22-1}{1.22}} = 1.648$$

or $$T_1 = \frac{T_4}{1.648} = \frac{1540.8}{1.648} = 934.9 \text{ K or } 661.9°\text{C}.$$

(*i*) **Work done per kg of gas, W :**

Heat supplied $= c_v (T_4 - T_3)$

$= 0.75 (1540.8 - 385.2) = 866.7$ kJ/kg

Heat rejected $= c_p (T_1 - T_2) = 0.92(934.9 - 300)$

$= 584.1$ kJ/kg

Work done / kg of gas,

W = Heat supplied – heat rejected

$= 866.7 - 584.1 = 282.6$ kJ/kg

= 282600 Nm/kg. **(Ans.)**

(*ii*) **Efficiency of the cycle :**

Efficiency, $$\eta = \frac{\text{Work done}}{\text{Heat supplied}} = \frac{282.6}{866.7}$$

= 0.326 or **32.6%.** **(Ans.)**

21.9. ERICSSON CYCLE

It is so named as it was invented by Ericsson. Fig. 21.32 shows *p-v* diagram of this cycle.

It comprises of the following *operations* :

(*i*) 1–2—*Rejection of heat at constant pressure*

(*ii*) 2–3—*Isothermal compression*

(*iii*) 3–4—*Addition of heat at constant pressure*

(*iv*) 4–1—*Isothermal expansion.*

Considering 1 kg of air.

Volume ratio, $r = \dfrac{v_2}{v_3} = \dfrac{v_1}{v_4}$

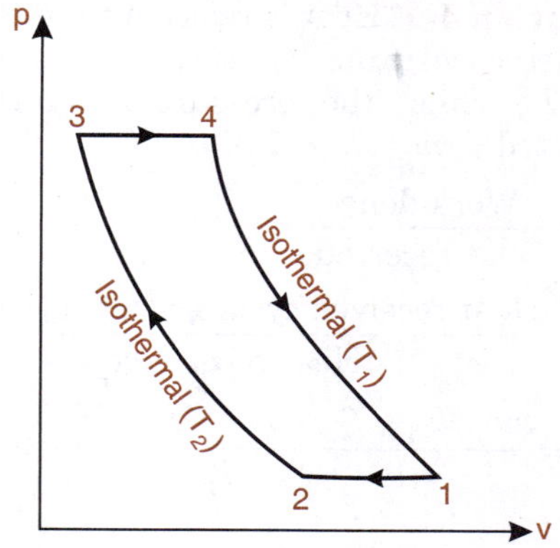

Fig. 21.32

Heat supplied to air from an external source

= Heat supplied during the isothermal expansion 4–1

$= RT_1 \log_e r$

Heat rejected by air to an external source

$= RT_2 \, . \log_e r$

Work done = Heat supplied – heat rejected

$= RT_1 \, . \log_e r - RT_2 \, . \log_e r = R \log_e r \,(T_1 - T_2)$

$$\eta = \frac{\text{Work done}}{\text{Heat supplied}} = \frac{R \log_e r \,(T_1 - T_2)}{RT_1 \, . \log_e r}$$

$$= \frac{T_1 - T_2}{T_1} \quad ...(21.15)$$

which is the same as Carnot cycle.

Note. For **'Stirling cycle', Miller cycle and Lenoir cycle** please refer the Author's popular book on "I.C. Engines".

21.10. BRAYTON CYCLE

Brayton cycle is *a constant pressure cycle for a perfect gas*. It is also called **Joule cycle.** The heat transfers are achieved in reversible constant pressure heat exchangers. An ideal gas turbine plant would perform the processes that make up a Brayton cycle. The cycle is shown in the Fig. 21.33 (*a*) and it is represented on *p-v* and *T-s* diagrams as shown in Fig. 21.33 (*b, c*).

The various operations are as follows :

Operation 1–2. The air is compressed isentropically from the lower pressure p_1 to the upper pressure p_2, the temperature rising from T_1 to T_2. No heat flow occurs.

Operation 2–3. Heat flows into the system increasing the volume from V_2 to V_3 and temperature from T_2 to T_3 whilst the pressure remains constant at p_2. Heat received $= mc_p (T_3 - T_2)$.

Operation 3–4. The air is expanded isentropically from p_2 to p_1, the temperature falling from T_3 to T_4. No heat flow occurs.

Operation 4–1. Heat is rejected from the system as the volume decreases from V_4 to V_1 and the temperature from T_4 to T_1 whilst the pressure remains constant at p_1. Heat rejected $= mc_p\,(T_4 - T_1)$.

$$\eta_{\text{air-standard}} = \frac{\text{Work done}}{\text{Heat received}}$$

$$= \frac{\text{Heat received/cycle} - \text{Heat rejected/cycle}}{\text{Heat received/cycle}}$$

$$= \frac{mc_p\,(T_3 - T_2) - mc_p\,(T_4 - T_1)}{mc_p\,(T_3 - T_2)} = 1 - \frac{T_4 - T_1}{T_3 - T_2}$$

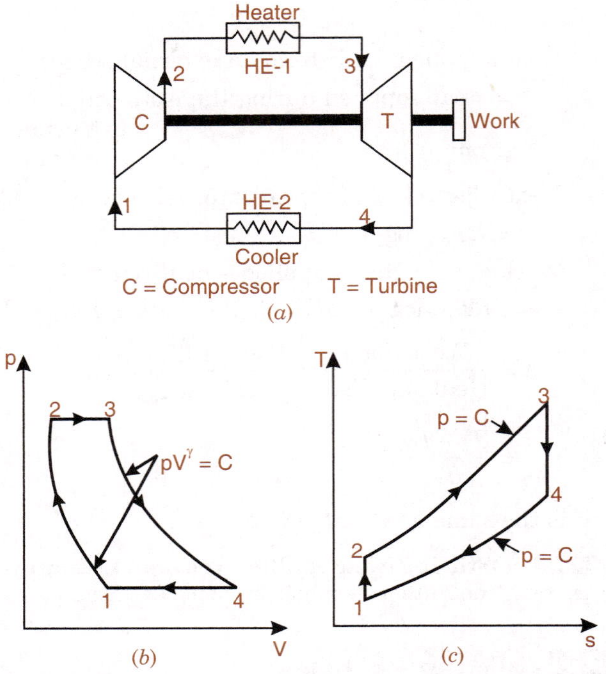

Fig. 21.33. Brayton cycle : (*a*) Basic components of a gas turbine power plant (*b*) *p-V* diagram (*c*) *T-s* diagram.

Now, from isentropic expansion,

$$\frac{T_2}{T_1} = \left(\frac{p_2}{p_1}\right)^{\frac{\gamma-1}{\gamma}}$$

$$T_2 = T_1\,(r_p)^{\frac{\gamma-1}{\gamma}}, \text{ where } r_p = \text{pressure ratio.}$$

Similarly $\dfrac{T_3}{T_4} = \left(\dfrac{p_2}{p_1}\right)^{\frac{\gamma-1}{\gamma}}$ or $T_3 = T_4\,(r_p)^{\frac{\gamma-1}{\gamma}}$

$$\therefore \quad \eta_{\text{air-standard}} = 1 - \frac{T_4 - T_1}{T_4(r_p)^{\frac{\gamma-1}{\gamma}} - T_1(r_p)^{\frac{\gamma-1}{\gamma}}}$$

$$= 1 - \frac{1}{(r_p)^{\frac{\gamma-1}{\gamma}}} \qquad \text{...(21.16)}$$

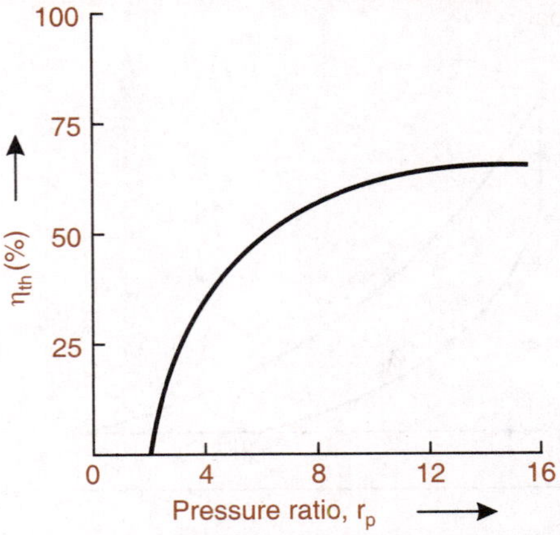

Fig. 21.34. Effect of pressure ratio on the efficiency of Brayton cycle.

The eqn. (21.16) shows that the *efficiency of the ideal joule cycle increases with the pressure ratio. The absolute limit of upper pressure is determined by the limiting temperature of the material of the turbine at the point at which this temperature is reached by the compression process alone, no further heating of the gas in the combustion chamber would be permissible and the work of expansion would ideally just balance the work of compression so that no excess work would be available for external use.*

Pressure Ratio for maximum work :

Now we shall prove that the *pressure ratio for maximum work is a function of the limiting temperature ratio.*

Work output during the cycle

$$= \text{Heat received/cycle} - \text{heat rejected/cycle}$$

$$= mc_p\,(T_3 - T_2) - mc_p\,(T_4 - T_1)$$

$$= mc_p\,(T_3 - T_4) - mc_p\,(T_2 - T_1)$$

$$= mc_p\,T_3\left(1 - \frac{T_4}{T_3}\right) - T_1\left(\frac{T_2}{T_1} - 1\right)$$

In case of a given turbine the minimum temperature T_1 and the maximum temperature T_3 are prescribed, T_1 being the temperature of the atmosphere and T_3 the maximum temperature which the metals of turbine would withstand. Consider the specific heat at constant pressure c_p to be constant. Then,

Since, $\dfrac{T_3}{T_4} = (r_p)^{\frac{\gamma-1}{\gamma}} = \dfrac{T_2}{T_1}$

Using the constant

'z' $= \dfrac{\gamma-1}{\gamma}$, we have, work output/cycle

$$W = K\left[T_3\left(1 - \frac{1}{r_p^{\,z}}\right) - T_1\,(r_p^{\,z} - 1)\right]$$

Differentiating with respect to r_p

$$\frac{dW}{dr_p} = K\left[T_3 \times \frac{z}{r_p(z+1)} - T_1\, z r_p^{(z-1)}\right]$$

$= 0$ for a maximum

$$\therefore \quad \frac{zT_3}{r_p^{(z+1)}} = T_1 z (r_p)^{(z-1)}$$

$$\therefore \quad r_p^{2z} = \frac{T_3}{T_1}$$

$$\therefore \quad r_p = (T_3/T_1)^{1/2z} \quad i.e., \quad r_p = (T_3/T_1)^{\frac{\gamma}{2(\gamma-1)}} \quad ...(21.17)$$

Thus, the *pressure ratio for maximum work is a function of the limiting temperature ratio.*

Work Ratio

Work ratio is defined as the *ratio of net work output to the work done by the turbine.*

$$\therefore \quad \text{Work ratio} = \frac{W_T - W_C}{W_T}$$

[where, W_T = Work obtained from this turbine,
and W_C = Work supplied to the compressor.]

$$= \frac{mc_p(T_3 - T_4) - mc_p(T_2 - T_1)}{mc_p(T_3 - T_4)}$$

$$= 1 - \frac{T_2 - T_1}{T_3 - T_4}$$

$$= 1 - \frac{T_1}{T_3}\left[\frac{(r_p)^{\frac{\gamma-1}{\gamma}} - 1}{1 - \dfrac{1}{(r_p)^{\frac{\gamma-1}{\gamma}}}}\right] = 1 - \frac{T_1}{T_3}(r_p)^{\frac{\gamma-1}{\gamma}} \quad ...(21.18)$$

Example 21.32. *Air enters the compressor of a gas turbine plant operating on Brayton cycle at 101.325 kPa, 27°C. The pressure ratio in the cycle is 6. Calculate the maximum temperature in the cycle and the cycle efficiency. Assume $W_T = 2.5\ W_C$, where W_T and W_C are the turbine and the compressor work respectively. Take $\gamma = 1.4$.* **(P.U.)**

Solution. Pressure of intake air, $p_1 = 101.325$ kPa

Temperature of intake air, $T_1 = 27 + 273 = 300$ K

The pressure ratio in the cycle, $r_p = 6$

(*i*) **Maximum temperature in the cycle, T_3 :**

Refer to Fig. 21.35.

$$\frac{T_2}{T_1} = \left(\frac{p_2}{p_1}\right)^{\frac{\gamma-1}{\gamma}} = (r_p)^{\frac{\gamma-1}{\gamma}} = (6)^{\frac{1.4-1}{1.4}} = 1.668$$

$$\therefore \quad T_2 = 1.668\ T_1 = 1.668 \times 300 = 500.4 \text{ K}$$

$$\text{Also,} \quad \frac{T_3}{T_4} = (r_p)^{\frac{\gamma-1}{\gamma}} = (6)^{\frac{1.4-1}{1.4}} = 1.668$$

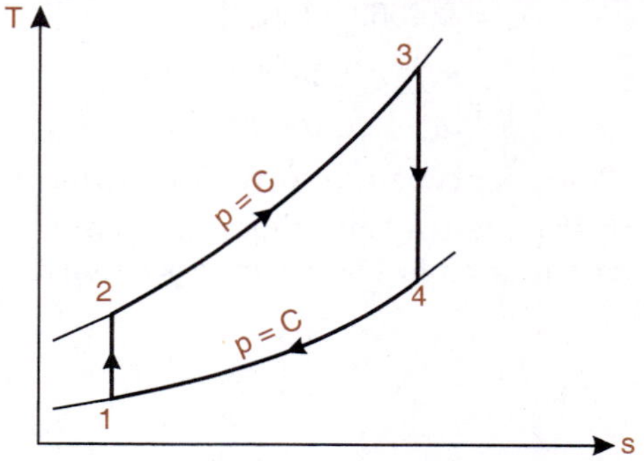

Fig. 21.35

$$\therefore \quad T_4 = \frac{T_3}{1.668}$$

But $\quad W_T = 2.5\ W_C$ (given)

$$\therefore \quad mc_p(T_3 - T_4) = 2.5\ mc_p(T_2 - T_1)$$

or $\quad T_3 - \dfrac{T_3}{1.668} = 2.5\,(500.4 - 300) = 501$

or $\quad T_3\left(1 - \dfrac{1}{1.668}\right) = 501$

$$\therefore \quad T_3 = \frac{501}{\left(1 - \dfrac{1}{1.668}\right)}$$

$= \mathbf{1251\ K}$ or $\mathbf{978°C.}$ **(Ans.)**

(*ii*) **Cycle efficiency, η_{cycle} :**

Now, $\quad T_4 = \dfrac{T_3}{1.668} = \dfrac{1251}{1.668} = 750$ K

$$\therefore \quad \eta_{cycle} = \frac{\text{Net work}}{\text{Heat added}}$$

$$= \frac{mc_p(T_3 - T_4) - mc_p(T_2 - T_1)}{mc_p(T_3 - T_2)}$$

$$= \frac{(1251 - 750) - (500.4 - 300)}{(1251 - 500.4)}$$

$= \mathbf{0.4}$ or $\mathbf{40\%.}$ **(Ans.)**

[***Check*** ; $\eta_{cycle} = 1 - \dfrac{1}{(r_p)^{\left(\frac{\gamma-1}{\gamma}\right)}} = 1 - \dfrac{1}{(6)^{\frac{1.4-1}{1.4}}}$ **(Ans.)**
$= \mathbf{0.4}$ or $\mathbf{40\%.}$]

Example 21.33. *A gas turbine is supplied with gas at 5 bar and 1000 K and expands it adiabatically to 1 bar. The mean specific heat at constant pressure and constant volume are 1.0425 kJ/kg K and 0.7662 kJ/kg K respectively.*

(i) Draw the temperature-entropy diagram to represent the processes of the simple gas turbine system.

(ii) Calculate the power developed in kW per kg of gas per second and the exhaust gas temperature. **(GATE)**

Solution. *Given* : $p_1 = 1$ bar ; $p_2 = 5$ bar ; $T_3 = 1000$ K ;
$c_p = 1.0425$ kJ/kg K ;

$c_v = 0.7662$ kJ/kg K

$$\gamma = \frac{c_p}{c_v} = \frac{1.0425}{0.7662} = 1.36$$

(*i*) Temperature-entropy (T-s) diagram :

Temperature-entropy diagram representing the processes of the simple gas turbine system is shown in Fig. 21.36.

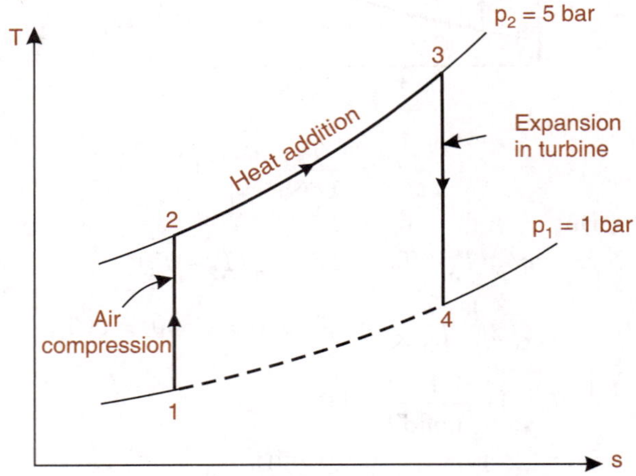

Fig. 21.36

(*ii*) Power required :

$$\frac{T_4}{T_3} = \left(\frac{p_1}{p_2}\right)^{\frac{\gamma-1}{\gamma}} = \left(\frac{1}{5}\right)^{\frac{1.36-1}{1.36}} = 0.653$$

$\therefore \quad T_4 = 1000 \times 0.653 = 653$ K

Power developed per kg of gas per second

$= c_p (T_3 - T_4)$

$= 1.0425 (1000 - 653) =$ **361.7 kW.** **(Ans.)**

Example 21.34. *An isentropic air turbine is used to supply 0.1 kg/s of air at 0.1 MN/m² and at 285 K to a cabin. The pressure at inlet to the turbine is 0.4 MN/m². Determine the temperature at turbine inlet and the power developed by the turbine. Assume c_p = 1.0 kJ/kg K.* **(GATE)**

Solution. *Given* : $\dot{m}_a$ = 0.1 kg/s ; p_1 = 0.1 MN/m² = 1 bar, T_4 = 285 K ; p_2 = 0.4 MN/m² = 4 bar ; c_p = 1.0 kJ/kg K.

Temperature at turbine inlet, T_3 :

$$\frac{T_3}{T_4} = \left(\frac{p_2}{p_1}\right)^{\frac{\gamma-1}{\gamma}} = \left(\frac{4}{1}\right)^{\frac{1.4-1}{1.4}} = 1.486$$

$\therefore$ **T_3 = 285 × 1.486 = 423.5 K.** **(Ans.)**

Power developed, P :

$P = \dot{m}_a c_p (T_3 - T_4)$

$= 0.1 \times 1.0 (423.5 - 285)$

= 13.85 kW. **(Ans.)**

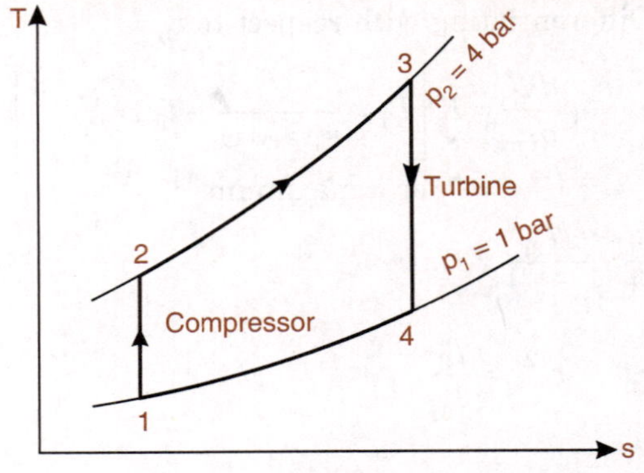

Fig. 21.37

Example 21.35. *Consider an air standard cycle in which the air enters the compressor at 1.0 bar and 20°C. The pressure of air leaving the compressor is 3.5 bar and the temperature at turbine inlet is 600°C. Determine per kg of air :*

(*i*) *Efficiency of the cycle,* (*ii*) *Heat supplied to air,*

(*iii*) *Work available at the shaft,*

(*iv*) *Heat rejected in the cooler, and*

(*v*) *Temperature of air leaving the turbine.*

For air γ = 1.4 and c_p = 1.005 kJ/kg K.

Solution. Refer to Fig. 21.35.

Pressure of air entering the compressor,

$p_1 = 1.0$ bar

Temperature at the inlet of compressor,

$T_1 = 20 + 273 = 293$ K

Pressure of air leaving the compressor,

$p_2 = 3.5$ bar

Temperature of air at turbine inlet,

$T_3 = 600 + 273 = 873$ K

(*i*) Efficiency of the cycle, η_{cycle} :

$$\eta_{cycle} = 1 - \frac{1}{(r_p)^{\frac{\gamma-1}{\gamma}}}$$

$$= 1 - \frac{1}{(3.5)^{\frac{1.4-1}{1.4}}} = \textbf{0.30 or 30\%.} \quad \textbf{(Ans.)}$$

$$\left(\because \; r_p = \frac{p_2}{p_1} = \frac{3.5}{1.0} = 3.5\right)$$

(*ii*) Heat supplied to air :

For compression process 1–2, we have

$$\frac{T_2}{T_1} = \left(\frac{p_2}{p_1}\right)^{\frac{\gamma-1}{\gamma}} = \left(\frac{3.5}{1}\right)^{\frac{1.4-1}{1.4}} = 1.43$$

$\therefore \quad T_2 = T_1 \times 1.43 = 293 \times 1.43 \simeq 419$ K

$\therefore$ Heat supplied to air,

$$Q_1 = c_p (T_3 - T_2) = 1.005\ (873 - 419)$$
$$= \mathbf{456.27\ kJ/kg.} \quad \textbf{(Ans.)}$$

(*iii*) **Work available at the shaft, W :**

We know that, $\eta_{cycle} = \dfrac{\text{Work output } (W)}{\text{Heat input } (Q_1)}$

or $\quad 0.30 = \dfrac{W}{456.27}$ or $W = 0.3 \times 456.27$

$$= 136.88 \text{ kJ/kg}$$

(*iv*) **Heat rejected in the cooler, Q_2 :**

Work output (W)

= Heat supplied (Q_1) – heat rejected (Q_2)

$\therefore \quad Q_2 = Q_1 - W = 456.27 - 136.88$

$= \mathbf{319.39\ kJ/kg.}$ **(Ans.)**

(*v*) **Temperature of air leaving the turbine, T_4 :**

For *expansion (isentropic) process 3–4*, we have

$$\frac{T_3}{T_4} = (r_p)^{\frac{\gamma-1}{\gamma}} = (3.5)^{\frac{1.4-1}{1.4}} = 1.43$$

$$\therefore \quad T_4 = \frac{T_3}{1.43} = \frac{873}{1.43} = \mathbf{610.5\ K.} \quad \textbf{(Ans.)}$$

[**Check** : Heat rejected in the air cooler at constant pressure during the process 4–1 can also be calculated as : Heat rejected = $m \times c_p (T_4 - T_1) = 1 \times 1.005 \times (610.5 - 293)$ = 319.1 kJ/kg]

Example 21.36. *Air enters the compressor of a gas turbine plant operating on Brayton cycle at 1 bar, 27°C. The pressure ratio in the cycle is 6. If $W_T = 2.5\ W_C$, where W_T and W_C are the turbine and compressor work respectively, calculate the maximum temperature and the cycle efficiency.* **(GATE)**

Solution. *Given* : $p_1 = 1$ bar ; $T_1 = 27 + 273 = 300$ K ;

$$\frac{p_2}{p_1} = 6 \ ; \ W_T = 2.5\ W_C$$

Maximum temperature, T_3 :

Now, $\quad \dfrac{T_2}{T_1} = \left(\dfrac{p_2}{p_1}\right)^{\frac{\gamma-1}{\gamma}} = (6)^{\frac{1.4-1}{1.4}} = (6)^{\frac{0.4}{1.4}} = 1.668$

$\therefore \quad T_2 = 300 \times 1.668 = 500.4$ K

Also, $\quad \dfrac{T_3}{T_4} = \left(\dfrac{p_2}{p_1}\right)^{\frac{\gamma-1}{\gamma}} = (6)^{\frac{0.4}{1.4}} = 1.668$

or $\quad T_3 = 1.668\ T_4$

Now, compressor work

$$W_C = mc_p (T_2 - T_1),$$

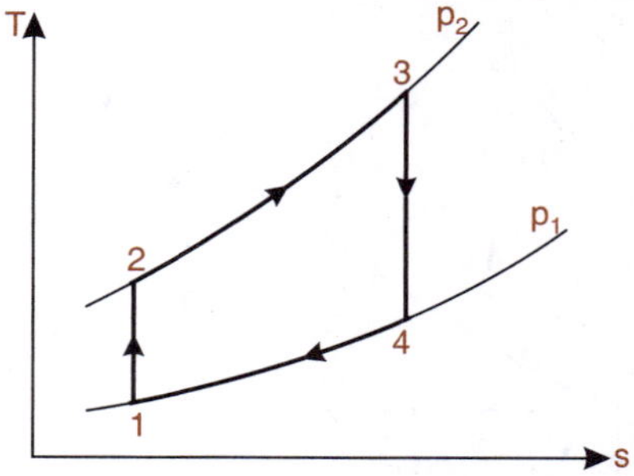

Fig. 21.38

and $\quad$ turbine work,

$$W_T = mc_p (T_3 - T_4)$$

Since $\quad W_T = 2.5\ W_C \quad$ ......(Given)

$\therefore \quad mc_p(T_3 - T_4) = 2.5 \times mc_p(T_2 - T_1)$

$$\left(T_3 - \frac{T_3}{1.668}\right) = 2.5\ (500.4 - 300)$$

or $\quad T_3 \left(1 - \dfrac{1}{1.668}\right) = 501$

$\therefore \quad \mathbf{T_3 = 1251\ K.}$ **(Ans.)**

Cycle efficiency, η_{cycle} :

$$\eta_{cycle} = \frac{W_T - W_C}{mc_p(T_3 - T_2)} = \frac{mc_p(T_3 - T_4) - mc_p(T_2 - T_1)}{mc_p(T_3 - T_2)}$$

$$= \frac{\left(1251 - \dfrac{1251}{1.668}\right) - (500.4 - 300)}{(1251 - 500.4)}$$

$$= \frac{501 - 200.4}{750.6} = \mathbf{0.40 \text{ or } 40\%.} \quad \textbf{(Ans.)}$$

Example 21.37. *A closed cycle ideal gas turbine plant operates between temperature limits of 800°C and 30°C and produces a power of 100 kW. The plant is designed such that there is no need for a regenerator. A fuel of calorific 45000 kJ/kg is used. Calculate the mass flow rate of air through the plant and rate of fuel consumption.* **(GATE)**

Assume $c_p = 1$ kJ/kg K and $\gamma = 1.4$.

Solution. *Given* : $T_1 = 30 + 273 = 303$ K ;

$T_3 = 800 + 273 = 1073$ K ;

$C = 45000$ kJ/kg ; $c_p = 1$ kJ/kg K ;

$\gamma = 1.4$; $W_{turbine} - W_{compressor} = 100$ kW.

$\dot{m}_a$, $\dot{m}_f$:

Since no regenerator is used we can assume the turbine expands the gases upto T_4 in such a way that the exhaust gas temperature from the turbine is equal to the temperature of air coming out of the compressor *i.e.*, $T_2 = T_4$

$$\frac{p_2}{p_1} = \frac{p_3}{p_4}, \ \frac{p_2}{p_1} = \left(\frac{T_2}{T_1}\right)^{\frac{\gamma}{\gamma-1}} \text{ and } \frac{p_3}{p_4} = \left(\frac{T_3}{T_4}\right)^{\frac{\gamma}{\gamma-1}}$$

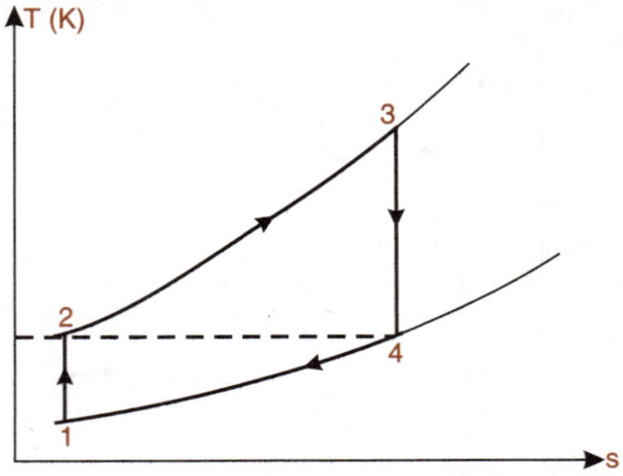

Fig. 21.39

$$\therefore \quad \frac{T_2}{T_1} = \frac{T_3}{T_4} = \frac{T_3}{T_2} \quad (\because \; T_2 = T_4 \text{ ......assumed})$$

or $$T_2^{\,2} = T_1 T_3 \quad \text{or} \quad T_2 = \sqrt{T_1 T_3}$$

or $$T_2 = \sqrt{303 \times 1073} = 570.2 \text{ K}$$

Now, $$W_{turbine} - W_{compressor} = \dot{m}_f \times C \times \eta$$

or $$100 = \dot{m}_f \times 45000 \times \left[1 - \frac{T_4 - T_1}{T_3 - T_2}\right]$$

$$= \dot{m}_f \times 45000 \left[1 - \frac{570.2 - 303}{1073 - 570.2}\right]$$

$$= \dot{m}_f \times 21085.9$$

or $$\dot{\mathbf{m}}_\mathbf{f} = \frac{100}{21085.9} = \mathbf{4.74 \times 10^{-3}\ kg/s.} \quad \textbf{(Ans.)}$$

Again, $$W_{turbine} - W_{compressor} = 100 \text{ kW}$$

$$(\dot{m}_a + \dot{m}_f)(T_3 - T_4) - \dot{m}_a \times 1 \times (T_2 - T_1) = 100$$

or $$(\dot{m}_a + 0.00474)(1073 - 570.2) - \dot{m}_a (570.2 - 303) = 100$$

or $$(\dot{m}_a + 0.00474) \times 502.8 - 267.2\, \dot{m}_a = 100$$

or $$502.8\, \dot{m}_a + 2.383 - 267.2\, \dot{m}_a = 100$$

or $$235.6\, \dot{m}_a = 97.617$$

$$\therefore \quad \dot{\mathbf{m}}_\mathbf{a} = \mathbf{0.414\ kg/s.} \quad \textbf{(Ans.)}$$

Example 21.38. *In a gas turbine plant working on Brayton cycle, the air at inlet is 27°C, 0.1 MP$_a$. The pressure ratio is 6.25 and the maximum temperature is 800°C. The turbine and compressor efficiencies are each 80%. Find compressor work, turbine work, heat supplied, cycle efficiency and turbine exhaust temperature. Mass of air may be considered as 1 kg. Draw T-s diagram.* **(N.U.)**

Solution. Refer to Fig. 21.40.

Given : $T_1 = 27 + 273 = 300$ K ;

$p_1 = 0.1$ MP$_a$; $r_p = 6.25$,

$T_3 = 800 + 273 = 1073$ K ;

$\eta_{comp.} = \eta_{turbine} = 0.8$.

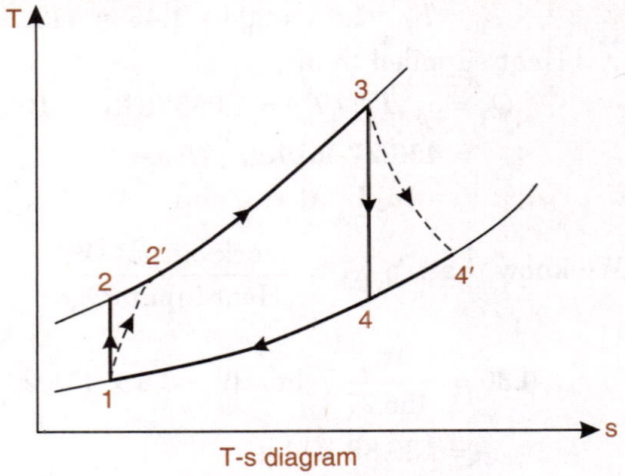

Fig. 21.40

For the *compression process 1–2,* we have

$$\frac{T_2}{T_1} = \left(\frac{p_2}{p_1}\right)^{\frac{\gamma-1}{\gamma}} = (r_p)^{\frac{\gamma-1}{\gamma}} = (6.25)^{\frac{1.4-1}{1.4}} = 1.688$$

or $$T_2 = 300 \times 1.688 = 506.4 \text{ K}$$

Also, $$\eta_{comp.} = \frac{T_2 - T_1}{T_2' - T_1}$$

or $$0.8 = \frac{506.4 - 300}{T_2' - 300}$$

or $$T_2' = \frac{506.4 - 300}{0.8} + 300 = 558 \text{ K}$$

∴ **Compressor work,**

$$W_{comp.} = 1 \times c_p \times (T_2' - T_1)$$
$$= 1 \times 1.005\,(558 - 300)$$
$$= \mathbf{259.29\ kJ/kg.} \quad \textbf{(Ans.)}$$

For *expansion process 3–4,* we have

$$\frac{T_3}{T_4} = \left(\frac{p_3}{p_4}\right)^{\frac{\gamma-1}{\gamma}} = (r_p)^{\frac{\gamma-1}{\gamma}} = (6.25)^{\frac{1.4-1}{1.4}} = 1.688$$

or $$T_4 = \frac{T_3}{1.688} = \frac{1073}{1.688} = 635.66 \text{ K}$$

Also, $$\eta_{turbine} = \frac{T_3 - T_4'}{T_3 - T_4}$$

or $$0.8 = \frac{1073 - T_4'}{1073 - 635.66}$$

or $$T_4' = 1073 - 0.8\,(1073 - 635.66) = 723.13 \text{ K}$$

∴ **Turbine work,**

$$W_{turbine} = 1 \times c_p \times (T_3 - T_4') \text{ (neglecting fuel mass)}$$
$$= 1 \times 1.005\,(1073 - 723.13)$$
$$= \mathbf{351.6\ kJ/kg.} \quad \textbf{(Ans.)}$$

Net work output,

$$W_{net} = W_{turbine} - W_{comp.}$$
$$= 351.6 - 259.29 = 92.31 \text{ kJ/kg}$$

Heat supplied,

$$Q_s = 1 \times c_p \times (T_3 - T_2')$$
$$= 1 \times 1.005 \times (1073 - 558)$$
$$= \mathbf{517.57\ kJ/kg.} \quad \textbf{(Ans.)}$$

Cycle efficiency,

$$\eta_{cycle} = \frac{W_{net}}{Q_s} = \frac{92.31}{517.57}$$
$$= \mathbf{0.1783} \text{ or } \mathbf{17.83\%}. \quad \textbf{(Ans.)}$$

Turbine exhaust temperature, $T_4' =$ **723.13 K** or **450.13°C. (Ans.)**

The *T-s* diagram is shown in Fig. 21.40.

Example 21.39. *Find the required air-fuel ratio in a gas turbine whose turbine and compressor efficiencies are 85% and 80%, respectively. Maximum cycle temperature is 875°C. The working fluid can be taken as air (c_p = 1.0 kJ/kg K, γ = 1.4) which enters the compressor at 1 bar and 27°C. The pressure ratio is 4. The fuel used has calorific value of 42000 kJ/kg. There is a loss of 10% of calorific value in the combustion chamber.* **(GATE)**

Solution. *Given :* $\eta_{turbine}$ = 85% ; $\eta_{compressor}$ = 80% ; T_3 = 273 + 875 = 1148 K ; T_1 = 27 + 273 = 300 K ; c_p = 1.0 kJ/kg K ; γ = 1.4 ; p_1 = 1 bar; p_2 = 4 bar (since pressure ratio is 4) ; C = 42000 kJ/kg K, η_{cc} = 90% (since loss in the combustion chamber is 10%)

For isentropic compression 1–2, we have

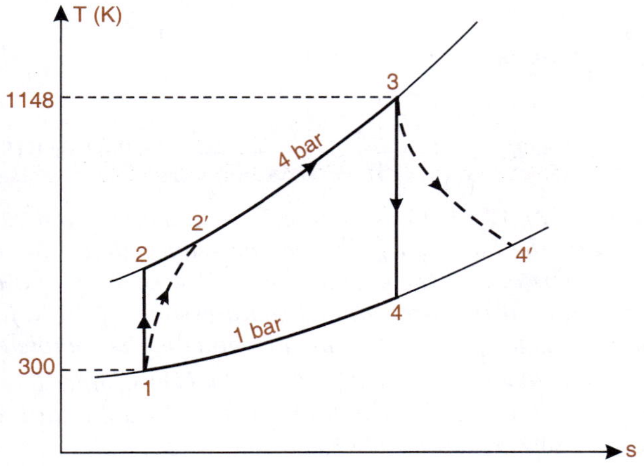

Fig. 21.41

$$\frac{T_2}{T_1} = \left(\frac{p_2}{p_1}\right)^{\frac{\gamma-1}{\gamma}} = (4)^{\frac{1.4-1}{1.4}} = 1.486$$

$$\therefore \quad T_2 = 300 \times 1.486 = 445.8 \text{ K}$$

$$\eta_{compressor} = \frac{T_2 - T_1}{T_2' - T_1}$$

or
$$0.8 = \frac{445.8 - 300}{T_2' - 300}$$

or
$$T_2' = \frac{445.8 - 300}{0.8} + 300 = 482.2 \text{ K}$$

Now, heat supplied by the fuel = heat taken by the burning gases

$$0.9 \times m_f \times C = (m_a + m_f) \times c_p \times (T_3 - T_2')$$

$$\therefore \quad C = \left(\frac{m_a + m_f}{m_f}\right) \times \frac{c_p(T_3 - T_2')}{0.9}$$

$$= \left(\frac{m_a}{m_f} + 1\right) \times \frac{c_p(T_3 - T_2')}{0.9}$$

or
$$42000 = \left(\frac{m_a}{m_f} + 1\right) \times \frac{1.00(1148 - 482.27)}{0.9}$$

$$= 739.78 \left(\frac{m_a}{m_f} + 1\right)$$

$$\therefore \quad \frac{m_a}{m_f} = \frac{42000}{739.78} - 1 = 55.77 \text{ say } 56$$

∴ **A/F ratio = 56 : 1. (Ans.)**

ADDITIONAL/TYPICAL EXAMPLES

Example 21.40. *In an engine working on the Otto cycle between given lower and upper limits of absolute temperature, T_1 and T_2 respectively, show that for maximum work to be done per kg, the ratio of compression is given by :*

$$r = \left(\frac{T_3}{T_1}\right)^{1.25}$$

where, γ = Ratio of specific heats = 1.4

Solution. Fig. 21.42, shows the cycle for the Otto cycle.

$$T_2 = T_1 \times (r)^{\gamma-1} ; \; T_4 = T_3 \times \left(\frac{1}{r}\right)^{\gamma-1}$$

Work done, $W = c_v(T_3 - T_2) - c_v(T_4 - T_1)$

$$= c_v(T_3 - T_1 \times r^{\gamma-1}) - c_v \left\{T_3\left(\frac{1}{r}\right)^{\gamma-1} - T_1\right\}$$

For maximum work, *W* is differentiated with the variable *r* and equated to 0.

i.e.,
$$\frac{dW}{dr} = [0 - (\gamma - 1)T_1(r)^{\gamma-2}]$$
$$- [- T_3 \times (\gamma - 1)r^{-\gamma} - 0] = 0$$

or,
$$(\gamma - 1)\, T_1\, (r)^{\gamma-2} = (\gamma - 1)\, T_3\, r^{-\gamma}$$

or,
$$\frac{T_3}{T_1} = \frac{(r)^{\gamma-2}}{(r)^{-\gamma}} = (r)^{2(\gamma-1)}$$

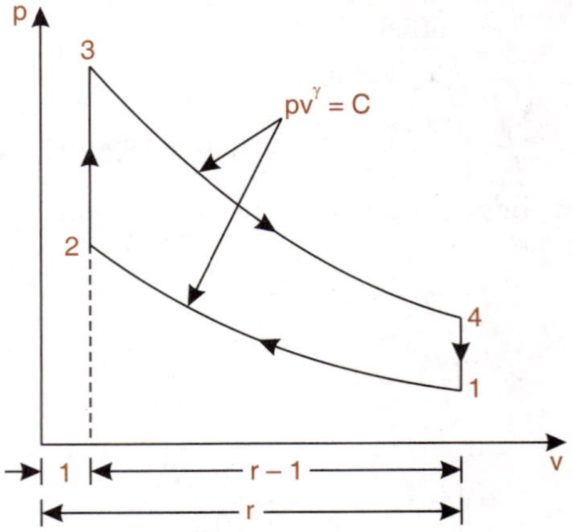

Fig. 21.42

$$\therefore \quad \mathbf{r} = \left(\frac{T_3}{T_1}\right)^{\frac{1}{2(\gamma-1)}} = \left(\frac{T_3}{T_1}\right)^{\frac{1}{2(1.4-1)}} = \left(\frac{T_3}{T_1}\right)^{1.25}$$

...**Proved.**

Example 21.41. *A diesel engine contains 0.1 m³ of air at 0.98 bar and 30°C at the beginning of compression. The compression ratio is 15 and the volume at cut-off is 0.0125 m³. Determine for the corresponding air standard cycle :*

(i) The cut-off ratio ;

(ii) The per cent clearance ;

(iii) The work done ;

(iv) The air standard efficiency.

Take c_p = 1.005 kJ/kg K and γ = 1.4. **(GATE)**

Solution. Fig. 21.43 shows the diesel cycle on *p-v* diagram.

Given : $V_1 = 0.1$ m³ ; $T_1 = 30 + 273 = 303$ K ; $p_1 = 0.98$ bar.

Compression ratio, $\quad r = \dfrac{V_1}{V_2} = 15$

$\therefore$ Clearance volume, $V_2 = \dfrac{V_1}{15} = \dfrac{0.1}{15} = 0.006667$ m³

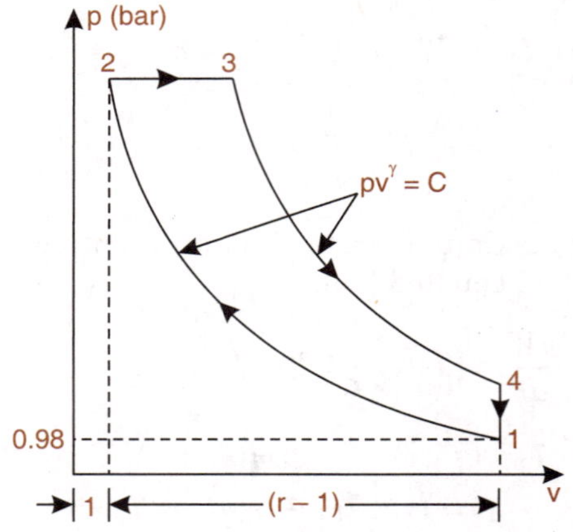

Fig. 21.43

(i) **Cut-off ratio,** $\rho = \dfrac{V_3}{V_2}$

$= \dfrac{0.0125}{0.006667} = $ **1.875. (Ans.)**

(ii) **The per cent clearance** $= \dfrac{1}{15}$

= **0.06667** or **6.667%. (Ans.)**

(iii) **The work done, W :**

$T_2 = T_1 \times (r)^{\gamma-1} = 303 \times (15)^{1.4-1} = 895.1$ K

$T_3 = \rho \times T_2 = 1.875 \times 895.1 = 1678.3$ K

$T_4 = T_3 \times \left(\dfrac{\rho}{r}\right)^{\gamma-1} = 1678.3 \times \left(\dfrac{1.875}{15}\right)^{1.4-1}$

$= 730.5$ K

Heat supplied,

$Q_{2-3} = c_p(T_3 - T_2) = 1.005(1678.3 - 895.1)$
$= 787.1$ kJ/kg

Heat rejected,

$Q_{4-1} = c_v\,(T_4 - T_1) = 0.718\,(730.5 - 303)$

$= 306.9$ kJ/kg $\quad \left(\because \; c_v = \dfrac{c_p}{\gamma} = \dfrac{1.005}{1.4} = 0.718\right)$

$\therefore$ Work done, $\quad W = Q_{2-3} - Q_{4-1} = 787.1 - 306.9$

= **480.2 kJ/kg. (Ans.)**

(iv) **The air standard efficiency,** $\eta_{\text{air-standard}}$ **:**

$\therefore \quad \eta_{air\text{-}standard} = \dfrac{W}{Q_{2-3}} = \dfrac{480.2}{787.1}$

= **0.61** or **61%. (Ans.)**

[**Check :** $\quad \eta_{diesel} = 1 - \dfrac{\rho^\gamma - 1}{\gamma(r)^{\gamma-1}(\rho - 1)}$

$= 1 - \dfrac{(1.875)^{1.4} - 1}{1.4 \times (15)^{1.4-1} \times (1.875 - 1)} = 0.61$ or 61%]

Example 21.42. *A diesel engine having compression ratio of 16 uses the fuel C_7H_{16}. The compression follows the law $pv^{1.4}$ = constant. If the engine uses 64% excess air and the temperature at the beginning of compression is 325 K, find the percentage of stroke at which combustion is completed.*

Assume $c_p = 1.0 + 21 \times 10^{-5}\,T$ kJ/kg K and T is in degrees' kelvin. C.V. = 44000 kJ/kg. Assume that air contains 23% by weight of O_2.

Solution. *Given :* $r = 16$; Excess air used = 64% ;

$T_1 = 325$ K ; C.V. = 44000 kJ/kg.

Fig. 21.44 shows the cycle of operation. The minimum amount of air required to burn 1 kg of fuel can be calculated by using the following chemical equation :

$$C_7H_{16} + 11O_2 = 7CO_2 + 8H_2O$$
$$100 + 352 = 308 + 144$$

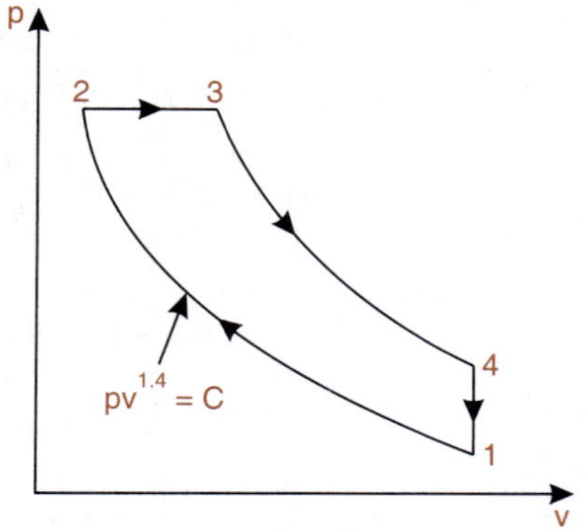

Fig. 21.44

$\therefore$ 1 kg of fuel requires $\frac{352}{100}$ kg of O_2 and

$\frac{352}{100} \times \frac{100}{23}$ kg of air = 15.3 kg

Actual air supplied = 15.3 × 1.64 = 25.092 kg/kg of fuel.

$$\% \text{ cut-off} = \frac{v_3 - v_2}{v_1 - v_2} = \frac{v_3 - v_2}{16v_2} = \frac{1}{16}\left(\frac{v_3}{v_2} - 1\right) \quad ...(i)$$

From *compression process 1–2*, we have

$$\left(\because \frac{v_1}{v_2} = r = 16\right)$$

$$\frac{T_2}{T_1} = \left(\frac{v_1}{v_2}\right)^{\gamma-1} = (16)^{1.4-1}$$

or, $T_2 = 325 \times (16)^{0.4} = 985$ K

The heat supplied per kg of fuel = Heat given to the gases during the process 2–3.

$$\therefore \quad 44000 = \int_{T_2}^{T_3} \left(\frac{m_a}{m_f} + 1\right) c_p \, dT$$

$$44000 = \int_{T_2}^{T_3} (25.092 + 1)(1.0 + 21 \times 10^{-5} T) \, dT$$

or, $$\frac{44000}{26.092} = \int_{985}^{T_3} (1.0 + 21 \times 10^{-5} \, T) \, dT$$

or, $$1686.34 = (T_3 - 985) + \frac{21 \times 10^{-5}}{2} [T_3^2 - (985)^2]$$

or, $$1686.34 = T_3 - 985 + 10.5 \times 10^{-5} \, T_3^2 - 101.87$$

or, $$10.5 \times 10^{-5} \, T_3^2 + T_3 - 2762.7 = 0$$

or, $$T_3 = \frac{-1 \pm \sqrt{1^2 + 4 \times 10.5 \times 10^{-5} \times 2762.7}}{2 \times 10.5 \times 10^{-5}}$$

$$= \frac{-1 + 1.47}{2 \times 10.5 \times 10^{-5}} = 2238 \text{ K.}$$

From *constant pressure process 2–3*, we have

$$\frac{v_2}{T_2} = \frac{v_3}{T_3} \quad \text{or} \quad \frac{v_3}{v_2} = \frac{T_3}{T_2} = \frac{2238}{985} = 2.272$$

Substituting this in eqn. (*i*), we have

$$\therefore \quad \% \text{ cut-off} = \frac{1}{16}(2.272 - 1)$$

$= \mathbf{0.0795}$ or **7.95%.** **(Ans.)**

Example 21.43. *In a hypothetical air cycle, consisting of three processes, an adiabatic compression is followed by an isothermal expansion to the initial volume of compression. Finally a heat rejection process completes the cycle.*

(*i*) *Draw the cycle on p-v and T-s diagrams and derive an expression for thermal efficiency of the cycle in terms of compression ratio r.*

(*ii*) *Also find the efficiency and m.e.p. of the cycle if the compression ratio is 14 and the induction conditions are 1 bar and 27°C. Take for air, c_p = 1.005 kJ/kg K and c_v = 0.718 kJ/kg K.* **(P.U.)**

Solution. *Given :* $r = 14$; $p_1 = 1$ bar ; $T_1 = 27 + 273 = 300$ K ;

$c_p = 1.005$ kJ/kg K ;

$c_v = 0.718$ kJ/kg K.

(*i*) The *p-v* and *T-s* diagrams of the cycle are shown in Fig. 21.45.

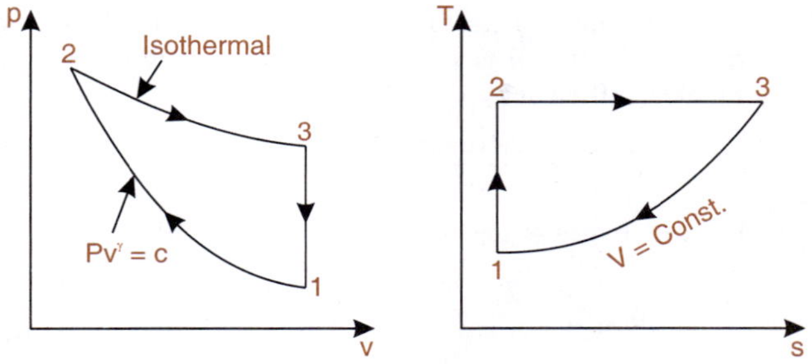

Fig. 21.45

Consider 1 kg of air.

Consider *adiabatic process 1–2 :*

$$\frac{T_2}{T_1} = \left(\frac{v_1}{v_2}\right)^{\gamma-1} = (r)^{\gamma-1}$$

$\therefore$ $T_2 = T_1 (r)^{\gamma-1}$; $Q_{1-2} = 0$

Consider *isothermal process 2–3 :*

$T_2 = T_3$

$Q_{2\text{-}3} = W = RT_2 \ln(r)$

Consider *heat-rejection (constant volume) process 3–1 :*

$Q_{3\text{-}1} = c_v (T_3 - T_1)$

Efficiency,

$$\eta = \frac{\text{Work done}}{\text{Heat added}} = \frac{Q_{2-3} - Q_{3-1}}{Q_{2-3}} = 1 - \frac{Q_{3-1}}{Q_{2-3}}$$

$$= 1 - \frac{c_v(T_3 - T_1)}{RT_2 \ln(r)} = 1 - \frac{c_v[T_1(r)^{\gamma-1} - T_1]}{RT_1(r)^{\gamma-1} \ln(r)}$$

$$= \mathbf{1 - \frac{c_v}{R \ln(r)}\left[1 - \frac{1}{(r)^{\gamma-1}}\right]}. \quad \textbf{(Ans.)}$$

(ii) $Q_{added} = Q_{2-3} = RT_2 \ln(r)$

where, $R = c_p - c_v = 1.005 - 0.718$

$= 0.287$ kJ/kg K,

$T_2 = T_1(r)^{\gamma-1} = 300 \times (14)^{1.4-1}$

$= 300 \times 2.874 = 862$ K

$$\left(\because \; \gamma = \frac{c_p}{c_v} = \frac{1.005}{0.718} = 1.4\right)$$

$\therefore$ $Q_{added} = 0.287 \times 862 \times \ln(14) = 652.88$ kJ/kg

$Q_{rejected} = c_v (T_3 - T_1) = 0.718(862 - 300)$

$= 403.52$ kJ/kg

$$\eta = 1 - \frac{Q_{rejected}}{Q_{added}} = 1 - \frac{403.52}{652.88}$$

$=$ **0.3819** or **38.19%.** **(Ans.)**

$$\left[\textit{Alternatively}: \eta = 1 - \frac{0.718}{0.287 \times \ln(14)}\left\{1 - \frac{1}{(14)^{1.4-1}}\right\} \right.$$

$$\left. = 1 - 0.9479 \times 0.652 = 0.3819 \text{ or } 38.19\% \right]$$

$W_{net} = Q_{added} - Q_{rejected}$

$= 652.88 - 403.52 = 249.36$ kJ/kg

Swept volume, $v_s = v_1 - v_2 = v_1 - \dfrac{v_1}{14} = \dfrac{13}{14} v_1$

or, $\dfrac{13}{14}\left(\dfrac{RT_1}{p_1}\right) = \dfrac{13}{14} \times \dfrac{(0.287 \times 1000)}{1 \times 10^5} \times 300$

$= 0.7995$ m³/kg

$\therefore$ **m.e.p.** $= \dfrac{W_{net}}{v_s} = \dfrac{249.36}{0.7995}$

$=$ **311.89 kN/m² or 3.1189 bar.** **(Ans.)**

HIGHLIGHTS

1. **A cycle** is defined as a repeated series of operations occurring in a certain order.
2. The efficiency of an engine using air as the working medium is known as an '**Air standard efficiency**'.
3. Relative efficiency,

$$\eta_{relative} = \frac{\text{Actual thermal efficiency}}{\text{Air standard efficiency}}.$$

4. **Carnot cycle** efficiency,

$$\eta_{Carnot} = \frac{T_1 - T_2}{T_1}.$$

5. **Otto cycle** efficiency,

$$\eta_{Otto} = 1 - \frac{1}{(r)^{\gamma - 1}}.$$

Mean effective pressure,

$$p_{m(Otto)} = \frac{p_1 r[(r^{\gamma-1} - 1)(r_p - 1)]}{(\gamma - 1)(r - 1)}.$$

6. **Diesel cycle** efficiency,

$$\eta_{Diesel} = 1 - \frac{1}{\gamma(r)^{\gamma - 1}}\left[\frac{\rho^\gamma - 1}{\rho - 1}\right]$$

Mean effective pressure,

$$p_{m(Diesel)} = \frac{p_1 r^\gamma [\gamma(\rho - 1) - r^{1-\gamma}(\rho^\gamma - 1)]}{(\gamma - 1)(r - 1)}.$$

7. **Dual cycle** efficiency,

$$\eta_{Dual} = 1 - \frac{1}{(r)^{\gamma - 1}}\left[\frac{(\beta.\rho^\gamma - 1)}{(\beta - 1) + \beta\gamma(\rho - 1)}\right]$$

Mean effective pressure,

$$p_{m(Dual)} = \frac{p_1 r^\gamma [\beta(\rho - 1) + (\beta - 1) - r^{1-\gamma}(\beta\rho^\gamma - 1)]}{(\gamma - 1)(r - 1)}.$$

8. Atkinson cycle efficiency,

$$\eta_{Atkinson} = 1 - \gamma . \frac{(r - \alpha)}{r^\gamma - \alpha^\gamma}$$

where α = Compression ratio, r = Expansion ratio.

9. Brayton cycle,

$$\eta_{Brayton} = 1 - \frac{1}{(r_p)^{\frac{\gamma-1}{\gamma}}}, \text{ where } r_p = \text{Pressure ratio.}$$

OBJECTIVE TYPE QUESTIONS

Choose the Correct Answer :

1. The air standard Otto cycle comprises
   (a) two constant pressure processes and two constant volume processes
   (b) two constant pressure processes and two constant entropy processes
   (c) two constant volume processes and two constant entropy processes
   (d) none of the above.
2. The air standard efficiency of Otto cycle is given by
   (a) $\eta = 1 + \dfrac{1}{(r)^{\gamma+1}}$ (b) $\eta = 1 - \dfrac{1}{(r)^{\gamma-1}}$
   (c) $\eta = 1 - \dfrac{1}{(r)^{\gamma+1}}$ (d) $\eta = 2 - \dfrac{1}{(r)^{\gamma-1}}$.
3. The thermal efficiency of theoretical Otto cycle
   (a) increases with increase in compression ratio
   (b) increases with increase in isentropic index γ
   (c) does not depend upon the pressure ratio
   (d) follows all the above.
4. The work output of theoretical Otto cycle
   (a) increases with increase in compression ratio

(*b*) increases with increase in pressure ratio

(*c*) increases with increase in adiabatic index γ

(*d*) follows all the above.

5. For same compression ratio

(*a*) thermal efficiency of Otto cycle is greater than that of Diesel cycle

(*b*) thermal efficiency of Otto cycle is less than that of Diesel cycle

(*c*) thermal efficiency of Otto cycle is same as that for Diesel cycle

(*d*) thermal efficiency of Otto cycle cannot be predicted.

6. In air standard Diesel cycle, at fixed compression ratio and fixed value of adiabatic index (γ)

(*a*) thermal efficiency increases with increase in heat addition cut-off ratio

(*b*) thermal efficiency decreases with increase in heat addition cut-off ratio

(*c*) thermal efficiency remains same with increase in heat addition cut-off ratio

(*d*) none of the above.

ANSWERS

1. (*b*)	**2.** (*b*)	**3.** (*d*)	**4.** (*d*)
5. (*a*)	**6.** (*b*).		

THEORETICAL QUESTIONS

1. What is a cycle ? What is the difference between an ideal and actual cycle ?

2. What is an air-standard efficiency ?

3. What is relative efficiency ?

4. Derive expressions of efficiency in the following cases :

(*i*) Carnot cycle

(*ii*) Diesel cycle

(*iii*) Dual combustion cycle.

5. Explain "Air standard analysis" which has been adopted for I.C. engine cycles. State the assumptions made for air standard cycles.

6. Derive an expression for 'Atkinson cycle'.

UNSOLVED EXAMPLES

1. A Carnot engine working between 377°C and 37°C produces 120 kJ of work. Determine :

(*i*) The heat added in kJ.

(*ii*) The entropy change during heat rejection process.

(*iii*) The engine thermal efficiency.

[**Ans.** (*i*) 229.5 kJ ; (*ii*) 0.353 kJ/K ; (*iii*) 52.3%]

2. Find the thermal efficiency of a Carnot engine whose hot and cold bodies have temperatures of 154°C and 15°C respectively. [**Ans.** 32.55%]

3. Derive an expression for change in efficiency for a change in compression ratio. If the compression ratio is increased from 6 to 8, what will be the percentage increase in efficiency ? [**Ans.** 8%]

4. The efficiency of an Otto cycle is 50% and γ is 1.5. What is the compression ratio ? [**Ans.** 4]

5. An engine working on Otto cycle has a volume of 0.5 m^3, pressure 1 bar and temperature 27°C at the commencement of compression stroke. At the end of compression stroke, the pressure is 10 bar. Heat added during the constant volume process is 200 kJ. Determine :

(*i*) Percentage clearance

(*ii*) Air standard efficiency

(*iii*) Mean effective pressure

(*iv*) Ideal power developed by the engine if the engine runs at 400 r.p.m. so that there are 200 complete cycles per minutes.

[**Ans.** (*i*) 23.76% ; (*ii*) 47.2% ; (*iii*) 2.37 bar ; (*iv*) 321 kW]

6. The compression ratio in an air-standard Otto cycle is 8. At the beginning of compression process, the pressure is 1 bar and the temperature is 300 K. The heat transfer to the air per cycle is 1900 kJ/kg of air. Calculate :

(*i*) Thermal efficiency

(*ii*) The mean effective pressure.

[**Ans.** (*i*) 56.47% ; (*ii*) 14.24 bar]

7. An engine 200 mm bore and 300 mm stroke works on Otto cycle. The clearance volume is 0.0016 m^3. The initial pressure and temperature are 1 bar and 60°C. If the maximum pressure is limited to 24 bar, find :

(*i*) The air-standard efficiency of the cycle

(*ii*) The mean effective pressure for the cycle.

Assume ideal conditions.

[**Ans.** (*i*) 54.08% ; (*ii*) 1.972 bar]

8. Calculate the air standard efficiency of a four stroke Otto cycle engine with the following data :

Piston diameter (bore) = 137 mm ; Length of stroke = 130 mm ;

Clearance volume 0.00028 m^3.

Express clearance as a percentage of swept volume.

[**Ans.** 56.1% ; 14.6%]

9. In an ideal Diesel cycle, the temperatures at the beginning of compression, at the end of compression and at the end of the heat addition are 97°C, 789°C and 1839°C. Find the efficiency of the cycle. [**Ans.** 59.6%]

10. An air-standard Diesel cycle has a compression ratio of 18, and the heat transferred to the working fluid per cycle is 1800 kJ/kg. At the beginning of the compression stroke, the pressure is 1 bar and the temperature is 300 K. Calculate : (*i*) Thermal efficiency, (*ii*) The mean effective pressure. [**Ans.** (*i*) 61% ; (*ii*) 13.58 bar]

11. 1 kg of air is taken through a Diesel cycle. Initially the air is at 15°C and 1 ata. The compression ratio is 15 and the heat added is 1850 kJ. Calculate : (*i*) The ideal cycle efficiency, (*ii*) The mean effective pressure.

[**Ans.** (*i*) 55.1% ; (*ii*) 13.4 bar]

12. What will be loss in the ideal efficiency of a Diesel engine with compression ratio 14 if the fuel cut-off is delayed from 6% to 9% ? [**Ans.** 2.1%]

13. The pressures on the compression curve of a diesel engine are at $\frac{1}{8}$th stroke 1.4 bar and at $\frac{7}{8}$th stroke

14 bar. Estimate the compression ratio. Calculate the air standard efficiency of the engine if the cut-off occurs at $\frac{1}{15}$th of the stroke. [**Ans.** 18.54 ; 63.7%]

14. A compression ignition engine has a stroke 270 mm, and a cylinder diameter of 165 mm. The clearance volume is 0.000434 m^3 and the fuel ignition takes place at constant pressure for 4.5 per cent of the stroke. Find the efficiency of the engine assuming it works on the Diesel cycle. [**Ans.** 61.7%]

15. The following data belong to a Diesel cycle :
Compression ratio = 16 : 1 ; Heat added = 2500 kJ/kg ; Lowest pressure in the cycle = 1 bar ; Lowest temperature in the cycle = 27°C.
Determine :
(*i*) Thermal efficiency of the cycle.
(*ii*) Mean effective pressure.
[**Ans.** (*i*) 45% ; (*ii*) 16.8 bar]

16. The compression ratio of an air-standard Dual cycle is 12 and the maximum pressure in the cycle is limited to 70 bar. The pressure and temperature of cycle at the beginning of compression process are 1 bar and 300 K. Calculate : (*i*) Thermal efficiency, (*ii*) Mean effective pressure.
Assume : cylinder bore = 250 mm, stroke length = 300 mm, c_p = 1.005, c_v = 0.718 and γ = 1.4.
[**Ans.** (*i*) 61.92% ; (*ii*) 9.847 bar]

17. The compression ratio of a Dual cycle is 10. The temperature and pressure at the beginning of the cycle are 1 bar and 27°C. The maximum pressure of the cycle is limited to 70 bar and heat supplied is limited to 675 kJ/kg of air. Find the thermal efficiency of the cycle. [**Ans.** 59.5%]

18. An air standard Dual cycle has a compression ratio of 16, and compression begins at 1 bar, 50°C. The maximum pressure is 70 bar. The heat transferred to air at constant pressure is equal to that at constant volume. Determine :
(*i*) The cycle efficiency.
(*ii*) The mean effective pressure of the cycle.
Take : c_p = 1.005 kJ/kg K, c_v = 0.718 kJ/kg K.
[**Ans.** (*i*) 66.5% ; (*ii*) 4.76 bar]

19. Compute the air standard efficiency of a Brayton cycle operating between a pressure of 1 bar and a final pressure of 12 bar. Take γ = 1.4. [**Ans.** 50.8%]

Variable Specific Heats

22

22.1. GENERAL ASPECTS

The ***specific heat*** *of any substance (as earlier stated) is the ratio of the heat required to raise the temperature of a unit mass of the substance through one degree centigrade.* Different substances have different values of specific heat. In case of gases, the temperature can be raised in two ways, *e.g.,* either at *constant pressure* or *constant volume.* Accordingly we have two specific heats c_p and c_v. It is often convenient to use specific heats for the mol of a substance. A mol is M kilograms, called kilogram-mol abbreviated as kg mol. Here M is the molecular weight of the substance.

Thus *molar specific heat*

$$C = M.\ c \text{ kJ/mol K or kcal/mol °K}$$

Similarly, $C_p = M.\ c_p$ kJ/mol K or kcal/mol °K

and $C_v = M.\ c_v$ kJ/mol K or kcal/mol °K

In general, the specific heats are not constant. *The specific heat varies largely with temperature but not very significantly with pressure except at very high pressure.* Thus in simple calculations, the variation in specific heat with pressure is neglected.

The specific heats of gases increase with the rise in temperature since the vibrational energy of the molecules increases with temperature. The effect of variable specific heats on the engine performance at higher temperature is considerable and it is, therefore, necessary to study these effects.

It is generally assumed that the specific heat is a linear function of temperature and the following relations hold good.

$$c_p = a + KT \quad ...(22.1)$$

$$c_v = b + KT \quad ...(22.2)$$

where a, b and K are constant.

$$R = (c_p - c_v) = (a + KT) - (b + KT)$$
$$= a - b \quad ...(22.3)$$

This linear relationship holds good upto about 1500 K. Above 1500 K, the increase is more rapid and the specific heats are given by :

$$c_p = a + K_1T + K_2T^2 \quad ...(22.4)$$

$$c_v = b + K_1T + K_2T^2 \quad ...(22.5)$$

If the internal energy of vibration is ignored the specific heats of gases may be written as :

$$R = c_p - c_v \text{ kJ/kg K or kcal/kg °K.}$$

For monoatomic gases :

$$c_v = \frac{3R}{2} \text{ kJ/kg K or kcal/kg °K}$$

$$Mc_v = \frac{3MR}{2} \text{ kJ/mol K or kcal/mol °K}$$

where, R = The characteristic gas constant expressed in kJ/kg K or kcal/kg °K ;

and MR = Molecular weight × characteristic gas constant
= Universal gas constant
= 8.315 kJ/kg mol K or 1.986 kcal/kg mol °K

c_v is expressed in kJ/kg K or kcal/kg °K

Also $$c_p = R + c_v = R + \frac{3R}{2} = \frac{5R}{2}$$

$$Mc_p = \frac{5}{2} MR$$

Thus Mc_p (monoatomic)
= 20.787 kJ/kg mol K or 4.965 kcal/kg mol °K

For diatomic gases :

$$c_v = \frac{5}{2} R \text{ or } Mc_v = \frac{5}{2} MR$$

and $$c_p = \frac{5}{2} R + R = \frac{7}{2} R$$

$\therefore$ Mc_p (diatomic)

= 29.1 kJ/kg mol K or 6.95 kcal/kg mol °K.

For polyatomic gases :

(Approx.) $c_v = 3R$ or $M_{cv} = 3MR$

$c_p = 4R$ or $M_{cp} = 4M_{cp}$

Thus M_{cp} (polyatomic) = 33.26 kJ/kg mol K or 7.944 kcal/kg mol °K.

It can be shown, from above relations, that the value of ratio $\frac{c_p}{c_v}$ for the gases is :

Gases	c_p/c_v
Monoatomic	5/3
Diatomic	7/5
Polyatomic	4/3

Table 22.1 gives the values of M_{cp} (*i.e.*, c_p) for some gases.

Table 22.1. Variable Specific Heats at Low Pressure

S. No.	*Gas (Temp. range)*	*Molecular weight (M)*	M_{cp} *in kcal/kg mol °K*
1.	Air (280 – 1500 K)	29	$6.36 + 1785 \times 10^{-4}\, T - 27.7 \times 10^{-8}\, T^2$
2.	Oxygen (280 – 2800 K)	32	$11.515 - 128/T^{1/2} + 840\, T^{-1}$
3.	Hydrogen (300 – 2200 K)	2.016	$5.76 + 10.4 \times 10^{-4}\, T + 14.9/T^{1/2}$
4.	Nitrogen (280 – 5000 K)	28.016	$9.47 - 1930/T + 35.7 + 10/T^2$
5.	Carbon monoxide (280 – 2800 K)	28	$9.46 - 1825/T + 33 \times 10^4/T^2$
6.	Carbon dioxide (280 – 3500 K)	44	$16.2 - 3620/T + 43.4 \times 10^4/T^2$
7.	Water (280 – 3000 K)	18.016	$19.86 - 445/T^{0.5} - 41.5/T$

1 kcal/kg mol °K = 4.1868 kJ/kg mol K.

—Fig. 22.1 shows the variation of M_{cp} for CO_2, O_2, CO and N_2 with temperature.

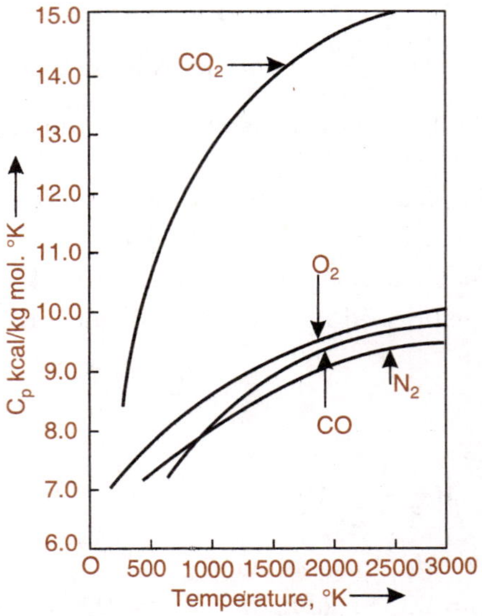

Fig. 22.1. Variation of molar specific heat with temperature.

—Fig. 22.2 shows the effect of humidity on properties of air. R is the characteristic gas constant ; c_p and c_v are the specific heats at constant pressure and constant volume respectively, γ = ratio of specific heats.

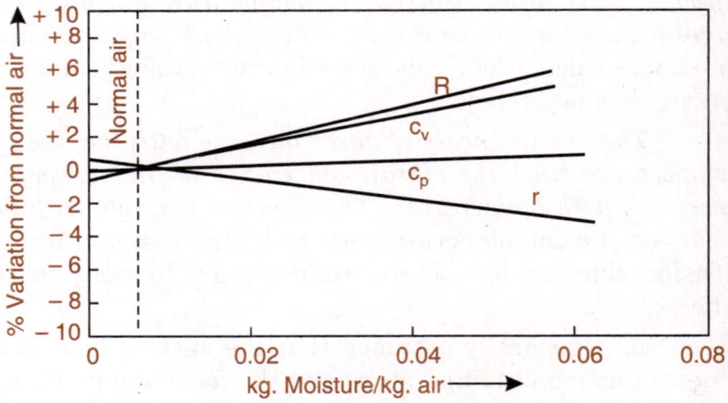

Fig. 22.2. Effect of humidity on properties of air.

22.2. CHANGE OF INTERNAL ENERGY AND ENTHALPY DURING A PROCESS WITH VARIABLE SPECIFIC HEATS

The small change in internal energy (du) of a *unit mass* of a gas for small change in temperature (dT) is given by :

$$du = c_v\, dT$$

$$\therefore \quad u_2 - u_1 = \int_{T_1}^{T_2} c_v\, dT = \int_{T_1}^{T_2} (b + KT)\, dT$$

$$= b(T_2 - T_1) + \frac{K}{2}(T_2^2 - T_1^2)$$

$$= (T_2 - T_1)\left[b + K\left(\frac{T_1 + T_2}{2}\right)\right]$$

$$= (T_2 - T_1)(b + KT_m)$$

where, T_m (mean temperature) $= \dfrac{T_1 + T_2}{2}$

$c_{vm} = b + KT_m$ (c_{vm} means mean specific heat at constant volume)

$$\therefore \quad u_2 - u_1 = c_{vm}(T_2 - T_1) \qquad \text{...(22.6)}$$

Similarly, change in enthalpy is given by :

$$dh = c_p\, dT$$

$$\therefore \quad h_2 - h_1 = c_{pm}(T_2 - T_1) \qquad \text{...(22.7)}$$

where, c_{pm} mean specific heat at constant pressure.

22.3. HEAT TRANSFER DURING A PROCESS WITH VARIABLE SPECIFIC HEATS

For a closed system, the heat-flow is expressed as :

$$dQ = du + dW \text{ (considering 1 kg of gas)}$$

or

$$dQ = c_v\, dT + pdv$$

$$\therefore \quad \frac{dQ}{dv} = c_v \cdot \frac{dT}{dv} + p \qquad \text{...(22.8)}$$

Also, $T = \dfrac{p \times v}{R}$ (According to gas equation)

$$\therefore \quad dT = \frac{1}{R}(pdv + vdp)$$

$$\therefore \quad \frac{dT}{dv} = \frac{1}{R}\left(p + v \cdot \frac{dp}{dv}\right)$$

Substituting this value in eqn. (22.8), we get

$$\frac{dQ}{dv} = \frac{c_v}{R}\left(p + \frac{vdp}{dv}\right) + p$$

Now inserting the values of

$$c_p = a + KT,\ c_v = b + KT \text{ and } R = a - b$$

in the above equation, we get

$$\frac{dQ}{dv} = \frac{b + KT}{(a - b)}\left(p + \frac{vdp}{dv}\right) + p$$

$$= \frac{1}{(a - b)}\left[bp + \frac{bvdp}{dv} + KTp + \frac{KTvdp}{dv} + ap - bp\right]$$

$$= \frac{b}{(a - b)}\left[v \cdot \frac{dp}{dv} + \frac{KTp}{b} + \frac{KTv}{b} \cdot \frac{dp}{dv} + \frac{a}{b}p\right]$$

$$= \frac{1}{\left(\dfrac{a}{b} - 1\right)}\left(\frac{a}{b}p + v \cdot \frac{dp}{dv}\right) + \frac{KT}{(a - b)}\left(p + v \cdot \frac{dp}{dv}\right)$$

$$\frac{a}{b} = \frac{c_p - KT}{c_v - KT} = \gamma'$$

$$\therefore \quad \frac{dQ}{dv} = \frac{1}{(\gamma' - 1)}\left(\gamma' p + v \cdot \frac{dp}{dv}\right) + \frac{KT}{(a - b)}\left(p + v \cdot \frac{dp}{dv}\right) \qquad \text{...(22.9)}$$

If the expansion or compression is polytropic, then

pv^n = constant, where $n < \gamma'$.

Differentiating the above equation,

$$p \cdot nv^{n-1} \cdot dv + v^n \cdot dp = 0$$

$$v \cdot \frac{dp}{dv} = -p \cdot n$$

Inserting this value in eqn. (22.9) the heat exchange in a polytropic process is given by,

$$\frac{dQ}{dv} = \frac{1}{(\gamma' - 1)}(\gamma' p - p \cdot n) + \frac{KT}{(a - b)}(p - p \cdot n)$$

$$= \frac{(\gamma' - n)}{(\gamma - 1)} \cdot p + \frac{KT}{(a - b)}(1 - n)p$$

$$= \left[\frac{\gamma' - n}{\gamma' - 1} - \frac{(n - 1)}{(a - b)}KT\right]p$$

$$\therefore \quad dQ = \left[\frac{(\gamma' - n)}{(\gamma' - 1)} - \frac{(n - 1)}{(a - b)}KT\right]pdv$$

$$= \left[\left(\frac{\gamma' - n}{\gamma' - 1}\right) - \left(\frac{n - 1}{a - b}\right)KT\right]dW \qquad \text{...(22.10)}$$

If $K = 0$

$$dQ = \frac{\gamma' - n}{\gamma' - 1} \cdot dW.$$

22.4. ISENTROPIC EXPANSION WITH VARIABLE SPECIFIC HEATS

The heat transfer to a system is expressed as :

$$dQ = du + dW$$

or $\quad dQ = c_v\, dT + pdv$ (considering one kg of air)

For isentropic process, $\quad dQ = 0$

$$\therefore \quad c_v\, dT + pdv = 0 \text{ or } c_v \frac{dT}{T} + \frac{pdv}{T} = 0$$

$$\therefore \quad c_v \frac{dT}{T} + R \cdot \frac{dv}{v} = 0 \qquad \left(\because \frac{p}{T} = \frac{R}{v}\right)$$

Inserting the value of R and c_v in the above equation, we get

$$(b + KT) \frac{dT}{T} + (a - b) \cdot \frac{dv}{v} = 0$$

$$\therefore \quad (b + KT) \frac{dT}{T} + (a - b) \frac{dv}{v} = 0$$

Integrating both sides, we get

$$b \log_e T + KT + (a - b) \log_e v = \text{constant}$$

or $\quad \log_e (T)^b + \log_e (e)^{KT} + \log_e (v)^{a-b} = \text{constant}$

or $\quad (T)^b\, (e)^{KT}\, (v)^{a-b} = \text{constant}$

or $$T.(e)^{\frac{K}{b}.T} \cdot (v)^{\left(\frac{a}{b} - 1\right)} = \text{constant} \qquad ...(22.11)$$

$$\therefore \quad \frac{T}{v} \cdot (e)^{\frac{K}{b}T} \cdot (v)^{\frac{a}{b}} = \text{constant}$$

But, $$\frac{T}{v} = \frac{p}{R} = \frac{p}{(a - b)}$$

Inserting the value of $\frac{T}{v} = \frac{p}{R}$ in the above equation, we get

$$p(e)^{\frac{KT}{b}} \cdot (v)^{\frac{a}{b}} = \text{constant}$$

or $$p(v)^{a/b} \cdot e^{KT/b} = \text{constant} \qquad ...(22.12)$$

22.5. 'ENTROPY CHANGE' DURING A PROCESS WITH VARIABLE SPECIFIC HEATS

Let p_1, v_1, T_1 and p_2, v_2, T_2 be the initial and final conditions of the gas ; then change in entropy is given by (specific heats changing linearly with the temperature),

$$ds = \frac{dQ}{T} = \frac{pdv}{T} + \frac{c_v dT}{T}$$

$$= R \cdot \frac{dv}{v} + c_v \cdot \frac{dT}{T} \qquad \left(\because \frac{p}{T} = \frac{R}{v}\right)$$

$$= (a - b) \frac{dv}{v} + (b + KT) \frac{dT}{T}$$

Integrating between the limits, we get

$$s_2 - s_1 = (a - b) \log_e (v_2/v_1) + b \log_e (T_2/T_1) + K(T_2 - T_1) \qquad ...(22.13)$$

To express the equation in terms of p and v

$$\frac{p_1 v_1}{T_1} = \frac{p_2 v_2}{T_2} \text{ or } \frac{T_2}{T_1} = \frac{p_2 v_2}{p_1 v_1}$$

or $\quad \log_e T_2/T_1 = \log_e (p_2/p_1) + \log_e (v_2/v_1)$

Inserting this value in eqn. (22.13), we get

$$s_2 - s_1 = (a - b) \log_e (v_2/v_1) + b[\log_e (p_2/p_1) + \log_e (v_2/v_1)] + K(T_2 - T_1)$$

$$= a \log_e v_2/v_1 - b \log_e v_2/v_1 + b \log_e p_2/p_1 + b \log_e (v_2/v_1) + K(T_2 - T_1)$$

$$\therefore \quad s_2 - s_1 = a \log_e (v_2/v_1) + b \log_e (p_2/p_1) + K(T_2 - T_1) \qquad ...(22.14)$$

To express the equation in terms of p and T

$$\log_e (v_2/v_1) = \log_e (T_2/T_1) - \log_e (p_2/p_1)$$

Inserting this value in eqn. (22.14), we get

$$s_2 - s_1 = a \log_e (T_2 - T_1) - (a - b) \log_e (p_2/p_1) + K\, (T_2 - T_1) \qquad ...(22.15)$$

22.6. EFFECTS OF VARIABLE SPECIFIC HEATS ON AIR STANDARD EFFICIENCIES OF OTTO AND DIESEL CYCLES

(i) **Otto Cycle.** The air standard efficiency of Otto cycle is given by

$$\eta = 1 - \frac{1}{(r)^{\gamma - 1}}$$

$$= 1 - (r)^{-\frac{R}{c_v}} \qquad \left(\because \gamma - 1 = \frac{R}{c_v}\right)$$

$$\therefore \quad 1 - \eta = (r)^{-\frac{R}{c_v}}$$

Taking log on both sides

$$\log_e (1 - \eta) = -\frac{R}{c_v} \log_e r$$

Differentiating the above equation,

$$-\frac{1}{1 - \eta} \cdot d\eta = -R \log_e r \left(-\frac{1}{c_v^2} dc_v\right)$$

$$= R \cdot \frac{dc_v}{c_v^2} \log_e r$$

The change in the efficiency with variation in specific heat is expressed as :

$$\frac{d\eta}{\eta} = -\frac{1 - \eta}{\eta} \cdot \frac{R}{c_v} \cdot \frac{dc_v}{c_v} \cdot \log_e r$$

$$\therefore \quad \frac{d\eta}{\eta} = -\frac{1 - \eta}{\eta} \cdot (\gamma - 1) \cdot \log_e r \cdot \frac{dc_v}{c_v} \qquad ...(22.16)$$

The negative sign indicates the decrease in efficiency with increase in c_v.

The eqn. (22.16) gives *the percentage variation in air standard efficiency of Otto cycle on account of percentage variation in* c_v.

(ii) **Diesel Cycle.** The air standard efficiency of the Diesel cycle is expressed as :

$$\eta = 1 - \frac{1}{(r)^{\gamma-1}} \left[\frac{\rho^\gamma - 1}{\gamma(\rho - 1)}\right]$$

$$\therefore \quad 1 - \eta = \frac{1}{(r)^{\gamma-1}} \left[\frac{\rho^\gamma - 1}{\gamma(\rho - 1)}\right]$$

Taking log on both sides, we get

$$\log_e (1 - \eta) = -(\gamma - 1) \log_e r + \log_e (\rho^\gamma - 1) - \log_e \gamma - \log_e (\rho - 1)$$

Differentiating the above equation with respect to γ

$$-\frac{1}{1-\eta} \cdot \frac{d\eta}{d\gamma} = -\log_e \gamma + \frac{\rho^\gamma . \log_e \rho}{\rho^\gamma - 1} - \frac{1}{\gamma}$$

$$\therefore \quad \frac{d\eta}{d\gamma} = (1 - \eta) \left[\log_e \gamma - \frac{\rho^\gamma . \log_e \rho}{\rho^\gamma - 1} + \frac{1}{\gamma}\right]$$

Multiplying the above equation by $\frac{d\gamma}{\eta}$

$$\frac{d\eta}{\eta} = \frac{1-\eta}{\eta} . d\gamma \left[\log_e r - \frac{\rho^\gamma . \log_e \rho}{\rho^\gamma - 1} + \frac{1}{\gamma}\right] \quad ...(22.17)$$

But $\quad c_p - c_v = R$

When R remains constant as the changes in c_p and c_v are considered with the same rate,

$$\therefore \quad \gamma - 1 = \frac{R}{c_v}$$

Defferentiating the above equation

$$d\gamma = -\frac{R}{{c_v}^2} . dc_v = -\frac{R}{c_v} \cdot \frac{dc_v}{c_v}$$

$$= -(\gamma - 1) \frac{dc_v}{c_v}$$

i.e.,
$$d\gamma = -(\gamma - 1) \frac{dc_v}{c_v} \quad ...(22.18)$$

Inserting the value of $d\gamma$ from eqn. (22.18) into eqn. (22.17), we get

$$\frac{d\eta}{\eta} = -\frac{1-\eta}{\eta} (\gamma - 1) \left[\log_e r - \frac{\rho^\gamma . \log_e \rho}{\rho^\gamma - 1} + \frac{1}{\gamma}\right] \frac{dc_v}{c_v} \quad ...(22.19)$$

The eqn. (22.19) *gives the percentage variation in air standard efficiency of Diesel Cycle on account of percentage variation in* c_v.

WORKED EXAMPLES

Example 22.1. *Deduce an expression for the change in entropy of 1 kg of gas in terms of the temperature T_1 and T_2 before and after compression, if the law of compression is pv^n = constant and the specific heats are of the form,*

$$c_p = a + KT$$
$$c_v = b + KT.$$

Solution. We know that, $dQ = du + dW$

$$= c_v \, dT + pdv$$

Dividing both sides by T, $\quad \frac{dQ}{T} = c_v . \frac{dT}{T} + \frac{pdv}{T}$

or
$$ds = c_v \frac{dT}{T} + \frac{P}{T} . dv$$

Inserting the value of c_v,

$$ds = b \frac{dT}{T} + KdT + \frac{p}{T} dv \quad ...(i)$$

Also,
$$pV = RT$$
$$\therefore \quad pdv + vdp = RdT$$

and
$$pv^n = C \quad ...(ii)$$
$$\therefore \quad p \, . \, n(v)^{n-1} . dv + v^n . dp = 0$$
$$npdv = -vdp \quad ...(iii)$$

From *(ii)* and *(iii)*, we get

$$pdv - npdv = RdT$$
$$pdv = \frac{RdT}{1-n}$$

Inserting in *(i)*, we get

$$ds = b \frac{dT}{T} + KdT + \frac{R}{(1-n)} \cdot \frac{dT}{T}$$
$$= \left[b - \frac{R}{(n-1)}\right] \frac{dT}{T} + KdT$$

Integrating, we get

$$s_2 - s_1 = \left[b - \frac{R}{(n-1)}\right] \log_e \frac{T_2}{T_1} + K(T_2 - T_1)$$

Substituting, $R = a - b$,

or
$$s_2 - s_1 = \left[\mathbf{b} - \frac{\mathbf{a - b}}{\mathbf{(n-1)}}\right] \mathbf{\log_e} \frac{\mathbf{T_2}}{\mathbf{T_1}} + \mathbf{K(T_2 - T_1)}.$$

(Ans.)

Example 22.2. *An Otto cycle engine with a compression ratio of 10 uses a petroleum fuel of calorific value 48000 kJ/kg. The air fuel ratio is 15 : 1. The temperature and pressure of the charge at the end of suction are 57°C and 1 bar respectively. Determine the maximum pressure in the cycle with mean index of compression as 1.36 and the value of specific heat at constant volume heat addition expressed as*

$c_v = 0.7117 + 2.1 \times 10^{-4}$ kJ/kg K, where T is the mean temperature.

If the value of c_v, remains constant at 0.7117 kJ/kg K, and also compression index is unaltered, how will the maximum pressure be affected ?

Solution. Compression ratio, $r = 10$

Calorific value of petroleum fuel = 48000 kJ/kg

Air-fuel ratio = 15 : 1

Temperature at the end of suction,

$$T_1 = 57 + 273 = 330 \text{ K}$$

Pressure at the end of suction, $p_1 = 1$ bar

Mean index of compression, $n = 1.36$.

We have $p_1V_1^{1.36} = p_2V_2^{1.36}$

$$\therefore \quad p_2 = p_1 \times \left(\frac{V_1}{V_2}\right)^{1.36} = 1 \times (10)^{1.36} = 22.91 \text{ bar}$$

$$\text{Also,} \quad \frac{T_2}{T_1} = \left(\frac{p_2}{p_1}\right)^{\frac{n-1}{n}} = \left(\frac{22.91}{1}\right)^{\frac{1.36-1}{1.36}} = 2.29$$

$$\therefore \quad T_2 = 330 \times 2.29 = 756 \text{ K}$$

Mean specific heat during constant volume heat addition

$$c_{v(\text{mean})} = 0.7117 + 2.1 \times 10^{-4}\left(\frac{T_3 + T_2}{2}\right)$$

Thus heat added at constant volume/kg of charge

$$= \frac{48000}{6} = 3000 \text{ kJ/kg of charge}$$

$$\text{And } 3000 = \left[0.7117 + 2.1 \times 10^{-4}\left(\frac{T_3 + 756}{2}\right)\right] \times (T_3 - 756)$$

$$\therefore \quad 3000 = 0.7117\, T_3 + \frac{2.1 \times 10^{-4}}{2}\, T_3^2 + \frac{2.1 \times 10^{-4} \times 756}{2}\, T_3 - 0.7117 \times 756$$

$$- \frac{2.1 \times 10^{-4} \times 756}{2}\, T_3 - \frac{2.1 \times 10^{-4}}{2} \times 756 \times 756$$

$$= 0.7117\, T_3 + 0.000105\, T_3^2 - 598$$

$$= 0.000105\, T_3^2 + 0.7117\, T_3 - 598$$

or $$0.000105\, T_3^2 + 0.7117\, T_3 - 3598 = 0$$

Solving for T_3, $T_3 = 3370$ K

$\therefore$ *Maximum pressure in the cycle,*

$$p_3 = p_2 \times \frac{T_3}{T_2} = 22.91 \times \frac{3370}{756} = \mathbf{102.1 \text{ bar.}} \quad \textbf{(Ans.)}$$

When c_v remains constant at 0.7117 kJ/kg K ;

$$3000 = 0.7117\,(T_3 - 756)$$

or $$T_3 = \frac{3000}{0.7117} + 756 = 4971 \text{ K}$$

$$\text{And} \quad p_3 = p_2\,\frac{T_3}{T_2} = 22.91 \times \frac{4971}{756}$$

$$= \mathbf{150.6 \text{ bar.}} \quad \textbf{(Ans.)}$$

Example 22.3. *Find the percentage change in efficiency of an Otto cycle for a compression ratio of 7 to 1 if the specific heat at constant volume increases by 3%.*

Solution. The change in efficiency with variation in specific heat is given by :

$$\frac{d\eta}{\eta} = -\frac{1-\eta}{\eta}\,(\gamma - 1)\log_e r\,\frac{dc_v}{c_v} \quad ...(i)$$

$$\text{Also,} \quad \eta = 1 - \frac{1}{(r)^{\gamma-1}} = 1 - \frac{1}{(7)^{1.4-1}}$$

$$= 0.541 \text{ or } 54.1\%$$

Inserting this value in the eqn. (*i*), we get

$$\frac{d\eta}{\eta} = -\left(\frac{1 - 0.541}{0.541}\right)(1.4 - 1)\log_e 7 \times 0.03$$

$$= \mathbf{-\,0.0198} \text{ or } \mathbf{-1.98\%.} \quad \textbf{(Ans.)}$$

Negative sign indicates decrease in efficiency.

Example 22.4. *The following particulars relate to a Diesel cycle :*

Compression ratio = 18, cut-off = 5% of stroke, mean specific heat c_v for cycle = 0.71 kJ/kg K, characteristic gas constant = 0.285 kJ/kg K.

Knowing that the specific heat increases with temperature, if the mean specific heat for the air standard cycle increases by 2 per cent determine the percentage change in the air standard efficiency.

Solution. Compression ratio, $r = 18$

Cut-off = 5% of stroke

Mean specific heat for cycle, $c_v = 0.71$ kJ/kg K

Characteristic gas constant, $R = 0.285$ kJ/kg K

Increase in c_v = 2%.

Change in air-standard efficiency :

The percentage variation in the air standard efficiency on account of the percentage variation in the value of c_v is given by

$$\frac{d\eta}{\eta} = -\left(\frac{1-\eta}{\eta}\right)(\gamma - 1)\left[\log_e r - \frac{\rho^\gamma \log_e \rho}{\rho^\gamma - 1} + \frac{1}{\gamma}\right]\frac{dc_v}{c_v} \quad ...(i)$$

[Article 22.6, Eqn. (22.19)]

$$\text{Now,} \quad \rho = \frac{\%\text{ cut-off}}{100}\,(r - 1) + 1$$

$$= \frac{5}{100}\,(18 - 1) + 1 = 1.85$$

$$\text{and} \quad \gamma = 1 + \frac{R}{c_v} = 1 + \frac{0.285}{0.71} = 1.4$$

$$\text{Also,} \quad \frac{dc_v}{c_v} = \frac{2}{100}$$

Efficiency of Diesel cycle is given by

$$\eta = 1 - \frac{1}{\gamma . (r)^{\gamma - 1}} \left[\frac{\rho^{\gamma - 1}}{\rho - 1}\right]$$

$$= 1 - \frac{1}{1.4\,(18)^{1.4 - 1}} \left[\frac{(1.85)^{1.4 - 1}}{1.85 - 1}\right]$$

$$= 1 - 0.2248 \left(\frac{1.366}{0.85}\right) = 0.6387 \text{ or } 63.87\%$$

Now substituting the values of η and other quantities in the eqn. (*i*), we get

$$\frac{d\eta}{\eta} = -\frac{(1 - 0.6387)}{0.6387} \times (1.4 - 1)$$

$$\left[\log_e 18 - \frac{(1.85)^{1.4} \log_e 1.85}{(1.85)^{1.4} - 1} + \frac{1}{1.4}\right] \times \frac{2}{100}$$

$$= -0.5657 \times 0.4 \left[2.89 - \frac{1.455}{1.366} + 0.714\right] \times \frac{2}{100}$$

$$= -0.226\,(2.89 - 1.065 + 0.714) \times \frac{2}{100}$$

$$= -0.01147 = -1.147\%$$

Negative sign means *decrease*.

Hence *percentage decrease in efficiency*

= **1.147%.** **(Ans.)**

HIGHLIGHTS

1. The specific heat varies largely with temperature but not very significantly with pressure except at very high pressure.
2. The specific heats of gases increase with the rise in temperature since the vibrational energy of molecules increases with temperature.
3. Change of *internal energy* (per kg) during a process with variable specific heats

$$u_2 - u_1 = c_{vm}\,(T_2 - T_1)$$

Change of *enthalpy*, $h_2 - h_1 = c_{pm}\,(T_2 - T_1)$.

4. Heat transfer during a process with variable specific heats,

$$dQ = \left[\left(\frac{\gamma' - n}{\gamma' - 1}\right) - \left(\frac{n - 1}{a - b}\right) KT\right] dW.$$

5. In case of isentropic expansion with variable specific heats

$$p(v)^{a/b} \,.\, e^{KT/b} = \text{constant}.$$

6. 'Entropy change' during a process with variable specific heats,

$$s_2 - s_1 = a \log_e \left(\frac{T_2}{T_1}\right) - (a - b) \log_e \left(\frac{p_2}{p_1}\right) + K(T_2 - T_1).$$

7. Percentage variation in air standard efficiency on account of percentage variation in c_v in case of :

(*i*) **Otto cycle :** $\dfrac{d\eta}{\eta} = -\dfrac{1 - \eta}{\eta}\,(\gamma - 1) \log_e r \times \dfrac{dc_v}{c_v}$.

(*ii*) **Diesel cycle :** $\dfrac{d\eta}{\eta} = -\dfrac{1 - \eta}{\eta}\,(\gamma - 1)$

$$\left[\log_e r - \frac{\rho^\gamma . \log_e \rho}{\rho^\gamma - 1} + \frac{1}{\gamma}\right] \frac{dc_v}{c_v}.$$

OBJECTIVE TYPE QUESTIONS

Choose the Correct Answer :

1. The value of c_v of a real gas
(*a*) increases with increase in temperature
(*b*) decreases with increase in temperature
(*c*) remains constant with increase in temperature
(*d*) first increases with increase in temperature and then decreases with further increase in temperature.
2. The value c_p of a real gas
(*a*) increases with increase in temperature
(*b*) decreases with increase in temperature
(*c*) remains constant with increase in temperature
(*d*) first increases with increase in temperature and then decreases with further increase in temperature.
3. The value of c_p
(*a*) decreases with increase in moisture content in air
(*b*) increases with increase in moisture content in air
(*c*) remain same irrespective of the moisture content in air
(*d*) increases with increase in moisture content and then decreases with further increase in moisture content in air.
4. The value of $\gamma\ (= c_p/c_v)$ for air
(*a*) decreases with increase in moisture content in air
(*b*) increase with increases in moisture content in air
(*c*) remains constant irrespective of the increase in moisture content in air
(*d*) increases first and decreases later with moisture content in air.
5. The value of $\gamma\ (= c_p/c_v)$ for real gas
(*a*) decreases with increase in temperature
(*b*) increases with increase in temperature
(*c*) remains same irrespective of increase in temperature
(*d*) increases first with increase in temperature and then decreases with further increase in temperature.
6. The change of internal energy during a process with variable specific heats is equal to
(*a*) $c_p\,(T_2 - T_1)$ (*b*) $c_{vm}\,(T_2 - T_1)$
(*c*) $c_v/(T_2 - T_1)$ (*d*) $c_{vm}/(T_2 - T_1)$.
7. The change of enthalpy during a process with variable specific heats is equal to
(*a*) $c_p\,(T_2 - T_1)$ (*b*) $c_{pm}\,(T_2 - T_1)$
(*c*) $c_p/(T_2 - T_1)$ (*d*) $c_{pm}/(T_2 - T_1)$.

8. Heat transfer during a process with variable specific heats is equal to

(a) $\left[\left(\frac{\gamma' - n}{\gamma' - 1}\right) - \left(\frac{n-1}{a-b}\right) KT\right] dW$

(b) $\left[\left(\frac{\gamma' - n^2}{\gamma' - 1}\right) - \left(\frac{n-2}{a-b}\right) KT\right] dW$

(c) $\left[\left(\frac{\gamma'^2 - n^2}{\gamma' - 1}\right) - \left(\frac{n-1}{a-b}\right) KT\right] dW$

(d) $\left[\left(\frac{\gamma' - n}{\gamma' + 1}\right) - \left(\frac{n-1}{a-b}\right) KT^2\right] dW.$

ANSWERS

1. (a)	**2.** (a)	**3.** (b)	**4.** (a)
5. (a)	**6.** (b)	**7.** (b)	**8.** (a).

THEORETICAL QUESTIONS

1. What are molar specific heats ?

2. Derive expressions for change of internal energy and enthalpy during a process with variable specific heats.

3. Prove that in case of isentropic expansion with variable specific heats,

$$p(v)^{a/b} \cdot e^{KT/b} = \text{constant}.$$

4. Derive an expression for 'entropy change' during a process with variable specific heats.

5. Taking variation in specific heat into account, derive the following expressions :

For **Otto cycle** :

$$\frac{d\eta}{\eta} = \left[-\frac{1-\eta}{\eta} \times (\gamma - 1) \times \log_e r\right] \frac{dc_v}{c_v}$$

For **Diesel cycle** :

$$\frac{d\eta}{\eta} = -\frac{1-\eta}{\eta}(\gamma - 1)\left[\log_e r - \frac{\rho^\gamma \log_e \rho}{\rho^\gamma - 1} + \frac{1}{\gamma}\right] \frac{dc_v}{c_v}.$$

UNSOLVED EXAMPLES

1. Find the heat added and change in entropy per kg of air when it is heated from 300°K to 400°K at constant pressure.

$$\left[\textbf{Hint.} = \int_{T_1}^{T_2} c_p dT \text{ and } s_2 - s_1 = \int_{T_1}^{T_2} c_p \cdot \frac{dT}{T}.\right]$$

[**Ans.** 24.25 kcal/kg, 0.0695 kcal/kg °K]

2. Find the change in enthalpy as 1 kg of oxygen is heated from 500°K to 2000°K. The specific heat at constant pressure is given by

$$c_p = 11.515 - \frac{172}{\sqrt{T}} + \frac{1530}{T} \text{ kcal/kg °K.}$$

[**Ans.** 366 kcal/kg]

3. Calculate the percentage change in air standard efficiency of a Diesel cycle which has the following particulars : Compression ratio = 20 : 1, cut-off = 5% of stroke, $c_v = 0.171$ kcal/kg °K ; $R = 29.27$ m kgf/kg °K ; increase in mean value of specific heat at constant volume = 2.5%.

[**Ans.** 0.66%]

4. A gas of 0.15 m^3 volume at a pressure of 50 kgf/cm^2 weighs 6 kg. It is expanded isentropically until the temperature falls to 600°K. The values of specific heats are given by $c_p = 0.225 + 0.00005T$ and $c_v = 0.155 + 0.00005T$. Calculate : (i) Work done ; (ii) Pressure at the end of expansion. [**Ans.** (i) 792 kcal, (ii) 3.07 kgf/cm^2]

5. The compression ratio of an engine working on Diesel cycle is 13.5 and air-fuel ratio used is 28 : 1. The C.V. of the fuel used is 10,000 kcal/kg. If the temperature at the end of compression is 800°C, find out the percentage of stroke at which the cut-off occurs. [**Ans.** 8.2% of stroke]

6. The temperatures at the five salient points of a Dual combustion cycles are 363°K, 988°K, 1294°K, 2135°K and 950°K. Find the thermal efficiency of the cycle.

Assume : $c_p = 0.2395 + 0.00003T$

and $c_v = 0.171 + 0.00003T$, where T is in °K.

[**Ans.** 50.5%]

Internal Combustion Engines

23

23.1. HEAT ENGINES

Any type of engine or machine which derives heat energy from the combustion of fuel or any other source and converts this energy into mechanical work is termed as a ***heat engine.***

Heat engines may be classified into two main classes as follows :

1. External Combustion Engines.
2. Internal Combustion Engines.

1. External Combustion Engines (E.C. Engines)

In this case, *combustion of fuel takes place outside the cylinder* as in case of *steam engines* where the heat of combustion is employed to generate steam which is used to move a piston in a cylinder. Other examples of external combustion engines are *hot air engines, steam turbine* and *closed cycle gas turbine*. These engines are generally used for driving locomotives, ships, generation of electric power etc.

2. Internal Combustion Engines (I.C. Engines)

In this case, *combustion of the fuel with oxygen of the air occurs within the cylinder* of the engine. The internal combustion engines group includes engines employing mixtures of combustible gases and air, known as *gas engines*, those *using lighter liquid fuel* or spirit known as *petrol engines* and those using heavier liquid fuels, known as *oil compression ignition* or *diesel engines*.

The *external combustion engines* claim the following *advantages over internal combustion engines :*

1. Starting torque is generally high.
2. Because of external combustion of fuel, cheaper fuels can be used. Even solid fuels can be used advantageously.
3. Due to external combustion of fuel it is possible to have flexibility in arrangement.
4. These units are self-starting with the working fluid whereas in case of internal combustion engines, some additional equipment or device is used for starting the engines.

Reciprocating internal combustion engines offer the following *advantages over external combustion engines :*

1. Overall efficiency is high.
2. Greater mechanical simplicity.
3. Weight to power ratio is generally low.

4. Generally lower initial cost.
5. Easy starting from cold conditions.
6. These units are compact and thus require less space.

23.2. DEVELOPMENT OF I.C. ENGINES

Many experimental engines were constructed around 1878. The first really successful engine did not appear, however until 1879, when a German engineer Dr. Otto built his famous Otto gas engine. The operating cycle of this engines was based upon principles first laid down in 1860 by a French engineer named Bea de Rochas. The majority of modern I.C. engines operate according to these principles.

The development of the wellknown Diesel engine began about 1883 by Rudoff Diesel. Although this differs in many important respects from the Otto engine, the operating cycle of modern high speed Diesel engines is thermodynamically very similar to the Otto cycle.

23.3. CLASSIFICATION OF I.C. ENGINES

Internal combustion engines may be *classified* as given below :

1. According to cycle of operation :

(*i*) Two-stroke cycle engines
(*ii*) Four-stroke cycle engines.

2. According to cycle of combustion :

(*i*) Otto cycle engine (combustion at constant volume)
(*ii*) Diesel cycle engine (combustion at constant pressure)
(*iii*) Dual-combustion or Semi-Diesel cycle engine (combustion partly at constant volume and partly at constant pressure).

3. According to arrangement of cylinder :

(*i*) Horizontal engine (*ii*) Vertical engine
(*iii*) *V*-type engine (*iv*) Radial engine etc.

4. According to their uses :

(*i*) Stationary engine (*ii*) Portable engine
(*iii*) Marine engine (*iv*) Automobile engine
(*v*) Aero engine etc.

5. According to the fuel employed and the method of fuel supply to the engine cylinder :

(*i*) Oil engine (*ii*) Petrol engine
(*iii*) Gas engine (*iv*) Kerosene engine
(*v*) Carburettor, hot bulb, solid injection and air injection engine.

6. According to the speed of the engine :

(*i*) Low speed engine
(*ii*) Medium speed engine
(*iii*) High speed engine.

7. According to method of ignition :

(*i*) Spark ignition (S.I.) engine
(*ii*) Compression ignition (C.I.) engine.

8. According to method of cooling the cylinder :

(*i*) Air-cooled engine
(*ii*) Water-cooled engine.

9. According to method of Governing :

(*i*) Hit and miss governed engine
(*ii*) Quality governed engine
(*iii*) Quantity governed engine.

10. According to valve arrangement :

(*i*) Overhead valve engine
(*ii*) *L*-head type engine
(*iii*) *T*-head type engine
(*iv*) *F*-head type engine.

11. According to number of cylinders :

(*i*) Single cylinder engine
(*ii*) Multi-cylinder engine.

23.4. APPLICATIONS OF I.C. ENGINES

The I.C. engines are generally used for :

(*i*) Road vehicles (*e.g.*, scooter, motorcycle, buses etc.)
(*ii*) Air craft
(*iii*) Locomotives
(*iv*) Construction in civil engineering equipment such as bull-dozer, scraper, power shovels etc.
(*v*) Pumping sets
(*vi*) Cinemas
(*vii*) Hospital
(*viii*) Several industrial applications.

Note. Prime movers in all *construction equipment* are invariable I.C. engines, unless of course, when drive is electric. Use of steam source for this equipment is almost absolete.

23.5. BASIC IDEA OF I.C. ENGINES

The basic idea of internal combustion engine is shown in Fig. 23.1. The cylinder which is closed at one end is filled with a mixture of fuel and air. As the crankshaft turns it pushes cylinder. The piston is forced up and compresses the mixture in the top of the cylinder. The mixture is set alight and, as it burns, it creates a gas pressure on the piston, forcing it down the cylinder. This motion is shown by arrow '1'. The piston pushes on the rod which pushes on the crank. The crank is given rotary (turning) motion as shown by the arrow '2'. The flywheel fitted on the end of the crankshaft stores energy and keeps the crank turning steadily.

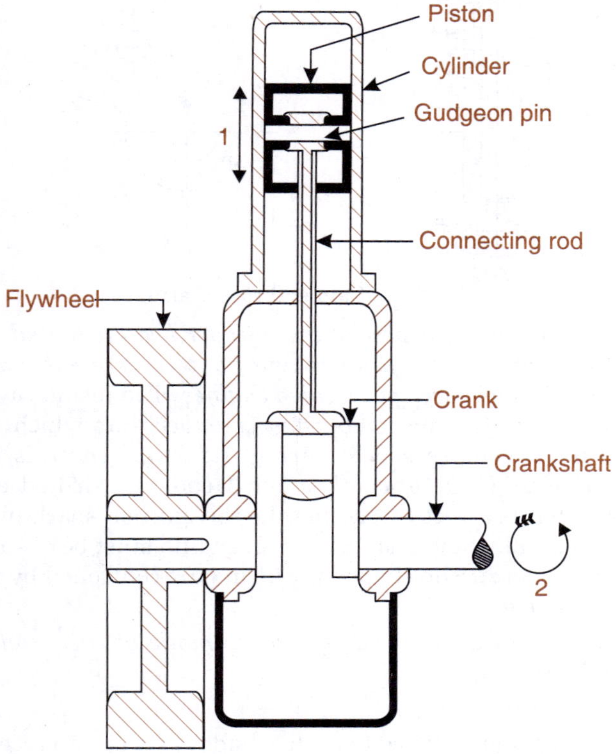

Fig. 23.1. Basic idea of I.C. engine.

23.6. DIFFERENT PARTS OF I.C. ENGINES

Here follows the detail of the various parts of an internal combustion engine.

A cross-section of an air-cooled I.C. engine with principal parts is shown in Fig. 23.2.

A. Parts common to both petrol and diesel engine :

1. Cylinder
2. Cylinder head
3. Piston
4. Piston rings
5. Gudgeon pin
6. Connecting rod
7. Crank
8. Crankshaft
9. Engine bearing
10. Crankcase
11. Flywheel
12. Governor
13. Valves and valve operating mechanism.

B. Parts for petrol engines only :

1. Spark plugs
2. Carburettor
3. Fuel pump.

C. Parts for Diesel engine only :

1. Fuel pump.
2. Injector.

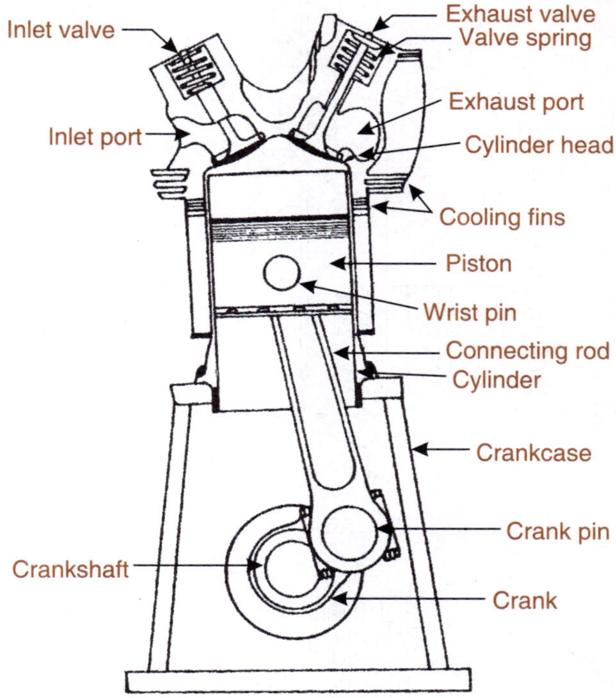

Fig. 23.2. Air-cooled I.C. engine.

A. Parts Common to Both Petrol and Diesel Engines

1. Cylinder

The cylinder contains gas under pressure and guides the piston. It is in direct contact with the products of combustion and it must be cooled. The ideal form consists of a plain cylindrical barrel in which the piston slides. The movement of the piston or stroke being in most cases, longer than the bore. This is known as the "*stroke bore ratio*". The upper end consists of a combustion or clearance space in which the ignition and combustion of the charge takes place. In practice, it is necessary to depart from the ideal hemispherical shape in order to accommodate the valves, sparking plugs etc. and to control the combustion. Sections of an air-cooled cylinder and a water-cooled cylinder are shown in Figs. 23.3 and 23.4 respectively. *The cylinder is made of hard grade cast iron and is usually cast in one piece.*

2. Cylinder head

One end of the cylinder is closed by means of a *removable cylinder head* (Fig. 23.3) which usually contains the inlet or admission valve [Fig. 23.5 (*a*)] for admitting the mixture of air and fuel and exhaust valve [Fig. 23.5 (*b*)] for discharging the product of combustion. Two valves are kept closed, by means of cams (Fig. 23.6) geared to the engine shaft. The passage in the cylinder head leading to and from the valves are called *ports*. The pipes which connect the inlet ports of the various cylinders to a common intake pipe for the engine is called the inlet *manifold*. If the exhaust ports are similarly connected to a common exhaust system, this system of piping is called *exhaust manifold*.

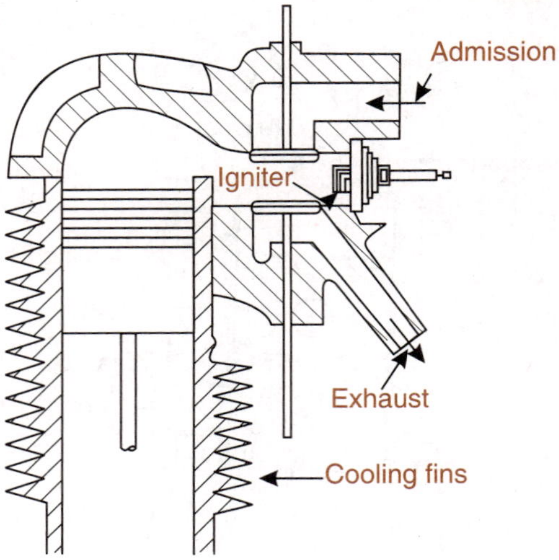

Fig. 23.3. Air-cooled cylinder.

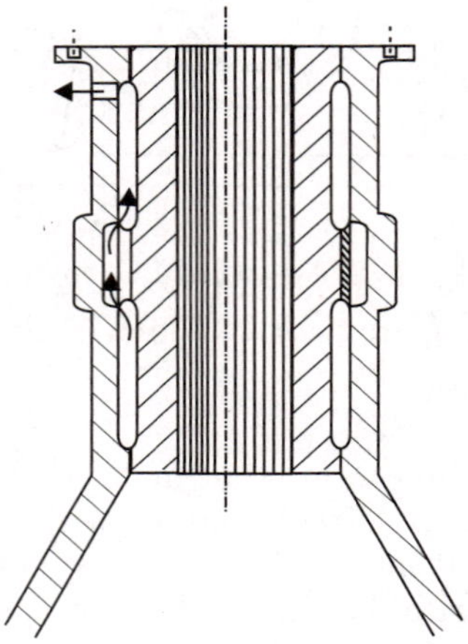

Fig. 23.4. Water-cooled cylinder.

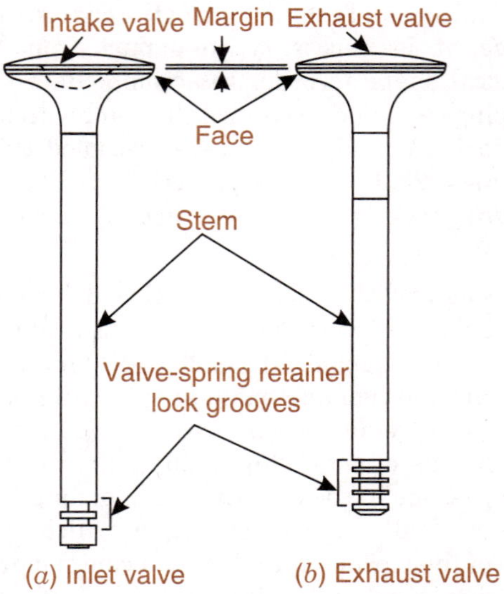

Fig. 23.5

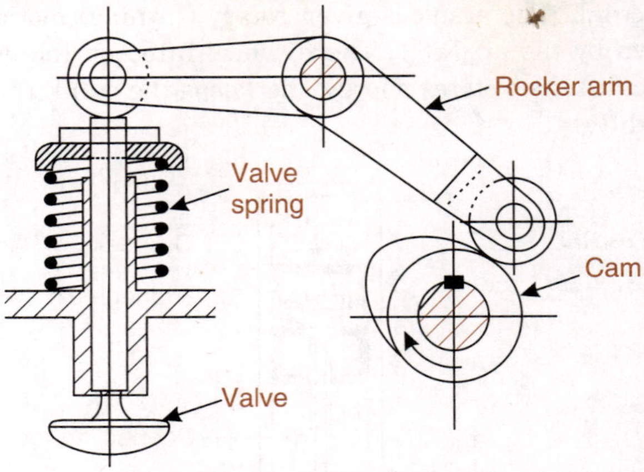

Fig. 23.6. Cam and rocker arm.

The main purpose of the cylinder head is to seal the working ends of the cylinders and not to permit entry and exit of gases on cover head valve engines. The inside cavity of head is called the *combustion chamber*, into which the mixture is compressed for firing. Its *shape controls the direction and rate of combustion*. Heads are drilled and tapped with correct thread to take the ignition spark plug. All the combustion chambers in an engine must be of same shape and size. The shape may be in part controlled by the piston shape.

The cylinder head is usually made of cast iron or aluminium.

3. Piston

A piston is fitted to each cylinder as a face to receive gas pressure and transmit the thrust to the connecting rod.

The piston must (*i*) give gas tight seal to the cylinder through bore, (*ii*) slide freely, (*iii*) be light and (*iv*) be strong. The thrust on the piston on the power stroke tries to tilt the piston as the connecting rod swings, sideways. The piston wall, called the *skirt* must be strong enough to stand upto this side thrust. *Pistons are made of cast iron or aluminium alloy for lightness*. Light alloy pistons expand more than cast iron one therefore they need large clearances to the bore, when cold, or special provision for expansion. Pistons may be solid skirt or split skirt. A section through a split skirt piston is shown in Fig. 23.7.

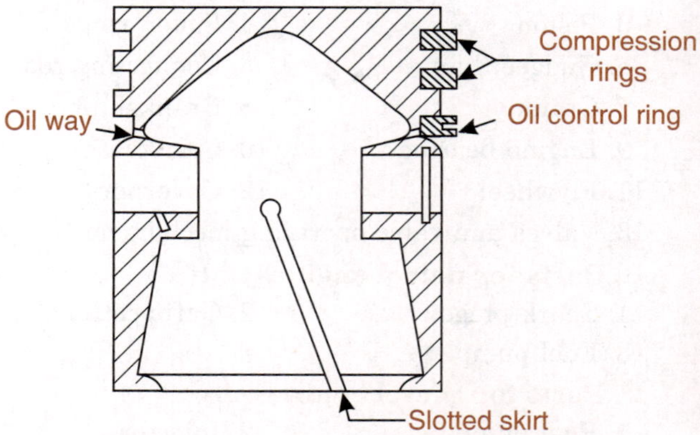

Fig. 23.7. Section through a split skirt piston.

4. Piston rings

The piston must be a fairly loose fit in the cylinder. If it were a tight fit, it would expand as it got hot and might stick tight in the cylinder. If a piston sticks it could ruin the engine. On the other hand, if there is too much clearance between the piston and cylinder walls, much of the pressure from the burning gasoline vapour will leak past the piston. This means, that the push on the piston will be much less effective. It is the push on the piston that delivers the power from the engines.

To provide a good sealing fit between the piston and cylinder, pistons are equipped with piston rings, as shown in Fig. 23.8. The rings are usually made of cast iron of fine grain and high elasticity which is not affected by the working heat. Some rings are of alloy spring steel. They are split at one point so that they can be expanded and slipped over the end of the piston and into ring grooves which have been cut in the piston. When the piston is installed in the cylinder the rings are compressed into ring grooves which have been cut in the piston. When the piston is installed in the cylinder, the rings are compressed into the ring grooves so that the split ends come almost together. The rings fit tightly against the cylinder wall and against the sides of the ring grooves in the piston. Thus, they form a good seal between the piston and the cylinder wall. The rings can expand or contract as they heat and cool and still make a good deal. Thus they are free to slide up and down the cylinder wall.

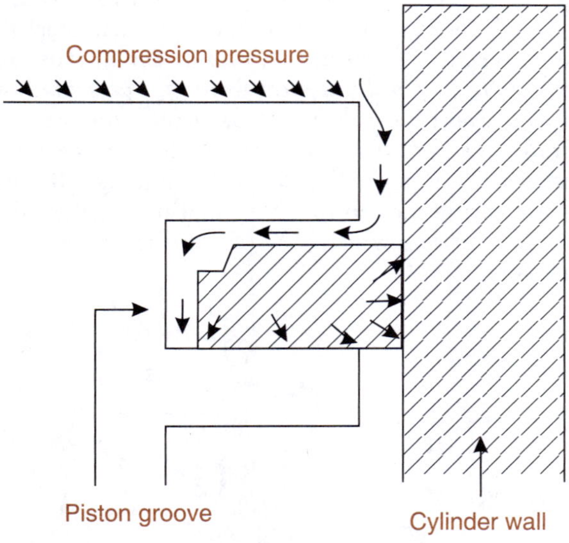

Fig. 23.8. Working of a piston ring.

Fig. 23.8 shows how the piston ring works to hold in the compression and combustion pressure. The arrows show the pressure above the piston passing through clearance between the piston and the cylinder wall. It presses down against the top and against the back of the piston rings as shown by the arrows. This pushes the piston ring firmly against the bottom of the piston ring groove. As a result there are good seals at both of these points. The higher the pressure in the combustion chamber, the better the seal.

Small two-stroke cycle engines have two rings on the piston. Both are compression rings (Fig. 23.9). Two rings are used to divide up the job of holding the compression and combustion pressure. This produces better sealing with less ring pressure against the cylinder wall.

Four-stroke cycle engines have an extra ring, called the oil control ring (Fig. 23.10). Four stroke cycle engines are so constructed that they get much more oil in the cylinder wall than do two stroke cycle engines. This additional oil must be scraped off to prevent it from getting up into the combustion chamber, where it would burn and cause trouble.

Refer to Figs. 23.9 and 23.10 the compression rings have a rectilinear cross-section and oil rings are provided with a groove in the middle and with through holes spaced at certain interval from each other. The oil collected from the cylinder walls flows through these holes into the piston groove whence through the holes in the body of the piston and down its inner walls into the engine crankcase.

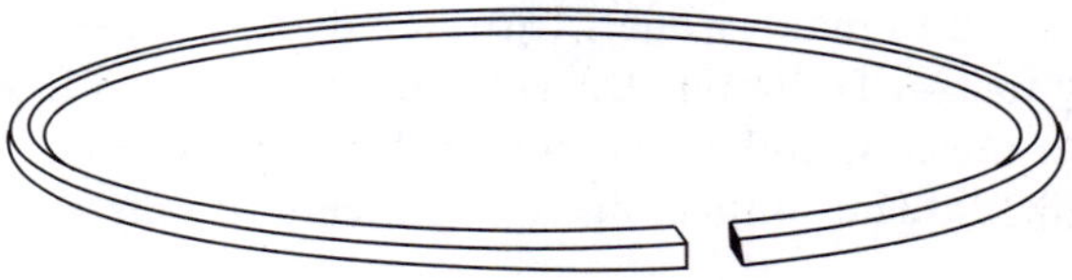

Fig. 23.9. Compression ring.

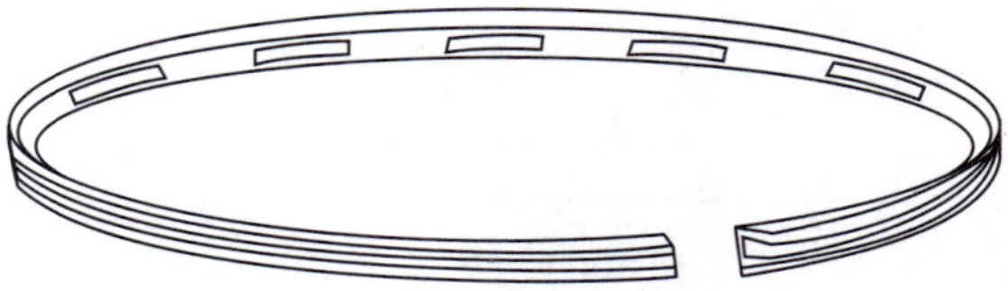

Fig. 23.10. Oil ring.

5. Gudgeon pin (or wrist pin or piston pin)

These are *hardened steel parallel spindles* fitted through the piston bosses and the small end bushes or eyes to allow the connecting rods to swivel. Gudgeon pins are a press fit in the piston bosses of light alloy pistons when cold. For removal or fitting, the piston should be dipped in hot water or hot oil, this expands the bosses and the pins can be removed or fitted freely without damage.

It is made hollow for lightness since it is a reciprocating part.

6. Connecting rod

Refer to Fig. 23.11 (on next page). The connecting rod transmits the piston load to the crank, causing the latter to turn, thus converting the reciprocating motion of the piston into a rotary motion of the crankshaft. The lower or "big end" of the connecting rod turns on "crank pins".

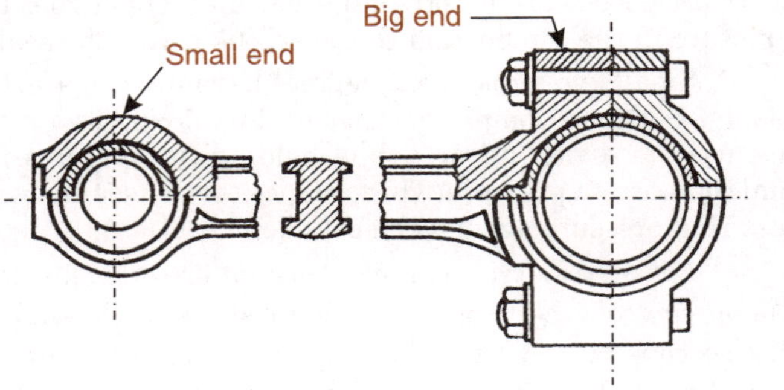

Fig. 23.11. Connecting rod.

The connecting rods are made of nickel, chrome and chrome vandium steels. For small engines the material may be aluminium.

7. Crank

The piston moves up and down in the cylinder. This up and down motion is called reciprocating motion. The piston moves in a straight line. The straight line motion must be changed to rotary, or turning motion, in most machines, before it can do any good. That is rotary motion is required to make wheels turn, a cutting blade spin or a pulley rotate. To change the reciprocating motion to rotary motion a crank and connecting rod are used. (Figs. 23.12 and 23.13). The connecting rod connects the piston to the crank.

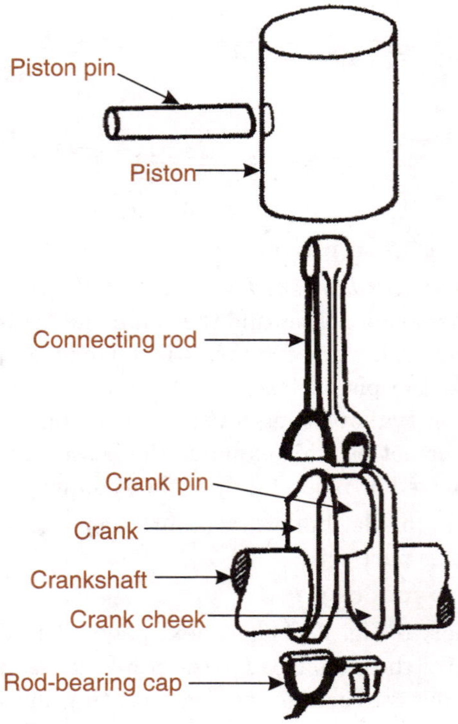

Fig. 23.12

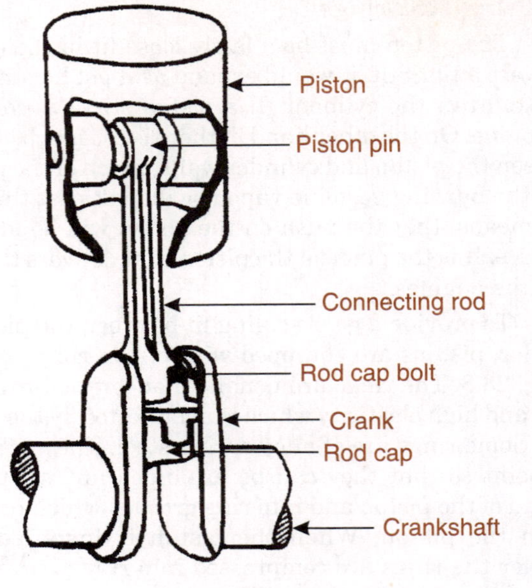

Fig. 23.13

Note. The crank end of the connecting rod is called rod "big end". The piston end of the connecting rod is called the rod "small end".

8. Crankshaft

The crank is part of the crankshaft. The crankshaft of an internal combustion engine receives via its cranks the efforts supplied by the pistons to the connecting rods. All the engines auxiliary mechanisms with mechanical transmission are geared in one way or the another to the crankshaft. *It is usually a steel forging, but some makers use special types of cast iron such as spheroidal graphitic or nickel alloy castings* which are cheaper to produce and have good service life. Refer to Fig. 23.14. The crankshaft converts the reciprocating motion to rotary motion. The crankshaft mounts in bearings which, encircle the journals so it can rotate freely.

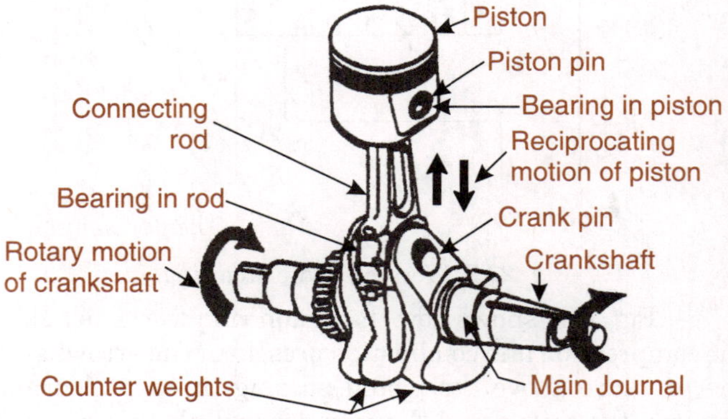

Fig. 23.14. Crankshaft and other parts.

The shape of the crankshaft *i.e.,* the mutual arrangement of the cranks depend on the number and arrangement of cylinders and the turning order of the engine. Fig. 23.15 shows a typical crankshaft layout for a four-cylinder engine.

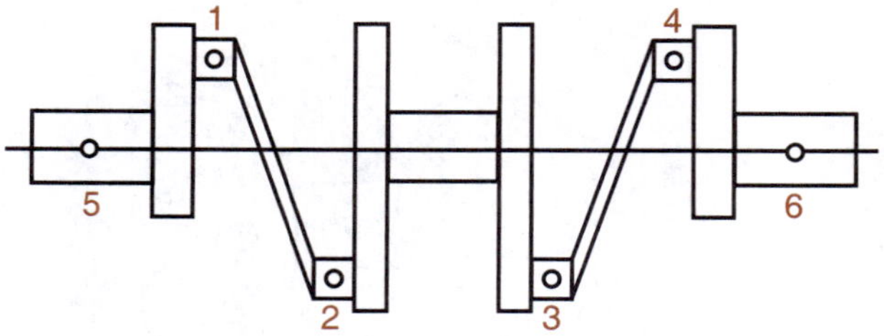

Fig. 23.15. Typical crankshaft layout.

9. Engine bearing

The crankshaft is supported by bearing. The connecting rod big end is attached to the crank pin on the crank of the crankshaft by a bearing. A piston pin at the rod small end is used to attach the rod to the piston. The piston pin rides in bearings. Everywhere there is rotary action in the engine, bearings are used to support the moving parts. The purpose of bearing is to reduce the friction and allow the parts to move easily. Bearings are lubricated with oil to make the relative motion easier.

Bearings used in engines are of two types : *sliding* or *rolling* (Fig. 23.16).

The sliding type of bearings are sometimes called bushings or sleeve bearings because they are in the shape of a sleeve that fits around the rotating journal or shaft. The sleeve-type connecting rod big end bearings usually called simply rod bearings and the crankshaft supporting bearings called the main bearings are of the split sleeve type. They must be split in order to permit their assembly into the engine. In the rod bearing, the upper half of the bearing is installed in the rod, the lower half is installed in the rod bearing cap. When the rod cap is fastened to the rod shown in Fig. 23.13 a complete sleeve bearing is formed. Likewise, the upper halves of the main bearings are assembled in the engine and then the main bearing caps, with the lower bearing halves are attached to the engine to complete the sleeve bearings supporting the crankshaft.

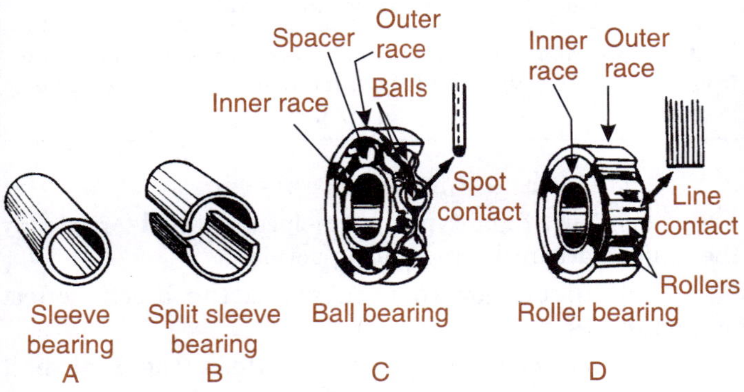

Fig. 23.16. Bearings.

The typical bearing half is made of steel or bronze back to which a lining of relatively soft bearing material is applied. Refer to Fig. 23.17. This relatively soft bearing material, which is made of several materials such as copper, lead, tin and other metals, has the ability to conform to slight irregularities of the shaft rotating against it. If wear does take place, it is the bearing that wears and the bearing can be replaced instead of much more expansive crankshaft or other engine part.

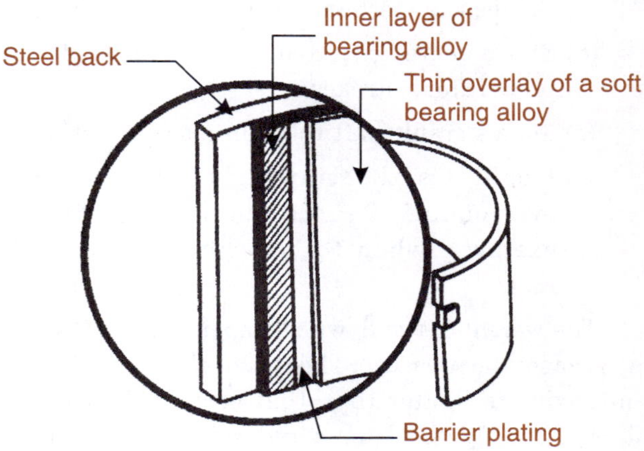

Fig. 23.17. Bearing half (details).

The rolling-type bearing uses balls or rollers between the stationary support and the rotating shaft. Refer to Fig. 23.16. Since the balls or rollers provide rolling contact, the frictional resistance to movement is much less. In some roller bearing, the rollers are so small that they are hardly bigger than needles. These bearings are called *needle bearings*. Also some rollers bearings have the rollers set at an angle to the races, the rollers roll in are tapered. These bearings are called *tapered roller bearings*. Some ball and roller bearings are sealed with their lubricant already in place. Such bearings require no other lubrication. Other do require lubrication from the oil in the gasoline (two-stroke cycle engines) or from the engine lubrication system (four-stroke cycle engines).

The type of bearing selected by the designers of the engine depends on the design of the engine and the use to which the engine will be put. *Generally, sleeve bearings, being less expensive and satisfactory for most engine applications, are used. In fact sleeve bearings are used almost universally in automobile engines. But you will find some engines with ball and roller bearings to support the crankshaft and for the connecting rod and piston-pin bearings.*

10. Crankcase

The main body of the engine to which the cylinders are attached and which contains the crankshaft and crankshaft bearing is called *crankcase*. This member also holds other parts in alignment and resists the explosion and inertia forces. It also protects the parts from dirt etc. and serves as a part of lubricating system.

11. Flywheel

Refer to Figs. 23.1 and 23.18. A flywheel (steel or cast iron disc) secured on the crankshaft performs the following *functions* :

(*a*) Brings the mechanism out of dead centres.

(*b*) Stores energy required to rotate the shaft during preparatory strokes.

(*c*) Makes crankshaft rotation more uniform.

(*d*) Facilitates the starting of the engine and overcoming of short time over loads as, for example, when the machine is started from rest.

The weight of the flywheel depends upon the nature of variation of the pressure. The flywheel for a double acting steam engine is lighter than that of a single-acting one. Similarly, the flywheel for a two-stroke cycle engine is lighter than a flywheel used for a four-stroke cycle engine. *Lighter flywheels are used for multi-cylinder engines.*

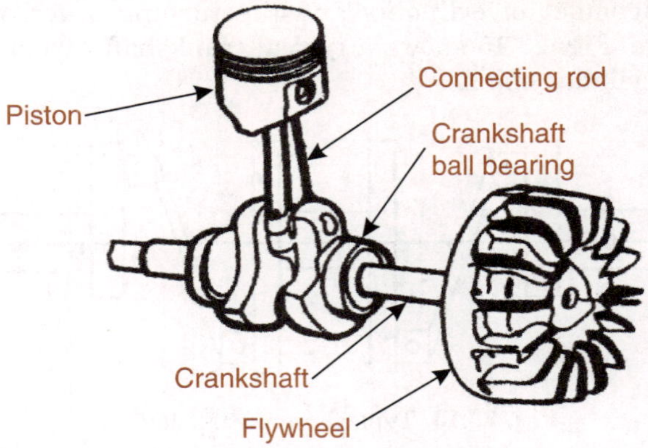

Fig. 23.18. Flywheel secured on crankshaft.

12. Governor

A governor may be defined *as a device for regulating automatically output of a machine by regulating the supply of working fluid*. When the speed decreases due to increase in load the supply valve is opened by mechanism operated by the governor and the engine therefore speeds up again to its original speed. If the speed increases due to a decreases of load the governor mechanism closes the supply valve sufficiently to slow the engine to its original speed. *Thus the function of a governor is to control the fluctuations of engine speed due to changes of load.*

Comparison between a Flywheel and a Governor

S. No.	*Flywheel*	*Governor*
1.	It is provided on engines and fabricating machines *viz.*, rolling mills, punching machines ; shear machines, presses etc.	It is provided on prime movers such as engines and turbines.
2.	Its function is to store the available mechanical energy when it is in excess of the load requirement and to part with the same when the available energy is less than that required by the load.	Its function is to regulate the supply of driving fluid producing energy, according to the load requirement so that at different loads almost a constant speed is maintained.
3.	It works continuously from cycle to cycle.	It works intermittently *i.e.,* only when there is change in load.
4.	In engines it takes care of fluctuations of speed during thermodynamic cycle.	It takes care of fluctuations of speed due to variation of load over long range of working engines and turbines.
5.	In fabrication machines it is very economical to use it in that it reduces capital investment on prime movers and their running expenses.	But for governor, there would have been unnecessarily more consumption of driving fluid. Thus it economies its consumption.

Types of governors :

Governors are classified as follows :

1. Centrifugal governor

(*a*) Gravity controlled, in which the centrifugal force due to the revolving masses is largely balanced by gravity.

(*b*) Spring controlled, in which the centrifugal force is largely balanced by springs.

2. Inertia and flywheel governors

(*a*) Centrifugal type, in which centrifugal forces play the major part in the regulating action.

(*b*) Inertia governor, in which the inertia effect predominates.

The *inertia type* governors are fitted to the crankshaft or flywheel of an engine and so differ radically in appearance from the centrifugal governors. The balls are so arranged that the inertia force caused by an angular acceleration or retardation of the shaft tends to alter their positions. The amount of displacement of governor balls is controlled by suitable springs and through the governor mechanism, alters the fuel supply to the engine. The inertia governor is more sensitive than centrifugal but it becomes very difficult to balance the revolving parts. For this reason *centrifugal governors are more frequently used*. We shall discuss centrifugal governors only.

Important centrifugal governors are :

1. Watt governor
2. Porter governor
3. Proell governor
4. Hartnell governor.

1. Watt governor

It is the primitive governor as used by Watt on some of his early steam engines. It is used for a very slow speed engine and this is why it has now become obsolete.

Refer to Fig. 23.19. Two arms are hinged at the top of the spindle and two revolving balls are fitted on the other ends of the arms. One end of each of the links are hinged with the arms, while the other ends are hinged with the sleeve, which may slide over the spindle. The speed of the crankshaft is transmitted to the spindle through a pair of bevel gears by means of a suitable arrangement. So the rotation of the spindle of the governor causes the weights to move away from the centre due to the centrifugal force. This makes the sleeve to move in the upward direction. This movement of the sleeve is transmitted by the lever to the throttle valve which partially closes or opens the steam pipe and reduces or increases the supply of steam to the engine. So the engine speed may be adjusted to a normal limit.

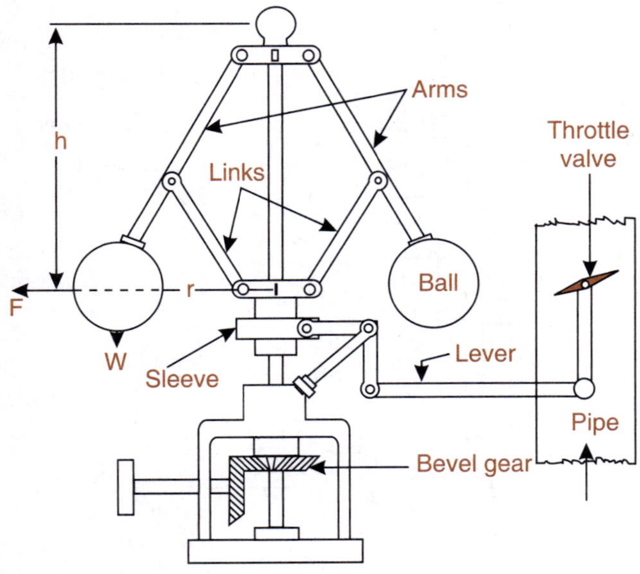

Fig. 23.19. Watt governor.

2. Porter governor

Fig. 23.20 shows diagrammatically a Porter governor where two or more masses called the governor balls rotate about the axis of the governor shaft which is driven through suitable gearing from the engine crankshaft. The governor balls are attached to the arms. The lower arms are attached to the *sleeve which acts as a central weight*. If the speed of the rotation of the balls increases owing to a decrease of load on the engine, the governor balls fly outwards and the sleeve moves upwards thus closing the fuel passage till the engine speed comes back to its designed speed. If the engine speed decreases owing to an increase of load, the governor balls fly inwards and the sleeve moves downwards thus opening the fuel passage more for oil till the engine speed comes back to its designed speed. The engine is said to be running at its designed speed when the outward inertia or centrifugal force is just balanced by the inward controlling force.

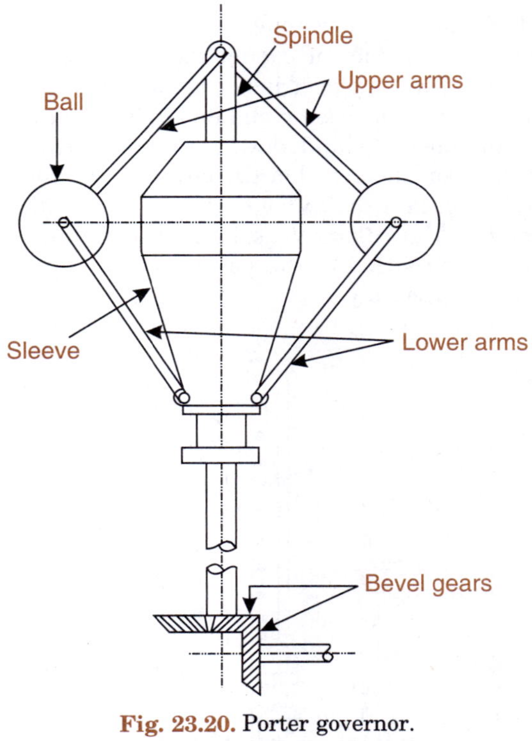

Fig. 23.20. Porter governor.

3. Proell governor

Refer to Fig. 23.21. It is a modification of porter governor. The governor balls are carried on an *extension of the lower arms*. For given values of weight of the ball, weight of the sleeve and height of the governor, a Proell governor runs at a *lower speed* than a Porter governor. *In order to give the same equilibrium speed a ball of smaller mass may be used in Proell governor*.

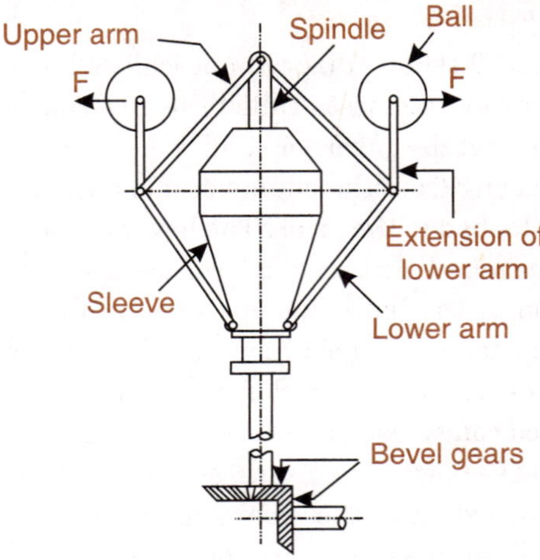

Fig. 23.21. Proell governor.

4. Hartnell governor

The Hartnell governor is a spring loaded governor in which the controlling force, to a great extent, is provided by the spring thrust.

Fig. 23.22 shows one of the types of Hartnell governors. It consists of casing fixed to the spindle. A compressed spring is placed inside the casing which presses against the top of the casing and on adjustable collars. The sleeve can move up and down on the vertical spindle depending upon the speed of the governor. Governor balls are carried on bell crank lever which are pivoted on the lower end of the casing. The balls will fly outwards or inwards as the speed of the governor shaft increases or decreases respectively.

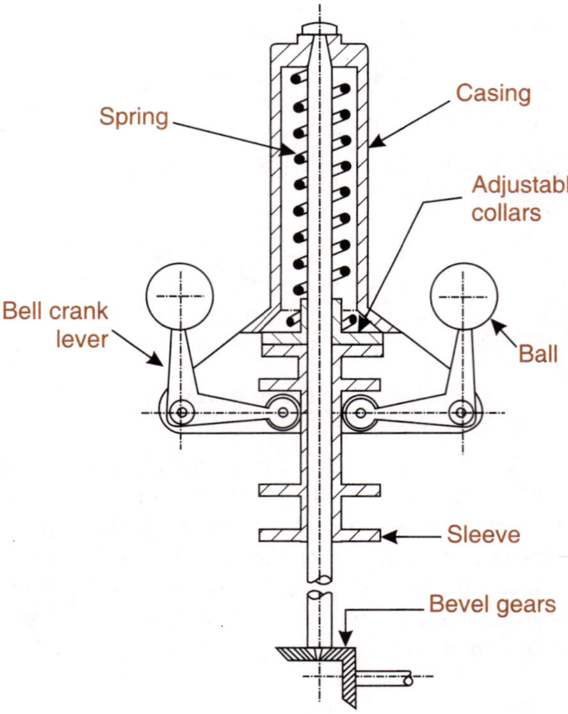

Fig. 23.22. Hartnell governor.

13. Valves and valve operating mechanism.

With few exceptions the inlet and exhaust of internal combustion engines are controlled by poppet valves. These valves are held to their seating by strong springs, and as the valves usually open inwards, the pressure in the cylinder helps to keep them closed. The valves are lifted from their seats and the ports opened either by cams having projecting portion designed to give the period of opening required or by eccentrics operating through link-work. Of these two methods the cam gear is more commonly used, but in either case it is necessary that the valve gear shaft of an engine should rotate but once from beginning to end of a complete cycle, however many strokes may be involved in the completion of that cycle. This is necessary to secure a continuous regulation of the valve gear as required. For this purpose the cams or eccentrics of four-stroke engines are mounted on shafts driven by gearing at half the speed of the crankshaft. The curves used for the acting faces of the cams depend on the speed of the engine and rapidity of valve opening desired.

Fig. 23.23 shows a valve gear for I.C. engine. It consists of poppet valve, the steam bushing or guide, valve spring, spring retainer, lifter or push rod, camshaft and half speed gear for a four-stroke engine. The poppet valve, in spite of its shortcomings of noise and difficulties of cooling is

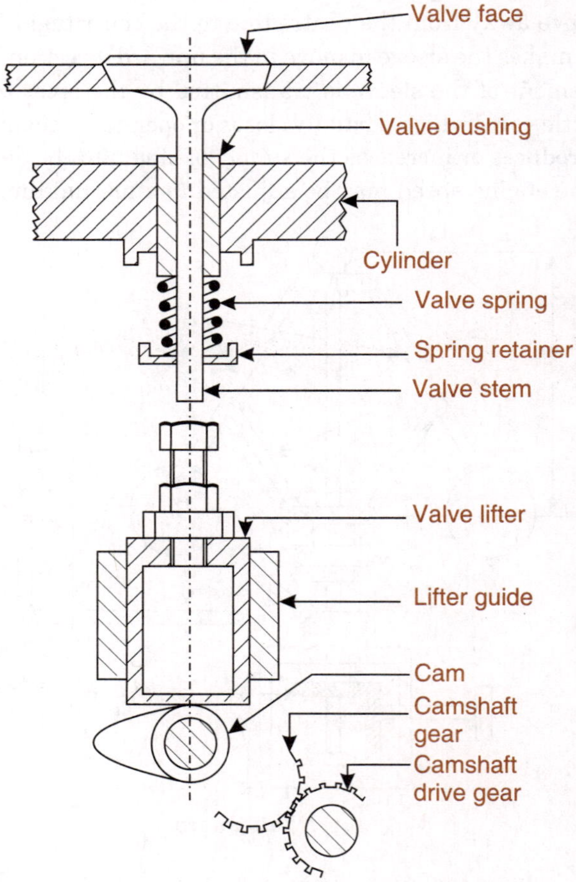

Fig. 23.23. Valve gear for I.C. engine.

commonly used due to its simplicity and capacity for effective sealing under all operating conditions. The valve is subjected to very heavy duty. It holds in combustion chamber and is exposed to high temperatures of burning gases. Exhaust valve itself may attain a high temperature while external cooling is not available. Special heat resisting alloys are therefore used in the construction of the exhaust valve and it may sometimes have a hollow construction filled with mineral salts to provide for heat dissipation. The salts become liquid when valve is working and transfer heat from the head to the stem from which it is carried through the stem guide to the cylinder block.

The timing of the valves *i.e.,* their opening and closing with respect to the travel of the piston is very important thing for efficient working of the engine. The drive of the camshaft is arranged through gears or chain and sprocket (called timing gear or timing chain). Any wearing of the gears or chain and sprocket would result in disturbing the precise timing of the valves. It is desirable, therefore, to avoid use of multiple gears of long chains in the camshaft drive.

Valve timing

Theoretically the valves open and close at top dead centre (T.D.C.) or at bottom dead centre (B.D.C.) but practically they do so sometime before or after the piston reaches the upper or lower limit of travel. There is a reason for this. Look at the inlet valve, for example. It normally opens several degrees of crankshaft-rotation before T.D.C. on the exhaust stroke. That is the intake valve begins to open before the exhaust stroke is finished. This gives the valve enough time to reach the fully open position before the intake stroke begins. Then, when the intake stroke starts, the intake valve is already wide open and air-fuel mixture can start to enter the cylinder, immediately. Likewise the intake valve remains open for quite a few degrees of crankshaft rotation after the piston has passed B.D.C. at the end of the intake stroke. This allows additional time for air-fuel mixture to continue to flow into the cylinder. The fact that the piston has already passed B.D.C. and is moving up or the compression stroke while the intake valve is still open does not effect the movement of air-fuel mixture into the cylinder. Actually air-fuel mixture is still flowing in as the intake valve starts to close.

This is due to the fact that air-fuel mixture has inertia. That is, it attempts to keep on flowing after it once starts through the carburettor and into the engine cylinder. The momentum of the mixture then keeps it flowing into the cylinder even though the piston has started up on the compression stroke. This packs more air-fuel mixture into the cylinder and results in a stronger power stroke. In other words, this improves *volumetric efficiency*.

For a somewhat similar reason, the exhaust valve opens well before the piston reaches B.D.C. on the power stroke. As the piston nears B.D.C., most of the push on the piston has ended and nothing is lost by opening the exhaust valve towards the end of the power stroke. This gives the exhaust gases additional time to start leaving the cylinder so that exhaust is well started by the time the piston passes B.D.C. and starts up on the exhaust stroke. The exhaust valve then starts opening for some degrees of crankshaft rotation after the piston has passed T.D.C. and intake stroke has started. This makes good use of momentum of exhaust gases. They are moving rapidly towards the exhaust port, and leaving the exhaust valve open for a few degrees after the intake stroke starts giving the exhaust gases some additional time to leave the cylinder. This allows more air-fuel mixture to enter on the intake stroke so that the stronger power stroke results. That is, it improves volumetric efficiency.

The actual timing of the valves varies with different four-stroke cycle engines, but the typical example for an engine is shown in Fig. 23.24. Note that the inlet valve opens 15° of crankshaft rotation before T.D.C. on the exhaust stroke and stays open until 50° of crankshaft rotation after B.D.C. on the compression stroke. The exhaust valve opens 50° before B.D.C. on the power stroke and stays open 15° after T.D.C. on the inlet stroke. This gives the two valves an overlap of 30° at the end of exhaust stroke and beginning of the *compression stroke*.

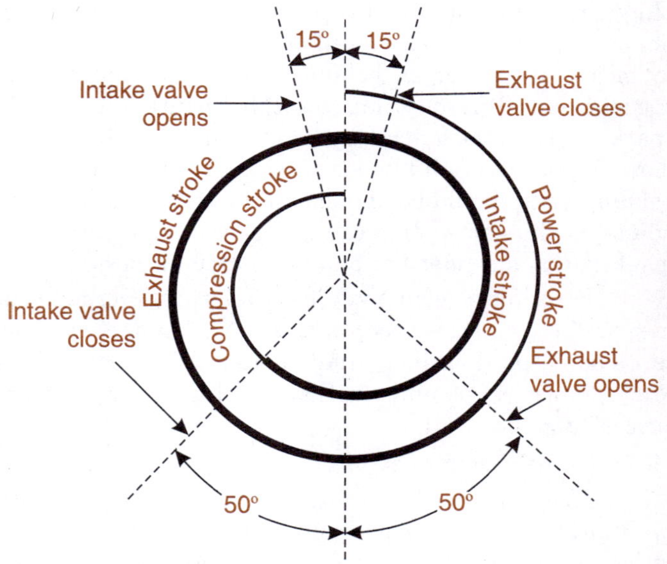

Fig. 23.24. Typical valve timing diagram.

B. Parts Common to Petrol Engine Only

Spark plug :

The main function of a spark plug is to conduct the high potential from the ignition system into the combustion chamber. It provides the proper gap across which spark is produced by applying high voltage, to ignite the combustion chamber.

A spark plug entails the following requirements :

(*i*) It must withstand peak pressures up to at least 55 bar.

(*ii*) It must provide suitable insulation between two electrodes to prevent short circuiting.

(*iii*) It must be capable of withstanding high temperatures to the tune of 2000°C to 2500°C over long periods of operation.

(*iv*) It must offer maximum resistance to erosion burning away of the spark points irrespective of the nature of fuel used.

(*v*) It must possess a high heat resistance so that the electrodes do not become sufficiently hot to cause the preignition of the charge within the engine cylinder.

(*vi*) The insulating material must withstand satisfactorily the chemical reaction effects of the fuel and hot products of combustion.

(*vii*) Gas tight joints between the insulator and metal parts are essential under all operating conditions.

Refer to Fig. 23.25. The spark plug consists of a metal shell having two electrodes which are insulated from each other with an air gap. High tension current jumping from the supply electrode produces the necessary spark. Plugs are sometimes identified by the heat range or the relative temperature obtained during operation. The correct type of plug with correct width of gap between the electrodes are important factors. The spark plug gap can be easily checked by means of a feeler gauge and set as per manufacturer's specifications. It is most important that while adjusting the spark plug it is the outer earthed electrode *i.e.*, tip which is moved in or out gradually for proper setting of the gap. No bending force should be applied on the centre-electrode for adjusting the gap as this can cause crack and fracture of insulation and sender the plug absolutely useless.

Porcelain is commonly used as insulating material in spark plugs, as it is cheap and easy to manufacture. Mica can also be used as insulating material for spark plugs. Mica, however, cannot withstand high temperatures successfully.

Simple carburettor :

The function of a carburettor is to atomise and metre the liquid fuel and mix it with the air as it enters the induction system of the engine, maintaining under all conditions of operation fuel-air proportions appropriate to those conditions.

All modern carburettors are based upon Bernoulli's theorem,

$$C^2 = 2gh$$

where C is the velocity in metres/sec, g is the acceleration due to gravity in metre/sec^2 and h is the head causing the flow expressed in metres of height of a column of the fluid.

The equation of mass rate of flow is given by,

$$m = \rho A\sqrt{2gh}$$

where ρ is the density of the fluid and A is the cross-sectional area of fluid stream.

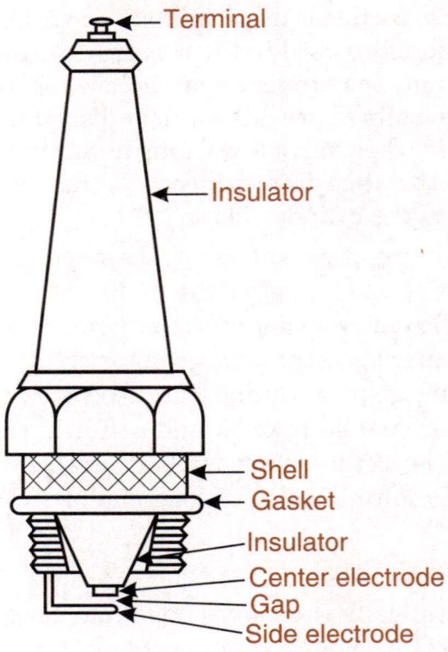

Fig. 23.25. Spark plug.

In Fig. 23.26 is shown simple carburettor. L is the float chamber for the storage of fuel. The fuel supplied under gravity action or by fuel pump enters the float chamber through the filter F. The arrangement is such that when the oil reaches a particular level the float valve M blocks the inlet passage and thus cuts off the fuel oil supply. On the fall of oil level, the float descends down, consequently intake passage opens and again the chamber is filled with oil. Then the float and the float valve maintains a constant fuel oil level in the float chamber. N is the jet from which the fuel is sprayed into the air stream as it enters the carburettor at the inlet S and passes through the throat or venturi R. The fuel level is slightly below the outlet of the jet when the carburettor is inoperative.

As the piston moves down in the engine cylinder, suction is produced in the cylinder as well as in the induction manifold Q as a result of which air flows through the carburettor. The velocity of air increases as it passes through the constriction at the venturi R and pressure decreases due to conversion of a portion of pressure head into kinetic energy. Due to decreased pressure at the venturi and hence by virtue of difference in pressure (between the float chamber and the venturi) the jet issues fuel oil into air stream. Since the jet has a very fine bore, the oil issuing from the jet is in the form of fine spray ; it vapourises quickly and mixes with the air. This air-fuel mixture enters the engine cylinder ; its quantity being controlled by varying the position of the throttle valve T.

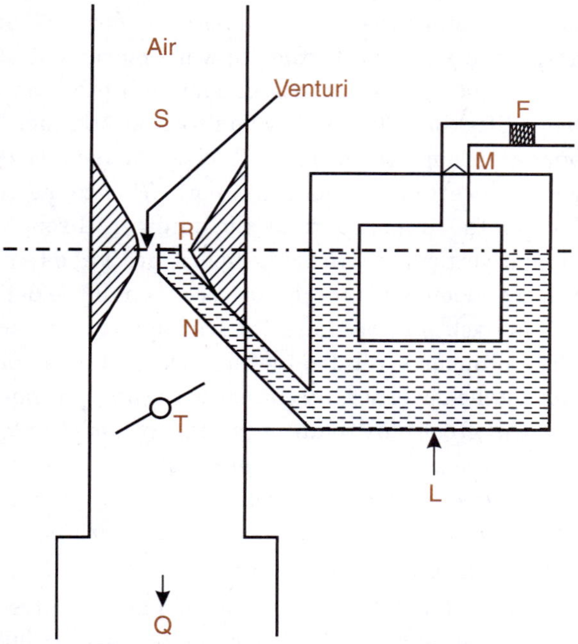

Fig. 23.26. Simple carburettor.

Limitations :

(*i*) Although theoretically the air-fuel ratio supplied by a simple (single jet) carburettor should remain constant as the throttle goes on opening, actually it provides increasingly richer mixture as the throttle is opened. This is because of the reason that the density of air tends to decrease as the rate of flow increases.

(*ii*) During idling, however, the nearly closed throttle causes a reduction in the mass of air flowing through the venturi. At such low rates of air flow, the pressure difference between the float chamber and the fuel discharge nozzle becomes very small. It is sufficient to cause fuel to flow through the jet.

(*iii*) Carburettor does not have arrangement for providing rich mixture during starting and warm up.

In order to *correct for faults* :

(*i*) number of compensating devices are used for (*ii*) an idling jet is used which helps in running the engine during idling. For (*iii*) choke arrangement is used.

Air-fuel mixtures

- The theoretically correct mixture of air and petrol is 15 : 1 (approximately). Thus, the uniform supply of such mixture would result in burning without leaving excess of air of fuel. But it is difficult to get such a mixture in actual practice. When there is too little air, some of the fuel goes unburnt or simply changed to carbon. When there is too much air in mixture, it burns slowly and erratically and there is a loss of power.

  There is however a *range of mixtures* between which combustion will take place.

  — The '*lower limit*' is approximately 7 : 1 to 10 : 1. This mixture is *barely explosive*.

  — The 'upper limit' is approximately 20 : 1. The mixture burns *irregularly*.

The above limits will also vary with the *characteristics* of *the fuel*, the *shape of the combustion space and the temperature and pressure in the combustion space.*

- *Mixture requirements of automotive engines :*

  — For "*average cruising speeds*" the air-fuel ratio is approximately from 15 : 1 to 7 : 1.

  — In order to obtain "*maximum power*" to be able to accelerate the engine quickly, a richer mixture of the ratio of about 12 : 1 is desirable. This is also called *maximum power ratio*. To start the engine *from cold* even a mixture richer than this but above the lower limit is obtained.

  — For "*maximum economy*", *i.e.,* less fuel consumption for unit power, the mixture ratio may be approximately 16 : 1 to 17 : 1. The ratio, however, *entails loss of power*.

Hence, it is essential that the carburettor should be so designed that proper proportions of air and fuel are obtained to meet these varying operating conditions.

Fuel pump (for carburettor-petrol engine).

Refer to Fig. 23.27. This type of pump is used in petrol engine for supply of fuel to the carburettor. Due to rotation of the crankshaft the cam pushes the lever in the upward direction. One end of the lever is hinged while the other end pulls the diaphragm rod with the *diaphragm*. So the diaphragm comes in the downward direction against the compression of the spring and thus a vacuum is produced in the pump chamber. This causes the fuel to enter into the pump chamber from the *glass bowl* through the *strainer* and the inlet valve, the impurities of the fuel ; if there is any, deposit at the bottom of the glass bowl. On the return stroke the spring pushes the diaphragm in the upward direction forcing the fuel from the pump chamber into the carburettor through the *outlet valve*.

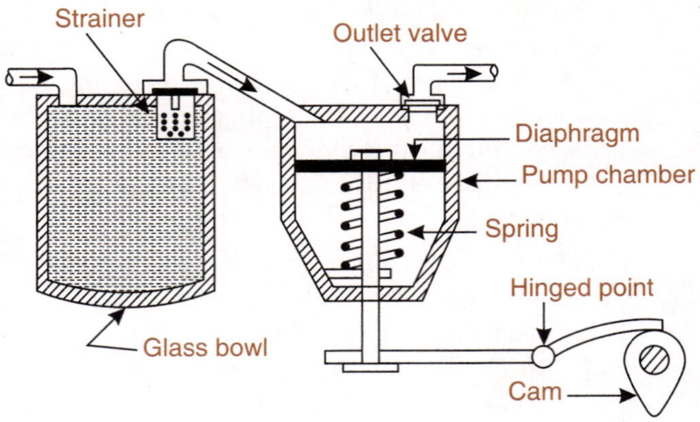

Fig. 23.27. Fuel pump for carburettor.

C. Parts for Diesel Engine Only

FUEL PUMP

Refer to Fig. 23.28. *L* is the plunger which is driven by a cam and tappet mechanism at the bottom (not shown in the figure) *B* is the barrel in which the plunger reciprocates. There is the rectangular vertical groove in the plunger which extends from top to another helical groove. *V* is the delivery valve which lifts off its seat under the liquid fuel pressure and against the spring force (*S*). The fuel pump is connected to fuel atomiser through the passage *P*, *SP* and *Y* are the spill and supply ports respectively. When the plunger is at its bottom stroke the ports *SP* and *Y* are uncovered (as shown in Fig. 23.28) and oil from low pressure pump (not shown) after being filtered is forced into the barrel.

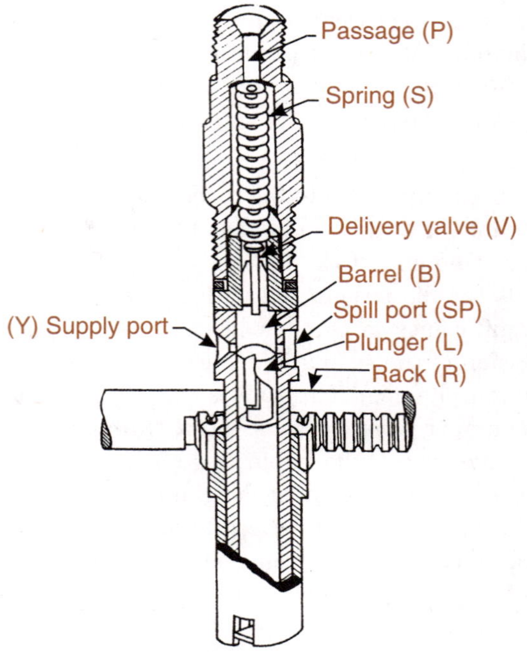

Fig. 23.28. Fuel pump.

When the plunger moves up due to cam and tappet mechanism, a stage reaches when both the ports *SP* and *Y* are closed and with the further upward movement of the plunger the fuel gets compressed. The high pressure thus developed lifts the delivery valve off its seat and fuel flows to atomiser through the passage *P*. With further rise of the plunger, at a certain moment, the port *SP* is connected to the fuel in the upper part of the plunger through the rectangular vertical groove by the helical groove ; as a result of which a sudden drop in pressure occurs and the delivery valve falls back and occupies its seat against the spring force. The plunger is rotated by the rack *R* which is moved in or out by the governor. *By changing the angular position of the helical groove (by rotating the plunger) of the plunger relative to the supply port, the length of stroke during which the oil is delivered can be varied and thereby quantity of fuel delivered to the engine is also varied accordingly.*

Fuel atomiser or injector

Refer to Fig. 23.29. It consists of a nozzle valve (*NV*) fitted in the nozzle body (*NB*). The nozzle valve is held on

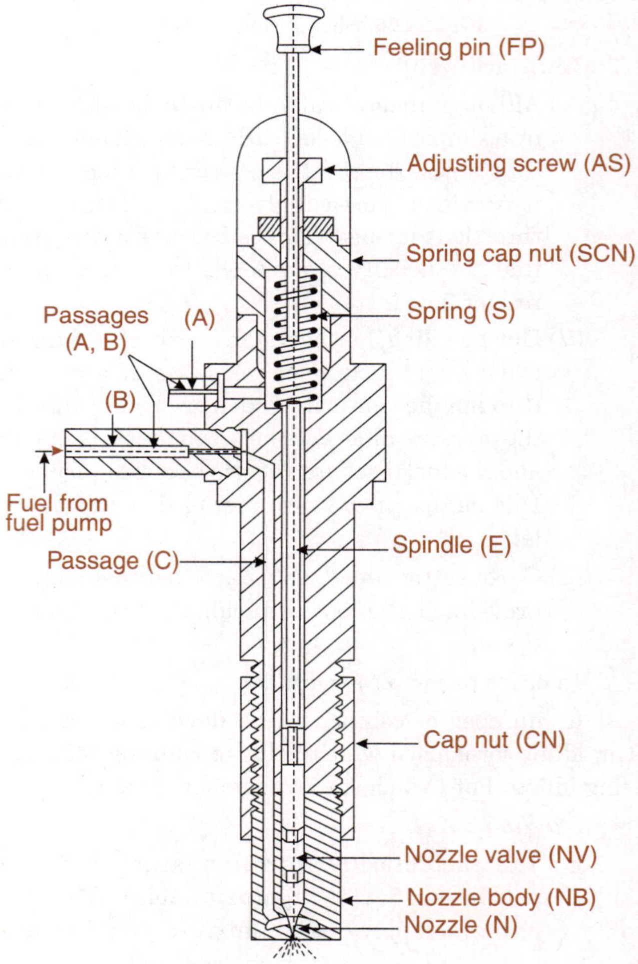

Fig. 23.29. Fuel atomiser or injector.

its seat by a spring '*S*' which exerts pressure through the spindle *E*. '*AS*' is the adjusting screw by which the nozzle valve lift can be adjusted. Usually the nozzle valve is set to lift at 135 to 170 bar pressure. *FP* is the feeling pin which indicates whether valve is working properly or not. The oil under pressure from the fuel pump enters the injector through the passages *B* and *C* and lifts the nozzle valve. The fuel travels down the nozzle *N* and injected into the engine cylinder in the form of fine sprays. When the pressure of the oil falls, the nozzle valve occupies its seat under the spring force and fuel supply is cut off. Any leakage of fuel accumulated above the valve is led to the fuel tank through the passage *A*. The leakage occurs when the nozzle valve is *worn out*.

List of engine parts, materials, method of manufacture and functions

S.No.	*Name of the part*	*Material*	*Function*	*Method of manufacture*
1.	*Cylinder*	Hard grade cast iron	Contains gas under pressure and guides the piston.	Casting.
2.	*Cylinder head*	Cast iron or aluminium	Main function is to seal the working end of the cylinder and not to permit entry and exit of gases on overhead valve engines.	Casting, forging.
3.	*Piston*	Cast iron or aluminium alloy	It acts as a face to receive gas pressure and transmits the thrust to the connecting rod.	Casting, forging.
4.	*Piston rings*	Cast iron	Their main function is to provide a good sealing fit between the piston and cylinder.	Casting.
5.	*Gudgeon pin*	Hardened steel	It supports and allows the connecting rod to swivel.	Forging.
6.	*Connecting rod*	Alloy steel ; for small engines the material may be aluminium	It transmits the piston load to the crank, causing the latter to turn, thus converting the reciprocating motion of the piston into rotary motion of the crankshaft.	Forging.
7.	*Crankshaft*	In general the crankshaft is made from a high tensile forging, but special cast irons are sometimes used to produce a light weight crank-shaft that does not require a lot of machining.	It converts the reciprocating motion of the piston into the rotary motion.	Forging.
8.	*Main bearings*	The typical bearing half is made of steel or bronze back to which a lining of relatively soft bearing material is applied.	The function of bearing is to reduce the friction and allow the parts to move easily.	Casting.
9.	*Flywheel*	Steel or cast iron.	In engines it takes care of fluctuations of speed during thermodynamic cycle.	Casting.
10.	*Inlet valve*	Silicon chrome steel with about 3% carbon.	Admits the air or mixture of air and fuel into engine cylinder.	Forging.
11.	*Exhaust valve*	Austenitic steel	Discharges the product of combustion.	Forging.

23.7. TERMS CONNECTED WITH I.C. ENGINES

Refer to Fig. 23.30.

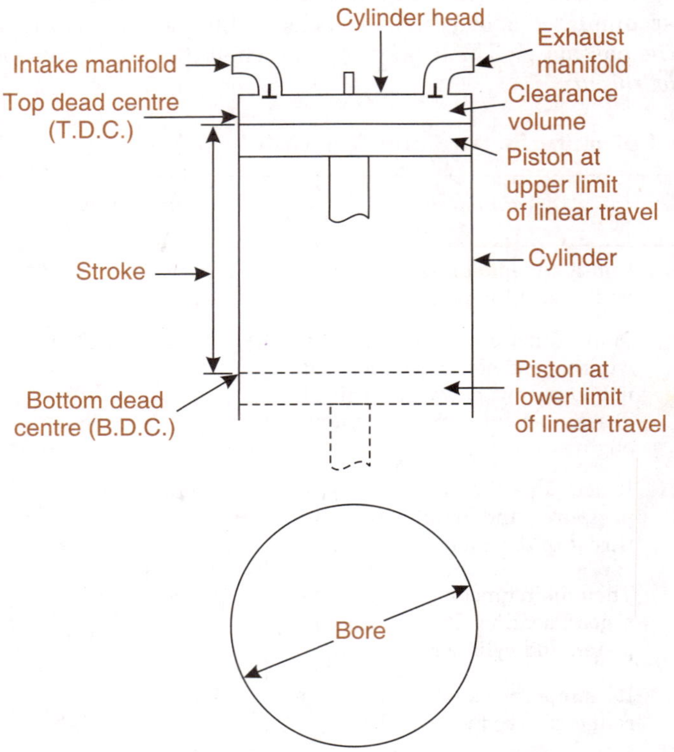

Fig. 23.30. Terms relating I.C. engines.

Bore. *The inside diameter of the cylinder is called **bore**.*

Stroke. As the piston reciprocates inside the engine cylinder, it has got limiting upper and lower positions beyond which it cannot move and reversal of motion takes place at these limiting positions.

*The linear distance along the cylinder axis between two limiting positions, is called **stroke**.*

Top Dead Centre (T.D.C.). *The top most position of the piston towards cover end side of the cylinder is called "top dead centre".* In case of horizontal engines, this is known as inner dead centre.

Bottom Dead Centre (B.D.C.). *The lowest position of the piston towards the crank end side of the cylinder is called "bottom dead centre".* In case of horizontal engines it is called outer dead centre.

Clearance volume. *The volume contained in the cylinder above the top of the piston, when the piston is at top dead centre, is called the **clearance volume**.*

Swept volume. *The volume swept through by the piston in moving between top dead centre and bottom dead centre, is called **swept volume** or **piston displacement**.* Thus, when piston is at bottom dead centre, total volume = swept volume + clearance volume.

Compression ratio. *It is ratio of total cylinder volume to clearance volume.*

Refer to Fig. 23.30. Compression ratio (r) is given by

$$r = \frac{V_s + V_c}{V_c}$$

where, V_s = Swept volume, and

V_c = Clearance volume.

The compression ratio varies from 5 : 1 to 11 : 1 (average value 7 : 1 to 9 : 1) in *S.I. engines* and from 12 : 1 to 24 : 1 (average value 15 : 1 to 18 : 1) in *C.I. engines*.

Piston speed. *The average speed of the piston is called **piston speed.***

Piston speed = 2 LN

where, L = Length of the stroke, and

N = Speed of the engine in r.p.m.

23.8. WORKING CYCLES

An internal combustion engine can work on any one of the following cycles :

(*a*) Constant volume or Otto cycle

(*b*) Constant pressure or Diesel cycle

(*c*) Dual combustion cycle.

These may be either *four-stroke cycle or two-stroke cycle engines*.

(*a*) **Constant volume or Otto cycle.** The cycle is so called because heat is supplied at constant volume. Petrol, gas and light oil work on this cycle. In the case of a petrol engine the proper mixing of petrol and air takes place in the carburettor which is situated outside the engine cylinder. The proportionate mixture is drawn into the cylinder during the suction stroke. In a gas engine also, air and gas is mixed outside the engine cylinder and this mixture enters the cylinder during the suction stroke. In light oil engines the fuel is converted to vapours by a vapouriser which receives heat from the exhaust gases of the engine and their mixture flows towards engine cylinder during suction stroke.

(*b*) **Constant pressure or Diesel cycle.** In this cycle only air is drawn in the engine cylinder during the suction stroke, this air gets compressed during the compression stroke and its pressure and temperature increase by a considerable amount. Just before the end of the stroke a metered quantity of fuel under pressure adequately more than that developed in the engine cylinder is injected in the form of fine sprays by means of a fuel injector. Due to very high pressure and temperature of the air the fuel ignites and hot gases thus produced throw the piston downwards and work is obtained. *Heavy oil engines make use of this cycle.*

(*c*) **Dual combustion cycle.** This cycle is also called *semi-diesel cycle*. It is so named because heat is added *partly at constant volume and partly at constant pressure*. In this cycle only air is drawn in the engine cylinder during suction stroke. The air is then compressed in hot combustion chamber at the end of the cylinder during the compression stroke to a pressure of about 26 bar. The heat of compressed air together with heat of combustion chamber ignites the fuel. The fuel is injected into the cylinder just before the end of compression stroke where it ignites immediately. The fuel injection is continued until the point of cut-off is reached. The burning of fuel at first takes place at constant volume and continues to burn at constant pressure during the first part of expansion or working stroke. The field of application of this cycle is *heavy oil engines*.

23.9. INDICATOR DIAGRAM

An "*indicator diagram*" is *a graph between pressure and volume* ; the former being taken on vertical axis and the latter on the horizontal axis. This is obtained by an instrument known as *indicator*. The indicator diagrams are of two types : (*a*) Theoretical or hypothetical, (*b*) Actual. The theoretical or hypothetical indicator diagram is always longer in size as compared to the actual one, since in the former losses are neglected. *The ratio of the area of the actual indicator diagram to the theoretical one is called* ***diagram factor.***

23.10. FOUR-STROKE CYCLE ENGINES

Here follows the description of the four-stroke otto and diesel-cycle engines.

Otto engines. The Otto four-stroke cycle refers to its use in petrol engines, gas engines, light oil engines and heavy oil engines in which the mixture of air and fuel are drawn in the engine cylinder. Since ignition in these engines is due to a spark, therefore they are also called *spark ignition engines*.

The various strokes of a four-stroke (Otto) cycle engine are detailed below.

Refer to Fig. 23.31.

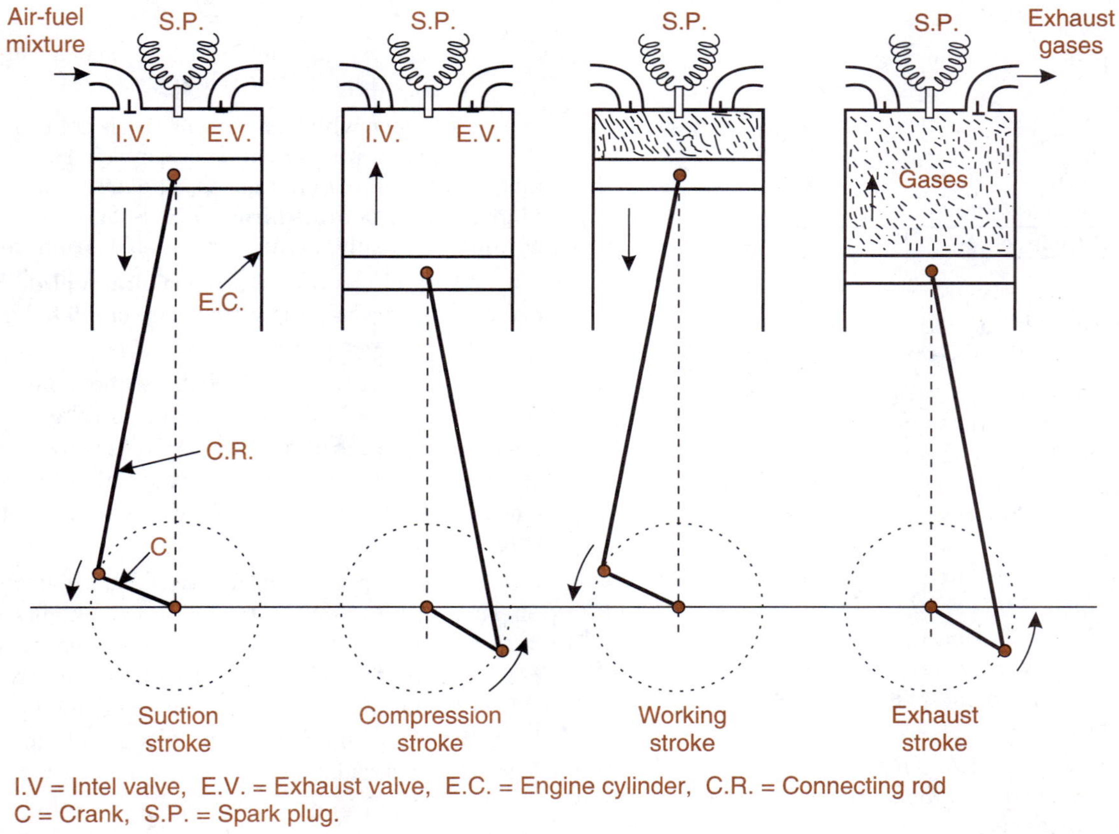

Fig. 23.31. Four-stroke Otto cycle engine.

1. Suction stroke. During this stroke (also known as *induction stroke*) the piston moves from top dead centre (T.D.C.) to bottom dead centre (B.D.C.) ; the inlet valve opens and proportionate fuel-air mixture is sucked in the engine cylinder. This operation is represented by the line 5-1 (Fig. 23.32). The exhaust valve remains closed through out the stroke.

2. Compression stroke. In this stroke, the piston moves (1-2) towards (T.D.C.) and compresses the enclosed fuel-air mixture drawn in the engine cylinder during suction. The pressure of the mixture rises in the cylinder to a value of about 8 bar. Just before the end of this stroke the operating-plug initiates a spark which ignites the mixture and combustion takes place at constant volume (line 2-3) (Fig. 23.32). Both the inlet and exhaust valves remain closed during the stroke.

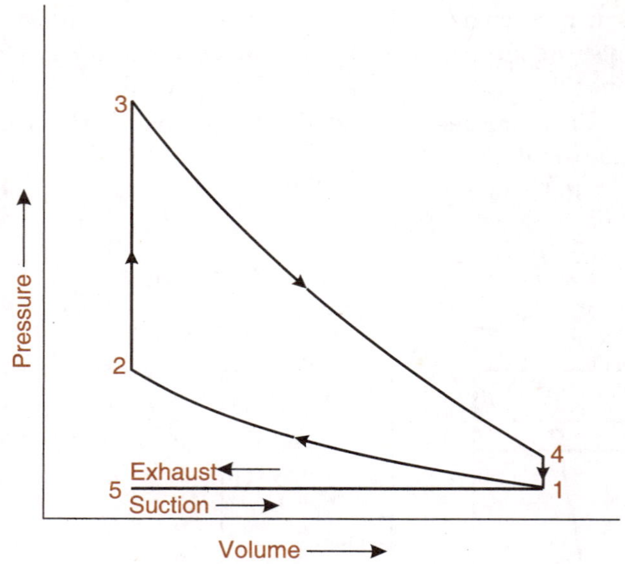

Fig. 23.32. Theoretical *p-V* diagram of a four-stroke Otto cycle engine.

3. Expansion or working stroke. When the mixture is ignited by the spark plug the hot gases are produced which drive or throw the piston from T.D.C. to B.D.C. and thus the work is obtained in this stroke. It is during this stroke when we get work from the engine ; the other three strokes namely suction, compression and exhaust being idle. *The flywheel mounted on the engine shaft stores energy during this stroke and supplies it during the idle strokes*. The expansion of the gases is shown by 3-4. (Fig. 23.32). Both the valves remain closed during the start of this stroke but when the piston just reaches the B.D.C. the exhaust valve opens.

4. Exhaust stroke. This is the last stroke of the cycle. Here the gases from which the work has been collected become useless after the completion of the expansion stroke and are made to escape through exhaust valve to the atmosphere. This removal of gas is accomplished during this stroke. The piston moves from B.D.C. to T.D.C. and the exhaust gases are driven out of the engine cylinder ; this is also called *scavenging*. This operation is represented by the line (1-5) (Fig. 23.32).

Fig. 23.33 shows the actual indicator diagram of four-stroke Otto cycle engine. It may be noted that line 5-1 is below the atmospheric pressure line. This is due to the fact that owing to restricted area of the inlet passages the entering fuel-air mixture cannot cope with the speed of the piston. The exhaust line 4-5 is slightly above the atmospheric pressure line. This is due to restricted exhaust passages which do not allow the exhaust gases to leave the engine-cylinder quickly.

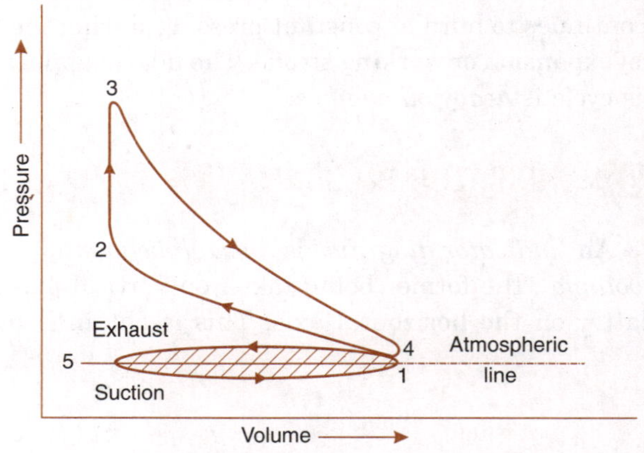

Fig. 23.33. Actual *p-V* diagram of a four-stroke Otto cycle engine.

The loop which has area 4-5-1 is called *negative loop* ; it gives the pumping loss due to admission of fuel air mixture and removal of exhaust gases. The area 1-2-3-4 is the total or gross work obtained from the piston and network can be obtained by subtracting area 4-5-1 from area 1-2-3-4.

Diesel engines (four-stroke cycle). As is the case of Otto four-stroke ; this cycle too is completed in *four-strokes* as follows. (Refer to Fig. 23.34).

1. Suction stroke. With the movement of the piston from T.D.C. to B.D.C. during this stroke, the inlet valve opens and the air at atmospheric pressure is drawn inside the engine cylinder ; the exhaust valve however remains closed. This operation is represented by the line 5-1 (Fig. 23.35).

2. Compression stroke. The air drawn at atmospheric pressure during the suction stroke is compressed to high pressure and temperature (to the value of 35 bar and 600°C respectively) as the piston moves from B.D.C. to T.D.C. This operation is represented by 1-2 (Fig. 23.35). Both the inlet and exhaust valves do not open during any part of this stroke.

3. Expansion or working stroke. As the piston starts moving from T.D.C. a matered quantity of fuel is injected into the hot compressed air in fine sprays by the fuel injector and it (fuel) starts burning at constant pressure

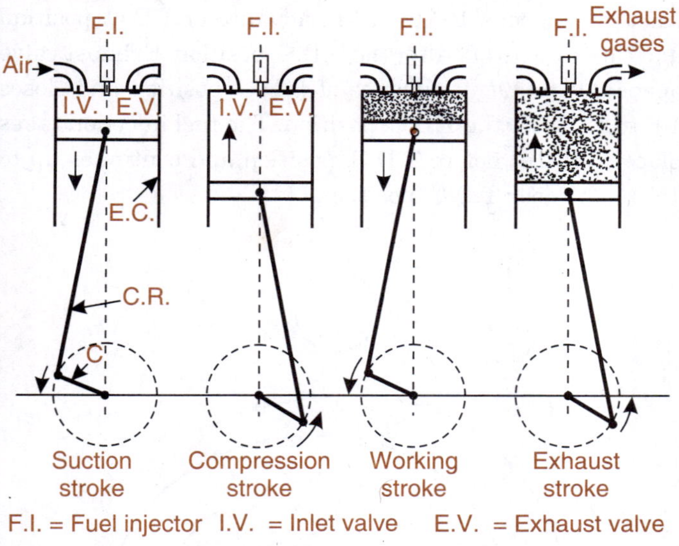

Fig. 23.34. Four-stroke Diesel cycle engine.

shown by the line 2-3. At the point 3 fuel supply is cut off. The fuel is injected at the end of compression stroke but in actual practice the ignition of the fuel starts before the end of the compression stroke. The hot gases of the cylinder expand adiabatically to point 4, thus doing work on the piston. The expansion is shown by 3-4 (Fig. 23.35).

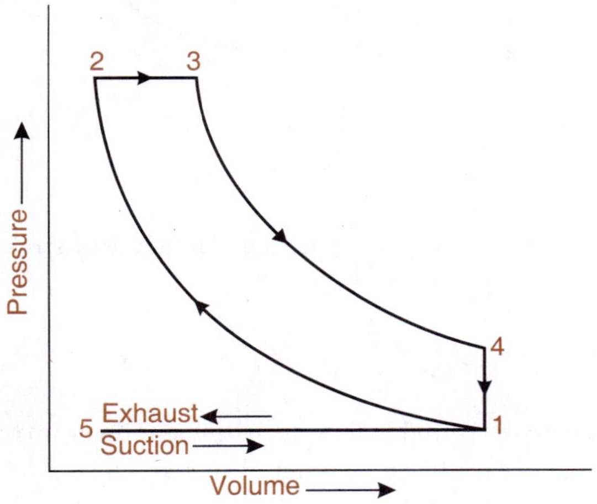

Fig. 23.35. Theoretical *p-V* diagram of a four-stroke Diesel cycle.

4. Exhaust stroke. The piston moves from the B.D.C. to T.D.C. and the exhaust gases escape to the atmosphere through the exhaust valve. When the piston reaches the T.D.C. the exhaust valve closes and the cycle is completed. This stroke is represented by the line 1-5 (Fig. 23.35).

Fig. 23.36 shows the actual indicator diagram for a four-stroke Diesel cycle engine. It may be noted that line 5-1 is below the atmospheric pressure line. This is due to the fact that owing to the restricted area of the inlet passages the entering air can't cope with the speed of the piston. The exhaust line 4-5 is slightly above the atmospheric line. This is because of the restricted exhaust passages which do not allow the exhaust gases to leave the engine cylinder quickly.

The loop of area 4-5-1 is called negative loop ; it gives the pumping loss due to admission of air and removal of exhaust gases. The area 1-2-3-4 is the total or gross work obtained from the piston and net work can be obtained by subtracting area 4-5-1 from area 1-2-3-4.

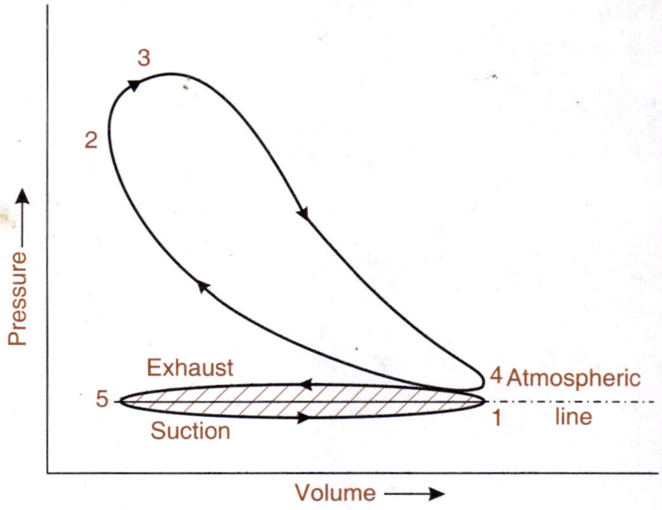

Fig. 23.36. Actual *p-V* diagram of four-stroke Diesel cycle.

Valve Timing Diagrams (Otto and Diesel engines)

1. Otto engine. Fig. 23.37 shows a theoretical valve timing diagram for *four-stroke "Otto cycle" engines* which is

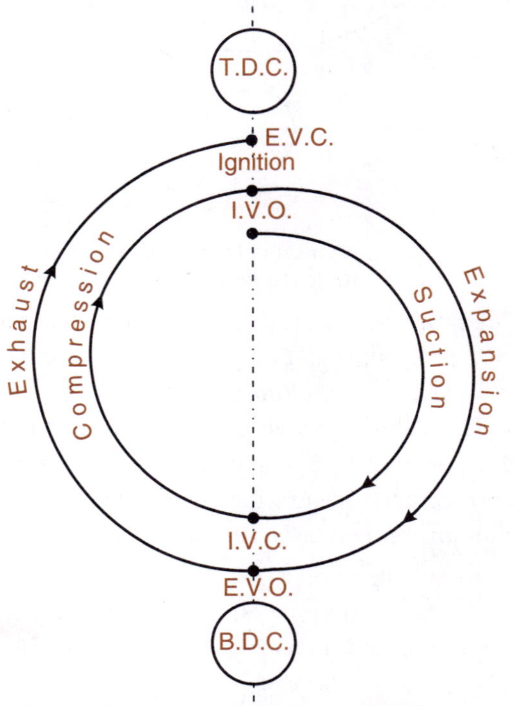

Fig. 23.37. Theoretical valve timing diagram (four-stroke Otto cycle engine).

self-explanatory. In actual practice, it is difficult to open and close the valve instantaneously ; so as to get better performance of the engine the valve timings are modified. In Fig. 23.38 is shown an actual valve timing diagram. The inlet valve is opened 10° to 30° in advance of the T.D.C. position to enable the fresh charge to enter the cylinder and to help the burnt gases at the same time, to escape to the atmosphere. The suction of the mixture continues up to 30°–40° or even 60° after B.D.C. position. The inlet valve closes and the compression of the entrapped mixture starts.

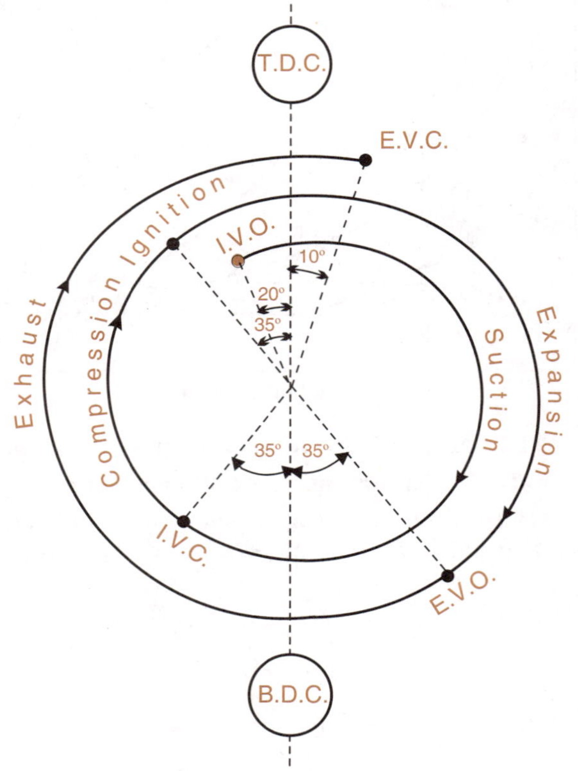

Fig. 23.38. Actual valve timing diagram (four-stroke Otto cycle engines).

The sparking plug produces a spark 30° to 40° before the T.D.C. position ; thus fuel gets more time to burn. The pressure becomes maximum nearly 10° past the T.D.C. position. The exhaust valve opens 30° to 60° before the B.D.C. position and the gases are driven out of the cylinder by piston during its upward movement. The exhaust valve closes when piston is nearly 10° past T.D.C. position.

2. Diesel engines. Fig. 23.39 shows the valve timing diagram of a *four-stroke "Diesel cycle" engine* (theoretical valve timing diagram, is however the same as Fig. 23.37). Inlet valve opens 10° to 25° in advance of T.D.C. position and closes 25° to 50° after the B.D.C. position. Exhaust valve opens 30° to 50° in advance of B.D.C. position and closes 10° to 15° after the T.D.C. position. The fuel injection takes place 5° to 10° before T.D.C. position and continues up to 15° to 25° near T.D.C. position.

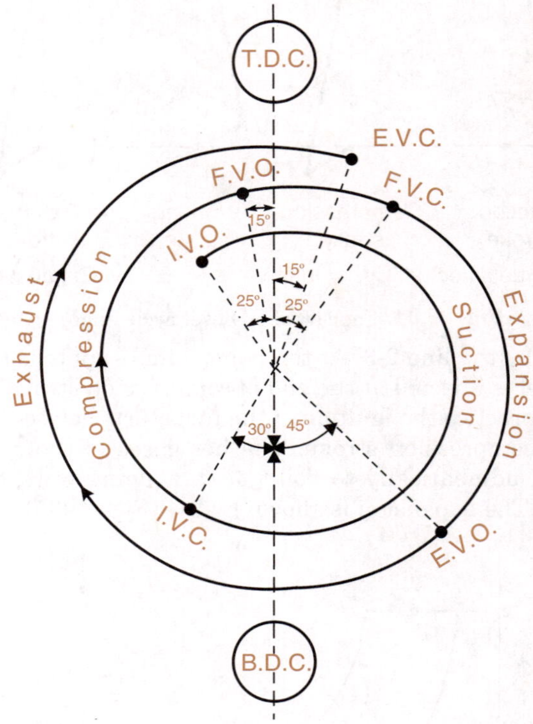

Fig. 23.39. Actual valve timing diagram (four-stroke Diesel cycle engines).

23.11. TWO-STROKE CYCLE ENGINES

In 1878, Dugald-clerk, a British engineer introduced a cycle which could be completed in two *strokes of piston rather than four strokes* as is the case with the four-stroke cycle engines. The engines using this cycle were called two-stroke cycle engines. In this engine suction and exhaust strokes are eliminated. Here *instead of valves, ports are used. The exhaust gases are driven out from engine cylinder by the fresh charge of fuel entering the cylinder nearly at the end of the working stroke.*

Fig. 23.40 shows a two-stroke petrol engine (used in scooters, motor cycles etc.). The cylinder L is connected to a closed crank chamber C.C. During the upward stroke of the piston M, the gases in L are compressed and at the same time fresh air and fuel (petrol) mixture enters the crank chamber through the valve V. When the piston moves downwards, V closes and the mixture in the crank chamber is compressed. Refer to Fig. 23.40 (*i*), the piston is moving upwards and is compressing an explosive change which has previously been supplied to L. Ignition takes place at the end of the stroke. The piston then travels downwards due to expansion of the gases (Fig. 23.40 (*ii*)) and near the end of this stroke the piston uncovers the exhaust port (E.P.) and the burnt exhaust gases escape through this port (Fig. 23.40 (*iii*)). The transfer port (T.P.) then is uncovered immediately, and the compressed charge from the crank chamber flows into the cylinder and is deflected upwards by the *hump* provided on the head of the piston. It may be noted that the incoming air-petrol mixture helps the removal of gases from the engine-cylinder ; if, in case these exhaust gases do not leave the cylinder, the fresh charge gets diluted and efficiency of the engine will decrease. The piston then again starts moving from B.D.C. to T.D.C. and the charge gets compressed when E.P. (exhaust port) and T.P. are covered by the piston ; thus the cycle is repeated.

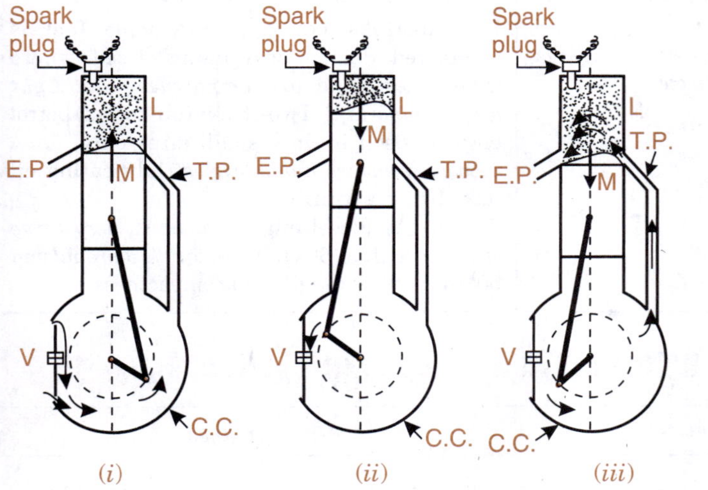

L = Cylinder, E.P. = Exhaust port, T.P. = Transfer port,
V = Valve, C.C = Crank chamber

Fig. 23.40. Two-stroke cycle engine.

Fig. 23.41 shows the *p-V* diagram for a two-stroke cycle engine. It is only for the main cylinder or the top side of the piston. Fig. 23.42 shows self-explanatory port timing diagram for a two-stroke cycle engine.

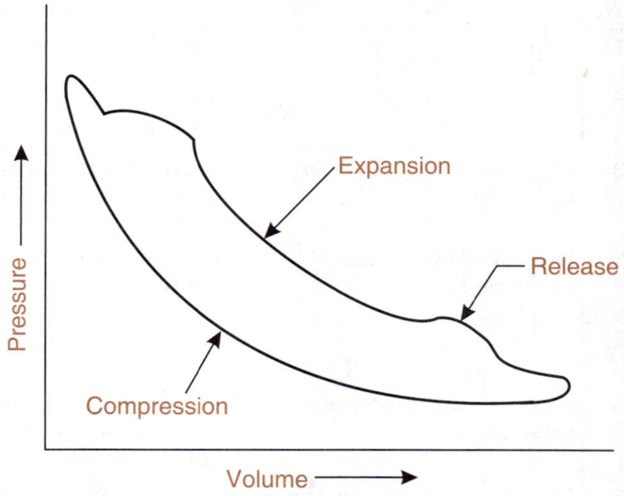

Fig. 23.41. *p-V* diagram for a two-stroke cycle engine.

In a two-stroke Diesel cycle engine all the operations are the same as in the spark ignition (Otto cycle) engine with the differences ; firstly in this case, only air is admitted into cylinder instead of air-fuel mixture and secondly fuel injector is fitted to supply the fuel instead of a sparking plug.

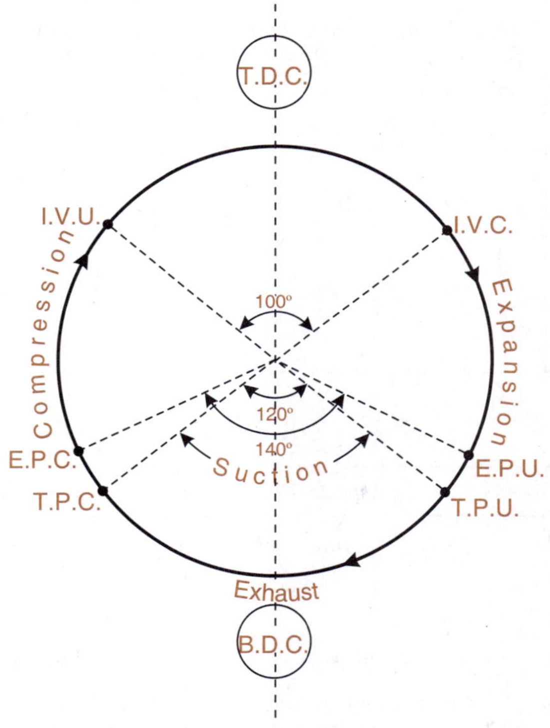

Fig. 23.42. Port timing diagram.

23.12. COMPARISON OF FOUR-STROKE AND TWO-STROKE CYCLE ENGINES

S. No.	*Aspects*	*Four-stroke cycle engines*	*Two-stroke cycle engines*
1.	*Completion of cycle*	The cycle is completed in *four strokes of the piston* or in *two revolutions of the crankshaft*. Thus one power stroke is obtained in every two revolutions of the crankshaft.	The cycle is completed in *two strokes of the piston* or in *one revolution of the crankshaft*. Thus one power stroke is obtained in each revolution of the crankshaft.
2.	*Flywheel required-heavier or lighter*	Because of (*i*) turning-movement is not so uniform and hence *heavier* flywheel is needed.	More uniform turning movement and hence *lighter* flywheel is needed.
3.	*Power produced for same size of engine*	Again because of one power stroke for two revolutions, power produced for same size of engine is *small* or for the same power the engine is heavy and bulky.	Because of one power stroke for one revolution, power produced for same size of engine in *more* (theoretically twice, actually about 1.8 times) or for the same power the engine is light and compact.
4.	*Cooling and lubrication requirements*	Because of one power stroke in two revolutions *lesser* cooling and lubrication requirements. Lesser rate of wear and tear.	Because of one power stroke in one revolution *greater* cooling and lubrication requirement. Great rate of wear and tear.
5.	*Value and valve mechanism*	The four-stroke engine *contains* valve and valve mechanism.	Two-stroke engines have *no* valves but only ports (some two-stroke engines are fitted with conventional exhaust valves).
6.	*Initial cost*	Because of the heavy weight and complication of valve mechanism, *higher* is the initial cost.	Because of light weight and simplicity due to absence of valve mechanism, *cheaper* in initial cost.
7.	*Volumetric efficiency*	Volumetric efficiency *more* due to more time of induction.	Volumetric efficiency *less* due to lesser time for induction.
8.	*Thermal and part-load efficiencies*	Thermal efficiency higher, part load efficiency better than two stroke cycle engine.	Thermal efficiency lower, part load efficiency lesser than four stroke cycle engine.
9.	*Applications*	Used where efficiency is important ; in *cars, buses, trucks tractors, industrial engines, aeroplane, power generators etc.*	In two-stroke petrol engine some fuel is exhausted during scavenging. Used where (*a*) *low cost*, and (*b*) *compactness and light weight important*. Two-stroke (air-cooled) petrol engines used in very small sizes only, *lawn movers, scooters motor cycles* (lubricating oil mixed with petrol). Two-stroke diesel engines used in *very large sizes* more than 60 cm bore, for *ship propulsion* because of low weight and compactness.

23.13. COMPARISON OF SPARK IGNITION (S.I.) AND COMBUSTION IGNITION (C.I.) ENGINES

S. No.	*Aspects*	*S.I. engines*	*C.I. engines*
1.	*Thermodynamic cycle*	Otto cycle	Diesel cycle ...... For slow speed engines Dual cycle ...... For high speed engines
2.	*Fuel used*	Petrol	Diesel.
3.	*Air-fuel ratio*	10 : 1 to 20 : 1	18 : 1 to 100 : 1.
4.	*Compression ratio*	Upto 11 ; Average value 7 to 9 ; Upper limit of compression ratio fixed by *anti-knock quality of fuel.*	12 to 24 ; Average value 15 to 18 ; Upper limit of compression ratio is limited by *thermal and mechanical stresses.*
5.	*Combustion*	Spark ignition	Compression ignition.
6.	*Fuel supply*	By carburettor ...... cheap method	By injection ...... explosive method.

7.	*Operating pressure* (*i*) Compression pressure (*ii*) Maximum pressure	7 bar to 15 bar 45 bar to 60 bar	30 bar to 50 bar 60 bar to 120 bar.
8.	*Operating speed*	High speed : 2000 to 6000 r.p.m.	Low speed : 400 r.p.m. Medium speed : 400 to 1200 r.p.m. High speed : 1200 to 3500 r.p.m.
9.	*Control of power*	Quantity governing ...... by throttle	Quality governing ...... by rack.
10.	*Calorific value*	44 MJ/kg	42 MJ/kg.
11.	*Cost of running*	High	Low.
12.	*Maintenance cost*	Minor maintenance required	Major overhaul required but less frequently.
13.	*Supercharging*	Limited by *detonation*. Used only in *aircraft engines*.	Limited by *blower power and mechanical and thermal stresses. Widely used.*
14.	*Two-stroke operation*	*Less suitable*, fuel loss in scavenging. But small two-stroke engines are used in mopeds, scooters and motorcycles due to their *simplicity and low cost.*	No fuel loss in scavenging. *More suitable.*
15.	*High powers*	No	Yes.
16.	*Uses*	Mopeds, scooters, motorcycles, simple engine passenger cars, air-crafts etc.	Buses, trucks locomotives, tractors, earth moving machinery and stationary generating plants.

23.14. COMPARISON BETWEEN A PETROL ENGINE AND A DIESEL ENGINE

S. No.	*Petrol engine*	*Diesel engine*
1.	Air-petrol mixture is sucked in the engine cylinder during suction stroke.	Only air is sucked during suction stroke.
2.	Spark plug is used.	Employs an injector.
3.	Power is produced by spark ignition.	Power is produced by compression ignition.
4.	Thermal efficiency up to 25%.	Thermal efficiency up to 40%.
5.	Occupies less space.	Occupies more space.
6.	More running cost.	Less running cost.
7.	Light in weight.	Heavy in weight.
8.	Fuel (Petrol) costlier.	Fuel (Diesel) cheaper.
9.	Petrol being volatile is dangerous.	Diesel is non-dangerous as it is non-volatile.
10.	Pre-ignition possible.	Pre-ignition not possible.
11.	Works on Otto cycle.	Works on Diesel cycle.
12.	Less dependable.	More dependable.
13.	Used in *cars* and *motor cycles*.	Used in heavy duty vehicles like *trucks, buses* and *heavy machinery*.

23.15. HOW TO TELL A TWO-STROKE CYCLE ENGINE FROM A FOUR-STROKE CYCLE ENGINE ?

S. No.	*Distinguishing features*	*Four-stroke cycle engine*	*Two-stroke cycle engine*
1.	*Oil sump and oil-filter plug*	It has an oil sump and oil-filter plug.	It does not have oil sump and oil-filter plug.
2.	*Oil drains etc.*	It requires oil drains and refills periodically, just an automobile do.	In this type of engine, the oil is added to the gasoline so that a mixture of gasoline and oil passes through the carburettor and enters the crankcase with the air.
3.	*Location of muffler (exhaust silencer)*	It is installed at the head end of the cylinder at the exhaust valve location.	It is installed towards the middle of the cylinder, at the exhaust port location.
4.	Name plate	If the name plate mentions the type of oil and the crankcase capacity, or similar data, it is a four-stroke cycle engine.	If the name plate tells to mix oil with the gasoline, it is a two-stroke cycle engine.

23.16. IGNITION SYSTEM

(Petrol Engines)

The operator of a spark ignition engine expects the ignition system to fire thousands of consecutive cycles in cylinders full of fuel-air mixture without a "*miss*" although the manifod pressure may vary from 0.35 to 2.7 bar, the fuel-air ratio may vary from 0.06 to 0.12 and the r.p.m. from 400 to 5000. In addition the ignition must occur at the proper crank angle so that the time losses are held at a minimum. In a view of this, the *requirements of ignition-systems* may be put down as given below :

1. A source of electric energy must be there.
2. A means for stepping up the voltage from the source to the very high potential required to produce a high tension arc across the spark plug gap that ignites the combustible mixture.
3. A means for timing and distributing the high voltage *i.e.,* supply of high voltage to each spark plug at the exact instant is required in every cycle in each cylinder.
4. Adjustment of spark advance with speed and load.

The two basic ignition systems in current use are :

1. Battery or coil ignition system.
2. Magneto ignition system.

Both the systems have been preferred to considerable extent and these systems fulfil the requirements for satisfactory operation.

1. Battery or coil-ignition system

Most of the modern spark ignition engines use battery ignition system. This system consists of the following components :

1. Battery (6 or 12 volts)
2. Ignition switch
3. Induction coil
4. Circuit/contact breaker
5. Condenser
6. Distributor.

Refer to Fig. 23.43.

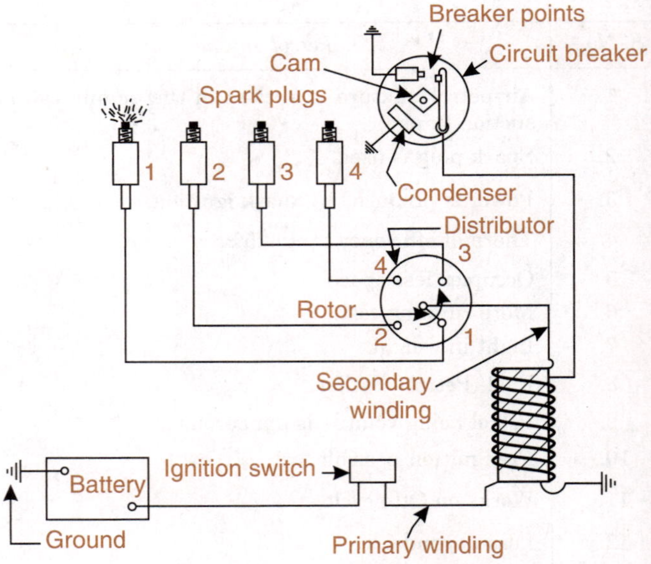

Fig. 23.43. Battery or coil-ignition system.

One terminal of the battery is ground to the frame of the engine, and other is connected through the ignition switch to one primary terminal of the ignition coil (consisting of a comparatively few turns of thick wire wound round an iron core). The other primary terminal is connected to one end of the contact points of the circuit breaker and through closed points to ground. The primary circuit of the ignition coil thus gets completed when contact points of the circuit breaker are together and switch is closed. The secondary terminal of the coil is connected to the central contact of the distributor and hence to distributor rotor. The secondary circuit consists of secondary winding (consisting of a large number of turns of fine wire) of the coil, distributor and four spark plugs. The contact breaker is driven by a cam whose speed is *half* the engine speed (for *four-stroke engines*) and breaks the primary circuit one for each cylinder during one complete cycle of the engine.

The breaker points are held on contact by a spring except when forced apart by lobs of the cam.

To start with, the ignition switch is made on and the engine is cranked *i.e.,* turned by hand when the contacts touch, the current flows from battery through the switch, primary winding of the induction coil to circuit breaker points and the circuit is completed through the ground. A condenser connected across the terminals of the contact breaker points prevent the sparking at these points. The rotating cam breaks open the contacts immediately and breaking of this primary circuit brings about a change of magnetic field ; due to which a very high voltage to the tune of 8000 to 12000 V is produced across the secondary terminals. (The number of turns in the secondary winding may be 50 to 100 times than in primary winding). Due to high voltage the spark jumps across the gap in the spark plug and air-fuel mixture is ignited in the cylinder.

On account of its combined cheapness, convenience of maintenance, attention and general suitability, it has been adopted universally on automobiles.

2. Magneto-ignition system

The magneto-ignition system is similar in principle to the battery system except that the magnetic field in the core of the primary and secondary windings is produced by a rotating permanent magnet (Fig. 23.44). As the magnet turns, the field is produced from a positive maximum to a negative maximum and back again. As this magnetic field falls from a positive maximum value, a voltage and current are induced in the primary winding. The primary current produces a magnetic field of its own which keeps the total magnetic field surrounding the primary and secondary winding approximately constant. When the permanent magnet has turned for enough so that its contribution to the total field is strongly negative, the breaker points are opened and the magnetic field about the secondary winding suddenly goes from a high positive value to a high negative value. This induces a high voltage in the secondary winding which is led to the proper spark plug by the distributor.

The *magneto is an efficient, reliable, self contained unit which is often preferred for aircraft engines because storage batteries are heavy and troublesome*. Special starting means are required, however, as the magneto will not furnish enough voltage for ignition at low speeds. Variation in ignition timing is more difficult with the magneto, since the breaker point must be opened when the rotating magnets are in the most favourable position. It is possible to change the engine crank angle at which the mangeto points open without disturbing the relationship between point opening and magnet position by designing the attachment pad so that the entire magneto body may be rotated a few degrees about its own shaft. Obviously this method is not as satisfactory as rotating a timer cam-plate.

Advantages of battery system :

1. It offers better sparks at low speeds, starting and for cranking purposes.
2. The initial cost of the system is low.
3. It is a reliable system and periodical maintenance required is negligible except for battery.
4. Items requiring attention can be easily located in more accessible position than those of magnetos.
5. The high speed engine drive is usually simpler than magneto drive.

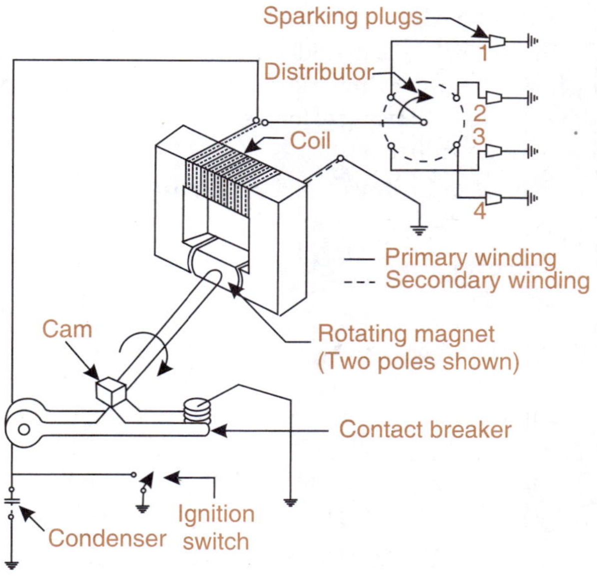

Fig. 23.44. Magneto-ignition system.

6. Adjustment of spark timing has no deterimental effect over the complete ignition timing range.

Disadvantages :

1. With the increasing speed, sparking voltage drops.

2. Battery, the only unreliable component of the system needs regular attention. In case battery runs down, the engine cannot be started as induction coil fails to operate.
3. Because of battery, bulk of the system is high.

Advantages of magneto system :

1. The system is more reliable as there is no battery or connecting cable.
2. The system is more suitable for medium and very high speed engines.
3. With the use of cobalt steel and nickel-aluminium magnet metals very light and compact units can be made which require very little room.
4. With recent development this system has become fairly reliable.

Disadvantages :

1. At low speeds and during cranking the voltage is very low. This has been overcome by suitable modifications in the circuit.
2. Adjustment of the spark timing *i.e.,* advance or retard, has deterimental effect upon the spark voltage or energy.
3. The powerful sparks at high engine speeds cause burning of the electrodes.

Firing order :

In case of multi-cylinder engines the order in which spark inside engine cylinders must occur is decided on the consideration of balancing of forces. Firing orders for various engines are given below :

No. of cylinders	Firing order
Two	1, 2
Three	1, 3, 2
Four	1, 4, 3, 2 or 1, 3, 4, 2
Six	1, 5, 3, 6, 2, 4 or 1, 4, 2, 6, 3, 5
Eight	1, 6, 2, 5, 8, 3, 7, 4 or 1, 8, 7, 3, 6, 5, 4, 2.

Eight (Vee) : Either of the following alternatives.

(A)	1*L*, 1*R*, 4*L*, 4*R*, 2*R*, 3*L*, 3*R*, 2*L*
(B)	1*L*, 4*R*, 2*R*, 2*L*, 3*R*, 3*L*, 4*L*, 1*R*
(C)	1*L*, 3*L*, 2*R*, 4*R*, 3*R*, 2*L*, 4*L*, 1*R*
(D)	1*L*, 4*R*, 4*L*, 2*L*, 3*R*, 3*L*, 2*R*, 1*R*
(E)	1*L*, 3*L*, 3*R*, 2*L*, 2*R*, 1*R*, 4*L*, 4*R*

L and *R* indicate cylinder on left and right hand side respectively. The firing order for a four-stroke engine with its cylinder numbered consecutively from 1 to n will be 1, 3, 5, 7, ... to n for one revolution of the crankshaft and 2, 4, 6, 8, ... to $(n - 1)$ for the next revolution.

23.17. FUEL INJECTION SYSTEM

(Diesel Engines) :

The *main functions of a fuel injection system are* :

1. Filter the fuel.
2. Metre or measure the correct quantity of fuel to be injected.
3. Time the fuel injection.
4. Control the rate of fuel injection.
5. Automise or break up the fuel to fine particles.
6. Properly distribute the fuel in the combustion chamber.

The injection systems are manufactured with great accuracy, especially the parts that actually metre and inject the fuel. Some of the tolerances between the moving parts are very small of the order of 1 micron. Such closely fitting parts require special attention during manufacture and hence the injection systems are costly.

In compression ignition engines (diesel and semi-diesel) two methods of fuel injection are used. These are

(*a*) Air injection

(*b*) Solid or airless injection.

(*a*) Air injection

In this method of fuel injection air is compressed in the compressor to a very high pressure (much higher than developed in the engine cylinder at the end of the compression stroke) and then injected through the fuel nozzle into the engine cylinder. The rate of fuel admission can be controlled by varying the pressure of injection air. Storage air bottles which are kept charged by an air compressor (driven by the engine) supply the high pressure air. This method is absolete these days.

(*b*) Solid or airless injection

It is also termed as *mechanical injection*. Here a fuel pump is used which supplies a measured quantity of fuel to the atomiser or injector which injects it (fuel) at a very high velocity into the engine cylinder in the form of sprays. The injection pressure varies from 100 to 145 bar (or even more in some cases). This pressure is produced by the fuel pump.

Fuel injection system consists of the following :

1. Fuel pump.
2. Injector or fuel atomiser.

For details, refer to Figs. 23.28 and 23.29.

23.18. ELECTRONIC FUEL INJECTION

Fig. 23.45 shows the fuel injection system-L-Jetronic with air flow metering (developed by Robert Bosch Corp.). It consists of the following units :

1. Fuel delivery system
2. Air induction system
3. Sensors and air-flow control system
4. Electronic control unit.

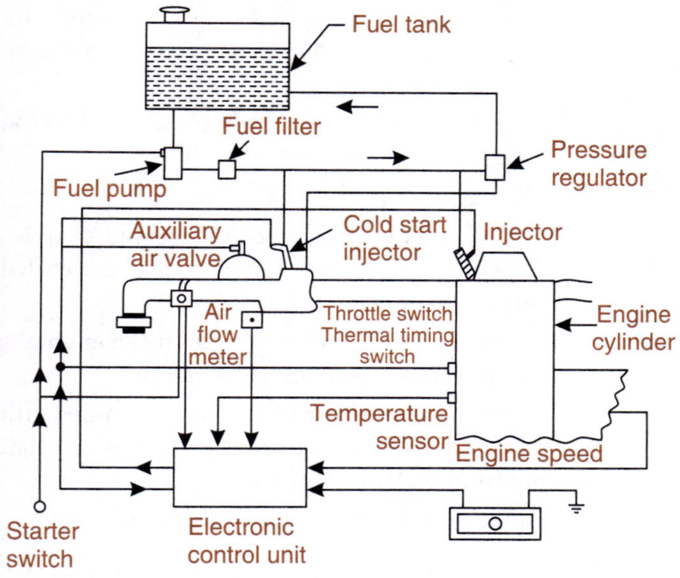

Fig. 23.45. Fuel injection system-L-Jetronic with air-flow metering.

1. **Fuel delivery system :**

- It consists of an electrically driven fuel pump which draws fuel from a *fuel tank*. The pump forces the oil through a *filter* into a line at the end of which is situated a *pressure regulator*, which in turn is connected is connected to intake manifold.
- The pressure regulator keeps the pressure difference between the fuel pressure and the manifold pressure constant, so that the quantity of fuel injected is dependent on the injector open time only.

2. **Air induction system :**

- After passing the air filter, the incoming air flows through an air flow meter, which generates a voltage signal (depending on the quantity of air flow).
- Just behind the *throttle valve* is fitted a cold start *magnetic injection valve*, which injects additional fuel for cold start. This valve also supplies the extra fuel needed during warm-up period.
- An *auxiliary valve* (which by-passes the throttle valve) supplies the extra air required for idling (in addition to rich air-fuel mixture). This extra air increases the engine speed after cold start to acceptable idling speed.
- To the throttle valve is attached a *throttle switch* equipped with a set of contacts which generate a sequence at voltage signals during the opening of throttle valve. The voltage signals result in injection of additional fuel required for acceleration.

3. **Electronic control unit :**

- The *sensors* are incorporated to measure the operating data at different locations. The data measured by the sensors are transmitted to the *electronic control unit which computes the amount of fuel injected during each engine cycle. The amount of fuel injected in varied by varying the injector opening time only.*
- The sensors used are :
  — Manifold pressure ;
  — Engine speed ;
  — Temperature at the intake manifold.

4. **Injection time :**

- For every revolution of the camshaft, the fuel is injected twice, each injection contributing half of a fuel quantity required for engine cycle.
- The injectors, at different phases of the operating cycle, are operated simultaneously.

Advantages and disadvantages of petrol injection :

Advantages :

1. Better starting and acceleration.
2. Owing to absence of any restriction (such as venturies and other metering elements in air passage) there is increased *volumetric efficiency and consequently increased power and torque*.
3. Higher compression ratios (higher by 1 to 1.5) can be employed (due to lower mixture temperatures in the engine cylinders).
4. Engines having fuel injection system can be used in any tilt position (which will cause surface trouble in carburettor).
5. Lower specific fuel consumption (since mixture to each cylinder is distributed more effectively).
6. Blowbracks and icing are eliminated.

Disadvantages :

1. High initial cost (due to precise and complicated component assemblies).
2. More noisy
3. Increased service problem
4. More bulky and heavy (than that of a carburettor).

23.19. COOLING SYSTEMS

In an I.C. engine, the temperature of the gases inside the engine cylinder may vary from 35°C or less to as high as 2750°C during the cycle. If an engine is allowed to run without external cooling, the cylinder walls, cylinder and pistons will tend to assume the average temperature of the gases to which they are exposed, which may be of the order of 1000 to 1500°C. Obviously at such high temperature ; the metals will loose their characteristics and piston will expand considerably and sieze the liner. Of course theoretically thermal efficiency of the engine will improve without cooling but actually the engine will sieze to run. If the cylinder wall temperature is allowed to rise above a certain limit, about 65°C, the lubricating oil will begin to evaporate rapidly and both cylinder and piston may be damaged. Also high temperature may cause excessive stress in some parts rendering them useless for further operation. In view of this, part of the heat generated inside the engine cylinder is allowed to be carried away by the cooling system. *Thus cooling system is provided on an engine for the following* **reasons** :

1. The even expansion of piston in the cylinder may result in seizure of the piston.
2. High temperatures reduce strength of piston and cylinder liner.
3. Overheated cylinder may lead to preignition of the charge, in case of spark ignition engine.
4. Physical and chemical changes may occur in lubricating oil which may cause sticking of piston rings and excessive wear of cylinder.

Almost 25 to 35 per cent of total heat supplied in the fuel is removed by the cooling medium. Heat carried away by lubricating oil and heat lost by radiation amounts 3 to 5 per cent of total heat supplied.

There are mainly two methods of cooling I.C. engine :

1. Air cooling.
2. Liquid cooling.

1. Air cooling

In this method, heat is carried away by the air flowing over and around the engine cylinder. It is used in scooters, motorcycles etc. Here fins are cast on the cylinder head and cylinder barrel which provide additional conductive and radiating surface. (Fig. 23.46). The fins are arranged in such a way that they are at right angles to the cylinder axis.

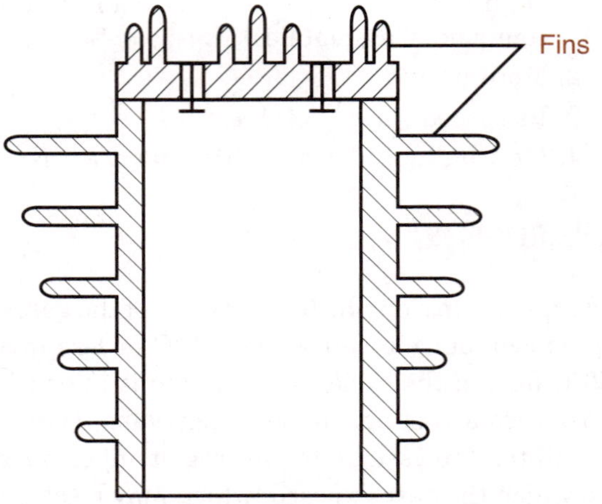

Fig. 23.46. Air cooling.

Advantages :

1. The design of the engine becomes simpler as no water jackets are required. The cylinder can be of identical dimensions and individually detachable and therefore cheaper to renew in case of accident etc.
2. Absence of cooling pipes, radiator etc. makes the cooling system simpler.
3. No danger of coolant leakage etc.
4. The engine is not subjected to freezing troubles etc. usually encountered in case of water-cooled engine.
5. The weight per B.H.P. of the air-cooled engine is less than that of water-cooled engine.
6. In this case engine is rather a self contained unit as it requires no external components *e.g.*, radiator, headers, tank etc.
7. Installation of air-cooled engines is easier.

Disadvantages :

1. Their movement is noisy.
2. Non-uniform cooling.
3. The output of air-cooled engine is less than that of a liquid cooled engine.
4. Maintenance is not easy.
5. Smaller useful compression ratio.

2. Liquid cooling

In this method of cooling engines, the cylinder walls and heads are provided with jackets through which the cooling liquid can circulate. The heat is transferred from cylinder walls to the liquid by convection and conduction. The liquid becomes heated in its passage through the jackets and is itself cooled by means of an air-cooled radiator system. The heat-from liquid in turn is transferred to air.

Various methods are used for circulating the water around the cylinder and cylinder head. These are :

1. Thermo-syphon cooling
2. Forced or pump cooling
3. Cooling with thermostatic regulator
4. Pressurised water cooling
5. Evaporative cooling.

1. Thermo-syphon cooling

The basis of this type of cooling is the fact that water becomes light on heating. Fig. 23.47 shows the thermo-syphon cooling arrangement. The top of radiator is connected to the top of water jacket by a pipe and bottom of the radiator to the bottom of the water jacket. Water travels down the radiator across which air is passed to cool it. The air-flow can take place due to vehicle motion or a fan can be provided for the purpose.

This system has the advantage that it is quite simple and automatic and is without any water pump unless there is leak, there is nothing to get out of order.

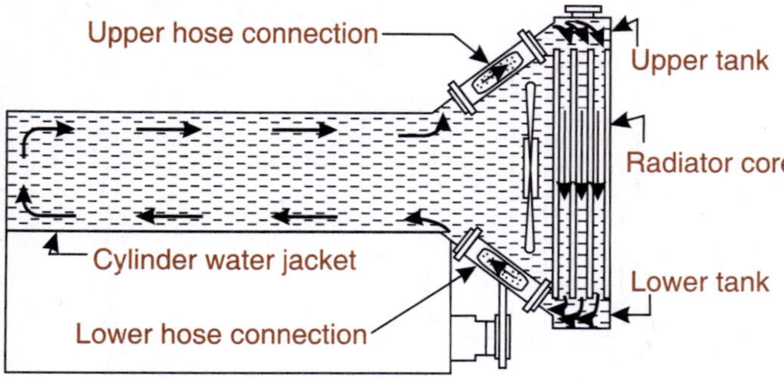

Fig. 23.47. Thermo-syphon cooling.

The major shortcoming of this system is that cooling depends only on the temperature and is independent of the engine speed. The rate of circulation is slow and insufficient. The circulation of water starts only after the engine has become hot enough to cause thermo-syphon action. This system requires that the radiator be above the engine for gravity flow of water to engine.

Thermo-syphon system is not widely used at present.

2. Forced or pump cooling

Refer to Fig. 23.48. In this system, a pump is used to cause positive circulation of water in the water jacket. Usually the pump is belt driven from the engine.

The advantage of forced system is that *cooling is ensured* under all conditions of operation.

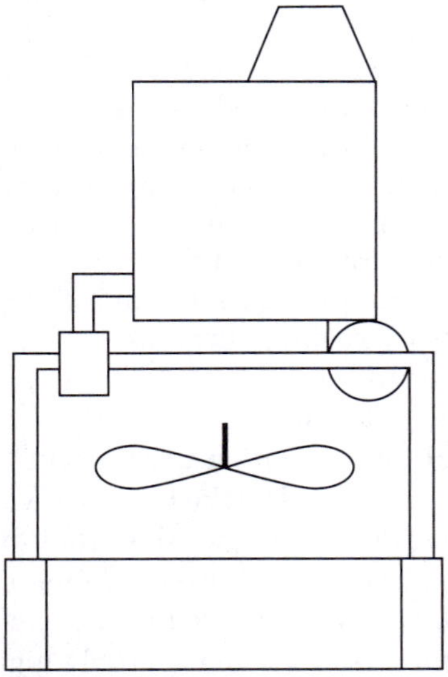

Fig. 23.48. Forced or pump system.

This system entails the following *demerits* :

(*i*) The cooling is independent of temperature. This may, under certain circumstances result in over cooling the engine.

(*ii*) While moving uphill the cooling requirement is increased because more fuel is burned. However, the coolant circulation is reduced which may result in over-heating the engine.

(*iii*) As soon as the engine is stopped the cooling also ceases. This is undesirable because cooling must continue till the temperatures are reduced to normal values.

3. Thermostat cooling

Too lower cylinder barrel temperature, may result in severe corrosion damage due to condensation of acids on the barrel wall. To avoid such a situation it is customary to use a thermostat (a temperature controlling device) to stop flow of coolant below a pre-set cylinder barrel temperature. Most modern cooling system employ a thermo-static device which prevents the water in the engine jacket from circulating through the radiator for cooling until its temperature has reached to a value suitable for efficient engine operation.

Fig. 23.49 shows a systematic diagram of a thermostatically controlled cooling system. Also shown is a typical thermostat (Fig. 23.50). It consists of bellows which are made of thin copper tubes, partially filled with a volatile

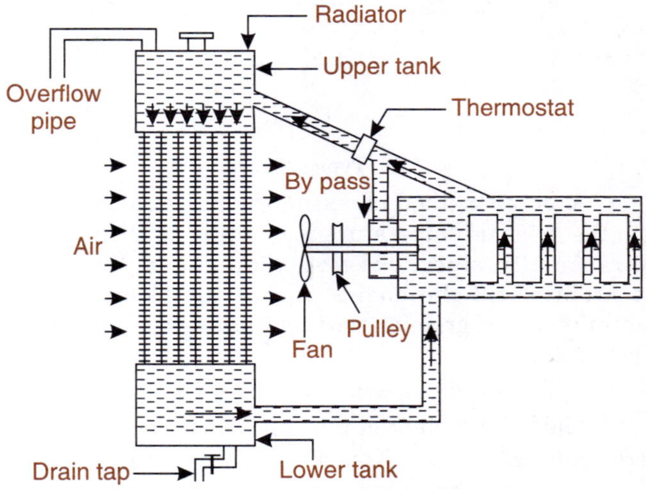

Fig. 23.49. Thermostatically controlled cooling system.

liquid like ether or methyl alcohol. The volatile liquid changes into vapour at the correct working temperature, thus creating enough pressure to expand the bellows. The temperature at which the thermostat operates is set by the manufacturers and cannot be altered. The movement of the bellows opens the main valve in the ratio of temperature rise, increasing or restricting the flow of water from engine to the radiator. Hence when the normal temperature of the engine has been reached the valve opens and circulation of water commences. When the unit is closed the gas condenses and so the pressure falls. The bellows collapse and the thermostat seats on its seat and circulation around thermostat stops. When the thermostat valve is not open and the engine is running the water being pumped rises in pressure and causes the pressure relief valve to open. Thus the water completes its circulation through the by-pass as shown in Figs. 23.49 and 23.50. Now when the temperature of water around the engine-cylinder rises upto a certain limit, it causes the thermostat valve to open. The pressure of water being pumped falls and pressure relief valve closes. So the flow of cooling water in the normal circuit commences through the radiator. This accelerates the rise of temperature of the cylinder walls and water and more power is developed in a few moments of the starting of the engine.

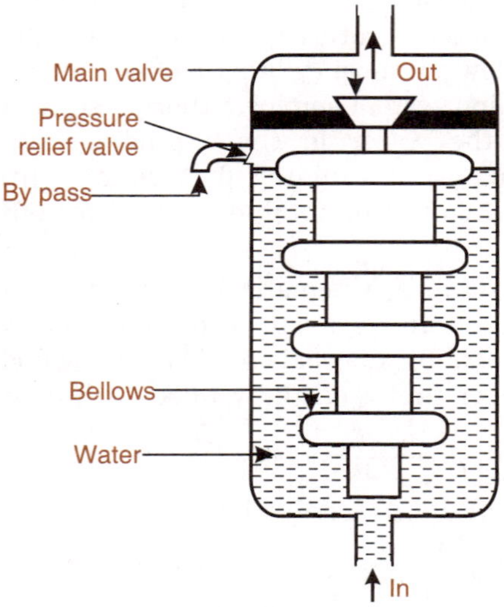

Fig. 23.50. Typical thermostat.

Another method of warming up the radiator water upto the normal temperature is by utilising the shutter in the radiator in order to restrict the incoming air through the radiator till the engine warms up. Thereafter, the shutter is opened gradually so that the desired rate of cooling is achieved.

4. Pressurised water cooling

The boiling point of the coolant can be increased by increasing its pressure. This allows a greater heat transfer to occur in the radiator due to a larger temperature differential. Usually the water pressure is kept between 1.5 bar to 2.0 bar. Use of pressurised water cooling requires an additional valve, called vacuum valve, to avoid formation of vacuum when the water is cooled after engine has stopped. A safety valve in the form of pressure relief valve is provided so that whenever cap is opened the pressure is immediately relieved.

5. Evaporative cooling

In this system, also called steam or vapour cooling, the temperature of the cooling water is allowed to reach a temperature of 100°C. This method of cooling utilises the high latent heat of vapourisation of water to obtain cooling with minimum of water. Fig. 23.51 shows such a system. The cooling circuit is such that coolant is always liquid but the steam formed is flashed off in the separate vessel. The make up water so formed is sent back for cooling. This system is used for cooling of many types of industrial engines.

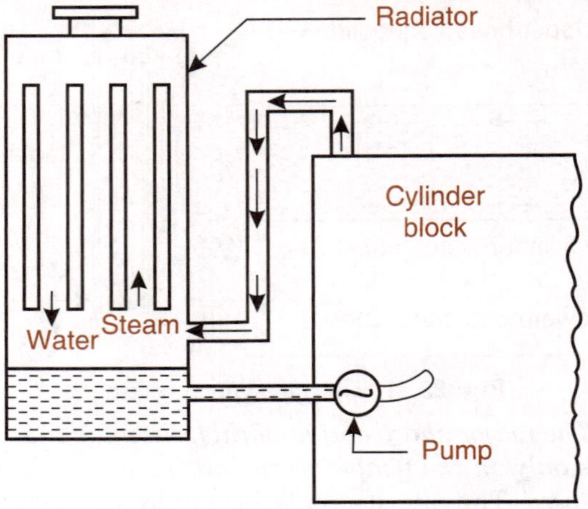

Fig. 23.51. Evaporating cooling.

Advantages of liquid cooling

1. Compact design of engine with appreciably smaller frontal area is possible.
2. The fuel consumption of high compression liquid-cooled engine is rather lower than for air-cooled one.
3. More even cooling of cylinder barrels and heads due to jacketing makes it easier to reduce the cylinder head and valve seating temperature.
4. In case of water-cooled engine installation is not necessary at the front of mobile vehicles, air crafts etc. as the cooling system can be conveniently located wherever required. This is not possible in case of air-cooled engines.
5. The size of the engine does not involve serious problem as far as design of cooling system is concerned. In case of air-cooled engines particularly in high horse power range difficulty is encountered in circulation of required quantity of air for cooling purposes.

Disadvantages

1. This is dependent system in which supply of water for circulation in the jacket is required.
2. Power absorbed by the pump for water circulation is considerably higher than that for cooling fans.
3. In the event of failure of cooling system serious damage may be caused to the engine.
4. Cost of system is considerably high.
5. System requires considerable attention for the maintenance of various parts of system.

23.20. LUBRICATION SYSTEMS

***Lubrication** is the admittance of oil between two surfaces having relative motion.* The purpose of lubrication may be one or more of the following :

1. Reduce friction and wear between the parts having relative motion.

2. Cool the surfaces by carrying away heat generated due to friction.
3. Seal a space adjoining the surfaces such as piston rings and cylinder liner.
4. Clean the surface by carrying away the carbon and metal particles caused by wear.
5. Absorb shock between bearings and other parts and consequently reduce noise.

Properties of lubricants :

The chief qualities to be considered in selecting oil for lubrication are :

1. Viscosity	2. Flash point
3. Fire point	4. Cloud point
5. Pour point	6. Oiliness
7. Corrosion	8. Emulsification
9. Physical stability	10. Chemical stability
11. Neutralisation number	12. Adhesiveness
13. Film strength	14. Specific gravity.

1. **Viscosity.** *It is the ability of the oil to resist internal deformation due to mechanical stresses and hence it is a measure of the ability of the oil film to carry a load.* A more viscous oil can carry a greater load, but it will offer greater friction to sliding movement of the one bearing surface over the other. Viscosity varies with the temperature and hence if a surface to be lubricated is normally at high temperature it should be supplied with oil of a higher viscosity than would be suitable for, say journal bearings.

2. **Flash point.** *It is defined as the lowest temperature at which the lubricating oil will flash when a small flame is passed across its surface.* The flash point of the oil should be sufficiently high so as to avoid flashing of oil vapours at the temperatures occurring in common use. High flash point oils are needed in air compressors.

3. **Fire point.** *It is the lowest temperature at which the oil burns continuously.* The fire point also must be high in a lubricating oil, so that oil does not burn in service.

4. **Cloud point.** When subject to low temperatures the oil changes from liquid stage to a plastic or solid state. In some cases the oil starts solidifying which makes it to appear cloudy. The temperature at which this takes place is called the *cloud point*.

5. **Pour point.** *Pour point is the lowest temperature at which the lubricating oil will pour.* It is an indication of its ability to move at low temperatures. This property must be considered because of its effect on starting an engine in cold weather and on free circulation of oil through exterior feed pipes when pressure is not applied.

6. **Oiliness.** *This is the property which enables oil to spread over and adhere to the surface of the bearing.* It is most important in boundary lubrication.

7. **Corrosion.** A lubricant should not corrode the working parts and it must retain its properties even in the presence of foreign matter and additives.

8. **Emulsification.** A lubricating oil, when mixed with water is emulsified and loses its lubricating property. The emulsification number is an index of the tendency of an oil to emulsify with water.

9. **Physical stability.** A lubricating oil must be stable physically at the lowest and highest temperatures between which the oil is to be used. At the lowest temperature there should not be any separation of solids, and at the highest temperature it should not vaporise beyond a certain limit.

10. **Chemical stability.** A lubricating oil should also be stable chemically. There should not be any tendency for oxide formation.

11. **Neutralisation number.** An oil may contain certain impurities that are not removed during refining. The neutralisation number test is a simple procedure to determine acidity or alkalinity of an oil. It is the weight in milligrams of potassium hydroxide required to neutralise the acid content of one gram of oil.

12. **Adhesiveness.** *It is the property of lubricating oil due to which the oil particles stick with the metal surfaces.*

13. **Film strength.** *It is the property of a lubricating oil due to which the oil retains a thin film between the two surfaces even at high speed and load.* The film does not break and the two surfaces do not come in direct contact. Adhesiveness and film strength cause the lubricant to enter the metal pores and cling to the surfaces of the bearings and journals keeping them wet when the journals are at rest and presenting metal to contact until the film of lubricant is built up.

14. **Specific gravity.** It is a measure of density of oil. It is an indication regarding the grade of lubricant by comparing one lubricant with other. It is determined by a hydrometre which flats in the oil, and the gravity is read on the scale of the hydrometer at the surface of the oil.

Main parts of an engine to be lubricated :

The main parts of an engine which need lubrication are as given under :

(*i*) Main crankshaft bearings.
(*ii*) Big-end bearings.
(*iii*) Small-end or gudgeon pin bearings.
(*iv*) Piston rings and cylinder walls.
(*v*) Timing gears.
(*vi*) Camshaft and camshaft bearings.
(*vii*) Valve mechanism.
(*viii*) Valve guides, valve tappets and rocker arms.

Various lubrication systems used for I.C. engines may be classified as :

1. Wet sump lubrication system.
2. Dry sump lubrication system.
3. Mist lubrication system.

1. **Wet Sump Lubrication System**

These systems employ a large capacity oil sump at the base of crank chamber, from which the oil is drawn by a low pressure oil pump and delivered to various parts. Oil there gradually returns back to the sump after serving the purpose.

(*a*) **Splash system.** Refer to Fig. 23.52. This system is used on some *small four-stroke stationary engines*. In this case the caps on the big ends bearings of connecting rods are provided with scoops which, when the connecting rod is in the lowest position, just dip into oil troughs and thus direct the oil through holes in the caps to the big end bearings. Due to splash of oil it reaches the lower portion of the cylinder walls, crankshaft and other parts requiring lubrication. Surplus oil eventually flows back to the oil sump. Oil level in the troughs is maintained by means of a oil pump which takes oil from sump, through a filter.

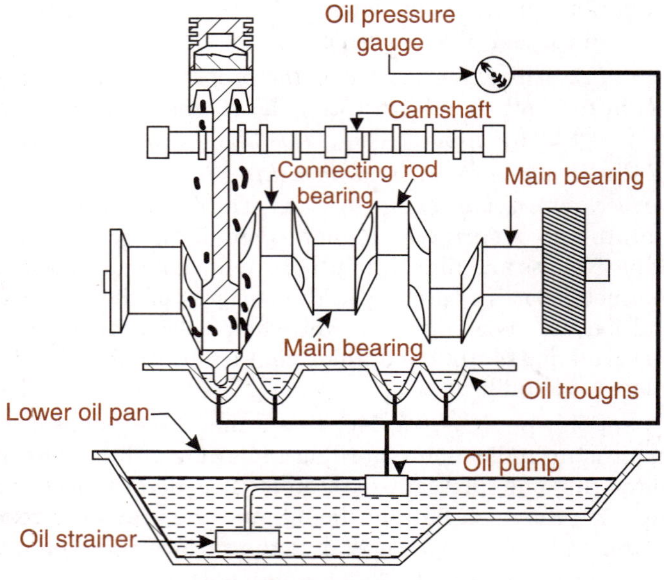

Fig. 23.52. Splash system.

Splash system is suitable for low and medium speed engines having moderate bearing load pressures. For high performance engines, which normally operate at high bearing pressures and rubbing speeds this system does not serve the purpose.

(*b*) **Semi-pressure system.** This method is a combination of splash and pressure systems. It incorporates the advantages of both. In this case main supply of oil is located in the base of crank chamber. Oil is drawn from the lower portion of the sump through a filter and is delivered by means of a gear pump at pressure of about 1 bar to the main bearings. The big end bearings are lubricated by means of a spray through nozzles. Thus oil also lubricates the cams, crankshaft bearings, cylinder walls and timing gears. An oil pressure gauge is provided to indicate satisfactory oil supply.

The system is less costly to install as compared to pressure system. It enables higher bearing loads and engine speeds to be employed as compared to splash system.

(*c*) **Full pressure system.** In this system, oil from oil sump is pumped under pressure to the various parts requiring lubrication. Refer to Fig. 23.53. The oil is drawn from the sump through filter and pumped by means of a gear pump. Oil is delivered by the pressure pump at pressure ranging from 1.5 to 4 bar. The oil under pressure is supplied to main bearings of crankshaft and camshaft. Holes drilled through the main crankshafts bearing journals, communicate oil to the big end bearings and also small end bearings through hole drilled in connecting rods. A pressure gauge is provided to confirm the circulation of oil to the various parts. A pressure regulating valve is also provided on the delivery side of this pump to prevent excessive pressure.

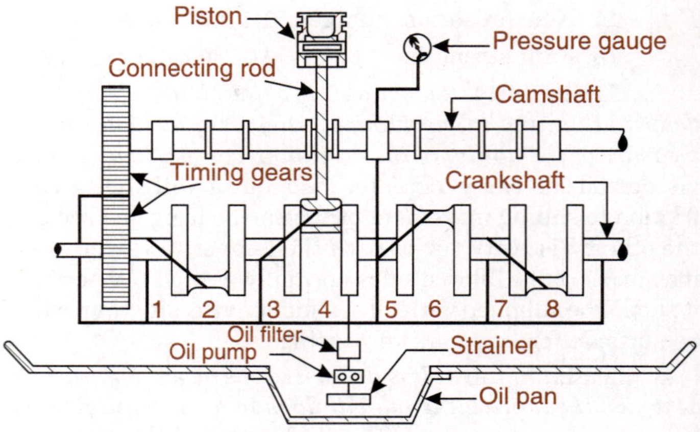

Fig. 23.53. Full pressure system.

- *This system finds favour from most of the engine manufacturers as it allows high bearing pressure and rubbing speeds.*

The general arrangement of ***wet sump lubrication system*** is shown in Fig. 23.54. In this case oil is always contained in the sump which is drawn by the pump through a strainer.

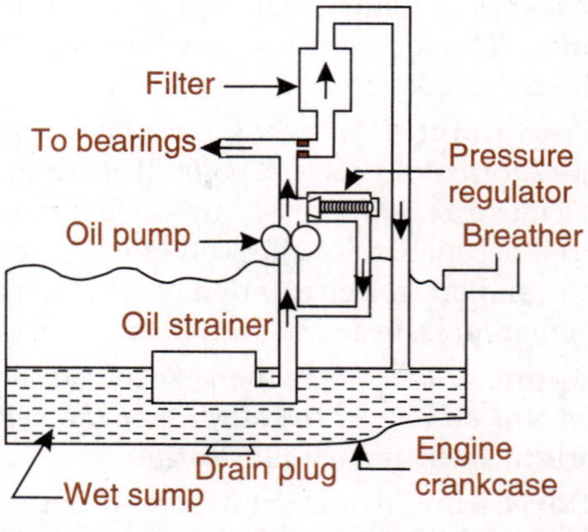

Fig. 23.54. Wet sump lubrication system.

2. **Dry Sump Lubrication System**

Refer to Fig. 23.55. In this system, the oil from the sump is carried to a separate storage tank outside the engine cylinder block. The oil from sump is pumped by means of a sump pump through filters to the storage tank. Oil from storage tank is pumped *to the engine cylinder* through oil cooler. Oil pressure may vary from 3 to 8 bar. *Dry sump lubrication system is generally adopted for high capacity engines.*

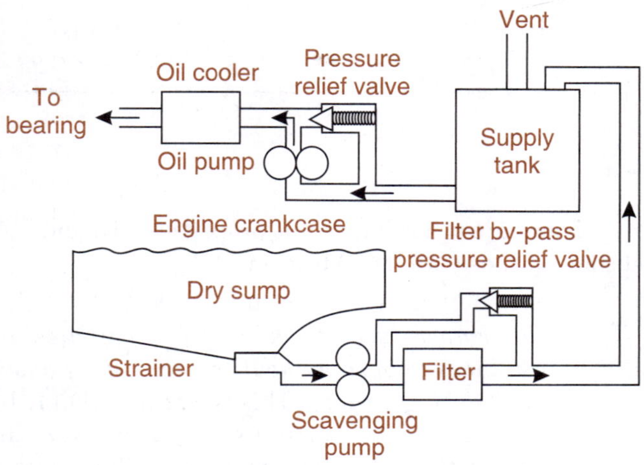

Fig. 23.55. Dry sump lubrication system.

3. **Mist Lubrication System**

This system is used for two-stroke cycle engines. Most of these engines are crank charged, *i.e.,* they employ crankcase compression and thus, are not suitable for crankcase lubrication. These engines are lubricated by adding 2 to 3 per cent lubricating oil in the fuel tank. The oil and fuel mixture is induced through the carburettor. The gasoline is vaporised ; and the oil in the form of mist, goes via crankcase into the cylinder. The oil which impinges on the crankcase walls lubricates the main and connecting rod bearings, and rest of the oil which passes on the cylinder during charging and scavenging periods, lubricates the piston, piston rings and the cylinder.

Advantages :

1. System is simple.
2. Low cost (because no oil pump, filter etc., are required.)

Disadvantages :

1. A portion of the lubricating oil invariably burns in combustion chamber. This bearing oil when burned, and still worse, when partially burned in combustion chamber leads to heavy exhaust emissions and formation of heavy deposit on piston crown, ring grooves and exhaust port which interferes with the efficient engine operation.
2. One of the main functions of lubricating oil is the protection of anti-friction bearings etc., against corrosion. Since the oil comes in close contact with acidic vapours produced during the combustion process, it rapidly loses its anti-corrosion properties resulting in corrosion damage of bearings.
3. For effective lubrication oil and fuel must be thoroughly mixed. This requires either separate mixing prior to use or use of some additive to give the oil good mixing characteristics.
4. Due to higher exhaust temperature and less efficient scavenging the crankcase oil is diluted. In addition some lubricating oil burns in combustion chamber. This results in 5 to 15 per cent higher lubricant consumption for two-stroke engine of similar size.
5. Since there is no control over the lubricating oil, once introduced with fuel, most of the two-stroke engines are over-oiled most of the time.

23.21. GOVERNING OF I.C. ENGINE

The function of a governor is to keep the speed of engine constant irrespective of the changes in load on the engine. The governor is usually of centrifugal type.

In petrol engine, the control is exercised by means of a throttle valve which is placed in intake manifold. The quantity of mixture entering the cylinder depends on the amount of opening of throttle valve. The position of throttle valve is controlled by the governor (centrifugal type). In diesel engines, the flow of fuel is controlled by centrifugal governor which actuates link rods which in turn operate some device on the fuel pump and consequently portion of the fuel by-passes. The governor in plunger type injection pump alters the relative angular position of the plunger.

Following are the methods of governing I.C. engines :

(*i*) Hit and miss method

(*ii*) Quality governing

(*iii*) Quantity governing

(*i*) **Hit and miss method.** Refer to Fig. 23.56. When the speed increases the permissible value the governor sleeve *S* gets lifted up, as a result of which the lever *A* lifts the distant piece *B*, so that the pecker *K* misses it. Thus the gas inlet valve *V* does not open and the usual charge does not enter the cylinder. This continues until the speed is reduced and *B* occupies its initial position. Explosions are thus missed intermittently but every charge is of normal strength. *This method is commonly used in gas engines.*

(*ii*) **Quality governing.** In this method of governing, the *mixture strength is altered*. In gas engine it is effected by reducing the amount of gas supplied to the engine. This is accomplished by varying the lift of the gas valve. In oil engines, quality governing is carried out by varying the quality of fuel oil entering the cylinder per cycle, it is done by changing the angular position of the helical groove of the pump plunger.

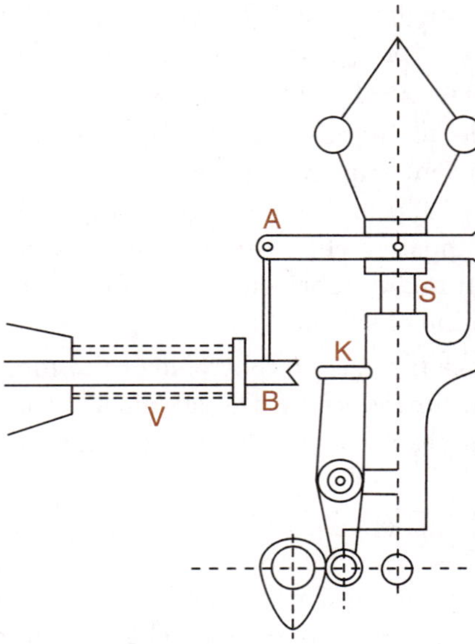

Fig. 23.56. Hit and miss governing.

In this type of governing, the ignition is not always satisfactory and thermal efficiency is reduced.

(*iii*) **Quantity governing.** Here *mixture strength remaining the same, the quantity of mixture entering the cylinder is altered*. When the speed is too high a lesser amount of charge is admitted into the cylinder. The compression ratio and air standard efficiency remain unchanged. The pressure after compression and during working stroke is lower, the less work thus obtained during the cycle reduces the speed. *This method is preferred for large engines*.

23.22. LIQUID FUELS FOR RECIPROCATING COMBUSTION ENGINES

The majority of fuels for I.C. engines are obtained by the distillation of crude petroleum. This may be followed by pressure and heat treatment to reform some of the products coupled with mixing of some additives.

The distribution of the boiling points of some fuels by specific gravity is shown in Fig. 23.57.

Petrol *may be defined as a distillate with a boiling point that does not exceed 200°C* and ***diesel fuel*** *as a mixture of gas oils and fuel oils.*

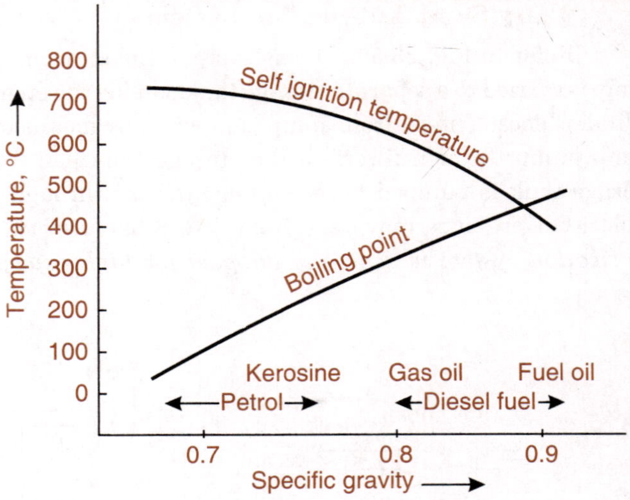

Fig. 23.57

There are other important properties of liquid fuels and some of these are listed below :

1. **The self ignition temperature (S.I.T.).** It *is the temperature at which the fuel ignites and continues to burn without the need of a flame to initiate the burning*. This is not a constant but depends on the *mixture straight*, *pressure* and *combustion chamber geometry*. The effect of pressure is most important and is shown in Fig. 23.58.

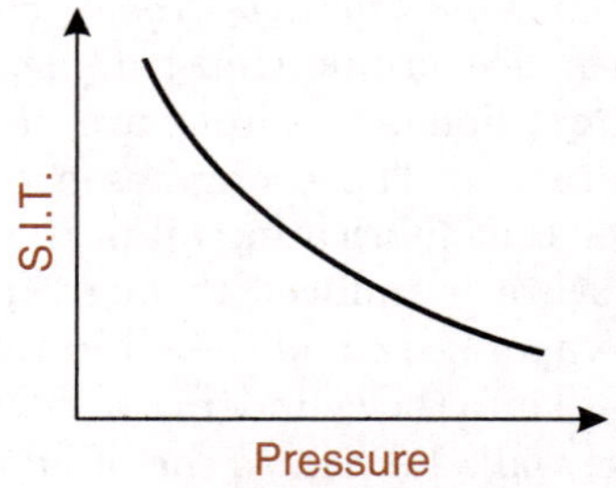

Fig. 23.58

2. **Calorific valve.** The calorific value influences the available power of the engine. When considering calorific valve, the effect of specific gravity on fuels marketed by volume should be taken into account.
3. **Enthalpy of vaporisation.** The enthalpy of vaporisation affects the temperature of the charge. A *fuel with a high enthalpy (viz. Alcohol) of vaporisation will give a cooler charge* and thus *a greater mass can be contained in the same volume.*
4. **Volatility.** The volatility of a fuel is of importance *for promoting fast reations. Low boiling point fuels are more volatile.*

The following points are worth noting :

- In a **S.I. engine** which uses petrol, the fuel should be volatile so that evaporation is quick enough for combustible mixture to be available for the spark to ignite. The self ignition temperature should be *high enough to ensure that ignition does not occur accidently*.
- For a **C.I. engine** which uses gas, diesel or fuel oil, the self ignition temperature must be *below the temperature of air after compression in order that combustion may be completed*.

Even when a reactive mixture is at a temperature that is greater than the self ignition temperature there is a time lag before the combustion appears to commence which is known as the *induction period*. The induction period is *proportional to the square of the self ignition temperature*. In order to promote the fast reactions necessary in C.I. it is found necessary that the *temperature of the air after compression must be greatly in excess of the self ignition temperature*.

23.23. COMBUSTION PHENOMENON IN S.I. ENGINES

Normal Combustion. In a spark-ignition engine, a single intensely high temperature spark passes across the electrodes, leaving behind a thin thread of flame. From this thin thread, combustion spreads to the envelope of mixture immediately surrounding it at a rate which depends primarily upon the temperature of the flame front itself and to a secondary degree, upon both the temperature and the density of the surrounding envelope. In the actual engine cylinder, the mixture is not at rest but is in highly turbulent condition. The turbulence breaks the filament of a flame into a ragged front, thus presenting a far greater area of surface from which heat is being radiated ; hence its advance is speeded up enormously.

According to Ricardo, the combustion process can be imagined as if developing in two stages, *one the growth and development of a self-propagating nucleus of flame (ignition lag), and the other the spread of that flame throughout the combustion chamber*. The former is a chemical process depending upon the nature of the fuel, temperature and pressure, the proportion of the exhaust gas and also upon the temperature co-efficient of the fuel, *i.e.*, the relationship between temperature and rate of acceleration of oxidation or burning.

Fig. 23.59 shows the p-θ diagram of a petrol engines, *LNQM* assumes compression curve having no ignition. In fact ignition takes place at about 35° before the top dead centre at point *N*. First stage of combustion, the ignition lag, starts from this point and no pressure rise is noticeable. *Q* is the point where the pressure rise can be detected. From this point it deviates from the simple compression (motoring) curve. The time lag between first igniting of fuel and the commencement of the main phase of combustion is called the *period of incubation* or is also known as ***ignition lag.*** The time is normally about 0.0015 seconds. The maximum pressure is reached at about 12° after top dead centre point. Although the point of maximum pressure marks the completion of flame travel, it does not mean that at this point the whole of the heat of fuel has been liberated, for even after the passage of the flame, some further chemical adjustments due to reassociation, etc., will continue to a greater or less degree throughout the expansion stroke. This is known as ***after burning.***

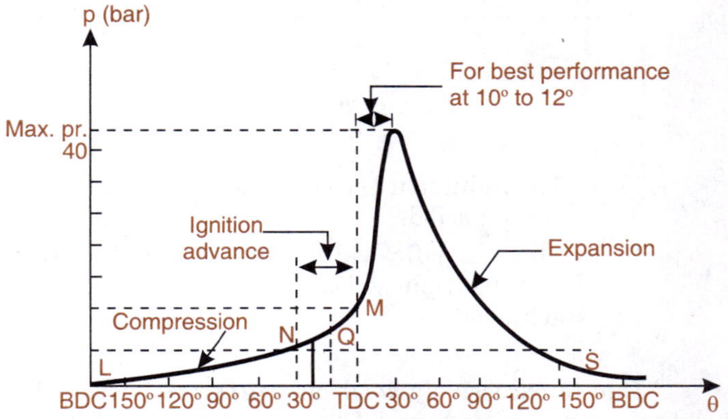

Fig. 23.59. Pressure-crank angle diagram of a petrol engine.

Factors affecting normal combustion in S.I. engines

The factors which affect normal combustion in S.I. engines are brief described below :

1. ***Induction pressure.*** As the pressure falls delay period increases and the ignition must be earlier at low pressures. A *vacuum control* may be incorporated.
2. ***Engine speed.*** As speed increases the constant time delay period needs more crank angle and ignition must be earlier. A *centrifugal control* may be employed.
3. ***Ignition timing.*** If ignition is too early the peak pressure will occur too early and work transfer falls. If ignition is too late the peak pressure will be low and work transfer falls. Combustion may not be complete by the time the exhaust valve opens and the valve may burn.
4. ***Mixture strength.*** Although the stoichiometric ratio should give the best results, the effect of dissociation shown in Fig. 23.60 is to make a slightly rich mixture necessary for maximum work transfer.
5. ***Compression ratio.*** An increase in compression ratio increases the maximum pressure and the work transfer.
6. ***Combustion chamber.*** The combustion chamber should be designed to *give a short flame path to avoid knock and it should promote optimum turbulence.*

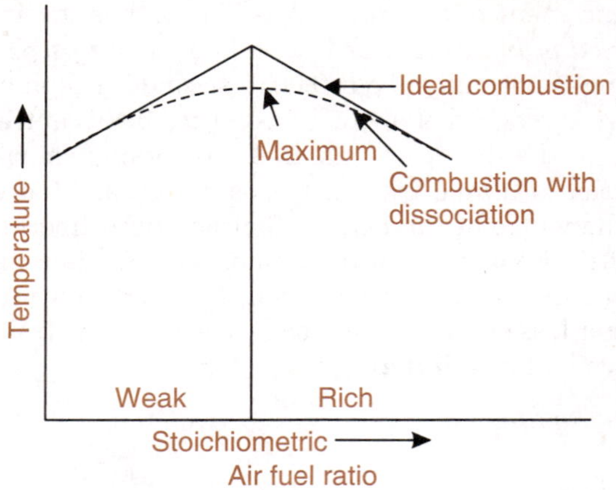

Fig. 23.60

7. ***Fuel choice.***

- The induction period of the fuel will affect the delay period.
- The calorific value and the enthalpy of vaporisation will affect the temperatures achieved.

Abnormal Combustion

Due to excessively weak mixtures combustion may be slow or may be mistimed. These are however obvious.

There are *two combustion abnormalities* which are less obvious :

- The *first* of these is **free** or **post ignition** of the mixture by *incandescent carbon particles in the chamber*. This will have the *effect of reducing the work transfer.*
- The *second* abnormality is generally known as **knock** and is a complex condition with many facets. A simple explanation shows that knock occurs when the unburnt portion of the gas in the combustion chamber is heated by combustion and radiation so that its *temperature becomes greater than the self ignition temperature.* If *normal progressive combustion is not completed before the end of the induction period* then a *simultaneous explosion of the unburnt gas will occur*. This explosion is accompanied by a detonation (pressure) wave which will be repeatedly rejected from the cylinder walls setting up a high frequency resonance which gives an audible noise. The *detonation wave causes excessive stress and also destroys the thermal boundary layer at the cylinder walls causing over heating.*

Note. Refer to Articles 23.24 and 23.25 for details of Pre-ignition and detonation.

23.24. PRE-IGNITION

Refer to Fig. 23.61. Some parts of combustion chamber like the exhaust valve or spark plug electrode may have very high temperature (700°C or more) before the spark is applied near the end of compression stroke. Carbon particles formed due to incomplete combustion deposit on the combustion chamber wall and have also very high temperature. The overheated parts or glowing carbon deposits may cause premature ignition of the fuel-air mixture during compression before the spark is applied and flame front begins to propagate from these surfaces. This *premature combustion which starts before application of spark is called* ***pre-ignition***. When the spark is applied it will produce a separate flame front. As a result multiple

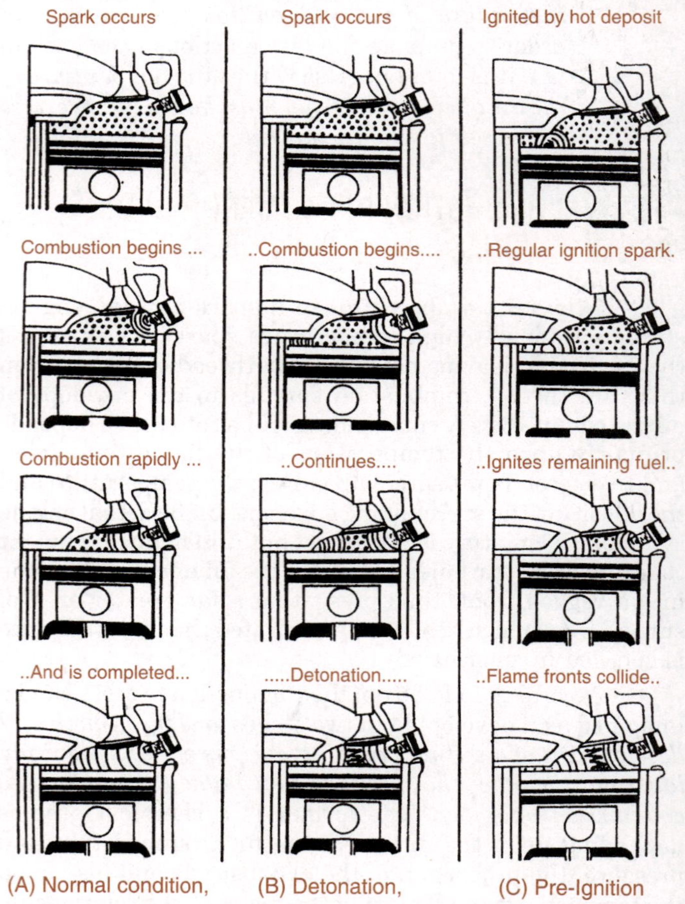

Fig. 23.61. Normal combustion, detonation and pre-ignition.

flame fronts will propagate in the mixture increasing the rate of pressure rise. If pre-ignition occurs sufficiently early in the compression stroke, the power output will be reduced due to large increase in compression work for compressing the combustion products. Pre-ignition causes much higher pressure and temperature of last part of the charge than normal ignition because of its earlier occurrence on the compression stroke and also due to appearance of multiple flame fronts after the spark. Therefore, it increases the tendency of detonation in engines. *Pre-ignition is a serious type of abnormal combustion*. It increases the heat transfer to the cylinder walls because high temperature gases remain in contact with the cylinder for a longer period. The load on the crankshaft during compression is abnormally high. This may cause crank failure. *Over-heated spark plugs and exhaust valves which are the main causes of pre-ignition should be carefully avoided in engines.*

Tests for pre-ignition

The standard test for pre-ignition is to shut off the ignition. If the engine still fires, it is assumed that pre-ignition was taking place when the ignition was on. Experience shows that this assumption is not always valid. *Sudden loss of power with no evidence of mechanical malfunctioning is fairly good evidence of pre-ignition.* The best proof of pre-ignition is the appearance of an indicator card taken with a high speed indicator of the balanced-pressure type.

23.25. DETONATION OR "PINKING"

At present the amount of power that can be developed in the cylinder of a petrol engine is fixed by the liability of a fuel to detonate, *i.e.*, just before the flame has completed its course across the combustion chamber and remaining unburnt charge fires throughout its mass spontaneously without external assistance.

The result is a tremendously rapid and local increase in pressure which sets up pressure waves that hit the cylinder walls with such violence that the walls emit a sound like a '*ping*'. It is the ping that manifests detonation. *Thus a very sudden rise of pressure during combustion accompanied by metallic hammer like sound is called* ***detonation***.

The region in which detonation occurs is farthest removed from the sparking plug, and is named the "detonation zone" and even with severe detonation this zone is rarely more than one quarter the clearance volume.

Process of detonation

After the passage of the spark there is a rise of temperature and pressure due to the combustion of the fuel ignited, and to a less extent by the upward motion of the piston. Both temperature and pressure combine to accelerate the velocity of the flame front in compressing the unburnt portion of the charge in the detonation zone. Ultimately the temperature in this zone reaches such a high value that chemical reaction proceeds at a far greater rate than that at which the flame is advancing. Here we have *combustion unaccompanied by flame, producing a very high rate of pressure rise*.

Theories of detonation

There are two general theories of knocking/ detonation :

(*i*) The auto-ignition theory

(*ii*) The detonation theory.

(*i*) **Auto-ignition theory.** Auto-ignition refers to initiation of combustion without the necessity of a flame. The auto-ignition theory of knock assumes that the flame velocity is *normal* before the onset-of auto-ignition and that gas vibrations are created by a number of end-gas elements auto-igniting almost simultaneously.

(*ii*) **Detonation theory**. In the auto-ignition theory, it is assumed that the flame velocity is normal before the onset of auto-ignition whereas in detonation theory a true *detonating wave* formed by preflame reactions has been proposed as the mechanism for explosive auto-ignition. Such a shock wave would travel through the chamber at about twice the sonic velocity and would compress the gases to pressures and temperatures where the reaction should be practically instantaneous.

In fact knocking or detonation is a complex phenomenon and no single explanation may be sufficient to explain it fully.

Effects of detonation :

1. Noise and roughness
2. Mechanical damage
3. Carbon deposits
4. Increase in heat transfer
5. Decrease in power output and efficiency
6. Pre-ignition.

Control of detonation :

The detonation can be *controlled or even stopped by the following methods* :

1. Increasing engine r.p.m.
2. Retarding spark.
3. Reducing pressure in the inlet manifold by throttling.
4. Making the ratio too lean or too rich, preferably latter.
5. ***Water injection.*** Water injection increases the delay period as well as reduces the flame temperature.
6. *Use of high octane fuel can eliminate detonation.* High octane fuels are obtained by adding additives known as dopes (such as tetra-ethyl of lead, benzol, xylene etc.), to petrol.

Fig. 23.61 shows normal combustion, detonation and pre-ignition.

23.26. FACTORS AFFECTING KNOCK

The *likelihood of knock is increased by any reduction in the induction period of combustion and any reduction in the progressive explosion flame velocity*. Particular factors are listed below :

1. **Fuel choice :** A low self ignition temperature promotes knock.
2. **Induction pressure :** Increase of pressure decreases the self ignition temperature and the induction period. Knock will tend to occur *at full throttle.*
3. **Engine speed :** Low engine speeds will give low turbulence and low flame velocities (combustion period is constant in angle) and knock may occur at low speed.
4. **Ignition timing :** *Advanced ignition timing* increases peak pressures and *promotes* knock.
5. **Mixture strength :** *Optimum mixture strength* gives high pressures and *promotes* knock.
6. **Compression ratio :** *High compression ratios* increase the cylinder pressures and *promote* knock.
7. **Combustion chamber design :** Poor design gives long flame paths, poor turbulence and insufficient cooling all of which *promote* knock.
8. **Cylinder cooling :** *Poor cooling* raises the mixture temperature and *promotes* knock.

23.27. PERFORMANCE NUMBER (PN)

Performance number *is a useful measure of detonation tendency*. It has been developed from the conception of knock limited indicated mean effective pressure (KLIMEP), when inlet pressure is used as the dependent variable.

$$\text{Performance number (PN)} = \frac{\text{KLIMEP of test fuel}}{\text{KLIMEP of iso-octane}}$$

The performance number is *obtained on specified engine*, under specified set of conditions by varying the inlet pressure.

Highest Useful Compression Ratio (HUCR)

The **highest useful compression ratio** *is the highest compression ratio employed at which a fuel can be used in a specified engine under specified set of operating conditions, at which detonation first becomes audible with both the ignition and mixture strength adjusted to give the highest efficiency.*

23.28. DESIRABLE CHARACTERISTICS OF COMBUSTION CHAMBER FOR S.I. ENGINES

From the point of view of *attaining highest resistance to detonation* under a given set of service conditions, and also *promote high power output ; high thermal efficiency, and smooth operation,* the following would appear to be *desirable characteristics for S.I. engines* :

1. Short combustion time.
2. Short ratio of flame path to bore.
3. Absence of hot surfaces in the end gas region.
4. Use of squish areas particularly in the end gas region.
5. High velocity through the inlet valve.
6. Large surface to volume ratio for the end gas (to have adequate cooling in the detonation zone).
7. Cooling of hot spots.

23.29. COMBUSTION CHAMBER DESIGN—S.I. ENGINES

Combustion chambers are usually designed with every possible attempt made to meet the following objectives :

1. To regulate the rate of pressure rise such that the greatest force is applied to the piston as closely after T.D.C. on the 'power stroke' as possible, with a gradual decrease in the force on the piston during the power stroke. The forces must be applied to the piston smoothly, however, thus placing a limit on the rate of the pressure rise, as well as the position of the peak pressure with respect to T.D.C.

2. To prevent the possibility of detonation at all times.

To obtain these objectives attempt is made to design the combustion chambers with the *following factors in mind :*

(*i*) To achieve the highest possible flame front velocity through the creation of high turbulence of the minute "swirl" type.

(*ii*) To burn the largest mass of charge as soon as possible after ignition (consistent with a smooth application of force), with progressive reduction in the mass of the charge burnt toward the end of the combustion.

(*iii*) To reduce the possibility of detonation by :

— Reducing the temperature of the last portion of the charge to burn, through the application of a high surface to volume ratio in that part of the combustion chamber where this portion burns. Such a ratio increases the heat transfer to the combustion chamber walls and thereby tends to reduce the temperature of the final unburned charge.

— Reducing the distance for the flame to travel by centrally locating the spark plug, or in some engines, by using dual spark plugs.

Fig. 23.62 shows a few representative types of combustion chambers, of which there are many more variation.

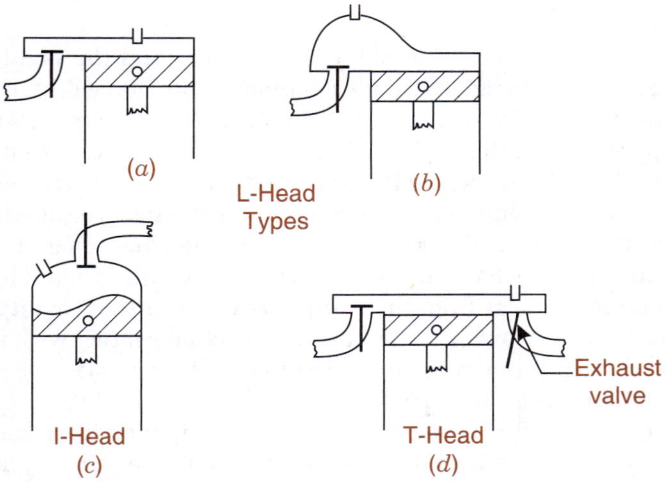

Fig. 23.62. Some typical combustion chambers (not to scale).

It may be noted that these chambers are designed to obtain the objectives outlined above, namely :

— A *high combustion rate at the start.*

— A *high surface to volume ratio near the end of burning.*

— A *rather centrally located spark plug.*

23.30. OCTANE NUMBER

The compression ratio which can be utilized depends on the fuel to be used and a scale has been developed against which the knock tendency of a fuel can be rated. The rating is given an *octane number.*

— The fuel under test is compared with a mixture of *iso-octane* (high rating) and *normal heptane* (low rating), by volume. *The octane number of the fuel is the percentage of octane in the reference mixture which knocks under the same conditions as the fuel.* The number obtained depends on the conditions of the test and the two main methods in use (the *research* and *motor* methods) give different ratings for the same fuel. The motor test is carried out at the higher temperature and gives the lower rating. The difference between the two is taken as the measure of the temperature sensitivity of the fuel.

— High octane fuels (upto 100) can be produced by refining techniques, but it is done more cheaply, and more frequently, by the use of anti-knock *additives*, such a *tetraethyl lead*. (An addition of 1.1 cm^3 of tetraethyl lead to one litre of 80 octane petrol increases the octane number to 90). Fuels have been developed which have a higher anti-knock rating than iso-octane and this has lead to an extension of the octane scale.

Advantages of high-octane fuel

The *advantages* of high-octane fuel are as follows :

1. The engine can be operated at high compression ratio and therefore, with high efficiency without detonation.
2. The engine can be supercharged to high output without detonation.
3. Optimum spark advance may be employed raising both power and efficiency.

23.31. TURBULENCE IN S.I. ENGINES

- Turbulence plays a very important role in combustion phenomenon in S.I. (as well C.I.) engines. The flame speed is very low in non-turbulent mixtures. *A turbulent motion of the mixture intensifies the processes of heat transfer and mixing of the burned and unburned portions in the flame front* (diffusion). These two factors cause the velocity of turbulent flame to increase practically in proportion to the turbulent velocity. The turbulence of the mixture is due to admission of fuel-air mixture through comparatively narrow sections of the intake pipe, valves etc., in the suction stroke. The turbulence can be increased at the end of the compression stroke by suitable design of combustion chamber which involves the geometry of cylinder head and piston crown.
- The degree of turbulence *increases directly with the piston speed.*

The *effects of turbulence* can be summed up as follows :

1. Turbulence accelerates chemical action by intimate mixing of fuel and oxygen. Thus weak mixtures can be burnt.
2. The increase of flame speed due to turbulence reduces the combustion time and hence minimises the tendency to detonate.
3. Turbulence increase the heat flow to the cylinder wall and in the limit excessive turbulence may extinguish the flame.
4. Excessive turbulence results in the more rapid pressure rise (though maximum pressure may be lowered) and the high pressure rise causes the

crankshaft to spring and rest of the engine to vibrate with high periodicity, resulting in rough and noisy running of the engine.

The following points are worth noting :

Swirl :

- *The main macro mass motion within the cylinder is rotational motion called* ***swirl***. It is generated by constructing the intake system to give a tangential component to the intake flow as it enters the cylinder. *This is done by shaping and contouring the intake manifold, valve ports and even the piston face.*
- Swirl *greatly enhances the mixing of air and fuel* to give a homogeneous mixture in the very short time available for this in modern high speed engines. It is also a *main mechanism for very rapid spreading of the flame front during the combustion process.*

Squish and Tumble :

- As the piston approaches T.D.C. at the end of compression stroke, the volume around the outer edges of the combustion chamber is suddenly reduced to a very small value. Many modern combustion chamber designs have most of the clearance volume near the centreline of the cylinder. As the piston approaches T.D.C. the gas mixture occupying the volume at the *outer radius of the cylinder is forced radially inward as this outer volumne is reduced to near zero. This radial inward motion of the gas mixture is called "squish".* It adds to other mass motions within the cylinder to mix the air and fuel and to quickly spread the flame front. Maximum squish velocity usually occurs at about 10° before T.D.C.
- As the piston nears T.D.C. *squish motion generates a secondary rotational flow* called "**tumble**". *This rotation occurs about a circumferential axis near the outer edge of the piston bowl.*

23.32. COMBUSTION PHENOMENON IN C.I. ENGINES

The process of combustion in the compression ignition (C.I.) engine is fundamentally different from that in a spark-ignition engine. In C.I. engine combustion occurs by the high temperature produced by the compression of the air, *i.e.,* it is an *auto-ignition*. For this a minimum compression ratio of 12 is required. The efficiency of the cycle increases with higher values of compression ratio but the maximum pressure reached in the cylinder also increases. This requires heavier construction. The upper limit of compression ratio in a C.I. engine is due to mechanical factor and is a compromise between high efficiency and low weight and cost. The normal compression ratios are in the range of 14 to 17, but may be upto 23. The air-fuel ratios used in the C.I. engine lie between 18 and 25 as against about 14 in the S.I. engine, and hence C.I. engines are bigger and heavier for the same power than S.I. engines.

In the C.I. engine, the intake is air alone and the fuel is injected at high pressure in the form of fine droplets near the end of compression. This leads to delay period in the C.I. engine, is greater than that in the S.I. engine. The *exact phenomenon of combustion in the C.I. engine* is explained below :

Each minute droplet of fuel as it enters the highly heated air of engine cylinder is quickly surrounded by an envelope of its own vapour and this, in turn and at an appreciable interval is inflamed at the surface of the envelope. To evaporate the liquid, latent heat is abstracted from the surrounding air which reduces the temperature of the thin layer of air surrounding the droplet, and some time must elapse before this temperature can be raised again by abstracting heat from the main bulk of air in this vicinity. As soon as this vapour and the air in actual contact with it reach a certain temperature, ignition will take place. Once ignition has been started and a flame established the heat required for further evaporation will be supplied from that released by combustion. The vapour would be burning as fast as it can find fresh oxygen, *i.e.*, it will depend upon the rate at which it is moving through the air or the air is moving past it.

In the C.I. engine, the fuel is not fed in at once but is spread over a definite period. The first arrivals meet air whose temperature is only a little above their self-ignition temperature and the delay is more or less prolonged. The later arrivals find air already heated to a far higher temperature by the burning of their predecessors and therefore light up much more quickly, almost as they issue from the injector nozzle, but their subsequent progress is handicapped for there is less oxygen to find.

If the air within the cylinder were motionless, only a small proportion of the fuel would find sufficient oxygen, for it is impossible to distribute the droplets uniformly throughout the combustion space. Therefore some air movement is absolutely essential, as in the S.I. engine. But there is a fundamental difference between the air movements in the two types of engines. In the S.I. engine we call it *turbulence* and mean a confusion of whirls and eddies with no general direction of flow (to break up the surface of the flame front, and to distribute the shreds of flame throughout an externally prepared combustible mixture). In the C.I. engine we call it *air swirl* and mean an orderly movement of the whole body of the air, with or without some eddying or turbulence, so as to bring a continuous supply of fresh air to each burning droplet and sweep away the products of combustion which otherwise tend to suffocate it.

Three phases of C.I. engine combustion

In the C.I. engine, combustion may be considered in *three distinct stages* as shown in Fig. 23.63.

1. Ignition delay period.
2. Period of rapid or uncontrolled combustion.
3. Period of controlled combustion.

The third phase is followed by *after burning* (or burning on the expansion stroke), which *may be called the fourth phase of combustion*.

1. **Ignition delay period.** The delay period is counted from the start of injection to the point where the p-θ combustion curve departs from air compression (or no ignition or motoring) curve. The delay period can be roughly sub-divided into ***physical delay*** and ***chemical delay***. *The period of physical delay is the time between the beginning of injection and the attainment of chemical reaction conditions*. In the physical delay period, the *fuel is atomized, vaporized, mixed with air, and raised in temperature*. In the *chemical delay period reaction starts slowly and then accelerates until inflammation or ignition takes place* (it may be noted that the ignition delay in the S.I. engine is essentially equivalent to the chemical delay in the C.I. engine).

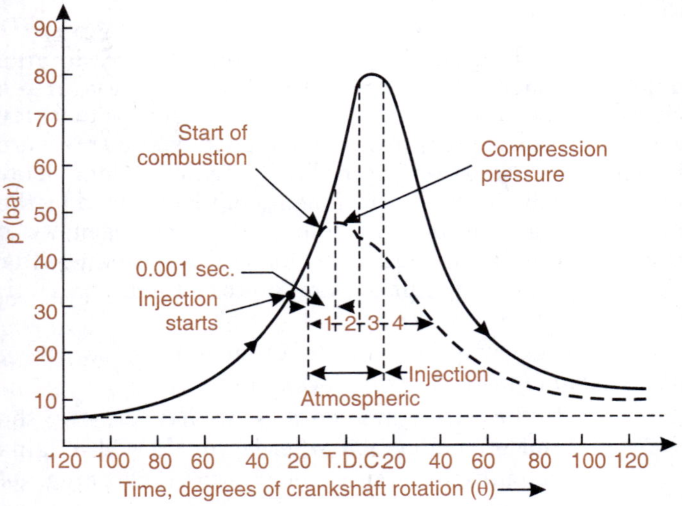

Fig. 23.63. Combustion phenomenon of C.I. engine.

The delay period exerts a great influence in the C.I. engine combustion phenomenon. It is clear that the pressure reached during the second stage will depend upon the duration of the delay period ; the longer the delay, the more rapid and higher the pressure rise, since more fuel will be present in the cylinder before the rate of burning comes under control. This causes rough running and may cause *diesel knock*. Therefore we must *aim to keep the delay period as short as possible, both for the sake of smooth running and in order to maintain control over the pressure changes. But some delay period is necessary otherwise the droplets would not be dispersed in the air for complete combustion*. However, the delay period imposed upon is greater than what is needed and the designer's efforts are to shorten it as much as possible.

2. **Period of rapid or uncontrolled combustion.** The second stage of combustion in C.I. engines, after the delay period, is the period of rapid or uncontrolled combustion. This period is counted from the end of the delay period to the point of maximum pressure on the indicator diagram. In this second stage of combustion, the rise of pressure is rapid because during the delay period the droplets of fuel have had time to spread themselves out over a wide area and they have fresh air all around them. *About one-third of heat is evolved during this process*.

The rate of pressure rise depends on the amount of fuel present at the end of delay period, degree of turbulence, fineness of atomization and spray pattern.

3. **Period of controlled combustion.** At the end of second stage of combustion, the temperature and pressure, are so high that the fuel droplets injected in the third stage burn almost as they enter and any further pressure rise can be controlled by purely mechanical means, *i.e.*, by the injection rate. The period of controlled combustion is assumed to end at maximum cycle temperature. *The heat evolved by the end of controlled combustion is about 70 to 80 per cent.*

4. **After burning.** The combustion continues even after the fuel injection is over, because of poor distribution of fuel particles. This burning may continue in the expansion stroke upto 70° to 80° of crank travel from T.D.C. This continued burning, called the *after burning*, may be considered as the fourth stage of the combustion. The total heat evolved by the end of entire combustion process is 95 to 97% ; 3 to 5% of heat goes as unburned fuel in exhaust.

In the p-V diagram, the stages of combustion are not seen because of little movement of piston with crank angle at the end and reversal of stroke. So for studying the combustion stages, therefore, a pressure-crank angle or time, p-θ or p-t diagram is invariably used. In the actual diagram, the various stages of combustion look merged, yet the individual stage is distinguishable.

Factors Affecting Combustion in C.I. Engines :

In compression ignition combustion, the length of the delay period plays a vital role. This period serves a useful purpose in that it allows the fuel jet to penetrate well into the combustion space. If there were no delay the fuel would burn at the injector and there would be an oxygen deficiency around the injector resulting in incomplete combustion. If delay is too long the amount of fuel available for stimultaneous explosion is too great and the resulting pressure rise is too rapid :

Delay is *reduced* by the following :

— High charge temperature ;
— High fuel temperature ;
— Good turbulence ;
— A fuel with a short induction period.

Fuel choice : The choice of fuel has a large effect on combustion since induction period of the fuel contributes a major part of the delay period.

Combustion chamber design : The shape of the chamber is critical so that the fuel is introduced with swirl and the pistons are also shaped to produce a '*squish*' effect, the whole concept of design being to give controlled motion of the air and fuel rather than rely on haphazard turbulence.

Engine load : In most engines injection starts at a fixed crank angle and continues as long as the rack setting allows. At light loads all the fuel will be injected in the delay and simultaneous explosion periods, or at very light loads when temperatures are lower and the delay period longer injection may be completed in the delay period.

Engine speed : The effect of speed is allied to the design of the combustion chamber and varies with individual designs. Delay may be reduced or increased.

Abnormal combustion in C.I. engines :

In C.I. engines abnormal combustion is not as great a problem as in S.I. engines. The only abnormality is "**diesel knock**". *This occurs when the delay period is excessively long so that there is a large amount of fuel in the cylinder for the simultaneous explosion phase.* The rate of pressure rise per degree of crank angle is then so great than an *audible knocking sound occurs.* Running is rough and if allowed to become extreme the *increase in mechanical and thermal stresses may damage the engine.* **Knock** *is thus a function of the fuel chosen and may be avoided by choosing a fuel with characteristics that do not give too long a delay period.*

23.33. DELAY PERIOD (OR IGNITION LAG) IN C.I. ENGINES

— In C.I. (compression ignition) engine, the fuel which is in atomised form is considerably colder than the hot compressed air in the cylinder. Although the actual ignition is almost instantaneous, an appreciable time elapses before the combustion is in full progress. This time occupied is called the *delay period* or *ignition lag*. *It is the time immediately following injection of the fuel during which the ignition process is being initiated and the pressure does not rise beyond the value it would have due to compression of air* :

— The delay period extends for about 13°, movement of the crank. The time for which it occurs decreases with increase in engine speed.

The *delay period depends upon the following :*

(*i*) Temperature and pressure in the cylinder at the time of injection.

(*ii*) Nature of the fuel mixture strength.

(*iii*) Relative velocity between the fuel injection and air turbulence.

(*iv*) Presence of residual gases.

(*v*) Rate of fuel injection.

(*vi*) To small extent the finess of the fuel spray.

The delay period increases with load but is not much affected by injection pressure.

— *The delay period should be as short as possible since a long delay period gives a more rapid rise in pressure and thus causes knocking.*

23.34. DIESEL KNOCK

If the delay period in C.I. engines is long a large amount of fuel will be injected and accumulated in the chamber. The auto-ignition of this large amount of fuel may cause high rate of pressure rise and high maximum pressure which may cause *knocking* in diesel engines. A long delay period not only increases the amount of fuel injected by the moment of ignition but also improves the homogeneity of the fuel-air mixture and its chemical preparedness for explosion type self-ignition similar to detonation is S.I. engines.

The following are the ***differences*** in *the knocking phenomena of the S.I. and C.I. engines* :

1. In the S.I. engine, the detonation occurs near the end of combustion whereas in the C.I. engine detonation occurs near the beginning of combustion.
2. The detonation in the S.I. engine is of a homogeneous charge causing very high rate of pressure rise and very high maximum pressure. In the C.I. engine, the fuel and air are imperfectly mixed and hence the rate of pressure rise is normally lower than that in the detonating part of the charge in the S.I. engine.
3. In the C.I. engine the fuel is injected into the cylinder only at the end of the compression stroke, there is no question of pre-ignition as in S.I. engine.
4. In the S.I. engine, it is relatively easy to distinguish between knocking and non-knocking operation as the human ear easily finds the distinction.

23.35. CETANE NUMBER

- The cetane rating of a diesel fuel is a *measure of its ability to autoignite quickly* when it is injected into the compressed and heated air in the engine. Though ignition delay is affected by several engine design parameters such as compression ratio, injection rate, injection time, inlet air temperature etc., it is also dependent on hydrocarbon composition of the fuel and to some extent on its volatility characteristic. The cetane number is a numerical measure of the influence the diesel fuel has in determining the ignition delay. *Higher the cetane rating of the fuel lesser is the propensity for diesel knock.*
- The procedure for obtaining octane number is similar to that for obtaining the octane number of petrols. Reference mixtures of *cetane* ($C_{16}H_{34}$) (high ignitability), and *α-methyl-naphthalene* ($C_{11}H_{10}$) (low ignitability), are used. The mixture is made by volume and the ignitability of the test fuel is quoted as the *percentage of cetane in the reference mixture which has the same ignitability.*
- For higher speed engines, the cetane number required is about 50, for medium speed engines about 40, and for slow speed engines about 30.
- Cetane number is the most important single fuel property which affects the exhaust emissions, noise and startability of a diesel engine. In general, *lower the cetane number higher are the hydrocarbon emissions and noise levels.* Low centane fuels increase ignition delay so that start of combustion is near to top dead centre. This is similar to retarding of injection timing which is also known to result in higher hydrocarbon levels.

— In general, *a high octane value implies a low cetane value.*

The relation between cetane number and delay period for a particular engine at a particular set of running conditions is illustrated in Fig. 23.64.

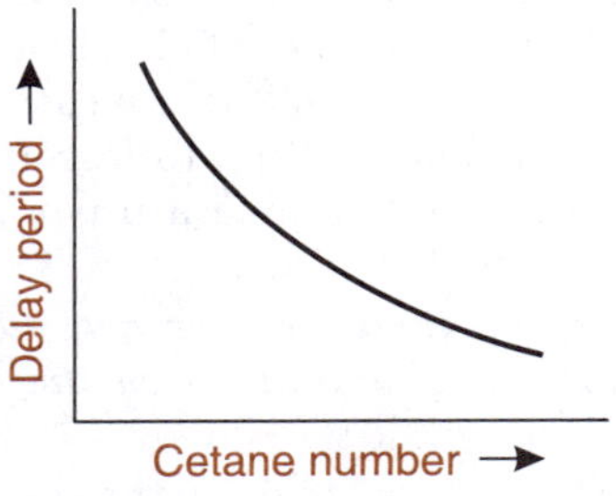

Fig. 23.64

23.36. BASIC DESIGNS OF C.I. ENGINE COMBUSTION CHAMBERS

In C.I. engines several types of combustion chambers are used. Each of these has its own peculiarities, and desirable, as well as undesirable features. Any one of these combustion chambers may produce good results in one field of application, but less desirable, or even poor results in another. No one combustion chamber design has yet been developed which will produce the best result is all types of engines. The particular design chosen, then, must be that which accomplishes the best performance for the application desired.

Four specific designs which find wide use in C.I. engines are discussed below :

1. The non-turbulent type
   Open combustion chamber.
2. The turbulent type
   (*i*) Turbulent chamber
   (*ii*) Precombustion chamber
   (*iii*) Energy cell.

1. Non-turbulent type

Fig. 23.65 (*a*) illustrates the usual design of *open combustion chamber*, which is representative of non-turbulent type. The fuel is injected directly into the upper portion of the cylinder, which acts as the combustion chamber. This type depends little on turbulence to perform the mixing. Consequently, the heat loss to the chamber walls is relatively low, and easier starting results. In order to obtain proper penetration and dispersal of the fuel necessary for mixing with the air, however, high injection pressures and multi-orifice nozzles are required. This necessitates small nozzle openings and results in more frequent clogging or diversion of the fuel spray by accumulated carbon particles, with consequent higher maintenance costs.

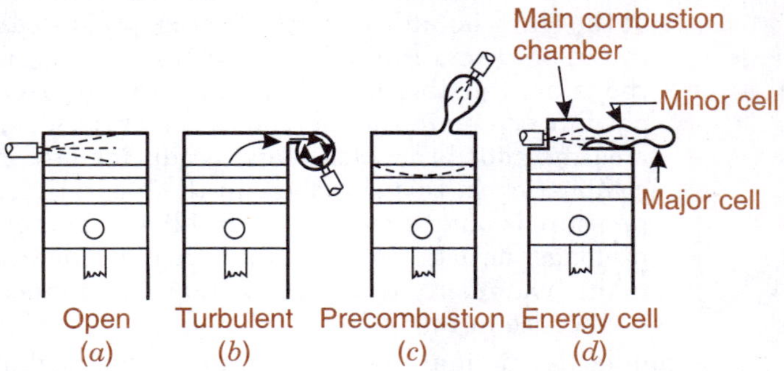

Fig. 23.65. Some commonly used C.I. engine combustion chambers.

This type of chamber is ordinarily used on low speed engines, where injection is spread through a greater period of time and thus ignition delay is a relatively less important factor. Consequently, less costly fuels with longer ignition delay may be used.

2. **Turbulent type**

The turbulent chamber, precombustion chamber and energy cell are variations of the turbulent type of chamber and are illustrated in Fig. 23.65 (*b*), (*c*) and (*d*).

— In the '*turbulent chamber*', Fig. 23.65 (*b*), the upward moving piston forces a flow of air into a small antechamber, thus imparting a rotary motion to the air passing the pintle type nozzle. As the fuel is injected into the rotating air, it is partially mixed with this air, and commences to burn. The pressure built up in the antechamber by the expanding burning gases force the burning and unburned fuel and air mixtures back into the main chamber, again imparting high turbulence and further assisting combustion.

— In the '*precombustion chamber*', Fig. 23.65 (*c*), the upward moving piston forces part of the air into a side chamber, called the precombustion chamber. Fuel is injected into the air in the precombustion chamber by a pintle type nozzle. The combustion of the fuel and air produces high pressures in the precombustion chamber, thus creating high turbulence and producing good mixing and combustion.

— The '*energy cell*' is more complex than the precombustion chamber. It is illustrated in Fig. 23.65 (*d*). As the piston moves up on the compression stroke, some of the air is forced into the major and minor chambers of the energy cell. When the fuel is injected through the pintle type nozzle, part of the fuel passes across the main combustion chamber and enters the minor cell, where it is mixed with the entering air. Combustion first commences in the main combustion chamber where the temperature is higher, but the rate of burning is slower in this location, due to insufficient mixing of the fuel and air. The burning in the minor cell is slower at the start, but due to better mixing, progresses at a more rapid rate. The pressures built up in the minor cell, therefore, force the burning gases out into the main combustion chamber, thereby creating added turbulence and producing better combustion in this chamber. In the mean time, pressure is built up in the major cell, which then prolongs the action of the jet stream entering the main chamber, thus continuing to induce turbulence in the main chamber.

Summarily it may be said that a particular combustion chamber design must be chosen to perform a given job. No one combustion chamber can produce an ultimate of performance in all tasks. As most engineering work, the design of the chamber must be based on a compromise, after full considerations of the following factors : (*i*) Heat lost to combustion chamber walls, (*ii*) Injection pressure, (*iii*) Nozzle design, (*iv*) Maintenance, (*v*) Ease of starting, (*vi*) Fuel requirement, (*vii*) Utilisation of air, (*viii*) Weight relation of engine to power output, (*ix*) Capacity for variable speed operation.

23.37. SUPERCHARGING

The purpose of supercharging is to raise the volumetric efficiency above that value which can be obtained by normal aspiration.

The engine is an air pump. Increasing the air consumption permits greater quantities of fuel to be added, and results in a greater potential output. The indicated power produced is almost directly proportional to the engine air consumption. While brake power is not so closely related to air consumption, it is nevertheless, dependent upon the mass of air consumed. It is desirable, then, that the engine takes in greatest possible mass of air.

Three possible methods which might be utilized to increase the air consumption of an engine are :

1. *Increasing the piston displacement*, but this increases the size and weight of the engine, and *introduces additional cooling problems*.
2. *Running the engine at higher speeds,* which results in *increased fluid and mechanical friction losses, and imposes greater inertia stresses on engine parts*.
3. *Increasing the density of the charge,* such that a greater mass of charge is introduced into the same volume or same total piston displacement.

The last method of increasing the air capacity of an engine is widely used, and is termed *supercharging*.

The apparatus used to increase the air density is known as a *supercharger*. It is merely a *compressor* which provides a denser charge to the engine, thereby enabling the consumption of a greater mass of charge with the same total piston displacement. During the process of compressing the charge, the supercharger produces the following effects :

(*i*) *Provides better mixing of the air-fuel mixture*. The turbulent effect created by the supercharger assists in additional mixing of the fuel and air particles. The arrangement of certain types of superchargers, particularly the *centrifugal type*, also encourages more even distribution of the charge to the cylinders.

(*ii*) *The temperature of the charge is raised as it is compressed, resulting in a higher temperature within the cylinders*. This is partially beneficial in that it helps to produce better vapourisation of fuel (in case of S.I. engines) but deterimental in that it tends to lessen the density of the charge. The increase in temperature of the charge also effects the detonation of the fuel.

Supercharging tends to increase the possibility of detonation in a S.I. engine and lessen the possibility in a C.I. engine.

(*iii*) Power is required to drive the supercharger. This is usually taken from the engine and thereby removes, from over-all engine output, some of the gain in power obtained through supercharging.

Compressors used are of the following three types :

(*i*) **Positive displacement type** used with many reciprocating engines in stationary plants, vehicles and marine installations.

(*ii*) **Axial flow type** seldom used to supercharge reciprocating engines, it is widely used as the compressor unit of the *gas turbines.*

(*iii*) **Centrifugal type** *widely used* as the *supercharger for reciprocating engines, as well as compressor for gas turbines*. It is almost exclusively used as the supercharger with reciprocating power plants for *aircraft because it is relatively light and compact, and produces continuous flow rather than pulsating flow as in some positive displacement types*.

Supercharging of S.I. engines

The schematic arrangement for supercharging S.I. engine and an ideal *p-V* diagram for a supercharged constant volume cycle are shown in Fig. 23.66 (*a*), (*b*).

Apart from increases in pressures over the un-supercharged cycle the main difference is that *pumping loop* (1–5–6–7–1) is now a positive one. The net indicated power (I.P.) is obtained by adding the contribution from pumping loop to that of the *power loop* (1–2–3–4–1). The gas is exhausted from the cylinder at the pressure of point 4 on the diagram. This value is high compared with atmospheric pressure and a considerable amount of energy is lost by allowing the gas to blow down in this way.

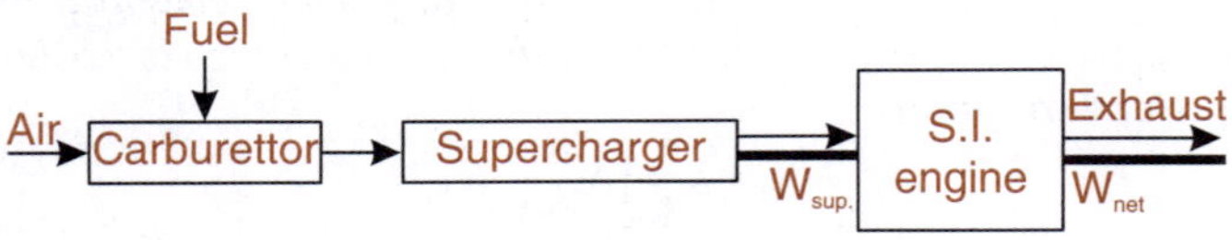

(*a*)

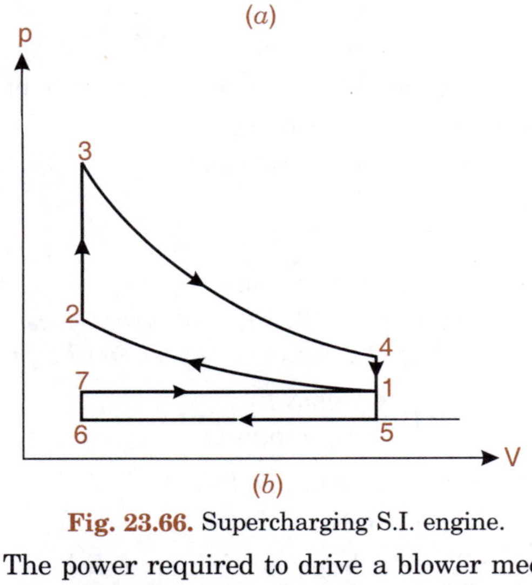

(*b*)

Fig. 23.66. Supercharging S.I. engine.

The power required to drive a blower mechanically connected to the engine must be subtracted from the engine output to obtain the net brake power (B.P.) of the supercharged engine.

Then, $$I_{\text{m.e.p.}} = \left(\frac{\text{area } 12341 + \text{area } 15671}{\text{length of diagram}}\right) \times \text{spring number} \quad ...(23.1)$$

and brake power (B.P.) = ($\eta_{\text{mech.}}$ × I.P.) – (power required to drive blower)

(for mechanically driven blowers only) ...(23.2)

(*i*) As far as S.I. engines are concerned, supercharging is employed *only for 'aircraft' and 'racing car engine'*. This is because the *increase in supercharging pressure increases the tendency to detonate*.

(*ii*) Supercharging of petrol engines, because of its poor fuel economy, is not very popular and is used only when a large amount of power is needed or when more power is needed *to compensate altitude loss*.

Supercharging of C.I. engines

Fig. 23.67 shows the schematic arrangement for supercharging C.I. engine and an ideal *p-V* diagram for a supercharged constant pressure (diesel) cycle.

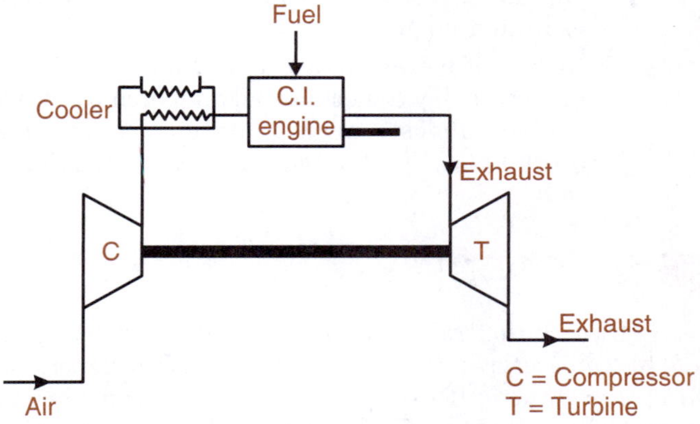

(*a*)

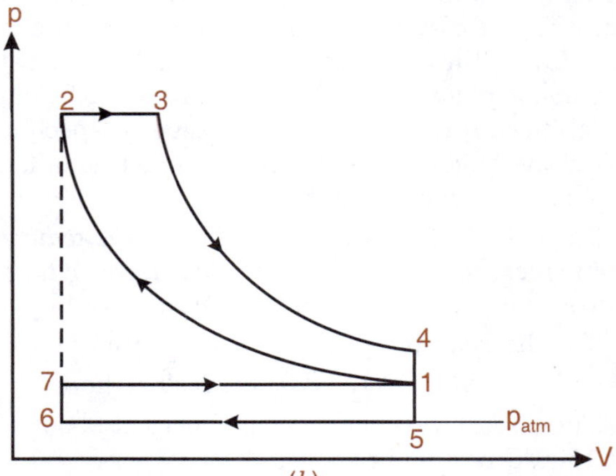

(*b*)

Fig. 23.67. Supercharging C.I. engine.

Unlike S.I. engines supercharging does not result in any combustion problem, *rather it improves combustion in diesel engine*. Increase in pressure and temperature of the intake air reduces significantly delay and hence the rate of pressure rise resulting in a *better, quieter and smoother combustion*. This improvement in combustion allows a poor quality fuel to be used in a diesel engine and it is also not sensitive to the type of fuel used. *The increase in intake temperature reduces volumetric and thermal efficiency but increase in density due to pressure compensates for this and intercooling is not necessary except for highly supercharged engines.*

If an unsupercharged engine is supercharged it will increase the reliability and durability of the engine due to smoother combustion and lower exhaust temperatures. The degree of supercharging is limited by thermal and mechanical load on the engine and strongly depends on the type of supercharger used and design of the engine.

Effects of supercharging on performance of the engine :

1. The *'power output'* of a supercharged engine is *higher* than its naturally aspirated counterpart.
2. The *'mechanical efficiencies'* of supercharged engines are *slightly better* than the naturally aspirated engines.
3. Inspite of better mixing and combustion due to reduced delay a mechanically supercharged otto engine almost always have *'specific fuel consumption' higher than a naturally aspirated engine*.

23.38. DISSOCIATION

Dissociation refers to *disintegration of burnt gases at high temperatures. It is a reversible process and increases with temperature*. During dissociation a considerable amount of heat is absorbed. This heat will be liberated when the elements recombine as the temperature falls. Thus the general effect of dissociation is a suppression of a part of the heat during the combustion period and the liberation of it as expansion proceeds, a condition which is really identical with the effects produced by the change in specific heat. However, the effect of dissociation is much smaller than that of change of specific heat.

The dissociation, in general, lowers the temperature and, consequently, the pressure at the beginning of the stroke, this causes a loss of power and efficiency.

The dissociation mainly is of CO_2 into CO and O_2

$$2CO + O_2 \rightleftharpoons 2CO_2 + \text{heat}$$

The dissociation of CO_2 commences at about 1000°C and at 1500°C it amounts to 1%.

There is very little dissociation of H_2O and is observed to take place at about 1400°C.

$$2H_2 + O_2 \rightleftharpoons 2H_2O + \text{heat.}$$

Dissociation is more severe in the chemically correct mixture. If the mixture is weaker, it gives temperatures lower than those required for dissociation to take place while if it is richer, during combustion it will give out CO and O_2 both of which suppress the dissociation of CO_2.

The *dissociation has a more pronounced effect in S.I. engines*. In the C.I. engines, the heterogeneous mixture of air and fuel tend to lower the temperature and hence the dissociation.

23.39 PERFORMANCE OF I.C. ENGINES

Engine performance is an indication of the degree of success with which it does its assigned job *i.e.*, conversion of chemical energy contained in the fuel into the useful mechanical work.

In evaluation of engine performance certain basic parameters are chosen and the effect of various operating conditions, design concepts and modifications on these parameters are studied. The *basic performance parameters* are numerated and discussed below :

1. Power and mechanical efficiency
2. Mean effective pressure and torque
3. Specific output
4. Volumetric efficiency
5. Fuel-air ratio
6. Specific fuel consumption
7. Thermal efficiency and heat balance
8. Exhaust smoke and other emissions
9. Specific weight.

1. Power and mechanical efficiency

(*i*) **Indicated power.** *The total power developed by combustion of fuel in the combustion chamber is called **indicated power**.*

$$\text{I.P.} = \frac{np_{mi}\, LANk \times 10}{6} \text{ kW} \qquad ...(23.3)$$

where, n = Number of cylinders,

p_{mi} = Indicated mean effective pressure, bar,

L = Length of stroke, m,

A = Area of piston, m^2, and

$k = \frac{1}{2}$ for 4-stroke engine

= 1 for 2-stroke engine.

(*ii*) **Brake power (B.P.).** *The power developed by an engine at the output shaft is called the **brake power**.*

$$\text{B.P.} = \frac{2\pi NT}{60 \times 1000} \text{ kW} \qquad ...(23.4)$$

where, N = Speed in r.p.m., and

T = Torque in Nm.

The difference between I.P. and B.P. is called *frictional power*, F.P.

i.e., F.P. = I.P. – B.P. ...(23.5)

*The ratio of B.P. to I.P. is called **mechanical efficiency*** *i.e.,* *Mechanical efficiency,* $\eta_{mech.} = \dfrac{B.P.}{I.P.}$ ...(23.6)

2. Mean effective pressure and torque

***Mean effective pressure** is defined as hypothetical pressure which is thought to be acting on the piston throughout the power stroke.* If it is based on I.P. it is called *indicated mean effective pressure* ($I_{m.e.p.}$ or p_{mi}) and if based on B.P. it is called brake mean effective pressure ($B_{m.e.p.}$ or p_{mb}). Similarly, frictional mean effective pressure ($F_{m.e.p.}$ or p_{mf}) can be defined as :

$$F_{m.e.p.} = I_{m.e.p.} - B_{m.e.p.} \quad ...(23.7)$$

The torque and mean effective pressure are related by the engine size.

Since the power (P) of an engine is dependent on its size and speed, therefore it is not possible to compare engine on the basis of either power or torque. *Mean effective pressure is the true indication of the relative performance of different engines.*

3. Specific output

It is defined as the *brake output per unit of piston displacement* and is given by :

$$\text{Specific output} = \frac{B.P.}{A \times L}$$

$$= \text{Constant} \times p_{mb} \times \text{r.p.m.} \quad ...(23.8)$$

For the same piston displacement and brake mean effective pressure (p_{mb}) an engine running at higher speed will give more output.

4. Volumetric efficiency

It is defined as the ratio of actual volume (reduced to N.T.P.) of the charge drawn in during the suction stroke to the swept volume of the piston.

The average value of this efficiency is from 70 to 80 per cent but in case of *supercharged engine* it may be *more* than 100 per cent, if air at about atmospheric pressure is forced into the cylinder at a pressure greater than that of air surrounding the engine.

5. Fuel-air ratio

It is the ratio of the mass of fuel to the mass of air in the fuel-air mixture.

"Relative fuel air ratio" is defined as the *ratio of the actual fuel air ratio to that of stoichiometric fuel-air ratio required to burn the fuel supplied.*

6. Specific fuel consumption (s.f.c.)

It is the mass of fuel consumed per kW developed per hour, and is a criterion of economical power production.

i.e., $$\text{s.f.c.} = \frac{\dot{m}_f}{B.P.} \text{ kg/kWh.}$$

7. Thermal efficiency and heat balance

Thermal efficiency. It *is the ratio of indicated work done to energy supplied by the fuel.*

If $\dot{m}_f$ = Mass of fuel used in kg/sec., and

C = Calorific value of fuel (lower),

Then *indicated thermal efficiency* (based on I.P.),

$$\eta_{th.\,(I)} = \frac{I.P.}{\dot{m}_f \times C} \quad ...(23.9)$$

and *brake thermal efficiency* (based on B.P.)

$$\eta_{th.\,(B)} = \frac{B.P.}{\dot{m}_f \times C} \quad ...(23.10)$$

Heat balance sheet

The performance of an engine is generally given by heat balance sheet.

To draw a heat balance sheet for I.C. engine, it is run at constant load. Indicator diagram is obtained with the help of an indicator. The quantity of fuel used in a given time and its calorific value, the amount, inlet and outlet temperatures of cooling water and the weight of exhaust gases are recorded. After calculating I.P. and B.P. the heat in different items is found as follows :

Heat supplied by fuel

For *petrol and oil engines*, heat supplied = $m_f \times C$, where m_f and C are mass used per minute (kg) and lower calorific value (kJ or kcal) of the fuel respectively.

For *gas engines*, heat supplied = $V \times C$, where V and C is volume at N.T.P. (m³/min.) and lower calorific value of gas respectively.

(*i*) ***Heat absorbed in I.P.***

Heat equivalent of I.P. (per minute) = I.P. × 60 kJ ...(23.11)

(*ii*) ***Heat taken away by cooling water***

If m_w = Mass of cooling water used per minute,

t_1 = Initial temperature of cooling water, and

t_2 = Final temperature of cooling water,

Then, heat taken away by water

$$= m_w \times c_w \times (t_2 - t_1) \quad ...(23.12)$$

where c_w = Specific heat of water.

(*iii*) ***Heat taken away by exhaust gases***

If, m_e = Mass of exhaust gases (kg/min),

c_{pg} = Mean specific heat at constant pressure,

t_e = Temperature of exhaust gases, and

t_r = Room (or boiler house) temperature,

Then heat carried away by exhaust gases

$$= m_e \times c_{pg}(t_e - t_r) \quad ...(23.13)$$

Note. The mass of exhaust gases can be obtained by adding together mass of fuel supplied and mass of air supplied.

The heat balance sheet from the above data can be drawn as follows :

Item	kJ	Per cent
Heat supplied by fuel	...	...
(*i*) Heat absorbed in I.P.	...	...
(*ii*) Heat taken away by cooling water	...	...
(*iii*) Heat carried away by exhaust gases	...	...
(*iv*) Heat unaccounted for (by difference)	...	...
Total	...	...

8. Exhaust smoke and other emissions

Smoke is an indication of incomplete combustion. It limits the output of an engine if air pollution control is the consideration. *Exhaust emissions* have of late become a matter of grave concern and with the enforcement of legislation on air pollution in many countries, it has become necessary to view them as performance parameters.

9. Specific weight

It is defined as the weight of the engine in kg for each B.P. developed. It is an indication of the engine bulk.

Basic Measurements :

To evaluate the performance of an engine following *basic measurements* are usually undertaken :

1. Speed
2. Fuel consumption
3. Air consumption
4. Smoke density
5. Exhaust gas analysis
6. Brake power
7. Indicated power and friction power
8. Heat going to cooling water
9. Heat going to exhaust.

1. Measurement of speed

The speed may be measured by :

(*i*) Revolution counters

(*ii*) Mechanical tachometer

(*iii*) Electrical tachometer.

2. Fuel measurement

The fuel consumed by an engine can be measured by the following methods :

(*i*) Fuel flow method

(*ii*) Gravemetric method

(*iii*) Continuous flow meters.

3. Measurement of air consumption

The air consumption can be measured by the following methods :

(*i*) Air box method

(*ii*) Viscous-flow air meter.

(*i*) *Air box method*

Fig. 23.68 shows the arrangement of the system. It consists of air-tight chamber fitted with a sharp edged orifice of known co-efficient of discharge. The orifice is located away from the suction connection to the engine. Due to the suction of engine, there is a pressure depression in the air box or chamber which causes the flow through the orifice. For obtaining a steady flow, the volume of chamber should be sufficiently large compared with the swept volume of the cylinder, generally 500 to 600 times the swept volume. It is assumed that the intermittent suction of the engine will not effect the air pressure in the air box as volume of the box is sufficiently large, and pressure in the box remains same.

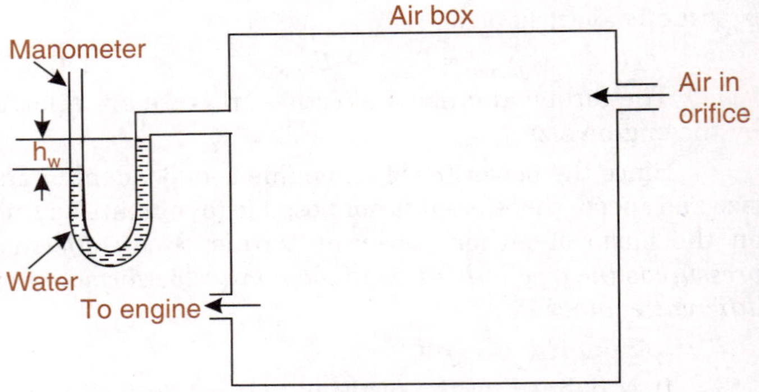

Fig. 23.68. Air-box method for measuring air.

A water manometer is used to measure the pressure difference causing the flow through the orifice. The depression across the orifice should not exceed 100 to 150 mm of water.

Let A = Area of orifice, m^2,

d = Diameter of orifice, cm,

h_w = Head of water in cm causing the flow,

C_d = Co-efficient of discharge for orifice,

ρ_a = Density of air in kg/m^3 under atmospheric conditions, and

ρ_w = Density of water in kg/m^3.

Head in meters of air (H) is given by :

$$H.\rho_a = \frac{h_w}{100}\rho_w$$

$$\therefore \quad H = \frac{h_w}{100} \times \frac{\rho_w}{\rho_a} = \frac{h_w}{100} \times \frac{1000}{\rho_a} = \frac{10h_w}{\rho_a} \text{ m of air}$$

The velocity of air passing through the orifice is given by,

$$C_a = \sqrt{2gH} \text{ m/s} = \sqrt{2g\frac{10h_w}{\rho_a}} \text{ m/s}$$

The volume of air passing through the orifice,

$$V_a = C_d \times A \times C_a = C_d A\sqrt{2g\frac{10h_w}{\rho_a}}$$

$$= 14\, AC_d\sqrt{\frac{h_w}{\rho_a}} \text{ m}^3\text{/s}$$

$$= 840\, AC_d\sqrt{\frac{h_w}{\rho_a}} \text{ m}^3\text{/min.}$$

Mass of air passing through the orifice is given by

$$m_a = V_a\rho_a = 14 \times \frac{\pi d^2}{4 \times 100^2} \times C_d \sqrt{\frac{h_w}{\rho_a}} \times \rho_a$$

$$= 0.0011\, C_d \times d^2 \sqrt{h_w \rho_a} \text{ kg/s}$$

$$= 0.066\, C_a \times d^2 \sqrt{h_w \rho_a} \text{ kg/min.} \quad ...(23.14)$$

(*ii*) ***Viscous-flow air meter***

Alcock viscous-flow air meter is another design of air meter. It is not subjected to the errors of the simple types of flow meters. With the air-box the flow is proportional to the square root of the pressure difference across the orifice. With the Alcock meter the air flows through a form of honey-comb so that flow is viscous. The resistance of the element is directly proportional to the air velocity and is measured by means of an inclined manometer. Felt pads are fitted in the manometer connections to damp out fluctuations. The meter is shown in Fig. 23.69.

The accuracy is improved by fitting a damping vessel between the meter and the engine to reduce the effect of pulsations.

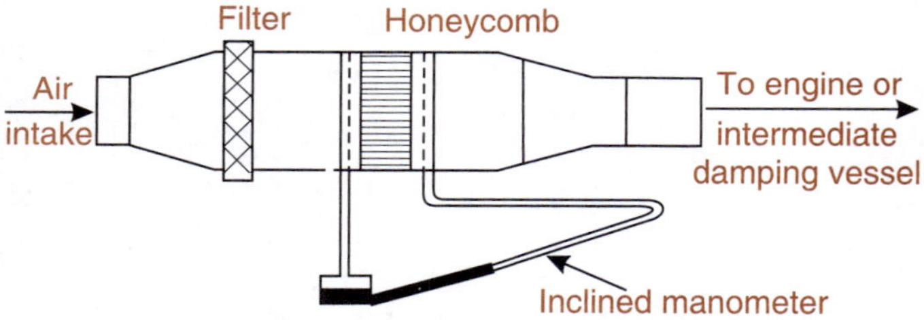

Fig. 23.69. Alcock viscous-flow air meter.

4. Measurement of exhaust smoke

The following smoke meters are used :

(*i*) Bosch smoke meter

(*ii*) Hatridge smoke meter

(*iii*) PHS smoke meter.

5. Measurement of exhaust emission

Substances which are emitted to the atmosphere from any opening down stream of the exhaust part of the engine are termed as exhaust emissions. Some of the more commonly used instruments for measuring exhaust components are given below :

(*i*) Flame ionisation detector

(*ii*) Spectroscopic analysers

(*iii*) Gas chromatography.

6. Measurement of B.P.

The B.P. of an engine can be determined by a brake of some kind applied to the brake pulley of the engine. The arrangement for determination of B.P. of the engine is known as *dynamometer*. The dynamometers are classified into following two classes :

(*i*) Absorption dynamometers

(*ii*) Transmission dynamometers.

(*i*) **Absorption dynamometers.** *Absorption dynamometers are those that absorb the power to be measured by friction.* The power absorbed in friction is finally dissipated in the form of heat energy.

Common forms of absorption dynamometers are :

- Prony brake
- Rope brake
- Hydraulic brake
- Fan brake
- Electrical brake dynamometers
  — Eddy current dynamometer
  — Swinging field d.c. dynamometer.

(*ii*) **Transmission dynamometers.** These are also called *torquemeters*. These are very accurate and are used where continuous transmission of load is necessary. There are used mainly in automatic units.

Here we shall discuss *Rope brake dynamometer* only :

Rope brake dynamometer

Refer to Fig. 23.70. A rope is wound round the circumference of the brake wheel. To prevent the rope from slipping small wooden blocks (not shown in the Fig. 23.70) are laced to rope. To one end of the rope is attached a spring balance (S) and the other end carries the load (W). The speed of the engine is noted from the tachometer (revolution counter).

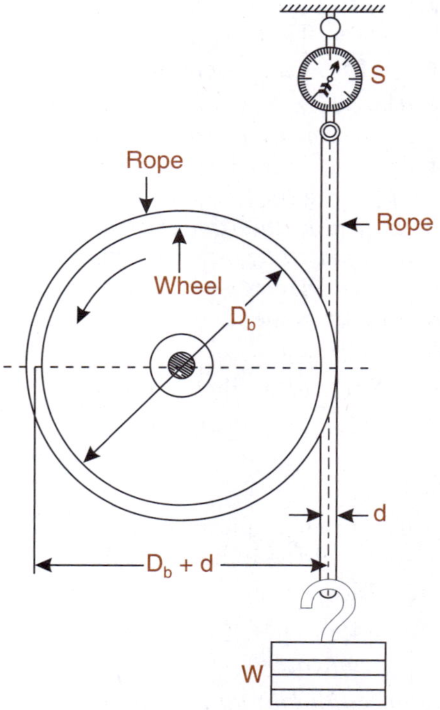

Fig. 23.70. Rope brake dynamometer.

If, W = weight at the end of the rope, N,
S = spring balance reading, N,
N = engine speed, r.p.m.,
D_b = diameter of the brake wheel, m,
d = diameter of the rope, m, and
$(D_b + d)$ = effective diameter of the brake wheel,

Then work/revolution

$$= \text{Torque} \times \text{angle turned per revolution}$$

$$= (W - S) \times \left(\frac{D_b + d}{2}\right) \times 2\pi$$

$$= (W - S)(D_b + d) \times \pi$$

Work done/min $= (W - S)\, \pi\, (D_b + d)\, N$

Work done/sec $= \dfrac{(W - S)\, \pi\, (D_b + d)\, N}{60}$

$$\therefore \quad \text{B.P.} = \frac{(W - S)\, \pi\, (D_b + d)\, N}{60 \times 1000} \text{ kW} \qquad ...(23.15)$$

$$= \frac{(W - S)\, \pi\, D_b N}{60 \times 1000} \quad \text{...... if } d \text{ is neglected}$$

or

$$\left(= \frac{T \times 2\pi N}{60 \times 1000} \text{ kW} \right) \qquad ...(23.16)$$

Rope brake is cheap and easily constructed but not very accurate because of changes in friction co-efficient of the rope with temperature.

Measurement of Indicated power (I.P.)

The power developed in the engine cylinder or at the piston is necessarily greater than that at the crankshaft due to engine losses. Thus,

I.P. = B.P. + engine losses.

Indicated power is usually determined with the help of a p-V diagram taken with the help of an indicator. In case indicated power cannot be measured directly, it is made possible by measuring the brake power and also the engine losses. If the *indicator diagram* is available, the indicated power may be *computed by measuring the area of* diagram, either with a *planimeter* or by *ordinate method*, and dividing by the stroke measurement in order to obtain the mean effective pressure (m.e.p.).

i.e.,
$$p_{mi} = \frac{\text{Net area of diagram in mm}^2}{\text{Length of diagram in mm}} \times \text{Spring constant}$$

where p_{mi} is in bar. ...(23.17)

(The spring constant is given in bar per mm of vertical movement of the indicator stylus.)

Engine indicators

The main types of engine indicators are :

1. *Piston indicator*
2. *Balanced diaphragm type indicator*

(*i*) The Farnborough balanced engine indicator
(*ii*) Dickinson-Newell indicator
(*iii*) MIT balanced pressure indicator
(*iv*) Capacitance-type balance pressure indicator.

3. *Electrical indicators*

In addition to this, *optical indicators* are also used.

Calculation of indicated power (I.P.) :

If, p_{mi} = Indicated mean effective pressure, bar,
A = Area of piston, m^2,
L = Length of stroke, m,
N = Speed of the engine, r.p.m.,
$k = \frac{1}{2}$ for 4-stroke engine,
= 1 for 2-stroke engine.

Then, Force on the piston $= p_{mi} \times A \times 10^5$ N

Work done per working stroke

$$= \text{Force} \times \text{length of stroke}$$

$$= p_{mi} \times A \times 10^5 \times L \text{ Nm}$$

Work done per second = Work done per stroke × number of working stroke per second

$$= p_{mi} \times L \times A \times 10^5 \times \frac{N}{60} \times k \text{ Nm/s or J/s}$$

$$= \frac{p_{mi} \times LANk \times 10^5}{60 \times 1000} \text{ kW}$$

i.e., Indicated power,

$$\text{I.P.} = \frac{p_{mi}\, LANk \times 10}{6} \text{ kW}$$

If n is the number of cylinders, then

$$\text{I.P.} = \frac{np_{mi}\, LANk \times 10}{6} \text{ kW} \qquad ...(23.18)$$

Morse test

This test is only applicable to *multi-cylinder engines.*

The engine is run at the required speed and the torque is measured. One cylinder is cut out, by shorting the plug if an S.I. engine is under test, or by disconnecting an injector if a C.I. engine is under test. *The speed falls because of the loss of power with one cylinder cut out, but is restored by reducing the load.* The torque is measured again when the speed has reached its original value. If the values of I.P. of the cylinders are denoted by I_1, I_2, I_3 and I_4 (considering a four-cylinder engine), and the power losses in each cylinder are denoted by L_1, L_2, L_3 and L_4, then the value of B.P., B at the test speed with all cylinders firing is given by

$$B = (I_1 - L_1) + (I_2 - L_2) + (I_3 - L_3) + (I_4 - L_4) \qquad ...(i)$$

If *number 1 cylinder is cut out,* then the contribution I_1 is lost ; and if the losses due to that cylinder remain the same as when it is firing, then the B.P., B_1, now obtained at the same speed is

$$B_1 = (0 - L_1) + (I_2 - L_2) + (I_3 - L_3) + (I_4 - L_4) \quad ...(ii)$$

Subtracting eqn. (*ii*) from eqn. (*i*), we get

$$B - B_1 = I_1 \quad ...(23.19)$$

Similarly,

$B - B_2 = I_2$ when cylinder number 2 is cut out,

and $B - B_3 = I_3$ when cylinder number 3 is cut out,

and $B - B_4 = I_4$ when cylinder number 4 is cut out

Then, for the engine,

$$I = I_1 + I_2 + I_3 + I_4 \quad ...(23.20)$$

Measurement of frictional power (F.P.) :

The frictional power of an engine can be determined by the following methods :

1. Willan's line method (used for C.I. engines only)
2. Morse test
3. Motoring test
4. Difference between I.P. and B.P.

1. ***Willan's line method***

At a constant engine speed the load is reduced in increments and the corresponding B.P. and gross fuel consumption readings are taken. A graph is then drawn of fuel consumption against B.P. as in Fig. 23.71. The graph drawn in called the Willan's line (analogous to Willan's line for a steam engine), and is extrapolated back to cut the B.P. axis at the point *L*. The reading *OL* is taken as the power loss of the engine at that speed. The fuel consumption at zero B.P. is given by *OM* ; and if the relationship between fuel consumption and B.P. is assumed to be linear, then a fuel consumption *OM* is equivalent to a power loss of *OL*.

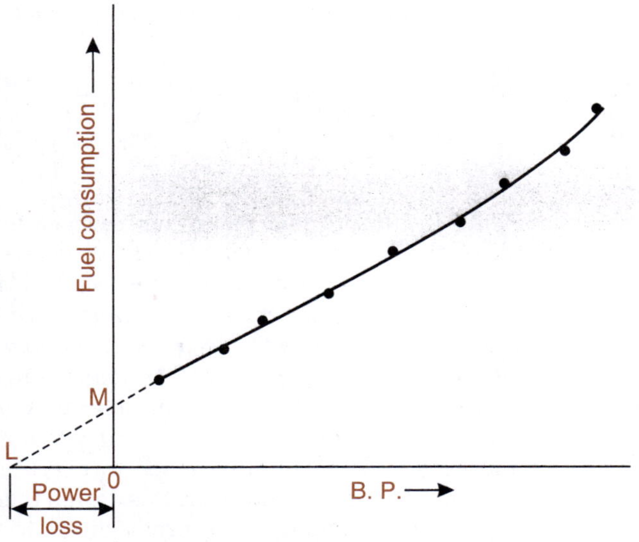

Fig. 23.71. Willan's line method.

2. ***Morse test***

In 'Morse test' (already discussed), frictional power can be found by subtracting $(B.P.)_n$ from $(I.P.)_n$

i.e., $\quad F.P. = (I.P.)_n - (B.P.)_n$

where *n* is the number of cylinders.

3. ***Motoring test***

In this test the engine is first run upto the desired speed by its own power and allowed to remain under the given speed and load conditions for sometime so that oil, water and engine component temperatures reach stable conditions. The power of the engine during this period is absorbed by a dynamometer (usually of electrical type). The fuel supply is then cut off and by suitable electric switching devices the dynamometer is converted to run as a motor to drive or 'motor' the engine at the same speed at which it was previously running. The power supply to the motor is measured which is a measure of F.P. of the engine.

4. ***Difference between I.P. and B.P.***

The method of finding the F.P. by finding the difference between I.P. as obtained from an indicator diagram, and B.P. as obtained by a dynamometer is the ideal method. However, due to difficulties in obtaining accurate indicator diagrams, especially at high engine speeds, this method is usually only used in research laboratories and its use at commercial level is very limited.

23.40. ENGINE PERFORMANCE CURVES

- The following *parameters* are of interest (to engineers) :
  - — Torque
  - — Power
  - — Specific fuel consumption and its inverse
  - — I.C. engine efficiency
  - — Brake mean effective pressure ; this is defined as : Brake power/swept volume × cyclic efficiency (in this relation swept volume is kept constant). Also brake power = torque × angular speed, thus at constant speed, the b.e.m.p. is directly proportional to torque and either may be used. The ratio of brake mean effective pressure to the indicated mean effective pressure (from indicator diagram) may be seen to be equal to the ratio of brake power to the indicated power which is defined as mechanical efficiency. Thus indicator diagram and the output torque may be connected with a suitable allowance for mechanical efficiency.
- *The torque-speed relations* (Fig. 23.72) exhibit a curve even though it would seen reasonable to expect the mean effective pressure to be constant at all speeds. However, at low speeds leakage through valves etc., becomes of greater significance so that the m.e.p. falls and at high speeds the volumetric efficiency causes the induced mass to fall with a parallel fall in m.e.p. *The power curves are the product of the torque curves with speed.*

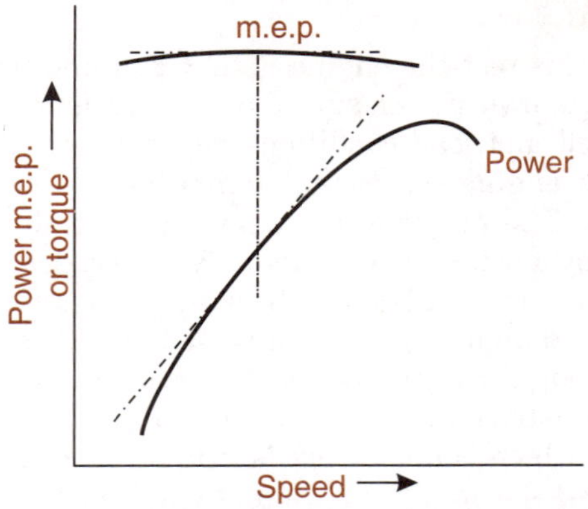

Fig. 23.72. Power-speed and torque-speed curves for the I.C. engine.

Specific fuel consumption relations :

S.I. engines : Refer to Fig. 23.73. The curves are plotted for *constant throttle opening*, *constant speed and constant ignition setting*. The *only variable is the air-fuel ratio*. The effect of ignition setting or speed may by ascertained by producing a new family of curves. An alternative method of plotting these parameters is to use the air-fuel ratio as the abscissa. Here it can be seen that maximum economy occurs with a slightly weak mixture. This means that there is excess air and combustion is complete. Maximum power occurs with a slightly rich mixture when all the available oxygen is used. The I.C. engine efficiency is the inverse of the specific fuel consumption with the constant calorific value as a factor. Thus the curves of specific fuel consumption (s.f.c.) also represent efficiency. The maximum value of brake I.C. engine efficiency for S.I. and C.I. engines are of the order 35% and 40% respectively.

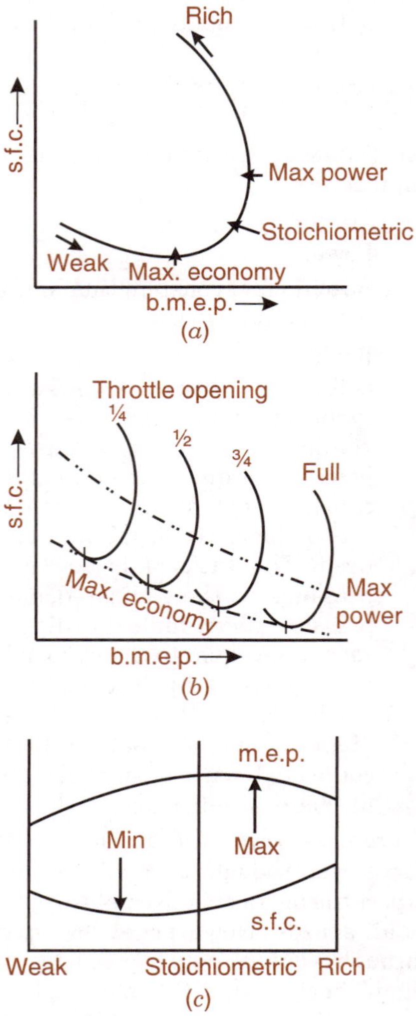

Fig. 23.73. Specific fuel consumption-brake mean effective pressure curves for the S.I. engine.

C.I. engine. The flat curve of Fig. 23.74 illustrates that at *part load the compression ignition engine is more economical than the spark ignition engine.* This is the benefit of quality control rather than quantity control of power.

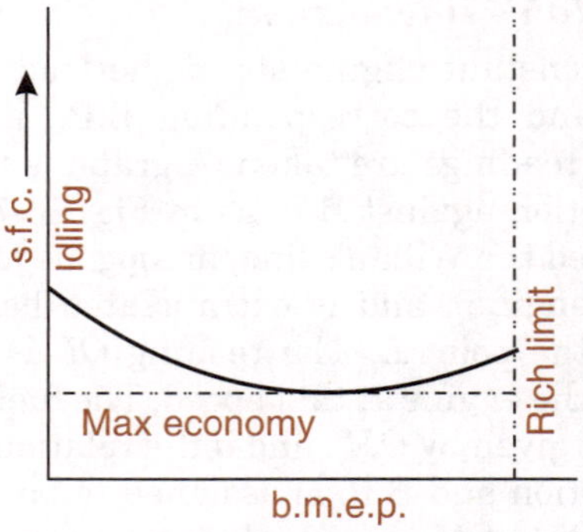

Fig. 23.74. Specific fuel consumption-brake mean effective pressure curve for the C.I. engine.

23.41. THE WANKEL ROTARY COMBUSTION (RC) ENGINE

Introduction

- Scheffel in 1952, to get a patent for *rotary engine*, utilised the *principle that oval or elliptical rotors can be designed to maintain contact ; while turning about fixed centres, and that three or more rotors can be run to enclose between them a continuously varying volume*. The four volumes between the rotors with the suitable arrangement of ports, ignition system and adequate compression ratio, could be made to execute a *four-phase Otto cycle*. However, this design failed due to its complexity and great difficulties/ problems involved in its manufacture.
- Felix Wankel (German inventor), in 1954, got a patent for design of four-phase rotary engine working on the Otto cycle principle.
- Later Dr. Frocede made certain modifications and an engine was developed, called as KKM

(Kreiskolbenmotor) : now popularly known as Wankel Rotating Combustion (RC) engine.

Construction and Working

Refer to Fig. 23.75

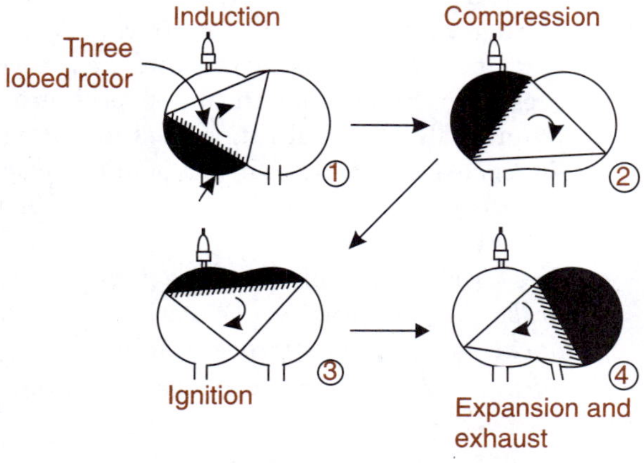

Fig. 23.75. The Wankel engine.

Construction : It consists of the following parts :

(*i*) ***Rotor*** (three lobed).

(*ii*) ***Eccentric or output shaft*** with its integral eccentric. No connecting rod is required as the rotor rotates directly on the eccentric shaft. The *output torque is transmitted to the shaft through eccentric.*

(*iii*) ***Internal and external timing gears.*** They maintain the phase relationship between the rotation of the rotor and the eccentric shaft and eventually *control the orbital motion of the motor.*

Working :

- The Wankel engine works on the *four-phase principle.* (The word phase corresponds to stroke of the reciprocating engine).
- The engine having three lobed rotor is driven eccentrically in a casing in such a way that there are *three separate volumes trapped between the rotor and the casing. These three volumes perform "induction", "compression",* "combustion", *"expansion"* and *"exhaust"* processes in sequences. There are three power impulses for each revolution of the rotor, and since the eccentric or output shaft rotates at three times the speed of the rotor, there *is only one power impulse for each revolution of the output shaft of a single bank rotary engine.*

One complete *thermodynamic cycle* is completed over 360° rotation of the rotor ; the section phase takes 90° of rotor movement and so also the other three phases. One *thermodynamic phase* is completed every 270° rotation of the output shaft, since the output shaft makes three revolutions for every single rotation of the rotor.

Features

1. Simple construction, less mechanical loss, smooth motion and does not require a cranking mechanism.
2. *Good power volume ratio.*
3. *No reciprocating parts* and hence no balancing problem and complicated engine vibrations eliminated.
4. Due to the absence of intake-exhaust valve mechanism, the correct timings for opening and closing (the ports) can be maintained even at high speeds.
5. Low torque fluctuation.

There are problems in the design, notably of sealing and of heat transfer but these have been overcome sufficiently well for spark ignition engine to be marketed.

23.42. STRATIFIED CHARGE ENGINES AND DUEL-FUEL ENGINES

Stratified Charge Engines

- Whereas several S.I. engines are designed to have a homogeneous air-fuel mixture throughout the combustion chamber, some modern stratified charge engines are designed to have a *different air-fuel ratio at different locations within the combustion chamber. A rich mixture that ignites readily is desired around the spark plug, while the major volume of the combustion chamber is filled with a very lean mixture that gives good fuel economy. Special intake systems* are necessary to supply this non-homogeneous mixture. Combination of multiple valves and multiple fuel injectors, alongwith flexible valve and injection timing are used to accomplish the desired results.
- Some stratified charge SI engines are operated with no throttle, which raises the volumetric efficiency. Speed is controlled by proper timing and quantity of fuel input.

Duel Fuel Engines

- Owing to various technical and financial reasons, some engines are designed to operate *using a combination of two fuels.* For instance, in some third-world countries *dual fuel engines are used because of high cost of diesel fuel.*
- Large C.I. engines are run on a *combination of methane and diesel oil.* Methan is the main fuel because it is more cheaply available. However, methane is not a good C.I. fuel by itself because it does not readily self-ignite (due to its high octane number). *A small amount of diesel oil is injected at the proper cycle time. This ignites in a normal manner and initiates combustion in the methane-air mixture filling the cylinder.*

- On these types of engines, *combination of fuel input systems are needed.*

Note. For more details on I.C. engines please refer the Author's book on **"Internal Combustion Engines."**

WORKED EXAMPLES

Example 23.1. *The factors that tend to increase detonation is S.I. engine tend to reduce knocking in C.I. engine. Discuss this statement with reference to the following influencing factors :*

(*i*) *Compression ratio*

(*ii*) *Inlet temperature*

(*iii*) *Self-ignition temperature of fuel*

(*iv*) *Time-lag of ignition temperature of fuel, and*

(*v*) *Combustion chamber wall temperature.*

(U.P.S.C.)

Solution. The *factors that tend to increase detonation is S.I. engine tend to reduce knocking in C.I. engine. This is justified by the following factors* :

(*i*) **Compression ratio :**

If the compression ratio is increased then air temperature and pressure at the end of compression also increases, thereby decreasing the delay period. So *the tendency for detonation increases by increasing the compression.*

(*ii*) **Inlet temperature :**

Any increase in the inlet temperature increases the temperature at the end of compression thereby increasing the tendency to knocking in case of S.I. engines.

(*iii*) **Self-ignition temperature :**

The self ignition temperature is the temperature of auto-ignition of charge. This causes detonation in S.I. engines. However, in case of C.I. engines, the early auto-ignition is necessary to avoid knocking.

(*iv*) **Time-lag of ignition of fuel :**

The time-lag of ignition of fuel should be short to avoid knocking in C.I. engines. However short ignition time causes knocking in S.I. engines.

(*v*) **Combustion chamber wall temperature :**

If the temperature of combustion chamber wall is high then auto-ignition of the charge takes place, causing detonation early in the S.I. engine. On the other hand quick auto-ignition due to high wall temperature helps in reducing knocking in C.I. engine.

Example 23.2. *Discuss the air-fuel ratio requirements of a petrol engine automobile for starting and warm up, idling and low load, normal power range, maximum power range and acceleration. State how these requirements are achieved.*

(U.P.S.C)

Solution. Air-fuel ratio requirements of a petrol engine :

1. ***For idling and low speed (From no-load to about 20% rated power) :***

- No-load running of the engine is called "*idling*". During idling throttle is nearly closed and the suction pressure is very low, *i.e.,* exhaust pressure is higher than the intake pressure. This requires that air-fuel ratios used for idling and low speeds say upto about 20% of full-load *should be rich for smooth engine operation* (F/A ratio 0.8 or A/F = 12.5 : 1)
- The richening of mixture *increases the probability of contact between fuel and air particles and thus improves combustion.*

2. ***For maximum power range (From about 75% to 100% rated power) :***

It requires A/F ratio about 14 : 1. *Besides providing maximum power, rich mixture also prevents overheating of exhaust valve at high load to inhibit detonation.* At high loads, there is a greater heat transfer to engine parts.

3. ***For starting and warm up :***

Starting from cold the speed as well as engine temperatures are low hence much of 'heating ends', supplied by the carburettor do not vapourise and remain in liquid form. Therefore during starting a very rich mixture must be supplied *i.e.,* A/F ratio = 3 to 1.5 : 1.

4. ***Acceleration requirements :***

Acceleration requirements refer to an increase in engine speed resulting from opening the throttle. However, the main purpose of opening the throttle is to provide an increase in torque and whether or not an increase in speed follows depends an nature of load.

Example 23.3. *Discuss the formation of exhaust emissions in petrol engines. How do these emissions vary with air-fuel ratio ?*

Describe with a sketch how a three-way catalytic converter controls pollution in petrol cars. **(U.P.S.C.)**

Solution. Petrol consists of a mixture of various hydrocarbons. Perfect combustion gives only CO_2 and water vapours plus air whereas incomplete combustion gives CO and unburn hydrocarbons also in exhaust. Nitrogen (N_2) also at 1100°C reacts with oxygen and forms oxides (NO_2) which is very harmful.

(*i*) **Carbon monoxide (CO) :**

Due to insufficient amount of air in air-fuel mixture exhaust can be made free of CO if A/F ratio is 16 : 1 ; more percentage of CO increases in idle range and decreases with speed.

(*ii*) **Hydrocarbons :**

These are direct result of incomplete combustion. Hydrocarbon levels are higher during high speed deceleration, idling and low speed operations.

(*iii*) **Particulate matter and partial oxidation product :**

Organic and inorganic compounds of higher molecular weight and lead compounds resulting from the use of TEL are exhausted in the form of very small size particles of the order 0.02 to 0.06 μ.

(*iv*) **Oxides of nitrogen (NO_x).** High temperature and availability of oxygen are two main reasons of formation of NO_x.

The variation of emission with A/F ratio is shown in Fig. 23.76

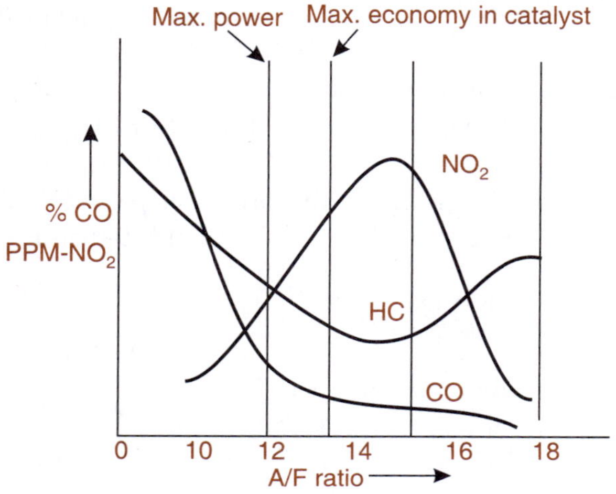

Fig. 23.76. Variation of emissions with A/F.

Fig. 23.77 shows a catalytic counter package.

Example 23.4. (*a*) *What is the reason that two-stroke engine is not used in car even though it develops theoretically twice power than that of four-stroke engine ?*

(*b*) *'Air fuel ratio in a S.I. engine varies from 8 to 16 approximately while such variation in a C.I. engine is from 100 at no-load to 20 at full load.' Explain.*

(*c*) *Explain the difference between :*

(*i*) *Pre-ignition ;* (*ii*) *auto-ignition ;* (*iii*) *detonation.*

Solution. (*a*) A majority of cars are fitted with SI engines *due to light weight and good pick up*. The two-stroke SI engine is *not used in cars as it suffers from two big disadvantage-**fuel loss** and **idling difficulty***.

In S.I. engines using carburettor, the scavenging is done with fuel-air mixture, and only the fuel mixed with the retained air is used for combustion. Thus a *part of fuel is lost with scavenging air, giving poor fuel economy.*

The two-stroke S.I. engine *runs irregularly and even may stop at low speeds* when m.e.p. is reduced to about 2 bar. This is due to large amount of residual gas (more than in four-stroke engine) mixing with small amount of charge. *At low speeds there may be back firing due to slow burning rate.*

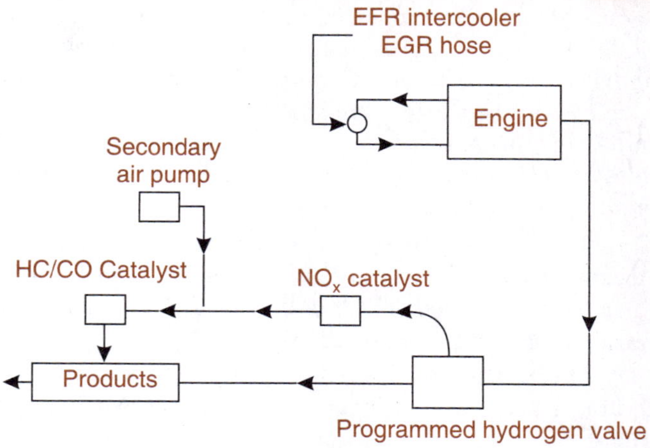

Fig. 23.77. Catalytic converter.

Both the above *drawbacks* may be avoided by using fuel injection. But this makes the system complicated, and the maintenance cost is also increased (fuel injection pump is the first to give trouble), and hence not suitable for car engine.

(*b*) In **S.I. engines,** *the combustion is homogeneous.* A flame nucleus is formed at the spark plug electrodes, and the flame propagates in a more or less homogeneous mixture of air and fuel. The ignition limits of air-fuel ratio are narrow, between about 8 : 1 to 18 : 1. The mixture proportion in S.I. engines *should be within this limit for the initiation and sustaining of the flame.* The carburettor supplies the mixture of air and fuel between this limit, depending upon the engine requirements of starting from cold, idling, normal running (maximum economy mixture), and maximum power (rich mixture).

On the other hand, in C.I. *engines the combustion is heterogeneous,* and the load control is by varying the quantity of fuel. There is no throttling of inlet air. A very small quantity of fuel is supplied by injector at starting and no load. As the load increases, the quantity of fuel is increased. At full load, for smoke free exhaust and best utilization of air, the air-fuel ratio reduces to a about 20 : 1, depending upon the type of combustion chamber.

The ignition starts at several points in the combustion chamber at locations where the local mixture of air-fuel formed is between the ignition limits (irrespective of the overall air-fuel ratio in the cylinder being much higher than the limits of ignition).

(*c*) (*i*) **Pre-ignition.** In SI engines, the combustion during the normal working is initiated by an electric spark. The spark is timed to occur at a definite point just before the end of the compression stroke. The ignition of the charge should not occur before the spark is introduced in the cylinder. If the ignition starts due to any other source when the piston is still moving on the compression stroke, it is known as ***pre-ignition***. *Pre-ignition will develop excessive pressure before the end of compression stroke tending to push* *the piston opposite to the direction in which it is moving, resulting in loss of power, violent thumping, stopping the engine or even mechanical damage*. Pre-ignition *may occur on account of persistent detonation, overheated spark plug points, overheated exhaust valve, incandescent carbon deposits on the surface of the cylinder or spark plug, or faulty timings of the spark plug.*

(*ii*) **Auto-ignition.** It is one of the theories of knocking in SI engines. *Auto-ignition* refers to the *initiation of combustion without the necessity of a flame*. The auto-ignition theory of knocking assumes that the flame velocity is normal before the onset of auto-ignition, and that gas vibrations are initiated by a number of end gas elements auto-igniting almost simultaneously. Auto-ignition does not occur immediately as the self-ignition temperature is reached. Some ignition delay period is required before the reaction becomes explosive. During the delay period some preflame reactions occur before giving rise to a flame. The exact method of formation of the preflame reactions is not known.

(*iii*) **Detonation.** The second theory of knocking in the SI engines is detonation. It is *the name given to the violent waves produced within the cylinder of an SI engine.* The noise produced is like that produced by a sharp ringing blow upon the metal of a cylinder.

The region in which the detonation occurs is far away from the spark plug, and is knows as the *detonation zone*. After a spark is produced, there is a rise of pressure and temperature due to the combustion of the ignited fuel. This rise in temperature and pressure both combine to increase the velocity of the flame, compressing the unburnt portion of the charge of the detonation zone. Finally, the temperature in the detonation zone reaches such a high value that *chemical reaction occurs at a far greater rate than in the advancing flame*. *Before the flame completes its course* across the combustion chamber, the whole mass of remaining unburnt charge *ignites instantaneously*. This spontaneous ignition of a portion of the charge sets *rapidly moving high pressure waves* that hit cylinder walls with such violence that the cylinder walls gives out a loud pinking noise. *It is this noise that expresses or indicates detonation.*

Example 23.5. *A two-stroke cycle internal combustion engine has a mean effective pressure of 6 bar. The speed of the engine is 1000 r.p.m. If the diameter of piston and stroke are 110 mm and 140 mm respectively, find the indicated power developed.*

Solution. Mean effective pressure (indicated),

$p_{mi} = 6$ bar

Engine speed, $N = 1000$ r.p.m.

Diameter of the piston, $D = 110$ mm $= 0.11$ m

Stroke length, $L = 140$ mm $= 0.14$ m

Indicated power developed, P :

$$\text{Indicated power, I.P.} = \frac{np_{mi}\, LANk \times 10}{6} \text{ kW}$$

Here, n = No. of cylinders = 1

and $k = 1$ ..... for 2-stroke cycle engine

$$\therefore \quad \text{I.P.} = \frac{1 \times 6 \times 0.14 \times \frac{\pi}{4} \times (0.11)^2 \times 1000 \times 1 \times 10}{6}$$

$= \textbf{13.3 kW.}$ **(Ans.)**

Example 23.6. *A 4-cylinder four-stroke petrol engine develops 14.7 kW at 1000 r.p.m. The mean effective pressure is 5.5 bar. Calculate the bore and stroke of the engine, if the length of stroke is 1.5 times the bore.*

Solution. Number of cylinders, $n = 4$

Power developed, $P = 14.7$ kW

Engine speed, $N = 1000$ r.p.m.

Indicated mean effective pressure, $p_{mi} = 5.5$ bar

Length of stroke, $L = 1.5\, D$ (bore)

For four stroke cycle, $k = \frac{1}{2}$

$\boldsymbol{L = ?, D = ?}$

Indicated power developed,

$$\text{I.P.} = \frac{np_{mi}\, LANk \times 10}{6} \text{ kW}$$

$$14.7 = \frac{4 \times 5.5 \times 1.5D \times \pi / 4D^2 \times 1000 \times \frac{1}{2} \times 10}{6}$$

$$\therefore \quad D^3 = \frac{14.7 \times 6 \times 4 \times 2}{4 \times 5.5 \times 1.5 \times \pi \times 1000 \times 10} = 0.0006806$$

$\therefore \quad D = \textbf{0.0879}$ or **87.9 mm. (Ans.)**

and $\quad L = \textbf{1.5} \times \textbf{87.9} = \textbf{131.8 mm.}$ **(Ans.)**

Example 23.7. *A single-cylinder, four-stroke cycle oil engine is fitted with a rope brake. The diameter of the brake wheel is 600 mm and the rope diameter is 26 mm. The dead load on the brake is 200 N and the spring balance reads 30 N. If the engine runs at 450 r.p.m., what will be the brake power of the engine ?*

Solution. Diameter of the brake wheel,

$D_b = 600$ mm $= 0.6$ m

Rope diameter, $d = 26$ mm $= 0.026$ m

Dead load on the brake, $W = 200$ N

Spring balance reading, $S = 30$ N

Engine speed, $N = 450$ r.p.m.

Brake power, B.P. :

Brake power is given by, B.P.

$$= \frac{(W - S)\,\pi\,(D_b + d)N}{60 \times 1000} \text{ kW}$$

$$= \frac{(200 - 30)\pi(0.6 + 0.026) \times 450}{60 \times 1000} = \textbf{2.5 kW.}$$ **(Ans.)**

Example 23.8. *A four-cylinder four-stroke, spark-ignition engine develops a maximum brake torque of 160 Nm at 3000 r.p.m. Calculate the engine displacement, bore and stroke. The brake mean effective pressure at the maximum engine torque point is 960 kPa. Assume bore is equal to stroke.* **(GATE)**

Solution. *Given* $n = 4$; $k = \frac{1}{2}$ (engine being 4-stroke cycle),

$T_b = 160$ Nm,
$N = 3000$ r.p.m., $p_{mb} = 960$ kPa
$= 960 \times 10^3$ N/m^2 = 9.6 bar ; D = L

D, L, displacement :

$$\text{Power developed} = \frac{2\pi N T_b}{60 \times 1000} \text{ kW}$$

$$= \frac{p_m LANk \times 10}{6} \text{ kW}$$

i.e.,
$$\frac{2\pi N T_b}{60 \times 1000} = \frac{p_{mb} LANk \times 10}{6}$$

Substituting the values, we get

$$\frac{2\pi \times 3000 \times 160}{60 \times 1000} = \frac{9.6 \times D \times \frac{\pi}{4} \times D^2 \times 3000 \times \frac{1}{2} \times 10}{6}$$

$$50.265 = 18849.6\ D^3$$

$$\therefore \quad D = \left(\frac{50.265}{18849.6}\right)^{1/3}$$

$$= \mathbf{0.1387\ m} \text{ or } \mathbf{138.7\ mm.} \quad \textbf{(Ans.)}$$

$$\therefore \quad \mathbf{L = D = 138.7\ mm.} \quad \textbf{(Ans.)}$$

$$\text{Displacement} = \frac{\pi}{4} D^2 \times L = \frac{\pi}{4} \times 0.1387^2 \times 0.1387$$

$$= \mathbf{0.002095\ m^3.} \quad \textbf{(Ans.)}$$

Example 23.9. *A turbo-charged six-cylinder diesel engine has the following performance details :*

(i) Work done during compression and expansion = 820 kW

(ii) Work done during intake and exhaust = 50 kW

(iii) Rubbing friction in the engine = 150 kW

(iv) Net work done by turbine = 40 kW

If the brake mean effective pressure is 0.6 MPa, determine the bore and stroke of the engine taking the ratio of bore to stroke as 1 and engine speed as 1000 r.p.m. **(GATE 1998)**

Solution. *Given :* $p_{mb} = 0.6$ MP$_a$ = 6 bar ; $\frac{D}{L} = 1$;
$N = 1000$ r.p.m.

D, L :

Net work available = 820 – (50 + 150 + 40) = 580 kW

$$\text{B.P} = \frac{n \times p_{mb}\ LANk \times 10}{6}$$

$$580 = \frac{6 \times 6 \times D \times \frac{\pi}{4} D^2 \times 1000 \times \frac{1}{2} \times 60}{6}$$

$$= 23562\ D^3$$

$$D = \left(\frac{580}{23562}\right)^{1/3}$$

$$= 0.2908 \text{ m} \quad \text{or} \quad 290.8 \text{ mm}$$

Hence **D = L = 290.8 mm. (Ans.)**

Example 23.10. *A spark-ignition engine, designed to run on octane* ($C_8 H_{18}$) *fuel, is operated on methane* (CH_4). *Estimate the ratio of the power input of the engine with methane fuel to that with octane. In both cases the fuel ratio is stoichiometric, the mixture is supplied to the engine at the same conditions, the engine runs at the same speed, and has the same volumetric and thermal efficiencies. The heating value of methane is 50150 kJ/kg while that of octane in 44880 kJ/kg.* **(U.P.S.C.)**

Solution. In spark-ignition engine, the air standard efficiency is given as :

$$\eta_{\text{air-standard}} = 1 - \frac{1}{(r)^{\gamma - 1}},$$

where r is the compression ratio which depends on engine parameters and has no relevance to fuel *e.g.*, quantity, type, calorific value etc.

Indicated power of engine is given by

$$\text{I.P.} = \frac{p_m\ LANk \times 10}{6} \text{ kW}$$

when L (length), A (area) and N (r.p.m.) depend on engine construction.

Only p_m (mean effective pressure) depends on engine operation.

For air-standard Otto cycle,

$$p_m = \frac{\text{Net work}}{\text{Displacement}} = \frac{Q_s \times \text{efficiency}}{\text{Displacement}}$$

where Q_s is the energy supplied.

Since Q_s in proportional to the mass of the fuel supplied times calorific value of fuel, therefore

$$\frac{(\text{Power})_{\text{methane}}}{(\text{Power})_{\text{octane}}} = \frac{C_{\text{methane}}}{C_{\text{octane}}}$$

$$= \frac{50150}{44880} = \mathbf{1.117.} \quad \textbf{(Ans.)}$$

Example 23.11. *A large four-stroke cycle diesel engine runs at 2000 r.p.m. The engine has a displacement of 25 litres and a brake mean effective pressure of 0.6 MN/m². It consumes 0.018 kg/s of fuel (calorific value = 42000 kJ/kg). Determine the brake power and brake thermal efficiency.* **(GATE)**

Solution. *Given* : N = 2000 r.p.m. ;

$$k = \frac{1}{2} \text{ (4-stroke cycle engine) ;}$$

Displacement ($A \times L$) = 25 litres = 25 × 10⁻³

= 0.025 m³ ; p_{mb} = 0.6 MN/m² = 6 bar

$\dot{m}_f$ = 0.018 kg/s ; C = 42000 kJ/kg

Brake power, B.P. :

$$\text{B.P.} = \frac{p_m LANk \times 10}{6}$$

$$= \frac{6 \times 0.025 \times 2000 \times \frac{1}{2} \times 10}{6} = \textbf{250 kW.} \quad \textbf{(Ans.)}$$

Brake thermal efficiency,

$$\eta_{\text{th (B)}} = \frac{\text{B.P.}}{\dot{m}_f \times \text{C}}$$

$$= \frac{250}{0.018 \times 42000}$$

= **0.3307** or **33.07%.** **(Ans.)**

Example 23.12. *A rope brake was used to measure the brake power of a single-cylinder, four-stroke cycle petrol engine. It was found that the torque due to brake load is 175 Nm and the engine makes 500 r.p.m. Determine the brake power developed by the engine.*

Solution. Torque due to brake load, T = 175 Nm

Engine speed, N = 500 r.p.m.

Brake power, B.P. :

$$\text{Brake power, B.P.} = \frac{2\pi NT}{60 \times 1000} = \frac{2\pi \times 500 \times 175}{60 \times 1000}$$

= **9.16 kW.** **(Ans.)**

Example 23.13. *Following observations were recorded during a test on a single-cylinder oil engine :*

Bore = 300 mm ; stroke = 450 mm ; speed = 300 r.p.m. ; i.m.e.p. = 6 bar ; net brake load = 1.5 kN ; brake drum diameter = 1.8 metres ; brake rope diameter = 2 cm.

Calculate : (i) Indicater power ; (ii) Brake power ; (iii) Mechanical efficiency. **(AMIE Winter, 2000)**

Solution. Bore of engine cylinder, D = 300 mm = 0.3 m

Stroke length, L = 450 mm = 0.45 m

Engine speed, N = 300 r.p.m.

Indicated mean effective pressure, p_{mi} = 6 bar

Net brake load, $(W - S)$ = 1.5 kN

Diameter of brake drum, D_b = 1.8 m

Brake rope diameter, d = 2 cm = 0.02 m

(i) **Indicated power, I.P. :**

$$\text{I.P.} = \frac{n\, p_{mi}\, LANk \times 10}{6} \quad \left[\text{where } k = \frac{1}{2} \text{ ...... four-stroke engine and } n = \text{no. of cylinders}\right]$$

$$= \frac{1 \times 6 \times 0.45 \times \frac{\pi}{4} \times 0.3^2 \times 300 \times \frac{1}{2} \times 10}{6}$$

= **47.71 kW.** **(Ans.)**

(ii) **Brake power, B.P. :**

$$\text{B.P.} = \frac{(W - S)\, \pi\, (D_b + d)N}{60}$$

$$= \frac{1.5 \times \pi(1.8 + 0.02) \times 300}{60}$$

= **42.88 kW.** **(Ans.)**

(iii) **Mechanical efficiency, $\eta_{\text{mech.}}$:**

$$\eta_{\text{mech.}} = \frac{\text{B.P.}}{\text{I.P.}} = \frac{42.88}{47.71}$$

= **0.8987** or **89.87%.** **(Ans.)**

Example 23.14. *The power output of an I.C. engine is measured by a rope brake dynamometer. The diameter of the brake pulley is 700 mm and the rope diameter is 25 mm. The load on the light side of the rope is 50 kg mass and spring balance reads 50 N. The engine running at 900 r.p.m. consumes fuel of calorific value of 44000 kJ/kg, at a rate of 4 kg/h.*

Assume g = 9.81 m/s². Calculate :

(i) Brake specific fuel consumption

(ii) Brake thermal efficiency. **(GATE)**

Solution. *Given* : D_b = 700 mm = 0.7 m ,

d = 25 mm = 0.025 m,

W = 50 kg, S = 50 N ; N = 900 r.p.m. ;

C = 44000 kJ/kg, m_f = 4 kg/h

(i) **Brake specific fuel consumption, b.s.f.c :**

Brake power,

$$\text{B.P} = \frac{(W - S)\, \pi\, (D_b + d) \times N}{60 \times 1000} \text{ kW}$$

$$= \frac{(50 \times 9.81 - 50) \times \pi\, (0.7 + 0.025) \times 900}{60 \times 1000}$$

= 15.05 kW

Brake specific fuel consumption,

$$\text{b.s.f.c.} = \frac{m_f \text{ (kg/h)}}{\text{B.P.(kW)}}$$

$$= \frac{4}{15.05} = \textbf{0.266 kg/kWh.} \quad \textbf{(Ans.)}$$

(*ii*) **Brake thermal efficiency, $\eta_{th(B)}$:**

$$\eta_{th(B)} = \frac{B.P.}{\dot{m}_f \times C} \quad \text{(where } \dot{m}_f = \text{fuel used, in kg/s)}$$

$$= \frac{15.05}{(4/3600) \times 44000}$$

$= \mathbf{0.3078}$ or **30.78%.** **(Ans.)**

Example 23.15. *A four cylinder four-stroke S.I. engine has a compression ratio of 8 and bore of 100 mm, with stroke equal to the bore. The volumetric efficiency of each cylinder is equal to 75%. The engine operates at a speed of 4800 r.p.m. with an air-fuel ratio 15.*

Given that the calorific value of fuel = 42 MJ/kg, atmospheric density = 1.12 kg/m³, mean effective pressure in the cylinder = 10 bar and mechanical efficiency of the engine = 80%, determine the indicated thermal efficiency and the brake power. **(GATE)**

Solution. *Given* : Number of cylinders, $n = 4$; $k = \frac{1}{2}$ (engine being four-stroke) ; $r = 8$; D = 100 mm = 0.1 m ; $L = D = 0.1$ m ; $\eta_{vol} = 75\%$; $N = 4800$ r.p.m. ; Air-fuel ratio = 15 ; $C = 42$ MJ/kg ; $\rho = 1.12$ kg/m³ ; $p_{mi} = 10$ bar, $\eta_{mech} = 80\%$.

Indicated thermal efficiency, $\eta_{th.(I)}$:

Indicated power,

$$I.P. = \frac{n\, p_{mi}\, LANk \times 10}{6}$$

$$= \frac{4 \times 10 \times 0.1 \times \left(\frac{\pi}{4} \times 0.1^2\right) \times 4800 \times \frac{1}{2} \times 10}{6} \text{ kW}$$

$= 125.66$ kW

$$\text{Air consumption} = n \times \frac{\pi}{4} D^2 \times L \times \frac{N}{2} \times \eta_{vol.}$$

$$= 4 \times \frac{\pi}{4} \times 0.1^2 \times 0.1 \times \frac{4800}{2} \times 0.75$$

$= 5.655$ m³/min $= 0.09425$ m³/s

Mass flow of air, $\dot{m}_a = 0.09425 \times 1.12 = 0.1056$ kg/s

$$\text{Fuel consumption, } \dot{m}_f = \frac{0.1056}{\text{Air-fuel ratio}} = \frac{0.1056}{15}$$

$= 0.00704$ kg/s

$$\therefore \quad \eta_{th(I)} = \frac{I.P.}{\dot{m}_f \times C}$$

$$= \frac{125.66 \times 10^3}{0.00704 \times 42 \times 10^6}$$

$= \mathbf{0.425}$ or **42.5%.** **(Ans.)**

Brake power, B.P. :

Brake power = Indicated power × η_{mech}.

$= 125.66 \times 0.8 = \mathbf{100.53\ kW.}$ **(Ans.)**

Example 23.16. *A single-cylinder four-stroke diesel engine running at 1800 r.p.m. has a bore of 85 mm and a stroke of 110 mm. It takes 0.56 kg of air per minute and develops a brake power output of 6 kW while the air-fuel ratio is 20 : 1. The calorific value of the fuel used is 42550 kJ/kg, and the ambient air density is 1.18 kg/m³. Calculate :*

(i) The volumetric efficiency, and

(ii) Brake specific fuel consumption. **(GATE)**

Solution. *Given* : $N = 1800$ r.p.m. ; $D = 85$ mm $= 0.085$ m ; $L = 110$ mm $= 0.11$ m ;

Air flow rate, $m = 0.56$ kg/min. ; B.P. = 6 kW ;

Air-fuel ratio = 20 : 1 ;

$C = 42550$ kJ/kg ; $\rho = 1.18$ kg/m³.

(*i*) **The volumetric efficiency, $\eta_{vol.}$:**

$$\text{Volume displacement} = \frac{\pi}{4} D^2 \times L \times \frac{N}{2}$$

$$= \frac{\pi}{4} \times (0.085)^2 \times 0.11 \times \frac{1800}{2}$$

$= 0.5617$ m³/min

Mass of air $= 0.5617 \times 1.18 = 0.663$ kg/min

∴ Volumetric efficiency

$$= \frac{0.56}{0.663} = \mathbf{0.845} \text{ or } \mathbf{84.5\%.} \quad \textbf{(Ans.)}$$

(*ii*) **Brake specific fuel consumption (b.s.f.c.)**

Fuel consumption

$$= \frac{0.56}{\text{Air-fuel ratio}} = \frac{0.56}{20} = 0.028 \text{ kg/min.}$$

∴ Brake specific fuel consumption

$$= \frac{\text{Fuel used / hour}}{B.P.} \text{ kg/kWh}$$

$$= \frac{0.028 \times 60}{6} = \mathbf{0.28\ kg/kWh.} \quad \textbf{(Ans.)}$$

☞**Example 23.17.** *Following data refer to a four-stroke double-acting diesel engine having cylinder diameter 200 mm and piston stroke 350 mm.*

m.e.p. on cover side	*= 6.5 bar*
m.e.p. on crank side	*= 7 bar*
Speed	*= 420 r.p.m.*
Diameter of piston rod	*= 20 mm*
Dead load on the brake	*= 1370 N*
Spring balance reading	*= 145 N*
Brake wheel diameter	*= 1.2 m*
Brake rope diameter	*= 20 mm*

Calculate the mechanical efficiency of the engine.

Solution. *Given* : $P_{mi(cover)} = 6.5$ bar,

$p_{mi(crank)} = 7$ bar, $D = 0.2$ m,

$L = 0.35$ m,

$N = 420$ r.p.m., $d_{rod} = 20$ mm = 0.02 m,
$W = 1370$ N, $S = 145$ N,

$D_b = 1.2$ m, $d = 0.02$ m, $k = \frac{1}{2}$ ... 4-stroke cycle engine

Mechanical efficiency ; $\eta_{mech.}$ **:**

Area of cylinder on cover end side,

$$A_{cover} = \pi/4\ D^2 = (\pi/4) \times (0.2)^2 = 0.03141 \text{ m}^2$$

Effective area of cylinder on crank end side,

$$A_{crank} = \pi/4\ (D^2 - d_{rod}^2)$$
$$= \pi/4\ (0.2^2 - 0.02^2) = 0.0311 \text{ m}^2$$

Indicated power on cover end side,

$$\text{I.P.}_{\cdot(\text{cover})} = \frac{p_{mi\,(\text{cover})} \times LANk \times 10}{6}$$

$$= \frac{6.5 \times 0.35 \times 0.03141 \times 420 \times \frac{1}{2} \times 10}{6}$$

$$= 25 \text{ kW}$$

Indicated power on crank end side,

$$\text{I.P.}_{\cdot(\text{crank})} = \frac{p_{mi\,(\text{crank})} \times LANk \times 10}{6}$$

$$= \frac{7 \times 0.35 \times 0.0311 \times 420 \times \frac{1}{2} \times 10}{6}$$

$$= 26.67 \text{ kW}$$

Total I.P. = 25 + 26.67 = 51.67 kW

Now, brake power,

$$\text{B.P.} = \frac{(W - S)\,\pi\,(D_b + d)N}{60 \times 1000}$$

$$= \frac{(1370 - 145)\pi(1.2 + 0.02) \times 420}{60 \times 1000} \text{ kW}$$

$$= 32.86 \text{ kW}$$

Mechanical efficiency,

$$\eta_{mech.} = \frac{\text{B.P.}}{\text{I.P.}} = \frac{32.86}{51.67} = 0.6359 = \mathbf{63.59\%}. \quad \textbf{(Ans.)}$$

Example 23.18. *The following data refer to an oil engine working on Otto four-stroke cycle :*

Brake power	*= 14.7 kW*
Suction pressure	*= 0.9 bar*
Mechanical efficiency	*= 80%*
Ratio of compression	*= 5*
Index of compression curve	*= 1.35*
Index of expansion curve	*= 1.3*
Maximum explosion pressure	*= 24 bar*
Engine speed	*= 1000 r.p.m.*
Ratio of stroke : bore	*= 1.5*

Find the diameter and stroke of the piston.

Solution. Refer to Fig. 23.78.

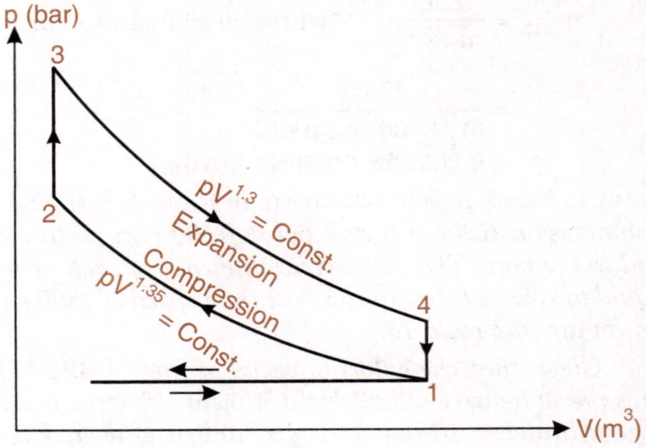

Fig. 23.78

B.P. = 14.7 kW, $p_1 = 0.9$ bar,

$\eta_{mech.} = 80\%$, $r = 5$, $p_3 = 24$ bar

$N = 1000$ r.p.m., $\frac{L}{D} = 1.5$

D = ?, L = ?

Compression ratio, $r = \frac{V_1}{V_2} = \frac{V_4}{V_3}$

To find p_2, considering *compression process 1–2*, we have

$$p_1 V_1^{1.35} = p_2 V_2^{1.35}$$

or $$\frac{p_2}{p_1} = \left(\frac{V_1}{V_2}\right)^{1.35} = (5)^{1.35} = 8.78$$

∴ $$p_2 = p_1 \times 8.78 = 0.9 \times 8.78 = 7.9 \text{ bar}$$

To find p_4, considering *expansion process 3–4,* we have

$$p_3 V_3^{1.3} = p_4 V_4^{1.3}$$

or $$\frac{p_3}{p_4} = \left(\frac{V_4}{V_3}\right)^{1.3} = (5)^{1.3} = 8.1$$

∴ $$p_4 = \frac{p_3}{8.1} = \frac{24}{8.1} = 2.96 \text{ bar}$$

Work done/cycle = Area 1–2–3–4 = Area under the curve 3–4 – area under the curve 1–2

$$= \frac{p_3V_3 - p_4V_4}{1.3 - 1} - \frac{p_2V_2 - p_1V_1}{1.35 - 1}$$

$$= \frac{p_3V_3 - p_4V_4}{0.3} - \frac{p_2V_3 - p_1V_4}{0.35}$$

$[\because\ V_1 = V_4 \text{ and } V_2 = V_3]$

$$= \frac{10^5(24V_3 - 2.9V_4)}{0.3} - \frac{10^5(7.9V_3 - 0.9V_4)}{0.35}$$
$$= 10^5\ [(80V_3 - 9.67V_4) - (22.57V_3 - 2.57V_4)]$$

$= 10^5(80V_3 - 9.67V_4 - 22.57V_3 + 2.57V_4)$

$= 10^5(57.43V_3 - 7.1V_4)$

$= 10^5(57.43V_3 - 7.1 \times 5V_3) \quad \left[\because \frac{V_4}{V_3} = 5\right]$

$= 10^5 \times 21.93\ V_3$ Nm.

Mean effective pressure (theoretical),

$$p_m = \frac{\text{Work done/cycle}}{\text{Stroke volume } (V_s)}$$

$$= \frac{10^5 \times 21.93V_3}{(V_4 - V_3)} = \frac{10^5 \times 21.93V_3}{5V_3 - V_3}$$

$= 10^5 \times 5.48$ N/m² or 5.48 bar.

Now, $\eta_{mech.} = \frac{\text{B.P.}}{\text{I.P.}}$

$\therefore \quad \text{I.P.} = \frac{\text{B.P.}}{\eta_{mech.}} = \frac{14.7}{0.8} = 18.37$ kW.

To find D and L :

$$\text{I.P.} = \frac{p_{mi} LANk \times 10}{6} \text{ kW}$$

$$14.7 = \frac{5.48 \times 1.5D \times \pi / 4 \times D^2 \times 1000 \times \frac{1}{2} \times 10}{6}$$

$$\therefore D^3 = \frac{14.7 \times 6 \times 4 \times 2}{5.48 \times 1.5 \times \pi \times 1000 \times 10} = 0.002855$$

or $D = 0.1418$ m $\simeq$ **0.1398 m** or **139.8 mm. (Ans.)**

and $L = 1.5D = 1.5 \times 139.8 =$ **209.7 mm. (Ans.)**

Example 23.19. *The following results refer to a test on a petrol engine :*

Indicated power = 30 kW ;

brake power = 26 kW ;

Engine speed = 1000 r.p.m.;

fuel per brake-power hour

= 0.35 kg ;

Calorific value of the fuel used = 43900 kJ/kg.

Calculate : (i) The indicated thermal efficiency,

(ii) The brake thermal efficiency, and

(iii) The mechanical efficiency.

Solution. Indicated power, I.P. = 30 kW

Brake power, B.P. = 26 kW

Engine speed, $N = 1000$ r.p.m.

Fuel per brake-power hour = 0.35 kg/B.P.-h

Calorific value of the fuel used, $C = 43900$ kJ/kg

Now, fuel consumption per hour = 0.35 × 26 = 9.1 kg/h.

(i) Indicated thermal efficiency,

$$\eta_{th(I)} = \frac{\text{I.P.}}{\dot{m}_f \times C} = \frac{30}{\left(\frac{9.1}{3600}\right) \times 43900}$$

= **0.27** or **27%. (Ans.)**

(ii) Brake thermal efficiency,

$$\eta_{th(B)} = \frac{\text{B.P.}}{\dot{m}_f \times C} = \frac{26}{\frac{9.1}{3600} \times 43900}$$

= **0.234** or **23.4%. (Ans.)**

(iii) Mechanical efficiency,

$$\eta_{mech.} = \frac{\text{B.P.}}{\text{I.P.}} = \frac{26}{30} = \mathbf{0.866} \text{ or } \mathbf{86.6\%}. \quad \textbf{(Ans.)}$$

Example 23.20. *The output of an I.C. engine is measured by a rope brake dynamometer. The diameter of the brake pulley is 750 mm and rope diameter is 50 mm. The dead load on the tight side of the rope is 400 N and the spring balance reading is 50 N. The engine consumes 4.2 kg/h of fuel at rated speed of 1000 r.p.m. The calorific value of fuel is 43900 kJ/kg. Calculate :*

(i) Brake specific fuel consumption, and

(ii) Brake thermal efficiency.

Solution. Diameter of brake pulley,

$D_b = 750$ mm = 0.75 m

Rope diameter, $d = 50$ mm = 0.05 m

Dead load, $W = 400$ N

Spring balance reading, $S = 50$ N

Consumption of fuel = 4.2 kg/h

Rated speed, $N = 1000$ r.p.m.

Calorific value of fuel, $C = 43900$ kJ/kg

Brake power,

$$\text{B.P.} = \frac{(W - S)\,\pi\,(D_b + d)N}{60 \times 1000} \text{ kW}$$

$$= \frac{(400 - 50)\pi(0.75 + 0.05) \times 1000}{60 \times 1000} = 14.66 \text{ kW.}$$

(i) Brake specific fuel consumption,

$$\text{s.f.c. (brake)} = \frac{4.2}{14.66} = \mathbf{0.286 \text{ kg/kWh.}} \quad \textbf{(Ans.)}$$

(ii) Brake thermal efficiency,

$$\eta_{th(B)} = \frac{\text{B.P.}}{\dot{m}_f \times C} = \frac{14.66}{\frac{4.2}{3600} \times 43900}$$

($\dot{m}_f$ = Fuel used in kg/s)

= **0.286** or **28.6%. (Ans.)**

Example 23.21. *A six-cylinder, four-stroke 'Petrol engine' having a bore of 90 mm and stroke of 100 mm has a compression ratio of 7. The relative efficiency with reference to indicated thermal efficiency is 55% when the indicated specific fuel consumption is 0.3 kg/kWh. Estimate the calorific value of the fuel and fuel consumption (in kg/h), given that the imep is 8.5 bar and speed is 2500 r.p.m.*

(AMIE)

Solution. Number of cylinders, $n = 6$

Bore of each cylinder, $D = 90$ mm $= 0.09$ m

Stroke length, $L = 100$ mm $= 0.1$ m

Compression ratio, $r = 7$

Relative efficience, $\eta_{relative} = 55\%$

Indicates specific fuel consumption = 0.3 kg/kWh

Indicated mean effective pressure, i.m.e.p. = 8.6 bar

Engine speed, $N = 2500$ r.p.m.

Calorific value of fuel (C) and fuel consumption (in kg/h) :

$$\eta_{air\ standard} = 1 - \frac{1}{(r)^{\gamma-1}} = 1 - \frac{1}{(7)^{1.4-1}} = 0.5408$$

$$\text{Now, } \eta_{relative} = \frac{\eta_{th(I)}}{\eta_{air\ standard}}$$

or $\eta_{th(I)} = \eta_{relative} \times \eta_{air\ standard}$

or Indicated thermal efficiency,

$$\eta_{th(I)} = 0.55 \times 0.5408 = 0.297$$

$$\text{But} \quad \eta_{th(I)} = \frac{\text{I.P.}}{\dot{m}_f \times C} = \frac{1}{(0.3/3600) \times C}$$

$$\text{or} \quad 0.297 = \frac{3600}{0.3 \times C} \quad \text{or } C = \frac{3600}{0.297 \times 0.3}$$

$= \mathbf{40404\ kJ/kg.}$ **(Ans.)**

$$\text{Now, indicated power, I.P.} = \frac{np_{mi}LANk \times 10}{6}$$

[where $k = \frac{1}{2}$ ...... four-stroke cycle engine]

$$= \frac{6 \times 8.6 \times 0.1 \times \frac{\pi}{4} \times 0.09^2 \times 2500 \times \frac{1}{2} \times 10}{6}$$

$= 68.39$ kW

$\therefore$ Fuel consumption

$= 0.3 \times 68.39 = \mathbf{20.52\ kg/h.}$ **(Ans.)**

☞**Example 23.22.** *A 4-cylinder two-stroke cycle petrol engine develops 30 kW at 2500 r.p.m. The mean effective pressure on each piston is 8 bar and mechanical efficiency is 80%. Calculate the diameter and stroke of each cylinder of stroke to bore ratio 1.5. Also calculate the fuel consumption of the engine, if brake thermal efficiency is 28%. The calorific value of the fuel is 43900 kJ/kg.*

Solution. Number of cylinder, $n = 4$

Brake power, B.P. = 30 kW

Engine speed, $N = 2500$ r.p.m.

Mean effective pressure $p_{mi} = 8$ bar

Mechanical efficiency, $\eta_{mech.} = 80\%$

Length of stroke, $L = 1.5\ D$ (bore)

Brake thermal efficiency, $\eta_{th(B)} = 28\%$

Calorific value of the fuel, $C = 43900$ kJ/kg

$k = 1$ ...... for 2-stroke cycle engine

(*i*) **L = ?, D = ?**

$$\eta_{mech.} = \frac{\text{B.P.}}{\text{I.P.}}$$

$$0.8 = \frac{30}{\text{I.P.}}$$

$$\therefore \quad \text{I.P.} = \frac{30}{0.8} = 37.5 \text{ kW}$$

$$\text{Also,} \quad \text{I.P.} = \frac{np_{mi}LANk \times 10}{6}$$

$$37.5 = \frac{4 \times 8 \times 1.5D \times \pi/4D^2 \times 2500 \times 1 \times 10}{6}$$

$$\therefore D^3 = \frac{37.5 \times 6 \times 4}{4 \times 8 \times 1.5 \times \pi \times 2500 \times 10} = 0.0002387$$

or $D = \mathbf{0.062\ m}$ or **62 mm.** **(Ans.)**

and $L = 62 \times 1.5 = \mathbf{93\ mm.}$ **(Ans.)**

(*ii*) **Fuel consumption :**

Brake thermal efficiency, $\eta_{th(B)} = \dfrac{\text{B.P.}}{\dot{m}_j \times C}$ ($\dot{m}_f$ = Fuel used in kg/s)

$$0.28 = \frac{30}{\dot{m}_f \times 43900}$$

$$\therefore \quad \dot{m}_f = \frac{30}{0.28 \times 43900}$$

$= \mathbf{0.00244\ kg/s}$ or **8.78 kg/h.** **(Ans.)**

Example 23.23. *A six cylinder, 4-stroke SI engine having a piston displacement of 700 cm³ per cylinder developed 78 kW at 3200 r.p.m. and consumed 27 kg of petrol per hour. The calorific value of petrol is 44 MJ/kg. Estimate :*

(*i*) *The volumetric efficiency of the engine if the air-fuel ratio is 12 and intake air is at 0.9 bar, 32°C.*

(*ii*) *The brake thermal efficiency, and*

(*iii*) *The brake torque.*

For air, R = 0.287 kJ/kg K. **(N.U.)**

Solution. Number of cylinders = 6

Piston displacement per cylinder

$= 700$ cm^3 or 700×10^{-6} m^3

Power developed, $P = 78$ kW

Speed of the engine, $N = 3200$ r.p.m.

Mass of fuel used, $m_f = 27$ kg/h

Calorific value of fuel, $C = 44$ MJ/kg

Air-fuel ratio = 12

Intake air pressure, $p_1 (= p_a) = 0.9$ bar

Intake air temperature, $T_1 = (T_a) = 32 + 273 = 305$ K

For air, $R = 0.287$ kJ/kg K

(*i*) **Volumetric efficiency of the engine, η_{vol} :**

Mass of air, m_a = Air-fuel ratio × mass of fuel

$= 12 \times 27 = 324$ kg/h

Also, $p_a V_a = m_a R_a T_a$ or $V_a = \dfrac{m_a R_a T_a}{P_a}$

$\therefore$ Volume of intake air,

$$V_a = \frac{324 \times 0.287 \times 305}{0.9 \times 10^2} = 315.126 \text{ m}^3/\text{h}$$

Swept volume per hour = Piston displacement per cylinder × no. of cylinder × $\dfrac{N}{2}$ × 60

$$= 700 \times 10^{-6} \times 6 \times \frac{3200}{2} \times 60$$

$$= 403.2 \text{ m}^3/\text{h}$$

$\therefore$ Volumetric efficiency,

$$\eta_{vol.} = \frac{\text{Volume of intake air}}{\text{Swept volume}}$$

$$= \frac{315.126}{403.2} = \mathbf{0.781} \text{ or } \mathbf{78.1\%}. \quad \textbf{(Ans.)}$$

(ii) **Brake thermal efficiency η_{BT} :**

$$\eta_{BT} = \frac{\text{Brake work}}{\text{Heat supplied by fuel}} = \frac{78}{27 \times \dfrac{44 \times 10^3}{3600}}$$

$$= \frac{78 \times 3600}{27 \times (44 \times 10^3)}$$

$$= \mathbf{0.2364} \text{ or } \mathbf{23.64\%}. \quad \textbf{(Ans.)}$$

(iii) **The brake torque, T_B :**

$$P = \frac{2\pi N T_B}{60} \quad \text{or} \quad 78 = \frac{2\pi \times 3200 \times T_B}{60}$$

$$\therefore \quad T_B = \frac{78 \times 60}{2\pi \times 3200} = \mathbf{0.2328 \text{ kNm.}} \quad \textbf{(Ans.)}$$

Example 23.24. *A 6-cylinder, four-stroke gas engine with a stroke volume of 1.75 litres develops 26.3 kW at 504 r.p.m. The m.e.p. is 6 bar. Find the average number of times each cylinder misfires in one minute.*

Solution. Number of cylinders, $n = 6$

Stroke volume, $V_s = 1.75$ litres $= 1.75 \times 10^{-3}$ m^3

Indicated power, I.P. = 26.3 kW

Engine speed, $N = 504$ r.p.m.

Mean effective pressure, $p_{mi} = 6$ bar

$k = 1/2$ ..... for 4-stroke engine.

Average number of times each cylinder misfires/min :

$$\text{I.P.} = \frac{n p_{mi} LANk \times 10}{6} \text{ kW}$$

$$26.3 = \frac{6 \times 6 \times 1.75 \times 10^{-3} \times N \times 1/2 \times 10}{6}$$

$[\because \; LA = V_s = 1.75 \times 10^{-3} \text{ m}^3]$

$$N = \frac{26.3 \times 6 \times 2 \times 1000}{6 \times 6 \times 1.75 \times 10} = 500 \text{ r.p.m.}$$

Actual number of fires in one minute

$$= \frac{500}{2} \times 6 = 1500$$

Expected number of fires in one minute

$$= \frac{504}{2} \times 6 = 1512$$

Number of misfires/min. = 1512 – 1500 = 12.

Average number of times each cylinder misfires in one minute $= \dfrac{12}{6} = \mathbf{2.}$ **(Ans.)**

☞**Example 23.25.** *The following data refer to a car engine having 4 cylinders.*

Bore = 75 mm ; stroke = 90 mm ; engine to rear axle ratio 39 : 8 ; wheel diameter with tyre fully inflated 650 mm. The petrol consumption for a distance of 3.2 km when car was moving at a speed of 48 km per hour was found to be 0.227 kg.

If the mean effective pressure is 5.625 bar, determine the indicated power and thermal efficiency. Calorific value of the petrol may be taken as 43470 kJ/kg.

Solution. Bore, $D = 75$ mm or 0.075 m

Stroke length, $L = 90$ mm or 0.09 m

Number of cylinders, $n = 4$

Engine to rear axle ratio = 39 : 8

Wheel diameter with tyre fully inflated = 650 mm or 0.65 m

Petrol consumption for a distance of 3.2 km at a speed of 48 km/h = 0.227 kg

Mean effective pressure, $p_{mi} = 5.625$ bar

Calorific value of petrol, $C = 43470$ kJ/kg

$k = 1/2$ ...... for 4-stroke engine.

Indicated power. I.P. :

Speed of the car = 48 km/h = $\dfrac{48 \times 1000}{60}$ = 800 m/min.

In N_t are the revolutions made by the tyre per minute, then $\pi D N_t = 800$

$$\therefore \quad N_t = \frac{800}{\pi \times 0.650} = 392 \text{ r.p.m.}$$

As the rear axle ratio is 39 : 8,

$$\therefore N_e \text{ (speed of the engine shaft)} = \frac{392 \times 39}{8} = 1911 \text{ r.p.m.}$$

$$\text{I.P.} = \frac{n p_{mi} LANk \times 10}{6} \text{ kW}$$

$$= \frac{4 \times 5.625 \times 0.09 \times \pi/4 \times 0.075^2 \times 1911 \times \dfrac{1}{2} \times 10}{6}$$

$$= \mathbf{14.25 \text{ kW.}} \quad \textbf{(Ans.)}$$

Indicated thermal efficiency :

To find indicated thermal efficiency, let us find $\dot{m}_f$ first :

Speed of the car $= \frac{48}{60} = 0.8$ km/min.

Time for covering 3.2 km $= \frac{3.2}{0.8} = 4$ min.

Amount of fuel consumed in 4 min. = 0.227 kg

$\therefore$ Fuel consumed/sec $= \frac{0.227}{4 \times 60} = 0.000946$ kg/s

Now, *Indicated thermal efficiency,*

$$\eta_{th(I)} = \frac{\text{I.P.}}{\dot{m}_f \times C} = \frac{14.25}{0.000946 \times 43470}$$

$= \mathbf{0.346}$ or **34.6%.** **(Ans.)**

Example 23.26. *The following readings were taken during the test of a single-cylinder four stroke oil engine :*

Cylinder diameter	*= 250 mm*
Stroke length	*= 400 mm*
Gross m.e.p.	*= 7 bar*
Pumping m.e.p.	*= 0.5 bar*
Engine speed	*= 250 r.p.m.*
Net load on the brake	*= 1080 N*
Effective diameter of the brake	*= 1.5 metres*
Fuel used per hour	*= 10 kg*
Calorific value of fuel	*= 44300 kJ/kg*

Calculate : (i) Indicated power ;

(ii) Brake power ;

(iii) Mechanical efficiency ;

(iv) Indicated thermal efficiency.

Solution. $D = 250$ mm $= 0.25$ m, $L = 400$ mm $= 0.4$ m,

$p_{mg} = 7$ bar,

$p_{mp} = 0.5$ bar, $N = 250$ r.p.m.,

$D_b = 1.5$ m, $\dot{m}_f = \frac{10}{3600} = 0.00277$ kg/s

$C = 44300$ kJ/kg, $n = 1$, $(W - S) = 1080$ N

Net, $p_m = p_{mg} - p_{mp} = 7 - 0.5 = 6.5$ bar.

***(i)* Indicated power, I.P. :**

$$\text{I.P.} = \frac{np_m LANk \times 10}{6}$$

$$= \frac{1 \times 6.5 \times 0.4 \times \pi / 4 \times 0.25^2 \times 250 \times \frac{1}{2} \times 10}{6} \text{ kW}$$

$= \mathbf{26.59}$ **kW. (Ans.)**

***(ii)* Brake power, B.P. :**

$$\text{B.P.} = \frac{(W - S)\pi D_b N}{60 \times 1000} \text{ kW} = \frac{1080 \times \pi \times 1.5 \times 250}{60 \times 1000}$$

$= \mathbf{21.2}$ **kW. (Ans.)**

***(iii)* Mechanical efficiency, $\eta_{mech.}$:**

$$\eta_{mech.} = \frac{\text{B.P.}}{\text{I.P.}} = \frac{21.2}{26.59} = \mathbf{0.797} \text{ or } \mathbf{79.7\%}. \quad \textbf{(Ans.)}$$

***(iv)* Indicated thermal efficiency, $\eta_{th(I)}$:**

$$\eta_{th(I)} = \frac{\text{I.P.}}{\dot{m}_f \times C} = \frac{26.59}{0.00277 \times 44300}$$

$= \mathbf{0.216}$ or **21.6%.** **(Ans.)**

Example 23.27. *The brake thermal efficiency of a diesel engine is 30 per cent. If the air to fuel ratio by weight is 20 and the calorific value of the fuel used is 41800 kJ/kg, what brake mean effective pressure may be expected at S.T.P. conditions ?*

Solution. Brake thermal efficiency, $\eta_{th(B)} = 30\%$

Air-fuel ratio by weight $= 20$

Calorific value of fuel used, $C = 41800$ kJ/kg

Brake mean effective pressure, p_{mb} :

$$\text{Brake thermal efficiency} = \frac{\text{Work produced}}{\text{Heat supplied}}$$

$$0.3 = \frac{\text{Work produced}}{41800}$$

$\therefore$ Work produced per kg of fuel $= 0.3 \times 41800$
$= 12540$ kJ

Mass of air used per kg of fuel = 20 kg

S.T.P. conditions refer to 1.0132 bar and 15°C

$$\text{Volume of air used} = \frac{mRT}{p} = \frac{20 \times 287 \times (273 + 15)}{1.0132 \times 10^5}$$

$= 16.31$ m^3

Brake mean effective pressure,

$$p_{mb} = \frac{\text{Work done}}{\text{Cylinder volume}} = \frac{12540 \times 1000}{16.31 \times 10^5}$$

$= \mathbf{7.69}$ **bar. (Ans.)**

Example 23.28. *In a test on single-cylinder four-stroke cycle gas engine with explosion in every cycle, the gas consumption given by the metre was 0.216 m³ per minute ; the pressure and temperature of the gas being 75 mm of water and 17°C respectively. Air consumption was 2.84 kg/min., the temperature being 17°C and barometer reading 745 mm of mercury. The bore of the engine was 250 mm and stroke 475 mm and r.p.m. 240.*

Find volumetric efficiency of the engine referred to volume of charge at N.T.P. Assume R for air as 287 Nm/kg K.

Solution. Gas consumption, $V_1 = 0.216$ m³/min.

Pressure of the gas $= 75$ mm of water

Temperature of gas, $T_1 = 17 + 273 = 290$ K

Air consumption $= 2.84$ kg/min

Temperature of air $= 17 + 273 = 290$ K

Barometer reading $= 745$ mm Hg

Bore of the engine, $D = 250$ mm $= 0.25$ m

Stroke of engine, $L = 475$ mm = 0.475 m

Engine speed, $N = 240$ r.p.m.

R for air = 287 Nm/kg K.

Volumetric efficiency, $\eta_{vol.}$:

Pressure of the gas, $p_1 = 745 + \dfrac{75}{13.6} = 750.5$ mm of mercury.

At N.T.P.

$$p_2 = 760 \text{ mm of mercury}$$
$$T_2 = 0 + 273 = 273 \text{ K}$$
$$V_2 = ?$$

To find volume of gas used at N.T.P. (V_2), using the relation :

$$\frac{p_1V_1}{T_1} = \frac{p_2V_2}{T_2}$$

$$\frac{750.5 \times 0.216}{290} = \frac{760 \times V_2}{273}$$

$$\therefore \quad V_2 = \frac{750.5 \times 0.216 \times 273}{760 \times 290} = 0.201 \text{ m}^3$$

Gas used per stroke $= \dfrac{0.201}{240/2} = 0.001675$ m^3

Volume occupied by air at N.T.P. (V) :

$$pV = mRT$$

or

$$V = \frac{mRT}{p} = \frac{2.84 \times 287 \times 273}{1.0132 \times 10^5}$$
$$= 2.196 \text{ m}^3/\text{min}$$

Air used per stroke $= \dfrac{2.196}{240/2} = 0.0183$ m^3 at N.T.P.

Mixture of gas and air used per stroke

$$= 0.001675 + 0.0183 = 0.0199 \text{ m}^3$$

Volumetric efficiency,

$$\eta_{vol.} = \frac{\text{Actual volume of mixture drawn per stroke at N.T.P.}}{\text{Swept volume of system}}$$

$$= \frac{0.0199}{\pi/4 \times 0.25^2 \times 0.475} = \mathbf{0.853} \text{ or } \mathbf{85.3\%}. \quad \textbf{(Ans.)}$$

☞**Example 23.29.** *The following particulars were obtained in a trial on a 4-stroke gas engine :*

Duration of trial	*= 1 hour*
Revolutions	*= 14000*
Number of missed cycle	*= 500*
Net brake load	*= 1470 N*
Mean effective pressure	*= 7.5 bar*
Gas consumption	*= 20000 litres*
L.C.V. of gas at supply condition	*= 21 kJ/litre*
Cylinder diameter	*= 250 mm*
Stroke	*= 400 mm*
Effective brake circumference	*= 4 m*
Compression ratio	*= 6.5 : 1*

Calculate : (*i*) *Indicated power*
(*ii*) *Brake power*
(*iii*) *Mechanical efficiency*
(*iv*) *Indicated thermal efficiency*
(*v*) *Relative efficiency.*

Solution. $N = \dfrac{14000}{60} = \dfrac{700}{3}$ r.p.m. ; $W–S = 1470$ N,

$p_{mi} = 7.5$ bar ; $V_g = \dfrac{20000}{3600} = 5.55$ litres/s,

$D = 250$ mm = 0.25 m, $L = 400$ mm = 0.4 m

$\pi(D_b + d) = 4$ m, $r = 6.5$, $n = 1$.

(*i*) **Indicated power, I.P. :**

$$\text{I.P.} = \frac{np_{mi}LANk \times 10}{6}$$

$$Nk = \left(\frac{14000}{2} - 500\right)\Big/ 60$$

$$= \frac{6500}{60} \text{ working cycles/min.}$$

$$\therefore \text{I.P.} = \frac{1 \times 7.5 \times 0.4 \times \pi/4 \times 0.25^2 \times (6500/60) \times 10}{6}$$

$$= \mathbf{26.59 \text{ kW}}. \quad \textbf{(Ans.)}$$

(*ii*) **Brake power, B.P. :**

$$\text{B.P.} = \frac{(W - S)\,\pi\,(D_b + d)N}{60 \times 1000} = \frac{1470 \times 4 \times (700/3)}{60 \times 1000}$$

$$= \mathbf{22.86 \text{ kW}}. \quad \textbf{(Ans.)}$$

(*iii*) **Mechanical efficiency, $\eta_{mech.}$:**

$$\eta_{mech.} = \frac{\text{B.P.}}{\text{I.P.}} = \frac{22.86}{26.59} = \mathbf{0.859} \text{ or } \mathbf{85.9\%}. \quad \textbf{(Ans.)}$$

(*iv*) **Indicated thermal efficiency, $\eta_{th.(I)}$:**

$$\eta_{th(I)} = \frac{\text{I.P.}}{V_g \times C} = \frac{26.59}{5.5 \times 21} = \mathbf{0.23} \text{ or } \mathbf{23\%}. \quad \textbf{(Ans.)}$$

(*v*) **Relative efficiency, $\eta_{relative}$:**

$$\eta_{relative} = \frac{\eta_{thermal}}{\eta_{air\text{-}standard}}$$

$$\eta_{air\text{-}standard} = 1 - \frac{1}{(r)^{\gamma-1}} = 1 - \frac{1}{(6.5)^{1.4-1}}$$
$$= 0.527 \text{ or } 52.7\%$$

$$\therefore \quad \eta_{relative} = \frac{0.23}{0.527}$$
$$= \mathbf{0.436} \text{ or } \mathbf{43.6\%}. \quad \textbf{(Ans.)}$$

Example 23.30. *The compression curve on the indicator diagram for a gas engine follows the law $pV^{1.3}$ = constant. At two points on the curve at $\frac{1}{4}$ stroke and $\frac{3}{4}$ stroke the pressures are 1.4 bar and 3.6 bar respectively. Determine the compression ratio of the engine. Calculate the thermal efficiency and the gas consumption per I.P. hour, if the relative efficiency is 0.4 and the gas has the calorific value of 18800 kJ/m³.*

Solution. Refer to Fig. 23.79.

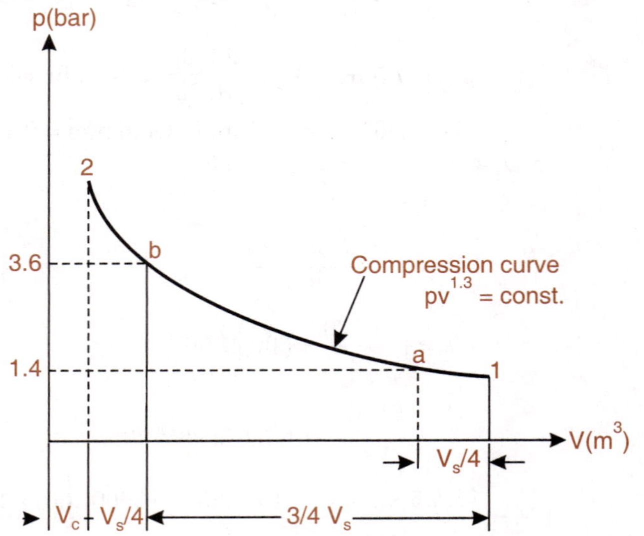

Fig. 23.79

Compression law,	$pV^{1.3}$ = constant
Pressure at 'a',	p_a = 1.4 bar
Pressure at 'b',	p_b = 3.6 bar
Volume at 'a',	$V_a = V_c + 0.75V_s$
Volume at 'b',	$V_b = V_c + 0.25V_s$
Also,	$p_aV_a^{1.3} = p_bV_b^{1.3}$

or $$\frac{V_a}{V_b} = \left(\frac{p_b}{p_a}\right)^{1/1.3} = \left(\frac{3.6}{1.4}\right)^{1/1.3} = 2.067$$

Also, $$\frac{V_a}{V_b} = \frac{V_c + 0.75V_s}{V_c + 0.25V_s} = 2.067$$

or $$(V_c + 0.75V_s) = 2.067\ (V_c + 0.25V_s)$$

or $$V_c + 0.75V_s = 2.067V_c + 0.516V_s$$

or $$0.234V_s = 1.067V_c \text{ or } \frac{V_s}{V_c} = 4.56.$$

$$\textit{Compression ratio} = \frac{V_s + V_c}{V_c} = \frac{V_s}{V_c} + 1$$

$$= 4.56 + 1 = \mathbf{5.56.}\ \textbf{(Ans.)}$$

Air standard efficiency,

$$\eta_{\text{air-standard}} = 1 - \frac{1}{(r)^{\gamma-1}} = 1 - \frac{1}{(5.56)^{1.4-1}}$$

$$= 0.496 \text{ or } 49.6\%$$

$$\eta_{\text{relative}} = \frac{\eta_{\text{thermal}}}{\eta_{\text{air-standard}}}$$

$$0.4 = \frac{\eta_{\text{thermal}}}{0.496}$$

$$\therefore\ \eta_{\text{thermal}} = 0.4 \times 0.496 = \mathbf{0.198} \text{ or } \mathbf{19.8\%}.\ \textbf{(Ans.)}$$

But, $$\eta_{\text{thermal}} = \frac{\text{I.P.}}{V_g \times C}$$

(V_g = Volume of gas used in m³/s)

$$0.198 = \frac{1}{V_g \times 18800}$$

$$\therefore\ V_g = \frac{1}{0.198 \times 18800} \text{ m}^3\text{/s}$$

$$= \frac{1}{0.198 \times 18800} \times 3600$$

$$= \mathbf{0.967\ m^3/I.P.\ hour.}\ \textbf{(Ans.)}$$

Example 23.31. *A 6-cylinder petrol engine has a volume compression ratio of 5 : 1. The clearance volume of each cylinder is 0.000115 m³. The engine consumes 10.5 kg of fuel per hour whose calorific value is 41800 kJ/kg. The engine runs at 2500 r.p.m. and the efficiency ratio is 0.65.*

Calculate the average indicated mean effective pressure developed.

Solution. The ideal cycle referred to the petrol engine working is Otto cycle.

Number of cylinder,	$n = 6$
Compression ratio,	$r = 5$
Clearance volume of each cylinder	= 0.000115 m³
Fuel consumed	= 10.5 kg/h
Calorific value of fuel,	C = 41800 kJ/kg
Engine speed,	N = 2500 r.p.m.
Efficiency ratio	= 0.65.

Mean effective pressure developed, p_m :

Air-standard efficiency in case of Otto cycle is given by

$$\eta_{\text{air-standard}} = 1 - \frac{1}{(r)^{\gamma-1}} = 1 - \frac{1}{(5)^{1.4-1}}$$

$$= 0.457 \text{ or } 47.5\%$$

Also, $$\eta_{\text{ratio}} = \frac{\eta_{\text{thermal}}}{\eta_{\text{air-standard}}}$$

$$\therefore\ \eta_{\text{thermal}} = \eta_{\text{ratio}} \times \eta_{\text{air-standard}}$$

$$= 0.65 \times 0.475 = 0.308$$

But, $$\eta_{\text{thermal(I)}} = \frac{\text{I.P.}}{\dot{m}_f \times C}$$

or $$0.308 = \frac{\text{I.P.}}{\frac{10.5}{3600} \times 41800}$$

$$\therefore \quad \text{I.P.} = \frac{0.308 \times 41800 \times 10.5}{3600}$$

$= 37.55$ kW $= 37.5 \times 10^3$ Nm/s

∴ Net work from one cycle per cylinder

$$= \frac{37.5 \times 10^3 \times 60}{6 \times (2500/2)} = 300 \text{ Nm}$$

Also, $$r = \frac{V_s + V_c}{V_c} = 5$$

$$\therefore \quad V_s + V_c = 5V_c$$

or $$V_s = 4V_c = 4 \times 0.000115 = 0.00046 \text{ m}^3$$

∴ *Mean affective pressure developed*

$$p_m = \frac{\text{W}_{\text{net}} \text{ per cycle}}{V_s} = \frac{300}{0.00046 \times 10^5} \text{ bar}$$

$=$ **6.52 bar.** **(Ans.)**

Example 23.32. *An engine is required to develop 100 kW, the mechanical efficiency of the engine is 86% and the engine uses 55 kg/h of fuel. Due to improvement in the design and operating conditions, there is reduction in engine friction to the extent of 4.8 kW. If the indicated thermal efficiency remains the same, determine the saving in fuel in kg/h.*

(AMIE)

Solution. Required brake power = 100 kW

$$\text{Indicated power} = \frac{100}{0.86} = 116.28 \text{ kW}$$

$$(s.f.c.)_I = \frac{55}{116.25} = 0.473 \text{ kg/kWh}$$

Friction power = 116.28 – 100 = 16.28 kW.

Given that indicated thermal efficiency remains same after improvement, the $(s.f.c.)_I$ also remains the same.

After improvement :

Brake power	= 100 kW
Friction power	= 16.28 – 4.8 = 11.48 kW
Indicated power	= 100 + 11.48 = 111.48 kW
Fuel consumption	= 111.48 × 0.473 = 52.73 kg/h
Saving in fuel	= 55 – 52.73 = 2.27 kg/h

or $$\frac{2.27}{55} \times 100 = \mathbf{4.127\%}. \quad \textbf{(Ans.)}$$

☞**Example 23.33.** *A 2-cylinder C.I. engine with a compression ratio 13 : 1 and cylinder dimensions of 200 mm × 250 mm works on two-stroke cycle and consumes 14 kg/h of fuel while running at 300 r.p.m. The relative and mechanical efficiencies of engine are 65% and 76% respectively. The fuel injection is effected upto 5% of stroke. If the calorific value of the fuel used is given as 41800 kJ/kg, calculate the mean effective pressure developed.*

Solution. Refer to Fig. 23.80.

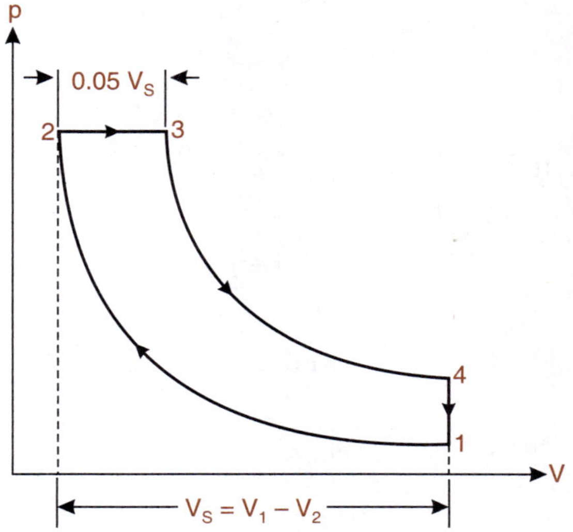

Fig. 23.80

Diameter of cylinder,	$D = 200$ mm $= 0.2$ m
Stroke length,	$L = 250$ mm $= 0.25$ m
Number of cylinders,	$n = 2$
Compression ratio,	$r = 13$
Fuel consumption	= 14 kg/h
Engine speed,	$N = 300$ r.p.m.
Relative efficiency,	$\eta_{relative} = 65\%$
Mechanical efficiency,	$\eta_{mech.} = 76\%$
Cut-off	= 5% of stroke
Calorific value of fuel,	$C = 41800$ kJ/kg

$k = 1$ ...... for two-stroke cycle engine

Cut-off ratio, $$\rho = \frac{V_3}{V_2}$$

Also, $$V_3 - V_2 = 0.05V_s = 0.05(V_1 - V_2)$$

or $$V_3 - V_2 = 0.05\,(13V_2 - V_2) \quad \left[\because \frac{V_1}{V_2} = r = 13\right]$$

or $$V_3 - V_2 = 0.05 \times 12V_2 = 0.6V_2.$$

$$\therefore \quad \frac{V_3}{V_2} = \rho = 1.6$$

$$\eta_{\text{air-standard}} = 1 - \frac{1}{\gamma(r)^{\gamma-1}}\left[\frac{\rho^\gamma - 1}{\rho - 1}\right]$$

$$= 1 - \frac{1}{1.4(14)^{1.4-1}}\left[\frac{1.6^{1.4} - 1}{1.6 - 1}\right]$$

$= 1 - 0.248 \times 1.55$

$= 0.615\%$ or 61.5%

Also, $$\eta_{relative} = \frac{\eta_{thermal}}{\eta_{air\text{-}standard}}$$

$$0.65 = \frac{\eta_{thermal}}{0.615}$$

$\therefore$ $$\eta_{thermal} = 0.65 \times 0.615 = 0.4$$

But, $$\eta_{thermal\ (I)} = \frac{\text{I.P.}}{\dot{m}_f \times C}$$

$$0.4 = \frac{\text{I.P.}}{\frac{14}{3600} \times 41800}$$

$\therefore$ $$\text{I.P.} = \frac{0.4 \times 14 \times 41800}{3600} = 65 \text{ kW}$$

Now, $$\eta_{mech.} = \frac{\text{B.P.}}{\text{I.P.}}$$

$$0.76 = \frac{\text{B.P.}}{65}$$

$\therefore$ $$\text{B.P.} = 0.76 \times 65 = 49.4 \text{ kW}$$

Mean effective pressure can be calculated based on I.P. or B.P. of the engine.

$$\text{I.P.} = \frac{np_{mi}LANk \times 10}{6}$$

where, p_{mi} = Indicated mean effective pressure,

$$65 = \frac{2 \times p_{mi} \times 0.25 \times \pi / 4 \times 0.2^2 \times 300 \times 1 \times 10}{6}$$

$$\therefore p_{mi} = \frac{65 \times 6 \times 4}{2 \times 0.25 \times \pi \times 0.2^2 \times 300 \times 10}$$

= **8.27 bar. (Ans.)**

and brake mean effective pressure,

p_{mb} = 0.76 × 8.27 = **6.28 bar. (Ans.)**

☞Example 23.34. *Following data relate to 4-cylinder four-stroke petrol engine. Air-fuel ratio by weight = 16 : 1, calorific value of the fuel = 45200 kJ / kg, mechanical efficiency = 82%, air-standard efficiency = 52%, relative efficiency = 70%, volumetric efficiency = 78%, stroke / bore ratio = 1.25, suction conditions = 1 bar 25°C, r.p.m. = 2400 and power at brakes = 72 kW.*

Calculate : (i) Compression ratio

(ii) Indicated thermal efficiency

(iii) Brake specific fuel consumption

(iv) Bore and stroke.

Solution. Air fuel ratio by weight = 16 : 1

No. of cylinders, $n = 4$

Calorific value of fuel, C = 45200 kJ/kg

Mechanical efficiency, $\eta_{mech.}$ = 82%

Air-standard efficiency, $\eta_{air\text{-}standard}$ = 52%

Relative efficiency, $\eta_{relative}$ = 70%

Volumetric efficiency, $\eta_{vol.}$ = 78%

Stroke/bore ratio, = 1.25

Engine speed, N = 2400 r.p.m.

Suction conditions p = 1 bar,

T = 25 + 273 = 298 K

Stroke/bore ratio = 1.25

Brake power, B.P. = 72 kW.

(i) Compression ratio, r :

For petrol engine, air standard efficiency is given by :

$$\eta_{air\text{-}standard} = 1 - \frac{1}{(r)^{\gamma-1}}$$

$$0.52 = 1 - \frac{1}{(r)^{1.4-1}} \text{ or } \frac{1}{(r)^{1.4-1}} = 0.48$$

or $$(r)^{0.4} = \frac{1}{0.48} = 2.08$$

or $$r = (2.08)^{1/0.4} = (2.08)^{2.5} = 6.2$$

i.e., Compression ratio = **6.2. (Ans.)**

(ii) Indicated thermal efficiency, $\eta_{th(I)}$:

$$\eta_{relative} = \frac{\eta_{thermal}}{\eta_{air\text{-}standard}} \text{ or } 0.7 = \frac{\eta_{thermal}}{0.52}$$

$\therefore$ $\eta_{thermal\ (I)}$ = 0.7 × 0.52 = 0.364 or 36.4%

i.e., Indicated thermal efficiency = **36.4%. (Ans.)**

(iii) Brake specific fuel consumption (b.s.f.c.) :

$$\text{Indicated power, I.P.} = \frac{\text{B.P.}}{\eta_{mech.}} = \frac{72}{0.82} = 87.8 \text{ kW}$$

Also, $$\eta_{th(I)} = \frac{\text{I.P.}}{\dot{m}_f \times C}$$

$$0.364 = \frac{87.8}{\dot{m}_f \times 45200},$$

where $\dot{m}_f$ = Fuel used in kg/s

$\therefore$ $$\dot{m}_f = \frac{87.8}{0.364 \times 45200} = 0.00533 \text{ kg/s}$$

Brake specific fuel consumption,

$$\text{b.s.f.c.} = \frac{\text{Fuel used / sec.}}{\text{B.P.}} = \frac{0.00533}{72} \text{ kg/kWs}$$

$$= \frac{0.00533}{72} \times 3600 \text{ kg/kWh}$$

= **0.2665 kg/kWh. (Ans.)**

(iv) Bore and stroke :

Mass of air-fuel mixture = 1 + 16 = 17 kg/kg of fuel

$\therefore$ For 0.00533 kg/s of fuel supplied to engine the mass of air-fuel mixture

= 17 × 0.00533 = 0.0906 kg/s

$\therefore$ Volume of air-fuel mixture supplied to the engine per sec.

$$\frac{mRT}{P} = \frac{0.0906 \times 287 \times (25 + 273)}{1 \times 10^5} = 0.07748 \text{ m}^3\text{/s}$$

$$\eta_{vol.} = \frac{\text{Mass of mixture supplied / sec.}}{\text{Swept volume}}$$

$$0.78 = \frac{0.07748}{\text{Swept volume}}$$

∴ Swept volume

$$= \frac{0.07748}{0.78} = 0.0993 \text{ m}^3\text{/s}$$

But swept volume/sec.

$$= \left(\frac{\pi}{4} D^2 \times L\right) \times \text{no. of cylinders} \times \frac{\text{r.p.m.}}{2 \times 60}$$

$$= \frac{\pi}{4} \times D^2 \times 1.25D \times 4 \times \frac{2400}{2 \times 60} = 0.0993$$

$$\therefore \quad D^3 = \frac{0.0993 \times 4 \times 2 \times 60}{\pi \times 1.25 \times 4 \times 2400} = 0.001264$$

∴ D = **0.108 m** or **108 mm.** **(Ans.)**

and $L = 108 \times 1.25$ = **135 mm.** **(Ans.)**

☞**Example 23.35.** *A single-cylinder four-stroke gas engine has a bore of 180 mm and stroke of 340 mm and is governed on hit-and-miss principle. When running at 400 r.p.m. at full load, indicator cards are taken which give a working loop mean effective pressure of 6.4 bar, and a pumping loop mean effective pressure of 0.36 bar. Diagrams from the dead cycle give a mean effective pressure of 0.64 bar. The engine was run light at the same speed (i.e., with no load), and a mechanical counter recorded 46 firing strokes per minute.*

Calculate : (i) Full load brake power.

(ii) Mechanical efficiency of the engine. **(P.U.)**

Solution. Number of cylinders, $n = 1$

Bore, D = 180 mm = 0.18 m

Stroke, L = 340 mm = 0.34 m

Engine speed, N = 400 r.p.m.

Working loop mean effective pressure = 6.4 bar

Pumping loop mean effective pressure = 0.36 bar

Mean effective pressure (dead cycle) = 0.64 bar

Firing strokes/min. = 46

Refer to Fig. 23.81.

(*i*) Full load brake power, B.P. :

Net indicated mean effective pressure,

$p_{mi\ (net)}$ = Working (or power) loop mean effective pressure – pumping loop mean effective pressure

= 6.4 – 0.36 = 6.04 bar

Also, working cycles/min = 46

and, dead cycles/min $= \left(\frac{400}{2} - 46\right) = 154$

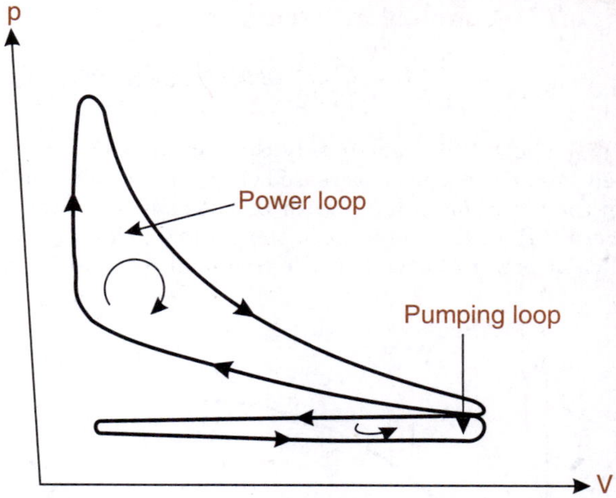

Fig. 23.81

Therefore, since there is no brake power output,

Frictional power, F.P. = (net I.P.) – (pumping power of dead cycles) ...(*i*)

Now,

$$\text{Net I.P.} = \frac{np_{mi}(\text{net}) \times LANk \times 10}{6}$$

$$= \frac{1 \times 6.04 \times 0.34 \times \pi/4 \times 0.18^2 \times 46 \times 10}{6} \quad [\because \ Nk = 46]$$

= 4 kW

Pumping power of dead cycles

$$= \frac{np_{mi(d)}\, LANk \times 10}{6}$$

$$= \frac{1 \times 0.64 \times 0.34 \times \pi/4 \times 0.18^2 \times 154 \times 10}{6} \quad [\because \ Nk = 154]$$

= 1.42 kW

Substituting the above values in eqn. (*i*), we get

F.P. = 4 – 1.42 = 2.58 kW

At full load the engine fires regularly every two revolutions, and there are $\frac{400}{2} = 200$ firing strokes per minute.

$$\therefore \text{ I.P.} = \frac{n\, p_{mi(net)}\, LANk \times 10}{6}$$

$$= \frac{1 \times 6.04 \times 0.34 \times \pi/4 \times 0.18^2 \times 200 \times 10}{6} \quad \left[Nk = 400 \times \frac{1}{2} = 200\right]$$

= 17.42 kW

Hence brake power,

B.P. = (I.P. – F.P.) = 17.42 – 2.58 = **14.84 kW.** **(Ans.)**

(*ii*) **Mechanical efficiency, $\eta_{mech.}$:**

$$\eta_{mech.} = \frac{\text{B.P.}}{\text{I.P.}} = \frac{14.84}{17.42} = \mathbf{0.852} \text{ or } \mathbf{85.2\%}. \quad \textbf{(Ans.)}$$

Note. The F.P. is very nearly constant at a given engine speed ; and if the load is decreased giving lower values of B.P., then the variation in $h_{mech.}$ with B.P. is shown in Fig. 23.82. At zero B.P. at the same speed the engine is developing just sufficient power to overcome the frictional resistance.

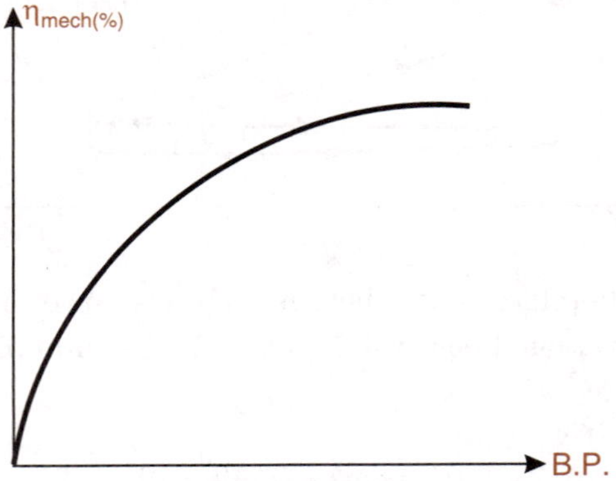

Fig. 23.82

Example 23.36. *During the trial of a gas engine following observations were recorded :*

Bore	*= 320 mm*
Stroke	*= 420 mm*
Speed	*= 200 r.p.m.*
Number of explosions/min.	*= 90*
Gas used	*= 11.68 m³/h*

Pressure of gas = 170 mm of water above atmospheric pressure

Barometer	*= 755 mm (mercury)*
Mean effective pressure	*= 6.2 bar*
Calorific value of gas used	*= 21600 kJ/kg at N.T.P.*
Net load on brake	*= 2040 N*
Brake drum diameter	*= 1.2 m*
Ambient temperature	*= 25°C*

Calculate : (i) Mechanical efficiency, and (ii) Brake thermal efficiency.

Solution. *Given :* $n = 1$; $D = 0.32$ m ;

$L = 0.42$ m ; $N = 200$ r.p.m.,

$$Nk = 90 \;;\; V_g = \frac{11.68}{3600} = 0.00324 \text{ m}^3/\text{s}$$

$$\text{Pressure of gas} = 755 + \frac{170}{13.6} = 767.5 \text{ mm Hg}$$

$p_{mi} = 6.2$ bar, $C = 21600$ kJ/kg at N.T.P.

$(W - S) = 1840$ N, $D_b = 1$ m.

(*i*) **Mechanical efficiency :**

As the number of explosions per minute is given as 90 per minute and r.p.m. of engine is 200 it shows that the *engine is operating on four-stroke cycle.*

Indicated power (I.P.) is given by the relation :

$$\text{I.P.} = \frac{n p_{mi} LANk \times 10}{6}$$

$$= \frac{1 \times 6.2 \times 0.42 \times \pi / 4 \times 0.32^2 \times 90 \times 10}{6} \qquad [\because \; Nk = 90]$$

$= 31.4$ kW

$$\text{B.P.} = \frac{(W - S)\, \pi\, D_b N}{60 \times 1000} = \frac{2040 \times \pi \times 1.2 \times 200}{60 \times 1000}$$

$= 25.6$ kW

∴ *Mechanical efficiency,*

$$\eta_{mech.} = \frac{\text{B.P.}}{\text{I.P.}} = \frac{25.6}{31.4} = \mathbf{0.815} \text{ or } \mathbf{81.5\%}. \quad \textbf{(Ans.)}$$

(*ii*) **Brake thermal efficiency :**

Volume of gas at N.T.P. :

$p_1 = 767.5$ mm Hg, $V_1 = 11.68$ m³/h,

$T_1 = 25 + 273 = 298$ K

$p_2 = 760$ mm Hg, $T_2 = 0 + 273 = 273$ K

Now, to find V_2, using the relation :

$$\frac{p_1 V_1}{T_1} = \frac{p_2 V_2}{T_2}$$

$$V_2 = \frac{p_1 V_1 T_2}{p_2 T_1} = \frac{767.5 \times 11.68 \times 273}{760 \times 298}$$

$= 10.8$ m³/h

Brake thermal efficiency,

$$\eta_{th.(B)} = \frac{\text{B.P.}}{V_g \times C} \qquad [V_g = \text{Volume of gas in m}^3/\text{s}]$$

$$= \frac{25.6}{\frac{10.8}{3600} \times 21600} = \mathbf{0.395} \text{ or } \mathbf{39.5\%}. \quad \textbf{(Ans.)}$$

☞**Example 23.37.** *Air consumption for a four-stroke petrol engine is measured by means of a circular orifice of diameter 3.2 cm. The co-efficient of discharge for the orifice is 0.62 and the pressure across the orifice is 150 mm of water. The barometer reads 760 mm of Hg. Temperature of air in the room is 20°C. The piston displacement volume is 0.00178 m³. The compression ratio is 6.5. The fuel consumption is 0.135 kg/min of calorific value 43900 kJ/kg. The brake power developed at 2500 r.p.m. is 28 kW. Determine :*

(*i*) *The volumetric efficiency on the basis of air alone.*

(*ii*) *The air-fuel ratio.*

(*iii*) *The brake mean effective pressure.*

(*iv*) *The relative efficiency on the brake thermal efficiency basis.*

Solution. Diameter of circular orifice,

$d = 3.2$ cm $= 0.032$ m

Co-efficient of discharge, $C_d = 0.62$

Pressure across orifice, h_w = 150 mm of water
Temperature of air in the room = 20°C
Piston displacement = 0.00178 m^3
Compression ratio, $r = 6.5$
Fuel consumption = 0.135 kg/min
Calorific value of fuel, C = 43900 kJ/kg
Brake power, B.P. = 28 kW
Speed = 2500 r.p.m.

$$k = \frac{1}{2} \text{ .... for 4-stroke cycle, engine.}$$

(*i*) Volumetric efficiency on the basis of air alone :

Characteristic gas equation is written as,

$$pV = mRT$$

or

$$\frac{m}{V} = \frac{p}{RT} = \frac{1.0132 \times 10^5}{287 \times (20 + 273)} = 1.2 \text{ kg/m}^3$$

Also, 150 mm of $H_2O = \frac{150}{1000} \times 1000 = 150 \text{ kg/m}^2$

Thus head of air column causing flow,

$$H = \frac{150}{1.2} = 125 \text{ m}$$

Thus air flow through the orifice

$$= \text{Air consumption} = C_d \times A \times \sqrt{2gH}$$

$$= 0.62 \times \frac{\pi}{4} \times (0.032)^2 \times \sqrt{2 \times 9.81 \times 125}$$

$$= 0.0247 \text{ m}^3\text{/s}$$

Therefore, air consumption per stroke

$$= \frac{0.0247 \times 60}{\left(\frac{2500}{2}\right)} = 0.001185 \text{ m}^3$$

∴ Volumetric efficiency,

$$\eta_{vol.} = \frac{\text{Air consumption of stroke}}{\text{Piston displacement}}$$

$$= \frac{0.001185}{0.00178} = \mathbf{0.665 \text{ or } 66.5\%.} \quad \textbf{(Ans.)}$$

(*ii*) Air-fuel ratio :

Mass of air drawn into the cylinder per min.

$$= 0.0247 \times 60 \times 1.2 = 1.778 \text{ kg/min}$$

$$\therefore \quad \text{Air-fuel ratio} = \frac{1.778}{0.135} = \mathbf{13.67 : 1.} \quad \textbf{(Ans.)}$$

(*iii*) Brake mean effective pressure, p_{mb} :

$$\text{B.P.} = \frac{n \times p_{mb} LANk \times 10}{6}$$

$$28 = \frac{1 \times p_{mb} \times 0.00178 \times 2500 \times \frac{1}{2} \times 10}{6}$$

$$[\because \quad LA = 0.00178 \text{ m}^3]$$

$$\therefore \quad p_{mb} = \frac{28 \times 6 \times 2}{0.00178 \times 2500 \times 10} = \mathbf{7.55 \text{ bar.}} \quad \textbf{(Ans.)}$$

(*iv*) Relative efficiency :

$$\eta_{\text{air-standard}} = 1 - \frac{1}{(r)^{\gamma-1}} = 1 - \frac{1}{(6.5)^{1.4-1}}$$

$$= 0.527 \text{ or } 52.7\%$$

Brake thermal efficiency,

$$\eta_{th.(B)} = \frac{\text{B.P.}}{\dot{m}_f \times C} = \frac{28}{\frac{0.135}{60} \times 43900}$$

$$= 0.2835 \text{ or } 28.35\%$$

$$\therefore \quad \eta_{\text{relative}} = \frac{\eta_{\text{thermal (B)}}}{\eta_{\text{air-standard}}} = \frac{0.2835}{0.527}$$

$$= \mathbf{0.5379 \text{ or } 53.79\%.} \quad \textbf{(Ans.)}$$

Example 23.38. *A six-cylinder, four-stroke CI engine is tested against a water brake dynamometer for which B.P. = WN/17 × 10^3 in kW, where W is the brake load in newton and N is the speed of the engine in the r.p.m. The air consumption was measured by means of a sharp edged orifice. During the test following observations were taken :*

Bore = 10 cm
Stroke = 14 cm
Speed = 2500 r.p.m.
Brake load = 480 N
Barometer reading = 76 cm of Hg
Orifice diameter = 3.3 cm
Co-efficient of discharge of orifice = 0.62
Pressure drop across orifice = 14 cm of Hg
Room temperature = 25°C
Fuel consumption = 0.32 kg/min.

Calculate the following :

(i) The volumetric efficiency ; (ii) The brake mean effective pressure (b.m.e.p.) ; (iii) The engine torque ; (iv) The brake specific fuel consumption (b.s.f.c.).

Solution. (*i*) Volumetric efficiency, η_{vol} :

V_s = Swept volume,

$$= \frac{\pi}{4} D^2 L \times \frac{N}{60 \times 2} \times \text{No. of cylinders,}$$

for 4-stroke. (where N = r.p.m.)

$$= \frac{\pi}{4} (0.1)^2 \times 0.14 \times \frac{2500}{60 \times 2} \times 6 = 0.137 \text{ m}^3\text{/s.}$$

Barometer

$$= 76 \text{ cm Hg} = \left[\frac{76}{100} \times 13.6 \times 10^3 \times 9.81\right] \times 10^{-3}$$

$$= 101.4 \text{ kN/m}^2$$

$$\rho_a = \frac{p}{R_a T} = \frac{101.3}{0.287\,(273 + 25)} = 1.1844 \text{ kg/m}^3$$

$$\Delta p = 14 \text{ cm of Hg} = \frac{14}{100} \times 13.6 \times 10^3 \times 9.81$$
$$= 18.678 \times 10^3 \text{ N/m}^2$$
$$\Delta p = \rho_a \times 9.81 \times h_a,$$

where, h_a = Head, m of air, causing flow

or $$h_a = \frac{18.678 \times 10^3}{1.1844 \times 9.81} = 1607.5 \text{ m of air}$$

V_a = Volume flow rate of air, at free air conditions

$$= C_d \frac{\pi}{4}(d_0)^2 \sqrt{2gh_a}$$
$$= 0.62 \times \frac{\pi}{4}\left(\frac{3.3}{100}\right)^2 \sqrt{2 \times 9.81 \times 1607.5}$$
$$= 0.094 \text{ m}^3/\text{s}$$

$$\%\eta_{vol} = \frac{V_a}{V_s} \times 100 = \frac{0.094}{0.137} \times 100 = \mathbf{68.6\%.} \quad \textbf{(Ans.)}$$

(*ii*) **The brake mean effective pressure p_{mb} :**

$$B.P. = \frac{WN}{17} \times 10^{-3} \text{ kW}$$
$$= \frac{480 \times 2500}{17} \times 10^{-3} = 70.588 \text{ kW}$$
$$= p_{mb} LA \times \frac{N}{60} \times \frac{1}{2} \times 6, \text{ for six-cylinder, four-stroke engine}$$

or $$p_{mb} = \frac{70.588 \times 60 \times 2}{0.14 \times \frac{\pi}{4}(0.1)^2 \times 2500 \times 6}$$
$$= \mathbf{513.57 \text{ kN/m}^2.} \quad \textbf{(Ans.)}$$

(*iii*) **Engine torque, T :**

$$\text{B.P.} = 2\pi NT$$

or $$\text{Torque, } (T) = \frac{\text{B.P.}}{2\pi N} = \frac{70.588 \times 10^3}{2\pi \times \frac{2500}{60}}$$
$$= \mathbf{269.63 \text{ Nm.}} \quad \textbf{(Ans.)}$$

(*iv*) **Brake specific fuel consumption, b.s.f.c. :**

$$\text{b.s.f.c.} = \frac{m_f \text{ (kg/h)}}{\text{B.P.}} = \frac{0.32 \times 60}{70.588}$$
$$= \mathbf{0.272 \text{ kg/kWh.}} \quad \textbf{(Ans.)}$$

☞**Example 23.39.** *A single-cylinder 4-stroke diesel engine gave the following results while running on full load :*

Area of indicator card	*= 300 mm²*
Length of diagram	*= 40 mm*
Spring constant	*= 1 bar/mm*
Speed of the engine	*= 400 r.p.m.*
Load on the brake	*= 370 N*
Spring balance reading	*= 50 N*
Diameter of brake drum	*= 1.2 m*
Fuel consumption	*= 2.8 kg/h*
Calorific value of fuel	*= 41800 kJ/kg*
Diameter of the cylinder	*= 160 mm*
Stroke of the piston	*= 200 mm*

Calculate : (i) Indicated mean effective pressure ; (ii) Brake power and brake mean effective pressure ; (iii) Brake specific fuel consumption, brake thermal and indicated thermal efficiencies.

Solution. *Given :* $N = 400$ r.p.m. ; $W = 370$ N, ; $S = 50$ N ; $D_b = 1.2$ m ; $m_f = 2.8$ kg/h ; $C = 41800$ kJ/kg ; $D = 0.16$ m ; $L = 0.2$ m ;

$$k = \frac{1}{2} \text{ ...... for 4-stroke cycle engine.}$$

(*i*) **Indicated mean effective pressure, p_{mi} :**

$$p_{mi} = \frac{\text{Area of indicator diagram or card} \times \text{spring costant}}{\text{Length of diagram}}$$
$$= \frac{300 \times 1}{40} = \mathbf{7.5 \text{ bar.}} \quad \textbf{(Ans.)}$$

Indicated power,

$$\text{I.P.} = \frac{np_{mi}LANk \times 10}{6}$$
$$= \frac{1 \times 7.5 \times 0.2 \times \pi/4 \times 0.16^2 \times 400 \times \frac{1}{2} \times 10}{6}$$
$$= 10.05 \text{ kW.}$$

(*ii*) **B.P. : , p_{mb} :**

$$\text{Brake power, B.P.} = \frac{(W - S)\,\pi\, D_b N}{60 \times 1000}$$
$$= \frac{(370 - 50)\pi \times 1.2 \times 400}{60 \times 1000}$$
$$= \mathbf{8.04 \text{ kW.}} \quad \textbf{(Ans.)}$$

Also, $$\text{B.P.} = \frac{np_{mb} \times LANk \times 10}{6}$$

$$8.04 = \frac{1 \times p_{mb} \times 0.2 \times \pi/4 \times 0.16^2 \times 400 \times \frac{1}{2} \times 10}{6}$$

$$\therefore \quad p_{mb} = \frac{8.04 \times 6 \times 4 \times 2}{0.2 \times \pi \times 0.16^2 \times 400 \times 10} = \mathbf{6 \text{ bar.}} \quad \textbf{(Ans.)}$$

(*iii*) **b.s.f.c. :, $\eta_{th(B)}$:, $\eta_{th.(I)}$:**

Brake specific fuel consumption,

b.s.f.c = Fuel consumption per B.P. hour

$$= \frac{2.8}{8.04} = \mathbf{0.348 \text{ kg/B.P. hour.}} \quad \textbf{(Ans.)}$$

Brake thermal efficiency,

$$\eta_{th.(B)} = \frac{\text{B.P.}}{\dot{m}_f \times C} = \frac{8.04}{\frac{2.8}{3600} \times 41800}$$
$$= \mathbf{0.2473} \text{ or } \mathbf{24.73\%.} \quad \textbf{(Ans.)}$$

Indicated thermal efficiency,

$$\eta_{th.(I)} = \frac{B.P.}{\dot{m}_f \times C} = \frac{10.05}{\frac{2.8}{3600} \times 41800}$$

$= \mathbf{0.3091}$ or **30.91%. (Ans.)**

Example 23.40. *A 4-cylinder, four-stroke cycle engine, 82.5 mm bore × 130 mm stroke develops 28 kW while running at 1500 r.p.m. and using a 20 per cent rich mixture. If the volume of the air in the cylinder when measured at 15.5°C and 762 mm of mercury is 70 per cent of the swept volume, the theoretical air-fuel ratio is 14.8, heating value of petrol used is 45980 kJ/kg and the mechanical efficiency of the engine is 90%, find :*

(i) The indicated thermal efficiency.

(ii) The brake mean effective pressure.

Take R = 287 Nm/kg K.

Solution. Number of cylinders, $n = 4$

Engine bore,	$D = 0.0825$ m
Stroke length,	$L = 0.13$ m
Brake power,	B.P. = 28 kW
Engine speed,	$N = 1500$ r.p.m.
Theoretical air-fuel ratio	= 14.8
Calorific value of fuel,	$C = 45980$ kJ/kg
Mechanical efficiency,	$\eta_{mech.} = 90\%$.

(i) Indicated thermal efficiency, $\eta_{th.(I)}$:

Swept volume, $V_s = \pi/4\, D^2L = \pi/4 \times 0.0825^2 \times 0.13$

$= 0.000695 \text{ m}^3$

$$\text{Volume of air drawn in} = \frac{70}{100} \times 0.000695$$

$= 0.0004865 \text{ m}^3$

Given : $p = \frac{762}{760} \times 1.0132 = 1.015$ bar

$V = 0.0004865 \text{ m}^3$ (calculated above)

$R = 287$ Nm/kg K

$T = 15.5 + 273 = 288.5$ K

∴ m = Mass of air/stroke/cylinder

$$= \frac{pV}{RT} = \frac{1.015 \times 10^5 \times 0.0004865}{287 \times 288.5}$$

$= 0.000596$ kg

Theoretical mass of air used per minute

$$= 0.000596 \times \frac{1500}{2} \times 4 = 1.788 \text{ kg}$$

Theoretical air-fuel ratio = 14.8

∴ Theoretical mass of fuel used/min

$$= \frac{1.788}{14.8} = 0.1208 \text{ kg/min}$$

When using 20% rich mixture, then

$\dot{m}_f$ = Mass of fuel burnt/sec

$$= \frac{0.1208}{60} \times \frac{120}{100} = 0.002416 \text{ kg/s}$$

$$\text{Now, } \eta_{mech.} = \frac{B.P.}{I.P.} = \frac{28}{I.P.}$$

$$0.9 = \frac{28}{I.P.}$$

$$\therefore \quad I.P. = \frac{28}{0.9} = 31.11 \text{ kW.}$$

Indicated thermal efficiency,

$$\eta_{th.(I)} = \frac{I.P.}{\dot{m}_f \times C} = \frac{31.11}{0.002416 \times 45980}$$

$= \mathbf{0.28}$ or **28%. (Ans.)**

(ii) Brake mean effective pressure, p_{mb} :

$$B.P. = \frac{n \times p_{mb} LANk \times 10}{6}$$

$$28 = \frac{4 \times p_{mb} \times 0.13 \times \pi/4 \times 0.0825^2 \times 1500 \times \frac{1}{2} \times 10}{6}$$

$$\therefore \quad p_{mb} = \frac{28 \times 6 \times 4 \times 2}{4 \times 0.13 \times \pi \times 0.0825^2 \times 1500 \times 10}$$

$= \mathbf{8.06}$ **bar. (Ans.)**

Example 23.41. *During the test of 40 minutes on a single-cylinder gas engine of 200 mm cylinder bore and 400 mm stroke, working on the four-stroke cycle and governed by hit and miss method of governing, the following readings were taken :*

Total number of revolutions	*= 9400*
Total number of explosions	*= 4200*
Area of indicator diagram	*= 550 mm²*
Length of indicator diagram	*= 72 mm*
Spring number	*= 0.8 bar/mm*
Brake load	*= 540 N*
Brake wheel diameter	*= 1.6 m*
Brake rope diameter	*= 2 cm*
Gas used	*= 8.5 m³*
Calorific value of gas	*= 15900 kJ/m³*

Calculate :

(i) Indicated power,

(ii) Brake power, and

(iii) Indicated and brake thermal efficiencies.

Solution. *Given :* $D = 0.2$ m ; $L = 0.4$ m ;

$$N_t = 9400 \text{ r.p.m.} ; Nk = \frac{4200}{40} = 105 ;$$

$(W - S) = 540$ N ; $D_b = 1.6$ m ; $d = 0.02$ m ;

$$V_g = \frac{8.5}{40 \times 60} = 0.00354 \text{ m}^3/\text{s}.$$

(*i*) Indicated power, I.P. :

Indicated mean effective pressure.

$$p_{mi} = \frac{\text{Area of indicator diagram} \times \text{spring number}}{\text{Length of the diagram}}$$

$$= \frac{550 \times 0.8}{72} = 6.11 \text{ bar}$$

$$\text{I.P.} = \frac{np_{mi}LANk \times 10}{6}$$

$$= \frac{1 \times 6.11 \times 0.4 \times \pi/4 \times 0.2^2 \times 105 \times 10}{6}$$

= **13.4 kW. (Ans.)**

(*ii*) Brake power, B.P. :

$$\text{B.P.} = \frac{(W - S)\,\pi\,(D_b + d)N}{60 \times 1000}$$

$$= \frac{540 \times \pi\,(1.6 + 0.02) \times \left(\frac{9400}{40}\right)}{60 \times 1000}$$

= **10.76 kW. (Ans.)**

(*iii*) Indicated thermal efficiency :

$$\eta_{th.(I)} = \frac{\text{I.P.}}{V_g \times C} = \frac{13.4}{0.00354 \times 15900}$$

= **0.238** or **23.8%. (Ans.)**

Brake thermal efficiency,

$$\eta_{th.(B)} = \frac{\text{B.P.}}{V_g \times C} = \frac{10.76}{0.00354 \times 15900}$$

= **0.191** or **19.1%. (Ans.)**

☞**Example 23.42.** *The following observations were recorded during the test on a 6-cylinder, 4-stroke Diesel engine :*

Bore	*= 125 mm*
Stroke	*= 125 mm*
Engine speed	*= 2400 r.p.m.*
Load on dynamometer	*= 490 N*
Dynamometer constant	*= 16100*
Air orifice diameter	*= 55 mm*
Co-efficient of discharge	*= 0.66*
Head causing flow through orifice	*= 310 mm of water*
Barometer reading	*= 760 mm Hg*
Ambient temperature	*= 25°C*
Fuel consumption	*= 22.1 kg/h*
Calorific value of fuel	*= 45100 kJ/kg*
Per cent carbon in the fuel	*= 85%*
Per cent hydrogen in the fuel	*= 15%*
Pressure of air at the end of suction stroke	*= 1.013 bar*
Temperature at the end of suction stroke	*= 25°C*

Calculate :

(i) Brake mean effective pressure,

(ii) Specific fuel consumption,

(iii) Brake thermal efficiency,

(iv) Volumetric efficiency, and

(v) Percentage of excess air supplied.

Solution. $n = 6$, $D = 0.125$ m, $L = 0.125$ m,

$N = 2400$ r.p.m.

$W = 490$ N, C_D = dynamometer constant = 16100

d_0 = orifice diameter = 0.055 m, $C_d = 0.66$,

$h_w = 310$ mm

$$\dot{m}_f = \frac{22.1}{3600} = 0.00614 \text{ kg/s}, \; C = 45100 \text{ kJ/kg},$$

$$k = \frac{1}{2} \text{ ..... for 4-stroke cycle engine.}$$

(*i*) Brake mean effective pressure, p_{mb} :

$$\text{Brake power, B.P.} = \frac{W \times N}{C_D} = \frac{490 \times 2400}{16100} = 73 \text{ kW}$$

Also,

$$\text{B.P.} = \frac{np_{mb}LANk \times 10}{6}$$

$$73 = \frac{6 \times p_{mb} \times 0.125 \times \pi/4 \times 0.125^2 \times 2400 \times \frac{1}{2} \times 10}{6}$$

$$\therefore \quad p_{mb} = \frac{76 \times 6 \times 4 \times 2}{6 \times 0.125 \times \pi \times 0.125^2 \times 2400 \times 10}$$

= **3.96 bar. (Ans.)**

(*ii*) Specific fuel consumption, b.s.f.c. :

$$\text{b.s.f.c.} = \frac{22.1}{73} = \mathbf{0.3027 \text{ kg/kWh.}} \quad \textbf{(Ans.)}$$

(*iii*) Brake thermal efficiency, $\eta_{th.(B)}$:

$$\eta_{th.(B)} = \frac{\text{B.P.}}{\dot{m}_f \times C} = \frac{73}{0.00614 \times 45100}$$

= **0.2636** or **26.36%. (Ans.)**

(*iv*) Volumetric efficiency, $\eta_{vol.}$:

Stroke volume of cylinder = $\pi/4\ D^2 \times L$

$= \pi/4 \times 0.125^2 \times 0.125 = 0.00153$ m^3

The volume of air passing through the orifice of the air box per minute is given by,

$$V_a = 840\, A_0\, C_d \sqrt{\frac{h_w}{\rho_a}}$$

where, C_d = Discharge co-efficient of orifice = 0.66,

A_0 = Area of cross-section of orifice,

$= \pi/4\ d_0^2 = \pi/4 \times (0.055)^2 = 0.00237$ m^2,

h_w = Head causing flow through orifice in cm of water,

$$= \frac{310}{10} = 31 \text{ cm, and}$$

ρ_a = Density of air at 1.013 bar and 25°C

$$= \frac{P}{RT} = \frac{1.013 \times 10^5}{287 \times (25 + 273)} = 1.18 \text{ kg/m}^3.$$

$$\therefore \quad \text{Volume of air, } V_a = 840 \times 0.00237 \times 0.66 \sqrt{\frac{31}{1.18}}$$

$$= 6.73 \text{ m}^3\text{/min}$$

∴ Actual volume of air per cylinder

$$= \frac{6.73}{n} = \frac{6.73}{6} = 1.12 \text{ m}^3\text{/min}$$

∴ Air supplied per stroke per cylinder

$$= \frac{1.12}{(2400/2)} = 0.000933 \text{ m}^3$$

$$\therefore \quad \eta_{vol.} = \frac{\text{Volume of air actually supplied}}{\text{Volume of air theoretically required}}$$

$$= \frac{0.000933}{0.00153} = \mathbf{0.609} \text{ or } \mathbf{60.9\%}. \quad \textbf{(Ans.)}$$

(*v*) **Percentage of excess air supplied :**

Quantity of air required per kg of fuel for complete combustion

$$= \frac{100}{23}\left[C \times \frac{8}{3} + H_2 \times \frac{8}{1}\right]$$

where C is the fraction of carbon and H_2 is the fraction of hydrogen present in the fuel respectively.

$$= \frac{100}{23}\left[0.85 \times \frac{8}{3} + 0.15 \times 8\right]$$

= 15.07 kg/kg of fuel

Actual quantity of air supplied per kg of fuel

$$= \frac{V_a \times \rho_a \times 60}{22.1} = \frac{6.73 \times 1.18 \times 60}{22.1}$$

= 21.56 kg

$$\therefore \quad \textit{Percentage excess air} = \frac{21.56 - 15.07}{15.07} \times 100$$

= **43.06%. (Ans.)**

HEAT BALANCE SHEET

Example 23.43. *The following observations were recorded in a test of one hour duration on a single-cylinder oil engine working on four-stroke cycle.*

Bore	*= 300 mm*
Stroke	*= 450 mm*
Fuel used	*= 8.8 kg*
Calorific value of fuel	*= 41800 kJ/kg*
Average speed	*= 200 r.p.m.*
m.e.p.	*= 5.8 bar*
Brake friction load	*= 1860 N*
Quantity of cooling water	*= 650 kg*
Temperature rise	*= 22°C*
Diameter of the brake wheel	*= 1.22 m*

Calculate : (i) Mechanical efficiency, and (ii) Brake thermal efficiency.

Draw the heat balance sheet.

Solution. *Given :* $n = 1, D = 0.3 \text{ m}, L = 0.45 \text{ m},$

$m_f = 8.8$ kg/h, $C = 41800$ kJ/kg,

$N = 200$ r.p.m., $p_{mi} = 5.8$ bar,

$(W - S) = 795$ N, $D_b = 1.22$ m,

$k = \frac{1}{2}$ ..... for 4-stroke cycle engine

$m_w = 650$ kg and $t_{w_2} - t_{w_1} = 22$°C.

(*i*) **Mechanical efficient, $\eta_{mech.}$:**

Indicated power,

$$\text{I.P.} = \frac{np_{mi}LANk \times 10}{6}$$

$$= \frac{1 \times 5.8 \times 0.45 \times \pi/4 \times 0.3^2 \times 200 \times \frac{1}{2} \times 10}{6}$$

= 30.7 kW

Brake power,

$$\text{B.P.} = \frac{(W - S)\,\pi DN}{60 \times 1000} = \frac{1860 \times \pi \times 1.22 \times 200}{60 \times 1000}$$

= 23.76 kW

$$\eta_{mech.} = \frac{\text{B.P.}}{\text{I.P.}} = \frac{23.76}{30.7} = \mathbf{0.773} \text{ or } \mathbf{77.3\%}. \quad \textbf{(Ans.)}$$

(*ii*) **Brake thermal efficiency, $\eta_{th.(B)}$:**

$$\eta_{th.(B)} = \frac{\text{B.P.}}{\dot{m}_f \times C} = \frac{23.76}{\frac{8.8}{3600} \times 41800}$$

= **0.232** or **23.2%. (Ans.)**

Heat supplied = 8.8 × 41800 = 367840 kJ/h.

(*i*) Heat equivalent of I.P.

= I.P. × 3600 kJ/h

= 30.7 × 3600 = 110520 kJ/h.

(*ii*) Heat carried away by cooling water

$= m_w \times c_{pw} \times (t_{w_2} - t_{w_1})$

= 650 × 4.18 × 22 = 59774 kJ/h.

Heat balance sheet (hourly basis) :

Item	kJ	Per cent
Heat supplied by fuel	**367840**	**100**
(*i*) Heat absorbed in I.P.	110520	30.05
(*ii*) Heat taken away by cooling water	59774	16.25
(*iii*) Heat carried away by exhaust gases, radiation etc. (by difference)	197546	53.70
Total	367840	100

Example 23.44. *In a trial of a single-cylinder oil engine working on dual cycle, the following observations were made :*

Compression ratio = *15*
Oil consumption = *10.2 kg/h*
Calorific value of fuel = *43890 kJ/kg*
Air consumption = *3.8 kg/min*
Speed = *1900 r.p.m.*
Torque on the brake drum = *186 Nm*
Quantity of cooling water used = *15.5 kg/min*
Temperature rise = *36°C*
Exhaust gas temperature = *410°C*
Room temperature = *20°C*
c_p *for exhaust gases* = *1.17 kJ/kg K*

Calculate : (*i*) *Brake power,*
(*ii*) *Brake specific fuel consumption, and*
(*iii*) *Brake thermal efficiency.*
Draw heat balance sheet on minute basis.

Solution. *Given :* $n = 1$, $r = 15$, $m_f = 10.2$ kg/h,
$C = 43890$ kJ/kg, $m_a = 3.8$ kg/min.,
$N = 1900$ r.p.m., $T = 186$ Nm,
$m_w = 15.5$ kg/min, $t_{w_2} - t_{w_1} = 36°C$, $t_g = 410°C$,
$t_r = 20°C$ and $c_p = 1.17$.

(*i*) **Brake power, B.P. :**

$$\text{B.P.} = \frac{2\pi NT}{60 \times 1000} = \frac{2\pi \times 1900 \times 186}{60 \times 1000}$$

= **37 kW. (Ans.)**

(*ii*) **Brake specific fuel consumption, b.s.f.c. :**

$$\text{b.s.f.c.} = \frac{10.2}{37} = \mathbf{0.2756\ kg/kWh.}\ \textbf{(Ans.)}$$

(*iii*) **Brake thermal efficiency,**

$$\eta_{th(B)} = \frac{\text{B.P.}}{m_f \times C} = \frac{37}{\frac{10.2}{3600} \times 43890}$$

= **0.2975** or **29.75%. (Ans.)**

—Heat supplied by the fuel per minute

$$= \frac{10.2}{60} \times 43890 = 7461 \text{ kJ/min}$$

(*i*) Heat equivalent of B.P.
= B.P. × 60 = 37 × 60 = 2220 kJ/min.

(*ii*) Heat carried away by cooling water

$$= m_w \times c_{pw} (t_{w_2} - t_{w_1}) = 15.5 \times 4.18 \times 36$$

= 2332 kJ/min.

(*iii*) Heat carried away by exhaust gases

$$= m_g \times c_{pg} \times (t_g - t_r)$$

$$= \left(\frac{10.2}{60} + 3.8\right) \times 1.17 \times (410 - 20)$$

= 1811 kJ/min.

Heat balance sheet (minute basis)

Item	kJ	Per cent
Heat supplied by fuel	**7461**	**100**
(*i*) Heat absorbed in B.P.	2220	29.8
(*ii*) Heat taken away by cooling water	2332	31.2
(*iii*) Heat carried away by exhaust gases	1811	24.3
(*iv*) Heat unaccounted for (by difference)	1098	14.7
Total	7461	100

☞**Example 23.45.** *From the data given below, calculate indicated power, brake power and draw a heat balance sheet for a two-stroke diesel engine run for 20 minutes at full load :*

r.p.m. = *350*
m.e.p. = *3.1 bar*
Net brake load = *640 N*
Fuel consumption = *1.52 kg*
Cooling water = *162 kg*
Water inlet temperature = *30°C*
Water outlet temperature = *55°C*
Air used/kg of fuel = *32 kg*
Room temperature = *25°C*
Exhaust temperature = *305°C*
Cylinder bore = *200 mm*
Cylinder stroke = *280 mm*
Brake diameter = *1 metre*
Calorific value of fuel = *43900 kJ/kg*
Steam formed per kg of fuel in the exhaust = *1.4 kg*
Specific heat of steam in exhaust = *2.09 kJ/kg K*
Specific heat of dry exhaust gases = *1.0 kJ/kg K.*

Solution. *Given :* N = 350 r.p.m., p_{mi} = 3.1 bar, $(W\text{-}S)$ = 640 N, m_f = 1.52 kg, m_w = 162 kg, t_{w_1} = 30°C, t_{w_2} = 55°C, m_a = 32 kg/kg of fuel, t_r = 25°C, t_g = 305°C, D = 0.2 m, L = 0.28 m, D_b = 1 m, C = 43900 kJ/kg, c_{ps} = 2.09, c_{pq} = 1.0 and k = 1 ..... for two-stroke cycle engine.

(*i*) **Indicated power, I.P. :**

$$\text{I.P.} = \frac{np_{mi}LANk \times 10}{6}$$

$$= \frac{1 \times 3.1 \times 0.28 \times \pi/4 \times 0.2^2 \times 350 \times 1 \times 10}{6}$$

= **15.9 kW. (Ans.)**

(*ii*) **Brake power, B.P. :**

$$\text{B.P.} = \frac{(W-S)\,\pi\, D_b N}{60 \times 1000} = \frac{640 \times \pi \times 1 \times 350}{60 \times 1000}$$

= **11.73 kW. (Ans.)**

—Heat supplied in *20 minutes*

= 1.52 × 43900 = 66728 kJ

(*i*) Heat equivalent of I.P. in 20 minutes

= I.P. × 60 × 20 = 15.9 × 60 × 20 = 19080 kJ

(*ii*) Heat carried away by cooling water

$= m_w \times c_{pw} \times (t_{w_2} - t_{w_1})$

= 162 × 4.18 × (55 – 30) = 16929 kJ

Total mass of air = 32 × 1.52 = 48.64 kg

Total mass of exhaust gases

= Mass of fuel + mass of air

= 1.52 + 48.64 = 50.16 kg

Mass of steam formed = 1.4 × 1.52 = 2.13 kg

∴ Mass of dry exhaust gases

= 50.16 – 2.13 = 48.03 kg

(*iii*) Heat carried away by dry exhaust gases

$= m_g \times c_{pg} \times (t_g - t_r)$

= 48.03 × 1.0 × (305 – 25) = 13448 kJ

(*iv*) Heat carried away by steam

$= 2.13[h_f + h_{fg} + c_{ps}\,(t_{sup} - t_s)]$

At 1.013 bar pressure (atmospheric pressure assumed) :

h_f = 417.5 kJ/kg, h_{fg} = 2257.9 kJ/kg

= 2.13 [417.5 + 2257.9 + 2.09 (305 – 99.6)]

= 6613 kJ/kg ..... neglecting sensible heat of water at room temperature.

Heat balance sheet (20 minute basis) :

Item	kJ	Per cent
Heat supplied by fuel	**66728**	**100**
(*i*) Heat equivalent of I.P.	19080	28.60
(*ii*) Heat carried away by cooling water	16929	25.40
(*iii*) Heat carried away by dry exhaust gases	13448	20.10
(*iv*) Heat carried away by steam in exhaust gases	6613	9.90
(*v*) Heat unaccounted for (by difference)	10658	16.00
Total	66728	100.00

Example 23.46. *During the trial of a single-acting oil engine, cylinder diameter 200 mm, stroke 280 mm, working on two-stroke cycle and firing every cycle, the following observations were made :*

Duration of trial	*= 1 hour*
Total fuel used	*= 4.22 kg*
Calorific value	*= 44670 kJ/kg*
Proportion of hydrogen in fuel	*= 15%*
Total number of revolutions	*= 21000*
Mean effective pressure	*= 2.74 bar*
Net brake load applied to a drum of 1 m diameter	*= 600 N*
Total mass of cooling water circulated	*= 495 kg*
Inlet temperature of cooling water	*= 13°C*
Outlet temperature of cooling water	*= 38°C*
Air used	*= 135 kg*
Temperature of air in test room	*= 20°C*
Temperature of exhaust gases	*= 370°C*

Assume : c_p (gases) = 1.005 kJ/kg K ; c_p (steam) at atmospheric pressure = 2.093 kJ/kg K.

Calculate the thermal efficiency and draw up the heat balance. **(U.P.S.C.)**

Solution. *Given :* D = 200 mm = 0.2 m ;

L = 280 mm = 0.28 m ; m_f = 4.22 kg/h ;

$$C = 44670 \text{ kJ/kg} ; \text{r.p.m.} = \frac{21000}{60} = 350 ;$$

p_{mi} = 2.74 bar ; D_b = 1 m ; $(W - S)$ = 600 N ;

m_w = 495 kg/h ; t_{w_1} = 13°C, t_{w_2} = 38°C ;

m_a = 135 kg/h ; t_r = 20°C ; t_g = 370°C ;

C_{pg} = 1.005 kJ/kg K ;

c_{ps} = 2.093 kJ/kg K

Thermal efficiency, $\eta_{th.}$:

Indicated power,

$$\text{I.P.} = \frac{p_{mi} LANk \times 10}{6}$$

$$= \frac{2.74 \times 0.28 \times \frac{\pi}{4} \times 0.2^2 \times 350 \times 1 \times 10}{6}$$

$= 14.06$ kW

($k = 1$, engine being 2-stroke cycle)

Thermal efficiency (indicated),

$$\eta_{th.(I)} = \frac{\text{I.P.}}{\dot{m}_f \times C}$$

$$= \frac{14.06}{\frac{4.22}{3600} \times 44670}$$

$= \mathbf{0.268}$ or **26.8%.** **(Ans.)**

Brake power,

$$\text{B.P.} = \frac{(W - S)\,\pi\, D_b N}{60 \times 1000} \text{ kW}$$

$$= \frac{600 \times \pi \times 1 \times 350}{60 \times 1000} = 10.99 \text{ kW}$$

Heat balance sheet (minute basis) :

$$\text{Heat input} = \frac{4.22}{60} \times 44670 = 3141.8 \text{ kJ/ min.}$$

(*i*) Heat equivalent of

B.P. = 10.99 × 60 = 659.4 kJ/min.

(*ii*) Heat lost to cooling water $= m_m \times c_{pw} \times (t_{w_2} - t_{w_1})$

$$= \frac{495}{60} \times 4.186 \times (38 - 13) = 863.4 \text{ kJ/min.}$$

Mass of exhaust gases (wet) = mass of air/min + mass of fuel/min

$$= \frac{135}{60} + \frac{4.22}{60} = 2.32 \text{ kg/min}$$

Steam in exhaust gases = 9 × H_2 × mass of fuel used/min

$$= 9 \times \frac{15}{100} \times \frac{4.22}{60} = 0.095 \text{ kg/min}$$

Mass of dry exhaust gases/min,

m_g = Mass of exhaust gas (wet) – mass of H_2O produced/min

= 2.32 – 0.095 = 2.225 kg/min.

(*iii*) Heat carried away be dry exhaust gases

$= m_g \times c_{pg} \times (t_g - t_r)$

= 2.225 × 1.005 × (370 – 20) = 782.6 kJ/min

(*iv*) Assuming that steam in exhaust gases exists as superheated steam at atmospheric pressure and exhaust gas temperature, the enthalpy of 1 kg of steam at atmospheric pressure 1.013 tar ≃ 1 bar and 370°C

$= h_{sup} - h$

(where, h = sensible heat of water at room temperature).

$= [h_f + h_{fg} + c_{ps}\,(t_{sup} - t_s) - h]$

$= (417.5 + 2257.9 + 2.093\,(370 - 99.6) - 1 \times 4.18 \times (20 - 0)]$

= 3157.7 kJ/min

Heat carried away by steam

= 0.095 × 3157.7 ≃ 300 kJ/min

Heat balance sheet (minute basis) :

Item	kJ	Per cent
Heat supplied by fuel	**3141.8**	**100**
(*i*) Heat equivalent of B.P.	659.4	20.99
(*ii*) Heat carried away by cooling water	863.4	27.48
(*iii*) Heat carried away by dry exhaust gases	782.6	24.91
(*iv*) Heat carried away by steam	300	9.55
(*v*) Heat unaccounted for (by difference)	536.4	17.07
Total	3141.8	100

☞**Example 23.47.** *During a test on a two-stroke oil engine on full load the following observations were recorded :*

Speed = *350 r.p.m.*
Net brake load = *590 N*
Mean effective pressure = *2.8 bar*
Oil consumption = *4.3 kg/h*
Jacket cooling water = *500 kg/h*
Temperature of jacket water at inlet and outlet = 25°C and 50°C respectively
Air used per kg of oil = *33 kg*
Temperature of air in test room = *25°C*
Temperature of exhaust gases = *400°C*
Cylinder diameter = *220 mm*
Stroke length = *280 mm*
Effective brake diameter = *1 metre*
Calorific value of oil = *43900 kJ/kg*
Proportion of hydrogen in fuel oil = *15%*
Mean specific heat of dry exhaust gases = 1.0 kJ/kg K
Specific heat of steam = *2.09 kJ/kg K*

Calculate : (i) Indicated power, and (ii) Brake power. Also draw up heat balance sheet on minute basis.

Solution. *Given :* $n = 1$, $N = 350$ r.p.m., $(W–S) = 590$ N,

$p_{mi} = 2.8$ bar $m_f = 4.3$ kg/h, $m_w = 500$ kg/h,

$t_{w_1} = 25°C$, $t_{w_2} = 50°C$ $m_a = 33$ kg/kg of oil,

$t_r = 25°C$, $t_g = 400°C$, $D = 0.22$ m

$L = 0.28$ m, $D_b = 1$ m, $C = 43900$ kJ/kg,

$c_{pg} = 1.0$, $c_{ps} = 2.09$

$k = 1$ ..... for two-stroke cycle engine.

(*i*) **Indicated power, I.P. :**

$$\text{I.P.} = \frac{np_{mi}LANk \times 10}{6}$$

$$= \frac{1 \times 2.8 \times 0.28 \times \pi/4 \times 0.22^2 \times 350 \times 1 \times 10}{6}$$

= **17.38 kW. (Ans.)**

(*ii*) **Brake power, B.P. :**

$$\text{B.P.} = \frac{(W - S)\,\pi\, D_b N}{60 \times 1000} = \frac{590 \times \pi \times 1 \times 350}{60 \times 1000}$$

= **10.81 kW. (Ans.)**

Heat supplied per minute

$$= \frac{4.3}{60} \times 43900 = 3146 \text{ kJ/min.}$$

(*i*) Heat equivalent of

I.P. = 17.38 × 60 = 1042.8 kJ/min.

(*ii*) Heat lost to cooling water

$$= m_w \times c_{pw} \times (t_{w_2} - t_{w_1})$$

$$= \frac{500}{60} \times 4.18 \times (50 - 25) = 870.8 \text{ kJ/min}$$

Now, $2H_2 + O_2 = 2H_2O$

4 32 36

1 8 9

i.e., 1 kg of H_2 produces 9 kg of H_2O

∴ Mass of H_2O produced per kg of fuel burnt

= 9 × H_2 × mass of fuel used/min.

$$= 9 \times 0.15 \times \frac{4.3}{60} = 0.0967 \text{ kg/min.}$$

Total mass of exhaust gases (wet)/min.

= Mass of air/min. + mass of fuel/min.

$$= \frac{(33 + 1) \times 4.3}{60} = 2.436 \text{ kg/min.}$$

Mass of dry exhaust gases/min.

= Mass of wet exhaust gases/min

– Mass of H_2O produced/min.

= 2.436 – 0.0967 = 2.339 kg/min.

(*iii*) Heat lost to dry exhaust gases

$$= m_g \times c_{pg} \times (t_g - t_r)$$

$$= 2.339 \times 1.0 \times (400 - 25) = 887 \text{ kJ/min.}$$

(*iv*) Assuming that steam in exhaust gases exists as superheated steam at atmospheric pressure and exhaust gas temperature, the enthalpy of 1 kg of steam at atmospheric pressure 1.013 ≃ 1 bar and 400°C

$= h_{sup} - h$ (where h is the sensible heat of water at room temperature)

$$= [h_f + h_{fg} + c_{ps}\,(t_{sup} - t_s)] - 1 \times 4.18 \times (25 - 0)$$

$$= [417.5 + 2257.9 + 2.09\,(400 - 99.6)] - 104.5$$

= 3355 kJ/min.

∴ Heat carried away by steam

= 0.0967 × 3355 = 320.6 kJ/min,

Heat balance sheet (minute basis) :

Item	kJ	Per cent
Heat supplied by fuel	**3146**	**100**
(*i*) Heat equivalent of I.P.	1042.8	33.15
(*ii*) Heat carried away by cooling water	870.8	27.70
(*iii*) Heat carried away by dry gases	887	28.15
(*iv*) Heat carried away by steam	320.6	10.20
(*v*) Heat unaccounted for (by difference)	24.8	0.80
Total	314.6	100

Example 23.48. *During a test on a Diesel engine the following observations were made :*

The power developed by the engine is used for driving a D.C. generator. The output of the generator was 210 A at 200 V ; the efficiency of generator being 82%. The quantity of fuel supplied to the engine was 11.2 kg/h ; calorific value of fuel being 42600 kJ/kg. The air-fuel ratio was 18 : 1.

The exhaust gases were passed through a exhaust gas calorimeter for which the observations were as follows : Water circulated through exhaust gas calorimeter = 580 litres/h. Temperature rise of water through calorimeter = 36°C. Temperature of exhaust gases at exit from calorimeter = 98°C. Ambient temperature = 20°C.

Heat lost to jacket cooling water is 32% of the total heat supplied.

If the specific heat of exhaust gases be 1.05 kJ/kg K draw up the heat balance sheet on minute basis.

Solution. Output of generator : 210 A at 200 V

Generator efficiency = 82%

Fuel used = 11.2 kg/h

Calorific value of fuel = 42600 kJ/kg

Air-fuel ratio = 18 : 1

Mass of water circulated through calorimeter,

m_c = 580 litres or 580 kg/h

Temperature rise of water, $t_{w_2} - t_{w_1} = 36°C$

Temperature of exhaust gases at exit from calorimeter = 98°C

Ambient temperature = 20°C

Heat lost to jacket cooling water

= 32% of the total heat supplied

Specific heat of exhaust gases = 1.05 kJ/kg K

Total power generated $= VI = 200 \times 210$

$= 42000$ W = 42 kW

Power available at the brakes of the engine,

$$\text{B.P.} = \frac{42}{0.82} = 51.22 \text{ kW}$$

Total heat supplied to the engine = Fuel supplied per min. × Calorific value of fuel

$$= \frac{11.2}{60} \times 42600 = 7952 \text{ kJ/min.}$$

(*i*) Heat equivalent of B.P. = 51.22 × 60 = 3073 kJ/min

Mass of exhaust gases formed per minute

$$= \text{Fuel supplied/min.} \left(\frac{A}{F} \text{ ratio} + 1\right) ; \quad \left[\frac{A}{F} \text{ ratio means air-fuel ratio}\right]$$

$$= \frac{11.2}{60} (18 + 1) = 3.55 \text{ kg/min.}$$

(*ii*) Heat carried away by exhaust gases/min.

= Heat gained by water in exhaust gas calorimeter from exhaust gases + Heat in exhaust gases at exit from exhaust gas calorimeter above room temperature.

$$= m_w \times c_{pw} \times (t_{w_2} - t_{w_1}) + m_g \times c_{pg}(t_g - t_r)$$

$$= \frac{580}{60} \times 4.18 \times 36 + 3.55 \times 1.05 \,(98 - 20)$$

= 1454.6 + 290.7 = 1745.3 kg/min.

(*iii*) Heat lost to jacket cooling water

= 0.32 × 7952 = 2544.6 kJ/min.

Heat balance sheet (minute basis) :

Item	kJ	Per cent
Heat supplied	**7952**	**100**
(*i*) Heat equivalent of B.P.	3073	38.7
(*ii*) Heat carried away by exhaust gases	1745.3	21.9
(*iii*) Heat lost to jacket cooling water	2544.6	32.0
(*iv*) Heat unaccounted for (by difference)	589.1	7.4
Total	7952	100

☞**Example 23.49.** *During a trial of a single-cylinder, 4-stroke diesel engine the following observations were recorded :*

Bore	*= 340 mm*
Stroke	*= 440 mm*
r.p.m.	*= 400*
Area of indicator diagram	*= 465 mm²*
Length of diagram	*= 60 mm*
Spring constant	*= 0.6 bar/mm*
Load on hydraulic dynamometer	*= 950 N*
Dynamometer constant	*= 7460*
Fuel used	*= 10.6 kg/h*
Calorific value of fuel	*= 49500 kJ/kg*
Cooling water circulated	*= 25 kg/min*
Rise in temperature of cooling water	*= 25°C*

The mass analysis of fuel is :

Carbon	*= 84%*
Hydrogen	*= 15%*
Incombustible	*= 1%*

The volume analysis of exhaust gases is :

Carbon dioxide	*= 9%*
Oxygen	*= 10%*
Nitrogen	*= 81%*
Temperature of exhaust gases	*= 400°C*
Specific heat of exhaust gases	*= 1.05 kJ/kg°C*
Ambient temperature	*= 25°C*

Partial pressure of steam in exhaust gases

= 0.030 bar

Specific heat of superheated steam = 2.1 kJ/kg°C.

Draw up heat balance sheet on minute basis.

Solution. *Given :* $n = 1, D = 0.34$ m, $L = 0.44$ m, $N = 400$ r.p.m., $W = 950$ N,

C_d (dynamometer constant) = 7460, $m_f = 10.6$ kg/h,

$C = 49500$ kJ/kg, $m_w = 25$ kg/min, $(t_{w_2} - t_{w_1}) = 25$°C,

$t_g = 400$°C, $c_{pg} = 1.05$ kJ/kg°C, $c_{ps} = 2.1$ kJ/kg°C.

Mean effective pressure,

$$p_{mi} = \frac{\text{Area of indicator diagram} \times \text{spring constant}}{\text{Length of indicator diagram}}$$

$$= \frac{465 \times 0.6}{60} = 4.65 \text{ bar}$$

Indicated power,

$$\text{I.P.} = \frac{n p_{mi} LANk \times 10}{6}$$

$$= \frac{1 \times 4.65 \times 0.44 \times \pi / 4 \times 0.34^2 \times 400 \times \frac{1}{2} \times 10}{6}$$

= 61.9 kW

Brake power,

$$\text{B.P.} = \frac{W \times N}{C_d} = \frac{950 \times 400}{7460} = 50.9 \text{ kW}$$

Frictional power,

F.P. = I.P. – B.P. = 61.9 – 50.9 = 11 kW

Heat supplied per minute

= Fuel used per min. × Calorific value

$$= \frac{10.6}{60} \times 49500 = 8745 \text{ kJ/min.}$$

(*i*) Heat equivalent of

B.P. = B.P. × 60 = 50.9 × 60 = 3054 kJ/min.

(*ii*) Heat lost in friction

= F.P. × 60 = 11 × 60 = 660 kJ/min.

(*iii*) Heat carried away by cooling water

$$= m_w \times c_{pw} \times (t_{w_2} - t_{w_1})$$

= 25 × 4.18 × 25 = 2612.5 kJ/min.

Mass of air supplied per kg of fuel

$$= \frac{N \times C}{33(CO + CO_2)} = \frac{81 \times 84}{33(0 + 9)} = 22.9 \text{ kg}$$

Mass of exhaust gases formed per kg of fuel

= 22.9 + 1 = 23.9 kg

Mass of exhaust gases formed/min.

$$= 23.9 \times \frac{10.6}{60} = 4.22 \text{ kg}$$

Mass of steam formed per kg of fuel

= 9 × 0.15 = 1.35 kg

∴ Mass of steam formed per min.

$$= 1.35 \times \frac{10.6}{60} = 0.238 \text{ kg/min.}$$

Mass of dry exhaust gases formed per min.

= 4.22 – 0.238 = 3.982 kg.

(*iv*) Heat carried away by dry exhaust gases/min.

$= m_g \times c_{pg} \times (t_g - t_r)$

= 3.982 × 1.05 × (400 – 25) = 1568 kJ/min.

Steam is carried away by exhaust gases. The temperature of steam is also the same as that of exhaust gases *e.g.*, 400°C.

At partial pressure of steam 0.03 bar, the saturation temperature is 24.1°C. Therefore, steam is superheated.

Enthalpy of steam

$= h_g + c_{ps}(t_{sup} - t_s)$

= 2545.5 + 2.1 (400 – 24.1) = 3334.89 kJ/kg.

(*v*) ∴ Heat carried by steam in exhaust gases

= 3334.89 × 0.238 = 793.7 kJ/min.

(*vi*) Heat unaccounted for = Total heat supplied – Heat equivalent of B.P. – heat lost in friction – heat carried away by cooling water – heat carried away by dry exhaust gases – heat carried away by steam in exhaust gases

= 8745 – (3054 + 660 + 2612.5 + 1568 + 793.7)

= 56.8 kJ/min.

Heat balance sheet on minute basis :

Item	kJ	Per cent
Heat supplied	**8745**	**100**
(*i*) Heat equivalent of B.P.	3054	34.92
(*ii*) Heat lost in friction	660	7.55
(*iii*) Heat carried away by cooling water	2612.5	29.87
(*iv*) Heat carried away by dry exhaust gases	1568	17.93
(*v*) Heat carried away by steam in exhaust gases	793.7	9.07
(*vi*) Heat unaccounted for	56.8	0.66
Total	8745	100

MORSE TEST

Example 23.50. *In a test of a 4-cylinder, 4-stroke engine 75 mm bore and 100 mm stroke, the following results were obtained at full throttle at a particular constant speed and with fixed setting of fuel supply of 6.0 kg/h.*

B.P. with all cylinder working = 15.6 kW

B.P. with cylinder no. 1 cut-out = 11.1 kW

B.P. with cylinder no. 2 cut-out = 11.03 kW

B.P. with cylinder no. 3 cut-out = 10.88 kW

B.P. with cylinder no. 4 cut-out = 10.66 kW

If the calorific value of the fuel is 83600 kJ/kg and clearance volume is 0.0001 m^3*, calculate :*

(*i*) *Mechanical efficiency,*

(*ii*) *Indicated thermal efficiency, and*

(*iii*) *Air-standard efficiency.*

Solution. B.P. = I.P.–F.P.

Assuming that the engine is running at constant speed the *frictional and pumping losses remain constant.* Now if one cylinder is cut-out it will not produced any power but the frictional losses and power lost in operating the valves will remain the same as the speed of the engine is constant. The B.P. reduction at the crankshaft due to one cylinder cut out will be *exactly equal to the I.P. produced by that cylinder.*

Therefore,

I.P. produced in cylinder 1, $IP_1 = BP - BP_1 = 15.6 - 11.1 = 4.5$ kW

I.P. produced in cylinder 2, $IP_2 = BP - BP_2 = 15.6 - 11.03 = 4.57$ kW

I.P. produced in cylinder 3, $IP_3 = BP - BP_3 = 15.6 - 10.88 = 4.72$ kW

I.P. produced in cylinder 4, $IP_4 = BP - BP_4 = 15.6 - 10.66 = 4.94$ kW

Total I.P. produced $= IP_1 + IP_2 + IP_3 + IP_4$

I.P. = 4.5 + 4.57 + 4.72 + 4.94 = 18.73 kW

(*i*) **Mechanical efficiency, $\eta_{mech.}$:**

$$\eta_{mech.} = \frac{\text{B.P.}}{\text{I.P.}} = \frac{15.6}{18.73}$$

= **0.833** or **83.3%.** **(Ans.)**

(*ii*) **Indicated thermal efficiency, $\eta_{th(I)}$:**

$$\eta_{th.(I)} = \frac{\text{I.P.}}{\dot{m}_f \times C} = \frac{18.73}{\frac{6}{3600} \times 83600}$$

= **0.1344** or **13.44%.** **(Ans.)**

(*iii*) **Air-standard efficiency, $\eta_{air\text{-}standard}$:**

Stroke volume,

$$V_s = \frac{\pi}{4} D^2 L = \frac{\pi}{4} \times 0.075^2 \times 0.1 = 0.0004417 \text{ m}^3$$

Clearance volume, $V_c = 0.0001 \text{ m}^3$

Compression ratio,

$$r = \frac{V_s + V_c}{V_c} = \frac{0.0004417 + 0.0001}{0.0001} = 5.4$$

$$\therefore \quad \eta_{air\text{-}standard} = 1 - \frac{1}{(r)^{\gamma-1}} = 1 - \frac{1}{(5.4)^{1.4-1}}$$

= **0.49** or **49%.** **(Ans.)**

Example 23.51. *A 4-cylinder petrol engine has a bore of 60 mm and a stroke of 90 mm. Its rated speed is 2800 r.p.m. and it is tested at this speed against brake which has a torque arm of 0.37 m. The net brake load is 160 N and the fuel consumption is 8.986 litres/h. The specific gravity of petrol used is 0.74 and it has a lower calorific value of 44100 kJ/kg. A Morse test is carried out and the cylinders are cut out in the order 1, 2, 3, 4 with corresponding brake loads of 110, 107, 104 and 110 N respectively. Calculate for this speed :*

(*i*) *The engine torque,*

(*ii*) *The brake mean effective pressure,*

(*iii*) *The brake thermal efficiency,*

(*iv*) *The specific fuel consumption,*

(*v*) *Mechanical efficiency, and*

(*vi*) *Indicated mean effective pressure.*

Solution. Number of cylinders, $n = 4$

Bore,	D = 60 mm = 0.06 m
Stroke,	L = 90 mm = 0.09 m
Speed,	N = 2800 r.p.m.
Torque arm	= 0.37 m
Not brake lead	= 160 N
Specific gravity of petrol	= 0.74
Fuel consumption	= 8.986 litres/h
	= 8.986 × 1 × 0.74 kg/h
Calorific value	= 44100 kJ/kg

(*i*) **Engine torque, T :**

Engine torque,

T = Net brake load × Torque arm

= 160 × 0.37 = **59.2 Nm.** **(Ans.)**

Brake power,

$$\text{B.P.} = \frac{2\pi NT}{60 \times 1000} = \frac{2\pi \times 2800 \times 59.2}{60 \times 1000} = 17.36 \text{ kW}$$

(*ii*) **Brake mean effective pressure, p_{mb} :**

$$\text{B.P.} = \frac{n p_{mb} LANk \times 10}{6}$$

$$17.36 = \frac{4 \times p_{mb} \times 0.09 \times \frac{\pi}{4} \times (0.06)^2 \times 2800 \times \frac{1}{2} \times 10}{6}$$

$$\therefore p_{mb} = \frac{17.36 \times 6 \times 4 \times 2}{4 \times 0.09 \times \pi \times (0.06)^2 \times 2800 \times 10}$$

= **7.31 bar.** **(Ans.)**

(*iii*) **Brake thermal efficiency, $\eta_{th\ (B)}$:**

$$\eta_{th(B).} = \frac{\text{B.P.}}{\dot{m}_f \times C} = \frac{17.36}{\frac{(8.986 \times 1 \times 0.74)}{3600} \times 44100}$$

= **0.213** or **21.3%.** **(Ans.)**

(*iv*) **Specific fuel consumption, s.f.c. :**

$$\text{s.f.c.} = \frac{m_f}{\text{B.P.}} = \frac{6.65}{17.36} = \mathbf{0.383\ kg/kWh.} \quad \textbf{(Ans.)}$$

(*v*) **Mechanical efficiency, $\eta_{mech.}$:**

Since the speed is constant, substituting the brake loads instead of the values of B.P. as follows :

$$IP_1 = BP - BP_1 = 160 - 110 = 50 \text{ N}$$
$$IP_2 = BP - BP_2 = 160 - 107 = 53 \text{ N}$$
$$IP_3 = BP - BP_3 = 160 - 104 = 56 \text{ N}$$
$$IP_4 = BP - BP_4 = 160 - 110 = 50 \text{ N}$$

Hence for the engine, the indicated load is given by

$$IP = IP_1 + IP_2 + IP_3 + IP_4$$
$$= 50 + 53 + 56 + 50 = 209 \text{ N}$$

$$\therefore \quad \eta_{mech.} = \frac{BP}{IP} = \frac{160}{209} = \mathbf{0.765} \text{ or } \mathbf{76.5\%.} \quad \textbf{(Ans.)}$$

(*vi*) **Indicated mean effective pressure, p_{mi} :**

$$\eta_{mech.} = \frac{p_{mb}}{p_{mi}}$$

$$\therefore \quad p_{mi} = \frac{p_{mb}}{\eta_{mech.}} = \frac{7.31}{0.765} = \mathbf{9.55\ bar.} \quad \textbf{(Ans.)}$$

SUPERCHARGING

☞**Example 23.52.** *The average indicated power developed in a C.I. engine is 13 kW/m³ of free air induced per minute. The engine is a three-litre four-stroke engine running at 3500 r.p.m., and has a volumetric efficiency of 81%, referred to free air conditions of 1.013 bar and 15°C. It is proposed to fit a blower, driven mechanically from the engine. The blower has an isentropic efficiency of 72% and works through a* *pressure ratio of 1.72. Assume that at the end of induction the cylinders contain a volume of charge equal to the swept volume, at the pressure and temperature of the delivery from the blower. Calculate the increase in brake power to be expected from the engine.*

Take all mechanical efficiencies as 78%.

Solution. Capacity of the engine = 3 litres = 0.003 m^3

$$\text{Swept volume} = \frac{3500}{2} \times 0.003 = 5.25 \text{ m}^3/\text{min.}$$

Unsupercharged induced volume = 5.25 × $\eta_{vol.}$

= 5.25 × 0.81 = 4.25 m^3

Blower delivery pressure = 1.72 × 1.013 = 1.74 bar

Temperature after isentropic compression

$$= 288 \times (1.72)^{(1.4-1)/1.4} = 336.3 \text{ K} \quad \left[\because \frac{T_2}{T_1} = \left(\frac{p_2}{p_1}\right)^{\frac{\gamma-1}{\gamma}}\right]$$

∴ Blower delivery temperature

$$= 288 + \left(\frac{336.3 - 288}{0.72}\right) = 355 \text{ K}$$

$$\left[\because \eta_{isen.} = \frac{T_2 - T_1}{T_{2'} - T_1} \text{ or } T_{2'} = T_1 + \frac{T_2 - T_1}{\eta_{isen.}}\right]$$

The blower delivers 5.25 m^3/min at 1.74 bar and 355 K.

Equivalent volume at 1.013 bar and 15°C

$$= \frac{5.25 \times 1.74 \times 288}{1.013 \times 355} = 7.31 \text{ m}^3/\text{min.}$$

∴ Increase in induced volume

= 7.31 – 4.25 = 3.06 m^3/min.

∴ Increase in indicated power from air induced

= 13 × 3.06 = 39.78 kW

Increase in I.P. due to the increased induction pressure

$$= \frac{(1.74 - 1.013) \times 10^5 \times 5.25}{10^3 \times 60} = 6.36 \text{ kW}$$

i.e., Total increase in

I.P. = 39.78 + 6.36 = 46.14 kW

∴ Increase in engine

B.P. = $\eta_{mech.}$ × 46.14 = 0.78 × 46.14 = 35.98 kW

From this must be *deducted the power required to drive the blower*

Mass of air delivered by blower

$$= \frac{1.74 \times 10^5 \times 5.25}{60 \times 287 \times 355} = 0.149 \text{ kg/s}$$

Work input to blower $= \dot{m} c_p (355 - 288)$

= 0.149 × 1.005 × 67

$$\therefore \text{Power required} = \frac{0.149 \times 1.005 \times 67}{0.78}$$

= 12.86 kW

∴ **Net increase in B.P.**

= 35.98 – 12.86 = **23.12 kW. (Ans.)**

HIGHLIGHTS

1. Any type of engine or machine which derives heat energy from the combustion of fuel or any other source and converts this energy into mechanical work is termed as a **heat engine**.
2. The function of a carburettor is to atomise and meter the liquid fuel and mix it with air as it enters the injection system of the engine maintaining under all conditions of operation fuel-air proportion approximate to those conditions.
3. The two basic ignition systems in current use are :
   (*i*) Battery or coil-ignition system
   (*ii*) Magneto-ignition system.
4. Following are the methods of governing I.C. engines :
   (*i*) Hit and miss method (*ii*) Quality governing
   (*iii*) Quantity governing.
5. Pre-ignition is the premature combustion which starts before the application of spark. Overheated spark plugs and exhaust valves which are the main causes of pre-ignition should be carefully avoided in engines.
6. A very sudden rise in pressure during combustion accompanied by metallic hammer like sound is called **detonation**. The region in which detonation occurs is farthest removed from the sparking plug, and is named the 'detonation zone' and even with severe detonation this zone is rarely more than that one quarter the clearance volume.
7. The **octane number** is the percentage of octane in the mixture of iso-octane (high rating) and normal heptane (low rating), by volume] which knocks under the same conditions as the fuel.
8. **Delay period or ignition lag** is the time immediately following injection of fuel during which the ignition process is being initiated and the pressure does not rise beyond the value it would have due to compression of air.
9. Higher the cetane rating of the fuel lesser is the propensity for diesel knock. In general a high octane value implies a low cetane value.

10. The purpose of **supercharging** is to raise the volumetric efficiency above that value which can be obtained by normal aspiration. Supercharging of petrol engines, because of its poor fuel economy, is not very popular and is used only when a large amount of power is needed or when more power is needed to compensate altitude loss.
11. **Dissociation** refers to disintegration of burnt gases at high temperatures. It is a reversible process and increases with temperature. Dissociation, in general, causes a loss of power and efficiency.
12. **Performance of I.C. engines.** Some important relation :

(*i*) Indicated power (I.P.) = $\dfrac{np_{mi}LANk \times 10}{6}$ kW

(*ii*) Brake (B.P.) = $\dfrac{(W - S)\,\pi\,(D_b + d)N}{60 \times 1000}$ kW

or $\left(= \dfrac{2\pi NT}{60 \times 1000}\,\text{kW}\right)$

(*iii*) Mechanical efficiency, $\eta_{mech.} = \dfrac{\text{B.P.}}{\text{I.P.}}$

(*iv*) Thermal efficiency (indicated), $\eta_{th.(I)} = \dfrac{\text{I.P.}}{\dot{m}_f \times C}$

and thermal efficiency (brake), $\eta_{th.(B)} = \dfrac{\text{B.P.}}{\dot{m}_f \times C}$

where $\dot{m}_f$ = Mass of fuel used in kg/sec.

(*v*) $\eta_{relative} = \dfrac{\eta_{thermal}}{\eta_{air\text{-}standard}}$

(*vi*) Measurement of air consumption by **air box method** : Volume of air passing through the orifice,

$$V_a = 840\, AC_d \sqrt{\frac{h_w}{\rho_a}}$$

and mass of air passing through the orifice,

$$m_a = 0.066\, C_d \times d^2 \sqrt{h_w \rho_a} \text{ kg/min}$$

where, A = Area of orifice, m^2,
d = Diameter of orifice, cm,
h_w = Head of water in 'cm' causing the flow, and
ρ_a = Density of air in kg/m^3 under atmospheric conditions.

OBJECTIVE TYPE QUESTIONS

Choose the Correct Answer :

1. In a four-stroke cycle engine, the four operations namely suction, compression, expansion and exhaust are completed in the number of revolutions of crankshaft equal to
(*a*) four (*b*) three
(*c*) two (*d*) one.
2. In a two-stroke cycle engine, the operations namely suction, compression, expansion and exhaust are completed in the number of revolutions of crankshaft equal to
(*a*) four (*b*) three
(*c*) two (*d*) one.
3. In a four-stroke cycle S.I. engine the camshaft runs
(*a*) at the same speed as crankshaft
(*b*) at half the speed of crankshaft
(*c*) at twice the speed of crankshaft
(*d*) at any speed irrespective of crankshaft speed.
4. The following is an S.I. engine
(*a*) Diesel engine (*b*) Petrol engine
(*c*) Gas engine (*d*) none of the above.
5. The following is C.I. engine
(*a*) Diesel engine (*b*) Petrol engine
(*c*) Gas engine (*d*) none of the above.
6. In a four-stroke cycle petrol engine, during suction stroke
(*a*) only air is sucked in
(*b*) only petrol is sucked in
(*c*) mixture of petrol and air is sucked in
(*d*) none of the above.
7. In a four-stroke cycle diesel engine, during suction stroke
(*a*) only air is sucked in
(*b*) only fuel is sucked in
(*c*) mixture of fuel and air is sucked in
(*d*) none on the above.
8. The two-stroke cycle engine has
(*a*) one suction valve and one exhaust valve operated by one cam
(*b*) one suction valve and one exhaust valve operated by two cams
(*c*) only ports covered and uncovered by piston to effect charging and exhausting
(*d*) none of the above.
9. For same output, same speed and same compression ratio the thermal efficiency of a two-stroke cycle petrol engine as compared to that for four-stroke cycle petrol engine is
(*a*) more
(*b*) less
(*c*) same as long as compression ratio is same
(*d*) same as long as output is same.
10. The ratio of brake power to indicated power of an I.C. engine is called
(*a*) mechanical efficiency (*b*) thermal efficiency
(*c*) volumetric efficiency (*d*) relative efficiency.
11. The specific fuel consumption of a diesel engine as compared to that for petrol engines is
(*a*) lower (*b*) higher
(*c*) same for same output (*d*) none of the above.
12. The thermal efficiency of petrol engine as compared to diesel engine is
(*a*) lower
(*b*) higher
(*c*) same for same power output
(*d*) same for same speed.

13. Compression ratio of petrol engines is in the range of
(*a*) 2 to 3 (*b*) 7 to 10
(*c*) 16 to 20 (*d*) none of the above.

14. Compression ratio of diesel engines may have a range
(*a*) 8 to 10 (*b*) 10 to 15
(*c*) 16 to 20 (*d*) none of the above.

15. The thermal efficiency of good I.C. engine at the rated load is in the range of
(*a*) 80 to 90% (*b*) 60 to 70%
(*c*) 30 to 35% (*d*) 10 to 20%.

16. In case of S.I. engine, to have best thermal efficiency the fuel-air mixture ratio should be
(*a*) lean (*b*) rich
(*c*) may be lean or rich (*d*) chemically correct.

17. The fuel-air ratio, for maximum power of S.I. engines, should be
(*a*) lean (*b*) rich
(*c*) may be lean or rich (*d*) chemically correct.

18. In case of petrol engine, at starting
(*a*) rich fuel-air ratio is needed
(*b*) weak fuel-air ratio is needed
(*c*) chemically correct fuel-air ratio is needed
(*d*) any fuel-air ratio will do.

19. Carburettor is used for
(*a*) S.I. engines (*b*) Gas engines
(*c*) C.I. engines (*d*) none of the above.

20. Fuel injector is used in
(*a*) S.I. engines (*b*) Gas engines
(*c*) C.I. engines (*d*) none of the above.

21. Very high speed engines are generally
(*a*) Gas engines (*b*) S.I. engines
(*c*) C.I. engines (*d*) Steam engines.

22. In S.I. engine, to develop high voltage for spark plug
(*a*) battery is installed (*b*) distributor is installed
(*c*) carburettor is installed (*d*) ignition coil is installed.

23. In S.I. engine, to obtain required firing order
(*a*) battery is installed (*b*) distributor is installed
(*c*) carburettor is installed (*d*) ignition coil is installed.

24. For petrol engines, the method of governing employed is
(*a*) quantity governing (*b*) quality governing
(*c*) hit and miss governing (*d*) none of the above.

25. For diesel engines, the method of governing employed is
(*a*) quantity governing (*b*) quality governing
(*c*) hit and miss governing (*d*) none of the above.

26. Voltage developed to strike spark in the spark plug is in the range
(*a*) 6 to 12 volts (*b*) 1000 to 2000 volts
(*c*) 20000 to 25000 volts (*d*) none of the above.

27. In a 4-cylinder petrol engine the standard firing order is
(*a*) 1-2-3-4 (*b*) 1-4-3-2
(*c*) 1-3-2-4 (*d*) 1-3-4-2.

28. The torque developed by the engine is maximum
(*a*) at minimum speed of engine
(*b*) at maximum speed of engine
(*c*) at maximum volumetric efficiency speed of engine
(*d*) at maximum power speed of engine

29. Iso-octane content in a fuel for S.I. engines
(*a*) retards auto-ignition
(*b*) accelerates auto-ignition
(*c*) does not affect auto-ignition
(*d*) none of the above.

30. Normal heptane content in fuel for S.I. engines
(*a*) retards auto-ignition
(*b*) accelerates auto-ignition
(*c*) does not affect auto-ignition
(*d*) none of the above.

31. The knocking in S.I. engines increases with
(*a*) increase in inlet air temperature
(*b*) increase in compression ratio
(*c*) increase in cooling water temperature
(*d*) all of the above.

32. The knocking in S.I. engines gets reduced
(*a*) by increasing the compression ratio
(*b*) by retarding the spark advance
(*c*) by increasing inlet air temperature
(*d*) by increasing the cooling water temperature.

33. Increasing the compression ratio in S.I. engines
(*a*) increases the tendency for knocking
(*b*) decreases tendency for knocking
(*c*) does not affect knocking
(*d*) none of the above.

34. The knocking tendency in petrol engines will increase when
(*a*) speed is decreased
(*b*) speed is increased
(*c*) fuel-air ratio is made rich
(*d*) fuel-air ratio is made lean.

35. The ignition quality of fuels for S.I. engines is determined by
(*a*) cetane number rating (*b*) octane number rating
(*c*) calorific value rating (*d*) volatility of the fuel.

36. Petrol commercially available in India for Indian passenger cars has octane number in the range
(*a*) 40 to 50 (*b*) 60 to 70
(*c*) 80 to 85 (*d*) 95 to 100.

37. Octane number of the fuel used commercially for diesel engine in India is in the range
(*a*) 80 to 90 (*b*) 60 to 80
(*c*) 60 to 70 (*d*) 40 to 45.

38. The knocking tendency in C.I. engines increases with
(*a*) decrease of compression ratio
(*b*) increase of compression ratio
(*c*) increasing the temperature of inlet air
(*d*) increasing cooling water temperature.

39. Desirable characteristic of combustion chamber for S.I. engines to avoid knock is
(*a*) small bore
(*b*) short ratio of flame path to bore
(*c*) absence of hot surfaces in the end region of gas
(*d*) all of the above.

ANSWERS

1. (*c*)	**2.** (*d*)	**3.** (*b*)	**4.** (*b*)
5. (*a*)	**6.** (*c*)	**7.** (*a*)	**8.** (*c*)
9. (*b*)	**10.** (*a*)	**11.** (*a*)	**12.** (*a*)
13. (*b*)	**14.** (*c*)	**15.** (*c*)	**16.** (*a*)
17. (*b*)	**18.** (*a*)	**19.** (*a*)	**20.** (*c*)
21. (*b*)	**22.** (*d*)	**23.** (*b*)	**24.** (*a*)
25. (*b*)	**26.** (*c*)	**27.** (*d*)	**28.** (*c*)
29. (*a*)	**30.** (*b*)	**31.** (*d*)	**32.** (*b*)
33. (*a*)	**34.** (*a*)	**35.** (*b*)	**36.** (*c*)
37. (*d*)	**38.** (*a*)	**39.** (*d*).	

THEORETICAL QUESTIONS

1. Name the two general classes of combustion engines and state how do they basically differ in principle ?

2. Discuss the relative advantages and disadvantages of internal combustion and external combustion engines.

3. What are the two basic types of internal combustion engines ? What are the fundamental differences between the two ?

4. What is the function of a governor ? Enumerate the types of governors and discuss with a neat sketch the Porter governor.

5. Differentiate between a flywheel and a governor.

6. (*a*) State the function of a carburettor in a petrol engine.
(*b*) Describe a simple carburettor with a neat sketch and also state its limitations.

7. Explain with neat sketches the construction and working of the following :
(*i*) Fuel pump (*ii*) Injector.

8. Explain the following terms as applied to I.C. engines :
Bore, stroke, T.D.C., B.D.C., clearance volume, swept volume, compression ratio and piston speed.

9. Explain with suitable sketches the working of a four-stroke otto engine.

10. Discuss the difference between ideal and actual valve timing diagrams of a petrol engine.

11. In what respects four-stroke diesel cycle (compression ignition) engine differs from four-stroke cycle spark ignition engine ?

12. Discuss the difference between theoretical and actual valve timing diagrams of a diesel engine.

13. What promotes the development of two-stroke engines ? What are the two main types of two-stroke engines.

14. Describe with a suitable sketch the two-stroke cycle spark ignition (SI) engine. How its indicator diagram differs from that of four-stroke cycle engine ?

15. Compare the relative advantages and disadvantages of four-stroke and two-stroke cycle engines.

16. Discuss with suitable sketches the following ignition systems used in petrol engines :
(*a*) Coil or battery-ignition system
(*b*) Magneto-ignition system.

17. State the relative advantages and disadvantages of battery and magneto-ignition systems.

18. (*a*) Why do we feel the necessity of cooling an I.C. engine ?
(*b*) Explain briefly the following methods of cooling I.C. engines :
(*i*) Air cooling
(*ii*) Liquid cooling.
Also state their relative advantages and disadvantages.

19. (*a*) State the purposes of lubrication.
(*b*) Discuss with the help of suitable sketches the following :
(*i*) Wet pump lubrication
(*ii*) Dry pump lubrication.

20. Discuss briefly the following methods of governing I.C. engines :
(*a*) Hit and miss method
(*b*) Quality governing
(*c*) Quantity governing.

21. What do you mean by pre-ignition ? How can it be detected ?

22. Describe the phenomenon of detonation or knocking in S.I. engines. How can it be controlled ?

23. (*a*) What do you mean by performance of I.C. engine ?
(*b*) Discuss briefly the basic performance parameters.
(*c*) Discuss with suitable sketch the brake rope dynamometer.

24. Describe how the I.P. of a multi-cylinder engine is measured ?

25. Describe the method commonly used in laboratory for measuring the air supplied to an I.C. engine.

26. Derive the formula used for finding the mass of air supplied to an engine using an orifice tank.

27. Explain the phenomenon of auto-ignition. Explain how auto-ignition is responsible for knocking in S.I. engines.

28. Explain the phenomena of knocking in S.I. engine. What are the different factors which influence the knocking ? Describe the methods used to suppress it.

29. Explain the difference between
(*i*) pre-ignition,
(*ii*) auto-ignition and
(*iii*) detonation.

30. What is meant by ignition delay ?

31. What are causes of knock in C.I. engines ?

32. What are the different methods used in C.I. engines to create turbulence in the mixture ?
Explain its effect on power output and thermal efficiency of the engine.

33. What do you mean by 'octane number' and 'cetane number' of fuels ? How are they determined ?

UNSOLVED EXAMPLES

1. A single cylinder petrol engine working on two-stroke cycle develops indicated power of 5 kW. If the mean effective pressure is 7.0 bar and the piston diameter is 100 mm, calculate the average speed of the piston.
[**Hint.** Average piston speed = 2 *LN*.] [**Ans.** 109.1 m/s]

2. A 4-cylinder petrol engine works on a mean effective pressure of 5 bar and engine speed of 1250 r.p.m. Find the indicated power developed by the engine if the bore is 100 mm and stroke 150 mm. [**Ans.** 6.11 kW]

3. A 4-cylinder four-stroke S.I. engine is designed to develop 44 kW indicated power at a speed of 3000 r.p.m. The compression ratio used is 6. The law of compression and expansion is $pV^{1.3}$ = constant and heat addition and rejection takes place at constant volume. The pressure and temperature at the beginning of compression stroke are 1 bar and 50°C. The maximum pressure of the cycle is limited to 30 bar. Calculate the diameter and stroke of each cylinder assuming all cylinders have equal dimensions.

Assume diagram factor = 0.8 and ratio of stroke/bore = 15. [**Ans.** D = 95 mm, L = 142.5 mm]

4. During the trial of a four-stroke diesel engine, the following observations were recorded :

Area of indicator diagram = 475 mm^2, length of indicator diagram = 62 mm, spring number = 1.1 bar/mm, diameter of piston = 100 mm, length of stroke = 150 mm, engine r.p.m. = 375.

Determine : (*i*) Indicated mean effective pressure
(*ii*) Indicated power.

[**Ans.** (*i*) 8.43 bar ; (*ii*) 3.1 kW]

5. A 4-cylinder, four-stroke diesel engine runs at 1000 r.p.m. The bore and stroke of each cylinder are 100 mm and 160 mm respectively. The cut-off is 6.62% of the stroke. Assuming that the initial condition of air inside the cylinder are 1 bar and 20°C, mechanical efficiency of 75%, calculate the air-standard efficiency and brake power developed by the engine.

Also, calculate the brake specific fuel consumption if the air/fuel ratio is 20 : 1. Take R for air as 0.287 kJ/kg K and clearance volume as 0.000084 m^3.

[**Ans.** 61.4%, 21.75 kW, 0.4396 kg/kWh]

6. During a trial of a two-stroke diesel engine the following observations were recorded :

Engine speed = 1500 r.p.m., load on brakes = 120 kg, length of brake arm = 875 mm.

Determine :

(*i*) Brake torque (*ii*) Brake power.

[**Ans.** (*i*) 1030 Nm ; (*ii*) 161.8 kW]

7. A four-stroke gas engine developes 4.2 kW at 180 r.p.m. and at full load. Assuming the following data, calculate the relative efficiency based on indicated power and air-fuel ratio used. Volumetric efficiency = 87%, mechanical efficiency = 74%, clearance volume = 2100 cm^3, swept volume = 9000 cm^3, fuel consumption = 5 m^3/h, calorific value of fuel = 16750 kJ/m^3. [**Ans.** 50.2%, 7.456 : 1]

8. During the trial of a four-stroke cycle gas engine the following data were recorded :

Area of indicator diagram	=	565.8 mm^2
Length of indicator diagram	=	74.8 mm
Spring index	=	0.9 bar/mm
Cylinder diameter	=	220 mm
Stroke length	=	430 mm
Number of explosions/min	=	100

Determine : (*i*) Indicated mean effective pressure.
(*ii*) Indicated power.

[**Ans.** (*i*) 6.8 bar ; (*ii*) 18.5 kW]

9. The following observations were recorded during a trial of a four-stroke engine with rope brake dynamometer :

Engine speed = 650 r.p.m., diameter of brake drum = 600 mm, diameter of rope = 50 mm, dead load on the brake drum = 32 kg, spring balance reading = 4.75 kg.

Calculate the brake power. [**Ans.** 5.9 kW]

10. The following data refer to a four-stroke petrol engine :

Engine speed = 2000 r.p.m., ideal thermal efficiency = 35%, relative efficiency = 80%, mechanical efficiency = 85%, volumetric efficiency = 70%.

If the engine develops 29.42 kW brake power calculate the cylinder swept volume. [**Ans.** 0.00185 m^3]

11. A single-cylinder four-stroke gas engine has a bore of 178 mm and a stroke of 330 mm and is governed by hit and miss principle. When running at 400 r.p.m. at full load, indicator cards are taken which give a working loop mean effective pressure of 6.2 bar, and a pumping loop mean effective pressure of 0.35 bar. Diagrams from the dead cycle give a mean effective pressure of 0.62 bar. The engine was run light at the same speed (*i.e.,* with no load), and a mechanical counter recorded 47 firing strokes per minute. Calculate :

(*i*) Full load brake power (*ii*) Mechanical efficiency of the engine. [**Ans.** (*i*) 13.54 kW ; (*ii*) 84.7%]

12. During a 60 minutes trial of a single-cylinder four-stroke engine the following observations were recorded :

Bore = 0.3 m, stroke = 0.45 m, fuel consumption = 11.4 kg, calorific value of fuel = 42000 kJ/kg, brake mean effective pressure = 6.0 bar, net load on brakes = 1500 N, r.p.m. = 300, brake drum diameter = 1.8 m, brake rope diameter = 20 mm, quantity of jacket cooling water = 600 kg, temperature rise of jacket water = 55°C, quantity of air as measured = 250 kg, exhaust gas temperature = 420°C, c_p for exhaust gases = 1 kJ/kg K, ambient temperature = 20°C.

Calculate :

(*i*) Indicated power ;
(*ii*) Brake power ;
(*iii*) Mechanical efficiency ;
(*iv*) Indicated thermal efficiency.

Draw up a heat balance sheet on minute basis.

[**Ans.** (*i*) 47.7 kW, (*ii*) 42.9 kW, (*iii*) 89.9%, (*iv*) 35.86%]

13. A quality governed four-stroke, single-cylinder gas engine has a bore of 146 mm and a stroke of 280 mm. At 475 r.p.m. and full load the net load on the friction brake is 433 N, and the torque arm is 0.45 m. The indicator diagram gives a net area of 578 mm^2 and a length of 70 mm with a spring rating of 0.815 bar/mm.

Calculate :

(*i*) The indicated power (*ii*) Brake power
(*iii*) Mechanical efficiency.

[**Ans.** (*i*) 12.5 kW ; (*ii*) 9.69 kW ; (*iii*) 77.5%]

14. A two-cylinder four-stroke gas engine has a bore of 380 mm and a stroke of 585 mm. At 240 r.p.m. the torque developed is 5.16 kNm.

Calculate : (*i*) Brake power (*ii*) Mean piston speed in m/s
(*iii*) Brake mean effective pressure.

[**Ans.** (*i*) 129.8 kW ; (*ii*) 4.68 m/s ; (*iii*) 4.89 bar]

15. The engine of Problem 14 is supplied with a mixture of coal gas and air in the proportion of 1 to 7 by volume. The estimated volumetric efficiency is 85% and the calorific value of the coal gas is 16800 kJ/m^3. Calculate the brake thermal efficiency of the engine. [**Ans.** 27.4%]

16. A 4-cylinder, four-stroke diesel engine has a bore of 212 mm and a stroke of 292 mm. At full load at 720 r.p.m., the b.m.e.p. is 5.93 bar and the specific fuel consumption is 0.226 kg/kWh. The air/fuel ratio as determined by exhaust gas analysis is 25 : 1. Calculate the brake thermal efficiency and volumetric efficiency of the engine.

Atmospheric conditions are 1.01 bar and 15°C and calorific value for the fuel may be taken as 44200 kJ/kg. [**Ans.** 36% ; 76.5%]

17. A 4-cylinder petrol engine has an output of 5 kW at 2000 r.p.m. A Morse test is carried out and the brake torque readings are 177, 170, 168 and 174 Nm respectively. For normal running at this speed the specific fuel consumption is 0.364 kg/kWh. The calorific value of fuel is 44200 kJ/kg.

Calculate : (*i*) Mechanical efficiency
(*ii*) Brake thermal efficiency of the engine.
[**Ans.** (*i*) 82% ; (*ii*) 22.4%]

18. A V-8 four-stroke petrol engine is required to give 186.5 kW at 440 r.p.m. The brake thermal efficiency can be assumed to be 32% at the compression ratio of 9 : 1. The air/fuel ratio is 12 : 1 and the volumetric efficiency at this speed is 69%. If the stroke to bore ratio is 0.8, determine the engine displacement required and the dimensions of the bore and stroke. The calorific value of the fuel is 44200 kJ/kg, and the free air conditions are 1.013 bar and 15°C.
[**Ans.** 5.12 litres ; 100.6 mm ; 80.5 mm]

19. During the trial (60 minutes) on a single-cylinder oil engine having cylinder diameter 300 mm, stroke 450 mm and working on the four-stroke cycle, the following observations were made :

Total fuel used = 9.6 litres, calorific value of fuel = 45000 kJ/kg, total number of revolutions = 12624, gross indicated mean effective pressure = 7.24 bar, pumping i.m.e.p. = 0.34 bar, net load on the brake = 3150 N, diameter of brake wheel drum = 1.78 m, diameter of the rope = 40 mm, cooling water circulated = 545 litres, cooling water temperature rise = 25°C, specific gravity of oil = 0.8.

Determine : (*i*) Indicated power.
(*ii*) Brake power.
(*iii*) Mechanical efficiency.

Draw up the heat balance sheet on minute basis.
[**Ans.** (*i*) 77 kW ; (*ii*) 61.77 kW ; (*iii*) 80.22%]

20. The following results were obtained on full load during a trial on a two stroke oil engine :

Engine speed	= 350 r.p.m.
Net brake load	= 600 N
m.e.p.	= 2.75 bar
Oil consumption	= 4.25 kg/h
Temperature rise of jacket cooling water	= 25°C
Air used per kg of oil	= 31.5 kg
Temperature of air in test room	= 20°C
Temperature of exhaust gases	= 390°C

Following data also apply to the above test :

Cylinder diameter	= 220 mm
Stroke	= 280 mm
Effective brake diameter	= 1 metre
Calorific value of oil	= 45000 kJ/kg
Proportion of hydrogen in fuel oil	= 15%
Partial pressure of steam in exhaust gases	= 0.04 bar
Mean specific heat of exhaust gases	= 1.0 kJ/kg K
Specific heat of superheated steam	= 2.1 kJ/kg K
Specific heat of water	= 4.186 kJ/kg K

Determine : (*i*) Indicated power.
(*ii*) Brake power.
(*iii*) Mechanical efficiency.

Draw up heat balance sheet for the test.
[**Ans.** (*i*) 17.1 kW ; (*ii*) 11 kW ; (*iii*) 64.33%]

21. A 4-cylinder, four-stroke diesel engine developes 83.5 kW at 1800 r.p.m. with specific fuel consumption of 0.231 kg/kWh, and air/fuel ratio of 23 : 1. The analysis of fuel is 87% carbon and 13% hydrogen, and the calorific value of the fuel is 43500 kJ/kg. The jacket cooling water flows at 0.246 kg/s and its temperature rise is 50 K. The exhaust temperature is 316°C. Draw up an energy balance for the engine. Take R = 0.302 kJ/kg K and c_p = 1.09 kJ/kg K for the dry exhaust gases and c_p = 1.86 kJ/kg K for superheated steam. The temperature in the test house is 17.8°C, and the exhaust gas pressure is 1.013 bar.
[**Ans.** B.P. = 35.8%, cooling water = 22.1%, exhaust = 24%, radiation and unaccounted = 16.7%]

22. During the trial of a single-cylinder, 4-stroke, diesel engine the following observations were recorded :

Bore = 350 mm, stroke = 450 mm, r.p.m. = 400, area of indicator diagram = 472 mm^2, length of indicator diagram = 62 mm, spring constant = 0.59 bar/mm, load on hydraulic dynamometer = 970 N, dynamometer constant = 7500, fuel used = 10.78 kg/h, calorific value of fuel = 50000 kJ/kg, cooling water circulated = 24 litres/min, rise in temperature of cooling water = 24°C. The main analysis of fuel is : carbon = 85%, hydrogen = 14%, incombustibles = 1%. The volume analysis of exhaust gases is : carbon dioxide = 8%, oxygen = 11%, nitrogen = 81%. Temperature of exhaust gases = 380°C, specific heat of exhaust gases = 1.05 kJ/kg°C, ambient temperature = 20°C, partial pressure of steam in exhaust gases = 0.03 bar, specific heat of superheated steam = 2.1 kJ/kg°C.

Draw up the heat balance sheet on minute basis.
[**Ans.** (*i*) Heat equivalent of B.P. = 34.55%,
(*ii*) heat lost in friction = 8.73%,
(*iii*) Heat carried away by cooling water = 26.84%,
(*iv*) Heat in dry exhaust gases = 19.54%,
(*v*) Heat carried away by steam in exhaust gases = 7.24%,
(*vi*) Heat unaccounted for = 3.10%]

23. A 4-cylinder petrol engine has a bore of 57 mm and a stroke of 90 mm. Its rated speed is 2800 r.p.m. and it is tested at this speed against a brake, which has a torque arm of 0.356 m. The net brake load is 155 N and the fuel consumption is 6.74 litres/h. The specific gravity of petrol used is 0.735 and it has a lower calorific value of

44200 kJ/kg. A Morse test is carried out and the cylinders are cut out in order 1, 2, 3, 4 with corresponding brake loads 111, 106.5, 104.2 and 111 N, respectively. Calculate for this speed :

(*i*) The engine torque.
(*ii*) The brake mean effective pressure.
(*iii*) The brake thermal efficiency.
(*iv*) The specific fuel consumption.
(*v*) The mechanical efficiency.
(*vi*) The indicated mean effective pressure.

[**Ans.** (*i*) 55.2 Nm ; (*ii*) 7.55 bar ; (*iii*) 26.6% ; (*iv*) 0.306 kg/kWh ; (*v*) 82.8% ; (*vi*) 9.12 bar]

24. The average indicated power developed in C.I. engine is 12.9 kW/m^3 of free air induced per minute. The engine is three-litre four-stroke engine running at 3500 r.p.m., and has a volumetric efficiency of 80%, referred to free conditions of 1.013 bar and 15°C. It is proposed to fit a blower, driven mechanically from the engine. The blower has an isentropic efficiency of 75% and works through a pressure ratio of 1.7. Assume that at the end of induction the cylinders contain a volume of charge equal to the swept volume, at the pressure and temperature of the delivery from the blower. Calculate the increase in B.P. to be expected from the engine.
Take all mechanical efficiencies as 80%. [**Ans.** 25.3 kW]

Air Compressors

24

24.1. General aspects. 24.2. Classification of air compressors. 24.3. Reciprocating compressors—Construction and working of a reciprocating compressor (single stage)—Single-stage compressor—equation for work (neglecting clearance)—Equation for work (with clearance volume)—Volumetric efficiency—Actual *p-V* diagram for single-stage compressor— Multi-stage compression—Efficiency of compressor—How to increase isothermal efficiency ?—Clearance in compressors—Effect of clearance volume—Free air delivered and displacement—Compressor performance—Effect of atmospheric conditions on the output of a compressor—Control of compressors—Arrangement of reciprocating compressors—Intercooler—Compressed air motors—Reciprocating air motor—Rotary type air motor. **24.4. Rotary compressors**—Classification—Displacement compressors—Steady flow compressors. **24.5. Comparison between reciprocating and centrifugal compressors. 24.6. Comparison between reciprocating and rotary air compressors. 24.7. Comparison between centrifugal and axial flow compressors.** *Highlights—Objective Type Questions—Theoretical Questions—Unsolved Examples.*

24.1. GENERAL ASPECTS

The compressed air finds application in the following *fields* :

1. It is widely employed for powering small engines, generally those of portable nature. Compressed air is used in such diversified fields as :

(*i*) operating tools in factories ;

(*ii*) operating drills and hammers in road building ;

(*iii*) excavating ;

(*iv*) tunneling and mining ;

(*v*) starting diesel engines ; and

(*vi*) operating brakes on buses, trucks and trains.

2. A large quantity of air at moderate pressure is used in smelting of various metals such as melting iron, in blowing converters, and cupola work.

3. Large quantities of air are used in the air-conditioning, drying, and ventilation fields. In many of these cases, there is little resistance to the flow of air ; and hence it does not have to be compressed (*i.e.,* measurably decreased in volume). For such cases fans serve the purpose of moving the air to the desired location. In other cases, particularly in drying work, there is appreciable resistance to the flow of air and a compressor of some sort is required to build up sufficient pressure to overcome the resistance to flow.

The function of a compressor is to take a definite quantity of fluid (usually gas, and most often air) *and deliver it at a required pressure.*

An air compressor takes in atmospheric air, compresses it and delivers the high-pressure air to a storage vessel from which it may be conveyed by the pipeline to wherever the supply of compressed air is required. Since the process of compressing the gas requires that work should be done upon it, it will be clear that a compressor must be driven by some form of prime mover. Of the energy received by the compressor from the prime mover, some will be absorbed in work done against friction, some will be lost to radiation and any coolant which might be employed to cool the machine, and the rest will be maintained within the air itself. The prime-mover converts only a fraction of the heat it receives from the source into work, and so far as the compressor alone is concerned, the energy which it receives is that which is available at the shaft of the prime mover.

The general arrangement of a compressor set is shown diagrammatically in Fig. 24.1.

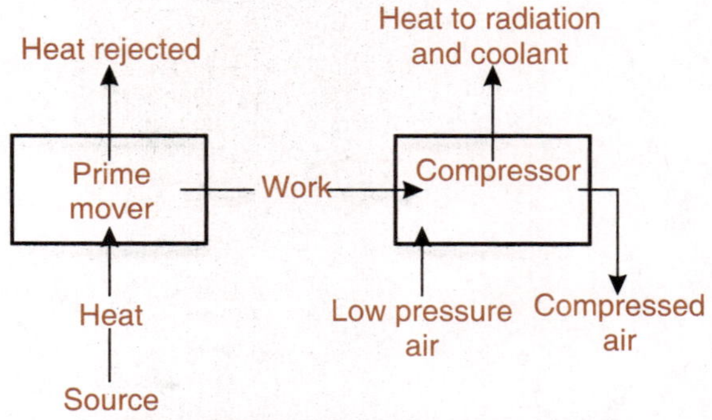

Fig. 24.1. General arrangement of a compressor set.

24.2. CLASSIFICATION OF AIR COMPRESSORS

Air and gas compressors are *classified* into *two main types* :

1. Reciprocating compressors ; and
2. Rotary compressors.

— According to whether or not the process of compressing is carried out in one unit or in several similar units in the one machine, a compressor may be *single-stage*, or *multi-stage.*

— Again, in case of reciprocating compressors, the air may be compressed in the cylinder on one side of the piston only, or use may be made of both piston faces. Such compressors are *single-acting* and *double-acting,* respectively.

— ***Centrifugal compressors,*** which are of the rotary type, may be single or double entry, which means that the compressor is filled with either one or two air intakes according to whether it is of the former or latter type when compression takes place in one or two units, respectively.

Air compressors may be classified in another manner, this time from an aspect of the *use to which they are put.*

— For example, air pumps and exhausters are used to produce vacua, their job being to remove air from a particular system to create a low pressure therein.

— *Blowers* and *superchargers* are essentially air compressors, but the *increase in pressure* which they produce is only small, and upto, say *0.7 to 1.05 bar.*

— A *booster is an air or gas compressor which is employed to raise the pressure of air/gas which has already been compressed.* It is where a slightly higher pressure is required, or where a loss of pressure has occurred in a long delivery line.

24.3. RECIPROCATING COMPRESSORS

24.3.1. Construction and Working of a Reciprocating Compressor (Single-stage)

Fig. 24.2 (*a*) shows a sectional view of a single-stage reciprocating compressor. It consists of a piston which reciprocates in a cylinder, driven through a connecting rod and crank mounted in a crankcase. There are inlet and delivery valves mounted in the head of the cylinder. These valves are usually of the pressure differential type, meaning that they will operate as the result of the difference of pressures across the valve. The working of this type of compressor is as follows :

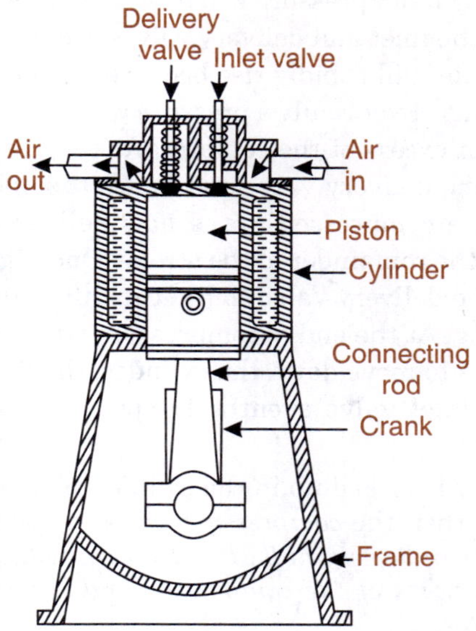

Fig. 24.2. (*a*) Sectional view of a single-stage reciprocating compressor.

As shown in Fig. 24.2 (*b*), the piston is moving down the cylinder and any residual compressed air left in the cylinder after the previous compression will expand and will eventually reach a pressure slightly below intake pressure early or in the stroke. This means that the pressure outside the inlet valve is now higher than on the inside and hence the inlet valve will lift off its seat. A stop is provided to limit its lift and to retain it in its valve seating. Thus a fresh charge of air will be aspirated into the cylinder for the remainder of the *induction stroke,* as it is called. During this stroke the delivery valve will remain closed, since the compressed air on the outside of this valve is at a much higher pressure than the induction stroke.

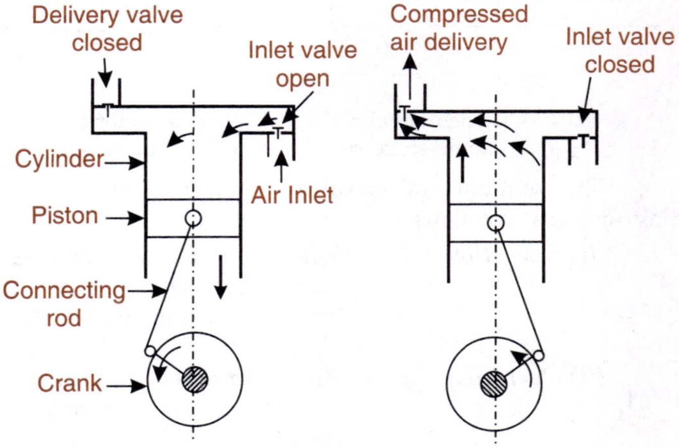

Fig. 24.2 (*b*) **Fig. 24.2** (*c*)

As shown in Fig. 24.2 (*c*) the piston is now moving upwards. At the beginning of this upward stroke, a slight increase in cylinder pressure will have closed the inlet valve. Since both the inlet and delivery valves are now closed, the pressure of air will rapidly rise because it is now locked up in the cylinder. Eventually a pressure will be reached which is slightly in excess of the compressed air pressure on the outside of the delivery valve and hence the delivery valve will lift. The compressed air is now delivered from the cylinder in the remainder of the stroke. Once again there is a stop on the delivery valve to limit its lift and to retain it in its seating. At the end of compression stroke piston once again begins to move down the cylinder, the delivery valve closes; the inlet valve eventually opens and the cycle is repeated.

As air is locked up in the cylinder of a reciprocating compressor then the *compression pressure* for this type of compressor *can be very high. It is limited by the strength of the various parts of the compressor and the power of the driving motor.*

It may be noted that there is *intermittent flow of air in a reciprocating air compressor.*

24.3.2. Single-stage Compressor : Equation for Work (*neglecting clearance*)

In Fig. 24.3 is shown a theoretical *p-V* diagram for a single-stage reciprocating air compressor, neglecting clearance.

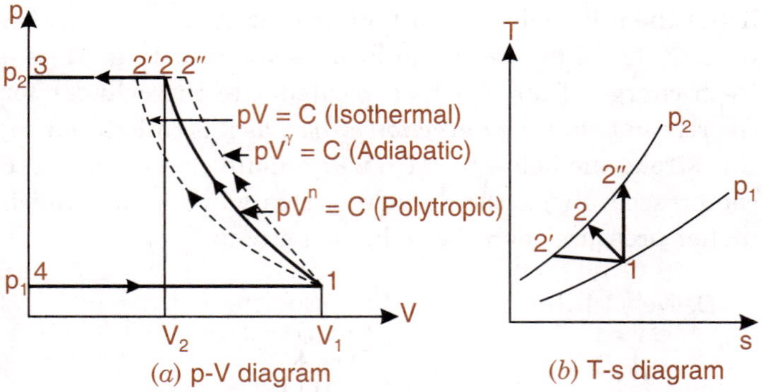

Fig. 24.3. Theoretical *p-V* and *T-s* diagrams for a single-stage reciprocating air compressor.

The sequence of operations as represented on the diagram, are as follows :

(*i*) *Operation 4-1* : Volume of air V_1 aspirated into the compressor at pressure p_1 and temperature T_1.

(*ii*) *Operation 1-2* : Air compressed according to the law $pV^n = C$ from p_1 to pressure p_2. Volume decreases from V_1 to V_2. Temperature increases from T_1 to T_2.

(*iii*) *Operation 2-3* : Compressed air of volume V_2 and at pressure p_2 with temperature T_2 delivered from the compressor.

During compression, due to its excess temperature above the compressor surroundings, the air will lose some heat. Thus, neglecting the internal effect of friction which is small in the case of the reciprocating compressor, the index *n is less than* γ, the adiabatic index. Since work must be put into an air compressor to run it, every effort is made to reduce this amount of work input. Inspection of *p-V* diagram shows the frictionless adiabatic as 1-2″ and that if compression were along the isothermal 1-2′ instead of polytropic 1-2 then the work done, given by the area of the diagram, would be reduced and, infact, would then be '*minimum*'. *Isothermal compression cannot be achieved in practice but an attempt is made to approach the isothermal case by cooling the compressor either by addition of cooling fins or a water jacket to the compressor cylinder.* For a reciprocating compressor, a comparison between the actual work done during compression and the ideal isothermal work done is made by means of the ***isothermal efficiency.***

This is defined as,

$$\text{Isothermal efficiency} = \frac{\text{Isothermal work done}}{\text{Actual work done}}$$

Thus, the higher the isothermal efficiency, the more nearly has the actual compression approached the ideal isothermal compression.

Total shaft work done/cycle, W = Area 41234

or W = Area under 4-1 – Area under 1-2 – Area under 2-3

$$= p_1V_1 - \frac{p_2V_2 - p_1V_1}{n-1} - p_2V_2$$

$$= (p_1V_1 - p_2V_2) - \left(\frac{p_2V_2 - p_1V_1}{n-1}\right)$$

$$= (p_1V_1 - p_2V_2) + \left(\frac{p_1V_1 - p_2V_2}{n-1}\right)$$

$$= \left(1 + \frac{1}{n-1}\right)(p_1V_1 - p_2V_2)$$

$$\therefore \quad W = \left(\frac{n}{n-1}\right)(p_1V_1 - p_2V_2) \quad \text{...(24.1)}$$

This equation can be modified as follows :

$$W = \frac{n}{n-1}(p_1V_1 - p_2V_2)$$

$$= \frac{n}{n-1} \cdot p_1V_1\left(1 - \frac{p_2V_2}{p_1V_1}\right) \quad \text{...(24.2)}$$

Now, $p_1V_1^n = p_2V_2^n$

$$\therefore \quad \frac{V_2}{V_1} = \left(\frac{p_1}{p_2}\right)^{1/n}$$

and substituting this into eqn. (24.2), we have

$$W = \frac{n}{n-1} p_1V_1 \left\{1 - \frac{p_2}{p_1}\left(\frac{p_1}{p_2}\right)^{1/n}\right\}$$

$$= \frac{n}{n-1} p_1V_1 \left\{1 - \frac{p_2}{p_1}\left(\frac{p_2}{p_1}\right)^{-\frac{1}{n}}\right\}$$

$$= \frac{n}{n-1} p_1V_1 \left\{1 - \left(\frac{p_2}{p_1}\right)^{1-\frac{1}{n}}\right\}$$

$$= \frac{n}{n-1} p_1V_1 \left\{1 - \left(\frac{p_2}{p_1}\right)^{\frac{n-1}{n}}\right\} \quad ...(24.3)$$

The solution to this equation will always come out *negative* showing that work must be done *on* the compressor. Since only the *magnitude* of the work done is required from the expression then it is often written,

$$W = \frac{n}{n-1} p_1V_1 \left\{\left(\frac{p_2}{p_1}\right)^{\frac{n-1}{n}} - 1\right\} \quad ...(24.4)$$

$$= \frac{n}{n-1} mRT_1 \left\{\left(\frac{p_2}{p_1}\right)^{\frac{n-1}{n}} - 1\right\} \quad ...(24.5)$$

If the air delivery temperature T_2 is required then this can be obtained by using this equation :

$$\frac{T_2}{T_1} = \left(\frac{p_2}{p_1}\right)^{\frac{n-1}{n}} \quad \text{or} \quad T_2 = T_1\left(\frac{p_2}{p_1}\right)^{\frac{n-1}{n}} \quad ...(24.6)$$

24.3.3. Equation for Work (with clearance volume)

In practice, all reciprocating compressors will have a *clearance volume.* The clearance volume is *that volume which remains in the cylinder after the piston has reached the end of its inward stroke.*

Refer to Fig. 24.4. At point 1, the cylinder is full of intake air, volume V_1 and the piston is about to commence its compression stroke. The air is compressed polytropically according to some law $pV^n = C$ to delivery pressure p_2 and volume V_2. At 2 the delivery valve theoretically opens and for the remainder of the stroke, 2 to 3, the compressed air is delivered from the cylinder. At 3 the piston has reached the end of its inward stroke and so on, delivery of compressed air ceases at 3. V_3 is the clearance volume and is filled at this stage with compressed air. As the piston begins the intake stroke *this residual air will expand,* according to some polytropic law $pV^n = C$, and it is not *until the pressure has reduced to intake pressure* at 4 that the inlet valve will begin to open thus permitting the intake of a fresh charge of air. For the remainder of the intake stroke a fresh charge is taken into the cylinder. This volume $(V_1 - V_4)$ is *effective swept volume.*

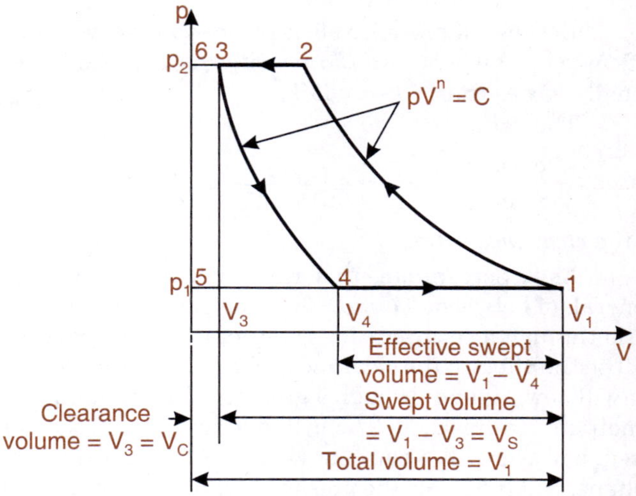

Fig. 24.4

Work done/cycle,

W = Net area 12341 = Area 51265 – Area 54365

Assuming the polytropic index to be same for both compression and clearance expansion, then,

$$W = \frac{n}{n-1} p_1V_1 \left\{\left(\frac{p_2}{p_1}\right)^{\frac{n-1}{n}} - 1\right\} - \frac{n}{n-1} p_4V_4 \left\{\left(\frac{p_3}{p_4}\right)^{\frac{n-1}{n}} - 1\right\} \quad ...(24.7)$$

But $p_4 = p_1$ and $p_3 = p_2$, then eqn. (24.7) becomes,

$$W = \frac{n}{n-1} p_1V_1 \left\{\left(\frac{p_2}{p_1}\right)^{\frac{n-1}{n}} - 1\right\} - \frac{n}{n-1} p_1V_4 \left\{\left(\frac{p_2}{p_1}\right)^{\frac{n-1}{n}} - 1\right\}$$

$$= \frac{n}{n-1} p_1(V_1 - V_4) \left\{\left(\frac{p_2}{p_1}\right)^{\frac{n-1}{n}} - 1\right\} \quad ...(24.8)$$

24.3.4. Volumetric Efficiency

Refer to Fig. 24.4. *The **volumetric efficiency** of a compressor is the ratio of free air delivered to the displacement of the compressor.* It is also the *ratio of effective swept volume to the swept volume.*

i.e., Volumetric efficiency

$$= \frac{\text{Effective swept volume}}{\text{Swept volume}} = \frac{V_1 - V_4}{V_1 - V_3} \quad ...(24.9)$$

Because of presence of clearance volume, volumetric efficiency is always *less than unity.* As a percentage, it usually varies from *60% to 85%.*

The ratio,

$$\frac{\text{Clearance volume}}{\text{Swept volume}} = \frac{V_3}{V_1 - V_3} = \frac{V_c}{V_s} = k \quad ...(24.10)$$

is the *clearance ratio.*

As a percentage, this ratio will have a value, in general, of between *4% and 10%.* The greater the pressure ratio through a reciprocating compressor, then the greater will be the effect of the clearance volume since the clearance air will now expand through a greater volume before intake conditions are reached. The cylinder size and stroke being fixed, however will mean that $(V_1 - V_4)$, the effective swept volume, will reduce as the pressure ratio increases and thus the volumetric efficiency reduces.

Volumetric efficiency,

$$\eta_{vol.} = \frac{V_1 - V_4}{V_1 - V_3} = \frac{(V_1 - V_3) + (V_3 - V_4)}{(V_1 - V_3)}$$

$$= 1 + \frac{V_3}{V_1 - V_3} - \frac{V_4}{V_1 - V_3}$$

$$= 1 + \frac{V_3}{V_1 - V_3} - \frac{V_4}{V_1 - V_3} \cdot \frac{V_3}{V_3}$$

$$= 1 + \frac{V_3}{V_1 - V_3} - \frac{V_3}{V_1 - V_3} \cdot \frac{V_4}{V_3}$$

$$= 1 + k - k \cdot \frac{V_4}{V_3}$$

$$= 1 + k - k \left(\frac{p_3}{p_4}\right)^{1/n} \qquad \left| \begin{array}{l} p_3 V_3^{\,n} = p_4 V_4^{\,n} \\ \dfrac{V_4}{V_3} = \left(\dfrac{p_3}{p_4}\right)^{1/n} \end{array} \right|$$

or $$\eta_{vol.} = 1 + k - k \left(\frac{p_2}{p_1}\right)^{1/n} \qquad (\because \; p_3 = p_2, p_4 = p_1) \quad ...(24.11)$$

or $$\eta_{vol.} = 1 + k - k \left(\frac{V_1}{V_2}\right) \quad ...(24.12)$$

The above equations are valid if the index of expansion and compression is same. However, it may be noted that the *clearance volumetric efficiency is dependent on only the index of expansion of the clearance volume from V_3 to V_4.* Thus, if the index of compression = n_c and index of expansion = n_e, the volumetric efficiency is given by

$$\eta_{vol.} = 1 + k - k \left(\frac{p_3}{p_4}\right)^{1/n_e} \quad ...(24.13)$$

$$= 1 + k - k \left(\frac{p_2}{p_1}\right)^{1/n_e} \quad ...(24.14)$$

$$= 1 + k - k \left(\frac{V_4}{V_3}\right) \quad ...(24.15)$$

In this case volumetric efficiency $= 1 + k - k \left(\dfrac{V_1}{V_2}\right)$.

In practice the air that is sucked in during the induction (suction) stroke gets heated up while passing through the hot valves and coming in contact with hot cylinder walls. There is wire drawing effect through the valves resulting in drop in pressure. Thus the ambient conditions are different from conditions obtained at state 1 in Fig. 24.4.

Let $p_{amb.}$ = Pressure of ambient air, and

$T_{amb.}$ = Temperature of ambient air

$$\therefore \quad \frac{p_{amb.} V_{amb.}}{T_{amb.}} = \frac{p_1 (V_1 - V_4)}{T_1}$$

Thus, $$V_{amb.} = \frac{p_1 \times T_{amb.}}{T_1 \times p_{amb.}} \times (V_1 - V_4)$$

Thus volumetric efficiency referred to ambient conditions may be written as

$$\eta_{vol.\,(amb.)} = \frac{V_{amb.}}{V_1 - V_3} = \frac{p_1 \times T_{amb.}}{T_1 \times p_{amb.}} \times \frac{V_1 - V_4}{V_1 - V_3}$$

But from eqn. (24.11)

$$\frac{V_1 - V_4}{V_1 - V_3} = 1 + k - k \left(\frac{p_2}{p_1}\right)^{1/n}$$

$$\therefore \quad \eta_{vol.\,(amb.)} = \frac{p_1 \times T_{amb.}}{T_1 \times p_{amb.}} \left[1 + k - k \left(\frac{p_2}{p_1}\right)^{1/n}\right] \quad ...(24.16)$$

$$= \frac{p_1 \times T_{amb.}}{T_1 \times p_{amb.}} \left[1 + k - k \left(\frac{V_2}{V_1}\right)\right] \quad ...(24.17)$$

(This efficiency should not be used for finding out the dimensions of the cylinder. For finding out the dimensions of the cylinder, the volumetric efficiency based on suction condition only should be used).

Fig. 24.5 shows the manner in which the volumetric efficiency varies with delivery pressure. Theoretically, the volumetric efficiency is 100% when the delivery pressure equals that of the surroundings, and in fact no compression takes place at all. It decreases rapidly with increase in delivery pressure at first, and then more slowly for increase in delivery pressure at higher pressure.

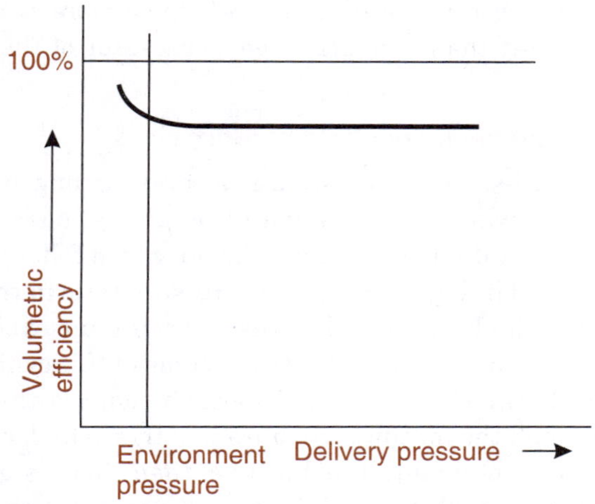

Fig. 24.5. Variation of volumetric efficiency with delivery pressure.

The *volumetric efficiency is lowered* by any of the following conditions :

(*i*) Very high speed
(*ii*) Leakage past the piston
(*iii*) Too large a clearance volume
(*iv*) Obstruction at inlet valves
(*v*) Overheating of air by contact with hot cylinder walls.
(*vi*) Inertia effect of air in suction pipe.

By paying careful attention in the design of the compressor to these causes of loss, an improvement in volumetric efficiency can be obtained.

24.3.5. Actual p-V (indicator) Diagram for Single-stage Compressor

Fig. 24.6 shows an actual compressor diagram. 1234 is the theoretical *p*-V diagram (already discussed). At point 4, when the clearance air has reduced to atmospheric pressure. The inlet valve in practice will not open. There are two main reasons for this : (*i*) there must be a pressure difference across the inlet valve in order to move it and (*ii*) inlet valve inertia. Thus, the pressure drops away until the valve is forced off its seat. Some *valve bounce* will then set in, as shown by the wavy line, and eventually intake will become near enough-steady at some pressure below atmospheric pressure. This negative pressure difference, called the *intake depression* settles naturally, showing that what is called suction is really the atmospheric air forcing its way into the cylinder against reduced pressure. A similar situation occurs at 2, at the beginning of compressed air delivery. There is a constant pressure rise, followed by valve bounce and the pressure then settles at some pressure above external delivery pressure. Compressed air is usually delivered into a tank called the receiver and hence external delivery pressure is sometimes called the receiver pressure.

Other small effects at inlet and delivery would be *gas inertia* and *turbulence.*

The practical effects (discussed above) are responsible for the addition of the two small shaded *negative work areas* shown in Fig. 24.6.

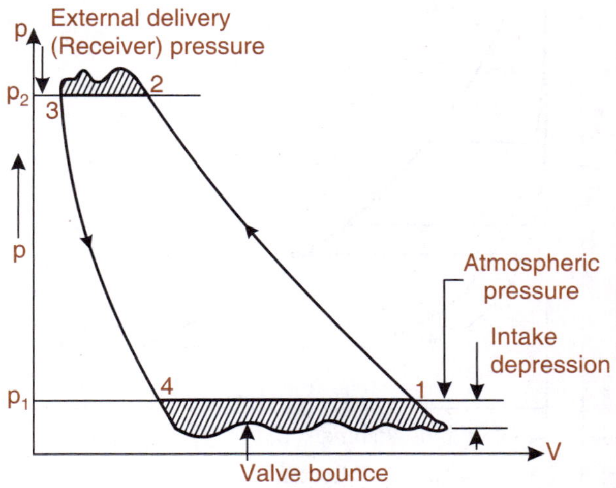

Fig. 24.6. Actual compressor *p-V* diagram.

24.3.6. Multi-stage Compression

In a single-stage reciprocating compressor if the delivery is restricted the delivery pressure will increase. If the delivery pressure is increased too far, however, certain disadvantages appear. Referring to Fig. 24.7 assume that the single-stage compressor is compressing to pressure p_2, the complete cycle to 1234. Clearance air expansion will be 3-4 and mass flow through the compressor will be controlled by the effective swept volume ($V_1 - V_4$). Assume now that a restriction is now placed on delivery. The delivery pressure becomes p_5 say, and the cycle is then 1567, clearance expansion being 6-7. The mass flow through the compressor is now controlled by effective swept volume ($V_1 - V_7$), which is less than ($V_1 - V_4$). In the limit, assuming the compressor to be strong enough, the compression would take place 1-8, where V_s = clearance volume, in which case there would be no delivery. It is seen, therefore, that as the *delivery pressure for a single-stage, reciprocating compressor is increased so the mass flow through the compressor decreases.* Note, also that as *the delivery pressure is increased, so also will the delivery temperature increase.* Referring to Fig. 24.7, $T_8 > T_5 > T_2$. *If high temperature air is not a requirement of the compressed air delivered, then, any increase in temperature represents an energy loss.*

If high pressure is to be delivered by a single-stage machine then it will require heavy working parts in order to accommodate the high pressure ratio through the machine. This will increase the balancing problem and the high torque fluctuation will require a heavier flywheel installation. Such disadvantages can be overcome by multi-stage compression.

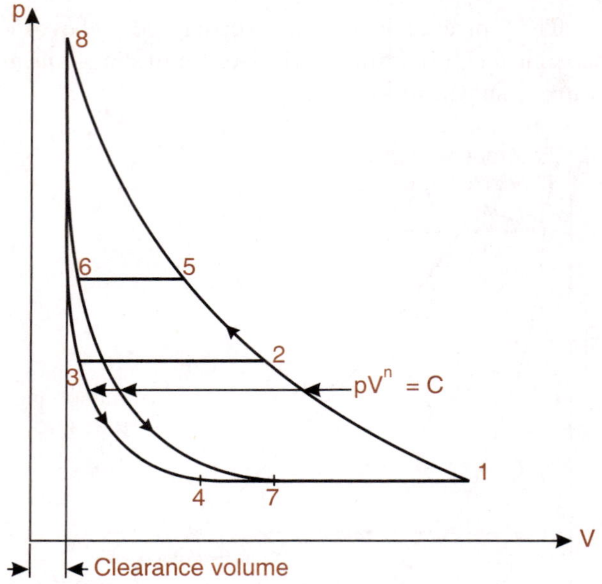

Fig. 24.7

Multi-stage compression is *very efficient* and is now-a-days almost universally adopted except for compressors where the overall pressure rise required is small. The method is not only advantageous from a thermodynamic point of view, but also has mechanical advantages over single-stage compression.

Advantages :

The important *advantages of multi-stage compression* can be summed up as follows :

1. The *air can be cooled* at pressures intermediate between intake and delivery pressures.
2. The *power* required to drive a multi-stage machine is *less* than would be required by a single-stage machine delivering the same quantity of air at the same delivery pressure.
3. Multi-stage machines have *better mechanical balance*.
4. The pressure range (and hence also the temperature range) may be kept within desirable limits. This results in (*i*) *reduced losses due to air leakage* (*ii*) *improved lubrication,* due to lower temperatures and (*iii*) *improved volumetric efficiency*.
5. The cylinder, in a single-stage machine, must be robust enough to withstand the delivery pressure. The down pressure cylinders of a multi-stage machine may be *lighter in construction* since the maximum pressure therein is low.

Disadvantages :

Inspite of all these advantages, a multi-stage compressor with intercoolers is likely to be more expensive in initial cost than a single-stage compressor of the same capacity.

Multi-stage reciprocating compressors

Multi-stage compression is a series arrangement of cylinders in which compressed air from the cylinder before, becomes the intake air for the cylinder which follows. This is illustrated in Fig. 24.8. The low pressure ratio in the low-pressure cylinder means that the clearance air expansion is reduced and the effective swept volume of this cylinder is increased. Since in this cylinder which controls the mass flow through the machine, because it is this cylinder which introduces the air into the machine, then there is greater mass flow through the multi-stage arrangement than the single-stage machine.

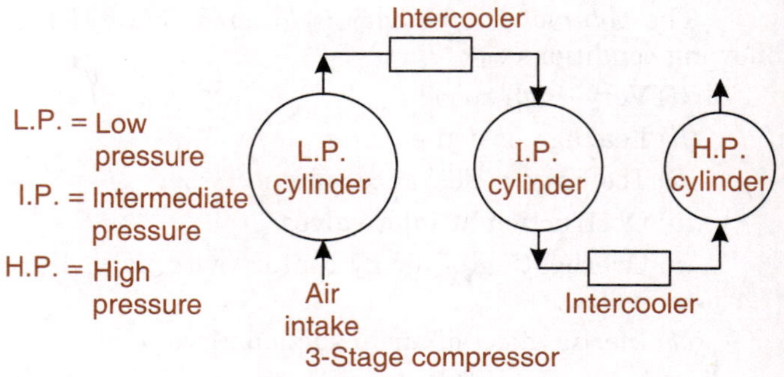

Fig. 24.8

If an *intercooler* is installed between cylinders, in which the compressed air is cooled between cylinders, then the *final delivery temperature is reduced.* This reduction in temperature means a reduction in internal energy of the delivered air, and since this energy must have come from the input energy required to drive the machine, this results in a *decrease in input work requirement for a given mass of delivered air*.

It is common to find machines with either two or three stages of compression. *The complexity of the machinery limits the number of stages.*

Refer to Fig. 24.8. The cylinders are shown with diameters which decrease as the pressure increases. This is because, as the pressure increases, so the volume of a given mass of gas decreases. There is continuity of mass flow through a compressor and hence each following cylinder will require a smaller volume due to its increased pressure range. This reduction in volume is usually accomplished by reducing the cylinder diameter.

Fig. 24.9 shows cycle arrangements in the development of the *ideal conditions* required for multi-stage compression. For simplicity, clearance is neglected.

Referring to Fig. 24.9, the overall pressure range is p_1 to p_3. Cycle 8156 is that of single- stage compressor. Cycles 8147 and 7456 are that of a two-stage compressor without intercooling between cylinders. Cycles 8147 and 7236 are that of a two-stage compressor with *perfect intercooling* between cylinders.

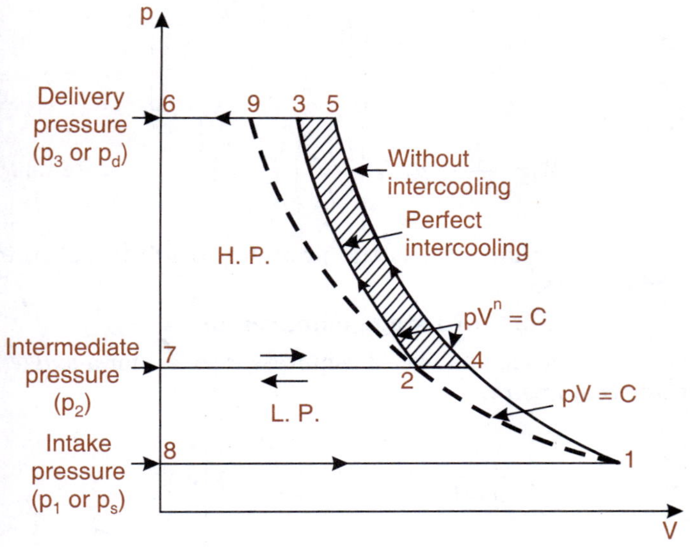

Fig. 24.9

'Perfect intercooling' means that after the initial compression in the L.P. cylinder, with its consequent temperature rise, the air is cooled in an intercooler back to its original temperature. This means, referring to Fig. 21.9, $T_2 = T_1$, in which case point 2 lies on isothermal through point 1. This shows that multi-stage compression, with perfect intercooling, approaches more closely the ideal isothermal compression than in the case with single-stage compression.

Ideal conditions for multi-stage compressors :

Case 1. Single-stage compressor :

As earlier stated cycle 8156 is that of a single-stage compressor, neglecting clearance. For this cycle,

$$W = \frac{n}{n-1} p_1V_1 \left[\left(\frac{p_5}{p_1} \right)^{\frac{n-1}{n}} - 1 \right] \quad ...(24.18)$$

Delivery temperature, $T_5 = T_1 \left(\frac{p_5}{p_1} \right)^{\frac{n-1}{n}}$ ...(24.19)

Case 2. Two-stage compressor :

(*i*) ***Without intercooling***

8147 ...... Low pressure cycle

7456 ...... High pressure cycle.

For this arrangement work done,

$$W = \frac{n}{n-1} p_1V_1 \left[\left(\frac{p_4}{p_1} \right)^{\frac{n-1}{n}} - 1 \right] + \frac{n}{n-1} \cdot p_4V_4 \left[\left(\frac{p_5}{p_4} \right)^{\frac{n-1}{n}} - 1 \right] \quad ...(24.20)$$

This will give the same result as that of eqn. (24.18). The final delivery temperature will also be given by eqn. (24.19), because there is no intercooling.

(*ii*) ***With perfect intercooling***

8147 ...... Low pressure cycle

7236 ...... High pressure cycle.

For this arrangement,

$$W = \frac{n}{n-1} p_1V_1 \left[\left(\frac{p_4}{p_1} \right)^{\frac{n-1}{n}} - 1 \right] + \frac{n}{n-1} p_2V_2 \left[\left(\frac{p_3}{p_2} \right)^{\frac{n-1}{n}} - 1 \right] \quad ...(24.21)$$

Delivery temperature is given by

$$T_3 = T_2 \left(\frac{p_3}{p_2} \right)^{\frac{n-1}{n}} = T_1 \left(\frac{p_3}{p_2} \right)^{\frac{n-1}{n}}, \text{ Since } T_2 = T_1 \quad ...(24.22)$$

Now, since $T_2 = T_1$, then

$$p_2V_2 = p_1V_1 \quad ...(24.23)$$

Also $p_4 = p_2$ ...(24.24)

Inserting eqns. (24.23) and (24.24) in eqn. (24.21)

$$W = \frac{n}{n-1} p_1V_1 \left[\left(\frac{p_2}{p_1} \right)^{\frac{n-1}{n}} + \left(\frac{p_3}{p_2} \right)^{\frac{n-1}{n}} - 2 \right] \quad ...(24.25)$$

Now, inspection of Fig. 24.9 *shows the shaded area 2453 which is the work saving which occurs as a result of using an intercooler.*

Conditions for minimum work :

It will be observed from the Fig. 24.9 that as intermediate pressure $p_2 \to p_1$, then area 2453 → 0. Also as $p_2 \to p_3$, then area 2453 → 0. This means, therefore, that an intermediate pressure p_2 exists which makes area 2453 a maximum. This is the condition when W is a minimum.

Inspection of eqn. 24.25 shows that for minimum W, $[(p_2/p_1)^{n-1/n} + (p_3/p_2)^{n-1/n}]$ must be minimum, all other parts of the equation being constant in this consideration and p_2 is the variable.

Hence, for *minimum*, W,

$$\frac{dW}{dp_2} = \frac{d\left[\left(\frac{p_2}{p_1}\right)^{\frac{n-1}{n}} + \left(\frac{p_3}{p_2}\right)^{\frac{n-1}{n}}\right]}{dp_2} = 0$$

Differentiating, with respect to p_2,

$$\frac{1}{(p_1)^{\frac{n-1}{n}}} \times \left(\frac{n-1}{n}\right) p_2^{\left(\frac{n-1}{n}\right)-1} + (p_3)^{\frac{n-1}{n}} \times -\left(\frac{n-1}{n}\right)(p_2)^{-\left(\frac{n-1}{n}\right)-1} = 0$$

or $$\frac{1}{p_1^{\frac{n-1}{n}}} \times \left(\frac{n-1}{n}\right)(p_2)^{-1/n} = (p_3)^{\frac{n-1}{n}} \times \left(\frac{n-1}{n}\right) p_2^{\frac{-2n+1}{n}}$$

or $$\frac{p_2^{-1/n}}{p_2^{\left(\frac{-2n+1}{n}\right)}} = (p_1p_3)^{\frac{n-1}{n}}$$

or $$p_2^{-1/n} \times p_2^{-\left(\frac{-2n+1}{n}\right)} = (p_1p_3)^{\frac{n-1}{n}}$$

or $$p_2^{-1/n}\, p_2^{\frac{2n-1}{n}} = (p_1p_3)^{\frac{n-1}{n}}$$

or $$p_2^{\frac{2n-2}{n}} = (p_1p_3)^{\frac{n-1}{n}}$$

$\therefore$ $$p_2^2 = p_1p_3 \qquad \text{...(24.26)}$$

or $$p_2 = \sqrt{p_1p_3} \qquad \text{...(24.27)}$$

and $$\frac{p_2}{p_1} = \frac{p_3}{p_2} \qquad \text{...(24.28)}$$

or *pressure ratio per stage is equal.*

p_2 obtained from eqn. (24.27) will give ideal intermediate pressure which, with perfect intercooling, will give the minimum W.

With these ideal conditions, inserting equations (24.23), (24.24) and (24.28) into eqn. (24.21) shows that *there is equal work per cylinder.*

Hence, $$W = \frac{2n}{n-1} p_1V_1 \left[\left(\frac{p_2}{p_1}\right)^{\frac{n-1}{n}} - 1\right] \qquad \text{...(24.29)}$$

Inserting eqn. (24.27) in eqn. (24.29), we get

$$W = \frac{2n}{n-1} p_1V_1 \left[\left\{\frac{(p_1p_3)^{1/2}}{p_1}\right\}^{\frac{n-1}{n}} - 1\right]$$

$$= \frac{2n}{n-1} p_1V_1 \left[\left\{\left(\frac{p_3}{p_1}\right)^{1/2}\right\}^{\frac{n-1}{n}} - 1\right]$$

$$\therefore \quad W = \frac{2n}{n-1} p_1V_1 \left[\left(\frac{p_3}{p_1}\right)^{\frac{n-1}{2n}} - 1\right] \qquad \text{...(24.30)}$$

Note that p_3/p_1 is the pressure ratio through the compressors.

Case 3. Multi-stage compressor :

From the analysis of compressor so far, for a *single-stage* compressor,

$$W = \frac{n}{n-1} p_1V_1 \left[\left(\frac{p_2}{p_1}\right)^{\frac{n-1}{n}} - 1\right]$$

For a *two-stage* compressor,

$$W = \frac{2n}{n-1} p_1V_1 \left[\left(\frac{p_3}{p_1}\right)^{\frac{n-1}{2n}} - 1\right]$$

It seems reasonable to assume, therefore, that for a *three-stage* machine,

$$W = \frac{3n}{n-1} p_1V_1 \left[\left(\frac{p_4}{p_1}\right)^{\frac{n-1}{3n}} - 1\right]$$

and for *x-stage* compressor,

$$W = \frac{xn}{n-1} p_1V_1 \left[\left(\frac{p_{(x+1)}}{p_1}\right)^{\frac{n-1}{xn}} - 1\right] \qquad \text{...(24.31)}$$

This equation is very important, since it *applies to any type of compressor or motor, and even to vapour engines, provided n = or < γ.*

Note that $\frac{p_{(x+1)}}{p_1}$ is the pressure ratio through the compressor, in each case.

Note, also, that since for an ideal compressor there is equal work per cylinder, for an x-stage compressor

$$W = \frac{xn}{n-1} p_1V_1 \left[\left(\frac{p_2}{p_1}\right)^{\frac{n-1}{n}} - 1\right] \qquad \text{...(24.32)}$$

To determine the intermediate pressures for an *x-stage* machine running under ideal conditions, use is made of eqn. (24.28).

Here it is shown that the pressure ratio per stage is equal.

Hence, for an *x-stage* machine,

$$\frac{p_2}{p_1} = \frac{p_3}{p_2} = \frac{p_4}{p_3} \; ..... \; \frac{p_{(x+1)}}{p_x} = Z, \text{ say} \quad ...(24.33)$$

From this,

$$p_2 = Zp_1$$
$$p_3 = Zp_2 = Z^2p_1$$
$$p_4 = Zp_3 = Z^3p_1$$
$$\vdots$$
$$p_{x+1} = Zp_x = Z^xp_1$$

$$\therefore \quad Z^x = \frac{p_{x+1}}{p_1}$$

or
$$Z = \sqrt[x]{\frac{p_{(x+1)}}{p_1}}$$

$$= \sqrt[x]{\text{(Pressure ratio through compressor)}} \quad ...(24.34)$$

Inserting the value of Z in eqn. (24.33) will determine the intermediate pressures.

In the event of *intercooling being imperfect we must treat each stage as a separate compressor*, in which case 'x' in eqn. (24.31) will be unity. With this special value of 'x' the power per stage can be calculated, and finally the total power is the sum of the powers per stage :

$$W = \frac{n_1}{n_1 - 1} p_1V_1 \left[\left(\frac{p_2}{p_1}\right)^{\frac{n_1-1}{n_1}} - 1\right]$$
$$+ \frac{n_2}{n_2 - 1} p_2V_2 \left[\left(\frac{p_3}{p_2}\right)^{\frac{n_2-1}{n_2}} - 1\right] + ...$$

Heat rejection per stage per kg of air :

If the air is cooled to its initial temperature the whole of the work done in compression must be rejected to the cooling medium.

Hence for a *single-stage* the heat rejected is given by,

Heat rejected

$$W = \frac{n}{n-1} p_1V_1 \left[\left(\frac{p_2}{p_1}\right)^{\frac{n-1}{n}} - 1\right] \quad ...(24.35)$$

and since for 1 kg of air, $p_1V_1 = RT_1$ and $\frac{T_2}{T_1} = \left(\frac{p_2}{p_1}\right)^{\frac{n-1}{n}}$,

then eqn. (24.35) may be written as

$$W = \frac{n}{n-1} RT_1 \left[\frac{T_2}{T_1} - 1\right] \text{ per kg of air}$$

$$= \frac{n}{n-1} \frac{R}{J} (T_2 - T_1) \text{ heat units}$$

$$\begin{bmatrix} J = 1 \text{......S.I. units} \\ J = 427 \text{...... M.K.S. units} \end{bmatrix}$$

But
$$\frac{R}{J} = c_p - c_v$$

$$W = \frac{n}{n-1} (c_p - c_v)(T_2 - T_1) \quad ...(24.36)$$

$\therefore$ ***Heat rejected with perfect intercooling***

$$= \left[c_p + c_v\left(\frac{\gamma - n}{n-1}\right)\right] (T_2 - T_1) \text{ per kg of air} \quad ...(24.37)$$

$$\left[\textbf{Note. } \frac{n}{n-1}(c_p - c_v) = c_p + \frac{c_v(\gamma - n)}{n-1}\right]$$

The first term in eqn. (24.37) represents the *heat rejected at constant pressure in the intercooler* ; whilst the second term represents the *heat rejected during compression alone* ; and writing $c_v = R/J\,(\gamma - 1)$ it may be reduced to the form

$$\frac{\gamma - n}{\gamma - 1} \times \text{work done in heat units}$$

To find the change in entropy (s) during the first stage of compression, we have, from the definition of entropy,

$$ds = \frac{dW}{T} = \frac{dQ}{T} \quad (\because \text{ Work done = Heat rejected})$$

Differentiating eqn. (24.37) and dividing by T,

$$ds = \frac{dW}{T} = \left[c_p + c_v\left(\frac{\gamma - n}{n-1}\right)\right] \frac{dT}{T}$$

Integration gives the change in s as

$$(s_2 - s_1) = \left[c_p + c_v\left(\frac{\gamma - n}{n-1}\right)\right] \log_e T_2/T_1$$

$$= \frac{n}{n-1} (c_p - c_v) \log_e T_2/T_1$$

$$\left(\text{Substituting, } \gamma = \frac{c_p}{c_v}\right) \quad ...(24.38)$$

For the complete isothermal *two-stage* compression the change in entropy,

$$(s_2 - s_1) = \frac{R}{J} \log_e p_3/p_1 = (c_p - c_v) \log_e p_3/p_1 \quad ...(24.39)$$

But if the work done in *stage compression is* to be *minimum*

$$\frac{T_2}{T_1} = \left(\frac{p_2}{p_1}\right)^{\frac{n-1}{n}} = \left(\frac{p_3}{p_1}\right)^{\frac{n-1}{2n}} \quad ...(24.40)$$

By inserting eqn. (24.40) in eqn. (24.38)

$$(s_2 - s_1) = \frac{n}{n-1}(c_p - c_v)\log_e\left(\frac{p_3}{p_1}\right)^{\frac{n-1}{2n}}$$

$$= \left(\frac{c_p - c_v}{2}\right)\log_e p_3/p_1 \quad \text{...(24.41)}$$

Comparing eqn. (24.39) and eqn. (24.41) it will be seen that the one is half the value of the other ; hence the work done per stage is minimum when increase in entropy per stage

$$= \frac{\text{Isothermal increase in entropy for whole compression}}{\text{Number of stages}}$$

and the *maximum temperature per stage is constant* and equal to T_2.

Actual p-V (indicator) diagram for two-stage compressor

The actual indicator diagram for a two-stage compressor is shown in Fig. 24.10. The wavy lines during induction and delivery strokes are due to "Flatter" of the disc valves. The L.P. and H.P. diagrams *overlap due to pressure drop in the intercooler,* and, of course, clearance effects are plainly visible in the actual indicator diagram. *The inertia and friction effects which result in valve flatter increase the area of the diagram slightly, and hence their effect is to increase the total work of compression.*

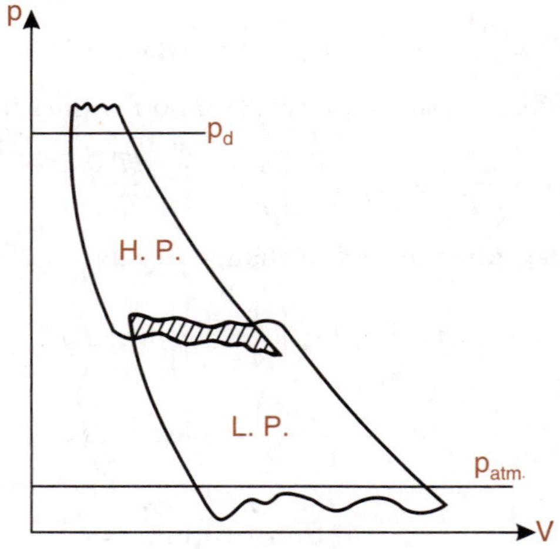

Fig. 24.10

24.3.7. Efficiency of Compressor

The theoretical horse power of a compressor is calculated on the assumption that the compression curve of *p-V* diagram is an isothermal. Then,

Isothermal work done/cycle = Area of *p-V* diagram

$$= p_1 V_1 \log_e r$$

$$\text{Isothermal power} = \frac{p_1 V_1 \log_e r \times N}{60 \times 1000} \text{ kW} \quad \text{...(24.42)}$$

In M.K.S. units

$$\text{Isothermal horse power} = \frac{p_1 V_1 \log_e r \times N}{4500}$$

where, N = number of cycles/min.

...[24.42 (*a*)]

The indicated power of a compressor is the power obtained from the actual indicator card taken during a test on the compressor,

$$\textit{Compressor efficiency} = \frac{\text{isothermal horse power}}{\text{indicated horse power}}$$

$$\textit{Isothermal efficiency} = \frac{\text{isothermal horse power}}{\text{shaft horse power}}$$

where the shaft horse power is the brake horse power required to drive the compressor. A usual value of isothermal efficiency is about 70 per cent.

The '*adiabatic efficiency*' of an air compressor is the ratio of the horse power required to drive the compressor compared with the area of the hypothetical indicator diagram assuming adiabatic compression.

or Adiabatic efficiency,

$$\eta_{adiabatic} = \frac{\left(\frac{\gamma}{\gamma-1}\right)p_1V_1\left[\left(\frac{p_2}{p_1}\right)^{\frac{\gamma-1}{\gamma}} - 1\right] \times \frac{N}{4500}}{\text{B.H.P. required to drive the compressor}} \quad \text{...(24.43)}$$

24.3.8. How to Increase Isothermal Efficiency ?

The following methods are employed to achieve nearly isothermal compression for high speed compressors :

(1) **Spray injection.** This method (used some years ago) assimilates the practice of injecting water into the compressor cylinder towards the compression stroke with the object of cooling the air. It entails the following *demerits* :

(*i*) It necessitates the use of a *special gear for injection* ;

(*ii*) The injected water *interferes with the cylinder lubrication and attacks cylinder walls and valves*, and

(*iii*) The water mixed with air should be *separated before using the air.*

(2) **Water jacketing.** It consists in circulating water around the cylinder through the water jacket which helps to cool the air during compression. *This method is commonly used in all types of reciprocating air compressors.*

(3) **Inter-cooling.** When the speed of the compressor is high and pressure ratio required is also high with single-stage compression water jacketing proves to be less effective. The use of inter-cooling is restored to in addition to the water jacketing by dividing the compression into two or more stages. The air compressed in the first stage is cooled in an intercooler (heat exchanger) to its original temperature before passing it on to the following (second) stage.

(4) **External fins.** The small capacity air compressors can be effectively cooled by using fins on their external surfaces.

(5) **By a suitable choice of cylinder proportions.** By providing a short stroke and a large bore in conjunction with sleeve valves, a much greater surface is available for cooling, and the surface of the cylinder head is far more effective in this respect than the surface of the barrel, because the periodic motion of the piston does not allow the barrel to be exposed to the air for a sufficient time for heat to flow away. Moreover the air is compressed against the cylinder cover. Unfortunately clearance increases as the square of the bore, but in the Broom-Wade compressor this increase is compensated for by the mechanically operated valve.

Mechanical efficiency. In general, the mechanical efficiency is the ratio of the mechanical output to the mechanical input. For an air compressor,

$$\textit{Mechanical efficiency, } \eta_{mech.} = \frac{\text{I.H.P. of compressor}}{\text{Shaft horse power}}.$$

24.3.9. Clearance in Compressors

The *clearance volume* consists of the following *two spaces* :

(*i*) The space between the cylinder end and the piston to allow for wear and to give mechanical freedom and

(*ii*) The space for the reception of valves.

In high-class H.P. compressors the clearance volume may be as little as 3 per cent of the swept volume, lead fuse wire being used to determine the actual width of the gap between the cylinder end and the piston, whilst in cheap L.P. compressors the clearance may be 6 per cent of the swept volume, in which the thickness of a flattened ball of putty is a measure of gap.

The direct effect of clearance is to make the volume taken in per stroke less than the swept volume, and because of the necessary increase in size of the compressor (to maintain the output) the power to drive the compressor is slightly increased. *The maximum compression pressure is also controlled by the clearance volume.*

The *value of clearance* may be expressed as follows :

(*i*) Since less precision is required in machining and erection, a large clearance cheapens a compressor and tends to increase its reliability.

(*ii*) A variable clearance is a convenient and safe way of controlling the output of a constant speed compressor.

(*iii*) Increasing the clearance in one stage throws more work on the stage below. In this way the temperature rise in the higher stages, consequent on controlling the output by throttling the L.P. suction, may be limited.

24.3.10. Effect of Clearance Volume

The ***clearance volume*** is *the volume within the cylinder when the piston it at the end of its inward travel plus the volume within the passages leading to the valves. The effect of clearance volume is to reduce the volume actually aspirated.* Therefore clearance volume should be as small as possible, but it cannot be reduced to zero since, for mechanical reasons, the piston face cannot be allowed to come into contact with cylinder head. The clearance volume of the compressor is given as *a percentage of stroke volume*.

The *p-V* diagram for a single-stage and a single-acting air-compressor with clearance volume is shown in Fig. 24.11. At the end of delivery stroke, the high pressure air is left in the clearance volume and *suction for the second cycle starts only when the air pressure falls to the atmospheric pressure*. This is represented by the expansion curve 3-4. Assuming the compression and expansion of the air follow the same law, the work done per cycle is given by the area 1-2-3-4-1 on *p-V* diagram.

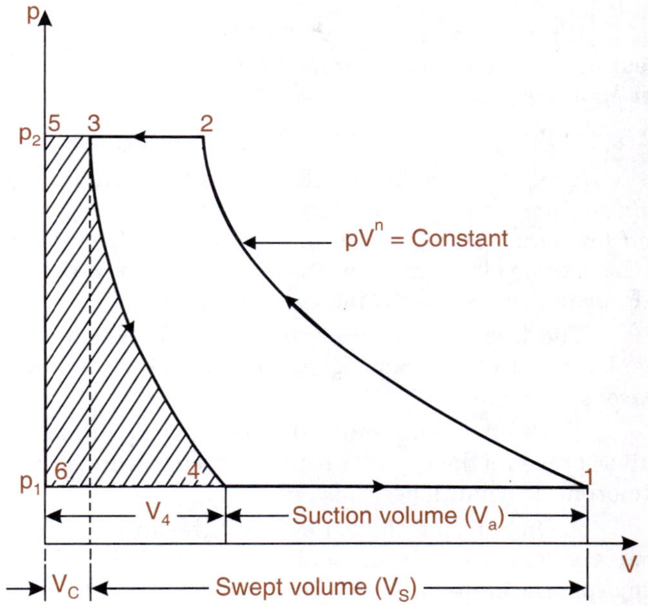

Fig. 24.11

$$\therefore \quad W(\text{area 1-2-3-4-1}) = W_{comp.}\ (\text{area 1-2-5-6-1}) - W_{exp.}\ (\text{area 3-4-6-5-3})$$

$$\therefore \quad W = \frac{n}{n-1} p_1 V_1 \left[\left(\frac{p_2}{p_1} \right)^{\frac{n-1}{n}} - 1 \right] - \frac{n}{n-1} p_4 V_4 \left[\left(\frac{p_3}{p_4} \right)^{\frac{n-1}{n}} - 1 \right]$$

As $p_3 = p_2$ and $p_4 = p_1$

$$\therefore \quad W = \frac{n}{n-1} p_1(V_1 - V_4)\left[\left(\frac{p_2}{p_4}\right)^{\frac{n-1}{n}} - 1\right] \quad ...(24.44)$$

$$= \frac{n}{n-1} p_1 V_a \left[\left(\frac{p_2}{p_1}\right)^{\frac{n-1}{n}} - 1\right] \quad ...(24.45)$$

where $V_a = V_1 - V_4$ is the actual volume of free-air delivered per cycle.

$$\therefore \quad W = \frac{n}{n-1} m_1 R T_1 \left[\left(\frac{p_2}{p_1}\right)^{\frac{n-1}{n}} - 1\right] \quad ...(24.46)$$

where m_1 is the actual mass of air delivered per cycle.

$\therefore$ Work delivered per kg of air delivered

$$= \frac{n}{n-1} R T_1 \left[\left(\frac{p_2}{p_1}\right)^{\frac{n-1}{n}} - 1\right] \quad ...(24.47)$$

[From eqns. (24.35) and (24.47) it is obvious that the clearance volume *does not affect the work of compression per kg of air*.]

24.3.11. Free Air Delivered (F.A.D.) and Displacement

The *free air delivered* (F.A.D.) is the *actual volume delivered at the stated pressure reduced to intake temperature and pressure*, and expressed in m^3/min. The displacement is the actual volume in m^3/min swept out per minute by the L.P. piston or pistons during the suction strokes.

The free air delivered per minute is *less* than the displacement of the compressor because of the following *reasons* :

1. The fluid resistance through the air intake, and valves prevents the cylinder being fully charged with air at atmospheric conditions.

2. On entering the hot cylinder the air expands ; so that the mass of air present (compared with that at atmospheric temperature) is reduced in the ratio : (Absolute atmospheric temperature)/(Absolute temperature of the air in the cylinder).

3. The high-pressure air trapped in the clearance space, must expand to a pressure below atmospheric before the automatic suction valves can open ; a portion of the suction stroke is therefore wasted in effecting this expansion.

4. A certain loss is caused by the leakage.

24.3.12. Compressor Performance

By compressor performance, we generally mean the mass of air delivered per minute per B.P. (or B.H.P.) on the machine.

For a machine of given capacity and numerial pressure the *performance of a compressor* is influenced by the following *factors* :

(*i*) The pressure range per cylinder.
(*ii*) The number of stages employed.
(*iii*) The clearance volume.
(*iv*) The speed of the machine.
(*v*) The cooling efficiency.
(*vi*) The air intake piping.
(*vii*) The type and disposition of the valves.

24.3.13. Effect of Atmospheric Conditions on the Output of a Compressor

A low barometer and a high temperature (as encountered at considerable elevations during day time in tropical countries) is responsible for an appreciable diminution in the mass output of compressors which have to operate under these conditions. The volumetric efficiency (when referred to a standard atmosphere) falls by about 3% per 300 mm increase in elevation, and 1% per 5°C increase in temperature. As a result of the considerable reduction in temperature after sun-down, and accompanying humidity, power plant in tropical climates runs considerably better at night.

24.3.14. Control of Compressors

Compressor control may be carried out in many different ways, depending on the circumstances in which they are used ; *e.g.*,

1. A compressor, directly driven by a steam engine, may be controlled by a combined centrifugal governor on the steam engine and an air-pressure regulator, the control consisting in an adjustment of the speed to suit the load. The mechanism operates either the steam throttle or varies the cut-off. This is suitable where the prime-mover may be run at reduced speeds without too great drop in efficiency.

2. Where the drive is by means of electrical motors it will usually be necessary to keep the speed constant (it may be inevitable with synchronous motors), and then some unloading device may be used to blow low-pressure air off to the atmosphere. By artificially obstructing the low-pressure intake and thus lowering the intake pressure in addition to the mass aspired, the temperature of delivery may be raised to a dangerous value due to the higher pressure ratio, and this method is therefore *not to be recommended.*

3. A method, commendable because it affords some control over the volumetric efficiency, *is to provide* ***variable clearance control.*** This is achieved by having *air pockets adjacent to the cylinder*, which are brought into communication with the cylinder by automatically operated valves.

4. With mechanically operated valves it is usual to hold the suction valve open for part of the compression stroke.

In many cases a combination of these, and other, controls may actually be used.

24.3.15. Arrangement of Reciprocating Compressors

As earlier discussed the reciprocating air compressors may be classified into single and multi-stage machines, and they may be single or double-acting. In the latter case, air is compressed alternately on either side of the piston, and consequently there are two compression strokes per revolution per cylinder. In some machines compression in the various stages take place in separate cylinders, the pistons in which are independently actuated from separate cranks upon the crankshaft. In others a compound cylinder, which forms either two or even three stages, fitted with a single piston, is employed and again, two pistons may be connected together each reciprocating in its own cylinder.

Fig. 24.12 shows three arrangements in diagram form :

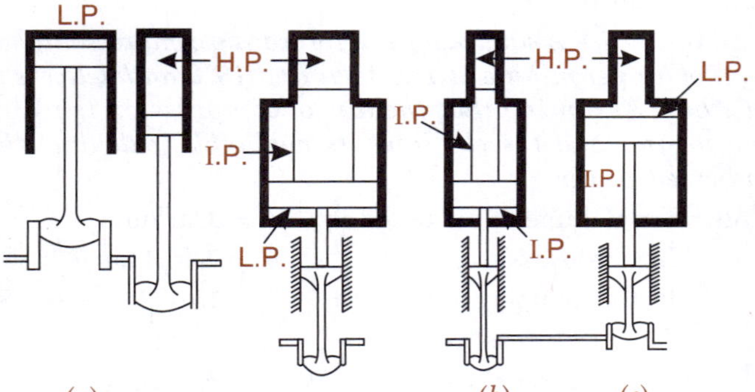

Fig. 24.12. Arrangement of reciprocating compressors.

Fig. 24.12 (*a*) shows a two-crank, two-stage single-acting air compressor. In this arrangement each piston is actuated independently from a separate crank. The cranks are kept 180° out of phase.

Fig. 24.12 (*b*) shows a single-crank, three-stage single-acting air compressor. Here compression of the air in the low pressure cylinder takes place on the down-stroke of the piston, whilst on the up-stroke compression in the intermediate and high pressure stages occurs simultaneously.

Fig. 24.12 (*c*) shows a two-crank, three-stage air compressor having double-acting low and intermediate pressure cylinders and two single-acting high pressure cylinders, compression in these high pressure cylinders taking place only during the up-stroke of the pistons.

Many such arrangements are possible but their discussion is beyond the scope of this book.

24.3.16. Intercooler

The cooler which is placed in between stages is called ***Intercooler.*** With the object of removing moisture, coolers are sometimes fitted after the last stage, and for this reason are called ***'Aftercoolers'***, but it should be understood that *aftercoolers cannot influence the work done in compression.*

Intercoolers and aftercoolers are simple heat exchangers in which heat is removed from air which has been compressed and its temperature has risen as a result of compression. A simplified section through an intercooler is shown in Fig. 24.13. The cooling water passes through the tubes which are secured between two tube plates, and the air circulates over the tubes through a system of baffles.

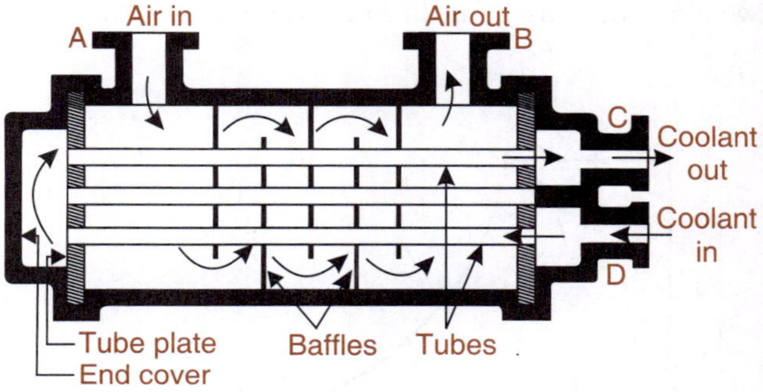

Fig. 24.13. Intercooler.

24.3.17. Compressed Air Motors

Compressed air is employed in a wide variety of applications in industry. For some purposes air-operated motors are the most suitable forms of power, especially where there are safety requirements to be met as in mining applications. Pneumatic breakers, picks, spades, rammers, vibrators, riveters, etc. form a range of hand tools which have wide applications in constructional work. They are light in construction and suitable for operation in remote situations for which other forms of power tools may not be suitable. The action required of such tools, with the associated simplicity and robustness of construction, is obtained with an air-operated design.

The most general types of motors are :

(1) Piston type or Reciprocating type.

(2) Rotary type.

24.3.18. Reciprocating Air Motor

The cycle in the reciprocating type air motor is reverse of that in the reciprocating compressor. Air is supplied to the air motor from an air receiver in which the air is at approximately ambient temperature. There is a pressure drop in the air line between the receiver and the motor. The air expands in a motor cylinder to atmospheric pressure in a manner which is polytropic (*i.e.*, the expansion is internally reversible and the law of expansion is

pV^n = constant, where $n < \gamma$, and is usually about 1.3). If the air is initially at ambient temperature, then this form of expansion will bring about a reduction in air temperature as lower pressures are reached. The temperatures reached may be sufficiently low to be below the dew point of the moisture in the air, this may be condensed, and the water formed may even be cooled to its freezing point. This may lead to the formation of ice in the cylinder with the consequence of blocked valves. To prevent this condition it may be necessary to *pre-heat the air* to an initial temperature which is high enough *to prevent the ice formation.* This heating of air causes an increase in volume at supply pressure and reduces the demand from the compressor. Further, the temperature at which the heat transfer is required is low, and a low grade supply of heat or 'waste heat' may be utilized for the purpose.

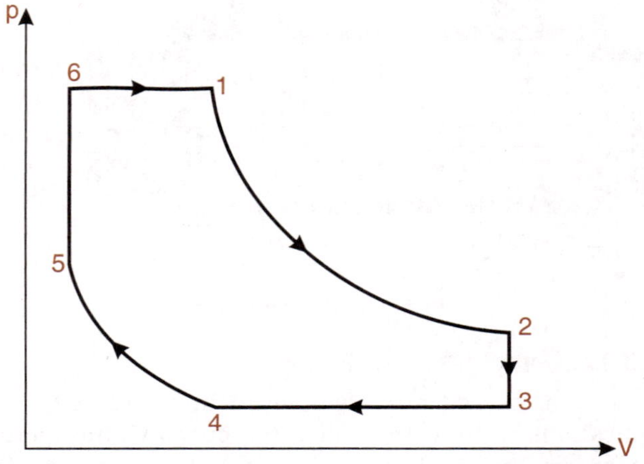

Fig. 24.14. Hypothetical diagram for a reciprocating air motor.

Fig. 24.14 shows a hypothetical diagram for a reciprocating air motor. The sequence of operations is as follows :

(*i*) **Operation 1–2 :** The air expands from 1 (p_1) to 2 (p_2) at the end of the stroke (according to the law $pV^n = C$).

(*ii*) **Operation 2–3. :** Blow down (release) of air from 2 to 3 (at constant volume).

(*iii*) **Operation 3–4 :** Air is exhausted from 3 to 4, and at 4 compression of the trapped or cushion air begins.

(*iv*) **Operation 4–5–6 :** Air at supply pressure p_6, is admitted to the cylinder at the point 5 where it mixes irreversibly with the cushion air. The pressure in the cylinder is rapidly brought upto the inlet valve, p_6.

(*v*) **Operation 6–1 :** The supply of air is made at constant pressure behind the moving piston to the point of cut-off at 1. The cut-off ratio is given by

$$\text{Cut-off ratio} = \frac{V_1 - V_6}{V_3 - V_6}$$

The effect of cushion air is to give a smoother running motor. The position of the point 5 depends on the point of initial compression 4, and on the law of compression pV^n = constant. The conditions may be such that the points 5 and 6 coincide. The analysis of such a diagram is best carried out from basic principles.

Note. For a given power, a reciprocating air compressor consumes less air than the rotary form, because of the reduced leakage and of the greater expansion that is possible. This however is only secured at the expense of a heavier, and more costly mechanism.

24.3.19. Rotary Type Air Motor

The *air turbine* is valveless, small in size, light in weight, and requires no internal lubrication, but air friction is high, and any dampness in the air causes rapid deterioration of the blading at low temperatures.

The sliding blade eccentric drum type requires internal lubrication, and even so the slots, in which the blades move, wear rapidly.

The toothed wheel type has a smaller friction and can expand damp air without internal deterioration. In the "herringbone" type expansion is possible together with a high starting torque and extreme mechanical simplicity. This commends the turbine for colliery work inspite of its extra vagance on air.

Example 24.1. *A single-stage reciprocating compressor takes 1 m³ of air per minute at 1.013 bar and 15°C and delivers it at 7 bar. Assuming that the law of compression is $pV^{1.35}$ = constant, and the clearance is negligible, calculate the indicated power.*

Solution. Volume of air taken in, $V_1 = 1 \text{ m}^3/\text{min}$

Intake pressure, $p_1 = 1.013$ bar

Initial temperature, $T_1 = 15 + 273 = 288$ K

Delivery pressure, $p_2 = 7$ bar

Law of compression : $pV^{1.35}$ = constant

Indicated power I.P. :

Mass of air delivered per min.,

$$m = \frac{p_1V_1}{RT_1} = \frac{1.013 \times 10^5 \times 1}{287 \times 288} = 1.226 \text{ kg/min}$$

$$\text{Delivery temperature, } T_2 = T_1 \left(\frac{p_2}{p_1}\right)^{(n-1)/n}$$

$$= 288 \left(\frac{7}{1.013}\right)^{(1.35-1)/1.35} = 475.2 \text{ K}$$

$$\text{Indicated work} = \frac{n}{n-1} mR (T_2 - T_1) \text{ kJ/min}$$

$$= \frac{1.35}{1.35 - 1} \times 1.226 \times 0.287 (475.2 - 288)$$

$$= 254 \text{ kJ/min}$$

i.e., *Indicated power I.P* $= \dfrac{254}{60}$ = **4.23 kW. (Ans.)**

Example 24.2. *If the compressor of example 21.1 is driven at 300 r.p.m. and is a single-acting, single-cylinder machine, calculate the cylinder bore required, assuming a stroke to bore ratio of 1.5 : 1. Calculate the power of the motor required to drive the compressor if the mechanical efficiency of the compressor is 85% and that of the motor transmission is 90%.*

Solution. Volume dealt with per minute at inlet

$= 1 \text{ m}^3/\text{min}.$

$\therefore$ Volume drawn in per cycle

$$= \frac{1}{300} = 0.00333 \text{ m}^3/\text{cycle}$$

i.e., Cylinder volume = 0.00333 m^3

$$\therefore \quad \frac{\pi}{4} D^2 L = 0.00333$$

(where D = bore, L = stroke)

$$i.e., \quad \frac{\pi}{4} D^2 (1.5 \times D) = 0.00333 \quad \text{or} \quad D^2 = \frac{0.00333 \times 4}{\pi \times 1.5}$$

i.e., *Cylinder bore*, D = **0.1414 m** or **141.4 mm.** **(Ans.)**

$$\text{Power input to the compressor} = \frac{4.23}{0.85} = 4.98 \text{ kW}$$

$$\therefore \quad \textit{Motor power} = \frac{4.98}{0.9} = \textbf{5.53 kW.} \quad \textbf{(Ans.)}$$

Example 24.3. *An air compressor takes in air at 1 bar and 20°C and compresses it according to law $pv^{1.2}$ = constant. It is then delivered to a receiver at a constant pressure of 10 bar. R = 0.287 kJ/kg K. Determine :*

(*i*) *Temperature at the end of compression ;*

(*ii*) *Work done and heat transferred during compression per kg of air.*

Solution. Refer to Fig. 24.15.

$T_1 = 20 + 273 = 293$ K ; p_1 = 1 bar ; p_2 = 10 bar

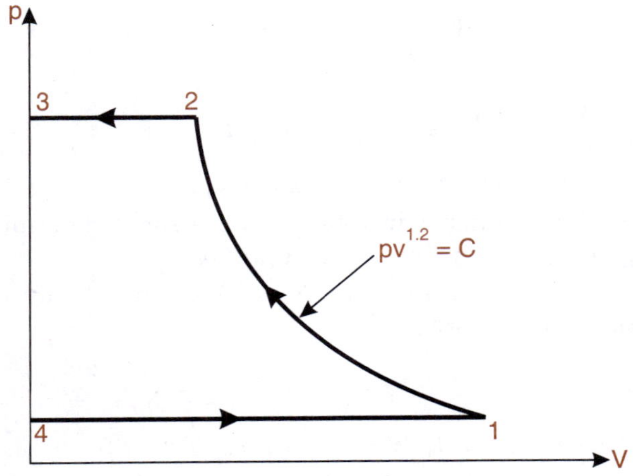

Fig. 24.15

Law of compression : $pv^{1.2} = C$; $R = 0.287$ J kJ/kg K

(*i*) **Temperature at the end of compression, T_2 :**

For *compression process 1–2*, we have

$$\frac{T_2}{T_1} = \left(\frac{p_2}{p_1}\right)^{\frac{n-1}{n}} = \left(\frac{10}{1}\right)^{\frac{1.2-1}{1.2}} = 1.468$$

or $\quad T_2 = T_1 \times 1.468 = 293 \times 1.468$

$= \textbf{430 K}$ or **157°C.** **(Ans.)**

(*ii*) **Work done and heat transferred during compression per kg of air :**

Work done,

$$W = mRT_1 \frac{n}{n-1}\left[\left(\frac{p_2}{p_1}\right)^{\frac{n-1}{n}} - 1\right] \quad \text{...[Eqn. (24.5)]}$$

$$= 1 \times 0.287 \times 293 \times \left(\frac{1.2}{1.2-1}\right)\left[\left(\frac{10}{1}\right)^{\frac{1.2-1}{1.2}} - 1\right]$$

= **236.13 kJ/kg of air.** **(Ans.)**

Heat transferred during compression,

$$Q = W + \Delta U$$

$$= \frac{p_1v_1 - p_2v_2}{n-1} + c_v(T_2 - T_1)$$

$$= \frac{R(T_1 - T_2)}{n-1} + c_v(T_2 - T_1) = (T_2 - T_1)\left[c_v - \frac{R}{n-1}\right]$$

$$= (430 - 293)\left[0.718 - \frac{0.287}{1.2-1}\right]$$

= **– 98.23 kJ/kg.** **(Ans.)**

Negative sign indicates heat rejection.

Example 24.4. *Following data relate to a performance test of a single-acting 14 cm × 10 cm reciprocating compressor :*

Suction pressure	*= 1 bar*
Suction temperature	*= 20°C*
Discharge pressure	*= 6 bar*
Discharge temperature	*= 180°C*
Speed of compressor	*= 1200 r.p.m.*
Shaft power	*= 6.25 kW*
Mass of air delivered	*= 1.7 kg/min*

Calculate the following :

(*i*) *The actual volumetric efficiency ;*

(*ii*) *The indicated power ;*

(*iii*) *The isothermal efficiency ;*

(*iv*) *The mechanical efficiency ;*

(*v*) *The overall isothermal efficiency.*

(D.U.)

Solution. *Given :* p_1 = 1 bar ; T_1 = 20 + 273 = 293 K ;
p_2 = 6 bar ; T_2 = 180 + 273 = 453 K ;
N = 1200 r.p.m., P_{shaft} = 6.25 kW ;
m_a = 1.7 kg/min.

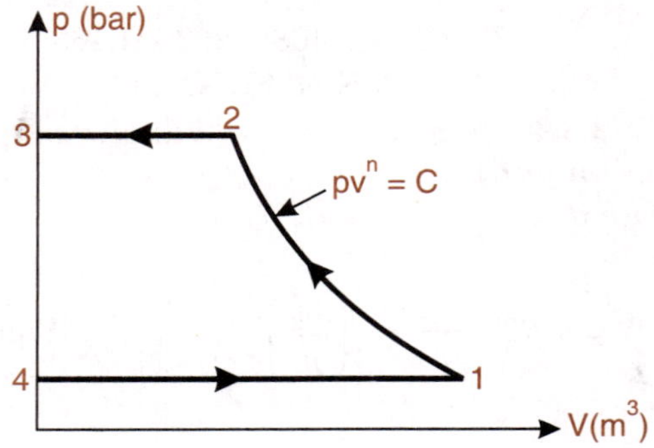

Fig. 24.16

(*i*) **The actual volumetric efficiency, $\eta_{vol.}$:**

Displacement volume (m³/min)

$$V_d = \frac{\pi}{4} D^2 L \times N \quad \text{(for single-acting compressor)}$$

$$= \frac{\pi}{4} \times \left(\frac{14}{100}\right)^2 \times \left(\frac{10}{100}\right) \times 1200 = 1.8473 \text{ m}^3\text{/min}$$

$$\text{F.A.D.} = \frac{mRT_1}{p_1} = \frac{1.7 \times (0.287 \times 1000) \times 293}{1 \times 10^5}$$

$$= 1.4295 \text{ m}^3\text{/min.}$$

$$\therefore \quad \eta_{vol.} = \frac{\text{F.A.D.}}{V_d} \times 100$$

$$= \frac{1.4295}{1.8473} \times 100 = \mathbf{77.38\%.} \quad \textbf{(Ans.)}$$

(*ii*) **The indicated power, I.P. :**

$$\frac{T_2}{T_1} = \left(\frac{p_2}{p_1}\right)^{\frac{n-1}{n}} \quad \text{or} \quad \frac{n-1}{n} = \frac{\ln (T_2/T_1)}{\ln (p_2 p_1)}$$

or
$$\frac{1}{n} = 1 - \frac{\ln (453/293)}{\ln (6/1)}$$

or
$$n = 1.32$$

Hence, index of compression, n = 1.32

∴ Indicated power,

$$\text{I.P.} = \frac{n}{n-1} mRT_1 \left\{\left(\frac{p_2}{p_1}\right)^{\frac{n-1}{n}} - 1\right\}$$

$$= \frac{1.32}{1.32 - 1} \times \frac{1.7}{60} \times 0.287 \times 293 \left\{\left(\frac{6}{1}\right)^{\frac{1.32-1}{1.32}} - 1\right\}$$

$$= 5.346 \text{ kJ/s or kW}$$

i.e., **I.P. = 5.346 kW. (Ans.)**

(*iii*) **Isothermal efficiency, η_{iso} :**

Isothermal power = $mRT_1 \ln (p_2/p_1)$

$$= \frac{1.7}{60} \times 0.287 \times 293 \ln (6/1)$$

$$= 4.269 \text{ kJ/s or kW}$$

$$\therefore \quad \eta_{iso} = \frac{4.269}{5.346} \times 100 = \mathbf{79.85\%.} \quad \textbf{(Ans.)}$$

(*iv*) **The mechanical efficiency, η_{mech} :**

$$\eta_{mech} = \frac{\text{Indicated power}}{\text{Shaft power}} \times 100$$

$$= \frac{5.346}{6.25} \times 100 = \mathbf{85.5\%.} \quad \textbf{(Ans.)}$$

(*v*) **The overall isothermal efficiency, $\eta_{overall\ (iso)}$:**

$$\eta_{overall\ (iso)} = \frac{\text{Isothermal power}}{\text{Shaft power}} \times 100$$

$$= \frac{4.269}{6.25} \times 100 = \mathbf{68.3\%.} \quad \textbf{(Ans.)}$$

Example 24.5. (*a*) *Show that the compressor work obtained from the analysis of a conventional card with clearance and polytropic processes for the reciprocating compressor is identical to that obtained from the analysis of a reversible steady flow rotary compressor where in certain mass of a gas is compressed from the initial condition of pressure p_1 and t_1 respectively to the final pressure p_2 in accordance to $pv^n = C$.*

(*b*) *A low pressure, water jacketed steady flow rotary compressor compresses polytropically 6.75 kg/min of air from 1 atm. and 21°C to 0.35 bar gauge and 43°C. Neglecting the change in kinetic energy find the work and mass of water circulated if the temperature rise of the cooling water is 3.3°C. Take c_p (for air) = 1.003 kJ/kg K.* **(P.U.)**

Solution. (*a*) In reciprocating compressors the work required is

$$W = \frac{n}{n-1} p_1 V_1 \left[\left(\frac{p_2}{p_1}\right)^{\frac{n-1}{n}} - 1\right]$$

$$= \frac{n}{n-1} mR (T_2 - T_1) \quad \text{...(1)}$$

When compression is adiabatic, $n = \gamma$

Work transfer in rotary compressor is determined by *applying steady flow energy equation.*

(*i*) **With isentropic flow.** Applying steady flow energy equation, we get

$$h_1 + \frac{C_1^2}{2g} + W = h_2' + \frac{C_2^2}{2g}$$

For $C_1 = C_2$, $W = h_2' - h_1 = c_p(T_2' - T_1)$

where T_2' is the temperature after isentropic compression

Since $$c_p = \frac{\gamma R}{\gamma - 1}$$

$$W = \frac{\gamma}{\gamma - 1} R(T_2' - T_1) \qquad ...(2)$$

Eqn. (1) is similar to eqn. (2) for unit mass.

(ii) **With compression polytropic.** In actual practice due to internal heating there is increase of work done above isentropic work, and work done is

$$W = c_p(T_2 - T_1) = \frac{\gamma}{\gamma - 1} R(T_2 - T_1)$$

where T_2 is the actual temperature, *i.e.*, obtained by using the relationship $pv^n = C$.

(iii) **With cooled compression.** Some heat is being taken away by cooling of compressor and so

$$W = c_p(T_2 - T_1) + Q$$

(*b*) **Work,** $W = m\, c_p\, (T_2 - T_1)$

$= 6.75 \times 1.003\ (43 - 21)$

$= \mathbf{148.94\ min.}$ **(Ans.)**

If the compression would have been isentropic

$$T_2' = T_1\, (r_p)^{\frac{\gamma - 1}{\gamma}} = (21 + 273) \left(\frac{1.35}{1}\right)^{\frac{1.4-1}{1.4}}$$

$= 320.3$°K or 47.3°C

Heat rejected to cooling water

$= m\, c_p(T_2' - T_2)$

$= 6.75 \times 1.003\ (47.3 - 43) = 29.11$ kJ

Mass of cooling water, $\mathbf{m_w} = \dfrac{29.11}{c_{pw} \times (t_{w2} - t_{w1})}$

$$= \frac{29.11}{4.18 \times 3.3} = \mathbf{2.11\ kg/min.} \quad \textbf{(Ans.)}$$

Example 24.6. *A single-stage double-acting air compressor is required to deliver 14 m³ of air per minute measured at 1.013 bar and 15°C. The delivery pressure is 7 bar and the speed 300 r.p.m. Take the clearance volume as 5% of the swept volume with the compression and expansion index of n = 1.3. Calculate :*

(i) Swept volume of the cylinder;

(ii) The delivery temperature;

(iii) Indicated power.

Solution. Quantity of air to be delivered = 14 m³/min

Intake pressure and temperature $p_1 = 1.013$ bar,

$T_1 = 15 + 273 = 288$ K

Delivery pressure, $p_2 = 7$ bar

Compressor speed, $N = 300$ r.p.m.

Clearance volume, $V_c = 0.05\, V_s$

Compression and expansion index, $n = 1.3$

(i) **Swept volume of the cylinder, V_s :**

Swept volume, $V_s = V_1 - V_3 = V_1 - V_c = V_1 - 0.05\, V_s$

$\therefore \quad V_1 = 1.05\, V_s$

Volume induced per cycle = $(V_1 - V_4)$

and $$V_1 - V_4 = \frac{14}{300 \times 2} = 0.0233 \text{ m}^3$$

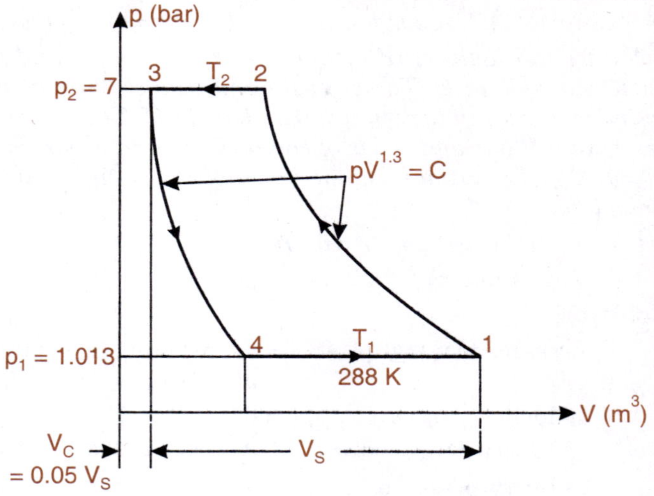

Fig. 24.17

Now, $\quad V_1 = 1.05\, V_s$ and $\dfrac{V_4}{V_3} = \left(\dfrac{p_2}{p_1}\right)^{1/n}$

$$= \left(\frac{7}{1.013}\right)^{1/1.3} = 4.423$$

i.e., $\quad V_4 = 4.423\, V_3 = 4.423 \times 0.05\, V_s = 0.221\, V_s$

$\therefore \quad (V_1 - V_4) = 1.05\, V_s - 0.221\, V_s = 0.0233$

$$\therefore \quad V_s = \frac{0.0233}{(1.05 - 0.221)} = 0.0281 \text{ m}^3$$

i.e., *Swept volume of the cylinder* $= \mathbf{0.0281\ m^3.}$ **(Ans.)**

(ii) **The delivery temperature, T_2 :**

Using the relation, $\dfrac{T_2}{T_1} = \left(\dfrac{p_2}{p_1}\right)^{\frac{n-1}{n}}$

$$\therefore \; T_2 = T_1 \times \left(\frac{p_2}{p_1}\right)^{\frac{n-1}{n}} = 288 \times \left(\frac{7}{1.013}\right)^{\frac{1.3-1}{1.3}} = 450 \text{ K}$$

$\therefore$ *Delivery temperature*

$= 450 - 273 = \mathbf{177°C.}$ **(Ans.)**

(iii) **Indicated power :**

Indicated power

$$= \frac{n}{n-1} p_1(V_1 - V_4) \left\{ \left(\frac{p_2}{p_1}\right)^{\frac{n-1}{n}} - 1 \right\}$$

$$= \frac{1.3}{1.3 - 1} \times \frac{1.013 \times 10^5 \times 14}{10^3 \times 60} \left\{ \left(\frac{7}{1.013}\right)^{\frac{1.3-1}{1.3}} - 1 \right\} \text{ kW}$$

= 57.56 kW.

i.e., *Indicated power* $= \mathbf{57.56\ kW.}$ **(Ans.)**

☞**Example 24.7.** *A single-stage, double-acting compressor has a free air delivery (F.A.D.) of 14 $m^3/min.$ measured at 1.013 bar and 15°C. The pressure and temperature in the cylinder during induction are 0.95 bar 32°C. The delivery pressure is 7 bar and index of compression and expansion, n = 1.3. The clearance volume is 5% of the swept volume. Calculate :*

(*i*) *Indicated power required ;*

(*ii*) *Volumetric efficiency.*

Solution.

Free air delivery, F.A.D. = 14 m^3/min. (measured at 1.013 bar and 15°C)

Induction pressure, $p_1 = 0.95$ bar

Induction temperature, $T_1 = 32 + 273 = 305$ K

Delivery pressure, $p_2 = 7$ bar

Index of compression and expansion, $n = 1.3$

Clearance volume, $V_3 = V_c = 0.05\ V_s$

(*i*) **Indicated power :**

Mass delivered per minute

$$m = \frac{pV}{RT} = \frac{1.013 \times 10^3 \times 14}{0.287 \times 288 \times 10^3} = 17.16 \text{ kg/min.}$$

[where F.A.D. per minute is V at p(= 1.013 bar) and T(15 + 273 = 288 K)]

To find T_2, using the equation,

$$\frac{T_2}{T_1} = \left(\frac{p_2}{p_1}\right)^{\frac{n-1}{n}}$$

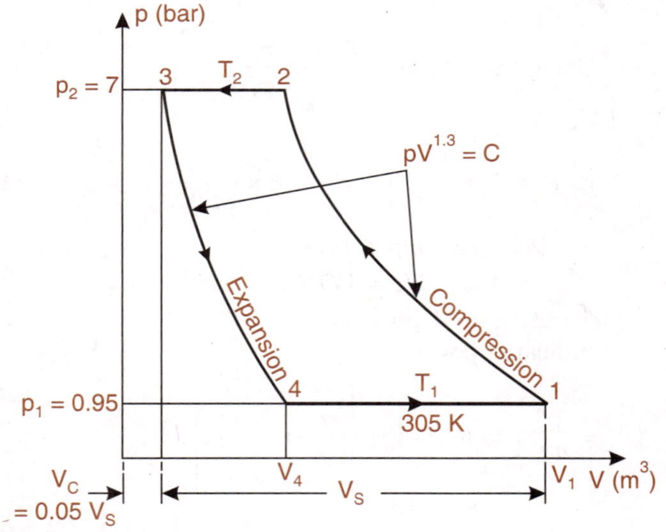

Fig. 24.18

$$\therefore \quad T_2 = T_1 \left(\frac{p_2}{p_1}\right)^{\frac{n-1}{n}} = 305 \times \left(\frac{7}{0.95}\right)^{\frac{1.3-1}{1.3}} = 483.5 \text{ K.}$$

Indicated power

$$= \frac{n}{n-1}\, mR(T_2 - T_1)$$

$$= \frac{1.3}{1.3-1} \times 17.16 \times 0.287\ (483.5 - 305)$$

$$= 3809.4 \text{ kJ/min.}$$

$\therefore$ **Indicated power**

$$= \frac{3809.4}{60} = \mathbf{63.49\ kW.} \quad \textbf{(Ans.)}$$

(*ii*) **Volumetric efficiency :**

Using the relation,

$$\frac{V_4}{V_3} = \left(\frac{p_3}{p_4}\right)^{1/n} = \left(\frac{p_2}{p_1}\right)^{1/1.3} = \left(\frac{7}{0.95}\right)^{1/1.3} = 4.65$$

$$\therefore \quad V_4 = 4.65 \times V_3 = 4.65 \times 0.05\ V_s = 0.233\ V_s$$

$$\therefore V_1 - V_4 = V_1 - 0.233 V_s = 1.05\ V_s - 0.233\ V_s$$

$$= 0.817\ V_s$$

Now, $$m = \frac{pV}{RT} = \frac{p_1 (V_1 - V_4)}{RT_1}$$

i.e., $$\text{F.A.D./cycle, } V = (V_1 - V_4)\ \frac{T}{T_1} \cdot \frac{p_1}{p}$$

(where p_1 and T_1 are the suction conditions)

$$\therefore \quad V = 0.817\ V_s \times \frac{288}{305} \times \frac{0.95}{1.013} = 0.723\ V_s$$

$\therefore$ Volumetric efficiency,

$$\eta_{vol.} = \frac{V}{V_s} = \frac{0.723\ V_s}{V_s} = \mathbf{0.723 \text{ or } 72.3\%.} \quad \textbf{(Ans.)}$$

Example 24.8. (*a*) *What is meant by volumetric efficiency of a reciprocating compressor ? How is it affected by (i) speed of the compressor ; (ii) delivery pressure ; and (iii) throttling across the valves.*

(*b*) *State at least six uses of compressed air.*

(*c*) *The free air delivered by a single-stage double-acting reciprocating compressor, measured at 1 bar and 15°C of free air, is 16 m^3/min. The pressure and temperature of air inside the cylinder during suction are 0.96 bar and 30°C respectively and delivery pressure is 6 bar. The compressor has a clearance of 4% of the swept volume and the mean piston speed is limited to 300 m/min. Determine :*

(*i*) *Power input to the compressor if mechanical efficiency is 90% and compression efficiency 85% ;*

(*ii*) *Stroke and bore if the compressor runs at 500 r.p.m.*

Take index of compression and expansion = 1.3.

(M.U.)

Solution. (*a*) ***Volumetric efficiency*** *of a compressor is defined as the ratio of Free Air Delivered (FAD) to the swept volume.* ***FAD*** is the *volume of air delivered by the compressor measured at some reference condition* (which may be the ambient condition or the standard sea level condition). FAD is *less* than the swept volume due to the following *reasons* :

(*i*) Throttling and pressure drop at inlet valve and passages ;

(*ii*) Heating of inlet air by coming in contact with hot cylinder walls ; and

(*iii*) Re-expansion of compressed air retained in the clearance volume.

Effects of parameters on volumetric efficiency :

(*i*) ***Speed of compressor.*** As the speed is increased the pressure drop in the inlet passage and the inlet valve increases. Further the air temperature during intake also increases due to less available for cooling. Both of these factors *reduce* volumetric efficiency of compressor *with increase of its speed.*

(*ii*) ***Delivery pressure.*** Refer to Fig. 24.19. With increase of delivery pressure the pressure ratio increases and hence during inward stroke *a-b*, the effective swept volume is reduced. The volume of air delivered (FAD) is reduced from V_d to V_d' when the delivery pressure in increased from p_d to p_d'. Thus the volumetric efficiency is *decreased* when the *delivery pressure is increased.*

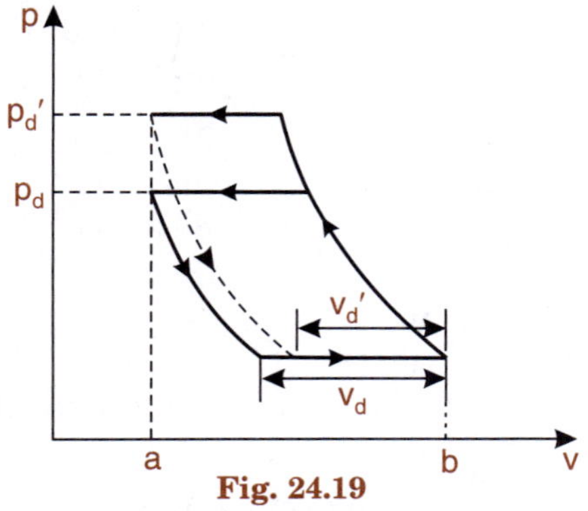

Fig. 24.19

(*iii*) ***Throttling across the valves.*** Throttling across the inlet valve reduces the pressure in the cylinder at the end of the inlet stroke. Further, throttling at the inlet and delivery valves increases the pressure ratio. Both of these effects would *reduce the FAD, and hence the volumetric efficiency of the compressor.*

(*b*) **Uses of compressed air :**

(*i*) Driving a compressed air engine (air motor).

(*ii*) Driving pneumatic tools.

(*iii*) Spray painting.

(*iv*) Cleaning surfaces by air blast.

(*v*) Conveying solid and powdered materials in pipelines.

(*vi*) Blast furnaces and boiler furnaces.

(*vii*) Pneumatic controls.

(*viii*) Inflating tyres of automobiles and tractors.

(*c*) Refer to Fig. 24.20.

Given : F.A.D. = 16 m³/min (measured at 1 bar and 15°C) ; $p_1 = 0.96$ bar,

$$T_1 = 30 + 273 = 303 \text{ K}, n = 1.3 ;$$
$$V_3 = V_c = 0.04\, V_s ;\ p_2 = 6 \text{ bar} ;$$
$$\eta_{mech.} = 90\% ;\ \eta_{comp.} = 85\% ;$$

Piston speed = 300 m/min ; N = 500 r.p.m.

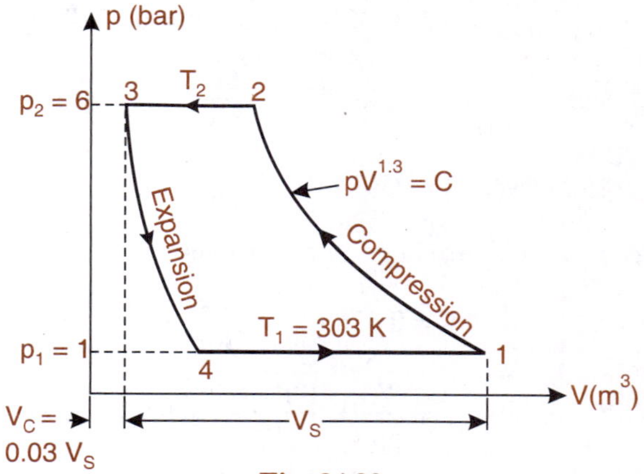

Fig. 24.20

(*i*) **Power input to compressor :**

Mass flow rate of compressor

$$m = \frac{pV}{RT} = \frac{(1 \times 10^5) \times 16}{287 \times 288} = 19.36 \text{ kg/min.}$$

[where F.A.D. per minute is V at p (= 1 bar) and T (15 + 273 = 288 K)]

To find T_2, using the equation,

$$\frac{T_2}{T_1} = \left(\frac{p_2}{p_1}\right)^{\frac{n-1}{n}}$$

$$\therefore \quad T_2 = T_1 \times \left(\frac{p_2}{p_1}\right)^{\frac{n-1}{n}} = 303 \times \left(\frac{6}{0.96}\right)^{\frac{1.3-1}{1.3}}$$

$$= 462.4 \text{ K}$$

Power input to compressor

$$= \left[\frac{n}{n-1} mR\,(T_2 - T_1)\right] \times \frac{1}{\eta_{mech.} \times \eta_{comp.}}$$

$$= \left[\left(\frac{1.3}{1.3-1}\right) \times 19.36 \times 0.287\,(462.4 - 303)\right] \times \frac{1}{0.9 \times 0.85}$$

= 5016.9 kJ/min = **83.6 kJ/s (kW). (Ans.)**

(*ii*) **Stroke (L) and bore (D) :**

Piston speed = $2LN$

$\therefore$ 300 = 2 × L × 500

or **L = 0.3 m** or **300 mm. (Ans.)**

$$\text{F.A.D.} = \frac{\pi}{4} D^2L \times 2N \times \eta_{vol.} \text{ ...... for double-acting air compressor} \quad ...(1)$$

To find $\eta_{vol.}$ proceed as follows :

$$\frac{V_4}{V_3} = \left(\frac{p_3}{p_4}\right)^{1/n} = \left(\frac{p_2}{p_1}\right)^{1/1.3} = \left(\frac{6}{0.96}\right)^{1/1.3} = 4.094$$

$\therefore \quad V_4 = 4.094 \times V_3 = 4.094 \times 0.04\, V_s = 0.1637\, V_s$

$\therefore \quad V_1 - V_4 = V_1 - 0.1637\, V_s$

$= 1.04\, V_s - 0.1637\, V_s = 0.8763\, V_s$

Now $\quad m = \dfrac{pV}{RT} = \dfrac{p_1(V_1 - V_4)}{RT_1}$

i.e., F.A.D./cycle, $V = (V_1 - V_4)\dfrac{T}{T_1} \cdot \dfrac{p_1}{p}$

(where p_1 and T_1 are suction conditions)

$$\therefore \quad V = 0.8763\, V_s \times \frac{288}{303} \times \frac{0.96}{1} = 0.799\, V_s$$

$$\therefore \quad \eta_{vol.} = \frac{V}{V_s} = \frac{0.799 V_s}{V_s} = 0.799$$

Substituting the values in eqn. (1), we get

$$16 = \frac{\pi}{4} D^2 \times 0.3 \times 2 \times 500 \times 0.799$$

$$\therefore \quad \mathbf{D} = \left(\frac{16 \times 4}{\pi \times 0.3 \times 2 \times 500 \times 0.799}\right)^{1/2}$$

$= \mathbf{0.29\ m}$ or $\mathbf{290\ mm}$. **(Ans.)**

Example 24.9. *A single-stage single-acting air compressor delivers 0.6 kg of air per minute at 6 bar. The temperature and pressure at the end of suction stroke are 30°C and 1 bar. The bore and stroke of the compressor are 100 mm and 150 mm respectively. The clearance is 3% of the swept volume. Assuming the index of compression and expansion to be 1.3, find :*

(*i*) *Volumetric efficiency of the compressor,*

(*ii*) *Power required if the mechanical efficiency is 85%, and*

(*iii*) *Speed of the compressor (r.p.m).* **(N.U.)**

Solution. Refer to Fig. 24.21.

Mass of air delivered, $m = 0.6$ kg/min.

Delivery pressure, $p_2 = 6$ bar

Induction pressure, $p_1 = 1$ bar

Induction temperature, $T_1 = 30 + 273 = 303$ K

Bore, $D = 100$ mm $= 0.1$ m

Stroke length, $L = 150$ mm $= 0.15$ m

Clearance volume, $V_c = 0.03\, V_s$

Mechanical efficiency, $\eta_{mech.} = 85\%$

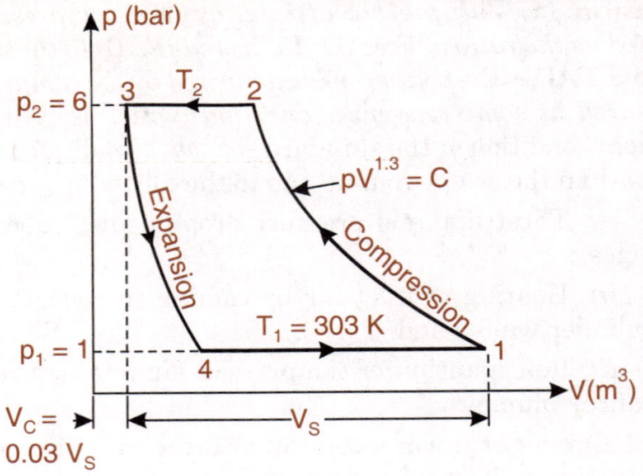

Fig. 24.21

(*i*) Volumetric efficiency of compressor, $\eta_{vol.}$:

$$\eta_{vol.} = 1 + k - k\left(\frac{p_2}{p_1}\right)^{1/n} \quad \text{...[Eqn. (24.12)]}$$

where $\quad k = \dfrac{V_c}{V_s} = 0.03$

$$\therefore \quad \eta_{vol.} = 1 + 0.03 - 0.03\left(\frac{6}{1}\right)^{\frac{1}{1.3}}$$

$= \mathbf{0.91096}$ or $\mathbf{91.096\%}$. **(Ans.)**

(*ii*) Power required to drive the compressor :

$$\text{Indicated power} = \frac{n}{n-1} mRT_1 \left[\left(\frac{p_2}{p_1}\right)^{1/n} - 1\right]$$

$$= \frac{1.3}{1.3 - 1} \times \frac{0.6}{60} \times 0.287 \times 303 \left[\left(\frac{6}{1}\right)^{\frac{1.3-1}{1.3}} - 1\right]$$

$= 1.93$ kW

$\therefore \quad$ Power required to drive the compressor $= \dfrac{1.93}{\eta_{mech.}}$

$= \dfrac{1.93}{0.85} = \mathbf{2.27\ kW.}$ **(Ans.)**

(*iii*) Speed of the compressor, N (r.p.m.) :

Free air delivery, F.A.D.

$$= \frac{mRT_1}{p_1} = \frac{0.6 \times 0.287 \times 1000 \times 303}{1 \times 10^5}$$

$= 0.5218$ m³/min.

Displacement volume

$$= \frac{\text{F.A.D.}}{\eta_{vol.}} = \frac{0.5218}{0.91096} = 0.5728 \text{ m}^3\text{/min.}$$

Also, $\quad 0.5728 = \dfrac{\pi}{4} D^2L \times N$ (for single-acting compressor)

or $\quad 0.5728 = \dfrac{\pi}{4} \times 0.1^2 \times 0.15 \times N$

∴ Speed of compressor

$$N = \frac{0.5728 \times 4}{\pi \times 0.1^2 \times 0.15} = \mathbf{486.2\ r.p.m.}\quad \textbf{(Ans.)}$$

☞ **Example 24.10.** *An air compressor having stroke length of 88 cm and clearance volume of 2 per cent of the swept volume delivers air at a pressure of 8.2 bar. In order to study the effect of clearance on free air delivery and work expended, the compressor was overhauled and a distance piece of 0.55 cm was fitted between the cylinder head and the cylinder. The compressor was then commissioned under the changed clearance. Calculate :*

(*i*) *Percentage change in the volume of free air delivered, and*

(*ii*) *Percentage change in power expended.*

Before and after overhauling the piston had a suction pressure 1.025 bar and the index of compression and expansion was 1.3. **(M.U.)**

Solution. Stroke length of the air compressor, L = 88 cm

Clearance volume, $V_c = \frac{2}{100} \times V_s = 0.02V_s$, where V_s is the swept volume

The pressure at which air is delivered, p_2 = 8.2 bar

Index of compression and expression, n = 1.3

(*i*) Percentage change in the volume of free air delivered :

Before overhauling :

Stroke volume = Area of the piston × Stroke length

$= A$ (cm²) × 88 (cm) = 88 A cm³

Clearance volume,

$$V_c = 0.02 \times 88\,A = 1.76\,A \text{ cm}^3 = V_3$$

Cylinder volume,

$$V_1 = V_s + V_c = 88\,A + 1.76\,A = 89.76\,A \text{ cm}^3$$

Considering *expansion process 3-4*, we have

$$p_3 V_3^n = p_4 V_4^n$$

or $$V_4 = V_3 \left(\frac{p_3}{p_4}\right)^{1/n}$$

or $$V_4 = 1.76\,A \left(\frac{8.2}{1.025}\right)^{\frac{1}{1.3}} = 8.712\,A \text{ cm}^3$$

∴ Volume of free air sucked in $= V_1 - V_4 = (89.76\,A - 8.712\,A) = 81.048\,A$ cm³

After overhauling :

When the distance piece is inserted the clearance space is increased, consequently for the same stroke volume, the cylinder volume would increase.

Clearance volume, $V_3 = (1.76 + 0.55)A = 2.31\,A$ cm³

Cylinder volume, $V_1 = 88\,A + 2.31\,A = 90.31\,A$ cm³

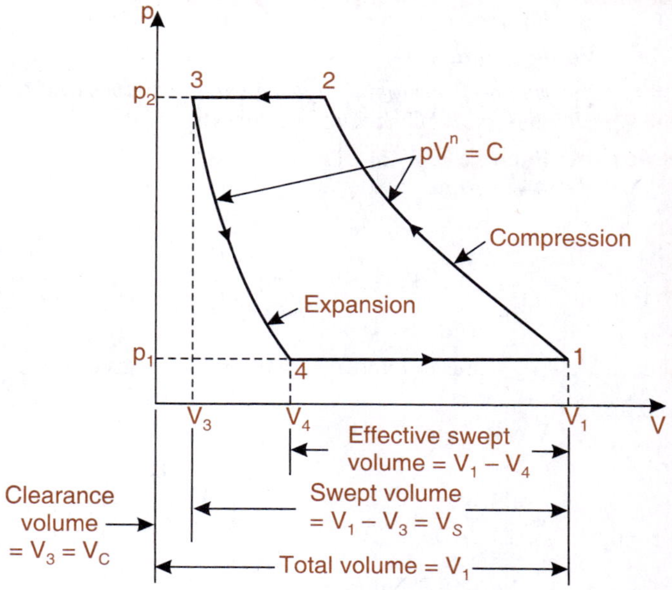

Fig. 24.22

From *expansion curve 3-4*, we have

$$p_3 V_3^n = p_4 V_4^n$$

or $$\frac{V_4}{V_3} = \left(\frac{p_3}{p_4}\right)^{1/n} = \left(\frac{p_2}{p_1}\right)^{\frac{1}{n}} = \left(\frac{8.2}{1.028}\right)^{\frac{1}{1.3}} = 4.95$$

or $$V_4 = V_3 \times 4.95 = 2.31\,A \times 4.95 = 11.43\,A$$

∴ Volume of free air sucked in $= V_1 - V_4$

$= 90.31\,A - 11.43\,A = 78.88\,A$ cm³

Percentage change in free air delivery

$$= \frac{81.048A - 78.88A}{81.048A} \times 100 = \mathbf{2.67\%}.\quad \textbf{(Ans.)}$$

(*ii*) Percentage change in power expended :

$$p = \frac{n}{n-1} \times p_1 \times \text{volume of free air sucked in} \left[\left(\frac{p_2}{p_1}\right)^{\frac{n-1}{n}} - 1\right]$$

From the above expression it is evident that with index '*n*' and pressures p_1 and p_2 remaining unchanged, the power required to run the compressor is *directly proportional to free air sucked in.*

∴ *Percentage change in power expended* = **2.67% (decrease). (Ans.)**

Example 24.11. *A 4-cylinder double-acting compressor is required to compress 30 m³/min of air at 1 bar and 27°C to a pressure of 16 bar. Determine the size of motor required and cylinder dimensions if the following data is given :*

Speed of the compressor,	*N = 320 r.p.m.*
Clearance volume,	V_c *= 4%*
Stroke to bore ratio,	*L/D = 1.2*

Mechanical efficiency, $\eta_{mech.} = 82\%$

Value of index, $n = 1.32.$

Assume no pressure change in suction valves and the air gets heated by 12°C during suction stroke.

Solution. Refer to Fig. 24.11.

Net work done

$$= \frac{n}{n-1} \times p_1 \times (V_1 - V_4)\left[\left(\frac{p_2}{p_1}\right)^{\frac{n-1}{n}} - 1\right]$$

where $V_1 - V_4$ = suction volume = 30 m³/min (given) = $\frac{30}{60}$ = 0.5 m³/s

∴ Work done

$$= \frac{1.32}{1.32-1} \times 1 \times 10^5 \times 0.5\left[\left(\frac{16}{1}\right)^{\frac{1.32-1}{1.32}} - 1\right]$$

= 197648.9 Nm/s

$$\text{Theoretical power} = \frac{197648.9}{1000} = 197.64 \text{ kW}$$

∴ **Motor power**

$$= \frac{197.64}{\eta_{mech.}} = \frac{197.64}{0.82} = \textbf{241 kW.} \quad \textbf{(Ans.)}$$

Volumetric efficiency,

$$\eta_{vol.} = \left[1 + k - k\left(\frac{p_2}{p_1}\right)^{1/n}\right] \times \frac{p_i T_a}{p_a T_i}$$

(Suffix '*i*' and '*a*' stand for inside and atmospheric conditions)

i.e.,
$$\eta_{vol.} = \left[1 + 0.04 - 0.04\left(\frac{16}{1}\right)^{1/1.32}\right] \times \frac{1 \times (273+27)}{1 \times (273+39)}$$

= 0.686 or 68.6%

Now swept volume of one cylinder

$$= \frac{30}{4} \times \frac{1}{2 \times 320} \times \frac{1}{0.686} = 0.01708 \text{ m}^3$$

$$\therefore \quad \frac{\pi}{4} D^2 L = 0.01708 \quad \text{or} \quad \frac{\pi}{4} D^2 \times 1.2\, D = 0.01708$$

$$\therefore \quad D^3 = \frac{0.01708 \times 4}{\pi \times 1.2}$$

∴ **D = 0.263 m** or **263 mm. (Ans.)**

and **L** = 1.2 × 263 = **315.6 mm. (Ans.)**

Example 24.12. *A two-cylinder single-acting air compressor is to deliver 16 kg of air per minute at 7 bar from suction conditions 1 bar and 15°C. Clearance may be taken as 4% of stroke volume and the index for both compression and re-expansion as 1.3. Compressor is directly coupled to a four-cylinder four-stroke petrol engine which runs at 2000 r.p.m. with a brake mean effective pressure of 5.5 bar. Assuming a stroke-bore ratio of 1.2 for both engine and compressor and a mechanical efficiency of 82% for compressor, calculate the required cylinder dimensions.*

Solution. Refer to Fig. 24.11.

$$\text{Amount of air delivered per cylinder} = \frac{16}{2} = 8 \text{ kg/min.}$$

Suction conditions : p_1 = 1 bar, T_1 = 15 + 273 = 288 K

From the gas equation, $p_1(V_1 - V_4) = mRT_1$

or
$$V_1 - V_4 = \frac{mRT_1}{p_1} = \frac{8 \times 287 \times 288}{1 \times 10^5} = 6.61 \text{ m}^3\text{/min.}$$

$$= \frac{6.61}{2000} = 0.003305 \text{ m}^3\text{/stroke} \; .....$$

compressor being single-acting

From expansion curve,

$$\frac{V_4}{V_3} = \left(\frac{p_3}{p_4}\right)^{1/n} = \left(\frac{7}{1}\right)^{1/1.3} = 4.467$$

$$\therefore \quad V_4 = 4.467\, V_3 = 0.04\, V_s \times 4.467 = 0.1787\, V_s$$

(where V_s is the swept volume)

Since $V_1 = V_3 + V_s = 0.04\, V_s + V_s = 1.04\, V_s$

∴ $V_1 - V_4 = 1.04\, V_s - 0.1787\, V_s = 0.8613\, V_s$

But $V_1 - V_4$ is also equal to 0.003305

∴ $0.8613\, V_s = 0.003305$

i.e.,
$$V_s = \frac{0.003305}{0.8613} = 0.003837 \text{ m}^3$$

If L_c = length of stroke of compressor, and

D_c = diameter of the cylinder of the compressor,

then

$L_c = 1.2\, D_c$ ....(given)

$$\therefore \quad \frac{\pi}{4} D_c^2 \times L_c = V_s$$

or
$$\frac{\pi}{4} D_c^2 \times 1.2\, D_c = 0.003837$$

or
$$D_c^3 = \frac{0.003837 \times 4}{\pi \times 1.2}$$

i.e., $\mathbf{D_c}$ = **0.1596 m** or **159.6 mm. (Ans.)**

$\mathbf{L_c}$ = 159.6 × 1.2 = **191.5 mm. (Ans.)**

Now indicated power of the compressor

$$= \frac{n}{n-1} \times mRT_1\left[\left(\frac{p_2}{p_1}\right)^{\frac{n-1}{n}} - 1\right]$$

$$= \frac{1.3}{1.3-1} \times \frac{16}{60} \times 0.287 \times 288\left[\left(\frac{7}{1}\right)^{\frac{1.3-1}{1.3}} - 1\right]$$

= 54.1 kW

Brake power of the engine

$$= \frac{54.1}{\eta_{mech.}} = \frac{54.1}{0.82} = 65.97 \text{ kW}$$

Now, $65.97 = \dfrac{n_e\, p_{mb}\, L_e\, AN}{60}$

If D_e = diameter of the engine cylinder,

L_e = length of the stroke of engine = 1.2 D_e,

n_e = number of engine cylinders,

Then,

$$65.97 = \frac{4 \times (5.5 \times 10^5) \times 1.2\, D_e \times \frac{\pi}{4} \times D_e^2 \times 2000}{60 \times 10^3}$$

$$\therefore \quad D_e^3 = \frac{65.97 \times 60 \times 10^3 \times 4}{4 \times 10^5 \times 5.5 \times 1.2 \times \pi \times 2000}$$

$= 0.0009545 \text{ m}^3$

i.e., $\mathbf{D_e = 0.0984}$ **m** or **98.4 mm. (Ans.)**

$\mathbf{L_e}$ = 1.2 × 98.4 = **118.1 mm. (Ans.)**

Example 24.13. *A single-stage double-acting air compressor delivers air at 7.5 bar. The pressure and temperature at the end of suction stroke are 1 bar and 25°C. It delivers 2.2 m³ of free air per minute when the compressor is running at 310 r.p.m. The clearance volume is 5% of stroke volume. The pressure and temperature of ambient air are 1.03 and 20°C.*

Determine :

(i) Volumetric efficiency of the compressor ;

(ii) Diameter and stroke of the cylinder if both are equal, and

(iii) I.P. of the compressor and B.P. if the mechanical efficiency is 85%.

Take : Index of compression = 1.25, and Index of expansion = 1.3.

Solution. Refer to Fig. 24.23.

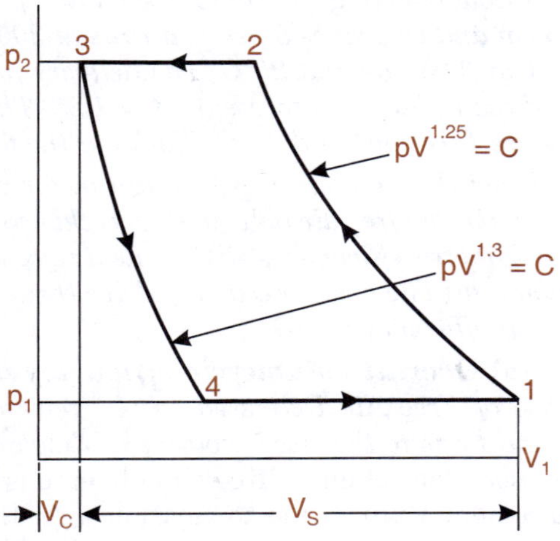

Fig. 24.23

Given : p_1 = 1 bar, p_2 = 7.5 bar,

T_1 = 25 + 273 = 298 K, $V_{amb.}$ = 2.2 m³

N = 310 r.p.m., $V_c = 0.05\, V_s$, $p_{amb.}$ = 1.03 bar,

$T_{amb.}$ = 20 + 273 = 293 K, $\eta_{mech.}$ = 85%.

(*i*) **Volumetric efficiency of the compressor $\eta_{vol.}$:**

$$\text{Volumetric efficiency} = \frac{V_1 - V_4}{V_s}$$

From the *expansion curve 3-4,*

$$p_2 V_3^{1.3} = p_4 V_4^{1.3}$$

or, $$p_2 V_c^{1.3} = p_1 V_4^{1.3} \quad (\because \; V_3 = V_c \text{ and } p_4 = p_1)$$

$$V_4 = V_c \left(\frac{p_2}{p_1}\right)^{\frac{1}{1.3}} = V_c \left(\frac{7.5}{1}\right)^{1/1.3}$$

$$= V_c\, (7.5)^{0.769} = 4.71\, V_c$$

But, $V_c = 0.05\, V_s$

$\therefore \quad V_4 = 4.71 \times 0.05\, V_s = 0.2355\, V_s$

$$\therefore \quad \eta_{vol.} = \frac{(V_s + V_c) - 0.2355\, V_s}{V_s}$$

$$= \left(\frac{V_s + 0.05\, V_s - 0.2355\, V_s}{V_s}\right)$$

= **0.814** or **81.4%. (Ans.)**

(*ii*) **Diameter and stroke of the cylinder D and L :**

The volume of air delivered at suction condition is given by

$$V_1 = \frac{p_{amb}\, V_{amb}\, T_1}{p_1\, T_{amb}} = \frac{1.03 \times 2.2 \times 298}{1 \times 293}$$

= 2.305 m³/min.

The volume delivered per minute is given by

$V_1 = (V_s \times \eta_{vol.} \times \text{r.p.m.}) \times 2$

$\therefore \quad 2.305 = V_s \times 0.814 \times 310 \times 2$

$$\therefore \quad V_s = \frac{2.305}{0.814 \times 310 \times 2}$$

= 0.00456 m³ or 4560 cm³

$\therefore \quad \pi/4\, D^2 L = 4560$

$\pi/4\, D^3 = 4560 \qquad (\because \; D = L)$

$$\therefore \quad D = \left(\frac{4560 \times 4}{\pi}\right)^{1/3} = 17.97 \text{ cm} \simeq \mathbf{18 \text{ cm.}} \;\; \textbf{(Ans.)}$$

$\therefore$ **L = 18 cm. (Ans.)**

(*iii*) **I.P. of compressor and B.P :**

The work done per cycle of operation for *double-acting* compressor is given by :

$$W = 2 \times \left[\frac{n_1}{n_1 - 1} \cdot p_1V_1 \left\{ \left(\frac{p_2}{p_1} \right)^{\frac{n_1-1}{n_1}} - 1 \right\} \right.$$

$$\left. - \frac{n_2}{n_2 - 1} \cdot p_4V_4 \left\{ \left(\frac{p_3}{p_4} \right)^{\frac{n_2-1}{n_2}} - 1 \right\} \right]$$

as $\quad p_3 = p_2$ and $p_4 = p_1$

$$= 2 \left[\frac{n_1}{n_1 - 1} \cdot p_1V_1 \left\{ \left(\frac{p_2}{p_1} \right)^{\frac{n_1-1}{n_1}} - 1 \right\} \right.$$

$$\left. - \frac{n_2}{n_2 - 1} \cdot p_1V_4 \left\{ \left(\frac{p_2}{p_1} \right)^{\frac{n_2-1}{n_2}} - 1 \right\} \right] \quad ...(i)$$

where, n_1 = Index of compression,

n_2 = Index of expansion,

$V_1 = V_c + V_s = 0.05\ V_s + V_s = 1.05\ V_s$, and

$V_4 = 0.235\ V_s$.

Inserting these values in the eqn. (*i*), we get

$$\text{W/cycle} = 2 \times p_1V_s \left[\frac{n_1}{n_1 - 1} \times 1.05 \left\{ \left(\frac{p_2}{p_1} \right)^{\frac{n_1-1}{n_1}} - 1 \right\} \right.$$

$$\left. - \frac{n_2}{n_2 - 1} \times 0.2355 \left\{ \left(\frac{p_2}{p_1} \right)^{\frac{n_2-1}{n_2}} - 1 \right\} \right]$$

$$= 2 \times 1 \times 10^5 \times 0.00456$$

$$\left[\frac{1.25}{1.25 - 1} \times 1.05 \left\{ (7.5)^{0.25/1.25} - 1 \right\} \right.$$

$$\left. - \frac{1.3}{1.3 - 1} \times 0.2355 \left\{ (7.5)^{0.3/1.3} - 1 \right\} \right]$$

$$= 912\ [(5 \times 1.05 \times 0.496) - 1.0205 \times 0.59]$$

$$= 1825.7 \text{ Nm/cycle}$$

$$\therefore \text{ I.P.} = \frac{\text{Work done/cycle} \times \text{r.p.m.}}{60 \times 1000}$$

$$\frac{1825.7 \times 310}{60 \times 1000} = 9.43 \text{ kW}$$

$$\therefore \text{ B.P.} = \frac{\text{I.P.}}{\eta_{mech.}} = \frac{9.43}{0.85} = \mathbf{11.09\ kW.} \quad \textbf{(Ans.)}$$

Example 24.14. *A single-stage single-acting compressor delivers 14 m^3 of free air per minute from 1 bar to 7 bar. The speed of compressor is 310 r.p.m. Assuming that compression and expansion follow the law $pV^{1.35}$ = constant and clearance is 5% of the swept volume, find the diameter and stroke of the compressor. Take L = 1.5 D. The temperature and pressure of air at the suction are same as atmospheric air.*

Solution. We know that,

$$n_{\text{vol.}} = 1 + k - k \left(\frac{p_2}{p_1} \right)^{1/n}$$

$$= 1 + 0.05 - 0.05 \left(\frac{7}{1} \right)^{1/1.35}$$

$$\left(\because \ k = \frac{V_c}{V_s} = \frac{0.05\ V_s}{V_s} = 0.05 \right)$$

$$= 0.839 \text{ or } 83.9\%$$

The free air delivered per minute is given by,

$$V_s \times \eta_{\text{vol.}} \times 310 = 14$$

$$\therefore \quad V_s = \frac{14}{\eta_{\text{vol.}} \times 310} = \frac{14}{0.839 \times 310} = 0.0538 \text{ m}^3$$

But, $\quad V_s = \dfrac{\pi}{4} D^2L$

or $\quad 0.0538 = \dfrac{\pi}{4} D^2 \times 1.5D$

or $\quad D = \left(\dfrac{0.538 \times 4}{\pi \times 1.5} \right)^{1/3} = \mathbf{0.77\ m}$ or $\mathbf{77\ cm.}$ **(Ans.)**

$L = 1.5\ D = 1.5 \times 77 = \mathbf{115.5\ cm.}$ **(Ans.)**

Example 24.15. (*a*) *Define the term 'overall volumetric efficiency' with reference to a reciprocating compressor. Discuss the parameters in brief which affect it.*

(*b*) *Show the effect of increase in compression ratio in a single-stage reciprocating compressor on p-v diagram and give its physical explanation.*

(*c*) *A double-acting air compressor having size (D × L) 33 × 35 cm and clearance 5 per cent runs at 300 r.p.m. It takes in air at 0.95 bar and 25°C. The delivery pressure is 4.5 bar and the index of compression, n = 1.25. The free air conditions are 1.013 bar and 20°C. Work out the following :*

(*i*) *Show the process on p-V diagram for cover and crank ends ; (ii) The free air delivered as reckoned from the apparent volumetric efficiency ; (iii) The heat rejected during compression ; (iv) The power needed to drive the compressor if mechanical efficiency is 80%.* **(AMIE)**

Solution. (*a*) ***Overall volumetric efficiency*** *is defined as the ratio of Free Air Delivered (FAD) referred to the ambient conditions to the swept volume or displacement of the compressor*. The volume of free air delivered is less than the displacement volume due to clearance. At the end of the delivery stroke the clearance space is filled with compressed air. On the inward stroke, the air will be admitted only after the clearance air is expanded to the inlet conditions. This would reduce the effective swept volume. Further, in practice the air that is sucked in during the induction stroke gets heated up while passing through the hot valves and coming in contact with hot cylinder walls. There is also wire drawing effect through the valves resulting in drop in pressure. Thus the conditions obtained at the end of induction stroke (p_1, T_1) are different from the ambient conditions (p_{amb}, T_{amb}). Therefore the overall volumetric efficiency is

$$\eta_{v,\text{ overall}} = \frac{p_1}{T_1} \times \frac{T_{amb}}{p_{amb}} \left[1 - C\left\{\left(\frac{p_2}{p_1}\right)^{1/n} - 1\right\}\right]$$

where, C = Clearance ratio (ratio of clearance volume to swept volume), and

n = Index of re-expansion.

The parameters which affect the overall volumetric efficiency are as given below :

(*i*) *Engine speed.* Higher the engine speed, greater is the wire drawing effect, and also higher the temperature of cylinder walls, lower the efficiency.

(*ii*) Leakage past the piston lowers the volumetric efficiency.

(*iii*) Too large a clearance volume will lower the clearance volumetric efficiency, and hence the overall volumetric efficiency.

(*iv*) Obstructions at inlet valves will increase wire drawing, and hence lower the volumetric efficiency.

(*v*) Inertia effects of air in suction pipe may lower the volumetric efficiency.

(*b*) The ***effect of increase in compression ratio on p-V diagram*** is as shown in Fig. 24.24.

By increasing the delivery from p_2 to p_2',

(*i*) The pressure ratio is increased from $\left(\frac{p_2}{p_1}\right)$ to $\left(\frac{p_2'}{p_1}\right)$.

(*ii*) The compression work per kg is increased from 1-2-3-4 to 1-2′-3′-4′.

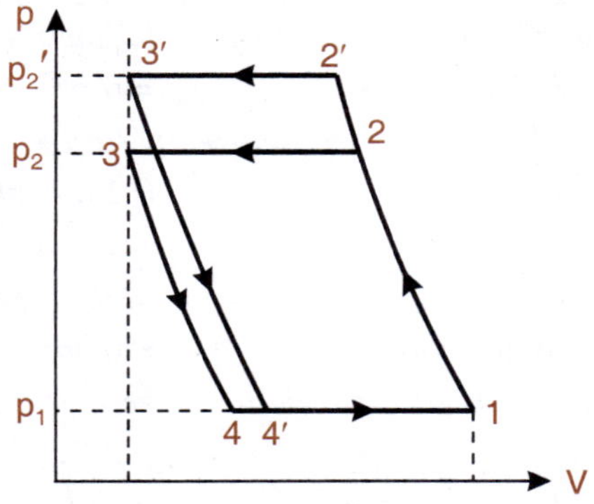

Fig. 24.24

(*iii*) The effective swept volume is decreased from ($V_1 - V_4$) to ($V_1 - V_4'$). This is due to the fact that by increasing the pressure ratio the pressure of air in the clearance volume, at the end of delivery stroke, is increased. The re-expansion 3′–4′ occupies a large fraction of swept volume (as compared to 3–4). This reduces the effective suction stroke with increase in pressure ratio.

(*c*) (*i*) **Process on p-V diagram :**

Neglecting the effect of piston rod the processes for cover and crank ends are shown on the *p-V* diagram as indicated in Fig. 24.25.

(*ii*) **The free air delivered :**

$$V_S = \frac{\pi}{4} D^2 L \times 2N \text{ for double-acting}$$

$$\frac{\pi}{4}\left(\frac{33}{100}\right)^2 \times \frac{35}{100} \times 2 \times 300 = 17.96 \text{ m}^3/\text{min.}$$

$\eta_{vol,\,C}$ = clearance volumetric efficiency

$$= 1 - C\left[\left(\frac{p_2}{p_1}\right)^{1/n} - 1\right]$$

$$= 1 - 0.05\left[\left(\frac{4.5}{0.95}\right)^{1/1.25} - 1\right] = 0.8765$$

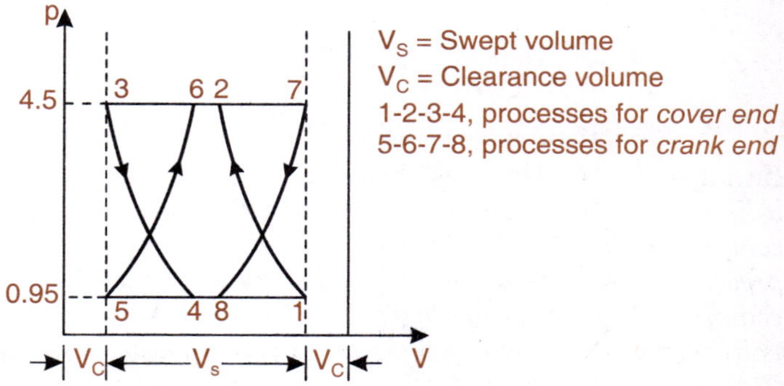

Fig. 24.25

The actual air drawn in per min.

$$= (V_1 - V_4) = V_a = \eta_{vol,\,C} \times V_s$$

$$= 0.8765 \times 17.96 = 15.74 \text{ m}^3/\text{min.}$$

This volume of air drawn is measured at 0.95 bar and 25°C. The free air conditions are 1.013 bar and 20°C.

Therefore the free air delivered, in m³/min is

$$= \frac{p_{suc}}{p_{amb}} \times \frac{T_{amb}}{T_{suc}} \times 15.74$$

$$= \frac{0.95}{1.013} \times \frac{273 + 20}{273 + 25} \times 15.74$$

$$= \mathbf{14.51\ m^3/min.} \quad \textbf{(Ans.)}$$

(*iii*) **Heat rejected during compression :**

Mass of air delivered per min.

$$\dot{m}_a = \frac{(0.95 \times 10^5) \times 15.74}{287 \times 298} = 17.48 \text{ kg/min.}$$

$$T_2 = T_1 \left(\frac{p_2}{p_1}\right)^{n-1/n} = 298 \left(\frac{4.5}{0.95}\right)^{0.25/1.25} = 406.7 \text{ K}$$

Heat rejected during compression,

$$Q_2 = mc_v \frac{\gamma - n}{n - 1} (T_2 - T_1)$$

$$= 17.48 \times 0.717 \times \frac{1.4 - 1.25}{1.25 - 1} (406.7 - 298)$$

$$= \mathbf{817.4 \text{ kJ/min.}} \quad \textbf{(Ans.)}$$

(*iv*) **Power needed to drive the compressor P :**

$$P = \frac{1}{\eta_{\text{mech.}}} \left[\frac{n}{n-1} mRT_1 \left\{\left(\frac{p_2}{p_1}\right)^{\frac{n-1}{n}} - 1\right\}\right]$$

$$= \frac{1}{0.8} \left[\frac{1.25}{0.25} \times \frac{17.48}{60} \times 0.287 \times 298 \left\{\left(\frac{4.5}{0.95}\right)^{\frac{0.25}{1.25}} - 1\right\}\right] \text{ kJ/S}$$

$$= \mathbf{56.82 \text{ kJ/s}} \text{ or } \mathbf{kW.} \quad \textbf{(Ans.)}$$

Example 24.16. *Air at 103 kPa and 27°C is drawn in L.P. cylinder of a two-stage air compressor and is isentropically compressed to 700 kPa. The air is then cooled at constant pressure to 37°C in an intercooler and is then again compressed isentropically to 4 MPa in the H.P. cylinder, and is delivered at this pressure. Determine the power required to run the compressor if it has to deliver 30 m³ of air per hour measured at inlet conditions.* **(M.U.)**

Solution. Refer to Fig. 24.26.

Pressure of intake air (L.P. cylinder),

$p_1 = 103$ kPa

Temperature of intake air,

$T_1 = 27 + 273 = 300$ K

Pressure of air entering H.P. cylinder,

$p_2 = 700$ kPa

Temperature of air entering H.P. cylinder,

$T_2 = 37 + 273 = 310$ K

Pressure of air after compression in H.P. cylinder,

$p_3 = 4$ MPa or 4000 kPa

Volume of air delivered = 30 m³/h

Power required to run the compressor, P :

Mass of air compressed,

$$m = \frac{(103 \times 10^3) \times 30}{(0.287 \times 1000) \times 300} = 35.89 \text{ kg/h}$$

For the *compression process 1-2′*, we have

$$\frac{T_2'}{T_1} = \left(\frac{p_2}{p_1}\right)^{\frac{\gamma-1}{\gamma}} = \left(\frac{700}{103}\right)^{\frac{1.4-1}{1.4}}$$

$$= 1.7289 \text{ or } T_2' = 300 \times 1.7289 = 518.7 \text{ K}$$

Similarly for the *compression process 2–3*, we have

$$\frac{T_3}{T_2} = \left(\frac{p_3}{p_2}\right)^{\frac{\gamma-1}{\gamma}} = \left(\frac{4000}{700}\right)^{\frac{1.4-1}{1.4}}$$

$$= 1.6454 \text{ or } T_3 = 310 \times 1.645 = 510.1 \text{ K}$$

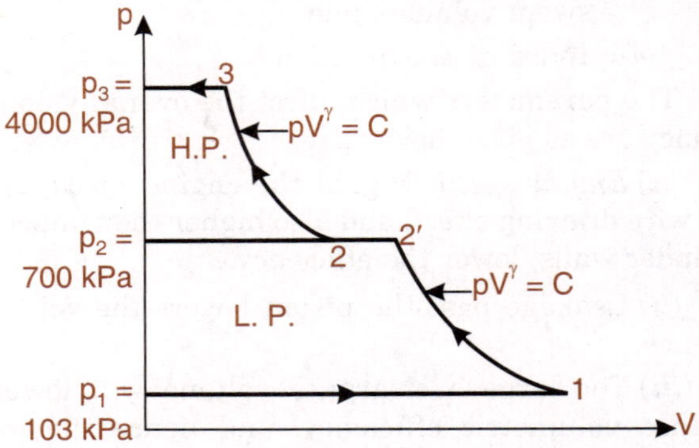

Fig. 24.26

∴ Work required to run the compressor,

$$W = \frac{\gamma}{\gamma - 1} [mR (T_2' - T_1) + mR (T_3 - T_2)]$$

$$= \frac{\gamma}{\gamma - 1} \times mR [(T_2' - T_1) + (T_3 - T_2)]$$

$$= \frac{1.4}{1.4 - 1} \times \frac{35.89}{3600} \times 0.287 [(518.7 - 300) + (510.1 - 310)] = 4.194 \text{ kN m/s}$$

Hence power required to run the compressor

$= \mathbf{4.194 \text{ kW.}}$ **(Ans.)**

Example 24.17. *A trial on a two-stage single-acting reciprocating air compressor gave the following data :*

Free air delivered	*= 6 m³/min*
Atmospheric pressure and temperature	*= 1 bar and 27°C*
Delivery pressure	*= 40 bar*
Speed	*= 400 r.p.m.*
Intermediate pressure	*= 6 bar*
Temperature at the inlet to the second stage	*= 27°C*
Law of compression	*= $pV^{1.3}$ = constant*

Mechanical efficiency = *80%*

Stroke of L.P. = *diameter of L.P.*

= *stroke of H.P.*

Calculate :

(*i*) *Cylinder-diameters ;*

(*ii*) *Power required, neglect clearance.*

(D.U.)

Solution. Let, $D_{L.P.}$ = dia of L.P. cylinder, and

$D_{H.P.}$ = dia of H.P. cylinder.

L = Stroke = $D_{L.P.}$

Assuming a volumetric efficiency of 100%,

$$\text{F.A.D.} = \frac{\pi}{4} D^2_{L.P.} \times \text{r.p.m. (for single-acting)}$$

or

$$6 = \frac{\pi}{4} D^2_{L.P.} \times D_{L.P.} \times 400 \text{ m}^3/\text{min}$$

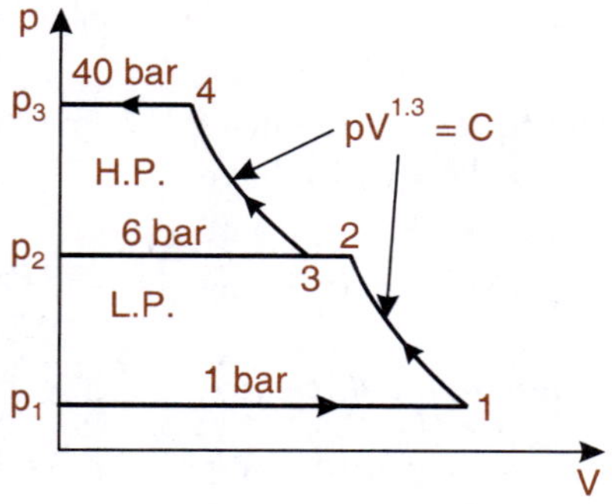

Fig. 24.27

$\therefore$ $D_{L.P.} = 0.2673$ m = **267.3 mm = L. (Ans.)**

Swept volume of H.P. cylinder

= Volume of air at 27°C and 6 bar

$$= 1 \text{ m}^3/\text{min} = \frac{\pi}{4} D^2_{H.P.} \times L \times \text{r.p.m.}$$

$$= \frac{\pi}{4} D^2_{H.P.} \times 0.2673 \times 400$$

$[\because \; p_1 V_{L.P.} = p_2 V_{H.P.}$ or $1 \times 6 = 6 \times V_{H.P.}$ or $V_{H.P.} = 1 \text{ m}^3/\text{min}]$

$\therefore$ $D_{H.P.}$ = **0.109 m or 109 mm. (Ans.)**

$$\text{Indicated work} = \frac{n}{n-1} p_1 V_1 \left[\left(\frac{p_2}{p_1} \right)^{\frac{n-1}{n}} - 1 \right] + \frac{n}{n-1} p_2 V_3 \left[\left(\frac{p_3}{p_2} \right)^{\frac{n-1}{n}} - 1 \right]$$

$$= \frac{n}{n-1} mRT_1 \left[\left(\frac{p_2}{p_1} \right)^{\frac{n-1}{n}} - 1 \right] + \frac{n}{n-1} mRT_3 \left[\left(\frac{p_3}{p_2} \right)^{\frac{n-1}{n}} - 1 \right]$$

$$= \frac{n}{n-1} mRT_1 \left[\left(\frac{p_2}{p_1} \right)^{\frac{n-1}{n}} + \left(\frac{p_3}{p_2} \right)^{\frac{n-1}{n}} - 2 \right],$$

as $T_1 = T_3$.

Also, $m = \dfrac{p_1 V_1}{RT_1}$

$$= 100 \times \frac{6}{60} \times \frac{1}{0.287\,(273 + 27)}$$

= 0.116 kg/s

$\therefore$ Indicated work

$$= \frac{1.3}{0.3} \times 0.116 \times 0.287 \times 300 \left[\left(\frac{6}{1} \right)^{\frac{0.3}{1.3}} + \left(\frac{40}{6} \right)^{\frac{0.3}{1.3}} - 2 \right]$$

= 45.94 kJ/s (kW)

$$\text{Power required} = \frac{45.99}{0.8} = \mathbf{58.42 \text{ kW}}. \quad \textbf{(Ans.)}$$

Example 24.18. *A two-stage single-acting reciprocating compressor takes in air at the rate of 0.2 m³/s. The intake pressure and temperature of air are 0.1 MPa and 16°C. The air is compressed to a final pressure of 0.7 MPa. The intermediate pressure is ideal and intercooling is perfect. The compression index in both the stages is 1.25 and the compressor runs at 600 r.p.m. Neglecting clearance, determine :*

(*i*) *The intermediate pressure,*

(*ii*) *The total volume of each cylinder,*

(*iii*) *The power required to drive the compressor, and*

(*iv*) *The rate of heat rejection in the intercooler.*

Take c_p *= 1.005 kJ/kg K and R = 0.287 kJ/kg K.*

(D.U.)

Solution. Intake volume, $V_1 = 0.2 \text{ m}^3/\text{s}$

Intake pressure, $p_1 = 0.1$ MPa

Intake temperature, $T_1 = 16 + 273 = 289$ K

Final pressure, $p_3 = 0.7$ MPa

Compression index in both stages, $n_1 = n_2 = n = 1.25$

Speed of the compressor, $N = 600$ r.p.m.

$c_p = 1.005$ kJ/kg K ; $R = 0.287$ kJ/kg K

(*i*) **The intermediate pressure, p_2 :**

$$p_2 = \sqrt{p_1 p_3} = \sqrt{0.1 \times 0.7} = 0.2646 \text{ MPa}$$

......(Perfect intercooling)

(*ii*) **The total volume of each cylinder, V_{s_1}, V_{s_2} :**

We know that, $V_{s_1} \times \frac{N}{60} = V_1$ or $V_{s_1} \times \frac{600}{60} = 0.2$

$\therefore$ V_{s_1} (Volume of L.P. cylinder) $= \frac{60 \times 0.2}{600}$

$= \mathbf{0.02\ m^3}$. **(Ans.)**

Also, $p_1 V_{s_1} = p_2 V_{s_2}$ or $V_{s_2} = \frac{p_1 V_{s_1}}{p_2}$

$\therefore$ V_{s_2} (Volume of H.P. cylinder) $= \frac{0.1 \times 0.02}{0.2646}$

$= \mathbf{0.00756\ m^3}$. **(Ans.)**

(*iii*) **The power required to drive the compressor, P :**

$$P = \frac{2n}{n-1} p_1 V_1 \left[\left(\frac{p_3}{p_1} \right)^{\frac{n-1}{2n}} - 1 \right]$$

......(Perfect intercooling)

$$= \frac{2 \times 1.25}{1.25 - 1} \times (0.1 \times 10^3) \times 0.2 \left[\left(\frac{0.7}{0.1} \right)^{\frac{1.25-1}{2 \times 1.25}} - 1 \right]$$

$= \mathbf{42.96\ kW}$. **(Ans.)**

(*iv*) **The rate of heat rejection in the intercooler :**

Mass of air handled, $m = \frac{p_1 V_1}{R T_1} = \frac{(0.1 \times 10^3) \times 0.2}{0.287 \times 289}$

$= 0.241$ kg/s

Also, $$\frac{T_2}{T_1} = \left(\frac{p_2}{p_1} \right)^{\frac{n-1}{n}}$$

or $$\frac{T_2}{289} = \left(\frac{0.2646}{0.1} \right)^{\frac{1.25-1}{1.25}}$$

or $T_2 = 351.1$ K

$\therefore$ Heat rejected in the intercooler

$= m \times c_p \times (T_2 - T_1)$

$= 0.241 \times 1.005 \times (351.1 - 289)$

$= \mathbf{15.04\ kJ/s}$ or $\mathbf{15.04\ kW}$. **(Ans.)**

Example 24.19. *A two-stage air compressor with complete intercooling delivers air to the mains at a pressure of 30 bar, the suction conditions being 1 bar and 15°C. If both cylinders have the same stroke, find the ratio of cylinders diameters, for the efficiency of compression to be a maximum. Assume the index of compression to be 1.3.*

Solution. Refer to Fig. 24.28.

Volume of L.P. cylinder $= V_1$

Volume of H.P. cylinder $= V_2$

If $D_{L.P.}$ and $D_{H.P.}$ are the diameters of the low pressure cylinder and high pressure cylinder respectively, then

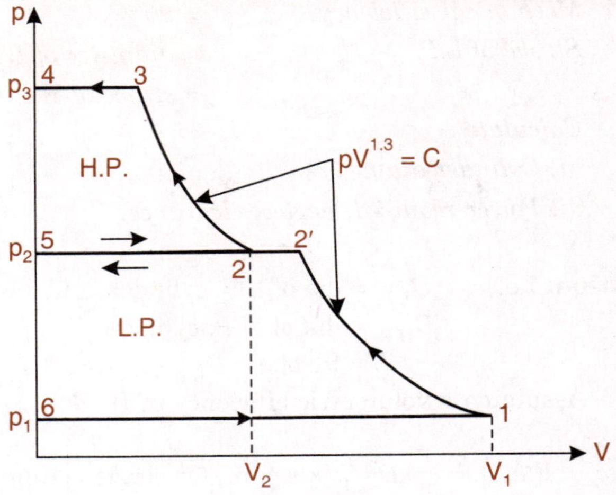

Fig. 24.28

$$\frac{V_1}{V_2} = \frac{\frac{\pi}{4} D^2_{L.P.}}{\frac{\pi}{4} D^2_{H.P.}} \quad \text{or} \quad \frac{D_{L.P.}}{D_{H.P.}} = \sqrt{\frac{V_1}{V_2}}$$

From the *curve 1-2′* following the law ; $pV^{1.3} = C$, we have

$$p_1 V_1^{1.3} = p_2 V_2'^{1.3} \quad \text{or} \quad \frac{V_1}{V_2'} = \left(\frac{p_2}{p_1} \right)^{1/1.3}$$

But for *maximum efficiency*

$$p_2 = \sqrt{p_1 p_3} = \sqrt{1 \times 30} = 5.48 \text{ bar}$$

$$\therefore \quad \frac{V_1}{V_2'} = (5.48)^{1/1.3} = 3.7 \quad ...(i)$$

Now temperature T_2' at the end of compression in the low pressure cylinder

$$T_2' = T_2 \times \left(\frac{p_2}{p_1} \right)^{\frac{n-1}{n}} = (15 + 273) \left(\frac{5.48}{1.0} \right)^{\frac{1.3-1}{1.3}}$$

$= 426.4$ K.

From *constant pressure process of 2′-2*

$$\frac{V_2}{T_2} = \frac{V_2'}{T_2'}$$

i.e., $$\frac{V_2'}{V_2} = \frac{T_2'}{T_2} = \frac{426.4}{(15 + 273)} = 1.48 \quad ...(ii)$$

From (*i*) and (*ii*),

$$\frac{V_1}{V_2} = \frac{V_1}{V_2'} \times \frac{V_2'}{V_2} = 3.7 \times 1.48 = 5.476$$

$$\therefore \quad \frac{D_{L.P.}}{D_{H.P.}} = \sqrt{\frac{V_1}{V_2}} = \sqrt{5.476} = 2.34.$$

i.e., **Ratio of cylinder diameters = 2.34. (Ans.)**

Example 24.20. *In a two-stage air compressor the pressures are atmospheric 1.0 bar ; intercooler 7.4 bar ; delivery 42.6 bar. Assuming complete intercooling to the original temperature of 15°C and compression index n = 1.3, find the work done in compressing 1 kg of air.*

If both cylinders have the same stroke and the piston diameters are 9 cm and 3 cm and the volumetric efficiency of the compressor is 90 per cent, will the intercooler pressure be steady or will it rise or fall as the compressor continues working ? Justify your answer. **(P.U.)**

Solution. With *complete intercooling, work done,*

$$W = \frac{n}{n-1} RT_1 \left[\left(\frac{p_2}{p_1}\right)^{\frac{n-1}{n}} + \left(\frac{p_3}{p_2}\right)^{\frac{n-1}{n}} - 2 \right]$$

$$= \frac{1.3}{1.3-1} \times 0.287 \times 288 \left[\left(\frac{7.4}{1.0}\right)^{\frac{1.3-1}{1.3}} + \left(\frac{42.6}{7.4}\right)^{\frac{1.3-1}{1.3}} - 2 \right]$$

$= 358.17\ (1.587 + 1.497 - 2) =$ **388.25 kJ. (Ans.)**

Since *intercooling is perfect,*

$$p_1V_1 = p_2'V_2' \quad \text{or} \quad \frac{V_1}{V_2'} = \frac{7.4}{1.0} = 7.4$$

Ratio of effective cylinder volumes

$$= \frac{\text{Effective volume of L.P. cylinder}}{\text{Volume of H.P. cylinder}}$$

$$= \frac{\pi/4 \times (0.09)^2 \times l \times 0.9}{\pi/4 \times (0.03)^2 \times l} = 8.1$$

As the ratio of effective cylinder volumes is more than the ratio of the volumes obtained from *p-V* diagram, *more air is supplied to high pressure cylinder than can hold, therefore H.P. cylinder would suck less air from intercooler than received from L.P. cylinder. Thus pressure in the intercooler will rise.*

It may be noted that the effective volume of L.P. cylinder neglecting clearance is actual volume × volumetric efficiency. It applies to the *L.P. cylinder only as the air is sucked in this cylinder.*

Example 24.21. *A single-acting two-stage air compressor deals with 4 m³/min of air under atmospheric conditions of 1.016 bar and 15°C with a speed of 250 r.p.m. The delivery pressure is 78.65 bar. Assuming complete intercooling find the minimum power required by the compressor and the bore and stroke of the compressor. Assume a piston speed of 3 m/s, mechanical efficiency of 75% and volumetric efficiency of 80% per stage. Assume the polytropic index of compression in both the stages to be n = 1.25 and neglect clearance.* **(N.U.)**

Solution. In the Fig. 24.29. 1–2 shows compression in L.P. cylinder, 2–3 shows cooling before compression in H.P. cylinder and 3–4 compression in H.P. cylinder.

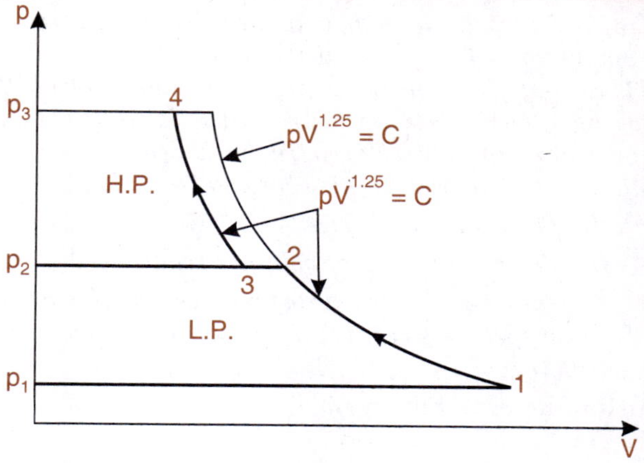

Fig. 24.29

Piston speed $= 2lN$

$\therefore$ **Stroke length,** $l = \dfrac{\text{Piston speed}}{2N} = \dfrac{3 \times 60}{2 \times 250}$

= **0.36 m. (Ans.)**

Volume handled $= \pi/4\ d^2 l \times N \times \eta_{vol.}$

$4 = \pi/4\ d^2 \times 0.36 \times 250 \times 0.8$

$\therefore$ **Diameter,** $d = \left(\dfrac{4 \times 4}{\pi \times 0.36 \times 250 \times 0.8}\right)^{1/2}$

= **0.266 m. (Ans.)**

Mass handled by compressor,

$$m = \frac{pV}{RT} = \frac{1.016 \times 10^5 \times 4}{287 \times 288}$$

= 4.916 kg/min. = 0.0819 kg/s.

Intermediate pressure,

$$p_2 = \sqrt{p_3 p_1} = \sqrt{78.65 \times 1.016} = 8.94 \text{ bar}$$

Temperature at the end of first stage compression,

$$T_2 = T_1 \times (r_p)^{\frac{n-1}{n}} = 288 \times \left(\frac{8.94}{1.016}\right)^{\frac{1.25-1}{1.25}}$$

$= 444.9$ K $\quad \left[\text{where } r_p = \dfrac{p_2}{p_1}\right]$

Work required

$$= 2 \times \frac{n}{n-1} \times mR\,(T_2 - T_1) \times \frac{1}{\eta_{mech.}}$$

$$= 2 \times \frac{1.25}{1.25-1} \times 0.0819 \times 287\,(444.9 - 288) \times \frac{1}{0.75} \times \frac{1}{1000}$$

= **49.17 kW. (Ans.)**

Example 24.22. *In a single-acting two-stage reciprocating air compressor 4.5 kg of air per min. are compressed from 1.013 bar and 15°C through a pressure ratio of 9 to 1. Both stages have the same pressure ratio, and the law of compression and expansion in both stages is $pV^{1.3}$ = constant. If the intercooling is complete, calculate :*

(*i*) *The indicated power*

(*ii*) *The cylinder swept volumes required.*

Assume that the clearance volumes of both stages are 5% of their respective swept volumes and that the compressor runs at 300 r.p.m.

Solution. Refer to Fig. 24.30.

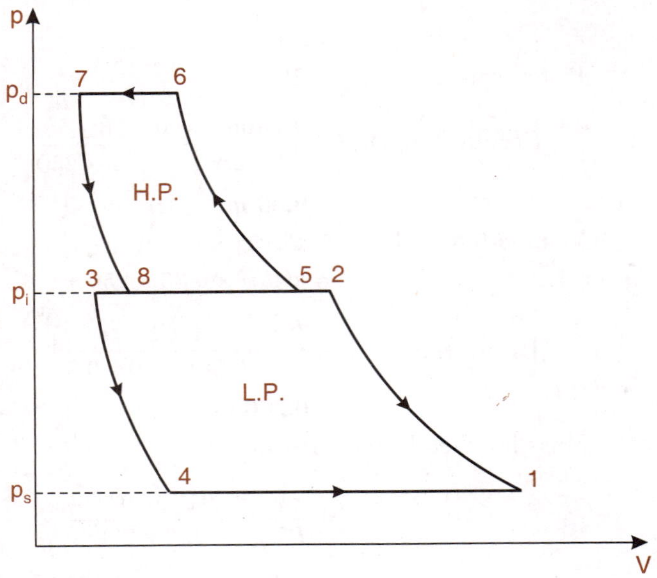

Fig. 24.30

Amount of air compressed, m = 4.5 kg/min.

Suction conditions, p_s = 1.013 bar,

T_s = 15 + 273 = 288 K

Pressure ratio, $\dfrac{p_d}{p_s} = 9$

Also, $\dfrac{p_i}{p_s} = \dfrac{p_d}{p_i}$ ...(Given)

Compression, expansion index, n = 1.3

Clearance volume in each stage = 5% of swept volume

Speed of the compressor, N = 300 r.p.m.

(*i*) **Indicated power :**

$$\therefore \quad \frac{p_i}{p_s} = \frac{p_d}{p_i}$$

$$\therefore \quad p_i^2 = p_s \times p_d = p_s \times 9p_s = 9p_s^2$$

$$\therefore \quad p_i = 3\,p_s \quad i.e., \quad \frac{p_i}{p_s} = 3$$

Now using the equation, $\dfrac{T_i}{T_s} = \left(\dfrac{p_i}{p_s}\right)^{\frac{n-1}{n}} = (3)^{\frac{1.3-1}{1.3}}$

$$\therefore \quad T_i = T_s \times (3)^{0.3/1.3} = 288 \times (3)^{0.3/1.3} = 371 \text{ K}$$

Now as n, m and temperature difference are the same for both stages, then the *work done in each stage is the same*.

Total work required per min.

$$= 2 \times \frac{n}{n-1}\, mR\,(T_i - T_s)$$

$$= 2 \times \frac{1.3}{1.3-1} \times 4.5 \times 0.287\,(371 - 288) = 929 \text{ kJ/min.}$$

$\therefore$ **Indicated power** $= \dfrac{929}{60}$ = **15.48 kW. (Ans.)**

(*ii*) **The cylinder swept volumes required :**

The mass induced per cycle,

$$m = \frac{4.5}{300} = 0.015 \text{ kg/cycle.}$$

This mass is passed through each stage in turn

For the L.P. pressure cylinder (Fig. 24.30)

$$V_1 - V_4 = \frac{mRT_s}{p_s} = \frac{0.015 \times 287 \times 288}{1.013 \times 10^5}$$

$$= 0.0122 \text{ m}^3\text{/cycle}$$

$$\eta_{vol.} = \frac{V_1 - V_4}{V_s} = 1 + k - k\left(\frac{p_i}{p_s}\right)^{1/n}$$

$$= 1 + 0.05 - 0.05\,(3)^{1/1.3}$$

$$= 0.934 \qquad \left(k = \frac{V_c}{V_s} = 0.05\right)$$

i.e., $\eta_{vol.} = 0.934$

$$\therefore \quad V_s = \frac{V_1 - V_4}{\eta_{vol.}} = \frac{V_1 - V_4}{0.934} = \frac{0.0122}{0.934}$$

$$= 0.0131 \text{ m}^3\text{/cycle}$$

i.e., **Swept volume of L.P. cylinder $V_{s(L.P.)}$ = 0.0131 m³. (Ans.)**

For the high pressure stage, a mass of 0.015 kg/cycle is drawn in at 15°C and a pressure of p_i = 3 × 1.013 = 3.039 bar

i.e., Volume drawn in $= \dfrac{0.015 \times 287 \times 288}{3.039 \times 10^5}$

$$= 0.00408 \text{ m}^3\text{/cycle}$$

$$\eta_{vol.} = 1 + k - k\left(\frac{p_d}{p_i}\right)^{1/n}$$

and since $\dfrac{V_c}{V_s}$ (= k) is the same as for the low pressure stage

and also $\dfrac{p_d}{p_i} = \dfrac{p_i}{p_s}$ then $\eta_{vol.}$ is 0.934 as above.

$\therefore$ **Swept volume of H.P. stage,**

$$V_{s(H.P.)} = \frac{0.00408}{0.934} = \mathbf{0.004367\ m^3.}\ \textbf{(Ans.)}$$

It may be noted that the clearance ratio $\left(k = \dfrac{V_c}{V_s}\right)$ is the same in each cylinder, and the suction temperatures are the same since *intercooling is complete*, therefore the swept volumes are in the ratio of the suction pressures,

i.e., $$V_{s(H.P.)} = \frac{V_{L.P.}}{3} = \frac{0.0131}{3} = 0.00437 \text{ m}^3.$$

Example 24.23. *A two-stage compressor delivers 2.2 m³ free air per minute. The pressure and temperature of air at the suction are 1 bar and 25°C respectively. The pressure at the delivery is 55 bar. The clearance in L.P. cylinder is 5% and also in H.P. cylinder is 5% of the stroke. Assuming perfect intercooling between the two stages, find the minimum power required to run the compressor at 210 r.p.m.*

If the strokes of both the cylinders are equal to the diameter of the L.P. cylinder, find the diameters and strokes. What is the ratio of the cylinder volumes ; Law of compression and re-expansion in both the cylinders is $pV^{1.3}$ = constant.

Solution. Refer to Fig. 24.31.

Free air delivered, $V_1 = 2.2$ m³/min, $p_1 (= p_s) = 1$ bar,

$T_1 = 25 + 273 = 298$ K ;

$p_d = 55$ bar ; $N = 210$ r.p.m. ;

Law of compression and expansion ;

$pV^{1.3}$ = constant.

Clearance in each of L.P. and H.P. cylinders = 5% of the stroke

Minimum power required to run the compression, P :

For *two-stage compressor* (with *perfect intercooling*) work done is given by,

$$W = \frac{2n}{n-1} p_1V_1 \left[\left(\frac{p_d}{p_s}\right)^{\frac{n-1}{2n}} - 1\right]$$

$$= \frac{2n}{n-1} mRT_1 \left[\left(\frac{p_d}{p_s}\right)^{\frac{n-1}{2n}} - 1\right]$$

$(\because \; p_1V_1 = mRT_1)$

But, $$m = \frac{p_1V_1}{RT_1} = \frac{1 \times 10^5 \times 2.2}{287 \times 298} = 2.57 \text{ kg/min}$$

$$\therefore \quad W = \frac{2 \times 1.3}{(1.3 - 1)} \times 2.57 \times 287 \times 298 \left[\left(\frac{55}{1}\right)^{\frac{(1.3-1)}{2 \times 1.3}} - 1\right]$$

$= 1119919$ Nm/min.

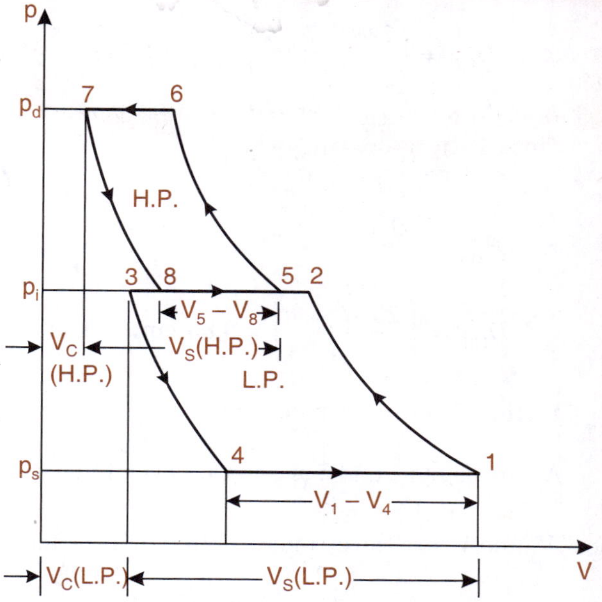

Fig. 24.31

$$\therefore \quad P = \frac{1119919}{60 \times 1000} = \textbf{18.66 kW.} \quad \textbf{(Ans.)}$$

Diameters and strokes :

We know that, $p_i = \sqrt{p_s p_d}$

(p_i = Intermediate pressure)

$= \sqrt{1 \times 55} = 7.4$ bar

The volumetric efficiency of L.P. cylinder is given by

$$n_{vol.(1)} = 1 + k - k\left(\frac{p_2}{p_1}\right)^{1/n} \qquad (p_2 = p_i)$$

$$= 1 + 0.05 - 0.05\left(\frac{7.4}{1}\right)^{1/1.3}$$

$= 0.817$ or 81.7%

The volumetric efficiency of the H.P. cylinder is same, because

$$\frac{p_d}{p_i} = \frac{p_i}{p_s} \text{ and } \frac{V_{c(H.P.)}}{V_{s(H.P.)}} = \frac{V_{c(L.P.)}}{V_{s(L.P.)}}$$

Free air or air at suction condition delivered per minute is given by,

Free air $= (V_1 - V_4) \times$ r.p.m.

$= V_{s(L.P.)}\, \eta_{vol.(L.P.)} \times$ r.p.m.

$= V_{s(L.P.)} \times 0.817 \times 210 = 2.2$

$$\therefore \; V_{s(L.P.)} = \frac{2.2}{0.817 \times 210} \text{ m}^3 = \frac{2.2 \times 10^6}{0.817 \times 210}$$

$= 12822.7$ cm³

$\pi/4\, D^2_{L.P.}\, L_{L.P.} = 12822.7$

$\pi/4\, D^3_{L.P.} = 12822.7$

$[\because \; L_{L.P.} = L_{H.P.} = D_{L.P.}]$ (Given)

$$\therefore \quad D_{L.P.} = \left(\frac{12822.7 \times 4}{\pi}\right)^{1/3} = \mathbf{25.37\ cm.} \quad \textbf{(Ans.)}$$

$$\therefore \quad L_{L.P.} = L_{H.P.} = \mathbf{25.37\ cm.} \quad \textbf{(Ans.)}$$

Since, $\eta_{vol.(L.P.)} = \eta_{vol.(H.P.)}$

$$\therefore \quad D^2_{L.P.}\, p_1 = D^2_{H.P.}\, p_i$$

$$\therefore \quad D^2_{H.P.} = \frac{D^2_{L.P.}\, p_s}{p_i} \qquad (\because \; p_1 = p_s)$$

$$D_{H.P.} = \sqrt{\left(\frac{25.37^2 \times 1}{7.4}\right)} = \mathbf{9.33\ cm.} \quad \textbf{(Ans.)}$$

Ratio of cylinder volumes, $\frac{V_1}{V_s}$:

As the points 1 and 5 are on isothermal line we have

$$p_1 V_1 = p_5 V_5$$

(where V_1 and V_5 are the cylinder volumes of L.P. and H.P.)

$$\frac{V_1}{V_5} = \frac{p_5}{p_1} = \frac{p_i}{p_s} = \frac{7.4}{1} = \mathbf{7.4.} \quad \textbf{(Ans.)}$$

Example 24.24. *A two-stage double-acting air compressor, operating at 220 r.p.m. takes in air at 1.0 bar and 27°C. The size of the L.P. cylinder is 360 × 400 mm ; the stroke of H.P. cylinder is the same as that of the L.P. cylinder and the clearance of both the cylinders is 4%. The L.P. cylinder discharges the air at a pressure of 4.0 bar. The air passes through the intercooler so that it enters the H.P. cylinder at 27°C and 3.80 bar, finally it is discharged from the compressor at 15.2 bar. The value of n in both the cylinders is 1.3, c_p = 1.0035 kJ/kg K and R = 0.287 kJ/kg K.*

Calculate : (i) The heat rejected in the intercooler ;

(ii) Diameter of H.P. cylinder ;

(iii) The power required to drive H.P. cylinder.

Solution. Refer to Fig. 24.32.

Speed of compressor, N = 220 r.p.m.

p_1 = 1.0 bar, p_5 (= p_8) = 3.8 bar,

p_2 (= p_3) = 4.0 bar,

p_6 (= p_7) = 15.2 bar

T_1 = 27 + 273 = 300 K,

T_5 = 27 + 273 = 300 K,

Clearance of L.P. and H.P. cylinders = 4%

Value of 'n' for both the cylinders = 1.3

c_p = 1.0035 kJ/kg K ; R = 0.287 kJ/kg K

Diameter of L.P. cylinder, $D_{L.P.}$ = 360 mm = 0.36 m

.Stroke of L.P. cylinder, $L_{L.P.}$ = 400 mm = 0.4 m

Swept volume of L.P. cylinder/min

$= \pi/4\, D^2_{L.P.} \times L_{L.P.} \times (220 \times 2)$

$= \pi/4 \times 0.36^2 \times 0.4 \times (220 \times 2)$

= 17.91 m³/min

...... since compressor is double-acting

Volumetric efficiency referred to condition at 1

$$= 1 + k - k\left(\frac{p_2}{p_1}\right)^{1/n} = 1 + 0.04 - 0.04\left(\frac{4.0}{1.0}\right)^{1/1.3}$$

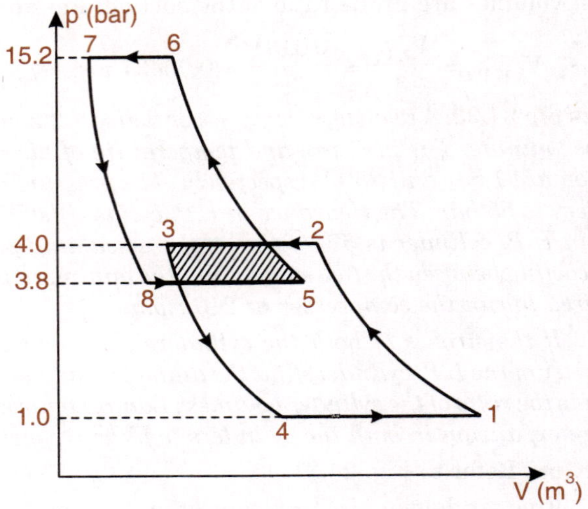

Fig. 24.32

$$= 0.9238 \text{ (or } 92.38\%) \qquad \left[\because \; k = \frac{V_c}{V_s} = \frac{4}{100} = 0.04\right]$$

$\therefore$ Volume of air drawn, referred to condition at 1

= 0.9238 × 17.91 = 16.54 m³/min

Thus, mass of air/min

$$m = \frac{p_1 V_1}{RT_1} = \frac{1.0 \times 10^5 \times 16.54}{0.287 \times 300 \times 10^3} = 19.21 \text{ kg/min}$$

$$\text{Also,} \quad T_2 = T_1 \times \left(\frac{p_2}{p_1}\right)^{\frac{n-1}{p_n}} = 300\left(\frac{4}{1}\right)^{\frac{1.3-1}{1.3}} = 413 \text{ K}$$

(*i*) Heat rejected in the intercooler :

Heat rejected in the intercooler

$= mc_p(T_2 - T_5) = 19.21 \times 1.0035\,(413 - 300)$

= **2178.3 kJ/min.** **(Ans.)**

(*ii*) Diameter of H.P. cylinder :

Volume of air drawn in H.P. cylinder per minute

$$V_s = \frac{mRT_5}{p_5} = \frac{19.21 \times 0.287 \times 300 \times 10^3}{3.8 \times 10^5}$$

= 4.352 m³/min.

Since the pressure ratio and clearance percentage in H.P. and L.P. cylinders are the same, therefore, the volumetric efficiency of both cylinders is same referred to condition at the start of compression.

$\therefore$ Swept volume of H.P. cylinder

$$= \frac{4.352}{0.9238} = 4.71 \text{ m}^3\text{/min}$$

i.e., $\pi/4\, D^2_{H.P.} \times L_{H.P.} \times (2 \times 220) = 4.71$

$$\pi/4\, D^2_{H.P.} \times 0.4 \times (2 \times 220) = 4.71 \quad [\because \ L_{H.P.} = L_{L.P.} = 0.4 \text{ m}]$$

$$\therefore \quad D_{H.P.} = \left(\frac{4.71 \times 4}{\pi \times 0.4 \times 2 \times 220}\right)^{1/2}$$

$$= 0.1846 \text{ m or } 184.6 \text{ mm} \quad \text{(given)}$$

i.e., *Diameter of H.P. cylinder* = **184.6 mm. (Ans.)**

(iii) **Power required to drive H.P. cylinder :**

Since the initial temperature and pressure ratio in the L.P. and the H.P. cylinders are the same, $T_6 = T_2$

∴ Power required for H.P. cylinder

$$= \frac{n}{n-1} mR(T_2 - T_1)$$

$$= \frac{1.3}{1.3-1} \times \frac{19.12}{60} \times 0.287\,(413 - 300)$$

= **45 kW. (Ans.)**

Example 24.25. *A single-acting two-stage air compressor delivers air at 18 bar. The temperature and pressure of the air before the compression in L.P. cylinder are 25°C and 1 bar. The discharge pressure of L.P. cylinder is 4.2 bar. The pressure of air leaving the intercooler is 4 bar and the air is cooled to 25°C. The diameter and stroke of L.P. cylinder are 40 cm and 50 cm respectively. The clearance volume is 5% stroke in both cylinders. The speed of the compressor is 200 r.p.m. Assuming the index of compression and re-expansion in both cylinders as 1.25, c_p for air = 1.004 kJ/kg K, find :*

(i) Power required to run the compressor, and

(ii) Heat rejected in intercooler/min.

Solution. Fig. 24.33 (*a*) and (*b*) shows the process on *p-V* and *T-s* diagrams

Given : $p_s = 1$ bar, $p_i = 4.2$ bar, $p_i' = 4.0$ bar,

$p_d = 18$ bar, $T_5 = T_1 = 25 + 273 = 298$ K

$D_{L.P.} = 0.4$ m, $L_{L.P.} = 0.5$ m

$N = 200$ r.p.m., $V_c = 5\%$ of stroke in both the cylinders, and

$c_p = 1.004$ kJ/kg K

$$V_{s(L.P.)} = \frac{\pi}{4} D^2_{L.P.} L_{L.P.} = \frac{\pi}{4} \times 0.4^2 \times 0.5$$

$$= 0.0628 \text{ m}^3$$

The volumetric efficiency of L.P. cylinder is given by,

$$\eta_{vol.\,(L.P.)} = 1 + k - k\left(\frac{p_i}{p_s}\right)^{1/n}$$

$$= 1 + 0.05 - 0.05\left(\frac{4.2}{1}\right)^{1/1.25} \quad \left(k = \frac{V_c}{V_s} = 0.05\right)$$

$$= 0.892 \text{ or } 89.2\%$$

$$\therefore \quad (V_1 - V_4) = \eta_{vol.\,(L.P.)} \times V_{s_1} = 0.892 \times 0.0628$$

$$= 0.056 \text{ m}^3$$

Considering the *compression curve 1–2*

$$\frac{T_2}{T_1} = \left(\frac{p_i}{p_s}\right)^{\frac{n-1}{n}}$$

$$T_2 = T_1\left(\frac{p_i}{p_s}\right)^{\frac{n-1}{n}} = 298\left(\frac{4.2}{1}\right)^{\frac{1.25-1}{1.25}} = 397 \text{ K}$$

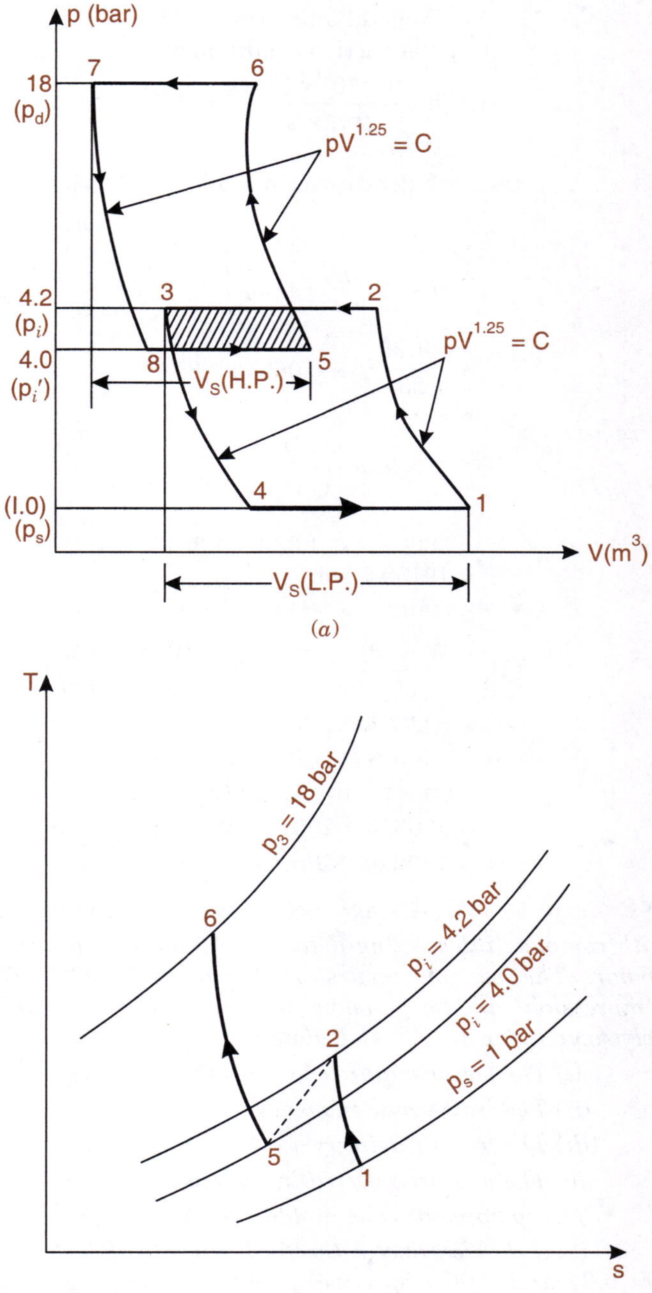

Fig. 24.33

$$W_{(L.P.)}/\text{cycle} = \frac{n}{n-1}\, mRT_1 \left[\left(\frac{p_i}{p_s}\right)^{\frac{n-1}{n}} - 1\right]$$

$$W_{(H.P.)}/\text{cycle} = \frac{n}{n-1}\, mRT_5 \left[\left(\frac{p_d}{p_i}\right)^{\frac{n-1}{n}} - 1\right]$$

As $T_5 = T_1$ (given)

$$\left[\begin{array}{l} m = \text{Mass of air of volume } (V_1 - V_4) \\ \quad\text{at suction condition of L.P. cylinder} \\ \therefore\ m = \dfrac{1 \times 10^5 \times 0.056}{287 \times 298} = 0.0655 \text{ kg/stroke} \end{array}\right]$$

∴ Total work done/cycle, $W = W_{(L.P.)} + W_{(H.P.)}$

$$= \frac{n}{n-1}\, mRT_1 \left[\left(\frac{p_i}{p_s}\right)^{\frac{n-1}{n}} + \left(\frac{p_d}{p_i{'}}\right)^{\frac{n-1}{n}} - 2\right]$$

$$= \frac{1.25}{1.25-1} \times 0.0655 \times 287$$

$$\times\, 298 \left[\left(\frac{4.2}{1}\right)^{\frac{1.25-1}{1.25}} + \left(\frac{18}{4.0}\right)^{\frac{1.25-1}{1.25}} - 2\right]$$

$= 28009.8 \times (1.332 + 1.351 - 2)$
$= 19130.7$ Nm

(*i*) **Power required to run the compressor**

$$\text{I.P.} = \frac{W \times N\,(\text{r.p.m})}{60 \times 1000}\ \text{kW} = \frac{19130.7 \times 200}{60 \times 1000}$$

= **63.77 kW. (Ans.)**

(*ii*) **Heat rejected in intercooler/min**

$= (m \times \text{r.p.m.}) \times c_p \times (T_2 - T_1)$
$= (0.0655 \times 200) \times 1.004\,(397 - 298)$
= **1302.08 kJ/min. (Ans.)**

☞Example 24.26. *A single-acting two-stage compressor with complete intercooling delivers 10.5 kg/min of air at 16 bar. The suction occurs at 1 bar and 27°C. The compression and expansion processes are reversible, polytropic index n = 1.3. Calculate :*

(*i*) *The power required to drive the compressor*
(*ii*) *The isothermal efficiency*
(*iii*) *The free air delivery*
(*iv*) *The heat transferred in intercooler*
The compressor runs at 440 r.p.m.

(*v*) *If the clearance ratios for L.P. and H.P. cylinders are 0.04 and 0.06 respectively, calculate the swept and clearance volumes for each cylinder.*

Solution. *Given :* $p_1 = 1.0$ bar, $p_2 = 4.0$ bar,
$p_3 = 16.0$ bar
$T_1 = 27 + 273 = 300$ K, $n = 1.3$,

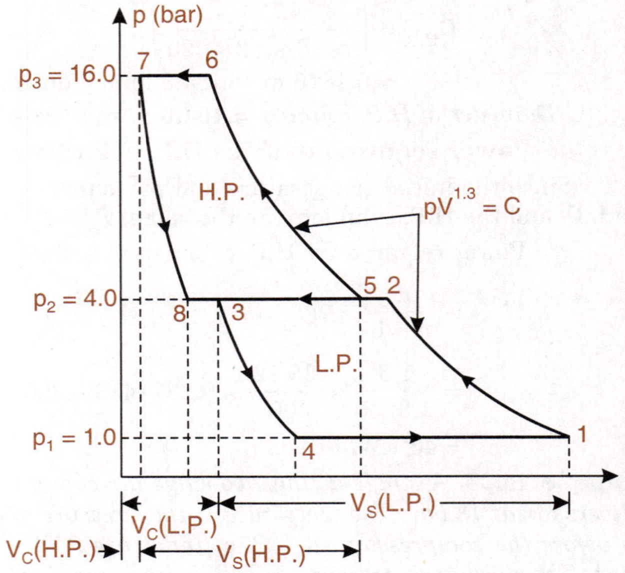

Fig. 24.34

Amount of air delivered = 10.5 kg/min
Clearance ratio for H.P. cylinder, $k = 0.04$
Clearance ratio for H.P. cylinder, $k = 0.06$

The pressure ratio per stage $= \sqrt{p_1 p_3} = \sqrt{1 \times 16} = 4$

(*i*) **Power required :**

Work done in two stages with *perfect intercooling*

$$= \frac{2n}{n-1}\, mRT_1 \left[\left(\frac{p_3}{p_1}\right)^{\frac{n-1}{2n}} - 1\right]$$

$$= \frac{2 \times 1.3}{1.3-1} \times \frac{10.5}{60} \times 0.287 \times 300 \left[(16)^{\frac{1.3-1}{2 \times 1.3}} - 1\right]$$

= **49.21 kW. (Ans.)**

(*ii*) **Isothermal efficiency :**

Isothermal work

$$= mRT_1 \log_e \left(\frac{p_3}{p_1}\right) = \frac{10.5}{60} \times 0.287 \times 300 \times \log_e \left(\frac{16}{1}\right)$$

= **41.77 kW. (Ans.)**

Isothermal efficiency

$$= \frac{41.77}{49.21} = \mathbf{0.8488 \text{ or } 84.88\%.}\ \textbf{(Ans.)}$$

(*iii*) **Free air delivery (F.A.D.) :**

Free air delivery,

$$V = \frac{mRT_1}{p_1} = \frac{10.5 \times 0.287 \times 300 \times 10^3}{1 \times 10^5}$$

= **9.04 m³/min. (Ans.)**

(*iv*) **Heat transferred in intercooler :**

Temperature at the end of compression

$$T_2 = T_1 \times \left(\frac{p_2}{p_1}\right)^{\frac{n-1}{n}} = 300\left(\frac{4}{1}\right)^{\frac{1.3-1}{1.3}} = 413 \text{ K}$$

Heat transferred in intercooler

$$= m \times c_p \times (T_2 - T_5) = m \times c_p\,(T_2 - T_1) \quad [\because \; T_5 = T_1]$$

$$= \frac{10.5}{60} \times 1.005\,(413 - 300) = \mathbf{19.87 \text{ kW}}. \quad \textbf{(Ans.)}$$

(*v*) **Swept and clearance volumes :**

Volumetric efficiency for L.P. stage,

$$\eta_{vol.(L.P.)} = 1 + k - k\left(\frac{p_2}{p_1}\right)^{\frac{1}{n}}$$

$$= 1 + 0.04 - 0.04(4)^{1/1.3} = 0.9238 \text{ or } 92.38\%$$

Similarly volumetric efficiency of H.P. stage,

$$\eta_{vol.\,(H.P.)} = 1 + 0.06 - 0.06(4)^{1/1.3}$$

$$= \mathbf{0.8857} \text{ or } \mathbf{88.57\%}. \quad \textbf{(Ans.)}$$

L.P. Stage :

(*i*) **Swept volume,**

$$\mathbf{V_{s(L.P.)}} = (V_1 - V_3) = \frac{\text{Free air delivery}}{\text{Speed} \times \eta_{vol.\,(L.P.)}}.$$

$$= \frac{9.04}{440 \times 0.9238} = \mathbf{0.0222 \text{ m}^3}. \quad \textbf{(Ans.)}$$

(*ii*) **Clearance volume,**

$$\mathbf{V_{c\,(L.P.)}} = 0.04\,(V_1 - V_3) \text{ or } 0.04\,V_{s(L.P.)}$$

$$= 0.04 \times 0.0222 = \mathbf{0.000888 \text{ m}^3}. \quad \textbf{(Ans.)}$$

H.P. Stage :

(*i*) **Swept volume**

$$\mathbf{V_{s\,(H.P.)}} = (V_5 - V_7)$$

$$= \frac{\text{Free air delivery}}{\text{Stage pressure ratio} \times \text{speed} \times \eta_{vol.\,(H.P.)}}$$

$$= \frac{9.04}{4 \times 440 \times 0.8857} = \mathbf{0.0058 \text{ m}^3}. \quad \textbf{(Ans.)}$$

(*ii*) **Clearance volume,**

$$\mathbf{V_{c\,(H.P.)}} = 0.06 \times (V_5 - V_7) \text{ or } 0.06\,V_{s(H.P.)}$$

$$= 0.06 \times 0.0058 = \mathbf{0.000348 \text{ m}^3}. \quad \textbf{(Ans.)}$$

Example 24.27. *The pressure limits of a 3-stage compressor are 1.05 bar and 40 bar. The compressor supplies 3 m³ of air per minute. The law of compression is $pV^{1.25}$ = Constant. Calculate on one minute basis :*

(*i*) *Indicated work done assuming conditions to be those for maximum efficiency ;*

(*ii*) *Isothermal work between the same pressure limits ;*

(*iii*) *Isothermal efficiency ;*

(*iv*) *Indicated work if the machine were of one-stage only ;*

(*v*) *Percentage saving in work done to using three stages instead of one.*

Solution.

(*i*) **Work done/min.**

$$= \frac{3n}{n-1}\,p_1V_1\left[\left(\frac{p_d}{p_s}\right)^{\frac{n-1}{3n}} - 1\right]$$

$$= \frac{3 \times 1.25}{1.25 - 1} \times 1.05 \times 10^5 \times 3\left[\left(\frac{40}{1.05}\right)^{\frac{1.25-1}{3 \times 1.25}} - 1\right]$$

$$= \mathbf{1297725.7 \text{ Nm/min.}} \quad \textbf{(Ans.)}$$

(*ii*) **Isothermal work done/min**

$$= 10^5\,p_1V_1 \log_e \frac{p_d}{p_s}$$

$$= 10^5 \times 1.05 \times 3 \times \log_e \frac{40}{1.05}$$

$$= \mathbf{1146628.1 \text{ Nm}}. \quad \textbf{(Ans.)}$$

(*iii*) **Isothermal efficiency**

$$= \frac{1146628.1}{1297725.7} = \mathbf{0.883} \text{ or } \mathbf{88.3\%}. \quad \textbf{(Ans.)}$$

(*iv*) **Single-stage, work done/min.**

$$= \frac{n}{n-1}\,p_1V_1\left[\left(\frac{p_d}{p_s}\right)^{\frac{n-1}{n}} - 1\right]$$

$$= \frac{1.25}{1.25 - 1} \times 1.05 \times 10^5 \times 3\left[\left(\frac{40}{1.05}\right)^{\frac{1.25-1}{1.25}} - 1\right]$$

$$= \mathbf{1686780.2 \text{ Nm}}. \quad \textbf{(Ans.)}$$

(*v*) **% Work saved**

$$= \frac{1686780.2 - 1297725.7}{1686780.2}$$

$$= \mathbf{0.23} \text{ or } \mathbf{23\%}. \quad \textbf{(Ans.)}$$

Example 24.28. *A 3-stage compressor is used to compress air from 1.0 bar to 36 bar. The compression in all stages follows the law $pV^{1.25}$ = C. The temperature of air at the inlet of compressor is 300 K. Neglecting the clearance and assuming perfect intercooling, find out the indicated power required in kW to deliver 15 m³ of air per minute measured at inlet conditions and intermediate pressures also. Take R = 0.287 kJ/kg K.*

Solution. *Given :* $p_2 = 1.0$ bar, $p_4 = 36$ bar,

$n = 1.25$, $R = 0.287$ kJ/kg K

$T_1 = 300$ K

As there is *perfect intercooling,*

$$\frac{p_2}{p_1} = \frac{p_3}{p_2} = \frac{p_4}{p_3} = \left(\frac{p_4}{p_1}\right)^{1/3} = \left(\frac{36}{1}\right)^{1/3} = 3.302$$

$\therefore$ **Intermediate pressures,**

$\mathbf{p_2} = 3.302\, p_1 = 3.302 \times 1 = \mathbf{3.302\ bar.}$ **(Ans.)**

$\mathbf{p_3} = 3.302\, p_2 = 3.302 \times 3.302 = \mathbf{10.9\ bar.}$ **(Ans.)**

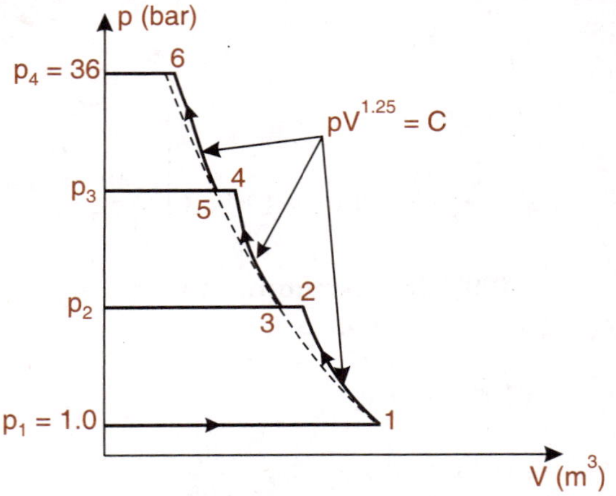

Fig. 24.35

Now $T_2 = T_1 \left(\frac{p_4}{p_1}\right)^{\frac{n-1}{n} \times \frac{1}{x}}$

[where x = no. of stages = 3]

$\therefore \quad T_2 = 300 \left(\frac{36}{1}\right)^{\left(\frac{1.25-1}{1.25}\right) \times \frac{1}{3}} = 380.9 \text{ K}$

Mass of air handled per minute

$$= \frac{p_1 V_1}{RT_1} = \frac{1.0 \times 10^5 \times 15}{0.287 \times 1000 \times 300}$$

= 17.42 kg/min.

Total work done in three stages (in kJ/s)

$$= \left[\frac{n}{n-1} . mR(T_2 - T_1)\right] \times x$$

$$= \left[\frac{1.25}{(1.25-1)} \times \frac{17.42}{60} \times 0.287 \times (380.9 - 300)\right] \times 3$$

= 101.11 kJ/s or 101.11 kW

i.e., **Indicated power required = 101.11 kW. (Ans.)**

☞**Example 24.29.** *A 3-stage double-acting compressor, operating at 200 r.p.m. takes in air at 1.0 bar and 20°C. The low pressure cylinder size is 350 mm × 400 mm. The intermediate pressure cylinder and the high pressure cylinder have the same stroke as the low pressure cylinder. The discharge pressures from the first stage and second stage are 4.0 bar and 16.0 bar and the air is finally delivered at 64.0 bar. The air is cooled to initial temperature in the intercooler after each stage and there is a drop of pressure of 0.2 bar in each of the intercoolers. The clearance volume in each cylinder is 4% of the stroke volume, but the compression indices are 1.2, 1.25 and 1.3 for compression and expansion in each of the 1st, 2nd and 3rd stages respectively. Neglect the effect of piston rods and assume R = 0.287 kJ/kg K, and* c_p *= 1.00 kJ/kg K. Determine :*

(i) Heat rejected in each of the intercoolers and also during compression process in each stage. Also find the heat rejected in the after-cooler if the delivered air is cooled to initial temperature.

(ii) The diameter of the intermediate pressure and the high pressure stage cylinders.

(iii) The shaft power required to drive the compressor with mechanical efficiency of 80%.

Take c_p *= 1.005 kJ/kg K.*

Solution. Fig. 24.36 [(*a*) and (*b*)] shows the *p-V* and *T-s* diagrams.

The swept volume of low pressure cylinder per minute,

$$V_{S(L.P.)} = \frac{\pi}{4} D^2_{L.P.} L \times (\text{r.p.m.}) \times 2$$

$$= \frac{\pi \times 0.35^2 \times 0.4 \times 200 \times 2}{4} = 15.394 \text{ m}^3/\text{min.}$$

[r.p.m. is multiplied by 2 since the compressor is *double-acting*]

Volumetric efficiency referred to conditions at 1, *i.e.,* 1 bar and 20°C are :

$$\eta_{vol.\ (1st\ stage)} = 1 + k - k \left(\frac{p_2}{p_1}\right)^{\frac{1}{n}}$$

$$= 1 + 0.04 - 0.04 \left(\frac{4}{1}\right)^{\frac{1}{1.2}}$$

= 0.913 or 91.3%

$$\eta_{vol.\ (2nd\ stage)} = 1 + 0.04 - 0.04 \left(\frac{p_6}{p_5}\right)^{\frac{1}{1.25}}$$

$$= 1.04 - 0.04 \left(\frac{16}{3.8}\right)^{\frac{1}{1.25}}$$

= 0.9136 or 91.36%

$$\eta_{vol.\ (3rd\ stage)} = 1 + 0.04 - 0.04 \left(\frac{p_{10}}{p_9}\right)^{\frac{1}{1.3}}$$

$$= 1.04 - 0.04 \left(\frac{64}{15.8}\right)^{\frac{1}{1.3}}$$

= 0.9227 or 92.27%

Volume of air taken in at 1 bar 20°C

$(V_1 - V_4)$/min = $V_{S(L.P.)}$/min. × $\eta_{vol.\ (1st\ stage)}$

= 15.394 × 0.913 = 14.05 m³/min.

And mass of air/min.,

$$m = \frac{p_1(V_1 - V_4)}{RT_1} = \frac{1 \times 10^5 \times 14.05}{0.287 \times 293 \times 10^3}$$
$$= 16.708 \text{ kg/min.}$$

Also $T_2 = T_1 \left(\frac{p_2}{p_1}\right)^{\frac{n-1}{n}} = 293 \left(\frac{4}{1}\right)^{\frac{1.2-1}{1.2}} = 369 \text{ K}$

and $T_6 = T_5 \left(\frac{p_6}{p_5}\right)^{\frac{n-1}{n}} = 293 \left(\frac{16}{3.8}\right)^{\frac{1.25-1}{1.25}}$
$= 390.6$ K.

and $T_{10} = T_9 \left(\frac{p_{10}}{p_9}\right)^{\frac{n-1}{n}} = 293 \left(\frac{64}{15.8}\right)^{\frac{1.3-1}{1.3}}$
$= 404.6$ K

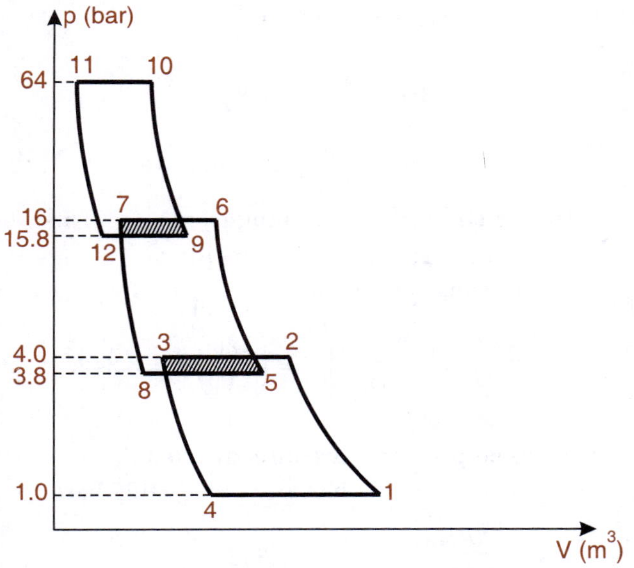

(a)

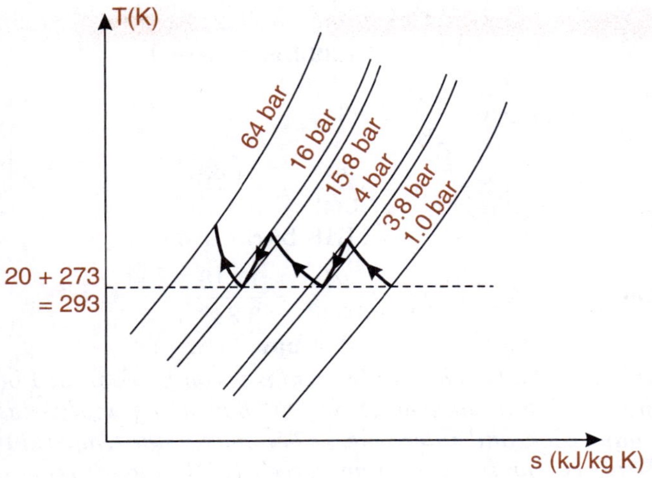

(b)

Fig. 24.36

(*i*) **Heat rejected in each cooler :**

Heat rejected in intercooler *after 1st stage*

$= mc_p (T_2 - T_5) = 16.708 \times 1.005 (369 - 293)$

$=$ **1276.16 kJ/min. (Ans.)**

And heat rejected in intercooler *after 2nd stage*

$= mc_p(T_6 - T_9) = 16.708 \times 1.005 \times (390.6 - 293)$

$=$ **1638.85 kJ/min. (Ans.)**

And heat rejected in the *after-cooler*

$= mc_p(T_{10} - T_1) = 16.708 \times 1.005 (404.6 - 293)$

$=$ **1873.93 kJ/min. (Ans.)**

(*ii*) **$D_{I.P.}$; $D_{H.P.}$:**

Volume drawn in intermediate pressure cylinder/min.

$$(V_5 - V_8)/\text{min.} = \frac{mRT_5}{p_5} = \frac{16.708 \times 0.287 \times 293 \times 10^3}{3.8 \times 10^5}$$
$$= 3.69 \text{ m}^3/\text{min.}$$

∴ Swept volume of intermediate cylinder/min

$$V_{s(\text{intermediate})}/\text{min} = \frac{3.69}{\eta_{vol.(2\text{nd stage})}} = \frac{3.69}{0.9136}$$
$$= 4.039 \text{ m}^3/\text{min.}$$

or $\frac{\pi}{4} D^2_{\text{I.P.}} L \times (r.p.m.) \times 2 = 4.039$

$$\therefore \quad D_{\text{I.P.}} = \left[\frac{4.039 \times 4}{\pi \times L \times (r.p.m.) \times 2}\right]^{1/2}$$
$$= \left[\frac{4.039 \times 4}{\pi \times 0.4 \times 200 \times 2}\right]^{1/2}$$
$$= 0.179 \text{ m or } 179 \text{ mm}$$

i.e., $D_{\text{I.P.}} =$ **179 mm. (Ans.)**

Volume drawn in high pressure cylinder/min

$$(V_9 - V_{12})/\text{min} = \frac{mRT_9}{p_9} = \frac{16.708 \times 0.287 \times 293 \times 10^3}{15.8 \times 10^5}$$
$$= 0.889 \text{ m}^3/\text{min.}$$

∴ Swept volume of high pressure cylinder,

$$V_{s(\text{H.P.})}/\text{min} = \frac{0.889}{\eta_{vol.\,(2\text{nd stage})}} = \frac{0.889}{0.9227}$$
$$= 0.9364 \text{ m}^3/\text{min.}$$

or $\frac{\pi}{4} D^2_{\text{H.P.}} L \times (r.p.m.) \times 2 = 0.9364$

$$\therefore \quad D_{\text{H.P.}} = \left(\frac{0.9364 \times 4}{\pi \times 0.4 \times 200 \times 2}\right)^{1/2}$$
$$= 0.0863 \text{ m or } 86.3 \text{ mm}$$

i.e., $D_{\text{H.P.}} =$ **86.3 mm. (Ans.)**

(iii) **Shaft power :**

Shaft power is given by :

$$\left[\frac{1.2}{1.2-1} mR\,(T_2 - T_1) + \frac{1.25}{1.25-1} mR\,(T_6 - T_5) + \frac{1.3}{1.3-1} mR\,(T_{10} - T_9)\right] \times \frac{1}{60 \times \eta_{mech.}}$$

$$= \frac{1}{60 \times 0.80}\, mR \left[\frac{1.2}{0.2} \times (369 - 293) + \frac{1.25}{0.25}(390.6 - 293) + \frac{1.3}{0.3} \times (404.6 - 293)\right]$$

$$= \frac{16.708 \times 0.287}{60 \times 0.8}\,(456 + 488 + 483.6) = 142.6 \text{ kW}$$

i.e., **Shaft power = 142.6 kW. (Ans.)**

Heat rejected during compression process during a stage above may not be confused with heat rejected per stage, which includes heat transferred during suction and delivery, if any. Thus heat rejected during compression process 1–2.

$$= \left[c_v \left(\frac{\gamma - n}{\gamma - 1}\right)(T_2 - T_1)\right] \times m$$

Also $c_p = c_v + R,$

$\therefore$ $c_v = c_p - R = 1.005 - 0.287 = 0.718$

and $\gamma = \frac{c_p}{c_v} = \frac{1.005}{0.718} = 1.4$

Therefore, heat transfer

$$= \left[0.718 \left(\frac{1.4 - 1.2}{1.4 - 1}\right)(369 - 293)\right] \times 16.708$$

= 455.86 kJ/min. (Ans.)

Similarly for **2nd stage** compression process 5–6

Heat transfer

$$= \left[0.718 \left(\frac{1.4 - 1.2.5}{1.4 - 1}\right)(390.6 - 293)\right] \times 16.708$$

= 439.07 kJ/min. (Ans.)

Similarly for **3rd stage** compression process 9-10

Heat transfer

$$= \left[0.718 \left(\frac{1.4 - 1.3}{1.4 - 1}\right)(404.6 - 293)\right] \times 16.708$$

= 334.69 kJ/min. (Ans.)

Example 24.30. *A multi-stage air compressor is to be designed to elevate the pressure from 1 bar to 125 bar such that stage pressure ratio will not exceed 4. Determine :*

(i) Number of stages

(ii) Exact stage-pressure ratios

(iii) Intermediate pressures.

Solution. *(i)* **Number of stages, x :**

Assuming *perfect intercooling,* the condition for *minimum* work of compression in multi-stage compression is

$$\frac{(p)_{x+1}}{(p)_x} = \left[\frac{(p)_{x+1}}{(p)_1}\right]^{1/x}$$

In this case $\frac{(p)_{x+1}}{(p)_x}$ is restricted to 4 and

$$\frac{(p)_{x+1}}{(p)_1} = \frac{125}{1} = 125$$

Hence $4 = (125)^{1/x}$

$$\log_e 4 = \frac{1}{x} \log_e 125$$

$$1.386 = \frac{1}{x} \times 4.828$$

$$\therefore \quad x = \frac{4.828}{1.386} = 3.48 \text{ say } 4$$

Hence the number of stages, x = 4. (Ans.)

(ii) **Exact stage-pressure ratios :**

Again using the relation

$$\frac{(p)_{x+1}}{(p)_x} = \left[\frac{(p)_{x+1}}{(p)_1}\right]^{1/x} = \left(\frac{125}{1}\right)^{1/4} = \mathbf{3.343.} \quad \textbf{(Ans.)}$$

(iii) **Intermediate pressures :**

Extreme pressures are already fixed

i.e., $p_1 = 1$ bar, $p_{4+1} = p_5 = 125$ bar

Also $\frac{(p)_{4+1}}{(p)_4} = \frac{p_5}{p_4} = 3.343$

$$\therefore \quad p_4 = \frac{p_5}{3.343} = \frac{1.25}{3.343}$$

= 37.39 bar. (Ans.)

Similarly $\frac{p_4}{p_3} = 3.343$

$$\therefore \quad p_3 = \frac{p_4}{3.343} = \frac{37.39}{3.343}$$

= 11.18 bar. (Ans.)

and $p_2 = \frac{p_3}{3.343} = \frac{11.18}{3.343}$

= 3.344 bar. (Ans.)

Example 24.31. *The requirement is to compress air at 1 bar and 25°C and deliver it at 160 bar using multi-stage compression and intercoolers. The maximum temperature during compression must not exceed 125°C and also cooling in the intercooler is done so as not to drop the temperature below 40°C. The law of compression followed is $pV^{1.25}$ = constant for all stages.*

Calculate : (*i*) *Number of stages required,*

(*ii*) *Work input per kg of air, and*

(*iii*) *Heat rejected in the intercoolers.*

Take $R = 0.287$ kJ/kg K, $c_v = 0.71$ kJ/kg K.

Solution. Let, p_s = Suction pressure = 1 bar,

T_s = Suction temperature = 25 + 273 = 298 K,

p_1 = Delivery pressure from 1st stage = entry pressure to 2nd stage,

T_1 = Delivery temperature after every stage,

p_d = Delivery pressure from (*N* + 1)th stage,

x = Number of stages after the 1st stage, and

T_{ratio} = Temperature ratio for 2nd stage, 3rd stage ..... *n*th stage.

(*i*) Number of stages :

Thus,
$$\frac{p_1}{p_s} = \left(\frac{T_1}{T_s}\right)^{\frac{n}{n-1}}$$

$$= \left(\frac{125+273}{25+273}\right)^{\frac{1.25}{1.25-1}} = \left(\frac{398}{298}\right)^5 = 4.25$$

$\therefore$ $p_1 = p_s \times 4.25 = 1.0 \times 4.25 = 4.25$ bar

Also
$$\frac{T_2}{T_1} = \frac{T_3}{T_2} = \frac{T_4}{T_3} = \ldots\ldots \frac{T_x}{T_{x-1}}$$

or
$$\frac{p_x}{p_1} = (T_{ratio})^{\frac{xn}{n-1}}$$

and
$$T_{ratio} = \frac{(125+273)}{(40+273)} = \frac{398}{313} = 1.2715$$

or
$$\frac{160}{4.25} = (1.2175)^{\frac{x \times 1.25}{(1.25-1)}}$$

or
$$37.65 = (1.2715)^{5x}$$

$$\log_e 37.65 = 5x \times \log_e 1.2715$$

$$3.628 = 5x \times 0.24$$

$\therefore$
$$x = \frac{3.628}{5 \times 0.24} = 3.02 \text{ or say } 3$$

Hence, number of stages = 3 + 1 = **4. (Ans.)**

(*ii*) Work done per kg of air :

Pressure ratio in 1st stage = 4.25

Pressure ratio in the following stage

$$= \left(\frac{160}{4.25}\right)^{1/3} = 3.351 \text{ bar}$$

Temperature leaving 2nd, 3rd and 4th stages

= 125 + 273 = 398 K (Given)

Work done/kg in 1st stage

$$= \frac{n}{n-1} RT_s \left[\left(\frac{p_1}{p_s}\right)^{\frac{n-1}{n}} - 1\right]$$

$$= \frac{1.25}{1.25-1} \times 0.287 \times 298 \left[(4.25)^{\frac{1.25-1}{1.25}} - 1\right]$$

= 143.5 kJ

Work done in the following 3 stages

$$= 3 \times \frac{1.25}{1.25-1} \times 0.287 \times 313 \left[(3.351)^{\frac{1.25-1}{1.25}} - 1\right]$$

= 368.6 kJ

Total work done/kg

= 143.5 + 368.6 = **512.1 kJ/kg. (Ans.)**

(*iii*) Heat rejected in the intercoolers :

Heat rejected in the intercoolers

$= 3c_p\,(398 - 313)$

$= 3 \times 0.997 \times 85$

$[\because \; c_p = c_v + R = 0.71 + 0.287 = 0.997$ kJ/kg$]$

= **254.23 kJ/kg. (Ans.)**

Example 24.32. *A 3-stage air compressor supplying air blast for an oil engine has a rated capacity of 10.5 m³ of free air/min, and is driven by the main engine at 100 r.p.m. The pressure at the suction to L.P. cylinder is 1 bar and at delivery from H.P. cylinder is 95 bar. The fractional clearances are 0.04 and 0.07 for I.P. and H.P. cylinders. Assuming the temperature at the end of suction in all cylinders is 25°C i.e., perfect intercooling, stage pressures in geometric progression and the law of compression $pV^{1.25}$ = constant, calculate the swept volume of each cylinder. The free air conditions are 1.013 bar and 15°C.*

Solution. Refer to Fig. 24.37. Since the pressures are in geometric progression (given),

$$\frac{p_d}{p_{i2}} = \frac{p_{i2}}{p_{i1}} = \frac{p_{i1}}{p_s} = z \text{ (Pressure ratio).}$$

$$p_{i1} = zp_s$$

$$p_{i2} = p_{i1}z = z^2 p_s$$

$$p_d = p_{i2}z = z^3 p_s$$

[p_d = Delivery pressure (H.P. cylinder)

p_{i2} = Intermediate pressure-2

p_{i1} = Intermediate pressure-1

p_s = Suction pressure (L.P. cylinder)]

$$\therefore \quad z = \left(\frac{p_d}{p_s}\right)^{1/3}$$

$$= \left(\frac{95}{1}\right)^{1/3} = 4.56$$

$\therefore$ $p_{i1} = zp_s = 4.56 \times 1 = 4.56$ bar

$p_{i2} = zp_{i1} = 4.56 \times 4.56 \times 20.8$ bar

$$\eta_{vol.(L.P.)} = 1 + k - k\,(z)^{1/n}$$

$$= 1 + 0.04 - 0.04\,(4.56)^{1/1.25}$$

= 0.905 or 90.5%

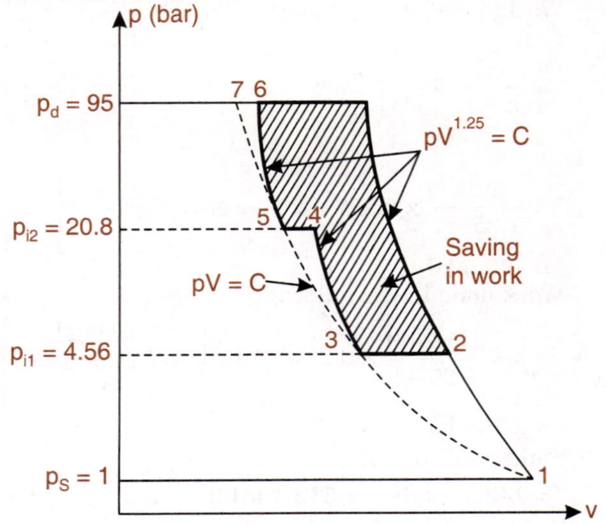

Fig. 24.37

Similarly, $\eta_{vol.\ (L.P.)} = \eta_{vol.\ (H.P.)}$

$$= 1 + 0.07 - 0.07\ (4.56)^{1/1.25}$$

$$= 0.834 \text{ or } 83.4\%$$

Swept volume of each cylinder

Volume of free air reduced to suction condition of L.P. cylinder is given by

$$\frac{p_1 V_1}{T_1} = \frac{p_{amb.} V_{amb.}}{T_{amb.}}$$

$$V_1 = \frac{p_{amb.} V_{amb.} T_1}{T_{amb.}\ p_1}$$

$$= \frac{1.013 \times 10^5 \times 10.5 \times (273 + 25)}{(273 + 15) \times 1 \times 10^5}$$

$= 11.0$ m³/min.

∴ Swept capacity of L.P. cylinder

$$= \frac{11.0}{\eta_{vol.(L.P)} \times \text{r.p.m.}}$$

$$= \frac{11.0}{0.905 \times 100} = \mathbf{0.1215\ m^3.}\quad \textbf{(Ans.)}$$

Again volume of free air reduced to suction conditions of I.P. cylinder

$$= \frac{1.013 \times 10^5 \times 10.5 \times 298}{288 \times 4.56 \times 10^5} = 2.41 \text{ m}^3/\text{min}$$

Swept capacity of I.P. cylinder

$$= \frac{2.41}{0.834 \times 100} = \mathbf{0.089\ m^3/min.}\quad \textbf{(Ans.)}$$

Again volume of free air reduced to suction conditions of H.P. cylinder

$$= \frac{1.013 \times 10^5 \times 10.5 \times 298}{288 \times 20.8 \times 10^5} = 0.529 \text{ m}^3/\text{min}$$

Swept capacity of H.P. cylinder

$$= \frac{0.529}{0.834 \times 100} = \mathbf{0.00634\ m^3.}\quad \textbf{(Ans.)}$$

Example 24.33. *Show that the heat rejected per stage per kg of air in a reciprocating compressor with perfect intercooling is given by*

$$\left[c_p + c_v \left(\frac{\gamma - n}{n - 1}\right)\right] (T_2 - T_1)$$

where $(T_2 - T_1)$ = *Temperature rise during compression,*
n = *Polytropic index,*
γ = *Adiabatic index, and*
c_p, c_v = *Two specific heats of air.*

Solution. In a compressor heat is rejected in two stages :

(*i*) *During compression* when heat is rejected to the cylinder walls ;

(*ii*) *During intercooling*.

Heat rejected during *compression*

$$= \frac{\gamma - n}{\gamma - 1} \times \text{compression work}$$

$$= \frac{\gamma - n}{\gamma - 1}\left(\frac{p_2 v_2 - p_1 v_1}{n - 1}\right)$$

$$= \frac{\gamma - n}{\gamma - 1}\left(\frac{T_2 - T_1}{n - 1}\right) R \text{ ...... per kg of air}$$

For perfect intercooling the temperature of air after compression T_2 must be reduced to initial temperature T_1.

Heat rejected in *intercooling*

$$= c_p\ (T_2 - T_1) \text{ ..... per kg of air}$$

∴ Total heat rejected per kg of air

$$= \frac{\gamma - n}{\gamma - 1} \times \frac{R}{n - 1}\ (T_2 - T_1) + c_p\ (T_2 - T_1)$$

$$= \left[c_p + \frac{R}{\gamma - 1}\left(\frac{\gamma - n}{n - 1}\right)\right] (T_2 - T_1)$$

$$= \left[c_p + c_v\left(\frac{\gamma - n}{n - 1}\right)\right] (T_2 - T_1).\quad \textbf{Proved.}$$

Example 24.34. *The cylinder of an "Air motor" has a bore of 6.35 cm and a stroke of 11.4 cm. The supply pressure is 6.3 bar, the supply temperature 24°C, and exhaust pressure is 1.013 bar. The clearance volume is 5% of the swept volume and the cut-off ratio is 0.5. The air is compressed by the returning piston after it has travelled through 0.95 of its stroke. The law of compression and expansion is* $pV^{1.3}$ = *constant. Calculate the temperature at the end of expansion and the indicated power of the motor which runs at 300 r.p.m.*

Also calculate the air supplied per minute. Take R = 0.287 kJ/kg K.

Solution. Refer to Fig. 24.14.

$$\text{Swept volume} = \frac{\pi \times (0.0635)^2 \times 0.114}{4} = 0.000361 \text{ m}^3$$

Clearance volume = $V_6 = V_5 = 0.05 \times 0.000361$
$= 0.000018\ m^3$

$$V_1 = \frac{0.000361}{2} + 0.000018 = 0.000198\ m^3$$

$$V_2 = 0.000361 + 0.000018 = 0.000379\ m^3$$

$$V_4 - V_5 = 0.05 \times 0.000361 \simeq 0.000018\ m^3$$

$$\therefore \quad V_4 = 0.000018 + 0.000018 = 0.000036\ m^3$$

$$p_1 V_1^n = p_2 V_2^n$$

$$\therefore \quad p_2 = p_1 \left(\frac{V_1}{V_2}\right)^n = 6.3 \left(\frac{0.000198}{0.000379}\right)^{1.3} = 2.71\ \text{bar}$$

Temperature at the end of expansion :

$$T_2 = T_1 \left(\frac{V_1}{V_2}\right)^{n-1} = 297 \left(\frac{0.000198}{0.000379}\right)^{0.3} = 244.4\ \text{K}$$

i.e., Temperature after expansion

$= 244.4 - 273 = -$ **28.6°C.** **(Ans.)**

Indicated power of the motor :

$$p_5 = p_4 \left(\frac{V_4}{V_5}\right)^n = 1.013 \left(\frac{0.000036}{0.000018}\right)^{1.3} = 2.494\ \text{bar}$$

Work done per cycle = Area 1234561

i.e., Work done

$$= p_1(V_1 - V_6) + \frac{(p_1V_1 - p_2V_2)}{n-1} - p_3(V_3 - V_4) - \left(\frac{p_5V_5 - p_4V_4}{n-1}\right)$$

$\therefore$ Work done/cycle

$$= 10^5 \times 6.3\ (0.000198 - 0.000018) + \frac{10^5\ (6.3 \times 0.000198 - 2.71 \times 0.000379)}{(1.3 - 1)}$$

$$- 10^5 \times 1.013\ (0.000379 - 0.000036) - \frac{10^5\ (2.494 \times 0.000018 - 1.013 \times 0.000036)}{(1.3 - 1)}$$

$$= 113.4 + 73.44 - 34.74 - 2.808 = 149.29\ \text{Nm}$$

$\therefore$ I.P. $= \dfrac{149.29 \times 300}{60 \times 1000} =$ **0.746 kW.** **(Ans.)**

Air supplied per minute :

The mass induced per cycle is given by $(m_1 - m_4)$. It is necessary to determine the *temperature of air at 4*, which *can be taken as equal to that at 3*. It is assumed that the air in the cylinder at the point 2 expands isentropically to the exhaust pressure.

$$\therefore \quad T_3 = T_2 \left(\frac{p_3}{p_2}\right)^{\frac{\gamma - 1}{\gamma}} = 244.4 \left(\frac{1.013}{2.71}\right)^{0.4/1.4} = 184.4\ \text{K}$$

i.e., $m_4 = \dfrac{p_4 \times V_4}{RT_4} = \dfrac{1.013 \times 10^5 \times 0.000036}{287 \times 184.4}$

$= 0.0000689$ kg

Also, $m_1 = \dfrac{p_1 V_1}{RT_1} = \dfrac{6.3 \times 10^5 \times 0.000198}{287 \times 297}$

$= 0.001463$ kg

$\therefore$ Induced mass/cycle = (0.001463 – 0.0000689) kg

$\therefore$ Mass of air supplied/min

= (0.001463 – 0.0000689) × 300

= **0.418 kg/min.** **(Ans.)**

24.4. ROTARY COMPRESSORS

- **Rotary compressors** *are the machines which develop pressure and have a rotor as their primary element when compared with the piston sliding mechanism of the reciprocating compressor.*
- *Whenever large quantities of air or gas are required at relatively low pressure* **rotary compressors** *are employed.*

24.4.1. Classification

The rotary compressors are *classified* as follows :

1. *Displacement (positive) compressors :*

(*i*) Roots blower

(*ii*) Sliding vane compressor

(*iii*) Lysholm compressor

(*iv*) Screw compressor.

2. *Steady-flow (or Non-positive displacement) compressors :*

(*i*) Centrifugal (or radial) compressor

(*ii*) Axial flow compressor.

24.4.2. Displacement Compressors

'Displacement Compressors' are those compressors in which air is compressed by being trapped in the reduced space formed by two sets of engaging surfaces.

24.4.2.1. Roots Blower

The two lobe type is shown in Fig. 24.38, but three and four lobe versions are in use for higher pressure ratios. One of the rotors is connected to the drive and the second rotor is gear driven from the first. In this way the rotors rotate in phase and the profile of the lobes is of cycloidal or involute form giving correct making of the lobes to seal the delivery side from the inlet side. This sealing continues untill delivery commences. There must be some clearance between the lobes, and between the casing and the lobes, to reduce wear, this clearance forms a leakage path which has an increasingly adverse effect on efficiency as the pressure ratio increases.

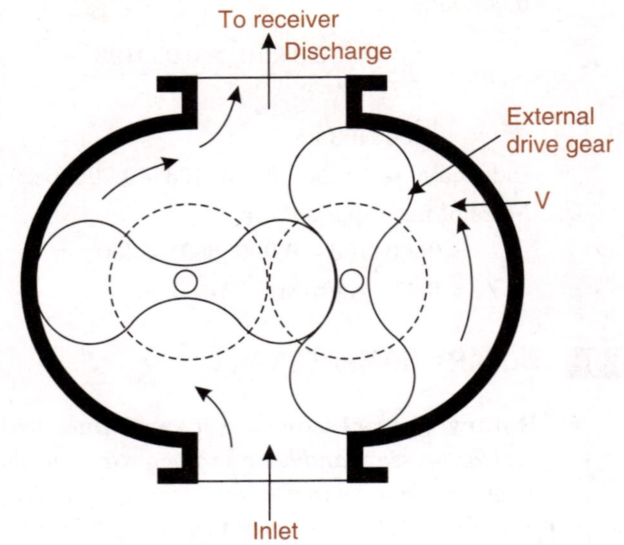

Fig. 24.38. Roots blower, two lobe rotors.

As each side of the each lobe faces its side of the casing a volume of gas V, at pressure p_1, is displaced towards the delivery side at constant pressure. A further rotation of the rotor opens this volume to the receiver, and the gas flows back from the receiver, since this gas is at a-higher pressure. The gas induced is compressed irreversibly by that from the receiver, to the pressure p_2 and then delivery begins. This process is carried out *four times* per revolution of the driving shaft.

For this machine the *p-V* diagram is shown in Fig. 24.39, in which the pressure rise from p_1 to p_2 is shown as an irreversible process at constant volume.

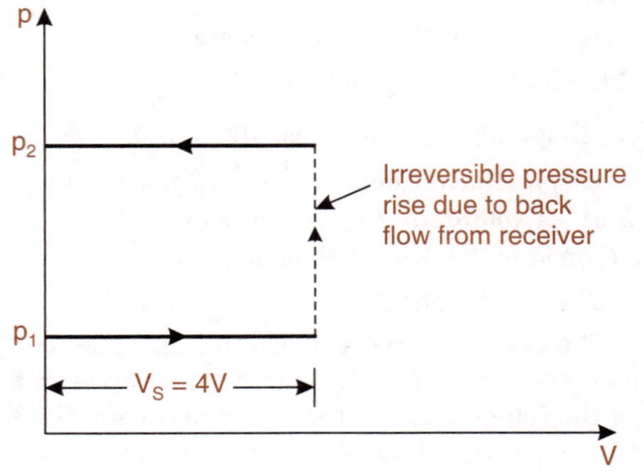

Fig. 24.39. *p-V* diagram for roots blower.

Work done per cycle $= (p_2 - p_1)V$

$\therefore$ Work done per revolution

$$= 4(p_2 - p_1)V \quad ...(24.48)$$

If V_s is the volume dealt with per minute at p_1 and T_1 then

$$\text{Work done/min.} = (p_2 - p_1)V_s \quad ...[24.48\ (a)]$$

The ideal compression process from p_1 to p_2 is a reversible adiabatic (*i.e.*, isentropic) process. The work done per minute ideally is given by,

$$\text{Work done/min.} = \frac{\gamma}{\gamma-1}\, p_1 V_s \left\{\left(\frac{p_2}{p_1}\right)^{\frac{\gamma-1}{\gamma}} - 1\right\}$$

Then a comparison may be made on the basis of a roots efficiency,

i.e., $$\text{Roots efficiency} = \frac{\text{Work done isentropically}}{\text{Actual work done}}$$

$$= \frac{\frac{\gamma}{\gamma-1}\, p_1 V_s \left\{\left(\frac{p_2}{p_1}\right)^{\frac{\gamma-1}{\gamma}} - 1\right\}}{V_s (p_2 - p_1)}$$

$$= \frac{\frac{\gamma}{\gamma-1}\, p_1 V_s \left\{(r)^{\frac{\gamma-1}{\gamma}} - 1\right\}}{p_1 V_s (r-1)}$$

$$\left(\text{where } r = \frac{p_2}{p_1} = \text{pressure ratio}\right)$$

Also, $$\frac{\gamma}{\gamma-1} = \frac{c_p}{R}$$

$$\therefore \quad \text{Roots efficiency} = \frac{c_p}{R}\left[\frac{(r)^{\frac{\gamma-1}{\gamma}} - 1}{(r-1)}\right] \quad ...(24.49)$$

In case of a Roots air blower values of pressure ratio, r of 1.2, 1.6, and 2 give values for the Roots efficiency of 0.945, 0.84 and 0.765 respectively. These values show that the *efficiency decreases as the pressure ratio increases.*

This machine has a number of *imperfections but is well suited to such tasks as the scavenging and supercharging of I.C. engines.*

Roots blowers are built for capacities from 0.14 m^3/min. to 1400 m^3/min, and pressure ratios of the order of 2 to 1 for a single-stage machine and 3 to 1 for a two-stage machine.

24.4.2.2. Vane Type Blower

Refer to Fig. 24.40. A vane type blower consists of a rotor mounted eccentrically in the body, and supported by ball and roller bearings in the end covers of the body. The rotor is slotted to take the blades which are of a non-metallic material, usually fibre or carbon. As each blade moves past the inlet passage, compression begins due to decreasing volume between the rotor and casing. Delivery begins with the arrival of each blade at the delivery passage. This type of compression differs from that of the Roots blower in that some or all of the compression is obtained before the trapped volume is opened to delivery. Further compression can be obtained by the back-flow of air from the receiver which occurs in an irreversible manner.

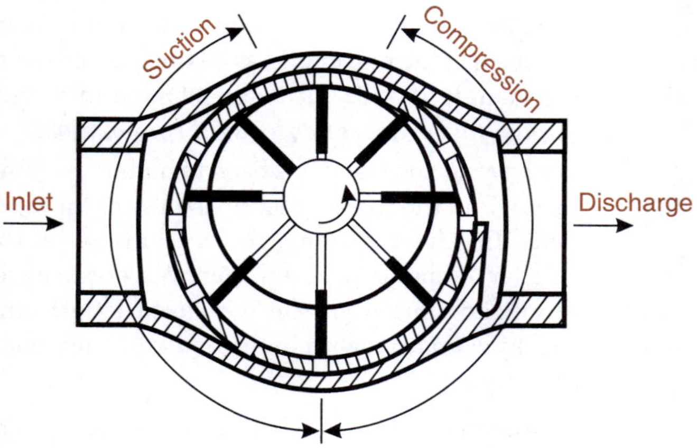

Fig. 24.40. Vane type blower.

Fig. 24.41 shows the *p-V* diagram. V_s is the induced volume at pressure p_1 and temperature T_1. Compression occurs to the pressure p_i, the ideal form for an uncooled machine being isentropic. At this pressure the displaced gas is opened to the receiver and gas flowing back from the

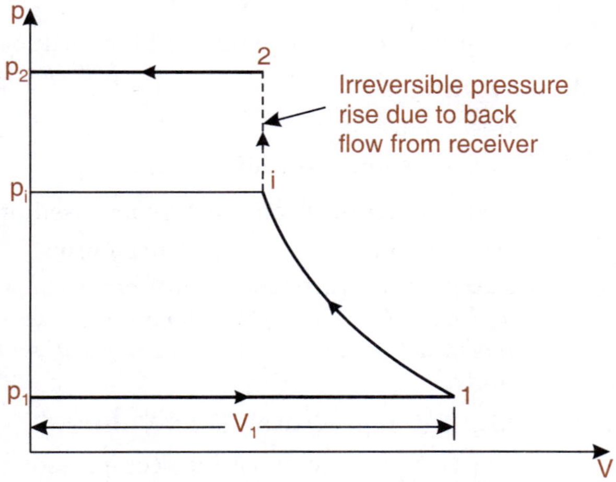

Fig. 24.41. *p-V* diagram for vane blower.

receiver raises the pressure irreversibly to p_2. The work done per revolution with *N* vanes is given by the following expression :

$$W = N\frac{\gamma}{\gamma-1}p_1V_1\left[\left(\frac{p_i}{p_1}\right)^{\frac{\gamma-1}{\gamma}}-1\right] + N(p_2 - p_i)V_i \quad \text{...(24.50)}$$

- The vane blowers *require less work* compared to roots blower for the same capacity and pressure rise.
- They are commonly used to deliver upto 150 m^3 of air per minute at pressure ratio upto 8.5.
- The speed limit of a vane blower is 3000 r.p.m.

Example 24.35. *Compare the work inputs required for a Roots blower and a Vane type compressor having the same induced volume of 0.03 m^3/rev., the inlet pressure being 1.013 bar and the pressure ratio 1.5 to 1. For the Vane type assume that internal compression takes place through half the pressure range.*

Solution. Inlet pressure, $p_1 = 1.013$ bar

Pressure ratio, $\dfrac{p_2}{p_1} = 1.5$

$\therefore \quad p_2 = 1.5\,p_1 = 1.5 \times 1.013 = 1.52$ bar.

For the **Roots blower**, refer to Fig. 24.42.

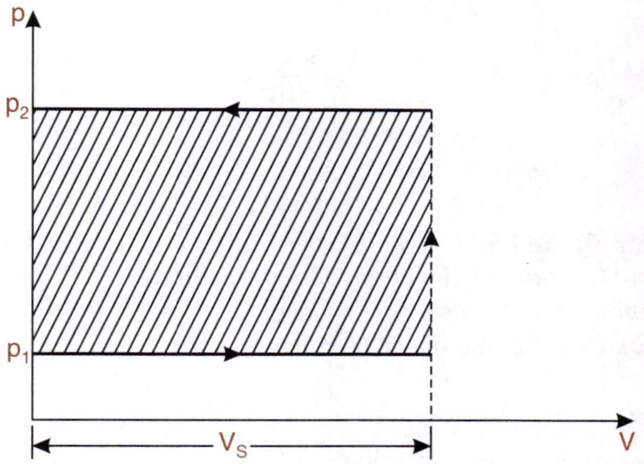

Fig. 24.42

Work done/rev. $= (p_2 - p_1)V_s$

$$= (1.52 - 1.013) \times \frac{10^5 \times 0.03}{10^3} = \mathbf{1.52\ kJ.} \quad \textbf{(Ans.)}$$

For the **Vane type**, $p_i = \dfrac{1.52 + 1.013}{2} = 1.266$ bar

Refer to Fig. 24.43.

Work required = (Area *A* + Area *B*)

Now, $\quad$ Area $A = \dfrac{\gamma}{\gamma-1}p_1V_s\left[\left(\dfrac{p_i}{p_1}\right)^{(\gamma-1)/\gamma}-1\right]$

$$= \frac{1.4}{1.4-1} \times \frac{1.013 \times 10^5 \times 0.03}{10^3}\left[\left(\frac{1.266}{1.013}\right)^{\frac{1.4-1}{1.4}}-1\right] \text{ kJ/rev.}$$

$= 3.5 \times 1.013 \times 100 \times 0.03 \times 0.066$

$= 0.702$ kJ/rev.

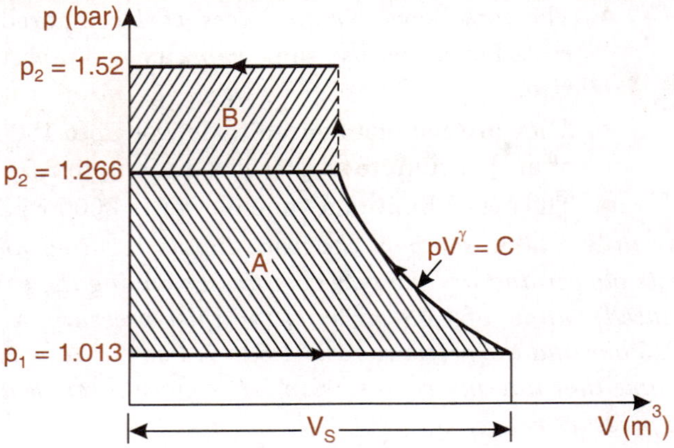

Fig. 24.43

Area $B = (p_2 - p_1)V_b$

Now, $V_b = V_s \left(\frac{p_1}{p_2}\right)^{1/\gamma} = 0.03 \left(\frac{1.013}{1.266}\right)^{1/1.4}$

$= 0.0256 \text{ m}^3$

Area $B = \frac{(1.52 - 1.266) \times 10^5 \times 0.0256}{10^3}$

$= 0.65$ kJ/rev.

∴ Work required

$= 0.702 + 0.65 =$ **1.352 kJ/rev.** **(Ans.)**

Example 24.36. *A roots blower compresses 0.08 m³ of air from 1.0 bar to 1.5 bar per revolution. Calculate the compressor efficiency.*

Solution. Volume of air to be compressed,

$V = 0.08 \text{ m}^3$

Intake pressure, $p_1 = 1.0$ bar

Pressure after compression, $p_2 = 1.5$ bar

Actual work done,

$W_{actual} = (p_2 - p_1) V = 10^5(1.5 - 1.0) \times 0.08 = 4000$ Nm.

Also ideal work done per revolution is given by,

$$W_{ideal} = \frac{\gamma}{\gamma - 1} p_1 V_1 \left[\left(\frac{p_2}{p_1}\right)^{(\gamma-1)/\gamma} - 1\right]$$

$$= \frac{1.4}{1.4 - 1} \times 1.0 \times 10^5 \times 0.08 \left[\left(\frac{1.5}{1.0}\right)^{\frac{1.4-1}{1.4}} - 1\right]$$

$= 3438.89$ Nm.

$$\therefore \eta_{compressor} = \frac{W_{ideal}}{W_{actual}} = \frac{3438.89}{4000}$$

$=$ **0.8597** or **85.97%.** **(Ans.)**

24.4.3. Steady-flow Compressors

The compressors in which compression occurs by transfer of kinetic energy from a rotor are called ***Steady-flow compressors.***

The centrifugal type of compressor was used in the earliest gas turbine units for aircraft.

- For low pressure ratios (no greater than about 4 : 1) the centrifugal compressor is *lighter* and is able to *operate effectively over a wide range of mass flows at any one speed,* than its axial-flow counterpart.
- For *larger units with higher pressure ratios the axial flow compressor is more efficient and is usually preferred.* For industrial and large marine gas turbine plants axial flow compressors are usually used, although some units *may employ two or more centrifugal compressors.*
- — For *aircraft* the trend has been to higher pressure ratios, and the compressor is usually of the *axial flow.* In aircraft units the advantage of the smaller diameter axial flow compressor can offset the disadvantage of the increased length and weight compared with an equivalent centrifugal compressor.

Advantages *of centrifugal compressors over axial flow compressors :*

1. Smaller length.
2. Contaminated atmosphere does not deteriorate the performance.
3. Can perform efficiently over wide range of mass flows at any speed.
4. Cheaper to produce.
5. More robust.
6. Less prone to icing troubles at high altitudes.

Disadvantages :

1. Large frontal area.
2. Lower maximum efficiency.

Uses : The centrifugal compressors are used in :

(*i*) Superchargers (*ii*) Turbo-prop.

- **Centrifugal compressors** *are preferred where simplicity, light weight, ruggedness are more important than maximum efficiency and smaller diameter.*

24.4.3.1. Static and Total Head Values

As compared to reciprocating compressors the *velocities encountered in centrifugal compressors are very large and therefore total head quantities should be considered* while analysing centrifugal compressors. The *total head quantities take into account the kinetic energy of the air passing through the compressor.*

Let us consider, a horizontal passage of varying area through which air is flowing (Fig. 24.48). Applying steady flow energy equation to the system for 1 kg of air flow (assuming no external heat transfer and work transfer to the system), we get

$$u_1 + p_1v_1 + \frac{C_1^2}{2} = u_2 + p_2v_2 + \frac{C_2^2}{2}$$

$$\left[u_1 + \frac{p_1v_1}{J} + \frac{C_1^2}{2gJ} = u_1 + \frac{p_2v_2}{J} + \frac{C_2^2}{2gJ} \quad \text{...... in MKS units}\right]$$

$$h_1 + \frac{C_1^2}{2} = h_2 + \frac{C_2^2}{2}$$

$$\therefore \quad c_pT_1 + \frac{C_1^2}{2} = c_pT_2 + \frac{C_2^2}{2}$$

$$\therefore \quad c_pT + \frac{C^2}{2} = \text{constant} \quad \text{...(24.51)}$$

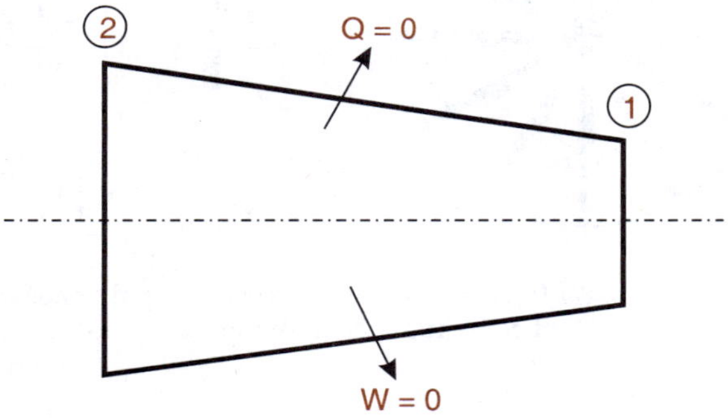

Fig. 24.44

Temperature 'T' is called the ***static temperature*** or the *temperature* of the *air measured by the thermometer when the thermometer is moving at the air velocity*. If the moving air is brought to rest under reversible conditions, the total kinetic energy of the air is converted into heat energy, increasing the temperature and pressure of the air. This temperature and pressure of the air is known as ***"stagnation"*** or ***"total head"*** *temperature and pressure*. The total head temperature and pressure are denoted by a suffix notation '0'.

$$\therefore \quad c_pT + \frac{C^2}{2} = c_pT_0 \quad \text{...(24.52)}$$

where T_0 is known as *total head or stagnation temperature*

$$\therefore \quad T_0 - T = \frac{C^2}{2c_p} \quad \text{...(24.53)}$$

or

$$h_0 - h = \frac{C^2}{2} \quad \text{...(24.54)}$$

For finding the total head pressures, use the equation,

$$\frac{p_0}{p} = \left(\frac{T_0}{T}\right)^{\frac{\gamma}{\gamma - 1}} \quad \text{...(24.55)}$$

where, p = Static pressure,

T = Static temperature,

p_0 = Stagnation pressure, and

T_0 = Stagnation temperature.

Adiabatic process and isentropic process :

- In **adiabatic process** the system does not exchange any heat with the surroundings *i.e.*, no heat enters or leaves the working fluid externally. The *ideal reversible adiabatic process* is called **isentropic process** *and in this process entropy remains constant.*

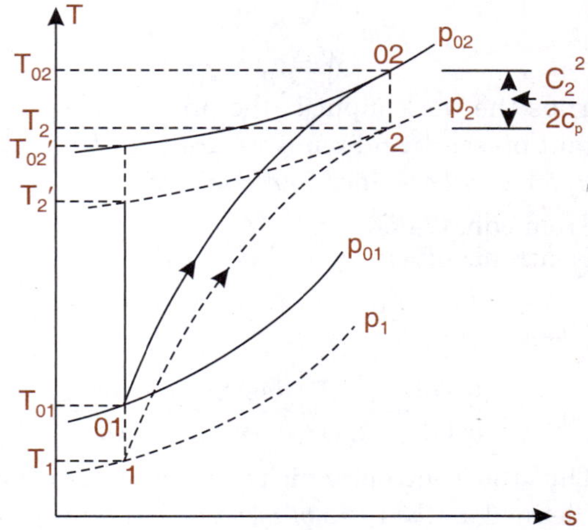

Fig. 24.45. Total head and static quantities on T-s diagram.

- During adiabatic compression in a rotary compressor there is friction between molecules of air and between air and blade passages, eddies formation and shocks at entry and exit. These factors *cause internal generation of heat and consequently the maximum temperature reached would be more than that for adiabatic compression.* This *results in a progressive increase in entropy.* Such a process *though adiabatic is not isentropic.* The heat generated by friction etc., may be removed continuously with the result that the process might not involve any entropy change. The process *would then be isentropic but not adiabatic as heat has been transferred.*

Isentropic efficiency. *"Isentropic efficiency" of rotary compressor* may be defined as *the ratio of isentropic temperature rise to actual temperature rise.*

With reference to Fig. 24.45, which represents a combined diagram for static and stagnation (total head) values, we have

$$\text{Isentropic efficiency} = \frac{\text{Isentropic temperature rise}}{\text{Actual temperature rise}}$$

$$\eta_{isen} = \frac{T_{02}' - T_{01}}{T_{02} - T_{01}}$$

...based on total values ...[24.56 (*a*)]

$$= \frac{T_2' - T_1}{T_2 - T_1}$$

...based on static values ...[24.56 (*b*)]

During compression process work has to be imparted to the impeller. The energy balance equation would then yield :

$$c_pT_1 + \frac{C_1^2}{2} = c_pT_2 + \frac{C_2^2}{2} - W$$

or $$c_pT_{01} = c_pT_{02} - W$$

or $$W = c_p(T_{02} - T_{01}) \quad ...(24.57)$$

Thus the work input is the product of specific heat at constant pressure and temperature rise. *This relation is true both for adiabatic and isentropic processes.*

From eqn. (24.56),
Isentropic efficiency,

$$\eta_{isen} = \frac{T_{02}' - T_{01}}{T_{02} - T_{01}}$$

or $$\eta_{isen} = \frac{c_p(T_{02}' - T_{01})}{c_p(T_{02} - T_{01})} = \frac{\text{Isentropic work}}{\text{Actual work}} \quad ...(24.58)$$

Thus the isentropic efficiency of a rotary compressor may be defined *as the ratio of isentropic compression work to actual compression work.*

24.4.3.2. Centrifugal Compressor

Fig. 24.46 shows a centrifugal compressor (with double sided impeller). It consists of the following *parts* :

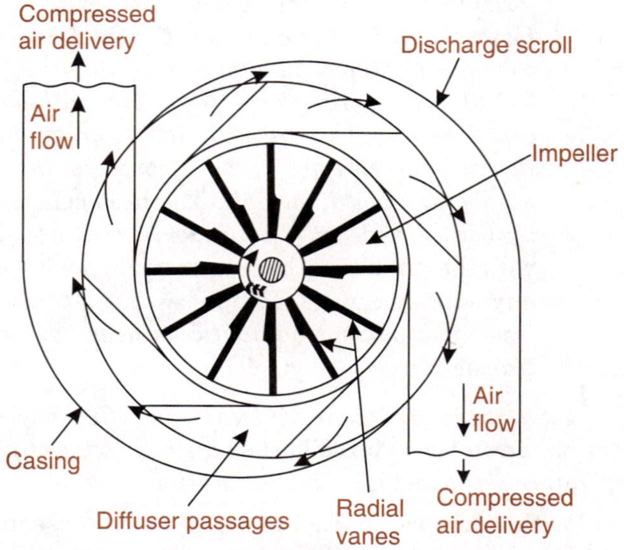

Fig. 24.46. Centrifugal compressor.

1. *Curved radial vanes.* A series of curved radial vanes are attached to and rotate with the shaft.

2. *Impeller.* The impeller is a disc fitted with radial vanes. The impeller is *generally forged or die-casted of low silicon aluminium alloy.*

The impeller may be single-eyed or double sided (having eye on either side of the compressor, so that air is drawn in on both sides) as shown in Figs. 24.47 and 24.48 respectively. The advantage of double sided impeller is that the *impeller is subjected to approximately equal forces in an axial direction.*

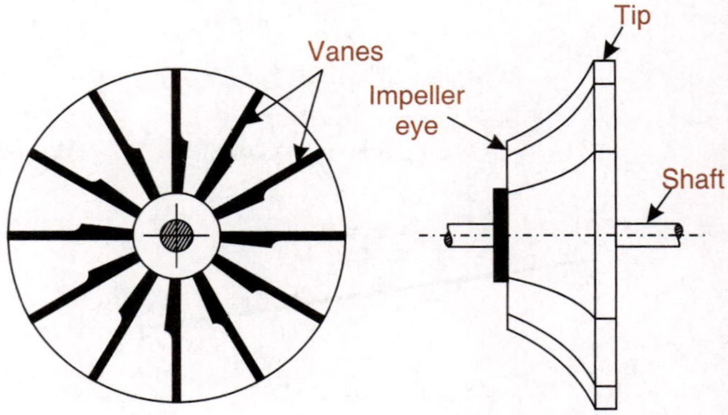

Fig. 24.47. Impeller (single-eyed) and radial vanes of centrifugal compressor.

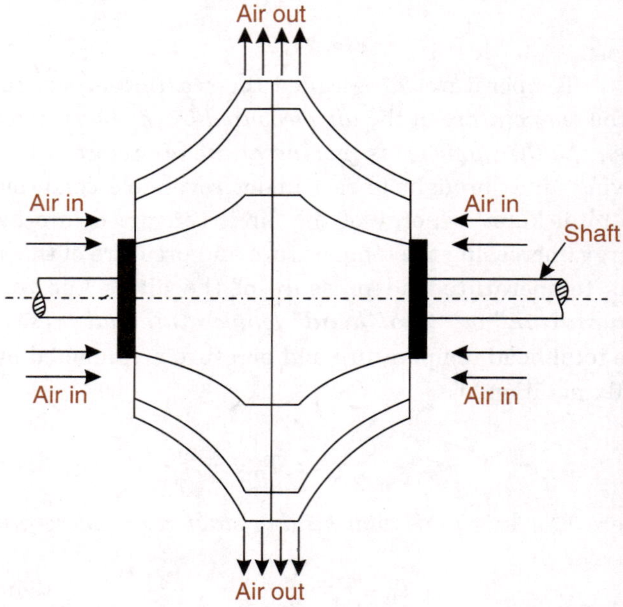

Fig. 24.48. Double eyed impeller.

3. *Casing.* The casing surrounds the rotating impeller.

4. *Diffuser.* The diffuser is housed in a radial portion of the casing.

Working :

- Air *enters the eye of the "impeller"* at a mean radius r_m with a low velocity C_1 and atmospheric pressure p_1. Depending upon the centrifugal action of the impeller, the air moves radially outwards and during its movement is guided by the impeller vanes. The impeller transfers the energy of the drive to the air causing a rise both in static pressure and temperature, and increase in velocity. Let the increased pressure and velocity be p_2 and C_2 respectively. The *work input equals the rise in total temperature.*
- The air now *enters the diverging passage* called *"diffuser"* where it is efficiently slowed down. The *kinetic energy is converted into pressure energy* with the result that *there is a further rise in static pressure*. Let the increased pressure and the reduced velocity be p_3 and C_3. The changes of pressure and velocity of air passing through the impeller and diffuser are shown in Fig. 24.49.

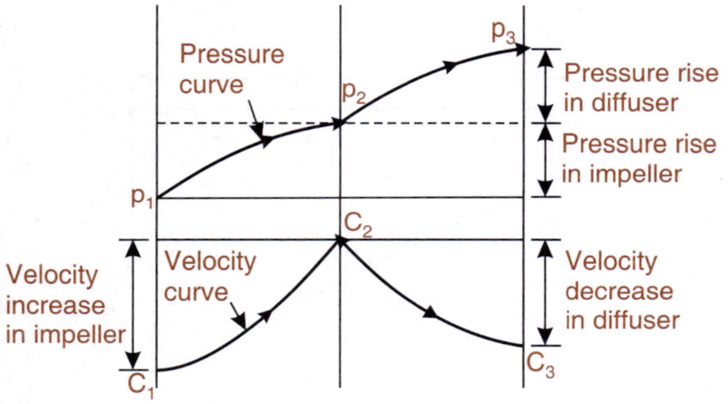

Fig. 24.49. Variations of pressure and velocity of air passing through impeller and diffuser.

- In practice *nearly half the total pressure is achieved in impeller and remaining half in the diffuser.* A pressure ratio of 4.5 : 1 can be achieved with *single-stage* centrifugal compressor. For *higher ratios,* multi-stage compressors are used. In multi-stage compressors, the outlet of the first stage is passed to the second stage and so on. A pressure ratio of 12 : 1 is possible with multi-stage centrifugal compressors.

24.4.3.2.1. Velocity Diagrams and Theory of Operation of Centrifugal Compressors

Let, C_{bl_1} = Mean blade velocity at entrance,

C_{bl_2} = Mean blade velocity at exit,

C_1 = Absolute velocity at inlet to the rotor,

C_2 = Absolute velocity at outlet to the rotor,

C_{r_1} = Relative velocity of air at entry of rotor,

C_{r_2} = Relative velocity of air at exit of rotor,

C_{w_1} = Velocity of whirl at inlet,

C_{w_2} = Velocity of whirl at outlet,

C_{f_1} = Velocity of flow at inlet,

C_{f_2} = Velocity of flow at outlet,

α_1 = Exit angle from the guide vane or inlet angle of the guide vane,

β_1 = Inlet angle to the rotor or impeller,

β_2 = Outlet angle from the rotor or impeller, and

α_2 = Inlet angle to the diffuser.

Fig. 24.50 shows the velocity diagrams for the inlet and outlet of the impeller. It is *assumed that the entry of the air is 'axial',* therefore the *whirl component at the inlet* (C_{w_1}) is *zero* and therefore $C_1 = C_{f_1}$. The enlarged views

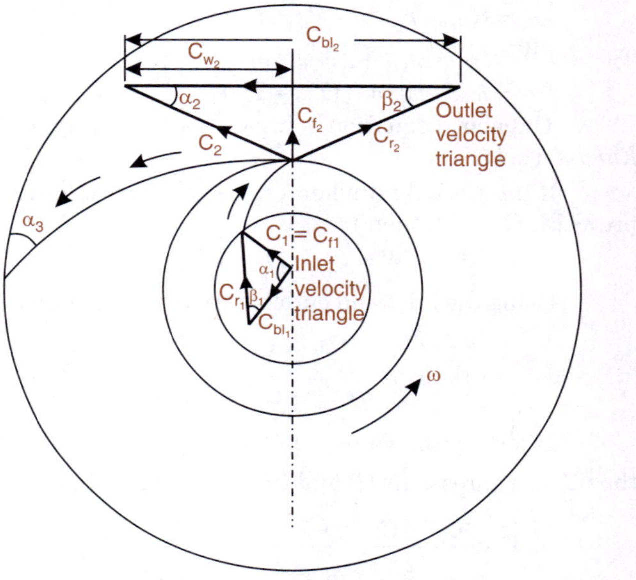

Fig. 24.50

of inlet and outlet velocity diagrams are shown in Fig. 24.51 (*a*) and 24.51 (*b*). To *avoid shock at entry and exit the blade must be parallel to the relative velocity* of *air at inlet or outlet* and therefore β_1 and β_2 are the *impeller* blade angles at the inlet and outlet. The *diffuser* blade angle *must be parallel to the absolute velocity of air from the impeller* (C_2), therefore α_2 is the diffuser blade angle at the inlet and α_3 is the diffuser blade angle at the outlet. If the *discharge from the diffuser is circumferential,* then its blade angle at outlet (α_3) should be *as small as possible.*

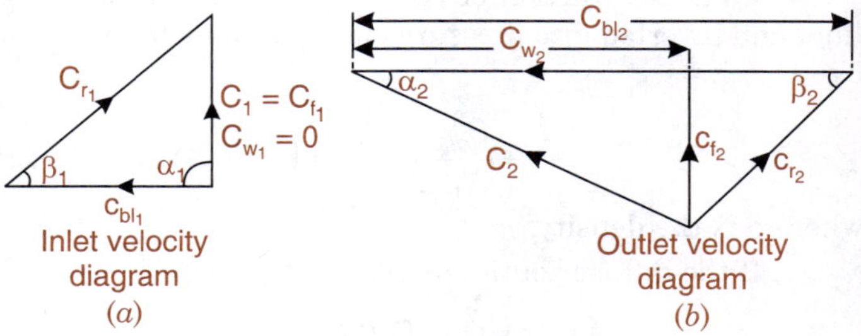

Fig. 24.51. Velocity diagrams.

Work done by impeller (Euler's work) :

The work supplied to a fluid in a stage of compressor may be found by applying the moment of momentum theorem. Consider 1 kg of working fluid passing through the impeller. The theoretical torque which must be supplied to the impeller will be equal to the rate of change of moment of momentum experienced by the working fluid.

The theoretical torque = $(C_{w2} \cdot r_2 - C_{w1}r_1)$, where r_1 and r_2 are the radii at the inlet and outlet of the impeller respectively.

If ω is the angular velocity in rad/s, the work done on 1 kg of fluid will be,

$$W = \text{Theoretical torque} \times \text{angular velocity}$$
$$= (C_{w2} \cdot r_2 - C_{w1} \cdot r_1)\,\omega$$
$$= (C_{w2} \cdot r_2 \cdot \omega - C_{w1} \cdot r_1 \cdot \omega)$$

or,
$$W = C_{w2}C_{bl_2} - C_{w1}C_{bl_1}$$
$$= h_{o2} - h_{o1} = c_p(T_{o2} - T_{o1}) \quad ...(24.59)$$

The above equation is known as *Euler's equation* or *Euler's work*.

If the working fluid enters radially *i.e.,* if there is no prewhirl, $C_{w1} = 0$, then

$$W = C_{w2} \,.\, C_{bl_2} \text{ J/kg} \quad ...(24.60)$$

Using the inlet and outlet velocity triangles, we have

$$C_{r1}{}^2 = C_{bl_1}^2 + C_1{}^2 - 2C_{bl_1}C_{w1} \quad ...(i)$$
$$C_{r_2}^2 = C_{bl_2}^2 + C_2{}^2 - 2C_{bl_2}C_{w2} \quad ...(ii)$$

Inserting the values of $C_{w2} \,.\, C_{bl_2}$ and $C_{w1} \,.\, C_{bl_1}$ from the above expressions (*i*) and (*ii*) in eqn. (24.59), we get

$$W = \underbrace{\frac{C_2{}^2 - C_1{}^2}{2}}_{\text{First term}} + \underbrace{\frac{C_{r_1}^2 - C_{r_2}^2}{2}}_{\text{Second term}} + \underbrace{\frac{C_{bl_2}^2 - C_{bl_1}^2}{2}}_{\text{Third term}} \quad ...(24.61)$$

— The *first term* shows the *increase in K.E. of 1 kg of working fluid in the impeller that has to converted into the pressure energy in the 'diffuser'.*

— The *second term* shows the *pressure rise in the impeller due* to *'diffusion action'* (as the relative velocity decreases from inlet to outlet).

— The *third term* shows the pressure rise in the impeller due to *'centrifugal action'* (as the working fluid enters at a lower diameter and comes out at a higher diameter).

Thus the fraction of K.E. imparted to the working fluid and inverted into pressure energy in impeller is given by

$$\frac{C_{r_1}^2 - C_{r_2}^2}{2} + \frac{C_{bl_2}^2 - C_{bl_1}^2}{2} = \int_1^2 \frac{dp}{\rho} \quad ...(24.62)$$

where ρ is the density.

If the diffuser outlet velocity is C_4, then

$$\frac{C_2{}^2 - C_4{}^2}{2} = \int_2^4 \frac{dp}{\rho} + \Delta h_{\text{dif. loss}}$$

i.e., the difference of K.E. at the impeller outlet and diffuser outlet introduces the work partly utilised for pressure increase, and partly irreversibly inverted into heat due to losses in diffuser.

- **Power required per impeller** for $\dot{m}$ kg of air flow in one second,

$$P = \frac{\dot{m}\, C_{w_2}\, C_{bl_2}}{1000} \text{ kW, using eqn. (24.60)} \quad ...(24.63)$$

— If the blade is radial (ideal case), then the velocity diagram at the outlet of the impeller is as shown in Fig. 24.52. As $C_{w_2} = C_{bl_2}$, the *work done per kg of air flow per second* is given by

$$W = C_2^2 \quad ...(24.64)$$

Since the air cannot leave the impeller at a velocity greater than the impeller tip velocity, the maximum work supplied per kg of air flow per second is given by eqn. (24.64).

— Now consider the steady flow at the inlet and outlet of the impeller, assuming the heat transfer during the flow of air through the impeller is zero.

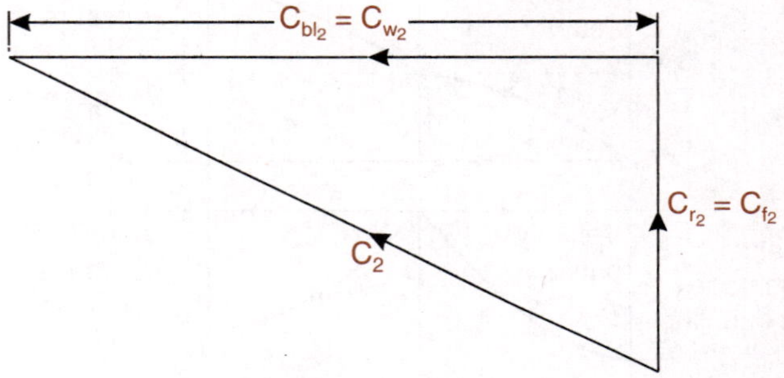

Fig. 24.52

$$h_1 + \frac{C_1^2}{2} + W = h_2 + \frac{C_2^2}{2}$$

$$W = \left(h_2 + \frac{C_2^2}{2}\right) - \left(h_1 + \frac{C_1^2}{2}\right)$$
$$= c_p\left(T_2 + \frac{C_2^2}{2c_p}\right) - c_p\left(T_1 + \frac{C_1^2}{2c_p}\right)$$
$$= c_p\,T_{02} - c_pT_{01} = c_p(T_{02} - T_{01}) \quad ...(24.65)$$

$$\therefore \quad W = c_pT_{01}\left[\frac{T_{02}}{T_{01}} - 1\right] = c_pT_{01}\left[\frac{T_2\left(\dfrac{p_{02}}{p_2}\right)^{\frac{\gamma-1}{\gamma}}}{T_1\left(\dfrac{p_{01}}{p_1}\right)^{\frac{\gamma-1}{\gamma}}} - 1\right]$$

$$= c_pT_{01}\left[\left(\frac{p_{02}}{p_{01}}\right)^{\frac{\gamma-1}{\gamma}} - 1\right] = c_pT_{01}\left[(r_{p0})^{\frac{\gamma-1}{\gamma}} - 1\right] \quad ...(24.66)$$

where r_{p0} is the *pressure ratio based on stagnation pressures.*

In most practical problems, $C_1 = C_2$, then equations (24.65) and (24.66) are reduced to

$$W = c_p (T_2 - T_1) \quad ...(24.67)$$

$$= c_p T_1 \left[(r_p)^{\frac{\gamma-1}{\gamma}} - 1 \right] \quad ...(24.68)$$

Now from eqns. (24.64) and (24.66)

$$C_2^2 = c_p T_{01} \left[\left(\frac{p_{02}}{p_{01}} \right)^{\frac{\gamma-1}{\gamma}} - 1 \right]$$

$$\therefore \quad \frac{p_{02}}{p_{01}} = \left[\frac{C_2^2}{c_p T_{01}} + 1 \right]^{\frac{\gamma}{\gamma-1}} \quad ...(24.69)$$

If $\quad C_1 = C_2$

$$\frac{p_2}{p_1} = \left[\frac{C_2^2}{c_p T_1} + 1 \right]^{\frac{\gamma}{\gamma-1}} \quad ...(24.70)$$

Thus as per eqn. (24.66) the *power input to the compressor* depends upon the following *factors* :

(*i*) Mass flow of air through the compressor.

(*ii*) Total temperature at the inlet of the compressor.

(*iii*) Total pressure ratio of the compressor which depends upon the square of the impeller tip velocity.

In the above analysis :

p_{01} = Stagnation pressure at inlet of the compressor ;
T_{01} = Stagnation temperature at inlet ;
p_{02} = Stagnation pressure at outlet ;
T_{02} = Stagnation temperature at outlet ;
p_1 = Static pressure at inlet ;
T_1 = Static temperature at inlet ;
p_2 = Static pressure at outlet ;
T_2 = Static temperature at outlet.

24.4.3.2.2. Width of Blades of Impeller and Diffuser

If the mass of the air flowing per second is constant and is known, then the width of blades of impeller and diffuser can be calculated as follows :

Let, $\dot{m}$ = Mass of air flowing per second,
b_1 = Width (or height) of impeller at inlet,
C_{f_1} = Velocity of flow at inlet of the impeller,
v_1 = Volume of 1 kg of air at the inlet,
r_1 = Radius of impeller at the inlet,

Then, $$\dot{m} = \frac{\text{Volume of air flowing per second}}{\text{Volume of 1 kg of air}}$$

$$= \frac{2\pi r_1 b_1 \times C_{f_1}}{v_1}$$

But as the air is trapped radially,

$$C_{f_1} = C_1$$

$$\therefore \quad \dot{m} = \frac{2\pi r_1 b_1 \times C_1}{v_1} \quad ...(24.71)$$

i.e., $$b_1 = \frac{\dot{m} v_1}{2\pi_1 r_1 C_1} \quad ...[24.71\ (a)]$$

Similarly the width of impeller blade at the outlet can be found by using suffix 2 in eqn. (24.71)

$$\dot{m} = \frac{2\pi r_2 b_2 \times C_{f_2}}{v_2} \quad ...(24.72)$$

The width or height of the impeller blades at the outlet and height of diffuser blade at the inlet should be same theoretically.

The width or height of the diffuser blades at the outlet, is given by

$$\dot{m} = \frac{2\pi r_d b_d \times C_{fd}}{v_d} \quad ...(24.73)$$

where suffix '*d*' represents the quantities at the *outlet of the diffuser*.

If, n = Number of blades on the impeller, and
t = Thickness of the blade,

then eqns. (24.71), (24.72) and (24.73) are expressed as follows :

$$\dot{m} = \frac{(2\pi r_1 - nt) b_1 C_{f_1}}{v_1} \quad ...(24.74)$$

$$\dot{m} = \frac{(2\pi r_2 - nt) b_2 C_{f_2}}{v_2} \quad ...(24.75)$$

$$\dot{m} = \frac{(2\pi r_d - nt) b_d C_{f_d}}{v_d} \quad ...(24.76)$$

24.4.3.2.3. Isentropic Efficiency of the Compressor

The following *losses* occur when air flows through the impeller :

(*i*) Friction between the air layers moving with relative velocities and friction between the air and flow passages

(*ii*) Shock at entry

(*iii*) Turbulence caused in air.

The losses mentioned above cause an increase in enthalpy of the air without increase of pressure therefore the *actual temperature* of air coming out from the compressor is *more* than the temperature of air if it is compressed isentropically. The *actual work* required for the same increase in pressure ratio is *more* due to *irreversibilities.* The actual and isentropic compression for the same pressure ratio is shown in Fig. 24.53.

The isentropic efficiency is given by the relation,

$$\eta_{isen} = \frac{\text{Isentropic work}}{\text{Actual work}} = \frac{h'_{02} - h_{01}}{h_{02} - h_{01}}$$

$$= \frac{T'_{02} - T_{01}}{T_{02} - T_{01}} \quad ...(24.77)$$

provided specific heat at constant pressure (c_p) remains constant.

If $C_1 = C_2$, then $\eta_{isen} = \frac{T'_2 - T_1}{T_2 - T_1}$ ...(24.78)

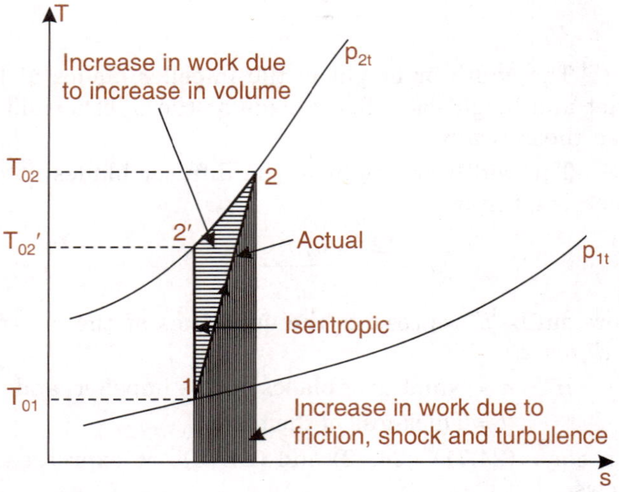

Fig. 24.53. Actual and isentropic compression on *T-s* diagram.

$$= \frac{\frac{T'_2}{T_1} - 1}{\frac{T_2}{T_1} - 1} = \frac{\left(\frac{p_2}{p_1}\right)^{\frac{\gamma - 1}{\gamma}} - 1}{\left(\frac{p_2}{p_1}\right)^{\frac{n-1}{n}} - 1}, \; n > \gamma.$$

24.4.3.2.4. Slip Factor and Pressure Co-efficient

In the earlier analysis it was assumed that $C_{w_2} = C_{bl_2}$ but this condition is not satisfied in actual practice *due to secondary flow effects* and therefore in actual compressors $C_{w_2} < C_{bl_2}$.

The difference between ($C_{bl_2} - C_{w_2}$) is known as **slip.**

Slip factor (ϕ_s). It is defined as the *ratio of actual whirl component* (C_{w_2}) *and the ideal whirl component* (C_{bl_2})

$$\therefore \quad \phi_s = \frac{C_{w_2}}{C_{bl_2}} = 1 \text{ if } C_{w_2} = C_{bl_2} \quad ...(24.79)$$

The stagnation (total head) pressure ratio is given by

$$\frac{p_{02}}{p_{01}} = \left(\frac{T'_{02}}{T_{01}}\right)^{\frac{\gamma-1}{\gamma}} = \left(1 + \frac{T'_{02} - T_{01}}{T_{01}}\right)^{\frac{\gamma-1}{\gamma}}$$

Substituting the value of $(T'_{02} - T_{01})$ from eqn. (24.77), we get

$$\frac{p_{02}}{p_{01}} = \left[1 + \frac{\eta_{isen}\,(T_{02} - T_{01})}{T_{01}}\right]^{\frac{\gamma-1}{\gamma}} \quad ...(24.80)$$

As per eqn. [24.60 (*a*)], the actual work done per kg of air is given by

$$c_p(T_{02} - T_{01}) = C_{bl_2} - C_{w_2}$$

The actual work done per kg of air by the compressor is always greater than $C_{bl_2} C_{w_2}$ due to fluid friction and windage losses, therefore the actual work is obtained by multiplying $C_{bl_2} C_{w_2}$ by a factor ϕ_w known as **work factor** or **power input factor.**

$$\therefore \quad c_p(T_{02} - T_{01}) = \phi_w C_{bl_2} C_{w_2}$$

$$\therefore \quad T_{02} - T_{01} = \frac{\phi_w C_{bl_2} C_{w_2}}{c_p} \quad ...(24.81)$$

Now substituting the value of eqn. (24.81) into eqn. (24.80), we have

$$\frac{p_{02}}{p_{01}} = \left[1 + \frac{\eta_{isen}\, \phi_w C_{bl_2} C_{w_2}}{c_p\, T_{01}}\right]^{\frac{\gamma-1}{\gamma}}$$

Now substituting the value of C_{w_2} from eqn. (24.79), we have

$$\frac{p_{02}}{p_{01}} = \left[1 + \frac{\eta_{isen}\, \phi_w\, \phi_s\, C_{bl}{}^2}{c_p\, T_{01}}\right]^{\frac{\gamma-1}{\gamma}} \quad ...(24.82)$$

Pressure Co-efficient (ϕ_p). It is defined as the *ratio of isentropic work to Euler work.*

$$\therefore \quad \phi_p = \frac{\text{Isentropic work}}{\text{Euler work}} = \frac{c_p\,(T_{02}' - T_{01})}{C_{bl_2}\, C_{w_2}}$$

Using eqn. (24.77) and assuming the vanes of the impeller are radial and counter flow of fluid is neglected ($C_{w_2} = C_{bl_2}$)

$$\phi_p = \frac{c_p\, \eta_{isen}\,(T_{02} - T_{01})}{C_{bl_2}{}^2}$$

Now using eqn. (24.81) which is

$$c_p(T_{02} - T_{01}) = \phi_w\, C_{bl_2}\, C_{w_2} = \phi_w \phi_s C_{bl_2}{}^2$$

$$\phi_p = \frac{\phi_w\, \phi_s\, C^2_{bl_2}\, \eta_{isen.}}{C^2_{bl_2}} = \phi_w \phi_s\, \eta_{isen.}$$

...(24.83)

24.4.3.2.5. The Effect of Impeller Blade Shape on Performance

The following shapes of blades are utilized in the impellers of centrifugal compressors :

1. *Backward-curved blades* ($\beta_2 < 90°$)

2. *Radial-curved blades* ($\beta_2 = 90°$)

3. *Forward-curved blades* ($\beta_2 > 90°$)

Fig. 24.54, shows the relative performance of these blades. Centrifugal effects on the curved blades *create a bending moment and produce increased stresses which reduce the maximum speed at which the impeller can run.*

- Normally backward blades/vanes with β_2 between 20–25° are employed except in cases where high head is the major consideration.
- Sometimes compromise is made between the low energy transfer (backward-curved vanes) and high outlet velocity (forward-curved vanes) by using *radial vanes.*

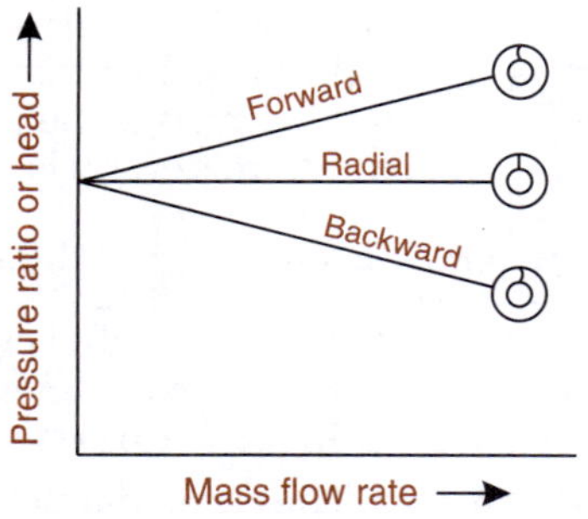

Fig. 24.54. Characteristics of backward-curved, radial-curved, and forward-curved vanes.

Advantages of radial-blade impellers :

1. Can be manufactured *easily.*
2. Lowest unit blade stress for a given diameter and rotational speed, hence *highest weight.*
3. Free from complex bending stresses.
4. Equal energy conversion in impeller and diffuser, *giving high pressure ratios with good efficiency*.

In view of the above reasons, the *impeller with radial blades has been the logic choice of the designers of aircraft centrifugal compressors.*

24.4.3.2.6. Diffuser System

In a centrifugal compressor, the diffuser plays a significant role in overall compression process. Whereas the *impeller* is designed to impart energy to the air by increasing its velocity as efficiently as possible, the '**diffuser**' *converts this imparted kinetic energy to pressure-rise.* For a radial-bladed impeller, the diffuser contributes about one-half of the overall static pressure-rise.

In the *vaned diffuser* the vanes are used to remove the whirl of the fluid at a higher rate than is possible by a simple increase in radius, thereby reducing the length of flow path and diameter. The *vaned diffuser is advantageous where small size is important as in the case of air craft engines.*

Fig. 24.55 shows a typical vaned diffuser. There is a clearance between the impeller and vane leading edges amounting to about 10 to 20% of the diameter for compressors. This *space* constitutes a vaneless diffuser and its functions are :

(*i*) To smooth out velocity variation between the impeller tip and vanes.

(*ii*) To reduce the circumferential pressure gradient at impeller tip.

(*iii*) To reduce the Mach number at entry to the vanes.

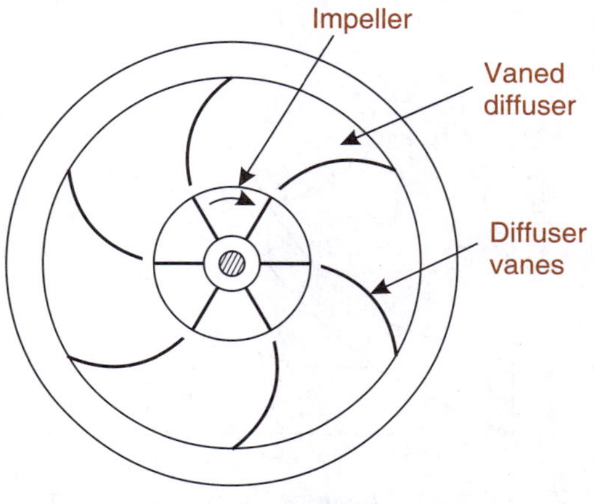

Fig. 24.55. Typical vaned diffuser.

The flow follows an approximately *logarithmic spiral path* to the vanes after which it is constrained by the diffuser channels. For rapid diffusion the axis of the channel is straight and tangential to the spiral. The design of the passages is usually based on the simple channel theory with an equivalent degree of divergence ranging between 8 to 12° to control separation.

- *The number of diffuser vanes has a direct bearing on the size and the efficiency of the diffuser.*
- *When the number of diffuser passages is less than the number of impeller passages a more uniform total flow results.*

Diffuser efficiency *is defined as,*

$$\eta_d = \frac{(p_2 - p_1)}{\frac{w}{2g}(C_1^2 - C_2^2)} \qquad \text{...(24.84)}$$

Suffices 1 and 2 denote upstream and downstream conditions of diffuser and w is the weight density.

In order to achieve higher diffuser efficiency, the following points need be considered at the time of design :

1. The entrance blade angle of the diffuser must be such that *air impinges on it with a small angle of attack.*

2. Sudden changes in flow are to be avoided. After the air leaves the diffuser, it must be turned through 90° to

flow in an axial direction to the combustion chamber. This turning may be achieved by vanes installed in the diffuser below.

3. The area of the flow passage must be large enough to handle the air and it must expand within certain maximum and reasonable limits.

24.4.3.2.7. Losses in Centrifugal Compressors

The losses in a centrifugal compressor may be categorized as follows :

1. **Friction losses.** These losses are proportional to C^2 and hence proportional to m^2.

2. **Incidence loss.** These losses in terms of drag co-efficient C_D is proportional to C_DC^2.

Fig. 24.56 shows the variation of losses with respect to the mass flow rate.

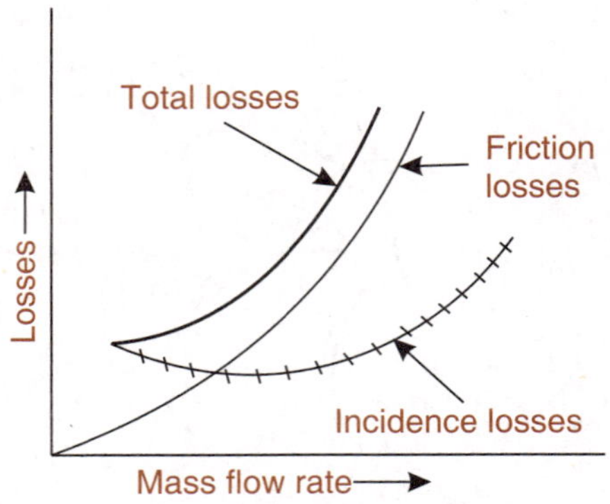

Fig. 24.56. Variation of losses with respect to mass flow rate.

24.4.3.2.8. Selection of Compressors Geometrics

The various compressors' geometrics are selected on the basis of the following :

1. Number of blades in impeller
2. Blade angles
3. Impeller diameters
4. Impeller widths
5. Impeller material
6. Vaneless diffuser
7. Vaned diffuser.

1. **Number of blades in impeller.** The optimum number of blades which gives the best efficiency can be chosen by experience for a particular requirement. A number of empirical relations are available for calculating the optimum number of blades.

- *Vincent* suggests that the optimum blade number varies from 18 to 22 for radial bladed impeller having diameter from 25 cm to 36 cm.

2. **Blade angles :**

- Among the outlet and inlet blade angles, the former influence the latter to a great extent.
- The inlet blade angle, within the reasonable limit, does not affect the performance and, its optimum value is about 30° to 35°.

3. **Impeller diameters.** For *radial blades,* the tip impeller diameter is calculated by the following equations :

$$\text{Actual work input} = \phi_w \phi_s\, C_{bl}^2 \text{ J/kg} \quad ...(i)$$

and
$$C_{bl_2} = \frac{\pi D_2 N}{60} \quad ...(ii)$$

After knowing the tip diameter, the inlet diameter is calculated from the value of diameter ratio. Generally $\frac{d_2}{d_1}$ varies from 1.6 to 2.

4. **Impeller width.** If b_1 and b_2 are the blade width at inlet and outlet of impeller, then neglecting the thickness of the blades it is calculated by the equation

$$m = \pi d_1 b_1\, C_{f_1}\rho_1 = \pi d_2 b_2\, C_{f_2}\rho_2$$

Generally $C_{f_1} = C_{f_2}$.

5. **Impeller material.** The impeller of the centrifugal compressor is generally forged or die casted of low silicon aluminium alloy.

6. **Vaneless diffuser.** The function of vaneless diffuser or space is to stabilize the flow for shockless entry into the bladed diffuser and to invert some portions of K.E. into pressure energy. The diameter ratio of vaneless to impeller tip diameter varies from $\frac{D_3}{D_2}$ = 1/0.06 to 1.12.

Since the flow in the vaneless diffuser is assumed to be logarithmic spiral, hence $\alpha_2 = \alpha_3$. Generally $b_2 = b_3$ = width of vaneless diffuser. In some cases $b_3 > b_2$.

7. **Vaned diffuser.** The outlet diameter of the vaned diffuser depends upon the choice for the velocity desired from the outlet of the compressor.

By the *use of bladed diffuser the dimensions of the machine can be reduced due to reduction in diffuser size.* The outlet diameter is calculated by the equation,

$$m = \pi d_4 b_4\, C_4 \rho p_4 \sin \alpha_4$$

where $b_4 = b_2 = b_3$.

— The number of vanes may vary from 10 to 30 but *should not coincide with number of vanes in impeller* to avoid resistance.

— $\frac{D_4}{D_3}$ may vary from 1.25 to 1.6 and the maximum diffusion angle is around 10°.

- *Isentropic efficiency of vaned diffuser at design condition is higher than that of vaneless but its off-design performance is poor than that of vaneless.*

24.4.3.2.9. Compressor Characteristics

When flow is taking place in an impeller channel, there are certain inlet losses, friction and separation losses and discharge losses in the diffuser. If these losses and the effect of slip and non-uniform distribution of radial velocity around the periphery of the impeller are taken into account, the head-capacity characteristic for the backward-curved vanes would take the form *LM* as shown in Fig. 24.57.

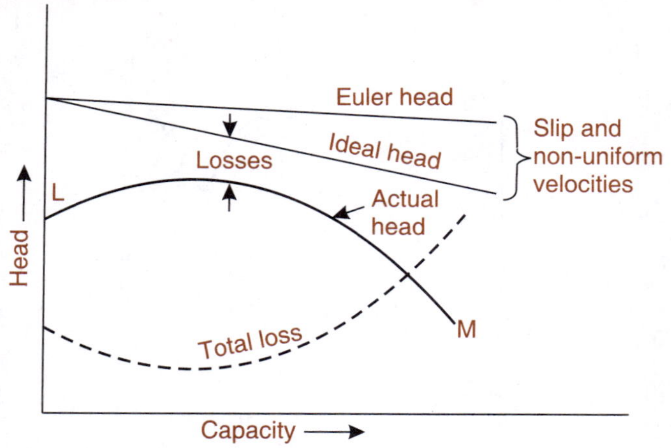

Fig. 24.57. Actual characteristics of a centrifugal compressor.

24.4.3.2.10. Surging and Choking

Fig. 24.58 shows a typical characteristics of a centrifugal compressor at one particular speed. *Consider that the compressor is running at point N* :

- If now the *resistance to flow is increased* (say, by, closing the valve provided at the delivery line of the compressor), the equilibrium point moves to *M*.

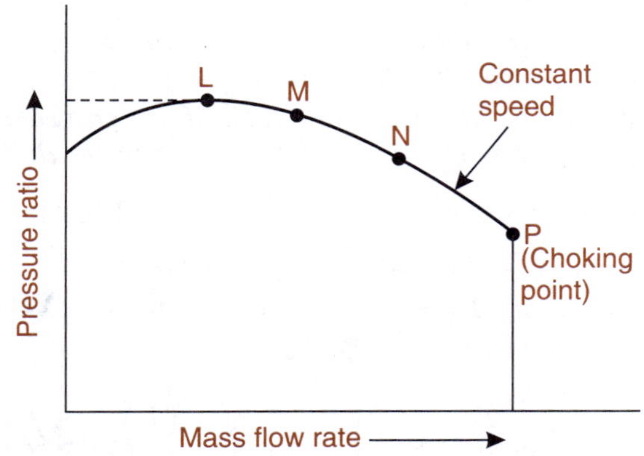

Fig. 24.58. A typical characteristic at one particular speed.

- Any further restriction to the flow will cause the operating point to shift to the left, ultimately arriving at point *L*. At this point *maximum pressure ratio is obtained.* If the flow is still reduced from this point then the pressure ratio will *reduce.* At this moment, there is a higher pressure in the downstream system than the compressor delivery and the *flow stops* and *may even reverse in direction.*
- *After a short interval of time compressor again starts to deliver fluid.* The *pressure starts to increase from a very low value and operating point moves from right to left.* If the downflow conditions are unchanged then once again the flow will break down after point *L* and the cycle will be repeated with *a high frequency.* This phenomenon is known as **surging** or **pumping.**

  This *instability will be severe in compressors producing high pressure ratios,* which may ultimately *lead to physical damage due to impact loads and high-frequency vibration.*
- Owing to this particular phenomenon of surging or pumping at *low mass flow rates the compressor cannot be operated at any point left of the maximum pressure ratio point.* That is, it cannot be operated on the positive slope of the characteristic.

  On the characteristic, the following *situation occurs at higher mass flow rate points* :

At a constant rotor speed the tangential velocity component at the impeller tip *remains constant.* With the *increase in mass flow the pressure ratio decreases* and hence the density is *decreased.* Consequently, the radial velocity is increased considerably, which increases the absolute velocity and incidence angle at the diffuser vane tip. Thus, there is rapid progression towards a *choking state.* The slope of the characteristic therefore steepens and finally after point '*P*' mass flow *cannot be increased any further.* The characteristic finally becomes *vertical.* The point '*P*' on the characteristic curve is called **choking** *point.*

When the compressor is used with gas turbine then the characteristics must be matched properly otherwise troubles will be experienced either due to surging or due to low efficiency.

Note. The *choking mass flow can be varied by changing the impeller rotational speed.*

24.4.3.2.11. Performance of Centrifugal Compressors

- Fig. 24.59 (*a*) shows the relationship between pressure ratio, power and efficiency curves *versus* flow rate for various values of speeds such as N_1, N_2 etc. At *a certain speed, efficiency increases as the flow rate increases and reaches a maximum value after which it decreases.* Accordingly as *the flow rate increases the power consumed also increases.*
- Fig. 24.59 (*b*) shows the performance and constant efficiency curves.

  Such a plot does not take into account the varying inlet temperature and pressure. In addition to this, these plots cannot show the comparison of performance for similar compressors of different sizes. To account for all these, the *performance curves are plotted with 'dimensionless parameters'.* These dimensionless parameters are : Pressure ratio, $\frac{p_2}{p_1}$; speed parameter, $\frac{N_1}{\sqrt{T_1}}$ and flow parameter $\frac{m\sqrt{T_1}}{p_1}$ [Fig. 24.59 (*c*) and (*d*)].

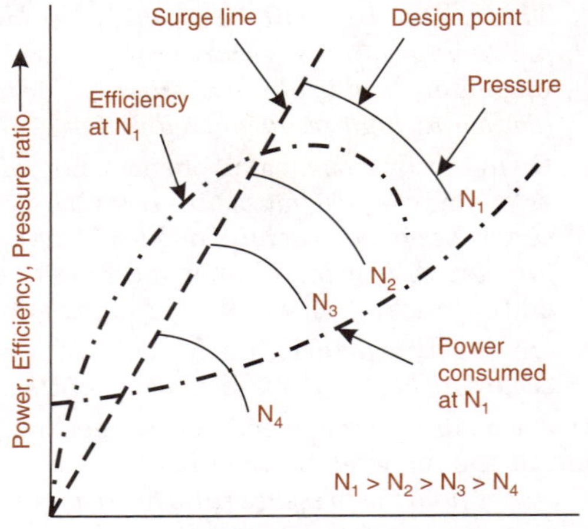

(*a*)

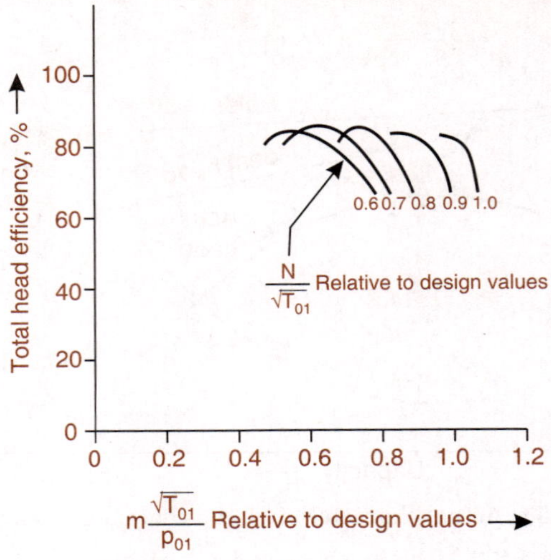

(*d*)

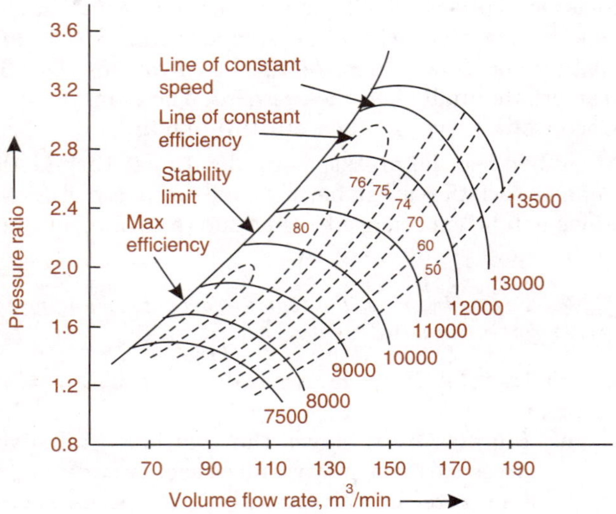

(*b*)

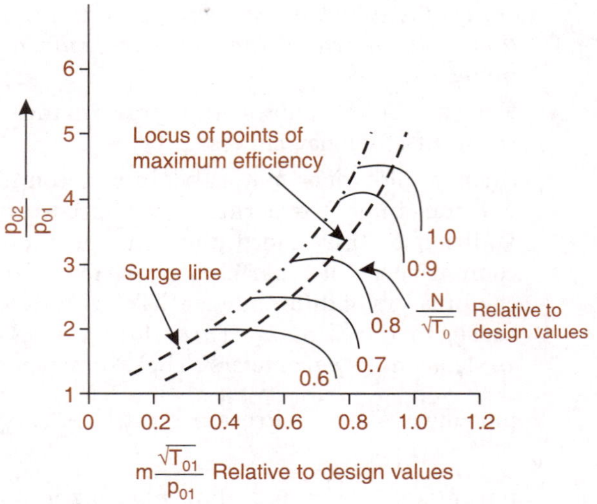

(*c*)

Fig. 24.59. Performance curves of centrifugal compressor.

Example 24.37. *A centrifugal compressor used as a supercharger for aero-engines handles 150 kg/min. of air. The suction pressure and temperature are 1 bar and 290 K. The suction velocity is 80 m/s. After compression in the impeller the conditions are 1.5 bar 345 K and 220 m/s. Calculate :*

(*i*) *Isentropic efficiency.*

(*ii*) *Power required to drive the compressor.*

(*iii*) *The overall efficiency of the unit.*

It may be assumed that K.E. of air gained in the impeller is entirely converted into pressure in the diffuser.

Solution. *Given :* $\dot{m} = \dfrac{150}{60} = 2.5$ kg/s ; $p_1 = 1$ bar ;

$T_1 = 290$ K ; $C_1 = 80$ m/s ;

$p_2 = 1.5$ bar ; $T_2 = 345$ K ; $C_2 = 220$ m/s.

(*i*) **Isentropic efficiency, $\eta_{isen.}$:**

$$\frac{T_2'}{T_1} = \left(\frac{p_2}{p_1}\right)^{\frac{\gamma-1}{\gamma}} = \left(\frac{1.5}{1}\right)^{\frac{1.4-1}{1.4}} = 1.1228$$

or $\quad T_2' = 290 \times 1.1228 = 325.6$ K

$\therefore$ Isentropic work done

$$= c_p(T_2' - T_1) + \frac{C_2^2 - C_1^2}{2 \times 1000}$$

$$= 1.005(325.6 - 290) + \frac{(220)^2 - (80)^2}{2 \times 1000}$$

$$= 35.778 + 21 = 56.78 \text{ kJ/kg.}$$

Work done in the impeller

$$= c_p(T_2 - T_1) + \frac{(220)^2 - (80)^2}{2 \times 1000}$$

$$= 1.005(345 - 290) + \frac{(220)^2 - (80)^2}{2 \times 1000}$$

$$= 55.275 + 21 = 76.27 \text{ kJ/kg}$$

$$\therefore \quad \eta_{isen} = \frac{\text{Isentropic work}}{\text{Actual work}} = \frac{56.78}{76.27}$$

$= \mathbf{0.7445}$ or **74.45%. (Ans.)**

(ii) **Power required to drive the compressor, P :**

$P = \dot{m} \times$ Work done in the impeller (kJ/kg)

$= 2.5 \times 76.27 =$ **190.67 kW. (Ans.)**

(iii) **The overall efficiency of the unit, $\eta_{overall}$:**

As K.E. gained in the impeller is converted into pressure, hence

$$c_p(T_3 - T_2) = \frac{C_2^2 - C_1^2}{2 \times 1000}$$

$$1.005(T_3 - 345) = \frac{(220)^2 - (80)^2}{2 \times 1000}$$

or $$T_3 = 365.9 \text{ K.}$$

The *pressure of air after leaving the diffuser, p_3* :

$$\frac{T_3}{T_2} = \left(\frac{p_3}{p_2}\right)^{\frac{\gamma-1}{\gamma}}$$

or $$\frac{p_3}{p_2} = \left(\frac{T_3}{T_2}\right)^{\frac{\gamma}{\gamma-1}} = \left(\frac{365.9}{345}\right)^{\frac{1.4}{1.4-1}} = 1.2286$$

$\therefore \quad p_3 = 1.5 \times 1.2286 = 1.843$ bar.

After isentropic compression, the *delivery temperature from diffuser, T_3'* :

$$\frac{T_3'}{T_1} = \left(\frac{p_3}{p_1}\right)^{\frac{\gamma-1}{\gamma}} = \left(\frac{1.843}{1}\right)^{\frac{1.4-1}{1.4}} = 1.191$$

or $$T_3' = 290 \times 1.191 = 345.39 \text{ K}$$

$$\therefore \eta_{overall} = \frac{T_3' - T_1}{T_3 - T_1} = \frac{345.39 - 290}{365.9 - 290}$$

$= \mathbf{0.7298}$ or **72.98%. (Ans.)**

Example 24.38. *A single inlet-type centrifugal compressor handles 528 kg/min. of air. The ambient air conditions are 1 bar and 20°C. The compressor runs at 20000 r.p.m. with isentropic efficiency of 80 percent. The air is compressed in the compressor from 1 bar static pressure to 4.0 bar total pressure. The air enters the impeller eye with a velocity of 145 m/s with no prewhirl. Assuming that the ratio of whirl speed to tip speed is 0.9, calculate :*

(i) Rise in total temperature during compression if the change in K.E. is negligible.

(ii) The tip diameter of the impeller.

(iii) Power required.

(iv) Eye diameter if the hub diameter is 12 cm.

Solution. *Given* : $\dot{m} = \frac{528}{60} = 8.8$ kg/s ;

$p_1 = 1$ bar, $T_1 = 20 + 273 = 293$ K ;

$N = 20000$ r.p.m. ; $\eta_{isen} = 80\%$;

$p_{01} = 1$ bar ; $p_{02} = 4.0$ bar ;

$C_1 = 145$ m/s ; $\frac{C_{w2}}{C_{bl_2}} = 0.9$; $d_h = 10$ cm $= 0.1$ m.

(i) **Rise in total temperature during compression if the change in K.E. is negligible :**

Refer to Fig. 24.60. The suffix '0' indicates the total values.

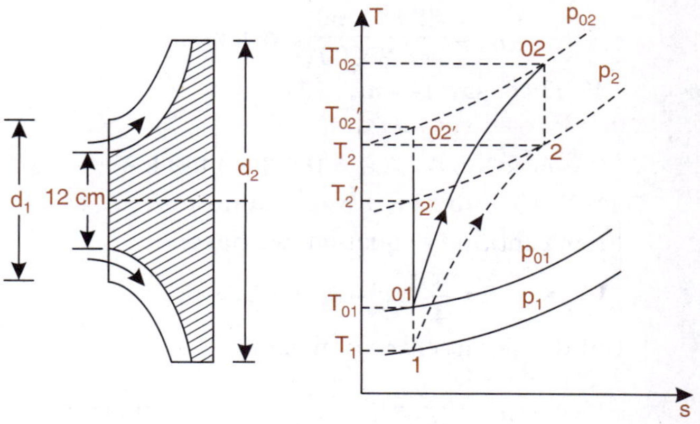

Fig. 24.60

The stagnation temperature at inlet to the machine,

$$T_{01} = T_1 + \frac{C_1^2}{2c_p} = 293 + \frac{(145)^2}{2 \times 1.005 \times 1000}$$

$= 303.5$ K

Now, $$\frac{T_{01}}{T_1} = \left(\frac{p_{01}}{p_1}\right)^{\frac{\gamma-1}{\gamma}} \text{ or } p_{01} = p_1 \times \left(\frac{T_{01}}{T_1}\right)^{\frac{\gamma}{\gamma-1}}$$

or $$p_{01} = 1 \times \left(\frac{303.5}{293}\right)^{\frac{1.4}{1.4-1}} = 1.131 \text{ bar}$$

$$\frac{T_{02}'}{T_{01}} = \left(\frac{p_{02}}{p_{01}}\right)^{\frac{\gamma-1}{\gamma}} = \left(\frac{4.0}{1.131}\right)^{\frac{1.4-1}{1.4}} = 1.435$$

$\therefore \quad T_{02}' = 303.5 \times 1.435 = 435.5$ K.

$\therefore$ ***Isentropic rise in total temperature***

$= 435.5 - 303.5 = 132$°C

Hence, *Actual rise in total temperature*

$$= \frac{132}{\eta_{isen}} = \frac{132}{0.8} = \mathbf{165°C.} \quad \textbf{(Ans.)}$$

(ii) **The tip diameter of the impeller, d_2 :**

Work consumed by the compressor $= c_p \times (\Delta t)_{actual}$

$= 1.005 \times 165 = 165.8$ kJ/kg

Work consumed by the compressor is also given by Euler's equation without prewhirl as :

$$W = \frac{C_{w_2} \times C_{bl_2}}{1000} \text{ kJ/kg} = 165.8 \text{ kJ/kg}$$

But $$\frac{C_{w_2}}{C_{bl_2}} = 0.9 \quad \therefore \quad C_{w_2} = 0.9 C_{bl_2}$$

$$\therefore \quad 165.8 = \frac{C_{bl_2}^2 \times 0.9}{1000}$$

or $$C_{bl_2} = \left(\frac{165.8 \times 1000}{0.9}\right)^{1/2} = 429.2 \text{ m/s}$$

But $$C_{bl_2} = 429.2 = \frac{\pi d_2 N}{60} = \frac{\pi d_2 \times 20000}{60}$$

$$\therefore \quad d_2 = \frac{429.2 \times 60}{\pi \times 20000} = 0.4098 \text{ m}$$

or 40.98 cm say **41 cm. (Ans.)**

(iii) **Power required, P :**

P = $\dot{m}$ × 165.8 = 8.8 × 165.8 = **1459 kW. (Ans.)**

(iv) **Eye diameter if hub diameter is 12 cm, d_1 :**

From continuity equation, we have

$$\dot{m} = \frac{\pi}{4}(d_1^2 - d_h^2) \times C_1 \times \rho_1$$

But density at entry is given by,

$$\rho_1 = \frac{p_1}{RT_1} = \frac{1 \times 10^5}{287 \times 293} = 1.189 \text{ kg/m}^3$$

$$\therefore \quad 8.8 = \frac{\pi}{4}(d_1^2 - 0.12^2) \times 145 \times 1.189$$

or $$d_1^2 = \frac{8.8 \times 4}{\pi \times 145 \times 1.189} + 0.12^2 = 0.07939 \text{ m}^2$$

$$\therefore \quad d_1 = \mathbf{0.2818 \text{ m or } 28.2 \text{ cm.}} \quad \textbf{(Ans.)}$$

Example 24.39. *A centrifugal compressor running at 10000 r.p.m. delivers 660 m³/min. of free air. The air is compressed from 1 bar and 20°C to a pressure ratio of 4 with an isentropic efficiency of 82%. Blades are radial at outlet of impeller and flow velocity of 62 m/s may be assumed throughout constant. The outer radius of impeller is twice the inner and the slip factor may be assumed as 0.9. The blade area co-efficient may be assumed 0.9 at inlet. Calculate :*

(i) Final temperature of air.
(ii) Theoretical power.
(iii) Impeller diameters at inlet and outlet.
(iv) Breadth of impeller at inlet.
(v) Impeller blade angle at inlet.
(vi) Diffuser blade angle at inlet.

Solution. Given : N = 10000 r.p.m. ; Volume of air delivered, V = 660 m³/min. ;

p_1 = 1 bar, T_1 = 20 + 273 = 293 K ;

$r_p = 4$, $\eta_{isen} = 0.82$; $C_{f_2} = 62$ m/s ;

$r_2 = 2r_1$; $\phi_s = 0.9$; Blade area co-efficient, $k_a = 0.9$.

(i) **Final temperature of air, T_2 :**

$$\frac{T_2'}{T_1} = (r_p)^{\gamma-1/\gamma} = (4)^{\frac{1.4-1}{1.4}} = 1.486$$

$$\therefore \quad T_2' = 293 \times 1.486 = 435.4 \text{ K}$$

Now, $$\eta_{isen} = \frac{T_2' - T_1}{T_2 - T_1}$$

or $$T_2 = T_1 + \frac{T_2' - T_1}{\eta_{isen}} = 293 + \frac{435.4 - 293}{0.82}$$

= **466.7 K. (Ans.)**

(ii) **Theoretical power, P :**

Mass flow rate,

$$\dot{m} = \frac{PV}{RT} = \frac{1 \times 10^5 \times (660/60)}{287 \times 293} = 13.08 \text{ m}^3\text{/s}$$

$$\therefore \quad P = \dot{m} c_p (T_2 - T_1)$$
$$= 13.08 \times 1.005(466.7 - 293)$$
$$= \mathbf{2283.3 \text{ kW.}} \quad \textbf{(Ans.)}$$

(iii) **Impeller diameters at inlet and outlet, d_1, d_2 :**

For radial blades, work input to the compressor is given by,

$$\text{Work done} = \frac{\phi_s C_{bl_2}^2}{1000} = c_p(T_2 - T_1)$$

Here T_2 is the final temperature of air from the exit of compressor.

or $$C_{bl_2} = \left[\frac{1000 \times c_p(T_2 - T_1)}{\phi_s}\right]^{1/2}$$

$$= \left[\frac{1000 \times 1.005(466.7 - 293)}{0.9}\right]^{1/2}$$

= 440.4 m/s

Also, $$C_{bl_2} = \frac{\pi d_2 N}{60} = 440.4$$

or $$d_2 = \frac{60 \times 440.4}{\pi \times 10000}$$

= **0.8411 m** or **84.11 cm. (Ans.)**

$$\therefore \quad d_1 = \frac{d_2}{2} = \frac{84.11}{2} = \mathbf{42.06 \text{ cm.}} \quad \textbf{(Ans.)}$$

(iv) **Breadth of impeller at inlet, b_1 :**

Volume flow rate = $2\pi r_1 b_1 C_{f_1} k_a$, where k_a is the blade area coefficient

$$\therefore \quad b_1 = \frac{\text{Volume flow rate}}{2\pi r_1 \cdot C_{f_1} \cdot k_a}$$

$$= \frac{(660/60)}{2\pi \times (0.4206/2) \times 62 \times 0.9}$$

= **0.1492 m** or **14.92 cm. (Ans.)**

(v) **Impeller blade angle at inlet β_1 :**

$$\tan \beta_1 = \frac{C_{f_1}}{C_{bl_1}} = \frac{62}{(440.4/2)} = 0.2816$$

$$\therefore \quad \beta_1 = \tan^{-1}(0.2816) = \mathbf{15.73°.} \quad \textbf{(Ans.)}$$

(vi) **Diffuser blade angle at inlet, α_2 :**

$$\tan \alpha_2 = \frac{C_{f_2}}{\phi_s \cdot C_{bl_2}} = \frac{62}{0.9 \times 440.4} = 0.1564$$

$$\therefore \quad \alpha_2 = \tan^{-1}(0.1564) = \mathbf{8.9°.} \quad \textbf{(Ans.)}$$

Example 24.40. *A centrifugal blower compresses 4.8 m³/s of air from 1 bar and 20°C to 1.5 bar. The index of compression n is 1.5. The flow velocity at inlet and outlet of the machine is the same and equal to 65 m/s. The inlet and outlet impeller diameters are 0.32 m and 0.62 m respectively. The blower rotates at 8000 r.p.m. Calculate :*

(*i*) *The blade angles at inlet and outlet of the impeller.*
(*ii*) *The absolute angle at the tip of the impeller.*
(*iii*) *The breadth of blade at inlet and outlet.*

It may be assumed that no diffuser is employed and the whole pressure increase takes place in the impeller and the blades have negligible thickness.

Solution. *Given* : $V_1 = 4.8$ m³/s ; $p_1 = 1$ bar ;
$T_1 = 20 + 273 = 293$ K ; $n = 1.5$
$C_{f1} = C_{f2} = 65$ m/s ; $d_1 = 0.32$ m ;
$d_2 = 0.62$ m ; $N = 8000$ r.p.m.

The temperature at the outlet of the compressor,

$$\frac{T_2'}{T_1} = \left(\frac{p_2}{p_1}\right)^{\frac{\gamma-1}{\gamma}} = \left(\frac{1.5}{1}\right)^{\frac{1.5-1}{1.5}} = 1.1447$$

$\therefore \quad T_2' = 293 \times 1.1447 = 335.4$ K

The peripheral velocity at inlet,

$$C_{bl_1} = \frac{\pi d_1 N}{60} = \frac{\pi \times 0.32 \times 8000}{60} = 134 \text{ m/s}$$

The tip peripheral velocity at outlet,

$$C_{bl_2} = \frac{\pi d_2 N}{60} = \frac{\pi \times 0.62 \times 8000}{60} = 259.7 \text{ m/s}$$

$$\text{Work done } = c_p(T_2' - T_1) = \frac{C_{bl_2} C_{w_2}}{1000}$$

$$\therefore \quad C_{w_2} = \frac{1.005(335.4 - 293) \times 1000}{259.7} = 164.1 \text{ m/s}$$

(*i*) The blade angles at inlet and outlet of the impeller, β_1, β_2 :

$$\tan \beta_1 = \frac{C_{f1}}{C_{bl_1}} = \frac{65}{134} \quad \therefore \quad \beta_1 = \mathbf{25.88°}. \quad \textbf{(Ans.)}$$

$$\tan \beta_2 = \frac{C_{f2}}{C_{bl_2} - C_{w2}} = \frac{65}{259.7 - 164.1}$$

$= 0.6799 \quad \therefore \quad \beta_2 = \mathbf{34.2°}.$ **(Ans.)**

(*ii*) The absolute angle at the tip of the impeller, α_2 :

$$\tan \alpha_2 = \frac{C_{f2}}{C_{w2}} = \frac{65}{164.1} = 0.3961$$

$\therefore \quad \alpha_2 = \mathbf{21.6°}.$ **(Ans.)**

(*iii*) The breadth of blade at inlet and outlet, b_1, b_2 :

Discharge at the inlet, $V_1 = 2\pi r_1 b_1 C_{f1}$

$4.8 = 2\pi \times 0.16 \times b_1 \times 65$

$$\therefore \quad b_1 = \frac{4.8}{2\pi \times 0.16 \times 65}$$

$= \mathbf{0.0734 \text{ m}}$ or **7.34 cm.** **(Ans.)**

Let V_2 be the discharge at the outlet, then

$$V_2 = \frac{p_1 V_1}{T_1} \times \frac{T_2}{p_2} \qquad \left[\because \frac{p_1 V_1}{T_1} = \frac{p_2 V_2}{T_2} \quad \text{Taking } T_2' = T_2\right]$$

$$= \frac{1 \times 10^5 \times 4.8}{293} \times \frac{335.4}{1.5 \times 10^5} = 3.66 \text{ m}^3\text{/s}$$

$\therefore \quad V_2 = 3.66 = 2\pi r_2 \,.\, b_2 \,.\, C_{f2} = 2\pi \times 0.31 \times b_2 \times 65$

or $$b_2 = \frac{3.66}{2\pi \times 0.31 \times 65}$$

$= \mathbf{0.0289 \text{ m}}$ or **2.89 cm.** **(Ans.)**

Example 24.41. *A centrifugal compressor delivers 16.5 kg/s of air with a total head pressure ratio of 4 : 1. The speed of the compressor is 15000 r.p.m. Inlet total head temperature is 20°C, slip factor 0.9, power input factor 1.04 and 80% isentropic efficiency. Calculate :*

(*i*) *Overall diameter of the impeller.*
(*ii*) *Power input.*

Solution. *Given* : $\dot{m} = 16.5$ kg/s ; Pressure ratio, $r_{0p} = 4$;
$N = 15000$ r.p.m. ;
$T_{01} = 20 + 273 = 293$ K
$\phi_s = 0.9$; $\phi_w = 1.04$; $\eta_{isen} = 80\%$.

(*i*) The overall diameter of the impeller, D :

$$\frac{T_{02}'}{T_{01}} = \left(\frac{p_{02}}{p_{01}}\right)^{\frac{\gamma-1}{\gamma}} = (4)^{\frac{1.4-1}{1.4}} = 1.486$$

$\therefore \quad T_{02}' = T_{01} \times 1.486 = 293 \times 1.486 = 435.4$ K

$$\eta_{isen} = \frac{T_{02}' - T_{01}}{T_{02} - T_{01}}$$

or $$0.8 = \frac{435.4 - 293}{T_{02} - T_{01}}$$

or $$T_{02} - T_{01} = \frac{435.4 - 293}{0.8} = 178 \text{ K}$$

$$\therefore \text{ Work done} = \frac{\phi_w \phi_s C_{bl_2}^2}{1000} = c_p(T_{02} - T_{01})$$

or $$C_{bl_2} = \left[\frac{1000 \times c_p(T_{02} - T_{01})}{\phi_w \phi_s}\right]^{1/2}$$

$$= \left[\frac{1000 \times 1.005 \times 178}{1.04 \times 0.9}\right]^{1/2} = 437.2 \text{ m/s}$$

Also, $$C_{bl_2} = \frac{\pi D N}{60} = 437.2$$

$$\therefore \quad D_2 = \frac{437.2 \times 60}{\pi \times 15000}$$

$= \mathbf{0.5567 \text{ m}}$ or **55.67 cm.** **(Ans.)**

(*ii*) Power input, P :

$P = \dot{m} c_p (T_{02} - T_{01})$
$= 16.5 \times 1.005 \times 178 = \mathbf{2951.7 \text{ kW}}.$ **(Ans.)**

☞**Example 24.42.** *The following data pertain to a centrifugal compressor :*

Total pressure ratio	*= 3.6 : 1*
Diameter of inlet eye of compressor impeller	*= 35 cm*
Axial velocity at inlet	*= 140 m/s*
Mass flow	*= 12 kg/s*
The velocity in the delivery duct	*= 120 m/s*
The tip speed of impeller	*= 460 m/s*
Speed of impeller	*= 16000 r.p.m.*
Total head isentropic efficiency	*= 80%*
Pressure co-efficient	*= 0.73*

Ambient conditions = 1.013 bar and 15°C.

Calculate :

(*i*) *The static pressure and temperature at inlet and outlet of compressor.*

(*ii*) *The static pressure ratio.*

(*iii*) *Work of compressor per kg of air.*

(*iv*) *The theoretical power required.*

Solution. *Given :* r_p (pressure ratio) = 3.6

$T_{01} = 15 + 273 = 288$ K ; $\eta_{isen} = 0.8$

$\dot{m} = 12$ kg/s ; ϕ_p (Pressure co-efficient) = 0.73 ;

$C_{bl_2} = 120$ m/s

(*i*) The static pressure and temperature at inlet and outlet of compressor :

Total head temperature rise

$$\Delta T_0 = \frac{T_{01}\left(r_p^{\frac{\gamma-1}{\gamma}} - 1\right)}{\eta_{isen}} = \frac{288\left[(3.6)^{\frac{1.4-1}{1.4}} - 1\right]}{0.8}$$

$= 159.1°C$

$\therefore$ $\mathbf{T_{02}} = T_{01} + \Delta T_0 = 288 + 159.1 = \mathbf{447.1\ K.}$ **(Ans.)**

The static temperature at exist is

$$\mathbf{T_2} = T_{02} - \frac{C_{bl_2}^2}{2c_p} = 447.1 - \frac{120^2}{2 \times 1.005 \times 10^3}$$

$= 447.1 - 7.1 = \mathbf{440\ K.}$ **(Ans.)**

The *static pressure at exit is*

$$p_2 = p_{02} - \frac{\rho_2 C_{bl_2}^2}{2} \qquad \left(\text{where } \rho_2 = \frac{p_2}{RT_2}\right)$$

i.e.,
$$p_2 = p_{02} - \frac{p_2 \times 120^2}{2 \times 0.287 \times 440 \times 10^3}$$

$= p_{02} - 0.057\, p_2$

But $p_{02} = 1.013 \times 3.6 = 3.65$ bar

$\therefore$ $p_2 = 3.65 - 0.057\, p_2$ or $1.057\, p_2 = 3.65$

$\therefore$ $\mathbf{p_2 = 3.45\ bar.}$ **(Ans.)**

To calculate *static conditions at inlet :*

$$T_{01} = T_1 + \frac{140^2}{2c_p} = T_1 + \frac{140^2}{2 \times 1.005 \times 10^3} = T_1 + 9.75$$

$\therefore$ $\mathbf{T_1} = 288 - 9.75 = \mathbf{278.25\ K.}$ **(Ans.)**

$$p_1 = p_{01} - \frac{p_1 \times 140^2}{2 \times 0.287 \times 278.25 \times 10^3}$$

$= 1.013 - 0.123\, p_1$

$\therefore$ $\mathbf{p_1 = 0.9\ bar.}$ **(Ans.)**

(*ii*) Static pressure ratio

$$= \frac{p_2}{p_1} = \frac{3.45}{0.9} = \mathbf{3.83.} \quad \textbf{(Ans.)}$$

(*iii*) Work done on air

$= c_p \times \Delta T_0$

$= 1.005 \times 159.1 = $ **159.89 kJ/kg of air. (Ans.)**

(*iv*) Theoretical power required to drive compressor

$= \dot{m}\, c_p\, \Delta T_0 = 12 \times 1.005 \times 159.1$

$=$ **1918.7 kW. (Ans.)**

Example 24.43. *Air at a temperature of 300 K flows in a centrifugal compressor running at 18000 r.p.m. The other data given is as follows :*

Isentropic total head efficiency	*= 0.76*
Outer diameter of blade tip	*= 550 mm*
Slip factor	*= 0.82*

Calculate :

(*i*) *The temperature rise of air passing through the compressor.*

(*ii*) *The static pressure ratio.*

Assume that the absolute velocities of air at inlet and exit of the compressor are same. Take c_p = 1.005 kJ/kg K.

Solution. $C_{bl_2} = \dfrac{\pi DN}{60} = \dfrac{\pi \times (550/1000) \times 18000}{60}$

$= 518.36$ m/s.

Work done per kg of air $= (C_{w_2} . C_{bl_2} - C_{w_1} . C_{bl_1})$

But $\qquad C_{w_1} = 0$ and $\phi_s = \dfrac{C_{w_2}}{C_{bl_2}}$

$\therefore$ Work done per kg of air

$= C_{bl_2}^2 \times \phi_s = (518.36)^2 \times 0.82 = 220331$ W

$= 220.331$ kW ...(*i*)

(*i*) Temperature rise of air, $\mathbf{T_1 - T_2}$:

Also work done $= c_p(T_2 - T_1)$ ...(*ii*)

Equating (*i*) and (*ii*), we get

$$\mathbf{T_2 - T_1} = \frac{220.331}{c_p} = \frac{220.331}{1.005} = \mathbf{219.23°C.} \quad \textbf{(Ans.)}$$

$$\eta_{isen} = \frac{T_2' - T_1}{T_2 - T_1}$$

i.e.,
$$0.76 = \frac{T_2' - 300}{219.23} \quad \text{or} \quad T_2' = 466.6 \text{ K.}$$

(*ii*) **The static pressure ratio :**

The static pressure ratio is given by

$$\frac{p_2}{p_1} = \left(\frac{T_2'}{T_1}\right)^{\frac{\gamma}{\gamma-1}} = \left(\frac{466.6}{300}\right)^{\frac{1.4}{1.4-1}} = \mathbf{4.69.} \quad \textbf{(Ans.)}$$

24.4.3.3. Axial Flow Compressor

24.4.3.3.1. Construction and Working

In an axial flow compressor, the *flow proceeds* throughout the compressor in a direction essentially *parallel to the axis of the machine.*

Construction. Refer to Fig. 24.61.

- An axial flow compressor consists of adjacent rows of rotor (moving) blades and stator (fixed) blades. The rotor blades are mounted on the rotating drum and stator blades are fixed to the casing stator. *One stage* of the machine comprises a row of rotor blades followed by a row of stator blades.
- For efficient operation the blades are of air foil section based on aerodynamic theory. The blades are so designed that wasteful losses due to shock and turbulence are prevented and the blades are free from *stalling troubles.* (The blades are said to *be stalled when the air stream fails to follow the blade contour*). Whereas the compressor blades have aerofoil section, the turbine blades have profiles formed by a number of circular arcs. This is so because the acceleration process being carried out in the converging blade passages of a reaction turbine is much more efficient and stable process as compared with the diffusing or decelerating process being carried out in the diverging passage between the blades of an axial flow compressor.

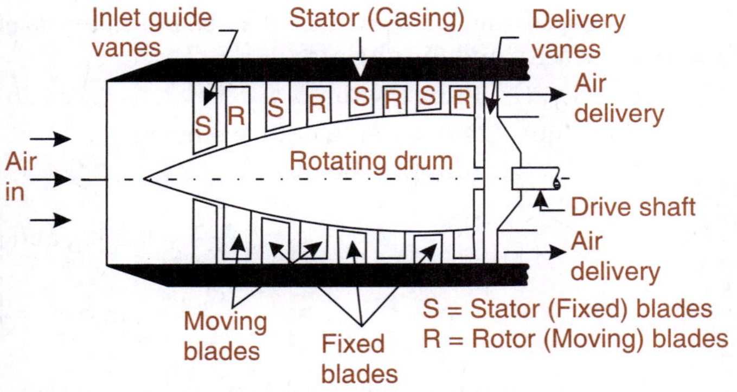

Fig. 24.61. Axial flow compressor.

- The annular area is usually reduced from inlet to outlet of the compressor. This is to *keep the flow velocity constant throughout the compressor length*. In the diverging passages of the moving blades, there is rise in temperature due to diffusion. The absolute velocity is also increased due to work input.

The "*fixed blades*" serve the following two *purposes* :

(*i*) Convert a *part of the K.E. of the fluid into pressure energy. This conversion is achieved by diffusion process carried out in the diverge blade passages.*

(*ii*) *Guide and redirect the fluid flow so that entry to the next stage is without shock.*

Working

Basically, the compression is performed in a similar manner to that of the centrifugal type. The work input to the rotor shaft is transferred by the moving blades to the air, thus accelerating it. The blades are so arranged that the *spaces between the blades form diffuser passages, and hence the velocity of the air relative to the blades is decreased as the air passes through them, and there is a rise in pressure. The air is then further diffused in the stator blades, which are also arranged to form diffuser passages.* In the fixed stator blades the *air is turned through an angle so that its direction is such that it can be allowed to pass to a second row of moving rotor blades. It is usual to have a relatively large number of stages and to maintain a constant work input per stage* (*e.g.*, from 5 to 14 stages have been used).

— The necessary reduction in volume may be allowed by flaring the stator or by flaring the rotor. It is more common to use a flared rotor, and this type is shown diagrammatically in Fig. 24.61.

— It is usually arranged to have an *equal temperature rise in the moving and the fixed blades, and to keep the axial velocity of air constant throughout the compressor.* Thus, *each stage of the compression is exactly similar with regard to air velocity and blade inlet and outlet angles.*

— A diffusing flow is less stable than a converging flow, and for this reason the blade shape and profile is much more important for a compressor than for a reaction turbine. The *design of compressor blades is based on aerodynamic theory and an aerofoil shape* is used.

Note. Two forms of rotors have been used namely the *drum* and *disc types*. The **disc type** is used where consideration of *low weight* is more important than cost as in *aircraft applications*. The drum type is more suitable for *static industrial applications*. In some applications, combination of both types has been used.

Materials. The following materials are used for the various components of an axial flow compressor :

1. ***Rotor bladings.*** The materials listed below are in the *increasing* order of weight and their ability to withstand high temperature :

(*i*) Fibrous composites (*ii*) Aluminium
(*iii*) Titanium (*iv*) Steel
(*v*) Nickel alloy.

2. *Rotor :*

- For *rotor shafts* and disc .... *"steel"*.
- *Aircraft engines may use titanium at the front stages and "nickel alloy" in the rest.*

3. *Stator bladings :*

- Same materials as that of rotor but *steel is the most common.*

4. *Castings.* These may be of *cast magnesium, aluminium, steel or iron or fabricated from titanium or steel.*

— NC (Numerically controlled) machines make dies and the blades are manufactured by precision forging. Blades are also machined by CNC copying machines.

24.4.3.3.2. Velocity Diagrams and Work Done of a stage of Axial Flow Compressors.

Fig. 24.62 shows the velocity triangles for one stage of an axial flow compressor. All angles are measured from the axial direction and the blade velocity C_{bl} is taken to be *same* at blade entry and exist. This is because the air enters and leaves the blades at *almost equal radii.*

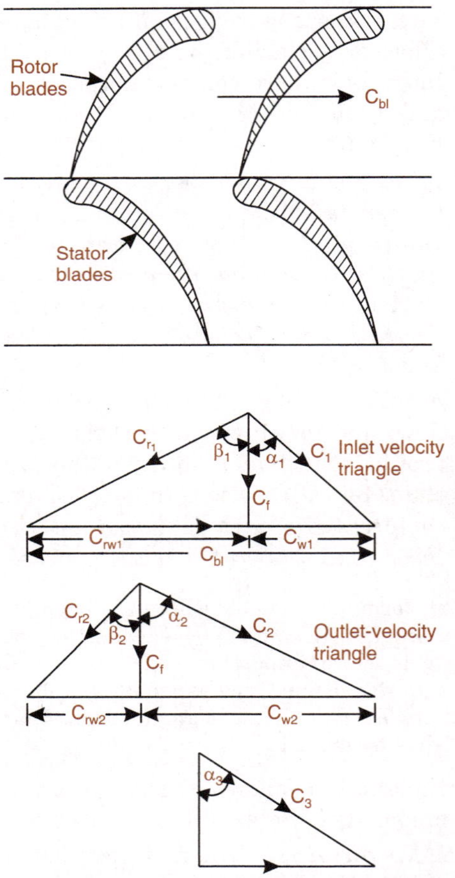

Fig. 24.62. Velocity diagrams for axial flow compressor.

- Air approaches the rotor blade with absolute velocity C_1 and at an angle α_1. The relative velocity C_{r1}, obtained by the vectorial addition of absolute velocity C_1 and blade velocity C_{bl}, has the inclination β_1 with the axial direction.
- Due to diffusion in the diverging passages formed by rotor blades, there is some pressure rise. This is at the expense of relative velocity and so the relative velocity decreases from C_{r1} to C_{r2}. Since work is being done on the air by rotor blades, the air would ultimately leave the rotor with increased absolute velocity C_2.
- The air then enters the stator blades and the diffusion and deceleration takes place in the diverging passage of stator blades. Finally the air leaves the stator blades with velocity C_3 at an angle α_3 and is redirected to the *next stage*. Generally it is *assumed that absolute velocity C_3 leaving the compressor stage equals the approach velocity C_1.*

From the velocity triangles, we have

$$\frac{C_{bl}}{C_f} = \tan \alpha_1 + \tan \beta_1 \quad ...(24.85)$$

and,

$$\frac{C_{bl}}{C_f} = \tan \beta_2 + \tan \alpha_2 \quad ...(24.86)$$

Assume *1 kg of flow of air* through the *compressor stage*.

From Newton's second law of motion,

Tangential force per kg = $C_{w2} - C_{w1}$

Work absorbed by the stage per kg of air,

$$W_{st} = C_{bl}(C_{w2} - C_{w1}) = c_p(T_{02} - T_{01})_{act} \quad ...(24.87)$$

where C_{w1} and C_{w2} are the whirl components of absolute velocities at inlet and exit of rotating blades.

(It may be noted that here whirl component at the entrance of the compressor is not *zero* because of the fact that air flows *axially and not radially.*)

The expression for work done may be put in terms of flow/axial velocity and air angles.

$$W_{st} = C_{bl}\, C_f \quad (\tan \alpha_2 - \tan \alpha_1) \quad ...(24.88)$$

From eqns. (24.85) and (24.86), we have

$$W_{st} = C_{bl}\, C_f \quad (\tan \beta_1 - \tan \beta_2) \quad ...(24.89)$$

$$\left[\because \frac{C_{w1}}{C_f} = \tan \alpha_1, \text{ and } \quad \frac{C_{w2}}{C_f} = \tan \alpha_2 \right]$$

The work input to the axial flow compressor may also be obtained by the Euler's equation. For each kg of air delivered, we have

$$W_{st} = C_{bl_2}\, C_{w2} - C_{bl_1}\, C_{w1}$$

Since $C_{w2} - C_{w1} = [(C_{bl} - C_{rw2}) - (C_{bl} - C_{rw1})]$

$$= C_{rw1} - C_{rw2}$$

Further $C_{bl_1} = C_{bl_2} = C_{bl}$, the above equation is modified as,

$$W_{st} = C_{bl}(C_{rw1} - C_{rw2})$$
$$= C_{bl}\, \Delta C_{rw} = c_p\, (\Delta T_0)_{act}$$

By use of velocity triangle and cosine theorem,

$$W_{st} = \underset{\text{(First term)}}{\frac{C_{r1}^2 - C_{r2}^2}{2}} + \underset{\text{(Second term)}}{\frac{C_2^2 - C_3^2}{2}} \quad ...(24.90)$$

Here $\quad C_3 = C_1$

— The *first term* on the right side of the above equation introduces the part of the work supplied by a rotating cascade, which is converted into pressure due to *diffusion action* in rotating cascade itself.

— The second term represents the increment of K.E. in rotating cascade that has to be converted into pressure energy in stationary cascade. Comparing this equation to the work input to centrifugal compressor, we find that the term centrifugal action $\left[\frac{C_{bl_2}^2 - C_{bl_1}^2}{2}\right]$ is missing in axial flow compressors. Due to this reason the *pressure ratio per stage in axial flow compressor is much less than that of centrifugal compressor.*

The stage temperature rise, regardless of efficiency of compression, will be given by the equation

$$(\Delta T)_{act} = \frac{C_{bl}C_f}{c_p}(\tan \beta_1 - \tan \beta_2) \quad ...(24.91)$$

Pressure rise in isentropic flow through a cascade :

Consider the incompressible isentropic and steady flow through a cascade from uniform condition 1 to uniform condition 2. From Bernoulli's equation, we have

$$\frac{p_1}{\rho} + \frac{C_1^2}{2} = \frac{p_2}{\rho} + \frac{C_2^2}{2}$$

Thus the static isentropic pressure rise may be expressed in terms of the inlet dynamic head.

$$(\Delta p)_{isen} = (p_2)_{isen} - \rho\left(\frac{C_1^2 - C_2^2}{2}\right)$$

$$= (p_2)_{isen} - \frac{\rho}{2}[(C_{f1}^2 + C_{w1}^2) - (C_{f2}^2 + C_{w2}^2)]$$

Since $\quad C_{f1} = C_{f2} = C_f$

$$\therefore \quad (\Delta p)_{isen} = \frac{\rho}{2}(C_{w1}^2 - C_{w2}^2)$$

$$= \frac{\rho}{2}\, C_f^2 (\tan^2 \alpha_1 - \tan^2 \alpha_2) \quad ...(24.92)$$

$[\because \; C_{w1} = C_f \tan \alpha_1, \text{ and } C_{w2} = C_f \tan \alpha_2]$

24.4.3.3.3. Degree of Reaction

Degree of reaction (R_d) is defined as *the ratio of pressure rise in the compressor stage.*

i.e., $\quad R_d = \dfrac{\text{Pressure rise in the rotor blades}}{\text{Pressure rise in the stage}}$

Pressure rise in the compressor stage equals work input per stage and is

$$= C_{bl}\,(C_{w_2} - C_{w_1})$$

Pressure rise in the rotor blades is at the expense of K.E. and is

$$= \frac{C_{r_1}^2 - C_{r_2}^2}{2}$$

$$\therefore R_d = \frac{C_{r_1}^2 - C_{r_2}^2}{2C_{bl}\,(C_{w_2} - C_{w_1})} \quad ...(24.93)$$

Refer inlet and outlet velocity triangles :

$$C_{w_2} = C_{bl} - C_f \tan \beta_2$$
$$C_{w_1} = C_{bl} - C_f \tan \beta_1$$
$$\therefore \quad C_{w_2} - C_{w_1} = C_f (\tan \beta_1 - \tan \beta_2) \quad ...(24.94)$$

Similarly from velocity triangles,

$$C_{r_1}^2 = (C_f)^2 + (C_f \tan \beta_1)^2$$
$$C_{r_2}^2 = (C_f)^2 + (C_f \tan \beta_2)^2$$
$$\therefore C_{r_1}^2 - C_{r_2}^2 = C_f^2\,(\tan^2 \beta_1 - \tan^2 \beta_2) \quad ...(24.95)$$

So $\quad R_d = \dfrac{C_f^2(\tan^2 \beta_1 - \tan^2 \beta_2)}{2C_{bl}.C_f\,(\tan \beta_1 - \tan \beta_2)}$

$$= \frac{1}{2}\frac{C_f}{C_{bl}}(\tan \beta_1 + \tan \beta_2) \quad ...(24.96)$$

Degree of reaction is usually kept as 0.5,

$$\therefore \quad 0.5 = \frac{1}{2} \cdot \frac{C_f}{C_{bl}}(\tan \beta_1 + \tan \beta_2)$$

or $\quad \dfrac{C_{bl}}{C_f} = \tan \beta_1 + \tan \beta_2$

But $\quad \dfrac{C_{bl}}{C_f} = \tan \alpha_1 + \tan \beta_1 = \tan \alpha_2 + \tan \beta_2$

(from velocity triangles)

$$\therefore \quad \frac{C_{bl}}{C_f} = \tan \beta_1 + \tan \beta_2$$
$$= \tan \alpha_1 + \tan \beta_1 = \tan \alpha_2 + \tan \beta_2$$

From this $\alpha_1 = \beta_2$; $\alpha_2 = \beta_1$

So with 50% reaction blading, the compressors have symmetrical blades and with this type of set-up losses in flow path are amply reduced.

In *symmetrical blades, the tip clearance and fluid friction losses are minimum.*

24.4.3.3.4. Polytropic Efficiency

The work input to a compressor, with usual notations,

$$W = c_p(T_{02} - T_{01})$$
$$= c_p(T_{02}' - T_{01}) \times \left(\frac{T_{02} - T_{01}}{T_{02}' - T_{01}}\right)$$
$$= c_p\left(\frac{T_{02} - T_{01}}{T_{02}' - T_{01}}\right) \times (T_{02'} - T_{01})$$
$$= c_p\, \eta_{isen}\, T_{01}\left(\frac{T_{02}'}{T_{01}} - 1\right)$$
$$= c_p\, \eta_{isen}\, T_{01}\left[\left(\frac{p_{02}}{p_{01}}\right) - 1\right] \quad ...(24.97)$$

Eqn. (24.97) indicates that for the same isentropic efficiency η_{isen} and pressure ratio $\frac{p_{02}}{p_{01}}$, the *work input is proportional to the initial temperature*. Thus in a compressor consisting of several stages of equal isentropic efficiency, each succeeding stage will have to perform more work because it has to deal with a fluid of increased temperature delivered to it from the preceding stage. Thus the overall isentropic efficiency which is a useful measure of the overall performance of the machine cannot be used to compare two compressors having different pressure ratios. *Overall efficiency will be less for a compressor operating at a higher pressure ratio.*

For comparing the performance of compressors with different stages, the concept of polytropic efficiency is introduced.

Polytropic efficiency *is the isentropic efficiency of one stage of a multi-stage compressor.* This small stage efficiency is *constant for all stages of a compressor with infinite number of stages.*

Let us consider the compression process of a multi-stage compressor on *T-s* plot of Fig. 24.63.

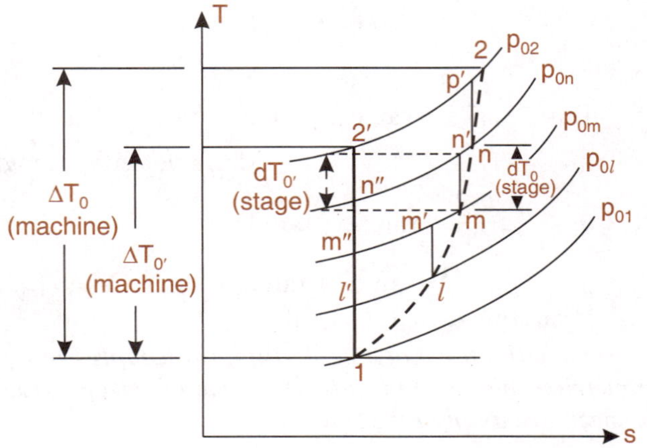

Fig. 24.63. Concept of polytropic efficiency.

The gas is being compressed from pressure p_{01} to p_{02} in *four stages of equal pressure ratio.* p_{ol}, p_{om} and p_{on} are the intermediate pressures.

Now by definition :

Overall isentropic efficiency (stagnation) for the *machine (m/c),*

$$\eta_{isen}(m/c) = \frac{\Delta T_0'}{\Delta T_0}$$

Stagnation isentropic efficiency for a *stage, (st)*

$$\eta_{isen(st)} = \frac{dT_0'}{dT_0}$$

The total actual temperature rise ΔT_0 can be represented as :

$$\Delta T_0 = \frac{(\Delta T_o')_{machine}}{\eta_{isen\,(m/c)}} \quad ...(i)$$

or

$$\Delta T_0 = \frac{\Sigma\,(dT_0')_{stage}}{\eta_{isen\,(st)}} \quad ...(ii)$$

Equating expressions (*i*) and (*ii*), we get

$$\frac{(\Delta T_0')_{machine}}{\eta_{isen\,(m/c)}} = \frac{\Sigma\,(dT_0')_{stage}}{\eta_{isen\,(st)}}$$

or

$$\frac{\eta_{isen\,(m/c)}}{\eta_{isen\,(st)}} = \frac{(\Delta T_0')_{machine}}{\Sigma\,(dT_0')_{stage}} \quad ...(24.98)$$

From Fig. 24.63, we have :

$$(\Delta T_0')_{machine} = (1\text{-}l') + (l'\text{-}m'') + (m''\text{-}n'') + (n''\text{-}2')$$
$$\Sigma(dT_0')_{stage} = (1\text{-}l') + (l\text{-}m') + (m\text{-}n') + (n\text{-}p')$$

On *T-s* plot, the constant pressure lines '*diverge*' *towards right*

i.e., $l\text{-}m' > l'\text{-}m''$

and $m\text{-}n' > m''\text{-}n''$

So we can say that $\Sigma(dT_0')_{stage} > (\Delta T_0')_{machine}$

Thus from eqn. (24.98), $\eta_{isen(st)} > \eta_{isen\,(m/c)}$

The *small stage efficiency* $\eta_{isen(st)}$ *which is constant for all stages is called polytropic efficiency and is designated by* η_p.

"Polytropic efficiency" in terms of entry and delivery pressure and temperatures and the ratio of specific heats :

Refer to Fig. 24.63. The actual compression path 1-2 is *irreversible, but the end states 1 and 2 are in equilibrium and lie on the same polytropic path characterised by :*

$$p_0 v_0^{\,n} = \text{constant } z_1$$

Let the corresponding reversible isentropic path 1-2′ with end states 1 and 2′ characterised by :

$$p_0 v^{\gamma} = \text{constant } z_2$$

For irreversible path, we may write

$$p_0 = z_1 \rho_0^{\,n}$$

On differentiation, we get

$$dp_0 = z_1 n \rho_0^{\,n-1}.d\rho_0 = n\,\frac{p_0}{\rho_0}\,d\rho_0 \quad ...(24.99)$$

Now, the characteristic gas equation may be written as :

$$\frac{p_0}{\rho_0} = RT_0 \quad \text{or} \quad \rho_0 = \frac{p_0}{RT_0}$$

On differentiation, we get :

$$d\rho_0 = \frac{1}{R}\left[\frac{T_0.dp_0 - p_0.dT_0}{T_0^2}\right] \quad ...(24.100)$$

From eqns. (24.99) and (24.100), we have

$$dp_0 = n\frac{p_0}{\rho_0}\frac{1}{R}\left[\frac{T_0.dp_0 - p_0.dT_0}{T_0^2}\right]$$

$$= n\,.\,RT_0.\frac{1}{R}\left[\frac{T_0.dp_0 - p_0.dT_0}{T_0^2}\right]$$

$$= n\left[\frac{T_0.dp_0 - p_0.dT_0}{T_0}\right]$$

$$dp_0 = n.dp_0 - \frac{n.p_0.dT_0}{T_0}$$

or $$\frac{n.p_0.dT_0}{T_0} = n.dp_0 - dp_0 = dp_0(n-1)$$

or Actual stage temperature,

$$dT_0 = dp_0\left(\frac{n-1}{n}\right)\frac{T_0}{p_0} \quad ...(24.101)$$

A similar treatment to the ideal compression process following the law $p_0 v_0^{\gamma} = z_2$ would give the stage isentropic temperature dT_0' expressed as

$$dT_0' = dp_0\left(\frac{\gamma-1}{\gamma}\right)\frac{T_0}{p_0} \quad ...(24.102)$$

Thus, by the definition of polytropic or small stage efficiency, we get :

$$\eta_p = \frac{dT_0'}{dT_0} = \frac{dp_0\left(\frac{\gamma-1}{\gamma}\right)\frac{T_0}{p_0}}{dp_0\left(\frac{n-1}{n}\right)\frac{T_0}{p_0}}$$

or $$\eta_p = \left(\frac{\gamma-1}{\gamma}\right)\left(\frac{n}{n-1}\right) \quad ...(24.103)$$

Eqn. (24.103) gives the value of polytropic efficiency in terms of exponent n and the adiabatic exponent γ.

Substituting the value of $\frac{n}{n-1}$ from eqn. (24.101) into eqn. (24.103) we have :

$$\eta_p = \frac{\gamma-1}{\gamma}.\frac{dp_0}{p_0}.\frac{T_0}{dT_0}$$

or $$\eta_p\frac{dT_0}{T_0} = \frac{\gamma-1}{\gamma}.\frac{dp_0}{p_0}$$

Integrating between the two end states 1 and 2, we get

$$\eta_p \ln\left(\frac{T_{02}}{T_{01}}\right) = \frac{\gamma-1}{\gamma}\ln\left(\frac{p_{02}}{p_{01}}\right)$$

or $$\eta_p = \frac{\frac{\gamma-1}{\gamma}\ln\left(\frac{p_{02}}{p_{01}}\right)}{\ln\left(\frac{T_{02}}{T_{01}}\right)} = \frac{\ln\left(\frac{p_{02}}{p_{01}}\right)^{\frac{\gamma-1}{\gamma}}}{\ln\left(\frac{T_{02}}{T_{01}}\right)} \quad ...(24.104)$$

The eqn. (24.104) gives the polytropic efficiency in terms of *pressure ratio* $\frac{p_{02}}{p_{01}}$, *temperature ratio* $\frac{T_{02}}{T_{01}}$ and the *ratio of specific heat* γ.

24.4.3.3.5. Flow Coefficient, Head or Work Coefficient, Deflection coefficient and Pressure Coefficient

1. Flow coefficient (ϕ_f). The flow coefficient of axial flow compressor stage is defined as,

$$\phi_f = \frac{C_{f1}}{C_{bl}}$$

Since $C_{bl} = C_{rw1} + C_{w1} = C_{f1}(\tan\beta_1 + \tan\alpha_1)$

$$\phi_f = \frac{C_{f1}}{C_{f1}(\tan\beta_1 + \tan\alpha_1)}$$

$$= \frac{1}{\tan\beta_1 + \tan\alpha_1} \quad ...(24.105)$$

Also, $C_{f1} = C_{f2} = C_f$

$$\therefore \quad \phi_f = \frac{C_{f2}}{C_{f2}(\tan\beta_2 + \tan\alpha_2)}$$

$$= \frac{1}{\tan\beta_2 + \tan\alpha_2} \quad ...(24.106)$$

2. Head or work coefficient (ϕ_h). *It is defined as the ratio of actual work done to the kinetic energy corresponding to the mean peripheral velocity.*

Thus,

$$\phi_h = \frac{c_p\,\Delta T}{C_{bl}^2/2} = \frac{2(C_{w2} - C_{w1})}{C_{bl}} = 2\left(\frac{\tan\alpha_2 - \tan\alpha_1}{\tan\beta_2 + \tan\alpha_2}\right) \quad ...(24.107)$$

3. Deflection co-efficient (ϕ_{def}). It is defined as,

$$\phi_{def} = \frac{C_{bl}(C_{w2} - C_{w1})}{C_{bl}^2} = \frac{C_{w2} - C_{w1}}{C_{bl}}$$

or $$\phi_h = 2\,\phi_{def} \quad ...(24.108)$$

4. Pressure co-efficient (ϕ_p). *It is defined as the ratio of isentropic work done to kinetic energy corresponding to the peripheral velocity.*

Thus, $$\phi_p = \frac{c_p\,\Delta T_{isen}}{C_{bl}^2/2} = \eta_{isen}\,\phi_h \quad ...(24.109)$$

24.4.3.3.6. Pressure Increase in a Stage of an Axial Flow Compressor and Number of Stages

The pressure ratio is expressed as

$$\frac{p_2}{p_1} = \left[1 + \eta_{isen} \frac{T_2 - T_1}{T_1}\right]^{\gamma/(\gamma-1)} \qquad ...(24.110)$$

Let T_1 and T_2 denote the temperature of the working fluid at inlet and outlet of *rotating blades.* Hence the temperature increase is

$$\Delta T_R = T_2 - T_1 = \frac{C_{r1}^2 - C_{r2}^2}{2 \times c_p}$$

$$\therefore \quad \frac{p_2}{p_1} = \left[1 + \eta_R \frac{\Delta T_R}{T_1}\right]^{\gamma/(\gamma-1)} \qquad ...(24.111)$$

The temperature rise in *stationary blades* is given by,

$$\Delta T_s = T_3 - T_2 = \frac{C_2^2 - C_3^2}{2 \times c_p}$$

$$\therefore \quad \frac{p_3}{p_2} = \left[1 + \eta_s \frac{\Delta T_s}{T_2}\right]^{\gamma/(\gamma-1)}$$

Hence pressure rise in the stationary blades is

$$\Delta p_s = p_3 - p_2 = p_2 \left[\left\{1 + \eta_s \frac{\Delta T_s}{T_2}\right\}^{\gamma/\gamma - 1} - 1\right]$$

The pressure increase in a stage is

$$\Delta p_{st} = \Delta p_R + \Delta p_s$$

and $$\Delta T_{st} = \Delta T_R + \Delta T_s$$

The stagnation pressure ratio is given by

$$\frac{p_{02}}{p_{01}} = \left[1 + \eta_{0\,isen} \frac{\Delta T_0}{T_{01}}\right]^{\gamma/(\gamma-1)}$$

If the work done per stage is assumed to be the same, then the number of stages (N) is given by,

$$N = \frac{\Delta T_0'}{(\Delta T_0')_{stage}} \qquad ...(24.112)$$

If the pressure ratio per stage be the same, then

$$(r_p)_{stage} = \frac{p_{02}}{p_{01}} = \frac{p_{03}}{p_{02}} = \frac{p_{0(N+1)}}{p_{0N}}$$

The overall pressure ratio is given by

$$r_p = \left[(r_p)_{stage}\right]^N \qquad ...(24.113)$$

or $$N = \frac{\ln (r_p)}{\ln\left[(r_p)_{stage}\right]} \qquad ...(24.114)$$

where $(r_p)_{stage}$ varies from 1.12 to 1.2.

24.4.3.3.7. Losses in Axial Flow Compressor Stage

In actual practice, various losses occur while the fluid flows through a compressor stage. The total pressure loss arises in three ways :

1. Profile losses on the surface of the blades.
2. Skin friction on the annulus walls.
3. Secondary flow losses.

The various losses represented on graph between stage efficiency and flow coefficient is shown in Fig. 24.64.

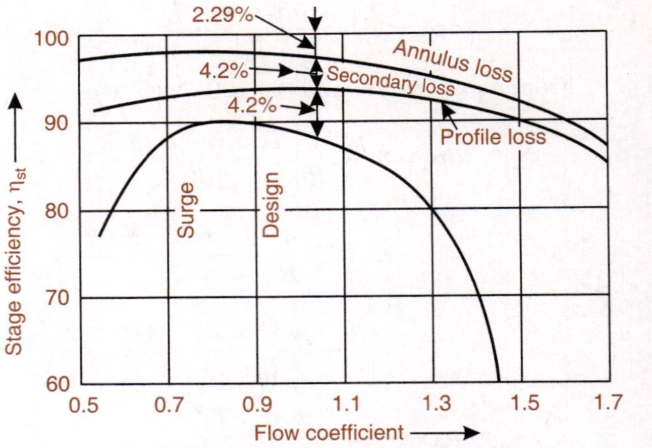

Fig. 24.64. Losses in compressor stage.

1. Profile losses on the surface of the blades :

- By profile losses, we mean the total pressure loss of two dimensional rectilinear cascade arising from the skin friction on the *surface* and due to the mixing of flow particles after the blades.
- These losses are usually determined experimentally.

2. Skin friction loss on the annulus walls :

- The wall friction total pressure losses arising from the skin friction on the annulus walls and the secondary losses are *difficult to analyse as boundary layer growth on these walls is a complex three-dimensional phenomena.*
- Empirical relations (by Howell, Haller) are available for calculating drag co-efficient.

3. Secondary flow losses :

- In an axial flow compressor blade channels, certain secondary flows are *produced by combined effects of curvature and boundary layer.*
- *Secondary flow is produced when a streamwise components of velocity is developed from the deflection of an initially sheared flow.* Such secondary flow occurs when a developed pipe flow enters a bend, when a sheared flow passes over an aerofoil of finite thickness or an aerofoil of finite lift or when a boundary level meets an obstacle normal to the surface over which it is flowing (*e.g.*, a wind blowing past a telegraph pole).
- One of most important engineering aspects of secondary flow occurs in *axial turbo-machinery aerodynamics where boundary layers growing on the casing and hub walls of the machines are deflected by rows of blades-stationary and rotating.*

24.4.3.3.8. Surging, Choking and Stalling—Compressor Characteristics

Surging. In axial flow and centrifugal compressors *"surging" is an unstable limit of operation.* ***Surging*** *is caused due to unsteady, periodic and reversal of flow through the compressor when the compressor has to operate at less mass flow rate than a predetermined value (a value corresponding to maximum pressure).* As the flow is drastically reduced than this predetermined value, this surge can reach such a magnitude as to endanger the compressor and in many cases *mechanical failures may result.* The alternating stresses to which the rotor of the machine is subjected during this irregular working condition, *may damage compressor bearings, rotor blading and scales.* Severe surge have been known to *bend the rotor shaft.*

Choking. When the *pressure ratio is unity (i.e., there is no compression), theoretically mass flow rate becomes maximum.* This generally occurs when the Mach number corresponding to relative velocity at inlet becomes *sonic. The maximum mass flow rate possible in compressor is known as* ***choking flow****. "Choking" means fixed mass flow rate regardless of pressure ratio (i.e., characteristic becomes vertical).*

Fig. 24.65, shows the *compressor characteristics.*

In the compressor where the flow is against the pressure gradient the incidence loss due to incorrect fluid angles relative to the blades becomes more important and pressure ratio (r_p) falls sharply at conditions away from the design point. This loss, added to the friction loss which will increase with increase of mass flow rate, gives a pressure ratio-mass flow rate relation as shown in Fig. 24.65.

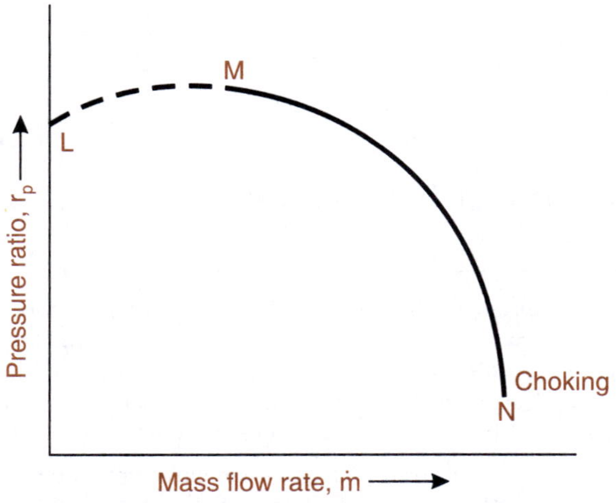

Fig. 24.65. The compressor characteristics.

- At point N, the compressor is *choked* and is *passing the maximum mass flow rate.*
- On the section MN of the curve the *flow is stable.* A fall in mass flow rate will result in a rise in pressure ratio which will tend to restore the fall.
- On the section LM of the curve, the flow is *not stable.* A fall in mass flow rate will be accompanied by a fall in pressure ratio. In this situation any small disturbance causing a check in mass flow will cause a fall in pressure ratio and the *flow may reverse at some point.* When the temporary disturbance is removed, the flow will pick up and it is found that small disturbances cause the flow to *oscillate rapidly.* The oscillations is noisy and can, if allowed to continue, cause structural damage in the compressor. It is called '*surge*' and the point M on the curve marks the limit of useful operation of the compressor. If a compressor is running normally at the point where surge usually commences it is possible to induce surge merely by passing the hand across the inlet. It is found that *compressor efficiency is highest at point adjacent to* M and it is therefore advisable to able to *operate as close to* M *as possible.*

Stalling. *"Stalling" of a stage of axial flow compressor is defined as the aerodynamic stall or the breakway of the flow from suction side of the blade aerofoil.* It may be due to *lesser flow rate* than designed value or due to *non-uniformity in the blade profile.* Thus stalling is *ahead phenomenon of surging.*

A multi-stage compressor may operate stable in the unsurged region with one or more of the stages stalled and rest of the stages unstalled. In other words, *stalling is a local phenomenon whereas surging is a complete system phenomena.*

24.4.3.3.9. Performance of Axial Flow Compressor

- Fig. 24.66 (*a*) shows the relationship between pressure ratio, power and efficiency *versus* flow rate for various values of speeds such as N_1, N_2, etc. At a *certain speed, efficiency increases as the flow rate increases and reaches a maximum value after which it decreases.* Accordingly as *the flow rate increases the power consumed also increases.*
- Fig. 24.66 (*b*) shows the performance and constant efficiency curves.

— Such a plot does not take into account the varying inlet temperature and pressure. In addition to this, these plots cannot show the comparison of performance for similar compressors of different sizes. To, account for all these, the *performance curves are plotted with 'dimensionless parameters'*. These dimensionless parameters are : Pressure ratio $\frac{p_2}{p_1}$; speed parameter, $\frac{N_1}{\sqrt{T_1}}$ and flow parameter $\frac{m\sqrt{T_1}}{p_1}$,

Refer to Fig. 24.67 (*a*) and (*b*).

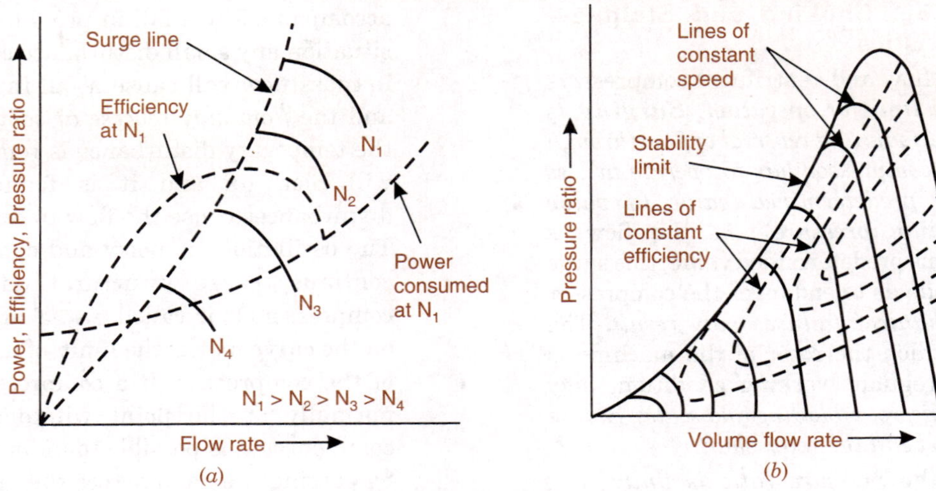

Fig. 24.66. Performance curves of axial flow compressor.

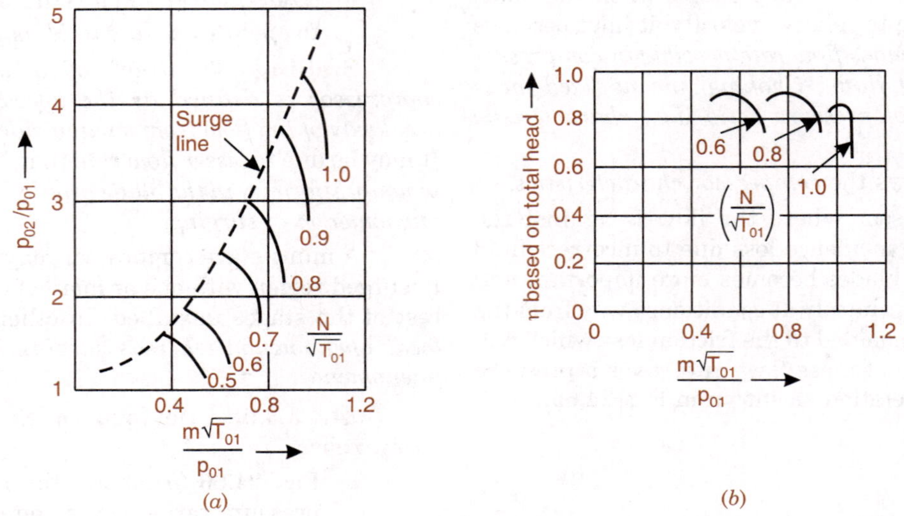

Fig. 24.67

24.5. COMPARISON BETWEEN RECIPROCATING AND CENTRIFUGAL COMPRESSORS

S. No.	*Aspects*	*Reciprocating compressors*	*Centrifugal compressors*
1.	*Vibration problems*	Greater vibration problems (due to the presence of reciprocating parts the machine is poorly balanced)	Less vibrational problems since the machine does not have reciprocating parts.
2.	*Mechanical efficiency*	Lower (due to the presence of several sliding or bearing members)	Higher comparatively (due to the absence of numerous sliding or bearing members)
3.	*Installed first-cost*	Higher	Lower (where pressure and volume conditions are favourable.)
4.	*Pressure ratio per stage*	About 5 to 8	About 3 to 4.5.

5.	*Capability to deliver pressure*	High pressure (By multistaging, high delivery pressure upto 5000 atm. may be achieved).	Medium pressure (By multi-staging, the delivery pressure upto 400 atm. may be achieved).
6.	*Capability of delivering volume of air/gas*	Small (By using multicylinders, the volume may be increased).	Greater (per unit of building space).
7.	*Flexibility in capacity and pressure range*	Greater	No flexibility in capacity and pressure range.
8.	*Maintenance expenses*	Higher	Lower
9.	*Continuity of service*	Lesser	Greater
10.	*Compression efficiency*	Higher, at compression ratio above 2.	Higher, at compression ratio less than 2.
11.	*Adaptability*	Adaptability to low speed drive	Adaptability to high speed, low maintenance cost drivers such as turbines
12.	*Operating attention*	More	Less
13.	*Mixing of working fluid with lubricating oil*	Always a chance	No chance
14.	*Suitability*	For low, medium and high pressures and low and medium gas volumes.	For low and medium pressures and large gas volumes.

24.6. COMPARISON BETWEEN RECIPROCATING AND ROTARY AIR COMPRESSORS

S. No.	*Aspects*	*Reciprocating air compressors*	*Rotary air compressors*
1.	*Suitability*	Suitable for low discharge of air at high pressure	Suitable for handling large volumes of air at low pressures.
2.	*Operational speed*	Low	Usually high
3.	*Air supply*	Pulsating	Continuous
4.	*Balancing*	Cyclic vibrations occur	Less vibrations
5.	*Lubricating system*	Generally complicated	Generally simple lubrication systems are required
6.	*Quality of air delivered*	Generally contaminated with oil	Air delivered is relatively more clean.
7.	*Air compressor size*	Large for the given discharge	Small for same discharge
8.	*Free air handled*	250–300 m^3/min	2000–3000 m^3/min
9.	*Delivery pressure*	High	Low
10.	*Usual standard of compression*	Isothermal compression	Isentropic compression

24.7. COMPARISON BETWEEN CENTRIFUGAL AND AXIAL FLOW COMPRESSORS

S. No.	*Aspects*	*Centrifugal compressors*	*Axial flow compressors*
1.	*Type of flow*	Axial (Parallel to the direction of axis of the machine)	Radial
2.	*Pressure ratio per stage*	High, about 4.5 : 1. Thus unit is *compact* — In supersonic compressors, the pressure ratio is about 10 but at the cost of efficiency. Operation is *not* so difficult and risky.	Low, about 1.2 : 1. This is due to absence of centrifugal action. To achieve the pressure ratio equal to that per stage in centrifugal compressor 10 to 20 stages are required. Thus, the unit is *less compact* and *less rugged.*

3.	*Isothermal efficiency*	About 80 to 82%	About 86 to 88% (with modern aerofoil blades)
4.	*Frontal area*	Larger	Smaller (This makes the axial flow compressors *more suitable for jet engines due to less drag*).
5.	*Flexibility of operation*	More (due to adjustable pre-whirl and diffuser vanes)	Less
6.	*Part load performance*	Better	Poor
7.	*Effect of deposit formation on the surface of impeller rotor*	*Performance not adversely affected*	*Performance adversely affected*
8.	*Starting torque required*	Low	High
9.	*Suitability for multistaging*	Slightly difficult	More suitable for multi-staging
10.	*Delivery pressure possible*	Upto 400 bar	upto 20 bar
11.	*Applications*	Used in blowing engines in steel mills, low pressure refrigeration, big central air conditioning plants, fertiliser and industry, supercharging I.C. engines, gas pumping in long distance pipe lines etc. — Previously it was used in jet engines	Mostly used in jet engines (due to higher efficiency and smaller frontal area). Also preferred in power plant gas turbines and steel mills.
12.	*Efficiency vs. speed curve*	More flat (Fig. 24.68)	Less flat comparatively Fig. 24.68

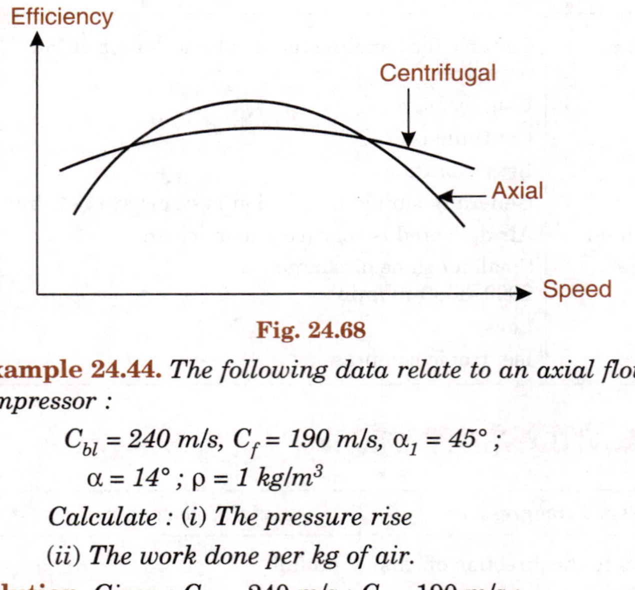

Fig. 24.68

Example 24.44. *The following data relate to an axial flow compressor :*

$C_{bl} = 240$ *m/s*, $C_f = 190$ *m/s*, $\alpha_1 = 45°$;

$\alpha = 14°$; $\rho = 1$ *kg/m*^{3}

Calculate : (i) The pressure rise

(ii) The work done per kg of air.

Solution. *Given* : $C_{bl} = 240$ m/s ; $C_f = 190$ m/s ;

$\alpha_1 = 45°$; $\alpha_2 = 14°$, $\rho = 1$ kg/m^3

***(i)* The pressure rise, Δp :**

The pressure rise through a ring of rotating blades,

$$\Delta p = \frac{\rho}{2} C_f^2 (\tan^2 \alpha_1 - \tan^2 \alpha_2) \qquad \text{...[Eqn. (24.92)]}$$

$$= \frac{1}{2} \times \frac{(190)^2}{10^5} [(\tan 45°)^2 - (\tan 14°)^2]$$

$$= \mathbf{0.169\ bar.} \quad \textbf{(Ans.)}$$

***(ii)* The work done per kg of air, W :**

$$W = C_{bl}\, C_f (\tan \alpha_1 - \tan \alpha_2)$$

$$= \frac{240 \times 190}{10^3} (\tan 45° - \tan 14°)$$

$$= \mathbf{34.23\ kW.} \quad \textbf{(Ans.)}$$

Example 24.45. *An axial flow compressor having eight stages and with 50% reaction design compresses air in the pressure ratio of 4 : 1. The air enters the compressor at 20°C and flows through it with a constant speed of 90 m/s. The rotating blades of compressor rotate with a mean speed of 180 m/s. Isentropic efficiency of the compressor may be taken as 82%. Calculate :*

(i) Work done by the machine

(ii) Blades angles.

Assume $\gamma = 1.4$ *and* $c_p = 1.005$ *kJ/kg K.*

Solution.
$$\eta_{isen} = \frac{T_2' - T_1}{T_2 - T_1}$$

Also
$$\frac{T_2'}{T_1'} = \left(\frac{p_2}{p_1}\right)^{\frac{\gamma - 1}{\gamma}} = (4)^{\frac{1.4-1}{1.4}} = (4)^{\frac{0.4}{1.4}} = 1.486$$

$\therefore \quad T_2' = (20 + 273) \times 1.486 = 435.4$ K

$\therefore \quad \eta_{isen} = \dfrac{435.4 - 293}{T_2 - 293}$ or $0.82 = \dfrac{142.4}{T_2 - 293}$

$\therefore \quad T_2 = \dfrac{142.4}{0.82} + 293 = 466.6$ K

Work required/kg

$= c_p(T_2 - T_1) = 1.005(466.6 - 293)$

$=$ **174.47 kJ/kg. (Ans.)**

Now, work done/kg

$=$ Number of stages $\times C_{bl}(C_{w_2} - C_{w_1})$

$174.47 = 8 \times C_{bl}\, C_f(\tan \alpha_2 - \tan \alpha_1)$

[Refer to Fig. 24.62]

$\therefore \quad \tan \alpha_2 - \tan \alpha_1 = \dfrac{174.47 \times 1000}{8 \times 180 \times 90} = 1.346$

For 50% reaction blading, $\alpha_2 = \beta_1$ and $\alpha_1 = \beta_2$

$\therefore \quad 1.346 = \tan \beta_1 - \tan \alpha_1$

Now, $\quad \tan \alpha_1 + \tan \beta_1 = \dfrac{C_{bl}}{C_f} = \dfrac{180}{90} = 2$

i.e., $\quad \tan \beta_1 - \tan \alpha_1 = 1.346 \quad$ ...(*i*)

$\tan \beta_1 + \tan \alpha_1 = 2 \quad$ ...(*ii*)

From (*i*) and (*ii*), we get

$2 \tan \beta_1 = 3.346$

$\therefore \quad \beta_1 =$ **59.1°** $= \alpha_2$. **(Ans.)**

and $\quad \alpha_1 =$ **18.1°** $= \beta_2$. **(Ans.)**

Example 24.46. *An axial flow compressor with an overall isentropic efficiency of 85% draws air at 20°C and compresses it in the pressure ratio of 4 : 1. The mean blade speed and flow velocity are constant throughout the compressor. Assuming 50% reaction blading and taking blade velocity as 180 m/s and work input factor as 0.82, calculate :*

(*i*) *Flow velocity*

(*ii*) *Number of stages*

Take $\alpha_1 = 12°$, $\beta_1 = 42°$.

Solution. Given : $\eta_{isen} = 85\%$, $T_1 = 20 + 273 = 293$ K

Pressure ratio, $\quad \dfrac{p_2}{p_1} = 4$, $C_{bl} = 180$ m/s

Work input factor $= 0.82$

$$\frac{T_2'}{T_1} = \left(\frac{p_2}{p_1}\right)^{\frac{\gamma-1}{\gamma}} = (4)^{\frac{1.4-1}{1.4}} = 1.486$$

$\therefore \quad T_2' = 293 \times 1.486 = 435.4$ K

Now $\quad \eta_{isen} = \dfrac{T_2' - T_1}{T_2 - T_1}$

$0.85 = \dfrac{435.4 - 293}{T_2 - 293}$

$\therefore \quad T_2 = 460.5$ K

Theoretical work required per kg

$= c_p(T_2 - T_1) = 1.005(460.5 - 293)$

$= 168.33$ kJ/kg

From velocity Δs (Fig. 24.62) :

$\dfrac{C_{bl}}{C_f} = \tan \alpha_1 + \tan \beta_1$

$= \tan 12° + \tan 42° = 0.212 + 0.9 = 1.112$

$\therefore \quad C_f = \dfrac{C_{bl}}{1.112} = \dfrac{180}{1.112} =$ **161.8 m/s. (Ans.)**

Work done per stage $= C_{bl}(C_{w_2} - C_{w_1}) \times$ work input factor

Now, $C_{w_2} = C_f \tan \alpha_2 = 161.8 \tan 42° = 145.7$ m/s

$(\because \alpha_2 = \beta_1)$

and $\quad C_{w_1} = C_f \tan \alpha_1 = 161.8 \tan 12° = 34.4$ m/s

$\therefore$ Work done per stage $= 180(145.7 - 34.4) \times 0.82 \times 10^{-3}$ kJ/kg $= 16.4$ kJ/kg

$\therefore \quad$ Number of stages $= \dfrac{168.33}{16.43} \simeq 10$

i.e., **Number of stages = 10. (Ans.)**

Example 24.47. *In an eight stage axial flow compressor, the overall stagnation pressure ratio achieved is 5 : 1 with an overall isentropic efficiency of 92 per cent. The inlet stagnation temperature and pressure at inlet are 290 K and 1 bar. The work is divided equally between the stages. The mean blade speed is 160 m/s and 50% reaction design is used. The axial velocity through the compressor is constant and is equal to 90 m/s. Calculate :*

(*i*) *The blade angles.* (*ii*) *The power required.*

Solution. *Given :* $N = 8$; $r_p = 5 : 1$; $\eta_{isen} = 92\%$; $T_{01} = 290$ K ; $p_{01} = 1$ bar ; $C_{bl} = 160$ m/s ; Degree of reaction = 50% ; $C_f = 90$ m/s.

(*i*) The blade angles, $\alpha_1, \beta_1, \alpha_2, \beta_2$:

Refer to Fig. 24.69 for velocity diagrams. Since the degree of reaction is 50% reaction, the *blades are symmetrical* and hence the velocity diagrams are *identical.* Thus

$\alpha_1 = \beta_2$ and $\alpha_2 = \beta_1$

Let suffix N denotes the *number of stages.*

With isentropic compression the temperature of air leaving the compressor stage is

$$T_{0N}' = T_{01}\left(\frac{p_{0N}}{p_{01}}\right)^{(\gamma-1/\gamma)} = 290 \times (5)^{\frac{1.4-1}{1.4}}$$

$= 459.3$ K

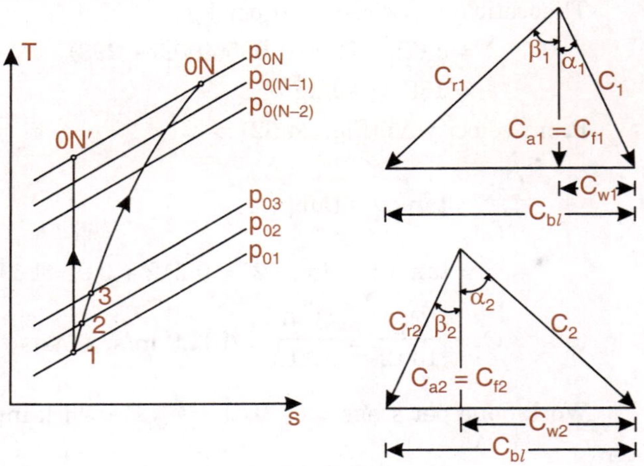

Fig. 24.69

But $\quad \eta_{isen} = \dfrac{T_{ON}' - T_{01}}{T_{ON} - T_{01}}$

$$0.92 = \frac{459.3 - 290}{T_{oN} - 290}$$

$$\therefore \quad T_{ON} = \frac{459.3 - 290}{0.92} + 290 = 474 \text{ K}$$

The work consumed by the compressor

$$= c_p(T_{ON} - T_{01}) = (C_{w2} - C_{w1})\, C_{bl} \times \text{N}$$

or $\quad c_p(T_{ON} - T_{01}) = C_f\,(\tan \alpha_2 - \tan \alpha_1)\, C_{bl}\,.\, N$

$\therefore \quad \tan \alpha_2 - \tan \alpha_1$

$$= \frac{c_p(T_{ON} - T_{01})}{C_f . C_{bl} . N} = \frac{1.005\,(474 - 290) \times 10^3}{90 \times 160 \times 8}$$

$$= 1.605 \qquad ...(i)$$

From velocity triangles, we have

$$\frac{C_{bl}}{C_f} = \tan \alpha_1 + \tan \beta_1 = \frac{160}{90} = 1.778 \qquad ...(ii)$$

Adding (*i*) and (*ii*), we get

$$\tan \beta_1 = \frac{1.605 + 1.778}{2} = 1.6915 \qquad (\because \alpha_1 = \beta_1)$$

or $\quad \beta_1 = \tan^{-1}(1.6915) = 59.4°$

$\therefore \quad$ **$\beta_1 = \alpha_2 = 59.4°$. (Ans.)**

Putting the value of $\tan \beta_1$ in (*ii*), we have

$$\tan \alpha_1 + 1.6915 = 1.778$$

or $\quad \tan \alpha_1 = 0.0865$

or $\quad \alpha_1 = \tan^{-1}(0.0865) = 4.94°$

$\therefore \quad$ **$\alpha_1 = \beta_2 = 4.94°$. (Ans.)**

(*ii*) The power required by compressor P :

$$P = \dot{m}\, c_p(T_{ON} - T_{01})$$
$$= 1 \times 1.005(474 - 290)$$
$$= \mathbf{184.9 \text{ kW}}. \quad \textbf{(Ans.)}$$

Example 24.48. *In an axial flow compressor, the overall stagnation pressure ratio achieved is 4 with overall stagnation isentropic efficiency 85 per cent. The inlet stagnation pressure and temperature are 1 bar and 300 K. The mean blade speed is 180 m/s. The degree of reaction is 0.5 at the mean radius with relative air angles of 12° and 32° at the rotor inlet and outlet respectively. The work done factor is 0.9. Calculate :*

(*i*) *Stagnation polytropic efficiency.*
(*ii*) *Number of stages.*
(*iii*) *Inlet temperature and pressure.*
(*iv*) *Blade height in the first stage if the hub-tip ratio is 0.42, mass flow rate 19.5 kg/s.*

Solution. Given : $r_p = \dfrac{p_{0N}}{p_{01}} = 4$; $\eta_{isen} = 85\%$;

$$p_{01} = 1 \text{ bar} ; T_{01} = 300 \text{ K}$$

$C_u = 180$ m/s ; Degree of reaction, $R_d = 0.5$. Work done factor = 0.9 ; Hub-tip ratio = 0.42 ; $\dot{m}$ = 19.5 kg/s.

Refer to Fig. 24.70. For 50% reaction, the inlet and outlet velocity diagrams are identical. Hence $\alpha_1 = \beta_2 = 12°$; $\alpha_2 = \beta_1 = 32°$.

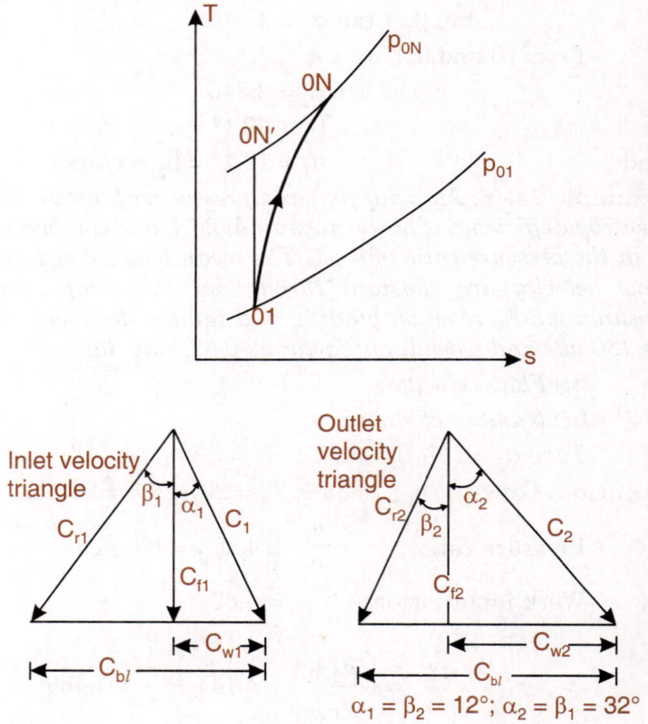

Fig. 24.70

(*i*) Stagnation polytropic efficiency, η_p :

The temperature at the end of the compression stage due to isentropic expansion is

$$T_{ON}' = T_{01}\left(\frac{p_{0N}'}{p_{01}}\right)^{\frac{\gamma - 1}{\gamma}} = 300\,(4)^{\frac{1.4 - 1}{1.4}} = 445.8 \text{ K}$$

$$\eta_{0\,isen} = 0.85 = \frac{T_{0N}' - T_{01}}{T_{0N} - T_{01}} = \frac{445.8 - 300}{T_{0N} - 300}$$

$$\therefore \quad T_{0N} = \frac{445.8 - 300}{0.85} + 300 = 471.5 \text{ K}$$

Now
$$\eta_p = \frac{\ln\left(\dfrac{p_{0N}}{p_{01}}\right)^{\frac{\gamma-1}{\gamma}}}{\ln\left(\dfrac{T_{0N}}{T_{01}}\right)} \quad \text{...[Eqn. (24.104)]}$$

or
$$\eta_p = \frac{\ln(4)^{\frac{1.4-1}{1.4}}}{\ln\left(\dfrac{471.5}{300}\right)} = \frac{0.396}{0.452}$$

$= \mathbf{0.8761}$ or **87.61%. (Ans.)**

(*ii*) **Number of stages, N :**

From the configuration of velocity triangles,

$$\frac{C_{bl}}{C_f} = \tan \alpha_1 + \tan \beta_1 = \tan 12° + \tan 32° = 0.8374$$

$$\therefore \quad C_f = \frac{C_{bl}}{0.8374} = \frac{180}{0.8374} = 214.9 \text{ m/s}$$

Now, $C_{w1} = C_f \tan \alpha_1 = 214.9 \times \tan 12° = 45.68$ m/s

$C_{w2} = C_f \tan \alpha_2 = 214.9 \times \tan 32° = 134.28$ m/s

Work consumed per stage

$= C_{bl}(C_{w2} - C_{w1}) \times$ work done factor ...[Eqn. (24.87)]

$$= \frac{180\,(134.28 - 45.68) \times 0.9}{1000} = 14.35 \text{ kJ/kg}$$

Total work consumed by the compressor

$= c_p(T_{0N} - T_{01}) = 1.005(471.5 - 300)$
$= 172.36$ kJ/kg

$\therefore$ Number of stages,

$$N = \frac{172.36}{14.35} = \mathbf{12\ stages.} \quad \textbf{(Ans.)}$$

(*iii*) **Inlet temperature and pressure, T_1, p_1 :**

The absolute velocity C_1 at exit from the guide vanes and approaching to moving blades of first stage,

$$C_1 = \frac{C_f}{\cos \alpha_1} = \frac{214.9}{\cos 12°} = 219.7 \text{ m/s}$$

Temperature,

$$T_1 = T_{01} - \frac{C_1^2}{2c_p} = 300 - \frac{(219.7)^2}{2 \times 1.005 \times 1000}$$

$= \mathbf{276\ K.}$ **(Ans.)**

Assuming reversible flow through the guide vanes put ahead of the first stage,

$$\frac{p_1}{p_{01}} = \left(\frac{T_1}{T_{01}}\right)^{\frac{\gamma}{\gamma-1}}$$

or
$$p_1 = 1 \times \left(\frac{276}{300}\right)^{\frac{1.4}{1.4-1}} = \mathbf{0.7469\ bar.} \quad \textbf{(Ans.)}$$

(*iv*) **Blade height in the first stage, l :**

The density of air approaching to first stage,

$$\rho_1 = \frac{p_1}{RT_1} = \frac{0.7469 \times 10^5}{287 \times 276} = 0.9429 \text{ kg/m}^3$$

From the continuity equation,

$\rho_1 A_1 C_f = \dot{m} = 19.5$

$0.9429 \times \pi r_1^2[1 - (0.42)^2] \times 241.9 = 19.5$

or
$$r_1 = \left[\frac{19.5}{0.9429 \times \pi \times [1 - (0.42)^2] \times 241.9}\right]^{1/2}$$

$= 0.1818$ m or 18.18 cm

But $\dfrac{r_h}{r_1} = 0.42 \quad \therefore \quad r_h = 18.18 \times 0.42 = 7.636$ cm

Hence height of the blade in the first stage,

$l = r_1 - r_h = 18.18 - 7.636 = \mathbf{10.544\ cm.}$ **(Ans.)**

Example 24.49. *A multi-stage axial flow compressor delivers 20 kg/s of air. The inlet stagnation condition is 1 bar and 17°C. The power consumed by the compressor is 4350 kW.*

Calculate :

(*i*) *Delivery pressure.*

(*ii*) *Number of stages.*

(*iii*) *Overall isentropic efficiency of the compressor.*

Assume temperature rise in the first stage is 15°C, the polytropic efficiency of compression is 88% and the stage stagnation pressure ratio is constant.

Solution. *Given :* $\dot{m} = 20$ kg/s ; $p_{01} = 1$ bar ;

$T_{01} = 17 + 273 = 290$ K ;

$P = 4350$ kW ; $T_{02} = 15 + 290 = 305$ K ;

$\eta_p = 88\%$.

(*i*) **Delivery pressure, P_{0N} :**

Now,
$$\eta_p = \frac{\ln\left(\dfrac{p_{02}}{p_{01}}\right)^{\frac{\gamma-1}{\gamma}}}{\ln\left(\dfrac{T_{02}}{T_{01}}\right)}$$

or
$$0.88 = \frac{\ln\left(\dfrac{p_{02}}{p_{01}}\right)^{0.2857}}{\ln\left(\dfrac{305}{290}\right)}$$

or
$$\ln\left(\frac{p_{02}}{p_{01}}\right)^{0.2857} = 0.88 \times \ln\left(\frac{305}{290}\right) = 0.04438$$

or $$\left(\frac{p_{02}}{p_{01}}\right)^{0.2857} = e^{0.04438} = 1.0454$$

or $$\frac{p_{02}}{p_{01}} = (1.0454)^{\frac{1}{0.2857}} = 1.168$$

The power required by the compressor,

$$P = \dot{m} \times c_p \times (T_{0N} - T_{01})$$

$$4350 = 20 \times 1.005 \times (T_{0N} - 290)$$

or $$T_{0N} = \frac{4350}{20 \times 1.005} + 290 = 506.4 \text{ K}$$

Also, the polytropic efficiency of the compressor is given as

$$\eta_p = \left(\frac{\gamma - 1}{\gamma}\right)\left(\frac{n}{n-1}\right) \quad \text{...[Eqn. (24.103)]}$$

or $$0.88 = \left(\frac{1.4-1}{1.4}\right)\left(\frac{n}{n-1}\right)$$

or $$\frac{n}{n-1} = \frac{0.88 \times 1.4}{(1.4 - 1)} = 3.08$$

During polytropic compression from first stage to last, we have

$$\frac{p_{0N}}{p_{01}} = \left(\frac{T_{0N}}{T_{01}}\right)^{\frac{n}{n-1}} = \left(\frac{506.4}{290}\right)^{3.08} = 5.567$$

$\therefore$ $p_{0N} = 1 \times 5.567 =$ **5.567 bar. (Ans.)**

(*ii*) **Number of stages, N :**

As the pressure ratio for each stage is the same hence

$$\frac{p_{02}}{p_{01}} = \frac{p_{03}}{p_{02}} = \frac{p_{04}}{p_{03}} = \ldots\ldots \frac{p_{0N}}{p_{0(N-1)}}$$

where suffix N indicates the number of stages.

$$\therefore \quad \left(\frac{p_{02}}{p_{01}}\right)^N = \frac{p_{0N}}{p_{01}}$$

Taking log on both the sides, we get

$$N = \ln\left(\frac{p_{02}}{p_{01}}\right) = \ln\left(\frac{p_{0N}}{p_{01}}\right)$$

or $$\text{N} = \frac{\ln\left(\frac{p_{0N}}{p_{01}}\right)}{\ln\left(\frac{p_{02}}{p_{01}}\right)} = \frac{\ln(5.567)}{\ln(1.168)} = \mathbf{11.} \quad \textbf{(Ans.)}$$

(*iii*) **Overall isentropic efficiency of the compressor, $(\eta_{overall})_{isen}$:**

With isentropic compression the temperature of air leaving the compressor is,

$$T_{0N}' = T_{01}\left(\frac{p_{0N}}{p_{01}}\right)^{\frac{\gamma-1}{\gamma}} = 290\,(5.567)^{\frac{1.4-1}{1.4}} = 473.6 \text{ K}$$

$$\therefore \quad (\eta_{overall})_{isen} = \frac{T_{0N}' - T_{01}}{T_{0N} - T_{01}} = \frac{473.6 - 290}{506.4 - 290}$$

$= \mathbf{0.8484}$ or **84.84%. (Ans.)**

HIGHLIGHTS

1. An air compressor takes in atmospheric air, compresses it and delivers the high pressure air to a storage vessel from which it may be conveyed by a pipeline to wherever the supply of compressed air is required.
2. Air and gas compressors are classified into two main types :
   (*i*) Reciprocating compressor
   (*ii*) Rotary compressors.
3. **Single-stage compressor**
   Equation for work (neglecting clearance),
   $$W = \frac{n}{n-1} mRT_1 \left\{\left(\frac{p_2}{p_1}\right)^{\frac{n-1}{n}} - 1\right\}$$
   Equation for work (with clearance volume),
   $$W = \frac{n}{n-1} p_1(V_1 - V_4)\left\{\left(\frac{p_2}{p_1}\right)^{\frac{n-1}{n}} - 1\right\}.$$
4. **'Volumetric efficiency'** of a compressor is the ratio of free air delivered to the displacement of the compressor.
   $$\eta_{vol.} = 1 + k - k\left(\frac{p_3}{p_4}\right)^{\frac{1}{n_e}} = 1 + k - k\left(\frac{p_2}{p_1}\right)^{\frac{1}{n_e}}$$
   $$= 1 + k - k\left(\frac{p_4}{p_3}\right) = 1 + k - k\,(V_1/V_2)$$
   where, $k = \dfrac{V_c \text{ (clearance volume)}}{V_s \text{ (swept volume)}}$ = Clearance ratio.
5. Multi-stage compression is very efficient and is now-a-days almost universally adopted except for compressors where the overall pressure rise required is small.
6. In a two-stage compressor efficiency will be maximum when
   $$\frac{p_2}{p_1} = \frac{p_3}{p_2} \text{ or } p_2 = \sqrt{p_1 p_3}$$
   For x-stage compressor the work to be supplied,
   $$W = \frac{xn}{n-1} p_1 V_1 \left[\left(\frac{p_{x+1}}{p_1}\right)^{\frac{n-1}{xn}} - 1\right]$$

This equation is very important, since it applies to any type of compressor or motor, and even to vapour engines, provided n = or < γ.

7. An **air-turbine** is valveless, small in size, light in weight, and requires no internal lubrication, but air friction is high, and any dampness in the air causes rapid deterioration of the blading at low temperatures.
8. **Displacement compressors** are those compressors in which air is compressed by being trapped in the reduced space formed by two sets of engaging surfaces.
9. **Steady-flow compressors** are those compressors in which compression occurs by transfer of kinetic energy from a rotor.
10. **Slip factor** is defined as the ratio of actual whirl component and ideal whirl component.
11. **Pressure co-efficient** is defined as the ratio of isentropic work to Euler work.
12. **Degree of reaction (R)** is defined as the ratio of pressure rise in the compressor stage

i.e., $$R_d = \frac{\text{Pressure rise in the rotor blades}}{\text{Pressure rise in the stage}}.$$

OBJECTIVE TYPE QUESTIONS

Choose the Correct Answer :

1. The work input to air compressor is minimum if the compression law followed is
   (*a*) $pV^{1.35} = C$ (*b*) Isothermal $pV = C$
   (*c*) Isentropic $pV^{\gamma} = C$ (*d*) $pV^{1.2} = C$.
2. For reciprocating air compressor the law of compression desired is isothermal and that may be possible by
   (*a*) very low speeds
   (*b*) very high speeds
   (*c*) any speed as speed does not affect the compression law
   (*d*) none of the above.
3. Work input to the air compressor with 'n' as index of compression
   (*a*) increases with increase in value of n
   (*b*) decreases with increase in value of n
   (*c*) remains same whatever the value of n
   (*d*) first increases and then decreases with increase of value of n.
4. Work done in a single-stage, single-acting air compressor without clearance per kg of air delivered when the compression process is isothermal is given by
   (*a*) $\frac{n}{n-1} p_1 v_1 \left[\left(\frac{p_2}{p_1} \right)^{\frac{n-1}{n}} - 1 \right]$
   (*b*) $\frac{\gamma}{\gamma-1} p_1 v_1 \left[\left(\frac{p_2}{p_1} \right)^{\frac{\gamma-1}{\gamma}} - 1 \right]$
   (*c*) $p_1 v_1 \log_e \frac{p_2}{p_1}$
   (*d*) $\frac{n}{n-1} p_1 v_1 \log_e \frac{p_2}{p_1}$.
5. The clearance volume in reciprocating air compressors is provided to
   (*a*) to reduce the work done per kg of air delivered
   (*b*) to increase the volumetric efficiency of the compressor
   (*c*) to accommodate valves in the head of the compressor
   (*d*) to create turbulence in the air to be delivered.
6. With increase in clearance volume, the ideal work of compressing 1 kg of air
   (*a*) increases
   (*b*) decreases
   (*c*) remains same
   (*d*) first increases and then decreases.
7. In reciprocating air compressors the clearance ratio is given by
   (*a*) $\frac{\text{Total volume of cylinder}}{\text{Clearance volume}}$
   (*b*) $\frac{\text{Swept volume of cylinder}}{\text{Clearance volume}}$
   (*c*) $\frac{\text{Clearance volume}}{\text{Swept volume of cylinder}}$
   (*d*) $\frac{\text{Clearance volume}}{\text{Total volume of cylinder}}$.
8. With suction pressure being atmospheric, increase in delivery pressure with fixed clearance volume
   (*a*) increase volumetric efficiency
   (*b*) decreases volumetric efficiency
   (*c*) does not change volumetric efficiency
   (*d*) first increases volumetric efficiency and then decreases it.
9. Mechanical efficiency of reciprocating air compressor is expressed as
   (*a*) $\frac{\text{B.P.}}{\text{I.P.}}$ (*b*) $\frac{\text{I.P.}}{\text{B.P.}}$
   (*c*) $\frac{\text{F.P.}}{\text{B.P.}}$ (*d*) $\frac{\text{F.P.}}{\text{I.P.}}$.
10. For the same overall pressure ratio, the leakage of air past the piston for multi-stage compression as compared to single-stage compression, is
    (*a*) more (*b*) less
    (*c*) constant (*d*) may be more or less.

11. Work done in a two stage reciprocating air compressor with imperfect cooling is given by

(*a*) $\frac{n}{n-1}p_1V_1\left[\left(\frac{p_2}{p_1}\right)^{\frac{n-1}{n}}-1\right]+\frac{n}{n-1}p_2V_2\left[\left(\frac{p_3}{p_2}\right)^{\frac{n-1}{n}}-1\right]$

(*b*) $\frac{2n}{n-1}p_1V_1\left[\left(\frac{p_2}{p_1}\right)^{\frac{n-1}{n}}-1\right]$

(*c*) $\frac{2n}{n-1}p_1V_1\left[\left(\frac{p_3}{p_1}\right)^{\frac{n-1}{2n}}-1\right]$

(*d*) $\frac{n-1}{2n}p_1V_1\left[\left(\frac{p_3}{p_1}\right)^{\frac{n-1}{2n}}-1\right]$.

12. Work done in a two-stage reciprocating air compressor with perfect intercooling is given by

(*a*) $\frac{n}{n-1}p_1V_1\left[\left(\frac{p_2}{p_1}\right)^{\frac{n-1}{n}}+\left(\frac{p_3}{p_1}\right)^{\frac{n-1}{n}}-2\right]$

(*b*) $\frac{2n}{n-1}p_1V_1\left[\left(\frac{p_3}{p_1}\right)^{\frac{n-1}{n}}-1\right]$

(*c*) $\frac{n}{n-1}p_1V_1\left[\left(\frac{p_3}{p_1}\right)^{\frac{n-1}{n}}-1\right]$

(*d*) $\frac{n}{n-1}p_1V_1\left[\left(\frac{p_2}{p_1}\right)^{\frac{n-1}{n}}\left(\frac{p_3}{p_2}\right)^{\frac{n-1}{n}}-2\right]$.

13. In reciprocating air compressor the method of controlling the quantity of air delivered is done by
(*a*) throttle control (*b*) blow-off control
(*c*) clearance control (*d*) all of the above.

14. The efficiency of Vane type air compressor as compared to Roots air compressor for the same pressure ratio is
(*a*) more (*b*) less
(*c*) same (*d*) may be more or less.

15. In centrifugal air compressor the pressure developed depends on
(*a*) impeller tip velocity (*b*) inlet-temperature
(*c*) compression index (*d*) all of the above.

16. In a centrifugal air compressor the pressure ratio is increased by
(*a*) increasing the speed of impeller keeping its diameter fixed
(*b*) increasing the diameter of the impeller keeping its speed constant
(*c*) reducing inlet temperature, keeping impeller diameter and speed fixed
(*d*) all of the above.

ANSWERS

1. (*b*)	**2.** (*a*)	**3.** (*a*)	**4.** (*c*)
5. (*c*)	**6.** (*c*)	**7.** (*c*)	**8.** (*b*)
9. (*b*)	**10.** (*b*)	**11.** (*a*)	**12.** (*b*)
13. (*d*)	**14.** (*a*)	**15.** (*d*)	**16.** (*d*).

THEORETICAL QUESTIONS

1. Enumerate the applications of compressed air.
2. State how are the air compressors classified ?
3. Describe with a neat sketch the construction and working of a single-stage single-acting reciprocating air compressor.
4. Prove that the work done/kg of air in a compressor is given by

$$W = RT_1\frac{n}{n-1}\left[(r_p)^{\frac{n-1}{n}}-1\right],\ \text{where } r_p = \text{pressure ratio.}$$

5. Prove that the volumetric efficiency of a single-stage compressor is given by

$$\eta_{vol.} = 1 + k - k\left(\frac{p_2}{p_1}\right)^{\frac{1}{n}},\ \text{where } k = \frac{V_c}{V_s}.$$

6. State the conditions which lower the volumetric efficiency.
7. Explain with a neat sketch actual *p-V* diagram for a single-stage compressor.
8. What do you mean by multi-stage compression ? State its advantages.
9. What is the effect of atmospheric conditions on the output of a compressor ?
10. Explain briefly with a neat sketch a reciprocating air motor.
11. Write a short note on 'rotary type air motor'.
12. What is a rotary compressor ? How are rotary compressors classified ?
13. What is a centrifugal compressor ? How does it differ from an axial flow compressor ?
14. What is Euler's work ?
15. What is a slip factor and a pressure co-efficient ?
16. Describe briefly an axial-flow compressor.
17. Draw the velocity diagrams of an axial-flow compressor.
18. What do you mean by 'surging' and 'choking' ?
19. Prove that the work done in two-stage compressor per kg of air delivered with perfect intercooling is given by

$$W/\text{kg} = \frac{2n}{n-1}RT_1\left[\left(\frac{p_d}{p_s}\right)^{\frac{n-1}{2n}}-1\right]$$

Using the above equation, prove that the work done/kg of air in 'x' stages with perfect intercooling is given by

$$W/\text{kg} = \frac{xn}{n-1} RT_1 \left[\left(\frac{p_d}{p_s} \right)^{\frac{n-1}{xn}} - 1 \right].$$

20. Prove that the heat rejected (per kg of air) with perfect intercooling

$$= \left[c_p + c_v \left(\frac{\gamma - n}{n-1} \right) \right] (T_2 - T_1).$$

21. Explain with a neat sketch the actual p-V diagram for a two-stage compressor.

22. Define the following efficiencies as applied to reciprocating air compressor :

(*i*) Compressor efficiency. (*ii*) Isothermal efficiency.
(*iii*) Adiabatic efficiency. (*iv*) Mechanical efficiency.

23. Write short notes on any three of the following :

(*i*) Clearance in compressors
(*ii*) Free air delivered (F.A.D.) and displacement.
(*iii*) Compressor performance.
(*iv*) Control of compressors.
(*v*) Arrangement of reciprocating compressors.
(*vi*) Intercooler.
(*vii*) Compressed air motors.

UNSOLVED EXAMPLES

1. Air is to be compressed in a single-stage reciprocating compressor from 1.013 bar and 15°C to 7 bar. Calculate the indicated power required for a free air delivery of 0.3 m³/min, when the compression process is :

(*i*) Isentropic
(*ii*) Reversible isothermal
(*iii*) Polytropic, with $n = 1.25$.

[**Ans.** (*i*) 1.31 kW ; (*ii*) 0.98 kW ; (*iii*) 1.19 kW]

2. The compressor of the above example is to run at 1000 r.p.m. If the compressor is single-acting and has a stroke/bore ratio of 1.2/1, calculate the bore size required.

[**Ans.** 68.3 mm]

3. A single-stage, single-acting air compressor running at 1000 r.p.m. delivers air at 25 bar. For this purpose the induction and free air conditions can be taken as 1.013 bar and 15°C, and the free air delivery as 0.25 m³/min. The clearance volume is 3% of the swept volume and the stroke/bore ratio is 1.2 : 1. Calculate the bore and stroke and the volumetric efficiency of this machine. Take the index of compression and expansion as 1.3. Calculate also the indicated power and the isothermal efficiency.

[**Ans.** 73.2 mm ; 87.84 mm ; 67.6% ; 2 kW ; 67.5%]

4. A single-acting compressor is required to deliver air at 70 bar from an induction pressure of 1 bar, at the rate of 2.4 m³/min measured at free-air conditions of 1.013 bar and 15°C. The temperature at the end of the induction stroke is 32°C. Calculate the indicated power required if the compression is carried out in two stages with an ideal intermediate pressure and complete intercooling. The index of compression and expansion for both stages is 1.25. What is the saving in power over single-stage compression ?

If the clearance volume is 3% of the swept volume in each cylinder, calculate the swept volumes of the cylinders. The speed of the compressor is 750 r.p.m.

If the mechanical efficiency of the compressor is 85%, calculate the power output in kilowatts of the motor required.

[**Ans.** 22.7 kW ; 6 kW ; 0.00396 m³, 0.000474 m³, 26.75 kW]

5. A single-cylinder, single-acting air compressor running at 300 r.p.m. is driven by a 23 kW electric motor. The mechanical efficiency of the drive between motor and compressor is 87%. The air inlet conditions are 1.013 bar and 15°C and the delivery pressure is 8 bar. Calculate the free-air delivery in m³/min, the volumetric efficiency, and the bore and stroke of the compressor. Assume that the index of compression and expansion is $n = 1.3$, that the clearance volume is 7% of the swept volume and that the bore is equal to the stroke.

[**Ans.** 4.47 m³/min ; 73% ; 296 mm]

6. A two-stage air compressor consists of three cylinders having the same bore and stroke. The delivery pressure is 7 bar and the free air delivery is 4.2 m³/min. Air is drawn in at 1.013 bar, 15°C and an intercooler cools the air to 38°C. The index of compression is 1.3 for all the three cylinders. Neglecting clearance calculate :

(*i*) The intermediate pressure
(*ii*) The power required to drive the compressor
(*iii*) The isothermal efficiency.

[**Ans.** (*i*) 2.19 bar; (*ii*) 16.3 kW; (*iii*) 84.5%]

7. A two-stage double-acting air compressor, operating at 200 r.p.m., takes in air at 1.013 bar and 27°C. The size of the L.P. cylinder is 350 × 380 mm ; the stroke of H.P. cylinder is the same as that of the L.P. cylinder and the clearance of both the cylinders is 4%. The L.P. cylinder discharges the air at a pressure of 4.052 bar. The air passes through the intercooler so that it enters the H.P. cylinder at 27°C and 3.85 bar, finally it is discharged from the compressor at 15.4 bar. The value of 'n' in both cylinders is 1.3. $c_p = 1.0035$ kJ/kg K and $R = 0.287$ kJ/kg K.

Calculate : (*i*) The heat rejected in the intercooler
(*ii*) The diameter of H.P. cylinder
(*iii*) The power required to drive the H.P. cylinder.

[**Ans.** (*i*) 1805.68 kg/min; (*ii*) 179.5 mm; (*iii*) 37.3 kW]

8. A single-acting two-stage compressor with complete intercooling delivers 10 kg/min of air at 16 bar. The suction occurs at 1 bar and 15°C. The expansion and compression processes are reversible polytropic with polytropic index $n = 1.25$. Calculate :

(*i*) The power required.
(*ii*) The thermal efficiency.
(*iii*) The free air delivery.
(*iv*) Heat transferred in intercooler.

(*v*) If the clearance ratios for L.P. and H.P. cylinders are 0.04 and 0.06 respectively, calculate the swept and clearance volumes for each cylinder.

The speed of the compressor is 400 r.p.m.

[**Ans.** (*i*) 44 kW; (*ii*) 86.8 %; (*iii*) 8.266 m^3/min; (*iv*) 15.41 kW; (*v*) 0.0225 m^3, 0.0009 m^3 ; 0.00588 m^3, 0.000353 m^3]

9. A two-stage air compressor with complete intercooling delivers air to the mains at a pressure of 30 bar, the suction conditions being 1 bar and 27°C. If both cylinders have the same stroke, find the ratio of the cylinders diameters, for the efficiency of compression to be a maximum. Assume the index of compression to be 1.3. [**Ans.** 2.337]

10. A 4-cylinder double-acting compressor is required to compress 25 m^3/min of air at 1 bar and 25°C to a pressure of 15 bar. Determine the size of the motor required and the cylinder dimensions if the following additional data is given :

Clearance volume = 5%, L/D = 1.2, r.p.m. = 300, Mechanical efficiency = 80%, Value of index, n = 1.35.

Assume no pressure change in suction valves and the air gets heated by 10°C during suction stroke.

[**Ans.** D = 550 mm, L = 680 mm]

11. A three-stage compressor is used to compress hydrogen from 1.04 bar to 35 bar. The compression in all stages follows the law $pV^{1.25} = C$. The temperature of hydrogen at inlet of compressor is 288 K. Neglecting clearance and assuming perfect intercooling, find :

(*i*) Indicated power required to deliver 14 m^3 of hydrogen per minute measured at inlet conditions.

(*ii*) Intermediate pressures.

Take R = 4125 J/kg K.

[**Ans.** (*i*) 96.2 kW ; (*ii*) 3.354 bar, 10.815 bar]

12. A three-stage reciprocating air compressor compresses air from 1 bar and 17°C to 35 bar. The law of compression is $pV^{1.25}$ = C and is same for all the stages of compression. Assuming perfect intercooling and neglecting clearance and valve resistance, find the minimum power required to compress 15 m^3/min of free air. Also find the intermediate pressures.

[**Ans.** 100.3 kW, 3.271 bar, 10.7 bar]

13. A three-stage single-acting air compressor running in an atmosphere at 1.013 bar and 15°C has a free air delivery of 2.83 m^3/min. The suction pressure and temperature are 0.98 bar and 32°C respectively. Calculate the indicated power required, assuming complete intercooling, n = 1.3, and that the machine is designed for minimum work. The delivery pressure is to be 70 bar.

[**Ans.** 24.2 kW]

14. Using the data of example 13 determine heat loss to the cylinder jacket cooling water and heat loss to the intercooler circulating water, per minute.

[**Ans.** 90 kJ/min, 375 kJ/min.]

15. A four-stage compressor works between limits of 1 bar and 112 bar. The index of compression in each stage is 1.28, the temperature at the start of compression in each stage is 32°C and the intermediate pressures are so chosen that the work is divided equally among the stages. Neglecting clearance find :

(*i*) The volume of free air delivered per kWh at 1.013 bar and 15°C.

(*ii*) The temperature at delivery from each stage.

(*iii*) The isothermal efficiency.

[**Ans.** (*i*) 6.24 m^3/kWh; (*ii*) 122°C; (*iii*) 87.87%]

16. A multi-stage air compressor is to be designed to elevate the pressure from 1 bar to 120 bar such that stage pressure ratio will not exceed 4. Determine :

(*i*) Number of stages.

(*ii*) Exact stage pressure ratio.

(*iii*) Intermediate pressures.

[**Ans.** (*i*) 4; (*ii*) 3.31; (*iii*) 36.25 bar, 10.95 bar, 3.31 bar]

17. In an ideal four-stage reciprocating air compressor, the inlet pressure is 96 kN/m^2 and inlet temperature is 300 K. The air is delivered at a pressure of 27.6 MN/m^2. The compressor is designed for the minimum power requirement and has perfect intercooling. The reversible compression and expansion processes both conform to the relation $pV^{1.2} = C$. Determine :

(*i*) Intermediate pressures.

(*ii*) The air delivery temperature.

(*iii*) The ideal isothermal efficiency.

For air, which may be assumed to be a perfect gas, the specific gas constant is 0.28702 kJ/kg K.

[**Ans.** (*i*) 395 kN/m^2; 1628 kN/m^2 and 7 is N/m^2; (*ii*) 380 K; (*iii*) 88.6%]

Rotary compressors

18. Air at 1.013 bar and 15°C is to be compressed at the rate of 5.6 m^3/min to 11.75 bar. Two machines are considered : (*i*) the roots blower ; and (*ii*) a sliding vane rotary compressor. Compare the powers required, assuming for the vane type that internal compression takes place through 75% of the pressure rise before delivery takes place, and that the compressor is an ideal uncooled machine. [**Ans.** 6.88 kW, 5.75 kW]

19. Air is compressed in a two-stage vane type compressor from 1.013 bar to 8.75 bar. Assuming equal pressure ratio in each stage, calculate the power required. Assume that in each compression is complete and that intercooling between stages is 75% complete. Calculate also the capacity of the high pressure stage in cubic metres per minute for a free air delivery of 42 m^3/min measured at 1.013 bar and 15°C. The machine is uncooled except for the intercooler and operates in an ideal manner.

[**Ans.** 187 kW ; 15.6 m^3/min]

20. A roots blower compresses 0.06 m^3 of air from 1.0 bar to 1.45 bar per revolution. Calculate the compressor efficiency. [**Ans.** 87.11%]

21. Free air of 30 m^3/min is compressed from 1.013 bar to 2.23 bar. Calculate the power required (*i*) if the compression is carried out in roots blower, (*ii*) if the compression is carried out in vane blower. Assume that there is 25% reduction in volume before the back-flow occurs, and (*iii*) the isentropic efficiency in each case.

[**Ans.** (*i*) 60.85 kW; (*ii*) 48.46 kW; (*iii*) 73.69%, 92.53%]

22. A centrifugal compressor is designed to have a pressure ratio of 3.5 : 1. The inlet eye of the compressor impeller is 30 cm in diameter. The axial velocity at inlet is 130 m/s and the mass flow is 10 kg/s. The velocity in the delivery duct is 115 m/s. The tip speed of the impeller is 450 m/s and runs at 16000 r.p.m. with total head isentropic efficiency of 78% and pressure co-efficient of 0.72. The ambient conditions are 1.013 bar and 15°C.

Calculate :

(*i*) The static pressure ratio

(*ii*) The static pressure and temperature at inlet and outlet of compressor

(*iii*) Work of compressor per kg of air, and

(*iv*) The theoretical power required.

[**Ans.** (*i*) 4.21 ; (*ii*) 0.917 bar, 279.6 K, 3.86 bar, 461.07 K ; (*iii*) 180.29 kJ/kg of air ; (*iv*) 1802.9 kW]

23. Air at a temperature of 290 K flows in a centrifugal compressor running at 20000 r.p.m. The other data given is as follows :

Slip factor = 0.80 ; Isentropic total head efficiency = 0.75 ; Outer diameter of blade tip = 500 mm. Determine :

(*i*) The temperature rise of air passing through the compressor.

(*ii*) The static pressure ratio.

Assume that the velocities of air at inlet and exit of the compressor are same. [**Ans.** (*i*) 218.62°C ; (*ii*) 4.8]

24. An axial flow air compressor of 50% reaction design has blades with inlet and outlet angles of 45° and 10° respectively. The compressor is to produce a pressure ratio of 6 : 1 with an overall isentropic efficiency of 0.85 when the air inlet temperature is 40°C. The blade speed and axial velocity are constant throughout the compressor. Assuming a value of 200 m/s for the blade speed, find the number of stages required when the work factor is (*i*) unity (*ii*) 0.89 for all stages. [**Ans.** (*i*) 9 ; (*ii*) 10]

25. A centrifugal compressor running at 9000 r.p.m. delivers 600 m^3/min of free air. The air is compressed from 1 bar and 20°C to a pressure ratio of 4 with an isentropic efficiency of 0.82. Blades are radial at outlet of impeller and the flow velocity of 62 m/s may be assumed throughout constant. The outer radius of the impeller is twice the inner and the slip factor may be assumed as 0.9. The blade area co-efficient may be assumed as 0.9 at the inlet. Calculate :

(*i*) Final temperature of air.

(*ii*) Theoretical power.

(*iii*) Impeller diameters at inlet and outlet.

(*iv*) Breadth of the impeller at inlet.

(*v*) Impeller blade angle at inlet.

(*vi*) Diffuse blade angle at inlet.

[**Ans.** (*i*) 466.85 K ; (*ii*) 2077.7 kW ; (*iii*) 46.745 cm, 94.9 cm ; (*iv*) 12.2 cm; (*v*) 15.7°; (*vi*) 8.9°]

26. A single inlet type centrifugal compressor handles 8 kg/s of air. The ambient air conditions are 1 bar and 20°C. The compressor runs at 22000 r.p.m. with isentropic efficiency of 82%. The air is compressed in the compressor from 1 bar static pressure to 4.2 bar total pressure. The air enters the impeller eye with a velocity of 150 m/s with no prewhirl. Assuming that the ratio of whirl speed to tip speed is 0.9, calculate :

(*i*) Rise in total temperature during compression if the change in K.E. is negligible.

(*ii*) The tip diameter of the impeller.

(*iii*) Power required.

(*iv*) Eye diameter if the hub diameter is 10 cm.

[**Ans.** (*i*) 167.67°C ; (*ii*) 37.56 cm ; (*iii*) 1348 kW ; (*iv*) 25.9 cm]

27. In an axial flow compressor, the overall stagnation pressure ratio achieved is 4 with overall stagnation isentropic efficiency 86 per cent. The inlet stagnation pressure and temperature are 1 bar and 320 K. The mean blade speed is 190 m/s. The degree of reaction is 0.5 at the mean radius with relative air angles of 10° and 30° at rotor inlet and outlet respectively. The work done factor is 0.9. Calculate :

(*i*) Stagnation polytropic efficiency.

(*ii*) Number of stages.

(*iii*) Inlet temperature and pressure.

(*iv*) Blade height in the first stage if the hub-tip ratio is 0.4, mass flow rate is 20 kg/s.

[**Ans.** (*i*) 88.4% ; (*ii*) 11 ; (*iii*) 287.39 K, 0.6864 bar; (*iv*) 11.4 cm]

28. A multi-stage axial flow compressor delivers 18 kg/s of air. The inlet stagnation condition is 1 bar and 20°C. The power consumed by the compressor is 4260 kW. Calculate :

(*i*) Delivery pressure.

(*ii*) Number of stages.

(*iii*) Overall isentropic efficiency of the compressor.

Assume temperature rise in the first-stage is 18°C. The polytropic efficiency of compression is 0.9 and the stage stagnation pressure ratio is constant.

[**Ans.** (*i*) 6.41 bar ; (*ii*) 10 ; (*iii*) 87.24%]

Gas Turbines and Jet Propulsion

25

25.1. GAS TURBINES—GENERAL ASPECTS

Probably a wind-mill was the first turbine to produce useful work, wherein the there is no pre-compression and no combustion. The characteristic features of a gas turbine as we think of the name today include a *compression process* and a *heat addition* (or combustion) *process*. The gas turbine represents perhaps the most satisfactory way of producing very large quantities of power in a self-contained and compact unit. The gas turbine may have a future use in conjunction with the oil engine. For smaller gas turbine units, the inefficiencies in compression and expansion processes become greater and to improve the thermal efficiency it is necessary to use a heat exchanger. In order that a small gas turbine may compete for economy with the small oil engine or petrol engine it is necessary that a compact effective heat exchanger be used in the gas turbine cycle. The thermal efficiency of the gas turbine alone is still quite modest 20 to 30% compared with that of a modern steam turbine plant 38 to 40%. It is possible to construct combined plants whose efficiencies are of order of 45% or more. Higher efficiencies might be attained in future.

The following *are* the major fields of application of gas turbines :

1. Aviation	2. Power generation
3. Oil and gas industry	4. Marine propulsion.

The efficiency of a gas turbine is not the criteria for the choice of this plant. A gas turbine is used in aviation and marine fields because it is *self contained, light weight not requiring cooling water and generally fit into the overall shape of the structure*. It is selected for power generation because of its *simplicity, lack of cooling water, needs quick installation and quick starting*. It is used in oil and gas industry because of *cheaper supply of fuel and low installation cost*.

The gas turbines have the following limitations : *(i) They are not self starting ; (ii) low efficiencies at part loads ; (iii) non-reversibility ; (iv) higher rotor speeds; and (v) overall efficiency of the plant low*.

25.2. CLASSIFICATION OF GAS TURBINES

The gas turbines are mainly divided into two groups :

1. Constant pressure combustion gas turbine

(*a*) Open cycle constant pressure gas turbine

(*b*) Closed cycle constant pressure gas turbine.

2. Constant volume combustion gas turbine

In almost *all the fields open cycle gas turbine plants are used. Closed cycle plants were introduced at one stage because of their ability to burn cheap fuel*. In between their progress remained slow because of availability of cheap oil and natural gas. Because of rising oil prices, now again, the attention is being paid to closed cycle plants.

25.3. MERITS OF GAS TURBINES

(*i*) Merits over I.C. engines :

1. The mechanical efficiency of a gas turbine (95%) is quite high as compared with I.C. engine (85%) since the I.C. engine has a large number of sliding parts.
2. A gas turbine does not require a flywheel as the torque on the shaft is continuous and uniform. Whereas a flywheel is a must in case of an I.C. engine.
3. The weight of gas turbine per H.P. developed is less than that of an I.C. engine.

4. The gas turbine can be driven at a very high speeds (40000 r.p.m.) whereas this is not possible with I.C. engines.
5. The work developed by a gas turbine per kg of air is more as compared to an I.C. engine. This is due to the fact that gases can be expanded upto atmospheric pressure in case of a gas turbine whereas in an I.C. engine expansion upto atmospheric pressure is not possible.
6. The components of the gas turbine can be made lighter since the pressures used in it are very low, say 5 bar compared with I.C. engine, say 60 bar.
7. In the gas turbine the ignition and lubrication systems are much simpler as compared with I.C. engines.
8. Cheaper fuels such as paraffine type, residue oils or powdered coal can be used whereas special grade fuels are employed in petrol engine to check knocking or pinking.
9. The exhaust from gas turbine is less polluting comparatively since excess air is used for combustion.
10. Because of low specific weight the gas turbines are particularly suitable for use in aircrafts.

Demerits of gas turbines

1. The thermal efficiency of a simple turbine cycle is low (15 to 20%) as compared with I.C. engines (25 to 30%).
2. With wide operating speeds the fuel control is comparatively difficult.
3. Due to higher operating speeds of the turbine, it is imperative to have a speed reduction device.
4. It is difficult to start a gas turbine as compared to an I.C. engine.
5. The gas turbine blades need a special cooling system.
6. One of the main demerits of a gas turbine is its *very poor thermal efficiency at part loads,* as the quantity of air remains same irrespective of load, and output is reduced by reducing the quantity of fuel supplied.
7. Owing to the use of nickel-chromium alloy, the manufacture of the blades is difficult and costly.
8. For the same output the gas turbine produces five times exhaust gases than I.C. engine.
9. Because of prevalence of high temperature (1000 K for blades and 2500 K for combustion chamber) and centrifugal force the life of the combustion chamber and blades is short/small.

(*ii*) Merits over steam turbines :

The gas turbine entails the following *advantages over steam turbines* :

1. Capital and running cost less.
2. For the same output the space required is far less.
3. Starting is more easy and quick.
4. Weight per H.P. is far less.
5. Can be installed anywhere.
6. Control of gas turbine is much easier.
7. Boiler along with accessories not required.

25.4. CONSTANT PRESSURE COMBUSTION GAS TURBINES

25.4.1. Open Cycle Gas Turbines

Refer to Fig. 25.1. The fundamental gas turbine unit is one operating on the open cycle in which a rotary compressor and a turbine are mounted on a common shaft. Air is drawn into the compressor and after compression passes to a combustion chamber. Energy is supplied in the combustion chamber by spraying fuel into the air stream, and the resulting hot gases expand through the turbine to the atmosphere. In order to achieve network output from the unit, the turbine must develop more gross work output than is required to drive the compressor and to overcome mechanical losses in the drive. The products of combustion coming out from the turbine are exhausted to the atmosphere as they cannot be used any more. The working fluids (air and fuel) must be replaced continuously as they are exhausted into the atmosphere.

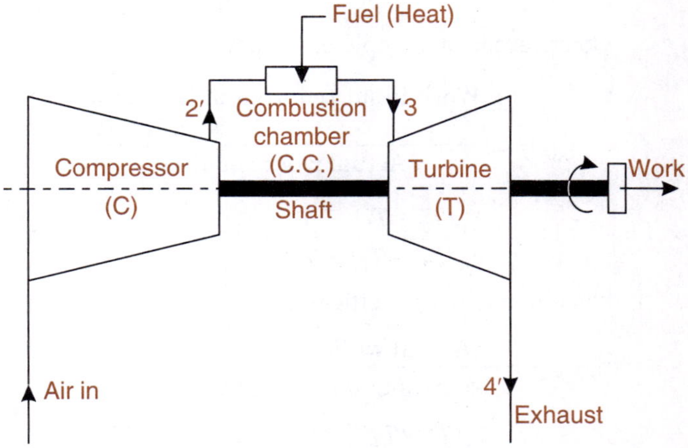

Fig. 25.1. Open cycle gas turbine.

If pressure loss in the combustion chamber is neglected, this cycle may be drawn on a *T-s* diagram as shown in Fig. 25.2.

- 1-2′ represents : *Irreversible adiabatic compression.*

- 2′-3 represents : *Constant pressure heat supply in the combustion chamber.*
- 3-4′ represents : *Irreversible adiabatic expansion.*
- 1–2 represents : *Ideal isentropic compression.*
- 3–4 represents : *Ideal isentropic expansion.*

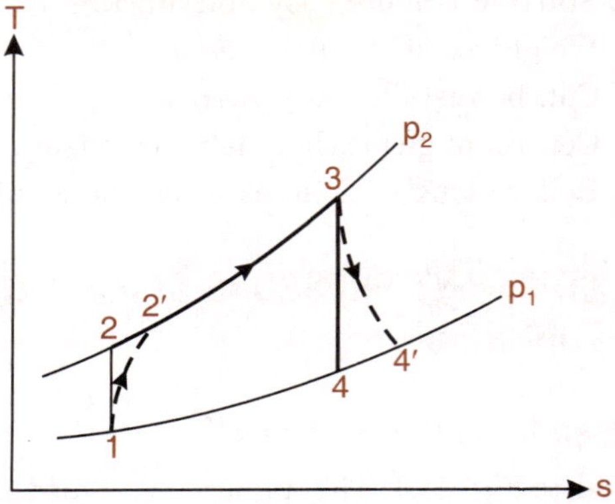

Fig. 25.2

Assuming change in kinetic energy between the various points in the cycle to be negligibly small compared with enthalpy changes and then applying the flow equation to each part of cycle, for unit mass, we have

Work input (compressor) $= c_p\,(T_2' - T_1)$

Heat supplied (combustion chamber) $= c_p\,(T_3 - T_2')$

Work output (turbine) $= c_p\,(T_3 - T_4')$

$\therefore$ Network output = Work output – Work input

$$= c_p\,(T_3 - T_4') - c_p(T_2' - T_1)$$

and

$$\eta_{thermal} = \frac{\text{Network output}}{\text{Heat supplied}} = \frac{c_p\,(T_3 - T_4') - c_p\,(T_2' - T_1)}{c_p(T_3 - T_2')}$$

Compressor isentropic efficiency,

$$\eta_{comp} = \frac{\text{Work input required in isentropic compression}}{\text{Actual work required}}$$

$$= \frac{c_p(T_2 - T_1)}{c_p(T_2' - T_1)} = \frac{T_2 - T_1}{T_2' - T_1} \quad ...(25.1)$$

Turbine isentropic efficiency,

$$\eta_{turbine} = \frac{\text{Actual work output}}{\text{Isentropic work output}}$$

$$= \frac{c_p\,(T_3 - T_4')}{c_p(T_3 - T_4)} = \frac{T_3 - T_4'}{T_3 - T_4} \quad ...(25.2)$$

Note. With the variation in temperature, the value of the specific heat of a real gas varies, and also in the open cycle, the specific heat of the gases in the combustion chamber and in turbine is different from that in the compressor because fuel has been added and a chemical change has taken place. Curves showing the variation of c_p with temperature and air/fuel ratio can be used, and a suitable mean value of c_p and hence γ can be found out. It is usual in gas turbine practice to assume fixed mean value of c_p and γ for the expansion process, and fixed mean values of c_p and γ for the compression process. In an open cycle gas turbine unit the mass flow of gases in turbine is greater than that in compressor due to mass of fuel burned, but it is possible to neglect mass of fuel, since the air/fuel ratios used are large. Also, in many cases, air is bled from the compressor for cooling purposes, or in the case of air-craft at high altitudes, bled air is used for de-icing and cabin air-conditioning. This amount of air bled is approximately the same as the mass of fuel injected therein.

25.4.2. Methods for Improvement of Thermal Efficiency of Open Cycle Gas Turbine Plant

The following methods are employed *to increase the specific output and thermal efficiency of the plant :*

1. Intercooling 2. Reheating
3. Regeneration.

1. Intercooling. A compressor in a gas turbine cycle utilises the major percentage of power developed by the gas turbine. The work required by the compressor can be reduced by compressing the air in two stages and *incorporating an intercooler* between the two as shown in Fig. 25.3. The corresponding *T-s* diagram for the unit is shown in Fig. 25.4. The actual processes take place as follows :

1-2′	...	L.P. (Low pressure) compression
2′-3	...	Intercooling
3-4′	...	H.P. (High pressure) compression
4′-5	...	C.C. (Combustion chamber)-heating
5-6′	...	T (Turbine)-expansion

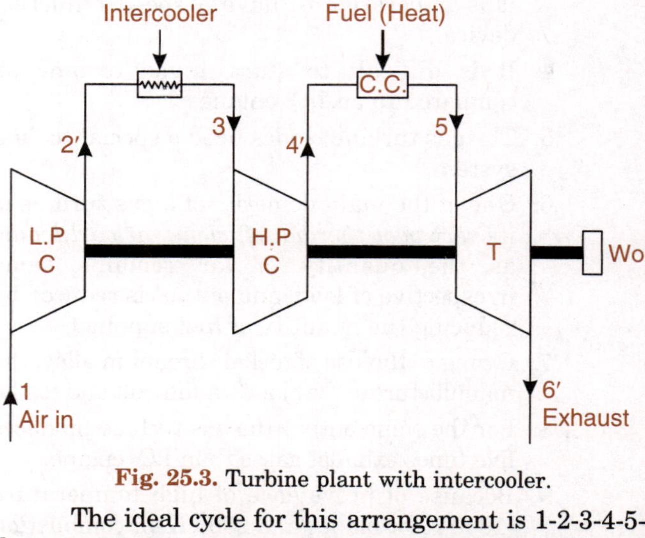

Fig. 25.3. Turbine plant with intercooler.

The ideal cycle for this arrangement is 1-2-3-4-5-6 ; the compression process without intercooling is shown as 1-*L*′ in the actual case, and 1-*L* in the ideal isentropic case.

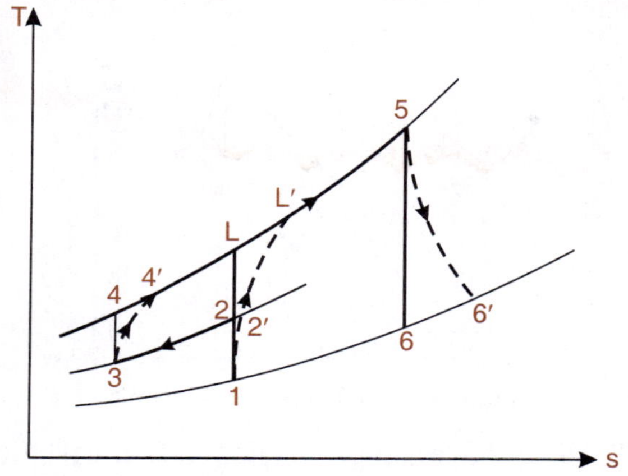

Fig. 25.4. *T-s* diagram for the unit.

Now,

Work input (*with intercooling*)

$= c_p(T_2' - T_1) + c_p(T_4' - T_3)$ ...(25.3)

Work input (*without intercooling*)

$= c_p(T_L' - T_1) = c_p(T_2' - T_1) + c_p(T_L' - T_2')$ ...(25.4)

By comparing equation (25.4) with equation (25.3) it can be observed that the *work input with intercooling is less than the work input with no intercooling*, when $c_p (T_4' - T_3)$ is *less* than $c_p(T_L' - T_2')$. This is so if it is assumed that isentropic efficiencies of the two compressors, operating separately, are each equal to the isentropic efficiency of the single compressor which would be required if no intercooling were used. Then $(T_4' - T_3) < (T_L' - T_2')$ since the pressure lines diverge on the *T-s* diagram from left to the right.

$$\text{Again, work ratio} = \frac{\text{Network output}}{\text{Gross work output}}$$

$$= \frac{\text{Work of expansion} - \text{Work of compression}}{\text{Work of expansion}}$$

From this we may conclude that *when the compressor work input is reduced then the work ratio is increased.*

However, the heat supplied in the combustion chamber when intercooling is used in the cycle, is given by,

Heat supplied *with intercooling* $= c_p(T_5 - T_4')$

Also the heat supplied when intercooling is not used, with the same maximum cycle temperature T_5, is given by

Heat supplied *without intercooling* $= c_p (T_5 - T_L')$

Thus, the *heat supplied when intercooling is used is greater than with no intercooling. Although the network output is increased by intercooling it is found in general that the increase in heat to be supplied causes the thermal efficiency to decrease.* When intercooling is used a supply of cooling water must be readily available. The additional bulk of the unit may offset the advantage to be gained by increasing the work ratio.

2. **Reheating.** The output of a gas turbine can be amply improved by expanding the gases in two stages with a *reheater* between the two as shown in Fig. 25.5. The H.P. turbine drives the compressor and the L.P. turbine provides the useful power output. The corresponding *T-s* diagram is shown in Fig. 25.6. The line 4′-*L*′ represents the expansion in the L.P. turbine if reheating is *not* employed.

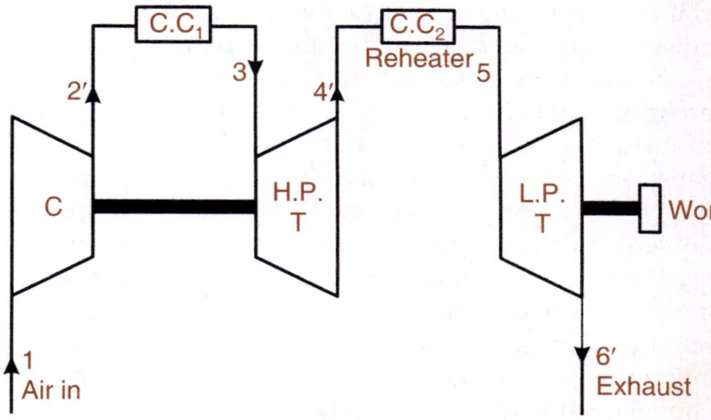

Fig. 25.5. Gas turbine with reheater.

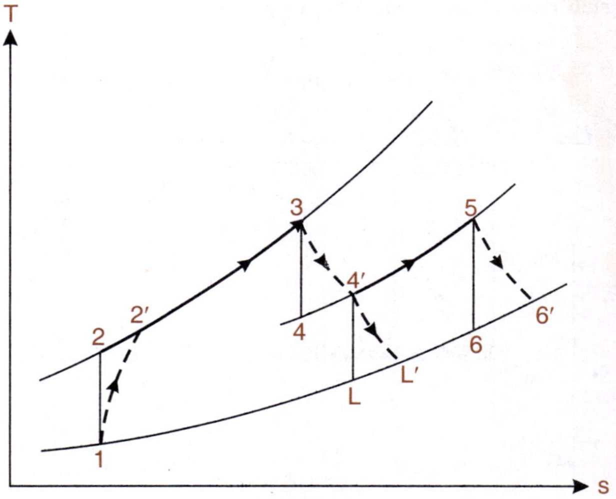

Fig. 25.6. *T-s* diagram for the unit.

Neglecting mechanical losses the work output of the H.P. turbine must be exactly equal to the work input required for the compressor i.e., $c_{pa} (T_2' - T_1) = c_{pg} (T_3 - T_4')$

The work output (net output) of L.P. turbine is given by,

Network output (with reheating) $= c_{pg} (T_5 - T_6')$

and Network output (without reheating) $= c_{pg} (T_4' - T_L')$

Since the pressure lines diverge to the right on *T-s* diagram it can be seen that the temperature difference $(T_5 - T_6')$ is always *greater* than $(T_4' - T_L')$, so that *reheating increases the network output.*

Although network is increased by reheating the heat to be supplied is also increased, and the net effect *can be to reduce the thermal efficiency*

Heat supplied $= c_{pg}(T_3 - T_2') + c_{pg}(T_5 - T_4')$.

Note. c_{pa} and c_{pg} stand for specific heats of air and gas respectively at constant pressure.

3. **Regeneration.** The exhaust gases from a gas turbine carry a large quantity of heat with them since their temperature is far above the ambient temperature. They can be used to heat the air coming from the compressor thereby reducing the mass of fuel supplied in the combustion chamber. Fig. 25.7 shows a gas turbine plant with a regenerator. The corresponding *T-s* diagram is shown in Fig. 25.8. 2′–3 represents the heat flow into the compressed air during its passage through the heat exchanger and 3–4 represents the heat taken in from the combustion of fuel. Point 6 represents the temperature of exhaust gases at discharge from the heat exchanger. The maximum temperature to which the air could be heated in the heat exchanger is ideally that of exhaust gases, but less than this is obtained in practice because a temperature gradient must exist for an unassisted transfer of energy. The *effectiveness* of the heat exchanger is given by :

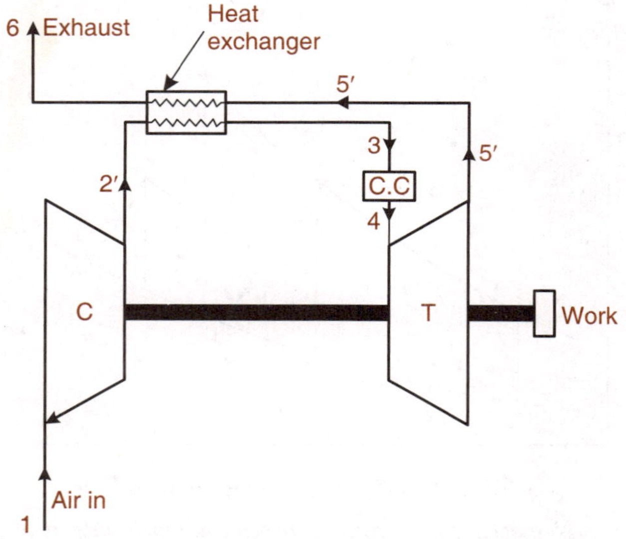

Fig. 25.7. Gas turbine with regenerator.

Effectiveness,

$$\varepsilon = \frac{\text{Increase in enthalpy per kg of air}}{\text{Available increase in enthalpy per kg of air}}$$

$$= \frac{(T_3 - T_2')}{(T_5' - T_2')} \qquad \text{...(25.5)}$$

(assuming c_{pa} and c_{pg} to be equal)

A heat exchanger is usually used in large gas turbine units for marine propulsion or industrial power.

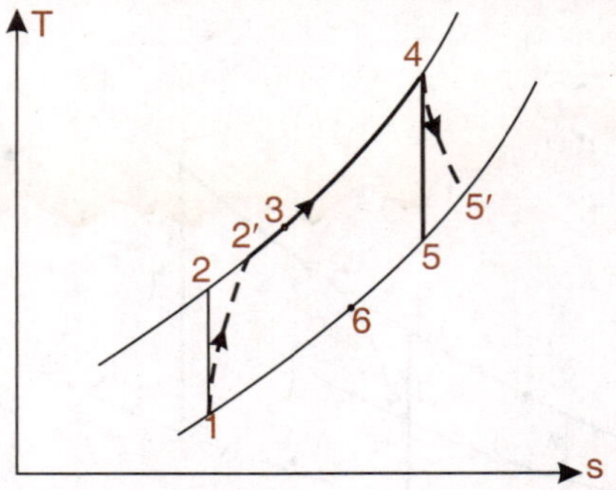

Fig. 25.8. *T-s* diagram for the unit.

25.4.3. Effect of Operating Variables on Thermal Efficiency

The thermal efficiency of *actual open cycle* depends on the following thermodynamic variables :

(*i*) Pressure ratio

(*ii*) Turbine inlet temperature (T_3)

(*iii*) Compressor inlet temperature (T_1)

(*iv*) Efficiency of the turbine ($\eta_{turbine}$)

(*v*) Efficiency of the compressor (η_{comp}).

Effect of turbine inlet temperature and pressure ratio :

If the permissible turbine inlet-temperature (with the other variables being constant) of an *open cycle gas turbine power plant* is increased its *thermal efficiency* is *amply improved*. A practical limitation to increasing the turbine inlet temperature, however, is the ability of the material available for the turbine blading to *withstand the high rotative and thermal stresses.*

Refer to Fig. 25.9. For *a given turbine inlet temperature, as the pressure ratio increases, the heat supplied as well as the heat rejected are reduced. But the ratio of change of heat supplied is not the same as the ratio of change heat rejected.* As a consequence, there *exists an optimum pressure ratio producing maximum thermal efficiency for a given turbine inlet temperature.*

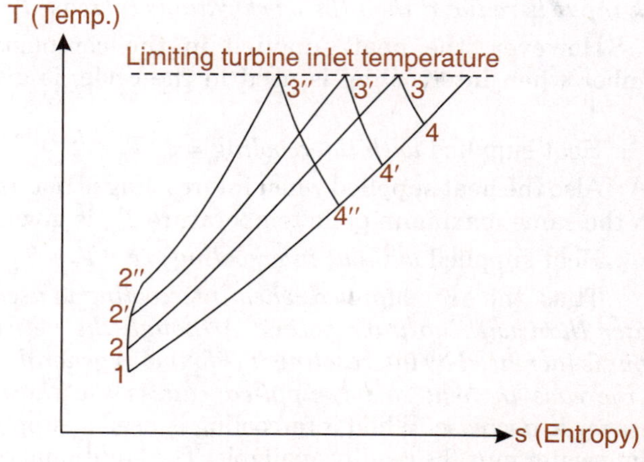

Fig. 25.9

As the pressure ratio increases, the thermal efficiency also increases until it becomes maximum and then it drops off with a further increase in pressure ratio (Fig. 25.10). Further, as the turbine inlet temperature increases, the peaks of the curves flatten out giving a greater range of ratios of pressure optimum efficiency.

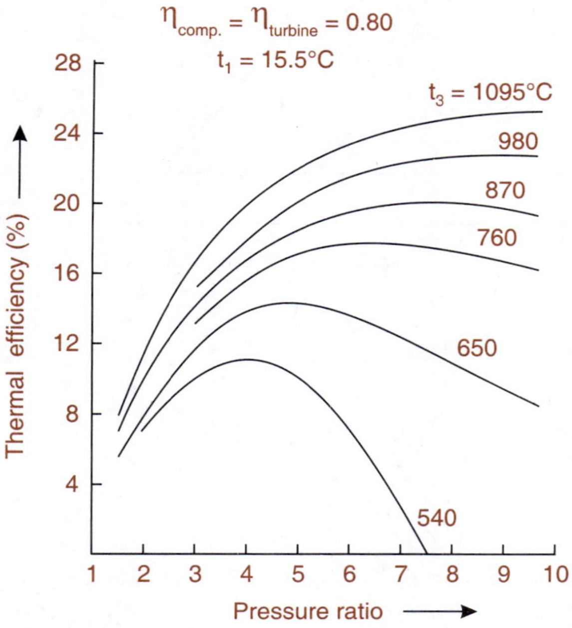

Fig. 25.10. Effect of pressure ratio and turbine inlet temperature.

Following particulars are worthnoting :

Gas temperatures	*Efficiency (gas turbine)*
550 to 600°C	20 to 22%
900 to 1000°C	32 to 35%
Above 1300°C	more than 50%

Effect of turbine and compressor efficiencies :

Refer to Fig. 25.11. The thermal efficiency of the actual gas turbine cycle is very sensitive to variations in the efficiencies of the compressor and turbine. There is a particular pressure ratio at which maximum efficiencies occur. For lower efficiencies, the peak of the thermal efficiency occurs at lower pressure ratios and vice versa.

Effect of compressor inlet temperature :

Refer to Fig. 25.12 (on next page). *With the decrease in the compressor inlet temperature there is increase in thermal efficiency of the plant.* Also the peaks of thermal efficiency occur at high pressure ratios and the curves become flatter giving thermal efficiency over a wider pressure ratio range.

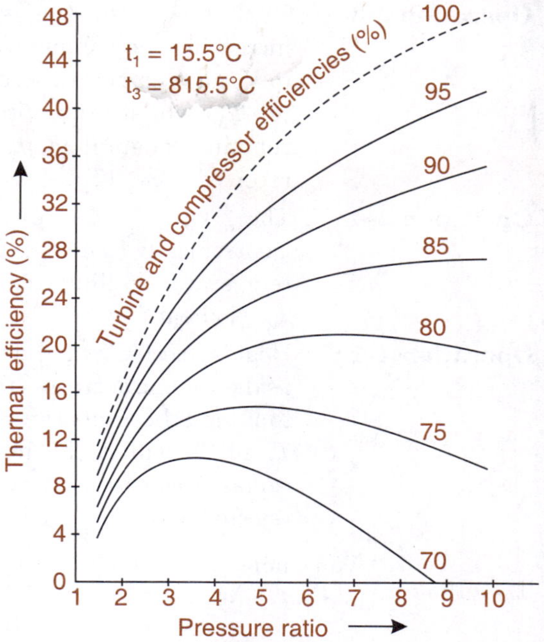

Fig. 25.11. Effect of compressor efficiency.

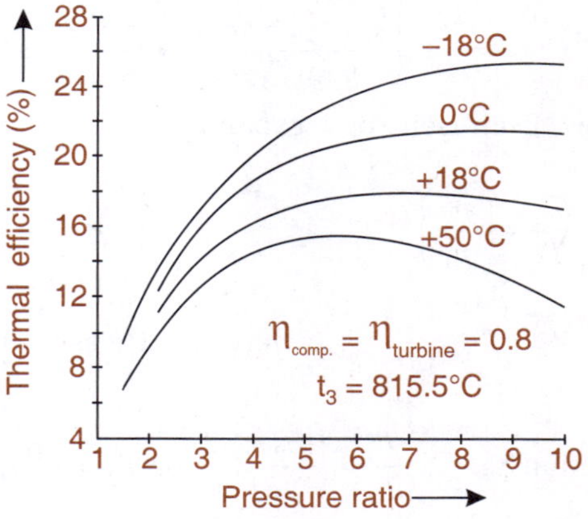

Fig. 25.12

25.4.4. Closed Cycle Gas Turbine (Constant Pressure or Joule Cycle)

Fig. 25.13 shows a gas turbine operating on a constant pressure cycle in which the closed system consists of air behaving as an ideal gas. The various operations are as follows : Refer Figs. 25.14 and 25.15.

Operation 1–2 : The air is compressed isentropically from the lower pressure p_1 to the upper pressure p_2, the temperature rising from T_1 to T_2. No heat flow occurs.

Operation 2–3 : Heat flow into the system increasing the volume from V_2 to V_3 and temperature from T_2 to T_3 whilst the pressure remains constant at p_2. Heat received = $mc_p (T_3 - T_2)$.

Operation 3–4 : The air is expanded isentropically from p_2 to p_1, the temperature falling from T_3 to T_4. No heat flow occurs.

Operation 4–1 : Heat is rejected from the system as the volume decreases from V_4 to V_1 and the temperature from T_4 to T_1 whilst the pressure remains constant at p_1. Heat rejected = $mc_p (T_4 - T_1)$

$$\eta_{air\text{-}standard} = \frac{\text{Work done}}{\text{Heat received}}$$

$$= \frac{\text{Heat received/cycle} - \text{Heat rejected/cycle}}{\text{Heat received/cycle}}$$

$$= \frac{mc_p(T_3 - T_2) - mc_p(T_4 - T_1)}{mc_p(T_3 - T_2)} = 1 - \frac{T_4 - T_1}{T_3 - T_2}$$

Now, from isentropic expansion

$$\frac{T_2}{T_1} = \left(\frac{p_2}{p_1}\right)^{\frac{\gamma-1}{\gamma}}$$

$$T_2 = T_1 \, (r_p)^{\frac{\gamma-1}{\gamma}}, \text{ where } r_p = \text{Pressure ratio}$$

Similarly, $\frac{T_3}{T_4} = \left(\frac{p_2}{p_1}\right)^{\frac{\gamma-1}{\gamma}}$ or $T_3 = T_4 \, (r_p)^{\frac{\gamma-1}{\gamma}}$

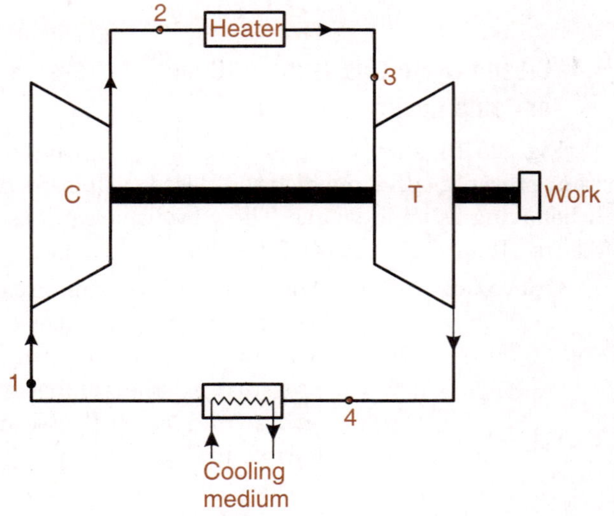

Fig. 25.13. Closed cycle gas turbine.

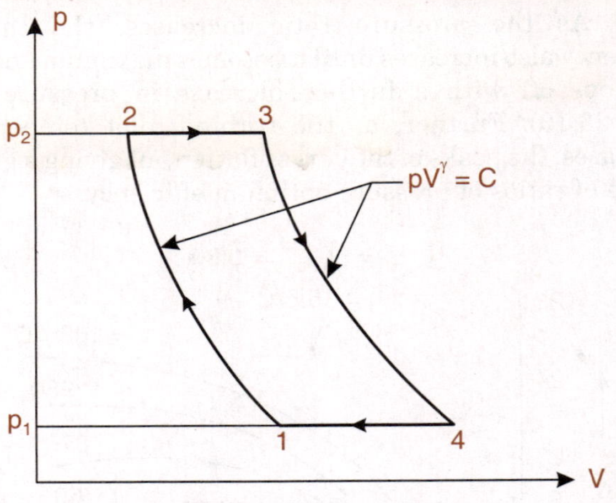

Fig. 25.14. *p-V* diagram.

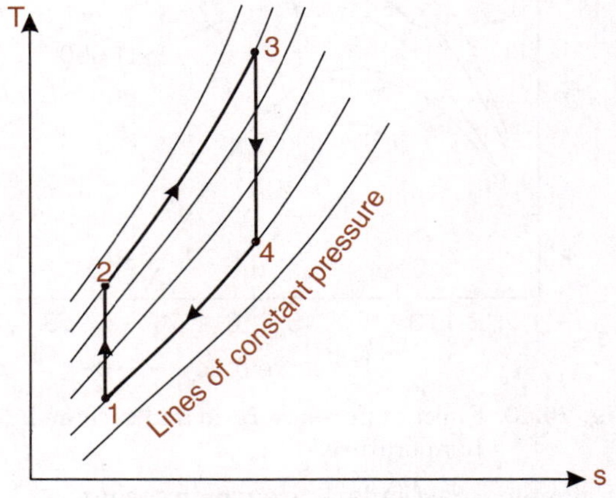

Fig. 25.15. *T-s* diagram.

$$\therefore \quad \eta_{air\text{-}standard} = 1 - \frac{T_4 - T_1}{T_4 (r_p)^{\frac{\gamma-1}{\gamma}} - T_1 (r_p)^{\frac{\gamma-1}{\gamma}}}$$

$$= 1 - \frac{1}{(r_p)^{\frac{\gamma-1}{\gamma}}} \quad \quad ...(25.6)$$

The expression shows that the *efficiency of the ideal joule cycle increases with the pressure ratio. The absolute limit of pressure is determined by the limiting temperature of the material of the turbine at the point at which this temperature is reached by the compression process alone, no further heating of the gas in the combustion chamber would be permissible and the work of expansion would ideally just balance the work of compression so that no excess work would be available for external use.*

Now we shall prove that the *pressure ratio for maximum work is a function of the limiting temperature ratio.*

Work output during the cycle

= Heat received/cycle – Heat rejected/cycle

= $mc_p (T_3 - T_2) - mc_p (T_4 - T_1)$

$$= mc_p\,(T_3 - T_4) - mc_p\,(T_2 - T_1)$$

$$= mc_pT_3\left(1 - \frac{T_4}{T_3}\right) - T_1\left(\frac{T_2}{T_1} - 1\right)$$

In case of a given turbine the minimum temperature T_1 and the maximum temperature T_3 are prescribed, T_1 being the temperature of the atmosphere and T_3 the maximum temperature which the metals of turbine would withstand. Consider the specific heat at constant pressure c_p to be constant. Then,

$$\text{Since,}\quad \frac{T_3}{T_4} = (r_p)^{\frac{\gamma - 1}{\gamma}} = \frac{T_2}{T_1}$$

Using the constant 'z' $= \dfrac{\gamma - 1}{\gamma}$, we have, work output/cycle

$$W = K\left[T_3\left(1 - \frac{1}{r_p^{\,z}}\right) - T_1(r_p^{\,z} - 1)\right]$$

Differentiating with respect to r_p

$$\frac{dW}{dr_p} = K\left[T_3 \times \frac{z}{r_p(z+1)} - T_1 z r_p^{(z-1)}\right]$$

$$= 0 \text{ for a maximum}$$

$$\therefore\quad \frac{zT_3}{r_p^{(z+1)}} = T_1 z (r_p)^{(z-1)}$$

$$\therefore\quad r_p^{\,2z} = \frac{T_3}{T_1}$$

$$\therefore\quad r_p = (T_3/T_1)^{1/2z} \text{ i.e., } r_p = (T_3/T_1)^{\frac{\gamma}{2(\gamma - 1)}}$$

Thus the *pressure ratio for maximum work is a function of the limiting temperature ratio.*

Fig. 25.16 shows an arrangement of closed cycle stationary gas turbine plant in which air is continuously circulated. This ensures that the air is not polluted by the addition of combustion waste product, since the heating of air is carried out in the form of heat exchanger shown in the diagram as air heater. The air exhausted from the power turbine is cooled before readmission to L.P. compressor. The various operations as indicated on T-s diagram (Fig. 25.17) are as follows :

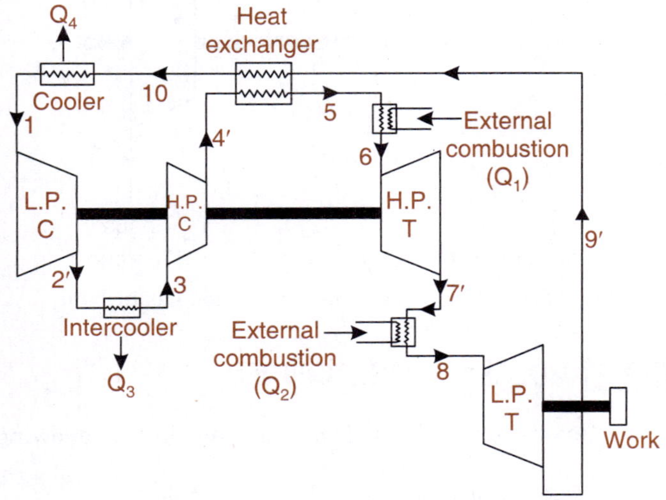

Fig. 25.16. Closed cycle gas turbine plant.

Operation 1-2′ : Air is compressed from p_1 to p_x in the L.P. compressor.

Operation 2′-3 : Air is cooled in the intercooler at constant pressure p_x.

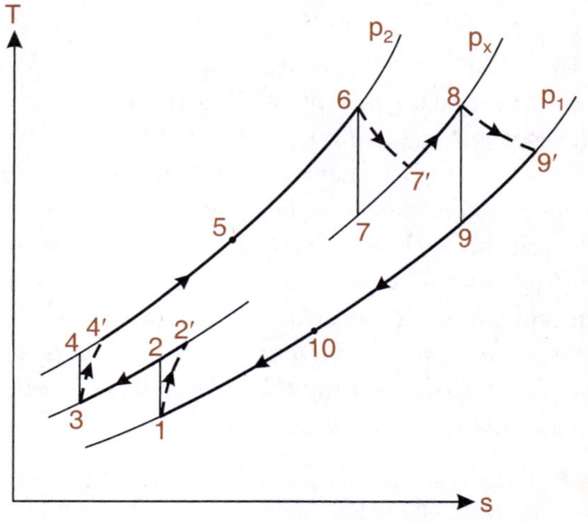

Fig. 25.17. T-s diagram for the plant.

Operation 3–4′ : Air is compressed in the H.P. compressor from p_x to p_2.

Operation 4′–5 : High pressure air is heated at constant pressure by exhaust gases from power turbine in the heat exchanger to T_5.

Operation 5–6 : High pressure air further heated at constant pressure to the maximum temperature T_6 by an air heater (through external combustion).

Operation 6–7′ : The air is expanded in the H.P. turbine from p_2 to p_x producing work to drive the compressor.

Operation 7′–8 : Exhaust air from the H.P. turbine is heated at constant pressure in the air heater (through external combustion) to the maximum temperature $T_8(= T_6)$.

Operation 8–9′ : The air is expanded in the L.P. turbine from p_x to p_1, producing energy for a flow of work externally.

Operation 9′–10 : Air from L.P. turbine is passed to the heat exchanger where energy is transferred to the air delivered from the H.P. compressor. The temperature of air leaving the heat exchanger and entering the cooler is T_{10}.

Operation 10–11 : Air cooled to T_1 by the cooler before entering the L.P. compressor.

The energy balance for the whole plant is as follows :

$$Q_1 + Q_2 - Q_3 - Q_4 = W$$

In a closed cycle plant, in practice, the control of power output is achieved by varying the mass flow by the use of a reservoir in the circuit. The *reservoir maintains the design pressure and temperature and therefore achieves an approximately constant level of efficiency for varying* loads. In this cycle since it is closed, *gases other than air with favourable properties can be used* ; furthermore it is possible to burn solid fuels in the combustion heaters. *The major factor responsible for inefficiency in this cycle is the large irreversible temperature drop which occurs in the air heaters between the furnace and circulating gas.*

Note **1.** In a closed cycle gas turbines, although air has been extensively used, the use of 'helium' which though of a lower density, has been inviting the attention of manufacturers for its use, for large output gas turbine units. *The specific heat of helium at constant pressure is about 'five times' that of air, therefore for each kg mass flow the heat drop and hence energy dealt within helium machines is nearly five times of those in case of air.* The *surface area of the heat exchanger for helium can be kept as low as 1/3* of that *required for gas turbine plant using air as working medium.* For the same temperature ratio and for the plants of the same output the *cross-sectional area required for helium is much less than that for air.* It may therefore be concluded that the *size of helium unit is considerably small comparatively.*

2. Some gas turbine plants work on a combination of two cycles the open cycle and the closed cycle. Such a combination is called the *semi-closed cycle. Here a part of the working fluid is confined within the plant and another part flows from and to atmosphere.*

25.4.5. Merits and Demerits of Closed Cycle Gas Turbine Over Open Cycle Gas Turbine

Merits of closed cycle :

1. Higher thermal efficiency
2. Reduced size
3. No contamination
4. Improved heat transmission
5. Improved part load efficiency
6. Lesser fluid friction
7. No loss of working medium
8. Greater output
9. Inexpensive fuel.

Demerits of closed cycle :

1. Complexity
2. Large amount of cooling water is required. This limits its use to stationary installation or marine use where water is available in abundance.
3. Dependent system.
4. The weight of the system per H.P. developed is high comparatively, therefore not economical for moving vehicles.
5. Requires the use of a very large air heater.

25.5. CONSTANT VOLUME COMBUSTION TURBINES

Refer to Fig. 25.18. In a constant volume combustion turbine, the compressed air from an air compressor C is admitted into the combustion chamber D through the valve A. When the valve A is closed, the fuel is admitted into the chamber by means of a fuel pump P. Then the mixture is ignited by means of a spark plug S. The combustion takes place at constant volume with increase of pressure. The valve B opens and the hot gases flow to the turbine T, and finally, they are discharged, into atmosphere. The energy of the hot gases is thereby converted into mechanical energy. For continuous running of the turbine these operations are repeated.

The main demerit associated with this type of turbine is that the *pressure difference and velocities of hot gases are not constant ; so the turbine speed fluctuates.*

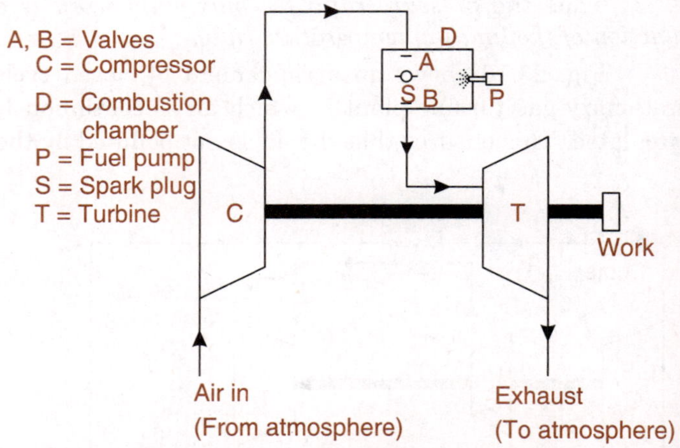

Fig. 25.18. Constant volume combustion gas turbine.

25.6. USES OF GAS TURBINES

Gas turbines find wide *applications* in the following *fields* :

1. Supercharging
2. Turbo-jet and turbo-propeller engines

3. Marine field
4. Railway
5. Road transport
6. Electric power generation
7. Industry.

25.7. GAS TURBINE FUELS

The various fuels used in gas turbines are enumerated and discussed below :

1. Gaseous fuels
2. Liquid fuels
3. Solid fuels

1. **Gaseous fuels.** *Natural gas is the ideal fuel for gas turbines*, but this is not available everywhere.

Blast furnace and *producer gases* may also be used for gas turbine power plants.

2. **Liquid fuels.** Liquid fuels of petroleum origin such as distillate oils or residual oils are most commonly used for gas turbine plant. The essential qualities of these fuels include *proper volatility, viscosity and calorific value.* At the same time it *should be* free *from any contents of moisture* and *suspended impurities that would log the small passages of the nozzles and damage valves and plungers of the fuel pumps.*

Minerals like *sodium, vanadium* and *calcium* prove *very harmful* for the *turbine blading* as these build deposits or corrode the blades. The sodium in ash should be less than 30% of the vanadium content as otherwise the ratio tends to be critical. The actual sodium content may be between 5 ppm to 10 ppm (part per million). If the vanadium is over 2 ppm, the magnesium in ash tends to become critical. *It is necessary that the magnesium in ash is at least three times the quantity of vanadium.* The content of calcium and lead should not be over 10 ppm and 5 ppm respectively.

Sodium is removed from residual oils by mixing with 5% of water and then double centrifuging when sodium leaves with water. Magnesium is added to the washed oil in the form of epsom salts, before the oil is sent into the combustor. This checks the corrosive action of vanadium. Residual oils burn with less ease than distillate oils and the latter are often used to start the unit from cold, after which the residual oils are fed in the combustor. In cold conditions residual oils need to be preheated.

3. **Solid fuels.** The use of solid fuels such as coal in pulverised form in gas turbines presents several difficulties most of which have been only partially overcome yet. The pulverising plant for coal in gas turbines applications is much lighter and small than its counterpart in steam generators. *Introduction of fuel in the combustion chamber of a gas turbine is required to be done against a high pressure whereas the pressure in the furnace of a steam plant is atmospheric.* Furthermore, the *degree of completeness of combustion in gas turbine applications has to be very high as otherwise soot and dust in gas would deposit on the turbine blading.*

Some practical applications of solid fuel burning in turbine combustors have been commercially, made available in recent years. In one such design finely crushed coal is used instead of pulverised fuel. This fuel is carried in stream of air tangentially into one end of a cylindrical furnace while gas comes out at the centre of opposite end. As the fuel particles roll around the circumference of the furnace they are burnt and a high temperature of about 1650°C is maintained which causes the mineral matter of fuel to be converted into a liquid slag. The slag covers the walls of the furnace and runs out through a top hole in the bottom. The result is that fly-ash is reduced to a very small content in the gases. In *another design* a regenerator is used to transfer the heat to air, the combustion chamber being located on the outlet of the turbine, and the combustion is carried out in the turbine exhaust stream. The advantage is that only clean air is handled by the turbine.

Example 25.1. *The air enters the compressor of an open cycle constant pressure gas turbine at a pressure of 1 bar and temperature of 20°C. The pressure of the air after compression is 4 bar. The isentropic efficiencies of compressor and turbine are 80% and 85% respectively. The air-fuel ratio used is 90 : 1. If flow rate of air is 3.0 kg/s, find :*

(*i*) *Power developed.*

(*ii*) *Thermal efficiency of the cycle.*

Assume $c_p = 1.0$ *kJ/kg K and* $\gamma = 1.4$ *of air and gases*

Calorific value of fuel = 41800 kJ/kg.

Solution. $p_1 = 1$ bar ; $T_1 = 20 + 273 = 293$ K

$p_2 = 4$ bar ; $\eta_{compressor} = 80\%$;

$\eta_{turbine} = 85\%$

Air-fuel ratio = 90 : 1 ; Air flow rate, $m_a = 3.0$ kg/s

(*i*) Power developed, P :

Refer to Fig. 25.19 (*b*)

$$\frac{T_2}{T_1} = \left(\frac{p_2}{p_1}\right)^{\frac{\gamma-1}{\gamma}} = \left(\frac{4}{1}\right)^{\frac{1.4-1}{1.4}} = 1.486$$

$$\therefore \quad T_2 = (20 + 273) \times 1.486 = 435.4 \text{ K}$$

$$\eta_{compressor} = \frac{T_2 - T_1}{T_2' - T_1}$$

$$0.8 = \frac{435.4 - 293}{T_2' - 293}$$

$$\therefore \quad T_2' = \frac{435.4 - 293}{0.8} + 293 = 471 \text{ K}$$

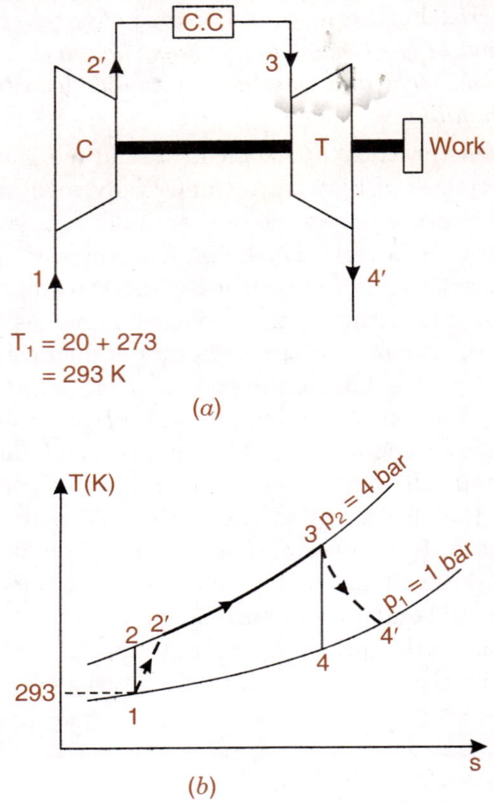

Fig. 25.19

Heat supplied by fuel = Heat taken by burning gases

$$m_f \times C = (m_a + m_f)\, c_p(T_3 - T_2')$$

(where m_a = mass of air, m_f = mass of fuel)

$$\therefore \quad C = \left(\frac{m_a}{m_f} + 1\right) c_p\,(T_3 - T_2')$$

$$\therefore \quad 41800 = (90 + 1) \times 1.0 \times (T_3 - 471)$$

i.e.,
$$T_3 = \frac{41800}{91} + 471 = 930 \text{ K}$$

Again,
$$\frac{T_4}{T_3} = \left(\frac{p_4}{p_3}\right)^{\frac{\gamma-1}{\gamma}} = \left(\frac{1}{4}\right)^{0.4/1.4} = 0.672$$

$$\therefore \quad T_4 = 930 \times 0.672 = 624.9 \text{ K}$$

$$\eta_{turbine} = \frac{T_3 - T_4'}{T_3 - T_4}$$

$$0.85 = \frac{930 - T_4'}{930 - 824.9}$$

$$\therefore \quad T_4' = 930 - 0.85\,(930 - 624.9) = 670.6 \text{ K}$$

$$W_{turbine} = m_g \times c_p \times (T_3 - T_4')$$

(where m_g is the mass of hot gases formed per kg of air)

$$\therefore \quad W_{turbine} = \left(\frac{90+1}{90}\right) \times 1.0 \times (930 - 670.6)$$
$$= 262.28 \text{ kJ/kg of air.}$$

$$W_{compressor} = m_a \times c_p \times (T_2' - T_1)$$
$$= 1 \times 1.0 \times (471 - 293) = 178 \text{ kJ/kg of air}$$

$$W_{net} = W_{turbine} - W_{compressor}$$
$$= 262.28 - 178 = 84.28 \text{ kJ/kg of air.}$$

Hence, *power developed,*

$P = 84.28 \times 3 =$ **252.84 kW/kg of air.** **(Ans.)**

(*ii*) Thermal efficiency of cycle, $\eta_{thermal}$:

Heat supplied per kg of air passing through combustion chamber

$$= \frac{1}{90} \times 41800 = 464.44 \text{ kJ/kg of air}$$

$$\therefore \quad \eta_{thermal} = \frac{\text{Work output}}{\text{Heat supplied}} = \frac{84.28}{464.44}$$

= **0.1814** or **18.14%.** **(Ans.)**

Example 25.2. *A gas turbine unit has a pressure ratio of 6 : 1 and maximum cycle temperature of 610°C. The isentropic efficiencies of the compressor and turbine are 0.80 and 0.82 respectively. Calculate the power output in kilowatts of an electric generator geared to the turbine when the air enters the compressor at 15°C at the rate of 16 kg/s.*

Take c_p = 1.005 kJ/kg K and γ = 1.4 for the compression process, and take c_p = 1.11 kJ/kg K and γ = 1.333 for the expansion process.

Solution. $T_1 = 15 + 273 = 288$ K ;

$$T_3 = 610 + 273 = 883 \text{ K} ; \quad \frac{p_2}{p_1} = 6,$$

$$\eta_{compressor} = 0.80 ; \ \eta_{turbine} = 0.82 ;$$

Air flow rate = 16 kg/s

For compression process : c_p = 1.005 kJ/kg K, γ = 1.4

For expansion process : c_p = 1.11 kJ/kg K, γ = 1.333

In order to evaluate the network output it is necessary to calculate temperatures T_2' and T_4'. To calculate T_2' we must first calculate T_2 and then use the isentropic efficiency.

For an isentropic process,

$$\frac{T_2}{T_1} = \left(\frac{p_2}{p_1}\right)^{\frac{\gamma-1}{\gamma}} = (6)^{\frac{1.4-1}{1.4}} = 1.67$$

$$\therefore \quad T_2 = 288 \times 1.67 = 481 \text{ K}$$

Also,
$$\eta_{compressor} = \frac{T_2 - T_1}{T_2' - T_1}$$

$$0.8 = \frac{481 - 288}{T_2' - T_1}$$

$$\therefore \quad T_2' = \frac{481 - 288}{0.8} + 288 = 529 \text{ K}$$

Similarly for the turbine,

$$\frac{T_3}{T_4} = \left(\frac{p_3}{p_4}\right)^{\frac{\gamma-1}{\gamma}} = \left(\frac{p_2}{p_1}\right)^{\frac{\gamma-1}{\gamma}} = (6)^{\frac{1.333-1}{1.333}} = 1.565$$

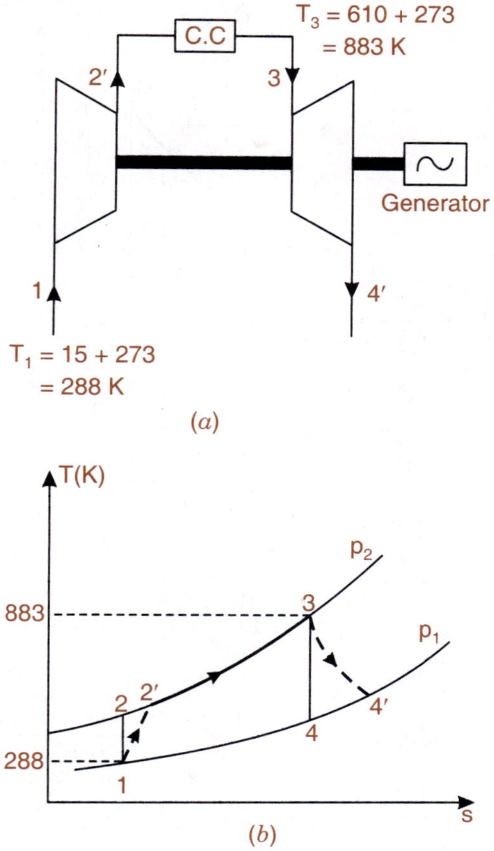

Fig. 25.20

$$\therefore \quad T_4 = \frac{T_3}{1.565} = \frac{883}{1.565} = 564 \text{ K}$$

Also,
$$\eta_{turbine} = \frac{T_3 - T_4'}{T_3 - T_4} = \frac{883 - T_4'}{883 - 564}$$

$$\therefore \quad 0.82 = \frac{883 - T_4'}{883 - 564}$$

$$\therefore \quad T_4' = 883 - 0.82\,(883 - 564) = 621.4 \text{ K}$$

Hence,

Compressor work input,
$$W_{compressor} = c_p\,(T_2' - T_1)$$
$$= 1.005\,(529 - 288) = 242.2 \text{ kJ/kg}$$

Turbine work output,
$$W_{turbine} = c_p\,(T_3 - T_4')$$
$$= 1.11\,(883 - 621.4) = 290.4 \text{ kJ/kg}$$

∴ Network output,
$$W_{net} = W_{turbine} - W_{compressor}$$
$$= 290.4 - 242.2 = 48.2 \text{ kJ/kg}$$

Power in kilowatts

= 48.2 × 16 = **771.2 kW.** **(Ans.)**

Example 25.3. *A gas turbine unit receives air at 1 bar and 300 K and compresses it adiabatically to 6.2 bar. The compressor efficiency is 88%. The fuel has a heating valve of 44186 kJ/kg and the fuel-air ratio is 0.017 kJ/kg of air. The turbine internal efficiency is 90%. Calculate the work of turbine and compressor per kg of air compressed and thermal efficiency.*

For products of combustion, c_p = 1.147 kJ/kg K and γ = 1.333. **(U.P.S.C. 1997)**

Solution. *Given* : p_1 (= p_4) = 1 bar, T_1 = 300 K ;

p_2(= p_3) = 6.2 bar ; $\eta_{compressor}$ = 88% ;

C = 44186 kJ/kg ;

Fuel-air ratio = 0.017 kJ/kg of air, $\eta_{turbine}$ = 90% ;

c_p = 1.147 kJ/kg K ; γ = 1.333.

For isentropic compression process 1–2 :

$$\frac{T_2}{T_1} = \left(\frac{p_2}{p_1}\right)^{\frac{\gamma-1}{\gamma}} = \left(\frac{6.2}{1}\right)^{\frac{1.4-1}{1.4}} = 1.684$$

$$\therefore \quad T_2 = 300 \times 1.684 = 505.2 \text{ K}$$

Now,
$$\eta_{compressor} = \frac{T_2 - T_1}{T_2' - T_1}$$

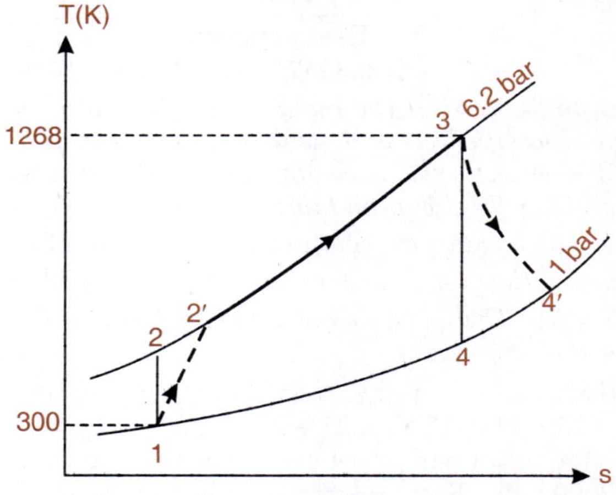

Fig. 25.21

$$0.88 = \frac{505.2 - 300}{T_2' - 300}$$

$$T_2' = \left(\frac{505.2 - 300}{0.88} + 300\right) = 533.2 \text{ K}$$

Heat supplied $= (m_a + m_f) \times c_p(T_3 - T_2') = m_f \times C$

or
$$\left(1 + \frac{m_f}{m_a}\right) \times c_p(T_3 - T_2') = \frac{m_f}{m_a} \times C$$

or
$$(1 + 0.017) \times 1.005(T_3 - 533.2) = 0.017 \times 44186$$

$$\therefore \quad T_3 = \frac{0.017 \times 44186}{(1 + 0.017) \times 1.005} + 533.2 = 1268 \text{ K}$$

For isentropic expansion process 3–4 :

$$\frac{T_4}{T_3} = \left(\frac{p_4}{p_3}\right)^{\frac{\gamma-1}{\gamma}} = \left(\frac{1}{6.2}\right)^{\frac{1.333-1}{1.333}} = 0.634$$

$\therefore \quad T_4 = 1268 \times 0.634 = 803.9 \text{ K}$

$(\because \ \gamma_g = 1.333 \ \ldots\ldots\text{Given})$

Now, $\eta_{turbine} = \dfrac{T_3 - T_4'}{T_3 - T_4}$

$$0.9 = \frac{1268 - T_4'}{1268 - 803.9}$$

$\therefore \quad T_4' = 1268 - 0.9(1268 - 803.9) = 850.3 \text{ K}$

$W_{compresssor} = c_p(T_2' - T_1) = 1.005(533.2 - 300)$
$= 234.4 \text{ kJ/kg}$

$W_{turbine} = c_{pg}(T_3 - T_4') = 1.147(1268 - 850.3)$
$= 479.1 \text{ kJ/kg}$

Network $= W_{turbine} - W_{compressor}$
$= 479.1 - 234.4 = 244.7 \text{ kJ/kg}$

Heat supplied per kg of air
$= 0.017 \times 44186 = 751.2 \text{ kJ/kg}$

$\therefore$ Thermal efficiency,

$$\eta_{th.} = \frac{\text{Network}}{\text{Heat supplied}} = \frac{244.7}{751.2}$$

$= \mathbf{0.3257}$ or **32.57%. (Ans.)**

Example 25.4. *Find the required air-fuel ratio in a gas turbine whose turbine and compressor efficiencies are 85% and 80%, respectively. Maximum cycle temperature is 875°C. The working fluid can be taken as air (c_p = 1.0 kJ/kg K, γ = 1.4) which enters the compressor at 1 bar and 27°C. The pressure ratio is 4. The fuel used has calorific value of 42000 kJ/kg. There is a loss of 10% of calorific value in the combustion chamber.* **(GATE)**

Solution. *Given :* $\eta_{turbine}$ = 85% ; $\eta_{compressor}$ = 80% ; T_3 = 273 + 875 = 1148 K, T_1 = 27 + 273 = 300 K ; c_p = 1.0 kJ/kg K ; γ = 1.4, p_1 = 1 bar, p_2 = 4 bar (Since pressure ratio is 4) ; C = 42000 kJ/kg K, η_{cc} = 90% (since loss in the combustion chamber is 10%)

For isentropic compression 1–2 :

$$\frac{T_2}{T_1} = \left(\frac{p_2}{p_1}\right)^{\frac{\gamma-1}{\gamma}} = (4)^{\frac{1.4-1}{1.4}} = 1.486$$

$\therefore \quad T_2 = 300 \times 1.486 = 445.8 \text{ K}$

$$\eta_{compressor} = \frac{T_2 - T_1}{T_2' - T_1}$$

or $$0.8 = \frac{445.8 - 300}{T_2' - 300}$$

or $$T_2' = \frac{445.8 - 300}{0.8} + 300 = 482.2 \text{ K}$$

Now, heat supplied by the fuel = heat taken by the burning gases

$$0.9 \times m_f \times C = (m_a + m_f) \times c_p \times (T_3 - T_2')$$

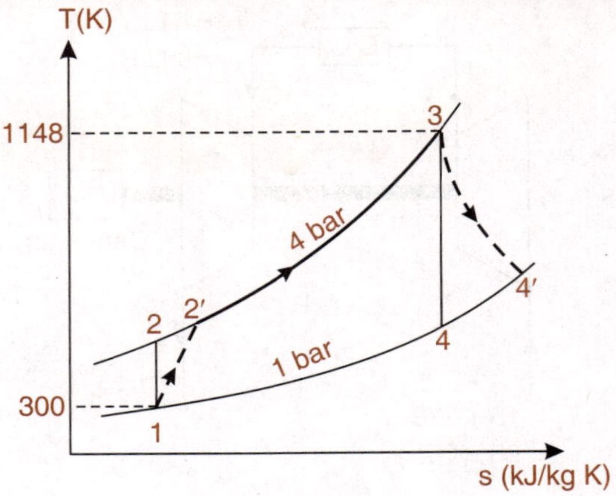

Fig. 25.22

$$\therefore \quad C = \left(\frac{m_a + m_f}{m_f}\right) \times \frac{c_p(T_3 - T_2')}{0.9}$$

$$= \left(\frac{m_a}{m_f} + 1\right) \times \frac{c_p(T_3 - T_2')}{0.9}$$

or $$42000 = \left(\frac{m_a}{m_f} + 1\right) \times \frac{1.00\,(1148 - 482.2)}{0.9}$$

$$= 739.78\left(\frac{m_a}{m_f} + 1\right)$$

$$\therefore \quad \frac{m_a}{m_f} = \frac{42000}{739.78} - 1 = 55.77 \text{ say } 56$$

$\therefore$ **A/F ratio = 56 : 1. (Ans.)**

Example 25.5. *Calculate the thermal efficiency and work ratio of the plant is example 25.4, assuming that c_p for the combustion process is 1.11 kJ/kg K.*

Solution. Heat supplied $= c_p(T_3 - T_2')$
$= 1.11\,(883 - 529) = 392.9 \text{ kJ/kg}$

$$\eta_{thermal} = \frac{\text{Network output}}{\text{Heat supplied}} = \frac{48.2}{392.9}$$

$= \mathbf{0.1226}$ or **12.26%. (Ans.)**

Now, *Work ratio* $= \dfrac{\text{Network output}}{\text{Gross work output}} = \dfrac{48.2}{W_{turbine}}$

$$= \frac{48.2}{290.4} = \mathbf{0.166.} \quad \textbf{(Ans.)}$$

Example 25.6. *In a constant pressure open cycle gas turbine air enters at 1 bar and 20°C and leaves the compressor at 5 bar. Using the following data : Temperature of gases entering the turbine = 680°C, pressure loss in the combustion chamber = 0.1 bar, $\eta_{compressor}$ = 85%, $\eta_{turbine}$ = 80%, $\eta_{combustion}$ = 85%, γ = 1.4 and c_p = 1.024 kJ/kg K for air and gas, find*

(*i*) *The quantity of air circulation if the plant develops 1065 kW.*

(*ii*) *Heat supplied per kg of air circulation.*

(*iii*) *The thermal efficiency of the cycle.*

Mass of the fuel may be neglected. **(M.U.)**

Solution. *Given :* $p_1 = 1$ bar, $p_2 = 5$ bar,

$p_3 = 5 - 0.1 = 4.9$ bar, $p_4 = 1$ bar,

$T_1 = 20 + 273 = 293$ K,

$T_3 = 680 + 273 = 953$ K,

$\eta_{compressor} = 85\%$, $\eta_{turbine} = 80\%$, $\eta_{combustion} = 85\%$,

For air and gases : $c_p = 1.024$ kJ/kg K, $\gamma = 1.4$

Power developed by the plant, $P = 1065$ kW.

(*i*) The quantity of air circulation, m_a :

For *isentropic compression 1-2,*

$$\frac{T_2}{T_1} = \left(\frac{p_2}{p_1}\right)^{\frac{\gamma-1}{\gamma}} = \left(\frac{5}{1}\right)^{\frac{1.4-1}{1.4}} = 1.584$$

$$\therefore \quad T_2 = 293 \times 1.584 = 464 \text{ K}$$

Now, $\eta_{compressor} = \dfrac{T_2 - T_1}{T_2' - T_1}$ *i.e.,* $0.85 = \dfrac{464 - 293}{T_2' - 293}$

$$\therefore \quad T_2' = \frac{464 - 293}{0.85} + 293 = 494 \text{ K}$$

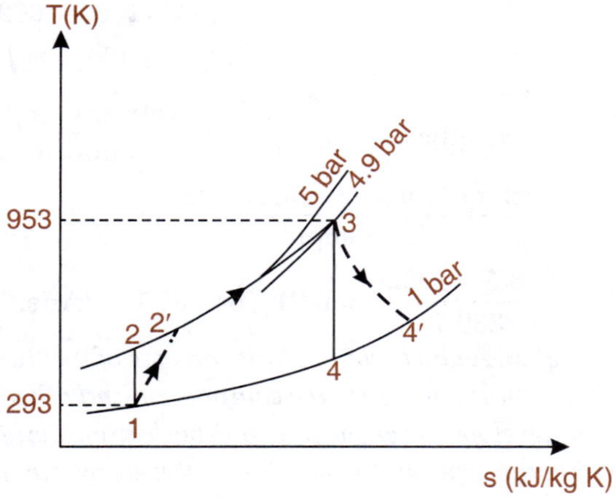

Fig. 25.23

For *isentropic expansion process 3-4,*

$$\frac{T_4}{T_3} = \left(\frac{p_4}{p_3}\right)^{\frac{\gamma-1}{\gamma}} = \left(\frac{1}{4.9}\right)^{\frac{1.4-1}{1.4}} = 0.635$$

$$\therefore \quad T_4 = 953 \times 0.635 = 605 \text{ K}$$

Now, $\eta_{turbine} = \dfrac{T_3 - T_4'}{T_3 - T_4}$

$$0.8 = \frac{953 - T_4'}{953 - 605}$$

$$\therefore \quad T_4' = 953 - 0.8(953 - 605) = 674.6 \text{ K}$$

$$W_{compressor} = c_p (T_2' - T_1) = 1.024 (494 - 293) = 205.8 \text{ kJ/kg}$$

$$W_{turbine} = c_p (T_3 - T_4') = 1.024 (953 - 674.6) = 285.1 \text{ kJ/kg.}$$

$$\therefore \quad W_{net} = W_{turbine} - W_{compressor} = 285.1 - 205.8 = 79.3 \text{ kJ/kg of air}$$

If the mass of air is flowing is m_a kg/s, the power developed by the plant is given by

$$P = m_a \times W_{net} \text{ kW}$$

$$1065 = m_a \times 79.3$$

$$\therefore \quad m_a = \frac{1065}{79.3} = 13.43 \text{ kg.}$$

i.e., **Quantity of air circulation = 13.43 kg. (Ans.)**

(*ii*) Heat supplied per kg of air circulation :

Actual heat supplied per kg of air circulation

$$= \frac{c_p (T_3 - T_2')}{\eta_{combustion}} = \frac{1.024 (953 - 494)}{0.85} = 552.9 \text{ kJ/kg.}$$

(*iii*) Thermal efficiency of the cycle, $\eta_{thermal}$:

$$\eta_{thermal} = \frac{\text{Work output}}{\text{Heat supplied}} = \frac{79.3}{552.9}$$

$$= \mathbf{0.1434 \text{ or } 14.34\%.} \quad \textbf{(Ans.)}$$

Example 25.7. *In a gas turbine the compressor is driven by the high pressure turbine. The exhaust from the high pressure turbine goes to a free low pressure turbine which runs the load. The air flow rate is 20 kg/s and the minimum and maximum temperatures are respectively 300 K and 1000 K. The compressor pressure ratio is 4. Calculate the pressure ratio of the low pressure turbine and the temperature of exhaust gases from the unit. The compressor and turbine are isentropic. c_p of air and exhaust gases = 1 kJ/kg K and $\gamma = 1.4$.* **(GATE)**

Solution. *Given :* $\dot{m}_a = 20$ kg/s ; $T_1 = 300$ K ;

$T_3 = 1000$ K, $\dfrac{p_2}{p_1} = 4$;

$c_p = 1$ kJ/kg K ; $\gamma = 1.4$,

Pressure ratio of low pressure turbine, $\dfrac{p_4}{p_5}$:

Since the compressor is driven by high pressure turbine,

$$\therefore \quad \frac{T_2}{T_1} = \left(\frac{p_2}{p_1}\right)^{\frac{\gamma-1}{\gamma}} = (4)^{\frac{0.4}{1.4}} = 1.486$$

or $\quad T_2 = 300 \times 1.486 = 445.8$ K

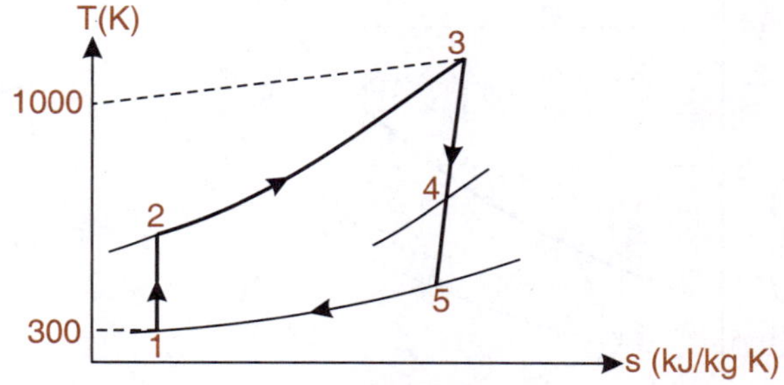

Fig. 25.24

Also, $\dot{m}_a c_p (T_2 - T_1) = \dot{m}_a c_p (T_3 - T_4)$ (neglecting mass of fuel)

or $T_2 - T_1 = T_3 - T_4$

$445.8 - 300 = 1000 - T_4$, or $T_4 = 854.2$ K

For process 3–4 :

$$\frac{T_3}{T_4} = \left(\frac{p_3}{p_4}\right)^{\frac{\gamma-1}{\gamma}}, \text{ or } \frac{p_3}{p_4} = \left(\frac{T_3}{T_4}\right)^{\frac{1.4}{0.4}}$$

or $$\frac{p_3}{p_4} = \left(\frac{1000}{854.2}\right)^{3.5} = 1.736$$

Now, $$\frac{p_3}{p_4} = \frac{p_3}{p_5} \times \frac{p_5}{p_4} = 4 \times \frac{p_5}{p_4}$$

$$\left(\because \frac{p_3}{p_5} = \frac{p_2}{p_1} = 4\right)$$

$\therefore$ $$\frac{p_5}{p_4} = \frac{1}{4}\left(\frac{p_3}{p_4}\right) = \frac{1}{4} \times 1.736 = 0.434$$

Hence pressure ratio of low pressure turbine

$$= \frac{p_4}{p_5} = \frac{1}{0.434} = \mathbf{2.3.} \quad \textbf{(Ans.)}$$

Temperature of the exhaust from the unit T_5 :

$$\frac{T_4}{T_5} = \left(\frac{p_4}{p_5}\right)^{\frac{\gamma-1}{\gamma}} = (2.3)^{\frac{1.4-1}{1.4}} = 1.269$$

$\therefore$ $$\mathbf{T_5} = \frac{T_4}{1.269} = \frac{854.2}{1.269} = \mathbf{673\ K.}$$

Example 25.8. *In an air-standard regenerative gas turbine cycle the pressure ratio is 5. Air enters the compressor at 1 bar, 300 K and leaves at 490 K. The maximum temperature in the cycle is 1000 K. Calculate the cycle efficiency, given that the efficiency of the regenerator and the adiabatic efficiency of the turbine are each 80%. Assume for air, the ratio of specific heats is 1.4. Also, show the cycle on a T-s diagram.*

(GATE)

Solution. *Given :* $p_1 = 1$ bar ; $T_1 = 300$ K, $T_2' = 490$ K ; $T_3 = 1000$ K

$$\frac{p_2}{p_1} = 5, \eta_{turbine} = 80\%, \varepsilon = 80\% = 0.8 \ ; \gamma = 1.4$$

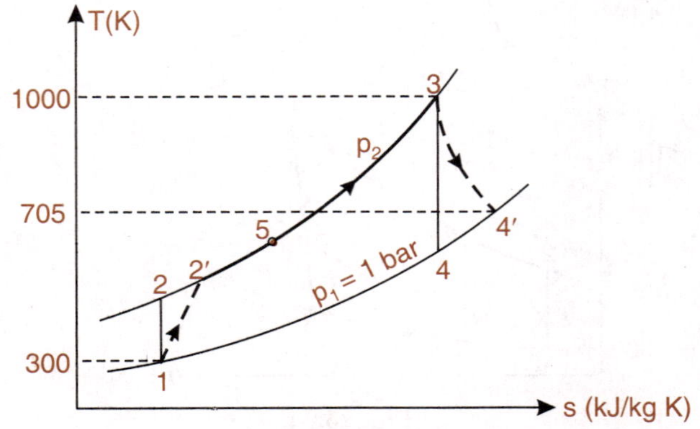

Fig. 25.25

Now, $$\frac{T_3}{T_4} = \left(\frac{p_2}{p_1}\right)^{\frac{\gamma-1}{\gamma}} = (5)^{\frac{1.4-1}{1.4}} = 1.5838$$

$\therefore$ $$T_4 = \frac{T_3}{1.5838} = \frac{1000}{1.5838} = 631.4 \text{ K}$$

Also, $$\eta_{turbine} = \frac{T_3 - T_4'}{T_3 - T_4}$$

or $$0.8 = \frac{1000 - T_4'}{1000 - 631.4}$$

$\therefore$ $T_4' = 1000 - 0.8\,(1000 - 631.4) = 705$ K

Effectiveness of heat exchanger, $$\varepsilon = \frac{T_5 - T_2'}{T_4' - T_2'}$$

or $$0.8 = \frac{T_5 - 490}{705 - 490}$$

$\therefore$ $T_5 = 0.8\,(705 - 490) + 490 = 662$ K

Work consumed by compressor $= c_p\,(T_2' - T_1)$
$= 1.005\,(490 - 300) = 190.9$ kJ/kg

Work done by turbine $= c_p\,(T_3 - T_4')$
$= 1.005\,(1000 - 705)$
$= 296.5$ kJ/kg.

Heat supplied $= c_p\,(T_3 - T_5)$
$= 1.005\,(1000 - 662)$
$= 339.7$ kJ/kg

$\therefore$ Cycle efficiency, $$\eta_{cycle} = \frac{\text{Network}}{\text{Heat supplied}}$$

$$= \frac{\text{Turbine work} - \text{Compressor work}}{\text{Heat supplied}}$$

$$= \frac{296.5 - 190.9}{339.7} = \mathbf{0.31} \text{ or } \mathbf{31\%.} \quad \textbf{(Ans.)}$$

Example 25.9. *A gas turbine plant consists of two turbines. One compressor turbine to drive compressor and other power turbine to develop power output and both are having their own combustion chambers which are served by air directly from the compressor. Air enters the compressor at 1 bar and 288 K and is compressed to 8 bar with an isentropic efficiency of 76%. Due to heat added in the combustion chamber, the inlet temperature of gas to both turbines is 900°C. The isentropic efficiency of turbines is 86% and the mass flow rate of air at the compressor is 23 kg/s. The calorific value of fuel is 4200 kJ/kg. Calculate the output of the plant and the thermal efficiency if mechanical efficiency is 95% and generator efficiency is 96%. Take c_p = 1.005 kJ/kg K and γ = 1.4 for air and c_{pg} = 1.128 kJ/kg K and γ = 1.34 for gases.*

(AMIE)

Solution. *Given :* $p_1 = 1$ bar ; $T_1 = 288$ K ; $p_2 = 8$ bar, $\eta_{(isen)} = 76\%$; $T_3 = 900°C$ or 1173 K, $\eta_{T(isen.)} = 86\%$, $m_a = 23$ kg/s ; C.V. = 4200 kJ/kg ; $\eta_{mech.} = 95\%$; $\eta_{gen.} = 96\%$; $c_p = 1.005$ kJ/kg ; $\gamma_a = 1.4$; $c_{pg} = 1.128$ kJ/kg K ; $\gamma_g = 1.34$.

The arrangement of the plant and the corresponding *T-s* diagram are shown in Fig. 25.26 (*a*), (*b*) respectively.

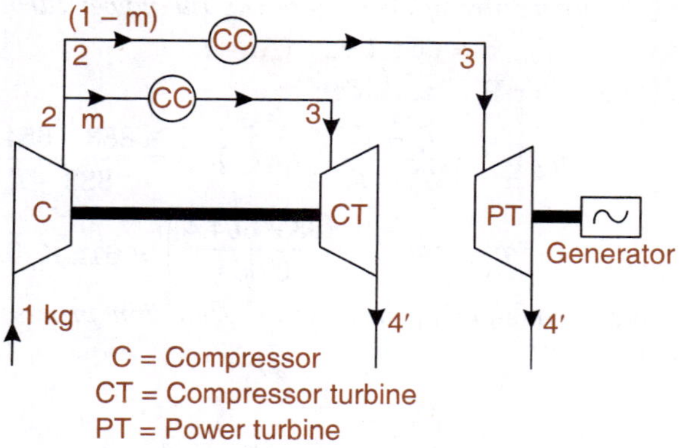

(*a*)

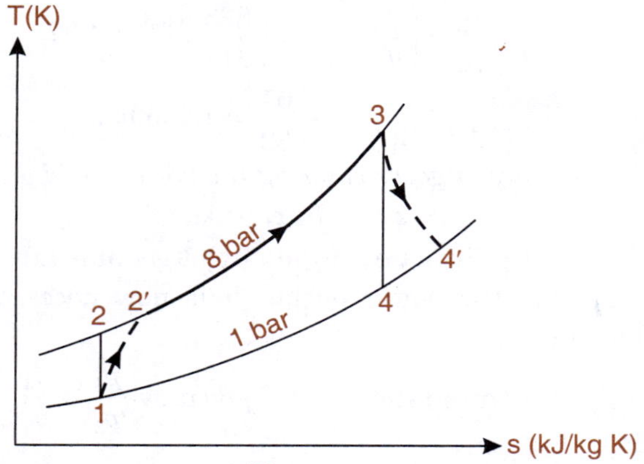

(*b*)

Fig. 25.26

Considering *isentropic compression process 1–2,* we have

$$\frac{T_2}{T_1} = \left(\frac{p_2}{p_1}\right)^{\frac{\gamma-1}{\gamma}} = \left(\frac{8}{1}\right)^{\frac{1.4-1}{1.4}} = 1.811$$

$\therefore \quad T_2 = 288 \times 1.811 = 521.6$ K

Also, $\eta_{C(isen.)} = \dfrac{T_2 - T_1}{T_2' - T_1}$

or $\quad 0.76 = \dfrac{521.6 - 288}{T_2' - 288}$

or $\quad T_2' = \dfrac{521.6 - 288}{0.76} + 288 = 595.4$ K

Considering *isentropic expansion process 3–4,* we have

$$\frac{T_4}{T_3} = \left(\frac{p_4}{p_3}\right)^{\frac{\gamma-1}{\gamma}} = \left(\frac{1}{8}\right)^{\frac{1.34-1}{1.34}} = 0.59$$

$\therefore \quad T_4 = 1173 \times 0.59 = 692.1$ K

Also, $\eta_{T(isen.)} = \dfrac{T_3 - T_4'}{T_3 - T_4}$

or $\quad 0.86 = \dfrac{1173 - T_4'}{1173 - 692.1}$

$\therefore \quad T_4' = 1173 - 0.86\,(1173 - 692.1) = 759.4$ K

Consider 1 kg of air-flow through compressor

$W_{compressor} = c_p\,(T_2' - T_1')$
$= 1.005\,(595.4 - 288) = 308.9$ kJ

This is equal to work of compressor turbine.

$\therefore \quad 308.9 = m_1 \times c_{pg}\,(T_3 - T_4')$, neglecting fuel mass

or $\quad m_1 = \dfrac{308.9}{1.128\,(1173 - 759.4)} = 0.662$ kg

and flow through the power turbine $= 1 - m = 1 - 0.662$
$= 0.338$ kg

$\therefore \quad W_{PT} = (1 - m) \times c_{pg}(T_3 - T_4')$
$= 0.338 \times 1.128\,(1173 - 759.4) = 157.7$ kJ

$\therefore$ **Power output** $= 23 \times 157.7 \times \eta_{mech.} \times \eta_{gen.}$
$= 23 \times 157.7 \times 0.95 \times 0.96$
$=$ **3307.9 kJ. (Ans.)**

$Q_{input} = c_{pg}T_3 - c_{pa}T_2'$
$= 1.128 \times 1173 - 1.005 \times 595.4$
$= 724.7$ kJ/kg of air

Thermal efficiency,

$$\eta_{th} = \frac{157.7}{724.7} \times 100 = \mathbf{21.76\%.} \quad \textbf{(Ans.)}$$

☞**Example 25.10.** *Air is drawn in a gas turbine unit at 15°C and 1.01 bar and pressure ratio is 7 : 1. The compressor is driven by the H.P. turbine and L.P. turbine drives a separate power shaft. The isentropic efficiencies of compressor, and the H.P. and L.P. turbines are 0.82, 0.85 and 0.85 respectively. If the maximum cycle temperature is 610°C, calculate :*

(*i*) *The pressure and temperature of the gases entering the power turbine.*

(*ii*) *The net power developed by the unit per kg/s mass flow.*

(*iii*) *The work ratio.*

(*iv*) *The thermal efficiency of the unit.*

Neglect the mass of fuel and assume the following :

For compression process :

$c_{pa} = 1.005$ *kJ/kg K and* $\gamma = 1.4$

For combustion and expansion processes :

$c_{pg} = 1.15$ *kJ/kg and* $\gamma = 1.333$.

Solution. *Given :* $T_1 = 15 + 273 = 288$ K,

$p_1 = 1.01$ bar, Pressure ratio $= \dfrac{p_2}{p_1} = 7$,

$\eta_{compressor} = 0.82$, $\eta_{turbine\,(H.P.)} = 0.85$,

$\eta_{turbine\,(L.P.)} = 0.85$,

Maximum cycle temperature,

$T_3 = 610 + 273 = 883$ K

(*i*) **Pressure and temperature of the gases entering the power turbine, p_4' and T_4' :**

Considering *isentropic compression 1–2*, we have

$$\frac{T_2}{T_1} = \left(\frac{p_2}{p_1}\right)^{\frac{\gamma-1}{\gamma}} = (7)^{\frac{1.4-1}{1.4}} = 1.745$$

$\therefore \quad T_2 = 288 \times 1.745 = 502.5$ K

Also $\quad \eta_{compressor} = \dfrac{T_2 - T_1}{T_2' - T_1}$

$$0.82 = \frac{502.5 - 288}{T_2' - 288}$$

$$\therefore \quad T_2' = \frac{502.5 - 288}{0.82} + 288 = 549.6 \text{ K}$$

$$W_{compressor} = c_{pa}(T_2' - T_1) = 1.005 \times (549.6 - 288) = 262.9 \text{ kJ/kg}$$

Now, the *work output of H.P. turbine = Work input to compressor*

$\therefore \quad c_{pg}(T_3 - T_4') = 262.9$

i.e., $\quad 1.15(883 - T_4') = 262.9$

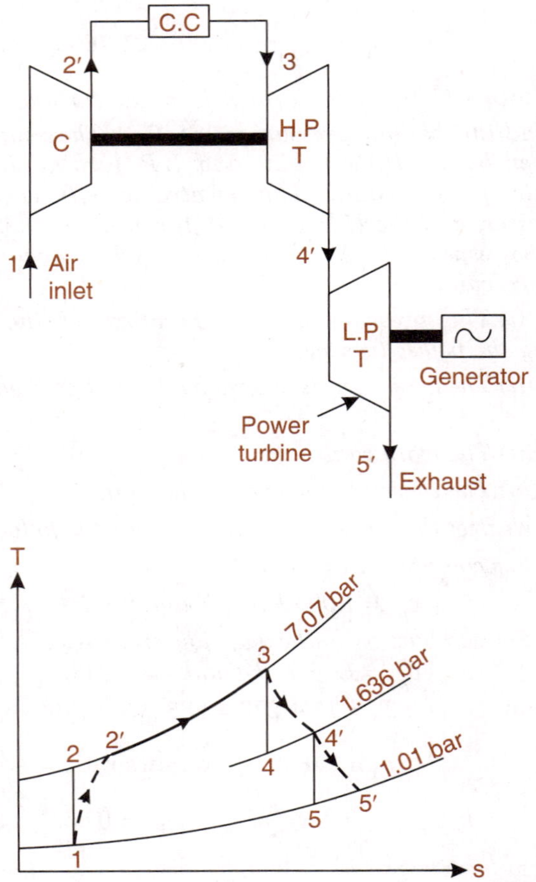

Fig. 25.27

$$\therefore \quad T_4' = 883 - \frac{262.9}{1.15} = 654.4 \text{ K}$$

i.e., *Temperature of gases entering the power turbine*

= **654.4 K. (Ans.)**

Again, for H.P. turbine :

$$\eta_{turbine} = \frac{T_3 - T_4'}{T_3 - T_4} \quad i.e., \quad 0.85 = \frac{883 - 654.4}{883 - T_4}$$

$$\therefore \quad T_4 = 883 - \left(\frac{883 - 654.4}{0.85}\right) = 614 \text{ K}$$

Now, considering *isentropic expansion process 3-4*, we have

$$\frac{T_3}{T_4} = \left(\frac{p_3}{p_4}\right)^{\frac{\gamma-1}{\gamma}}$$

or $$\frac{p_3}{p_4} = \left(\frac{T_3}{T_4}\right)^{\frac{\gamma}{\gamma-1}} = \left(\frac{883}{614}\right)^{\frac{1.33}{0.33}} = 4.32$$

i.e., $$p_4 = \frac{p_3}{4.32} = \frac{7.07}{4.32} = 1.636 \text{ bar}$$

i.e., *Pressure of gases entering the power turbine*

= **1.636 bar. (Ans.)**

(*ii*) **Net power developed per kg/s mass flow, P :**

To find the power output it is now necessary to calculate T_5'.

The pressure ratio, $\dfrac{p_4}{p_5}$, is given by $\dfrac{p_4}{p_3} \times \dfrac{p_3}{p_5}$

i.e., $$\frac{p_4}{p_5} = \frac{p_4}{p_3} \times \frac{p_2}{p_1} = \frac{7}{4.32} = 1.62$$

$(\because \quad p_2 = p_3 \text{ and } p_5 = p_1)$

Then,

$$\frac{T_4'}{T_5} = \left(\frac{p_4}{p_5}\right)^{\frac{\gamma-1}{\gamma}} = (1.62)^{\frac{0.33}{1.33}} = 1.127$$

$$\therefore \quad T_5 = \frac{T_4'}{1.127} = \frac{654.4}{1.127} = 580.6 \text{ K.}$$

Again, for L.P. turbine

$$\eta_{turbine} = \frac{T_4' - T_5'}{T_4' - T_5}$$

i.e., $$0.85 = \frac{654.4 - T_5'}{654.4 - 580.6}$$

$$\therefore \quad T_5' = 654.4 - 0.85\,(654.4 - 580.6) = 591.7 \text{ K}$$

$$W_{L.P.\ turbine} = c_{pg}\,(T_4' - T_5') = 1.15\,(654.4 - 591.7) = 72.1 \text{ kJ/kg}$$

Hence *net power output* (per kg/s mass flow)

= **72.1 kW. (Ans.)**

(*iii*) **Work ratio :**

$$\textit{Work ratio} = \frac{\text{Network output}}{\text{Gross work output}} = \frac{72.1}{72.1 + 262.9}$$

= **0.215. (Ans.)**

(*iv*) **Thermal efficiency of the unit, $\eta_{thermal}$:**

Heat supplied $= c_{pg}(T_3 - T_2') = 1.15\ (883 - 549.6)$
$= 383.4$ kJ/kg

$$\therefore \quad \eta_{thermal} = \frac{\text{Network output}}{\text{Heat supplied}} = \frac{72.1}{383.4}$$

$= \mathbf{0.188}$ or **18.8%. (Ans.)**

Example 25.11. *The pressure ratio of an open-cycle gas turbine power plant is 5.6. Air is taken at 30°C and 1 bar. The compression is carried out in two stages with perfect intercooling in between. The maximum temperature of the cycle is limited to 700°C. Assuming the isentropic efficiency of each compressor stage as 85% and that of turbine as 90%, determine the power developed and efficiency of the power plant, if the air-flow is 1.2 kg/s. The mass of fuel may be neglected, and it may be assumed that $c_p = 1.02$ kJ/kg K and $\gamma = 1.41$.* **(P.U.)**

Solution. Refer to Fig. 25.28.

Pressure ratio of the open-cycle gas turbine = 5.6

Temperature of intake air, $T_1 = 30 + 273 = 303$ K

Pressure of intake air, $p_1 = 1$ bar

Maximum temperature of the cycle, $T_5 = 700 + 273 = 973$ K

Isentropic efficiency of each compressor, $\eta_{comp.} = 85\%$

Isentropic efficiency of turbine, $\eta_{turbine} = 90\%$

Rate of air-flow, $\dot{m}_a = 1.2$ kg/s

$c_p = 1.02$ kJ/kg K and $\gamma = 1.41$.

Power developed and efficiency of the power plant :

Assuming that the pressure ratio in each stage is same, we have

$$\frac{p_2}{p_1} = \frac{p_4}{p_3} = \sqrt{\frac{p_4}{p_1}} = \sqrt{5.6} = 2.366$$

Since the pressure ratio and the isentropic efficiency of each compressor is the same then the *work input required for each compressor is the same* since both the compressors have the same inlet temperature (*perfect intercooling*) *i.e.*, $T_1 = T_3$ and $T_2' = T_4'$:

$$\text{Now,} \quad \frac{T_2}{T_1} = \left(\frac{p_2}{p_1}\right)^{\frac{\gamma-1}{\gamma}} = (2.366)^{\frac{1.41-1}{1.41}} = 1.2846$$

or $\quad T_2 = 303 \times 1.2846 = 389.23$ K

$$\text{Also,} \quad \eta_{comp.} = \frac{T_2 - T_1}{T_2' - T_1} \quad \text{or} \quad 0.85 = \frac{389.23 - 303}{T_2' - 303}$$

$$\text{or} \quad T_2' = \frac{389.23 - 303}{0.85} + 303 = 404.44 \text{ K}$$

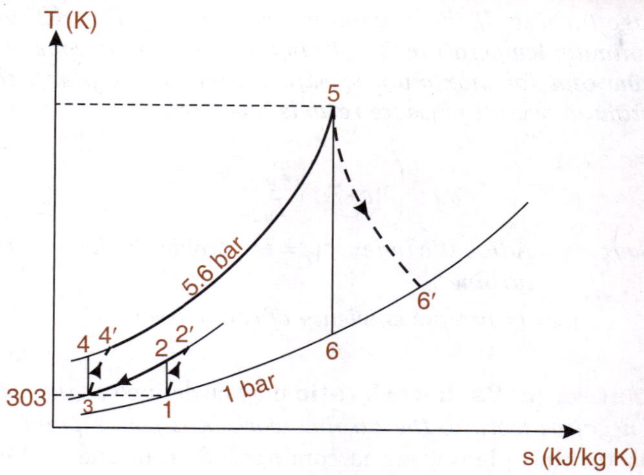

Fig. 25.28

Work input to 2-stage compressor,

$$W_{comp.} = 2 \times m \times c_p(T_2' - T_1)$$
$$= 2 \times 1.2 \times 1.02\ (404.44 - 303) = 248.32 \text{ kJ/s}$$

For turbine, we have

$$\frac{T_5}{T_6} = \left(\frac{p_5}{p_6}\right)^{\frac{\gamma-1}{\gamma}} = (5.6)^{\frac{1.41-1}{1.41}} = 1.65$$

$$\text{or} \quad T_6 = \frac{T_5}{1.65} = \frac{973}{1.65} = 589.7 \text{ K}$$

$$\text{Also,} \quad \eta_{turbine} = \frac{T_5 - T_6'}{T_5 - T_6}$$

$$\text{or} \quad 0.9 = \frac{973 - T_6'}{973 - 589.7}$$

or $\quad T_6' = 973 - 0.9\ (973 - 589.7) = 628$ K

$\therefore$ Work output of turbine,

$$W_{turbine} = m \times c_p(T_5 - T_6') = 1.2 \times 1.02\ (973 - 628)$$
$$= 422.28 \text{ kJ/s}$$

Network output,

$$W_{net} = W_{turbine} - W_{comp.}$$
$$= 422.28 - 248.32 = 173.96 \text{ kJ/s or kW}$$

Hence *power developed* $=$ **173.96 kW. (Ans.)**

Heat supplied,

$$Q_s = m \times c_p \times (T_5 - T_4')$$
$$= 1.2 \times 1.02 \times (973 - 404.44) = 695.92 \text{ kJ/s}$$

$\therefore$ Power plant efficiency,

$$\eta_{th} = \frac{W_{net}}{Q_s} = \frac{173.96}{695.92} = \mathbf{0.25} \text{ or } \mathbf{25\%}. \ \textbf{(Ans.)}$$

Example 25.12. (*a*) *Why are the back work ratios relatively high in gas turbine plants compared to those of steam power plants ?*

(*b*) *In a gas turbine plant compression is carried out in two stages with perfect intercooling and expansion in one* *stage turbine. If the maximum temperature (T_{max} K) and minimum temperature (T_{min}K) in the cycle remain constant, show that for maximum specific output of the plant, the optimum overall pressure ratio is given by*

$$r_{opt} = \left(\eta_T \cdot \eta_C \cdot \frac{T_{max}}{T_{min}}\right)^{\frac{2\gamma}{3(\gamma-1)}}$$

where γ = *Adiabatic index ;* η_T = *Isentropic efficiency of the turbine.*

η_C = *Isentropic efficiency of compressor.*

(P.U.)

Solution. (*a*) **Back work ratio** may be defined as *the ratio of negative work to the turbine work in a power plant.* In gas turbine plants, air is compressed from the turbine exhaust pressure to the combustion chamber pressure. This work is given by $-\int v dp$. As the specific volume of air is very high (even in closed cycle gas turbine plants), the compressor work required is very high, and also *bulky compressor is required.* In steam power plants, the turbine exhaust is changed to liquid phase in the condenser. The pressure of condensate is raised to boiler pressure by condensate extraction pump and boiler feed pump in series since the specific volume of water is very small as compared to that of air, the pump work ($-\int v dp$), is also very small. From the above reasons, the back work ratio

$$= \frac{-\int v dp}{\text{Turbine work}}$$

for gas *turbine plants is relatively high compared to that for steam power plants.*

(*b*) Refer to Fig. 23.29.

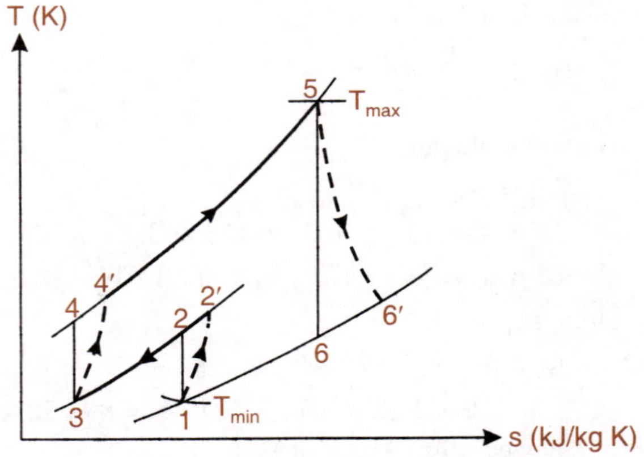

Fig. 25.29

Assuming optimum pressure ratio in each stage of the compressors $\sqrt{r}$,

$$\frac{T_2}{T_1} = \left(\frac{p_2}{p_1}\right)^{\frac{\gamma-1}{\gamma}}$$

or $$T_2 = T_{min} \times (r)^{\frac{\gamma-1}{2\gamma}}$$

$$W_{compressor} = 2[c_p (T_2' - T_1)] \text{ for both compressors}$$

$$= 2c_p \frac{T_2 - T_1}{\eta_C} = \frac{2c_p}{\eta_C} T_{min}\left[(r)^{\frac{\gamma-1}{2\gamma}} - 1\right], \text{ as}$$

$$T_1 = T_{min}$$

Also, $$\frac{T_5}{T_6} = \left(\frac{p_2}{p_1}\right)^{\frac{\gamma-1}{\gamma}} = (r)^{\frac{\gamma-1}{\gamma}}$$

$$\therefore \quad T_6 = \frac{T_5}{(r)^{\gamma-1/\gamma}} = \frac{T_{max}}{(r)^{\gamma-1/\gamma}}, \text{ as } T_5 = T_{max}$$

$$W_{turbine} = c_p (T_5 - T_6') = c_p \left[T_{max} - \frac{T_{max}}{(r)^{\gamma-1/\gamma}}\right] \eta_T, \text{ as}$$

$$\eta_T = \frac{T_5 - T_6'}{T_5 - T_6}$$

$$= c_p T_{max}\left[1 - \frac{1}{(r)^{\gamma-1/\gamma}}\right]\eta_T$$

$$W_{net} = W_{turbine} - W_{compressor}$$

$$= c_p \eta_T T_{max}\left[1 - \frac{1}{(r)^{\gamma-1/\gamma}}\right] - \frac{2c_p}{\eta_C} T_{min} [(r)^{\frac{\gamma-1}{2\gamma}} - 1]$$

For maximum work output,

$$\frac{dW_{net}}{dr} = 0$$

or $$-c_p \eta_T T_{max}\left(-\frac{\gamma-1}{\gamma}\right)(r)^{-\left(\frac{\gamma-1}{\gamma}\right)-1} - \frac{2c_p}{\eta_C} T_{min}\left(\frac{\gamma-1}{2\gamma}\right)(r)^{\frac{\gamma-1}{2\gamma}-1} = 0$$

or $$\eta_T \eta_C \frac{T_{max}}{T_{min}} = (r)^{3(\gamma-1)/2\gamma}, \text{ on simplification.}$$

Hence, the optimum pressure ratio is

$$r_{opt} = \left[\eta_T \cdot \eta_C \cdot \frac{T_{max}}{T_{min}}\right]^{\frac{2\gamma}{3(\gamma-1)}}. \quad ..... \textbf{ Proved.}$$

Example 25.13. *In a gas turbine the compressor takes in air at a temperature of 15°C and compresses it to four times the initial pressure with an isentropic efficiency of 82%. The air is then passed through a heat exchanger heated by the turbine exhaust before reaching the combustion chamber. In the heat exchanger 78% of the available heat is given to the air. The maximum temperature after constant pressure combustion is 600°C, and the efficiency of the turbine is 70%. Neglecting all losses except those mentioned, and assuming the working fluid throughout the cycle to have the characteristic of air find the efficiency of the cycle.*

Assume R = 0.287 kJ/kg K and γ = *1.4 for air and constant specific heats throughout.*

Solution. *Given :* $T_1 = 15 + 273 = 288$ K, Pressure ratio,

$$\frac{p_2}{p_1} = \frac{p_3}{p_4} = 4, \ \eta_{compressor} = 82\%.$$

Effectiveness of the heat exchanger, $\varepsilon = 0.78$,

$\eta_{turbine} = 70\%$, Maximum temperature,

$T_3 = 600 + 273 = 873$ K.

Efficiency of the cycle, η_{cycle} :

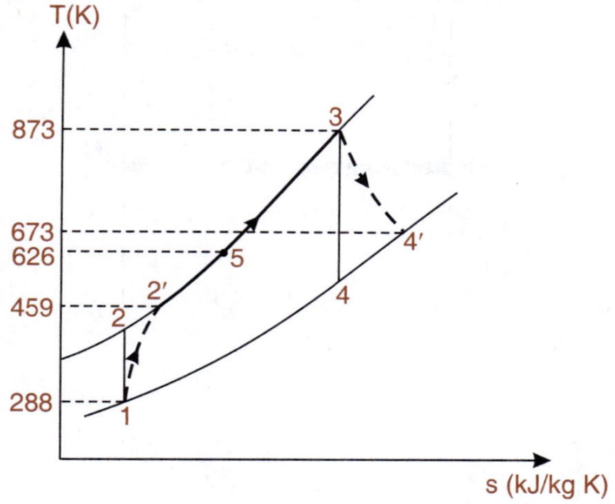

Fig. 25.30

Considering the *isentropic compression 1–2*, we have

$$\frac{T_2}{T_1} = \left(\frac{p_2}{p_1}\right)^{\frac{\gamma-1}{\gamma}} = (4)^{\frac{1.4-1}{1.4}} = 1.486$$

$\therefore \quad T_2 = 288 \times 1.486 = 428$ K

Now, $\quad \eta_{compressor} = \dfrac{T_2 - T_1}{T_2' - T_1}$

i.e., $\quad 0.82 = \dfrac{428 - 288}{T_2' - 288}$

$\therefore \quad T_2' = \dfrac{428 - 288}{0.82} + 288 = 459$ K

Considering the *isentropic expansion process 3–4*, we have

$$\frac{T_3}{T_4} = \left(\frac{p_3}{p_4}\right)^{\frac{\gamma-1}{\gamma}} = (4)^{\frac{1.4-1}{1.4}} = 1.486$$

$\therefore \quad T_4 = \dfrac{T_3}{1.486} = \dfrac{873}{1.486} = 587.5$ K.

Again, $\quad \eta_{turbine} = \dfrac{T_3 - T_4'}{T_3 - T_4} = \dfrac{873 - T_4'}{873 - 587.5}$

i.e., $\quad 0.70 = \dfrac{873 - T_4'}{873 - 587.5}$

$\therefore \quad T_4' = 873 - 0.7\,(873 - 587.5)$
$= 673$ K

$W_{compressor} = c_p(T_2' - T_1)$

But $\quad c_p = R \times \dfrac{\gamma}{\gamma - 1} = 0.287 \times \dfrac{1.4}{1.4 - 1}$
$= 1.0045$ kJ/kg K

$\therefore \quad W_{compressor} = 1.0045\,(459 - 288)$
$= 171.7$ kJ/kg

$W_{turbine} = c_p(T_3 - T_4') = 1.0045\,(873 - 673)$
$= 200.9$ kJ/kg

$\therefore$ Network $= W_{turbine} - W_{compressor}$
$= 200.9 - 171.7 = 29.2$ kJ/kg.

Effectiveness for heat exchanger, $\varepsilon = \dfrac{T_5 - T_2'}{T_4' - T_2'}$

i.e., $\quad 0.78 = \dfrac{T_5 - 459}{673 - 459}$

$\therefore \quad T_5 = (673 - 459) \times 0.78 + 459 = 626$ K

$\therefore$ Heat supplied by fuel per kg
$= c_p(T_3 - T_5) = 1.0045\,(873 - 626)$
$= 248.1$ kJ/kg

$\therefore \quad \eta_{cycle} = \dfrac{\text{Network done}}{\text{Heat supplied by the fuel}} = \dfrac{29.2}{248.1}$
$=$ **0.117 or 11.7%. (Ans.)**

Example 25.14. *A gas turbine employs a heat exchanger with a thermal ratio of 72%. The turbine operates between the pressures of 1.01 bar and 4.04 bar and ambient temperature is 20°C. Isentropic efficiencies of compressor and turbine are 80% and 85% respectively. The pressure drop on each side of the heat exchanger is 0.05 bar and in the combustion chamber 0.14 bar. Assume combustion efficiency to be unity and calorific value of the fuel to be 41800 kJ/kg.*

Calculate the increase in efficiency due to heat exchanger over that for simple cycle.

Assume c_p is constant throughout and is equal to 1.024 kJ/kg K, and assume $\gamma = 1.4$.

For simple cycle the air-fuel ratio is 90 : 1, and for the heat exchange cycle the turbine entry temperature is the same as for a simple cycle.

Solution. Simple Cycle. Refer to Fig. 25.31.

$$\frac{T_2}{T_1} = \left(\frac{p_2}{p_1}\right)^{\frac{\gamma-1}{\gamma}} = \left(\frac{4.40}{1.01}\right)^{\frac{1.4-1}{1.4}} = 1.486$$

$\therefore \quad T_2 = 293 \times 1.486 = 435.4$

Also, $\quad \eta_{compressor} = \dfrac{T_2 - T_1}{T_2' - T_1}$

$0.8 = \dfrac{435.4 - 293}{T_2' - 293}$

$\therefore \quad T_2' = \dfrac{435.4 - 293}{0.8} + 293 = 471$ K

Now, $\quad m_f \times C = (m_a + m_f) \times c_p \times (T_3 - T_2')$

[m_a = mass of air, m_f = mass of fuel]

$$\therefore \quad T_3 = \frac{m_f \times C}{c_p\,(m_a + m_f)} + T_2'$$

$$= \frac{1 \times 41800}{1.024\,(90 + 1)} + 471 = 919.5 \text{ K}$$

Also, $\quad \dfrac{T_4}{T_3} = \left(\dfrac{p_4}{p_3}\right)^{\frac{\gamma - 1}{\gamma}}$

or $\quad T_4 = T_3 \times \left(\dfrac{p_4}{p_3}\right)^{\frac{\gamma - 1}{\gamma}} = 919.5 \times \left(\dfrac{1.01}{3.9}\right)^{\frac{1.4 - 1}{1.4}}$

$$= 625 \text{ K}$$

Again, $\quad \eta_{turbine} = \dfrac{T_3 - T_4'}{T_3 - T_4}$

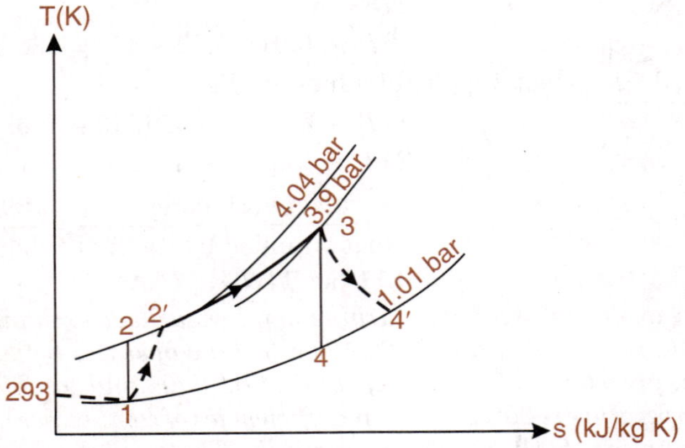

Fig. 25.31

$$\therefore \quad 0.85 = \frac{919.5 - T_4'}{919.5 - 625}$$

$$\therefore \quad T_4' = 919.5 - 0.85(919.5 - 625) = 669 \text{ K}$$

$$\eta_{thermal} = \frac{(T_3 - T_4') - (T_2' - T_1)}{(T_3 - T_2')}$$

$$= \frac{(919.5 - 669) - (471 - 293)}{(919.5 - 471)} = \frac{72.5}{448.5}$$

$= \mathbf{0.1616}$ or **16.16%.** **(Ans.)**

Heat Exchanger Cycle. Refer to Fig. 25.32 (*a*), (*b*)

$T_2' = 471$ K (as for simple cycle) ;

$T_3 = 919.5$ K (as for simple cycle)

To find T_4' :

$p_3 = 4.04 - 0.14 - 0.05 = 3.85$ bar ;

$p_4 = 1.01 + 0.05 = 1.06$ bar

$$\therefore \quad \frac{T_4}{T_3} = \left(\frac{p_4}{p_3}\right)^{\frac{\gamma - 1}{\gamma}} = \left(\frac{1.06}{3.85}\right)^{\frac{1.4 - 1}{1.4}} = 0.69$$

i.e., $\quad T_4 = 919.5 \times 0.69 = 634$ K

$$\eta_{turbine} = \frac{T_3 - T_4'}{T_3 - T_4} \; ; \quad 0.85 = \frac{919.5 - T_4'}{919.5 - 634}$$

$$\therefore \quad T_4' = 919.5 - 0.85\,(919.5 - 634) = 677 \text{ K}$$

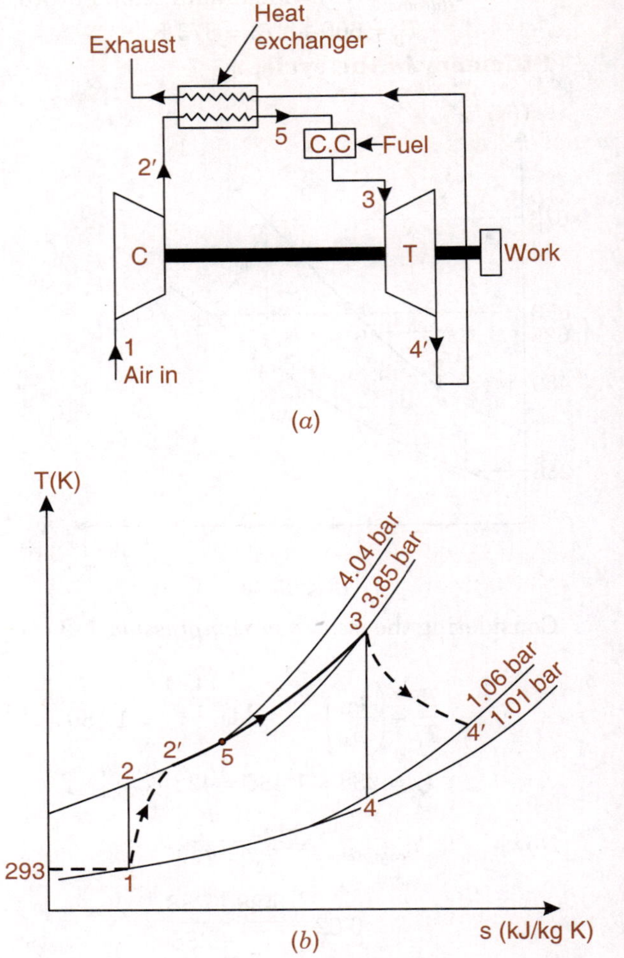

Fig. 25.32

To find T_5 :

Thermal ratio (or effectiveness),

$$\varepsilon = \frac{T_5 - T_2'}{T_4' - T_2'} \qquad \therefore \quad 0.72 = \frac{T_5 - 471}{677 - 471}$$

$$\therefore \; T_5 = 0.72\,(677 - 471) + 471 = 619 \text{ K}$$

$$\eta_{thermal} = \frac{(T_3 - T_4') - (T_2' - T_1)}{(T_3 - T_5)}$$

$$= \frac{(919.5 - 677) - (471 - 293)}{(919.5 - 619)} = \frac{64.5}{300.5}$$

$= \mathbf{0.2146}$ or **21.46%**

$\therefore$ **Increase in thermal efficiency**

$= 21.46 - 16.16 =$ **5.3%.** **(Ans.)**

Example 25.15. *A 4500 kW gas turbine generating set operates with two compressor stages ; the overall pressure ratio is 9 : 1. A high pressure turbine is used to drive the compressors, and a low-pressure turbine drives the generator. The temperature of the gases at entry to the high pressure turbine is 625°C and the gases are reheated to 625°C after expansion in the first turbine. The exhaust gases leaving the low-pressure turbine are passed through a heat exchanger to heat air leaving the high pressure stage compressor. The compressors have equal pressure ratios and intercooling is complete between the stages. The air inlet temperature to the unit is 20°C. The isentropic efficiency of each compressor stage is 0.8, and the isentropic efficiency of each turbine stage is 0.85, the heat exchanger thermal ratio is 0.8. A mechanical efficiency of 95% can be assumed for both the power shaft and compressor turbine shaft. Neglecting all pressure losses and changes in kinetic energy calculate :*

(*i*) *The thermal efficiency ;*

(*ii*) *Work ratio of the plant ;*

(*iii*) *The mass flow in kg/s.*

Neglect the mass of the fuel and assume the following :

For air : c_{pa} = 1.005 kJ/kg K and γ = 1.4

For gases in the combustion chamber and in turbines and heat exchanger, c_{pg} = 1.15 kJ/kg K and γ = 1.333.

Solution. Refer to Fig. 25.33

Given : $T_1 = 20 + 273 + 293$ K,

$T_6 = T_8 = 625 + 273 = 898$ K

Efficiency of each compressor stage = 0.8

Efficiency of each turbine stage = 0.85,

$\eta_{mech.} = 0.95, \varepsilon = 0.8$

(*i*) **Thermal efficiency, $\eta_{thermal}$:**

Since the pressure ratio and the isentropic efficiency of each compressor is the same then the work input required for each compressor is the same since both compressors have the same air inlet temperature *i.e.*, $T_1 = T_3$ and $T_2' = T_4'$.

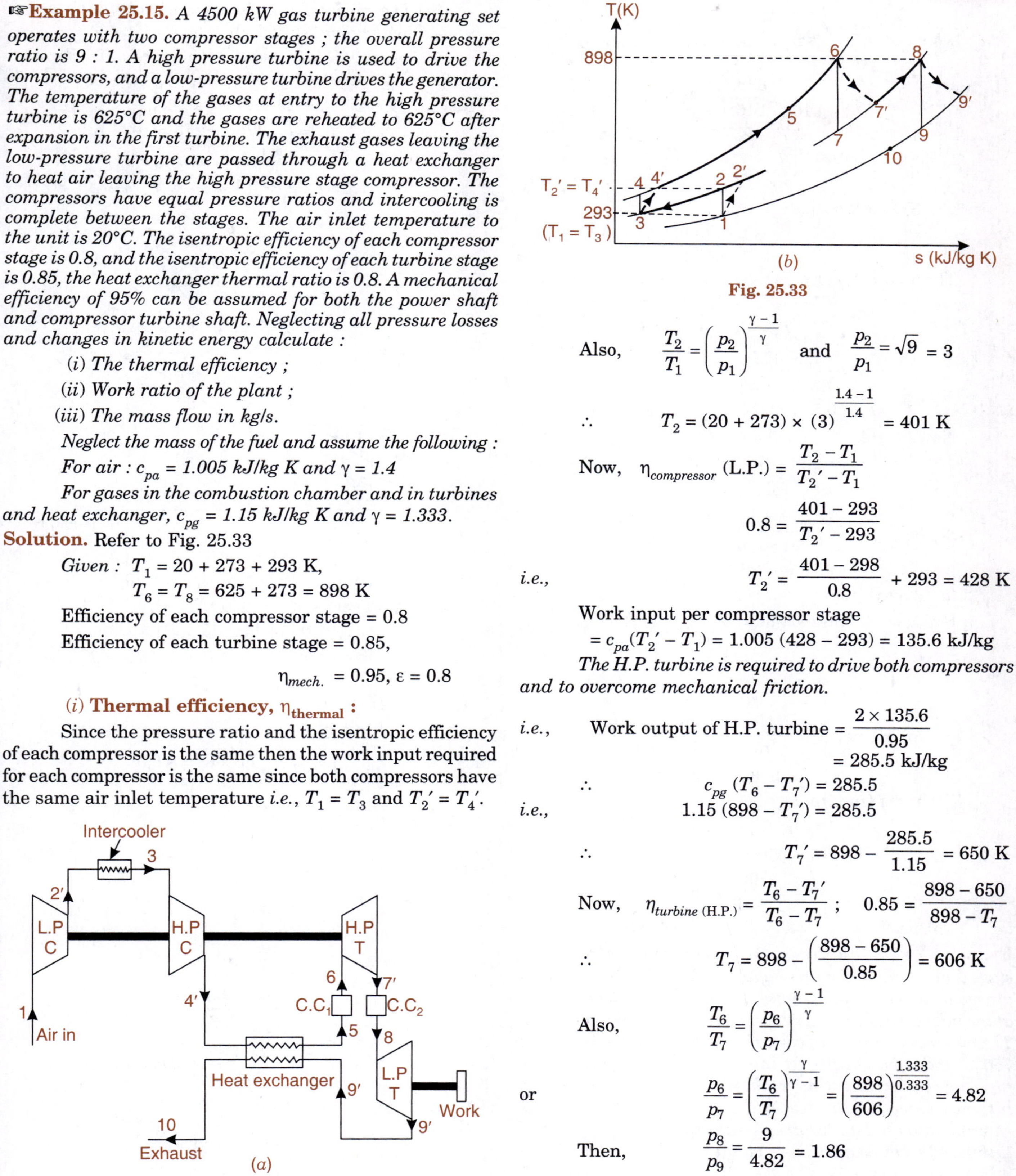

Fig. 25.33

Also, $\dfrac{T_2}{T_1} = \left(\dfrac{p_2}{p_1}\right)^{\frac{\gamma-1}{\gamma}}$ and $\dfrac{p_2}{p_1} = \sqrt{9} = 3$

$\therefore \quad T_2 = (20 + 273) \times (3)^{\frac{1.4-1}{1.4}} = 401$ K

Now, $\eta_{compressor}$ (L.P.) $= \dfrac{T_2 - T_1}{T_2' - T_1}$

$0.8 = \dfrac{401 - 293}{T_2' - 293}$

i.e., $T_2' = \dfrac{401 - 298}{0.8} + 293 = 428$ K

Work input per compressor stage

$= c_{pa}(T_2' - T_1) = 1.005\ (428 - 293) = 135.6$ kJ/kg

The H.P. turbine is required to drive both compressors and to overcome mechanical friction.

i.e., Work output of H.P. turbine $= \dfrac{2 \times 135.6}{0.95}$

$= 285.5$ kJ/kg

$\therefore \quad c_{pg}\ (T_6 - T_7') = 285.5$

i.e., $1.15\ (898 - T_7') = 285.5$

$\therefore \quad T_7' = 898 - \dfrac{285.5}{1.15} = 650$ K

Now, $\eta_{turbine\ (H.P.)} = \dfrac{T_6 - T_7'}{T_6 - T_7}$; $\quad 0.85 = \dfrac{898 - 650}{898 - T_7}$

$\therefore \quad T_7 = 898 - \left(\dfrac{898 - 650}{0.85}\right) = 606$ K

Also, $\dfrac{T_6}{T_7} = \left(\dfrac{p_6}{p_7}\right)^{\frac{\gamma-1}{\gamma}}$

or $\dfrac{p_6}{p_7} = \left(\dfrac{T_6}{T_7}\right)^{\frac{\gamma}{\gamma-1}} = \left(\dfrac{898}{606}\right)^{\frac{1.333}{0.333}} = 4.82$

Then, $\dfrac{p_8}{p_9} = \dfrac{9}{4.82} = 1.86$

Again, $\dfrac{T_8}{T_9} = \left(\dfrac{p_8}{p_9}\right)^{\frac{\gamma-1}{\gamma}} = (1.86)^{\frac{1.333-1}{1.333}} = 1.16$

$\therefore \quad T_9 = \dfrac{T_8}{1.16} = \dfrac{898}{1.16} = 774 \text{ K}$

Also, $\eta_{turbine\ (L.P.)} = \dfrac{T_8 - T_9'}{T_8 - T_9}$; $0.85 = \dfrac{898 - T_9'}{898 - 774}$

$\therefore \quad T_9' = 898 - 0.85\ (898 - 774) = 792.6 \text{ K}$

$\therefore$ Network output $= c_{pg}(T_8 - T_9') \times 0.95$

$= 1.15\ (898 - 792.6) \times 0.95$

$= 115.15$ kJ/kg

Thermal ratio or effectiveness of heat exchanger,

$$\varepsilon = \frac{T_5 - T_4'}{T_9' - T_4'} = \frac{T_5 - 428}{792.6 - 428}$$

i.e., $\quad 0.8 = \dfrac{T_5 - 428}{792.6 - 428}$

$\therefore \quad T_5 = 0.8\ (792.6 - 428) + 428 = 719.7 \text{ K}$

Now, Heat supplied $= c_{pg}(T_6 - T_5) + c_{pg}(T_8 - T_7')$

$= 1.15\ (898 - 719.7) + 1.15\ (898 - 650)$

$= 490.2$ kJ/kg

$\therefore \quad \eta_{thermal} = \dfrac{\text{Network output}}{\text{Heat supplied}} = \dfrac{115.15}{490.2}$

$=$ **0.235** or **23.5%.** **(Ans.)**

(ii) **Work ratio :**

Gross work of the plant $= W_{turbine\ (H.P.)} + W_{turbine\ (L.P.)}$

$= 285.5 + \dfrac{115.15}{0.95}$

$= 406.7$ kJ/kg

$\therefore$ **Work ratio** $= \dfrac{\text{Network output}}{\text{Gross work output}} = \dfrac{115.15}{406.7}$

$=$ **0.283.** **(Ans.)**

(iii) **Mass flow in $\dot{m}$:**

Let the mass flow be $\dot{m}$, then

$\dot{m} \times 115.15 = 4500$

$\therefore \quad \dot{m} = \dfrac{4500}{115.15} = 39.08$ kg/s

i.e., **Mass flow** $=$ **39.08 kg/s.** **(Ans.)**

Example 25.16. *In a closed cycle gas turbine there is two-stage compressor and a two-stage turbine. All the components are mounted on the same shaft. The pressure and temperature at the inlet of the first-stage compressor are 1.5 bar and 20°C. The maximum cycle temperature and pressure are limited to 750°C and 6 bar. A perfect intercooler is used between the two-stage compressors and a reheater is used between the two turbines. Gases are heated in the reheater to 750°C before entering into the L.P. turbine. Assuming the compressor and turbine efficiencies as 0.82, calculate :*

(i) The efficiency of the cycle without regenerator.

(ii) The efficiency of the cycle with a regenerator whose effectiveness is 0.70.

(iii) The mass of the fluid circulated if the power developed by the plant is 350 kW,

The working fluid used in the cycle is air. For air : $\gamma = 1.4$ and $c_p = 1.005$ kJ/kg K.

Solution. *Given :* $T_1 = 20 + 273 = 293$ K,

$T_5 = T_7 = 750 + 273 = 1023$ K,

$p_1 = 1.5$ bar,

$p_2 = 6$ bar, $\eta_{compressor} = \eta_{turbine} = 0.82.$

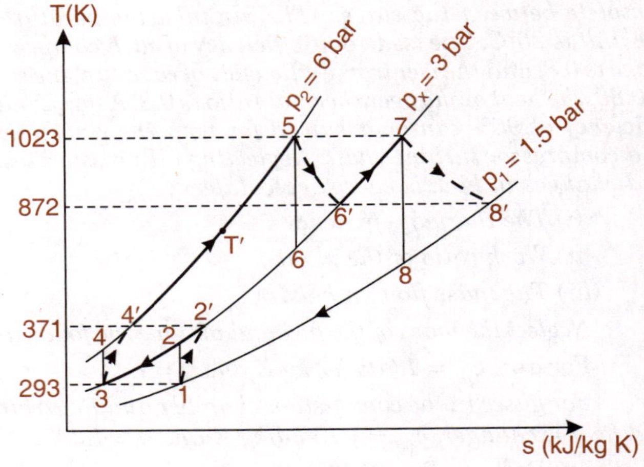

Fig. 25.34

Effectiveness of regenerator, $\varepsilon = 0.70$,

Power developed, $P = 350$ kW.

For air : $c_p = 1.005$ kJ/kg, K, $\gamma = 1.4$

As per given conditions : $T_1 = T_3$, $T_2' = T_4'$

$$\frac{T_2}{T_1} = \left(\frac{p_2}{p_1}\right)^{\frac{\gamma-1}{\gamma}}$$

and $\quad p_x = \sqrt{p_1 p_2} = \sqrt{1.5 \times 6} = 3$ bar

Now, $T_2 = T_1 \times \left(\dfrac{p_2}{T_1}\right)^{\frac{\gamma-1}{\gamma}}$

$= 293 \times \left(\dfrac{3}{1.5}\right)^{\frac{1.4-1}{1.4}} = 357$ K

$\eta_{compressor\ (L.P.)} = \dfrac{T_2 - T_1}{T_2' - T_1}$

$0.82 = \dfrac{357 - 293}{T_2' - 293}$

$\therefore \quad T_2' = \dfrac{357 - 293}{0.82} + 293 = 371$ K

i.e., $\quad T_2' = T_4' = 371$ K

Now, $\quad \dfrac{T_5}{T_6} = \left(\dfrac{p_5}{p_6}\right)^{\frac{\gamma-1}{\gamma}} = \left(\dfrac{p_2}{p_x}\right)^{\frac{1.4-1}{1.4}}$ $\quad \left[\because \begin{array}{l} p_5 = p_2 \\ p_6 = p_x \end{array}\right]$

$$\frac{1023}{T_6} = \left(\frac{6}{3}\right)^{0.286} = 1.219$$

$$\therefore \quad T_6 = \frac{1023}{1.219} = 839 \text{ K}$$

$$\eta_{turbine\,(H.P.)} = \frac{T_5 - T_6'}{T_5 - T_6}$$

$$0.82 = \frac{1023 - T_6'}{1023 - 839}$$

$$\therefore \quad T_6' = 1023 - 0.82\,(1023 - 839) = 872 \text{ K}$$

$T_8' = T_6' = 872$ K as $\eta_{turbine\,(H.P.)} = \eta_{turbine\,(L.P.)}$

and $T_7 = T_5 = 1023$ K

Effectiveness of regenerator, $\varepsilon = \dfrac{T' - T_4'}{T_8' - T_4'}$

where T' is the temperature of air coming out of regenerator

$$\therefore \quad 0.70 = \frac{T' - 371}{872 - 371}$$

i.e., $T' = 0.70\,(872 - 371) + 371 = 722$ K

Network available,

$$W_{net} = [W_{T(H.P.)} + W_{T(L.P.)}] - [W_{C(H.P.)} + W_{C(L.P.)}]$$

$= 2\,[W_{T(L.P.)} - W_{C(L.P.)}]$ as the *work developed by each turbine is same and work absorbed by each compressor is same.*

$$\therefore W_{net} = 2c_p\,[(T_5 - T_6') - (T_2' - T_1)]$$
$$= 2 \times 1.005\,[(1023 - 872) - (371 - 293)]$$
$$= 146.73 \text{ kJ/kg of air}$$

Heat supplied per kg of air *without regenerator*

$$= c_p(T_5 - T_4') + c_p(T_7 - T_6')$$
$$= 1.005\,[(1023 - 371) + (1023 - 872)]$$
$$= 807 \text{ kJ/kg of air}$$

Heat supplied per kg of air *with regenerator*

$$= c_p(T_5 - T') + c_p(T_7 - T_6')$$
$$= 1.005\,[(1023 - 722) + (1023 - 872)]$$
$$= 454.3 \text{ kJ/kg}$$

(*i*) $\eta_{\textbf{thermal (without regenerator)}}$

$= \dfrac{146.73}{807} =$ **0.182** or **18.2%. (Ans.)**

(*ii*) $\eta_{\textbf{thermal (with regenerator)}}$

$= \dfrac{146.73}{454.3} =$ **0.323** or **32.3%. (Ans.)**

(*iii*) **Mass of fluid circulated, $\dot{m}$:**

Power developed, $P = 146.73 \times \dot{m}$ kW

$\therefore \quad 350 = 146.73 \times \dot{m}$

i.e., $\dot{m} = \dfrac{350}{146.73} = 2.38$ kg/s

i.e., *Mass of fluid circulated* **= 2.38 kg/s. (Ans.)**

☞**Example 25.17.** *The air in a gas turbine plant is taken in L.P. compressor at 293 K and 1.05 bar and after compression it is passed through intercooler where its temperature is reduced to 300 K. The cooled air is further compressed in H.P. unit and then passed in the combustion chamber where its temperature is increased to 750°C by burning the fuel. The combustion products expand in H.P. turbine which runs the compressors and further expansion is continued in L.P. turbine which runs the alternator. The gases coming out from L.P. turbine are used for heating the incoming air from H.P. compressor and then expanded to atmosphere.*

Pressure ratio of each compressor = 2, isentropic efficiency of each compressor stage = 82%, isentropic efficiency of each turbine stage = 82%, effectiveness of heat exchanger = 0.72, air flow = 16 kg/s, calorific value of fuel = 42000 kJ/kg, c_v (for gas) = 1.0 kJ/kg K, c_p (gas) = 1.15 kJ/kg K, γ (for air) = 1.4, γ (for gas) = 1.33.

Neglecting the mechanical, pressure and heat losses of the system and fuel mass also determine the following :

(*i*) *The power output.*

(*ii*) *Thermal efficiency.*

(*iii*) *Specific fuel consumption.*

Solution. *Given :* $T_1 = 293$ K, $T_3 = 300$ K, $\dfrac{p_2}{p_1} = \dfrac{p_4}{p_3} = 2$,

$T_6 = 750 + 273 = 1023$ K,

$\eta_{compressor} = 82\%$, $\eta_{turbine} = 82\%$, $\varepsilon = 0.72$,

$\dot{m}_a = 16$ kg/s, $C = 42000$ kJ/kg,

$c_{pa} = 1.0$ kJ/kg K, $c_{pg} = 1.15$ kJ/kg K,

γ (for air) = 1.4, γ (for gas) 1.33.

$$\frac{T_2}{T_1} = \left(\frac{p_2}{p_1}\right)^{\frac{\gamma-1}{\gamma}} = (2)^{\frac{1.4-1}{1.4}} = 1.219$$

$$\therefore \quad T_2 = 293 \times 1.219 = 357 \text{ K}$$

Also, $\eta_{compressor} = \dfrac{T_2 - T_1}{T_2' - T_1}$

$$\therefore \quad 0.82 = \frac{357 - 293}{T_2' - 293}$$

$$\therefore \quad T_2' = \left(\frac{357 - 293}{0.82}\right) + 293 = 371 \text{ K}$$

Similarly, $\dfrac{T_4}{T_3} = \left(\dfrac{p_4}{p_3}\right)^{\frac{\gamma-1}{\gamma}} = (2)^{\frac{1.4-1}{1.4}} = 1.219$

$$\therefore \quad T_4 = 300 \times 1.219 = 365.7 \text{ K}$$

and $\eta_{compressor} = \dfrac{T_4 - T_3}{T_4' - T_3}$

$$\therefore \quad 0.82 = \frac{365.7 - 300}{T_4' - 300}$$

i.e., $T_4' = \left(\dfrac{365.7 - 300}{0.82}\right) + 300 = 380$ K

Work output of H.P. turbine = Work input to compressor.

Neglecting mass of fuel we can write

$$c_{pg}(T_6 - T_7') = c_{pa}[(T_2' - T_1) + (T_4' - T_3)]$$

$$1.15\,(1023 - T_7') = 1.0\,[(371 - 293) + (380 - 300)]$$

or $$1.15\,(1023 - T_7') = 158$$

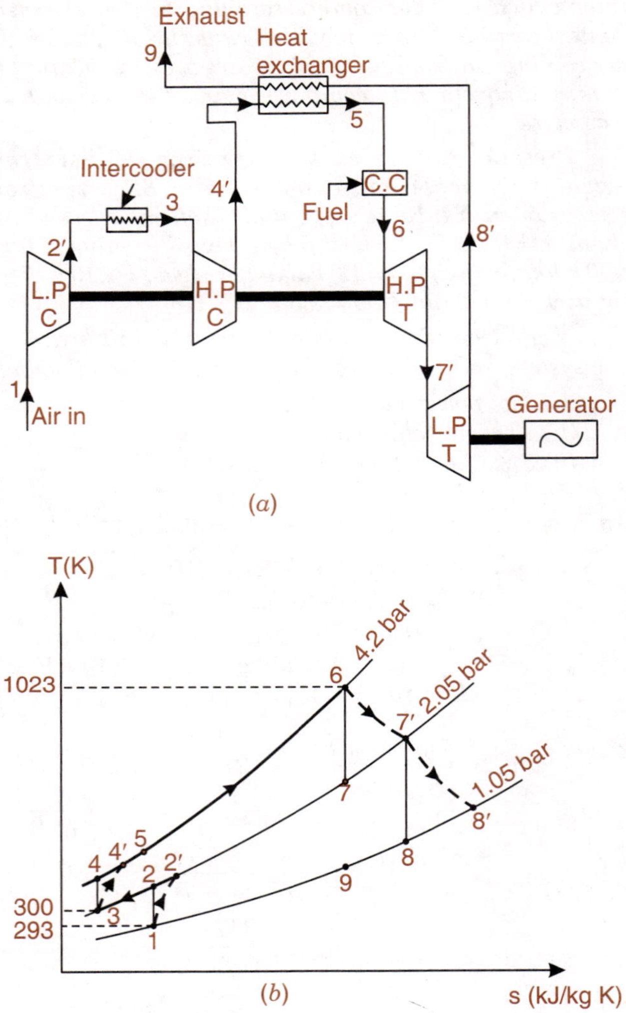

Fig. 25.35

$\therefore$ $$T_7' = 1023 - \frac{15.8}{1.15} = 886\text{ K}$$

Also, $$\eta_{turbine\,(H.P.)} = \frac{T_6 - T_7'}{T_6 - T_7}$$

i.e., $$0.82 = \frac{1023 - 886}{1023 - T_7}$$

$\therefore$ $$T_7 = 1023 - \left(\frac{1023 - 886}{0.82}\right) = 856\text{ K}$$

Now, $$\frac{T_6}{T_7} = \left(\frac{p_6}{p_7}\right)^{\frac{\gamma - 1}{\gamma}}$$

$\therefore$ $$\frac{p_6}{p_7} = \left(\frac{T_6}{T_7}\right)^{\frac{\gamma}{\gamma - 1}} = \left(\frac{1023}{856}\right)^{\frac{1.33}{1.33 - 1}} = 2.05$$

i.e., $$p_7 = \frac{p_6}{2.05} = \frac{4.2}{2.05} = 2.05\text{ bar}$$

$[\because\ p_6 = 1.05 \times 4 = 4.2\text{ bar}]$

$$\frac{T_7'}{T_8} = \left(\frac{p_7}{p_8}\right)^{\frac{\gamma - 1}{\gamma}} = \left(\frac{2.05}{1.05}\right)^{\frac{1.33 - 1}{1.33}} = 1.18$$

$$T_8 = \frac{T_7'}{1.18} = \frac{886}{1.18} = 751\text{ K}$$

Again, $$\eta_{turbine\,(L.P.)} = \frac{T_7' - T_8'}{T_7' - T_8}$$

$$0.82 = \frac{886 - T_8'}{886 - 751}$$

$\therefore$ $$T_8' = 886 - 0.82\,(886 - 751) = 775\text{ K}$$

(*i*) Power output :

Net power output $= c_{pg}(T_7' - T_8')$

$= 1.15\,(886 - 775) = 127.6$ kJ/kg

$\therefore$ Net output per second $= \dot{m} \times 127.6$

$= 16 \times 127.6 = 2041.6$ kJ/s = **2041.6 kW. (Ans.)**

(*ii*) Thermal efficiency :

Effectiveness of heat exchanger,

$$\varepsilon = \frac{T_5 - T_4'}{T_8' - T_4'}$$

i.e., $$0.72 = \frac{T_5 - 380}{778 - 380}$$

$\therefore$ $$T_5 = 0.72\,(775 - 380) + 380 = 664\text{ K}$$

Heat supplied in combustion chamber per second

$= \dot{m}_a\, c_{pg}\,(T_6 - T_5)$

$= 16 \times 1.15\,(1023 - 664) = 6605.6$ kJ/s

$\therefore$ $$\eta_{thermal} = \frac{2041.6}{6605.6} = \mathbf{0.309 \text{ or } 30.9\%.}$$ **(Ans.)**

(*iii*) Specific fuel consumption :

If m_f is the mass of fuel supplied per kg of air, then

$$m_f \times 42000 = 1.15\,(1023 - 664)$$

$\therefore$ $$\frac{1}{m_f} = \frac{42000}{1.15\,(1023 - 664)} = \frac{101.7}{1}$$

$\therefore$ Air-fuel ratio = 101.7 : 1

$\therefore$ Fuel supplied per hour

$$= \frac{16 \times 3600}{101.7} = 566.37\text{ kg/h}$$

$\therefore$ *Specific fuel consumption*

$$= \frac{566.37}{2041.6} = \mathbf{0.277\text{ kg/kWh.}}$$ **(Ans.)**

☞**Example 25.18.** *Air is taken in a gas turbine plant at 1.1 bar 20°C. The plant comprises of L.P. and H.P. compressors and L.P. and H.P. turbines. The compression in L.P. stage is upto 3.3 bar followed by intercooling to 27°C. The pressure of air after H.P. compressor is 9.45 bar. Loss in pressure during intercooling is 0.15 bar. Air from H.P. compressor is transferred to heat exchanger of effectiveness 0.65 where it is heated by the gases from L.P. turbine. After heat exchanger the air passes through combustion chamber. The temperature of gases supplied to H.P. turbine is 700°C. The gases expand in H.P. turbine to 3.62 bar and air then reheated to 670°C before expanding in L.P. turbine. The loss of pressure in reheater is 0.12 bar. Determine :*

(i) The overall efficiency

(ii) The work ratio

(iii) Mass flow rate when the power generated is 6000 kW.

Assume : Isentropic efficiency of compression in both stages = 0.82.

Isentropic efficiency of expansion in turbines = 0.85.

For air : $c_p = 1.005$ *kJ/kg K,* $\gamma = 1.4$.

For gases : $c_p = 1.15$ *kJ/kg K,* $\gamma = 1.33$.

Neglect the mass of fuel.

Solution. *Given :* $T_1 = 20 + 273 = 293$ K, $p_1 = 1.1$ bar,

$p_2 = 3.3$ bar, $T_3 = 27 + 273 = 300$ K,

$p_3 = 3.3 - 0.15 = 3.15$ bar,

$p_4 = p_6 = 9.45$ bar, $T_6 = 973$ K,

$T_8 = 670 + 273 = 943$ K, $p_8 = 3.5$ bar,

$\eta_{compressors} = 82\%$, $\eta_{turbines} = 85\%$,

Power generated = 6000 kW,

Effectiveness, $\varepsilon = 0.65$, $c_{pa} = 1.005$ kJ/kg K,

$\gamma_{air} = 1.44$, $c_{pg} = 1.15$ kJ/kg K

and $\gamma_{gases} = 1.33$.

Refer to Fig. 25.36.

Now,
$$\frac{T_2}{T_1} = \left(\frac{p_2}{p_1}\right)^{\frac{\gamma-1}{\gamma}} = \left(\frac{3.3}{1.1}\right)^{\frac{1.4-1}{1.4}} = 1.369$$

$\therefore$
$$T_2 = 293 \times 1.369 = 401 \text{ K}$$

$$\eta_{compressor\ (\text{L.P.})} = 0.82 = \frac{T_2 - T_1}{T_2' - T_1} = \frac{401 - 293}{T_2' - 293}$$

$\therefore$
$$T_2' = \left(\frac{401 - 293}{0.82}\right) + 293 = 425 \text{ K}$$

Again,
$$\frac{T_4}{T_3} = \left(\frac{p_4}{p_3}\right)^{\frac{\gamma-1}{\gamma}} = \left(\frac{9.45}{3.15}\right)^{\frac{1.4-1}{1.4}} = 1.369$$

$\therefore$
$$T_4 = 300 \times 1.369 = 411 \text{ K}$$

Now,
$$\eta_{compressor\ (\text{H.P.})} = \frac{T_4 - T_3}{T_4' - T_3}$$

$$0.82 = \frac{411 - 300}{T_4' - 300}$$

$\therefore$
$$T_4' = \left(\frac{411 - 300}{0.82}\right) + 300 = 435 \text{ K}$$

Similarly,
$$\frac{T_6}{T_7} = \left(\frac{p_6}{p_7}\right)^{\frac{\gamma-1}{\gamma}} = \left(\frac{9.45}{3.62}\right)^{\frac{1.33-1}{1.33}} = 1.268$$

$\therefore$
$$T_7 = \frac{T_6}{1.268} = \frac{973}{1.268} = 767 \text{ K}$$

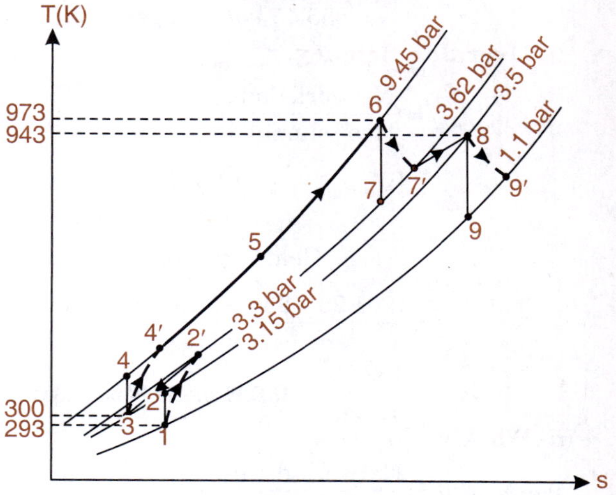

Fig. 25.36

Also,
$$\eta_{turbine\ (\text{H.P.})} = \frac{T_6 - T_7'}{T_6 - T_7}$$

$$0.85 = \frac{973 - T_7'}{973 - 767}$$

$\therefore$
$$T_7' = 973 - 0.85\,(973 - 767) = 798 \text{ K}$$

Again,
$$\frac{T_8}{T_9} = \left(\frac{p_8}{p_9}\right)^{\frac{\gamma-1}{\gamma}} = \left(\frac{3.5}{1.1}\right)^{\frac{1.33-1}{1.33}} = 1.332$$

$\therefore$
$$T_9 = \frac{T_8}{1.332} = \frac{943}{1.332} = 708 \text{ K}$$

$$\eta_{turbine\ (\text{L.P.})} = \frac{T_8 - T_9'}{T_8 - T_9}$$

$$0.85 = \frac{943 - T_9'}{943 - 708}$$

$\therefore$
$$T_9' = 943 - 0.85\,(943 - 708) = 743 \text{ K.}$$

Effectiveness of heat exchanger,

$$\varepsilon = 0.65 = \frac{T_5 - T_4'}{T_9' - T_4'}$$

i.e.,
$$0.65 = \frac{T_5 - 435}{743 - 435}$$

$\therefore$
$$T_5 = 0.65\,(743 - 435) + 435 = 635 \text{ K}$$

$$W_{turbine\ (\text{H.P.})} = c_{pg}\,(T_6 - T_7') = 1.15\,(973 - 798)$$
$$= 201.25 \text{ kJ/kg of gas}$$

$$W_{turbine\ (\text{L.P.})} = c_{pg}(T_8 - T_9') = 1.15\,(943 - 743)$$
$$= 230 \text{ kJ/kg of gas}$$

$$W_{compressor\ (L.P.)} = c_{pa}(T_2' - T_1) = 1.005\ (425 - 293)$$
$$= 132.66 \text{ kJ/kg of air}$$
$$W_{compressor\ (H.P.)} = c_{pa}(T_4' - T_3) = 1.005\ (435 - 300)$$
$$= 135.67 \text{ kJ/kg of air}$$
$$\text{Heat supplied} = c_{pg}(T_6 - T_5) + c_{pg}(T_8 - T_7')$$
$$= 1.15\ (973 - 635) + 1.15\ (943 - 798)$$
$$= 555.45 \text{ kJ/kg of gas}$$

(*i*) **Overall efficiency, $\eta_{overall}$:**

$$\eta_{overall} = \frac{\text{Network done}}{\text{Heat supplied}}$$
$$= \frac{[W_{turbine\ (H.P.)} + W_{turbine\ (L.P.)}] - [W_{comp.\ (L.P.)} + W_{comp.\ (H.P.)}]}{\text{Heat supplied}}$$
$$= \frac{(201.25 + 230) - (132.66 + 135.67)}{555.45}$$
$$= \frac{162.92}{555.45} = \mathbf{0.293 \text{ or } 29.3\%.} \quad \textbf{(Ans.)}$$

(*ii*) **Work ratio :**

$$\text{Work ratio} = \frac{\text{Network done}}{\text{Turbine work}}$$
$$= \frac{[W_{turbine\ (H.P.)} + W_{turbine\ (L.P.)}] - [W_{comp.\ (L.P.)} + W_{comp.\ (H.P.)}]}{[W_{turbine\ (H.P.)} + W_{turbine\ (L.P.)}]}$$
$$= \frac{(201.25 + 230) - (132.66 + 135.67)}{(201.25 + 230)}$$
$$= \frac{162.92}{431.25} = 0.377.$$

i.e., *Work ratio* = **0.377. (Ans.)**

(*iii*) **Mass flow rate, $\dot{m}$:**

Network done = 162.92 kJ/kg.

Since mass of fuel is neglected, for 6000 kW, mass flow rate,

$$\dot{m} = \frac{6000}{162.92} = 36.83 \text{ kg/s}$$

i.e., *Mass flow rate* = **36.83 kg/s. (Ans.)**

25.8. JET PROPULSION

The *principle of jet propulsion* involves *imparting momentum to a mass of fluid in such a manner that the reaction of imparted momentum provides a propulsive force*. It may be achieved by expanding the gas, which is at high temperature and pressure, through a nozzle due to which a high velocity jet of hot gases is produced (in the atmosphere) that gives a propulsive force (in opposite direction due to its reaction). For jet propulsion the open cycle gas turbine is *most suitable*.

The propulsion system may be classified as follows :

1. **Air stream jet engines,** (Air-breathing engines)

(*a*) *Steady combustion systems ;* continuous air flow

(*i*) Turbo-jet (*ii*) Turbo-prop

(*iii*) Ram jet

(*b*) *Intermittent combustion system ;* intermittent flow

(*i*) Pulse jet or flying bomb.

2. **Self contained rocket engines** (Non-air breathing engines)

(*i*) Liquid propellent (*ii*) Solid propellent.

In **air stream jet engines** the *oxygen necessary for the combustion is taken from the surrounding atmosphere* whereas in a **rocket engine** *the fuel and the oxidiser are contained in the body of the unit which is to be propelled.*

Note. The turbo-jet and turbo-prop are modified forms of simple open cycle gas turbine. The ram jet and pulse jet are *athodyds* (aero-thermo-dynamic ducts) *i.e.,* straight duct type of jet engines having *no compressor and turbine wheels.*

In the past air propulsion was achieved by a **"Screw propeller"**. In this system the total power developed by the turbine (full expansion) is used to drive the compressor and propeller. Fig. 25.37 shows the power plant for screw propeller. By controlling the supply of fuel in the combustion chamber the power supplied to the propeller can be controlled. *The rate of increase of efficiency of screw propeller is higher at lower speeds but its efficiency falls rapidly at higher speeds above the sonic velocity.*

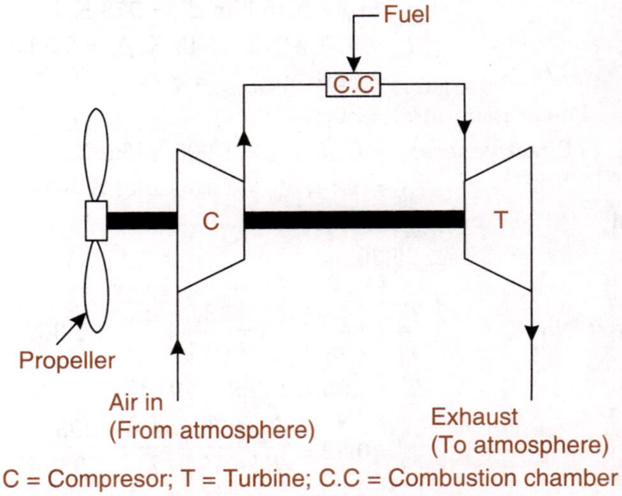

Fig. 25.37. Power plant for screw propeller.

25.8.1. Turbo-Jet

25.8.1.1. Description

Fig. 25.38 shows a turbo-jet unit.

- It consists of *diffuser* at entrance which slows down the air (entering at velocity equal to the plane speed) and part of the kinetic energy of the air stream is converted into pressure ; this type of compression is called as *ram compression*.

- The air is further compressed to a pressure of 3 to 4 bar in a rotary compressor (usually of axial flow type).
- The compressed air then enters the combustion chamber (C.C.) where fuel is added. The combustion of fuel takes place at sensibly constant pressure and subsequently temperature rises rapidly.
- The hot gases then enter the gas turbine where *partial expansion* takes place. The *power produced is just sufficient to drive the compressor, fuel pump and other auxiliaries.*
- The exhaust gases from the gas turbine which are at a higher pressure than atmosphere are expended in a nozzle and a very high velocity jet is produced which provides a forward motion to the aircraft by the jet reaction (Newton's third law of motion).

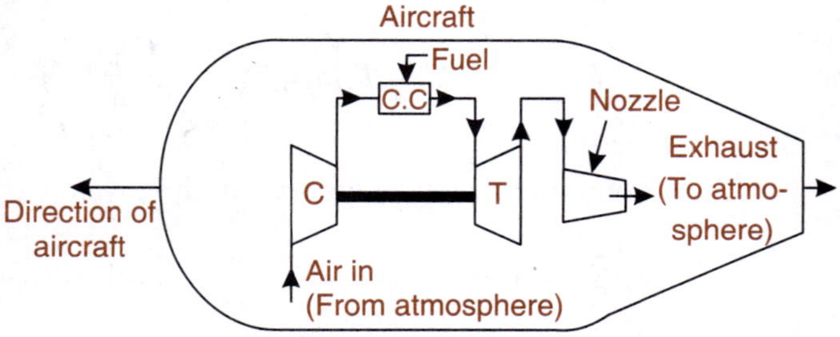

Fig. 25.38. Gas turbine plant for turbo-jet.

At higher speeds the turbo-jet gives higher propulsion efficiency. The turbo-jets are most suited to the aircrafts travelling *above 800 km/h.*

The overall efficiency of a turbo-jet is the product of the thermal efficiency of the gas turbine plant and the propulsive efficiency of the jet (nozzle).

Advantages of Turbo-jet engines

1. Construction much simpler (as compared to multi-cylinder piston engine of comparable power).
2. Engine vibrations absent.
3. Much higher speeds possible (more than 3000 km/h achieved).
4. Power supply is uninterrupted and smooth.
5. Weight to power ratios superior (as compared to that of reciprocating type of aero-engine).
6. Rate of climb higher.
7. Requirement of major overhauls less frequent.
8. Radio interference much less.
9. Maximum altitude ceiling as compared to turbo-prop and conventional piston type engines.
10. Frontal area smaller.
11. Fuel can be burnt over a large range of mixture strength.

Disadvantages of turbo-jet engines

1. Less efficient.
2. Life of the unit comparatively shorter.
3. The turbo-jet becomes rapidly inefficient below 550 km/h.
4. More noisy (than a reciprocating engine).
5. Materials required are quite expensive.
6. Require longer strip since length of take-off is too much.
7. At take-off the thrust is low, this effect is overcome by boosting.

25.8.1.2. Basic Cycle for Turbo-jet Engine

The basic cycle for the turbo-jet engine is the *Joule or Brayton cycle* as shown is Fig. 25.39. The various processes are as follows :

Process 1-2 : The air entering from atmosphere is *diffused isentropically* from velocity C_1 down to zero (*i.e.,* C_2 = 0). This indicates that the diffuser has an efficiency of 100%, this is termed as *ram compression.*

Process 1-2′ is the *actual process.*

Process 2-3 : *Isentropic* compression of air.

Process 2′-3′ shows the *actual* compression of air.

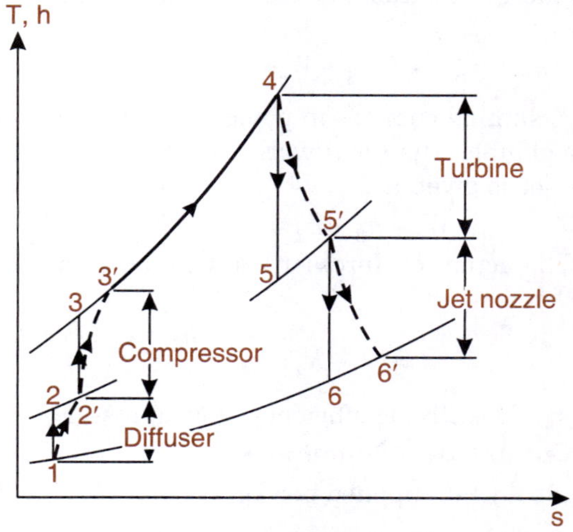

Fig. 25.39. *T-s* diagram of turbo-jet.

Process 3-4 : *Ideal* addition of heat at constant pressure $p_3 = p_4$

Process 3′-4 shows the *actual* addition of heat at constant process $p_3 = p_4$.

Process 4-5 : *Isentropic expansion* of gas in the *turbine*.

Process 4-5′ shows the *actual* expansion in the turbine.

Process 5-6 : *Isentropic expansion* of gas in the *nozzle*.

Process 5′-6′ shows the *actual expansion* of gas in the *nozzle*.

Consider *1 kg of working fluid flowing through the system.*

Diffuser :

Between states 1 and 2, the energy equation is given by :

$$\frac{C_a^2}{2} + h_1 + Q_{1-2} = \frac{C_2^2}{2} + h_2 + W_{1-2}$$

where C_a (= C_1) = Velocity of entering air from atmosphere.

In an ideal diffuser $C_2 = 0$, $Q_{1\text{-}2} = 0$ and $W_{1\text{-}2} = 0$.

$\therefore$ Enthalpy at state 2 is, $h_2 = h_1 + \dfrac{C_a^2}{2}$ kJ/kg

or $$T_2 = T_1 + \frac{C_a^2}{2.c_p} \quad (\because \; h = c_p.T) \quad \text{...(25.7)}$$

Process 1-2′ shows actual process in diffuser.

Diffuser efficiency, $$\eta_d = \frac{h_2 - h_1}{h_2' - h_1} = \frac{T_2 - T_1}{T_2' - T_1}$$

or $$h_2' = h_1 + \frac{C_a^2}{2\,\eta_d}$$

or $$T_2' = T_1 + \frac{C_a^2}{2\,.\,c_p\,.\,\eta_d} \quad \text{...(25.8)}$$

Compressor :

Energy equation between states 2 and 3 gives

$$h_2 + \frac{C_2^2}{2} + Q_{2-3} + W_c = h_3 + \frac{C_3^2}{2}$$

Assuming changes in potential and kinetic energies to be negligible, the ideal work expended in running the compressor is given as,

$$W_c = h_3 - h_2 = c_p(T_3 - T_2)$$

The actual compressor work (to be supplied by the turbine)

$$= h_3' - h_2 = \frac{h_3 - h_2}{\eta_c} = \frac{c_p\,(T_3 - T_2)}{\eta_c}$$

(where η_c = Isentropic efficiency of compressor)

Combustion chamber :

Ideal heat supplied per kg, $Q = h_4 - h_3 = c_p(T_4 - T_3)$

$$\text{Actual heat supplied} = \left(1 + \frac{m_f}{m_a}\right) h_4 - h_3'$$

or $$Q_a = c_{pg}\left(1 + \frac{m_f}{m_a}\right) T_4 - c_{pa}\,.\,T_3'$$

(where c_{pg} and c_{pa} are specific heats of gases and air at constant pressure respectively)

Turbine :

Between states 4 and 5, the energy equation is given by :

$$h_4 + \frac{C_4^2}{2} + Q_{4-5} = h_5 + \frac{C_5^2}{2} + W_t$$

If $Q_{4\text{–}5} = 0$, then turbine work,

$$W_t = (h_4 - h_5) + \frac{(C_4^2 - C_5^2)}{2}$$

If the change in kinetic energy is neglected, we have

$$W_t = (h_4 - h_5) = c_p(T_4 - T_5)$$

Actual turbine work $= h_4 - h_5' = c_p(T_4 - T_5')$

$$= c_p(T_4 - T_5) \times \eta_t$$

(where η_t = Isentropic efficiency of turbine)

For the simplification, turbine work = compressor work

or $$c_p(T_4 - T_5') = c_p(T_4 - T_5)\,\eta_t = \frac{c_p\,(T_3 - T_2)}{\eta_c}$$

or $$T_5' = T_4 - (T_4 - T_5)\eta_t = T_4 - \frac{c_p\,(T_3 - T_2)}{\eta_c}$$

Jet nozzel :

Energy equation between states 5 and 6 gives

$$h_5 + \frac{C_5^2}{2} = h_6 + \frac{C_6^2}{2} \quad \text{...Ideal case}$$

$$h_5' + \frac{C_5'^2}{2} = h_6' + \frac{C_6'^2}{2} \quad \text{...Actual case}$$

If $C_5'^{\,2}$ is very less as compared to $C_6'^{\,2}$, we have

$$h_5' = h_6' + \frac{C_6'^2}{2}$$

or $$C_6' = \sqrt{2\,(h_5' - h_6')} = \sqrt{2\,\eta_n\,(h_5' - h_6)}$$

or $$C_6' = \sqrt{2\,\eta_n\,c_p\,(T_5' - T_6)} \quad \text{...(25.9)}$$

(where η_n = Nozzle efficiency)

Thermal efficiency (η_{th}) is given by :

$$\eta_{th} = \frac{(h_4 - h_6') - (h_3' - h_1)}{(h_4 - h_3')}$$

$$= \frac{(T_4 - T_6') - (T_3' - T_1)}{(T_4 - T_3')} \quad \text{...(25.10)}$$

25.8.1.3. Thrust, Thrust-power, Propulsive Efficiency and Thermal Efficiency.

Thrust (T)

Let C_a = Forward velocity of aircraft through air, m/s. Assuming the atmospheric air to be still the velocity of air, relative to the aircraft, at entry to the aircraft will be C_a. It is called *velocity of approach of air*.

C_j = Velocity of jet (gases) relative to the exit nozzle/ aircraft ; m/s.

$\left[1+\frac{\text{fuel }(m_f)}{\text{air }(m_a)}\right]$ = Mass of products leaving the nozzle for 1 kg of air.

Thrust is the force *produced due to change of momentum.*

Now, absolute velocity of gases leaving aircraft

$$= (C_j - C_a)$$

Absolute velocity of air entering the aircraft = 0

$$\therefore \quad \text{Change of momentum} = \left(1+\frac{m_f}{m_a}\right)(C_j - C_a)$$

Hence, thrust,

$$T = \left(1+\frac{m_f}{m_a}\right)(C_j - C_a) \text{ N/kg of air/s} \quad ...(25.11)$$

$$T = (C_j - C_a) \text{ N/kg of air/s, } \textit{neglecting mass of fuel} \quad ...(25.12)$$

Thrust power (T.P.) :

It is defined as the rate at which work must be developed by the engine if the aircraft is to be kept moving at a constant velocity C_a against friction force or drag.

$\therefore$ Thrust power = Forward thrust × speed of aircraft

$$\text{or} \quad \text{T.P.} = \left[\left(1+\frac{m_f}{m_a}\right)(C_j - C_a)\right] C_a \text{ W/kg of air} \quad ...(25.13)$$

$$= (C_j - C_a)\, C_a \text{ W/kg of air if } \textit{mass of fuel is neglected}$$

$$= \frac{(C_j - C_a)\, C_a}{1000} \text{ kW/kg of air} \quad ...(25.14)$$

Propulsive power (P.P.) :

The *energy required to change the momentum of the mass flow of gas represents the propulsive power.* It is expressed as the *difference between the rate of kinetic energies of the entering air and exit gases.*

Mathematically,

$$\text{P.P.} = \Delta\text{K.E.} = \frac{\left(1+\frac{m_f}{m_a}\right)C_j^2}{2} - \frac{C_a^2}{2} \text{ W/kg} \quad ...(25.15)$$

$$= \frac{C_j^2 - C_a^2}{2} \text{ W/kg, } \textit{neglecting mass of fuel}$$

$$= \frac{C_j^2 - C_a^2}{2 \times 1000} \text{ kW/kg of air} \quad ...(25.16)$$

Propulsive efficiency ($\eta_{prop.}$) :

The *ratio of thrust power to propulsive power is called the propulsive efficiency of the propulsive unit.*

$$\eta_{prop.} = \frac{\text{Thrust power}}{\text{Propulsive power}} = \frac{\left[\left(1+\frac{m_f}{m_a}\right)(C_j - C_a)\right] C_a}{\left[\frac{\left(1+\frac{m_f}{m_a}\right)C_j^2}{2} - \frac{C_a^2}{2}\right]}$$

$$= \frac{2\left[\left(1+\frac{m_f}{m_a}\right)(C_j - C_a)\right] C_a}{\left[\left(1+\frac{m_f}{m_a}\right)C_j^2 - C_a^2\right]} \quad ...(25.17)$$

Neglecting the mass of fuel,

$$\eta_{prop.} = \frac{2(C_j - C_a)\, C_a}{C_j^2 - C_a^2} = \frac{2(C_j - C_a)\, C_a}{(C_j + C_a)(C_j - C_a)}$$

$$\text{or} \quad \eta_{prop.} = \frac{2C_a}{C_j + C_a} \quad \text{or} \quad \frac{2}{\left(\frac{C_j}{C_a} + 1\right)} \quad ...(25.18)$$

From eqn. (25.18) it is evident that the *propulsive efficiency increases with an increase in aircraft velocity C_a.* $\eta_{prop.}$ becomes 100% when C_a approaches C_j ; thrust reduces to zero (Eqn. 25.12).

Thermal efficiency (η_{th}) :

It is defined as the *ratio of propulsive work and the energy released by the combustion of fuel.*

$$\text{or,} \quad \eta_{th} = \frac{\text{Propulsive work}}{\text{Heat released by the combustion of fuel}}$$

$$= \frac{\text{Increase in kinetic energy of the gases}}{\text{Heat released by the combustion of fuel}}$$

$$\text{or,} \quad \eta_{th} = \frac{\left(1+\frac{m_f}{m_a}\right)C_j^2 - C_a^2}{2\left[\frac{m_f}{m_a} \times \text{calorific value}\right]} \quad ...(25.19)$$

$$\simeq \frac{(C_j^2 - C_a^2)}{2 \times \left(\frac{m_f}{m_a}\right) \times \text{calorific value}} \quad ...(25.20)$$

Overall efficiency (η_0) is given by :

$$\eta_0 = \eta_{th} \times \eta_{prop.} = \frac{(C_j^2 - C_a^2)}{2 \times \left(\frac{m_f}{m_a}\right) \times \text{calorific value}} \times \frac{2C_a}{C_j + C_a}$$

$$= \frac{(C_j - C_a)\, C_a}{\left(\frac{m_f}{m_a}\right) \times \text{calorific value}} \quad ...(25.21)$$

For *maximum overall efficiency the aircraft velocity C_a is one half of the jet velocity C_j.*

The jet *efficiency* (η_{jet}) is defined as :

$$\eta_{jet} = \frac{\text{Final kinetic energy in the jet}}{\text{Isentropic heat drop in the jet pipe + Carry over from the turbine}}.$$

Example 25.19. *A turbo-jet engine consumes air at the rate of 60.2 kg/s when flying at a speed of 1000 km/h. Calculate :*

(i) Exit velocity of the jet when the enthalpy change for the nozzle is 230 kJ/kg and velocity co-efficient is 0.96.

(ii) Fuel flow rate in kg/s when air-fuel ratio is 70 : 1

(iii) Thrust specific fuel consumption

(iv) Thermal efficiency of the plant when the combustion efficiency is 92% and calorific value of the fuel used is 42000 kJ/kg.

(v) Propulsive power

(vi) Propulsive efficiency

(vii) Overall efficiency.

Solution. Rate of air consumption, $\dot{m}_a = 60.2$ kg/s

Enthalpy change for nozzle, $\Delta h = 230$ kJ/kg

Velocity coefficient, $z = 0.96$

Air-fuel ratio $= 70 : 1$

Combustion efficiency, $\eta_{combustion} = 92\%$

Calorific value of fuel, C.V. $= 42000$ kJ/kg

Aircraft velocity, $C_a = \frac{1000 \times 1000}{60 \times 60} = 277.8$ m/s

(i) Exit velocity of jet, C_j :

$$C_j = z\sqrt{2\,\Delta h \times 1000}\text{, where } \Delta h \text{ is in kJ}$$

$$= 0.96\sqrt{2 \times 230 \times 1000} = 651 \text{ m/s.}$$

i.e., **Exit velocity of jet = 651 m/s. (Ans.)**

(ii) Fuel flow rate :

Rate of fuel consumption,

$$\dot{m}_f = \frac{\text{Rate of air consumption}}{\text{Air-fuel ratio}}$$

$$= \frac{60.2}{70} = \mathbf{0.86\ kg/s.}\quad \textbf{(Ans.)}$$

(iii) Thrust specific fuel consumption :

Thrust is the force produced due to change of momentum.

Thrust produced $= \dot{m}_a\,(C_j - C_a)$, neglecting mass of fuel.

$$= 60.2\,(651 - 277.8) = 22466.6 \text{ N.}$$

∴ Thrust specific fuel consumption

$$= \frac{\text{Fuel consumption}}{\text{Thrust}} = \frac{0.86}{22466.6}$$

$$= \mathbf{3.828 \times 10^{-5}\ kg/N\ of\ thrust/s.}\quad \textbf{(Ans.)}$$

(iv) Thermal efficiency, $\eta_{thermal}$:

$$\eta_{thermal} = \frac{\text{Work output}}{\text{Heat supplied}}$$

$$= \frac{\text{Gain in kinetic energy per kg of air}}{\text{Heat supplied by fuel per kg of air}}$$

$$= \frac{(C_j^2 - C_a^2)}{\left(\frac{m_f}{m_a}\right) \times \text{C.V.} \times \eta_{combustion} \times 1000}$$

$$= \frac{(651^2 - 277.8^2)}{2 \times \frac{1}{70} \times 42000 \times 0.92 \times 1000}$$

$$= 0.3139 \text{ or } 31.39\%$$

i.e., Thermal efficiency $=$ **31.39%. (Ans.)**

(v) Propulsive power :

Propulsive power

$$= \dot{m}_a \times \left(\frac{C_j^2 - C_a^2}{2}\right) = \frac{60.2}{1000} \times \left(\frac{651^2 - 277.8^2}{2}\right) \text{ kW}$$

$$= \mathbf{10433.5\ kW.}\quad \textbf{(Ans.)}$$

(vi) Propulsive efficiency, $\eta_{prop.}$:

$$\eta_{prop.} = \frac{\text{Thrust power}}{\text{Propulsive power}} = \frac{2\,C_a}{C_j + C_a} \quad \text{...[Eqn. (25.18)]}$$

$$= \frac{2 \times 277.8}{651 + 277.8} = 0.598 \text{ or } \mathbf{59.8\%.}\quad \textbf{(Ans.)}$$

(vii) Overall efficiency, η_0 :

$$\eta_0 = \frac{\text{Thrust work}}{\text{Heat supplied by fuel}}$$

$$= \frac{(C_j - C_a)C_a}{\left(\frac{m_f}{m_a}\right) \times \text{C.V.} \times \eta_{combustion}} \quad \text{...(25.22)}$$

$$= \frac{(651 - 277.8) \times 277.8}{\frac{1}{70} \times 42000 \times 0.92 \times 1000}$$

$$= \mathbf{0.1878 \text{ or } 18.78\%.}\quad \textbf{(Ans.)}$$

Example 25.20. *The following data pertain to a turbo-jet flying at an altitude of 9500 m :*

Speed of the turbo-jet = 800 km/h

Propulsive efficiency = 55%

Overall efficiency of the turbine plant = 17%

Density of air at 9500 m altitude = 0.17 kg/m³

Drag on the plane = 6100 N

Assuming calorific value of the fuels used as 46000 kJ/kg,

Calculate :

(i) Absolute velocity of the jet.

(*ii*) *Volume of air compressed per min.*
(*iii*) *Diameter of the jet.*
(*iv*) *Power output of the unit.*
(*v*) *Air-fuel ratio.*

Solution. *Given :* Altitude = 9500 m, $C_a = \frac{800 \times 1000}{60 \times 60}$ = 222.2 m/s,

$\eta_{propulsive}$ = 55%, $\eta_{overall}$ = 17% ; density of air at 9500 m altitude = 0.17 kg/m^3 ; drag on the plane = 6100 N.

(*i*) **Absolute velocity of the jet, $(C_j - C_a)$:**

$$\eta_{propulsive} = 0.55 = \frac{2C_a}{C_j + C_a}$$

where, C_j = Velocity of gases at nozzle exit relative to the aircraft, and

C_a = Velocity of the turbo-jet/aircraft.

$$\therefore \quad 0.55 = \frac{2 \times 222.2}{C_j + 222.2}$$

i.e.,
$$C_j = \frac{2 \times 222.2}{0.55} - 222.2 = 585.8 \text{ m/s}$$

$\therefore$ Absolute velocity of jet = $C_j - C_a$ = 585.8 – 222.2 = 363.6 m/s.

(*ii*) **Volume of air compressed/min. :**

Propulsive force $= \dot{m}_a (C_j - C_a)$

$$6100 = \dot{m}_a (585.8 - 222.2)$$

$$\therefore \quad \dot{m}_a = 16.77 \text{ kg/s}$$

$\therefore$ Volume of air compressed/min.

$$= \frac{16.77}{0.17} \times 60 = \textbf{5918.8 kg/min.} \quad \text{(Ans.)}$$

(*iii*) **Diameter of the jet, d :**

Now,
$$\frac{\pi}{4} d^2 \times C_j = 5918.8$$

i.e.,
$$\frac{\pi}{4} d^2 \times 585.8 = (5918.8/60)$$

$$\therefore \quad d = \left(\frac{5918.8 \times 4}{60 \times \pi \times 585.8}\right)^{1/2}$$

$$= 0.463 \text{ m} = 463 \text{ mm}$$

i.e., **Diameter of the jet = 463 mm. (Ans.)**

(*iv*) **Power output of the unit :**

Thrust power = Drag force × velocity of turbo-jet
= 6100 × 222.2 N-m/s

$$= \frac{6100 \times 222.2}{1000} = 1355.4 \text{ kW}$$

$$\textbf{Turbine output} = \frac{\text{Thrust power}}{\text{Propulsive efficiency}} = \frac{1355.4}{0.55}$$

= **2464.4 kW. (Ans.)**

(*v*) **Overall efficiency, η_0 :**

$$\eta_0 = \frac{\text{Heat equivalent of output}}{\dot{m}_f \times \text{C.V.}}$$

i.e.,
$$0.17 = \frac{2464.4}{\dot{m}_f \times 46000}$$

$$\therefore \quad \dot{m}_f = \frac{2464.4}{0.17 \times 46000} = 0.315 \text{ kg/s}$$

$$\therefore \quad \text{Air-fuel ratio} = \frac{\text{Air used (in kg/s)}}{\text{Fuel used (in kg/s)}} = \frac{16.77}{0.315} = 53.24$$

i.e., *Air-fuel ratio* = **53.24 : 1. (Ans.)**

Example 25.21. *In a jet propulsion unit air is drawn into the rotary compressor at 15°C and 1.01 bar and delivered at 4.04 bar. The isentropic efficiency of compression is 82% and the compression is uncooled. After delivery the air is heated at constant pressure until the temperature reaches 750°C. The air then passes through a turbine unit which drives the compressor only and has an isentropic efficiency of 78% before passing through the nozzle and expanding to atmospheric pressure of 1.01 bar with an efficiency of 88%. Neglecting any mass increase due to the weight of the fuel and assuming that R and γ are unchanged by combustion, determine :*

(*i*) *The power required to drive the compressor.*

(*ii*) *The air-fuel ratio if the fuel has a calorific value of 42000 kJ/kg.*

(*iii*) *The pressure of the gases leaving the turbine.*

(*iv*) *The thrust per kg of air per second.*

Neglect any effect of the velocity of approach.

Assume for air : R = 0.287 kJ/kg K, γ = 1.4.

Solution. *Given :* T_1 = 15 + 273 = 288 K,
p_1 = 1.01 bar, p_2 = 4.04 bar,
T_3 = 750 + 273 = 1023 K,
$\eta_{compressor}$ = 82%, $\eta_{turbine}$ = 78%,
η_{nozzle} = 88%,
R_{air} = 0.287 kJ/kg K, γ_{air} = 1.4.

Refer to Fig. 25.40.

$$\frac{T_2}{T_1} = \left(\frac{p_2}{p_1}\right)^{\frac{\gamma-1}{\gamma}} = \left(\frac{4.04}{1.01}\right)^{\frac{1.4-1}{1.4}} = 1.486$$

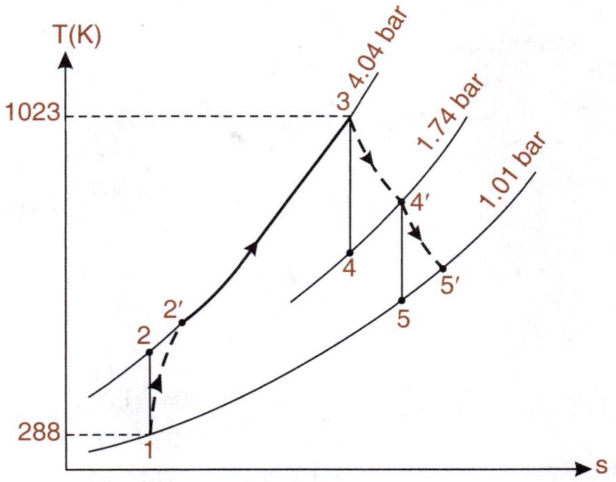

Fig. 25.40

$\therefore \quad T_2 = 2.88 \times 1.486 = 428 \text{ K}$

$$\eta_{compressor} = \frac{T_2 - T_1}{T_2' - T_1} \quad i.e., \quad 0.82 = \frac{428 - 288}{T_2' - 288}$$

$$\therefore \quad T_2' = \left(\frac{428 - 288}{0.82}\right) + 288 = 458.7 \text{ K}$$

$$c_p = R \times \left(\frac{\gamma}{\gamma - 1}\right) = 0.287 \times \frac{1.4}{(1.4 - 1)}$$

$$= 1.004 \text{ kJ/kg K}$$

(*i*) Power required to drive the compressor :

Power required to drive the compressor (per kg of air/sec.)

$$= c_p(T_2' - T_1) = 1.004(458.7 - 288)$$

$= \mathbf{171.38\ kW.}$ **(Ans.)**

(*ii*) Air-fuel ratio :

$$m_f \times C = (m_a + m_f) \times c_p \times (T_3 - T_2')$$

where, m_a = Mass of air per kg of fuel, and
= Air-fuel ratio.

$$\therefore \quad m_a = \frac{m_f \times C}{c_p\,(T_3 - T_2')} - m_f$$

$$\therefore \quad \frac{m_a}{m_f} = \frac{C}{c_p\,(T_3 - T_2')} - 1$$

$$= \frac{42000}{1.004(1023 - 458.7)} - 1 = 73.1$$

i.e., **Air-fuel ratio = 73.1 : 1. (Ans.)**

(*iii*) Pressure of the gases leaving the turbine, p_4 :

Neglecting effect of fuel on mass flow,

Actual turbine work = actual compressor work

i.e., $c_p(T_2' - T_1) = c_p(T_3 - T_4')$

or $T_2' - T_1 = T_3 - T_4'$

$\therefore \quad 458.7 - 288 = 1023 - T_4'$

$\therefore \quad T_4' = 852.3 \text{ K}$

Also, $$\eta_{turbine} = \frac{T_3 - T_4'}{T_3 - T_4}$$

$$0.78 = \frac{1023 - 852.3}{1023 - T_4}$$

$$\therefore \quad T_4 = 1023 - \left(\frac{1023 - 852.3}{0.78}\right) = 804 \text{ K}$$

$$\therefore \quad \frac{T_4}{T_3} = \left(\frac{p_4}{p_3}\right)^{\frac{\gamma - 1}{\gamma}}$$

or $$\frac{p_4}{p_3} = \left(\frac{T_4}{T_3}\right)^{\frac{\gamma}{\gamma - 1}} = \left(\frac{804}{1023}\right)^{\frac{1.4}{1.4 - 1}} = 0.43$$

or $p_4 = 4.04 \times 0.43 = \mathbf{1.74\ bar.}$ **(Ans.)**

(*iv*) Thrust per kg of air per second :

$$\frac{T_4'}{T_5} = \left(\frac{p_4}{p_5}\right)^{\frac{\gamma - 1}{\gamma}} = \left(\frac{1.74}{1.01}\right)^{\frac{1.4 - 1}{1.4}} = 1.168$$

$$\therefore \quad T_5 = \frac{T_4'}{1.168} = \frac{852.3}{1.168} = 729.7 \text{ K}$$

$$\eta_{nozzle} = \frac{T_4' - T_5'}{T_4' - T_5}$$

$$0.88 = \frac{852 - T_5'}{852.3 - 729.7}$$

$$\therefore \quad T_5' = 852.3 - 0.88\,(852.3 - 729.7) = 744.4 \text{ K}$$

If C_j is the jet velocity, then

$$\frac{C_j^2}{2} = c_p(T_4' - T_5')$$

$$\therefore \quad C_j = \sqrt{2 \times c_p\,(T_4' - T_5')}.$$

$$= \sqrt{2 \times 1.004(852.3 - 744.4) \times 1000}$$

$$= 465.5 \text{ m/s}$$

$\therefore$ Thrust per kg per second

$= 1 \times 465.5 = \mathbf{465.5\ N.}$ **(Ans.)**

Example 25.22. *A turbo-jet engine travels at 216 m/s in air at 0.78 bar and – 7.2°C. Air first enters diffuser in which it is brought to rest relative to the unit and it is then compressed in a compressor through a pressure ratio of 5.8 and fed to a turbine at 1110°C. The gases expand through the turbine and then through the nozzle to atmospheric pressure (i.e., 0.78 bar). The efficiencies of diffuser, nozzle and compressor are each 90%. The efficiency of turbine is 80%. Pressure drop in the combustion chamber is 0.168 bar. Determine :*

(i) Air-fuel ratio ;

(ii) Specific thrust of the unit ;

(iii) Total thrust, if the inlet cross-section of diffuser is 0.12 m².

Assume calorific value of fuel as 44150 kJ/kg of fuel.

Solution. Refer to Fig. 25.41.

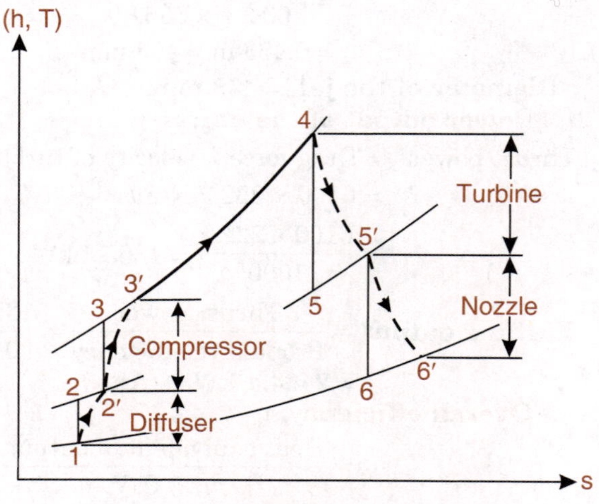

Fig. 25.41

Speed of the aircraft, $C_a = 216$ m/s

Intake air temperature, $T_1 = -7.2 + 273 = 265.8$ K

Intake air pressure, $p_1 = 0.78$ bar

Pressure ratio in the compressor, $r_p = 5.8$

Temperature of gases entering the gas turbine,

$$T_4 = 1110 + 273 = 1383 \text{ K}$$

Pressure drop in combustion chamber = 0.168 bar

$$\eta_d = \eta_n\ ;\ \eta_c = 90\%\ ;\ \eta_t = 80\%.$$

Calorific value of fuel, $C.V. = 44150$ kJ/kg of coal

(*i*) Air-fuel ratio :

For ideal diffuser (*i.e.*, process 1-2) the energy equation is given by :

$$h_2 = h_1 + \frac{C_a^2}{2} \quad \text{or} \quad h_2 - h_1 = \frac{C_a^2}{2}$$

or $$T_2 - T_1 = \frac{C_a^2}{2c_p}$$

or $$T_2 = T_1 + \frac{C_a^2}{2c_p}$$

$$= 265.8 + \frac{216^2}{2 \times 1.005 \times 1000} = 289 \text{ K}$$

For actual diffuser (*i.e.*, process 1-2′),

$$\eta_d = \left(\frac{h_2 - h_1}{h_2' - h_1}\right) \quad \text{or} \quad h_2' - h_1 = \frac{h_2 - h_1}{\eta_d}$$

or $$h_2' = h_1 + \frac{h_2 - h_1}{\eta_d} = h_1 + \frac{C_a^2}{2\,\eta_d}$$

or $$T_2' = T_1 + \frac{C_a^2}{2\,c_p\,\eta_d}$$

$$= 265.8 + \frac{216^2}{2 \times 1.005 \times 1000 \times 0.9} = 291.6 \text{ K}$$

Now, $$\frac{T_2}{T_1} = \left(\frac{p_2}{p_1}\right)^{\frac{\gamma-1}{\gamma}} \quad \text{or} \quad \frac{289}{265.8} = \left(\frac{p_2}{0.78}\right)^{\frac{1.4-1}{1.4}}$$

or $$(1.087)^{3.5} = \left(\frac{p_2}{0.78}\right)$$

or $$p_2 = 0.78 \times (1.087)^{3.5} = 1.044 \text{ bar}$$

Again, $$\frac{T_3}{T_2'} = (r_p)^{\frac{\gamma-1}{\gamma}} = (5.8)^{\frac{1.4-1}{1.4}} = 1.652$$

or $$T_3 = 291.6 \times 1.652 = 481.7 \text{ K}$$

Also, $$\eta_c = \frac{T_3 - T_2'}{T_3' - T_2'} \quad \text{or} \quad T_3' = T_2' + \frac{T_3 - T_2'}{\eta_c}$$

$$= 291.6 + \frac{481.7 - 291.6}{0.9} = 502.8 \text{ K}$$

Assume $c_{pg} = c_{pa} = c_p$

Heat supplied $= (m_a + m_f)\,c_p\,T_4 - m_a c_p\,T_3' = m_f \times C$

or $$m_a\,c_p\,T_4 + m_f\,c_p\,T_4 - m_a\,c_p\,T_3' = m_f \times C$$

or $$m_a c_p (T_4 - T_3') = m_f (C - c_p\,T_4)$$

or $$\frac{m_a}{m_f} = \frac{C - c_p\,T_4}{c_p\,(T_4 - T_3')} = \frac{44150 - 1.005 \times 1383}{1.005(1383 - 502.8)} = 48.34$$

∴ Air-fuel ratio = **48.34. (Ans.)**

(*ii*) Specific thrust of the unit :

$$p_4 = p_3 - 0.168 = 5.8 \times 1.044 - 0.168 = 5.88 \text{ bar}$$

Assume that the turbine drives *compressor only* (and not accessories also as is the usual case)

∴ $$c_p(T_3' - T_2') = c_p(T_4 - T_5')$$

or $$T_3' - T_2' = T_4 - T_5' \quad \text{or} \quad T_5' = T_4 - (T_3' - T_2')$$

$$= 1383 - (502.8 - 291.6) = 1171.8 \text{ K}$$

Also, $$\eta_t = \frac{T_4 - T_5'}{T_4 - T_5}$$

or $$T_5 = T_4 - \frac{T_4 - T_5'}{\eta_t}$$

$$= 1383 - \frac{1383 - 1171.8}{0.8} = 1119 \text{ K}$$

Now, $$\frac{T_4}{T_5} = \left(\frac{p_4}{p_5}\right)^{\frac{\gamma-1}{\gamma}} = \left(\frac{5.88}{p_5}\right)^{\frac{1.4-1}{1.4}}$$

or $$\left(\frac{1383}{1119}\right)^{3.5} = \frac{5.88}{p_5} \quad \text{or } p_5 = 2.8 \text{ bar}$$

Again, $$\frac{T_5'}{T_6} = \left(\frac{p_5}{p_6}\right)^{\frac{\gamma-1}{\gamma}} = \left(\frac{2.8}{0.78}\right)^{\frac{1.4-1}{1.4}} = 1.44$$

or $$T_6 = \frac{T_5'}{1.44} = \frac{1171.8}{1.44} = 813.75 \text{ K}$$

and $$\eta_n = \frac{T_5' - T_6'}{T_5' - T_6}$$

or $$T_6' = T_5' - \eta_n\,(T_5' - T_6)$$

$$= 1171.8 - 0.9\,(1171.8 - 813.75)$$

$$= 849.5 \text{ K}$$

Velocity at the exit of the nozzle,

$$C_j = 44.72\sqrt{h_5' - h_6'} = 44.72\sqrt{c_p\,(T_5' - T_6')}$$

$$= 44.72\sqrt{1.005(1171.8 - 849.5)} = 804.8 \text{ m/s}$$

$$\text{Specific thrust} = (1 + m_f) \times C_j = \left(1 + \frac{1}{48.34}\right) \times 804.8$$

= **821.45 N/kg of air/s. (Ans.)**

(*iii*) Total Thrust :

Volume of flowing air, $V_1 = 0.12 \times 216 = 25.92$ m³/s

Mass flow,

$$m_a = \frac{p_1 V_1}{RT_1} = \frac{0.78 \times 10^5 \times 25.92}{(0.287 \times 1000) \times 265.8} = 26.5 \text{ kg/s}$$

∴ Total thrust

= 26.5 × 821.45 = **21768.4 N. (Ans.)**

☞**Example 25.23.** *The following data pertain to a jet engine flying at an altitude of 9000 metres with a speed of 215 m/s.*

Thrust power developed	*750 kW*
Inlet pressure and temperature	*0.32 bar, – 42°C*
Temperature of gases leaving the combustion chamber	*690°C*
Pressure ratio	*5.2*
Calorific value of fuel	*42500 kJ/kg*
Velocity in ducts (constant)	*195 m/s*
Internal efficiency of turbine	*86%*
Efficiency of compressor	*86%*
Efficiency of jet tube	*90%*

For air : $c_p = 1.005$, $\gamma = 1.4$, $R = 0.287$

For combustion gases, $c_p = 1.087$

For gases during expansion, $\gamma = 1.33$.

Calculate the following :

(i) Overall thermal efficiency of the unit ;

(ii) Rate of air consumption ;

(iii) Power developed by the turbine ;

(iv) The outlet area of jet tube ;

(v) Specific fuel consumption is kg per kg of thrust.

Solution. Refer to Fig. 25.42.

Given : T.P. = 750 kW ; $p_1 = 0.32$ bar, $T_1 = -42 + 273 = 231$ K ; $T_3 = 690 + 273 = 963$ K ; $r_{pc} = 5.2$; $C = 42500$ kJ/kg ; $C_a = 215$ m/s, $C_4' = 19.5$ m/s, $\eta_c = 0.86$; $\eta_t = 0.86$; $\eta_{jt} = 0.9$.

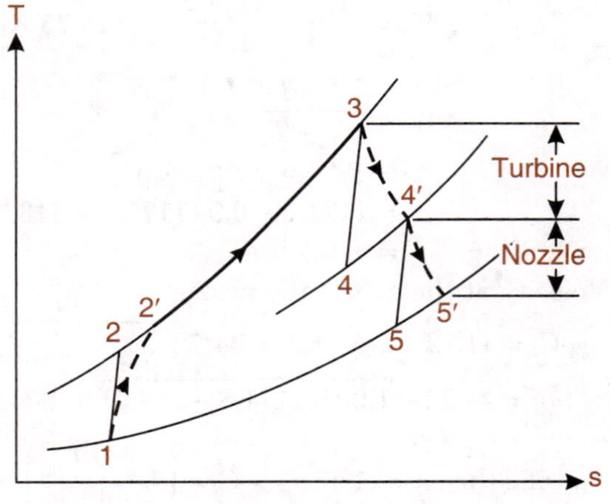

Fig. 25.42

Refer to Fig. 25.42.

Let m_f = kg of fuel required per kg of air

Then, heat supplied per kg of air

$$= 42500\, m_f = (1 + m_f) \times 1.087(T_3 - T_2') \quad ...(i)$$

Now, $$\frac{T_2}{T_1} = \left(\frac{p_2}{p_1}\right)^{\frac{\gamma-1}{\gamma}} = (5.2)^{\frac{1.4-1}{1.4}} = (5.2)^{0.2857} = 1.60$$

or $$T_2 = 231 \times 1.60 = 369.6 \text{ K}$$

Also, $$\eta_c = \frac{T_2 - T_1}{T_2' - T_1} \quad \text{or} \quad T_2' = T_1 + \frac{T_2 - T_1}{\eta_c}$$

$$= 231 + \frac{369.6 - 231}{0.86} = 392.2 \text{ K}$$

Substituting the value of T_2' in eqn. (*i*), we get

$$42500\, m_f = (1 + m_f) \times 1.087\,(963 - 392.2)$$
$$= 620.46\,(1 + m_f)$$

or $$42500\, m_f = 620.46 + 620.46\, m_f$$

or $$m_f = \frac{620.46}{(42500 - 620.46)}$$
$$= 0.0148 = \text{fuel-air ratio}$$

$\therefore$ Air-fuel ratio $= \dfrac{1}{0.0148} = 67.56 : 1$

The discharge velocity $C_j = C_5'$ cannot be determined from the thrust equation because the rate of air-flow is not known. It may be determined from the expression of jet efficiency.

Jet efficiency,

$$\eta_{jet} = \frac{\text{Final kinetic energy in the jet}}{\text{Isentropic heat drop in the jet pipe + Carry-over from the turbine}}$$

or $$\eta_{jet} = \frac{C_j^2/2}{c_{pg}(T_4' - T_5) + C_4'^2/2} \quad \text{(where } C_4' = 195 \text{ m/s)} \quad ...(ii)$$

Since the turbine's work is to drive the compressor only, therefore,

$$c_{pa}(T_2' - T_1) = c_{pg}\left(1 + \frac{m_f}{m_a}\right)(T_3 - T_4')$$

or $$1.005\,(392.2 - 231) = 1.087(1 + 0.0148)\,(963 - T_4')$$

or $$T_4' = 963 - \frac{1.005(392.2 - 231)}{1.087(1 + 0.0148)}$$
$$= 816.13 \text{ K}$$

Let r_{pt} = expansion pressure ratio in turbine

i.e., $$r_{pt} = \frac{p_3}{p_4}$$

r_{pj} = expansion pressure ratio in jet tube

i.e., $$r_{pj} = \frac{p_4}{p_5}$$

$$\therefore \quad r_{pt} \times r_{pj} = \frac{p_3}{p_4} \times \frac{p_4}{p_5} \simeq 5.2$$

Now, $$\eta_t = \frac{T_3 - T_4'}{T_3 - T_4} \quad \text{or} \quad T_4 = T_3 - \frac{T_3 - T_4'}{\eta_t}$$

$$= 963 - \frac{963 - 816.13}{0.86} = 792.2 \text{ K}$$

Also, $$\frac{T_3}{T_4} = \left(\frac{p_3}{p_4}\right)^{\frac{\gamma-1}{\gamma}} = \left(\frac{p_3}{p_4}\right)^{\frac{1.33-1}{1.33}}$$

or $$\frac{963}{792.2} = (r_{pt})^{0.248}$$

or $$r_{pt} = \left(\frac{963}{792.2}\right)^{\frac{1}{0.248}} = 2.197$$

$\therefore$ $$r_{pj} = \frac{p_4}{p_5} = \frac{5.2}{2.197} = 2.366$$

Thus, $$\frac{T_4'}{T_5} = (r_{pj})^{\frac{\gamma-1}{\gamma}} = (2.366)^{\frac{1.33-1}{1.33}} = 1.238$$

or $$T_5 = \frac{T_4'}{1.238} = \frac{816.13}{1.238} = 659.23 \text{ K}$$

Substituting the values in eqn. (*ii*), we get

$$0.9 = \frac{C_j^2/2}{1.087 \times 1000(816.13 - 659.23) + 195^2/2}$$

$$= \frac{C_j^2/2}{189562.8}$$

$$\therefore \quad C_j = \sqrt{0.9 \times 189562.8 \times 2} = 584.13 \text{ m/s}$$

(*i*) **Overall efficiency, η_0 :**

$$\eta_0 = \frac{\left[\left(1 + \frac{m_f}{m_a}\right)C_j - C_a\right]C_a}{\left(\frac{m_f}{m_a}\right) \times C}$$

$$= \frac{[(1 + 0.0148) \times 584.13 - 215]215}{1000 \times 0.0148 \times 42500}$$

= **0.1291** or **12.91%. (Ans.)**

(*ii*) **Rate of air consumption, $\dot{m}_a$:**

Thrust power = Thrust × Velocity of the unit

$$750 = \left[\frac{\left\{\left(1 + \frac{m_f}{m_a}\right)C_j - C_a\right\}\dot{m}_a}{1000}\right]C_a$$

or $$750 = \frac{\{(1 + 0.0148) \times 584.13 - 215\} \times \dot{m}_a}{1000} \times 215$$

$$= 81.22\ \dot{m}_a$$

or $$\dot{m}_a = \frac{750}{81.22} = \mathbf{9.234\ kg/s.} \quad \textbf{(Ans.)}$$

(*iii*) **Power developed by the turbine, P_t :**

$$P_t = \dot{m}_a\left(1 + \frac{m_f}{m_a}\right)c_{pg}(T_3 - T_4')$$

$$= 9.234(1 + 0.0148) \times 1.087(963 - 816.13)$$

= **1496 kW. (Ans.)**

(*iv*) **The outlet area of jet tube, A_{jt} :**

Now, $$\frac{C_j^2 - C_4'^2}{2} = c_{pg}(T_4' - T_5')$$

or $$T_5' = T_4' - \frac{C_j^2 - C_4'^2}{2 \times c_{pg}}$$

$$= 816.13 - \frac{(584.13^2 - 195^2)}{2 \times 1.087 \times 1000} = 676.67 \text{ K}$$

Assume the exit pressure of the gases be equal to atmospheric pressure *i.e.*, 0.32 bar.

Density of exhaust gases,

$$\rho = \frac{p_5'}{RT_5'} = \frac{0.32 \times 10^5}{0.29 \times 1000 \times 676.67} = 0.163 \text{ m}^3/\text{kg}$$

(Assuming R = 0.29 for the gases)

Also, discharge of jet area

$$= A_{jt} \times C_j \times \rho = \dot{m}_a\left(1 + \frac{m_f}{m_a}\right)$$

or $$A_{jt} \times 584.13 \times 0.163 = 9.234\ (1 + 0.0148)$$

or $$A_{jt} = \mathbf{0.0984\ m^2.} \quad \textbf{(Ans.)}$$

(*v*) **Specific fuel consumption in kg per kg of thrust :**

Specific fuel consumption

$$= \frac{0.0148 \times 9.234 \times 3600}{1000 \times (750/215)}$$

= **0.141 kg/thrust-hour. (Ans.)**

25.8.2. Turbo-prop

Fig. 25.43 shows a turbo-prop system employed in aircrafts. Here the expansion of gases takes place *partly in turbine* (80%) and *partly* (20%) *in the nozzle*. The power developed by the turbine is consumed in running the compressor and the propeller. The propeller and jet produced by the nozzle give forward motion to the aircraft. The turbo-prob entails the advantages of turbo-jet (*i.e., low specific weight and simplicity in design*) and propeller (*i.e., high power for take-off and high propulsion efficiency at speeds below 600 km/h*). The overall efficiency of the turbo-prop is improved by providing the diffuser before the compressor as shown. The pressure rise takes place in the diffuser. This pressure rise takes place due to *conversion of kinetic energy of the incoming air* (equal to aircraft velocity) *into pressure energy by the diffuser*. This type of compression is known as ***"ram effect"***.

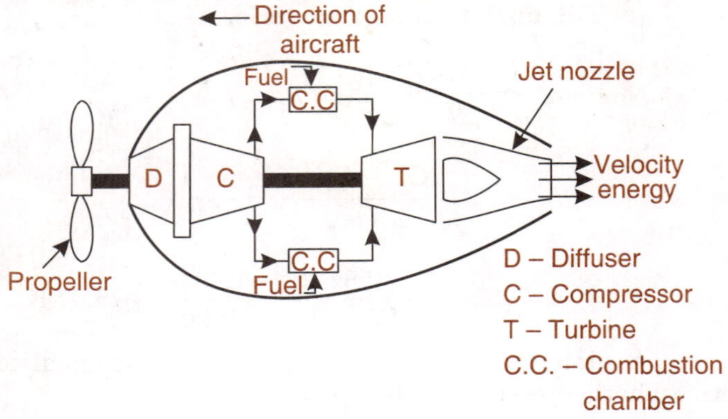

Fig. 25.43. Turbo-prop.

25.8.3. Ram-jet

Ram-jet is also called *athodyd, Lorin tube* or *flying stovepipe*. Ram-jet engines have the capability to fly at *supersonic speeds*. Fig. 25.44 shows a schematic diagram of a ram-jet engine (compressor and turbine are not necessary as the entire compression depends only on the ram compression).

- The ram-jet engine consists of a *diffuser* (used for compression), *combustion chamber* and *nozzle*.
- The air enters the ram-jet plant with *supersonic speed* and is slowed down to *sonic velocity* in the *supersonic diffuser*, consequently the pressure suddenly increases in the supersonic diffuser to the formation of shock wave. The pressure of air is further increased in the subsonic diffuser *increasing the temperature of the air above the ignition temperature*.
- In the combustion chamber, the fuel is injected through injection nozzles. The fuel air mixture is then ignited by means of a spark plug and combustion temperatures of the order of 2000 K are attained. The expansion of gases towards the diffuser entrance is restricted by pressure barrier at the after end of the diffuser and as a result the hot gases are constrained to move towards the nozzle and undergo expansion ; the pressure energy is converted into the kinetic energy. The high velocity gases leaving the nozzle provide forward thrust to the unit.

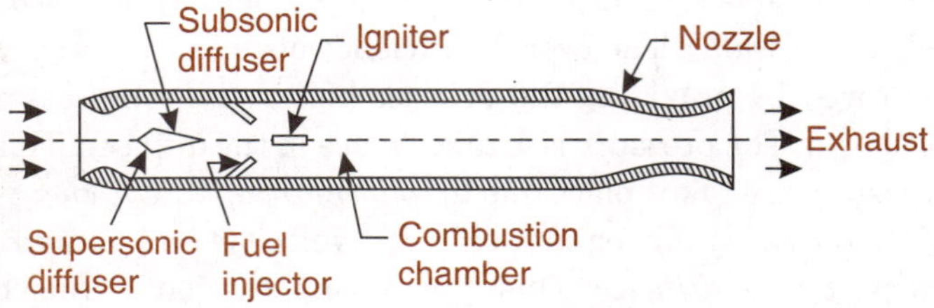

Fig. 25.44. Schematic diagram of a Ram-jet propulsion unit.

The best performance of ram-jet engine is obtained at flight speed of 1700 km/h to 2000 km/h.

Advantages of ram-jet engine

The ram-jet engine possesses the following *advantages* over other types of jet engines :

1. No moving parts.
2. Light in weight.
3. Wide variety of fuels may be used.

Shortcomings/Limitations

1. It *cannot be started of its own*. It *has to be accelerated to a certain flight velocity by some launching device*. A ram-jet is *always equipped with a small turbo-jet which starts the ram-jet*.

2. The fuel consumption is too large at low and moderate speeds.

3. For successful operation, the diffuser needs to be designed carefully so that kinetic energy associated with high entrance velocities is efficiently converted into pressure.

4. To obtain steady combustion, certain elaborate devices in form of flame holders or pilot flame are required.

25.8.4. Pulse-jet Engine

A *pulse-jet engine* is an intermittent combustion engine and it operates on a cycle similar to a reciprocating engine, whereas the turbo-jet and ram-jet engines are continuous in operation and are based on Brayton cycle. A pulse-jet engine like an athodyd, develops thrust by a high velocity of jet of exhaust gases without the aid of compressor or turbine. Its development is primarily due to the inability of the ram-jet to be self starting. Fig. 25.45 shows a schematic arrangement of a pulse-jet propulsion unit.

- The incoming air is compressed by ram effect in the diffuser section and the grid passages which are opened and closed by V-shaped non-return valves.

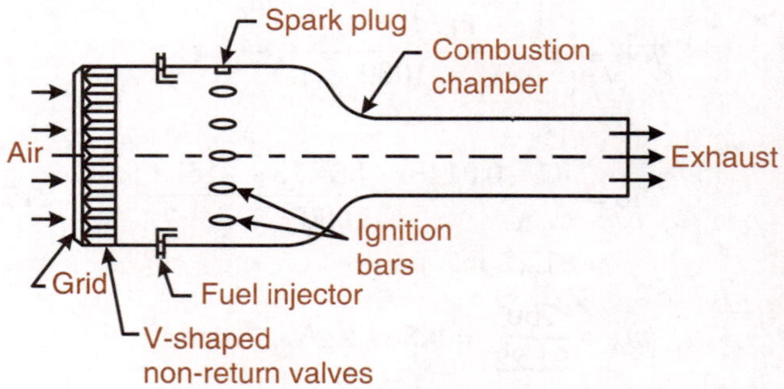

Fig. 25.45. Pulse-jet engine.

- The fuel is then injected into the combustion chamber by fuel injectors (worked from the air pressure from the compressed air bottles). The combustion is then initiated by a spark plug (once the engine is operating normally, the spark is turned off and the residual flame in the combustion chamber is used for combustion).

- As a result of combustion (of mixture of air and fuel) the temperature and pressure of combustion products increase. Because the combustion pressure is higher than the ram pressure, the non-return valves get closed and consequently the hot gases flow out of the tail pipe with a high velocity and in doing so give a forward thrust to the unit.
- With the escape of gases to the atmosphere, the static pressure in the chamber falls and the high pressure air in the diffuser forces the valves to open and fresh air is admitted for combustion during a new cycle.

Advantages :

1. Simple in construction and very inexpensive as compared to turbo-jet engine. Well adapted to pilotless aircraft.
2. Capable of producing static thrust and thrust in excess of drag at much low speeds.

Shortcomings :

1. High intensity of noise.
2. Severe vibrations.
3. High rate of fuel consumption and low thermodynamic efficiency.
4. Intermittent combustion as compared to continuous combustion in a turbo-jet engine.
5. The operating altitude is limited by air density consideration.
6. Serious limitation to mechanical valve arrangement.

25.8.5. Rocket Engines

Similar to jet propulsion, the thrust required for rocket propulsion is produced by the high velocity jet of gases passing through the nozzle. But the main difference is that in case of *jet propulsion the oxygen required for combustion is taken from the atmosphere and fuel is stored whereas for rocket engine, the fuel and oxidiser both are contained in a propelling body and as such it can function in vacuum.*

The rockets may be classified as follows :

1. According to the type of propellents :

(*i*) Solid propellent rocket

(*ii*) Liquid propellent rocket.

2. According to the number of motors :

(*i*) Single-stage rocket (consists of one rocket motor)

(*ii*) Multi-stage rocket (consists of more than one rocket motor).

Fig. 25.46 shows a simple type single stage liquid propellent (the fuel and the oxidiser are commonly known as propellents) rocket. It consists of a fuel tank *FT*, an oxidiser tank *O*, two pumps P_1, P_2, a steam turbine *ST* and a combustion chamber C.C. The fuel tank contains alcohol and oxidiser tank contains liquid oxygen. The fuel and the oxidiser are supplied by the pumps to the combustion chamber where the fuel is ignited by electrical means. The pumps are driven with the help of a steam turbine. Here the steam is produced by mixing a very concentrated hydrogen-peroxide with potassium permanganate. The products of combustion are discharged from the combustion chamber through the nozzle *N*. So the rocket moves in the opposite direction. In some modified form, this type of rocket may be used in missiles.

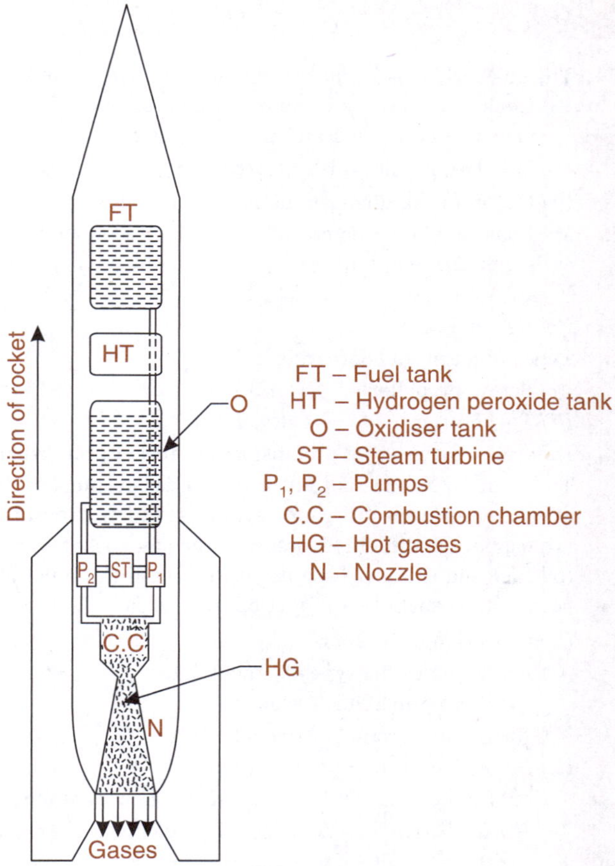

Fig. 25.46. Rocket.

25.8.5.1. Requirements of an ideal rocket propellent

An ideal rocket propellent should have the following *characteristics/properties :*

1. High heat value
2. Reliable smooth ignition
3. Stability and ease of handling and storing

4. Low toxicity and corrosiveness
5. Highest possible density so that it occupies less space.

25.8.5.2. Applications of rockets

The fields of application of rockets are as follows :

1. Long range artillary
2. Lethal weapons
3. Signalling and firework display
4. Jet assisted take-off
5. For satellites
6. For space ships
7. Research.

25.8.5.3. Thrust work, propulsive work and propulsive efficiency

In rocket propulsion, since air is self contained, the entry velocity relative to aircraft is zero. Neglecting the friction and other losses, we have the following formulae.

Thrust work $= C_j C_a$

Propulsive work $= C_j C_a + \dfrac{(C_j - C_a)^2}{2} = \dfrac{C_j^2 + C_a^2}{2}$

Rocket propulsive efficiency

$$= \frac{C_j C_a}{(C_j^2 + C_a^2)/2} = \frac{2\,C_j C_a}{C_j^2 + C_a^2} = \frac{2\left(\dfrac{C_a}{C_j}\right)}{1 + \left(\dfrac{C_a}{C_j}\right)^2} \quad \text{...(25.23)}$$

HIGHLIGHTS

1. The gas turbines are mainly divided into two groups :
   (*i*) Constant pressure combustion gas turbine
   (*a*) Open cycle constant pressure gas turbine
   (*b*) Closed cycle constant pressure gas turbine.
   (*ii*) Constant volume combustion gas turbine.
2. Methods for improvement of thermal efficiency of open cycle gas turbine plant :
   (*i*) Intercooling (*ii*) Reheating
   (*iii*) Regeneration.
3. Types of jet propulsion systems :
   (*i*) Screw propeller (*ii*) Turbo-jet
   (*iii*) Turbo-prop (*iv*) Ram-jet.
4. Difference between jet propulsion and rocket propulsion :
   The main difference is that in case of jet propulsion the oxygen required for combustion is taken from the atmosphere and fuel is stored whereas for rocket engine the fuel and oxidiser both are contained in a propelling body and as such it can function in vacuum.
5. Classification of rockets :
   (*i*) According to the type of propellents :
   (*a*) Solid propellent rocket
   (*b*) Liquid propellent rocket.
   (*ii*) According to the number of motors :
   (*a*) Single-stage rocket (consists of one rocket motor)
   (*b*) Multi-stage rocket (consists of more than one rocket motor).

OBJECTIVE TYPE QUESTIONS

Choose the Correct Answer :

1. Thermal efficiency of a gas turbine plant as compared to Diesel engine plant is
   (*a*) higher (*b*) lower
   (*c*) same (*d*) may be higher or lower.
2. Mechanical efficiency of a gas turbine as compared to internal combustion reciprocating engine is
   (*a*) higher (*b*) lower
   (*c*) same (*d*) unpredictable.
3. For a gas turbine the pressure ratio may be in the range
   (*a*) 2 to 3 (*b*) 3 to 5
   (*c*) 16 to 18 (*d*) 18 to 22.
4. The air standard efficiency of closed gas turbine cycle is given by (r = pressure ratio for the compressor and turbine)
   (*a*) $\eta = 1 - \dfrac{1}{(r_p)^{\gamma - 1}}$ (*b*) $\eta = 1 - (r_p)^{\gamma - 1}$
   (*c*) $\eta = 1 - \left(\dfrac{1}{r_p}\right)^{\frac{\gamma - 1}{\gamma}}$ (*d*) $\eta = (r_p)^{\frac{\gamma - 1}{\gamma}} - 1$.
5. The work ratio of closed cycle gas turbine plant depends upon
   (*a*) pressure ratio of the cycle and specific heat ratio
   (*b*) temperature ratio of the cycle and specific heat ratio
   (*c*) pressure ratio, temperature ratio and specific heat ratio
   (*d*) only on pressure ratio.
6. Thermal efficiency of closed cycle gas turbine plant increases by
   (*a*) reheating (*b*) intercooling
   (*c*) regenerator (*d*) all of the above.
7. With the increase in pressure ratio thermal efficiency of a simple gas turbine plant with fixed turbine inlet temperature
   (*a*) decreases
   (*b*) increases
   (*c*) first increases and then decreases
   (*d*) first decreases and then increases.
8. The thermal efficiency of a gas turbine cycle with ideal regenerative heat exchanger is
   (*a*) equal to work ratio (*b*) less than work ratio
   (*c*) more than work ratio (*d*) unpredictable.

9. In a two-stage gas turbine plant reheating after first stage
(*a*) decreases thermal efficiency
(*b*) increases thermal efficiency
(*c*) does not effect thermal efficiency
(*d*) none of the above.

10. In a two-stage gas turbine plant, reheating after first stage
(*a*) increases work ratio
(*b*) decreases work ratio
(*c*) does not affect work ratio
(*d*) none of the above.

11. In a two-stage gas turbine plant, with intercooling and reheating
(*a*) both work ratio and thermal efficiency improve
(*b*) work ratio improves but thermal efficiency decreases
(*c*) thermal efficiency improves but work ratio decreases
(*d*) both work ratio and thermal efficiency decrease.

12. For a jet-propulsion unit, ideally the compressor work and turbine work are
(*a*) equal (*b*) unequal
(*c*) not related to each other (*d*) unpredictable.

13. Greater the difference between jet velocity and aeroplane velocity
(*a*) greater the propulsive efficiency
(*b*) less the propulsive efficiency
(*c*) unaffected is the propulsive efficiency
(*d*) none of the above.

ANSWERS

1. (*b*)	**2.** (*a*)	**3.** (*c*)	**4.** (*c*)
5. (*c*)	**6.** (*d*)	**7.** (*c*)	**8.** (*a*)
9. (*a*)	**10.** (*a*)	**11.** (*b*)	**12.** (*a*)
13. (*b*).			

THEORETICAL QUESTIONS

1. What do you mean by the term 'gas turbine' ? How are gas turbines classified ?
2. State the merits of gas turbines over I.C. engines and steam turbines. Discuss also the demerits over gas turbines.
3. Describe with neat sketches the working of a simple constant pressure open cycle gas turbine.
4. Discuss briefly the methods employed for improvement of thermal efficiency of open cycle gas turbine plant.
5. Describe with neat diagram a closed cycle gas turbine. State also its merits and demerits.
6. Explain with a neat sketch the working of a constant volume combustion turbine.
7. Enumerate the various uses of gas turbines.
8. Write a short note on fuels used for gas turbines.
9. Explain the working difference between propeller-jet, turbo-jet and turbo-prop.
10. State the fundamental differences between the jet propulsion and rocket propulsion.

UNSOLVED EXAMPLES

1. In an air standard gas turbine engine, air at a temperature of 15°C and a pressure of 1.01 bar enters the compressor, where it is compressed through a pressure ratio of 5. Air enters the turbine at a temperature of 815°C and expands to original pressure of 1.01 bar. Determine the ratio of turbine work to compressor work and the thermal efficiency when the engine operates on ideal Brayton cycle.

   Take : $\gamma = 1.4$, $c_p = 1.005$ kJ/kg K. [**Ans.** 2.393 ; 37.03%]

2. In an open cycle constant pressure gas turbine air enters the compressor at 1 bar and 300 K. The pressure of air after the compression is 4 bar. The isentropic efficiencies of compressor and turbine are 78% and 85% respectively. The air-fuel ratio is 80 : 1. Calculate the power developed and thermal efficiency of the cycle if the flow rate of air is 2.5 kg/s.

   Take c_p = 1.005 kJ/kg K and $\gamma = 1.4$ for air and c_{pg} = 1.147 kJ/kg K and $\gamma = 1.33$ for gases. R = 0.287 kJ/kg K. Calorific value of fuel = 42000 kJ/kg.

   [**Ans.** 204.03 kW/kg of air ; 15.54%]

3. A gas turbine has a pressure ratio of 6/1 and a maximum cycle temperature of 600°C. The isentropic efficiencies of the compressor and turbine are 0.82 and 0.85 respectively. Calculate the power output in kilowatts of an electric generator geared to the turbine when the air enters the compressor at 15°C at the rate of 15 kg/s.

   Take : c_p = 1.005 kJ/kg K and $\gamma = 1.4$ for the compression process, and take c_p = 1.11 kJ/kg K and $\gamma = 1.333$ for the expansion process. [**Ans.** 920 kW]

4. Calculate the thermal efficiency and the work ratio of the plant in example 3 (above), assuming that c for the combustion process is 1.11 kJ/kg K. [**Ans.** 15.8% ; 0.206]

5. The gas turbine has an overall pressure ratio of 5 : 1 and a maximum cycle temperature of 550°C. The turbine drives the compressor and an electric generator, the mechanical efficiency of the drive being 97%. The ambient temperature is 20°C and the isentropic efficiencies for the compressor and turbine are 0.8 and 0.83 respectively. Calculate the power output in kilowatts for an air flow of 15 kg/s. Calculate also the thermal efficiency and the work ratio. Neglect changes is kinetic energy, and the loss of pressure in combustion chamber. [**Ans.** 655 kW ; 12% ; 0.168]

6. Air is drawn in a gas turbine unit at 17°C and 1.01 bar and the pressure ratio is 8 : 1. The compressor is driven by the H.P. turbine and the L.P. turbine drives a separate power shaft. The isentropic efficiencies of the compressor, and the H.P. and L.P. turbines are 0.8, 0.85 and 0.83, respectively. Calculate the pressure and temperature of the gases entering the power turbine, the net power developed by the unit per kg/s of mass flow, the work ratio and the thermal efficiency of the unit. The maximum cycle temperature is 650°C.

   For the compression process take c_p = 1.005 kJ/kg K and $\gamma = 1.4$

For the combustion process and expansion process, take

$c_p = 1.15$ kJ/kg K and $\gamma = 1.333$

Neglect the mass of fuel.

[**Ans.** 1.65 bar, 393°C ; 74.5 kW ; 0.201 ; 19.1%]

7. In a gas turbine plant, air is compressed through a pressure ratio of 6 : 1 from 15°C. It is then heated to the maximum permissible temperature of 750°C and expanded in two stages each of expansion ratio $\sqrt{6}$, the air being reheated between the stages to 750°C. A heat exchanger allows the heating of the compressed gases through 75 per cent of the maximum range possible. Calculate : (*i*) The cycle efficiency (*ii*) The work ratio (*iii*) The work per kg of air.

The isentropic efficiencies of the compressor and turbine are 0.8 and 0.85 respectively.

[**Ans.** (*i*) 32.75% ; (*ii*) 0.3852 ; (*iii*) 152 kJ/kg]

8. At the design speed the following data apply to a gas turbine set employing the heat exchanger : Isentropic efficiency of compressor = 75%, isentropic efficiency of the turbine = 85%, mechanical transmission efficiency = 99%, combustion efficiency = 98%, mass flow = 22.7 kg/s, pressure ratio = 6 : 1, heat exchanger effectiveness = 75%, maximum cycle temperature = 1000 K.

The ambient air temperature and pressure are 15°C and 1.013 bar respectively. Calculate :

(*i*) The net power output (*ii*) Specific fuel consumption (*iii*) Thermal efficiency of the cycle.

Take the lower calorific value of fuel as 43125 kJ/kg and assume no pressure-loss in heat exchanger and combustion chamber.

[**Ans.** (*i*) 2019 kW ; (*ii*) 0.4799 kg/kWh ; (*iii*) 16.7%]

9. In a gas turbine plant air at 10°C and 1.01 bar is compressed through a pressure ratio of 4 : 1. In a heat exchanger and combustion chamber the air is heated to 700°C while its pressure drops 0.14 bar. After expansion through the turbine the air passes through a heat exchanger which cools the air through 75% of maximum range possible, while the pressure drops 0.14 bar, and the air is finally exhausted to atmosphere. The isentropic efficiency of the compressor is 0.80 and that of turbine 0.85.

Calculate the efficiency of the plant. [**Ans.** 22.76%]

10. In a marine gas turbine unit a high-pressure stage turbine drives the compressor, and a low-pressure stage turbine drives the propeller through suitable gearing. The overall pressure ratio is 4 : 1, and the maximum temperature is 650°C. The isentropic efficiencies of the compressor, H.P. turbine, and L.P. turbine are 0.8, 0.83, and 0.85 respectively, and the mechanical efficiency of both shafts is 98%. Calculate the pressure between turbine stages when the air intake conditions are 1.01 bar and 25°C. Calculate also the thermal efficiency and the shaft power when the mass flow is 60 kg/s. Neglect kinetic energy changes, and pressure loss in combustion.

[**Ans.** 1.57 bar ; 14.9% ; 4560 kW]

11. In a gas turbine unit comprising L.P. and H.P. compressors, air is taken at 1.01 bar 27°C. Compression in L.P. stage is upto 3.03 bar followed by intercooling to 30°C. The pressure of air after H.P. compressor is 58.7 bar. Loss in pressure during intercooling is 0.13 bar. Air from H.P. compressor is transferred to heat exchanger of effectiveness 0.60 where it is heated by gases from L.P. turbine. The temperature of gases supplied to H.P. turbine is 750°C. The gases expand in H.P. turbine to 3.25 bar and are then reheated to 700°C before expanding in L.P. turbine. The loss of pressure in reheater is 0.1 bar. If isentropic efficiency of compression in both stages is 0.80 and isentropic efficiency of expansion in turbine is 0.85, calculate : (*i*) Overall efficiency (*ii*) Work ratio (*iii*) Mass flow rate when the gas power generated is 6500 kW. Neglect the mass of fuel.

Take, for air : $c_p = 1.005$ kJ/kg K, $\gamma = 1.4$

for gases : $c_{pg} = 1.15$ kJ/kg K, $\gamma = 1.3$.

[**Ans.** (*i*) 16.17% ; (*ii*) 0.2215 ; (*iii*) 69.33 kg of air/sec.]

12. In a gas turbine installation, air is taken in L.P. compressor at 15°C 1.1 bar and after compression it is passed through intercooler where its temperature is reduced to 22°C. The cooled air is further compressed in H.P. unit and then passed in the combustion chamber where its temperature is increased to 677°C by burning the fuel. The combustion products expand in H.P. turbine which runs the compressor and further expansion is continued in the L.P. turbine which runs the alternator. The gases coming out from L.P. turbine are used for heating the incoming air from H.P. compressor and then exhausted to atmosphere.

Taking the following data determine : (*i*) power output (*ii*) specific fuel consumption (*iii*) Thermal efficiency :

Pressure ratio of each compressor = 2, isentropic efficiency of each compressor stage = 85%, isentropic efficiency of each turbine stage = 85%, effectiveness of heat exchanger = 0.75, air-flow = 15 kg/sec., calorific value of fuel = 45000 kJ/kg, c_p(for gas) = 1 kJ/kg K, c_p (for gas) = 1.15 kJ/kg K, γ (for air) = 1.4, γ (for gas) = 1.33.

Neglect the mechanical, pressure and heat losses of the system and fuel mass also.

[**Ans.** (*i*) 1849.2 kW ; (*ii*) 0.241 kg/kWh ; (*iii*) 33.17%]

13. A turbo-jet engine flying at a speed of 960 km/h consumes air at the rate of 54.5 kg/s. Calculate : (*i*) Exit velocity of jet when the enthalpy change for the nozzle is 200 kJ/kg and velocity co-efficient is 0.97, (*ii*) fuel flow rate in kg/s when air-fuel ratio is 75 : 1 (*iii*) Thrust specific fuel consumption (*iv*) Thermal efficiency of the plant when the combustion efficiency is 93% and calorific value of the fuel is 45000 kJ/kg, (*v*) Propulsive power (*vi*) Propulsive efficiency (*vii*) Overall efficiency.

[**Ans.** (*i*) 613.5 m/s ; (*ii*) 0.7267 kg/s ; (*iii*) 4.3×10^{-5} kg/N of thrust/s ; (*iv*) 25.44% ; (*v*) 8318 kW ; (*vi*) 60.6% ; (*vii*) 16.58%]

14. A turbo-jet has a speed of 750 km/h while flying at an altitude of 10000 m. The propulsive efficiency of the jet is 50% and overall efficiency of the turbine plant is 16%. The density of air at 10000 m altitude is 0.173 kg/m. The drag on the plant is 6250 N. The calorific value of the fuel is 48000 kJ/kg. Calculate : (*i*) Absolute velocity of the jet (*ii*) Volume of air compressed per minute (*iii*) Diameter of the jet (*iv*) Power output of the unit in kW (*v*) Air-fuel ratio. [**Ans.** (*i*) 417.3 m/s ; (*ii*) 5194 m/min ; (*iii*) 415 mm ; (*iv*) 2500 kW ; (*v*) 46.01]

Refrigeration

26

26.1. FUNDAMENTALS OF REFRIGERATION

26.1.1. Introduction

Refrigeration *is the science of producing and maintaining temperatures below that of the surrounding atmosphere.* This means the removing of heat from a substance to be cooled. Heat always passes downhill, from a warm body to a cooler one, until both bodies are at the same temperature. Maintaining perishables at their required temperatures is done by refrigeration. Not only perishables but today many human work spaces in offices and factory buildings are air-conditioned and a refrigeration unit is the heart of the system.

Before the advent of mechanical refrigeration water was kept cool by storing it in semi-porous jugs so that the water could seep through and evaporate. The evaporation carried away heat and cooled the water. This system was used by the Egyptians and by Indians in the Southwest. Natural ice from lakes and rivers was often cut during winter and stored in caves, straw-lined pits, and later in sawdust-insulated buildings to be used as required. The Romans carried pack trains of snow from Alps to Rome for cooling the Emperor's drinks. Though these methods of cooling all make use of natural phenomena, they were used to maintain a lower temperature in a space or product and may properly be called refrigeration.

In simple, *refrigeration means the cooling of or removal of heat from a system.* The equipment employed to maintain the system at a low temperature is termed as *refrigerating system* and the system which is kept at lower temperature is called *refrigerated system.* Refrigeration is generally produced in one of the following *three ways* :

(*i*) By *melting of a solid.*

(*ii*) By *sublimation of a solid.*

(*iii*) By *evaporation of a liquid.*

Most of the commercial refrigeration is produced by the evaporation of a liquid called *refrigerant. Mechanical refrigeration* depends upon the evaporation of liquid refrigerant and its circuit includes the equipment naming ***evaporator, compressor, condenser*** and ***expansion valve.*** It is used for preservation of food, manufacture of ice, solid carbon dioxide and control of air temperature and humidity in the air-conditioning system.

Important refrigeration applications :

1. Ice making
2. Transportation of foods above and below freezing
3. Industrial air-conditioning
4. Comfort air-conditioning
5. Chemical and related industries
6. Medical and surgical aids
7. Processing food products and beverages
8. Oil refining and synthetic rubber manufacturing
9. Manufacturing and treatment of metals
10. Freezing food products
11. Miscellaneous applications :
    (*i*) Extremely low temperatures
    (*ii*) Plumbing
    (*iii*) Building construction, etc.

26.1.2. Elements of Refrigeration Systems

All refrigeration systems must include atleast *four basic units* as given below :

(*i*) *A low temperature thermal "sink" to which heat will flow from the space to be cooled.*

(*ii*) *Means of extracting energy from the sink, raising the temperature level of this energy, and delivering it to a heat receiver.*

(*iii*) *A receiver to which heat will be transferred from the high temperature high-pressure refrigerant.*

(*iv*) *Means of reducing of pressure and temperature of the refrigerant as it returns from the receiver to the "sink".*

26.1.3. Refrigeration Systems

The various refrigeration systems may be enumerated as below :

1. Ice refrigeration
2. Air refrigeration system
3. Vapour compression refrigeration system
4. Vapour absorption refrigeration system
5. Special refrigeration systems
   (*i*) Adsorption refrigeration system
   (*ii*) Cascade refrigeration system
   (*iii*) Mixed refrigeration system
   (*iv*) Vortex tube refrigeration system
   (*v*) Thermoelectric refrigeration
   (*vi*) Steam jet refrigeration system.

26.1.4. Co-efficient of Performance (C.O.P.)

The performance of a refrigeration system is expressed by a term known as the *"co-efficient of performance"*, which is defined as the *ratio of heat absorbed by the refrigerant while passing through the evaporator to the work input required to compress the refrigerant in the compressor* ; in short it is the *ratio between heat extracted and work done* (in heat units).

If, R_n = Net refrigerating effect,

W = Work expanded in by the machine during the same interval of time,

Then, $$\text{C.O.P.} = \frac{R_n}{W}$$

and $$\text{Relative C.O.P.} = \frac{\text{Actual C.O.P.}}{\text{Theoretical C.O.P.}}$$

where, Actual C.O.P. = Ratio of R_n and W *actually measured during a test*

and, Theoretical C.O.P. = Ratio of *theoretical values* of R_n and W *obtained by applying laws of thermodynamics to the refrigeration cycle.*

26.1.5. Standard Rating of a Refrigeration Machine

The rating of a refrigeration machine is obtained by refrigerating effect or amount of heat extracted in a given time from a body. The rating of the refrigeration machine is given by a unit of refrigeration known as ***"standard commercial tonne of refrigeration"*** which is defined as *the refrigerating effect produced by the melting of 1 tonne of ice from and at 0°C in 24 hours.* Since, the latent heat of fusion of ice is 336 kJ/kg, the refrigerating effect of 336 × 1000 kJ in 24 hours is rated as one *tonne*, *i.e.*,

$$1 \text{ tonne of refrigeration (TR)} = \frac{336 \times 1000}{24} = 14000 \text{ kJ/h.}$$

Note. ***Ton of refrigeration (TR).*** A ton of refrigeration is basically an American unit of refrigerating effect (R.E.). It originated from the rate at which heat is required to be removed to freeze *one ton of water from and at 0°C*. Using American units this is equal to removal of 200 BTU of heat per minute, and MKS unit it is adopted as 50 kcal/min or 3000 kcal/hour. *In S.I. units its conversion is rounded of to 3.5 kJ/s (kW) or 210 kJ/min.*

In this book we shall be adopting,

1 tonne of refrigeration = 14000 kJ/h (1 ton = 0.9 tonne).

26.2. AIR REFRIGERATION SYSTEM

26.2.1. Introduction

Air cycle refrigeration is one of the earliest methods of cooling developed. It became obsolete for several years because of its low co-efficient of performance (C.O.P.) and high operating costs. It has, however, been applied to aircraft refrigeration systems, where with low equipment weight, it can utilise a portion of the cabin air according to the supercharger capacity. The main characteristic feature of *air refrigeration system, is that throughout the cycle the refrigerant remains in gaseous state.*

The air refrigeration system can be divided in two systems :

(*i*) Closed system (*ii*) Open system.

In ***closed*** (or *dense air*) ***system*** the *air refrigerant is contained within the piping or components parts of the system at all times and refrigerator with usually pressures above atmospheric pressure.*

In the ***open system*** the refrigerator is replaced by the *actual space to be cooled with the air expanded to atmospheric pressure, circulated through the cold room and then compressed to the cooler pressure.* The pressure of operation in this system is inherently limited to operation *at atmospheric pressure* in the refrigerator.

A closed system claims the following *advantages* over open system : (*i*) In a closed system the suction to compressor may be at high pressure. The sizes of expander and compressor can be kept within reasonable limits by using dense air ; (*ii*) In open air system, the air picks up moisture from the products kept in the refrigerated chamber ; the moisture may freeze during expansion and is likely to choke the valves whereas it does not happen in closed system and (*iii*) In open system, the expansion of the refrigerant can be carried only upto atmospheric pressure prevailing in the cold chamber but for a closed system there is no such restriction.

26.2.2. Reversed Carnot Cycle

If a machine working on reversed Carnot cycle is driven from an external source, it will work or function as a refrigerator. The *production of such a machine has not been possible practically because the adiabatic portion of the stroke would need a high speed while during isothermal portion of stroke a very low speed will be necessary. This variation of speed during the stroke, however is not practicable.*

p-V and *T-s* diagrams of reversed Carnot cycle are shown in Fig. 26.1 (*a*) and (*b*). Starting from point *l*, the clearance space of the cylinder is full of air, the air is then expanded adiabatically to point *p* during which its temperature falls from T_1 to T_2, the cylinder is put in contact with a cold body at temperature T_2. The air is then expanded isothermally to the point *n*, as a result of which heat is extracted from the cold body at temperature T_2. Now the cold body is removed ; from *n* to *m* air undergoes adiabatic compression with the assistance of some external power and temperature rises to T_1. A hot body at temperature T_1 is put in contact with the cylinder. Finally the air is compressed isothermally during which process heat is rejected to the hot body.

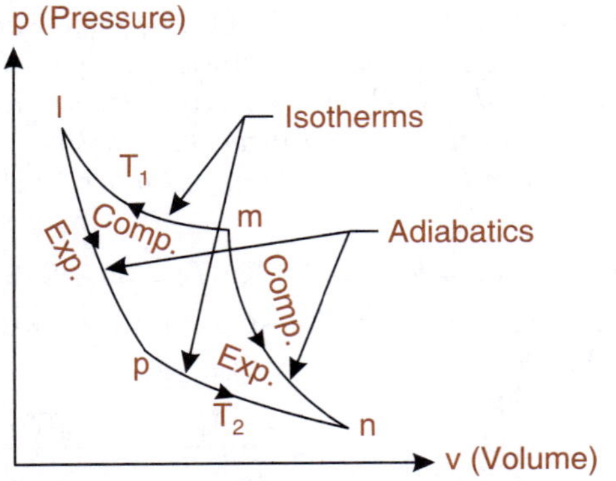

Fig. 26.1 (*a*) *p-V* diagram for reversed Carnot cycle.

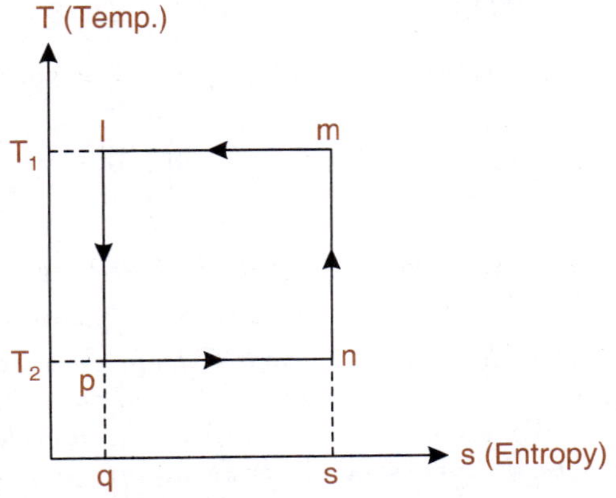

Fig. 26.1 (*b*) *T-s* diagram for a reversed Carnot cycle.

Refer to Fig. 26.1 (*b*)

Heat abstracted from the cold body = Area '*npqs*'

$= T_2 \times pn$

Work done per cycle = Area '*lpnm*'

$= (T_1 - T_2) \times pn$

Co-efficient of performance,

$$\text{C.O.P.} = \frac{\text{Heat extracted from the cold body}}{\text{Work done per cycle}}$$

$$= \frac{\text{Area '}npqs\text{'}}{\text{Area '}lpnm\text{'}} = \frac{T_2 \times pn}{(T_1 - T_2) \times pn} = \frac{T_2}{T_1 - T_2} \quad ...(26.1)$$

Since, the co-efficient of performance (C.O.P.) means the ratio of the desired effect in kJ/kg to the energy supplied in kJ/kg, therefore C.O.P. in case of Carnot cycle run either as a refrigerating machine or a heat pump or as a heat engine will be as given below :

(*i*) **For a reversed Carnot cycle 'refrigerating machine' :**

$$\text{C.O.P.}_{(\text{ref.})} = \frac{\text{Heat extracted from the cold body/cycle}}{\text{Work done per cycle}}$$

$$= \frac{T_2 \times pn}{(T_1 - T_2) \times pn} = \frac{T_2}{T_1 - T_2} \quad ...(26.2)$$

(*ii*) **For a Carnot cycle 'heat pump' :**

$$\text{C.O.P.}_{(\text{heat pump})} = \frac{\text{Heat rejected to the hot body/cycle}}{\text{Work done per cycle}}$$

$$= \frac{T_1 \times lm}{(T_1 - T_2) \times pn} = \frac{T_1 \times pn}{(T_1 - T_2) \times pn} \quad (\because \; lm = pn)$$

$$= \frac{T_1}{T_1 - T_2} \quad ...(26.3)$$

$$= 1 + \frac{T_2}{T_1 - T_2} \quad ...(26.4)$$

This indicates that *C.O.P. of heat pump is greater than that of a refrigerator working on reversed Carnot cycle between* the same *temperature* limits T_1 and T_2 by *unity.*

(*iii*) **For a Carnot cycle 'heat engine' :**

$$\text{C.O.P.}_{(\text{heat engine})} = \frac{\text{Work obtained/cycle}}{\text{Heat supplied/cycle}}$$

$$= \frac{(T_1 - T_2) \times pn}{T_1 \times lm}$$

$$= \frac{(T_1 - T_2) \times pn}{T_1 \times pn} \quad (\because \; lm = pn)$$

$$= \frac{T_1 - T_2}{T_1} \quad ...(26.5)$$

Example 26.1. *A Carnot refrigerator requires 1.3 kW per tonne of refrigeration to maintain a region at low temperature of – 38°C. Determine :*

(i) C.O.P. of Carnot refrigerator

(ii) Higher temperature of the cycle

(iii) The heat delivered and C.O.P. when this device is used as heat pump.

Solution. $T_2 = 273 - 38 = 235$ K

Power required per tonne of refrigeration = 1.3 kW

(*i*) C.O.P. of Carnot refrigerator :

$$\text{C.O.P.}_{\text{(Carnot ref.)}} = \frac{\text{Heat absorbed}}{\text{Work done}}$$

$$= \frac{1 \text{ tonne}}{1.3} = \frac{14000 \text{ kJ/h}}{1.3 \times 60 \times 60 \text{ kJ/h}}$$

$$= \mathbf{2.99.} \quad \textbf{(Ans.)}$$

(*ii*) Higher temperature of the cycle, T_1 :

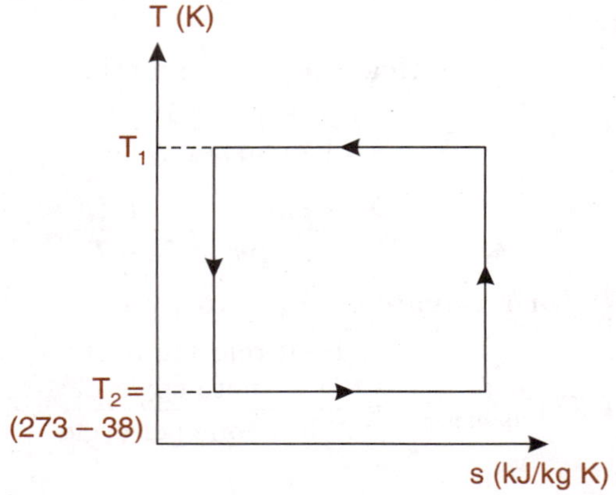

Fig. 26.2

$$\text{C.O.P.}_{\text{(Carnot ref.)}} = \frac{T_2}{T_1 - T_2}$$

i.e.,
$$2.99 = \frac{235}{T_1 - 235}$$

$$\therefore \quad T_1 = \frac{235}{2.99} + 235 = 313.6 \text{ K}$$

$$= 313.6 - 273 = \mathbf{40.6°C.} \quad \textbf{(Ans.)}$$

(*iii*) Heat delivered as heat pump

= Heat absorbed + Work done

$$= \frac{14000}{60} + 1.3 \times 60 = \mathbf{311.3 \ kJ/min.} \quad \textbf{(Ans.)}$$

$$\textbf{C.O.P.}_{\text{(heat pump)}} = \frac{\text{Heat delivered}}{\text{Work done}} = \frac{311.3}{1.3 \times 60}$$

$$= \mathbf{3.99.} \quad \textbf{(Ans.)}$$

Example 26.2. *A refrigerating system operates on the reversed Carnot cycle. The higher temperature of the refrigerant in the system is 35°C and the lower temperature is – 15°C. The capacity is to be 12 tonnes. Neglect all losses. Determine :*

(i) Co-efficient of performance.

(ii) Heat rejected from the system per hour.

(iii) Power required.

Solution. (*i*)
$$T_1 = 273 + 35 = 308 \text{ K}$$
$$T_2 = 273 - 15 = 258 \text{ K}$$

Capacity = 12 tonne

$$\textbf{C.O.P.} = \frac{T_2}{T_1 - T_2} = \frac{258}{308 - 258} = \mathbf{5.16.} \quad \textbf{(Ans.)}$$

(*ii*) Heat rejected from the system per hour :

$$\text{C.O.P.} = \frac{\text{Refrigerating effect}}{\text{Work input}}$$

$$5.16 = \frac{12 \times 14000 \text{ kJ/h}}{\text{Work input}}$$

$$\therefore \quad \text{Work input} = \frac{12 \times 14000}{5.16} = 32558 \text{ kJ/h.}$$

Thus, *heat rejected/hour*

= Refrigerating effect/hour + Work input/hour

= 12 × 14000 + 32558 = **200558 kJ/h.** **(Ans.)**

(*iii*) Power required :

$$\textit{Power required} = \frac{\text{Work input/hour}}{60 \times 60} = \frac{32558}{60 \times 60}$$

$$= \mathbf{9.04 \ kW.} \quad \textbf{(Ans.)}$$

Example 26.3. *A cold storage is to be maintained at – 5°C while the surroundings are at 35°C. The heat leakage from the surroundings into the cold storage is estimated to be 29 kW. The actual C.O.P. of the refrigeration plant used is one third that of an ideal plant working between the same temperatures. Find the power required to drive the plant.*

(AMIE)

Solution. $T_2 = -5 + 273 = 268$ K ; $T_1 = 35 + 273 = 308$ K

Heat leakage from the surroundings into the cold storage = 29 kW

$$\text{Ideal C.O.P.} \quad = \frac{T_2}{T_1 - T_2} = \frac{268}{308 - 268} = 6.7$$

$$\text{Actual C.O.P.} \quad = \frac{1}{3} \times 6.7 = 2.233 = \frac{R_n}{W}$$

(where R_n = net refrigerating effect, and W = work done)

or
$$2.233 = \frac{29}{W} \quad \text{or} \quad W = \frac{29}{2.233}$$

$$= 12.98 \text{ kJ/s}$$

Hence, power required to drive the plant

$$= \mathbf{12.98 \ kW.} \quad \textbf{(Ans.)}$$

Example 26.4. *Ice is formed at 0°C from water at 20°C. The temperature of the brine is – 8°C. Find out the kg of ice formed per kWh. Assume that the refrigeration cycle used is perfect reversed Carnot cycle. Take latent heat of ice as 335 kJ/kg.*

Solution. Latent heat of ice = 335 kJ/kg

$$T_1 = 20 + 273 = 293 \text{ K}$$
$$T_2 = -8 + 273 = 265 \text{ K}$$
$$\text{C.O.P.} = \frac{T_2}{T_1 - T_2} = \frac{265}{293 - 265} = 9.46$$

Heat to be extracted per kg of water (to form ice at 0°C *i.e.*, 273 K), R_n

$$= 1 \times c_{pw} \times (293 - 273) + \text{latent heat of ice}$$
$$= 1 \times 4.18 \times 20 + 335 = 418.6 \text{ kJ/kg}$$

Also, 1 kWh = 1 × 3600 = 3600 kJ

Also, $\text{C.O.P.} = \frac{R_n}{W} = \frac{\text{Refrigerating effect in kJ/kg}}{\text{Work done in kJ}}$

$$\therefore \quad 9.46 = \frac{m_{ice} \times 418.6}{3600}$$

i.e., $$m_{ice} = \frac{9.46 \times 3600}{418.6} = 81.35 \text{ kg}$$

Hence, ice formed per kWh = 81.35 kg. (Ans.)

Example 26.5. *Find the least power of a perfect reversed heat engine that makes 400 kg of ice per hour at – 8°C from feed water at 18°C. Assume specific heat of ice as 2.09 kJ/kg K and latent heat 334 kJ/kg.*

Solution.
$$T_1 = 18 + 273 = 291 \text{ K}$$
$$T_2 = -8 + 273 = 265 \text{ K}$$
$$\text{C.O.P.} = \frac{T_2}{T_1 - T_2} = \frac{265}{291 - 265} = 10.19$$

Heat absorbed per kg of water (to form ice at – 8°C *i.e.*, 265 K),

$$R_n = 1 \times 4.18\,(291 - 273) + 334 + 1 \times 2.09 \times (273 - 265)$$
$$= 425.96 \text{ kJ/kg}$$

Also, $\text{C.O.P.} = \frac{\text{Net refrigerating effect}}{\text{Work done}} = \frac{R_n}{W}$

i.e., $$10.19 = \frac{425.96 \times 400}{W}$$

$$\therefore \quad W = \frac{425.96 \times 400}{10.19} = 16720.7 \text{ kJ/h}$$
$$= 4.64 \text{ kJ/s or } 4.64 \text{ kW}$$

Hence, least power = 4.64 kW. (Ans.)

☞**Example 26.6.** *The capacity of the refrigerator (working on reversed Carnot cycle) is 280 tonnes when operating between – 10°C and 25°C. Determine :*

(*i*) *Quantity of ice produced within 24 hours when water is supplied at 20°C.*

(*ii*) *Minimum power (in kW) required.*

Solution. (*i*) **Quantity of ice produced :**

Heat to be extracted per kg of water (to form ice at 0°C)

$$= 4.18 \times 20 + 335 = 418.6 \text{ kJ/kg}$$

Heat extraction capacity of the refrigerator

$$= 280 \text{ tonnes}$$
$$= 280 \times 14000$$
$$= 3920000 \text{ kJ/h}$$

∴ Quantity of ice produced in 24 hours,

$$m_{ice} = \frac{3920000 \times 24}{418.6 \times 1000} = \mathbf{224.75 \text{ tonnes.}} \quad \textbf{(Ans.)}$$

(*ii*) **Minimum power required :**

$$T_1 = 25 + 273 = 298 \text{ K}$$
$$T_2 = -10 + 273 = 263 \text{ K}$$
$$\text{C.O.P.} = \frac{T_2}{T_1 - T_2} = \frac{263}{298 - 263} = 7.51$$

Also, $\text{C.O.P.} = \frac{\text{Net refrigerating effect}}{\text{Work done /min}} = \frac{R_n}{W}$

i.e., $$7.51 = \frac{3920000}{W}$$

$$\therefore \quad W = \frac{3920000}{7.51} \text{ kJ/h} = 145 \text{ kJ/s}$$

∴ **Power required = 145 kW. (Ans.)**

Example 26.7. *A cold storage plant is required to store 20 tonnes of fish. The temperature of the fish when supplied = 25°C ; storage temperature of fish required = – 8°C ; specific heat of fish above freezing point = 2.93 kJ/kg °C ; specific heat of fish below freezing point = 1.25 kJ/kg °C ; freezing point of fish = – 3°C. Latent heat of fish = 232 kJ/kg.*

If the cooling is achieved within 8 hours ; find out :

(*i*) *Capacity of the refrigerating plant.*

(*ii*) *Carnot cycle C.O.P. between this temperature range.*

(*iii*) *If the actual C.O.P. is $\frac{1}{3}$rd of the Carnot C.O.P. find out the power required to run the plant.*

Solution. Heat removed in 8 hours from each kg of fish

$$= 1 \times 2.93 \times [25 - (-3)] + 232 + 1 \times 1.25\,[-3 - (-8)]$$
$$= 82.04 + 232 + 6.25 = 320.29 \text{ kJ/kg}$$

Heat removed by the plant/min

$$= \frac{320.29 \times 20 \times 1000}{8} = 800725 \text{ kJ/h}$$

(*i*) **Capacity of the refrigerating plant**

$$= \frac{800725}{14000} = \mathbf{57.19 \text{ tonnes.}} \quad \textbf{(Ans.)}$$

(*ii*)
$$T_1 = 25 + 273 = 298 \text{ K}$$
$$T_2 = -8 + 273 = 265 \text{ K}$$

∴ **C.O.P. of reversed Carnot cycle**

$$= \frac{T_2}{T_1 - T_2} = \frac{265}{298 - 265} = \mathbf{8.03.} \quad \textbf{(Ans.)}$$

(*iii*) **Power required :**

$$\text{Actual C.O.P.} = \frac{1}{3} \times \text{Carnot C.O.P.}$$

$$= \frac{1}{3} \times 8.03 = 2.67$$

But actual C.O.P.

$$= \frac{\text{Net refrigerating effect/min}}{\text{Work done /min}} = \frac{R_n}{W}$$

$$2.67 = \frac{800725}{W} \text{ kJ/h}$$

$$\therefore \quad W = \frac{800725}{2.67} = 299897 \text{ kJ/h} = 83.3 \text{ kJ/s}$$

$\therefore$ *Power required to run the plant*

= 83.3 kW. (Ans.)

Example 26.8. *A heat pump is used for heating the interior of a house in cold climate. The ambient temperature is – 5°C and the desired interior temperature is 25°C. The compressor of heat pump is to be driven by a heat engine working between 1000°C and 25°C. Treating both cycles as reversible, calculate the ratio in which the heat pump and heat engine share the heating load.* **(P.U.)**

Solution. Refer to Fig. 26.3.

Given : $T_1 = 1000 + 273 = 1273$ K ; $T_2 = 25 + 273 = 298$ K ;

$T_3 = -5 + 273 = 268$ K ; $T_4 = 25 + 273 = 298$ K

The ratio in which the heat pump and heat engine share the heating load, $\frac{Q_4}{Q_1}$ **:**

Since both the cycles are *reversible*, therefore,

$$\frac{Q_3}{Q_4} = \frac{T_3}{T_4} \quad \text{and} \quad \frac{Q_2}{Q_1} = \frac{T_2}{T_1}$$

or $$\frac{Q_3}{Q_4} = \frac{268}{298} \quad \text{or} \quad Q_3 = \frac{268}{298} Q_4 \text{ and } \frac{Q_2}{Q_1} = \frac{298}{1273}$$

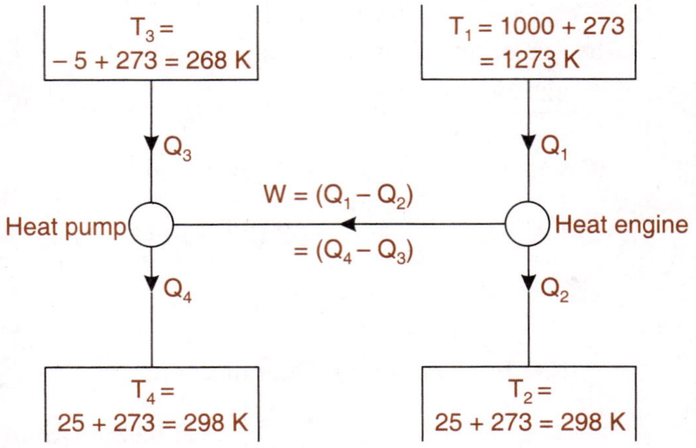

Fig. 26.3

Heat engine drives the heat pump,

$$\therefore \quad W = (Q_1 - Q_2) = Q_4 - Q_3$$

Dividing both sides by Q_1, we have

$$1 - \frac{Q_2}{Q_1} = \frac{Q_4 - Q_3}{Q_1}$$

$$1 - \frac{298}{1273} = \frac{Q_4 - \frac{268}{298} Q_4}{Q_1}$$

$$\frac{975}{1273} = \frac{30}{298} \times \frac{Q_4}{Q_1}$$

$$\therefore \quad \frac{Q_4}{Q_1} = \frac{975}{1273} \times \frac{298}{30} = \mathbf{7.608.} \quad \textbf{(Ans.)}$$

26.2.3. Reversed Brayton Cycle

Fig. 26.4 shows a schematic diagram of an air refrigeration system working on reversed Brayton cycle. *Elements* of this systems are :

1. Compressor
2. Cooler (Heat exchanger)
3. Expander
4. Refrigerator.

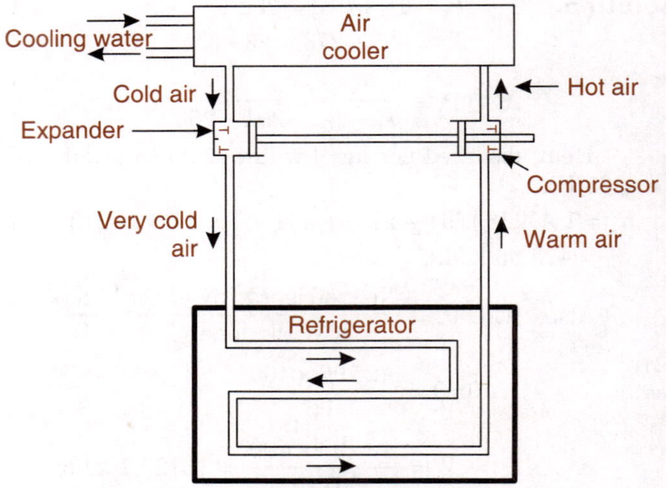

Fig. 26.4. Air refrigeration system.

In this system, work gained from expander is employed for compression of air, consequently less external work is needed for operation of the system. In practice it may or may not be done *e.g.*, in some aircraft refrigeration systems which employ air refrigeration cycle the expansion work may be used for driving other devices.

This system uses reversed *Brayton cycle* which is described below :

Fig. 26.5 (*a* and *b*) shows *p-V* and *T-s* diagrams for a reversed Brayton cycle. Here it is *assumed* that (*i*) *absorption and rejection of heat are constant pressure processes and* (*ii*) Compression and expansion are *isentropic processes*.

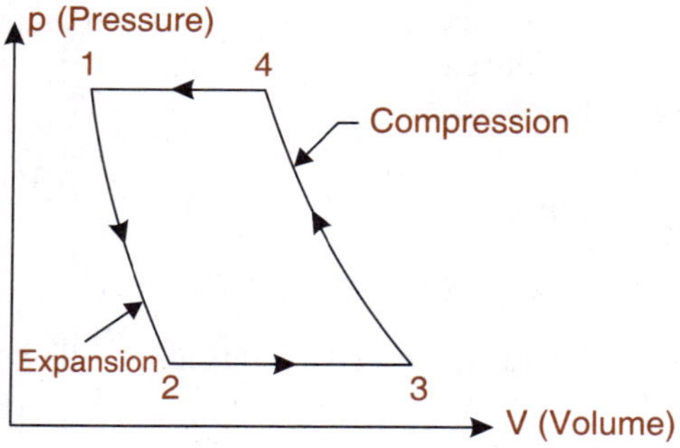

Fig. 26.5. (*a*) *p-V* diagram.

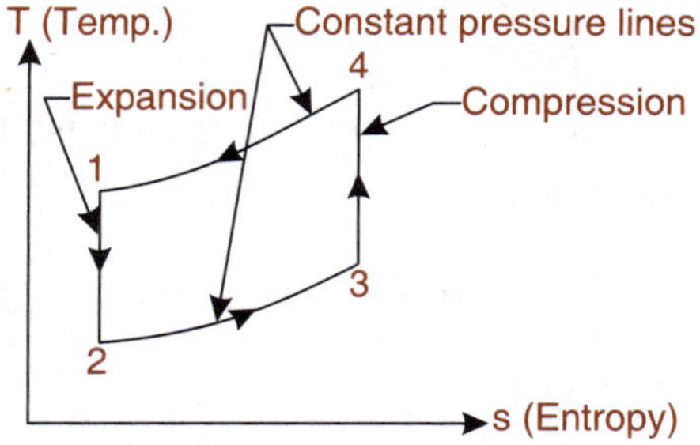

Fig. 26.5. (*b*) *T-s* diagram.

Considering m kg of air :

Heat absorbed in refrigerator,

$$Q_{added} = m \times c_p \times (T_3 - T_2)$$

Heat rejected is cooler,

$$Q_{rejected} = m \times c_p \times (T_4 - T_1)$$

If the process is considered to be polytropic, the *steady flow work of compression is* given by,

$$W_{comp} = \frac{n}{n-1}(p_4V_4 - p_3V_3) \quad ...(26.6)$$

Similarly work of expansion is given by,

$$W_{exp.} = \frac{n}{n-1}(p_1V_1 - p_2V_2) \quad ...(26.7)$$

Equations (26.6) and (26.7) may easily be reduced to the theoretical isentropic process shown in Fig. 26.5 (*b*) by substituting $\gamma = n$ and the known relationship.

$$R = c_p\left(\frac{\gamma-1}{\gamma}\right) \text{J}$$

The net external work required for operation of the cycle

= Steady flow work of compression

– Steady flow work of expansion

$$= W_{comp.} - W_{exp.}$$

$$= \left(\frac{n}{n-1}\right)(p_4V_4 - p_3V_3 - p_1V_1 + p_2V_2)$$

$$= \left(\frac{n}{n-1}\right)mR(T_4 - T_3 - T_1 + T_2) \quad \left[\because \begin{array}{l} p_1V_1 = mRT_1 \\ p_2V_2 = mRT_2 \\ p_3V_3 = mRT_3 \\ p_4V_4 = mRT_4 \end{array}\right]$$

$$= \left(\frac{n}{n-1}\right)\frac{mR}{J}(T_4 - T_3 - T_1 + T_2)$$

in heat units.

But

$$R = c_p\left(\frac{\gamma-1}{\gamma}\right) J$$

(J = 1 in S.I. units)

$$\therefore W_{comp.} - W_{exp.} = \left(\frac{n}{n-1}\right)\left(\frac{\gamma-1}{\gamma}\right) mc_p(T_4 - T_3 + T_2 - T_1) \quad ...(26.8)$$

For *isentropic compression* and *expansion*,

$$W_{net} = mc_p(T_4 - T_3 + T_2 - T_1)$$

Now according to law of conservation of energy the network on the gas must be equivalent to the net heat rejected.

$$\text{Now,} \quad \text{C.O.P.} = \frac{W_{added}}{Q_{rejected} - Q_{added}} = \frac{Q_{added}}{W_{net}}$$

For the air cycle assuming polytropic compression and expansion, co-efficient of performance is :

$$\text{C.O.P.} = \frac{m \times c_p \times (T_3 - T_2)}{\left(\frac{n}{n-1}\right)\left(\frac{\gamma-1}{\gamma}\right) m \times c_p \times (T_4 - T_3 + T_2 - T_1)}$$

$$= \frac{(T_3 - T_2)}{\left(\frac{n}{n-1}\right)\left(\frac{\gamma-1}{\gamma}\right)(T_4 - T_3 + T_2 - T_1)} \quad ...(26.9)$$

Note. The reversed Brayton cycle is same as the Bell-Coleman cycle. Conventionally Bell-Coleman cycle refers to a closed cycle with expansion and compression taking place in reciprocating expander and compressor respectively, and heat rejection and heat absorption taking place in condenser and evaporator respectively.

With the development of efficient centrifugal compressors and gas turbines, the processes of compression and expansion can be carried out in centrifugal compressors and gas turbines respectively. Thus, the shortcoming encountered with conventional reciprocating expander and compressor is overcome. Reversed Brayton cycle finds its application for air-conditioning of aeroplanes where air is used as refrigerant.

26.2.4. Merits and Demerits of Air refrigeration System

Merits

1. Since air is non-flammable, therefore there is no risk of fire as in the machine using NH_3 as the refrigerant.

2. It is cheaper as air is easily available as compared to the other refrigerants.

3. As compared to the other refrigeration systems the weight *of air refrigeration system per tonne of refrigeration is quite low, because of this reason this system is employed in aircrafts.*

Demerits

1. The C.O.P. of this system is very low in comparison to other systems.

2. The weight of air required to be circulated is more compared with refrigerants used in other systems. This is

due to the fact that heat is carried by air in the form of *sensible heat*.

Example 26.9. *A Bell-Coleman refrigerator operates between pressure limits of 1 bar and 8 bar. Air is drawn from the cold chamber at 9°C, compressed and then it is cooled to 29°C before entering the expansion cylinder. Expansion and compression follow the law $pv^{1.35}$ = constant. Calculate the theoretical C.O.P. of the system.*

For air take $\gamma = 1.4$, $c_p = 1.003$ kJ/kg K.

Solution. Fig. 26.6 shows the working cycle of the refrigerator.

Given : $p_2 = 1.0$ bar ;
$p_1 = 8.0$ bar ;
$T_3 = 9 + 273 = 282$ K ;
$T_4 = 29 + 273 = 302$ K.

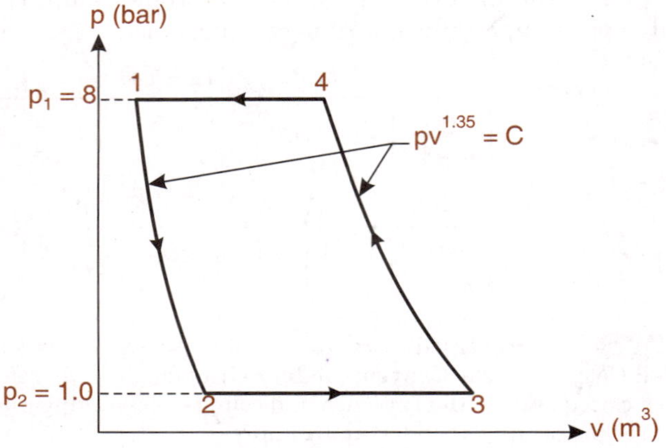

Fig. 26.6

Considering *polytropic compression 3–4*, we have

$$\frac{T_4}{T_3} = \left(\frac{p_1}{p_2}\right)^{\frac{n-1}{n}} = \left(\frac{8}{1}\right)^{\frac{1.35-1}{1.35}} = (8)^{0.259} = 1.71$$

or $T_4 = T_3 \times 1.71 = 282 \times 1.71 = 482.2$ K

Again, considering *polytropic expansion 1–2*, we have

$$\frac{T_1}{T_2} = \left(\frac{p_1}{p_2}\right)^{\frac{n-1}{n}} = \left(\frac{8}{1}\right)^{\frac{1.35-1}{1.35}} = 1.71$$

$$T_2 = \frac{T_1}{1.71} = \frac{302}{1.71} = 176.6 \text{ K}$$

Heat extracted from cold chamber per kg of air

$= c_p (T_3 - T_2) = 1.003 (282 - 176.6)$
$= 105.7$ kJ/kg.

Heat rejected in the cooling chamber per kg of air

$= c_p (T_4 - T_1) = 1.003 (482.2 - 302) = 180.7$ kJ/kg.

Since, the compression and expansion are not isentropic, difference between heat rejected and heat absorbed is not equal to the work done because there are heat transfers to the surroundings and from the surroundings during compression and expansion.

To find the work done, the area of the diagram '1-2-3-4' is to be considered :

Work done

$$= \frac{n}{n-1}(p_4V_4 - p_3V_3) - \frac{n}{n-1}(p_1V_1 - p_2V_2)$$

$$= \frac{n}{n-1} R[(T_4 - T_3) - (T_1 - T_2)]$$

The value of R can be calculated as follows

$$\frac{c_p}{c_v} = \gamma$$

$$\therefore \quad c_v = \frac{c_p}{\gamma} = \frac{1.003}{1.4} = 0.716$$

$R = (c_p - c_v) = 1.003 - 0.716 = 0.287$ kJ/kg K.

$\therefore$ Work done

$$= \frac{1.35}{0.35} \times 0.287 \, [(482.2 - 282) - (302 - 176.6)]$$

$= 82.8$ kJ/kg.

$$\therefore \quad \textbf{C.O.P.} = \frac{\text{Heat abstracted}}{\text{Work done}} = \frac{105.7}{82.4}$$

$=$ **1.27. (Ans.)**

Example 26.10. *An air refrigeration open system operating between 1 MPa and 100 kPa is required to produce a cooling effect of 2000 kJ/min. Temperature of the air leaving the cold chamber is – 5°C and at leaving the cooler is 30°C. Neglect losses and clearance in the compressor and expander. Determine :*

(*i*) *Mass of air circulated per min.* ;

(*ii*) *Compressor work, expander work, cycle work* ;

(*iii*) *C.O.P. and power in kW required.*

Solution. Refer to Fig. 26.7.

Pressure, $p_1 = 1$ MPa $= 1000$ kPa ; $p_2 = 100$ kPa

Refrigerating effect produced = 2000 kJ/min

Temperature of air leaving the cold chamber,

$T_3 = -5 + 273 = 268$ K

Temperature of air leaving the cooler,

$T_1 = 30 + 273 = 303$ K

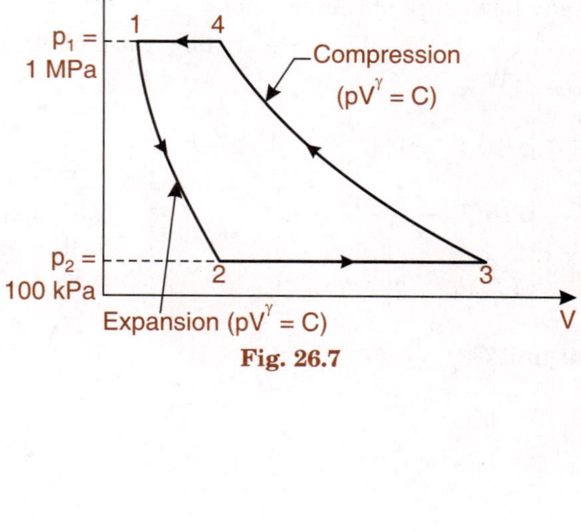

Fig. 26.7

(*i*) **Mass of air circulated per minute, m :**

For the *expansion process 1–2*, we have

$$\frac{T_1}{T_2} = \left(\frac{p_1}{p_2}\right)^{\frac{\gamma-1}{\gamma}} = \left(\frac{1000}{100}\right)^{\frac{1.4-1}{1.4}} = 1.9306$$

or $$T_2 = \frac{T_1}{1.9306} = \frac{303}{1.9306} = 156.9 \text{ K}$$

Refrigerating effect per kg $= 1 \times c_p (T_3 - T_2) = 1.005 (268 - 156.9) = 111.66$ kJ/kg

∴ Mass of air circulated per minute

$$= \frac{\text{Refrigerating effect}}{\text{Refrigerating effect per kg}}$$

$$= \frac{2000}{111.66} = \mathbf{17.91 \text{ kg/min.}} \quad \text{(Ans.)}$$

(*ii*) **Compressor work ($W_{comp.}$), expander work ($W_{exp.}$) and cycle work (W_{cycle}) :**

For *compression process 3–4*, we have

$$\frac{T_4}{T_3} = \left(\frac{p_4}{p_3}\right)^{\frac{\gamma-1}{\gamma}} = \left(\frac{1000}{10}\right)^{\frac{1.4-1}{1.4}} = 1.9306$$

or $$T_4 = 268 \times 1.9306 = \mathbf{517.4 \text{ K.}} \quad \text{(Ans.)}$$

Compressor work, $W_{comp.}$: $\dfrac{\gamma}{\gamma - 1} mR (T_4 - T_3)$

$$= \frac{1.4}{1.4 - 1} \times 17.91 \times 0.287 (517.4 - 268)$$

$$= \mathbf{4486.85 \text{ kJ/min.}} \quad \text{(Ans.)}$$

Expander work, $W_{exp.}$: $\dfrac{\gamma}{\gamma - 1} mR (T_1 - T_2)$

$$= \frac{1.4}{1.4 - 1} \times 17.91 \times 0.287 (303 - 156.9)$$

$$= \mathbf{2628.42 \text{ kJ/min.}} \quad \text{(Ans.)}$$

Cycle work, W_{cycle} : $W_{comp.} - W_{exp.}$

$= 4486.85 - 2628.42$

$= \mathbf{1858.43 \text{ kJ/min.}}$ **(Ans.)**

(*iii*) **C.O.P. and power required (P) :**

$$\text{C.O.P.} = \frac{\text{Refrigerating effect}}{\text{Work required}} = \frac{2000}{1858.43}$$

$= \mathbf{1.076.}$ **(Ans.)**

Power required,

$$P = \text{Work per second} = \frac{1858.43}{60} \text{ kJ/s or kW}$$

$= \mathbf{30.97 \text{ kW.}}$ **(Ans.)**

☞**Example 26.11.** *A refrigerating machine of 6 tonnes capacity working on Bell-Coleman cycle has an upper limit of pressure of 5.2 bar. The pressure and temperature at the start of the compression are 1.0 bar and 16°C respectively. The compressed air cooled at constant pressure to a temperature of 41°C enters the expansion cylinder. Assuming both expansion and compression processes to be adiabatic with γ = 1.4, calculate :*

(*i*) *Co-efficient of performance.*

(*ii*) *Quantity of air in circulation per minute.*

(*iii*) *Piston displacement of compressor and expander.*

(*iv*) *Bore of compressor and expansion cylinders. The unit runs at 240 r.p.m. and is double- acting. Stroke length = 200 mm.*

(*v*) *Power required to drive the unit*

For air take γ = 1.4 and c_p = 1.003 kJ/kg K.

Solution. Refer to Fig. 26.8.

$T_3 = 16 + 273 = 289$ K ; $T_1 = 41 + 273 = 314$ K

$p_1 = 5.2$ bar ; $p_2 = 1.0$ bar.

Considering the *adiabatic compression 3–4*, we have

$$\frac{T_4}{T_3} = \left(\frac{p_1}{p_2}\right)^{\frac{\gamma-1}{\gamma}} = \left(\frac{5.2}{1}\right)^{\frac{1.4-1}{1.4}} = (5.2)^{0.286} = 1.6$$

∴ $T_4 = 1.6$; $T_3 = 1.6 \times 289 = 462.4$ K

Considering the *adiabatic expansion 1–2*, we have

$$\frac{T_1}{T_2} = \left(\frac{p_1}{p_2}\right)^{\frac{\gamma-1}{\gamma}}, \quad \frac{314}{T_2} = \left(\frac{5.2}{1}\right)^{\frac{0.4}{1.4}} = 1.6$$

or $$T_2 = \frac{314}{1.6} = 196.25 \text{ K.}$$

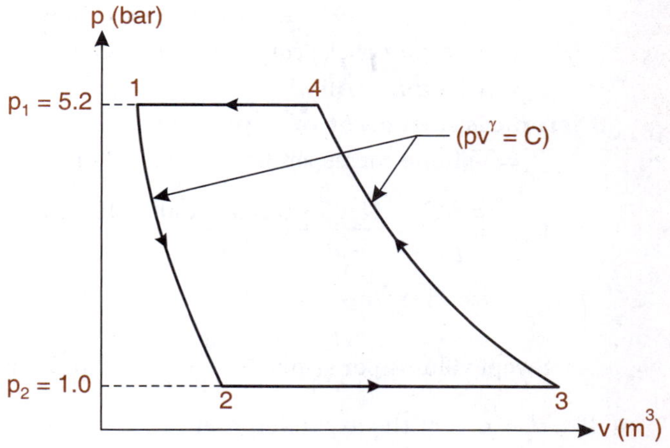

Fig. 26.8

(*i*) **C.O.P. :**

Since both the compression and expansion processes are isentropic/adiabatic reversible,

$$\therefore \text{ C.O.P. of the machine} = \frac{T_2}{T_1 - T_2} = \frac{196.25}{314 - 196.25}$$

$= \mathbf{1.67.}$ **(Ans.)**

(*ii*) **Mass of air in circulation :**

Refrigerating effect per kg of air

$= c_p (T_3 - T_2) = 1.003 (289 - 196.25)$

$= 93.03$ kJ/kg.

Refrigerating effect produced by the refrigerating machine

$= 6 \times 14000 = 84000$ kJ/h.

Hence mass of air in circulation

$$= \frac{84000}{93.03 \times 60} = \textbf{15.05 kg/min.} \quad \textbf{(Ans.)}$$

(*iii*), (*iv*) **Piston displacement of compressor :**

= Volume corresponding to point 3 *i.e.*, V_3

$$\therefore \quad V_3 = \frac{mRT_3}{p_2} = \frac{15.05 \times 0.287 \times 1000 \times 289}{1.0 \times 10^5}$$

$= \textbf{12.48 m}^3\textbf{/min.}$ **(Ans.)**

∴ Swept volume per stroke

$$= \frac{12.48}{2 \times 240} = 0.026 \text{ m}^3$$

If, d_c = Dia. of compressor cylinder, and

l = Length of stroke,

then $\frac{\pi}{4} d_c^2 \times l = 0.026$

or $$\frac{\pi}{4} d_c^2 \times \left(\frac{200}{1000}\right) = 0.026$$

$$\therefore \quad d_c = \left(\frac{0.026 \times 1000 \times 4}{\pi \times 200}\right)^{1/2}$$

$= 0.407$ m or 407 mm

i.e., *Diameter or bore of the compressor cylinder*

= 407 mm. (Ans.)

Piston displacement of expander

= Volume corresponding to point 2 *i.e.*, V_2

$$\therefore \quad V_2 = \frac{mRT_2}{p_2} = \frac{15.05 \times 0.287 \times 1000 \times 196.25}{1 \times 10^5}$$

$= \textbf{8.476 m}^3\textbf{/min.}$ **(Ans.)**

$$\therefore \quad \text{Swept volume per stroke} = \frac{8.476}{2 \times 240} = 0.0176 \text{ m}^3.$$

If d_e = dia. of the expander, and

l = length of stroke,

then $$\frac{\pi}{4} d_e^2 \times l = 0.0176$$

or $$\frac{\pi}{4} d_e^2 \times \left(\frac{200}{1000}\right) = 0.0176$$

$$\therefore \quad d_e = \left(\frac{0.0176 \times 1000 \times 4}{\pi \times 200}\right)^{1/2}$$

$= 0.335$ m or 335 mm

i.e., *Diameter or bore of the expander cylinder*

= 335 mm. (Ans.)

(*v*) **Power required to drive the unit :**

$$\text{C.O.P.} = \frac{\text{Refrigerating effect}}{\text{Work done}} = \frac{R_n}{W}$$

$$1.67 = \frac{6 \times 14000}{W}$$

$$W = \frac{6 \times 14000}{1.67} = 50299.4 \text{ kJ/h} = 13.97 \text{ kJ/s}.$$

Hence power required **= 13.97 kW. (Ans.)**

26.3. SIMPLE VAPOUR COMPRESSION SYSTEM

26.3.1. Introduction

Out of all refrigeration systems, the vapour compression system is the most important system from the view point of *commercial* and *domestic utility*. It is the most practical form of refrigeration. In this system the *working fluid is a vapour*. It readily evaporates and condenses or changes alternately between the vapour and liquid phases without leaving the refrigerating plant. During evaporation, it absorbs heat from the cold body. This heat is used as its latent heat for converting it from the liquid to vapour. In condensing or cooling or liquifying, it rejects heat to external body, thus creating a cooling effect in the working fluid. This refrigeration system thus acts as a latent heat pump since it pumps its latent heat from the cold body or brine and rejects it or delivers it to the external hot body or cooling medium. The principle upon which the vapour compression system works apply to all the vapours for which tables of Thermodynamic properties are available.

26.3.2. Simple Vapour Compression Cycle

In a simple vapour compression system fundamental processes are completed in one cycle.

These are :

1. Compression 2. Condensation
3. Expansion 4. Vaporisation.

The flow diagram of such a cycle is shown in Fig. 26.9.

The *vapour at low temperature and pressure* (state '2') enters the "*compressor*" where it is compressed isentropically and subsequently its temperature and pressure increase considerably (state '3'). This vapour after leaving the compressor enters the "*condenser*" where it is condensed into *high pressure liquid* (state '4') and is collected in a "*receiver tank*". From receiver tank it passes through the "*expansion valve*", here it is *throttled down to a lower pressure* and has a low temperature (state '1'). After finding its way through expansion "valve" it finally passes on to "*evaporator*" where it *extracts heat from the surroundings* or *circulating fluid being refrigerated and vapourises to low pressure vapour (state '2')*.

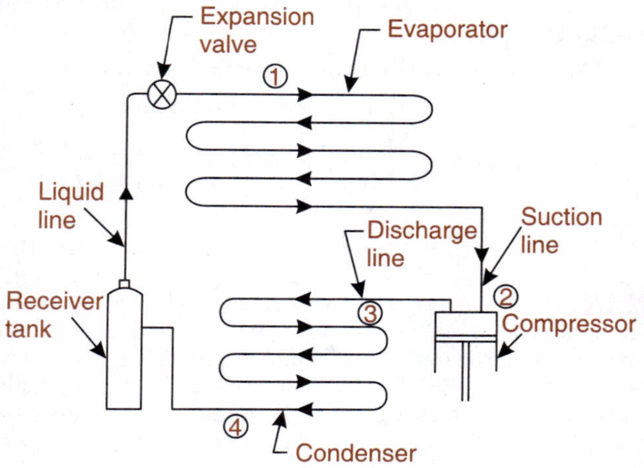

Fig. 26.9. Vapour compression system.

Merits and demerits of vapour compression system over Air refrigeration system :

Merits :

1. C.O.P. is quite high as the working of the cycle is very near to that of reversed Carnot cycle.

2. When used on ground level the running cost of vapour-compression refrigeration system is only 1/5th of air refrigeration system.

3. For the same refrigerating effect the size of the evaporator is smaller.

4. The required temperature of the evaporator can be achieved simply by adjusting the throttle valve of the same unit.

Demerits :

1. Initial cost is high.

2. The major disadvantages are *inflammability*, *leakage of vapours* and *toxity*. These have been overcome to a great extent by improvement in design.

26.3.3. Functions of Parts of a Simple Vapour Compression System

Here follows the brief description of various parts of a simple vapour compression system shown in Fig. 26.9.

1. Compressor. The function of a compressor is to remove the *vapour* from the evaporator, and to *raise its temperature and pressure to a point such that it* (vapour) *can be condensed with available condensing media*.

2. Discharge line (or **hot gas line**). A hot gas or discharge line *delivers the high-pressure, high-temperature vapour from the discharge of the compressor to the condenser*.

3. Condenser. The function of a condenser is to *provide a heat transfer surface through which heat passes from the hot refrigerant vapour to the condensing medium*.

4. Receiver tank. A receiver tank is used to provide *storage for a condensed liquid* so that a constant supply of liquid is available to the evaporator as required.

5. Liquid line. A liquid line carries the liquid refrigerant from the receiver tank to the refrigerant flow control.

6. Expansion valve (refrigerant flow control). Its function is to *meter the proper amount of refrigerant to the evaporator and to reduce the pressure of liquid entering the evaporator so that liquid will vapourize in the evaporator at the desired low temperature and take out sufficient amount of heat*.

7. Evaporator. An evaporator *provides a heat transfer surface through which heat can pass from the refrigerated space into the vapourizing refrigerant*.

8. Suction line. The suction line *conveys the low pressure vapour from the evaporator to the suction inlet of the compressor*.

26.3.4. Vapour Compression Cycle on Temperature-Entropy (T-s) Diagram

We shall consider the following three cases :

1. When the vapour is dry and saturated at the end of compression. Fig. 26.10 represents the vapour compression cycle, on *T-s* diagram the points 1, 2, 3 and 4 correspond to the state points 1, 2, 3 and 4 in Fig. 26.9.

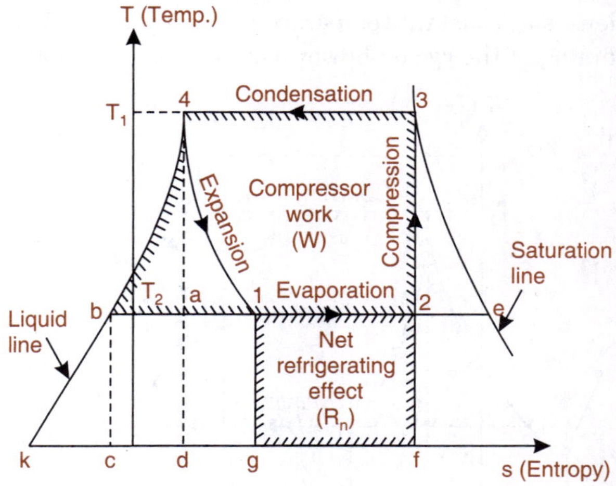

Fig. 26.10. *T-s* diagram.

At point '2' the vapour which is at low temperature (T_2) and low pressure enters the compressor's cylinder and is compressed adiabatically to '3' when its temperature increases to the temperature T_1. It is then condensed in the condenser (line 3–4) where it gives up its latent heat to the condensing medium. It then undergoes throttling expansion while passing through the expansion valve and its again reduces to T_2, it is represented by the line 4–1. From the *T-s* diagram it may be noted that due to this expansion the liquid partially evaporates, as its dryness fraction is represented by the ratio $\frac{b_1}{b_2}$. At '1' it enters the evaporator where it is further evaporated at constant pressure and constant temperature to the point '2' and the cycle is completed.

Work done by the compressor = W = Area '2-3-4-b-2'

Heat absorbed = Area '2-1-g-f-2'

$$\therefore \quad \text{C.O.P.} = \frac{\text{Heat extracted or refrigerating effect}}{\text{Work done}}$$

$$= \frac{\text{Area '2-1-}g\text{-}f\text{-2'}}{\text{Area '2-3-4-}b\text{-2'}}$$

or
$$\text{C.O.P.} = \frac{h_2 - h_1}{h_3 - h_2} \quad ...[26.10\ (a)]$$

$$= \frac{h_2 - h_4}{h_3 - h_2} \quad ...[26.10\ (b)]$$

($\because$ $h_1 = h_4$, *since during the throttling expansion 4–1 the total heat content remains unchanged*).

2. When the vapour is superheated after compression. If the compression of the vapour is continued after it has become dry, the vapour will be superheated, its effect on *T-s* diagram is shown in Fig. 26.11. The vapour enters the compressor at condition '2' and is compressed to '3' where it is superheated to temperature T_{sup}. Then it enters the condenser. Here firstly superheated vapour cools to temperature T_1 (represented by line 3-3′) and then it condenses at constant temperature along the line 3′-4 ; the remaining of the cycle ; however is the same as before.

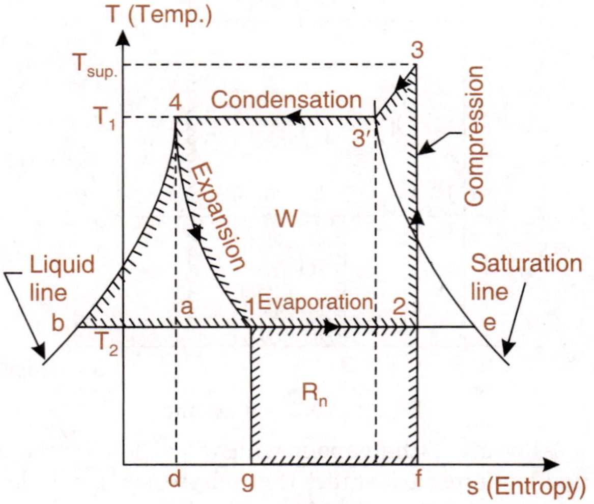

Fig. 26.11. *T-s* diagram.

Now, Work done = Area '2-3-3′-4-b-2'

and Heat extracted/absorbed = Area '2-1-g-f-2'

$$\therefore \quad \text{C.O.P.} = \frac{\text{Heat extracted}}{\text{Work done}} = \frac{\text{Area '2-1-}g\text{-}f\text{-2'}}{\text{Area '2-3-3'-4-}b\text{-2'}}$$

$$= \frac{h_2 - h_1}{h_3 - h_2} \quad ...[26.10\ (c)]$$

In this case $h_3 = h_3' + c_p\,(T_{sup.} - T_{sat.})$ and h_3' = total heat of dry and saturated vapour at the point '3″.

3. When the vapour is wet after compression. Refer to Fig. 26.12.

Work done by the compressor = Area '2-3-4-b-2'

Heat extracted = Area '2-1-g-f-2'

$$\therefore \quad \text{C.O.P.} = \frac{\text{Heat extracted}}{\text{Work done}} = \frac{\text{Area '2-1-}g\text{-}f\text{-2'}}{\text{Area '2-3-4-}b\text{-2'}}$$

$$= \frac{h_2 - h_1}{h_3 - h_2} \quad ...[26.10(d)]$$

Note. If the vapour is not superheated after compression, the operation is called 'WET COMPRESSION' and if the vapour is superheated at the end of compression, it is known as 'DRY COMPRESSION'. Dry compression, in actual practice is always preferred as it gives *higher volumetric efficiency* and *mechanical efficiency* and there are *less chances of compressor damage*.

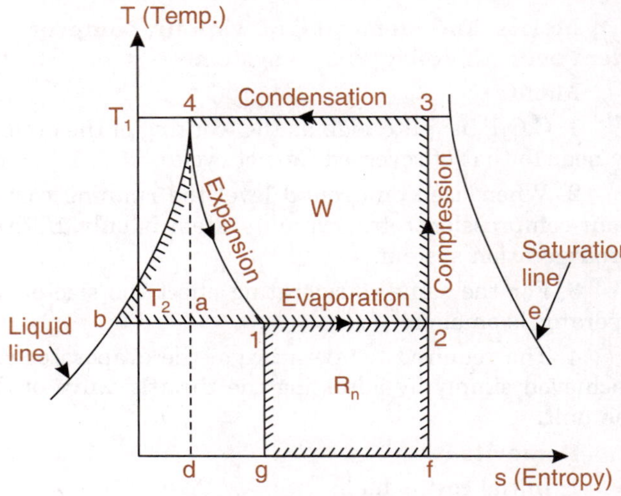

Fig. 26.12. *T-s* diagram.

26.3.5. Pressure-Enthalpy (p-h) Chart

The diagram commonly used in the analysis of the refrigeration cycle are :

(*i*) Pressure-enthalpy (*p-h*) chart

(*ii*) Temperature-entropy (*T-s*) chart.

Of the two, the pressure-enthalpy diagram seems to be the more useful.

The condition of the refrigerant in any thermodynamic state can be represented as a point on the *p-h* chart. The point on the *p-h* chart that represents the condition of the refrigerant in any one particular thermodynamic state may be located if any two properties of the refrigerant for that state are known, the other properties of the refrigerant for that state can be determined directly from the chart for studying the performance of the machines.

Refer to Fig. 26.13. The chart is dividing into three areas that are separated from each other by the saturated liquid and saturated vapour lines. The region on the chart to the *left* of the saturated liquid line is called the *sub-cooled region.* At any point in the sub-cooled region the refrigerant is in the liquid phase and its temperature is below the saturation temperature corresponding to its pressure. The area to the *right* of the saturated vapour line is superheated region and the refrigerant is in the form of a *superheated vapour.* The section of the chart between the saturated liquid and saturated vapour lines is the two phase region and represents the change in phase of the refrigerant between liquid and vapour phases. At any point between two saturation lines the refrigerant is in the form of a liquid vapour mixture. *The distance between the two lines along any constant pressure line, as read on the enthalpy scale at the bottom of the chart, is the latent heat of vaporisation of the refrigerant at that pressure.*

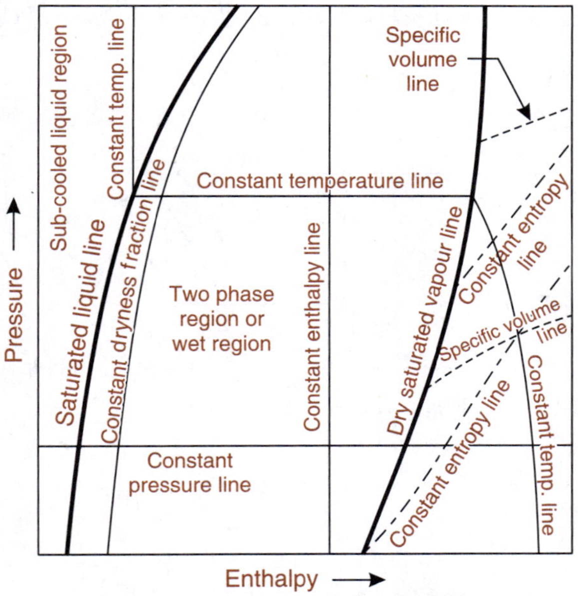

Fig. 26.13. Pressure enthalpy (*p-h*) chart.

The horizontal lines extending across the chart are lines of 'constant pressure' and the vertical lines are lines of constant enthalpy. The lines of 'constant temperature' in the sub-cooled region are almost vertical on the chart and parallel to the lines of constant enthalpy. In the centre section, since the refrigerant changes state at a constant temperature and pressure, the lines of constant temperature are parallel to and coincide with the lines of constant pressure. At the saturated vapour line the lines of constant temperature change direction again and, in the superheated vapour region, fall of sharply toward the bottom of the chart.

The straight lines which extend diagonally and almost vertically across the superheated vapour region are lines of constant entropy. The curved, nearly horizontal lines crossing the superheated vapour region are lines of constant volume.

p-h chart gives directly the changes in enthalpy and pressure during a process for thermodynamic analysis.

26.3.6. Simple Vapour Compression Cycle on p-h Chart

Fig. 26.14 shows a simple vapour compression cycle on a *p-h* chart. The points 1, 2, 3 and 4 correspond to the points marked in Fig. 26.9.

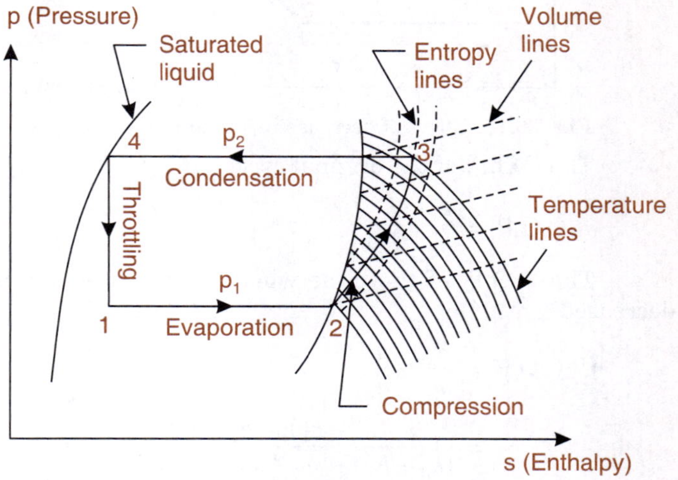

Fig. 26.14. Simple vapour compression cycle on *p-h* chart.

The dry saturated vapour (at state 2) is drawn by the compressor from evaporator at lower pressure p_1 and then it (vapour) is compressed isentropically to the upper pressure p_2. The isentropic compression is shown by the line 2-3. Since the vapour is dry and saturated at the start of compression it becomes superheated at the end of compression as given by point 3. The process of *condensation which takes place at constant pressure* is given by the line 3-4. The vapour now reduced to saturated liquid is throttled through the expansion valve and the process is shown by the line 4-1. At the point 1 a mixture of vapour and liquid enters the evaporator where it gets dry saturated as shown by the point 2. The cycle is thus completed.

Heat extracted (or refrigerating effect produced),

$$R_n = h_2 - h_1$$

Work done, $$W = h_3 - h_2$$

$$\therefore \quad \text{C.O.P.} = \frac{R_n}{W} = \frac{h_2 - h_1}{h_3 - h_2}$$

The values of h_1, h_2 and h_3 can be directly read from *p-h* chart.

26.3.7. Factors Affecting the Performance of a Vapour Compression System

The factors which affect the performance of a vapour compression system are given below :

1. **Effect of suction pressure.** The effect of *decrease* in suction pressure is shown in Fig. 26.15.

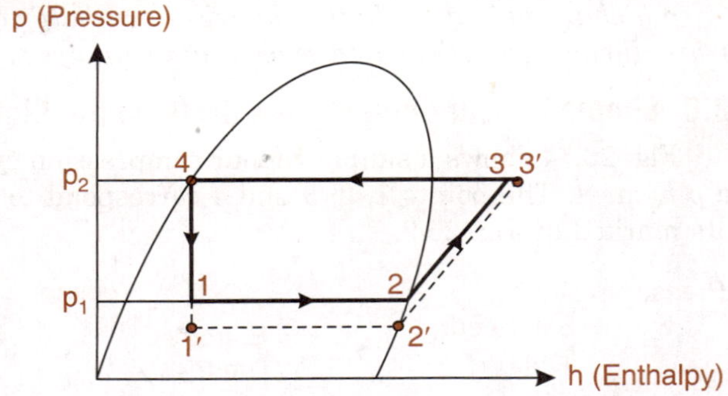

Fig. 26.15. Effect of decrease in suction pressure.

The C.O.P. of the original cycle,

$$\text{C.O.P.} = \frac{h_2 - h_1}{h_3 - h_2}$$

The C.O.P. of the cycle when suction pressure is decreased,

$$\text{C.O.P.} = \frac{h_2' - h_1'}{h_3' - h_2'}$$

$$= \frac{(h_2 - h_1) - (h_2 - h_2')}{(h_3 - h_2) + (h_2 - h_2') + (h_3' - h_3)}$$

$$(\because\ h_1 = h_1')$$

This shows that the *refrigerating effect is decreased and work required is increased. The net effect is to reduce the refrigerating capacity of the system (with the same amount of refrigerant flow) and the C.O.P.*

2. **Effect of delivery pressure.** Fig. 26.16 shows the effect of *increase in delivery pressure*.

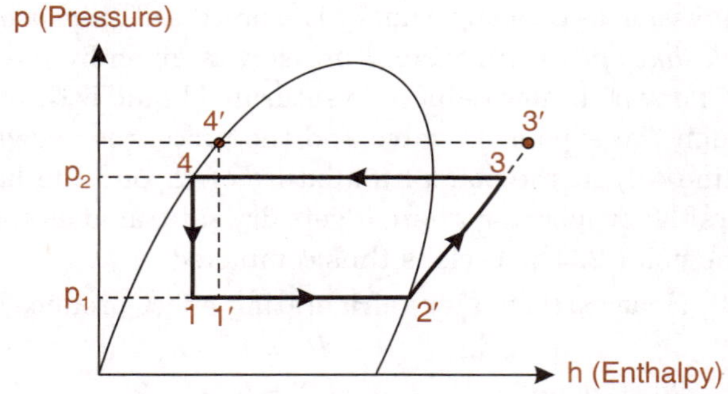

Fig. 26.16. Effect of increase in delivery pressure.

C.O.P. of the original cycle,

$$\text{C.O.P.} = \frac{h_2 - h_1}{h_3 - h_2}$$

C.O.P. of the cycle when delivery pressure is increased,

$$\text{C.O.P.} = \frac{h_2 - h_1'}{h_3' - h_2} = \frac{(h_2 - h_1) - (h_1' - h_1)}{(h_3 - h_2) + (h_3' - h_3)}$$

The effect of increasing the delivery/discharge pressure is just similar to the effect of decreasing the suction pressure. *The only difference is that the effect of decreasing the suction pressure is more predominant than the effect of increasing the discharge pressure.*

The following points may be noted :

(*i*) *As the discharge temperature required in the summer is more as compared with winter, the same machine will give less refrigerating effect (load capacity decreased) at a higher cost.*

(*ii*) *The increase in discharge pressure is necessary for high condensing temperatures and decrease in suction pressure is necessary to maintain low temperature in the evaporator.*

3. **Effect of superheating.** As may be seen from the Fig. 26.17 the effect of superheating is to increase the refrigerating effect but this increase in refrigerating effect is at the cost of increase in amount of work spent to attain the upper pressure limit. *Since the increase in work is more as compared to increase in refrigerating effect, therefore overall effect of superheating is to give a low value of C.O.P.*

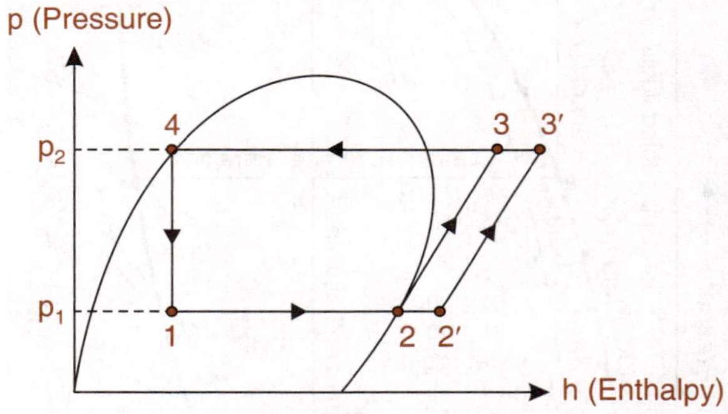

Fig. 26.17. Effect of superheating.

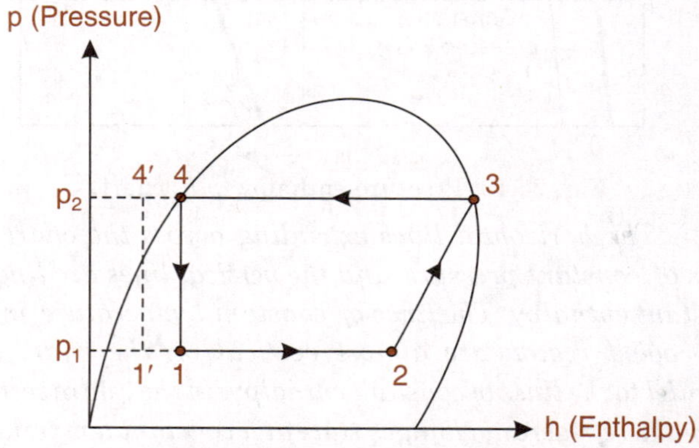

Fig. 26.18. Effect of sub-cooling of liquid.

4. **Effect of sub-cooling of liquid.** *'Sub-cooling' is the process of cooling the liquid refrigerant below the condensing temperature for a given pressure.* In Fig. 26.18 the process of sub-cooling is shown by 4-4′. As is evident from the figure the *effect of sub-cooling is to increase the refrigerating effect. Thus sub-cooling results in increase of C.O.P. provided that no further energy has to be spent to obtain the extra cold coolant required.*

The *sub-cooling* or *undercooling* may be done by any of the following methods :

(*i*) Inserting a special coil between the condenser and the expansion valve.

(*ii*) Circulating greater quantity of cooling water through the condenser.

(*iii*) Using water cooler than main circulating water.

5. Effect of suction temperature and condenser temperature. The performance of the vapour compression refrigerating cycle varies considerably with both vaporising and condensing temperatures. Of the two, the *vaporising temperature has far the greater effect*. It is seen that *the capacity and performance of the refrigerating system improve as the vapourising temperature increases and the condensing temperature decreases*. Thus refrigerating system *should always be designed to operate at the highest possible vapourising temperature and lowest possible condensing temperature*, of course, keeping in view the requirements of the application.

26.3.8. Actual Vapour Compression Cycle

The actual vapour compression cycle differs from the theoretical cycle in several ways because of the following *reasons* :

(*i*) Frequently the liquid refrigerant is sub-cooled before it is allowed to enter the expansion valve, and usually the gas leaving the evaporator is superheated a few degrees before it enters the compressor. This superheating may occur as a result of the type of expansion control used or through a pick up of heat in the suction line between the evaporator and compressor.

(*ii*) Compression, although usually assumed to be isentropic, may actually prove to be neither isentropic nor polytropic.

(*iii*) Both the compressor suction and discharge valves are actuated by pressure difference and this process requires the actual suction pressure inside the compressor to be slightly below that of the evaporator and the discharge pressure to be above that of condenser.

(*iv*) Although isentropic compression assumes no transfer of heat between the refrigerant and the cylinder walls, actually the cylinder walls are hotter than the incoming gases from the evaporator and colder than the compressed gases discharged to the condenser.

(*v*) Pressure drop in long suction and liquid line piping and any vertical differences in head created by locating the evaporator and condenser at different elevations.

Fig. 26.19 shows the actual vapour compression cycle on *T-s* diagram. The various processes are discussed as follows :

Process 1-2-3. This process represents passage of refrigerant through the evaporator, with *1-2* indicating *gain of latent heat of vapourisation, and 2-3, the gain of superheat before entrance to compressor*. Both of these processes approach very closely to the constant pressure conditions (assumed in theory).

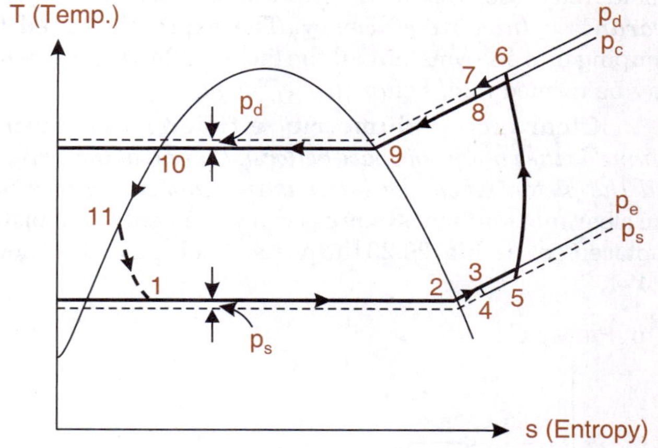

Fig. 26.19. Actual vapour compression cycle (*T-s* diagram).

Process 3-4-5-6-7-8. This path/process represents the passage of the vapour refrigerant from entrance to the discharge of the compressor. Path 3-4 represents the throttling action that occurs during passage through the suction valves, and path 7-8 represents the throttling during passage through exhaust valves. Both of these actions are accompanied by an entropy increase and a slight drop in temperature.

Compression of the refrigerant occurs along path 5-6, which is actually neither isentropic nor polytropic. The heat transfers indicated by path 4-5 and 6-7 occur essentially at constant pressure.

Process 8-9-10-11. This process represents the passage of refrigerant through the condenser with 8-9 indicating removal of superheat, 9-10 the removal of latent heat, and 10-11 removal of heat of liquid or sub-cooling.

Process 11-1. This process represents passage of the refrigerant through the expansion valve, both theoretically and practically an irreversible adiabatic path.

26.3.9. Volumetric Efficiency

A compressor which is theoretically perfect would have neither clearance nor losses of any type and would pump on each stroke a quantity of refrigerant equal to piston displacement. No actual compressor is able to do this, since it is impossible to construct a compressor without clearance or one that will have no wire drawing through the suction and discharge valves, no superheating of the suction gases upon contact with the cylinder walls, or no leakage of gas past the piston or the valves. All these factors effect the volume of gas pumped or the capacity of the compressor, some of them affect the H.P. requirements per tonne of refrigeration developed.

'Volumetric efficiency' is defined as the *ratio of actual volume of gas drawn into the compressor* (at evaporator temperature and pressure) on *each stroke to the piston displacement*. If the effect of *clearance alone* is considered, the resulting expression may be termed *clearance volumetric efficiency*. The expression used for grouping into one constant all the factors affecting efficiency may be termed *total volumetric efficiency*.

Clearance volumetric efficiency. *'Clearance volume' is the volume of space between the end of the cylinder and the piston when the latter is in dead centre position.* The clearance volume is expressed as a percentage of piston displacement. In Fig. 26.20 the piston displacement is shown as 4′-1.

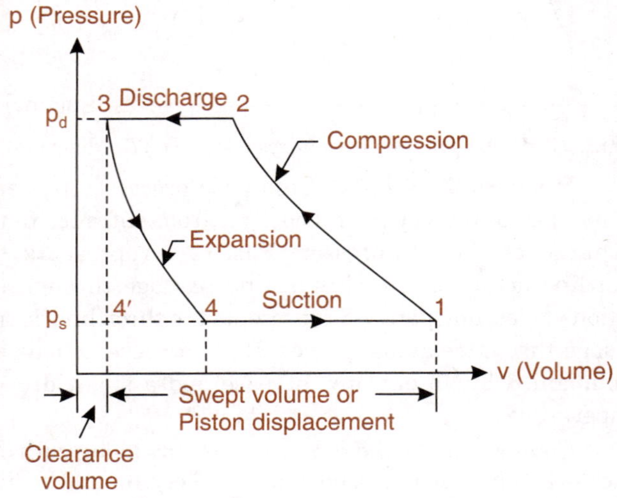

Fig. 26.20

During the suction stroke 4′-1, the vapour filled in clearance space at a discharge pressure p_d expands along 3-4 and the suction valve opens only when pressure has dropped to suction pressure p_s, therefore actual volume sucked will be $(v_1 - v_4)$ while the swept volume is $(v_1 - v_4')$. The *ratio of actual volume of vapour sucked to the piston displacement is defined as clearance volumetric efficiency*.

Thus,

Clearance volumetric efficiency,

$$\eta_{cv} = \frac{v_1 - v_4}{v_1 - v_4'} = \frac{v_1 - v_4}{v_1 - v_3} \qquad (\because \; v_4' = v_3)$$

Considering *polytropic expansion process 3-4*, we have

$$p_s v_4^n = p_d v_3^n$$

or
$$\frac{p_d}{p_s} = \left(\frac{v_4}{v_3}\right)^n \quad \text{or} \quad v_4 = v_3 \cdot \left(\frac{p_d}{p_s}\right)^{1/n}$$

If the clearance ratio,

$$C = \frac{v_3}{v_1 - v_3} = \frac{\text{Clearance volume}}{\text{Swept volume}}$$

Thus
$$\eta_{cv} = \frac{v_1 - v_4}{v_1 - v_3} = \frac{(v_1 - v_4') - (v_4 - v_4')}{(v_1 - v_3)}$$

$$= \frac{(v_1 - v_3) - (v_4 - v_3)}{(v_1 - v_3)} \qquad (\because \; v_4' = v_3)$$

$$= 1 - \frac{v_4 - v_3}{v_1 - v_3}$$

$$= 1 - \frac{v_3\left(\frac{p_d}{p_s}\right)^{1/n} - v_3}{v_1 - v_3}$$

$$= 1 + \frac{v_3}{v_1 - v_3}\left[1 - \left(\frac{p_d}{p_s}\right)^{1/n}\right]$$

$$= 1 + C - C\left(\frac{p_d}{p_s}\right)^{1/n}$$

Hence clearance volumetric efficiency,

$$\eta_{cv} = 1 + C - C\left(\frac{p_d}{p_s}\right)^{1/n} \qquad \text{...(26.11)}$$

Total volumetric efficiency. The total volumetric efficiency (η_{tv}) of a compressor is best obtained by actual *laboratory measurements of the amount of refrigerant compressed and delivered to the condenser*. It is very difficult to predict the effects of wire-drawing, cylinder wall heating, and piston leakage to allow any degree of accuracy in most cases. The total volumetric efficiency can be approximately calculated if the pressure drop through the suction valves and the temperature of the gases at the end of the suction stroke are known and if it is assumed that there is no leakage past the piston during compression, it can be calculated [by modifying the eqn. (26.11)] by using the following equation :

$$\eta_{tv} = \left[1 + C - C\left(\frac{p_d}{p_s}\right)^{1/n}\right] \times \frac{p_c}{p_s} \times \frac{T_s}{T_c} \qquad \text{...(26.12)}$$

where the subscript '*c*' refers to compressor cylinder and '*s*' refers to the evaporator or the suction line just adjacent to the compressor.

26.3.10. Mathematical Analysis of Vapour Compression Refrigeration

(*i*) **Refrigerating effect.** *Refrigerating effect is the amount of heat absorbed by the refrigerant in its travel through the evaporator*. In Fig. 26.10 this effect is represented by the expression.

$$Q_{evap.} = (h_2 - h_1) \text{ kJ/kg} \qquad \text{...(26.13)}$$

In addition to the latent heat of vaporisation it may include any heat of superheat absorbed in the evaporator.

(*ii*) **Mass of refrigerant.** Mass of refrigerant circulated (per second per tonne of refrigeration) may be calculated by *dividing the amount of heat by the refrigerating effect*.

∴ Mass of refrigerant circulated,

$$m = \frac{14000}{3600\,(h_2 - h_1)} \text{ kg/s-tonne} \quad ...(26.14)$$

because one tonne of refrigeration means cooling effect of 14000 kJ/h.

(iii) **Theoretical piston displacement.** Theoretical piston displacement (per tonne of refrigeration per minute) may be found by *multiplying the mass of refrigerant to be circulated (per tonne of refrigeration per sec.) by the specific volume of the refrigerant gas, $(v_g)_2$*, at its entrance of compressor. Thus,

Piston displacement$_{(Theoretical)}$

$$= \frac{14000}{3600\,(h_2 - h_1)}(v_g)_2 \text{ m}^3\text{/s-tonne} \quad ...(26.15)$$

(iv) **Power (Theoretical) required.** Theoretical power per tonne of refrigeration is the power, *theoretically required to compress the refrigerant*. Here volumetric and mechanical efficiencies are not taken into consideration. Power required may be calculated as follows :

(a) ***When compression is isentropic :***

Work of compression $= h_3 - h_2$ ...(26.16)

Power required $= m(h_3 - h_2)$ kW

where, m = Mass of refrigerant circulated in kg/s.

(b) ***When compression follows the general law pV^n = constant :***

$$\text{Work of compression} = \frac{n}{n-1}(p_3v_3 - p_2v_2) \text{ Nm/kg}$$

Power required

$$= m \times \frac{n}{n-1}(p_3v_3 - p_2v_2) \times \frac{1}{10^3} \text{ kW } (p \text{ is in N/m}^2) \quad ...(26.17)$$

(v) **Heat rejected to compressor cooling water.** If the compressor cylinders are jacketed, an appreciable amount of heat may be rejected to the cooling water during compression. If the suction and discharge compression conditions are known, this heat can be determined as follows :

Heat rejected to compressor cooling water

$$= \left[\frac{n}{(n-1)}\left(\frac{p_3v_3 - p_2v_2}{1000}\right) - (h_3 - h_2)\right] \text{ kJ/kg } (p \text{ is in N/m}^2) \quad ...(26.18)$$

(vi) **Heat removed through condenser.** Heat removed through condenser includes all heat removed through the condenser, either as latent heat, heat of superheat, or heat of liquid. This is *equivalent to the heat absorbed in the evaporator plus the work of compression*.

∴ Heat removed through condenser $= m(h_3 - h_4)$ kJ/s (m = mass of refrigerant circulated in kg/s) ...(26.19)

26.4. VAPOUR ABSORPTION SYSTEM

26.4.1. Introduction

In a *vapour absorption system the refrigerant is absorbed on leaving the evaporator, the absorbing medium being a solid or liquid.* In order that the sequence of events should be continuous it is necessary for the refrigerant to be separated from the absorbent and subsequently condensed before being returned to the evaporator. The separation is accomplished by the application of direct heat in a *'generator'*. The solubility of the refrigerant and absorbent must be suitable and the plant which uses *ammonia as the refrigerant and water as absorbent* will be described.

26.4.2. Simple Vapour Absorption System

Refer to Fig. 26.21 for a simple absorption system. The solubility of ammonia in water at low temperatures and pressures is higher than it is at higher temperatures and pressures. The ammonia vapour leaving the evaporator at

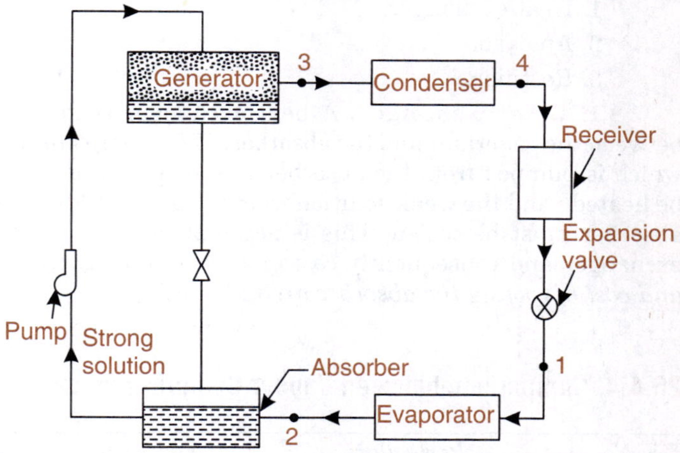

Fig. 26.21. *(a)* Simple vapour absorption system.

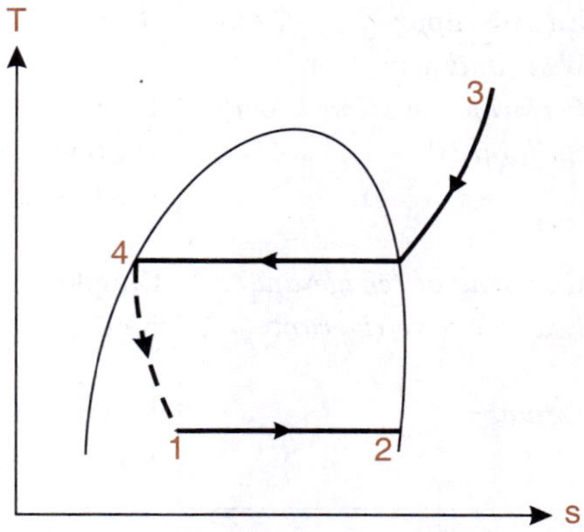

Fig. 26.21. *(b)* Simple vapour absorption system—*T-s* diagram.

point 2 is readily absorbed in the low temperature hot solution in the absorber. This process is accompanied by the rejection of heat. The ammonia in water solution is pumped to the higher pressure and is heated in the generator. *Due to reduced solubility of ammonia in water at the higher pressure and temperature, the vapour is removed from the solution.* The vapour then passes to the condenser and the weakened ammonia in water solution is returned to the absorber.

In this system the *work done on compression is less than in vapour compression cycle* (since pumping a liquid requires much less work than compressing a vapour between the same pressures) but a heat input to the generator is required. The heat may be supplied by any convenient form *e.g.*, steam or gas heating.

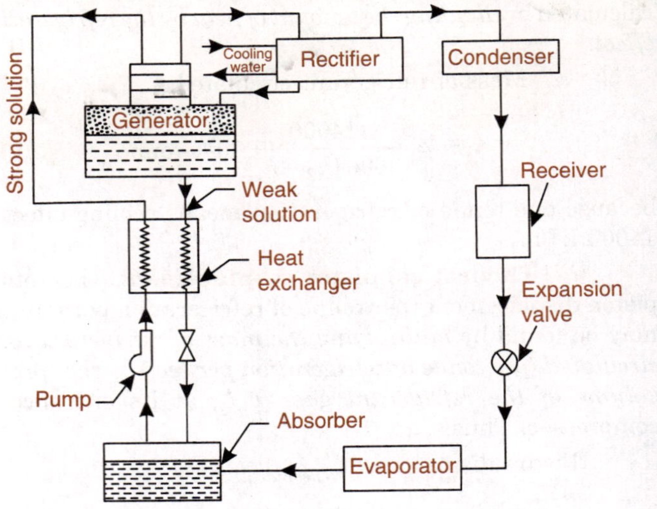

Fig. 26.22

26.4.3. Practical Vapour Absorption System

Refer to Fig. 26.22. Although a simple vapour absorption system can provide refrigeration *yet its operating efficiency is low*. The following *accessories* are fitted to make the system more practical and improve the performance and working of the plant.

1. Heat exchanger.
2. Analyser.
3. Rectifier.

1. **Heat exchanger.** A heat exchanger is located between the generator and the absorber. The strong solution which is pumped from the absorber to the generator must be heated ; and the weak solution from the generator to the absorber must be cooled. This is accomplished by a heat exchanger and consequently *cost of heating the generator and cost of cooling the absorber are reduced.*

2. **Analyser.** An analyser consists of a series of trays mounted above the generator. Its *main function is to remove partly some of the unwanted water particles associated with ammonia vapour going to condenser. If these water vapours are permitted to enter condenser mthey may enter the expansion valve and freeze ; as a result the pipe line may get choked.*

3. **Rectifier.** A rectifier is a water-cooled heat exchanger *which condenses water vapour and some ammonia and sends back to the generator*. Thus *final reduction or elemination of the percentage of water vapour takes place in a rectifier*.

The co-efficient of performance (C.O.P.) of this system is given by :

$$\text{C.O.P.} = \frac{\text{Heat extracted from the evaporator}}{\text{Heat supplied in the generator} + \text{Work done by the liquid pump}}.$$

26.4.4. Comparison between Vapour Compression and Vapour Absorption Systems

S. No.	*Particulars*	*Vapour compression system*	*Vapour absorption system*
1.	*Type of energy supplied*	Mechanical—a high grade energy	Mainly heat—a low grade energy
2.	*Energy supply*	Low	High
3.	*Wear and tear*	More	Less
4.	*Performance at part loads*	Poor	System not affected by variations of loads.
5.	*Suitability*	Used where high grade mechanical energy is available	Can also be used at remote places as it can work even with a simple kerosene lamp (of course in small capacities)
6.	*Charging of refrigerant*	Simple	Difficult
7.	*Leakage of refrigerant*	More chances	No chance as there is no compressor or any reciprocating component to cause leakage.
8.	*Damage*	Liquid traces in suction line may damage the compressor	Liquid traces of refregerant present in piping at the exit of evaporator constitute no danger.

Example 26.12. *A refrigeration machine is required to produce i.e., at 0°C from water at 20°C. The machine has a condenser temperature of 298 K while the evaporator temperature is 268 K. The relative efficiency of the machine is 50% and 6 kg of Freon-12 refrigerant is circulated through the system per minute. The refrigerant enters the compressor with a dryness fraction of 0.6. Specific heat of water is 4.187 kJ/kg K and the latent heat of ice is 335 kJ/kg. Calculate the amount of ice produced on 24 hours. The table of properties of Freon-12 is given below :*

Temperature K	*Liquid heat kJ/kg*	*Latent heat kJ/g*	*Entropy of liquid kJ/kg*
298	59.7	138.0	0.2232
268	31.4	154.0	0.1251

(U.P.S.C.)

Solution. *Given* : m = 6 kg/min. ; $\eta_{relative}$ = 50% ; x_2 = 0.6 ; c_{pw} = 4.187 kJ/kg K ; Latent heat of ice = 335 kJ/kg.

Refer to Fig. 26.23

h_{f_2} = 31.4 kJ/kg ; h_{fg_2} = 154.0 kJ/kg ;

h_{f_3} = 59.7 kJ/kg ;

h_{fg_3} = 138 kJ/kg ; h_{f_4} = 59.7 kJ/kg

...From the table given above

$$h_2 = h_{f_2} + x_2 h_{fg_2}$$
$$= 31.4 + 0.6 \times 154 = 123.8 \text{ kJ/kg}$$

For *isentropic compression 2-3*, we have

$$s_3 = s_2$$

$$s_{f_3} + x_3 \frac{h_{fg_3}}{T_3} = s_{f_2} + x_2 \frac{h_{fg_2}}{T_2}$$

$$0.2232 + x_3 \times \frac{138}{298} = 0.1251 + 0.6 \times \frac{154}{268}$$
$$= 0.4698$$

$$\therefore \quad x_3 = (0.4698 - 0.2232) \times \frac{298}{138}$$
$$= 0.5325$$

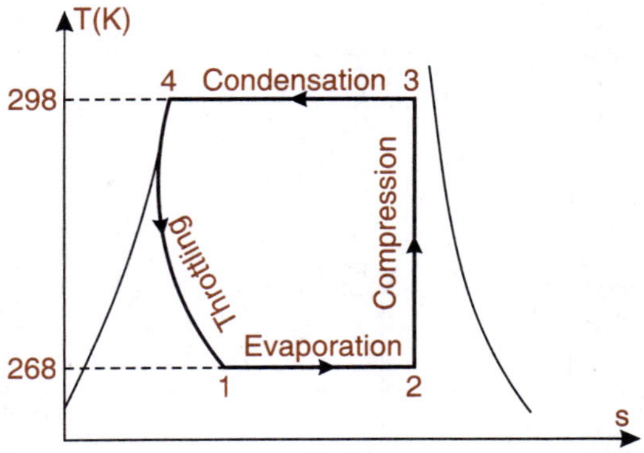

Fig. 26.23

Now, $h_3 = h_{f_3} + x_3 h_{fg_3}$ = 59.7 + 0.5325 × 138
= 133.2 kJ/kg

Also, $h_1 = h_{f_4}$ = 59.7 kJ/kg

$$\text{Theoretical C.O.P.} = \frac{R_n}{W}$$

$$= \frac{h_2 - h_1}{h_3 - h_2} = \frac{123.8 - 59.7}{133.2 - 123.8} = 6.82$$

Actual C.O.P. = $\eta_{relative}$ × (C.O.P.)$_{theoretical}$
= 0.5 × 6.82 = 3.41

Heat extracted from 1 kg of water at 20°C for the formation of 1 kg of ice at 0°C

= 1 × 4.187 × (20 – 0) + 335
= 418.74 kJ/kg

Let m_{ice} = Mass of ice formed in kg/min.

$$(\text{C.O.P.})_{actual} = 3.41 = \frac{R_n(\text{actual})}{W}$$

$$= \frac{m_{ice} \times 418.74}{m(h_3 - h_2)} = \frac{m_{ice} \times 418.74 \text{ (kJ/min)}}{6(133.2 - 123.8) \text{ (kJ/min)}}$$

$$\therefore \quad m_{ice} = \frac{6(133.2 - 123.8) \times 3.41}{418.74} = 0.459 \text{ kg/min}$$

$$= \frac{0.459 \times 60 \times 24}{1000} \text{ tonnes (in 24 hours)}$$

= **0.661 tonne. (Ans.)**

Example 26.13. *28 tonnes of ice from and at 0°C is produced per day in an ammonia refrigerator. The temperature range in the compressor is from 25°C to – 15°C. The vapour is dry and saturated at the end of compression and an expansion valve is used. Assuming a co-efficient of performance of 62% of the theoretical, calculate the power required to drive the compressor.*

Temp. °C	*Enthalpy (kJ/kg)*		*Entropy of liquid (kJ/kg K)*	*Entropy of vapour kJ/kg K*
	Liquid	*Vapour*		
25	100.04	1319.22	0.3473	4.4852
– 15	– 54.56	1304.99	– 2.1338	5.0585

Take latent heat of ice = 335 kJ/kg.

Solution. Theoretical C.O.P. = $\dfrac{h_2 - h_1}{h_3 - h_2}$

Here, h_3 = 1319.22 kJ/kg ;
$h_1 = h_4$ (*i.e.*, h_{f4}) = 100.04 kJ/kg
...From the table above.

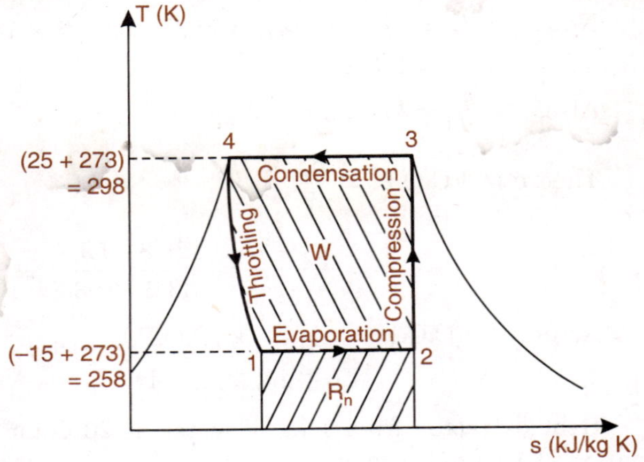

Fig. 26.24

To find h_2, let us first find dryness at point 2.

Entropy at '2' = Entropy at '3' (Process 2-3 being isentropic)

$$s_{f_2} + x_2\, s_{fg_2} = s_{g_3}$$

$$-2.1338 + x_2 \times [5.0585 - (-2.1338)] = 4.4852$$

$$\therefore \quad x_2 = \frac{4.4852 + 2.1338}{5.0585 + 2.1338} = 0.92$$

$$\therefore \quad h_2 = h_{f_2} + x_2\, h_{fg_2}$$

$$= -54.56 + 0.92 \times [1304.99 - (-54.56)]$$

$$= 1196.23 \text{ kJ/kg.}$$

$$\therefore \quad \text{C.O.P.}_{\text{(theoretical)}} = \frac{1196.23 - 100.04}{1319.22 - 1196.23} = 8.91.$$

$$\therefore \quad \text{C.O.P.}_{\text{(actual)}} = 0.62 \times \text{C.O.P.}_{\text{(theoretical)}} \quad \text{...Given}$$

i.e., $\quad \text{C.O.P.}_{\text{(actual)}} = 0.62 \times 8.91 = 5.52$

Actual refrigerating effect per kg

$= \text{C.O.P.}_{\text{(actual)}} \times$ work done

$= 5.52 \times (h_3 - h_2)$

$= 5.52 \times (1319.22 - 1196.23)$

$= 678.9$ kJ/kg

Heat to be extracted per hour

$$= \frac{28 \times 1000 \times 335}{24} = 390833.33 \text{ kJ}$$

Heat to be extracted per second

$$= \frac{390833.33}{3600} = 108.56 \text{ kJ/s.}$$

∴ Mass of refrigerant circulated per second

$$= \frac{108.56}{678.9} = 0.1599 \text{ kg}$$

Total work done by the compressor per second

$= 0.1599 \times (h_3 - h_2)$

$= 0.1599\,(1319.22 - 1196.23)$

$= 19.67$ kJ/s

i.e., **Power required to drive the compressor = 19.67 kW. (Ans.)**

Example 26.14. *A refrigerating plant works between temperature limits of – 5°C and 25°C. The working fluid ammonia has a dryness fraction of 0.62 at entry to compressor. If the machine has a relative efficiency of 55%, calculate the amount of ice formed during a period of 24 hours. The ice is to be formed at 0°C from water at 15°C and 6.4 kg of ammonia is circulated per minute. Specific heat of water is 4.187 kJ/kg and latent heat of ice is 335 kJ/kg.*

Properties of NH_3 (datum – 40°C).

Temp. °C	*Liquid heat kJ/kg*	*Latent heat kJ/kg*	*Entropy of liquid kJ/kg K*
25	298.9	1167.1	1.124
– 5	158.2	1280.8	0.630

Solution. Fig. 26.25 shows the *T-s* diagram of the cycle.

Enthalpy at point '2',

$$h_2 = h_{f_2} + x_2\, h_{fg_2} = 158.2 + 0.62 \times 1280.8$$

$$= 952.3 \text{ kJ/kg}$$

Enthalpy at point '1', $\quad h_1 = h_{f_4} = 298.9$ kJ/kg

Also, entropy at point '2' = entropy at point '3'

i.e., $\quad s_2 = s_3$

$$s_{f_2} + x_2\, s_{fg_2} = s_{f_3} + x_3\, s_{fg_3}$$

$$0.630 + 0.62 \times \frac{1280.8}{(-5 + 273)} = 1.124 + x_2 \times \frac{1167.1}{(25 + 273)}$$

i.e., $\quad x_3 = 0.63$

∴ Enthalpy at point '3', $h_3 = h_{f_3} + x_3\, h_{fg_3}$

$= 298.9 + 0.63 \times 1167.1$

$= 1034.17$ kJ/kg

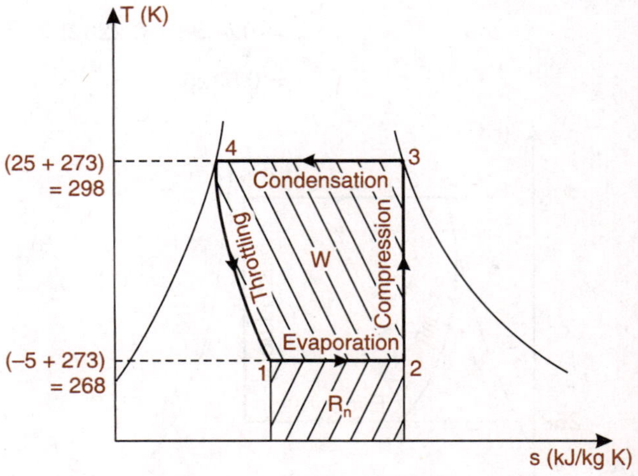

Fig. 26.25

$$\text{C.O.P.}_{\text{(theoretical)}} = \frac{h_2 - h_1}{h_3 - h_2} = \frac{952.3 - 298.9}{1034.17 - 952.3}$$

$$= \frac{653.4}{81.87} = 7.98.$$

$$\text{C.O.P.}_{\text{(actual)}} = 0.55 \times 7.98 = 4.39$$

Work done per kg of refrigerant

$$= h_3 - h_2 = 1034.17 - 952.3$$
$$= 81.87 \text{ kJ/kg}$$

Refrigerant in circulation, m = 6.4 kg/min.

∴ Work done per second

$$= 81.87 \times \frac{6.4}{60} = 8.73 \text{ kJ/s}$$

Heat extracted per kg of ice formed

$$= 15 \times 4.187 + 335 = 397.8 \text{ kJ.}$$

Amount of ice formed in 24 hours,

$$m_{\text{ice}} = \frac{8.73 \times 3600 \times 24}{397.8} = \mathbf{1896.1 \text{ kg.}} \quad \textbf{(Ans.)}$$

☞**Example 26.15.** *A simple vapour compression plant produces 5 tonnes of refrigeration. The enthalpy values at inlet to compressor, at exit from the compressor, and at exit from the condenser are 183.19, 209.41 and 74.59 kJ/kg respectively. Estimate :*

(*i*) *The refrigerant flow rate,*

(*ii*) *The C.O.P.,*

(*iii*) *The power required to drive the compressor, and*

(*iv*) *The rate of heat rejection to the condenser.*

Solution. Total refrigeration effect produced = 5 *TR* (tonnes of refrigeration)

$$= 5 \times 14000 = 70000 \text{ kJ/h or } 19.44 \text{ kJ/s}$$
$$(\because 1 \; TR = 14000 \text{ kJ/h})$$

Refer to Fig. 26.26.

Given : h_2 = 183.19 kJ/kg ; h_3 = 209.41 kJ/kg ;
$h_4 (= h_1)$ = 74.59 kJ/kg (Throttling process)

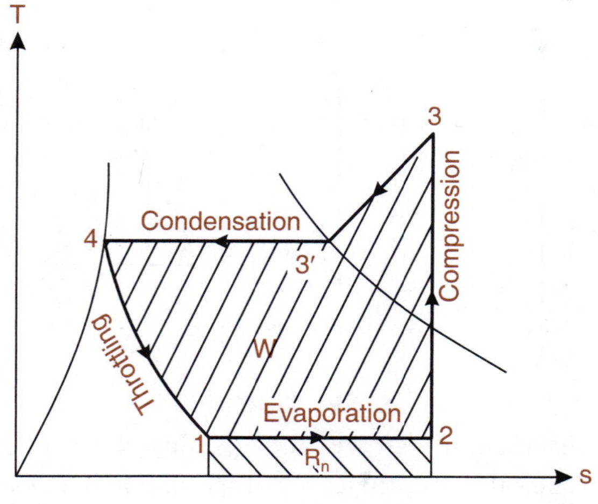

Fig. 26.26

(*i*) **The refrigerant flow rate, $\dot{m}$:**

Net refrigerating effect produced per kg = $h_2 - h_1$

$$= 183.19 - 74.59 = 108.6 \text{ kJ/kg}$$

∴ Refrigerant flow rate,

$$\dot{m} = \frac{19.44}{108.6} = \mathbf{0.179 \text{ kg/s.}} \quad \textbf{(Ans.)}$$

(*ii*) **The C.O.P. :**

$$\text{C.O.P.} = \frac{R_n}{W} = \frac{h_2 - h_1}{h_3 - h_2} = \frac{183.19 - 74.59}{209.41 - 183.19}$$

$$= \mathbf{4.142.} \quad \textbf{(Ans.)}$$

(*iii*) **The power required to drive the compressor, P :**

$$P = \dot{m} \, (h_3 - h_2) = 0.179 \, (209.41 - 183.19)$$
$$= \mathbf{4.69 \text{ kW.}} \quad \textbf{(Ans.)}$$

(*iv*) **The rate of heat rejection to the condenser :**

The rate of heat rejection to the condenser

$$= \dot{m} \, (h_3 - h_4)$$
$$= 0.179 \, (209.41 - 74.59)$$
$$= \mathbf{24.13 \text{ kW.}} \quad \textbf{(Ans.)}$$

Example 26.16. (*i*) *What are the advantages of using an expansion valve instead of an expander in a vapour compression refrigeration cycle ?*

(*ii*) *Give a comparison between centrifugal and reciprocating compressors.*

(*iii*) *An ice-making machine operates on ideal vapour compression refrigeration cycle using refrigerant R-12. The refrigerant enters the compressor as dry saturated vapour at – 15°C and leaves the condenser as saturated liquid at 30°C. Water enters the machine at 15°C and leaves as ice at – 5°C. For an ice production rate of 2400 kg in a day, determine the power required to run the unit. Find also the C.O.P. of the machine. Use refrigerant table only to solve the problem. Take the latent heat of fusion for water as 335 kJ/kg.*

Solution. (*i*) If an expansion cylinder is used in a vapour compression system, *the work recovered would be extremely small, in fact not even sufficient to overcome the mechanical friction*. It will not be possible to gain any work. Further, the *expansion cylinder is bulky*. On the other hand the *expansion valve is a very simple and handy device, much cheaper than the expansion cylinder*. It *does not need installation, lubrication or maintenance*.

The expansion valve *also controls the refrigerant flow rate according to the requirement, in addition to serving the function of reducting the pressure of the refrigerant.*

(*ii*) The comparison between centrifugal and reciprocating compressors :

The comparison between centrifugal and reciprocating compressors is given in the table below :

S. No.	*Particulars*	*Centrifugal compressor*	*Reciprocating compressor*
1.	*Suitability*	Suitable for handling large volumes of air at low pressures	Suitable for low discharges of air at high pressure.
2.	*Operational speeds*	Usually high	Low
3.	*Air supply*	Continuous	Pulsating
4.	*Balancing*	Less vibrations	Cyclic vibrations occur
5.	*Lubrication system*	Generally simple lubrication systems are required.	Generally complicated
6.	*Quality of air delivered*	Air delivered is relatively more clean	Generally contaminated with oil.
7.	*Air compressor size*	Small for given discharge	Large for same discharge
8.	*Free air handled*	2000–3000 m³/min	250–300 m³/min
9.	*Delivery pressure*	Normally below 10 bar	500 to 800 bar
10.	*Usual standard of compression*	Isentropic compression	Isothermal compression
11.	*Action of compressor*	Dynamic action	Positive displacement.

(*iii*) Using property table of R-12 :

$$h_2 = 344.927 \text{ kJ/kg}$$
$$h_4 = h_1 = 228.538 \text{ kJ/kg}$$
$$(c_p)_v = 0.611 \text{ kJ/kg°C}$$
$$s_2 = s_3$$

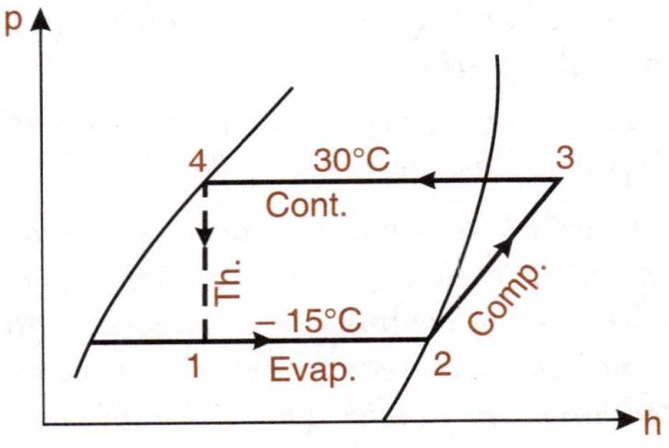

Fig. 26.27

or $$1.56323 = 1.5434 + 0.611 \log_e \left[\frac{t_3 + 273}{30 + 273}\right]$$

or $$t_3 = 39.995°\text{C}$$

$$h_3 = 363.575 + 0.611(39.995 - 30) = 369.68 \text{ kJ/kg.}$$

$$R_n/\text{kg} = h_2 - h_1 = 344.927 - 228.538 = 116.389 \text{ kJ/kg}$$

$$W/\text{kg} = h_3 - h_2 = 369.68 - 344.927 = 24.753$$

$$\textbf{C.O.P.} = \frac{R_n}{W} = \frac{116.389}{24.753} = \mathbf{4.702.} \quad \textbf{(Ans.)}$$

Assuming c_p for ice = 2.0935 kJ/kg°C

Heat to be removed to produce ice

$$= \frac{2400}{24 \times 3600} [4.187(15 - 0) + 335 + 2.0935(0 - (-5))]$$

= 11.3409 kJ/s = Work required, kJ/s (kW) × C.O.P.

$$\therefore \quad \text{Work required (Power)} = \frac{11.3409}{4.702} = \mathbf{2.4 \ kW.} \quad \textbf{(Ans.)}$$

Example 26.17. *A R-12 refrigerator works between the temperature limits of – 10°C and + 30°C. The compressor employed is of 20 cm × 15 cm, twin cylinder, single-acting compressor having a volumetric efficiency of 85%. The compressor runs at 500 r.p.m. The refrigerant is sub-cooled and it enters at 22°C in the expansion valve. The vapour is superheated and enters the compressor at – 2°C. Work out the following :*

(i) Show the process on T-s and p-h diagrams ; (ii) The amount of refrigerant circulated per minute ; (iii) The tonnes of refrigeration ; (iv) The C.O.P. of the system. **(M.U.)**

Solution. (*i*) **Process on T-s and p-h diagrams :**

The processes on *T-s* and *p-h* diagrams are shown in Fig. 26.28.

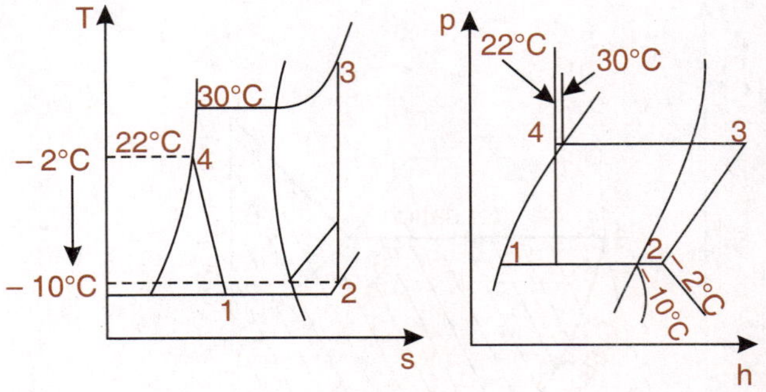

Fig. 26.28

(*ii*) **Mass of refrigerant circulated per minute :**

The value of enthalpies and specific volume read from *p-h* diagram are as under :

$h_2 = 352$ kJ/kg ; $h_3 = 374$ kJ/kg
$h_4 = h_1 = 221$ kJ/kg ; $v_2 = 0.08$ m³/kg

Refrigerants effect per kg

$= h_2 - h_1 = 352 - 221 = 131$ kJ/kg

Volume of refrigerant admitted per min.

$= \frac{\pi}{4} D^2 L \times$ r.p.m. $\times 2 \times \eta_{vol}$, for twin cylinder, single acting

$= \frac{\pi}{4} (0.2)^2 \times 0.15 \times 500 \times 2 \times 0.85$

$= 4$ m³/min

Mass of refrigerant per min

$= \frac{4}{0.08}$ = **50 kg/min. (Ans.)**

(iii) Cooling capacity in tonnes of refrigeration :

Cooling capacity $= 50(h_2 - h_1) = 50 \times 131$

= 6550 kJ/min or 393000 kJ/h

or $= \frac{393000}{14000}$ = **28.07 TR. (Ans.)**

($\because$ 1 tonne of refrigeration TR = 14000 kJ/h)

(iv) The C.O.P. of the system :

Work per kg $= (h_2 - h_1) = 374 - 352 = 22$ kJ/kg

C.O.P. $= \frac{131}{22}$ = **5.95. (Ans.)**

Example 26.18. *In a standard vapour compression refrigeration cycle, operating between an evaporator temperature of – 10°C and a condenser temperature of 40°C, the enthalpy of the refrigerant, Freon-12, at the end of compression is 220 kJ/kg. Show the cycle diagram on T-s plane. Calculate :*

(i) The C.O.P. of the cycle.

(ii) The refrigerating capacity and the compressor power assuming a refrigerant flow rate of 1 kg/min. You may use the extract of Freon-12 property table given below :

t(°C)	*p(MPa)*	*h_f(kJ/kg)*	*h_g(kJ/kg)*
– 10	0.2191	26.85	183.1
40	0.9607	74.53	203.1

(GATE)

Solution. The cycle is shown on *T-s* diagram in Fig. 26.29.

Given : Evaporator temperature = – 10°C
Condenser temperature = 40°C

Enthalpy at the end of compression, $h_3 = 220$ kJ/kg
From the table given, we have

$h_2 = 183.1$ kJ/kg ; $h_1 = h_{f_4} = 26.85$ kJ/kg

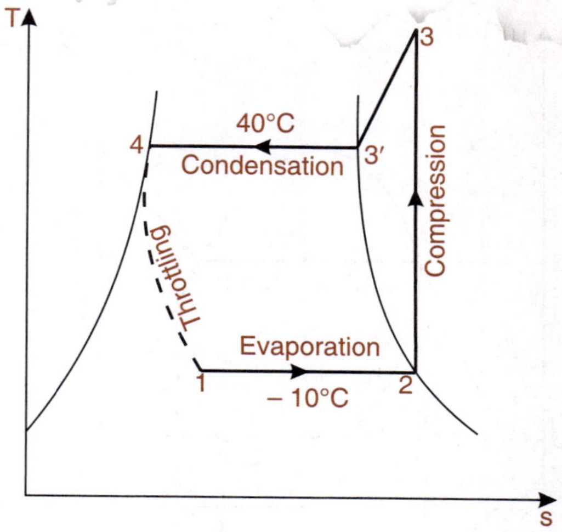

Fig. 26.29

(i) The C.O.P. the cycle :

$$\text{C.O.P.} = \frac{R_n}{W} = \frac{h_2 - h_1}{h_3 - h_2}$$

$$= \frac{183.1 - 74.53}{220 - 183.1} = \textbf{2.94.} \quad \textbf{(Ans.)}$$

(ii) Refrigerating capacity :

Refrigerating capacity $= m(h_2 - h_1)$

[where m = mass flow rate of refrigerant = 1 kg/min ...(Given)]

$= 1 \times (183.1 - 74.53)$ = **108.57 kJ/min. (Ans.)**

Compressor power :

Compressor power $m(h_3 - h_2)$

$= 1 \times (220 - 183.1) = 36.9$ kJ/min or 0.615 kJ/s

= **0.615 kW. (Ans.)**

Example 26.19. *A Freon-12 refrigerator producing a cooling effect of 20 kJ/s operates on a simple cycle with pressure limits of 1.509 bar and 9.607 bar. The vapour leaves the evaporator dry saturated and there is no undercooling. Determine the power required by the machine.*

If the compressor operaters at 300 r.p.m. and has a clearance volume of 3% of stroke volume, determine the piston displacement of the compressor. For compressor assume that the expansion following the law $pv^{1.13}$ = constant.

Given :

Temperature °C	*p_s bar*	*v_g m³/kg*	*Enthalpy h_f*	*kJ/kg h_g*	*Entropy s_f*	*kJ/kg K s_g*	*Specific heat kJ/kg K*
– 20	1.509	0.1088	17.8	178.61	0.073	0.7082	—
40	9.607	—	74.53	203.05	0.2716	0.682	0.747

(U.P.S.C.)

Solution. *Given* : (From the table above) :

$h_2 = 178.61$ kJ/kg ; $h_3' = 203.05$ kJ/kg ;

$h_{f_4} = 74.53$ kJ/kg $= h_1$

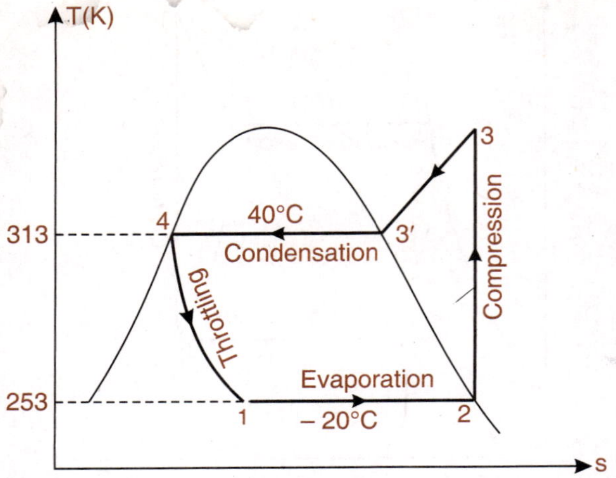

Fig. 26.30

Now, cooling effect $= \dot{m}(h_2 - h_1)$

$$20 = \dot{m}(178.61 - 74.53)$$

$$\therefore \quad \dot{m} = \frac{20}{(178.61 - 74.53)} = 0.192 \text{ kg/s}$$

Also, $s_3 = s_2$

$$s_3' + c_p \ln\left(\frac{T_3}{T_3'}\right) = 0.7082$$

$$0.682 + 0.747 \ln\left(\frac{T_3}{313}\right) = 0.7082$$

or $$\ln\left(\frac{T_3}{313}\right) = \frac{0.7082 - 0.682}{0.747} = 0.03507$$

or $$\frac{T_3}{313} = e^{0.03507} = 1.0357$$

$\therefore \quad T_3 = 313 \times 1.0357 = 324.2$ K

Now, $h_3 = h_3' + c_p(324.2 - 303)$

$= 203.05 + 0.747\,(324.2 - 313)$

$= 211.4$ kJ/kg

Power required :

Power required by the machine $= \dot{m}(h_3 - h_2)$

$= 0.192\,(211.4 - 178.61) =$ **6.29 kW. (Ans.)**

Piston displacement, V :

Volumetric efficiency,

$$\eta_{vol.} = 1 + C - C\left(\frac{p_d}{p_s}\right)^{1/n}$$

$$= 1 + 0.03 - 0.03\left(\frac{9.607}{1.509}\right)^{\frac{1}{1.13}} = 0.876 \text{ or } 87.6\%$$

The volume of refrigerant at the intake conditions is

$$\dot{m} \times v_g = 0.192 \times 0.1088 = 0.02089 \text{ m}^3\text{/s}$$

Hence the swept volume

$$= \frac{0.02089}{\eta_{vol.}} = \frac{0.02089}{0.876} = 0.02385 \text{ m}^3\text{/s}$$

$$\therefore \quad \mathbf{V} = \frac{0.02385 \times 60}{300} = \mathbf{0.00477\ m^3.} \quad \textbf{(Ans.)}$$

Example 26.20. *A food storage locker requires a refrigeration capacity of 50 kW. It works between a condenser temperature of 35°C and an evaporator temperature of –10°C. The refrigerant is ammonia. It is sub-cooled by 5°C before entering the expansion valve by the dry saturated vapour leaving the evaporator. Assuming a single cylinder, single-acting compressor operating at 1000 r.p.m. with stroke equal to 1.2 times the bore.*

Determine : (i) The power required, and

(ii) The cylinder dimensions.

Properties of ammonia are :

Saturation temperature, °C	*Pressure bar*	*Enthalpy (kJ/kg)*		*Entropy (kJ/kg K)*		*Specific volume (m³/kg)*		*Specific heat (kJ/kg K)*	
		Liquid	*Vapour*	*Liquid*	*Vapour*	*Liquid*	*Vapour*	*Liquid*	*Vapour*
– 10	*2.9157*	*154.056*	*1450.22*	*0.82965*	*5.7550*	—	*0.417477*	—	*2.492*
35	*13.522*	*366.072*	*1488.57*	*1.56605*	*5.2086*	*1.7023*	*0.095629*	*4.556*	*2.903*

(U.P.S.C.)

Solution. *Given* : (From the table above)

$h_2 = 1450.22$ kJ/kg ; $h_3' = 1488.57$ kJ/kg ;

$h_{f_4} = 366.072$ kJ/kg ;

$h_{f_4'} = h_1 = h_{f_4} - 4.556\,(308 - 303)$

$= 366.07 - 4.556\,(308 - 303) = 343.29$ kJ/kg

Also $s_3 = s_2$

or $$s_3' + c_p \ln\left(\frac{T_3}{T_3'}\right) = 5.755$$

$$5.2086 + 2.903 \ln\left(\frac{T_3}{308}\right) = 5.755$$

or $\ln\left(\frac{T_3}{308}\right) = \frac{5.755 - 5.2086}{2.903} = 0.1882$

$$\frac{T_3}{308} = e^{0.1882} = 371.8 \text{ K}$$

Now, $h_3 = h_3' + c_p(T_3 - T_3')$
$= 1488.57 + 2.903\ (371.8 - 308)$
$= 1673.8$ kJ/kg

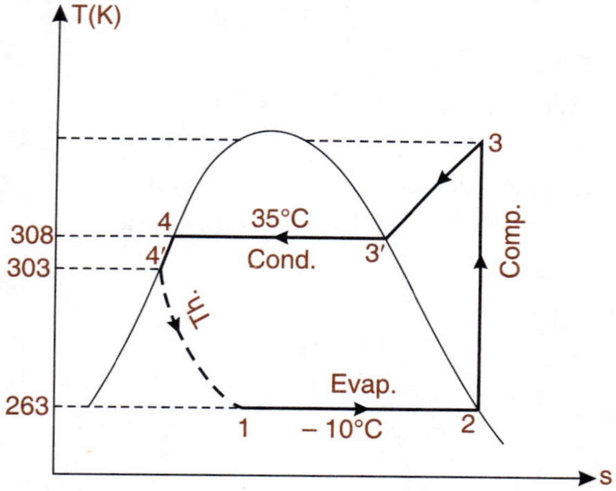

Fig. 26.31

Mass of refrigerant,

$$\dot{m} = \frac{50}{h_2 - h_1} = \frac{50}{1450.22 - 343.29}$$
$$= 0.04517 \text{ kJ/s}$$

(*i*) Power required :

Power required $= \dot{m}\ (h_3 - h_2)$
$= 0.04517\ (1673.8 - 1450.22)$
$= \mathbf{10.1\ kW.}$ **(Ans.)**

(*ii*) Cylinder dimensions :

$$\dot{m} = \frac{\pi}{4} D^2 \times L \times \frac{N}{60} \times 0.417477$$
$$= 0.04517 \text{ (calculated above)}$$

or $\frac{\pi}{4} D^2 \times 1.2D \times \frac{1000}{60} \times 0.417477 = 0.04517$

or $D^3 = \frac{0.04517 \times 4 \times 60}{\pi \times 1.2 \times 1000 \times 0.417477} = 0.006888$

∴ Diameter of cylinder,

$\mathbf{D} = (0.006888)^{1/3} = \mathbf{0.19\ m.}$ **(Ans.)**

and, Length of the cylinder,

$L = 1.2D = 1.2 \times 0.19 = \mathbf{0.228\ m.}$ **(Ans.)**

Example 26.21. *A refrigeration cycle uses Freon-12 as the working fluid. The temperature of the refrigerant in the evaporator is – 10°C. The condensing temperature is 40°C. The cooling load is 150 W and the volumetric efficiency of the compressor is 80%. The speed of the compressor is 720 r.p.m. Calculate the mass flow rate of the refrigerant and the displacement volume of the compressor.*

Properties of Freon-12

Temperature (°C)	*Saturation pressure (MPa)*	*Enthalpy (kJ/kg)*		*Specific volume (m³/kg) Saturated vapour*
		Liquid	*Vapour*	
– 10	*0.22*	*26.8*	*183.0*	*0.08*
40	*0.96*	*74.5*	*203.1*	*0.02*

(GATE)

Solution. *Given* : Cooling load = 150 W ; $\eta_{vol.} = 0.8$;
N = 720 r.p.m.

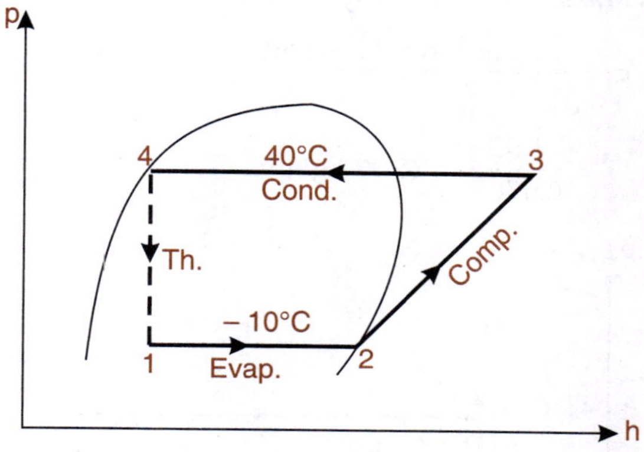

Fig. 26.32

Mass flow rate of the refrigerant ṁ :

Refrigerating effect $= h_2 - h_1$
$= 183 - 74.5 = 108.5$ kJ/kg

Cooling load $= \dot{m} \times (108.5 \times 1000) = 150$

or $\dot{\mathbf{m}} = \frac{150}{108.5 \times 1000}$
$= \mathbf{0.001382\ kJ/s.}$ **(Ans.)**

Displacement volume of the compressor :

Specific volume at entry to compressor,

$v_2 = 0.08$ m³/kg (From table)

∴ Displacement volume of compressor

$$= \frac{\dot{m}v_2}{\eta_{vol.}} = \frac{0.001382 \times 0.08}{0.8}$$
$$= \mathbf{0.0001382\ m^3/s.}$$ **(Ans.)**

Example 26.22. *In a simple vapour compression cycle, following are the properties of the refrigerant R-12 at various points :*

Compressor inlet :	$h_2 = 183.2$ *kJ/kg*
	$v_2 = 0.0767$ *m³/kg*
Compressor discharge :	$h_3 = 222.6$ *kJ/kg*
	$v_3 = 0.0164$ *m³/kg*
Compressor exit :	$h_4 = 84.9$ *kJ/kg*
	$v_4 = 0.00083$ *m³/kg*

The piston displacement volume for compressor is 1.5 litres per stroke and its volumetric efficiency is 80%. The speed of the compressor is 1600 r.p.m.

Find : (i) Power rating of the compressor (kW) ;

(ii) Refrigerating effect (kW). **(GATE 1996)**

Solution. Piston displacement volume

$$= \frac{\pi}{4} d^2 \times l = 1.5 \text{ litres}$$

$$= 1.5 \times 1000 \times 10^{-6} \text{ m}^3\text{/stroke}$$

$$= 0.0015 \text{ m}^3\text{/revolution.}$$

(*i*) Power rating of the compressor (kW) :

Compressor discharge

$$= 0.0015 \times 1600 \times 0.8\ (\eta_{vol.}) = 1.92 \text{ m}^3\text{/min.}$$

Mass flow rate of compressor,

$$m = \frac{\text{Compressor discharge}}{v_2}$$

$$= \frac{1.92}{0.0767} = 25.03 \text{ kg/min.}$$

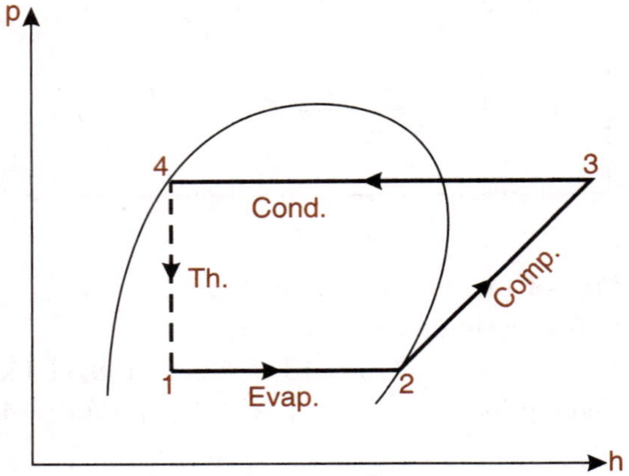

Fig. 26.33

Power rating of the compressor $= \dot{m}(h_3 - h_2)$

$$= \frac{25.03}{60} (222.6 - 183.2) = \mathbf{16.44\ kW.} \quad \textbf{(Ans.)}$$

(*ii*) Refrigerating effect (kW) :

Refrigerating effect $= \dot{m}\,(h_2 - h_1) = \dot{m}\,(h_2 - h_4)$ $\quad (\because\ h_1 = h_4)$

$$= \frac{25.03}{60} (183.2 - 84.9)$$

$$= \mathbf{41\ kW.} \quad \textbf{(Ans.)}$$

Example 26.23. *A refrigerator operating on standard vapour compression cycle has a co-efficiency performance of 6.5 and is driven by a 50 kW compressor. The enthalpies of saturated liquid and saturated vapour refrigerant at the operating condensing temperature of 35°C are 62.55 kJ/kg and 201.45 kJ/kg respectively. The saturated refrigerant vapour leaving evaporator has an enthalpy of 187.53 kJ/kg. Find the refrigerant temperature at compressor discharge. The c_p of refrigerant vapour may be taken to be 0.6155 kJ/kg°C.* **(GATE)**

Solution. *Given* : C.O.P. = 6.5 ; W = 50 kW,

$$h_3' = 201.45 \text{ kJ/kg},$$

$$h_{f_4} = h_1 = 69.55 \text{ kJ/kg} ;\ h_2 = 187.53 \text{ kJ/kg}$$

$$c_p = 0.6155 \text{ kJ/kg K}$$

Temperature, t_3 :

Refrigerating capacity $= 50 \times \text{C.O.P.}$

$= 50 \times 6.5 = 325$ kW

Heat extracted per kg of refrigerant

$= 187.53 - 69.55 = 117.98$ kJ/kg

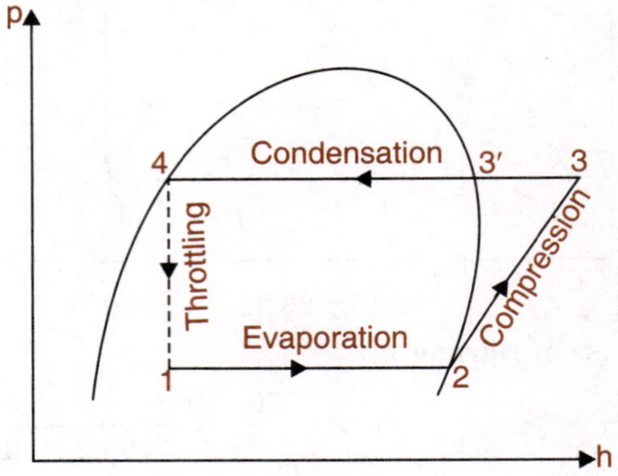

Fig. 26.34

Refrigerant flow rate $= \dfrac{325}{117.98} = 2.755$ kg/s

Compressor power = 50 kW

$\therefore$ Heat input per kg $= \dfrac{50}{2.755} = 18.15$ kJ/kg

Enthalpy of vapour after compression

$= h_2 + 18.15 = 187.53 + 18.15$

$= 205.68$ kJ/kg

Superheat $= 205.68 - h_3' = 205.68 - 201.45$

$= 4.23$ kJ/kg

But $\quad 4.23 = 1 \times c_p\,(t_3 - t_3') = 1 \times 0.6155 \times (t_3 - 35)$

$$\therefore \quad t_3 = \frac{4.23}{0.6155} + 35 = \mathbf{41.87°C.} \quad \textbf{(Ans.)}$$

Note. The compressor rating of 50 kW is assumed to be the enthalpy of compression, in the absence of any data on the efficiency of compressor.

Example 26.24. *A vapour compression heat pump is driven by a power cycle having a thermal efficiency of 25%. For the heat pump, refrigerant-12 is compressed from saturated vapour at 2.0 bar to the condenser pressure of 12 bar. The isentropic efficiency of the compressor is 80%. Saturated liquid enters the expansion valve at 12 bar. For the power* *cycle 80% of the heat rejected by it is transferred to the heated space which has a total heating requirement of 500 kJ / min. Determine the power input to the heat pump compressor. The following data for refrigerant-12 may be used :*

Pressure, bar	*Temperature, °C*	*Enthalpy, kJ / kg*		*Entropy, kJ / kg K*	
		Liquid	*Vapour*	*Liquid*	*Vapour*
2.0	– 12.53	24.57	182.07	0.0992	0.7035
12.0	49.31	84.21	206.24	0.3015	0.6799

Vapour specific heat at constant pressure = 0.7 kJ/kg K. **(U.P.S.C.)**

Solution. Heat rejected by the cycle $= \dfrac{500}{0.8} = 625$ kJ/min.

Assuming isentropic compression of refrigerant, we have

Entropy of dry saturated vapour at 2 bar

= Entropy of superheated vapour at 12 bar

$$0.7035 = 0.6799 + c_p \ln \frac{T}{(49.31 + 273)}$$

$$= 0.6799 + 0.7 \times \ln\left(\frac{T}{322.31}\right)$$

or $$\ln\left(\frac{T}{322.31}\right) = \frac{0.7035 - 0.6799}{0.7} = 0.03371$$

or $$T = 322.31\,(e)^{0.03371} = 333.4 \text{ K}$$

$\therefore$ Enthalpy of superheated vapour at 12 bar

$= 206.24 + 0.7(333.4 - 322.31)$

$= 214$ kJ/kg

Heat rejected per cycle = 214 – 84.21 = 129.88 kJ/kg

Mass flow rate of refrigerant

$$= \frac{625}{129.88} = 4.812 \text{ kg/min}$$

Work done on compressor

$= 4.812\,(214 - 182.07)$

$= 153.65$ kJ/min = 2.56 kW

Actual work of compresson

$$= \frac{2.56}{\eta_{compressor}} = \frac{2.56}{0.8} = 3.2 \text{ kW}$$

Hence power input to the heat pump compressor

= **3.2 kW. (Ans.)**

☞**Example 26.25.** *A food storage locker requires a refrigeration system of 2400 kJ/min. capacity at an evaporator temperature of 263 K and a condenser temperature of 303 K. The refrigerant used is freon-12 and is subcooled by 6°C before entering the expansion valve and vapour is superheated by 7°C before leaving the evaporator coil. The compression of refrigerant is reversible adiabatic. The refrigeration compressor is two-cylinder single-acting with stroke equal to 1.25 times the bore and operates at 1000 r.p.m.*

Properties of Freon-12

Saturation temp, K	*Absolute pressure, bar*	*Specific volume of vapour, m³ / kg*	*Enthalpy, kJ/kg*		*Entropy, kJ/kg K*	
			Liquid	*Vapour*	*Liquid*	*Vapour*
263	2.19	0.0767	26.9	183.2	0.1080	0.7020
303	7.45	0.0235	64.6	199.6	0.2399	0.6854

Take : Liquid specific heat = 1.235 kJ/kg K ;

Vapour specific heat = 0.733 kJ/kg K. Determine :

(i) Refrigerating effect per kg.

(ii) Mass of refrigerant to be circulated per minute.

(iii) Theoretical piston displacement per minute.

(iv) Theoretical power required to run the compressor, in kW.

(v) Heat removed through condenser per min.

(vi) Theoretical bore and stroke of compressor.

Solution. The cycle of refrigeration is represented on *T-s* diagram on Fig. 26.35.

Enthalpy at '2', $\quad h_2 = h_2' + c_p\,(T_2 - T_2')$

From the given table :

$$h_2' = 183.2 \text{ kJ/kg}$$

$(T_2 - T_2')$ = Degree of superheat as the vapour enters the compressor = 7°C

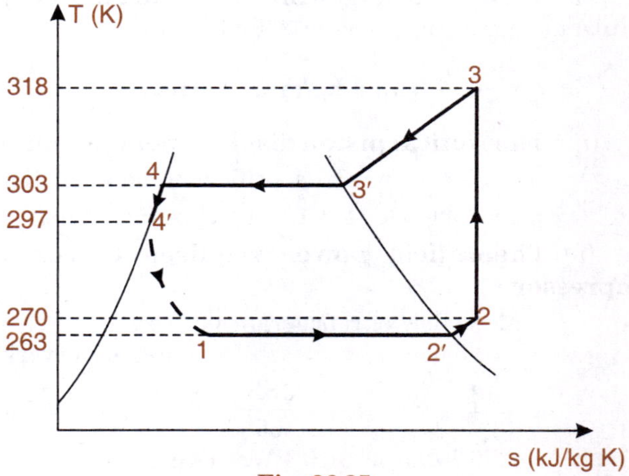

Fig. 26.35

$\therefore \quad h_2 = 183.2 + 0.733 \times 7 = 188.33$ kJ/kg

Also, entropy at '2',

$$s_2 = s_2' + c_p \log_e \frac{T_2}{T_2'}$$

$$= 0.7020 + 0.733 \log_e \left(\frac{270}{263}\right)$$

$$= 0.7212 \text{ kJ/kg K}$$

For isentropic process 2-3

Entropy at '2' = Entropy at '3'

$$0.7212 = s_3' + c_p \log_e \left(\frac{T_3}{T_3'}\right)$$

$$= 0.6854 + 0.733 \log_e \left(\frac{T_3}{303}\right)$$

$$\therefore \quad \log_e \left(\frac{T_3}{303}\right) = 0.0488$$

i.e., $T_3 = 318$ K

Now, enthalpy at '3', $h_3 = h_3' + c_p (T_3 - T_3')$

$= 199.6 + 0.733 (318 - 303) = 210.6$ kJ/kg.

Also, enthalpy at 4′,

$$h_{f_4'} = h_{f_4} - (c_p)_{\text{liquid}} (T_4 - T_4')$$

$$= 64.6 - 1.235 \times 6 = 57.19 \text{ kJ/kg}$$

For the *process 4′-1,*

Enthalpy at 4′ = enthalpy at 1 = 57.19 kJ/kg

For specific volume at 2,

$$\frac{v_2'}{T_2'} = \frac{v_2}{T_2}$$

$$\therefore \quad v_2 = \frac{v_2'}{T_2'} \times T_2 = 0.0767 \times \frac{270}{263} = 0.07874 \text{ m}^3\text{/kg}$$

(*i*) **Refrigerating effect per kg**

$= h_2 - h_1 = 188.33 - 57.19 =$ **131.14 kJ/kg. (Ans.)**

(*ii*) **Mass of refrigerant to be circulated per minute** for producing effect of 2400 kJ/min.

$= \frac{2400}{131.14} =$ **18.3 kg/min. (Ans.)**

(*iii*) **Theoretical piston displacement per minute**

= Mass flow/min. × specific volume at suction

$= 18.3 \times 0.07874 =$ **1.441 m³/min. (Ans.)**

(*iv*) **Theoretical power required to run the compressor**

= Mass flow of refrigerant per sec. × compressor work/kg

$= \frac{18.3}{60} \times (h_3 - h_2) = \frac{18.3}{60} (210.6 - 188.33)$ kJ/s

= **6.79 kJ/s** or **6.79 kW. (Ans.)**

(*v*) **Heat removed through the condenser per min.**

= Mass flow of refrigerant × heat removed per kg of refrigerant

$= 18.3 (h_3 - h_{f_4'}) = 18.3 (210.6 - 57.19)$

= **2807.4 kJ/min. (Ans.)**

(*vi*) **Theoretical bore (d) and stroke (l) :**

Theoretical piston displacement per cylinder

$$= \frac{\text{Total displacement per minute}}{\text{Number of cylinder}} = \frac{1.441}{2}$$

$= 0.7205$ m³/min.

Also, length of stroke = 1.25 × diameter of piston

Hence, $0.7205 = \pi/4 \, d^2 \times (1.25 \, d) \times 1000$

i.e., $d =$ **0.09 m** or **90 mm. (Ans.)**

and $l = 1.25 \, d = 1.25 \times 90 =$ **112.5 mm. (Ans.)**

☞**Example 26.26.** *A refrigeration system of 10.5 tonnes capacity at an evaporator temperature of – 12°C and a condenser temperature of 27°C is needed in a food storage locker. The refrigerant ammonia is sub-cooled by 6°C before entering the expansion valve. The vapour is 0.95 dry as it leaves the evaporator coil. The compression in the compressor is of adiabatic type.*

Using p-h chart find :

(*i*) *Condition of volume at outlet of the compressor*

(*ii*) *Condition of vapour at entrance to evaporator*

(*iii*) *C.O.P.*

(*iv*) *Power required, in kW.*

Neglect valve throttling and clearance effect.

Solution. Refer to Fig. 26.36.

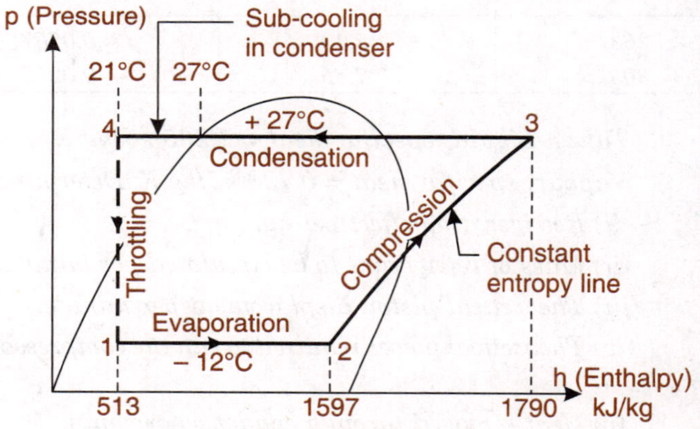

Fig. 26.36

Using *p-h* chart for ammonia,

- Locate point '2' where – 12°C cuts 0.95 dryness fraction line.
- From point '2' move along constant entropy line and locate point '3' where it cuts constant pressure line corresponding to + 27°C temperature.

- From point '3' follow constant pressure line till it cuts + 21°C temperature line to get point '4'.
- From point '4' drop a vertical line to cut constant pressure line corresponding to – 12°C and get the point '5'.

The values as read from the chart are :

$$h_2 = 1597 \text{ kJ/kg}$$
$$h_3 = 1790 \text{ kJ/kg}$$
$$h_4 = h_1 = 513 \text{ kJ/kg}$$
$$t_3 = 58°\text{C}$$
$$x_1 = 0.13.$$

(*i*) **Condition of the vapour at the outlet of the compressor**

= 58 – 27 = **31°C superheat. (Ans.)**

(*ii*) **Condition of vapour at entrance to evaporator,**

x_1 = **0.13. (Ans.)**

(*iii*) **C.O.P.** $= \dfrac{h_2 - h_1}{h_3 - h_2} = \dfrac{1597 - 513}{1790 - 1597}$

= **5.6. (Ans.)**

(*iv*) **Power required :**

$$\text{C.O.P.} = \frac{\text{Net refrigerating effect}}{\text{Work done}} = \frac{R_n}{W}$$

$$5.6 = \frac{10.5 \times 14000}{W \times 60}$$

$$\therefore \quad W = \frac{10.5 \times 14000}{5.6 \times 60} \text{ kJ/min} = 437.5 \text{ kJ/min.}$$

= 7.29 kJ/s.

i.e., **Power required = 7.29 kW. (Ans.)**

☞**Example 26.27.** *The evaporator and condenser temperatures of 20 tonnes capacity freezer are – 28°C and 23°C respectively. The refrigerant – 22 is sub-cooled by 3°C before it enters the expansion valve and is superheated to 8°C before leaving the evaporator. The compression is isentropic. A six-cylinder single-acting compressor with stroke equal to bore running at 250 r.p.m. is used. Determine :*

(*i*) *Refrigerating effect/kg.*
(*ii*) *Mass of refrigerant to be circulated per minute.*
(*iii*) *Theoretical piston displacement per minute.*
(*iv*) *Theoretical power.*
(*v*) *C.O.P.*
(*vi*) *Heat removed through condenser.*
(*vii*) *Theoretical bore and stroke of the compressor.*

Neglect valve throttling and clearance effect.

Solution. Refer to Fig. 26.37. Following the procedure as given in the previous example plot the points 1, 2, 3 and 4 on *p-h* chart for freon-22. The following values are obtained :

$$h_2 = 615 \text{ kJ/kg}$$
$$h_3 = 664 \text{ kJ/kg}$$
$$h_4 = h_1 = 446 \text{ kJ/kg}$$
$$v_2 = 0.14 \text{ m}^3\text{/kg.}$$

(*i*) **Refrigerating effect per kg**

$= h_2 - h_1 = 615 - 446 =$ **169 kJ/kg. (Ans.)**

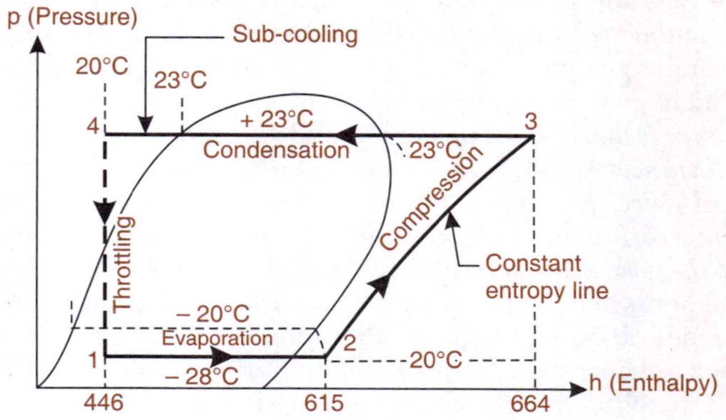

Fig. 26.37

(*ii*) **Mass of refrigerant to be circulated per minute,**

$$m = \frac{20 \times 14000}{169 \times 60} = \textbf{27.6 kg/min.} \quad \textbf{(Ans.)}$$

(*iii*) **Theoretical piston displacement**

= Specific volume at suction × Mass of refrigerant used/min

= 0.14 × 27.6 = **3.864 m³/min (Ans.)**

(*iv*) **Theoretical power**

$$= m \times (h_3 - h_2) = \frac{27.6}{60}(664 - 615) = 22.54 \text{ kJ/s}$$

= **22.54 kW. (Ans.)**

(*v*) **C.O.P.** $= \dfrac{h_2 - h_1}{h_3 - h_2} = \dfrac{615 - 446}{664 - 615} =$ **3.45. (Ans.)**

(*vi*) **Heat removed through the condenser**

$= m\,(h_3 - h_4) = 27.6\,(664 - 446)$

= **6016.8 kJ/min. (Ans.)**

(*vii*) **Theoretical displacement per minute per cylinder**

$$= \frac{\text{Total displacement/ min.}}{\text{Number of cylinders}} = \frac{3.864}{6}$$

= 0.644 m³/min

Let diameter of the cylinder = d

Then, stroke length, $l = d$

Now, $$\frac{\pi}{4} d^2 \times l = \frac{0.644}{950}$$

or $$\frac{\pi}{4} d^2 \times d = \frac{0.644}{950}$$

i.e., **d** = 0.0952 m or **95.2 mm. (Ans.)**

and **l = 95.2 mm. (Ans.)**

26.5. REFRIGERANTS

A ***'refrigerant'*** *is defined as any substance that absorbs heat through expansion or vaporisation and loses it through condensation in a refrigeration system.* The term 'refrigerant' in the broadest sense is also applied to such *secondary cooling mediums* as cold water or brine, solutions. Usually refrigerants include only those working mediums which pass through the cycle of *evaporation, recovery, compression, condensation* and *liquification. These substances absorb heat at one place at low temperature level and reject the same at some other place having higher temperature and pressure. The rejection of heat takes place at the cost of some mechanical work.* Thus circulating cold mediums and cooling mediums (such as ice and solid carbon dioxide) are not primary refrigerants. In the early days only four refrigerants, *Air, ammonia* (NH_3), *Carbon dioxide* (CO_2), *Sulphur dioxide* (SO_2), possessing chemical, physical and thermodynamic properties permitting their efficient application and service in the practical design of refrigeration equipment were used. All the refrigerants change from liquid state to vapour state during the process.

26.5.1. Classification of Refrigerants

The refrigerants are classified as follows :

1. Primary refrigerants.
2. Secondary refrigerants.

1. **Primary refrigerants** *are those working mediums or heat carriers which directly take part in the refrigeration system and cool the substance by the absorption of latent heat, e.g., Ammonia, Carbon dioxide, Sulphur dioxide, Methyl chloride, Methylene chloride, Ethyl chloride and Freon group etc.*

2. **Secondary refrigerants** *are those circulating substances which are first cooled with the help of the primary refrigerants and are then employed for cooling purposes, e.g., ice, solid carbon dioxide* etc. These refrigerants cool substances by absorption of their sensible heat.

The primary refrigerants are grouped as follows :

(*i*) **Halocarbon compounds.** In 1928, Charles Kettening and Dr. Thomas Mighey invented and developed this group of refrigerant. In this group are included refrigerants which contain one or more of three halogens, chlorine and bromine and they are sold in the market under the names as *Freon, Genetron, Isotron*, and *Areton*. Since the refrigerants belonging to this group have outstanding merits over the other group's refrigerants, therefore they find wide field of application in domestic, commercial and industrial purposes.

The list of the halocarbon-refrigerants commonly used is given below :

R-10 — Carbon tetrachloride (CCl_4)
R-11 — Trichloro-monofluoro methane (CCl_3F)
R-12 — Dichloro-difluoro methane (CCl_2F_2)
R-13 — Mono-bromotrifluoro methane ($CBrF_3$)
R-21 — Dichloro monofluoro methane ($CHCl_2F$)
R-22 — Mono chloro difluoro methane ($CHClF_2$)
R-30 — Methylene-chloride (CH_2Cl_2)
R-40 — Methyle chloride (CH_3Cl)
R-41 — Methyle fluoride (CH_3F)
R-100 — Ethyl chloride (C_2H_5Cl)
R-113 — Trichloro trifluoroethane ($C_2F_3Cl_3$)
R-114 — Tetra-fluoro dichloroethane ($Cl_2F_4Cl_2$)
R-152 — Difluoro-ethane ($C_2H_6F_2$)

(*ii*) **Azeotropes.** The refrigerants belonging to this group consists of mixtures of different substances. These substances cannot be separated into components by distillations. They possess fixed thermodynamic properties and do not undergo any separation with changes in temperature and pressure. An azeotrope behaves like a simple substance.

Example. R-500. It contains 73.8% of (R-12) and 26.2% of (R-152).

(*iii*) **Hydrocarbons.** Most of the refrigerants of this group are organic compounds. Several hydrocarbons are used successfully in commercial and industrial installations. Most of them possess satisfactory thermodynamic properties but are highly inflammable. Some of the important refrigerants of this group are :

R-50 — Methane (CH_4)
R-170 — Ethane (C_2H_6)
R-290 — Propane (C_2H_8)
R-600 — Butane (C_4H_{10})
R-601 — Isobutane [$CH(CH_3)_3$]

(*iv*) **Inorganic compounds.** Before the introduction of hydrocarbon group these refrigerants were most commonly used for all purposes.

The important refrigerants of this group are :

R-717 — Ammonia (NH_3)
R-718 — Water (H_2O)
R-729 — Air (mixture of O_2, N_2, CO_2 etc.)
R-744 — Carbon dioxide (CO_2)
R-764 — Sulphur dioxide (SO_2)

(*v*) **Unsaturated organic compound.** The refrigerants belonging to this group possess ethylene or propylene as their constituents. They are :

R-1120 — Trichloroethylene ($C_3H_4Cl_3$)
R-1130 — Dichloroethylene ($C_2H_4Cl_2$)
R-1150 — Ethylene (C_3H_6)
R-1270 — Propylene.

26.5.2. Desirable Properties of an Ideal Refrigerant

An ideal refrigerant should possess the following properties :

1. **Thermodynamic properties :**

(*i*) Low boiling point

(*ii*) Low freezing point

(*iii*) Positive pressures (but not very high) in condenser and evaporator.

(*iv*) High saturation temperature

(*v*) High latent heat of vaporisation.

2. **Chemical properties :**

(*i*) Non-toxicity

(*ii*) Non-flammable and non-explosive

(*iii*) Non-corrosiveness

(*iv*) Chemical stability in reacting

(*v*) No effect on the quality of stored (food and other) products like flowers, with other materials *i.e.*, furs and fabrics.

(*vi*) Non-irritating and odourless.

3. **Physical properties :**

(*i*) Low specific volume of vapour

(*ii*) Low specific heat

(*iii*) High thermal conductivity

(*iv*) Low viscosity

(*v*) High electrical insulation.

4. **Other properties :**

(*i*) Ease of leakage location

(*ii*) Availability and low cost

(*iii*) Ease of handling

(*iv*) High C.O.P.

(*v*) Low power consumption per tonne of refrigeration.

(*vi*) Low pressure ratio and pressure difference.

Some important properties (mentioned above) are discussed below :

Freezing point. As the refrigerant must operate in the cycle above its freezing point, it is evident that the same for the refrigerant *must be lower than system temperatures.* It is found that except in the case of water for which the freezing point is 0°C, other refrigerants have reasonably low values. Water, therefore, can be used only in air-conditioning applications which are above 0°C.

Condenser and evaporator pressures. The evaporating pressure should be as near atmospheric as possible. If it is *too low,* it would result in *a large volume of the suction vapour.* If it is *too high, overall high pressures including condenser pressure would result necessitating stronger equipment and consequently higher cost. A positive pressure is required in order to eliminate the possibility of the entry of air and moisture into the system. The normal boiling point of the refrigerant should, therefore, be lower than the refrigerant temperature.*

Critical temperature and pressure. Generally, for high C.O.P. the *critical temperature should be very high so that the condenser temperature line on p-h diagram is far removed from the critical point.* This ensures reasonable refrigerating effect as it is very small with the state of liquid before expansion near the critical point.

The *critical pressure should be low so as to give low condensing pressure.*

Latent heat of vaporisation. It should be *as large as possible to reduce the weight of the refrigerant to be circulated in the system. This reduces initial cost of the refrigerant. The size of the system will also be small and hence low initial cost.*

Toxicity. Taking into consideration comparative hazard to life due to gases and vapours underwriters Laboratories have divided the compounds into *six groups.* Group six contains compounds with a very low degree of toxicity. It includes R_{12}, R_{114}, R_{13}, etc. Group one, at the other end of the scale, includes the most toxic substances such as SO_2.

Ammonia is not used in comfort air-conditioning and in domestic refrigeration because of inflammability and toxicity.

Inflammability. Hydrocarbons (*e.g.*, methane, ethane etc.) are highly explosive and inflammable. Fluorocarbons are neither explosive nor inflammable. *Ammonia is explosive* in a mixture with air in concentration of 16 to 25% by volume of ammonia.

Volume of suction vapour. The *size of the compressor depends on the volume of suction vapour per unit* (say per tonne) *of refrigeration. Reciprocating compressors are used with refrigerants with high pressures and small volumes of the suction vapour. Centrifugal or turbo-compressors are used with refrigerants with low pressures and large volumes of the suction vapour.* A high volume flow rate for a given capacity is required for centrifugal compressors to permit flow passages of sufficient width to minimise drag and obtain high efficiency.

Thermal conductivity. *For a high heat transfer co-efficient a high thermal conductivity is desirable.* R_{22} has better heat transfer characteristics than R_{12} ; R_{21} is still better, R_{13} has poor heat transfer characteristics.

Viscosity. For *a high heat transfer co-efficient a low viscosity is desirable.*

Leak tendency. The *refrigerants should have low leak tendency.* The greatest drawback of *fluorocarbons* is the fact that they are *odourless.* This, at times, results in a complete loss of costly gas from leaks without being detected.

An ammonia leak can be very easily detected by pungent odour.

Refrigerant cost. The cost factor is only relevant to the extent of the price of the initial charge of the refrigerant which is very small compared to the total cost of the plant and its installation. The cost of losses due to leakage is also important. In small-capacity units requiring only a small charge of the refrigerant, the cost of refrigerant is immaterial.

The *cheapest refrigerant is Ammonia.* R_{12} *is slightly cheaper than* R_{22}. R_{12} and R_{22} have replaced ammonia in the dairy and frozen food industry (and even in cold storages) *because of the tendency of ammonia to attack some food products.*

Co-efficient of performance and power per tonne. Practically all common refrigerants have approximately same C.O.P. and power requirement.

Table 26.1 gives the values of C.O.P. for some important refrigerants.

Table 26.1. C.O.P. of some important refrigerants

S. No.	*Refrigerant*	*C.O.P.*
1.	Carnot value	5.74
2.	R_{11}	5.09
3.	R_{113}	4.92
4.	Ammonia	4.76
5.	R_{12}	4.70
6.	R_{22}	4.66
7.	R_{144}	4.49
	CO_2	2.56

Action with oil. No chemical reaction between refrigerant and lubricating oil of the compressor should take place. Miscibility of the oil is quite important as some oil should be carried out of the compressor crankcase with the hot refrigerant vapour to lubricate the pistons and discharge valves properly.

Reaction with materials of construction. While selecting a material to contain the refrigerant this material should be given a due consideration. Some metals are attacked by the refrigerants ; *e.g., ammonia reacts with copper, brass or other cuprous alloys in the presence of water*, therefore in *ammonia systems the common metals used are iron and steel. Freon group* does not react with steel, copper, brass, zinc, tin and aluminium but is corrosive to magnesium and aluminium having magnesium more than 2%. Freon group refrigerants tend to dissolve natural rubber in packing and gaskets but synthetic rubber such as neoprene are entirely suitable. The hydrogenerated hydrocarbons may react with zinc but not with copper, aluminium, iron and steel.

26.5.3. Properties and Uses of Commonly Used Refrigerants

1. Air

Properties :

(*i*) No cost involved ; easily available.

(*ii*) Completely non-toxic.

(*iii*) Completely safe.

(*iv*) The C.O.P. of air cycle operating between temperatures of 80°C and – 15°C is 1.67.

Uses :

(*i*) Air is one of the earliest refrigerants and was widely used even as late as World War I wherever a completely non-toxic medium was needed.

(*ii*) Because of low C.O.P., it is used only where *operating efficiency is secondary* as in *air-craft refrigeration.*

2. Ammonia (NH_3)

Properties :

(*i*) It is highly toxic and flammable.

(*ii*) It has the excellent thermal properties.

(*iii*) It has the *highest refrigerating effect per kg of refrigerant.*

(*iv*) Low volumetric displacement.

(*v*) Low cost.

(*vi*) Low weight of liquid circulated per tonne of refrigeration.

(*vii*) High efficiency.

(*viii*) The evaporator and condenser pressures are 3.5 bar abs. and 13 bar abs. (app.) respectively at standard conditions of – 15°C and 30°C.

Uses :

(*i*) It is widely used in large industrial and commercial reciprocating compression systems where high toxicity is secondary.
It is extensively used in *ice plants, packing plants, large cold storages and skating rinks* etc.

(*ii*) It is widely used as the refrigerant in *absorption systems.*

The following points are worth noting :

- Ammonia *should never be used with copper, brass and other copper alloys ; iron and steel should be* used in *ammonia systems instead.*
- In ammonia systems, to *detect the leakage a sulphur candle is used which gives off a dense white smoke when ammonia vapour is present.*

3. Sulphur dioxide (SO_2)

Properties :

(*i*) It is a colourless gas or liquid.

(*ii*) It is extremely toxic and has a pungent irritating odour.

(*iii*) It is non-explosive and non-flammable.

(*iv*) It has a liquid specific gravity of 1.36.

(*v*) Works at low pressures.

(*vi*) Possesses small latent heat of vaporisation.

Uses :

It finds little use these days. However, its use was made in small machines in early days.

• *The leakage of sulphur dioxide may be detected by bringing aqueous ammonia near the leak, this gives off a white smoke.*

4. Carbon dioxide (CO_2)

Properties :

(*i*) It is a colourless and odourless gas, and is heavier than air.

(*ii*) It has liquid specific gravity of 1.56.

(*iii*) It is non-toxic and non-flammable.

(*iv*) It is non-explosive and non-corrosive.

(*v*) It has *extremely high operating pressures.*

(*vi*) It gives *very low refrigerating effect.*

Uses :

This refrigerant has received only limited use because of the high power requirements per tonne of refrigeration and the high operating pressures. In *former years* it was selected for *marine refrigeration,* for *theater air-conditioning systems,* and for *hotel and institutional refrigeration* instead of ammonia because it is non-toxic.

At the present-time its use is limited primarily to the *manufacture of dry ice* (solid carbon dioxide).

• The *leak detection of CO_2* is done by *soap solution.*

5. Methyl chloride (CH_3Cl)

Properties :

(*i*) It is a colourless liquid with a faint, sweet, non-irritating odour.

(*ii*) It has liquid specific gravity of 1.002 at atmospheric pressure.

(*iii*) It is neither flammable nor toxic.

Uses :

It has been used in the past in both domestic and commercial applications. It should never be used with aluminium.

6. R-11 (Trichloro monofluoro methane)

Properties :

(*i*) It is composed of one carbon, three chlorine and one fluorine atoms (or parts by weight) and is *non-corrosive, non-toxic* and *non-flammable.*

(*ii*) It dissolves natural rubber.

(*iii*) It has a boiling point of – 24°C.

(*iv*) It mixes completely with mineral lubricating oil under all conditions.

Uses :

It is employed for 50 tonnes capacity and over in small office buildings and factories. A centrifugal compressor is used in the plants employing this refrigerant.

• *Its leakage is detected by a halide torch.*

7. R-12 (Dichloro-difluoro methane) or Freon-12

Properties :

(*i*) It is *non-toxic, non-flammable,* and *non-explosive,* therefore it is *most suitable refrigerant.*

(*ii*) It is fully oil miscible therefore it simplifies the problem of oil return.

(*iii*) The operating pressures of R-12 in evaporator and condenser under *standard tonne of refrigeration* are 1.9 bar abs. and 7.6 bar abs. (app.).

(*iv*) Its latent heat at – 15°C is 161.6 kJ/kg.

(*v*) C.O.P. = 4.61.

(*vi*) It does not break even under the extreme operating conditions.

(*vii*) It condenses at moderate pressure and under atmospheric conditions.

Uses :

1. It is suitable for high, medium and low temperature applications.

2. It is used for domestic applications.

3. It is *excellent electric insulator therefore it is universally used in sealed type compressors.*

8. R-22 (Monochloro-difluro methane) or Freon-22

R-22 refrigerant is superior to R-12 in many respects. It has the following properties and uses :

Properties :

(*i*) The compressor displacement per tonne of refrigeration with R-22 is 60% less than the compressor displacement with R-12 as refrigerant.

(*ii*) R-22 is miscible with oil at condenser temperature but tries to separate at evaporator temperature when the system is used for very low temperature applications (– 90°C). Oil separators must be incorporated to return the oil from the evaporator when the system is used for such low temperature applications.

(*iii*) The pressures in the evaporator and condenser at standard tonne of refrigeration are 2.9 bar abs. and 11.9 bar abs. (app.).

(*iv*) The latent heat at – 15°C is low and is 89 kJ/kg.

The *major disadvantage of R-22 compared with R-12 is the high discharge temperature which requires water cooling of the compressor head and cylinder.*

Uses :

R-22 is universally used in commercial and industrial low temperature systems.

HIGHLIGHTS

1. Refrigeration is the science of producing and maintaining temperatures below that of the surrounding atmosphere.
2. Refrigeration is generally produced in one of the following three ways :
   (*i*) By melting a solid ;
   (*ii*) By sublimation of a solid ;
   (*iii*) By evaporation of a liquid.
3. Co-efficient of performance (C.O.P.) is defined as the ratio of heat absorbed by the refrigerant while passing through the evaporator to the work input required to compress the refrigerant in the compressor ; in short it is the ratio between heat extracted and work done (in heat units).
4. Relative C.O.P. $= \dfrac{\text{Actual C.O.P}}{\text{Theoretical C.O.P}}$.
5. 1 tonne of refrigeration = 14000 kJ/h.
6. The main characteristic feature of air refrigeration system is that throughout the cycle the refrigerant remains in *gaseous state*.
   The air refrigeration system may be of two types :
   (*i*) Closed system and (*ii*) Open system.
7. Co-efficient of performance of a 'refrigerator' working on a reversal Carnot cycle
   $$= \frac{T_2}{T_1 - T_2}$$
   For a Carnot cycle 'heat pump' C.O.P. $= \dfrac{T_1}{T_1 - T_2}$
   For a Carnot cycle 'heat engine' C.O.P. $= \dfrac{T_1 - T_2}{T_1}$.
8. For air refrigeration system working on reversed Brayton cycle.
   $$\text{C.O.P.} = \frac{(T_3 - T_2)}{\left(\frac{n}{n-1}\right)\left(\frac{\gamma - 1}{\gamma}\right)(T_4 - T_3 + T_2 - 1)} .$$
9. The following air refrigeration systems are used in aeroplanes :
   (*i*) Simple cooling system (*ii*) Boot strap system
   (*iii*) Regenerative cooling system.
10. In a simple vapour compression cycle the following processes are completed :
    (*i*) Compression (*ii*) Condensation
    (*iii*) Expansion (*iv*) Vaporisation.
11. The various parts of a simple vapour compression cycle are : Compressor, Discharge line (or hot gas line), Condenser, Receiver tank, Liquid line, Expansion valve, Evaporator and Suction line.
12. If the vapour is not superheated after compression, the operation is called 'Wet compression' and if the vapour is superheated at the end of compression, it is known as 'Dry compression'. Dry compression, in actual practice is always preferred as it gives higher volumetric efficiency and mechanical efficiency and there are less chances of compressor damage.
13. *p-h* chart gives directly the changes in enthalpy and pressure during a process for thermodynamic analysis.
14. When suction pressure is decreased, the refrigerating effect is decreased and work required is increased. The net effect is to reduce the refrigerating capacity of the system and the C.O.P.
15. The overall effect of superheating is to give a low value of C.O.P.
16. 'Sub-cooling' results in increase of C.O.P. provided that no further energy has to be spent to obtain the extra cold coolant required.
17. The refrigerating system should always be designed to operate at the highest possible vaporising temperature and lowest possible condensing temperature, of course, keeping in view the requirements of the application.
18. 'Volumetric efficiency' is defined as the ratio of the actual volume of gas drawn into the compressor (at evaporator temperature and pressure) on each stroke to the piston displacement. If the effect of clearance alone is considered, the resulting expression may be termed 'clearance volumetric efficiency'. The expression used for grouping into one constant all the factors affecting efficiency may be termed 'total volumetric efficiency'.
19. Clearance volumetric efficiency, $\eta_{cv} = 1 + C - C\left(\dfrac{p_d}{p_s}\right)^{1/n}$
    where, $C = \dfrac{\text{Clearance volume}}{\text{Swept volume}}$
    p_d = Displacement pressure
    p_s = Suction pressure.
20. Total volumetric efficiency,
    $$\eta_{tv} = \left[1 + C - C\left(\frac{p_d}{p_s}\right)^{1/n}\right] \times \frac{p_c}{p_s} \times \frac{T_s}{T_c}$$
    where subscript '*c*' refers to compressor cylinder and '*s*' refers to the evaporator on the suction line just adjacent to the compressor.

OBJECTIVE TYPE QUESTIONS

Fill in the blanks :

1. ...... means the cooling of or removal of heat from a system.
2. Most of the commercial refrigeration is produced by the evaporation of a liquid ......
3. ...... is the ratio between the heat extracted and the work done.
4. ...... $= \dfrac{\text{Actual C.O.P}}{\text{Theoretical C.O.P.}}$.
5. The C.O.P. for Carnot refrigerator is equal to ......
6. The C.O.P. for a Carnot heat pump is equal to .....
7. The C.O.P. for a Carnot refrigerator is ...... than that of Carnot heat pump.

8. The C.O.P. of an air refrigeration system is ...... than a vapour compression system.
9. In a refrigeration system the heat rejected at higher temperature = ... + ...
10. Out of all the refrigeration systems, the ...... system is the most important system from the stand point of commercial and domestic utility.
11. The function of a ...... is to remove the vapour from the evaporator and to raise its temperature and pressure to a point such that it (vapour) can be condensed with normally available condensing media.
12. The function of a ...... is to provide a heat transfer surface through which a heat passes from the hot refrigerant vapour to the condensing medium.
13. The function of ...... is to meter the proper amount of refrigerant to the evaporator and to reduce the pressure of liquid entering the evaporator so that liquid will vaporise in the evaporator at the desired low temperature.
14. ...... provides a heat transfer surface through which heat can pass from the refrigerated space or product into the vaporising refrigerant.
15. If the vapour is not superheated after compression, the operation is called .....
16. If the vapour is superheated at the end of compression, the operation is called ......
17. When the suction pressure decreases the refrigerating effect and C.O.P. are ......
18. ...... results in increase of C.O.P. provided that no further energy has to be spent to obtain the extra cold coolant required.
19. ...... efficiency is defined as the ratio of actual volume of gas drawn into the compressor (at evaporator temperature and pressure) on each stroke to the piston displacement.

ANSWERS

1. Refrigeration 2. Refrigerant 3. C.O.P.
4. Relative C.O.P. 5. $\frac{T_2}{T_1 - T_2}$ 6. $\frac{T_1}{T_1 - T_2}$
7. Less 8. Less
9. Refrigeration effect + work done
10. Vapour compression 11. Compressor
12. Condenser
13. Expansion valve 14. Evaporator
15. Wet compression 16. Dry compression 17. Reduced
18. Sub-cooling 19. Volumetric.

THEORETICAL QUESTIONS

1. Define the following :
   (*i*) Refrigeration (*ii*) Refrigerating system
   (*iii*) Refrigerated system.
2. Enumerate different ways of producing refrigeration.
3. Enumerate important refrigeration applications.
4. State elements of refrigeration systems.
5. Enumerate systems of refrigeration.
6. Define the following :
   (*i*) Actual C.O.P. (*ii*) Theoretical C.O.P.
   (*iii*) Relative C.O.P.
7. What is a standard rating of a refrigeration machine ?
8. What is main characteristic feature of an air refrigeration system ?
9. Differentiate clearly between open and closed air refrigeration systems.
10. Explain briefly an air refrigerator working on a reversed Carnot cycle. Derive expression for its C.O.P.
11. Derive an expression for C.O.P. for an air refrigeration system working on reversed Brayton cycle.
12. State merits and demerits of an air refrigeration system.
13. Describe a simple vapour compression cycle giving clearly its flow diagram.
14. State merits and demerits of 'Vapour compression system' over 'Air refrigeration system'.
15. State the functions of the following parts of a simple vapour compression system :
   (*i*) Compressor, (*ii*) Condenser,
   (*iii*) Expansion valve, and (*iv*) Evaporator.
16. Show the vapour compression cycle on 'Temperature-Entropy' (*T-s*) diagram for the following cases :
   (*i*) When the vapour is dry and saturated at the end of compression.
   (*ii*) When the vapour is superheated after compression.
   (*iii*) When the vapour is wet after compression.
17. What is the difference between 'Wet compression' and 'Dry compression' ?
18. Write a short note on 'Pressure Enthalpy (*p-h*) chart'.
19. Show the simple vapour compression cycle on a *p-h* chart.
20. Discuss the effect of the following on the performance of a vapour compression system :
   (*i*) Effect of suction pressure
   (*ii*) Effect of delivery pressure
   (*iii*) Effect of superheating
   (*iv*) Effect of sub-cooling of liquid
   (*v*) Effect of suction temperature and condenser temperature.
21. Show with the help of diagrams, the difference between theoretical and actual vapour compression cycles.
22. Define the terms 'Volumetric efficiency' and 'Clearance volumetric efficiency'.
23. Derive an expression for 'Clearance volumetric efficiency'.
24. Explain briefly the term 'Total volumetric efficiency'.
25. Explain briefly simple vapour absorption system.
26. Give the comparison between a vapour compression system and a vapour absorption system.

UNSOLVED EXAMPLES

1. The co-efficient of performance of a Carnot refrigerator, when it extracts 8350 kJ/min from a heat source, is 5. Find power required to run the compressor.
[**Ans.** 27.83 kW]

2. A reversed cycle has refrigerating C.O.P. of 4,

(*i*) Determine the ratio T_1/T_2 ; and

(*ii*) If this cycle is used as heat pump, determine the C.O.P. and heat delivered.

[**Ans.** (*i*) 1.25 ; (*ii*) 50 kW, 5]

3. An ice plant produces 10 tonnes of ice per day at 0°C, using water at room temperature of 20°C. Estimate the power rating of the compressor motor if the C.O.P. of the plant is 2.5 and overall electromechanical efficiency is 0.9.

Take latent heat of freezing for water = 335 kJ/kg

Specific heat of water = 4.18 kJ/kg. [**Ans.** 21.44 kW]

4. An air refrigeration system operating on Bell Coleman cycle, takes in air from cold room at 268 K and compresses it from 1.0 bar to 5.5 bar. The index of compression being 1.25. The compressed air is cooled to 300 K. The ambient temperature is 20°C. Air expands in an expander where the index of expansion is 1.35.

Calculate : (*i*) C.O.P. of the system (*ii*) Quantity of air circulated per minute for production of 1500 kg of ice per day at 0°C from water at 20°C. (*iii*) Capacity of the plant in terms of kJ/s.

Take c_p = 4.18 kJ/kg K for water, c_p = 1.005 kJ/kg K for air

Latent heat of ice = 335 kJ/kg.

[**Ans.** (*i*) 1.974 ; (*ii*) 5.814 kg/min ; (*iii*) 7.27 kJ/s]

5. The temperature in a refrigerator coil is 267 K and that in the condenser coil is 295 K. Assuming that the machine operates on the reversed Carnot cycle, calculate :

(*i*) C.O.P.$_{\text{(ref.)}}$

(*ii*) The refrigerating effect per kW of input work.

(*iii*) The heat rejected to the condenser.

[**Ans.** (*i*) 9.54 ; (*ii*) 9.54 kW ; (*iii*) 10.54 kW]

6. An ammonia vapour-compression refrigerator operates between an evaporator pressure of 2.077 bar and a condenser pressure of 12.37 bar. The following cycles are to be compared ; in each case there is no undercooling in the condenser, and isentropic compression may be assumed :

(*i*) The vapour has a dryness fraction of 0.9 at entry to the compressor.

(*ii*) The vapour is dry saturated at entry to the compressor.

(*iii*) The vapour has 5 K of superheat at entry to the compressor.

In each case calculate the C.O.P.$_{\text{(ref.)}}$ and the refrigerating effect per kg.

What would be the C.O.P.$_{\text{(ref.)}}$ of a reversed Carnot cycle operating between the same saturation temperatures ?

[**Ans.** 4.5 ; 957.5 kJ/kg ; 4.13 ; 1089.9 kJ/kg ; 4.1 ; 1101.4 kJ/kg]

7. A refrigerator using Freon-12 operates between saturation temperatures of – 10°C and 60°C, at which temperatures the latent heats are 156.32 kJ/kg and 113.52 kJ/kg respectively. The refrigerant is dry saturated at entry to the compressor and the liquid is not undercooled in the condenser. The specific heat of liquid freon is 0.970 kJ/kg K and that of the superheated freon vapour is 0.865 kJ/kg K. The vapour is compressed isentropically in the compressor. Using no other information than that given, calculate the temperature at the compressor delivery, and the refrigerating effect per kg of Freon. [**Ans.** 69.6°C ; 88.42 kJ/kg]

8. A heat pump using ammonia as the refrigerant operates between saturation temperatures of 6°C and 38°C. The refrigerant is compressed isentropically from dry saturation and there is 6 K of undercooling in the condenser. Calculate :

(*i*) C.O.P.$_{\text{(heat pump)}}$

(*ii*) The mass flow of refrigerant

(*iii*) The heat available per kilowatt input.

[**Ans.** (*i*) 8.8 ; (*ii*) 25.06 kg/h ; (*iii*) 8.8 kW]

9. An ammonia vapour-compression refrigerator has a single-stage, single-acting reciprocating compressor which has a bore of 127 mm, a stroke of 152 mm and a speed of 240 r.p.m. The pressure in the evaporator is 1.588 bar and that in the condenser is 13.89 bar. The volumetric efficiency of the compressor is 80% and its mechanical efficiency is 90%. The vapour is dry saturated on leaving the evaporator and the liquid leaves the condenser at 32°C. Calculate the mass flow of refrigerant, the refrigerating effect, and the power ideally required to drive the compressor.

[**Ans.** 0.502 kg/min ; 9.04 kW ; 2.73 kW]

10. An ammonia refrigerator operates between evaporating and condensing temperatures of – 16°C and 50°C respectively. The vapour is dry saturated at the compressor inlet, the compression process is isentropic and there is no undercooling of the condensate.

Calculate :

(*i*) The refrigerating effect per kg,

(*ii*) The mass flow and power input per kW of refrigeration, and

(*iii*) The C.O.P.$_{\text{(ref.)}}$.

[**Ans.** (*i*) 1003.4 kJ/kg ; (*ii*) 3.59 kg/h ; 0.338 kW ; (*iii*) 2.96]

11. 30 tonnes of ice from and at 0°C is produced in a day of 24 hours by an ammonia refrigerator. The temperature range in the compressor is from 298 K to 258 K. The vapour is dry saturated at the end of compression and expansion valve is used. Assume a co-efficient of performance of 60% of the theoretical and calculate the power in kW required to drive the compressor. Latent heat of ice is 334.72 kJ/kg.

Temp. K	Enthalpy (kJ/kg)		Entropy of liquid (kJ/kg K)	Entropy of vapour (kJ/kg)
	Liquid	Vapour		
298	100.04	1319.22	0.3473	4.4852
258	– 54.56	1304.99	– 2.1338	5.0585

[**Ans.** 21.59 kW]

12. A refrigerant plant works between temperature limits of – 5°C (in the evaporator) and 25°C (in the condenser). The working fluid ammonia has a dryness fraction of 0.6 at entry to the compressor. If the machine has a relative efficiency of 50%, calculate the amount of ice formed during a period of 24 hours. The ice is to be formed at 0°C from water at 20°C and 6 kg of ammonia is circulated per minute. Specific heat of water is 4.187 kJ/kg and latent heat of ice is 335 kJ/kg.

Properties of ammonia (datum – 40°C) :

Temp. K	Liquid heat kJ/kg	Latent heat kJ/kg	Entropy of liquid kJ/kg°C
298	298.9	1167.1	1.124
268	158.2	1280.8	0.630

[**Ans.** 1640.5 kg]

13. A food storage locker requires a refrigeration system of 2500 kJ/min capacity at an evaporator temperature of – 10°C and a condenser temperature of 30°C. The refrigerant used is Freon-12 and sub-cooled by 5°C before entering the expansion valve and vapour is superheated by 6°C before leaving the evaporator coil. The compression of refrigerant is reversible adiabatic. The refrigeration compressor is two-cylinder single- acting with stroke equal to 1.3 times the bore and operates at 975 r.p.m. Determine (using thermodynamic tables of properties for Freon-12) :

(*i*) Refrigerating effect per kg.
(*ii*) Mass of refrigerant to be circulated per minute.
(*iii*) Theoretical piston displacement per minute.
(*iv*) Theoretical power required to run the compressor, in kW.
(*v*) Heat removed through the condenser per minute.
(*vi*) Theoretical bore and stroke of compressor.

Properties of Freon-12 :

Saturation temp. °C	Absolute pressure	Specific volume of vapour m^3/kg	Enthalpy		Entropy	
			Liquid kJ/kg	Vapour kJ/kg	Liquid kJ/kg K	Vapour kJ/kg K
– 10°C	2.19	0.0767	26.9	183.2	0.1080	0.7020
30°C	7.45	0.0235	64.6	199.6	0.2399	0.6854

Take : Liquid specific heat = 1.235 kJ/kg K
Vapour specific heat = 0.735 kJ/kg K.

[**Ans.** (*i*) 129.17 kJ/kg ; (*ii*) 19.355 kg/min ; (*iii*) 1.518 m^3/min ; (*iv*) 7.2 kW ; (*v*) 2931 kJ/min ; (*vi*) 91 mm, 118 mm]

14. A vapour compression refrigerator uses methyl chloride and works in the pressure range of 11.9 bar and 5.67 bar. At the beginning of the compression, the refrigerant is 0.96 dry and at the end of isentropic compression, it has a temperature of 55°C. The refrigerant liquid leaving the condenser is saturated. If the mass flow of refrigerant is 1.8 kg/min. Determine :

(*i*) Co-efficient of performance.
(*ii*) The rise in temperature of condenser cooling water if the water flow rate is 16 kg/min.
(*iii*) The ice produced in the evaporator in kg/hour from water at 15°C and ice at 0°C.

Properties of methyl chloride :

Saturation temp. (°C)	Pressure (bar)	Enthalpy (kJ/kg)		Entropy (kJ/kg K)	
		h_f	h_g	s_f	s_g
– 20	11.9	30.1	455.2	0.124	1.803
25	5.67	100.5	476.8	0.379	1.642

Take : Specific enthalpy of fusion of ice = 336 kJ/kg

Specific heat of water = 4.187 kJ/kg. [**Ans.** 4.97, 10.9°C, 91.3 kg]

15. A vapour compression refrigerator circulates 4.5 kg of NH_3 per hour. Condensation take place at 30°C and evaporation at – 15°C. There is no under-cooling of the refrigerant. The temperature after isentropic compression is 75°C and specific heat of superheated vapour is 2.82 kJ/kg K. Determine :

(*i*) Co-efficient of performance.

(*ii*) Ice produced in kg per hour in the evaporator from water at 20°C and ice at 0°C. Take : Enthalpy of fusion of ice = 336 kJ/kg, specific heat of water = 4.187 kJ/kg.

(*iii*) The effective swept volume of the compressor in m³/min.

Properties of ammonia :

Sat. temp. (K)	*Enthalpy (kJ/kg)*		*Entropy (kJ/kg K)*		*Volume (m³/kg)*	
	h_f	h_g	s_f	s_g	v_f	v_g
303	323.1	1469	1.204	4.984	0.00168	0.111
258	112.3	1426	0.457	5.549	0.00152	0.509

[**Ans.** 4.95, 682 kg/h, 2.2 m³/min]

Air-conditioning

27

27.1. INTRODUCTION

"Air-conditioning" *is the simultaneous control of temperature, humidity, motion and purity of the atmosphere in confined space.* Thus the important factors which are involved in a complete air-conditioning installation are (*i*) Temperature control ; (*ii*) Humidity control ; (*iii*) Air movement and circulation and (*iv*) Air filtering, cleaning and purification. Complete air-conditioning provides simultaneous control of these factors for both summer and winter. In addition to comfort phases of air-conditioning many industries have found that air-conditioning of their plants has made possible *more complete control of manufacturing processes and material and improves the quality of the finished products.*

The development of the central heating plant was an early step towards modern air-conditioning. Another step was *development of automatic control* for *regulating the heating plant and providing* the proper humidity. The third step was the *development of automatic* refrigeration devices which could be employed for summer cooling and dehumidifying the air.

Note. **For "Psychrometrics" please refer Chapter 9.**

27.2. AIR-CONDITIONING SYSTEMS

27.2.1. Introduction

An ***air-conditioning system*** *is defined as an assembly of different parts of the system used to produce a specified condition of air within a required space or building.*

The basic elements of air-conditioning systems (of whatever form) are :

1. **Fans** for moving air.
2. **Filters** for cleaning air, either fresh, recirculated or both.
3. **Refrigerating plant** connected to heat exchange surface, such as finned coils or chilled water sprays.
4. **Means for warming** the air, such as hot water or steam heated coils or electrical elements.
5. **Means for humidification ; and or dehumidification**.
6. **Control system** to regulate automatically the amount of cooling or warming.

27.2.2. Air-conditioning Cycle

Refer to Fig. 27.1. An air-conditioning cycle comprises the following *steps* :

- The fan forces air into duct work which is connected to the openings in the room. These openings are commonly called *outlets* or *terminals*.
- The duct work directs the air to the room through the outlets.
- The air enters the room and either heats or cools as required. Dust particles from the room enter the air stream and are carried along with it.
- Air then flows from the room through a second outlet (sometimes called the return outlet) and enters the return duct work, where dust particles are removed by a *filter*.

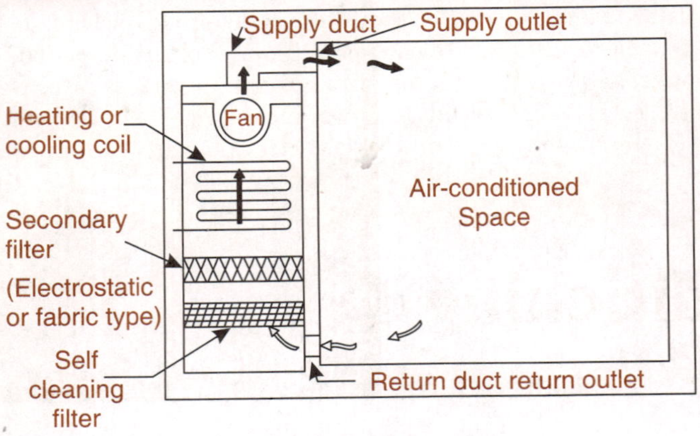

Fig. 27.1. Air-conditioning cycle.

- After the air is cleaned, it is either heated or cooled depending upon the condition in the room. If cool air is required, the air is passed over the surface of a *cooling coil* ; if warm air required, the air is passed through a *combustion chamber* or over the surface of a *heating coil*.
- Finally the air flows back to the fan, and the cycle is completed.

The main parts of the equipment in the air-conditioning cycle are :

(*i*) Fan (*ii*) Supply ducts
(*iii*) Supply outlets (*iv*) Space to be conditioned
(*v*) Return outlets (*vi*) Return ducts
(*vii*) Filter, and
(*viii*) Heating chamber or cooling coil.

1. **Fan.** The primary function of a fan is to *move air to and from the room.* The air that a fan moves in air conditioning system is made up of :

(*i*) All outdoor air ;
(*ii*) All indoor room air (called recirculated air)
(*iii*) A combination of outdoor and indoor air.

The fans *pulls* air from outdoors or from the room but in most systems it *pulls* air from both sources at the same time. The amount of air supplied by the fan must be *regulated* since drafts in the room cause discomfort, and poor air movement slows the heat rejection processes. This can be achieved by (*i*) chosing a fan that can deliver the correct amount of air, and (*ii*) by controlling the speed of the fan so that air stream in the room provides good circulation but does not cause drafts. Of course, the fan is only one of the pieces of equipment that contributes body comfort.

2. **Supply duct.** The function of a supply duct is *to direct the air from fan to the room.* In order that air may flow freely it should be as short as possible and have minimum number of turns.

3. **Supply outlets.** The function of supply outlets is *to distribute the air evenly in a room.* These outlets may (*i*) fan the air, (*ii*) direct air in a jet stream and (*iii*) may do a combination of both. Since supply outlets can either fan or jet the air stream, therefore they are able to exert some control on the direction of air delivered by the fan. This direction control plus the location and the number of outlets in the room contribute a great deal to comfort or discomfort effect of air pattern.

4. **Space.** It is very important to have an enclosed space (*i.e.*, room) since if it does not exist it would be impossible to complete the air cycle since contained air from supply outlets would flow into the atmosphere.

5. **Return outlets.** These are the openings in the room surface. They are employed *to allow room air to enter the return duct* (*i.e.*, return outlets allow air to pass *from* the room). They are usually located at opposite extreme of a wall or room from the supply outlet.

6. **Filters.** A filter is primarily used to *clean the air by removing dust and dirt particles*. They are usually located at some point in the return air duct. They are made of many materials from *spun glass to composite plastic*. Other types operate on electrostatic principle.

7. **Cooling coil and heating coil or combustion chamber.** The cooling coil, and heating coil or combustion chamber can be located either ahead or after the fan, *but should always be located after the filter.* A filter ahead of the coil is necessary to prevent the excessive dirt, dust and dirt particles from covering the coil surface.

Summer operation. The air-conditioning cycle *cools the air* during summer operation. Return air from the room passes over the surface of *cooling coil*, and the air is cooled to the required temperature. If there is too much moisture present, it is removed automatically as the air is cooled by the coil.

Winter operation. The air-conditioning cycle *adds heat to the air* during winter operation. This is achieved by passing the return air from the room over the surface of a *heating coil* etc.

27.2.3. Air-conditioning Systems

The air-conditioning systems are mainly *classified* as :

1. Central systems 2. Zoned systems
3. Unitary systems 4. Unitary-central systems.

Another method of *classification* of air-conditioning system is as follows :

1. Single-air systems 2. Dual-air systems
3. Primary-air systems 4. Unit systems
5. Panel systems.

27.2.3.1. Central System

This type of system is suitable for air-conditioning large spaces such as *theatres, cinemas restaurants, exhibition halls, or big factory spaces where no sub-division exists.* The central systems, are generally employed for the loads above 25 TR and 2500 m^3/min of conditioned air. The unitary systems can be more economically employed for low capacity (below 25 TR) units.

In central system, the equipment such as fans, coils, filters and their encasement are designed for assembly in the field. A central system serves different rooms, requires individual control of each room. The condenser, compressor, dampers, heating, cooling and humidifying coils and fan are located at one place say basement. The conditioned air is carried to the different rooms by means of supply ducts and returned back to the control plant through return ducts. Part of the supply air to the rooms may be exhausted outdoors. Outdoor air enters from a intake which should be situated on that side of the building least exposed to solar heat. It should not be close to the ground or to dust collected roof. The air after passing through damper passes through filters. The filters may be of a mechanical cleaned type, replaceable-cell type or may be electrostatic. The cleaned air then passes to the conditioning equipment in the following order : Tempering (or preheater) coil, cooling coil, humidifier (Air washer), heating coil and finally fan. Fig. 27.2, shows a schematic diagram of complete (year round) air-conditioning system.

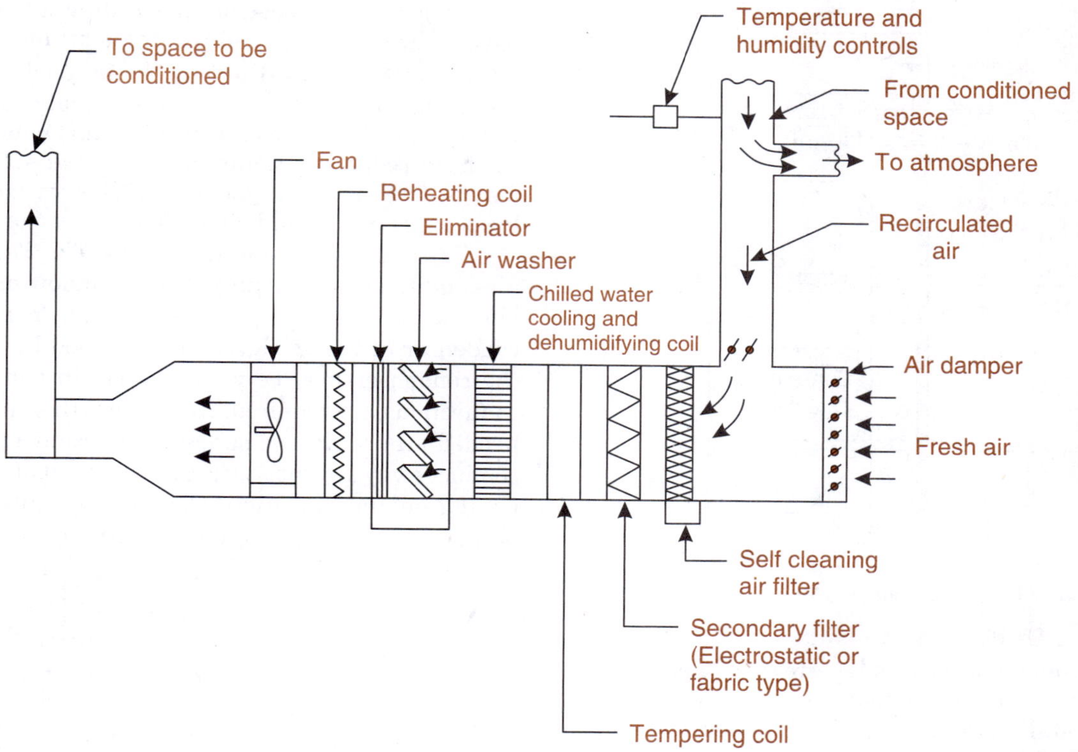

Fig. 27.2. Central system.

Advantages of central system :

1. Low investment cost as compared to total cost of separate unit.
2. Space occupied is unimportant as compared to a room unit conditioner which must be placed in the room.
3. Better accessibility for maintenance.
4. The running cost is less per unit of refrigeration.
5. Noise and vibration troubles are less to the people in air-conditioned places as the air- conditioning plant is far away from the air-conditioned places.
6. The exhaust air can be returned and partly reused with obvious saving in heating and refrigeration.

Disadvantages of central system :

The disadvantage of the central system is that it results in *large size ducts which are costly and occupy large space.*

27.2.3.2. Zoned System

When in a building several rooms or floors are to be served, it is necessary to consider means by which the varying heat gains in the different departments can be dealt with. Some rooms may have some, and others not ; some may be crowded and others empty ; again some may contain heat-producing equipment. Variations in requirements of this kind are the most common case with which air-conditioning has to deal, and for this central system is unsuitable. *Zoned system is one approach.*

In zoned system the building is divided into zones such that as nearly as possible conditions may be expected to exist. Each zone is provided with its own local recirculating fan and booster cooler or heater, and this unit receives fresh air supply conditioned to some average temperature and corrected for humidity by means of what is in effect a central plant. Such an arrangement is shown in Fig. 27.3. In this case the central plant is on the roof. The circulating units are fixed overhead adjacent to the corridor, taking then return air there from, and distributing ducts are run above the corridor false ceiling delivering into the various rooms, floor by floor each floor, constituting a separate zone. The return air from the room passes through grilles into the corridor which acts as the returning air collecting duct. The cooling or heating booster coils could be served from circulating water mains, each coil being controlled locally according to the requirements of the zone served.

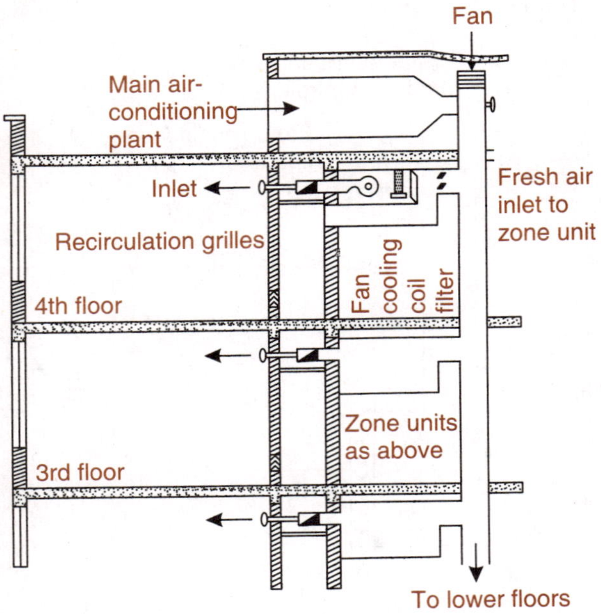

Fig. 27.3. Zoned system.

27.2.3.3. Unitary Systems

The components of unitary air-conditioned system are assembled in the factory itself. These assembled units are usually installed in or immediately adjacent to a zone or space to be conditioned. The package units are available in the size ranges of greatest usage to obtain economics of factory production.

Unitary system is commonly preferred for 15 tonnes capacity or above 200 m^3/min. of flow. The units of even 100 tonnes capacity have also been manufactured.

Various *factory assembled units* available are :

(*i*) Attic (or exhaust) fan.

(*ii*) Remote units.

(*iii*) Self contained units.

(*iv*) Room air-conditioners.

(*v*) Unit air coolers.

(*i*) **Attic fans.** An attic or exhaust fan is a cooling unit without any heat transfer element such as a cooling coil. When the sun sets the temperature of outdoor air reduces to cool levels whereas the indoor temperatures are high. To reduce the inside temperature an attic fan is placed in the attic. It *draws outdoor air into several rooms of the building through various doors and windows and finally discharges from attic to outdoors.* Consequently *a circulation of cool outdoor air is set up in the building. A propeller type fan* is usually recommended as it can handle large volume of air at low pressure efficiently.

(*ii*) **Remote units.** A system in which air handling unit is separated from the condensing unit is called a *remote system*. The conditioning or air handling unit is called *remote unit*. It consists of a fan (either propeller or centrifugal type) with its driving motor, cooling coil, heating coil, filters, drip pan, louvers etc., with or without the duct connections at the outlet. Remote units are available in capacities ranging from 2 to 100 TR. These units are available for floor mounting or for suspension from ceiling. Some remote units have air-washing and coil wetting features.

(*iii*) **Self contained units.** In a self contained unit the condensing unit and other functional elements (such as coil and fans) are encased in the same cabinet. Fresh air can be introduced if required. The discharge from the casing may be free pressure type (*i.e.,* with or without duct work). Proper means should be adopted to cool the compressor.

(*iv*) **Room air-conditioners.** Fig. 27.4 shows a unit air-conditioner for mounting in a window or wall bracket. Unit air-conditioners of small size generally have the condenser of the refrigerator air cooled but in large sizes the condenser may be water-cooled, in which case piping connections are required. Apart from this the only services needed are an electric supply and a connection to drain to conduct away any moisture condensed out of the atmosphere during dehumidification. Compressors in most units are *hermetic* and *therefore quiet in running.*

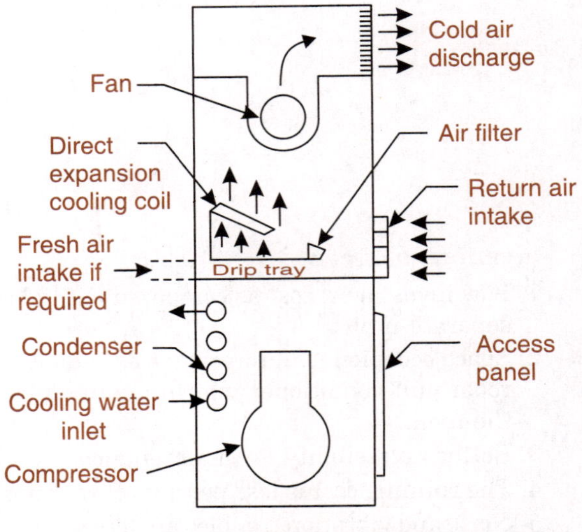

Fig. 27.4. Unit air-conditioner.

Units of considerable size are suitable for industrial applications, in which case ducting may be connected for distribution.

(*v*) **Unit air coolers.** A unit air cooler is a special application of remote units. It primarily reduces the temperature in insulated and sealed storage rooms. The rating of these coolers is the basic rating expressed in kJ per hour per C-degree temperature differential between the refrigerant and the air. A defrosting coil is necessary when the temperature is below 2°C. It may be mounted on the floor or wall or suspended from the ceiling.

Advantages of unitary system. The unitary system commands following *advantages over the central system.*

1. There is saving in the installation and assembly labour charges.
2. Zoning and duct work eliminated.
3. In unitary system exact requirement of each separate room is met whereas in central system the individual needs of separate rooms cannot the met.
4. Failure of the unit puts off conditioning in only one room whereas the failure of the central plant off-sets all the rooms to be served.
5. Only those rooms which need cooling will have their units running, whereas the central plant will have to run all the time for the sake of only a few rooms.
6. *The specific feature of a unitary system is that there in individual room-temperature control.*

27.2.3.4. Unitary-Central Systems

In a unitary-central system *each room is provided with a room unit which gets a supply of conditioned air from a central system.* The *main aim of such systems it to either decrease the size of the ducts or to eliminate them completely.*

The following three unitary-central systems are in common use :

(*i*) Induction units.
(*ii*) All-air high velocity systems.
(*iii*) Fan coil units.

(*i*) **Induction units.** An induction unit *uses the principle of induction as a means of recirculation.* Each induction unit receives primary conditioned air under pressure from a central plant.

Fig. 27.5 shows an induction unit (*generally placed below the window*). The primary air, in summer, is treated in a surface or spray dehumidifier and dehumidified to sufficiently low dew point temperature to compensate for the latent heat gain. The primary air, in winter is treated in a humidifier to a sufficiently high dew point to take care of latent heat losses. The water to the cooling coil in the induction room unit is supplied from a central plant by means of a centrifugal pump. In summer this water is chilled in a central water cooler and in winter it is heated by means of a water heater centrally located. The quantity of water supplied to the induction coil can be controlled by means of a room thermostat for controlling the room temperature.

The high velocity of primary air jets induces a secondary air stream to flow from the room through the coil of the induction unit. The room air is therefore treated to a proper temperature while it flows through the coil. The mixture of primary (20%) and secondary (80%) air is then delivered freely at the outlet of the unit. The quantity of primary air required is greatly reduced since the secondary room air takes up part of the room load. The primary air is usually sufficient for ventilation purposes and consists wholly of outdoor air. Consequently there is no need to have return air ducts and zoning of primary air and water circuits.

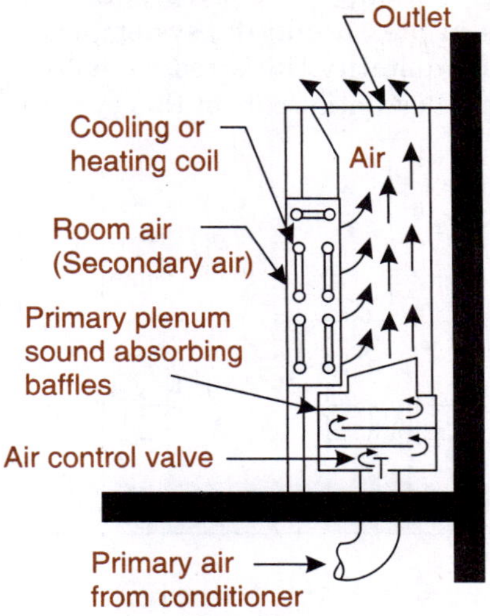

Fig. 27.5. Induction unit.

(*ii*) **All-air high velocity system.** In these systems special control and acaustic equipment are employed since they use high static pressures and high duct velocities. The ducts used are round and are designed carefully in order to reduce friction and noise.

All air high velocity systems are of the following two types :

1. Single duct system. 2. Dual duct system.

1. **Single duct system.** Fig. 27.6 shows a schematic diagram of a single duct system. Each room in a zone has a volume control on the room outlet box which varies the volume of air supplied to each individual room. From the outlet box, the high velocity air is fed into a diffuser. The box itself contains some sound control equipment.

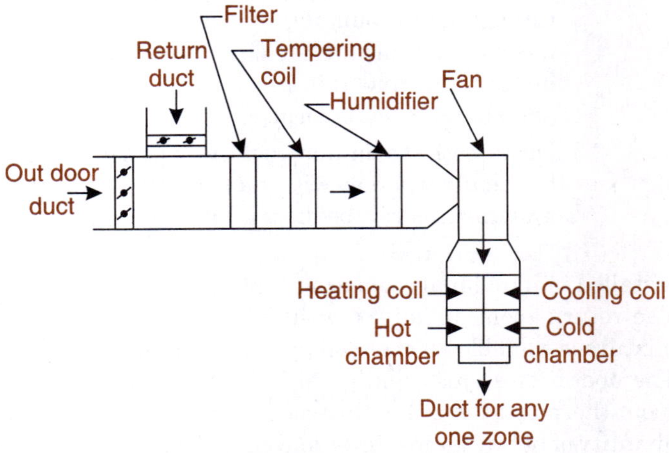

Fig. 27.6. Single duct system.

2. **Dual duct system.** In this system use is made of two ducts, *one conveying warm air* and *one conveying cool air* and *each room contains a blender* so arranged with air valves or dampers that all warm, all cool or some mixture of both is delivered into the room. Fig. 27.7 illustrates such a unit, incorporated there in being a means for automatically recirculating the total air-delivery, such that, regardless of variations in pressure in the system, each unit delivers its correct air quantity. Referred to as 'constant volume control', it is an essential part of this system.

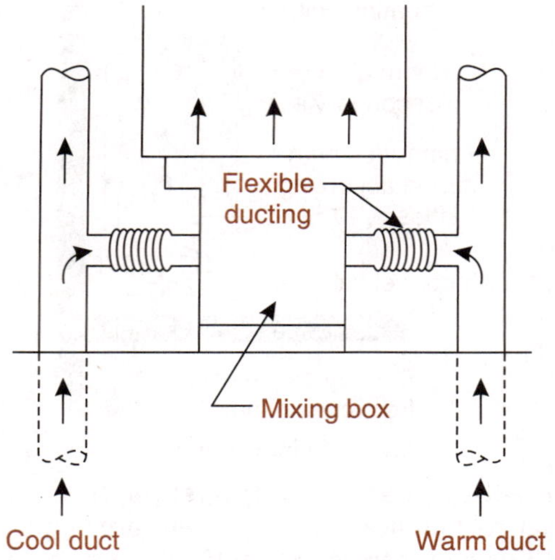

Fig. 27.7. Dual duct system.

Advantages of duct system :

1. Any room may be warmed or cooled according to a need without zoning or any problem of change-over thermostats.
2. Core area of a building, or rooms requiring high rates of ventilation may equally be served off the same systems, no separate plants being necessary.

Disadvantages of duct system :

1. Ducts are larger and fan blowers greater.
2. Owing to the greater quantity of air in circulation throughout the building, dual duct system usually incorporates means for circulation back to the main plant, and this involves return air ducts and shafts in some form.
3. Ducts both warm and cool, need to be installed thermally whereas with some forms of induction systems this can be dispensed with.

(iii) **Fan coil units.** The fan coil units which are installed in individual room are supplied with primary air. The return room air mixes with the primary air and the mixture is then blown over coil by a fan located in the unit. The operation is just similar to induction coil units. The room thermostat is of heating-cooling type. It controls the quantity of hot water in winter and chilled water is supplied to the room coil.

These units, like induction units, are located around the periphery of a building, usually under windows.

27.2.3.5. Ice System of Air-conditioning

In this system cooling is done by melting ice. This is a practical air-conditioning method for churches, theatres and public halls that have short operating hours and relatively high peak loads. Since the investment in mechanical refrigeration equipment is expensive for short periods, ice fills the need. A small quantity of ice in a water cooling tank can release refrigeration at a rapid rate.

27.2.3.6. Selection of System

The selection of system type is determined by a great variety design requisites, and the final choice may vary from a minimum cost, summer-relief cooling installation to a closely controlled, year-round air-conditioning system. The more important factors to be considered in the selection of a system may be classified as follows :

1. General :

(*i*) New or existing building.
(*ii*) Nature of building occupancy.
(*iii*) Type and construction of building.
(*iv*) Relative magnitudes of cooling air heating loads, particularly including the highly variable effect of solar load.
(*v*) Ratio of interior to peripheral zones.

2. Functional :

(*i*) Independent control of temperature and humidity.
(*ii*) Zone or individual room control.
(*iii*) Simultaneous heating and cooling capacity for reversal of intermediate season loads on exterior zones.
(*iv*) Air-distribution in space, including effect of control method.
(*v*) Ventilation-air supply, including flushing action for concentrated loads or contamination sources.
(*vi*) Mixing of air between spaces.
(*vii*) Cleanliness of supply air.
(*viii*) Noise and vibration.
(*ix*) Adaptability to partial-load operation.
(*x*) Local code requirements.

3. Initial Cost :

(*i*) Required refrigeration capacity.
(*ii*) Apparatus and equipment requirements.
(*iii*) Cost of duct work, piping, and miscellaneous services.
(*iv*) Possible use of existing heating system, ventilation ducts, piping etc.
(*v*) Cost of required alterations to existing building.
(*vi*) Cost of design engineering.

4. Operating and owning cost :

(*i*) Relative costs of power, steam, fuel and water.

(*ii*) Possible use of 100% outside air for cooling without refrigeration during intermediate seasons.

(*iii*) Amount of cancellation of cooling effect by heating for control purposes.

(*iv*) Service and maintenance requirements.

(*v*) Rental value of space occupied by air-conditioning equipment, ducts, and pipes.

(*vi*) Rate of depreciation and obsolescence of equipment.

(*vii*) Insurance, taxes and financing costs.

5. Space requirements :

(*i*) Equipment and apparatus space sizes and locations.

(*ii*) Duct and pipe sizes and locations.

(*iii*) Possible use of marginal spaces, such as abandoned elevator shafts, closets etc.

6. Installation problems :

(*i*) Disturbance of occupants of existing building.

(*ii*) Adaptability to progressive or partial installation.

(*iii*) Adaptability to installation of equipment in isolated or remote areas.

(*iv*) Skills required by installation mechanics.

7. Miscellaneous :

(*i*) Flexibility for architectural and partition changes.

(*ii*) Simplicity of operation and maintenance.

(*iii*) Appearance.

27.2.3.7. Applications of Air-conditioning

1. Industrial applications :

(*i*) Food industry. (*ii*) Photographic industry.

(*iii*) Textile industry. (*iv*) Printing industry.

(*v*) Machine tool industry.

2. Theatres air-conditioning.

3. Departmental-store air-conditioning.

4. Transport air-conditioning :

(*i*) Automobile air-conditioning.

(*ii*) Train air-conditioning.

(*iii*) Aircraft air-conditioning.

(*iv*) Ship air-conditioning.

5. Air-conditioning for television centre.

6. Air-conditioning for computer centre.

7. Air-conditioning of automatic telephone exchange building.

8. Museum air-conditioning.

9. Hospital air-conditioning.

27.3. AIR-CONDITIONING EQUIPMENT, COMPONENTS AND CONTROLS

27.3.1. Air-conditioning Equipment

The air-conditioning equipment can be broadly *classified* in two groups :

1. Package units 2. Central units.

Package units include window room air-conditioners and all such installations where functional components are packed all together.

Central units refer to the installation of the different components at different points of the building.

As for function-wise distribution is concerned both types will fall in the following categories : (*i*) Only cooling type system, (*ii*) Combination cooling and heating type system and (*iii*) Heat pump system.

27.3.1.1. Package Units

Package units are of two types :

1. Window type. 2. Console type.

1. **Window type package units.** A *package unit* is a *self contained unit* because the complete unit including evaporator and condensing unit is *all incorporated in a common enclosure.* The normal capacity of such a unit is 1 and 1.5 TR. There are window mounting models which are normally capable of cooling, heating, cleaning and circulating the air. The air distribution is met by a grill arrangement which also allows fresh air through dampers. The dampers are also provided for exhaust purposes. A window type air-conditioner is basically designed for cooling of room where it is installed. The entire systems consists of following sub-assemblies :

Sub-assembly	*Parts*
1. System assembly :	(*i*) Evaporator (*ii*) Capillary (*iii*) Condenser (*iv*) Strainer (*v*) Compressor
2. Motor, fan and blower assembly	(*i*) Fan (*ii*) Blower motor (*iii*) Motor mounting brackets
3. Cabinet and grill assembly	(*i*) Cabinet (*ii*) Grill
4. Switch board panel assembly	(*i*) Selector switch (*ii*) Relay (*iii*) Thermostat (*iv*) Fan motor capacitor (*v*) Running capacitor for compressor motor and starting capacitor for compressor motor.

Fig. 27.8 shows a schematic diagram of a *window type air-conditioner*.

2. **Console type package units.** These units are usually packaged in decorative cabinets and are mounted on the floor directly. Normally water-cooled condensers are employed with such units. The capacities are normally 3 to 20 TR. For proper delivery of conditioned air to the space a short length of duct work is attached or built-in grilles are placed on the top. For heating cycle, if desired, electric strip heaters can be installed. These heaters are inter-locked with the blower motor using in addition a heating thermostat for safety and temperature control purposes.

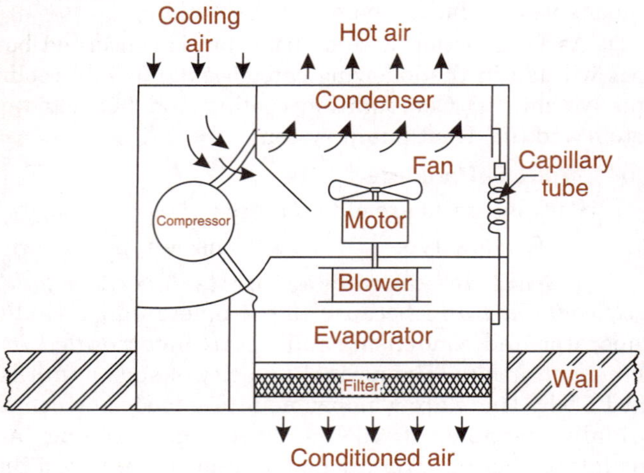

Fig. 27.8. Window type air-conditioner.

27.3.1.2. Central Units

In central units (central air-conditioning system) air handling unit is generally separated from the condensing unit. Such units are available in two types : *Horizontal* and *vertical* depending upon the position of filter and the drain pan. The cooling coil in these units may be either *direct expansion* (DX) or *chilled water type*. The functional elements of a central unit are : (*i*) Equipment for air flow like blower and a motor to drive it ; (*ii*) Equipment for cooling and/or dehumidification like direct expansion (DX) coil or chilled water coil ; (*iii*) Equipment for cleaning of air like filters ; (*iv*) Equipment for dehydration ; and (*v*) Equipment for operational and safety controls.

Fig. 27.9 shows a typical control air-conditioning unit.

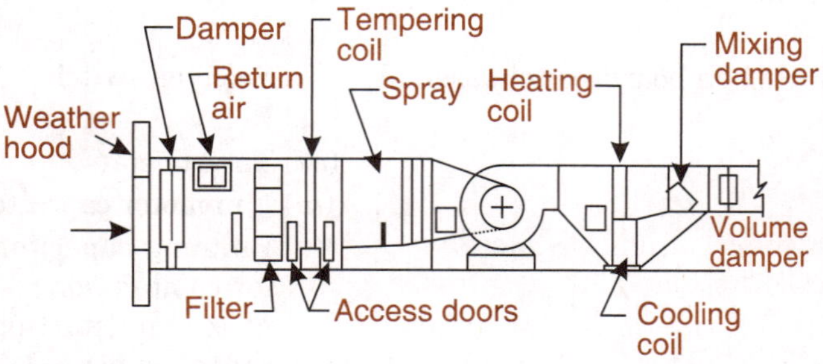

Fig. 27.9. Typical control air-conditioning unit.

27.3.2. Air-conditioning Components

Some important components used in air-conditioning equipment (systems) are discussed in the following articles.

27.3.2.1. Filters

The function of filters is to *arrest* the solid impurities such as soot, ash, lint, smoke and fumes and even living organisms such as virus, bacteria and fungus spores. These are placed *ahead of heating or cooling coils.* The filters are broadly classified as follows :

1. Dry air filters
2. Viscous impingement filters
3. Electrostatic filters.

1. **Dry air filter.** In these filters the filtering medium consists of fabrics made out of wool felt, cotton batting, cellulose fibre etc. The medium is supported in pockets of wire frame or V-shaped plates in order to increase the area of medium exposed to air. Such filters have high dust arrestance including lint. The cleanable filtering medium in permanent frames must be cleaned by application of a vaccum cleaner on the windward side. The *throw away* filters have frame of card board or fibre ; these must be replaced by a new one, after collection of full dust load.

2. **Viscous impingement filters.** In these filters the filtering medium consists of relatively coarse fibres packed between two panels of expanded metal screen enclosed in a frame. The coarse fibres may be of glass, steel, wool, wire screen, animal hair, hemp fibres, metal stamping or shavings. The *fibres are dipped in oil or grease called adhesive intended to retain dust particles which come in contact with it.* There are three types of such filters :

(*i*) Throw away or replaceable type
(*ii*) Manually cleaned type
(*iii*) Automatic or self cleaning filter.

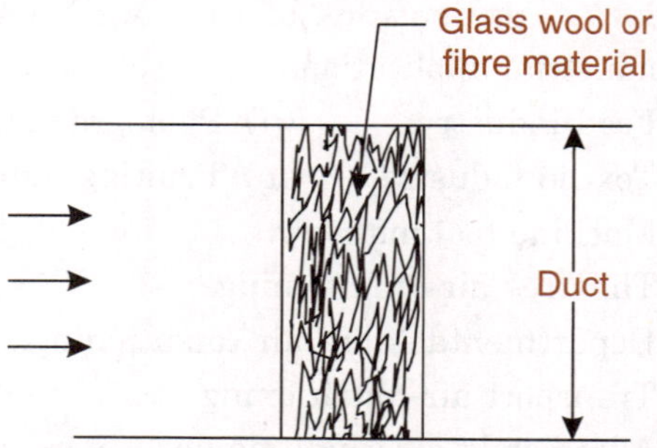

Fig. 27.10. Throw away or replaceable type filter.

A *throw away* or *replaceable* type (Fig. 27.10) filter contains fibrous material such as glass wool having *card-board frames* and retaining grilles. It is designed to be discarded after one period of use. The field of application of such filters is unit air-conditioners.

The *manually cleaned type of filter* has *metal frame*. This type of filter must be cleaned periodically in order to limit the pressure which increases as the filter gets clogged. Air jet, water jet, steam jet, kerosene or oil may be used for cleaning.

The automatic viscous filter (Fig. 27.11) consists of a continuous roll of material coated with oil and is motor driven across the air stream. The roll passes over rollers and moves alternately through a though of oil and the air stream. The *through serves dual purposes of washing off the dirt and recoating the fabric of the roll with relatively clean oil.*

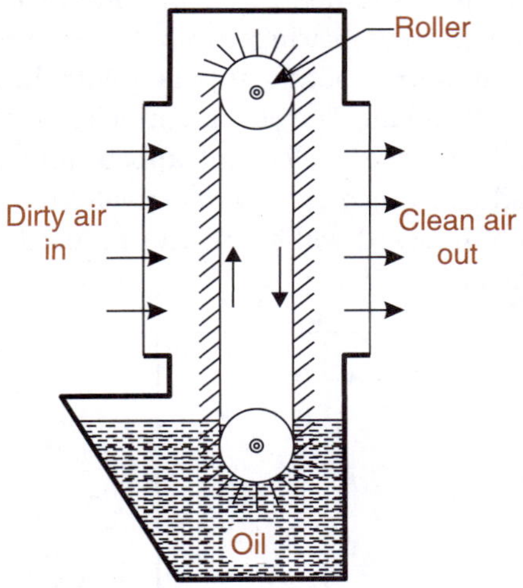

Fig. 27.11. Automatic viscous filter.

3. **Electrostatic filters.** These filters make use of the *principle of electrostatics.* The dirty stream is made to pass through a series of charging plates where dirt particles get electrically charged. There this air stream passes through a bank of collector plates which are given an opposite electrical charge. Consequently the charged dirt particles adhere to the collector plates of opposite polarity which are removable for cleaning.

The tobacco smoke can be cleaned by electrostatic method where very high D.C. voltage (about 15000 V) is employed. Fig. 27.12 shows a typical electrostatic filter.

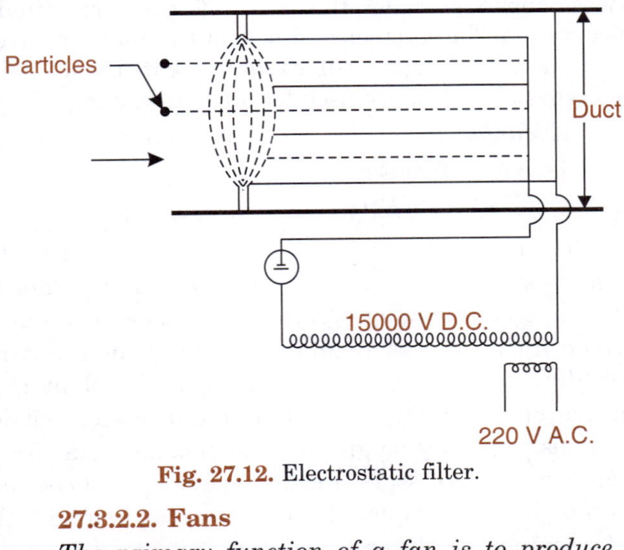

Fig. 27.12. Electrostatic filter.

27.3.2.2. Fans

The primary function of a fan is to produce air movements through heating, ventilating and air-conditioning apparatus. It essentially consists of a rotating wheel which is surrounded by a stationary member known as a housing. The fans can be classified in two general groups.

1. Axial flow fans.
2. Centrifugal or radial-flow fans.

1. **Axial flow fans.** These fans *produce a flow of air in a direction parallel to the axis of rotation*. These are *suitable for handling large air volumes* and can be used where noise-level considerations are not important. They are therefore, used in industrial ventilation and air-conditioning systems. The majority of fans are of centrifugal or radial-flow type. Fig. 27.13 shows a schematic representation of an axial flow fan. The flow of air is parallel to the axis of the impeller. Blades are of aerofoil section. The tips run with as fine a clearance as practicable in a cylindrical casing. Air enters in the axial direction and leaves with a rotational component due to the work done. The absolute velocity of the leaving air is higher than the axial velocity.

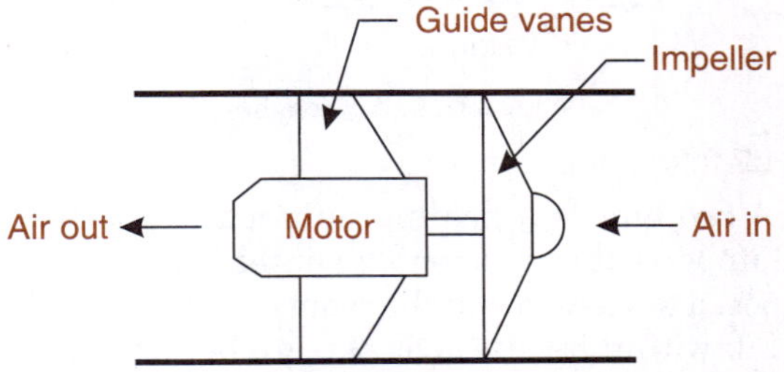

Fig. 27.13. Axial flow fan.

The axial flow fan most commonly used, the *disc or propeller fan,* is best adapted to applications requiring flow against low resistances. When an axial-flow fan is placed in a cylindrical drum-type housing equipped with satisfactory directional vanes, it is called a *vane-axial-flow fan.* If vanes are not included, the term *tube-axial* is used. These fans can be applied in air-conditioning work and operate at good efficiencies and satisfactory noise-level if properly selected for air delivery and system resistance.

2. **Centrifugal or radial-flow fans.** In case a *system has duct work, centrifugal fans have to be used as the static pressure drop is considerable.* But when there is no duct work, propellers or axial flow fans can be used. Nevertheless, *in window type and package units drum-type centrifugal fans only are used, whereas most exhaust fans are of axial type.* A centrifugal fan possesses the following *advantages* :

1. Quiet and efficient operation at high pressures.
2. It is easy to connect centrifugal fan inlet to large apparatus sections and its outlet to smaller supply duct sections.

A centrifugal or a radial flow fan consists of an impeller running in a casing, normally of a volute shape. The air enters axially and is discharged at the periphery.

27.3.2.3. Air Washer

Fig. 27.14 shows a schematic diagram of an air washer. Here air flows through a spray of water and during this flow air may be cooled to heated, humidified or dehumidified, or simply adiabatically saturated, depending on the mean surface temperature of water. The water is, accordingly, externally cooled or heated or simply recirculated by a pump. In case of humidification of air loss of water is compensated by make-up water. In order to minimise loss of water droplets eliminator plates are used.

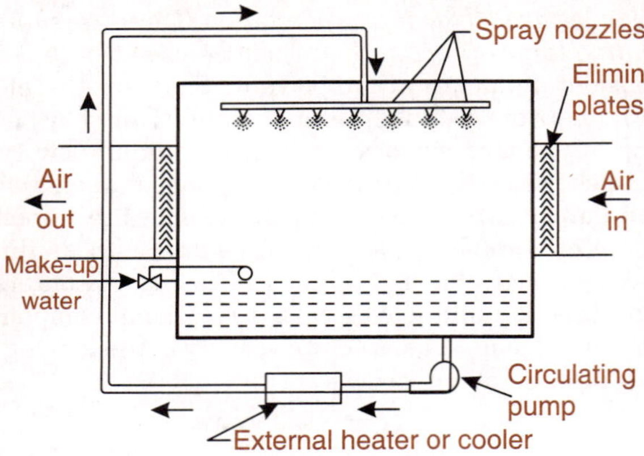

Fig. 27.14. Air washer.

27.3.2.4. Radiator

A *radiator* is a heating unit/heat transfer element. It is an important component of winter heating equipment. It is exposed to view within the room to be heated. All objects which fall within its visible range are heated by *radiation*. It also imparts heat by conduction to surrounding air which in turn circulates the heat by *natural convection currents*. Thus heat is transferred to room by the radiator both by radiation and convection. When a radiator is placed or covered in such a way that it is not visible in the room, it is said to be a *concealed radiator*. This type of radiator heats the room by *natural convection air current*. *'Base board radiators* are installed along the bottom of the walls in places of the conventional board. These are supplied with steam, hot water or electric energy. They operate with gravity circulated (natural convection) room air and have a substantial portion of their surface directly exposed to the room. The **'finned tube units'** consist of metal tubing with metallic fins bonded to the tube. The tubing is supplied with hot water or steam. They may be completely exposed to room or enclosed with top, front or inclined outlets. They operate with gravity or natural circulation of room air. **Pipe coils** are sometimes used in factory buildings and are usually placed under windows or along exposed wall. The output of these units is expressed in kJ/hr. or in sq. metre of equivalent direct radiation (EDR).

Preferably a radiator should be installed near floor and should deliver a horizontal natural current of heated air so that it mixes readily with room air and thereby reduces stratification *i.e.*, difference of temperature of air at floor and ceiling.

27.3.2.5. Convector

A convector is also a heating unit. It may or may not be concealed and is placed in the room to be heated. It transfers heat to the room mainly by convection. The convection currents may be natural or forced. It may be supplied (like radiator) with steam, hot water or electrical energy. Fig. 27.15 shows a schematic representation of a convector. Cold room air enters at the bottom while the heated air is discharged at the outlet grille by natural currents set up due to heating. Convectors with electrical resistors usually have ratings from 1 kW to 8 kW. Base board type units are usually supplied with hot water but sometimes with steam. They do not interfere with furniture placement. They give a pleasant appearance and can be easily installed. *Finned tube units (convectors)* are placed in recesses of exposed walls and can be used with either water or steam.

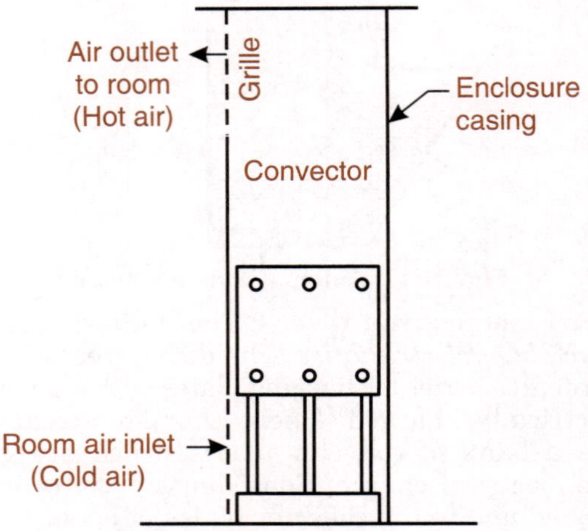

Fig. 27.15. Convector.

27.3.3. Air-conditioning Controls

The field of application of air-conditioning control devices depends upon the size of the area under consideration, the number and size of the units involved, etc. Depending upon particular requirements the following three types of control are used for air conditioning :

1. *Manual* ; 2. *Automatic* ;
3. *Semi-automatic.*

27.3.3.1. Manually Controlled System

The manual control can be an *ordinary knife switch which must be operated by hand when the actuating mechanism is started or stopped.* This method requires *personal attention* and regulation at frequent intervals, *especially when the load varies.* Even if the plant is in operation only at certain times of the day, inside and outside conditions may vary so greatly that constant changes in adjustment are necessary. In this type of control *constant attention* is required by the operator to maintain the predetermined conditions, and the *cost of operation is usually found to be excessive.* If a manual control is provided, the user may permit the unit to operate until the room is too cold, or else he may forget to start the machine until after the room has reached an uncomfortable warm temperature. In either case, the user will not be satisfied, and therefore it is usually wise to install an automatic control.

27.3.3.2. Automatic System

To meet the *varying load* almost constantly completely automatic system may be employed. The various instruments used in this type of control are : *Time clocks, thermostats, humidity controllers*, and *automatic dampers*. Automatically operated controls provide all the features that are considered most desirable for temperature and humidity regulation as well as economy of operation. In this system *operator or owner is relieved of all responsibility* for *maintaining the predetermined conditions*.

27.3.3.3. Semi-automatic Control System

This system makes use of a close-differential thermostat and in most cases it is entirely satisfactory. Time clocks are also very desirable on installations where definite shut-down periods are required. The fresh air intake from outdoors should be equipped with adjustable louvers, so that maximum amount of air can be regulated for both summer cooling and winter heating. On any unit below 25 TR capacity it is desirable to keep the control system as simple and free from service liabilities as possible.

The control devices that may be encountered are briefly outlined as follows :

Thermostats. *A thermostat is a device that operates by changes in room temperature.* In some thermostats a bimetal strip is used to make and break a set of contacts. In another design, the bimetal strip is coiled into a loose helix, with outer coil rigidly fixed to the case. The inner coil is free to move in accordance with temperature changes, and to this free end is secured a glass capsule containing an ample charge of mercury. Two electric wires are seated in one end of the capsule, their base ends projecting into the capsule. When the capsule is tilted in one direction, the mercury flows to the lowest end ; if this happens to the plain end ; the circuit is broken. The electrical circuit is completed only when the capsule is tilted in the other direction, so that mercury flows to the end containing the two wire ends. The mercury bridges the space between the two wires and acts as a metallic conductor.

27.3.3.4. Automatic Humidity Control

Humidistat is an instrument which controls the humidity automatically. In the instrument the element is some type of hygroscopic material, such as wood, human hair, paper and similar materials, when the material absorbs water from the air, expansion takes place. It is this action that is used to make and break the circuit controlling humidifying or dehumidifying apparatus.

27.3.3.5. Air Movement Control

Movement of air is usually determined by factors *viz.*, size of the blowers, the size of the ducts and supply grilles, and the speed of the blower motor. These factors are established when the system is installed and can only be varied by changing the motor-speeds or by the manipulation of suitable dampers.

27.3.3.6. Automatic Temperature Control

The automatic device that controls the temperature of the unit is the *thermostat*. It contains an element that is sensitive to changes in temperature. This element is usually a bimetal strip or a metal bellows filled with a volatile liquid. An automatic temperature control system is generally operated by making and breaking an electric circuit or by opening and closing a compressed air line. Electrically operated systems are used by manufacturers for practically all installations whereas compressed air operated systems are employed in extremely large central and multiple installations.

27.3.3.7. Limit Switches

The use of limit switches is made in heating installations to prevent the fan motor from circulating cold air when the heat is off. These are also employed to prevent the humidifier from operating when the heat is off, or to cut-off refrigeration when heat is on.

27.3.3.8. Time Clocks

Timers are merely clock-operated switches that can be placed either with the thermostat or manual switch, thus automatically preventing operation during certain periods.

27.4. AIR DISTRIBUTION

27.4.1. Definitions

Intake : An opening through which air is *returned* or *exhausted* from the space.

Outlet : An opening through which air is *supplied* to the treated space.

Grille : A functional or *decorative covering for an outlet or intake.*

Register : A grille provided with a damper.

Diffuser : An outlet grille (or appurtenance) designed to guide the direction of the air.

Throw : The horizontal or vertical axial distance that an air stream travels on leaving the outlet.

Drop : The vertical distance the lower edge of the air stream drops

	between the time it leaves the outlet and the time it reaches the end of its throw.
Primary air	: The air coming out of the outlet.
Secondary air	: The room air picked up by the primary air by entrainment.
Total air	: The mixture of primary and secondary air.
Aspect ratio	: Grille-dimension ratio, length to width.

27.4.2. Principles of Air Distribution

1. Air should be distributed in the room so that the required temperatures, humidity and air velocity are maintained in the occupied zone of about 1.8 m above the floor.
2. Air stratification, temperature difference, dead pockets, high drafts, stagnation layers, convection currents and spot ceiling or heating must be avoided.
3. There should be through mixing of the conditioned air discharged into the room through outlets with the air inside the room.
4. The temperature differential in the room should not be larger than 1°C.
5. The velocity should be in the range of 7 m/min to 17 m/min to avoid low or high drafts. Down flow and flow directed to the faces of people is preferred to the upward flow and flow directed to backs or sides of people.
6. The exhaust and inlet points must be so arranged that fresh air is available in all parts of the room.

27.4.3. Air Handling System

The air handling system consists of :

1. **Air distribution system**—comprising various *inlet* for recirculated air and ducts for the supply air.
2. **Duct system**—including the *return duct, supply duct* and *air-conditioning apparatus* comprising *dampers, filters, coil* or *air washer*.
3. **Fan**—provides necessary energy to move the air.

Fig. 27.16 shows schematic air-flow diagram for an air-conditioning system. It may be seen that for circulation of air a closed loop is formed. In this loop the reference point is room itself which can be considered at atmospheric pressure. Through the inlets the air enters and it continues to drop in pressure until is reaches the fan. The fan raises the pressure, and this pressure thereafter starts dropping again until the air enters the space. Therefore, the pressure on the discharge side of the fan is positive and on suction side negative.

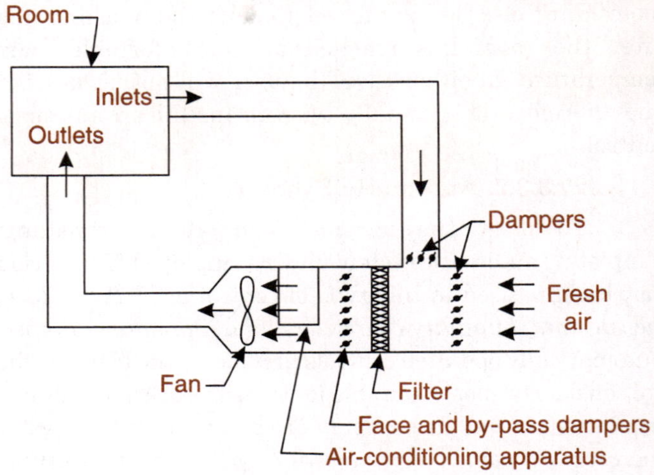

Fig. 27.16. Air-flow diagram.

27.4.4. Room Air Distribution

27.4.4.1. Requirements of Good Room Air Distribution

The primary requirement of good room air distribution is to create a proper combination of temperature, humidity and air motion in the occupied zone which is normally at 1.8 m above the floor level. The maximum variation in temperature in a single room should not be more then 1°C, and within rooms, 2°C. The *desirable air velocity* is 7.5 to 9 m/min.

27.4.4.2. Draft

It is defined as *any localized feeling of coolness or warmth of any portion of the body due to both air movement and air temperature with humidity and radiation, considered constant.* The warmth or coolness of a draft is measured above or below the controlled room condition of 24.4°C DBT at the centre of the room and air moving at approximately 9 m/min.

27.4.4.3. Types of Supply Air Outlet

Number of outlets depends upon the total quantity of air to be distributed, size and type of outlets selected. The *location* of the outlet is determined by height, shape, structural features, window and door openings of the room. Several types of outlets are available and the *selection* of the type of the outlet depends on the required throw and the air quantity to be thrown by each outlet.

Outlets may be divided into four types :

1. Grille outlets.
2. Slot diffuser outlets.
3. Ceiling diffuser outlets.
4. Perforated ceiling panels.

1. **Grille outlets.** Grilles with *fixed bars* can give direction at the outlet and gives a *constant capacity* at the outlet. This can be used where air capacity can be

determined accurately and is not likely to change. *Adjustable bar grilles* provide for *varying the air quantity* if required. Horizontal and vertical, both adjustments, if provided make control easier.

2. **Slot diffuser outlet.** Slotted outlets have performance similar to bar grilles. Long narrow slots give better induction of room air and mixing is perfected in a short throw. They suit mostly to window outlets and linear design of architecture. These should not be installed in ceilings.

3. **Ceiling diffuser outlets.** These outlets supply air in multiple layers and induction is rapid at a short distance after the air leaves the outlets. They may also be fitted with dampers.

4. **Perforated ceiling panels.** These are particularly suited to low ceilings. It offers unobstructive appearance and sound absorbing device.

27.4.5. Duct Systems

Ducts are employed to supply conditioned air (from air-conditioning plant) to the outlets which further distribute air in the room or occupied zone of space. The ducts which supply air to the room/space are called *supply ducts* and those (ducts) which extract the air from conditioned space and send back to the air-conditioning plant are known as *return ducts*.

The supply ducts may be arranged in the following ways :

1. Loop perimeter duct system.
2. Radial perimeter duct system.
3. Extended plenum duct system.

In *loop perimeter duct system* (Fig. 27.17) the conditioned air is carried in several feeder ducts to a common continuous closed loop duct around the perimeter of the building. The outlets then supply air to the room. The system is well adopted to *concrete slab construction* on ground, for down draft and cold floors. To avoid waste of heat, edge insulation of the slab is essential.

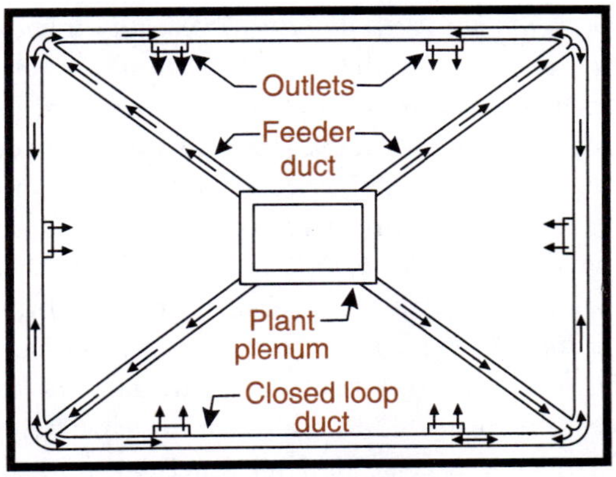

Fig. 27.17. Loop perimeter duct system.

In *radial perimeter duct system* separate ducts supply conditioned air to the respective supply outlets from the plant plenum as shown in Fig. 27.18. This system is recommended for *crawl-space houses.* It can also be used in *slab floors,* but will not be effective as the loop system in warming the perimeter of the floor.

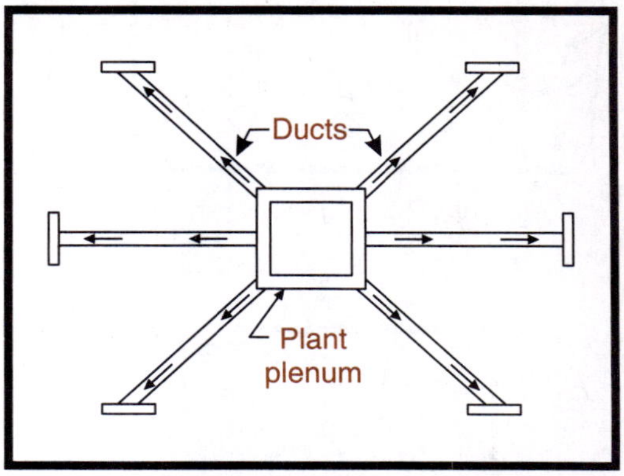

Fig. 27.18. Radial perimeter duct system.

In *extended plenum duct system* the plenum is usually extended to one or both sides of the air-conditioning plant. Each supply outlet is fed with conditioned air from separate duct as shown in Fig. 27.19. This system is recommended for *houses with basement or crawl space.*

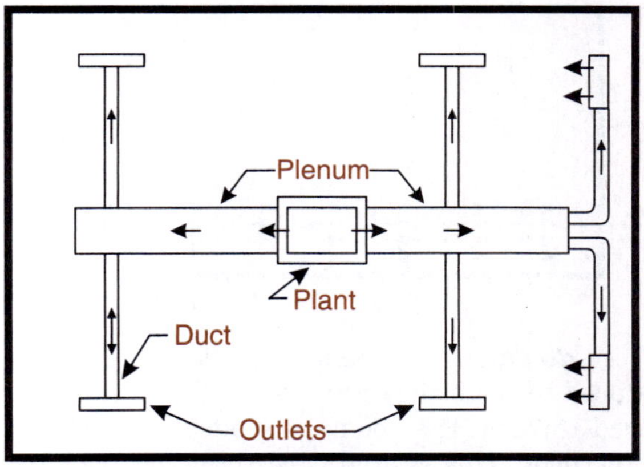

Fig. 27.19. Extended plenum duct system.

27.4.6. Air Distribution Systems

Conditioned air from air-conditioning plant may be fed to a room through supply by any of the following air distribution systems :

1. Ejector system
2. Downward system
3. Upward system.

In an ***ejector system*** the inlet grille ejects the air into the room and induces sufficient velocity for circulation. Figs. 27.20 and 27.21 show important systems of this type. Former is cheaper and very simple in construction. Latter system is known as *pan-type* arrangement. It provides more uniform distribution of air. Fig. 27.22 shows an arrangement which is recommended for place such as recreation places and gymnasiums where the persons occupying the room are moving or doing some exercises.

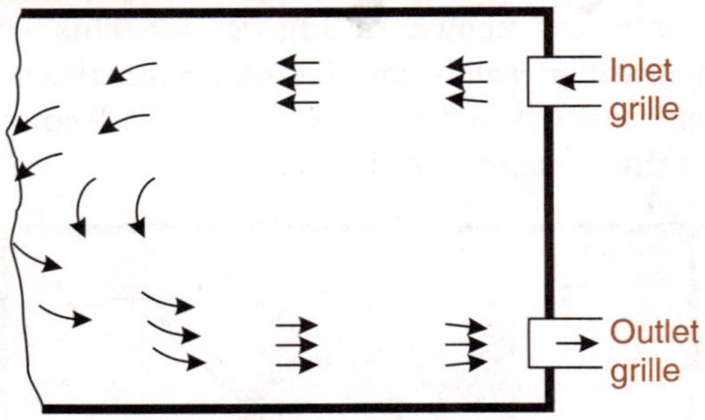

Fig. 27.20. Ejector system.

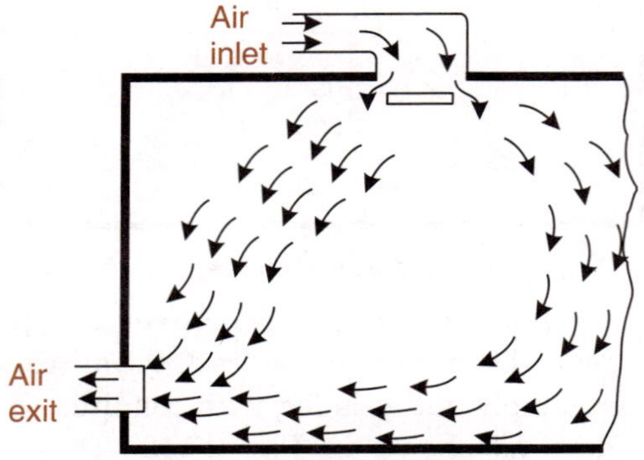

Fig. 27.21. Ejector system (Pan type arrangement)

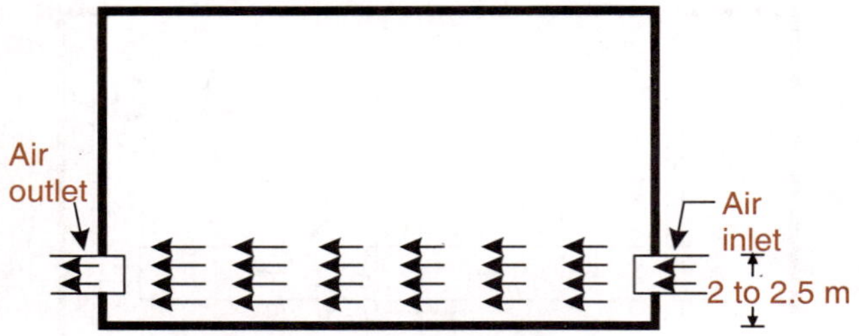

Fig. 27.22

In ***downward system*** of air flow (Fig. 27.23) air is introduced through the openings located in the ceiling and removed through the openings made in floor or in walls near the floor. This system is costly because of the use of perforated ceiling. It may be employed in schools, offices, theatres and auditoriums.

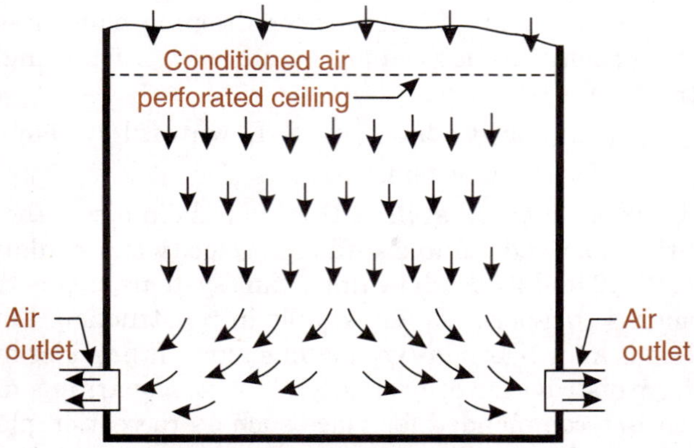

Fig. 27.23. Downward System.

In an ***upward system*** of flow the air outlets are located in walls near the ceiling or in the ceiling. This system is employed in the situations when the air in the room occupied by persons rises carrying with it foul smell/odour from their bodies.

27.4.7. Duct Design Methods

The major consideration in duct design is that ducts should carry the necessary volume of conditioned air from fan outlet to the conditioned space *with minimum frictional and dynamic pressure losses and still be economical in size.* The duct layout must be made so as to reach the outlet with least number of bends, obstructions and changes in area. With rectangular ducts the aspect ratio of 4 : 1 and less is desirable but in no case must be made greater than 8 : 1. For given cross-sectional area minimum sheet is required with square cross-section. The *velocities in the ducts must be high enough to reduce the size of the ducts but low enough to reduce the noise and the pressure losses, to economise requirements.*

After deciding the layout and calculating the requirements of air quantities at various outlets the size of the ducts may be determined by using any one of the following methods. There are mainly *three methods of sizing the ducts* :

1. Velocity reduction
2. Equal friction (or constant pressure loss)
3. Static regain.

1. Velocity reduction method. This method is the easiest in sizing ducts but is entirely empirical and takes no account of the relative pressures losses in various branches. Balancing is obtained by dampering. It is usually *adopted for vary simple systems* and is not recommended for vigorous design of any less simple systems. Considerable experience and judgement is required in selecting velocities such as to make the system optimum in economy and power.

2. Equal friction method. A well-known design method for supply systems employs *equal friction per unit of length.* This is *superior to the velocity reduction method since it prevents one section of duct from having an excessive resistance compared with another in a system having symmetrical branch and outlet runs.* However, its use in the design of complicated systems, particularly those with several runs of widely different lengths and resistances, or with long runs with many grilles, is not recommended. Such a system will be *difficult to balance,* since the method makes no provision for equalising pressure drops in branches or for providing the same static pressure behind each outlet. In sizing a rectangular run with a number of outlets in branches, it is desirable to maintain an aspect ratio (ratio of side lengths) as close to 1 as practical, and to change only one of the dimensions at a time. *When there is an appreciable difference in equivalent length between branch ducts, the longest would be sized by equal friction and the* *total pressure drop for the runs should be computed. The pressure drop between the fan and point of take-off for each of the other runs should then the subtracted from this total to give the desired total pressure drop for each branch.* If the same friction loss per unit length is used for all runs, dampering will be required at the take-offs to shorter runs. With package equipment where the fan pressure is set, it may be necessary to estimate the total equivalent length of the longest run and divide it into the available pressure to determine the unit pressure drop.

3. Static regain method. In order to achieve perfect balancing, the pressure losses in various branches should be made equal, so that the pressure at all outlets is the same. This is possible if the friction loss in each branch is made equal to the gain in pressure due to reduction in velocity. The gain in pressure is taken as *static pressure regain*. For complete regain, it may not be possible to design economically very long branches and the branches very near to the fan. In such cases, it may be *sufficient to design the main duct for complete regain and provide same pressure at all outlets from the main duct for branches. Partial regain may be considered a good practice for a first few outlets from the main duct so that same pressure loss is allowed in the beginning. At the rest of the outlets all frictional losses may be regained by reducing the velocity.*

27.4.8. Leakage of Air and Maintenance of Ducts

Depending on air pressure, type of construction, and workmanship air leakage varies over wide limits. Leakages from 5 to 30% have been observed during actual tests on typical supply systems. Corner holes normally account for only a small portion. The largest usual source is at transverse seams located against the wall or ceiling in such manner that light joints are almost impossible. Allowances should be made for leakage, depending on job conditions. For supply systems with static pressures in excess of 2.5 cm, calking, felting or soldering is recommended. Very little maintenance is required for ventilating and air-conditioning ducts. When they are dry, the deterioration by corrosion is usually negligible. *Periodic cleaning is important* because even with comparatively efficient air-cleaning devices, dirt accumulates over a period of time. A shift in dampers will frequently blow a cloud of dirt into the room. To allow accessability for cleaning ducts should be provided with access doors.

27.5. LOAD ESTIMATION

27.5.1. Introduction

The primary requirement of cooling or heating equipment is that these must be able to remove or add heat at the rate at which it is produced or removed, and maintain the given comfort conditions in the room. The proportion of sensible heat to latent heat decides the slope of '*sensible heat ratio*' or '*enthalpy humidity difference ratio*' and hence also the condition at which the conditioned air must enter the room because it lies on the line drawn from the room condition at the slope of above ratios. The cost of the equipment will be quite high if it is designed for maximum heating or cooling loads. The owner may be prepared to tolerate some uncomfort for a short time to save capital cost investment. If the equipment is designed for average loads, there may be long periods of uncomfort and the equipment may soon become unpopular. Hence the *capacity of equipment must be estimated at a value, which wisely accounts for the physical and economical comfort of the owner*. It is only the exact and detailed estimation of heating and cooling loads which helps in this decision.

The *estimation of load* involves the following *variables* :

1. Magnitude and direction of wind velocity.
2. Outside humidity and temperature.
3. Nature of construction, materials used.
4. Orientation of openings, windows and doors.
5. Periods of occupancy and the number of persons in the room, activities of the persons etc.

One has to be satisfied with the estimation under assumed conditions keeping the variables fixed and allowances must be made for, when the actual conditions are different from those assumed. Thus to make a precise load calculation is unfortunately not simple and some of the sources of the load are difficult to predict and evaluate. For a practising engineer it is not possible to make detailed calculations and therefore he must consult latest editions of 'Heating ventilation and air-conditioning guide' and make quick easy estimates.

27.5.2. Cooling Load Estimate

For air-conditioning the cooling load can be classified as follows :

1. Room load—which falls on the *room directly*.

2. Total load—which falls on the *air-conditioning apparatus*.

1. Room load

(*a*) **Room sensible heat (RSH)**

(*i*) Solar and transmission heat gain through walls, roof etc.

(*ii*) Solar and transmission heat gain through glass.

(*iii*) Transmission gain through partition walls, ceiling, floor etc.

(*iv*) Infiltration

(*v*) Internal heat gain from people, power lights, appliances etc.

(*vi*) Additional heat gain not accounted above, safety factor etc.

(*vii*) Supply duct heat gain, supply duct leakage loss and fan horse power.

(*b*) **Room latent heat (RLH)**

(*i*) Infiltration

(*ii*) Internal heat gain from people, steam appliances etc.

(*iii*) Vapour transmission

(*iv*) Additional heat gain not accounted above, safety factor, etc.

(*v*) Supply duct leakage loss.

It may be noted that if we add *'by-passed air load'* to room sensible heat, we shall get *'effective room sensible heat'* (ERSH). Similarly adding 'by-passed air load' to room latent heat we shall get *'effective room latent heat'* (ERLH).

2. Grand total load (on air-conditioning apparatus)

(*a*) **Sensible heat**

(*i*) Effective room sensible heat

(*ii*) Sensible heat of the outside air that is not by-passed.

(*iii*) Return duct heat gain, return duct leakage gain, dehumidifier pump horse power and dehumidifier and piping losses.

Total sensible heat (TSH) is obtained by adding items (*i*) to (*iii*).

(*b*) **Latent-heat**

(*i*) Effective room latent heat.

(*ii*) Latent heat of outside air which is not by-passed.

(*iii*) Return duct leakage gain.

Total latent heat (TLH) is obtained by adding items (*i*) to (*iii*).

Grand total heat (GTH) = Total sensible heat (TSH) + Total latent heat (TLH)

i.e., $$\text{GTH} = \text{TSH} + \text{TLH} \quad ...(27.1)$$

27.5.3. Heating Load Estimate

Heating-load estimate is prepared on the basis of 'maximum probable heat loss' of the room or space to be heated. Following points should be considered while making heat-load calculations :

1. Transmission heat loss. The transmission heat loss from walls, roof, etc., is calculated on the basis of just the outside and inside temperature difference.

2. Solar radiation. Normally there is no solar radiation present and hence no solar heat gain at the time of the peak load which normally occurs in the early morning hours.

3. Internal heat gains. The heating requirement is reduced due to internal heat gains from occupants, lights motors and machinery etc.

27.5.4. Solar Radiation

The *solar radiation intensity* normal to the sun's rays incident upon a plane surface situated in the outer limits of the earth's atmosphere, varies with time of the year as the distance of the earth from the sun changes. Its value when the earth is at its mean distance from the sun is called the '*solar constant*'. The normal value of the solar constant is assumed as 5040 kJ/m^2-hr. Radiation received at the surface of the earth is must less because, much of it, while passing through earth's atmosphere, is scattered and absorbed by dust and vapour particles and the gases in the atmosphere.

The solar heat reaches any part of the earth's surface in the form of two radiations :

1. Beam or direct radiation. The part of the sun's radiation which travels through the atmosphere and reaches the earth's surface directly is called *Beam* or *direct radiation. It is maximum when the surface is normal to the sun's rays.* The intensity of the radiation can be increased or decreased by *changing the orientation of the surface.*

2. Diffuse or sky radiation. A large part of the sun's radiation is scattered, reflected back into space and absorbed by the earth's atmosphere. A part of this radiation is re-radiated and reaches the earth's surface uniformly from all directions. It is called *diffuse or sky radiation*. It *does not normally change with orientation of surface.*

The total solar radiation reaching a surface is equal to the *sum of the direct and diffuse radiation.*

27.5.5. Solar Heat Gain Through Glass

Glass which is a major cladding material of most buildings, provides the most direct route for entry of solar radiation. For these reasons, the proper estimation of heat gain through glass is necessary.

Heat transmitted through a glass surface depends on the wavelength of radiation and physical and chemical characteristics of the glass. Part of the radiation is *absorbed*, part is *reflected* and the rest is *transmitted.* Glass is opaque to the radiant energy emitted from sources below 200°C. Thus glass has high *transmittivity for short wavelength and low transmittivity for long wavelength radiation.*

Direct solar heat gain can be *reduced* by using different types of glass, glass construction and shades as given below :

(*i*) Double pane glass reduces the solar heat by 10% to 20%.

(*ii*) Special heat absorbing glass reduces the solar heat by 25%.

(*iii*) Stained glass can reduces it upto 65% depending upon its colour.

(*iv*) Shading devices installed on the outside of windows reduce sun load upto 15%.

(*v*) Ventilation blinds and curtain shades reduce it by 30 to 35%.

27.5.6. Heat Flow Through Building Structures (Thermal Barriers)

Refer to Fig. 27.24. Transfer of heat by conduction (Q) from outside to inside under steady state due to the temperature difference is given by the equation :

$$Q = UA\,(t_0 - t_i) \qquad \text{...(27.2)}$$

where, Q = Heat transfer rate kJ/h,

U = Overall co-efficient of heat transfer,

A = Area, m^2,

t_0 = Outside temperature,

t_i = Inside temperature,

and, *overall thermal resistance*,

$$\frac{1}{U} = \frac{1}{f_0} + \frac{x_1}{K_1} + \frac{x_2}{K_2} + ...+ \frac{1}{f_i} \qquad \text{...(27.3)}$$

where, f_0 = Film co-efficient on outside wall,

x = Thickness of material,

K = Conductivity of material, and

f_i = Film co-efficient on inside wall.

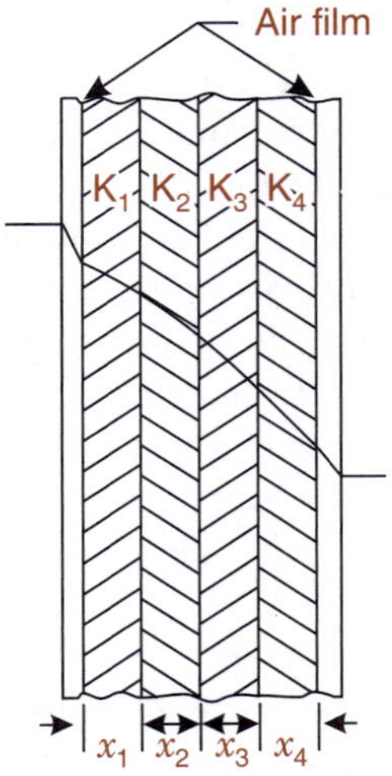

Fig. 27.24

Subscripts 1, 2 etc., refer to successive layers of materials which make up the walls.

The film *co-efficient varies directly with wind velocity*.

While considering the steady state conduction, the outdoor air temperature should be assumed to be the mean average during the day. This assumption of steady state is justified only in cases where the temperature difference inside and outside is very large and fluctuation of outside air temperature is small fraction of the temperature difference between inside and outside. For the cases where fluctuation in periodic variation is large compared to temperature difference inside and outside, the equivalent temperature difference should be used.

Loss or gain of heat from adjacent rooms through partitions and from the ground through floor may be calculated by using the following equation :

$$Q = UA\,(t' - t_i) \qquad \text{...(27.4)}$$

where U and A refer to the separating partition or the floor and t' refers to the air temperature of adjacent room or ground.

27.5.7. Infiltration

Both sensible and latent heat losses result from the *leakage* of outside air into the building through cracks and opening, causing displacement of an equal volume of room air. The infiltration depends on the length and width of cracks around the windows, doors, walls, partitions and floors. The nature of cracks depends upon the building construction workmanship and the building age and deterioration.

In order to estimate air leakage through *window* and *door cracks* the following two methods may be used.

1. Air change method 2. Crack length method.

27.5.7.1. Air Change Method

In this method, the infiltration is based on the volume of the room and is given by the number of times air changes to fill the volume of the room. The sensible heat loss/cooling load due to such air changes and other infiltration can be calculated as follows :

$$Q = \rho Vc\,(t_0 - t_i) \qquad \text{...(27.5)}$$

where, ρ = Density of air, kg/m^3,

V = Rate of air infiltration or changes in m^3/h,

c = Specific heat of air,

t_0 = Outside air temperature, and

t_i = Inside air temperature.

27.5.7.2. Crack Length Method

Crack length method is *more accurate and logical*. This method gives volume of air infiltrated per metre of crack length.

Air change method is purely arbitrary and based on the engineering practice. It assumes that infiltration air replaces the air in the room in a certain time. It should be used only when it is not possible to use crack method due to uncertainty of the crack length. Air change method is convenient for reception halls and vestibules, where air change is very frequent.

Note. The crack method is used in case of windows and the air change method is more convenient to use in the case of doors.

27.5.7.3. Chimney or Stack Effect

Due to differences between temperatures and humidities, differences in the densities of air between the outside and inside of the buildings are produced. Consequently pressure differences occur causing flow of air known as *chimney* or *stack effect*. When the inside temperature is lower than the outside, the stack effect produces positive inside pressure at lower levels and negative inside pressure at higher levels. As a result, the outward flow of air takes place at lower levels and the inward flow at higher levels, with the neutral zone in the middle. The reverse is true when the inside is at a higher temperature than the outside. Thus, in *'summer' infiltration is at the top and exfiltration at the bottom.* Similarly in *'winter' infiltration is at the bottom and exfiltration at the top*.

27.5.8. Internal Heat Gains

The sensible and latent heat gains due to occupants, lights, appliances, machines, piping etc., within the conditioned space form the components of *internal heat gains*.

27.5.8.1. Heat Load of Occupants

The occupants in a conditioned space give out heat at a metabolic rate that more or less depends on their rate of working. The relative proportion of the sensible and latent heats given out, however, depends on the ambient dry bulb temperature (DBT). The lower the *DBT*, the greater the heat given out as sensible heat. The usual problem in calculating the occupancy load lies in the *estimation of the exact number of people present.*

27.5.8.2. Electrical Load

Electrical load consists of lights, heaters, rheostats and other heat dissipating electrical devices.

Lighting load. Electrical lights generate a sensible heat equal to the amount of the electric power consumed. Most of the energy is liberated as heat and the rest as light which also eventually becomes heat after multiple reflections. As a rough calculations one may use the lighting load equal to 33.5 W/m^2 to produce a lighting standard of 540 lumens/m^2 in an office space. After the wattage is known the heat gain may be calculated as follows :

Incandescent : Q = total watts,

Fluorescent : $Q = 1.25 \times$ total watts.

Appliance load. Both sensible and latent heats are contributed by most appliances. The latent heat produced depends on the function the appliances perform, such as drying, cooking etc. *Gas appliances* produce additional moisture as a product of combustion. Such loads can be considerably reduced by providing properly designed loads with a positive exhaust system of suction over the appliances. *Electric motors* contribute sensible heat to the conditioned space. A part of the power input is directly converted into heat due to the inefficiency of the motor and is dissipated through the frame of the motor. This power is (Input) (1-motor efficiency). The rest of the power input is utilised by the driven mechanism for doing work which may or may not result in a heat gain to the space. It depends whether the energy input goes outside or to the conditioned space.

27.5.8.3. Product Load

The term *'product'* as used here means anything whose temperature is reduced by the refrigerating equipment. It includes perishable commodities such as food stuff and other items such as welding rods, plastics, medicines and liquids of the all types. The quantity of heat to be given up by the product to the refrigerating space depends upon : (*i*) Initial state of the product at the time of entering ; (*ii*) Final state of product it attains after storing ; (*iii*) Nature of product ; (*iv*) Weight of product and its specific heat above and below freezing ; (*v*) Freezing temperature ; and (*vi*) Latent heat of fusion.

27.5.8.4. Process Load

For each industrial air-conditioning process there is a specific procedure for calculating the cooling and heating load. The requirements for the process may involve the control of one or more of the following factors : (*i*) Regain of moisture content by hygroscopic materials, such as cotton, silk, tobacco, etc., and the accompanying heat liberated ; (*ii*) Drying load ; (*iii*) Rate of chemical and biochemical reactions ; (*iv*) Rate of crystallization, freezing, freez-drying etc., and (*v*) Sensible cooling load.

27.5.9. System Heat Gains

The system heat gain is the heat gain (or loss) of an air-conditioning system. It includes the following :

1. Supply air duct heat gain and leakage loss.
2. Heat gain from air-conditioning fan.
3. Return air duct heat and leakage gain.
4. Heat gain from dehumidifier pump and piping.

The sensible heat gain should be initially estimated and then included in the total heat load for the air-conditioning plant. The same should be checked after the whole plant has been designed.

Note. Safety factor' is strictly a factor of probable error in the estimation of the load. For the purpose, additional 5 per cent heat should be added to the room sensible and latent heats.

HIGHLIGHTS

1. An air-conditioning system is defined as an assembly of different parts of the system used to produce specified condition of air within a required space or building.
2. The basic elements of an air-conditioning system are : Fans, filters, refrigerating plant, means for warming the air, means for humidification and/or dehumidification ; and control system.
3. The main parts of the equipment in the air-conditioning cycle are :
   Fan, supply ducts, supply outlets, space to be conditioned, return outlets, return ducts, filter and heating chamber or cooling coil.
4. The air-conditioning systems are mainly classified as follows :
   (*i*) Central systems (*ii*) Zoned systems
   (*iii*) Unitary systems (*iv*) Unitary-central systems.
5. Air-conditioning equipment primarily include package units and central units. Package unit may be of window or console type.
6. A package units is a self-contained unit because the complete unit including evaporator and condensing unit is all incorporated in a common enclosure.
7. Central units are available in two types : horizontal and vertical depending upon position of filter and the drain pan.
8. Air-conditioning components : Filters, fans, air washer, radiator, convector etc.
9. Air-conditioning controls may be manual, automatic and semi-automatic.
10. Some important control devices are : Thermostats, automatic humidity control, air movement control, automatic temperature control, limit switches and time clocks.
11. Air distribution system comprises of :
    (*i*) Air distribution system (*ii*) Duct system
    (*iii*) Fan.
12. *Draft* is defined as any localized feeling of coolness or warmth of any portion of the body due to both air movement and air temperature with humidity and radiation considered constant.
13. Outlets may be classified as follows :
    (*i*) Grille outlets
    (*ii*) Slot diffuser outlets
    (*iii*) Ceiling diffuser outlets
    (*iv*) Perforated ceiling panels.
14. The supply ducts may be arranged in the following ways :
    (*i*) Loop perimeter duct system
    (*ii*) Radial perimeter duct system
    (*iii*) Extended plenum duct system.
15. Air distribution systems may be divided into three types :
    (*i*) Ejector system (*ii*) Downward system
    (*iii*) Upward system.
16. Three methods of sizing the ducts are :
    (*i*) Velocity reduction method
    (*ii*) Equal friction method
    (*iii*) Static regain method.
17. The estimation of load involves the following variables :
    (*i*) Magnitude and direction of wind velocity.
    (*ii*) Outside humidity and temperature.
    (*iii*) Nature of construction, materials used.
    (*iv*) Orientation of openings, windows and doors.
    (*v*) Periods of occupancy and number of persons in the room, activities of the persons etc.
18. For air-conditioning the cooling load can be classified as follows :
    (*i*) Room load—which falls on the *room directly*
    (*ii*) Total load—which falls on the *air-conditioning apparatus.*
19. The solar heat reaches any part of the earth's surface in the form of two radiations :
    (*i*) Beam or direct radiation
    (*ii*) Diffuse or sky radiation.
20. Direct solar heat gain can be reduced by using different types of glasses, glass construction and shades.
21. Transfer of heat by conduction (Q) from outside to inside under steady state due to the temperature difference is given by the equation :
    $$Q = UA(t_0 - t_i)$$
    and 'overall thermal resistance',
    $$\frac{1}{U} = \frac{1}{f_0} + \frac{x_1}{K_1} + \frac{x_2}{K_2} + ... + \frac{1}{f_i}.$$
22. In order to estimate air leakage through *window* and *door cracks* the following two methods may be used :
    (*i*) Air change method (*ii*) Crack length method.
    Air change method is purely arbitrary and based on the engineering practice.
    The crack method is used in case of windows and the air change method is more convenient to use in the case of doors.

OBJECTIVE TYPE QUESTIONS

Fill in the Blanks :

1. The simultaneous control of temperature, humidity, motion and purity of the atmosphere in a confined space is called ......
2. The art of measuring the moisture content of air is termed ......
3. ...... is the temperature of air as registered by an ordinary thermometer.
4. The difference between the dry bulb and wet bulb temperature is called ....
5. ...... is the ratio of the partial pressure of water vapour in the mixture to the saturated partial pressure at the dry bulb temperature, expressed as percentage.
6. The heat that does not affect the temperature but changes the state of a substance when added to or subtracted from it is called ......

7. The ratio of mass of water vapour associated with unit mass of water vapour associated with saturated unit mass of dry air is called ......
8. Air undergoes sensible ...... whenever it passes over a surface that is at a temperature less than the dry bulb temperature of the air but greater than dew point temperature.
9. The ratio of the room sensible heat to the sum of room sensible heat and room latent heat is called ......
10. ...... is the ratio of total sensible heat to the grand total heat that the cooling or the conditioning apparatus should handle.
11. An assembly of different parts of the system used to produce specified condition of air within a required space or building is called ......
12. ...... are used for moving air.
13. ...... are employed for cleaning air.
14. The function of a ...... is to direct the air from fan to the room.
15. ...... distribute the air evenly in a room.
16. ...... is a cooling unit without any heat transfer element such as a cooling coil.
17. A system in which air handling unit is separated from condensing unit is called a ...... system.
18. ...... unit uses the principle of induction as a means of recirculation.
19. In ...... system zoning and duct work is eliminated.
20. In ...... system each room is provided with a room unit which gets a supply of conditioned air from a central system.
21. An opening through which air is supplied to the treated space is called ......
22. ...... is a grille provided with a damper.
23. ...... is an outlet grille designed to guide the direction of the air.
24. ...... air is the air coming out of the outlet.
25. ..... air is the air picked up by the primary air by entrainment.
26. Grille-dimensions ratio is called ...... ratio.
27. An opening through which air is returned or exhausted from the space is called ......
28. ...... provides necessary energy to move the air.
29. The various duct systems are ......, and ......
30. Three methods of sizing the ducts are ....., and ......
31. The primary requirement of ...... or ...... equipment is that these must be able to remove or add heat at the rate at which it produced or removed, and maintain the given comfort conditions in the room.
32. The ...... of equipment must be estimated at a value, which wisely accounts for the physical and economical comfort of the owner.
33. For air-conditioning the cooling load can be classified as ...... load and ...... load.
34. The normal value of solar constant is assumed as ...... kJ/m-hr.
35. The solar heat reaches any part of the earth's surface in the form of two radiations ...... and ......
36. Complete the equation :

$$Q = \text{......} A(\text{...} - \text{...}).$$

37. The film co-efficient varies directly with ......
38. Estimation of air leakage through window and door cracks can be done by two methods ...... and ......
39. ...... method is purely arbitrary and based on engineering practice.
40. ...... method is used in case windows.
41. Package units may be of ...... or ...... type.
42. The cooling coil in central units may be either ...... or ...... type.
43. The function of a...... is to arrest the solid impurities, such as soot, ash, lint, smoke and fumes and even living organisms such as virus, bacteria and fungus spores.
44. The manually cleaned type of filter has ...... frame.
45. The primary function of a ...... is to produce air movements through heating, ventilating and air-conditioning apparatus.
46. ...... fans produce a flow of air in a direction parallel to the axis of rotation.
47. A ...... fan consists of an impeller running in a casing, normally of a volute shape.
48. Heat is transferred to the room by the radiator both by ...... and ......
49. A convector transfers heat to the room mainly by ......
50. Three types of air-conditioning controls are ......, ... and ......

ANSWERS

1.	Air-conditioning ;	**2.**	Psychrometry ;
3.	DBT ;	**4.**	Wet bulb depression ;
5.	Relative humidity ;	**6.**	Latent heat ;
7.	Degree of saturation ;	**8.**	Cooling ;
9.	RSHF ;	**10.**	GSHF ;
11.	Air-conditioning system ;	**12.**	Fans ;
13.	Filters ;	**14.**	Supply duct ;
15.	Supply outlets ;	**16.**	Attic fan ;
17.	Remote ;	**18.**	Induction ;
19.	Unitary ;	**20.**	Unitary-central system ;
21.	Outlet ;	**22.**	Register ;
23.	Diffuser ;	**24.**	Primary ;
25.	Secondary ;	**26.**	Aspect ;
27.	Intake ;	**28.**	Fan ;
29.	Loop perimeter, radial perimeter and extended plenum ;		
30.	Velocity-reduction, equal friction and static regain ;		
31.	Cooling, heating ;		
32.	Capacity ;	**33.**	Room total ;
34.	5040 ;	**35.**	Beam, diffuse ;
36.	U, t_0, t_i ;	**37.**	Wind velocity ;
38.	Air change, crack length ;		
39.	Air change ;	**40.**	Crack length ;
41.	Window, console ;		

42. Direct expansion, chilled water ;
43. Filter ; 44. Metal ;
45. Fan ; 46. Axial flow ;
47. Centrifugal or a radial flow ;
48. Radiation and convection ;
49. Convection ;
50. Manual, automatic and semi-automatic.

THEORETICAL QUESTIONS

1. Define the term 'air-conditioning'.
2. Define an 'air-conditioning system'. Name its basic elements.
3. Explain with a neat sketch an 'air-conditioning cycle'.
4. Enumerate the main parts of the equipment in the air-conditioning cycle.
5. How are air-conditioning systems classified ?
6. Explain with neat diagram the working of central system of air-conditioning.
7. Write a short note on zone system of air-conditioning.
8. Define a 'Unitary system'. Where is it commonly preferred ? Explain a room air-conditioner with a neat sketch.
9. State the advantages of central system over unitary system of air-conditioning.
10. Discuss briefly any two of the following :
    (*i*) Attic fans (*ii*) Remote units
    (*iii*) Self contained units (*iv*) Unit air coolers.
11. Explain any two of the following unitary central systems :
    (*i*) Induction units
    (*ii*) All-air high velocity systems
    (*iii*) Fan coil units.
12. Write a short note on 'Ice system of air-conditioning'.
13. State the factors which should be taken into consideration while selecting a system of air-conditioning.
14. How are air-conditioning equipment classified ? In what ways package units differ from central units ?
15. Describe briefly with a neat sketch a window type air-conditioner.
16. Write a short note on 'console type' package units.
17. Enumerate the functional elements of a central unit.
18. What is the function of a filter ? How are filters classified ? Explain briefly an automatic or self cleaning filter.
19. What purpose is served by a fan ? How are fans classified ?
20. State the field of application of 'Axial flow' and 'Centrifugal' fans.
21. Write short notes on any two of the following :
    (*i*) Air washer (*ii*) Radiator
    (*iii*) Convector.
22. Enumerate air-conditioning controls. Describe briefly any two of them.
23. Explain briefly any two of the following control devices :
    (*i*) Thermostats
    (*ii*) Automatic humidity control
    (*iii*) Air movement control
    (*iv*) Automatic temperature control.
24. Define the following as applied to 'air distribution'. Intake, Outlet, Grille, Register Diffuser, Throw, Drop, Primary air.
25. State the principles of air distribution.
26. What is an air handling system ? Give air-flow diagram for an air-conditioning system.
27. What are the requirements of good room air distribution ?
28. Describe briefly the following :
    (*i*) Grille outlets
    (*ii*) Slot diffuser outlets
    (*iii*) Ceiling diffuse outlets
    (*iv*) Perforated ceiling panels.
29. What is the function of ducts ? Enumerate duct systems and explain anyone of them with a neat sketch.
30. Explain any two of the following air distribution systems :
    (*i*) Ejector system (*ii*) Downward system
    (*iii*) Upward system.
31. Enumerate methods of duct design and explain in detail the 'equal friction method'.
32. Write a short note on 'air leakage and duct maintenance'.
33. List the variables which are involved in the estimation of load.
34. Enumerate the components of cooling-load estimate.
35. What points should be considered while making heat load calculations ?
36. Write a short note on 'solar radiation'.
37. Write down the procedure for calculating heat gains through building structures.
38. What do you mean by the term 'infiltration' ? Explain briefly how air leakage through window and door cracks can be estimated ?
39. Write a short note on 'chimney or stack effect'.
40. Enumerate and explain the components of internal heat gains.
41. List the components of system heat gains.

28 Heat Transfer

28.1. Modes of heat transfer. 28.2. Heat transfer by conduction—Fourier's law of heat conduction—Thermal conductivity of materials—Thermal resistance (R_{th})—General heat conduction equation in Cartesian coordinates—Heat conduction through plane and composite walls—The overall heat-transfer coefficient—Heat conduction through hollow and composite cylinders—Heat conduction through hollow and composite spheres—Critical thickness of insulation. **28.3. Heat transfer by convection. 28.4. Heat exchangers**—Introduction—Types of heat exchangers—Heat exchanger analysis—Logarithmic mean temperature difference (LMTD)—**28.5. Heat transfer by radiation**—Introduction—Surface emission properties—Absorptivity, reflectivity and transmissivity—Concept of a black body—The Stefan-Boltzmann law—Kirchhoff 's law—Planck's law—Wien's displacement law—Intensity of radiation and Lambert's cosine law—Radiation exchange between black bodies separated by a non-absorbing medium. *Highlights—Objective Type Questions—Theoretical Questions—Unsolved Examples.*

28.1. MODES OF HEAT TRANSFER

"*Heat transfer*" which is defined as the *transmission of energy from one region to another as a result of temperature gradient* takes place by the following *three modes :*

(*i*) Conduction ; (*ii*) Convection ;

(*iii*) Radiation.

Heat transmission, in majority of real situations, occurs as a result of combinations of these modes of heat transfer. *Example :* The water in a boiler shell receives its heat from the fire-bed by conducted, convected and radiated heat from the fire to the shell, conducted heat through the shell and conducted and convected heat from the inner shell wall, to the water. *Heat always flows in the direction of lower temperature.*

The above three modes are similar in that a temperature differential must exist and the heat exchange is in the direction of decreasing temperature ; each method, however, has different controlling laws.

(*i*) **Conduction.** '*Conduction*' is the *transfer of heat from one part of a substance to another part of the same substance, or from one substance to another in physical contact with it, without appreciable displacement of molecules forming the substance.*

In *solids*, the heat is conducted by the following *two mechanisms :*

(*a*) ***By lattice vibration*** (The faster moving molecules or atoms in the hottest part of a body transfer heat by impacts some of their energy to adjacent molecules).

(*b*) ***By transport of free electrons*** (Free electrons provide an energy flux in the direction of decreasing temperature—For metals, especially good electrical conductors, the electronic mechanism is responsible for the major portion of the heat flux except at low temperature).

In case of *gases*, the mechanism of heat conduction is simple. The kinetic energy of a molecule is a function of temperature. These molecules are in a continuous random motion exchanging energy and momentum. When a molecule from the high temperature region collides with a molecule from the low temperature region, it loses energy by collisions.

In liquids, the mechanism of heat is nearer to that of gases. However, the molecules are more closely spaced and intermolecular forces come into play.

(*ii*) **Convection.** '*Convection' is the transfer of heat within a fluid by mixing of one portion of the fluid with another*.

- Convection is possible only in a fluid medium and is *directly linked* with the *transport of medium itself.*
- Convection constitutes the *macroform* of the heat transfer since macroscopic particles of a fluid moving in space cause the heat exchange.
- The effectiveness of heat transfer by convection depends largely upon the mixing motion of the fluid.

This mode of heat transfer is met with in situations where energy is transferred as heat to a flowing fluid at any surface over which flow occurs. This mode is *basically conduction in a very thin fluid layer at the surface and then mixing caused by the flow.* The heat flow depends on the properties of fluid and is independent of the properties of the material of the surface. However, the shape of the surface will influence the flow and hence the heat transfer.

Free or natural convection. *Free or natural convection occurs where the fluid circulates by virtue of the natural differences in densities of hot and cold fluids ; the denser portions of the fluid move downward because of the greater force of gravity, as compared with the force on the less dense.*

Forced convection. *When the work is done to blow or pump the fluid, it is said to be forced convection.*

(*iii*) **Radiation.** *'Radiation' is the transfer of heat through space or matter by means other than conduction or convection.*

Radiation heat is thought of as *electromagnetic waves or quanta* (as convenient) an emanation of the same nature as light and radio waves. *All bodies radiate heat ; so a transfer of heat by radiation occurs because hot body emits more heat than it receives and a cold body receives more heat than it emits.* Radiant *energy* (being electromagnetic radiation) *requires no medium for propagation and will pass through a vacuum.*

Note. The rapidly oscillating molecules of the hot body produce electromagnetic waves in hypothetical medium called *ether.* These waves are identical with light waves, radio waves and X-rays, differ from them only in *wavelength* and travel with an approximate velocity of 3×10^8 m/s. These waves carry energy with them and *transfer it to the relatively slow-moving molecules of the cold body* on which they happen to fall. The molecular energy of the later increases and results in a rise of its temperature. Heat travelling by radiation is known as *radiant heat.*

The properties of radiant heat in general, are similar to those of light. Some of the properties are :

(*i*) It does not require the presence of a material medium for its transmission.

(*ii*) Radiant heat can be reflected from the surfaces and obeys the ordinary laws of reflection.

(*iii*) It travels with velocity of light.

(*iv*) Like light, it shows interference, diffraction and polarisation etc.

(*v*) It follows the law of inverse square.

The wavelength of heat radiations is longer than that of light waves, hence they are invisible to the eye.

28.2. HEAT TRANSFER BY CONDUCTION

28.2.1. Fourier's Law of Heat Conduction

Fourier's law of heat conduction is an empirical law based on observation and states as follows :

"The rate of flow of heat through a simple homogeneous solid is directly proportional to the area of the section at right angles to the direction of heat flow, and to change of temperature with respect to the length of the path of the heat flow".

Mathematically, it can be represented by the equation :

$$Q \propto A \cdot \frac{dt}{dx}$$

where, Q = Heat flow through a body per unit time (in watts), W,

A = Surface area of heat flow (*perpendicular to the direction of flow*), m^2,

dt = Temperature difference of the faces of block (homogeneous solid) of thickness 'dx' through which heat flows,°C or K, and

dx = Thickness of body in the direction of flow, m.

Thus, $$Q = -k \cdot A \frac{dt}{dx} \quad ...(28.1)$$

where, k = Constant of proportionality and is known as *thermal conductivity of the body.*

The –ve sign of k [eqn. (28.1)] is to take care of the decreasing temperature alongwith the direction of increasing thickness or the direction of heat flow. The temperature gradient $\frac{dt}{dx}$ is *always negative along positive* x direction and therefore the value of Q *becomes* +ve.

Assumptions :

The following are the *assumptions* on which Fourier's law is based :

1. Conduction of heat takes place under *steady state conditions.*
2. The heat flow is unidirectional.
3. The temperatures gradient is *constant* and the temperature profile is *linear.*
4. There is no internal heat generation.
5. The bounding surfaces are isothermal in character.
6. The material is homogeneous and isotropic (*i.e.*, the value of thermal conductivity is *constant in all directions*).

Some essential features of Fourier's Law :

Following are some *essential features of Fourier's law* :

1. It is applicable to all matter (may be solid, liquid or gas).
2. It is based on experimental evidence and cannot be derived from first principle.
3. It is a vector expression indicating that heat flow rate is in the direction of decreasing temperature and is normal to an isotherm.
4. It helps to define thermal conductivity 'k' (transport property) of the medium through which heat is conducted.

28.2.2. Thermal Conductivity of Materials

From eqn. (28.1), we have $k = \frac{Q}{A} \cdot \frac{dx}{dt}$

The value of $k = 1$ when $Q = 1, A = 1$ and $\frac{dt}{dx} = 1$

Now $k = \frac{Q}{1} \cdot \frac{dx}{dt}$ (unit of k : $\text{W} \times \frac{1}{\text{m}^2} \times \frac{\text{m}}{\text{K (or °C)}}$
= W/m K or W/m°C)

Thus, the *thermal conductivity of a material is defined* as follows :

"The amount of energy conducted through a body of unit area, and unit thickness in unit time when the difference in temperature between the faces causing heat flow is unit temperature difference".

It follows from eqn. (28.1) that materials with high thermal conductivities are good conductors of heat, whereas materials with low thermal conductives are good thermal insulator. *Conduction of heat occurs most readily in pure metals, less so in alloys, and much less readily in non-metals.* The very low thermal conductivities of certain thermal insulators *e.g.*, cork is due to their porosity, the air trapped within the material acting as an insulator.

Thermal conductivity (a property of material) depends essentially upon the following factors :

(*i*) Material structure

(*ii*) Moisture content

(*iii*) Density of the material

(*iv*) Pressure and temperature (operating conditions)

Thermal conductivities (average values at normal pressure and temperature) of some common materials are as under :

Material	*Thermal conductivity (k) (W/mK)*	*Material*	*Thermal conductivity (k) (W/mK)*
1. Silver	410	8. Asbestos sheet	0.17
2. Copper	385	9. Ash	0.12
3. Aluminum	225	10. Cork, felt	0.05–0.10
4. Cast-iron	55–65	11. Saw dust	0.07
5. Steel	20–45	12. Glass wool	0.03
6. Concrete	1.20	13. Water	0.55–0.7
7. Glass (window)	0.75	14. Freon	0.0083

Following points *regarding thermal conductivity—its variation for different materials and under different conditions* are worth noting :

1. Thermal conductivity of a material is due to flow of free electrons (in case of *metals*) and lattice vibrational waves (in case of *fluids*).
2. Thermal conductivity in case of *pure metals* is the highest (k = 10 to 400 W/m°C). It *decreases with increase in impurity*.

   The range of k for other materials is as follows :

   Alloys : k = 12 to 120 W/m°C

   Heat insulating and building materials :

   k = 0.023 to 2.9 W/m°C

   Liquids : k = 0.2 to 0.5 W/m°C

   Gases and vapours : k = 0.006 to 0.05 W/m°C.
3. Thermal conductivity of a metal varies considerably when it (metal) is heat treated or mechanically processed/formed.
4. Thermal conductivity of *most metals decreases with the increase in temperature* (*aluminium* and *uranium* being the *exceptions*).
   — In most of *liquids* the value of thermal conductivity tends to decrease with temperature (water being an exception) due to decrease in density with increase in temperature.
   — In case of gases the value of thermal conductivity *increases with temperature.* Gases with higher molecular weights have smaller thermal conductivities than with lower molecular weights. This is because the mean molecular path of gas molecules decreases with increase in density and k is directly proportional to the mean free path of the molecule.
5. The dependence of thermal conductivity (k) on temperature, for most materials is almost linear ;

$$k = k_0 (1 + \beta t) \quad ...(28.2)$$

   where, k_0 = Thermal conductivity at 0°C, and

   β = Temperature coefficient of thermal conductivity, 1/°C (It is usually *positive for non-metals* and *insulating materials* (magnesite bricks being the exception) and *negative for metallic conductors* (aluminium and certain non-ferrous alloys are the exceptions).
6. In case of solids and liquids, thermal conductivity (k) is only very weakly dependent on pressure ; in case of gases the value of k is independent of pressure (near standard atmospheric).
7. In case of non-metallic solids :
   — Thermal conductivity of porous materials depends upon the type of gas or liquid present in the voids.
   — Thermal conductivity of a *damp material is considerably higher than that of the dry material and water taken individually.*

— Thermal conductivity *increases with increase in density*.

8. The Wiedemann and Franz law (based on experiment results), regarding thermal and electrical conductivities of a material, states as follows :

"*The ratio of the thermal and electrical conductivities is the same for all metals at the same temperature ; and that the ratio is directly proportional to the absolute temperature of the metal.*"

Mathematically, $\frac{k}{\sigma} \propto T$

or $$\frac{k}{\sigma T} = C \quad \text{...(28.3)}$$

where, k = Thermal conductivity of metal at temperature T(K),

σ = Electrical conductivity of metal at temperature T(K), and

C = Constant (for all metals) is referred to as Lorenz number

($= 2.45 \times 10^{-8}$ WΩ/K^2 ; Ω stands for ohms).

This law conveys that the materials which are *good conductors of electricity are also good conductors of heat.*

28.2.3. Thermal Resistance (R_{th})

When two physical systems are described by similar equations and have similar boundary conditions, these are said to be *analogous*. The heat transfer processes may be compared by *analogy* with the flow of electricity in an electrical resistance. As the flow of electric current in the electrical resistance is directly proportional to potential difference (dV) ; similarly heat flow rate, Q, is directly proportional to temperature difference (dt), the driving force for heat conduction through a medium.

As per Ohm's law (in electric-circuit theory), we have

$$\text{Current } (I) = \frac{\text{Potential difference } (dV)}{\text{Electrical resistance } (R)} \quad \text{...(28.4)}$$

By analogy, the heat flow equation (Fourier's equation) may be written as

Heat flow rate

$$(Q) = \frac{\text{Temperature difference } (dt)}{\left(\frac{dx}{kA}\right)} \quad \text{...(28.5)}$$

By comparing eqns. (28.4) and (28.5), we find that I is analogous to Q, dV is analogous to dt and R is analogous to the quantity $\left(\frac{dx}{kA}\right)$. The quantity $\frac{dx}{kA}$ is called *thermal conduction resistance* $(\mathbf{R_{th}})_{cond.}$

i.e., $$(R_{th})_{cond.} = \frac{dx}{kA}$$

The reciprocal of the thermal resistance is called *thermal conductance*.

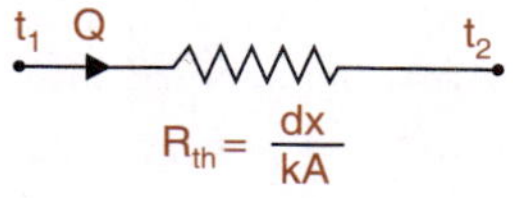

Fig. 28.1

It may be noted that *rules for combining electrical resistances in series and parallel apply equally well to thermal resistances.*

The concept of thermal resistance is quite helpful while making calculations for flow of heat.

28.2.4. General Heat Conduction Equation in Cartesian Coordinates

Consider an infinitesimal rectangular parallelopiped (volume element) of sides dx, dy and dz parallel, respectively, to the three axes (X, Y, Z) in a medium in which temperature is varying with location and time as shown in Fig. 28.2.

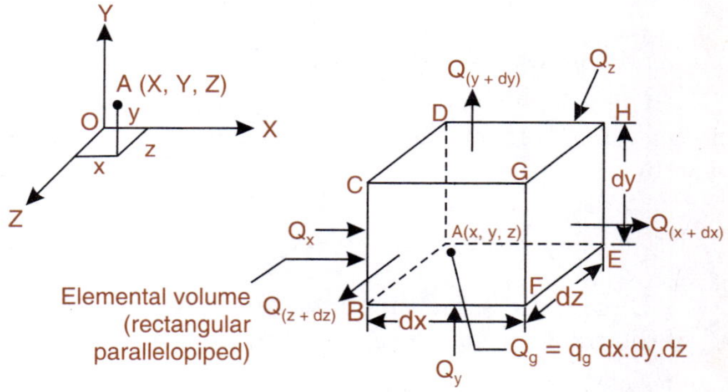

Fig. 28.2. Elemental volume for three-dimensional heat conduction analysis—Cartesian coordinates.

Let, t = Temperature at the left face $ABCD$; this temperature may be assumed uniform over the entire surface, since the area of this face can be made arbitrarily *small*.

$\frac{dt}{dx}$ = Temperature changes and rate of change along X-direction.

Then, $\left(\frac{\partial t}{\partial x}\right) dx$ = Change of temperature through distance dx, and

$t + \left(\frac{\partial t}{\partial x}\right) dx$ = Temperature on the right face $EFGH$ (at distance dx from the left face $ABCD$).

Further, let,

k_x, k_y, k_z = Thermal conductivities (direction characteristics of the material) along X, Y and Z-axes.

If the directional characteristics of a material are equal/same, it is called an *"Isotropic material"* and if unequal/different *'Anisotropic material'*.

q_g = Heat generated per unit volume per unit time.

Inside the control volume there may be heat sources due to flow of electric current in electric motors and generators, nuclear fission etc.

[Note. q_g may be function of position or time, or both].

ρ = Mass density of material.

c = Specific heat of the material.

Energy balance/equation for volume element :

Net heat accumulated in the element due to conduction of heat from all the coordinate directions considered (A) + heat generated within the element (B) = Energy stored in the element (C) ...(1)

Let, Q = Rate of *heat flow* in a direction, and

$Q' = (Q.d\tau)$ = *Total* heat flow (flux) in that direction (in time $d\tau$).

A. *Net heat accumulated in the element due to conduction of heat from all the directions considered :*

Quantity of heat flowing into the element from the left face $ABCD$ during the time interval $d\tau$ in X-direction is given by :

$$\text{Heat influx. } Q_x' = -k_x(dy.dz)\frac{\partial t}{\partial x}.d\tau \quad ...(i)$$

During the same time interval $d\tau$ the heat flowing out of the right face of control volume ($EFGH$) will be :

$$\text{Heat efflux. } Q'_{(x+dx)} = Q_x' + \frac{\partial}{\partial x}(Q_x')\,dx \quad ...(ii)$$

$\therefore$ Heat accumulation in the element due to heat flow in X-direction,

$$dQ_x' = Q_x' - \left[Q_x' + \frac{\partial}{\partial x}(Q_x')\,dx\right]$$

[Subtracting (*ii*) from (*i*)]

$$= -\frac{\partial}{\partial x}(Q_x')\,dx = -\frac{\partial}{\partial x}\left[-k_x(dy.dz)\frac{\partial t}{\partial x}.d\tau\right]dx$$

$$= \frac{\partial}{\partial x}\left[k_x\frac{\partial t}{\partial x}\right]dx.dy.dz.d\tau \quad ...(28.6)$$

Similarly the heat accumulated due to heat flow by conduction along Y and Z directions in time $d\tau$ will be :

$$dQ_y' = \frac{\partial}{\partial x}\left[k_y\frac{\partial t}{\partial y}\right]dx.dy.dz.d\tau \quad ...(28.7)$$

$$dQ_z' = \frac{\partial}{\partial z}\left[k_z\frac{\partial t}{\partial z}\right]dx.dy.dz.d\tau \quad ...(28.8)$$

$\therefore$ *Net heat accumulated* in the element due to conduction of heat from all the coordinate directions considered

$$= \frac{\partial}{\partial x}\left[k_x\frac{\partial t}{\partial x}\right]dx.dy.dz.d\tau + \frac{\partial}{\partial y}\left[k_y\frac{\partial t}{\partial y}\right]dx.dy.dz.d\tau + \frac{\partial}{\partial z}\left[k_z\frac{\partial t}{\partial z}\right]dx.dy.dz.d\tau$$

$$= \left[\frac{\partial}{\partial x}\left(k_x\frac{\partial t}{\partial x}\right) + \frac{\partial}{\partial y}\left(k_y\frac{\partial t}{\partial y}\right) + \frac{\partial}{\partial z}\left(k_z\frac{\partial t}{\partial z}\right)\right]dx.dy.dz.d\tau \quad ...(28.9)$$

B. *Total heat generated within the element* (Q_g') :

The total heat generated in the element is given by :

$$Q_g' = q_g(dx.dy.dz)\,d\tau \quad ...(28.10)$$

C. *Energy stored in the element :*

The total heat accumulated in the element due to heat flow along coordinate axes [eqn. (28.9)] and the heat generated within the element [eqn. (28.10)] together serve to increase the thermal energy of the element/lattice. This increase in thermal energy is given by :

$$\rho(dx.dy.dz)c.\frac{\partial t}{\partial \tau}.d\tau \quad ...(28.11)$$

[$\because$ Heat stored in the body = Mass of the body × specific heat of the body material × rise in the temperature of body].

Now, substituting eqns. (28.9), (28.10), (28.11), in the eqn. (1), we have

$$\left[\frac{\partial}{\partial x}\left(k_x\frac{\partial t}{\partial x}\right) + \frac{\partial}{\partial y}\left(k_y\frac{\partial t}{\partial y}\right) + \frac{\partial}{\partial z}\left(k_z\frac{\partial t}{\partial z}\right)\right]dx.dy.dz.d\tau + q_g(dx.dy.dz)d\tau = \rho(dx.dy.dz)c.\frac{\partial t}{\partial \tau}.d\tau$$

Dividing both sides by $dx.dy.dz.d\tau$, we have

$$\frac{\partial}{\partial x}\left(k_x\frac{\partial t}{\partial x}\right) + \frac{\partial}{\partial y}\left(k_y\frac{\partial t}{\partial y}\right) + \frac{\partial}{\partial z}\left(k_z\frac{\partial t}{\partial z}\right) + q_g = \rho.c.\frac{\partial t}{\partial \tau} \quad ...(28.12)$$

or, using the vector operator ∇, we get

$$\nabla.(k\nabla t) + q_g = \rho.c.\frac{\partial t}{\partial \tau}$$

This is known as the **general heat conduction equation** for **'non-homogeneous material', self heat generating'** and **'unsteady three-dimensional flow'**. This equation establishes in *differential form the relationship between the time and space variation of temperature at any point of solid through which heat flow by conduction takes place.*

General heat conduction equation for constant thermal conductivity :

In case of *homogeneous* (in which properties *e.g.*, specific heat, density, thermal conductivity etc. are same everywhere in the material) and *isotropic* (in which properties are independent of surface orientation) material, $k_x = k_y = k_z = k$ and diffusion equation (28.12) becomes

$$\frac{\partial^2 t}{\partial x^2} + \frac{\partial^2 t}{\partial y^2} + \frac{\partial^2 t}{\partial z^2} + \frac{q_g}{k} = \frac{\rho.c}{k} \cdot \frac{\partial t}{\partial \tau} = \frac{1}{\alpha} \cdot \frac{\partial t}{\partial \tau} \quad ...(28.13)$$

where $\alpha = \frac{k}{\rho.c} = \frac{\text{Thermal conductivity}}{\text{Thermal capacity}}$

The quantity, $\alpha = \frac{k}{\rho.c}$ is known as **thermal diffusivity.**

— The larger the value of α, the faster will the heat diffuse through the material and its temperature will change with time. This will result either due to a high value of thermal conductivity *k* or a low value of heat capacity *ρ.c*. A low value of heat capacity means the less amount of heat entering the element would be absorbed and used to raise its temperature and more would be available for onward transmission. Metals and gases have relatively high value of α and their response to temperature changes is quite rapid. The non-metallic solids and liquids respond slowly to temperature changes because of their relatively small value of thermal diffusivity.

— Thermal diffusivity is an important characteristic quantity for *unsteady conduction situations.*

Eqn. (28.13) by using Laplacian ∇^2, may be written as :

$$\nabla^2 t + \frac{q_g}{k} = \frac{1}{\alpha} \cdot \frac{\partial t}{\partial \tau} \quad ...[28.13\ (a)]$$

Eqn. (28.13), governs the temperature distribution under unsteady heat flow through a material which is homogeneous and isotropic.

Other simplified forms of heat conduction equation in cartesian coordinates :

(*i*) For the case when *no internal source of heat generation is present.* Eqn. (28.13) reduces to

$\frac{\partial^2 t}{\partial x^2} + \frac{\partial^2 t}{\partial y^2} + \frac{\partial^2 t}{\partial z^2} = \frac{1}{\alpha} \cdot \frac{\partial t}{\partial \tau}$ [Unsteady state $\left(\frac{\partial t}{\partial \tau} \neq 0\right)$ heat flow with no internal heat generation] or

$$\nabla^2 t = \frac{1}{\alpha} \cdot \frac{\partial t}{\partial \tau} \quad \textbf{...(Fourier's equation)} \quad ...(28.14)$$

(*ii*) Under the situations when temperature does not depend on time, the conduction then takes place in steady state $\left(i.e., \frac{\partial t}{\partial \tau} = 0\right)$ and the eqn. (28.13) reduces to

$$\frac{\partial^2 t}{\partial x^2} + \frac{\partial^2 t}{\partial y^2} + \frac{\partial^2 t}{\partial z^2} + \frac{q_g}{k} = 0$$

or

$$\nabla^2 t + \frac{q_g}{k} = 0$$

...(Poisson's equation) ...(28.15)

In the absence of internal heat generation, eqn. (28.15) reduces to

$$\frac{\partial^2 t}{\partial x^2} + \frac{\partial^2 t}{\partial y^2} + \frac{\partial^2 t}{\partial z^2} = 0$$

or $\nabla^2 t = 0$ **...(Laplace equation)** ...(28.16)

(*iii*) *Steady state and one-dimensional heat transfer*

$$\frac{\partial^2 t}{\partial x^2} + \frac{q_g}{k} = 0 \quad ...(28.17)$$

(*iv*) *Steady state, one-dimensional, without internal heat generation*

$$\frac{\partial^2 t}{\partial x^2} = 0 \quad ...(28.18)$$

(*v*) *Steady state, two dimensional, without internal heat generation*

$$\frac{\partial^2 t}{\partial x^2} + \frac{\partial^2 t}{\partial y^2} = 0 \quad ...(28.19)$$

(*vi*) *Unsteady state, one dimensional, without internal heat generation*

$$\frac{\partial^2 t}{\partial x^2} = \frac{1}{\alpha} \cdot \frac{\partial t}{\partial \tau} \quad ...(28.20)$$

28.2.5. Heat Conduction Through Plane and Composite Walls

28.2.5.1. Heat Conduction Through a Plane Wall

Refer to Fig. 28.3 (*a*). Consider a plane wall of homogeneous material through which heat is flowing *only in x-direction.*

Let, L = Thickness of the plane wall,
A = Cross-sectional area of the wall,
k = Thermal conductivity of the wall material, and
t_1, t_2 = Temperatures maintained at the two faces 1 and 2 of the wall, respectively.

The general heat conduction equation in cartesian coordinates is given by :

$$\frac{\partial^2 t}{\partial x^2} + \frac{\partial^2 t}{\partial y^2} + \frac{\partial^2 t}{\partial z^2} + \frac{q_g}{k} = \frac{1}{\alpha} \cdot \frac{\partial t}{\partial \tau} \quad ...[\text{Eqn. } (28.13)]$$

If the heat conduction takes place under the conditions, steady state $\left(\frac{\partial t}{\partial \tau} = 0\right)$, one-dimensional

$\left[\frac{\partial^2 t}{\partial y^2} = \frac{\partial^2 t}{\partial z^2} = 0\right]$ and with no internal heat generation $\left(\frac{q_g}{k} = 0\right)$ then the above equation is reduced to :

$$\frac{\partial^2 t}{\partial x^2} = 0, \text{ or } \frac{d^2 t}{dx^2} = 0 \qquad ...(28.21)$$

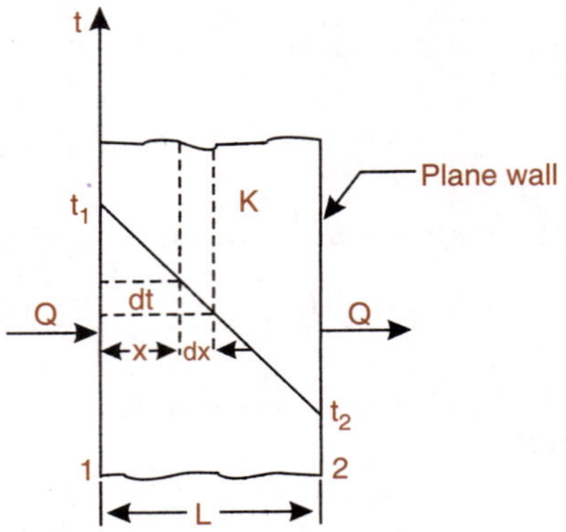

(a)

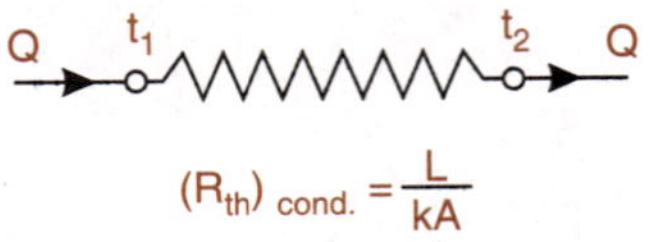

(b)

Fig. 28.3. Heat conduction through a plane wall.

By integrating the above differential twice, we have

$$\frac{\partial t}{\partial x} = C_1 \text{ and } t = C_1 x + C_2 \qquad ...(28.22)$$

where C_1 and C_2 are the arbitrary constants. The values of these constants may be calculated from the known boundary conditions as follows :

At $x = 0$ $\quad t = t_1$

At $x = L$ $\quad t = t_2$

Substituting the values in the eqn. (28.22), we get

$$t_1 = O + C_2 \text{ and } t_2 = C_1 L + C_2$$

After simplification, we have,

$$C_2 = t_1 \text{ and } C_1 = \frac{t_2 - t_1}{L}$$

Thus, the eqn. (28.22) reduces to :

$$t = \left(\frac{t_2 - t_1}{L}\right) x + t_1 \qquad ...(28.23)$$

The eqn. (28.23) indicates that *temperature distribution across a wall is linear* and is *independent of thermal conductivity*. Now heat through the plane wall can be found by using Fourier's equation as follows :

$$Q = -kA \frac{dt}{dx}, \left(\text{where } \frac{dt}{dx} = \text{temperature gradient}\right)$$

...[Eqn. (28.1)]

But, $\frac{dt}{dx} = \frac{d}{dx}\left[\left(\frac{t_2 - t_1}{L}\right) x + t_1\right] = \frac{t_2 - t_1}{L}$

$$\therefore \quad Q = -kA \frac{(t_2 - t_1)}{L} = \frac{kA\,(t_1 - t_2)}{L} \qquad ...(28.24)$$

Eqn. (28.24) can be written as :

$$Q = \frac{(t_1 - t_2)}{(L/kA)} = \frac{(t_1 - t_2)}{(R_{th})_{cond.}} \qquad ...(28.25)$$

where, $(R_{th})_{cond.}$ = Thermal resistance to heat conduction. Fig. 28.3 (*b*) shows the *equivalent thermal circuit* for heat flow through the plane wall.

Let us now find out the condition when instead of space, weight is the main criterion for selection of the insulation of a plane wall.

Thermal resistance (conduction) of the wall,

$$(R_{th})_{cond.} = \frac{L}{kA} \qquad ...(i)$$

Weight of the wall, $\quad W = \rho A L \qquad ...(ii)$

Eliminating L from (*i*) and (*ii*), we get

$$W = \rho A.(R_{th})_{cond.}\, kA = (\rho.k)A^2.(R_{th})_{cond.} \qquad ...(28.26)$$

The eqn. (28.26) stipulates the condition that, for a specified thermal resistance, the *lightest insulation will be one which has the smallest product of density* (ρ) *and thermal conductivity* (*k*).

28.2.5.2. Heat Conduction Through a Composite Wall

Refer to Fig. 28.4 (*a*). Consider the transmission of heat through a composite wall consisting of a number of slabs.

Let L_A, L_B, L_C = Thicknesses of slabs *A*, *B* and *C* respectively (also called path lengths),

k_A, k_B, k_C = Thermal conductivities of the slabs *A*, *B* and *C* respectively,

$t_1, t_4 (t_1 > t_4)$ = Temperatures at the wall surfaces 1 and 4 respectively, and

t_2, t_3 = Temperatures at the interfaces 2 and 3 respectively.

Since the quantity of heat transmitted per unit time through each slab/layer is same, we have

$$Q = \frac{k_A . A\,(t_1 - t_2)}{L_A} = \frac{k_B . A\,(t_2 - t_3)}{L_B} = \frac{k_C . A\,(t_3 - t_4)}{L_C}$$

(Assuming that there is a perfect contact between the layers and no temperature drop occurs across the interface between the materials).

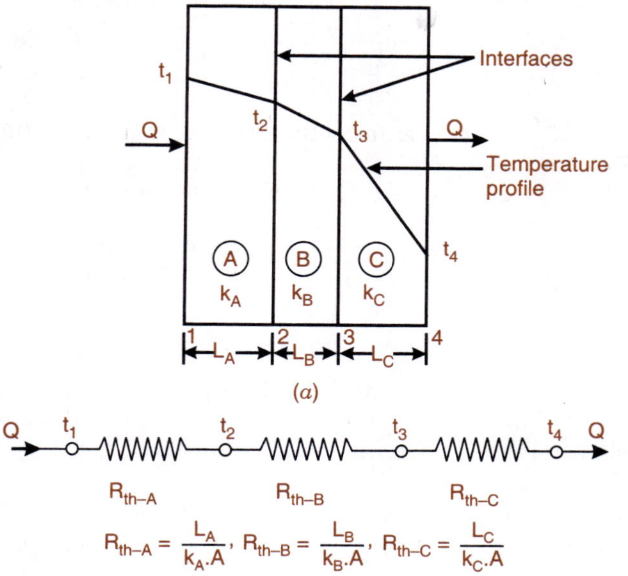

Fig. 28.4. Steady state conduction through a composite wall.

Rearranging the above expression, we get

$$t_1 - t_2 = \frac{Q\,.\,L_A}{k_A\,.\,A} \qquad ...(i)$$

$$t_2 - t_3 = \frac{Q\,.\,L_B}{k_B\,.\,A} \qquad ...(ii)$$

$$t_3 - t_4 = \frac{Q\,.\,L_C}{k_C\,.\,A} \qquad ...(iii)$$

Adding (*i*), (*ii*) and (*iii*), we have

$$(t_1 - t_4) = Q\left[\frac{L_A}{k_A\,.\,A} + \frac{L_B}{k_B\,.\,A} + \frac{L_C}{k_C\,.\,A}\right]$$

or

$$Q = \frac{A\,(t_1 - t_4)}{\left[\frac{L_A}{k_A} + \frac{L_B}{k_B} + \frac{L_C}{k_C}\right]} \qquad ...(28.27)$$

or

$$Q = \frac{(t_1 - t_4)}{\left[\frac{L_A}{k_A\,.\,A} + \frac{L_B}{k_B\,.\,A} + \frac{L_C}{k_C\,.\,A}\right]}$$

$$= \frac{(t_1 - t_4)}{[R_{th\text{-}A} + R_{th\text{-}B} + R_{th\text{-}C}]} \qquad ...(28.28)$$

If the composite wall consists of *n* slabs/layers, then

$$Q = \frac{[t_1 - t_{(n+1)}]}{\sum_{1}^{n} \frac{L}{kA}} \qquad ...(28.29)$$

In order to solve more complex problems involving both series and parallel thermal resistances, the electrical analogy may be used. A typical problem and its analogous electric circuit are shown in Fig. 28.5.

$$Q = \frac{\Delta\, t_{overall}}{\Sigma\, R_{th}} \qquad ...(28.30)$$

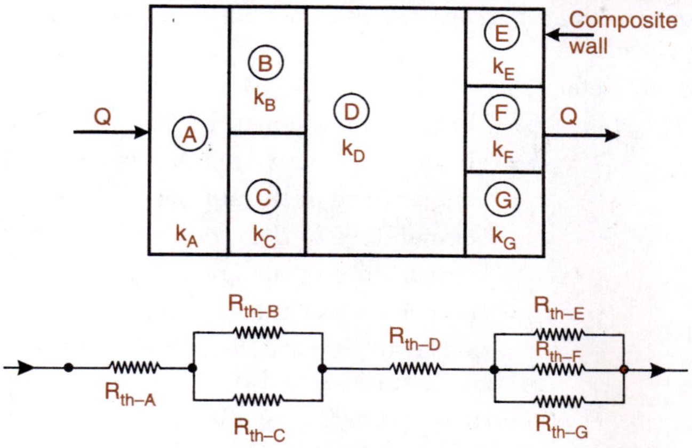

Fig. 28.5. Series and parallel one-dimensional heat transfer through a composite wall and electrical analog.

Thermal contact resistance. In a composite (multi-layer) wall, the calculations of heat flow are made on the assumptions : (*i*) The contact between the adjacent layers is perfect, (*ii*) At the interface there is no fall of temperature, and (*iii*) At the interface the temperature is continuous, although there is discontinuity in temperature gradient. In real systems, however, due to surface roughness and void spaces (usually filled with air) the contact surfaces *touch only at discrete locations.* Thus there is not a single plane of contact, which means that the area available for the flow of heat at the interface will be small compared to geometric face area. Due to this reduced area and presence of air voids, *a large resistance to heat flow at the interface occurs.* This resistance is known as *thermal contact resistance* and it *causes temperature drop between two materials at the interface* as shown in Fig. 28.6.

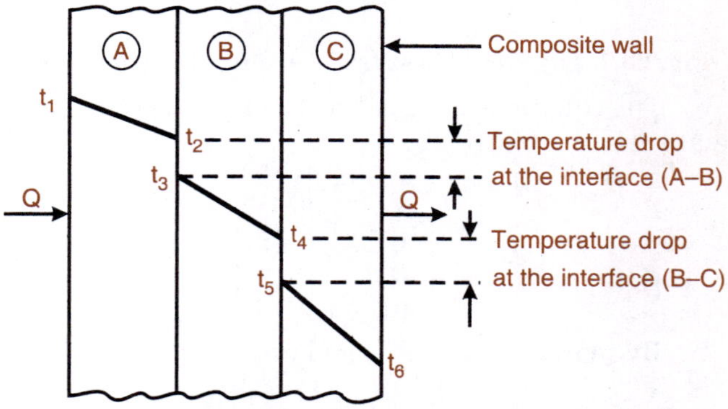

Fig. 28.6. Temperature drops at the interfaces.

Refer to Fig. 28.6. The contact resistances are given by

$$(R_{th\text{-}AB})_{\text{cond.}} = \frac{(t_2 - t_3)}{Q/A} \text{ and } (R_{th\text{-}BC})_{\text{cont.}} = \frac{(t_4 - t_5)}{Q/A}.$$

28.2.6. The Overall Heat-transfer Coefficient

While dealing with the problems of fluid to fluid heat transfer across a metal boundary, it is usual to adopt an overall heat transfer coefficient *U* which *gives the heat transmitted per unit area per unit time per degree temperature difference between the bulk fluids on each side of the metal.*

Refer to Fig. 28.7.

Let, L = Thickness of the metal wall,

k = Thermal conductivity of the wall material,

t_1 = Temperature of the surface-1,

t_2 = Temperature of the surface-2,

t_{hf} = Temperature of the hot fluid,

t_{cf} = Temperature of the cold fluid,

h_{hf} = Heat transfer coefficient from hot fluid to metal surface, and

h_{cf} = Heat transfer coefficient from metal surface to cold fluid.

(The suffices *hf* and *cf* stand for hot fluid and cold fluid respectively.)

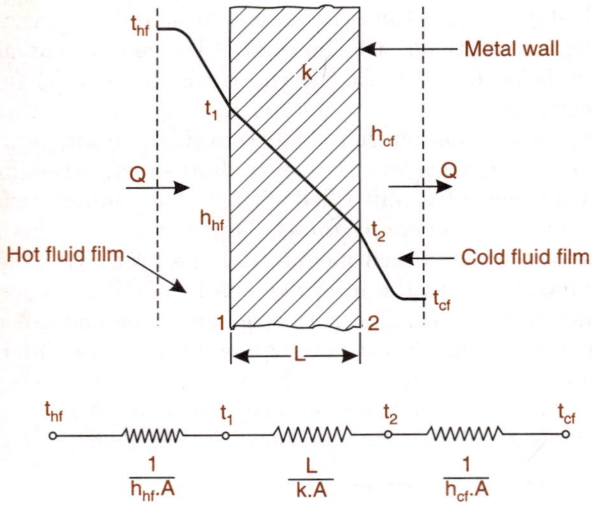

Fig. 28.7. The overall heat transfer through a plane wall.

The equations of heat flow through the fluid and the metal surface are given by

$$Q = h_{hf} \, . \, A(t_{hf} - t_1) \quad \text{...}(i)$$

$$Q = \frac{k_A \, (t_1 - t_2)}{L} \quad \text{...}(ii)$$

$$Q = h_{cf} \, . \, A(t_2 - t_{cf}) \quad \text{...}(iii)$$

By rearranging (*i*), (*ii*) and (*iii*), we get

$$t_{hf} - t_1 = \frac{Q}{h_{hf} . A} \quad \text{...}(iv)$$

$$t_1 - t_2 = \frac{QL}{k \, . \, A} \quad \text{...}(v)$$

$$t_2 - t_{cf} = \frac{Q}{h_{cf} \, . \, A} \quad \text{...}(vi)$$

Adding (*iv*), (*v*) and (*vi*), we get

$$t_{hf} - t_{cf} = Q \left[\frac{1}{h_{hf} \, . \, A} + \frac{L}{k \, . \, A} + \frac{1}{h_{cf} \, . \, A} \right]$$

or,
$$Q = \frac{A \, (t_{hf} - t_{cf})}{\frac{1}{h_{hf}} + \frac{L}{k} + \frac{1}{h_{cf}}} \quad \text{...(28.31)}$$

If U is the overall coefficient of heat transfer, then

$$Q = U.A \, (t_{hf} - t_{cf}) = \frac{A \, (t_{hf} - t_{cf})}{\frac{1}{h_{hf}} + \frac{L}{k} + \frac{1}{h_{cf}}}$$

or,
$$U = \frac{1}{\frac{1}{h_{hf}} + \frac{L}{k} + \frac{1}{h_{cf}}} \quad \text{...(28.32)}$$

It may be noticed from the above equation that if the individual coefficients differ greatly in magnitude only a change in the *least* will have significant effect on the rate of heat transfer.

Example 28.1. *The inner surface of a plane brick wall is at 60°C and the outer surface is at 35°C. Calculate the rate of heat transfer per m^2 of surface area of the wall, which is 220 mm thick. The thermal conductivity of the brick is 0.51 W/m°C.*

Solution. Temperature of the inner surface of the wall

$$t_1 = 60°\text{C}.$$

Temperature of the outer surface of the wall,

$$t_2 = 35°\text{C}$$

The thickness of the wall, L = 220 mm = 0.22 m

Thermal conductivity of the brick,

$$k = 0.51 \text{ W/m°C}$$

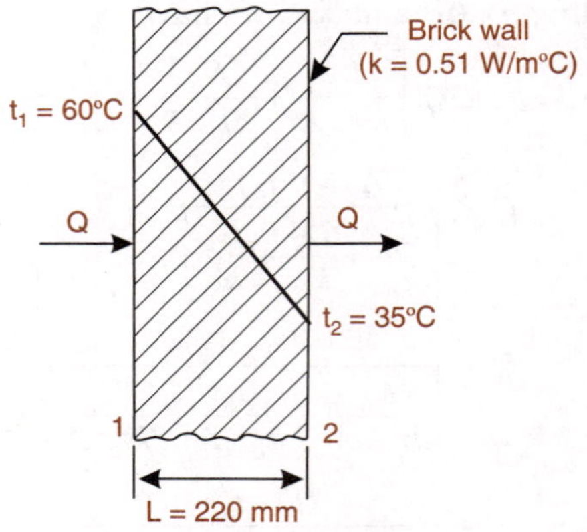

Fig. 28.8

Rate of heat transfer per m²: q

Rate of heat transfer per unit area,

$$q = \frac{Q}{A} = \frac{k(t_1 - t_2)}{L}$$

or,
$$q = \frac{0.51 \times (60 - 35)}{0.22} = \mathbf{57.95 \ W/m^2}. \quad \textbf{(Ans.)}$$

Example 28.2. *A reactor's wall 320 mm thick, is made up of an inner layer of fire brick (k = 0.84 W/m°C) covered with a layer of insulation (k = 0.16 W/m°C). The reactor operates* *at a temperature of 1325°C and the ambient temperature is 25°C.*

(i) Determine the thickness of fire brick and insulation which gives minimum heat loss.

(ii) Calculate the heat loss presuming that the insulating material has a maximum temperature of 1200°C.

If the calculated heat loss is not acceptable, then state whether addition of another layer of insulation would provide a satisfactory solution.

Solution. Refer to Fig. 28.9.

Given : $t_1 = 1325°C$; $t_2 = 1200°C$, $t_3 = 25°C$;

$L_A + L_B = L = 320$ mm or 0.32 m

$\therefore \quad L_B = (0.32 - L_A)$; ...(i)

$k_A = 0.84$ W/m°C ;

$k_B = 0.16$ W/m°C.

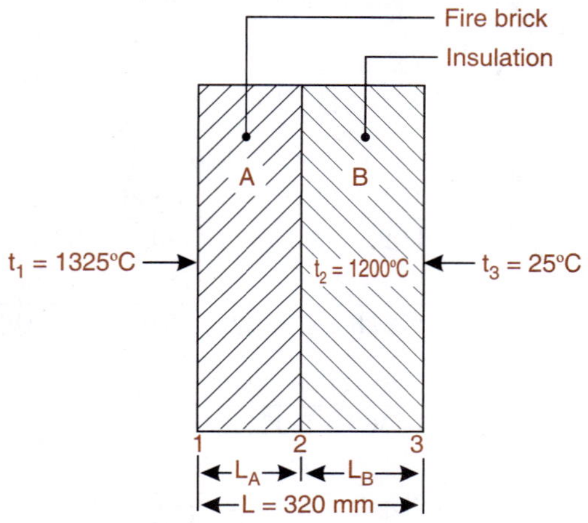

Fig. 28.9

(i) **L_A : ; L_B :**

The heat flux, under steady state conditions, is constant throughout the wall and is same for each layer. Then for *unit area* of wall,

$$q = \frac{t_1 - t_3}{L_A/k_A + L_B/k_B} = \frac{t_1 - t_2}{L_A/k_A} = \frac{t_2 - t_3}{L_B/k_B}$$

Considering first two quantities, we have

$$\frac{(1325 - 25)}{L_A/0.84 + L_B/0.16} = \frac{(1325 - 1200)}{L_A/0.84}$$

or $$\frac{1300}{1.190\, L_A + 6.25(0.32 - L_A)} = \frac{105}{L_A}$$

or $$\frac{1300}{1.190\, L_A + 2 - 6.25\, L_A} = \frac{105}{L_A}$$

or $$\frac{1300}{2 - 5.06\, L_A} = \frac{105}{L_A}$$

or $$1300 L_A = 105\,(2 - 5.06\, L_A)$$

or $$1300\, L_A = 210 - 531.3\, L_A$$

or $$L_A = \frac{210}{(1300 + 531.3)}$$

$= \mathbf{0.1146\ m}$ or **114.6 mm. (Ans.)**

$\therefore$ Thickness of insulation

$L_B = 320 - 114.6 = \mathbf{205.4\ mm.}$ **(Ans.)**

(ii) **Heat loss per unit area,** q :

Heat loss per unit area, $q = \dfrac{t_1 - t_2}{L_A/k_A} = \dfrac{1325 - 1200}{0.1146/0.84}$

$= \mathbf{916.23\ W/m^2}$. **(Ans.)**

If another layer of insulating material is added, the heat loss from the wall will reduce ; consequently the temperature drop across the fire brick lining will drop and the interface temperature t_2 will *rise*. As the interface temperature is *already fixed*. Therefore, a *satisfactory solution will not be available* by *adding layer of insulation.*

Example 28.3. *An exterior wall of a house may be approximated by a 0.1 m layer of common brick (k = 0.7 W/m°C) followed by a 0.04 m layer of gypsum plaster (k = 0.48 W/m°C). What thickness of loosely packed rock wool insulation (k = 0.065 W/m°C) should be added to reduce the heat loss or (gain) through the wall by 80 per cent ?*

(N.U.)

Solution. Refer to Fig. 28.10.

Thickness of common brick, $L_A = 0.1$ m

Thickness of gypsum plaster, $L_B = 0.04$ m

Thickness of rock wool, $L_C = x$ (in m) = ?

Thermal conductivities :

Common brick, $k_A = 0.7$ W/m°C

Gypsum plaster, $k_B = 0.48$ W/m°C

Rock wool, $k_C = 0.065$ W/m°C

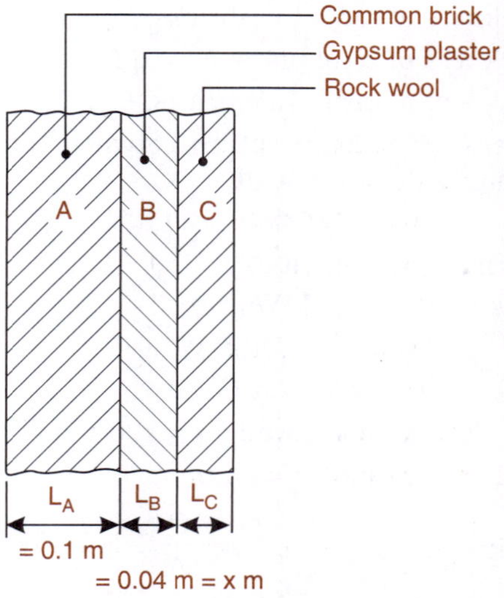

Fig. 28.10

Case I. *Rock wool insulation* **not** *used :*

$$Q_1 = \frac{A\,(\Delta t)}{\dfrac{L_A}{k_A} + \dfrac{L_B}{k_B}} = \frac{A(\Delta t)}{\dfrac{0.1}{0.7} + \dfrac{0.04}{0.48}} \qquad ...(i)$$

Case II. *Rock wool insulation used :*

$$Q_2 = \frac{A\,(\Delta t)}{\frac{L_A}{k_A} + \frac{L_B}{k_B} + \frac{L_C}{k_C}} = \frac{A(\Delta t)}{\frac{0.1}{0.7} + \frac{0.04}{0.48} + \frac{x}{0.065}} \quad ...(ii)$$

But, $Q_2 = (1 - 0.8)Q_1 = 0.2\ Q_1$ ...(Given)

$$\therefore \quad \frac{A(\Delta t)}{\frac{0.1}{0.7} + \frac{0.04}{0.48} + \frac{x}{0.065}} = 0.2 \times \frac{A(\Delta t)}{\frac{0.1}{0.7} + \frac{0.04}{0.48}}$$

or, $$\frac{0.1}{0.7} + \frac{0.04}{0.48} = 0.2\left[\frac{0.1}{0.7} + \frac{0.04}{0.48} + \frac{x}{0.065}\right]$$

or, $$0.1428 + 0.0833 = 0.2\ [0.1428 + 0.0833 + 15.385\ x]$$

or, $$0.2261 = 0.2\ (0.2261 + 15.385\ x)$$

or, $$x = 0.0588 \text{ m or } 58.8 \text{ mm}$$

Thus, the *thickness of rock wool insulation should be* **58.8 mm. (Ans.)**

Example 28.4. *A furnace wall consists of 200 mm layer of refractory bricks, 6 mm layer of steel plate and a 100 mm layer of insulation bricks. The maximum temperature of the wall is 1150°C on the furnace side and the minimum temperature is 40°C on the outermost side of the wall. An accurate energy balance over the furnace shows that the heat loss from the wall is 400 W/m². It is known that there is a thin layer of air between the layers of refractory bricks and steel plate. Thermal conductivities for the three layers are 1.52, 45 and 0.138 W/m°C respectively. Find :*

(i) To how many millimetres of insulation brick is the air layer equivalent ?

(ii) What is the temperature of the outer surface of the steel plate ?

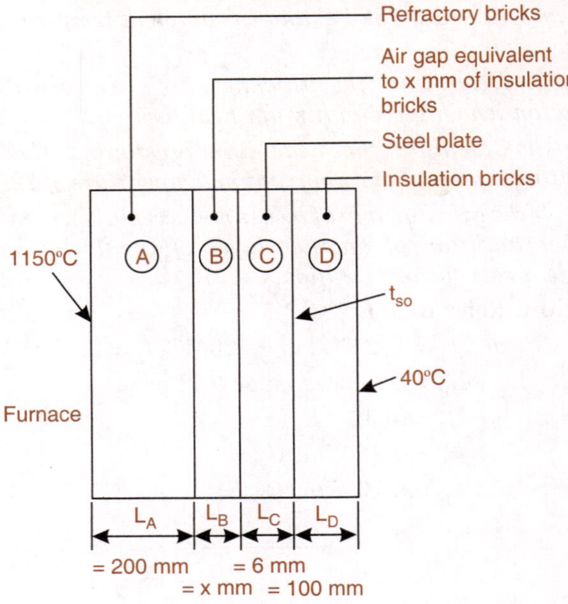

Fig. 28.11

Solution. Refer to Fig. 28.11.

Thickness of refractory bricks, $L_A = 200$ mm $= 0.2$ m

Thickness of steel plate, $L_C = 6$ mm $= 0.006$ m

Thickness of insulation bricks, $L_D = 100$ mm $= 0.1$ m

Difference of temperature between the innermost and outermost side of the wall,

$$\Delta t = 1150 - 40 = 1110°\text{C}$$

Thermal conductivities :

$$k_A = 1.52 \text{ W/m°C} ;$$
$$k_B = k_D = 0.138 \text{ W/m°C} ;$$
$$k_C = 45 \text{ W/m°C}$$

Heat loss from the wall, $q = 400$ W/m²

(i) **The value of x = (L_C) :**

We know, $$Q = \frac{A\,.\,\Delta t}{\Sigma \frac{L}{k}} \text{ or } \frac{Q}{A} = q = \frac{\Delta t}{\Sigma \frac{L}{k}}$$

or $$400 = \frac{1110}{\frac{L_A}{k_A} + \frac{L_B}{k_B} + \frac{L_C}{k_C} + \frac{L_D}{k_D}}$$

or $$400 = \frac{1110}{\frac{0.2}{1.52} + \frac{(x/1000)}{0.138} + \frac{0.006}{45} + \frac{0.1}{0.138}}$$

$$= \frac{1110}{0.1316 + 0.0072x + 0.00013 + 0.7246}$$

$$= \frac{1110}{0.8563 + 0.0072x}$$

or $$0.8563 + 0.0072\ x = \frac{1110}{400} = 2.775$$

or $$x = \frac{2.775 - 0.8563}{0.0072}$$

$$= \mathbf{266.5\ mm.} \quad \textbf{(Ans.)}$$

(ii) **Temperature of the outer surface of the steel plate t_{so} :**

$$q = 400 = \frac{(t_{so} - 40)}{L_D/k_D}$$

or $$400 = \frac{(t_{so} - 40)}{(0.1/0.138)} = 1.38(t_{so} - 40)$$

or $$t_{so} = \frac{400}{1.38} + 40 = \mathbf{329.8°C.} \quad \textbf{(Ans.)}$$

Example 28.5. *Find the heat flow rate through the composite wall as shown in Fig. 28.12. Assume one dimensional flow.*

$k_A = 150$ W/m°C,

$k_B = 30$ W/m°C,

$k_C = 65$ W/m°C and

$k_D = 50$ W/m°C. **(M.U.)**

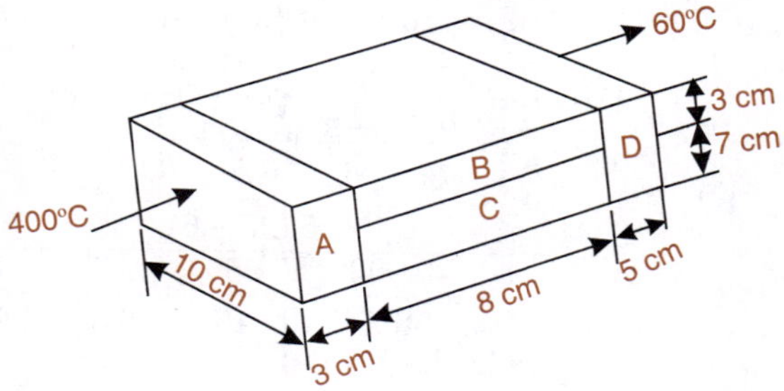

Fig. 28.12

Solution. The thermal circuit for heat flow in the given composite system (shown in Fig. 28.12) has been illustrated in Fig. 28.13.

Thickness :

$L_A = 3 \text{ cm} = 0.03 \text{ m}$;
$L_B = L_C = 8 \text{ cm} = 0.08 \text{ m}$;
$L_D = 5 \text{ cm} = 0.05 \text{ m}$

Areas :

$A_A = 0.1 \times 0.1 = 0.01 \text{ m}^2$;
$A_B = 0.1 \times 0.03 = 0.003 \text{ m}^2$
$A_C = 0.1 \times 0.07 = 0.007 \text{ m}^2$;
$A_D = 0.1 \times 0.1 = 0.01 \text{ m}^2$

Heat flow rate, Q :

The thermal resistances are given by

$$R_{th\text{-}A} = \frac{L_A}{k_A A_A} = \frac{0.03}{150 \times 0.01} = 0.02$$

$$R_{th\text{-}B} = \frac{L_B}{k_B A_B} = \frac{0.08}{30 \times 0.003} = 0.89$$

$$R_{th\text{-}C} = \frac{L_C}{k_C A_C} = \frac{0.08}{65 \times 0.007} = 0.176$$

$$R_{th\text{-}D} = \frac{L_D}{k_D A_D} = \frac{0.05}{50 \times 0.01} = 0.1$$

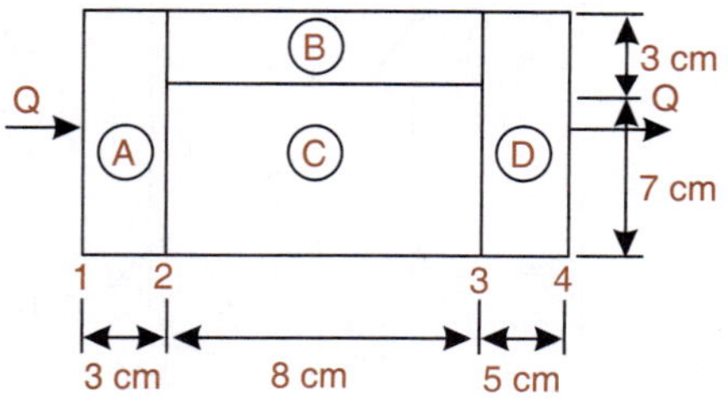

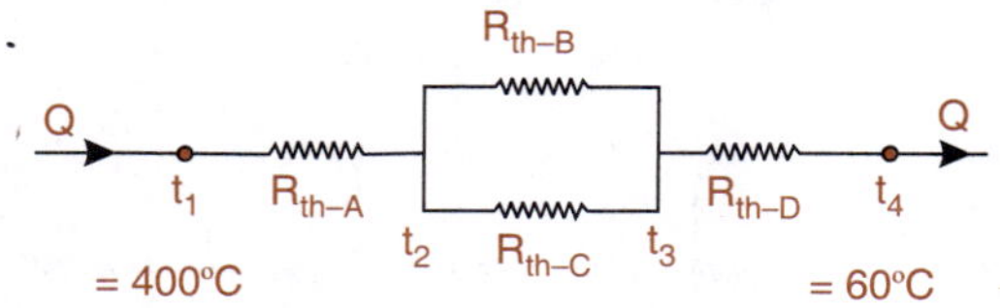

Fig. 28.13. Thermal circuit for heat flow in the composite system.

The equivalent thermal resistance for the parallel thermal resistances $R_{th\text{-}B}$ and $R_{th\text{-}C}$ is given by :

$$\frac{1}{(R_{th})_{eq.}} = \frac{1}{R_{th-B}} + \frac{1}{R_{th-C}} = \frac{1}{0.89} + \frac{1}{0.176} = 6.805$$

$$\therefore \quad (R_{th})_{eq.} = \frac{1}{6.805} = 0.147$$

Now, the total thermal resistance is given by

$$(R_{th})_{total} = R_{th\text{-}A} + (R_{th})_{eq.} + R_{th\text{-}D}$$
$$= 0.02 + 0.147 + 0.1 = 0.267$$

$$\therefore \quad Q = \frac{(\Delta t)_{overall}}{(R_{th})_{total}} = \frac{(400 - 60)}{0.267}$$

$= \mathbf{1273.4\ W.}$ **(Ans.)**

Example 28.6. *A mild steel tank of wall thickness 12 mm contains water at 95°C. The thermal conductivity of mild steel is 50 W/m°C, and the heat transfer coefficients for the inside and outside the tank are 2850 and 10 W/m²°C, respectively. If the atmospheric temperature is 15°C, calculate :*

(*i*) *The rate of heat loss per m² of the tank surface area ;*

(*ii*) *The temperature of the outside surface of the tank.*

Solution. Refer to Fig. 28.14.

Thickness of mild steel tank wall

$L = 12 \text{ mm} = 0.012 \text{ m}$

Temperature of water, $t_{hf} = 95°\text{C}$
Temperature of air, $t_{cf} = 15°\text{C}$
Thermal conductivity of mild steel, $k = 50 \text{ W/m°C}$
Heat transfer coefficients :
Hot fluid (water), $h_{hf} = 2850 \text{ W/m}^2\text{°C}$
Cold fluid (air), $h_{cf} = 10 \text{ W/m}^2\text{°C}$

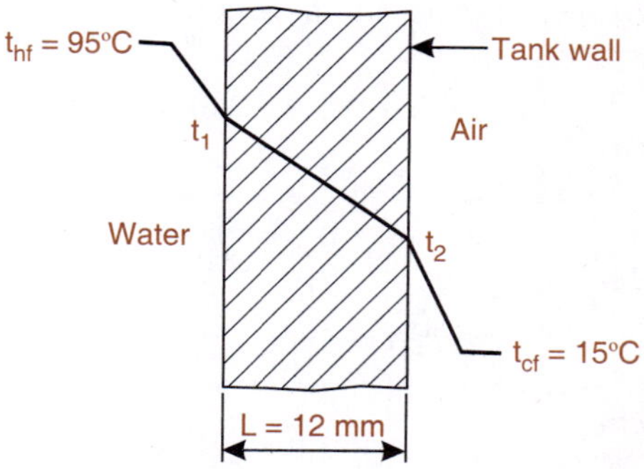

Fig. 28.14

(*i*) **Rate of heat loss per m² of the tank surface area, q :**

Rate of heat loss per m² of tank surface,

$$q = UA(t_{hf} - t_{cf})$$

The overall heat transfer coefficient, U is found from the relation ;

$$\frac{1}{U} = \frac{1}{h_{hf}} + \frac{L}{k} + \frac{1}{h_{cf}} = \frac{1}{2850} + \frac{0.012}{50} + \frac{1}{10}$$
$$= 0.0003508 + 0.00024 + 0.1 = 0.1006$$

$\therefore \quad U = \dfrac{1}{0.1006} = 9.94 \text{ W/m}^2\text{°C}$

$\therefore \quad q = 9.94 \times 1 \times (95 - 15) = $ **795.2 W/m²**. **(Ans.)**

(*ii*) **Temperature of the outside surface of the tank, t_2 :**

We know that, $q = h_{cf} \times 1 \times (t_2 - t_{cf})$

or $\quad 795.2 = 10(t_2 - 15)$

or $\quad t_2 = \dfrac{795.2}{10} + 15 = $ **94.52°C. (Ans.)**

Example 28.7. *The interior of a refrigerator having inside dimensions of 0.5 m × 0.5 m base area and 1 m height, is to be maintained at 6°C. The walls of the refrigerator are constructed of two mild steel sheets 3 mm thick (k = 46.5 W/m°C) with 50 mm of glass wool insulation (k = 0.046 W/m°C) between them. If the average heat transfer coefficients at the inner and outer surfaces are 11.6 W/m²°C and 14.5 W/m²°C respectively, calculate :*

(*i*) *The rate at which heat must be removed from the interior to maintain the specified temperature in the kitchen at 25°C, and*

(*ii*) *The temperature on the outer surface of the metal sheet.*

Solution. Refer to Fig. 28.15

Given : $L_A = L_C = 3 \text{ mm} = 0.003 \text{ m}$;

$L_B = 50 \text{ mm} = 0.05 \text{ m}$;

$k_A = k_C = 46.5 \text{ W/m°C}$; $k_B = 0.046 \text{ W/m°C}$;

$h_0 = 11.6 \text{ W/m}^2\text{°C}$; $h_i = 14.5 \text{ W/m}^2\text{°C}$;

$t_0 = 25\text{°C}$; $t_i = 6\text{°C}$.

The total area through which heat is coming into the refrigerator

$$A = 0.5 \times 0.5 \times 2 + 0.5 \times 1 \times 4 = 2.5 \text{ m}^2$$

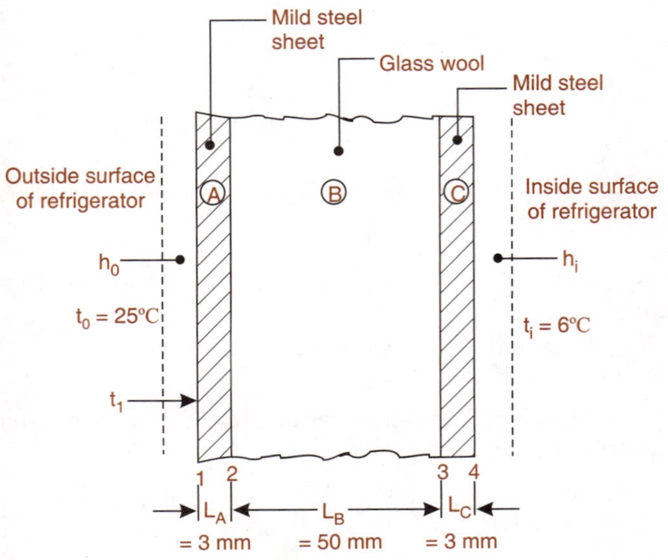

Fig. 28.15

(*i*) **The rate of removal of heat, Q :**

$$Q = \frac{A(t_0 - t_i)}{\dfrac{1}{h_o} + \dfrac{L_A}{k_A} + \dfrac{L_B}{k_B} + \dfrac{L_C}{k_C} + \dfrac{1}{h_i}}$$

$$= \frac{2.5\,(25 - 6)}{\dfrac{1}{11.6} + \dfrac{0.003}{46.5} + \dfrac{0.05}{0.046} + \dfrac{0.003}{46.5} + \dfrac{1}{14.5}}$$

= **38.2 W. (Ans.)**

(*ii*) **The temperature at the outer surface of the metal sheet, t_1 :**

$$Q = h_o A(25 - t_1)$$

or $\quad 38.2 = 11.6 \times 2.5\,(25 - t_1)$

or $\quad t_1 = 25 - \dfrac{38.2}{11.6 \times 2.5} = $ **23.68°C. (Ans.)**

Example 28.8. *A furnace wall is made up of three layers of thicknesses 250 mm, 100 mm and 150 mm with thermal conductivities of 1.65, k and 9.2 W/m°C respectively. The inside is exposed to gases at 1250°C with a convection coefficient of 25 W/m²°C and the inside surface is at 1100°C, the outside surface is exposed air at 25°C with convection coefficient of 12 W/m²°C. Determine :*

(*i*) *The unknown thermal conductivity 'k' ;*

(*ii*) *The overall heat transfer coefficient ;*

(*iii*) *All surface temperatures.*

Solution. $L_A = 250 \text{ mm} = 0.25 \text{ m}$; $L_B = 100 \text{ mm} = 0.1 \text{ m}$;

$L_C = 150 \text{ mm} = 0.15 \text{ m}$; $k_A = 1.65 \text{ W/m°C}$;

$k_C = 9.2 \text{ W/m°C}$; $\quad t_{hf} = 1250\text{°C}$;

$t_1 = 1100\text{°C}$

$h_{hf} = 25 \text{ W/m}^2\text{°C}$; $\quad h_{cf} = 12 \text{ W/m}^2\text{°C}$

(*i*) **Thermal conductivity, k (= k_B) :**

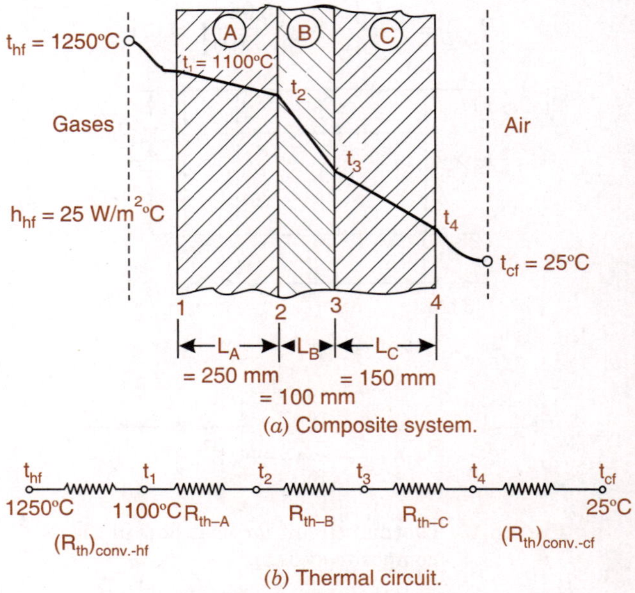

Fig. 28.16

The rate of heat transfer per unit area of the furnace wall,

$$q = h_{hf}(t_{hf} - t_1)$$
$$= 25(1250 - 1100) = 3750 \text{ W/m}^2$$

Also, $$q = \frac{(\Delta t)_{overall}}{(R_{th})_{total}}$$

or $$q = \frac{(t_{hf} - t_{cf})}{(R_{th})_{conv-hf} - R_{th-A} + R_{th-B} + R_{th-C} + (R_{th})_{conv-cf}}$$

or $$3750 = \frac{(1250 - 25)}{\frac{1}{h_{hf}} + \frac{L_A}{k_A} + \frac{L_B}{k_B} + \frac{L_C}{k_C} + \frac{1}{h_{cf}}}$$

or $$3750 = \frac{1225}{\frac{1}{25} + \frac{0.25}{1.65} + \frac{0.1}{k_B} + \frac{0.15}{9.2} + \frac{1}{12}}$$

$$= \frac{1225}{0.04 + 0.1515 + \frac{0.1}{k_B} + 0.0163 + 0.0833}$$

$$= \frac{1225}{0.2911 + \frac{0.1}{k_B}}$$

or $$3750\left(0.289 + \frac{0.1}{k_B}\right) = 1225$$

or $$\frac{0.1}{k_B} = \frac{1225}{3750} - 0.2911 = 0.0355$$

$\therefore$ $$k_B = k = \frac{0.1}{0.0355} = \mathbf{2.817\ W/m^2°C.}$$ **(Ans.)**

***(ii)* The overall transfer coefficient, U :**

The overall heat transfer coefficient, $U = \frac{1}{(R_{th})_{total}}$

Now,

$$(R_{th})_{total} = \frac{1}{25} + \frac{0.25}{1.65} + \frac{0.1}{2.817} + \frac{0.15}{9.2} + \frac{1}{12}$$
$$= 0.04 + 0.1515 + 0.0355 + 0.0163 + 0.0833$$
$$= 0.3266°\text{C m}^2/\text{W}$$

$\therefore$ $$U = \frac{1}{(R_{th})_{total}} = \frac{1}{0.3266}$$
$$= \mathbf{3.06\ W/m^2°C.}$$ **(Ans.)**

***(iii)* All surface temperatures ; t_1, t_2, t_3, t_4 :**

$$q = q_A = q_B = q_C$$

or $$3750 = \frac{(t_1 - t_2)}{L_A/k_A} = \frac{(t_2 - t_3)}{L_B/k_B} = \frac{(t_3 - t_4)}{L_C/k_C}$$

or $$3750 = \frac{(1110 - t_2)}{0.25/1.65}$$

or $$t_2 = 1100 - 3750 \times \frac{0.25}{1.65} = \mathbf{531.8°C.}$$ **(Ans.)**

Similarly, $$3750 = \frac{(531.8 - t_3)}{0.1/2.817}$$

or $$t_3 = 531.8 - 3750 \times \frac{0.1}{2.817} = \mathbf{398.6°C.}$$ **(Ans.)**

and $$3750 = \frac{(398.6 - t_4)}{(0.15/9.2)}$$

or $$t_4 = 398.6 - 3750 \times \frac{0.5}{9.2} = \mathbf{337.5°C.}$$ **(Ans.)**

[Check using outside convection,

$$q = \frac{(337.5 - 25)}{1/h_{cf}} = \frac{(337.5 - 25)}{1/12} = 3750 \text{ W/m}^2]$$

28.2.7. Heat Conduction Through Hollow and Composite Cylinders

28.2.7.1. Heat Conduction through a Hollow Cylinder

Refer to Fig. 28.17. Consider a hollow cylinder made of material having constant thermal conductivity and insulated at both ends.

Let r_1, r_2 = Inner and outer radii ;

t_1, t_2 = Temperature of inner and outer surfaces, and

k = Constant thermal conductivity within the given temperature range.

Consider an element at radius '*r*' and thickness '*dr*' for a length of the hollow cylinder through which heat is transmitted. Let *dt* be the temperature drop over the element.

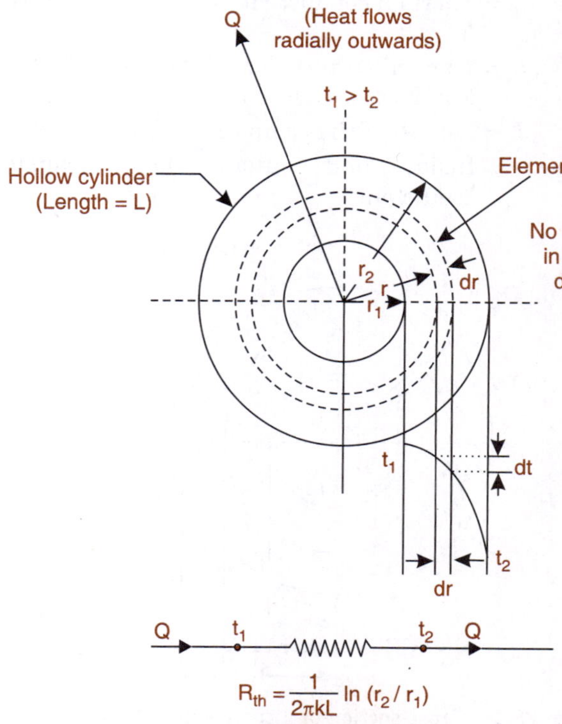

Fig. 28.17

Area through which heat is transmitted. $A = 2\pi r . L$.

Path length = dr (over which the temperature fall is dt)

$$\therefore \quad Q = -kA . \left(\frac{dt}{dr}\right) = -k . 2\pi r . L \frac{dt}{dr} \text{ per unit time}$$

or $$Q . \frac{dr}{r} = -k . 2\pi L.dt$$

Integrating both sides, we get

$$Q \int_{r_1}^{t_2} \frac{dr}{r} = -k.2\pi L \int_{t_1}^{t_2} dt$$

or $$Q \left[\ln (r)\right]_{r_1}^{r_2} = k.2\pi L \left[t\right]_{t_1}^{t_2}$$

or $$Q.\ln(r_2/r_1) = k.2\pi L(t_2 - t_1) = k.2\pi L(t_1 - t_2)$$

$$\therefore \quad Q = \frac{k.2\pi L(t_1 - t_2)}{\ln (r_2/r_1)} = \frac{(t_1 - t_2)}{\left[\frac{\ln(r_2/r_1)}{2\pi k L}\right]} \quad ...(28.33)$$

28.2.7.2. Heat Conduction Through a Composite Cylinder

Consider flow of heat through a composite cylinder as shown in Fig. 28.18.

Let t_{hf} = The temperature of the hot fluid flowing inside the cylinder,

t_{cf} = The temperature of the cold fluid (atmospheric air),

k_A = Thermal conductivity of the inside layer A,

k_B = Thermal conductivity of the outside layer B,

t_1, t_2, t_3 = Temperature at the points 1, 2 and 3 (see Fig. 28.18),

L = Length of the composite cylinder, and

h_{hf}, h_{cf} = Inside and outside heat transfer coefficients.

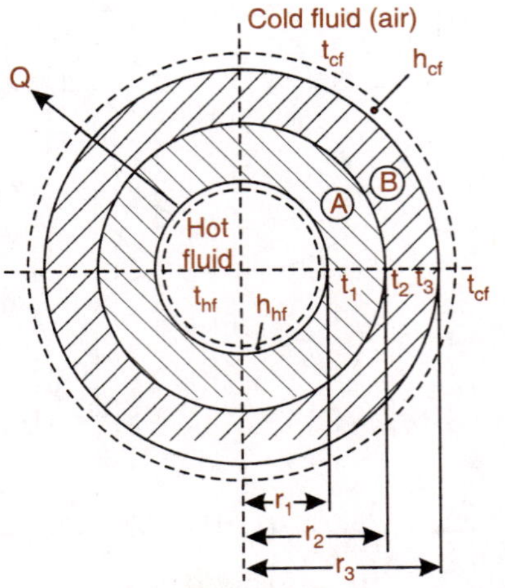

Fig. 28.18. Cross-section of a composite cylinder.

The rate of heat transfer is given by,

$$Q = h_{hf} . 2\pi r_1 . L(t_{hf} - t_1) = \frac{k_A . 2\pi L(t_1 - t_2)}{\ln (r_2/r_1)}$$

$$= \frac{k_B . 2\pi L (t_2 - t_3)}{\ln (r_3/r_2)} = h_{cf} . 2\pi r_3 . L(t_3 - t_{cf})$$

Rearranging the above expression, we get

$$t_{hf} - t_1 = \frac{Q}{h_{hf} . r_1 . 2\pi L} \quad ...(i)$$

$$t_1 - t_2 = \frac{Q}{\frac{k_A . 2\pi L}{\ln (r_2/r_1)}} \quad ...(ii)$$

$$t_2 - t_3 = \frac{Q}{\frac{k_B . 2\pi L}{\ln (r_3/r_2)}} \quad ...(iii)$$

$$t_3 - t_{cf} = \frac{Q}{h_{cf} . r_3 . 2\pi L} \quad ...(iv)$$

Adding (*i*), (*ii*), (*iii*) and (*iv*), we have

$$\frac{Q}{2\pi L}\left[\frac{1}{h_{hf} . r_1} + \frac{1}{\frac{k_A}{\ln (r_2/r_1)}} + \frac{1}{\frac{k_B}{\ln (r_3/r_2)}} + \frac{1}{h_{cf} . r_3}\right] = t_{hf} - t_{cf}$$

$$\therefore \quad Q = \frac{2\pi L (t_{hf} - t_{cf})}{\left[\frac{1}{h_{hf} . r_1} + \frac{1}{\frac{k_A}{\ln (r_2/r_1)}} + \frac{1}{\frac{k_B}{\ln (r_3/r_2)}} + \frac{1}{h_{cf} . r_3}\right]}$$

$$\therefore \quad Q = \frac{2\pi L (t_{hf} - t_{cf})}{\left[\frac{1}{h_{hf} . r_1} + \frac{\ln (r_2/r_1)}{k_A} + \frac{\ln(r_3/r_2)}{k_B} + \frac{1}{h_{cf} / r_3}\right]} \quad ...(28.34)$$

If there are '*n*' concentric cylinders, then

$$Q = \frac{2\pi L (t_{hf} - t_{cf})}{\left[\frac{1}{h_{hf} . r_1} + \sum_{n=1}^{n=n} \frac{1}{k_n} \ln \{r_{(n+1)}/r_n\} + \frac{1}{h_{cf} . r_{(n+1)}}\right]} \quad ...(28.35)$$

If inside and outside heat transfer coefficients are *not* considered then the above equation can be written as

$$Q = \frac{2\pi L [t_1 - t_{(n+1)}]}{\sum_{n=1}^{n=n} \frac{1}{k_n} \ln [r_{(n+1)}/r_n]} \quad ...(28.36)$$

Example 28.9. *A thick walled tube of stainless steel with 20 mm inner diameter and 40 mm outer diameter is covered with a 30 mm layer of asbestos insulation (k = 0.2 W/m°C). If the inside wall temperature of the pipe is maintained at 600°C and the outside insulation at 1000°C, calculate the heat loss per metre of length.*

(Punjab University)

Solution. Refer to Fig. 28.19,

Given, $r_1 = \frac{20}{2} = 10 \text{ mm} = 0.01 \text{ m}$

$r_2 = \frac{40}{2} = 20 \text{ mm} = 0.02 \text{ m}$

$r_3 = 20 + 30 = 50 \text{ mm} = 0.05 \text{ m}$

$t_1 = 600°C, t_3 = 1000°C, k_B = 0.2 \text{ W/m°C}$

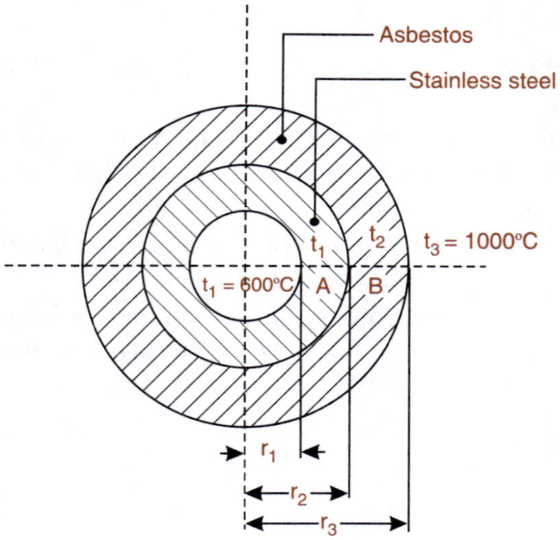

Fig. 28.19

Heat transfer per metre of length, Q/L :

$$Q = \frac{2\pi L\,(t_1 - t_3)}{\frac{\ln(r_2/r_1)}{k_A} + \frac{\ln(r_3/r_2)}{k_B}}$$

Since the thermal conductivity of stainless steel is not given therefore, neglecting the resistance offered by stainless steel to heat transfer across the tube, we have

$$\frac{Q}{L} = \frac{2\pi(t_1 - t_3)}{\frac{\ln(r_3/r_2)}{k_B}} = \frac{2\pi(600 - 1000)}{\frac{\ln(0.05/0.02)}{0.2}}$$

$= -$ **548.57 W/m. (Ans.)**

Negative sign indicates that the heat transfer takes place *radially inward.*

Example 28.10. *Hot air at a temperature of 65°C is flowing through a steel pipe of 120 mm diameter. The pipe is covered with two layers of different insulating materials of thickness 60 mm and 40 mm, and their corresponding thermal conductivities are 0.24 and 0.4 W/m°C. The inside and outside heat transfer coefficients are 60 and 12 W/m°C. The atmosphere is at 20°C. Find the rate of heat loss from 60 m length of pipe.*

Solution. Refer to Fig. 28.20.

Given : $r_1 = \frac{120}{2} = 60 \text{ mm} = 0.06 \text{ m}$

$r_2 = 60 + 60 = 120 \text{ mm} = 0.12 \text{ m}$

$r_3 = 60 + 60 + 40 = 160 \text{ mm} = 0.16 \text{ m}$

$k_A = 0.24 \text{ W/m°C}\ ;\quad k_B = 0.4 \text{ W/m°C}$

$h_{hf} = 60 \text{ W/m}^2\text{°C}\ ;\quad h_{cf} = 12 \text{ W/m}^2\text{°C}$

$t_{hf} = 65°C\ ;\quad t_{cf} = 20°C$

Length of pipe, $L = 60$ m

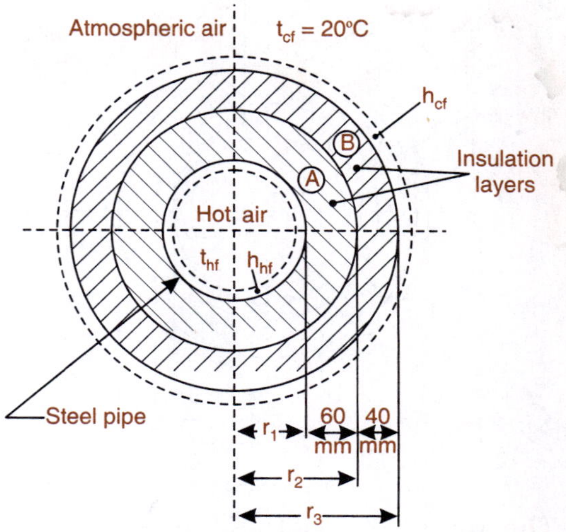

Fig. 28.20

Rate of heat loss, Q :

Rate of heat loss is given by

$$Q = \frac{2\pi L\,(t_{hf} - t_{cf})}{\left[\frac{1}{h_{hf}\cdot r_1} + \frac{\ln(r_2/r_1)}{k_A} + \frac{\ln(r_3/r_2)}{k_B} + \frac{1}{h_{cf}\cdot r_3}\right]}$$ [Eqn. (28.34)]

$$= \frac{2\pi \times 60(65 - 20)}{\left[\frac{1}{60 \times 0.06} + \frac{\ln(0.12/0.06)}{0.24} + \frac{\ln(0.16/0.12)}{0.4} + \frac{1}{12 \times 0.16}\right]}$$

$$= \frac{16964.6}{0.2777 + 2.8881 + 0.7192 + 0.5208} = 3850.5 \text{ W}$$

i.e., Rate of heat loss = **3850.5 W. (Ans.)**

Example 28.11. *A 150 mm steam pipe has inside diameter of 120 mm and outside diameter of 160 mm. It is insulated at the outside with asbestos. The steam temperature is 150°C and the air temperature is 20°C. h (steam side) = 100 W/m²°C, h (air side) = 30 W/m²°C, k (asbestos) = 0.8 W/m°C and k (steel) = 42 W/m°C. How thick should the asbestos be provided in order to limit the heat losses to 2.1 kW/m² ?* **(N.U.)**

Solution. Refer to Fig. 28.21.

Given : $r_1 = \frac{120}{2} = 60 \text{ mm} = 0.06 \text{ m}$

$r_2 = \frac{160}{2} = 80 \text{ mm} = 0.08 \text{ m}$

$k_A = 42$ W/m°C ; $k_B = 0.8$ W/m°C

$t_{hf} = 150$°C ; $t_{cf} = 20$°C

$h_{hf} = 100$ W/m²°C ; $h_{cf} = 30$ W/m²°C

Heat loss = 2.1 kW/m²

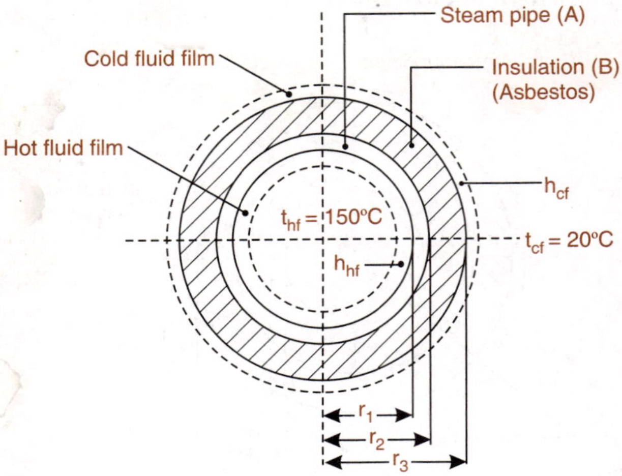

Fig. 28.21

Thickness of insulation (asbestos), $(r_3 - r_2)$:

Area for heat transfer

$= 2\pi\, r\, L$ (where L = length of the pipe)

$\therefore$ Heat loss

$= 2.1 \times 2\pi\, r\, L$ kW

$= 2.1 \times 2\pi \times 0.075 \times L = 0.989\, L$ kW

$= 0.989$ L $\times 10^3$ watts

$$\left(\begin{aligned}&\text{where } r, \text{ mean radius} = \frac{150}{2}\\ &= 75 \text{ mm or } 0.075 \text{ m ... Given}\end{aligned}\right)$$

Heat transfer rate in such a case is given by

$$Q = \frac{2\pi L\,(t_{hf} - t_{cf})}{\left[\dfrac{1}{h_{hf}\,.\,r_1} + \dfrac{\ln(r_2/r_1)}{k_A} + \dfrac{\ln(r_3/r_2)}{k_B} + \dfrac{1}{h_{cf}\,.\,r_3}\right]}$$

...[Eqn. (28.34)]

$$0.989\, L \times 10^3 = \frac{2\pi L\,(150 - 20)}{\left[\dfrac{1}{100 \times 0.06} + \dfrac{\ln(0.08/0.06)}{42} + \dfrac{\ln(r_3/0.08)}{0.8} + \dfrac{1}{30 \times r_3}\right]}$$

$$0.989 \times 10^3 = \frac{816.81}{\left[0.16666 + 0.00685 + \dfrac{\ln(r_3/0.08)}{0.8} + \dfrac{1}{30 r_3}\right]}$$

or $$\frac{\ln(r_3/0.08)}{0.8} + \frac{1}{30\, r_3} = \frac{816.81}{0.989 \times 10^3} - (0.16666 + 0.00685) = 0.6524$$

or $$1.25 \ln(r_3/0.08) + \frac{1}{30\, r_3} - 0.6524 = 0$$

Solving by hit and trial, we get

$r_3 \simeq 0.105$ m or 105 mm

$\therefore$ Thickness of insulation

$= r_3 - r_2 = 105 - 80 =$ **25 mm. (Ans.)**

28.2.8. Heat Conduction Through Hollow and Composite Spheres

28.2.8.1. Heat Conduction through Hollow Sphere

Refer to Fig. 28.22. Consider a hollow sphere made of material having constant thermal conductivity.

Let r_1, r_2 = Inner and outer radii,

t_1, t_2 = Temperature of inner and outer surfaces, and

k = Constant thermal conductivity of the material with the given temperature range.

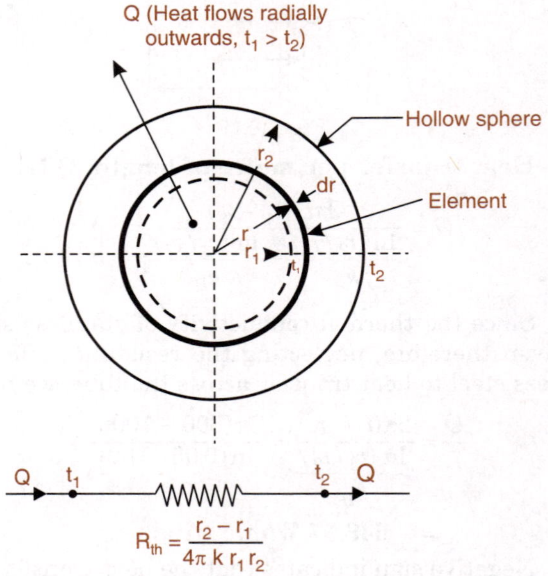

Fig. 28.22. Steady state conduction through a hollow sphere.

Consider a small element of thickness dr at any radius r.

Area through which the heat is transmitted, $A = 4\pi r^2$

$$\therefore \quad Q = -\,k\,.\,4\pi r^2\,.\,\frac{dt}{dr}$$

Rearranging and integrating the above equation, we obtain

$$Q\int_{r_1}^{r_2}\frac{dr}{r^2} = -\,4\pi k\int_{t_1}^{t_2} dt$$

or $$Q\left[\frac{r^{-2+1}}{-2+1}\right]_{r_1}^{r_2} = -\,4\pi k\,\Big[\,t\,\Big]_{t_1}^{t_2}$$

or $$-Q\left(\frac{1}{r_2}-\frac{1}{r_1}\right) = -4\pi k(t_2 - t_1)$$

or $$\frac{Q(r_2 - r_1)}{r_1 r_2} = 4\pi k\,(t_1 - t_2)$$

or $$Q = \frac{4\pi k r_1 r_2 (t_1 - t_2)}{(r_2 - r_1)} = \frac{t_1 - t_2}{\left[\dfrac{(r_2 - r_1)}{(4\pi k r_1 r_2)}\right]} \quad ...(28.37)$$

28.2.8.2. Heat Conduction through a Composite Sphere

Considering Fig. 28.23 as cross-section of a composite sphere, the heat flow equation can be written as follows :

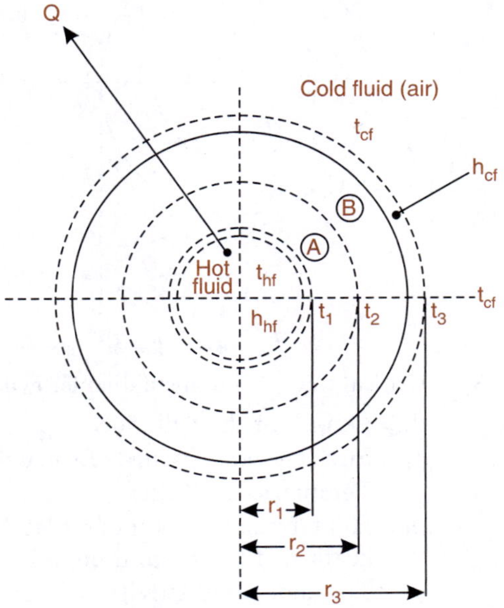

Fig. 28.23. Steady state conduction through a composite sphere.

$$Q = h_{hf}\,.\,4\pi r_1^2\,(t_{hf} - t_1)$$
$$= \frac{4\pi k_A r_1 r_2 (t_1 - t_2)}{(r_2 - r_1)} = \frac{4\pi k_B r_2 r_3\,(t_1 - t_3)}{(r_3 - r_2)}$$
$$= h_{cf}\,.\,4\pi r_3^2\,(t_3 - t_{cf})$$

By rearranging the above equation, we have

$$t_{hf} - t_1 = \frac{Q}{h_{hf}\,.\,4\pi r_1^2} \quad ...(i)$$

$$t_1 - t_2 = \frac{Q\,(r_2 - r_1)}{4\pi k_A\,.\,r_1 r_2} \quad ...(ii)$$

$$t_2 - t_3 = \frac{Q\,(r_3 - r_2)}{4\pi k_B\,.\,r_2 r_3} \quad ...(iii)$$

$$t_3 - t_{cf} = \frac{Q}{h_{cf}\,.\,4\pi r_3^2} \quad ...(iv)$$

Adding (*i*), (*ii*), (*iii*) and (*iv*), we get

$$\frac{Q}{4\pi}\left[\frac{1}{h_{hf}\,.\,r_1^2} + \frac{(r_2 - r_1)}{k_A\,.\,r_1 r_2} + \frac{(r_3 - r_2)}{k_B\,.\,r_2 r_3} + \frac{1}{h_{cf}\,.\,r_3^2}\right] = t_{hf} - t_{cf}$$

$$\therefore\ Q = \frac{4\pi(t_{hf} - t_{cf})}{\left[\dfrac{1}{h_{hf}\,.\,r_1^2} + \dfrac{(r_2 - r_1)}{k_A\,.\,r_1 r_2} + \dfrac{(r_3 - r_2)}{k_B\,.\,r_2 r_3} + \dfrac{1}{h_{cf}\,.\,r_3^2}\right]} \quad ...(28.38)$$

If there are n concentric spheres then the equation (28.38) can be written as follows :

$$Q = \frac{4\pi(t_{hf} - t_{cf})}{\left[\dfrac{1}{h_{hf}\,.\,r_1^2} + \sum_{n=1}^{n=n}\left\{\dfrac{r_{(n+1)} - r_n}{k_n\,.\,r_n\,.\,r_{(n+1)}}\right\} + \dfrac{1}{h_{cf}\,.\,r^2_{(n+1)}}\right]} \quad ...(28.39)$$

If inside and outside heat transfer coefficients are considered, then the above equation can be written as follows :

$$Q = \frac{4\pi\,[t_1 - t_{(n+1)}]}{\sum_{n=1}^{n=n}\left[\dfrac{r_{(n+1)} - r_n}{k_n\,.\,r_n\,.\,r_{(n+1)}}\right]} \quad ...(28.40)$$

Example 28.12. *A spherical shaped vessel of 1.4 m diameter is 90 mm thick. Find the rate of heat leakage, if the temperature difference between the inner and outer surfaces is 220°C. Thermal conductivity of the material of the sphere is 0.083 W/m°C.*

Solution. Refer to Fig. 28.24.

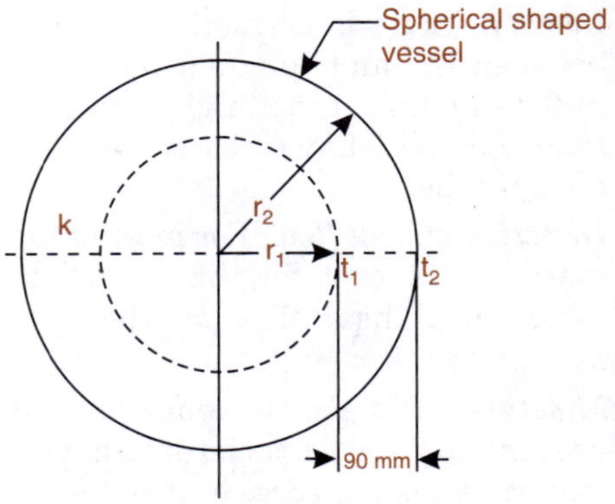

Fig. 28.24

Given : $r_2 = \frac{1.4}{2} = 0.7$ m ;

$r_1 = 0.7 - \frac{90}{1000} = 0.61$ m ;

$t_1 - t_2 = 220$°C ; $k = 0.083$ W/m°C

The rate of heat transfer/leakage is given by

$$Q = \frac{(t_1 - t_2)}{\left[\dfrac{(r_2 - r_1)}{4\pi k r_1 r_2}\right]} \quad ...[\text{Eqn. } (28.37)]$$

$$= \frac{220}{\left[\dfrac{(0.7 - 0.61)}{4\pi \times 0.083 \times 0.61 \times 0.7}\right]} = 1088.67 \text{ W}$$

i.e., Rate of heat leakage = **1088.67 W. (Ans.)**

28.2.9. Critical Thickness of Insulation

28.2.9.1. Insulation-General aspects

Definition. *A material which retards the flow of heat with reasonable effectiveness* is known as *'Insulation'*. Insulation serves the following *two purposes :*

(*i*) It prevents the heat flow from the system to the surroundings ;

(*ii*) It prevents the heat flow from the surroundings to the system.

Applications :

The *fields of application* of insulations are :

(*i*) Boilers and steam pipes

(*ii*) Air-conditioning systems

(*iii*) Food preserving stores and refrigerators

(*iv*) Insulating bricks (employed in various types of furnaces)

(*v*) Preservation of liquid gases etc.

Factors affecting thermal conductivity

Some of the important factors which affect thermal conductivity (k) of the insulators (the value of k should be always low to reduce the rate of heat flow) are as follows :

1. *Temperature.* For most of the insulating materials, the value of k *increases with increase in temperature.*
2. *Density.* There is no mathematical relationship between k and ρ (density). The common understanding that high density insulating materials will have higher values of k in not always true.
3. *Direction of heat flow.* For most of the insulating materials (except few like wood) the effect of direction of heat flow on the values of k is negligible.
4. *Moisture.* It is always considered necessary to prevent ingress of moisture in the insulating materials during service, it is however difficult to find the effect of moisture on the values of k of different insulating materials.
5. *Air pressure.* It has been found that the value of *k decreases with decrease in pressure.*
6. *Convection in insulators.* The value of k increases due to the phenomenon of convection in insulators.

28.2.9.2. Critical Thickness of Insulation

The addition of insulation always increases the conductive thermal resistance. But when the total thermal resistance is made of conductive thermal resistance [$(R_{th})_{cond.}$] and convective thermal resistance [$(R_{th})_{conv.}$], the addition of insulation in *some cases* may *reduce the convective thermal resistance due to increase in surface area,* as in the case of a cylinder and a sphere, and the total thermal resistance may actually decrease resulating in increased heat flow. It may be shown that the thermal resistance actually decreases and then increases in some cases.

"The thickness upto which heat flow increases and after which heat flow decreases is termed as ***Critical thickness.*** In case of cylinders and spheres it is called ***'Critical radius'.***

A. Critical thickness of insulation for cylinder :

Consider a solid cylinder of radius r_1 insulated with an insulation of thickness $(r_2 - r_1)$ as shown in Fig. 28.25.

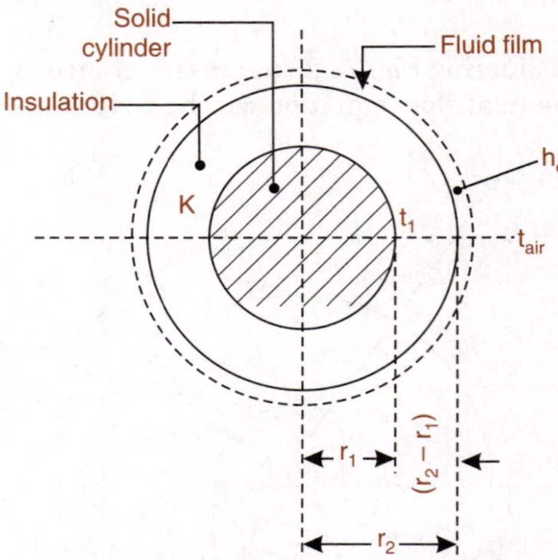

Fig. 28.25. Critical thickness of insulation for cylinder.

Let, L = Length of the cylinder,

t_1 = Surface temperature of the cylinder,

t_{air} = Temperature of air,

h = Heat transfer coefficient at the outer surface of the insulation, and

k = Thermal conductivity of insul-ating material.

Then the rate of heat transfer from the surface of the solid cylinder to the surroundings is given by

$$Q = \frac{2\pi L\,(t_1 - t_{air})}{\dfrac{\ln (r_2/r_1)}{k} + \dfrac{1}{h_o \cdot r_2}} \qquad ...(28.41)$$

From eqn. (28.41) it is evident that as r_2 increases, the factor $\frac{\ln (r_2/r_1)}{k}$ increases but the factor $\frac{1}{h_o \cdot r_2}$ decreases. Thus Q becomes *maximum* when the *denominator* $\left[\frac{\ln (r_2/r_1)}{k} + \frac{1}{h_o \cdot r_2}\right]$ becomes *minimum.* The required condition is

$$\frac{d}{dr_2}\left[\frac{\ln (r_2/r_1)}{k} + \frac{1}{h_o \cdot r_2}\right] = 0$$

(r_2 being the only variable)

$$\therefore \quad \frac{1}{k} \cdot \frac{1}{r_2} + \frac{1}{h_o}\left(-\frac{1}{r_2^2}\right) = 0$$

or $$\frac{1}{k} - \frac{1}{h_o \cdot r_2} = 0$$

or $$h_o \cdot r_2 = k$$

or $$r_2 \, (= r_c) = \frac{k}{h_o} \quad \text{...(28.42)}$$

The above relation represents the *condition for minimum resistance and consequently *maximum heat flow rate.* The *insulation radius at which resistance to heat flow is minimum* is called the *'critical radius'* (r_c). The critical radius r_c is dependent of the thermal quantities k and h_o and is independent of r_1 (*i.e.,* cylinder radius).

*It may be noted that if the second derivative of the denominator is evaluated, it will come out to be *positive.* This would verify that heat flow rate will be maximum, when $r_2 = r_c$.

In eqn. (28.41) ln $(r_2/r_1)/k$ is the conduction (insulation) thermal resistance which *increases* with increasing r_2 and $1/h_o.r_2$ is convective thermal resistance which *decreases* with increasing r_2. At $r_2 = r_c$ *the rate of increase of conductive resistance of insulation is equal to the rate of decrease of convective resistance thus giving a minimum value for the sum of thermal resistances.*

In the physical sense we may arrive at the following conclusions :

(*i*) For cylindrical bodies with $r_1 < r_c$, the heat transfer increases by adding insulation till $r_2 = r_c$ as shown in Fig. [28.26 (*a*)]. If insulation thickness is further increased, the rate of heat loss will decrease from this peak value, but until a certain amount of insulation denoted by r_2' at b is added, the heat loss rate is still greater for the solid cylinder. This happens when r_1 is small and r_c is large, *viz.,* the thermal conductivity of the insulation k is high (poor insulating material) and h_o is low. A *practical application would be the insulation of electric cables which should be good insulator for current but poor for heat.*

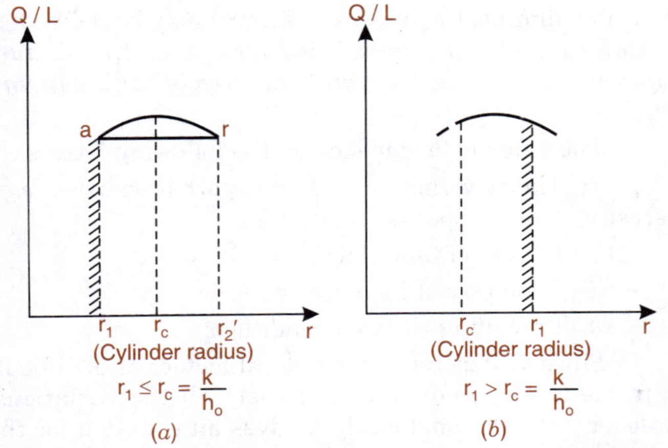

Fig. 28.26. Dependence of heat loss on insulation thickness.

(*ii*) For cylindrical bodies with $r_1 > r_c$, the heat transfer decreases by adding insulation [Fig. 28.26 (*b*)]. This happens when r_1 is large and r_c is small, *viz.,* a good insulating material is used with low k and h_o is high. In *steam and refrigeration pipes* heat insulation is the main objective. For insulation to be properly effective in restricting heat transmission, the *outer radius must be greater than or equal to the critical radius.*

B. Critical thickness of insulation for sphere :

Refer to Fig. 28.27. The equation of heat flow through a sphere with insulation is given as

$$Q = \frac{(t_1 - t_{air})}{\left[\frac{r_2 - r_1}{4\pi k r_1 r_2}\right] + \frac{1}{4\pi r_2^2 h_o}}$$

Adopting the same procedure as that of a cylinder, we have

$$\frac{d}{dr_2}\left[\frac{r_2 - r_1}{4\pi k r_1 r_2} + \frac{1}{4\pi r_2^2 . h_o}\right] = 0$$

or $$\frac{d}{dr_2}\left[\frac{1}{kr_1} - \frac{1}{kr_2} + \frac{1}{r_2^2 h_o}\right] = 0$$

or $$\frac{1}{kr_2^2} - \frac{2}{r_2^3 h_o} = 0$$

or $$r_2^3 h_o = 2kr_2^2$$

or $$r_2 (= r_c) = \frac{2k}{h_o} \quad \text{...(28.43)}$$

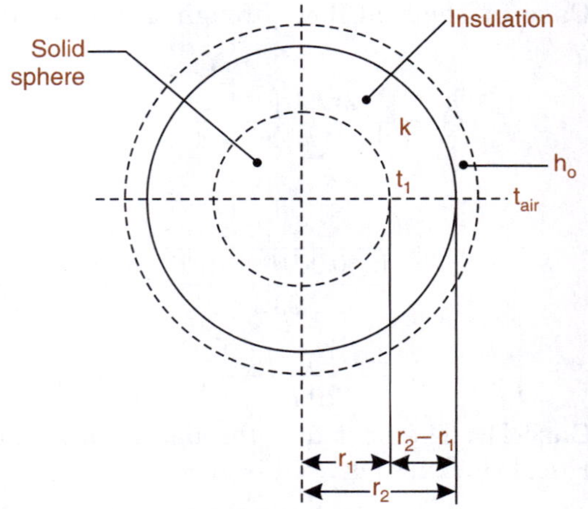

Fig. 28.27

Example 28.13. *A small electric heating application uses wire of 2 mm diameter with 0.8 mm thick insulation (k = 0.12 W/m°C). The heat transfer coefficient (h_o) on the insulated surface is 35 W/m²°C. Determine the critical thickness of insulation in this case and the percentage change in the heat transfer rate if the critical thickness is used, assuming the temperature difference between the surface of the wire and surrounding air remains unchanged.*

Solution. Refer to Fig. 28.28.

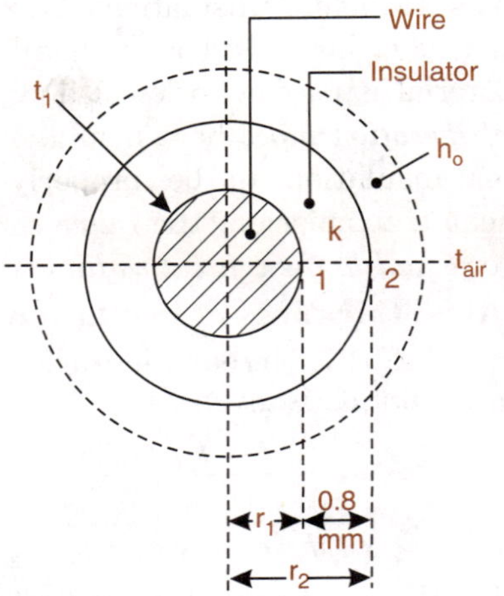

Fig. 28.28

Given : $r_1 = \frac{2}{2} = 1$ mm $= 0.001$ m

$r_2 = 1 + 0.8 = 1.8$ mm $= 0.0018$ m

$k = 0.12$ W/m°C, $h_o = 35$ W/m²°C

Critical thickness of insulation :

The critical radius of insulation is given by

$$r_c = \frac{k}{h_o} = \frac{0.12}{35} = 3.43 \times 10^{-3} \text{ m or } 3.43 \text{ mm.}$$

$\therefore$ Critical thickness of insulation

$= r_c - r_1 = 3.43 - 1 =$ **2.43 mm. (Ans.)**

Percentage change in heat transfer rate :

Case I : The heat flow through an insulated wire is given by

$$Q_1 = \frac{2\pi L\,(t_1 - t_{air})}{\dfrac{\ln(r_2/r_1)}{k} + \dfrac{1}{h_o \cdot r_2}}$$

$$= \frac{2\pi L\,(t_1 - t_{air})}{\dfrac{\ln(0.0018/0.001)}{0.12} + \dfrac{1}{35 \times 0.0018}}$$

$$= \frac{2\pi L\,(t_1 - t_{air})}{20.77} \qquad ...(i)$$

Case II : The heat flow through an insulated wire when critical thickness is used is given by

$$Q_2 = \frac{2\pi L\,(t_1 - t_{air})}{\dfrac{\ln(r_c/r_1)}{k} + \dfrac{1}{h_o \cdot r_c}}$$

$$= \frac{2\pi L\,(t_1 - t_{air})}{\dfrac{\ln(0.00343/0.001)}{0.12} + \dfrac{1}{35 \times 0.00343}}$$

$$= \frac{2\pi L\,(t_1 - t_{air})}{18.6} \qquad ...(ii)$$

$\therefore$ Percentage increases in heat flow by using critical thickness of insulation

$$= \frac{Q_2 - Q_1}{Q_1} \times 100 = \frac{\dfrac{1}{18.6} - \dfrac{1}{20.77}}{\dfrac{1}{20.77}} \times 100$$

= **11.6%. (Ans.)**

28.3. HEAT TRANSFER BY CONVECTION

- The rate equation for the convective heat transfer (regardless of particular nature) between a surface and an adjacent fluid is prescribed by *Newton's law of cooling* (Refer to Fig. 28.29)

$$Q = hA(t_s - t_f) \qquad ...(28.44)$$

where, Q = Rate of conductive heat transfer,

A = Area exposed to heat transfer,

t_s = Surface temperature,

t_f = Fluid temperature, and

h = Co-efficient of conductive heat transfer.

The units of h are,

$$h = \frac{Q}{A\,(t_s - t_f)} = \frac{\text{W}}{\text{m}^2\,°\text{C}} \text{ or W/m}^2\text{°C or W/m}^2\text{K}$$

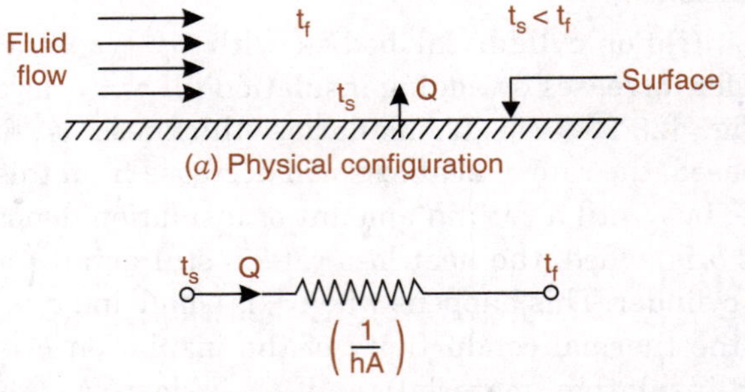

(*a*) Physical configuration

(*b*) Equivalent circuit

Fig. 28.29. Convective heat-transfer.

The coefficient of convective heat transfer '*h*' (also known as film heat transfer coefficient) may be defined as "*the amount of heat transmitted for a unit temperature difference between the fluid and unit area of surface in unit time.*"

The value of '*h*' depends on the following factors :

(*i*) Thermodynamic and transport properties (*e.g.*, viscosity, density, specific heat etc.) ;

(*ii*) Nature of fluid flow ;

(*iii*) Geometry of the surface ;

(*iv*) Prevailing thermal conditions.

Since '*h*' depends upon several factors, it is difficult to frame a single equation to satisfy all the variations, however a dimensional analysis gives an equation for the purpose which is given as under :

$$\frac{h_D}{k} = Z\left(\frac{\rho CD}{\pi}\right)^a \left(\frac{c_p\mu}{k}\right)^b \left(\frac{D}{L}\right)^c \quad ...(28.45)$$

or $$Nu = Z\,(Re)^a\,(Pr)^b \left(\frac{D}{L}\right)^c$$

where, Nu = Nusselt number $\left(\frac{h_D}{k}\right)$,

Re = Reynolds number $\left(\frac{\rho \bar{u} D}{\mu}\right)$,

Pr = Prandtl number $\left(\frac{c_p\mu}{k}\right)$,

$\frac{D}{L}$ = Diameter to length ratio,

Z = A constant to be determined experimentally,

ρ = Density,

μ = Dynamic viscosity, and

C = Velocity.

The *mechanisms of convection* in which phase changes are involved lead to the important *fields of boiling and condensation*. Refer to Fig. 28.29 (*b*). The quantity $\frac{1}{hA}\left[Q = \frac{t_s - t_f}{(1/hA)} \text{ ... Eqn (28.44)}\right]$ is called *convection thermal resistance* $[(R_{th})_{conv.}]$ *to heat flow.*

- **Dimensionless numbers :**

Reynolds numbers, $Re = \frac{VL}{\nu}$

Prandtl number, $Pr = \frac{\mu c_p}{k} = \frac{\nu}{\alpha}$

Nusselt number, $Nu = \frac{hL}{k}$

Stanton number, $St = \frac{h}{\rho V c_p} = \frac{Nu}{Re \times Pr}$

Peclet number, $Pe = \frac{LV}{\alpha} (= Re.Pr)$

Graetz number, $G = Pe\left(\frac{\pi D}{4}\right)$

Grashoff number, $Gr = \frac{\rho^2 \beta g \Delta t L^3}{\mu^2}$.

Example 28.14. *A hot plate 1 m × 1.5 m is maintained at 300°C. Air at 25°C blows over the plate. If the convective heat transfer coefficient is 20 W/m²°C, calculate the rate of heat transfer.*

Solution. Area of the plate exposed to heat transfer,

$$A = 1 \times 1.5 = 1.5 \text{ m}^2$$

Plate surface temperature, $t_s = 300°C$

Temperature of air (fluid), $t_f = 20°C$

Convective heat-transfer coeffficient,

$$h = 20 \text{ W/m}^2°C$$

Rate of heat transfer, Q :

From Netwon's law of cooling,

$$Q = hA\,(t_s - t_f)$$
$$= 20 \times 1.5(300 - 20) = \mathbf{8400\ W\ or\ 8.4\ kW.} \quad \textbf{(Ans.)}$$

Example 28.15. *A wire 1.5 mm in diameter and 150 mm long is submerged in water at atmospheric pressure. An electric current is passed through the wire and is increased until the water boils at 100°C. Under the condition if convective heat transfer coefficient is 4500 W/m²°C find how much electric power must be supplied to the wire to maintain the wire surface at 120°C ?*

Solution. Diameter of the wire, d = 1.5 mm = 0.0015 m

Length of the wire, l = 150 mm = 0.15 m

∴ Surface area of the wire (exposed to heat transfer),

$$A = \pi\, d\, l = \pi \times 0.0015 \times 0.15 = 7.068 \times 10^{-4} \text{ m}^2$$

Wire surface temperature, t_s = 120°C

Water temperature, t_f = 100°C

Convective heat transfer coefficient, h = 4500 W/m²°C

Electric power to be supplied :

Electric power which must be supplied = Total convection loss (Q)

$$\therefore \quad Q = hA(t_s - t_f) = 4500 \times 7.068 \times 10^{-4}\,(120 - 100)$$
$$= \mathbf{63.6\ W.} \quad \textbf{(Ans.)}$$

Example 28.16. *Water flows inside a tube 45 mm in diameter and 3.2 m long at a velocity of 0.78 m/s. Determine the heat transfer co-efficient and the rate of heat transfer if the mean water temperature is 50°C and the wall is isothermal at 70°C. For water at 50°C take k = 0.66 W/mK, ν = 0.478 × 10⁻⁶ m²/s and Prandtl number = 2.98.*

Solution. Diameter of the tube, D = 45 mm = 0.045 m

Length of the tube, l = 3.2 m

Velocity of water, $\bar{u}$ = 0.78 m/s

For water at 60°C, k = 0.66 W/mK

Kinematic viscosity, $\nu = 0.478 \times 10^{-6}$ m²/s

Pr = 2.98

Reynolds number is given by

$$Re = \frac{D\bar{u}}{\nu} = \frac{0.045 \times 0.78}{0.478 \times 10^{-6}} = 73431$$

From Dittus and Boelter equation, Nusselt number,

$$Nu = 0.023\,(Re)^{0.8}\,(Pr)^{0.4}$$

$$\frac{hD}{k} = 0.023\,(73431)^{0.8}\,(2.98)^{0.4}$$

$$\frac{h \times 0.045}{0.66} = 0.023 \times 7810.9 \times 1.547$$

$$\therefore \quad h = 4076 \text{ W/m}^2 \text{ K}$$

i.e., Heat transfer co-efficient = **4076 W/m² K (Ans).**

$$Q = hA\,(t_w - t_f) = 4076 \times \pi DL\,(70 - 50)$$
$$= 4076 \times \pi \times 0.045 \times 3.2 \times 20$$
$$= 36878 \text{ or } 36.878 \text{ kW}$$

i.e., Rate of heat transfer = **36.878 kW. (Ans).**

Example 28.17. *When 0.5 kg of water per minute is passed through a tube of 20 mm diameter, it is found to be heated from 20°C to 50°C. The heating is accomplished by condensing steam on the surface of the tube and subsequently the surface temperature of the tube is maintained at 85°C. Determine the length of the tube required for developed flow.*

Take the thermo-physical properties of water at 60°C as :

$$\rho = 983.2 \text{ kg/m}^2,\ c_p = 4.178 \text{ kJ/kg K},$$
$$k = 0.659 \text{ W/m°C},\ \nu = 0.478 \times 10^{-6} \text{ m}^2\text{/s}.$$

Solution. *Given :* m = 0.5 kg/min, D = 20 mm = 0.02 m,

t_i = 20°C, t_o = 50°C

Length of the tube required for fully developed flow, L :

The mean film temperature,

$$t_f = \frac{1}{2}\left(85 + \frac{20+50}{2}\right) = 60°\text{C}$$

Let us first determine the type of the flow

$$m = \rho A \bar{u} = 983.2 \times \frac{\pi}{4} \times (0.02)^2 \times \bar{u} = \frac{0.5}{60} \text{ (kg/s)}$$

or
$$\bar{u} = \frac{0.5}{60} \times \frac{4}{\pi} \times \frac{1}{983.2 \times (0.02)^2} = 0.0269 \text{ m/s}$$

Reynolds number,

$$Re = \frac{D.\bar{u}}{\nu} = \frac{0.02 \times 0.0269}{0.478 \times 10^{-6}} = 1125.5$$

Since Re < 2000, hence the flow is *laminar.*

With *constant wall temperature* having *fully developed flow,*

$$Nu = \frac{hD}{k} = 3.65$$

or
$$h = \frac{3.65\,k}{D} = \frac{3.65 \times 0.659}{0.02} = 120.26 \text{ W/m}^2\text{°C}$$

The rate of heat transfer,

$$Q = A_s\, h\, (t_s - t_\infty) = m\, c_p\, (t_o - t_i)$$

Here, $$t_\infty = \frac{20+50}{2} = 35°\text{C} = t_b$$

$$\therefore\ (\pi \times 0.02 \times L) \times 120.26 \times (85 - 35)$$
$$= \frac{0.5}{60} \times (4.178 \times 10^3)(50 - 20)$$

or $$377.8\, L = 1044.5$$

or $$L = \frac{1044.5}{377.8} = \mathbf{2.76\ m.}\quad \textbf{(Ans.)}$$

28.4. HEAT EXCHANGERS

28.4.1. Introduction

A **'heat exchanger'** may be defined *as an equipment which transfers the energy from a hot fluid to a cold fluid, with maximum rate and minimum investment and running costs.*

In heat exchangers the temperature of each fluid changes as it passes through the exchangers, and hence the temperature of the dividing wall between the fluids also changes along the length of the exchanger.

Examples of heat exchangers :

(*i*) Intercoolers and preheaters ;
(*ii*) Condensers and boilers in steam plant ;
(*iii*) Condensers and evaporators in refrigeration units ;
(*iv*) Regenerators ;
(*v*) Automobile radiators ;
(*vi*) Oil coolers of heat engine ;
(*vii*) Milk chiller of a pasteurising plant ;
(*viii*) Several other industrial processes.

28.4.2. Types of Heat Exchangers

In order to meet the widely varying applications, several types of heat exchangers have been developed which are classified on the basis of *nature of heat exchange process, relative direction of fluid motion, design and constructional features,* and *physical state of fluids.*

1. Nature of heat exchange process

Heat exchangers, on the basis of nature of heat exchange process, are classified as follows :

(*i*) Direct contact (or open) heat exchangers.
(*ii*) Indirect contact heat exchangers.
(*a*) Regenerators. (*b*) Recuperators.

(*i*) ***Direct contact heat exchangers.*** In a direct contact or open heat exchanger the exchange of heat takes place by direct mixing of hot and cold fluids and transfer of heat and mass takes place simultaneously. The use of such units is made under conditions where mixing of two fluids is either harmless or desirable. *Examples* : (*i*) Cooling towers ; (*ii*) Jet condensers ; (*iii*) Direct contact feed heaters.

Fig. 28.30 shows a direct contact heat exchanger in which steam mixes with cold water, gives its latent heat to water and gets condensed. Hot water and non-condensable gases leave the container as shown in the figure.

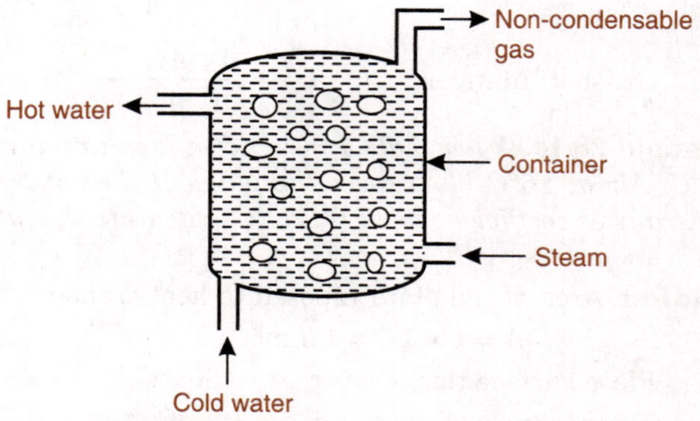

Fig. 28.30. Direct contact or open heat exchanger.

(*ii*) ***Indirect contact heat exchangers.*** In this type of heat exchanger, the heat transfer between two fluids could be carried out by transmission through wall which separates the two fluids. This type includes the following :

(*a*) Regenerators.

(*b*) Recuperators or surface exchangers.

(*a*) *Regenerators* : In a *regenerator* type of heat exchanger the *hot and cold fluids pass alternately through a space containing solid particles (matrix), these particles providing alternately a sink and a source for heat flow.*

Examples : (*i*) I.C. engines and gas turbines ; (*ii*) Open hearth and glass melting furnaces ; (*iii*) Air heaters of blast furnaces.

A *regenerator* generally operates periodically (the solid matrix alternately stores heat extracted from the hot fluid and then delivers it to the cold fluid). However, in some regenerators the matrix is made to rotate through the fluid passages arranged side by side which makes the heat exchange process *continuous.*

The performance of these regenerators is affected by the following *parameters :*

(*i*) Heat capacity of regenerating material,

(*ii*) The rate of absorption, and

(*iii*) The release of heat.

Advantages :

1. Higher heat transfer coefficient ;
2. Less weight per kW of the plant ;
3. Minimum pressure loss ;
4. Quick response to load variation ;
5. Small bulk weight ;
6. Efficiency quite high.

Disadvantages :

1. Costlier compared to recuperative heat exchangers.
2. Leakage is the main trouble, therefore, perfect sealing is required.

(*b*) *Recuperators : 'Recuperator'* is the most important type of heat exchanger in which the flowing fluids exchanging heat are on either side of *dividing wall* (in the form of pipes or tubes generally). These heat exchangers are used when two fluids cannot be allowed to mix *i.e.,* when the mixing is undesirable.

Examples : (*i*) Automobile radiators, (*ii*) Oil coolers, intercoolers, *air preheaters,* economisers, superheaters, condensers and surface feed heaters of a steam power plant, (*iii*) Milk chiller of pasteurising plant, (*iv*) Evaporator of an ice plant :

Advantages :

1. Easy construction ;
2. More economical ;
3. More surface area for heat transfer ;
4. Much suitable for stationary plants.

Disadvantages :

1. Less heat transfer coefficient ;
2. Less generating capacity ;
3. Heavy and sooting problems.

The flow through *direct heat exchangers and recuperators* may be treated as *steady state* while through regenerators the *flow is essentially transient.*

2. Relative direction of fluid motion

According to the relative directions of two fluid streams the heat exchangers are classified into the following *three* categories :

(*i*) Parallel-flow or unidirection flow

(*ii*) Counter-flow

(*iii*) Cross-flow.

(*i*) ***Parallel-flow heat exchangers.*** In a *parallel-flow exchanger,* as the name suggests, the two fluid streams (hot and cold) travel in the *same direction.* The two streams enter at one end and leave at the other end. The flow arrangement and variation of temperatures of the fluid streams in case of parallel flow heat exchangers, are shown in Fig. 28.31. It is evident from the Fig. 28.31. (*b*) that the *temperature difference* between the hot and cold fluids goes on *decreasing* from inlet to outlet. Since this type of heat exchanger needs a large area of heat transfer, therefore, it is *rarely used in practice.*

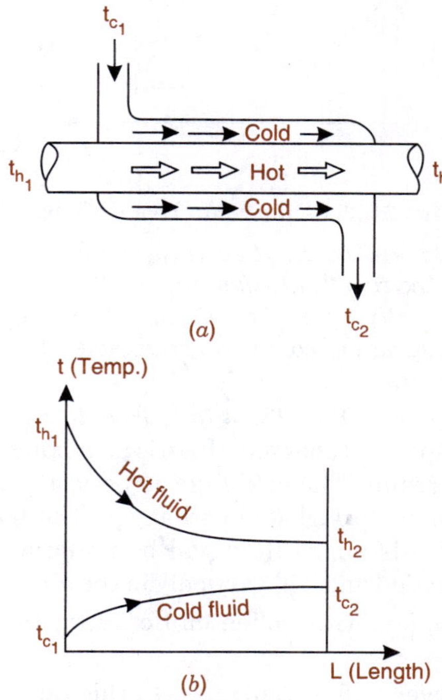

Fig. 28.31. Parallel-flow heat exchanger.

Examples : Oil coolers, oil heaters, water heaters etc.

As the two fluids are separated by a wall, this type of heat exchanger may be called *parallel- flow recuperator or surface heat exchanger.*

(*ii*) ***Counter-flow heat exchangers.*** In a counter-flow heat exchanger, the two fluids flow in *opposite directions*. The hot and cold fluids enter at the opposite ends. The flow arrangement and temperature distribution for such a heat exchanger are shown schematically in Fig. 28.32. The *temperature difference* between the two fluids remains more or less *nearly constant*. This type of heat exchanger, due to counter flow, gives *maximum rate of heat transfer for a given surface area*. Hence such heat exchangers are *most favoured* for heating and cooling of fluids.

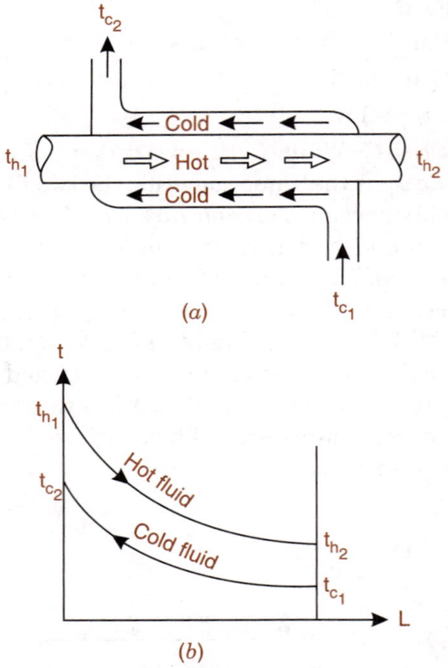

Fig. 28.32. Counter-flow heat exchanger.

(*iii*) ***Cross-flow heat exchangers.*** In cross-flow heat exchangers, the *two fluids (hot and cold) cross one another in space, usually at right angles*. Fig. 28.32 shows a schematic diagram of common arrangements of cross-flow heat exchangers.

- Refer to Fig. 28.33 (*a*) : *Hot fluid* flows in the separate tubes and there is no mixing of the fluid streams. The *cold fluid* is perfectly mixed as it flows through the exchanger. The temperature of this mixed fluid will be uniform across any section and will vary only in the direction of flow.

Examples : The cooling unit of refrigeration system etc.

- Refer to Fig. 28.33 (*b*) : In this case each of the fluids follows a prescribed path and is unmixed as it flows through heat exchanger. Hence the temperature of the fluid leaving the heater section is not uniform.

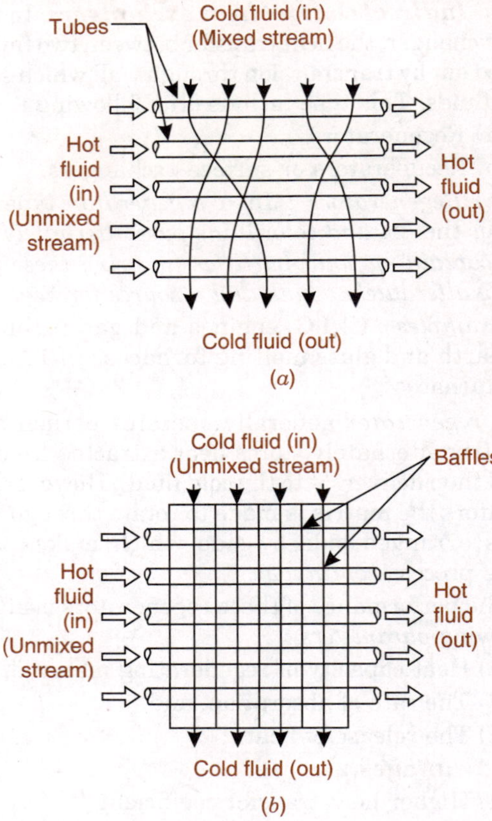

Fig. 28.33. Cross-flow heat exchangers.

Examples : Automobile radiator etc.

- In yet another arrangement, *both the fluids are mixed* while they travel through the exchanger ; consequently the temperature of both the fluids is uniform across the section and varies only in the direction in which flow takes place.

3. Design and constructional features

On the basis of design and constructional features, the heat exchangers are classified as under :

(*i*) ***Concentric tubes.*** In this type, two concentric tubes are used, each carrying one of the fluids. This direction of flow may be parallel or counter as depicted in Fig. 28.31 (*a*) and Fig. 28.32 (*a*). The effectiveness of the heat exchanger is increased by using swirling flow.

(*ii*) ***Shell and tube.*** In this type of heat exchanger one of the fluids flows through a bundle of tubes enclosed by a shell. The other fluid is forced through the shell and it flows over the outside surface of the tubes. Such an arrangement is employed where *reliability* and *heat transfer effectiveness are important.* With the use of multiple tubes heat transfer rate is amply improved due to increased surface area.

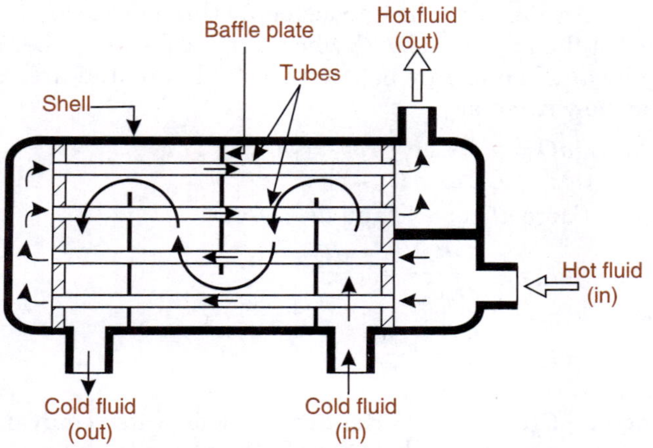

(*a*) One-shell pass and two-tube pass heat exchanger.

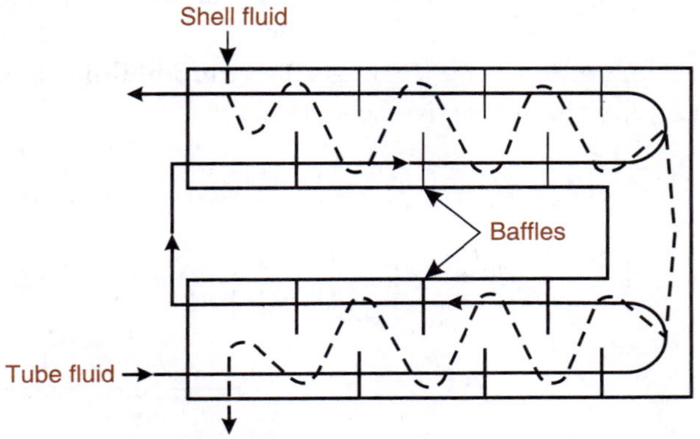

(*b*) Two-shell pass and Four-tube pass heat exchanger

Fig. 28.34. Shell and tube heat exchangers.

(*iii*) ***Multiple shell and tube passes.*** Multiple shell and tube passes are used for *enhancing the overall heat transfer. Multiple shell pass* is possible where the fluid flowing through the shell is re-routed. The shell side fluid is forced to flow back and forth across the tubes by baffles. *Multiple tube pass* exchangers are those which re-route the fluid through tubes in the opposite direction.

(*iv*) ***Compact heat exchangers.*** These are special purpose heat exchangers and have a very large transfer surface area per unit volume of the exchanger. They are generally employed when convective heat transfer sufficient associated with one of the fluids is much smaller than that associated with the other fluid.

Example : Plate-fin, flattened fin tube exchangers etc.

4. Physical state of fluids

Depending upon the physical state of fluids the heat exchangers are classified as follows :

(*i*) Condensers (*ii*) Evaporators

(*i*) ***Condensers.*** In a condenser, the condensing fluid remains at constant temperature throughout the exchanger while the temperature of the colder fluid gradually increases from inlet to outlet. The hot fluid loses latent part of heat which is accepted by the cold fluid (Refer to Fig. 28.35).

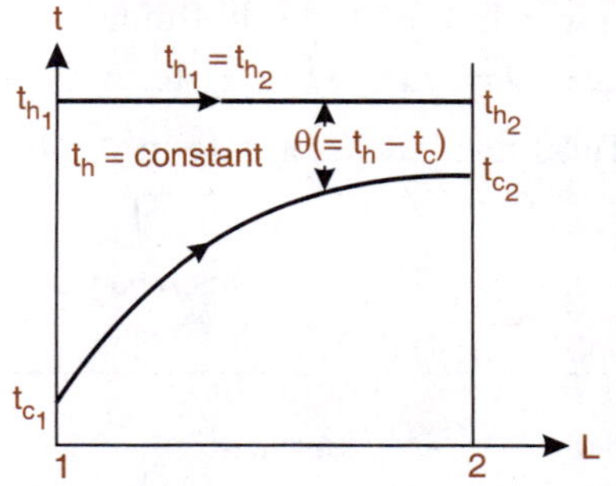

Fig. 28.35. Temperature distribution in a condenser.

(*ii*) ***Evaporators.*** In this case, the boiling fluid (cold fluid) remains at constant temperature while the temperature of hot fluid gradually decreases from inlet to outlet (Refer to Fig. 28.36).

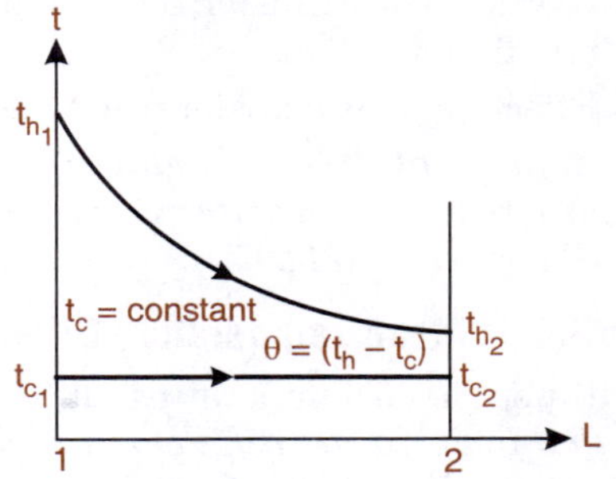

Fig. 28.36. Temperature distribution in an evaporator.

28.4.3. Heat Exchanger Analysis

For designing or predicting the performance of a heat exchanger it is necessary that the total heat transfer may be related with its governing parameters : (*i*) U (overall heat transfer coefficient due to various modes of heat transfer), (*ii*) A total surface area of the heat transfer, and (*iii*) t_1, t_2 (the inlet and outlet fluid temperatures). Fig. 28.37 shows the overall energy balance in a heat exchanger.

Let, $\dot{m}$ = Mass flow rate, kg/s,

c_p = Specific heat of fluid at constant pressure J/kg°C,

t = Temperature of fluid, °C, and

Δt = Temperature drop or rise of a fluid across the heat exchanger.

Subscripts h and c refer to the *hot* and *cold* fluids respectively ; subscripts 1 and 2 correspond to the *inlet* and *outlet* conditions respectively.

Assuming that there is no heat loss to the surroundings and potential and kinetic energy changes are negligible, from the energy balance in a heat exchanger, we have :

Heat given up by the hot fluid,

$$Q = m_h \, c_{ph} \, (t_{h_1} - t_{h_2}) \quad ...(28.46)$$

Heat picked up by the cold fluid,

$$Q = m_c\, c_{pc}\, (t_{c_2} - t_{c_1}) \quad \text{...(28.47)}$$

Total heat transfer rate in the heat exchanger,

$$Q = UA\, \theta_m \quad \text{...(28.48)}$$

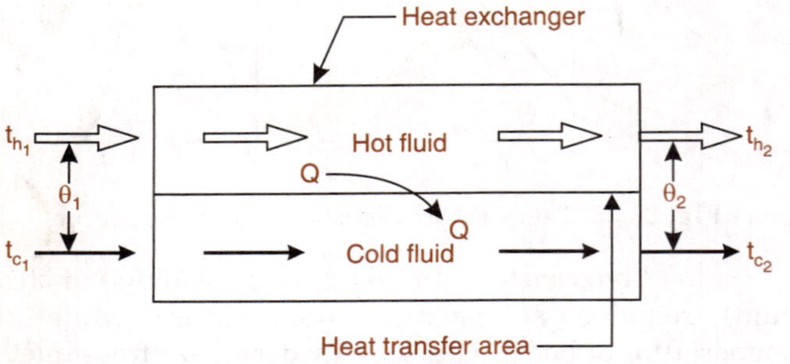

Fig. 28.37. Overall energy balance in a heat exchanger.

where, U = Overall heat transfer coefficient between the two fluids,

A = Effective heat transfer area, and

θ_m = *Appropriate mean value of temperature difference or logarithmic mean temperature difference (LMTD).*

28.4.4. Logarithmic Mean Temperature Difference (LMTD)

Logarithmic mean temperature difference (LMTD) is defined as *that temperature difference which, if constant, would give the same rate of heat transfer as actually occurs under variable conditions of temperature difference.*

In order to derive expression for *LMTD* for various types of heat exchangers, the following ***assumptions*** are made :

1. The overall heat transfer coefficient U is constant.
2. The flow conditions are steady.
3. The specific heats and mass flow rates of both fluids are constant.
4. There is no loss of heat to the surroundings, due to the heat exchanger being perfectly insulated.
5. There is no change of phase either of the fluids during the heat transfer.
6. The changes in potential and kinetic energies are negligible.
7. Axial conduction along the tubes of the heat exchanger is negligible.

28.4.4.1. Logarithmic Mean Temperature Difference for "Parallel-flow"

Refer to Fig. 28.38, which shows the flow arrangement and distribution of temperature in a single-pass parallel-flow heat exchanger.

Let us consider an elementary area dA of the heat exchanger. The rate of flow of heat through this elementary area is given by

$$dQ = U\, dA\, (t_h - t_c) = U \,.\, dA \,.\, \Delta t$$

As a result of heat transfer dQ through the area dA, the hot fluid is cooled by dh whereas the cold fluid is heated up by dt_c. The energy balance over a differential area dA may be written as

$$dQ = -\,\dot{m}_h \,.\, c_{ph} \,.\, dt_h = \dot{m}_c \,.\, c_{pc} \,.\, dt_c$$
$$= U \,.\, dA \,.\, (t_h - t_c) \quad \text{...(28.49)}$$

(Here dt_h is – ve and dt_c is +ve)

or $$dt_h = -\frac{dQ}{\dot{m}_h c_{ph}} = -\frac{dQ}{C_h}$$

and $$dt_c = \frac{dQ}{\dot{m}_c c_{pc}} = \frac{dQ}{C_c}$$

where, $C_h = \dot{m}_h\, c_{ph}$ = Heat capacity or water equivalent of hot fluid, and

$C_c = \dot{m}_c\, c_{pc}$ = Heat capacity or water equivalent of cold fluid.

$\dot{m}_h$ and $\dot{m}_c$ are the mass flow rates of fluids and c_{ph} and c_{pc} are the respective specific heats.

$$\therefore \quad dt_h - dt_c = -\,dQ\left[\frac{1}{C_h} + \frac{1}{C_c}\right]$$

$$d\theta = -\,dQ\left[\frac{1}{C_h} + \frac{1}{C_c}\right] \quad \text{...(28.50)}$$

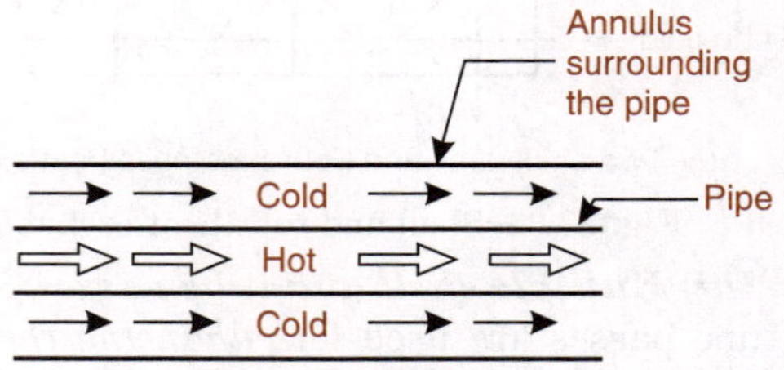

(*a*) Flow arrangement.

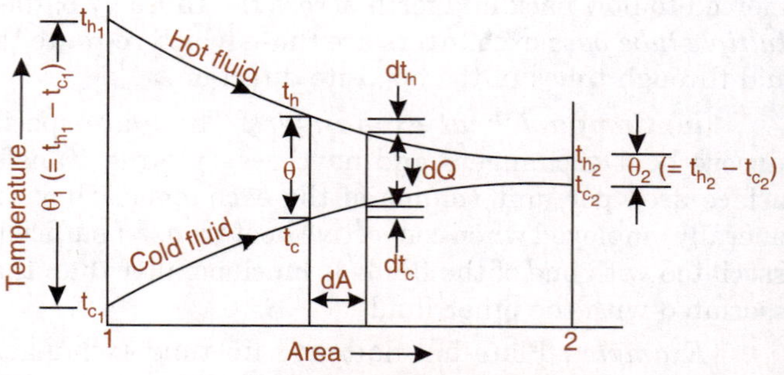

(*b*) Temperature distribution.

Subscripts *h*, *c* refer to *hot* and *cold* fluids
Subscripts 1, 2 refer to *inlet* and *outlet* conditions.

Fig. 28.38. Calculation of LMTD for a parallel-flow heat exchanger.

Substituting the value of dQ from eqn. (28.4) the above equation becomes

$$d\theta = -\,U \,.\, dA\, (t_h - t_c)\left[\frac{1}{C_h} + \frac{1}{C_c}\right]$$

or $$d\theta = -U\,.\,dA\,.\,\theta\left[\frac{1}{C_h}+\frac{1}{C_c}\right]$$

or $$\frac{d\theta}{\theta} = -U.\,dA\left[\frac{1}{C_h}+\frac{1}{C_c}\right]$$

Integrating between inlet and outlet conditions (*i.e.*, from $A = 0$ to $A = A$), we get

$$\int_1^2 \frac{d\theta}{\theta} = -\left[\frac{1}{C_h}+\frac{1}{C_c}\right]\int_{A=0}^{A=A} U\,.\,dA$$

or $$\ln(\theta_2/\theta_1) = -UA\left[\frac{1}{C_h}+\frac{1}{C_c}\right] \quad ...(28.51)$$

Now, the total heat transfer rate between the two fluids is given by

$$Q = C_h\,(t_{h_1} - t_{h_2}) = C_c\,(t_{c_2} - t_{c_1}) \quad ...(28.52)$$

or $$\frac{1}{C_h} = \frac{t_{h_1} - t_{h_2}}{Q} \quad ...[28.52\ (a)]$$

$$\frac{1}{C_c} = \frac{t_{c_2} - t_{c_1}}{Q} \quad ...[28.52\ (b)]$$

Substituting the values of $\frac{1}{C_h}$ and $\frac{1}{C_c}$ into eqn. (28.51), we get

$$\ln(\theta_2/\theta_1) = -UA\left[\frac{t_{h_1} - t_{h_2}}{Q} + \frac{t_{c_2} - t_{c_1}}{Q}\right]$$

$$= \frac{UA}{Q}\,[(t_{h_2} - t_{c_2}) - (t_{h_1} - t_{c_1})]$$

$$= \frac{UA}{Q}\,(\theta_2 - \theta_1)$$

$$Q = \frac{UA\,(\theta_1 - \theta_1)}{\ln(\theta_2/\theta_1)}$$

The above equation may be written as

$$Q = UA\,\theta_m \quad ...(28.53)$$

where $$\theta_m = \frac{\theta_2 - \theta_1}{\ln(\theta_2/\theta_1)} = \frac{\theta_1 - \theta_2}{\ln(\theta_1/\theta_2)} \quad ...(28.54)$$

θ_m is called the *logarithmic mean temperature difference (LMTD)*.

28.4.4.2. Logarithmic Mean Temperature Difference for "Counter-flow"

Refer to Fig. 28.39, which shows the flow arrangement and temperature distribution in a single-pass counter-flow heat exchanger.

Let us consider an elementary area dA of the heat exchanger. The rate of flow of heat through this elementary area is given by

$$dQ = U\,.\,dA\,(t_h - t_c) = U\,.\,dA\,.\,\Delta t \quad ...(28.55)$$

In this case also, due to heat transfer dQ through the area dA, the hot fluid is cooled down by dt_h whereas the cold fluid is heated by dt_c. The energy balance over a differential area dA may be written as

$$dQ = -\dot{m}_h\,.\,c_{ph}\,.\,dt_h = -\dot{m}_c\,.\,c_{pc}\,.\,dt_c \quad ...(28.56)$$

In a counter-flow system, the temperatures of both the fluids *decrease* in the direction of heat exchanger length, hence the –ve signs.

$$\therefore \quad dt_h = -\frac{dQ}{\dot{m}_h c_{ph}} = -\frac{dQ}{C_h}$$

and $$dt_c = -\frac{dQ}{\dot{m}_c c_{pc}} = -\frac{dQ}{C_c}$$

$$\therefore \quad dt_h - dt_c = -dQ\left[\frac{1}{C_h} - \frac{1}{C_c}\right]$$

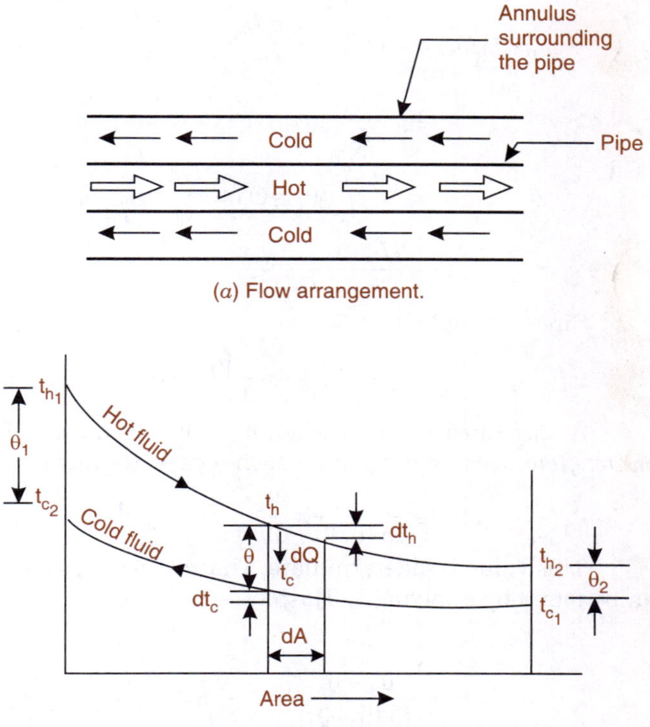

(*a*) Flow arrangement.

(*b*) Temperature distribution.

Fig. 28.39. Calculation of LMTD for a counter-flow heat exchanger.

or $$d\theta = -dQ\left[\frac{1}{C_h} - \frac{1}{C_c}\right] \quad ...(28.57)$$

Inserting the value of dQ from eqn. (28.55), we get

$$d\theta = -U\,dA\,(t_h - t_c)\left[\frac{1}{C_h} - \frac{1}{C_c}\right]$$

$$= -U\,dA\,.\,\theta\left[\frac{1}{C_h} - \frac{1}{C_c}\right]$$

or $$\frac{d\theta}{\theta} = -U\,dA\left[\frac{1}{C_h} - \frac{1}{C_c}\right]$$

Integrating the above equation from $A = 0$ to $A = A$, we get

$$\ln(\theta_2/\theta_1) = -U.A\left[\frac{1}{C_h} - \frac{1}{C_c}\right] \quad ...(28.58)$$

Now, the total heat transfer rate between the two fluids is given by

$$\theta = C_h (t_{h_1} - t_{h_2}) = C_c (t_{c_2} - t_{c_1}) \quad ...(28.59)$$

or $$\frac{1}{C_h} = \frac{t_{h_1} - t_{h_2}}{Q} \quad ...[28.60\ (a)]$$

or $$\frac{1}{C_c} = \frac{t_{c_2} - t_{c_1}}{Q} \quad ...[28.60\ (b)]$$

Substituting the values of $\frac{1}{C_h}$ and $\frac{1}{C_c}$ into eqn. (28.58), we get

$$\ln(\theta_2/\theta_1) = -UA\left[\frac{t_{h_1} - t_{h_2}}{Q} - \frac{t_{c_2} - t_{c_1}}{Q}\right]$$

$$= -\frac{UA}{Q}\,[(t_{h_1} - t_{c_2}) - (t_{h_2} - t_{c_1})]$$

$$= -\frac{UA}{Q}(\theta_1 - \theta_2) = \frac{UA}{Q}(\theta_2 - \theta_1)$$

or $$Q = \frac{UA(\theta_2 - \theta_1)}{\ln(\theta_2/\theta_1)}$$

Since $Q = UA\,\theta_m$

$$\therefore \quad \theta_m = \frac{\theta_2 - \theta_1}{\ln(\theta_2/\theta_1)} = \frac{\theta_1 - \theta_2}{\ln(\theta_1/\theta_2)} \quad ...(28.61)$$

A special case arises when $\theta_1 = \theta_2 = \theta$ in case of a *counter-flow* heat exchanger. In such a case, we have

$$\theta_m = \frac{\theta - \theta}{\ln(\theta/\theta)} = \frac{0}{0}$$

This value is indeterminate. The value of θ_m for such can be found by applying L′ Hospital's rule :

$$\lim_{\theta_2 \to \theta_1} \frac{\theta_2 - \theta_1}{\ln(\theta_2/\theta_1)} = \lim_{(\theta_2/\theta_1) \to 1} \frac{\theta_1\left[\frac{\theta_2}{\theta_1} - 1\right]}{\ln(\theta_2/\theta_1)}$$

Let $(\theta_2/\theta_1) = R$. Therefore, the above expression can be written as

$$\lim_{R \to 1} \frac{\theta(R-1)}{\ln(R)}$$

Differentiating the numerator and denominator with respect to R and taking limits, we get

$$\lim_{(R \to 1)} \frac{\theta}{(1/R)} = \theta$$

Hence when $\theta_1 = \theta_2$ eqn. (28.3) becomes

$$Q = UA\,\theta$$

θ_m *(LMTD) for a counter-flow unit is always greater than that for a parallel-flow unit ;* hence counter-flow heat exchanger can transfer *more* heat than parallel-flow one ; in other words a counter-flow heat exchanger needs a *smaller heating surface for the same rate of heat transfer.* For this reason, the counter-flow arrangement is usually used.

When the temperature variations of the fluids are relatively small, then temperature variation curves are approximately straight lines and adequately accurate results are obtained by taking the *arithmetic mean temperature difference (AMTD).*

$$AMTD = \frac{t_{h_1} + t_{h_2}}{2} - \frac{t_{c_1} + t_{c_2}}{2} = \frac{(t_{h_1} - t_{c_1}) + (t_{h_2} - t_{c_2})}{2}$$

$$= \frac{\theta_1 + \theta_2}{2} \quad ...(28.62)$$

However, practical considerations suggest that the logarithmic mean temperature difference (θ_m) should be invariably used when $\frac{\theta_1}{\theta_2} > 1.7$.

Example 28.18. *The flow rates of hot and cold water streams running through a parallel-flow heat exchanger are 0.2 kg/s and 0.5 kg/s respectively. The inlet temperatures on the hot and cold sides are 75°C and 20°C respectively. The exit temperature of hot water is 45°C. If the individual heat transfer coefficients on both sides are 650 W/m²°C, calculate the area of the heat exchanger.*

Solution. *Given :* $\dot{m}_h = 0.2$ kg/s ; $\dot{m}_c = 0.5$ kg/s ; $t_{h_1} = 75°C$;

$t_{h_2} = 45°C$; $t_{c_1} = 20°C$; $h_i = h_o = 650$ W/m²°C.

The area of heat exchanger, A :

The heat exchanger is shown diagrammatically in Fig. 28.40.

The heat transfer rate,

$$Q = m_h \times c_{ph} \times (t_{h_1} - t_{h_2})$$

$$= 0.2 \times 4.187 \times (75 - 45) = 25.122 \text{ kJ/s}$$

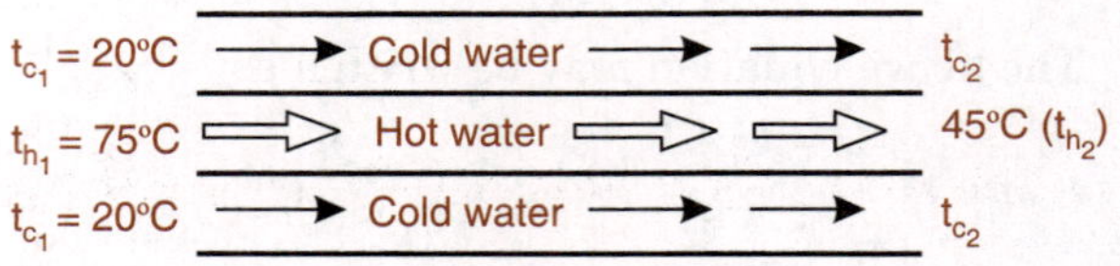

(*a*) Flow arrangement.

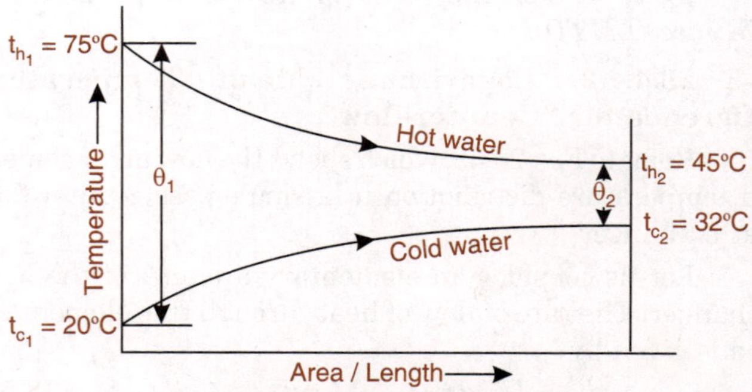

(*b*) Temperature distribution.

Fig. 28.40. Parallel-flow heat exchanger.

Heat lost by hot water = Heat gained by cold water

$$\dot{m}_h \times c_{ph} \times (t_{h_1} - t_{h_2}) = \dot{m}_c \times c_{pc} \times (t_{c_2} - t_{c_1})$$

$$0.2 \times 4.187 \times (75 - 45) = 0.5 \times 4.187 \times (t_{c_2} - 20)$$

$$\therefore \quad t_{c_2} = 32°C$$

Logarithmic mean temperature difference (*LMTD*) is given by,

$$\theta_m = \frac{\theta_1 - \theta_2}{\ln(\theta_1/\theta_2)} \quad ...[\text{Eqn. (28.54)}]$$

or

$$\theta_m = \frac{(t_{h_1} - t_{c_1}) - (t_{h_2} - t_{c_2})}{\ln[(t_{h_1} - t_{c_1})/(t_{h_2} - t_{c_2})]}$$

$$= \frac{(75 - 20) - (45 - 32)}{\ln[(75 - 20)/(45 - 32)]} = \frac{55 - 33}{\ln(55/13)}$$

$$= 29.12°C$$

Overall heat transfer coefficient *U* is calculated from the relation

$$\frac{1}{U} = \frac{1}{h_i} + \frac{1}{h_o} = \frac{1}{650} + \frac{1}{650} = \frac{1}{325}$$

$$\therefore \quad U = 325 \text{ W/m}^2\text{°C}$$

Also, $Q = U A \theta_m$

or

$$A = \frac{Q}{U\theta_m} = \frac{25.122 \times 1000}{325 \times 29.12}$$

$= \mathbf{2.65\ m^2}$. **(Ans.)**

Example 28.19. *In a counter-flow double pipe heat exchanger, water is heated from 25°C to 65°C by an oil with a specific heat of 1.45 kJ/kg K and mass flow rate of 0.9 kg/s. The oil is cooled from 230°C to 160°C. If the overall heat transfer coefficient is 420 W/m²°C, calculate the following :*

(*i*) *The rate of heat transfer,*

(*ii*) *The mass flow rate of water, and*

(*iii*) *The surface area of the heat exchanger.*

Solution. *Given :* $t_{c_1} = 25°C$; $t_{c_2} = 65°C$, $c_{ph} = 1.45$ kJ/kg K ;

$\dot{m}_h = 0.9$ kg/s ; $t_{h_1} = 230°C$;

$t_{h_2} = 160°C$, $U = 420$ W/m²°C.

(*i*) **The rate of heat transfer, *Q* :**

$$Q = \dot{m}_h \times c_{ph} \times (t_{h_1} - t_{h_2})$$

or $Q = 0.9 \times (1.45) \times (230 - 160) = \mathbf{91.35\ kJ/s.}$ **(Ans.)**

(*ii*) **The mass flow rate of water, $\dot{m}_c$:**

Heat lost by oil (hot fluid) = Heat gained by water (cold fluid)

$$\dot{m}_h \times c_{hp} \times (t_{h_1} - t_{h_2}) = \dot{m}_c \times c_{pc} \times (t_{c_2} - t_{c_1})$$

$$91.35 = \dot{m}_c \times 4.187\,(65 - 25)$$

$$\therefore \quad \dot{m}_c = \frac{91.35}{4.187 \times (65 - 25)}$$

$= \mathbf{0.545\ kg/s.}$ **(Ans.)**

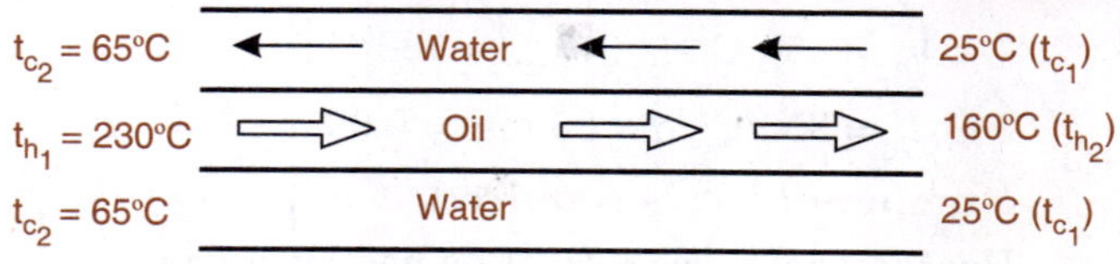

(*a*) Flow arrangement.

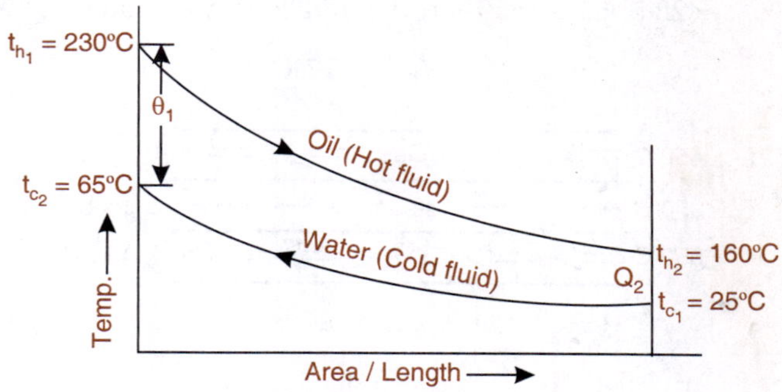

(*b*) Temperature distribution.

Fig. 28.41. Counter-flow heat exchanger.

(*iii*) **The surface area of heat exchanger, *A* :**

Logarithmic mean temperature difference (LMTD) is given by

$$\theta_m = \frac{\theta_1 - \theta_2}{\ln(\theta_1/\theta_2)}$$

$$= \frac{(t_{h_1} - t_{c_2}) - (t_{h_2} - t_{c_1})}{\ln[(t_{h_1} - t_{c_2})/(t_{h_2} - t_{c_1})]}$$

$$= \frac{(230 - 65) - (160 - 25)}{\ln[(230 - 65)/(160 - 25)]}$$

or

$$\theta_m = \frac{165 - 135}{\ln[(165/135)]} = 149.5°C$$

Also, $Q = U A \theta_m$

or

$$A = \frac{Q}{U\theta_m} = \frac{91.35 \times 10^3}{420 \times 149.5} = \mathbf{1.45\ m^2.}$$ **(Ans.)**

Example 28.20. *Steam enters a counter-flow heat exchanger, dry saturated at 10 bar and leaves at 35°C. The mass flow of steam is 800 kg/min. The gas enters the heat exchanger at 650°C and mass flow rate is 1350 kg/min. If the tubes are 30 mm diameter and 3 m long, determine the number of tubes required. Neglect the resistance offered by metallic tubes. Use the following data :*

For steam : $t_{sat} = 180°C$ *(at 10 bar) ;*

$c_{ps} = 2.71$ *kJ/kg°C ;* $h_s = 600$ *W/m²°C*

For gas : $c_{pg} = 1$ *kJ/kg°C ;* $h_g = 250$ *W/m²°C*

(P.U.)

Solution. *Given :* $\dot{m}_s = \dot{m}_c = \frac{800}{60} = 13.33$ kg/s ;

$$\dot{m}_g = \dot{m}_h = \frac{1350}{60} = 22.5 \text{ kg/s} ;$$

$t_{h_1} = 650°C$; $t_{c_1} (= t_{sat}) = 180°C$;

$t_{c_2} = 350°C$; $d = 30$ mm $= 0.03$ m ; $L = 3$ m.

Number of tubes required, *N* :

Heat lost by gases = Heat gained by steam

$$\dot{m}_h \times c_{ph} \times (t_{h_1} - t_{h_2}) = \dot{m}_c \times c_{pc} \times (t_{c_2} - t_{c_1})$$

$$22.5 \times 1 \times (650 - t_{h_2}) = 13.33 \times 2.71 \times (350 - 180)$$

$$\therefore \quad t_{h_2} = 377°C$$

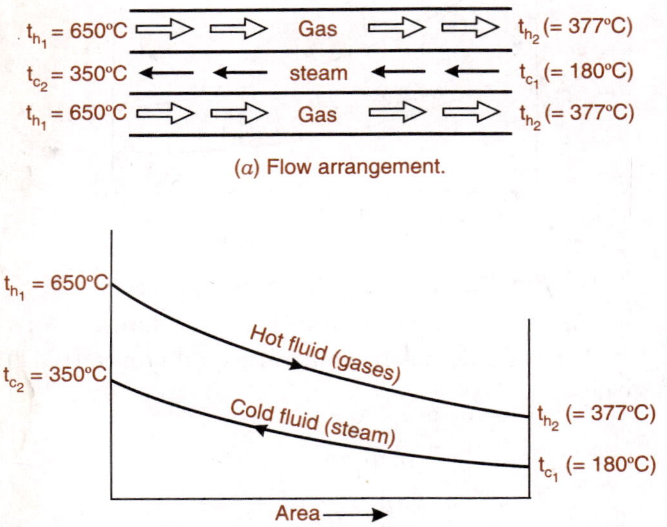

(*a*) Flow arrangement.

(*b*) Temperature distribution.

Fig. 28.42. Counter-flow heat exchanger.

Overall heat transfer coefficient is given by

$$\frac{1}{U} = \frac{1}{h_g} + \frac{d_o}{d_i}\frac{1}{h_s} = \frac{1}{h_g} + \frac{1}{h_s} \text{ as } d_i \approx d_o$$

...(given)

or $$U = \frac{h_g \times h_s}{h_g + h_s} = \frac{250 \times 600}{250 + 600} = 176.5 \text{ W/m}^2\text{°C}$$

Total heat transfer rate is given by

$$Q = U A \theta_m \qquad ...(i)$$

where $A = N \times (\pi d L) = N \times \pi \times 0.03 \times 3 = 0.2827 \text{ Nm}^2$

$$Q = 22.5 \times (1 \times 10^3) \times (650 - 377) = 6142.5 \times 10^3 \text{ W}$$

$$\theta_m = \frac{\theta_1 - \theta_2}{\ln(\theta_1/\theta_2)} = \frac{(t_{h1} - t_{c2}) - (t_{h2} - t_{c1})}{\ln[(t_{h1} - t_{c2})/(t_{h2} - t_{c1})]}$$

$$= \frac{(650 - 350) - (377 - 180)}{\ln[(650 - 300)/(377 - 180)]} = \frac{300 - 197}{\ln(300/197)}$$

$$= 244.9°C$$

Substituting the values in eqn. (*i*), we get

$$6142.5 \times 10^3 = 176.5 \times 0.2827 \text{ N} \times 244.9$$

or $$N = \frac{6142.5 \times 10^3}{176.5 \times 0.2827 \times 244.9}$$

= 503 tubes. (Ans.)

Example 28.21. *A two-pass surface condenser is required to handle the exhaust from a turbine developing 15 MW with specific steam consumption of 5 kg/kWh. The condenser vacuum is 660 mm of Hg when the barometer reads 760 mm of Hg. The mean velocity of water is 3 m/s, water inlet temperature is 24°C. The condensate is saturated water and outlet temperature of cooling water in 4°C less than the condensate temperature. The quality of exhaust steam is 0.9 dry. The overall heat transfer coefficient based on outer area of tubes is 4000 W/m²°C. The water tubes are 38.4 mm in outer diameter and 29.6 mm in inner diameter. Calculate the following :*

(*i*) *Mass of cooling water circulated in kg/min,*

(*ii*) *Condenser surface area,*

(*iii*) *Number of tubes required per pass, and*

(*iv*) *Tube length.* **(P.U.)**

Solution. *Given* : $d_i = 29.6$ mm $= 0.0296$ m ;

$d_o = 38.4$ mm $= 0.0384$ m ;

$U = 4000$ W/m²°C ; $V = 3$ m/s ;

$t_{c_1} = 24°C$; x (dryness fraction) $= 0.9$.

The pressure of the steam in the condenser,

$$p_s = \frac{760 - 660}{760} \times 1.0133 = 0.133 \text{ bar}$$

The properties of steam at $p_s = 0.133$ bar, from steam table, are :

$$t_{sat} = 51°C \; ; \; h_{fg} = 2592 \text{ kJ/kg}$$

$$\therefore \quad t_{c_1} = 51 - 4 = 47°C$$

The steam condensed per minute,

$$\dot{m}_s (= \dot{m}_h) = \frac{(15 \times 1000) \times 5}{60} = 1250 \text{ kg/min}$$

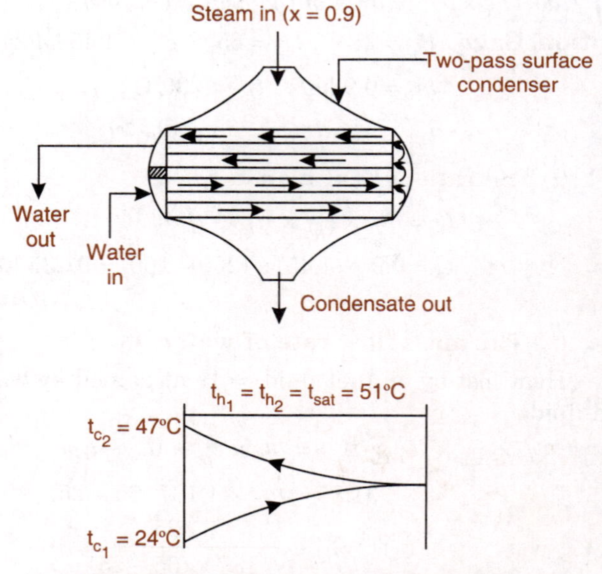

Fig. 28.43. A two-pass surface condenser.

(*i*) **Mass of cooling water circulated per minute, $\dot{m}_w$ (= $\dot{m}_c$) :**

Heat lost by steam = Heat gained by water

$$\dot{m}_h \times (x \,.\, h_{fg}) = \dot{m}_c \times c_{pc} \times (t_{c_2} - t_{c_1})$$

$$1250 \times (0.9 \times 2592) = \dot{m}_c \times 4.187\,(47 - 24)$$

$$\therefore \qquad \dot{m}_c\ (= \dot{m}_w) = \mathbf{30280\ kg/min.} \quad \textbf{(Ans.)}$$

(*ii*) **Condenser surface area, A :**

$$Q = \frac{\dot{m}_s \times (x \,.\, h_{fg})}{60} = U A\, \theta_m \qquad ...(i)$$

where,

$$\theta_m = \frac{\theta_1 - \theta_2}{\ln(\theta_1/\theta_2)} = \frac{(t_{h_1} - t_{c_1}) - (t_{h_2} - t_{c_2})}{\ln[(t_{h_1} - t_{c_1})/(t_{h_2} - t_{c_2})]}$$

$$= \frac{(51 - 24) - (51 - 47)}{\ln[(51-24)/(51-47)]} = \frac{(27 - 4)}{\ln(27/4)}$$

$$= 12.04°C$$

Substituting the values in eqn. (*i*), we get

$$\frac{1250}{60} \times (0.9 \times 2592 \times 10^3) = 4000 \times A \times 12.04$$

or

$$A = \mathbf{1009.1\ m^2.} \quad \textbf{(Ans.)}$$

(*iii*) **Number of tubes required per pass, N_p :**

$$\dot{m}_w = \left(\frac{\pi}{4}\, d_i^2 \times V \times \rho\right) \times N_p$$

$$\frac{30280}{60} = \frac{\pi}{4} \times (0.0296)^2 \times 3 \times 1000 \times N_p$$

or

$$N_p = \frac{30280 \times 4}{60 \times \pi \times (0.0296)^2 \times 3 \times 1000}$$

$$= 244.46 \text{ say } \mathbf{245.} \quad \textbf{(Ans.)}$$

(Total number of tubes required,

$$N = 2N_p = 2 \times 245 = 490)$$

(*iv*) **Tube length, L :**

$$A = (\pi d_o L) \times (2N_p)$$

$$1009.1 = (\pi \times 0.0384 \times L) \times (2 \times 245)$$

or

$$L = \frac{1009.1}{\pi \times 0.0384 \times 2 \times 245} \simeq \mathbf{17.1\ m.} \quad \textbf{(Ans.)}$$

Example 28.22. *A feed water heater which supplies hot water to a boiler comprises a shell and tube heat exchanger with one-shell pass and two-tube passes. One hundred thin-walled tubes each of 20 mm diameter and length of 2 m per pass are used. Under normal operating conditions, water enters the tubes at 10 kg/s and 17°C and is heated by condensing saturated steam at 1 atm. on the outer surface of the tubes. The convection coefficient of the saturated steam is 10 kW/m²°C. Determine the water exit temperature.*

Use the following properties of water :

c_p = 4.18 kJ/kg°C ; μ = 0.596 × 10^{-3} Ns/m² ;
k = 0.635 W/m°C and Pr = 3.93. **(M.U.)**

Solution. *Given* : *p* (number of tube passes) = 2, *N* (total number of tubes) = 200,

d = 20 mm = 0.02 m ; (length per pass) = 2 m,

$\dot{m}_w = \dot{m}_c$ = 10 kg/s, t_{c_1} = 17°C.

Water exit temperature, t_{c_2} :

$$\dot{m}_c = \frac{\pi}{4}\, d^2 \times V \times \rho \times N_p$$

$$\left[\begin{array}{l}\text{where } V = \text{velocity of water ;}\\ N_p = \text{number of tubes}\\ \text{per pass} = \dfrac{N}{p} = \dfrac{200}{2} = 100\end{array}\right]$$

or

$$10 = \frac{\pi}{4} \times 0.02^2 \times V \times 1000 \times 100$$

$$\therefore \qquad V = \frac{10 \times 4}{\pi \times 0.02^2 \times 1000 \times 100} = 0.318 \text{ m/s}$$

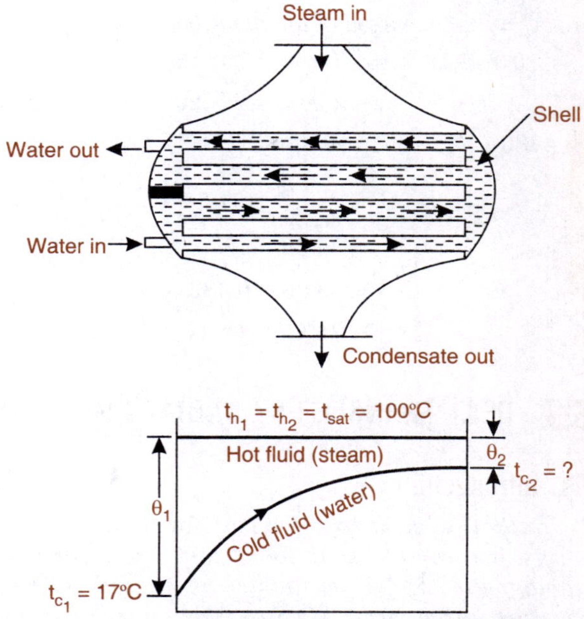

Fig. 28.44. One-shell pass and two-tube passes condenser.

Using non-dimensional heat transfer equation to water side, we get

$$Nu = \frac{h_i d}{k} = 0.023\,(Re)^{0.8}\,(Pr)^{0.33}$$

or

$$h_i = \frac{k}{d} \times 0.023\,(Re)^{0.8}\,(Pr)^{0.33} \qquad ...(i)$$

$$Re = \frac{\rho V d}{\mu} = \frac{1000 \times 0.318 \times 0.02}{0.596 \times 10^{-3}} = 10671$$

Substituting the values in eqn. (*i*), we get

$$h_i = \frac{0.635}{0.02} \times 0.023\,(10671)^{0.8}\,(3.93)^{0.33}$$

$$= 1915 \text{ W/m}^2{}°\text{C}$$

The overall heat transfer coefficient is given by the relation,

$$\frac{1}{U} = \frac{1}{h_i} + \frac{1}{h_o}$$

$$\frac{1}{U} = \frac{1}{1915} + \frac{1}{10 \times 10^3} = 0.000622$$

$$\therefore \quad U = \frac{1}{0.000622} = 1607.7 \text{ W/m}^2\text{°C}$$

Further, $\theta_1 = t_{h_1} - t_{c_1} = 100 - 17 = 83\text{°C}$

$$\theta_2 = t_{h_2} - t_{c_2} = 100 - t_{c_2}$$

$\therefore$ Arithmetic mean temperature difference,

$$AMTD = \frac{\theta_1 + \theta_2}{2} = \frac{83 + (100 - t_{c_2})}{2} = 91.5 - 0.5\, t_{c_2}$$

The heat transfer rate is given by,

$$Q = \dot{m}_c\, c_{pc}\, (t_{c_2} - t_{c_1}) = U A_s\, (AMTD)$$

$$= U \times (\pi\, d\, L \times N)\, (AMTD)$$

(where A_s = Surface area of *all the tubes in both passes*)

or $\quad 10 \times (4.18 \times 10^3)\, (t_{c_2} - 17) = 1607.7$

$$\times (\pi \times 0.02 \times 2 \times 200) \times (91.5 - 0.5\, t_{c_2})$$

$$41800\, (t_{c_2} - 17) = 40406\, (91.5 - 5\, t_{c_2})$$

or $\quad t_{c_2} - 17 = \dfrac{40406}{41800}\, (91.5 - 0.5\, t_{c_2})$

$$= 0.966\, (91.5 - 0.5\, t_{c_2})$$

$$= 88.39 - 0.483\, t_{c_2}$$

or $\quad t_{c_2}$ = **71°C. (Ans.)**

28.5 HEAT TRANSFER BY RADIATION

28.5.1. Introduction

'Radiation' heat transfer is defined as *"the transfer of energy across a system boundary by means of an electromagnetic mechanism which is caused solely by a temperature difference."* Whereas the heat transfer by conduction and convection takes place only in the *presence of medium, radiation heat transfer does not require a medium.* Radiation exchange, in fact, occurs most effectively in *vacuum.* Further, the rate of heat transfer by *conduction and convection* varies as the temperature difference to the *first power,* whereas the *radiant heat exchange* between two bodies depends on the difference between their temperature to the *'fourth power'.* Both the amount of radiation and the quality of radiation depend upon temperature. The *dissipation from the filament of a vacuum tube* or the *heat leakage through the evacuated walls of a thermos flask* are some familiar *examples* of heat transfer by radiation.

The contribution of *radiation* to heat transfer is very significant at high absolute temperature levels such as those prevailing in *furnaces, combustion chambers, nuclear explosions* and in *space applications.* The *solar energy* incident upon the earth is also governed by the *laws of radiation.*

The energy which a radiating surface releases is *not continuous* but is in the form of successive and separate (discrete) packet or quanta of energy called *photons.* The photons are propagated through space as rays ; the movement of swarm of photons is described as *electromagnetic waves.* The photons travel (with speed equal to that of light) in straight paths with unchanged frequency ; when they approach the receiving surface, there occurs reconversion of wave motion into *thermal energy* which is partly absorbed, reflected or transmitted through the receiving surface (the magnitude of each fraction depends, upon the nature of the surface that receives the *thermal radiation*).

All types of electromagnetic waves are classified in terms of *wavelength* and are propagated at the speed of light (c) *i.e.,* 3×10^8 m/s. The electromagnetic spectrum is shown in Fig. 28.45. The distinction between one form of radiation and another lies only in its frequency (f) and wavelength (λ) which are related by

$$c = \lambda \times f \quad \text{...(28.63)}$$

The emission of thermal radiation (range lies between wavelength of 10^{-7} m and 10^{-4} m) *depends upon the nature, temperature and state of the emitting surface.* However, with *gases the dependence is also upon the thickness of the emitting layer and the gas pressure.*

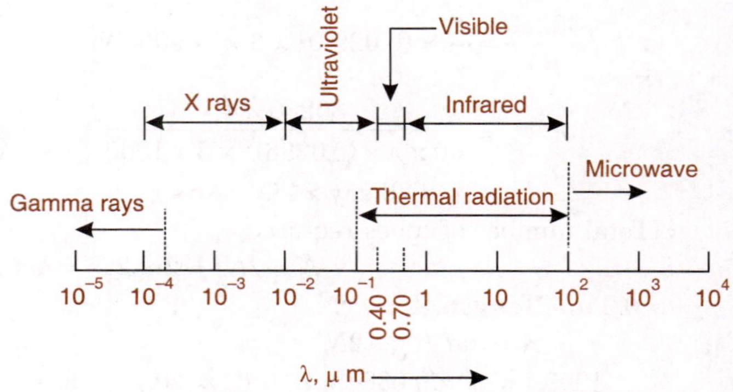

Fig. 28.45. Spectrum of electromagnetic radiation.

Thermal radiations exhibit characteristics similar to those of *visible light,* and *follow optical laws.* These can be *reflected, refracted* and are *subject to scattering and absorption when they pass through a media.* They get *polarised* and *weakened* in strength with inverse square of radial distance from the radiating surface.

28.5.2. Surface Emission Properties

The rate of emission of radiation by a body depends upon the following *factors* :

(*i*) The temperature of the surface,

(*ii*) The nature of the surface, and

(*iii*) The wavelength or frequency of radiation.

The parameters which deal with the surface emission properties are given below :

(i) **Total emissive power (E).** The *emissive power* is defined as the *total amount of radiation emitted by a body per unit area and time.* It is expressed in W/m². The *emissive power of a black body,* according to Stefan- Boltzmann, is *proportional to absolute temperature to the fourth power.*

$$E_b = \sigma\, T^4 \text{ W/m}^2 \qquad \text{...(28.64)}$$

$$E_b = \sigma A\, T^4 \text{ W} \qquad \text{...[28.64 (a)]}$$

where σ = Stefan-Boltzmann constant

$= 5.67 \times 10^{-8}$ W/m² K⁴.

(ii) **Monochromatic (spectral) emissive power (E_1).** It is often necessary to determine the spectral distribution of the energy radiated by a surface. At any given temperature the amount of radiation emitted per unit wavelength varies at different wavelengths. For this purpose the *monochromatic emissive power* E_λ of the surface is used. It is defined as *the rate of energy radiated per unit area of the surface per unit wavelength.*

The total emissive power is given by

$$E = \int_o^\infty E_\lambda\, d\lambda \text{ W/m}^2 \qquad \text{...(28.65)}$$

(iii) **Emission from real surface-emissivity.** The emissive power from a real surface is given by

$$E = \varepsilon\, \sigma\, A T^4 \text{ W} \qquad \text{...(28.66)}$$

where ε = emissivity of the material.

Emissivity(ε). It is defined as the *ability of the surface of a body to radiate heat.* It is also defined as the *ratio of the emissive power of any body to the emissive power of a black body of equal temperature* $\left(i.e., \varepsilon = \dfrac{E}{E_b}\right)$. Its values varies for different substances ranging from 0 to 1. For a black body ε = 1, for a white body surface ε = 0 and for gray bodies it lies between 0 and 1. It may vary with temperature or wavelength.

(iv) Intensity of radiation.

(v) Radiation density and pressure.

(vi) Radiosity (J). It refers to *all of the radiant energy leaving a surface.*

(vii) Interrelationship between surface emission and irradiation properties.

28.5.3. Absorptivity, Reflectivity and Transmissivity

When incident radiation also called **irradiation** (defined as the *total incident radiation on a surface from all directions per unit time and per unit area of surface),* expressed in W/m² and denoted by (*G*) impinges on a surface, three things happens ; a part is *reflected* back (G_r), a part is *transmitted* through (G_t), and the remainder is *absorbed* (G_a) depending upon the characteristics of the body, as shown in Fig. 28.46.

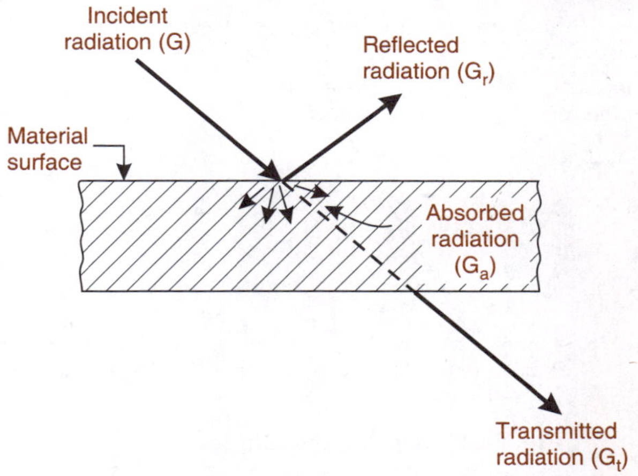

Fig. 28.46. Absorption, reflection and transmission of radiation.

By the conservation of energy principle,

$$G_a + G_r + G_t = G$$

Dividing both sides by *G*, we get

$$\frac{G_a}{G} + \frac{G_r}{G} + \frac{G_t}{G} = \frac{G}{G}$$

$$\alpha + \rho + \tau = 1 \qquad \text{...(28.67)}$$

where, α = *Absorptivity* (or fraction of incident radiation absorbed),

ρ = *Reflectivity* (or fraction of incident radiation reflected), and

τ = *Transmissivity* (or fraction of incident radiation transmitted).

When the incident radiation is absorbed, it is converted into internal energy.

Black body. For perfectly absorbing body, α = 1, ρ = 0, τ = 0. Such a body is called a *'black body' (i.e., a black body is one which neither reflects nor transmits any part of the incident radiation but absorbs all of it).* In practice, a perfect black body (α = 1) does not exist. However, its concept is very important.

Opaque body. When no incident radiation is transmitted through the body, it is called an *'opaque body'.*

For the opaque body τ = 0, and eqn. (28.67) reduces to

$$\alpha + \rho = 1 \qquad \text{...(28.68)}$$

Solids generally do not transmit unless the material is of very thin section. Metals absorb radiation within a fraction of a micrometre, and insulators within a fraction of millimetre. Glasses and liquids are, therefore, generally considered as opaque.

White body. If all the incident radiation falling on the body are reflected, it is called a *'white body'.*

For a white body, ρ = 1, α = 0 and τ = 0.

Gases such as hydrogen, oxygen and nitrogen (and their mixture such as air) have a transmissivity of practically *unity.*

Reflections are of two types : Refer to Fig. 28.47.

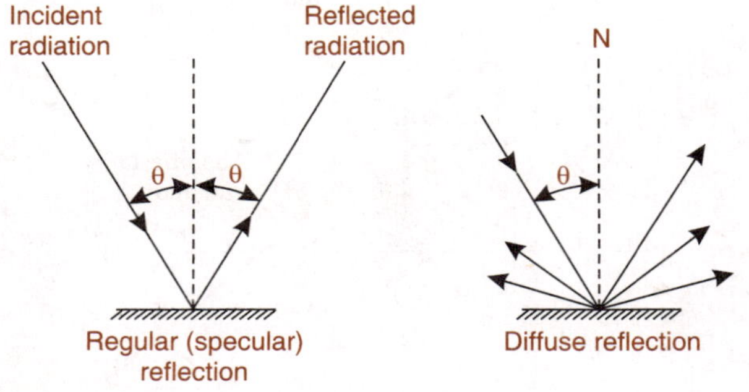

Fig. 28.47. Regular and diffuse reflections.

1. Regular (specular) reflection
2. Diffuse reflection.

Regular reflection implies that angle between the reflected beam and the normal to the surface *equals* the angle made by the incident radiation with the same normal. Reflection from highly polished and smooth surfaces approaches specular characteristics.

In a *diffused reflection,* the incident beam is reflected in *all directions*. Most of the engineering materials have rough surfaces, and these rough surfaces give diffused reflections.

Gray body. If the radiative properties, α, ρ τ of a body are assumed to be uniform over the entire wavelength spectrum, then such a body is called *gray body*. A *gray body* is also defined as one *whose absorptivity of a surface does not vary with temperature and wavelength of the incident radiation* [$\alpha = (\alpha)_\lambda$ = constant.]

A *coloured body* is one whose absorptivity of a surface *varies* with the wavelength of radiation [$\alpha \neq (\alpha)_\lambda$].

28.5.4. Concept of a Black Body

A black body is an object *that absorbs all the radiant energy reaching its surface* (for a black body $\alpha = 1$, $\rho = 0$, $\tau = 0$). No actual body is perfectly black ; the concept of a black body is an idealization with which the radiation characteristics of real bodies can be conveniently compared.

A *black body* has the following *properties* :

(*i*) It absorbs all the incident radiation falling on it and does not transmit or reflect regardless of wavelength and direction.

(*ii*) It emits maximum amount of thermal radiations at all wavelengths at any specified temperature.

(*iii*) It is a *diffuse emitter* (*i.e.,* the radiation emitted by a black body is independent of direction).

Consider a hollow enclosure with a very small hole for the passage of incident radiation as shown in Fig. 28.48. Incident radiant energy passes through the small opening ; some of this energy is absorbed by the inside surface and some is reflected. However, most of this energy is absorbed on a second incidence. Again, a small fraction is reflected. After a number of such reflections the amount unabsorbed is exceedingly small and very little of the original incident energy is reflected back out of the opening. A small hole leading into a cavity (Hohlraum) thus acts very nearly as a black body because all the radiant energy entering through it gets absorbed.

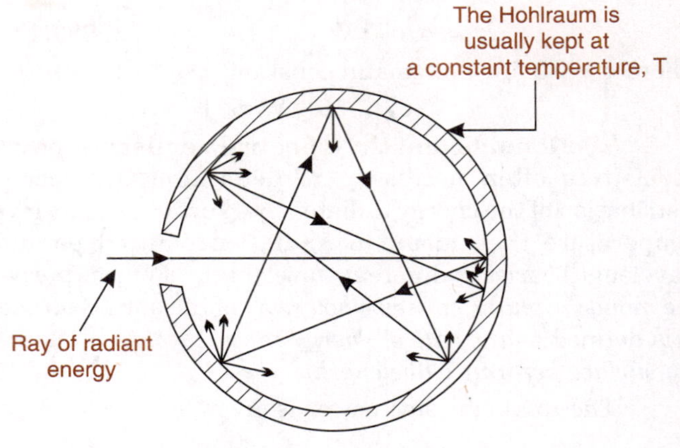

Fig. 28.48. Concept of a black body.

Isothermal furnaces, with small apertures, approximate a black body and are frequently used to *calibrate* heat flux gauges, thermometers and other radiometric devices.

28.5.5. The Stefan-Boltzmann Law

The law states that *the emissive power of a black body is directly proportional to the fourth power of its absolute temperature.*

i.e.,
$$E_b = \sigma T^4 \quad ...(28.69)$$

where, E_b = Emissive power of a black body, and

σ = Stefan-Boltzmann constant

$= 5.67 \times 10^{-8}$ W/m² K⁴.

Equation (28.69) can be rewritten as :

$$E_b = 5.67 \left(\frac{T}{100}\right)^4 \quad ...(28.70)$$

28.5.6. Kirchhoff's Law

The law states that *at any temperature the ratio of total emissive power E to the total absorptivity α is a constant for all substances which are in thermal equilibrium with their environment.*

Let us consider a large radiating body of surface area A which encloses a small body (1) of surface area A_1 (as shown in Fig. 28.49). Let the energy fall on the unit surface of the body at the rate E_b. Of this energy, generally, a fraction α, will be absorbed by the small body. Thus this energy absorbed by the small body (1) is $\alpha_1 A_1 E_b$, in which α_1 is the absorptivity of the body. When thermal equilibrium is attained, the *energy absorbed* by the body (1) must be equal to the *energy emitted*, say, E_1 per unit surface. Thus, at equilibrium, we may write

$$A_1E_1 = \alpha_1 A_1 E_b \quad ...(28.71)$$

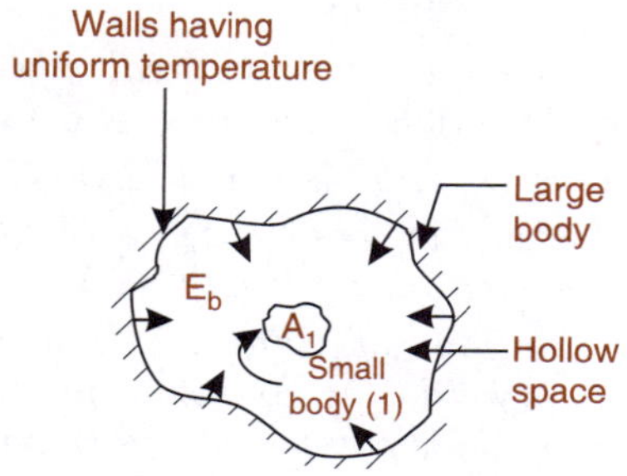

Fig. 28.49. Derivation of Kirchhoff's law.

Now we remove body (1) and replace it by body (2) having absorptivity α_2. The radiative energy impinging on the surface of this body is again E_b. In this case, we may write

$$A_2E_2 = \alpha_2 A_2 E_b \quad ...(28.72)$$

By considering generality of bodies, we obtain

$$E_b = \frac{E_1}{\alpha_1} = \frac{E_2}{\alpha_2} = \frac{E}{\alpha} \quad ...(28.73)$$

Also, as per definition of emissivity ε, we have

$$\varepsilon = \frac{E}{E_b}$$

or
$$E_b = \frac{E}{\varepsilon} \quad ...(28.74)$$

By comparing eqns. (28.73) and (28.74), we obtain

$$\varepsilon = \alpha \quad ...(28.75)$$

(α is always *smaller* than 1. Therefore, the emissive power E is always *smaller* than the emissive power of a black body at equal temperature).

Thus, kirchhoff's law also states that *the emissivity of a body is equal to its absorptivity when the body remains in thermal equilibrium with its surroundings*.

28.5.7. Planck's Law

In 1900 Max Planck showed by quantum arguments that the spectral distribution of the radiation intensity of a black body is given by

$$(E_\lambda)_b = \frac{2\pi c^2 h \lambda^{-5}}{\exp\left(\frac{ch}{\lambda k T}\right) - 1} \quad ...(\text{Planck's law}) \quad ...(28.76)$$

where, $(E_\lambda)_b$ = Monochromatic (single wavelength) emissive power of a black body,

c = Velocity of light in vacuum, $2.998 \times 10^8 \simeq 3 \times 10^8$ m/s,

h = Planck's constant = 6.625×10^{-34} js,

λ = Wavelength, μm,

k = Boltzmann constant = 1.3805×10^{-23} J/K,

and T = Absolute temperature, K.

Hence the unit of $(E_\lambda)_b$ is W/m². μm

Quite often the Planck's law is written as

$$(E_\lambda)_b = \frac{C_1\lambda^{-5}}{\exp\left[\frac{C_2}{\lambda T}\right] - 1} \quad ...(28.77)$$

where, $C_1 = 2\pi c^2 h = 3.742 \times 10^8$ W.μm⁴/m² ;

$$C_2 = \frac{ch}{k} = 1.4388 \times 10^4 \text{ μmK}$$

Equation (28.76) is of great importance as it provides quantitative results for the radiation from a black body.

The quantity $(E_\lambda)_b$, *monochromatic emissive power*, is defined as the *energy emitted by the black surface in all directions at a given wavelength λ per unit wavelength interval around λ* ; that is, the rate of energy emission in the interval $d\lambda$ is equal to $(E_\lambda)_b\, d\lambda$. The total emissive power and monochromatic emissive power are related by the equation

$$E_b = \int_0^\infty (E_\lambda)_b\, d\lambda \quad ...(28.78)$$

A plot of $(E_\lambda)_b$ as a function of temperature and wavelength is given in Fig. 28.50.

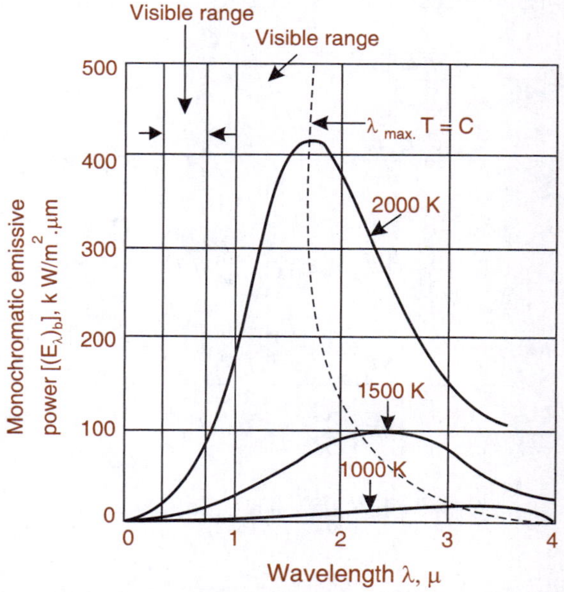

Fig. 28.50. Variation of emissive power with wavelength.

The plot shows that the following distinct characteristics of black body radiations :

1. *The energy emitted at all wavelengths increases with rise in temperature.*
2. *The peak spectral emissive power shifts towards a smaller wavelength at higher temperatures.* This shift signifies that at elevated temperature, much

of the energy is emitted in a narrow band ranging on both sides of wavelength at which the monochromatic power is maximum.

3. The *area* under the monochromatic emissive power versus wavelength, at any temperature, gives the *rate of radiant energy emitted* within the wavelength interval $d\lambda$. Thus,

$$dE_b = (E_\lambda)_b \, d\lambda$$

or

$$E_b = \int_{\lambda=0}^{\lambda=\infty} (E_\lambda)_b \, d\lambda$$

... over the entire range of length.

The integral represents the total emissive power per unit area radiated from a black body.

28.5.8. Wien's Displacement Law

In 1893 Wien established a relationship between the temperature of a black body and the wavelength at which the maximum value of monochromatic emissive power occurs. A peak monochromatic emissive power occurs at a particular wavelength. Wien's displacement law states that the *product of λ_{max} and T is constant, i.e.,*

$$\lambda_{max} \, T = \text{constant} \qquad ...(28.79)$$

$$(E_\lambda)_b = \frac{C_1 \lambda^{-5}}{\exp\left(\frac{C_2}{\lambda T}\right) - 1}$$

$(E_\lambda)_b$ becomes maximum (if T remains constant) when

$$\frac{d(E_\lambda)_b}{d\lambda} = 0$$

i.e.,

$$\frac{d(E_\lambda)_b}{d\lambda} = \frac{d}{d\lambda}\left[\frac{C_1 \lambda^{-5}}{\exp\left(\frac{C_2}{\lambda T}\right) - 1}\right] = 0$$

or

$$\frac{\left[\exp\left(\frac{C_2}{\lambda T}\right) - 1\right](-5 C_1 \lambda^{-6}) - C_1 \lambda^{-5}\left\{\exp\left(\frac{C_2}{\lambda T}\right)\frac{C_2}{T}\left(-\frac{1}{\lambda^2}\right)\right\}}{\left[\exp\left(\frac{C_2}{\lambda T}\right) - 1\right]^2} = 0$$

or

$$-5 C_1 \lambda^{-6} \exp\left(\frac{C_2}{\lambda T}\right) + 5 C_1 \lambda^{-6} + C_1 C_2 \lambda^{-5} \frac{1}{\lambda^2 T} \exp\left(\frac{C_2}{\lambda T}\right) = 0$$

Dividing both side by $5C_1 \lambda^{-6}$, we get

$$-\exp\left(\frac{C_2}{\lambda T}\right) + 1 + \frac{1}{5} C_2 \frac{1}{\lambda T} \exp\left(\frac{C_2}{\lambda T}\right) = 0$$

Solving this equation by trial and error method, we get

$$\frac{C_2}{\lambda T} = \frac{C_2}{\lambda_{max} T} = 4.965$$

$$\therefore \quad \lambda_{max} \, T = \frac{C_2}{4.965} = \frac{1.439 \times 10^4}{4.965} \ \mu\text{mK}$$
$$= 2898 \ \mu\text{mK} \ (\simeq 2900 \ \mu\text{mK})$$

i.e.,

$$\lambda_{max} \, T = 2898 \ \mu\text{mK} \qquad ...(28.80)$$

This law holds true for more *real substances* ; there is however some deviation in the case of a metallic conductor where the product (λ_{max} .T) is found to vary with absolute temperature. It is used in *predicting a very high temperature through measurement of wavelength.*

A combination of Planck's law and Wien's displacement law yields the condition for the maximum monochromatic emissive power for a black body.

$$(E_{\lambda b})_{max} = \frac{C_1 (\lambda_{max})^{-5}}{\exp\left[\frac{C_2}{\lambda_{max} T}\right] - 1}$$

$$= \frac{0.374 \times 10^{-15}\left(\frac{2.898 \times 10^{-3}}{T}\right)^{-5}}{\exp\left[\frac{1.4388 \times 10^{-2}}{2.898 \times 10^{-3}}\right] - 1}$$

or $(E_{\lambda b})_{max} = 1.285 \times 10^{-5} \, T^5$ W/m² per metre wavelength ...(28.81)

28.5.9. Intensity of Radiation and Lambert's Cosine Law

28.5.9.1. Intensity of Radiation

When a surface element emits radiation, all of it will be intercepted by a hemispherical surface placed over the element. The **intensity of radiation** (I) is defined as the *rate of energy leaving a surface in a given direction per unit solid angle per unit area of the emitting surface normal to the mean direction in space.* A **solid angle** is defined as a *portion of the space inside a sphere enclosed by a conical surface with the vertex of the cone at the centre of the sphere.* It is measured by the ratio of the spherical surface enclosed by the cone to the square of the radius of the sphere ; it unit is steradian (sr). The solid angle subtended by the complete hemisphere is given by : $\frac{2\pi r^2}{r^2} = 2\pi$.

Fig. 28.51 (a) shows a small black surface of area dA (emitter) emitting radiation in different directions. A black body radiation collector through which the radiation pass is located at an angular position characterised by *zenith angle* θ towards the surface normal and angle ϕ of a spherical coordinate system. Further the collector subtends a solid angle $d\omega$ when viewed from a point on the emitter.

Let us now consider radiation from the elementary area dA_1 at the centre of a sphere as shown in Fig. 28.51. Suppose this radiation is absorbed by a second elemental area dA_2, a portion of the hemispherical surface.

The projected area of dA_1 on a plane perpendicular to the line joining dA_1 and dA_2 = $dA_1 \cos\theta$.

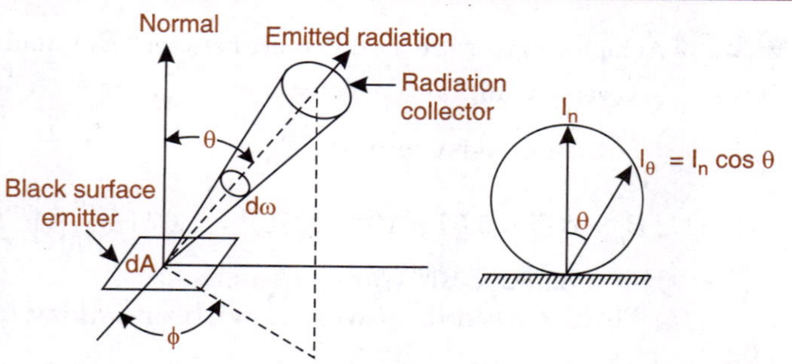

(*a*) Special distribution of radiations emitted from a surface.

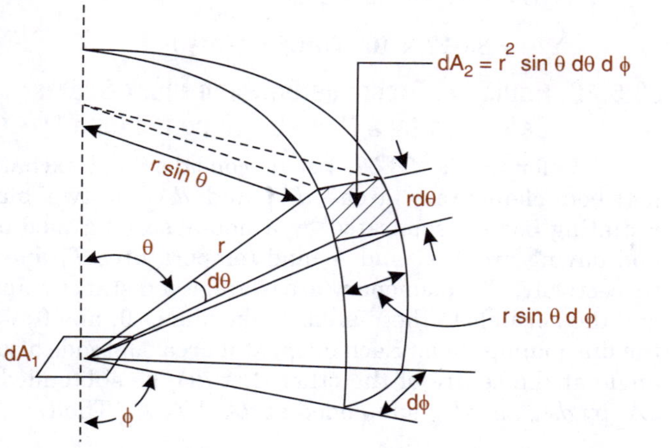

(*b*) Illustration for evaluating area dA_2

Fig. 28.51. Radiation from an elementary surface.

The solid angle subtended by $dA_2 = \dfrac{dA_2}{r^2}$

∴ The intensity of radiation,

$$I = \frac{dQ_{1-2}}{dA_1 \cos\theta \times \dfrac{dA_2}{r^2}} \qquad ...(28.82)$$

where dQ_{1-2} is the rate of radiation heat transfer from dA_1 to dA_2.

It is evident from Fig. 28.51 (*b*) that,

$$dA_2 = r\, d\theta\, (r \sin\theta\, d\phi)$$

or

$$dA_2 = r^2 \sin\theta . d\theta . d\phi \qquad ...(28.83)$$

From eqns. (28.82) and (28.83), we obtain

$$dQ_{1-2} = I\, dA_1 \,.\, \sin\theta \,.\, \cos\theta \,.\, d\theta \,.\, d\phi$$

The total radiation through the hemisphere is given by

$$Q = I\, dA_1 \int_{\theta=0}^{\theta=\frac{\pi}{2}} \int_{\phi=0}^{\phi=2\pi} \sin\theta \cos\theta\, d\theta\, d\phi$$

$$= 2\pi I\, dA_1 \int_{\theta=0}^{\theta=\frac{\pi}{2}} \sin\theta \cos\theta\, d\theta$$

$$= \pi I\, dA_1 \int_{\theta=0}^{\frac{\pi}{2}} 2 \sin\theta \cos\theta\, d\theta$$

$$= \pi I\, dA_1 \int_{\theta=0}^{\frac{\pi}{2}} \sin 2\theta\, d\theta$$

or

$$Q = \pi I\, dA_1 \qquad ...(28.84)$$

Also $\qquad Q = E.dA_1$

$\therefore \qquad E\, dA_1 = \pi I\, dA_1$

or $\qquad E = \pi I$

i.e., The *total emissive power of a diffuse surface is equal to π times its intensity of radiation.*

28.5.9.2. Lambert's Cosine Law

The law states that the *total emissive power E_θ from a radiating plane surface in any direction is directly proportional to the cosine of the angle of emission.* The angle of emission θ is the angle subtended by the normal to the radiating surface and the direction vector of emission of the receiving surface. If E_n be the total emissive power of the radiating surface in the direction of its normal, then

$$E_\theta = E_n \cos\theta \qquad ...(28.85)$$

The above equation is true only for diffuse radiation surface. The radiation emanating from a point on a surface is termed diffused if the intensity, I is constant. This law is also known as *Lambert's law of diffuse radiation.*

Example 28.23. *The effective temperature of a body having an area of 0.12 m² is 527°C. Calculate the following :*

(*i*) *The total rate of energy emission.*

(*ii*) *The intensity of normal radiation, and*

(*iii*) *The wavelength of maximum monochromatic emissive power.*

Solution. *Given* : $A = 0.12\ m^2$; $T = 527 + 273 = 800$ K

(*i*) **The total rate of energy emission, E_b :**

$$E_b = \sigma A T^4 \text{ W (watts)} \qquad ...[\text{Eqn. (28.64 (a))}]$$

$$= 5.67 \times 10^{-8} \times 0.12 \times (800)^4$$

$$= 5.67 \times 0.12 \times \left(\frac{800}{100}\right)^4 = \mathbf{2786.9\ W.} \quad \textbf{(Ans.)}$$

(*ii*) **The intensity of normal radiation, I_{bn} :**

$$I_{bn} = \frac{E_b}{\pi}, \text{ where } E_b \text{ is in W/m}^2 \text{ K}^4$$

$$= \frac{\sigma T^4}{\pi} = \frac{5.67 \times \left(\dfrac{800}{100}\right)^4}{\pi}$$

$$= \mathbf{7392.5\ W/m^2.sr.} \quad \textbf{(Ans.)}$$

(*iii*) **The wavelength of maximum monochromatic emissive power, λ_{max} :**

From Wien's displacement law,

$$\lambda_{max}.T = 2898\ \mu\text{m K} \qquad ...[\text{Eqn. (28.80)}]$$

or

$$\lambda_{max} = \frac{2898}{T} = \frac{2898}{800} = \mathbf{3.622\ \mu m.} \quad \textbf{(Ans.)}$$

Example 28.24. *Assuming the sun to be a black body emitting radiation with maximum intensity at λ = 0.49 μm, calculate the following :*

(*i*) *The surface temperature of the sun, and*

(*ii*) *The heat flux at surface of the sun.*

Solution. Given : λ_{max} = 0.49 μm

(*i*) **The surface temperature of the sun, T :**

According to Wien's displacement law,

$$\lambda_{max} \cdot T = 2898 \text{ μmK}$$

$$\therefore \quad T = \frac{2898}{\lambda_{max}} = \frac{2898}{0.48} = \mathbf{5914\ K.} \quad \textbf{(Ans.)}$$

(*ii*) **The heat flux at the surface of the sun, $(E)_{sun}$:**

$$(E)_{sun} = \sigma T^4 = 5.67 \times 10^{-8}\, T^4 = 5.67 \left(\frac{T}{100}\right)^4$$

$$= 5.67 \times \left(\frac{5914}{100}\right)^4$$

$$= \mathbf{6.936 \times 10^7\ W/m^2.} \quad \textbf{(Ans.)}$$

Example 28.25. *Calculate the following for an industrial furnance in the form of a black body and emitting radiation of 2500°C :*

(*i*) *Monochromatic emissive power at 1.2 μm length,*

(*ii*) *Wavelength at which the emission is maximum,*

(*iii*) *Maximum emissive power,*

(*iv*) *Total emissive power, and*

(*v*) *Total emissive power of the furnance if it is assumed as a real surface with emissivity equal to 0.9.*

Solution. *Given :* $T = 2500 + 273 = 2773$ K ;

$\lambda = 12$ μm, $\varepsilon = 0.9$

(*i*) **Monochromatic emissive power at 1.2 μm length, $(E_\lambda)_b$:**

According to Planck's law,

$$(E_\lambda)_b = \frac{C_1 \lambda^{-5}}{\exp\left(\frac{C_2}{\lambda T}\right) - 1} \quad \text{...[Eqn. (28.77)]}$$

where, $C_1 = 3.742 \times 10^8$ W. μm^4/m^2

$= 0.3742 \times 10^{-15}$ W.m^4/m^2 and

$C_2 = 1.4388 \times 10^{-2}$ mK

Substituting the values, we get

$$(E_\lambda)_b = \frac{0.3742 \times 10^{-15} \times (1.2 \times 10^{-6})^{-5}}{\exp\left(\frac{1.4388 \times 10^{-4}}{1.2 \times 10^{-6} \times 2773}\right) - 1} = \frac{1.5 \times 10^{14}}{74.48}$$

$$= \mathbf{2.014 \times 10^{12}\ W/m^2.} \quad \textbf{(Ans.)}$$

(*ii*) **Wavelength at which the emission is maximum, λ_{max} :**

According to Wien's displacement law,

$$\lambda_{max} = \frac{2898}{T} = \frac{2898}{2773} = \mathbf{1.045\ μm.} \quad \textbf{(Ans.)}$$

(*iii*) **Maximum emissive power, $(E_{\lambda b})_{max}$:**

$$(E_{\lambda b})_{max} = 1.285 \times 10^{-5}\, T^5 \text{ W/m}^2 \text{ per metre length} \quad \text{...[Eqn. (28.81)]}$$

$$= 1.285 \times 10^{-5} \times (2773)^5$$

$$= \mathbf{2.1 \times 10^{12}\ W/m^2\ per\ metre\ length.} \quad \textbf{(Ans.)}$$

[**Note.** At high temperature the difference between $(E_\lambda)_b$ and $(E_{\lambda b})_{max}$ is very small].

(*iv*) **Total emissive power, E_b :**

$$E_b = \sigma T^4 = 5.67 \times 10^{-8} (2273)^4 = 5.67 \left(\frac{2773}{100}\right)^4$$

$$= \mathbf{3.352 \times 10^6\ W/m^2.} \quad \textbf{(Ans.)}$$

(*v*) **Total emissive power, E with emisivity (ε) = 0.9**

$$E = \varepsilon\, \sigma T^4 = 0.9 \times 5.67 \left(\frac{2773}{100}\right)^4$$

$$= \mathbf{3.017 \times 10^6\ W/m^2.} \quad \textbf{(Ans.)}$$

28.5.10. Radiation Exchange Between Black Bodies Separated by a Non-absorbing Medium

Refer to Fig. 28.52. Let us consider heat exchange between elementary areas dA_1 and dA_2 of two black radiating bodies, separated by a non-absorbing medium, and having areas A_1 and A_2 and temperatures T_1 and T_2 respectively. The elemenary areas are at a distance r apart and the normals to these areas make angles θ_1 and θ_2 with the line joining them. Each elemental area subtends a solid angle at the centre of the other. Let $d\omega_1$ be subtended at dA_1 by dA_2 and $d\omega_2$ subtended at dA_2 by dA_1. Then

$$d\omega_1 = \frac{dA_2 \cos\theta_2}{r^2}, \text{ and } d\omega_2 = \frac{dA_1 \cos\theta_1}{r^2} \quad \text{...(28.86)}$$

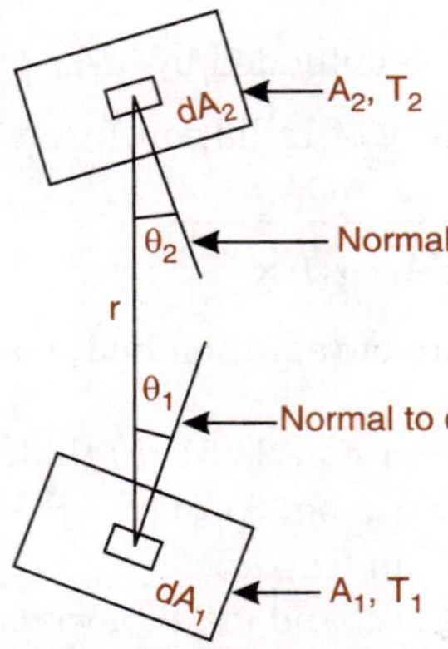

Fig. 28.52. Radiation heat exchange between two black surfaces.

The energy leaving dA_1 in the direction given by the angle per unit solid angle = $I_{b_1}\, dA_1 \cos\theta_1$.

where, I_b = Black body intensity, and

$dA_1 \cos\theta_1$ = Projection of dA_1 on the line between the centres.

The rate of radiant energy leaving dA_1 and striking on dA_2 is given by

$$dQ_{1-2} = I_{b_1}\, dA_1 \cos\theta_1 .\, d\omega_1$$

$$= \frac{I_{b_1} \cos\theta_1 \cos\theta_2\, dA_1\, dA_2}{r^2} \quad \text{...(28.87)}$$

This energy is absorbed by the elementary area dA_2, since both the surfaces are black. The quantity of energy radiated by dA_2 and absorbed by dA_1 is given by

$$dQ_{2-1} = \frac{I_{b_2} \cos\theta_2 \cos\theta_1 \, dA_2 \, dA_1}{r^2} \quad ...(28.88)$$

The net rate of transfer of energy between dA_1 and dA_2 is

$$dQ_{12} = dQ_{1-2} - dQ_{2-1}$$

$$= \frac{dA_1 \, dA_2 \cos\theta_1 \cos\theta_2}{r^2} \left(I_{b_1} - I_{b_2}\right)$$

But $I_{b_1} = \frac{E_{b_1}}{\pi}$ and $I_{b_2} = \frac{E_{b_2}}{\pi}$ ...[Eqn. (28.88)]

$$\therefore \quad dQ_{12} = \frac{dA_1 \, dA_2 \cos\theta_1 \cos\theta_2}{\pi r^2} (E_{b_1} - E_{b_2}) \quad ...(28.89)$$

or

$$dQ_{12} = \frac{\sigma dA_1 \, dA_2 \cos\theta_1 \cos\theta_2}{\pi r^2} (T_1^2 - T_2^4) \quad ...(28.90)$$

The rate of total net heat transfer for the total areas A_1 and A_2 is given by

$$Q_{12} = \int dQ_{12} = \sigma (T_1^2 - T_2^4) \int_{A_1}\int_{A_2} \frac{\cos\theta_1 \cos\theta_2 \, dA_1 \, dA_2}{r^2} \quad ...(28.91)$$

The rate of radiant energy emitted by A_1 that falls on A_2, from eqn. (28.87), is given by

$$Q_{1-2} = I_{b_1} \int_{A_1}\int_{A_2} \frac{\cos\theta_1 \cos\theta_2 \, dA_1 \, dA_2}{r^2}$$

$$Q_{1-2} = \sigma T_1^4 \int_{A_1}\int_{A_2} \frac{\cos\theta_1 \cos\theta_2 \, dA_1 \, dA_2}{\pi r^2} \quad ...(28.92)$$

The rate of total energy radiated by A_1 is given by,

$$Q_1 = A_1 \sigma T_1^4$$

Hence the fraction of the rate of energy leaving area A_1 and impinging on area A_2 is given by

$$\frac{Q_{1-2}}{Q_1} = \frac{1}{A_1} \int_{A_1}\int_{A_2} \frac{\cos\theta_1 \cos\theta_2 \, dA_1 \, dA_2}{\pi r^2} \quad ...(28.93)$$

or

$$\frac{Q_{1-2}}{Q_1} = F_{1-2} \quad ...[29.93\ (a)]$$

F_{1-2} is known as '**configuration factor**' or '**surface factor**' or '**view factor**' between the two radiating surfaces and is a function of geometry only.

Thus, the ***shape factor*** may be defined as *"The fraction of radiative energy that is diffused from one surface element and strikes the other surface directly with no intervening reflections."*

Further, $Q_{1-2} = F_{1-2} A_1 \sigma T_1^4$ ...(28.94)

Similarly, the rate of radiant energy by A_2 that falls on A_1, from eqn. (28.88), is given by

$$Q_{2-1} = \sigma T_2^4 \int_{A_1}\int_{A_2} \frac{\cos\theta_1 \cos\theta_2 \, dA_1 \, dA_2}{\pi r^2}$$

The rate of total energy radiated by A_2 is given by

$$Q_2 = A_2 \sigma T_2^4$$

Hence the fraction of the rate of energy leaving area A_2 and impinging on area A_1 is given by

$$\frac{Q_{2-1}}{Q_2} = \frac{1}{A_2} \int_{A_1}\int_{A_2} \frac{\cos\theta_1 \cos\theta_2 \, dA_1 \, dA_2}{\pi r^2} \quad ...(28.95)$$

or

$$\frac{Q_{2-1}}{Q_2} = F_{2-1}$$

F_{2-1} is the shape factor of A_2 with respect to A_1.

$$Q_{2-1} = F_{2-1} A_2 \sigma T_2^4 \quad ...(28.96)$$

From eqns. (28.93) and (28.95), we get

$$A_1 F_{1-2} = A_2 F_{2-1} \quad ...(28.97)$$

The above result is known as **reciprocity theorem.** It indicates that the net radiant interchange may be evaluated by computing one way configuration factor from either surface to the other. Thus the net rate of heat transfer between two surfaces A_1 and A_2 is given by

$$Q_{12} = A_1 F_{1-2} \sigma (T_1^4 - T_2^4)$$

$$= A_2 F_{2-1} \sigma (T_1^4 - T_2^4) \quad ...(28.98)$$

It may be noted that eqn. (28.98) is applicable to *black surfaces only and must not be used* for surfaces having emissivities very different from unity.

Example 28.26. *A body at 1000°C in black surroundings at 500°C has an emissivity of 0.42 at 1000°C and an emissivity of 0.72 at 500°C. Calculate the rate of heat loss by radiation per m².*

(i) When the body is assumed to be gray with ε = 0.42.

(ii) When the body is not gray.

Assume that the absorptivity is independent of the surface temperature.

Solution. *(i)* **When the body is gray with ε = 0.42 :**

$T_1 = 1000 + 273 = 1273$ K

$T_2 = 500 + 273 = 773$ K

ε at 1000°C = 0.42

ε at 500°C = 0.72

$\sigma = 5.67 \times 10^{-8}$

Heat loss per m² by radiation,

$$q = \varepsilon\sigma (T_1^4 - T_2^4)$$

$$= 0.42 \times 5.67 \times 10^{-8} [(1273)^4 - 773)^4]$$

$$= 54893 \text{ W}$$

i.e., *Heat loss per m² by radiation* = **54.893 kW. (Ans.)**

(ii) **When the body is not gray :**

Absorptivity when source is at 500°C = Emissivity when body is at 500°C

i.e., absorptivity, α = 0.72

Then, energy emitted

$$= \varepsilon\sigma T_1^4 = 0.42 \times 5.67 \times 10^{-8} \times (1273)^4$$

and, Energy absorbed

$= \alpha\sigma T_2^4 = 0.72 \times 5.67 \times 10^{-8} \times (773)^4$

i.e., q = Energy emitted – Energy absorbed

$= 0.42 \times 5.67 \times 10^{-8} \times (1273)^4 - 0.72 \times 5.67 \times 10^{-8} \times (773)^4$

$= 62538 - 14576 = 47962$ W

i.e., Heat loss per m^2 *by radiation* = **47.962 kW. (Ans.)**

Example 28.27. *A long steel rod, 22 mm in diameter, is to be heated from 420°C to 540°C. It is placed concentrically in a long cylindrical furnace which has an inside diameter of 180 mm. The inner surface of the furnace is at a temperature of 1100°C, and has an emissivity of 0.82. If the surface of the rod has an emissivity of 0.62, find the time required for the heating operation.*

Take for steel : c = 0.67 kJ/kg K, ρ= 7845 kg/m³.

Solution. Refer to Fig. 28.53.

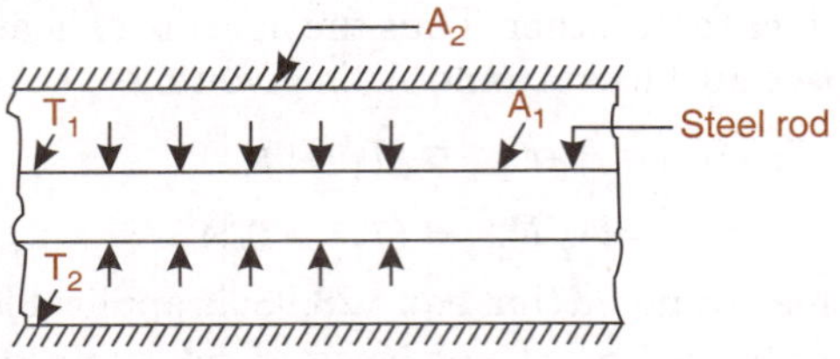

Fig. 28.53

Diameter of the steel rod = 22 mm = 0.022 m

Inside diameter of the furnace = 180 mm = 0.18 m

Emissivity $\varepsilon_1 = 0.62$

Emissivity $\varepsilon_2 = 0.82$

Specific heat of steel, $c = 0.67$ kJ/kg K

Density of steel, $\rho = 7845$ kg/m³

$T_1 = 420 + 273 = 693$ K ..... 1st case

and $= 540 + 273 = 813$ K ..... 2nd case

$T_2 = 1100 + 273 = 1373$ K

The surface area of the rod, $A_1 = \pi \times 0.022 \times l$ m²

The surface area of the furnace, $A_2 = \pi \times 0.18 \times l$ m²

Time required for the heating operation, t_h :

Initial rate of heat absorption by radiation, when the rod is at 420°C or 693 K

$$Q_i = \frac{A_1\sigma(T_1^4 - T_2^4)}{\frac{1}{\varepsilon_1} + \frac{A_1}{A_2}\left(\frac{1}{\varepsilon_2} - 1\right)}$$

$$= \frac{\pi \times 0.022 \times 1 \times 5.67 \times 10^{-8}\,(693^4 - 1373^4)}{\frac{1}{0.62} + \left(\frac{\pi \times 0.022 \times l}{\pi \times 0.18 \times l}\right)\left[\frac{1}{0.82} - 1\right]}$$

$$= \frac{-13022.5}{1.64} = -7940.5 \text{ W/m}$$

[∵ $l = 1$ m...assumed]

Rate of heat absorption at the end of the heating process, when the rod is at 540°C or 813 K

$$Q_e = \frac{A_2\,\sigma(T_1^4 - T_2^4)}{\frac{1}{\varepsilon_1} + \frac{A_1}{A_2}\left(\frac{1}{\varepsilon_2} - 1\right)}$$

$$= \frac{\pi \times 0.022 \times 1 \times 5.67 \times 10^{-8}\,(813^4 - 1373^4)}{\frac{1}{0.62} + \frac{\pi \times 0.022 \times l}{\pi \times 0.18 \times l}\left[\frac{1}{0.82} - 1\right]}$$

$$= \frac{-12214.3}{1.64} = -7447.7 \text{ W/m}$$

∴ Average rate of heat absorption during the heating process

$$Q_{av.} = \frac{7940.5 + 7447.7}{2} = 7694.1 \text{ W/m}$$

Time required for heating, t_h is obtained from the equation

$$mc_p\,\Delta T = Q_{av.} \times t_h$$

$$\therefore \quad t_h = \frac{[\pi/4 \times (0.022)^2 \times 1 \times 7845] \times 0.67 \times (540 - 420) \times 1000}{7694.1}$$

= **31.16 s. (Ans.)**

Example 28.28. *Calculate the heat transfer rate per* m^2 *area by radiation between the surfaces of two long cylinders having radii 100 mm and 50 mm respectively. The smaller cylinder being in the larger cylinder. The axes of the cylinders are parallel to each other and separated by a distance of 20 mm. The surfaces of inner and outer cylinders are maintained at 127°C and 27°C respectively. The emissivity of both the surfaces is 0.5.*

Assume the medium between the two cylinders is non-absorbing. **(P.U.)**

Solution. *Given* : $r_1 = 50$ mm $= 0.05$ m ;

$r_2 = 100$ mm $= 0.1$ m,

$T_1 = 127 + 273 = 400$ K,

$T_2 = 27 + 273 = 300$ K, $\varepsilon_1 = \varepsilon_2 = 0.5$

The heat transfer between two concentric or eccentric cylinders is given by

$$(Q_{12})_{net} = \frac{A_1\,\sigma(T_1^4 - T_2^4)}{\left(\frac{1-\varepsilon_1}{\varepsilon_1}\right) + \frac{1}{F_{1-2}} + \left(\frac{1-\varepsilon_2}{\varepsilon_2}\right)\frac{A_1}{A_2}}$$

Here $F_{1-2} = 1$ and $\frac{A_1}{A_2} = \frac{2\pi r_1 L}{2\,\pi r_2 L} = \frac{r_1}{r_2}$

Substituting the values, we have

$$(Q_{12})_{net} = \frac{1 \times 5.67\left[\left(\frac{400}{100}\right)^4 - \left(\frac{300}{100}\right)^4\right]}{\left(\frac{1-0.5}{0.5}\right) + 1\left(\frac{1-0.5}{0.5}\right) \times \frac{0.05}{0.1}} = \frac{992.25}{2.5}$$

= **396.9 W/m². (Ans.)**

Example 28.29. *Three thin walled infinitely long hollow cylinders of radii 5 cm, 10 cm and 15 cm are arranged concentrically as shown in Fig. 28.54. T_1 = 1000 K and T_3 = 300 K.*

Assuming $\varepsilon_1 = \varepsilon_2 = \varepsilon_3 = 0.05$ and vacuum in the spaces between the cylinders, calculate the steady state temperature of cylinder surface 2 and heat flow per m² area of cylinder 1. **(P.U.)**

Solution. *Given :* $r_1 = 5$ cm ; $r_2 = 10$ cm ; $r_3 = 15$ cm ;

$T_1 = 1000$ K ; $T_3 = 300$ K

$\varepsilon_1 = \varepsilon_2 = \varepsilon_3 = 0.05.$

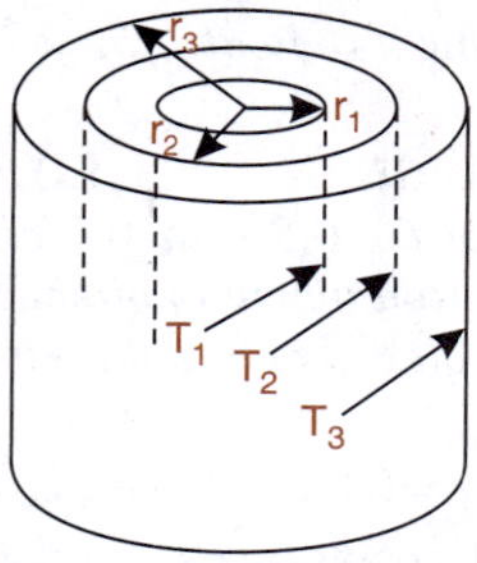

Fig. 28.54

For steady state heat flow,

$$Q_{12} = Q_{23}$$

or
$$\frac{A_1\sigma(T_1^4 - T_2^4)}{\left(\frac{1-\varepsilon_1}{\varepsilon_1}\right) + \frac{1}{F_{1-2}} + \left(\frac{1-\varepsilon_2}{\varepsilon_2}\right)\frac{A_1}{A_2}} = \frac{A_2\,\sigma(T_2^4 - T_3^4)}{\left(\frac{1-\varepsilon_2}{\varepsilon_2}\right) + \frac{1}{F_{2-3}} + \left(\frac{1-\varepsilon_3}{\varepsilon_3}\right)\frac{A_2}{A_3}}$$

In this case $F_{1-2} = F_{2-3} = 1$; and

$$\frac{A_1}{A_2} = \frac{r_1}{r_2} = \frac{5}{10} = 0.5$$

$$\frac{A_2}{A_3} = \frac{r_2}{r_3} = \frac{10}{15} = 0.67$$

$$\therefore \quad \frac{2\pi r_1 L\left[\left(\frac{1000}{100}\right)^4 - \left(\frac{T_2}{100}\right)^4\right]}{\left(\frac{1-0.05}{0.05}\right) + 1 + \left(\frac{1-0.05}{0.05}\right)\times 0.5} = \frac{2\pi r_2 L\left[\left(\frac{T_2}{100}\right)^4 - \left(\frac{300}{100}\right)^4\right]}{\left(\frac{1-0.05}{0.05}\right) + 1 + \left(\frac{1-0.05}{0.05}\right)\times 0.67}$$

or
$$\frac{0.05(10000 - x^4)}{29.4} = \frac{0.1(x^4 - 81)}{32.73}$$

or
$$(1000 - x^4) = \frac{29.5 \times 0.1}{32.73 \times 0.05}(x^4 - 81) = 1.8(x^4 - 81)$$

or
$$2.8x^4 = 10000 - 145.8 = 9854.2$$

or
$$x = \left(\frac{9854.2}{2.8}\right)^{1/4} = 7.7$$

or
$$\frac{T_2}{100} = 7.7 \text{ or } T_2 = 770 \text{ K}$$

∴ Heat flow per m² area of cylinder 1,

$$Q_{12} = \frac{A_1\,\sigma(T_1^4 - T_2^4)}{\left(\frac{1-\varepsilon_1}{\varepsilon_1}\right) + 1 + \left(\frac{1-\varepsilon_2}{\varepsilon_2}\right)\frac{A_1}{A_2}}$$

$$= \frac{1 \times 5.67\left[\left(\frac{1000}{100}\right)^4 - \left(\frac{770}{100}\right)^4\right]}{\left(\frac{1-0.05}{0.05}\right) + 1 + \left(\frac{1-0.05}{0.05}\right)\times 0.5}$$

$$= \frac{5.67 \times (10000 - 3515.3)}{29.5}$$

= **1246.4 W. (Ans.)**

Example 28.30. *Two concentric spheres 210 mm and 300 mm diameters with the space between them evacuated are to be used to store liquid air (– 153°C) in a room at 27°C. The surfaces of the spheres are flushed with aluminium (ε = 0.03) and latent heat of vaporisation of liquid air is 209.35 kJ/kg. Calculate the rate of evaporation of liquid air.* **(M.U.)**

Solution. *Given :* $r_1 = \frac{210}{2} = 105$ mm $= 0.105$ m ;

$r_2 = \frac{300}{2} = 150$ mm $= 0.15$ m ;

$T_1 = -153 + 273 = 120$ K ;

$T_2 = 27 + 273 = 300$ K ;

$\varepsilon_1 = \varepsilon_2 = 0.03$, $h_{fg} = 209.35$ kJ/kg.

Rate of evaporation of liquid air :

The heat flow from the inner sphere surface to outer sphere surface is given by,

$$Q_{12} = \frac{A_1\sigma(T_1^4 - T_2^4)}{\left(\frac{1-\varepsilon_1}{\varepsilon_1}\right) + \frac{1}{F_{1-2}} + \left(\frac{1-\varepsilon_2}{\varepsilon_2}\right)\frac{A_1}{A_2}}$$

$$= \frac{4\pi r_1^2\sigma(T_1^4 - T_2^4)}{\left(\frac{1-\varepsilon_1}{\varepsilon_1}\right) + 1 + \left(\frac{1-\varepsilon_2}{\varepsilon_2}\right)\frac{r_1^2}{r_2^2}}$$

$$= \frac{4\pi \times (0.105)^2 \times 5.67\left[\left(\frac{120}{100}\right)^4 - \left(\frac{300}{100}\right)^4\right]}{\left(\frac{1-0.03}{0.03}\right) + 1 + \left(\frac{1-0.03}{0.03}\right)\times\left(\frac{0.105}{0.15}\right)^2}$$

$$= \frac{0.7855\,(2.07 - 81)}{32.33 + 1 + 15.84} = \frac{-61.99}{49.17} = -1.26 \text{ W}$$

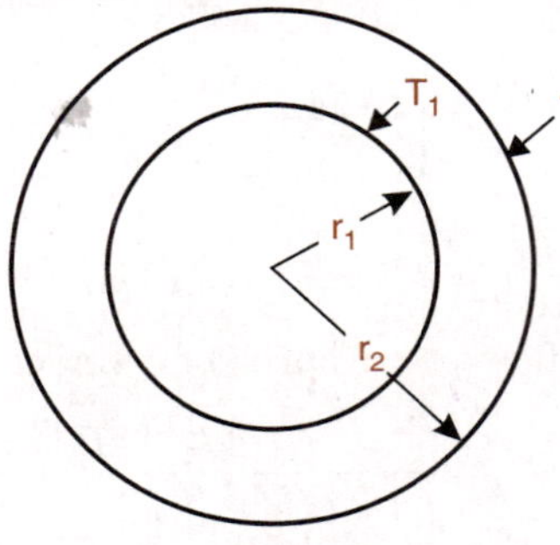

Fig. 28.55

– ve sign indicates that heat is gained by the surface 1, means, heat is flowing from outside surface to inside surface.

∴ The rate of evaporation

$$= \frac{1.26 \times 3600}{209.35 \times 1000} = \mathbf{0.0217\ kg/h.} \quad \textbf{(Ans.)}$$

Example 28.31. *Liquid oxygen (boiling temperature = – 182°C) is to be stored in spherical container of 30 cm diameter. The system is insulated by an evacuated space between inner sphere and surrounding 45 cm inner diameter concentric sphere. For both spheres ε = 0.03 and temperature of the outer sphere is 30°C. Estimate the rate of heat flow by radiation to the oxygen in the container.*

Solution. *Given* : $T_1 = -182 + 273 = 91$ K,

$T_2 = 30 + 273 = 303$ K, $\varepsilon_1 = \varepsilon_2 = 0.03$

$d_1 = 30$ cm $= 0.3$ m, $d_2 = 45$ cm $= 0.45$ m.

Rate of heat flow, Q_{12} :

The heat flow between the two concentric spheres by radiation is given by

$$Q_{12} = \frac{A_1 \sigma (T_1^4 - T_2^4)}{\frac{1-\varepsilon_1}{\varepsilon_1} + \frac{1}{F_{1-2}} \left(\frac{1-\varepsilon_2}{\varepsilon_2}\right)\frac{A_1}{A_2}}$$

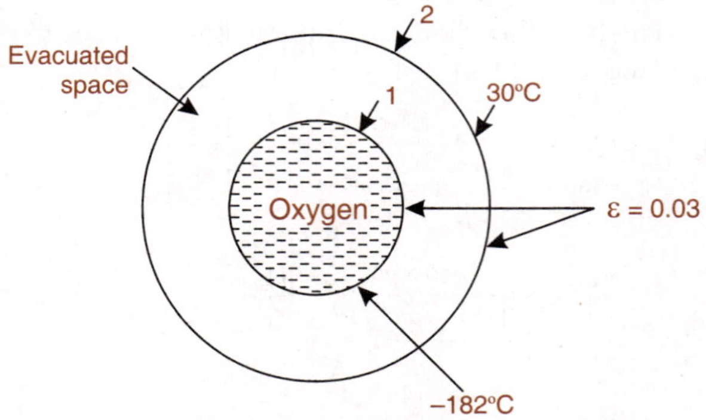

Fig. 28.56

For concentric spheres

$$F_{1-2} = 1$$

and
$$\frac{A_1}{A_2} = \left(\frac{d_1}{d_2}\right)^2 = \left(\frac{0.3}{0.45}\right)^2 = 0.4444$$

$$A_1 = 4\pi r_1^2 = 4\pi \times \left(\frac{0.3}{2}\right)^2 = 0.283 \text{ m}^2$$

Now substituting the values in the above equation, we get

$$Q_{12} = \frac{0.283 \times 5.67 \left[\left(\frac{91}{100}\right)^4 - \left(\frac{303}{100}\right)^4\right]}{\left(\frac{1-0.03}{0.03}\right) + 1 + \left(\frac{1-0.03}{0.03}\right) \times 0.4444}$$

$$= \frac{0.283 \times 5.67(0.686 - 84.289)}{32.33 + 1 + 14.37} = \mathbf{-2.81\ W}$$

–ve sign shows *heat flows from outside to inside.* **(Ans.)**

Example 28.32 (Radiation shield). *The large parallel planes with emissivities 0.3 and 0.8 exchange heat. Find the percentage reduction when a polished aluminium shield of emissivity 0.04 is placed between them. Use the method of electrical analogy.*

Solution. *Given :* $\varepsilon_1 = 0.3$; $\varepsilon_2 = 0.8$; $\varepsilon_3 = 0.04$

Consider all resistances (surface resistances and space resistances) per unit surface area.

For steady state heat flow,

$$\frac{E_{b_1} - E_{b_3}}{\left(\frac{1-\varepsilon_1}{\varepsilon_1}\right) + 1 + \left(\frac{1-\varepsilon_3}{\varepsilon_3}\right)} = \frac{E_{b_3} - E_{b_2}}{\left(\frac{1-\varepsilon_3}{\varepsilon_3}\right) + 1 + \left(\frac{1-\varepsilon_2}{\varepsilon_2}\right)}$$

$[\because A_1 = A_2 = A_3 = 1 \text{ m}^2 \text{ and } F_{1-3}, F_{3-2} = 1]$

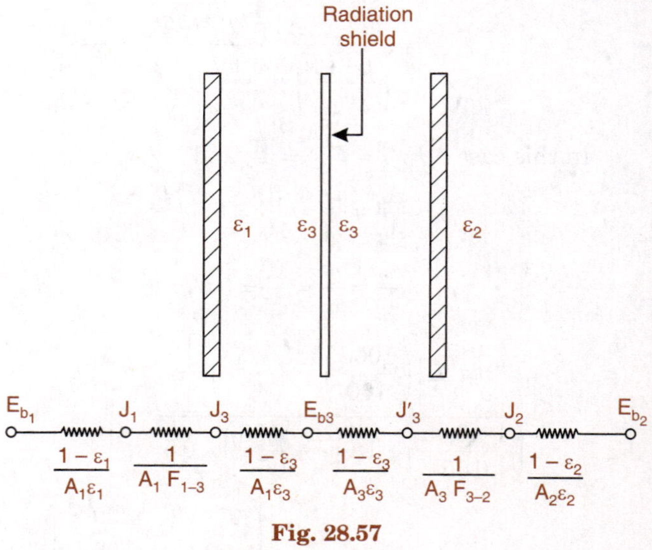

Fig. 28.57

or
$$\frac{\sigma(T_1^4 - T_3^4)}{\frac{1}{\varepsilon_1} + \frac{1}{\varepsilon_3} - 1} = \frac{\sigma(T_3^4 - T_2^4)}{\frac{1}{\varepsilon_3} + \frac{1}{\varepsilon_2} - 1}$$

or
$$\frac{T_1^4 - T_3^4}{\frac{1}{0.3} + \frac{1}{0.04} - 1} = \frac{T_3^4 - T_2^4}{\frac{1}{0.04} + \frac{1}{0.8} - 1}$$

or $$\frac{T_1^4 - T_3^4}{27.33} = \frac{T_3^4 - T_2^4}{25.25}$$

or $$T_1^4 - T_3^4 = \frac{27.33}{25.25}(T_3^4 - T_2^4)$$
$$= 1.08\ (T_3^4 - T_2^4)$$
$$= 1.08\ T_3^4 - 1.08\ T_2^4$$

or $$2.08\ T_3^4 = T_1^4 + 1.08\ T_2^4$$

or $$T_3^4 = \frac{1}{2.08}(T_1^4 + 1.08\ T_2^4)$$
$$= 0.48\ (T_1^4 + 1.08\ T_2^4) \quad ...(i)$$

Q_{12} (heat flow **without shield**)

$$= \frac{\sigma(T_1^4 - T_2^4)}{\frac{1}{\varepsilon_1} + \frac{1}{\varepsilon_2} - 1} = \frac{\sigma(T_1^4 - T_2^4)}{\frac{1}{0.3} + \frac{1}{0.8} - 1} = \frac{\sigma(T_1^4 - T_2^4)}{3.58} \quad ...(ii)$$

Q_{13} (heat flow **with shield**)

$$= \frac{\sigma(T_1^4 - T_3^4)}{\frac{1}{\varepsilon_1} + \frac{1}{\varepsilon_3} - 1} = \frac{\sigma(T_1^4 - T_3^4)}{\frac{1}{0.3} + \frac{1}{0.4} - 1} = \frac{\sigma(T_1^4 - T_3^4)}{27.33} \quad ...(iii)$$

∴ **Percentage reduction** in heat flow due to shield

$$= \frac{Q_{12} - Q_{13}}{Q_{12}}$$
$$= 1 - \frac{Q_{13}}{Q_{12}} = 1 - \frac{\sigma(T_1^4 - T_3^4)/27.33}{\sigma(T_1^4 - T_2^4)/3.58}$$
$$= 1 - \frac{3.58}{27.33}\left[\frac{T_1^4 - T_3^4}{T_1^4 - T_2^4}\right]$$
$$= 1 - 0.131\left[\frac{T_1^4 - 0.48\ (T_1^4 + 1.08\ T_2^4)}{T_1^4 - T_2^4}\right]$$
$$= 1 - 0.131\left[\frac{T_1^4 - 0.48\ T_1^4 + 0.52\ T_2^4}{T_1^4 - T_2^4}\right]$$
$$= 1 - 0.131\left[\frac{0.52\ (T_1^4 - T_2^4)}{(T_1^4 - T_2^4)}\right]$$
$$= 1 - 0.131 \times 0.52$$
= **0.932** or **93.2%.** **(Ans.)**

HIGHLIGHTS

1. Heat transfer may be defined as the transmission of energy from one region to another as a result of temperature gradient and it takes place by three modes : conduction, convection and radiation.
2. Fourier's law of conduction : $Q = -kA\frac{dt}{dx}$
   where, Q = Heat flow through a body per unit time,
   A = Surface area of heat flow (perpendicular to the direction of flow),
   dt = Temperature difference,
   dx = Thickness of body in the direction of flow, and
   k = Thermal conductivity of the body.
3. Conduction of heat flow through slabs :
   $$Q = \frac{A(t_1 - t_2)}{\left(\frac{x_1}{k_1} + \frac{x_2}{k_2} + \frac{x_3}{k_3}\right)} \quad ......3 \text{ slabs.}$$
4. Conduction of heat flow through pipe walls and lagging :
   $$Q = \frac{(t_1 - t_2)}{\frac{1}{2\pi l}\left[\frac{1}{k}\log_e\frac{r_2}{r_1}\right]}.$$
5. Conduction of heat through a hollow sphere :
   $$Q = \frac{t_1 - t_2}{\left(\frac{r_2 - r_1}{4\pi k r_1 r_2}\right)}.$$
6. Heat transfer by convection :
   $$Q = hA\ (t_1 - t_2).$$
7. Overall heat transfer co-efficient :
   $$U = \frac{1}{\frac{1}{h_{hf}} + \sum\frac{L}{k} + \frac{1}{h_{cf}}}.$$
8. Heat transfer between two fluids separated by the walls of a composite tube of solid material :
   $$Q = \frac{(t_{hf} - t_{cf})}{\frac{1}{2\pi L}\left[\frac{1}{r_1 h_{hf}} + \frac{1}{k_1}\log_e\frac{r_2}{r_1} + \frac{1}{k_2}\log_e\frac{r_3}{r_2} + \frac{1}{r_4 h_{cf}}\right]}.$$
9. A heat exchanger may be defined as an equipment which transfers the energy from a hot fluid to a cold fluid, with maximum rate and minimum investment and running cost.
10. The net heat transfer in case of gray bodies with emissivities ε_1 and ε_2 is given by :
    $$Q = \frac{\sigma A(T_1^4 - T_2^4)}{\left(\frac{1}{\varepsilon_1} + \frac{1}{\varepsilon_2}\right) - 1}$$
    In case of concentric or long co-axial cylinder,
    $$Q = \frac{A_1\sigma(T_1^4 - T_2^4)}{\frac{1}{\varepsilon_1} + \frac{A_1}{A_2}\left(\frac{1}{\varepsilon_2} - 1\right)}.$$

OBJECTIVE TYPE QUESTIONS

Choose the Correct Answer :

1. The Fourier's law of heat transfer by conduction is expressed as
   (a) $Q = kA^2\frac{dt}{dx}$ (b) $Q = kA\frac{dt}{dx}$
   (c) $Q = k^2A\frac{dx}{dt}$ (d) $Q = k^3A\frac{dx}{dt}$.

2. The heat transfer is constant when
   (a) temperature remains constant with time
   (b) temperature decreases with time
   (c) temperature increases with time
   (d) any of these.
3. The co-efficient of thermal conductivity is defined as
   (a) Quantity of heat transfer per unit area per one degree drop in temperature
   (b) Quantity of heat transfer per one degree temperature drop per unit area
   (c) Quantity of heat transfer per unit time per unit area
   (d) Quantity of heat transfer per unit time per unit area per one degree temperature drop per unit length.
4. The thermal conductivity is expressed as
   (a) W/mK (b) W/m^{2}K
   (c) W/hmK (d) W/h^{2}m^{2}K.
5. Heat transfer from higher temperature to low temperature takes place according to
   (a) Fourier law
   (b) First law of thermodynamics
   (c) Second law of thermodynamics
   (d) Zeroth law of thermodynamics.
6. Conduction through flat composite wall is given by :

   (a) $Q = \dfrac{t_1 - t_4}{\dfrac{x_1}{k_1A} + \dfrac{x_2}{k_2A} + \dfrac{x_3}{k_3A}}$

   (b) $Q = \dfrac{t_1 - t_4}{\dfrac{k_1A}{x_1} + \dfrac{k_2A}{x_2} + \dfrac{k_3A}{x_3}}$

   (c) $Q = \dfrac{(t_1 - t_4)A}{\dfrac{k_1}{x_1} + \dfrac{k_2}{x_2} + \dfrac{k_3}{x_3}}$

   (d) $Q = \dfrac{\dfrac{k_1A}{x_1} + \dfrac{k_2A}{x_2} + \dfrac{k_3A}{x_3}}{(t_1 - t_4)}$

   where Q = heat transfer, t_1, t_2, t_3 and t_4 temperatures on surfaces of composite wall, x_1, x_2, x_3, x_4 thicknesses of different composite wall layers.
7. Conduction through hollow, radial one dimensional heat transfer is expressed as

   (a) $Q = \dfrac{2\pi L(t_1 - t_2)k}{\log_e r_2/r_1}$ (b) $Q = \dfrac{2\pi L(t_1 - t_2)}{k(r_2 - r_1)}$

   (c) $Q = \dfrac{2\pi L \log_e (t_1/t_2)}{(r_2 - r_1)k}$ (d) $Q = \dfrac{2\pi L(t_1 - t_2)k}{\log_e r_2/r_1}$.
8. The radial heat transfer rate through hollow cylinder increases as the ratio of outer radius to inner radius
   (a) decreases (b) increases
   (c) constant (d) none of the above.
9. Stefan-Boltzmann law is expressed as
   (a) $Q = \sigma AT^4$ (b) $Q = \sigma A^2T^4$
   (c) $Q = \sigma AT^2$ (d) $Q = AT^4$.
10. The quantity of heat radiation is dependent on
   (a) area of the body only
   (b) shape of the body only
   (c) temperature of the body only
   (d) on all (a), (b) and (c).

ANSWERS

1. (b)	2. (a)	3. (d)	4. (a)
5. (c)	6. (a)	7. (a)	8. (a)
9. (a)	10. (c).		

THEORETICAL QUESTIONS

1. Enumerate the three modes by which heat can be transferred from one place to another. Which is the slowest of all ?
2. How do you define the thermal conductivity of a material ?
3. What do you understand by the terms 'convective heat transfer co-efficient' and 'overall heat transfer co-efficient'.
4. Derive an expression for heat loss in kJ/m^2-hr through a composite wall of layers (i) without considering convective heat transfer co-efficients and (ii) considering the convective heat transfer co-efficients.
5. Classify the heat exchangers according to the flow directions of fluid and give few examples of each in actual field of application.
6. Prove that the mean temperature difference in a parallel-flow heat exchanger is given by

$$LMTD\ (t_m) = \frac{t_1 - t_2}{\log_e \dfrac{t_1}{t_2}}.$$

UNSOLVED EXAMPLES

1. The inner surface of a plane brick wall is at 40°C and the outer surface is at 20°C. Calculate the rate of heat transfer per m^2 of surface area of the wall, which is 250 mm thick. The thermal conductivity of the brick is 0.52 W/mK. [**Ans.** 41.6 W/m^2]
2. Determine the rate of heat flow through the boiler wall made of 2 cm thick steel and covered with an insulating material of 0.5 cm thick. The temperatures at the inner and outer surfaces of the wall are 300°C and 50°C respectively.
   k (steel) = 58 W/mK
   k (insulation) = 0.116 W/mK. [**Ans.** 5.8 kW/m^2]
3. A mild steel tank of wall thickness 10 mm contains water at 90°C. Calculate the rate of heat loss per m^2 of tank surface area when the atmospheric temperature is 15°C.

The thermal conductivity of mild steel is 50 W/mK, and the heat transfer co-efficients for inside and outside the tank are 2800 and 11 W/m^2 K, respectively. Calculate also the temperature of the outside surface of the tank.

[Ans. 820 W/m^2, 89.6°C]

4. A cold storage room has walls made of 0.23 m of brick on the outside, 0.08 m of plastic foam, and finally 15 mm of wood on the inside. The outside and inside air temperatures are 22°C and – 2°C respectively. If the inside and outside heat transfer co-efficients are respectively 29 and 12 W/m^2 K and the thermal conductivities of brick, foam and wood are 0.98, 0.02 and 0.17 W/mK respectively determine (*i*) the rate of heat removal by refrigeration if the total wall area is 90 m^2, and (*ii*) the temperature of the inside surface of the brick.

[Ans. (*i*) 486.4 W; (*ii*) 20.28°C]

5. The wall of a refrigerated van is of 1.5 mm of steel sheet at outer surface, 10 mm plywood at the inner surface and 2 cm of glasswool in between. Calculate the rate of heat flow, if the temperatures of the inside and outside surfaces are – 15°C and 24°C.

Take : k (steel) = 23.2 W/mK,

k (glass-wool) = 0.014 W/mK

and k (plywood) = 0.052 W/mK. **[Ans.** 6 kW/m^2]

6. Sheets of brass and steel, each 10 mm thick, are placed in contact. The outer surface of brass is kept at 100°C and outer surface of steel is kept at 0°C. What is the temperature of the common interface ? The thermal conductivities of brass and steel are in the ratio of 2 : 1.

[Ans. 66.7°C]

7. The wall of a furnace is made up of 250 mm of fire brick, k = 1.05 W/mK ; 120 mm of insulation brick, k = 0.85 W/mK, and 200 mm of red brick, k = 0.85 W/mK. The inner and outer surface temperatures of the walls are 850°C and 65°C respectively. Calculate the temperatures at the contact surfaces.

Neglect the resistance of mortar joints.

[Ans. 703°C, 210°C]

8. Calculate the heat flowing through a furnace wall 0.23 m thick, the inside and outside surface temperatures of which are 1000°C and 200°C respectively. Assume that the mean thermal conductivity of the wall material is 1.1 W/mK. Assuming that 7 mm of insulation (k = 0.075 W/mK) is added to the outside surface of the wall and reduces the heat loss 20% ; calculate the outside surface temperature of the wall. If the cost of the insulation is Rs. 70 per sq m what time will be required to pay for the insulation ? Base the calculations on the 24 hours operation per day and 199 days per year. Heat energy may be valued at ₹ 10 per 1000 kWh.

[Ans. 3826 W/h-m^2 ; 74.3°C ; 1.916 years]

9. A flat wall of a furnace is composed of two layers of different materials having thicknesses of 0.115 m and 0.6 m with thermal conductivities of 0.16 W/m K and 10.6 W/m K respectively. If 1 kW/h of heat passes through every sq m area, estimate the drop in temperature at the contact between the two walls. The temperature inside the furnace is 1000°C and that at outside layer is 150°C.

[Ans. 74°C]

10. A furnace wall consists of 250 mm fire brick, 125 mm insulating brick, and 250 mm building brick. The inside wall is at temperature of 600°C and the atmospheric temperature is 20°C. Calculate the heat loss per m^2 of wall area and the temperature of the outside wall surface of the furnace. The heat transfer co-efficient for the outside surface is 10 W/m^2 K, and the thermal conductivities of the fire brick, insulating brick and building brick are 1.4, 0.2 and 0.7 W/m K respectively. Neglect radiation. **[Ans.** 0.46 kW/m^2 ; 66°C]

11. Hot air at a temperature of 60°C is flowing through a steel pipe of 100 mm diameter. The pipe is covered with two layers of different insulating materials of thicknesses 50 mm and 30 mm, and their corresponding thermal conductivities are 0.23 and 0.37 W/m K. The inside and outside heat transfer co-efficients are 58 and 12 W/m^2 K. The atmosphere is at 25°C. Find the rate of heat loss from a 50 m length of pipe. Neglect the resistance of the steel pipe. **[Ans.** 2.334 kW]

12. A steel pipe of 100 mm bore and 7 mm wall thickness, carrying steam at 260°C, is insulated with 40 mm of a high temperature diatomaceous earth covering. This covering is in turn insulated with 60 mm of asbestos felt. If the atmospheric temperature is 15°C, calculate the rate at which heat is lost by the steam per m length of the pipe. The heat transfer co-efficients for the inside and outside surfaces are 550 and 15 W/m^2 K, respectively and the thermal conductivities of steel, diatomaceous earth and asbestos felt are 50, 10.09 and 0.07 W/m K respectively. Calculate also the temperature of the outside surface. **[Ans.** 116 W ; 22.8°C]

13. A 250 mm steam main, 225 metres long is covered with 50 mm of high temperature insulation (k = 0.095 W/m K) and 40 mm of low temperature insulation (k = 0.065 W/m K). The inner and outer surface temperatures as measured are 400°C and 50°C respectively. Calculate :

(*i*) The total heat loss per hour.

(*ii*) The total heat loss per m^2 of outer surface.

(*iii*) The heat loss per m^2 of pipe surface.

(*iv*) The temperature between the two layers of insulation.

Neglect heat conduction through pipe material.

[Ans. (*i*) 265514 kJ/h; (*ii*) 873.5 kJ/h; (*iii*) 1502.5 kJ/h; (*iv*) 215°C]

14. A steam pipe of 160 mm inside diameter and 170 mm outside diameter (k = 58 W/m K) is covered with first layer of insulating material of 30 mm thickness (k = 0.17 W/m K) and second layer of insulating material of 50 mm thickness (k = 0.093 W/m K). The temperature of steam passing through the pipe is 300°C and ambient air temperature surrounding the pipe is 30°C. Taking inner and outer heat transfer co-efficients 30 and 5.8 W/m^2 K respectively, find the heat lost per metre length of pipe. **[Ans.** 216 W/m]

15. A small hemispherical oven is built of an inner layer of insulating fire brick 125 mm thick, and an outer covering of 85% magnesia 40 mm thick. The inner surface of the oven is at 800°C and the heat transfer co-efficient for the outside surface is 10 W/m^2 K, the room temperature is 20°C. Calculate the heat loss through the hemisphere if the inside radius is 0.6 m. Take the thermal conductivities of fire brick and 85% magnesia as 0.31 and 0.05 W/mK, respectively. [**Ans.** 1.93 kW]

16. A spherical shaped vessel of 1.2 m diameter is 100 mm thick. Find the rate of heat leakage, if the temperature difference between the inner and outer surfaces is 200°C. Thermal conductivity of the material is 0.3 kJ/m-h-°C. [**Ans.** 2262 kJ/h]

17. Exhaust gases flowing through a tubular heat exchanger at the rate of 0.3 kg/s are cooled from 400°C to 120°C by water initially at 10°C. The specific heat of exhaust gases and water may be taken as 1.13 and 4.19 kJ/kg K respectively, and overall heat transfer co-efficient from gases to water is 140 W/m^2 K. Calculate the surface area required when the cooling water flow is 0.4 kg/s.
(*i*) For parallel-flow ; (*ii*) For counter-flow.
[**Ans.** (*i*) 4.0 m^2; (*ii*) 3.37 m^2]

18. Water flows inside a tube 50 mm in diameter and 3 m long at a velocity of 0.8 m/s. Determine the heat transfer co-efficient and the rate of heat transfer if the mean water temperature is 50°C and the wall is isothermal at 70°C. For water at 60°C, take k = 0.66 W/m K, ν (kinematic viscosity) = 0.478 × 10^{-6} m^2/s, and Prandtl number = 2.98. [**Ans.** 4075 W/m^{2}K ; 38.39 kW]

19. Liquid air at – 153°C is stored in the space of two concentric spheres of 21 cm and 30 cm diameters. The surface emissivities are 0.03. Assume the outer surface temperature is 27°C. Considering only radiation heat transfer and taking the latent heat of liquid air of 209 kJ/kg, find the rate of evaporation. Take σ = 2.04 × 10^{-4} kJ/h-m^2 K^4. [**Ans.** 21.7 kg/h]

20. A body at 1100°C in black surroundings at 550°C has an emissivity of 0.4 at 1100°C and an emissivity of 0.7 at 550°C. Calculate the ratio of heat loss by radiation per m^2,
(*i*) when the body is assumed to be grey with ε = 0.4
(*ii*) when the body is not grey.
[**Ans.** (*i*) 70.22 kW; (*ii*) 62.42 kW]

21. A long steel rod, 20 mm in diameter, is to be heated from 427°C to 538°C. It is placed concentrically in a long cylindrical furnace which has an inside diameter of 160 mm. The inner surface of the furnace is at a temperature of 1093°C, and has an emissivity of 0.85. If the surface of the rod has an emissivity of 0.6, find the time required for the heating operation.
Take for steel : ρ = 7845 kg/m^3 and c = 0.67 kJ/kg K. [**Ans.** 29.88 s]

Non-Conventional Power Generation

29

29.1. **Introduction.** 29.2. **Wind power plants**—Introduction—Wind availability and measurement—Types of wind mills—Wind-electric generating power plant—Types of wind machines—Wind-powered battery chargers—Wind electricity in small independent grids—Wind electricity economics—Problems in operating large wind power generators. **29.3. Tidal power plants**—Introduction—Components of tidal power plants—Classification and operation of tidal power plants—Advantages and limitations of tidal power generation. **29.4. Solar power plants**—Solar energy—General terms and introduction—Collectors in various ranges and applications—Flat plate collectors—Focusing (or concentrating) collectors—Solar pond technology—Low temperature thermal power generation—Medium temperature systems using focusing collectors—High temperature systems—Solar farm and solar tower plant. **29.5. Geothermal power plants**—Geothermal energy—Geothermal sources—Geothermal power estimates—Environmental problems—Applications of geothermal energy—Advantages and disadvantages of geothermal energy over other energy forms—Geothermal energy in India and abroad. **29.6. Bio-gas plants.** *Highlights—Theoretical Questions.*

29.1. INTRODUCTION

A plenty of energy is needed to sustain industrial growth and agricultural production. The existing sources of energy such as coal, oil uranium etc., may not be adequate to meet the ever increasing energy demands. These conventional sources of energy are also depleting and may be exhausted at the end of the century or beginning of the next century. Consequently, sincere and untiring efforts shall have to be made by the scientists and engineers in exploring the possibilities of harnessing energy from several non-conventional energy sources. The various non-conventional energy sources are as follows :

(*i*) Solar energy
(*ii*) Wind energy
(*iii*) Energy from biomass and biogas
(*iv*) Ocean thermal energy conversion
(*v*) Tidal energy
(*vi*) Geothermal energy
(*vii*) Hydrogen energy
(*viii*) Fuel cells
(*ix*) Magneto-hydrodynamics generator
(*x*) Thermionic converter
(*xi*) Thermo-electric power.

Advantages of non-conventional energy sources

The leading advantages of non-conventional energy sources are :

1. They do not pollute the atmosphere.
2. They are available in large quantities.
3. They are well-suited for decentralised use.

According to energy experts the non-conventional energy sources can be used with advantage for *power generation as well as other applications* in a large number of locations and situations in our country.

In this chapter, power generation through solar energy, geothermal energy and biomass and biogas energy will be discussed.

29.2. WIND POWER PLANTS

29.2.1. Introduction

The electrical energy can be generated by wind energy. The wind energy, which is an indirect source of energy, can be used to run a wind will which in turn drives a generator to produce electricity. Although wind mills have been used for more than a dozen centuries for grinding grain and pumping water, interest in large scale power generation has developed over the past 50 years. A largest wind generator built in the past was 800 kW unit operated in France from 1958–60. The flexible 3 blades propeller was about 35 m in diameter and produced the rated power in a 60 km/hour wind with a rotation speed of 47 r.p.m. The maximum power developed was 12 MW. In India the interest in the wind mills was shown in the last fifties and early sixties. Apart from importing a few from outside, new designs were also developed, but it was not sustained. It is only in last 10–15 years that development work is going on in many institutions. An important reason for this lack of interest in wind energy must be that wind, in India is relatively low and vary appreciably with seasons. These low and seasonal winds imply a high cost of exploitation of wind energy. In our country high wind speeds are, however, available in coastal areas of Sourashtra, Western Rajsthan and some parts of central India. In these areas there could be a possibility of using medium and large sized wind mills for generation of electricity.

Characteristics of wind energy

1. Wind-power systems do not pollute the atmosphere.
2. Fuel provision and transport are not required in wind-power systems.
3. Wind energy is a renewable source of energy.
4. Wind energy when produced on small scale is cheaper, but competitive with conventional power generating systems when produced on a large scale.

Wind energy entails following *shortcomings/problems* :

1. It is fluctuating in nature.
2. Due to its irregularity it needs storage devices.
3. Wind power generating systems produce ample *noise*.

29.2.2. Wind Availability and Measurement

Wind energy can only be economical in areas of good wind availability. Wind energy differs with region and season and also, possibly to an even greater degree with local terrain and vegetation. Although wind speeds generally increase with height, varying speeds are found over different kinds of terrain. Observations of wind speed are carried out at meteorological stations, airports and lighthouses and are recorded regularly with ten minute mean values being taken every three hours at a height of 10 m. But airports, sometimes are in valleys and many wind speed metres are situated low and combinations of various, other factors mean that reading can be misleading. It is difficult, therefore, to determine the real wind speed of a certain place without actual *in situ* measurements.

The World Meteorological Organization (WMO) has accept four methods of wind recording :

(*i*) Human observation and log book.

(*ii*) Mechanical cup-counter anemometers.

(*iii*) Data logger.

(*iv*) Continuous record of velocity and direction.

1. **Human observation and log book.** This involves using the Beaufort Scale of wind strengths which defines visible 'symptoms' attributable to different wind speeds. The method is *cheap* and *easily implemented* but is *often unreliable*. The best that can be said of such records is that they are better than nothing.

2. **Mechanical cup-counter anemometers.** The majority of meteorological stations use mechanical cup-counter anemometers. By taking the readings twice or three times a day, it is possible to estimate the mean wind speed. This is a *low cost method*, but is *only relatively reliable*. The instrument has to be in good working order, it has to be correctly sited and should be reliably read at least daily.

3. **Data logger.** The equipment summarizes velocity frequency and direction. It is *more expensive and prone to technical failures but gives accurate data*. The method is tailored to the production of readily interpretable data of relevance to wind energy assessment. It does not keep a time series record but *presents the data in processed form*.

4. **Continuous record of velocity and direction.** This is how data is recorded at major airports of permanently manned meteorological stations. The *equipment is expensive and technically complex*, but it retains a detailed times-series record (second-by-second) of wind direction and wind speed. Results are given in copious quantities of data which require lengthy and expensive analysis.

29.2.3. Types of Wind Mills

The various types of wind mills which are practically useful are shown in Fig. 29.1.

1. **Multiple blade type.** It is the most widely used wind mill. It has 15 to 20 blades made from metal sheets. The *sail type* has three blades made by stitching out triangular pieces of convass cloth. Both these types run at low speeds of 60 to 80 r.p.m.

2. **Savonius type.** This type of wind mill has hollow circular cylinder sliced in half and the halves are mounted on vertical shaft with a gap in between. Torque is produced by the pressure difference between the two sides of the half facing the wind. This is quite efficient but needs a large surface area.

3. **Darrieus type.** This wind mill needs much less surface area. It is shaped like an egg beater and has two or three blades shaped like aerofoils.

It may be noted that :

— Both the Savonius and Darrieus types are mounted on a *vertical axis* and hence they *can run independently of the direction of wind*.

— The horizontal axis mills *have to face the direction of the wind* in order to generate power.

Performance of wind mills

The performance of a wind mill is defined as 'Co-efficient of performance' (K_p).

$$K_p = \frac{\text{Power delivered by the rotor}}{\text{Maximum power available in the wind}}$$

or

$$K_p = \frac{P}{P_{max}} = \frac{P}{\frac{1}{2}\rho A U_w^3}$$

where ρ = Density of air,
A = Swept area, and
U_w = Velocity of wind.

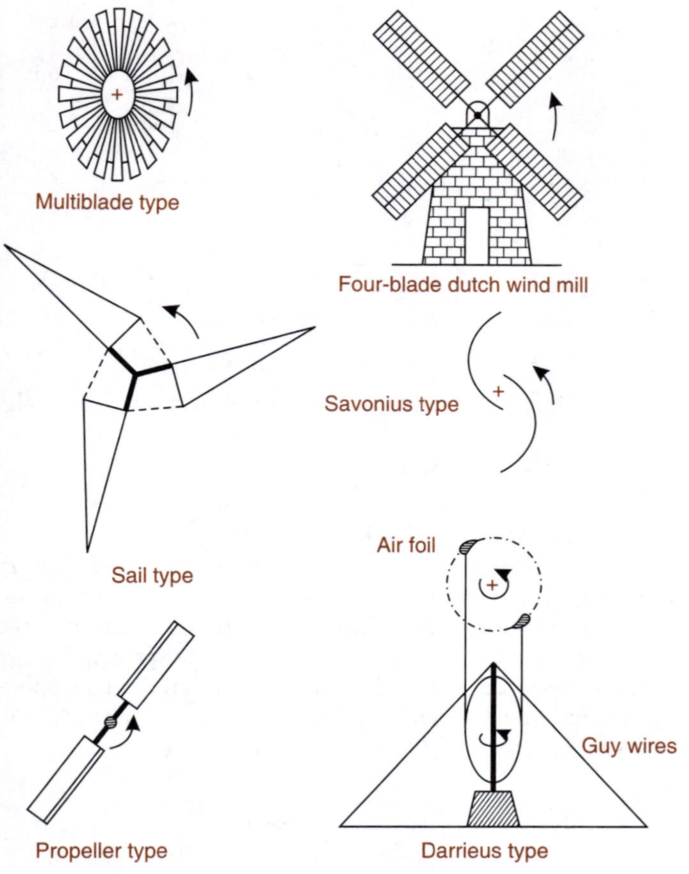

Fig. 29.1. Types of wind mills.

Fig. 29.2 shows a plot between K_p and tip speed ratio U_{bt}/U_w where U_{bt} = speed of blade tip.

It can be seen that K_p is the *lowest of Savonius* and *Dutch types* whereas the *propeller types have the highest value*.

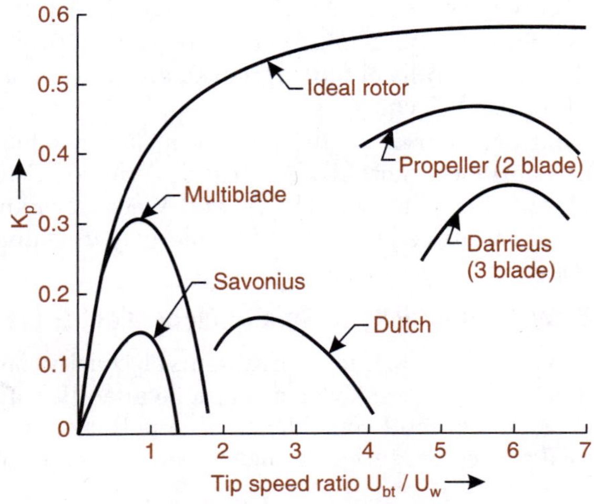

Fig. 29.2. K_p of wind mills.

In the designing of wind mills, it is uppermost to keep the power to weight ratio at the lowest possible level.

29.2.4. Wind-electric Generating Power Plant

Fig. 29.3 shows the various parts of a wind-electric generating power plant. These are :

1. Wind turbine or rotor
2. Wind mill head—it houses speed increaser, drive shaft, clutch, coupling etc.

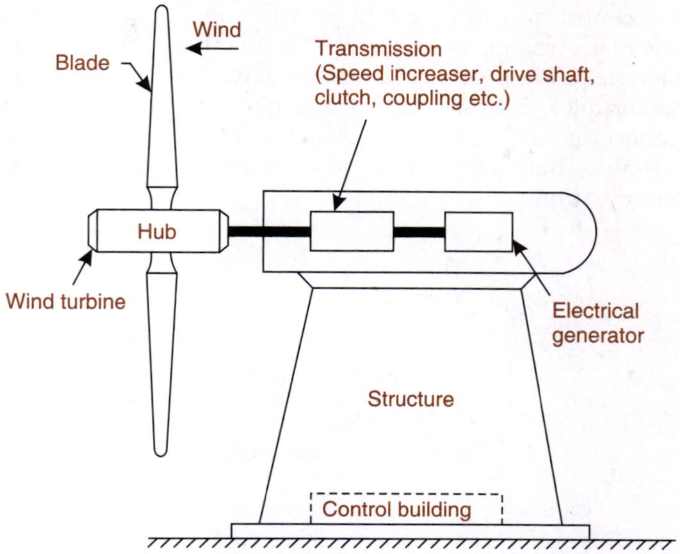

Fig. 29.3. Wind-electric generating power plant.

3. Electrical generator
4. Supporting structure

— The most important component is the **rotor.** For an effective utilisation, all components should be properly designed and matched with the rest of the components.

— The wind mill head performs the following functions :

(*i*) It supports the rotor housing and the rotor bearings.

(*ii*) It also houses any control mechanism incorporated like changing the pitch of the blades for safety devices and tail vane to orient the rotor to face the wind, the latter is facilitated by mounting it on the top of the supporting structure on suitable bearings.

— The wind turbine may be located either *unwind* or *downwind* of the power. In the unwind location the wind encounters the turbine before reaching the tower. *Downwind rotors are generally preferred especially for the large aerogenerators*.

— The **supporting structure** is designed to withstand the wind load during gusts. Its type and height is related to cost and transmission system incorporated. Horizontal axis wind turbines are mounted on towers so as to be above the level of turbulence and other ground related effects.

29.2.5. Types of Wind Machines

Wind machines (aerogenerators) are generally classified as follows :

1. Horizontal axis wind machines
2. Vertical axis wind machines.

1. Horizontal axis wind machines. Fig. 29.4 shows a schematic arrangement of a horizontal axis wind machine. Although the common wind turbine with a horizontal axis is simple in principle, yet the design of a complete system, especially a large one that would produce electric power economically, is complex. It is of paramount importance that the components like rotor, transmission, generator and tower should not only be as efficient as possible but they must also function effectively in combination.

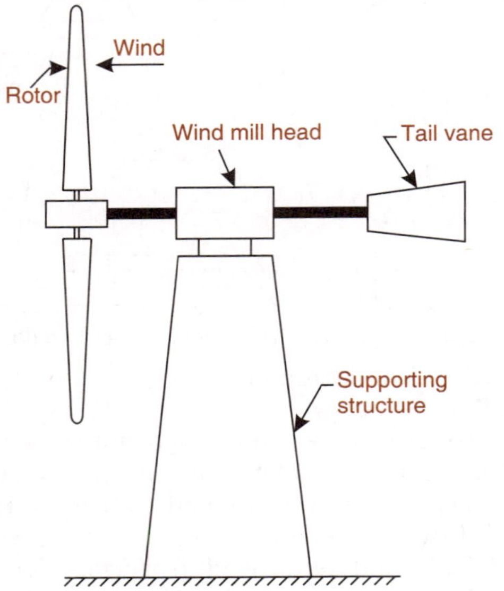

Fig. 29.4. Horizontal axis wind machine.

2. Vertical axis wind machines. Fig. 29.5 shows vertical axis type wind machine. One of the main advantages of vertical axis rotors is that they do not have to be turned into the windstream as the wind direction changes. Because their operation is independent of wind direction, vertical axis wind machines are called *panemones.*

29.2.6. Wind-Powered Battery Chargers

One application of wind energy systems which is of considerable potential importance (to developing countries) is the use of small wind generators to charge batteries for powering lighting, radio communication and hospital equipment. Wind generators have been in use in Europe and North America since the 1920s, although their use declined considerably.

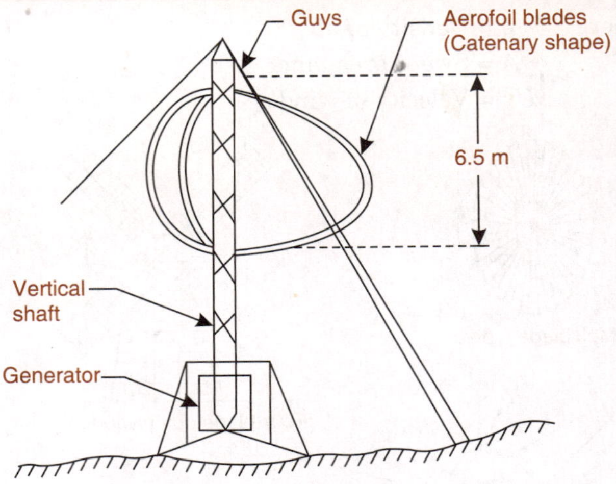

Fig. 29.5. Vertical axis wind machine.

A battery charging system has to include the following :

(*i*) A wind powered generator
(*ii*) A converter
(*iii*) A container for the batteries.

Fig. 29.6 shows a set-up of wind powered battery charging system. It is worthnoting that 12 volt batteries, which are rechargeable using wind generators, can be used to power fluorescent tube lighting which is six times more efficient than tungsten filament lamps. Such lighting opens up a number of important development opportunities in areas which normally have no lighting.

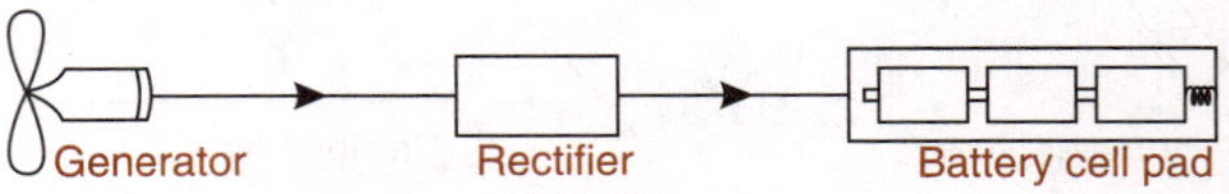

Fig. 29.6

For small wind generators the total system efficiency is made up as follows :

Wind regime matching efficiency	60% (approx.)
Rotor efficiency	35% (approx.)
Generator and wiring efficiency	70% (approx.)
Battery charge/discharge efficiency	70% (approx.)

Cumulatively, a total energy capture efficiency of about 10% is generally obtained from small wind generators utilized for battery charging.

Battery charging wind generators are produced in several countries, notably Australia, France, Sweden, Switzerland, the U.K., the U.S.A. and West Germany. In developing countries production is underway in China and has started in India.

29.2.7. Wind Electricity in Small Independent Grids

Refer to Fig. 29.7. In such systems electricity consumption fluctuates constantly as does the availability of wind energy. The degree of coincidence of supply and demand can be calculated by statistical means and it has been found that electricity supply with an acceptable degrees of reliability cannot be based solely on wind energy. If an extensive grid does not exist, electricity storage (batteries) or a back-up system (diesel) is required. Loads for remote systems of upto 6 kWh/day equivalent to an average power consumption of 250 W with a duty cycle of 24 hours, can be provided with battery storage.

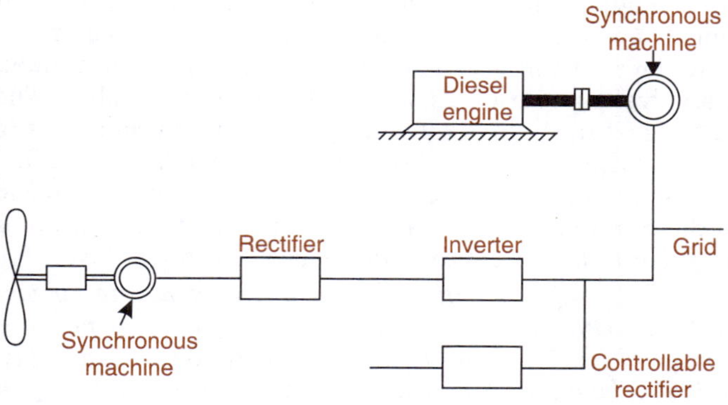

Fig. 29.7. Principle of combined wind/diesel power generation.

If a diesel and wind generator are used in conjunction with a grid, the *diesel generator should only be used when wind energy is absent*. Problems can occur, however, when the diesel generator is called on to change its output frequently as wind energy availability fluctuates. Besides decreasing the oil saving, diesel generation on this basis leads to more frequent overhauls of the generator. Both factors will increase costs. Several methods of overcoming these problems have been tried but there is not yet an established solution. Some development work has still to be done before wind generators can be run in parallel with diesel on a routine basis.

29.2.8. Wind Electricity Economics

Wind generator power costs are heavily linked to the characteristics of a wind resource in a specific location. The cost of supplied power declines as wind speeds increase, and the power supplied increases in proportion to the *cube of the wind speed.*

Matching available energy and load requirements is also important in wind energy economics. The correct size of wind generator must be chosen together with some kind of storage or co-generation with an engine or a grid to obtain the best economy. The ideal application is a task that can utilize a variable power supply, *e.g.,* ice making or water purification.

Regarding the *economics*, the choice of interest rate obviously has a major effect on the overall energy cost. With low interest rates, capital intensive power sources such as solar and wind are favoured. Other factors bearing a strong influence on the economics of wind electricity are the standard of maintenance and service facilities and the cost of alternative energy supplies in the particular area.

29.2.9. Problems in Operating Large Wind Power Generators

The operation of large wind power generators entails the following problems :

1. **Location of site.** The most important factor is locating a site big enough which has a reasonable average high wind velocity.

Sourastra and Coastal Regions in India are promising areas.

2. **Constant angular velocity.** A constant angular velocity is a must for generating A.C. (alternating current) power and this means very sensitive governing.

3. **Variation in wind velocity.** The wind velocity varies with time and varies in direction and also varies from the bottom to top of a large rotor (some rotors are as long as 50 metres). This causes fatigue in blades.

4. **Need of a storage system.** At zero velocity conditions, the power generated will be zero and this means some storage system will have to be incorporated along with the wind mill.

5. **Strong supporting structure.** Since the wind mill generator will have to be located at a height, the supporting structure will have to be designed to withstand high wind velocity and impacts. This will add to the initial costs of the wind mill.

6. **Occupation of large areas of land.** Large areas of land will become unavailable due to wind mill gardens (places where many wind mills are located). The whole area will have to be protected to avoid accidents.

Inspite of all these difficulties, interest to develop wind mills is there since this is a clean source of energy.

29.3. TIDAL POWER PLANTS

29.3.1. Introduction

The periodic rise and fall of the water level of sea which are carried by the action of the sun and moon on water of the earth is called the '**tide**'. Tidal energy can furnish a significant portion of all such energies which are renewable in nature. The large scale up and down movement of sea water represents an unlimited source of energy. If some part of this vast energy can be converted into electrical energy, it would be an important source of hydropower.

The *main feature of the tidal cycle is the difference in water surface elevations at the high tide and at the low tide.* If this differential head could be utilized in operating a hydraulic turbine, the tidal energy could be converted into electrical energy by means of an attached generator.

29.3.2. Components of Tidal Power Plants

The following are the components of a tidal power plant :

1. The dam or dyke (low wall) to form the pool or basin.

2. Sluice ways from the basins to the sea and *vice versa*.
3. The power house.

Dam or dyke. The function of dam or dyke is to form a barrier between the sea and the basin or between one basin and the other in case of multiple basins.

Sluice ways. These are used to fill the basin during the high tide or empty the basin during the low tide, as per operational requirement. These devices are controlled through gates.

Power house. A power house houses turbines, electric generators and other auxiliary equipment. As far as possible the power house and sluice ways should be in *alignment* with the dam or dyke.

29.3.3. Classification and Operation of Tidal Power Plants

Tidal power plants are classified as follows :

1. Single basin arrangement

(*i*) Single ebb-cycle system

(*ii*) Single tide-cycle system

(*iii*) Double cycle system.

2. Double basin arrangement

In a *single basin arrangement* power can be generated only *intermittently*. In this arrangement only one basin interacts with the sea. The two are separated by a dam or dyke and the flow between them is through sluice ways located conveniently along the dam. The rise and fall of tidal water levels provide the potential head.

Fig. 29.8 shows a general arrangement of single basin tidal power plant (double cycle system). Such plants generally use *reversible water turbines* so that power is generated on low tide as well high tide. The operation of the plant is as follows :

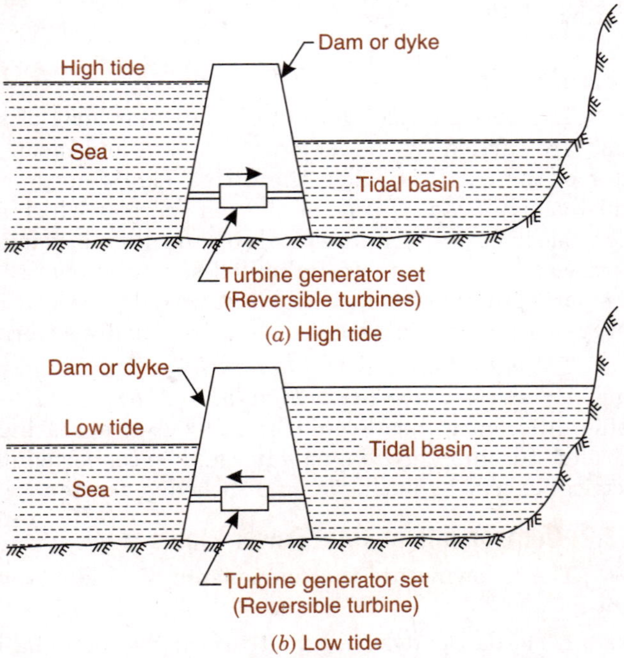

Fig. 29.8. General arrangement of tidal power plant.

When the incoming tide sea level and tidal-basin level are equal, the turbine conduit is closed. When the sea level rises, and about half way to high tide the turbine valves are opened and the sea water flows into the basin through the turbine runner generating power. This also raises the level of water in the basin. The turbine continues to generate power until the tide passes through its high point and begins to drop. The water head then quickly diminishes till it is not enough to supply the no-load losses. By pass valve then quickly opens to let water into the basin to gain maximum water level. When sea and basin water level are again equal, the valves are closed as well as the turbine conduit. The basin level then stays constant while the tide continues to go out. After sufficient head has developed, the turbine valves are agained opened and water now flows from basin to the sea, thereby generating power. The plant continues to generate power till the tide reaches its lowest level.

A *single basin plant cannot generate power continuously, though it might do so by using a pumped storage plant if the load it supplies fluctuates considerably*. A *double basin scheme* can provide power *continuously* or on demand, which is a *great advantage*. The drawback is that the civil works become more extensive. In the simplest *double-basin scheme* there must be a dam between each basin and the sea and also a dam between the basins, containing the power house. One basin is maintained always at a lower level than the other. The lower reservoir empties at low tide, the upper reservoir is replinshed at high tide. If the generating capacity is to be large, the reservoirs must be large which means that long dams would be required.

Fig. 29.9 shows a *tidal power plant-double basin operation*.

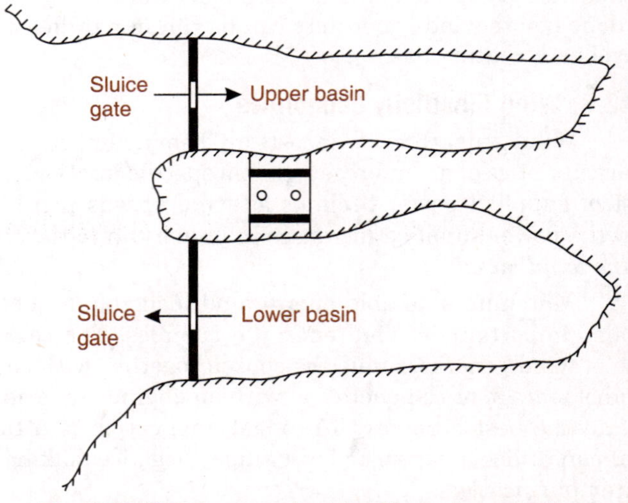

Fig. 29.9. Tidal power plant-double basin operation.

29.3.4. Advantages and Limitations of Tidal Power Generation

Advantages :

1. Tidal power is completely independent of the precipitation (rain) and its uncertainty, besides being inexhaustible.

2. Large area of valuable land is not required.
3. When a tidal power plant works in combination with thermal or hydroelectric system peak power demand can be effectively met with.
4. Tidal power generation is free from pollution.

Limitations :

1. Due to variation in tidal range the output is not uniform.
2. Since the turbines have to work on a wide range of head variation (due to variable tidal range) the plant efficiency is affected.
3. There is a fear of machinery being corroded due to corrosive sea water.
4. It is difficult to carry out construction in sea.
5. As compared to other sources of energy, the tidal power plant is costly.
6. Sedimentation and silteration of basins are the problems associated with tidal power plants.
7. The power transmission cost is high because the tidal power plants are located away from load centres.

— The first commercial tidal power station in the World was constructed in France in 1965 across the mouth of La Rance Estuary. It has a high capacity of 240 MW. The average tidal range at La Rance is 8.4 m and the dam built across the estuary encloses an area of 22 km^2.

29.4. SOLAR POWER PLANTS

29.4.1. Solar Energy—General Terms and Introduction

General Terms

Solar constant. Solar constant is the energy from the sun, per unit time, received on a unit area of surface perpendicular to the radiation, in space, at the earth's mean distance from the sun. According to Thekaekara and Drummond (1971) the value of the solar constant is 1353 W/M^2 (1.940 cal/cm^2 min or 4871 kJ/m^2 hr).

Beam radiation. *The solar radiation received from the sun without change of direction is called beam radiation.*

Diffuse radiation. *It is the solar radiation received from the sun after its direction has been changed by reflection and scattering by the atmosphere.*

Air mass. *It is the path length of radiation through the atmosphere, considering the vertical path at sea level as unity.*

Zenith angle. *It is angle between the beam for the sun and the vertical.*

Solar altitude. *It is the angle between the beam from the sun and horizontal i.e.,* (90-Zenith angle).

Solar or short-wave radiation. It is the radiation originating from the sun, at a source temperature of about 6000 K and in the wavelength range of 0.3 to 3.0 μm.

Long-wave radiation. Radiation originating from source at temperatures near ordinary ambient temperatures and thus substantially all at wavelength greater than 3 μm.

Declination. *It is the angular position of the sun at solar moon with respect to the plane of the equator (north positive).*

Solar Energy—Introduction

The surface of the earth receives from the sun about 10^{14} kW of solar energy which is approximately five order of magnitude greater than currently being consumed from all resources. It is evident that sun will last for 10^{11} years. Even through the sunlight is filtered by the atmosphere one square metre of the land exposed to direct sunlight receives the energy equivalent of about 1 H.P. or 1 kW. However, this last amount of solar energy reaching earth is not easily convertible and certainly is not "free".

There are two obvious obstacles to harnessing solar energy. Firstly it is not constantly available on earth. Thus, some form of storage is needed to sustain a solar power system through the night and during periods when local weather conditions obscure the sun. Secondly the solar energy is diffused. Although the total amount of energy is enormous, the collection and conservation of solar energy into useful forms must be carried out over a large area which entails a large capital investment for the conversion apparatus.

Solar energy, therefore, most likely will be developed not because it is cheaper than alternative energy sources but because these alternative sources sooner or later (*i*) will be exhausted, (*ii*) will become increasingly more expensive, (*iii*) will continue to political and economical control by the nations possessing them, and (*iv*) will produce undesirable yet incompletely understood environmental consequences, especially on large scale that will be required to meet projected demands even with controlled growth.

Solar energy has some good advantages in comparison to the other sources of power. *Solar radiation does not contaminate environment or endanger ecological balance. It avoids major problems like exploration, extraction and transportation.*

29.4.2. Collectors in Various Ranges and Applications

Following list gives the thermal applications of solar energy and possible temperature ranges :

1. Low temperature (t = 100°C) :

(*i*) Water heating
(*ii*) Space heating
(*iii*) Space cooling
(*iv*) Drying

} ... Flat plate

2. Medium temperature (100 to 200°C) :

(*i*) Vapour engines and turbines
(*ii*) Process heating
(*iii*) Refrigeration
(*iv*) Cooking

... Cylindrical parabola

3. High temperature (> 200°C) :

(*i*) Steam engines and turbines
(*ii*) Stirling engine
(*iii*) Thermoelectric generator

... Parabolloid mirror arrays

The above classification of low, medium and high temperature ranges is somewhat arbitrary. Heating water for domestic applications, space heating and cooling and drying of agricultural products (and industrial products) is generally at temperature below 100°C, achieved using flat plate collectors with one or two glass plate covers. Refrigeration for preservation of food products, heating for certain industrial processes, and operation of engines and turbines using low boiling organic vapours is possible at somewhat higher temperature of 100 to 200°C and may be achieved using focusing collectors with cylindrical parabola reflectors requiring only one directional diurnal tracking. Conventional steam engines and turbines, stirling hot air engines, and thermoelectric generators require the solar collectors to operate at high temperatures. Solar collectors operating at temperature above 200°C generally consist of parabolloid reflector as an array of mirrors reflecting to a central target, and requiring two directional diurnal tracking.

The concentrators or focusing type collectors can give high temperatures than flat plate collectors, but they entail the following *shortcomings/limitations* :

1. Non-availability and high cost of materials required. These materials must be easily shapeable, yet have a long life ; they must be light weight and capable of retaining their brightness in tropical weather. Anodised aluminium and stainless steel are two such materials but they are expensive and not readily available in sufficient quantities.
2. They require direct light and are not operative when the sun is even partly covered with clouds.
3. They need tracking systems and reflecting surfaces undergo deterioration with the passage of time.
4. These devices are also subject to similar vibration and movement problems as radar antenna dishes.

29.4.3. Flat Plate Collectors

29.4.3.1. Description

Fig. 29.10 shows a flat plate collector which consists of *four essential components* :

1. ***An absorber plate.*** It intercepts and absorbs solar radiation. This plate is usually metallic (copper, aluminium or steel), although plastics have been used in some low temperature applications. In most cases it is coated with a material to enhance the absorption of solar radiation. The coating may also be tailored to minimise the amount of infrared radiation emitted.

A heat transport fluid (usually air or water) is needed to extract the energy collected and passes over, under or through passages which form an integral part of the plate.

2. ***Transparent covers.*** These are one or more sheets of solar radiation transmitting materials and are placed above the absorber plate. They allow solar energy to reach the absorber plate while reducing convection, conduction and re-radiation heat losses.

3. ***Insulation beneath the absorber plate.*** It minimises and protects the absorbing surface from heat losses.

4. ***Box like structure.*** It contains the above components and keeps them in position.

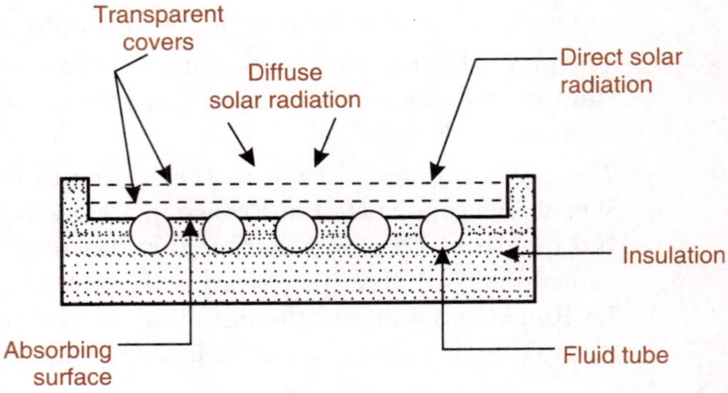

Fig. 29.10. Flat plate solar collector.

Various types of flat plate collectors have been designed and studied. These include tube in plate, corrugated type, spiral wound type etc. Other criteria is single exposure, double exposure or exposure and reflector type. The collector utilizes sheets of any of the highly conducting material *viz.* copper, aluminium or galvanized iron. The sheets are painted dead black for increasing the absorbtivity. The sheets are provided with one or more glass or plastic covers with air gap in between to reduce the heat transfer losses. The sides which are not exposed to solar radiation are well-insulated. The whole assembly is fixed in air tight wooden box which is mounted on simple device to give the desired angle of inclination. The dimensions of collectors should be such as to make their handling easy. The collector will absorb the sun energy (direct as well as diffused) and transfer it to the fluid (air, water or oil) flowing within the collector. Basically, a flat plate collector is effective most of time, reliable for good many years and also inexpensive.

Use of flat mirrors in flat plate collectors improves the output, permitting higher temperatures of operation. Side mirrors are used either at north and south edges or at east and west edges of the collector or a combination both. The mirrors may be of reversible or non-reversible type.

29.4.3.2. Analysis

Consider an object exposed to sun radiations of intensity P, per unit area at the surface of the body. These radiations, will partly be absorbed by the body, while the remaining will be partly transmitted and rest reflected. If we take the incident radiations equal to unity, then the absorbed, reflected, and transmitted parts of energy will add up to unity. These parts are called absorption co-efficient, reflection co-efficient and transmission co-efficient and represented by the symbols α, ρ and τ respectively.

Using the above symbols, we can write

$$\alpha + \rho + \tau = 1 \qquad ...(i)$$

The absorbed part of the solar radiations, which is equal to α is, responsible for increasing the temperature of the body. However, the body also losses energy by conduction, convection and radiation. The equilibrium temperature of the body will be that at which the heat losses from the body are equal to the absorbed radiations. For analysis, purposes, if we represent the body, by a flat plate and assume that the convection and conduction losses are negligible to begin with, then at equilibrium temperature, the absorbed solar radiations should be equal to the radiations losses from the flat plate. The radiation losses are equal to $\varepsilon\sigma T^4$, where ε and T are the emission co-efficient and absolute temperature respectively of flat plate and σ is the Boltzman's constant.

Therefore, at equilibrium,

$$\alpha P = \varepsilon\sigma T^4 \qquad ...(ii)$$

or

$$\frac{\alpha P}{\varepsilon} = \sigma T^4 \qquad ...(iii)$$

From equation (*iii*), it is evident that comparatively higher equilibrium temperature will be obtained where the quantity *i.e.*, the ratio of absorption co-efficient to emission co-efficient of the flat plate is more. However, this has been demonstrated by an equation obtained under idealised condition. In the realistic conditions too, its nature will remain the same, but it will get modified by other influencing factors.

The collectors for which $\frac{\alpha}{\varepsilon}$ ratio is *equal to unity* are called **'Neutral Collectors'** and those for which the ratio is *greater than unity* are called **'Selective Collectors'**.

The amount of energy collected, however, does not depend on $\frac{\alpha}{\varepsilon}$ ratio. It primarily depends on higher value of α. So to obtain higher energy collection, one should use such flat plate whose absorption co-efficient is as high as possible, since the solar radiations are similar to those emitted by a body at a temperature of 5500°C and is as low as possible in the temperature range, normally attained by the collectors.

A flat plate painted black is placed on a well insulated base. If it is exposed to solar radiations where $P = 800$ W/m^2, a typical summer value for a tropical region, we obtain from equation (*iii*) the equilibrium temperature 70°C. Inspite of the simplifications here, it is a fair estimate of the temperature reached by a black plate left for a time in the tropical sun.

This method can be refined by including the convection losses and the energy gain as a result of absorption of diffused radiations by the flat plate.

If P' is the intensity of the diffused radiations and α' the absorption co-efficient, then equation (*ii*) becomes

$$\alpha P + \alpha' P' = h_c (T - T_a) = \varepsilon\sigma T^4 \qquad ...(iv)$$

This is valid, where the base is insulated, hence conduction losses are neglected. Here T_a is the atmospheric temperature and h_c is the convection heat transfer co-efficient.

29.4.4. Focusing (or Concentrating) Collectors

The main types of focusing or concentrating collectors are as follows :

1. Parabolic trough collector
2. Mirror strip collector
3. Fresnel lens collector
4. Flat plate collector with adjustable mirrors
5. Compound parabolic concentrator.

Fig. 29.11 (*a*) shows the principle of the parabolic trough collector, which is often used in focusing collectors. Solar radiation coming from the particular direction is collected over the area of reflecting surface and is concentrated

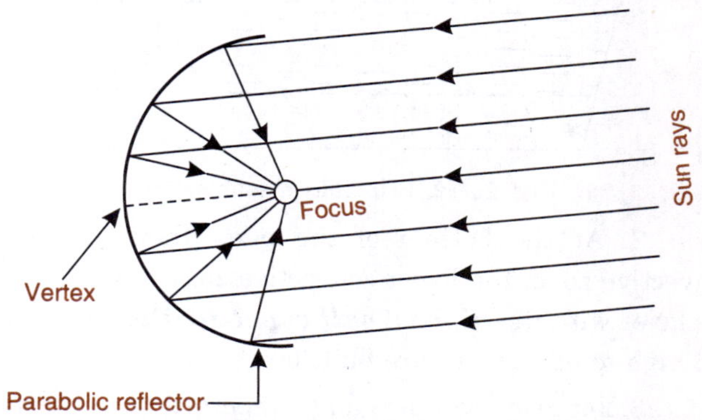

Fig. 29.11. (*a*) Cross-section of parabolic trough collector.

at the focus of the parabola, if the reflector is in the form of a trough with parabolic cross-section, the solar radiation is focused along a line. Mostly *cylindrical parabolic concentrators* are used in which absorber is placed along focus axis [Fig. 29.11 (*b*)].

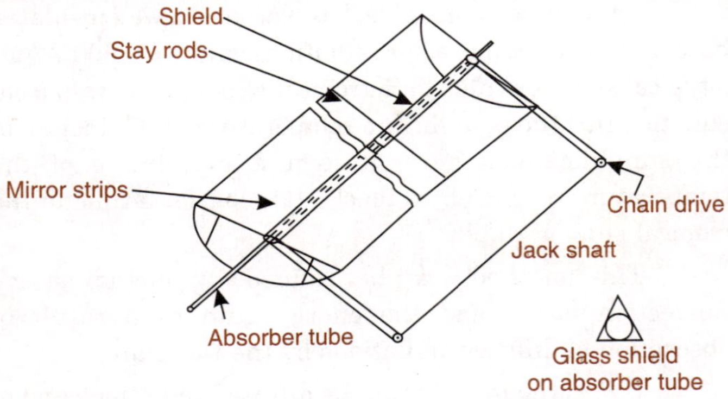

Fig. 29.11. (*b*) Cylindrical parabolic system.

29.4.5. Solar Pond Technology

Refer to Fig. 29.12. The vertical configuration of salt-gradient solar pond normally consists of following *three zones*.

1. Adjacent to the surface there is a *homogeneous convective zone* that serves as a buffer zone between environmental fluctuations at the surface and conductive heat transport from the layer below. This is the ***upper convective zone (UCZ)***.

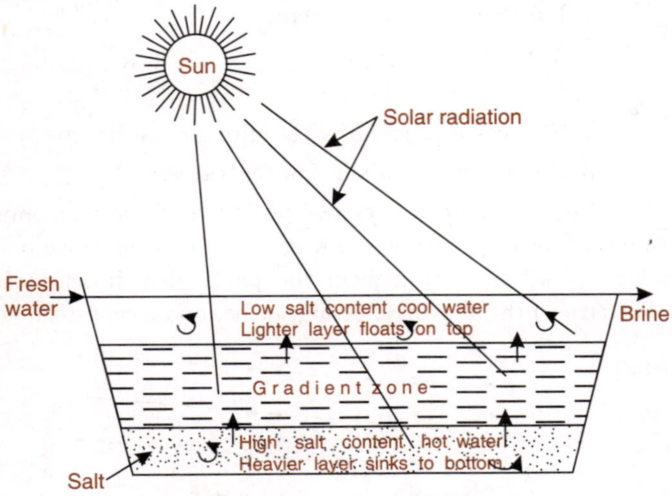

Fig. 29.12. Principle of solar pond.

2. At the bottom of the pond there is another convective zone, the lower convective zone of LCZ. This is the layer with the ***highest salt concentration*** and where the high temperatures are built up.

3. For given salinites and temperatures in the upper and lower convective zones, there exists a stable **intermediate gradient zone.** This zone keeps the two convective zones apart and gives the solar pond its unique thermal performance. This intermediate zone provides excellent insulation for the storage layer, while simultaneously transmitting the solar radiation. To maintain a solar pond in this non-equilibrium stationary state, it is necessary to replace the amount of salt that it transported by molecular diffusion from the LCZ to the UCZ. This means that salt must be added to the LCZ, and fresh water to the UCZ whilst brine is removed. The brine can be recycled, divided into water and salt (by solar distillation) and returned to the pond.

The major heat loss occurs from the surface of the solar pond. This heat loss can be prevented by spreading a plastic grid over the pond's surface to prevent disturbance by the wind. Disturbed water tends to lose heat transfer faster than when calm.

Due to the excessively high salt concentration of the LCZ, "*a plastic liner*" or impermeable soil must be used to "*prevent infiltration*" into the nearby ground water or soil. The liner is a factor that increases the cost of a solar pond. A site where the soil is naturally impermeable, such as the base of a natural pond or lake, or can be made impermeable by compaction or other means, will allow considerably lower power costs.

The optical transmission properties and related collection efficiency vary greatly and depend on the following :

(*i*) Salt concentration ;

(*ii*) The quantity of suspended dust or other particles ;

(*iii*) Surface impurities like leaves or debris, biological material like bacteria and algae ; and

(*iv*) The type of salt.

It becomes obvious that much higher efficiencies and storage can be achieved *through the utilization of refined or pure salt* whenever possible, as this *maximizes optical transmission*.

The solar pond is an effective collector of diffuse, as well as direct radiation, and will gather useful heat even on cloudy or overcast days. Under ideal conditions, the pond's absorption efficiency can reach 50% of incoming solar radiation, although actual efficiencies average about 20% due to heat losses. Once the lower layers of the pond reaches over 60°C the heat generated can be drawn off through a heat exchanger and used to drive a low temperature organic Rankine cycle (ORC) turbine. This harnesses the pressure differentials created when a low boiling point organic fluid or gas is boiled by heat from the pond *via*. a heat exchanger and cooled by a condenser to drive a turbine to generate electricity. The conversion efficiency of an organic Rankine cycle turbine driving an electric generator is 5–8% (which means 1–3% from insolation to electricity output).

29.4.6. Low Temperature Thermal Power Generation

29.4.6.1. Solar Pond Electric Power Plant

A low temperature thermal electric power production scheme using solar pond is shown schematically in Fig. 29.13. The energy obtained from a solar pond is used to drive a Rankine cycle heat engine. Hot water from the bottom level of the pond is pumped to the evaporator where the organic working fluid is vapourized. The vapour then flows under high pressure to the turbine where it expands and work thus obtained runs an electric generator producing electricity. The exhaust vapour is then condensed in a condenser and the liquid is pumped back to the evaporator and the cycle is repeated.

In Australia, a 2000 sq m solar pond equipped with a 20 kW engine has been installed.

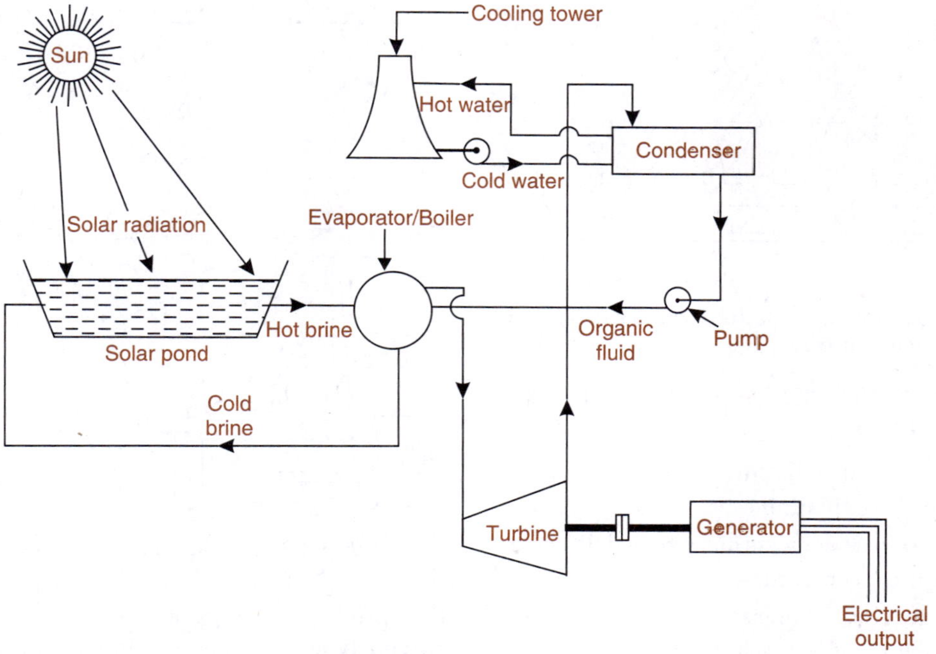

Fig. 29.13. Solar pond electric power plant.

29.4.6.2. Low Temperature Solar Power Plant

Fig. 29.14 shows a schematic diagram of a low temperature solar power plant. In this system an array of flat plate collectors is used to heat water to about 70°C and this heat is used to boil butane in a heat exchanger. The high pressure butane vapour thus obtained runs a butane turbine which in turn operates a hydraulic pump. The pump pumps the water from well which is used for irrigation purposes. The exhaust butane vapour (from butane turbine) is condensed with the help of water which is pumped by the pump and the condensate is returned to the heat exchanger (or boiler).

29.4.7. Medium Temperature Systems Using Focusing Collectors

A circular or rectangular parabolic mirror can collect the radiation and focus it on to a small area, a mechanism for moving the collector to follow the sun being necessary. Such devices are used for *metallurgical research* where *high purity and high temperatures are essential,* an example being a 55 m diameter collector giving about 1 MW at Mont Louis in Pyrenees. Smaller units having 20 m diameter

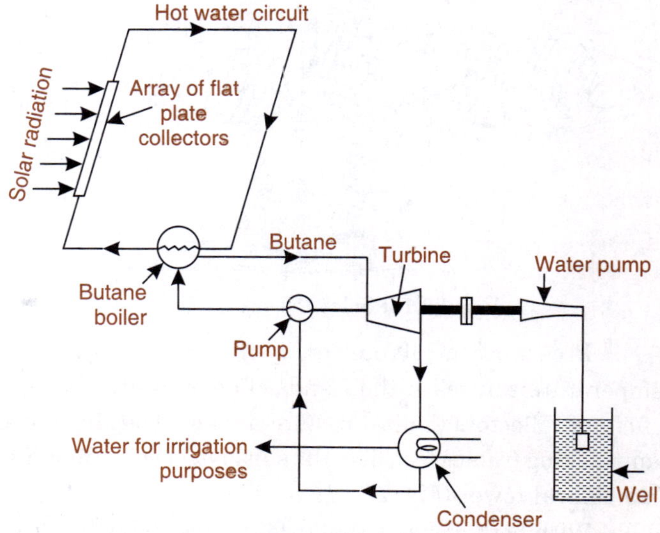

Fig. 29.14. Low temperature solar power plant.

reflector can give temperatures of 300°C over an area of about 50 m^2. The collector efficiency is about 50%. On a small scale, units about 1 m diameter giving temperatures of about 300°C have been used for cooking purposes.

Fig. 29.15 shows a concave solar energy collector focusing sun's rays on boiler at a focal point. Generation of steam at 250°C could give turbine efficiencies up to 20–25 per cent.

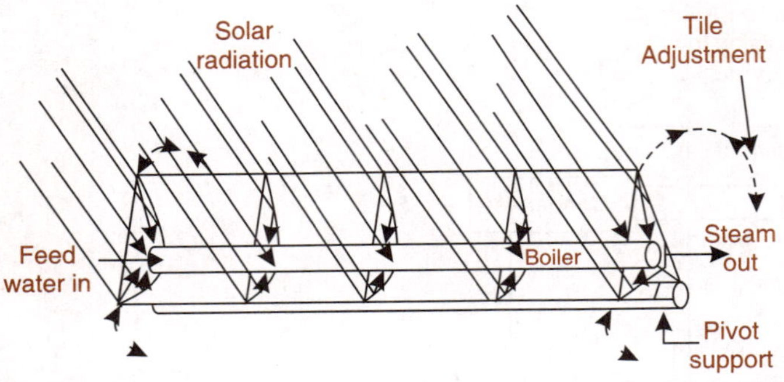

Fig. 29.15. Concave solar energy collector focuses sun's rays on boiler at focal point.

29.4.8. High Temperature Systems—Solar Farm and Solar Tower Plant

For a large scale production of process-heat the following two concepts are available :

1. **The solar farm.** It consists of a whole field covered with parabolic trough concentrators.

2. **The solar tower.** It consists of a central receiver on a tower and a whole field of tracking.

In case of a *'solar farm'* temperature at the point of focus can reach several hundred degrees celsius. Fig. 29.16 shows a solar tower system.

Fig. 29.16. Solar tower system.

In case of central receiver *"solar tower"* concentrators, temperature can reach thousands of degrees celsius, since a field of reflectors (heliostats) are arranged separately on sun-tracking frames to reflect the sun on to a boiler mounted on a central tower (Fig. 29.17).

With both systems ('solar farm' and 'solar tower'), *a heat transfer fluid or gas is passed through the point or line of insolation concentration to collect the heat and transfer it to the point of use. Such heat can be used either directly in industrial or commercial processes or indirectly in electricity production via. steam and a turbine.*

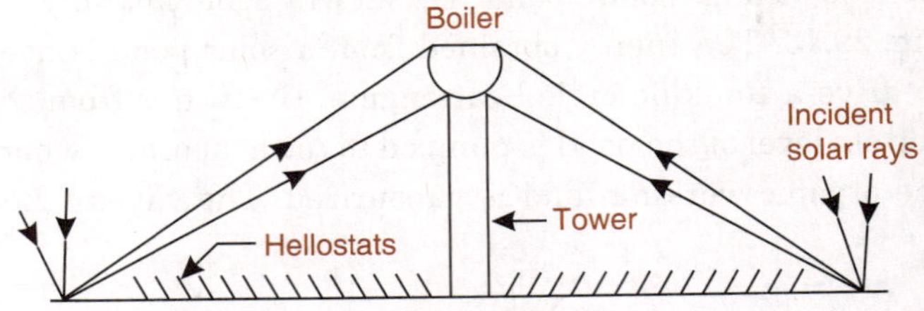

Fig. 29.17

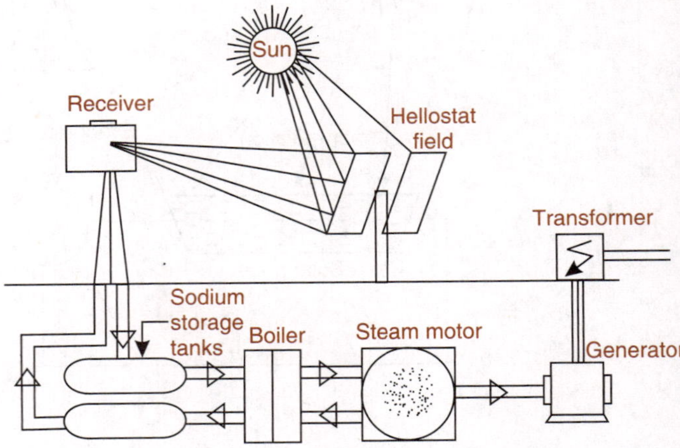

Fig. 29.18. Diagram of solar tower power plant.

The solar technologies such as the above two systems that produce very hot water or steam (*i.e.*, one 177°C) are currently still under development and, in general, these technologies are not cost competitive with conventional power sources such as oil or gas.

29.5. GEOTHERMAL POWER PLANTS

29.5.1. Geothermal Energy

Geothermal energy is the heat from high pressure steam coming from within the earth. It is a renewable source of energy derived from the rain water in the earth heated to over 180°C by subterranean hot rocks. Geothermal sources which show up hot water springs with temperature of steam upto 150°C are widely distributed over the globe.

Fig. 29.19 shows a schematic diagram depicting how hot springs are produced through hot magma (molten mass), the fractured crystalline rocks, the permeable rocks and percolating ground water.

29.5.2. Geothermal Sources

The following five general categories of geothermal sources have been identified :

1. Hydrothermal convective systems

(*i*) Vapour-dominated or dry steam fields

(*ii*) Liquid-dominated system or wet steam fields
(*iii*) Hot-water fields.
2. Geopressure resources
3. Petro-thermal or hot dry rocks (HDR)
4. Magma resources
5. Volcanoes.

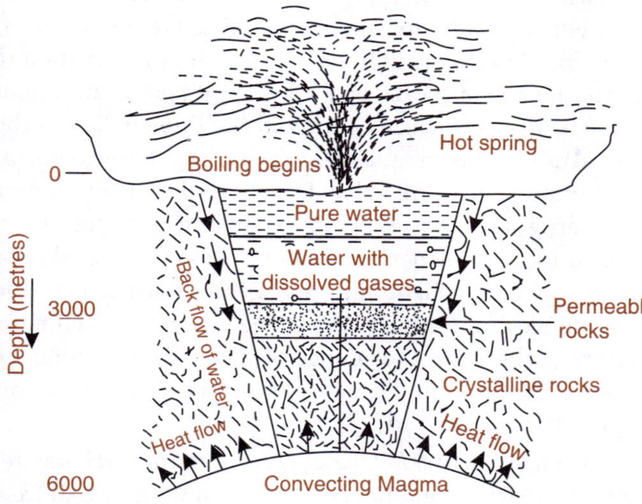

Fig. 29.19. Hot spring system structure.

The hydrothermal convective systems are best resources for geothermal energy exploitation at present. Hot dry rock is also being considered.

Fig. 29.20 shows a dry-steam open system used in Larderello (Italy) and Greyser (U.S.A.).

Fig. 29.21 shows a flash steam open type used in Cerro Prierol Mexico, Octake (Japan).

Fig. 29.22 shows a hot water closed (Binary) installed in Russia and U.S.A.

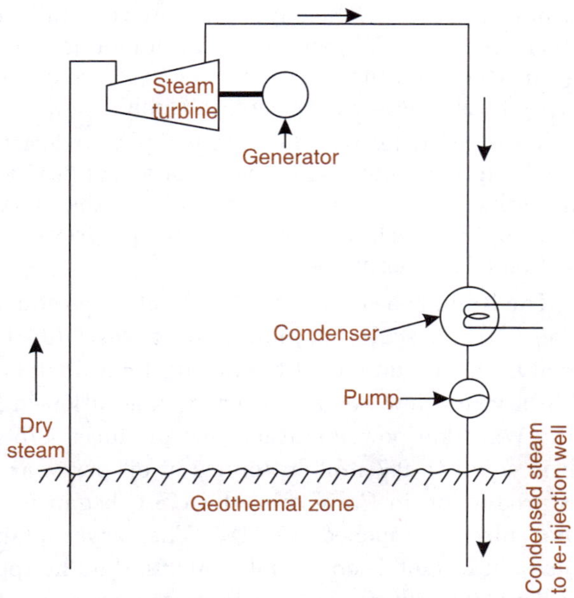

Fig. 29.20. Dry-steam open system.

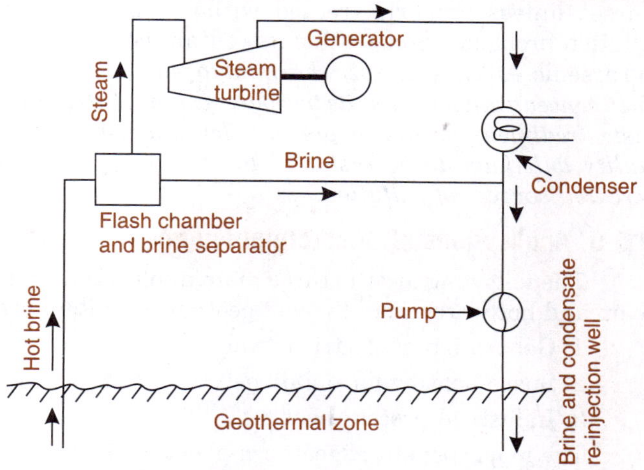

Fig. 29.21. Flash steam open type.

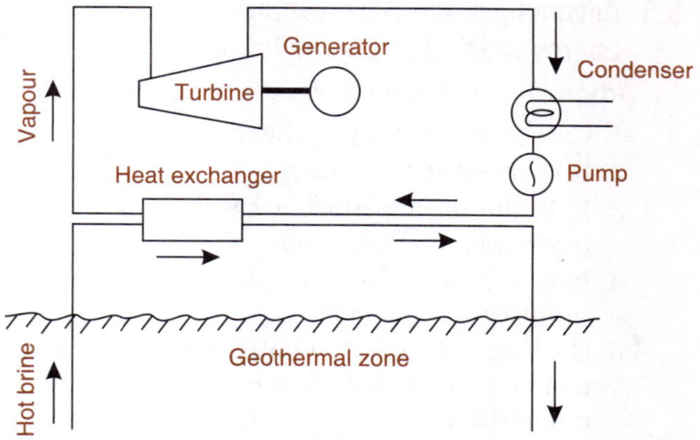

Fig. 29.22. Hot water closed (Binary).

29.5.3. Geothermal Power Estimates

Although geothermal power estimates vary very widely yet rough estimate is given below :

Depth	*Total stored energy (approximately)*
3 km	8×10^{21} joules
10 km	4×10^{22} joules

— The energy stored in hot springs = 10 per cent of the above quantities.

— If the above energy is extracted from a 3 km belt, with 1% thermal energy recovery factor, at a uniform rate of over 50 year period, a thermal power of 50 GW is obtained. With a thermal electric conversion efficiency of 20%, there will be a yield of only 10 GW of electric power.

29.5.4. Environmental Problems

Geothermal power plants create some environmental problems which are peculiar to them alone. The effluent will be salty and may contain sodium and potassium compounds. Additionally, in some cases lithium, fluorine, boron and arsenic compounds may be present. Such effluents cannot be discharged into the existing water courses unless properly treated without risking severe pollution problems. Some effluents contain boron, fluorine and arsenic. All these are *very harmful to plants and animal life in concentrations as low as two parts per million. Suitable waste treatment plants to prevent degradation of water quality will have to be installed to treat these new and increased sources of pollution.*

29.5.5. Applications of Geothermal Energy

The following are the three main applications of the steam and hot water from the wet geothermal reservoirs :

1. Generation of electric power
2. Space heating for buildings
3. Industrial process heat.

The major benefit of geothermal energy is its *varied application and versatility*.

29.5.6. Advantages and Disadvantages of Geothermal Energy over other Energy Forms

Advantages of geothermal energy :

1. Geothermal energy is cheaper.
2. It is versatile in its use.
3. It is the least polluting as compared to other conventional energy sources.
4. It is available for multiple uses from a single resource.
5. Geothermal power plants have the highest annual load factors of 85 per cent to 90 per cent compared 45 per cent to 50 per cent for fossil fuel plants.
6. It delivers greater amount of net energy from its system as compared to other alternative or conventional systems.
7. Geothermal energy from the earth's interior is almost as inexhaustible as solar or wind energy, so long as its sources are actively sought and economically tapped.

Disadvantages :

1. Low overall power production efficiency, (about 15% as compared to 35 to 40% for fossil fuel plants).
2. Drilling operation is noisy.
3. Large areas are needed for exploitation of geothermal energy as much of it is diffuse.
4. The withdrawl of large amounts of steam or water from a hydrothermal reservoir may result in surface subsidence or settlement.

29.5.7. Geothermal Energy in India and Abroad

Some progress is being made in India on tapping geothermal energy on a commercial scale. Engineers from the Geological survey of India have drilled about 50 shallow wells for steam in the Puga valley of the Ladakh region in Jammu and Kashmir. It may be possible to operate a 5-MW power station at the site. The Puga valley at an altitude of 4500 metres above sea level has the most promising geothermal field. The area extends to about 40 square kilometres out of which 5 sq km is active. A combination of wet and dry steam to the tune of 170 tonnes of hot water per hour and 20 tonnes/hour of dry steam (superheated steam suitable for running steam turbines) is available. This is enough to run a small power station to light the homes of local population. The geothermal heat can also be used for space heating in the Puga valley as the temperature in this area, especially during winter months, goes down to 35 degrees below freezing point. There are no other energy sources in Ladakh region and coal, petroleum etc. have to be transported from Srinagar. It can also be used for poultry farming, mushroom cultivation and pushmina wool processing which needs a warmer climate. In addition, there are good deposits of borax and sulphur in this area. Sulphur in elemental form is found only this region in the whole of India.

There are many *hot water springs* in India. Hot water springs represent heat energy coming out of the earth from a large body of molten rock that has been pushed up into the upper crust of the earth by geological forces. In North, they occur in Ladakh and Himachal Pradesh. In Western parts they are found in the Cambay region of Gujarat and Maharashtra. They are also found in the Singhbhum region of Bihar while there are some in Assam. The water from a hot spring at Garampani, near Jowai, in Assam is so hot in summer that rice kept in muslin bag gets cooked in on time.

The Geological Survey of India has so far identified about 350 hot springs sites which can be explored as sources of geothermal energy. The engineers have commissioned an experimental 1 kW generator running on geothermal energy in the Puga area. This is the first production of electricity from a hot water spring in India.

Many countries with hot springs in their territories have realised their potential for power and heat production. Countries like Italy, Iceland, New Zealand, the USA and the Russia have achieved remarkable progress in the application of geothermal energy.

The Italian power plant at Larderello was started in 1904 on a small scale but now it produces 540 MW of electricity. This is equivalent to burning 1.5 million tonnes of oil in a year. New Zealand started exploration in 1950 and the Wairakei power station now produces 175 MW, which is equal to 0.7 million tonnes of oil per year. The power production in California, the USA, began in 1960 and has already touched 50 MW. The power plant at Reykjavik in Iceland is on a small scale. In the Phillippines the drillers struck high pressure, high temperature steam at about 200 m only at Tiwi, a tiny sleepy village nestled at the base of the volcano Malino in 1967. By January, 1976, the first geothermal power plant at Tiwi produced 55 MW. A geothermal plant with a capacity of 11 MW has been in operation for nearly 20 years in Russia. The construction of another power plant at Mutnovsky with a capacity of 200 MW has been completed.

29.6. BIO-GAS PLANTS

In our country, there is a great potential for application of biomass as an **alternate source of energy.** We have plenty of agricultural and forest resources for reproduction of biomass.

The following are the biomass resources :

1. Concentrated wastes

(*i*) Municipal solid (*ii*) Sewage wood products
(*iii*) Industrial waste (*iv*) Manure at large lots.

2. Dispersed waste residue

(*i*) Crop residue (*ii*) Logging residue
(*iii*) Disposed manure.

3. Harvested bio-mass

(*i*) Standing biomass
(*ii*) Biomass energy plantations.

Bio-gas

The main source for production of bio-gas is wet **cowdung.** Some of the other sources are :

(*i*) Sewage (*ii*) Crop residue
(*iii*) Vegetable wates (*iv*) Water hyacinth
(*v*) Poultry droppings (*vi*) Pig-manures
(*vii*) Algae (*viii*) Ocean kelp.

Bio-gas, a mixture *containing 55-65% methane, 30-40% carbon dioxide* and the rest being the impurities hydrogen (hydrogen sulphide and some nitrogen), can *be produced from the decomposition of animal, plant and human waste.* It is a clean but slow burning gas and usually has a calorific value 5000 to 5500 kcal/kg. It can be used directly in cooking, reducing the demand for firewood. Moreover, the material from which the bio-gas is produced retains its value as a fertilizer and can be returned to soil.

Bio-gas is produced by digestion, pyrolysis or hydrogasification. *Digestion* is a biological process that occurs in the absence of oxygen and in the presence of anaerobic organisms at ambient pressures and temperatures of 35–70°C. The container in which this digestion takes place is known as the *digester*.

Bio-gas plants have been built in various designs. Janata model gobar gas plant is described below :

Janata Model Gobar Gas Plant

Constructional Features. A Janata model gobar gas plant comprises of following parts :

Refer to Fig. 29.23 :

1. Foundation 2. Digester
3. Dome 4. Inlet chamber
5. Outlet chamber 6. Mixing tank
7. Gas outlet pipe.

1. Foundation. The foundation is the amply compacted base of the digester made of cement concrete and brick ballast. Its construction is so carried out that it may provide a stable foundation to the digester walls and bear full load of slurry filled in the digester. It should be water proof so that no percolation or water leakage takes place.

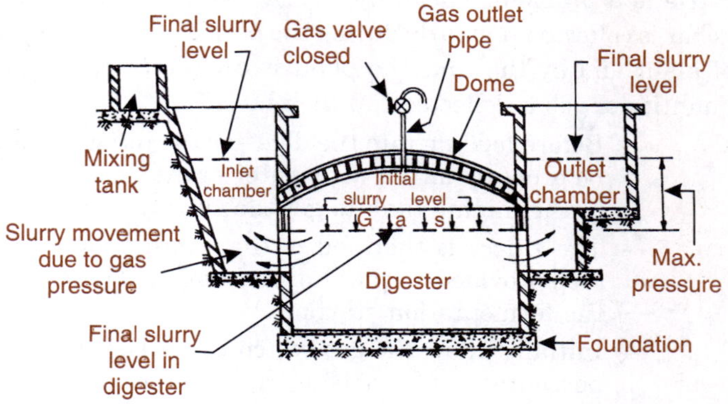

Fig. 29.23. Janata model gobar gas plant.

2. Digester. It is underground cylindrical wall portion made of bricks, sand and cement. It is this place where fermentation of dung takes place. It is also sometimes called 'fermentation tank'. Two rectangular openings facing each other are provided for inflow and outflow at almost middle of its height.

3. Dome. It is a hemispherical roof of the digester ; has a fixed height and forms the *critical part* in the construction of Janata gobar gas plant. The gas gets collected in the space of the dome and exerts pressure on the slurry in the digester.

4. Inlet chamber. An inlet chamber has a bell mouth shape and is made of bricks, cement and sand. It has its top opening at the ground level. Its outlet wall is made inclined/slopy to enable to daily cattle dung feed to move easily into the digester.

5. Outlet chamber. It is that part of the plant through which digested slurry moves out of the digester at a predetemined height. It has a small rectangular cross-section and above this it becomes larger to a defined height. For easy cleaning of the digester two steps are provided in it which enable a man to climb down. Its top opening is also at the ground level. Just near the top opening is provided a small outlet through which the digested/spent slurry flows to a compost pit.

6. Mixing tank. It is this tank where gobar and water are mixed properly in the ratio of 1 : 1 to make slurry which is then poured into the inlet chamber.

7. Gas outlet pipe. It is a small piece of G.I. pipe which is fitted at the top of the dome for conveying the gas to the points of use. A valve is fitted at its end to regulate the flow of gas to the gas connections.

Initial loading of the plant. After completion of the plant it is to be charged with cattle dung and water in the ratio of 1 : 1. Since Janata plant is designed for 50 days retention period, therefore initial loading will comprise of 50 times of gobar needed for daily feed. Naturally to arrange such a large quantity of gobar by farmers who have a few cattle is a problem. This difficulty can be overcome if the gobar is collected regularly during the period of construction of plant and by the time the plant is completed the large quantity of gobar collected can then be fed.

— Before feeding into the digester the dung should be properly mixed with water in the ratio of 1 : 1 in the mixing tank to get the slurry.

— The slurry is then fed to the digester upto the step provided on the Outlet chamber. Thereafter the fermentation starts.

— Initially allow the gas which comprises of large percentage of carbon dioxide, oxygen and hydrogen sulphide to escape since it will not burn.

— The gas obtained after this release will have a right combination of methane (60%) and carbon dioxide (40%) and burn with odourless blue flame in the gas chulha/burner/stove.

— Now, at this stage start feeding the plant with dung slurry.

— Regular, proper and careful feeding will ensure flawless gas service.

How does the Plant Function ?

— The cattle dung and water are mixed properly in the ratio 1 : 1 to form slurry which is then filled in the digester upto the height of its cylindrical portion.

— As soon as the fermentation of dung picks up the generated gas starts accumulating in the dome. This gas then exerts pressure on the slurry and displaces it into the Inlet and Outlet chambers. Consequently, the slurry level in the digester falls and rises in the Outlet chamber.

— This fall in the level of slurry continues (if the outlet pipe valve is closed) till the slurry level reaches the upper ends deges of the Inlet and Outlet gates.

— After this stage any further accumulation of gas would exert pressure and start escaping through these gates into the atmosphere. This condition is indicated by *bubbling* and froth formation on the surface of the slurry in the inlet and outlet chambers.

— The quantity of usable gas can be determined by calculating the increase in slurry volume in the inlet and outlet chambers.

— In the event of using gas, the slurry level in the digester rises while that in the inlet and outlet comes down.

— The increase in gas pressure due to increased generation in comparison to its utilisation will push/displace the slurry up in the inlet and outlet chambers while the decrease in pressure is balanced by the return flow of slury back in the digester. During the process approximately equivalent quantity of spent/digested slurry (top layer slurry) is discharged from the outlet chamber through the outlet opening ; this is so because fresh/undigested slurry which is fed into the inlet chamber is heavier than the digested/spent slurry and settles down in the digester.

Chemistry of Gas Generation

In gobar gas plant, the feed which comprises mainly of cattle dung is subjected to anaerobic fermentation as a result of which is produced combustible gas and fully matured organic manure which is superior to green manure available from dung otherwise. The whole process of feeding fresh slurry of dung and water and extraction of spent slurry in gobar gas plant is continuous one.

The cattle dung (and other fermentable materials like night soil, poultry or piggery droppings etc.) when confined in a place where there is no air gives rise to mainly two types of bacteria *viz. Acid forming bacteria* and *Gasifying bacteria.*

Acid forming bacteria convert carbohydrates, proteins, fats into vilatile acids and carbon dioxide is produced during the process. This phase is also known as liquification phase and is brought about by a set of *saprophytic bacteria* of means of extracellurenzine. These bacteria are less sensitive and can exist, develop and multiply in wide range of conditions.

Liquification phase is followed by gasification phase which is actively carried out by *methane bacteria*. They work upon volatile acid produced during the previous phase, with the help of intracellular enzime and convert it into methane and carbon dioxide.

The whole process of gobar gas generation is governed by the factors *viz.* Temperature of substrate, Loading rate, Solid concentration, Detention period, pH value, Nutrients concentration and Toxic substance etc.

The composition of the gas produced varies with the type of fermentable material used. In case of cattle dung on an average the gas produced consists of 55 to 60% methane and 40 to 45% carbon dioxide with little quantity of hydrogen, hydrogen sulphide. With night soil the percentage may be : Methane = 65%, Carbon dioxide = 34%, hydrogen sulphide = 0.6% and other gas 0.4%.

Diphasic Anaerobic Digestion

To tide over the problems generally associated with maintaining optimum fermentation conditions inside the digester, diphasic anaerobic digestion seems to be probable solution. In this system two separate digesters are used which separate the hydrogen and acid formation phase from methane formation phase ; this separation of phases is brought out by kinetic control. The acid phase is operated at low retention times so that only the fast growing acidogens of first hydrolysis and acid forming stages are retained in the digester and methogens are washed out. The methane phase takes place in a separate digester with the acid phase effluent as the feed material.

This system facilitates provision of optimum environment conditions for growth of two physiologically different types of bacteria which in turn ensures *higher efficiency of the process with subsequent reduction in the size of digester and its cost*. In this process, system monitoring is also rendered easier.

HIGHLIGHTS

1. A flat plate collector consists of the following four essential components :
   (*i*) An absorber plate
   (*ii*) Transparent covers
   (*iii*) Insulation beneath the absorber plate
   (*iv*) Box like structure.
2. Geothermal energy is the heat from high pressure steam coming from within the earth. The hydrothermal convective systems are heat resources for geothermal energy exploitation at present. Hot dry rock is also being considered.
3. The following are the biomass resources :
   (*i*) Concentrated wastes (*ii*) Dispensed waste residue
   (*iii*) Harvested biomass.
4. A Janata model gobar gas plant consists of the following parts :
   (*i*) Foundation (*ii*) Digester
   (*iii*) Dome (*iv*) Inlet chamber
   (*v*) Outlet chamber (*vi*) Mixing tank
   (*vii*) Gas outlet pipe.

THEORETICAL QUESTIONS

1. List the various non-conventional energy sources.
2. Explain with the help of a neat diagram a solar pond electric power plant.
3. Describe the working of a 'Solar tower plant'.
4. Give the working of a geothermal plant.
5. Explain with a neat sketch the construction and working of a Janata model gobar gas plant.

'GATE' and 'UPSC' Examinations' Questions (Latest-selected) with Answers/Solutions

30

(A) Conventional Questions with Answers/Solutions

(B) Multiple-choice Questions with Answers and "Explanations"

A. Conventional Questions with Answers/Solutions

Thermodynamics

Q. 30.1. *(a) State the Clausius inequality in words. (b) An inventor claims that he has developed a heat engine which absorbs 1200 kJ and 800 kJ from reservoirs at 800 K and 600 K respectively and rejects 600 kJ and 200 kJ as heat to reservoirs at 400 K and 300 K respectively. It delivers 1200 kJ work. Determine whether the heat engine is theoretically possible.* **(GATE)**

Solution. (*a*) In words, the "Clausius inequality" may be expressed as follows:

"When a system performs a reversible cycle, then

$$\sum_{cycle} \left(\frac{\delta Q}{T}\right) = 0,$$

but when the cycle is not reversible

$$\sum_{cycle} \left(\frac{\delta Q}{T}\right) < 0"$$

(*b*) *Given:* Q_1 = 1200 kJ; T_1 = 800 K; Q_2 = 800 kJ; T_2 = 600 K; Q_3 = – 600 kJ; T_3= 400 K; Q_4 = – 200 kJ; T_4 = 300 K.

Work delivered, W = 1200 kJ

Now as per 'Clausius inequality':

$$\sum_{cycle} \frac{\delta Q}{T} = \frac{Q_1}{T_1} + \frac{Q_2}{T_2} + \frac{Q_3}{T_3} + \frac{Q_4}{T_4}$$

$$= \frac{1200}{800} + \frac{800}{600} + \left(-\frac{600}{400}\right) + \left(-\frac{200}{300}\right)$$

$$= 0.666.$$

Since $\sum_{cycle} \frac{\delta Q}{T}$ = 0.666, is greater than zero, therefore, heat engine is theoretically *not possible*. **(Ans.)**

Q. 30.2. *A mass of m kg of fluid at temperature T_1 is mixed with an equal mass of same fluid at temperature T_2. Prove that the resultant change of entropy of universe is*

$$2mc \ln \frac{\left(\frac{T_1 + T_2}{2}\right)}{\sqrt{T_1 \times T_2}}$$

where, c = specific heat of fluid. **(IES)**

Solution. The resultant temperature (T) after mixing,

$$T = \frac{m_1 T_1 + m_2 T_2}{(m_1 + m_2)} = \frac{T_1 + T_2}{2}$$

$(\because \; m_1 = m_2 \text{ ...given})$

Assuming no loss, resultant change of entropy of universe,

$$\Delta S = mc \int_{T_1}^{T} \frac{dT}{T} + mc \int_{T_2}^{T} \frac{dT}{T}$$

$$= mc \ln\left(\frac{T}{T_1}\right) + mc \ln\left(\frac{T}{T_2}\right)$$

$$= mc \ln\left(\frac{T \times T}{T_1 T_2}\right) = mc \ln\left(\frac{T^2}{T_1 T_2}\right)$$

$$= mc \ln\left(\frac{T}{\sqrt{T_1 T_2}}\right)^2$$

Putting $T = \frac{T_1 + T_2}{2}$, in the above expression, we get:

$$\Delta S = 2mc \ln \frac{(T_1 + T_2)}{2\sqrt{T_1 T_2}} \text{ ...}$$ **Proved.**

Q. 30.3. *A heat source at 627°C transfer heat at the rate of 3000 kJ/min. to a system maintained at 287°C. A heat sink is available at 27°C. Assuming these temperatures to remain constant, find:*

(*i*) *Change in entropy of source.*

(*ii*) *Entropy production accompanying heat transfer.*

(iii) The original available energy.

(iv) The available energy after heat transfer. **(GATE)**

Solution. *Given :* $T_1 = 627 + 273 = 900$ K; $Q_1 = 3000$ kJ/min; $T_2 = 287 + 273 = 560$ K; $T_0 = 27 + 273 = 300$ K

(*i*) **Change in entropy of source, ΔS:**

$Q_1 = T_1 \cdot \Delta S$ or $3000 = 900 \times \Delta S$ or $\Delta S = 3.33$ kJ/K min (–).

Negative sign indicates the *decrease* in the entropy of source. **(Ans.)**

(*ii*) **Entropy production accompanying heat transfer, $\Delta S'$:**

$$T_1 \cdot \Delta S = T_2 \cdot \Delta S'$$

or,
$$\Delta S' = \frac{T_1 \cdot \Delta S}{T_2}$$

$$= \frac{3000}{560} = \mathbf{5.36\ kJ/K\ min.} \quad \textbf{(Ans.)}$$

(*iii*) **The original available energy:**

The original available energy

$$= \left(1 - \frac{T_0}{T_1}\right) Q_1$$

$$= \left(1 - \frac{300}{900}\right) \times 3000$$

$$= \mathbf{2000\ kJ/min.} \quad \textbf{(Ans.)}$$

(*iv*) **The available energy after heat transfer:**

The available energy after heat transfer

= Available energy before heat transfer – unavailable energy due to heat transfer

$= 2000 - T_0(\Delta S' - \Delta S)$

$= 2000 - 300(5.36 - 3.33)$

$= \mathbf{1391\ kJ/min.}$ **(Ans.)**

Q. 30.4. *An insulated tank of 1 m³ volume contains air at 0.1 MPa and 300 K. The tank is connected to high pressure line in which air at 1 MPa and 600 K flows. The tank is quickly filled with air by opening the valve between the tank and high pressure line. If the air pressure finally in tank is 1 MPa, determine the mass of air which enters the tank and entropy change associated with filling process.* **(IES)**

Solution. *Given:* $V_1 = 1$ m³; $p_1 = 0.1$ MPa = 1 bar; $T_1 = 300$ K; $p_p = 1$ MPa = 10 bar; $T_p = 600$ K

Mass of air which enters the tank, $(m_2 - m_1)$:

Initial mass of air in the tank, $m_1 = \dfrac{p_1 V_1}{R T_1}$

$$= \frac{(1 \times 10^5) \times 1}{287 \times 300} = 1.161 \text{ kg}$$

Final mass of air in the tank, $m_2 = \dfrac{p_2 V_2}{R T_2}$

or,
$$m_2 T_2 = \frac{(10 \times 10^5) \times 1}{287} = 3484.32$$

Now,
$$m_2 u_2 = m_2 \times (c_2 T_2)$$
$$= 3484.32 \times (0.718) = 2501.74$$

Internal energy of gas,

$$E_1 = m_1 u_1 + (m_2 - m_1)(u_p + 0)$$
$$E_2 = m_2 u_2$$

Change in internal energy,

$$\Delta E = m_2 u_2 - m_1 u_1 - (m_2 - m_1)(u_p)$$

Also, work done in filling process,

$$W = p_p (V_2 - V_1) = p_p[-(m_2 - m_1) v_p]$$
$$= -(m_2 - m_1) p_p v_p$$

$\therefore$ Heat involved, $Q = \Delta E + W$

$$= [m_2 u_2 - m_1 u_1 - (m_2 - m_1) u_p] - (m_2 - m_1) p_p v_p$$
$$0 = (m_2 u_2 - m_1 u_1) - (m_2 - m_1)(u_p + p_p v_p)$$
$$0 = (m_2 u_2 - m_1 u_1) - (m_2 - m_1)(h_p)$$

or,
$$0 = [2501.74 - (1.161 \times 0.718 \times 300)] - [(m_2 - 1.161)(1.005 \times 600)]$$

or,
$$2251.66 = 603 m_2 - 700.1$$

or,
$$m_2 = \frac{2251.66 + 700.1}{603} = 4.895 \text{ kg}$$

Mass of air which enters the tank $= m_2 - m_1$

$= 4.895 - 1.161 = \mathbf{3.734\ kg}$ **(Ans.)**

Entropy change, $S_2 - S_1$:

Temperature of the gases in tank, $T_2 = \dfrac{m_2 T_2}{m_2}$

$$= \frac{3484.32}{4.895} = 711.8 \text{ K}$$

Entropy change associated with filling process is given by

$$T\,dS = c_p\,dT - v \cdot dp$$

or,
$$S_2 - S_1 = \Delta m \left[c_p \ln\left(\frac{T_2}{T_1}\right) - R \ln\left(\frac{p_2}{p_1}\right)\right]$$

$$= \Delta m c_p \ln\left(\frac{T_2}{T_1}\right) \qquad (\because \ p_2 = p_1)$$

$$= 3.734 \times 1.005 \ln\left(\frac{711.8}{600}\right)$$

$$= \mathbf{0.6452\ kJ/K} \quad \textbf{(Ans.)}$$

Q. 30.5. *A reversible engine 'A' absorbs energy from a reservoir at temperature T_1 and rejects energy to a reservoir at temperature T_2. A second engine 'B' absorbs the same amount of energy as rejected by engine 'A', from the same reservoir at temperature T_2 and rejects energy to a reservoir at temperature T_3. What will be the relation between T_1, T_2 and T_3 if*

(*i*) *the efficiencies of both the engines 'A' and 'B' are the same and*

(*ii*) *the work delivered by both the engines is the same?* **(IES)**

Solution. Refer to Fig. 30.1.

Relation between T_1, T_2 and T_3 :

$$\text{Efficiency of engine } A,\ \eta_A = \frac{Q_1 - Q_2}{Q_1}$$

$$= \frac{T_1 - T_2}{T_1} \quad \text{... for reversible engine}$$

and $$W_A = Q_1 - Q_2$$

$$\text{Similarly, } \eta_B = \frac{Q_2 - Q_3}{Q_2} = \frac{T_2 - T_3}{T_2}$$

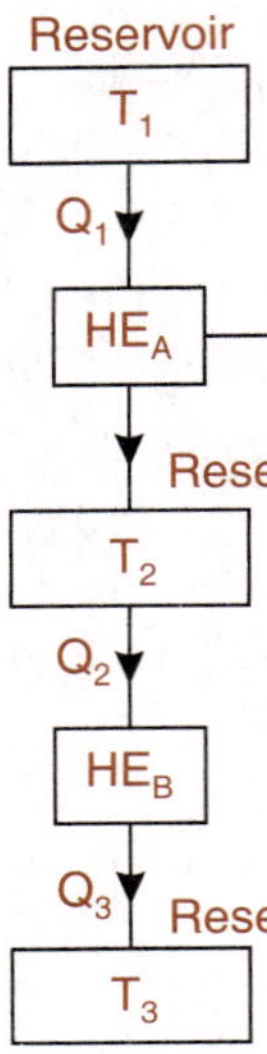

Fig. 30.1

(i) When the efficiencies of both the engines are equal:

$$\frac{T_1 - T_2}{T_1} = \frac{T_2 - T_3}{T_2}$$

or, $$T_1T_2 - T_2^2 = T_1T_2 - T_1T_3$$

or, $$T_2 = \sqrt{\mathbf{T_1T_3}} \quad \textbf{(Ans.)}$$

(ii) When the works delivered by both the engines is same:

$$W_A = W_B$$

or, $$Q_1 - Q_2 = Q_2 - Q_3$$

or, $$2Q_2 = Q_1 + Q_3$$

or, $$Q_2 = \frac{Q_1 + Q_3}{2}$$

For reversible case, using absolute temperature scale $\left(\frac{Q_1}{Q_2} = \frac{T_1}{T_2}\right)$, we have:

$$T_2 = \frac{\mathbf{T_1 + T_3}}{\mathbf{2}} \quad \textbf{(Ans.)}$$

Q. 30.6. *An ideal gas is heated at constant volume until its temperature is 3 times the original temperature, then it is expanded isothermally till it reaches its original pressure. The gas is then cooled at constant pressure till it is restored to the original state. Determine the net work done per kg of gas if the initial temperature is 350 K.* ***(GATE)***

Solution. *Given :* $T_1 = 350$ K; $T_2 = 3T_1 = 3 \times 350 = 1050$ K

Refer to Fig. 30.2.

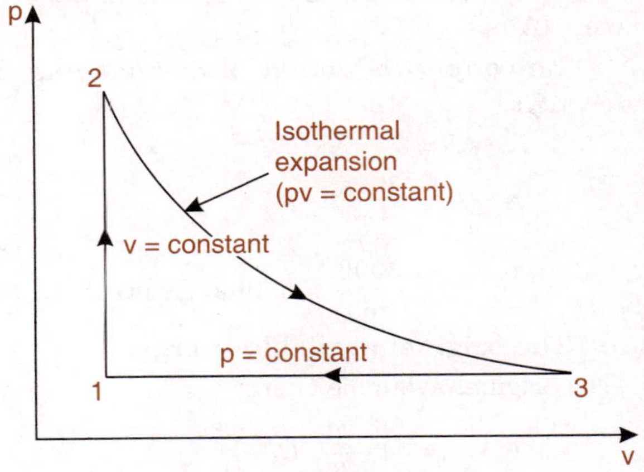

Fig. 30.2

Network done:

'Process 1–2' : *Constant volume process:*

$$\frac{p_1v_1}{T_1} = \frac{p_2v_2}{T_2} \text{ or } \frac{p_1}{T_1} = \frac{p_2}{T_2} \qquad (\because \ v_1 = v_2)$$

or, $$\frac{p_2}{p_1} = \frac{T_2}{T_1} = 3$$

Work done, $W_{1-2} = 0$, since there is no change in volume.

'Process 2–3' : *Isothermal expansion:*

$$p_2v_2 = p_3v_3 \qquad (\because \ T_2 = T_3)$$

or, $$p_2v_2 = p_1v_3 \qquad (\because \ p_3 = p_1)$$

or, $$\frac{v_2}{v_3} = \frac{p_1}{p_2} = \frac{1}{3}$$

$$\text{Work done, } W_{2-3} = \int_2^3 p\,dv$$

$$= \int_2^3 \frac{RT}{v}dv = RT_2 \ln\left(\frac{v_3}{v_2}\right)$$

$$= 287 \times 1050 \ln(3) \times 10^{-3}$$

$$= 331.1 \text{ kJ/kg}$$

'Process 3–1' : *Constant pressure process:*

$$\text{Work done, } W_{3-1} = \int_3^1 p\,dv = p(v_1 - v_3)$$

$$= p_1v_1 - p_3v_3$$

$$= R(T_1 - T_3)$$

$$= 287(350 - 1050) \times 10^{-3}$$

$$= -\,200.9 \text{ kJ/kg}$$

$$\text{Network done} = W_{1-2} + W_{2-3} + W_{3-1}$$

$$= 0 + 331.1 + (-\,200.9)$$

$$= \mathbf{130.2\ kJ/kg} \quad \textbf{(Ans.)}$$

Q. 30.7. *4 kg of water at 40°C are mixed with 6 kg of water at 100°C in a steady flow process. Calculate:*

(i) The temperature of resulting mixture,

(ii) The change in entropy, and

(iii) The unavailable energy with respect to the energy receiving water at 40°C. **(IES)**

Solution. *Given:* $m_1 = 4$ kg; $T_1 = 40 + 273 = 313$ K; $m_2 = 6$ kg; $T_2 = 100 + 273 = 373$ K

(*i*) **The temperature of resulting mixture, T_f:**

By energy balance, we get:

$$m_1 \times c_p \times (T_f - T_1) = m_2 \times c_p \times (T_2 - T_f)$$

or, $$4 \times (T_f - 313) = 6 \times (373 - T_f)$$

or, $$4T_f - 1252 = 2238 - 6T_f$$

or, $$T_f = \frac{2238 + 1252}{10}$$

$$= 349 \text{ K or } \mathbf{76°C} \quad \textbf{(Ans.)}$$

(*ii*) **Change in entropy, ΔS :**

$$\Delta S = m_1 c_p \ln\left(\frac{T_f}{T_1}\right) + m_2 c_p \ln\left(\frac{T_f}{T_2}\right)$$

$$= 4 \times 4.18 \ln\left(\frac{349}{313}\right) + 6 \times 4.18 \times \ln\left(\frac{349}{373}\right)$$

$$= 1.82 - 1.668$$

$$= \mathbf{0.152\ kJ/K} \quad \textbf{(Ans.)}$$

(*iii*) **Unavailable energy w.r.t. the energy receiving water at 40°C, $(\text{U.E.})_{40°C}$:**

Assuming room temperature as 20°C, for water at 40°C, the unavailable energy,

$$(\text{U.E.})_{40°C} = T_0(\Delta S)_{40°C}$$

$$= (20 + 273) \times mc_p \ln\left(\frac{349}{313}\right)$$

$$= 293 \times 4 \times 4.18 \times \ln\left(\frac{349}{313}\right)$$

$$= \mathbf{533.34\ kJ} \quad \textbf{(Ans.)}$$

Q. 30.8. *A house is to be maintained at 25°C in summer as well as in winter. For this purpose, it is proposed to use a reversible device as a refrigerator in summer and as a heat pump in winter. The ambient temperature is 40°C in summer and 3°C in winter. The energy losses as heat from the roof and the wall is estimated at 5 kW per degree Celsius temperature difference between the room and the ambient conditions. Calculate the power required to operate the device in summer and in winter.* **(IES)**

Solution. *Summer season :* $T_1 = 25 + 273 = 298$ K;

Ambient temperature, $T_2 = 40 + 273 = 313$ K;

Energy losses as heat from the roof and the wall = 5 kW/°C.

Refer to Figs. 30.3 and 30.4.

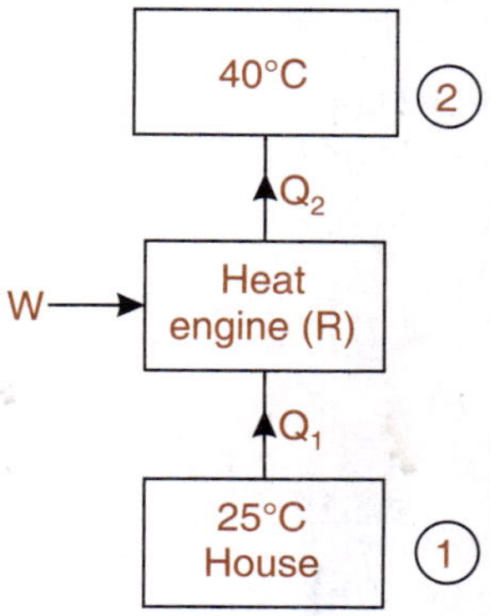

Fig. 30.3. Summer season.

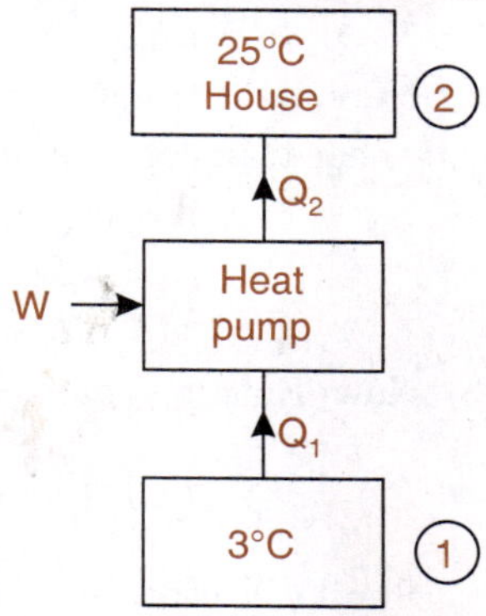

Fig. 30.4. Winter season.

Power required to operate the device in "Summer season", P :

For a reversible heat engine :

$$\frac{Q_1}{Q_2} = \frac{T_1}{T_2}$$

$$Q_1 = (\text{Energy loss/°C} \times \Delta T) \text{ kW}$$

$$= 5 \times (40 - 25) = 75 \text{ kW}$$

$\therefore$ $$\frac{75}{Q_2} = \frac{298}{313}$$

or, $$Q_2 = \frac{75 \times 313}{298} = 78.78 \text{ kW}$$

Hence, power required,

$$P = Q_2 - Q_1 = 78.78 - 75$$

$$= \mathbf{3.78\ kW} \quad \textbf{(Ans.)}$$

"Winter season" :

$$T_1 = 3 + 273 = 276 \text{ K};$$

$$T_2 = 25 + 273 = 298 \text{ K}$$

Now, $$Q_2 = 5 \times (25 - 3) = 110 \text{ kW}$$

Again, $$\frac{Q_1}{Q_2} = \frac{T_1}{T_2} = \frac{276}{298}$$

or, $$\frac{Q_1}{110} = \frac{276}{298} \text{ or } Q_1 = 101.88 \text{ kW}$$

Hence, power required,

$$P = Q_2 - Q_1$$

$$= 110 - 101.88 = \mathbf{8.12\ kW} \quad \textbf{(Ans.)}$$

Q. 30.9. *A heat pump operates between two identical bodies. In the beginning, both the bodies are at the same temperature T_1 but operation as of heat pump cools down one of the body to temperature T_2. Show that for the operation of heat pump the minimum work input needed by the heat pump for unit mass is given* $W_{min} = c\left[\frac{T_1^2}{T_2} + T_2 - 2T_1\right]$, *where c is the specific heat of bodies.* **(IES)**

Solution. Let, T_f = Final temperature attained by the body to which heat is rejected (*B*).

Now, work input, $W = Q_B - Q_A$

Here, heat transfer from body $A = c(T_1 - T_2)$

and heat transfer from body $B = c(T_f - T_1)$

$$\therefore \quad W = c(T_f - T_1) - c(T_1 - T_2)$$
$$= c[(T_f - T_1) - (T_1 - T_2)]$$
$$= c(T_f + T_2 - 2T_1) \quad ...(i)$$

Now, Entropy change of A,

$$\Delta S_A = \int_{T_1}^{T_2} c \cdot \frac{dT}{T} = c \cdot \ln\left(\frac{T_2}{T_1}\right)$$

Similarly, Entropy change of B,

$$\Delta S_B = \int_{T_1}^{T_f} c \cdot \frac{dT}{T} = c \cdot \ln\left(\frac{T_f}{T_1}\right)$$

For a *reversible* system, entropy change of universe = 0

$$\therefore \quad \Delta S_A + \Delta S_B = 0$$

or,
$$c \cdot \ln\left(\frac{T_2}{T_1}\right) + c \cdot \ln\left(\frac{T_f}{T_1}\right) = 0$$

or,
$$c \cdot \ln\left(\frac{T_2 T_f}{T_1^2}\right) = 0 \quad \text{or} \quad \frac{T_2 T_f}{T_1^2} = 1$$

or,
$$T_f = \frac{T_1^2}{T_2} \quad ...(ii)$$

Substituting eqn. (*ii*) in eqn. (*i*), we get :

$$W_{min} = c\left[\frac{T_1^2}{T_2} + T_2 - 2T_1\right] \ldots \quad \textbf{Proved.}$$

Q. 30.10. *An air compressor is used to fill rapidly a 3 m³ tank at 20°C and 1 atm. The filling process is governed by the law $pv^{1.4}$ = constant. The kinetic and potential energy effects are negligible. The ratio of the final to the initial mass of air in the tank is 4. Work out the following:*

(*i*) *Sketch the system and list the assumptions made.*

(*ii*) *Work input to the compressor.* **(IES)**

Solution. *Given:* V_f = 3 m³; T_1 = 20 + 273 = 293 K; p_1 = 1 atm = 1.0132 bar

Law : $pV^{1.4}$ = constant; $\frac{m_2}{m_1} = 4$

(*i*) **Sketch of the system and assumption made :**

Fig. 30.5 shows the system.

Assumption made :

1. Air enters at ambient conditions (T_{amb} = 293 K, p = 1 atm).
2. No heat exchange in the system.
3. K.E. and P.E. effects are negligible.

(*ii*) **Work input to the compressor, W_{input} :**

For adiabatic process:

$$p_1 V_1^{\gamma} = p_2 V_2^{\gamma}$$

or,
$$p_2 = p_1\left(\frac{V_1}{V_2}\right)^{\gamma} = p_1\left(\frac{m_2}{m_1}\right)^{\gamma}$$

$$\left[\begin{array}{l} \because \quad m_1 = \dfrac{V_1}{v_1}, \text{ or } v_1 = \dfrac{V_1}{m_1} \\ \quad m_2 = \dfrac{V_2}{v_2} \text{ or } v_2 = \dfrac{V_2}{m_2} \\ \therefore \quad \dfrac{v_1}{v_2} = \dfrac{V_1/m_1}{V_2/m_2} = \dfrac{m_2}{m_1} \\ \text{where, } v = \text{specific volume} \end{array}\right]$$

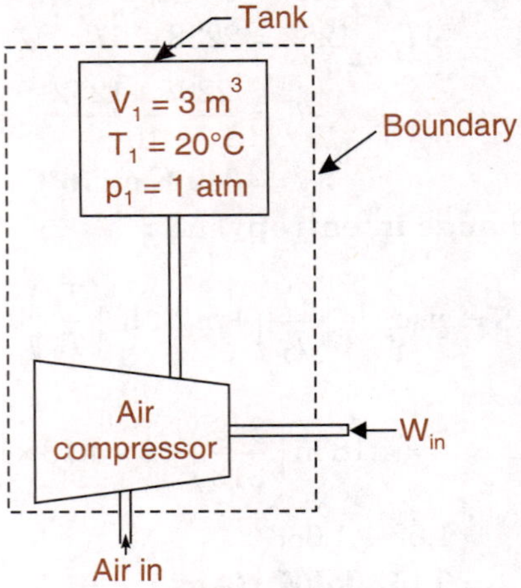

Fig. 30.5

or,
$$p_2 = 1.0132 \times 10^5 \times (4)^{1.4}$$
$$= 7.056 \times 10^5 \text{ N/m}^2$$

By applying energy equation, we get :

$$\frac{dE}{dt} = -\frac{dW}{dt} + \sum m_i(h_i)$$

On integration, we have:

$$\Delta E = -\Delta W + \Delta m(h_1)$$

or,
$$(m_2u_2 - m_1u_1) = -\Delta W + (m_2 - m_1)\, c_pT_1$$

or,
$$\Delta W = -(m_2u_2 - m_1u_1) + (m_2 - m_1)(c_pT_1)$$

or,
$$\Delta W = (-m_2c_vT_2 + m_1c_vT_1) + (m_2 - m_1)(c_pT_1) \ldots(i)$$

Also,
$$\frac{T_2}{T_1} = \left(\frac{p_2}{p_1}\right)^{\frac{\gamma-1}{\gamma}}$$

or,
$$T_2 = 293 \times \left(\frac{7.056 \times 10^5}{1.0132 \times 10^5}\right)^{\frac{1.4-1}{1.4}} = 510 \text{ K}$$

Initial mass in the tank, $m_1 = \dfrac{pV}{RT}$

$$= \frac{(1.0132 \times 10^5) \times 3}{287 \times 293} = 3.61 \text{ kg}$$

$$m_2 = 4\, m_1 = 4 \times 3.61 = 14.44 \text{ kg}$$

Inserting the values in eqn. (*i*), we get:

$$W_{in} (= \Delta W) = [(-14.44 \times 0.71 \times 510) + (3.61 \times 0.71 \times 293)] + [(14.44 - 3.61) \times 1.005 \times 293]$$
$$= -4477.7 + 3189.1$$
$$= \mathbf{1288.6\ kJ\ (-)}\ \textbf{(Ans.)}$$

(–ve sign indicates work done on the air).

Q. 30.11. *Steam enters a turbine at an enthalpy of 3300 kJ/kg and a velocity of 180 m/s. The steam comes out of turbine at an enthalpy of 2700 kJ/kg with a velocity of 120 m/s. At the condition of steady state, the turbine develops work equal to 550 kJ/kg of steam flowing through the turbine. The heat transfer between the turbine and its surroundings occurs at an average temperature of 370 K. The entropy of steam at inlet and exist of turbine are 6.932 kJ/kg K and 7.361 kJ/kg K, respectively. Neglecting the changes in potential energy between inlet and outlet, work out the following :*

Calculate the rate at which entropy is produced within the turbine per kg of steam flowing. **(IES)**

Solution. *Given :*

At inlet, h_1 = 3300 kJ/kg; C_1 = 180 m/s; s_1 = 6.932 kJ/kg K;

At exit, h_2 = 2700 kJ/kg; C_2 = 120 m/s; s_2 = 7.361 kJ/kg K; Turbine work, W = 550 kJ/kg.

Rate at which entropy is produced/generated, s_{gen} :

Refer to Fig. 30.6.

Applying energy equation for steady flow at inlet and exit, we get:

$$Q + \left(h_1 + \frac{C_1^2}{2}\right) = W + \left(h_2 + \frac{C_2^2}{2}\right)$$

or,
$$Q = W + (h_2 - h_1) + \left(\frac{C_2^2}{2} - \frac{C_1^2}{2}\right)$$

$$= 550 + (2700 - 3300) + \left[\left(\frac{120^2}{2} - \frac{180^2}{2}\right)10^{-3}\right]$$

$$= 550 - 600 - 9 = -59 \text{ kJ/kg-s}$$

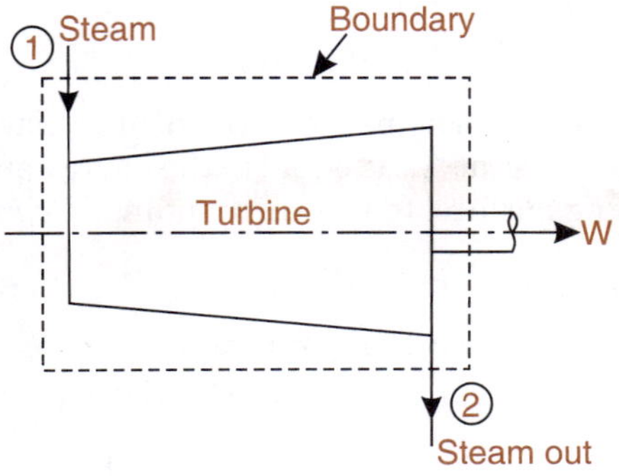

Fig. 30.6

Applying second law to the steady flow, we have:

$$\frac{ds}{dt} = \frac{Q}{T} + \sum m_i s_i - \sum m_0 s_0 + s_{gen}$$

For steady flow, $\frac{ds}{dt} = 0$

$$\therefore \quad s_{gen} = m_2 s_2 - m_1 s_1 - \frac{Q}{T}$$

$$= (7.361 - 6.932) - \left(-\frac{59}{370}\right)$$

$$= \mathbf{0.588\ kJ/kg\ K\text{-}s} \quad \textbf{(Ans.)}$$

Q. 30.12. *A finite thermal system having heat capacity, C = 0.04 T² J/K initially at 600 K. Estimate the maximum work obtainable from the thermal system if the surroundings is at 300 K.* **(GATE)**

Solution. *Given :* Heat capacity, $C = 0.04\ T^2$; T_1 = 600 K; T_2 = 300 K.

Maximum work obtainable, W:

For finite thermal system, $dQ = CdT$

or,
$$Q = \int_{300}^{600} 0.04T^2\, dT$$

$$= \left[\frac{0.04T^3}{3}\right]_{600}^{300} = \frac{1}{75}(600^3 - 300^3) \times 10^{-3}$$

$$= 2520 \text{ kJ}$$

Change in the entropy of the *body* = $\int \frac{dQ}{T}$

Change in entropy of *sink* = $\frac{Q - W}{T}$

For maximum work, considering process to be reversible, $\Delta S = 0$.

$$\text{Now, } \int_{600}^{300} C.\frac{dT}{T} = \int_{600}^{300} 0.04T\, dT$$

$$= \frac{0.04}{2}[(300)^2 - (600)^2] \times 10^{-3}$$

$$= -5.4 \text{ kJ for finite thermal system}$$

and
$$\frac{Q - W}{T} = \frac{2520 - W}{T}$$

$$\therefore \quad -5.4 + \frac{2520 - W}{300} = 0$$

or,
$$\frac{2520 - W}{300} = 5.4$$

or,
$$W = \mathbf{900\ kJ} \quad \textbf{(Ans.)}$$

Q. 30.13. *(a) State Gouy-Stodola theorem of irreversibility.*

(b) Estimate the irreversibility associated with the expansion process of air through a very small constriction in a pipe from pressure and temperature, respectively of 8 bar and 600 K to pressure of 1.2 bar. Assume air to be an ideal gas. The temperature of surroundings is 25°C. **(IES)**

Solution. (*a*) **Gouy-Stodola theorem of irreversibility:**

"It states that rate of loss of available energy or energy in a process is proportional to the rate of entropy generation (S_{gen})".

(Irreversibility, $I = T_0 \Delta S_{univ} = T_0 S_{gen.}$)

(*b*) *Given :* p_1 = 8 bar; T_1 = 600 K; p_2 = 1.2 bar; T_0 = 25 + 273 = 298 K.

Irreversibility, *I*:

Since the process is *similar to throttling*, therefore, the final temperature, $T_2 = T_1$ = 600 K and $\Delta Q = 0$

Extropy change in process:

$$(\Delta S)_{system} = -R \ln \left(\frac{p_2}{p_1}\right)$$

$$= -287 \times \ln \left(\frac{1.2}{8}\right) \times 10^{-3}$$

$$= -0.5445 \text{ kJ/kg K}$$

As, there is not heat transfer, $(\Delta S)_{surr.} = 0$

$$(\Delta S)_{universe} = (\Delta S)_{system} + (\Delta S)_{surr.}$$

$$= -5.445 + 0 = -0.5445 \text{ kJ/kg K}$$

∴ Irreversibility (for unit mass flow rate), $I = T_0 \Delta_{univ.}$

$$= 298 \times (-0.5445)$$

$$= \mathbf{162.26\ (-)\ kJ/kg\text{-}s}$$

(–ve sign means loss of available energy). **(Ans.)**

Q. 30.14. *The heat capacity at constant pressure of a certain system is a function of temperature only and may be expressed as*

$$C_p = 2.093 + \frac{41.87}{t + 100} \text{ J/°C}$$

where t is the temperature in °C. The system is heated while it is maintained at a pressure of 1 atmosphere until its volume increases from 2000 cm³ to 2400 cm³ and its temperature increases from 0°C to 100°C.

(i) Find the magnitude of heat interaction.

(ii) How much does the internal energy of the system increase? **(GATE)**

Solution. *Given :* $C_p = 2.093 + \dfrac{41.87}{t+100}$ J/°C; p = 1 atm. = 1.0132 bar; V_1 = 2000 cm³; V_2 = 2400 cm³; t_1 = 0°C; t_2 = 100°C.

(*i*) **Magnitude of heat interaction, ΔQ:**

For constant pressure process, $dQ = \int_{t_1}^{t_2} C_p \, dt$

or,
$$\Delta Q = \int_{t_1}^{t_2} \left(2.093 + \frac{41.87}{t+100}\right) dt$$

$$= 2.093\,(t_2 - t_1) + 41.87 \ln \left(\frac{t_2 + 100}{t_1 + 100}\right)$$

$$= 2.093\,(100 - 0) + 41.87 \ln \left(\frac{200}{100}\right)$$

$$= \mathbf{238.32\ kJ.} \quad \textbf{(Ans.)}$$

(*ii*) **Increase in internal energy, Δ*u*:**

Work done, $\Delta W = \int_1^2 p dV = p\int_1^2 dV = p(V_2 - V_1)$

$$= (1.0132 \times 10^5)\,(2400 - 2000) \times 10^{-6}$$

$$= 40.53 \text{ J}$$

∴ $\Delta u = u_2 - u_1 = \Delta Q - \Delta W = 238.32 - 40.53$

$$= \mathbf{197.79\ kJ.} \quad \textbf{(Ans.)}$$

Q. 30.15. *Consider an engine in outer space which operates on the Carnot cycle. The only way in which heat can be transferred from the engine is by radiation. The rate at which heat is radiated is proportional to the fourth power of the absolute temperature and to the area of the radiating surface. Show that for a given power output and a given T_1, the area of the radiator will be a minimum when $\dfrac{T_2}{T_1} = \dfrac{3}{4}$.* **(IES)**

Solution. Refer to Fig. 30.7.

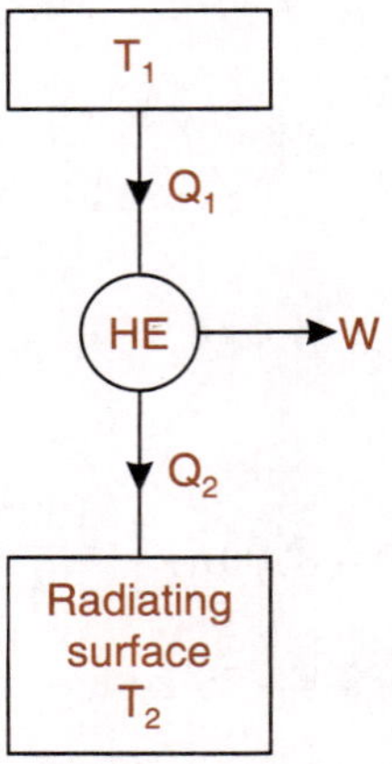

Fig. 30.7

For the heat engine (*HE*), the heat rejected Q_2 to the 'radiating surface' at 'T_2' is equal to the energy emitted from the 'radiating surface' to the surroundings by *radiation*.

$$Q_2 \propto AT_2^4 = kAT_2^4 .$$

where, k = Constant, and

A = Area of the 'radiating surface'.

Now, $$\eta = \frac{W}{Q_1} = \frac{T_1 - T_2}{T_1}$$

or, $$\frac{W}{T_1 - T_2} = \frac{Q_1}{T_1} = \frac{Q_2}{T_2} = \frac{kAT_2^4}{T_2} = kAT_2^3$$

or, $$A = \frac{W}{kT_2^3 (T_1 - T_2)} = \frac{W}{k (T_1 T_2^3 - T_2^4)}$$

For a given W and T_1, A will be minimum when,

$$\frac{dA}{dT_2} = -\frac{W}{k}\left[\frac{3T_1T_2^2 - 4T_2^3}{(T_1T_2^3 - T_2^4)^2}\right] = 0$$

Since $(T_1 T_2^3 - T_2^4) \neq 0$

$$\therefore \quad 3T_1T_2^2 - 4T_2^3 = 0 \text{ or } 3T_1 T_2^2 = 4T_2^3$$

or, $$\frac{T_2}{T_1} = \frac{\mathbf{3}}{\mathbf{4}}.$$ **(Ans.)**

Q. 30.16. *For an isentropic expansion of a gas with $c_p = a + kT$, $c_v = b + kT$ and $c_p - c_v = R$, show that $T^b v^{a-b} e^{kT}$ = constant.*

Solution. *Given :* $c_p = a + kT$; $c_v = b + kT$; $c_p - c_v = R$

Now, $c_p - c_v = (a + kT) - (b + kT) = (a - b) = R$

For an ideal gas:

$$\Delta S = c_v \frac{dT}{T} + R\frac{dV}{V}$$

$$= (b + kT)\frac{dT}{T} + (a - b)\frac{dV}{V}$$

$$= b\frac{dT}{T} + kdT + (a - b)\frac{dV}{V}$$

Since the expansion is 'isentropic', therefore,

$$b\frac{dT}{T} + kdT + (a - b)\frac{dV}{V} = 0$$

On integration, we get:

$$b \ln (T) + k(T) + (a - b) \ln (V) = \text{constant}$$

or, $$\ln T^b + \ln e^{kT} + \ln (V)^{a-b} = \text{constant}$$

or, $$T^b e^{kT} V^{a-b} = \text{constant.}$$ **Proved**

Applied Thermodynamics

Q. 30.17 (*Steam power plant*). *A steam power plant generating 500 MW of electrical power employs a natural circulation boiler which supplies steam at a pressure of 150 bars and temperature of 550°C. The condenser pressure is 0.05 bar. The turbine, mechanical and generator efficiencies are 87%, 98% and 99% respectively. The boiler uses pulverized coal having a calorific value of 26 MJ/kg and yields 92% efficiency. The feedwater passing through the feed heaters enters the boiler at 160°C. The risers of the furnace are 55 m high while the downcomers are placed outside the furnace for producing natural circulation. The quality of steam at the top of the riser is 12% and a minimum exit velocity of mixture leaving the risers and entering the drum is 1.4 m/s. The dimensions of the riser tubes are 60 mm OD and 3 mm wall thickness while the dimensions of downcomers are 185 mm OD and 8 mm thick. Assume no pressure drop and heat loss to the risers. Work out the following:*

(*i*) *Show the process on T-s diagram.*

(*ii*) *The generation of rate of steam ignoring the amount of steam bled off to feed heaters in kg/s.*

(*iii*) *The rate of fuel flow required in kg/s.*

(*iv*) *The evaporation factor.*

(*v*) *The circulation ratio.*

(*vi*) *The number of riser tubes.*

(*vii*) *The number of downcomers.*

Take the following properties:

$$h_1 = 3448.6 \text{ kJ/kg}, \; h_2 = 1987 \text{ kJ/kg}$$
$$h_{f_3} = 137.8 \text{ kJ/kg}, \; h_{f_5} = 675.5 \text{ kJ/kg}$$
$$\rho_{riser,\,top} = 396 \text{ kg/m}^3$$
$$\rho_{downcomer,\,inlet} = 603 \text{ kg/m}^3$$

where suffixes 1, 2, 3 and 5 denote boiler exit (or turbine inlet), condenser exit, pump inlet and boiler inlet respectively.

(IES)

Solution. *Given:* Power generated by the plant = 500 MW; p_1 = 150 bars; t_1 = 550°C; $p_{cond.}$ = 0.05 bar; η_t = 87%; η_m = 98%; η_g = 99%; C.V. = 26 MJ/kg; η_{boiler} = 92%

(*i*) Fig. 30.8 shows the **process on T-s diagram**

(*ii*) **Rate of steam generation, $\dot{m}_s$:**

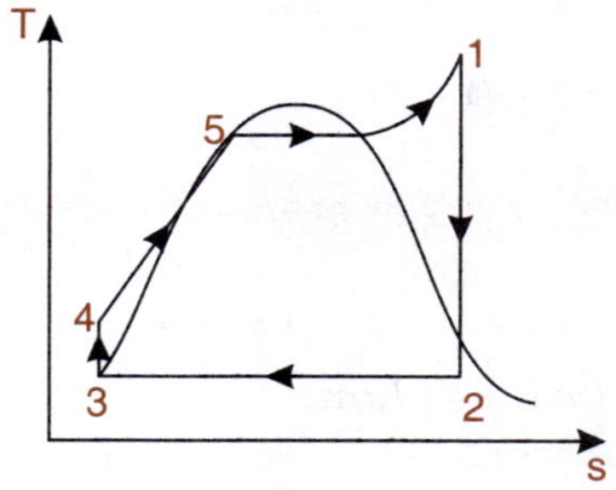

Fig. 30.8. T-s diagram.

Assuming isentropic expansion in the turbine and neglecting other losses, we have:

$$\dot{m}_s(h_1 - h_2)\,\eta_t\eta_m\eta_g = 500 \times 10^3$$

or $$\dot{m}_s = \frac{500 \times 10^3}{(3448.6 - 1987) \times 0.87 \times 0.98 \times 0.99}$$

$$= \mathbf{405.28 \text{ kg/s.}}$$ **(Ans.)**

(*iii*) **Rate of fuel flow, $\dot{m}_{fuel}$:**

$$\dot{m}_{fuel}(C.V.)_{fuel} \cdot \eta_{boiler} = \dot{m}_s(h_1 - h_{f5})$$

$$\dot{m}_{fuel} \times (26 \times 10^3) \times 0.92 = 405.28(3448.6 - 675.57)$$

or, $\dot{m}_{fuel}$ = **46.98 kg/s. (Ans.)**

(*iv*) **Evaporation factor:**

$$\text{Evaporation factor} = \frac{\dot{m}_s}{\dot{m}_{fuel}} = \frac{405.28}{46.98}$$

= **8.627 (Ans.)**

(*v*) **Circulation ratio, CR:**

$$CR = \frac{1}{\text{Quantity of steam at the top of the riser}(x_{top})}$$

$$= \frac{1}{x_{top}} = \frac{1}{0.12} = \mathbf{8.333}$$ **(Ans.)**

(*vi*) **Number of riser tubes:**

Let us first calculate rate of steam formation in one riser, $\dot{m}_{rs}$.

$$\dot{m}_{rs} = A_i \rho_{top} C_i x_{top} \quad \text{(where, } C \text{ stands for velocity)}$$

$$= \frac{\pi}{4}[(0.06 - 0.006)^2] \times 396 \times 1.4 \times 0.12$$

$$= 0.1524 \text{ kg/s}$$

$$\therefore \text{ Number of riser tubes} = \frac{\dot{m}_s}{\dot{m}_{rs}} = \frac{405.28}{0.1524}$$

= **2660 tubes (Ans.)**

(*vii*) **Number of downcomers:**

Let us first calculate the mass flow rate of feed water through one downcomer, $\dot{m}_{dw}$.

$$\dot{m}_{dw} = A_i \rho_i C_i = \frac{\pi}{4}[(0.185 - 0.016)^2] \times 603 \times 1.4$$

$$= 18.94 \text{ kg/s}$$

$$\therefore \text{ Number of downcomers} = \frac{\dot{m}_s}{\dot{m}_{dw}} = \frac{405.28}{18.94}$$

= 21.4 say **22 (Ans.)**

Q. 30.18 ***(Gas turbine plant).*** *In an open-cycle gas turbine plant, the air enters at 15°C and 1 bar, and is compressed in a compressor to a pressure ratio of 15. The air from the exit of compressor is first heated in a heat exchanger which is 75% efficient by turbine exhaust gas and then in a combustor to a temperature of 1600 K. The same gas expands in a two stage turbine such that the expansion work is maximum. The exhaust gas from h.p. turbine is reheated to 1500 K and then expands to l.p. turbine. The isentropic efficiencies of compressor and turbine may be taken as 86% and 88% respectively. The mechanical efficiencies for compressor and turbine are 97% each. The alternator efficiency is 98%. The output of turbo-alternator is 250 MW. Work out the following:*

(*i*) *Show the process on T-s diagram.*

(*ii*) *The cycle thermal efficiency.*

(*iii*) *The work ratio.*

(*iv*) *The specific power output.*

(*v*) *The mass flow rate of air.* **(IES)**

Solution. *Given:* T_1 = 15 + 273 = 288 K; p_1 = 1 bar; $\frac{p_2}{p_1}$ = 15; Effectiveness of heat exchange, ε = 0.75; T_3 = 1600 K; T_5 = 1500 K; $(\eta_{isen})_C$ = 86% ; $(\eta_{isen})_T$ = 88% ; $(\eta_{mech})_C$ = $(\eta_{mech})_T$ = 97%.

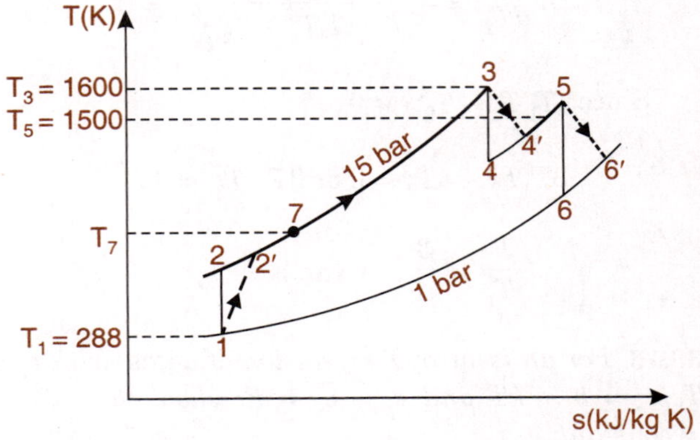

Fig. 30.9. T-s diagram.

(*i*) Fig. 30.9 shows the **process on *T-s* diagram.**

(*ii*) **Cycle thermal efficiency, $\eta_{Thermal}$:**

$$\frac{T_2}{T_1} = \left(\frac{p_2}{p_1}\right)^{\frac{\gamma-1}{\gamma}} = (15)^{\frac{1.4-1}{1.4}}$$

$$= 2.1678$$

or, $$T_2 = 288 \times 2.1678 = 624.33 \text{ K}$$

$$(\eta_{isen})_C = 0.86 = \frac{T_2 - T_1}{T_2' - T_1}$$

$$= \frac{624.33 - 288}{T_2' - 288}$$

or, $$T_2' = \left(\frac{624.33 - 288}{0.86}\right) + 288$$

$$= 679.1 \text{ K}$$

Now, $$p_4 = p_5 = \sqrt{p_1 p_2} = \sqrt{1 \times 15}$$

$$= 3.87 \text{ bar}$$

Again, $$\frac{T_4}{T_3} = \left(\frac{p_4}{p_3}\right)^{\frac{\gamma-1}{\gamma}} \quad \text{or} \quad \frac{T_4}{1600} = \left(\frac{3.87}{15}\right)^{\frac{1.4-1}{1.4}}$$

or, $$T_4 = 1086.5 \text{ K}$$

Similarity, $$\frac{T_6}{T_5} = \left(\frac{p_6}{p_5}\right)^{\frac{\gamma-1}{\gamma}} = \frac{T_6}{1500} = \left(\frac{1}{3.87}\right)^{\frac{1.4-1}{1.4}}$$

or, $$T_6 = 1018.9 \text{ K}$$

At $(\eta_{isen})_T = 88\%$, we have:

$$0.88 = \frac{T_3 - T_4'}{T_3 - T_4} = \frac{1600 - T_4'}{1600 - 1086.5}$$

or, $T_4' = 1600 - 0.88(1600 - 1086.5)$
$= 1148.1$ K

Again, $0.88 = \frac{T_5 - T_6'}{T_5 - T_6} = \frac{1500 - T_6'}{1500 - 1018.9}$

or, $T_6' = 1500 - 0.88(1500 - 1018.9)$
$= 1076$ K

Effectiveness, $\varepsilon = 0.75 = \frac{T_7 - T_2'}{T_6' - T_2'} = \frac{T_7 - 679.1}{1076.6 - 679.1}$

or, $T_7 = 0.75(1076.6 - 679.1) + 679.1$
$= 977.2$ K

$W_{turbine} = [c_p(T_3 - T_4') + c_p(T_5 - T_6')] \times \eta_{mech}$
$= [1.005(1600 - 1148.1)$
$+ 1.005(1500 - 1076.6)] \times 0.97$
$= 853.27$ kJ/kg

$W_{compressor} = \frac{c_p(T_2' - T_1)}{\eta_{mech}} = \frac{1.005(679.1 - 288)}{0.97}$
$= 405.21$ kJ/kg

$Q_{supplied}$, $Q_s = c_p(T_3 - T_7)$
$= 1.005(1600 - 977.2)$
$= 625.9$ kJ/kg

Hence, $\eta_{thermal} = \frac{W_{turbine} - W_{compressor}}{Q_s}$
$= \frac{853.29 - 405.21}{625.9}$
$= 0.7159$ or **71.59%** **(Ans.)**

(*iii*) **Work ratio:**

Work ratio $= \frac{W_{turbine} - W_{compressor}}{W_{turbine}}$
$= \frac{853.29 - 405.21}{853.29}$
$= 0.525$ or **52.5%** **(Ans.)**

(*iv*) **Specific power output:**

Specific power output = net work output
$= W_{turbine} - W_{compressor}$
$= 853.29 - 405.21$
$=$ **448.08 kJ/kg** **(Ans.)**

(*v*) **Mass flow rate of air, $\dot{m}_a$:**

$$\dot{m}_a \times \text{Net work output} = \frac{250 \times 10^3}{\eta_{alternator}} = \frac{250 \times 10^3}{0.98}$$

or, $\dot{m}_a \times 448.08 = \frac{250 \times 10^3}{0.98}$

or, $\dot{m}_a = \frac{250 \times 10^3}{448.08 \times 0.98}$
$=$ **569.32 kg/s** **(Ans.)**

Q. 30.19 ***(Gas turbine plant).*** *A two stage compression with intercooling in between stages and a single stage turbine with regeneration is employed in an open cycle gas turbine plant. Air at 1 bar and 15°C enters the compressor and the maximum pressure ratio is 5 and the maximum temperature in the cycle is 800°C. The rate of air flow through the cycle is 250 kg/sec and the calorific value of the fuel used is 42 MJ/kg. The isentropic efficiencies of both the compressors is 0.8 and the effectiveness of the regenerator is 0.7, and the isentropic efficiency of the turbine is 0.9. The combustion efficiency is 0.95, the mechanical efficiency is 0.96 and the generator is 0.75. Take c_p of air = 1.005 kJ/kg K and $\gamma = 1.4$ and for gases c_p = 1.08 kJ/kg K and $\gamma = 1.33$.*

Assuming perfect intercooling and neglecting pressure and heat losses, determine:

(*i*) *The air fuel ratio.*

(*ii*) *The cycle efficiency.*

(*iii*) *The specific fuel cosumption.* ***(IES)***

Solution. *Given:* $p_1 = 1$ bar; $T_1 = 15 + 273 = 288$ K; Max. pressure ratio, $\frac{p_5}{p_6} = 5$; $T_5 = 273 + 800 = 1073$ K; $\dot{m}_a = 250$ kg/s; C.V. = 42 MJ/kg or 42×10^3 kJ/kg; Isentropic efficiency of each compressor, $\eta_{comp.} = 0.8$; Effectiveness, $\varepsilon = 0.7$; Isentropic efficiency of turbine, $\eta_{turbine} = 0.9$; $\eta_{combustion} = 0.95$; $\eta_{mech.} = 0.96$; $\eta_{gen} = 0.75$, c_{pa} of air = 1.005 kJ/kg K; $\gamma_a = 1.4$ and for gases $c_{pg} = 1.08$ kJ/kg K and $\gamma_g = 1.33$.

Refer to Fig. 30.10

(*i*) **Air fuel ratio,** $\frac{\dot{m}_a}{\dot{m}_f}$ **:**

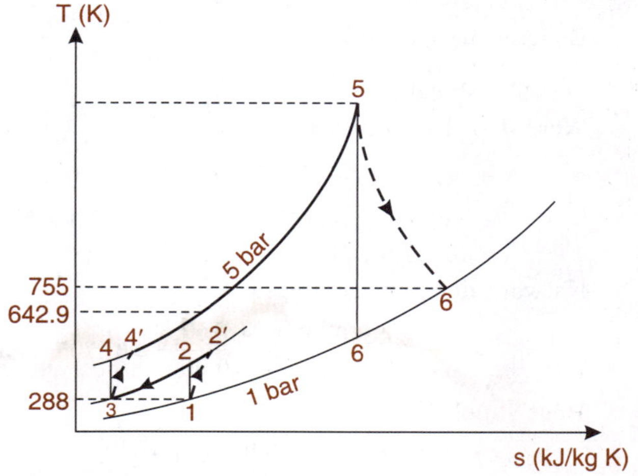

Fig. 30.10

Due to perfect intercooling, $p_2 = p_3 = \sqrt{p_1 p_4} = \sqrt{1 \times 5}$ = 2.23 bar.

Since the pressure ratio and isentropic efficiency of each compressor is same then the *work input required for each compressor is the same* since both the compressors have the same inlet temperature *(perfect intercooling), i.e.* $T_1 = T_3$ and $T_2' = T_4'$.

For compressors:

$$\frac{T_2}{T_1} = \left(\frac{p_2}{p_1}\right)^{\frac{\gamma_a - 1}{\gamma_a}} \quad \text{or} \quad \frac{T_2}{288} = (2.23)^{\frac{1.4-1}{1.4}}$$

or, $T_2 = 362.2$ K

Also, $(\eta_{isen})_{C\text{-}I} = \dfrac{T_2 - T_1}{T_2' - T_1}$ or $0.8 = \dfrac{362.2 - 288}{T_2' - 288}$

or, $T_2' = 380.7$ K

Similarly, $\dfrac{T_4}{T_3} = \left(\dfrac{p_4}{p_3}\right)^{\frac{\gamma_a - 1}{\gamma_a}}$ or $\dfrac{T_4}{288} = \left(\dfrac{5}{2.23}\right)^{\frac{1.4-1}{1.4}}$

or, $T_4 = 362.7$ K

Again, $(\eta_{isen})_{C\text{-}II} = \dfrac{T_4 - T_3}{T_4' - T_3}$

or, $0.8 = \dfrac{362.7 - 288}{T_4' - 288}$

or, $T_4' = 381.4$ K

For turbine:

$$\frac{T_5}{T_6} = \left(\frac{p_5}{p_6}\right)^{\frac{\gamma_g - 1}{\gamma_g}} \quad \text{or} \quad \frac{1073}{T_6} = \left(\frac{5}{1}\right)^{\frac{1.33-1}{1.33}}$$

$T_6 = 719.7$ K

$$(\eta_{isen})_T = 0.9 = \frac{T_5 - T_6'}{T_5 - T_6} = \frac{1073 - T_6'}{1073 - 719.7}$$

or, $T_6' = 755$ K

Effectiveness, $\varepsilon = 0.7 = \dfrac{T_7 - T_4'}{T_6' - T_4'} = \dfrac{T_7 - 381.4}{755 - 381.4}$

or, $T_7 = 642.9$ K

Work done by compressor = $W_{C\text{-}I} + W_{C\text{-}II}$

$= \dot{m}_a\, c_{pa}(T_2' - T_1) + \dot{m}_a\, c_{pa}(T_4' - T_3)$

$= 250 \times 1.005[(380.7 - 288) + (381.4 - 288)]$

$= 46757.6$ kJ/s

Network done by compressor,

$$= \frac{46757.6}{\eta_{mech}} = \frac{46757.6}{0.96} = 48705.8 \text{ kJ/s}$$

Heat supplied,

$Q_s = \dot{m}_f \times \text{C.V.} \times \eta_{combustion}$

$= \dot{m}_f \times (42 \times 10^6) \times 0.95$

$= 39.9 \times 10^3\, \dot{m}_f$

By energy balance equation, we have:

$Q_s = (\dot{m}_a + \dot{m}_f)\, c_{pg}\, T_5 - \dot{m}_a\, c_{pa}\, T_7$

$= (250 + \dot{m}_f) \times 1.08 \times 1073 - 250 \times 1.005 \times 642.9$

or, $Q_s = (1158.8\, \dot{m}_f + 128181.4)$

$\therefore$ $1158.8\, \dot{m}_f + 128181.4 = 39.9 \times 10^3\, m_f$

or, $\dot{m}_f = \dfrac{128181.4}{[(39.9 \times 10^3) - 1158.8]} = 3.31$ kg/s

Hence, $\dfrac{\dot{m}_a}{\dot{m}_f} = \dfrac{250}{3.31} = $ **75.53** **(Ans.)**

(*ii*) **Cycle efficiency, η_{cycle}:**

$W_{net} = W_{turbine} - W_{compressor}$

$W_{turbine} = (\dot{m}_a + \dot{m}_f)\, c_{pg}(T_5 - T_6')$

$= (250 + 3.31) \times 1.08\ (1073 - 755)$

$= 86996.8$ kW

Network done by turbine $= 86996.8 \times \eta_{gen}$

$= 86996.8 \times 0.75 = 65247.6$ kW

$\therefore$ $W_{net} = 65247.6 - 46757.6 = 18490$ kW

Hence, $\eta_{cycle} = \dfrac{W_{net}}{Q_s} = \dfrac{18490}{\dot{m}_f \times \text{C.V.} \times \eta_{combustion}}$

$$= \frac{18490}{3.31 \times (42 \times 10^3) \times 0.95}$$

= **0.14 or 14%** **(Ans.)**

(*iii*) **Specific fuel consumption:**

Specific fuel consumption $= \dfrac{\dot{m}_f \times 3600}{18490}$

$= \dfrac{3.31 \times 3600}{18490} =$ **0.644 kg/kWh** **(Ans.)**

Q. 30.20 ***(Gas turbine plant).*** *Air is taken in a gas turbine plant at 1.1 bar 20°C. The plant comprises L.P. and H.P. compressors and L.P. and H.P. turbines. The compression in L.P. stage is upto 3.3 bar followed by intercooling to 27°C. The pressure of air after H.P. compressor is 9.45 bar. Loss in pressure during intercooling is 0.15 bar. Air from H.P. compressor is transferred to heat exchanger of effectiveness 0.65 where it is heated by the gases from L.P. turbine. After heat exchanger the air passes through combustion chamber. The temperature of gases supplied to H.P. turbine is 700°C. The gases expand in H.P. turbine to 3.62 bar and air then reheated to 670°C before expanding in L.P. turbine. The loss of pressure in reheater is 0.12 bar. Determine :*

(i) The overall efficiency.

(ii) The work ratio.

(iii) Mass flow rate when the power generated is 6000 kW.

Assume: Isentropic efficiency of compression in both stages = 0.82.

Isentropic efficiency of expansion in turbines = 0.85.

For air : c_p = 1.005 kJ/kg K, γ = 1.4.

For gases : c_p = 1.15 kJ/kg K, γ = 1.33.

Neglect the mass of fuel. ***(IES)***

Solution. *Given :* $T_1 = 20 + 273 = 293$ K; $p_1 = 1.1$ bar;

$p_2 = 3.3$ bar; $T_3 = 27 + 273 = 300$ K;

$p_3 = 3.3 - 0.15 = 3.15$ bar; $p_4 = p_6 = 9.45$ bar;

$T_6 = 973$ K;

$T_8 = 670 + 273 = 943$ K; $p_8 = 3.5$ bar;

$\eta_{compressors} = 82\%$;

$\eta_{turbines} = 85\%$; Power generated = 6000 kW;

Effectiveness, $\varepsilon = 0.65$; $c_{pa} = 1.005$ kJ/kg K;

$\gamma_{air} = 1.44$; $c_{pg} = 1.15$ kJ/kg K and $\gamma_{gases} = 1.33$.

Refer to Fig. 30.11.

$$\frac{T_2}{T_1} = \left(\frac{p_2}{p_1}\right)^{\frac{\gamma-1}{\gamma}} = \left(\frac{3.3}{1.1}\right)^{\frac{1.4-1}{1.4}} = 1.369$$

$$\therefore \quad T_2 = 293 \times 1.369 = 401 \text{ K}$$

$$\eta_{compressor\ (L.P.)} = 0.82 = \frac{T_2 - T_1}{T_2' - T_1} = \frac{401 - 293}{T_2' - 293}$$

$$\therefore \quad T_2' = \left(\frac{401 - 293}{0.82}\right) + 293 = 425 \text{ K}$$

Again,
$$\frac{T_4}{T_3} = \left(\frac{p_4}{p_3}\right)^{\frac{\gamma-1}{\gamma}} = \left(\frac{9.45}{3.15}\right)^{\frac{1.4-1}{1.4}} = 1.369$$

$$\therefore \quad T_4 = 300 \times 1.369 = 411 \text{ K}$$

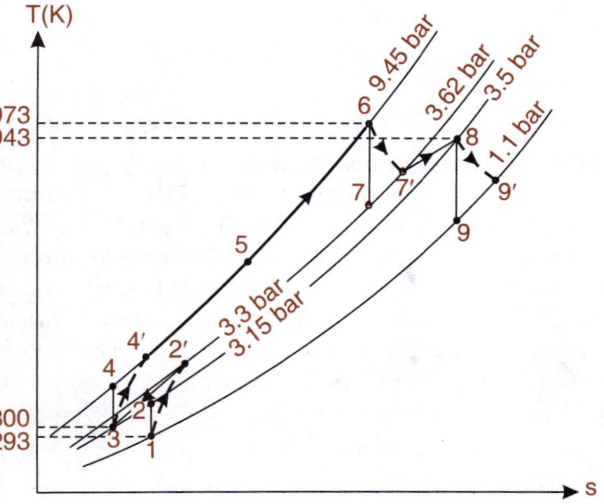

Fig. 30.11

Now,
$$\eta_{compressor\ (H.P.)} = \frac{T_4 - T_3}{T_4' - T_3}$$

$$0.82 = \frac{411 - 300}{T_4' - 300}$$

$$\therefore \quad T_4' = \left(\frac{411 - 300}{0.82}\right) + 300 = 435 \text{ K}$$

Similarly,
$$\frac{T_6}{T_7} = \left(\frac{p_6}{p_7}\right)^{\frac{\gamma-1}{\gamma}} = \left(\frac{9.45}{3.62}\right)^{\frac{1.33-1}{1.33}} = 1.268$$

$$\therefore \quad T_7 = \frac{T_6}{1.268} = \frac{973}{1.268} = 767 \text{ K}$$

Also,
$$\eta_{turbine\ (H.P.)} = \frac{T_6 - T_7'}{T_6 - T_7}$$

$$0.85 = \frac{973 - T_7'}{973 - 767}$$

$$\therefore \quad T_7' = 973 - 0.85(973 - 767) = 798 \text{ K}$$

Again,
$$\frac{T_8}{T_9} = \left(\frac{p_8}{p_9}\right)^{\frac{\gamma-1}{\gamma}} = \left(\frac{3.5}{1.1}\right)^{\frac{1.33-1}{1.33}} = 1.332$$

$$\therefore \quad T_9 = \frac{T_8}{1.332} = \frac{943}{1.332} = 708 \text{ K}$$

$$\eta_{turbine\ (L.P.)} = \frac{T_8 - T_9'}{T_8 - T_9}$$

$$0.85 = \frac{943 - T_9'}{943 - 708}$$

$$\therefore \quad T_9' = 943 - 0.85(943 - 708) = 743 \text{ K.}$$

Effectiveness of heat exchanger,

$$\varepsilon = 0.65 = \frac{T_5 - T_4'}{T_9' - T_4'}$$

i.e.,
$$0.65 = \frac{T_5 - 435}{743 - 435}$$

$$\therefore \quad T_5 = 0.65(743 - 435) + 435 = 635 \text{ K}$$

$W_{turbine\ (H.P.)} = c_{pg}\,(T_6 - T_7')$

$= 1.15(973 - 798) = 201.25$ kJ/kg of gas

$W_{turbine\ (L.P.)} = c_{pg}(T_8 - T_9')$

$= 1.15(943 - 743) = 230$ kJ/kg of gas

$W_{compressor\ (L.P.)} = c_{pa}(T_2' - T_1)$

$= 1.005(425 - 293) = 132.66$ kJ/kg of air

$W_{compressor\ (H.P.)} = c_{pa}(T_4' - T_3)$

$= 1.005(435 - 300) = 135.67$ kJ/kg of air

Heat supplied $= c_{pg}(T_6 - T_5) + c_{pg}(T_8 - T_7')$

$= 1.15(973 - 635) + 1.15(943 - 798)$

$= 555.45$ kJ/kg of gas

(i) **Overall efficiency, $\eta_{overall}$:**

$$\eta_{overall} = \frac{\text{Net work done}}{\text{Heat supplied}}$$

$$= \frac{[W_{turbine\,(H.P.)} + W_{turbine\,(L.P.)}] - [W_{comp.\,(L.P.)} + W_{comp.\,(H.P.)}]}{\text{Heat supplied}}$$

$$= \frac{(201.25 + 230) - (132.66 + 135.67)}{555.45}$$

$$= \frac{162.92}{555.45}$$

= 0.293 or **29.3%.** **(Ans.)**

(*ii*) **Work ratio:**

$$\text{Work ratio} = \frac{\text{Network done}}{\text{Turbine work}}$$

$$= \frac{[W_{turbine(H.P.)} + W_{turbine(L.P.)}] - [W_{comp.(L.P.)} + W_{comp.(H.P.)}]}{[W_{turbine(H.P.)} + W_{turbine(L.P.)}]}$$

$$= \frac{(201.25 + 230) - (132.66 + 135.67)}{(201.25 + 230)}$$

$$= \frac{162.92}{431.25} = 0.377$$

i.e., *Work ratio* = **0.377** **(Ans.)**

(*iii*) **Mass flow rate, $\dot{m}$:**

Network done = 162.92 kJ/kg.

Since mass of fuel is neglected, for 6000 kW, mass flow rate,

$$\dot{m} = \frac{6000}{162.92} = 36.83 \text{ kg/s}$$

i.e., *Mass flow rate* = **36.83 kg/s.** **(Ans.)**

Q. 30.21 ***(Gas turbine).*** *A gas turbine utilizes two-stages centrifugal compressor. The pressure ratios for the first and second stages are 2.5 and 2.1 respectively. The flow of air is 10 kg/s, this air being drawn at 1.013 bar and 20°C. If the temperature drop in the intercooler is 60°C and the isentropic efficiency is 90% for each stage, calculate:*

(i) The actual temperature at the end of each stage.

(ii) The total compressor power.

Assume $\gamma = 1.4$ and $c_p = 1.005$ kJ/kg K for air.

(*IES*)

Solution. *Given:* $\frac{p_2}{p_1} = 2.5$; $\frac{p_3}{p_2} = 2.1$; $\dot{m} = 10$ kg/s; $p_1 = 1.013$ bar; $T_1 = 20 + 273 = 293$ K; Temperature drop in intercooler $T_2 - T_3 = 60°C$; $\eta_{isen} = 90\%$; $\gamma = 1.4$;

$c_p = 1.005$ kJ/kg K for air.

(*i*) **Actual temperature at the end of each stage, T_2, T_3:**

For isentropic compression:

$$\frac{T_2'}{T_1} = \left(\frac{p_2}{p_1}\right)^{\frac{\gamma-1}{\gamma}} \quad \text{or} \quad \frac{T_2'}{293} = (2.5)^{\frac{1.4-1}{1.4}}$$

or, $T_2' = 380.68$ K

Also, $\eta_{isen} = \frac{T_2' - T_1}{T_2 - T_1}$ or $0.9 = \frac{380.68 - 293}{T_2 - 293}$

or, $0.9(T_2 - 293) = 380.68 - 293 = 87.68$

or, $T_2 = \frac{87.68}{0.9} + 293$ = **390.42 K** **(Ans.)**

and, $T_3 = 390.42 - 60$ = **330.42 K** **(Ans.)**

(*ii*) **Total compressor power:**

Compressor power

$$= \dot{m}c_p[(T_2 - T_1) + (T_4 - T_3)] \quad \ldots(i)$$

Now, $\frac{T_4'}{T_3} = \left(\frac{p_3}{p_2}\right)^{\frac{\gamma-1}{\gamma}}$

or, $\frac{T_4'}{330.42} = (2.1)^{\frac{1.4-1}{1.4}}$

or, $T_4' = 330.42\,(2.1)^{\frac{0.4}{1.4}} = 408.44$ K

Also, $0.9 = \frac{T_4' - T_3}{T_4 - T_3} = \frac{408.44 - 330.42}{T_4 - 330.42}$

$$= \frac{78.02}{T_4 - 330.42}$$

or, $0.9(T_4 - 330.42) = 78.02$ or $T_4 = 417.11$ K

Insolving the values in eqn. (*i*), we get:

Compressor power

$= 10 \times 1.005[(390.42 - 293) + (417.11 - 330.42)]$

= **1850.3 kW** **(Ans.)**

Q. 30.22 ***(Centrifugal compressor).*** *A centrifugal compressor running at 16000 rpm takes in air at 17°C and compresses it through a pressure ratio of 4 with an isentropic efficiency of 82 per cent. The blades are radially inclined and the slip factor is 0.85. Guide vanes at inlet give the air an angle of prewhirl of 20° to the axial direction. The mean diameter of the impeller eye is 200 mm and the absolute air velocity at inlet is 120 m/s. Calculate the impeller tip diameter.* **(*IES*)**

Solution. *Given:* $N = 16000$ rpm; $T_1 = 17 + 223 = 290$ K; Pressure ratio, $r_p = 4$; $\eta_{isen} = 82\%$; Slip factor, $\phi_s = 0.85$,

Inlet prewhirl angle = 20°(to axis direction); Mean impeller diameter, $D_1 = 200$ mm = 0.2 m; Absolute air velocity at inlet, $C_1 = 120$ m/s

Refer to Fig. 30.12.

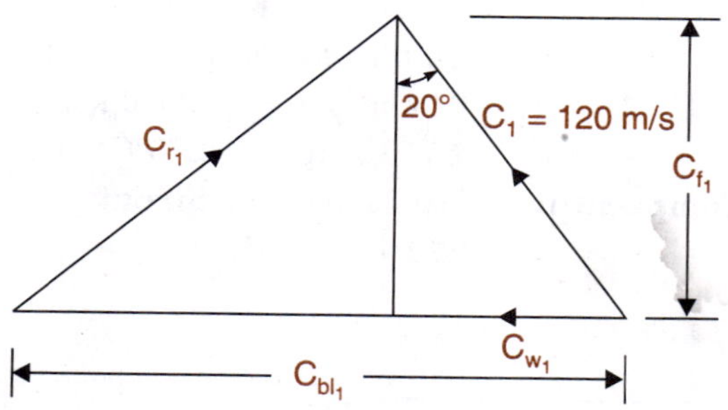

Fig. 30.12

Impeller up diameter D_1:

Velocity of flow at inlet,

$$C_{f_1} = 120 \cos 20° = 112.76 \text{ m/s}$$

Velocity of whirl at inlet,

$$C_{w_1} = 120 \sin 20° = 41.04 \text{ m/s}$$

Peripheral velocity at inlet, $C_{bl_1} = \dfrac{\pi D_1 N}{60}$

$$= \frac{\pi \times 0.2 \times 16000}{60}$$

$$= 167.55 \text{ m/s}$$

Slip factor, $\phi_s = 0.85 = \dfrac{C_{w_2}}{C_{bl_2}}$

or, $\quad C_{w_2} = 0.85\, C_{bl_2}$

Now, $\quad \dfrac{T_2'}{T} = \left(\dfrac{p_2}{p_1}\right)^{\frac{\gamma-1}{\gamma}}$ or $\dfrac{T_2'}{290} = (4)^{\frac{1.4-1}{0.4}}$

or, $\quad T_2' = 430.9 \text{ K}$

$$\eta_{isen} = 0.82 = \frac{T_2' - T_1}{T_2 - T_1} = \frac{430.9 - 290}{T_2 - 290}$$

or, $\quad T_2 = 290 + \dfrac{430.9 - 290}{0.82} = 461.8 \text{ K}$

Work done $= c_p(T_2 - T_1) = 1.005(461.8 - 290)$

$= 172.66$ kJ/kg $\quad$ ...(*i*)

Also, $\quad$ Work done $= (C_{w_2} C_{bl_2} - C_{w_1} C_{bl_1})$

$= (0.85\, C_{bl_2} \times C_{bl_2} - 41.04 \times 167.55)$

$= 0.85\, C_{bl_2}^2 - 6876.25$ $\quad$ ...(*ii*)

Equating equations (*i*) and (*ii*), we get:

$$0.85\, C_{bl_2}^2 - 6876.25 = 172.66 \times 10^3$$

$$C_{bl_2} = \left(\frac{6876.25 + 172.66 \times 10^3}{0.85}\right)^{1/2}$$

$= 459.6$ m/s

But, $\quad C_{bl_2} = \dfrac{\pi D_2 N}{60}$ or $459.6 = \dfrac{\pi D_2 \times 16000}{60}$

or, $\quad D_2 = \dfrac{459.6 \times 60}{\pi \times 16000} = 0.548$ m

or, $\quad$ **= 548 mm $\quad$ (Ans.)**

Q. 30.23 ***(Reciprocating compressor).*** *A single-stage, double-acting compressor has a free air delivery (F.A.D.) of 14 m³/min. measured at 1.013 bar and 15°C. The pressure and temperature in the cylinder during induction are 0.95 bar 32°C. The delivery pressure is 7 bar and index of compression and expansion, n = 1.3. The clearance volume is 5% of the swept volume. Calculate :*

(*i*) *Indicated power required ;*

(*ii*) *Volumetric efficiency.* **(GATE)**

Solution. Free air delivery (F.A.D.) = 14 m³/min. (measured at 1.013 bar and 15°C)

Induction pressure, $p_1 = 0.95$ bar

Induction temperature, $T_1 = 32 + 273 = 305$ K

Delivery pressure, $p_2 = 7$ bar

Index of compression and expansion, $n = 1.3$

Clearance volume, $V_3 = V_c = 0.05\, V_s$

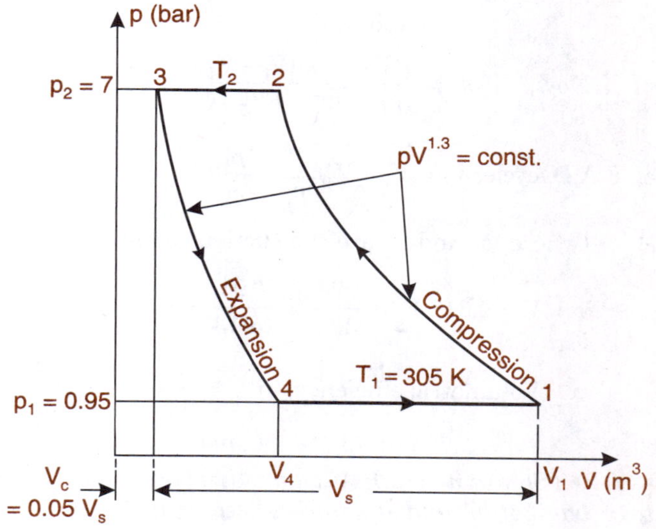

Fig. 30.13

Refer to Fig. 30.13.

(*i*) **Indicated power :**

Mass delivered per minute,

$$m = \frac{pV}{RT} = \frac{1.013 \times 10^3 \times 14}{0.287 \times 288 \times 10^3} = 17.16 \text{ kg/min.}$$

[where F.A.D. per minute is V at p(= 1.013 bar) and T(15 + 273 = 288 K)]

To find T_2, using the equation,

$$\frac{T_2}{T_1} = \left(\frac{p_2}{p_1}\right)^{\frac{n-1}{n}}$$

$$\therefore \quad T_2 = T_1\left(\frac{p_2}{p_1}\right)^{\frac{n-1}{n}}$$

$$= 305 \times \left(\frac{7}{0.95}\right)^{\frac{1.3-1}{1.3}} = 483.5 \text{ K}$$

$$\text{Indicated power} = \frac{n}{n-1}\, mR(T_2 - T_1)$$

$$= \frac{1.3}{1.3-1} \times 17.16 \times 0.287\,(483.5 - 305)$$

$$= 3809.4 \text{ kJ/min.}$$

$$\therefore \text{ Indicated power} = \frac{3809.4}{60} = \mathbf{63.49\ kW} \quad \textbf{(Ans.)}$$

(*ii*) **Volumetric efficiency:**

Using the relation,

$$\frac{V_4}{V_3} = \left(\frac{p_3}{p_4}\right)^{1/n} = \left(\frac{p_2}{p_1}\right)^{1/1.3} = \left(\frac{7}{0.95}\right)^{1/1.3} = 4.65$$

$$\therefore \quad V_4 = 4.65 \times V_3 = 4.65 \times 0.05\, V_s = 0.233\, V_s$$

$$\therefore \quad V_1 - V_4 = V_1 - 0.233 V_s = 1.05\, V_s - 0.233\, V_s = 0.817\, V_s$$

Now, $$m = \frac{pV}{RT} = \frac{p_1(V_1 - V_4)}{RT_1}$$

i.e., F.A.D./cycle, $V = (V_1 - V_4)\dfrac{T}{T_1} \cdot \dfrac{p_1}{p}$

(where, p_1 and T_1 are the suction conditions)

$$\therefore \quad V = 0.817\, V_s \times \frac{288}{305} \times \frac{0.95}{1.013} = 0.723\, V_s$$

$$\therefore \quad \text{Volumetric efficiency, } \eta_{vol.} = \frac{V}{V_s} = \frac{0.723\, V_s}{V_s}$$

$$= 0.723 \text{ or } \mathbf{72.3\%}. \quad \textbf{(Ans.)}$$

Q. 30.24 ***(Steam nozzle).*** *A steam nozzle is supplied steam at 15 bar 350°C and discharges steam at 1 bar. If the diverging portion of the nozzle is 80 mm long and the throat diameter is 6 mm, determine the cone angle of the divergent portion. Assume 12% of the total available enthalpy drop is lost in friction in the divergent portion. Also determine the velocity and temperature of the steam at throat.* ***(IES)***

Solution. p_1 = 15 bar, 350°C; p_3 = 1 bar; $k = 1 - 0.12 = 0.88$

When steam supplied to the nozzle is *superheated*, the pressure at throat,

$$p_2 = 0.546\, p_1 = 0.546 \times 15 = 8.19 \text{ bar}$$

Refer to Fig. 30.14

From Mollier chart :

$$h_1 = 3150 \text{ kJ/kg},\ h_2 = 2992 \text{ kJ/kg}$$
$$v_2 = 0.24 \text{ m}^3\text{/kg},\ h_3 = 2580 \text{ kJ/kg}$$
$$v_3' = 1.75 \text{ m}^3\text{/kg},\ t_2 = 270°\text{C}$$

i.e., **Temperature of the steam at throat = 270°C. (Ans.)**

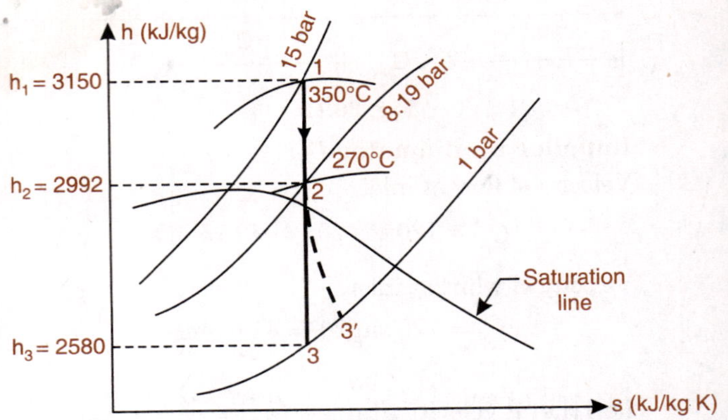

Fig. 30.14

Velocity of steam at throat:

$$C_2 = 44.72\sqrt{h_d} = 44.72\sqrt{(h_1 - h_2)}$$

$$= 44.72\sqrt{(3150 - 2992)}$$

$$= \mathbf{562.12\ m/s} \quad \textbf{(Ans.)}$$

From the conditions at nozzle throat, mass flow rate,

$$\dot{m} = \frac{A_2 C_2}{v_2} = \frac{\pi/4\,(6/1000)^2 \times 562.12}{0.24}$$

$$= 0.0662 \text{ kg/s}$$

At exit :

$$C_3' = 44.72\sqrt{kh_d'}$$

$$= 44.72\sqrt{0.88 \times (h_1 - h_3)}$$

$$= 44.72\sqrt{0.88 \times (3150 - 2580)}$$

$$= 1001.5 \text{ m/s}$$

$$\text{Exit area of the nozzle, } A_3 = \frac{\dot{m} \times v_3'}{C_3'}$$

$$= \frac{0.0662 \times 1.75}{1001.5} = 0.0001156 \text{ m}^2$$

i.e., $$\frac{\pi}{4} D_3^2 = 0.0001156$$

$$\therefore \quad D_3 = \left(\frac{4 \times 0.0001156}{\pi}\right)^{1/2}$$

$$= 0.012 \text{ m or } 12.1 \text{ mm}$$

If θ be the cone angle of nozzle,

$$\tan\theta = \frac{(12.1 - 6)}{2 \times 80} = 0.03812 \quad \text{or} \quad \theta = 2°\ 11'$$

Thus, **cone angle** = 2 × 2° 11′ = **4° 22′**. **(Ans.)**

Q. 30.25 ***(Impulse turbine).*** *The first stage of a turbine is a two-row velocity compounded impulse wheel. The steam velocity at inlet is 600 m/s and the mean blade velocity is 120 m/s. The nozzle angle is 16° and the exit angles for the first-row of moving blades, the fixed blades, and the second row of moving blades are 18°, 21° and 35° respectively.*

(i) Calculate the blade inlet angles for each row.

(ii) Calculate also for each row of moving blades, the driving force and the axial thrust on the wheel for a mass flow of 1 kg/s.

(iii) Calculate the diagram efficiency for the wheel and the diagram power per kg/s steam flow.

(iv) What would be the maximum possible diagram efficiency for the given steam inlet velocity and nozzle angle ?

Take the blade velocity co-efficient as 0.9 for all blades. **(IES)**

Solution. Refer to Fig. 30.15.

$$\alpha = 16°,\ \phi = 18°,\ C_1 = 600 \text{ m/s},\ C_{bl} = 120 \text{ m/s}$$

$$\alpha' = 21°,\ \phi' = 35°,\ \dot{m}_s = 1 \text{ kg/s},$$

Blade velocity co-efficient, $K = 0.9$

With the above data velocity triangles can be drawn.

From the diagram (*by measurement*)

$$C_w = C_{w_1} + C_{w_0} = 875 \text{ m/s};$$

$$C_w' = C'_{w_1} + C'_{w_0} = 294 \text{ m/s}$$

$$C_{f_1} = 168 \text{ m/s},\ C_{f_0} = 135 \text{ m/s};$$

$$C'_{f_1} = 106 \text{ m/s},\ C'_{f_0} = 97 \text{ m/s}$$

(*i*) **Blade inlet angles:**

First row: $\theta =$ **20°** (*moving blade*)

$\beta =$ **24.5°** (*fixed blade*)

Second row: $\theta' =$ **34.5°** (*moving blade*). **(Ans.)**

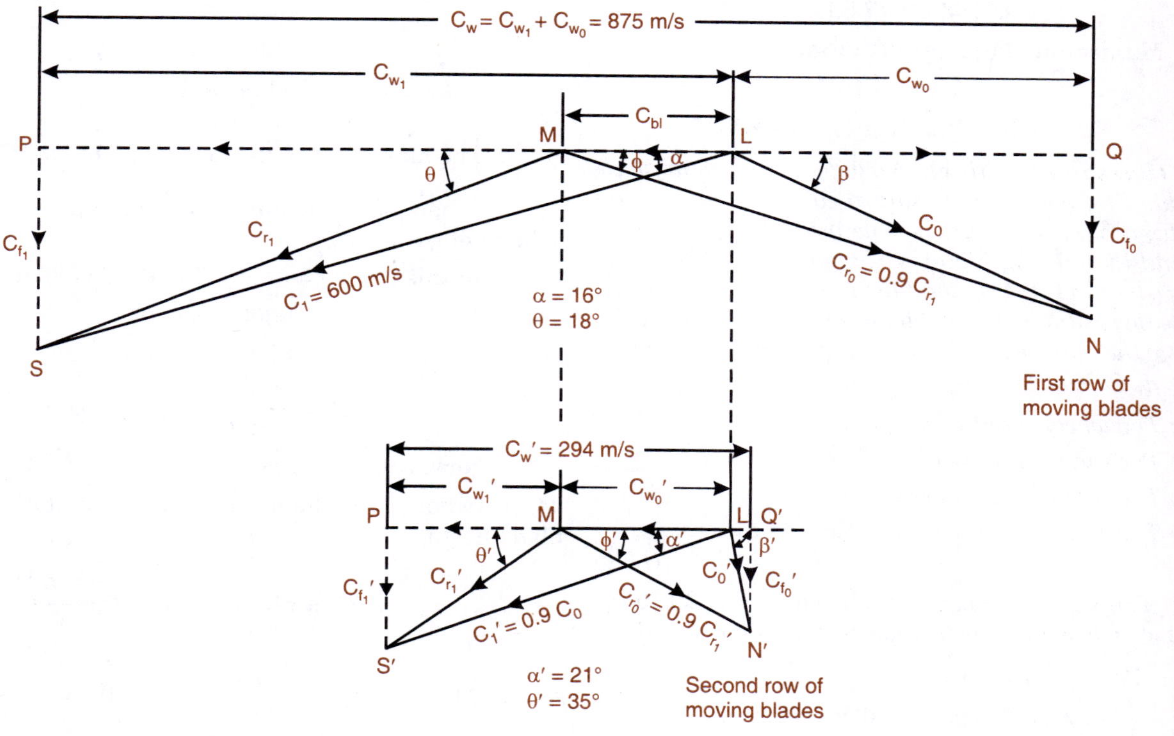

Fig. 30.15

(*ii*) **Driving force:**

First row of moving blades

$$= \dot{m}_s(C_{w_1} + C_{w_0}) = \dot{m}_s C_w$$

$$= 1 \times 875 = \mathbf{875\ N} \quad \textbf{(Ans.)}$$

Second row of moving blades

$$= \dot{m}_s(C_{w_1}' + C_{w_0}') = \dot{m}_s C_w'$$

$$= 1 \times 294 = \mathbf{294\ N} \quad \textbf{(Ans.)}$$

Axial thrust :

First row of moving blades

$= \dot{m}_s(C_{f_1}' + C_{f_0}') = 1 \times (168 - 135)$

$= 33$ N

Second row of moving blades

$= \dot{m}_s(C_{f_1}' - C_{f_0}') = 1 \times (106 - 97)$

$= 9$ N

Total axial thrust = 33 + 9 = **42 N per kg/s** **(Ans.)**

(*iii*) **Power developed** $= \dfrac{\dot{m}_s(C_w + C_w')\,C_{bl}}{1000}$

$= \dfrac{1 \times (875 + 294) \times 120}{1000}$

= **140.28 kW per kg/s.**

Diagram efficiency $= \dfrac{2C_{bl}(C_w + C_w')}{C_1^2}$

$= \dfrac{2 \times 120\,(875 + 294)}{(600)^2}$

= **0.7793 or 77.93%** **(Ans.)**

(*iv*) **Maximum diagram efficiency**

$= \cos^2 \alpha = \cos^2 16°$

= 0.924 or **92.4%.** **(Ans.)**

Q. 30.26 ***(Parson's turbine).*** *A stage of a turbine with Parson's blading delivers dry saturated steam at 2.7 bar from the fixed blades at 90 m/s. The mean blade height is 40 mm, and the moving blade exit angle is 20°. The axial velocity of steam is 3/4 of the blade velocity at the mean radius. Steam is supplied to the stage at the rate of 9000 kg/h. The effect of the blade tip thickness on the annulus area can be neglected. Calculate:*

(*i*) *The wheel speed in r.p.m..*

(*ii*) *The diagram power.*

(*iii*) *The diagram efficiency.*

(*iv*) *The enthalpy drop of the steam in this stage.*

(IES)

Solution. The velocity diagram is shown in Fig. 30.16 (*a*) and the blade wheel annulus is represented in Fig. 30.16 (*b*).

Pressure = 2.7 bar, $x = 1$,

$C_1 = 90$ m/s, $h = 40$ mm = 0.04 m

$\alpha = \phi = 20°$, $C_{f_1} = C_{f_0} = 3/4\ C_{bl}$

Rate of steam supply = 9000 kg/h.

(*i*) **Wheel speed, *N* :**

$C_f = 3/4\ C_{bl} = C_1 \sin 20°$

$= 90 \sin 20° = 30.78$ m/s

$\therefore \quad C_{bl} = 30.78 \times 4/3 = 41.04$ m/s

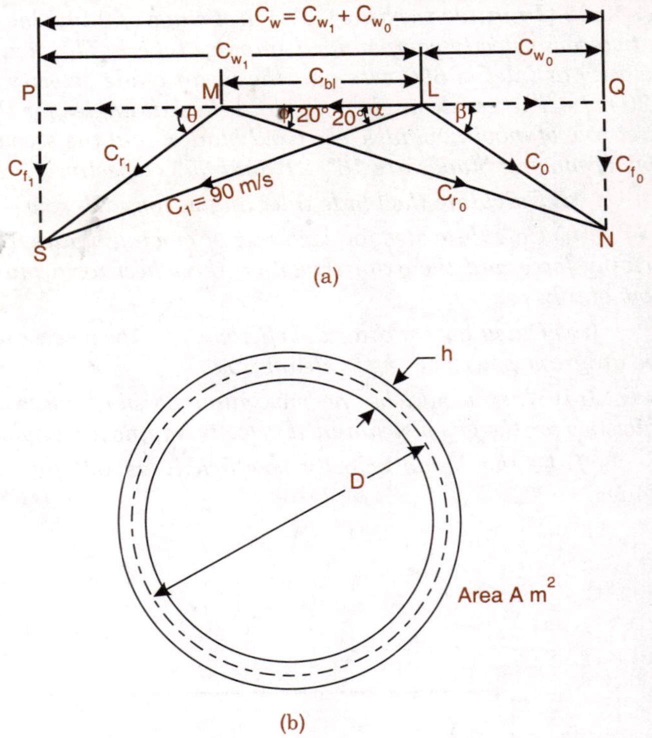

(a)

(b)

Fig. 30.16

The mass flow of steam is given by: $\dot{m}_s = \dfrac{C_f A}{v}$

(where, A is the annulus area, and v is the specific volume of the steam).

In this case, $v = v_g$ at 2.7 bar = 0.6686 m³/kg

$$\therefore \quad \dot{m}_s = \frac{9000}{3600} = \frac{30.78}{0.6686}$$

or, $$A = \frac{9000 \times 0.6686}{3600 \times 30.78} = 0.054 \text{ m}^2$$

Now, annulus area, $A = \pi D h$

(where, D is the mean diameter, and h is the mean blade height)

$$\therefore \quad 0.054 = \pi D \times 0.04 \quad \text{or} \quad D = \frac{0.054}{\pi \times 0.04} = 0.43 \text{ m}$$

Also, $$C_{bl} = \frac{\pi D N}{60} \quad \text{or} \quad 41.04 = \frac{\pi \times 0.43 \times N}{60}$$

or, $$N = \frac{41.04 \times 60}{\pi \times 0.43} = \textbf{1823 rpm}$$ **(Ans.)**

(*ii*) **The diagram power:**

Diagram power $= \dot{m}_s\ C_w\ C_{bl}$

Now, $C_w = 2C_1 \cos \alpha - C_{bl}$

$= 2 \times 90 \times \cos 20° - 41.04$

$= 128.1$ m/s

$\therefore$ *Diagram power* $= \dfrac{9000 \times 128.1 \times 41.04}{3600 \times 1000}$

$= \mathbf{13.14\ kW}$ **(Ans.)**

(*iii*) **The diagram efficiency:**

Rate of doing work per kg/s $= C_w\, C_{bl} = 128.1 \times 41.04$ N m/s

Also, energy input to the moving blades per stage

$$= \frac{C_1^2}{2} + \frac{C_{r_0}^2 - C_{r_1}^2}{2} = \frac{C_1^2}{2} + \frac{C_1^2 - C_{r_1}^2}{2}$$

$$= C_1^2 - \frac{C_{r_1}^2}{2} \qquad (\because\ C_{r_0} = C_1)$$

Referring to Fig. 30.16, we have:

$$C_{r_1}^2 = C_1^2 + C_{bl}^2 - 2C_1\, C_{bl} \cos \alpha$$

$= 90^2 + 41.04^2 - 2 \times 90 \times 41.04 \times \cos 20°$

$= 8100 + 1684.28 - 6941.69$

or, $C_{r_1} = 53.3$ m/s

Energy input $= 90^2 - \dfrac{53.3^2}{2} = 6679.5$ Nm per kg/s

$\therefore$ Diagram efficiency $= \dfrac{128.1 \times 41.04}{6679.5}$

$= 0.787$ or **78.7%.** **(Ans.)**

(*iv*) **Enthalpy drop in the stage:**

Enthalpy drop in the moving blades

$$= \frac{C_{r_0}^2 - C_{r_1}^2}{2} = \frac{90^2 - 53.3^2}{2 \times 1000}$$

$= 2.63$ kJ/kg $\qquad (\because\ C_{r_0} = C_1)$

$\therefore$ Total enthalpy drop per stage $= 2 \times 2.63$

$= \mathbf{5.26\ kJ/kg}$ **(Ans.)**

Q. 30.27. *(a) How do 'fire tube boilers' differ from water tube boilers?*

(b) How are the impurities in feed water removed from the boiler? ***(GATE)***

Solution (*a*) The comparison between the 'fire tube boilers' and 'water tube boiler' is give below.

S.No.	*Particulars*	*Fire-tube boilers*	*Water-tube boilers*
1.	*Position of water and hot gases*	Hot gases inside the tubes and water outside the tubes.	Water inside the tubes and hot gases outside the tubes.
2.	*Mode of firing*	Generally internally fired.	Externally fired.
3.	*Operating pressure*	Operating pressure limited to 16 bar.	Can work under as high pressure as 100 bar.
4.	*Rate of steam production*	Lower	Higher.
5.	*Suitability*	Not suitable for large power plants.	Suitable for large power plants.
6.	*Risk on bursting*	Involves lesser risk on explosion due to lower pressure.	Involves more risk on bursting due to high pressure.
7.	*Floor area*	For a given power it occupies more floor area.	For a given power it occupies less floor-area.
8.	*Construction*	Difficult	Simple
9.	*Transportation*	Difficult	Simple
10.	*Shell diameter*	Large for same power	Small for same power
11.	*Chances of explosion*	Less	More
12.	*Treatment of water*	Not so necessary	More necessary
13.	*Accessibility of various parts*	Various parts not so easily accessible for cleaning, repair and inspection.	Various parts are more accessible.
14.	*Requirement of skill*	Require less skill for efficient and economic working.	Require more skill and careful attention.

(*b*) Before feeding the water (as make-up water) to the boiler it is *adequately treated.* Still, it has some impurities in the form of dissolved solids (TDS). These solids are expressed in ppm(parts per million). In order to *maintain a certain ppm in the drum, 'blow down' is necessary. Trisodium phosphate* or *tanin* is injected in suitable dose periodically to help precipate the salts at the bottom of the drum.

Q. 30.28. *Defind air rate, specific power and the cycle work ratio in a gas turbine. What is the significance of these parameters?* **(IES)**

Solution. (*i*) **Air rate.** It is defined as *air supplied per second in a gas turbine cycle for combustion of fuel.*

(*ii*) **Specific power.** The *power output per unit mass of air supplied per second* is termed as "specific power".

(*iii*) **Cycle work ratio.** The 'work ratio' in a gas turbine cycle is defined as the *ratio of net work output to the gross work output.*

$$\text{In other words, Work ratio} = \frac{\text{Network output}}{\text{Gross work output}}$$

$$= \frac{\text{Work of expansion } (W_T) - \text{Work of compression } (W_C)}{\text{Work of expansion } (W_T)}$$

or, $$\text{Work ratio} = \frac{W_T - W_C}{W_T}$$

Significance of the parameters:

(*i*) "*Air rate*" is significant in deciding the *size of the compressor*

(*ii*) "*Specific power*" decides the *efficiency of the combustion process* and also conveys whether complete combustion is taking place or not.

(*iii*) "*Work ratio*" is significant in deciding the ratio between $(W_T - W_C)$ and (W_T) and hence *efficiency of the cycle.*

Q. 30.29. *Using a layout diagram describe a gas turbine plant for turbo-jet.* **(IES)**

Solution. Fig. 30.17 shows a turbo-jet unit.

- It consists of *diffuser* at entrance which slows down the air (entering at velocity equal to the plane speed) and part of the kinetic energy of the air stream is converted into pressure ; this type of compression is called as *ram compression*.
- The air is further compressed to a pressure of 3 to 4 bar in a rotary compressor (usually of axial flow type).
- The compressed air then enters the combustion chamber (C.C.) where fuel is added. The combustion of fuel takes place at sensibly constant pressure and subsequently temperature rises rapidly.
- The hot gases then enter the gas turbine where *partial expansion* takes place. The *power produced is just sufficient to drive the compressor, fuel pump and other auxiliaries*.
- The exhaust gases from the gas turbine which are at a higher pressure than atmosphere are expended in a nozzle and a very high velocity jet is produced which provides a forward motion to the air-craft by the jet reaction (Newton's third law of motion).

At higher speeds the turbo-jet gives higher propulsion efficiency. The turbo-jets are most suited to the air-crafts travelling *above 800 km/h*.

The overall efficiency of a turbo-jet is the product of the thermal efficiency of the gas turbine plant and the propulsive efficiency of the jet (nozzle).

Advantages of Turbo-jet engines

1. Construction much simpler (as compared to multi-cylinder piston engine of comparable power).
2. Engine vibrations absent.
3. Much higher speeds possible (more than 3000 km/h achieved).
4. Power supply is uninterrupted and smooth.
5. Weight to power ratios superior (as compared to that of reciprocating type of aero-engine).

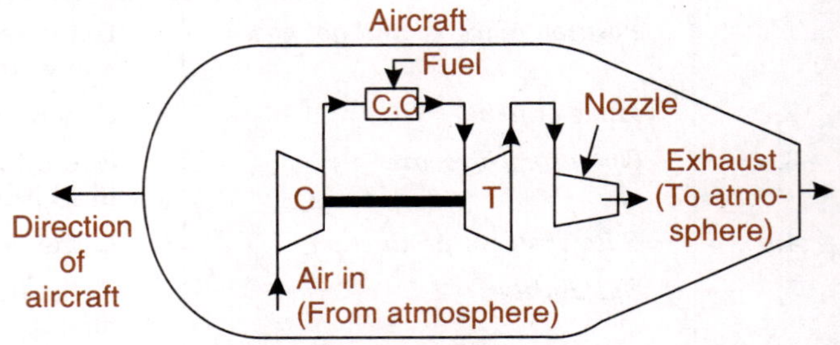

Fig. 30.17. Gas turbine plant for turbo-jet.

6. Rate of climb higher.
7. Requirement of major overhauls less frequent.
8. Radio interference much less.
9. Maximum altitude ceiling as compared to turbo-prop and conventional piston type engines.
10. Frontal area smaller.
11. Fuel can be burnt over a large range of mixture strength.

Disadvantages of turbo-jet engines

1. Less efficient.
2. Life of the unit comparatively shorter.
3. The turbo-jet becomes rapidly inefficient below 550 km/h.
4. More noisy (than a reciprocating engine).
5. Materials required are quite expensive.
6. Require longer strip since length of take-off is too much.
7. At take-off the thrust is low, this effect is overcome by boosting.

Q. 30.30 (*a*) *What is the objective of steam turbine governing. Enumerate the various methods of governing. Give the comparison of 'Throttle and Nozzel control governing'.*

(*b*) *What is hydro-mechanical speed governing system (prime control).* **(IES)**

Solution. (*a*) The *objective of governing is to keep the turbine speed fairly constant irrespective of load. The principal methods* of steam turbine governing are as follows:

1. Throttle governing
2. Nozzle governing
3. By-pass governing
4. Combination of 1 and 2 and 1 and 3.

Comparison of Throttle and Nozzle control governing

S.No	Aspects	Throttle control	Nozzle control
1.	*Throttling losses*	Severe	No throttling losses (Actually there are a little throttling losses in Nozzles valves which are partially open).
2.	*Partial admission losses*	Low	High.
3.	*Heat drop available*	Lesser	Larger
4.	*Use*	Used in impulse and reaction turbines both.	Used in impulse and also in reaction (if initial stage impulse) turbines.
5.	*Suitability*	Small turbines	Medium and larger turbines.

(*b*) The 'hydro-mechanical speed governing system (prime control)' uses a centrifugal governor (mechanical) to sense the speed of the turbine and the reference is set for rated speed of the turbine. The governor senses the speed through the accessory drive shaft and fly weights. Initially, the governor hydraulically amplifies the error signal and provides an actual output position that can be linked to a fuel valve. The fuel valve either throttles or by-passes the metered fuel from the main fuel control. The speed sensing device may be electrical or hydraulic.

Q. 30.31. *What is the effect of impeller blade shape on performance of centrifugal compressors.* **(IES)**

Solution. The following shapes of blades are utilized in the impellers of centrifugal compressors:

1. Backward-curved blades ($\beta_2 < 90°$)

2. Radial-curved blades ($\beta_2 = 90°$)

3. Forward-curved blades ($\beta_2 > 90°$)

Fig. 30.18 shows the relative performance of these blades. Centrifugal effects on the curved blades create a bending moment and produce increased stresses which reduce the maximum speed at which the impeller can run.

- Normally backward blades/vanes with β_2 between 20–25° are employed except in cases where high head is the major consideration.
- Sometimes compromise is made between the low energy transfer (backward-curved vanes) and high outlet velocity (forward-curved vanes) by using radial vanes.

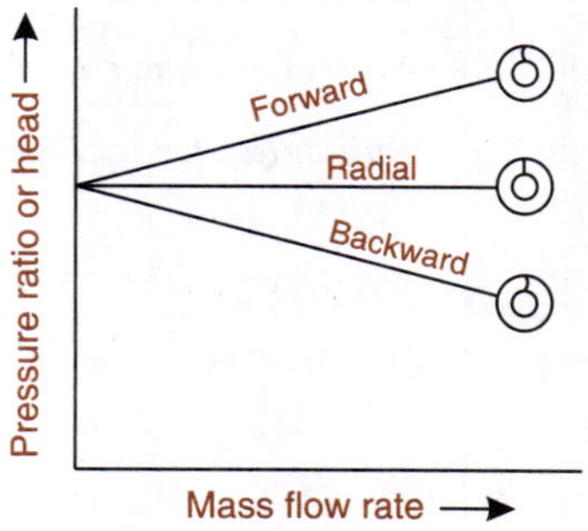

Fig. 30.18. Characteristics of backward-curved, radial-curved and forward-curved vanes.

Advantages of radial-blade impellers:

1. Can be manufactured easily.
2. Lowest unit blade stress for a given diameter and rotational speed, hence highest weight.
3. Free from complex bending stresses.
4. Equal energy conversion in impeller and diffuser, giving high pressure ratios with good efficiency.

In view of the above reasons, the impeller with radial blades has been the logic choice of designers of aircraft centrifugal compressors.

Q. 30.32 *Discuss the relative merits and demerits of axial flow compressors over centrifugal compressors.* ***(GATE)***

Solution. The relative merits and demerits of axial flow compressors over centrifugal compressors are given in a tabular form below:

S.No.	Aspects	Centrifugal Compressors	Axial Flow Compressors
1.	*Type of flow*	Axial (Parallel to the direction of axis of the machine)	Radial
2.	*Pressure ratio per stage*	High, about 4.5 : 1. Thus unit is *compact* — In supersonic compressors, the pressure ratio is about 10 but at the cost of efficiency. Operation is *not* so difficult and risky.	Low, about 1.2 : 1. This is due to absence of centrifugal action. To achieve the pressure ratio equal to that per stage in centrifugal compressor 10 to 20 stages are required. Thus the unit is *less compact* and *less rugged.*
3.	*Isothermal efficiency*	About 80 to 82%	About 86 to 88% (with modern aerofoil blades)
4.	*Frontal area*	Larger	Smaller (This makes the axial flow compressors *more suitable for jet engines due to less drag*).
5.	*Flexibility of operation*	More (due to adjustable pre-whirl and diffuser vanes)	Less
6.	*Part load performance*	Better	Poor
7.	*Effect of deposit formation on the surface of impeller rotor*	Performance not adversely affected	Performance adversely affected
8.	*Starting torque required*	Low	High
9.	*Suitability for multistaging*	Slightly difficult	More suitable for multistaging
10.	*Delivery pressure possible*	Upto 400 bar	Upto 20 bar
11.	*Applications*	Used in blowing engines in steel mills, low pressure refrigeration, big central air conditioning plants, fertiliser and industry, supercharging I.C. engines, gas pumping in long distance pipe lines etc. — Previously it was used in jet engines	Mostly used in jet engines (due to higher efficiency and smaller frontal area). Also preferred in power plant gas turbines and steel mills.
12.	*Efficiency vs speed curve*	More flat	Less flat comparatively

Q. 30.33 *What are the constructional features of an axial flow compressor? How air is compressed and what is the method of getting higher compression ratio in such compressor?* ***(IES)***

Solution. Axial flow compressor:

In an axial flow compressor, the flow proceeds throughout the compressor in a direction essentially parallel to the axis of the machine.

Construction. Refer to Fig. 30.19

- An axial flow compressor consists of adjacent rows of rotor (moving) blades and stator (fixed) blades. The rotor blades are mounted on the rotating drum and stator blades are fixed to the casing stator. *One stage* of the machine comprises a row of rotor blades followed by a row of stator blades.
- For efficient operation the blades are of air foil section based on aerodynamic theory. The blades are so designed that wasteful losses due to shock and turbulence are prevented and the blades are free from *stalling troubles* (The blades are said to *be stalled when the air stream fails to follow the blade contour*). Whereas the compressor blades have aerofoil section, the turbine blades have profiles formed by a number of circular arcs. This is so because the acceleration process being carried out in the converging blade passages of a reaction turbine is much more efficient and stable process as compared with the diffusing or decelerating process being carried out in the diverging passage between the blades of an axial flow compressor.

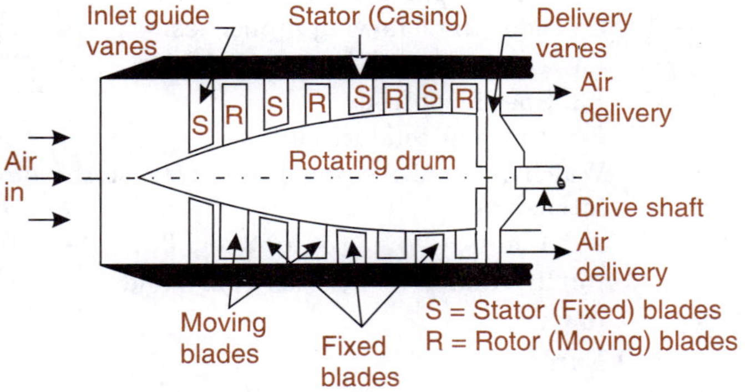

Fig. 30.19. Axial flow compressor.

- The annular area is usually reduced from inlet to outlet of the compressor. This is to *keep the flow velocity constant throughout the compressor length*. In the diverging passages of the moving blades, there is rise in temperature due to diffusion. The absolute velocity is also increased due to work input.

The *"fixed blades"* serve the following two *purposes* :

(*i*) Convert a *part of the K.E. of the fluid into pressure energy. This conversion is achieved by diffusion process carried out in the diverge blade passages.*

(*ii*) *Guide and redirect the fluid flow so that entry to the next stage is without shock.*

Working

Basically, the compression is performed in a similar manner to that of the centrifugal type. The work input to the rotor shaft is transferred by the moving blades to the air, thus accelerating it. The blades are so arranged that the *spaces between the blades form diffuser passages, and hence the velocity of the air relative to the blades is decreased as the air passes through them, and there is a rise in pressure. The air is then further diffused in the stator blades, which are also arranged to form diffuser passages.* In the fixed stator blades the *air is turned through an angle so that its direction is such that it can be allowed to pass to a second row of moving rotor blades. It is usual to have a relatively large number of stages and to maintain a constant work input per stage* (*e.g.*, from 5 to 14 stages have been used).

- The necessary reduction in volume may be allowed by flaring the stator or by flaring the rotor. It is more common to use a flared rotor, and this type is shown diagrammatically in Fig. 30.19.
- It is usually arranged to have an *equal temperature rise in the moving and the fixed blades, and to keep the axial velocity of air constant throughout the compressor.* Thus *each stage of the compression is exactly similar with regard to air velocity and blade inlet and outlet angles*.
- A diffusing flow is less stable than a converging flow, and for this reason the blade shape and profile is much more important for a compressor than for a reaction turbine. The *design of compressor blades is based on aerodynamic theory and an aerofoil shape* is used.

Q. 30.34 *What do you mean by supersaturated flow through steam nozzles? Discuss the causes of supersaturation phenomena.* **(GATE)**

Solution. Refer to Article **18.4.**

Q. 30.35 *Why are downcomers fewer in number and bigger in diameter, while risers are more in number and smaller in diameter in a steam generator.* **(IES)**

Solution. In a steam generator, $\Delta p = \frac{\rho f L V^2}{2D}$; the downcomers are meant to move the water by *gravity, 'bigger'* the diameter (D) 'less' the pressure drop (Δp) due to *friction,* since $\Delta p \propto \frac{1}{D}$. Hence, the downcomers are made bigger in diameter, which may vary from 150 mm to 200 mm or even higher.

Q. 30.36 *Why are steam turbines compounded? What are methods of compounding?* **(GATE)**

Solution. Refer to Article **19.5.**

B. Multiple-Choice Questions with 'Answers' and 'Explanations'

THERMODYNAMICS

Q. 30.1. A thermal electric power plant produces 1000 MW of power. If the coal releases 900 × 10^7 kJ/h of energy, then the rate at which heat is rejected from the power plant is

(*a*) 500 MW (*b*) 1000 MW
(*c*) 1500 MW (*d*) 2000 MW.

Ans. (*c*)

Explanation:

Heat rejected from the plant = Energy released by the coal – energy produced by the plant

$$= \frac{900 \times 10^7}{3600} \times 10^{-3} - 1000 = 1500 \text{ MW}$$

Q. 30.2. A heat engine is supplied with 250 kJ/s of heat at a constant fixed temperature of 227°C; the heat is rejected at 27°C, the cycle is reversible, then the amount of heat rejected is

(*a*) 250 kJ/s (*b*) 200 kJ/s
(*c*) 180 kJ/s (*d*) 150 kJ/s.

Ans. (*d*)

Explanation:

Heat supplied by the heat engine, Q_1 = 250 kJ/s;
Source temperature, T_1 = 227 + 273 = 500 K;
Sink temperature, T_2 = 27 + 273 = 300 K

For a reversible cycle: $\frac{Q_1}{T_1} = \frac{Q_2}{T_2}$ (Q_2 = Heat rejected)

or, $\frac{250}{500} = \frac{Q_2}{300}$

$\therefore \quad Q_2 = \frac{250 \times 300}{500} = 150$ kJ/s

Q. 30.3. 0.70 kg/s of air enters with a specific enthalpy of 290 kJ and leaves it with 450 kJ of specific enthalpy. Velocities at inlet and exit are 6 m/s and 2 m/s respectively. Assuming adiabatic process, what is power input to the compressor?

(*a*) 120 kW (*b*) 118 kW
(*c*) 115 kW (*d*) 112 kW.

Ans. (*d*)

Explanation:

$$\text{Power input to the compressor} = \dot{m}\left[\left(h_2 + \frac{C_2^2}{2}\right) - \left(h_1 + \frac{C_1^2}{2}\right)\right]$$

$$= 0.7\left[\left(450 + \frac{6^2}{2 \times 1000}\right) - \left(290 + \frac{2^2}{2 \times 1000}\right)\right]$$

= 112 kN

Q. 30.4. Which cycle consists of two reversible isotherms and two reversible isobars?

(*a*) Carnot cycle (*b*) Stirling cycle
(*c*) Ericsson cycle (*d*) Brayton cycle.

Ans. (*c*)

Explanation:

Ericsson cycle consists of following *operations*:

(*i*) Rejection of heat at constant pressure (*i.e.* isobar)
(*ii*) Isothermal compression (*i.e.* isotherm)
(*iii*) Addition of heat at constant pressure (isobar)
(*iv*) Isothermal expansion (isotherm).

Hence (*c*) is the correct answer.

Q. 30.5. Consider the following properties:

1. Temperature 2. Viscosity
3. Specific entropy
4. Thermal conductivity

Which of the above properties of a system is/are intensive?

(*a*) 1 only (*b*) 2 and 3
(*c*) 2, 3 and 4 (*d*) 1, 2, 3 and 4.

Ans. (*d*)

Explanation:

A propety which is *independent of mass* is termed as 'intensive propety'. Here all the given properties are independent of mass, hence (*d*) is the correct answer.

Q. 30.6. A reversible heat engine rejects 50 per cent of the heat supplied during a cycle of operation. If this engine is reversed and operates as a heat pump, then what is its coefficient of performance?

(*a*) 1.0 (*b*) 1.5
(*c*) 2.0 (*d*) 2.5.

Ans. (*c*)

Explanation:

For a reversible engine: $\frac{Q_1}{T_1} = \frac{Q_2}{T_2}$ or $\frac{Q_1}{T_1} = \frac{0.5\,Q_1}{T_2}$

or, $\frac{T_2}{T_1} = 0.5$

If the engine reversed and operates as a heat pump, then coefficient of performance (COP) heat pump

$$= \frac{T_1}{T_1 - T_2} = \frac{1}{1-(T_2/T_1)} = \frac{1}{1-0.5} = 2$$

Q. 30.7. The thermal efficiency of a Carnot heat engine is 30%. If the engine is reversed in operation to work as a heat pump with operating conditions unchanged, then what will be the COP for heat pump?

(*a*) 0.30 (*b*) 2.33
(*c*) 3.33 (*d*) cannot be calculated.

Ans. (*c*)

Explanation.

$$\eta_{carnot} = 30\%; \ (COP)_{heat\ pump} = ?$$

$$\eta_{carnot} = 0.3 = \frac{Q_1 - Q_2}{Q_1} = 1 - \frac{Q_2}{Q_1} \text{ or } \frac{Q_2}{Q_1} = 1 - 0.3 = 0.7$$

$$(COP)_{heat\ pump} = \frac{Q_1}{Q_1 - Q_2} = \frac{1}{1-(Q_2/Q_1)} = \frac{1}{1-0.7} = 3.33$$

Q. 30.8. A Carnot engine rejects 30% of absorbed heat to a sink at 30°C. The temperature of the heat source is

(*a*) 100°C (*b*) 433°C
(*c*) 737°C (*d*) 1010°C

Ans. (*c*)

Explanation:

$$\eta_{carnot} = \frac{Q_1 - Q_2}{Q_1} = \frac{Q_1 - 0.3Q_1}{Q_1} = 0.7$$

$$= \frac{T_1 - T_2}{T_1} \quad \text{or,} \quad 0.7\,T_1 = T_1 - T_2$$

or, $0.3\,T_1 = T_2$

or, $$T_1 = \frac{273+30}{0.3} = 1010 \text{ K or } 737°\text{C}$$

Q. 30.9. An engine operates between temperature limits of 900 K and T_2 and another between T_2 and 400 K. For both to be equally efficient, the value of T_2 will be

(*a*) 700 K (*b*) 600 K
(*c*) 750 K (*d*) 650 K.

Ans. (*b*)

Explanation:

Given: $T_1 = 900$ K; $T_3 = 400$ K; $T_2 = ?$

When both engines are to be equally efficient, then

$$T_2 = \sqrt{T_1 T_3} = \sqrt{900 \times 400} = 600 \text{ K}$$

Q. 30.10. An inventor states that his new engine rejects to the sink 40% of heat absorbed from the source while the source and sink temperatures are 327°C and 27°C respectively. His engine is therefore equivalent to

(*a*) Joule engine (*b*) Striling engine
(*c*) Impossible engine (*d*) Carnot engine.

Ans. (*c*)

Explanation:

Given: $T_1 = 327 + 273 = 600$ K;
$T_2 = 27 + 273 = 300$

The maximum possible efficiency of the engine is the Carnot efficiency, given by

$$\eta_{carnot} = \frac{T_1 - T_2}{T_1} = \frac{600-300}{300} = 0.5 \text{ or } 50\%$$

Efficiency of the inventor's engine,

$$\eta = \frac{Q_1 - Q_2}{Q_1} = \frac{Q_1 - 0.4Q}{Q_1} = 0.6 \text{ or } 60\%$$

Since efficiency claimed is greater than η_{carnot}, hence the inventor's engine is equivalent to *impossible engine*.

Q. 30.11. A closed system receives 60 kJ heat but its internal energy decreases by 30 kJ. Then the work done by the system is

(*a*) 90 kJ (*b*) 30 kJ
(*c*) – 30 kJ (*d*) – 90 kJ.

Ans. (*a*)

Explanation:

$dQ = 60$ kJ; $du = -30$ kJ

According to the first law of thermodynamics,

$$dQ = du + dW$$

or, $60 = -30 + dW$

or, $dW = 90$ kJ

Q. 30.12. The value of $\oint \frac{dQ}{T}$ for an irreversible cycle is

(*a*) Equal to zero (*b*) Greater than zero
(*c*) Less than zero (*d*) Unity.

Ans. (*c*)

Explanation:

Gausius inequality may be expressed as follows: "When a system performs a reversible cycle,

$$\oint \frac{dQ}{T} = 0$$

but when the cycle is *irreversible:*

$$\oint \frac{dQ}{T} < 0\text{"}$$

Hence (*c*) is the correct answer.

Q. 30.13. Increase in entropy of a system represents

(*a*) Increase in availability of energy

(*b*) Increase in temperature

(*c*) Decreases in pressure

(*d*) Degradation of energy.

Ans. **(*d*)**

Explanation:

Entropy is a measure of degree of disorderness. *Increase in entropy* of a system represents *degradation of energy*.

Q. 30.14. Work transfer between the system and the surroundings

(*a*) is a point function

(*b*) is always given by $\int p\, dv$

(*c*) is a function of pressure only

(*d*) depends on the path followed by the system.

Ans. **(*d*)**

Explanation

There are certain quantities which cannot be located on a graph by a '*point*' but are given by the '*area*' or so on that graph. In that case, the area on the graph pertaining to the particular process, is a *function of path of the process*. Such quantities are called "*path functions*".

Examples: Heat, "*work*" etc.

Hence, (*d*) is the correct answer.

Q. 30.15. A closed system is one in which

(*a*) Mass does not cross boundaries of the system, though energy may do so

(*b*) Mass crosses the boundary but not the energy

(*c*) Neither mass nor energy cross the boundary of the system

(*d*) Both energy and mass cross the boundaries of the system.

Ans. **(*a*)**

Explanation:

If the boundary of a system is *impervious to the flow of matter,* it is called a 'closed system'. In this *system energy may cross the boundaries* of the system.

Q. 30.16. A thermodynamic system is considered to be an isolated one if

(*a*) mass transfer and entropy change are zero

(*b*) entropy change and energy transfer are zero

(*c*) energy transfer and mass transfer are zero

(*d*) mass transfer and volume change are zero.

Ans. **(*c*)**

Explanation:

An isolated system' is that system which *exchanges neither energy nor matter* with any other system or with environment.

Q. 30.17. The constant volume gas thermometer works on the principle that

(*a*) at low pressure, the temperature of the gas is independent of its pressure at constant volume

(*b*) at high pressure, the temperature of the gas is independent of its pressure at constant volume

(*c*) at low pressure, the temperature of the gas is proportional to its pressure at constant volume.

(*d*) at high pressure, the temperature of the gas is proportional to its pressure at constant volume.

Ans. **(*c*)**

Explanation:

The constant volume gas thermometer works on the principle that at low pressure the temperature of the gas is proportional to its pressure at constant volume.

At low pressure the gas behaves as an ideal gas and follows the ideal gas equation: $pV = mRT$; for V = constant m, R also being constant $p \propto T$.

Q. 30.18. There is no work transfer involved in this process

(*a*) Adiabatic expansion

(*b*) Isothermal expansion

(*c*) Polytropic expansion

(*d*) Free expansion.

Ans. **(*d*)**

Explanation:

In "free expansion process," *no work is done* on or by the fluid, since the boundary of the system does not move. No heat flows to or from the fluid since the system is well lagged. The process is therefore, adiabatic but irreversible.

Q. 30.19. This process is one in which there is only work interaction between the system and the surroundings

(*a*) Diabetic process

(*b*) Adiabatic process

(*c*) Isothermal process

(*d*) Quasi-static process.

Ans. **(*b*)**

Explanation:

An "*adiabatic process*" is one in which there is only *work interaction* between the system and surroundings; *no heat is transferred* to or from the fluid during the process.

- For an adiabatic process to take place, perfect thermal in evolution for the system must be available.

Q. 30.20. For a given temperature T_1, as the difference between T_1 and T_2 increases, the COP of a Carnot heat pump

(*a*) increases

(*b*) decreases

(*c*) does not change

(*d*) first decreases then increases.

Ans. **(*b*)**

Explanation:

$$(\text{COP})_{\text{heat pump}} = \frac{T_1}{T_1 - T_2}$$

When T_1 is constant, and $T_1 - T_2$ increases, the COP will decrease.

Q. 30.21. A heat engine is supplied with 2512 kJ/min of heat at 650°C. Heat ejection with 900 kJ/min takes place at 100°C. This type of heat engine is

(*a*) ideal (*b*) irreversible

(*c*) impossible (*d*) practical.

Ans. **(*c*)**

Explanation:

Given : $Q_1 = 2512$ kJ/min; $T_1 = 650 + 273 = 923$ K;
$Q_2 = 900$ kJ/min;
$T_2 = 100 + 273 = 373$ K

$$\eta_{\text{carnot}} = \frac{T_1 - T_2}{T_1} = 1 - \frac{T_2}{T_1} = 1 - \frac{373}{923} = 0.596$$

or 59.6%

$$\eta_{\text{actual}} = \frac{Q_1 - Q_2}{Q_1} = \frac{2512 - 900}{2512} = 0.642 \text{ or } 64.2\%$$

Since $\eta_{\text{actual}} > \eta_{\text{carnot}}$, therefore, this heat engine is *impossible*.

Q. 30.22. If the work done on a closed system is 20 kJ/kg, and 40 kJ/kg heat is rejected from the system, its internal energy decreases by

(*a*) 20 kJ/kg (*b*) 60 kJ/kg

(*c*) –20 kJ/kg (*d*) –60 kJ/kg.

Ans. **(*c*)**

Explanation:

Given: $W = -20$ kJ/kg;
$Q = -40$ kJ/kg; $(u_2 - u_1) = ?$
$Q = (u_2 - u_1) + W$ or -40
$= (u_2 - u_1) - 20$

or, $u_2 - u_1 = -20$ kJ/kg

Q. 30.23. A Carnot heat pump is used to heat a house. The outside temperature is –3°C and the indoor temperature is 27°C. If the heat loss from the house is 40 kW the power required to operate the heat pump is

(*a*) 1 kW (*b*) 2 kW

(*c*) 3 kW (*d*) 4 kW.

Ans. **(*d*)**

Explanation:

Given: $T_1 = 27 + 273 = 300$ K; $T_2 = -3 + 273 = 270$ K; $Q = 40$ kW; $W = ?$

$$\text{COP} = \frac{T_1}{T_1 - T_2} = \frac{300}{(300 - 270)} = 10$$

$$\text{Now, COP} = \frac{Q}{W} \text{ or } W = \frac{Q}{\text{COP}} = \frac{40}{10} = 4 \text{ kW}$$

Q. 30.24. The enthalpy drop for flow through convergent horizontal nozzle is 100 kJ/kg.

If the velocity of approach at inlet to the nozzle is negligible, the exit velocity of the fluid is

(*a*) 20 m/s (*b*) 400 m/s

(*c*) 447.2 m/s (*d*) 520.8 m/s.

Ans. **(*c*)**

Explanation:

Enthalpy drop = 100 kJ/kg, Velocity of approach, $C_1 = 0$. By applying steady flow energy equation at inlet and exit of nozzle and neglecting change in datum and assuming no heat and work transfer, we get:

$$h_1 + \frac{C_1^2}{2} = h_2 + \frac{C_2^2}{2}$$

or, $(h_1 - h_2) = \frac{C_2^2}{2}$ or $100 \times 10^3 = \frac{C_2^2}{2}$

or, $C_2 = 447.2$ m/s.

Q. 30.25. The difference between constant pressure specific heat c_p and constant volume specific heat c_v for pure substance

(*a*) approaches zero at triple point

(*b*) approaches zero as the absolute temperature approaches zero

(*c*) is always equal to the gas constant R

(*d*) approaches zero at critical point.

Ans. **(*b*)**

Explanation:

The following relation holds good for a *pure substance*:

$$c_p - c_v = \frac{TV\beta^2}{K_T}$$, where T = Absolute temperature,

V = Volume, β = Co-efficient of volume expensivity; K_T = Isothermal compressibility.

Thus, when T approaches zero, $c_p - c_v$ approaches zero. Hence (*b*) is the correct answer.

Q. 30.26. Joule-Thomson coefficient is the slope of

(*a*) constant enthalpy lines on *T-s* diagram

(*b*) constant enthalpy lines on *T-p* diagram

(*c*) inversion curve on *T-s* diagram

(*d*) inversion curve on *T-p* diagram.

Ans. **(*b*)**

Explanation:

The *slope* of a constant enthalpy line on a *T-p diagram* at a particular state may be *positive*, *zero* or *negative* value. The shape is called "*Joule-Thomson coefficient*", μ and is given by:

$$\mu = \left(\frac{\partial T}{\partial p}\right)_h = 0 \text{ for an ideal gas.}$$

Q. 30.27. Which gas shows a heating effect in the Joule-Thomson experiment, while undergoing throttling process through a porous plug of cotton wool?

(*a*) Carbon dioxide (*b*) Hydrogen

(*c*) Oxygen (*d*) Nitrogen.

Ans. **(*b*)**

Explanation:

Hydrogen (H_2) gas shows a *heating effect* in the Joule-Thomson experiment, while undergoing throttling process through a porous plug of cotton wool. All other gases except hydrogen, helium and neon *cool* upon expansion by the Joule-Thomson process.

Q. 30.28. An ideal gas at 27°C is heated at constant pressure till the volume becomes three times. The temperature of the gas will then be

(*a*) 81°C (*b*) 900°C

(*c*) 627°C (*d*) 927°C.

Ans. **(*c*)**

Explanation:

Given : $T_1 = 27 + 273 = 300$ K; $V_2 = 3V_1$

As the ideal gas is heated at constant pressure

(p = constant), therefore, $\frac{T_1}{T_2} = \frac{V_1}{V_2}$ or $\frac{300}{T_2} = \frac{V_1}{3V_1}$

or $T_2 = 300 \times 3 = 900$ K or 627°C.

Q. 30.29. A heat engine using lake water at 12°C as source and the surrounding atmosphere at 2°C as sink executes 1080 cycles per min. If the amount of heat drawn per cycle is 57 J, then the output of the engine will be

(*a*) 66 W (*b*) 56 W

(*c*) 46 W (*d*) 36 W.

Ans. **(*d*)**

Explanation:

$$T_1 = 12 + 273 = 285 \text{ K},$$
$$T_2 = 2 + 273 = 275 \text{ K}$$

Number of cycles executed per second $= \frac{1080}{60} = 18$ cycles/sec.

Amount of heat drawn per cycle = 57 J

$$\eta = \frac{T_1 - T_2}{T_1} = \frac{285 - 275}{285} = 0.0351$$

Work done per cycle = 57 × η = 57 × 0.351 = 2 J

$$\left(\because \quad \eta = \frac{W}{Q}\right)$$

∴ Work done per second = 2 × 18 = 36 J/s or 36 W

Thus correct answer is (*d*).

Q. 30.30. The loss of available energy associated with the transfer of 1000 kJ of heat from a constant temperature system at 600 K to another at 400 K when the environmental temperature is 300 K is

(*a*) 166.67 kJ (*b*) 250 kJ

(*c*) 500 kJ (*d*) 750 kJ.

Ans. **(*a*)**

Explanation:

Heat rejected to 400 K system is given as:

$$\frac{600 - 400}{600} = \frac{Q_{600} - Q_{400}}{1000}$$

or, $$\frac{1}{3} = \frac{1000 - Q_{400}}{1000}$$

∴ $$Q_{400} = 1000 - \frac{1}{3} \times 1000 = 666.67 \text{ kJ}$$

Heat rejected to 300 K system is given as

$$\frac{600 - 300}{600} = \frac{1000 - Q_{300}}{1000}$$

or $$\frac{1}{2} = \frac{1000 - Q_{300}}{1000}$$

∴ $$Q_{300} = 1000 - \frac{1}{2} \times 1000 = 500 \text{ kJ}$$

Hence loss of available energy = 666.67 – 500 = 166.67 kJ

Thus (*a*) is the correct choice.

Q. 30.31. One reversible heat engine operates between 1600 K and T_2K, and another reversible heat engine operates between T_2K and 400 K. If both the engines have the same heat input and output, then the temperature T_2 must be equal to

(*a*) 1000 K (*b*) 1200 K

(*c*) 1400 K (*d*) 800 K.

Ans. (*d*)

Explanation:

Since both the reversible engines have the same heat input and output, therefore, $\frac{T_1 - T_2}{T_1}$ is same for both the engines, therefore,

$$\frac{1600 - T_2}{1600} = \frac{T_2 - 400}{T_2}$$

or, $$1 - \frac{T_2}{1600} = 1 - \frac{400}{T_2}$$

or, $$T_2^2 = 1600 \times 400 \text{ or } T_2 = 800 \text{ K}$$

Thus (*d*) is the correct choice.

Q. 30.32. The efficiency of a reversible cyclic process undergone by a substance as shown in Fig. 30.20 is

(*a*) 0.40 (*b*) 0.55

(*c*) 0.66 (*d*) 0.80.

Ans. (*c*)

Explanation:

$$\eta = \frac{T_1 - T_2}{T_1} = \frac{1500 - 500}{1500} = 0.66$$

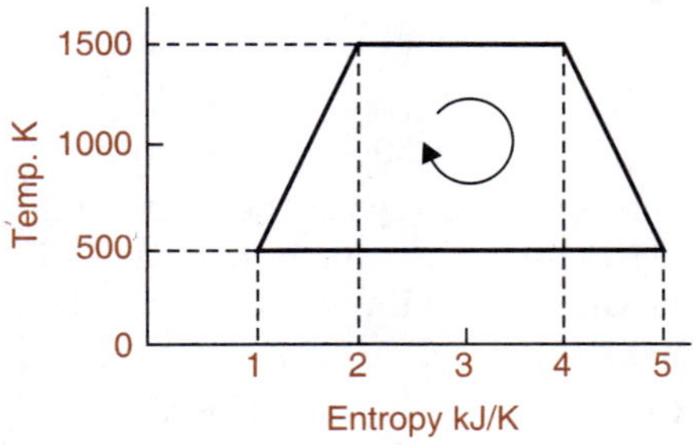

Fig. 30.20

Thus the correct choice is (*c*).

Q. 30.33. Otto cycle efficiency is higher than Diesel cycle efficiency for the same compression ratio and heat input, because in Otto cycle

(*a*) combustion is at constant volume

(*b*) expansion and compression are isentropic

(*c*) maximum temperature is higher

(*d*) heat rejected is lower.

Ans. (*d*)

Explanation:

In case of Otto cycle, for the same heat input and same compression ratio, efficiency is higher (compared to Diesel Cycle) because heat rejected is less.

Q. 30.34. Match List I with List II and select the correct using the codes given below the lists:

List I (*Cycle operating between fixed temperature limits*)	**List II** (*Characteristic*)
A. Carnot cycle	1. Efficiency depends upon cut-off ratio and compression ratio
B. Brayton cycle	2. Efficiency depends upon volume compression ratio
C. Otto cycle	3. Efficiency depends only upon pressure ratio
D. Diesel cycle	4. Efficiency depends only upon temperature limits
	5. Efficiency depends upon specific heats of the working substance

Codes:	A	B	C	D
(*a*)	3	4	2	5
(*b*)	4	2	1	5
(*c*)	4	3	2	1
(*d*)	3	4	5	1

Ans. (*c*)

Explanation:

A matches with 4 since in the case of Carnot cycle,

$\eta = \frac{T_1 - T_2}{T_1}$, *i.e.,* η depends upon temperature limits. *B matches* with 3 since in the case of Brayton cycle, $\eta = \frac{1}{(r_p)^{\frac{\gamma-1}{\gamma}}}$, *i.e.,* η depends upon only pressure ratio (r_p), *C* matches with 2 since efficiency of Otto cycle depends on volume compression work. *D* matches with 1 since efficiency of Diesel Cycle depends upon cut-off ratio and compression ratio.

Thus (*c*) is the correct choice.

Q. 30.35. A standard vapour is compressed to half its volume without changing its temperature. The result is that:

(*a*) All the vapour condenses to liquid

(*b*) Some of the liquid evaporates and the pressure does not change

(*c*) The pressure is double its initial value

(*d*) Some of the vapour condenses and the pressure does not change.

Ans. (*c*)

Explanation:

$$T_1 = T_2, V_2 = \frac{V_1}{2} \quad \text{... Given}$$

From, $p_1 V_1 = p_2 V_2$,

$$p_1 V_1 = p_2 \times \frac{V_1}{2} \text{ or } p_2 = 2p_1.$$

Q. 30.36. The minimum weight of air required per kg of fuel for complete combustion of a fuel is given by

(*a*) $11.6\text{ C} + 34.8\left(H_2 - \frac{O_2}{8}\right) + 4.35\text{ S}$

(*b*) $11.6\text{ C} + 34.8\left(H_2 + \frac{O_2}{8}\right) + 4.35\text{ S}$

(*c*) $11.6\text{ C} + 34.8\left(H_2 - \frac{O_2}{8}\right) + 6.35\text{ S}$

(*d*) $11.6\text{ C} + 64.8\left(H_2 + \frac{O_2}{8}\right) + 4.35\text{ S}$

Ans. **(*a*)**

Explanation:

In a fuel, the oxygen required for C, H_2, S and O_2 is given by

$$\frac{8}{3}\text{C} + 8H_2 + \text{S} - O_2 \text{ or } \frac{8}{3}\text{C} + 8\left(H_2 - \frac{O_2}{8}\right) + \text{S}$$

Since atmospheric air contains 23 per cent oxygen by weight, therefore, minimum air required per kg of fuel

$$= \frac{100}{23}\left[\frac{8}{3}\text{C} + 8\left(H_2 - \frac{O_2}{8}\right) + \text{S}\right]$$

$$= 11.6\text{ C} + 34.8\left(H_2 - \frac{O_2}{8}\right) + 4.35\text{ S}$$

Thus correct choice is (*a*).

Q. 30.37. Given that along the path 1-2-3, a system absorbs 100 kJ as heat and does 60 kJ work while along the path 1-4-3 it does 20 kJ work (Fig. 30.21). The heat absorbed during the cycle 1-4-3 is

(*a*) – 140 kJ (*b*) – 80 kJ

(*c*) – 40 kJ (*d*) + 60 kJ.

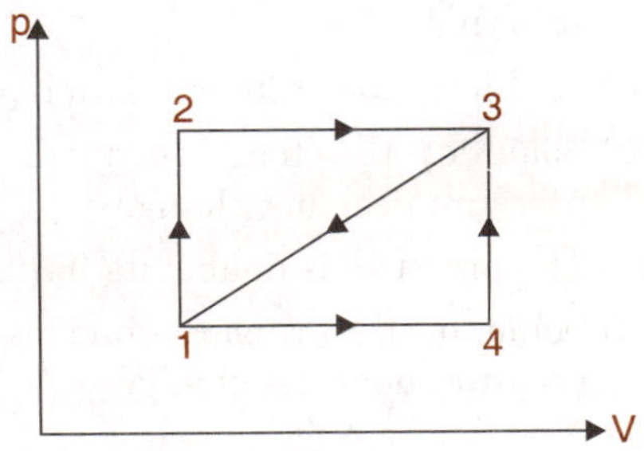

Fig. 30.21

Ans. **(*d*)**

Explanation:

$$Q_{123} = U_{13} + W_{123}$$

or, $100 = U_{13} + 60$

$\therefore$ $U_{13} = 40$ kJ

$$Q_{143} = U_{13} + W_{143}$$
$$= 40 + 20 = 60 \text{ kJ}$$

Q. 30.38. In a cyclic heat engine operating between a source temperature of 600°C and a sink temperature of 20°C, the *least* rate of heat rejection per kW net output of the engine is

(*a*) 0.460 kW (*b*) 0.505 kW

(*c*) 0.588 kW (*d*) 0.650 kW.

Ans. **(*b*)**

Explanation:

$$h_{max} = \frac{T_1 - T_2}{T_1} = \frac{(600 + 273) - (20 + 273)}{(600 + 273)}$$

$$= \frac{580}{873} = 0.6644$$

$$\text{Also,} 0.6644 = \frac{W \text{ (net output)}}{Q_{added}} = \frac{1}{Q_{added}}$$

or, $Q_{added} = 1.505$ kW

$\therefore$ Least rate of heat rejection,

$$Q_{rejected} = Q_{added} - W$$
$$= 1.505 - 1 = 0.505 \text{ kW}$$

Q. 30.39. In a reversible cycle, the source temperature is 227°C and the sink temperature is 27°C. The maximum available work for heat input of 100 kJ will be

(*a*) 100 kJ (*b*) 60 kJ

(*c*) 40 kJ (*d*) 88 kJ.

Ans. **(*c*)**

Explanation:

$$\eta_{max} = \frac{T_1 - T_2}{T_1} = \frac{(227 + 273) - (27 + 273)}{(227 + 273)}$$

$$= \frac{500 - 300}{500} = 0.4$$

$$\text{Also, } 0.4 = \frac{W}{Q}, \text{ or } 0.4 = \frac{W}{100}$$

$\therefore$ $W = 0.4 \times 100 = 40$ kJ

Q. 30.40. A regenerative steam cycle renders

(*a*) increased work output per unit mass of steam

(*b*) decreased work output per unit mass of steam

(*c*) increases thermal efficiency

(*d*) decreased work output per unit mass of steam as well as increased thermal efficiency.

Ans. (*c*)

Explanation:

In a regenerative steam cycle, a part of steam is extracted from turbine and used for heating the condensate. By doing so, some work is lost per unit mass of steam (corresponding to steam extracted out), but its heat is not wasted to cooling water and is conserved within the cycle; as a consequence of which thermal efficiency is increased.

Q. 30.41. Heat and work are

(*a*) intensive properties
(*b*) extensive properties
(*c*) point functions
(*d*) path functions.

Ans. (*d*)

Explanation:

Heat and work are path functions; these are inexact differentials.

Q. 30.42. The values of enthalpy of steam at the inlet and outlet of a steam of a steam turbine in a Rankine cycle are 2800 kJ/kg and 1800 kJ/kg respectively. Neglecting pump work, the specific steam consumption in kg/kWh is

(*a*) 3.60 (*b*) 0.36
(*c*) 0.06 (*d*) 0.01.

Ans. (*a*)

Explanation:

Specific steam consumption (kg/kWh)

$$= \frac{3600}{\text{Network done}} = \frac{3600}{(2800 - 1800)}$$

$$= 3.6 \text{ kg/kWh}$$

Q. 30.43. An ideal Brayton cycle, operating between the pressure limits of 1 bar and 6 bar, minimum and maximum temperature of 300 K and 1500 K. The ratio of specific heat of the working fluid is 1.4. The approximate final temperature in Kelvin at the end of the compression and expansion processes are respectively.

(*a*) 500 and 900 (*b*) 900 and 500
(*c*) 500 and 5400 (*d*) 900 and 950.

Ans. (*a*)

Explanation:

Refer to Fig. 30.22

$p_1 = 1$ bar, $T_{min} = 300$ K
$p_2 = 6$ bar, $T_{max} = 1500$ K
$\gamma = 1.4$

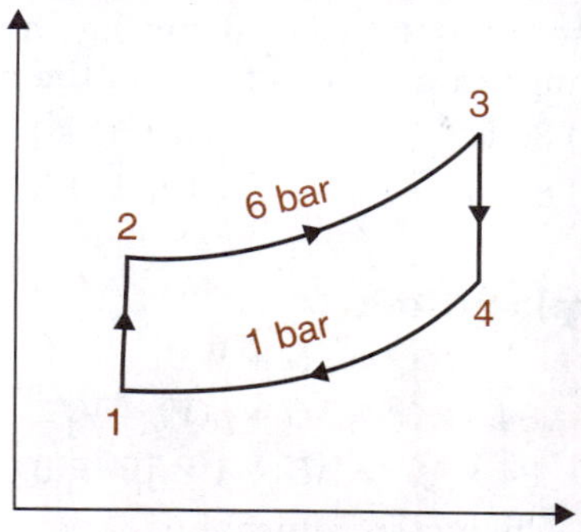

Fig. 30.22

$$\frac{T_2}{T_1} = \left(\frac{p_2}{p_1}\right)^{\frac{\gamma-1}{\gamma}}$$

$$\frac{T_2}{300} = \left(\frac{6}{1}\right)^{\frac{0.4}{1.4}} = 1.668$$

or, $T_2 = 500$ K

$$\frac{T_3}{T_4} = \left(\frac{6}{1}\right)^{\frac{0.4}{1.4}}$$

or, $\frac{1500}{T_4} = 1.668$

or, $T_4 \simeq 900$ K

Q. 30.44. Which one of the following *pV-T* diagrams (Fig. 30.23) correctly represents that of an ideal gas?

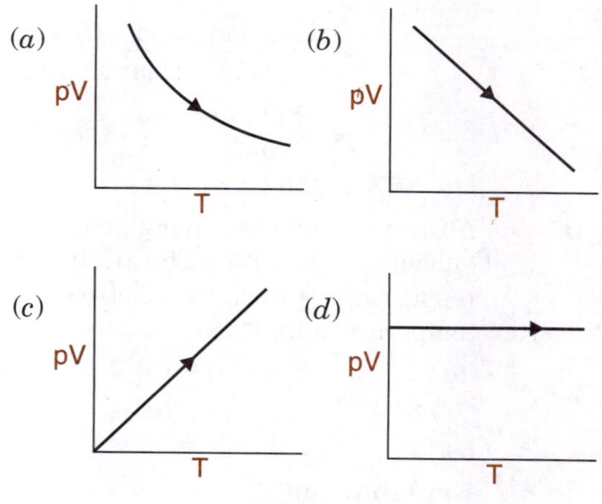

Fig. 30.23

Ans. (*c*)

Explanation:

In case of an ideal gas $pV = mRT$, which means that there is direct straight line relationship between pV and T; it is shown in Fig. 30.23 (*c*). Hence (*c*) is the correct choice.

Q. 30.45. A mixture of gases expands from 0.03 m^3 to 0.06 m^3 at a constant pressure of 1 MPa and absorbs 84 kJ of heat during the process. The change in internal energy of the mixture is

(*a*) 30 kJ (*b*) 54 kJ
(*c*) 84 kJ (*d*) 114 kJ.

Ans. (*b*)

Explanation:

$$Q = \Delta U + W$$

or, $84 \times 10^3 = \Delta U + p(V_2 - V_1)$

or, $84 \times 10^3 = \Delta U + 1 \times 10^6 \times 0.03$

∴ Change in internal energy,

$$\Delta U = 84 \times 10^3 - 0.03 \times 10^6$$
$$= 54000 \text{ J} \quad \text{or} \quad 54 \text{ kJ}$$

Q. 30.46. A heat engine is supplied with 250 kJ/s of heat at a constant fixed temperature of 227°C. The heat is rejected at 27°C. The cycle is reversible, if the amount of heat rejected is

(*a*) 273 kJ/s (*b*) 200 kJ/s
(*c*) 180 kJ/s (*d*) 150 kJ/s.

Ans. (*d*)

Explanation:

$T_1 = 227 + 273 = 500$ K; $T_2 = 27 + 273 = 300$ K;

Heat supplied = 250 kJ/s

Efficiency of reversible engine

$$= \frac{T_1 - T_2}{T_1} = \frac{500 - 300}{500} = 0.4$$

Also, $0.4 = \dfrac{\text{Work done}}{\text{Heat supplied}}$

$$= \frac{\text{Heat supplied} - \text{Heat rejected}(Q_R)}{\text{Heat supplied}}$$

$$= \frac{250 - Q_R}{250}$$

or, $Q_R = 150$ kJ/s

Q. 30.47. A Carnot engine receiving heat at 400 K has an efficiency of 25%. The COP of a Carnot refrigerator working between the same temperature limits is

(*a*) 1 (*b*) 2
(*c*) 3 (*d*) 4.

Ans. (*c*)

Explanation:

$$T_1 = 400, \ \eta_{carnot} = 25\% \quad \text{COP} = ?$$

$$\eta_{carnot} = \frac{T_1 - T_2}{T_1} \quad \text{or,} \quad 0.25 = \frac{400 - T_2}{400}$$

or, $T_2 = 300$ K

$$\text{COP} = \frac{T_2}{T_1 - T_2} = \frac{300}{400 - 300} = 3$$

Q. 30.48. Which one of the following diagram (Fig. 30.24) represents Otto cycle on temperature (*T*)-entropy (*s*) plane?

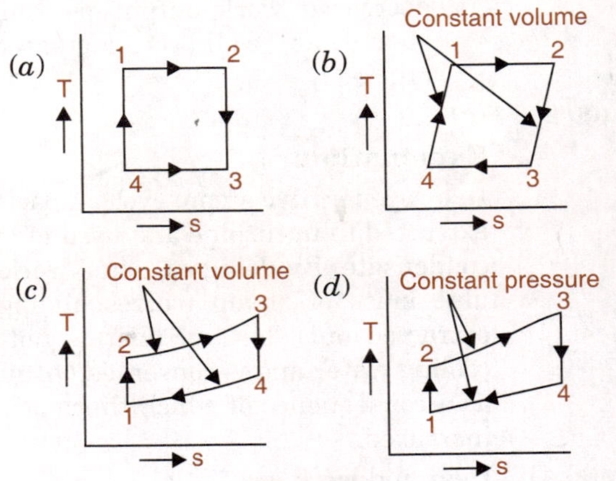

Fig. 30.24

Ans. (*c*)

Explanation:

In an Otto-cycle there are two isentropics and two constant volume processes.

Q. 30.49. A new temperature scale in degrees N is to be defined. The boiling and freezing points on this scale are 400°N and 100°N respectively.

What will be the reading on new scale corresponding to 60°C?

(*a*) 120°N (*b*) 180°N
(*c*) 220°N (*d*) 280°N.

Ans. (*d*)

Explanation:

On new scale, the boiling and freezing points are 400°N and 100°N respectively, *i.e.*, range, corresponding to 100°C, is 300°N. Therefore reading on new scale corresponding to 60°C

$$= 100 + \frac{300}{100} \times 60 = 280°\text{N}$$

Q. 30.50. A mono-atomic ideal gas (γ = 1.67, molecular weight = 40) is compressed adiabatically from 0.1 MPa, 300 K to 0.2 MPa. The universal gas constant is 8.314 kJ-K mol^{-1} K^{-1}. The work of compression of the gas (in kJ kg^{-1}) is

(*a*) 29.7 (*b*) 19.9
(*c*) 13.3 (*d*) 0.

Ans. (*a*)

Explanation:

Gas constant, $R = \dfrac{R_0}{M} = \dfrac{8.314}{40}$

or, $R = 0.20785$ kJ/kg K

$$p_1 = 0.1 \text{ MPa} = 100 \text{ kPa}$$
$$T_1 = 300 \text{ K}$$
$$p_2 = 0.2 \text{ MPa} = 200 \text{ kPa}$$

$$\text{Work done} = \frac{p_1V_1 - p_2V_2}{\gamma - 1}$$

$$= \frac{mR(T_1 - T_2)}{\gamma - 1}$$

$$\frac{T_2}{T_1} = \left(\frac{p_2}{p_1}\right)^{\gamma - 1/\gamma}$$

$$= \left(\frac{200}{100}\right)^{\frac{1.67-1}{1.67}} = 1.32$$

or, $T_2 = 300 \times 1.32 = 396$ K

$$\text{Work done/kg} = \frac{0.20785\,(300 - 396)}{0.67}$$

$$= -\,29.78 \text{ kJ/kg}$$

Negative sign shows compression work.

Q. 30.51. Consider the following two processes;

1. A heat source at 1200 K loses 2500 kJ of heat to a sink at 800 K
2. A heat source at 800 K loses 2000 kJ of heat to a sink at 500 K

Which of the following statements is true?

(*a*) Process I is more irreversible than Process II

(*b*) Process II is more irreversible than Process I

(*c*) Irreversibility associated in both the processes are equal

(*d*) Both the processes are reversible.

Ans. **(*b*)**

Explanation:

More irreversible process is associated with more entropy increase

$$(\Delta S)_a = \frac{-2500}{1200} + \frac{2500}{800}$$

$$= 1.041 \text{ kJ/K}$$

$$(\Delta S)_b = \frac{-2000}{800} + \frac{2000}{500}$$

$$= 1.5 \text{ kJ/K}$$

Thus, as $(\Delta S)_b > (\Delta S)_a$, process *b* is more irreversible.

Q. 30.52. In Rankine cycle, regeneration results in higher efficiency because

(*a*) pressure inside the boiler increases

(*b*) heat is added before steam enters the low pressure turbine

(*c*) average temperature of heat addition in the boiler increases

(*d*) total work delivered by the turbine increases.

Ans. **(*c*)**

Explanation: Refer to Fig. 30.25.

The object of the regenerative feed heating cycle is to supply the working fluid to the boiler at same state between 2 and 2′ (rather than at state 2) thereby increasing the average temperature of heat addition to the cycle.

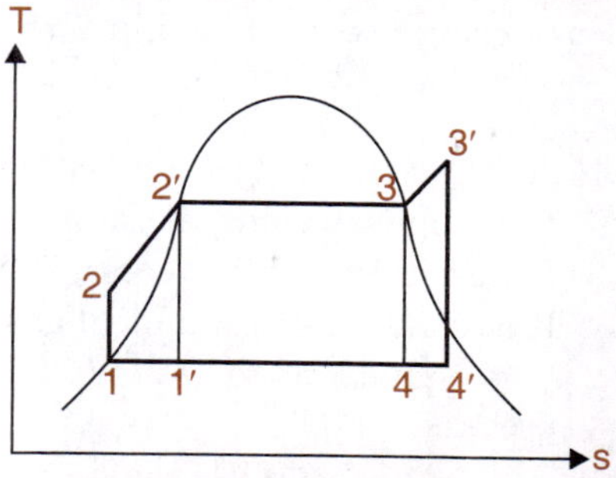

Fig. 30.25

Q. 30.53. Considering the variation of static pressure and absolute velocity in an impulse steam turbine across one row of moving blades

(*a*) both pressure and velocity decrease

(*b*) pressure decreases but velocity increases

(*c*) pressure remains constant, while velocity increases

(*d*) pressure remains constant while velocity decreases

Ans. **(*d*)**

Explanation: Refer to Fig. 30.26

In impulse turbine, the steam pressure remains constant while it flows through the moving blades, where only the kinetic energy converts into mechanical energy.

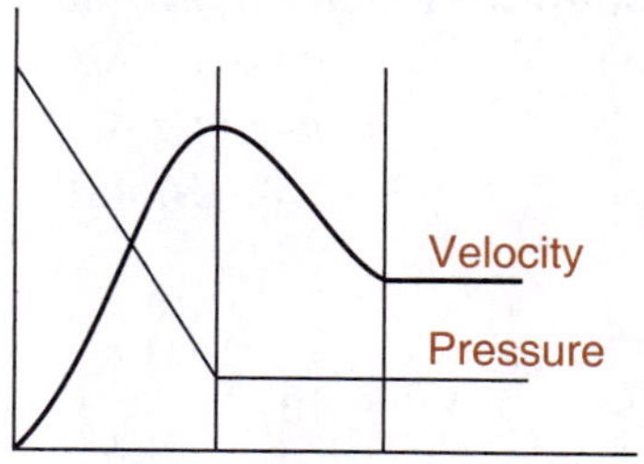

Fig. 30.26

Q. 30.54. A 2 kW, 40 litres water heater is switched on for 20 minutes. The heat capacity c_p for water is 4.2 kJ/kg K. Assuming all the electrical energy has gone into heating the water, increase of water temperature in degree centigrade is

(*a*) 2.7 (*b*) 4.0

(*c*) 14.3 (*d*) 25.25.

Ans. **(*c*)**

Explanation:

Heat supplied,

$$Q = \text{power} \times \text{time}$$

$$= 2 \times 1000 \times 20 \times 60$$

$$= 2400 \times 1000 \text{ J} \quad \text{or} \quad 2400 \text{ kJ}$$

But heat required to increase the water temperature,

$$Q = mc_p\Delta T$$

$$2400 = (0.04 \times 1000) \times 4.2 \times \Delta T$$

$$\Delta T = 14.3°\text{C}.$$

Q. 30.55. Considering the relationship $TdS = dU + pdV$ between the entropy (S), internal energy (U), pressure (p), temperature (T) and volume (V), which of the following statements is correct?

(a) It is applicable only for a reversible process

(b) For an irreversible process, $TdS > dU + pdV$

(c) It is valid only for an ideal gas

(d) It is equivalent to 1st law, for a reversible process.

Ans. **(d)**

Explanation:

$$TdS = dU + pdV$$

This equation holds good for any process reversible or irreversible, undergone by a closed system, since it is a relation among properties which are independent of the both.

Q. 30.56. In a gas turbine, hot combustion products with the specific heat c_p = 0.98 kJ/kg K, and c_v = 0.7538 kJ/kg K enter the turbine at 20 bar, 1500 K and exit at 1 bar. The isentropic efficiency of the turbine is 0.94. The work developed by the turbine per kg of gas flow is

(a) 689.66 kJ/kg (b) 794.66 kJ/kg

(c) 1009.72 kJ/kg (d) 1312.00 kJ/kg.

Ans. **(a)**

Explanation: Refer to Fig. 30.27.

Given: T_3 = 1500 K

c_p = 0.98 kJ/kg K

c_v = 0.7538 kJ/kg K

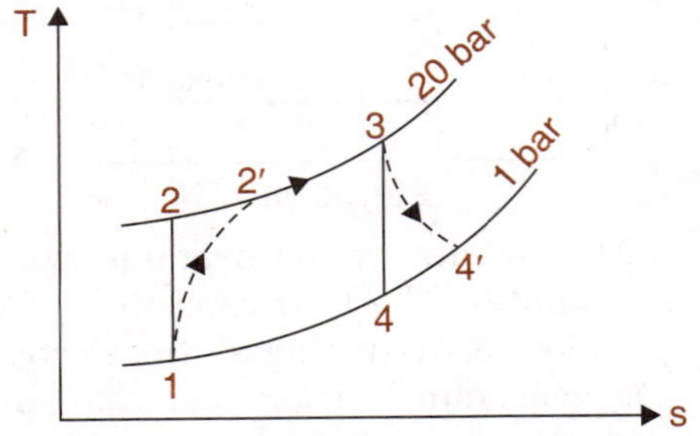

Fig. 30.27

$$\therefore \quad \gamma = \frac{c_p}{c_v} = 1.3$$

Now, $$\frac{T_4}{T_3} = \left(\frac{p_4}{p_3}\right)^{\frac{\gamma-1}{\gamma}}$$

or, $$\frac{T_4}{1500} = \left(\frac{1}{20}\right)^{\frac{0.3}{1.3}}$$

or, $$T_4 = 751.37 \text{ K}$$

Now turbine efficiency,

$$\eta = \frac{T_3 - T_4'}{T_3 - T_4}$$

or, $$0.94 = \frac{1500 - T_4'}{1500 - 751.34}$$

$$\therefore \quad T_4' = 796.26 \text{ K}$$

Turbine work,

$$W_t = c_p(T_3 - T_4')$$
$$= 0.98(1500 - 796.26)$$
$$= 689.66 \text{ kJ/kg.}$$

Q. 30.57. For an engine operating on air standard Otto cycle, the clearance volume is 10% of the swept volume. The specific heat ratio of air is 1.4. The air standard cycle efficiency is

(a) 38.3% (b) 39.8%

(c) 60.2% (d) 61.7%.

Ans. **(d)**

Explanation:

$$\eta_{otto} = 1 - \left(\frac{V_c}{V_1}\right)^{\gamma-1}$$

$$= 1 - \left(\frac{V_c}{V_c + V_s}\right)^{\gamma-1}$$

where, V_c = Clearance vol.,

= 10% of V_s = 0.1 V_s, and

V_s = Swept volume

$$\therefore \quad \eta_{otto} = 1 - \left(\frac{0.1V_s}{0.1V_s + V_s}\right)^{\gamma-1}$$

$$= 1 - \left(\frac{0.1}{1.1}\right)^{1.4-1}$$

$$= 0.617 = 61.7\%$$

Q. 30.58. A gas contained in a cylinder is compressed, the work required for compression being 5000 kJ. During the process, heat interaction of 2000 kJ causes the surroundings to be heated. The change in internal energy of the gas during the process is

(a) – 7000 kJ (b) – 3000 kJ

(c) + 3000 kJ (d) + 7000 kJ.

Ans. **(c)**

Explanation:

$$\Delta Q = W + \Delta U$$

$$-2000 \text{ kJ} = -5000 \text{ kJ} + \Delta U$$

$$\Delta U = 3000 \text{ kJ.}$$

Q. 30.59. Which of the following are boiler mountings?

1. Fusible plug　　2. Blow-off cock
3. Steam trap　　4. Feed check valve

Select the correct answer using the code given below:

(*a*) 1, 2 and 3　　(*b*) 2, 3 and 4
(*c*) 1, 3 and 4　　(*d*) 1, 2 and 4.

Ans. (*d*)

Explanation:

"Boiler mountings" are different fittings and devices which are *necessary* for *the operation and safety of boilers*. Usually these devices are mounted over the shell.

In accordance with the Indian boiler regulation the following mountings should be fitted to the boiler.

Two safety valves, two water level indicators, A pressure gauge, A steam stop valve, A feed check valve, A blow-off cock, An-attachment for inspector's test gauge, A man hole, Mud holes or sight holer.

- Boilers of Lancashire and Cornish type should be fitted with a high pressure and low water safety valve.

All land boilers should have a 'fusible' plug in each furnace. Thus steam trap is not a mounting. Hence (*d*) is the correct answer.

Q. 30. 60. If the cross-section of a nozzle is increasing in the direction of flow in supersonic flow, then in the downstream direction

(*a*) both pressure and velocity will increase
(*b*) both pressure and velocity will decrease
(*c*) pressure will increase but velocity will decrease
(*d*) pressure will decrease but velocity will increase.

Ans. (*d*)

Explanation:

In case of a supersonic flow, Mach number, $M > 1$

We know that: $\dfrac{dA}{A} = \dfrac{1}{\gamma}\cdot\dfrac{dp}{p}\left(\dfrac{1-M^2}{M^2}\right)$

Here $\dfrac{dA}{A}$ = +ve, $\dfrac{dp}{p}$ = –ve

Hence, pressure will decrease and velocity will increase.

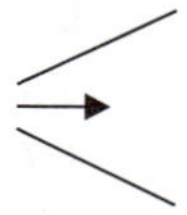

Fig. 30.28. Divergent nozzle.

Q. 30.61. The compounding of steam turbines done

(*a*) to improve efficiency
(*b*) to reduce the speed of rotor
(*c*) to reduce exit losses
(*d*) to increase the turbine output.

Ans. (*b*)

Explanation:

The *compounding* of steam turbines is done to *reduce* of speed of the rotor.

- If the steam is expanded from the boiler pressure to condenser pressure in one stage (as in simple impulse turbine) the speed of the rotor becomes *tremendously high* which crops up practical complicacies. There are several methods of reducing this speed to lower value; the different methods of compounding are:

(*i*) Velocity compounding
(*ii*) Pressure compounding
(*iii*) Pressure velocity compounding
(*iv*) Reaction turbine.

Q. 30.62. In a Brayton cycle, the value of optimum pressure ratio for maximum network done between temperatures T_1 and T_3, where T_3 is the maximum temperature and T_1 is the minimum temperature is

(*a*) $r_p = \left(\dfrac{T_3}{T_1}\right)^{\frac{\gamma}{\gamma-1}}$　　(*b*) $r_p = \left(\dfrac{T_3}{T_1}\right)^{\frac{\gamma-1}{2\gamma}}$

(*c*) $r_p = \left(\dfrac{T_3}{T_1}\right)^{\frac{\gamma}{2(\gamma-1)}}$　　(*d*) $r_p = \left(\dfrac{T_3}{T_1}\right)^{\frac{2(\gamma-1)}{\gamma}}$.

Ans. (*c*)

Explanation:

In case of a Brayton cycle pressure ratio (r_p) for maximum network done is given as:

$r_p = \left(\dfrac{T_3}{T_1}\right)^{\frac{\gamma}{2(\gamma-1)}}$ (*i.e.*, r_p *is a function of the limiting temperature ratio.*)

Q. 30.63. In a radial blade centrifugal compressor, the velocity of blade tip is 400 m/s and slip factor is 0.9. Assuming the absolute velocity at inlet to be axial, the work done per kg of flow is

(*a*) 36 kJ　　(*b*) 72 kJ
(*c*) 144 kJ　　(*d*) 360 kJ.

Ans. (*c*)

Explanation:

Given: Velocity of blade tip, C_{bl} = 400 m/s; Slip factor ϕ_s = 0.9

Work done = $\phi_s\ C_{bl}^2$ = 0.9 × 400^2 ×10^{-3} kJ = 144 kJ.

Q. 30.64. Which one of the following is used to bring down the speed of an impulse steam turbine to practical limits?

(*a*) A centrifugal governor

(*b*) Compounding of the turbine

(*c*) A large flywheel

(*d*) A gear box.

Ans. (*b*)

Explanation:

If the steam is expanded from boiler pressure to condenser pressure in a simple impulse turbine, the rotor speed becomes tremendously high which crops up practical complicacies. To control this high speed *"compounding"* of the turbine is required (All the methods of reducing rotor speed utilise multiple system of rotor in series, keyed on a common shaft, involve and the steam pressure or jet velocity is *absorbed in stages* as the steam flows over the blades.)

Q. 30.65. Consider the following for a gas turbine plant:

1. Intercooling 2. Regeneration
3. Reheat

Which of the above features in a simple gas turbine cycle increase the work ratio?

(*a*) 1, 2 and 3 (*b*) Only 1 and 2

(*c*) Only 2 and 3 (*d*) Only 1 and 3.

Ans. (*d*)

Explanation:

$$\text{Work ratio} = \frac{\text{Net work output}}{\text{Gross work output}}$$

$$= \frac{\text{Work of expansion} - \text{Work of compression}}{\text{Work of expansion}}$$

or, $\dfrac{W_T - W_C}{W_T}$ (*i.e.*, Intercooling and reheating increase the work ratio.)

- An heat exchanger (Regenerating feature) is usually used in large gas turbine units for marine propulsion or industrial power.

Hence (*d*) is the correct answer.

Q. 30.66. What is the ratio of the isentropic work to Euler's work known as?

(*a*) Pressure coefficient

(*b*) Slip factor

(*c*) Work factor

(*d*) Degree of reaction.

Ans. (*c*)

Explanation:

The '*work factor*' is the ratio of isentropic work to Euler's work.

$$i.e., \text{Work factor} = \frac{\text{Isentropic work}}{\text{Euler's work}}$$

The value of input factor is kept nearer to unity, the actual value is about 1.04.

- 'Work factor' is introduced to take care of losses due to friction between the casing and the air carried by the vanes etc; thus actual work input is greater than the theoretical value.

Q. 30.67. Consider the following statements:

For a large aviation gas turbine, an axial flow compressor is usually preferred over centrifugal compressor because

1. the maximum efficiency is higher
2. the frontal area is lower
3. the pressure rise per stage is more
4. the cost is lower

Which of these statements are correct?

(*a*) 1 and 4 (*b*) 1 and 2

(*c*) 1, 2 and 3 (*d*) 2, 3 and 4.

Ans. (*b*)

Explanation:

For a large aviation gas turbine, an axial flow compressor is usually preferred over centrifugal compressor due to the reasons that it has *lower frontal area and higher maximum efficiency.*

- Axial flow compressor is also preferred in power plant gas turbines and steel mills.

Q. 30.68. In an axial flow compressor, when the degree of reaction is 50%, it implies that

(*a*) Work done in compression will be the least

(*b*) 50% stages of the compression will be ineffective

(*c*) Pressure after compression will be optimum

(*d*) The compressor will have symmetrical blades.

Ans. (*d*)

Explanation:

In a reaction turbine if the degree of reaction is 50%, $\theta = \beta$ and $\phi = \alpha$ which means that moving blade and fixed blade have the *same shape* (*i.e.*, symmetrical). This condition gives *symmetrical* velocity diagrams. This type of turbine is known as 'Parson's reaction turbine'.

Q. 30.69. The power required to drive a turbo-compressor for a given pressure ratio decreases when

(*a*) Air is heated at entry

(*b*) Air is cooled at entry

(*c*) Air is cooled at exit

(*d*) Air is heated at exit.

Ans. (*b*)

Explanation:

The power required to drive a turbo-compressor for a given pressure ratio decreases when air is cooled at entry, this is so because *density of air at low temperature is high.*

Q. 30.70. Consider the following statements indicating a comparison between rocket and jet propulsion systems:

1. Both rocket and jet engines carry the fuel and oxidant
2. Rockets do not employ compressor or propeller
3. Rockets can operate in vacuum also
4. Rockets can use solid fuels and oxidants.

Which of these statements are correct?

(*a*) 1, 2, 3 and 4 (*b*) Only 1 and 2

(*c*) Only 2, 3 and 4 (*d*) Only 1, 3 and 4

Ans. (*c*)

Explanation:

In case of *'jet propulsion'* the oxygen required for combustion is taken from the atmosphere and fuel is stored whereas in *'Rockets-engine'*, the fuel and oxidiser *both* are contained in a propelling body and as such it can function in *vacuum.*

Q. 30.71. What does application of centrifugal air compressors lead to?

(*a*) Large frontal area of aircraft

(*b*) Higher flow rate through the engine

(*c*) Higher aircraft speed

(*d*) Lower frontal area of the aircraft.

Ans. (*a*)

Explanation:

The application of centrifugal air compressor leads to large frontal area of aircraft and consequently drag increases. That is why axial flow compressors are used in aircraft.

Q. 30.72. Consider the following:

1. Injector 2. Economizer
3. Blow-off cock 4. Steam stop valve

Which of the above is/are not boiler mountings?

(*a*) Only 1 (*b*) Only 1 and 2

(*c*) 1, 2 and 3 (*d*) 2 and 4.

Ans. (*b*)

Explanation:

Injector and economiser are some of the important boiler accessories.

Q. 30.73. Wilson line is associated with which one of the following?

(*a*) Total steam consumption with respect to power output

(*b*) Supersonic flow of steam through a nozzle

(*c*) Nozzle flow with friction

(*d*) Supersaturated flow of steam through a nozzle

Ans. (*d*)

Explanation:

Wilson line is associated with the supersaturated flow of steam through a nozzle.

The limiting condition of under-cooling at which condensation commences and is assumed to restore conditions of normal thermal equilibrium is called the *"Wilson line".*

Q. 30.74. Which one of the following expresses the maximum blade efficiency of a Parson's turbine?

(*a*) $\dfrac{2\cos^2\alpha}{1+\cos^2\alpha}$ (*b*) $\dfrac{\cos^2\alpha}{1+2\cos^2\alpha}$

(*c*) $\dfrac{\cos\alpha}{1+\cos^2\alpha}$ (*d*) $\dfrac{\cos\alpha}{2}$.

Ans. (*b*)

Explanation:

The expression for maximum blade efficiency of Parson's turbine is given as: $(\eta_{bl})_{max}$

$= \dfrac{2\cos^2\alpha}{1+\cos^2\alpha}$, where α is the angle of discharging tip. The condition for *maximum efficiency* of a reaction turbine is derived by the following assumptions:

(*i*) The degree of reaction is 50%

(*ii*) The moving and fixed blades are *symmetrical*

(*iii*) The velocity of steam at exit from the preceding stage is same as velocity of steam at the entrance to the succeeding stage.

Q. 30.75. In a reaction turbine the enthalpy drop in a stage is 60 units, the enthalpy drop in the moving blades is 32 units. The degree of reaction is

(*a*) 0.533 (*b*) 0.284

(*c*) 0.466 (*d*) 1.875.

Ans. (*a*)

Explanation:

The *'degree of reaction'* (R_d) of reaction turbine stage is defined as the ratio of heat drop over the moving blades (Δh_m) to the total heat drop ($\Delta h_f + \Delta h_m$) in the stage.

or, $$R_d = \frac{\Delta h_m}{\Delta h_f + \Delta h_m} = \frac{32}{60} = 0.533$$

Q. 30.76. A converging diverging nozzle is connected to a gas pipeline. As the inlet of the nozzle (converging section) the Mach number is 2. It is observed that there is a shock in the diverging section. The value of the Mach number at the throat is

(*a*) < 1 (*b*) Equal to 1
(*c*) > 1 (*d*) ≥ 1.

Ans. **(*b*)**

Explanation:

In a converging-doverging nozzle, the value of Mach number at the throat is always unity, *i.e.* $M = 1$.

Q. 30.77. What is the function of heavy water in a nuclear reactor?

(*a*) It serves as a coolant
(*b*) It serves as moderator
(*c*) It serves as a coolant as well as a moderator
(*d*) It serves as a neutron absorber.

Ans. **(*c*)**

Explanation:

As a moderator, heavy water (D_2O) in a nuclear reactor is the best material available [other commonly used moderators are H_2O, He (gas), Be and C(graphite)]. D_2O can be used as a *coolant* as well.

- In a nuclear reactor the function of a *'moderate'* is *slow down* the neutrons *from the high velocities but not absorb them.*
- The function of *'coolant'* is to remove the intense *heat produced* in the reactor and to bring out for further utilisation.

Q. 30.78. Consider the following statements:

1. The gases measured directly by Orsat apparatus from a flue gas sample are CO_2, O_2 and N_2
2. Bomb calorimeter measures higher calorific value of fuel at constant pressure
3. For burning 1 kg of fuel (carbon) to carbon monoxide, the stoichiometric quantity of air required is 8/3 kg

Which of these statements is/are correct?

(*a*) Only 1 (*b*) Only 2
(*c*) Only 3 (*d*) 1, 2 and 3.

Ans. **(*c*)**

Explanation:

- The gases measured directly by 'Orsat apparatus' from a flue gas sample are CO_2, O_2 and CO (Not N_2).
- A 'bomb calorimeter' measures the higher or gross calorific value because the fuel sample is burnt at a *constant volume* (not constant pressure).

Q. 30.79. Blade erosion in steam turbines takes place

(*a*) due to high temperature steam
(*b*) due to droplets in steam
(*c*) due to high rotational speed
(*d*) due to high flow rate.

Ans. **(*b*)**

Explanation:

In steam turbines the blade erosion takes place due to *droplets in steam.* When the speed is high and moisture *exceeds 10 per cent* the effect of moisture is *most prominent*, the most effected portion is the *back* of the inlet edge of the blade, where either grooves are formed or even some portion breaks away. Due to centrifugal force the water particles tend to concentrate in the outer annulus and their tip speed is greater than the root speed, hence errosion effect is *most on tips.*

Q. 30.80. Consider the following statements in respect of axial flow air compressors:

1. An axial flow air compressor is often described as a reversed reaction turbine.
2. With 50% degree of reaction, the velocity diagrams are symmetrical.

Which of these statements is/are correct?

(*a*) 1 only (*b*) 2 only
(*c*) Both 1 and 2 (*d*) Neither 1 nor 2.

Ans. **(*c*)**

Explanation:

(*c*) is the correct answer because the working principle of an axial flow compressor is opposite to that of a reaction turbine and the velocity diagrams of a reaction turbine are symmetrical when the degree of reaction is 50 per cent.

Q. 30.81. Which one of the following boiler accessories does not need 'flue-gas' for its operation?

(*a*) Economizer (*b*) Preheater
(*c*) Injector (*d*) Superheater.

Ans. **(*c*)**

Explanation:

The function of an *'injector'* is to feed water into the boiler. It does *not* need 'fuel-gas' for its operation.

– It is commonly employed for vertical and locomotive boilers and does not find application in large capacity high pressure boilers. It is also used where the space is not available for the installation of a feed pump.

Q. 30.82. Which one of the following fittings is mounted on the boiler to put off the fire in the furnace, when water level falls to an unsafe limit?

(*a*) Feed check valve (*b*) Safety valve
(*c*) Fusible plug (*d*) Blow-off cock.

Ans. **(*c*)**

Explanation:
The function of a *'fusible plug'* is to protect the boiler against damage due to overheating for low water level. It is mounted on the fire box crown plate to put off the fire in the furnace when the water level falls to an unsafe limit.

Q. 30.83. Which one of the following statements is correct about the Fanno flow?
(*a*) For an initially subsonic flow, the effect of friction is to decrease the Mach number towards unity
(*b*) For an initially supersonic flow, the effect of friction is to increase the Mach number towards unity
(*c*) At the point of maximum entropy, the Mach number is unity
(*d*) Stagnation pressure always increases along the Fanno line.

Ans. **(*c*)**

Explanation:
In case of 'Fanno flow' the Mach number (M) is unity at the point of maximum entropy.

Q. 30.84. The turbine of the turbo-prop engine as compared to that of turbo-jet engine is
(*a*) similar (*b*) smaller
(*c*) bigger (*d*) unpredictable.

Ans. **(*c*)**

Explanation:
The turbine of the turbo-prop engine as compared to that of a turbo-jet engine is *bigger* because in the former case turbine expansion process is *greater* than that of the latter one.

Q. 30.85. Consider the following statements in respect of gas turbines:
A gas turbine plant with reheater leads to a
1. considerable improvement in the work output
2. considerable improvement in the thermal efficiency

Which of these statements is/are correct?
(*a*) 1 only (*b*) 2 only
(*c*) Both 1 and 2 (*d*) Neither 1 nor 2.

Ans. **(*a*)**

Explanation:
In a gas turbine 'reheating' increases the work output but the thermal efficiency may be increased or decreased depending upon the mean temperature of heat addition.

Q. 30.86. An open cycle constant pressure gas turbine uses a fuel of calorific value 40,000 kJ/kg with air fuel ratio 80 : 1 and develops a net output of 80 kJ/kg of air. The thermal efficiency of the cycle is
(*a*) 61% (*b*) 16%
(*c*) 18% (*d*) None of these.

Ans. **(*b*)**

Explanation:
Given: Calorific value = 40,000 kJ/kg

$$\dot{m}_a / \dot{m}_f = 80 : 1$$
$$W_{out} = 80 \text{ kJ/kg of air}$$
$$\eta_{thermal} = \frac{\text{Work output } (W_{out})}{\text{Heat supplied } (Q_{in})} = \frac{\dot{m}_a \times 80}{\dot{m}_f \times 40{,}000}$$
$$= \frac{80 \times 80}{40000} = 0.16 \text{ or } 16\%$$

Q. 30.87. Uranium 238 is represented as $_{92}U^{238}$. What does it imply?
(*a*) It has 92 protons and 146 neutrons
(*b*) It has 146 protons and 92 electrons
(*c*) It has 92 protons and 238 neutrons
(*d*) It has 92 neutrons and 238 protons.

Ans. **(*a*)**

Explanation:
Uranium 238 is represented as $_{92}U^{238}$
i.e. Number of proton = atomic number = 92,
and Number of neutron = 238 – 92 = 146

Q. 30.88. Water-tube boilers are preferred for
(*a*) high pressure and high output
(*b*) high pressure and low output
(*c*) low pressure and high output
(*d*) low pressure and low output.

Ans. **(*a*)**

Explanation:
Water-tube boilers are preferred for high pressure and high output.
- These boilers are used exclusively when pressureabove 10 bar and capacity in excess of 7000 kg of steam per hour is required. 'Babcock and Wilcox water-tube boiler' is example of horizontal straight tube boiler and may be designed for stationary or marine purposes.

Q. 30.89. Given,
$$\eta_s = \text{stage efficiency}$$
$$\eta_n = \text{nozzle efficiency}$$
$$\eta_b = \text{blade efficiency}$$
Which one of the following is correct?
(*a*) $\eta_n = \eta_b \eta_s$ (*b*) $\eta_b = \eta_s \eta_n$
(*c*) $\eta_b \times \eta_n \times \eta_s = 1$ (*d*) $\eta_s = \eta_b \eta_n$.

Ans. **(*d*)**

Explanation:
Stage efficiency = Blade efficiency × nozzle efficiency
i.e. $\eta_s = \eta_b \times \eta_n$

Q. 30.90. A centrifugal compressor is suitable for which of the following?
(*a*) High pressure ratio, low mass flow
(*b*) Low pressure ratio, low mass flow
(*c*) High pressure ratio, high mass flow
(*d*) Low pressure ratio, high mass flow.

Ans. (*d*)

Explanation:

A centrifugal pressure is suitable for low and medium pressures (pressure ratio about 3 to 4.5) and high mass flow (large gas volumes).

Q. 30.91. The suction pressure is 1 bar and delivery pressure is 125 bar. What is the ideal intermediate pressure at the end of first stage for a 3-stage air compressor?

(*a*) 25 bar (*b*) 5 bar
(*c*) 10 bar (*d*) 20 bar.

Ans. (*b*)

Explanation:

Suction pressure = 1 bar

Ideal intermediate pressure at the end of 1st stage = $(5)^1$ = 5 bar

[Pressure after 2nd stage = $(5)^2$ = 25 bar;
Pressure after 3rd stage = $(5)^3$ = 125 bar]

Q. 30.92. A nuclear reactor is said to be critical when the neutron population in the reactor core is

(*a*) rapidly increasing leading to the point of explosion
(*b*) decreasing from a specific value
(*c*) reduced to zero
(*d*) constant.

Ans. (*d*)

Explanation:

A nuclear reactor is said to be critical when the neutron population in the reactor core is '*Constant*'.

Q. 30.93. A gas turbine plant working on Joule cycle produces 4000 kW of power. If its work ratio is 40% the power consumed by the compressor is

(*a*) 2000 kW (*b*) 4000 kW
(*c*) 6000 kW (*d*) 8000 kW.

Ans. (*c*)

Explanation:

Given: $W_T - W_C$ = 4000 kW; Work ratio = 0.4

$$\text{Work ratio} = \frac{\text{Network output}}{\text{Gross work output}} = \frac{W_T - W_C}{W_T}$$

$$\text{or,} \quad 0.4 = \frac{4000}{W_T}$$

$$\text{or,} \quad W_T = 10000 \text{ kW}$$

∴ Power consumed by compressor

$$W_T = 10000 - 4000 = 6000 \text{ kW}$$

Q. 30.94. The ratio of the static enthalpy rise in the rotor to the static enthalpy rise in the stage of an axial flow compressor defined as

(*a*) Power input factor
(*b*) Flow coefficient
(*c*) Temperature coefficient
(*d*) Degree of reaction.

Ans. (*d*)

Explanation:

In an axial flow compressor, degree of reaction (R_d) is defined as the ratio of pressure rise in the compressor stage. It is equal to the ratio of static enthalpy rise in the rotor to the static enthalpy rise in the stage of an axial flow compressor.

Q. 30.95. In a gas turbine cycle with regeneration

(*a*) pressure ratio increases
(*b*) work output decreases
(*c*) thermal efficiency increases
(*d*) heat input increases.

Ans. (*c*)

Explanation:

In a gas turbine cycle with '*regeneration' thermal efficiency increases* (The exhaust gases from a gas turbine carry a large quantity of heat with them since their temperature is far above the ambient temperature, so they can be used to heat the air coming from the compressor thereby *reducing the mass of* fuel supplied in the combustion chamber).

Q. 30.96. In a normal shock wave in one dimensional flow

(*a*) the entropy remains constant
(*b*) the entropy increases across the shock
(*c*) the entropy decreases across the shock
(*d*) the velocity, pressure and density increase across the shock.

Ans. (*b*)

Explanation:

In a normal shock wave in one dimensional flow, there is always a change from supersonic to subsonic speed which involves pressure rise and never the reverse. As per second law of thermodynamics entropy always increases during irreversible adiabatic change, thus *entropy increases* across the shock.

MISCELLANEOUS OBJECTIVE TYPE QUESTIONS

(With Answers)

MISCELLANEOUS OBJECTIVE TYPE QUESTIONS

(With Answers)

SET-I

1. Choose the appropriate answer :

(*i*) It is related to supersaturated flow in steam nozzles

(*a*) Reheat factor (*b*) Wilson line

(*c*) State point locus (*d*) Mach number.

(*ii*) The ideal refrigeration cycle in aircraft is

(*a*) vapour compression cycle

(*b*) vapour absorption cycle

(*c*) steam jet refrigeration

(*d*) reversed Brayton cycle.

(*iii*) For supersonic flow the converging duct is a

(*a*) nozzle (*b*) diffuser

(*c*) venturi

(*d*) duct in which velocity remains constant.

(*iv*) Morse test measures the indicated power of a

(*a*) SI engine (*b*) CI engine

(*c*) Steam engine (*d*) Steam turbine.

(*v*) It has zero cetane number

(*a*) Normal heptane

(*b*) Alpha-methyl naphthalene

(*c*) Cetane (*d*) Iso-octane.

2. State whether the statements are correct or wrong :

(*i*) An isentropic process is always adiabatic.

(*ii*) There is no pressure drop across a reaction turbine stage.

(*iii*) The maximum velocity that can be achieved with an incompressible fluid is the sonic velocity.

(*iv*) The function of a governor is to maintain the shaft speed constant as the load varies.

(*v*) A convergent nozzle is used to obtain supersonic velocity.

(*vi*) The integral $\int pdv$ gives work for any process.

3. Fill in the blanks :

(*i*) COP of a refrigerator is defined as ....

(*ii*) Mach number is defined as ....

(*iii*) Equivalent evaporation of a boiler is defined as ....

(*iv*) A boiler is specified by ....

(*v*) Isothermal efficiency of a compressor is defined as ....

(*vi*) Degree of reaction is defined as ....

4. Match the sets :

Set A	*Set B*
1. Fusible plug	(*a*) SI engine
2. Pre-whirl	(*b*) CI engine
3. Fuel injection	(*c*) Steam turbine
4. Capillary tube	(*d*) Fire tube boiler
5. Otto cycle	(*e*) Centrifugal compressor
6. Labyrinth glands	(*f*) Vapour compression refrigeration plant.

ANSWERS

1. (*i*) Wilson line (*ii*) Reversed Brayton cycle
(*iii*) Diffuser (*iv*) SI engine
(*v*) Alpha-methyl naphthalene.

2. (*i*) Correct (*ii*) Wrong
(*iii*) Correct (*iv*) Correct
(*v*) Wrong (*vi*) Wrong.

3. (*i*) COP of refrigerator

$$= \frac{\text{Net refrigerating effect } (R_n)}{\text{Work done } (W)}$$

(*ii*) *Mach number* is the *ratio of the actual velocity to the sonic velocity.*

(*iii*) *Equivalent evaporation of a boiler* is defined as the *amount of water evaporated from feed water at 100°C and formed into dry and saturated system at 100°C at normal pressure.* It is usually written as *"from and at 100°C".*

(*iv*) A boiler is *specified* by : (*a*) Pressure, temperature, rate of steam generation (*b*) Fire tube or water tube (*c*) Fuel used and method of firing.

(*v*) *Isothermal efficiency* is the *ratio of isothermal power to shaft power.*

(*vi*) *Degree of reaction* of a reaction turbine (R_d) is defined as *the ratio of heat drop over moving blades to the total heat drop in the stage.*

i.e., $R_d = \dfrac{\text{Heat drop in moving blades}}{\text{Heat drop in the stage}} = \dfrac{\Delta h_m}{\Delta h_f + \Delta h_m}$

(where surfaces *f* and *m* stand for fixed and moving blades respectively)

4. (1) (*d*) (2) (*e*) (3) (*b*) (4) (*f*) (5) (*a*) (6) (*c*).

SET-II

1. Choose the correct answer :

(*i*) Cycle efficiency of a modern thermal power plant is approximately

(*a*) 29% (*b*) 60%
(*c*) 80% (*d*) 44%
(*e*) none of the above.

(*ii*) Three similar diesel engines (stroke to bore ratio same) have diameters of 10 cm, 25 cm and 35 cm and they run at speeds of (1) 2000 r.p.m., (2) 700 r.p.m. and (3) 570 r.p.m. respectively. Compression ratio and all other conditions of operation of the engines to be the same, which engine will have more tendency to knocking ?

(*a*) 3 (*b*) 2
(*c*) 1 (*d*) None of the above.

(*iii*) In the above question (*ii*), which engine will have more tendency for smoky exhaust ?

(*a*) 1 (*b*) 2
(*c*) 3 (*d*) none of the above.

Justify in one sentence.

(*iv*) Amyl nitrite, ethyl nitrate or ether added in small quantities with diesel fuel alters the rate of preflame reaction

(*a*) decreases (*b*) increases
(*c*) remains the same (*d*) none of the above.

(*v*) Some metal organic compounds (tetraethyl lead or iron carbonyl) added to gasoline in some quantities alter the ignition delay of the mixture appreciably

(*a*) decreases (*b*) increases
(*c*) remains the same (*d*) none of the above.

(*vi*) Overall thermal efficiency of a modern thermal plant is approximately

(*a*) 50% (*b*) 44%
(*c*) 20% (*d*) 30%
(*e*) none of the above.

(*vii*) Otto cycle is efficient

(*a*) more than (*b*) less than
(*c*) equal to (*d*) none of the above.

The diesel cycle for the same compression ratio. Prove mathematically.

(*viii*) The work done per stage for two-row impulse turbine at optimum blade speed ratio and unit mass flow is

(*a*) mU^2 (*b*) $2mU^2$
(*c*) $4mU^2$ (*d*) $8mU^2$.

where m = mass flow, U = peripheral speed.

(*ix*) The cooling range of a cooling tower is the difference in

(*a*) temperature between the hot water entering the cooling tower and the cooled water leaving the tower
(*b*) temperature between the hot water entering the cooling tower and the wet bulb temperature
(*c*) temperature between the hot water entering the cooling tower and dry bulb temperature
(*d*) none of the above.

(*x*) The approach of cooling tower is the difference in temperatures between

(*a*) cold water leaving cooling tower and wet bulb temperature
(*b*) cold water leaving cooling tower and dry bulb temperature
(*c*) hot water coming to the cooling tower and dry bulb temperature
(*d*) hot water coming into the cooling tower and wet bulb temperature.

(*xi*) The turbine of a gas turbine plant uses

(*a*) simple impulse
(*b*) two-row velocity compounded impulse stage
(*c*) impulse stage followed by reaction stage
(*d*) reaction stage.

(*xii*) Three velocity diagrams are given below for optimum U/C_1 ratios, same values of U and α_1. Identity which is reaction stage

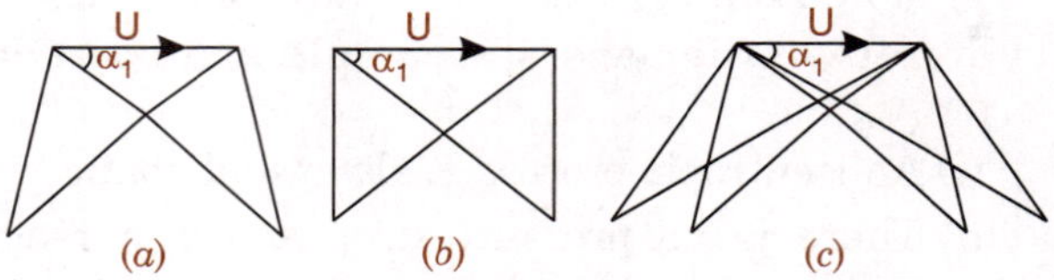

(*a*) (*b*) (*c*)

2. Answer the following :

(*i*) In steam turbine, there is 'reheat factor'. Why in turbine of a gas turbine engine, 'reheat factor' is omitted but accounted for otherwise ?

(*ii*) When a gas is compressed, the work done per unit mass flow may be given by

(*a*) $(h_2 - h_1)$ (*b*) $-\int_{v_1}^{v_2} v.dp$

State precisely in which cases (*a*) and in which cases (*b*) can be applied.

(*iii*) What is the difference between hot and cold insulation, *i.e.,* insulations for a steam pipe and for chilled brine line ?

(*iv*) The velocity diagrams (reaction stage) for blade speed ratio less than, equal to, and greater than the nominal optimum value are given below :

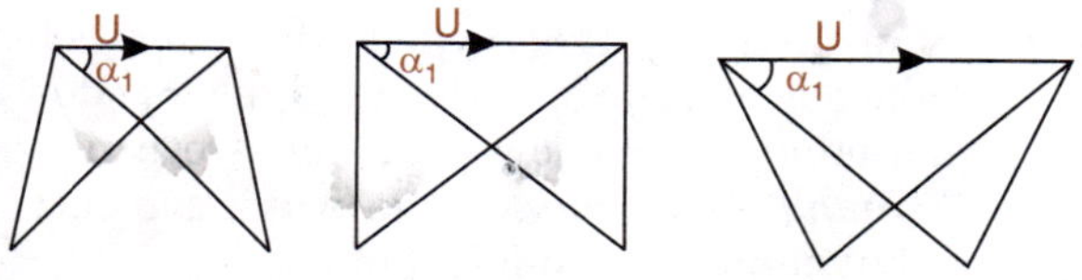

(*a*) (*b*) (*c*)

Mark each of them with any one of $\frac{U}{C_1} \gtreqless \cos \alpha_1$ with usual notations.

ANSWERS

1. (*i*) Cycle efficiency of a modern thermal power plant is 44% (app.).

(*ii*) (*a*) In *diesel engine, the knocking depends on slow preflame reaction.* At slower speeds these reactions are slow due to less swirl and knocking can occur.

(*iii*) (*c*) Larger diameter engines are *more smoky due to less penetration of fuel to find oxygen.* Slow speed will also give less turbulence.

(*iv*) (*b*) Diesel fuel additives reduce delay period by increasing the rate of preflame reactions.

(*v*) (*b*) Gasoline engine additives increase the delay period to avoid ignition of end charge before flame reaches there.

(*vi*) (*c*) Overall thermal efficiency of modern thermal plant is about 20%.

(*vii*) (*a*) The efficiency of diesel cycle is given by :

$$\eta_{diesel} = 1 - \frac{1}{\gamma (r)^{\gamma - 1}} \left[\frac{\rho^{\gamma} - 1}{\rho - 1} \right]$$

where, ρ = cut-off ratio and r = compression ratio.

The value of ρ more than one increases numerator faster than denominator and efficiency decreases. Hence for a *given compression ratio,* the *Otto cycle is more efficient.*

(*viii*) (*b*) $\frac{U}{C_1} = \frac{\cos \alpha}{2}$, Work done per stage $= 2U(C_1 \cos \alpha - U) \times m = 2U(2U - U) \times m = 2U^2 m$

(*ix*) (*b*) is the correct choice.

(*x*) (*a*) is true.

(*xi*) (*b*) Pressure ratio is sufficiently low (5) and two row velocity compounded impulse stage can be used to reduce speed.

(*xii*) (*b*) is the correct choice since in a reaction turbine, the velocity ratio $\frac{U}{C_1} = \cos \alpha$ is satisfied by Fig. (*b*).

2. (*i*) The term reheat factor does not come into effect in gas turbines due to the fact that pressure ratio is low and there is single (nozzle) expansion. However, in a turbo-jet engine it can come into existence. Adiabatic efficiency of turbine accounts for it.

(*ii*) (*a*) ......adiabatic flow process :

(*b*) ......isothermal process (pv = constant).

(*iii*) There should not be any change in the structure of insulating material at the temperatures involved.

(*iv*) (*a*) $\frac{U}{C_1} < \cos \alpha_1$ (*b*) $\frac{U}{C_1} = \cos \alpha_1$

(*c*) $\frac{U}{C_1} > \cos \alpha_1$.

SET-III

1. Are the following statement correct ? Rectify the incorrect statements changing only the wrong part of the sentence :

(*i*) Stoichiometric air is required for complete combustion.

(*ii*) The most important characteristics for effective combustion of liquid fuels are flame point and power point.

(*iii*) Alkali is mainly responsible for water side corrosion in boilers.

(*iv*) In large high pressure boilers, a high degree of steam purity is necessary to prevent deposits in the circuit.

(*v*) A high temperature of flue gases at boiler exit means a too rich mixture of fuel and air.

2. With respect to a steam plant, who do you understand by :

(*i*) Metallurgical limit

(*ii*) Attemperator

(*iii*) Non-return valve

(*iv*) Condensate pump

(*v*) Low temperature corrosion.

3. Differentiate between the following :

(*i*) Land based gas turbines and 'aircraft turbine engines' ;

(*ii*) Deaerator and demineralizer ;

(*iii*) Intercooling in 'reciprocating compressor' and in 'rotary compressor' ;

(*iv*) Ejector and injector ;

(*v*) Regeneration in Brayton cycle and in Rankine cycle.

ANSWERS

1. (*i*) Stoichiometric air is the *theoretical* air required for complete combustion.

(*ii*) The most important characteristic for effective combustion of liquid fuels are *ignition delay* and *auto-ignition temperature*.

(*iii*) *Acid* is mainly responsible for water side corrosion in boilers.

(*iv*) In large high pressure boilers, a high degree of *water* purity is necessary to prevent deposits in the circuit.

(*v*) A high temperature of fuel gases at boiler exit means a rich mixture of fuel and air.

2. (*i*) **Metallurgical limit.** The maximum/highest temperature which boiler and turbine can withstand is decided by the metallurgy of the material of which these components are made.

(*ii*) **Attemperator.** An attemperator (or desuperheater) is an apparatus employed to *reduce the temperature of superheat steam.* Normally attemperators are 'spray type'.

(*iii*) **Non-return valve.** A valve which allows flow to take place in one direction (and reverse flow does not-occur) is known as non-return valve. *Example* : Feed check valve (water can enter into it but cannot go from it).

(*iv*) **Condensate pump.** A pump which is used to extract condensate from the condenser (to the hot well) is known as condensate pump.

(*v*) **Low temperature corrosion.** Low temperature corrosion is the corrosion of condenser tubes and is due to the *deposits of alkali and iron sulphates* which slowly eat the tube metal and eventually cause failure. This type of corrosion is more dangerous than scale formation. It can be *minimised by :*

(*a*) using a liquid base additive ;

(*b*) expoxy coatings applied to the water surface ; and

(*c*) intermittent chlorination of water.

3. (*i*) **Land based gas turbines and aircraft turbine engines**

Land based gas turbines	*Aircraft turbine engines*
Used in rear oil fields where natural gas is available ; used in industry where gas is available as by-product ; used in electric generating stations (as base load and peak load plants).	The jet aircraft plant makes use of kinetic energy of gases by expansion of turbine exhaust gases to atmospheric pressure in a propelling nozzle (in an industrial gas turbine, kinetic energy of leaving the turbine is wasted).
In industrial gas turbine, a life of 1000000 working hours (without major overhaul) is common.	The life of aircraft gas turbines may be only 500 working hours.

(*ii*) **Deaerator and demineralizer**

Deaerator	*Demineralizer*
In a *deaerator water is heated for the boiler feed.* The entering water is sprayed through the nozzles in the atmosphere of steam filling the heater. The heating steam is the bled steam from the turbine.	In a *demineralizer, water is freed from its mineral content* by evaporation or by a series of cation and anion exchangers to produce essentially distilled water.
	Demineralizing process, perhaps is the most economical method of producing make up water for high pressure boilers.

(*iii*) **Intercooling in reciprocating and rotary compressors**

Reciprocating compressors	*Rotary compressors*
The compressed air from L.P. cylinder passes into the *intercooler where it gets cooled to initial temperature.*	Generally, cooling is not adopted in rotary compressors.
	In *centrifugal compressors* cooling results in very complicated construction. In *axial flow compressors,* the only method of cooling is to pass cooling fluid (generally air) at low temperature through hollow blades which results in costly construction.

(*iv*) **Ejector and Injector**

Ejector

Steam operated air ejectors find very wide field of application for the *production of high vacuum.*

The operation of ejector consists in utilizing the various drag of high velocity steam jet for the ejection of air and other incondensable gases from a chamber ; it is chiefly used for exhausting the air from steam condensers. The steam jet flows through an air chamber where it entrains the air and any other gases which are adjacent to its surface ; the kinetic energy of the resulting mixture is then converted to pressure energy by being pushed through a diverging cone, or diffuser. The increase of pressure thus obtained enables the mixture to be discharged against a pressure which is higher than that of the entertaining chamber. The entertaining operation is due to the viscous drag between the air and steam jet.

For steam plants where a high vacuum pressure is maintained it is imperative to use a multi-stage ejector.

Injector

A steam injector is employed to force water into the boiler under pressure.

It makes use of the principle of a steam nozzle by which it utilizes the kinetic energy of a steam jet for increasing the pressure and velocity of a corresponding quantity of water.

(*v*) **Regeneration in Brayton cycle and in Rankine cycle**

Brayton cycle

The temperature of the gases coming out from the turbine (in Brayton cycle) is quite high and this energy is utilized in a regenerator to heat the air coming from the compressor to the combustion chamber. This regeneration will require *less amount of heat supplied* for the same maximum temperature at turbine inlet thereby *improving the cycle efficiency.*

Rankine cycle

In a Rankine cycle, energy possessed in the steam itself is utilised for heating the feed water. The steam is extracted from the steam turbine during the expansion and its energy is utilized in heating the feed water in the feed water heaters. This energy regenerated requires less amount of heat in the boiler, thereby improving the efficiency of the cycle.

SET-IV

1. Choose the appropriate answers :

(*i*) In a reciprocating compressor, as the clearance volume increases
(*a*) work of compression increases
(*b*) amount of air delivered increases
(*c*) amount of air delivered decreases
(*d*) volumetric efficiency increases.

(*ii*) The working substance in Rankine cycle
(*a*) is gas (*b*) is vapour
(*c*) can be gas or vapour.

(*iii*) The diagram factor is the ratio of
(*a*) actual mep to inlet pressure
(*b*) actual mep to theoretical mep
(*c*) theoretical mep to actual mep
(*d*) theoretical mep to inlet pressure.

(*iv*) In reciprocating steam engines, the point of cut-off is the point at which
(*a*) expansion is stopped (*b*) cushioning begins
(*c*) cushioning is stopped
(*d*) steam supply is stopped.

(*v*) Willan's line gives a linear relationship between steam consumption and
(*a*) thermal efficiency
(*b*) mechanical efficiency
(*c*) brake power (*d*) indicated power.

(*vi*) Parson's turbine is
(*a*) an impulse turbine with degree of reaction = 0
(*b*) an impulse turbine with degree of reaction = 0.5
(*c*) an impulse turbine with degree of reaction = 0.75
(*d*) an impulse turbine with degree of reaction = 1.0.

2. Are the following statements correct ? Rectify the incorrect statements, changing only the wrong part :

(*i*) For the saturated air with water vapour, wet bulb temperature is equal to the dew-point temperature.

(*ii*) Work done per cycle/stroke volume is the expression for the mean effective pressure of Otto cycle.

(*iii*) COP of the reversed Carnot cycle decreases when the upper temperature is increased, the lower temperature is kept constant.

(*iv*) Theoretical air is stoichiometric air.

(*v*) Reversible constant temperature process is the process during which, heat is added in Rankine vapour power cycle.

ANSWERS

1. (*i*) (*c*) (*ii*) (*b*) (*iii*) (*b*)
(*iv*) (*d*) (*v*) (*d*) (*vi*) (*b*).

2. (*i*) For the saturated air with water vapour, the wet bulb temperature is equal to dry bulb temperature as well as dew point temperature.

(*ii*) 'Work done per cycle/stroke volume' is the expression for mean effective pressure of any cycle.

(*iii*) COP of the reversed Carnot cycle decreases when the upper temperature is increased, but the lower temperature is kept constant.

(*iv*) Theoretical air is stoichiometric air.

(*v*) Irreversible constant pressure process is the process during which heat is added in Rankine vapour power cycle.

SET-V

1. Examine the following statements, and find out whether statement is true or false :

(*i*) Compared with reciprocating compressors, rotary compressors are smaller in size for the same discharge.

(*ii*) Babcook and Wilcox boiler is a water-tube boiler, extensively used as a mobile boiler.

(*iii*) In expansive steam engine, steam pushes the piston at constant pressure when it is expanded for developing work.

(*iv*) Sonic velocity is obtained at the exit of a convergent steam nozzle flowing maximum mass per unit area.

(*v*) The steam velocity in a reaction turbine is very high, resulting in a relatively high speed of the turbine.

(*vi*) Edward air pump is a reciprocating type wet air pump, where both the air and condensate are removed by a single pump.

(*vii*) The overlapping of a 4-stroke petrol engine is quite low, as the higher value would allow more amount of petrol pass-out with the exhaust gases.

(*viii*) Higher danger of explosion during transport comes in the way of extensive use of gaseous fuel.

(*ix*) Excellent miscibility with lubricating oil of widely used refrigerant, Freon-12 helps in solving the lubrication problem.

(*x*) The acceleration of a rocket is highest at the take-off to enable the rocket to lift.

2. Find out the correct answer among the alternate options :

(*i*) The main reason for adopting the axial flow compressor instead of centrifugal compressor in aircraft turbines is that

(*a*) the starting torque for axial flow compressor is high

(*b*) the frontal area of axial flow compressor is considerably less

(*c*) the efficiency at middle speed range is higher.

(*ii*) Locomotive boiler is identified with the following outstanding feature :

(*a*) It is very compact

(*b*) It can generate steam in very short time

(*c*) It attains a high combustion efficiency as the combustion space is much larger.

(*iii*) In a throttle-governed reciprocating steam engine

(*a*) the steam pressure is maintained constant and the total quantity of steam admitted is reduced

(*b*) the pressure as well as the quantity of steam admitted to the engine is reduced to give better performance at all loads

(*c*) the speed is maintained constant and the pressure of the admitted steam is reduced.

(*iv*) The main effect of presence of friction in steam nozzles is

(*a*) to reduce the enthalpy drop

(*b*) to reduce the exit pressure

(*c*) to reduce the dryness fraction

(*v*) In a velocity-compounded impulse turbine

(*a*) there is only one set of nozzles and two or more rows of moving blades

(*b*) there is one row of fixed blades at the entry of each row of moving blades

(*c*) the velocity is increased after each stage.

(*vi*) In a surface condenser

(*a*) the cooling water and exhaust steam mix with each other and the steam is condensed

(*b*) the cooling water is passed through the tubes and the exhaust steam surrounds the tubes

(*c*) the cooling water is sprayed over the tubes through which exhaust steam is passed.

(*vii*) The statement that theoretically, a 2-stroke engine develops twice the power of a 4-stroke engine at the same speed, is derived from the fact that

(*a*) the working cycle is completed in one revolution of the crankshaft in a 2-stroke engine, whereas it takes two revolutions in a 4-stroke engine

(*b*) a 2-stroke engine runs at twice engine speed compared with a 4-stroke engine

(*c*) a 2-stroke engine has a set of two cylinders and pistons.

(*viii*) Coal gas is obtained

(*a*) by the partial combustion of coal in a stream of air and steam

(*b*) by passing air and a large amount of steam over waste coal

(*c*) as a by-product during the destructive distillation of coal.

(*ix*) Winter air-conditioning implies

(*a*) maintaining the temperature and humidity of the air

(*b*) increasing the temperature and humidity of the air

(*c*) increasing the temperature and maintaining the humidity of the air.

(*x*) The thermal efficiency of simple open cycle gas turbine plant is improved by regeneration, as this

(*a*) increases the work output from the same size of plant

(*b*) decreases the temperature of gases at the turbine inlet

(*c*) decreases the quantity of heat supplied.

ANSWERS

1.	(*i*) True	(*ii*) False	(*iii*) False	(*iv*) True
	(*v*) False	(*vi*) True		
	(*vii*) False	(*viii*) False	(*ix*) True	(*x*) True.
2.	(*i*) (*b*)	(*ii*) (*a*)	(*iii*) (*b*)	(*iv*) (*a*)
	(*v*) (*a*)	(*vi*) (*b*)	(*vii*) (*a*)	(*viii*) (*c*)
	(*ix*) (*b*)	(*x*) (*c*).		

SET-VI

1. Examine the following statements and find out whether the statement is true or false :

(*i*) A two-stroke engine occupies larger floor area than a four-stroke engine.

(*ii*) High speed compression ignition engines operate on dual combustion cycle.

(*iii*) The volume of air sucked by compressor during suction stroke is called swept volume.

(*iv*) An open cycle gas turbine works on same cycle as that of a closed cycle gas turbine.

(*v*) The highest temperature during cycle in vapour compression system occurs after evaporation.

(*vi*) Refrigerant freon-11 has the lowest freezing point.

(*vii*) The ratio of heat actually utilised in producing steam to the heat liberated in furnace is called factor of evaporation.

(*viii*) If the relative humidity is low, the rate of evaporation of water will be high.

(*ix*) A compound steam engine requires a heavier flywheel.

(*x*) The flow through a nozzle is regarded as adiabatic flow.

2. Find out the correct answer among the alternate options :

(*i*) The ratio of the volume charge admitted at NTP to the swept volume of the piston is called

(*a*) overall efficiency

(*b*) volumetric efficiency

(*c*) relative efficiency.

(*ii*) Rotary compressors are used for delivering

(*a*) small quantities of air at high pressures

(*b*) large quantities of air at high pressures

(*c*) large quantities of air at low pressures.

(*iii*) The efficiency of a jet engine is higher at

(*a*) high speeds (*b*) low altitudes

(*c*) high altitudes.

(*iv*) In SI units one tonne of refrigeration is equal to

(*a*) 210 kJ/min (*b*) 420 kJ/min

(*c*) 620 kJ/min.

(*v*) In electrolux refrigerator

(*a*) ammonia evaporates in hydrogen

(*b*) hydrogen evaporates in ammonia

(*c*) ammonia is absorbed in water.

(*vi*) When the speed of the crankshaft is between 100 r.p.m. and 250 r.p.m., the engine is said to be

(*a*) slow speed steam engine

(*b*) medium speed steam engine
(*c*) high speed steam engine.

(*vii*) The average value of diagram factor lies between
(*a*) 0.2 to 0.5 (*b*) 0.5 to 0.65
(*c*) 0.65 to 0.9.

(*viii*) De Laval turbines are mostly used
(*a*) for small power purposes and low speeds
(*b*) for small power purposes and high speeds
(*c*) for large power purposes.

(*ix*) The ratio of the temperature rise of cooling water to the vacuum temperature minus inlet cooling water temperature is called
(*a*) condenser efficiency (*b*) vacuum efficiency
(*c*) nozzle efficiency.

(*x*) A steam nozzle converts
(*a*) heat energy of steam into potential energy
(*b*) kinetic energy into heat energy of steam
(*c*) heat energy of steam into kinetic energy.

ANSWERS

1. (*i*) False (*ii*) True (*iii*) False (*iv*) True
(*v*) False (*vi*) False (*vii*) False (*viii*) True
(*ix*) False (*x*) True.

2. (*i*) (*b*) (*ii*) (*c*) (*iii*) (*a*) (*iv*) (*a*)
(*v*) (*c*) (*vi*) (*b*) (*vii*) (*c*) (*viii*) (*b*)
(*ix*) (*a*) (*x*) (*c*).

SET-VII

1. Fill in the blanks :

(*i*) If both mass and energy cross the boundary it is called ...... system.

(*ii*) Rankine cycle efficiency is given by expression ....

(*iii*) COP is defined as ....

(*iv*) The expression for work done during polytropic process is ....

(*v*) For perfect intercooling in a two-stage air compressor the relation between the pressures p_1, p_2 and p_3 is ...... for maximum efficiency.

2. Indicate whether the statements are correct or wrong :

(*i*) In throttling process entropy remains constant.

(*ii*) In jet condenser the coolant water and condensate water mix with each other.

(*iii*) In impulse turbine, the expansion of steam takes place in guide vanes.

(*iv*) The volumetric efficiency of a compressor increases with increase in compression ratio.

(*v*) In open cycle gas turbine the exhaust gas from the turbine is rejected into the atmosphere.

3. Match the set :

(*i*) Fuel injector	SI engine
(*ii*) Throttle valve	Condenser
(*iii*) Reheat factor	Diesel engine
(*iv*) Brayton cycle	Steam turbine
(*v*) Vacuum efficiency	Gas turbine

4. Given below are some statements which are applicable either to SI engine or Diesel engine. Classify them as applicable to SI or Diesel engines :

(*i*) Air-fuel mixture is sucked inside the cylinder during suction stroke.

(*ii*) Compression ratio is 20 : 1.

(*iii*) Mechanical efficiency is high.

(*iv*) Chances of pre-ignition are existing.

(*v*) Running cost is low.

ANSWERS

1. (*i*) open (*ii*) $\dfrac{(h_1 - h_2)}{(h_1 - h_{f2})}$

(*iii*) $\dfrac{\text{Refrigerating effect}}{\text{Work supplied}}$ (*iv*) $\dfrac{(p_1v_1 - p_2v_2)}{(n-1)}$

(*v*) $p_2 = \sqrt{p_1 p_3}$

2. (*i*) Wrong (*ii*) Correct
(*iii*) Correct (*iv*) Wrong
(*v*) Correct

3. (*i*) Fuel injector, Diesel engine
(*ii*) Throttle valve, SI engine
(*iii*) Reheat cycle, Steam turbine
(*iv*) Brayton cycle, gas turbine.
(*v*) Vacuum efficiency, condenser

4. (*i*) SI (*ii*) Diesel
(*iii*) Diesel (*iv*) SI
(*v*) Diesel.

SET-VIII

1. Indicate whether the following statements are correct or wrong :

(*i*) The two constant entropy lines cannot intersect.

(*ii*) Once-through boilers never have a steam drum.

(*iii*) The clearance volume of a reciprocating compressor is given as a percentage of stroke volume.

(*iv*) In a centrifugal compressor the axial component of the fluid velocity produces rotational effect on the rotor.

(*v*) The suction pipe diameter of a refrigeration plant should be less than the discharge pipe diameter.

2. Fill in the blanks:

(*i*) The stoichiometric air fuel ratio in a gas turbine cycle is ....

(*ii*) In a ...... process the change of entropy is equal to δ Q/T.

(*iii*) If the thermal efficiency of a heat engine becomes 100% it violates ...... law of thermodynamics.

(*iv*) In an ideal throttling process ...... remains constant.

(*v*) In order to have a high COP the pressure range of the compressor should be...... .

3. Match the set:

(*i*) Scavenging	(I) Steam turbine
(*ii*) Velocity compounding	(II) SI engine
(*iii*) Rankine cycle	(III) Two-stroke engine
(*iv*) Octane number	(IV) Nozzle
(*v*) Accelerating flow	(V) Curtis stage.

4. Select the most appropriate answer:

(*i*) The internal energy of a system is a function of temperature only, this statement holds good:

(*a*) for any pure substance
(*b*) for a perfect gas only
(*c*) for any gas (*d*) none of the above.

(*ii*) The critical pressure of the steam is :

(*a*) 10 bar (*b*) 221.1 bar
(*c*) 100 bar (*d*) 2212 bar.

(*iii*) In a Freon plant the following material should be used for pipes :

(*a*) copper (*b*) brass
(*c*) aluminium (*d*) steel.

(*iv*) In a gas turbine plant reheating is done mainly to

(*a*) increase the power output
(*b*) decrease the peak temperature
(*c*) reduce the size of turbine
(*d*) increase the outlet temperature.

(*v*) The speed of the camshaft of a 4-stroke IC engine is:

(*a*) equal to the crankshaft speed
(*b*) half of the crankshaft speed
(*c*) twice the crankshaft speed
(*d*) none of the above.

ANSWERS

1. (*i*) Correct (*ii*) Correct
(*iii*) Correct (*iv*) Wrong
(*v*) Wrong.

2. (*i*) never used (*ii*) reversible
(*iii*) second (*iv*) enthalpy
(*v*) low.

3. (*i*) Scavenging, two-stroke engine
(*ii*) velocity compounding, curtis stage
(*iii*) Rankine cycle, steam turbine
(*iv*) Octane number, SI engine
(*v*) Accelerating flow, nozzle.

4. (*i*) *b* (*ii*) *b*
(*iii*) this is wrong, as all are appropriate,
(*iv*) *a* (*v*) *b*.

SET-IX

1. Give the reasons for the following facts:

(*i*) Nozzles are more efficient than diffusers.

(*ii*) Impulse reaction steam turbine blading is more efficient than impulse bladings.

(*iii*) Deaerator is invariably used in a steam power plant.

(*iv*) Pressure rise per stage in a centrifugal compressor is more than that of an axial flow compressor.

(*v*) Thermal efficiency of a jet engine increases with decreases in inlet air temperature.

2. Choose the correct answer:

(*i*) Volumetric efficiency of a reciprocating compressor increases due to

(*a*) increase in clearance ratio and pressure ratio
(*b*) decrease in clearance ratio and pressure ratio
(*c*) decrease in clearance ratio, pressure ratio and exponent of re-expansion
(*d*) decrease in clearance ratio and pressure ratio and increase in exponent of re-expansion.

(*ii*) Brake specific fuel consumption of an SI engine is around

(*a*) 0.1 kg/kWh (*b*) 0.3 kg/kWh
(*c*) 0.5 kg/kWh (*d*) 0.7 kg/kWh.

(*iii*) Specific steam consumption in a modern steam power plant is around

(*a*) 3 kg/kWh (*b*) 6 kg/kWh
(*c*) 9 kg/kWh (*d*) 12 kg/kWh.

(*iv*) Two-stroke engine is not used in a passenger car mainly due to

(*a*) noisy exhaust
(*b*) poor brake specific fuel consumption
(*c*) need of more lubrication
(*d*) overheating of piston.

(*v*) In the gasoline fuel, to produce anti-knock effect, one of the following is added

(*a*) tetra-ethyl lead

(*b*) tetra-ethyl lead and ethylene dibromide

(*c*) benzyl amino phenol

(*d*) ethylene dibromide and benzyl amino phenol.

ANSWERS

1. (*i*) In diffusers the pressure gradient exists in the direction opposite to that of flow. Also stream lines tend to break away from the diverging walls reversing their direction, in order to flow with a favourable pressure gradient. This results in turbulence. This situation does not arise in nozzles. Hence *nozzles are more efficient than diffusers.*

(*ii*) In impulse-reaction turbines the enthalpy drop in each stage is less, and hence the velocities are low. The blade speed can therefore be kept optimum. Further due to low velocities, the blade friction losses are also less. The velocities in impulse bladings are relatively high. Therefore the *impulse-reaction blading is more efficient than impulse blading.*

(*iii*) De-aerator is used in steam power plants to reduce the oxygen content of feed water to zero, or practically zero. This is necessary as the oxygen is present in feed water as corrosive agent.

(*iv*) In centrifugal compressor, the pressure rise is due to three effects: *impulsive effect* (change in absolute kinetic energy), *reaction effect* (change is relative velocity), and *centrifugal effect*. In axial flow compressor the last effect is absent. This last effect makes a significant contribution to the pressure rise in centrifugal compressors. Also, the average velocities are much higher in centrifugal compressors than in axial flow compressors. Hence, the pressure rise per stage in these compressors is more than that in axial flow compressors.

(*v*) With decrease in inlet air temperature, the density of inlet air increases, and hence the mass flow rate too. More fuel could be burned. Both of these effects increase the power and also the thrust produced by the engine. But this will have little effect on the thermal efficiency.

2. The correct answers are :

(*i*) (*d*) (*ii*) (*b*) (*iii*) (*b*)

(*iv*) (*b*) (*v*) (*a*).

Index

A

B

C

TABLE V

Conversion Factors

Force

1 newton	=	1 kg-m/sec^2
	=	0.012 kgf
1 kgf	=	9.81 N

Pressure

1 bar	=	750.06 mm Hg
	=	0.9869 atm
	=	10^5 N/m^2
	=	10^3 kg/m-sec^2
1 N/m^2	=	1 pascal
	=	10^{-5} bar
	=	10^{-2} kg/m-sec^2
1 atm	=	760 mm Hg
	=	1.03 kgf/cm^2 = 1.01325 bar
	=	1.01325×10^5 N/m^2

Work, Energy or Heat

1 joule	=	1 newton metre
	=	1 watt-sec
	=	2.7778×10^{-7} kWh
	=	0.239 cal
	=	0.239×10^{-3} kcal
1 cal	=	4.184 joule
	=	1.1622×10^{-6} kWh
1 kcal	=	4.184×10^3 joule
	=	427 kgfm
	=	1.1622×10^{-3} kWh
1 kWh	=	8.6×10^5 cal
	=	860 kcal
	=	3.6×10^6 joule
1 kgfm	=	$\left(\frac{1}{427}\right)$ kcal = 9.81 joules

Power

1 watt	=	1 joule/sec = 0.86 kcal/h
1 h.p.	=	75 mkgf/sec = 0.1757 kcal/sec
	=	735.3 watt
1 kW	=	1000 watts
	=	860 kcal/h

Specific heat

1 kcal/kg °K	=	4.18 kJ/kg K

Thermal conductivity

1 watt/m-K	=	0.8598 kcal/h-m °C
1 kcal/h-m °C	=	1.16123 watt/m-K
	=	1.16123 joules/s-m-K

Heat transfer co-efficient

1 watt/m^2-K	=	0.86 kcal/m^2-h °C
1 kcal/m^2-h °C	=	1.163 watt/m^2-K

IMPORTANT ENGINEERING CONSTANTS AND EXPRESSIONS IN SI UNITS

S.No.	*Engineering constants and expressions*	*M.K.S. system*	*S.I. units*
1.	Value of g_0	9.81 kg-m/kgf-sec^2	1 kg-m/N-sec^2
2.	Universal gas constant	848 kgf-m/kg mole °K	848 × 9.81 = 8314 J/kg-mole °K (∵ 1 kgf-m = 9.81 joules)
3.	Gas constant (R)	29.27 kgf m/kg °K for air	$\frac{8314}{29}$ = 287 joules/kg K for air
4.	Specific heats (for air)	c_v = 0.17 kcal/kg °K c_p = 0.24 kcal/kg °K	c_v = 0.17 × 4.184 = 0.71128 kJ/kg K c_p = 0.24 × 4.184 = 1 kJ/kg K
5.	Flow through nozzle-exit velocity (C_2)	91.5 $\sqrt{U}$ where U is in kcal	44.7 $\sqrt{U}$ where U is in kJ
6.	Refrigeration 1 ton	= 50 kcal/min	= 210 kJ/min
7.	**Heat transfer** The Stefan Boltzman Law is given by :	$Q = \sigma T^4$ kcal/m^2-h when $\sigma = 4.9 \times 10^{-8}$ kcal/h-m^2 °K^4	$Q = \sigma T^4$ watts/m^2-h when $\sigma = 5.67 \times 10^{-8}$ W/m^2K^4